ORGANIC PHOSPHORUS COMPOUNDS

Volume 6

G. M. KOSOLAPOFF
Auburn University

and

L. MAIER
Monsanto Research S. A.

WILEY-INTERSCIENCE, a Division of John Wiley & Sons, Inc.
New York • London • Sydney • Toronto

Library of Congress Cataloging in Publication Data:

Kosolapoff, Gennady M
Organic phosphorus compounds.

1950 ed. published under title: Organophosphorus compounds.
Includes bibliographies.
1. Organophosphorus compounds. I. Maier, L., joint author. II. Title.

QD412.P1K55 1973 547'.07 72-1359
ISBN 0-471-50445-9 (v. 6)

Printed in the United States of America

10 9 8 7 6 5 4 3 2 1

Contents

Chapter 14. Phosphinic Acids and Derivatives

P. C. CROFTS

University of Manchester Institute of Science and Technology, Great Britain

This chapter surveys phosphinic acids RR'P(O)OH, and the several related classes of acid in which one or both of the oxygen atoms of phosphinic acids are replaced by sulfur or by selenium. Their salts, esters, anhydrides, acyl derivatives, amides, imides, acylamino derivatives, and hydrazides are included as are corresponding derivatives, RR'P(NR")X, of phosphinimidic acids (the acids themselves having no free existence since they are labile tautomers of phosphinic amides). Phosphinimidic halides are included, but trihalophosphoranes, and phosphinic halides and pseudohalides, which are all closely related, structurally and synthetically, to the subjects of this chapter, are described in Chapters 5B and 9.

No extensive review of all this material has appeared since the publication in 1950 of Kosolapoff's previous book, which describes the approximately 100 phosphinic acids (there called secondary phosphonic acids) and their derivatives which had been reported at that time. The literature has now been searched comprehensively using the subject index of Chemical Abstracts through volume 69 (to December 1968) and by searching the pages of section 29 in volumes 70-72 (to June 1970), for which subject indexes were not then available. The primary literature has thus been covered up to about the end of 1969, although possibly with a few gaps on non-preparative topics from mid-1968, and has yielded information about more than 1000 acids and their derivatives.

General methods for the preparation and interconversion of phosphinic (phosphinothioic, etc.) acids and derivatives are classified and discussed in 22 sections, most of which are subdivided. Sections I-XIV deal with methods for preparing compounds containing the fundamental phosphinic structure

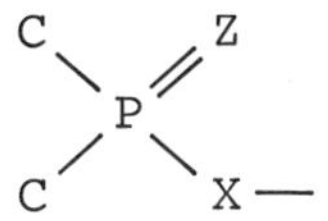

(X, Z = O, S, Se, NR)

from starting materials of other structural types. The remaining preparative sections deal with methods for converting one phosphinic compound into another and hence also cover most of the important reactions of phosphinic acids and their derivatives. The reactions discussed in Sections XX-XXV are those involving changes at phosphorus by which the various functional types of phosphinic compounds (acids, esters, etc.) are interconverted, while Sections XXX-XXXII are concerned with reactions which modify organic groups attached to phosphorus, oxygen, sulfur, or nitrogen, without affecting the phosphinic functional group.

Individual synthetic methods are distinguished in the list of compounds by Roman numerals and lower-case letters corresponding to the section and subsection in which the method is discussed. An asterisk against a symbol for a synthetic method in the compound list indicates a variation which does not justify a separate heading but which is mentioned in the text under the unasterisked method. Miscellaneous reactions, for each of which only a few examples have been reported, are described in the entries for compounds for whose preparation they have been used.

A. METHODS OF PREPARATION OF PHOSPHINIC ACIDS AND DERIVATIVES, AND THEIR THIO, SELENO, AND IMIDO ANALOGS, FROM COMPOUNDS CONTAINING PHOSPHORUS IN OTHER OXIDATION STATES OR HAVING OTHER THAN TWO PHOSPHORUS-CARBON BONDS

I. ROUTES FROM SECONDARY PHOSPHINES

Ia. OXIDATION OF SECONDARY PHOSPHINES

$$R_2PH \xrightarrow{[O]} R_2PO_2H$$

The first recorded preparation of a phosphinic acid was by A. W. Hofmann, who in 1872 oxidized dimethylphosphine with fuming nitric acid and isolated dimethylphosphinic acid as a crystalline solid.[293] The method lay almost unused during the next eighty years because of the inaccessibility of secondary phosphines, but as new routes to these have been developed (see Chapter 1) it has been used occasionally, with hydrogen peroxide in acetic acid,[292] in acetone,[317] or in methanol,[106] or with water and sodium peroxide[318,321] as oxidizing agents.

Ib. ADDITION OF SULFUR TO SECONDARY PHOSPHINES

$$R_2PH + 2S \longrightarrow R_2PS_2H$$

This reaction, like the previous one, is historically important having been used for the first recorded preparation of a phosphinodithioic acid in 1892.[294] It also passed through a period of disuse but is now of more importance for the preparation of phosphinodithioic acids than oxidation is for phosphinic acids. The reaction is exothermic and is most commonly carried out in benzene,[292,310,315] but acetone,[586] acetonitrile,[586] aqueous ammonia,[649] carbon disulfide,[314] ether,[294] and pyridine[18] have also been used as reaction media.

Ic. REACTIONS OF SECONDARY PHOSPHINES WITH ORGANIC AZIDES

$$R_2PH + 2R'N_3 \longrightarrow R_2P(:NR')NHR' + 2N_2$$

See Section IVc.

Id. REACTIONS OF SECONDARY PHOSPHINES WITH DISULFIDES

$$R_2PH + R'S{\cdot}SR' \longrightarrow R_2P(=S)SR' + R'H$$

This reaction[260] is carried out in benzene under reflux and gives excellent yields. Free radicals are apparently involved, since if hydroquinone is present other products are obtained, including a phosphinothoate if aqueous 80% t-butanol is the reaction solvent.

Ie. REACTIONS OF SECONDARY PHOSPHINES WITH ARENESULFONYL CHLORIDES

$$2R_2PH + ArSO_2Cl \longrightarrow R_2P(=O)SAr + R_2POH + HCl$$

This may involve mutual oxidation and reduction to give a phosphinic chloride and a thiophenol which then react. Good yields are reported using diphenylphosphine or an arylalkylphosphine, with pyridine as base and benzene as solvent.[697] Di-(2-cyanoethyl)phosphine also reacts in this way, but dibutylphosphine gave instead the phosphinic acid and a thiophenol or diaryl disulfide.[103]

II. ROUTES FROM SECONDARY PHOSPHINE OXIDES AND SECONDARY PHOSPHINE SULFIDES

IIa. OXIDATION OF SECONDARY PHOSPHINE OXIDES

$$R_2P(O)H \xrightarrow{[O]} R_2PO_2H$$

A number of secondary phosphine oxides have been oxidized to phosphinic acids following their isolation in a pure state from reactions of dialkyl phosphites[146,303,482,533,789] or of monoalkyl phosphonites[179] with Grignard reagents, or in other ways (see Chapter 11).[100,101,180,416,682,724] Air or oxygen were used for the earlier oxidations,[100,724] but hydrogen peroxide is now commonly used, and bromine water or acid potassium permanganate,[303] and alkaline potassium ferricyanide,[146] have also proved satisfactory.

IIb. HALOGENATION OF SECONDARY PHOSPHINE OXIDES AND HYDROLYSIS OF THE UNISOLATED PHOSPHINIC HALIDES

$$R_2P(O)H \longrightarrow [R_2P(O)X] \longrightarrow R_2PO_2H$$

This method, using phosphorus pentachloride for the chlorination stage, is reported to be more effective than oxidation with hydrogen peroxide for conversion of long-chain dialkylphosphine oxides to phosphinic acids.[789]

IIc. HALOGENATION OF SECONDARY PHOSPHINE OXIDES AND REACTION OF THE UNISOLATED PHOSPHINIC HALIDES WITH ALCOHOLS OR AMINES

$$R_2P(O)H \longrightarrow [R_2P(O)X] \begin{cases} \longrightarrow R_2P(O)OR' \\ \longrightarrow R_2P(O)NHR' \end{cases}$$

Crude phosphinic bromides have been prepared from secondary phosphine oxides by reaction with bromine,[734] or with carbon tetrachloride[647] or bromotrichloromethane,[142] and converted to esters and amides by reactions XXIe and XXIi.

IId. ADDITION OF SULFUR TO SECONDARY PHOSPHINE OXIDES, OR TO SECONDARY PHOSPHINE SULFIDES

$$R_2P(O)H + S \longrightarrow R_2POSH$$

$$R_2P(S)H + S \longrightarrow R_2PS_2H$$

Phosphinothioic acids may be conveniently prepared from secondary phosphine oxides, either crude (Section VIg) or isolated and purified,[292,351] but the relative inaccessibility of secondary phosphine sulfides makes these less useful as a source of phosphinodithioic acids.[586]

IIe. REACTIONS OF SECONDARY PHOSPHINE SULFIDES WITH SULFENYL HALIDES[134]

$$R_2P(S)H + ArSCl \longrightarrow R_2P(=S)SAr + HCl$$

IIf. REACTIONS OF SECONDARY PHOSPHINE OXIDES WITH ORGANIC DISULFIDES

$$R_2P(O)H + R'S \cdot SR' \longrightarrow R_2P(=O)SR' + R'SH$$

Reactions of this type were first carried out by heating dioctylphosphine oxide with several dialkyl disulfides and catalytic amounts of sodium to 110-140°,[592] but in a subsequent investigation of reactions between eight diarylphosphine oxides and diphenyl disulfide, they were performed in t-butyl alcohol without added base at 30°.[261] Although products were isolated, the interest in this case was primarily in the kinetics of the reaction, and a mechanism, involving nucleophilic attack on the disulfide by the phosphinous acid tautomer of the phosphine oxide, was suggested. Either ionic or radical mechanisms can operate in cognate reactions of disulfides with secondary phosphines,[260] and with methyl diphenylphosphinite.[154]

IIg. REACTIONS OF SECONDARY PHOSPHINE OXIDES WITH SOME CARBONYL GROUPS

$$R_2P(O)H + R'CO \cdot CO_2Et \longrightarrow R_2P(O) \cdot CR'(OH) \cdot CO_2Et \longrightarrow$$

$$R_2P(O) \cdot O \cdot CHR \cdot CO_2Et$$

Secondary phosphine oxides react with aldehydes and ketones to give tertiary phosphine oxides containing an α-hydroxyalkyl group (see Chapter 6), but when there is another electron attracting group attached to the carbonyl group, as in α-ketocarboxylic and α-ketophosphonic esters, the initial exothermically formed adduct rearranges on heating to give a phosphinate.[617]

IIh.   REACTIONS OF SECONDARY PHOSPHINE SULFIDES WITH
       MORPHOLINOSULFUR CHLORIDE[17]

$$R_2P(S)H \;+\; \underset{O}{\underset{|}{\underset{N}{\underset{|}{SCl}}}} \quad \xrightarrow{\text{Base}} \quad R_2P{\overset{S}{\diagup}}{\underset{S\cdot N}{\diagdown}} \diagdown O$$

$$\downarrow \; HCl \;+\; R_2P(S)H$$

$$R_2P(S)\cdot S\cdot P(S)R_2$$

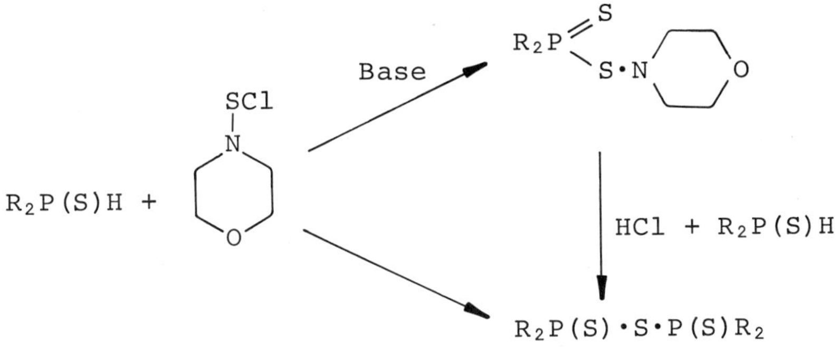

III.   ROUTES FROM BIPHOSPHINE DISULFIDES AND OTHER
       COMPOUNDS CONTAINING PHOSPHORUS-PHOSPHORUS
       BONDS

IIIa. OXIDATION OF BIPHOSPHINE DISULFIDES

$$R_2P(S)\cdot P(S)R_2 \xrightarrow{[O]} \begin{array}{l} R_2PO_2H \\[1mm] R_2P(O)\cdot O\cdot P(O)R_2 \end{array}$$

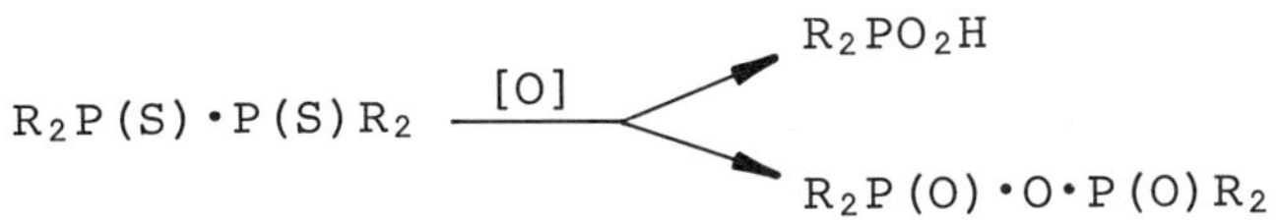

Biphosphine disulfides, obtained by reactions of thiophosphoryl chloride or phosphonothioic dihalides and Grignard reagents [see Chapter 2], provide by their oxidation
a useful source of a number of lower symmetrical,[447] unsymmetrical,[475] and cyclic dialkylphosphinic acids.[706]
Because of the insolubility and consequent ease of isolation of tetramethylbiphosphine disulfide, this route is
particularly convenient for the preparation of dimethylphosphinic acid,[424,677] which is otherwise a rather in-
accessible compound.
     The oxidation to phosphinic acids may be effected with
nitric acid,[424,706] hydrogen peroxide,[677] or mercuric
oxide in benzene.[447,475] Phosphinic anhydrides are obtained using mercuric oxide in benzene under strictly anhydrous conditions.[447]

     IIIb.   OXIDATION OR ADDITION OF SULFUR TO BIPHOSPHINES
OR POLYPHOSPHINES. Bonds between two tervalent phosphorus
atoms are readily cleaved by oxygen or sulfur with accompanying conversion of the phosphorus atoms to the quinquevalent state. Several reactions have been described in
which products were variously a phosphinic acid,[676] a
phosphinic anhydride (radical mechanism suggested),[741] and

a phosphinothioic thioanhydride, $R_2P(S){\cdot}S{\cdot}P(S)R_2$.[313]

Oxidation of the cyclic triphosphine 1,2,3-triphenyl-1,2,3-triphosphaindane with potassium permanganate in acetone yields o-phenylenedi(phenylphosphinic acid).[483]

IIIc. HYDROLYSIS OF BIPHOSPHINE DISULFIDES

$$R_2P(S){\cdot}P(S)R_2 \longrightarrow R_2POSH + R_2P(S)H$$

This reaction,[133] carried out with 10% aqueous sodium hydroxide at 90° for one hour, appears to be a convenient route to lower aliphatic phosphinothioic acids, particularly if the yields could be improved by controlled oxidation of the secondary phosphine sulfides which are also formed in the reaction.

IIId. FUSION OF BIPHOSPHINE DISULFIDES WITH SODIUM SULFIDE AND SULFUR OR WITH SODIUM SELENIDE AND SELENIUM

$$R_2P(S){\cdot}P(S)R_2 + Na_2S + S \longrightarrow 2R_2PS_2^- \; Na^+$$

The reaction formulated above (R = Et, Pr, Bu) gave excellent yields at temperatures of 140-185°,[444,448] but the parallel reaction of tetraethylbiphosphine disulfide with sodium selenide and selenium proved more difficult, because of the high melting points of the inorganic reactants, and gave only a 35% yield of sodium diethylphosphinoselenothioate.[443]

IV. ROUTES FROM DERIVATIVES OF PHOSPHINOUS, PHOSPHINOTHIOUS, AND PHOSPHINOSELENOUS ACIDS

IVa. OXIDATION OF, OR ADDITION OF SULFUR OR SELENIUM TO, PHOSPHINITE, PHOSPHINOTHIOITE, OR PHOSPHINOSELENOITE ESTERS

$$R_2P{\cdot}OR' \longrightarrow R_2P(O)OR'$$

$$R_2P{\cdot}SR' \longrightarrow R_2P(O)SR'$$

$$R_2P{\cdot}SeR' \longrightarrow R_2P(O)SeR'$$

$$R_2P{\cdot}OR' \longrightarrow R_2P(S)OR'$$

$$R_2P{\cdot}SR' \longrightarrow R_2P(S)SR'$$

$$R_2P{\cdot}OR' \longrightarrow R_2P(Se)OR'$$

The six types of reaction formulated above all proceed readily, give good yields, and are uncomplicated by side reactions, but their synthetic value is limited by the relative difficulty of preparing the initial tervalent phosphorus esters [see Chapter 11].

The reactions are generally exothermic, and many occur spontaneously at room temperature, this tendency being increased if R and R' are alkyl and diminished if they are aryl groups. From qualitative comparisons of pairs of esters it appears that ease of formation increases in the following order:

$$R_2P(Se)OR' < R_2P(S)OR' < R_2P(O)OR' \sim R_2P(S)SR' <$$

$$R_2P(O)SR' < R_2P(O)SeR'$$

although comparisons of oxidation on the one hand, and additions of sulfur and selenium on the other, are made difficult by the different states of the reactants.

Phosphinites may be oxidized by dry air,[776] oxygen,[595,599] or nitrogen oxides,[596,598] in reactions which are often exothermic at room temperature. Oxidation of phosphinothioites require moderation by cooling, dilution of nitrogen dioxide with nitrogen, and the use of a solvent,[14] while oxidation of phenyl diphenylphosphinoselenoite proceeds very readily[514,588] in spite of the moderating effect of the phenyl groups.

Additions of sulfur are often carried out in benzene. They are exothermic both for phosphinothioites,[10,11,755b] and for phosphinites,[527,596,749,755b] although for the latter it is common to heat under reflux for several hours to complete the reaction.[202,598,599]

Addition of sulfur to an acyl phosphinite (a phosphinous-carboxylic anhydride) apparently proceeds more slowly than to esters.[344]

The only recorded addition of selenium to a phosphinite required gentle heating.[527]

Oxidation and addition of sulfur are valuable methods for preparing phosphinate and phosphinothioate esters of acid-sensitive alcohols, whose phosphinites can be obtained by transesterification reactions.[595,599] In cases where there is little risk of hydrolysis, oxidations may be carried out with hydrogen peroxide.[76,755b] They may also be effected indirectly with halogens and water or ethanol,[208,527,749] but there are no obvious general advantages in this.

IVb. OXIDATION, OR ADDITION OF SULFUR TO PHOSPHINOUS AMIDES

$$R_2P\cdot NR'R'' \longrightarrow R_2P(O)\cdot NR'R''$$

$$R_2P\cdot NR'R'' \longrightarrow R_2P(S)\cdot NR'R''$$

Oxidation of diphenylphosphinous diethylamide by air occurs during one week in decane at room temperature but is much slower in dimethylformamide or butanol even on heating.[559] A number of six-membered cyclic amides of aminobiphenylylphosphinous acids have been oxidized to the corresponding cyclic phosphinic amides with hydrogen peroxide,[118] and this reagent has been used for oxidation of dimethylamides of acetylenic phosphinous acids,[122] and for oxidation of diphenylphosphinous phenylsulfonyl imide.[731] Secondary and tertiary amides of diphenylphosphinous acid have been oxidized by refluxing with activated manganese dioxide in benzene,[739] and have been converted to phosphinothioic amides by heating with sulfur in benzene,[739] and in carbon disulfide.[736] Sulfur has been added to bis-(trifluoromethyl) phosphonous amide and dimethylamide by heating in sealed tubes.[170]

IVc. REACTIONS OF CHLORIDES, ESTERS, AND AMIDES OF PHOSPHINOUS ACIDS WITH ORGANIC AZIDES

$$R_2PX + R'N_3 \longrightarrow R_2P(:NR')X + N_2$$

These reactions, together with the related reaction of secondary phosphines with azides (Section Ic),[159,582] provide an important approach to a variety of phosphinimidic compounds. Except for the reaction of diphenylphosphine with triphenylsilyl azide (90 hr at 110-115°),[582] and also two points noted below, the reactions proceed smoothly in either ether or benzene at 20-50° and give yields, after moderate reaction times of about 45-80%.

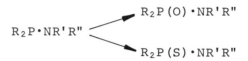

The largest group of reactions which have been reported are those of aryl azides with diethylphosphinous tertiary amides,[246,479] and with some dialkyl- and diarylphosphinous secondary amides.[239,245,787] A few reactions have been described of acyl azides with phosphinites,[338] and with phosphinic chlorides (at 80-90°),[159] and of phosphinic azides with phosphinites,[247] while one paper reports single examples of reactions of an alkyl azide with a phosphinous amide, of an aryl azide with a phosphinite and with a phosphinous toluene-p-sulfonyl amide (8% yield) and of toluene-p-sulfonyl azide with a phosphinite and with a phosphinic chloride. Reactions of phosphinous amides with toluene-p-sulfonyl azide proceed anomalously and yield stable adducts, $R_2P(:N\cdot N:N\cdot Tos)NR'R''$.[94]

IVd. REACTIONS OF PHOSPHINOUS CHLORIDES WITH N,N-DICHLOROSULFONAMIDES

$$R_2PCl + ArSO_2\cdot NCl_2 \longrightarrow R_2P(=N\cdot SO_2\cdot Ar)(Cl) + Cl_2$$

This reaction[729] is performed in boiling benzene, and hence under much milder conditions than that of trichlorophosphoranes with sulfonamides (Section XXc) which is another route to the same products. It is stated to occur in two stages:

$$2R_2PCl + ArSO_2\cdot NCl_2 \longrightarrow R_2P(Cl):N\cdot SO_2\cdot Ar + R_2PCl_3$$

$$R_2PCl_3 + ArSO_2\cdot NCl_2 \longrightarrow R_2P(Cl):N\cdot SO_2\cdot Ar + 2Cl_2$$

IVe. COMBINED OXIDATION AND HYDROLYSIS OF PHOSPHINOUS CHLORIDES

$$R_2PCl \longrightarrow R_2PO_2H$$

This process has been used many times since diphenylphosphinic acid was first prepared in 1877 by the reaction of diphenylphosphinous chloride with dilute nitric acid.[526] Other methods of combining the two reactions are by chlorination and hydrolysis,[156,519] by using hydrogen peroxide alone[298] or in alkaline solution,[119,169,579] or by using alkaline potassium permanganate.[645] Yields are usually excellent.

Treatment of diphenylphosphinous chloride with t-butyl hydroperoxide gives t-butyl diphenylphosphinate in a combined oxidation and conversion of an acid chloride to an ester.[740]

IVf. REACTIONS OF PHOSPHINOUS CHLORIDES WITH SODIUM ARENESULFINATES

$$R_2PCl + ArSO_2^- \, Na^+ \longrightarrow R_2P(=O)(SAr)$$

This combined oxidation-reduction and esterification is carried out in dimethylformamide.[698]

IVg. REACTIONS OF PHOSPHINOUS CHLORIDES WITH CHLOROAMINE, AND MOISTURE OR AMMONIA[244,735a]

$$Ph_2PCl + NH_2Cl \longrightarrow Ph_2P(NH_2)Cl \cdot NH \cdot PPh_2Cl_2$$

Moist air:

$$\longrightarrow Ph_2P(O) \cdot NH \cdot P(O)Ph_2 + Ph_2P(Cl){:}N \cdot P(O)Ph_2$$

NH_3:

$$\longrightarrow Ph_2P(NH_2){:}N \cdot P({:}NH_2^+)Ph_2\ Cl^-$$

IVh. REACTIONS OF PHOSPHINITE ESTERS WITH α-HALOKETONES

$$R_2P \cdot OR + R' \cdot CO \cdot CR''_2X \longrightarrow R_2P(O) \cdot O \cdot CR'{:}CR''_2$$

Several phosphinate esters have been prepared by Perkov reactions of dialkyl- and diarylphosphinites. Reactions with chloral are exothermic and are carried out by mixing with ice-cooling and then gently warming to complete the reaction.[6,357,387] Reactions of diphenylphosphinites with chloroacetone give unsaturated esters of diphenylphosphinic acid by mixing at room temperature followed by heating to 150° for 4 hr; but slow mixing of the reactants at 170° results in normal Arbuzov reactions and formation of tertiary phosphine oxides.[515]

Exothermic formation of an unsaturated ester occurred on mixing propyl dipropylphosphinite and S-ethyl chloro(thioacetate).[794]

V. ROUTES FROM DERIVATIVES OF PHOSPHONOUS, PHOSPHONOTHIOUS AND PHOSPHONODITHIOUS ACIDS

Va. ARBUZOV REACTIONS OF DIALKYL PHOSPHONITES OR DIALKYL PHOSPHONODITHIOITES

$$RP(OR')_2 + R''X \longrightarrow RR''P(O)OR'$$

This is a major route to esters of alkylaryl- and unsymmetrical dialkyl-phosphinic acids, and hence by hydrolysis (Section XXIIIa) to the acids themselves, and over 12% of the literature references to this chapter include reactions of this type. These reactions are typically carried out by mixing the phosphonite and the halide at room temperature, heating slowly, moderating any overvigorous reaction, and finally maintaining the mixture at 120-160° for several hours. Yields are normally good to very good but may be reduced by unnecessary overheating resulting in some pyrolysis of ester to acid.[662]

Most of the reactions which have been described

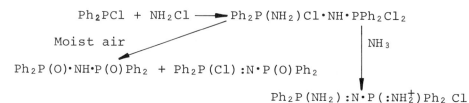

involve unsubstituted, saturated or unsaturated, aliphatic or side-chain aromatic halides, and unsubstituted phosphonites and hence yield products which contain only hydrocarbon groups. Provided, however, that the structure of the halide is consistent with its undergoing S_N2 reactions, a variety of substituents may be present in either of the reactants. Reactions of dihalides may yield esters either of haloalkylphosphinic acids,[280,394] or of diphosphinic acids.[150,493,743] Diphosphinates may also be prepared from diphosphonites and alkyl halides,[60] whilst reactions of diphosphonites with dihalides yield polyphosphinates.[689] Esters of cyclic phosphinic acids are obtained from intramolecular reactions of ω-haloalkylphosphonite esters.[15,279]

Other substituents in phosphonites involved in Arbuzov reactions have been limited to aromatic halide,[560] dialkylamino,[138] and ketone and ester groups,[71] but this probably merely reflects the unavailability of substituted phosphonites. In halides, substituents such as hydroxy, thiol, carboxylic acid and primary or secondary amino groups, which would themselves react with phosphonites, may not be present, but halides used have included alkoxy,[31,56,266] alkylthio,[431] acyloxy,[56] carboxylic ester[26,266,283] and tertiary amide,[38,190,266] carbothiolic ester,[794] nitrile,[90] and tertiary amino groups.[266]

Acyl halides react normally, but at much lower temperatures because of their high reactivity, to give esters of acylphosphinic acids.[308,650] Bromoacetone gives predominantly 2-ketopropylphosphinates,[47,607] with minor amounts of isopropenyl phosphonates which are the sole product from chloroacetone.[47] Vinyl esters of phosphonic acids would be obtained from α-haloaldehydes, but their acetals and also those of β-haloaldehydes react normally and give acetals or enol esters,[666,667] which can be converted (see Section XXXg) without difficulty to aldehydoalkylphosphinates.

More unusually substituted halides, which nevertheless give normal reactions, are those with a phthalimido group,[603] with a variety of silicon-containing substituents,[126] with sulfonyl fluoride,[253] and sulfonate ester groups,[288] and with butyl di-(chloromethyl)phosphinate which yields a triphosphinate ester.[478] Nitro groups are usually reduced by tervalent phosphorus esters, but o-dinitrobenzene reacts with diethylmethylphosphonite to give an o-nitrophenylphosphinate.[114]

There are many examples of Arbuzov reactions of phosphonites with allyl and benzyl-type halides, but very few involving secondary or tertiary halides of any kind, isolated cases being reactions of isopropyl iodide,[30] diphenylmethyl chloride and bromide,[223,283] triphenylmethyl bromide,[27] and a few substituted halides.[90,266,283,667]

Reactions of vinyl halides are limited to those of perfluoropropene[330,400] which reacts at the 1-position, of 1-chloroperfluorocyclopentene which reacts with replacement of the 2-fluorine atom,[207] of 1,2-dichloroperfluorocyclopentene in which both chlorine atoms are replaced to give a diphosphinate,[206] and of 1-cyanovinyl bromide (α-bromoacrylonitrile),[644] and benzylsulfonyl- and phenylsulfonyl-vinyl bromides[365] which, whether 1- or trans-2-substituted, react to give trans-2- substituted products.

There is one case of an Arbuzov-type reaction with a diarylphosphonite:[524] formation of a quasi-phosphonium intermediate occurs readily but alkaline hydrolysis is then necessary to obtain the aryl phosphinate.

Reactions of trialkyl phosphites with carbon tetrachloride, which occur at lower temperatures than ordinary Arbuzov reactions, are known to proceed by a free-radical mechanism. This apparently operates also in reactions of phosphonites with carbon tetrachloride which are reported to take place at room temperature,[354] during twenty minutes on a steam bath,[38] or on refluxing (at 70-80°).[358,388]

Arbuzov reactions of normal heterolytic type may occur, without added halide but probably catalyzed by impurities, during distillation of crude phosphonites.[25,356] True uncatalyzed isomerizations of ordinary pure phosphonites require temperatures of 240° or more,[633] but there are two classes of phosphonite which rearrange intramolecularly under less extreme conditions. 2-Chloroalkyl esters of phosphonous acids are isomerized to 2-chloroalkylphosphinates at the temperatures (∿150°) of normal Arbuzov reactions:[236,337,343] evidence for a cyclic quasi-phosphonium intermediate is provided by the conversion of a 1,3-dichloroprop-2-yl phosphonite into a 2,3-dichloroprop-1-yl-phosphinate.[680] Allyl phosphonites may isomerize to allylphosphinates during distillation,[358,360] or on heating to 110-140°: the interconversion of 1-methylallyl and but-2-enyl groups[605] indicates a cyclic transition state in these reactions and also in the rearrangements at room temperature of crude alk-1-ynyl phosphonites to alka-1,2-dienylphosphinates.[96,124,125,305,606]

Reactions of alkyl phosphonodithioites with halides at 130-150° give a variety of products, but alkyl phosphinodithioates predominate and a number of these have been prepared in this way.[10,29,41,429]

Vb. ARBUZOV REACTIONS OF DIALKYL PHOSPHONITES, AND CONVERSION OF THE CRUDE PHOSPHINATE TO A PHOSPHINIC ACID

$$RP(OR')_2 + R''X \longrightarrow [RR''P(O)OR'] \longrightarrow RR''PO_2H$$

Phosphinates, which are particularly high boiling,[150,499,501] or whose purification is otherwise difficult,[124,207] may be hydrolyzed in the crude state and the resultant acids purified by recrystallization. Also included in this section are some Arbuzov reactions in which esters were pyrolyzed because of over-vigorous conditions in sealed-tube reactions.[27]

Vc. ARBUZOV REACTIONS OF ALKYL PHOSPHONAMIDOTHIONITES[13]

$$RP(SR')NR_2 + R''X \longrightarrow RR''P(S)NR_2$$

Vd. REACTIONS OF PHOSPHONOUS DI-(DIALKYLAMIDES) WITH ALKYL HALIDES, AND HYDROLYSIS OF THE PRODUCTS

$$RP(NR'_2)_2 + R''X \longrightarrow [RR''P^+(NR'_2)_2\ X^-] \longrightarrow$$

$$[RR''P^+(NR'_2)_2\ OH^-] \longrightarrow RR''PO_2H$$

There are several reports of this route to phosphinic acids.[521,523,602] The first stage occurs readily, giving crystalling quasi-phosphonium halides, but the number of steps involved in conversion into hydroxides, pyrolysis of these, and purification of the crude phosphinic acids through their silver salts, explains why this method has fallen into disuse.

Ve. ALKYLATION OF SODIUM DERIVATIVES OF MONOALKYL PHOSPHONITES OR MONOALKYL PHOSPHONOTHIOITES

$$R{-}P(\overset{\ominus}{=}O)(OR')\ Na^+ + R''X \longrightarrow RR''P(O)OR' + NaX$$

The active hydrogen atom in monoesters of phosphonous and phosphonothious acids enables them to form alkali-metal derivatives which are soluble in inert solvents (ether, dioxan, toluene) and react with alkyl halides, etc., to give phosphinates. This alternative to the Arbuzov reaction has been comparatively little used. It has the advantage of using reactants which are more easily prepared and less air-sensitive than dialkyl phosphonites, but the disadvantages of requiring a reaction solvent and more manipulation.

Both alkyl-[39,40] and aryl-[9,139,760] phosphonites have been used, together with alkyl iodides from ethyl to nonyl,[40] s-octyl toluene-p-sulfonate (with partial retention of optical activity),[139] several α,ω-alkylene dihalides,[9] and ethyl chloroacetate.[283]

Unusual halides which reacted normally, are N,Ndiethylchloroformamide,[39] and sodium chloroacetate[760] which would not be suitable for an Arbuzov reaction. Chloroacetone gave an epoxy phosphinate.[47]

Reactions of O-ethyl phenylphosphonothioite with five halides containing alkylthio or ethoxycarbonyl groups, including one secondary halide, are described in a patent.[709]

Vf. ALKYLATION OF SODIUM DERIVATIVES OF MONOALKYL PHOSPHONITES, FOLLOWED BY HYDROLYSIS OF THE RE-SULTING PHOSPHINATE ESTERS

$$R-P\overset{O}{\underset{OR'}{\diagdown}} Na^+ + R''X \xrightarrow[-NaX]{} [RR''P(O)OR'] \longrightarrow RR''PO_2H$$

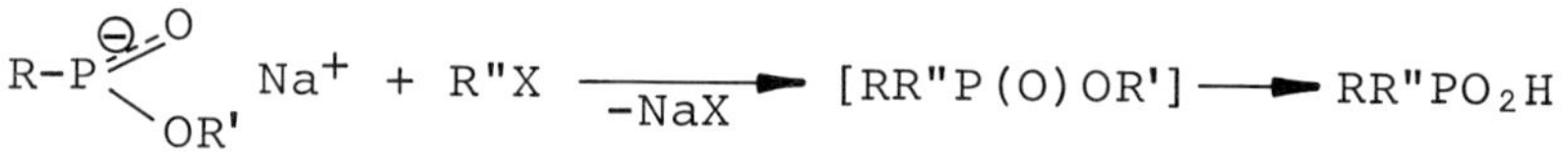

It has sometimes been convenient to hydrolyze a crude ester prepared by the reaction described in the previous section.[409]

VI. ROUTES IN WHICH PHOSPHORUS-CARBON BONDS ARE FORMED BY REACTIONS OF ORGANOMETALLIC COMPOUNDS WITH HALIDES AND OTHER DERIVATIVES OF PHOSPHORUS ACIDS

Halides and esters containing a phosphoryl group or a tervalent phosphorus atom generally react readily with Grignard reagents and other organometallic compounds. Such reactions are equally suitable for the attachment of either aliphatic or aromatic groups to phosphorus, and thus provide routes to a wide range of phosphinic acids.

Control of the substitution so that the product contains only two organic groups attached to phosphorus may be approximate, or may be achieved more precisely by using halides or esters with only one or two replacable groups. Hydrolysis, and oxidation if the phosphorus is not already fully quinquevalent, is necessary to convert the initial product into a phosphinic acid.

VIa. REACTIONS OF ORGANOMETALLIC COMPOUNDS WITH PHOS-PHORUS OXYCHLORIDE OR WITH PHOSPHONIC CHLORIDES, FOLLOWED BY HYDROLYSIS OF THE RESULTING PHOS-PHINIC CHLORIDES

$$POCl_3 + 2RMgX \xrightarrow[-2MgXCl]{} [R_2P(O)Cl] \longrightarrow R_2PO_2H$$

This method was first described soon after the discovery of Grignard reagents (see Chapter 9). Addition of phosphorus oxychloride to solutions of benzyl- and of 1-naphthyl-magnesium halides each afforded a mixture of phosphinic acid and tertiary phosphine oxide,[691] and similar formation of both di- and tri- substituted products is the usual result of other such reactions.[225,410,538,590] Although separation of these mixtures usually presents no problem, the overall yields of phosphinic acids are only moderate.

Factors which have been reported to improve yields are slow reversed addition, i.e., Grignard reagent into a solution of phosphorus oxychloride, allowing gentle boiling of the ether, and fairly dilute reagents so that the final concentration is about 0.2 molar.[406] The effect of different reactant ratios has been investigated,[590] and improved yields have been reported when heterocyclic bases were previously added to the phosphorus oxychloride.[254,410]

Reactions with Grignard reagents containing a p-tosyloxy,[543] or a p-trimethylsilyl substituent,[225] or derived from pyrrole or indoles,[538] with p-dimethylaminophenyl-lithium,[682] and with Ivanov reagents (organomagnesium compounds with an α-carboxylic acid or amide group),[86] have been described as well as those with unsubstituted aliphatic and aromatic Grignard reagents. Reactions of phosphonic dichlorides have been largely confined to examples containing bulky groups: Phosphinic acids were the only products isolated (in moderate yield) from some of these,[143,145] but in others the use of t-alkylmagnesium halides led to predominant formation of secondary phosphine oxides.[101,416]

Reactions of thiophosphoryl and thiophosphonic halides with alkylmagnesium halides often give phosphine sulfides (see Chapter 7) and biphosphine disulfides (see Chapter 2) and do not provide a general route to sulfur-containing acids. Exceptions are the reaction of thiophosphoryl chloride and benzylmagnesium chloride which gives dibenzylphosphinothioic acid as well as the tertiary phosphine sulfide,[753] and reactions of s-butyl- and t-butyl-magnesium halides from which small yields of phosphinodithioic acids have been obtained.[127,272]

VIb. REACTIONS OF ORGANOMETALLIC COMPOUNDS WITH ESTER-CHLORIDES OF PHOSPHORIC OR PHOSPHONIC ACIDS

$$Cl_2P(O)OR' + 2RMgX \longrightarrow R_2P(O)OR' + 2MgXCl$$

Although alkoxy and aryloxy groups attached to phosphorus can be replaced in reactions with organometallic compounds, Grignard reagents discriminate sufficiently well between chlorine atoms and medium-sized alkoxy groups for this reaction to have been used for preparing butyl and isopentyl dialkylphosphinates.[590,696]

A related reaction is that of phenylmagnesium bromide with the cyclic phosphorochloridate of ethane-1,2-diol, in which the Grignard reagent attacks the phosphorus-chlorine bond and one of the two phosphorus-oxygen single bonds to give 2-hydroxyethyl diphenylphosphinate.[177]

VIc. REACTIONS OF ORGANOMETALLIC COMPOUNDS WITH AMIDE-CHLORIDES OF PHOSPHORIC OR PHOSPHONIC ACIDS, USUALLY FOLLOWED BY HYDROLYSIS OF THE RESULTING PHOSPHINIC AMIDE

$$Cl_2P(O)NR'_2 + 2RM \xrightarrow[-2MX]{} R_2P(O)NR'_2 \longrightarrow R_2PO_2H$$

Whereas ester groups are fairly reactive towards organometallic compounds, dialkylamide groups provide a very satisfactory way of limiting substitution in phosphorus chlorides and are easily removed by acid hydrolysis.

The reaction of sodium with a mixture of N,N-diethylphosphoroamidic dichloride and bromobenzene[525] may be regarded as the earliest reaction in this category. All subsequent examples have used preformed organometallic compounds, which have usually been aliphatic[143,408,528] or aromatic[408,528,738] Grignard reagents, although aryllithium compounds[58,59,534,757] have also been used. Difunctional Grignard reagents and organolithium compounds have been used to prepare cyclic phosphinic acids,[414,421] and diphosphinic acids.[58,60]

The blocking groups have most usually been dimethylamino in phosphonamidic chlorides,[58,143] and diethylamino in phosphoramidic dichlorides,[408,738] although piperidino groups were first introduced for this purpose,[528] and di-isopropylamino,[542] anilino,[527] and benzoylimino and substituted benzoylimino groups[160] have also been used.

The phosphinic amides obtained have occasionally been isolated and purified,[160,525,542,757] but more usually have been hydrolyzed directly to phosphinic acids, by heating with hydrochloric acid.[143,408,528]

VId. REACTIONS OF ORGANOMETALLIC COMPOUNDS WITH PHOSPHORUS TRICHLORIDE OR WITH PHOSPHONOUS CHLORIDES, FOLLOWED BY OXIDATION AND HYDROLYSIS OF THE RESULTING PHOSPHINOUS CHLORIDES

$$PCl_3 + 2RM \longrightarrow [R_2PCl] \longrightarrow R_2PO_2H$$

Reactions of this type, giving a crude phosphinous dichloride which is then oxidized and hydrolyzed, as described in Section IVe, have been reported.[86,541,580] Yields are generally fairly low,[119,148,474] and unless the extra reactivity of a tervalent phosphorus halide is necessary in order to promote the reaction with the organometallic reagent the method is not particularly attractive.

VIe. REACTION OF AN ORGANOMETALLIC COMPOUND WITH AN ESTER-CHLORIDE OF A PHOSPHONOUS ACID, FOLLOWED BY ADDITION OF SULFUR TO THE RESULTING PHOSPHINITE ESTER

$$RP(OR')Cl + R''M \longrightarrow [RR''P\cdot OR] \longrightarrow RR''P(S)OR'$$

The only example of this reaction involves an α-keto triethylstannane whose low reactivity makes it sufficiently discriminating to attack the phosphorus-chlorine bond without replacing the ester group.[574]

VIf. REACTION OF AN ORGANOMETALLIC COMPOUND WITH A PHOSPHONAMIDOUS CHLORIDE, FOLLOWED BY OXIDATION AND HYDROLYSIS OF THE RESULTING PHOSPHINOUS AMIDE

$$RP(NR'_2)Cl + R''M \longrightarrow [RR''P\cdot NR'_2] \longrightarrow RR''PO_2H$$

The only example[287] of this type of reaction yielded an acid which has also been satisfactorily prepared from the corresponding phosphonamidic chloride. It seems likely, however, in view of the deactivating effect of an amide group on adjacent phosphorus-chlorine bonds, that the extra reactivity of a tervalent phosphorus chloride could, in other cases, favor this route over that of Section VIc.

VIg. REACTIONS OF ORGANOMETALLIC COMPOUNDS WITH DIALKYL PHOSPHITES OR DIALKYL PHOSPHOROTHIOITES, FOLLOWED BY OXIDATION OR ADDITION OF SULFUR OR SELENIUM

$$(RO)_2P(O)H + 3R'MgX \longrightarrow [R'_2P(O)MgX] \longrightarrow [R'_2P(O)H] \longrightarrow R'_2PO_2H$$

This method, together with similar sequences in which the secondary phosphine oxide is isolated in a pure state and oxidized in a separate stage, provides one of the best general routes to many symmetrical unsubstituted dialkyl- and diaryl- phosphinic acids.

Unsubstituted primary[423,504,591,751] and secondary[145,488,504] and aryl-[422,600] magnesium halides, and also 3,3,3-trifluoropropyl magnesium chloride[452] have been used. The requirement for at least three equivalents of Grignard reagent could be disadvantageous but can be circumvented by using the sodium derivative of the dialkyl phosphite.[591,600]

Oxidation of the crude secondary phosphine oxide is effected with hydrogen peroxide,[423,591,755a] or otherwise (see Sections IIa and IIb), while sulfur[452,504] or selenium[490] are added by heating in an inert solvent under reflux.

Reactions of dialkyl phosphorothioites (which are much less readily available than dialkyl phosphites) with Grignard reagents and subsequent treatment with sulfur give phosphinodithioic acids in some cases.[505]

VIh. REACTIONS OF ORGANOMETALLIC COMPOUNDS WITH ANHYDRIDES OF PHOSPHONIC OR PHOSPHONOTHIOIC ACIDS

$$(RPO_2)_n + R'MgX \longrightarrow RR'PO_2H$$

$$(RPOS)_2 + R'MgX \longrightarrow RR'POSH$$

These reactions[182] are slow (2 hr at 100°) and give poor yields in comparison with most of the others in this group. They are of interest, but not of synthetic value.

VII. REACTIONS OF PHOSPHORUS TRIHALIDES AND OF PHOSPHONOUS DICHLORIDES WITH ANHYDROUS DIAZONIUM SALTS

$$RPCl_2 + ArN_2^+ BF_4^- \longrightarrow [ArRP^+Cl_2 \; BF_4^-] \xrightarrow{H_2O} ArRPO_2H$$

Reactions of dry aryldiazonium fluoroborates with phosphorus trichloride in ethyl acetate in presence of cuprous chloride, followed by addition of water and removal of the solvent by steam distillation, yield arylphosphonic acids together with small and variable amounts of symmetrical diarylphosphinic acids.[166] Isopropyl acetate, butyl acetate, and dioxan may also be used as reaction solvents, and copper bronze and cuprous bromide are alternative catalysts. Diazonium fluorosilicates give less vigorous reactions but similar yields.[210]

Yields of phosphinic acids are improved by using phosphorus tribromide and cuprous bromide,[168,214,221] but increasing the proportion of diazonium salt to phosphorus trichloride was not markedly effective.[157] Generally better, but still very variable, yields of unsymmetrical diarylphosphinic acids,[209,290,484] and alkylarylphosphinic acids,[209,495,496] are obtained by using aryl- or alkylphosphonous dichlorides, and this method has been preferred over the dialkyl phosphite/Grignard reagent route (Section VIg) for the preparation of diphenylphosphinic acid.[482]

The particular value of this reaction lies in the wide variety of substituted aromatic phosphinic acids, of known orientation, to which it can lead. There are many examples with alkyl, alkoxy, halogen, and nitro substituents in benzene rings, and it has also been used for the preparation of (substituted biphenylyl)phenylphosphinic acids.[119]

VIII. ROUTES IN WHICH PHOSPHORUS-CARBON BONDS ARE FORMED BY ELECTROPHILIC AROMATIC SUBSTITUTION REACTIONS

VIIIa. REACTIONS BETWEEN PHOSPHORUS OXYCHLORIDE AND VERY REACTIVE AROMATIC COMPOUNDS

$$ArH + POCl_3 \longrightarrow [Ar_2P(O)Cl] \longrightarrow Ar_2PO_2H$$

Reactions of phosphorus trichloride with aromatic compounds can be assisted by aluminum chloride, but this is not possible with phosphorus oxychloride which forms a stable non-ionic aluminum chloride complex. Aromatic compounds therefore require to be highly activated towards electrophilic attack in order to react with phosphorus oxychloride, and such reactions have been described only with dimethylaniline and diethylaniline giving moderate yields, after hydrolytic working up, of the corresponding di-(p-dialkamino-phenyl)phosphinic acids.[100,682]

VIIIb. FRIEDEL-CRAFTS REACTIONS OF PHOSPHORUS TRICHLORIDE, FOLLOWED BY OXIDATION AND HYDROLYSIS OR ALCOHOLYSIS OF THE RESULTING DIARYLPHOSPHINOUS CHLORIDES

$$ArH + PCl_3 + AlCl_3 \longrightarrow [Ar_2PCl\cdot AlCl_3] \longrightarrow Ar_2PO_2H$$

$$[Ar_2PCl\cdot AlCl_3] \downarrow [Ar_2PCl_3] \longrightarrow Ar_2P(O)OEt$$

Diarylphosphinous chlorides, formed by prolonged reactions between aromatic compounds, phosphorus trichloride, and aluminum chloride, cannot be extracted, because of complex formation with the aluminum chloride, but may be converted into diarylphosphinic acids by hydrolysis and atmospheric oxidation,[522] or into their ethyl esters by chlorination and treatment of the crude trichlorophosphorane with ethanol.[407,411,413,417]

Aluminum chloride induced isomerization of the aryl groups[417] limits the value of this method, but it has been used to prepare cyclic phosphinic acids from diaryl ethers and thioethers,[216,219,458] shorter reaction times being sufficient in these cases.

VIIIc.   ALUMINUM CHLORIDE PROMOTED REACTIONS OF PHOS-
PHORUS PENTASULFIDE OR OF PHOSPHONODITHIOIC
ANHYDRIDES WITH AROMATIC HYDROCARBONS

$$P_4S_{10} + 8ArH \longrightarrow 4Ar_2PS_2H + 2H_2S$$

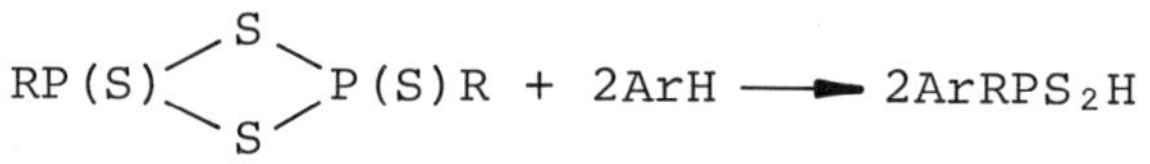

$$RP(S) \begin{array}{c} S \\ \diagup \diagdown \\ \diagdown \diagup \\ S \end{array} P(S)R + 2ArH \longrightarrow 2ArRPS_2H$$

The reaction of phosphorus pentasulfide with benzene
and aluminum chloride provides an excellent route to diphenylphosphinodithioic acid,[289] and hence to many other
diphenylphosphinic and diphenylphosphinous compounds.[755b]
The isomerizing effect of aluminum chloride discourages
the use of other aromatic compounds in this reaction,
although toluene has been reported to give a crystalline
phosphinodithioic acid.[531]

The reaction of phosphonodithioic anhydrides, prepared
from phosphonothioic dichlorides and hydrogen sulfide,
with benzene and aluminum chloride gives 70-90% yields of
phenylalkylphosphinodithioic acids.[563]

IX.   REACTIONS OF ACIDS AND ESTERS CONTAINING PHOS-
PHORUS-HYDROGEN BONDS AND OF PHOSPHONOUS DI-
HALIDES, WITH CARBONYL AND IMINO GROUPS AND WITH
ORTHOCARBOXYLIC ESTERS

This group of reactions involves nucleophilic attack
on a carbonyl or similar group, by a phosphorus atom which
is tervalent or pseudo-tervalent [i.e., in the group
$>PH(O)$ which is potentially tautomeric with $>P \cdot OH$].   The
products contain a new phosphorus-carbon bond with a
hydroxyl or other substituent on the α-carbon atom.

IXa.   REACTIONS OF HYPOPHOSPHOROUS ACID OR PHOSPHON-
OUS ACIDS WITH ALDEHYDES OR KETONES

$$H_3PO_2 + 2RR'CO \longrightarrow [RR'C(OH)]_2PO_2H$$

$$\begin{array}{c} R \\ \diagdown \\ \diagup \\ H \end{array} PO_2H + R'R''CO \longrightarrow \begin{array}{c} R \\ \diagdown \\ R'R''C \diagup \\ \diagdown OH \end{array} PO_2H$$

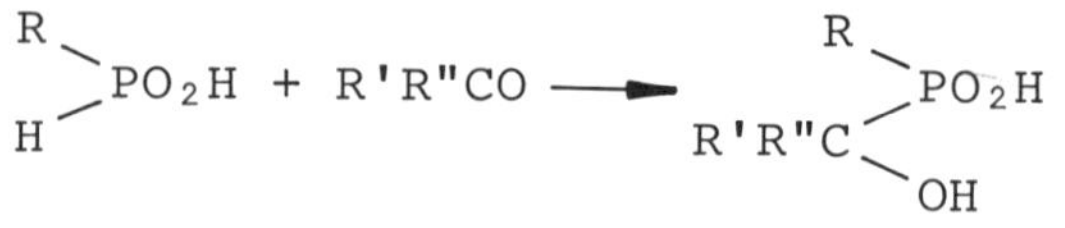

Reactions of this type were carried out before and
soon after the beginning of this century,[487,615] by heating hypophosphorous acid with aldehydes or ketones on a

waterbath for several days. Subsequent examples have used essentially the same procedure except for some shortening of the reaction time. The kinetics of reactions of hypophosphorous acid with formaldehyde and with benzaldehyde have been investigated.[296]

In some cases α-hydroxyalkylphosphonous acids[487,573] (from addition of one of the P-H bonds of hypophosphorous acid to a carbonyl compound), or an α-aminoalkylphosphonous acid[433] (from hypophosphorous acid and a Schiff's base), have been isolated and then treated with a different carbonyl compound, and reactions between aldehydes and unsubstituted phosphonous acids have also been described.[594,640]

Addition of arylphosphonous acids to the ketonic carbonyl group of methyl and ethyl pyruvate,[615] gave crystalline phosphinic acids which were isomerized to undistillable mono-esters of phosphonic acids, on heating to 150-180°.

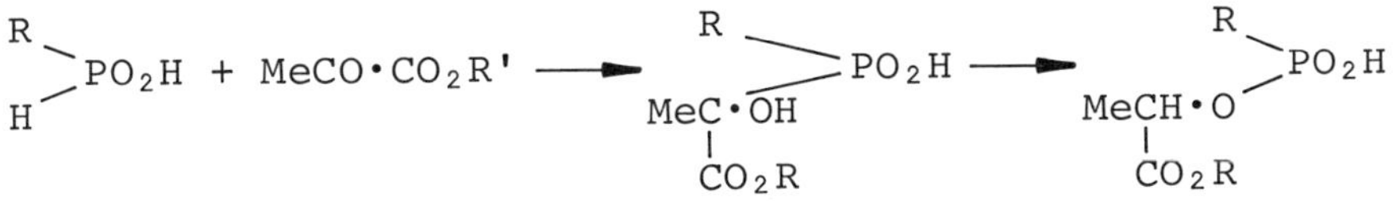

IXb. REACTIONS OF HYPOPHOSPHOROUS ACID OR PHOSPHONOUS ACIDS WITH AZOMETHINES OR RELATED COMPOUNDS

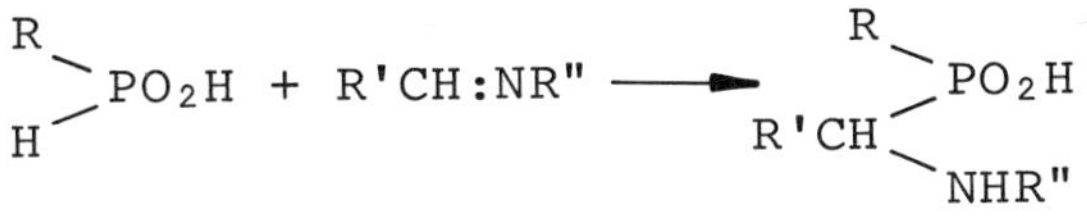

Reactions as formulated above have been carried out between phenylphosphonous acid and several Schiff's bases during 8 hr in boiling benzene.[640] Related reactions are those of a phosphonous acid with aryl isocyanates,[205] and with bis(diethylamino)methane,[594]

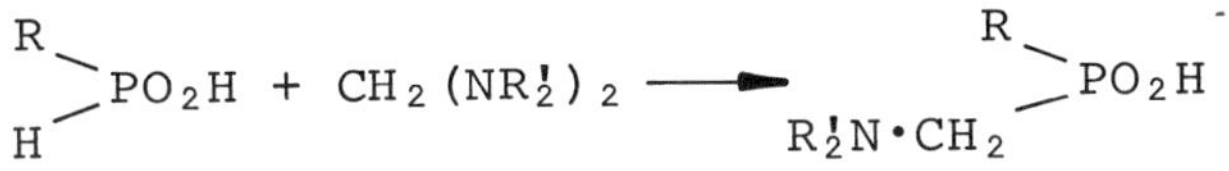

and the reaction of hypophosphorous acid with formaldehyde and secondary amines, which in presence of hydrochloric acid involves both P-H bonds.[477]

$$H_3PO_2 + 2CH_2O + 2R_2NH \longrightarrow (R_2N\cdot CH_2)_2PO_2H + 2H_2O$$

IXc. REACTIONS OF ALKYL HYPOPHOSPHITES, OR OF MONO-ALKYL PHOSPHONITES OR PHOSPHONOTHIOITES WITH ALDEHYDES OR KETONES

$$H_2P(O)OR + 2R'CHO \longrightarrow [R'CH(OH)]_2P(O)OR$$

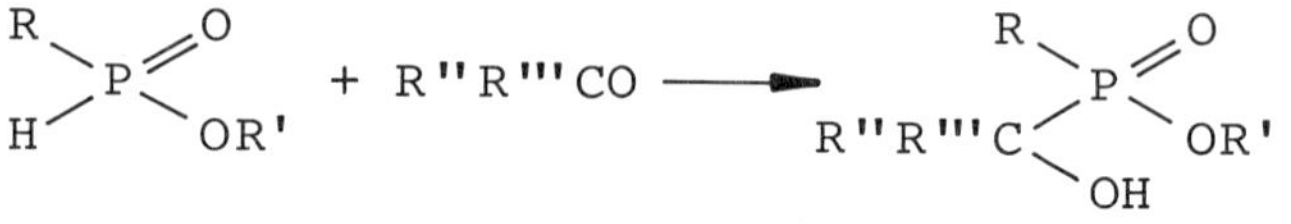

Reactions between monoalkyl phosphonites and unsubstituted aldehydes or ketones occur during several days at room temperature or more rapidly on gentle heating,[2,7,8] either alone or with a trace of a sodium alkoxide added as a catalyst.[99,609,610,612] Reactions with chloral,[3,61,70] and with a ketone containing inductively electron-attracting groups[793] are initially exothermic and are heated for completion.

Some products from unsubstituted alkyl aryl ketones,[612] but more particularly those from α-keto-esters,[618] and from other compounds with a mesomerically electron-attracting group adjacent to the reacting carbonyl group,[613,619,629] are rearranged to phosphonates if heated to 110-120°:

$$RP(O)(OR')\cdot CR''R'''\cdot OH \longrightarrow RP(O)(OR')\cdot O\cdot CHR''R'''$$

The rearrangement, the kinetics of which have been investigated,[614] is catalyzed by sodium alkoxides,[613,629] and in some cases the rearranged product is obtained directly if sodium alkoxide is present.[611]

Additions of phosphonites to vinyl acetate are catalyzed by sodium alkoxides and occur exothermically. As reactions of a derivative of acetaldehyde, they are probably best considered here rather than with reactions of other unsaturated compounds in Section IXf. The products were originally formulated as β-acetoxyethylphosphinates,[632] but an α-acetoxyphosphinate structure was later suggested.[628]

Reactions of unsubstituted[4,530] and substituted[618] phosphonothioites, with aldehydes and ketones, proceed similarly to those of phosphonites. The rather difficultly accessible alkyl hypophosphites react with chloral in petrol at 50° to give fairly good yields of esters of bis(2,2,2-trichloro-1-hydroxyethyl)phosphinic acid.[325]

IXd. REACTIONS OF MONOALKYL PHOSPHONITES WITH AZO-
METHINES, AND RELATED REACTIONS

$$\underset{H}{\overset{R}{>}}P\overset{\displaystyle O}{\underset{OR'}{\diagup}} + R''CH:NR''' \longrightarrow \underset{R''CH}{\overset{R}{>}}P\overset{\displaystyle O}{\underset{\substack{OR' \\ NHR'''}}{}}$$

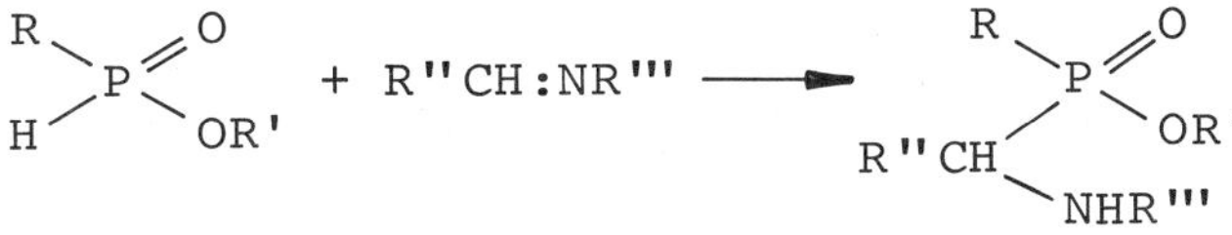

Monoalkyl phosphonites react with Schiff's bases on mixing at room temperature and addition of a catalytic amount of a sodium alkoxide.[604,636,769] Other preparations of phosphinic esters by addition of phosphorus-hydrogen bonds across carbon-nitrogen double bonds are the reaction of butyl ethylphosphonite with ethyl isocyanate,[632] and reactions of several S-alkyl alkylphosphono-(phenylhydrazono)thioites, RPH(:N·NPh)SR', with Schiff's bases.[123]

Reactions of monoalkyl phosphonites with aldehydes and ketones, in presence of ammonia, at 100°, afford phosphinates containing α-(primary amino) alkyl groups,[346] whilst a similar reaction between ethyl phenylphosphonite, acetone, and ethylenediamine gave the expected ethylene-bis(α-aminoalkylphosphinate).[347]

Reactions of esters containing phosphorus-hydrogen bonds with dimethylamine derivatives of formaldehyde, include those of alkyl hypophosphites with bis(diethyl-amino)methane and with diethylaminomethoxymethane,[324] and of several O-alkyl phosphonothioites with bis(diethyl-amino)methane.[536,589]

The preparation of several esters of α-aminoalkyl-phosphinic acids by reactions of three-covalent ester-amides or ester-hydrazides of phosphonous acids with aldehydes or ketones is included in this section because of the similarity of the products to those discussed above. The reactions, which bear a formal similarity to the formation of an α-chloroalkylphosphinic chloride from a phosphonous dichloride and an aldehyde or ketone are:

PhP(OPr)NEt$_2$ + RCHO $\longrightarrow$ $\underset{RCH}{\overset{Ph}{>}}P\overset{\displaystyle O}{\underset{\substack{OPr \\ NEt_2}}{}}$ (ref. 537)

EtP(SR)NH·NHPh + R'R''CO $\longrightarrow$ $\underset{R'R''C}{\overset{Et}{>}}P\overset{\displaystyle O}{\underset{\substack{SR \\ NH·NHPh}}{}}$ (ref. 5)

[R,R' = alkyl; R'' = alkyl or H]

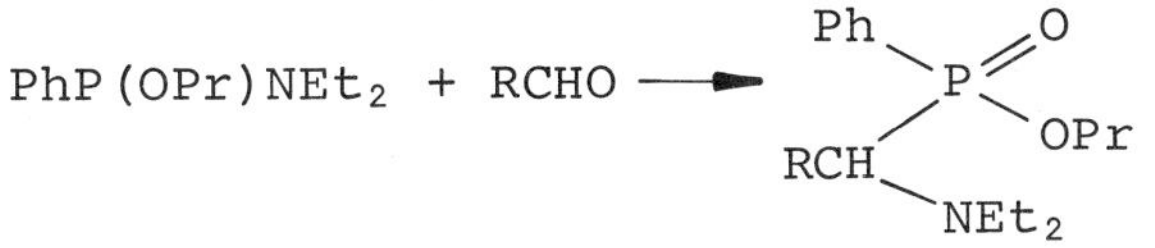

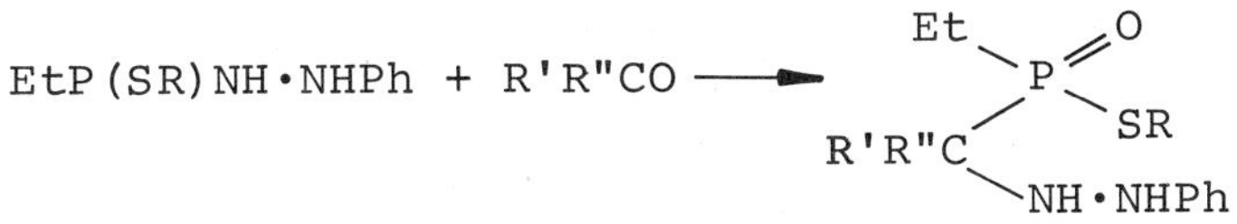

The second reaction could involve a prior conversion to an S-alkyl ethylphosphonothioite and the phenylhydrazone of the carbonyl compound, which then react by addition of the newly formed phosphorus-hydrogen bond across the new carbon-nitrogen double bond. This possibility is not available for the first reaction which could perhaps best be tentatively formulated with a five-covalent intermediate.

IXe.   REACTIONS OF MONOALKYL PHOSPHONITES WITH ORTHO-
        CARBOXYLIC ESTERS

$$\underset{H}{\overset{R}{>}}P\overset{O}{\underset{OR'}{=}} + R''C(OEt)_3 \longrightarrow (EtO)_2CR'' \underset{OR'}{\overset{R}{>}}P\overset{O}{=}$$

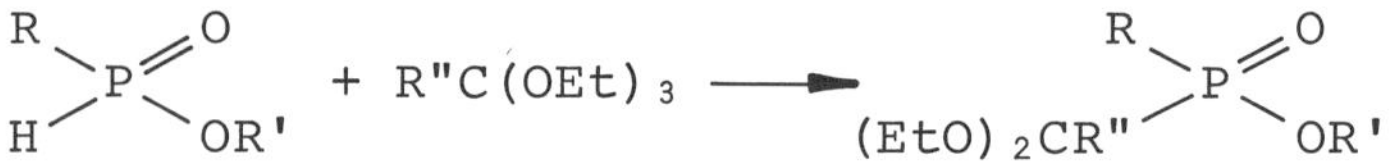

Acetals of α-aldehydophosphinates are obtained, as formulated above (R" = H) when monoalkyl phosphonites are heated with trialkyl orthoformates to 180°.[660]

Analogously, heating monoalkyl phosphonites with triethyl orthoacetate to 180°,[544] or with triethylamine and ketene diethyl acetal to 100°,[256] yielded acetals of α-ketophosphonates (R" = Me). These eliminated ethanol, in one case during heating to 180°, and in another case only when sodium ethoxide was present during heating, to give the corresponding enol ethers.[544]

IXf.   REACTIONS OF MONOALKYL PHOSPHONITES AND PHOS-
        PHONOTHIOITES WITH α,β-UNSATURATED ESTERS,
        NITRILES, AND RELATED COMPOUNDS

$$RPH(O)OR' + \underset{}{>}C=C-C\overset{O}{<} \longrightarrow -\underset{\underset{RP(O)OR'}{|}}{C}-CH-C\overset{O}{<}$$

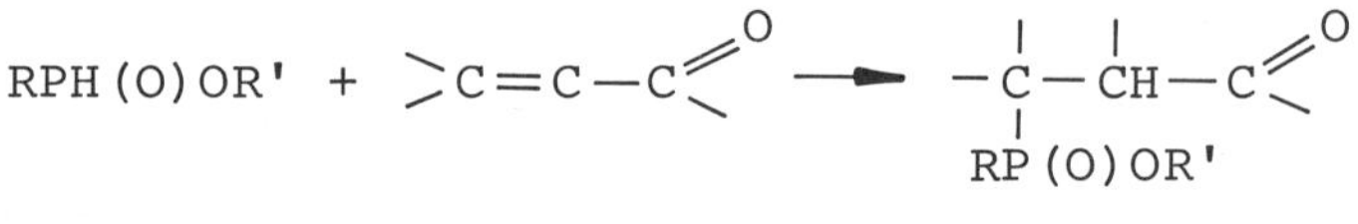

Carbonyl groups, and others which attract electrons mesomerically, activate carbon-carbon multiple bonds, conjugated with them, for nucleophilic addition of phosphonites. Such reactions, all catalyzed by sodium alkoxides, have been observed between monoalkyl phosphonites and carbon-carbon double bonds conjugated with ketone,[637] ester,[90,636,641,642] nitrile,[90,635] sulfone,[639] and 2-pyridyl[492] groups, and between a monoalkyl phosphonothioite and acrylonitrile.[530] Monoaddition of ethyl ethylphosphonite occurs to the triple bond of ethyl phenylpropiolate,[642] but it is noteworthy that reaction of methyl and ethyl ethylphosphonites with propynal occurs at the carbonyl group rather than at the triple bond,[609] and α,β-olefinic aldehydes would probably react similarly.

IXg. REACTIONS OF PHOSPHONOUS DIHALIDES WITH ALDEHYDES OR KETONES, AND SUBSEQUENT HYDROLYSIS

$$RPCl_2 + R'R''CO \longrightarrow (R'R''CCl)RP(O)Cl \longrightarrow$$

$$(R'R''CCl)RPO_2H \text{ or } [R'R''C(OH)]RPO_2H$$

Crude α-chloroalkylphosphinic chlorides, obtained from reactions of phosphonous dichlorides and unconjugated aldehydes or ketones, have been hydrolyzed to phosphinic acids with or without accompanying hydrolysis of the α-chloro substituent.

Examples include reactions of phenylphosphonous dichloride with acetaldehyde,[521] benzaldehyde,[521] and o-chlorobenzaldehyde,[224] from which α-hydroxyphosphinic acids were obtained, and those with cyclohexanone,[224] and with benzophenone,[222] which gave α-chloro acids.

Since phosphinic anhydrides can be prepared by heating phosphinic halides with paraformaldehyde (Section XXIg), reactions of phosphonous dihalides with paraformaldehyde may give anhydrides of halomethylphosphinic acids as well, or instead of, phosphinic halides.

$$RPX_2 + (CH_2O)_n \longrightarrow (XCH_2)RP(O)X \longrightarrow$$

$$(XCH_2)RP(O)\cdot O\cdot P(O)R(CH_2X)$$

IXh. REACTION OF A PHOSPHONOUS DICHLORIDE WITH AN ORTHOCARBOXYLIC ESTER[162]

$$PhPCl_2 + CH(OEt)_3 \longrightarrow \begin{array}{c} Ph \\ (EtO)_2CH \end{array} \begin{array}{c} O \\ \diagdown P \diagup \\ \diagup \quad \diagdown \\ OEt \end{array}$$

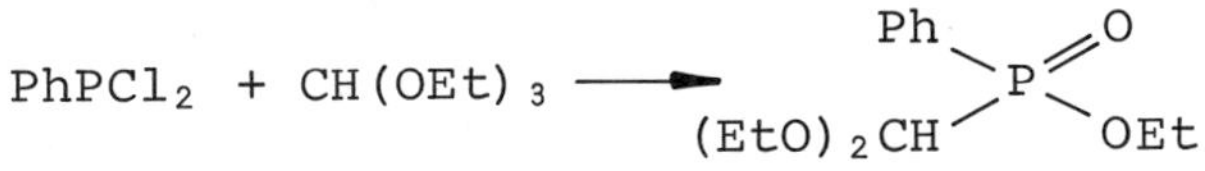

X. REACTIONS OF HALIDES AND ESTERS OF PHOSPHOROUS AND PHOSPHONOUS ACIDS WITH CONJUGATED DIENES, α,β-UNSATURATED KETONES, ACIDS, AND AMIDES, AND α-DIKETONES

Derivatives of phosphinic acids are obtained by reactions of several types of tervalent phosphorus compound with compounds containing appropriate arrangements of conjugated double bonds. Many of these, giving 5-membered cyclic products, are clearly cheletropic 1,4-cycloadditions, and similarly formed cyclic intermediates are also very probable, as was suggested more than half a century ago,[137] in those reactions which yield acyclic products.

Although relatively few examples of the reactions in Sections Xb to Xe have been described, these reactions, and analogous preparations in some cases of similarly substituted phosphinic chlorides, have considerable potential value in that they provide routes to phosphinic acids which also contain other synthetically useful functional groups.

Xa. REACTIONS OF ETHYLENE PHOSPHOROHALIDITES, OR RELATED COMPOUNDS, WITH CONJUGATED DIENES

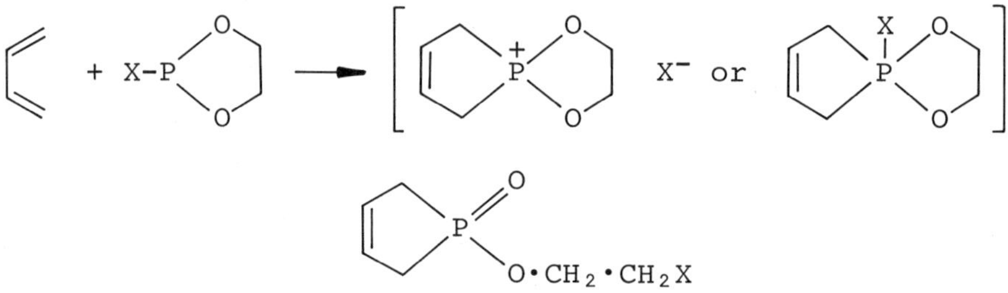

Cheletropic reactions of phosphorus trihalides with 1,3-dienes afford crystalline cyclic trihalophosphoranes, from which 5-membered cyclic phosphinic acids and their derivatives are readily prepared.[278,302]

Similar reactions of diesters of phosphorous acid, in which the third bond from phosphorus is joined to a fluorine,[670] chlorine,[44,45,425,672,673,674] or bromine,[781] atom, or to a thiocyanate,[185] alkoxy,[184,426] alkylthio,[186] or acyloxy[188] group, yield esters of cyclic phosphinic acids. If, as formulated above, the diester is derived from a glycol (or a mono or dithio analog), which has mostly been the case because of the greater ease of preparation of these cyclic esters, the third group originally bonded to phosphorus usually becomes attached to the other end of the alkylene group in the product. An exception to this is provided by reactions of O,S-ethylene O-methyl phosphorothioite,[426] which yield methyl phosphinates with elimination of ethylene sulfide or its equivalent. The cyclic phosphorochloridites, etc., have mostly been derived from 1,2-diols, 1-hydroxy-2-thiols,[426] or 1,2-dithiols,[425,673] with a few from 1,3-diols.[674,781] Diethyl phosphorofluoridite is the only acyclic aliphatic ester whose use in this reaction has been reported.[670]

All the reactions with aliphatic esters have been carried out by heating in sealed tubes at temperatures ranging between 80° and 150° for periods of from 3 to 50 hr. Diaryl phosphorochloridites and aryl phosphorodichloridites, for which hydrolysis must occur during working up in order to yield phosphinates, react very slowly,[43] but o-phenylene phosphorohalidites give crystalline phosphorane-type adducts after one day at room temperature and heating with water is necessary to convert these into phosphinate esters.[669]

Xb. REACTIONS OF PHOSPHONOUS DICHLORIDES WITH α,β-UNSATURATED KETONES

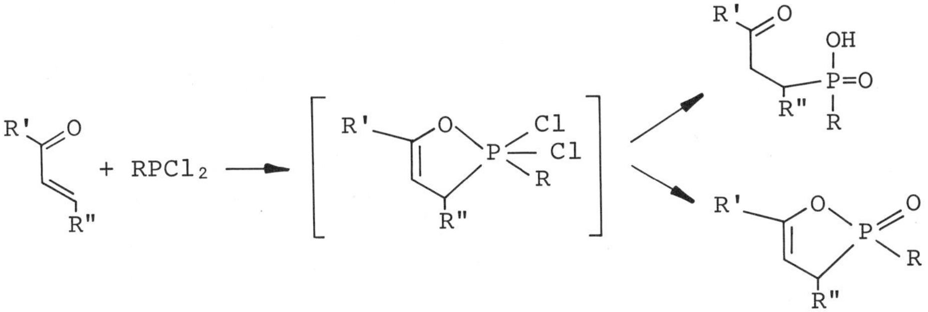

There are several variants of this reaction. In the earliest of these, mesityl oxide was produced by addition of phosphorus pentoxide to a mixture of acetone and a phosphonous dichloride, with which it reacted, and after hydrolysis yielded the phosphinic acid (wrongly formulated with the phosphorus α to the carbonyl group).[520] Similar reactions have been carried out, without phosphorus pentoxide but heating the mixture to 80-100° for 10-20 hr, and have yielded cyclic enol esters.[577]

More usually, an unsaturated ketone (mesityl oxide benzalacetophenone or cinnamalacetophenone) and a phosphonous dichloride have been mixed in acetic acid to give a phosphinic acid,[136,137] or in acetic anhydride to give a cyclic enol ester which can be isolated by distillation,[33,73] or hydrolyzed to the acid.[173]

A related reaction is that between diethyl phenylphosphonite and acrylaldehyde to give the ethyl enol ether of ethyl(3-oxopropyl) phenylphosphinate.[361]

Xc. REACTIONS OF PHOSPHONOUS DICHLORIDES, OR PHOSPHONOCHLORIDITE, PHOSPHONOCHLORIDOTHIOITE, OR PHOSPHONITE ESTERS, WITH α,β-UNSATURATED CARBOXYLIC ACIDS

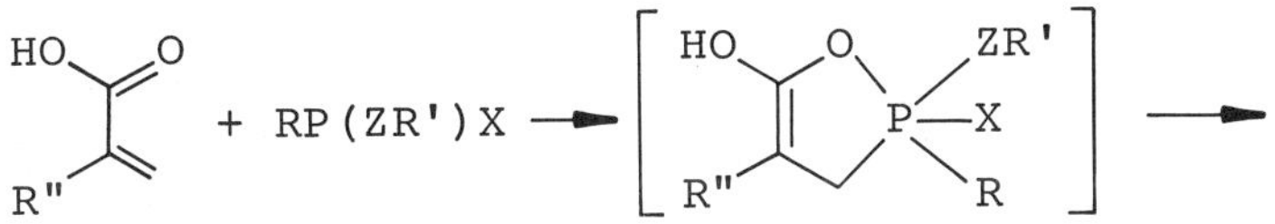

[Z, X = O, Cl; S, Cl; or O, OR']

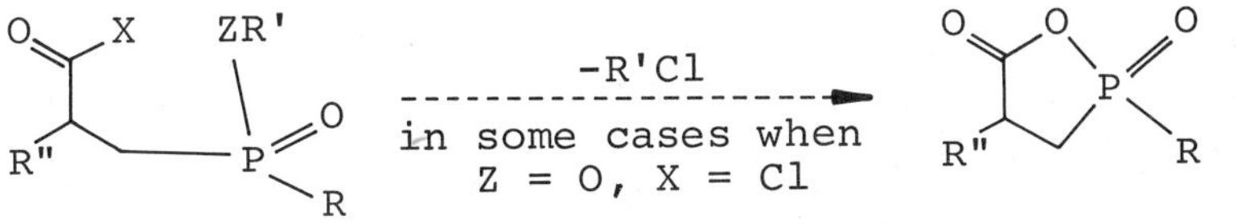

Reactions of phosphonous dichlorides with α,β-unsaturated acids yield di(acid chlorides) of β-carboxyphosphinic acids.[369,370,371,379] These dichlorides are easily converted by acetic anhydride into cyclic phosphinic-carboxylic anhydrides which are also sometimes isolated from the above reactions as well[378] or instead[380] of dichlorides, presumably by the action on these of excess of the carboxylic acid.

In analogous reactions, formulated above, alkyl phosphonochloridites and α,β-unsaturated acids either yield cyclic phosphinic-carboxylic anhydrides,[383] or if the esterifying groups are sufficiently chlorine-substituted to resist their elimination by reaction with the carboxylic chloride group, (C-acid chlorides P-esters) of β-carboxyphosphinic acids may be obtained.[386,624] These chloride-esters are converted into cyclic anhydrides by further heating,[386,773] but this does not happen to (C-acid chlorides) of S-alkyl β-carboxyphosphinothioates which are formed in high yields from alkyl phosphonochloridothionites and α,β-unsaturated acids.[12]

Dialkyl phenylphosphonites react exothermically with α,β-unsaturated acids to give diesters of (β-carboxyalkyl)phenylphosphinic acids.[361]

Xd. REACTIONS OF PHOSPHONOCHLORIDITE ESTERS WITH α,β-UNSATURATED CARBOXYLIC AMIDES

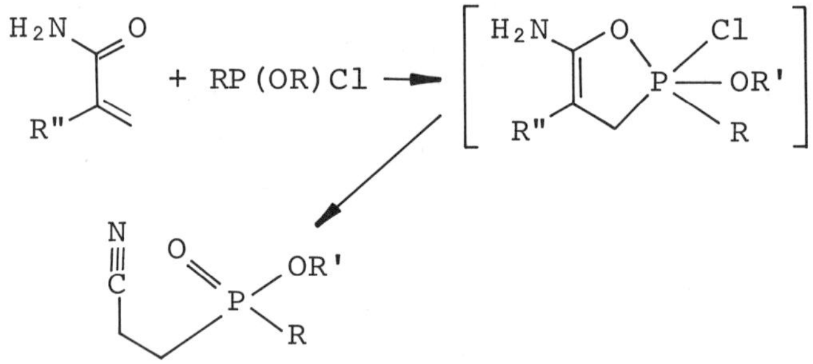

Reactions of acrylamide and methacrylamide with ethyl, phenyl-, and p-tolyl-phosphonochloridites, in which the esterifying groups are chlorine-substituted and therefore not readily lost, are exothermic and, after heating to complete the reaction, yield esters of (β-cyanoalkyl) phosphinic acids.[384,625] Similar reactions of phosphonous dichlorides give (β-cyanoalkyl)phosphinic chlorides.[372,374,377,382,620,621]

Xe. REACTIONS OF PHOSPHONOUS DICHLORIDES WITH α-DIKETONES

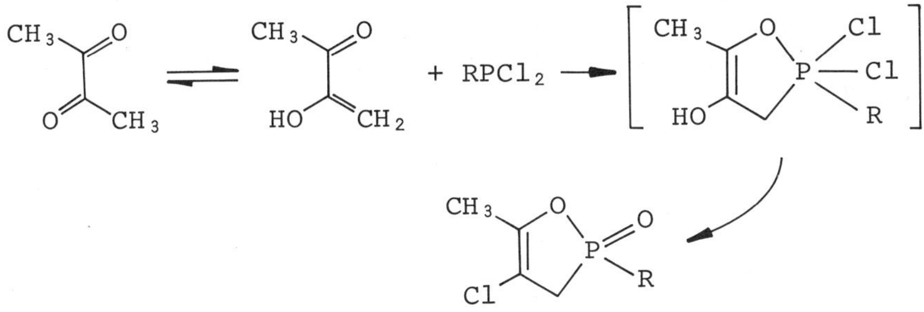

Reactions of 2,3-butanedione with methyl-, ethyl-, and phenyl-phosphonous dichlorides constitute the only published examples of this reaction. It is exothermic on mixing and, after standing overnight before distillation, affords poor yields of cyclic enol esters of (2-chloro-3-oxobutyl)phosphinic acids.[622,623]

XI. ROUTES INVOLVING CLEAVAGE OF PHOSPHORUS-CARBON BONDS

XIa. REACTIONS OF TERTIARY PHOSPHINE OXIDES WITH HYDROXIDES

$$RR'R''PO + OH^- \longrightarrow R'R''PO_2H + R^-$$

(where R^- is the most stable carbanion)

Good yields of phosphinic acids are obtained by fusion, at 200-300°, of tertiary phosphine oxides with sodium or potassium hydroxides.[300,491,797] When there is a choice of leaving groups, the most electronegative one is preferentially expelled and a single phosphinic acid, rather than a mixture is obtained.[189,300]

Boiling aqueous alkali is sufficient for removal of one 2-thienyl,[491] phenylethynyl,[199] or hydroxymethyl[291] group from their tertiary phosphine oxides, whilst one trifluoromethyl group is removed when tertiary phosphine oxides (or their dihalophosphorane precursors) containing two or three such groups react with water or an amine at room temperature[63,110,178] (see also Chapter 6).

XIb. REACTIONS OF TERTIARY PHOSPHINE OXIDES WITH SODIUM HYDRIDE, OR OF TERTIARY PHOSPHINES WITH POTASSIUM, AND SUBSEQUENT OXIDATION

$$R_3PO + NaH \longrightarrow [R_2P(O)H] \longrightarrow R_2PO_2H$$

$$R_3P + K \longrightarrow [R_2PH] \longrightarrow R_2PO_2H$$

Cleavage of tertiary phosphine oxides by fusion with sodium hydride at 160° follows a similar pattern to the sodium hydroxide reaction discussed in the previous section, with benzyl groups removed in preference to phenyl groups, and alkyl groups not removed.[297]

The reactions of tertiary phosphines with potassium were carried out during 8 hr in refluxing dioxan.[319] Those of type $R_2R'P$ gave single phosphinic acids but the choice of group removed was less regular than from tertiary phosphine oxides (see also Chapter 1).

XIc. AUTOXIDATION OF TERTIARY PHOSPHINES

$$R_3P \longrightarrow R_3PO + R_2P(O)OR$$

Passage of air or oxygen through 10% solutions of tributyl- or tricyclohexyl-phosphine in hexane gave roughly equal amounts of tertiary phosphine oxide and of phosphinate.[102]

XII. FREE RADICAL ADDITIONS OF ACIDS AND ESTERS CONTAINING PHOSPHORUS-HYDROGEN BONDS TO UNSATURATED HYDROCARBONS

XIIa. ADDITION OF HYPOPHOSPHOROUS ACID TO OLEFINS

$$H_2PO_2H + 2R\cdot CH{:}CH_2 \longrightarrow (R\cdot CH_2\cdot CH_2)_2PO_2H$$

Peroxide initiated reactions of terminal olefins with 50% aqueous hypophosphorus acid at 135°,[789] or in alcohol under reflux,[585] for 12-24 hr, or with sodium hypophosphite in methanol,[594] give di-n-alkylphosphinic acids. Dicyclohexylphosphinic acid is similarly obtained from cyclohexene,[584,594] and addition of phenylphosphonous acid to cyclohexene has also been reported.[584]

XIIb. ADDITION OF MONO-ESTERS OF PHOSPHONOUS AND PHOSPHONOTHIOUS ACIDS TO OLEFINS AND ACETYLENES

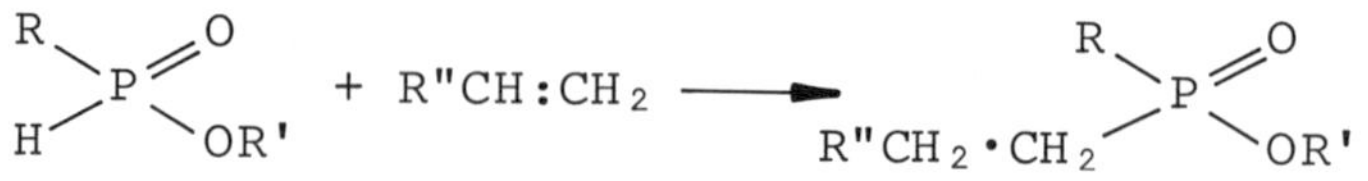

Reactions of this type, initiated by dibenzoyl peroxide or ultraviolet radiation, have been carried out at 80-90° with cyclohexene and with several acyclic terminal olefins,[627] and at 120-130° with hexa-1,5-diene which gave both mono- and di-addition, and with butyl vinyl ether.[631]

Reactions with acetylenes result in mono-addition at 80-95°,[630] and in di-addition (with one phosphinic acid group attached to each of the initially acetylenic carbon atoms) at 140-180°.[638]

XIII. PHOTOCHEMICAL REACTION OF PHOSPHORUS-HYDROGEN BONDS IN PRESENCE OF CHLORINE, OR OF PHOSPHORUS-CHLORINE BONDS, WITH A SATURATED HYDROCARBON

$$RPH(O)OR' + R''H + Cl_2 \longrightarrow RR''P(O)OR'$$

$$Cl_2P(O)OR + 2R'H \longrightarrow R'_2P(O)OR$$

$$RPCl(O)OR' + R''H \longrightarrow RR''P(O)OR'$$

Reactions have been carried out between cyclohexane, several monoesters of phosphonous acids and chlorine (high-pressure mercury lamp), and also between cyclohexane and phosphorodichloridate or phosphonochloridate esters (deuterium lamp).[545] Yields of phosphinates averaged over 60%, but less symmetrical hydrocarbons presumably would give mixtures of isomeric products.

B. METHODS FOR INTERCONVERSION OF PHOSPHINIC ACIDS, HALIDES AND OTHER FUNCTIONAL DERIVATIVES, AND THEIR THIO, SELENO, AND IMIDO ANALOGS

XX. REACTIONS OF TRIHALOPHOSPHORANES

XXa. HYDROLYSIS OF TRIHALOPHOSPHORANES

$$R_2PX_3 + 2H_2O \longrightarrow R_2PO_2H + 3HX$$

Trihalophosphoranes are readily and very exothermically hydrolyzed to phosphinic acids, either in one stage or using sulfur dioxide to give the intermediate phosphinic chloride, which can be purified before hydrolyzing further.[196] The trihalophosphoranes may be obtained in various ways, e.g., by reaction of phosphorus pentachloride or of tetrachlorophosphoranes with olefins,[196] or by chlorination of phosphinous dichlorides,[519] phosphinothioic chlorides,[452] phosphinodithioic acids,[755b] or tetra-alkylbiphosphine disulfides[436] (see also Chapter 5B). One important group are the unsaturated cyclic trihalophosphoranes, which are obtained by cheletropic reactions of conjugated dienes with phosphorus trichloride or phosphorus tribromide. Hydrolyses and other reactions of these usually involve migration of the double bond from the 3,4- to the 2,3- position if the halogen is chlorine, but no rearrangement if it is bromine.[52,278]

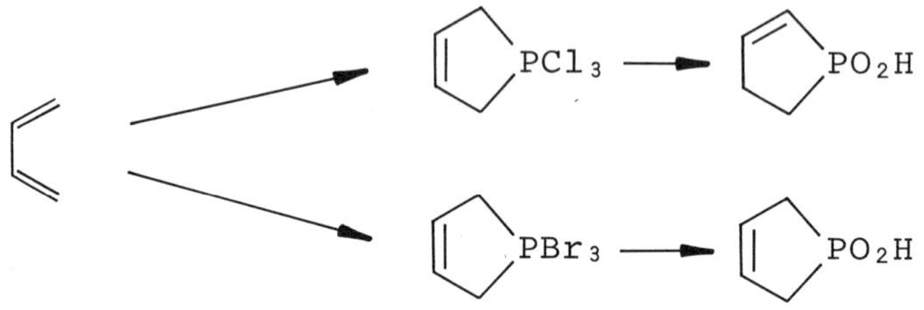

XXb. ALCOHOLYSIS OF TRIHALOPHOSPHORANES

$$R_2PX_3 + 2R'OH \longrightarrow R_2P(O)OR' + R'X + 2HX$$

The most important of these reactions, as of the hydrolyses described in the previous section, involve cyclic unsaturated trihalophosphoranes and, like the hydrolyses, result in double-bond migration in reactions of trichlorophosphoranes but not in those of tribromophosphoranes.[52,302]

Another group of alcoholyses are of crude trichlorophosphoranes obtained by chlorination of aluminum chloride complexes of diarylphosphinous chlorides, from aluminum chloride assisted reactions of phosphorus trichloride with aromatic compounds.[407,417]

$$Ar_2PCl \cdot AlCl_3 \longrightarrow Ar_2PCl_3 \longrightarrow Ar_2P(O)OEt$$

XXc. REACTIONS OF TRICHLOROPHOSPHORANES WITH AMINES AND AMIDES

$$R_2PCl_3 + R'NH_2 \longrightarrow R_2P(:NR')Cl$$

The only N-unsubstituted phosphinimidic chloride which has been reported is the P,P-bis(trichloromethyl) compound which was obtained as a low-melting, distillable, crystalline solid by reaction of the trichlorophosphorane and ammonia at -10°.[428] Other phosphinimidic chlorides have required higher reaction temperatures in their preparations by this route: diphenyltrichlorophosphorane and arylamine hydrochlorides were heated in carbon tetrachloride under reflux, and the resulting quasi-phosphonium compounds, $Ph_2P^+Cl \cdot NHAr\ Cl^-$ were then heated with pyridine in benzene;[795] diphenyltrichlorophosphorane and dichloro- and trichloro-acetamides or methyl dichloro- and trichloro-acetimidates were heated in carbon tetrachloride or benzene;[733] while reactions of various diaryltrichlorophosphoranes and arylsulfonamides required 2-3 hr at 120-130°.[729,730,732] The products were viscous oils or glasses except for N,P,P-triphenylphosphinimidic chloride,[795] and a range of N-arylsulfonyl-P,P-diphenylphosphinimidic chlorides,[731] which were obtained as crystalline solids.

XXd.   REACTIONS OF A TRICHLOROPHOSPHORANE WITH AMMONIUM CHLORIDE, FOLLOWED BY HYDROLYSIS OF THE PRODUCT

$$2Ph_2PCl_3 + NH_4Cl \longrightarrow Ph_2P \underset{N}{\overset{Cl\quad Cl}{\diagup\diagdown}} P^+Ph_2 \quad Cl^- \longrightarrow$$

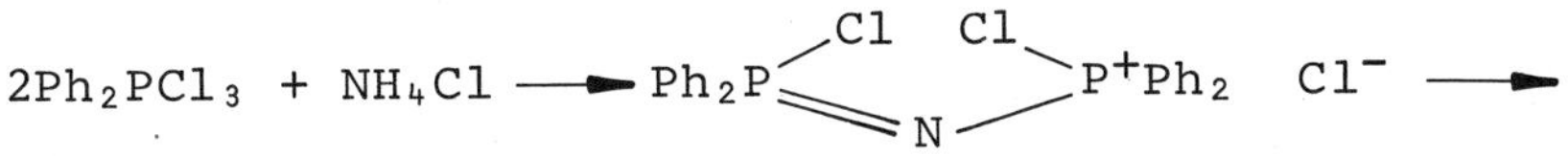

$$Ph_2P(O)\cdot NH\cdot P(O)Ph_2$$

The first stage was carried out in 1,2-dichloroethane at 130° and the second in chloroform with added formic acid, under reflux.[204]

XXI.   REACTIONS OF PHOSPHINIC, PHOSPHINOTHIOIC, PHOSPHINOSELENOIC, AND PHOSPHINOIMIDIC HALIDES

Phosphinic halides, and their thio, seleno, and imido analogs, are valuable precursors of many of the corresponding acids and their derivatives. The chemistry of phosphinic (etc.) halides is fully dealt with in Chapter 9, but for present purposes a distinction may usefully be made between their syntheses from compounds of non-phosphinic type, e.g., the preparation of phosphinic chlorides by careful hydrolysis of the reaction products of phosphonous dichlorides, alkyl chlorides and aluminum chloride,

$$RPCl_2 + R'Cl + AlCl_3 \longrightarrow [RR'P^+Cl_2\ AlCl_4^-] \xrightarrow{\ H_2O\ } RR'P(O)Cl$$

or of phosphinothioic bromides by controlled bromination of tetra-alkylbiphosphine disulfides,

$$R_2P(S)\cdot P(S)R_2 + Br_2 \longrightarrow 2R_2P(S)Br$$

and, on the other hand, their preparations from compounds which form the subject of this chapter, and particularly from phosphinic (etc.) acids (Section XXIIk).

The former methods, when combined with hydrolysis, alcoholysis, etc., of the halides, provide routes to phosphinic acids comparable with those of Sections I-XIII in starting from compounds of quite different structural types. Reactions of the second type are essential first stages in many interconversions of phosphinic (etc.) acids and their derivatives, the second stages of which are described below.

XXIa. HYDROLYSIS OF PHOSPHINIC, PHOSPHINOTHIOIC, OR PHOSPHINOSELENOIC HALIDES

$$R_2P(O)Cl + H_2O \longrightarrow R_2PO_2H$$

$$R_2P(S)Cl + HO^- \longrightarrow R_2POSH$$

$$R_2P(Se)Cl + HO^- \longrightarrow R_2POSeH$$

Most phosphinic chlorides are hydrolyzed about as easily as is phosphorus oxychloride, and therefore require no more than stirring with ice or water, followed by evaporation of the resulting solution, to obtain the phosphinic acid. More difficulty is encountered if the organic groups are bulky and shield the phosphorus atom from attack but in such cases heating with dilute nitric acid,[84, 791] or aqueous alkali,[746,798] is usually adequate. Di-t-butylphosphinic chloride is an extreme case and required two days boiling with aqueous sodium hydroxide.[147]

Phosphinothioic chlorides are generally more sluggish in their reactions with nucleophilic reagents, but are hydrolyzed satisfactorily to phosphinothioic acids by heating with aqueous or alcoholic sodium hydroxide,[98,289, 506] and this procedure has also been used to hydrolyze a phosphinoselenoic chloride to the corresponding acid.[490]

XXIb. REACTIONS OF PHOSPHINIC, PHOSPHINOTHIOIC, OR PHOSPHINOSELENOIC HALIDES WITH SODIUM HYDROSULFIDE

$$R_2P(O)Cl + HS^- \longrightarrow R_2POSH$$

$$R_2P(S)Cl + HS^- \longrightarrow R_2PS_2H$$

$$R_2P(Se)Cl + HS^- \longrightarrow R_2PSSeH$$

Phosphinothioic,[506,646] phosphinodithioic,[345,505] and phosphinoselenothioic[443] acids have all been obtained as sodium salts, and hence by acidification as free acids, by treatment of appropriate chlorides with sodium hydrosulfide in methanol or ethanol. The reactions are normally fairly rapid but t-butylmethylphosphinothioic bromide required 15 hr under reflux in dioxan.[437]

XXIc. REACTIONS OF PHOSPHINOTHIOIC, OR PHOSPHINOSELENOIC HALIDES WITH SODIUM HYDROSELENIDE

$$R_2P(S)Cl + HSe^- \longrightarrow R_2PSSeH$$

$$R_2P(Se)Cl + HSe^- \longrightarrow R_2PSe_2H$$

Reactions of sodium hydroselenide with phosphinothioic or phosphinoselenoic chlorides in ethanol gave sodium salts of diethylphosphinoselenothioic and diethylphosphinodiselenoic acids.[442,443]

XXId. HYDROLYSIS OF PHOSPHINIMIDIC HALIDES

$$R_2P(:NR')Cl + H_2O \longrightarrow R_2P(O)NHR'$$

N-Unsubstituted,[428] N-aryl,[795] N-acyl,[159,733] and N-arylsulfonyl[731,732] phosphinimidic chlorides are all readily converted by moisture into the amides or mixed imides which are the stable tautomers of phosphinimidic acids.

XXIe. REACTIONS OF PHOSPHINIC, PHOSPHINOTHIOIC, OR PHOSPHINIMIDIC HALIDES WITH ALCOHOLS OR PHENOLS, OR THEIR ANIONS OR THOSE OF THEIR SULFUR OR SELENIUM ANALOGS. This group of reactions provides routes of major importance to phosphinate, phosphinothioate, phosphinodithioate, phosphinoselenoate and phosphinimidate esters.

$$R_2P(O)Cl \longrightarrow R_2P(O)OR'$$

A great many phosphinates have been made in this way: only a few representative references are quoted. Reactions of phosphinic chlorides with alcohols can be carried out simply by mixing and removing hydrogen chloride under reduced pressure, or by heating under reflux in an inert solvent,[157,243] while aryl esters can be prepared by heating mixtures of phosphinic chlorides and phenols fairly strongly for several hours.[232,527] More commonly, however, either a sodium alkoxide in the corresponding alcohol,[419,423] or in benzene,[677] or a sodium aryloxide in benzene or ether,[664] is used, or the reaction is carried out using the alcohol or phenol together with a tertiary base (usually triethylamine or pyridine) in ether or benzene.[75,264,482,683]

A related reaction of phosphinic chlorides is that with epoxides to give 2-chloroalkyl esters.[236,362]

$$R_2P(O)Cl \longrightarrow R_2P(O)SR'$$

Relatively few phosphinothioate S-esters have been

prepared by reactions of phosphinic chlorides with thiols, in presence of triethylamine,[370,385,656] or with sodium arylsulfides.[697] This is probably not because of any particular difficulties in these reactions, but is a result of the generally greater convenience of the sodium phosphinothioate-alkyl halide route (Section XXIIa).

$$R_2P(S)Cl \longrightarrow R_2P(S)OR'$$

There are, on the other hand, many accounts, particularly in the patent literature, of preparations of phosphinothioate O-esters by reactions of phosphinothioic chlorides with alcohols or phenols and a tertiary base, or more usually with a sodium alkoxide or aryloxide.[57,316,436,718]

$$R_2P(S)Cl \longrightarrow R_2P(S)SR'$$

Phosphinodithioates have been prepared by reactions of sodium thiolates,[506,671] and also of thiols and triethylamine,[278] with phosphinothioic chlorides. Related reactions are those of phosphinothioic chlorides with S-alkyl esters of dithiocarbonic and dithiocarboxylic acids,[163] and with potassium ethyl trithiocarbonate.[710]

$$R_2P(O)Cl \longrightarrow R_2P(O)SeR'$$

The phenyl Se-ester of diphenylphosphinoselenoic acid has been prepared from diphenylphosphinic chloride and the bromomagnesium derivative of selenophenol.[587]

$$R_2P(:NR')Cl \longrightarrow R_2P(:NR')OR''$$

Reactions of phosphinimidic chlorides with sodium alkoxides or aryloxides yield phosphinimidates.[729,730]

XXIf. REACTIONS OF PHOSPHINIC, PHOSPHINOTHIOIC, OR PHOSPHINOSELENOIC CHLORIDES WITH PHOSPHINIC (ETC.) ACIDS, OR WITH A HALF-EQUIVALENT OF WATER, OR WITH HYDROGEN SULFIDE

$$R_2P(O)Cl + R_2PO_2^- \longrightarrow R_2P(O)\cdot O\cdot P(O)R_2$$

Phosphinic anhydrides may be prepared by heating the phosphinic chloride with the sodium phosphinate in an inert solvent,[434] but a commoner method has been to treat the phosphinic chloride in benzene or an ether with triethylamine and a limited amount of water.[377,620,767]

The stable isomers of monothio- and dithio-(phosphinic anhydrides) have the sulfur atoms in the thiophosphoryl groups rather than in bridging positions. Unsymmetric phosphinic-phosphinothioic anhydrides, $R_2P(O)\cdot O\cdot P(S)R_2$, are consequently obtained from reactions of phosphinic chlorides with phosphinothioic acids,[98] or by treatment of a phosphinic chloride with hydrogen sulfide,[529] as well as from a phosphinothioic chloride and a phosphinic acid;[476] while phosphinothioic anhydrides, $R_2P(S)\cdot O\cdot P(S)R_2$, are the products of hydrolysis,[28,524] or of treatment with dry silver oxide,[447] of phosphinothioic halides, or of reactions of phosphinothioic chlorides with the corresponding potassium phosphinothioate,[502] or with the phosphinothioic acid and pyridine.[476]

Phosphinothioic thioanhydrides, including an unsymmetrical compound, $Pr_2P(S)\cdot S\cdot P(S)Et_2$, which disproportionated to symmetrical compounds on redistillation, have been prepared by heating phosphinothioic bromides with sodium phosphinodithioates,[447] and an analogous preparation of a phosphinoselenoic selenoanhydride has been reported.[442]

XXIg. REACTIONS OF PHOSPHINIC CHLORIDES WITH PARAFORMALDEHYDE[539]

$$R_2P(O)Cl + CH_2O \longrightarrow R_2P(O)\cdot O\cdot P(O)R_2$$

XXIh. REACTIONS OF PHOSPHINIC CHLORIDES WITH PHOSPHINATE ESTERS

$$R_2P(O)Cl + R_2P(O)OEt \longrightarrow R_2P(O)\cdot O\cdot P(O)R_2 + EtCl$$

This is a very satisfactory route to anhydrides of aliphatic[423,424] and aromatic[142] phosphinic anhydrides and to their mono- and di-thio analogs.[476] Attempted preparation of unsymmetrical anhydrides, $R_2P(O)\cdot O\cdot P(O)R'_2$, gave mixtures of two symmetrical anhydrides.[412] The reactions are carried out without solvent at 150-220°, and a mechanism has been suggested involving nucleophilic attack on the phosphinic chloride by the phosphoryl oxygen of the ester.[752]

XXIi. REACTIONS OF PHOSPHINIC, PHOSPHINOTHIOIC, OR PHOSPHINIMIDIC HALIDES WITH AMMONIA, AMINES, OR HYDRAZINES

$$R_2P(Z)Cl + R'R''NH \longrightarrow R_2P(Z)NR'R''$$

$$[Z = O,\ S,\ N\cdot SO_2\cdot C_6H_4\text{-p-Me}]$$

A large number of phosphinic and phosphinothioic N-substituted and N,N-disubstituted amides have been prepared by reactions of chlorides with excess of primary or secondary amines, or with primary or secondary amines in presence of triethylamine or pyridine, in ether, benzene, hexane, or carbon tetrachloride. The reactions, which are normally carried out with ice-cooling during mixing, followed by heating under reflux for about 2 hr, are scattered throughout the literature with only a few papers[198, 267,546,576,664,757] describing the preparation of more than a single diethylamide or anilide. Satisfactory results were only obtained from aminoheterocycles if they were converted into N-trimethylsilyl derivatives before reactions with phosphinic chlorides.[282]

A few primary amides have been prepared by reactions of phosphinic and phosphinothioic chlorides with aqueous ammonia.[316,547,746] Phosphinic and phosphinothioic hydrazides have been obtained by addition of the chlorides to an excess of hydrazine,[92,744] or substituted hydrazines,[28,32,91,567] and N,N'-di(phosphinic) and N,N'-di(phosphinothioic) hydrazides have also been prepared.[92, 744]

Several N-tosylphosphinimidic amides and hydrazides have been prepared by reactions of this type.[787]

XXIj. REACTIONS OF PHOSPHINIC OR PHOSPHINOTHIOIC CHLORIDES WITH ANIONS OF PHOSPHINIC OR PHOSPHINOTHIOIC AMIDES[268,703]

$$R_2P(Z)Cl + R_2P(Z)NR'^- \longrightarrow R_2P(Z)\cdot NR'\cdot P(Z)R_2$$

$$[Z = O, S]$$

XXIk. REACTIONS OF DICHLORIDES OF CARBOXYLIC-PHOSPHINIC ACIDS

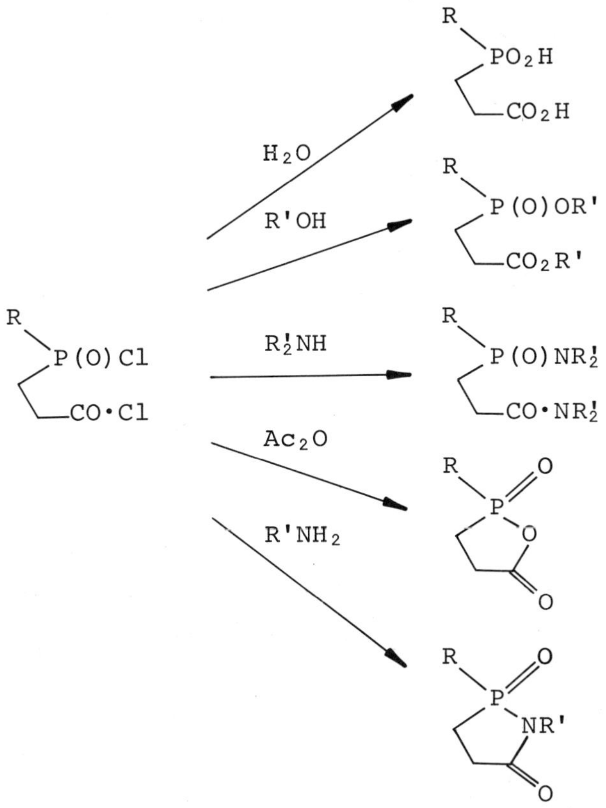

Di(acid chlorides) of β-carboxyphosphinic acids, obtained from phosphonous dichlorides and α,β-unsaturated carboxylic acids (Section Xc), react with water, alcohols, secondary amines, or an excess of ammonia, under the same conditions as the corresponding reactions of monofunctional acid chlorides, to give di-acids, di-esters, or diamides. Of more interest are the reactions of the di(acid chlorides) which give 5-membered cyclic phosphinic-carboxylic anhydrides or cyclic imides. The former are carried out by heating with acetic anhydride to about 60° for 1 hr,[369,370,373,378,381,608] or by heating with one equivalent of acetic acid,[371] while in the latter a mixture of a primary amine and triethylamine is added to a solution of the di(acid chloride) at 0°, followed by heating to 60-80° for 1-2 hr.[375,626]

XXII. REACTIONS OF ACIDS

XXIIa. ESTERIFICATION OF PHOSPHINIC, PHOSPHINOTHIOIC, PHOSPHINODITHIOIC, PHOSPHINOSELENOTHIOIC, AND PHOSPHINODISELENOIC ACIDS BY ALKYLATION OF THEIR ANIONS

$$R_2POS^- \ Na^+ + R'X \longrightarrow R_2P(O)SR'$$

Apart from a few early preparations of phosphinates from alkyl iodides and silver salts of phosphinic acids,[487] almost all these reactions have involved sodium or potassium salts and aliphatic (often substituted) chlorides or bromides. The method has been little used for phosphinates, which require high temperatures,[19] or are obtained in lower yields than by other methods,[273] except for preparations of dialkylaminoalkyl phosphinates in which either the potassium salt,[259] or the free acid,[738] may be used and which may proceed through cyclic ammonium ions.

In contrast, many alkyl phosphinothioates,[502] and phosphinodithioates,[316,448,505] and all four of the known phosphinoselenothioates and phosphinodiselenoates,[443] have been prepared by this route, which is particularly well represented in the patent literature.[470,716,717,718] The reaction is facilitated by the higher nucleophilicity of selenium and sulfur than that of oxygen and in reactions where the anion is formally hetero-ambident, S-esters are formed rather than O-esters, and Se-esters rather than S-esters. Acetone, acetonitrile, dioxan, and ethanol have been used as reaction solvents, and good yields are usually obtained after three hours at around 60°. The kinetics of reactions of benzyl chloride with four dialkylphosphinodithioic acids and two diarylphosphinodithioic acids have been investigated and found all to have pseudounimolecular rate constants of about 5×10^{-4} sec^{-1} at 25° in 95% ethanol.[341]

XXIIb. ESTERIFICATION OF PHOSPHINIC ACIDS BY TREATMENT WITH TRIALKYL PHOSPHITES

$$R_2PO_2H + (R'O)_3P \longrightarrow R_2P(O)OR' + (R'O)_2P(O)H$$

Trialkyl phosphites are effective mild esterifying agents and have been used to prepare esters of di-(allyloxymethyl)-,[72] di-(allylthiomethyl)-,[328] and other phosphinic acids,[328,548] by heating the reactants in sealed tubes to about 140° for several hours.

XXIIc. ESTERIFICATION OF PHOSPHINIC ACIDS USING DIAZOMETHANE

$$R_2PO_2H + CH_2N_2 \longrightarrow R_2P(O)OCH_3 + N_2$$

Diazomethane is a convenient reagent for small-scale preparation of methyl esters of phosphinic acids and has been used on several occasions.[86,179,273,309,800]

XXIId. ESTERIFICATION OF PHOSPHINODITHIOIC ACIDS BY REACTION WITH HYDROXY COMPOUNDS

$$R_2PS_2H + R'OH \longrightarrow R_2P(S)SR'$$

Direct esterification of phosphinic acids has not been reported, but diphenylphosphinodithioate esters have been prepared by heating the acid with primary alcohols from methyl to dodecyl.[295] The reaction failed with isopropyl and s-butyl alcohols, but proceeded satisfactorily with t-butyl alcohol and with octan-2-ol although olefins may have been intermediates in these cases.

The formation of an amidomethyl ester by a Mannich-type reaction between dimethylphosphinodithioic acid, formaldehyde, and benzo-1,2,3-triazin-4-one,[194] falls logically into this section if the N-methylol derivative of the heterocycle is assumed to be an intermediate.

XXIIe. ESTERIFICATION OF PHOSPHINOTHIOIC AND PHOSPHINODITHIOIC ACIDS BY THEIR ADDITION TO CARBON-CARBON DOUBLE BONDS. Diphenylphosphinodithioic acids undergo base-catalyzed nucleophilic addition to α,β-unsaturated aldehydes, ketones, nitriles, esters, and to maleic anhydride.[430] Uncatalyzed additions of phosphinodithioic acids to olefins,[295] to vinyl ethers [exothermic reactions, product formulated as $R_2P(S)S{\cdot}CHMe{\cdot}OR'$],[353,713] and to several rather complicated α,β-unsaturated carbonyl compounds,[556,557,558] have also been described.

Dialkylphosphinothioic acids and dialkylphosphinodithioic acids also add to divinyl ether:[708] the products are formulated as $R_2P(Z){\cdot}S{\cdot}CH_2{\cdot}CH_2{\cdot}O{\cdot}CH{:}CH_2$.

XXIIf. OXIDATION AND/OR HYDROLYSIS OF PHOSPHINOTHIOIC AND PHOSPHINODITHIOIC ACIDS

$$R_2PS_2H \longrightarrow R_2PO_2H$$

Phosphinic acids are readily obtained from phosphinothioic and phosphinodithioic acids by air-oxidation,[310] or treatment with steam,[289] hydrogen peroxide,[448] or 6N nitric acid.[289] The reaction is of preparative value for diphenylphosphinic acid because of the ready synthesis of the corresponding phosphinodithioic acid from benzene, phosphorus pentasulfide, and aluminum chloride.[289]

XXIIg. CONVERSION OF ACIDS TO ANHYDRIDES OR THIOANHYDRIDES BY STRONG HEATING. Conversion of a dithiophosphinic acid to its thioanhydride, $R_2P(S)\cdot S\cdot P(S)R_2$ occurs at a much lower temperature (170°)[295] than that required (410°) for dehydration of a phosphinic acid.[474]

XXIIh. CONVERSION OF ACIDS TO ANHYDRIDES BY TREATMENT WITH CARBODIIMIDES[535]

$$2Et_2POSH + C_6H_{11}\cdot N{:}C{:}N\cdot C_6H_{11} \longrightarrow$$

$$Et_2P(O)\cdot O\cdot P(S)Et_2 + C_6H_{11}\cdot NH\cdot CS\cdot NH\cdot C_6H_{11}$$

XXIIi. ACYLATION OF PHOSPHINATE AND PHOSPHINODITHIOATE ANIONS BY REACTIONS WITH CARBOXYLIC OR SULFONIC CHLORIDES OR WITH CARBOXYLIC ANHYDRIDES

$$R_2PS_2^- Na^+ + R'\cdot COX \longrightarrow R_2P(S)\cdot S\cdot CO\cdot R'$$

More of these reactions, like the analogous ester preparations described in Section XXIIa, involve phosphinodithioates than involve phosphinates: none involving phosphinothioates have been reported, although they would be expected to proceed satisfactorily. The preparations of mixed anhydrides of phosphinic acids include reactions of diphenylphosphinic acid with acetic anhydride,[774] and of its silver salt with arylsulfonic chlorides.[697] Reactions of sodium salts of phosphinodithioic acids with aliphatic and aromatic acyl chlorides,[249,294,316] with S-alkyl chlorothioformates,[286] and with N,N-dialkylcarbamoyl chlorides,[286,366] have been reported.

XXIIj. OXIDATION OF PHOSPHINODITHIOATE, PHOSPHINOSELENOTHIOATE, AND PHOSPHINODISELENOATE ANIONS

$$R_2PS_2^- Na^+ \longrightarrow R_2P(S)\cdot S\cdot S\cdot P(S)R_2$$

Phosphinodithioate anions are oxidized to phosphinothioic disulfide by iodine in aqueous potassium iodide,[444,448] or in methanol;[586] or by cupric ions[444,552] with accompanying formation of cuprous phosphinodithioates.

Iodine in aqueous potassium iodide similarly oxidized sodium diethylphosphinoselenothioate to diethylphosphinothioic diselenide,[304,443] while iodine in carbon tetrachloride gave this together with a small amount of the triselenide. Diethylphosphinoselenoic triselenide was obtained from sodium diethylphosphinodiselenoate and iodine in aqueous potassium iodide.[443]

XXIIk. CONVERSION OF ACIDS INTO PHOSPHINIC AND PHOSPHINOTHIOIC CHLORIDES

$$R_2PO_2H \longrightarrow R_2P(O)Cl$$

Phosphinic and phosphinothioic chlorides are discussed fully in Chapter 9, but their formation from acids are important reactions and merit a brief discussion here.

The preparation of phosphinic chlorides from phosphinic acids can be effected satisfactorily by heating the acids with either phosphorus pentachloride or thionyl chloride, and the literature contains many examples of the use of each of these reagents.

Thionyl chloride usually removes sulfur or selenium in preference to oxygen, and hence affords phosphinic chlorides from phosphinothioic[98,343] and phosphinoselenoic[489] acids. Conversion of a phosphinothioic acid into a phosphinothioic chloride by thionyl chloride has, however, been reported[452] but phosphorus pentachloride is a more reliable reagent for this purpose.[505] Phosphorus pentachloride,[289,316] thionyl chloride,[586] and dry hydrogen chloride[289] have all been used to prepare phosphinothioic chlorides from phosphinodithioic acids.

XXIII. REACTIONS OF ESTERS

XXIIIa. HYDROLYSIS OF PHOSPHINATE ESTERS

$$R_2P(O)OR' \longrightarrow R_2PO_2H$$

Many phosphinic acids (but not phosphinothioic or phosphinodithioic acids which would be desulfurized) have been prepared by hydrolyses of their lower alkyl esters, but the experimental conditions are so generally uniform that little discussion is required. The normal procedure is to heat a mixture of the ester and concentrated, or constant-boiling ($\sim$20%), hydrochloric acid under reflux for several hours and then evaporate to obtain the crude acid. A few alkaline hydrolyses have been reported but possess no general advantages to outweigh the extra stages necessary to isolate the products.

There have been a number of kinetic investigations of various aspects of phosphinate ester hydrolysis, mostly in alkaline media.[15,65,66,67,158,227,248,270,399,453]

XXIIIb. PYROLYSIS OF ESTERS. A number of esters of diphenylphosphinic acid have been pyrolyzed in mechanistic studies,[74,79,500] and other individual cases of phosphinate[358] and phosphinothioate[437] pyrolyses have also been reported.

XXIIIc. TRANSESTERIFICATION

$$R_2P(O)OR' \longrightarrow R_2P(O)OR''$$

Reactions of phosphinates with alcohols have been little used, reported examples being the preparation of cyclohexyl methylphenylphosphinates by 12-hr heating of methyl and ethyl esters with cyclohexanol,[128] several transesterifications of 1-alkoxyphospholene 1-oxides,[675] and a conversion of a phenyl ester into the ethyl ester as the last step of a series of reactions.[801] Reactions of S-alkyl phosphinothioates with alkoxide ions have yielded phosphinates in stereochemical studies.[69,263]

Of more general synthetic value are reactions of phosphinates with alkyl halides.

$$R_2P(O)OR' + R''X \longrightarrow R_2P(O)OR'' + R'X$$

These have given benzyl and long-chain alkyl esters, in reactions catalyzed by triphenylphosphine,[284] ethylene di-esters from 2-chloroethyl esters,[237] and cyclic esters of ω-hydroxyalkylphosphinic acids by intramolecular reactions of unisolated haloalkylphosphinate esters resulting from Arbuzov reactions of α,ω-dihalides with phosphinites.[229]

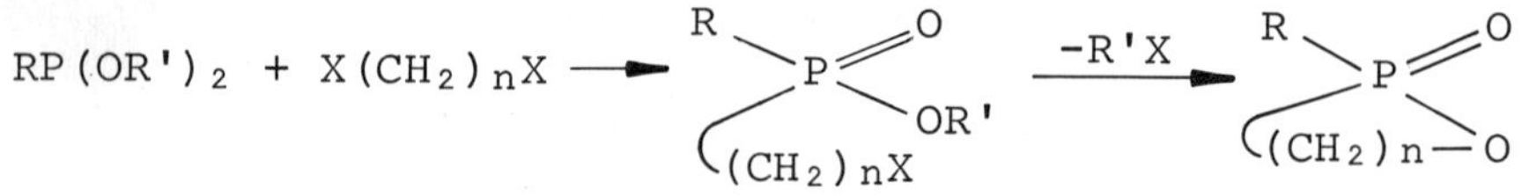

Analogous intramolecular eliminations of ethyl acetate from ethyl ω-acetoxyphosphinates have also been described.[56]

XXIIId. HYDROLYSIS AND ALCOHOLYSIS OF CYCLIC ENOL PHOSPHINATES

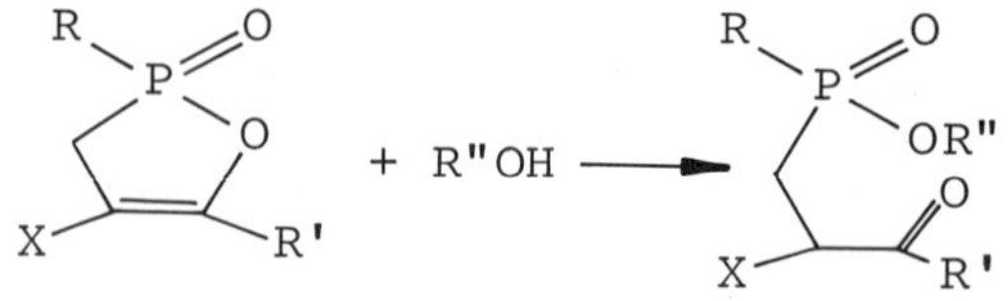

Ring-opening of these cyclic esters to give 3-oxoalkylphosphinic acids or their esters, occurs exothermically with water or alcohols for the enol esters (X =

Cl)[622,623] obtained from phosphonous dichlorides and an α-diketone (Section Xe), but the products from phosphonous dichlorides and α,β-unsaturated ketones (Section Xb) require heating with water[73,577] or alkali[50] to give acids, or with alcohols[577] or sodium alkoxides[50] to give esters.

XXIIIe. O-ALKYL → S-ALKYL ISOMERIZATION OF PHOSPHINOTHIOATE ESTERS

$$R_2P(S)OR' \longrightarrow R_2P(O)SR'$$

Reactions of this type have long been known for phosphorothioates but only one phosphinothioate example has been described.[502] The reaction was carried out by heating the O-alkyl ester (R = Ph, R' = Et) with ethyl bromide in a sealed tube.

There is also one example of a similar isomerization brought about by a catalytic amount of methyl iodide during 30 hr at room temperature, and involving a 1,5- rather than a 1,3- migration:[705]

$$Ph_2P(OMe){:}N{\cdot}P(S)Ph_2 \longrightarrow Ph_2P(O){\cdot}N{:}P(SMe)Ph_2$$

XXIIIf. REACTIONS OF ESTERS WITH AMINES OR HYDRAZINE. Phosphinic amides apparently cannot be prepared by aminolysis of phosphinates because of the lower nucleophilicity of amines towards phosphinic groups than towards carbon (compare Section XXIVe). This generalization is supported by reactions of haloalkylphosphinates with amines, in which halogen atoms are replaced by amino groups without affecting the phosphinate ester group,[306,330] and by reactions of alkyl aryl(trichloromethyl)phosphinates with primary and secondary amines which afforded salts of the phosphinic acids and alkylated amines:[391,392]

$$Ar(CCl_3)P(O)OR + 2R'R''NH \longrightarrow Ar(CCl_3)PO_2^- \; R'R''N^+H_2 + RR'R''N$$

Heating diesters of phosphinic-carboxylic diacids with an excess of hydrazine or semicarbazide results in reaction at the carboxylic ester group but not at the phosphinic ester group.[665] This suggests a similar resistance of phosphinates towards hydrazine as towards amines, but this is apparently much less pronounced since methyl methylphenylphosphinate has been reported to react slightly exothermically with hydrazine to give a good yield of methylphenylphosphinic hydrazide.[744]

XXIIIg. HYDROLYSIS OF PHOSPHINIMIDATE ESTERS

$$R_2P({:}NR')OR'' + H_2O \longrightarrow R_2P(O)NHR' + R''OH$$

The only phosphinimidic esters which have been described have acyl,[338] arylsulfonyl,[730] or phosphinyl[247] groups on the imidic nitrogen atom. They are stable to water but are cleaved by dilute aqueous hydrochloric acid or aqueous alcoholic alkali,[730] or by dry hydrogen chloride,[247,338] to N-acylated or N-sulfonylated phosphinic amides, or to phosphinic imides.

XXIIIh. CONVERSION OF ESTERS TO ANHYDRIDES, BY REACTION WITH GEM-DIHALIDES

$$2R_2P(O)OR' + RCHX_2 \longrightarrow R_2P(O)\cdot O\cdot P(O)R_2 + RCHO + 2R'X$$

This reaction has been carried out using methyl esters of several phosphinic acids and either benzylidene dichloride or methylene iodide.[276] Yields were 62-77%.

XXIIIi. CONVERSION OF PHOSPHINATES INTO PHOSPHINIC CHLORIDES

$$R_2P(O)OR' + PCl_5 \longrightarrow R_2P(O)Cl + R'Cl + POCl_3$$

There are several reports of this reaction between phosphinates and phosphorus pentachloride being used preparatively. Reactions of halogen-containing esters have required fairly strong heating,[236,791] but milder conditions have been used in other cases.[350,504,656]

XXIV. REACTIONS OF ANHYDRIDES

The heat of hydrolysis of the P-O-P linkage of two phosphinic anhydrides has been found to be considerably more than that for the corresponding reaction of pyrophosphates.[418] This is reflected in the ready cleavage of phosphinic anhydrides by water, alcohols, and amines, but not, in the one case reported, by a Grignard reagent.[142]

Analogous reactions of acyclic phosphinic-carboxylic and phosphinic-sulfonic mixed anhydrides have not apparently been investigated, but a number of such reactions of cyclic anhydrides of dibasic phosphinic-carboxylic acids have been reported. Alcoholysis and aminolysis proceed dissimilarly, indicating a difference between the relative nucleophilicities of alcohols and of amines towards phosphinic and carboxylic groups.

XXIVa. HYDROLYSIS OF ANHYDRIDES

$$R_2P(O)\cdot O\cdot P(O)R_2 + H_2O \longrightarrow 2R_2PO_2H$$

Phosphinic anhydrides are rapidly hydrolyzed by water,[423] but phosphinothioic anhydrides require alkali,[28] or prolonged high temperatures,[524] for hydrolysis. The hydrolysis of a P-S-P link apparently occurs rather more easily than that of a P-O-P link, since dipropylphosphinothioic anhydride and thioanhydride require about 21 and 14 hr respectively, for 50% hydrolysis in aqueous dioxan at room temperature,[447] and the relative stability of diphenylphosphinothioic anhydride is demonstrated by its preparation by the exothermic reaction of the corresponding thioanhydride with 6N nitric acid.[295]

Bistrifluoromethylphosphinothioic thioanhydride was cleaved to the phosphinodithioic acid by hydrogen sulfide during 4 hr at 100°.[258]

XXIVb. ALCOHOLYSIS OF ANHYDRIDES

$$R_2P(O){\cdot}O{\cdot}P(O)R_2 + R'OH \longrightarrow R_2P(O)OR' + R_2PO_2H$$

Although all anhydrides would be expected to undergo this reaction, it is only synthetically useful when anhydrides are easily prepared from non-phosphinic precursors.[374,377,539]

XXIVc. AMINOLYSIS OF ANHYDRIDES

$$R_2P(O){\cdot}O{\cdot}P(O)R_2 + R'NH_2 \longrightarrow R_2P(O)NHR' + R_2PO_2H$$

The same remarks apply here as to the previous reaction, and examples of its use are even less common.[539]

XXIVd. ALCOHOLYSIS OR HYDROLYSIS OF CYCLIC PHOSPHINIC-CARBOXYLIC ANHYDRIDES

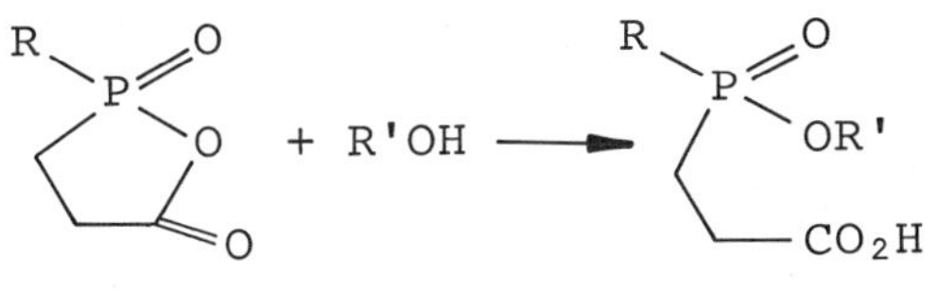

These anhydrides may be hydrolyzed to β-carboxyalkylphosphinic acids,[369,370] but most such acids have been prepared from phosphinic-carboxylic dichlorides (Section XXIk). Alcoholysis of anhydrides, in which the exocyclic phosphorus-bound group is methyl,[369,373,381] ethyl,[380,386,608] or chloromethyl,[378] and in which the two carbon atoms between the phosphinic and carboxylic groups are unsubstituted,[378,381,386] have a methyl group attached to the α-carbon,[369] or to the β-carbon atom,[373,380,386] or are joined by a double bond,[608] all result in phosphorus-oxygen bond cleavage and formation of an alkyl (β-carboxyalkyl)phosphinate. Exceptionally, the cyclic anhydride having a phenyl group attached to phosphorus and an olefinic bond in the ring, undergoes acyl-oxygen bond cleavage by ethanol and formation of (2-ethoxycarbonylvinyl)-phenylphosphinic acid.[370]

XXIVe. AMINOLYSIS OF CYCLIC PHOSPHINIC-CARBOXYLIC ANHYDRIDES

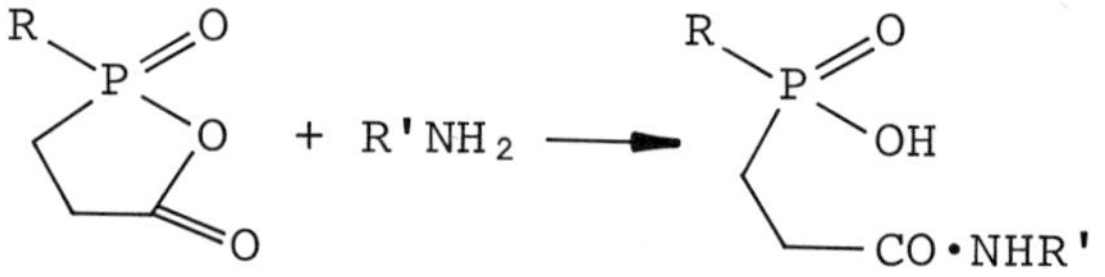

Twenty-four individual reactions of primary aromatic amines, and one of phenothiazine,[380] with cyclic anhydrides having a methyl, ethyl, chloromethyl, or phenyl group attached to phosphorus, and an unsubstituted, α- or β-methylated, or unsaturated ring, all resulted in acyl-oxygen bond cleavage and formation of (arylcarbamoylalkyl) phosphinic acids.[369,373,381,386,608] The reactants were heated at 80-90° for 7-8 hr,[380] or in sealed tubes at 120° for 1-4 hr.[371,378]

XXV. REACTIONS OF AMIDES AND OF IMIDES

XXVa. CONVERSION OF AMIDES TO IMIDES BY STRONG HEATING

$$RR'P(Z)NHR'' \longrightarrow RR'P(Z)\cdot NR''\cdot P(Z)RR'$$

This reaction has been observed with dibutylphosphinic amide,[547] with a diamide of an ethylenedi(phosphinic acid) which gave a 5-membered cyclic imide,[498] with dimethylphosphinothioic amide,[702] and with the amide and N-phenylamide of diphenylphosphinothioic acid, both of which also gave other, cyclic, products.[725]

XXVb. CHLORINATION OR BROMINATION OF PHOSPHINOTHIOIC IMIDES[702,705]

$$R_2P(S)\cdot NH\cdot P(S)R'_2 \longrightarrow R_2P(X){:}N\cdot P^+(X)R'_2\ X^-$$

XXVc.   ALKYLATION OF AMIDES AND IMIDES.   Phosphinic amides have been alkylated, either with alkyl halides after conversion of the amide into its anion,[268] or with a large excess of diazomethane as a further reaction after formation of an N-substituted amide by a 1,3-dipolar addition of diazomethane to a phosphinic isothiocyanate.[759] The overall result in each case is the replacement of a hydrogen atom attached to nitrogen and the retention of the phosphinic amide structure.

$$R_2P(O)NHR' \longrightarrow R_2P(O)NR'R''$$

In contrast, diphenylphosphinothioic imide is methylated on sulfur.   Diazomethane gives a neutral monomethyl derivative, and this on prolonged treatment with dimethyl sulfate affords an ionic dimethylated product, which is also obtained from the anion of the original imide and dimethyl sulfate.[703]

$$Ph_2P(S) \cdot NH \cdot P(S)Ph_2 \xrightarrow{CH_2N_2} Ph_2P(SMe):N \cdot P(S)Ph_2$$

$$\downarrow Li \qquad\qquad\qquad\qquad\qquad \downarrow Me_2SO_4$$

$$Li^+[Ph_2P(S) \cdot N \cdot P(S)Ph_2]^- \xrightarrow{Me_2SO_4} Ph_2P(SMe):N \cdot P^+(SMe)Ph_2\ X^-$$

$$[X^- \text{ is a complex anion } (SbCl_6^-, \text{ or } BPh_4^-)$$
added to assist isolation]

Analogous behavior is exhibited by the mixed imide of diphenylphosphinothioic acid and N,N-dimethylthiocarbamic acid, which reacts with diazomethane to give a mixture of two S-methylated products, without any methylation of the central nitrogen atom.[704]

$$Ph_2P(S) \cdot NH \cdot CS \cdot NMe_2 + CH_2N_2 \longrightarrow$$

$$Ph_2P(S) \cdot N:C(SMe)NMe_2 + Ph_2P(SMe):N \cdot CS \cdot NMe_2$$

$$74\% \qquad\qquad\qquad\qquad 26\%$$

Phosphinimidic tertiary amides,[246] or anions of phosphinimidic secondary amides,[245] are alkylated on both nitrogen atoms to yield quasi-phosphonium halides.

$$Et_2P(:NPh)NMeAr + MeI \longrightarrow Et_2P^+(NMePh)NMeAr\ I^-$$

$$Et_2P(:NPh)N^-Ph + 2MeI \longrightarrow Et_2P^+(NMePh)_2\ I^-$$

XXVd. HYDROLYSIS OF AMIDES

$$R_2P(O)NHR' + H_2O \longrightarrow R_2PO_2H$$

Phosphinic primary amides,[547,796] and dimethylamides,[796] are easily hydrolyzed, but more vigorous conditions are required for hydrolysis of phosphinic arylamides,[118,757,796] and of all phosphinothioic amides.[796] The kinetics of hydrolysis of a series of di(p-substituted aryl)phosphinic anilides have been investigated.[757]

XXVe. ALCOHOLYSIS OR HYDROLYSIS OF CYCLIC PHOSPHINIC-CARBOXYLIC IMIDES

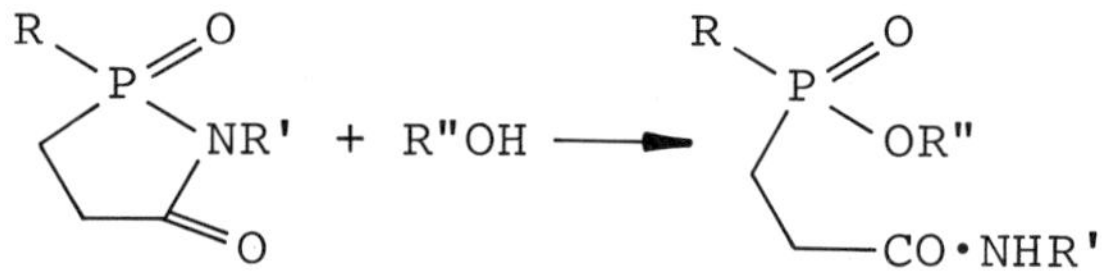

Treatment with water, or heating with ethanol under reflux, both result in cleavage of the phosphorus-nitrogen bond and formation of an N-substituted β-carbamoylalkylphosphinic acid or its ethyl ester.[626]

C. MODIFICATION OF ORGANIC GROUPS IN PHOSPHINIC ACIDS, THEIR DERIVATIVES, AND ANALOGS

XXX. MODIFICATION OF GROUPS ATTACHED TO PHOSPHORUS

Over fifty reactions have been used to alter the phosphorus-attached groups of phosphinic acids and their derivatives either with or, more usually, without functional change at phosphorus. Only a very few examples have been described of many of these reactions, e.g., hydrogenation of olefinic bonds,[278,302,494,672] and of benzene rings,[214,220,396] allylic prototropic rearrangements,[155,606] halogenation of methyl groups,[677] and α-halogenation of ketones,[137] ozonolysis,[136] and nucleophilic addition of thiols,[601] and of alcohols,[364] to olefinic bonds, hydration of acetylenic bonds,[395] replacement of hydroxyl by halogen,[121,326] ring-opening of epoxides,[36] conversion of carbonyl groups to acetals,[231] and to enol-ethers,[556,634] decarboxylation,[801] esterification of carboxyl groups,[600,801] hydrolysis,[89,90,281,801] and transesterification,[600] of carboxylate esters, hydrolysis of nitriles,[89,90,450,647] and addition of hydrogen sulfide to a nitrile group to give a carbothionic amide.[648] Similarly, there have been reports of the hydrogenolysis of benzylamino to primary amino groups,[769] of diazotization, and diazo-coupling,[175] or diazo-replacement,[167,217] of quaternization of tertiary amino groups,[138,771] of Hofmann elimination from quaternary ammonium salts,[771] of oxidation of thioethers to sulfones,[216,328,431] and of oxidation or addition of sulfur to peripheral phosphine groups.[59] A few reactions have also been recorded in which carbon-carbon bonds are formed by Diels-Alder reactions,[373,398,771] or by reactions depending on carbanion formation, either in another reactant which then reacts with a halogen-substituted phosphinate,[493] or more often at a carbon atom adjacent only to a phosphoryl group,[507,616] or flanked by a phosphoryl group and another electron-attracting group.[203,431,643]

Other types of reaction which have been more extensively used are classified below, either individually or in groups of related reactions.

XXXa. ADDITION OF HALOGEN OR EPOXIDATION OF OLEFINIC BONDS. Chlorine, in carbon tetrachloride below 0°, has been added to several esters of allylphosphinic acids,[461,465,655] and bromine, in acetic acid, carbon tetrachloride or chloroform has been added to unsaturated phosphinic acids.[136,196,494]

Peracetic and monoperphthalic acid have been used to form epoxides from esters of phospholene acids,[37] and of allylphosphinic acids.[466]

XXXb. OXIDATION OF METHYL OR METHYLENE GROUPS ATTACHED TO BENZENE RINGS. Eight permanganate oxidations of methyl groups, all in symmetrical diarylphosphinic acids, have been described,[214,215,216,217,522,524,600] including examples where the oxidized groups were ortho,[215] meta,[216] and para,[214,217,600] to the phosphinic acid group. In one case, only one of the two p-methyl groups was oxidized,[524] and in another case, where the orientation is not explicit, only one out of three methyl groups in each ring was oxidized. The preferred reaction solvent is aqueous pyridine.[214]

There is one report of oxidation of a methylene group between two benzene rings.[222]

XXXc. NITRATION. Eight nitrations of phosphinic acids, giving well-characterized products,[62,120,172,214,215,216] and three less definite ones,[524] have been reported. The reactions, except one where the rings bore carboxyl substituents,[216] were carried out at temperatures between 0° and 35°. Mixed acids or fuming nitric acid were used for arylphosphinic acids, in which the nitro group entered meta to the phosphinic acid group, but concentrated nitric acid sufficed for benzyl[62,120] and

2-phenylethyl[62] acids, which were nitrated in the para-position.

XXXd. REPLACEMENT OF CHLORINE BY IODINE, OR REPLACEMENT OF HALOGEN BY HYDROXYL, THIOL, ALKOXY, ARYLOXY, AMINO OR SUBSTITUTED AMINO GROUPS. Halogen atoms are activated for nucleophilic replacement when attached to carbon adjacent to phosphoryl, and most of the replacement reactions discussed in this section involve α-chloroalkylphosphinic acids or their derivatives. There are, however, some examples of replacement of halogen atoms more remote from phosphorus, both in alkyl,[16,330] and in aryl groups,[168,217,407,682] the aromatic substitutions being catalyzed by copper powder or cuprous oxide.

Both chlorine atoms of di-(chloromethyl)phosphinic acid and of several of its esters and amides,[549] and also those of an O-ester of di-(chloromethyl)phosphinothioic acid,[583] have been replaced by iodine by heating with potassium iodide in dimethylformamide, and one single substitution is also recorded.[549] Replacement of an α-chlorine by hydroxyl during hydrolysis of a phosphinic chloride has been described,[222] and similar reactions occur during aqueous treatment of the crude products of reaction IXg from phosphonous dichlorides and aldehydes.[521] Replacement of α-chlorine by thiol groups is apparently best effected by using thiourea and hydrolyzing the resultant di-(S-thiouronium) compound,[327] a more complex reaction with formation of a 1,3-thiaphosphetan ring occurring during alkaline hydrolysis of di-(chloromethyl)-phosphinothioic acid.[329]

Reactions of sodium alkoxides with esters of di-(chloromethyl) phosphinic acid yield esters of di-(alkoxymethyl) phosphinic acids only if the alkyl groups (identical in the starting ester and in the alkoxide to avoid transesterification) are propyl, butyl, or pentyl; ether formation occurring if they are methyl or ethyl.[548] Replacement of halogen by aryloxy groups has been described both for di-(chloromethyl)phosphinic acid and its ethyl ester, in dimethylformamide,[550] and for di-p-bromophenylphosphinic acid.[217]

The largest number of replacement reactions of this type are those involving ammonia or amines. Aliphatic substitutions include those of di(chloromethyl)phosphinic acid with aniline and with primary aliphatic amines,[307] of several of its esters (again not methyl or ethyl) and amides with dimethylamine,[306] of an ester and an amide of di-(3-chloropropyl)phosphinic acid with diethylamine,[16] and the replacement of the 2-fluorine atom in esters of methyl(perfluoropropenyl)phosphinic acid in reactions with diethylamine, morpholine and piperidine.[330] In diarylphosphinic acids, replacement of one or two p-chlorine atoms by amino,[407,682] or of one or two m-bromine atoms by methylamino groups,[168] proceeded satisfactorily but products could not be obtained from various primary amines and either di-o- or di-p- bromophenylphosphinic acids.[168]

XXXe. ELIMINATION OF HYDROGEN HALIDE GIVING OLEFINIC OR ACETYLENIC BONDS. In contrast to the assistance to nucleophilic replacement of α-halogen atoms which a phosphoryl group provides, its effect on β-haloalkyl groups is to facilitate dehydrohalogenation by bases.

Vinyl-,[494] and phenylethynyl-,[198] phosphinic acids are thus obtained from 2-bromoethyl- or 2-chlorostyryl- acids by the action of hot alcoholic potassium hydroxide. Esters of β-halogenophosphinic acids may be dehydrohalogenated using alcoholic potassium hydroxide[234] or triethylamine[234,337] to esters of unsaturated acids, or with accompanying hydrolysis at phosphorus to give the unsaturated acids.[197] Phenylethynylphosphinic acids have also been obtained from 2-chlorostyryl tertiary phosphine oxides by dehydrohalogenation and accompanying cleavage of one phosphorus-carbon bond by reaction XIa.[199]

A striking demonstration of the resistance of an α-halovinyl group to elimination or substitution is afforded by ethyl (1-bromovinyl)-ethylphosphinate which with hot alcoholic potassium hydroxide undergoes a Michael addition to the double bond.[364]

XXXf. ALKYLATION OR ACYLATION OF HYDROXYL, THIOL, OR AMINO GROUPS. Ethyl, allyl, and benzyl halides and chloroacetic acid react with di(hydroxymethyl)phosphinic acid,[72,548] or with di(mercaptomethyl)-phosphinic acid[328] in aqueous sodium hydroxide to give acids containing ether or thioether groups. 1-Hydroxyalkylphosphinic acids,[487,572,573,775] and their esters,[571,775] have been acylated with acid chlorides or anhydrides, and di(hydroxymethyl)-phosphinic acid has been converted into its di-(phenylcarbamoyl) derivative by reaction with phenyl isocyanate.[323] Acylation by acid chlorides of mercapto[328] and of secondary amino[307] groups in phosphinic acids have also been reported.

XXXg. HYDROLYSIS OF ACETALS. Arbuzov reactions cannot be carried out on halogenated aldehydes, but normal reactions occur if the aldehydo group is protected as an acetal. The products are easily converted by gentle acid hydrolysis into esters of aldehydophosphinic acids.[659,666,667]

XXXh. CONVERSION OF CARBOXYLIC CHLORIDES TO CARBOXYLIC ACIDS, CARBOXYLATE ESTERS, OR CARBOTHIOLATE ESTERS. Reactions of di-acid chlorides of carboxylic-phosphinic chlorides are discussed in Section XXIm. Other reactions involving carboxylic chloride groups are those of phosphinate[383,386,624,773] and phosphinothiolate[12] esters containing carboxylic chloride substituents obtained by reaction Xc. Most of these reactions, which are not numerous, are with alcohols to give carboxylate ester groups,[383,386,624,773] but hydrolysis to a carboxylic acid,[624] and reactions with thiols,[12,773] are also described.

XXXi. REDUCTION OF NITRO GROUPS. Neutral or alkaline conditions are used for reduction of nitroarylphosphinic acids in order to dissolve the acid. Low pressure hydrogenation over Raney nickel[167,209,214] gives the best yields, but ferrous sulfate in aqueous ammonia,[486,496] and sodium dithionite[175] have also been used as reducing agents.

XXXj. SYNTHESIS OF HETEROCYCLIC GROUPS. Esters of phosphinic acids containing heterocyclic groups, linked to phosphorus through carbon atoms, have been prepared by reactions of phosphinate esters containing acetal groups. Such reactions with aliphatic 1,3-,[654] and 1,4- and 1,5-,[269] diols give saturated 6-, 7-, and 8-membered rings containing two oxygen atoms. Reactions with o-aminophenols,[653,657] o-aminothiophenol,[658] o-phenylenediamine,[652] and phenylhydrazine,[651] yield benzoxazoles, benzthiazolines, benzimidazoles, and indoles, respectively, the Fischer indole syntheses involving a 3 → 2 migration of the phosphorus-containing group.

XXXI. MODIFICATION OF GROUPS ATTACHED TO OXYGEN OR SULFUR

The only reported reactions involving changes to O-ester groups are those of ethoxysilylalkyl phosphinates with a silanol.[20]

$$Me_2P(O){\cdot}O{\cdot}CH_2{\cdot}SiMe_n(OEt)_{3-n} + HO{\cdot}SiR_3 \longrightarrow$$

$$Me_2P(O){\cdot}O{\cdot}CH_2{\cdot}SiMe_n(O{\cdot}SiR_3)_{3-n}$$

$$[n = 1 \text{ or } 2]$$

Modifications of S-ester groups (all in phosphinodithioates) include the reaction of chloromethyl phosphinodithioates with sodium N,N-dimethyldithiocarbamate,[707,718]

$$Me_2P(S) \cdot S \cdot CH_2Cl + Me_2N \cdot CS_2^- \ Na^+ \longrightarrow$$

$$Me_2P(S) \cdot S \cdot CH_2 \cdot S \cdot CS \cdot NMe_2 + NaCl$$

a conversion of a carboxyamide substituent into a carbothionic amide by heating with phosphorus pentasulfide,[715] and methylations with diazomethane of a phenolic hydroxyl group,[558] and of a carbonyl group to its enol ether.[556]

XXXII. MODIFICATION OF GROUPS ATTACHED TO NITROGEN

Only a few examples have been reported of any of the following reactions. Amides have been alkylated with diazomethane,[759] or by reaction of their anions with alkyl halides.[268] Phosphinic hydrazides have been converted to hydrazones,[434,745,756] and have been acylated,[92,744] at the terminal nitrogen atom. Oxidation of several phosphinic, phosphinothioic, and phosphinimidic hydrazides has given the corresponding strongly colored azo-compounds,[91,92,93] while a p-phenylene di-(phosphinic amide) has been oxidized to the quinone di-(phosphinic imide).[555]

Two reactions which form heterocycles are those of phosphinic hydrazides with β-keto-esters,[756]

$$Ph_2P(O) \cdot NH \cdot NH_2 + R \cdot CO \cdot CH_2 \cdot CO_2R' \longrightarrow Ph_2P(O)-N\underset{O}{\overset{N}{\diagdown}}\diagup R$$

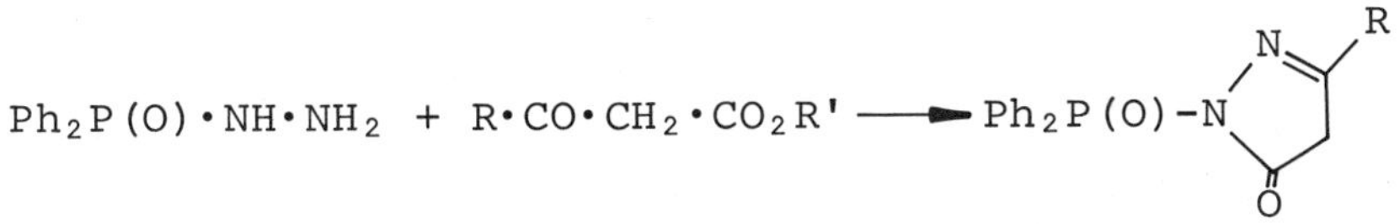

and of phosphinic isothiocyanates with aliphatic diazo-compounds.[759]

$$R_2P(O) \cdot N:C:S + R'CHN_2 \longrightarrow R_2P(O)-NH \diagup \underset{R'}{\overset{S}{\diagdown}} \overset{N}{\underset{N}{\parallel}}$$

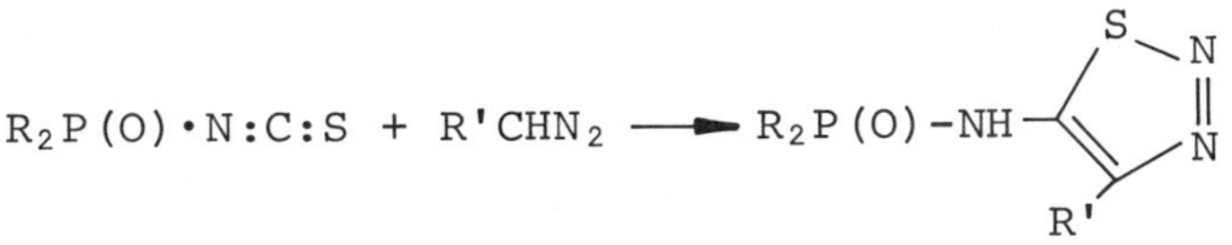

XXXIIa. PREPARATION OF PHOSPHINIC AND PHOSPHINOTHIOIC

UREAS. About twenty N-phosphinic ureas have been prepared by reaction of diphenylphosphinic isocyanate with primary and secondary amines in benzene.[772]

$$Ph_2P(O) \cdot NCO + RNH_2 \longrightarrow Ph_2P(O) \cdot NH \cdot CO \cdot NHR$$

A comparable number of N-phosphinic thioureas have been similarly prepared from diphenylphosphinic isothiocyanate,[758] and a few N-phosphinothioic thioureas, $Ph_2P(S) \cdot NH \cdot CS \cdot NRR'$,[704] and an N-phosphinic dithiocarbamate, $Ph_2P(O)\cdot NH\cdot CS\cdot SBu$,[758] have been obtained by analogous methods.

Several other, less generally convenient, routes have also been used to a few N-phosphinic ureas,[772] while N,N'-di(diphenylphosphinic) urea has been obtained by hydration of the corresponding carbodiimide.[783]

D. PHYSICAL PROPERTIES OF PHOSPHONIC ACIDS, AND OF THEIR THIO, SELENO, AND IMIDO ANALOGS

Apart from a few of the lower dialkylphosphinic acids which are liquids at room temperature, the great majority of phosphinic acids are colorless crystalline solids. The smallest acids are very soluble in cold water, but aqueous solubility decreases rapidly with increasing molecular weight while solubility in organic solvents tends to decrease, subject of course to the solubilizing or desolubilizing influences of substituents. Aqueous ethanol, toluene, and ligroin are commonly used recrystallization solvents. Unsubstituted aliphatic and aromatic phosphinic acids mostly have pK_a values in water between about 3.0 and 3.5, with variations from this range because of steric effects,[145,183,421,446] and higher values in aqueous ethanol.[503,584]

Phosphinothioic and phosphinodithioic acids usually have progressively lower melthing points than the corresponding phosphinic acids and are generally less soluble in water and somewhat stronger acids.[336,341,345,446] The colors of phosphinoselenothioic and phosphinodiselenoic acids are presumably due to separation of selenium since their esters are colorless.

Alkali-metal salts of phosphinic, phosphinothioic, and phosphinodithioic acids are water-soluble, and when isolated often form hydrated crystals. Crystalline salts, which may be used for characterization of the acids, are also formed with organic bases. Salts of phosphinic acids with some divalent metals (those of zinc and cobalt having been most investigated) are polymeric materials, which often have appreciable solubilities in organic solvents, and which in some cases can be drawn into fibers.[88,140,685] Structures in which metal atoms are connected alternately through single and triple phosphinate bridges have been indicated by x-ray studies.[250,251] Divalent metal salts of phosphinodithioic acids are well crystalline compounds, which melt comparitively sharply without decomposition and are monomeric or dimeric both in solution[445] and in the solid state.[115,335]

Melting points, boiling points, densities, and refractive indices of individual compounds are recorded in the compound list. Infrared and ^1H NMR spectra which have been run routinely on many of the more recently prepared compounds are not included, but values of ^{31}P chemical shifts are listed, and references are given to any detailed studies of infrared or NMR spectra as well as to those of ultraviolet and mass spectra and other, less commonly investigated, physical properties.

E. LIST OF COMPOUNDS

Derivatives are listed under their parent acid, except for phosphinimidic derivatives where each one forms a separate entry. Compounds are divided into five major types (phosphinic; phosphinothioic; phosphinodithioic; phosphinoselenoic, etc.; and phosphinimidic derivatives) and these primary divisions, which are very unequal in size, are subdivided into 31 structural classes, arranged so as to bring together acids with similar chemistry and prepared by similar methods. Since routes to aliphatic phosphinic acids are generally more dependent on their substituents than are those to aromatic phosphinic acids, the classification has been based not only on the general nature of the two organic groups attached to phosphorus but also on the types of substituents present in aliphatic groups.

Within each structural class, acids are listed in order of increasing seniority of the first of the two groups attached to phosphorus and then in order of increasing seniority of the second. In classes based on the presence of a substituent, the substituted group is placed first. Increasing seniority for ordering purposes depends on the following factors in the order listed: (a) increase in the number of carbon atoms linked together, i.e., not beyond a hetero-atom in a functional group, (b) increase in unsaturation or substitution, (c) increasing branching, and increasing proximity of branching to the phosphorus atom, and (d) increasing proximity of unsaturation, or of substituents, to phosphorus. The seniority of a group is not increased by the formation of functional derivatives of a substituent: acids containing functionally modified substituents in the first group are listed after the unmodified acid but before acids containing the unmodified first group and a different second group.

Each entry for an individual acid and its derivatives follows the following order:

Acid and its salts
O-esters
S-esters
Anhydride, and mixed anhydrides with other phosphinic acids (entry with the first of the two acids to be

listed)
Mixed anhydrides with carboxylic and sulfonic acids
Amides
Imides (mixed phosphinic imides listed under first acid)
N-acylated amides, and ureas
Hydrazides
Other derivatives

Within each part of the list concerned with esters or amides of a particular acid, those with unsubstituted aliphatic esterifying or N-substituting groups are listed first in order of increasing size, branching, and unsaturation, then those with substituted aliphatic groups, similarly arranged, then phenyl and substituted phenyl derivatives in order of increasing number of substituents, followed by those with naphthyl and other condensed aromatic groups.

The entry for each compound, i.e., an acid or a derivative, indicates the methods by which it has been prepared, records values of common physical constants, and gives literature references to these and to more complex physical methods. References are quoted to all recorded preparations (preliminary publications being omitted whenever a full paper covering all the same material has also appeared) except for a few of the very commonest compounds, whose preparations by the same route have been described many times. For physical constants, one value has been selected whenever, as is often the case, several numerically similar values are recorded in the literature. For such compounds, any odd widely different values have been ignored, but if only two sources are available and give significantly divergent values then both have been quoted. Since the listing of references to preparations is comprehensive, a complete list of recorded values of physical constants may fairly easily be obtained from them, if needed.

E.1. Phosphinic Acids and Their Derivatives

E.1.1. Unsubstituted Aliphatic Phosphinic Acids

Me_2PO_2H. Ia.[111,293] IIIa.[424,677] XXIVa.[257] m. 87.5-88.5°,[145] vapor pressure given by $\log_{10}p=7.313-2880/T$, b.(extrapolated) 377°, Trouton constant 20.2 cal./deg./mole,[111] x-ray structure determination,[252] IR,[367] IR and Raman spectra,[240,578] ^{31}P -48.6 ppm,[539,700] ^{31}P of anion -37.5 ppm,[700] mass spectrum,[271] pK_a 3.08,[145] 3.13 (in 7% EtOH), 5.15 (in 80% EtOH),[503] Be salt,[737] TiO, $Ti(OMe)_2$, and $Ti(OBu)_2$ derivs.,[22,151] Cr^3 salt,[690] Co salt, blue needles,[129] m. 342-343°,[686] Zn salt, decomp.

440°,[685] Me_2Al deriv., m. 43°,[130] Me_2Ga deriv., m. 54°,[130] Me_2In deriv., m. 75-76°,[130] Pr_3Si deriv., b_5 127-128°,[723] $HO{\cdot}CH_2CH_2N^+Me_3$ salt, m. 112.5°.[334]

Me ester. XXIe.[334,677] b_{10} 65-66°,[334] n_D^{12} 1.4350,[334] n_D^{22} 1.4299,[677] mass spectrum.[271]

Et ester. XXIe.[424] d_4^{25} 1.0278,[424] n_D^{25} 1.4261,[424] IR,[367] ^{31}P -50.3 ppm,[554] mass spectrum.[271]

$CH_2{:}CH{\cdot}CH_2$ ester. XXIe. b_{14} 97.5-98.5°, n_D^{22} 1.4456.[677]

$CH{\vdots}C{\cdot}CH_2$ ester. XXIe. $b_{2.5}$ 89-89.5°, n_D^{22} 1.4608.[677]

$MeCH{:}CH{\cdot}CH_2$ ester. XXIe. b_{11} 108°, n_D^{21} 1.4543.[677]

$CH_2{:}CMe{\cdot}CH_2$ ester. XXIe. $b_{13.5}$ 103-104°, n_D^{20} 1.4507.[677]

$MeC{\vdots}C{\cdot}CH_2$ ester. XXIe. $b_{2.5}$ 81-81.5°, n_D^{22} 1.4534.[677]

$(CH_2)_4$ di-ester. XXIe. m. 77-78°.[754]

$ClCH_2{\cdot}CH_2$ ester. XXIe. n_D^{20} 1.4581.[677]

$Me_2N{\cdot}CH_2{\cdot}CH_2$ ester. XXIe. b_3 78-82°, n_D^{13} 1.4509.[334]

$EtO{\cdot}SiMe_2{\cdot}CH_2$ ester. XXIIa. b_4 106-108°, d_4^{20} 1.0040, n_D^{20} 1.4355.[19]

$Et_3Si{\cdot}O{\cdot}SiMe_2{\cdot}CH_2$ ester. XXXI. b_2 117-118°, d_4^{20} 0.9609, n_D^{20} 1.4409.[20]

$Me_2PhSi{\cdot}O{\cdot}SiMe_2{\cdot}CH_2$ ester. XXXI. b_2 155-157°, d_4^{20} 1.0329, n_D^{20} 1.4820.[20]

$MePh_2Si{\cdot}O{\cdot}SiMe_2{\cdot}CH_2$ ester. XXXI. $b_{0.15}$ 172-174°, d_4^{20} 1.0780, n_D^{20} 1.5275.[20]

$(EtO)_2SiMe{\cdot}CH_2$ ester. XXIIa. b_2 114-115°, d_4^{20} 1.0390, n_D^{20} 1.4325.[19]

$(Et_3Si{\cdot}O)_2SiMe{\cdot}CH_2$ ester. XXXI. b_1 153-156°, d_4^{20} 0.9634, n_D^{20} 1.4442.[20]

$(Me_2PhSi{\cdot}O)_2SiMe{\cdot}CH_2$ ester. XXXI. $b_{0.15}$ 147-150°, d_4^{20} 1.0615, n_D^{20} 1.5034.[20]

$(MePh_2Si{\cdot}O)_2SiMe{\cdot}CH_2$ ester. XXXI. $b_{0.0001}$ 250°, d_4^{20} 1.1178, n_D^{20} 1.5580.[20]

$CH_2{\cdot}SiMe_2{\cdot}O{\cdot}SiMe_2{\cdot}CH_2$ di-ester. XXIIa. b_5 208-210°, d_4^{20} 1.0869, n_D^{20} 1.4550.[19]

$Me_2P(O){\cdot}O{\cdot}P(O)Me_2$. XXIg.[539] XXIh.[424] XXIIIh.[276] By thermal reorganization (4 1/2 hr at 450-480°) of $MePO_3HNa$,[257] m. 119-121°,[424] $b_{0.2}$ 135°,[539] b_{15} 190-192°,[424] ^{31}P -52.6 ppm,[539] ΔH(hydrolysis) -17.5 kcal $mole^{-1}$.[418]

1,2,3-thiadiazol-5-ylamide. From $Me_2P(O){\cdot}NCS + CH_2N_2$. m. 161-163°.[759]

benzimidazol-2-ylamide. XXIi. m. 194-196°.[282]

$Me_2P(O){\cdot}NH{\cdot}P(O)Me_2$. XXVb + hydrolysis,[702] ^{31}P -43.4 ppm,[565] K salt, needles, m. 253-254°.[702]

$Ph_2P(O){\cdot}NH{\cdot}P(O)Me_2$. XXVb + hydrolysis, needles, m. 217°, ^{31}P -44.4 and -21.3 ppm.[702]

$H_2N{\cdot}NH{\cdot}P(O)Me_2$. XXIi. m. 158-161°.[744]

$PhNH{\cdot}NH{\cdot}P(O)Me_2$. XXIi. m. 73-77°.[91]

$PhN{:}N{\cdot}P(O)Me_2$. From phenylhydrazide + HgO, dark red liquid, decomp. at room temperature.[91]

$EtMePO_2H$. IIIa,[475] VIc.[143] m. 7-8°,[143] $b_{1.5}$ 122°,[475] b_{11} 170-172°,[143] n_D^{20} 1.4514.[143]

Et ester. XXIe. b_{17} 94-98°.[476]

Bu ester. Ve,[40] XXIe.[799] $b_{1.5}$ 73-74°,[40] b_3 85°,[799] d_4^{20} 0.9689,[40] n_D^{20} 1.4350.[40]

i-Bu ester. XXIe. b_3 74-76°, d_4^{20} 0.9719, n_D^{20} 1.4308.[799]

C_6H_{13} ester. Ve. b_{12} 132-133°, d_4^{20} 0.9483, n_D^{20} 1.4381.[40]

$CH_2 \cdot CH_2$ di-ester. IVa. $b_{0.5}$ 117-118°, n_D^{20} 1.4712.[598]

$p\text{-}O_2N \cdot C_6H_4$ ester. XXIe. m. 65.5-67°.[476]

EtMeP(O)·O·P(O)EtMe. XXIh. b_{15} 189-193°, n_D^{20} 1.4701.[476]

EtMeP(S)·O·P(O)EtMe. Diastereoisomers: $b_{0.2}$ 124-127°, n_D^{20} 1.5051; $b_{0.2}$ 137-142°, n_D^{20} 1.5041; ^{31}P -98.1 and -54.0 ppm.[476]

Et_2PO_2H. IIIa.[447] VIg.[422] XIb.[297] m. 19°,[446,447] $b_{0.08}$ 92°,[145] $b_{1.3}$ 126°,[446] b_{21} 194-195°,[422] n_D^{20} 1.4551,[446] mass spectrum,[271] pK_a 3.29.[145]

Me ester. XXIe.[661] b_{12} 86°,[661] d_0^{20} 1.0261,[661] n_D^{20} 1.4392,[661] mass spectrum.[271]

Et ester. Va.[38,661,662] XXIe.[424] b_{10} 88-88.5°,[662] d_4^{20} 0.9908,[662] n_D^{20} 1.4437,[662] n_D^{25} 1.4375,[424] kinetics of hydrolysis.[15,67,453]

Pr ester. XXIe. b_{13} 103-104°, d_0^{20} 0.9916, n_D^{20} 1.4356.[661]

i-Pr ester. XXIe.[661] b_{14} 93.5°,[661] d_0^{20} 0.9666,[661] n_D^{20} 1.4337,[661] mass spectrum.[271]

Bu ester. Ve.[40] XXIe.[661] $b_{0.5}$ 80-81°,[40] b_{12} 115°,[661] d_0^{20} 0.9596,[661] n_D^{20} 1.4375.[661]

i-Bu ester. XXIe. b_{13} 112°, d_0^{20} 0.9530, n_D^{20} 1.4355.[661]

C_6H_{13} ester. XXIe. b_{13} 141-142.5°, d_0^{20} 0.9431, n_D^{20} 1.4420.[661]

C_7H_{15} ester. Ve.[40] XXIe.[661] $b_{0.5}$ 106-107°,[40] b_4 129-130°,[661] d_0^{20} 0.9364,[661] n_D^{20} 1.4440.[661]

$PhCH_2$ ester. XXIIIc. $b_{0.5}$ 133-135°.[284]

C_8H_{17} ester. XXIe. $b_{1.25}$ 125-126°, d_0^{20} 0.9306, n_D^{20} 1.4458.[661]

$(CH_2)_4$ di-ester. XXIe. $b_{0.1}$ 172°.[754]

$HO \cdot CH_2 \cdot CH_2$ ester. XXIe. $b_{0.1}$ 90-94°, n_D^{25} 1.4582, kinetics of hydrolysis.[453]

$MeO \cdot CH_2 \cdot CH_2$ ester. XXIe. $b_{0.05}$ 70-75°, n_D^{25} 1.4431, kinetics of hydrolysis.[453]

CCl_2:CH ester. IVh.[6,681] $b_{0.5}$ 81-83°,[681] d_4^{20} 1.2465,[681] n_D^{20} 1.4810.[681]

Ph ester. XXIe. $b_{0.6}$ 132-133°, d_4^{20} 1.0941, n_D^{20} 1.5164.[664]

$o\text{-}Me \cdot C_6H_4$ ester. XXIe. $b_{0.5}$ 138.5-139°, d_4^{20} 1.0778, n_D^{20} 1.5161.[664]

p-Me·C_6H_4 ester. XXIe. m. 53.5-54°, $b_{0.3}$ 154-155°.[664]

o-Cl·C_6H_4 ester. XXIe. $b_{0.25}$ 133°, d_4^{20} 1.1869, n_D^{20} 1.5242.[664]

p-Cl·C_6H_4 ester. XXIe. m. 35-38°, $b_{0.25}$ 134-135°, d_4^{20} 1.1963, n_D^{20} 1.5257.[664]

o-O_2N·C_6H_4 ester. XXIe. $b_{0.6}$ 168-170°, d_4^{20} 1.2312, n_D^{20} 1.5324.[664]

m-O_2N·C_6H_4 ester. XXIe. $b_{0.3}$ 171-171.5°, d_4^{20} 1.2288, n_D^{20} 1.5326.[664]

p-O_2N·C_6H_4 ester. XXIe.[664,683] $b_{0.18}$ 164.5-166°,[664] d_4^{20} 1.2356,[664] n_D^{20} 1.5432,[664] kinetics of alkaline hydrolysis,[248] rates of hydrolysis and of reaction with fly-brain cholinesterase.[227]

$Et_2P(O)$·O·$P(O)Et_2$. IIIa.[447] XXIh.[424,752] $b_{0.5}$ 124-125°,[447] b_{14} 186-188°,[424] d_4^{23} 1.1053,[424] n_D^{20} 1.4227,[752] n_D^{23} 1.4720.[424]

$Et_2P(S)$·O·$P(O)Et_2$. XXIf.[98,529] XXIIh.[535] $b_{0.05}$ 93-94°,[98] $b_{0.5}$ 146-147°,[529] n_D^{20} 1.5037,[98] n_D^{25} 1.5056.[529]

amide. XXIi. m. 62-64°.[547]

Et amide. XXIi. $b_{0.5}$ 120.5-122°, d_4^{20} 0.9890, n_D^{20} 1.4603.[664]

Et_2 amide. XXIi. b_{16} 134-135°, d_4^{20} 0.9493, n_D^{20} 1.4564.[664]

(CH_2:CH)MePO_2H.

Bu ester. Va. b_2 64-64.2°, d_4^{20} 0.9862, n_D^{20} 1.4452.[340]

(CH_2:CH)EtPO_2H.

Me ester. XXIe. b_{11} 80°, d_4^{20} 1.0500, n_D^{20} 1.4518.[362]

Et ester. XXIe. b_{14} 86.5°, d_4^{20} 1.0210, n_D^{20} 1.4495.[362]

Pr ester. XXIe. b_{10} 94-95°, d_4^{20} 0.9851, n_D^{20} 1.4468.[362]

i-Pr ester. XXIe. b_{10} 83°, d_4^{20} 0.9777, n_D^{20} 1.4440.[362]

CH_2:CH·CH_2 ester. XXIe. b_{10} 99-100°, d_4^{20} 1.0160, n_D^{20} 1.4640.[362]

Bu ester. Va.[340] XXIe.[362] b_2 68.5-69°,[340] b_{10} 106°,[362] d_4^{20} 0.9716,[340] n_D^{20} 1.4472.[340]

i-Bu ester. XXIe. b_{10} 99-100°, d_4^{20} 0.9674, n_D^{20} 1.4448.[362]

Me_2CH·CH_2·CH_2 ester. XXIe. b_{10} 112°, d_4^{20} 0.9561, n_D^{20} 1.4466.[362]

$ClCH_2CH_2$ ester. XXIe*. b_7 107°, d_4^{20} 1.1610, n_D^{20} 1.4695.[362]

Ph ester. XXIe. b_3 114-115°, d_4^{20} 1.1140, n_D 1.5240.[762]

$(CH_2:CH)_2PO_2H$. XXIa. $b_{0.002}$ 130-132°, d_4^{20} 1.1874, n_D^{20} 1.4880.[343]

$(CH⋮C)_2PO_2H$.

Me_2 amide. IVb. VIc*. m. 83°.[122]

$PrMePO_2H$. IIIa. $b_{0.0001}$ 122°, n_D^{20} 1.4518.[475]

$PrEtPO_2H$.

Pr ester. Va. b_8 104-105°, d_4^{20} 0.9657, n_D^{20} 1.4370.[38]

Bu ester. Ve. $b_{0.5}$ 83-84°, d_4^{20} 0.9461, n_D^{20} 1.4381.[40]

C_7H_{15} ester. Ve. $b_{0.5}$ 111-113°, d_4^{20} 0.9262, n_D^{20} 1.4438.[40]

p-$O_2N{\cdot}C_6H_4$ ester. XXIe. $b_{0.1}$ 140°, n_D^{30} 1.5411, rates of hydrolysis and of reaction with fly-brain cholinesterase.[227]

Pr_2PO_2H. IIIa.[436,447] VIg.[423] XXIIf.[448] m. 59.5°,[447] pK 3.46,[145] La, ZrO_2^{2+}, Fe^{3+}, Co^{3+}, Sn^{4+}, Ce^{3+}, and UO_2^{2+} salts.[447]

Et ester. XXIe. b_{14} 110-112°, d_4^{21} 0.9567, n_D^{21} 1.4369.[423]

Pr ester. IVa.[596] Va.[662] $b_{0.4}$ 95-97°,[662] d_4^{20} 0.9447,[662] n_D^{20} 1.4386.[662]

Bu ester. Va.[504] Ve.[352] b_1 78-79°,[504] $b_{2.5}$ 90.5-91.5°,[352] d_4^{20} 0.9343,[504] n_D^{20} 1.4389.[504]

CCl_2:CH ester. IVh. b_2 104-106°, d_4^{20} 1.1728, n_D^{20} 1.4733.[6]

CH_2:C(SEt) ester. IVh. b_2 115°, d_4^{20} 1.0476, n_D^{20} 1.4910.[794]

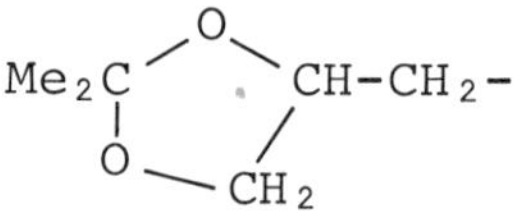

ester. IVa. $b_{0.2}$ 143-144°, d_4^{20} 1.0376, n_D^{20} 1.4530.[596]

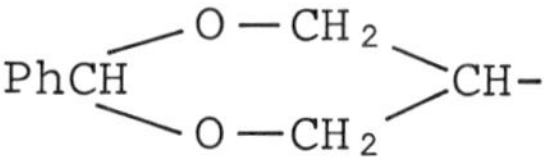

ester. IVa. $b_{0.0001}$ 117-118°, n_D^{20} 1.5190.[596]

NC·CH_2·CH_2 ester. IVa. d_4^{20} 1.0251, n_D^{20} 1.4722.[595]

EtO_2C·CH:CMe ester. XXIe. b_3 146-147°.[718]

Furfuryl ester. IVa. d_4^{20} 1.0440, n_D^{20} 1.4918.[595]

1,4:3,6 - dianhydro -D - glucitol di-ester. IVa. $b_{0.01}$ 190-195°, n_D^{20} 1.4780, $[\alpha]_D^{20}$ +79.8°.[599]

1,4:3,6- dianhydro -D -glucitol di-ester. IVa. $b_{0.01}$ 200-210°, n_D^{20} 1.4750, $[\alpha]_D^{20}$ +47.6°.[599]

o-$O_2N{\cdot}C_6H_4$ ester. XXIe. $b_{0.12}$ 158.5-160°, d_4^{20} 1.1697, n_D^{20} 1.5224.[664]

m-$O_2N{\cdot}C_6H_4$ ester. XXIe. $b_{0.16}$ 174-176°, d_4^{20} 1.1710, n_D^{20} 1.5256.[664]

p-$O_2N{\cdot}C_6H_4$ ester. XXIe.[664,683] $b_{0.28}$ 184-187°, d_4^{20} 1.1816, n_D^{20} 1.5360,[664] kinetics of alkaline hydrolysis,[248] rates of hydrolysis and of reaction with fly-brain cholinesterase.[227]

$Pr_2P(O){\cdot}O{\cdot}P(O)Pr_2$. IIIa.[447] XXIh.[423] m. 28-30°, b_1 138-140°, d_4^{21} 1.0299, n_D^{21} 1.4629.[423]

Amide. XXIi. m. 68-69.5°.[547]

Et_2 amide. XXIi. $b_{0.22}$ 104°, d_4^{20} 0.9287, n_D^{20} 1.4566.[664]

Morpholide. XXIi. $b_{1.2}$ 145-146.5°, d_4^{20} 1.0421, n_D^{20}

1.4806.[546]
Piperidide. XXIi. $b_{0.25}$ 117-118°, d_4^{20} 0.9888, n_D^{20} 1.4800.[546]

i-PrMePO$_2$H. VIc. $b_{0.05}$ 96-98°, n_D^{24} 1.4502.[143]

i-PrEtPO$_2$H. VIc. $b_{0.017}$ 89.5°, n_D^{25} 1.4567.[143]
Bu ester. Ve. $b_{0.5}$ 70-72°, d_4^{20} 0.9452, n_D^{20} 1.4365.[40]
p-$O_2N \cdot C_6H_4$ ester. XXIe. $b_{0.1}$ 130°, n_D^{30} 1.5384, rates of hydrolysis and of reaction with fly-brain cholinesterase.[227]

i-Pr$_2$PO$_2$H. VIg. m. 47.5-49.5°, $b_{0.02}$ 84-86°,[145] mass spectrum,[271] pK_a 3.56.[145]
Me ester. XXIe. Mass spectrum.[271]
CCl_2:CH ester. IVh. b_4 90-91°, d_4^{20} 1.1641, n_D^{20} 1.4700.[6]
p-$O_2N \cdot C_6H_4$ ester. XXIe.[664,683] m. 78.5-79.5°,[664] $b_{0.002}$ 155-158°,[683] n_D^{20} 1.5358,[683] kinetics of alkaline hydrolysis,[248] rates of hydrolysis and of reaction with fly-brain cholinesterase.[227]
Morpholide. XXIi. m. 35-36.5°, $b_{0.7}$ 132.5-134°, d_4^{35} 1.0456, n_D^{35} 1.4812.[546]
Piperidide. XXIi. m. 29-30°, $b_{0.7}$ 123-124.5°, d_4^{30} 0.9877, n_D^{30} 1.4810.[546]

(CH_2:CH·CH_2)MePO$_2$H.
i-Pr ester. Va. b_7 92°, d_4^{20} 0.9833, n_D^{20} 1.4435.[461]
i-Bu ester. Va. b_{11} 109-111°, d_4^{20} 0.9711, n_D^{20} 1.4483.[462]
C_5H_{11} ester. Va. b_8 120-122°, d_4^{20} 0.9640, n_D^{20} 1.4508.[461]
$Me_3C \cdot CH_2$ ester. Va. $b_{0.08}$ 76-78°, d_4^{20} 0.9500, n_D^{20} 1.4415.[462]

(CH_2:CH·CH_2)EtPO$_2$H.
Me ester. Va. b_8 92-94°, d_4^{20} 1.1564, n_D^{20} 1.4621.[466]
Et ester. Va. b_{15} 107°, d_4^{20} 0.9974, n_D^{20} 1.4519.[461]
Pr ester. Va. b_{10} 111-112.5°, d_4^{20} 0.9792, n_D^{20} 1.4480.[362]
i-Pr ester. Va. b_{10} 98-99°, d_4^{20} 0.9701, n_D^{20} 1.4450.[362]
CH_2:CH·CH_2 ester. Va.[362,605] b_2 86.5-88°,[605] b_{12} 115-117°,[362] d_4^{20} 1.0100,[362] n_D^{20} 1.4672.[362]
Bu ester. Va. b_{10} 120-122°, d_4^{20} 0.9670, n_D^{20} 1.4495.[362]
i-Bu ester. Va. b_{10} 116-118°, d_4^{20} 0.9596, n_D^{20} 1.4476.[362]
$Me_2CH \cdot CH_2 \cdot CH_2$ ester. Va. b_{10} 129-130°, d_4^{20} 0.9506, n_D^{20} 1.4485.[362]

(CH_2:CH·CH_2)(CH_2:CH)PO$_2$H.
Bu ester. Va. $b_{1.5}$ 78-78.5°, d_4^{20} 0.9790, n_D^{20} 1.4629.[340]

(CH_2:CH·CH_2)PrPO$_2$H.
Pr ester. Va. b_{12} 120-121°, d_4^{20} 0.9686, n_D^{20} 1.4478.[465]

i-Pr ester. Va. b_7 103-104°, d_4^{20} 0.9796, n_D^{20} 1.4608.[461]
C_5H_{11} ester. Va. b_9 140-141°, d_4^{20} 0.9615, n_D^{20} 1.4630.[461]

$(CH_2{:}CH{\cdot}CH_2)$i-PrPO_2H.
i-Pr ester. Va. b_{10} 100-102°, d_4^{20} 0.9577, n_D^{20} 1.4459.[461]

$(CH_2{:}CH{\cdot}CH_2)_2PO_2H$.
Et ester. Va. b_{15} 118-120°, d_4^{20} 1.0048, n_D^{20} 1.4676.[465]
$CH_2{:}CH$ ester. XXIe (using Me·CHO + Et_3N). $b_{0.03}$ 100-103°, d_4^{20} 1.0240, n_D^{20} 1.4840.[656]
Pr ester. Va. b_9 116-118°, d_4^{20} 0.9862, n_D^{20} 1.4640.[465]
Bu ester. Va.[340] VIb.[656] $b_{0.04}$ 89-91°,[656] b_2 66-66.5°,[340] d_4^{20} 0.9741,[340] n_D^{20} 1.4629.[340]
i-Bu ester. Va. b_{11} 120-122°, d_4^{20} 0.9682, n_D^{20} 1.4585.[462]
MeCH:CH·CH_2 ester. XXIe. $b_{0.04}$ 91-93°, d_4^{20} 1.0009, n_D^{20} 1.4810.[656]
$Me_3C{\cdot}CH_2$ ester. Va. $b_{0.15}$ 98-100°, d_4^{20} 0.9490, n_D^{20} 1.4540.[462]
Cyclo-C_6H_{11} ester. XXIe. $b_{0.07}$ 100-104°, d_4^{20} 1.0207, n_D^{20} 1.4860.[656]
$Me_2CH{\cdot}(CH_2)_4$ ester. XXIe. $b_{0.07}$ 118-120°, d_4^{20} 0.9445, n_D^{20} 1.460.[656]
2,5-Me_2-cyclohexyl ester. XXIe. $b_{0.03}$ 115-117°, d_4^{20} 1.0008, n_D^{20} 1.4850.[656]
2,5,5-Me_3-cyclohexyl ester. XXIe. $b_{0.07}$ 112-113°, d_4^{20} 0.9797, n_D^{20} 1.4770.[656]
PhCH:CH·CH_2 ester. XXIe. $b_{0.03}$ 146-147°, d_4^{20} 1.0681, n_D^{20} 1.5410.[656]
$ClCH_2{\cdot}CH_2$ ester. XXIe. $b_{0.04}$ 104-107°, d_4^{20} 1.1357, n_D^{20} 1.4860.[656]
$CCl_3{\cdot}CH_2$ ester. XXIe. $b_{0.04}$ 99-100°, d_4^{20} 1.2866, n_D^{20} 1.4970.[656]
$ClCH_2$·CHMe ester. XXIe. $b_{0.02}$ 99-100°, d_4^{20} 1.1048, n_D^{20} 1.4850.[656]
$Et_2N{\cdot}CH_2{\cdot}CH_2$ ester. XXIe. $b_{0.07}$ 116-118°, d_4^{20} 0.9785, n_D^{20} 1.4710, ethiodide, m. 74.5-76°.[656]
$O_2N{\cdot}CMe_2{\cdot}CH_2$ ester. XXIe. $b_{0.02}$ 121-123°, d_4^{20} 1.1192, n_D^{20} 1.4785.[656]

$(CH_2{:}C{:}CH)EtPO_2H$.
CH⋮C·CH_2 ester. Va*. $b_{2.5}$ 113.5-115°, d_4^{20} 1.0836, n_D^{20} 1.5056.[606]

$(CH_3{\cdot}C{\vdots}C)EtPO_2H$.
CH⋮C·CH_2 ester by base-catalyzed rearrangement of previous ester. b_3 121-124°, d_4^{20} 1.0845, n_D^{20} 1.4894.[606]

$(CH_3{\cdot}C{\vdots}C)_2PO_2H$.
Me_2 amide. IVb. VIc*. $b_{0.3}$ 140°.[122]

$BuMePO_2H$. IIIa. m. 36-37°, $b_{0.0001}$ 142°, n_D^{40} 1.4439.[475]
Bu ester. VIb. b_4 111-112°, d_4^{20} 0.9411, n_D^{20} 1.4409.[590]
$Me_2CH{\cdot}CH_2{\cdot}CH_2$ ester. VIb. b_7 132-133°, d_4^{20} 0.9343, n_D^{20} 1.4422.[590]

$BuEtPO_2H$.
Et ester. Va. b_{10} 106-108°, d_4^{20} 0.9932, n_D^{20} 1.4382.[38]
Pr ester. Ve. b_{10} 119-120°, d_4^{20} 0.9483, n_D^{20} 1.4395.[40]
Bu ester. Ve. $b_{0.5}$ 89-90°, d_4^{20} 0.9372, n_D^{20} 1.4400.[40]
C_8H_{17} ester. Ve. b_1 133-134°, d_4^{20} 0.9159, n_D^{20} 1.4467.[40]
$p\text{-}O_2N{\cdot}C_6H_4$ ester. XXIe. $b_{0.1}$ 150°, n_D^{30} 1.5351, rates of hydrolysis and of reaction with fly-brain cholinesterase.[227]
$BuEtP(S){\cdot}O{\cdot}P(O)BuEt$. XXIf. $b_{0.7}$ 164-166°, n_D^{25} 1.4922.[529]

$Bu(CH_2{:}CH)PO_2H$.
$ClCH_2{\cdot}CH_2$ ester. XXXe. From ester of 2-chloroethyl acid, b_2 106-107°, d_4^{20} 1.1048, n_D^{20} 1.4694.[337]
Bu ester. Va. b_2 90-90.5°, d_4^{20} 0.9486, n_D^{20} 1.4490.[340]

$Bu(CH_2{:}CH{\cdot}CH_2)PO_2H$.
Bu ester. Va. b_{12} 139-140°, d_4^{20} 0.9480, n_D^{20} 1.4522.[655]

Bu_2PO_2H. Ia.[750] IIIa.[447] VIa.[410,590] VIg.[422,423,591,755] m. 70-71°,[410] IR,[161] pK_a 3.41,[145] 3.50 (in 7% EtOH), 5.63 (in 80% EtOH);[503] Be salt;[238,737] Mg salt, needles, solubilities in H_2O, EtOH, C_6H_6;[174] Ca salt, needles, solubilities in H_2O, EtOH, C_6H_6;[174] Ba salt, solubilities in H_2O, EtOH, C_6H_6;[174] Co^{2+} salt, blue;[140,242] Co^{2+} mixed salt of Bu_2PO_2H and $(C_8H_{17})_2PO_2H$;[684] Cu^{2+} salt, green needles, solubilities in H_2O, EtOH, C_6H_6;[174] Zn salt,[140,242,684] x-ray crystal structure;[250] Zn mixed salt of Bu_2PO_2H and $(C_6H_{13})_2PO_2H$, x-ray crystal structure;[241] Zn mixed salt of Bu_2PO_2H and $(C_8H_{17})_2PO_2H$;[684] Zn mixed salt of Bu_2PO_2H and $(C_{10}H_{21})_2PO_2H$;[241] Bu_3Sn derivative, $b_{0.0001}$ 200-235°;[320] Pb^{2+} salt, needles, solubilities in H_2O, EtOH, C_6H_6.[174]
Me ester. XXIe.[113] b_3 95°,[113] d_4^{25} 0.9501,[569] n_D^{25} 1.4441,[569] viscosity (at 25°) 43.5 millipoise,[113] completely miscible with water below 35.5° and above 191°,[569] extractant for U and Pu.[112]
Et ester. XXIe.[102,113,423] b_2 103°,[113] b_{18} 149-151°,[423] d_4^{20} 0.9394,[569] n_D^{20} 1.4420,[569] n_D^{25} 1.4404,[113] viscosity (at 25°) 40.8 millipoise,[113] completely miscible with water below 14° and above 226°,[569] solubility at 25° of ester in water 13g/1,[113] solubility at 25° of water in ester

416g/l,[113] extractant for U and Pu.[112]

Pr ester. d_4^{20} 0.9298, n_D^{20} 1.4439, completely miscible with water above 243°.[569]

Bu ester. Va.[662,780] XIc.[102] XXIe.[102,113,419,435] b_1 126°,[113] d_4^{20} 0.9251,[511] d_4^{25} 0.9179,[569] n_D^{25} 1.4442,[569] magnetic rotation,[451,780] diamagnetic susceptibility,[779] dipole moment 3.29D,[510,511] viscosity (at 25°) 53.9 millipoise,[113] ΔH (formation) -183.6 kcal/mole,[742] ΔH (combustion) -2070.2 kcal/mole,[742] heat of mixing with $CHCl_3$,[419] completely miscible with water above 253°,[569] solubility at 25° of ester in water 4.5g/l,[113] solubility at 25° of water in ester 160g/l,[113] phase equilibria in system ester $-H_2O$ $-HNO_3$,[568] phase equilibria in system ester $-H_2O$ $-H_2SO_4$,[570] phase equilibria in system ester $-H_2O$ $-UO_2SO_4$,[570] extractant for U and Pu,[112] extractant for Zr,[770] complex $UO_2(NO_3)_2 \cdot 2Bu_2P(O)OBu$ m. 56-57°.[112]

C_5H_{11} ester. VIb. b_1 146-148°, d_4^{20} 0.9124, n_D^{20} 1.4443.[590]

cyclo-C_6H_{11} ester. XXIe. $b_{0.7}$ 139°, n_D^{25} 1.4650.[102]

C_9H_{19} ester. XXIIIc. $b_{0.002}$ 149°.[284]

$(CH_2)_4$ di-ester. XXIe. $b_{0.2}$ 219°.[754]

CCl_2:CH ester. IVh. b_6 143-144°, d_4^{20} 1.1259, n_D^{20} 1.4710.[6]

$BuO_2C \cdot CH_2$ ester. IIg. b_2 150-152°, d_4^{20} 1.0052, n_D^{20} 1.4530.[617]

$EtO_2C \cdot CHMe$ ester. IIg. b_1 155-156°, d_4^{20} 1.0243, n_D^{20} 1.4510.[617]

$EtO_2C \cdot CHPh$ ester. IIg. b_2 165-167°, d_4^{20} 1.0342, n_D^{20} 1.4800.[617]

$(EtO)_2P(O) \cdot CHMe$ ester. IIg. b_2 168-170°, d_4^{20} 1.0728, n_D^{20} 1.4625.[617]

$(EtO)_2P(O) \cdot CHPh$ ester. IIg. b_2 168-170°, d_4^{20} 1.0455, n_D^{20} 1.4720.[617]

o-$O_2N \cdot C_6H_4$ ester. XXIe. $b_{0.35}$ 184-186°, d_4^{20} 1.1250, n_D^{20} 1.5151.[664]

m-$O_2N \cdot C_6H_4$ ester. XXIe. $b_{0.2}$ 190-191°, d_4^{20} 1.1253, n_D^{20} 1.5168.[664]

p-$O_2N \cdot C_6H_4$ ester. XXIe.[664,683] $b_{0.28}$ 205-206°, d_4^{20} 1.1320, n_D^{20} 1.5264,[664] kinetics of alkaline hydrolysis,[248] rates of hydrolysis and of reaction with fly-brain cholinesterase.[227]

m-$CF_3 \cdot C_6H_4$ ester. XXIe. $b_{0.1}$ 118°, n_D^{25} 1.4634.[152]

3,5-$(CF_3)_2C_6H_3$ ester. XXIe. $b_{0.04}$ 100-103°, n_D^{25} 1.4375.[152]

4-HO-3,5-(t-Bu)$_2C_6H_2$ ester. XXIe. m. 86.5-88°.[679]

$Bu_2P(O) \cdot O \cdot P(O)Bu_2$. XXIh. b_2 170-171°, d_4^{21} 0.9908, n_D^{20} 1.4628.[423]

Et amide. XXIi. $b_{0.37}$ 157-158°, d_4^{20} 0.9363, n_D^{20} 1.4598.[664]

Et$_2$ amide. XXIi.[542,664] $b_{0.2}$ 126-127°, d_4^{20} 0.9173, n_D^{20} 1.4581.[664]
i-Pr$_2$ amide. VIc*. m. 65-71.5°, $b_{0.3}$ 124°.[542]
Morpholide. XXIi. m. 41.5-42.5°, $b_{0.7}$ 168-170°.[546]
Piperidide. XXIi. $b_{0.3}$ 140-141°, d_4^{20} 0.9638, n_D^{20} 1.4754.[546]
$Bu_2P(O){\cdot}NH{\cdot}P(O)Bu_2$. (XXIi + XXVa),[547] XXIIIg.[247] m. 121-122°,[547] 136.5-137.5°.[247]

i-Bu_2PO_2H. Ia.[750] VIa.[590] VId.[580] XXIIIa.[663] m. 43-45°,[580] $b_{0.4}$ 140-141°,[580] b_4 190-192°,[590] pK_a 3.70 (in 7% EtOH), 5.63 (in 80% EtOH).[503]
i-Bu ester, Va. b_3 99-102°, d_0^{20} 0.9151, n_D^{21} 1.4311.[663]
CCl_2:CH ester. IVh. b_3 107-108°, d_4^{20} 1.1268, n_D^{20} 1.4723.[6]
p-$O_2N{\cdot}C_6H_4$ ester. XXIe.[664,683] m. 41-43°,[664] $b_{0.15}$ 176.5-177°,[664] n_D^{20} 1.5269,[683] kinetics of alkaline hydrolysis.[248]
Amide. XXIi. m. 76-77°.[547]
Piperidide. XXIi. m. 55-56.5°, $b_{1.5}$ 140-140.5°.[546]
$CF_3{\cdot}CO{\cdot}NH$ $\cdot$P(O)(i-Bu)$_2$. XXIIIg. m. 87.5-88.5°.[338]
$CCl_3{\cdot}CO{\cdot}NH{\cdot}P(O)$(i-Bu)$_2$. XXIIIg. m. 181.5-183°.[338]

s-Bu_2PO_2H. VIg. $b_{0.03}$ 120°, n_D^{20} 1.4610.[141]
p-$O_2N{\cdot}C_6H_4$ ester. XXIe. $b_{0.01}$ 170-173°, n_D^{20} 1.5352,[683] kinetics of alkaline hydrolysis.[248]

t-$BuMePO_2H$. VIa.[101] XXIa.[147] m. 88-89°.[101]

t-$BuEtPO_2H$. VIa.[191,416] VIc.[227] XXIa.[147] m. 100-101°.[147]
p-$O_2N{\cdot}C_6H_4$ ester. XXIe. $b_{0.2}$ 145°, n_D^{30} 1.5381, rates of hydrolysis and of reaction with fly-brain cholinesterase.[227]

t-Bu(i-Pr)PO_2H. VIa.[143] XXIa.[147] m. 84°.[147]

t-$BuBuPO_2H$. VIa. Low-melting solid, $b_{0.02}$ 138°.[101]

t-Bu_2PO_2H. Ia.[292] VIa.[145] VId.[145] XXIa.[147] m. 208-210°,[145] pK_a 4.24.[145]
Ph ester. IVa. $b_{0.2}$ 130-140° (bath temp.).[749]

(MeCH:CH$\cdot CH_2$)$MePO_2H$.
Pr ester. Va. $b_{0.03}$ 72-74°, d_4^{20} 0.9767, n_D^{20} 1.4512.[463]
Bu ester. Va. b_8 126-129°, d_4^{20} 0.9635, n_D^{20} 1.4500.[463]
i-Bu ester. Va. $b_{0.03}$ 80-82°, d_4^{20} 0.9586, n_D^{20} 1.4490.[463]

(MeCH:CH$\cdot CH_2$)$EtPO_2H$.
Me ester. Va. b_1 65-67°, d_4^{20} 0.9943, n_D^{20} 1.4522.[464]
Et ester. Va. $b_{0.03}$ 76-79°, d_4^{20} 0.9828, n_D^{20} 1.4550.[464]
CH_2:CH$\cdot$CHMe ester. Va*. $b_{2.5}$ 97-98°, d_4^{20} 0.9714, n_D^{20} 1.4620.[605]

(MeCH:CH$\cdot CH_2$)i-$PrPO_2H$.
Me ester. Va. $b_{0.1}$ 64-66°, d_4^{20} 0.9691, n_D^{20} 1.4503.[464]

CH_2:CH·CH_2 ester. Va. b_{14} 128-129°, d_4^{20} 0.9679, n_D^{20} 1.4600.[464]

(MeCH:CH·CH_2)(CH_2:CH·CH_2)PO_2H.

Me ester. Va. $b_{0.15}$ 76-78°, d_4^{20} 1.0135, n_D^{20} 1.4670.[464]

Pr ester. Va. $b_{0.04}$ 90-92°, d_4^{20} 0.9730, n_D^{20} 1.4648.[463]

CH_2:CH·CH_2 ester. Va. $b_{0.2}$ 93-95°, d_4^{20} 1.0024, n_D^{20} 1.4787.[464]

Bu ester. Va. $b_{0.05}$ 114-115°, d_4^{20} 0.9630, n_D^{20} 1.4630.[463]

i-Bu ester. Va. $b_{0.03}$ 94-96°, d_4^{20} 0.9553, n_D^{20} 1.4620.[463]

(CH_2:CH·CHMe)EtPO_2H.

MeCH:CH·CH_2 ester. Va*. b_2 90-90.5°, d_4^{20} 0.9825, n_D^{20} 1.4698.[605]

(Me_2C:CH)EtPO_2H.

Et ester. Va. b_{10} 100-103°, d_4^{20} 1.0023, n_D^{20} 1.4560.[363]

(C_5H_{11})EtPO_2H.

p-O_2N·C_6H_4 ester. XXIe. $b_{0.1}$ 160°, n_D^{30} 1.5365, rates of hydrolysis and of reaction with fly-brain cholinesterase.[227]

(C_5H_{11})(CH_2:CH·CH_2)PO_2H.

C_5H_{11} ester. Va. b_8 156-158°, d_4^{20} 0.9329, n_D^{20} 1.4535.[461]

$(C_5H_{11})_2PO_2H$. IIb. m. 68-69°,[789] Be salt.[737]

C_5H_{11} ester. Va. b_1 119°, d_4^{20} 0.9154, n_D^{20} 1.446,[780] magnetic rotation,[451,780] diamagnetic susceptibility,[779] dipole moment 3.22D.[510,511]

p-O_2N·C_6H_4 ester. XXIe. $b_{0.03}$ 173-176°, n_D^{20} 1.5135,[683] kinetics of alkaline hydrolysis.[248]

(Me_2CH·CH_2·CH_2)MePO_2H.

Me_2CH·CH_2·CH_2 ester. VIb. b_4 140-141°, d_4^{20} 0.9266, n_D^{20} 1.4398.[590]

$(Me_2CH{\cdot}CH_2{\cdot}CH_2)_2PO_2H$. VIg. m. 37-39°, b_3 230-233°.[591]

Bu ester. XXIe. $b_{0.4}$ 121-122°, d_4^{22} 0.9076, n_D^{20} 1.4440.[435]

($EtCMe_2$)t-BuPO_2H. VId. m. 154-158°.[148]

$(EtCMe_2)_2PO_2H$. IVe. m. 132-136°.[148]

(cyclo-C_5H_9)MePO_2H. XXIa. XXIIIa. b_2 171-172.6°, d_4^{20} 1.1390, n_D^{20} 1.4900; p-toluidine salt m. 70-71°.[350]

Bu ester. Va. Also by hydrogenation of ester of cyclopentadienyl acid, b_1 90-91°, d_4^{20} 1.0026, n_D^{20} 1.4628.[350]

(cyclopentadienyl)MePO_2H.

Bu ester. Va. b_1 91-92°, d_4^{20} 1.0512, n_D^{20} 1.4918.[350]

(C_6H_{13})MePO_2H. VIa. b_1 216-218°.[590]

Me_2CH·CH_2·CH_2 ester. VIb. b_2 139-141°, d_4^{20} 0.9143, n_D^{20} 1.4432.[590]

$(C_6H_{13})EtPO_2H$.
Me ester. XIIb. b_{11} 126-127°, d_4^{20} 1.0471, n_D^{20} 1.4755.[627]
Et ester. Ve. $b_{0.5}$ 93°, d_4^{20} 0.9302, n_D^{20} 1.4413.[40]
i-Bu ester. Ve. $b_{0.5}$ 106-107°, d_4^{20} 0.9235, n_D^{20} 1.4413.[40]
p-$O_2N{\cdot}C_6H_4$ ester. XXIe. $b_{0.1}$ 160°, n_D^{30} 1.5241, rates of hydrolysis and of reaction with fly-brain cholinesterase.[227]

$(C_6H_{13})(CH_2{:}CH{\cdot}CH_2)PO_2H$.
Bu ester. Va. b_2 132-133°, d_4^{20} 0.9349, n_D^{20} 1.4530.[465]

$(C_6H_{13})_2PO_2H$. IIb.[789] VIa.[410] VIg.[591,755] XIIa.[789] m. 78-79°;[789] Co salt;[242] Zn salt.[242]
Et ester. XXIe.[113] $b_{1.5}$ 143°,[113] d_4^{25} 0.9102,[113] n_D^{25} 1.4460,[113] viscosity (at 25°) 75.7 millipoise,[113] solubility at 25° of ester in water < 0.1 g/l,[113] solubility at 25° of water in ester 150 g/l,[113] extractant for U and Pu.[112]
Bu amide. IIc. $b_{0.05}$ 154-155°, n_D^{25} 1.4597.[734]

$[CH_2{:}CH{\cdot}(CH_2)_4]EtPO_2H$.
Et ester. XIIb. b_2 102-103°, d_4^{20} 1.0004, n_D^{20} 1.4712.[631]

(cyclo-C_6H_{11})$MePO_2H$.
Me ester. Va. $b_{1.2}$ 90°.[402]
(-)-menthyl esters. XXIe and fractional crystallization.[402] Diastereoisomers: m. 110-111°, $[\alpha]_D^{\sim 25}$ -54° (c~2,C_6H_6), and m. 80-81°, $[\alpha]_D^{\sim 25}$ -59° (c~2,C_6H_6),[402] NMR.[460]

(cyclo-C_6H_{11})$EtPO_2H$. XXIIIa. m. 59-60.5°.[627]
Et ester. XIIb.[627] XIII.[545] $b_{0.8}$ 86-87°,[545] b_{11} 142-143°,[627] d_4^{20} 1.0241,[627] n_D^{20} 1.4705.[627]

(cyclo-C_6H_{11})$_2PO_2H$. Ia.[750] VIa.[309] XIIa.[584 594] Also by hydrogenation of Ph_2PO_2H,[220] m. 143°,[309] pK_a 5.92 (in 75% EtOH), 6.64 (in 95% EtOH).[503]
Me ester. XXIIc. m. 58-59°.[309]
Et ester. XIII. b_{11} 162-164°, n_D^{20} 1.4402.[545]
Bu ester. XXIe. $b_{0.3}$ 134°, n_D^{24} 1.4900.[102]
cyclo-C_6H_{11} ester. XIc. XXIe. m. 85-86°.[102]
Ph ester. XIII. $b_{0.7}$ 121-124°, n_D^{20} 1.4438.[545]
Hydrazide. XXIi. m. 150-155°.[92]

$(C_7H_{15})EtPO_2H$.
Et ester. XIIb. b_{10} 147-148°, d_4^{20} 0.9350, n_D^{20} 1.4440.[627]
Bu ester. XXIIb. b_{13} 170-171°, d_4^{20} 0.9230, n_D^{20} 1.4460.[627]
p-$O_2N{\cdot}C_6H_4$ ester. XXIe. $b_{0.1}$ 165°, n_D^{30} 1.5212, rates of hydrolysis and of reaction with fly-brain cholinesterase.[227]

$(C_7H_{15})_2PO_2H$. IIb.[789] VIg.[591,755] XIIa.[594] m. 77-78°;[789] Na salt, IR;[578] Be salt.[737]
C_7H_{15} ester. Va.[780] $b_{0.5}$ 164°,[780] d_4^{20} 0.8955,[780]

n_D^{20} 1.452,[780] magnetic rotation,[451,780] diamagnetic susceptibility,[779] dipole moment 3.16 D.[510,511]

$(C_5H_{11}\cdot CH{:}CH)EtPO_2H$.
Et ester. XIIb. b_3 124-125°, d_4^{20} 0.9572, n_D^{20} 1.4560.[630]
i-Pr ester. XIIb. b_{12} 145-147°, d_4^{20} 0.9423, n_D^{20} 1.4530.[630]

$(cyclo\text{-}C_6H_{11}\cdot CH_2)_2PO_2H$. By hydrogenation of $(PhCH_2)_2PO_2H$, m. 100-101°.[396]

$(PhCH_2)MePO_2H$. IIIa. m. 121-122°.[475]
Et ester. Va. $b_{0.4}$ 120-125°.[743]

$(PhCH_2)EtPO_2H$. XXIIIa. m. 71-72°.[62]
Et ester. Ve. b_1 115-115.5°, d_4^{20} 1.0793, n_D^{20} 1.522.[40]
i-Pr ester. Va. b_6 153-156°.[62]
Bu ester. Ve. $b_{0.5}$ 124-125°, d_4^{20} 1.0397, n_D^{20} 1.5050.[40]
C_9H_{19} ester. Ve. $b_{0.5}$ 164-166°, d_4^{20} 0.9814, n_D^{20} 1.4948.[40]

$(PhCH_2)BuPO_2H$. IIb.[179] XXIIIa.[688] m. 95.5-96.5° from H_2O, 109-110° from Me_2CO-hexane.[179]
Bu ester. Va. b_2 145°, n_D^{20} 1.5013.[688]

$(PhCH_2)_2PO_2H$. IIa.[533] VIa.[691] VIc.[528] XIa.[797] XXIIf.[98] XXIIIa.[352] m. 192°;[454] $NH_4(7H_2O)$, $Na(7H_2O)$, $K(7H_2O)$, $Mg(3H_2O)$, $Ba(8H_2O)$, Cu, Ag, and Cd salts.[454]
Me ester. XXIIa. m. 75°.[467]
Bu ester. IXc. m. 77.5-80°.[352]
(-)-menthyl ester. XXIe. m. 143-144°, $[\alpha]_D^{20}$ -110° (c 5, EtOH), pyrolysis.[500]

$(C_8H_{17})EtPO_2H$.
$Me_2CH\cdot CH_2\cdot CH_2$ ester. VIb. b_3 158-160°, d_4^{20} 0.9088, n_D^{20} 1.4443.[590]
$p\text{-}O_2N\cdot C_6H_4$ ester. XXIe. $b_{0.1}$ 170°, n_D^{30} 1.5200, rates of hydrolysis and of reaction with fly-brain cholinesterase.[227]

$(C_8H_{17})_2PO_2H$. IIb.[789] VIg.[591,755] XIIa.[585,789] Also by thermal disproportionation of $(C_8H_{17})_2POH$,[532] m. 85°,[789] b_1 325-327°,[591] extractant for Y^{3+}, Ce^{3+}, Eu^{3+}, Tm^{3+}, Yb^{3+}, UO_2^{2+}, Am^{3+}, and Cm^{3+};[585] Na salt, IR;[578] Zn salt.[684]
Bu ester. IIc. $b_{0.15}$ 158°, n_D^{25} 1.4515.[734]
$BuO_2\cdot CH_2$ ester. IIg. b_2 178-180°, d_4^{20} 0.9436, n_D^{20} 1.4540.[617]
$EtO_2C\cdot CHMe$ ester. IIg. b_2 198-200°, d_4^{20} 0.9568, n_D^{20} 1.4540.[617]
$(EtO)_2P(O)\cdot CHMe$ ester. IIg. b_2 192-194°, d_4^{20} 0.9875, n_D^{20} 1.4520.[617]
Bu amide. IIc. $b_{0.1}$ 170°, n_D^{25} 1.4610.[734]

$(C_6H_{13}\cdot CH{:}CH)EtPO_2H$.
Et ester. XIIb. $b_{15.5}$ 157-158°, d_4^{20} 0.9568, n_D^{20}

1.4585.[630]
$(PhCH_2 \cdot CH_2)_2PO_2H$. VIg. m. 95-97°.[62]
$(o\text{-}Me \cdot C_6H_4 \cdot CH_2)_2PO_2H$. XIa. m. 153-154°.[797]
$(m\text{-}Me \cdot C_6H_4 \cdot CH_2)_2PO_2H$. XIa. m. 110-112°.[797]
$(p\text{-}Me \cdot C_6H_4 \cdot CH_2)_2PO_2H$. XIa. m. 173-174°.[797]
$(PhCH{:}CH)EtPO_2H$.
Et ester. Va. $b_{0.03}$ 118-119°, d_4^{20} 1.0821, n_D^{20} 1.5525,[455] polarography.[456]
$(PhCH{:}CH)BuPO_2H$. Va. $b_{0.04}$ 162-163°, d_4^{20} 1.0355, n_D^{20} 1.5268, polarography.[456]
$(PhCH{:}CH)_2PO_2H$. XIII. m. 157-158°.[199]
$(PhC{\vdots}C)(PhCH{:}CH)PO_2H$. From $(PhCCl{:}CH)(PhCH{:}CH)PO_2H$ + alcoholic KOH. m. 84-86°.[198]
$(PhC{\vdots}C)_2PO_2H$. XIa.[199] From $(PhCCl{:}CH)_2PO_2H$ + alcoholic KOH.[198] m. 146-147°.[198]
Et ester. $b_{0.06}$ 190-193°.[395]
Ph ester. XXIe. m. 83-84°.[198]
$p\text{-}Cl \cdot C_6H_4$ ester. XXIe. m. 96-97°.[198]
$m\text{-}O_2N \cdot C_6H_4$ ester. XXIe. m. 105-106.5°.[198]
$p\text{-}O_2N \cdot C_6H_4$ ester. XXIe. m. 103-104°.[198]
Ph amide. XXIi. m. 179-179.5°.[198]
$p\text{-}ClC_6H_4$ amide. XXIi. m. 140-142°.[198]
$2,4\text{-}Cl_2C_6H_3$ amide. XXIi. m. 121.5-122°.[198]
$p\text{-}BrC_6H_4$ amide. XXIi. m. 143-145°.[198]
$p\text{-}MeO \cdot C_6H_4$ amide. XXIi. m. 132-133°.[198]
$p\text{-}MeC_6H_4$ amide. XXIi. m. 176-177°.[198]
$(C_9H_{19})EtPO_2H$. XXIIIa. m. 35.5-36°.[627]
Et ester. XIIb. b_{12} 172°, d_4^{20} 0.9310, n_D^{20} 1.4460.[627]
$(C_9H_{19})_2PO_2H$. IIb. m. 84-85°.[789]
$(PhCH_2 \cdot CH_2 \cdot CH_2)_2PO_2H$. VIg. m. 99-101°.[152]
$m\text{-}CF_3 \cdot C_6H_4$ ester. XXIe. $b_{0.05}$ 210°, n_D^{25} 1.5382.[152]
$(p\text{-}Et \cdot C_6H_4 \cdot CH_2)_2PO_2H$. XIa. m. 167-168°.[796]
$(C_{10}H_{21})MePO_2H$. VIb. m. 28°, b_4 173-175°, d_4^{30} 0.8830, n_D^{30} 1.4412.[590]
$(C_{10}H_{21})EtPO_2H$.
Et ester. Ve. b_1 130-131°, d_4^{20} 0.9160, n_D^{20} 1.4480.[40]
Pr ester. Ve. b_1 140-141°, d_4^{20} 0.9129, n_D^{20} 1.4478.[40]
Bu ester. Ve. $b_{1.5}$ 148-150°, d_4^{20} 0.9083, n_D^{20} 1.4489.[40]
C_9H_{19} ester. Ve. $b_{0.5}$ 181-182°, d_4^{20} 0.8947, n_D^{20} 1.4521.[40]
$(C_{10}H_{21})_2PO_2H$. IIb. XIIa. m. 87.7-88.3°;[789] Mg salt, needles, m. 82-83°, solubilities in H_2O, EtOH, C_6H_6;[174] Ca salt, waxy needles, solubilities in H_2O, EtOH, C_6H_6;[174] Ba salt, needles, m. < 100°, solubilities in H_2O, EtOH, C_6H_6;[174] Co salt;[242] Cu^{2+} salt, blue needles, m. 94°, solubilities in H_2O, EtOH, C_6H_6;[174] Zn salt;[242] Pb^{2+} salt, solubilities in H_2O, EtOH, C_6H_6.[174]
$(p\text{-}PrC_6H_4 \cdot CH_2)_2PO_2H$. XIa. m. 136-137°.[797]
$(p\text{-}i\text{-}PrC_6H_4 \cdot CH_2)_2PO_2H$. XIa. m. 157-158°.[797]

(PhCH:CH·CH:CH)(PhCH:CH)PO_2H. XXIa. m. 150-151°.[196]
Ph ester. XXIe. m. 67-70°.[196]
Ph amide. XXIi. m. 183-185°.[196]
$(PhCH{:}CH{\cdot}CH{:}CH)_2PO_2H$. XXIa. m. 201-202°.[196]
Ph ester. XXIe. m. 144-145.5°.[196]
Ph amide. XXIi. m. 196-197°.[196]
$(C_{11}H_{23})_2PO_2H$. IIb. m. 89-90°.[789]
$(C_{12}H_{25})_2PO_2H$. IIb.[789] VIg.[755] XIIa.[789] m. 93.8-94.8°.[789]
$(C_{14}H_{29})_2PO_2H$. IIb. XIIa. m. 97-98°.[789]
$(C_{16}H_{33})_2PO_2H$. IIb. XIIa. m. 102.5-103.5°.[789]
$(C_{18}H_{37})_2PO_2H$. IIb. XIIa. m. 105.3-106°.[789]

E.1.2. Aliphatic/Aromatic Phosphinic Acids with Unsubstituted Aliphatic Groups (and Unsubstituted or Substituted Aromatic Groups)

$PhMePO_2H$. Ia.[183] IIIa.[475] IIIb.[676] Vd.[521,602] Vf.[409] XIa.[300] XIb.[297] XXIa.[84] XXIIIa.[25,128] m. 133-134°,[409,475,602,676] ^{31}P -40.1 ppm,[539] pK_a 2.96;[183] Be salt;[737] TiO and $Ti(OEt)_2$ derivs.;[151] Co salt, blue needles,[129] m. 210-211°;[686] Zn salt,[88,685] softens 200°, decomp. 415°;[88] (-)-hydrindamine salt, m. 175° decomp., $[\alpha]_D^{17}$ -19.65°;[602] (±)-hydroxyhydrindamine salt, m. 170° decomp.;[602] (-)-menthylamine salt, m. 188-189°, $[\alpha]_D^{15}$ -22.35°;[602] cinchonine salt ($2H_2O$), m. 170-172°, $[\alpha]_D^{17}$ +126.7°;[602] cinchonidine salt ($4H_2O$), m. 154°,[602,676] $[\alpha]_D^{17.5}$ -70.5° (EtOH);[602] quinine salt ($4H_2O$), m. 164-166°, $[\alpha]_D^{15}$ -113.45° (EtOH).[602]
Me ester. Va.[31,128,633] XXIe.[243,262] $b_{0.1}$ 83-85°,[262] $b_{0.6}$ 104-106°,[264] b_{14} 142°,[31] d_4^{20} 1.1561, n_D^{20} 1.5244,[633] mass spectrum.[107]
Et ester. Va.[636] XXIe.[128,243] b_{10} 135°,[636] b_{20} 159°,[128] d_4^{20} 1.1040,[636] n_D^{20} 1.5169.[636]
$CH_2{:}CH$ ester. XXIe (using MeCHO + Et_3N). b_2 102-104°, d_4^{20} 1.1166, n_D^{20} 1.5224.[233]
i-Pr ester. Va.[636] VIb.[69] $b_{0.6}$ 95°,[69] b_{10} 146°,[636] d_4^{20} 1.0764, n_D^{20} 1.5097.[636]
$CH_2{:}CH{\cdot}CH_2$ ester. Va. b_1 112-113°, d_0^{20} 1.1110, n_D^{20} 1.5208.[360]
cyclo-C_6H_{11} ester. XXIe. XXIIIc. $b_{0.5}$ 130°, n_D^{22} 1.5259.[128]
$PhCH_2$ ester. XXIe. $b_{0.5}$ 152-154°.[264]
(-)-$C_6H_{13}{\cdot}CHMe$ ester. XXIe. $b_{0.6}$ 113°, n_D^{20} 1.4945, $[\alpha]_D^{20}$ +4.46°.[264]
trans-2-i-Pr-cyclohexyl esters. XXIe and fractional crystallization. m. 67-70° and 95-97°, NMR.[460]
(-)-menthyl (R)-ester. XXIe and fractional crystallization. m. 89°, $[\alpha]_D^{\sim 25}$ -16° (c∿2, C_6H_6),[402]

NMR,[460] ORD.[459]
(−)-menthyl (S)-ester. XXIe and fractional crystallization.[402,575] $[\alpha]_D^{\sim 25}$ −94° (c∿2, C_6H_6),[402] $[\alpha]_D^{25}$ −93.8° (c 1.45, $CHCl_3$),[575] NMR,[460] ORD.[459]
cholesteryl (S)-ester. XXIe and fractional crystallization. m. 134-134.5°, $[\alpha]_D^{25}$ −81.4° (c 4.53, $CHCl_3$).[575]
$ClCH_2 \cdot CH_2$ ester. Va. b_2 153-156°, d_4^{20} 1.2582, n_D^{20} 1.5360.[235]
$CCl_3 \cdot CMe_2$ ester. Va. $b_{0.08}$ 115-117°, d_4^{20} 1.3333, n_D^{20} 1.5361.[449]
3,5-(t-Bu)$_2$-4-HO·C_6H_2 ester. XXIe. m. 128.5-130°.[679]
PhMeP(O)·O·P(O)PhMe. XXIg.[539] XXIIIh.[276] m. 94°,[539] b_2 200°,[539] ^{31}P −43.7 ppm.[539]
PhMeP(S)·O·P(O)PhMe. XXIf. $b_{0.01}$ 188-192°.[476]
(S,R)-PhMeCH·NH·P(O)PhMe. XXIi + chromatography. m. 118.5-119.7°, $[\alpha]_D^{25}$ −16.1° (c 1.38, $CHCl_3$).[575]
(S,S)-PhMeCH·NH·P(O)PhMe. XXIi + chromatography. Also formed with inversion at P by reactions of menthyl (S)-ester and of cholesteryl (S)-ester with (S)-PhMeCH·NHLi, m. 133-134°, $[\alpha]_D^{25}$ −64.6°. (c 1.96, $CHCl_3$).[575]
(±)-PhNH·P(O)PhMe. XXIi. m. 142°.[243]
(S)-PhNH·P(O)PhMe. Formed with retention of configuration by reaction of the carbanion of (R)-PhMeP(O)·CH_2Ph with PhCH:NPh,[301] also formed with inversion at P by reactions of menthyl (S)-ester and of cholesteryl (S)-ester with PhNHLi,[575] m. 164°,[301] $[\alpha]_D^{25}$ −26.1° (c 0.755, MeOH).[575]
$H_2N \cdot NH \cdot P(O)PhMe$. XXIIIf. m. 128-132°.[744]
PhMeP(O)·NH·NH·P(O)PhMe. XXIi. m. 198-204°.[744]

$PhEtPO_2H$. VII.[495] XIa.[300] XIb.[319] XXIIIa.[24] Also by hydrogenation of Ph(CH_2:CH)PO_2H,[494] m. 79-80°.[24]
Me ester. XXIIIc*. $b_{0.8}$ 106-107°, n_D^{23} 1.5218, also prepared optically active but not optically pure,[263] mass spectrum.[107]
Et ester. Va.[24,633,636] $b_{0.4}$ 97-98°,[666] $b_{2.5}$ 122-123°,[633] b_{10} 140°,[636] b_{16} 162-164°,[24] d_4^{20} 1.0829,[636] n_D^{20} 1.5129,[636] polarography.[456]
i-Pr ester. Va. b_9 146.5°, d_4^{20} 1.0573, n_D^{20} 1.5078.[636]
CH_2:CH·CH_2 ester. Va. b_3 124-125°, d_4^{20} 1.0854, n_D^{20} 1.5155.[360]
$ClCH_2 \cdot CH_2$ ester. Va. b_2 178-180°, d_4^{20} 1.2472, n_D^{20} 1.5400.[235]
CCl_2:CH ester. IVh. b_5 168-170°, d_4^{20} 1.2678, n_D^{20} 1.5437.[387]
p-$O_2N \cdot C_6H_4$ ester. XXIe. m. 80-81°, rates of hydrolysis and of reaction with fly-brain cholinesterase.[227]
PhEtP(S)·O·P(O)PhEt. XXIf. $b_{0.2}$ 160-165°, n_D^{30} 1.5902.[529]

$Ph(CH_2:CH)PO_2H$. XXIa.[234] XXXe.[343,494] m. 120-121°.[343]
Me ester. XXIe. b_1 93-95°, d_4^{20} 1.1880, n_D^{20} 1.5336.[234]
Et ester. XXIe. b_1 116-118°, d_4^{20} 1.1150, n_D^{20} 1.5249.[234]
Pr ester. XXIe. b_1 118-120°, d_4^{20} 1.1120, n_D^{20} 1.5170.[234]
$CH_2:CH{\cdot}CH_2$ ester. XXIe. b_1 118-120°, d_4^{20} 1.1295, n_D^{20} 1.5278.[234]
Bu ester. IVa.[340] XXIe.[234] b_1 120-122°,[234] d_4^{20} 1.0899,[234] n_D^{20} 1.5119.[234]
$CH_2{\cdot}CH_2$ di-ester. XXIIIc. $b_{0.04}$ 150-153°, d_4^{20} 1.2300, n_D^{20} 1.5635.[237]
$ClCH_2{\cdot}CH_2$ ester. XXXe. b_1 123-125°, d_4^{20} 1.2392, n_D^{20} 1.5395.[234]
Ph ester. XXIe. b_1 130-133°, d_4^{20} 1.1025, n_D^{20} 1.5391.[234]
Et_2 amide. XXIi. m. 41-42°, $b_{0.06}$ 108-109°.[576]
$CH_2:CH{\cdot}CH_2$ amide. XXIi. m. 91-92°.[576]
Aziridide. XXIi. m. 31-33°, $b_{0.09}$ 94-95°.[576]
Morpholide. XXIi. m. 63-64°.[576]
Piperidide. XXIi. $b_{0.05}$ 114-115°.[576]

$PhPrPO_2H$
Pr ester. Va. b_{14} 163°, d_0^0 1.0591, d_0^{16} 1.0463, $n_D^{22.5}$ 1.4979.[31]
(-)-menthyl (R)-ester. XXIe + fractional crystallization. m. 86°, $[\alpha]_D^{\sim 25}$ -14° (c∿2, C_6H_6),[402] NMR.[460]
(-)-menthyl (S)-ester. XXIe + fractional crystallization. m. 40°, $[\alpha]_D^{\sim 25}$ -81° (c∿2, C_6H_6),[402] NMR.[460]

$Ph(i\text{-}Pr)PO_2H$. IIa.[416] XXIa. ($AlCl_3$- catalyzed isomerization of Pr to i-Pr having evidently occurred during the preparation of the phosphinic chloride),[84] XXIIIb (low-melting product).[30] m. 88-89°.[416]
i-Pr ester. Va. b_{11} 146-147°, d_0^{17} 1.0813, n_D^{20} 1.4929.[30]

$Ph(CH_2:CH{\cdot}CH_2)PO_2H$. XXIIIa. A viscous oil;[155] cyclo-$C_6H_{11}NH_2$ salt, m. 192°.[155]
$CH_2:CH{\cdot}CH_2$ ester. Va.[155,360,605] $b_{0.5}$ 129-130°,[605] d_4^{20} 1.0960,[605] n_D^{20} 1.5250.[605]
$ClCH_2{\cdot}CH_2$ ester. Va. $b_{1.5}$ 169-170°, d_4^{20} 1.2145, n_D^{20} 1.5351.[235]
$CCl_3{\cdot}CMe_2$ ester. Va. $b_{0.18}$ 141-142°, d_4^{20} 1.2899, n_D^{20} 1.5355°.[449]
1-CCl_3-cyclopentyl ester. Va. m. 55-56°, $b_{0.22}$ 162°, d_4^{20} 1.3073, n_D^{20} 1.5500.[368]

$Ph(CH_3{\cdot}CH:CH)PO_2H$. By isomerization of previous acid by heating with KOH in MeOH. m. 65°;[155] cyclohexylamine salt, m. variable but > 200° on rapid heating.[155]

$Ph(CH_2:C:CH)PO_2H$. XXIIIa.[125] m. 94-96°,[125] dipole moment 2.50D,[124] mass spectrum.[124]

 CH⋮C·CH_2 ester. Va*. $b_{0.3}$ 128-132°.[125]

$PhBuPO_2H$. IIb.[179] VIh.[182] Vf.[409] Oil, $b_{0.0001}$ 240-250°;[182] crystalline hydrate;[179] dicyclohexylamine salt, m. 145.5-147°;[182] Zn salt, x-ray crystal structure.[251]

Ph(i-Bu)PO_2H. Vb. m. 64-65°.[27]

 i-Bu ester. Va. $b_{0.5}$ 115-117°.[280]

Ph(t-Bu)PO_2H. Ia.[292] VIa.[101,416] IIa.[101,292,416] m. 160-161°.[101,416]

$Ph(EtCMe_2)PO_2H$. IIa. m. 99-101°.[101]

$Ph(Me_2:C:CH)PO_2H$. XXIIIa.[125] m. 79-80°,[125] dipole moment 2.70D.[124]

 CH⋮C·CMe_2 ester. Va*. Undistillable,[96,125] n_D^{20} 1.5344.[96]

Ph(cyclo-C_6H_{11})PO_2H. XIa.[300] XIb.[319] XIIa.[584] XXIa.[798] XXIIIa.[627] m. 99-100.5°,[627] 121-122°,[319] pK_a 5.02 (in 75% EtOH), 5.60 (in 95% EtOH).[584]

 Et ester. XIIb.[627] XIII.[545] b_2 131-133°,[545] b_{10} 182-183°,[627] d_4^{20} 1.1010,[627] n_D^{20} 1.4705,[545] 1.5288.[627]

Ph(PrCH:C:CH)PO_2H.

 CH⋮C·CHPr ester. Va*. Undistillable,[96,124] n_D^{20} 1.5325.[96]

$Ph(C_7H_{15})PO_2H$. XXIIIa. m. 40.5-41°.[627]

 Et ester. Ve. XIIb. b_{10} 192°, d_4^{20} 1.0085, n_D^{20} 1.4989.[627]

$Ph(PhCH_2)PO_2H$. IIa.[179] VId.[86] VId*.[86] m. 182.5-183.5°.[179]

 Me ester. XXIIc. m. 89-92°.[86]

 Et ester. Va. m. 62.5-63.5°, $b_{0.2}$ 180-182°.[299]

$Ph(C_8H_{17})PO_2H$. XXIIIa. m. 40°.[627]

 Et ester. XIIb. b_2 174-175°, d_4^{20} 1.0101, n_D^{20} 1.5000.[627]

(+)-Ph(C_6H_{13}·CHMe)PO_2H. XXIIIa. m. 34-36°, $[\alpha]_{546}^{27}$ +20.15° (c 10, $CHCl_3$).[139]

 Bu ester. Ve using optically pure s-octyl tosylate, some racemization during reaction, product purified by chromatography. $b_{0.21}$ 122-127°, $[\alpha]_{546}^{27}$ +6.62° (c 19, $CHCl_3$).[139]

Ph[$\overline{CH_2·(CH_2)_4·C}$:C:CH]$PO_2H$. Vb*.[125] m. 109-110°.[125] Dipole moment 2.62D,[124] mass spectrum.[124]

Ph(m-MeC_6H_4·CH_2)PO_2H.

 Et ester. Va. b_1 170-173°, d_4^{20} 1.1141, n_D^{20} 1.5580.[516]

Ph(p-MeC_6H_4·CH_2)PO_2H.

 Et ester. Va. m. 68-69°, $b_{0.7}$ 158-161°.[516]

Ph(PhCH:CH)PO_2H. XXIa. m. 150-151°.[197]

 Me ester. XXIe. m. 114-115°.[197]

 Et ester. XXIe. b_3 205-209°, d_4^{25} 1.1461, n_D^{25}

1.5990.[197]
Pr ester. XXIe. b_1 196-199°, d_4^{25} 1.1174, n_D^{25} 1.5890.[197]
Bu ester. XXIe. b_2 210-212°, d_4^{25} 1.1038, n_D^{25} 1.5825.[197]
Ph ester. XXIe. m. 98-100°.[197]
p-ClC_6H_4 ester. XXIe. m. 111-111.5°.[197]
Ph(PhC⋮C)PO_2H. XIa.[199] (XXIIIa + XXXe) from Ph(PhCCl:CH)P(O)OBu,[197] m. 126-127°.[197,199]
Ph(C_9H_{19})PO_2H. XXIIIa. m. 39°.[627]
Me ester. XIIb. b_2 180-182°, d_4^{20} 1.0163, n_D^{20} 1.5000.[627]
Et ester. XIIb. b_2 187-188.5°, d_4^{20} 0.9922, n_D^{20} 1.4970.[627]
Ph($C_{10}H_{21}$)PO_2H. XXIIIa. m. 55.5-56°.[627]
Et ester. XIIb. $b_{5.5}$ 199-200°, d_4^{20} 0.9851, n_D^{20} 1.4920.[627]
Ph(PhCH:CH·CH:CH)PO_2H. XXIa. m. 151-152°.[196]
Ph ester. XXIe. m. 98-101°.[196]
Ph amide. XXIi. m. 195-196°.[196]
Ph(PhC⋮C·C⋮C)PO_2H. XXXe. m. 141-141.5°.[196]
Ph($C_{12}H_{25}$)PO_2H. XIa. m. 66-67°.[300]
Ph(Ph_2CH)PO_2H. Vb.[223] m. 244-246°.[222,223]
Et ester. Va. m. 235-237°, not hydrolyzed by heating 12 hr with conc. HCl.[283]
Ph($PhCH_2·CH_2·CH_2$·CHPh)PO_2H. XIa from 1,3,5-triphenylphospholan-1-oxide. m. 143-147°.[189]
Ph($C_{18}H_{37}$)PO_2H. XIa. m. 67-68°.[300]
Ph(Ph_3C)PO_2H. Vb. m. 287-288°.[27]
(p-ClC_6H_4)$MePO_2H$. XXIa. m. 94°, pK_a 2.39.[560]
Me ester. Va. b_5 144°, d_4^{20} 1.2696, n_D^{20} 1.5363.[560]
$CH_2:CH_2·CH_2$ ester. Va. b_2 129-130°, d_4^{20} 1.1960, n_D^{20} 1.5308.[360]
(p-ClC_6H_4)$EtPO_2H$.
Et ester. Va. b_9 153-155°, d_4^{20} 1.1614, n_D^{20} 1.5162.[633]
(p-ClC_6H_4)($CH_2:CH·CH_2$)PO_2H.
$CH_2:CH·CH_2$ ester. Va.[360,605] b_2 148-149°,[360,605] d_4^{20} 1.1203,[360] 1.1820,[605] n_D^{20} 1.5338,[360] 1.5410.[605]
(o-BrC_6H_4)$EtPO_2H$. VII.[209] m. 98-101°,[209] UV.[333]
(p-BrC_6H_4)$EtPO_2H$. VII. m. 68-69°.[495]
(p-HO·C_6H_4)$EtPO_2H$. By alkaline hydrolysis of following acid. m. 160-161°.[543]
[p-(p-$MeC_6H_4·SO_2$·O·)C_6H_4]$EtPO_2H$. VIa. m. 153-153.5°.[543]
[p-(p-$MeC_6H_4·SO_2$·O)C_6H_4]$BuPO_2H$. VIa. m. 141-142.5°.[543]
(p-$H_2N·C_6H_4$)$MePO_2H$. XXXi. m. 167-168°.[496]
(p-$H_2N·C_6H_4$)$EtPO_2H$. XXXi. m. 172-173°.[209]
(p-$H_2N·C_6H_4$)$PrPO_2H$. XXXi. m. 168-169°.[496]
(p-$H_2N·C_6H_4$)$BuPO_2H$. XXXi. m. 172-173°.[496]

$(p\text{-}H_2N{\cdot}C_6H_4)(C_8H_{17})PO_2H$. XXXi. m. 173-174°.[496]
$(p\text{-}Me_2N{\cdot}C_6H_4)MePO_2H$.
Me ester. Va.[138,262] m. 81.0-82.2°,[138] $b_{0.4}$ 170-172°;[262] methiodide m. 161° decomp., resolved via dibenzoyltartrate into (+)-form, m. 155.6-156.4°, $[\alpha]_D^{25}$ +28°, and (-)-form, m. 155.8-156.4°, $[\alpha]_D^{25}$ -29°;[138] methopicrate m. 176.0-176.6°.[138]
$(p\text{-}Me_2N{\cdot}C_6H_4)(CH_2{:}CH{\cdot}CH_2)PO_2H$.
$CH_2{:}CH{\cdot}CH_2$ ester. Va. $b_{0.03}$ 146-147°, d_4^{20} 1.0998, n_D^{20} 1.5695.[605]
$(o\text{-}O_2N{\cdot}C_6H_4)MePO_2H$. XXIIIa. m. 154°.[114]
Me ester. Va*. $b_{0.01}$ 130°.[114]
$(m\text{-}O_2N{\cdot}C_6H_4)EtPO_2H$. VII. m. 122-123°.[495]
$(p\text{-}O_2N{\cdot}C_6H_4)MePO_2H$. VII. m. 190-192°.[496]
$(p\text{-}O_2N{\cdot}C_6H_4)EtPO_2H$. VII.[209,495] m. 127-130°,[209] UV.[333]
$(p\text{-}O_2N{\cdot}C_6H_4)PrPO_2H$. VII. m. 76-78°.[496]
$(p\text{-}O_2N{\cdot}C_6H_4)BuPO_2H$. VII. m. 96-98°.[496]
$(p\text{-}O_2N{\cdot}C_6H_4)(C_8H_{17})PO_2H$. VII.[496] $PhNH_2$ salt, m. 115-117°.[496]
$(p\text{-}MeC_6H_4)MePO_2H$. Vd.[523] XXIIIa.[393] m. 120°.[523]
Me ester. Va. b_{13} 151-152°, d_0^{15} 1.1204, n_D^{15} 1.5280.[393]
Pr ester. Va. b_{12} 167°, d_0^{16} 1.0650, n_D^{16} 1.5185.[393]
Bu ester. Va. b_{11} 178°, d_0^{16} 1.0583, n_D^{16} 1.5092.[393]
$(p\text{-}MeC_6H_4)EtPO_2H$. VII. m. 88-89°.[495]
Et ester. Va.[393,633,666] $b_{0.02}$ 112-115°,[666] b_{13} 161°,[393] d_4^{20} 1.0580,[666] n_D^{20} 1.5072.[666]
$(p\text{-}MeC_6H_4)(i\text{-}Pr)PO_2H$.
i-Pr ester. Va. b_{12} 147-149°, d_4^{20} 1.0531, n_D^{20} 1.5048.[356]
$(p\text{-}MeC_6H_4)(CH_2{:}CH{\cdot}CH_2)PO_2H$.
$CH_2{:}CH{\cdot}CH_2$ ester. Va. $b_{0.3}$ 122-123°, d_4^{20} 1.0769, n_D^{20} 1.5314.[605]
$(MeC_6H_4)(PhCH_2)PO_2H$. [orientation of MeC_6H_4 unstated but probably largely or all p] XXIIIa. m. 145°.[524]
Ph ester. Va*. m. 120°.[524]
$(p\text{-}EtC_6H_4)EtPO_2H$.
Et ester. Va. $b_{0.015}$ 96-98°, d_4^{20} 1.0580, n_D^{20} 1.5072.[666]
$(p\text{-}EtC_6H_4)(i\text{-}Pr)PO_2H$.
Et ester. Va. $b_{0.012}$ 103-105°, d_4^{20} 1.0359, n_D^{20} 1.5050.[666]
(1-indenyl)$MePO_2H$.
Et ester. Va. $b_{1.5}$ 127.5-128.5°, d_4^{20} 1.1434, n_D^{20} 1.5579.[350]
(1-naphthyl)$MePO_2H$.
Me ester. Va. $b_{0.1}$ 128-130°.[262]
(1-naphthyl)$(CH_2{:}CH{\cdot}CH_2)PO_2H$.
$CH_2{:}CH{\cdot}CH_2$ ester. Va. b_4 179.5-180°, d_4^{50} 1.1239, n_D^{50} 1.5832.[358]
(2-naphthyl)$EtPO_2H$. VII. m. 127-128°.[495]

(9-fluorenyl)$MePO_2H$.
Et ester. Va. m. 97-97.5°, b_2 164-166°.[350]

E.1.3. Aliphatic and Aliphatic/Aromatic Phosphinic Acids with Halogen or Nitro Substituents in Aliphatic Groups

$(CH_2F)PhPO_2H$. From $PhPCl_4$ + CH_2N_2 with subsequent hydrolysis, m. 94-95°.[790]
$(CF_3)_2PO_2H$. XIa*. XXa. b_{238} 137-138°, b_{760} 182° $pK_a < 1$.[178]
Me amide. XIa*. m. -21°, b_{760} (extrapolated) 154.6°.[110]
$(CF_3)(CH_2{:}C{:}CH)PO_2H$.
Et ester. Va*. b_1 67°, d_4^{20} 1.3100, n_D^{20} 1.4240.[305]
$CH{\vdots}C{\cdot}CH_2$ ester. Va*. $b_{0.009}$ 53-58°, d_4^{20} 1.2910, n_D^{20} 1.4400.[305]
$(CF_3)PhPO_2H$. XIa*.[63] m. 84-86°;[63] Ag salt, m. 294-296°.[63]
$(ClCH_2)MePO_2H$. XXIVa. Liquid,[539] ^{31}P -50.7 ppm.[539,540]
$(ClCH_2)MeP(O){\cdot}O{\cdot}P(O)Me(CH_2Cl)$. IXg*. m. 63°,[539] ^{31}P -45.9 ppm.[539,540]
$(ClCH_2)_2PO_2H$. XXIa.[121,326] m. 80.5 81.5°,[121] ^{31}P -32.0 ppm.[539,540]
Me ester. XXIe.[326,478] $b_{0.03}$ 79-80°,[326] d_4^{20} 1.4589,[326] n_D^{20} 1.4885,[326] ^{31}P -40.9 ppm,[478] kinetics of hydrolysis.[65,66]
Et ester. XXIe.[326,478,539] $b_{0.07}$ 89-90°,[326] $b_{1.5}$ 115-116°,[539] d_4^{20} 1.3688,[326] n_D^{20} 1.4806,[326] ^{31}P -39.7 ppm,[478,540] kinetics of hydrolysis.[65,66]
Pr ester. XXIe. $b_{0.03}$ 86-87°, d_4^{20} 1.3046, n_D^{20} 1.4768,[326] kinetics of hydrolysis.[65,66]
i-Pr ester. XXIe.[326,478] $b_{0.07}$ 92-93°,[326] d_4^{20} 1.3022,[326] n_D^{20} 1.4770,[326] ^{31}P -37.34 ppm,[478] kinetics of hydrolysis.[65,66]
$CH_2{:}CH{\cdot}CH_2$ ester. XXIe. $b_{0.01}$ 105°, d_4^{20} 1.3452, n_D^{20} 1.4955,[326] kinetics of hydrolysis.[65]
Bu ester. XXIe.[326,478] $b_{0.05}$ 85-93°,[478] d_4^{20} 1.2492,[326] n_D^{20} 1.4759,[326] ^{31}P -37.8 ppm,[478] kinetics of hydrolysis.[65,66]
i-Bu ester. XXIe. $b_{0.07}$ 97-99°, d_4^{20} 1.2513, n_D^{20} 1.4728,[326] kinetics of hydrolysis.[65,66]
C_5H_{11} ester. XXIe. $b_{0.03}$ 114-116°, d_4^{20} 1.2116, n_D^{20} 1.4732,[326] kinetics of hydrolysis.[65,66]
$Me_3C{\cdot}CH_2$ ester. XXIe. $b_{0.04}$ 106°, d_4^{20} 1.2099, n_D^{20} 1.4691,[326] kinetics of hydrolysis.[65,66]
(4,6-Me_2-2-oxopyrimidinyl)$CH_2{\cdot}CHMe$ ester. XXIe. Hydrochloride; m. 150.5-151°; picrate m. 121.5-122.5°.[678]
(4,6-Me_2-2-oxopyrimidinyl)$CH_2{\cdot}CH(CH_2OH)$ ester. XXIe. Hydrochloride; m. 162.5-163.5°.[678]
$CH_2{\cdot}CH_2$ di-ester. XXIe. m. 55°.[326]

$(CH_2)_4$ di-ester. XXIe. m. 111-112°.[326]
$(CH_2)_5$ di-ester. XXIe. m. 82-83°.[326]
$(CH_2)_6$ di-ester. XXIe. m. 83-84°.[326]
Ph ester. XXIe. $b_{0.02}$ 117-120°, d_4^{20} 1.3868, n_D^{20} 1.5485,[326] kinetics of hydrolysis.[65]
$(ClCH_2)_2P(O)\cdot O\cdot P(O)(CH_2Cl)_2$. IXg*.[539] XXIg.[539] m. 74°,[539] ^{31}P -37.3 ppm.[540]
Me amide. XXIi. m. 67-68°.[326]
Me_2 amide. XXIi. m. 80-81°.[326]
Et_2 amide. XXIi. m. 64-65°.[326]
Pr amide. XXIi. m. 75-76°.[326]
$CH_2:CH\cdot CH_2$ amide. XXIi. m. 92-93°.[326]
$(CH_2:CH\cdot CH_2)_2$ amide. XXIi. m. 66-67°.[326]
i-Bu amide. XXIi. m. 97-98°.[326]
$PhCH_2$ amide. XXIi. m. 97-99°.[326]
Aziridide. XXIi. m. 76-77°.[326]
Ph amide. XXIi. m. 142-143°.[326]

$(ClCH_2)EtPO_2H$.
Et ester. XXIe. b_3 78-78.5°, d_4^{20} 1.1632, n_D^{20} 1.4570.[348]

$(ClCH_2)PhPO_2H$. XXIa. m. 93-93.5°,[348] ^{31}P -36.3 ppm.[539]
Me ester. XXIe.[348,761] $b_{2.5}$ 125-126°,[348] d_4^{20} 1.2654,[348] n_D^{20} 1.5409.[348]
Et ester. XXIe. $b_{2.5}$ 132-134°, d_4^{20} 1.2329, n_D^{20} 1.5275.[348]
Et_2 amide. XXIVc. b_5 158°, ^{31}P -30.5.[539]
Ph amide. XXIi. m. 171-172°.[564]
$(ClCH_2)Ph\cdot P(O)\cdot O\cdot P(O)Ph(CH_2Cl)$. XXIg.[539] IXg*.[539] b_1 210°,[539] ^{31}P -32.3 ppm,[539] -30.9 ppm.[540]

$(CCl_3)MePO_2H$. By chlorination of Me_2PO_2H. m. 161-161.5°.[677]
Et ester. XXb. m. 40°.[165]
i-Pr ester. XXb. m. 45°.[165]

$(CCl_3)_2PO_2H$. XXa. (1 1/2 H_2O), m. 140-142°.[427]
Amide. XXId. m. 194-195°.[428]

$(CCl_3)EtPO_2H$.
Me ester. Va. $b_{2.5}$ 80-81°, d_4^{20} 1.4174, n_D^{20} 1.4889,[38] dipole moment 3.20D.[46]
Et ester. Va. $b_{1.5}$ 83-84°, d_4^{20} 1.3476, n_D^{20} 1.4819,[38] dipole moment 3.15D.[46]
Pr ester. Va. $b_{1.5}$ 88-89°, d_4^{20} 1.2790, n_D^{20} 1.4674,[38] dipole moment 3.29D.[46]

$(CCl_3)PhPO_2H$. XXIa.[84,791] XXIIIa.[391] m. 163-164°;[791] Et_2NH salt, m. 164.5-165.5°;[391,392] Pr_2NH salt, m. 144.5-145.5°;[391] $BuNH_2$ salt, m. 210°;[392] $BuEt_3N^+$ salt, m. 155.5-156.5°;[391] Bu_2NH salt, m. 104.5-105.5°;[391] $PhNH_2$ salt, m. 215.5-216°.[391,392]
Me ester. Va. m. 108°.[354]
Et ester. Va. m. 79°, b_1 147-148°.[354]
Pr ester. Va. b_1 150-152°, d_0^{14} 1.2918, n_D^{19} 1.4945.[354]

Bu ester. Va. b_3 149.5-150.5°, d_0^{20} 1.293, n_D^{20} 1.5315.[359]
i-Bu ester. Va. b_1 155-156°, d_0^{13} 1.2697, n_D^{18} 1.4993.[354]
$ClCH_2 \cdot CH_2$ ester. Va. b_5 185-186°, d_4^{20} 1.4690, n_D^{20} 1.5573.[357]
$(CCl_3)PhP(O) \cdot O \cdot P(O)Ph(CCl_3)$. XXIh. m. 170-172°.[791]
$CCl_3(p\text{-}ClC_6H_4)PO_2H$. XXIIIa. m. 198-199°;[358] NH_4^+ salt, m. 225-226°;[392] Et_2NH salt, m. 185-186°;[391] Pr_2NH salt, m. 139.5-140.5°;[391] Bu_2NH salt, m. 137-138°;[391] $PhNH_2$ salt, m. 212-213°.[391]
Et ester. Va. b_3 155-156.5°, d_4^{20} 1.4590, n_D^{20} 1.5572.[358]
Pr ester. Va. b_2 159°, d_4^{20} 1.3530, n_D^{20} 1.5378.[358]
Bu ester. Va. b_2 153.5°, d_4^{20} 1.3290, n_D^{20} 1.5390.[358]
$CCl_3(p\text{-}BrC_6H_4)PO_2H$. XXIIIa. m. 220.5-221°;[388] Et_2NH salt, m. 214-214.5°;[391] $MeEt_3N^+$ salt, m. 171.5-172.5°;[391] $PhNH_2$ salt, m. 225.5-226.5° decomp.[391]
Me ester. Va. m. 89.5-90°, b_6 175-178°, n_D^{20} 1.5772 (supercooled).[388]
Et ester. Va. b_1 176-178°, d_4^{20} 1.6218, n_D^{20} 1.5716.[388]
Pr ester. Va. b_4 175°, d_4^{20} 1.5582, n_D^{20} 1.5609.[388]
Bu ester. Va. b_4 186-187°, d_4^{20} 1.4900, n_D^{20} 1.5515.[388]
C_5H_{11} ester. Va. b_4 208°, d_4^{20} 1.4558, n_D^{20} 1.5490.[388]
$(CCl_3(p\text{-}MeO \cdot C_6H_4)PO_2H$.
Et ester. Va. b_4 145-147°, d_0^{16} 1.2650, n_D^{16} 1.5068.[355]
$CCl_3(p\text{-}MeC_6H_4)PO_2H$. XXIIIa. m. 184.5-185°.[393]
Me ester. Va. b_1 128-130°, d_0^{20} 1.2140, n_D^{20} 1.5312.[393]
Et ester. Va. b_2 157°, d_0^{16} 1.3103, n_D^{16} 1.5428.[393]
Pr ester. Va. b_3 169-170°, d_0^{16} 1.2944, n_D^{16} 1.5370.[393]
Bu ester. Va. b_2 180-181°, d_0^{16} 1.2143, n_D^{16} 1.5267.[393]
i-Bu ester. Va. b_4 178-180°, d_0^{16} 1.2250, n_D^{16} 1.5294.[393]
C_5H_{11} ester. Va. b_2 195-198°, d_4^{20} 1.1685, n_D^{20} 1.5165.[356]
C_6H_{13} ester. Va. b_2 203-204°, d_4^{20} 1.1570, n_D^{20} 1.5128.[356]
$ClCH_2 \cdot CH_2$ ester. Va. b_2 184-185°, d_4^{20} 1.3881, n_D^{25} 1.5462.[356]
$CCl_3[p\text{-}(i\text{-}Pr)C_6H_4]PO_2H$.
Et ester. Va. b_2 145.5-147°, d_4^{20} 1.2740, n_D^{20} 1.5350.[358]
Pr ester. Va. b_3 170°, d_4^{20} 1.1550, n_D^{20} 1.5127.[358]
$CCl_3(1\text{-naphthyl})PO_2H$. XXIIIb. m. 217-218°;[358] Et_2NH salt, m. 198-199°;[391] $PhNH_2$ salt, m. 199.5-200.5°

decomp.[391]
Me ester. Va. m. 101.5-102°, b_5 205-207°.[358]
Et ester. Va. m. 80-80.5°, b_3 179-180°.[358]
Pr ester. Va. b_1 186-187°, d_4^{20} 1.3460, n_D^{20} 1.5991.[358]
i-Pr ester. Va. b_4 230° decomp., d_4^{20} 1.3300, n_D^{20} 1.4930.[358]
Bu ester. Va. b_5 300°, n_D^{20} 1.5895.[358]
$(BrCH_2)_2PO_2H$. XXIVa. Cyclo-$C_6H_{11}NH_2$ salt, m. 151-152°.[176]
cyclo-C_6H_{11} amide. XXIi. m. 153°.[176]
$(BrCH_2)_2P(O)\cdot O\cdot P(O)(CH_2Br)_2$. IXg*. m. 51-52°, $b_{0.1}$ 220-226°.[176]
$(ICH_2)(ClCH_2)PO_2H$.
Et ester. XXXd. m. 33°, $b_{0.045}$ 102-104°, d_4^{50} 1.8264, n_D^{50} 1.5345.[549]
$(ICH_2)_2PO_2H$. XXXd. m. 139-140°.[549]
Et ester. XXXd. m. 99.5-100°.[549]
Bu ester. XXXd. m. 78-79°.[549]
Ph ester. XXXd. m. 86-87°.[549]
Et_2 amide. XXXd. m. 88-89°.[549]
$(CH_2{:}CH\cdot CH_2)_2$ amide. XXXd. m. 80-81°.[549]
$(ICH_2)PhPO_2H$.
Et ester. Va.[280,394] b_1 137-138°,[394] b_1 153°.[280]
i-Bu ester. Va. $b_{0.5}$ 157-160°.[280]
$(ClCH_2\cdot CH_2)EtPO_2H$.
Et ester. XXIe. b_3 86-88°, d_4^{20} 1.1300, n_D^{20} 1.4584.[362]
$ClCH_2\cdot CH_2$ ester. Va. b_4 132-134°, d_4^{20} 1.2690, n_D^{20} 1.4832.[362]
$(ClCH_2\cdot CH_2)(CH_2{:}CH)PO_2H$.
$ClCH_2\cdot CH_2$ ester. Va. b_2 120-122°, d_4^{20} 1.2260, n_D^{20} 1.4930.[343]
$(ClCH_2\cdot CH_2)BuPO_2H$.
$ClCH_2\cdot CH_2$ ester. Va. b_2 133-134°, d_4^{20} 1.2000, n_D^{20} 1.4798.[337]
$(ClCH_2\cdot CH_2)PhPO_2H$.
Me ester. XXIe. b_2 140-141°, d_4^{20} 1.2201, n_D^{20} 1.5355.[232]
Et ester. XXIe. b_1 130°, d_4^{20} 1.1896, n_D^{20} 1.5239.[232]
Pr ester. XXIe. b_2 148-150°, d_4^{20} 1.1680, n_D^{20} 1.5180.[232]
Bu ester. XXIe. b_2 154-156°, d_4^{20} 1.1200, n_D^{20} 1.5128.[232]
$ClCH_2\cdot CH_2$ ester. Va*. $b_{0.02}$ 135-142°, d_4^{20} 1.3017, n_D^{20} 1.5441.[236]
Ph ester. XXIe. $b_{2.5}$ 185-188°, n_D^{20} 1.5711.[232]
Aziridide. XXIi. m. 44-45°.[576]
$(ClCH_2\cdot CH_2)(p\text{-}MeC_6H_4)PO_2H$.
$ClCH_2\cdot CH_2$ ester. Va*. b_1 153-155°, d_4^{20} 1.2383, n_D^{20} 1.5452.[236]

$(CH_3\cdot CHCl)_2PO_2H$. From PCl_5 + $MeCHN_2$ with subsequent hydrolysis, m. 107°; Ag salt; $PhNH_2$ salt, decomp. 160°.[792]

$(ClCH_2\cdot CHCl)EtPO_2H$.
Et ester. XXIe. b_3 105-107°, d_4^{20} 1.2410, n_D^{20} 1.4765.[762]

$(ClCH:CH)EtPO_2H$.
Me ester. XXIe. b_{15} 111°, d_4^{20} 1.2010, n_D^{20} 1.4791.[766]
Et ester. XXIe. b_{15} 115°, d_4^{20} 1.1520, n_D^{20} 1.4710.[765]
$CH_2:CH\cdot CH_2$ ester. XXIe. b_{14} 127-129°, d_4^{20} 1.1450, n_D^{20} 1.4832.[766]

$(ClCH:CH)PhPO_2H$.
Me ester. XXIe. b_{11} 170-171°, d_4^{20} 1.2590, n_D^{20} 1.5535.[764]
Et ester. XXIe. b_9 166-167°, d_4^{20} 1.2114, n_D^{20} 1.5412.[764]
Pr ester. XXIe. b_{11} 180-181°, d_4^{20} 1.1790, n_D^{20} 1.5358.[764]
Bu ester. XXIe. b_8 181-182°, d_4^{20} 1.1500, n_D^{20} 1.5281.[764]
Ph ester. XXIe. m. 74°, b_8 198°.[764]

$(CH_2:CCl)EtPO_2H$.
Me ester. XXIe. b_{12} 82-86°, d_4^{20} 1.1870, n_D^{20} 1.4650.[762,765]
Et ester. XXIe. b_{10} 85-89°, d_4^{20} 1.1260, n_D^{20} 1.4625.[762,765]
Pr ester. XXIe. b_{12} 106-112°, d_4^{20} 1.1100, n_D^{20} 1.4600.[762,765]
$CH_2:CH\cdot CH_2$ ester. XXIe. b_3 93-94°, d_4^{20} 1.1360, n_D^{20} 1.4725.[762,765]
Bu ester. XXIe. b_{11} 112-115°, d_4^{20} 1.0760, n_D^{20} 1.4600.[762,765]
i-Bu ester. XXIe. b_{12} 103-106°, d_4^{20} 1.0870, n_D^{20} 1.4590.[762,765]
Ph ester. XXIe. b_3 122-132°, d_4^{20} 1.1900, n_D^{20} 1.5305.[762,765]

$(BrCH_2\cdot CH_2)EtPO_2H$.
Et ester. Va. b_2 105°.[493]

$(CH_2:CBr)EtPO_2H$.
Me ester. XXIe. b_{13} 95-100°, d_4^{20} 1.421, n_D^{20} 1.4845.[364]
Et ester. XXIe. b_{14} 100-102°, d_4^{20} 1.373, n_D^{20} 1.4840.[364]
Pr ester. XXIe. b_{15} 113-115°, d_4^{20} 1.289, n_D^{20} 1.4750.[364]
Bu ester. XXIe. b_{12} 126-128°, d_4^{20} 1.279, n_D^{20} 1.4780.[364]
i-Bu ester. XXIe. b_{10} 118-120°, d_4^{20} 1.273, n_D^{20} 1.4740.[364]

$(BrCH_2\cdot CHBr)PhPO_2H$. XXXa. m. 63-66°.[494]

$(CF_3\cdot CH_2\cdot CH_2)_2PO_2H$. XXa. b_5 170-171°.[452]

$(CF_3{\cdot}CF{:}CF)MePO_2H$.
Et ester. Va. b_8 69-71°, d_4^{20} 1.3154, n_D^{20} 1.3780.[330]
i-Pr ester. Va. b_2 72°, d_4^{20} 1.299, n_D^{20} 1.3795.[400]
i-Bu ester. Va. b_8 80-82°, d_4^{20} 1.2582, n_D^{20} 1.3855.[330]
Aziridide. XXIi. $b_{1.5}$ 60-62°, d_4^{20} 1.2815, n_D^{20} 1.4195.[330]
$(ClCH_2{\cdot}CH_2{\cdot}CH_2)_2PO_2H$.
Et_2 amide. XXIi. $b_{0.04}$ 158-159°, d_4^{20} 1.1678, n_D^{20} 1.5075.[16]
$CH_2{:}CH{\cdot}CH_2$ amide. XXIi. m. 167-169°.[16]
Aziridide. XXIi. $b_{0.008}$ 135-136°, d_4^{20} 1.2699, n_D^{20} 1.5255.[16]
Ph amide. XXIi. $b_{0.004}$ 170-171°.[16]
$(ClCH_2{\cdot}CHCl{\cdot}CH_2)MePO_2H$.
Pr ester. XXXa. $b_{0.0001}$ 115-117°, d_4^{20} 1.2280, n_D^{20} 1.4795.[465]
i-Pr ester. XXXa. m. 55-56°.[461]
C_5H_{11} ester. XXXa. $b_{0.002}$ 126-129°, d_4^{20} 1.1744, n_D^{20} 1.4772.[461]
$(ClCH_2{\cdot}CHCl{\cdot}CH_2)EtPO_2H$.
$(ClCH_2)_2CH$ ester. Va*. $b_{0.05}$ 149-151°, d_4^{20} 1.3768, n_D^{20} 1.5055.[680]
$(ClCH_2{\cdot}CHCl{\cdot}CH_2)BuPO_2H$.
Bu ester. XXXa. $b_{0.3}$ 145-148°, d_4^{20} 1.1209, n_D^{20} 1.4765.[655]
C_6H_{13} ester. XXXa. $b_{0.6}$ 154°, d_4^{20} 1.0934, n_D^{20} 1.4728.[465]
$(ClCH_2{\cdot}CHCl{\cdot}CH_2)PhPO_2H$.
$(ClCH_2)_2CH$ ester. Va*. $b_{0.02}$ 179-181°, n_D^{20} 1.5528.[680]
$(CF_3{\cdot}CH{:}CCl)MePO_2H$.
i-Pr ester. IXf. b_2 82-84°, n_D^{20} 1.4110.[230]
$(CF_3{\cdot}CH{:}CBr)MePO_2H$.
i-Pr ester. IXf. $b_{2.5}$ 108-112°, d_4^{20} 1.3320, n_D^{20} 1.4330.[230]
$(\overline{CF_2{\cdot}CF_2{\cdot}CF_2{\cdot}CCl{:}C})PhPO_2H$. XXIIIa. m. 171-172°.[207]
Et ester. Va. m. 30-42°, could not be distilled or recrystallized.[207]
(1-Cl-cyclohexyl)$PhPO_2H$. IXg. m. 149-150°.[224]
$(p\text{-}ClC_6H_4{\cdot}CH_2)BuPO_2H$. IIa. m. 121.5-122.5°.[179]
$(p\text{-}BrC_6H_4{\cdot}CH_2)PhPO_2H$. IIa. m. 214.5-215.5°.[179]
$(p\text{-}BrC_6H_4{\cdot}CH_2)_2PO_2H$. XIa. m. 192-193°.[797]
$(Ph{\cdot}CCl{:}CH)PhPO_2H$. XXIa. m. 151-152°.[197]
Me ester. XXIe. b_3 207-210°, d_4^{25} 1.2632, n_D^{25} 1.6112.[197]
Et ester. XXIe. b_3 210-214°, d_4^{25} 1.2178, n_D^{25} 1.5970.[197]
Bu ester. XXIe. b_4 225-227°, d_4^{25} 1.1789, n_D^{25} 1.5853.[197]
$(PhCCl{:}CH)(PhCH{:}CH)PO_2H$. XXIa. m. 125-126°.[198]

Ph ester. XXIe. m. 100-101°.[198]
p-ClC_6H_4 ester. XXIe. m. 117-118°.[198]
m-$O_2N\cdot C_6H_4$ ester. XXIe. m. 144-145°.[198]
p-$O_2N\cdot C_6H_4$ ester. XXIe. m. 137-138°.[198]
1-naphthyl ester. XXIe. m. 115-116°.[198]
Ph amide. XXIi. m. 196-197°.[198]
p-$MeO\cdot C_6H_4$ amide. XXIi. m. 180-181°.[198]
p-MeC_6H_4 amide. XXIi. m. 215-217°.[198]

$(Ph\cdot CHBr\cdot CHBr\cdot CHBr\cdot CHBr)PhPO_2H$. XXXa. m. 191° decomp.[196]

$(Ph_2CCl)PhPO_2H$.
Me ester. XXIe. m. 100-101°.[222]

(p-$O_2N\cdot C_6H_4\cdot CH_2$)$EtPO_2H$. XXXc. m. 142-144°.[62]

(p-$O_2N\cdot C_6H_4\cdot CH_2$)$_2PO_2H$. XXXc. m. 225-226°.[120]

(p-$O_2N\cdot C_6H_4\cdot CH_2\cdot CH_2$)$_2PO_2H$. XXXc. m. 196-199°.[62]

E.1.4. Aliphatic and Aliphatic/Aromatic Phosphinic Acids with Hydroxyl, or Functionally Modified Hydroxyl, Substituents in Aliphatic Groups

(o-$HO\cdot C_6H_4\cdot O\cdot CH_2$)$MePO_2H$.
Cyclic ester. From $(ClCH_2)MeP(O)Cl$ + o-$C_6H_4(OH)_2$. m. 52-57°, $b_{0.001}$ 114°.[786]

$(PhO\cdot CH_2)(ClCH_2)PO_2H$. XXXd. Yellow oil.[550]
Et ester. XXIIb. $b_{0.02}$ 140-141°, d_4^{20} 1.2398, n_D^{20} 1.5183.[550]

$(HO\cdot CH_2)_2PO_2H$. IXa[322] (kinetics of H_3PO_2-CH_2O reaction[296]). XIa.[291] Sirupy liquid;[322] Na salt, unmelted at 400°;[322] Na salt ($10H_2O$), m. 72-82°;[322] Ba salt;[291] cyclo-$C_6H_{11}NH_2$ salt, m. 180°;[176] $PhNH_2$ salt, m. 132-133.5°.[322]

$(EtO\cdot CH_2)_2PO_2H$. XXXf. $b_{0.0001}$ 220°.[548]
Et ester. XXIIb. $b_{0.04}$ 78-80°, d_4^{20} 1.0506, n_D^{20} 1.4353.[548]

$(PrO\cdot CH_2)_2PO_2H$.
Pr ester. XXXd. $b_{0.02}$ 78-79°, d_4^{20} 0.9994, n_D^{20} 1.4382.[548]

$(CH_2{:}CH\cdot CH_2\cdot O\cdot CH_2)_2PO_2H$. XXXf. $b_{0.001}$ 250°, d_4^{20} 1.1535, n_D^{20} 1.4790.[171]
Me ester. XXIIb. $b_{0.03}$ 90-92°, d_4^{20} 1.0940, n_D^{20} 1.4659.[171]
Et ester. XXIIb. $b_{0.03}$ 93-95°, d_4^{20} 1.0670, n_D^{20} 1.4620.[171]
Pr ester. XXIIb. $b_{0.03}$ 98-100°, d_4^{20} 1.0450, n_D^{20} 1.4613.[171]
$CH_2{:}CH\cdot CH_2$ ester. XXIIb. $b_{0.025}$ 98-100°, d_4^{20} 1.0627, n_D^{20} 1.4710.[171]
Bu ester. XXIIb. $b_{0.03}$ 105-110°, d_4^{20} 1.0301, n_D^{20} 1.4594.[171]

$(BuO\cdot CH_2)_2PO_2H$.
Bu ester. XXXd. $b_{0.03}$ 108-110°, d_4^{20} 0.9682, n_D^{20} 1.4415.[548]

$(C_5H_{11}\cdot O\cdot CH_2)_2PO_2H$.
C_5H_{11} ester. XXXd. $b_{0.0001}$ 125-126°, d_4^{20} 0.9554, n_D^{20} 1.4430.[548]
$(PhO\cdot CH_2)_2PO_2H$. XXXd. m. 132-132.5°.[550]
Et ester. XXIIb. $b_{0.0001}$ 170°, d_4^{20} 1.1894, n_D^{20} 1.5548.[801]
$(o\text{-}ClC_6H_4\cdot O\cdot CH_2)_2PO_2H$.
Et ester. [XXXd + XXIIb]. m. 81-82°.[550]
$(m\text{-}ClC_6H_4\cdot O\cdot CH_2)_2PO_2H$.
Et ester. [XXXd + XXIIb]. $b_{0.0001}$ 212°, d_4^{20} 1.3198, n_D^{20} 1.5630.[550]
$(p\text{-}ClC_6H_4\cdot O\cdot CH_2)_2PO_2H$. XXXd. m. 163.5-164°.[550]
$(p\text{-}BrC_6H_4\cdot O\cdot CH_2)_2PO_2H$. XXXd. m. > 350°.[550]
$(o\text{-}MeC_6H_4\cdot O\cdot CH_2)_2PO_2H$.
Et ester. [XXXd + XXIIb]. m. 80°.[550]
$(m\text{-}MeC_6H_4\cdot O\cdot CH_2)_2PO_2H$.
Et ester. [XXXd + XXIIb]. $b_{0.0001}$ 188-190°, d_4^{20} 1.1457, n_D^{20} 1.5475.[550]
$(p\text{-}MeC_6H_4\cdot O\cdot CH_2)_2PO_2H$. XXXd. m. 136-137°.[550]
$(AcO\cdot CH_2)(HO\cdot CH_2)PO_2H$. XXXf. $PhNH_2$ salt, m. 143-146°.[322]
$(AcO\cdot CH_2)_2PO_2H$. XXXf. $PhNH_2$ salt, m. 109-114°.[322]
$(PhNH\cdot CO\cdot O\cdot CH_2)_2PO_2H$. From $(HO\cdot CH_2)_2PO_2^-\ Na^+$ + PhNCO. Na salt, m. 76-77°.[323]
Ph amide. From $(HO\cdot CH_2)_2PO_2H$ + PhNCO. m. 98-100°.[323]
$(HO\cdot CH_2)EtPO_2H$.
Et ester. IXc. $b_{0.02}$ 108-110°, d_4^{20} 1.1202, n_D^{20} 1.4532,[610] loses EtOH at ∿200° giving viscous oils, $-[CH_2\cdot P(O)Et\cdot O]-_n$ (M.W. 750-2000) which on prolonged standing yield the dimeric cyclic ester, $\overline{O\cdot CH_2\cdot P(O)Et\cdot O\cdot CH_2\cdot P}(O)Et$, m. 210-212°.[610]
$(MeO\cdot CH_2)EtPO_2H$. XXIa. b_3 184-185°, d_4^{20} 1.1639, n_D^{20} 1.4458.[767]
Me ester. XXIe. b_{14} 115-116°, d_4^{20} 1.1152, n_D^{20} 1.4508.[767]
Et ester. XXIe. b_{15} 120-121°, d_4^{20} 1.0570, n_D^{20} 1.4432.[767]
Pr ester. XXIe. b_{14} 121-122°, d_4^{20} 1.0385, n_D^{20} 1.4431.[767]
Bu ester. XXIe. b_{14} 124-125°, d_4^{20} 1.0118, n_D^{20} 1.4412.[767]
Et_2 amide. XXIi. b_{16} 137-138°, d_4^{20} 1.0119, n_D^{20} 1.4637.[767]
$(MeO\cdot CH_2)EtP(O)\cdot O\cdot P(O)Et(CH_2\cdot OMe)$. XXIf. b_1 155-156°, d_4^{20} 1.1762, n_D^{20} 1.4680.[767]
$(EtO\cdot CH_2)EtPO_2H$. XXIa. b_1 170-171°, d_4^{20} 1.1077, n_D^{20} 1.4520.[767]
Me ester. XXIe. b_7 100-101°, d_4^{20} 1.0524, n_D^{20} 1.4385.[767]
Et ester. XXIe. b_2 92-93°, d_4^{20} 1.0232, n_D^{20} 1.4364.[767]
Pr ester. XXIe. b_1 93-94°, d_4^{20} 1.0004, n_D^{20}

1.4372.[767]
Bu ester. XXIe. b_{15} 144-146°, d_4^{20} 0.9894, n_D^{20} 1.4395.[767]
Et_2 amide. XXIi. b_1 96-97°, d_4^{20} 0.9920, n_D^{20} 1.4590.[767]
$(EtO{\cdot}CH_2)EtP(O){\cdot}O{\cdot}P(O)Et(CH_2{\cdot}OEt)$. XXIf. b_1 160-161°, d_4^{20} 1.1202, n_D^{20} 1.4620.[767]

$(PrO{\cdot}CH_2)EtPO_2H$. XXIa. b_1 183-184°, d_4^{20} 1.0772, n_D^{20} 1.4508.[767]
Me ester. XXIe. b_{19} 126-127°, d_4^{20} 1.0417, n_D^{20} 1.4438.[767]
Et ester. XXIe. b_{14} 114-115°, d_4^{20} 1.0091, n_D^{20} 1.4410.[767]
Pr ester. XXIe. b_{15} 125-126°, d_4^{20} 0.9892, n_D^{20} 1.4402.[767]
Bu ester. XXIe. b_{14} 134-135°, d_4^{20} 0.9798, n_D^{20} 1.4412.[767]
Et_2 amide. XXIi. b_2 107-108°, d_4^{20} 0.9803, n_D^{20} 1.4582.[767]
$(PrO{\cdot}CH_2)EtP(O){\cdot}O{\cdot}P(O)Et(CH_2{\cdot}OPr)$. XXIf. b_2 167-168°, d_4^{20} 1.0908, n_D^{20} 1.4592.[767]

$(BuO{\cdot}CH_2)EtPO_2H$. XXIa. b_2 199-200°, d_4^{20} 1.0450, n_D^{20} 1.4523.[767]
Me ester. XXIe. b_2 100-101°, d_4^{20} 1.0141, n_D^{20} 1.4426.[767]
Et ester. XXIe. b_2 104-105°, d_4^{20} 0.9874, n_D^{20} 1.4393.[767]
Pr ester. XXIe. b_1 106-107°, d_4^{20} 0.9746, n_D^{20} 1.4402.[767]
Bu ester. XXIe. b_2 115-116°, d_4^{20} 0.9649, n_D^{20} 1.4414.[767]
Et_2 amide. XXIi. b_2 120-121°, d_4^{20} 0.9661, n_D^{20} 1.4580.[767]
$(BuO{\cdot}CH_2)EtP(O){\cdot}O{\cdot}P(O)Et(CH_2{\cdot}OBu)$. XXIf. b_2 180-181°, d_4^{20} 1.0563, n_D^{20} 1.4603.[767]

$(HO{\cdot}CH_2{\cdot}CH_2{\cdot}O{\cdot}CH_2)EtPO_2H$.
Cyclic ester. By heating the Et ester of the next acid with 2% H_3PO_4, b_1 115-116°, d_4^{20} 1.2317, n_D^{20} 1.4750.[56]

$(AcO{\cdot}CH_2{\cdot}CH_2{\cdot}O{\cdot}CH_2)EtPO_2H$.
Et ester. Va. $b_{0.5}$ 124-125°, d_4^{20} 1.1274, n_D^{20} 1.4470.[56]

$(HO{\cdot}CH_2)PhPO_2H$.
Et ester. IXc. $b_{0.015}$ 185°, d_4^{20} 1.1927, n_D^{20} 1.5373,[610] loses EtOH at ∿200° giving a solid, $-[CH_2{\cdot}P(O)Ph{\cdot}O]-_n$ (M.W. ∿5000).[610]

$(MeO{\cdot}CH_2)PhPO_2H$.
Et ester. Va. b_2 138-139°, d_0^0 1.1543, $n_D^{20.5}$ 1.4891.[31]

$(HO{\cdot}CH_2)(C_8H_{17})PO_2H$. IIa. m. 77-78°.[105]

$(BuO{\cdot}CH_2{\cdot}CH_2)EtPO_2H$.

Et ester. XIIb. $b_{2.5}$ 116°, d_4^{20} 0.9956, n_D^{20} given as 1.14440, presumably should be 1.4440.[631]
$(AcO\cdot CH_2\cdot CH_2)EtPO_2H$.
Et ester. IXc*. b_4 110-111°, d_4^{20} 1.0794, n_D^{20} 1.4361,[632] alternative α-AcO structure suggested.[628]
Bu ester. IXc*. b_4 123°, d_4^{20} 1.0408, n_D^{20} 1.4412.[632]
$(EtO\cdot CH_2\cdot CH_2)PhPO_2H$.
Et ester. Va. $b_{2.5}$ 149.5°, d_0^{25} 1.0811, n_D^{25} 1.4910.[31]
$(AcO\cdot CH_2\cdot CH_2)PhPO_2H$.
Et ester. IXc*. b_4 149-150°, d_4^{20} 1.1329, n_D^{20} 1.4990.[632] See ref. 628 for revised structure of an analogous ester.
[MeCH(OAc)]$EtPO_2H$.
Et ester suggested as revised structure for Et ester of $(AcO\cdot CH_2\cdot CH_2)EtPO_2H$.[628]
$[MeCH(OH)]_2PO_2H$.
Cyclic acetal with MeCHO, see cyclic phosphinic acids.
$[MeCH(OH)](CCl_3\cdot CMe_2)PO_2H$.
1-CCl_3-cyclohexyl ester. IXc. m. 179-181°.[8]
[MeCH(OH)]$PhPO_2H$. IXg. m. 104°.[521]
$[CHCl_2\cdot CH(OH)]_2PO_2H$. Ia. IXa. m. 195°;[106] $PhNH_2$ salt, m. 175° decomp.[106]
$[CCl_3\cdot CH(OH)]MePO_2H$.
Me ester. IXc. m. 157-158°.[61]
Et ester. IXc. m. 142-143°.[61]
Pr ester. IXc. m. 150-152°.[61]
i-Pr ester. IXc. m. 129-130°.[61]
Bu ester. IXc. m. 127-128°.[61]
i-Bu ester. IXc. m. 100-101°.[61]
C_5H_{11} ester. IXc. m. 153-154°.[61]
PrCHMe ester. IXc. m. 112-113°.[61]
$ClCH_2\cdot CH_2$ ester. IXc. m. 86-87°.[61]
o-$ClC_6H_4\cdot CH_2$ ester. IXc. m. 134-135°.[61]
Ph ester. IXc. m. 150-151°.[61]
m-ClC_6H_4 ester. IXc. m. 144-145°.[61]
2,4-$Cl_2C_6H_3$ ester. IXc. m. 106-107°.[61]
2,4,6-$Cl_3C_6H_2$ ester. IXc. m. 114-115°.[61]
p-$O_2N\cdot C_6H_4$ ester. IXc. m. 121-122°.[61]
$[CCl_3\cdot CH(OH)]EtPO_2H$.
Me ester. IXc.[61,571] m. 120-121°.[61]
Et ester. IXc.[61,571] m. 150-151°.[61]
Pr ester. IXc. m. 155-156°.[61]
i-Pr ester. IXc. m. 140-141°.[61]
Bu ester. IXc. m. 162-163°.[61]
i-Bu ester. IXc. m. 130-131°.[61]
C_5H_{11} ester. IXc. waxy, m. 137-147°.[61]
PrCHMe ester. IXc. m. 152-153°.[61]
$PhCH_2$ ester. IXc. m. 132-133°.[61]
PhCHMe ester. IXc. m. 149-150°.[61]

$ClCH_2 \cdot CH_2$ ester. IXc. m. 88.5-89.5°.[61]
o-$ClC_6H_4 \cdot CH_2$ ester. IXc. m. 154-155°.[61]
[$CCl_3 \cdot$CH(OAc)]$EtPO_2H$.
Me ester. XXXf. $b_{0.006}$ 91-92°, d_4^{20} 1.408, n_D^{20} 1.4870.[571]
Et ester. XXXf. $b_{0.01}$ 93-94°, d_4^{20} 1.355, n_D^{20} 1.4833.[571]
Pr ester. XXXf. $b_{0.01}$ 108-112°, d_4^{20} 1.317, n_D^{20} 1.4790.[571]
i-Pr ester. XXXf. $b_{0.01}$ 103-104°, d_4^{20} 1.309, n_D^{20} 1.4752.[571]
Bu ester. XXXf. $b_{0.01}$ 121-123°, d_4^{20} 1.286, n_D^{20} 1.4778.[571]
i-Bu ester. XXXf. $b_{0.01}$ 121-123°, d_4^{20} 1.281, n_D^{20} 1.4760.[571]
[$CCl_3 \cdot$CH(OH)]$_2PO_2H$. IXa.[181,572] m. 205-213°;[181] Et_3N salt, unstable, m. 112-115°.[572]
Me ester. IXc. m. 173-174°.[325]
Pr ester. IXc. m. 175-176°.[325]
Bu ester. IXc. m. 173-174°.[325]
C_5H_{11} ester. IXc. m. 170-171°.[325]
C_6H_{13} ester. IXc. m. 166-167°.[325]
C_9H_{19} ester. IXc. m. 152-153°.[325]
[$CCl_3 \cdot$CH(OAc)]$_2PO_2H$. XXXf. m. 191-193°;[572] Et_3N salt, m. 124-126°.[572]
[$CCl_3 \cdot$CH(O·CO·Et)]$_2PO_2H$. XXXf. m. 183-184°.[572]
[$CCl_3 \cdot$CH(O·CO·Pr)]$_2PO_2H$. XXXf. m. 150-151.5°.[572]
[$CCl_3 \cdot$CH(O·CO·$CHMe_2$)]$_2PO_2H$. XXXf. m. 203°.[572]
[$CCl_3 \cdot$CH(O·CO·Bu)]$_2PO_2H$. XXXf. m. 130-131°.[572]
[$CCl_3 \cdot$CH(O·CO·$CH_2 \cdot CHMe_2$)]$_2PO_2H$. XXXf. m. 184-185°.[572]
[$CCl_3 \cdot$CH(O·CO·C_5H_{11})]$_2PO_2H$. XXXf. m. 98-100°.[572]
[$CCl_3 \cdot$CH(OH)]$BuPO_2H$.
Me ester. IXc. m. 104-105°.[3]
Et ester. IXc. m. 147-148°.[3]
Pr ester. IXc. m. 115-116°.[3]
i-Pr ester. IXc. m. 135-136°.[3]
Bu ester. IXc. m. 82-83°.[3]
i-Bu ester. IXc. m. 120-121°.[3]
s-Bu ester. IXc. m. 95-96°.[3]
C_5H_{11} ester. IXc. m. 137-138°.[3]
Et_2CH ester. IXc. m. 142-143°.[3]
$ClCH_2 \cdot CH_2$ ester. IXc. m. 80-81°.[3]
[$CCl_3 \cdot$CH(OH)]($CCl_3 \cdot CMe_2$)PO_2H.
1-CCl_3-cyclohexyl ester. IXc. m. 162.5°.[8]
[$CCl_3 \cdot$CH(OH)]$PhPO_2H$. IXa. m. 166-167°.[640]
Me ester. IXc. m. 122-123°,[7] mass spectrum.[107]
Et ester. Mass spectrum.[107]
1-CCl_3-cyclopentyl ester. IXc. m. 151°.[368]
[$CCl_3 \cdot$CH(OH)](C_7H_{15})PO_2H. IXa. m. 153-154°.[594]
[HO·(CH_2)$_3$]$EtPO_2H$.
Cyclic ester. From EtP(OEt)$_2$ + Br(CH_2)$_3$Br, b_2 83°.[229]

[EtCH(OH)](CCl_3·CMe_2)PO_2H.

1-CCl_3-cyclohexyl ester. IXc. m. 178-180°.[8]

[Me_2C(OH)][MeCH(OH)]PO_2H. IXa. m. 132° (rapid heating).[487]

[Me_2C(OH)][CCl_3·CH(OH)]PO_2H. IXa. m. 156-157° decomp.[573]

[Me_2C(OAc)][CCl_3·CH(OAc)]PO_2H. XXXf. m. 170-171°.[573]

[Me_2C(O·CO·Et)][CCl_3·CH(O·CO·Et)]PO_2H. XXXf. m. 145-146°.[573]

[Me_2C(O·CO·Pr)][CCl_3·CH(O·CO·Pr)]PO_2H. XXXf. m. 122-123°.[573]

[Me_2C(OH)]i-$PrPO_2H$. IIa. m. 160-161°.[104]

[Me_2C(OH)]$_2PO_2H$. IXa. m. 185°;[487] Na, Ag, La, and Pb salts.[487]

Me ester. XXIIa. m. 92°.[487]

Et ester. XXIIa. m. 95°.[487]

[Me_2C(OAc)]$_2PO_2H$. XXXf. m. 117°.[487]

[Me_2C(O·CO·Ph)]$_2PO_2H$. XXXf. m. 195-196°.[487]

[Me_2C(OH)](CCl_3·CMe_2)PO_2H.

1-CCl_3-cyclohexyl ester. IXc. m. 145-147°.[8]

[Me_2C(OH)]$PhPO_2H$. XXIIIa. m. 163°.[346]

Et ester. IXc. m. 94-96°.[346]

i-Bu ester. IXc. m. 94-95.5°.[7]

1-CCl_3-cyclopentyl ester. IXc. m. 130-131°.[368]

[CH⋮C·CH(OH)]$EtPO_2H$.

Me ester. IXc. m. 84°.[609]

Et ester. IXc. b_1 126-127°, d_4^{20} 1.1320, n_D^{20} 1.4750.[609]

[$ClCH_2$·CMe(OH)]$_2PO_2H$. IXa. m. 153.5-154.5°.[121]

[($ClCH_2$)$_2$C(OH)]$_2PO_2H$. IXa. m. 141-142°.[121]

[($ClCH_2$)$_2$C(OH)](CCl_3·CMe_2)PO_2H.

1-CCl_3-cyclohexyl ester. IXc. m. 168-168.5°.[8]

[HO(CH_2)$_4$]$EtPO_2H$.

Cyclic ester. By heating the Et ester of the next acid with 2% H_3PO_4. $b_{0.5}$ 87-88°, d_4^{20} 1.1211, n_D^{20} 1.4730.[56]

[AcO(CH_2)$_4$]$EtPO_2H$.

Et ester. Va. $b_{0.5}$ 126-128°, d_4^{20} 1.0683, n_D^{20} 1.4490.[56]

[HO(CH_2)$_4$](i-Bu)PO_2H.

Cyclic ester. From i-BuP$(OEt)_2$ + Cl$(CH_2)_4$Cl. b_1 103°.[229]

[PrCH(OH)](CCl_3·CMe_2)PO_2H.

1-CCl_3-cyclohexyl ester. m. 156-158°.[8]

[PrCH(OH)]$PhPO_2H$. IXc. m. 122-123.5°.[7]

[i-PrCH(OH)]$_2PO_2H$. By hydrolysis of cyclic acetal. m. 168-169°.[106]

Cyclic acetal with i-PrCHO, see cyclic phosphinic acids.

[EtMeC(OH)][CCl_3·CH(OH)]PO_2H. IXa. m. 153-154°.[573]

[MeCHCl·CMe(OH)]$EtPO_2H$.

Et ester. IXc. d_4^{20} 1.1631, n_D^{20} 1.4645.[2]

Pr ester. IXc. d_4^{20} 1.1230, n_D^{20} 1.4605.[2]

$[HO(CH_2)_5]PhPO_2H$.
Cyclic ester. From $PhP(OPh)_2 + Br(CH_2)_5Br$. b_1 120°.[229]
$[i\text{-}BuCH(OH)](CCl_3 \cdot CMe_2)PO_2H$.
1-CCl_3-cyclohexyl ester. IXc. m. 177-178°.[8]
$[i\text{-}BuCH(OH)]_2PO_2H$. IXa.[487,775] m. 230° (from 160° on slow heating);[487] K and Ba salts.[775]
(1-HO-cyclopentyl)$(CCl_3 \cdot CMe_2)PO_2H$.
1-CCl_3-cyclohexyl ester. IXc. m. 190° decomp.[8]
(1-HO-cyclopentyl)(cyclo-$C_5H_9)PO_2H$. IIa. m. 190-192°.[104]
(1-HO-cyclopentyl)$PhPO_2H$.
1-CCl_3-cyclopentyl ester. IXc. m. 142-143°.[368]
(1-HO-cyclohexyl)$(CCl_3 \cdot CMe_2)PO_2H$.
1-CCl_3-cyclohexyl ester. IXc. m. 166-167.5°.[8]
(1-HO-cyclohexyl)(cyclo-$C_6H_{11})PO_2H$. IIa. m. 200-201°.[104]
(1-HO-cyclohexyl)$PhPO_2H$. IXa. m. 184-185°.[640]
(1-HO-cyclohexyl)(p-$ClC_6H_4)PO_2H$. IXa. m. 194°.[640]
(1-HO-cyclohexyl)(p-$BrC_6H_4)PO_2H$. IXa. m. 189°.[640]
(1-HO-cyclohexyl)(p-$MeO \cdot C_6H_4)PO_2H$. IXa. m. 191°.[640]
(1-HO-cyclohexyl)(p-$MeC_6H_4)PO_2H$. IXa. m. 200-201°.[640]
$[C_6H_{13} \cdot CH(OH)][Me_2C(OH)]PO_2H$. IXa. m. 131° (rapid heating).[487]
$[C_6H_{13} \cdot CH(OH)][EtMeC(OH)]PO_2H$. IXa. m. 147° (rapid heating).[487]
$[C_6H_{13} \cdot CH(OH)]_2PO_2H$. IXa. m. 160°;[775] K and Ba salts.[775]
$[C_6H_{13} \cdot CH(OAc)]_2PO_2H$. XXXf. m. 94°.[775]
(p-$MeO \cdot C_6H_4 \cdot CH_2)PhPO_2H$. Va. b_{11} 236-238°.[299]
$[PhCH(OH)](HO \cdot CH_2)PO_2H$. IXa. m. 154° (rapid heating).[487]
$[PhCH(OH)][CCl_3 \cdot CH(OH)]PO_2H$. IXa. m. 172°.[573]
$[PhCH(OAc)][CCl_3 \cdot CH(OAc)]PO_2H$. XXXf. m. 181-183°.[573]
$[PhCH(OH)][Me_2C(OH)]PO_2H$. IXa. m. 182° (rapid heating).[487]
$[PhCH(OH)][CCl_3 \cdot CMe_2]PO_2H$.
1-CCl_3-cyclohexyl ester. IXc. m. 144-146°.[8]
$[PhCH(OH)][i\text{-}BuCH(OH)]PO_2H$. IXa. m. 203-205° (rapid heating).[487]
$[PhCH(OH)][PrMeC(OH)]PO_2H$. IXa. m. 170° (rapid heating).[487]
$[PhCH(OH)][Et_2C(OH)]PO_2H$. IXa. m. 192° (rapid heating).[487]
$[PhCH(OH)]PhPO_2H$. IXa.[640] IXg.[521] m. 112-114°,[521] 195°.[640]
Et ester. IXc.[7,346] m. 75°,[346] 182-182.5°.[7]
Bu ester. IXc. m. 76-78°.[7]
i-Bu ester. IXc. m. 100-101°.[7]
1-CCl_3-cyclopentyl ester. IXc. m. 165-167°.[368]
$[PhCH(OH)](p\text{-}ClC_6H_4)PO_2H$. IXa. m. 198°.[640]
$[PhCH(OH)](p\text{-}BrC_6H_4)PO_2H$. IXa. m. 191°.[640]
$[PhCH(OH)](p\text{-}MeO \cdot C_6H_4)PO_2H$. IXa. m. 176-177°.[640]
$[PhCH(OH)](p\text{-}MeC_6H_4)PO_2H$. IXa. m. 153-154°.[640]
$[PhCH(OH)]_2PO_2H$. IXa.[487,775,777] m. 230° (from 165° on slow heating);[487,777] K, Ba, Ag, and Hg^+ salts.[775]

Et ester. XXIIa.[775]
[PhCH(OEt)]$_2$PO$_2$H. Ia. IIa. m. 130°; (1H$_2$O), m. 104-110°.[180]
[PhCH(O·CHMe$_2$)]$_2$PO$_2$H. Ia. IIa. m. 113°; (1H$_2$O), m. 98-100°.[180]
[PhCH(OAc)]$_2$PO$_2$H. XXXf. m. ∿100°.[775]
[o-FC$_6$H$_4$·CH(OH)]EtPO$_2$H.
Bu ester. IXc. d_4^{20} 1.2670, n_D^{20} 1.5005.[2]
ClCH$_2$·CH$_2$ ester. IXc. d_4^{20} 1.4450, n_D^{20} 1.5235.[2]
[p-FC$_6$H$_4$·CH(OH)]EtPO$_2$H.
Bu ester. IXc. d_4^{20} 1.2700, n_D^{20} 1.5020.[2]
ClCH$_2$·CH$_2$ ester. IXc. d_4^{20} 1.4220, n_D^{20} 1.5145.[2]
[p-FC$_6$H$_4$·CH(OH)]PhPO$_2$H. IXa. m. 203°.[640]
[o-ClC$_6$H$_4$·CH(OH)]PhPO$_2$H. IXg. m. 283-284°.[224]
[p-ClC$_6$H$_4$·CH(OH)]PhPO$_2$H. IXa. m. 202°.[640]
[p-BrC$_6$H$_4$·CH(OH)]PhPO$_2$H. IXa. m. 183-184°.[640]
[o-O$_2$N·C$_6$H$_4$·CH(OH)]EtPO$_2$H.
Bu ester. IXc. m. 89-90°.[2]
ClCH$_2$·CH$_2$ ester. IXc. m. 120-121°.[2]
[m-O$_2$N·C$_6$H$_4$·CH(OH)]EtPO$_2$H.
Pr ester. IXc. m. 144-145°.[2]
Bu ester. IXc. m. 151-152°.[2]
ClCH$_2$·CH$_2$ ester. IXc. m. 110-111°.[2]
[m-O$_2$N·C$_6$H$_4$·CH(OH)](CCl$_3$·CMe$_2$)PO$_2$H.
1-CCl$_3$-cyclohexyl ester. IXc. m. 154-156°.[8]
[m-O$_2$N·C$_6$H$_4$·CH(OH)]PhPO$_2$H.
Me ester. IXc. m. 155-156°.[7]
Et ester. IXc. m. 120-121°.[7]
Bu ester. IXc. m. 120-121.5°.[7]
i-Bu ester. IXc. m. 142-143°.[7]
[p-O$_2$N·C$_6$H$_4$·CH(OH)]EtPO$_2$H.
Bu ester. IXc. m. 163-164°.[2]
ClCH$_2$·CH$_2$ ester. IXc. m. 128-129°.[2]
[p-O$_2$N·C$_6$H$_4$·CH(OH)]PhPO$_2$H. IXa. m. 211-212°.[640]
1-CCl$_3$-cyclopentyl ester. IXc. m. 171-173°.[368]
[Me$_2$CH·CH(OH)·CMe$_2$·CH(OH)]$_2$PO$_2$H. IXa (using Me$_2$CH·CHO which forms the aldol before reacting with H$_3$PO$_2$). m. 140-142°.[775]
[p-MeC$_6$H$_4$·CH(OH)](CCl$_3$·CMe$_2$)PO$_2$H.
1-CCl$_3$-cyclohexyl ester. IXc. m. 150-152°.[8]
[PhCMe(OH)][MeCH(OH)]PO$_2$H. IXa. m. 192° (rapid heating).[487]
[PhCMe(OH)]PhPO$_2$H.
i-Bu ester. IXc. m. 150-151°.[7]
[PhCMe(OH)](PhCHMe)PO$_2$H. IIa. m. 174-176°.[104]
[p-MeC$_6$H$_4$·CMe(OH)]EtPO$_2$H.
Me ester. IXc. m. 75°.[612]
Et ester. IXc. m. 83°, 112°.[612]
Pr ester. IXc. m. 106°.[612]
[PhCEt(OH)]EtPO$_2$H.
i-Pr ester. IXc. m. 99°.[612]

[p-(i-Pr)C_6H_4·CH(OH)]$_2PO_2H$. IXa. m. ∿140°; Ba salt.[775]
[1-naphthyl·CH(OH)]$PhPO_2H$.
Et ester. IXc. m. 152-153°.[99]
[2-naphthyl·CH(OH)]$PhPO_2H$.
Et ester. IXc. m. 171-173°.[99]
[2-naphthyl·CMe(OH)]$PhPO_2H$.
Et ester. IXc. m. 153-155°.[99]
[Ph_2C(OH)]$PhPO_2H$. [XXIa + XXXd]. m. 191-192°;[222] benzylisothiouronium salt, m. 170-171°.[222]

E.1.5. Aliphatic and Aliphatic/Aromatic Phosphinic Acids with Carbonyl, or Functionally Modified Carbonyl (Acetal, Enol Ether, Enol Ester), Substituents in Aliphatic Groups

[(EtO)$_2$CH]$PhPO_2H$.
Et ester. IXh. $b_{0.3}$ 130-132°, n_D^{20} 1.4964.[162]
[(EtO)$_2$CH](p-ClC_6H_4)PO_2H.
Et ester. IXe. $b_{0.35}$ 137-140°, d_4^{20} 1.1863, n_D^{20} 1.5115.[660]
(H·CO·CH_2)$EtPO_2H$.
Et ester. XXXg. $b_{0.22}$ 84-85°, d_4^{20} 1.0751, n_D^{20} 1.4431;[659] polarography;[668] 2,4-dinitrophenylhydrazone, m. 146-147°.[659]
[(MeO)$_2$CH·CH_2]$EtPO_2H$.
Me ester. IXe. b_{16} 102-105°, d_4^{20} 1.0787, n_D^{20} 1.4382.[660]
Et ester. IXe. b_{14} 111-114°, d_4^{20} 1.0563, n_D^{20} 1.4355.[660]
[(EtO)$_2$CH·CH_2]$EtPO_2H$.
Me ester. IXe. b_{12} 118-120°, d_4^{20} 1.0302, n_D^{20} 1.4350.[660]
Et ester. Va.[659] IXe.[660] $b_{0.5}$ 103-105°,[659] b_{13} 120-123°,[660] d_4^{20} 1.0135,[660] 1.0369,[659] n_D^{20} 1.4300,[660] 1.4375.[659]
(BuO·CH:CH)$EtPO_2H$.
Et ester. XXIe. b_3 117-119°, d_4^{20} 1.0040, n_D^{20} 1.4564.[766]
(ClCH_2·CH_2·O·CH:CH)$EtPO_2H$.
Et ester. XXIe. b_2 149°, d_4^{20} 1.0170, n_D^{20} 1.4840.[766]
(EtO·CH:CH)$_2PO_2H$.
Me ester. XXIe. b_4 154-156°, d_4^{25} 1.1085, n_D^{25} 1.4838.[195]
Et ester. XXIe. b_3 142-144°, d_4^{25} 1.0828, n_D^{25} 1.4815.[195]
Pr ester. XXIe. b_3 155-157°, d_4^{25} 1.0600, n_D^{25} 1.4770.[195]
Bu ester. XXIe. b_3 163-165°, d_4^{25} 1.0646, n_D^{25} 1.4775.[195]
C_7H_{15} ester. XXIe. b_2 183-185°, d_4^{25} 1.0090, n_D^{25} 1.4545.[195]

CH_2:CH·CH_2 ester. XXIe. b_3 158-159°, d_4^{25} 1.0844, n_D^{25} 1.4919.[195]
Ph ester. XXIe. b_2 184-186°, d_4^{25} 1.1667, n_D^{25} 1.5370.[195]

(BuO·CH:CH)$_2PO_2$H.
Me ester. XXIe. b_2 191-193°, d_4^{20} 1.0389, n_D^{20} 1.4085.[195]
Et ester. XXIe. b_3 180-181°, d_4^{20} 1.0235, n_D^{20} 1.4770.[195]
Pr ester. XXIe. b_2 174-175°, d_4^{20} 1.0256, n_D^{20} 1.4755.[195]
Bu ester. XXIe. b_2 182-183°, d_4^{20} 1.0381, n_D^{20} 1.4775.[195]

(H·CO·CH_2)PhPO_2H.
Et ester. XXXg. $b_{0.02}$ 137-140°, d_4^{20} 1.1313, n_D^{20} 1.5163,[666] polarography;[668] 2,4-dinitrophenylhydrazone, m. 133-134°.[666]

[(EtO)$_2$CH·CH_2]PhPO_2H.
Et ester. Va. $b_{0.2}$ 142-143°, d_4^{20} 1.0805, n_D^{20} 1.4943.[666]

(EtO·CH:CH)PhPO_2H.
Me ester. XXIe. b_2 144°, d_4^{20} 1.1425, n_D^{20} 1.5328.[23]
Et ester. XXIe. b_2 156°, d_4^{20} 1.1032, n_D^{20} 1.5220.[23]
Pr ester. XXIe. b_2 170-171°, d_4^{20} 1.0765, n_D^{20} 1.5195.[23]
Bu ester. XXIe. b_2 177-178°, d_4^{20} 1.0549, n_D^{20} 1.5173.[23]

(CH_2:CH·O·CH:CH)PhPO_2H.
Et ester. XXIe. b_2 145-146°, d_4^{20} 1.1480, n_D^{20} 1.5430.[763]

(H·CO·CH_2)(p-ClC_6H_4)PO_2H.
Et ester. XXXg. $b_{0.02}$ 144-146°, d_4^{20} 1.2435, n_D^{20} 1.5347,[667] polarography;[668] 2,4-dinitrophenylhydrazone, m. 175-176°.[667]

[(EtO)$_2$CH·CH_2](p-ClC_6H_4)PO_2H.
Et ester. Va. $b_{0.07}$ 143-145°, d_4^{20} 1.1678, n_D^{20} 1.5065.[667]

(H·CO·CH_2)(p-MeC_6H_4)PO_2H.
Et ester. XXXg. $b_{0.02}$ 133-136°, d_4^{20} 1.1359, n_D^{20} 1.5290,[666] polarography;[668] 2,4-dinitrophenylhydrazone, m. 166-167°.[666]

[(EtO)$_2$CH·CH_2](p-MeC_6H_4)PO_2H.
Et ester. Va. $b_{0.015}$ 134-137°, d_4^{20} 1.0719, n_D^{20} 1.4956.[666]

(H·CO·CH_2)(p-EtC_6H_4)PO_2H.
Et ester. XXXg. $b_{0.02}$ 134-137°, d_4^{25} 1.0927, n_D^{25} 1.5125,[666] polarography;[668] 2,4-dinitrophenylhydrazone, m. 160-162°.[666]

[(EtO)$_2$CH·CH_2](p-EtC_6H_4)PO_2H.
Et ester. Va. $b_{0.01}$ 142-146°, d_4^{20} 1.0587, n_D^{20} 1.4948.[666]

$(H\cdot CO\cdot CH_2)[p\text{-}(i\text{-}Pr)C_6H_4]PO_2H$.
Et ester. XXXg. $b_{0.02}$ 140-143°, d_4^{25} 1.0878, n_D^{25} 1.5148,[666] polarography;[668] 2,4-dinitrophenylhydrazone, m. 125-127°.[666]
$[(EtO)_2CH\cdot CH_2][p\text{-}(i\text{-}Pr)C_6H_4]PO_2H$.
Et ester. Va. $b_{0.015}$ 126-129°, d_4^{20} 1.0460, n_D^{20} 1.4940.[666]
$[(EtO)_2CMe](ClCH_2)PO_2H$.
Me ester. From $ClCH_2\cdot PH(O)(OMe)$ + $CH_2:C(OEt)_2$, $b_{0.01}$ 90-95°.[256]
$(Me\cdot CO)EtPO_2H$.
Me ester. Va. b_9 89-90°, d_4^{20} 1.1225, n_D^{20} 1.4405.[650]
Et ester. Va. b_{11} 117-118°, d_4^{20} 1.0997, n_D^{20} 1.4415.[650]
Bu ester. Major product from $EtP(OBu)_2$ + $CH_2:CO$, b_{10} 134-136°, d_4^{20} 1.0598, n_D^{20} 1.4471.[634]
$[CH_2:C(OAc)]EtPO_2H$.
Et ester. From $EtP(OEt)_2$ + $CH_2:CO$, also from $(Me\cdot CO)EtP(O)OEt$ + $CH_2:CO$, b_{10} 127°, d_4^{20} 1.1010, n_D^{20} 1.4510.[634]
Bu ester. Minor product from $EtP(OBu)_2$ + $CH_2:CO$. Also from $(Me\cdot CO)EtP(O)OBu$ + $CH_2:CO$, b_{10} 145-146°, d_4^{20} 1.0737, n_D^{20} 1.4512.[634]
$(Me\cdot CO)BuPO_2H$.
Bu ester. Va. $b_{0.05}$ 80°, n_D^{20} 1.4381.[308]
$[(p\text{-}HO\cdot C_6H_4\cdot O)_2CMe]BuPO_2H$.
Bu ester. From Bu ester of previous acid + $p\text{-}C_6H_4(OH)_2$ + BF_3, m. 172°.[308]
$(Me\cdot CO)PhPO_2H$.
Me ester. Va.[308,650] Also stated to be the major product from $PhP(OMe)_2$ + $CH_2:CO$,[634] $b_{0.03}$ 98-99°,[650] d_4^{20} 1.1935,[650] n_D^{20} 1.5285.[650]
Et ester. Va.[308] Also stated to be the major product from $PhP(OEt)_2$ + $CH_2:CO$, $b_{0.03}$ 97-98°,[308] n_D^{20} 1.5168.[308]
$[(EtO)_2CMe]PhPO_2H$.
Et ester. IXe. $b_{0.006}$ 117-118°, d_4^{20} 1.0960, n_D^{20} 1.4983.[544]
Pr ester. From Pr ester of previous acid + $CH(OEt)_3$, $b_{0.1}$ 151-153°, d_4^{20} 1.0712, n_D^{20} 1.4915.[231]
$[(p\text{-}HO\cdot C_6H_4\cdot O)_2CMe]PhPO_2H$.
Me ester. From $(Me\cdot CO)PhP(O)OMe$ + $p\text{-}C_6H_4(OH)_2$ + BF_3, m. 210-210.5°.[308]
Et ester. From $(Me\cdot CO)PhP(O)OEt$ + $p\text{-}C_6H_4(OH)_2$ + BF_3, m. 209.5-210°.[308]
$[(3\text{-}Cl\text{-}4\text{-}HO\cdot C_6H_3\cdot O)_2CMe]PhPO_2H$.
Me ester. From $(Me\cdot CO)PhP(O)OMe$ + $2\text{-}Cl\text{-}1,4\text{-}(HO)_2C_6H_3$ + BF_3, m. 222.5-223°.[308]
$[CH_2:C(OAc)]PhPO_2H$.
Me ester. From product from $PhP(OMe)_2$ + $CH_2:CO$. Also from $(Me\cdot CO)PhP(O)OMe$ + $CH_2:CO$, b_{10} 165-166°, d_4^{20}

1.1802, n_D^{20} 1.5001.[634]
Et ester minor product from $PhP(OEt)_2 + CH_2:CO$. Also from $(Me\cdot CO)PhP(O)OEt + CH_2:CO$, b_{10} 172-173°, d_4^{20} 1.1621, n_D^{20} 1.5078.[634]

$(Me\cdot CO)(p\text{-}ClC_6H_4)PO_2H$.
Me ester. Va. $b_{0.04}$ 130°, d_4^{20} 1.2974, n_D^{20} 1.5410.[650]

$[CH_2:C(OEt)](p\text{-}ClC_6H_4)PO_2H$.
Et ester. IXe. $b_{0.05}$ 144-146°, n_D^{20} 1.5028.[544]

$[(EtO)_2CMe](p\text{-}MeC_6H_4)PO_2H$.
Et ester. IXe. $b_{0.07}$ 126-127°, d_4^{20} 1.0747, n_D^{20} 1.5025.[544]

$[CH_2:C(OEt)](p\text{-}MeC_6H_4)PO_2H$.
Et ester. IXe. $b_{0.045}$ 122-125°, d_4^{20} 1.0873, n_D^{20} 1.5102.[544]

$(H\cdot CO\cdot CH_2\cdot CH_2)EtPO_2H$.
Et ester. XXXg. $b_{0.25}$ 96-98°, d_4^{20} 1.0625, n_D^{20} 1.4555,[659] polarography;[668] 2,4-dinitrophenylhydrazone, m. 161-162°.[659]

$[(EtO)_2CH\cdot CH_2\cdot CH_2]EtPO_2H$.
Et ester. Va. $b_{0.18}$ 103-105°, d_4^{20} 1.0100, n_D^{20} 1.4415.[659]

$(H\cdot CO\cdot CH_2\cdot CH_2)PhPO_2H$.
Et ester. XXXg. $b_{0.05}$ 136-138°, d_4^{20} 1.1521, n_D^{20} 1.5203,[667] polarography;[668] 2,4-dinitrophenylhydrazone, m. 149-150°.

$[(EtO)_2CH\cdot CH_2\cdot CH_2]PhPO_2H$.
Et ester. Va. $b_{0.25}$ 151-153°, d_4^{20} 1.0758, n_D^{20} 1.4930.[666]

$(EtO\cdot CH:CH\cdot CH_2)PhPO_2H$.
Et ester. Xb*.[361] Also by thermal elimination of EtOH from Et ester of previous acid,[666] $b_{0.015}$ 139-141°,[666] $b_{0.5}$ 150-152°,[361] d_4^{20} 1.0417,[666] n_D^{20} 1.5160.[666]

$(H\cdot CO\cdot CH_2CH_2)(p\text{-}ClC_6H_4)PO_2H$.
Et ester. XXXg. $b_{0.025}$ 161-163°, d_4^{20} 1.2498, n_D^{20} 1.5332,[667] polarography;[668] 2,4-dinitrophenylhydrazone, m. 164-165°.[667]

$[(EtO)_2CH\cdot CH_2\cdot CH_2](p\text{-}ClC_6H_4)PO_2H$.
Et ester. Va. $b_{0.025}$ 156-157°, d_4^{20} 1.1480, n_D^{20} 1.5030.[667]

$(H\cdot CO\cdot CH_2\cdot CH_2)(p\text{-}MeC_6H_4)PO_2H$.
Et ester. XXXg. $b_{0.004}$ 150-152°, d_4^{20} 1.1562, n_D^{20} 1.5290,[667] polarography.[668]

$[(EtO)_2CH\cdot CH_2\cdot CH_2](p\text{-}MeC_6H_4)PO_2H$.
Et ester. Va. $b_{0.06}$ 146-148°, d_4^{20} 1.0588, n_D^{20} 1.4914.[667]

$(Me\cdot CO\cdot CH_2)EtPO_2H$.
Et ester. Va. b_1 75-78°, d_4^{20} 1.0833, n_D^{20} 1.4502,[47] mass spectrum.[153]

$(Me\cdot CO\cdot CH_2)PhPO_2H$.
Me ester. Va. b_8 168-170°, d_4^{20} 1.1965, n_D^{20} 1.5220.[607]

(H·CO·CHMe)PhPO$_2$H.
Et ester. XXXg*. $b_{0.02}$ 116-120°, d_4^{20} 1.1173, n_D^{20} 1.5182.[667]

(EtO·CH:CHMe)PhPO$_2$H.
Et ester. Va using (EtO)$_2$CH·CHMeBr with thermal elimination of EtOH. $b_{0.15}$ 132-134°, d_4^{20} 1.0911, n_D^{20} 1.5130.[667]

(H·CO·CHMe)(p-MeC$_6$H$_4$)PO$_2$H.
Et ester. XXXg*. $b_{0.1}$ 137-140°, d_4^{20} 1.1070, n_D^{20} 1.5210.[667]

(EtO·CH:CHMe)(p-MeC$_6$H$_4$).
Et ester. Va using (EtO)$_2$CH·CHMeBr with thermal elimination of EtOH $b_{0.3}$ 152-154°, d_4^{20} 1.0840, n_D^{20} 1.5102.[667]

(Me·CO·CHCl·CH$_2$)MePO$_2$H.
Me ester. XXIIId. $b_{0.01}$ 97°, d_4^{20} 1.2443, n_D^{20} 1.4762.[623]

Cyclic enol ester $\overline{\text{O·CMe:CCl·CH}_2\text{·P}}$(O)Me. Xe. $b_{0.006}$ 90-92°, d_4^{20} 1.3074, n_D^{20} 1.5010.[623]

(Me·CO·CHCl·CH$_2$)EtPO$_2$H.
Me ester. XXIIId. $b_{0.005}$ 88-89°, d_4^{20} 1.2059, n_D^{20} 1.4720.[623]
Et ester. XXIIId. $b_{0.009}$ 100°, d_4^{20} 1.1590, n_D^{20} 1.4675.[623]
i-Pr ester. XXIIId. $b_{0.003}$ 82-83°, d_4^{20} 1.1198, n_D^{20} 1.4605.[623]

Cyclic enol ester $\overline{\text{O·CMe:CCl·CH}_2\text{·P}}$(O)Et. Xe. $b_{0.002}$ 86°, d_4^{20} 1.2568, n_D^{20} 1.4958.[623]

(Me·CO·CHCl·CH$_2$)PhPO$_2$H. XXIIId. d_4^{20} 1.2976, n_D^{20} 1.5463.[622]
Me ester. XXIIId. $b_{0.007}$ 121-124°, d_4^{20} 1.2120, n_D^{20} 1.5210.[622]
Et ester. XXIIId. $b_{0.01}$ 142-144°, d_4^{20} 1.1921, n_D^{20} 1.5205.[622]

Cyclic enol ester $\overline{\text{O·CMe:CCl·CH}_2\text{·P}}$(O)Ph. Xe. $b_{0.01}$ 129-131°, d_4^{20} 1.2841, n_D^{20} 1.5608.[622]

[Me·CO·CMe(OH)]EtPO$_2$H.
Et ester. IXc.[613] b_1 97-98°, d_4^{20} 1.1156, n_D^{20} 1.4563,[613,614] kinetics of P-C → P-O rearrangement.[614]
Pr ester. IXc. b_1 107-108°, d_4^{20} 1.0911, n_D^{20} 1.4570.[613]
i-Pr ester. IXc. b_1 98-99°, d_4^{20} 1.0827, n_D^{20} 1.4513.[613]

[Me·CO·CMe(OH)]PhPO$_2$H.
Et ester. IXc. m. 94-95°.[619]

(Me·CO·CH$_2$·CHMe)EtPO$_2$H.
Et ester. IXf. b_9 139°, d_4^{20} 1.0435, n_D^{20} 1.4561.[637]
Bu ester. IXf. b_5 156-157°, d_4^{20} 1.0121, n_D^{20} 1.4563.[637]

$(Me \cdot CO \cdot CH_2 \cdot CMe_2)MePO_2H$.
    ester. XXIIId. $b_{0.04}$ 73-74°.[49]
$(Me \cdot CO \cdot CH_2 \cdot CMe_2)EtPO_2H$. XXIIId. m. 112-113°.[49,50,577]
    Me ester. XXIIId.[49,50,577] $b_{0.1}$ 79-80°,[50] b_3 112-
    113°,[577] d_4^{20} 1.0705,[50] n_D^{20} 1.4642.[50]
    Et ester. IXf.[637] XXIIId.[49,50,577] $b_{0.02}$ 73-75°,[50]
    b_{14} 152.5°,[637] d_4^{20} 1.0351,[50] 1.0392,[637] n_D^{20}
    1.4577,[637] 1.4602.[50]
    Pr ester. XXIIId. b_5 131-133°, d_4^{20} 1.0300, n_D^{20}
    1.4640.[577]
    i-Pr ester. XXIIId. m. 82-84°.[577]
    Bu ester. IXf.[637] XXIIId.[577] $b_{0.09}$ 96-98°,[577] b_6
    154-155.5°,[637] d_4^{20} 1.0066,[637] 1.0250,[577] n_D^{20}
    1.4588,[637] 1.4631.[577]
    C_5H_{11} ester. XXIIId. $b_{0.09}$ 122-124°, d_4^{20} 0.9837,
    n_D^{20} 1.4502.[577]

    Cyclic enol ester $\overline{O \cdot CMe:CH \cdot CMe_2 \cdot P}(O)Et$. Xb. $b_{0.1}$
    69-71°, d_4^{20} 1.0565, n_D^{20} 1.4757.[33]
$(Me \cdot CO \cdot CH_2 \cdot CMe_2)PhPO_2H$. Xb.[520] XXIIId.[73,577] m.
    93.5°,[73] 121°;[577] ($1H_2O$), m. 86°;[520] $PhNH_2$ salt,
    m. 124°.[577]
    Me ester. XXIIId. $b_{0.3}$ 149-152°, d_4^{20} 1.1260, n_D^{20}
    1.5255.[577]
    Et ester. XXIIId. $b_{0.3}$ 154-156°, d_4^{20} 1.1250, n_D^{20}
    1.5235;[577] 2,4-dinitrophenylhydrazone, m. 213°.[577]
    Pr ester. XXIIId. $b_{0.2}$ 141-143°, d_4^{20} 1.0870, n_D^{20}
    1.5137;[577] 2,4-dinitrophenylhydrazone, m. 203°.[577]
    i-Pr ester. XXIIId. $b_{0.1}$ 131-133°, d_4^{20} 1.0968, n_D^{20}
    1.5150.[577]
    Bu ester. XXIIId. $b_{0.15}$ 146-148°, d_4^{20} 1.0697, n_D^{20}
    1.5110.[577]
    i-Bu ester. XXIIId. $b_{0.85}$ 158-160°, d_4^{20} 1.0741, n_D^{20}
    1.5115.[577]
    C_5H_{11} ester. XXIIId. $b_{0.8}$ 166-170°, d_4^{20} 1.0613, n_D^{20}
    1.5068.[577]
    $Me_2CH \cdot CH_2 \cdot CH_2$ ester. XXIIId. $b_{0.08}$ 138-140°, d_4^{20}
    1.0630, n_D^{20} 1.5075.[577]

    Cyclic enol ester $\overline{O \cdot CMe:CH \cdot CMe_2 \cdot P}(O)Ph$. Xb.[73,577]
    $b_{0.09}$ 136°,[577] b_{10} 170-172°,[73] d_4^{20} 1.0310,[577] n_D^{20}
    1.5415,[577] 1.5449.[73]
$(Me \cdot CO \cdot CH_2 \cdot CMe_2)(p-MeC_6H_4)PO_2H$. Xb. m. 102-103°.[520]
$(Ph \cdot CO)EtPO_2H$.
    Me ester. Va. $b_{0.52}$ 116.5-118°, d_4^{20} 1.1764, n_D^{20}
    1.5208; 2,4-dinitrophenylhydrazone, m. 175°.[650]
    Et ester. Va. $b_{0.38}$ 117.5-119°, d_4^{20} 1.1439, n_D^{20}
    1.5215; 2,4-dinitrophenylhydrazone, m. 116-
    117°.[650]
    Pr ester. Va. $b_{0.12}$ 123-124°, d_4^{20} 1.1135, n_D^{20}
    1.5148; 2,4-dinitrophenylhydrazone, m. 110-
    111°.[650]

Bu ester. Va. $b_{0.08}$ 125-126°, d_4^{20} 1.0961, n_D^{20} 1.5106, 2,4-dinitrophenylhydrazone, m. 90°.[650]
C_5H_{11} ester. Va. $b_{0.28}$ 135-141°, d_4^{20} 1.0754, n_D^{20} 1.5062, 2,4-dinitrophenylhydrazone, m. 78°.[650]
C_6H_{13} ester. Va. $b_{0.09}$ 136-138°, d_4^{20} 1.0623, n_D^{20} 1.5060.[650]

(Ph·CO)$PhPO_2H$.
Me ester. Va. m. 78.5°, $b_{0.25}$ 147-147.5°.[650]
Et ester. Va. $b_{0.08}$ 148.5-150°, d_4^{20} 1.1862, n_D^{20} 1.5760.[650]
Pr ester. Va. $b_{0.10}$ 158-159.5°, d_4^{20} 1.1630, n_D^{20} 1.5710.[650]
Bu ester. Va. $b_{0.09}$ 163-165°, d_4^{20} 1.1393, n_D^{20} 1.5617.[650]

(Ph·CO)(p-ClC_6H_4)PO_2H.
Me ester. Va. $b_{0.025}$ 158-159°, d_4^{20} 1.3024, n_D^{20} 1.5980.[650]
Et ester. Va. $b_{0.018}$ 168-169°, d_4^{20} 1.2706, n_D^{20} 1.5884.[650]

(Ph·CO)(p-MeC_6H_4)PO_2H.
Me ester. Va. $b_{0.02}$ 149.5°, d_4^{20} 1.1980, n_D^{20} 1.5846.[650]
Et ester. Va. $b_{0.015}$ 152-153.5°, d_4^{20} 1.1658, n_D^{20} 1.5746.[650]
Pr ester. Va. $b_{0.04}$ 158-159°, d_4^{20} 1.1430, n_D^{20} 1.5650.[650]

(o-ClC_6H_4·CO)$PhPO_2H$.
Me ester. Va. $b_{0.04}$ 158-160°, d_4^{20} 1.2981, n_D^{20} 1.5885.[650]
Et ester. Va. $b_{0.035}$ 174°, d_4^{20} 1.2534, n_D^{20} 1.5747.[650]
Pr ester. Va. $b_{0.09}$ 186-187°, d_4^{20} 1.2333, n_D^{20} 1.5693.[650]
Bu ester. Va. $b_{0.05}$ 180°, d_4^{20} 1.2070, n_D^{20} 1.5640.[650]
i-Bu ester. Va. $b_{0.05}$ 190°, d_4^{20} 1.2075, n_D^{20} 1.5624.[650]

(o-ClC_6H_4·CO)(p-ClC_6H_4)PO_2H.
Me ester. Va. $b_{0.04}$ 167°, d_4^{20} 1.3757, n_D^{20} 1.5982.[650]
Et ester. Va. $b_{0.04}$ 180-181°, d_4^{20} 1.3262, n_D^{20} 1.5815.[650]

(o-ClC_6H_4·CO)(p-MeC_6H_4)PO_2H.
Me ester. Va. $b_{0.04}$ 166-167°, d_4^{20} 1.2708, n_D^{20} 1.5850.[650]
Et ester. Va. $b_{0.04}$ 174-175°, d_4^{20} 1.2274, n_D^{20} 1.5717.[650]
Pr ester. Va. $b_{0.04}$ 191-192°, d_4^{20} 1.2086, n_D^{20} 1.5674.[650]

(Et·CO·CH_2·CMeEt)$EtPO_2H$.
Bu ester. IXf. b_{11} 176-177°, d_4^{20} 1.0002, n_D^{20} 1.4641.[637]

(Ph·CO·CH_2)$PhPO_2H$. From (PhC⋮C)$PhPO_2H$ + conc.H_2SO_4, poured on ice. m. 128-129°.[395]

(Ph·CO·CH$_2$)$_2$PO$_2$H. From (PhC⋮C)$_2$PO$_2$H + conc.H$_2$SO$_4$, poured
    on ice. m. 124-125°.[395]

   Et ester. From (PhC⋮C)$_2$P(O)OEt + conc.H$_2$SO$_4$, poured on
   ice. m. 63-64°; 2,4-dinitrophenylhydrazone, m.
   237-238°.[395]

(HO·CH:CPh)(PhCH$_2$)PO$_2$H.   (Enol structure of esters is
    indicated by IR and by violet color with FeCl$_3$.)

   Me ester. From (PhCH$_2$)$_2$P(O)H + H·CO$_2$Me.   m. 139-140°.[34]

   Et ester. From (PhCH$_2$)$_2$P(O)H + H·CO$_2$Et.   m. 106-108°.[34]

(p-MeC$_6$H$_4$·CO)(p-ClC$_6$H$_4$)PO$_2$H.

   Me ester.   Va.   b$_{0.03}$ 186-188°, d$_4^{20}$ 1.2728, n$_D^{20}$
    1.5950.[650]

   Et ester.   Va.   b$_{0.03}$ 198-200°, d$_4^{20}$ 1.2410, n$_D^{20}$
    1.5842.[650]

(p-MeC$_6$H$_4$·CO)(p-MeC$_6$H$_4$)PO$_2$H.

   Me ester.   Va.   b$_{0.03}$ 180-182°, d$_4^{20}$ 1.1731, n$_D^{20}$
    1.5856.[650]

   Et ester.   Va.   b$_{0.02}$ 183-185°, d$_4^{20}$ 1.1482, n$_D^{20}$
    1.5722.[650]

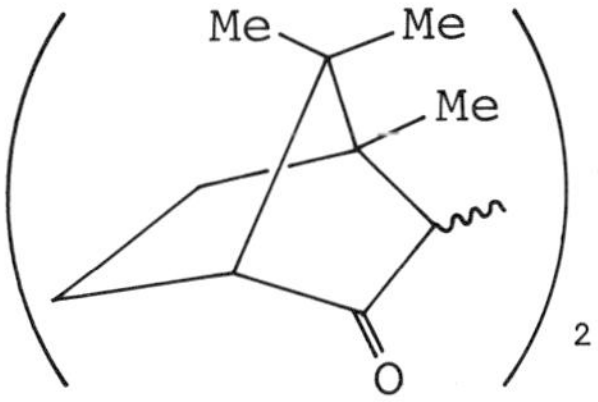

PO$_2$H.   VId.   m. 283° decomp. $[\alpha]_D^{20}$
151.1° (c 2.4, CHCl$_3$).[541]

(Me·CO·CH$_2$·CHPh)EtPO$_2$H.

   Et ester.   IXf.   b$_8$ 178-179°, d$_4^{20}$ 1.1084, n$_D^{20}$
    1.5150.[637]

   Bu ester.   IXf.   b$_6$ 194-196°, d$_4^{20}$ 1.0906, n$_D^{20}$
    1.5188.[637]

(Et·CO·CH$_2$·CHPh)EtPO$_2$H.

   Et ester.   IXf.   b$_8$ 178-180°, d$_4^{20}$ 1.1008, n$_D^{20}$
    1.5096.[637]

   Bu ester.   IXf.   b$_7$ 196-198°, d$_4^{20}$ 1.0664, n$_D^{20}$
    1.5109.[637]

[Ph·CO·CPh(OH)]PhPO$_2$H.

   Et ester.   IXc.   m. 166°.[619]

(Ph·CO·CH$_2$·CHPh)BuPO$_2$H.   Xb.   m. 191-193°.[173]

(Ph·CO·CH$_2$·CHPh)PhPO$_2$H.   XXIIId.[73,137]   m. 243-244°.[73]

   Cyclic enol ester $\overline{\text{O·CPh:CH·CHPh·P}}$(O)Ph.   Xb.[73,137]
   m. 162-163°.[73]

(Ph·CO·CHBr·CHPh)PhPO$_2$H. By bromination of previous acid.
   Also by addition of Br$_2$ to its cyclic enol ether
   and hydrolysis of the adduct.[137]   Separable into
   two forms, m. 195° decomp. (reacts slowly with
   aqueous Na$_2$CO$_3$ giving Ph·CO·CH:CHPh) and m. 150°
   (reacts rapidly with aqueous Na$_2$CO$_3$, and converted
   by sunlight into the higher-melting form).[137]

(PhCH:CH·CO·CH_2·CHPh)$PhPO_2H$. Xb. m. 235-236°.[136]
(PhCHBr·CHBr·CO·CH_2·CHPh)$PhPO_2H$. XXXa. m. 195° decomp.[136]
(PhCH:CBr·CO·CH_2·CHPh)$PhPO_2H$. XXXe. m. 200°.[136]
[Ph·CO·CH_2·CH(CH:CHPh)]$PhPO_2H$. Xb. m. 200°.[136]

E.1.6. Aliphatic and Aliphatic/Aromatic Phosphinic Acids with Carboxyl, or Functionally Modified Carboxyl, Substituents in Aliphatic Groups

(EtNH·CO)$EtPO_2H$.
Bu ester. IXd. b_{16} 177°, d_4^{20} 1.0589, n_D^{20} 1.4581.[632]
(Et_2N·CO)$EtPO_2H$.
Et ester. Va.[38] Ve.[39] b_2 107-108°, d_4^{20} 1.0538, n_D^{20} 1.4630.[39]
Pr ester. Va.[38] Ve.[39] b_2 112-113°, d_4^{20} 1.0346, n_D^{20} 1.4622.[39]
i-Pr ester. Ve. b_9 133-134°, d_4^{20} 1.0286, n_D^{20} 1.4595.[39]
Bu ester. Ve. $b_{0.5}$ 111-112°, d_4^{20} 1.0208, n_D^{20} 1.4625.[39]
i-Bu ester. Ve. b_9 145-146.5°, d_4^{20} 1.0170, n_D^{20} 1.4610.[39]
C_6H_{13} ester. Ve. $b_{2.5}$ 131-132°, d_4^{20} 0.9725, n_D^{20} 1.4500.[39]
C_7H_{15} ester. Ve. b_2 143-145°, d_4^{20} 0.9817, n_D^{20} 1.4565.[39]
C_8H_{17} ester. Ve. $b_{0.5}$ 137-138°, d_4^{20} 0.9782, n_D^{20} 1.4618.[39]
C_9H_{19} ester. Ve. $b_{1.5}$ 145-147°, d_4^{20} 0.9702, n_D^{20} 1.4612.[39]
$C_{10}H_{21}$ ester. Ve. $b_{1.5}$ 160-162°, d_4^{20} 0.9630, n_D^{20} 1.4600.[39]
(BuNH·CO)$PhPO_2H$. IXb. ($1H_2O$), m. 95°, resolidifies and then m. 116°.[205]
(PhNH·CO)$PhPO_2H$. IXb. m. 153-153.5° decomp.; Et_3N salt, m. 165-167° decomp.[205]
(p-ClC_6H_4·NH·CO)$PhPO_2H$. IXb. m. 161.5-162° decomp.; Et_3N salt, m. 170-170.5° decomp.[205]
(2,5-$Cl_2C_6H_3$·NH·CO)$PhPO_2H$. IXb. m. 153°.[205]
(EtO_2C·CH_2)$EtPO_2H$.
Et ester. Va. b_1 96-96.5°, d_4^{20} 1.0935, n_D^{20} 1.4443.[38]
Pr ester. Va. b_1 104-105°, d_4^{20} 1.0649, n_D^{20} 1.4420.[38]
(H_2N·NH·CO·CH_2)$EtPO_2H$. From Et ester of previous acid + N_2H_4. m. 200-202°.[665]
(CH_2:CH·CH_2·O·CO·CH_2)$_2PO_2H$.
Et ester. IVa. $b_{0.03}$ 135-136°, d_4^{20} 1.1662, n_D^{20} 1.4718.[766]
(HO_2C·CH_2)$PhPO_2H$. XXIIIa.[26,281] m. 121.5-122.5°.[26]
Me ester. Ve. m. 91.5-92.5°.[760]
Et ester. Ve. m. 92.5-93.5°.[760]

$(EtO_2C\cdot CH_2)PhPO_2H$.
Et ester. Va. $b_{0.9}$ 142-148°, d_4^{20} 1.1529, n_D^{20} 1.5078.[517]
i-Bu ester. Va.[26,283] Ve.[283] $b_{1.5}$ 160°,[283] b_7 195-198°,[26] d_0^0 1.1223.[26]

$(PhO_2C\cdot CH_2)PhPO_2H$.
Et ester. Va. $b_{1.1}$ 170-173°, d_4^{20} 1.1570, n_D^{20} 1.5255.[517]

$(p\text{-}ClC_6H_4\cdot O\cdot CO\cdot CH_2)PhPO_2H$.
Et ester. Va. $b_{0.8}$ 164-170°, d_4^{20} 1.1877, n_D^{20} 1.5334.[517]

$(EtS\cdot CO\cdot CH_2)PhPO_2H$.
Et ester. Va. $b_{0.4}$ 158-161°, d_4^{20} 1.1651, n_D^{20} 1.5270.[517]

$(p\text{-}ClC_6H_4\cdot S\cdot CO\cdot CH_2)PhPO_2H$.
Et ester. Va. $b_{0.6}$ 198-200°, d_4^{20} 1.2361, n_D^{20} 1.5635.[517]

$(Et_2N\cdot CO\cdot CH_2)PhPO_2H$.
Et ester. Va. $b_{0.55}$ 155-160°, d_4^{20} 1.1791, n_D^{20} 1.5179.[517]

$(H_2N\cdot NH\cdot CO\cdot CH_2)PhPO_2H$. By-product in preparation of Et ester, m. 272-274°.[665]
Et ester. From $(EtO_2C\cdot CH_2)PhP(O)OEt + N_2H_4$. m. 72-75°.[665]
Bu ester. From $(EtO_2C\cdot CH_2)PhP(O)OBu + N_2H_4$. m. 82-84°.[665]

$(Ph\cdot CO\cdot NH\cdot NH\cdot CO\cdot CH_2)PhPO_2H$.
Et ester. From $(EtO_2C\cdot CH_2)PhP(O)OEt + Ph\cdot CO\cdot NH\cdot NH_2$. m. 70-72°.[665]

$[BuO_2C\cdot CH(OH)]EtPO_2H$.
Pr ester. IXc. b_4 146°, d_4^{20} 1.0802, n_D^{20} 1.4490.[611]

$[MeO_2C\cdot CH(OMe)]BuPO_2H$.
Et ester. Va. $b_{0.04}$ 85-88°, n_D^{19} 1.4502.[266]

$[\overline{CH_2\cdot(CH_2)_4\cdot N}\cdot CO\cdot CH(OMe)]BuPO_2H$.
Et ester. Va. $b_{0.02}$ 113-115°.[266]

$[MeO_2C\cdot CH(OMe)]PhPO_2H$.
Et ester. Va. $b_{0.06}$ 140-142°, n_D^{19} 1.5135, ^{31}P -46.0 ppm.[226]

$[\overline{CH_2\cdot(CH_2)_4\cdot N}\cdot CO\cdot CH(OMe)]PhPO_2H$.
Et ester. Va. Two forms, m. 90-91° and m. 116-117°, ^{31}P -45.0 ppm.[266]

$(HO_2C\cdot CH_2\cdot CH_2)MePO_2H$. XXIk. m. 94°.[381]
Et ester. XXIVd. d_4^{20} 1.2095, n_D^{20} 1.4656.[381]
Cyclic anhydride $\overline{O\cdot CO\cdot CH_2\cdot CH_2\cdot P}(O)Me$. XXIk. m. 97-98°, $b_{0.5}$ 149-150°.[381]

$(MeO_2C\cdot CH_2\cdot CH_2)MePO_2H$.
Me ester. XXIk. $b_{0.22}$ 113-114°, d_4^{20} 1.1803, n_D^{20} 1.4520.[381]

$(EtO_2C\cdot CH_2\cdot CH_2)MePO_2H$.
Et ester. XXIk. $b_{0.005}$ 92-93°, d_4^{20} 1.0982, n_D^{20}

1.4470.[381]

$(PrO_2C\cdot CH_2\cdot CH_2)MePO_2H$.
Pr ester. XXIk. $b_{0.004}$ 112-113°, d_4^{20} 1.0468, n_D^{20} 1.4452.[381]

$(CH_2{:}CH\cdot CH_2\cdot O\cdot CO\cdot CH_2\cdot CH_2)MePO_2H$.
$CH_2{:}CH\cdot CH_2$ ester. XXIk. $b_{0.005}$ 121-122°, d_4^{20} 1.1326, n_D^{20} 1.4750.[381]

$(Ph\cdot NH\cdot CO\cdot CH_2\cdot CH_2)MePO_2H$. XXIVe. m. 147-148°.[381]

$(NC\cdot CH_2\cdot CH_2)MePO_2H$. XXIVb. $b_{0.01}$ 235°, d_4^{20} 1.2611, n_D^{20} 1.4850.[372]
Me ester. XXIe. XXIVb. $b_{0.2}$ 107°, d_4^{20} 1.1549, n_D^{20} 1.4582.[372]
Et ester. XXIe. $b_{0.05}$ 105°, d_4^{20} 1.1100, n_D^{20} 1.4554.[372]
Pr ester. XXIe. $b_{0.1}$ 121°, d_4^{20} 1.0744, n_D^{20} 1.4543.[372]
Bu ester. XXIe. $b_{0.06}$ 120°, d_4^{20} 1.0443, n_D^{20} 1.4533.[372]
Et_2 amide. XXIi. $b_{0.01}$ 132°, d_4^{20} 1.0530, n_D^{20} 1.4718.[372]
$(NC\cdot CH_2\cdot CH_2)MeP(O)\cdot O\cdot P(O)Me(CH_2\cdot CH_2\cdot CN)$. XXIf. $b_{0.06}$ 215°, d_4^{20} 1.2693, n_D^{20} 1.4935.[372]

$(HO_2C\cdot CH_2\cdot CH_2)(ClCH_2)PO_2H$. XXIk. m. 92-93°.[378]
Et ester. XXIVd. Undistillable, d_4^{20} 1.2937, n_D^{20} 1.4765.[378]
Cyclic anhydride $\overline{O\cdot CO\cdot CH_2\cdot CH_2\cdot P}(O)CH_2Cl$. Xc (by-product). m. 122-124°, $b_{0.006}$ 151-153°.[378]

$(EtO_2C\cdot CH_2\cdot CH_2)(ClCH_2)PO_2H$.
Et ester. XXIk. $b_{0.0045}$ 122-123°, d_4^{20} 1.2073, n_D^{20} 1.4642.[378]
$(EtO_2C\cdot CH_2\cdot CH_2)(ClCH_2)P(O)\cdot O\cdot P(O)(CH_2Cl)(CH_2\cdot CH_2\cdot CO_2Et)$ By-product in preparation of Et ester, $b_{0.005}$ 183-185°, d_4^{20} 1.3436, n_D^{20} 1.4840.[378]

$(PhNH\cdot CO\cdot CH_2\cdot CH_2)(ClCH_2)PO_2H$. XXIVe. m. 133-134°.[378]

$(HO_2C\cdot CH_2\cdot CH_2)EtPO_2H$. XXIk. m. 99.5°;[379] $H_2N(CH_2)_6NH_2$ salt, m. 58-60°.[379]
Me ester. XXIVd. d_4^{20} 1.2371, n_D^{20} 1.4740.[386]
Pr ester. XXIVd. d_4^{20} 1.1456, n_D^{20} 1.4650.[386]
Cyclic anhydride $\overline{O\cdot CO\cdot CH_2\cdot CH_2\cdot P}(O)Et$. Xc.[383] Pyrolysis of C-ester-P-chloride,[385] pyrolysis of C-chloride-P-ester,[386] $b_{0.003}$ 137-138°,[386] d_4^{20} 1.2912,[386] n_D^{20} 1.4860.[385,386]
Pr cyclic imide $Pr\overline{N\cdot CO\cdot CH_2\cdot CH_2\cdot P}(O)Et$. XXIk. $b_{0.01}$ 165°, d_4^{20} 1.1261, n_D^{20} 1.4915.[375]
Bu cyclic imide. XXIk. $b_{0.009}$ 137-140°, d_4^{20} 1.028, n_D^{20} 1.4920.[375]
Ph cyclic imide. XXIk. $b_{0.006}$ 169-172°, d_4^{20} 1.2310, n_D^{20} 1.5650.[375]
p-$MeO\cdot C_6H_4$ cyclic imide. XXIk. $b_{0.015}$ 218°, n_D^{20} 1.5610.[375]

m-$O_2N\cdot C_6H_4$ cyclic imide. XXIk. m. 102°.[375]
2,4-$(O_2N)_2C_6H_3$ cyclic imide. XXIk. m. 156°.[375]
o-MeC_6H_4 cyclic imide. XXIk. $b_{0.003}$ 170°, n_D^{20} 1.5515.[375]
m-MeC_6H_4 cyclic imide. XXIk. $b_{0.001}$ 165°, n_D^{20} 1.5525.[375]
p-MeC_6H_4 cyclic imide. XXIk. $b_{0.006}$ 182°, n_D^{20} 1.5568.[375]
1-naphthyl cyclic imide. XXIk. $b_{0.001}$ 201°.[375]

$(Cl\cdot CO\cdot CH_2\cdot CH_2)EtPO_2H$.
1-CCl_3-cyclopentyl ester. Xc. $b_{0.002}$ 195°, d_4^{20} 1.3861, n_D^{20} 1.5100.[386]

$(MeO_2C\cdot CH_2\cdot CH_2)EtPO_2H$.
Me ester. XXIk. $b_{0.22}$ 103-104°, d_4^{20} 1.1480, n_D^{20} 1.4510.[379]
1-CCl_3-cyclopentyl ester. XXXh. $b_{0.007}$ 155°, d_4^{20} 1.3304, n_D^{20} 1.5005.[386]

$(EtO_2C\cdot CH_2\cdot CH_2)EtPO_2H$. XXIa. d_4^{20} 1.2062, n_D^{20} 1.4720.[385]
Et ester. XXIa.[385] XXIk.[379,385] XXXh.[383] $b_{0.22}$ 116-118°,[379,385] d_4^{20} 1.0761,[379] n_D^{20} 1.4550.[379,385]
Bu ester. XXIa. $b_{0.001}$ 114°, d_4^{20} 1.0366, n_D^{20} 1.4495.[385]
Ph ester. XXIa. $b_{0.001}$ 148-150°, d_4^{20} 1.1455, n_D^{20} 1.5070.[385]
$(EtO_2C\cdot CH_2\cdot CH_2)EtP(O)\cdot O\cdot P(O)Et(CH_2\cdot CH_2\cdot CO_2Et)$. XXIf. Also as by-product in preparation of Et ester, $b_{0.7}$ 201°, d_4^{20} 1.1835, n_D^{20} 1.4690.[385]
Et_2 amide. XXIi. $b_{0.001}$ 127-128°, d_4^{20} 1.0443, n_D^{20} 1.4650.[385]

$(PrO_2C\cdot CH_2\cdot CH_2)EtPO_2H$.
1-CCl_3-cyclopentyl ester. XXXh. $b_{0.002}$ 156°, d_4^{20} 1.2411, n_D^{20} 1.4910.[386]

$(BuO_2C\cdot CH_2\cdot CH_2)EtPO_2H$.
Bu ester. IXf. b_1 141°, d_4^{20} 1.0146, n_D^{20} 1.4490.[641]

$[(ClCH_2)_2CH\cdot O\cdot CO\cdot CH_2\cdot CH_2]EtPO_2H$.
$(ClCH_2)_2CH$ ester. XXIk (with $\overline{O\cdot CH_2\cdot C}H\cdot CH_2Cl$). $b_{0.002}$ 237°, d_4^{20} 1.3669, n_D^{20} 1.4970.[379]

$(p$-$MeO\cdot C_6H_4\cdot NH\cdot CO\cdot CH_2\cdot CH_2)EtPO_2H$. XXIVe. m. 156-157°.[386]
$(m$-$O_2N\cdot C_6H_4\cdot NH\cdot CO\cdot CH_2\cdot CH_2)EtPO_2H$. XXIVe. m. 172-173°.[386]
$(p$-$MeC_6H_4\cdot NH\cdot CO\cdot CH_2\cdot CH_2)EtPO_2H$. XXIVe. m. 154-155°, 171-172°.[386]
(1-naphthyl$\cdot NH\cdot CO\cdot CH_2\cdot CH_2)EtPO_2H$. XXIVe. m. 160-161°.[386]

$(NC\cdot CH_2\cdot CH_2)EtPO_2H$.
Me ester. XXIe. $b_{0.2}$ 118°, d_4^{20} 1.1262, n_D^{20} 1.4625.[621]
Et ester. IXf.[621,635] XXIe.[621] $b_{0.03}$ 110°,[621] b_{12} 171-171.5°,[635] d_4^{20} 1.0821,[621] n_D^{20} 1.4528.[621]
Pr ester. XXIe. $b_{0.03}$ 124°, d_4^{20} 1.0539, n_D^{20} 1.4558.[621]
Bu ester. IXf.[635] XXIe.[621] $b_{0.03}$ 126°,[621] b_{13} 192°,[635] d_4^{20} 1.0329,[621] n_D^{20} 1.4554.[621]

i-Bu ester. XXIe. $b_{0.025}$ 118°, d_4^{20} 1.0289, n_D^{20} 1.4545.[621]
$CCl_3 \cdot CMe_2$ ester. Xd. m. 49°, $b_{0.05}$ 150-152°.[625]
1-CCl_3-cyclopentyl ester. Xd. $b_{0.05}$ 155-160°, d_4^{20} 1.3437, n_D^{20} 1.5135.[625]
Ph ester. XXIe. $b_{0.025}$ 152°, d_4^{20} 1.1602, n_D^{20} 1.5250.[621]

$(HO_2C \cdot CH_2 \cdot CH_2)_2PO_2H$. IIa.[647] Also by acid hydrolysis of $(NC \cdot CH_2 \cdot CH_2)_2P(O)O(i\text{-}Pr)$,[647] and by alkaline hydrolysis, with involuntary oxidation, of $(NC \cdot CH_2 \cdot CH_2)_2P(O)H$ [450] m. 159-160°.[450,647]

$(H_2N \cdot CS \cdot CH_2 \cdot CH_2)_2PO_2H$. From $(NC \cdot CH_2 \cdot CH_2)_2PH$ + S + aqueous NH_3. Also from $(NC \cdot CH_2 \cdot CH_2)_2PO_2H$ + H_2S + Et_3N in pyridine. m. 179° decomp.[648]

$(NC \cdot CH_2 \cdot CH_2)(H_2N \cdot CS \cdot CH_2 \cdot CH_2)PO_2H$. From $(NC \cdot CH_2 \cdot CH_2)_2P(O)H$ + S in moist i-PrOH. m. 142°.[648]

$(NC \cdot CH_2 \cdot CH_2)_2PO_2H$. Ia. m. 124-125°.[648]
i-Pr ester. IIc. m. 79-80°.[647]

$(HO_2C \cdot CH_2 \cdot CH_2)PhPO_2H$. By hydrolyses of $(MeO_2C \cdot CH_2 \cdot CH_2)PhP$-(O)OBu and of $(NC \cdot CH_2 \cdot CH_2)PhP(O)OBu$.[89,90] Also by (XXIIIa + XXXh).[624] m. 156-157°.[89]
Cyclic anhydride $\overline{O \cdot CO \cdot CH_2 \cdot CH_2 \cdot P}(O)Ph$. Xc. m. 74-75°, $b_{0.6}$ 185°.[624]
$CH_2{:}CH \cdot CH_2$ cyclic imide $CH_2{:}CH \cdot CH_2 \cdot \overline{N \cdot CO \cdot CH_2 \cdot CH_2 \cdot P}(O)Ph$. XXIk. m. 106-108°, $b_{0.05}$ 198°.[626]
Bu cyclic imide. XXIk. $b_{0.05}$ 159°, d_4^{20} 1.1530, n_D^{20} 1.5420.[626]
Ph cyclic imide. XXIk. m. 115-116°, $b_{0.06}$ 234°.[626]
o-MeC_6H_4 cyclic imide. XXIk. m. 140-142°, $b_{0.06}$ 225°.[626]
p-MeC_6H_4 cyclic imide. XXIk. m. 184-185°.[626]

$(Cl \cdot CO \cdot CH_2 \cdot CH_2)PhPO_2H$.
$(ClCH_2)_2CH$ ester. Xc. $b_{0.002}$ 200°, d_4^{20} 1.3615, n_D^{20} 1.5475.[624]
$CCl_3 \cdot CMe_2$ ester. Xc. $b_{0.002}$ 187°, d_4^{20} 1.3850, n_D^{20} 1.5540.[624]
1-CCl_3-cyclopentyl ester. Xc. $b_{0.002}$ 200°, d_4^{20} 1.3885, n_D^{20} 1.5560.[624]

$(MeO_2C \cdot CH_2 \cdot CH_2)PhPO_2H$.
Me ester. Xc. b_5 159-160°, d_4^{20} 1.1591, n_D^{20} 1.5062.[361]
Et ester. IXf. b_2 159-160°, d_4^{20} 1.1619, n_D^{20} 1.5081.[641]
Bu ester. IXf.[89,90] $b_{0.02}$ 108°,[89,90] $b_{0.9}$ 162°,[90] d_4^{20} 1.1156,[90] n_D^{20} 1.5028.[90]

$(EtO_2C \cdot CH_2 \cdot CH_2)PhPO_2H$.
Et ester. Xc. b_3 178-179°, d_4^{20} 1.1532, n_D^{20} 1.5040.[361]
$(ClCH_2)_2CH$ ester. XXXh. $b_{0.001}$ 205°, d_4^{20} 1.3000, n_D^{20} 1.5225.[624]

$CCl_3 \cdot CMe_2$ ester. XXXh. $b_{0.001}$ 240°, d_4^{20} 1.2464, n_D^{20} 1.5170.[624]

$(PrO_2C \cdot CH_2 \cdot CH_2)PhPO_2H$. Xc. b_5 183-184°, d_4^{20} 1.0691, n_D^{20} 1.4995.[361]

$(PhNH \cdot CO \cdot CH_2 \cdot CH_2)PhPO_2H$. XXVe. m. 155-156°.[626]

Et ester. XXVe. m. 164-165°.[626]

$(NC \cdot CH_2 \cdot CH_2)PhPO_2H$. XXIa. $b_{0.0045}$ 205-210°, n_D^{20} 1.5595.[382]

Me ester. XXIe. $b_{0.005}$ 155-156°, d_4^{20} 1.1882, n_D^{20} 1.5340.[382]

Et ester. IXf.[635] XXIe.[382] $b_{0.005}$ 156-157°,[382] d_4^{20} 1.1457,[382] n_D^{20} 1.5211.[382]

Pr ester. XXIe. $b_{0.0045}$ 158-159°, d_4^{20} 1.1204, n_D^{20} 1.5199.[382]

Bu ester. IXf.[89,90] $b_{0.01}$ 111-114°,[90] d_4^{20} 1.0981,[90] n_D^{20} 1.5139.[90]

$ClCH_2 \cdot CH_2$ ester. Xd. $b_{0.001}$ 179-180°, d_4^{20} 1.2745, n_D^{20} 1.5410.[384]

$CCl_3 \cdot CMe_2$ ester. Xd. m. 66-67°, $b_{0.001}$ 198-199°, n_D^{20} 1.5400.[384]

$CCl_3(CH_2)_4$ ester. Xd. m. 64.5°, $b_{0.001}$ 192°, n_D^{20} 1.5400.[384]

Ph ester. XXIe. m. 73-75°, $b_{0.005}$ 186-187°, d_4^{20} 1.2057, n_D^{20} 1.5732.[382]

$(NC \cdot CH_2 \cdot CH_2)PhP(O) \cdot O \cdot P(O)Ph(CH_2 \cdot CH_2 \cdot CN)$. XXIf. $b_{0.0045}$ 255-260°, n_D^{20} 1.5730.[382]

Et_2 amide. XXIi. $b_{0.004}$ 146-147°, d_4^{20} 1.1027, n_D^{20} 1.5362.[382]

$(Cl \cdot CO \cdot CH_2 \cdot CH_2)(p\text{-}MeC_6H_4)PO_2H$.

$CCl_3 \cdot CMe_2$ ester. Xc. $b_{0.002}$ 218°, d_4^{20} 1.3604, n_D^{20} 1.5435.[624]

$(MeO_2C \cdot CH_2 \cdot CH_2)(p\text{-}MeC_6H_4)PO_2H$.

Et ester. IXf. b_1 153-155°, d_4^{20} 1.1441, n_D^{20} 1.5131.[636]

$(NC \cdot CH_2 \cdot CH_2)(p\text{-}MeC_6H_4)PO_2H$. XXIa. $b_{0.004}$ 225-230°, d_4^{20} 1.2236, n_D^{20} 1.5575.[382]

Me ester. XXIe. $b_{0.004}$ 155-156°, d_4^{20} 1.1636, n_D^{20} 1.5320.[382]

Et ester. IXf.[636] XXIe.[382] $b_{0.004}$ 155-156°,[382] b_2 182-183°,[636] d_4^{20} 1.1216,[382] n_D^{20} 1.5220.[382]

Pr ester. XXIe. $b_{0.004}$ 156-157°, d_4^{20} 1.1053, n_D^{20} 1.5200.[382]

Bu ester. XXIe. $b_{0.004}$ 168-169°, d_4^{20} 1.0835, n_D^{20} 1.5158.[382]

$CCl_3 \cdot CMe_2$ ester. Xd. m. 72-74°, $b_{0.001}$ 197-198°.[384]

Ph ester. XXIe. m. 54-56°, $b_{0.004}$ 185-186°, d_4^{20} 1.1807, n_D^{20} 1.5710.[382]

$(NC \cdot CH_2 \cdot CH_2)(p\text{-}MeC_6H_4)P(O) \cdot O \cdot P(O)(C_6H_4\text{-}p\text{-}Me)(CH_2 \cdot CH_2 \cdot CN)$. XXIf. $b_{0.004}$ 270-275°, d_4^{20} 1.2274, n_D^{20} 1.5676.[382]

Et amide. XXIi. $b_{0.005}$ 156-157°, d_4^{20} 1.0851, n_D^{20} 1.5352.[382]

$(HO_2C \cdot CHMe)PhPO_2H$. By hydrolysis of ester of next

acid,[26,281] m. 168-169°.[26]

$(EtO_2C\cdot CHMe)PhPO_2H$.

i-Bu ester. Va.[26,283] Also from Na + $(EtO_2C\cdot CH_2)$-PhP(O)O(i-Bu) + MeI,[283] b_1 165°,[283] b_7 191-193°,[26] d_0^0 1.10535.[26]

$(EtO_2C\cdot CH{:}CH)MePO_2H$.

Et ester. XXIk. $b_{0.06}$ 104°, d_4^{20} 1.1330, n_D^{20} 1.4590.[373]

$(PrO_2C\cdot CH{:}CH)MePO_2H$.

Pr ester. XXIk. $b_{0.04}$ 112°, d_4^{20} 1.0701, n_D^{20} 1.4612.[373]

$(HO_2C\cdot CH{:}CH)EtPO_2H$.

Et ester. XXIVd. d_4^{20} 1.2004, n_D^{20} 1.4832.[608]

Cyclic anhydride O·CO·CH:CH·P(O)Et. XXIk. $b_{0.055}$ 110°, d_4^{20} 1.3391, n_D^{20} 1.4950.[608]

$(EtO_2C\cdot CH{:}CH)EtPO_2H$.

$CCl_3\cdot CMe_2$ ester. [Xc + XXXh]. $b_{0.04}$ 142-143°, d_4^{20} 1.3039, n_D^{20} 1.4920.[773]

1-CCl_3-cyclopentyl ester. [Xc + XXXh]. $b_{0.04}$ 176-178°, d_4^{20} 1.3108, n_D^{20} 1.5050.[773]

$(PhNH\cdot CO\cdot CH{:}CH)EtPO_2H$. XXIVe. m. 142-143°.[608]

$(BuS\cdot CO\cdot CH{:}CH)EtPO_2H$.

$CCl_3\cdot CMe_2$ ester. [Xc + XXXh]. $b_{0.04}$ 177-179°, d_4^{20} 1.2695, n_D^{20} 1.5160.[773]

$(NC\cdot CH{:}CH)EtPO_2H$.

Et ester. Va*. b_1 119-120°, d_4^{20} 1.1040, n_D^{20} 1.4530.[644]

Pr ester. Va*. $b_{0.5}$ 125-126°, d_4^{20} 1.0638, n_D^{20} 1.4530.[644]

$(HO_2C\cdot CH{:}CH)PhPO_2H$. XXIVd. m. 76-78°.[370]

Cyclic anhydride O·CO·CH:CH·P(O)Ph. Xc.[773] XXIk.[370] $b_{0.04}$ 165°,[370] d_4^{20} 1.3464,[370] n_D^{20} 1.5720.[370]

$(MeO_2C\cdot CH{:}CH)PhPO_2H$.

Me ester. [Xc + XXIk]. $b_{0.008}$ 134°, d_4^{20} 1.2209, n_D^{20} 1.5380.[371]

$(EtO_2C\cdot CH{:}CH)PhPO_2H$. XXIa. XXIVd*. m. 138-140°.[370]

Et ester. [Xc + XXIk]. $b_{0.008}$ 143°, d_4^{20} 1.1555, n_D^{20} 1.5240.[371]

$CCl_3\cdot CMe_2$ ester. [Xc + XXXh]. $b_{0.04}$ 190-192°, d_4^{20} 1.3249, n_D^{20} 1.5400.[773]

$(PrO_2C\cdot CH{:}CH)PhPO_2H$.

Pr ester. [Xc + XXIk]. $b_{0.006}$ 151-152°, d_4^{20} 1.1160, n_D^{20} 1.5158.[371]

$(BuO_2C\cdot CH{:}CH)PhPO_2H$.

Bu ester. [Xc + XXIk]. $b_{0.04}$ 166-167°, d_4^{20} 1.1075, n_D^{20} 1.5165.[371]

$(BuS\cdot CO\cdot CH{:}CH)PhPO_2H$.

$CCl_3\cdot CMe_2$ ester. [Xc + XXXh]. $b_{0.04}$ 208-210°, d_4^{20} 1.2931, n_D^{20} 1.5550.[773]

$(EtO_2C\cdot CH{:}CH)(p\text{-}MeC_6H_4)PO_2H$.

$CCl_3 \cdot CMe_2$ ester. [Xc + XXXh]. $b_{0.04}$ 192-194°, d_4^{20} 1.3010, n_D^{20} 1.5360.[773]

$[EtO_2C \cdot CMe(OH)]EtPO_2H$.
Et ester. IXc. d_4^{20} 1.1309, n_D^{20} 1.4478, kinetics of thermal (P-C → P-O) isomerization.[774]

$[NC \cdot CMe(OH)]EtPO_2H$.
Et ester. IXc. d_4^{20} 1.0920, n_D^{20} 1.4420, kinetics of thermal (P-C → P-O) isomerization.[774]

$[MeO_2C \cdot CMe(OH)]PhPO_2H$. IXa. m. 93-94°.[615]

$[EtO_2C \cdot CMe(OH)]PhPO_2H$. IXa. m. 93°.[615]
Et ester. IXc. m. 61-62°.[619]

$[EtO_2C \cdot CMe(OH)](p\text{-}ClC_6H_4)PO_2H$. IXa. m. 123°.[615]

$[EtO_2C \cdot CMe(OH)](p\text{-}BrC_6H_4)PO_2H$. IXa. m. 133°.[615]

$[EtO_2C \cdot CMe(OH)](p\text{-}MeO \cdot C_6H_4)PO_2H$. IXa. m. 118°.[615]

$[(EtO_2C)_2C(OH)]EtPO_2H$.
Et ester. IXc. d_4^{20} 1.1592, n_D^{20} 1.4430, isomerizes (P-C → P-O) on attempted distillation.[629]
i-Pr ester. IXc. d_4^{20} 1.1335, n_D^{20} 1.4400, isomerizes on attempted distillation.[629]

$[i\text{-}Pr_2N \cdot CO \cdot (CH_2)_3]BuPO_2H$.
Et ester. Va. $b_{0.005}$ 149-151°, n_D^{20} 1.4670.[190]

$(HO_2C \cdot CHMe \cdot CH_2)MePO_2H$.
Et ester. XXIVd. m. 99-100°.[373]
Cyclic anhydride $\overline{O \cdot CO \cdot CHMe \cdot CH_2 \cdot P}(O)Me$. XXIk. m. 63°, $b_{0.04}$ 128°.[373]

$(EtO_2C \cdot CHMe \cdot CH_2)MePO_2H$.
Et ester. XXIk. $b_{0.2}$ 108°, d_4^{20} 1.0759, n_D^{20} 1.4430.[373]

$(PhNH \cdot CO \cdot CHMe \cdot CH_2)MePO_2H$. XXIVe. m. 165-166°.[373]

$(NC \cdot CHMe \cdot CH_2)MePO_2H$. XXIVa. $b_{0.01}$ 197°, d_4^{20} 1.2250, n_D^{20} 1.4863.[374]
Me ester. XXIVa. m. 53-54°, $b_{0.65}$ 110°.[374]
Et ester. XXIVa. $b_{0.08}$ 112°, d_4^{20} 1.0720, n_D^{20} 1.4528.[374]
Pr ester. XXIe. $b_{0.07}$ 117°, d_4^{20} 1.0440, n_D^{20} 1.4525.[374]
Bu ester. XXIe. $b_{0.08}$ 120°, d_4^{20} 1.0261, n_D^{20} 1.4532.[374]
Me_2 amide. XXIi. $b_{0.15}$ 128°, d_4^{20} 1.0248, n_D^{20} 1.4780.[374]
$(NC \cdot CHMe \cdot CH_2)MeP(O) \cdot O \cdot P(O)Me(CH_2 \cdot CHMe \cdot CN)$. By-product in preparation of phosphinic chloride by Xc. $b_{0.01}$ 205°, d_4^{20} 1.1948, n_D^{20} 1.4810.[374]

$(HO_2C \cdot CHMe \cdot CH_2)EtPO_2H$. XXIk. m. 40-43°.[379]
Me ester. XXIVd. d_4^{20} 1.1620, n_D^{20} 1.4600.[386]
Et ester. XXIVd. d_4^{20} 1.0903, n_D^{20} 1.4505.[386]
Cyclic anhydride $\overline{O \cdot CO \cdot CHMe \cdot CH_2 \cdot P}(O)Et$. Xc.[380,386] XXIk.[380] m. 61-62°,[380,386] $b_{0.008}$ 133°,[380,386] d_4^{50} 1.1993,[380] n_D^{50} 1.4660.[380]
Bu cyclic imide $Bu\overline{N \cdot CO \cdot CHMe \cdot CH_2 \cdot P}(O)Et$. XXIk. $b_{0.004}$

134°, d_4^{20} 1.0691, n_D^{20} 1.4833.[375]
Ph cyclic imide. XXIk. $b_{0.003}$ 150°, n_D^{20} 1.5435.[375]
p-MeO·C_6H_4 cyclic imide. XXIk. $b_{0.001}$ 154°, n_D^{20} 1.5472.[375]
o-MeC_6H_4 cyclic imide. XXIk. $b_{0.006}$ 169°, n_D^{20} 1.5295.[375]
m-MeC_6H_4 cyclic imide. XXIk. $b_{0.006}$ 178°, n_D^{20} 1.5320.[375]
p-MeC_6H_4 cyclic imide. XXIk. $b_{0.006}$ 171°, n_D^{20} 1.5315.[375]
1-naphthyl cyclic imide. XXIk. $b_{0.006}$ 205°.[375]
(Cl·CO·CHMe·CH_2)$EtPO_2H$.
1-CCl_3-cyclopentyl ester. Xc. $b_{0.002}$ 200°, d_4^{20} 1.3489, n_D^{20} 1.5070.[386]
(EtO_2C·CHMe·CH_2)$EtPO_2H$.
Et ester. XXIk.[379] [XXIVd + continued heating with EtOH].[380] $b_{0.07}$ 86-87°,[379] d_4^{20} 1.0536,[379] n_D^{20} 1.4440.[379]
(BuO_2C·CHMe·CH_2)$EtPO_2H$.
Bu ester. IXf. b_1 137-138°, d_4^{20} 1.0025, n_D^{20} 1.4490.[641]
(PhNH·CO·CHMe·CH_2)$EtPO_2H$. XXIVe. m. 141-142°.[380]
(p-MeO·C_6H_4·NH·CO·CHMe·CH_2)$EtPO_2H$. XXIVe. m. 135-136°.[380]
(m-O_2N·C_6H_4·NH·CO·CHMe·CH_2)$EtPO_2H$. XXIVe. m. 138-139°.[380]
(m-MeC_6H_4·NH·CO·CHMe·CH_2)$EtPO_2H$. XXIVe. m. 136-137°.[380]
(p-MeC_6H_4·NH·CO·CHMe·CH_2)$EtPO_2H$. XXIVe. m. 121-122°.[380]
(4-O_2N-2-MeC_6H_3·NH·CO·CHMe·CH_2)$EtPO_2H$. XXIVe. m. 163-164°.[380]
(1-naphthyl·NH·CO·CHMe·CH_2)$EtPO_2H$. XXIVe. m. 115-117°.[380]
(phenothiazin-10-yl·NH·CO·CHMe·CH_2)$EtPO_2H$. XXIVe. m. 174°.[380]
(NC·CHMe·CH_2)$EtPO_2H$. XXIa. m. 68-70°, $b_{0.006}$ 195-196°, d_4^{20} 1.1666, n_D^{20} 1.4862.[377]
Me ester. XXIVb. $b_{0.009}$ 121°, d_4^{20} 1.0864, n_D^{20} 1.4582.[377]
Et ester. IXf.[635] XXIe.[377] $b_{0.005}$ 107-108°,[377] b_{17} 183-185°,[635] d_4^{20} 1.0048,[635] 1.0482,[377] n_D^{20} 1.4470,[635] 1.4550.[377]
Pr ester. XXIe. $b_{0.007}$ 125-126°, d_4^{20} 1.0289, n_D^{20} 1.4550.[377]
Bu ester. XXIe. $b_{0.05}$ 118°, d_4^{20} 1.0136, n_D^{20} 1.4545.[377]
CCl_3·CMe_2 ester. Xd. m. 67°, $b_{0.05}$ 140-145°.[625]
1-CCl_3·cyclo-C_5H_9 ester. Xd. $b_{0.05}$ 158°, d_4^{20} 1.2981, n_D^{20} 1.5080.[625]
Ph ester. XXIe. $b_{0.006}$ 146-147°, d_4^{20} 1.1254, n_D^{20} 1.5182.[377]
Et_2 amide. XXIi. $b_{0.006}$ 135°, d_4^{20} 1.0080, n_D^{20}

1.4710.[377]
(NC·CHMe·CH$_2$)EtP(O)·O·P(O)Et(CH$_2$·CHMe·CN). XXIf. $b_{0.006}$ 211-212°, d_4^{20} 1.1525, n_D^{20} 1.4840.[377]
(HO$_2$C·CHMe·CH$_2$)PhPO$_2$H. XXIk. m. 149-150.5°.[371]
Cyclic anhydride O·CO·CHMe·CH$_2$·P(O)Ph (ring). XXIk. m. 84-87°, $b_{0.002}$ 156-157°.[371]
(Cl·CO·CHMe·CH$_2$)PhPO$_2$H.
CCl$_3$·CMe$_2$ ester. Xc. $b_{0.002}$ 205°, d_4^{20} 1.3622, n_D^{20} 1.5410.[624]
1-CCl$_3$-cyclopentyl ester. Xc. $b_{0.002}$ 220°, d_4^{20} 1.3717, n_D^{20} 1.5520.[624]
(MeO$_2$C·CHMe·CH$_2$)PhPO$_2$H.
Me ester. Xc.[361] XXIk.[371] $b_{0.001}$ 123-124°,[371] b_5 163-165°,[361] d_0^{20} 1.1297,[361] n_D^{20} 1.5046.[361]
Et ester. IXf. $b_{0.05}$ 141-143°, d_4^{20} 1.1390, n_D^{20} 1.5064.[641]
i-Bu ester. IXf. b_1 156°, d_4^{20} 1.0962, n_D^{20} 1.4965.[641]
C_6H_{13} ester. IXf. b_1 176-177°, d_4^{20} 1.0589, n_D^{20} 1.4908.[641]
(EtO$_2$C·CHMe·CH$_2$)PhPO$_2$H.
Et ester. IXf.[641] Xc.[361] XXIk.[371] $b_{0.003}$ 125-126°,[371] b_1 159-160°,[641] b_8 185-186°,[361] d_4^{20} 1.1192,[371] n_D^{20} 1.4985.[371]
Pr ester. IXf. $b_{0.6}$ 149°, d_4^{20} 1.0988, n_D^{20} 1.4962.[641]
(PrO$_2$C·CHMe·CH$_2$)PhPO$_2$H.
Pr ester. Xc.[361] XXIk.[371] $b_{0.004}$ 139-141°,[371] b_8 188-190°,[361] d_4^{20} 1.1063,[371] n_D^{20} 1.5049.[371]
(PhNH·CO·CHMe·CH$_2$)PhPO$_2$H. XXIVe. m. 148-149°.[371]
(p-MeO·C$_6$H$_4$·NH·CO·CHMe·CH$_2$)PhPO$_2$H. XXIVe. m. 111-112.5°.[371]
(m-O$_2$N·C$_6$H$_4$·NH·CO·CHMe·CH$_2$)PhPO$_2$H. XXIVe. m. 170-171°.[371]
(m-MeC$_6$H$_4$·NH·CO·CHMe·CH$_2$)PhPO$_2$H. XXIVe. m. 140-141°.[371]
(p-MeC$_6$H$_4$·NH·CO·CHMe·CH$_2$)PhPO$_2$H. XXIVe. m. 155-156.5°.[371]
(4-O$_2$N-2-MeC$_6$H$_4$·NH·CO·CHMe·CH$_2$)PhPO$_2$H. XXIVe. m. 200-201°.[371]
(1-naphthyl·NH·CO·CHMe·CH$_2$)PhPO$_2$H. XXIVe. m. 162-163°.[371]
(NC·CHMe·CH$_2$)PhPO$_2$H.
(ClCH$_2$)$_2$CH ester. Xd. $b_{0.05}$ 157-161°, d_4^{20} 1.2952, n_D^{20} 1.5420.[625]
CCl$_3$·CMe$_2$ ester. Xd. m. 88°, $b_{0.05}$ 170-172°.[625]
1-CCl$_3$-cyclopentyl ester. Xd. $b_{0.05}$ 188°, d_4^{20} 1.3166, n_D^{20} 1.5430.[625]
(Cl·CO·CHMe·CH$_2$)(p-MeC$_6$H$_4$)PO$_2$H.
CCl$_3$CMe$_2$ ester. Xc. $b_{0.001}$ 193°, d_4^{20} 1.3416, n_D^{20} 1.5410.[624]
(MeO$_2$C·CHMe·CH$_2$)(p-MeC$_6$H$_4$)PO$_2$H.
Et ester. IXf. $b_{0.08}$ 148-149°, d_4^{20} 1.1226, n_D^{20} 1.5090.[636]

Pr ester. IXf. $b_{0.08}$ 149-150°, d_4^{20} 1.0898, n_D^{20} 1.5002.[636]

$(NC\cdot CHMe\cdot CH_2)(p\text{-}MeC_6H_4)PO_2H$.

Et ester. XXIe. $b_{0.005}$ 141-142°, d_4^{20} 1.1067, n_D^{20} 1.5135.[620]

Bu ester. XXIe. $b_{0.005}$ 153-154°, d_4^{20} 1.0604, n_D^{20} 1.5085.[620]

$CCl_3\cdot CMe_2$ ester. Xd. $b_{0.05}$ 180-183°, d_4^{20} 1.2827, n_D^{20} 1.5340.[625]

Ph ester. XXIe. $b_{0.004}$ 180-181°, d_4^{20} 1.1518, n_D^{20} 1.5612.[620]

$(NC\cdot CHMe\cdot CH_2)(p\text{-}MeC_6H_4)P(O)\cdot O\cdot P(O)(C_6H_4\text{-}p\text{-}Me)\text{-}(CH_2\cdot CHMe\cdot CN)$. XXIf. $b_{0.005}$ 225-227°, n_D^{20} 1.5545.[620]

Et_2 amide. XXIi. $b_{0.004}$ 165-166°, d_4^{20} 1.0725, n_D^{20} 1.5280.[620]

$(HO_2C\cdot CH_2\cdot CHMe)MePO_2H$. XXIk. d_4^{20} 1.2810, n_D^{20} 1.4750.[369]

Et ester. XXIVd. d_4^{20} 1.1594, n_D^{20} 1.4610.[369]

Cyclic anhydride $\overline{O\cdot CO\cdot CH_2\cdot CHMe\cdot P}(O)Me$. XXIk. Also by-product from $(Cl\cdot CO\cdot CH_2\cdot CHMe)EtP(O)Cl$ + EtOH. Crystals, deliquescent in air, $b_{0.06}$ 125°, d_4^{20} 1.2548, n_D^{20} 1.4800.[369]

$(EtO_2C\cdot CH_2\cdot CHMe)MePO_2H$.

Et ester. XXIk. $b_{0.06}$ 110°, d_4^{20} 1.0845, n_D^{20} 1.4490.[369]

$(PhNH\cdot CO\cdot CH_2\cdot CHMe)MePO_2H$. XXIVe. m. 143-144°.[369]

$(HO_2C\cdot CH_2\cdot CHMe)EtPO_2H$. XXIk. d_4^{20} 1.2600, n_D^{20} 1.4830.[376]

$(MeO_2C\cdot CH_2\cdot CHMe)EtPO_2H$.

Me ester. XXIk. $b_{0.15}$ 90-91°, d_4^{20} 1.1058, n_D^{20} 1.4532.[376]

$(EtO_2C\cdot CH_2\cdot CHMe)EtPO_2H$.

Et ester. IXf.[642] XXIk.[376] $b_{0.08}$ 87°,[376] $b_{1.5}$ 118-120°,[642] d_4^{20} 1.0540,[376] n_D^{20} 1.4475.[376]

$(PrO_2C\cdot CH_2\cdot CHMe)EtPO_2H$.

Pr ester. XXIk. $b_{0.004}$ 107-108°, d_4^{20} 1.0351, n_D^{20} 1.4486.[376]

$(CH_2{:}CH\cdot CH_2\cdot O\cdot CO\cdot CH_2\cdot CHMe)EtPO_2H$.

$CH_2{:}CH\cdot CH_2$ ester. XXIk. $b_{0.004}$ 112°, d_4^{20} 1.0675, n_D^{20} 1.4697.[376]

$(Et_2N\cdot CO\cdot CH_2\cdot CHMe)EtPO_2H$.

Et_2 amide. XXIk. $b_{0.0004}$ 152-153°, d_4^{20} 1.0085, n_D^{20} 1.4818.[376]

$(NC\cdot CH_2\cdot CHMe)EtPO_2H$.

Et ester. IXf. b_7 158-159.5°, d_4^{20} 1.0597, n_D^{20} 1.4550.[635]

$(HO_2C\cdot CH_2\cdot CHMe)PhPO_2H$. By hydrolyses of esters of next acid.[90,642] m. 82-82.5°.[642]

$(EtO_2C\cdot CH_2\cdot CHMe)PhPO_2H$.

Et ester. IXf. $b_{1.5}$ 154-155°, d_4^{20} 1.1128, n_D^{20} 1.5025.[642]

Bu ester. IXf. $b_{0.005}$ 106-108°, d_4^{20} 1.0781, n_D^{20}

1.4971.[90]

$(HO_2C\cdot CHEt)PhPO_2H$. By hydrolysis of ester of next acid. m. 124-125°.[90]

$(EtO_2C\cdot CHEt)PhPO_2H$.
Et ester. Va. $b_{0.1}$ 130-131°, d_4^{20} 1.1481, n_D^{22} 1.5010.[90]

$(HO_2C\cdot CMe_2)EtPO_2H$. By hydrolysis of ester of next acid. m. 171°.[281]
cyclic diphenylguanide (no structural formula given). XXIk (with 1,3-diphenylguanidine). m. 199-200°.[281]

$(EtO_2C\cdot CMe_2)EtPO_2H$.
i-Pr ester. From Na + $(EtO\cdot CO\cdot CHMe)EtP(O)O(i\text{-}Pr)$ + MeI. Also from 2Na + $(EtO_2C\cdot CH_2)EtP(O)O(i\text{-}Pr)$ + 2MeI. $b_{1.7}$ 113°.[281]

$(p\text{-}MeC_6H_4\cdot NH\cdot CO\cdot CMe_2)EtPO_2H$.
$p\text{-}MeC_6H_4$ amide. XXIk. m. 120-122°.[281]

$(HO_2C\cdot CMe_2)PhPO_2H$. By hydrolysis of ester of next acid. m. 123-125°;[281] NH_4^+ salt, m. 242-245° decomp.;[281] Na and Ba salts.[281]

$(EtO_2C\cdot CMe_2)PhPO_2H$.
i-Bu ester. Ve. Also from Na + $(EtO_2C\cdot CHMe)PhP(O)O\text{-}(i\text{-}Bu)$ + MeI, b_{10} 194-198°.[283]

$[HO_2C\cdot CH_2\cdot CH(CO_2H)]EtPO_2H$. By hydrolysis of Pr ester of next acid. m. 131-131.5°.[642]

$[EtO_2C\cdot CH_2\cdot CH(CO_2Et)]EtPO_2H$.
Et ester. IXf. b_1 142°, d_4^{20} 1.1142, n_D^{20} 1.4481.[642]
Pr ester. IXf. $b_{0.5}$ 151-152.5°, d_4^{20} 1.1015, n_D^{20} 1.4510.[642]

$[\overline{CO\cdot O\cdot CH_2\cdot CH_2\cdot C}(CO\cdot Me)]EtPO_2H$.
Et ester. Ve (forming bond between large group and P). $b_{0.05}$ 147°.[108]

$[\overline{CO\cdot O\cdot CH_2\cdot CH_2\cdot C}(CO\cdot Me)]BuPO_2H$.
Bu ester. Ve (forming bond between large group and P). $b_{0.05}$ 146°.[108]

$[MeO_2C\cdot CH_2\cdot CH_2\cdot CMe(CO_2Et)]PhPO_2H$.
Et ester. By base-catalyzed addition of $(EtO_2C\cdot CHMe)\text{-}PhP(O)OEt$ to $CH_2{:}CH\cdot CO_2Me$. $b_{0.005}$ 134°, n_D^{20} 1.5048.[203]

$(HO_2C\cdot CMe_2\cdot CO\cdot CMe_2)MePO_2H$. By alkaline hydrolysis under reflux of ester of next acid. m. 71.4-72.8°.[71]
Me ester. By alkaline hydrolysis at room temperature of ester of next acid. m. 119.4-120°.[71]

$(MeO_2C\cdot CMe_2\cdot CO\cdot CMe_2)MePO_2H$.
Me ester. Va. $b_{0.03}$ 96-97°.[71]

$(p\text{-}MeO_2C\cdot C_6H_4\cdot CH_2)PhPO_2H$.
Et ester. Va. m. 77-78°.[516]

$(HO_2C\cdot CHPh)PhPO_2H$. Vb.[283] VIa.[86] m. 79-81°;[86] $(1H_2O)$, m. 120°.[283]
Piperidide. XXIi. m. 149-151°.[86]

$(MeO_2C{\cdot}CHPh)PhPO_2H$.
Me ester. XXIIc (from previous acid). m. 86-88°.[86]
$(Me_2N{\cdot}CO{\cdot}CHPh)PhPO_2H$. VIa. m. 202-204°.[86]
Me ester. XXIIc. m. 180-182°.[86]
$[EtO_2C{\cdot}CPh(OH)]EtPO_2H$.
Et ester. IXc. m. 87-88°.[618]
Pr ester. IXc. d_4^{20} 1.1419, n_D^{20} 1.4965.[618]
Bu ester. IXc. d_4^{20} 1.1187, n_D^{20} 1.4968.[618]
$[EtO_2C{\cdot}CPh(OH)]PhPO_2H$.
Me ester. IXc. m. 118°.[618]
Et ester. IXc. m. 107°.[618]
$(EtO_2C{\cdot}\overline{CH{\cdot}CH_2{\cdot}CHMe{:}CHMe{\cdot}CH_2{\cdot}CH})MePO_2H$.
Et ester. By Diels-Alder reaction. $b_{0.15}$ 130°, d_4^{20} 1.0956, n_D^{20} 1.4858.[373]
$(HO_2C{\cdot}CH{\cdot}CHPh)EtPO_2H$. By hydrolysis of Et ester of next acid. m. 138-138.5°.[642]
$(EtO_2C{\cdot}CH{\cdot}CHPh)EtPO_2H$.
Et ester. IXf. $b_{0.5}$ 157-158°, d_4^{20} 1.1048, n_D^{20} 1.5040.[642]
Pr ester. IXf. b_1 164-166°, d_4^{20} 1.0844, n_D^{20} 1.5012.[642]
$(HO_2C{\cdot}CH_2{\cdot}CHPh)PhPO_2H$. By ozonolysis of $(PhCH{:}CX{\cdot}CO{\cdot}CH_2{\cdot}$-$CHPh)PhPO_2H$ [X = H, Br]. m. 212°.[136]
$(EtO_2C{\cdot}CH{:}CPh)EtPO_2H$.
Et ester. IXf. $b_{0.5}$ 147-148°, d_4^{20} 1.1054, n_D^{20} 1.5082.[642]
$[PhCH{:}C(CN)]EtPO_2H$.
Et ester. From $(NC{\cdot}CH_2)EtP(O)OEt$ + PhCHO. b_1 151-152°, d_4^{20} 1.1179, n_D^{20} 1.5580.[643]

E.1.7. Aliphatic and Aliphatic/Aromatic Phosphinic Acids with Amino, or Functionally Modified Amino, Substituents in Aliphatic Groups

$(H_2N{\cdot}CH_2)MePO_2H$. By prolonged acid hydrolysis of Et ester of N-phthaloyl acid (two below). m. 296-298°.[603]
$(o\text{-}HO{\cdot}C_6H_4{\cdot}NH{\cdot}CH_2)MePO_2H$.
Cyclic ester. From $(ClCH_2)MeP(O)Cl$ + $o\text{-}H_2N{\cdot}C_6H_4{\cdot}OH$. m. 68-74°.[786]
$(phthaloyl{:}N{\cdot}CH_2)MePO_2H$. XXIIIa. m. 231-233°.[603]
Et ester. Va. m. 109-111°.[603]
$(Me_2N{\cdot}CH_2)_2PO_2H$.
Pr ester. XXXd. $b_{0.03}$ 83-84°, d_4^{20} 0.9776, n_D^{20} 1.4510.[306]
Bu ester. XXXd. $b_{0.03}$ 90-91°, d_4^{20} 0.9668, n_D^{20} 1.4522.[306]
Ph ester. XXXd. $b_{0.06}$ 142-143°, d_4^{20} 1.072, n_D^{20} 1.5122.[306]
Me_2 amide. XXXd. m. 65-66°.[306]
$(Et_2N{\cdot}CH_2)_2PO_2H$.
Pr ester. IXd*. $b_{0.002}$ 67-69°, n_D^{20} 1.4510.[324]

C_7H_{15} ester. IXd*. $b_{0.007}$ 120-125°, d_4^{20} 0.9328, n_D^{20} 1.4525.[324]
C_8H_{17} ester. IXd*. $b_{0.04}$ 155-158°, d_4^{20} 0.9243, n_D^{20} 1.4530.[324]
C_9H_{19} ester. IXd*. $b_{0.004}$ 134-138°, d_4^{20} 0.9231, n_D^{20} 1.4555.[324]
$C_{10}H_{21}$ ester. IXd*. $b_{0.007}$ 162-165°, d_4^{20} 0.9172, n_D^{20} 1.4570.[324]
Et_2 amide. XXXd. m. 41-42°, $b_{0.0001}$ 101-104°.[306]
$(PrNH{\cdot}CH_2)_2PO_2H$. XXXd. Hydrochloride, m. 238-239°.[307]
$(CH_2{:}CH{\cdot}CH_2{\cdot}NH{\cdot}CH_2)_2PO_2H$. XXXd. Hydrochloride, m. 200-202°.[307]
$[(CH_2{:}CH{\cdot}CH_2)_2N{\cdot}CH_2]_2PO_2H$.
(allyl)$_2$ amide. XXXd. $b_{0.0001}$ 116-120°, d_4^{20} 0.9679, n_D^{20} 1.5008.[306]
$(BuNH{\cdot}CH_2)_2PO_2H$. XXXd. Hydrochloride, m. 240-242°.[307]
(piperidino$\cdot CH_2)_2PO_2H$. IXb. m. 170-180°, ^{31}P -16.0 ppm.[477]
$(PhNH{\cdot}CH_2)_2PO_2H$. XXXd. m. 180-181°.[307]
$(CH_2{:}CMe{\cdot}CO{\cdot}NPh{\cdot}CH_2)_2PO_2H$. XXXf. m. 70-71°.[307]
$(PhCH_2{\cdot}NH{\cdot}CH_2)_2PO_2H$. XXXd. Hydrochloride, m. 220-222°.[307]
$[CH_2{:}CMe{\cdot}CO{\cdot}N(CH_2Ph)CH_2]_2PO_2H$. XXXf. m. 146-147°; Na salt, m. 272.5-273.5°.[307]
$(H_2N{\cdot}CH_2)PhPO_2H$. By prolonged acid hydrolysis of N-phthaloyl acid (three below) or any of its esters. m. 286-287°.[603]
$(Me_2N{\cdot}CH_2)PhPO_2H$. [XXXd + XXIIIa]. m. 206-207°.[394]
$(Et_2N{\cdot}CH_2)(C_7H_{15})PO_2H$. IXb. m. 127-130°.[594]
(phthaloyl:N$\cdot CH_2)PhPO_2H$. XXIIIa. m. 270-273°.[603]
Me ester. Va. m. 94-96°.[603]
Et ester. Va. m. 111-113°.[603]
phthaloyl:N$\cdot CH_2$ ester. From PhP$(OMe)_2$ + phthaloyl:-N$\cdot CH_2Br$ under more vigorous conditions than for preparation of Me ester. m. 168-169°.[603]
$(H_2N{\cdot}CH_2{\cdot}CH_2)PhPO_2H$. Ia. m. 265°.[317]
$(H_2N{\cdot}CHMe)PhPO_2H$. From PhP$(OMe)_2$ + phthaloyl:N CHMeCl and hydrolysis of phthaloyl and ester groups.[603] Also by hydrogenolysis of next acid.[769] m. 268-269°.[769]
$(PhCH_2{\cdot}NH{\cdot}CHMe)PhPO_2H$. [IXd + XXIIIa]. m. 232-233°.[769]
[piperidino$\cdot CH(CO_2Me)]BuPO_2H$.
Et ester. Va. $b_{0.04}$ 115-120°, n_D^{19} 1.4775.[266]
[piperidino$\cdot$CH(CO$\cdot$piperidino)]$BuPO_2H$.
Et ester. Va. $b_{0.04}$ 160-165°, n_D^{20} 1.4863.[266]
$[Et_2N{\cdot}(CH_2)_3]_2PO_2H$.
Et ester. XXXd. $b_{0.006}$ 162-163°, d_4^{20} 0.9628, n_D^{20} 1.4709.[16]
Et_2 amide. XXXd. $b_{0.002}$ 137-138°, d_4^{20} 0.9422, n_D^{20} 1.4778.[16]
$(H_2N{\cdot}CHEt)PhPO_2H$. By hydrogenolysis of next acid. m. 265-266°.[769]

$(PhCH_2 \cdot NH \cdot CHEt)PhPO_2H$. [IXd + XXIIIa]. m. 252-253°.[769]
$(H_2N \cdot CMe_2)PhPO_2H$. [IXd + XXIIIa]. m. 224° decomp.[346]
Et ester. By-product in preparation of acid; hydrochloride, m. 166° decomp.[346]
$(AcNH \cdot CMe_2)PhPO_2H$. XXXf. m. 197-198°.[346]
$(PhNH \cdot CMe_2)[PhCH(OH)]PO_2H$. IXa. m. 189-190°.[433]
$[Et_2N \cdot C(CF_3):CF]MePO_2H$.
Et ester. XXXd. b_1 72-73°, d_4^{20} 1.1776, n_D^{20} 1.4180.[330]
$[morpholino \cdot C(CF_3):CF]MePO_2H$.
Et ester. XXXd. b_1 85-87°, d_4^{20} 1.3092, n_D^{20} 1.4330.[330]
$[piperidino \cdot C(CF_3):CF]MePO_2H$.
Et ester. XXXd. b_8 100-112°, d_4^{20} 1.3010, n_D^{20} 1.4300.[330]
i-Bu ester. XXXd. b_6 100-103°, d_4^{20} 1.0990, n_D^{20} 1.4280.[330]
$(H_2N \cdot CHPr)EtPO_2H$. By hydrogenolysis of next acid. m. 232-233°.[769]
$(PhCH_2 \cdot NH \cdot CHPr)EtPO_2H$. [IXd + XXIIIa]. m. 164-165°.[769]
$(H_2N \cdot CHPr)PhPO_2H$. By hydrogenolysis of next acid. m. 245-247°.[769]
$(PhCH_2 \cdot NH \cdot CHPr)PhPO_2H$. [IXd + XXIIIa]. m. 204-205°.[769]
$(H_2N \cdot CMeEt)PhPO_2H$. XXIIIa. m. 197-198° decomp., very hygroscopic.[346]
Et ester. IXd. Hydrochloride, m. 156-157°.[346]
$[HO_2C \cdot CH(NH_2) \cdot CH_2 \cdot CH_2]EtPO_2H$. From $(BrCH_2 \cdot CH_2)EtP(O)OEt$ + $AcNH \cdot CH(CO_2Et)_2$ with subsequent hydrolysis and decarboxylation. m. 185-187° decomp.[493]
$[HO_2C \cdot CH(NH_2) \cdot CH_2 \cdot CH_2]PhPO_2H$. From $(BrCH_2 \cdot CH_2)PhP(O)OEt$ + $AcNH \cdot CH(CO_2Et)_2$ with subsequent hydrolysis and decarboxylation. m. 235-237°.[493]
$(H_2N \cdot CHBu)PhPO_2H$. By hydrogenolysis of next acid, m. 244-245°.[769]
$(PhCH_2 \cdot NH \cdot CHBu)PhPO_2H$. [IXd + XXIIIa]. m. 175-176°.[769]
$(1\text{-}H_2N\text{-}cyclohexyl)PhPO_2H$. XXIIIa. $(1H_2O)$, m. 212°.[346]
Et ester. IXd. Oil. Hydrochloride, m. 149° and 193°; picrate, m. 168°.[346]
$[Et_2N \cdot CH(C_6H_{13})]PhPO_2H$.
Pr ester. From $PhP(OPr)NEt_2$ + $C_6H_{13} \cdot CHO$. $b_{0.01}$ 98-99°, d_4^{20} 0.9859, n_D^{20} 1.4950.[537]
$(p\text{-}H_2N \cdot C_6H_4 \cdot CH_2)_2PO_2H$. XXXi. m. 303°.[486]
$(PhNH \cdot CHPh)EtPO_2H$.
Et ester. IXd. m. 85°.[604]
Bu ester. IXd. m. 100°.[604]
$(p\text{-}MeO \cdot C_6H_4 \cdot NH \cdot CHPh)EtPO_2H$.
Bu ester. IXd. m. 107-108°.[604]
$(PhCH_2 \cdot NH \cdot CHPh)EtPO_2H$. [IXd + XXIIIa]. m. 212-213°.[769]
$(m\text{-}MeC_6H_4 \cdot NH \cdot CHPh)EtPO_2H$.
Et ester. IXd. m. 111.5-112°.[604]
$(H_2N \cdot CHPh)PhPO_2H$. XXIIIa.[346] Also by hydrogenolysis of

N-benzyl acid (five below),[769] m. 246-247°;[769] ($1H_2O$), m. 218°.[346]
Et ester. IXd. Hydrochloride, m. 164-165°.[346]
($Et_2N{\cdot}CHPh$)$PhPO_2H$.
Pr ester. From PhP(OPr)NEt_2 + PhCHO, $b_{0.005}$ 95-96°, d_4^{20} 1.0760, n_D^{20} 1.5395.[537]
(PhNH·CHPh)$PhPO_2H$.
Et ester. IXd. m. 176°.[604]
(p-MeO·C_6H_4·NH·CHPh)$PhPO_2H$.
Et ester. IXd. m. 157-158°.[604]
(p-O_2N·C_6H_4·NH·CHPh)$PhPO_2H$. IXb. m. 178-180°.[640]
($PhCH_2$·NH·CHPh)$PhPO_2H$. [IXd + XXIIIa]. m. 252-253°.[769]
(p-MeC_6H_4·NH·CHPh)$PhPO_2H$.
Et ester. IXd. m. 174-175°.[604]
(PhNH·CHPh)(p-MeC_6H_4)PO_2H.
i-Pr ester. IXd. m. 158-159°.[636]
[p-Me_2N·C_6H_4·CH(OH)]EtPO_2H.
Bu ester. IXc. m. 130-131°.[2]
[PhNH·CH(o-HO·C_6H_4)]$PhPO_2H$. IXb. m. 177°.[640]
[PhNH·CH(p-Me_2N·C_6H_4)]EtPO_2H.
Et ester. IXd. m. 123°.[604]

E.1.8. Aliphatic and Aliphatic/Aromatic Phosphinic Acids with Sulfur-Containing Substituents in Aliphatic Groups

(o-H_2N·C_6H_4·S·CH_2)MePO_2H.
Cyclic amide. From ($ClCH_2$)MeP(O)Cl + o-H_2N·C_6H_4·SH. m. 192-196° decomp.[786]
(HS·CH_2)$_2PO_2H$. By hydrolysis of di-(S-thiouronium-methyl) phosphinic acid (seven below). Oil, characterized by analysis and by formation of S-alkylated products.[327]
(HO_2C·CH_2·S·CH_2)$_2PO_2H$. XXXf. Yellow sirup.[328]
(EtO_2C·CH_2·S·CH_2)$_2PO_2H$. By azeotropic esterification of previous acid. Yellow sirup.[328]
Et ester. XXIIb. $b_{0.001}$ 134-138°, d_4^{20} 1.2306, n_D^{20} 1.5030.[328]
(CH_2:CH·CH_2·S·CH_2)$_2PO_2H$. XXXf. Yellow sirup.[328]
Et ester. XXIIb. $b_{0.001}$ 114-116°, d_4^{20} 1.1249, n_D^{20} 1.5280.[328]
Pr ester. XXIIb. $b_{0.001}$ 119-120°, d_4^{20} 1.1082, n_D^{20} 1.5187.[328]
Bu ester. XXIIb. $b_{0.001}$ 129-133°, d_4^{20} 1.0933, n_D^{20} 1.5187.[328]
($PhCH_2$.S·CH_2)$_2PO_2H$. XXXf. m. 87-89°.[328]
($PhCH_2$·SO_2·CH_2)$_2PO_2H$. From previous acid + peracetic acid. m. 256-257°.[328]
(Ph·CO·S·CH_2)PO_2H. XXXf. m. 140-141°.[328]
[H_2N·C(:NH)·S·CH_2]$_2PO_2H$. XXXd. Hydrochloride m. 225°.[327]
(EtS·CH_2)$PhPO_2H$. XXIIIa. m. 99°.[431]

Et ester. Va. $b_{0.5}$ 144-145°, n_D^{16} 1.5452.[431]
i-Bu ester. Va. n_D^{20} 1.5258.[431]
$(EtSO_2 \cdot CH_2)PhPO_2H$. From previous acid + peracetic acid. m. 165°.[431]
i-Bu ester. From i-Bu ester of previous acid + peracetic acid. m. 109-110°.[431]
$(F \cdot SO_2 \cdot CH_2 \cdot CH_2)MePO_2H$.
Et ester. Va. $b_{0.001}$ 117-119°, d_4^{20} 1.2895, n_D^{20} 1.4455.[253]
$(F \cdot SO_2 \cdot CH:CH)MePO_2H$.
Et ester. Va. $b_{0.004}$ 123-130°, d_4^{20} 1.3150, n_D^{20} 1.4392.[253]
$(Et \cdot SO_2 \cdot CH_2 \cdot CH_2)EtPO_2H$.
Et ester. IXf*. Solid, $b_{2.5}$ 204°.[639]
(trans-$PhSO_2 \cdot CH:CH)EtPO_2H$.
Et ester. Va. m. 86°.[365]
[trans-$(PhCH_2 \cdot SO_2) \cdot CH:CH]EtPO_2H$.
Et ester. Va. m. 92-93°.[365]
(trans-$PhSO_2 \cdot CH:CH)PhPO_2H$.
Me ester. Va. m. 164-165°.[365]
Et ester. Va. m. 101-102°.[365]
$[C_6Cl_5 \cdot S \cdot CH_2 \cdot CMe(OH)]EtPO_2H$.
Me ester. IXc. m. 155-158°.[793]
Et ester. IXc. m. 166-167°.[793]
Pr ester. IXc. m. 149-151°.[793]
$[EtSO_2 \cdot CH(CH_2 \cdot CH_2 \cdot CN)]PhPO_2H$.
i-Bu ester. From $(EtSO_2 \cdot CH_2)PhP(O)O$(i-Bu) + $CH_2:CH \cdot CN$. m. 125°.[431]
[p-$(EtO \cdot SO_2) \cdot C_6H_4 \cdot CH_2]PhPO_2H$.
Et ester. Va. m. 110-110.5°.[288]

E.1.9. Aliphatic and Aliphatic/Aromatic Phosphinic Acids with Silicon-Containing Substituents in Aliphatic Groups

$(Me_3SiCH_2)PhPO_2H$.
Et ester. Va. b_1 93-94°, d_4^{20} 1.0288, n_D^{20} 1.5039.[126]
(i-$PrMe_2Si \cdot CH_2)PhPO_2H$.
Et ester. Va. $b_{1.5}$ 114-117°, d_4^{20} 1.0227, n_D^{20} 1.5053.[126]
$(PhMe_2Si \cdot CH_2)PhPO_2H$.
Et ester. Va. b_3 152-154°, d_4^{20} 1.0888, n_D^{20} 1.5508.[126]
$(EtO \cdot SiMe_2CH_2)PhPO_2H$.
Et ester. Va. b_2 121-123°, d_4^{20} 1.0553, n_D^{20} 1.4962.[126]
$[(EtO)_2SiMe \cdot CH_2]PhPO_2H$.
Et ester. Va. b_2 126-129°, d_4^{20} 1.0743, n_D^{20} 1.4868.[126]
$[(EtO)_3SiCH_2]PhPO_2H$.
Et ester. Va. b_3 151-153°, d_4^{20} 1.0875, n_D^{20}

1.4796.[126]

$[(EtO)_3Si \cdot CH_2 \cdot CH_2]PhPO_2H$.
Et ester. Va. b_3 163-167°, d_4^{20} 1.0783, n_D^{20} 1.4740.[126]

$[(EtO)_2SiMe \cdot (CH_2)_3]PhPO_2H$.
Et ester. Va. $b_{2.5}$ 167-171°, d_4^{20} 1.0534, n_D^{20} 1.4863.[126]

$[(EtO)_2SiEt \cdot (CH_2)_3]PhPO_2H$.
Et ester. Va. b_3 184-188°, d_4^{20} 1.0775, n_D^{20} 1.4890.[126]

$[(EtO)_3Si \cdot (CH_2)_3]PhPO_2H$.
Et ester. Va. $b_{2.5}$ 184-186°, d_4^{20} 1.0730, n_D^{20} 1.4792.[126]

E.1.10. Diarylphosphinic Acids

Ph_2PO_2H. IVe.[298,526,579] VIa.[254,406] VIc.[528,738] VIg.[142,303,755] VII.[209,210] XIa.[300] XIb.[297] XXa.[519] XXIIf.[289] XXVd.[525] m. 190°,[519] 195-196°,[579] UV,[211,782] IR,[161] ^{31}P -25.5 ppm,[539,700] ^{31}P of anion -19.5 ppm,[700] -23.6 ppm,[565] x-ray powder data,[204] x-ray structure determination,[768] ΔH (fusion) 4.5 kcal/mole,[415] pK_a 2.32 (in 7% EtOH),[503] 3.43 (in 50% EtOH),[503] 4.10 (in 75% EtOH),[584] 4.24 (in 80% EtOH),[503] 4.70 (in 95% EtOH),[584] monomeric in AcOH,[212] dimeric in napthalene;[420] NH_4^+ salt, x-ray powder data;[204] Be salt,[88,238,737] Mg salt, needles, solubilities in H_2O, EtOH, C_6H_6;[174] Ca salt, needles, solubilities in H_2O, EtOH, C_6H_6;[174] Ba salt (2 1/2 H_2O), needles, solubilities in H_2O, EtOH, C_6H_6;[174] Ti salt, softens 135°, m. 180°;[149,151] $Ti(OAlk)_2$ derivs., etc.;[151] Zr salt, m. 472-474°;[149] Cr^{3+} salt;[690] Cr^{3+} salt (acetylacetone complex), green plates, x-ray structure determination;[788] Mo salt, salmon-pink powder;[748] Co salt, blue needles,[129] infusible at 450°;[686] Cu salt, blue needles, solubilities in H_2O, EtOH, C_6H_6;[174] Zn salt, decomp. 490°;[88,685] Me_2Al deriv., m. 153-156°;[130] Me_2Ga deriv. m. 164°;[130] Me_3Si deriv., m. 77-78°, $b_{0.03}$ 132°;[1] Et_3Ge deriv., $b_{0.001}$ 160°;[383] Sn^{4+} salt, m. 168-174°;[722] PhSn deriv., m. 210-216°;[722] Me_2Sn deriv., m. > 250°;[722] Et_2Sn deriv., m. 370-372°;[117] Bu_2Sn deriv., decomp. 230°;[722] Ph_2Sn deriv., m. > 250°;[722] Me_3Sn deriv., softens 214-216° and 255-258°, m. > 360°;[117] Et_3Sn deriv., m. 248°;[117,320] Pr_3Sn deriv., m. 226°;[117] Bu_3Sn deriv., m. 218°;[117,320] Ph_3Sn deriv., m. 250°,[320] 360°,[721] > 360°;[117] Pb^{2+} salt, powder;[174] Pb^{2+} basic salt, needles.[174]

Me ester. IVa.[76] XXIe.[75] $b_{0.34}$ 139-140°,[76] $b_{2.4}$ 178°,[75] ^{31}P -32.2 ppm.[700]

Et ester. IVa.[208] XXIe.[417,482] m. 39-41°,[208] $b_{0.35}$ 148-148.5°,[482] $b_{0.4}$ 160°,[208] $b_{1.5}$ 173-175°,[411,417]

n_D^{20} 1.5720,[482] n_D^{25} 1.5632.[411]
Pr ester. XXIe. m. 89.5-91°;[81] complex with 1/3 MgI_2, m. 110-112°.[80]
i-Pr ester. XXIe. m. 97-99°;[75] complex with 1/3 MgI_2, m. 115-117°.[80]
CH_2:CH·CH_2 ester. XXIe. b_5 162-165°.[75]
CH_2:CHMe ester. IVh.[35,515] m. 87-88°.[35]
i-Bu ester. XXIe. m. 79.3-80.5°;[81] complex with 1/3 MgI_2, m. 125-128°.[80]
t-Bu ester. XXIe.[75] IVe*.[740] m. 97-99°,[75] 111-112°.[740]
Me_3C·CH_2 ester. XXIe. m. 86.5-87.5°;[81] complex with 1/3 MgI_2, m. 137-141°.[80]
C_6H_{13} ester. IVa. $b_{0.1}$ 161-165°, d_4^{20} 1.0761, n^{20} 1.5449.[755b]
cyclo-C_6H_{11} ester. IVa.[598] XXIe.[75] m. 120-121°.[75]
cis-2-Me-cyclohexyl ester. XXIe. m. 84-85°, pyrolysis.[79]
trans-2-Me-cyclohexyl ester. XXIe. m. 73-74°, pyrolysis.[79]
trans-4-Me-cyclohexyl ester. XXIe. m. 132-133°.[75]
C_8H_{17} ester. IVa. $b_{0.2}$ 177-182°, d_4^{20} 1.0384, n_D^{20} 1.5330.[755b]
$PhCH_2$·CH_2 ester. XXIe. m. 65-66°.[74]
$PhCH_2$·CH_2·CH_2 ester. XXIe. m. 69.5-70°, pyrolysis.[79]
(-)-menthyl ester. XXIe.[79,460,500] m. 72-73°,[79] $[\alpha]_D^{20}$ -72° (c 5, EtOH),[500] $[\alpha]_D^{25}$ -73.0° (c 8.2, $CHCl_3$),[79] pyrolysis.[79,500]
(+)-neomenthyl ester. XXIe. m. 74-75°, $[\alpha]_D^{20}$ +10° (c 5, EtOH), pyrolysis.[500]
Bornyl ester. XXIe. m. 74-75°, $[\alpha]_D^{25.5}$ -14.3 (c 5.9, $CHCl_3$), pyrolysis.[79]
$PhCH_2$·CHPh ester. XXIe. m. 142-143°, pyrolysis.[79]
Cholesteryl ester. XXIe. m. 153-154°, $[\alpha]_D^{24}$ -20.1° (in $CHCl_3$).[74]
$(CH_2)_4$ di-ester. XXIe. m. 117-118°.[754]
CCl_2:CH ester. IVh. m. 44-45°, $b_{6.5}$ 205-206.5°.[387]
HO·CH_2·CH_2 ester. VIb*. XXIe. m. 103.5-104.5°.[177]
Me_2N·CH_2·CH_2 ester. XXIIa*. Methiodide, m. 191-192° decomp.[738]
Et_2N·CH_2·CH_2 ester. XXIIa*. Methiodide, m. 193-194° decomp.[738]
HO$(CH_2)_3$ ester. XXIe. m. 82-84°.[177]
$Et_2N(CH_2)_3$ ester. XXIIa*. Hydrochloride, m. 133-134°.[738]
Ph ester. XXIe.[75,527] XXIe*.[758] m. 135-136°.[75,527]
p - ClC_6H_4 ester. XXIe*. m. 113-116°.[758]
p-O_2N·C_6H_4 ester. XXIe.[227,270] XXIe*.[758] m. 149-150°,[270] kinetics of hydrolysis,[270] rates of hydrolysis and of reaction with fly-brain cholinesterase.[227]

p-Me•C_6H_4 ester. XXIe*. m. 121°.[758]
4-HO-3,5(t-Bu)$_2C_6H_2$ ester. XXIe. m. 167-168°.[679]
1-naphthyl ester. XXIe. m. 122-124°.[75]
$Ph_2P(O)$•O•P(O)Ph_2. IIIb.[741] XXIf.[434] XXIg.[539] XXIh.[142] XXIIIh.[276] m. 143.5-144.5°,[142] $b_{0.2}$ 230°,[539] ^{31}P -33.1 ppm,[539] ΔH(hydrolysis) -17.6 kcal/mole.[418]
CH_3•CO•O•P(O)Ph_2. XXIIi. m. 93.0-97.3°, gives $Ph_2P(O)$•O•P(O)Ph_2 on further heating.[774]
Ph•SO_2•O•P(O)Ph_2. XXIIi. m. 162-164°.[697]
p-Me•C_6H_4•SO_2•O•P(O)Ph_2. XXIIi. m. 188-189°.[697]
Amide. XXIi.[434,796] m. 168°,[434] 190-192°,[796] ^{31}P -25.5 ppm.[565,700]
Me_2 amide. XXIi.[796] m. 103-105°,[796] ^{31}P -29.6 ppm,[700] x-ray structure determination.[512]
Et_2 amide. IVb.[274,559] VIc*.[525] XXIi.[267,274] m. 141-142°.[267]
Pr amide. XXIi. m. 90-93°.[267]
i-Pr amide. XXIi. m. 146-148°.[267]
Bu amide. XXIi. m. 93-95°.[267]
BuMe amide. XXIi. $b_{0.4}$ 160-164°.[739]
t-Bu amide. IVb.[739] XXIi.[267,739] m. 136-137°.[739]
cyclo-C_6H_{11} amide. XXIi. m. 197-197.5°.[267]
$PhCH_2$ amide. XXIi. m. 111-112°.[267]
$(PhCH_2)_2$ amide. IVb. m. 141-143°.[739]
Me_3C•CH_2•CMe_2 amide. IVb. m. 257-258°.[739]
$Ph_2P(O)$•NH•CH_2•NH•P(O)Ph_2. From $Ph_2P(O)NH_2$ + $(CH_2O)_n$, m. 208-209°.[434]
$Ph_2P(O)$•NH•$(CH_2)_6$•NH•P(O)Ph_2. IVb. m. 118.5-119.5°.[739]
Ph amide. XXId.[795] XXIi.[267,303] m. 242-244°,[267,303] ^{31}P -20.0 ppm,[566] forms K deriv. which can be alkylated.[268]
PhMe amide. XXIi.[267] Also from $Ph_2P(O)NPh^-$ K^+ + MeI,[268] m. 116-118°.[267,268]
PhBu amide. From $Ph_2P(O)NPh^-$ K^+ + BuBr, m. 144-145°.[268]
Ph_2 amide. XXIi.[267] m. 105-106°,[267] ^{31}P -18.8 ppm.[566]
o-ClC_6H_4 amide. XXId. m. 140-142°.[795]
m-ClC_6H_4 amide. XXIi. m. 252-253°.[267]
p-ClC_6H_4 amide. XXId.[795] XXIi.[267] m. 215-216°.[267]
2,4-$Cl_2C_6H_3$ amide. XXId. m. 177-179°.[795]
p-BrC_6H_4 amide. XXId. m. 233-235°.[795]
p-EtO•C_6H_4 amide. XXId. m. 205-207°.[795]
m-O_2N•C_6H_4 amide. XXId. m. 195-197°.[795]
p-O_2N•C_6H_4 amide. XXId. m. 255-257°.[795]
o-MeC_6H_4 amide. XXIi. m. 127-129°.[267]
m-MeC_6H_4 amide. XXIi. m. 250-250.5°.[267]
p-MeC_6H_4 amide. XXIi. m. 205-206°, sublimes 195°.[267]
1-naphthyl amide. XXIi. m. 188-190°.[267]
2-naphthyl amide. XXIi. m. 264-268°.[267]

1,3-thiazol-2-yl amide. XXIi. m. 224-238°.[282]
1,2,3-thiadiazol-5-yl amide. From $Ph_2P(O)NCS + CH_2N_2$. m. 216-221°.[759]
(1,2,3-thiadiazol-5-yl)Me amide. From $Ph_2P(O)NCS$ + more CH_2N_2 than for previous compound. m. 171-172°.[759]
4-Ph-1,2,3-thiadiazol-5-yl amide. From $Ph_2P(O)NCS + PhCHN_2$. m. 184-185°.[759]
1,2,4-triazol-3-yl amide. XXIi. m. 240-241°.[282]
2-pyridyl amide. XXIi. m. 177-180°.[267]
3-pyridyl amide. XXIi. m. 203-204°.[267]
4-pyridyl amide. XXIi. m. 173-174°.[267]
pyrimidin-2-yl amide. XXIi. m. 187-188°.[282]
benzthiazol-2-yl amide. XXIi. m. 246°.[282]
benzimidazol-2-yl amide. XXIi. m. 272-273°.[282]

$Ph_2P(O)\cdot NH$— 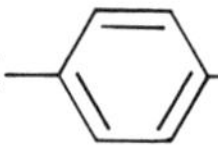 —$NH\cdot P(O)Ph_2$. XXIi. m. 198-200°.[555]

$Ph_2P(O)\cdot N$= 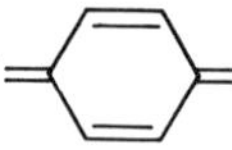 =$N\cdot P(O)Ph_2$. From previous compound + $Pb(OAc)_4$ (and reducible to previous compound with SO_2 or aqueous $NaHSO_3$). Orange crystals, m. 103-105°.[555]
$Ph_2P(O)\cdot NPh\cdot CH_2\cdot NPh\cdot P(O)Ph_2$. From $Ph_2P(O)NPh^- K^+ + CH_2I_2$.[268]
$Ph_2P(O)\cdot NH\cdot P(O)Ph_2$. IVg.[244] XXd.[204] XXIIIg.[247] m. 266.5°,[204] ^{31}P -14.5 ppm,[565] x-ray powder data,[204] not easily hydrolyzed, H is acidic and structure is alternatively written as $Ph_2P(OH):N\cdot P(O)Ph_2$;[204] NH_4^+ salt, x-ray powder data.[204]
$Ph_2P(O)\cdot NPh\cdot P(O)Ph_2$. XXIj. m. 211-213°.[268]
$Ph_2P(S)\cdot NH\cdot P(O)Ph_2$. XXId. XXIIIg. m. 172-174°, ^{31}P -54.2 and -22.1 ppm; K salt, m. 295-298°, ^{31}P -34.7 and -12.4 ppm; $2Ph_2P(S)\cdot NH\cdot P(O)Ph_2\cdot HSbCl_6$, m. 142-144°.[705]
$H\cdot CO\cdot NH\cdot P(O)Ph_2$. By atmospheric hydrolysis of $i\text{-}PrO\cdot CH:N\cdot P(O)Ph_2$ (formed, together with CH_2N_2, by reaction of $Ph_2P(O)\cdot N_3$ and $i\text{-}PrO\cdot CH:CH_2$). m. 178-180°.[78]
$CHCl_2\cdot CO\cdot NH\cdot P(O)Ph_2$. XXId. m. 237-238°.[733]
$CCl_3\cdot CO\cdot NH\cdot P(O)Ph_2$. XXId. m. 182-183°.[733]
$PhCO\cdot NH\cdot P(O)Ph_2$. VIc*.[160] XXId.[159] m. 149-151°,[160] 193-195°.[159]
$o\text{-}ClC_6H_4.CO\cdot NH\cdot P(O)Ph_2$. VIc*. m. 92-94°.[160]
$p\text{-}ClC_6H_4\cdot CO\cdot NH\cdot P(O)Ph_2$. VIc*. m. 204-205°.[160]
$o\text{-}BrC_6H_4\cdot CO\cdot NH\cdot P(O)Ph_2$. VIc*. m. 167-168°.[160]
$p\text{-}BrC_6H_4\cdot CO\cdot NH\cdot P(O)Ph_2$. VIc*. m. 208-209°.[160]
$m\text{-}O_2N\cdot C_6H_4\cdot CO\cdot NH\cdot P(O)Ph_2$. XXId. m. 213-215°.[159]
$p\text{-}O_2N\cdot C_6H_4\cdot CO\cdot NH\cdot P(O)Ph_2$. XXId. m. 227-228°.[159]
$i\text{-}PrC(NMe_2):N\cdot P(O)Ph_2$. From $Ph_2P(O)N_3 + Me_2C:CH\cdot NMe_2$.

m. 137.5-139°.[82]
$Ph_2P(O)\cdot N:C:N\cdot P(O)Ph_2$. XXIi [$Ph_2P(O)Cl + Ag_2CN_2$]. m. 109-110°.[783]
$H_2N\cdot CO\cdot NH\cdot P(O)Ph_2$. XXXIIa. m. 185-192°.[772]
$NC\cdot NH\cdot CO\cdot NH\cdot P(O)Ph_2$. XXXIIa. m. 138-141°.[772]
$Et_2N\cdot CO\cdot NH\cdot P(O)Ph_2$. XXIi. XXXIIa. m. 125-130°.[772]
$PrNH\cdot CO\cdot NH\cdot P(O)Ph_2$. XXXIIa. m. 195-200°.[772]
i-$PrNH\cdot CO\cdot NH\cdot P(O)Ph_2$. XXXIIa. m. 170-177°.[772]
$BuNH\cdot CO\cdot NH\cdot P(O)Ph_2$. XXXIIa. Also by reactions of previous compound, and of $MeO_2C\cdot NH\cdot P(O)Ph_2$ (no details of this compound) with a large excess of $BuNH_2$. m. 193-195°.[772]
$Bu_2N\cdot CO\cdot NH\cdot P(O)Ph_2$. XXXIIa. m. 255-258°.[772]
s-$BuNH\cdot CO\cdot NH\cdot P(O)Ph_2$. XXXIIa. m. 170-175°.[772]
t-$BuNH\cdot CO\cdot NH\cdot P(O)Ph_2$. XXXIIa. m. 180-188°.[772]
$C_5H_{11}\cdot NH\cdot CO\cdot NH\cdot P(O)Ph_2$. XXXIIa. m. 182-185°.[772]
$PhCH_2\cdot NH\cdot CO\cdot NH\cdot P(O)Ph_2$. XXXIIa. m. 210-212°.[772]
$PhNH\cdot CO\cdot NH\cdot P(O)Ph_2$. XXXIIa. Also from $Ph_2P(O)NH_2$ + PhNCO. m. 193-195°.[772]
$Ph_2N\cdot CO\cdot NH\cdot P(O)Ph_2$. XXXIIa. m. 117-120°.[772]
$C_6F_5\cdot NH\cdot CO\cdot NH\cdot P(O)Ph_2$. XXXIIa. m. 157-160°.[772]
p-$ClC_6H_4\cdot NH\cdot CO\cdot NH\cdot P(O)Ph_2$. XXXIIa. m. 205-210°.[772]
o-$MeC_6H_4\cdot NH\cdot CO\cdot NH\cdot P(O)Ph_2$. XXXIIa. m. 188-193°.[772]
m-$MeC_6H_4\cdot NH\cdot CO\cdot NH\cdot P(O)Ph_2$. XXXIIa. m. 195-198°.[772]
p-$MeC_6H_4\cdot NH\cdot CO\cdot NH\cdot P(O)Ph_2$. XXXIIa. m. 185-190°.[772]
2-pyridyl$\cdot NH\cdot CO\cdot NH\cdot P(O)Ph_2$. XXXIIa. m. 170-180°.[772]
4-pyridyl$\cdot NH\cdot CO\cdot NH\cdot P(O)Ph_2$. XXXIIa. m. 182-186°.[772]
$Ph_2P(O)\cdot NH\cdot CO\cdot NH\cdot P(O)Ph_2$. XXXIIa. Also from $Ph_2P(O)\cdot N:C:N\cdot P(O)Ph_2 + H_2O$. m. 178-179°.[772]
$PhNH\cdot NH\cdot CO\cdot NH\cdot P(O)Ph_2$. XXXIIa. m. 195-205°.[772]
$BuS\cdot CS\cdot NH\cdot P(O)Ph_2$. XXXIIa. m. 105-107°.[758]
$MeNH\cdot CS\cdot NH\cdot P(O)Ph_2$. XXXIIa. m. 184°.[758]
$EtNH\cdot CS\cdot NH\cdot P(O)Ph_2$. XXXIIa. m. 171-172°.[758]
$PrNH\cdot CS\cdot NH\cdot P(O)Ph_2$. XXXIIa. m. 162-163°.[758]
i-$PrNH\cdot CS\cdot NH\cdot P(O)Ph_2$. XXXIIa. m. 158-160°.[758]
$BuNH\cdot CS\cdot NH\cdot P(O)Ph_2$. XXXIIa. m. 148-151°.[758]
i-$BuNH\cdot CS\cdot NH\cdot P(O)Ph_2$. XXXIIa. m. 150-154°.[758]
cyclo-$C_6H_{11}\cdot NH\cdot CS\cdot NH\cdot P(O)Ph_2$. XXXIIa. m. 154-157°.[758]
$PhNH\cdot CS\cdot NH\cdot P(O)Ph_2$. XXXIIa. m. 141-142°.[758]
o-$ClC_6H_4\cdot NH\cdot CS\cdot NH\cdot P(O)Ph_2$. XXXIIa. m. 123-125°.[758]
p-$ClC_6H_4\cdot NH\cdot CS\cdot NH\cdot P(O)Ph_2$. XXXIIa. m. 124-127°.[758]
o-$MeO\cdot C_6H_4\cdot NH\cdot CS\cdot NH\cdot P(O)Ph_2$. XXXIIa. m. 119-120°.[758]
p-$MeO\cdot C_6H_4\cdot NH\cdot CS\cdot NH\cdot P(O)Ph_2$. XXXIIa. m. 129-131°.[758]
o-$MeC_6H_4\cdot NH\cdot CS\cdot NH\cdot P(O)Ph_2$. XXXIIa. m. 133-134°.[758]
p-$MeC_6H_4\cdot NH\cdot CS\cdot NH\cdot P(O)Ph_2$. XXXIIa. m. 131-132°.[758]
1-naphthyl$\cdot NH\cdot CS\cdot NH\cdot P(O)Ph_2$. XXXIIa. m. 106-110°.[758]
2-naphthyl$\cdot NH\cdot CS\cdot NH\cdot P(O)Ph_2$. XXXIIa. m. 137-138°.[758]
$NC\cdot NH\cdot C(:NH)\cdot NH\cdot P(O)Ph_2$. XXIi. m. 146-152°.[434]
$Ph\cdot SO_2\cdot NH\cdot P(O)Ph_2$. IVb. XXId. m. 205-206°.[731]
p-$ClC_6H_4\cdot SO_2\cdot NH\cdot P(O)Ph_2$. XXId. m. 220-221°.[731]
o-$O_2N\cdot C_6H_4\cdot SO_2\cdot NH\cdot P(O)Ph_2$. XXId. m. 199-200°.[731]

m-$O_2N{\cdot}C_6H_4{\cdot}SO_2{\cdot}NH{\cdot}P(O)Ph_2$. XXId. m. 192-193°.[731]
p-$O_2N{\cdot}C_6H_4{\cdot}SO_2{\cdot}NH{\cdot}P(O)Ph_2$. XXId. m. 204-205°.[731]
o-$MeC_6H_4{\cdot}SO_2{\cdot}NH{\cdot}P(O)Ph_2$. XXId. m. 183-184°.[731]
p-$MeC_6H_4{\cdot}SO_2{\cdot}NH{\cdot}P(O)Ph_2$. XXId. m. 214-215°.[731]
1-naphthyl$\cdot SO_2{\cdot}NH{\cdot}P(O)Ph_2$. XXId. m. 212-213°.[731]
2-naphthyl$\cdot SO_2{\cdot}NH{\cdot}P(O)Ph_2$. XXId. m. 213.5-214°.[731]
$HO{\cdot}NH{\cdot}P(O)Ph_2$. XXIi. m. 131° decomp.[434]
$H_2N{\cdot}NH{\cdot}P(O)Ph_2$. XXIi.[434,744] m. 210-212°.[744]
$Me_2N{\cdot}NH{\cdot}P(O)Ph_2$. IVb.[719] XXIi.[567,735] m. 167-168°.[567,735]
$Me_2N{\cdot}NMe{\cdot}P(O)Ph_2$. IVb. XXIi. m. 165-166.5°.[567]
$Me_2N{\cdot}NEt{\cdot}P(O)Ph_2$. IVb. XXIi. m. 140-141°.[567]
$PhNH{\cdot}NH{\cdot}P(O)Ph_2$. XXIi.[91,434] m. 187-189°,[91] m. 223° decomp.[434]
$Ph_2N{\cdot}NH{\cdot}P(O)Ph_2$. XXIi. m. 154-156°.[32]
$PhCH{:}N{\cdot}NH{\cdot}P(O)Ph_2$. From hydrazide + PhCHO. m. 193°.[434]
$EtO_2C{\cdot}CH_2{\cdot}CMe{:}N{\cdot}NH{\cdot}P(O)Ph_2$. From hydrazide + $Me{\cdot}CO{\cdot}CH_2{\cdot}$-$CO_2Et$. m. 123-124°.[756]
$\overline{CO{\cdot}CH_2{\cdot}CMe{:}N{\cdot}N}{\cdot}P(O)Ph_2$. By heating previous compound. m. 170-171°.[756]
$EtO_2C{\cdot}CH_2{\cdot}CPh{:}N{\cdot}NH{\cdot}P(O)Ph_2$. From hydrazide + $Ph{\cdot}CO{\cdot}CH_2{\cdot}$-$CO_2Et$. m. 156° decomp.[756]
$\overline{CO{\cdot}CH_2{\cdot}CPh{:}N{\cdot}N}{\cdot}P(O)Ph_2$. As previous compound but at higher temperature. m. 260-261°.[756]
$EtO_2C{\cdot}CH_2{\cdot}C(C_6H_4$-$p$-$NO_2){:}N{\cdot}NH{\cdot}P(O)Ph_2$. From hydrazide + p-$O_2N{\cdot}C_6H_4{\cdot}CO{\cdot}CH_2{\cdot}CO_2Et$. m. 185° decomp.[756]
$\overline{CO{\cdot}CH_2{\cdot}C(C_6H_4\text{-}p\text{-}NO_2){:}N{\cdot}N}{\cdot}P(O)Ph_2$. As previous compound but at higher temperature. m. 258°.[756]
$MeO_2C{\cdot}CH_2{\cdot}C(C_6H_4$-$p$-$O{\cdot}C_{18}H_{37})N{\cdot}NH{\cdot}P(O)Ph_2$. Minor product from hydrazide + p-$(C_{18}H_{37}{\cdot}O){\cdot}C_6H_4{\cdot}CO{\cdot}CH_2{\cdot}CO_2Me$. m. 108-109°.[756]
$\overline{CO{\cdot}CH_2{\cdot}C[C_6H_4\text{-}p\text{-}(O{\cdot}C_{18}H_{37})]{:}N{\cdot}N}{\cdot}P(O)Ph_2$. Major product of previous reaction. m. 124-125°.[756]
$Ph{\cdot}CO{\cdot}NH{\cdot}NH{\cdot}P(O)Ph_2$. XXIi.[744] Also from hydrazide + PhCOCl.[92,744] m. 199°.[92]
$PhC(OEt){:}N{\cdot}NH{\cdot}P(O)Ph_2$. From hydrazide + $PhC(OEt)_3$. m. 158-159°.[745]
$PhC{:}(NH){\cdot}NH{\cdot}NH{\cdot}P(O)Ph_2$. From hydrazide + PhC(:NH)OEt. m. 220-223°.[745]
$Ph_2P(O){\cdot}NH{\cdot}NH{\cdot}P(O)Ph_2$. XXIi from N_2H_4.[92] XXIi from $H_2N{\cdot}NH{\cdot}P(O)Ph_2$.[744] m. 253-255°.[92]
$Ph_2P(S){\cdot}NH{\cdot}NH{\cdot}P(O)Ph_2$. From hydrazide + $Ph_2P(S)Cl$. m. 206-212°.[744]
$PhN{:}N{\cdot}P(O)Ph_2$. From phenylhydrazide + N-bromosuccinimide. Also from PhN_2^+ BF_4^- + $Ph_2P(O)H$, dark red crystals. m. 105-106°.[93]
$Ph_2P(O){\cdot}N{:}N{\cdot}P(O)Ph_2$. From $Ph_2P(O)NH{\cdot}NH{\cdot}P(O)Ph_2$ + N-bromosuccinimide. Blue-violet crystals, m. 140-142° decomp. UV, ^{31}P -29.4 ppm.[92]

$(o\text{-}FC_6H_4)_2PO_2H$. VII. m. 220-222°.[221]
$(o\text{-}ClC_6H_4)PhPO_2H$. VII. m. 186-187°.[228]
Et ester. XXIe. m. 68-69°, $b_{0.02}$ 158-161°.[228]
$(o\text{-}ClC_6H_4)_2PO_2H$. VII. m. 233-236°,[166] UV.[332]
$(m\text{-}ClC_6H_4)PhPO_2H$. IVe.[645] VII.[209,210] m. 159-162°,[209] UV.[332]
Et ester. $b_{0.6}$ 170-174°, n_D^{25} 1.5701.[157]
$(m\text{-}ClC_6H_4)_2PO_2H$. VII.[166,210] m. 164-165°,[166] UV.[332]
Et ester. XXIe. $b_{0.7}$ 183-187°, n_D^{25} 1.5794.[157]
$(p\text{-}ClC_6H_4)PhPO_2H$. IVe.[645] VII.[482] m. 158.5-159.5°.[645]
Et ester. XXIe. $b_{0.3}$ 168-170°, $n_D^{20.5}$ 1.5740.[482]
$(p\text{-}ClC_6H_4)(m\text{-}ClC_6H_4)PO_2H$. IVe. m. 175° decomp.[645]
$(p\text{-}ClC_6H_4)_2PO_2H$. VII. m. 145-146°,[166] pK_a 1.68 (in 7% EtOH), 3.48 (in 80% EtOH).[503]
CCl_2:CH ester. IVh. m. 67-67.5°, $b_{0.5}$ 192°.[390]
Ph amide. XXIi. m. 209-209.5°.[757]
$(2,3\text{-}Cl_2C_6H_3)_2PO_2H$. VII. m. 278-280°.[157]
$(2,5\text{-}Cl_2C_6H_3)_2PO_2H$. VII.[157,221] m. 232-233°.[221]
$(3,5\text{-}Cl_2C_6H_3)_2PO_2H$. VII. m. 243-244.5°.[157]
$(2,3,6\text{-}Cl_3C_6H_2)_2PO_2H$. VII. m. 287-288.5°.[157]
$(2,4,5\text{-}Cl_3C_6H_2)_2PO_2H$. VII. m. 244.5-246°.[157]
$(o\text{-}BrC_6H_4)PhPO_2H$. VII.[209,210] m. 209-210°,[209] UV.[333]
Me ester. XXIe. $b_{0.1}$ 160-168°.[221]
Et ester. XXIe. $b_{0.1}$ 156-165°.[221]
i-Pr ester. XXIe. $b_{0.1}$ 183-191°.[221]
$(o\text{-}BrC_6H_4)_2PO_2H$. VII. 268.5-269.5°,[167] UV.[333]
Et ester. XXIe. m. 115-118°, $b_{0.1}$ 186-194°.[221]
$(m\text{-}BrC_6H_4)PhPO_2H$. VIc.[58] VII.[157] m. 161.5-162°.[157]
$(m\text{-}BrC_6H_4)_2PO_2H$. VII. m. 186.5-189°.[168]
$(p\text{-}BrC_6H_4)PhPO_2H$. IVe.[156,645] VIc.[58] m. 174.5°.[156]
$(p\text{-}BrC_6H_4)_2PO_2H$. VII. m. 170.5-172.5°.[168]
CCl_2:CH ester. IVh. m. 131-133°.[357]
Ph·CO·NH·P(O)$(C_6H_4\text{-}p\text{-}Br)_2$. VIc*. m. 187-188°.[160]
$p\text{-}ClC_6H_4$·CO·NH·P(O)$(C_6H_4\text{-}p\text{-}Br)_2$. VIc*. m. 205-207°.[160]
$p\text{-}BrC_6H_4$·CO·NH·P(O)$(C_6H_4\text{-}p\text{-}Br)_2$. VIc*. m. 204-206°.[160]
$(2,5\text{-}Br_2C_6H_3)_2PO_2H$. VII. m. 277-279°.[157]
$(o\text{-}IC_6H_4)PhPO_2H$. VII. m. 206-207.5°.[221]
$(m\text{-}IC_6H_4)_2PO_2H$. VII. m. 212.5-213.5°.[157]
$(p\text{-}IC_6H_4)_2PO_2H$. From di-p-$NH_2$ acid by diazotization and reaction with KI. m. 209-210°.[217]
$(o\text{-}HO\cdot C_6H_4)PhPO_2H$. By-product in $LiAlH_4$ reduction of o-MeO acid. m. 156-157.5°; benzylisothiouronium salt, m. 180°.[484]
$(m\text{-}HO\cdot C_6H_4)_2PO_2H$. From di-m-$NH_2$ acid by diazotization and hydrolysis. m. 226-229°.[167]
$(p\text{-}HO\cdot C_6H_4)_2PO_2H$. From di-p-$NH_2$ acid by diazotization and hydrolysis. m. 213-215°.[167]
$(o\text{-}MeO\cdot C_6H_4)PhPO_2H$. VII. m. 214-216°; benzylisothiouronium salt, m. 176°.[484]
$(o\text{-}MeO\cdot C_6H_4)_2PO_2H$. VIc.[408] VII.[221] m. 227-228°,[408]

234-238°.[221]

$(m\text{-}MeO\cdot C_6H_4)PhPO_2H$. VII.[290,484] m. 147-147.5°;[484] benzylisothiouronium salt, m. 181-183°.[290]

Me ester. XXIe. m. 45-49°.[290]

Et ester. XXIe. $b_{0.03}$ 158-164°, n_D^{21} 1.5704.[484]

$(p\text{-}MeO\cdot C_6H_4)PhPO_2H$. VII.[290,484] m. 184°;[290] benzylisothiouronium salt, m. 158°.[484]

$(p\text{-}MeO\cdot C_6H_4)_2PO_2H$. VIc.[408] VII.[64] XXIa.[343] m. 180-181°.[343]

Ph amide. XXIi. m. 209.5-211.5°.[757]

$(m\text{-}EtO\cdot C_6H_4)PhPO_2H$. VII. m. 131-134°; benzylisothiouronium salt, m. 176°.[290]

$(p\text{-}EtO\cdot C_6H_4)PhPO_2H$. VII.[290,484] m. 148-149°.[290]

$(o\text{-}PhO\cdot C_6H_4)PhPO_2H$. VII. m. 164-166°.[457]

$(p\text{-}PhO\cdot C_6H_4)_2PO_2H$. From di-p-Br acid + PhOH. m. 203-206°.[217]

$(o\text{-}H_2N\cdot C_6H_4)PhPO_2H$. XXXi. m. 166-167°.[175]

$(m\text{-}H_2N\cdot C_6H_4)PhPO_2H$. XXXi. m. 260-263° decomp.[209]

$(m\text{-}H_2N\cdot C_6H_4)_2PO_2H$. XXXi.[167,172] m. 287-289° decomp.,[167] UV.[214]

$(p\text{-}H_2N\cdot C_6H_4)PhPO_2H$. XXXi. m. 210-211.5°.[209]

$(p\text{-}H_2N\cdot C_6H_4)_2PO_2H$. XXXd.[682] XXXi.[167] m. 242-243°,[682] softens at 270°.[167]

$(3\text{-}H_2N\text{-}6\text{-}BrC_6H_3)_2PO_2H$. XXXi. Decomp. 300°.[214]

$(m\text{-}MeNH\cdot C_6H_4)(m\text{-}BrC_6H_4)PO_2H$. XXXd. m. 80-93°.[168]

$(m\text{-}MeNH\cdot C_6H_4)_2PO_2H$. XXXd. m. 232.5-233.5°.[168]

$(p\text{-}Me_2N\cdot C_6H_4)PhPO_2H$. XIa. m. > 370°.[300]

$(p\text{-}Me_2N\cdot C_6H_4)_2PO_2H$. IIa.[100,682] VIIIa.[100,682] m. 209-211°,[682] 249-250°.[100]

Ph amide. VIc*. m. 231-232.5°.[757]

$(p\text{-}Et_2N\cdot C_6H_4)_2PO_2H$. VIIIa. m. 253-254°.[100]

$[2\text{-}(2\text{-}HO\text{-}5\text{-}MeC_6H_3\cdot N{:}N\cdot)C_6H_4]PhPO_2H$. From $(o\text{-}H_2N\cdot C_6H_4)\text{-}PhPO_2H$ by diazotization and coupling with p-cresol. m. 173-175°.[175]

$[2(CMe{:}N\cdot NPh\cdot C(OH){:}C\cdot N{:}N\cdot)C_6H_4]PhPO_2H$. From $(o\text{-}H_2N\cdot C_6H_4)\text{-}PhPO_2H$ by diazotization and coupling with 1-Ph-3-Me-pyrazol-5-one. m. 216-218°.[175]

$(o\text{-}O_2N\cdot C_6H_4)PhPO_2H$. VII. m. 229-232°.[213]

$(m\text{-}O_2N\cdot C_6H_4)PhPO_2H$. VII.[209,210] m. 166-167°,[209] UV.[332]

$(m\text{-}O_2N\cdot C_6H_4)_2PO_2H$. VII.[166,210] XXXc.[172] m. 271-273°,[166,210] pK_a 2.37 (in 80% EtOH);[503] NH_4^+ salt, yellow prisms, m. 260°,[172] UV.[332]

$(p\text{-}O_2N\cdot C_6H_4)PhPO_2H$. VII.[209,210] m. 153-154°,[209] UV.[332]

$(p\text{-}O_2N\cdot C_6H_4)_2PO_2H$. VII.[166,210] m. 275-277°,[166] UV.[332]

Ph amide. XXIi. m. 229-231.5°.[757]

$(2\text{-}Cl\text{-}4\text{-}O_2N\cdot C_6H_3)_2PO_2H$. VII. m. 270-275° decomp.[221]

$(2\text{-}Br\text{-}3\text{-}O_2N\cdot C_6H_3)_2PO_2H$. VII. m. 296-297° decomp.[157]

$(2\text{-}Br\text{-}5\text{-}O_2N\cdot C_6H_3)_2PO_2H$. VII. m. > 300°.[221]

$(p\text{-}Ph_2P\cdot C_6H_4)PhPO_2H$. VId. m. 192.5-193.5°.[59]

$(p\text{-}Ph_2P\cdot C_6H_4)_2PO_2H$. VId. m. 258-261°.[59]

$[p\text{-}Ph_2P(O)\cdot C_6H_4]PhPO_2H$. From p-$Ph_2P$ acid + H_2O_2, m. 231-232°; monohydrate. m. 144°.[59]

$[p\text{-}Ph_2P(O)\cdot C_6H_4]_2PO_2H$. From di-p-$Ph_2P$ acid + H_2O_2. m. 192-196°.[59]
$[p\text{-}Ph_2P(S)\cdot C_6H_4]PhPO_2H$. From p-$Ph_2P$ acid + S. m. 268-270°.[59]
$[p\text{-}Ph_2P(S)\cdot C_6H_4]_2PO_2H$. From di-p-$Ph_2P$ acid + S. m. > 300°.[59]
$(p\text{-}Me_3Si\cdot C_6H_4)_2PO_2H$. VIa. m. 213-214.5°.[225]
$(m\text{-}MeC_6H_4)_2PO_2H$. VII.[157,214] m. 173.5-175°,[214] UV.[214]
$(p\text{-}MeC_6H_4)PhPO_2H$. XIa. m. 134-136°.[300]
  $Ph\cdot SO_2\cdot NH\cdot P(O)(C_6H_4\text{-}p\text{-}Me)Ph$. XXId. m. 187-188°.[732]
  $m\text{-}O_2N\cdot C_6H_4\cdot SO_2\cdot NH\cdot P(O)(C_6H_4\text{-}p\text{-}Me)Ph$. XXId. m. 142-143°.[732]
  $o\text{-}MeC_6H_4\cdot SO_2\cdot NH\cdot P(O)(C_6H_4\text{-}p\text{-}Me)Ph$. XXId. m. 135-136°.[732]
  $p\text{-}MeC_6H_4\cdot SO_2\cdot NH\cdot P(O)(C_6H_4\text{-}p\text{-}Me)Ph$. XXId. m. 175-176°.[732]
  2-naphthyl$\cdot SO_2\cdot NH\cdot P(O)(C_6H_4\text{-}p\text{-}Me)Ph$. XXId. m. 208-210°.[732]
$(p\text{-}MeC_6H_4)_2PO_2H$. VIc.[408,528] VIg.[600] VII.[214] XIb.[319] m. 131-132°,[408,600] 135-136°,[420] UV,[214,782] pKa 2.47 (in 7% EtOH), 3.66 (in 50% EtOH), 4.45 (in 80% EtOH).[503]
  $CCl_2{:}CH$ ester. IVh. m. 68.5-70°.[389]
  Ph amide. XXIi.[160,757] m. 160-162°,[160] 216-216.5°.[757]
  $Ph\cdot CO\cdot NH\cdot P(O)(C_6H_4\text{-}p\text{-}Me)_2$. VIc. m. 224-225°.[160]
  $p\text{-}ClC_6H_4\cdot CO\cdot NH\cdot P(O)(C_6H_4\text{-}p\text{-}Me)$. VIc. m. 208-209°.[160]
$(p\text{-}MeC_6H_4)(o\text{-}PhO\cdot C_6H_4)PO_2H$. VII. m. 174-177°.[457]
$(4\text{-}Cl\text{-}3\text{-}MeC_6H_3)_2PO_2H$. VII. m. 178-181°.[218]
$(2\text{-}Cl\text{-}5\text{-}MeC_6H_3)_2PO_2H$. VII. m. 235-237.5°.[214]
$(3\text{-}Cl\text{-}4\text{-}MeC_6H_3)_2PO_2H$. VII. m. 189-192°.[218]
$(2\text{-}Br\text{-}4\text{-}MeC_6H_3)_2PO_2H$. VII. m. 265-271°.[214,221]
$(2\text{-}Me\text{-}4\text{-}O_2N\cdot C_6H_3)_2PO_2H$. VII. m. 228-229°.[215]
$(2\text{-}Me\text{-}5\text{-}O_2N\cdot C_6H_3)_2PO_2H$. VII. m. 243-245°.[215]
$(3\text{-}Me\text{-}4\text{-}O_2N\cdot C_6H_3)_2PO_2H$. VII. m. 240-250°.[215]
$(4\text{-}Me\text{-}3\text{-}O_2N\cdot C_6H_3)_2PO_2H$. XXXc. m. 235-237.5°.[215]
$(m\text{-}CF_3\cdot C_6H_4)PhPO_2H$. IVe. m. 170-171°.[645]
$(o\text{-}HO_2C\cdot C_6H_4)PhPO_2H$. VII (from Me anthranilate). m. 161-164°.[213]
$(p\text{-}HO_2C\cdot C_6H_4)PhPO_2H$. IVe [from $(p\text{-}NC\cdot C_6H_4)PhPCl$]. m. 258-260°.[645]
$(p\text{-}HO_2C\cdot C_6H_4)_2PO_2H$. XXXb.[214,600] Unmelted at 330°.[600]
$(p\text{-}MeO_2C\cdot C_6H_4)_2PO_2H$. Esterification (MeOH + HCl) of previous acid. m. 191-192°.[600]
$(p\text{-}EtO_2C\cdot C_6H_4)_2PO_2H$.
  Et ester. From $(p\text{-}Cl\cdot CO\cdot C_6H_4)_2P(O)Cl$ + NaOEt. m. 162-164°.[600]
$(p\text{-}BuO_2C\cdot C_6H_4)_2PO_2H$. Transesterification of $(p\text{-}MeO_2C\cdot C_6H_4)_2PO_2H$. m. 124°.[600]
$(p\text{-}Et_2N\cdot CO\cdot C_6H_4)_2PO_2H$.
  Et_2 amide. From $(p\text{-}Cl\cdot CO\cdot C_6H_4)_2P(O)Cl$ + Et_2NH. m. 141-142°.[600]
$(2\text{-}Br\text{-}4\text{-}HO_2C\cdot C_6H_3)_2PO_2H$. XXXb. m. 278-282°.[217]
$(2\text{-}HO_2C\text{-}5\text{-}O_2N\cdot C_6H_3)_2PO_2H$. XXXb. m. 292-296°.[215]

$(2,5\text{-}Me_2C_6H_3)_2PO_2H$. IVe.[298] XIb.[319] m. 184°.[298]
$(2,4,5\text{-}Me_3C_6H_2)PhPO_2H$. IVe. m. 181°; $PhNH{\cdot}NH_2$ salt, m. 140.5°.[524]
$(2,4,5\text{-}Me_3C_6H_2)_2PO_2H$. VIIIb.[522,746] m. 205-206°.[746]
$(2,4,6\text{-}Me_3C_6H_2)_2PO_2H$. IIa.[146,747] IIIb.[746] VIa.[226] m. 210°,[226] UV.[747]
$(3\text{-}HO_2C\text{-}4,6\text{-}Me_2C_6H_2)_2PO_2H$. XXXb. m. 185°.[522]
$(2,3,4,6\text{-}Me_4C_6H)_2PO_2H$. IIa. m. 237°.[146]
$(p\text{-}t\text{-}BuC_6H_4)_2PO_2H$. VIIIb. m. 211-212°.[413]
(1-naphthyl)$PhPO_2H$. XIa.[300] XIb.[319] m. 185-187°.[300]
(1-naphthyl)$_2PO_2H$. VIa.[534] VIc.[691] VIg.[142] XIb.[319] m. 220°.[691]
Et ester. IIc. m. 138.5-139.5°.[142]
Ph amide. IIc. m. 289-291°.[142]
Anhydride. XXIf. m. 255-260°.[142]
(2-naphthyl)$PhPO_2H$. VIf. m. 165-166°.[402]
(-)-menthyl (R)-ester. XXIe + fractional crystallization.[402] m. 87-88°,[402] $[\alpha]_D^{\sim 25}$ -14° (C∿2, C_6H_6),[402] NMR.[460]
(-)-menthyl (S)-ester. XXIe + fractional crystallization.[402] m. 103-104°,[402] $[\alpha]_D^{\sim 25}$ -90° (c∿2, C_6H_6),[402] NMR.[460]
$(o\text{-}PhC_6H_4)PhPO_2H$. IVe.[119,474] VII.[119] m. 180-181°.[119]
Et ester. XXIe. m. 112-114°.[119]
Ph amide. XXIi.[119,474] m. 204-205°.[474]
$(p\text{-}PhC_6H_4)PhPO_2H$. IIa. m. 238-240°.[179]
Me ester. XXIIc. m. 100-101°.[179]
$(4\text{-}Br\text{-}2\text{-}PhC_6H_3)PhPO_2H$. VII. Needles, m. 193-194°.[119]
$[2\text{-}(2\text{-}H_2N{\cdot}C_6H_4)C_6H_4]PhPO_2H$. By hydrolysis of cyclic amide. Hydrochloride, m. 244° decomp.[118]
Cyclic amide. IVb. m. 289-290°.[118]
$[2\text{-}(2\text{-}H_2N{\cdot}C_6H_4)C_6H_4](p\text{-}BrC_6H_4)PO_2H$.
Cyclic amide. IVb. m. 223-226°.[118]
$[2\text{-}(2\text{-}H_2N{\cdot}C_6H_4)C_6H_4](p\text{-}Me_2N{\cdot}C_6H_4)PO_2H$.
Cyclic amide. IVb. m. 258-260°. Resolved via camphorsulfonates, giving (+)-isomer, m. 135-136°, $[\alpha]_D^{21}$ +152.8°; (-)-isomer, m. 135-136°, $[\alpha]_D^{22}$ -152.8°.[118]
$[2\text{-}(2\text{-}H_2N{\cdot}C_6H_4)C_6H_4](p\text{-}MeC_6H_4)PO_2H$.
Cyclic amide. IVb. m. 232-235°.[118]
$[2\text{-}(4\text{-}O_2N{\cdot}C_6H_4)C_6H_4]PhPO_2H$. VII. Buff needles, m. 234-237°.[119]
$(3\text{-}Me\text{-}6\text{-}PhC_6H_3)PhPO_2H$. VId. VII. m. 144-145°.[119]
$[2\text{-}(2\text{-}H_2N\text{-}4\text{-}MeC_6H_3)C_6H_4]PhPO_2H$.
Cyclic amide. IVb. m. 246-248°.[118]
$(p\text{-}PhCH_2C_6H_4)PhPO_2H$. By NaOH fusion of $[p\text{-}(Ph{\cdot}PO_2H{\cdot}CHPh)\text{-}C_6H_4]PhPO_2H$. m. 155-156°.[223]
$(p\text{-}Ph{\cdot}CO{\cdot}C_6H_4)PhPO_2H$. VII. XXXb. m. 183-185°.[222]

E.1.11. Phosphinic Acids Containing Heterocyclic Groups (Linked to Phosphorus Through Carbon Atoms)

$(\overline{O\cdot CH_2\cdot C}H\cdot CH_2)EtPO_2H$.
Me ester. XXXa. $b_{0.18}$ 98-100°, d_4^{20} 1.1564, n_D^{20} 1.4621.[466]
Pr ester. XXXa. $b_{0.06}$ 106-107°, d_4^{20} 1.0740, n_D^{20} 1.4540.[466]

$(\overline{O\cdot CH_2\cdot C}H\cdot CH_2)_2PO_2H$.
Pr ester. XXXa. $b_{0.015}$ 129.5-130°, d_4^{20} 1.1562, n_D^{20} 1.4650.[466]

$(\overline{O\cdot CH_2\cdot C}Me)EtPO_2H$.
Et ester. Ve*. b_1 81.5-82°, d_4^{20} 1.0914, n_D^{20} 1.4505.[47]

(5-H_2N-5-Me-1,3-dioxan-2-yl·CH_2)(p-ClC_6H_4)PO_2H.
Et ester. XXXj. m. 48-49°.[654]

(5-H_2N-5-Me-1,3-dioxan-2-yl·CH_2·CH_2)$PhPO_2H$.
Et ester. XXXj. m. 42-43°.[654]

(1,3-dioxepan-2-yl·CH_2)(p-MeC_6H_4)PO_2H.
Et ester. XXXj. $b_{0.09}$ 168-170°, d_4^{20} 1.1471, n_D^{20} 1.5145.[269]

(1,3-dioxocan-2-yl·CH_2)(p-MeC_6H_4)PO_2H.
Et ester. XXXj. $b_{0.06}$ 190-193°, d_4^{20} 1.0991, n_D^{20} 1.5052.[269]

(2-thienyl)[CCl_3·CH(OH)]PO_2H.
Me ester. IXc. m. 139-140°,[70] mass spectrum.[107]
Et ester. IXc. m. 104.5°,[70] mass spectrum.[107]
Pr ester. IXc. m. 96.7°.[70]
i-Pr ester. IXc. m. 139.5°.[70]
Bu ester. IXc. m. 69.7°.[70]

(2-thienyl)$_2PO_2H$. XIa. m. 192-193°.[491]

(indol-2-yl)$PhPO_2H$. XXIIIa. m. 147-148° decomp.[651]
Et ester. XXXj. m. 116-118°.[651]

(indol-3-yl)$_2PO_2H$. VIa. m. 190°.[538]

(indol-2-yl·CH_2)$PhPO_2H$. XXIIIa. m. 109-110°.[651]
Et ester. XXXj. m. 114-115° decomp.[651]

(2-Me-indol-3-yl)$_2PO_2H$. VIa. m. 159-160°.[538]

(2-pyridyl·CH_2·CH_2)$PhPO_2H$.
Et ester. XIf. $b_{0.1}$ 143-145°, d_4^{25} 1.1419, n_D^{25} 1.5560;[492] picrate, m. 158-159°.[492]

(benz-1,3-oxazol-2-yl·CH_2)$PhPO_2H$. XXIIIa. Na salt, m. 126-128°.[653]
Pr ester. XXXj. m. 106-108°.[653]

(benz-1,3-oxazol-2-yl·CH_2)(p-ClC_6H_4)PO_2H. XXIIIa. Na salt, m. 169-171°.[653]
Et ester. XXXj. m. 121-122°.[657]

(benz-1,3-oxazol-2-yl·CH_2)(p-MeC_6H_4)PO_2H. XXIIIa. m. 183-185°; K salt, m. 198-200°.[653]
Et ester. XXXj. m. 154-155° decomp.[653]

(6-O_2N-benz-1,3-oxazol-2-yl·CH_2)(p-ClC_6H_4)PO_2H.
Et ester. XXXj. m. 219-220°.[657]
(6-O_2N-benz-1,3-oxazol-2-yl·CH_2·CH_2)$PhPO_2H$.
Pr ester. XXXj. m. 184-185°.[657]
(benz-1,3-thiazolin-2-yl·CH_2)(p-MeC_6H_4)PO_2H.
Me ester. XXXj. m. 102-104°, $b_{0.2}$ 202-206°.[658]
(benz-1,3-thiazolin-2-yl·CH_2·CH_2)$PhPO_2H$.
Pr ester. XXXj. m. 134-136°, $b_{0.12}$ 232-234°.[658]
(benzimidazol-2-yl)(p-ClC_6H_4)PO_2H.
Et ester. XXXj. m. 305°.[652]
(benzimidazol-2-yl·CH_2)$PhPO_2H$.
Et ester. XXXj. m. 297-298°.[652]
Pr ester. XXXj. m. 196-208°.[652]
i-Bu ester. XXXj. m. 310-311°.[652]
(benzimidazol-2-yl·CH_2·CH_2)$PhPO_2H$.
Pr ester. XXXj. m. 283-285°.[652]

E.1.12. Cyclic Phosphinic Acids

 PO_2H. VIa (very low yield). m. 65° (unsharp) pK_a 3.07.[421]

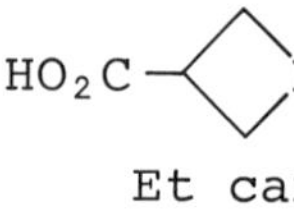 PO_2H. By hydrolysis and decarboxylation of Et ester of following acid. m. 136-138°.[801]
Et carboxylic ester. From acid + EtOH. m. 101-102°.[801]
di-Me ester. XXIIb. $b_{0.03}$ 85°, d_4^{20} 1.2744, n_D^{20} 1.4670.[801]
di-Et ester. XXIIb. $b_{0.01}$ 88°, d_4^{20} 1.1677, n_D^{20} 1.4603.[801]

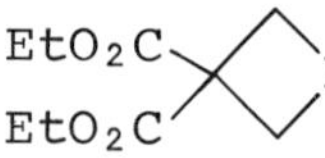 PO_2H.

Et ester. From $(ClCH_2)_2P(O)OPh$ + sodium derivative of diethyl malonate, followed by transesterification with EtOH. $b_{0.0001}$ 119°, d_4^{20} 1.2045, n_D^{20} 1.4720.[801]

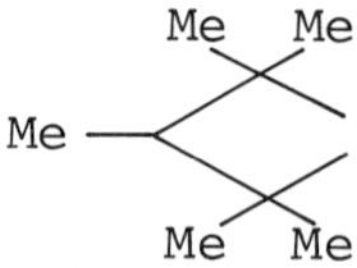 PO_2H. XXIa. m. 72-74°; dihydrate m. 54-58°.[513]
Me ester. XXIe. $b_{0.2}$ 66-69°, mixture of cis and trans isomers indicated by two close peaks on high resolution GLC.

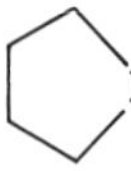 PO_2H. IIIa.[706] Cyclization by VIc (very low yield of product with m.p. 99-100°).[421] XXIIIa.[15, 279] m. 53-54.5°.[279]
Me ester. Hydrogenation of ester of 2,3-unsaturated acid, $b_{0.6}$ 74°, n_D^{20} 1.4702.[302]

Et ester. Cyclization by Va.[15,279] Also by hydrogenation of ester of 2,3-unsaturated acid.[278,302] b_{10} 108°,[15,302] n_D^{20} 1.4620,[15] kinetics of hydrolysis.[15,158]
Bu ester. Cyclization by Va. $b_{0.05}$ 76°.[279]
$(CH_2)_4$ di-ester. By-product during preparations of Et and Bu esters.[279] $b_{0.01}$ 218-220°.[279]

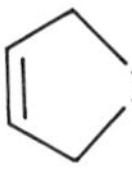PO_2H.

Me ester. XXb.[302] XXIe.[42] XXIIIc.[675] $b_{0.06}$ 55-60°,[302] b_{14} 113-114°,[42] d_4^{20} 1.1891,[42] n_D^{20} 1.4882.[42]
Et ester. Xa.[670] XXb.[302] XXIe.[42,48] XXIIIc.[675] $b_{0.2}$ 64-66°,[302] b_9 105°,[670] d_4^{20} 1.1361,[42,48] n_D^{20} 1.4801,[42,48] kinetics of hydrolysis.[158]
i-Pr ester. XXb. $b_{0.04}$ 53°, $b_{0.3}$ 66-70°, n_D^{20} 1.4647.[302]
CH⋮C·CH_2 ester. XXIe. b_9 144-145°, d_4^{20} 1.1801, n_D^{20} 1.5050.[675]
FCH_2·CH_2 ester. XXIe. m. 37°, b_1 95°.[187]
$ClCH_2$·CH_2 ester. Xa. b_{10} 154-155°, d_4^{20} 1.2428, n_D^{20} 1.5127.[672]
MeO·CH_2·CH_2 ester. Xa. $b_{0.5}$ 92°, d_4^{20} 1.1725, n_D^{20} 1.4770.[184]
CH_3·CO·O·CH_2·CH_2 ester. Xa. b_1 165°, d_4^{20} 1.2435, n_D^{20} 1.4790.[188]
CF_3·CO·O·CH_2·CH_2 ester. Xa. b_1 106°, d_4^{20} 1.3890, n_D^{20} 1.4215.[188]
MeS·CH_2·CH_2 ester. Xa. b_1 114°, d_4^{20} 1.2201, n_D^{20} 1.5295.[186]
NCS·CH_2·CH_2 ester. Xa. b_1 160°, d_4^{20} 1.3032, n_D^{20} 1.5330.[185]
SCN·CH_2·CH_2 ester. Minor product in preparation of previous ester. b_1 110°, d_4^{20} 1.2024, n_D^{20} 1.5070.[185]
$ClCH_2$·CHMe ester. Xa. b_{10} 147-148°, d_4^{20} 1.2369, n_D^{20} 1.4978.[672]
$ClCH_2$·CH_2·CHMe ester. Xa. $b_{0.1}$ 68-70°, d_4^{20} 1.2040, n_D^{20} 1.4892.[674]
ClCHMe·CHMe ester. Xa. $b_{0.05}$ 111-112°, n_D^{20} 1.4912.[302]
$ClCH_2$·CMe_2·CH_2 ester. Xa. $b_{0.2}$ 94-95°, d_4^{20} 1.1682, n_D^{20} 1.4930.[674]
Ph ester. Xa.[43] XXIe.[51] m. 48-49°,[51] 60-62° (after long standing),[43] $b_{0.04}$ 107-109°,[51] b_3 163-163.5°,[43] d_4^{20} 1.2065,[51] n_D^{20} 1.5568,[51] 1.5628.[43]
o-HO·C_6H_4 ester. Xa.[45,669] (Apparently different products), m. 128-129°,[45] $b_{0.5}$ 110°,[669] d_4^{20} 1.2601,[669] n_D^{20} 1.5375.[669]
p-O_2N·C_6H_4 ester. XXIe. m. 98-99°.[53,55]

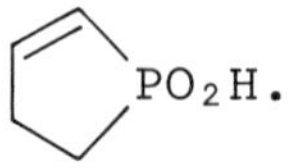

Me ester. XXb. $b_{0.08}$ given as 143-146°, appears to be an error, n_D^{20} 1.4918.[278]
Et ester. XXb. $b_{0.1}$ 75-78°, n_D^{20} 1.4836,[278] kinetics of hydrolysis.[158]
o-$O_2N{\cdot}C_6H_4$ ester. XXIe. $b_{0.037}$ 173-176°, d_4^{20} 1.3557, n_D^{20} 1.5760.[53,55]
m-$O_2N{\cdot}C_6H_4$ ester. XXIe. m. 84-86°.[55]
p-$O_2N{\cdot}C_6H_4$ ester. XXIe. m. 105-107°.[53,55]
2,4-$(O_2N)_2C_6H_3$ ester. XXIe. m. 105-107°.[55]

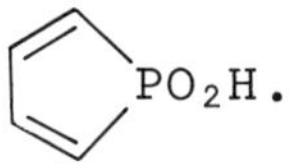

Et ester. By Hofmann elimination from 4-Me_3N^+-2,3-unsaturated ester[771]. By XXXe from 4-Br-2,3-unsaturated ester[771] or from 3,4-diBr ester.[398] λ_{max} 293nm (ε 1050),[771] rapidly dimerizes.[398,771]

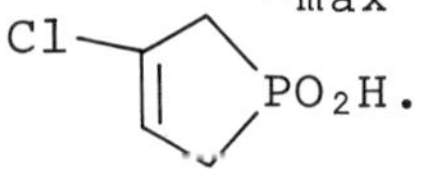

$ClCH_2{\cdot}CH_2$ ester. Xa. $b_{0.02}$ 131-134°, n_D^{20} 1.5142.[302]
$ClCH_2{\cdot}CHMe$ ester. Xa. b_{10} 169-170°, d_4^{20} 1.3327, n_D^{20} 1.5130.[672]

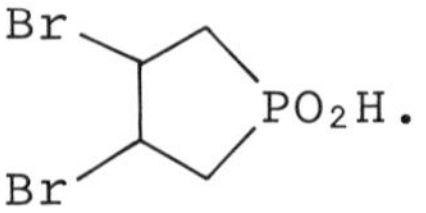

Ph ester. XXXa. m. 119-121°.[43]

Et ester. XXXa. $b_{0.03}$ 87-88°, d_4^{20} 1.2466, n_D^{20} 1.4820.[37]
Ph ester. XXa. Product is mixture of two stereoisomers, one (55%) m. 46-49°, $b_{0.04}$ 131-132°, d_4^{20} 1.2944, n_D^{20} 1.5578, and the other (31%) m. 65-66°.[37]

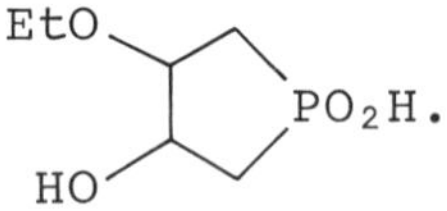

Et ester. From ester of 3,4-epoxy acid + EtOH. $b_{0.04}$ 148-150°, d_4^{20} 1.1726, n_D^{20} 1.4757.[36]

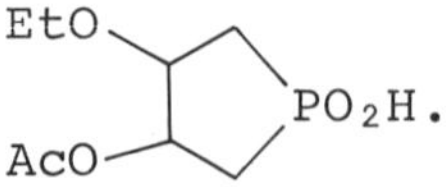

Et ester. From previous ester + Ac_2O. $b_{0.035}$ 110-111°, d_4^{20} 1.1601, n_D^{20} 1.4639.[36]

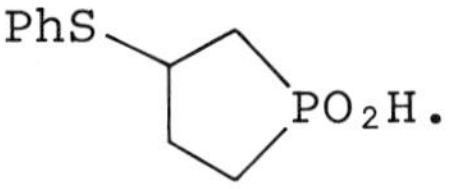

Et ester. From ester of 2,3-unsaturated acid + PhSH. $b_{0.005}$ 180°, d^{20} 1.1916, n_D^{20} 1.5697.[601]

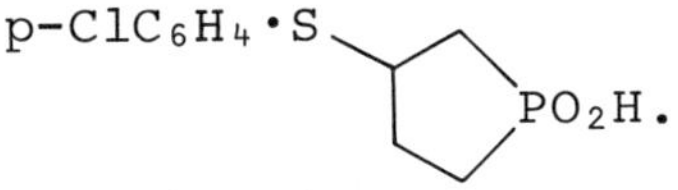

Et ester. From ester of 2,3-unsaturated acid + p-ClC_6H_4SH. $b_{0.01}$ 200°, d^{20} 1.2676, n_D^{20} 1.5750.[601]

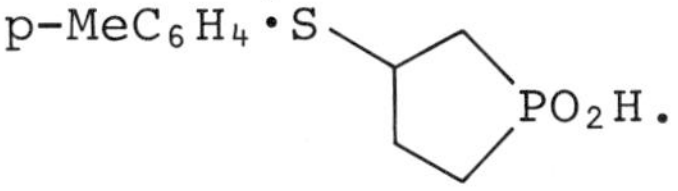

Et ester. From ester of 2,3-unsaturated acid + p-MeC_6H_4SH. $b_{0.005}$ 200°, d^{20} 1.1712, n_D^{20} 1.5611.[601]

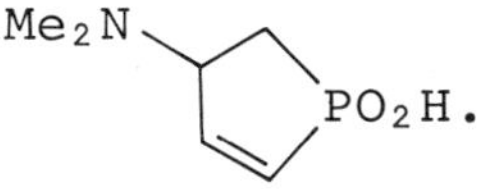

Et ester. XXXd. Oil; methiodide, m. 141-141.5°.[771]

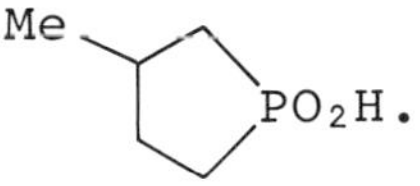

Et ester. Hydrogenation of ester of 3,4-unsaturated acid. $b_{0.7}$ 80-81°, n_D^{20} 1.4588.[278]

$ClCH_2 \cdot CHMe$ ester. Hydrogenation of ester of 3,4-unsaturated acid. b_{10} 149.5-150.5°, d_4^{20} 1.1490, n_D^{20} 1.4780.[672]

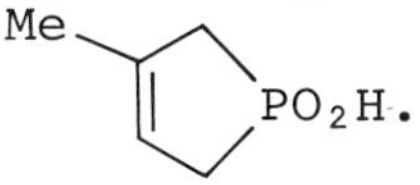

Free acid unknown, hydrolysis of chloride or bromide gives 2,3-unsaturated isomer.[42,778]

Me ester. Xa.[426] XXIe.[42,778] XXIIIc.[675] $b_{0.5}$ 86-87°,[426] b_{10} 117°,[42,778] d_4^{20} 1.1343,[42,778] n_D^{20} 1.4847.[42,778]

Et ester. Xa.[278,302,670] XXIe.[778] XXIIIc.[675] $b_{0.07}$ 74-76°,[302] b_9 118-119°,[778] d_4^{20} 1.0885,[778] n_D^{20} 1.4786.[778]

Pr ester. XXIe. b_8 123-124°, d_4^{20} 1.0584, n_D^{20} 1.4723.[778]

$CH_2{:}CH \cdot CH_2$ ester. XXIe. b_9 130-132°, d_4^{20} 1.0910, n_D^{20} 1.4919.[675]

Bu ester. XXIe. b_8 136-137°, d_4^{20} 1.0353, n_D^{20} 1.4732.[778]

C_5H_{11} ester. XXIe. b_8 146-147°, d_4^{20} 1.0201, n_D^{20} 1.4725.[778]

cyclo-C_6H_{11} ester. XXIe. $b_{0.02}$ 124-126°, d_4^{20} 1.0850, n_D^{20} 1.4990.[778]

$PhCH_2$ ester. XXIe. $b_{0.06}$ 138-139°, d_4^{20} 1.1465, n_D^{20} 1.5454.[778]
$ClCH_2 \cdot CH_2$ ester. Xa. b_{10} 161.5-162.5°, d_4^{20} 1.2390, n_D^{20} 1.4949.[672]
$MeO \cdot CH_2 \cdot CH_2$ ester. Xa. $b_{0.8}$ 109°, d_4^{20} 1.1329, n_D^{20} 1.4693.[184]
$EtS \cdot CH_2 \cdot CH_2$ ester. Xa. b_1 125°, d_4^{20} 1.1406, n_D^{20} 1.5145.[186]
$ClCH_2 \cdot CHMe$ ester. Xa. b_{10} 155-156°, d_4^{20} 1.1881, n_D^{20} 1.4948.[672]
$BrCH_2 \cdot CHMe$ ester. Xa. $b_{0.08}$ 110-112°, d_4^{20} 1.4204, n_D^{20} 1.5089.[781]
$ClCH_2 \cdot CH_2 \cdot CHMe$ ester. Xa. $b_{0.07}$ 68-70°, d_4^{20} 1.1723, n_D^{20} 1.4885.[674]
$BrCHMe \cdot CHMe$ ester. Xa. $b_{0.08}$ 109-111°, d_4^{20} 1.3426, n_D^{20} 1.5035.[781]
$ClCH_2 \cdot CMe_2 \cdot CH_2$ ester. Xa. $b_{0.2}$ 96-97°, d_4^{20} 1.1401, n_D^{20} 1.4900.[674]
$BrCH_2 \cdot CMeBu \cdot CH_2$ ester. Xa. $b_{0.08}$ 135-138°, d_4^{20} 1.2250, n_D^{20} 1.4985.[781]
Ph ester. Xa.[43] XXIe.[87] m. 71°,[87] b_2 152-154°,[87] d_4^{20} 1.1468,[87] n_D^{20} 1.5420.[87]
$o\text{-}O_2N \cdot C_6H_4$ ester. XXIe. $b_{0.06}$ 176-178°, d_4^{20} 1.3050, n_D^{20} 1.5670.[53,55]
$m\text{-}O_2N \cdot C_6H_4$ ester. XXIe. $b_{0.045}$ 172-175°, d_4^{20} 1.3075, n_D^{20} 1.5685.[53,55]
$p\text{-}O_2N \cdot C_6H_4$ ester. XXIe. $b_{0.015}$ 168°, n_D^{20} 1.5803.[55]
Bu_2 amide. XXIi. $b_{0.017}$ 116-120°, d_4^{20} 0.9745, n_D^{20} 1.4832.[54]
Piperidide. XXIi. $b_{0.013}$ 110-112°, d_4^{20} 1.0920, n_D^{20} 1.5200.[54]
Morpholide. XXIi. m. 52-53°, $b_{0.01}$ 140-141°, n_D^{20} 1.5528.[54]

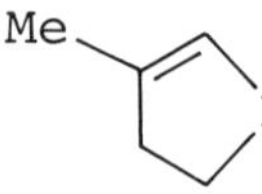
PO_2H. XXa.[278] XXIa.[42,778] m. 95-97°,[278] 116-117°,[42,778] $b_{0.02}$ 174-176°,[42,778] d_4^{20} 1.2165 (supercooled),[42,778] n_D^{20} 1.5129 (supercooled).[42,778]
Et ester. XXb. $b_{0.07}$ 94-95°, n_D^{20} 1.4862.[278]
Bu ester. XXb. $b_{0.2}$ 121-123°, n_D^{20} 1.4771.[278]
C_6H_{13} ester. XXb. $b_{0.02}$ 117-120°, n_D^{20} 1.4712.[278]
$C_{12}H_{25}$ ester. XXb. $b_{0.03}$ 156-157°, n_D^{20} 1.4714.[278]
Ph ester. XXIe. m. 57-59°, $b_{0.05}$ 145-155°, n_D^{20} 1.5560.[278]
Piperidide. XXIi. $b_{0.05}$ 130-134°, n_D^{20} 1.5502.[278]

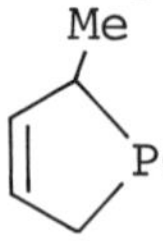
PO_2H.

Et ester. XXIe. b_1 87-90°, d_4^{20} 1.0869, n_D^{20} 1.4778.[87]

Bu ester. XXIe. $b_{1.5}$ 118-121°, d_4^{20} 1.0340, n_D^{20} 1.4725.[87]
MeO·CH$_2$·CH$_2$ ester. Xa. $b_{0.5}$ 105°, d_4^{20} 1.1315, n_D^{20} 1.4680.[184]
PhO·CH$_2$·CH$_2$ ester. XXIe. b_1 174-177°, d_4^{20} 1.1911, n_D^{20} 1.5358.[87]
p-ClC$_6$H$_4$·O·CH$_2$·CH$_2$ ester. XXIe. $b_{1.5}$ 184-187°, d_4^{20} 1.2709, n_D^{20} 1.5408.[87]
2,4-Cl$_2$C$_6$H$_3$·O·CH$_2$·CH$_2$ ester. XXIe. $b_{1.5}$ 200-204°, n_D^{20} 1.5529.[87]
3,4,5-Cl$_3$C$_6$H$_2$·O·CH$_2$·CH$_2$ ester. XXIe. $b_{1.5}$ 230-231°.[87]
3,5-Me$_2$C$_6$H$_3$·O·CH$_2$·CH$_2$ ester. XXIe. $b_{1.5}$ 179-181°, d_4^{20} 1.1190, n_D^{20} 1.5337.[87]
4-Cl-3,5-Me$_2$C$_6$H$_2$·O·CH$_2$·CH$_2$ ester. XXIe. b_1 185-190°, d_4^{20} 1.2079, n_D^{20} 1.5490.[87]
ClCH$_2$·CHMe ester. Xa. b_{10} 144-146°, d_4^{20} 1.1850, n_D^{20} 1.4908.[672]
Ph ester. Xa.[43] XXIe.[87] b_2 145-147°,[87] d_4^{20} 1.1709,[87] n_D^{20} 1.5471.[87]
p-ClC$_6$H$_4$ ester. XXIe. b_1 151-154°, d_4^{20} 1.2844, n_D^{20} 1.5547.[87]
2,4-Cl$_2$C$_6$H$_3$ ester. XXIe. $b_{0.5}$ 160-163°, d_4^{20} 1.3754, n_D^{20} 1.5630.[87]
2,4,5-Cl$_3$C$_6$H$_2$ ester. XXIe. b_1 172-176°.[87]
o-O$_2$N·C$_6$H$_4$ ester. XXIe. $b_{0.02}$ 160-162°, d_4^{20} 1.3014, n_D^{20} 1.5616.[55]
m-O$_2$N·C$_6$H$_4$ ester. XXIe. $b_{0.02}$ 177-179°, d_4^{20} 1.3049, n_D^{20} 1.5650.[55]
p-O$_2$N·C$_6$H$_4$ ester. XXIe. b_1 189-192°, n_D^{20} 1.5770.[87]
4-Cl-3,5-Me$_2$C$_6$H$_2$ ester. XXIe. b_2 171-175°, d_4^{20} 1.2096, n_D^{20} 1.5434.[87]
Et$_2$ amide. XXIi. $b_{1.5}$ 115-118°, d_4^{20} 1.0283, n_D^{20} 1.4919.[87]

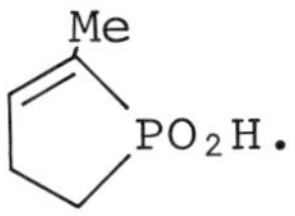

Et ester. XXb. $b_{0.07}$ 85-91°, n_D^{20} 1.4810.[278]

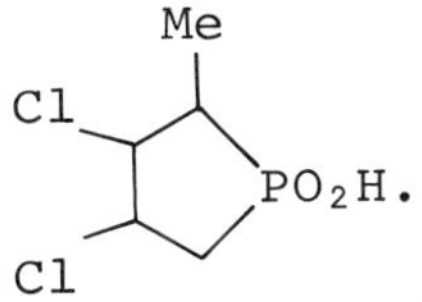

Bu ester. XXIe. b_3 143-148°, d_4^{20} 1.2658, n_D^{20} 1.4965.[87]
Ph ester. XXIe. b_2 160-162°, d_4^{20} 1.3210, n_D^{20} 1.5404.[87]

Me ester. XXXa. m. 53-56°, $b_{0.03}$ 87°, d_4^{20} 1.2432, n_D^{20} 1.4811.[37]
Et ester. XXXa. $b_{0.03}$ 94-95°, d_4^{20} 1.1768, n_D^{20} 1.4748.[37]
Pr ester. XXXa. $b_{0.023}$ 108-109°, d_4^{20} 1.1432, n_D^{20} 1.4721.[37]
i-Pr ester. XXXa. $b_{0.03}$ 91°, d_4^{20} 1.1375, n_D^{20} 1.4697.[37]
Bu ester. XXXa. $b_{0.03}$ 118°, d_4^{20} 1.1132, n_D^{20} 1.4701.[37]

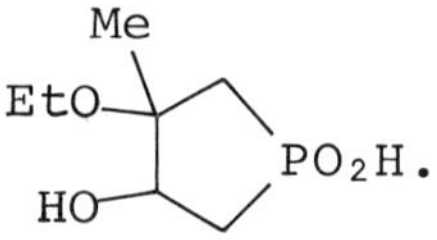

Et ester. From ester of 3,4-epoxy acid + EtOH. $b_{0.02}$ 139-139.5°, d_4^{20} 1.411, n_D^{20} 1.4733.[36]

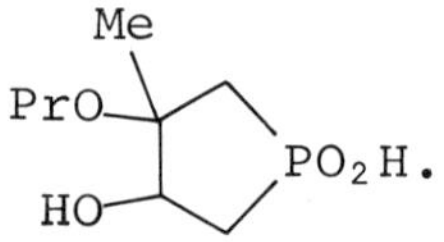

Pr ester. From ester of 3,4-epoxy acid + PrOH. $b_{0.05}$ 155-157°, d_4^{20} 1.0743, n_D^{20} 1.4657.[36]

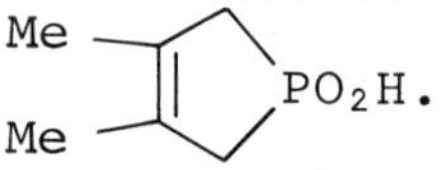 XXa.[52] XXIa.[43,778] XXIIIa.[44] m. 122-123°,[52,778] $b_{0.05}$ 168-169°.[52,778]

Me ester. Xa.[426] XXb.[302,508] XXIe.[48,52] $b_{0.1}$ 84-85°,[508] $b_{0.9}$ 99-100°,[508] b_{10} 131°,[48,52] d_4^{20} 1.1072,[48,52] n_D^{20} 1.4892.[48,52]
Et ester. XXb.[507] XXIe.[48] m. 42-43°,[48] $b_{0.2}$ 85-87°.[48]
Pr ester. XXIe. b_9 135-136°, d_4^{20} 1.0482, n_D^{20} 1.4778.[48]
$ClCH_2 \cdot CH_2$ ester. Xa.[44,302] b_1 120-121°, d_4^{20} 1.2021, n_D^{20} 1.5020.[44]
$MeO \cdot CH_2 \cdot CH_2$ ester. Xa. $b_{0.8}$ 130°, d_4^{20} 1.0960, n_D^{20} 1.4623.[184]
$NCS \cdot CH_2 \cdot CH_2$ ester. Xa. b_1 175°, d_4^{20} 1.2260, n_D^{20} 1.5280.[185]
$ClCH_2 \cdot CHCl \cdot CH_2$ ester. Xa. b_5 154-156°, d_4^{20} 1.2692, n_D^{20} 1.5083.[45]
$ClCHMe \cdot CH_2$ ester. Xa. $b_{3.5}$ 135-136°, d_4^{20} 1.1598, n_D^{20} 1.4960.[44]
$ClCH_2 \cdot CH_2 \cdot CHMe$ ester. Xa. $b_{0.1}$ 95-97°, d_4^{20} 1.1585, n_D^{20} 1.4892.[674]
$ClCHMe \cdot CHMe$ ester. Xa. b_3 144-145°, d_4^{20} 1.1383, n_D^{20} 1.4914.[45]

$ClCH_2 \cdot CMe_2 \cdot CH_2$ ester. Xa. $b_{0.5}$ 120-122°, d_4^{20} 1.1182, n_D^{20} 1.4922.[674]

Ph ester. Xa. $b_{2.5}$ 146-147°, d_4^{20} 1.1512, n_D^{20} 1.5480.[43]

$o\text{-}O_2N \cdot C_6H_4$ ester. XXIe. m. 86-88°.[53,55]

$p\text{-}O_2N \cdot C_6H_4$ ester. XXIe. m. 83-85°.[55]

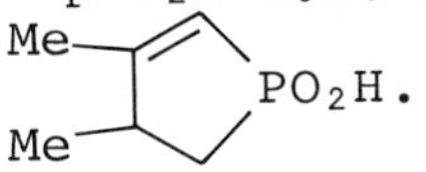

XXa.[52,278] XXIIIa.[52] m. 69-70°,[278] $b_{0.03}$ 195°,[52] n_D^{20} 1.5115.[52]

Me ester. XXb.[52,278] XXIIc.[278] $b_{0.03}$ 83-85°,[278] b_{10} 134-135°,[52] d_4^{20} 1.1062,[52] n_D^{20} 1.4882.[52]

Et ester. XXb. $b_{0.07}$ 97°, n_D^{20} 1.4800.[278]

C_6H_{13} ester. XXb. $b_{0.4}$ 122-126°, n_D^{20} 1.4710.[278]

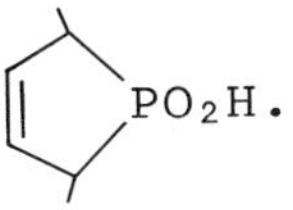

$ClCH_2 \cdot CHCl \cdot CH_2$ ester. Xa. b_6 134-136°, d_4^{20} 1.1912, n_D^{20} 1.4850.[45]

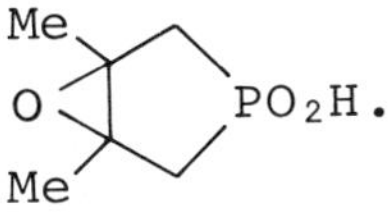

Et ester. XXXa. m. 79-80°.[37]

Bu ester. XXXa. $b_{0.025}$ 111-112°, d_4^{20} 1.0888, n_D^{20} 1.4681.[37]

Ph ester. XXXa. m. 93-95°, $b_{0.01}$ 140-142°.[37]

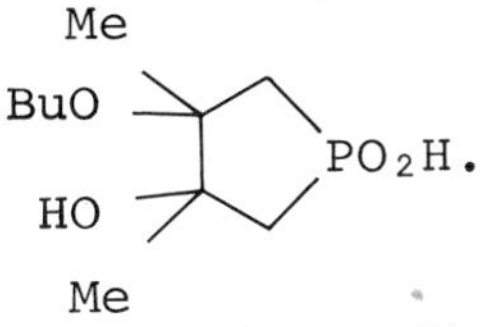

Bu ester. From ester of 3,4-epoxy acid + BuOH. $b_{0.045}$ 149-149.5°, d_4^{20} 1.0660, n_D^{20} 1.4751.[36]

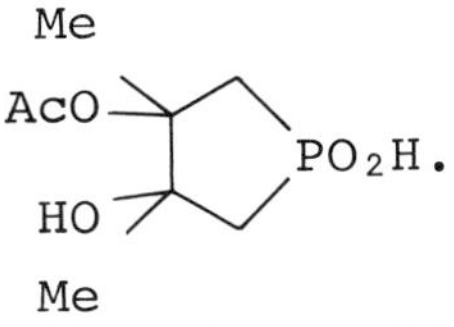

Et ester. From ester of 3,4-epoxy acid + AcOH. $b_{0.035}$ 162.5-163°, n_D^{20} 1.4720.[36]

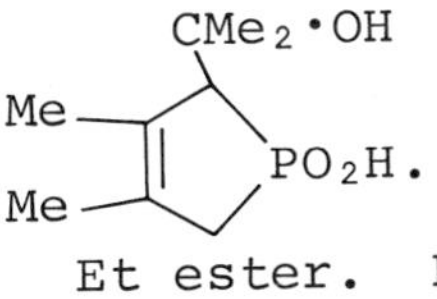

Et ester. From Li deriv. of ester of 3,4-diMe acid + Me_2CO. m. 80°.[507]

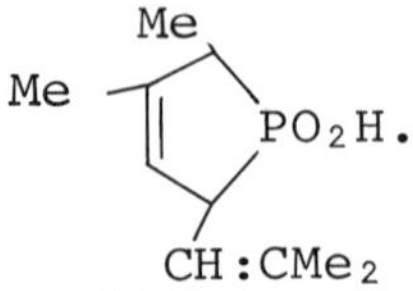

$ClCH_2{\cdot}CH_2$ ester. Xa. b_3 154-156°, d_4^{20} 1.1234, n_D^{20} 1.5026.[45]

$ClCH_2{\cdot}CHCl{\cdot}CH_2$ ester. Xa. m. 86-87°, b_3 164-165°, d_4^{20} 1.1666, n_D^{20} 1.5120.[45]

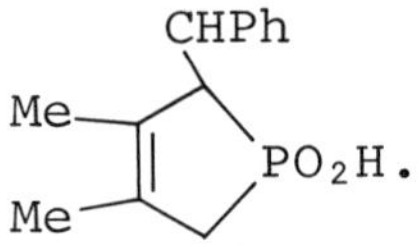

Et ester. From Li deriv. of ester of 3,4-diMe acid + $PhCH_2Br$. $b_{0.000025}$ 95-100°, n_D^{20} 1.5340.[507]

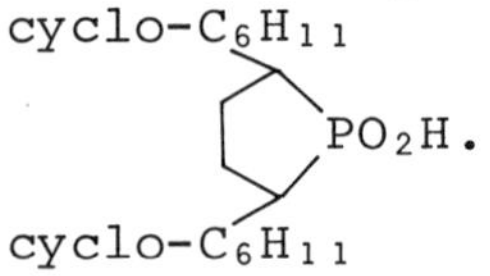

XIa. m. 272-276°.[189]

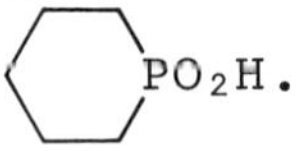

IIIa.[706] VIc.[414] XXIIIa.[15] m. 128-129°,[414,706] pK_a 2.73.[421]

Et ester. Va. b_{10} 109°, n_D^{20} 1.4489, kinetics of hydrolysis.[15]

Bu ester. XXIe. b_1 80°, d_4^{30} 0.9771, n_D^{30} 1.4405.[414]

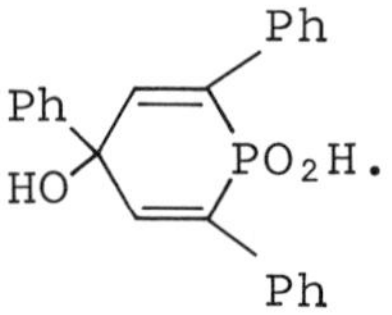

Me ester. By reaction of 1.1-dimethoxy-2,4,6-triphenylphospha(V)benzene with LiBr, Ac_2O, and H_2O_2, two stereoisomers. m. 194-198°, and m. 198-200°.[164]

Et ester. By Diels-Alder reaction. Oil, characterized by IR.[771]

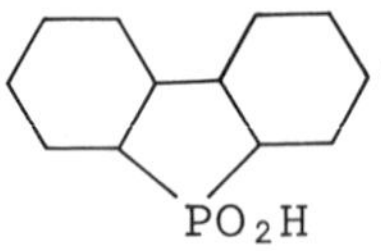

By hydrogenation of acid (below) having two benzenoid rings. m. 153-154.5°.[214]

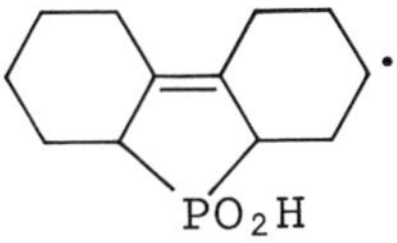

Et ester (C=C may be 2,3 in 5-membered ring). XXb. $b_{0.1}$ 162-167°, n_D^{20} 1.5170.[278]
$ClCH_2 \cdot CH_2$ ester. Xa. m. 186-188°.[45]
$ClCH_2 \cdot CHCl \cdot CH_2$ ester. Xa. m. 110-110.5°.[45]

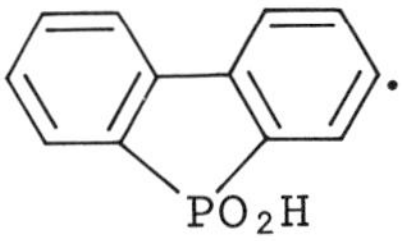

IVe.[169] Also by reduction of $(o\text{-}BrC_6H_4)_2$-PO_2H,[211] m. 253-257°,[211] UV.[211]

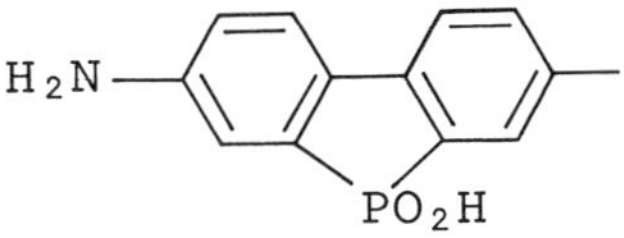

XXXi. m. > 300°, UV.[214]

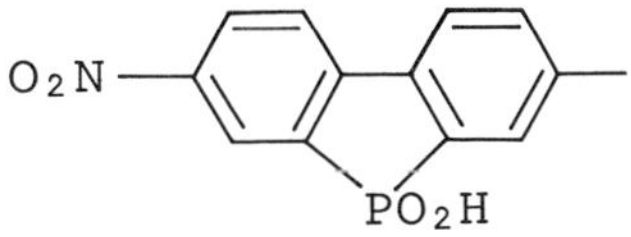

XXXc. Decomp. > 260°, UV.[214]

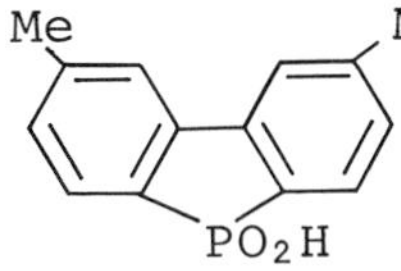

By reduction of $(2\text{-}Br\text{-}4\text{-}MeC_6H_3)_2PO_2H$. m. 303-305°, UV.[214]

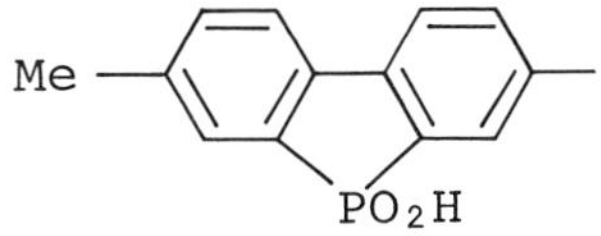

By reduction of $(2\text{-}Cl\text{-}5\text{-}MeC_6H_3)_2$-$PO_2H$. m. 325-328°, UV.[214]

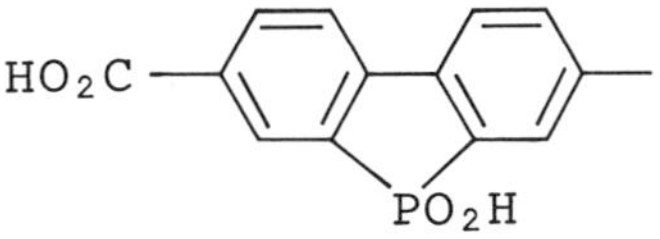

XXXb. m. > 300°, UV.[214]

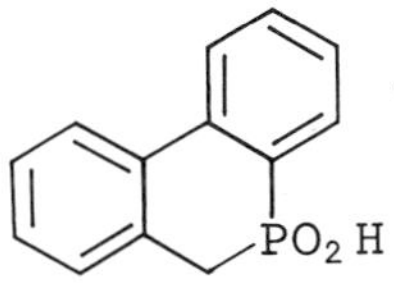

By thermal cyclization (350° for 6 hr) of $o\text{-}Ph \cdot C_6H_4 \cdot CH_2 \cdot PO_3H_2$. Melts indefinitely and resolidifies at 220-225° then m. 236-238°,[474] x-ray structure determination.[785]
Me ester. XXIe. m. 171.5-172.5°, UV.[474]
Ph ester. XXIe. m. 75-77°, $b_{0.1}$ 197-198°.[474]
Anhydride. XXIIg. m. 272-274°.[474]
Ph amide. XXIi. m. 202.5-204°.[474]

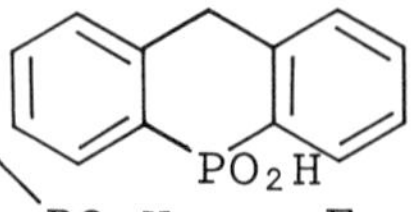 . IVe. Decomp. > 225°, UV.[169]

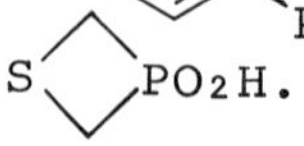 From $(ClCH_2)_2POSH$ + aqueous NaOH. m. 142°.[329]

Et ester. XXIIb. $b_{0.017}$ 60°, d_4^{20} 1.2551, n_D^{20} 1.5078.[329]

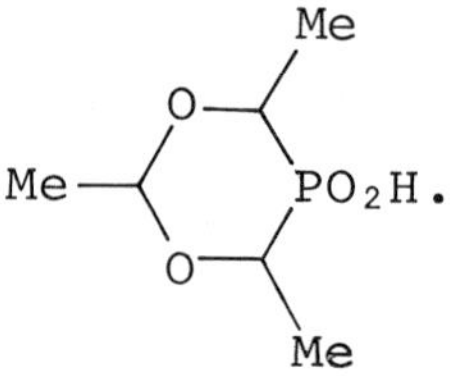 IXa. m. 163-165°.[800]

Me ester. XXIIc. $b_{0.003}$ 72°, d_4^{20} 1.2110, n_D^{20} 1.4663.[800]

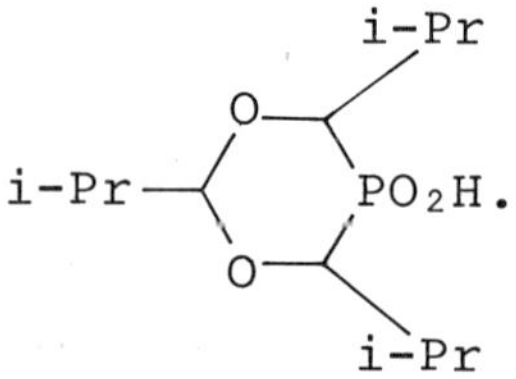 Ia. m. 159-160°.[106]

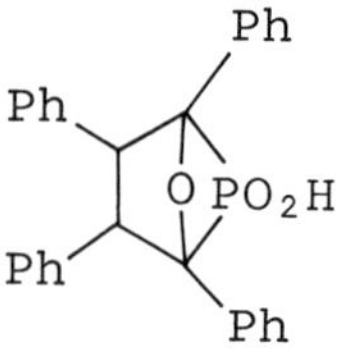

$ClCH_2 \cdot CH_2$ ester. Xa. m. 209-210°.[45]

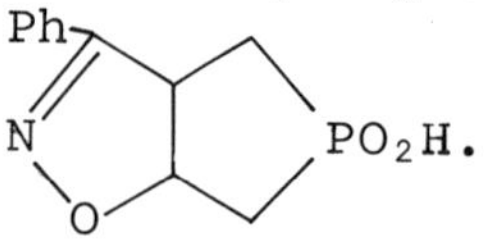

Ph ester. By addition of PhCNO to ester of 3,4-unsaturated monocyclic acid. m. 136-137°.[51]

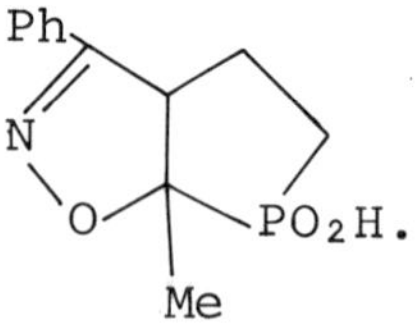

Ph ester. By addition of PhCNO to ester of 3-Me-2,3-unsaturated monocyclic acid. m. 134-135°, NMR indicates that OPh is exo to dihedral angle between rings.[51]

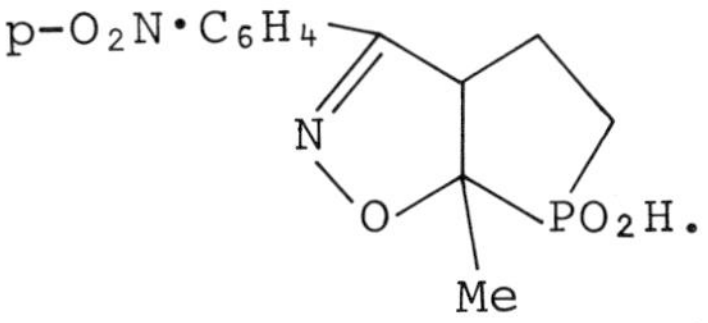

Ph ester. By addition of p-$O_2N \cdot C_6H_4 \cdot$CNO to ester of 3-Me-2,3-unsaturated monocyclic acid. m. 162-163°, NMR indicates that OPh is exo to dihedral angle between rings.[51]

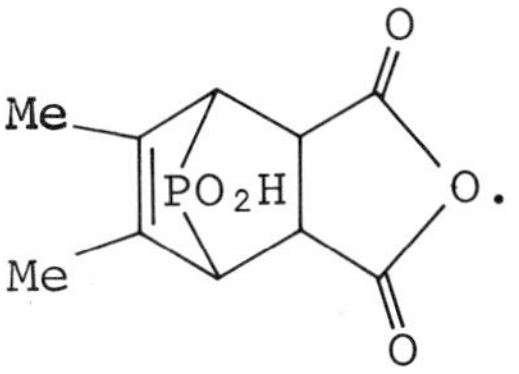 By Diels-Alder reaction between unisolated 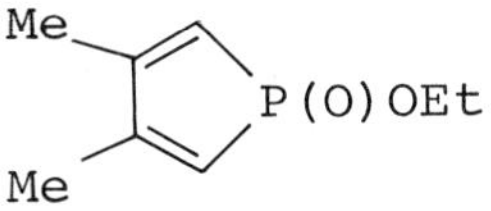 and maleic anhydride and spontaneous hydrolysis of the resultant ester. m. 261-263°.[398]

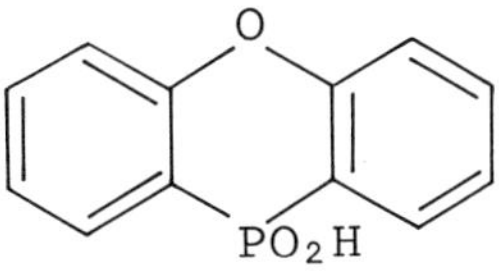 . IVe.[169,458] m. 231-234°,[169] UV.[169]

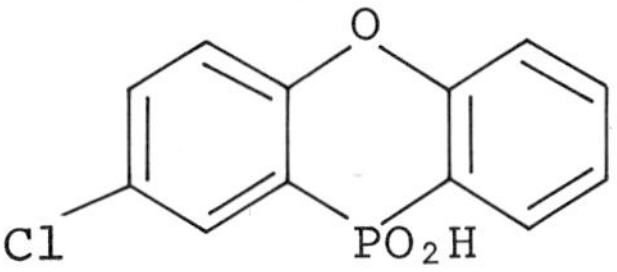 . IVe. m. 240-242°.[458]

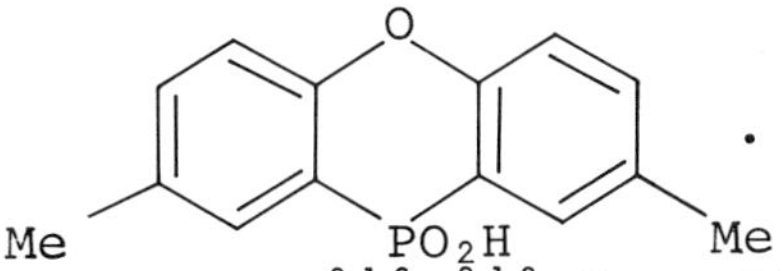 . IVe. On unisolated product from (p-$MeC_6H_4)_2O$, PCl_3 and $AlCl_3$.[219] m. > 300°,[219] UV;[216,219] Cr salt;[404] Co, Ni, and Cu salts;[403] Zn salt.[403,405]

$Et_2N \cdot CH_2 \cdot CH_2$ ester. XXIe. Hydrochloride, m. 186°.[259]

i-$Pr_2N \cdot CH_2 \cdot CH_2$ ester. XXIe. Hydrochloride, m. 158°.[259]

$Me_2N \cdot (CH_2)_3$ ester. XXIe. Hydrochloride, m. 207°.[259]

3-quinuclidinyl ester. XXIe. m. 148°; hydrochloride, m. 295°.[259]

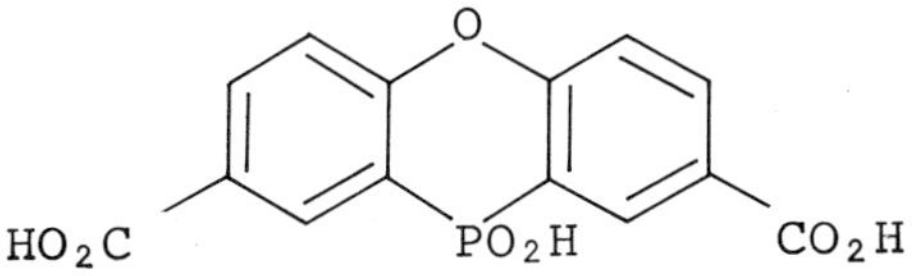 . XXXb. m. > 300°, UV.[216]

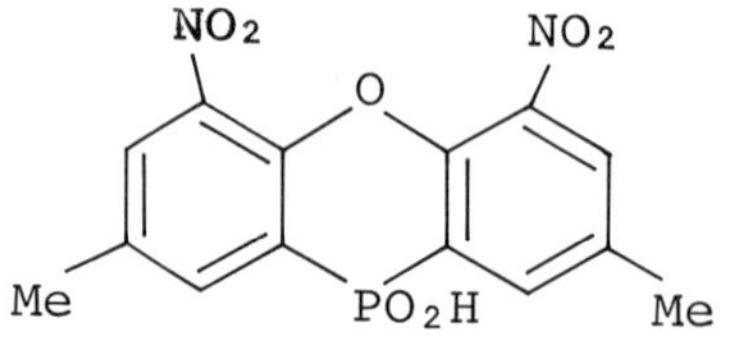

. XXXc (orientation not proved). m. > 300°, UV.[216]

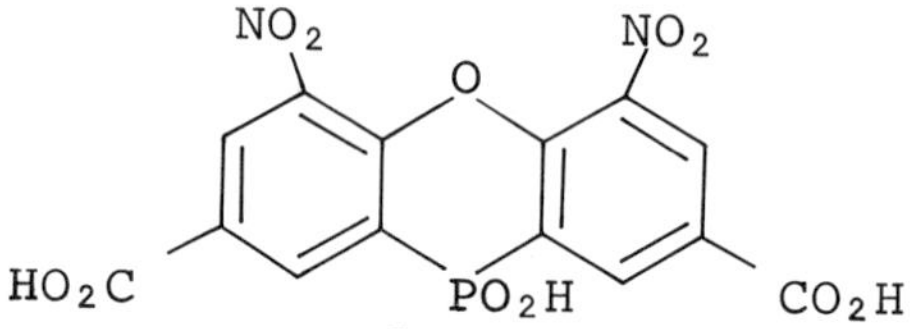

. XXXc. m. > 300°, UV.[216]

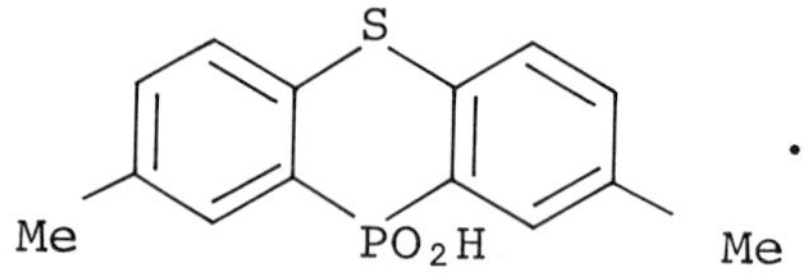

. IVe. On unisolated product from $(p\text{-}MeC_6H_4)_2S$, PCl_3, and $AlCl_3$. m. > 300°, UV.[216]

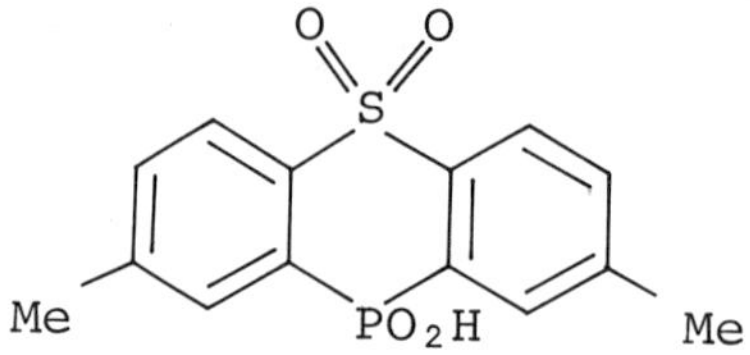

. From previous acid + H_2O_2, m. > 300°. UV.[216]

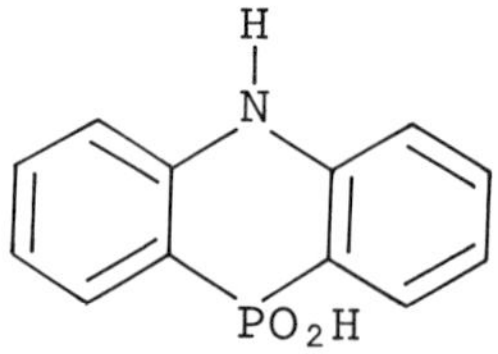

. IIa.[273,724] m. 277° (preheated block).[273]

Me ester. XXIe. XXIIa. XXIIc. m. 224-225°.[273]
Bu ester. XXIe. m. 139.5-141°.[273]
$PhCH_2$ ester. XXIIa. m. 179-180°.[273]
$Et_2N{\cdot}CH_2{\cdot}CH_2$ ester. XXIe. (2 1/2 H_2O), m. 80-83.5°; methiodide, m. 163-164°.[273]
$Et_2N{\cdot}(CH_2)_3$ ester. XXIe. ($2H_2O$), m. 42-43.5°.[273]
Anhydride. XXIf. Unmelted at 300°.[273]
Mixed anhydride with $PhCO_2H$. XXIIi. m. > 340°.[273]
$Et_2N{\cdot}CH_2{\cdot}CH_2$ amide. XXIi. ($1H_2O$), m. 191-194°; methiodide, m. 205°.[273]
$Me_2N{\cdot}(CH_2)_3$ amide. XXIi. ($1H_2O$), m. 185-186°.[273]
N-Me-piperazide. XXIi. m. 230-235°.[720]

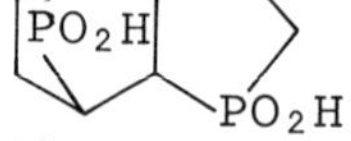

.

di-Et ester. By hydrogenation of the following ester. m. 80-81°,[398] kinetics of acid and alkaline hydrolyses.[399]

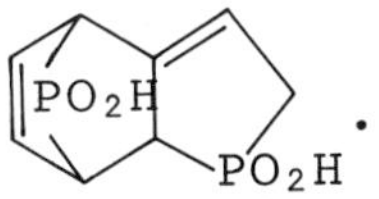

di-Et ester. By Diels-Alder dimerization of the ester of the monocyclic di-unsaturated acid. m. 125-126°,[398] kinetics of acid and alkaline hydrolyses.[399]

E.1.13. Acids Containing Two or More Phosphinic Acid Groups

$Et\cdot PO_2H\cdot CH_2\cdot PO_2H\cdot Et$.
Et_2 ester. Va. $b_{0.5}$ 125-127°.[285]
$Ph\cdot PO_2H\cdot CH_2\cdot PO_2H\cdot Ph$. Vb.[150,499] m. 233-234°.[150]
$Me\cdot PO_2H\cdot CH_2\cdot CH_2\cdot PO_2H\cdot Me$. XXIIIa. m. 189-191°.[497]
i-Pr_2 ester. Va. (±) and meso forms, m. 84-86° and m. 115-117°.[497]
$Et\cdot PO_2H\cdot CH_2\cdot CH_2\cdot PO_2H\cdot Et$. XXIIIa. m. 163-165°.[497]
i-Pr_2 ester. Va. (±) and meso forms, m. 50-52° and m. 83-85°.[497]
$Pr\cdot PO_2H\cdot CH_2\cdot CH_2\cdot PO_2H\cdot Pr$. XXIIIa. m. 181-183°.[497]
i-Pr_2 ester. Va. m. 55-56°.[497]
$Bu\cdot PO_2H\cdot CH_2\cdot CH_2\cdot PO_2H\cdot Bu$. XXIIIa. m. 194-196°.[497]
i-Pr ester. Va. m. 59-61°.[497]
$C_5H_{11}\cdot PO_2H\cdot CH_2\cdot CH_2\cdot PO_2H\cdot C_5H_{11}$. Vb. m. 189-190°.[501]
$C_6H_{13}\cdot PO_2H\cdot CH_2\cdot CH_2\cdot PO_2H\cdot C_6H_{13}$. Vb. m. 185-187°.[501]
cyclo-$C_6H_{11}\cdot PO_2H\cdot CH_2\cdot CH_2\cdot PO_2H\cdot$cyclo-$C_6H_{11}$. Ia. m. 266-271°.[321]
$Ph\cdot PO_2H\cdot CH_2\cdot CH_2\cdot PO_2H\cdot Ph$. Ia.[318,321] Vb.[275] XXIIIa.[9] m. 265-268°.[9]
Et_2 ester. Va. m. 68-70°.[493]
i-Pr_2 ester. Va.[498] Ve.[9] (±) and meso forms, m. 113-114° and m. 146-147°.[9]
Bu_2 ester. Ve. m. 87-88°.[9]
i-Bu_2 ester. Ve. m. 94-95°.[9]
Ph_2 ester. Ve. (±) and meso forms, m. 68-69° and m. 101-102°.[9]
di-amide. XXIi. m. 202-204°.[498]
di-(Et_2 amide). XXIi. m. 195-208°.[498]
di-(Bu amide). XXIi. m. 185-190°.[498]
di-(Ph amide). XXIi. m. 240-246°.[498]
Cyclic imide. XXVa. m. 240-243°.[498]
$PhCH_2\cdot PO_2H\cdot CH_2\cdot CH_2\cdot PO_2H\cdot CH_2Ph$. XXIIIa.[62]
i-Pr_2 ester. (±) and meso forms, m. 99-103° and m. 168-171°.[62]
$C_8H_{17}\cdot PO_2H\cdot CH_2\cdot CH_2\cdot PO_2H\cdot C_8H_{17}$. Vb. m. 177-179°.[501]
$C_{12}H_{25}\cdot PO_2H\cdot CH_2\cdot CH_2\cdot PO_2H\cdot C_{12}H_{25}$. Vb. m. 164-165°.[501]
trans-$Ph\cdot PO_2H\cdot CH{:}CH\cdot PO_2H\cdot Ph$.
Et_2 ester. Va from trans-$PhSO_2\cdot CH{:}CHBr$ or trans-$PhSO_2\cdot CH{:}CH\cdot PO_2Et\cdot Ph$. m. 135-136°.[365]

Ph·PO_2H·$(CH_2)_3$·PO_2H·Ph. Vb.[499] XXIIIa.[9] m. 157-159°.[499]
i-Pr_2 ester. Ve. m. 86-87°.[9]
Ph·PO_2H·$(CH_2)_4$·PO_2H·Ph. Vb.[499] XXIIIa.[9] m. 223-225°.[9,499]
i-Pr_2 ester. Ve. (±) and meso forms, m. 76.5-77.5° and m. 83-84°.[9]
i-Bu_2 ester. Ve. m. 88-89°.[9]
Ph·PO_2H·$(CH_2)_5$·PO_2H·Ph. Vb. m. 124-126°.[499]
Et·PO_2H·$(CH_2)_6$·PO_2H·Et.
Et_2 ester. XIIb. $b_{0.07}$ 190°, d_4^{20} 1.0512, n_D^{20} 1.4690.[631]
cyclo-C_6H_{11}·PO_2H·$(CH_2)_6$·PO_2H·cyclo-C_6H_{11}. XXIIf. m. 140-142°.[310]
Ph·PO_2H·$(CH_2)_6$·PO_2H·Ph. Vb. m. 159-161°.[499]
Et·PO_2H·CHBu·CH_2·PO_2H·Et. XIIb. $b_{0.04}$ 140-142°, d_4^{20} 1.0328, n_D^{20} 1.4675.[638]
Et·PO_2H·CH(CH:CH_2)·CHMe·CH_2·PO_2H·Et.
Et_2 ester. By base-catalyzed dimerization of (CH_2:CH·CH_2)EtP(O)OEt. $b_{1.5}$ 172-173°, d_4^{20} 1.0732, n_D^{20} 1.4782.[616]
Pr_2 ester. By base-catalyzed dimerization of (CH_2:CH·CH_2)EtP(O)OPr. b_2 192°, d_4^{20} 1.0471, n_D^{20} 1.4697.[616]
o-C_6H_4(PO_2H·Ph)$_2$. IIIb. m. 197-198°; mono-benzylisothiouronium salt, m. 201-202°.[483]
p-C_6H_4(PO_2H·Me)$_2$. VIf.[183] XXIIIa.[60] m. 233-235°,[60] pKa 2.67.[183]
Me_2 ester. Va. m. 107-109°.[60]
p-C_6H_4(PO_2H·Ph)$_2$. VIc.[58] VIf.[287] m. 213-215°,[58] 330-333°.[287]
p-C_6H_4(CH_2·PO_2H·Me)$_2$.
Et_2 ester. Va. m. 139°.[743]
p-C_6H_4(CH_2·PO_2H·Et)$_2$.
Et_2 ester. Va. m. 65°.[516]
p-C_6H_4(CH_2·PO_2H·Ph)$_2$.
Et_2 ester. Va.[516,751] m. 151-153.5°,[516] 186-187°.[751]
p-C_6H_4[CH(OH)·PO_2H·Ph]$_2$. IXa. m. 225-226°.[640]
Bu_2 ester. IXc. m. 204-206°.[85]
biphenyl-4,4'-(PO_2H·Me)$_2$. VIc. m. 307-310°.[60]
biphenyl-4,4'-(PO_2H·Ph)$_2$. VIc. m. 213-215°.[58]
Ph·PO_2H·CHPh·(p-C_6H_4)·PO_2H·Ph. From PhPCl_2, Ph_2CO, and moist $AlCl_3$.[222]
Me_2 ester. XXIe. m. 190-195°.[222]
Et·CHF·CH(PO_2H·Me)$_2$.
Et_2 ester. From MeP$(OEt)_2$ + ClCH:CH·SO_2F. $b_{0.00425}$ 142-147°, d_4^{20} 1.1802, n_D^{20} 1.4712.[253]

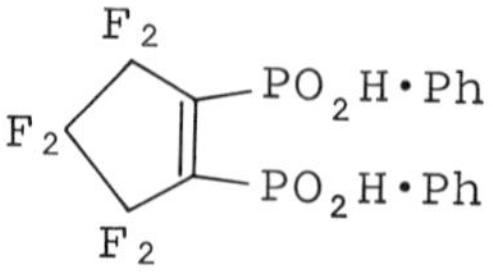

Et_2 ester. Va. $b_{0.45}$ 174-176°.[206]

$Ph\cdot PO_2H\cdot CMe_2\cdot NH\cdot CH_2\cdot CH_2\cdot NH\cdot CMe_2\cdot PO_2H\cdot Ph$.
    Et_2 ester. IXd. Dipicrate, m. 152° decomp.[347]
$p\text{-}C_6H_4(NH\cdot CHPh\cdot PO_2H\cdot C_6H_4\text{-}p\text{-}Me)_2$.
    Et_2 ester. IXd. m. 213-215°.[636]
    Pr_2 ester. IXd. m. 210-211°.[636]
$Ph\cdot PO_2H\cdot CH_2\cdot PO_2H\cdot CH_2\cdot PO_2H\cdot Ph$.
    Bu, Et, Bu ester. Va. Undistilled at 123°/0.05 torr, n_D^{20} 1.5318, ^{31}P -38.8 ppm (central P) and -33.6 ppm.[478]

Et esters:
$-[CH_2\cdot PO_2H\cdot (CH_2)_4\cdot PO_2H]-n$
$-[(CH_2)_4\cdot PO_2H\cdot (CH_2)_4\cdot PO_2H]-n$
$-[CH_2\cdot CH_2\cdot PO_2H\cdot (CH_2)_6\cdot PO_2H]-n$
$-[(CH_2)_4\cdot PO_2H\cdot (CH_2)_6\cdot PO_2H]-n$
$-[CH_2\cdot CH{:}CH\cdot CH_2\cdot PO_2H\cdot (CH_2)_6\cdot PO_2H]-n$
$-[CH_2\cdot (p\text{-}C_6H_4)\cdot CH_2\cdot PO_2H\cdot (CH_2)_6\cdot PO_2H]-n$
$-[CH_2\cdot CH_2\cdot O\cdot CH_2\cdot CH_2\cdot PO_2H\cdot (CH_2)_6\cdot PO_2H]-n$
$-[CH_2\cdot CH(OEt)\cdot PO_2H\cdot (CH_2)_6\cdot PO_2H]-n$
$-[CO\cdot (CH_2)_4\cdot CO\cdot PO_2H\cdot (CH_2)_6\cdot PO_2H]-n$
$-[CO\cdot (p\text{-}C_6H_4)\cdot CO\cdot PO_2H\cdot (CH_2)_6\cdot PO_2H]-n$
$-[SiMe_2\cdot PO_2H\cdot (CH_2)_6\cdot PO_2H]-n$
$-[P(O)Bu\cdot PO_2H\cdot (CH_2)_6\cdot PO_2H]-n$

Me, Et, Pr, i-Pr, and Bu esters:
$-[CH_2\cdot (p\text{-}C_6H_4)\cdot CH_2\cdot PO_2H\cdot (CH_2)_4\cdot PO_2H]-n$
$-[CH_2\cdot (p\text{-}C_6H_4)\cdot CH_2\cdot PO_2H\cdot (CH_2)_6\cdot PO_2H]-n$

The above esters were prepared by Arbusov reactions between dihalides corresponding to the first part of the formulae and tetra-alkyl butane-1,4- or hexane-1,6- diphosphonites. The products, with molecular weights up to ∿5000, were viscous liquids, waxes, or crosslinked solids.[689]

E.2. Phosphinothioic Acids and Their Derivatives

E.2.1. Unsubstituted Aliphatic Phosphinothioic Acids

Me_2POSH. IIIc.[141] XXIa.[506,718] m. 41.5-43°;[506] NH_4^+ salt, m. 152°;[718] Na salt, m. 147-148.5°;[506] IR spectra show that Na and Me_4N^+ salts are ionic whilst Zn, Cd, Mn, Ni, and praeseodidymium salts are inner complexes of predominantly thione structure (i.e., Me_2P(–O–)(=S→)metal),[342] analysis of IR spectra of anion and of Ni salt.[509]
    cyclo-C_6H_{11} O-ester. XXIe. m. 62°.[131,718]
    $EtS\cdot CH_2\cdot CH_2$ O-ester. XXIe.[506,718] $b_{0.01}$ 72°,[718] b_2 96-96.5°,[506] d_4^{20} 1.1036,[506] n_D^{20} 1.5334.[506]
    $Et_2N\cdot CH_2\cdot CH_2$ O-ester. XXIe. $b_{0.01}$ 70°.[718]
    $NC\cdot CH_2\cdot CH_2$ O-ester. XXIe. $b_{0.01}$ 84°.[718]
    $EtO_2C\cdot CH{:}CMe$ O-ester. XXIe. m. 58-60°, $b_{0.01}$ 60-61°.[472,718]

$Et_2N \cdot CO \cdot CH{:}CMe$ O-ester. XXIe. $b_{0.01}$ 80°.[472,718]
$\overline{CH_2 \cdot CMe_2 \cdot CH_2 \cdot CO \cdot CH{:}C}$ O-ester. XXIe. m. 90°.[472,718]
Ph O-ester. XXIe. m. 36-38°.[57,562]
o-MeC_6H_4 O-ester. XXIe. $b_{0.3}$ 99-102°.[57,562]
p-MeC_6H_4 O-ester. XXIe. $b_{0.2}$ 103°.[57,562]
2,6-$Me_2C_6H_3$ O-ester. XXIe. m. 94-96°.[57,562]
3,4-$Me_2C_6H_3$ O-ester. XXIe. m. 50-51°.[57,562]
3,5-$Me_2C_6H_3$ O-ester. XXIe.[57,562] m. 55-56°.[562]
p-t-BuC_6H_4 O-ester. XXIe. m. 59-61°.[57,562]
m-ClC_6H_4 O-ester. XXIe. m. 56-57°.[581]
p-ClC_6H_4 O-ester. XXIe.[57,131,562,718] m. 50°.[131,718]
p-$MeO \cdot C_6H_4$ O-ester. XXIe. m. 34-35°.[57]
p-$MeS \cdot C_6H_4$ O-ester. XXIe. m. 53°.[718]
p-$(NC \cdot CH_2 \cdot S)C_6H_4$ O-ester. XXIe. m. 78-80°.[693]
p-$(2,4\text{-}Cl_2C_6H_3 \cdot CH_2 \cdot S)C_6H_4$ O-ester. XXIe. m. 110°.[135]
o-$O_2N \cdot C_6H_4$ O-ester. XXIe. m. 60°.[718]
m-$O_2N \cdot C_6H_4$ O-ester. XXIe. m. 96°.[718]
p-$O_2N \cdot C_6H_4$ O-ester. XXIe.[57,562,718] m. 144-146°.[57,562]
m-$CF_3 \cdot C_6H_4$ O-ester. XXIe. m. 27-30°.[581]
p-$MeO_2C \cdot C_6H_4$ O-ester. XXIe. m. 115-117°.[57]
m-$(H_2N \cdot CO)C_6H_4$ O-ester. XXIe. m. 139°.[718]
o-$(MeNH \cdot CO)C_6H_4$ O-ester. XXIe. m. 96°.[192,712]
p-$NC \cdot C_6H_4$ O-ester. XXIe. m. 76°.[712]
2,4-$Cl_2C_6H_3$ O-ester. XXIe. m. 63-65°.[57,562]
3-Cl-4$(NC \cdot CH_2 \cdot S)C_6H_3$ O-ester. XXIe. m. 39-40°.[693]
2-Cl-4-$O_2N \cdot C_6H_3$ O-ester. XXIe. m. 100-101°.[57,562]
3-Cl-4-$O_2N \cdot C_6H_3$ O-ester. XXIe. m. 57.5-58.5°.[57,562]
2-MeO-4-$O_2N \cdot C_6H_3$ O-ester. XXIe. m. 96°.[193,473]
3-MeO-4-$O_2N \cdot C_6H_3$ O-ester. XXIe. m. 73°.[193,473]
3-Cl-4-$NC \cdot C_6H_3$ O-ester. XXIe. m. 96°.[485]
3-Me-4-$MeS \cdot C_6H_3$ O-ester. XXIe. m. 70°.[131,718]
2-Me-4-$O_2N \cdot C_6H_3$ O-ester. XXIe. m. 98°.[193,473]
3-Me-4-$O_2N \cdot C_6H_3$ O-ester. XXIe. m. 68°.[193,473]
4-NC-2-MeC_6H_3 O-ester. XXIe. m. 132°.[485]
4-NC-3-MeC_6H_3 O-ester. XXIe. m. 91°.[485]
2,4,5-$Cl_3C_6H_2$ O-ester. XXIe.[57,562,718] m. 79°.[718]
3,5-Cl_2-4-$O_2N \cdot C_6H_2$ O-ester. XXIe. m. 132-133°.[171]
4-Cl-3,5-$Me_2C_6H_2$ O-ester. XXIe. m. 62-63°.[57,562]
4-NC-2,6-$(i\text{-}Pr)_2C_6H_2$ O-ester. XXIe. m. 147°.[711]
4-NC-2,6-$(t\text{-}Bu)_2C_6H_2$ O-ester. XXIe. m. 118°.[711]
1-naphthyl O-ester. XXIe. m. 111°.[719]
2-naphthyl O-ester. XXIe. m. 88-89°.[719]
coumarin-4-yl O-ester. XXIe. m. 130-131°.[561]
coumarin-7-yl O-ester. XXIe. m. 155-157°.[561]
4-Me-coumarin-7-yl O-ester. XXIe. m. 132-135°.[561]
3-EtO_2C-coumarin-4-yl O-ester. XXIe. m. 131-134°.[561]
3-Cl-4-Me-coumarin-7-yl O-ester. XXIe. m. 203°.[131,718]
benz-1,2-thiazol-3-yl O-ester. XXIe. m. 106°.[469]
dihydro-3-oxobenz-1,4-thiazin-7-yl O-ester. XXIe.

m. 198-199°.[692]
1-Ph-3-Me-pyrazol-5-yl O-ester. XXIe. m. 78°.[472,718]
2-EtS-4-Me-pyrimidin-6-yl O-ester. XXIe. m. 56-57°.[472,718]
benzfurazan-5-yl O-ester. XXIe. m. 110-111°.[727]
-(p-C_6H_4)S•S(p-C_6H_4)- di O-ester. XXIe. m. 60°.[718]
p-MeC_6H_4•S•CH_2 S-ester. XXIIa. m. 68°.[716,718]
EtS•CH_2•CH_2 S-ester. XXIIa. $b_{0.01}$ 83°.[716,718]
CH_2:CH•S•CH_2•CH_2 S-ester. XXIIe. $b_{0.01}$ 84°.[708]
EtS•CH_2•CHMe S-ester. XXIIa. $b_{0.01}$ 94°.[716,718]
5-Me-2-MeS•C_6H_3•CH_2 S-ester. XXIIa. m. 98°.[716,718]
Amide. XXIi. m. 99-101°.[702]
Aziridide. XXIi. b_2 74-75°, d_4^{20} 1.0827, n_D^{20} 1.5330.[331]
p-MeC_6H_4 amide. XXIi. m. 119-120°.[144]
$Me_2P(S)$•NH•$P(S)Me_2$. XXIj. XXVa. m. 178°, ^{31}P -59.6 ppm;[702] Na salt ($1H_2O$), cubes, m. 103-104°; Na salt ($2H_2O$), needles, m. 112°, ^{31}P -43.8 ppm.[702]
$Ph_2P(S)$•NH•$P(S)Me_2$. XXIj. m. 156-157°, ^{31}P -64.1 and -52.5 ppm, Na salt ($1H_2O$), m. 187-189°, ^{31}P -43.4 and -37.6 ppm.[702]
Ph•CO•NH•NH•$P(S)Me_2$. XXIi. m. 182°.[744]

EtMePOSH. VIg. XXIa. XXIb. $b_{0.075}$ 62-64°, n_D^{23} 1.5325.[646]
$BuCMe_2$ O-ester. IVa. b_2 126-127°, d_4^{20} 0.9242, n_D^{20} 1.4682.[598]
C_6H_{13}•CHMe O-ester. IVa. b_2 147-148°, d_4^{20} 0.9477, n_D^{20} 1.4740.[598]
p-MeS•C_6H_4 O-ester. XXIe. $b_{0.02}$ 114-116°.[131]
p-O_2N•C_6H_4 O-ester. XXIe.[131,476] m. 69-70°.[476]

Et_2POSH. IIIc.[133] VIg.[345,504] XXIa.[98] m. 11.5°,[345,504] $b_{1.5}$ 88.5-89°,[345,504] n_D^{20} 1.5262,[345,504] d_4^{20} 1.1090,[345,504] pK_a 2.54,[446] 2.80 (in 7% EtOH), 4.88 (in 80% EtOH);[345] Na salt ($3H_2O$), m. 58-59°;[438] Co^{2+} salt, intensely blue crystals, m. 186-188°;[438] Co^{2+} salt (4-pyridine), pink, m. 63-64°, loses pyridine in air;[438] Zn salt, m. 186-188°;[438] Zn_4O $(OSPEt_2)_6$, m. 273°,[518] Cd salt, m. 119-122°;[438] In^{3+} salt, m. 133°;[438] Pb^{2+} salt, m. 99-102°;[438] cyclo-C_6H_{11}•NH_2 salt, m. 145-147°.[98]
i-Pr O-ester. XXIe. b_{11} 96.5-97°.[436]
cyclo-C_6H_{11} O-ester. XXIe. $b_{0.1}$ 74°.[718]
C_8H_{17} O-ester. IVa. b_3 129-130°, d_4^{20} 0.9448, n_D^{20} 1.4812.[349]
EtS•CH_2•CH_2 O-ester. XXIe. $b_{0.01}$ 84°.[131,718]
Et_2N•CH_2•CH_2 O-ester. XXIe. $b_{0.01}$ 78°.[131,718]
p-O_2N•C_6H_4 O-ester. XXIe. $b_{0.01}$ 121°.[718]
p-NC•C_6H_4 O-ester. XXIe. $b_{0.01}$ 119°.[192]
C_6Cl_5 O-ester. XXIe. m. 121°.[718]
2-Br-naphth-1-yl O-ester. XXIe. m. 38-41°.[719]
benzfurazan-5-yl O-ester. XXIe. m. 36.5-38.5°.[727]
Bu S ester. IVa. $b_{0.03}$ 68.5-69°, d_4^{20} 1.0068, n_D^{20}

1.4923.[14]

$EtS \cdot CH_2$ S-ester. XXIIa. $b_{0.01}$ 76°.[716,718]

$p\text{-}FC_6H_4 \cdot S \cdot CH_2$ S-ester. XXIIa. $b_{0.01}$ 110°.[191]

$EtS \cdot CH_2 \cdot CH_2$ S-ester. XXIe.[132,718] XXIIa.[716,718] $b_{0.01}$ 79°.[132,718]

$CH_2{:}CH \cdot S \cdot CH_2 \cdot CH_2$ S-ester. XXIIe. $b_{0.01}$ 88°.[708]

$Et_2N \cdot CH_2 \cdot CH_2$ S-ester. XXIIa. $b_{0.02}$ 81°.[716,718]

$EtO_2C \cdot CH_2$ S-ester. XXIIa. $b_{0.01}$ 70°.[716,718]

$MeNH \cdot CO \cdot CH_2$ S-ester. XXIIa. m. 80°.[716,718]

$EtS \cdot CH_2 \cdot CHMe$ S-ester. XXIIa. $b_{0.01}$ 92°.[716,718]

$EtO_2C \cdot CHMe$ S-ester. XXIIa. $b_{0.01}$ 80°.[716,718]

$3,4\text{-}Cl_2C_6H_3 \cdot CH_2$ S-ester. XXIIa. $b_{0.01}$ 117°.[716,718]

$p\text{-}ClC_6H_4$ S-ester. XXIe. $b_{0.01}$ 106°.[132,718]

$Et_2P(S) \cdot O \cdot P(S)Et_2$. XXIf. m. 42.5°, $b_{0.4}$ 146.5-147.5°.[447]

$Me \cdot CO \cdot O \cdot P(S)Et_2$. IVa*. b_2 72.5-73°, d_4^{20} 1.1059, n_D^{20} 1.4992.[344]

Et_2 amide. XXIi. $b_{0.0004}$ 71-72°, d_4^{20} 0.9854, n_D^{20} 1.5122.[13]

Aziridide. XXIi. $b_{0.06}$ 71-72°, d_4^{20} 1.0406, n_D^{20} 1.4700.[33]

Pr_2POSH. VIg.[345,504] XXIa.[504,718] m. 31°,[345,504] b_2 98.5-99°,[345,504] pK_a 2.58,[446] 2.83 (in 7% EtOH), 5.10 (in 80% EtOH);[345] NH_4^+ salt, m. 117-119°;[504] $p\text{-}C_6H_4(NH_2)_2$ salt, m. 125.5-127.5°.[351]

Me O-ester. XXIe. b_{12} 118-119°.[436]

Pr O-ester. IVa. $b_{0.5}$ 81-82°, d_4^{20} 0.9614, n_D^{20} 1.4778,[596] magnetic rotation.[451]

i-Pr O-ester. XXIe. b_{14} 125-125.5°.[436]

$\overline{O \cdot CMe_2 \cdot O \cdot CH_2 \cdot CH} \cdot CH_2$ O-ester. IVa. b_1 141-142°.[596]

1,4:3,6-dianhydro-D-mannitol di-O-ester. IVa. m. 64-65°, $[\alpha]_D^{20}$ +101.3°.[599]

1,4:3,6-dianhydro-D-glucitol di-O-ester. IVa. $b_{0.01}$ 195-200°, n_D^{20} 1.5225, $[\alpha]_D^{20}$ +64.3°.[599]

$p\text{-}MeS \cdot C_6H_4$ O-ester. XXIe. $b_{0.01}$ 117°.[718]

$p\text{-}O_2N \cdot C_6H_4$ O-ester. XXIe. m. 80-81°.[718]

$3\text{-}Me\text{-}4\text{-}MeS \cdot C_6H_3$ O-ester. XXIe. $b_{0.01}$ 127°.[718]

Bu S-ester. From $Pr_2P \cdot OPr$ + BuSCN, b_8 131-133°, d_4^{20} 0.9743, n_D^{20} 1.4630.[597]

$EtS \cdot CH_2 \cdot CH_2$ S-ester. XXIe. $b_{0.01}$ 89°.[132,718]

$Et_2N \cdot CH_2 \cdot CH_2$ S-ester. XXIe. $b_{0.01}$ 82°.[132,718]

$Pr_2P(S) \cdot O \cdot P(S)Pr_2$. XXIf. $b_{0.3}$ 141-142°, n_D^{20} 1.5244.[447]

i-Pr POSH. VIg. m. 76-77.5°,[345,504] pK_a 3.03 (in 7% EtOH), 5.46 (in 80% EtOH).[345]

$(CH_2{:}CH \cdot CH_2)_2POSH$.

Et S-ester. XXIe. $b_{0.04}$ 85-88°, d_4^{20} 1.0543, n_D^{20} 1.5230.[656]

$CH_2{:}CH \cdot CH_2$ S-ester. XXIe. $b_{0.04}$ 90-93°, d_4^{20} 1.0543, n_D^{20} 1.5350.[656]

BuMePOSH.

Et ester. XXIe. b_2 99-101°.[476]

$MeBuP(S) \cdot O \cdot P(S)BuMe$. XXIi. $b_{1.7}$ 191-192°, n_D^{20} 1.5052.[476]

Bu_2POSH. VIg.[345,504] m. 24-25°,[446] $b_{0.25}$ 81-81.5°,[345] pK_a 2.62,[446] 2.91 (in 7% EtOH), 5.14 (in 80% EtOH);[345] $Zn_4O(OSPBu_2)_6$, m. 104°, x-ray structure determination;[485] $PhNH_2$ salt, m. 70-71°.[586]
  $p-O_2N{\cdot}C_6H_4$ O-ester. XXIe. m. 44°.[718]
  $3-Me-4-MeS{\cdot}C_6H_3$ O-ester. XXIe. $b_{0.01}$ 129°.[718]
  Ph amide. XXIi. m. 66-67°.[586]
$i-Bu_2POSH$. VIg. m. 69.5-70.5°,[345,504] pK_a 3.17 (in 7% EtOH), 5.46 (in 80% EtOH);[345] $p-C_6H_4(NH_2)_2$ salt, m. 109-110.5°.[351]
  Bu O-ester. IVa. b_1 88-89°, d_4^{20} 0.9395, n_D^{20} 1.4790.[349]
$s-Bu_2POSH$. VIg. m. 55.5-57°, $b_{0.8}$ 102-102.5°, pK_a 3.10 (in 7% EtOH), 5.71 (in 80% EtOH).[345]
t-BuMePOSH. XXIa. m. 116.5°, ^{31}P -103.9 ppm;[437] Na salt, m. 334-336°, ^{31}P -80.6 ppm;[437] Zn salt, sinters with decomp. 300°, ^{31}P -90.2 ppm.[437]
  Me O-ester. XXIe. m. 90°, ^{31}P -112.0 ppm.[437]
  t-Bu O-ester. XXIe. $b_{0.25}$ 77-78°, ^{31}P -98.2 ppm.[437]
$t-Bu_2POSH$. VIg. m. 144-145°, pK_a 3.91 (in 7% EtOH), 6.09 (in 80% EtOH).[345]
  Ph O-ester. IVa. m. 53-54°.[749]
$(cyclo-C_6H_{11})MePOSH$.
  Ph amide. XXIi. m. 135.5-136°.[339]
$(cyclo-C_6H_{11})_2POSH$. x-ray structure determination.[312]
  Ph S-ester. Id*.[260] IIf.[261] m. 91-93°.[260]
$(C_5H_{11}{\cdot}CH{:}CH)i-PrPOSH$.
  i-Pr O-ester. XIIb. b_{13} 144-145°, d_4^{20} 0.9730, n_D^{20} 1.4720.[630]
$(PhCH_2)EtPOSH$.
  Et_2 amide. Vc. $b_{0.003}$ 134-135°, d_4^{20} 1.0669, n_D^{20} 1.5708.[13]
$(PhCH_2)_2POSH$. VIa*.[753] VIg.[98,504] m. 190.5-191.5°,[504] pK_a 4.64 (in 80% EtOH);[336] Na salt, m. 232-236°.[98]
$(C_8H_{17})_2POSH$.
  Et S-ester. IIf. m. 37°, b_3 200-202°.[592]
  Bu S-ester. From $(C_8H_{17})_2P(O)H + Et_2N{\cdot}SBu$,[593] IIf.[592] b_5 234-236°,[592] d_4^{20} 0.9148,[592] n_D^{20} 1.4770.[592]
  i-Bu S-ester. IIf. b_5 222-224°, d_4^{20} 0.9116, n_D^{20} 1.4777.[592]

E.2.2. Aliphatic/Aromatic Phosphinothioic Acids with Unsubstituted Aliphatic Groups

PhMePOSH. XXIb.[646] $b_{0.05}$ 82-85°,[646] n_D^{23} 1.5708,[646] resolved using brucine, phenylalanine, and quinine giving dicyclohexylamine salts, m. 152°, with $[\alpha]_{578}^{25}$ +9.18°, +9.25°, $[\alpha]_{546}^{25}$ +10.40°, +10.44°, and $[\alpha]_{578}^{25}$ -9.22°, $[\alpha]_{546}^{25}$ -10.42°;[68] cyclohexylamine salt, m. 157-159°;[646] dicyclohexylamine (±)-salt, m. 157°.[68]
  $p-O_2N{\cdot}C_6H_4$ O-ester. XXIe. m. 70-72°.[718]
  C_6H_{13} S-ester. XXIe. $b_{0.01}$ 96°.[132,718]
  $cyclo-C_6H_{11}$ S-ester. XXIe. m. 56°, $b_{0.01}$ 104°.[132,718]

EtS·CH_2·CH_2 S-ester. XXIIa. $b_{0.01}$ 101°.[716]
PhEtPOSH. IId.[646] XXIa.[263] XXIb.[646] XXIIIa*.[29] $b_{0.06}$ 111-112°,[646] n_D^{20} 1.5972,[646] resolved using quinine giving dicyclohexylamine salts with $[\alpha]_D^{20}$ +6.5°, and $[\alpha]_D^{20}$ -6.6°;[263] cyclohexylamine salt, m. 155-157°;[646] dicyclohexylamine (±)-salt, m. 158°.[263]
p-O_2N·C_6H_4 O-ester. XXIe. m. 56-58°.[718]
Me S-ester. XXIIa*. $b_{0.3}$ 120°, n_D^{20} 1.5755, $[\alpha]_D^{20}$ +10.4° from dicyclohexylamine salt with $[\alpha]_D^{20}$ +1.3°.[263]
Pr S-ester. IVa. $b_{0.08}$ 102-104°, d_4^{20} 1.1004, n_D^{20} 1.5594.[14]
PhBuPOSH. VIh. $b_{0.008}$ 93-94°, d_4^{20} 1.1234, n_D^{20} 1.5740.[182]
Ph(t-Bu)POSH. IId. m. 126-127°.[292]
Ph($PhCH_2$)POSH. XXIIIa*. m. 173-174°.[29]
p-MeC_6H_4 amide. XXIi. m. 190.5-191°.[144]
(p-MeC_6H_4)(i-Bu)POSH.
Ph S-ester. Ie. $b_{0.02}$ 113°.[697]
(3-phenanthryl)MePOSH.
$PhCH_2$ S-ester. [VIc + conversion to chloride + XXIe]. m. 69°, resolved by complex formation giving acids with $[\alpha]_D^{20}$ +17° and $[\alpha]_D^{20}$ -13.1°.[262]

E.2.3. Aliphatic and Aliphatic/Aromatic Phosphinothioic Acids with Substituents in Aliphatic Groups

$(CF_3)_2$POSH.
Amide. IVb.[170] XXIi.[258] m. -27.8 to -27.5°,[258] b. 131° (extrapolated).[170]
Me_2 amide. IVb.[170] XXIi.[258] m. -16.8 to -16.7°,[258] b. 139° (extrapolated).[170]
$(ClCH_2)_2$POSH.
Et O-ester. XXIe. $b_{0.03}$ 73°, d_4^{20} 1.3398, n_D^{20} 1.5342.[583]
Ph O-ester. XXIe. m. 56-57°.[583]
Et_2 amide. XXIi. $b_{0.001}$ 95°, d_4^{20} 1.2696, n_D^{20} 1.5508.[583]
Ph amide. XXIi. m. 102-103°.[583]
$(ICH_2)_2$POSH.
Ph O-ester. XXXd. m. 46-47°.[583]
(CF_3·CH_2·CH_2)$_2$POSH. VIg. b_8 149-151°.[452]
[CCl_3·CH(OH)]MePOSH.
Et O-ester. IXc. m. 173-175°.[4]
Pr O-ester. IXc. m. 165-166°.[4]
Bu O-ester. IXc. m. 161-162°.[4]
[CCl_3·CH(OH)]EtPOSH.
Bu O-ester. IXc. m. 156-158°.[4]
[CCl_3·CH(OH)]BuPOSH.
Pr O-ester. IXc. m. 138-140°.[4]
[Me_2C(OH)]MePOSH.

Et O-ester. IXc. m. 125-127°.[4]
[Ph·CH(OH)]MePOSH.
Et O-ester. IXc. m. 68-70°.[4]
Pr O-ester. IXc. m. 49-50°.[4]
[Ph·CH(OH)]EtPOSH.
Bu O-ester. IXc. m. 43-44°.[4]
Pr O-ester. IXc. m. 76-78°.[530]
[Ph·CH(OH)]BuPOSH.
Pr O-ester. IXc. m. 35-36°.[4]
$(Me·CO·CH_2)$BuPOSH.
Et O-ester. VIe. $b_{0.02}$ 136°, d_4^{20} 1.0380, n_D^{20} 1.4915.[574]
$(CH_2:CH·CH_2·O·CO·CH_2)_2$POSH.
Et O-ester. IVa. $b_{0.02}$ 126°, d_4^{20} 1.1564, n_D^{20} 1.5021.[776]
$(EtO_2C·CH_2)$PhPOSH.
Et O-ester. Ve. $b_{0.01}$ 78°.[709]
$(Cl·CO·CH_2·CH_2)$EtPOSH.
Et S-ester. Xc. $b_{0.003}$ 128-129°, d_4^{20} 1.2246, n_D^{20} 1.5247.[12]
Pr S-ester. Xc. $b_{0.003}$ 146-148°, d_4^{20} 1.1923, n_D^{20} 1.5194.[12]
i-Pr S-ester. Xc. $b_{0.003}$ 130-133°, d_4^{20} 1.1818, n_D^{20} 1.5145.[12]
Bu S-ester. Xc. $b_{0.003}$ 173-175°, d_4^{20} 1.1682, n_D^{20} 1.5168.[12]
i-Bu S-ester. Xc. $b_{0.003}$ 160-161°, d_4^{20} 1.1602, n_D^{20} 1.5131.[12]
$(EtO_2C·CH_2·CH_2)$EtPOSH.
Bu S-ester. XXIe. $b_{0.001}$ 124°, d_4^{20} 1.0806, n_D^{20} 1.4920.[385]
$(EtS·CO·CH_2·CH_2)$EtPOSH.
Et S-ester. XXXh. $b_{0.003}$ 180-182°, d_4^{20} 1.1345, n_D^{20} 1.5375.[12]
$(NC·CH_2·CH_2)$EtPOSH.
Pr O-ester. IXf. $b_{0.01}$ 91°, n_D^{25} 1.5013.[530]
$(NC·CH_2·CH_2)_2$POSH.
p-MeC_6H_4 S-ester. Ie. m. 120-122°.[103]
$(EtO_2C·CH_2·CH_2)$PhPOSH.
Et O-ester. Ve. $b_{0.01}$ 85°.[709]
$(EtO_2C·CHMe)$PhPOSH.
Et O-ester. Ve. $b_{0.01}$ 78°.[709]
$(EtO_2C·CH:CH)$PhPOSH.
Bu S-ester. XXIe. $b_{0.04}$ 180-182°, d_4^{20} 1.1339, n_D^{20} 1.5510.[370]
$[EtO_2C·CPh(OH)]$EtPOSH.
Et S-ester. IXc. $b_{0.04}$ 132°, d_4^{20} 1.1520, n_D^{20} 1.5348.[618]
$[EtO_2C·CPh(OH)]$PrPOSH.
Pr S-ester. IXc. $b_{0.04}$ 141°, d_4^{20} 1.1402, n_D^{20} 1.5071.[618]

$(o\text{-}HO \cdot C_6H_4 \cdot NH \cdot CH_2)MePOSH$.
Cyclic O-ester. From $(ClCH_2)MeP(S)Cl + o\text{-}H_2N \cdot C_6H_4 \cdot OH$. m. 96-100°, $b_{0.001}$ 166-169°.[786]
$(Et_2N \cdot CH_2)MePOSH$.
Pr O-ester. IXd. b_1 92-93°, d_4^{20} 0.9747, n_D^{20} 1.4830.[589]
Bu O-ester. IXd. b_1 102-104°, d_4^{20} 0.9662, n_D^{20} 1.4842.[589]
$(Et_2N \cdot CH_2)PhPOSH$.
Pr O-ester. IXd. $b_{0.03}$ 80-81°, n_D^{20} 1.5373.[536]
[EtMeC(NH·NHPh)]EtPOSH.
Et S-ester. IXd. d_4^{20} 1.0703, n_D^{20} 1.5642.[5]
[Pr·CH(NH·NHPh)]EtPOSH.
Et S-ester. IXd. $b_{0.6}$ 140°, d_4^{20} 1.0556, n_D^{20} 1.5552.[5]
Bu S-ester. IXd. d_4^{20} 1.0234, n_D^{20} 1.5432.[5]
[BuEtC(NH·NHPh)]EtPOSH.
Bu S-ester. IXd. d_4^{20} 0.9979, n_D^{20} 1.5303.[5]
$(EtSCH_2)PhPOSH$.
Et O-ester. Ve. $b_{0.01}$ 75°.[709]
$[2\text{-}HO\text{-}4(or5)\text{-}MeC_6H_3 \cdot S \cdot CH_2]MePOSH$.
Cyclic S-ester. From $(ClCH_2)MeP(O)Cl$ and 1-Me-3,4-$(HS)_2C_6H_3$. m. 76-78°, $b_{0.001}$ 176-177°.[786]
$(o\text{-}H_2N \cdot C_6H_4 \cdot S \cdot CH_2)MePOSH$.
Cyclic amide. From $(ClCH_2)MeP(S)Cl + o\text{-}H_2N \cdot C_6H_4 \cdot SH$. m. 102-109°. $b_{0.001}$ 174-178°.[786]
$(EtS \cdot CH_2 \cdot CH_2)PhPOSH$.
Et O-ester. Ve. $b_{0.01}$ 82°.[709]

E.2.4. Diarylphosphinothioic Acids

Ph_2POSH. XXIa.[289,345] XXIVa.[28] m. 141-143°,[289] pKa 1.88 (in 7% EtOH), 3.58 (in 80% EtOH);[345] K salt (2 1/2 H_2O), m. 213.5-215°;[502] MeHg deriv., m. 127-129°;[397] Me_2Al deriv., m. 227-228°;[130] Me_2Ga deriv., m. 203-204°.[13]
Me O-ester. XXIe.[502] Also by irradiation of $Ph_2P \cdot OMe + (PhCH_2)_2S_2$ in MeOH.[154] m. 84.5-85.5°.[502] x-ray structure determination.[687]
Et O-ester. XXIe. m. 42-43°.[502]
$CH_2{:}CH \cdot CH_2$ O-ester. IVa. $b_{0.035}$ 144-146°, d_4^{20} 1.1570, n_D^{20} 1.6160.[605]
C_6H_{13} O-ester. IVa. $b_{0.35}$ 180-183°, d_4^{20} 1.0856, n_D^{20} 1.5792.[755b]
$PhCH_2$ O-ester. By irradiation of $Ph_2P \cdot OMe + (PhCH_2)_2S_2$ in benzene. m. 85-86°.[154]
C_8H_{17} O-ester. IVa.[598,755b] $b_{0.01}$ 138.5-139°,[598] $b_{0.2}$ 180-185°,[755b] d_4^{20} 1.0798,[598] n_D^{20} 1.5710.[598]
$CH_2{:}CMe \cdot CO \cdot O \cdot CH_2 \cdot CH_2$ O-ester. XXIe. m. 62-63°.[277]
$EtO_2C \cdot CH{:}CMe$ O-ester. XXIe. m. 57-58°.[432]
4,6-Me_2-pyrimid-2-one-1-yl$\cdot CH_2 \cdot CH_2$ O-ester. XXIe. m. 108-109.5°.[678]

Ph O-ester. IVa.[527] XXIId*.[295] m. 124°.[527]
2-naphthyl O-ester. XXIe. m. 118°.[719]
Et S-ester. XXIIa. XXIIIe. m. 74.5-75°.[502]
C_6H_{13} S-ester. IVa. $b_{0.075}$ 169-171°, d_4^{20} 1.1003, n_D^{20} 1.5800.[755b]
$PhCH_2$ S-ester. XXIIa. m. 94-95°.[502]
C_8H_{17} S-ester. IVa. $b_{0.3}$ 197-200°, d_4^{20} 1.0702, n_D^{20} 1.5700.[755b]
Ph_2CH S-ester. XXIIa. m. 95-96°.[502]
p-$O_2N{\cdot}C_6H_4{\cdot}CH_2$ S-ester. XXIIa. m. 129-130°.[502]
cyclohex-1,3-ylene di-S-ester. Ie. m. 148-151°.[697]
Ph S-ester. Ie.[697] IIf.[261] IVa.[514] IVf.[698] Also from $Ph_2P{\cdot}OMe + Ph_2S_2$,[154] m. 90-91°.[154]
p-MeC_6H_4 S-ester. Ie.[697] IVf.[698] m. 112-113°.[698]
p-ClC_6H_4 S-ester. IVf. m. 111°.[698]
2-Cl-5-MeC_6H_3 S-ester. IVf. m. 89°.[698]
1-naphthyl S-ester. Ie. m. 110-112°.[697]
2-naphthyl S-ester. IVf. m. 113°.[698]
$Ph_2P(S){\cdot}O{\cdot}P(S)Ph_2$. XXIf.[28,476,502] m. 201.5-202°,[502] ^{31}P -73.9° ppm.[476]
$(p\text{-}ClC_6H_4)_2P(S){\cdot}O{\cdot}P(S)Ph_2$. XXIf. m. 84-85°.[502]
Amide. XXIi. m. 102-104°.[796]
Me amide. XXIi. m. 129-132°.[704]
Me_2 amide. XXIi. m. 88-90°.[704,796]
Aziridide. IVb. m. 70.5-71.5°.[739]
Bu amide. IVb. m. 56-57°.[739]
t-Bu amide. IVb. m. 120.5-121.5°.[736]
Piperidide. IVb. $b_{0.5}$ 160-164°.[736]
$\overline{CH_2{\cdot}(CH_2)_5{\cdot}N}{\cdot}P(S)Ph_2$. IVb. m. 90-91°.[739]
$\overline{CH_2{\cdot}(cyclohex\text{-}1,4\text{-}ylene){\cdot}CH_2{\cdot}N}{\cdot}P(S)Ph_2$. IVb. m. 128-129°.[739]
Ph amide. XXIi. m. 165-167°.[796]
$Et_2C_6H_3$ amide. IVb. m. 135-136°.[736]
2-pyridyl amide. IVb. m. 138-140.5°.[739]
$Ph_2P(S){\cdot}NH{\cdot}P(S)Ph_2$. XXIj.[703] XXVa.[725] m. 213.5-214.5°,[703] ^{31}P -55.1 ppm;[701,703] NH_4^+ salt, m. 210-212° decomp.;[703] K salt, m. 363-366°,[703] ^{31}P -35.6 ppm.[701]
$Ph_2P(S){\cdot}NPh{\cdot}P(S)Ph_2$. XXVa. m. 220°.[725]
$Ph_2P(S){\cdot}N{:}C{:}N{\cdot}P(S)Ph_2$. IVb. XXIi. m. 95-97°.[783]
$H_2N{\cdot}CS{\cdot}NH{\cdot}P(S)Ph_2$. XXXIIa. m. 143.5-144°.[704]
$MeNH{\cdot}CS{\cdot}NH{\cdot}P(S)Ph_2$. XXXIIa. m. 126-127°.[704]
$Me_2N{\cdot}CS{\cdot}NH{\cdot}P(S)Ph_2$. XXXIIa.[704] m. 95.5-96°,[704] ^{31}P -50.9 ppm;[701] K salt, m. 197-198°,[704] ^{31}P -40.7 ppm;[701] Me_2NH salt, m. 119-121°.[704] Reaction with CH_2N_2[704] gives a 3:1 mixture of $Me_2N{\cdot}C(SMe){:}N{\cdot}P(S)Ph_2$, m. 119-120°,[704] ^{31}P -41.6 ppm,[701] and $Me_2N{\cdot}CS{\cdot}N{:}P(SMe)Ph_2$, m. 150-151°,[704] ^{31}P -32.0,[701] both of which with Me_3O^+ $SbCl_6^-$ in H_2O give $Me_2N{\cdot}C(SMe){:}N{\cdot}P^+(SMe)Ph_2$ $SbCl_6^-$, m. 120-122°,

^{31}P -36.8 ppm.[701]

$Me_2N{\cdot}C(SMe){:}N{\cdot}P(S)Ph_2$. See previous compound.

$Me_2N{\cdot}NH{\cdot}P(S)Ph_2$. IVb. m. 95.5-97°.[567,735b]

$Me_2N{\cdot}NEt{\cdot}P(S)Ph_2$. IVb. m. 126.5-127.5°.[567]

$PhNH{\cdot}NH{\cdot}P(S)Ph_2$. XXIi. m. 133-134°.[91]

$Ph_2N{\cdot}NH{\cdot}P(S)Ph_2$. XXIi. m. 153-154°.[28]

$Ph_2P(S){\cdot}NH{\cdot}NH{\cdot}P(S)Ph_2$. XXIi. m. 204°.[92]

$PhCO{\cdot}NH{\cdot}NH{\cdot}P(S)Ph_2$. XXIi. m. 152-153°.[744]

$PhN{:}N{\cdot}P(S)Ph_2$. From $PhNH{\cdot}NH{\cdot}P(S)Ph_2$ + NaOBr. Red crystals, m. 72-73°.[91]

$Ph_2P(S){\cdot}N{:}N{\cdot}P(S)Ph_2$. From $Ph_2P(S){\cdot}NH{\cdot}NH{\cdot}P(S)Ph_2$ + N-bromosuccinimide. Brick-red powder, decomp. > -20°.[93]

$(p\text{-}ClC_6H_4)_2POSH$. XXIa. m. 103-104°, pKa 1.60 (in 7% EtOH), 3.00 (in 80% EtOH).[345]

Ph S-ester. IIf. m. 122.5-123°.[261]

$(p\text{-}BrC_6H_4)_2POSH$.

Ph S-ester. IIf. m. 120-122.5°.[261]

$(C_6F_5)_2POSH$.

Me O-ester. IVa. m. 90°, $b_{0.1}$ 106-108°.[202]

Et O-ester. IVa. m. 45°.[202]

Ph O-ester. IVa. $b_{0.2}$ 145-146°.[202]

$(p\text{-}MeO{\cdot}C_6H_4)_2POSH$.

Ph S-ester. IIf. m. 69.5-71.5°.[261]

$(p\text{-}MeC_6H_4)_2POSH$. m. 93-94°, pK_a 2.14 (in 7% EtOH), 4.12 (in 80% EtOH).[336]

Ph S-ester. IIf. m. 125-126°.[261]

$(MeC_6H_4)_2POSH$. [Orientation uncertain, since the starting $(MeC_6H_4)_2P(S)Cl$ was prepared from PhMe + $PSCl_3$ + $AlCl_3$, but probably mainly p].

Et O-ester. XXIe. m. 41-42°.[524]

Ph O-ester. XXIe. m. 135°.[524]

$(MeC_6H_4)_2P(S){\cdot}O{\cdot}P(S)(C_6H_4Me)_2$. XXIf. m. 165-166°.[524]

Amide. XXIi. m. 139°.[524]

Et_2 amide. XXIi. m. 177-178°.[524]

Piperidide. XXIi. m. 134°.[524]

Ph amide. XXIi. m. 152°.[524]

$PhNH{\cdot}NH{\cdot}P(S)(C_6H_4Me)_2$. XXIi. m. 135.5°.[524]

$(p\text{-}CF_3{\cdot}C_6H_4)_2POSH$.

Ph S-ester. IIf. m. 113-114°.[261]

$[3,5\text{-}(CF_3)_2C_6H_3]_2POSH$.

Ph S-ester. IIf. m. 115-116.5°.[261]

E.2.5. Cyclic Phosphinothioic Acids

POSH. XXIa. $b_{0.8}$ 120°, d_4^{20} 1.2851, n_D^{20} 1.6020.[671]

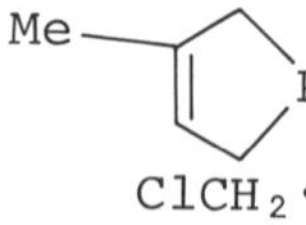
POSH. XXIa. $b_{0.5}$ 159-160°, d_4^{20} 1.2437, n_D^{20} 1.6034.[671]

$ClCH_2{\cdot}CH_2$ S-ester. Xa. b_5 160-163°, d_4^{20} 1.2520, n_D^{20} 1.5630.[673]

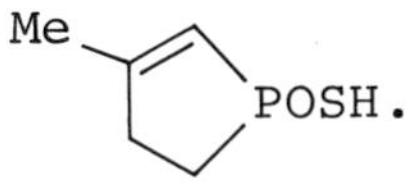

Et S-ester. XXIe. $b_{0.25}$ 110-111°.[278]
C_8H_{17} S-ester. XXIe. $b_{0.1}$ 158°, n_D^{20} 1.5198.[278]

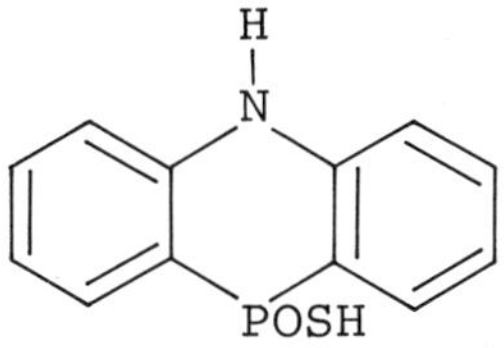

. IId. m. 213° decomp. (preheated block).[273]
Me O-ester. XXIe. m. 184-185.5°.[273]
Me S-ester. XXIIc. m. 266°.[273]

E.2.6. Acid Containing Two Phosphinothioic Acid Groups

Ph·POSH·$(CH_2)_4$·POSH·Ph.
Me_2 di-O-ester. XXIe. m. 77-80°.[316]
$(p\text{-}ClC_6H_4)_2$ di-O-ester. XXIe. m. 156°.[316]
$(p\text{-}O_2N{\cdot}C_6H_4)_2$ di-O-ester. XXIe. m. 168-176°.[316]

E.3. Phosphinodithioic Acids and Their Derivatives

E.3.1. Unsubstituted Aliphatic Phosphinodithioic Acids

Me_2PS_2H. Ib.[494] m. 47-50°;[494] Na salt m. 223-225° decomp.;[506] K salt, m. 130°;[718] Ni salt, x-ray structure determination;[335] Me_2Hg deriv., m. 120° decomp.;[397] Al salt, m. 186-188°;[130] Me_2Al deriv., m. 103-104°;[130] Me_2Ga deriv., m. 153-154°;[130] Me_2In deriv., m. 184-185°;[130] Me_2Tl deriv., decomp. > 185°.[130]
Et ester. XXIe*.[710] m. 86-87°,[710] b_1 78°.[718]
cyclo-C_6H_{11} ester. XXIe. m. 63°.[718]
$C_{12}H_{25}$ ester. XXIe. $b_{0.01}$ 84°.[718]
$ClCH_2$ ester. XXIIa.[707,718] m. 50°,[707] $b_{0.01}$ 60°.[707,718]
$CH_2{\cdot}O{\cdot}CH_2$ di-ester. XXIIa. m. 130°.[717,718]
$CH_2{\cdot}S{\cdot}CH_2$ di-ester. XXIIa. m. 88°.[717,718]
$Me_2N{\cdot}CS{\cdot}S{\cdot}CH_2$ ester. From $ClCH_2$ ester + $Me_2N{\cdot}CS{\cdot}S^-$ Na^+.[707,718]
EtS·CH_2 ester. XXIIa. $b_{0.01}$ 83°.[717,718]
p-$FC_6H_4{\cdot}S{\cdot}CH_2$ ester. XXIIa. m. 70-71°.[191]
p-$ClC_6H_4{\cdot}S{\cdot}CH_2$ ester. XXIIa. m. 60°.[717,718]
1-oxo-1,2-dihydrophthalazin-2-yl·CH_2 ester. XXIIa. m. 90°.[468]
3,4-dihydro-1,2,3-triazanaphthalene-3-yl·CH_2 ester. XXIIa.[717,718] XXIId*.[194] m. 166°.[717,718]

EtO·CHMe ester. XXIIe. $b_{0.01}$ 68°.[718]
EtS·CH_2·CH_2 ester. XXIe.[506] XXIIa.[717,718] $b_{0.01}$ 87°,[717,718] b_2 126.5-127°,[506] d_4^{20} 1.1408,[506] n_D^{20} 1.5923.[506]
CH_2:CH·S·CH_2·CH_2 ester. XXIIe. $b_{0.01}$ 76°.[708]
EtS·CHMe ester. XXIIe. $b_{0.01}$ 72°.[713,718]
EtO_2C·CH_2 ester. XXIIa. $b_{0.01}$ 75°.[717,718]
$MeSO_2$·NMe·CO·CH_2 ester. XXIIa. m. 105°.[714]
CH_2·CH_2·SO_2·CH_2·CH·NMe·CO·CH_2 ester. m. 104-105°.[726]
MeNH·CS·CH_2 ester from uncharacterized MeNH·CO·CH_2 ester[717,718] + P_4S_{10}. m. 66°.[715]
NC·CH_2 ester. XXIIa. $b_{0.01}$ 78°.[717,718]
EtS·$(CH_2)_3$ ester. XXIIa. $b_{0.01}$ 94°.[717,718]
EtS·CH_2·CHMe ester. XXIIa. $b_{0.01}$ 89°.[717,718]
$EtOC_2$·CHMe ester. XXIIa. m. 50°, $b_{0.01}$ 78°.[717,718]
NC·CH_2·CH_2 ester. XXIIa. $b_{0.01}$ 89°.[717,718]
EtO_2C·CMe(SMe)·CH_2 ester. XXIIa. $b_{0.06}$ 142°.[470]
MeO_2C·CH(SEt)·CHMe ester. XXIIa. $b_{0.1}$ 104°.[470]
EtO_2C·CH(SMe)·CHMe ester. XXIIa. $b_{1.3}$ 156°.[470]
EtO_2C·CH(SEt)·CHMe ester. XXIIa. $b_{0.8}$ 154°.[470]
H_2N·CO·CMe(SMe)·CH_2 ester. XXIIa. m. 114-115°.[695]
H_2N·CO·CMe(SEt)·CH_2 ester. XXIIa. m. 81°.[695]
-CH(CO·NH_2)·$(CH_2)_3$·CH($CONH_2$)- di-ester. XXIIa. m. 159-161°.[696]
-CH(CN)·$(CH_2)_3$·CH(CN)- di-ester. XXIIa. m. 162-163°.[696]
p-ClC_6H_4·CH_2 ester. XXIIa. m. 62°.[717,718]
p-MeS·C_6H_4·CH_2 ester. XXIIa. $b_{0.01}$ 112°.[717,718]
5-Me-2-MeS·C_6H_3·CH_2 ester. XXIIa. m. 51°.[717,718]
O·CH:C(OMe)·CO·CH:CH·CH_2 ester. XXIIa. m. 126°.[717,718]
benz-1,3-thiazol-2-yl·CH_2 ester. XXIIa. m. 75°.[200]
6-Me-benz-1,3-thiazol-2-yl·CH_2 ester. XXIIa. Bright red, m. 71°, $b_{0.01}$ 123°.[201]
Ph ester. XXIe. m. 68°.[718]
p-MeC_6H_4 ester. XXIe. m. 76°.[718]
o-ClC_6H_4 ester. IIe. m. 105°.[134]
p-ClC_6H_4 ester. XXIe. m. 109°.[718]
o-O_2N·C_6H_4 ester. IIe. m. 88.5°.[134]
m-O_2N·C_6H_4 ester. IIe. m. 94.5°.[134]
p-O_2N·C_6H_4 ester. IIe. m. 117.5°.[134]
$Me_2P(S)$·S·P(S)Me_2. IIh. m. 91°.[17]
Morpholino·S·P(S)Me_2. IIh. m. 99°.[17]
2,5-$Me_2C_6H_3$·S·S·P(S)Me_2. From previous compound + 2,5-$Me_2C_6H_3$·SH + HCl. m. 86°.[17]

$EtMePS_2H$.
Et ester. Va. $b_{0.03}$ 64-66°, d_4^{20} 1.0979, n_D^{20} 1.5650.[429]
Bu ester. Va. $b_{0.02}$ 76-77°, d_4^{20} 1.0408, n_D^{20} 1.5445.[429]

Et_2PS_2H. Ib.[294] IIId.[448] XXIb.[345,505] $b_{2.5}$ 68.5-69°,[505] b_{13} 130-131°,[446,448] d_4^{20} 1.1306,[505] n_D^{20} 1.5858,[505] IR,[109] pK_a 1.71 (in 7% EtOH), 2.53 (in 80% EtoH);[345] NH_4^+ salt, m. 193°;[294] Na salt, m. 153°;[448] Na salt ($2H_2O$), m. 124-125°;[448] Cr^{3+} salt, violet, m. 141°,[445] dipole moment 1.38D, μ_{eff} 3.85 BM;[439] Fe^{3+} salt, black, decomp. 80-85°, μ_{eff} 6.23 BM;[439] Co^{2+} salt, emerald green plates, m. 134.5°, dipole moment 1.43D,[440] μ_{eff} 4.46 BM,[439] x-ray structure determination;[115] Co^{2+} salt ($6NH_3$) flesh colored, m. ∿70°;[439] Co^{2+} salt (2 pyridine), blue-violet crystals, m. 110° decomp., μ_{eff} 5.03 BM;[439] Co^{3+} salt, coffee brown, decomp. ∿70°;[439] Ni salt, α and β forms described,[481] but only the α form is genuine,[728] violet needles, m. 142.5°,[440] dipole moment 1.49D,[440] diamagnetic;[439] Ni salt (2 thiophene), indigo-blue;[439] Ni salt (2 pyridine), green cubes, decomp. ∿125°, μ_{eff} 3.13 BM;[439] Cu^+ salt, m. 290° decomp.;[440] Cu^+ salt ($2Ph_3P$), m. 186° decomp.;[440] $Cu(OH)S_2PEt_2 \cdot 4$ pyridine, by passing O_2 through Cu^+ salt in pyridine, blue crystals, m. 223-225° decomp.;[440] Ag salt, m. 262° decomp.;[444] Au^+ salt, needles, m. 153°;[444] Zn salt, m. 166.5-167°,[445] dipole moment 1.37D,[440] x-ray determination;[115,116] Cd salt, m. 164.5-166°;[445] Hg^{2+} salt, m. 135-136°;[445] MeHg deriv., m. 40-43°;[397] In^{3+} salt, m. 127.5°;[445] Tl^+ salt, m. 139°;[444,445] Me_2Tl deriv., m. 224-225°;[97] Ph_2Tl deriv., m. 164°;[97] Cl_2Sn deriv., yellow, m. 179-180°,[97] dipole moment 7.71D;[440] Br_2Sn deriv., yellow, m. 175°;[97] Bu_2Sn deriv., plates, m. 40.5°, dipole moment 3.22D;[440] Ph_2Sn deriv., needles, m. 149.5°, dipole moment 3.32D;[440] Et_3Sn deriv., b_2 190°;[97] Ph_3Sn deriv., m. 88-89°;[97] Pb^{2+} salt, m. 114.5°,[445] dipole moment 1.94D;[440] Bi salt, bright yellow, m. 94.5°,[445] dipole moment 2.52D.[440]

Et ester. IVa.[10,11] Va.[10,429] XXIe.[718] $b_{0.05}$ 61-62°,[10,11] b_1 90°,[718] d_4^{20} 1.0594,[10] n_D^{20} 1.5559.[10]

Pr ester. IVa.[10,11] XXIIa.[448] $b_{0.05}$ 74-76°,[10,11] b_9 127°,[448] d_4^{20} 1.0363,[10,11] n_D^{20} 1.5542.[10,11]

i-Pr ester. IVa. $b_{0.04}$ 60-62°, d_4^{25} 1.0249, n_D^{25} 1.5385.[11]

Bu ester. IVa.[11] XXIIa.[448] $b_{0.05}$ 84-86°,[11] b_{11} 146.5-147°,[448] d_4^{25} 1.0174,[11] n_D^{20} 1.5411,[448] n_D^{25} 1.5372.[11]

i-Bu ester. IVa. $b_{0.06}$ 77-79°, d_4^{25} 1.0131, n_D^{25} 1.5343.[11]

cyclo-C_6H_{11} ester. XXIe. $b_{0.01}$ 86°.[718]

$PhCH_2$ ester. XXIIa. b_{11} 198-198.5°, d_4^{20} 1.122, n_D^{20} 1.6014.[448]

$C_{12}H_{25}$ ester. XXIe. $b_{0.01}$ 105°.[718]

CH_2 di-ester. XXIIa. m. 55°.[717]
$ClCH_2$ ester. XXIIa. $b_{0.01}$ 68°.[707,717,718]
$EtS{\cdot}CH_2$ ester. XXIIa. $b_{0.01}$ 82°.[717,718]
$EtO{\cdot}CHMe$ ester. XXIIe. $b_{0.01}$ 76°.[718]
$EtS{\cdot}CH_2{\cdot}CH_2$ ester. XXIIa. $b_{0.01}$ 96°.[717,718]
$CH_2{:}CH{\cdot}S{\cdot}CH_2{\cdot}CH_2$ ester. XXIIe. $b_{0.01}$ 92°.[708]
$EtS{\cdot}CHMe$ ester. XXIIa.[717] XXIIe.[713] $b_{0.01}$ 97°.[717]
$Et_2N{\cdot}CH_2{\cdot}CH_2$ ester. XXIIa. $b_{0.01}$ 90°.[717,718]
$EtO_2C{\cdot}CH_2$ ester. XXIIa. $b_{0.01}$ 91°.[717]
$NC{\cdot}CH_2$ ester. XXIIa. $b_{0.01}$ 88°.[717]
$EtS{\cdot}CH_2{\cdot}CHMe$ ester. XXIIa. $b_{0.01}$ 94°.[717,718]
$EtO_2C{\cdot}CHMe$ ester. XXIIa. $b_{0.01}$ 93°.[717,718]
$NC{\cdot}CH_2{\cdot}CH_2$ ester. XXIIa. $b_{0.01}$ 98°.[717]
$(MeNH{\cdot}CO)_2CH$ ester. XXIIa. m. 134°.[471]
$MeO_2C{\cdot}CH(SEt){\cdot}CHMe$ ester. XXIIa. $b_{0.3}$ 129°.[470]
$EtO_2C{\cdot}CH(SMe){\cdot}CHMe$ ester. XXIIa. $b_{0.4}$ 146°.[470]
$EtO_2C{\cdot}CH(SEt){\cdot}CHMe$ ester. XXIIa. $b_{0.6}$ 152°.[470]
$CH_2{\cdot}CO{\cdot}NH{\cdot}CO{\cdot}CH_2$ di-ester. XXIIa. m. 36-39°.[699]
p-ClC_6H_4 ester. XXIe. m. 62°.[718]
o-$O_2N{\cdot}C_6H_4$ ester. IIe. m. 57-58°.[134]
$Et_2P(S){\cdot}S{\cdot}P(S)Et_2$. IIh.[17] XXIf.[447] m. 52°,[17] $b_{0.5}$ 173-176°.[447]

$Pr_2P(S){\cdot}S{\cdot}P(S)Et_2$. XXIf. $b_{0.4}$ 175-177°, disproportionates on redistillation.
$Ph{\cdot}CO{\cdot}S{\cdot}P(S)Et_2$. XXIIi. m. 54°.[294]
$Me_2N{\cdot}CO{\cdot}S{\cdot}P(S)Et_2$. XXIIi. m. 70°.[717]
$Et_2P(S){\cdot}S{\cdot}S{\cdot}P(S)Et_2$. XXIIj.[444,448] Also from $CH_2{\cdot}CH_2{\cdot}O{\cdot}CH_2{\cdot}CH_2{\cdot}N{\cdot}S{\cdot}P(S)Et_2$ (second compound below) + Et_2PS_2H + HCl.[17] m. 56°.[17,448]
$Et_2P(S){\cdot}S{\cdot}S{\cdot}S{\cdot}P(S)Et_2$. By-product in preparation of Et_2PS_2H by Ib.[294] Also from Et_2PS_2H + SCl_2 or S_2Cl_2. m. 110.5°.[448]
Morpholino$\cdot S{\cdot}P(S)Et_2$. IIh. m. 102°.[17]
2,5-$Me_2C_6H_3{\cdot}S{\cdot}S{\cdot}P(S)Et_2$. From previous compound + 2,5-$Me_2C_6H_3{\cdot}SH$ + HCl. Also from Et_2PS_2H + 2,5-$Me_2C_6H_3{\cdot}SCl$. m. 48°.[17]

$PrEtPS_2H$.
Et ester. Va. $b_{0.1}$ 77-80°, d_4^{20} 1.0721, n_D^{20} 1.5515.[429]

Pr_2PS_2H. IIId.[448] VIg.[505] XXIb.[345,505] b_2 91-91.5°,[341,345,505] b_{12} 146-147°,[448] d_4^{20} 1.0691,[341,345,505] n_D^{20} 1.5632;[341,345,505] pK_a 1.84 (in 7% EtOH), 2.63 (in 80% EtOH);[345] Na salt, m. 177-178°;[448] Na salt($1H_2O$), m. 87-89°;[505] Na salt ($2H_2O$), m∿50-51°;[448] Co salt, emerald green prisms, m. 95.5°;[439] Ni salt, violet needles,[439] m. 149.5-150°;[505] Cu^+ salt, m. 240° decomp.;[444] $Cu(OH)S_2PPr_2{\cdot}4$ pyridine, by passing O_2 through Cu^+ salt in pyridine, blue crystals, m. 238° decomp.;[444] Ag salt, m. 285-290° decomp.;[444] Au^+ salt, needles, m. 175°;[444] Zn salt, m. 159°;[445] Cd salt, m. 179-

180°;[445] Hg^{2+} salt, m. 167-168°;[445] Pb^{2+} salt, m. 114-115°;[445] As salt, pale yellow, m. 133.5°;[445] Sb salt, bright yellow, m. 148.5°;[445] Bi salt, deep yellow, m. 201.5°.[445]
Bu ester. XXIIa. $b_{10.5}$ 158-159°, d_4^{20} 0.992, n_D^{20} 1.5156.[448]
$EtS \cdot CH_2 \cdot CH_2$ ester. XXIIa. $b_{0.1}$ 117-118°, d_4^{20} 1.0685, n_D^{20} 1.5623.[505]
$Pr_2P(S) \cdot S \cdot P(S)Pr_2$. IIh.[17] XXIf.[447] m. 47.5°,[447] $b_{0.7}$ 193.5-195°.[447]
$Pr_2P(S) \cdot S \cdot S \cdot P(S)Pr_2$. XXIIj.[448,505] m. 80-81°.[505]
Morpholino$\cdot S \cdot P(S)Pr_2$. IIh. m. 105°.[17]
2,5-$Me_2C_6H_3 \cdot S \cdot S \cdot P(S)Pr_2$. From previous compound + 2,5-$Me_2C_6H_3 \cdot SH$ + HCl.[17]

i-Pr_2PS_2H. XXIb.[345,505] Also minor product from i-PrMgHal + $PSCl_3$.[127] b_3 76-76.5°,[505] d_4^{20} 1.0877,[505] n_D^{20} 1.5745,[505] pK_a 1.64 (in 7% EtOH), 2.66 (in 80% EtOH);[345] Na salt, m. 153-154.5°;[505] Ni salt, intense blue,[481] m. 200°;[127]
$EtS \cdot CH_2 \cdot CH_2$ ester. XXIIa. $b_{0.1}$ 122-123°, d_4^{20} 1.0787, n_D^{20} 1.5661.[505]
$NC \cdot CH_2 \cdot CH_2$ ester. XXIIe. b_2 140-141°, d_4^{20} 1.0977, n_D^{20} 1.5566.[505]
i-$Pr_2P(S) \cdot S \cdot P(S)(i\text{-}Pr)_2$. XXIIj. m. 101.5-102.5°.[505]

$(CH_2{:}CH \cdot CH_2)EtPS_2H$.
Et ester. Va. $b_{0.05}$ 85-86°, d_4^{20} 1.0774, n_D^{20} 1.5675.[429]

Bu_2PS_2H. I-b.[649] IId.[586] IIId.[444] XXIb.[345,505] b_2 99-99.5°, d_4^{20} 1.0314, n_D^{20} 1.5481,[341,345,505] pK_a 1.79 (in 7% EtOH), 2.52 (in 80% EtOH);[345] NH_4^+ salt, m. 97-99°;[649] Na salt, m. 113-115°;[444] Ni salt, violet, m. 89-90°;[83,649] Ag salt, m. 122-125° decomp.;[444] Zn salt, m. 110°;[445] Cd salt, m. 147.5°;[445] $PhNH_2$ salt, m. 110-111°.[586]
$EtS \cdot CH_2 \cdot CH_2$ ester. XXIIa. $b_{0.1}$ 126-127°, d_4^{20} 1.0631, n_D^{20} 1.5516.[505]
$Bu_2P(S) \cdot S \cdot P(S)Bu_2$. IIh. $b_{0.1}$ 170-172°.[17]
$Bu_2P(S) \cdot S \cdot S \cdot P(S)Bu_2$. XXIIj. m. 62-63°.[586]
morpholino$\cdot S \cdot P(S)Bu_2$. IIh. m. 46°.[17]
2,5-$Me_2C_6H_3 \cdot S \cdot S \cdot P(S)Bu_2$. From previous compound + 2,5-$Me_2C_6H_3 \cdot SH$ + HCl. m. 49°.[17]

i-Bu_2PS_2H. Ib. m. 39-40°.[586]
Bu ester. Id. $b_{0.05}$ 97-103°.[260]
$EtO_2C \cdot CH_2 \cdot CH_2$ ester. Id. $b_{0.2}$ 152-157°.[260]

s-Bu_2PS_2H. One product from s-BuMgHal + $PSCl_3$, $b_{0.15}$ 90°, n_D^{20} 1.5552.[127]

t-$BuMePS_2H$. VIa*.[272] XXIb.[437] m. 220.5°,[272] ^{31}P -83.0 ppm;[272] Na salt, m. 334-338° decomp.;[437] Ni salt, blue-violet needles, m. 254°, ^{31}P -108.7 ppm;[272] Me_2N salt, m. 136.5°, ^{31}P -80.3 ppm.[272]

t-Bu_2PS_2H. Ib. Not obtained crystalline.[292]
t-$Bu_2P(S) \cdot S \cdot P(S)(t\text{-}Bu)_2$. IIIb. m. 182-184°.[313]

$(C_5H_{11})_2PS_2H$. Zn salt, m. 112.5°.[445]
(cyclo-$C_6H_{11})_2PS_2H$. Ib. m. 103-105°;[649] Ni salt, m. 270-275°;[83] Zn salt, m. 215-216°;[83] Pb^{2+} salt, m. 231-232°.[83]
Bu ester. Id. $b_{0.15}$ 167-175°.[260]
$(PhCH_2)EtPS_2H$.
Et ester. Va. b_2 136-138°, d_4^{20} 1.1272, n_D^{20} 1.6070.[41]
Bu ester. Va. $b_{0.05}$ 127-130°, d_4^{20} 1.0878, n_D^{20} 1.5820.[429]
$(PhCH_2)_2PS_2H$. VIg. XXIb. m. 132.5-133.5°.[505]
$PhCH_2$ ester. XXIIa. m. 111-112°.[505]
$NC \cdot CH_2 \cdot CH_2$ ester. XXIIe. m. 113-114°.[505]
$(PhCH_2)P(S) \cdot S \cdot P(S)(CH_2Ph)_2$. IIh. m. 188°.[17]
Morpholino$\cdot S \cdot P(S)(CH_2Ph)_2$. IIh. m. 96°.[17]
2,5-$Me_2C_6H_3 \cdot S \cdot S \cdot P(S)(CH_2Ph)_2$. From previous compound + 2,5-$Me_2C_6H_3 \cdot SH$ + HCl. m. 93°.[17]
$(C_8H_{17})_2PS_2H$. Ib.[83,649] Ni salt,[83,649] m. 49-51°.[83]
$(PhCH_2 \cdot CH_2)_2PS_2H$. Ib.[649] $NH_4{}^+$ salt, m. 201-203°;[649] Ni salt, m. 146-148°.[649]
$(C_{12}H_{25})_2PS_2H$. Ib.[649] Ni salt, m. 72-74°.[649]

E.3.2 Aliphatic/Aromatic Phosphinodithioic Acids with Unsubstituted Aliphatic Groups

$PhMePS_2H$. VIIIc. Liquid, decomp. on distillation, n_D^{20} 1.6787, ^{31}P -52.4 ppm; $NH_4{}^+$ salt, m. 160-161°; Ni salt, purple, m. 203-206°.[563]
$ClCH_2$ ester. XXIIa. $b_{0.01}$ 112°.[707,718]
$EtS \cdot CH_2 \cdot CH_2$ ester. XXIe. $b_{0.01}$ 110-115°.[718]
$PhEtPS_2H$. VIIIc. m. 64.5°, ^{31}P -62.8 ppm; NH_4^+ salt, m. 166-170°; Ni salt, purple, m. 164-165°.[563]
Et ester. IVa.[10] Va.[29] $b_{0.1}$ 112-115°,[10] $b_{3.5}$ 169-170°,[29] d_4^{20} 1.1392,[10] n_D^{20} 1.6140.[10]
Pr ester. IVa. $b_{0.05}$ 104-105°, d_4^{20} 1.1142, n_D^{20} 1.6032.[10]
i-Pr ester. IVa. $b_{0.05}$ 97-98°, d_4^{20} 1.0161, n_D^{20} 1.5993.[10]
Bu ester. IVa. $b_{0.06}$ 121-123°, d_4^{20} 1.0941, n_D^{20} 1.5925.[10]
i-Bu ester. IVa. $b_{0.1}$ 124-125°, d_4^{20} 1.0913, n_D^{20} 1.5916.[10]
Ph(i-Pr)PS_2H. VIIIc. $b_{0.001}$ 92°, n_D^{22} 1.6354, ^{31}P -74.6 ppm; Ni salt, purple, 206-207°.[563]
$PhBuPS_2H$. VIIIc. Liquid, decomp. on distillation, n_D^{22} 1.6222, ^{31}P -60.5 ppm; NH_4^+ salt, m. 120-122°; Ni salt, purple, m. 154-156°.[563]
Ph(t-Bu)PS_2H. Ib. m. 72-73°.[292]

E.3.3. Aliphatic and Aliphatic/Aromatic Phosphinodithioic Acids with Substituents in Aliphatic Groups

$(CF_3)_2PS_2H$. Ib.[170] IId.[170] XXIVa.[258] m. 14.0-14.5°,[258] b. 102° (extrapolated).[170]
Me ester. XXIe. m. -17.6 to -17.2°.[258]
$(NC\cdot CH_2\cdot CH_2)_2PS_2H$. Ib.[18,586] NH_4^+ salt, m. 163-165°;[586] Na salt, m. 104-106°;[18] pyridine salt, m. 95-96°;[18] cyclo-$C_6H_{11}NH_2$ salt, m. 123-124°.[586]
$H_2N\cdot CO\cdot NH\cdot CO\cdot CH_2$ ester. XXIIa. m. 132-134°.[18]
$(H_2N\cdot CH_2\cdot CH_2)EtPS_2H$. Ib. m. 248-251°.[315]
$(H_2N\cdot CH_2\cdot CH_2)(C_6H_{13})PS_2H$. Ib. m. 200-201°.[315]
$(H_2N\cdot CH_2\cdot CH_2)PhPS_2H$. Ib. m. 250-255°.[317]
$(EtNH\cdot CH_2\cdot CH_2)PhPS_2H$. Ib. m. 237-241°.[315]
$(Et_2N\cdot CH_2\cdot CH_2)PhPS_2H$. Ib. m. 192-194°.[315]
$(H_2N\cdot CH_2\cdot CH_2)(PhCH_2)PS_2H$. Ib. m. 253-257°.[315]
[2-HO-4(or 5)-$MeC_6H_3\cdot S\cdot CH_2$]$MePS_2H$.
Cyclic ester. From $(ClCH_2)MeP(S)Cl$ + 1-Me-3,4-$(HS)_2C_6H_3$. m. 92-96°, $b_{0.001}$ 150-153°.[786]

E.3.4. Diarylphosphinodithioic Acids

Ph_2PS_2H. Ib.[586] VIIIc.[289,563] XXIb.[444] m. 55-56°,[289] ^{31}P -52.3 ppm,[444] pK_a 1.75 (in 7% EtOH), 2.60 (in 80% EtOH);[345] Na salt, m. 255°;[444] MeMg deriv., m. 81-83°;[397] MeAl deriv., softens 180°, m. 260°;[130] Cl_2Ti deriv., red, m. 130°;[551] Cr^{3+} salt;[553] Mn salt, m. 250° decomp.;[553] Co^{2+} salt, lime green,[445] m. 280°;[552] Co^{2+} salt (2 pyridine), pink crystals;[552] Co^{2+} salt (2 $H\cdot CO\cdot NMe_2$), blue crystals;[552] Ni salt, blue-violet, m. > 270°;[445] Cu^{2+}salt, colorless, diamagnetic, m. 295-297°;[552] $Cu(OH)S_2PPh_2\cdot 4$ pyridine, by passing O_2 through Cu^{2+} salt in pyridine, blue crystals, m. 220-221°;[444] Ag salt, m. 150° decomp.;[553] Au^+ salt, colorless needles, m. 297° decomp;[444] Zn salt, unchanged at 200°;[445] Cd salt, stable to 250°;[553] Hg^{2+} salt, decomp. 240° without melting;[552] ClHg derivative, m. 192-193°;[353] Tl^+ salt, m. 189°;[444] Me_2Tl deriv., m. 153-155°;[97] Ph_2Tl deriv., m. 197°;[97] Cl_2Sn deriv., yellow, m. 179-180°;[97] Br_2Sn deriv., yellow, m. 175°;[97] Me_2Sn deriv., m. 161-162°;[97] Et_2Sn deriv., m. 162-163°;[97] Ph_2Sn deriv., m. 128-130°;[721] Ph_3Sn deriv., m. 128-130°;[721] Th^{4+} salt, m. > 250° decomp.;[553] UO_2^{2+} salt, m. 160° decomp.[553]
Me ester. XXIId. m. 82-83°.[295]
Et ester. XXIId. $b_{0.3}$ 166-167°, d_4^{20} 1.1919, n_D^{20} 1.6611.[295]
Bu ester. IVa.[163] XXIe*.[163] XXIId.[295] $b_{0.07}$ 162-164°,[295] $b_{0.5}$ 200-210°,[163] d_4^{20} 1.1471,[295]

n_D^{20} 1.6370.[163,295]
t-Bu ester. XXIId. XXIIe. m. 91-92°.[295]
C_6H_{13} ester. IVa.[755b] XXIId.[295] m. 46-47°,[755b] $b_{0.1}$ 190-192°.[295]
C_8H_{17} ester. IVa.[755b] XXIId.[295] $b_{0.2}$ 197-199°,[295] d_4^{20} 1.0845,[295] n_D^{20} 1.5972.[295]
C_6H_{13} CHMe ester. XXIId. XXIIe. $b_{0.015}$ 100-110°, d_4^{20} 1.0895, n_D^{20} 1.6050.[295]
$C_{10}H_{21}$ ester. XXIId. $b_{0.04}$ 155°, d_4^{20} 1.0640, n_D^{20} 1.5864.[295]
$C_{12}H_{25}$ ester. XXIId. $b_{0.2}$ 158°, d_4^{20} 1.0267, n_D^{20} 1.5726.[295]
PhO·CHMe ester. XXIIa. XXIIe. m. 56-57°.[353]
p-ClC_6H_4·O·CHMe ester. XXIIe. m. 59-60°.[353]
2,4-$Cl_2C_6H_3$·O·CHMe ester. XXIIe. m. 61.5-62.5°.[353]
p-O_2N·C_6H_4·O·CHMe ester. XXIIe. m. 110-111°.[353]
1-naphthyl·O·CHMe ester. XXIIe. m. 72-73°.[353]
2-naphthyl·O·CHMe ester. XXIIe. m. 97.7-98.7°.[353]
H·CO·CH_2·CH_2 ester. XXIIe. Liquid; p-nitrophenylhydrazone, m. 98-99°.[430]
Me·CO·CH_2·CH_2 ester. XXIIe. Liquid; p-nitrophenylhydrazone, m. 161-162°.[430]
Ph·CO·CH_2·CH_2 ester. XXIIe. m. 93°.[430]
CO·(o-C_6H_4)·CO·CH·CHPh ester. XXIIe. m. 115-117°.[557]
CO·(o-C_6H_4)·C(OMe):C·CHPh ester. From previous compound + CH_2N_2. m. 76-78°.[557]
CO·(o-C_6H_4)·CO·C(OH) ester. From Ph_2PS_2H + indanetrione. m. 114-116°.[557]
NC·CH_2·CH_2 ester. XXIIe. m. 63°.[430]
EtO_2C·CH:CMe ester. XXIe. m. 98-100°.[432]
HO_2C·CH_2·CH(CO_2H) ester. XXIIe. m. 175°.[430]
(MeO_2C)CH·CHPh ester. XXIIe. m. 89-90°.[430]
SO_2·(o-C_6H_4)·CO·CH·CHPh ester. XXIIe. m. 135°.[556]
SO_2·(o-C_6H_4)·C(OMe):C·CHPh ester. From previous compound + CH_2N_2. m. 115-117°.[556]
SO_2·[1,2-(4-MeC_6H_3)]·CO·CH·CHPh ester. XXIIe. m. 170°.[556]
EtO·$SiMe_2$·CH_2 ester. XXIIa. $b_{0.001}$ 210-230°, d_4^{20} 1.1481, n_D^{20} 1.6056.[21]
$(EtO)_2$SiMe·CH_2 ester. XXIIa. $b_{0.01}$ 230-235°, d_4^{20} 1.1545, n_D^{20} 1.5848.[21]
-CH_2·$SiMe_2$·O·$SiMe_2$·CH_2- di-ester. XXIIa. m. 126-127°.[21]
2,5-(Ph·CO·NH)$_2C_6H_3$ ester. XXIIe. m. 160°.[558]
2,5-(Ph·SO_2·NH)$_2C_6H_3$ ester. XXIIe. m. 140°.[558]
1-HO-4-(Ph·SO_2·NH)-naphth-2-yl ester. XXIIe. m. 185°.[558]

1-MeO-4-(Ph·SO_2·NH)-naphth-2-yl ester. From previous compound + CH_2N_2, m. 110°.[558]

$Ph_2P(S)$·S·$P(S)Ph_2$. IIh.[17] XXIIg.[295] m. 121.5°.[17]

Me·CO·S·$P(S)Ph_2$. XXIIi. m. 86-90°.[249]

Ph·CO·S·$P(S)Ph_2$. XXIIi. m. 120-121°.[249]

MeS·CO·$P(S)Ph_2$. XXIIi. m. 80°.[286]

PhS·CO·$P(S)Ph_2$. XXIIi. m. 62°.[286]

Et_2N·CO·$P(S)Ph_2$. XXIIi. m. 130-132.5°.[366]

$Ph_2P(S)$·S·S·$P(S)Ph_2$. XXIIj.[444,552] Also from Ph_2PH + excess S.[586] m. 146-147°.[586]

Morpholino·S·$P(S)Ph_2$. IIh. n_D^{20} 1.6280.[17]

p-FC_6H_4·S·S·$P(S)Ph_2$. From previous compound + p-FC_6H_4·SH + HCl. m. 145°.[17]

$(p\text{-}ClC_6H_4)_2PS_2H$. VIIIc. m. 82-83°, p$K_a$ 1.79 (in 7% EtOH), 2.69 (in 80% EtOH).[345]

$(p\text{-}MeC_6H_4)_2PS_2H$. VIIIc.[345,531] m. 80-81°,[341,345] pK_a 1.81 (in 7% EtOH), 2.65 (in 80% EtOH).[345]

Bu ester. XXIe*. $b_{0.02}$ 192-195°, n_D^{20} 1.6219.[163]

E.3.5 Cyclic Phosphinodithioic Acids

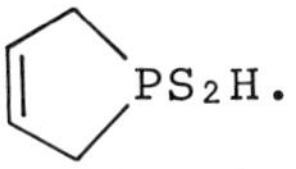

Et ester. XXIe. m. 32°, $b_{0.5}$ 98-101°.[671]

$ClCH_2$·CH_2 ester. Xa. $b_{0.5}$ 103°, d_4^{20} 1.3204, n_D^{20} 1.6205.[425]

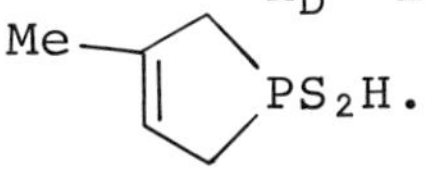

$ClCH_2$·CH_2 ester. Xa. b_5 176-178°, d_4^{20} 1.2860, n_D^{20} 1.6071.[673]

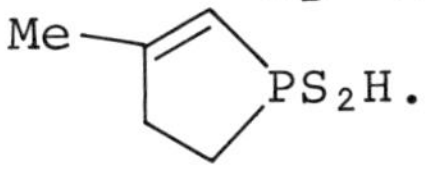

Et ester. XXIe. m. 30°, $b_{0.05}$ 121-124°, n_D^{20} 1.6055.[278]

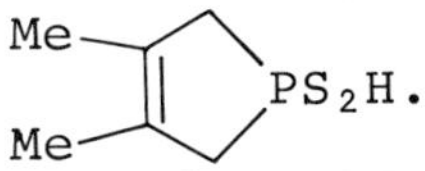

$ClCH_2$·CH_2 ester. Xa. $b_{0.5}$ 130°, d_4^{20} 1.2530, n_D^{20} 1.6030.[425]

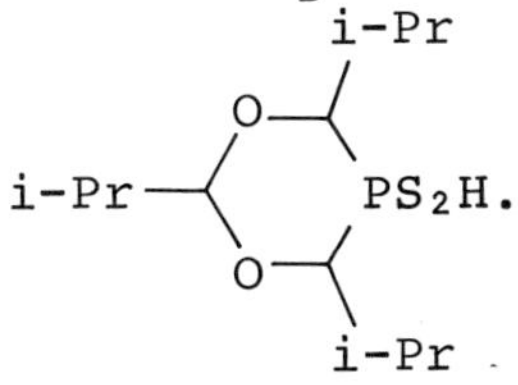

Ib. m. 49-52°; NH_4^+ salt, m. ~185° decomp.; Na salt, m. 286-287° decomp.[649]

E.3.6. Acids Containing Two Phosphinodithioic Acid Groups

Ph·PS_2H·PS_2H·Ph. Ib. Decomp. 225°.[314]
Bu·PS_2H·CH_2·CH_2·PS_2H·Bu.
    $(PhCH_2)_2$ ester. d_4^{20} 1.2006, n_D^{20} 1.6450.[265]
cyclo-C_6H_{11}·PS_2H·CH_2·CH_2·PS_2H·cyclo-C_6H_{11}.
    $(PhCH_2)_2$ ester. m. 109.5-110.5°.[265]
Ph·PS_2H·CH_2·CH_2·PS_2H·Ph. Ib. m. 142-144°.[321]
Ph·PS_2H·$(CH_2)_3$·PS_2H·Ph. Ib. Liquid.[314]
cyclo-C_6H_{11}·PS_2H·$(CH_2)_4$·PS_2H·cyclo-C_6H_{11}. Ib. m. 156-157°.[310]
Ph·PS_2H·$(CH_2)_4$·PS_2H·Ph. Ib.[314,316] m. 136.5-137.5°;[316] Na salt, decomp. 286°.[316]
    Me_2 ester. XXIIa. m. 105-109° (mixed diastereoisomers).[316]
    Et_2 ester. XXIIa. (±) and meso forms, m. from 72° and m. 115-119°, separated by crystallization.[316]
    $(PhCH_2)_2$ ester. XXIIa. m. 120-123°.[316]
    di-amide. XXIi. m. 180-182°.[316]
    di-(Bu amide). XXIi. m. 143-146°.[316]
    Ac_2 derivative. XXIIi. m. 166.5-168.5°.[316]
    $(Ph·CO)_2$ derivative. XXIIi. m. 187°.[316]
Et·PS_2H·$(CH_2)_5$·PS_2H·Et. Ib. Liquid; Ni salt, blue-violet crystals, decomp. 200°.[311]
Et·PS_2H·$(CH_2)_6$·PS_2H·Et. Ib. m. 75-77°; Co^{2+} salt, green crystals, decomp. 210-213°.[311]
cyclo-C_6H_{11}·PS_2H·$(CH_2)_6$·PS_2H·cyclo-C_6H_{11}. Ib. m. 108-110°.[310]

E.4. Phosphinoselenoic, Phosphinoselenothioic, and Phosphinodiselenoic Acids and Their Derivatives

Et_2POSeH. VIg. XXIa. m. 17°, $b_{0.05}$ 76-77°, d_4^{25} 1.4409, n_D^{25} 1.5624, n_D^{31} 1.5590.[490]
i-Pr_2POSeH. VIg. m. 88°.[488]
Bu_2POSeH. VIg. $b_{0.05}$ 93-94°, n_D^{20} 1.5260.[488]
s-Bu_2POSeH. VIg. $b_{0.05}$ 82-83°, n_D^{20} 1.5430.[488]
Ph_2POSeH.
    Ph O-ester. IVa. m. 114-115°.[527]
    Ph Se-ester. IVa.[514] IVa of unisolated Ph_2PSePh prepared by several routes.[588] XXIe.[587] Also from $Ph_2P(O)$·$P(O)Ph_2$ + PhSeBr,[588] m. 78-80°,[587,588] 95°.[514]
Et_2PSSeH. IIId.[441,443] XXIb.[443] XXIc.[441,443] The free acid is an unstable yellow oil; Na salt ($2H_2O$), m. 124-125°;[441,443] Zn salt, m. 157°;[441] Cd salt, m. 160°;[441] Ni salt, olive green, decomp. > 142°;[441] Pb^{2+} salt, green-yellow, m. 132°;[441] Bi salt, orange-red, m. 94°.[441]

Et Se-ester. XXIIa. Colorless, $b_{0.06}$ 71.5°, n_D^{20} 1.5854.[441,443]

Pr Se-ester. XXIIa. Colorless, $b_{0.06}$ 86-87°, n_D^{20} 1.5736.[441,443]

$Et_2P(S){\cdot}Se{\cdot}Se{\cdot}P(S)Et_2$. XXIIj. Yellow plates, m. 56.5°,[441,443] x-ray structure determination.[304]

$Et_2P(S){\cdot}Se{\cdot}Se{\cdot}Se{\cdot}P(S)Et_2$. From decomposition of acid. Also minor product of XXIIj. Ochre-colored crystals, m. 115-121°.[443]

Et_2PSe_2H. XXIc.[442,443] The free acid is an unstable orange oil; Na salt ($2H_2O$), m. 129-130°;[442,443] Pd^{2+} salt, brown-red, m. 197°;[442] Zn salt, m. 151°;[442] Cd salt, m. 158°;[442] In^{3+} salt, bright yellow, m. 146°;[442] Tl^+ salt, pale yellow, m. 129°;[442] Pb^{2+} salt, lemon yellow, m. 157-158°;[442] Bi salt, vermilion-red, m. 132°.[442]

Et ester. XXIIa. Colorless, $b_{0.13}$ 105°, n_D^{20} 1.6193.[443]

Pr ester. XXIIa. Colorless, $b_{0.1}$ 98°, n_D^{20} 1.6037.[443]

$Et_2P(Se){\cdot}Se{\cdot}P(Se)Et_2$. XXIf. Colorless plates, m. 65° [442,443]

$Et_2P(Se){\cdot}Se{\cdot}Se{\cdot}Se{\cdot}P(Se)Et_2$. XXIIj. Yellow needles, m. 135-137°.[443]

E.5. Derivatives of Phosphinimidic and Phosphinimidothioic Acids

Unsymmetrical phosphinimidic secondary amides are potentially tautomeric:

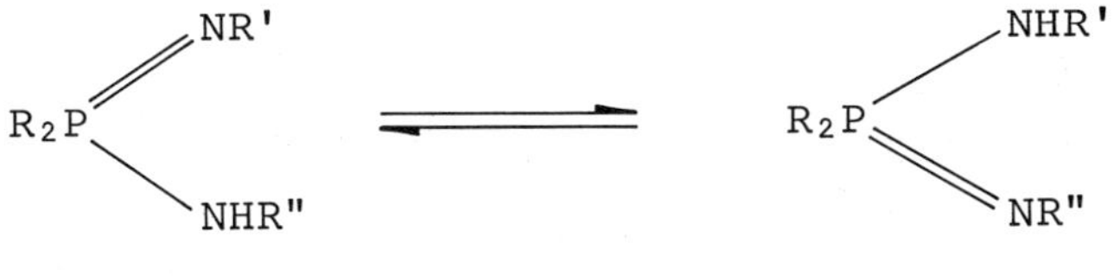

and individual compounds should be sought under both formulas. Those described have all been prepared by reaction of azides with phosphinous amides and are formulated with the group derived from the azide double-bonded to phosphorus.

E.5.1. Derivatives of Dialkylphosphinimidic Acids

$Et_2P(:NPh)NHPh$. IVc. m. 161-163°, treatment with NaH and then with MeI gives $Et_2P^+(NMePh)_2\ I^-$, m. 191.5-192°.[245]

$Et_2P(:NPh)NMePh$. IVc. $b_{0.0001}$ 103-105°, d_4^{20} 1.0902, n_D^{20} 1.6013.[246]

$Et_2P(:NPh)NMe{\cdot}C_6H_4$-m-F. IVc. $b_{0.0001}$ 110-112°, d_4^{20} 1.1382, n_D^{20} 1.5958.[479]

$Et_2P(:NPh)NMe{\cdot}C_6H_4$-m-Cl. IVc. $b_{0.0001}$ 125-126°, d_4^{20} 1.1628, n_D^{20} 1.6131.[479]
$Et_2P(:NPh)NMe{\cdot}C_6H_4$-p-Cl. IVc. $b_{0.0001}$ 120-121°, d_4^{20} 1.1620, n_D^{20} 1.6119.[479]
$Et_2P(:NPh)NMe{\cdot}C_6H_4$-p-Br. IVc. $b_{0.0001}$ 139-141°, d_4^{20} 1.3055, n_D^{20} 1.6265.[479]
$Et_2P(:NPh)NMe{\cdot}C_6H_4$-p-OMe. IVc. $b_{0.0001}$ 131-133°, d_4^{20} 1.1050, n_D^{20} 1.5989.[479]
$Et_2P(:NPh)NMe{\cdot}C_6H_4$-m-Me. IVc. $b_{0.0001}$ 110-112°, d_4^{20} 1.0688, n_D^{20} 1.5958.[479]
$Et_2P(:NPh)NMe{\cdot}C_6H_4$-p-Me. IVc. $b_{0.0001}$ 113-115°, d_4^{20} 1.2767, n_D^{20} 1.5967.[479]
$Et_2P(:NPh)NMe{\cdot}C_6H_4$-p-$CF_3$. IVc. $b_{0.0001}$ 108-110°, d_4^{20} 1.1916, n_D^{20} 1.5599.[479]
$Et_2P(:NPh)NMe{\cdot}C_6H_4$-p-CN. IVc. m. 93.5-95°.[479]
$Et_2P(:N{\cdot}C_6H_4$-m-F)NMePh. IVc. $b_{0.0001}$ 108-110°, d_4^{20} 1.1370, n_D^{20} 1.5879.[246]
$Et_2P(:N{\cdot}C_6H_4$-p-F)NMePh. IVc. $b_{0.0001}$ 114-115°, d_4^{20} 1.1340, n_D^{20} 1.5853.[246]
$Et_2P(:N{\cdot}C_6H_4$-m-Cl)NHPh. IVc. m. 143-144°.[245]
$Et_2P(N{\cdot}C_6H_4$-m-Cl)NMePh. IVc. $b_{0.0001}$ 123-125°, d_4^{20} 1.1555, n_D^{20} 1.6055.[246]
$Et_2P(:N{\cdot}C_6H_4$-p-Cl)NMePh. IVc. $b_{0.0001}$ 130-132°, d_4^{20} 1.1578, n_D^{20} 1.6065.[246]
$Et_2P(:N{\cdot}C_6H_4$-p-Cl)$NMe{\cdot}C_6H_4$-m-Cl. IVc. $b_{0.0001}$ 148-158°, n_D^{20} 1.6168.[479]
$Et_2P(:N{\cdot}C_6H_4$-p-OMe)NMePh. IVc. $b_{0.0001}$ 129-130°, d_4^{20} 1.1138, n_D^{20} 1.5985.[246]
$Et_2P(:N{\cdot}C_6H_4$-p-OMe)$NMe{\cdot}C_6H_4$-m-Cl. IVc. $b_{0.0001}$ 150-152°, n_D^{20} 1.6088.[479]
$Et_2P(:N{\cdot}C_6H_4$-p-NO_2)NMePh. IVc. $b_{0.0001}$ 147-150°, n_D^{20} 1.6150.[479]
$Et_2P(:N{\cdot}C_6H_4$-m-Me)NMePh. IVc. $b_{0.0001}$ 108-111°, d_4^{20} 1.0700, n_D^{20} 1.5958.[246]
$Et_2P(:N{\cdot}C_6H_4$-p-Me)NHPh. IVc. m. 140-148°.[245]
$Et_2P(:N{\cdot}C_6H_4$-p-Me)NMePh. IVc. $b_{0.0001}$ 110-112°, d_4^{20} 1.0708, n_D^{20} 1.5955.[246]
$Et_2P(:N{\cdot}C_6H_4$-p-CF_3)NMePh. IVc. $b_{0.0001}$ 110-112°, d_4^{20} 1.1899, n_D^{20} 1.5600.[246]
$Et_2P(:N{\cdot}C_6H_4$-p-CF_3)$NH{\cdot}C_6H_4$-p-OMe. IVc. m. 156.5-158°.[239]
$Et_2P(:N{\cdot}C_6H_4$-p-CN)NHPh. IVc. Very viscous, $b_{0.0001}$ 140-142°, n_D^{20} 1.6260.[246]
$Pr_2P(:N{\cdot}C_6H_4$-p-CF_3)$NH{\cdot}C_6H_4$-p-OMe. IVc. m. 129-131°.[239]
i-$Bu_2P(:N{\cdot}CO{\cdot}CF_3)$OBu. IVc. $b_{0.1}$ 95-98°, d_4^{20} 1.0640, n_D^{20} 1.4365.[338]
i-$Bu_2P(:N{\cdot}CO{\cdot}CCl_3)$OBu. IVc. $b_{0.005}$ 85°, d_4^{20} 1.1478, n_D^{20} 1.4877.[338]
i-$Bu_2P(:N{\cdot}C_6H_4$-p-CF_3)$NH{\cdot}C_6H_4$-p-OMe. IVc. m. 141-143°.[239]
$(CCl_3)_2P(:NH)Cl$. XXc. m. 41-42°, $b_{0.03}$ 68-69°.[428]

E.5.2. Derivatives of Diphenylphosphinimidic Acids

$Ph_2P(:NMe)NPr_2$. IVc. $b_{0.01}$ 155-160°.[787]

$Ph_2P(:N{\cdot}CO{\cdot}CHCl_2)Cl$. XXc. Oil.[733]

$Ph_2P(:N{\cdot}CO{\cdot}CCl_3)Cl$. XXc. m. 92-94°.[733]

$Ph_2P(:N{\cdot}CS{\cdot}NMe_2)SMe$. See $Ph_2P(S){\cdot}NH{\cdot}CS{\cdot}NMe_2$.

$Ph_2P{:}N{\cdot}CO{\cdot}NPh$ (ring). From $Ph_2PCl_2{\cdot}NCO + PhNH_3{}^+Cl^-$. m. 214-216°; hydrochloride, m. 142-144° decomp.[401]

$Ph_2P(:NPh)Cl$. IVc.[95] XXc.[795] m. 96-97°,[95] turns red in sunlight.[95]

$Ph_2P(:NPh)N_3$. From chloride + LiN_3. Yellow oil, slowly reddens in light, very easily hydrolyzed.[95]

$Ph_2P(:NPh)OEt$. IVc. m. 66-68°.[787]

$Ph_2P(:NPh)NPr_2$. IVc. m. 71-72°.[787]

$Ph_2P(:NPh)NHPh$. IVc. m. 178-179°.[787]

$Ph_2P(:NPh)NMe{\cdot}SO_2{\cdot}C_6H_4\text{-}p\text{-}Me$. IVc. m. 119-120°.[95,787]

$Ph_2P(:NPh)N{:}PPh_3$. From azide + Ph_3P. m. 192-193°.[95]

$Ph_2P(:N{\cdot}C_6H_4\text{-}p\text{-}CF_3)NH{\cdot}C_6H_4\text{-}p\text{-}OMe$. IVc. m. 121-122°.[239]

$Ph_2P(:N{\cdot}CO{\cdot}Ph)NH{\cdot}CO{\cdot}Ph$. Ic. m. 171-173°.[159]

$Ph_2P(:N{\cdot}CO{\cdot}C_6H_4\text{-}p\text{-}NO_2)NH{\cdot}CO{\cdot}C_6H_4\text{-}p\text{-}NO_2$. Ic. m. 223-224°.[159]

$Ph_2P(:N{\cdot}SO_2{\cdot}Ph)Cl$. XXc. m. 109-111°.[731]

$Ph_2P(:N{\cdot}SO_2{\cdot}Ph)OMe$. XXIe. m. 109-110°.[730]

$Ph_2P(:N{\cdot}SO_2{\cdot}Ph)OEt$. XXIe. m. 117-118°.[730]

$Ph_2P(:N{\cdot}SO_2{\cdot}Ph)OPh$. XXIe. m. 113-114°.[730]

$Ph_2P(:N{\cdot}SO_2{\cdot}C_6H_4\text{-}o\text{-}NO_2)Cl$. XXc. m. 159-161°.[731]

$Ph_2P(:N{\cdot}SO_2{\cdot}C_6H_4\text{-}m\text{-}NO_2)Cl$. XXc. m. 118-120°.[731]

$Ph_2P(:N{\cdot}SO_2{\cdot}C_6H_4\text{-}p\text{-}NO_2)Cl$. XXc. m. 164-168°.[731]

$Ph_2P(:N{\cdot}SO_2{\cdot}C_6H_4\text{-}p\text{-}NO_2)OMe$. XXIe. m. 170-172°.[730]

$Ph_2P(:N{\cdot}SO_2{\cdot}C_6H_4\text{-}p\text{-}NO_2)OEt$. XXIe. m. 49-50°.[730]

$Ph_2P(:N{\cdot}SO_2{\cdot}C_6H_4\text{-}o\text{-}Me)Cl$. XXc. m. 129-130°.[731]

$Ph_2P(:N{\cdot}SO_2{\cdot}C_6H_4\text{-}o\text{-}Me)OMe$. XXIe. m. 123-124°.[730]

$Ph_2P(:N{\cdot}SO_2{\cdot}C_6H_4\text{-}o\text{-}Me)OEt$. XXIe. m. 122-123°.[730]

$Ph_2P(:N{\cdot}SO_2{\cdot}C_6H_4\text{-}o\text{-}Me)OPh$. XXIe. m. 139-140°.[730]

$Ph_2P(:N{\cdot}SO_2{\cdot}C_6H_4\text{-}p\text{-}Me)Cl$. IVc.[91,95] XXc.[91,95,731] m. 99-101°,[731] 106-108°.[91,95]

$Ph_2P(:N{\cdot}SO_2{\cdot}C_6H_4\text{-}p\text{-}Me)N_3$. From chloride + NaN_3. m. 84-85°.[95]

$Ph_2P(:N{\cdot}SO_2{\cdot}C_6H_4\text{-}p\text{-}Me)OMe$. XXIe. m. 112-113°.[730]

$Ph_2P(:N{\cdot}SO_2{\cdot}C_6H_4\text{-}p\text{-}Me)OEt$. IVc.[787] XXIe.[730,787] m. 98-99°.[787]

$Ph_2P(:N{\cdot}SO_2{\cdot}C_6H_4\text{-}p\text{-}Me)OPh$. XXIe. m. 125-126°.[730]

$Ph_2P(:N{\cdot}SO_2{\cdot}C_6H_4\text{-}p\text{-}Me)NH_2$. XXIi. m. 139-140°.[787]

$Ph_2P(:N{\cdot}SO_2{\cdot}C_6H_4\text{-}p\text{-}Me)N{\cdot}CH_2{\cdot}CH_2$ (ring). XXIi. m. 153-154°.[95,787]

$Ph_2P(:N{\cdot}SO_2{\cdot}C_6H_4\text{-}p\text{-}Me)NHPh$. XXIi. m. 202-204°.[787]

$Ph_2P(:N \cdot SO_2 \cdot C_6H_4\text{-p-Me})N$ 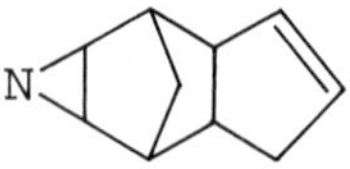 . From azide + dicyclopentadiene, m. 143-144°.[95]

$Ph_2P(:N \cdot SO_2 \cdot C_6H_4\text{-p-Me})NH \cdot SO_2 \cdot C_6H_4\text{-p-Me}$. XXIi. m. 203-205°.[787]

$Ph_2P(:N \cdot SO_2 \cdot C_6H_4\text{-p-Me})NH \cdot NH_2$. XXIi. m. 158-160°.[787]

$Ph_2P(:N \cdot SO_2 \cdot C_6H_4\text{-p-Me})NH \cdot NHPh$. XXIi. m. 215-217°.[91]

$Ph_2P(:N \cdot SO_2 \cdot C_6H_4\text{-p-Me})N:NPh$. From previous compound + N-bromosuccinimide. Red-violet crystals, m. 124-125°.[91]

$Ph_2P(:N \cdot SO_2 \cdot C_6H_4\text{-p-Me})N:PPh_3$. From azide + Ph_3P. m. 220-221°.[95]

$Ph_2P[:N \cdot SO_2 \cdot (1\text{-naphthyl})]Cl$. XXc. m. 114-116°.[731]

$Ph_2P[:N \cdot SO_2 \cdot (2\text{-naphthyl})]Cl$. XXc. m. 135-138°.[731]

$Ph_2P[:N \cdot SO_2 \cdot (2\text{-naphthyl})]OMe$. XXIe. m. 118-119°.[730]

$Ph_2P[:N \cdot SO_2 \cdot (2\text{-naphthyl})]OEt$. XXIe. m. 104-105°.[730]

$Ph_2P[:N \cdot SO_2 \cdot (2\text{-naphthyl})]OPh$. XXIe. m. 130-132°.[730]

$Ph_2P(:N \cdot N:N \cdot SO_2 \cdot C_6H_4\text{-p-Me})NPr_2$. IVc. Yellow, decomp. 72°.[94]

$Ph_2P(:N \cdot N:N \cdot SO_2 \cdot C_6H_4\text{-p-Me})NHPh$. IVc. Yellow, decomp. 80°.[94]

$[Ph_2P(:N \cdot N:N \cdot SO_2 \cdot C_6H_4\text{-p-Me})]_2NPh$. IVc. Yellow, decomp. 103°.[94]

$Ph_2P(:N \cdot SiPh_3)NH \cdot SiPh_3$. Ic. m. 161-162°, ^{31}P -43 ppm.[582]

E.5.3. Derivatives of Unsymmetrical Diarylphosphinimidic Acids

The phosphinimidic chlorides, prepared by IVd or XXc are viscous oils or brittle glasses. Reaction of the chlorides with sodium methoxide, phenoxide, or 1-naphthoxide, gives roughly equal mixtures of crystalline esters, listed below, and very viscous liquids apparently isomeric with them, the isomerism possibly resulting from restricted rotation about the P = N bond.

$(p\text{-}BrC_6H_4)PhP(:N \cdot SO_2 \cdot Ph)OMe$. XXIe. m. 81-82°.[729]

$(p\text{-}BrC_6H_4)PhP(:N \cdot SO_2 \cdot Ph)OPh$. XXIe. m. 187-188°.[729]

$(p\text{-}BrC_6H_4)PhP(:N \cdot SO_2 \cdot Ph)O(1\text{-naphthyl})$. XXIe. m. 158-159°.[729]

$(p\text{-}BrC_6H_4)PhP(:N \cdot SO_2 \cdot C_6H_4\text{-p-Cl})OPh$. XXIe. m. 138-139°.[729]

$(p\text{-}BrC_6H_4)PhP(:N \cdot SO_2 \cdot C_6H_4\text{-p-Cl})O(1\text{-naphthyl})$. XXIe. m. 126-127°.[729]

$(p\text{-}BrC_6H_4)PhP(:N \cdot SO_2 \cdot C_6H_4\text{-p-}NO_2)OMe$. XXIe. m. 86-87°.[729]

$(p\text{-}BrC_6H_4)PhP(:N \cdot SO_2 \cdot C_6H_4\text{-p-}NO_2)OPh$. XXIe. m. 167-168°.[729]

(p-BrC_6H_4)PhP(:N·SO_2·C_6H_4-p-NO_2)O(1-naphthyl). XXIe. m. 159-160°.[729]
(p-BrC_6H_4)PhP[:N·SO_2·(1-naphthyl)]OMe. XXIe. m. 162-163°.[729]
(p-BrC_6H_4)PhP[:N·SO_2·(1-naphthyl)]OPh. XXIe. m. 156-158°.[729]
(p-MeC_6H_4)PhP(:N·SO_2·Ph)OPh. XXIe. m. 168-170°.[729]
(p-MeC_6H_4)PhP(:N·SO_2·Ph)O(1-naphthyl). XXIe. m. 148-149°.[729]
(p-MeC_6H_4)PhP(:N·SO_2·C_6H_4-p-Cl)O(1-naphthyl). XXIe. m. 149-150°.[729]
(p-MeC_6H_4)PhP(:N·SO_2·C_6H_4-p-NO_2)O(1-naphthyl). XXIe. m. 152-153°.[729]
(p-MeC_6H_4)PhP[:N·SO_2·(1-naphthyl)]O(1-naphthyl). XXIe. m. 168-169°.[729]

E.5.4. Derivatives of Phosphinothioimidic Acids

(PhNH·CHPh)EtP(:N·NHPh)S(i-Pr). IXd. m. 145-147°.[123]
(p-O_2N·C_6H_4·NH·CHPh)MeP(:N·NHPh)SBu. IXd. m. 149-150°.[123]
(p-O_2N·C_6H_4·NH·CHPh)MeP(:N·NHPh)S·CH_2·CH_2·CH_2·$CHMe_2$. IXd. m. 147-148°.[123]
(m-MeC_6H_4·NH·CHPh)EtP(:N·NHPh)S(i-Pr). IXd. m. 152-154°.[123]
(p-MeC_6H_4·NH·CHPh)MeP(:N·NHPh)SBu. IXd. m. 151-152°.[123]

E.5.5. Pyrophosphinimidic and Pyrophosphinothioimidic Compounds

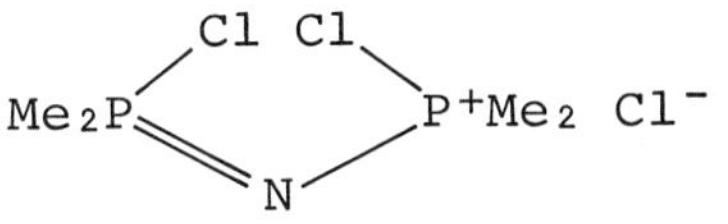

XXVb. Needles, m. 157°, decomp. 245-255°, ^{31}P -60.7 ppm.[702]

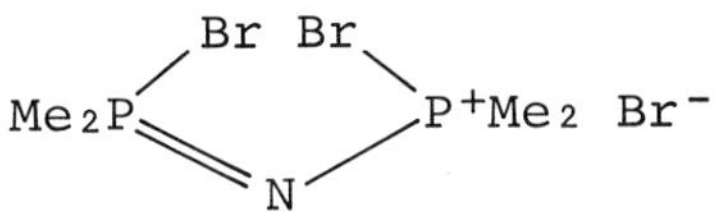

XXVb. Decomp. 83° without melting, ^{31}P -58.2 ppm.[702]

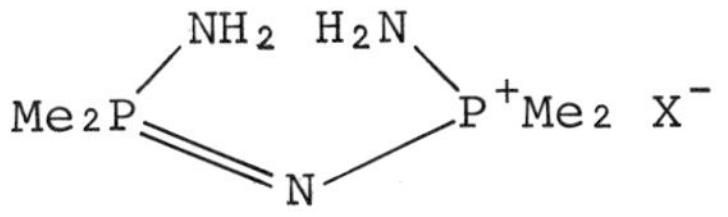

XXIi. X=Cl, long silky needles, hygroscopic, m. 108-109°; X=BPh_4, mother-of-pearl-like plates, m. 229°, ^{31}P -31.05 ppm.[702]

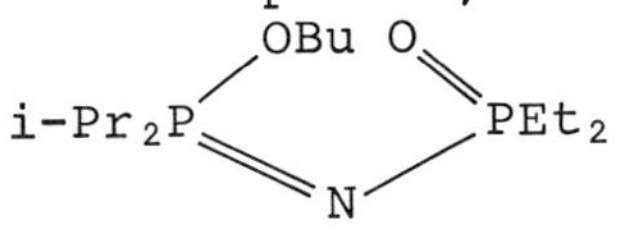

IVc. $b_{0.3}$ 132-134°, d_4^{20} 0.9526, n_D^{20} 1.4700.[247]

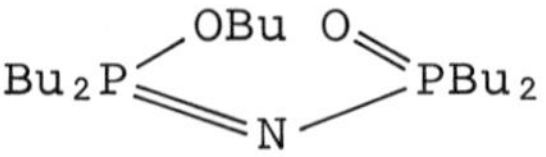

IVc. $b_{0.0003}$ 101-102°, d_4^{20} 1.0093, n_D^{20} 1.4540.[247]

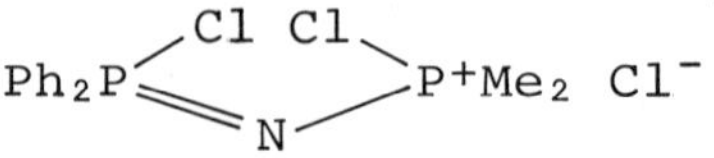

XXVb. Glassy, ^{31}P -64.3 and -39.2 ppm.[702]

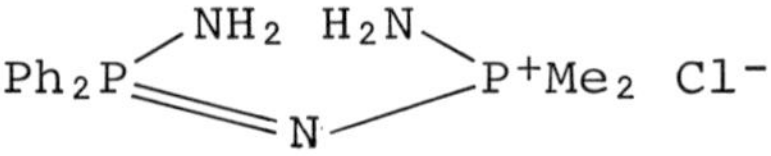

XXIi. Needles, m. 159-161°, ^{31}P -34.1 and -16.0 ppm.[702]

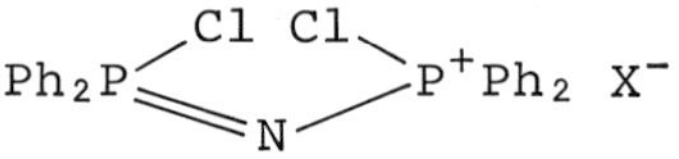

XXVb. X=Cl, m. 243-244°, ^{31}P -43.6 ppm.; X=PCl_6, crystalline, very sensitive to moisture, ^{31}P -43.3 ppm.; X=$SbCl_6$, m. 127-128°, ^{31}P -43.5 ppm.[705]

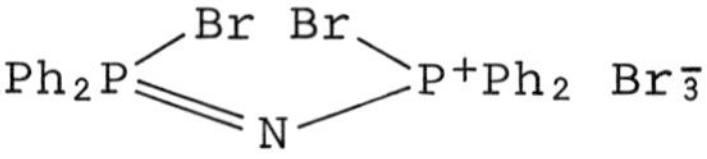

XXVb. Orange needles, loses Br_2 at room temperature.[705]

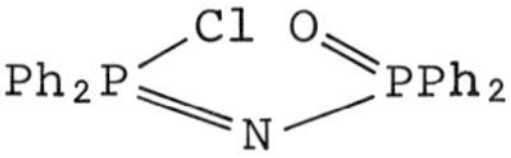

IVg. m. 282-284°.[244]

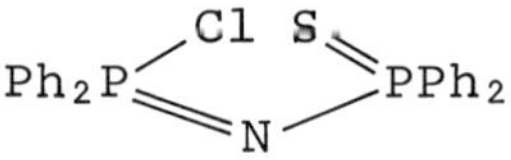

XXVb. m. 114-116°, ^{31}P -45.1 and -30.2 ppm.[705]

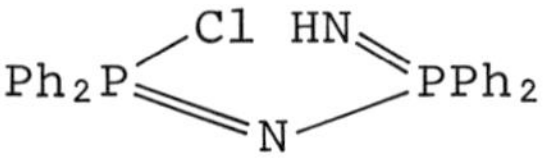

From Ph_2PCl + N_2H_4 (then heated), m. 268° decomp., ^{31}P -27.5 and -15.0 ppm.[735a]

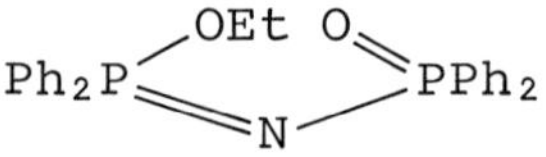

IVc. m. 98-99°.[247]

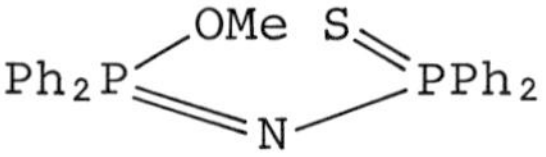

XXIe. Octahedral crystals, m. 151-152°, ^{31}P -41.8 and -28.9 ppm.[705]

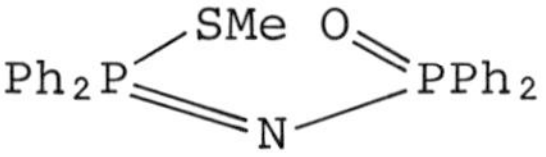

XXIIIe*. XXVc. m. 111-114°, ^{31}P -26.7 and -13.3 ppm.[705]

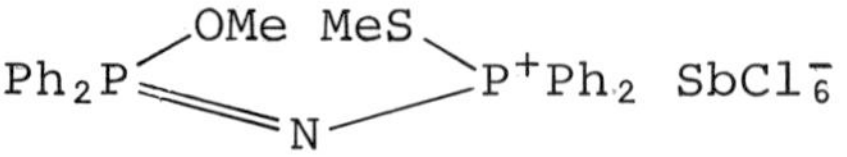

XXVc. Needles, m. 176-177°, ^{31}P -38.8 and -35.3 ppm.[705]

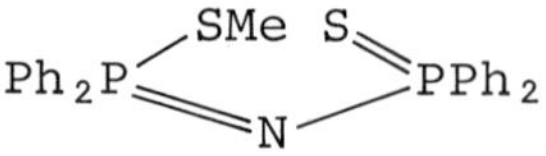

XXVc.[703] m. 190-191°,[703] ^{31}P -42.4 and -29.2 ppm.[701]

XXVc.[703] X=BPh_4, m. 138-139°;[703] X=PF_6, m. 168-170°;[703] X=$SbCl_6$, m. 187-188°,[703] ^{31}P -38.4 ppm.[701]

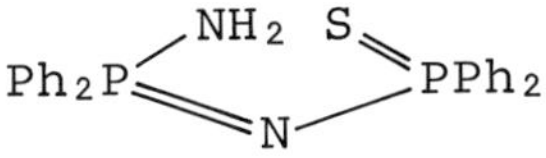

XXIi. m. 128-129° ^{31}P -42.4 and -20.6 ppm.; hydrochloride, m. 125° decomp., ^{31}P -55.9 and -40.7 ppm.[705]

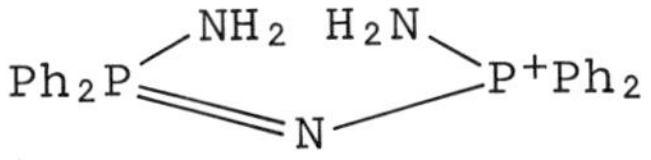

IVg.[244,735a] XXIi.[244] X=Cl, m. 245-246°;[735a] X=PF_6, m. 153-154°;[735a] X= picrate, m. 147°;[735a] X= anthraquinone-2-sulfonate, m. 218°.[735a]

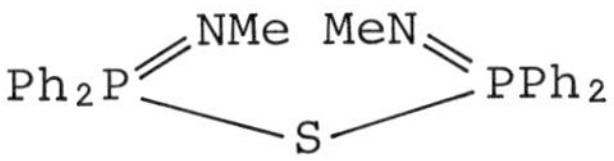

Minor product of heating $Ph_2P(S)NHMe$, m. 169-170°.[725]

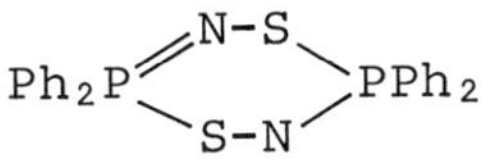

By strongly heating $Ph_2P(S)NHPh$, m. 265°.[725]

(Received December 3, 1971)

REFERENCES

1. Abel, E. W., and I. H. Sabherwal, J. Chem. Soc. (A) 1968, 1105.
2. Abramov, V. S., and V. I. Barabanov, Zh. Obshch. Khim., 36, 1830 (1966); C. A., 66, 55554 (1967).
3. Abramov, V. S. and V. I. Barabanov, Khim. Org. Soedin. Fosfora, Akad. Nauk SSSR, Otd. Obshch. Tekh. Khim., 1967, 135; C. A. 69, 67469 (1968).
4. Abramov, V. S., V. I. Barabanov, and Z. Ya. Sazonova, Zh. Obshch. Khim., 39, 1543 (1969); C. A. 71, 113045 (1969).
5. Abramov, V. S., R. Sh. Chenoborisov, and V. V. Markin, Zh. Obshch. Khim., 39, 464 (1969); C. A. 70, 115252 (1969).
6. Abramov., V. S., and N. I. Dyakonova, Zh. Obshch. Khim., 38, 1502 (1968); C. A. 69, 106805 (1968).
7. Abramov. V. S., and M. I. Kashirskii, Zh. Obshch. Khim., 28, 3056 (1958), C. A. 53, 10091 (1959).
8. Abramov, V. S., and V. K. Khairullin, Zh. Obshch. Khim., 27, 2387 (1957); C. A. 52, 7125 (1958).
9. Abramov, V. S., L. A. Tarasov, and F. G. Fatykhova, Zh. Obshch. Khim., 38, 1794 (1968); C. A. 70, 4224 (1969).
10. Akamsin, V. D., and N. I. Rizpolozhenskii, Izv. Akad. Nauk SSSR, Ser. Khim., 1966, 493; C. A. 65, 5480 (1966).

11. Akamsin, V. D., and N. I. Rizpolozhenskii, Dokl. Akad. Nauk SSSR, 168, 807 (1966); C. A. 65, 8953 (1966).
12. Akamsin, V. D., and N. I. Rizpolozhenskii, Izv. Akad. Nauk SSSR, Ser. Khim., 1967, 1976; C. A. 68, 29783 (1968).
13. Akamsin, V. D., and N. I. Rizpolozhenskii, Izv. Akad. Nauk SSSR, Ser. Khim., 1967, 1983; C. A. 68, 114692 (1968).
14. Akamsin, V. D., and N. I. Rizpolozhenskii, Izv. Akad. Nauk SSSR, Ser. Khim., 1967, 1987; C. A. 68, 114693 (1968).
15. Aksnes, G., and K. Bergesen, Acta Chem. Scand. 20, 2508 (1966).
16. Alexandrova, I. A., and G. M. Vinokurova, Izv. Akad. Nauk SSSR, Ser. Khim., 1969, 1163; C. A. 71, 50084 (1969).
17. Almasi, L., and L. Paskucz, Chem. Ber., 102, 1489 (1969).
18. American Cyanamid Co., Brit. 870,005; C. A. 56, 4797 (1962).
19. Andrianov, K. A., and I. K. Kuznetsova, Izv. Akad. Nauk SSSR, Otd. Khim. Nauk, 1961, 1454; C. A. 56, 501 (1962).
20. Andrianov, K. A., and I. K. Kuznetsova, Izv. Akad. Nauk SSSR, Otd. Khim. Nauk, 1961, 1792; C. A. 56, 7344 (1962).
21. Andrianov, K. A., and I. K. Kuznetsova, Izv. Akad. Nauk SSSR, Otd. Khim. Nauk, 1962, 456; C. A. 57, 15141 (1962).
22. Andrianov, K. A., and I. K. Kuznetsova, Izv. Akad. Nauk SSSR, Ser. Khim., 1964, 651; C. A. 61, 3143 (1964).
23. Anisimov, K. N., and N. E. Kolobova, Izv. Akad. Nauk SSSR, Otd. Khim. Nauk, 1962, 444; C. A. 57, 12529 (1962).
24. Arbuzov, A., J. Russ. Phys. Chem. Soc., 42, 395 (1910); C. A. 5, 1397 (1911).
25. Arbuzov, A. E., Zh. Obshch. Khim., 4, 898 (1934); C. A. 29, 2146 (1935).
26. Arbuzov, A. E., and B. A. Arbuzov, J. Russ. Phys. Chem. Soc., 61, 1599 (1929); C. A. 24, 5289 (1930).
27. Arbuzov, A. E., and I. A. Arbuzova, J. Russ. Phys. Chem. Soc., 61, 1905 (1929); C. A. 24, 5289 (1930).
28. Arbuzov, A. E., S. Yu. Baigil'dina, F. G. Valitova, and R. R. Shagidullin, Izv. Akad. Nauk SSSR, Ser. Khim., 1967, 1966; C. A. 68, 39722 (1968).
29. Arbuzov, A. E., and G. Kh. Kamai, J. Russ. Phys. Chem. Soc., 61, 2037 (1929); C. A. 24, 5736 (1930).

30. Arbuzov, A. E., G. Kh. Kamai, and O. N. Belorossova, Zh. Obshch. Khim., 15, 766 (1945); C. A. 41, 105 (1947).
31. Arbuzov, A. E., and A. I. Razumov, Izv. Akad. Nauk SSSR, Otd. Khim. Nauk, 1945, 167; C. A. 40, 3411 (1946).
32. Arbuzov, A. E., F. G. Valitova, A. V. Il'yasov, B. M. Kozyrev, and Yu. V. Yablokov, Dokl. Akad. Nauk SSSR, 147, 839 (1962); C. A. 58, 8535 (1963).
33. Arbuzov, B. A., V. E. Bel'skii, A. O. Vizel, K. M. Ivanovskaya, and G. Z. Motygullin, Dokl, Akad. Nauk SSSR, 176, 323 (1967); C. A. 68, 68130 (1968).
34. Arbuzov, B. A., G. G. Butenko, and E. G. Yarkova, Izv. Akad. Nauk SSSR, Ser. Khim., 1965, 1085; C. A. 63, 8397 (1965).
35. Arbuzov, B. A., N. A. Polezhaeva, V. S. Vinogradova, and A. K. Shamsutdinova, Izv. Akad. Nauk SSSR, Ser. Khim., 1965, 669; C. A. 63, 2998 (1965).
36. Arbuzov, B. A., A. P. Rakov, and A. O. Vizel, Izv. Akad. Nauk SSSR, Ser. Khim., 1969, 2230; C. A. 72 31931 (1970).
37. Arbuzov, B. A., A. P. Rakov, A. O. Vizel, L. A. Shapinskaya, and N. P. Kulikova, Izv. Akad. Nauk SSSR, Ser. Khim., 1968, 1313; C. A. 69, 87104 (1968).
38. Arbuzov, B. A., and N. I. Rizpolozhenskii, Izv. Akad. Nauk SSSR, Otd. Khim. Nauk, 1952, 854; C. A. 47, 9903 (1953).
39. Arbuzov, B. A., and N. I. Rizpolozhenskii, Izv. Akad. Nauk SSSR, Otd. Khim. Nauk, 1954, 631; C. A. 49, 11541 (1955).
40. Arbuzov, B. A., and N. I. Rizpolozhenskii, Izv. Akad. Nauk SSSR, Otd. Khim. Nauk, 1955, 253; C. A. 50, 3270 (1956).
41. Arbuzov, B. A., N. I. Rizpolozhenskii, and M. A. Zvereva, Izv. Akad. Nauk SSSR, Otd. Khim. Nauk, 1957, 179; C. A. 51, 11237 (1957).
42. Arbuzov, B. A., Yu. Yu. Samitov, A. O. Vizel, and T. V. Zykova, Dokl. Akad. Nauk SSSR, 159, 1062 (1964); C. A. 62, 6371 (1965).
43. Arbuzov, B. A., and L. A. Shapskinskaya, Izv. Akad. Nauk SSSR, Otd. Khim. Nauk, 1962, 65; C. A. 57, 13791 (1962).
44. Arbuzov, B. A., L. A. Shapshinskaya, and V. M. Erokhina, Izv. Akad. Nauk SSSR, Otd. Khim. Nauk, 1962, 2074; C. A. 58, 11396 (1963).
45. Arbuzov, B. A., L. A. Shapshinskaya, and V. M. Erokhina, Izv. Akad. Nauk SSSR, Ser. Khim., 1965, 1820; C. A. 64, 3588 (1966).
46. Arbuzov, B. A., and T. G. Shavsha-Tolkachaeva, Izv. Akad. Nauk SSSR, Otd. Khim. Nauk, 1954, 812; C. A. 49, 4352 (1955).

47. Arbuzov, B. A., V. S. Vinogradova, and M. A. Zvereva, Izv. Akad. Nauk SSSR, Otd. Khim. Nauk, 1960, 1772; C. A. 55, 16398 (1961).
48. Arbuzov, B. A., and A. O. Vizel, Dokl. Akad. Nauk SSSR, 158, 1105 (1964); C. A. 62, 2791 (1965).
49. Arbuzov, B. A., A. O. Vizel, and K. M. Ivanovskaya, Khim. Geterosikl. Soedin., 1967, 1130; C. A. 69, 67482 (1968).
50. Arbuzov, B. A., A. O. Vizel, K. M. Ivanovskaya, I. A. Studentsova, and R. S. Garaev, Dokl. Akad. Nauk SSSR, 182, 101 (1968); C. A. 70, 20170 (1969).
51. Arbuzov, B. A., A. O. Vizel, A. P. Pakov, and Yu. Yu. Samitov, Dokl. Akad. Nauk SSSR, 172, 1075 (1967); C. A. 66, 115630 (1967).
52. Arbuzov, B. A., A. O. Vizel, Yu. Yu. Samitov, and K. M. Ivanovskaya, Dokl. Akad. Nauk SSSR, 159, 582 (1964); C. A. 62, 6505 (1965).
53. Arbuzov, B. A., A. O. Vizel, and M. A. Zvereva, Izv. Akad. Nauk SSSR, Ser. Khim., 1967, 929; C. A. 68, 13074 (1968).
54. Arbuzov, B. A., A. O. Vizel, M. A. Zvereva, I. A. Studentsova, and R. S. Garaev, Izv. Akad. Nauk SSSR, Ser. Khim., 1966, 1848; C. A. 66, 95129 (1967).
55. Arbuzov, B. A., A. O. Vizel, M. A. Zvereva, I. A. Studentsova, R. S. Garaev, V. E. Bel'skii, M. G. Berim, S. D. Anisin, and L. A. Voitovich, Khim-Farm. Zh., 3, 9 (1969); C. A. 72, 55599 (1970).
56. Arbuzov, B. A., and D. Kh. Yarmukhametova, Izv. Akad. Nauk SSSR, Otd. Khim. Nauk, 1960, 1767; C. A. 55, 15507 (1961).
57. Baker, J. W., J. P. Chupp, and P. E. Newallis, U. S. 3,351,682; C. A. 68, 105359 (1968).
58. Baldwin, R. A., and M. T. Cheng, J. Org. Chem., 32, 1572 (1967).
59. Baldwin, R. A., M. T. Cheng, and D. G. Homer, J. Org. Chem., 32, 2176 (1967).
60. Baldwin, R. A., C. O. Wilson Jr., and R. I. Wagner, J. Org. Chem., 32, 2172 (1967).
61. Barabanov, V. I., and V. S. Abramov, Zh. Obshch. Khim., 35, 2225 (1965); C. A. 64, 11240 (1966).
62. Batkowski, T., P. Mastalerz, M. Michelewska, and B. Nitka, Roczniki Chem., 41, 471 (1967); C. A. 67, 32738 (1967).
63. Beg, M. A. A., and H. C. Clark, Can. J. Chem., 39, 564 (1961).
64. Bell, V. L. Jr., and G. M. Kosolapoff, J. Am. Chem. Soc., 75, 4901 (1953).
65. Bel'skii, V. E., M. V. Efremova, and I. M. Shermergorn, Izv. Akad. Nauk SSSR, Ser. Khim., 1966, 1654; C. A. 66, 64760 (1967).

66. Bel'skii, V. E., M. V. Efremova, and I. M. Shermergorn, Izv. Akad. Nauk SSSR, Ser. Khim., 1967, 923; C. A. 68, 12096 (1968).
67. Bel'skii, V. E., A. N. Pudovik, M. V. Efremova, V. N. Eliseenkov, and A. R. Panteleva, Dokl. Akad. Nauk SSSR, 180, 351 (1968); C. A. 69, 76091 (1968).
68. Benschop, H. P., and G. R. Van den Berg, Rec. Trav. Chim., 87, 362 (1968).
69. Benschop, H. P., G. R. Van den Berg, and H. L. Boter, Rec. Trav. Chim., 87, 387 (1968).
70. Bentov, M., L. David, and E. D. Bergmann, J. Chem. Soc., 1964, 4750.
71. Bentrude, W. G., and E. R. Witt, J. Am. Chem. Soc., 85, 2522 (1963).
72. Berezovskaya, I. V., L. A. Eliseeva, E. V. Kuznetsov, E. Kh. Mukhametzyanova, and I. M. Shermergorn, Izv. Akad. Nauk SSSR, Ser. Khim., 1968, 1369; C. A. 69, 87100 (1968).
73. Bergesen, K., Acta Chem. Scand., 19, 1784 (1965).
74. Berlin, K. D., and T. H. Austin, J. Org. Chem., 30, 2745 (1965).
75. Berlin, K. D., T. H. Austin, and M. Nagabushanam, J. Org. Chem., 30, 1267 (1965).
76. Berlin, K. D., T. H. Austin, and K. L. Stone, J. Am. Chem. Soc., 86, 1787 (1964).
77. Berlin, K. D., and M. A. R. Khayat, Tetrahedron, 22, 975 (1966).
78. Berlin, K. D., and M. A. R. Khayat, Tetrahedron, 22, 987 (1966).
79. Berlin, K. D., J. G. Morgan, M. E. Peterson, and W. C. Pivonka, J. Org. Chem., 34, 1266 (1969).
80. Berlin, K. D., and R. U. Pagilagan, Chem. Commun., 1966, 687.
81. Berlin, K. D., and R. U. Pagilagan, J. Org. Chem., 32, 129 (1967).
82. Berlin, K. D., and L. A. Wilson, Chem. Ind. (London), 1965, 1522.
83. Best, R. D., and R. C. Gordon, U. S. 3,325,444; C. A. 67, 65072 (1967).
84. Biddle, P., J. Kennedy, and J. L. Williams, Chem. Ind. (London), 1957, 1481.
85. Birum, G. H., and R. B. Clampitt, U. S. 3,372,209; C. A. 68, 95954 (1968).
86. Blicke, F. F., and S. Raines, J. Org. Chem., 29, 204 (1964).
87. Bliznyak, N. K., Z. N. Kvasha, and A. F. Kolomiets, Zh. Obshch. Khim., 37, 1811 (1967); C. A. 69, 2985 (1968).
88. Block, B. P., S. H. Rose, C. W. Schamm, E. S. Roth, and J. Simkin, J. Am. Chem. Soc., 84, 3200 (1962).

89. Bochwic, B., and J. Michalski, Nature, 167, 1035 (1951).
90. Bochwic, B., and J. Michalski, Roczniki Chem., 26, 593 (1952), C. A. 49, 2345 (1955).
91. Bock, H., and E. Baltin, Chem. Ber., 98, 2844 (1965).
92. Bock, H., and G. Rudolph, Chem. Ber., 98, 2273 (1965).
93. Bock, H., G. Rudolph, and E. Baltin, Chem. Ber., 98, 2054 (1965).
94. Bock, H., and W. Wiegräbe, Angew. Chem. Int. Ed., 2, 484 (1963).
95. Bock, H., and W. Wiegräbe, Chem. Ber., 99, 1068 (1966).
96. Boisselle, A. P., and N. A. Meinhardt, J. Org. Chem., 27, 1828 (1962).
97. Bonati, F., S. Cenini, and R. Ugo, J. Organometal. Chem., 9, 395 (1967).
98. Borecki, Cz., J. Michalski, and St. Musierowicz, J. Chem. Soc., 1958, 4081.
99. Borisov, G., Izv. Inst. Org. Khim., Bulg. Akad. Nauk, 3, 9 (1967); C. A. 68, 113741 (1968).
100. Bourneuf, M., Bull. Soc. Chim., 33, 1808 (1923).
101. Brown, A. D. Jr., and G. M. Kosolapoff, J. Chem. Soc. (C), 1968, 839.
102. Buckler, S. A., J. Am. Chem. Soc., 84, 3093 (1962).
103. Buckler, S. A., L. Doll, F. K. Lind, and M. Epstein, J. Org. Chem., 27, 794 (1962).
104. Buckler, S. A., and M. Epstein, Tetrahedron, 18, 1211 (1962).
105. Buckler, S. A., and M. Epstein, Tetrahedron, 18, 1221 (1962).
106. Buckler, S. A., and V. P. Wystrach, J. Am. Chem. Soc., 83, 168 (1961).
107. Budzikiewicz, H., and Z. Pelah, Monatsh. Chem., 96, 1739 (1965).
108. Büchel, K.-H., H. Röchling, and F. Korte, Liebig's Annalen, 685, 10 (1965).
109. Bulgakova, R. A., and R. R. Shagidullin, Izv. Akad. Nauk SSSR, Ser. Khim., 1968, 672; C. A. 69, 14306 (1968).
110. Burg, A. B., and A. J. Sarkis, J. Am. Chem. Soc., 87, 238 (1965).
111. Burg, A. B., and R. I. Wagner, J. Am. Chem. Soc., 75, 3872 (1953).
112. Burger, L. L., J. Phys. Chem., 62, 590 (1958).
113. Burger, L. L., and R. M. Wagner, Ind. and Eng. Chem., Chem. and Eng. Data Series, 3, 310 (1958).
114. Cadogan, J. I. G., D. J. Sears, and D. M. Smith, J. Chem. Soc. (C), 1969, 1314.

115. Calligaris, M., A. Ciana, and A. Ripamonti, Ric. Sci., 36, 1358 (1966); C. A. 66, 101234 (1967).
116. Calligaris, M., G. Nardin, and A. Ripamonti, Chem. Commun., 1968, 1014.
117. Campbell, I. G. M., G. W. A. Fowles, and L. A. Nixon, J. Chem. Soc., 1964, 1389.
118. Campbell, I. G. M., and J. K. Way, J. Chem. Soc., 1960, 5034.
119. Campbell, I. G. M., and J. K. Way, J. Chem. Soc., 1961, 2133.
120. Challenger, F., and A. T. Peters, J. Chem. Soc., 1929, 2610.
121. Chance, L. H., L. K. Leonard, and G. L. Drake, Ind. Eng. Chem., Product Res. Develop., 5, 252 (1966).
122. Charrier, C., and M. P. Simonnin, Compt. Rend., Ser. C, 264, 995 (1967).
123. Chenborisov, R. Sh., V. V. Markin, and F. G. Fatykhova, Zh. Obshch. Khim., 39, 927 (1969); C. A. 71, 50094 (1969).
124. Cherbuliez, E., A. Buchs, S. Jaccard, D. Janjic, and J. Rabinowitz, Helv. Chim. Acta, 49, 2395 (1966).
125. Cherbuliez, E., S. Jaccard, R. Prince, and J. Rabinowitz, Helv. Chim. Acta, 48, 632 (1965).
126. Chernyshev, E. A., E. F. Bugerenko, N. A. Nikolaeva, and A. D. Petrov, Dokl. Akad. Nauk SSSR, 147, 117 (1962); C. A. 58, 9120 (1963).
127. Christen, P. J., L. M. van der Linde, and F. N. Hooge, Rec. Trav. Chim., 78, 161 (1959).
128. Christol, H., and C. Marty, Compt. Rend., Ser. C, 262, 1722 (1966).
129. Coates, G. E., and D. S. Golightly, J. Chem. Soc., 1962, 2523.
130. Coates, G. E., and R. N. Mukherjee, J. Chem. Soc., 1964, 1295.
131. Cölln, R., and G. Schrader, Ger. 1,099,788; C. A. 55, 23918 (1961).
132. Cölln, R., and G. Schrader, Ger. 1,104,5C5; C. A. 56, 506 (1962).
133. Cölln, R., and G. Schrader, Ger. 1.138,771; C. A. 58, 12601 (1963).
134. Cölln, R., and G. Schrader, Ger. 1,141,990; C. A. 59, 1684 (1963).
135. Cölln, R., and G. Schrader, Ger. 1,179,197; C. A. 62, 483 (1965).
136. Conant, J. B., A H. Bump, and H. S. Holt, J. Am. Chem. Soc., 43, 1677 (1921).
137. Conant, J. B., and S. M. Pollack, J. Am. Chem. Soc., 43, 1665 (1921).

138. Coyne, D. M., W. E. McEwen, and C. A. VanderWerf, J. Am. Chem. Soc., 78, 3061 (1956).
139. Cram, D. J., R. D. Trepka, and P. St. Janiek, J. Am. Chem. Soc., 88, 2749 (1966).
140. Crescenzi, V., V. Giancotti, and A. Ripamonti, J. Am. Chem. Soc., 87, 391 (1965).
141. Crofts, P. C., unpublished.
142. Crofts, P. C., I. M. Downie, and K. Williamson, J. Chem. Soc., 1964, 1240.
143. Crofts, P. C., and I. S. Fox, J. Chem. Soc., 1958, 2995.
144. Crofts, P. C., and K. Gosling, J. Chem. Soc., 1964, 2486.
145. Crofts, P. C., and G. M. Kosolapoff, J. Am. Chem. Soc., 75, 3379 (1953).
146. Crofts, P. C., and D. W. B. Kydd, to be published.
147. Crofts, P. C., and D. M. Parker, J. Chem. Soc. (C), 1970, 332.
148. Crofts, P. C., and D. M. Parker, J. Chem. Soc. (C), 1970, 2529.
149. Dahl, G. H., U. S. 3,426,050; C. A. 70 68500 (1969).
150. Dahl, G. H., and B. P. Block, Inorg. Chem., 5, 1394 (1966).
151. Dahl, G. H., and B. P. Block, Inorg. Chem., 6, 1439 (1967).
152. Dalton, D. R., R. E. De Brunner, and E. S. Blake, U. S. 3,346,668; C. A. 69, 10536 (1968).
153. David, H., G. Martin, G. Mavel, and G. Sturtz, Bull. Soc. Chim. France, 1962, 1616.
154. Davidson, R. S., J. Chem. Soc. (C), 1967, 2131.
155. Davies, J. H., and P. Kirby, J. Chem. Soc., 1964, 3425.
156. Davies, W. C., and F. G. Mann, J. Chem. Soc., 1944, 276.
157. Denham, J. M., and R. K. Ingham, J. Org. Chem., 23, 1298 (1958).
158. Dennis, E. A., and F. H. Westheimer, J. Am. Chem. Soc., 88, 3431 (1966).
159. Derkach, G. I., and E. S. Gubnitskaya, Zh. Obshch. Khim., 34, 604 (1964); C. A. 60, 13268 (1964).
160. Derkach, G. I., E. S. Gubnitskaya, and A. V. Kirsanov, Zh. Obshch. Khim., 31, 3679 (1961); C. A. 57, 9876 (1962).
161. Detoni, S., and D. Hadzi, Spectrochim. Acta, 20, 949 (1964).
162. Dietsche, W., Liebig's Annalen, 712, 21 (1968).
163. Dietsche, W. H., Tetrahedron, 23, 3049 (1967).
164. Dimroth, K., and W. Stade, Angew. Chem. Int. Ed., 7, 881 (1968).

165. Dmitrieva, L. E., K. V. Karavanov, and S. Z. Ivin, Zh. Obshch. Khim., 38, 157 (1968); C. A. 69, 59350 (1968).
166. Doak, G. O., and L. D. Freedman, J. Am. Chem. Soc., 73, 5658 (1951).
167. Doak, G. O., and L. D. Freedman, J. Am. Chem. Soc., 74, 753 (1952).
168. Doak, G. O., and L. D. Freedman, J. Am. Chem. Soc., 75, 683 (1953).
169. Doak, G. O., L. D. Freedman, and J. B. Levy, J. Org. Chem., 29, 2382 (1964).
170. Dobbie, R. C., L. F. Doty, and R. G. Cavell, J. Am. Chem. Soc., 90, 2015 (1968).
171. Dörken, A., and G. Schrader, Brit. 941,631; C. A. 60, 6870 (1964).
172. Dörken, C., Ber., 21, 1505 (1888).
173. Drake, L. R., and C. S. Marvel, J. Org. Chem., 2, 387 (1937).
174. Drinkard, W. C., and G. M. Kosolapoff, J. Am. Chem. Soc., 74, 5520 (1952).
175. Dziomko, V. M., I. S. Markovich, and I. M. Yakhnis, Zh. Obshch. Khim., 37, 1897 (1967); C. A. 68, 29795 (1968).
176. Edmundson, R. S., and E. W. Mitchell, J. Chem. Soc. (C), 1966, 1096.
177. Edmundson, R. S., and J. O. L. Wrigley, Tetrahedron, 23, 283 (1967).
178. Emeleus, H. J., R. N. Haszeldine, and R. C. Paul, J. Chem. Soc., 1955, 563.
179. Emmick, T. L., and R. L. Letsinger, J. Am. Chem. Soc., 90, 3459 (1968).
180. Ettel, V., and J. Horak, Collection Czechoslav. Chem. Communs., 26, 1949 (1961).
181. Ettel, V., and J. Horak, Collection Czechoslav. Chem. Communs., 26, 2087 (1961).
182. Evdakov, V. P., and E. I. Alipova, Zh. Obshch. Khim., 37, 2508 (1967); C. A. 69, 27497 (1968).
183. Evleth, E. M. Jr., L. D. Freedman, and R. I. Wagner, J. Org. Chem., 27, 2192 (1962).
184. Evtikhov, Zh. L., N. A. Razumova, and A. A. Petrov, Dokl. Akad. Nauk SSSR, 181, 877 (1968); C. A. 70, 78084 (1969).
185. Evtikhov, Zh. L., N. A. Razumova, and A. A. Petrov, Zh. Obshch. Khim., 38, 196 (1968); C. A. 69, 77356 (1968).
186. Evtikhov, Zh. L., N. A. Razumova, and A. A. Petrov, Zh. Obshch. Khim., 38, 2341 (1968); C. A. 70, 37879 (1969).
187. Evtikhov, Zh. L., N. A. Razumova, and A. A. Petrov, Zh. Obshch. Khim., 39, 465 (1969); C. A. 70, 115236 (1969).

188. Evtikhov, Zh. L., N. A. Razumova, and A. A. Petrov, Zh. Obshch. Khim., 39, 2367 (1969); C. A. 72, 43791 (1970).
189. Ezzell, B. R., and L. D. Freedman, J. Org. Chem., 35, 241 (1970).
190. Falbe, J., R. Paatz, and F. Korte, Chem. Ber., 97, 2544 (1964).
191. Farbenfabriken Bayer A. G., Brit. 872,143; C. A. 56, 1482 (1962).
192. Farbenfabriken Bayer A. G., Brit. 950,145; C. A. 60, 14541 (1964).
193. Farbenfabriken Bayer A. G., Brit. 967,081; C. A. 62, 11853 (1965).
194. Farbenfabriken Bayer A. G., Fr. 1,528,547; C. A. 71, 3467 (1969).
195. Fedorova, G. K., and A. V. Kirsanov, Zh. Obshch. Khim., 35, 1483 (1965); C. A. 63, 14900 (1965).
196. Fedorova, G. K., and Ya. P. Shaturskii, Zh. Obshch. Khim., 36, 1262 (1966); C. A. 65, 15419 (1966).
197. Fedorova, G. K., Ya. P. Shaturskii, and A. V. Kirsanov, Probl. Organ. Sinteza, Akad. Nauk SSSR, Otd. Obshch. i Tekhn. Khim., 1965, 258; C. A. 64, 8228 (1966).
198. Fedorova, G. K., Ya. P. Shaturskii, and A. V. Kirsanov, Zh. Obshch. Khim., 35, 1984 (1965); C. A. 64, 6682 (1966).
199. Fedorova, G. K., Ya. P. Shaturskii, L. S. Moskalevskaya, Yu. S. Grushin, and A. V. Kirsanov, Zh. Obshch. Khim., 37, 2686 (1967); C. A. 69, 67490 (1968).
200. Fest, C., and W. Lorenz, Belg. 647,745; C. A. 63, 11616 (1965).
201. Fest, C., and G. Schrader, Ger. 1,161,275; C. A. 60, 10717 (1964).
202. Fild, M., Z. Anorg. Allgem. Chem., 358, 257 (1968).
203. Fiszer, B., and J. Michalski, Roczniki Chem., 26, 293 (1952); C. A. 50, 5550 (1956).
204. Fluck, E., and F. L. Goldmann, Chem. Ber., 96, 3091 (1963).
205. Fox, R. B., and W. J. Bailey, J. Org. Chem., 25, 1447 (1960).
206. Frank, A. W., J. Org. Chem., 30, 3663 (1965).
207. Frank, A. W., J. Org. Chem., 31, 1917 (1966).
208. Frank, A. W., and C. F. Baranaukas, J. Org. Chem., 31, 872 (1966).
209. Freedman, L. D., and G. O. Doak, J. Am. Chem. Soc., 74, 2884 (1952).
210. Freedman, L. D., and G. O. Doak, J. Am. Chem. Soc., 75, 4905 (1953).
211. Freedman, L. D., and G. O. Doak, J. Org. Chem., 21, 238 (1956).

212. Freedman, L. D., and G. O. Doak, J. Org. Chem., 21, 1533 (1956).
213. Freedman, L. D., and G. O. Doak, J. Org. Chem., 23, 769 (1958).
214. Freedman, L. D., and G. O. Doak, J. Org. Chem., 24, 638 (1959).
215. Freedman, L. D., and G. O. Doak, J. Org. Chem., 26, 2082 (1961).
216. Freedman, L. D., and G. O. Doak, J. Org. Chem., 29, 1983 (1964).
217. Freedman, L. D., and G. O. Doak, J. Med. Chem., 8, 891 (1965).
218. Freedman, L. D., and G. O. Doak, J. Org. Chem., 30, 1263 (1965).
219. Freedman, L. D., G. O. Doak, and J. R. Edmiston, J. Org. Chem., 26, 284 (1961).
220. Freedman, L. D., G. O. Doak, and E. L. Petit, J. Am. Chem. Soc., 77, 4262 (1955).
221. Freedman, L. D., H. Tauber, G. O. Doak, and H. J. Magnuson, J. Am. Chem. Soc., 75, 1379 (1953).
222. Freeman, K. L., and M. J. Gallagher, Aust. J. Chem., 19, 2025 (1966).
223. Freeman, K. L., and M. J. Gallagher, Aust. J. Chem., 19, 2159 (1966).
224. Freeman, K. L., and M. J. Gallagher, Aust. J. Chem., 21, 2297 (1968).
225. Frisch, K. C., and H. Lyons, J. Am. Chem. Soc., 75, 4078 (1953).
226. Fritzsche, H., U. Hasserodt, and F. Korte, Chem. Ber., 98, 1681 (1965).
227. Fukuto, T. R., R. L. Metcalf, and M. Y. Winton, J. Econ. Entomol., 54, 955 (1961).
228. Gallagher, M. J., E. C. Kirby, and F. G. Mann, J. Chem. Soc., 1963, 4846.
229. Garner, A. Y., U. S. 2,916,510; C. A. 54, 5571 (1960).
230. Gazieva, N. I., A. I. Shchekotikhin, and V. A. Ginsberg, Zh. Obshch. Khim., 38, 673 (1968); C. A. 69, 27491 (1968).
231. Gazizov, M. B., A. I. Razumov, and K. S. Gazizova, Zh. Obshch. Khim., 39, 709 (1969); C. A. 71, 39079 (1969).
232. Gefter, E. L., Zh. Obshch. Khim., 31, 952 (1961); C. A. 55, 23399 (1961).
233. Gefter, E. L., and M. I. Kabachnik, Dokl. Akad. Nauk SSSR, 114, 541 (1957); C. A. 52, 295 (1958).
234. Gefter, E. L., and I. A. Rogacheva, Zh. Obshch. Khim. SSSR, 31, 955 (1961); C. A. 55, 23399 (1961).
235. Gefter, E. L., and I. A. Rogacheva, Zh. Obshch. Khim SSSR, 33, 1177 (1963); C. A. 59, 10114 (1963).

236. Gefter, E. L., and I. A. Rogacheva, Zh. Obshch. Khim SSSR, 35, 1463 (1965); C. A. 63, 14900 (1965).
237. Gefter, E. L., and I. A. Rogacheva, Zh. Obshch. Khim SSSR, 36, 79 (1966); C. A. 64, 14208 (1966).
238. Gemiti, F., V. Giancotti, and A. Ripamonti, J. Chem. Soc. (A), 1968, 763.
239. Genkina, G. K., V. A. Gilyarov, and M. I. Kabachnik, Zh. Obshch. Khim., 38, 2513 (1968); C. A. 70, 57955 (1969).
240. Gerding, H., J. W. Maarsen, and D. H. Zijp, Rec. Trav. Chim., 77, 361 (1958).
241. Giancotti, V., F. Giordano, L. Randaccio, and A. Ripamonti, J. Chem. Soc. (A), 1968, 757.
242. Giancotti, V., and A. Ripamonti, Chim. Ind. (Milan), 48, 1065 (1966); C. A. 66, 29298 (1967).
243. Gibson, C. S., and D. J. A. Johnson, J. Chem. Soc., 1928, 92.
244. Gilson, I. T., and H. H. Sisler, Inorg. Chem., 4, 273 (1965).
245. Gilyarov, V. A., and M. I. Kabachnik, Zh. Obshch. Khim., 36, 282 (1966); C. A. 64, 15790 (1966).
246. Gilyarov, V. A., A. M. Maksudov, and M. I. Kabachnik, Zh. Obshch. Khim., 37, 2501 (1967); C. A. 69, 26442 (1968).
247. Gilyarov, V. A., E. N. Tsvetkov, and M. I. Kabachnik, Zh. Obshch. Khim., 36, 274 (1966); C. A. 64, 17408 (1966).
248. Ginjaar, L., and S. Blasse-Vel, Rec. Trav. Chim., 85, 694 (1966).
249. Giolito, S. L., U. S. 3,454,678; C. A. 71, 61543 (1969).
250. Giordano, F., L. Randaccio, and A. Ripamonti, Chem. Commun., 1967, 19.
251. Giordano, F., L. Randaccio, and A. Ripamonti, Chem. Commun., 1967, 1239.
252. Giordano, F., and A. Ripamonti, Acta Crystallog., 22, 678 (1967).
253. Gladshtein, B. M., B. I. Babkina, V. V. Fekotova, and L. Z. Soborovskii, Zh. Obshch. Khim., 34, 2897 (1964); C. A. 61, 16091 (1964).
254. Gleu, K., and A. Schubert, Ber., 73, 805 (1940).
255. Glockling, F., and K. A. Hooton, Proc. Chem. Soc., 1963, 146.
256. Goldwhite, H., and D. G. Rowsell, J. Am. Chem. Soc., 88, 3572 (1966).
257. Gordon, G. Ya., S. L. Varshavskii, and L. P. Kofman, Zh. Prikl. Khim. (Leningrad), 41, 1628 (1968); C. A. 69, 87113 (1968).
258. Gosling, K., and A. B. Burg, J. Am. Chem. Soc., 90, 2011 (1968).

259. Granoth, I., A. Kalir, and Z. Pelah, Israel J. Chem., 6, 651 (1968).
260. Grayson, M., and C. E. Farley, J. Org. Chem., 32, 236 (1967).
261. Grayson, M., C. E. Farley, and C. A. Streuli, Tetrahedron, 23, 1065 (1967).
262. Green, M., and R. F. Hudson, J. Chem. Soc., 1958, 3129.
263. Green, M., and R. F. Hudson, J. Chem. Soc., 1963, 540.
264. Green, M., and R. F. Hudson, J. Chem. Soc., 1963, 1004.
265. Grishina, O. N., and L. M. Kosova, Zh. Obshch. Khim., 37, 2276 (1967); C. A. 68, 87345 (1968).
266. Gross, H., G. Engelhardt, J. Freiberg, W. Bürger, and B. Costisella, Liebig's Annalen, 707, 35 (1967).
267. Gutmann, V., G. Mörtl, and K. Utvary, Monatsh. Chem., 93, 1114 (1962).
268. Gutmann, V., G. Mörtl, and K. Utvary, Monatsh. Chem., 94, 897 (1963).
269. Gurevich, P. A., N. I. Shelepova, and A. I. Razumov, Zh. Obshch. Khim., 38, 1905 (1968); C. A. 69, 106821 (1968).
270. Haake, P., and G. Hurst, J. Am. Chem. Soc., 88, 2544 (1966).
271. Haake, P., and P. S. Ossip, Tetrahedron, 24, 565 (1968).
272. Hägele, G., and W. Kuchen, Chem. Ber., 103, 2885 (1970).
273. Häring, M., Helv. Chim. Acta, 43, 1826 (1960).
274. Hart, W. A., and H. H. Sisler, Inorg. Chem., 3, 617 (1964).
275. Harwood, H. J., U. S. 3,157,694; C. A. 62, 4053 (1965).
276. Harwood, H. J., M. L. Becker, and R. R. Smith, J. Org. Chem., 32, 3882 (1967).
277. Hashimoto, S., and I. Furukawa, Kobunshi Kagaku, 24, 152 (1967); C. A. 68, 30193 (1968).
278. Hasserodt, U., K. Hunger, and F. Korte, Tetrahedron, 19, 1563 (1963).
279. Helferich, B., and E. Aufderhaar, Liebig's Annalen, 658, 100 (1962).
280. Henning, H. G., J. Prakt. Chem., 29, 93 (1965).
281. Henning, H. G., J. Prakt. Chem., 31, 304 (1966).
282. Henning, H. G., and G. Haack, Z. Chem., 6, 261 (1966).
283. Henning, H. G., and G. Hilgetag, J. Prakt. Chem., 29, 86 (1965).
284. Henning, H. G., G. Hilgetag, and G. Busse, J. Prakt. Chem., 33, 188 (1966).

285. Henning, H. G., and G. Petzold, Z. Chem., 5, 419 (1965).
286. Herbstman, S., U. S. 3,405,157; C. A. 70, 11811 (1969).
287. Herring, D. L., U. S. 3,341,576; C. A. 67, 108754 (1967).
288. Hieronymus, E., and R. Wirtz, Ger. 1,197,888; C. A. 63, 13318 (1965).
289. Higgins, W. A., P. W. Vogel, and W. G. Craig, J. Am. Chem. Soc., 77, 1864 (1955).
290. Hinton, R. C., F. G. Mann, and D. Todd, J. Chem. Soc., 1961, 5454.
291. Hoffman, A., J. Am. Chem. Soc., 52, 2995 (1930).
292. Hoffmann, H., and P. Schellenbeck, Chem. Ber., 99, 1134 (1966).
293. Hofmann, A. W., Ber., 5, 104 (1872).
294. Hofmann, A. W., and F. Mahla, Ber., 25, 2436 (1892).
295. Hopkins, T. R., and P. W. Vogel, J. Am. Chem. Soc., 78, 4447 (1956).
296. Horak, J., and V. Ettel, Collection Czechoslav. Chem. Communs., 26, 2410 (1961).
297. Horner, L., P. Beck, and V. G. Toscano, Chem. Ber., 94, 1317 (1961).
298. Horner, L., P. Beck, and V. G. Toscano, Chem. Ber., 94, 2122 (1961).
299. Horner, L., H. Hoffmann, W. Klink, H. Ertel, and V. G. Toscano, Chem. Ber., 95, 581 (1962).
300. Horner, L., H. Hoffmann, and H. G. Wippel, Chem. Ber., 91, 64 (1958).
301. Horner, L., and H. Winkler, Tetrahedron Letters, 1964, 3265.
302. Hunger, K., U. Hasserodt, and F. Korte, Tetrahedron, 20, 1593 (1964).
303. Hunt, B. B., and B. C. Saunders, J. Chem. Soc., 1957, 2413.
304. Husebye, S., Acta Chem. Scand., 20, 51 (1966).
305. Ignat'eva, G. V., Ya. S. Arbisman, Yu. A. Kondrat'ev, R. K. Bal'chenko, and S. Z. Ivin, Zh. Obshch. Khim., 38, 2816 (1968); C. A. 70, 78068 (1969).
306. Il'ina, M. K., and I. M. Shermergorn, Izv. Akad. Nauk SSSR, Ser. Khim., 1967, 1346; C. A. 67, 108702 (1967).
307. Il'ina, M. K., and I. M. Shermergorn, Izv. Akad. Nauk SSSR, Ser. Khim., 1968, 1860; C. A. 70, 37885 (1969).
308. Iliopulos, M. I., Chem. Ber., 99, 2410 (1966).
309. Issleib, K., and A. Brack, Z. Anorg. Allgem. Chem., 277, 258 (1954).
310. Issleib, K., and G. Döll, Chem. Ber., 94, 2664 (1961).

311. Issleib, K., and G. Döll, Z. Anorg. Allgem. Chem., 324, 259 (1963).
312. Issleib, K., and W. Gründler, Z. Krist., 119, 472 (1964).
313. Issleib, K., and M. Hoffmann, Chem. Ber., 99, 1320 (1966).
314. Issleib, K., and D. Jacob, Chem. Ber., 94, 107 (1961).
315. Issleib, K., R. Kümmel, H. Oehme, and I. Meissner, Chem. Ber., 101, 3612 (1968).
316. Issleib, K., and H. Oehme, Z. Anorg. Allgem. Chem., 343, 268 (1966).
317. Issleib, K., and H. Oehme, Chem. Ber., 100, 2685 (1967).
318. Issleib, K., and K. Standtke, Chem. Ber., 96, 279 (1963).
319. Issleib, K., and H. Völker, Chem. Ber., 94, 392 (1961).
320. Issleib, K., and B. Walther, J. Organometal. Chem., 10, 177 (1967).
321. Issleib, K., and H. Weichmann, Chem. Ber., 101, 2197 (1968).
322. Ivanov, B. E., and T. I. Karpova, Izv. Akad. Nauk SSSR, Ser. Khim., 1964, 1230; C. A. 61, 12030 (1964).
323. Ivanov, B. E., and T. I. Karpova, Izv. Akad. Nauk SSSR, Ser. Khim., 1967, 1851; C. A. 67, 108704 (1967).
324. Ivanov, B. E., and L. A. Kudryavtseva, Izv. Akad. Nauk SSSR, Ser. Khim., 1967, 1498; C. A. 68, 78359 (1968).
325. Ivanov, B. E., and L. A. Kudryavtseva, Izv. Akad. Nauk SSSR, Ser. Khim., 1968, 1633; C. A. 69, 87114 (1968).
326. Ivanov, B. E., A. R. Panteleeva, R. R. Shagidullin, and I. M. Shermergorn, Zh. Obshch. Khim., 37, 1856 (1967); C. A. 68, 29797 (1968).
327. Ivasyuk, N. V., E. Kh. Mukhametzyanova, and I. M. Shermergorn, Izv. Akad. Nauk SSSR, Ser. Khim., 1968, 1625; C. A. 69, 87109 (1968).
328. Ivasyuk, N. V., and I. M. Shermergorn, Izv. Akad. Nauk SSSR, Ser. Khim., 1968, 2388; C. A. 70, 29005 (1969).
329. Ivasyuk, N. V., and I. M. Shermergorn, Izv. Akad. Nauk SSSR, Ser. Khim., 1969, 481; C. A. 71, 13174 (1969).
330. Ivin, S. Z., V. K. Promonenkov, and E. A. Fokin, Zh. Obshch. Khim., 39, 1058 (1969); C. A. 71, 61485 (1969).
331. Ivin, S. Z., S. D. Shelakova, and V. K. Promonenkov, Zh. Obshch. Khim., 40, 561 (1970); C. A. 73, 14940 (1970).

332. Jaffe, H. H., and L. D. Freedman, J. Am. Chem. Soc., 74, 1069 (1952).
333. Jaffe, H. H., and L. D. Freedman, J. Am. Chem. Soc., 74, 2930 (1952).
334. Jean, H., Bull. Soc. Chim. France, 1957, 783.
335. Jones, P. E., G. B. Ansell, and L. Katz, Chem. Commun., 1968, 78.
336. Kabachnik, M. I., Acta Chim. Acad. Sci. Hung., 18, 407 (1959); C. A. 54, 354 (1960).
337. Kabachnik, M. I., G. A. Balueva, T. Ya. Medved, E. N. Tsvetkov, and Jung-Yu Chang, Kinetika i Kataliz, 6, 212 (1965); C. A. 63, 7039 (1965).
338. Kabachnik, M. I., V. A. Gilyarov, Cheng-Te Chang, and E. I. Matrosov, Izv. Akad. Nauk SSSR, Otd. Khim. Nauk, 1962, 1589; C. A. 58, 6673 (1963).
339. Kabachnik, M. I., and N. N. Godovikov, Dokl. Akad. Nauk SSSR, 110, 217 (1956); C. A. 51, 4982 (1957).
340. Kabachnik, M. I., Jung-Yu Chang, and E. N. Tsvetkov, Zh. Obshch. Khim., 32, 3351 (1962); C. A. 58, 9126 (1963).
341. Kabachnik, M. I., T. A. Mastryukova, G. A. Balueva, E. E. Kugucheva, A. E. Shipov, and T. A. Melent'eva, Zh. Obshch. Khim., 31, 140 (1961); C. A. 55, 22079 (1961).
342. Kabachnik, M. I., T. A. Mastryukova, E. I. Matrosov, and B. Fisher, Zh. Strukt. Khim., 6, 691 (1965); C. A. 64, 5945 (1966).
343. Kabachnik, M. I., T. A. Mastryukova, and T. A. Melent'eva, Zh. Obshch. Khim., 33, 382 (1963); C. A. 59, 1677 (1963).
344. Kabachnik, M. I., T. A. Mastryukova, and A. E. Shipov, Zh. Obshch. Khim., 33, 320 (1963); C. A. 59, 658 (1963).
345. Kabachnik, M. I., T. A. Mastryukova, A. E. Shipov, and T. A. Melentyeva, Tetrahedron, 9, 10 (1960).
346. Kabachnik, M. I., and T. Ya. Medved, Izv. Akad. Nauk SSSR, Otd. Khim. Nauk, 1954, 1024; C. A. 50, 219 (1956).
347. Kabachnik, M. I., T. Ya. Medved, G. K. Kozlova, V. S. Balabukha, E. A. Mironova, and L. I. Tikhonova, Izv. Akad. Nauk SSSR, Otd. Khim. Nauk, 1960, 651; C. A. 54, 22329 (1960).
348. Kabachnik, M. I., and E. S. Shepeleva, Izv. Akad. Nauk SSSR, Otd. Khim. Nauk, 1953, 862; C. A. 49, 843 (1955).
349. Kabachnik, M. I., and E. N. Tsvetkov, Dokl. Akad. Nauk SSSR, 135, 323 (1960); C. A. 55, 14288 (1961).
350. Kabachnik, M. I., and E. N. Tsvetkov, Zh. Obshch. Khim., 30, 3227 (1960); C. A. 55, 21067 (1961).

351. Kabachnik, M. I., and E. N. Tsvetkov, Izv. Akad. Nauk SSSR, Ser. Khim., 1963, 1227; C. A. 59, 12839 (1963).
352. Kabachnik, M. I., E. N. Tsvetkov, and Jung-Yu Chang, Dokl. Akad. Nauk SSSR, 125, 1260 (1959); C. A. 53, 21752 (1959).
353. Kalabina, A. V., Ming-Yin Liu, and N. V. Donskaya, Zh. Obshch. Khim., 34, 1117 (1964); C. A. 61, 1889 (1964).
354. Kamai, G., Zh. Obshch. Khim., 18, 443 (1948); C. A. 42, 7723 (1948).
355. Kamai, G., Dokl. Akad. Nauk SSSR, 66, 389 (1949); C. A. 44, 127 (1950).
356. Kamai, G., and R. K. Ismagilov, Zh. Obshch. Khim., 34, 439 (1964); C. A. 60, 13267 (1964).
357. Kamai, G., F. M. Kharrasova, G. I. Rakhimova, and R. B. Sultanova, Zh. Obshch. Khim., 39, 625 (1969); C. A. 71, 50104 (1969).
358. Kamai, G., F. M. Kharrasova, R. B. Sultanova, and S. Yu. Tukhvatullina, Zh. Obshch. Khim., 31, 3550 (1961); C. A. 57, 3477 (1962).
359. Kamai, G., F. M. Kharrasova, and S. Yu. Tukhvatullina, Tr. Kazansk.-Khim.-Tekhnol. Inst., 30, 18 (1962); C. A. 60, 5542 (1964).
360. Kamai, G., and V. A. Kukhtin, Zh. Obshch. Khim., 25, 1932 (1955); C. A. 50, 8502 (1956).
361. Kamai, G., and V. A. Kukhtin, Zh. Obshch. Khim., 28, 939 (1958); C. A. 52, 17162 (1958).
362. Kamai, G., and V. S. Tsivunin, Dokl. Akad. Nauk SSSR, 128, 543 (1959); C. A. 54, 7538 (1960).
363. Kamai, G., V. S. Tsivunin, and S. Kh. Nurtdinov, Zh. Obshch. Khim., 35, 1817 (1965); C. A. 64, 3587 (1966).
364. Kamai, G., V. S. Tsivunin, and L. A. Panina, Tr. Kazansk Khim.-Tekhnol. Inst., 30, 11 (1962); C. A. 60, 4180 (1964).
365. Kataev, E. G., F. R. Tantasheva, E. G. Yarkova, and E. A. Bernikov, Dokl. Akad. Nauk SSSR, 179, 862 (1968); C. A. 69, 77353 (1968).
366. Kennard, K. C., and D. M. Burness, J. Org. Chem., 24, 464 (1959).
367. Ketelaar, J. A. A., and H. R. Gersman, Rec. Trav. Chim., 78, 190 (1959).
368. Khairullin, V. K., Izv. Akad. Nauk SSSR, Ser. Khim., 1965, 1792; C. A. 64, 3590 (1966).
369. Khairullin, V. K., G. V. Dmitrieva, and A. N. Pudovik, Izv. Akad. Nauk SSSR, Ser. Khim., 1969, 1166; C. A. 71, 50121 (1969).
370. Khairullin, V. K., G. V. Dmitrieva, and A. N. Pudovik, Izv. Akad. Nauk SSSR, Ser. Khim., 1970, 468; C. A. 73, 3987 (1970).

371. Khairullin, V. K., V. N. Eliseenkov, and A. N. Pudovik, Zh. Obshch. Khim., 37, 871 (1967); C. A. 68, 105302 (1968).
372. Khairullin, V. K., R. M. Kondrat'eva, and A. N. Pudovik, Izv. Akad. Nauk SSSR, Ser. Khim., 1967, 2097; C. A. 68, 39721 (1968).
373. Khairullin, V. K., R. M. Kondrat'eva, and A. N. Pudovik, Zh. Obshch. Khim., 38, 288 (1968); C. A. 69, 106816 (1968).
374. Khairullin, V. K., R. M. Kondrat'eva, and A. N. Pudovik, Zh. Obshch. Khim., 38, 858 (1968); C. A. 69, 52218 (1968).
375. Khairullin, V. K., R. M. Kondrat'eva, and A. N. Pudovik, Izv. Akad. Nauk SSSR, Ser. Khim., 1968, 1375; C. A. 69, 87092 (1968).
376. Khairullin, V. K., and A. N. Pudovik, Zh. Obshch. Khim., 36, 494 (1966); C. A. 65, 738 (1966).
377. Khairullin, V. K., and A. N. Pudovik, Zh. Obshch. Khim., 37, 2742 (1967); C. A. 69, 67483 (1968).
378. Khairullin, V. K., A. N. Pudovik, and N. I. Kharitonova, Zh. Obshch. Khim., 39, 608 (1969); C. A. 71, 39082 (1969).
379. Khairullin, V. K., and R. R. Shagidullin, Zh. Obshch. Khim., 36, 289 (1966); C. A. 64, 15913 (1966).
380. Khairullin, V. K., R. R. Shagidullin, T. I. Sobchuk, and A. N. Pudovik, Khim. Org. Soedin. Fosfora, Akad. Nauk SSSR, Otd. Obshch. Tekh. Khim., 1967, 35; C. A. 69, 59329 (1968).
381. Khairullin, V. K., T. I. Sobchuk, and A. N. Pudovik, Zh. Obshch. Khim., 37, 710 (1967); C. A. 67, 54222 (1967).
382. Khairullin, V. K., T. I. Sobchuk, and A. N. Pudovik, Zh. Obshch. Khim., 38, 584 (1968); C. A. 69, 59347 (1968).
383. Khairullin, V. K., M. A. Vasyanina, and A. N. Pudovik, Izv. Akad. Nauk SSSR, Ser. Khim., 1967, 950; C. A. 67, 116918 (1967).
384. Khairullin, V. K., M. A. Vasyanina, and A. N. Pudovik, Izv. Akad. Nauk SSSR, Ser. Khim., 1967, 1603; C. A. 68, 13073 (1968).
385. Khairullin, V. K., M. A. Vasyanina, and A. N. Pudovik, Zh. Obshch. Khim., 39, 341 (1969); C. A. 71, 30547 (1969).
386. Khairullin, V. K., M. A. Vasyanina, A. N. Pudovik, and Yu. Yu. Samitov, Khim. Org. Soedin. Fosfora, Akad. Nauk SSSR, Otd. Obshch. Tekh. Khim., 1967, 29; C. A. 69, 59335 (1968).
387. Kharrasova, F. M., and G. Kamai, Zh. Obshch. Khim., 34, 2195 (1964); C. A. 61, 10705 (1964).

388. Kharrasova, F. M., G. Kamai, and R. R. Shagidullin, Zh. Obshch. Khim., 35, 1993 (1965); C. A. 64, 6680 (1966).
389. Kharrasova, F. M., G. Kamai, and R. R. Shagidullin, Zh. Obshch. Khim., 36, 1987 (1966); C. A. 66, 65592 (1967).
390. Kharrasova, F. M., G. Kh. Kamai, R. B. Sultanova, and R. R. Shagidullin, Zh. Obshch. Khim., 37, 687 (1967); C. A. 67, 53437 (1967).
391. Kharrasova, F. M., G. Kh. Kamai, R. B. Sultanova, and R. R. Shagidullin, Zh. Obshch. Khim., 37, 2532 (1967); C. A. 69, 36215 (1968).
392. Kharrasova, F. M., G. Kamai, R. B. Sultanova, and R. R. Shagidullin, Zh. Obshch. Khim., 39, 1274 (1969); C. A. 71, 101933 (1969).
393. Khisamova, Z. L., and G. Kamai, Zh. Obshch. Khim., 20, 1162 (1950) C. A. 45, 1531 (1951).
394. Khlebarov, N., Izv. Inst. Khim. Sredstva Selskoto Stopanstvo, Akad. Selskostopanskite Nauki Bulgar., 1, 13 (1962); C. A. 59, 14020 (1963).
395. Kirsanov, A. V., Ya. P. Shaturskii, and G. K. Fedorova, Zh. Obshch. Khim., 39, 2596 (1969); C. A. 72, 67044 (1970).
396. Kirsanov, A. V., L. P. Zhuravleva, M. I. Z'ola, G. L. Butova, and M. G. Suleimanova, Zh. Obshch. Khim., 37, 510 (1967); C. A. 67, 82253 (1967).
397. Kloes, H., and H. Shloer, Ger. 1,163,817; C. A. 61, 1894 (1964).
398. Kluger, R., F. Kerst, D. G. Lee, and F. H. Westheimer, J. Am. Chem. Soc., 89, 3919 (1967).
399. Kluger, R., F. Kerst, D. G. Lee, E. A. Dennis, and F. H. Westheimer, J. Am. Chem. Soc., 89, 3918 (1967).
400. Knunyants, I. L., R. N. Sterlin, V. V. Tyuleneva, and L. N. Pinkina, Izv. Akad. Nauk SSSR, Otd. Khim. Nauk, 1963, 1123; C. A. 59, 8784 (1963).
401. Kolotilo, M. V., and G. I. Derkach, Zh. Obshch. Khim., 39, 463 (1969); C. A. 70, 115261 (1969).
402. Korpiun, O., R. A. Lewis, J. Chickos, and K. Mislow, J. Am. Chem. Soc., 90, 4842 (1968).
403. Korshak, V. V., S. P. Krukovskii, and Jun-Hang Wang, Vysokomol. Soedin., Ser. B., 9, 583 (1967); C. A. 67, 100479 (1967).
404. Korshak, V. V., S. P. Krukovskii, Jun-Hang Wang, and B. V. Lokshin, Vysokomol. Soedin., Ser. B., 9, 628 (1967); C. A. 67, 100480 (1967).
405. Korshak, V. V., S. P. Krukovskii, V. E. Sheina, and V. G. Danilov, Vysokomol. Soedin., Ser. B., 10, 160 (1968); C. A. 69, 10820 (1968).
406. Kosolapoff, G. M., J. Am. Chem. Soc., 64, 2982 (1942).

407. Kosolapoff, G. M., J. Am. Chem. Soc., 70, 3465 (1948).
408. Kosolapoff, G. M., J. Am. Chem. Soc., 71, 369 (1949).
409. Kosolapoff, G. M., J. Am. Chem. Soc., 72, 4292 (1950).
410. Kosolapoff, G. M., J. Am. Chem. Soc., 72, 5508 (1950).
411. Kosolapoff, G. M., U. S. 2,594,453; C. A. 47, 1179 (1953).
412. Kosolapoff, G. M., J. Am. Chem. Soc., 74, 5520 (1952).
413. Kosolapoff, G. M., J. Am. Chem. Soc., 76, 3222 (1954).
414. Kosolapoff, G. M., J. Am. Chem. Soc., 77, 6658 (1955).
415. Kosolapoff, G. M., Dokl. Akad. Nauk SSSR, 167, 1303 (1966); C. A. 65, 3902 (1966).
416. Kosolapoff, G. M., and A. D. Brown Jr., J. Chem. Soc. (C), 1967, 1789.
417. Kosolapoff, G. M., and W. F. Huber, J. Am. Chem. Soc., 69, 2020 (1947).
418. Kosolapoff, G. M., and H. G. Kirksey, Dokl. Akad. Nauk SSSR, 176, 1339 (1967); C. A. 69, 18452 (1968).
419. Kosolapoff, G. M., and J. F. McCullough, J. Am. Chem. Soc., 73, 5392 (1951).
420. Kosolapoff, G. M., and J. S. Powell, J. Chem. Soc., 1950, 3535.
421. Kosolapoff, G. M., and R. F. Struck, J. Chem. Soc., 1957, 3739.
422. Kosolapoff, G. M., and R. F. Struck, J. Chem. Soc., 1959, 3950.
423. Kosolapoff, G. M., and R. M. Watson, J. Am. Chem. Soc., 73, 4101 (1951).
424. Kosolapoff, G. M., and R. M. Watson, J. Am. Chem. Soc., 73, 5466 (1951).
425. Kovolev, L. S., N. A. Razumova, and A. A. Petrov, Zh. Obshch. Khim., 38, 2277 (1968); C. A. 71, 13171 (1969).
426. Kovolev, L. S., N. A. Razumova, and A. A. Petrov, Zh. Obshch. Khim., 39, 869 (1969); C. A. 71, 50077 (1969).
427. Kozlov, E. S., and S. N. Gaidamaka, Zh. Obshch. Khim., 39, 933 (1969); C. A. 71, 50082 (1969).
428. Kozlov, E. S., S. N. Gaidamaka, and A. V. Kirsanov, Zh. Obshch. Khim., 39, 1648 (1969); C. A. 71, 91593 (1969).
429. Krasil'nikova, E. A., A. M. Potapov and A. I. Razumov, Zh. Obshch. Khim., 38, 609 (1968); C. A. 69, 43993 (1968).

430. Kreutzkamp, N., and J. Pluhatsch, Arch. Pharm., 291, 463 (1958).
431. Kreutzkamp, N., and J. Pluhatsch, Arch. Pharm., 292, 159 (1959).
432. Kreutzkamp, N., J. Pluhatsch, and H. Schindler, Arch. Pharm., 293, 900 (1960).
433. Kreutzkamp, N., C. Schimpfky, and K. Storck, Arch. Pharm., 300, 868 (1967).
434. Kreutzkamp, N., and H. Schindler, Arch. Pharm., 293, 296 (1960).
435. Kuchar, M., Chem. Prumysl., 13, 191 (1963); C. A. 59, 15305 (1963).
436. Kuchen, W., H. Buchwald, K. Strolenberg, and J. Metten, Liebig's Annalen, 652, 28 (1962).
437. Kuchen, W., and G. Hägele, Chem. Ber., 103, 2274 (1970).
438. Kuchen, W., and H. Hertel, Chem. Ber., 101, 1991 (1968).
439. Kuchen, W., and A. Judat, Chem. Ber., 100, 991 (1967).
440. Kuchen, W., A. Judat, and J. Metten, Chem. Ber., 98, 3981 (1965).
441. Kuchen, W., and B. Knop, Angew. Chem. Int. Ed., 3, 507 (1964).
442. Kuchen, W., and B. Knop, Angew. Chem. Int. Ed., 4, 244 (1965).
443. Kuchen, W., and B. Knop, Chem. Ber., 99, 1663 (1966).
444. Kuchen, W., and H. Mayatepek, Chem. Ber., 101, 3454 (1968).
445. Kuchen, W., J. Metten, and A. Judat, Chem. Ber., 97, 2306 (1964).
446. Kuchen, W., and H. Meyer, Z. Anorg. Allgem. Chem., 333, 71 (1964).
447. Kuchen, W., K. Strolenberg, and H. Buchwald, Chem. Ber., 95, 1703 (1962).
448. Kuchen, W., K. Strolenberg, and J. Metten, Chem. Ber., 96, 1733 (1963).
449. Kuryleva, M. A., and V. K. Khairullin, Izv. Akad. Nauk SSSR, Ser. Khim., 1965, 2133; C. A. 64, 11244 (1966).
450. Kuznetsov, E. V., R. K. Valetdinov, and Ts. Ya. Roitburd, Zh. Obshch. Khim., 33, 150 (1963); C. A. 59, 655 (1963).
451. Labarre, M. C., D. Voigt, and F. Gallais, Bull. Soc. Chim. France, 1967, 3328.
452. Larionova, M. A., A. L. Klebanskii, and V. A. Bartashev, Zh. Obshch. Khim., 33, 265 (1963); C. A. 59, 656 (1963).
453. Larsson, L., and G. Wallerberg, Acta Chem. Scand., 20, 1247 (1966).

454. Letts, E. A., and R. F. Blake, Trans. Roy. Soc. Edinburgh, 35, 527 (1889).
455. Levin, Ya. A., and V. S. Galeev, Zh. Obshch. Khim., 37, 1327 (1967); C. A. 68, 49688 (1968).
456. Levin, Ya. A., Yu. M. Kargin, V. S. Galeev, and V. I. Sannikova, Izv. Akad. Nauk SSSR, Ser. Khim., 1968, 411; C. A. 69, 7802 (1968).
457. Levy, J. B., G. O. Doak, and L. D. Freedman, J. Org. Chem., 30, 660 (1965).
458. Levy, J. B., L. D. Freedman, and G. O. Doak, J. Org. Chem., 33, 474 (1968).
459. Lewis, R. A., O. Korpiun, and K. Mislow, J. Am. Chem. Soc., 89, 4786 (1967).
460. Lewis, R. A., O. Korpiun, and K. Mislow, J. Am. Chem. Soc., 90, 4847 (1968).
461. Liorber, B. G., M. B. Gazizov, Z. M. Khammatova, and A. I. Razumov, Tr. Kazansk Khim.-Tekhnol. Inst., 33, 155 (1964); C. A. 65, 736 (1966).
462. Liorber, B. G., Z.M. Khammatova, I. V. Berezoskaya, and A. I. Razumov, Zh. Obshch. Khim., 38, 165 (1968); C. A. 69, 52225 (1968).
463. Liorber, B. G., Z. M. Khammatova, and A. I. Razumov, Zh. Obshch. Khim., 39, 1551 (1969); C. A. 71, 113046 (1969).
464. Liorber, B. G., Z.M. Khammatova, A. I. Razumov, T. V. Zykova, and T. B. Borisova, Zh. Obshch. Khim., 38, 878 (1968); C. A. 69, 67497 (1968).
465. Liorber, B. G., and A. I. Razumov, Zh. Obshch. Khim., 34, 1855 (1964); C. A. 61, 8334 (1964).
466. Liorber, B. G., and A. I. Razumov, Zh. Obshch. Khim., 36, 314 (1966); C. A. 64, 15914 (1966).
467. Litthauer, S., Ber., 22, 2144 (1889).
468. Lorenz, W., Fr. 1,335,759; C. A. 60, 558 (1964).
469. Lorenz, W., Ger. 1,160,440; C. A. 60, 10718 (1964).
470. Lorenz, W., H. G. Schicke, and G. Schrader, Ger. 1,116,657; C. A. 57, 666 (1962).
471. Lorenz, W., and G. Schrader, Belg. 609,155; C. A. 57, 16402 (1962).
472. Lorenz, W., and G. Schrader, Ger. 1,134,372; C. A. 58, 3353 (1963).
473. Lorenz, W., and G. Schrader, Belg. 609,802; C. A. 58, 11402 (1963).
474. Lynch, E. R., J. Chem. Soc., 1962, 3729.
475. Maier, L., Chem. Ber., 94, 3051 (1961).
476. Maier, L., Helv. Chim. Acta, 47, 1448 (1964).
477. Maier, L., Helv. Chim. Acta, 50, 1742 (1967).
478. Maier, L., Helv. Chim. Acta, 52, 827 (1969).
479. Maksudov, A. M., V. A. Gilyarov, and M. I. Kabachnik, Izv. Akad. Nauk SSSR, Ser. Khim., 1969, 871; C. A. 71, 39090 (1969).
480. Malatesta, L., Gazz. Chim. Ital., 77, 509 (1947).

481. Malatesta, L., and R. Pizzotti, Gazz. Chim. Ital., 76, 167 (1946).

482. Mallion, K. B., and F. G. Mann, J. Chem. Soc., 1964, 6121.

483. Mann, F. G., and M. J. Pragnell, J. Chem. Soc. (C), 1966, 916.

484. Mann, F. G., B. P. Tong, and V. P. Wystrach, J. Chem. Soc., 1963, 1155.

485. Mannes, K., G. Schrader, and K. Wedemeyer, Belg. 610,433; C. A. 58, 10238 (1963).

486. Marathe, K. G., N. S. Limaye, and B. V. Bhide, J. Sci. Ind. Research (India), 9B, 268 (1950); C. A. 45, 7543 (1951).

487. Marie, C., Ann. Chim. Phys. (8), 3, 335 (1904).

488. Markowska, A., Bull. Acad. Polon. Sci., Ser. Sci. Chim., 13, 149 (1965); C. A. 63, 9791 (1965).

489. Markowska, A., Bull. Acad. Polon. Sci., Ser. Sci. Chim., 15, 153 (1967); C. A. 67, 90891 (1967).

490. Markowska, A., and J. Michalski, Roczniki Chem., 34, 1675 (1960); C. A. 56, 7345 (1962).

491. Martin, K. R., and C. E. Griffin, J. Heterocyclic Chem., 3, 92 (1966).

492. Maruszewska-Wieczorkowska, E., and J. Michalski, Bull. Acad. Polon. Sci., Ser. Sci. Chim., 6, 19 (1958); C. A. 52, 16349 (1958).

493. Mastalerz, P., Roczniki Chem., 33, 985 (1959); C. A. 54, 6602 (1960).

494. Mastalerz, P., Roczniki Chem., 34, 1161 (1960); C. A. 55, 13357 (1961).

495. Mastalerz, P., Roczniki Chem., 36, 1093 (1962); C. A. 58, 5719 (1963).

496. Mastalerz, P., Roczniki Chem., 37, 187 (1963); C. A. 59, 6435 (1963).

497. Mastalerz, P., Roczniki Chem., 38, 61 (1964); C. A. 60, 14535 (1964).

498. Mastalerz, P., Roczniki Chem., 39, 33 (1965); C. A. 62, 16292 (1965).

499. Mastalerz, P., Roczniki Chem., 39, 1129 (1965); C. A. 64, 6684 (1966).

500. Mastalerz, P., and Z. E. Golubski, Roczniki Chem., 41, 1527 (1967); C. A. 68, 39717 (1968).

501. Mastalerz, P., and R. Tyka, Roczniki Chem., 38, 1529 (1964); C. A. 62, 9169 (1965).

502. Mastryukova, T. A., T. A. Melent'eva, and M. I. Kabachnik, Zh. Obshch. Khim., 35, 1197 (1965); C. A. 63, 11605 (1965).

503. Mastryukova, T. A., T. A. Melent'eva, A. E. Shipov, and M. I. Kabachnik, Zh. Obshch. Khim., 29, 2178 (1959); C. A. 54, 10463 (1960).

504. Mastryukova, T. A., A. E. Shipov, and M. I. Kabachnik, Zh. Obshch. Khim., 29, 1450 (1959); C. A. 54, 9729 (1960).
505. Mastryukova, T. A., A. E. Shipov, and M. I. Kabachnik, Zh. Obshch. Khim., 31, 507 (1961); C. A. 55, 22101 (1961).
506. Mastryukova, T. A., A. E. Shipov, and M. I. Kabachnik, Zh. Obshch. Khim., 32, 3579 (1962); C. A. 58, 11394 (1963).
507. Mathey, F., and J. P. Lampin, Compt. Rend., Ser. C, 270, 1531 (1970).
508. Mathey, F., and G. Mavel, Compt. Rend., Ser. C, 263, 855 (1966).
509. Matrosov, E. I., Izv. Akad. Nauk SSSR, Neorg. Mater., 3, 539 (1967); C. A. 67, 68940 (1967).
510. Mauret, P., J. P. Fayet, and M. C. Labarre, Compt. Rend., Ser. C, 265, 65 (1967).
511. Mauret, P., J. P. Fayet, D. Voigt, M. C. Labarre, and J. F. Labarre, J. Chim. Phys. Physico-Chim. Biol., 65, 549 (1968).
512. Mazhar-Ul-Haque, and C. N. Caughlan, Chem. Commun. 1966, 921.
513. McBride, J. J., E. Jungerman, J. V. Kilheffer, and R. J. Clutter, J. Org. Chem., 27, 1833 (1962).
514. McLean, R. A. N., Inorg. Nucl. Chem. Lett., 5, 745 (1969).
515. Medved, T. Ya., Yu. M. Polikarpov, K. S. Yudina, and M. I. Kabachnik, Izv. Akad. Nauk SSSR, Ser. Khim., 1965, 1707; C. A. 64, 3591 (1966).
516. Mel'nikov, N. N., Ya. A. Mandel'baum, and Z. M. Bakanova, Zh. Obshch. Khim., 31, 3953 (1961); C. A. 57, 11072 (1962).
517. Mel'nikov, N. N., Ya. A. Mandel'baum, V. I. Lomakina, and V. S. Livshits, Zh. Obshch. Khim., 31, 3949 (1961); C. A. 57, 11072 (1962).
518. Meriani, S., G. Nardin, and A. Ripamonti, Inorg. Chem., 6, 1931 (1967).
519. Michaelis, A., Ber., 10, 627 (1877).
520. Michaelis, A., Ber., 19, 1009 (1886).
521. Michaelis, A., Liebig's Annalen, 293, 193 (1896).
522. Michaelis, A., Liebig's Annalen, 294, 1 (1896).
523. Michaelis, A., Ber., 31, 1037 (1898).
524. Michaelis, A., Liebig's Annalen, 315, 43 (1901).
525. Michaelis, A., Liebig's Annalen, 326, 129 (1903).
526. Michaelis, A., and F. Graeff, Ber., 8, 1304 (1875).
527. Michaelis, A., and W. LaCoste, Ber., 18, 2109 (1885).
528. Michaelis, A., and F. Wegner, Ber., 48, 316 (1915).
529. Michalski, J., and A. Skowronska, Roczniki Chem., 34, 1381 (1960); C. A. 55, 19842 (1961).

530. Michalski, J., and Z. Tulimowski, Roczniki Chem., 36, 1781 (1962); C. A. 59, 10109 (1963).
531. Miller, C. O., and C. J. Dorer, U. S. 2,797,238; C. A. 51, 16537 (1957).
532. Miller, R. C., J. Org. Chem., 24, 2013 (1959).
533. Miller, R. C., J. S. Bradley, and L. A. Hamilton, J. Am. Chem. Soc., 78, 5299 (1956).
534. Mikhailov, B. M., and N. F. Kucherova, Dokl. Akad. Nauk SSSR, 74, 501 (1950); C. A. 45, 3343 (1951).
535. Mikolajczyk, M., Chem. Ber., 99, 2083 (1966).
536. Mizrakh, L. I., and V. P. Evdakov, Zh. Obshch. Khim., 36, 469 (1966); C. A. 65, 738 (1966).
537. Mizrakh, L. I., L. Yu Sandalova, and V. P. Evdakov, Zh. Obshch. Khim., 37, 1875 (1967); C. A. 68, 29792 (1968).
538. Mingoia, Q., Gazz. Chim. Ital., 62, 333 (1932).
539. Moedritzer, K., J. Am. Chem. Soc., 83, 4381 (1961).
540. Moedritzer, K., L. Maier, and L. C. D. Groenweghe, J. Chem. Eng. Data, 7, 307 (1962).
541. Morgan, G. T., and W. R. Moore, J. Chem. Soc., 97, 1697 (1910).
542. Morris, R. C., J. L. Van Winkle, and H. H. Wyrick, U. S. 2,714,064; C. A. 50, 10123 (1956).
543. Morrison, D. C., J. Am. Chem. Soc., 74, 3431 (1952).
544. Moskva, V. V., A. I. Maikova, and A. I. Razumov, Zh. Obshch. Khim., 39, 2451 (1969); C. A. 72, 79158 (1970).
545. Müller, E., and H. G. Padeken, Chem. Ber., 100, 521 (1967).
546. Mukhacheva, O. A., and A. I. Razumov, Zh. Obshch. Khim., 32, 2693 (1962); C. A. 58, 9119 (1963).
547. Mukhacheva, O. A., and A. I. Razumov, Zh. Obshch. Khim., 32, 2696 (1962); C. A. 58, 9119 (1963).
548. Mukhametzyanova, E. Ka., I. N. Faizullin, and I. M. Shermergorn, Izv. Akad. Nauk SSSR, Ser. Khim., 1969, 710; C. A. 71, 50079 (1969).
549. Mukhametzyanova, E. Kh., A. R. Panteleeva, and I. M. Shermergorn, Izv. Akad. Nauk SSSR, Ser. Khim., 1967, 1597; C. A. 68, 13075 (1968).
550. Mukhamzyanova, E. Kh., and I. M. Shermergorn, Izv. Akad. Nauk SSSR, Ser. Khim., 1969, 951; C. A. 71, 39096 (1969).
551. Mukherjee, R. N., A. K. Chatterjee, and J. Gupta, Indian J. Chem., 3, 514 (1965).
552. Mukherjee, R. N., V. V. Krishna Rao, and J. Gupta, Indian J. Chem., 4, 209 (1966).
553. Mukherjee, R. M., A. Y. Sonsale, and J. Gupta, Indian J. Chem., 4, 500 (1966).
554. Muller, N., P. C. Lauterbur, and J. Goldenson, J. Am. Chem. Soc., 78, 3557 (1956).

555. Mustafa, A., M. M. Sidky, and M. F. Zayed, Naturwissenschaften, 55, 83 (1968).
556. Mustafa, A., M. M. Sidky, S. M. A. D. Zayed, and W. M. Abdo, Tetrahedron, 24, 4725 (1968).
557. Mustafa, A., M. M. Sidky, S. M. A. D. Zayed, and M. R. Mahron, Liebig's Annalen, 712, 116 (1968).
558. Mustafa, A., M. M. Sidky, S. M. A. D. Zayed, and M. F. Zayed, Liebig's Annalen, 711, 198 (1968).
559. Myshkin, A. E., and V. P. Evdakov, Zh. Obshch. Khim., 38, 1776 (1968); C. A. 70, 4229 (1969).
560. Neimysheva, A. A., V. I. Savchuk, and I. L. Knunyants, Zh. Obshch. Khim., 36, 500 (1966); C. A. 65, 2103, (1966).
561. Newallis, P. E., J. P. Chupp, and J. W. Baker, U. S. 3,159,534; C. A. 62, 4053 (1965).
562. Newallis, P. E., J. P. Chupp, and J. W. Baker, U. S. 3,231,359; C. A. 64, 14218 (1966).
563. Newallis, P. E., J. P. Chupp, and L. C. D. Groenweghe, J. Org. Chem., 27, 3829 (1962).
564. Nielson, M. L., U. S. 3,284,497; C. A. 66, 28890 (1967).
565. Nielson, M. L., and J. V. Pustinger Jr., J. Phys. Chem., 68, 152 (1964).
566. Nielson, M. L., J. V. Pustinger Jr., and J. Strobell, J. Chem. Eng. Data, 9, 167 (1964).
567. Nielson, R. P., and H. H. Sisler, Inorg. Chem., 2, 753 (1963).
568. Nikolaev, A. V., Yu. A. Dyadin, and I. I. Yakovlev, Dokl. Akad. Nauk SSSR, 160, 363 (1965); C. A. 62, 11203 (1965).
569. Nikolaev, A. V., Yu. A. Dyadin, I. I. Yakovlev, V. B. Durasov, N. I. Yakovleva, and I. D. Khol'kina, Zh. Fiz. Khim., 40, 221 (1966); C. A. 64, 11937 (1966).
570. Nikolaev, A. V., I. M. Ivanov, and I. I. Yakovlev, Dokl. Akad. Nauk SSSR, 156, 888 (1964); C. A. 61, 7758 (1964).
571. Nikonorov, K. V., and E. A. Gurylev, Izv. Akad. Nauk SSSR, 1965, 2136; C. A. 64, 11245 (1966).
572. Nikonorov, K. V., E. A. Gurylev, F. F. Fakhrislamova, T. V. Raspopova, L. G. Urazaeva, M. G. Berim, K. B. Brudnaya, E. K. Naumova, and T. L. Nosreva, Dokl. Akad. Nauk SSSR, 172, 353 (1967); C. A. 67, 21968 (1967).
573. Nikonorov, K. V., E. A. Gurylev, L. G. Urazaeva, M. N. Nazypov, R. A. Asadov, and S. D. Anisin, Izv. Akad. Nauk SSSR, Ser. Khim., 1969, 2241; C. A. 72, 31914 (1970).
574. Novikova, Z. S., E. A. Efimova, and I. F. Lutsenko, Zh. Obshch. Khim., 38, 2345 (1968); C. A. 70, 29014 (1969).

575. Nudelman, A., and D. J. Cram, J. Am. Chem. Soc., 90, 3869 (1968).
576. Nuretdinov, I. A., R. R. Shagidullin, Yu. Ya. Shamonin, and N. P. Grechkin, Izv. Akad. Nauk SSSR, Ser. Khim., 1966, 839; C. A. 65, 5334 (1966).
577. Nurtdinov, S. Kh., V. S. Tsivunin, R. S. Khairullin, V. G. Kashtanova, and G. Kamai, Zh. Obshch. Khim., 40, 36 (1970); C. A. 72, 111569 (1970).
578. Nyquist, R. A., J. Mol. Struct., 2, 111 (1968).
579. Ocone, L. R., C. W. Schaumann, B. P. Block, and E. N. Walsh, Inorg. Synth., 8, 71 (1966).
580. Okhlobystin, O. Yu., and L. I. Zakharkin, Izv. Akad. Nauk SSSR, Otd. Khim. Nauk, 1958, 1006; C. A. 53, 1122 (1959).
581. Osborne, G. O., and J. Kidd, Brit. 991,590; C. A. 63, 4890 (1965).
582. Paciorek, K. L., and R. H. Kratzer, J. Org. Chem., 31, 2426 (1966).
583. Panteleeva, A. R., and I. M. Shermergorn, Izv. Akad. Nauk SSSR, Ser. Khim., 1968, 1644; C. A. 69, 87112 (1968).
584. Peppard, D. F., G. W. Mason, and C. M. Andrejasich, J. Inorg. Nucl. Chem., 27, 697 (1965).
585. Peppard, D. F., G. W. Mason, and S. Lewey, J. Inorg. Nucl. Chem., 27, 2065 (1965).
586. Peters, G., J. Org. Chem., 27, 2198 (1962).
587. Petragnani, N., and M. deMoura Campos, Chem. Ind. (London), 1965, 1076.
588. Petragnani, N., V. G. Toscano, and M. deMoura Campos, Chem. Ber., 101, 3070 (1968).
589. Petrov, K. A., A. A. Basyuk, V. P. Evdakov, and L. I. Mizrakh, Zh. Obshch. Khim., 34, 2226 (1964); C. A. 61, 12030 (1964).
590. Petrov, K. A., N. K. Bliznyuk, and V. P. Korotkova, Zh. Obshch. Khim., 30, 2995 (1960); C. A. 55, 18561 (1961).
591. Petrov, K. A., N. K. Bliznyuk, and T. N. Lysenko, Zh. Obshch. Khim., 30, 1964 (1960); C. A. 55, 6362 (1961).
592. Petrov, K. A., N. K. Bliznyuk, and I. Yu. Mansurov, Zh. Obshch. Khim., 31, 176 (1961); C. A. 55, 22097 (1961).
593. Petrov, K. A., N. K. Bliznyuk, and V. A. Savostenok, Zh. Obshch. Khim., 31, 1361 (1961); C. A. 55, 23317 (1961).
594. Petrov, K. A., T. N. Lysenko, B. Ya. Libman, V. V. Pozdnev, Khim. Org. Soedin. Fosfora, Akad. Nauk SSSR, Otd. Obshch. Tech. Khim., 1967, 181; C. A. 69, 67487 (1968).

595. Petrov, K. A., E. E. Nifant'ev, and R. G. Gol'tsova, Zh. Obshch. Khim., 32, 3716 (1962); C. A. 58, 12596 (1963).
596. Petrov, K. A., E. E. Nifant'ev, and L. V. Khorkhoyanu, Zh. Obshch. Khim., 31, 2889 (1961); C. A. 57, 860 (1962).
597. Petrov, K. A., E. E. Nifant'ev, and L. V. Khorkhoyanu, Zh. Obshch. Khim., 32, 3720 (1962); C. A. 59, 8844 (1963).
598. Petrov, K. A., E. E. Nifant'ev, L. V. Khorkhoyanu, and A. I. Trushkov, Zh. Obshch. Khim., 31, 3085 (1961); C. A. 57, 858 (1962).
599. Petrov, K. A., E. E. Nifant'ev, A. A. Shchegolev, and N. A. Khudyntsev, Zh. Obshch. Khim., 32, 3074 (1962); C. A. 58, 11456 (1963).
600. Petrov, K. A., V. A. Parshina, and G. L. Daruze, Zh. Obshch. Khim., 30, 3000 (1960); C. A. 55, 23399 (1961).
601. Pilgram, K., Tetrahedron Letters 1966, 3831.
602. Pope, W. J., and C. S. Gibson, J. Chem. Soc., 101, 740 (1912).
603. Popoff, I. C., L. K. Huber, B. P. Block, P. D. Morton, and R. P. Riordan, J. Org. Chem., 28, 2898 (1963).
604. Pudovik, A. N., Dokl. Akad. Nauk SSSR, 92, 773 (1953); C. A. 49, 3050 (1955).
605. Pudovik, A. N., I. M. Aladzheva, and L. V. Spirina, Zh. Obshch. Khim., 37, 700 (1967); C. A. 67, 21970 (1967).
606. Pudovik, A. N., I. M. Aladzheva, and L. N. Yakovenko, Zh. Obshch. Khim., 35, 1210 (1965); C. A. 63, 11609 (1965).
607. Pudovik, A. N., and V. P. Aver'yanova, Zh. Obshch. Khim., 26, 1426 (1956); C. A. 50, 14512 (1956).
608. Pudovik, A. N., G. V. Dmitrieva, Yu. Yu. Samitov, R. R. Shagidullin, and V. K. Khairullin, Zh. Obshch. Khim., 39, 2219 (1969); C. A. 72, 55581 (1970).
609. Pudovik, A. N., and O. S. Durova, Zh. Obshch. Khim., 36, 1460 (1966); C. A. 66, 11009 (1967).
610. Pudovik, A. N., and G. I. Evstaf'ev, Dokl. Akad. Nauk SSSR, 164, 1331 (1965); C. A. 64, 5219 (1966).
611. Pudovik, A. N., and I. V. Gur'yanova, Zh. Obshch. Khim., 37, 1649 (1967); C. A. 68, 13093 (1968).
612. Pudovik, A. N., I. V. Gur'yanova, and L. V. Banderova, Khim. Org. Soedin. Fosfora, Akad. Nauk SSSR, Otd. Obshch. Tech. Khim., 1967, 21; C. A. 69, 67485 (1968).
613. Pudovik, A. N., I. V. Gur'yanova, L. V. Banderova, and M. G. Limin, Zh. Obshch. Khim., 37, 876 (1967); C. A. 68, 2950 (1968).

614. Pudovik, A. N., I. V. Gur'yanova, L. V. Banderova, and G. V. Romanov, Zh. Obshch. Khim., 38, 143 (1968); C. A. 69, 96839 (1968).
615. Pudovik, A. N., I. V. Gur'yanova, and G. V. Romanov, Zh. Obshch. Khim., 39, 2418 (1969); C. A. 72, 79164 (1970).
616. Pudovik, A. N., I. V. Gur'yanova, and M. G. Zimin, Zh. Obshch. Khim., 37, 407 (1967); C. A. 67, 43887 (1967).
617. Pudovik, A. N., I. V. Gur'yanova, M. G. Zimin, and A. V. Durneva, Zh. Obshch. Khim., 39, 1081 (1969); C. A. 71, 61477 (1969).
618. Pudovik, A. N., I. V. Gur'yanova, M. G. Zimin, and O. E. Raevskaya, Zh. Obshch. Khim., 38, 1539 (1968); C. A. 70, 87915 (1969).
619. Pudovik, A. N., I. V. Gur'yanova, M. G. Zimin, and O. E. Raevskaya, Zh. Obshch. Khim., 39, 1021 (1969); C. A. 71, 61476 (1969).
620. Pudovik, A. N., and V. K. Khairullin, Zh. Obshch. Khim., 39, 1724 (1969); C. A. 71, 124579 (1969).
621. Pudovik, A. N., V. K. Khairullin, and G. V. Dmitrieva, Dokl. Akad. Nauk SSSR, 174, 372 (1967); C. A. 68, 2959 (1968).
622. Pudovik, A. N., V. K. Khairullin, and N. I. Kharitonova, Izv. Akad. Nauk SSSR, Ser. Khim., 1969, 466; C. A. 71, 13173 (1969).
623. Pudovik, A. N., V. K. Khairullin, Yu. Yu. Samitov, and R. R. Shagidullin, Zh. Obshch. Khim., 37, 865 (1967); C. A. 68, 105305 (1968).
624. Pudovik, A. N., V. K. Khairullin, and M. A. Vasyanina, Zh. Obshch. Khim., 37, 411 (1967); C. A. 67, 64487 (1967).
625. Pudovik, A. N., V. K. Khairullin, M. A. Vasyanina, and G. F. Novikova, Izv. Akad. Nauk SSSR, Ser. Khim., 1969, 2334; C. A. 72, 43787 (1970).
626. Pudovik, A. N., R. M. Kondrat'eva, and V. K. Khairullin, Izv. Akad. Nauk SSSR, Ser. Khim., 1969, 2076; C. A. 72, 12830 (1970).
627. Pudovik, A. N., and I. V. Konovalova, Zh. Obshch. Khim., 30, 2348 (1960); C. A. 55, 8326 (1961).
628. Pudovik, A. N., and I. V. Konovalova, Zh. Obshch. Khim., 32, 467 (1962); C. A. 58, 544 (1963).
629. Pudovik, A. N., I. V. Konovalova, and L. V. Banderova, Zh. Obshch. Khim., 35, 1206 (1965); C. A. 63, 13064 (1965).
630. Pudovik, A. N., I. V. Konovalova, and O. S. Durova, Zh. Obshch. Khim., 31, 2656 (1961); C. A. 56, 12925 (1962).
631. Pudovik, A. N., I. V. Konovalova, and A. A. Guryleva, Zh. Obshch. Khim., 33, 2924 (1963); C. A. 60, 5541 (1964).

632. Pudovik, A. N., I. V. Konovalova, and R. E. Krivonosova, Zh. Obshch. Khim., 26, 3110 (1956); C. A. 51, 8642 (1957).
633. Pudovik, A. N., and V. K. Krupnov, Zh. Obshch. Khim., 38, 1287 (1968); C. A. 69, 77340 (1968).
634. Pudovik, A. N., V. I. Nikitina, and G. P. Grupnov, Zh. Obshch. Khim., 29, 4019 (1959); C. A. 54, 20933 (1960).
635. Pudovik, A. N., and N. G. Poloznova, Zh. Obshch. Khim., 25, 778 (1955); C. A. 50, 2417 (1956).
636. Pudovik, A. N., and M. A. Pudovik, Zh. Obshch. Khim., 36, 1467 (1966); C. A. 66, 11001 (1967).
637. Pudovik, A. N., R. D. Sabirova, and T. A. Tener, Zh. Obshch. Khim., 24, 1026 (1954); C. A. 49, 8790 (1955).
638. Pudovik, A. N., and O. S. Shulyndina, Zh. Obshch. Khim., 39, 1014 (1969); C. A. 71, 61493 (1969).
639. Pudovik, A. N., and F. N. Sitdikova, Dokl. Akad. Nauk SSSR, 125, 826 (1959); C. A. 53, 19850 (1959).
640. Pudovik, A. N., L. V. Spirina, M. A. Pudovik, Yu. M. Kargin, and L. S. Andreeva, Zh. Obshch. Khim., 39, 1715 (1969); C. A. 71, 124597 (1969).
641. Pudovik, A. N., and D. Kh. Yarmukhametova, Izv. Akad. Nauk SSSR, Otd. Khim. Nauk, 1952, 902; C. A. 47, 10469 (1953).
642. Pudovik, A. N., and D. Kh. Yarmukhametova, Izv. Akad. Nauk SSSR, Otd. Khim. Nauk, 1954, 636; C. A. 49, 8789 (1955).
643. Pudovik, A. N., G. E. Yastrebova, L. M. Leont'eva, T. A. Zyablikova, and V. I. Nikitina, Zh. Obshch. Khim., 39, 1230 (1969); C. A. 71, 70690 (1969).
644. Pudovik, A. N., G. E. Yastrebova, V. E. Nikitina, and Yu. Yu. Samitov, Zh. Obshch. Khim., 38, 292 (1968); C. A. 69, 106815 (1968).
645. Quin, L. D., and R. E. Montgomery, J. Org. Chem., 28, 3315 (1963).
646. Ratajczak, A., Roczniki Chem., 36, 175 (1962); C. A. 57, 15147 (1962).
647. Rauhut, M. M., and H. A. Currier, J. Org. Chem., 26, 4628 (1961).
648. Rauhut, M. M., H. A. Currier, G. A. Peters, F. C. Schaefer, and V. P. Wystrach, J. Org. Chem., 26, 5135 (1961).
649. Rauhut, M. M., H. A. Currier, and V. P. Wystrach, J. Org. Chem., 26, 5133 (1961).
650. Razumov, A. I., and M. B. Gazizov, Zh. Obshch. Khim., 37, 2738 (1967); C. A. 69, 19263 (1968).
651. Razumov, A. I., and P. A. Gurevich, Zh. Obshch. Khim., 37, 1615 (1967); C. A. 68, 39730 (1968).
652. Razumov, A. I., and P. A. Gurevich, Zh. Obshch. Khim., 37, 1620 (1967); C. A. 68, 39731 (1968).

653. Razumov, A. I., P. A. Gurevich, B. G. Liorber, and T. B. Borisova, Zh. Obshch. Khim., 39, 392 (1969); C. A. 70, 115230 (1969).
654. Razumov, A. I., P. A. Gurevich, and V. V. Moskva, Zh. Obshch. Khim., 37, 961 (1967); C. A. 67, 100083 (1967).
655. Razumov, A. I., and B. G. Liorber, Dokl. Akad. Nauk SSSR, 135, 1150 (1960); C. A. 55, 13298 (1961).
656. Razumov, A. I., B. G. Liorber, I. V. Berezovskaya, and T. A. Tarzivolova, Khim.-Farm. Zh., 1, 41 (1967); C. A. 68, 69080 (1968).
657. Razumov, A. I., B. G. Liorber, and P. A. Gurevich, Zh. Obshch. Khim., 37, 2782 (1967); C. A. 69, 43977 (1968).
658. Razumov, A. I., B. G. Liorber, and P. A. Gurevich, Zh. Obshch. Khim., 38, 199 (1968); C. A. 69, 52216 (1968).
659. Razumov, A. I., and V. V. Moskva, Zh. Obshch. Khim., 34, 2589 (1964); C. A. 61, 14707 (1964).
660. Razumov, A. I., and V. V. Moskva, Zh. Obshch. Khim., 35, 1595 (1965); C. A. 63, 18144 (1965).
661. Razumov, A. I., and O. A. Mukhacheva, Dokl. Akad. Nauk SSSR, 91, 271 (1953); C. A. 48, 8725 (1954).
662. Razumov, A. I., and O. A. Mukhacheva, Zh. Obshch. Khim., 26, 2463 (1956); C. A. 51, 1822 (1957).
663. Razumov, A. I., O. A. Mukhacheva, and Sim-Do-Khen, Izv. Akad. Nauk SSSR, Otd. Khim. Nauk, 1952, 894; C. A. 47, 10466 (1953).
664. Razumov, A. I., O. A. Mukhacheva, and I. V. Zaikonnikova, Zh. Obshch. Khim., 27, 754 (1957); C. A. 51, 16332 (1957).
665. Razumov, A. I., R. L. Poznyak, K. B. Brudnaya, M. G. Berim, R. I. Slepova, Sh. Z. Tuktarova, and G. F. Rzhevskaya, Zh. Obshch. Khim., 37, 421 (1967); C. A. 67, 100208 (1967).
666. Razumov, A. I., and G. A. Savicheva, Zh. Obshch. Khim., 34, 2595 (1964); C. A. 61, 14707 (1964).
667. Razumov, A. I., and G. A. Savicheva, Zh. Obshch. Khim., 35, 2038 (1965); C. A. 64, 6679 (1966).
668. Razumov, A. I., G. A. Savicheva, and G. A. Budnikov, Zh. Obshch. Khim., 35, 1454 (1965); C. A. 63, 14681 (1965).
669. Razumova, N. A., Zh. L. Evtikhov, A. Kh. Voznesenskaya, and A. A. Petrov, Zh. Obshch. Khim., 39, 176 (1969); C. A. 70, 106608 (1969).
670. Razumova, N. A., Zh. L. Evtikhov, L. I. Zubtsova, and A. A. Petrov, Zh. Obshch. Khim., 38, 2342 (1968); C. A. 70, 47546 (1969).
671. Razumova, N. A., L. S. Kovalev, and L. I. Zubtsova, Zh. Obshch. Khim., 37, 1919 (1967); C. A. 68, 105295 (1968).

672. Razumova, N. A., and A. A. Petrov, Zh. Obshch. Khim., 33, 783 (1963); C. A. 59, 8783 (1963).
673. Razumova, N. A., and A. A. Petrov, Zh. Obshch. Khim., 34, 356 (1964); C. A. 60, 10710 (1964).
674. Razumova, N. A., and A. A. Petrov, Zh. Obshch. Khim., 34, 1886 (1964); C. A. 61, 8336 (1964).
675. Razumova, N. A., and I. M. Treskunova, Zh. Obshch. Khim., 34, 2949 (1964); C. A. 61, 16090 (1964).
676. Reesor, J. W. B., and G. F. Wright, J. Org. Chem., 22, 385 (1957).
677. Reinhardt, H., D. Bianchi, and D. Mölle, Chem. Ber., 90, 1656 (1957).
678. Reznik, V. S., and N. G. Pashkurov, Dokl. Akad. Nauk SSSR, 177, 604 (1967); C. A. 69, 27368 (1968).
679. Rieker, A., Z. Naturforsch., b, 21, 647 (1966).
680. Rizpolozhenskii, N. I., and A. A. Muslinkin, Izv. Akad. Nauk SSSR, Otd. Khim. Nauk, 1961, 1600; C. A. 56, 4791 (1962).
681. Rizpolozhenskii, N. I., and M. A. Zvereva, Izv. Akad. Nauk SSSR, Otd. Khim. Nauk, 1959, 358; C. A. 53, 17888 (1959).
682. Robins, R. K., and B. E. Christensen, J. Org. Chem., 16, 324 (1951).
683. de Roos, A. M., and H. J. Toet, Rec. Trav. Chim., 78, 271 (1959).
684. Rose, S. H., and B. P. Block, J. Am. Chem. Soc., 87, 2076 (1965).
685. Rose, S. H., and B. P. Block, J. Polymer Sci., Part A-I, 4, 573 (1966).
686. Rose, S. H., and B. P. Block, J. Polymer Sci., Part A-I, 4, 583 (1966).
687. de Saint-Giniez, D., A. Laurent, Nguyen-Thank-Thuong, C. Rerat, and P. Chabrier, Compt. Rend., Ser. C, 263, 1213 (1966).
688. Sander, M., Chem. Ber., 93, 1220 (1960).
689. Sander, M., Makromol. Chem., 55, 191 (1962).
690. Saraceno, A. J., and B. P. Block, Inorg. Chem., 3, 1699 (1964).
691. Sauvage, R., Compt. Rend., 139, 674 (1904).
692. Schicke, H. G., Belg. 652,189; C. A. 64, 8238 (1966).
693. Schicke, H. G., Ger. 1,192,202; C. A. 63, 9989 (1965).
694. Schicke, H. G., and A. Berger, Ger. 1,197,878; C. A. 63, 13319 (1965).
695. Schicke, H. G., and W. Lorenz, Ger. 1,144,264; C. A. 60, 4184 (1964).
696. Schicke, H. G., and G. Schrader, Ger. 1,122,060; C. A. 57, 12325 (1962).
697. Schindlbauer, H., Monatsh. Chem., 96, 2012 (1965).

698. Schindlbauer, H., and W. Prikosovich, Monatsh. Chem., 99, 1792 (1968).
699. Schloer, H. H., and G. Schrader, Belg. 616,335; C. A. 59, 1488 (1963).
700. Schmidpeter, A., and H. Brecht, Angew. Chem. Int. Ed., 6, 945 (1967).
701. Schmidpeter, A., H. Brecht, and H. Groeger, Chem. Ber., 100, 3063 (1967).
702. Schmidpeter, A., and J. Ebeling, Chem. Ber., 101, 815 (1968).
703. Schmidpeter, A., and H. Groeger, Z. Anorg. Allgem. Chem., 345, 106 (1966).
704. Schmidpeter, A., and H. Groeger, Chem. Ber., 100, 3052 (1967).
705. Schmidpeter, A., and H. Groeger, Chem. Ber., 100, 3979 (1967).
706. Schmutzler, R., Inorg. Chem., 3, 421 (1964).
707. Schrader, G., Ger. 1,078,125; C. A. 55, 14385 (1961).
708. Schrader, G., Ger. 1,102,138; C. A. 55, 10787 (1961).
709. Schrader, G., Ger. 1,115,248; C. A. 56, 14328 (1962).
710. Schrader, G., Ger. 1,136,334; C. A. 58, 1406 (1963).
711. Schrader, G., Belg. 615,668; C. A. 58, 11403 (1963).
712. Schrader, G., Belg. 615,669; C. A. 58, 11402 (1963).
713. Schrader, G., Ger. 1,143,513; C. A. 59, 1683 (1963).
714. Schrader, G., Ger. 1,152,407; C. A. 60, 557 (1964).
715. Schrader, G., Ger. 1,186,849; C. A. 62, 11689 (1965).
716. Schrader, G., and R. Cölln, Ger. 1,119,859; C. A. 57, 868 (1962).
717. Schrader, G., and W. Lorenz, Ger. 1,112,852; C. A. 56, 8747 (1962).
718. Schrader, G., W. Lorenz, R. Cölln, and H. Schlör, U. S. 3,232,830; C. A. 64, 15923 (1966).
719. Schrader, G., and H. Schlör, Ger. 1,138,048; C. A. 58, 6864 (1963).
720. Schroeder, D. C., P. O. Corcoran, C. L. Holden, and M. A. Mulligan, J. Org. Chem., 27, 1098 (1962).
721. Schumann, H., P. Jutzi, A. Roth, P. Schwabe, and E. Schauer, J. Organometal. Chem., 10, 71 (1967).
722. Schumann, H., H. Köpf, and M. Schmidt, J. Organometal. Chem., 2, 159 (1964).
723. Schwarz, R., and K. Schoeller, Chem. Ber., 91, 2103 (1958).
724. Sergeev, P. G., and D. G. Kudryashov, Zh. Obshch. Khim., 8, 266 (1938); C. A. 32, 5403 (1938).
725. Shaw, R. A., and E. H. M. Ibrahim, Angew. Chem. Int. Ed., 6, 556 (1967).

726. Shell International Research Maatschappij N. V., Neth. Appl. 6,504,109; C. A. 64, 14218 (1966).
727. Shell International Research Maatschappij N. V., Neth. Appl. 6,510,031; C. A. 64, 11216 (1966).
728. Shetty, P. S., P. Jose, and Q. Fernando, Chem. Commun., 1968, 788.
729. Shevchenko, V. I., and A. M. Pinchuk, Zh. Obshch. Khim., 35, 1492 (1965); C. A. 63, 14901 (1965).
730. Shevchenko, V. I., and V. T. Stratienko, Zh. Obshch. Khim., 30, 1958 (1960); C. A. 55, 6425 (1961).
731. Shevchenko, V. I., V. T. Stratienko, and A. M. Pinchuk, Zh. Obshch. Khim., 30, 1566 (1960); C. A. 55, 1490 (1961).
732. Shevchenko, V. I., V. T. Stratienko, and A. M. Pinchuk, Zh. Obshch. Khim., 35, 363 (1965); C. A. 62, 13173 (1965).
733. Shokol, V. A., G. I. Derkach, and A. V. Kirsanov, Zh. Obshch. Khim., 32, 166 (1962); C. A. 57, 16648 (1962).
734. Silver, H. B., J. Chem. Soc., 1967, 1326.
735a. Sisler, H. H., H. S. Ahuja, and N. L. Smith, Inorg. Chem., 1, 84 (1962).
735b. Sisler, H. H., R. P. Nielsen, T. H. Dexter, and D. J. Jaska, Inorg. Synth., 8, 74 (1966).
736. Sisler, H. H., and N. L. Smith, J. Org. Chem., 26, 611 (1961).
737. Slota, P. J. Jr., L. P. Freeman, and N. R. Fetter, J. Polymer Sci., Part A-I, 6, 1975 (1968).
738. Smith, B. E., and A. Burger, J. Am. Chem. Soc., 75, 5891 (1953).
739. Smith, N. L., and H. H. Sisler, J. Org. Chem., 26, 5145 (1961).
740. Sosnovsky, G. E., D. J. Rawlinson, and E. H. Zaret, Chem. Commun., 1966, 453.
741. Spanier, E. J., and F. E. Caropreso, J. Am. Chem. Soc., 92, 3348 (1970).
742. Starostin, A. D., A. V. Nikolaev, and Yu. A. Afanas'ev, Izv. Akad. Nauk SSSR, Ser. Khim., 1966, 1303; C. A. 66, 24923 (1967).
743. Steininger, E., Chem. Ber., 96, 3184 (1963).
744. Steininger, E., Monatsh. Chem., 97, 383 (1966).
745. Steininger, E., Monatsh. Chem., 97, 1193 (1966).
746. Stepanov, B. I., A. I. Bokanov, E. N. Karpova, and N. V. Danilova, Zh. Obshch. Khim., 40, 2217 (1970); C. A. 74, 141963 (1971).
747. Stepanov, B. I., E. N. Karpova, and A. I. Bokanov, Zh. Obshch. Khim., 39, 1544 (1969); C. A. 71, 113056 (1969).
748. Stephenson, T. A., E. Bannister, and G. Wilkinson, J. Chem. Soc., 1964, 2538.

749. Stewart, A. P., and S. Trippett, J. Chem. Soc. (C), 1970, 1263.
750. Stiles, A. R., F. F. Rust, and W. E. Vaughan, J. Am. Chem. Soc., 74, 3282 (1952).
751. Stilz, W., and H. Pommer, Ger. 1,122,065; C. A. 57, 3357 (1962).
752. Stoelzer, C., and A. Simon, Chem. Ber., 96, 896 (1963).
753. Strecker, W., and C. Grossmann, Ber., 49, 63 (1916).
754. Struck, R. F., J. Med. Chem., 9, 231 (1966).
755a. Strunin, B. N., O. Yu. Okholobystin, and L. I. Zakharkin, Izv. Akad. Nauk SSSR, Ser. Khim., 1963, 1373; C. A. 59, 14018 (1963).
755b. Stuebe, C., W. M. LeSuer, and G. R. Norman, J. Am. Chem. Soc., 77, 3526 (1955).
756. Tomachewski, G., and G. Geissler, Chem. Ber., 100, 919 (1967).
757. Tomachewski, G., and G. Kühn, J. Prakt. Chem., 38, 222 (1968).
758. Tomachewski, G., and D. Zanke, Z. Chem., 10, 117 (1970).
759. Tomachewski, G., and D. Zanke, Z. Chem., 10, 145 (1970).
760. Tsvetkov, E. N., R. A. Malevannaya, and M. I. Kabachnik, Zh. Obshch. Khim., 37, 695 (1967); C. A. 67, 21971 (1967).
761. Tsvetkov, E. N., R. A. Malevannaya, and M. I. Kabachnik, Zh. Obshch. Khim., 39, 1520 (1969); C. A. 71, 113054 (1969).
762. Tsivunin, V. S., and G. Kamai, Dokl. Akad. Nauk SSSR, 131, 1113 (1960); C. A. 54, 20843 (1960).
763. Tsivunin, V. S., G. Kh. Kamai, and S. V. Fridland, Zh. Obshch. Khim., 36, 436 (1966); C. A. 65, 741 (1966).
764. Tsivunin, V. S., G. Kamai, R. Sh. Khisamutdinova, and E. M. Smirnov, Zh. Obshch. Khim., 35, 1231 (1965); C. A. 63, 11608 (1965).
765. Tsivunin, V. S., G. Kamai, and G. K. Makeeva, Dokl. Akad. Nauk SSSR, 135, 1157 (1960); C. A. 55, 12271 (1961).
766. Tsivunin, V. S., G. Kamai, and D. B. Sultanova, Zh. Obshch. Khim., 33, 2149 (1963); C. A. 59, 12839 (1963).
767. Tsivunin, V. S., L. N. Krutskii, and G. Kamai, Zh. Obshch. Khim., 40, 597 (1970); C. A. 73, 25590 (1970).
768. Tung-Tsai Liang, and Kuo-Chen Chi, Hua Hsueh Hsueh Pao, 31, 155 (1965); C. A. 63, 12441 (1965) and 66, 6293 (1967).
769. Tyka, R., Tetrahedron Lett., 1970, 677.

770. Umezawa, H., and R. Hara, Anal. Chim. Acta, 25, 360 (1961).
771. Usher, D. A., and F. H. Westheimer, J. Am. Chem. Soc., 86, 4732 (1964).
772. Utvary, K., E. Freundlinger, and V. Gutmann, Monatsh. Chem., 97, 348 (1966).
773. Vasyanina, M. A., V. K. Khairullin, and A. N. Pudovik, Izv. Akad. Nauk SSSR, Ser. Khim., 1970, 452; C. A. 73, 3988 (1970).
774. Venezky, D. L., and C. F. Poranski Jr., J. Org. Chem., 32, 838 (1967).
775. Ville, J., Ann. Chim. Phys., (6), 23, 289 (1891).
776. Vinokurova, G. M., and S. G. Fattakov, Izv. Akad. Nauk SSSR, Ser., Khim., 1969, 1762; C. A. 72, 3538 (1970).
777. Viout, M. P., J. recherches centre natl. recherche sci., Labs. Bellevue (Paris), 28, 15 (1954); C. A. 50, 7077 (1956).
778. Vizel, A. O., M. A. Zvereva, K. M. Ivanovskaya, I. A. Studensova, V. G. Dumaev, and M. G. Berim, Dokl. Akad. Nauk SSSR, 160, 826 (1965); C. A. 62, 14721 (1965).
779. Voigt, D., M. C. Labarre, and L. Fournes, Compt. Rend., Ser. C, 262, 1113 (1966).
780. Voigt, D., M. C. Labarre, and J. P. Jaureguy, Bull. Soc. Chim. France, 1964, 3087.
781. Voznesenskaya, A. Kh., and N. A. Razumova, Zh. Obshch. Khim., 39, 387 (1969); C. A. 70, 115231 (1969).
782. Weil, T., Helv. Chim. Acta, 37, 654 (1954).
783. Weisz, A., and K. Utvary, Monatsh. Chem., 99, 2498 (1968).
784. Weitkamp, H., and F. Korte, Z. Anal. Chem., 204, 245 (1965).
785. Wheatley, P. J., J. Chem. Soc., 1962, 3733.
786. Wieber, M., and B. Eichorn, Monatsh. Chem., 99, 261 (1968).
787. Wiegrabe, W., and H. Bock, Chem. Ber., 101, 1414 (1968).
788. Wilkes, C. E., and R. A. Jacobson, Inorg. Chem., 4, 99 (1965).
789. Williams, R. H., and L. A. Hamilton, J. Am. Chem. Soc., 77, 3411 (1955).
790. Yagupol'skii, L. M., and Zh. M. Ivanova, Zh. Obshch. Khim., 30, 4026 (1960); C. A. 55, 22196 (1961).
791. Yagupol'skii, L. M., and P. A. Yufa, Zh. Obshch. Khim., 30, 1294 (1960); C. A. 55, 431 (1961).
792. Yakubovich, A. Ya., and V. A. Ginsburg, Zh. Obshch. Khim., 22, 1534 (1952); C. A. 47, 9256 (1953).

793. Yarmukhametova, D. Kh., and I. V. Cheplanova, Izv. Akad. Nauk SSSR, Ser. Khim., 1966, 1260; C. A. 65, 16995 (1966).
794. Zaishlova, I. A., I. L. Slosman, and Yu. G. Golobolov, Zh. Obshch. Khim., 36, 1838 (1966); C. A. 66, 55557 (1967).
795. Zhmurova, I. N., and A. V. Kirsanov, Zh. Obshch. Khim., 33, 1015 (1963); C. A. 59, 8782 (1963).
796. Zhmurova, I. N., I. Yu. Voitsekhovskaya, and A. V. Kirsanov, Zh. Obshch. Khim., 29, 2083 (1959); C. A. 54, 8681 (1960).
797. Zhuravleva, L. P., M. I. Z'ola, M. G. Suleimanova, and A. V. Kirsanov, Zh. Obshch. Khim., 38, 342 (1968); C. A. 69, 27484 (1968).
798. Zinov'ev, Yu. M., and L. Z. Soborovskii, Zh. Obshch. Khim., 26, 3030 (1956); C. A. 51, 8662 (1957).
799. Zinov'ev, Yu. M., and L. Z. Soborovskii, Zh. Obshch. Khim., 34, 929 (1964); C. A. 60, 15904 (1964).
800. Zyablikova, T. A., I. M. Magdeev, and I. M. Shermergorn, Izv. Akad. Nauk SSSR, Ser. Khim., 1968, 397; C. A. 69, 59345 (1968).
801. Zyablikova, T. A., A. R. Panteleeva, and I. M. Shermergorn, Izv. Akad. Nauk SSSR, Ser. Khim., 1969, 373; C. A. 70, 115228 (1969).

Chapter 15. Organic Derivatives (Esters and Organic Anhydro Acids) of Phosphoric and Polyphosphoric Acids

E. CHERBULIEZ

1211 Conches, Fossard 48, Switzerland

INTRODUCTION

The development of the chemistry of phosphoric acid esters has been extraordinarily vast in the last 15 years.

Besides a number of new methods, the number of definite chemical compounds in this field is so large that a complete bibliography would exceed the place assigned to this chapter in the present volume. In fact, it is relatively easy to find the bibliographic references pertaining to a determined substance with the help of the alphabetical indices published by the scientific journals and the different abstract publications. The same does not hold for methods: Until quite recently, the keywords of the indices were either incomplete, or the mention of the methods used in the published research work was by no means complete. Therefore, the author has aimed to be as exhaustive as possible for the methods: no pretention of completeness is made on behalf of compounds, all the more as the limit between individual chemical compounds and more or less definite substances is not easy to trace in the field of phosphoric esters; an example is given, for instance, by the nucleotides, the literature of which has enormously developed since the fundamental biological importance of this class of compounds has been understood.

GENERAL REMARKS

The substances which will be examined in this chapter are to be considered either as esters (with alcohols, phenols, or deriving from substances with hydrated carbonyl functions) or as mixed anhydrides with inorganic or organic oxyacids, or as containing both functions.

ESTERS

In contrast to the behavior of most carboxylic and mineral acids, the direct esterification of orthophosphoric acid or the acid functions of polyphosphoric acids is an extraordinary slow reaction which practically is not appropriate for the preparation of phosphoric esters. After 7 hr. in boiling ethanol, for instance, orthophosphoric acid gives practically no monoethyl phosphoric acid.[386a] This absence of reactivity of phosphoric acid is probably due to the fact that compounds in which the phosphorus is tetracoordinated with four electronegative atoms of medium volume (especially oxygen) are generally remarkably stable and trend neither to higher nor to lower coordinative saturation of the phosphorus atom. This feature is not exclusive for P; it is found equally but to a less degree in sulfur: Sulfonic acids yield practically no esters when heated with alcohols (though concentrated sulfuric acid does as it dissociates partially into water and the corresponding anhydride. This particularity has its counterpart in the fact that classical hydrolysis of phosphoric esters is equally slow so long as no particular factor intervenes, as will be shown in the discussion of the reactions of phosphoric esters. Nevertheless, it must be noted that this reaction of direct esterification has been reported recently to have been observed in certain cases.

Beck et al.,[218] for instance, report that uridine heated with phosphate in concentrated aqueous solution during 9 months at 65° yields uridine phosphate. Bliznyuk et al.[252] have patented the preparation of monoaryloxyethyl phosphoric acids by heating the corresponding alcohols with 85% phosphoric acid in an organic solvent with azeotropic distillation of water. These esterifications may be direct, but it is possible that in reality the reaction proceeds by slow production of polyphosphoric acids which are the phosphorylating agent. This course of the reaction intervenes most probably in "direct" phosphorylations by orthophosphoric acid at 140°,[232] by monosodium phosphate at 160°,[1125] or in the production of oligonucleotides by heating, e.g., uridine 2'(3')-phosphate with uridine at 160°.[1125] It is interesting to note that in this particular instance, only (3' → 5') and

(1' → 5') nucleotidic bonds are formed. In the "direct" esterification of phenols by phosphoric acid reported to occur at 280-300°,[789] one can admit that in reality the reaction proceeds via polyphosphoric acids (see A.I.1.β.1-1.2). Therefore we will not consider "direct esterification" as a method of preparing phosphoric esters.

A. METHODS OF PREPARATION

I. NON-CYCLIC ESTERS

In general, "direct" esterification of an alcohol with an acid function of a phosphoric acid derivative is realizable only with the help of particular condensing agents, such as disubstituted carbodiimides, and it is certain that reactive derivatives are produced first as intermediates. The production of monomolecular metaphosphoric acid or a derivative is considered as possible.[1344] It adds easily a hydroxyl grouping

$$RN{=}C{=}NR + H_3PO_4 \longrightarrow HPO_3; \qquad HPO_3 + ROH \longrightarrow ROPO_3H_2$$

Other possibilities may intervene, such as production of an additional product between carbodiimide and phosphoric acid, which is split by the alcoholic component into a substituted urea and a phosphoric ester, or with formation of a trimetaphosphate derivative which is the phosphorylating agent.[744,1358,1444a]

The general methods applied can be divided in two groups: use of reactive derivatives of phosphoric acid (anhydrides, halides, etc.) and use of phosphorus acids with coordinatively unsaturated P (mostly phosphorous acid) and which therefore are capable of direct esterification by alcohols, followed by ulterior transformation of the organic derivatives into molecules with tetracoordinated phosphorus (in the case of phosphorous acid derivatives, oxidation into the corresponding derivatives of phosphoric acid).

1. Use of P^V Derivatives

a. Anhydrides (Complete or Partial) of Orthophosphoric Acid

α. Reactions of Phosphorus Pentoxide with Hydroxy Compounds

This is one of the oldest (1848!) and cheapest methods for the synthesis of a mixture of primary and secondary phosphates.[330,1413,1414] But the old formulation of this reaction as

$$3ROH + P_2O_5 \longrightarrow (RO)PO_3H_2 + (RO)_2PO_2H$$

is quite insufficient and does not explain the fact that there appear always small quantities of free phosphoric acid and of tertiary ester. "Phosphorus pentoxide, P_2O_5" is not a monomeric molecule but has the structural unit of at least P_4O_{10}, which consists in essence of a closely knit network of phosphorus to oxygen to phosphorus bonds between the phosphoryl groups arranged approximately in a tetrahedron. The attack of an alcohol on such a structure proceeds by progressive cleavage of the anhydride bonds, which occurs essentially according to the laws of probability at any given instant, and in each P-O-P bond two "directions" of such cleavage can take place:[387]

RO┼H and H┼OR

A...P(O)┼O-P(O)...B A...P(O)-O┼P(O)...B

When such a process is followed through the destruction of all the anhydride linkages in the "pentoxide" unit, the formation of primary and secondary esters is the predominant process. However, there is a minor probability that tertiary esters and phosphoric acid will be formed; this possibility can be affected by factors inherent to the nature of the radical R (steric factors for instance).

Such a picture is verified by the observations that considerably greater molar ratios of hydroxy compounds are needed to bring the solid pentoxide into solution than the ratio shown in the equation given above.[1154] Controlled experimentation is rather lacking in the usual scientific literature on this reaction; a paper from 1945 may be used as a point of departure for specific cases.[387]

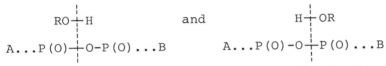

Since the primary and the secondary esters usually show decomposition symptoms at temperatures much above 100°, the reaction temperature is kept within such limits by suitable means. After completion of the actual dissolution of added pentoxide (usually four to five molar ratio of alcohol to the pentoxide is used), the mixture is quenched with water to hydrolyze any polyphosphate residues, and the individual primary and secondary esters are separated by fractional crystallization of suitable salts, when the radicals R are rather small. Barium salts are rather common, for the secondary esters yield barium salts that are much more water-soluble than those of the primary esters. Such procedures fail with the larger radicals, and the separations become extremely laborious if pure products are desired. In some cases use is made of the greater water solubility of the primary esters. The inorganic phosphate is usually removed beforehand by magnesia mixture or lithium hydroxide, followed by removal of residual acid by magnesium chloride in dilute ammonium hydroxide.[1428] Pyridine has been claimed to have beneficial effects on the conduct of this reaction.[1268]

As a serious limitation one should consider the fact that secondary alcohols may be partially dehydrated, yielding olefins, and that tertiary alcohols undergo quantitatively this type of reaction.

β. Reactions of Polyphosphoric Acids with Hydroxy Compounds

Polyphosphoric acids are partial anhydrides of orthophosphoric acid. These acids may affect a linear unbranched (type A) or branched (type B) chain or may present a cyclic structure which can present branched linear chains (types C and D respectively; the size of the cycle may vary).

$$\begin{array}{c} \text{OH} \quad\quad \text{OH} \quad\quad \text{OH} \\ | \quad\quad\quad | \quad\quad\quad | \\ \text{A:} \quad -P(O)-O-P(O)-O-P(O)- \end{array}$$

$$\begin{array}{c} \text{OH} \quad\quad\quad\quad \text{OH} \\ | \quad\quad\quad\quad\quad | \\ \text{B:} \quad -P(O)-O-P(O)-O-P(O)- \\ \quad\quad\quad\quad\quad | \\ \quad\quad\quad\quad\quad O \\ \quad\quad\quad\quad\quad | \\ \quad\quad\quad HO-P(O) \\ \quad\quad\quad\quad\quad | \end{array}$$

C:

$$HO-P(O) \underset{\underset{\underset{OH}{|}}{O\diagdown_{P}\diagup O}}{\overset{O}{\diagup \diagdown}} P(O)-OH$$

D:

$$HO-P(O) \underset{\underset{\underset{OH}{|}}{O\diagdown_{P}\diagup O}}{\overset{O}{\diagup \diagdown}} P(O) \overset{\overset{OH}{|}}{O-P(O)-}$$

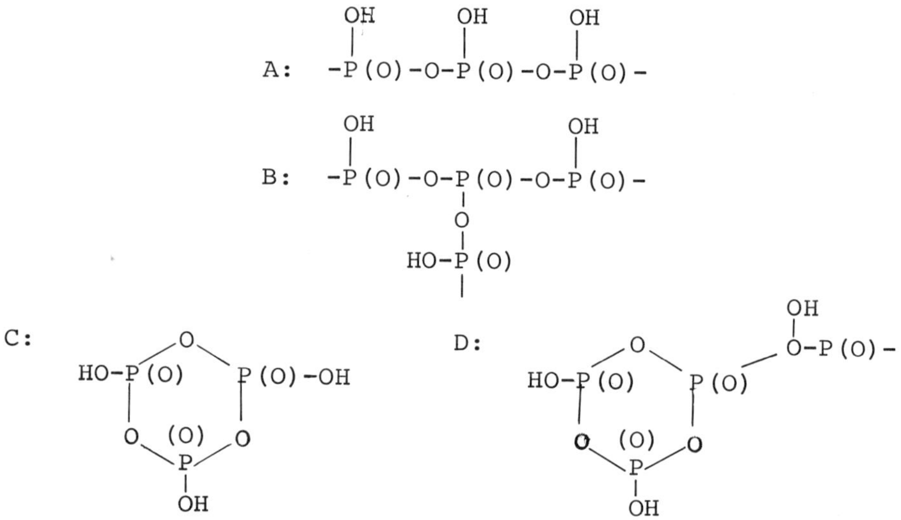

This type of compounds is easily obtained either by condensation of orthophosphoric acid by heat (150-250°C or even higher) or by reaction of phosphorus pentoxide with orthophosphoric acid. These acids can be prepared first and reacted afterwards with alcohols or phenols, or they may be obtained in situ, e.g., by heating glycerol with orthophosphoric acid at 150-200°.

When prepared by thermal condensation of H_3PO_4 at relatively low temperature (150-200°), the polyphosphoric acids produced present essentially unbranched chains with low condensation number (3 to 4).[387] The formation of the ester link proceeds as in the case of phosphorus pentoxide, and the cleavage of the P-O-P bonds may again proceed in two directions. The reaction is not purely statistical;

when using a linear unbranched polyphosphoric acid, the splitting of the chain by an alcohol yields preferentially primary esters (see chart for the reaction of triphosphoric acid with an alcohol).[388]

Reaction of triphosphoric acid ($H_5P_3O_{10}$) with an alcohol

$$\overset{\displaystyle OH}{(HO)_2P(O)-O-\overset{|}{P}(O)-O-P(O)(OH)_2} + HOR$$

$(HO)_2P(O)-O-P(O)(OH)_2 + (RO)PO_3H_2 \qquad (HO)_2P(O)-O-P(O)(OR)OH$

$$+ HOR \qquad\qquad\qquad\qquad\qquad\qquad\qquad\qquad + HOR \qquad + H_3PO_4$$

$$H_3PO_4 + (RO)PO_3H_2 \qquad\qquad 2(RO)PO_3H_2 \qquad\qquad H_3PO_4$$

$$+ (RO)_2PO_2H$$

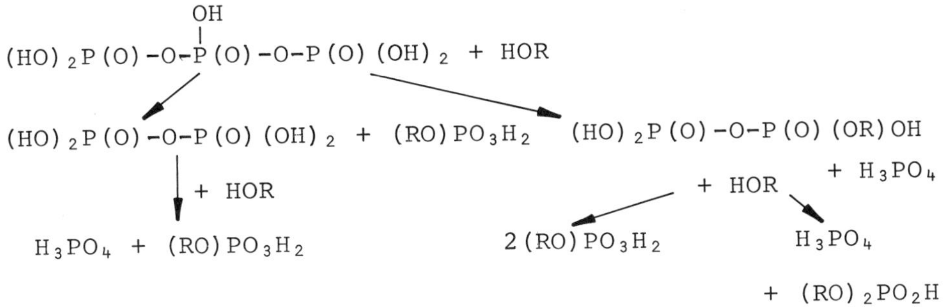

Overall reactions:

$$H_5P_3O_{10} + 2ROH = 2(RO)PO_3H_2 + H_3PO_4 \quad (more) \qquad or$$

$$H_5P_3O_{10} + 2ROH = (RO)_2PO_2H + 2H_3PO_4 \quad (less)$$

Therefore, the isolation of the primary esters presents in general no difficulties.

The reactions with polyphosphoric acid is generally realized by heating the alcohol in excess either pure or in ethereal solution with the acid on the water bath. After completion of the reaction (one to several hours) the mixture is quenched with water, the ether if present is eliminated, and the phosphoric esters are freed from phosphoric acid by neutralization with $Ba(OH)_2$ or $Mg(OH)_2$; the generally easily soluble Ba or Mg salts are crystallyzed from water (+ ethanol or a suitable organic solvent). The esterification of alcohols by "sirupy" phosphoric acid is due probably to the presence of polyphosphoric acid in this product generally prepared by concentrating commercial phosphoric acid at higher temperature and/or in vacuo.

Heating of hydroxy compounds with phosphoric acid to temperatures above 100° under reduced pressure yields mixtures of esters similar to the product mixture obtained by direct reaction with polyphosphoric acid (see above).

Polyhydroxy compounds yield complex mixtures and even resinous polymers. The complexity of the products is somewhat reduced by using acid salts (mono- or disodium phosphate). This may result from the fact that under these conditions the formation of higher polyphosphates is strongly reduced so that principally primary esters are formed (pyrophosphoric acid can yield only primary esters).

Solutions of phosphorus pentoxide in phosphoric acid, that is, partially hydrated phosphorus pentoxide mixtures (containing appreciable amounts of metaphosphoric acid) have been used for such reactions. Metaphosphoric acid being at least a cyclic triphosphoric acid, the tendency to form esters of condensed phosphoric acids--and mostly pyrophosphoric acid--is fairly high; the use of a moderate excess of the hydroxy compound helps to carry the reaction to a mixture of primary and secondary orthophosphoric esters.[460,1122] Such mixtures have been used for the esterification of hydroxy amino acids.[868]

The mechanism of the cleavage of the anhydride link is still under discussion. It seems to be likely that in a first step the highly polar phosphoryl groups form a hydrogen-bonded adduct with the hydroxyl compound; the adduct may shift to a quasi-phosphonium compound which then undergoes the observed cleavage.

b. Use of Haloderivatives of Pentavalent Phosphorus. Under this heading are considered derivatives with halogens or with halogens and oxygenated functions (O, OR). In all these compounds the halogens react with hydroxy compounds yielding POR functions, as do acyl halides. With phosphorus derivatives bearing more than one halogen atom halides of primary or secondary phosphoric esters are produced as intermediates; sometimes they can be isolated and will yield by hydrolysis of the very active acid halide function mono or diesterified phosphoric acids.

α. Phosphorus Pentahalides and Hydroxy Compounds

This possibility plays quite a minor role.

The reaction of phosphorus pentahalides with alcohols leads largely to the formation of alkyl halides, which is followed by the normal action of the resulting phosphorus oxyhalides. The effect is most pronounced in secondary and tertiary alcohols. The second reaction, as expected, is facilitated by the presence of pyridine and results in the formation of a spectrum of esters.[603,705] In some cases, monoalkyl dihalophosphates (not isolated) are formed which yield the corresponding monoesters by hydrolysis.[1482] Phenols react with the formation of quasi-phosphonium halides, which are usually not isolated as such, but the crude mixtures are treated with water or dilute aqueous alkali to form the corresponding esters. As a rule, a spectrum of esters is formed in practice, either from disproportionation, or from non-homogeneity of the intermediate products.[31]

$$PCl_5 + ROH \longrightarrow (H_2O) \longrightarrow ROP(O)(OH)_2$$

$$PCl_5 + 3ROH \longrightarrow (H_2O) \longrightarrow (RO)_3PO$$

$$PCl_5 + 4ROH \longrightarrow (H_2O) \longrightarrow (RO)_3PO$$

β. Phosphoryl Halides and Hydroxy Compounds

This reaction is generally realized in the presence of an organic tertiary base (B).[523]

In principle the following reactions can take place:

(a) $ROH + ROCl_3 + B \longrightarrow ROPOCl_2 + B{\cdot}HCl$

(b) $2ROH + POCl_3 + 2B \longrightarrow (RO)_2POCl + 2B{\cdot}HCl$

(c) $3ROH + POCl_3 + 3B \longrightarrow (RO)_3PO + 3B{\cdot}HCl$

The halides obtained by reactions (a) and (b) yield by hydrolysis of the reactive acid halide function the corresponding mono or diesterified phosphoric acids, $ROPO_3H_2$ and $(RO)_2PO_2H$, respectively.

The reactions with the phosphoryl halide are generally run in an inert solvent and at room temperature or below. The hydrolysis of the halides of primary or secondary phosphoric esters may be effected by water at room temperature or somewhat above it; the secondary esters require generally a higher temperature which favors partial hydrolysis of the ester groupings, especially in acid mediums. Therefore the hydrolysis of the halide function is best effected by dilute alkali. The free acids are obtained by acidification. Cyclic ester halides usually suffer ring opening during the hydrolysis.[1182]

This general procedure is the best laboratory procedure for the preparation of homogeneous tertiary phosphates of all categories. It has found a particular application in the field of phosphorylation of nucleosides yielding nucleotides (Yoshikawa et al.[1485,1486]): Isopropylidene derivatives of nucleosides are added at low temperatures (-10° or lower) to a solution of 2 mol-equiv. of phosphoryl chloride in 60-100 mol-equiv. of a trialkyl phosphate (generally trimethyl phosphate used)[1480,1487] under stirring continued until dissolution of the nucleoside derivative. The trialkyl phosphate acts as catalyst. The derivatives $R\text{-}OPOCl_2$ (R: isopropylidenenucleoside residue) are not isolated but hydrolyzed (chloridate and isopropylidene functions) yielding the nucleotides with good yields.

γ. Use of Halides of Primary or Secondary Phosphoric Esters

γ-1. Hydrolytic Transformation. The halides of partial esterified phosphoric acids are in theory hydrolyzable to the corresponding free acids.

$$(RO)P(O)Hal_2 + 2H_2O \longrightarrow (RO)_2PO_3H_2 + 2HHal$$

$$(RO)_2P(O)Hal + H_2O \longrightarrow (RO)_2PO_2H + HHal$$

This hydrolysis may be effected by water treatment at room temperature or somewhat above it. The secondary derivatives are usually fairly resistant and require higher temperatures; these conditions favor splitting of the ester functions. Often, the conversion of the halides to the salts of the derived acids is realized with dilute aqueous alkali. The acids are then obtained by acidification.

The halides are not always isolated: The crude reaction product of phosphorus oxychloride with the hydroxy compounds is hydrolyzed, and the acid esters are separated as salts with appropriate bases.[546,1482]

γ-2. Reaction with Hydroxy Compounds. The halophosphates react as acid halides with hydro-compounds or their metallic derivatives. In this manner $(RO)_2P(O)Cl$ with $R = C_2H_5$, $n\text{-}C_3H_7$, iso-C_3H_7, etc., have been used for the preparation of mixed tertiary phosphoric esters $(RO)_2$-$(R'O)PO$ in a very wide variety, following the general reaction:[177]

$$(RO)_2P(O)Cl + R'OH(R'OM) \longrightarrow (RO)_2(R'O)PO + HCl(MCl)$$

By use of a halophosphate with an R easily split off, the mixed tertiary esters will yield phosphoric acid monoesters. Phosphorylating agents of this sort used very often are the phenyl and benzyl derivatives (see below under 1.e.α-2). The phenyl ester function can be split by reflux with boiling $Ba(OH)_2$--the alkyl ester functions of phosphoric acid generally are not affected by this treatment--or by catalytic hydrogenolysis; the benzyl grouping can be eliminated by the last-mentioned procedure. Sometimes $(C_6H_5\text{-}O)P(O)Cl_2$ is used as phosphorylating agent; under appropriate conditions, the following reaction is privileged:[923]

$$ROH + (C_6H_5O)P(O)Cl_2 \longrightarrow (RO)(C_6H_5O)P(O)Cl + HCl$$

This halophosphate can be used as further phosphorylating agent, or it can be split (boiling $Ba(OH)_2$ for instance) with elimination of the acyl chloride and the phenyl ester functions, yielding the phosphorylated hydroxy compound $R\text{-}OPO_3H_2$.

The halophosphates of type $(RO)P(O)Cl_2$ can of course be used for the synthesis of tertiary mixed esters (RO)-$(R'O)_2PO$ which, in turn, with a suitable easily split off R can furnish the secondary phosphates $(R'O)_2PO_2H$.

c. Use of Anhydro Acid Derivatives. The anhydro function of this type of compounds is of course subject to attack by hydrolytic or alcoholytic means which yields less complex compounds. Alcoholysis in particular produces ester and acid functions in equimolecular amount.

From the viewpoint of phosphoric ester production one must distinguish acyl phosphates from derivatives with phosphoric acid anhydride functions, i.e., polyphosphoric acids. Acylphosphates are in general not phosphorylating agents but yield by alcoholysis carboxylic acid esters and phosphoric acid:

$$RCOOPO_3H_2 + R'OH \longrightarrow RCOOR' + H_3PO_4$$

so that they need not be considered here.

α. Alkyl Metaphosphates

An interesting reaction, even if its use is limited, is the attack of phosphorus pentoxide by ethers, principally by diethyl ether, yielding alkyl metaphosphates (Langheld, Steinhopf[842,843,845,1317]). These metaphosphates $(ROPO_2)_n$ are in reality a mixture of esters of cyclic polyphosphates. As such they may be hydrolyzed giving monoalkyl phosphoric acids (ethyl phosphoric acid); by alcoholysis at somewhat elevated temperature, mixtures of secondary and tertiary esters are produced which are easily separated. If the alkyl radicals of the metaphosphoric ester and the alcohol used for its degradation are different, the resulting mixture will be very complex due to the intervening disproportionation.[842,1122]

From alkyl metaphosphates tertiary alkyl esters may be produced by appropriate alkylating agents such as acetals[710] or even ethers.[711] This allows the direct preparation of triethyl phosphate by prolonged heating of P_2O_5 with diethyl ether (preferably in the presence of a halogenated hydrocarbon, generally chloroform).[711]

β. Pyrophosphates

Esters of pyrophosphoric acid can be used in two ways for the preparation of orthophosphoric esters, by controlled hydrolysis or as phosphorylating agents of hydroxy compound.

β-1. Controlled Hydrolysis of Pyrophosphoric Esters. Tetra-alkyl pyrophosphates give two equivalents of dialkyl phosphoric acid by treatment with water at moderately elevated temperature; the advantage of this procedure is the purity of the secondary ester practically devoid of primary ester.[1360]

In a similar manner the partial esters of pyrophosphoric acid yield the corresponding phosphoric esters or phosphoric acid, but evidently these esters are homogeneous only when originating from symmetric pyrophoric esters.

β-2. Pyrophosphoric Esters as Phosphorylating Agents. The reaction runs as follows:

$$(RO)_2P(O)O(O)P(OR)_2 + R'OH \longrightarrow (RO)_2(R'O)PO + (RO)_2PO_3H_2$$

By use of a radical R which is easily split off, the tertiary ester $(RO)_2(RO)PO$ will give the primary ester $R'OPO_3H_2$, e.g., Ref. 982.

This has been realized in particular with tetra-p-nitrophenyl pyrophosphate (prepared *in situ* from di-(p-nitrophenyl)-phosphoric acid and dicyclohexyl carbodiimide (see below), the p-nitrophenyl ester function being easily hydrolyzed at room temperature by 1N lithium hydroxide, which does not attack alcoholic ester functions under these conditions. The p-nitrophenyl residue may equally be removed by catalytic hydrogenolysis. This procedure has been applied to the phosphorylation of nucleosides (e.g., Khorana).[982]

γ. Hypophosphates

From the numerous types of anhydrides of phosphorus acids let us mention here the tetraalkyl hypophosphates $[(RO)_2PO]_2$ whose structure is still under discussion. These esters yield by alcoholysis products which support their assymetric structure.[63-65]

$$[(RO)_2PO]_2 + ROH \begin{cases} \longrightarrow (RO)_2PO_2H + (RO)_3P & (\sim 3/4) \\ \longrightarrow (RO)_3PO + (RO)_2POH & (\sim 1/4) \end{cases}$$

Therefore dialkyl phosphoric acids and trialkyl phosphates are produced by this reaction.

d. Esterification of Phosphoric Acid Functions by Other Agents

α. Reactions with Hydroxy Compounds with the Help of Condensing Agents

α-1. Disubstituted Carbodiimides. The direct esterification of phosphoric acid functions by alcohols is realizable only quite exceptionally (see Introduction), but the now classical disubstituted carbodiimides--generally dicyclohexylcarbodiimide--which have been successful in peptide synthesis--are used with excellent results for the esterification of phosphoric acid or its partial esters with hydroxy compounds. The technique is similar to that used in peptide synthesis: action of the substituted carbodiimide on the components in pyridine, aqueous pyridine, morpholine, etc., at ordinary temperature in the presence, or not, of an ion exchange resin, e.g., Dowex 50. Khorana and his group have used this procedure extensively since 1957 (e.g., Ref. 361). It is particularly useful in the field of nucleotide synthesis. It has been applied equally to the synthesis of cyclic phosphates from hydroxyalkyl phosphoric acids (see below under Cyclic Esters) and the phosphorylation of phosphoric acid mono- and diesters yielding pyrophosphoric and even higher condensed phosphoric acid derivatives (see below).

α-2. Arenesulfonyl Halides. Arenesulfonyl halides (chlorides used generally) activate the phosphorylation of alcoholic hydroxyl groups by phosphoric acid or a phosphoric monoester. The reaction proceeds smoothly at ordinary temperature with a technique similar to that used with the carbodiimides. This reaction has found interesting application in polynucleotide synthesis: e.g., thymidine with protected OH functions with the exception of 5'-OH is condensed by means of mesitylenesulfonyl chloride with similar protected 5'-O-acetylthymidine monophosphate. The resulting dinucleotide derivative is deacetylated and undergoes the same procedure giving a trinucleotide, etc.[744] The same reaction has been realized using an insoluble polymer support on which the nucleotide is fixed and condensed with the appropriate nucleotide derivative activated by an arenesulfonyl chloride (or picryl chloride).[1250] By repetition, polynucleotides as high as hexadecanucleotides have been synthesized.

Studying this reaction, Blackburn et al. consider the following mechanism as probable: The monoesterified phosphoric acid derivative yields a cyclic metaphosphate which reacts with the hydroxy compound by alcoholytic splitting of the P-O-P functions; therefore here again the general reaction treated in Section 1.a. seems to intervene.[245-247]

β. Reaction of Phosphoric Acid Functions with Diazoalkanes

The classical reaction of diazoalkanes with acidic compounds affords a very clean and mild esterification of phosphoric acid mono- and diesters. Its application is limited only by the difficulty of preparing the diazo derivatives of higher alkanes; diazomethane is mostly used, yielding methyl ester functions.[94]

γ. Reaction of Phosphoric Derivatives with Ethylene Oxides and Imines

γ-1. With Phosphorus Oxychloride. On warming in the presence of suitable catalysts such as iron filings[719,721] or anhydrous aluminum chloride,[459] epoxyolefins such as ethylene oxide react with phosphorus oxychloride yielding tertiary β-haloalkyl phosphates.

$$3\,\overset{\displaystyle O}{CH_2\!-\!CH_2} + POCl_3 \longrightarrow (ClCH_2CH_2O)_3PO$$

Lower rates of the oxide should yield presumably primary or secondary esters of this type.

γ-2. With Phosphoric Acid and Its Partial Esters. In the manner formulated above, phosphoric acid and its partial esters yield β-hydroxyalkyl phosphates by reaction with epoxides. Primary esters of this type are produced by reacting disodium phosphate with ethylene oxide.[96,164]

$$\overset{\displaystyle O}{CH_2\!-\!CH_2} + HO(O)P(ONa)_2 \longrightarrow HOCH_2CH_2OP(O)(ONa)_2$$

Ethylene imine warmed in deficient amount with sirupy phosphoric acid similarly forms monoaminoethyl phosphate.[395,1370]

$$\overset{\displaystyle NH}{CH_2\!-\!CH_2} + H_3PO_4 \longrightarrow H_2NCH_2CH_2OPO_3H_2$$

This reaction has been realized in acetone solution (6 hr reflux) with phenyl phosphoric acid, yielding the monophenyl ester of the above aminoethyl phosphate giving finally by hydrogenolysis (H_2/Pt in acetic acid) aminoethyl phosphoric acid.[1370] Similarly, DL-phosphoserine has been prepared (in acetone, 3 hr at 50°) starting from carboxyethylene imine.[1370]

δ. Reaction of Alkyl Halides with Salts of Phosphoric Acid Derivatives

In its oldest form (1854!) this reaction yields tertiary phosphates by heating for several hours silver phosphate with an excess of alkyl halide.[417]

$$Ag_3PO_4 + 3RX \longrightarrow (RO)_3PO + 3AgX \quad \text{(usually X = I)}$$

By use of monosilver phosphate (i.e., mixture of trisilver phosphate with phosphoric acid) primary esters may be prepared which can be isolated through the usual salts.[1501] The same result has been obtained by the use of the silver salt of trimetaphosphoric acid, yielding with alkyl iodide trialkyl trimetaphosphate; the latter gives monoalkyl phosphoric acid on hydrolysis, and a mixture of primary and secondary acid esters on alcoholysis. The secondary ester is homogeneous of course only if the radical R of the iodide and that of the alcohol are the same. As a result of disproportionation during these reactions, always some tertiary ester is formed.[842,1122]

Reactions of this type have been extensively used in the preparation of carbohydrate phosphates. The stereochemistry of the reaction is rather complicated insofar as the α- or β-configuration of the initial sugar derivative is some times conserved, and some times not. Thus α-acetobromoglucose treated with Ag_3PO_4 yields, after hydrolysis of the presumably tertiary ester, α-glucose 1-phosphate;[1469] on the other hand, for instance, 2,3,4,6-tetra-O-acetyl-D-glucopyranosyl bromide treated with Ag_3PO_4 yielded β-D-glucose 1-phosphate after deacylation with sodium methoxide in methanol.[1467]

Silver salts of secondary phosphates may be used to limit the reaction to production of primary phosphoric esters: diphenyl, dibenzyl, or phenyl benzyl phosphates are interesting agents because of the ease with which these groups are removed from the tertiary esters formed, either by hydrolysis or by catalytic hydrogenolysis.[1499] This technique has been shown to be very useful in the field of phosphorylated steroids (e.g., Ref. 985) and phosphatidic acids (e.g., Refs. 261, 985, 1263).

Other salts have been used in place of the silver salts.

Trisodium or tripotassium phosphate heated 10 to 16 hr in dimethylacetamide or dimethylsulfoxide yield with alkyl halides the corresponding tertiary phosphates, but the yields are not good: e.g., with n-butyl bromide 12% after 16 hr in dimethylacetamide at 140-150°.[1484]

Nucleotides and phosphorylated sugar derivatives have often been prepared by a very similar procedure in which the phosphorylating agent $(RO)_2PO_2H$ of the type mentioned above is used in the form of its salt with a strong tertiary amine, generally triethylamine;[1347] trimethylamine [1476] has also been used.

The tetramethylammonium salts $[(CH_3)_4N]OP(O)(OR)(OR')$ (resulting from demethylation of mixed tertiary phosphates with one methyl group by action of trimethylamine) refluxed 3-5 hr with bromomethyl-ketones $BrCH_2COX$ in chloroform or acetonitrile yield tertiary esters (RO)(R'O)PO-(OCH_2COX).[324]

ε. Reaction of Alkyl Sulfates with Metal Phosphates

This reaction seems to be less advantageous than the alkyl halide reaction. Trisodium phosphate or acid sodium phosphates alkylated by dialkyl sulfates have been used for the synthesis of primary esters (accompanied by some secondary esters).[166] The preparation of tetraethyl pyrophosphate by diethyl sulfate in boiling acetone is described in a patent.

ζ. Enzymatic Syntheses

A very interesting field which will certainly acquire great importance in the future especially in the domain of phosphoric esters of biological importance is offered by enzymatic synthesis. As examples let us mention the preparation of ^{14}C-labelled glucose 6-phosphate from U[^{14}C]glucose by hexokinase in the presence of adenosine triphosphate,[882] or the condensation of a 5'-protected uridine-2',3'-cyclophosphate with uridine (or other puric bases) by pancreatoribonuclease, yielding under ring opening the corresponding dinucleoside monophosphate.[1239]

e. Transformation of Phosphoric Esters. In this section we do not consider reactions modifying a radical R of a phosphoric ester (RO)(R'O)(R"O)PO (R',R" = org. radicals or H), for instance sulfonation of triphenylphosphate in the aromatic nuclei,[521] oxidation of adenosine monophosphate (ATP) into ATP N^1-oxide by action of hydrogen peroxide,[449] or dehydrohalogenation of β-chloroethyl dimethyl phosphate into vinyl dimethyl phosphate,[1390] but exclusively replacement reactions of one ester radical by another or by hydrogen. In the case of phosphoric esters, this last reaction should be termed "splitting" rather than hydrolysis; as it will be shown in the section on reaction mechanisms, the elimination of an ester group in aqueous solution is sometimes not a scission of POR by water (ester P⁝OR P-O⁝R) but an intramolecular reaction of the following type ... $P-O-CH_2-CH_2-R \rightarrow$ ... $POH + CH_2=CH-R$, depending on the nature of R.

α. Removal of Ester Groups

α-1. Partial Splitting of Secondary or Tertiary Phosphoric Esters. In general the stability of phosphoric esters decreases from primary to tertiary esters. Tertiary phosphates can be split readily to the secondary esters. Especially in the aliphatic series, such reactions can be used for synthetic purposes when alkaline solutions are used. The alkyl derivatives are attacked even at room temperature, provided that the radicals are moderately large; increased radical size rapidly increases hydrolysis resistance.[175,347,488,1122]

$$(RO)_3PO + NaOH \longrightarrow (RO)_2P(O)ONa + ROH$$

Acidic splitting is more difficult to control, and generally progressive degradation products, down to phosphoric acid, are formed.

Tertiary phosphates containing two phenyl or benzyl

groups may be hydrolyzed usually by mild treatment with dilute mineral acids at moderately elevated temperatures. Selective removal of one group may thus be performed, but general conditions must be found experimentally in each case.[132] Usually the secondary esters are quite stable to alkaline splitting and further attack occurs only under fairly drastic conditions. In acids the scission proceeds rather rapidly, but generally at a slower rate than shown in the degradation of the tertiary esters.[174,175,347,1122] Primary aliphatic esters are rather stable to alkaline splitting but are attacked by acidic media on heating.[168,169,173,462,1121,1406] The nature of the radicals is important. Alkyl groups usually show a double-inflection-rate curve, with essentially no scission at pH 8, a pronounced maximum about pH 4-5, a minimum at about pH 1, and a rapid rise in more acidic solutions.[414] Monoglycol phosphate shows a sharp maximum at pH 3-5, with the rest of the curve being similar to the above;[474] glycerophosphate derivatives show a shifted maximum at pH 3;[162] simple carbohydrate phosphates like diose phosphate, glucose 2-phosphate, and fructofuranose 6-phosphate show a minimum at pH 2-3, a maximum at pH 7, and a continued rise of hydrolysis rate in acidities beyond pH 1.[558] The hydrolytic treatment of glycerophosphates, in addition to the scission, also involves the transposition of the phosphoryl group. This is a reversible reaction with an equilibrium point at 87% 1-phosphate, attained readily by heating crude glycerophosphate in dilute mineral acids. This phenomenon is useful for the isolation of pure 1-isomer.[168,169,1406] A similar transposition occurs in xylose.[866]

α-2. Hydrogenolytic Removal of Phenyl or Benzyl Radicals. The phenyl and the benzyl group can generally be removed from the corresponding phosphates by catalytic hydrogenation.

Using the dibenzyl or diphenyl or benzyl phenyl chlorophosphates (RO)(R'O)P(O)Cl or the dichloridates $(RO)POCl_2$ ($R = C_6H_5CH_2$ or C_6H_5), respectively, the tertiary esters (XO)(RO)(RO)PO and $(XO)_2(RO)PO$ are prepared which by this mild procedure of hydrogenolysis yield the primary and secondary esters $XOPO_3H_2$ and $(XO)_2PO_2H$, respectively. The hydrogenolysis, however, does not operate smoothly in all cases, and no general conclusions on the occasions of failure may be made at this time. A very extensive study of conditions most favorable to success indicates the use of ethanol or dioxane for the solvent, preferably in the presence of a little water and a trace of acid (bases--Et_3N, morpholine, cyclohexylamine, $NaOCH_3$--inhibit hydrogenolytic splitting[410]). Shaking the solution with a preliminary amount of the catalyst is useful in removing catalyst poisons.[95]

Electrolytic splitting--which may be considered as a hydrogenolysis--of tribenzyl phosphate to dibenzyl phosphoric acid has been described (electrolyte: $(C_2H_5)_4NI$ in dimethyl formamide, Hg-cathode, potential needed to provide the limiting current on the phosphate polarogram).[930]

α-3. Partial Dealkylation by Halolysis. Debenzylation of phosphoric esters can be effected by heating with appropriate hydrohalogenic (or pseudohalogenic) salts (LiCl, NaBr, cyclohexylammonium iodide, etc., and even thiocyanates) in acetone or methyl ethyl ketone. When two benzyl groups are present, their removal proceeds in two steps.[113, 454, 974]

In this manner, monobenzylesters of phosphatidic acids have been prepared from their dibenzyl esters which were synthesized by using procedures described in Section 1.b.γ-2.

By refluxing with absolute dry calcium chloride in acetone, alkyl and benzyl groups may be removed, generally with good yield. When benzyl and small alkyl groups are simultaneously present, mixtures result, for instance[454]

$$(CH_3O)(C_6H_5CH_2O)_2PO \xrightarrow{CaI_2} [(CH_3O)(C_6H_5CH_2O)PO_2]_2Ca \text{ and } [(C_6H_5CH_2O)_2PO_2]_2Ca$$

(total yield 75%)

α-4. Partial Dealkylation of Tertiary Phosphoric Esters

By tertiary amines. Phosphoric esters are in general poor alkylating agents, with the exception of esters with methyl or benzyl groups. Heated with appropriate tertiary amines (Me_3N, N-methylmorpholine), these esters yield quaternary ammonium salts of diesterified phosphoric acids.[129, 324, 357, 391]

$$(RO)_3PO + X_3N \longrightarrow (RO)_2P(O)O[NX_3R]$$

From these salts, the corresponding acids $(RO)_2PO_2H$ (the two R being identical or different) may be obtained by known procedures, for instance by ion exchange resins. This dealkylation can be conducted by refluxing the reactants in dry aceton[357] or by refluxing in the amine used (e.g., methylmorpholine).[129]

A particular instance is the partial dealkylation of tertiary phosphates with a t-butylperoxy group by secondary amines (monoalkyl-cyclohexyl-amines) in which the butyloxy group is eliminated already at room temperature.[1274]

$$(Me_3COO)(RO)_2PO + C_6H_{11}NHR' \longrightarrow (RO)_2P(O)O[C_6H_{11}NHR']$$

By phosphorolysis. Phosphoric ester functions can be split by acidolysis. Carboxylic acids seem to react with difficulty and only in certain cases, but a recent Russian patent mentions the preparation of dialkyl phosphoric acids by reaction of trialkyl phosphates with phosphoric acid or monoalkyl phosphates, boric acid acting as a catalyst.[1100]

$$(RO)_3PO + H_3PO_4 \text{ (or } ROPO_3H_2) \longrightarrow (RO)_2PO_2H$$

No details are given in the abstract, and investigation of this reaction would be of interest.

β. Exchange Esterification of Tertiary Phosphates

Tertiary alkyl phosphates may be transesterified by heating with alcohols having a higher radical than that present in the ester.

In the presence of sodium alkoxide the reaction proceeds fairly readily and results in the formation of mixtures of mono- and diexchange products; the yields of individuals are rather poor. The first step can be represented as follows, the trialkyl phosphate behaving as alkylating agent and the detached group forming an ether with the alkoxide ion.[1219]

$$(RO)_3PO + R'OH + R'ONa \longrightarrow (RO)_2(R'O)PO + ROR'$$

The same reaction between tertiary aromatic phosphates and phenols is described in a patent with metal hydrides (particularly NaH) as catalyst, e.g., as follows (8-9 hr heating under nitrogen at 130-170° followed by distillation of the phenol liberated):[405]

$$(C_6H_5O)_3PO + 3\ 4\text{-}H_2NC_6H_4OH \longrightarrow (4\text{-}H_2NC_6H_4O)_3PO + 3C_6H_5OH$$

A similar exchange reaction has been observed between trialkyl phosphates and halides of higher alkyls.

Thus, triethyl phosphate heated some hours at higher temperature (e.g., 150°) with bromides of higher alkyls or the methyl esters of higher α-bromo aliphatic acids yields tertiary phosphates of the types $(C_2H_5O)_2(RO)PO$ with R = n-octyl and higher, and $(C_2H_5O)_2P(O)[OCH(R)COOCH_3]$ with R = n-octyl and higher.[918]

Tertiary phosphines [e.g., $(C_6H_5)_3P$] have been used as chain initiators, via the quaternary phosphonium halides $R(C_6H_5)_3PX$ (X = Cl or Br), of the transalkylation. The reaction is realized by heating equimolecular amounts of $(C_2H_5O)_3PO$ and Hal·X with 0.1 mol-equivalent of Ph_3P in an indifferent solvent (e.g., xylene) or without solvent at 120 to 180° until no more ethyl halide distilled (10 to 300 min.). More than one exchange may occur, but the product resulting from only one exchange is generally predominant.[672]

A patent[270] claims the preparation of mixed tertiary alkyl phosphates by exchange of radicals between tertiary phosphate and chlorosulfinate or chloroformate, as follows:

$$(RO)_3PO + (R'O)S(O)Cl \text{ [or } (R'O)C(O)Cl] \longrightarrow (RO)_2(R'O)PO + RCl + SO_2 \text{ (or } CO_2)$$

The reaction may occur, of course, more than once and mixtures are generally obtained. For instance equal molecular amounts of triethyl phosphate and butyl chlorosulfinate heated in 2-1/2 hr to 150° and maintained 1 hr at this temperature yield a mixture of diethyl butyl and ethyl dibutyl phosphate.

γ. Disproportionation of Partial Phosphoric Esters

Already in 1849 the disproportionation of lead diethyl phosphate into triethyl phosphate and lead monoethyl phosphate at 200° has been reported;[1414] barium salts behave similarly.[842] These reactions are of little practical interest but the analogous reaction with sodium salts is perhaps of more value even if it needs temperatures of approximately 300°.[1016,1472] It may be represented as follows.

$$\left.\begin{array}{c}(RO)_2(NaO)PO \\ + \\ 2(RO)_2(HO)PO\end{array}\right\} \quad 2(RO)_3PO + (NaO)(HO)_2PO$$

f. Miscellaneous. Under this heading we shall mention reactions which have not been, or cannot be, generalized. The enumeration is not exhaustive.

α. Desulfuration of Thiophosphates

Hydrolytic. Extensive hydrolytic treatment of compounds with the thiono group or a sulfhydryl group [$(RO)_3PS$, $(RO)_2POSH$, etc.] results in progressive loss of sulfur as hydrogen sulfide, with replacement by oxygen. The preparation of the thio derivative and its hydrolysis may be combined.[579,774] Of course the ester functions must be resistant to hydrolysis.

Oxidative. Thionophosphates (especially tertiary esters) are oxidized by strong oxidants, such as nitric acid, to the corresponding phosphates.[1111] In a particular case an alkylthiophosphoric acid which loses its thioester function by mild oxidation has been used. Condensation of 11-deoxycorticosterone with ethylthiophosphoric acid, $C_2H_5SPO_3H_2$ (pyridine, dicyclohexylcarbodiimide), followed by oxidation by iodine in aqueous acetone at room temperature yielded its 21-monophosphate.[1059] This procedure may be useful for acid or base sensitive products.

β. Addition of Phosphoric Acid to Olefins (Acetylenes)

In patents, the addition of phosphoric acid to olefins, yielding primary phosphates, is claimed by reaction at elevated temperatures and pressures in the presence of sulfuric acid, but no individual compounds are reported.[757] Similar addition of metaphosphoric acid has been described,[484] and the use of cuprous oxide or silver sulfate as catalysts has been proposed.[1014]

β-Cyanovinyl phosphate has been prepared by addition of phosphoric acid to cyanoacetylene.[538]

γ. Phosphorolysis of Glycosidic Bonds

In sugar chemistry acetylated glycosidic groups of O-acetylated sugars are phosphorylated by action of crystalline phosphoric acid, for instance at 50° for 2 hr.[297,373]

$$\begin{array}{ccc} OCOCH_3 & & OCOCH_3 \\ | & & | \\ -CH-CH\text{-}OCOCH_3 + H_3PO_4 \longrightarrow & -CH-CH\text{-}OPO_3H_2 \\ | & & | \\ -O & & -O \end{array}$$

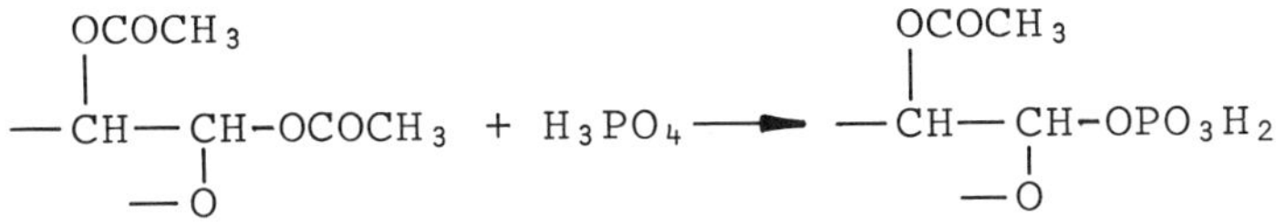

By deacetylation, the phosphorylated product yields the 1-phosphorylated sugar. In this manner D-glucopyranose 1,6-diphosphate has been prepared from 1,2,3,4-tetra-O-acetyl-β-D-glucopyranose 6-phosphate, the final product being isolated as tetra-cyclohexylammonium salt.[297]

A French patent claims the preparation--by a reaction of similar type--of D-glucose 6-phosphate (isolated as barium salt) by degradation of starch with phosphoric acid.[1270]

δ. Exchange of Radicals in Tertiary Alkyl Phosphates by Chlorosulfinate or Chloroformate Esters

A patent claims the replacement of radicals in tertiary alkylphosphates by heating with the chlorosulfinate or chloroformate ester of the alcohol whose radical is to be introduced; the reaction may be described as follows:

$$(RO)_3PO + R'OS(O)Cl \quad [\text{or } R'OC(O)Cl] \longrightarrow (RO)_2(R'O)PO + RCl + SO_2 \quad [\text{or } CO_2]$$

As the reaction involves each radical of the phosphate introduced, mixtures always result; with $(EtO)_3PO$ warmed with BuOS(O)Cl in 2.5 hr to 150° and maintained for 1 hr at that temperature, a mixture of diethyl butyl and ethyl dibutyl phosphates is obtained.[270]

ε. Tertiary Phosphates from Dialkyl Phosphoric Acids by Chloroformates

This reaction, described in patents,[255,257] proceeds as follows:

$$(RO)(R'O)PO_2Na + Et(or\ Me)OC(O)Cl \longrightarrow (RO)(R'O)(Et[or\ Me]O)PO + CO_2 + Et(or\ Me)Cl$$

It is realized by heating the components in higher ketones at 80° (or more) until no more carbon dioxide is produced.

ζ. Various Phosphorylation Procedures

ζ-1. Pyrophosphoryl Chloride. At sufficiently low temperatures, pyrophosphoryl chloride reacts with alcohols as an acid anhydride and yields alkyl phosphorodichloridates which may be hydrolyzed to the corresponding monoalkyl phosphoric acids or may be caused to react with alcohols to give tertiary phosphoric esters.

$$ROH + Cl_2P(O)OP(O)Cl_2 \longrightarrow Cl_2PO_2H + ROP(O)Cl_2$$

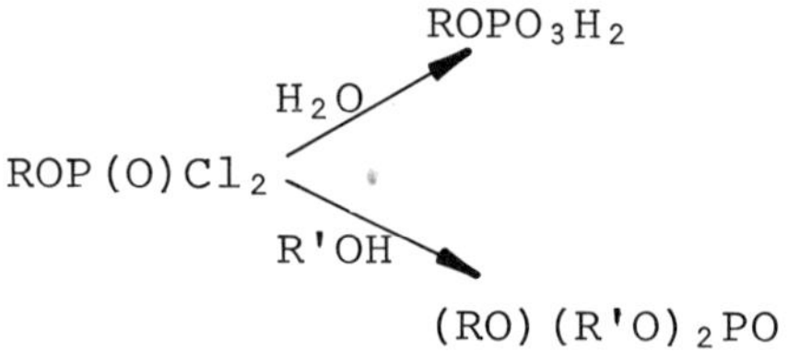

This reaction takes place when pyrophosphoryl chloride is added dropwise to a primary alcohol at -50°, the mixture is warmed carefully until reaction begins and then cooled again to -20°; this procedure is repeated before the final distillation.[636a]

With more complicated molecules like nucleosides (with protected secondary OH groups[816] or without[697]--the phosphorylation of primary OH groups is selective) or with 4,6-isopropylidene-ascorbic acid[1053] the reaction has been conducted in other solvents and at somewhat higher temperatures (0°), for instance in various organic solvents (nitriles, esters, etc.[733,1331]) or in pyridine-acetone.[1053]

ζ-2. Phosphoric Acid or Esters + Nitriles. Under the action of nitriles phosphoric acid functions can be esterified by alcoholic functions. Thus 2-cyanoethyl phosphate--itself an interesting agent for the preparation of primary phosphates (see next paragraph)--has been prepared by reacting anhydrous tri-triethylamine phosphate with 2-cyanoethanol and trichloroacetonitrile in the presence of triethylamine.[1101a] A series of monoterpenyl phosphates have been synthesized by a similar procedure.[449a] The production of pyrophosphates as byproducts has been reported.[1394] Mononucleotides yield similarly di- or oligonucleotides when the components are heated in dimethyl formamide some hours at 50° in the presence of trichloroacetonitrile.[442b] Malonitrile and acrylonitrile have equally been used.[1228]

A patent claims an exchange phosphorylation of 2' and 3' OH protected ribonucleosides to the corresponding 5' phosphate derivatives by heating with a tertiary phosphate, e.g., $(BuO)_3PO$, in the presence of malonitrile in dimethyl formamide.[1229]

ζ-3. Primary Phosphate Functions by Phosphorylation with Cyanoethyl Phosphoric Acid. The β-cyanoethyl ester function $CN\text{-}CH_2CH_2\text{-}O\text{-}P\cdots$ is very easily split in slightly alkaline medium but quite stable in neutral or acidic medium.[611] By condensation of an alcohol ROH with this derivative by action of an appropriate disubstituted carbodiimide, in most cases dicyclohexylcarbodiimide, e.g., in dry pyridine,[1344] the ester $ROP(OCH_2CH_2CN)O_2H$ is produced, which yields $ROPO_3H_2$ by mild alkaline treatment. This procedure has been used in different fields: phosphorylation of nucleosides (e.g., Ref. 462), simpler phosphoric esters like phenethyl phosphate[214] or phenolphthaleine monophosphate (by condensation of phenolphthaleine methyl ester with cyanoethylphosphate followed by mild alkaline and strong acid hydrolysis[936]).

ζ-4. Phosphorofluoridates for Selective Phosphorylaation of HO Groups. Phosphofluoridates $FPO(OR)_2$ are reported to phosphorylate in non aqueous solvents aminoalcohols exclusively at their hydroxyl groups, in contrast to agents like phosphorochloridates, tetraalkyl pyrophosphates, and N,N-dimethyl phosphoramido cyanates which react predominantly with NH_2 groups.[627]

ζ-5. Special Phosphorohalidates with Easily Eliminable Groups. Besides the halidates with phenyl or benzyl ester functions (see, for instance, Section 1.e.α-2) a series of other ester-halidates have been used in particular instances.

Di-cyclopropylcarbinyl phosphorochloridate, $(C_3H_5CH_2O)_2$-POCl, yields with alcohols mixed tertiary phosphates which are split to $(RO)PO_3H_2$ by 30 min reflux in 80% formic acid or 5 hr reflux in 10% acetic acid.[1299]

The analogous derivative of trichloroethanol, $(Cl_3CCH_2\text{-}O)_2POCl$, yields tertiary esters whose trichloroethyl groups are eliminated by action of Zn-Cu alloy in dimethyl formamide (50°, 1 hr) or zinc dust in 80% formic acid (room temp., 1 hr). This procedure has been useful for phosphorylation of protected nucleosides.[499]

Various amidophosphorohalidates are mentioned as phosphorylating agents, the amido function of the resulting ester is eliminated either by hydrolysis (e.g., morpholino groups by acid hydrolysis[1193] or contact with ion exchange resin with sulfonic acid ions at 60°[934] for morpholino phosphorodichloridate[724] and dimorpholino phosphorochloridate;[1193] alkaline hydrolysis for 2,6-lupetidino dichloridate)[725] or by special procedures (NHR' functions; e.g., methylamido groups can be removed not only by hydrolysis but equally well by nitrosation with nitrous acid (N_2O_3),[1181] or $NOSO_3H$ or nitrosyl chloride;[307a] $(RO)_2PONHR'$ yields with NaH the anion $(RO)_2PO\overline{N}R'$, split by CO_2, isocyanate, ketene, or aldehyde with liberation of $(RO)_2PO_2^-$[1427b]).

ζ-6. Alcoholysis of Amido Phosphoric Derivatives. The reaction proceeds as follows.

$$RR'NPO_3X_2 + R''OH \longrightarrow R''OPO_3X_2 + RR'NH$$

Examples: dibenzyl 1-imidazolylphosphate caused to react in dioxan at 40° with benzylidene tryptophanol yielded its dibenzyl phosphoric ester; splitting of the benzylidene group by acid hydrolysis and of the benzyl ester functions by hydrogenolysis yielded finally tryptophanyl phosphate.[670] Benzoylamidophosphoric acid heated 80 min with ethanol yielded ethyl phosphoric acid isolated as anilinium salt (79% yield).[1509]

ζ-7. Sugar Phosphates by Displacement of Sulfonyloxy Groups. Primary p-tolylsulfonyl groups in otherwise acetylated sugars or glycosides are displaced by the diphenyl phosphate group by refluxing with the lithium salt, $(C_6H_5O)_2PO_2Li$, in dimethyl formamide. Subsequent removal of the acetyl groups by aqueous ammonia and of the phenyl groups by hydrogenolysis yields the phosphorylated sugar or glycoside.[374]

ζ-8. Cyanovinyl Phosphate as Phosphorylating Agent. This monoester, $CNCH{=}CHOPO_3H_2$, suffers slow first order hydrolysis in water and is split by hydroxy compounds (and acids) as follows.

$$CNCH{=}CHOPO_3H_2 + HOR \longrightarrow ROPO_3H_2 + CNCH_2CHO$$

By this reaction nucleosides have been phosphorylated;[538] we shall find this reagent later for the phosphorylation of phosphates to pyrophosphates.

ζ-9.   Various Esterifying Agents.

Picryl chloride--similarly to arenesulfonyl halides
(see Section 1.d.α-2)--activates the esterification of an
alcoholic function by phosphoric acid or by a phosphoric
monoester;[1250] the reaction proceeds at room temperature
in the presence of a tertiary base in an inert solvent,
e.g., acetonitrile. It has been applied equally to the
preparation of nucleotides.

Dimethyl formamidinium chloride, $[(CH_3)_2N:CH(NH_2)]Cl$,
promotes the esterification of phosphoric monoesters by
an alcohol already at room temperature, e.g., 2 hr in
pyridine. Thus, for instance, phenyl benzyl phosphoric
acid has been prepared from phenyl phosphate and an excess
of benzyl alcohol. The procedure has been applied further
to nucleosides.[446]

Addition product of cyano bromo acetamide and triphenyl
phosphine, $HC(Br)(CN)-CONH_2 \cdot (C_6H_5)_3P$. The reaction proceeds in a solution of equimolecular amounts of the two
components (excess) and a phosphoric acid monoester in a
suitable solvent, e.g., dimethyl formamide, at room temperature overnight.[447,448] When using cyanoethyl phosphate,
the resulting diester is transformed into the monoester by
mild alkaline hydrolysis. The reaction proceeds probably
by a quasiphosphonium salt.[448]

$$\left(\begin{array}{c} NC-CH=C-NH_2 \\ | \\ O \\ \oplus| \\ P(C_6H_5)_3 \end{array} \right) Br^\ominus \xrightarrow[-HBr/-P(C_6H_5)_3]{ \begin{array}{c} O \\ \| \\ RO-P-OH \\ |\ominus \\ O \end{array} } \left(\begin{array}{c} NC-CH=C-NH_2 \\ | \\ O \\ | \\ RO-P=O \\ |\ominus \\ O \end{array} \right) \ +\ R'OH \longrightarrow$$

$$NCCH_2CONH_2$$
$$+ \quad *$$
$$\begin{array}{c} O \\ \| \\ RO-P-OR' \\ |\ominus \\ O \end{array}$$

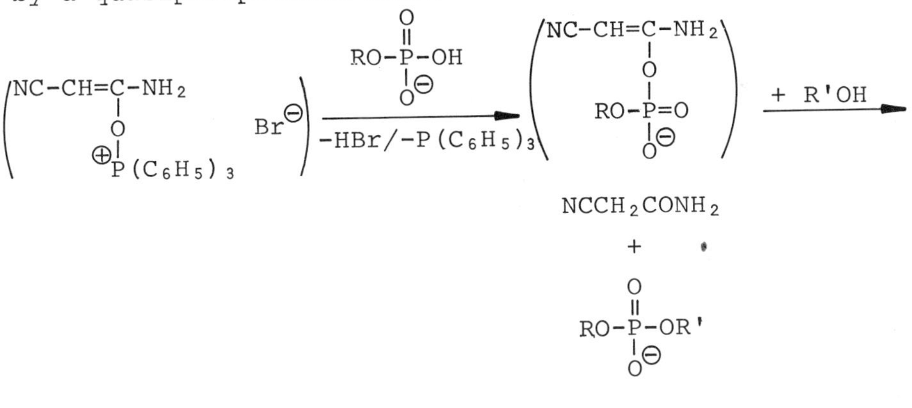

Mukaiyama et al.[1011] have shown that tetraethyl hypophosphate and aluminum ethoxide, treated with ethanol,
yield triethyl phosphate; the intervention of an addition
product of $Al(OC_2H_5)_3$ on the P atom of the phosphorous
part explains probably the orientation of the alcoholysis
of the P-O-P bond.

ζ-10.   Use of Cyclic Derivatives.

Cyclic phosphoric esters are in general easy to hydrolyze partially with opening of the cycle, yielding hydroxyalkyl phosphoric acids (starting from the free acids) or
their monoesters (starting from esters of the cyclic

secondary phosphates).[231,383a,833a,1379a]

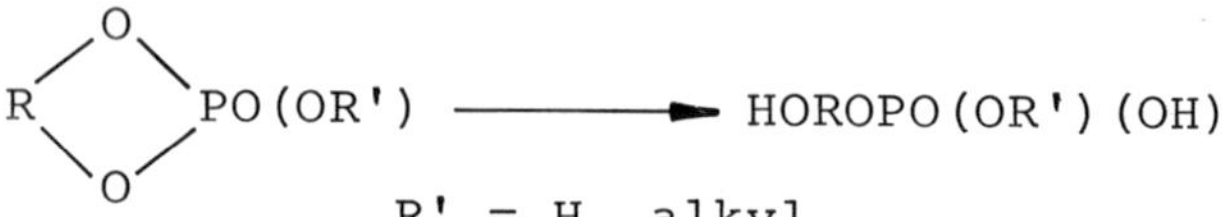

R' = H, alkyl

The hydrolysis can be conducted in acid or alkaline medium. The reaction is limited in its application because of the difficulty of preparing cyclic esters (see Section 2); one procedure is precisely the cyclization of mono-hydroxyalkyl phosphoric acids.

This procedure has been applied to derivatives whose cycle presents an acyl phosphate grouping which of course is rapidly split by water, for instance:[92]

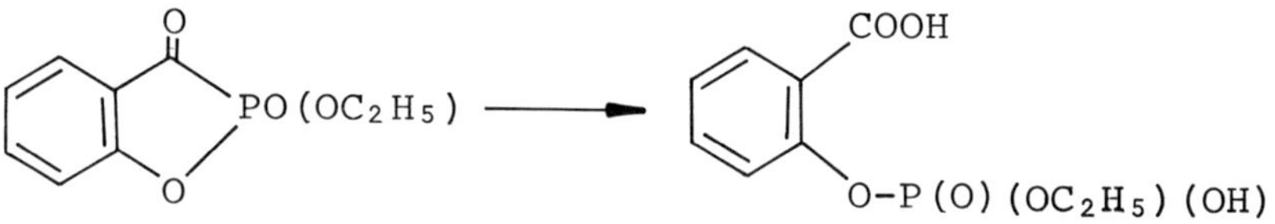

Anhydronucleosides treated with benzyl or dibenzyl phosphate yield 5'-benzyl (dibenzyl) phosphates;[977a] purine nucleosides with a 6-NH_2 group must first be acetylated.[977]

2-Ethoxy-2-oxo-3-methyl-1,3,2-oxazaphospholidine, a cyclic ester-amide, yields 2-methylamino-ethyl phosphate with appropriate acids, the amide function being split easily.[812]

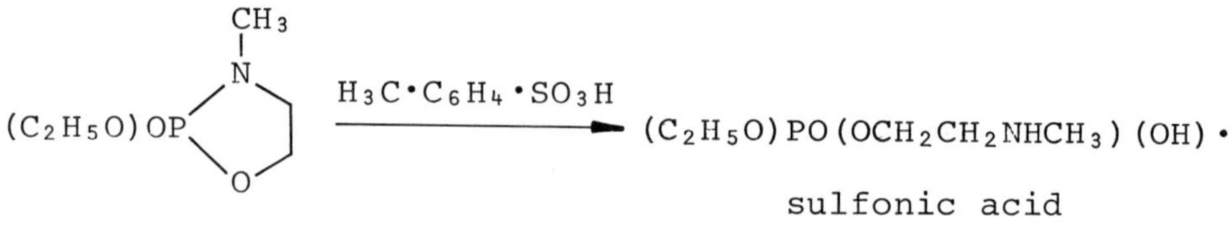

Double methylation of the methylamino group yields the corresponding choline derivative. The procedure is versatile insofar as the ethyl group can be replaced by other groups, for instance by a 1,2-protected glycerol residue for the synthesis of α-glycerophosphoric aminoalkyl esters.[812]

Hydrobenzoine cyclic phosphate adds under activation by diphenyl phosphorochloridate alcohols ROH, yielding hydrobenzoine phosphorodiesters; by hydrogenolysis, these diesters are split (hydrobenzoine is a substituted benzyl alcohol) to monoesters $ROPO_3H_2$. The reaction is described for butyl alcohol and nucleosides.[1387]

Silver trimetaphosphate treated with alkyl iodides

yields the corresponding triesters. By hydrolysis monoalkyl phosphoric acid is formed; alcoholysis is more complex and yields a mixture of monoalkyl phosphoric acid and pyrophosphate ester.[448b]

ζ-11. "Diazomethanolysis" of Glycerophosphate Derivatives. Diazomethane splits the bond between phosphatidic acid and the N-containing moiety in glycerophosphatides, yielding optically pure phosphatidic acids as dimethyl esters. The reaction is limited to glycerophosphatides with free amino group; lecithins are not cleaved.[152]

ζ-12. Adenosine Triphosphate (ATP). This important biological phosphate donator can act as such in a nonenzymatic reaction in solution in lower alcohols at room temperature.[504] Disodium ATP in methanol containing 0.25% HCl or H_2SO_4 transfers in 1 hr about 30% of its terminal phosphate, yielding monomethyl phosphoric acid. Presence of water (10%) inhibits the reaction. ADP or AMP do not react under these conditions.[504]

η. Rearrangement of N-Phosphoryl 1,2-Aminoalcohol to Aminoalkyl Phosphate

The rearrangement of N-acyl 1,2-aminoalcohols on heating with mineral acids to O-acyl 1,2-aminoalcohols has been extended to the N-diisopropylphosphoryl derivatives of aminoethanol, DL-serine, and DL-threonine, the isopropyl ester functions being hydrolyzed at the same time. For example:[1116]

$$HOCH_2\overset{\displaystyle NHPO(O\text{-}iC_3H_7)_2}{\underset{\displaystyle H}{C}}COOCH_3 \xrightarrow[\text{in 5-7\% HCl}]{\text{refluxed 4 hr}} H_2O_3POCH_2\overset{\displaystyle NH_2}{\underset{\displaystyle H}{C}}COOH$$

(yield 26%)

ι. From Phosphonic Acids

Phosphonic acids may serve in two ways for the preparation of phosphoric acid esters: by rearrangement or as donator of a PO_3H_2 group, depending on the nature of the radical fixed on the phosphorus atom.

Rearrangement has been observed first with dialkyl 2,2,2-trichloro-1-hydroxyethylphosphonate by dehydrohalogenation, yielding dialkyl 2-chlorovinyl phosphates as follows.[205]

$$CCl_3CH(OH)PO(OR)_2 \xrightarrow[NaOH]{-HCl} CCl_2{=}CHOPO(OR)_2$$

The dichloro derivative $CHCl_2CH(OH)PO(OC_2H_5)_2$ reacts in the same way.[205] The reaction may be generalized as long as in the $CR_3CH(OH)$-radical fixed on the P atom one R is capable of leaving easily as anion.[225]

When the resulting vinyl phosphate bears on C(2) two different substituents, e.g., H and Cl as in $CHCl{=}CHOPO(OC_2H_5)_2$, the vinyl group has probably trans configuration (absorption in the 10.4 nm region and none in the 14 nm region).[225]

Phosphorylating action has been observed with long chain β-chloroalkyl-phosphonic acids, for instance with β-chlorodecylphosphonic acid, treated at room temperature with an alcohol.

$$C_8H_{17}CH(Cl)CH_2PO_3H_2 + C_2H_5OH + c\text{-}C_6H_{11}NH_2 \longrightarrow C_8H_{17}CH{=}CH_2\ (89\%) + (C_2H_5O)PO_3H_2(c\text{-}C_6H_{11}NH_2)_2$$

It is interesting to note that phenol and even t-butyl alcohol react in the same way, yielding phenyl and t-butyl phosphoric acid, respectively.[926]

2. Use of P^{III} Derivatives (and Lower)

a. Monoalkyl Dichlorophosphates from Phosphorus Trichloride

α. Reaction with Peroxyalcohols

A patent[25] claims the production of alkyl dichlorophosphates by reacting phosphorus trichloride with peroxyalcohols (products not described!).

$$R(O_2H) + PCl_3 \longrightarrow (RO)P(O)Cl_2$$

The dichloridate may be used in ulterior classical transformations for producing the corresponding monoalkyl phosphoric acids or esters thereof.

β. Phosphorus Trichloride + Ethane + Oxygen

This procedure seems to be more interesting than the preceding because of the possibility of generalization. The components are caused to react first at -5° to 8° at 18 atm, then at 30° for 5 min. The result is a mixture of the dichloridates of ethanephosphonic and monoethyl phosphoric acids, both distillable. With the dichloridates the corresponding acid or higher esterified acids can be prepared by known methods. The following mechanism is proposed (hypothetical intermediaries in parentheses).

$$PCl_3 + O_2 \longrightarrow (PCl_3O_2) + PCl_3 \longrightarrow POCl_3$$

$$(PCl_3O_2) + EtH + PCl_3 \longrightarrow EtPOCl_2 + HCl + POCl_3$$

(phosphonic derivative)

$$(PCl_3O_2) + EtH \longrightarrow EtOPOCl_2 + HCl$$

(phosphoric derivative)

b. Oxidation of Phosphorous Derivatives. In principle derivatives with phosphorus in a lower oxidation state are easily oxidated, yielding derivatives of pentavalent phosphorus. In the case of phosphorous esters--the most important in this group--two difficulties arise: On one hand the preparation of organic phosphites is often costly particularly for the tertiary esters, and on the other most of the phosphorous esters are easily hydrolyzed so that "wet" agents should be avoided. Finally it must be remembered that phosphorous esters are very versatile in their oxidation reactions.

α. Reaction with Inorganic Oxidants

α-1. Oxygen and/or Sulfur Trioxide. Oxidation of tertiary phosphites by heating in a current of air[1505] or by treatment with air-sulfur trioxide[298] have been reported. More recent patents describe the oxidation by oxygen, air,[836] or "oxygen containing gases," under irradiation with ultraviolet light;[13,318,1400] as catalysts metals and metal oxides are proposed.[698,1249] In the photoöxidation, ozone is said to be no intermediate.[318] Aromatic phosphites are difficult to oxidize: triphenyl phosphite irradiated with a Hanovian lamp in dry air or oxygen yielded only 5% triphenyl phosphate after 18 hr![318] The same reaction can be realized by using sulfur trioxide in an appropriate solvent.[527] For instance to trimethyl phosphite in liquid sulfur dioxide (cooled in dry ice-methanol) SO_3 is added, then the temperature is raised to 50° and the solvent evaporated in vacuo; the yield is said to be 94% (patent!). Again aromatic phosphites are not oxidizable under these conditions.

α-2. Ozone, Hydrogen Peroxide. A Russian patent reports preparation of acid alkyl phosphates by treatment of the corresponding phosphites with ozone.[880] A similar possibility is reported for oxidation by 50% or 30% aqueous hydrogen peroxide.[668,851]

α-3. Sulfuryl Chloride, Phosgene. Secondary phosphites treated with $SOCl_2$ yield the corresponding dialkyl chlorophosphates.[94,526,530] The action of phosgene on tertiary phosphites yields equally secondary chlorophosphates, but this time by partial dealkylation.[756]

$$(RO)_3P + COCl_2 \longrightarrow (RO)_2POCl + RCl + CO$$

(and not as previously reported $(RO)_2P(O)COCl$)

From the chloridates the corresponding acids or by reaction with an alcohol or a phenol tertiary phosphates can be prepared.

α-4. Arsenic Anhydride. Bliznyuk et al.[250] have treated phosphorous acid with alcohols in the presence of As_2O_5 in an inert solvent (e.g., xylene) and claim to have obtained by simultaneous esterification and oxidation dialkyl phosphoric acids.

α-5. Nitrosyl Chloride, Nitrile Chloride, Nitrogen Oxides. In the hands of Arbuzov et al.[72] the first two compounds when treated with tertiary phosphites yielded the corresponding phosphates accompanied by small quantities of pyrophosphates. The authors suppose that this by-product results from partial conversion of $(RO)_3P$ into $(RO)_2POCl$ (not isolated) which reacts then with $(RO)_3PO$.

The findings concerning the action of nitrogen oxides are somewhat contradictory. Rizpolozhenskii et al.[1199] report that mixed N oxides at -20° in methylene chloride transform certain tertiary phosphites into the corresponding phosphates, but others, of the type $(RO)_2POCH_2CH_2COCH_3$ with R = Et or Pr, yield $(EtO)_2PO_2H$ and $(PrO)_2PO_2H$, respectively. N_2O_4 has been reported as yielding tertiary phosphates from tertiary phosphites,[775a,1100a] but this has been contradicted: the oxidation products are not phosphates $(RO)_3PO$ but pyrophosphates $R_4P_2O_7$.[637]

α-6. Mercury(II) Derivatives. Yellow mercuric oxide in acetone has been used for the oxidation of tertiary phosphites into phosphates; secondary phosphites did not react.[775a] In a procedure similar to that mentioned under Section α-4 above with arsenic anhydride, alkyl phosphoric acids have been prepared by reaction of phosphorous acid with an alcohol in the presence of mercuric salts and a tertiary base.[1065]

$$Hg(O_2CCH_3)_2 + PO_3H_3 + ROH \longrightarrow ROPO_3H_2 + Hg + 2CH_3COOH$$

The mercuric salt and phosphorous acid are heated under stirring in the presence of triethylamine in excess at 80° during 15 min. Besides the acetate, the chloride and the sulfate, and even Hg_2Cl_2, are effective; mercuric cyanide and dialkyl mercury are not.

α-7. Sulfur Dioxide. In a very complex reaction, sodium diethyl phosphite treated in benzene at ordinary temperature with sulfur dioxide yields a complicated mixture: a small quantity of sodium diethyl phosphate, accompanied by tetraethyl pyrophosphate and derivatives of thiophosphoric and thio- and dithio-pyrophosphoric esters.[1313] The mechanism of the reaction is obscure.

α-8. Halogens. The action of halogens is complex.
Chlorine. Secondary phosphites yield the corresponding secondary phosphochloridates.[632,894]

$$(RO)_2POH + Cl_2 \longrightarrow (RO)_2POCl + HCl$$

The phosphochloridates may be used for the preparation of dialkyl phosphoric acids (hydrolysis) or tertiary phosphates (reaction with R'OH). With tertiary phosphites the reaction can proceed in two ways:

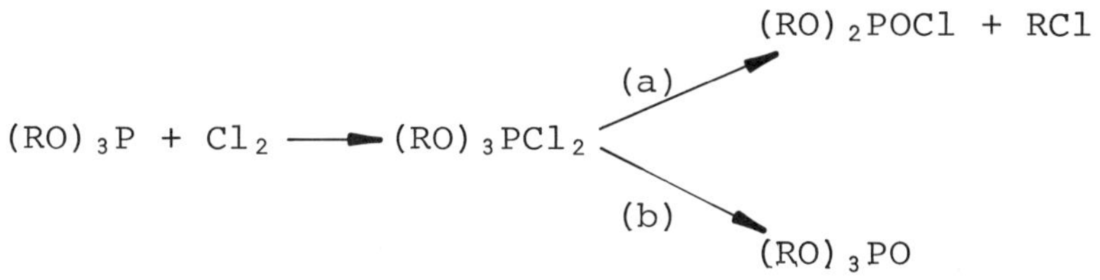

Type (a) applies to aliphatic phosphites. When esters with cyclic structure occupying two valences of phosphorus are used, the ring is cleaved and a haloalkyl ester function is formed in place of the alkyl halide:[1217]

The reaction is run at ordinary temperature.

A similar reaction starting from secondary halophosphites $(RO)_2PCl$ yields primary dihalophosphates.[1452]

Type (b) has been observed with the tertiary phosphite of 2,2,2-trichloroethanol[600] reacted with chlorine in pentane at -10°. The reaction may be due to the cumulative inductive effect of the nine Cl atoms reducing the nucleophilic activity of the lone pair of electrons on P. The type (b) reaction is favored by using the corresponding alcohol as solvent at 0 - 10°.[194,572] The same reaction has been realized starting from phosphorus trichloride by simultaneous esterification and oxidation by chlorine:[249]

$$ROH + 0.5\ PCl_3 + 0.5\ Cl_2 \longrightarrow (RO)_3PO$$

After 20 min at 35-40° HCl is blown off by a current of air, and the residue is distilled. With R = n-butyl, amyl, n-octyl yields of more than 80% were obtained.

Bromine behaves in general like chlorine.[601] Primary phosphites treated in the presence of an alcohol with bromine in dry pyridine yield mixed dialkyl phosphoric acids.[246]

$$ROPO_2H_2 + R'OH + Br_2 \longrightarrow (RO)(R'O)PO_2H + 2BrH$$

(3 days at room temperature)

It is interesting to note that in the absence of R'OH, the reaction gives no secondary phosphoric ester but symmetric dialkyl pyrophosphate.[246]

When iodine is used, the iodo derivatives $(RO)_3PI_2$ are too unstable to be isolated[430] in the case of aliphatic phosphites. With aromatic phosphites, polyiodides are produced,[567] which are hydrolyzed to the corresponding phosphates.

$$(PhO)_3P + \text{iodine} \longrightarrow (PhO)_3PI_4 \text{ and } (PhO)_3PI_9 \longrightarrow (PhO)_3PO$$

Hydrazides of phosphorous monoesters yield with iodine an intermediate (not isolated) which gives esters of phosphoric acid with alcohols, and pyrophosphates with phosphates:[287]

$$H(RO)P(O)NHNH_2 \xrightarrow{I_2} \text{intermediate (halophosphate?)}$$

→ + alcohol → phosphoric ester

→ + phosphate → pyrophosphate

With iodine in aqueous pyridine dibenzyl and diphenyl phosphorohydrazidites yielded the corresponding phosphoric acid diesters.[287]

$$(RO)_2PNHNH_2 \xrightarrow{\text{ox.}} (RO)_2PO_2H$$

α-9. Various Oxidants.

Potassium permanganate has been reported by a Japanese paper to allow the oxidation of riboflavine 5'-phosphite yielding the corresponding phosphate (no details given in the abstract).[1337]

By heating dialkyl phosphorous acids in carbon tetrachloride with anhydrous copper(II) chloride ($CuBr_2$ is less active) the corresponding phosphorochloridates are obtained from which the acids or tertiary esters may be prepared by classical procedures.

$$(RO)_2PHO + 2CuCl_2 \xrightarrow[100°]{\text{4 hr in dry air}} (RO)_2POCl + Cu_2Cl_2 + HCl$$

Lead tetraacetate produces instant oxidation of trialkyl

phosphites in benzene solution at room temperature; triaryl phosphites are similarly oxidized by silver(II) acetate in acetone.[480a]

α-10. Oxidation of Elemental Phosphorus in the Presence of Organic Substances. Red phosphorus heated under 40-150 atm with about 3 mol-equiv. of phenol in the presence of water several hours yields a mixture of phosphine, phosphonate, and phosphate, $(C_6H_5O)_3PO$, besides hydrogen (with a very small quantity of water no phosphate is formed).[1421]

A patent claims the preparation of a mixture of mono and dihydroxyalkyl phosphoric acids by heating white or yellow phosphorus in an oxygen containing gas with polyhydroxy compounds in the presence of a catalyst several hours at 95-98°, followed by stripping to 105°[195] (catalysts: Al, Sc, Ti, V, Cr, and their oxides). Thus with glycol a residue containing 25% of mono-hydroxyethyl and 53% dihydroxyethyl phosphoric acids is obtained.

β. Reaction with Organic Oxidants, without Condensation

β-1. Acyl Peroxides. Tertiary phosphites, vigorously shaken in an appropriate solvent ($CHCl_3$, CCl_4, ether) with an aroyl peroxide yield the corresponding phosphate and aroyl oxide (in absence of solvent an uncontrolled reaction takes place yielding phosphonates besides other products), e.g.,[312]

$$(C_2H_5O)_3P + (C_6H_5COO)_2 \longrightarrow (C_2H_5O)_3PO + (C_6H_5CO)_2O$$

A patent[354] describes the oxidation, with excellent yield, of tertiary phosphites by peracetic acid in solution in glacial acetic acid at 100°. With phosphites of unsaturated alcohols, e.g., trioleyl phosphite, the corresponding triepoxyalkyl phosphates are produced.

β-2. Tetrahalogenomethanes. Since the observation of Todd and co-workers[94-97] that in the presence of polyhalogenated alkanes--particularly $CBrCl_3$--secondary phosphites behave like secondary halophosphates with secondary or primary amines the very versatile reaction of phosphorous esters with tetrahalogenomethanes (CCl_4, CBr_4, $CBrCl_3$) has been much explored.

With dibenzyl phosphite and an alcohol in the presence of a tertiary amine (2,6-lutidine is particularly efficient) the yield of alkyl dibenzyl phosphates is modest.[94,97]

$$(PhCH_2O)_2PHO + ROH + CHal_4 + \text{base} \longrightarrow$$

$$(PhCH_2O)_2(RO)PO + CHHal_3 + \text{base·HHal}$$

Conducted with sodium hydroxide as base the reaction yields sodium dialkyl phosphate.[975]

$$(RO)_2PHO + CCl_4 + 2NaOH \longrightarrow (RO)_2PO_2Na + CHCl_3 + NaCl + H_2O$$

Dialkyl aryl phosphates have been prepared by action of carbon tetrachloride on dialkyl phosphite in the presence of an inorganic base (carbonate or hydroxide of an alkaline earth or an alkali, oxide of an alkaline earth), followed by reaction with a phenol, especially with nitro or halo groups on the ring.[421,1050]

Tertiary aliphatic amines, e.g., $(C_2H_5)_3N$, have been proposed[1315] and the reaction is conducted by refluxing equimolecular amounts of the reagents for some hours. The reaction can be realized with complicated alcohols, such as 4-methyl-7-hydroxycoumarin. Depending on the nature of the phosphite, the reaction proceeds more or less easily: with this coumarin derivative and di-2-chloroethyl phosphorous acid it is achieved after about 16 hr at 25-30°.[431]

Replacing the alcohol by ethylene oxide, the intermediary "halophosphate" splits the epoxy ring yielding a 2-haloethyl dialkyl phosphate,[517] e.g.:

$$(C_2H_5O)_2PHO + \underset{\diagdown O \diagup}{CH_2{-}CH_2} + CCl_4 \ (+3 \text{ drops } TiCl_4) \xrightarrow[\text{3 hr at } 100°]{\text{3 hr in ice}} (C_2H_5O)_2(ClCH_2CH_2O)PO + CHCl_3$$

With monoalkyl phosphorous acids, secondary phosphates result similarly.[246]

$$(RO)PO_2H_2 + R'OH + CCl_4 + NEt_3 \xrightarrow[\text{3 hr}]{\text{reflux}} (RO)(R'O)PO_2H + NEt_3 \cdot HCl + CHCl_3$$

With tertiary phosphites, the reaction of the ester with the tetrahalomethane in the presence of an alcohol yields the corresponding phosphate, with CBr_4 already below 0°, with CCl_4 after several hours of refluxing.[311]

$$(RO)_3P + CHal_4 \xrightarrow{\text{in } R'OH} (RO)_3PO + CHHal_3 + R'Hal$$

The reaction is considered to involve a radical chain process[311,761b] with solvolysis of a quasi-phosphonium intermediary,

$$[(RO)_3PCCl_3]^+Cl^- \xrightarrow{R'OH} [(RO)_3POR']^+Cl^- (+ CHCl_3) \longrightarrow (RO)_3PO + R'Cl$$

Kamaĭ et al.[761b] have studied the rate of the reaction depending on the nature of the radical of the phosphite and the temperature.

Tetraalkyl pyrophosphites are split by carbon tetrachloride (2 hr on the steam bath after the initial reaction) following the scheme given.

$$(RO)_2POP(OR)_2 \quad + \quad CCl_4 \left\langle \begin{array}{l} Cl_3CPO(OR)_2 \ + \ (RO)_2PCl \\[1em] Cl_3CP(OR)_2 \ + \ (RO)_2POCl \end{array} \right.$$

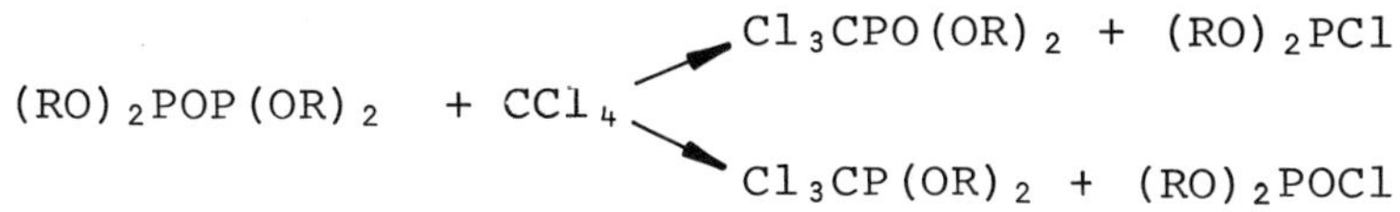

The proportion of these two types of reaction depends on the nature of the radical R; with R = ethyl or propyl, the halophosphate is obtained (proportion not stated in the abstract of this Russian paper), but not with R = butyl.

β-3. Epoxides. Trialkyl phosphites are oxidized smoothly at higher temperatures (about 175° for ethylene oxide) under pressure by epoxides, yielding phosphate and alkene, as follows.[1237]

$$(RO)_3P + \underset{\displaystyle \underset{O}{\diagdown\diagup}}{CH_2\text{-}CH_2} \longrightarrow (RO)_3PO + CH_2{=}CH_2$$

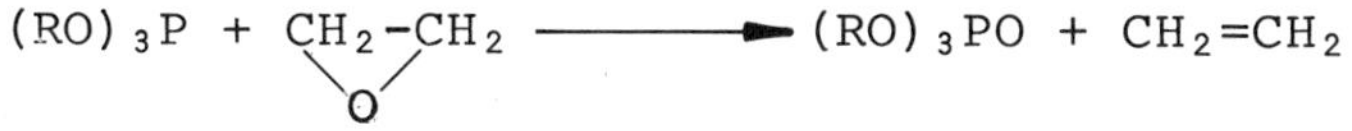

But it is interesting to note that an addition reaction between an epoxide and a tertiary phosphate has been reported (see below under Section γ-3).

β-4. Alpha Brominated Cyanoacetamides. Mukaiyama et al.[662,663,971,972,1011,1013] have shown that α-bromo and α-dibromocyanoacetamide when treated with phosphorous derivatives yield various phosphoric derivatives, depending on the conditions and the nature of the substrate. The mechanism of these reactions involves probably a (non-isolated) vinyl phosphate derivative and is therefore related to that of the reaction of phosphites with α-halo carbonyl compounds, yielding vinyl phosphates (see below, Section γ-2). In its effect it is either a simple oxidation of the phosphite or an oxidation combined with the phosphorylation of an alcohol by the acid function of the starting primary or secondary phosphite or by exchange of an ester group with the starting tertiary phosphite, represented by the following schemes.

$$(RO)_2PHO + CNCHBrCONH_2 \longrightarrow \left[\begin{array}{c} NCC{=}\underset{|}{C}\text{-}NH_2 \\ O \\ (RO)_2\overset{+}{P}\text{-}OH \end{array} \right] Br^- \xrightarrow{\quad C_6H_5CH_2OH \quad}$$

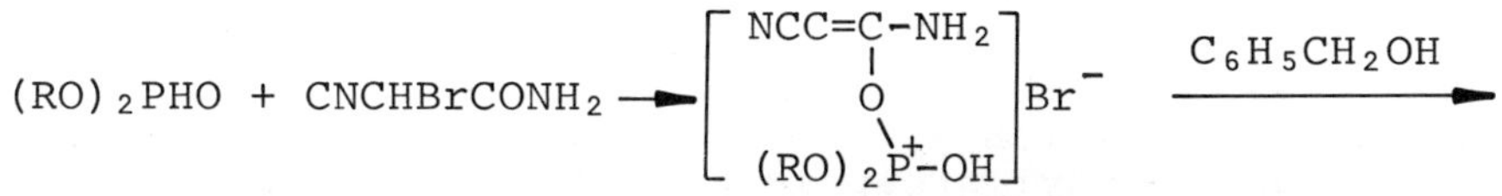

(not isolated)

$$
\left[
\begin{array}{c}
RO \quad\quad OCH_2C_6H_5 \\
\diagdown \overset{+}{} \diagup \\
P \\
\diagup \quad \diagdown \\
RO \quad\quad OH
\end{array}
\right] + CNCH_2CONH_2
$$

$$Br^-$$

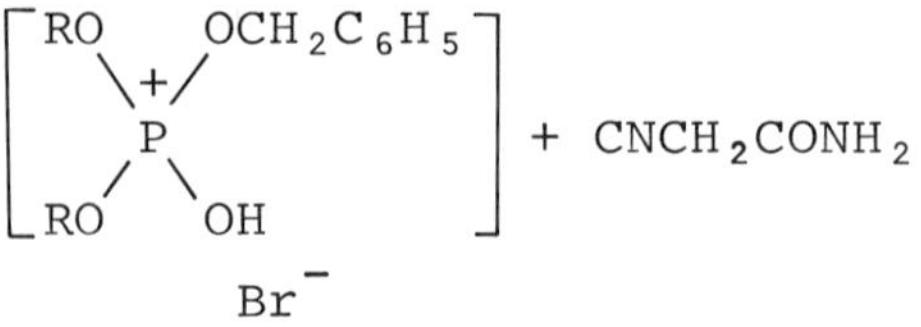

(not isolated)

$\downarrow$

$(RO)_2PO_2H + C_6H_5CH_2Br$

With R = C_2H_5 and with ethanol in the place of benzyl alcohol (Et is a less good leaving group than $PhCH_2$) the reaction yields a mixture:[971]

$$(EtO)_2PHO \xrightarrow{\quad CNCHBrCONH_2, \; EtOH \quad}$$

$$[(EtO)_2\overset{+}{P}(OEt)(OH)]Br^- \diagup\!\!\!\xrightarrow{\quad} \begin{array}{l}(EtO)_2PO_2H + EtBr \\ (EtO)_3PO + HBr\end{array}$$

(not isolated)

By the same mechanism a primary phosphorous ester $ROPO_2H_2$ when reacted with an alcohol R'OH yields the secondary ester $(RO)(R'O)PO_2H$. When using benzyl phosphorous acid the phosphoric monoester of the alcohol introduced, $R'OPO_3H_2$, is obtained by hydrogenolysis of the alkyl benzyl phosphoric acid.[971] This final result is obtained equally starting with ethyl phosphorodimorpholidite, $C_2H_5OP-(NC_4H_8O)_2$, this time with ester exchange.[972]

$$C_2H_5OP(NC_4H_8O)_2 + ROH \xrightarrow{\quad CNCHBrCONH_2 \quad} ROP(O)(NC_4H_8O)_2$$

$$+ CNCH_2CONH_2 + C_2H_5Br$$

By mild acid hydrolysis the dimorpholidate yields the alkyl phosphoric acid. Tertiary phosphites react with exchange of an ester group. The intermediate vinyl (enol) phosphate, analogous to that formulated above for secondary phosphites, is not isolated.[662,663]

$$(RO)_3P + CNCHBrCONH_2 + R'OH \xrightarrow[-40 \text{ to } -50°]{\text{ether}} (RO)_2(R'O)PO$$

$$+ RBr + CNCH_2CONH_2$$

When using benzyl diethyl phosphite and ethanol, propanol, or butanol the benzyl group is exchanged, as formulated for ethanol.[1012]

$(PhCH_2O)(EtO)_2P$ + $CNCHBrCONH_2$ + $EtOH$ $\longrightarrow$ $(EtO)_3PO$

$+ PhCH_2Br + CNCH_2CONH_2$

The dibromo derivative, $CNCBr_2CONH_2$, reacts similarly to the monobromo cyano acetamide and oxidizes two molecules of phosphite, the reaction with the first one yielding $CNCHBrCONH_2$ continuing the reaction with a second one.[662]

β-5. N-Halosuccinimides. N-Chlorosuccinimide oxidizes secondary phosphites yielding the corresponding phosphorochloridates which may be used for the preparation of the dialkyl phosphoric acids or its ethers.[1027]

$(C_6H_5CH_2O)_2PHO$ + $ClN\begin{matrix}\diagup COCH_2\\ |\\ \diagdown COCH_2\end{matrix}$ $\longrightarrow$ $(C_6H_5CH_2O)_2POCl$ + $HN\begin{matrix}\diagup COCH_2\\ |\\ \diagdown COCH_2\end{matrix}$

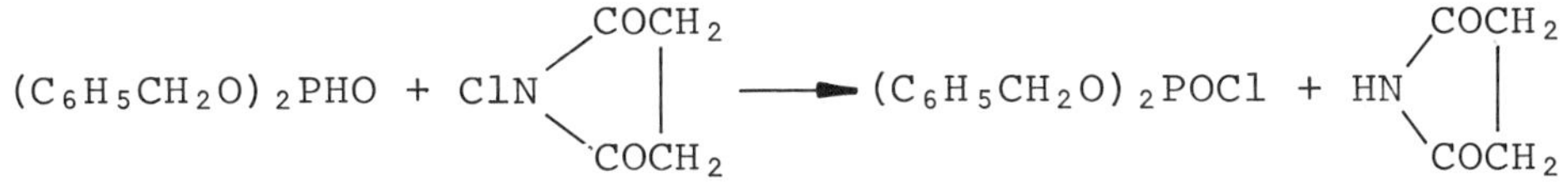

N-Bromosuccinimide reacts with benzyl diethyl phosphite in the presence of ethanol, propanol, or butanol similarly to α-bromocyanoacetamide (see above) yielding triethyl, diethyl propyl, and diethyl butyl phosphate, respectively.[1012]

β-6. Miscellaneous.
Primary nitroparaffins used as alkali salts of the acid form react with diethyl phosphorochloridite yielding the corresponding dialkyl phosphate and nitriles (2-nitropropane, a secondary nitro derivative, yields oximophosphates, see below under γ-4).[1011a]

$(C_2H_5O)_2PCl$ + $NaON(O)=CH-alkyl$ $\longrightarrow$ $(C_2H_5O)_2PO_2H$ + $NC-alkyl$

$+ NaCl$

Diethyl azodicarboxylate: the action of this compound on triethyl or tributyl phosphite, followed by reaction with benzyl or allyl alcohol, R'OH, produced the corresponding phosphates by alcoholysis of a non-isolated intermediary addition product.[973]

$(RO)_3P$ + $EtOC(O)-N=N-CO_2Et$ $\left[\begin{matrix} EtOC=N-\bar{N}-CO_2Et\\ |\\ O\\ \diagdown_+\\ P(OR)_3\end{matrix}\right]$ $\xrightarrow{R'OH}$

$[EtO_2C-NH-\bar{N}-CO_2Et, (RO)_3\overset{+}{P}-OR']$ $\longrightarrow$ $(RO)_3PO$ + $EtO_2C\overset{R'}{\underset{|}{N}}NHCO_2Et$

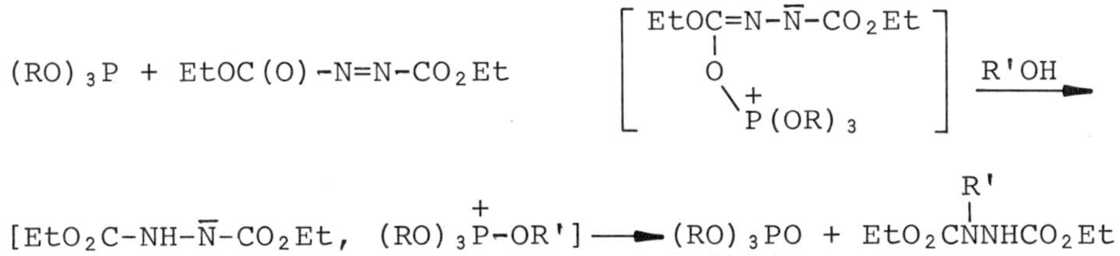

γ. Reaction with Organic Compounds by Addition or Condensation

γ-1. Quinones, α,β-Dicarbonyl Compounds. With methyl or ethyl phosphite, $(RO)_3P$, and p-quinones (p-benzoquinone,[1163] 1,4-naphthoquinone and its 2,3-dichloro derivative,[1159] anthraquinone[1159]) normal addition yields a mono O-alkyl hydroquinone phosphate, as shown for p-benzoquinone.

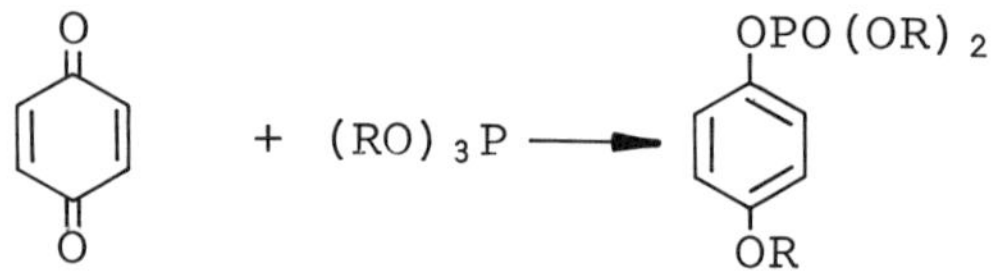

More complicated tertiary phosphites react in the same manner. For instance alkyl diglycyl phosphites, alkyl-$OP(OCH_2\text{-}CH\text{-}CH_2)_2$ (epoxide, O bridging CH and CH_2), add to p-benzoquinone with migration of a glycyl grouping, yielding:[1200]

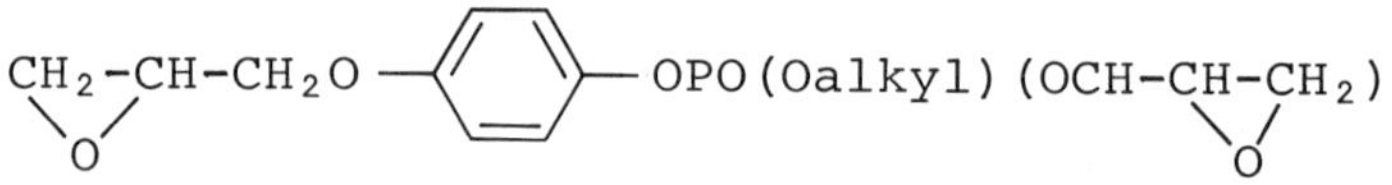

Chloranil reacts similarly with trimethyl and triethyl phosphite, but with triphenyl phosphite (after heating to 100°) no phosphorylated hydroquinone derivative was isolated, the mixture yielding by hydrolysis tetrachlorohydroquinone and triphenyl phosphate.[1165]

Dialkyl phosphorous acids react similarly. They add to p-benzoquinone[406a] or chloranil[1164] yielding the corresponding hydroquinone mono dialkylphosphates. Using dibenzyl phosphorous acid vitamin K_1 has been transformed into 1-dibenzylphosphoryl-dihydrovitamin K_1 which by hydrogenolysis afforded dihydrovitamin K_1 1-phosphate.[1202]

In all these reactions an addition product of the following type is probably involved as intermediary which undergoes a rearrangement with migration of one R radical.[1159]

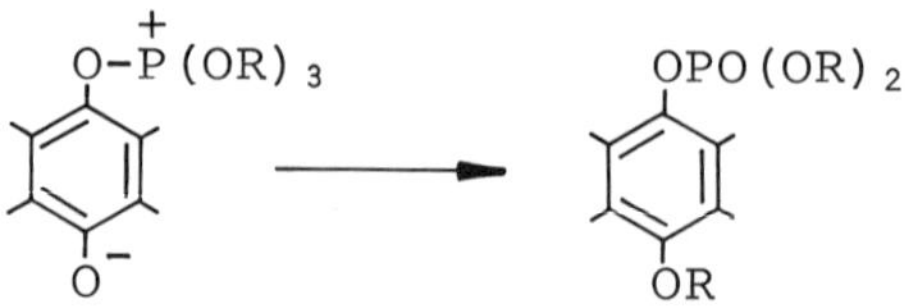

The reaction of o-quinones with phosphorous esters is in principle similar, but the intermediate addition product tends to cyclize yielding a cyclic pentaoxyphosphorane

derivative.[1159]   These phosphoranes are not always isolated;
they may undergo rearrangement or hydrolysis.

$$\text{C}=\text{O} + (RO)_3P \longrightarrow \overset{+}{\text{C}}\text{-O-P(OR)}_3 \longrightarrow \text{O-P(OR)}_3 \xrightarrow{\text{hydr.}} \begin{array}{c}\text{OPO(OR)}_2 \\ \text{OH}\end{array}$$

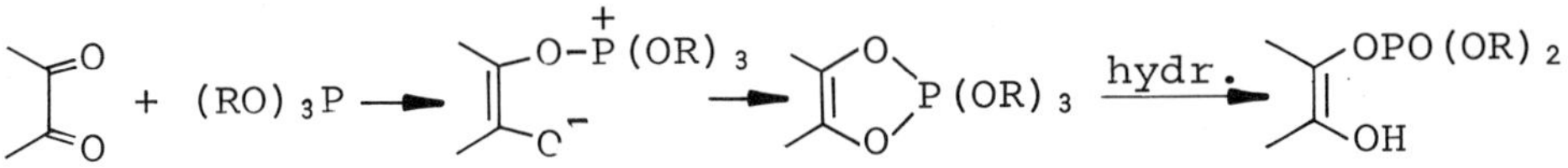

In this manner the corresponding derivatives have been prepared starting, e.g., from o-choranil,[1159] 1,2-naphtho-
quinone,[1159] phenanthrenequinone,[1165a] 4,5-pyrenequin-
one,[1159] 4,5,9,10-pyrenediquinone.[1159]

    α,β-Dicarbonyl compounds may behave similarly to o-
quinones.   Diacetyl reacted with trialkyl phosphites yielded
by rearrangement of the not isolated cyclopentaoxyphosphor-
ane a β-methoxyvinyl phosphoric ester, giving by mild hydrolysis an acetoin phosphate.[832]   Later the phosphorane
has been isolated (see Chapter 5A).[1165a]

$$\begin{array}{c}H_3CC=O \\ | \\ H_3CC=O\end{array} + (RO)_3P \xrightarrow{\text{fast}} \left[\begin{array}{c}H_3CC-O \\ \| \quad\quad P(OR)_3 \\ H_3CC-O\end{array}\right] \longrightarrow$$

$$(RO)_2PO[OCH(CH_3)=C(CH_3)OCH_3] \xrightarrow{\text{mild hydr.}} (RO)_2PO[OCH-$$

$$(CH_3)COCH_3]$$

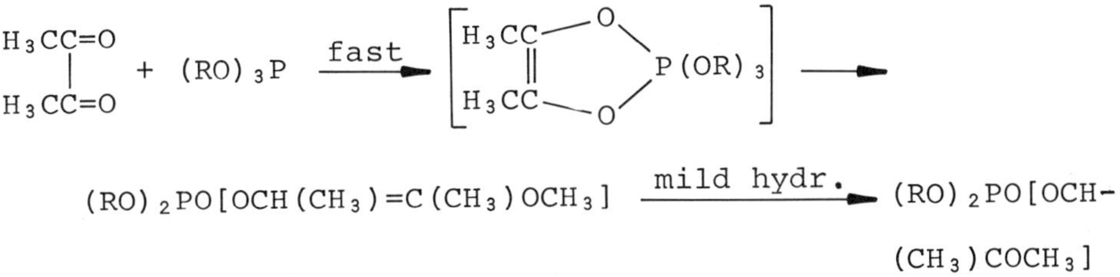

When using glyoxal the cyclic oxyphosphorane has been
isolated; by reaction with HCl or with acyl chlorides,
RCOCl (R = COCH₃, COCF₃, COC₆H₅, COCl), it yields derivatives of "sugarlike phosphates."[1167]

$$\begin{array}{c}HC=O \\ | \\ HC=O\end{array} + (CH_3O)_3P \longrightarrow \begin{array}{c}HC-O \\ \| \quad\quad P(OCH_3)_3 \\ HC-O\end{array}$$

$$\xrightarrow{\text{HCl}} \begin{array}{c}HCO-CH_2 \\ | \\ O-PO(OCH_3)_2\end{array} + CH_3Cl$$

$$\xrightarrow{\text{RCOCl}} \begin{array}{c}HCO-CH-COR \\ | \\ O-PO(OCH_3)_2\end{array} + CHCl_3$$

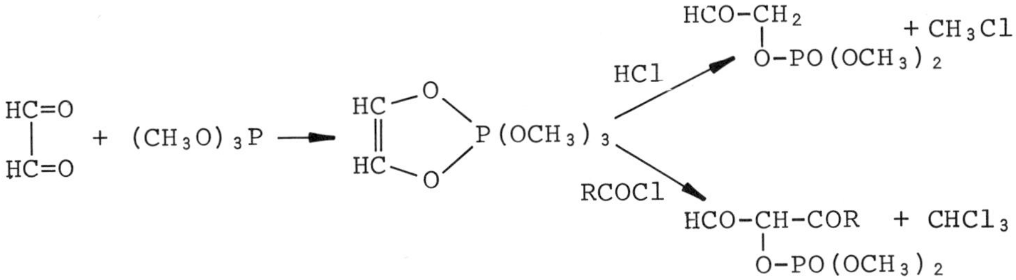

Depending on the conditions, α-diketones may react with
tertiary phosphites in the proportion 2:1 with formation
of new C-C bonds,[1166,1172] e.g., for diacetyl (see also
Chapter 5A):

$$
\begin{array}{c}
H_3CC=O \\
| \\
H_3CC=O
\end{array}
+
\begin{array}{c}
H_3CC\text{---}O \\
\| \\
H_3CC\text{---}O
\end{array}
P(OCH_3)_3
\xrightarrow{\text{slow}}
\begin{array}{c}
O\ \ CH_3 \\
\| \ \ | \\
H_3C\text{-}C\text{-}C \text{---} O \\
| \\
H_3C\text{-}C\text{-}C \\
\| \ \ | \ \ O \\
O\ \ CH_3
\end{array}
P(OCH_3)_3
$$

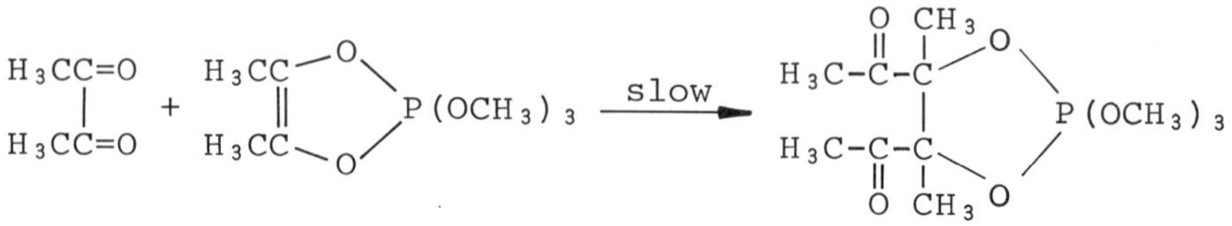

The component in the second step may be different from the
first, therefore this general reaction is extraordinary
versatile. Reacted, e.g., with methyl-pyruvate, the addition product of trimethyl phosphite to diacetyl, formulated
above, yields[614a]

$$
\begin{array}{c}
O\ \ CH_3 \\
\| \ \ | \\
H_3C\text{-}C\text{-}C \text{---} O \\
| \\
CH_3O\text{-}C\text{-}C \\
\| \ \ | \ \ O \\
O\ \ CH_3
\end{array}
P(OCH_3)_3
$$

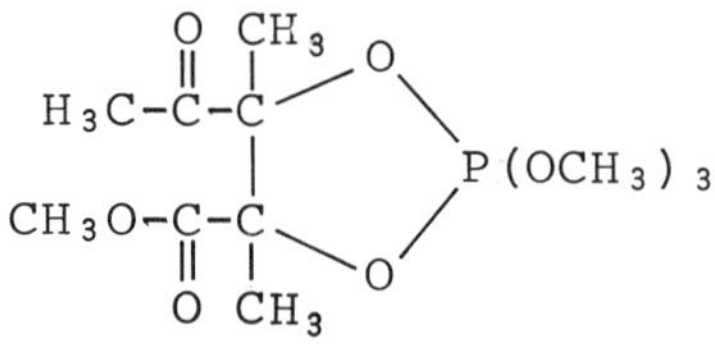

A review (from 1964) has been published by Ramirez.[1156a]
     Reacted with exactly 1 mole equ. of water under mild
conditions (e.g., in ether) the cyclic phosphoranes resulting from these reactions are generally hydrolyzed preferentially to cyclic ethylene or vinylen alkyl phosphates,
seldom to dialkyl substituted-alkyl phosphates by splitting
of the cycle.[614a] Under appropriate conditions, the monoalkyl cyclophosphates yield by hydrolysis the corresponding
cyclo acids (substituted 2-hydroxy-2-oxo-1,3,2-dioxaphos-
pholane).[614a]
     Alkyl pyruvate adds equally to trialkyl phosphites
(methyl and ethyl described) in two proportions, 2:1 and
1:1, depending on the conditions. In the cold, under
nitrogen, after some days, the 2:1 addition product (A)
is obtained; without cooling more of the 1:1 addition
product (B) results.[1148]

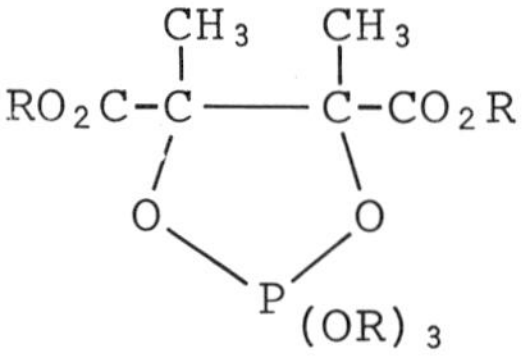

$$
\begin{array}{c}
CH_3\ \ \ CH_3 \\
| \ \ \ \ \ \ | \\
RO_2C\text{-}C \text{------} C\text{-}CO_2R \\
\diagup \ \ \ \ \ \diagdown \\
O \ \ \ \ \ \ \ \ \ O \\
\diagdown \ \ \diagup \\
P \\
(OR)_3
\end{array}
\quad (A)
\qquad
\begin{array}{c}
R \\
| \\
H_3CC\text{-}O\text{-}PO(OR)_2 \\
| \\
CO_2R
\end{array}
\quad (B)
$$

Dialkyl phosphorous acids add to ethyl mesoxalate in an
exothermic reaction yielding di(ethoxycarbonyl)methyl dialkyl phosphates, but when the reaction is run in the
cold, a phosphonate can be isolated which isomerizes at
100-120° (reaction catalyzed by C_2H_5ONa) to the phosphate.[1148a]

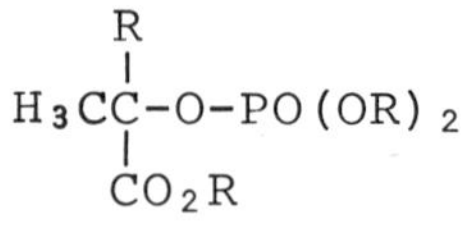

$$(C_2H_5O_2C)_2CO + OHP(OR)_2 \xrightarrow{\text{cold}} (C_2H_5O_2C)_2C(OH)PO(OR)_2$$

$$\xrightarrow{100\text{-}120°} (C_2H_5O_2C)_2CH\text{-}OPO(OR)_2$$

γ-2. α-Halocarbonyl Compounds. This most versatile reaction has been discovered in 1952 by Perkow.[1097a] Depending on the components and the conditions, the α-halocarbonyl grouping $-\underset{|}{\overset{}{C}}(Hal)-\overset{}{C}(=O)-$ --the carbonyl pertaining to an aldehyde, a ketone, or a carboxylic acid derivative-- reacts with tertiary phosphites yielding either phosphonates (Arbuzov reaction, the phosphite P atom adding onto the halogenated carbon which loses the halogen) or phosphates (Perkow reaction, the P atom adding onto the carbonyl oxygen, the halogen being lost again) as follows.

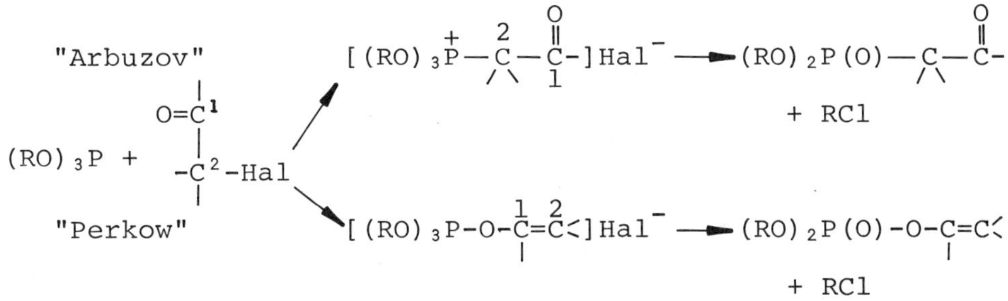

The Perkow type reaction in which we are interested here is favored by steric hindrance of the access to C-2 of the halocarbonyl group, marked positivation of carbonyl carbon atom C-1, difficulty of Hal to act as leaving group (Hal may be replaced by other leaving groups). Thus the production of vinyl phosphoric esters decreases in dependency on the halogen in the following order: Cl > Br > I (F). In the series of monochloro derivatives the order is: aldehydes > ketones > carboxylic esters; ethyl monochloroacetate yields only phosphonates, but monochloroacetyl chloride yields vinyl phosphates.[1145] Amides of α-halocarboxylic acids give vinyl phosphates with the exception of trichloroacetamides whose behavior is particular. Trichloroacetamide[1276a] and sometimes its N-alkylated derivatives yield trialkyl phosphates and trichlorovinylamines.[1276a]

$$(RO)_3P + Cl_3CC(O)NR'R'' \longrightarrow (RO)_3PO + R'R''NCCl{=}CCl_2$$

But N-acylated N-alkyl trichloroacetamides undergo with trialkyl phosphites the Perkow reaction yielding vinyl

phosphates. The acyl group may be cyclized on the alkyl
group, or replaced by a sulfonamide group. Therefore a
very large variety of vinyl phosphates is accessible by
this particular reaction:[1226a,1306]

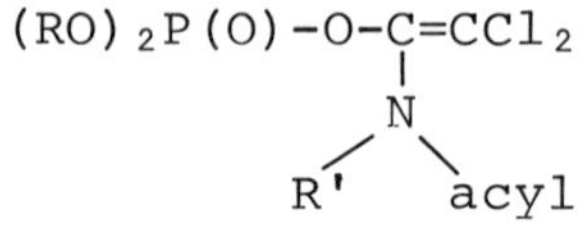

$$(RO)_2P(O)-O-C=CCl_2 \qquad R = alkyl \qquad R' = alkyl$$

$$
\begin{array}{c}
| \\
N \\
\diagup \quad \diagdown \\
R' \qquad acyl
\end{array}
$$

$$
\begin{aligned}
&acyl = CHO \qquad \left.\begin{array}{c} acyl \\ R' \end{array}\right\} = \begin{array}{c} CO \\ (CH_2)_{\overline{n}} \end{array}\!\!\diagdown X \\
& COalkyl \\
& CO_2alkyl \\
& SO_2N(alkyl)_2 \qquad X = CH_2, O, NH
\end{aligned}
$$

The general reaction of α-halocarbonyl compounds with tertiary phosphites has found very numerous applications under
various conditions (temperature, solvent, and period) in
the following fields.

α-Haloaldehydes, e.g., monochloroacetaldehyde,[20,564a]
dichloroacetaldehyde,[787,1097] chloral.[564,1097] Diethyl
propargyl phosphite, treated with chloral, yields diethyl
2,2-dichlorovinyl phosphate, $(C_2H_5O)_2(Cl_2C=CHO)PO$, by
elimination of the propargyl group of the phosphite.[1144]
The informations about the reaction of ethylene cyclic
phosphites with chloral are contradictory. Allen et al.[20]
have prepared methyl ethylene phosphate, $(CH_3O)-$
$\left(\begin{array}{c}CH_2O \\ CH_2O\end{array}\!\!\diagup\right)PO$, by reacting chloral with methyl ethylene phosphite, but a patent[1301] claims the production, starting
from chloral and various ethylene phosphites, $(RO)(C_2H_4O_2)-$
P, of 2-chloroethyl 2,2-dichlorovinyl phosphoric esters,
$(ClCH_2CH_2O)(Cl_2C=CHO)(RO)PO$ with R = methyl, hexyl, Cl_3-
CCH_2-, 4-chlorophenyl, and 2,4-dichlorophenyl. Other α-
haloaldehydes: bromal,[1096] 2,3-dichloropropionalde-
hyde,[1097] 2,3-dichlorobutyraldehyde.[1097]

α-Haloketones, e.g., simple haloketones like dichloro-
acetone $ClCH_2COCH_2Cl$,[1149b] fluorochloroacetones (chlorine
reacting), ClF_2COCCl_2F (chlorine of ClF_2C group reacting),[1289] ClF_2CCOCF_2Cl,[490,1456] $Cl_2FCOCFCl_2$,[1456] 1-chloro-
1-ethylthio-2-oxobutane and homologues [yielding 1-alkyl-
2-(alkylthio)vinyl dialkyl phosphates, e.g.,
$\left(\begin{array}{c}H_3C \\ H_3C\end{array}\!\!\diagdown CHO\right)_2PO(O-C-CH(SC_2H_5)]$,[254] α-chloro-α-phenylthioace-
$|$
C_2H_5
tone,[83] bromomethyl ethyl ketone (chloromethyl ethyl ketone
yields only phosphonate),[1149] ω-bromoacetophenone,[265,731]
ω,ω,ω-trichloroacetophenone,[787] different ω,ω-dichloro
and ω,ω,ω-trichloroacetophenones substituted in the phenyl
group (Cl, Br, OCH_3),[235,1103] 3,3-dibromocamphor,[20] yield
similar products.

Haloketones with other functional groups. Monobromo-
pyruvic acid reacted with tribenzyl phosphite yields di-
benzylphosphopyruvic acid, giving by hydrogenolysis phos-
phoenolpyruvic acid.[453]

$$BrCH_2COCOOH + (C_6H_5CH_2O)_3P \longrightarrow CH_2{=}C(CO_2H)OPO(OCH_2C_6H_5)_2$$

$$\longrightarrow CH_2{=}C(CO_2H)OPO_3H_2$$

Acetoacetate ester and amide derivatives: alkyl 2-chloro and 2,2-dichloroacetoacetates,[234,404] 2-chloroacetoacetate ester of mono-O-acetylglycol,[213] 2-chloro or bromo 2-alkylacetoacetate esters.[258] 2,2-Dichloroacetoacetic acid diethylamide (^{14}C labelled for the synthesis of labelled vinyl phosphates),[25a] 2-chloroacetoacetic acid methylamide,[1247] and 3-methoxypropylamide,[403] the N,O-dialkylhydroxylamides of the same acid.[234a]

α-Haloaliphatic derivatives: alkyl trichloroacetates,[564,596] trichloroacetic acid diethylamide,[564] trichloroacetyl chloride.[564,1145] As Pudovik et al.[1145] have found, this last compounds yields, depending on the experimental conditions, besides the phosphate a quantity (which may be superior) of a phosphonylated derivative: the dialkyl phosphoric ester of a phosphonylated vinyl alcohol, as follows:

$$2(C_2H_5O)_3P + CCl_3COCl \longrightarrow (C_2H_5O)_2P(O)\underset{\displaystyle PO(OC_2H_5)_2}{\underset{|}{C}}{=}CCl_2 + 2RCl$$

Monochloroacetyl chloride, α-bromoisobutyryl bromide, and monobromoacetyl bromide follow predominantly this anomalous route. Aliphatic and aromatic thioesters of α-chloromonothioalkanoic acids, alkyl-CHClCO(SR), yield in a normal Perkow reaction 1-alkyl(aryl)thiovinyl dialkylphosphates:[620,621,1437]

$$(RO)_2P(O)O{-}\underset{\displaystyle SR'}{\underset{|}{C}}{=}CH{-}R'' \qquad R = \text{alkyl} \qquad R'' = \text{H or alkyl}$$

$$R' = \text{alkyl or aryl}$$

Little mention is found on reaction between secondary phosphites and α-halocarbonyl compounds. The reaction is of a different type, by simple addition saturated tertiary phosphates may result, as follows.[272]

$$(CH_3O)_2PHO + CF_3COCF_3 \longrightarrow (CH_3O)_2P(O)OCH(CF_3)_2$$

But the Perkow reaction is claimed by another patent to result as follows:[1289]

$$(C_2H_5O)_2PHO + ClF_2CCOCFCl_2 \longrightarrow (C_2H_5O)_2P(O){-}O{-}\underset{\displaystyle CCl_2F}{\underset{|}{C}}{=}CF_2$$

γ-3. Vinyl Phosphates by Other Reagents than α-Halo-carbonyl Compounds. The halomercuri group may react similarly to the halogens. Trialkyl phosphites treated with 2-chloromercuriacetaldehyde or bromomercuri-acetone in non hydroxylic solvents yield vinyl phosphates as follows:[903]

$$(C_2H_5O)_3P + HalHgCH_2COR \longrightarrow (C_2H_5O)_2P(O)-O-CR=CH_2 +$$

$$C_2H_5Hal + Hg$$

$$R = H, CH_3$$

α-Oxo-toluensulfonates behave similarly.[468]

$$RCOCH_2OTs + (CH_3O)_3P \longrightarrow (CH_3O)_3P(O)-O-\underset{R}{C}=CH_2 + CH_3OTs$$

$$R = CH_3, C(CH_3)_3, C_6H_5$$

Monochloroethylene cyclocarbonate reacts slowly with trialkyl phosphites (4 days refluxing in benzene) partly by rearrangement into alkylphosphonate and partly yielding vinyl phosphate, as follows:[471a]

$$+ (RO)_3P \longrightarrow CH_2=CHOPO(OR)_2 + RCl + CO_2$$

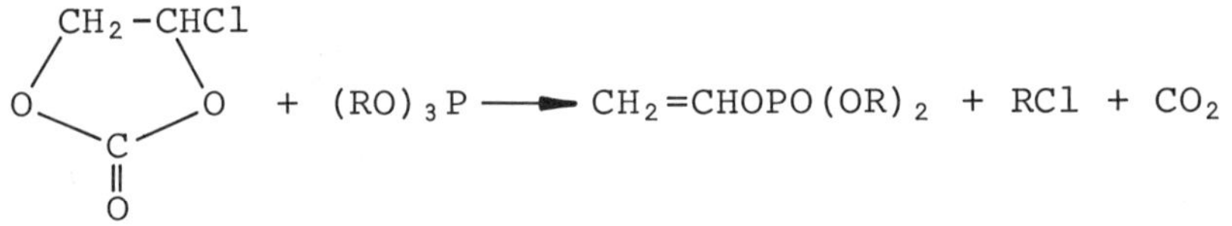

γ-4. Miscellaneous
Alloxan (but neither parabanic acid nor its N-1, N-3-dimethyl derivative) is phosphorylated by trimethyl phosphite like an o-quinone or an α,β-diketone.[1160]

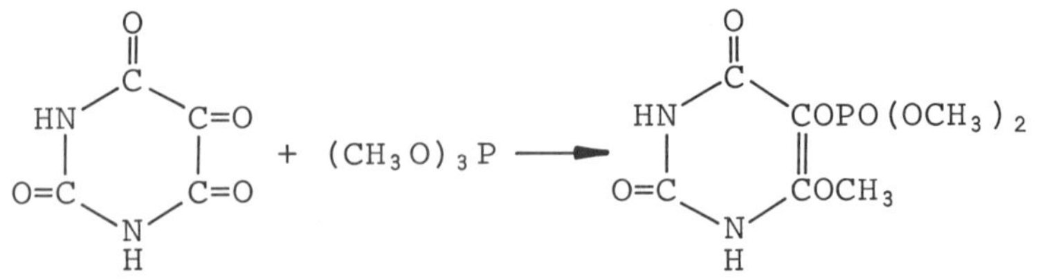

Using alloxan hydrate, the vinyl derivative (A) (30%) and the 5-dimethylphosphate of 5-hydroxybarbituric acid (40%) are isolated.

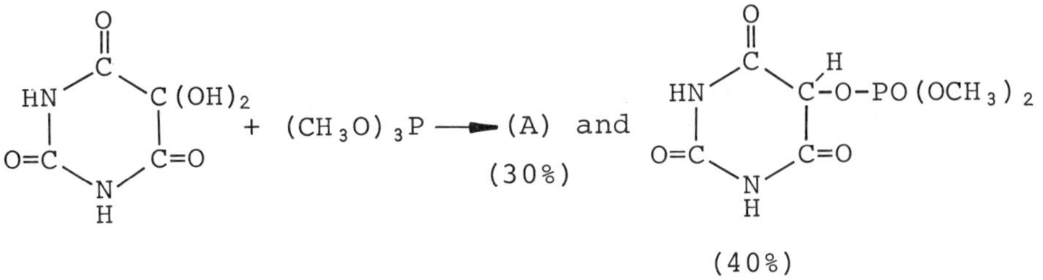

Alloxan hydrate reacted with dimethyl phosphorous acid yields only the latter product.

In contrast to the oxidation reaction of tertiary phosphites by epoxides (cf. above under β-3), a Russian patent[545] claims the production of a tertiary phosphate ester by an addition reaction between phosphorus trichloride and 3-vinylcyclohexene oxide. In the abstract no explanation is given of the oxidation mechanism.

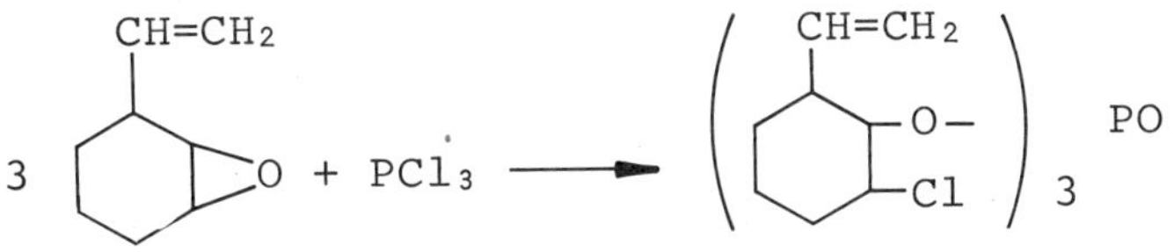

An interesting observation has been made by Denney et al.[469] When reacting $(C_2H_5O)_3P$ with ethyl peroxide in ether at room temperature, NMR-spectra show the presence of pentaethoxyphosphorus (not isolable) which decomposes in solution on standing at room temperature yielding finally after three weeks only triethyl phosphate. Trimethyl phosphite behaves similarly.

$$(C_2H_5O)P + (C_2H_5O)_2 \longrightarrow (C_2H_5O)_5P \longrightarrow (C_2H_5O)_3PO$$

+ decomposition products

Reactions similar to that of α-halocarbonyl compounds are offered by α-halo-α-nitrosoalkanes[607,917] and α-halo-α-nitro compounds $(CCl_3NO_2,$[18,19] $CH_2ClCHClNO_2,$[18,19] $CH_3-CClNO_2CH_3,$[18] $CH_3CBrNO_2CH_3,$[18,19] $HCClNO_2COOR$[251]) which yield with tertiary phosphites oximo esters instead of vinyl esters, the double bond being created with participation of the nitrogen atom.

The halonitroso derivatives react as follows.

$$\begin{matrix} Cl \\ ON \end{matrix}\!\!>\!\!C\!\!<\!\!\begin{matrix} R' \\ R'' \end{matrix} + P(OR)_3 \longrightarrow (RO)_2P(O)ON\!\!=\!\!C\!\!<\!\!\begin{matrix} R' \\ R'' \end{matrix} + RCl$$

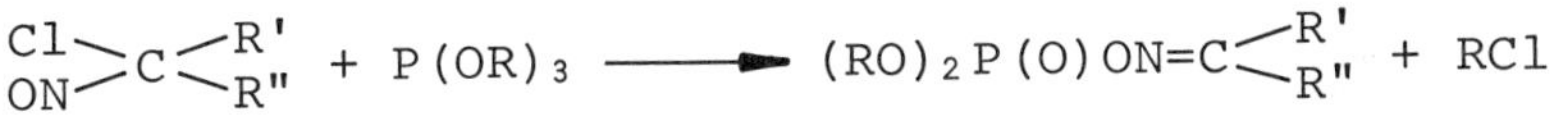

The halonitro derivatives react with two molecular equivalents of phosphite, one oxidized to phosphate and the other yielding the oximo ester

$$\begin{matrix} Hal \\ O_2N \end{matrix} > C < \begin{matrix} R' \\ R'' \end{matrix} + 2P(OR)_3 \longrightarrow (RO)_3PO + (RO)_2P(O)ON{=}C < \begin{matrix} R' \\ R'' \end{matrix} + RHal$$

II. CYCLIC ESTERS

General Remarks

Cyclic esters of phosphoric acid can be obtained from P^V compounds in two types of general reactions: (1) Creation, in one operation, of the double ester link between diol and phosphoric acid (or a phosphoric monoester); (2) Successive creation of the two ester links, i.e., cyclization of a derivative $...P^V \begin{matrix} \diagup OH \\ \diagdown ORX \end{matrix}$ with an ester grouping ORX presenting a functional group X capable of yielding an ester link with the phosphoryl group of the same molecule.

This classification is artificial inasmuch as the production of more than one ester link in one operation will be in reality the result of successive chemical reactions. But for practical purposes this classification will be adopted here.

An interesting source of phosphoric acid cyclic esters is represented by reactions of phosphorous compounds conducive either to phosphorous esters oxidizable to the corresponding phosphoric esters, or--by condensation with appropriate bifunctional compounds--to cyclic pentaoxyphosphorane derivatives yielding under certain conditions cyclic phosphates (the obtention of non cyclic phosphoric esters from these cyclic pentaoxyphosphoranes is mentioned above in Section I.2.γ-1).

Nomenclature

The systematic names of the cyclic phosphoric esters are to be build up in accord with the IUPAC rules. For the simpler cycles we have the following names:

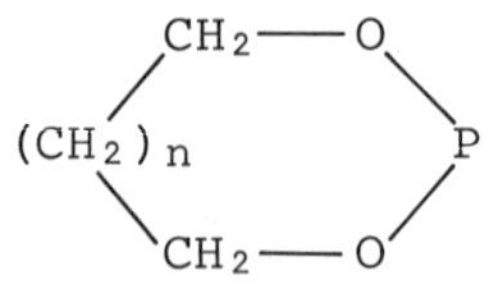

n = 0 : 1,3,2-dioxaphospholane

n = 1 : 1,3,2-dioxaphosphorinane

n = 2 : 1,3,2-dioxaphosphorepane

The cyclic phosphoric ester of 1,3-propanediol, e.g.,

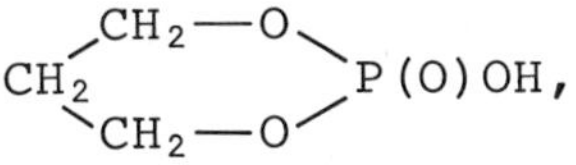, is 2-hydroxy-2-oxo-1,3,2-dioxaphosphorinane.

1. Reaction of Diols with Appropriate P^V Derivatives

a. With Phosphorus Oxychloride. Dihydroxy derivatives, mostly aliphatic and alicyclic diols, are reacted with phosphorus oxychloride in the presence or without an inert solvent. If necessary the reaction is moderated by cooling; a tertiary amine may be added as HCl acceptor or pyridine used as solvent.[130a,231,500,978,1318b,1327,1383] The reaction yields cyclic phosphorochloridates which afford the corresponding acids or their esters when treated in appropriate manner with water or an alcohol (the cyclic ester grouping is easily split in acid or alkaline medium).

Pentagonal and hexagonal cyclic esters are prepared readily by this procedure; higher cycles seem to be quite exceptional. Starting from 2,2'-dihydroxybiphenyl the obtainment of a dibenzophosphorepane system has been reported.[500a]

Thus, tris-hydroxymethyl-nitromethane has yielded 1-oxo-4-nitro-2,6,7-trioxa-1-phospha-bicyclo[2.2.2]octane (A),[1503] and cis-1,3,5-cyclohexanetriol has afforded 1-oxo-1-phospha-2,8,9-trioxaadamantane (B).[1318b]

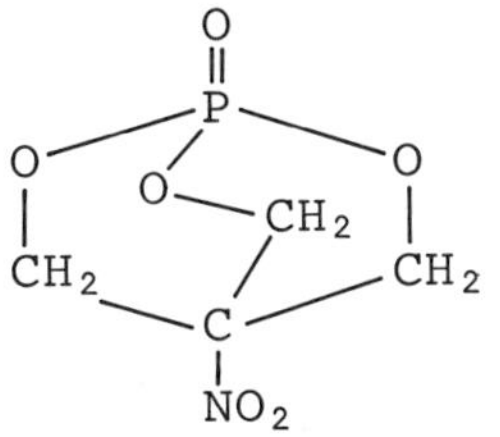

(A)

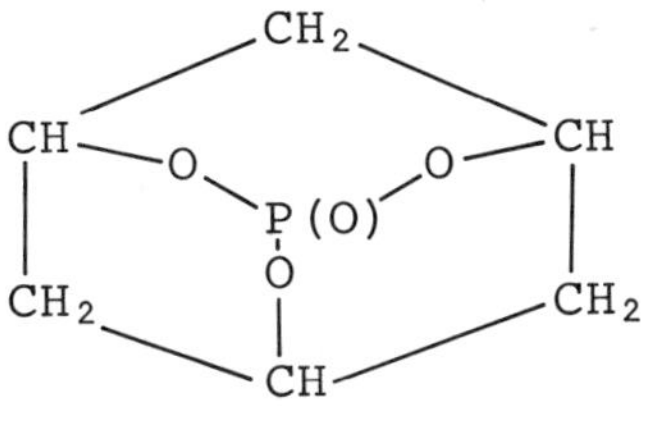

(B)

Pyrocatechol affords, depending on the ratio of the reactants, the phosphochloridate (C) or the neutral ester (D).[35a,280]

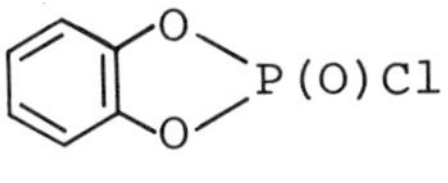

(C)

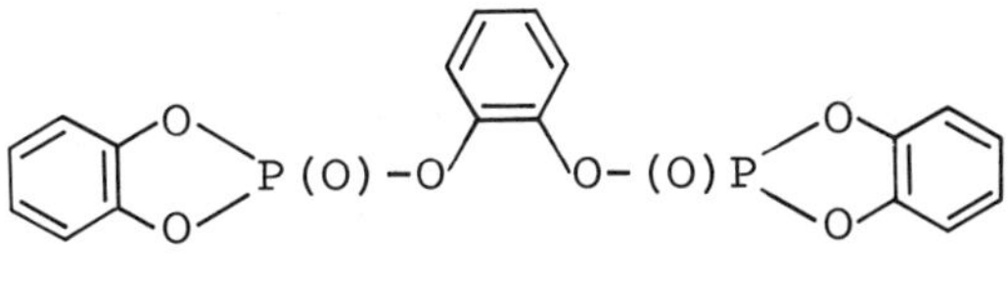

(D)

A cyclic ester-anhydride prepared in similar manner by

reacting salicylic acid with phosphorus oxychloride may be mentioned here: the phosphorochloridate (E) is capable of yielding the esters (F).[1107a]

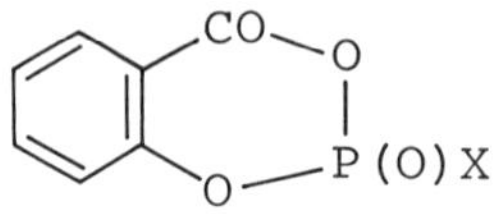

(E) : X = Cl (F) : X = OR

b. With Alkyl or Aryl Phosphorodichloridates ROP(O)-Cl_2. In a quite similar reaction diols with the hydroxyl groups in appropriate position yield the corresponding esters of cyclic x-methylene phosphoric acids, $RO(O)P{<}^{O}_{O}{>}R'$. Using a phosphorodichloridate with a radical R easily eliminated, the corresponding cyclic acids $R'{<}^{O}_{O}{>}PO_2H$ may be obtained.

Thus benzyl β-D-glucoside reacted with phenyl phosphorodichloridate afforded the corresponding 4,6-phenyl cyclophosphate yielding after dephenylation and debenzylation β-D-glucose 4,6-phosphate.[1221] Alkyl (and phenyl) phosphorodichloridates have been used to prepare 2-alkyloxy-2-oxo-1,3,2-benzodioxaphosphorinanes by reaction with orthohydroxybenzyl alcohol or its derivatives substituted in the nucleus.[1075,1281]

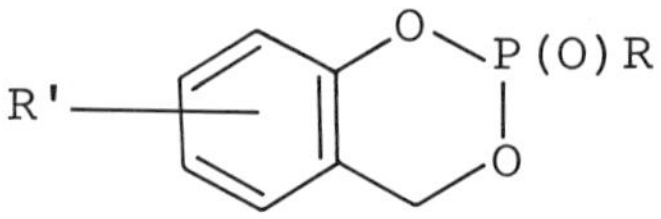

c. With Polyphosphoric Acid or Phosphorus Pentoxide. This direct procedure seems to be of very limited application. By heating pyrocatechol in excess with phosphorus pentoxide at 80-100° and distillation in vacuo o-phenylene cyclic phosphoric acid is obtained with an excellent yield calculated on P_2O_5, the excess pyrocatechol being recovered during distillation[599] (the nature of the distilled product had not been established by the author who isolated after crystallization from water o-hydroxyphenyl phosphoric acid resulting from the cyclic ester by hydrolysis). The same cyclic derivative is produced by heating the dihydric phenol with orthophosphoric acid at 200°: At this temperature polyphosphoric acid is produced which yields the monophosphoric ester of pyrocatechol, cyclized at the temperature used.[385]

Pyrophosphoric or polyphosphoric acid caused to react with hydroxy(or chloro)alkanols under appropriate conditions yields n-methylene cyclic phosphoric acids, the

intermediate hydroxy(or chloro)-alkyl phosphoric acid
being cyclized.

$$XCH_2(CH_2)_{n-2}CH_2OH \xrightarrow{\quad H_4P_2O_7 \quad} XCH_2(CH_2)_{n-2}CHOPO_3H_2 \longrightarrow$$

X = OH, Cl

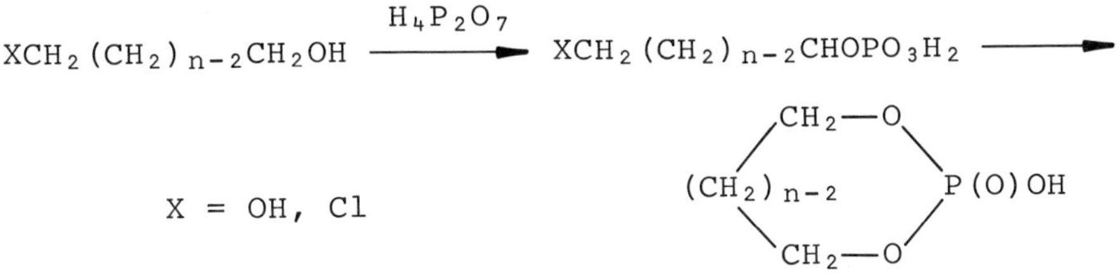

The yields are good with $ClCH_2CH_2OH$ (n = 2; phospholane)
and $ClCH_2CH_2CH_2OH$ (n = 3; phosphorinane) but much smaller
with the corresponding diols; the derivative with heptagonal cycle (phosphorepinane) is obtained only in very
small quantity and the octogonal derivative (n = 4) only
in traces (not isolated).[383a]

d.   With Phosphorus Pentachloride. This possibility
seems to have been explored only with aromatic derivatives. Pyrocatechol yields in addition to polymerized
products the trichloride (G)[37a] which is converted into
(C) by treatment with anhydrous oxalic acid;[35a] (C) can
be used for the preparation of the corresponding acid or
its esters.

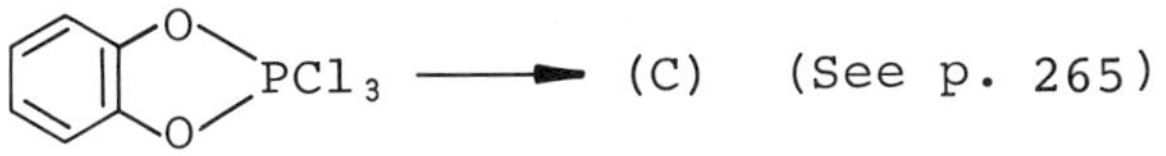

(G)

As early as 1859 Couper has examined the reaction between
methyl salicylate and PCl_5,[438] but the structure of the
resulting product (cyclic or not?) is not ascertained.
Much later (Pinkus[1107a]) it has been shown that salicylic
acid yields by this reaction a 2,2,2-trichloro-benzophos-
phorinane derivative which can be converted to the 2-
chloro-2-oxo derivative, useful as starting material for
the corresponding acid or its esters.

e.   By Transesterification with o-Phenylene Cyclic
Phosphoric Acid. This very particular reaction has been
applied to glycerol: heated at 90° with this cyclic
phenolic ester, glycerol yields 1,2-glycerol cyclic

phosphate, demonstrating once more the particular ease of formation of pentagonal cyclic esters.[1328] By a combined phosphorylation and transesterification, prednisolone, prednisone, and hydrocortisone reacted with $C_6H_4O_2POCl$ yield the corresponding o-hydroxyphenyl 17,21-cyclophosphates, the primary hydroxy group being attacked in the first step.[623a]

2. Cyclization of Phosphoric Acid Monoesters

a. Halogenoalkyl Phosphoric Acids. As emphasized under Section 1.c, the pentagonal and hexagonal cycles are most easily produced. This reaction can be realized by heating in neutral aqueous solution; examples:

cycle: pentagonal[833a] $BrCH_2CH_2OPO_3^{2-} \longrightarrow$

$$\begin{array}{c} CH_2-O \\ | \diagdown \\ CH_2-O \diagup \end{array} POO^- + Br^-$$

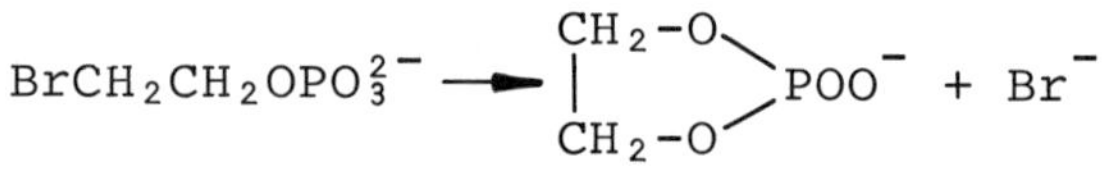

cycle: hexagonal[159a] $ClCH_2CH(OH)CH_2OPO_3^{2-} \longrightarrow$

$$(HO)HC\underset{CH_2-O}{\overset{CH_2-O}{\diagup\diagdown}} POO^- + Cl^-$$

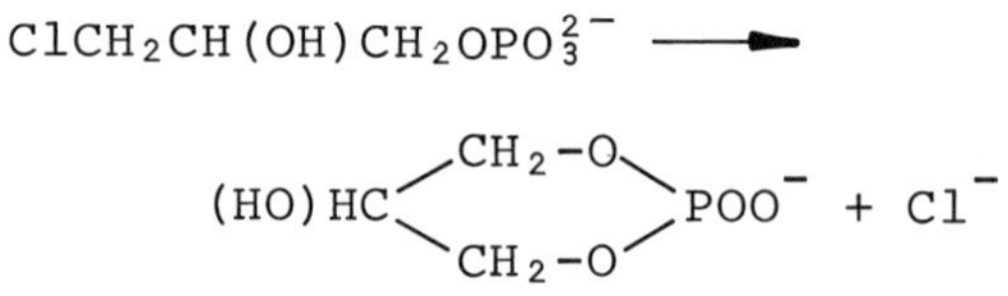

This cyclization results equally in anhydrous medium from the action of polyphosphoric acid at 100° during a sufficient period (20 hr) on the haloalkyl phosphoric acid.[383a]

b. Hydroxyalkyl Phosphoric Acids. Cyclizations of these acids, analogous to those mentioned above, have been realized:

By simple heating, e.g., 2-hydroxyethyl phosphoric acid heated in vacuo at 80° during several days yields ethylene cyclophosphoric acid, isolated as methyl ester;[383a] uridine 2'(3')-monophosphate heated in dimethylformamide at 150° yields some uridine 2',3'-cyclic phosphate.[991]

By action of carbodiimides. This procedure has proved particularly useful.[790,1383] Generally dicyclohexylcarbodiimide is used; the water-soluble cyclohexyl-β-(N-methylmorpholino)ethylcarbodiimide p-toluensulfonate has been proposed for ribonucleotide cyclization.[1354] Generally, five-membered rings are formed. With cyclophosphoric acids with higher rings, further action of the carbodiimide may lead to bi-cyclic pyrophosphates.[500a]

In the nucleotide field, by action of cyclohexylisocyanate,[458] N,N'-dicyclohexyl-4-morpholinocarboxamide,[580] P¹-diphenyl-P²-morpholinopyrophosphorochloridate[727] (all in pyridine or 2,6-lutidine solution).

By action of trifluoroacetic anhydride, used first for the preparation of 2',3'-cyclic phosphates,[296] and later for the cyclization of simpler compounds, e.g., 2-hydroxycyclohexyl phosphoric acid.[231]

By action of alkyl chlorocarbonates in the presence of a tertiary amine.[862,958] A patent claims the cyclization of ribonucleic acids in formamide plus liquid ammonia by action of a suspension of sodium amide in liquid ammonia.[1255]

By enzymic cyclization[520] (action of "active albumin fractions") saligenin cyclic phosphates have been prepared from saligenin mono-diarylphosphates.

3. Cyclic P^V Esters from Phosphorous Derivatives

a. From Cyclic Phosphites by Direct or Indirect Oxidation. Just as the phosphites, $(RO)_3P$, the cyclic phosphites, $(RO_2)P(OR')$, may yield phosphates by oxidation (cf. Section I.2.b.α). Oxidation is achieved preferentially by non-aqueous reagents since water tends to cleave the ring particularly if it is pentagonal. Besides MnO_2 and yellow mercuric oxide[122a,775a] N_2O_4 in methylene chloride or petrol ether has been shown to be successful.[442a,442b,775a,1100a] Bicyclic phosphites seem to be particularly resistant to cleavage. As examples the production of 4-alkyl-1-oxo-2,6,7-trioxa-1-phosphabicyclo[2.2.2]octanes by oxidation(oxygen or hydrogen peroxide) of bicyclophosphites can be cited.[700,1045a,1427a]

The cyclization-oxidation of 5-bromouridine 2'(3')-phosphite by hexachloroacetone, yielding 5-bromouridine 2',3'-cyclic phosphate, has been reported.[694]

The reaction of cyclic phosphates with chloral seems to be not quite unequivocal (cf. Section I.γ-2). In some instances the production of vinyl cyclophosphates has been described, especially with spirocyclic diphosphites.[20]

b. Via Pentaoxyphosphoranes

α. Trialkyl Phosphites and Vicinal Dicarbonyl Compounds
This very versatile reaction is due essentially to Ramirez and co-workers (since 1957);[1164] Russian authors have
published analogous reactions about the same time.[832]
These reactions will be mentioned here only as far as
they lead to cyclic phosphoric esters, the possibilities
they offer for creation of new C-C bonds being neglected.
(Reviews by Ramirez[1156a] and Chapter 5B.) The general
feature can be presented as follows. Compounds with
vicinal carbonyl functions add trialkyl phosphites (generally trimethyl phosphite used) yielding unsaturated
2,2,2-trialkoxy-1,3,2-dioxaphospholanes.

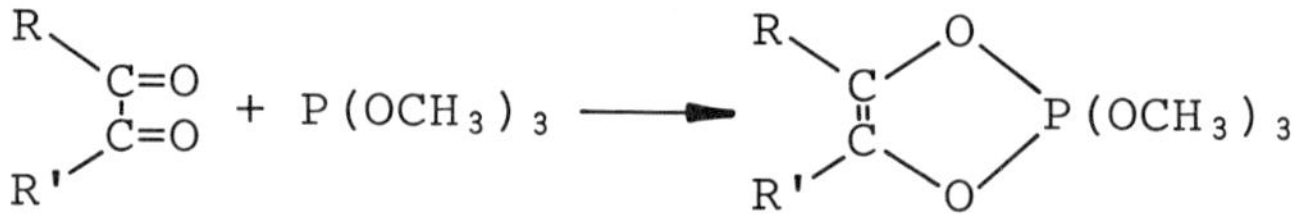

$$\begin{array}{c} R \\ \diagdown \\ C=O \\ C=O \\ \diagup \\ R' \end{array} + P(OCH_3)_3 \longrightarrow \begin{array}{c} R \diagdown O \\ C \diagdown \\ \| P(OCH_3)_3 \\ C \diagup \\ R' \diagup O \end{array} \qquad (A)$$

This reaction takes place at 60° (under nitrogen), followed by distillation. Exemplified in the case of diacetyl (R = R' = CH$_3$), the most studied diketone, the following easily proceeding reactions yield saturated tri-
alkyldioxaphospholanes:[1157,1161,1166,1170-1173]

With aliphatic or aromatic aldehydes, with rearrangement

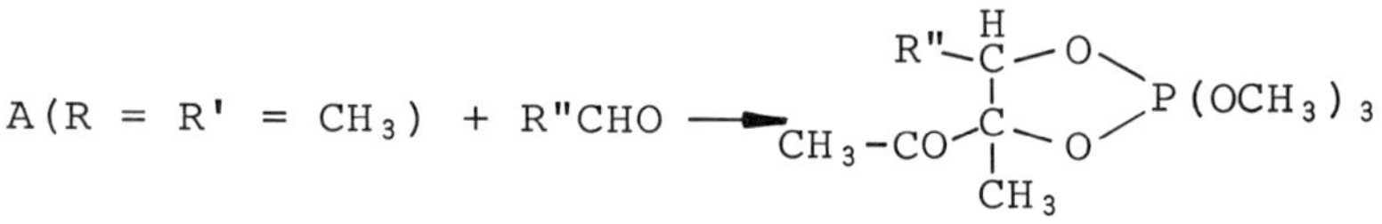

$$A(R = R' = CH_3) + R''CHO \longrightarrow \begin{array}{c} R''\!-\!\overset{H}{\underset{|}{C}}\!-\!O \diagdown \\ \diagdown P(OCH_3)_3 \\ CH_3-CO-\underset{|}{C} \diagup O \\ CH_3 \end{array}$$

(different stereoisomers)

With a second molecule of diacetyl

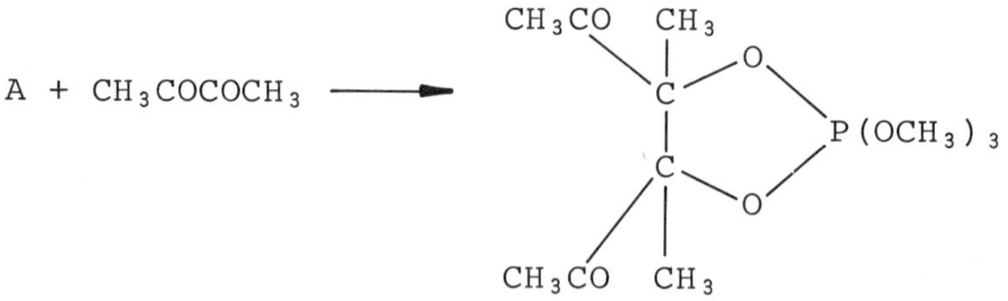

$$A + CH_3COCOCH_3 \longrightarrow \begin{array}{c} CH_3CO CH_3 \\ \diagdown | O \\ C \diagdown \\ | P(OCH_3)_3 \\ C \diagup \\ \diagup | O \\ CH_3CO CH_3 \end{array}$$

(different stereoisomers)

With ketones, with rearrangement; e.g.,

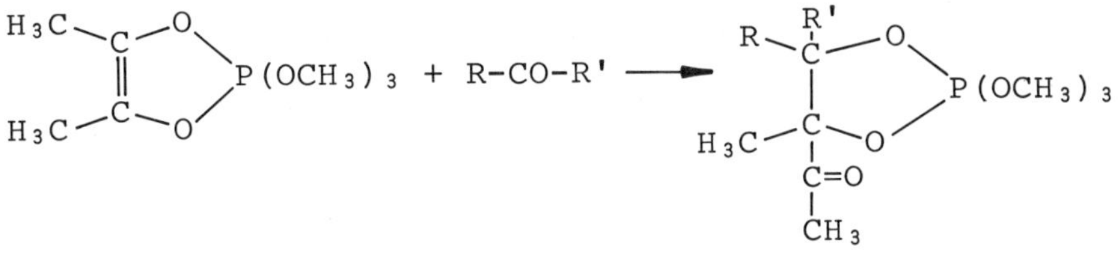

(different stereoisomers)

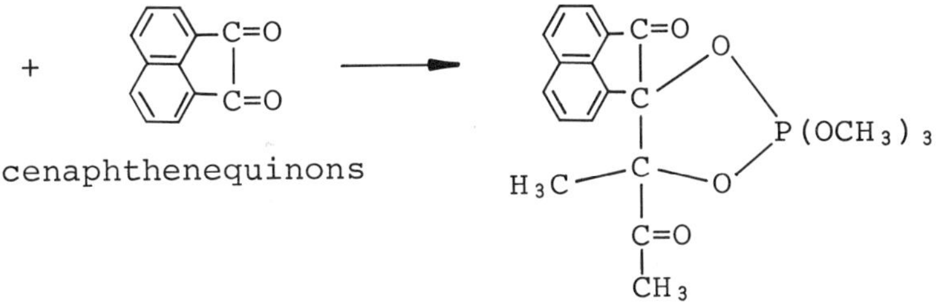

acenaphthenequinons

(stereoisomers)

With ketene, with rearrangement

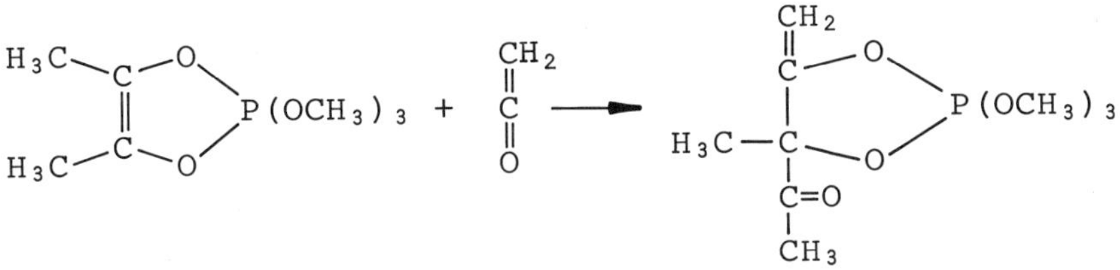

Similarly, Kirillova et al.[802] have prepared an unsaturated trialkoxy-dioxaphospholane by reacting methylphenylglyoxal with trialkylphosphites.

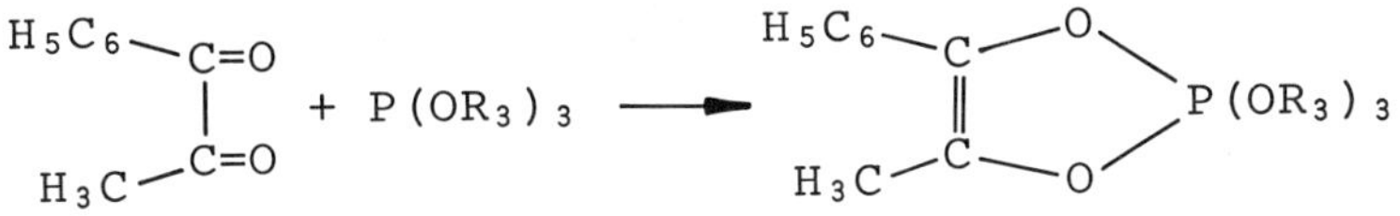

with R = CH_3, C_2H_5, n-C_3H_7, n-C_4H_9

1,2-Cyclohexadione yields the corresponding benzo derivative (Kuhtin et al.).[832] With amidodialkyl phosphites, Sandalova et al.[1223] have prepared similarly 2-dialkylamido-2,2-dialkoxy-1,3,2-dioxaphospholenes which yielded the corresponding trialkoxy derivatives when treated with the corresponding alcohol.

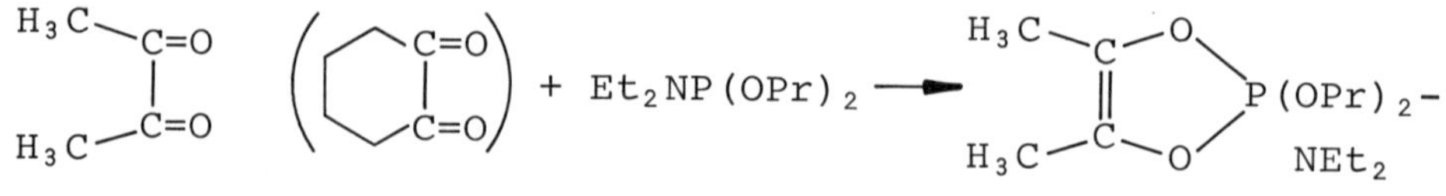

PrOH, steam bath, 30 min

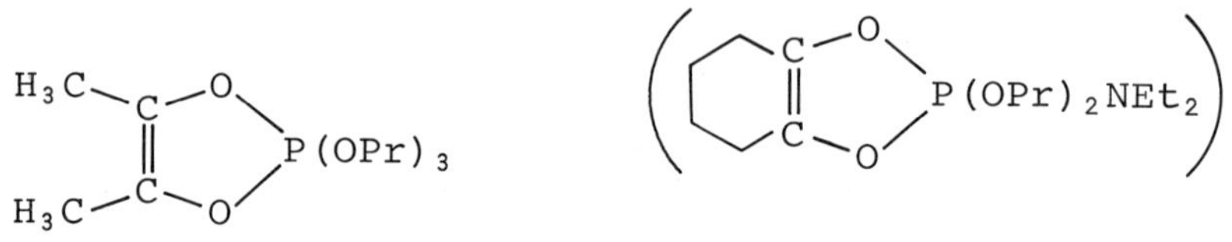

All these cyclic pentagonal pentaoxyphosphorus derivatives are of course very easily attacked by water or alcohols which tend to split the cycle yielding non cyclic phosphoric esters, and finally an organic phosphorus-free component besides simple alkyl or dialkyl phosphates, but under appropriate conditions--cautious treatment with one equivalent of water in aprotic solvents--the trialkoxy grouping on the P atom is degraded to an alkoxy oxo grouping, i.e., to a cyclic phosphoric ester:[1166]

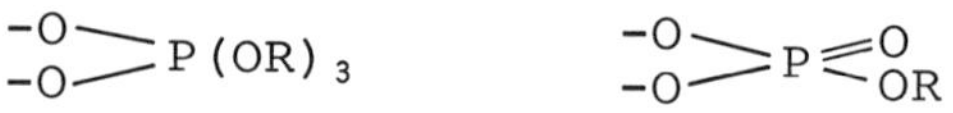

β. Condensation of Trialkyl Phosphites with o- and p-Nitrobenzaldehyde

Reaction of o- or p-nitrobenzaldehyde with trialkyl phosphites (methyl and ethyl used) yields already at ordinary temperature 1,3,2-dioxaphospholanes by simultaneous condensation between the carbonyl C atoms, e.g.:[1162]

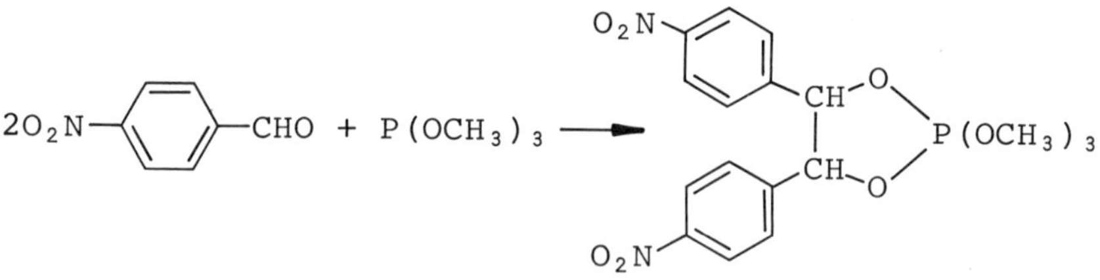

The resulting 2,2,2-trialkoxy derivative can be subjected to the mentioned degradation reactions.

This condensation seems to be rather limited: meta-nitrobenzaldehyde, chlorobenzaldehydes and in general aromatic aldehydes with electron-releasing groups do not react at 20°.

γ. Condensation of Triethyl Phosphite with Hexafluoroacetone

This reaction, similar to the preceding, yields the following cyclic pentaoxyphosphorane:[1207]

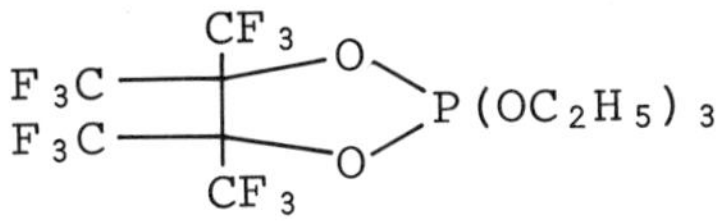

III. ESTERS OF ANHYDRO ACIDS WITH ANORGANIC OXYACIDS

1. Esters of Polyphosphoric Acids

Anhydro acids between phosphoric acid and phosphorus acids other than phosphoric acid seem to be known in a very limited manner and to be of small importance. The problem of phosphorous-phosphoric anhydride derivatives, $(RO)_2P-O-P(O)(OR)_2$ (distinct from hypophosphoric derivatives), is examined in Chapter 13. In the present section, these esters will be mentioned only insofar as they can serve as phosphorylating agents.

a. Creation of P-O-C Links

α. Direct Esterification

A very simple procedure is based on the classical reaction between a metal salt of the acid considered and an alkyl halide, realized by heating of the components. Generally silver salts are used.[350,417,647,1061,1213]

$$Ag_4P_2O_7 + 4RI \longrightarrow (RO)_2P(O)-O-P(O)(RO)_2 + 4AgI$$

The same procedure has been applied to silver[843,1439] or lead[321] metaphosphate. Due to the inhomogeneity of the starting material the resulting metaphosphoric esters are mixtures.

In a similar reaction bromosugar derivatives treated with triethyl ammonium tribenzyl pyrophosphate yield after hydrogenolytic debenzylation the corresponding sugar pyrophosphate.[1346]

Esterification by action of diazoalkanes seems not to have been studied (cf. the splitting action of diazomethane on certain phosphoric ester bonds; see Section I.1.f.ζ).

β. Cleavage of P-O-P Groupings

β-1. Polycondensed Phosphoric Acids and Alcohols (Phenols). The reaction of polycondensed phosphoric acids (polyphosphoric acid, metaphosphoric acid, phosphorus pentoxide) with hydroxy compounds yields in its final stage a mixture in which primary and secondary esters of orthophosphoric acid predominate (cf. Section I.1.a). Careful limited alcoholysis should give esters of lower condensated polyphosphoric acids but a definite individual compound has never been isolated.[6] The use of polyphosphoric acids as phosphorylating agents of certain phosphoric acid monoesters (e.g., nucleotides) shall be mentioned later (cf. Section b.ε).

β-2. Phosphorus Pentoxide and Ethers or Orthoesters. The reaction between phosphorus pentoxide and ethers leading to alkyl metaphosphates, first described by Langheld,[842,843] has been modified (heating, with agitation, in the presence of a halogenated hydrocarbon, e.g., chloroform),[1317] but its application is very limited: only ethyl ether seems to have been successful; diphenyl ether does not react and dibenzyl ether forms merely polymers.[1430] The nature of the resulting products seem to depend on the exact conditions of the experiment. Thilo et al.[1453] have found by chromatographic analysis in their product no cyclic polyphosphates but only open chain derivatives:

$C_2H_5O(HO)P(O)OPO_3H_2$; $(C_2H_5O)_3PO$; $C_2H_5O[P(O)(OH)O]_2P(O)-(OH)OC_2H_5$; $[HO(C_2H_5O)P(O)]_2O$; $HO[P(O)(OC_2H_5)O]_2P(O)(OH)-OC_2H_5$; $(C_2H_5O)_2PO_2H$; $(C_2H_5O)_2P(O)OP(O)(OC_2H_5)(OH)$; $[(C_2H_5O)_2P(O)]_2O$.

Burkhardt et al.[310] conclude that their "ethyl metaphosphate" was a mixture of bicyclic and monocyclic tetrametaphosphates, trimetaphosphate and linear tetraphosphate ethyl esters.

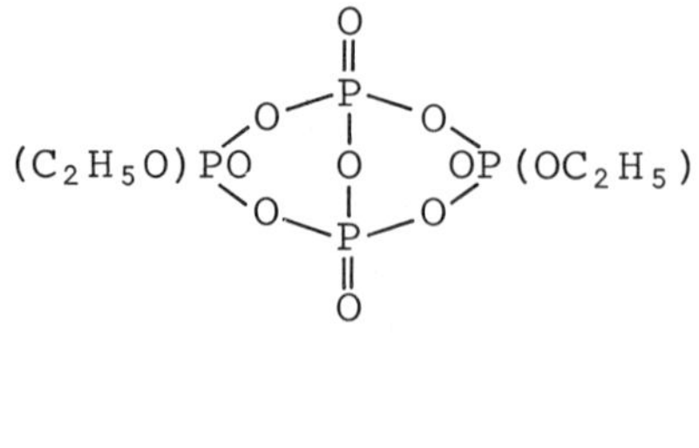

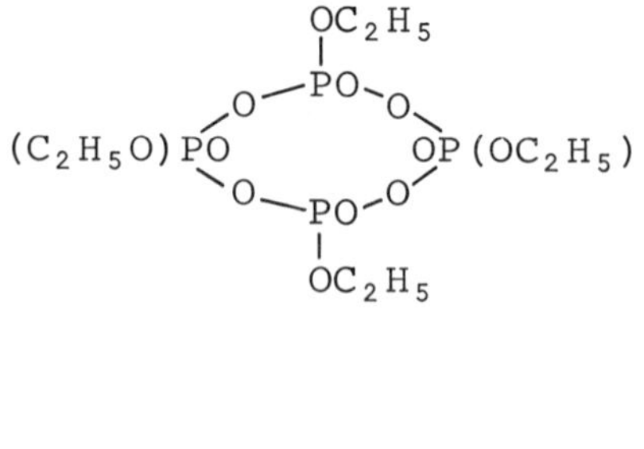

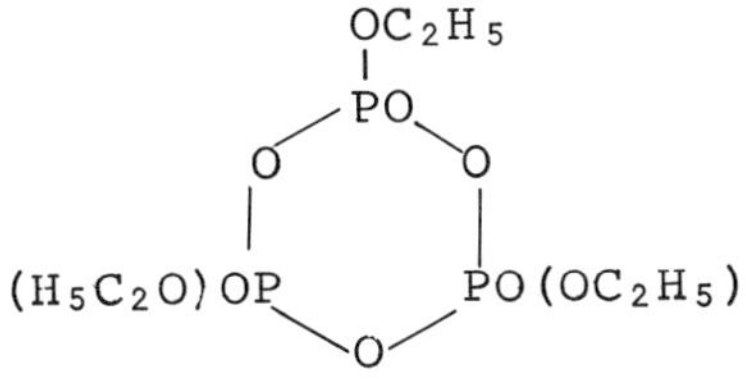

$(C_2H_5O)_2P(O)OP(O)(OC_2H_5)$-

$OP(O)(OC_2H_5)OP(O)(OC_2H_5)_2$

The reaction between phosphorus pentoxide and ethyl orthoformate in large excess (several hours at 125°) yields after distillation in vacuo a mixture containing besides ethyl formate and the excess of orthoformate some triethylphosphate and an important fraction of tetraethyl pyrophosphate.[269]

b. Creation of P-O-P Links

α. Direct Condensation ...POH + HOP... → P-O-P

α-1. Reaction with Disubstituted Carbodiimides. As mentioned above (see Section I.1.d.α-1) disubstituted carbodiimides have found a large application for condensation reactions between acid functions of various acids of phosphorus, in particular between acid functions of phosphoric acid mono and diesters. The reaction is conducted in the classical manner in anhydrous basic medium (often pyridine with or without a trialkyl amine, or acetone with a tertiary amine) at room temperature during a convenient period (several hours to 3 days), generally with dicyclohexylcarbodiimide. The reaction is realized either with one determined substance, and leads then to symmetric pyrophosphate diesters or tetraesters (e.g., Refs. 112, 114, 435, 695), or with two different components. In this latter case asymmetric pyrophosphate esters result: e.g., condensation of a nucleoside 5'-phosphate with orthophosphoric acid yields the corresponding 5'-pyrophosphate,[721a,1265] and with an excess of phosphoric acid 5'-triphosphate;[721a,1265] condensation of a nicotinamide mononucleotide with deoxyadenosine 5'-phosphate yields a diphosphopyridine nucleotide analog;[696] similarly nicotinamide-6-mercapto(or 6-methylthio)purine dinucleotides have been prepared.[1102]

α-2. Reaction with Other Organic Agents. A large number of organic agents have been shown to be capable of condensing two HO functions of partial phosphoric esters. In some cases intermediary products have been isolated but nevertheless we shall class these agents in the section of direct condensating agents.

Reaction of dialkyl phosphoric acids with ethyl vinyl ether, $C_2H_5OCH{=}CH_2$, (e.g., in ether in the presence of mercuric oxide with a small quantity of boron trifluoride as catalyst at room temperature during several days[661]) yields tetraalkyl pyrophosphates.[1011]

Diethyl phosphorochloridite, $(C_2H_5O)_2PCl$, and tetraethyl pyrophosphite, $(C_2H_5O)_2P\text{-}O\text{-}P(OC_2H_5)_2$, act similarly.[1011]

Salts of primary phosphoric esters (preferentially with tertiary organic bases) refluxed or heated to 80° in ether, acetonitrile, or dioxane with a keten dimer (generally the simplest one used) until no more CO_2 is evolved, yield the corresponding diprimary pyrophosphoric esters with good yields.[88,1011]

$$\begin{array}{l}CH_2{=}C{-}O\\ \quad\ |\quad\ \ |\\ \quad CH_2{-}C{=}O\end{array} + 2ROPO_3H_2 \xrightarrow{\text{base}} ROP(O)(HO)\text{-}O\text{-}P(O)(OH)OR + CH_3COCH_3 + CO_2$$

This reaction yields pyrophosphoric tetraesters when starting with secondary phosphates.[88] Nitriles are capable of provoking the transformation of primary phosphoric esters into symmetric diprimary pyrophosphates. With acetonitrile the reaction is slow and incomplete, but in the presence of a reactive halide the yield is good,[407] e.g.,

$$(C_6H_5O)PO_3Ag_2\,(\text{suspended in } CH_3CN + CH_3OCH_2Cl) \xrightarrow[\text{24 hr}]{\text{reflux}} (C_6H_5O)P(O)(OAg)\text{-}O\text{-}P(O)(OC_6H_5)(OAg)$$

With trichloroacetonitrile, Cl_3CCN, reaction of a phosphoric acid monoester is rapid and nearly quantitative.[407,442e] The probable intermediate is an imidoyl ester which phosphorylates the primary phosphate. (In section ε phosphorylation of phosphate by imidoyl phosphates will be examined.) Secondary phosphoric acid esters, $(RO)_2PO_2H$, do not react (no imidoyl ester produced?).[407]

$$RC{\equiv}N + HOP(OR')O_2H \longrightarrow \begin{array}{l}R\text{-}C{=}NH\\ \quad\ |\\ \quad O\text{-}P(OR')O_2H\end{array} \xrightarrow{R'OPO_3H_2} R'OP(O)(OH)\text{-}O\text{-}P(O)(OH)OR'$$

A similar reaction intervenes with cyanic acid esters, $ROC{\equiv}N$, as follows:[916]

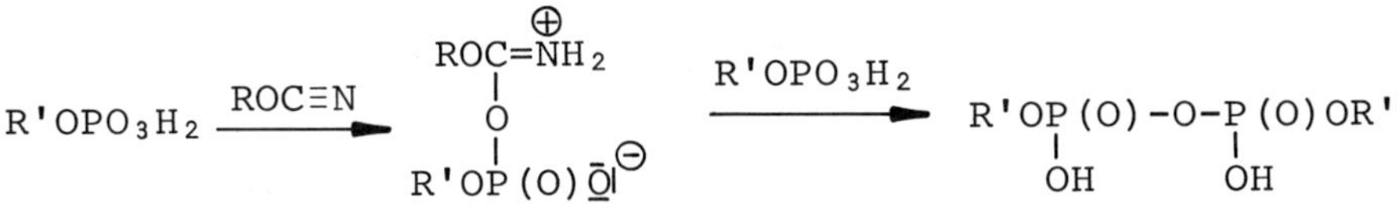

$$R'OPO_3H_2 \xrightarrow{ROC\equiv N} \begin{array}{c} \overset{\oplus}{ROC=NH_2} \\ | \\ O \\ | \\ R'OP(O)\bar{O}^{\ominus} \end{array} \xrightarrow{R'OPO_3H_2} \begin{array}{c} R'OP(O)-O-P(O)OR' \\ | \qquad\qquad | \\ OH \qquad\quad OH \end{array}$$

(not isolated)

With an excess of cyanic ester, in alcoholic solution, the pyrophosphate is activated and reacts with the alcohol, R"OH, yielding a secondary phosphate, (R'O)(R"O)-PO_2H.[916]

Dialkyl cyanamides (isomers of the carbodiimides) condense two molecules of a partial phosphoric acid ester yielding the corresponding symmetric pyrophosphate esters. Here again an imidoyl phosphate is the intermediary.[780a]

Ketene imides, $Ar_2C=C=NAr'$, may equally serve for the preparation of pyrophosphates, starting from phosphates, but the complexity of the synthesis of the reagent seems to limit its practical application.[454a]

Arenesulfonyl halides, already mentioned for the esterification of phosphoric acid (cf. Section I.1.d.α-2), may be used for the condensation of two molecules of phosphoric acid derivatives. Tetraphenyl and tetrabenzyl pyrophosphate have been obtained by action of p-nitrobenzene-sulfonyl chloride or p-toluenesulfonyl chloride on diphenyl and dibenzyl phosphoric acid, respectively.[433]

The condensation of primary phosphoric esters to symmetric diprimary pyrophosphates results from the action of 2,4-dinitrofluorobenzene in the ratio of 2:1 in pyridine at room temperature (10 days); a similar reaction results in the presence of a strong tertiary base (triethyl amine, e.g.). The supposed intermediates are in the first case a dinitrophenyl (DNP) alkyl phosphate ion, $DNPOP(O)(OR)O^-$, in the second case a fluorophosphate ion, $FP(O)(OR)O^-$; in each case the intermediary reacts with an ion $ROPO_3H^-$, giving the diprimary pyrophosphate ion ROP(O)(O^-)-O-P(O)(O^-)OR.[1465] The global reaction is represented as follows:

$$2ROPO_3H_2 + FC_6H_3(NO_2)_2 \longrightarrow ROP(O)(OH)\text{-}O\text{-}P(O)(OH)OR + HF + HOC_6H_3(NO_2)_2$$

Tetraphenyl and tetrabenzyl pyrophosphates have been prepared in excellent yield by reaction of carbon disulfide with the silver salts of the corresponding secondary phosphates [perhaps via the following intermediate: $(RO)_2P(O)\text{-}O\text{-}C(=S)\text{-}SAg$], as follows:[98]

$$2(RO)_2PO_2Ag + CS_2 \longrightarrow [(RO)_2PO]_2O + COS + Ag_2S$$

A singular reaction has been reported between a potassium alkyl methyl phosphate in excess and an O,O,S-trialkyl thiophosphate, leading by several hours of boiling in butyl ether to diprimary pyrophosphates.[685]

$$\underset{\text{(in excess)}}{(C_6H_5O)P(O)(OCH_3)OK} + (C_6H_5O)P(O)(OCH_3)SCH_3 \longrightarrow [(C_6H_5O)P(O)(OK)]_2O$$

Phosphorylation reactions via carbonium, phosphonium, and sulfonium intermediates, respectively, have been described (Todd).[775] The onium compound $\overset{+}{X}-\overline{Y}$ reacts as follows,

$$\overset{+}{X} + {}^-OP(O)(OR)OH \longrightarrow XOP(O)(OR)OH \xrightarrow{{}^-OP(O)(OR)OH} HO(RO)(O)P-O-P(O)(OR)OH$$

this last step and the further transformation of $\overset{+}{X}$ depending on its nature. With diphenyldichloromethane in a polar solvent the reaction proceeds as follows, with $[(C_6H_5)_2CCl]^+Cl^-$ as onium compound.

$$(C_6H_5)_2CCl_2 + 2(RO)PO_3H_2 \longrightarrow HO(RO)(O)P-O-P(O)(OR)OH + C_6H_5COC_6H_5 + 2HCl$$

(With an excess of diphenyldichloromethane the mixture phosphorylates an added alcohol.)

The phosphonium and sulfonium ions result respectively from triphenylphosphorus dichloride, $(C_6H_5)_3PCl_2 \longrightarrow [(C_6H_5)_3PCl]^+Cl^-$, or triphenylphosphine + polyhalomethane, (CCl_4, CBr_4), or diphenylsulfur dichloride + triphenylphosphine. Thus, e.g., the following reaction proceeds in 4 hr at 25° in acetonitrile or methyl nitrate:

$$(C_6H_5)PCl_2 + 2C_6H_5OPO_3H_2 \longrightarrow C_6H_5O(HO)(O)P-O-P(O)(OH)OC_6H_5$$

α-3. Reaction with Anorganic Agents. Here we find essentially sulfur compounds.

Dibenzyl phosphoric acid refluxed during 18 hr in benzene with thionyl chloride yielded tetrabenzyl pyrophosphate in 25% yield.[921] Treating 5'-O-trityl-thymidine + 3'-O-acetylthymidine 5'-monophosphate with 3-5 equivalents of thionyl chloride in dimethyl formamide at room temperature in view of the production of a 3' - 5' nucleotide link, the authors found after 22 hr besides the nucleotide 36-38% of P^1,P^2-di(3'-O-acetylthymidine) 5'-pyrophosphate.[730]

A patent claims the preparation of quaternary pyrophosphate esters by heating trialkyl phosphates with alkyl not higher than C_8 or lower triaryl phosphates with various sulfur halocompounds: $SOCl_2$, SO_2Cl_2, $SOBr_2$, SO_2Br_2, S_2Cl_2, e.g., as follows:[221]

$$2(RO)_3PO + SOCl_2 \longrightarrow R_4P_2O_7 + SO_2 + 2RCl$$

β. Rearrangement Reactions of P-O-P Links

β-1. Thermal Decomposition of Polyphosphates. Tetraalkyl pyrophosphates are decomposed or dissociated on heating into trialkyl phosphates and alkyl metaphosphates.[179] The claimed reversibility of this reaction has not been confirmed.

$$(RO)_2P(O)\text{-}O\text{-}P(O)(OR)_2 \longrightarrow (ROPO_2)_n + (RO)_3PO$$

The temperature necessary varies with the size of the alkyl group. Ethyl pyrophosphate dissociates rapidly at about 200°, the methyl derivative is cleaved at 130°.[179] As the metaphosphates produced decompose at such temperatures in olefins and metaphosphoric acid,[178,179] the reaction seems to be more interesting for the preparation of trialkyl phosphates from crude metaphosphate or pyrophosphate.

β-2. Reaction of Trialkyl Phosphates with Phosphorus Pentoxide. Phosphorus pentoxide heated with trialkyl phosphates until complete dissolution forms esters of polyphosphoric acids. The result is always a mixture. Depending on the proportions used crude products of the following types of neutral polyphosphoric acid esters may be expected.[6]

Pyrophosphates: $4(RO)_3PO + P_2O_5 \longrightarrow 3(RO)_4P_2O_3$
Triphosphates: $5(RO)_3PO + 2P_2O_5 \longrightarrow 3(RO)_5P_3O_5$
Tetraphosphates: $2(RO)_3PO + P_2O_5 \longrightarrow (RO)_6P_4O_7$
Metaphosphates: $(RO)_3PO + P_2O_5 \longrightarrow 3/n\ (ROPO_2)_n$

A patent describes a procedure yielding tetraethyl pyrophosphate sufficiently pure to be distilled: the reaction product of $2(C_2H_5O)_3PO + P_2O_5$ obtained at 50° (approximatively crude tetraphosphate) is heated 1 to 5 hr at 120-160° with triethyl phosphate, and then fractionated *in vacuo*.[657]

γ. Reactions with Halo-Oxygen Compounds of P^V

γ-1. Phosphorus Oxychloride. Phosphorylation of partial esters of phosphoric acid by phosphorus oxychloride seems to have been seldom used. A patent claims the synthesis of 3-creatinyl-5-methyl-5-(2-pyrophosphoryloxyethyl)-thiazoline by action of phosphorus oxychloride on the 5-(2-hydroxyethyl)-thiazoline derivative in glacial acetic acid (one week at 0°).[815]

Several patents describe the preparation of tetraalkyl pyrophosphates by reaction of trialkyl phosphates in excess with phosphorus oxychloride at 130-150°; these reactions yield mixtures with 30-45% tetra-alkyl pyrophosphate which can be separated by distillation in vacuo or molecular distillation.[656,658,1334]

γ-2. Halophosphoric Esters. Halophosphoric esters can yield pyrophosphates (and even higher condensed phosphates) in two ways: condensation with a mono or diesterified phosphoric acid (or a salt of this acid) (reaction A) or elimination of alkyl halide between the halophosphate and a trialkyl phosphate (reaction B).

(A) $(RO)(R'O)POCl + H(M)OPO(OR'')_2 \longrightarrow (RO)(R'O)P(O){-}O{-}P(O)(OR'')_2 + H(M)Cl$

(B) $(RO)_2POCl + (R'O)_3PO \longrightarrow (RO)_2P(O){-}O{-}P(O)(OR')_2 + R'Cl$

Reactions of type (A) have been realized by various procedures. Heating in pyridine or benzene + pyridine of diallyl phosphorochloridate with diallyl phosphoric acid yields tetraallyl pyrophosphate,[915] similarly P^1-diphenyl-P^2-dibenzyl pyrophosphate is prepared from diphenyl phosphorochloridate and dibenzyl phosphoric acid;[433] using the silver salt instead of the free acid the reaction is achieved in 10 min at 125°.[921] The reaction between the chloridate and the acid can be favored by the addition of a tertiary amine[1248,1365] or the use of tetramethyl ammonium salts.[1090] Monoalkyl phosphates react similarly; thus steroid monophosphates have been phosphorylated by action of diphenyl phosphorochloridate, yielding P^1-steroid P^2-diphenyl pyrophosphate.[1194]

This type of reaction has been applied to the synthesis of nucleoside pyrophosphates[130,132,1356] and, by repetition of the procedure, triphosphates.[130,628] With benzyl groups present as R' and R" debenzylation by hydrogenolysis (cf. Section I.1.e.α-2) or by reaction with N-methylmorpholine (cf. Section I.1.e.α-4) yields the monoesterified pyrophosphate or triphosphate.[628,954,1495a] Morpholine has equally been used as protecting group of phosphoric HO functions, the morpholino function being easily split by mild acid hydrolysis.[723,1495a] Triphosphates may be obtained from monophosphates by condensation of a phosphorochloridate with a pyrophosphoric derivative. In this manner a nucleoside monophosphate has been transformed into the triphosphate by condensation of the benzyl phosphorochloridate with a salt of P^1,P^2-dibenzyl pyrophosphate,[628] or by condensation of a nucleoside monophosphate with P^1-diphenyl P^2-morpholino pyrophosphorochloridate.[723] Uridine diphosphate glucose, identical with the natural coenzyme, has been synthesized by condensation of 2',3'-dibenzyluridine 5'-benzylphosphorochloridate with the trioctyl ammonium salt of α-D-glucose 1-phosphate in benzene in the presence of tributyl amine, followed by hydrogenolysis.[960a]

By reaction of monoethyl(phenyl) phosphorodichloridate with diethyl(phenyl) phosphoric acid in ether or benzene and pyridine, pentaethyl and pentaphenyl triphosphate have been prepared.[1261]

$$(RO)POCl_2 + 2(RO)_2PO_2H \longrightarrow (RO)_5P_3O_5$$

A particular application of reaction (A) is based on controlled hydrolysis of halophosphate esters. The free acid resulting from splitting of an HCl molecule condenses with a second molecule of halophosphate. The chlorophosphate in the presence of a tertiary amine is reacted with a limited amount of water (removal of the hydrogen chloride by evacuation is less favorable).[1360,1362] The replacement of the base by acrylonitrile has been proposed.[248] By treating an equimolecular mixture of two different dialkyl phosphorochloridates, the mixed pyrophosphates have been obtained in yields sometimes superior to 50%.[1049]

$$(RO)_2POCl + (R'O)(R''O)POCl + H_2O \xrightarrow[\text{at } -5°]{\text{ether, } N(C_2H_5)_3} (RO)_2P(O){-}O{-}P(O)(OR')(OR'')$$

Ethyl phosphorodichloridate, $C_2H_5OPOCl_2$, agitated with one mole-equiv. of water under light vacuum yielded a mixture containing besides ethyl polyphosphates some P^1,P^2-diethyl pyrophosphoric acid.[1020]

A similar reaction results from heating of aryl phosphorodichloridates, $ArOPOCl_2$, with anhydrous oxalic acid to 75-85°, yielding metaphosphates. The two halogens of the phosphorodichloridates are eliminated not as HCl but as oxalyl chloride.[28,53,94,437]

$$(RO)POCl_2 + (COOH)_2 \longrightarrow ROPO_2 + (COCl)_2$$

By this reaction o-chloroformylphenyl phosphorodichloridates yielded metaphosphates which could be purified by distillation.[28,53,437] This fact makes a revision of their constitution desirable.

Reactions of type (B). The elimination of alkyl halide between a P-Cl group and a phosphate ester group can take place by thermal decomposition of the dialkyl phosphorochloridates which contain the two groupings. This leads to metaphosphates.[179]

$$(RO)_2POCl \longrightarrow 1/n\ (ROPO_2)_n + RCl$$

Thermal decomposition of a mixture of dialkyl phosphorochloridate and trialkyl phosphate leads to pyrophosphates.[647,1302]

$$(RO)_2POCl + (RO)_3PO \xrightarrow{140\text{-}150^\circ} (RO)_2P(O)\text{-}O\text{-}P(O)(OR)_2 + RCl$$

Using triethyl phosphate and secondary phosphorochloridates with higher R (phenyl or 2-ethylcyclohexyl), the mixed pyrophosphate resulting from elimination of ethyl chloride is produced in good yield.[821]

δ. Reactions with Phosphamides

In a phosphoric acid or ester monoamide the P-NRR' group is liable to yield a P-O-P derivative by reaction with the HO function of a phosphoric acid derivative (Khorana, Todd).

$$\cdots \underset{\displaystyle OH}{P^V}(O)NRR' + HOP^V\cdots \longrightarrow \cdots\underset{\displaystyle OH}{P^V}(O)\text{-}O\text{-}P^V\cdots + HNRR'$$

The mechanism of this reaction is based perhaps on the following equilibrium formulated for benzyl phosphoramidic acid.[409]

$$C_6H_5CH_2OP(O){<}^{NH_2}_{OH} \rightleftharpoons C_6H_5CH_2OP(O){<}^{\overset{+}{N}H_3}_{O^-} \text{ (acylating agent)}$$

By autocondensation, benzyl phosphoramidic acid yields slowly dibenzyl pyrophosphate.

$$C_6H_5CH_2OP(O){<}^{\overset{+}{N}H_3}_{O^-} + {}^{HO}_{H_2N}{>}P(O)OCH_2C_6H_5 \longrightarrow C_6H_5CH_2O\underset{\displaystyle ONH_4}{P}(O)\text{-}O\text{-}$$

$$\underset{\displaystyle NH_2}{P}(O)OCH_2C_6H_5 \xrightarrow{H_2O} \text{diammonium } P^1,P^2\text{-dibenzyl pyrophosphate}$$

This very versatile reaction is realized by reacting a phosphoric ester monoamide with phosphoric acid or a phosphoric acid derivative with at least one acid function in solution or suspension in an appropriate solvent (dioxane, dimethyl formamide, pyridine, or a mixture) in the presence of a base (mostly tertiary amines) at room temperature or at 100° during the period necessary for the reaction, ranging from hours to days. The nature of the amido function can vary; bases used: ammonia,[360,362,409] hydrazine,[287] morpholine (very often, e.g., Refs. 300, 301, 902, 1211), cyclohexylamine, piperidine, p-anisidine,[983a] imidazole.[617a] As phosphoric acid component let us mention the following: phosphoric acid,[902,1488] phenyl phosphoric acid (possibility of ulterior elimination of the phenyl group),[983] sugar phosphates,[300,301,810,811,1211] pyrophosphoric acid (yielding triphosphates).[979,983a,1058,1496] Tri and higher phosphates result equally from treatment with a large excess of phosphorylating agent derived from orthophosphoric acid; e.g., adenosine 5'-phosphate treated by two mole-equiv. of triethyl ammonium phosphoroamidate at room temperature during 16 hr yielded 24% adenosine 5'-diphosphate but after treatment with 10 mole-equiv. of the same agent during 72 hr a mixture resulted containing 27% of diphosphate, 21% of triphosphate, and 17% of higher phosphates.[360] Sometimes, the reaction is very rapid: benzyl phosphoroguanidate heated 3 min in dioxane with phenyl phosphate yielded P^1-benzyl P^2-phenyl pyrophosphate (diammonium salt).[412]

The method has found wide application in the field of nucleotides (e.g., synthesis of ribo and deoxyribonucleoside 5'-triphosphates,[979] ADP and ATP,[360,362] cytidine 2'-phosphate 5'-diphosphate (via the 2',3' cyclic phosphate derivative by enzymatic scission of the cyclophosphate grouping),[1401] purine and pyrimidine 5'-diphosphate sugars[1211]). Steroid polyphosphates and nucleoside diphosphate steroids have equally been prepared.[1058] In the field of aliphatic compounds, one can mention the preparation of isopentenyl(3-methyl-3-butenyl) pyrophosphate (reaction of isopentenyl phosphoroamide with the addition product of phosphoric acid with dioxane in pyridine solution)[1488] and of DL-mevalonic acid pyrophosphate (reaction of 3-hydroxy-3-methyl-hex-5-enyl phosphoromorpholidate with phosphoric acid followed by periodic acid oxidation of the 5,6 double bond yielding the terminal carboxyl function).[902]

ε. Phosphorolysis of Vinyl and Imidoyl Phosphates

The common feature of these esters is the presence of a double bond on the carbon linked to the phosphoric oxygen: ···P-O-C=C and ···P-O-C=N. In each case the ester bond is split by action of an (alkyl or aryl) phosphate ion.

$$\cdots P{-}O{-}C{=} + {}^{-}O{-}P\cdots \longrightarrow \cdots P{-}O{-}P\cdots + {}^{-}O{-}C{=}$$

<u>Vinyl phosphates</u>. Cyanovinyl phosphoric acid yields pyrophosphates when treated with phosphoric acid (which may be partially esterified).[537,538]

$$CNCH{=}CHOPO_3H_2 + HOP(OR)O_2H \longrightarrow H_2O_3P{-}O{-}P(OR)O_2H + CNCH_2CHO$$

Similarly react 2,2-dichloro-1-ethoxyvinyl diethyl phosphate, $CCl_2{=}C(OC_2H_5)OP(O)(OC_2H_5)_2$, and 2-ethoxycarbonyl-1-ethoxyvinyl diethyl phosphate, $H_5C_2O_2CCH{=}C(OC_2H_5)OP(O)(OC_2H_5)_2$, e.g., as follows:[443,444]

$$CCl_2{=}C(OC_2H_5)OP(O)(OC_2H_5)_2 + HO_2P(OC_6H_5)_2 \longrightarrow (C_2H_5O)_2P(O){-}O{-}P(O)(OC_6H_5)_2 + HCCl_2CO_2C_2H_5$$

<u>Imidoyl phosphates</u>. In the first reaction of this type described N-phenylbenzimidoyl dibenzyl phosphate (prepared from N-phenylbenzimidoyl chloride and silver dibenzyl phosphate) was reacted with a secondary phosphoric acid ester,[95] e.g., as follows:

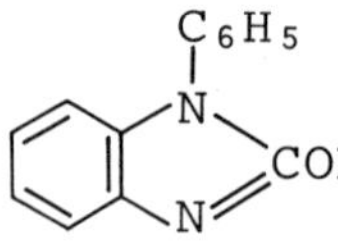

$$COP(O)(OCH_2C_6H_5)_2 + AgOP(O)(OCH_2C_6H_5)_2 \longrightarrow (C_6H_5CH_2)_4P_2O_7$$

Similarly P[1]-dibenzyl P[2]-diphenyl pyrophosphate (using diphenyl phosphoric acid) and uridine 5'-diphosphate (using tetrabutyl ammonium benzyl 2',3'-isopropylideneuridine 5'-phosphate followed by the elimination of the protecting groups) have been prepared.[94]

Imidazoyl phosphates have further been prepared from ketoximes by condensation with diaryl phosphoryl chlorides and Beckmann rearrangement.[780]

$$(CH_3)_2C{=}NOH + ClOP(OC_6H_5)_2 \longrightarrow CH_3N{=}C(CH_3)OPO(OC_6H_5)_2$$

In a simplified procedure the imidoyl phosphates are prepared <u>in situ</u> by Beckmann rearrangement of ketoxime arylsulfonates in the presence of diaryl phosphate anion. Cyclopentaneoxime p-nitrobenzene sulfonate was found to be particularly suited.[781] Unsymmetrical pyrophosphates can be prepared, e.g., by reacting the oxime sulfonate first with tetraethyl ammonium dibenzyl phosphate and then with diphenyl phosphoric acid. P^1-dibenzyl P^2-diphenyl pyrophosphate is produced.

ζ. Phosphorolysis of Phosphoric Anhydro Acids with Phosphorus and Carboxylic Acids

Phosphoric anhydro acids. In mixed phosphoric acid anhydrides the P-O-P bond is generally split by nucleophiles with liberation of the most stable anion (deriving from the strongest acid), i.e., the less positive P atom is attacked.[1356a] The following reactions are examples of phosphorolysis of tetraphenyl pyrophosphate,[433] realized by heating in acetonitrile in the presence of triethyl amine (no reaction without base).

$$2(C_6H_5CH_2O)_2PO_2H[(o\text{-}CH_3\text{-}C_6H_4O)_2PO_2H] + (C_6H_5O)_4P_2O_7 \longrightarrow$$

$$(C_6H_5CH_2O)_4P_2O_7[(o\text{-}CH_3\text{-}C_6H_4O)_4P_2O_7] + 2(C_6H_5O)_2PO_2H$$

The transphosphorylation presented by adenosine diphosphate in dimethyl sulfate in the presence of catalysts (Na^+, K^+, Rb^+, Cs^+ ; H^+, Li^+, NH_4^+, Mg^{2+}, Zn^{2+} are inactive) yielding adenosine monophosphate and triphosphate can be considered as resulting from the same type of transformation. The reaction is rather complex as adenosine triphosphate may phosphorylate another molecule of triphosphate yielding higher phosphates.[702]

Heated with polyphosphoric acid (orthophosphoric acid/70 - 90% phosphorus pentoxide) up to 150° thiamine yields thiamine mono and polyphosphates. A linear polyphosphoric acid is attacked by an alcohol preferentially at its ends (cf. Section I.1.a.β). Therefore it can be supposed that in a first step thiamine monophosphate is produced which in turn is phosphorylated by the polyphosphoric acid. A reaction product prepared with an 80% solution of phosphorus pentoxide in phosphoric acid contained 50% thiamine monophosphate, 35% diphosphate, and 15% triphosphate.[1083]

Mixed phosphoric/phosphorous anhydrides. In esters of this type of acid the nucleophile attacks the P^V atom of the $\cdots P^{III}\text{-}O\text{-}P^V\cdots$ group.[952a,1011]

$$(RO)_2P\text{-}O\text{-}P(O)(OR)_2 + HOPO(OR')_2 \longrightarrow$$

$$(RO)_2P(O)\text{-}O\text{-}P(O)(OR')_2 + (RO)_2PHO$$

Less nucleophilic agents like alcohols or amines attack P^{III} (using d orbitals to form donor complexes).[952a]

These tetraalkyl phosphoryl phosphites may be phosphorolyzed by phosphoric monesters, e.g., as follows, the mixed anhydride acting in the end as dehydrating agent when the P^V moiety consists of a $P(O)(OCH_2CH{=}CH_2)_2$ group.[919]

$$(C_2H_5O)_2P\text{-}O\text{-}P(O)(OCH_2CH{=}CH_2)_2 + 2H_2O_3P(OCH_2CH{=}CH_2) \xrightarrow[\text{3h at 50-55°}]{\text{pyridine-dioxane}} [(CH_2CH{=}CH_2O)\text{-}(HO)P(O)]_2O + (C_2H_5O)_2PHO + (CH_2CH{=}CH_2O)_2PO_2H$$

Similarly the same diallylphosphoryl diethyl phosphite, heated 1 hr in pyridine with phenyl phosphoric acid at 50-60° yielded P^1,P^2-diphenyl pyrophosphoric acid.

Acyl phosphates. Acyl phosphates are generally acylating agents. But acylphosphate esters derived from appropriate carboxylic acids (strong acids, acids with hindered carbonyl group) are phosphorylating. The following diethyl acyl phosphates, $(C_2H_5O)_2P(O)\text{-}OCOR$, are essentially phosphorylating: with R = C_6Cl_5, C_6HCl_4, $C_6H_2Cl_3$. Diethyl trifluoroacetyl phosphate disproportionates in 24 hr at 20° as follows:[840]

$$2(C_2H_5O)_2P(O)OCOCF_3 \longrightarrow (C_2H_5O)_4P_2O_3 + (CF_3CO)_2O$$

The following reaction between triethyl phosphate and acetic anhydride yielding tetraethyl pyrophosphate by heating at 140-205° in the presence of 0.001-0.6% of $BF_3\cdot(C_2H_5)_2O$ is perhaps based on the production of diethyl acetyl phosphate which disproportionates into tetraethyl pyrophosphate and acetic anhydride (a patent[479]).

A particular reaction is offered by bis(dibenzyl phosphoryl) oxalate which decomposes above its melting point (107-108°) to yield tetrabenzyl pyrophosphate.[921]

$$[COOP(O)(OCH_2C_6H_5)_2]_2 \longrightarrow (C_6H_5CH_2O)_4P_2O_3 + CO_2 + CO$$

η. Oxidative Phosphorylation

A series of very interesting experiments (Wieland) have shown that adenosine monophosphate and diphosphate are phosphorylated to di- and triphosphate, respectively, by inorganic phosphate during oxidation (bromine or tetrabromo-o-benzoquinone) in pyridine solution in the presence of organic compounds. The following have been used:[155,1454,1455,1455a,1455b]

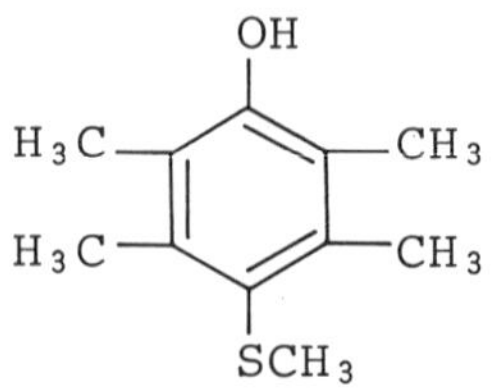

(I)

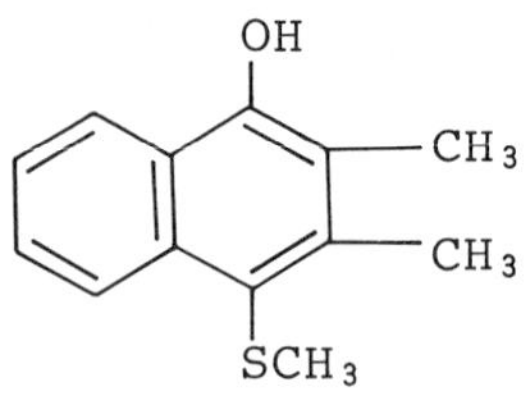

(II)

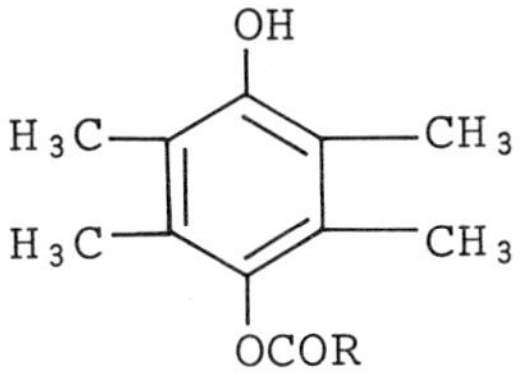

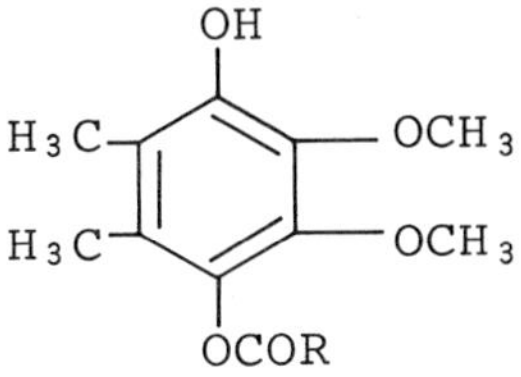

(III) (R = CH_3, $CH_2NHCOCH_3$ (IV)

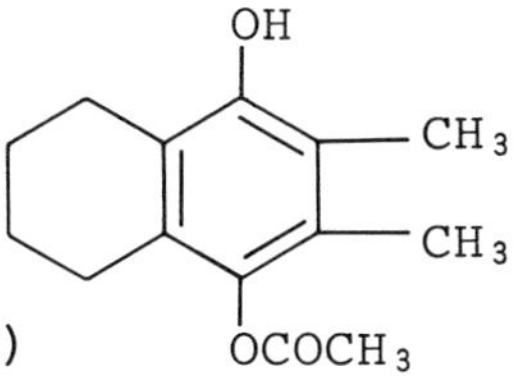

(V)

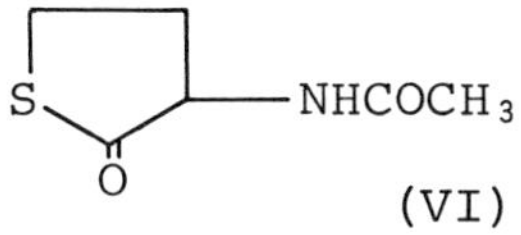

(VI)

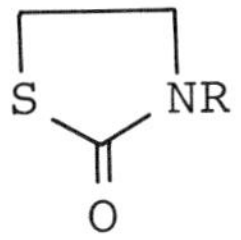

R = H, C_6H_5, CH_3

(VII)

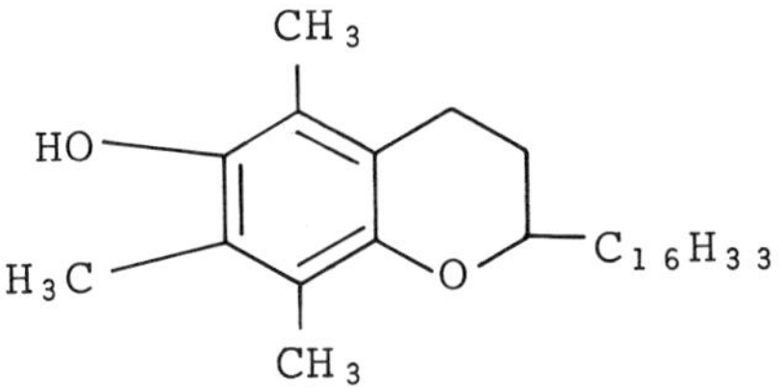

(VIII)

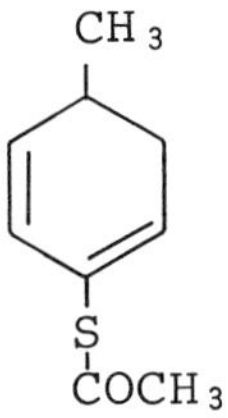

(IX)

The mechanisms are obscure. For the reaction with the compounds III-V, e.g., it is proposed that after electron elimination at the para HO group the activated acyl group yields acetyl phosphate as phosphorylating agent (but cf., e.g., the preceding paragraph on acyl phosphates).

The yields are variable and may attain 25% as in the case of the tetrabutyl ammonium salts of adenosine diphosphate and phosphoric acid oxidized with bromine and yielding adenosine triphosphate.[155]

Another interesting procedure of oxidative phosphorylation of phosphoric esters has been observed with hydroquinone monophosphates (quinol phosphates) when oxidized with bromine (Todd). In the course of this reaction the corresponding quinones are liberated with production of a phosphorylating agent which is thought to be monomolecular metaphosphate anion.[406a,b] The reaction is formulated as follows.

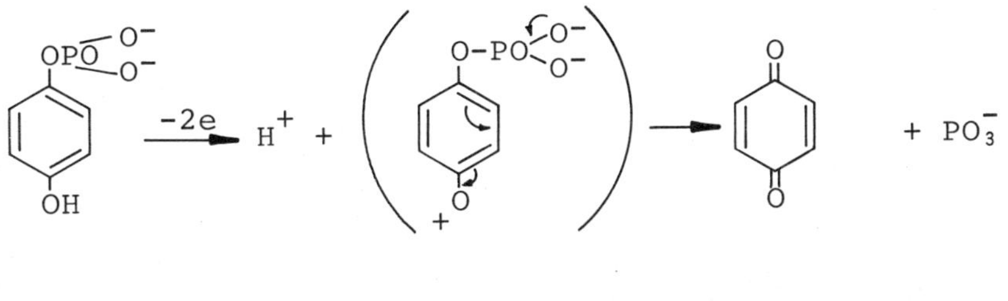

Similarly hydroquinone mono alkyl or dialkyl phosphates yield the corresponding alkyl or dialkyl metaphosphate anions, $^{-}OPO(OR)$ and $^{-}OP(OR)_2$, respectively.

In the presence of water or an alcohol phosphate or alkyl phosphate are formed, in the presence of a phosphate ester a pyrophosphate appears. Thus, e.g., 2,3-dimethyl-α-naphtohydroquinone monophosphate treated in dimethyl formamide with bromine in the presence of adenosine monophosphate yielded adenosine diphosphate.[406b]

Oxidative phosphorylation has been further observed with a number of compounds presenting a cyclic $C{=}N^+$ grouping.

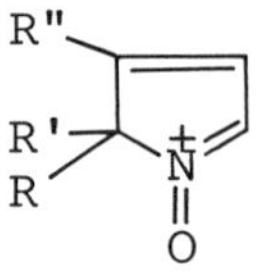
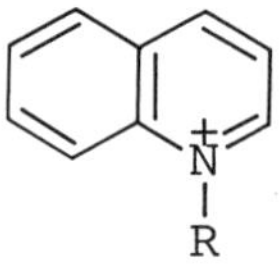
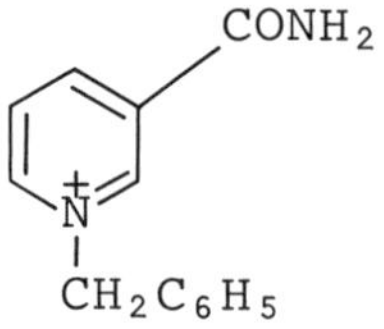

$R = R'' = H$; $R' = C_2H_5$ $\quad$ $R = O^-$

$R = R' = CH_3$; $R'' = H$ $\quad$ $R = CH_3$

$R = R' = R'' = CH_3$

These compounds form with tri(tetrabutyl ammonium) phosphate adducts which when oxidized in non aqueous solvents in the presence of 100% excess phosphate by bromine or potassium permanganate yield up to 25% of pyrophosphate.[408]

c. Oxidation of Compounds with Phosphorus of Lower Oxidation State. A patent claims the preparation of ethyl esters of polyphosphoric acids by oxidation of ethyl esters of polyphosphorus acids with P in lower state of oxidation by treatment with chlorine followed by ethanol. Thus tetraethyl hypophosphate or pyrophosphite yield tetraethyl pyrophosphate; the condensation product of sodium diethyl phosphite, $(C_2H_5O)_2PONa$, with ethyl phosphorodichloridate, $C_2H_5OPOCl_2$, yields finally pentaethyl triphosphate.[820]

Reaction between sodium dialkyl phosphites and dialkyl phosphorochloridates results in the formation of a complex mixture of tetraalkyl pyrophosphites, hypophosphates, and pyrophosphates.[60,62,65,966,1061] The origin of all these products resides probably in the initial formation of the dialkyl phosphite group, $(RO)_2PO$, which yields the different reaction products, respectively, by dimerization and oxidation-reduction exchange. The resolution of the mixture, e.g., by fractionated distillation, is deceiving, but oxidation of the crude mixture, e.g., by oxygen at elevated temperatures, yields the pyrophosphate esters.[1061] Ethyl chlorosulfonate, $C_2H_5OSO_2Cl$, can be used instead of the phosphorochloridate. The expected sulfophosphite decomposes during the reaction and yields essentially the same products besides trialkyl phosphate and sulfur dioxide, the phosphate resulting from the alkoxyl ion and the dialkyl phosphate group.[65,1063]

The oxidation of phosphite esters may result in the production of pyrophosphates, depending on the oxidant and the conditions.

With chlorine and bromine (and also with triarylmethyl bromides) sodium dialkyl phosphites lead to mixtures similar to those produced by phosphorochloridates.[60,62-65] The reaction is conducted in the cold with one atomic equivalent of halogen per mole of sodium salt with petroleum ether or ether as diluent-solvent.

In pyridine solution monoalkyl phosphorous acids are oxidized by bromine into the corresponding symmetric dialkyl pyrophosphoric acids.[246]

Nitrogen dioxide, N_2O_4, transforms dialkyl phosphites into tetraalkyl pyrophosphates.[442c,637]

In the course of the reaction between tetrachloromethane and secondary phosphite esters (leading to phosphorochloridates, cf. Section I.1.b.β-2) in the presence of a tertiary base, secondary reactions lead not only to hypophosphates (oxidizable into phosphates) but equally to pyrophosphates.[1314a]

Treating sodium adenosine 5'-phosphate and disodium phosphite dissolved in a minimum of water with bromine during 24 hr at 4° yields the corresponding di and triphosphates (yield about 13%); the yield may go up to 40% by using an organic solvent.[1464]

Mono and dibromomalonamide, $CHBr(CONH_2)_2$ and $CBr_2(CONH_2)_2$, respectively, reacted with trialkyl phosphites in the presence of dialkyl phosphoric acid yield tetralkyl pyrophosphates (in the presnce of an alcohol trialkyl phosphates are produced), e.g., as follows:[1013]

$$CHBr(CONH_2)_2 + (CH_3O)_3P + HOPO(OC_2H_5)_2 \longrightarrow$$

$$(CH_3O)_2P(O){-}O{-}P(O)(OC_2H_5)_2 + CH_2(CONH_2)_2 + CH_3Br$$

A particular reaction yielding a pyrophosphate ester starting with red phosphorus and bromine may be mentioned here. A patent claims the preparation of 3-creatinyl-4-methyl-5(2-hydroxyethyl)-thiazolyl pyrophosphate by reacting in glacial acetic acid creatine ethyl ester with red phosphorus, bromine, and 4-methyl-5(2-hydroxyethyl)-thiazole for 10 days, and saponifying finally the ester function by KOH in ethanol.[815] A phosphite ester seems to be probably an intermediary.

$$
\begin{array}{c}
\mathrm{CH_2CH_2OP(O)(OH)-O-P(O)(OH)_2} \\
|
\end{array}
$$

$$
S
\begin{array}{c}
\diagup\; \mathrm{C=CCH_3} \\
|\qquad\qquad\qquad \overset{\mathrm{NH}}{\underset{}{\parallel}} \\
\diagdown\; \mathrm{CH=N-CH_2N} \overset{\textstyle \mathrm{C-NH_2}}{\underset{\textstyle \mathrm{CH_2COOH}}{\diagup\diagdown}} \\
\qquad\quad \mathrm{Cl}
\end{array}
$$

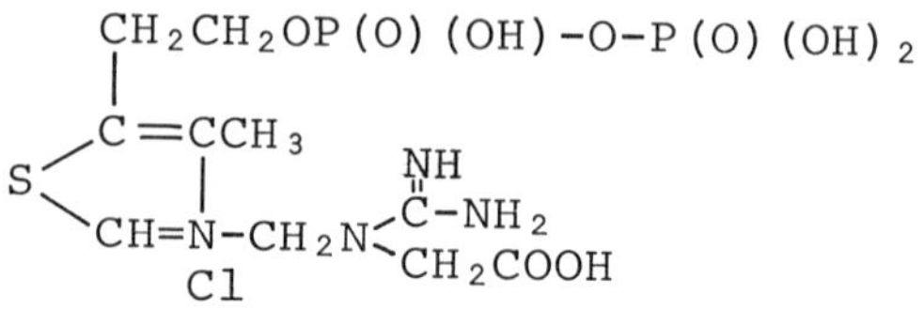

2. Esters of Anhydro Acids with Other Inorganic Oxy-
    acids

a. Sulfato and Sulfito Phosphates (Phosphosulfates,
Phosphosulfites). Adenosine 5'-phosphosulfate (sulfato-
phosphate), which is perhaps an intermediate in the enzymic activation of sulfate, has been synthesized from adenosine 5'-phosphate by the three following procedures.

By action of dicyclohexylcarbodiimide on the phosphate and concentrated sulfuric acid in aqueous pyridine;
the presence of adenosine 5'-phosphosulfate in the reaction products has been established by spectroscopic evidence.[1183]

By action on the adenosine phosphate of the sulfur
trioxide/pyridine complex at 40-50° in water in presence
of sodium hydrogen carbonate; the phosphosulfate has been
isolated as lithium salt.[125]

By acidolysis of l-ethoxyvinyl benzyl(or 2-cyanoethyl)
sulfate by adenosine 5'-phosphate (5 days at room temperature in the dark, in methylene chloride-dimethyl formamide solution) followed respectively by hydrogenolytic
debenzylation or mild alkaline decyanoethylation.[193]

A sulfito diphosphate ester has probably (no pure substance isolated) been obtained by reaction between thionyl
chloride (acting here as acyl chloride and not as dehydrating agent) and bis(p-nitrophenyl) phosphoric acid
(19 hr reflux in chloroform).[433]

$$
2(\mathrm{p\text{-}O_2N \cdot C_6H_4O})_2\mathrm{PO_2H} + \mathrm{SOCl_2} \longrightarrow
$$
$$
[(\mathrm{p\text{-}O_2N \cdot C_6H_4O})_2\mathrm{P(O)O}]_2\mathrm{SO} + 2\mathrm{HCl}
$$

The substance has been characterized by its reaction
products with cyclohexyl amine. This sort of derivatives
is perhaps an intermediate in the synthesis of pyrophosphates from secondary phosphate esters by thionyl chloride
(cf. Section III.1.b.α-3).

b. Silyl and Oxysilyl Phosphate Derivatives. This
kind of derivatives has been prepared from 4 types of
silicon compounds: alkyl(aryl) halosilanes, alkylated
alkoxysilanes, tetraalkylsilanes (orthosilicic esters),
and hexaalkyl disiloxanes.

From alkyl(aryl) halosilanes: Trialkyl halosilanes
treated with phosphoric acid,[1424] disodium hydrogen phosphate,[1211a] or sodium dialkyl phosphate[1211a] yield silicophosphoric derivatives, e.g., as follows:

$$3(CH_3)_3SiCl + H_3PO_4 \xrightarrow[6\ hr]{reflux} [(CH_3)_3SiO]_3PO + 3HCl$$

$$3(CH_3)_3SiCl + Na_2HPO_4 \xrightarrow[reflux,\ 2.5\ hr]{in\ ether} [(CH_3)_3SiO]_3PO + 2NaCl + HCl$$

$$(CH_3)_3SiCl + NaOPO(OCH_3)_2 \xrightarrow[20\ min,\ room\ temp.]{in\ ether,\ agit.} (CH_3)_3SiOPO(OCH_3)_2 + NaCl$$

Triethyl bromo(chloro)silane in excess caused to react with triethyl phosphate yields the corresponding trisilyl phosphate; for the less reactive chlorosilane derivative anhydrous ferric chloride is added as catalyst.[1424]

$$3(C_2H_5)_3SiBr + (C_2H_5O)_3PO \longrightarrow [(C_2H_5)_3SiO]_3PO + 3C_2H_5Br$$

Dimethyl(diphenyl) dichlorosilane treated respectively with phosphoric acid,[785] with sodium dimethyl phosphate,[1211a] or with triethyl phosphate[1211a] yields di-phosphoryloxysilane derivatives, e.g., as follows:

$$(C_6H_5)_2SiCl_2 + 2H_3PO_4 \xrightarrow{15\ hr\ at\ 80°} H_2O_3POSi(C_6H_5)_2OPO_3H_2 + 2HCl$$

$$(CH_3)_2SiCl_2 + 2NaOPO(OCH_3)_2 \xrightarrow[reflux,\ 1\ hr]{in\ ether} (CH_3O)_2P(O)OSi(CH_3)_2OP(O)(OCH_3)_2 + 2NaCl$$

$$(CH_3)_2SiCl_2 + 2(C_2H_5O)_3PO \xrightarrow{reflux,\ 2\ hr} (C_2H_5O)_2P(O)OSi(CH_3)_2OP(O)(OC_2H_5)_2 + 2C_2H_5Cl$$

In this last reaction the yield is modest. Besides the diphosphoryloxyderivative shown, higher condensed products are formed which contain the grouping $-P(O)(OC_2H_5)OSi(CH_3)_2O-$.

When reacted with one molecular equivalent of phosphoric acid, diphenyl dichlorosilane yields a bimolecular (cyclic ?) product, $[(C_6H_5)_2SiO_2PO(OH)]_2$.[785]

A mixture of dimethyl dichlorosilane and phosphoric acid in the ratio 3:2 heated several hours, finally *in vacuo*, yields a viscous product, $[(CH_3)_2Si]_3(PO_4)_2$, probably polymerized.[1424]

From alkylated alkoxysilanes: Trialkyl alkoxysilanes slowly distilled with phosphoric acid yield tris(trialkylsilyl) phosphates.[1211a,1424] The reaction seems to be reversible.[1211a] Example:

$$3(C_2H_5)_3SiOCH_3 + H_3PO_4 \longrightarrow [(C_2H_5)_3SiO]_3PO + 3CH_3OH$$

In a similar reaction but using phosphorous acid instead of phosphoric acid, the phosphate esters are obtained by refluxing the components in a stream of dry air during eight hours, e.g., as follows:[1423]

$$3(C_2H_5)_3SiOCH_3 + H_3PO_3 + O \text{ (dry air)} \longrightarrow [(C_2H_5)_3SiO]_3PO + 3CH_3OH$$

Phosphorus pentoxide refluxed several days in a large excess of trimethyl ethoxysilane dissolves slowly and adds about one molecule of the silane per phosphorus atom. The complex mixture produced contains not only $(CH_3)Si\text{-}O\text{-}P$ groups but still P-O-P groups. No defined products have been isolated.[1211a]

From tetraalkoxysilanes (orthosilicic acid esters): Tetraethoxysilane reacted with diethyl phosphorochloridate by refluxing during several hours gives in modest yield pentaethyl silicophosphate.[1211a]

$$Si(OC_2H_5)_4 + ClPO(OC_2H_5)_2 \longrightarrow (C_2H_5O)_3SiOPO(OC_2H_5)_2 + C_2H_5Cl$$

The same starting material yields with phosphoric acid in an exothermic reaction a mixture of more or less complex silicophosphoric esters, no definite product being isolated. Phosphorous pentoxide refluxed in the tetraethyl orthosilicic esters dissolves slowly; here again complex silicophosphoric esters with very few P-O-P groups are formed (no definite compound isolated).[1211a]

Hexaalkyl disiloxanes are claimed by a patent to yield trialkylsilyl dialkyl phosphates when heated up to about 210° with tetraalkyl pyrophosphates.[176]

$$R_3SiOSiR_3 + (RO)_2P(O)OP(O)(OR)_2 \longrightarrow 2(RO)_2P(O)OSiR_3$$

c. Alkyl(aryl) Stannyl and Germanyl Phosphoric Derivatives

Organotin(IV) phosphate derivatives have been prepared either from organotin halides or from organotin hydroxides or oxides.

The first type of reaction has been realized by treating the halide with a salt of a phosphoric (or polyphosphoric) acid which may be partially esterified.[399,401,828] The following reactions are shown as examples:

$$(C_6H_5)_3SnCl + NH_4OPO(OC_6H_5)_2 \xrightarrow[\text{reflux, 0.5 hr}]{\text{in } C_2H_5COCH_3} (C_6H_5)_3SnOPO(OC_6H_5)_2 + NH_4Cl$$ [828]

$$(C_6H_5)_2SnCl_2 + (NaO)_2P(O)OP(O)(OC_6H_{13})_2 \xrightarrow[\text{reflux}]{\text{in toluene}} 2NaCl + [(C_6H_5)_2SnO_2]P(O)OP(O)(OC_6H_{13})_2$$ [401]

Similarly, tributyl stannyl and triphenyl stannyl diethyl phosphates have been prepared.[828] A patent claims the preparation of tris(tributyltin) dibutyltin triphosphate by reaction between tributyltin chloride + dibutyltin dichloride and pentasodium triphosphate.[399]

The second type is represented by reactions between organotin oxides or hydroxides and phosphoric acid halide or anhydride, described in a patent.[399] With phosphorus oxychloride tributyltin hydroxide yields tris(tributyltin) phosphate.

$$3(C_4H_9)_3SnOH + POCl_3 \longrightarrow [(C_4H_9)_3SnO]_3PO + 3HCl$$

A mixture of diphenyltin oxide and phosphorus pentoxide refluxed in toluene yields diphenyltin pyrophosphate.

$$2(C_6H_5)_2SnO + P_2O_5 \longrightarrow [(C_6H_5)_2Sn]_2P_2O_7$$

In the field of germanium(IV) derivatives a patent claims the preparation of tris(trialkylgermanyl) phosphates by warming a mixture of hexaalkyl digermanoxane and phosphorus pentoxide in six hours up to 100°.[1438] Thus, e.g., tris(trimethylgermanyl) phosphate has been obtained.

$$3(CH_3)_3GeOGe(CH_3)_3 + P_2O_5 \longrightarrow 2[(CH_3)_3GeO]_3PO$$

d. Nitrosyl Phosphate Derivatives (Oximo Ester). The acid $ONOPO_3H_2$ (nitrosyl phosphate) is not known, but the "oximo phosphates", $RR'C{=}NOPO_3R''R'''$ may be considered as derivatives of N-substituted nitrosyl phosphoric acid or as O-phosphorylated isonitroso compounds. They have been obtained by methods mentioned above: reaction of oximes (resp. their Na derivatives) with dialkyl(diaryl) phosphorochloridates,[18,529,780,1375] (method, cf. Section I.1.b.γ-2); reaction of tertiary phosphites with α-halo-α-nitroso(or nitro)-alkanes[18,19,251,607,917,917a] (cf. Section I.2.b.γ-4); reaction of dialkyl phosphorochloridite with an aci-α-nitroalkane salt[1011a] (cf. Section I.2.b.β-6).

IV. ANHYDRO ACIDS WITH CARBOXYLIC ACIDS, AND THEIR ESTERS

1. Acylation of Phosphoric Acid (and Its Partial Esters) or Its Salts

a. By Acyl Halides. Acyl halides are very useful acylating agents for phosphoric acid which may be partially esterified. Generally salts are used.

Silver salts have found numerous applications. Trisilver phosphate, e.g., treated with an acyl chloride, is capable of yielding the mono, di, and trisubstituted products. Treated by an excess of acetyl chloride in cold ether it gives triacetyl phosphate.[886] The monoacyl derivatives can easily be prepared by this procedure in two ways: either use of monosilver phosphate or of silver dibenzyl phosphate (followed by hydrogenolytic debenzylation of the resulting acyl dibenzyl phosphate).

The monosilver salt has been prepared by trituration of the trisilver salt with phosphoric acid in ether.[368, 856,879] An alternative technique consists of stirring at room temperature trisilver phosphate and 85% phosphoric acid in a quantity corresponding to two mole-equivalents of H_3PO_4 with the acyl chloride in ether, e.g., as follows:[913]

$$3FCH_2COCl + Ag_3PO_4 + 2H_3PO_4 \longrightarrow 3FCH_2C(O)OPO_3H_2 + 3AgCl$$

(isolated as Li Salt)

An example of synthesis of an acyl phosphate with elimination of various protecting groups, using α-azidoacyl chlorides and silver dibenzyl phosphate for the preparation of α-aminoacyl phosphates is shown.[227]

$$CH_3CH(N_3)COCl + AgOPO(OCH_2C_6H_5)_2 \longrightarrow CH_3CH(N_3)C(O)OPO(OCH_2C_6H_5)_2 + AgCl$$

$$CH_3CH(N_3)C(O)OPO(OCH_2C_6H_5)_2 \xrightarrow{H_2/\text{Raney Ni}} CH_3CH(NH_2)C(O)OPO(OCH_2C_6H_5)_2 + N_2$$

$$CH_3CH(NH_2)C(O)OPO(OCH_2C_6H_5)_2 \xrightarrow{H_2/\text{Pd-C}} CH_3CH(NH_2)C(O)OPO_3H_2$$

(isolated as silver or barium salt)

In place of the silver salts other salts have been

proposed for particular applications. The higher secondary esters of phosphoric acid may be used as sodium salts since the stability of the corresponding acyl phosphates permits the use of the necessary higher temperatures.[1432] The attendant phosphate ion can be conveniently removed by freezing out of trisodium phosphate.[879] Alkoxycarbonyl phosphates have been prepared using sodium salts, e.g., as follows:[840]

$$C_2H_5OC(O)Cl + NaOPO(OC_2H_5)_2 \longrightarrow C_2H_5OC(O)OPO(OC_2H_5)_2 + NaCl$$

Formyl phosphate has been prepared similarly,[748] using formyl fluoride.

$$K_2HPO_4 + KHCO_3 + FC(O)H \xrightarrow[\text{(ice bath)}]{} HC(O)OPO_3K_2 + KF + H_2O + CO_2$$

After precipitation of potassium and phosphate ions by addition of lithium perchlorate and silver nitrate, the lithium salt of formyl phosphate has been isolated. The free acid (unstable even at 0°) has been obtained from the silver salt by means of methyl mercaptan.

The use of salts of tertiary amines (triethyl amine, N,N-diethylaniline, N-ethylpiperidine, N-methylmorpholine) is proposed by a patent for the acylation of secondary phosphoric acid esters. These ester-acids, the acyl chloride, and the tertiary base are stirred in ether during four hours.[432]

Even free acids may be used: Bis(dibenzyl phosphoryl) oxalate has been prepared by short refluxing of dibenzyl phosphoric acid in oxalyl chloride.[921]

$$2(C_6H_5CH_2O)_2PO_3H_2 + (COCl)_2 \longrightarrow (C_6H_5CH_2O)_2P(O)OC(O)C(O)OP(O)(OCH_2C_6H_5)_2 + 2HCl$$

As an example of the application of the silver salt/acyl chloride technique to condensed phosphoric acids the preparation of acetyl metaphosphate may be mentioned.[682]

$$AgPO_3 + CH_3COCl \longrightarrow CH_3COPO_3 + AgCl$$

b. By Ketenes. This useful method is limited in its application by the fact that only the simplest ketene, $CH_2{=}C{=}O$, is easily obtained. The synthesis of acetyl phosphate is realized as follows. Ketene is passed under good cooling in 85% phosphoric acid in ether. After treating the resulting mixture with ice water phosphate ion is removed by barium hydroxide, and the disilver acetyl phosphate is precipitated by silver nitrate.[228,715] The free acid is obtained from the silver salt by treatment with hydrogen sulfide. Acetyl dibenzyl phosphate is prepared similarly and yields acetyl phosphate by hydrogenolysis.[886] The direct acetylation of phosphoric acid is formulated as follows.

$$CH_2{=}C{=}O + HOPO_3H_2 \longrightarrow CH_3C(O)OPO_3H_2$$

c. By Vinyl Esters of Carboxylic Acids. Vinyl carboxylates lead to acyl phosphates by phosphorolysis. In a first example isopropenyl acetate was used. 85% phosphoric acid is added dropwise to cold isopropenyl acetate in excess, then a small quantity of concentrated sulfuric acid is added as catalyst, and the mixture is stirred during 35 min at 25°. The acetyl phosphate is isolated as lithium salt.[1309]

$$CH_2{=}C(CH_3)OC(O)CH_3 + H_3PO_4 \longrightarrow CH_3C(O)OPO_3H_2 + CH_3COCH_3$$

The same type of reaction has been realized with vinyl acetate, $CH_2{=}CHOC(O)CH_3$, in the presence of mercuric acetate/sodium ethylenediamine tetraacetate as catalyst (2-1/2 hr at 30°).[139]

2. Phosphorylation of Carboxylic Acids or Their Salts

a. By Phosphoryl Halides. Reactions of this type have been first carried out for the preparation of diacetyl dialkylamidophosphates.[241,1392] More recently this procedure has been used for the synthesis of trifluoroacetyl dialkyl phosphates,[840] and alkoxycarbonyl dialkyl

$$CF_3CO_2Na + ClPO(OC_2H_5)_2 \longrightarrow CF_3C(O)OPO(OC_2H_5)_2 + NaCl$$

phosphates (monoalkyl carbonic acid phosphoric acid anhydrides).[1245]

$$\underset{\text{(slight excess)}}{ROCO_2K} + ClPO(OR')_2 \xrightarrow[\text{3 hr, 40-50°}]{\text{dry acetone}} ROC(O)OPO(OR')_2 + KCl$$

b. By Phosphorus Acid Anhydrides: Pyrophosphates, Hypophosphates. The phosphorylation of carboxylic acids by means of pyrophosphates seems to be rather exceptional. The methylamide of N-benzoyl-DL-seryl pyrophosphate treated with acetic acid or glycine produces a very limited phosphorylation yielding 1-4% of the corresponding acyl phosphate, which was not isolated but only revealed by the formation of hydroxamic acids with hydroxylamine.[122]

Hypophosphates, i.e., esters of phosphorous-phosphoric anhydride, are capable of phosphorylation of carboxylic acids, the P^{III}-O-P^{V} group being split with liberation of the weaker P^{III} acid (Michalski).[861,919,952a]

$$XCO_2H + (RO)_2P\text{-}O\text{-}PO(OR)_2 \longrightarrow XC(O)OPO(OR)_2 + (RO)_2PHO$$

Thus, diallyl phosphoryl diethyl phosphite yields acetyl diallyl phosphate by reaction with acetic acid.[919]

$$CH_3CO_2H + (C_2H_5O)_2P\text{-}O\text{-}PO(OC_3H_5)_2 \longrightarrow CH_3C(O)OPO(OC_3H_5)_2 + (C_2H_5O)_2PHO$$

The reaction depends on sterical factors. With bulky X ($>C_3$) no reaction is observed; benzoic acid, e.g., does not react.

The acyl phosphates have been sometimes prepared *in situ* as acylating agents. Thus N-protected α-aminoacids treated successively with a mixed P^{III}/P^{V} ester anhydride and an α-aminoacid ester yield dipeptides, as follows.[861]

$$RCH(NHZ)CO_2H + (R'O)_2P\text{-}O\text{-}PO(OR')_2 \xrightarrow[\text{then ice cooled}]{\text{1 hr at }70°} RCH(NHZ)C(O)OPO(OR')_2, \text{ then}$$

$$+ H_2NCH(R)CO_2C_2H_5 + (CH_3)_3N \longrightarrow RCH(NHZ)C(O)NHCH(R)CO_2C_2H_5 + (CH_3)_3NHOPO(OC_2H_5)_2$$

c. By Vinyl Phosphates. Just as vinyl acetates are acetylating agents (cf. Section 1.c) vinyl phosphates (improperly called enol phosphates) are phosphorylating agents (cf. production of pyrophosphates, Section III.1.b.ε). Carboxylic acids yield acyl phosphates by reaction with vinyl phosphates (Cramer), as follows:

$$RCO_2H + R'CH{=}CH(R'')OPO_3H_2 \longrightarrow RC(O)OPO_3H_2 + R'CH_2C(R'')O$$

The reaction proceeds smoothly in appropriate solvents (benzene, acetone) at low or moderate temperature (0°-40°, depending on the components). Vinyl phosphates used: $(C_2H_5O)_2P(O)OC(OC_2H_5){=}CCl_2$,[443,444] $(C_2H_5O)_2P(O)OC(CO_2C_2H_5){=}CH_2$,[444] $(C_2H_5O)_2P(O)OC(OC_2H_5){=}CHCO_2C_2H_5$.[444,445]

A series of simple carboxylic acids as well as, e.g., N-protected α-aminoacids have thus been transformed into the corresponding acyl dialkyl phosphates.

3. Miscellaneous

a. Reaction between Salicylic Acid and Phosphorus Pentachloride, Oxychloride, and Alkyl Phosphorodichloridates. The reaction between phosphorus halides and salicylic acid or methyl salicylate is of particular complexity, and it is one of the earliest reactions studied leading to acyl phosphorus acid derivatives (mentioned already in Section II.1.d).

Describing in 1859 the reaction between phosphorus pentachloride and salicylic acid or its methyl ester, Couper assigned to the product obtained the structure (B) of a trichloro dioxyphosphorane derivative (but of course not under this name!). Couper prepared the same substance from (A), the reaction product of salicylic acid with phosphorus trichloride, by chlorination or by treatment with phosphorus pentachloride.[438]

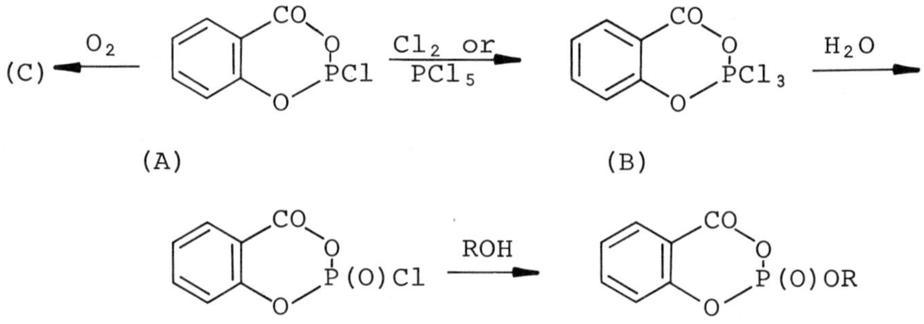

(A) (B)

(C) (D)

The structure of (B) was much discussed, but finally the constitution proposed by Couper was confirmed, as well as the structures of (A) and of (C). C is obtainable either from salicylic acid by phosphorus oxychloride or from (B) by careful treatment with water.[1207a]

The acyl chloride (C) gives with alcohols the esters (D) which are equally obtained by reaction of salicylic acid with alkyl phosphorodichloridates, $ROPOCl_2$.[92a,317a]

The above-mentioned reactions seem to be realizable also with substituted salicylic acids.

b. Oxidation of Acyl Phosphites. This quite plausible reaction seems to have been used only quite exceptionally. Thus the just mentioned 2-chloro-benzo-1,3,2-dioxaphospholane (A) is oxidized by air yielding the 2-chloro-2-oxo derivative (C).[317a]

The oxidation of monoacetyl pyrophosphite (whose structure is questionable!) by barium peroxide into monoacetyl pyrophosphate[933] is dubious.

c. Reaction between Secondary Diethylamidophosphate Esters and Anhydrides of Carboxylic Acids. This reaction proceeds smoothly, best below 30-40°, as follows.[525]

$$(RO)_2OPN(C_2H_5)_2 + (R'CO)_2O \longrightarrow R'C(O)OPO(OR)_2 + R'C(O)N(C_2H_5)_2$$

A series of diethylamidophosphate diesters have thus been treated with propionyl and butyryl anhydride.

V. ESTERS OF HALOPHOSPHORIC ACIDS

These substances are simultaneously esters and acyl halides and offer the double possibilities of reaction of the two functions. They are interesting foremost as intermediates for the synthesis of phosphoric ester derivatives resulting from the transformation of the acyl halide function (this function may yield ester, amide, or anhydride groupings by reactions during which the ester group(s) remain unchanged).

The phosphorohalidate esters most studied derive from chlorine and fluorine; bromine and iodine derivatives play quite a minor role. The fluoridates present a special interest due to their particular properties, chemical (relative stability, specific reactivity with HO groups in molecules containing H_2N function besides) as well as pharmacological (anticholinesterasic, myotic, insecticidal, etc.). For this reason the methods of preparation of phosphorofluoridates are listed apart.

Here again the list of halophosphoric esters given below does not pretend to be exhaustive.

1. Use of Phosphorus(V) Derivatives

a. Phosphorus Oxychloride or Oxybromide

α. Reactions with Hydroxy Compounds

This very simple and old method (first paper on the subject published in 1859) consists of the reaction of a phosphorus oxyhalide (generally ocychloride used) with alcohols or phenols in the absence or the presence of a tertiary amine (generally pyridine). The final product--$ROPOHal_2$ or $(RO)_2POHal$--depends roughly on the proportions of the reagents.[181,601,633,1236,1296,1434,1452] With an excess of the hydroxy compound and in the presence of a base, tertiary phosphates, $(RO)_3PO$, result of course (cf. Section I.1.b.β).

$$ROH + OPHal_3 \longrightarrow ROPOHal_2 + HHal$$

$$2ROH + OPHal_3 \longrightarrow (RO)_2POHal + 2HHal$$

The experimental conditions have been varied to a very large extent. Navech has presented in 1966 a review on this reaction with alcohols, which unhappily is not of an easy access.[1030a]

Primary aliphatic alcohols react satisfactorily at or below room temperature. For the accomplishment of the second step the elimination of the hydrogen halide is advantageous. This may be accomplished by carrying out the reaction under reduced pressure[416,633,1061] or by bubbling in an inert gas (e.g., nitrogen or carbon dioxide[601,1452]), or by addition of a base. In this case the best yields are obtained by using two moles of the base per mole of the oxyhalide.[100] Even higher alcohols react satisfactorily; with vitamin A alcohol, e.g., the reaction begins at -10° and is terminated at 0°.[737a] Halogenated alcohols need more drastic conditions: With glycol bromhydrine, $BrCH_2CH_2OH$, the phosphorodichloridate is obtained by heating with $POCl_3$ in carbon tetrachloride up to 60°.[749a] The fact that the elimination of HHal is advantageous for the second step yielding the phosphoromonohalidate seems to corroborate the assumption that in the first step a hydrogen-bonded adduct of the hydroxy compound with the phosphorus oxyhalide, or a quasi-phosphonium compound, $ROP(OH)Hal_3$, is produced which does not react further as such but only after loss of hydrogen halide, yielding the reactive phosphorodihalidate, $ROP(O)Hal_2$.[822]

A stepwise reaction using two different alcohols without isolation of the intermediary phosphorodihalidate, $ROPOHal_2$, is possible with the restriction that the second alcohol must have a higher alkyl radical than the first.[633]

Secondary and tertiary alcohols, generally, do not react smoothly, and alkyl halide formation may occur; methylphenyl carbinol, $(CH_3)(C_6H_5)CH(OH)$, e.g., is exclusively halogenated.[601,1296] An exception is offered by secondary alcohols with halogen substituted carbon atoms adjacent to the carbon bearing the hydroxyl group. Thus glycerol dichlorohydrine, $CH_2ClCH(OH)CH_2Cl$,[836a] and 1,1,1,3,3,3-hexachloropropane-2-ol, $CCl_3CH(OH)CCl_3$,[606a] yielded the corresponding di and monochloridates, $ROPOCl_2$ and $(RO)_2POCl$, either at higher temperature (2.5 hr at 130°[836a]) or after several days at room temperature.[606a]

Phosphorus oxybromide--which is rarely used--reacts more rapidly than phosphorus oxychloride: thus the hexachloropropane-2-ol just mentioned when reacted in hexane-pyridine at ordinary temperature yields with the oxychloride, respectively, $ROPOCl_2$ in three days, and $(RO)_2POCl$ in ten days whereas the oxybromide yields the secondary derivative, $(RO)_2POBr$, already in three days.[606a]

Phenols require in general more drastic conditions: temperatures up to the boiling point or the presence of a base. Under these conditions, the restriction to monosubstitution even when using a large excess of oxyhalidate is often not very effective.[514,689,746,964,1175,1215] Quite a number of metal salts are proposed as catalysts: $MgCl_2$,[208,259a,689,1492] $BaCl_2$ < NaCl < KCl < RbCl < CsCl < $FeCl_3$ (activity growing in this order),[770a] NaCl,[825a] LiCl.[12a] When using no solvent the reaction is achieved by heating up to 160-260° until no more hydrogen chloride is evolved.[1100c] Thus, orthocresol yields without difficulties the monosubstituted dichloridate and the disubstituted monochloridate, $(o-CH_3C_6H_4O)POCl_2$ and $(o-CH_3C_6H_4O)_2-POCl$, respectively;[835a] the two naphthols heated in benzene solution with phosphorus oxychloride and pyridine yield both the corresponding 1-naphthyl and 2-naphthyl phosphorodichloridates.[577] Complicated phenols such as tocopherol[1028a] or phenolic steroids (estrone, stilbestrol, etc.)[1099c] give the same reaction. 2,6-Dialkylphenols exhibit marked steric hindrance only with bulky radicals such as tert-butyl; reaction is then possible only in presence of Friedel-Crafts-type catalysts ($AlCl_3$, etc.) which provoke dealkylation or rearrangement of an orthosubstituent.[822a] Small substituents do not inhibit the reaction: pentafluorophenol yielded the mono as well as the disubstituted derivative, $C_6F_5OPOCl_2$ and $(C_6F_5O)_2-POCl$.[265a]

Di and polyhydroxy compounds may react in several ways: either polyphosphorylation (A) each hydroxy group yielding a phosphorodihalidate ester group, cyclization (B), or production of linear polycondensed derivatives (C), as shown for a dihydroxy compound and phosphorus oxychloride.

$$HOROH \longrightarrow Cl_2P(O)OROP(O)Cl_2, \text{ or}$$

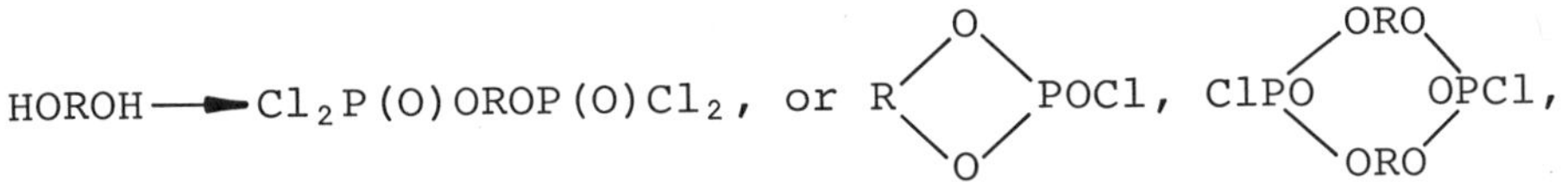

(A)

(B)

etc., or $HORO-P(O)Cl-ORO-P(O)Cl\cdots$

(C)

The reactions leading to products of types (B) and (C) presume as a first step the formation of a monosubstituted derivative, $HOROPCl_2$.

The reactions of glycol itself are complex and seem to yield products of type (B).[383a] But aliphatic vicinal glycols with longer chains, viz., propylene glycol, $CH_3CH(OH)CH_2OH$, and 3,4-dihydroxyhexane, $CH_3CH_2CH(OH)-CH(OH)CH_2CH_3$, when reacted with phosphorus oxychloride in pyridine at $-20°$ yield the corresponding phospholane derivatives (with five-membered cycle).[1383]

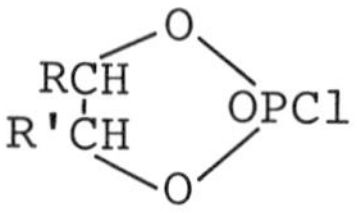

$$\begin{matrix} RCH \\ R'CH \end{matrix} \underset{O}{\overset{O}{\diagup}} OPCl \qquad (R = CH_3, \; R' = H; \; R = R' = CH_3CH_2)$$

Vicinal glycols with tertiary hydroxyl functions react in the same way, e.g., pinacol, $(H_3C)_2C(OH)-C(OH)(CH_3)_2$.[978] Phenylated glycols also lead to phospholane derivatives, e.g., meso-dihydrobenzoine, $C_6H_5CH(OH)-CH(OH)C_6H_5$.[1327],[1380a] Phosphorinane derivatives are easily obtained.[809a] Thus, 2,2-dimethyl-1,3-propanediol yields 2-chloro-2-oxo-5,5-dimethyl-1,3,2-dioxaphophosphorinane.[500]

$$(H_3C)_2C \begin{matrix} CH_2OH \\ \\ CH_2OH \end{matrix} \quad + \; POCl_3 \; \longrightarrow \; (H_3C)_2C \begin{matrix} CH_2-O \\ \\ CH_2-O \end{matrix} POCl$$

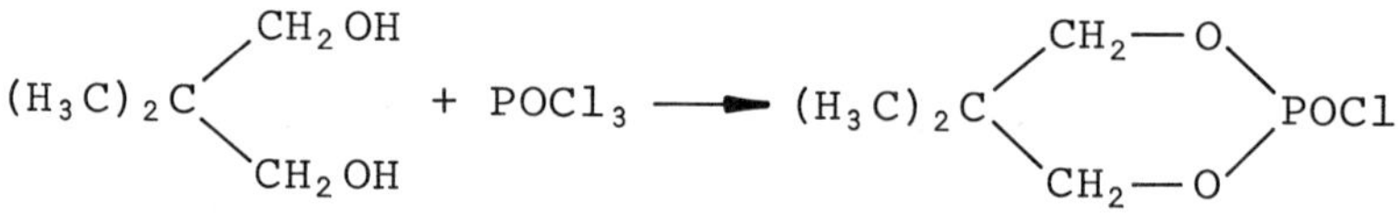

A number of spirocyclic derivatives have been obtained by this procedure: e.g., the compound (D) from 1,1-bis(hydroxymethyl)-cyclohexane,[848a] pentaerythritol di-phosphorochloridate (E),[9a],[1071a] compound (F) from 4,4-bis(hydroxymethyl)-1,2-dithiole.[915a]

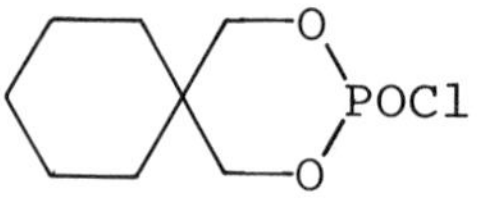

(D)   (E)

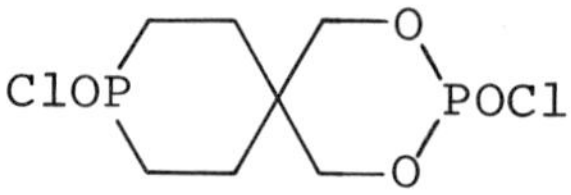

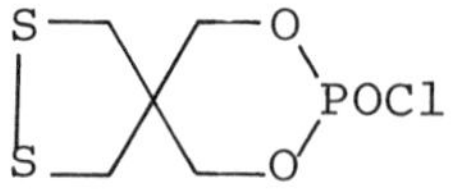

(F)

Even complicated ditertiary 1,3-glycols with appropriate configuration yield the corresponding cyclic phosphorohalidates, e.g., the compound (G) derived from a ditertiary dihydroxyterpenoid resulting from hydration-rearrangement of a cyperene epoxide.[885b]

POCl

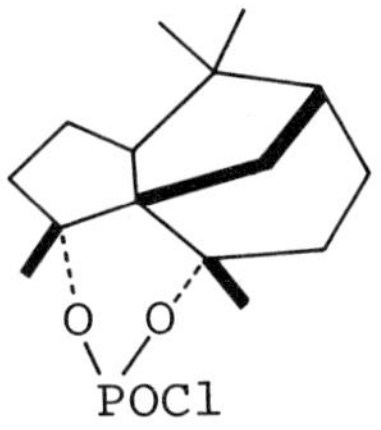

(G)

Dihydroxyphenols react very well. Ortho compounds (pyro-catechols) form the cyclic phenylene phosphorohali-dates.[1182] Meta and paradihydroxybenzenes react at both sites yielding di-phosphorodihalidates.[1182] Polyhydroxy phenols form resinous products, probably by polycondensa-tion.[241]

β. Reactions with Other than Hydroxy Compounds

Carbonyl compounds treated with phosphorus oxychloride in the presence of a tertiary amine are capable of reacting in their enolic forms, and lead to vinyl phosphorochlor-idates, but the yields are modest. Thus acetaldehyde shaken in an autoclave with phosphorus oxychloride and triethyl amine (exothermic reaction), then heated 1.5 hr at 100° yielded vinyl phosphorodichloridate, $(CH_2=CHO)-POCl_2$, and divinyl phosphorochloridate, $(CH_2=CHO)_2POCl$, depending on the ratio aldehyde:oxychloride (yield about 30%).[619] Similarly, isopropenyl phosphorodichloridate, $[CH_2=C(CH_3)O]POCl_2$, was obtained from acetone (two hr at 90°, autoclave), yield 5%![619]

A patent claims the addition of phosphorus oxychloride to epoxides (ethylene oxide, propylene oxide, phenylethyl-ene oxide) with ring opening and production of 2-chloro-alkyl phosphorodichloridates and di(2-chloroalkyl) phos-phorochloridates, as shown for propylene oxide.[814a] Other paper: Ref. 906a.

$$2CH_2\!\!-\!\!CHCH_3 + POCl_3 \longrightarrow (CH_3CHClCH_2O)_2POCl$$

The use of catalytic amounts of concentrated hydro-chloric acid permits the preparation of a number of these derivatives by letting the mixture containing one to three moles of 1,2-epoxide and one mole of oxychloride stand during several days at room temperature, or heating four to seven hours at 60-90°.[1146]

With an aliphatic ester the following reaction has been realized.[1153a]

$$CH_3COO\text{-}n\text{-}C_4H_9 + POCl_3 \xrightarrow{\text{4 hr, } H_3PO_4 \text{ as catalyst}} C_4H_9OPOCl_2 + CH_3COCl$$

Reacting phosphorus oxychloride with diazoalkanes in (moist ?) ether with anhydrous copper sulfate as catalyst, at -30°, Yakubovich et al. have observed the production, in small yield, of the corresponding alkyl phosphorodichloridates (diazomethane and diazoethane used). The intervention of moisture forming $HOPOCl_2$ *in situ* is admitted.[478a]

Tertiary phosphates in contact with phosphorus oxychloride are liable to undergo a disproportionation, as follows:

$$(RO)_3PO + POCl_3 \longrightarrow (RO)_2POCl + ROPOCl_2$$

This reaction is effective with lower esters at low temperatures and moderately effective with selected aromatic esters, principally the cyclic pyrocatechol derivatives, on heating to about 150°. Higher alkyl esters react not satisfactorily.[37,351,364,600,808] In comparison with other methods, this procedure seems not to be of great interest as the reactions are far from quantitative and the ratio of the reagents is of little influence on that of the products.[601] At higher temperatures, the same reagents give rise, in the aliphatic series, to polyphosphate esters (cf. Section III.1.b.γ-2).

Trimethylsilyloxy and trimethylgermanyloxy derivatives present interesting reactions with phosphorus oxyhalides.

Alkoxytrimethylsilanes, $RO(CH_3)_3Si$, react in two ways with phosphorus oxychloride (bromide), depending on the nature of the radical R:[539a,1297b]

(A) $(CH_3)_3SiOC_2H_5 + POCl_3 \longrightarrow C_2H_5OPOCl_2 + (CH_3)_3SiCl$

(B) $(CH_3)_3SiOCH(C_6H_5)_2 + POCl_3 \longrightarrow (CH_3)_3SiOPOCl_2 + (C_6H_5)_2CHCl$

2-Phenylpropyloxytrimethylsilane, $CH_3CH(C_6H_5)CH_2OSi(CH_3)_3$, reacts according to type (B).[539a]

Butyl phosphorodibromidate, $C_4H_9OPOBr_2$, which has been obtained from butoxytrimethylsilane by a reaction of type (A), is unstable.

Trimethylsilyl phosphorodichloridate has equally been prepared in a similar reaction between ethoxytrimethylsilane and hypophosphoryl chloride, $P_2O_2Cl_4$.[1297b]

The same silyl phosphorodichloridate is further

obtained from the reaction of hexamethyldisiloxane or tris(trimethylsilyl) phosphate with phosphorus oxychloride.[1297a]

$$(CH_3)_3SiOSi(CH_3)_3 + POCl_3 \longrightarrow (CH_3)_3SiOPOCl_2 + (CH_3)_3SiCl$$

$$[(CH_3)_3SiO]_3PO + 2POCl_3 \longrightarrow 3(CH_3)_3SiOPOCl_2$$

(yield about 30%)

This last reaction is somewhat similar to the disproportionation reaction between tertiary phosphates and phosphorus oxychloride, just mentioned.

Hexamethyldigermanoxane and (trimethylsilyloxy)trimethylgermane react similarly, as follows:[1297c]

$$(CH_3)_3GeOGe(CH_3)_3 + POCl_3 \longrightarrow (CH_3)_3GeOPOCl_2 + (CH_3)_3\text{-}GeCl$$

$$(CH_3)_3SiOGe(CH_3)_3 + POCl_3 \begin{cases} \longrightarrow (CH_3)_3SiOPOCl_2 + (CH_3)_3GeCl \quad \text{(about 80\%)} \\ \longrightarrow (CH_3)_3GeOPOCl_2 + (CH_3)_3SiCl \quad \text{(about 20\%)} \end{cases}$$

b. Phosphorus Pentachloride. Phosphorus pentachloride has been used in two ways for the preparation of halophosphoric esters.

α. With Hydroxy Compounds

These compounds react generally either by hydroxyl-chlorine exchange yielding alkyl chlorides, or by forming oxyhalophosphoranes of the types $ROPHal_4$ or $(RO)_2PHal_3$, but in a few cases production of phosphorohalidates, mostly of the type $(RO)_2POCl$, has been reported.[600,602,708,784,901] A mechanism has been proposed for this reaction, in which dissociation of the halo alkoxy phosphoranes, $(RO)_2PCl_3$ and $(RO)_3PCl_2$, into Cl^- and $(RO)_2\overset{+}{P}Cl_2$ and $(RO)_3\overset{+}{P}Cl$ intervenes, the phosphonium ions losing RCl and combining with Cl^-.[606a] The global reaction is as follows:

$$(RO)_2PCl_3 + ROH \longrightarrow (RO)_2POCl + RCl + HCl$$

In contrast to primary alcohols, secondary and tertiary alcohols are generally either dehydrated into olefins or halogenated into alkyl chlorides.[603,705] In both cases phosphorus oxychloride is formed which may undergo side reactions (cf. Section V.1.a.α).

The reaction with phenols requires generally heating,

which decomposes the oxyhalophosphoranes mentioned above with production of phosphorohalidates (see below under d).[99] The temperature necessary for the reaction depends on the nature of the phenol. Thus, p-nitrophenol treated with phosphorus pentachloride in chloroform during two hours with ice cooling yielded the secondary monochloridate, $(p\text{-}O_2NC_6H_4O)_2POCl$.[1382]

With compounds presenting in addition to the hydroxyl group a function liable to be halogenated, phosphorus pentachloride produces simultaneously chlorination of this last function and phosphorylation of the hydroxyl group. Thus, β,β,β-trichlorolactic acid yielded β,β,β-trichloro-α-(dichlorophosphoryl)-propionyl chloride, and tetrabenzoylhydroquinone, 1,4-bis(dichlorophosphoryl)-2,3,5,6-tetrakis(phenyldichloromethyl)benzene.

$$Cl_3CH(OH)COOH + PCl_5 \longrightarrow Cl_3CH(OPOCl_2)COCl + 2HCl$$ [857a]

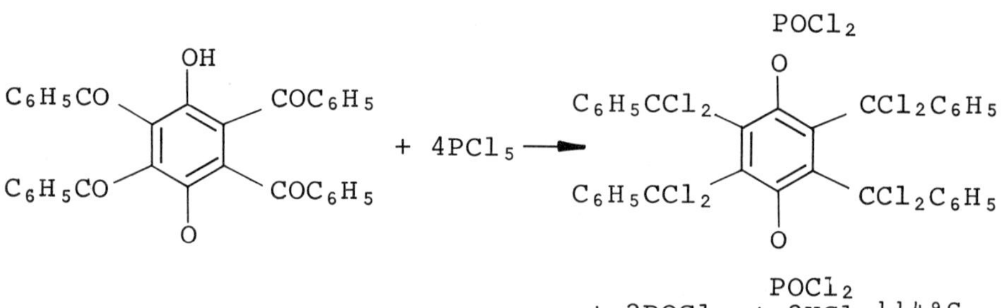

$$+ 2POCl_3 + 2HCl$$ [1149c]

Hydroxybenzoic acids react similarly. The carboxyl group is transformed into a chloroformyl function, and the hydroxyl is phosphorylated yielding a phosphorodichloridate ester.[27,46,438,1467,1468] A small excess of the pentachloride and gentle warming are proposed,[41] addition of a small amount of phosphorus oxychloride may be beneficial.[1467,1468] Heated further with phosphorus pentachloride the chloroformylaryl phosphorodichloridates are converted into derivatives with 5 labile chlorine atoms. Since these derivatives are distillable, the formulation as $Cl_3C\text{-}Ar\text{-}OPOCl_2$ seems favored in regard to the formulation as phosphorane derivatives, $ClOC\text{-}Ar\text{-}OPCl_4$.[33,46,51] The same products are obtained starting from the acyl chlorides of the hydrobenzoic acids, HO-Ar-COCl.[33,43,48,50]

$$HO_2C\text{-}Ar\text{-}OH + PCl_5 \longrightarrow 2HCl + ClOC\text{-}Ar\text{-}OPOCl_2 \xrightarrow[\text{heat}]{PCl_5} Cl_3C\text{-}Ar\text{-}OPOCl_2 + POCl_3$$

Hydroxyaryl sulfonic acids (sodium or potassium salts generally used) undergo with phosphorus pentachloride a similar reaction yielding the chlorosulfonyl dichlorophosphoryl derivatives, $ClO_2S-Ar-OPOCl_2$. The reaction needs usually temperature of at least 100°, and mixtures of pentachloride and oxychloride have been applied with similar results at temperatures up to 150-180°.[29,30,45,1507] The possibility has to be considered that at these temperatures the chlorosulfonyl group may be converted into a chlorine substituent.[30]

β. With Phosphoric Esters

Dealkylation of a cyclic ethyl phosphate by action of phosphorus pentachloride in benzene solution, 0.5 hr at 0-5°, has been reported. The phosphorochloridate produced has been isolated by distillation in vacuo.[623a]

$$
\begin{array}{c}
H_3C-C \\
\quad \parallel \\
H_3C-C
\end{array}
\begin{array}{c}
O \\
\diagdown \\
\diagup \\
O
\end{array}
PO(OC_2H_5) \; + \; PCl_5 \longrightarrow
\begin{array}{c}
H_3C-C \\
\quad \parallel \\
H_3C-C
\end{array}
\begin{array}{c}
O \\
\diagdown \\
\diagup \\
O
\end{array}
POCl
$$

$$+ \; C_2H_5Cl \; + \; POCl_3$$

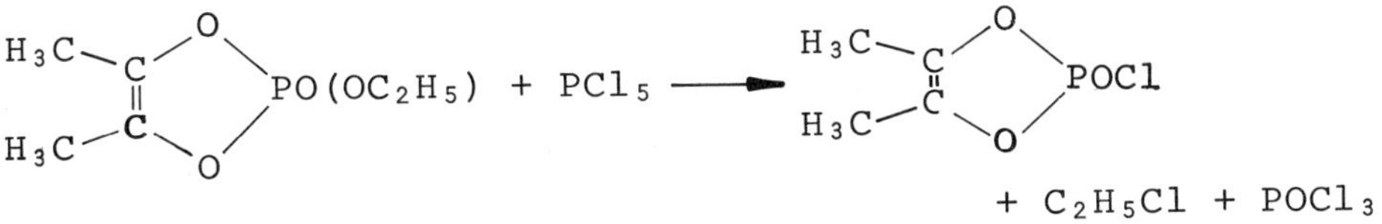

The conversion of partial phosphoric esters or their salts by reaction with phosphorus pentachloride has been used principally in the aromatic series. The reaction requires generally mild heating.[476,795,1499]

$$(RO)_2PO_2H \; + \; PCl_5 \longrightarrow (RO)_2POCl \; + \; POCl_3 \; + \; HCl$$

c. Pyrophosphoryl Chloride. At low temperatures (-50 to -20°)[636a,636b] or in the presence of appropriate solvents (metacresol, orthochlorophenol, acetonitrile, benzonitrile, etc.)[732-3] pyrophosphoryl chloride reacts with alcohols preferentially as anhydride and not as acyl chloride. The monoalkyl phosphorodichloridates produced may be used, even without isolation, for the preparation of the corresponding alkyl phosphoric acids (by hydrolysis) or of higher esters (by reaction with other hydroxyl compounds), as mentioned in Section I.f.ζ-1. Only primary hydroxyl functions seem to react well.[732] Effecting the reaction without solvent and beginning with good cooling, Grunze et al. have isolated by distillation of the resulting mixture a series of monoalkyl phosphorodichloridates deriving from primary aliphatic alcohols.[636a,636b]

Hexamethyldigermanoxane and (trimethylsilyloxy)trimethylgermane reacted with pyrophosphoryl chloride yield trimethylgermanyl phosphorodichloridate and trimethylsilyl phosphorodichloridate, as follows:[1297c]

$$(CH_3)_3GeOGe(CH_3)_3 \; + \; Cl_2P(O)OP(O)Cl_2 \longrightarrow 2(CH_3)_3GeOPOCl_2$$

$$(CH_3)_3SiOGe(CH_3)_3 + Cl_2P(O)OP(O)Cl_2 \longrightarrow (CH_3)_3SiOPOCl_2 + (CH_3)_3GeOPOCl_2$$

d. Conversion of Polyhalooxyphosphoranes (Quasi-Phosphonium Polyhalides). The reactions considered here involve the replacement of two halogen atoms by an oxygen atom on a phosphorus atom. Practically only aryl derivatives are used because of the limited stability of the polyhalo alkoxy phosphoranes.

The replacement may be realized by treatment with the theoretical amount of water or, better, by passing sulfur dioxide into the polyhalide. In this case the crude reaction product of phosphorus pentachloride with the hydroxy compound may be used.[40,494]

$$ROPHal_4 \text{ (or } (RO)_2PHal_3) + SO_2 \longrightarrow ROPOHal_2 \text{ (or } (RO)_2POHal) + SOHal_2$$

The secondary phosphorochloridates may be prepared by warming the trihalides with anhydrous oxalic acid,[31,37,631b] e.g.,

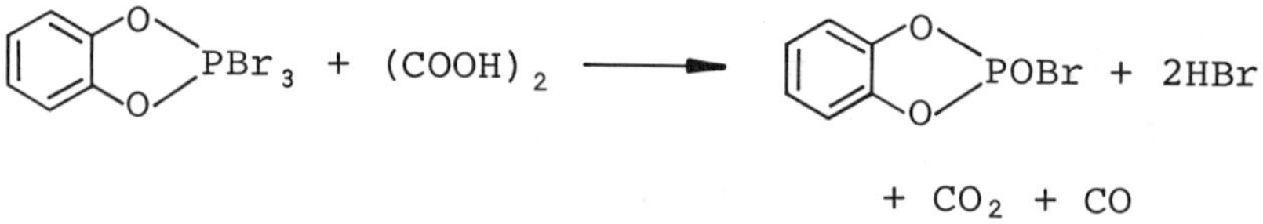

The cyclic pyrocatechol trichlorophosphorane derivative, $(C_6H_4O_2)PCl_3$, has yielded the corresponding phosphorochloridate by warming with acetone,[631b] or with a phos-

$$(C_6H_4O_2)PCl_3 + CH_3COCH_3 \longrightarrow (C_6H_4O_2)POCl + CH_3CCl_2CH_3$$

phinic acid which is chlorinated by the trihalide reacting similarly to phosphorus pentachloride;[670a] however, in this last case the reaction products are not easily separated.

$$(C_6H_4O_2)PCl_3 + RR'PO_2H \longrightarrow (C_6H_4O_2)POCl + RR'POCl + HCl$$

Diphenyl phosphorochloridate has been obtained by distillation of triphenoxyphosphorus dichloride.[1099a] This reaction represents the second step in the direct preparation of the dichloridate by reaction between phenol and phosphorus pentachloride (see above under section b.α).

$$(C_6H_5O)_3PCl_2 \longrightarrow (C_6H_5O)_2POCl + C_6H_5Cl$$

e. Secondary Phosphorochloridates from Primary Phosphorodichloridates. Primary phosphorodichloridates, $ROPOCl_2$, react normally with hydroxy compounds (in the presence, or not, of a tertiary base).

$$ROPOCl_2 + R'OH \longrightarrow (RO)(R'O)POCl + HCl$$

The reaction represents the second step in the direct preparation of secondary phosphorohalidates by treating phosphorus oxyhalide with hydroxy compounds without isolating the intermediate primary phosphorodihalidate, $ROPOHal_2$. But when using the isolated phosphorodichloridate the choice of the second hydroxy compound is not limited as it is for the "direct" preparation (see above under a.α). Thus, e.g., phenyl phosphorodichloridate treated with 2-chloroethanol yielded 2-chlorethyl phenyl phosphorochloridate $(ClCH_2CH_2O)(C_6H_5O)POCl$;[2] the same dichloridate reacted at -10° with 2,3-isopropylideneglycerol in the presence of quinoline furnished the corresponding phosphorochloridate,[149] and 2-fluoroethyl phos-

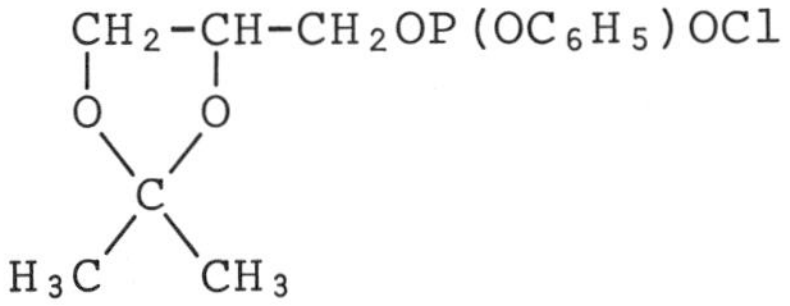

phorodichloridate treated in ether under cooling with methanol in the presence of triethyl amine gave methyl 2-fluoroethyl phosphorochloridate, $(CH_3O)(FCH_2CH_2O)POCl$.[756a]

It may be noted that aryl phosphorodichloridates, treated with diols, give no cyclic derivatives but tend to yield linear polycondensation products.[463a,1497a]

$$-(ArO)P(O)O-R-O(ArO)P(O)O-R-O-$$

f. Miscellaneous

α. Phosphorochloridate from Secondary Ester by Thionyl Chloride

Bis(paranitrophenyl) phosphoric acid refluxed two hours with thionyl chloride in dimethyl formamide yielded the corresponding phosphorochloridate.[85a]

$$(p\text{-}O_2NC_6H_4O)_2PO_2H + SOCl_2 \longrightarrow (p\text{-}O_2NC_6H_5O)_2POCl + SO_2 + HCl$$

β. Phosphorohalidates from Amides by Hydrogen Halides

An interesting observation reports the splitting of secondary phosphoroamidates by anhydrous hydrogen halides,[1262a] e.g., as shown:

$$(C_2H_5O)_2P(O)NRR' + 2HCl \longrightarrow (C_2H_5O)_2POCl + HNRR'HCl$$

This reaction is the reverse of the classical preparation of amides by the action of ammonia or primary and secondary amines on acyl halides. It is perhaps favored by a marked polarization of the P-N bond. It is irreversible as the detached NRR' group is transformed into an ammonium ion which does not react with the acyl halide function. The reaction is described for amides with R and R' = H, C_2H_5, and C_6H_5. The yield is good with hydrogen chloride; hydrogen bromide gives good yields with phenyl esters. With hydrogen iodide side reactions take place (no phosphoroiodate isolated); hydrogen fluoride does not react.

γ. Oxidation of Thionophosphorohalidates

Just as tertiary thionophosphoric esters may be converted into the corresponding tertiary phosphates (cf. Section I.1.f.α) diaryl thionophosphorochloridates have been reported to be oxidizable into the corresponding diaryl phosphorochloridates by nitric acid or hydrogen peroxide.[512] The hydrolyzability of the P-Hal function limits of course the field of application of this reaction.

δ. Thermic Decomposition of (Phenoxycarbonylimido)phosphorus Trichloride

This compound, heated *in vacuo* at 80°, is decomposed in a complex reaction yielding phosphorus oxychloride, triphenyl cyanurate, cyanuryl chloride, and about 46% of phenyl phosphorodichloridate.[802a] It is not stated if the reaction may be generalized.

$$C_6H_5OC(O)N{:}PCl_3 \longrightarrow C_6H_5OPOCl_2 \text{ (and other products)}$$

2. Use of Phosphorus(III) Derivatives

a. Phosphorus Trichloride. As mentioned already above (cf. Section I.2.a), alkyl phosphorodichloridates have been prepared in one operation from phosphorus trichloride by two ways: reaction with peroxyalcohols[25] (no definite products described in the patent), and reac-

$$R(OOH) + PCl_3 \longrightarrow ROPOCl_2 + HCl$$

tion of ethane and oxygen under pressure (18 atm) with phosphorus trichloride at temperatures beginning at -5° and rising until 30°. The resulting mixture contains ethanephosphonodichloridate, $C_2H_5POCl_2$, and ethyl phosphorodichloridate, $C_2H_5OPOCl_2$, separable by fractionated distillation *in vacuo*.[625] From the interaction of PCl_3, liquid ethylene and oxygen at -110° a mixture was obtained consisting of 17% $ClCH_2CH_2OPOCl_2$ and 83% $ClCH_2CH_2POCl_2$.[903a] A further reaction of this type has been reported by a Russian patent. Ethyl or butyl ether reacted at higher temperature with phosphorus trichloride in the presence of oxygen yield a complex mixture containing again phosphonodichloridate and phosphorodichloridate of the corresponding alkyl radical.[1270b]

b. Phosphorous Esters and Ester-Halides

α. Inorganic Oxydants

α-1. Halogens.

Chlorine is of course the most used halogen. Secondary and tertiary phosphites yield the corresponding phosphorochloridates as follows.

$$(RO)_2PHO + Cl_2 \longrightarrow (RO)_2POCl + HCl$$

$$(RO)_3P + Cl_2 \longrightarrow (RO)_2POCl + RCl$$

The techniques used are various: in carbon tetrachloride at 0-5° or higher,[936,1434a] without solvent,[471a,632,654a] etc.

The phosphite may be prepared and oxidized in one operation; e.g., dimethyl phosphorochloridate, $(CH_3O)_2$-POCl, has been prepared by treating phosphorus trichloride with methanol during 3.5 hr at -5 to -2° and reacting the resulting mixture with chlorine.[654a]

A dialkyl phosphorochloridite, $(RO)_2PCl$, treated with chlorine at low temperature, yields the corresponding alkyl phosphorodichloridate, as follows.[859a,1452]

$$(CF_2HCF_2CH_2O)_2PCl + Cl_2 \xrightarrow{-40°} CF_2HCF_2CH_2OPOCl_2 + CF_2HCF_2CH_2Cl$$ [859a]

Cyclic phosphorous esters of aliphatic diols, either as chloridites, 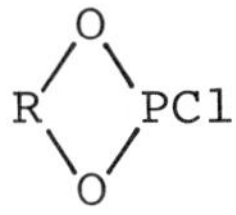 or as tertiary (alkyl) esters, 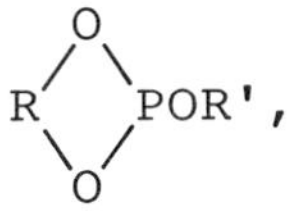 react similarly, with the difference that the ring is cleaved; in the final product a haloalkyl group appears in the place of the separated alkyl halide.[1217] Thus ethylene phosphorochloridite treated for 15 hr with chlorine at -18° yielded 2-chloroethyl phosphorodichloridate.

$$\begin{matrix} CH_2-O \\ | \qquad\quad \diagdown \\ \qquad\qquad\quad PCl + Cl_2 \longrightarrow ClCH_2CH_2OPOCl_2 \\ CH_2-O \diagup \end{matrix}$$

Similarly methyl ethylene phosphite, $(C_2H_4O)_2POCH_3$, gave methyl 2-chloroethyl phosphorochloridate, $(CH_3O)-(ClCH_2CH_2O)POCl$.[1217]

Cyclic phosphites deriving from pyrocatechol behave as the open-chained phosphites, i.e., without ring opening as shown with the methyl ester of the orthophenylene cyclic phosphite.[631b]

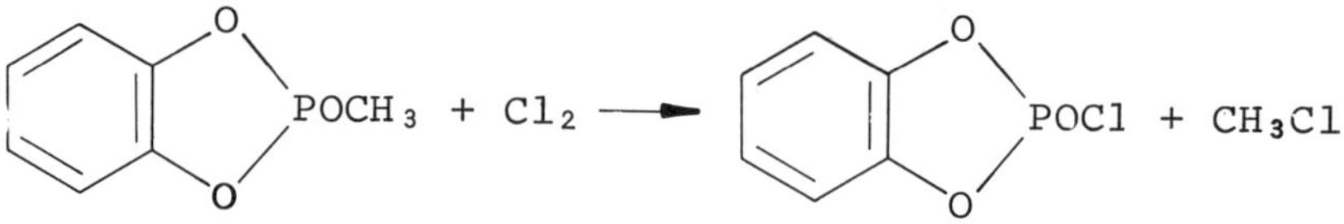

The derivative of salicylic acid behaves differently. The cyclic phosphorochloridite adds simply chlorine, and the resulting trichloride yields the cyclic phosphorochloridate by careful treatment with water.[1207a]

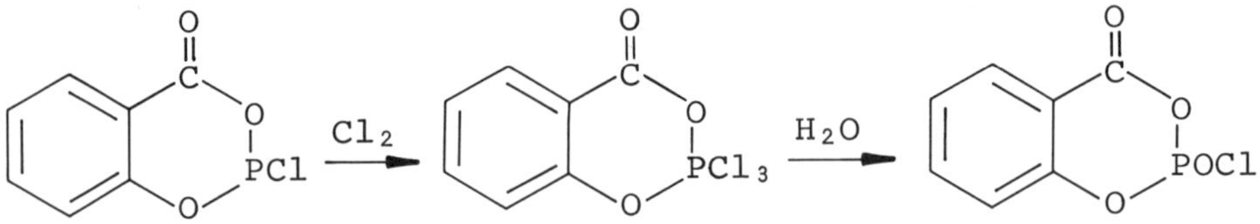

Bromine is not often used. It behaves like chlorine with secondary phosphites.[93b] With tertiary phosphites, normal reaction is generally observed[122a,606,631b] but sometimes an intermediate tribromide is isolated, e.g., as follows:[885a]

$[CH_2:C(CH_3)O]_2(C_6H_5O)P + Br_2 \longrightarrow [CH_2:C(CH_3)O]_2(C_6H_5O)-$

$PBr_2 \longrightarrow [CH_2:C(CH_3)O](C_6H_5O)POBr + CH_3C\equiv CH + HBr$

In this example of a bis(isopropenyl) phenyl phosphite, one of the unsaturated alphatic radical is finally eliminated. In the case of divinyl n-butyl phosphite, the saturated aliphatic radical disappears.[885a]

$(CH_2:CHO)_2(C_4H_9O)P + Br_2 \quad \dfrac{\text{in benzene, room}}{\text{temp. then at } 70\text{-}90°} \longrightarrow$

$(CH_2:CHO)_2POBr + C_4H_9Br$

The aliphatic iodine derivatives are too unstable to be

actually isolated. However, the reaction of a tertiary phosphite with iodine begins normally as is shown by the observation that methyl iodide is produced when methyl pyrocatechol phosphite is treated with iodine.[631b] Dimethyl phosphoroiodate, $(CH_3O)_2POI$, and diethyl phosphoroiodate, $(C_2H_5O)_2POI$, prepared from trimethyl phosphite and iodine in ether, are fairly stable in solution but have not been isolated; their presence results from the quantitative formation of anilide with aniline.[895] With aromatic phosphites, $(ArO)_3P$, finally polyiodide derivatives are obtained like $(C_6H_5O)_3PI_4$ and $(C_6H_5O)_3PI_9$ which, reacted with water, yield nearly quantitatively triphenyl phosphate.[567] Diphenyl phosphoroiodate has been prepared as follows:[567]

$$(C_6H_5O)_2(c\text{-}C_6H_{11}O)P + I_2 \longrightarrow (C_6H_5O)_2POI + c\text{-}C_6H_{11}I$$

α-2. Other Compounds

Phosphorus pentachloride has been used for the transformation of the cyclic phosphorochloridite derived from salicylic acid into the trichlorophosphorus compound just mentioned (cf. Section α-1, chlorine) which yields by moderate hydrolysis the corresponding phosphorochloridate,

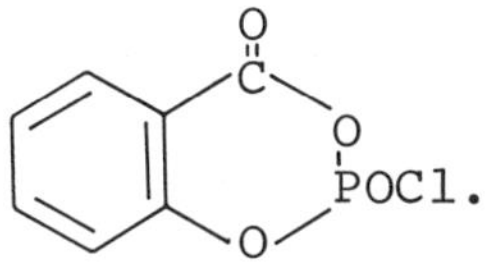
[1107b]

Sulfuryl chloride reacts similarly to chlorine but needs generally a somewhat higher temperature (about 40°). Use of an inert solvent is recommended.[94] Secondary as well as tertiary phosphites yield the corresponding phosphorochloridates.

$$(RO)_2PHO + SO_2Cl_2 \longrightarrow (RO)_2POCl + SO_2 + HCl$$ [94, 904a, 904b]

$$(RO)_3P + SO_2Cl_2 \longrightarrow (RO)_2POCl + SO_2 + RCl$$ [1127a]

The preparation of tertiary phosphites from alcohols and phosphorus trichloride and the subsequent oxidation by sulfuryl chloride has been performed without isolation of the phosphite, the two reactions taking place in the same benzene solution.[551a]

Methyl hypophosphate, treated with sulfuryl chloride in carbon tetrachloride under cooling, yielded dimethyl phosphorochloridate in modest yield.[1099b]

$$[(CH_3O)_2PO]_2 + SO_2Cl_2 \longrightarrow 2(CH_3O)_2POCl + SO_2$$

Thionyl chloride reacts similarly to sulfuryl chloride.[94, 1434a] However, the reaction is probably more complicated. With tertiary phosphites, the production of trialkyl thionophosphate, $(RO)_3PS$, trialkyl phosphate, $(RO)_3PO$, and alkyl chloride besides dialkyl phosphorochloridate, $(RO)_2POCl$, has been reported.[1128]

Disulfur dichloride, S_2Cl_2, has been used for the halogenation of secondary phosphites.[520a,562a]

With phosgene, $COCl_2$, aliphatic tertiary phosphites yield equally secondary phosphorochloridates.[1099b,1149a]

$$(RO)_3P + COCl_2 \longrightarrow (RO)_2POCl + CO + RCl$$

By means of dinitrogen tetroxide or yellow mercuric oxide diethyl phosphorochloridite has been oxidized into diethyl phosphorochloridate, $(C_2H_5O)_2POCl$.[775a]

Anhydrous copper(II) chloride oxidizes and halogenates dialkyl phosphites yielding dialkyl phosphorochloridates; copper(II) bromide yields the bromidate but is less active.[1266]

$$(RO)_2PHO + 2CuCl_2 \longrightarrow (RO)_2POCl + 2CuCl + HCl$$

Finally, simple treatment with oxygen transforms the cyclic phosphorochloridite derived from salicylic acid,

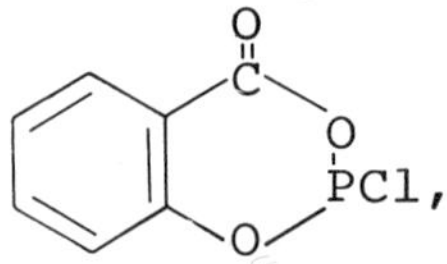

, into the corresponding phosphorochloridat

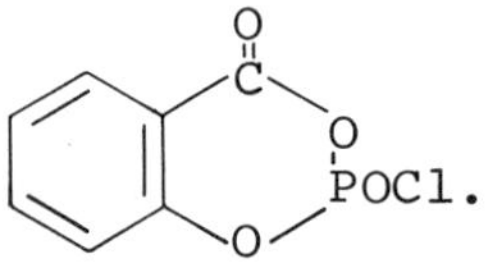

.[1107b]

β. Organic Compounds

β-1. Polyhaloalkanes. Since the work of Atherton et al. showing that mixtures of dialkyl phosphites with polyhalomethanes behave like secondary phosphorohalidates in their reaction with primary or secondary amines,[96,97] the preparation of phosphorohalidates by action of polyhaloalkanes (in the presence, or not, of tertiary amines) has been shown to be possible, starting not only from secondary phosphites but also from tertiary ones.

Thus, dialkyl phosphites reacted during 3 hr at 20° with carbon tetrachloride in the presence of triethyl amine yield phosphorochloridates.[1314a]

$$(RO)_2PHO + CCl_4 \longrightarrow (RO)_2POCl + HCCl_3$$

The reaction is possible likewise with tertiary phosphites, but it is then more complex. With carbon tetrachloride trialkyl phosphites yield a mixture of trichloromethylphosphonate, $Cl_3CPO(OR)_2$, and (less) dialkyl phosphorochloridate, $(RO)_2POCl$. Triethyl phosphite heated with bromoform up to 140° gave, in a reaction which was exothermic at the beginning, a mixture of diethyl phosphite, $(C_2H_5O)_2PHO$, and diethyl phosphorobromidate, $(C_2H_5O)_2POBr$.[760a]

The reaction between carbon tetrabromide and a tertiary phosphite with at least one aromatic radical, conducted in ether (exothermic in the beginning), leads to secondary phosphorobromidates, the eliminated radical being always aromatic. Thus diphenyl phosphorobromidate and methyl phenyl phosphorobromidate have been obtained from triphenyl phosphite and methyl diphenyl phosphite, respectively.[761a]

$$(RO)(R'O)(ArO)P + CBr_4 \longrightarrow (RO)(R'O)POBr \text{ (+ other products)}$$

Even pyrophosphites, reacted with carbon tetrachloride, lead to the corresponding phosphorochloridates, $(RO)_2$-$POCl$.[760] Example of use of hexachloroethane:[760b]

$$(RO)_3P + C_2Cl_6 \longrightarrow \underset{\text{(not isolated)}}{[(RO)_3PCl_2]} \longrightarrow (RO)_2POCl + RCl + C_2Cl_4$$

<u>β-2. Perchlorocycloalkenes(Bicycloalkenes)</u>. Perchlorocyclopentadiene reacts with triethyl phosphite at 0-5° as follows:[913a]

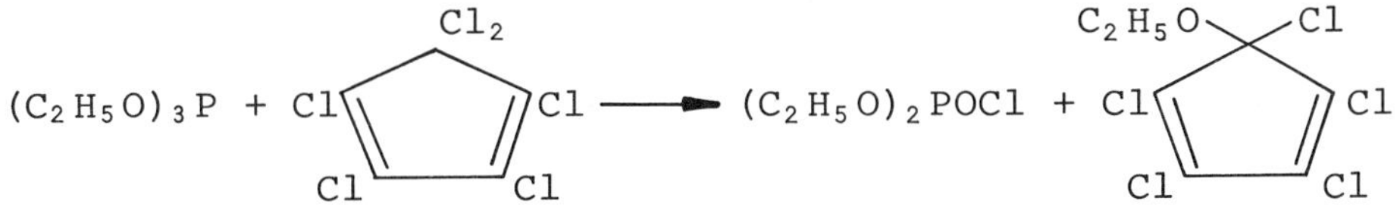

By a different mechanism--loss of two chlorine atoms--perchlorobis(2,4-cyclopentadiene-1-yl) transforms at 25° (in a slurry in pentane) trialkyl phosphites into the corresponding phosphorochloridates. Phosphites of unsaturated alcohols, e.g., tris(isopropenyl)phosphite, undergo this reaction without any attack on the double bonds.[913b]

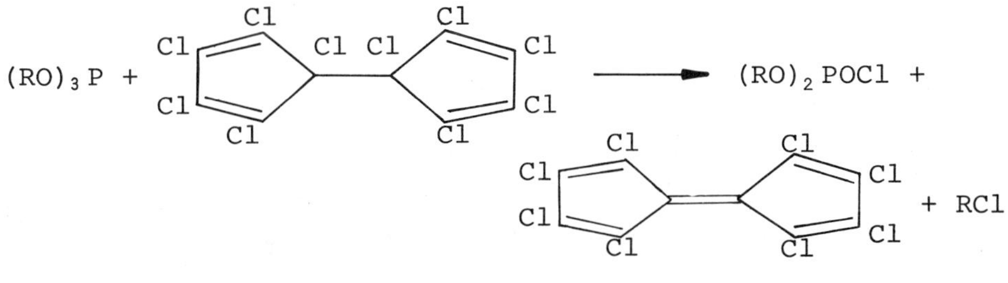

β-3. Derivatives of Carboxylic Acids

N-Chloro[395a,617c,1353a] and N-bromosuccinimide[617b] are useful agents transforming at moderate temperatures (up to 45-50°) secondary phosphites into phosphorohalidates. Unsaturated radicals in the phosphite, e.g., isopropenyl, remain unaffected.[617c] Even complicated derivatives, such as 2',3'isopropylidenadenosine 5'-(benzyl phosphite) react normally.[395a]

Alkyl formate when reacted with pyrocatechol phosphorobromidite is partially brominated, and partially oxidizes the bromidite into bromidate in a very complex reaction.[631a]

$$C_6H_4{<}^{O}_{O}{>}PBr + ROCHO \longrightarrow ROCHBr_2 \text{ and } C_6H_4{<}^{O}_{O}{>}POBr$$

Halonitriles such as 2,2,3-trichloro(or tribromo)-propionitrile are capable of halogenating trimethyl or triethyl phosphite into the corresponding dialkyl phosphorohalidate as shown for trimethyl phosphite and the bromine derivative.[1083a]

$$CH_2BrCBr_2CN + (CH_3O)_3P \longrightarrow (CH_3O)_2POBr + CH_2{:}CBrCN + CH_3Br$$

β-4. Sulfur Compounds: (Chlorothio)trichloromethane. This compound reacts with dialkyl phosphites yielding an O,O-dialkyl S-trichloromethyl thiophosphate and a dialkyl phosphorodichloridate, as follows:[520b]

$$(RO)_2PHO + ClSCCl_3 \begin{cases} \longrightarrow (RO)_2(Cl_3CS)PO + HCl \\ \longrightarrow (RO)_2POCl + CSCl_2 + HCl \end{cases}$$

Thus the (distillable) dialkyl phosphorochloridates with alkyl = methyl, ethyl, propyl, isopropyl, butyl, and isobutyl have been prepared.[520b]

γ. Metaphosphites. The "metaphosphites" considered here constitute a specific group of substances, the so-called chloroformyl metaphosphites resulting from the action of phosphorus trichloride on salicylic acid (or substituted salicylic acid). Anschütz et al., who discovered these substances, ascribed to them the dubious formula of monomolecular metaphosphites.[33,34,39,44,47,49] These compounds add two chlorine atoms by reaction with chlorine[33,40,43] or when heated with phosphorus pentachloride,[40] and yield the corresponding phosphorodichloridates, as shown for the derivative of salicylic acid:

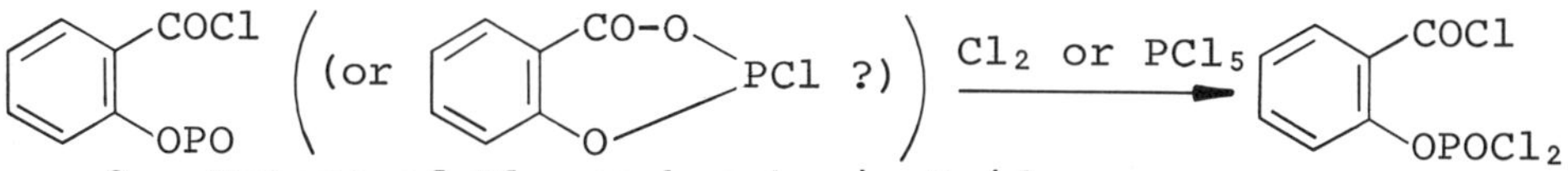

3. Esters of Fluorophosphoric Acids

a. Use of Phosphorus(V) Derivatives

α. Compounds with P-F Bonds

α-1. Phosphorus Oxyfluoride. Reacted under appropriate conditions (in pyridine or with bubbling in of ammonia) with two moles of an alcohol per mole of oxyfluoride, this compound yields dialkyl phosphorofluoridate.[1069b]

$$POF_3 + 2ROH + 2\ \text{base} \longrightarrow (RO)_2POF + 2\ \text{base}\cdot HF$$

α-2. Phosphorus Oxyfluorodichloride and Oxyfluorodibromide. Due to the relative stability of the P-F bond compared with the P-Cl or P-Br bond, the title compounds may yield monosubstituted phosphorofluoridohalidates, (RO)P(O)FHal, or disubstituted phosphorofluoridates, $(RO)_2POF$. By this procedure, e.g., methyl and ethyl phosphorofluoridochloridate, $CH_3OP(O)FCl$ and $CH_3CH_2OP(O)FCl$, respectively, as well as ethyl phosphorofluoridobromidate, $C_2H_5OP(O)FBr$, have been prepared.[1319a] Phenols react in the same way;[369,1069a] use of powdered potassium chloride as catalyst favors the production of the monosubstituted derivatives with mixed halogens, e.g., phenyl phosphorofluoridochloridate, $C_6H_5OP(O)FCl$.[264a] The possibilities of further transformation of the fluoridohalidates are evident: with a hydroxy compound they yield disubstituted phosphorofluoridates, (RO)(R'O)POF.[1069a] Controlled hydrolysis leads to monosubstituted hydrogen phosphorofluoridates, $ROP(F)O_2H$.[1319a]

α-3. Monoesters of Phosphorodifluoridic Acid. These compounds are characterized by the relative reactivity of one fluorine atom which allows the preparation of disubstituted phosphorofluoridates with two different radicals, e.g., as follows:[616a]

$$C_2H_5OPOF_2 + HOCH_2CH_2I \xrightarrow[\text{at } 0°]{1.5\ \text{hr}} (C_2H_5O)(ICH_2CH_2O)POF + HF$$

α-4. Transformations of Difluorophosphoric Acid and Fluorophosphoric Acid. The use of the silver salt of fluorophosphoric acid, $(AgO)_2POF$, for the synthesis of secondary phosphorofluoridates, $(RO)_2POF$, by reaction with alkyl iodides has been introduced already in 1932[841] but is still of interest.[430a,1235] The silver salt is heated with the alkyl iodide in benzene.

$$(AgO)_2POF + 2RI \longrightarrow (RO)_2POF + 2AgI$$

Just as said above, one of the fluorine atoms of difluorophosphoric acid is relatively easily exchanged. Thus reaction of difluorophosphoric acid with ethanol at 25° during 12 hr yields the monoethyl ester, $C_2H_5OP(F)-O_2H$.[698a]

Difluorophosphoric acid or its ammonium salt yield alkylsilyl derivatives by reaction with trialkylchlorosilane and dialkyldichlorosilane, as follows:

$$(CH_3)_3SiCl + HOPOF_2 \longrightarrow (CH_3)_3SiOPOF_2 + HCl$$ [239a]

$$R_3SiCl + NH_4OPOF_2 \xrightarrow[\text{ether, 0.5 hr}]{\text{refluxed in}} R_3SiOPOF_2 + NH_4Cl$$ [626a]

A large number of trisubstituted chlorosilanes have been used.[626a]

In a similar procedure, disubstituted dichlorosilanes, $RR'SiCl_2$, yielded dialkyl (or alkylaryl) (difluorophosphoryl)fluorosilane, the expected dialkyl (or alkylaryl) bis(difluorophosphoryl)silane, $RR'Si(OPOF_2)_2$, undergoing spontaneous decomposition.[626b]

$$RR'SiCl_2 + 2NH_4OPOF_2 \longrightarrow RR'Si(OPOF_2)F + 2NH_4Cl$$
$$+ \text{ decomposition products}$$

α-5. Pyrophosphoryl Fluoride. In a reaction similar to that which at low temperatures is given by pyrophosphorylchloride, pyrophosphorylfluoride is easily split by nucleophiles originating not only from hydroxy compounds but equally from trialkylchlorosilanes and trialkylchlorostannanes, yielding esters of difluorophosphoric acid, etc., e.g., as follows:

$$F_2P(O)\text{-}O\text{-}P(O)F_2 + C_2H_5OH \longrightarrow C_2H_5OPOF_2 + HOPOF_2$$

By this procedure the trimethylsilyl and tributylstannyl derivatives, $(CH_3)_3SiOPOF_2$ and $(C_4H_9)_3SnOPOF_2$, respectively, have been prepared.[1206a]

α-6. Synthesis of Fluoropyrophosphoric Esters. These syntheses are again based on the fact that the fluorine atom of an alkyl phosphorofluoridochloridate, ROPOFCl, does not participate easily in condensation reactions. Therefore the following reactions have been realized:[1319]

Condensation in ether-pyridine at 0°:

$$C_2H_5OPOFCl + HOPO(OC_2H_5)F \longrightarrow$$

$$(C_2H_5O)FP(O)\text{-}O\text{-}P(O)F(OC_2H_5) + HCl \cdot base$$

Condensation of a chloridate by controlled hydrolysis in presence of pyridine:

$$2C_2H_5OPOFCl + 1H_2O \text{ (in pyridine)} \longrightarrow$$

$$(C_2H_5O)FP(O)\text{-}O\text{-}P(O)F(OC_2H_5) + 2HCl \cdot base$$

Condensation of a chloridate with phosphorus oxychloride at 140°:

$$C_2H_5OPOFCl + (C_2H_5O)_3PO \longrightarrow$$

$$(C_2H_5O)FP(O)\text{-}O\text{-}P(O)(OC_2H_5)_2 + C_2H_5Cl$$

Condensation of a fluorophosphoric acid by action of dicyclohexylcarbodiimide (DCC) in ether at room temperature during 1 hr:

$$2C_2H_5OPOFOH + DCC \longrightarrow (C_2H_5O)FP(O)\text{-}O\text{-}P(O)F(OC_2H_5)$$

$$+ \text{ dicyclohexylurea}$$

Condensation at room temperature during 20 hr of diethyl vinyl phosphate with a fluorophosphoric acid:

$$C_2H_5OPOFOH + (CH_2{:}CHO)(C_2H_5O)_2PO \longrightarrow$$

$$(C_2H_5O)FP(O)\text{-}O\text{-}P(O)(OC_2H_5)_2 + CH_3CHO$$

The dissymmetric pyrophosphate esters are disproportionated during vacuum distillation.

β. Creation of the P-F Group in Phosphoric Ester Derivatives

β-1. Exchange Reactions with Phosphorochloridates. The chlorine-fluorine exchange is readily accomplished by heating the chloridate ester in a dry inert solvent (e.g., benzene) with sodium fluoride (e.g., Refs. 241, 819, 893, 1235, 1236), potassium fluoride,[193a,904b] ammonium fluoride,[564b,1353a] and potassium fluorosulfinate (without solvent, 0.5 hr at 120-130°).[1240a] The preparation of dialkyl phosphorofluoridates starting with phosphorus oxychloride without isolation of the intermediary dialkyl phosphorochloridate has been proposed in a patent: the alcohol (ethyl and isopropyl alcohol mentioned), phosphorus oxychloride and sodium fluoride are heated together in ethylene chloride.[517a] In another procedure the reaction mixture of phosphorus oxychloride with a sodium alkoxide is irradiated by ultraviolet light after addition of sodium fluoride.[1070] Primary dichloridates yield difluoridates (e.g., Ref. 616a).

The primary phosphorodichloridates may equally be used, but due to their hydrolyzability water is to be strictly avoided, which is not necessary for secondary phosphorochloridates.[241]

β-2. Exchange Reactions with Secondary Cyanophosphates. The title compounds undergo a similar exchange of the cyano group against fluorine by refluxing with potassium bifluoride in benzene.[241]

β-3. Polyphosphoric Esters and Hydrogen Fluoride. In the -P-O-P-O-P- chain of a polyphosphate ester anhydrous hydrogen fluoride splits the P-O bonds beginning at the end with the fluorine atom fixed on the terminal phosphorus atom. The reactions proceeds already at 20°.[698b,840a]

$$\text{-P-O-P-O-P-} + \text{HF} \longrightarrow \text{-PF} + \text{HOP-O-P-}$$

Tetraethyl pyrophosphate yields thus diethyl phosphorofluoridate, and the esters of higher polyphosphate react similarly, yielding not only dialkyl phosphorofluoridate but, as was to be expected, equally monoalkyl phosphoromonofluoridate. Ethyl metaphosphate yields ethyl fluorophosphoric acid, $C_2H_5OP(O)(OH)F$.

β-4. Fluorination of Monophosphoric Derivatives by Hydrogen Fluoride. Triphenyl phosphate is split by hydrogen fluoride with production of diphenyl phosphorofluoridate.[698b]

$$(C_6H_5O)_3PO + HF \longrightarrow (C_6H_5O)_2POF + C_6H_5F$$

A patent describes the condensation of monoalkyl or monoaryl phosphoric acid (or a salt of it) with hydrogen fluoride by means of trichloroacetonitrile or dicyclohexylcarbodiimide in an acid binding solvent. As examples are given methyl phosphorofluoridate, isolated as sodium salt, phenyl phosphorofluoridate (potassium salt), and sodium adenosyl 5-phosphorofluoridate.[933a]

β-5. Fluorination by Organic Fluorine Compounds. In the presence of strong tertiary amines 2,4-dinitrofluorobenzene and 2,4,6-trinitrofluorobenzene (picryl chloride) fluorinate phosphoric or thiophosphoric acid esters.[265c,1464a,1465]

The following mechanism is proposed:[1464a]

$$ROPO_3H_2 + F\text{-}\underset{O_2N}{\underset{|}{\bigcirc}}\text{-}NO_2 \xrightarrow{(C_2H_5)_3N} \begin{array}{c} ROP(O)O \\ -O \end{array} F\text{-}\underset{O_2N}{\bigcirc}\text{=}N\bar{O}_2$$

(hypothetic)

$$\longrightarrow ROP(O)F(OH) + HO(O_2N)_2C_6H_3$$

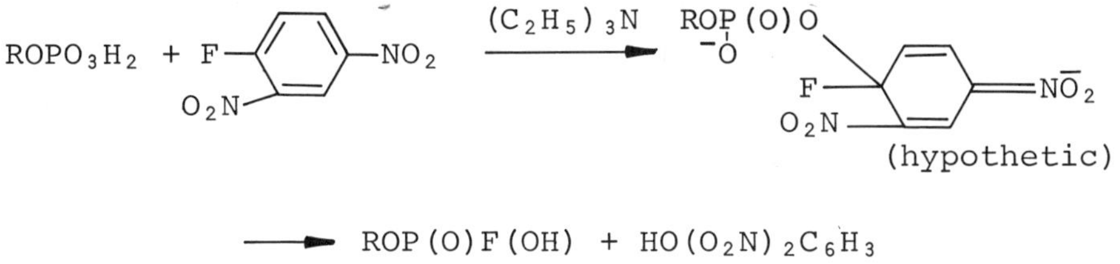

The use of a strong base is necessary; otherwise the intermediate product is split by a second molecule of the monosubstituted phosphoric acid with production of P^1,P^2-dialkyl(or diaryl) pyrophosphoric acid.[1464a] This procedure has been applied with success to nucleotides.[1465]

O,O-Dialkyl thiophosphoric acid, reacted as cyclohexylammonium salt with dinitrofluorobenzene or picrylchloride, yields the corresponding dialkyl phosphorofluoridate, $(RO)_2POF$.[265c]

Phenyl phosphoric acid reacted with <u>fluoranil</u> in the presence of a tertiary base yields phenyl hydrogen phosphorofluoridate.[410a] As intermediate a quinol addition product is postulated, as shown:

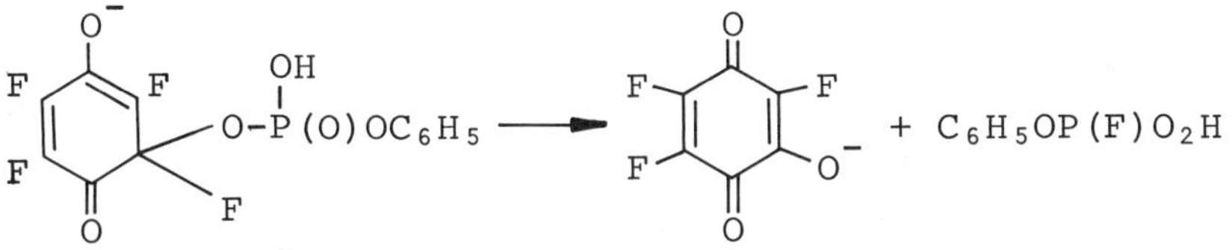

γ. Dialkyl Phosphorofluoridate from a Dialkyl Phosphonate
One instance of this sort of reaction has been reported recently, in the case of diethyl trichloromethylphosphonate heated with anhydrous potassium fluoride:[236a]

$$Cl_3CPO(OC_2H_5)_2 + KF \longrightarrow (C_2H_5O)_2POF + FCCl_3$$

b. Use of Phosphorus(III) Derivatives

α. Tertiary Phosphite + Phosphorus Pentafluoride
Trimethyl phosphite forms with phosphorus pentafluoride an adduct which, heated, breaks down in a complicated manner yielding some methyl phosphorodifluoridate, CH_3OPOF_2, besides CH_3OPF_4, $(CH_3O)_2PF$, and other products.[285a]

β. Secondary Phosphorofluoridites or Tertiary Phosphites
     and Halonitro, Halonitroso and Halofluoro Alkanes
The oxidation of secondary phosphorofluoridites, $(RO)_2PF$, by chloronitromethane or chloronitrosomethane, $Cl(O_2N)CH_2$ and $Cl(ON)CH_2$, respectively, in an inert solvent has been reported.[917b]

Reacted at -78° with 1,1,1,2,3,3,3-heptafluoro-2-nitrosopropane in difluorochloromethane, triethyl phosphite yields diethyl phosphorofluoridate.[498a] The reaction has been applied to a series of phosphites.

$$(CF_3)_2CF(NO) + (C_2H_5O)_3P \longrightarrow (C_2H_5O)_2POF + (CF_3)_2C{:}NOC_2H_5$$

Perhalogenated fluorochloroalkanes containing the grouping $-CF_2CCl_2-$, $-CF_2CFCl-$, or $-CFCl-CCl_2-$, the other positions being occupied by fluorine atoms, react with tertiary aliphatic phosphites (phenyl phosphite does not react[571a]) yielding the corresponding dialkyl phosphorofluoridates besides an unsaturated perhalogenated hydrocarbon with the grouping $-CCl{:}CCl-$, $-CF{:}CCl-$, or $-CF{:}CF-$, e.g., as follows:[194a]

$$CF_3CF_2CCl_2CF_3 + (CH_3O)_3P \xrightarrow[\text{45 min at 30°}]{\text{in benzene}} (CH_3O)_2POF + CF_3CF{:}CClCF_3 + CH_3Cl$$

The same reaction is afforded by $CF_3CFClCCl_2CF_3$ and by the saturated stable polymer $F(CF_2CFCl)_nF$.[571a] This last halogenated hydrocarbon is reacted with the phosphite in the ratio P:Cl = 1:1 at 100° where an exothermic reaction begins, as follows:

$$1/n\ F(CF_2CFCl)_nF + (RO)_3P \longrightarrow (RO)_2POF + 1/n\ F(CF{:}CCl)_nF + RCl$$

γ. Vinyl Phosphate from Tertiary Phosphite and Trichloroacetyl Chloride

This reaction is similar to the Perkow reaction (cf. Section I.2.b.γ-2) leading from tertiary phosphites to vinyl phosphates by means of α-chlorocarbonyl compounds. In the secondary phosphorofluoridite the F-P group remains unchanged during this reaction, but it is much more limited than with tertiary phosphites: only trichloroacetyl chloride yielded a phosphorofluoridate; acetyl chloride and benzoyl chloride were inactive. The reaction is conducted by heating the phosphorofluoridite with a large excess of trichloroacetylchloride during 5 hr at 80° under nitrogen. The yield was about 25% with the diethyl derivative.[1270a]

$$(C_2H_5O)_2PF + Cl_3CCOCl \longrightarrow (C_2H_5O)(Cl_2C{:}CClO)POF + C_2H_5Cl$$

VI. MISCELLANEOUS

1. Esters of Peroxyphosphoric Acid

Peroxyphosphoric acid is known since more than 50 years but only in 1959 the first representatives of its esters have been described which, in 1958, Larsson considered as probable intermediates in the Schönemann reac ion (oxidation of phosphonate derivatives by hydrogen peroxide).[851a] So far only tertiary esters of monoperoxyphosphoric acid with one t-butylperoxy grouping have been isolated.

They are prepared by reaction of phosphorochloridates with t-butyl peroxyalcohol in pyridine solution first at -10 to -20° then at room temperature[1192,1192a] or in a two-phase system (petrol ether and aqueous potassium hydroxide).[1275]

$$(RO)_2POCl + HOOC(CH_3)_3 \longrightarrow (RO)_2P[OOC(CH_3)_3]O + HCl$$

Esters with R = methyl, ethyl, propyl, isopropyl, butyl, isobutyl, phenyl, and benzyl have been prepared.[1275] Only esters with small aliphatic radicals are distillable in high vacuum, e.g., $(C_2H_5O)_2P(O_2\text{-t-}C_4H_9)O$ has b.p. 53-54°/0.003 Torr.[1192a] Bis(t-butylperoxy) esters were too unstable to be isolated.[1192a] Secondary esters of the type $(RO)_2PO_3H$ decompose in solution even at 0°.[1192a]

2. Phosphoric Acid Esters Labelled with Phosphorus Isotopes

Besides the stable isotope ^{31}P phosphorus has six isotopes (all radioactive) ranging from ^{28}P to ^{34}P, from which the species ^{32}P (half-life: 14.3 d) is most generally used for labelling purposes. The synthesis of ^{32}P-labelled phosphoric ester derivatives has been realized in three ways: use of inorganic ^{32}P compounds as starting material for classical syntheses; exchange reactions between non-labelled phosphoric derivatives and inorganic ^{32}P phosphate; creation of ^{32}P atoms by nuclear reactions.

a. Syntheses Starting from Inorganic ^{32}P Derivatives. As examples of these syntheses one can mention the use of phosphorus-32 oxychloride for the preparation of diphenyl phosphorochloridate labelled with ^{32}P for further syntheses,[657a] the use of labelled polyphosphoric acid prepared by dissolution of phosphorus pentoxide in labelled orthophosphoric acid (e.g., preparation of ^{32}P phosphorylcarnitine[968]), or the reaction of $^{32}PCl_3$ with hydroxy compounds followed by treatment of the labelled secondary phosphite by chlorine yielding the corresponding dialkyl phosphorochloridate which was further transformed into the fluoridate by reaction with sodium fluoride in carbon tetrachloride.[253a]

b. Exchange Reactions between Non-Labelled Phosphoric Derivatives and Inorganic ^{32}P Phosphate. This reaction has been realized by heating iron(III) ^{32}P phosphate with C_1 - C_6 trialkyl phosphates at 140° to 230°; at lower temperature there is no exchange, and above 230° decomposition takes place. As the presence of water accelerates the reaction, the exchange may result from intermediate hydrolysis. With triphenyl phosphate no exchange was observed even at 350°, and α-glyceryl phosphate exhibited in aqueous solution nearly no exchange.[1348,1349]

A rapid isotopic equilibrium has been reported for systems constituted by ^{32}P inorganic phosphate, adenosine triphosphate and added nucleotides in hemolysates of human erythrocytes previously incubated with pyruvic acid.[1399] The reaction products may be separated by high voltage electrophoresis in a pyridine-acetic acid-water buffer of pH 3.9.[1398a] Thus β and γ ^{32}P ribonucleoside triphosphate and β ^{32}P nucleoside diphosphate have been prepared.

c. Nuclear Reactions. ^{32}P may be prepared by neutron activation of ^{31}P. Thus the irradiation of orthophosphoric acid in glycerol produces ^{32}P phosphorous acid, hypophosphorous acid and glycerophosphorous acid (no P^V derivatives!); this last acid yields glycerophosphoric acid by oxidation with bromine. As in a mixture of labelled phosphorous acid, glycerol and ordinary phosphoric acid (or secondary potassium phosphate) no or nearly no glycerophosphorous acid was formed after two weeks at room temperature, the esterification is considered to result from the action of recoil atoms.[1478]

In a quite different reaction ^{32}P has been prepared by neutron irradiation of chlorine atoms under conditions which led to the production of ^{32}P phosphoric acid esters:

$$^{35}_{17}Cl + ^{1}_{0}n \longrightarrow ^{32}_{15}P + ^{4}_{2}He$$

Carbon tetrachloride with 1% of an alcohol, ROH, irradiated during 5 to 10 days with a flux of 1.17×10^6 n/s from a Po-Be source yielded a mixture containing besides inorganic phosphorus-32 compounds 12-16% of the ^{32}P produced as $(RO)_2{}^*PO_2H$ and respectively 28, 38, 44, 55, and 65% as $(RO)_3{}^*PO$ for R = methyl, ethyl, propyl, butyl, and amyl. Saturation of the tetrachloride with water prior to irradiation doubled approximatively the yield of labelled esters. The following mechanism is proposed. ^{32}P recoil atoms react with chlorine and other radicals formed by radiolysis of the substrate yielding phosphorus chlorides which by reaction with the alcohol and water present lead to the various P^V compounds revealed.[584,584a]

B. REACTIONS

A comprehensive review of structure and mechanisms in organic phosphorus chemistry has been published in 1965 (Hudson[709a]). The most important general features concerning the esters of phosphoric, polyphosphoric and halophosphoric acids as well as of acyl phosphates can be presented as follows.

I. PHOSPHORYLATION

Phosphorylation may proceed by bimolecular or by unimolecular reaction.

1. Bimolecular Phosphorylation

This is realized by reaction with an appropriate phosphorylating agent introduced either as such or prepared in situ.

In principle, phosphorohalidates, aromatic phosphates, and anhydrides react with bases (neutral or anionic) by direct displacement, e.g., as follows:

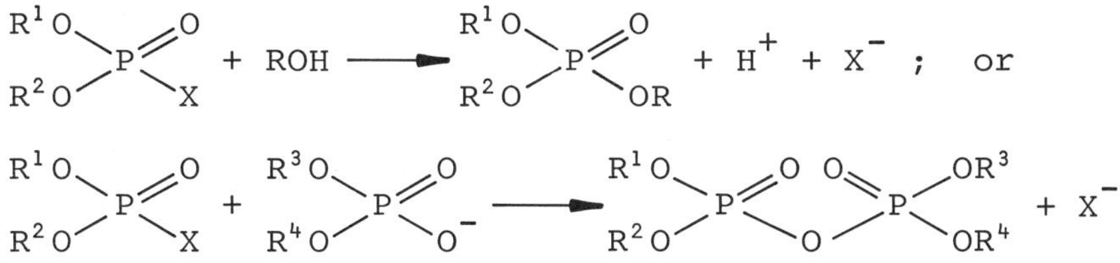

These reactions proceed with inversion of configuration (the molecules or ions of the type ABCP=O are tetrahedral).[625a] During hydrolysis in presence of $H_2{}^{18}O$ no ^{18}O exchange of the phosphoryl oxygen atom is observed in the case of chlorides,[485a] fluorides,[650a] or esters,[1314a] in confirmation of older considerations.[382,383] Therefore a short-lived addition product of diol type ($\gtrdot P <^{OH}_{OH}$) seems not to intervene (however, during alkaline hydrolysis or reaction with n-butylamine of the compounds (isoPropyl-O)$_2$P(O)X with X = F, Cl, CN, the formation of a transient penta-covalent intermediate has recently been considered as probable[709b]). These observations are compatible with the intervention of a transition state similar to that of a saturated carbon atom, i.e., with a S_N2 process.

In many cases the reactions schematized above lead to side reactions, and it is often advantageous to prepare the phosphorylating agent in situ. This is realized, e.g., by the use of agents such as carbodiimides, trichloroacetonitrile, etc. (cf. for the synthesis of esters in Section A.I.1 the paragraphs d.α and f.ζ-2, and for the production of the P-O-P bond Section A.III.1.b, α-1 and α-2). The protonated condensing agent yields the phosphorylating agent by reaction with the phosphate anion, e.g., as follows in the case of the production of a pyrophosphate:

$$(RO)_2PO_2H + R'N=C=NR'$$

proton transfer and substitution at the carbon atom

successive$\longrightarrow$ | simultaneous

$$R'N=C=\overset{+}{N}H \quad (RO)_2PO_2^-$$

$$R'N=\overset{|}{C}-NHR'$$

$$\begin{array}{c} | \\ O \\ | \\ PO(OR)_2 \end{array} + (RO)_2PO_2H \longrightarrow (RO)_2PO_2^- +$$

$$R'H\overset{+}{N}=\overset{|}{C}-NHR'$$

$$\begin{array}{c} | \\ O \\ | \\ PO(OR)_2 \end{array}$$

$$(RO)_2P(O)-O-P(O)(OR)_2 + R'HN-CO-NHR'$$

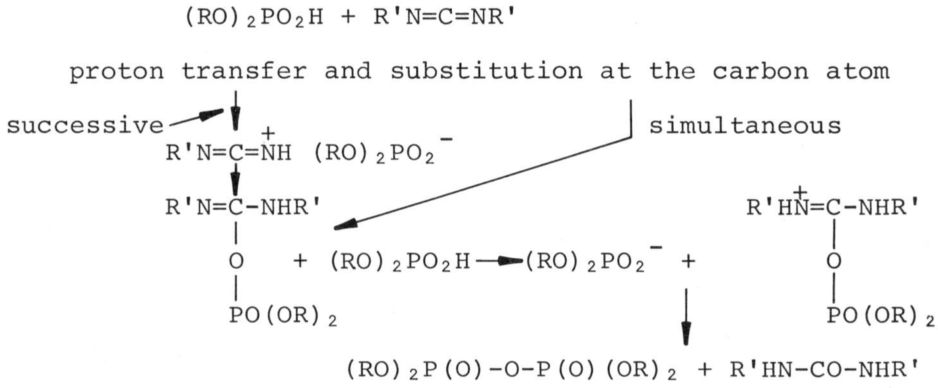

Similarly vinyl (enol) phosphates yield pyrophosphates by reaction with phosphoric acid or its primary or secondary esters (cf. Section A.III.1.b.ε).

In the oxidative phosphorylation, e.g., of adenosine diphosphate (yielding adenosine triphosphate) by action of bromine in the presence of phosphoric acid and various organic compounds capable of yielding quinonic derivatives, the phosphorylating agent is perhaps a quinone onium phosphate (cf. Section A.III.1.b.η), as follows in the case of α-tocopherol:[155]

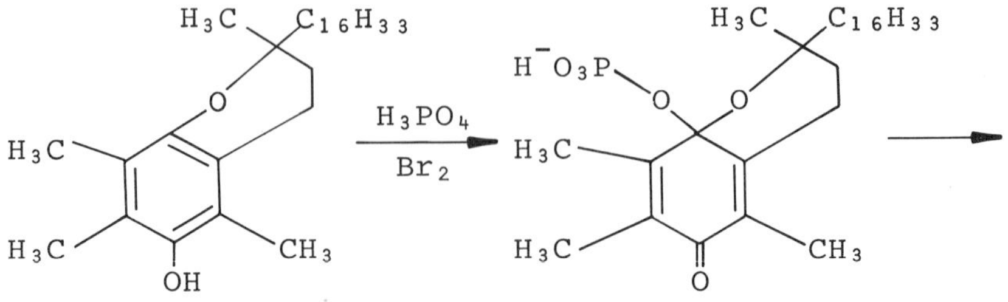

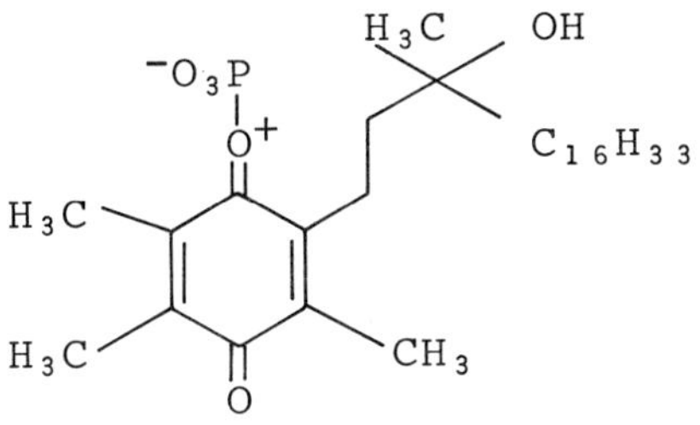

phosphorylating agent

The mechanism of phosphorylation by phosphorus oxychloride in particular has been studied by Hudson.[709b]

2. Unimolecular Phosphorylation[1355a,1449a]

By unimolecular heterolysis promoted by strongly electron-releasing groups a metaphosphate ion is produced as follows:

The metaphosphate ion yields esters by addition of alcohols, and pyrophosphates by addition of phosphoric acid (or its primary or secondary esters).

The high reactivity exhibited at a pH of about 4 by monoalkyl phosphates[162,474] may be due to a mechanism of this kind.

II. SCISSION OF ESTER FUNCTIONS

When speaking of hydrolysis one intends formally the replacement of the organic radical of an ester function by hydrogen. In the case of phosphoric esters, this reaction is in reality a very complex phenomenon.

$$(RO)(R'O)(R''O)P{=}O \longrightarrow (HO)(R'O)(R''O)P{=}O$$

R = hydrocarbon radical, R' and R" = H or hydrocarbon radicals; the hydrocarbon radicals may bear various functional groupings

This is shown by the extraordinary variety in the rate of this reaction, depending on the nature of the radicals R, R' and R", the pH of the medium, and the category of the ester.

1. "Hydrolysis"

In the series of phosphoric esters of simple aliphatic alcohols, the ease of hydrolysis in media of pH > 1 diminishes passing from tertiary to secondary and to primary esters. This is due to the electrostatic repulsion between the nucleophile (^-OH) and the ester in the anionic form.

For primary esters, e.g., methyl phosphoric acid, the curve of the rate of hydrolysis as a function of pH presents a minimum at about pH 1 and a maximum at pH 4-5 (e.g., Ref. 200a). This corresponds to the intervention of at least two mechanisms: at very low pH an acid catalyzed reaction (as for carboxylic esters) and at higher pH a reaction which is probably the unimolecular formation of metaphosphate anion.[365] The "hydrolysis" results from P-O fission (see, e.g., Ref. 200a); optically active 1-methoxyprop-2-yl phosphate, $CH_3OCH_2CH(CH_3)OPO_3H_2$, yields the alcohol with complete retention of configuration.[312]

The intermediary formation of metaphosphate is confirmed by the observation that in aqueous-alcoholic medium at pH ∿4.5 an exchange reaction takes place, as follows:

$$ROPO_3^{-} + R'OH \longrightarrow R'OPO_3^{-} + ROH$$

e.g., $R = C_2H_5$ and $R' = CH_3$

As under the conditions used, 100°, ∿120 hr) esterification of phosphoric acid by the aqueous alcohol is extremely slow, the primary ester $R'OPO_3H_2$ found cannot result from hydrolysis of $ROPO_3H_2$ followed by esterification of the liberated phosphoric acid by the alcohol R'OH; the formation of a metaphosphate intermediate explains the reaction.[378a]

Displacement reactions of this type are probably responsible for the exchange reactions between phosphoric esters and alcohols (phenols) (cf. Section A.I.1.e.β).

The nature of the organic radicals plays an important role.

Substitution of a phenyl group for an alkyl group produces a small increase in reactivity,[709d] enhanced by the presence of nitro groups.[123,605]

Vinyl (enol) phosphates are very reactive and hydrolyze in acidic as well as in alkaline medium. Alkaline hydrolysis results from nucleophilic attack of ^-OH on the electrophilic phosphorus atom with P-O fission;[1276,1306] acid hydrolysis proceeds probably by protonation of the β position of the vinyl group, followed by reaction with H_2O.[874a,1107,1306]

$$(RO)_2P(O)OC=C\diagdown \quad \xrightarrow{H^+} \quad \begin{array}{c} (RO)_2\overset{+}{P}(O)-O-\overset{|}{\underset{|}{C}}-\overset{/}{\underset{\diagdown}{CH}} \\ \updownarrow \\ (RO)_2\overset{+}{P}(O)-\overset{-}{O}=\overset{|}{\underset{|}{C}}-\overset{/}{\underset{\diagdown}{CH}} \end{array} \quad \xrightarrow{H_2O}$$

$$(RO)_2P(O)OH + O=\overset{|}{\underset{|}{C}}-\overset{/}{\underset{\diagdown}{CH}} + H^+$$

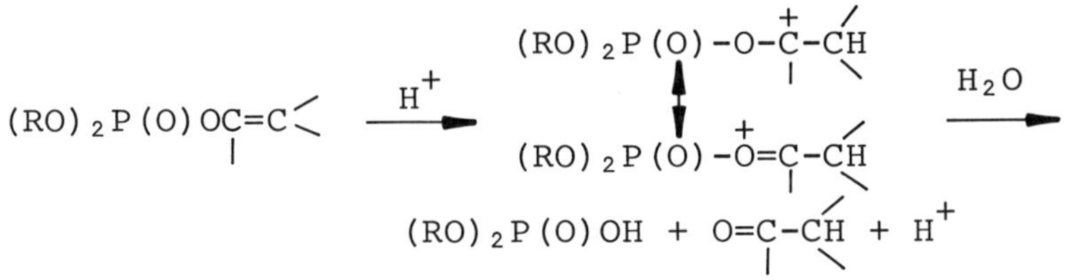

In very acidic medium ($HClO_4$, 1M and greater) the acid catalyzed hydrolysis with C-O fission predominates.[302,830]

Special kinetic studies are numerous: aliphatic esters,[200,801] dibutyl phosphate,[1489] ethyl phosphate,[652] hydrolysis of aliphatic esters promoted by $La(OH)_3$ gel,[313] by protenoids,[1074] methyl and neopentyl phosphate,[307] esters of phenol and substituted phenols (with the exception of nitrophenols),[199,200,305,366,381,699,941,942,1021] nitrated phenols,[123,303,304,306,472,665,914a] nucleotides.[1367,1495]

The geometry of the transition state of phosphate esters during hydrolysis has been studied by Dennis et al.,[471] the protolysis constants of phosphoric esters in dimethyl formamide have been measured by Potapov et al.[1133]

The hydrolysis of phosphoric esters of polyols presents particular features. As shown for the two glycerophosphoric acids, in acidic or basic medium an equilibrium between the two isomers is reached by intermediate formation of a cyclic derivative which is easily split, as shown:[370a,1198,1379,1411]

$$\begin{array}{c} CH_2OH \\ | \\ CH(OH) \\ | \\ CH_2OPO_3H_2 \end{array} \quad \underset{\xleftarrow{\hspace{1cm}}}{\overset{H^+}{\xrightarrow{\hspace{1cm}}}} \quad \begin{array}{c} CH_2OH \\ | \\ CHO \diagdown \\ \quad\quad\quad PO_2H \\ CH_2O \diagup \end{array} + H_2O \quad \underset{\xleftarrow{\hspace{1cm}}}{\overset{H^+}{\xrightarrow{\hspace{1cm}}}} \quad \begin{array}{c} CH_2OH \\ | \\ CHOPO_3H_2 \\ | \\ CH_2OH \end{array}$$

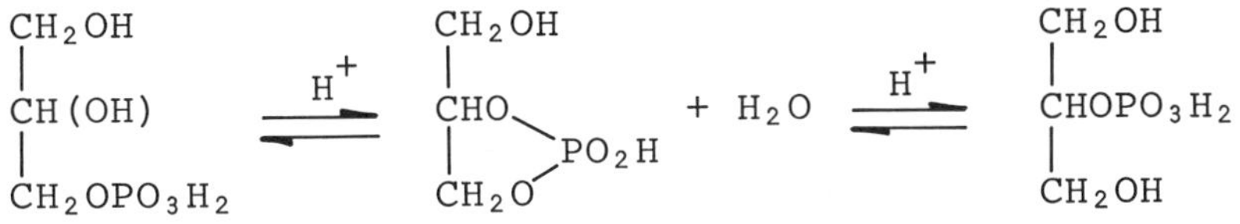

The pentagonal cyclic phosphate is indeed very rapidly hydrolyzed (see below).

Another instance of hydrolysis with formation of a cyclic intermediate is offered by esters of α-ketoalcohols with dialkyl phosphoric acids. These tertiary phosphates are very rapidly hydrolyzed, this time with production of a cyclic pentaoxyphosphorane derivative, in alkaline medium,[288,1168] as shown for dimethyl phosphoacetoin:

$$\begin{array}{c} CH_3-CHOPO(OCH_3)_2 \\ | \\ CH_3-C=O \end{array} + {}^-OH \quad \underset{\xleftarrow{\hspace{1cm}}}{\overset{}{\xrightarrow{\hspace{1cm}}}} \quad \begin{array}{c} CH_3-CHOPO(OCH_3)_2 \\ | \\ CH_3-\underset{|}{\overset{|}{C}}-O^- \\ OH \end{array} \quad \underset{\xleftarrow{\hspace{1cm}}}{\overset{}{\xrightarrow{\hspace{1cm}}}}$$

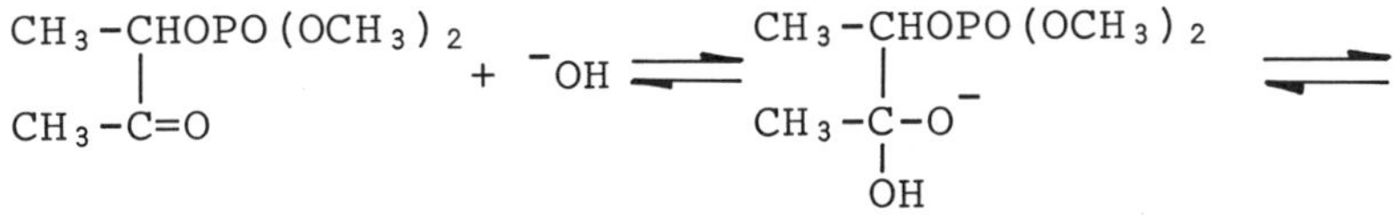

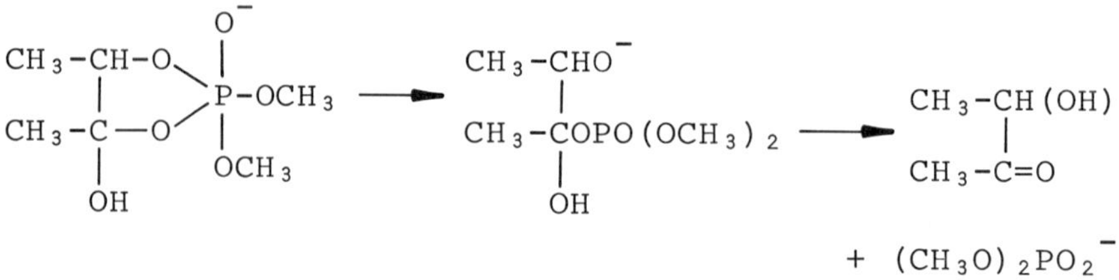

Kinetic data are to be found for esters of polyols and sugars: glycero dihydrogen phosphates and propane-1,2-diol 1-dihydrogen phosphate;[829] 2-hydroxycyclohexyl phosphates;[289,291] cyclohexyl 3-hydroxy-2-butyl hydrogen phosphate;[294] α-glucose 1-phosphate;[302] fructose 6-phosphate and 1,6-diphosphate;[1419] dephosphorylation of carbohydrate phosphates by ultraviolet light;[1366] 2(3 and 4)-pyridylmethyl dihydrogen phosphate;[1023] nucleotides,[1367,1495] diose phosphates;[559] glucose 4-phosphate.[497]

Besides acid catalysis many other catalytic effects have been observed. The rate of hydrolysis in alkaline medium is affected by the nature of the base in the case of nucleotides and glucose 1-phosphate,[1367] and of aromatic phosphates.[304] In acidic medium, it is dependent on the nature of the acid.[303] Therefore it is not surprising that, e.g., hydrolysis of salicyl phosphate, $HOOC-C_6H_4-OPO_3H_2$, presents a marked salt effect.[381] The promotion of the hydrolysis of various esters by lanthanum(III) hydroxide[313] has been mentioned above. In the catalysis of phosphate ester hydrolysis by "protenoids" (thermally prepared polycondensed aminoacids) these protenoids have been considered as "prebiotic enzymes"; their effect is inhibited by inhibitors of the natural enzymes.[1074]

Photochemical scission (VIS and UV) has been observed with nitrophenyl phosphates.[665] The hydrolysis of 2,4-dinitrophenyl phosphate dianion is strongly accelerated in the presence of dipolar aprotic solvents.[801]

In cyclic esters steric factors are very important. In general cyclic esters are split much more rapidly than similar non-cyclic derivatives. Thus the alkaline hydrolysis of ethylene phosphate is about 10 times more rapid than that of dimethyl phosphate, the first proceeding exclusively by P-O fission whereas the second exhibits only 90% P-O fission.[833b] The strain in the five-membered ring explains this difference. For orthophenylen phosphate and diphenyl phosphate the ratio of the rates is about the same, viz., $6 \cdot 10^6$.[759]

The importance of the strain is further illustrated by the fact that the hydrolysis of 1,2-dimethylvinylene alkyl phosphate (A) is more rapid than that of 1,2-dimethylethylene alkyl phosphate (B).[222,223]

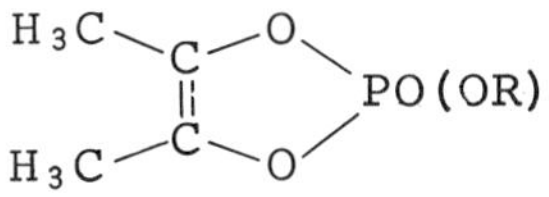

$$\underset{(A)}{\text{H}_3\text{C}-\overset{\overset{\text{C}}{\|}}{\underset{\text{C}}{\text{C}}}-\text{H}_3\text{C}} \quad \begin{array}{c}\text{O}\\ \text{O}\end{array}\text{PO(OR)}$$

(A)

$$\underset{(B)}{\text{H}_3\text{C}-\text{CH}-\text{H}_3\text{C}-\text{CH}} \quad \begin{array}{c}\text{O}\\ \text{O}\end{array}\text{PO(OR)}$$

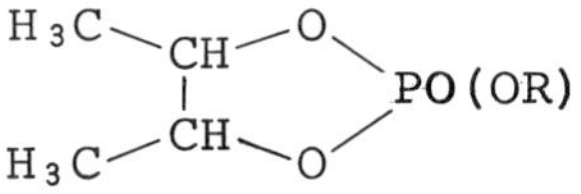

(B)

A detailed mechanism has been proposed for the alkaline hydrolysis of ethylene phosphates, based essentially on the assumption that the ring strain lowers the occupation of the phosphorus 3d orbitals and therefore deshields the P nucleus. This accounts for the high rate of nucleophilic attack in hydrolysis and O exchange.[267,1449a]

The alcoholysis of five-membered cyclic phosphates proceeds smoothly at ordinary temperature and yields the corresponding open-chained phosphoric esters.[624,1384] The reaction is acid catalyzed.[1384] Reaction of benzyl alcohol with 1,2-cyclic glycerol phosphate yields a mixture of about 1:1 α and β-glyceryl benzyl phosphoric acid; the cyclic ester of propane-1,2-diol reacts selectively with the P-O-C bond of the secondary alcohol function:[1384]

$$\underset{\text{CH}_2-\text{O}}{\overset{\text{CH}_3-\text{CH}-\text{O}}{|}} \text{PO}_2\text{H} + \text{HOCH}_2\text{C}_6\text{H}_5 \longrightarrow$$

$$\text{CH}_3-\text{CH(OH)}-\text{CH}_2\text{OP(OCH}_2\text{C}_6\text{H}_5)\text{O}_2\text{H}-$$

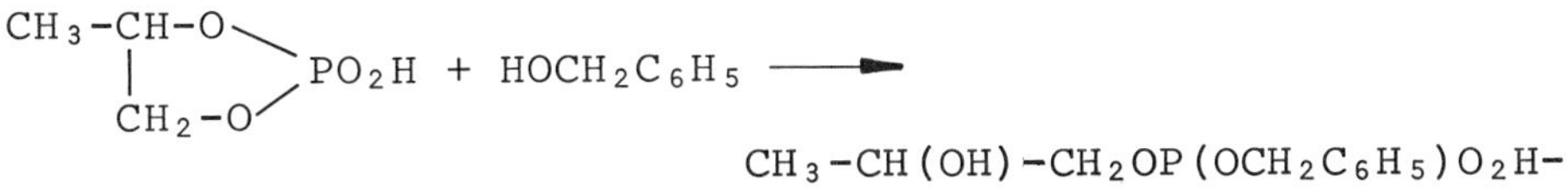

1,2-Dimethylvinylene ethyl phosphate reacts similarly, as follows:[624]

$$\underset{\text{CH}_3-\text{C}-\text{O}}{\overset{\text{CH}_3-\text{C}-\text{O}}{\|}} \text{PO(OC}_2\text{H}_5) + \text{HOC}_2\text{H}_5 \longrightarrow \text{CH}_3\text{CO-CH(CH}_3)-\text{OPO(OC}_2\text{H}_5)_2$$

Higher-membered cyclic phosphates--tetramethylen phosphate is the highest so far isolated[383a]--are much more stable. They are split in alkaline or acidic medium but--as the pentagonal esters[833a]--they are stable between pH 3.5 and 8.[383a,502a,1019]

A catalytic effect of the cation in alkaline hydrolysis has hereto been observed: trimethylene phosphates hydrolize with Ba²⁺ 35 times faster than with K⁺.[675]

With anhydrous hydrogen chloride in benzene solution the cyclic esters so far studied yield the corresponding haloalkyl open-chain phosphates.[1019]

$$\underset{\text{C}}{\overset{\text{C}}{|}} \begin{array}{c}\text{O}\\ \text{O}\end{array}\text{PO(OR)} + \text{HCl} \longrightarrow \text{ClC}\cdots\text{C-OPO(OR)OH}$$

R = alkyl or H

Ethylene ethyl phosphate treated with benzoic acid in benzene solution during eight days yielded a polyphosphoric derivative, as follows:

$$\begin{array}{c}\text{CH}_2\text{-O}\\ | \qquad\qquad \text{PO}(\text{OC}_2\text{H}_5)\\ \text{CH}_2\text{-O}\end{array} \xrightarrow{\quad \text{C}_6\text{H}_5\text{COOH} \quad} (\text{HOCH}_2\text{CH}_2\text{O})(\text{C}_2\text{H}_5\text{O})\text{PO-}$$

$$(\text{O}[\text{PO}(\text{OC}_2\text{H}_5)\text{OCH}_2\text{CH}_2\text{O}]_5\text{COC}_6\text{H}_5)$$

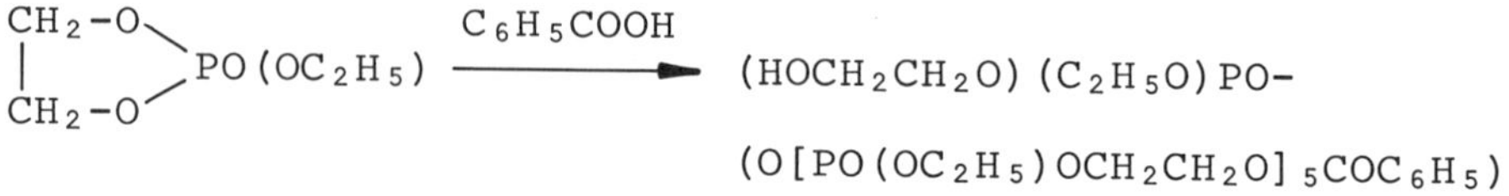

The authors suppose that electrophilic addition of benzoic acid on a C-O bond yields an acid ester, $(\text{C}_6\text{H}_5\text{COOCH}_2\text{CH}_2\text{O})\text{P}(\text{O})(\text{OC}_2\text{H}_5)\text{OH}$, which reacts similarly with another molecule of cyclic phosphate, yielding finally the polyphosphate ester found.[1190]

The relative stability of six-membered cyclic phosphate derivatives is demonstrated by the preparation of hexagonal cyclic phosphate esters with polyglycols by reaction of the corresponding phosphorochloridates with the polyglycols, described in a patent.[10]

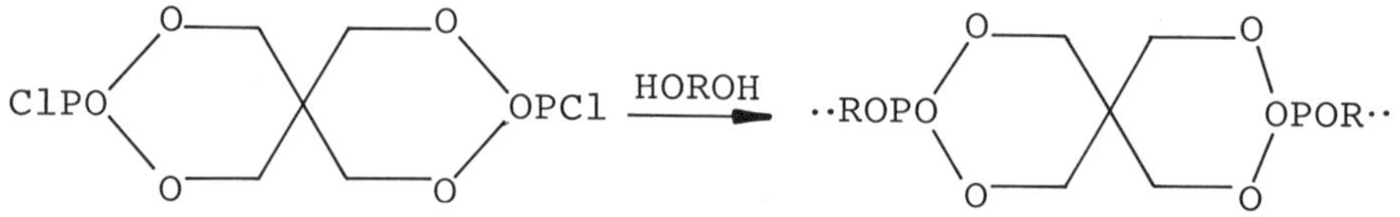

$$R = \text{HO}(\text{CH}_2\text{CH}_2\text{O})_n\text{H}$$

In relation with the insecticidal properties of p-nitrophenyl phosphates, e.g., paraoxon or diethyl p-nitrophenyl phosphate,[630] the behavior of five- and six-membered cyclic phosphate paranitrophenyl esters has been examined.

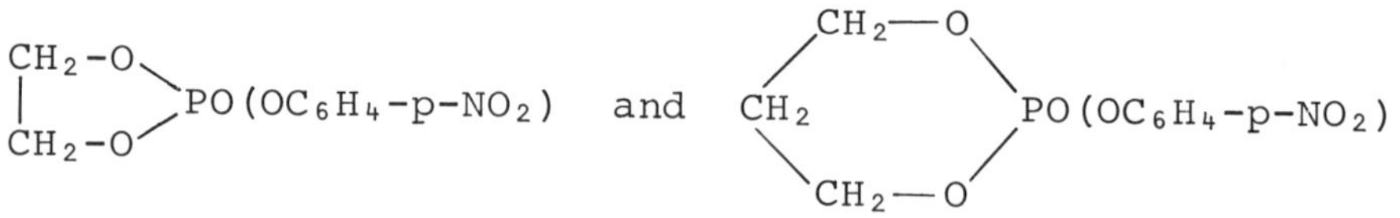

The six-membered derivative is split with liberation of paranitrophenol about at the same rate as paraoxon whereas the same reaction is almost instantaneous with the five-membered ester. Neither of them exhibits anticholesterase activity.[583]

2.   β-Elimination

The influence of the nature of the organic radical of phosphoric esters results not only in promoting either bimolecular or unimolecular scission of the ester function with liberation of the corresponding hydroxy compound; it can induce quite a different type of scission, i.e., β-elimination. This type of scission takes place preferentially in alkaline medium, in which monoalkyl phosphates are in general very stable. Rapid alkaline degradation of a phosphoric ester is therefore indicative of β-elimination.

Primary phosphoric esters in which the organic radical presents on the β carbon atom a conjugative group, e.g., -C≡N, -CO(NH₂), -CO(OR), -CO₂H, are characterized by rapid alkaline degradation.[384a] The ease of alkaline degradation of phosphoric esters of biological importance (e.g., glucose 2-phosphate, caseinogene) is long known.

The fact that the alkaline dephosphorylation of serine phosphate and the phosphopeptones resulting from partial hydrolysis of caseinogene is not a hydrolysis but a β-elimination has been established by several authors (e.g., Posternak[113,131a]).

This has been confirmed, e.g., by the production of acrylic acid by alkaline degradation of β-cyanoethyl phosphate during which the C-O bond is cleaved.[384a]

The conjugating group in β-position stabilizes the incipient carbanion, and in the alkaline medium a double negative charge results adjacent to the phosphorus atom, which both greatly reduce the rate of attack of base at the phosphorus atom.

$$N\equiv C-CH-CH_2-O-PO_3^{2-} \longrightarrow PO_4^{3-} + H_2O + NCC=CH_2 \xrightarrow{H_2O} HOOCC=CH_2$$

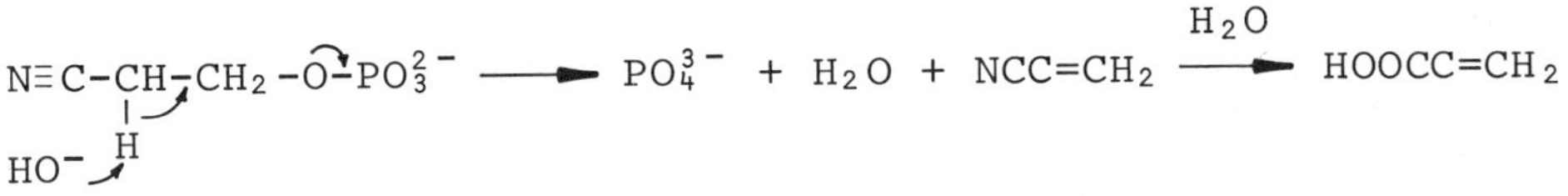

In confirmation of these findings, propargyl phosphate, $CH\equiv CCH_2OPO_3H_2$, and 2-cyano-2-methylpropyl phosphate, $(CH_3)_2C(CN)CH_2OPO_3H_2$, which bear no hydrogen on the β carbone atom, are very resistant to alkaline hydrolysis.[380b]

Kinetics of this type of degradation have been established for N-benzoyl-O-phosphoserine and its methylamide,[117,120] O-phosphoserine, O-phosphothreonine, phosphopeptones, and the catalytic effect of cerium and lanthanum ions,[192] glyceraldehyde and dihydroxyacetone phosphate,[290] glucose 3-phosphate.[825]

Another type of β-elimination is presented by β-aminoethyl phosphate which is capable of yielding aziridine.[292]

$$H_2NCH_2CH_2OPO(OC_2H_5)_2 \longrightarrow P(OC_2H_5)_2O_2^- + \overset{\overset{+}{NH_2}}{\underset{CH_2-CH_2}{\diagup\diagdown}}$$

3. Hydrolysis of Peroxyesters

The hydrolytic splitting of the dialkyl t-butyl phosphates attacks rapidly the P-O-O-C bond. Depending on the pH the scission takes place near the phosphorus or between the 2 oxygens. In water, the P-O fission is 80 to 90%, depending on the alkyl radical. For R = C_2H_5, O-O fission amounts to 26% in 0.1 N sodium hydroxide and to 90% in 2 N sodium hydroxide. The O-O fission yields methanol and acetone, as follows:

$$(RO)_2PO[OOC(CH_3)_3] \longrightarrow (RO)_2PO_2^- + {}^+OC(CH_3)_3 \xrightarrow{H_2O} H^+ + CH_3O\text{-}C(OH)(CH_3)\text{-}CH_3$$

$$\longrightarrow CH_3OH + CH_3COCH_3$$

4. Alkylating Reactions

Phosphoric acid esters are alkylating agents, stronger than carboxylic esters. Nucleophiles with high affinity toward the saturated carbon atom, e.g., I^-, react exclusively on the carbon (see e.g., Ref. 709c). This explains the dealkylation of phosphoric esters by halolysis with iodides, etc. (cf. Section A.I.1.e.α-3) realized by heating in acetone or methyl ethyl ketone, or by heating the ester directly with the salt (e.g., dealkylation of butyl phosphate by rare earth chlorides[1253]).

The dealkylation of phosphoric esters by appropriate tertiary amines (trimethylamine, N-methylmorpholine) results from alkylation of these amines (cf. Section A.I.1.-e.α-4). With diaryl methyl phosphates only methylation is observed.[1073] Pyrophosphates react similarly.[1448]

The alkylating action of phosphates and pyrophosphates is not limited to amines. 1 and 2-naphthylamine undergo nuclear alkylation, besides alkylation of the amino group, when heated with trimethyl phosphate at 250-275°,[753] e.g., as follows:

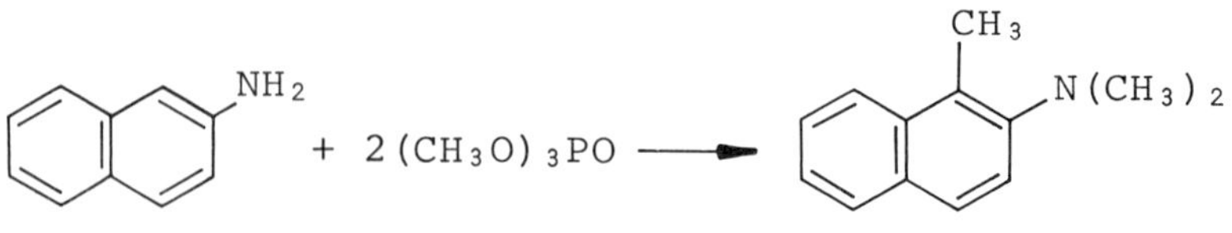

and

64%

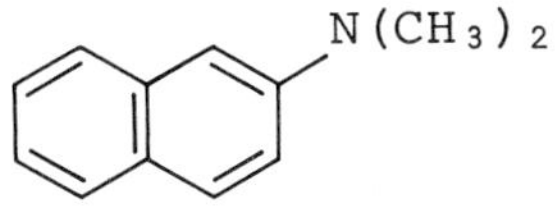

10%

Similarly tetrahydrocarbazole and 2,3-dimethylindole are methylated in β-position by heating with methyl polyphosphate at 160°.[1483]

A negative isotope effect has been observed with ^{14}C labelled trimethyl phosphate reacted with trimethylamine.[358]

In the presence of sterically hindered amines even alcohols may be alkylated, as, e.g., propanol which yields under these conditions benzyl propyl ether by reaction with tetrabenzyl pyrophosphate.[491]

5. Thermic Dealkylation

Aromatic esters are thermally very stable but the aliphatic ester functions tend to decompose at 400-500° (or below, depending on the nature of the radical) with production of olefines. Methyl groups are very stable.[215] For the tertiary alkyl aryl phosphates with 3 to 0 alkyls the stability grows in the following order: trialkyl < alkyl diphenyl < alkyl ditolyl and di(other aryl) < triaryl. Normal alkyl ester functions are more stable than branched ones but the absence of hydrogen on the β carbon atom enhances the stability of branched-alkyl esters.[590]

During pyrolysis isomerization of the olefines may occur. Thus mono, di, and tributyl phosphate yield a mixture of 1-butene and cis and trans 2-butene.[684]

The mono or diesters of phosphoric acid resulting from thermal degradation tend, of course, to yield polyphosphates. Self condensation may occur with appropriate ester groups. Thus mono β-chloroethyl phosphate undergoes the following thermal decomposition:[591]

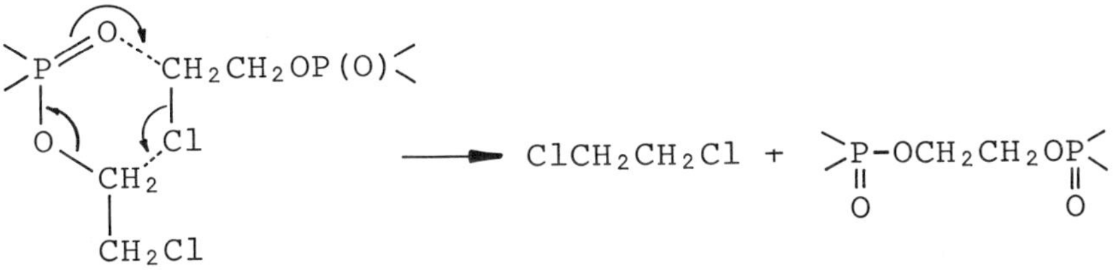

6. Acylating Reactions

As mentioned above (see subsection 4), phosphoric acid esters are in general alkylating agents. But depending on the nature of the ester radicals, they are capable of acting as acylating agents. This occurs with vinyl (enol) phosphates. Due to the positivation of the phosphorus atom they react with appropriate nucleophiles yielding the corresponding compound, nucleophile-OPO(OR)$_2$.

Vinyl phosphates in which the α carbon atom of the vinyl group bears an ethoxy radical are particularly apt to undergo this reaction, e.g., as follows.[443-5]

$$
\begin{array}{l}
\overset{\displaystyle OC_2H_5}{(C_2H_5O)_2P(O)-O-C=CHCO_2C_2H_5} \longrightarrow (C_2H_5O)_2P(O)OCOCH_3 \\
\quad + \quad -OCOCH_3 \quad H^+ \\
\qquad\qquad\qquad\qquad\qquad + C_2H_5O_2C-CH_2-CO_2C_2H_5
\end{array}
$$

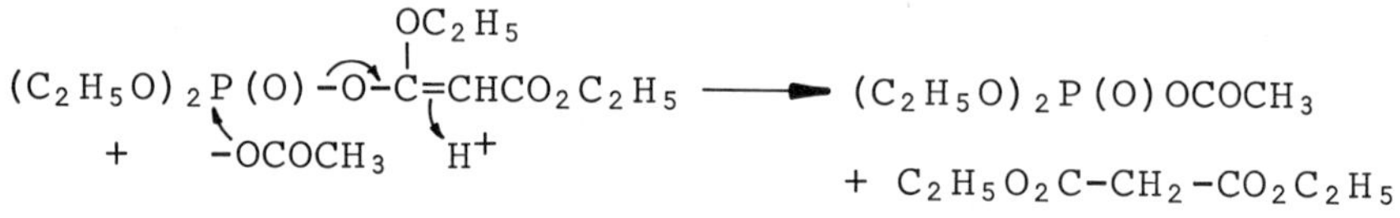

With di(hydrocarbyl) phosphoric acids the corresponding pyrophosphate tetraesters are formed.[443]

III.   MISCELLANEOUS REACTIONS OF ESTERS

Phosphates with unsaturated organic radicals are capable of _polymerization_ yielding high polymers and plastic masses. Triallyl and tris(2-methyl-allyl) phosphate polymerize at ~100° by initiation with benzoyl peroxide.[778] The polymerization (and copolymerization) of bis(methacryloyloxyethyl) phenyl or xylyl phosphate is induced at 20° by the same peroxide in the presence of N,N-dimethylaniline.[317]

On the other hand, diphenyl phosphorochloridate and tetraphenyl pyrophosphate reacted with cytidylic or pseudouridylic acid yield polymerized products, and with glycerol poly 3-(glyceryl-1-phosphoryl).[957,958]

The so-called Langheld ester (the reaction product of phosphorus pentoxide with ether, cf. Section A.III.1.a.β-2) has been used as _condensating agent_ in the Bischler-Napieralski reaction,[765] for the _esterification_ of phenol with carboxylic acids.[764]

Ethyl metaphosphate allows the preparation of isopropylidene derivatives of sugars by condensation of monoses with acetone.[1080]

Ethyl trimetaphosphate has been used for the synthesis of substituted benzimidazoles by condensation of o-phenylenediamine with carboxylic acids.[763]

By heating high molecular olefines with the complex formed between sulfur trioxide and an alkyl phosphate alkene sulfonates and hydroxyalkane sulfonates have been prepared.[1374]

Ethanolamine phosphatides catalyze the condensation of long chain aldehydes into 2,3-dialkyl acroleines. Choline phosphatides are less effective, simple amines are ineffective.[1297]

Bis(paranitrophenyl) phosphate, $(O_2NC_6H_4O)_2PO_2H$, has been used as a catalyst for nucleoside synthesis, e.g., for the condensation of theophylline with penta(O-acetyl) β-D-glucopyranose.[660]

Aromatic phosphates behave on substitution as expected. Thus triphenyl phosphate reacted with chlorine yields predominantly ortho and para tris(chlorophenyl) phosphate.[762]

The synthetic possibilities of cyclic pentaoxyphosphorane derivatives, discovered by Ramirez (see, e.g., Ref. 1169) and resulting in C-C condensation reactions, are to be considered in Chapter 5B.

IV. REACTIONS OF ANHYDRO ACID DERIVATIVES

1. Anhydro Acids with Inorganic Acids

a. With Oxyacids. The formation of these compounds obeys the general features presented above under I.

Practically, only the basic chemistry of anhydro acids with phosphorus acids has been explored. The anhydro function presents the general reactions of acid anhydrides, i.e., acylating reactions.

α. Polyphosphoric Esters

The P-O-P bond of inorganic polyphosphates is hydrolyzable rapidly only in acidic medium, in neutral or basic medium it is more or less stable. The same holds for this bond in the esters. As the phosphoric ester functions are generally hydrolyzed only very slowly, mild acidic hydrolysis of pyro and polyphosphate esters yields the corresponding orthophosphoric esters. Due to the stability of the P-O-P bond in alkaline medium pyrophosphoric esters with alkaline-labile ester functions yield inorganic pyrophosphate and the corresponding hydroxy compound by alkaline hydrolysis (e.g., scission of thiamine pyrophosphate by dilute aqueous barium hydroxide[1409]).

The hydrolysis of pyrophosphoric tetraesters, $(RO)_2$-$P(O)OP(O)(OR)_2$ has been investigated by Brock.[283] The author suggests a symmetric transition state for the P-O-P structure; the reaction is influenced chiefly by polar effects and very little--if at all--by steric effects.

The hydrolysis of the P-O-P bond is catalyzed by tertiary amines, in particular aromatic and heterocyclic amines of aromatic character (pyridines and imidazoles) (for a summary see, e.g., Ref. 444d). The catalytic activity of tertiary amines increases with their basicity.[450a]

Many polyphosphoric derivatives are of biological importance and their reactions have therefore been particularly examined.

The hydrolysis of seryl and diseryl pyrophosphate has been studied by Avaeva et al.[118,119,121]

Thiamine triphosphoric acid is split by water slowly into thiamine pyrophosphate and finally thiamine monophosphate; therefore the polyphosphoric chain is degraded in this medium beginning at the end with the free phosphoric acid residue. In alkaline medium the contrary takes place: the triphosphate as well as the pyrophosphate lose pyrophosphoric acid and yield thiamine monophosphate and thiamine, respectively.[1409]

Adenosine triphosphate (ATP) and adenosine diphosphate (ADP) have been extensively studied. The hydrolysis of ADP yielding adenylic acid and H_3PO_4 (i.e., here to degradation at the free phosphoric acid end) is of first order with respect to ADP and proceeds by nucleophilic attack by H_2O, as for the hydrolysis of acetyl phosphate and monoalkyl phosphates.[691]

In the metal ion catalyzed hydrolysis of ATP into ADP the reaction implies a special steric arrangement of metal ion, adenine ring, and phosphate groups in the respective nucleotide complexes. The catalytic (and possibly the enzymatic) enhancement of the degradation depends on the selective coordination of a metal ion or a proton on the non-terminal $-PO_2^-$ group, brought about by the occurrence of an additional ligand in juxtaposition to the $-PO_2^-$ group favoring, by a kind of chelate effect, the complex isomer in which the terminal P-O-P bond is labilized.[1298] ^{18}O from the aqueous medium enters the terminal phosphate group.[984a]

The acidic hydrolysis of uridine and N^3-methyluridine 5'(α-D-glucopyranosyl pyrophosphate) catalyzed by heterocyclic bases was found to be independent of the nature of the base.[301]

In aqueous <u>methylamine</u> the splitting of seryl pyrophosphate leads to O- and N-phosphoserine, due to interaction between phosphate and amino groups.[118,119]

With <u>alcohols</u> pyrophosphate esters react as does pyrophosphoric acid.

Between <u>carboxylic acids</u> and certain polyphosphate monoesters a transphosphorylation reaction may occur to a small extent, yielding acyl phosphates. Thus methylamido-N-benzoyl-DL-seryl pyrophosphate reacted with acetic acid or glycine yields respectively acetyl and glycyl phosphate (yield 1-4%); adenosine triphosphate and glycine in the presence of Be^{2+} at pH 5.2 gave in even better yield (maximum 22%) glycyl phosphate. The acyl phosphates have not been isolated but characterized by reaction with hydroxylamine yielding the corresponding organic hydroxamic acids.[122]

β. Phosphoric-Phosphorous Anhydro Acid Esters (Hypophosphates)

Michalski and co-workers have found that these derivatives may act as phosphorylating agents depending on the nucleophilicity of the partner. Alcohols and amines attack the phosphorus(III) atom, using a d-orbital for the formation of a donor complex. Acids attack the phosphorus(V) atom yielding acyl phosphates.[919,952a]

$$(RO)_2P(O)\text{-}O\text{-}P(OR)_2 + HOOCR' \longrightarrow R'COOPO(OR)_2 + (RO)_2PHO$$

With bulky R' (>C_3 or phenyl), however, no reaction takes place.[952a]

When using N-protected α-amino acids the corresponding mixed anhydrides result which may be used, without isolation, for the synthesis of dipeptides by reaction with a non-protected α-amino acid ester.[861]

Similarly primary or secondary phosphoric acid esters yield pyrophosphates.[919,952a]

γ. Other Anhydroacid Esters

Our knowledge in this field is very limited.

Sulfato phosphates are considered to be intermediates of enzymic activation of sulfate, therefore they should act as sulfating agents.[1183]

Silyl phosphates reacted with alcohols or phenol yield exclusively alkoxy and phenoxysilyl derivatives: they are silylating and not phosphorylating agents.

b. With Hydracids. The hydracids whose anhydro acids with partial phosphoric esters are of practical importance are hydrogen fluoride, chloride, and bromide. Phosphoroiodates are scarcely known (cf. Section A.V.2.b.-α-1). Just as in the inorganic chemistry of phosphoric acid derivatives, compounds bearing on the phosphorus simultaneously a halogen and a hydroxy group are unstable with the exception of fluorine. Therefore two types of phosphorochloridate and phosphorobromidate esters exist: $(RO)_2POHal$ and $(RO)PO(Hal)_2$; for fluoridates the type $(RO)P(O)(OH)F$ must be added. This difference between chlorine and bromine on one hand and fluorine on the other is due to the fact that the P-F bond is much more stable than the P-Cl or the P-Br bond. In fact, phosphorofluoridates present distinct chemical (e.g., selective phosphorylation of hydroxy groups in aminoalcohols with phosphorofluoridates) and biological (insecticidal activity) properties.

As set forth in Section A.V. the derivatives considered here are simultaneously esters and acyl halides, and are capable of exhibiting the reactions of both types of functions. In general the reactivity of the acyl halide grouping exceeds that of the ester function (exceptions for symmetric dialkyl difluoropyrophosphates, see below). Therefore the compounds considered may phosphorylate water (= hydrolysis yielding mono or diesters of phosphoric acid, alcohols and phenols (yielding esters), primary and secondary amines (yielding phosphoroamidates), salts of acids (yielding acyl phosphates and pyrophosphates). Phosphorodihalidates, $(RO)PO(Hal)_2$ react stepwise and can give rise to mixed phosphorochloridates or bromidates, (RO)(R'O)POHal. This type of reaction is particularly easy to realize with phosphorofluoridate-chloridate esters, (RO)POFCl, in which the chlorine atom is much more reactive than the fluorine atom.[1069a] These esters with mixed halogens are easily obtained by reaction of phosphorus oxydibromofluoride or oxydichlorofluoride, $POBr_2F$ and $POCl_2F$.[1069a,1319a]

The hydrolysis of phosphorochloridates, $(RO)_2POCl$, is considered to proceed by a one stage bimolecular displacement reaction of chlorine by water; in $H_2{}^{18}O$ there is no oxygen exchange.[485a] Dimethyl formamide is a very active catalyst of the hydrolysis.[1510] For solvolysis in general, see Ref. 485a.

The exchange reaction of the chlorine atom of compounds of the types $(RO)_2POCl$ and $(RO)POCl_2$ with $H^{36}Cl$ has been investigated by Drago et al.[485b]

Aksnes et al. have studied the hydrolysis of phosphorofluoridates, $(RO)_2POF$.[12c] In relation to the specific biological properties of this sort of compounds it is interesting to note that in animal plasma, red blood cells, and tissues an enzyme has been found that promotes this hydrolysis of the P-F group. This enzyme is unrelated to the enzymes catalyzing phosphatide hydrolysis.[926a]

In symmetric dialkyl difluoropyrophosphates, (RO)P-(F)(O)-O-(O)(F)P(OR), the alkyl-O-P bonds are activated. Carboxylic acids are transformed into esters and sodium iodide dealkylates these compounds; hydrogen chloride and iodide dealkylate equally. In each case, difluoro pyrophosphoric acid (or its salt), (HO)P(F)(O)-O-(O)(F)P(OH), is formed.[1319c]

2. Anhydro Acids with Carboxylic Acids (Acyl Phosphates)

The acyl phosphates are characterized by the acid anhydride group -C(O)-O-P···. In its reactivity it is placed between the P^V-O-P^V anhydride function of lesser reactivity and the carboxylic acid anhydride function, C(O)-O-C(O), generally of higher reactivity. The attention has been drawn to this class of compounds first by biochemists because of the importance of acetyl phosphate derivatives in biochemical acetylation.

<u>Hydrolysis</u> of acyl phosphates is as complex as that of monoalkyl phosphates as may be shown by the chemical hydrolysis of acetyl phosphate.

The non-enzymatic hydrolysis of this compound presents a pH-rate profile as complicated as that of methyl phosphate, e.g., and reveals the intervention of at least three mechanisms.[818a,818b] In very acidic medium (pH < 1) the non-dissociated substrate undergoes an acid catalyzed hydrolysis. At higher pH progressively the monoanion and finally the dianion predominate. Both are split probably by an unimolecular reaction with elimination of monomeric metaphosphate anion, as follows:[480b]

$$CH_3C \underset{O}{\overset{O}{\|}} \cdots \overset{H}{\underset{O}{\cdots}} \overset{O}{\underset{O}{\cdots}} P-O^- \longrightarrow CH_3CO_2H + PO_3^-; \quad CH_3\overset{O}{\overset{\|}{C}}-O-\overset{O^-}{\underset{O}{\overset{\|}{P}}}-O^- \longrightarrow$$

$$CH_3CO_2^- + PO_3^-$$

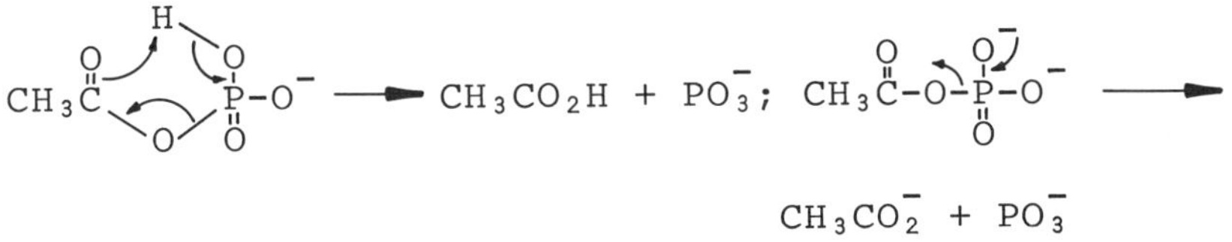

These conceptions are confirmed or partially contradicted by the following findings.

Simple acyl phosphates, in particular acetyl phosphate, were found to be cleaved in alkaline medium with C-O fission. Acidic and enzymatic hydrolysis proceed by P-O cleavage.[229]

Aliphatic acyl phosphates with bulky acyls exhibit a large ionic effect. Compared with trimethylacetyl and acetyl, it is strongest with 3,3-dimethylbutyryl for which C-O cleavage was again found.[1104]

Acetyl dibenzyl phosphate, $(CH_3COO)(C_6H_5CH_2O)_2PO$, presents general base catalysis in its hydrolysis which proceeds by C-O cleavage.[629]

The existence of at least two mechanisms results from the investigation of the hydrolysis of formyl phosphate: a proton catalyzed scission in acidic medium and, at pH > 5.5, ^-OH catalyzed hydrolysis. The kinetic study revealed two rate minima, at pH 5.5 and 7.2, respectively.[748]

The hydrolysis of acetyl phosphate is subject to metal ion catalysis which is most pronounced for the dianion, probably because this is the better chelating agent.[818b] In an investigation of the metal ion catalysis of the dianion, $CH_3COOP(O)O_2^{2-}$, between pH 5.8 and 8.8, the rate of hydrolysis was found to depend not only on temperature, pH, and catalyst (bivalent ions used: Zn, Co, Mn, Ni, Ca, Mg) but equally on the buffer. With Ca^{2+} and Mg^{2+} three types of reaction were found: (1) uncatalyzed, (2) attack on the intermediate metal complex by water, (3) attack on the complex by ^-OH.[1067]

The influence of the nature of the acyl radical on

stability and kind of hydrolysis is shown by the following examples.

Fluoracetyl phosphate, $FCH_2COOPO_3H_2$, is too unstable to be isolated pure. Its enzymatic hydrolysis is more rapid than that of acetyl phosphate.[913]

Ethoxycarbonyl phosphate could not be isolated by debenzylation of its dibenzyl ester, $(C_2H_5COO)(C_6H_5CH_2O)_2$-PO because spontaneous decomposition with evolution of carbon dioxide occurred.[849]

The two steps of the successive hydrolysis of dibenzoyl phosphate present each particular characteristic, as follows:[813]

$$(C_6H_5COO)_2PO_2H \xrightarrow[\text{catalysis}]{\text{base}} (C_6H_5COO)PO_3H_2 \xrightarrow[\text{neither acid nor base catalysis}]{} H_3PO_4$$

The cleavage occurs at the O-C bond in alkaline medium, and at the P-O bond in neutral and acidic medium.

In relation with the easy hydrolysis of acyl phosphates the following interpretation of the rapid hydrolysis of salicyl phosphate is of interest. This compound is split with a rate maximum at pH 5.3 (between 37 and 47°) with participation of the o-carboxyl group, as follows:[367]

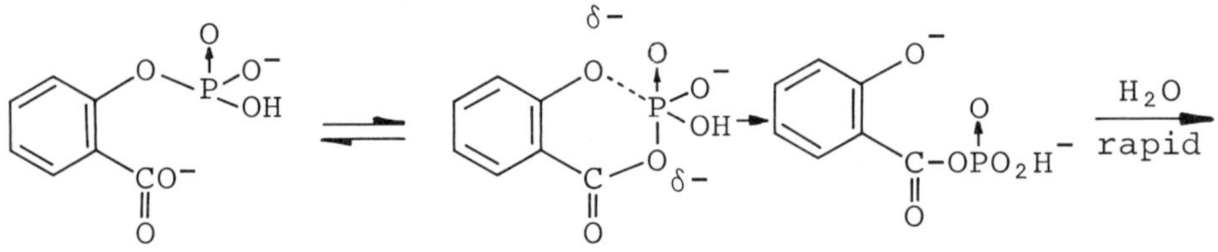

salicylic acid + H_3PO_4

Reacting with <u>alcohols</u> or <u>primary and secondary amines</u>, acyl phosphates are acylating, and not phosphorylating agents. Thus, e.g., carbamoyl phosphates may be used for transferring the carbamoyl group on the amino group of an amino acid, for instance with ornithine as follows,[883] the less steric hindered amino group being acylated:

$$HO_2CCH(NH_2)CH_2CH_2CH_2NH_2 + H_2NC(O)OPO_3H_2 \longrightarrow HO_2CCH(NH_2)CH_2CH_2CH_2NHC(O)NH_2 + H_3PO_4$$

But here too exceptions are found due to a large amount to sterical factors. Acyl groups with sterically hindered CO groups prevent the nucleophilic acylation and favor phosphorylation.

In the series of acyl diethyl phosphates, $(C_2H_5O)_2$-PO(OCOR), the compounds with R = CF_3, OC_2H_5, $C(CH_3)_3$ are acylating. The derivatives with R = C_6Cl_5, C_6HCl_4, $C_6H_2Cl_3$ are essentially phosphorylating. The acyl phosphate with R = $C(C_6H_5)_3$ treated with aniline yielded 29% of phosphoroanilidate, $(C_2H_5O)_2P(O)NHC_6H_5$, and 57% of N-triphenylacetyl anilide, $(C_6H_5)_3CCONHC_6H_5$.[840]

Trifluoroacetyl diethyl phosphate exhibits a particular reaction: it disproportionates in 24 hr at 20° into the corresponding symmetric acid anhydrides, as follows.[840]

$$2(C_2H_5O)_2PO(OCOCF_3) \longrightarrow (C_2H_5O)_2P(O){-}O{-}(O)P(OC_2H_5)_2 + (CF_3CO)_2O$$

C. GENERAL PHYSICAL PROPERTIES

I. ACIDITY

Esters of phosphoric, anhydrophosphoric, or halophosphoric acids which bear no hydroxyl group on the phosphorus atom(s) are neutral, with the exception, of course, of derivatives presenting organic radicals bearing free acidic or basic functions. In the case of anhydro or halophosphoric derivatives solvolysis generating hydroxy functions on a phosphorus atom must be excluded. Phosphorochloridates and bromidates undergo easy hydrolysis or alcoholysis producing acidic functions (iodidates are scarcely known--cf. Section A.V.2.b.α-1--but behave similarly). Anhydro acid derivatives with phosphorus acids are generally stable in neutral and often in alkaline medium but are easily split in acidic medium. Acyl phosphates are hydrolyzable at every pH and present generally a maximum of stability at about pH 4-5. Phosphorofluoridate groups are hydrolyzed only slowly. Cyclic esters with five-membered ring are split very rapidly, those with larger rings (as far as is known; cf. Section A.II.1.a) are relatively stable in neutral medium.

Partial esters of phosphoric and polyphosphoric acids are strong acids, di and monobasic, respectively. In the series of aliphatic esters the acidity, as compared with phosphoric acid, increases upon alkylation. This rise is in regression with the size of the alkyl group; a dialkyl ester is stronger than the corresponding monoalkyl ester.[834]

A general investigation of regularities in the dissociation constants of phosphorus acid derivatives has been published by Razumov et al.[1180] Its conclusion in respect to phosphoric acid derivatives is that the generic relations of organic phosphoric acids to inorganic phosphorus acids are best deduced by replacement of H by R rather than by that of OH by R.

The dissociation constants of dioctyl phosphoric acid in various aliphatic alcohols from C_4 to C_9 (unbranched chains), 3-methyl-propane-1-ol, t-butyl alcohol, cyclohexanol, and cyclohexane have been determined by Dulova et al.[492] The IR absorption spectra presented H bonding lines between 1220 and 1280 cm^{-1} and the P-O-C absorption in the 980-1100 cm^{-1} region.

The HO-P derivatives vary in their strength as a result of hydrogen bonding either to the OH oxygen (acidity decreasing) or to the P=O group (acidity increasing), but steric factors intervene equally.[834]

The progressive fall of acidity with increased size of the radicals is probably responsible for the formulation of monoesters with complex alcohols, e.g., cholesteryl phosphate, as "pyrophosphates" (Wagner-Jauregg[1432, 1432a,1433] and others[523,1494]). These compounds may be regarded as presenting dimeric or trimeric phosphate anions associated by hydrogen bonding (yielding ring structures?). The alkali metal salts of these acid esters appear to have similar ring structures.[540,714,1210,1432, 1432a]

II. HYDROGEN BONDING--DIMERIZATION

Determination of hydrogen bonding in tertiary phosphates shows a rapid rise of the "additive" factor in the oxygen of the phosphoryl group upon esterification.[99,920] As the halophosphates are probably intermediate, the entire series from phosphorus oxyhalides to tertiary esters may be expected to have a progressive increase of additive affinity at the oxygen atom, providing the driving force for reaction such as the formation of polyphosphates from the esters and phosphorohalidate derivatives.[822] The tertiary phosphates show the semipolar bonding of the oxygen in the phosphoryl group, P=O, in parachor and molecular refractivity measurements, rather than the double bond character.[73-75,612,1416]

By IR spectroscopy, the following order of hydrogen bonding ability has been established for a number of tertiary amides and esters of phosphoric and trithiolphosphoric acid: $[(CH_3)_2N]_3PO$ > $(C_6H_5O)_3PO$ > $(C_2H_5S)_3PO$ > $(C_6H_5S)_3PO$ > $(p\text{-}CH_3C_6H_4O)_3PO$ > $(m\text{-}CH_3C_6H_4O)_3PO$ > $(o\text{-}CH_3C_6H_4O)_3PO$ > $[(C_2H_5)_2N]_3PO$.[653]

Hydrogen bonding has been investigated in dialkyl propargyl phosphates.[1338]

Association (mostly dimerization) due to hydrogen

bonding was found to be stronger in partial esters of phosphoric acid than in carboxylic acids. In benzene and naphthalene, the following were found to be dimeric: bis-(2-ethylhexyl), diphenyl, bis(2,6-dimethylhept-4-yl), bis[p-(1,1,3,3-tetramethylbutyl)-phenyl], and dibutyl phosphoric acid whereas the following monoalkyl phosphoric acids presented higher association: (2-ethylhexyl) and (p-1,1,3,3-tetramethylbutyl)-phenyl phosphates.[1094]

Dimerization constants are given for bis(2-ethylhexyl) phosphoric acid in benzene and carbon tetrachloride (and interaction with tributyl phosphate in the same solvents) [1288] and for various trialkyl phosphates in cyclohexane, dodecane, and hexane.[1099]

III. MOLECULAR STRUCTURE

A comprehensive survey of structural chemistry of phosphorus compounds in general due to Corbridge was published in 1966.[433a] The author examines not only inorganic derivatives; numerous data are given concerning primary and secondary phosphoric esters either as free acids or as salts, and tertiary esters.

All four-connected compounds with a single phosphorus pentavalent phosphorus atom have a tetrahedral configuration. In the compounds with four oxygen atoms around the phosphorus, one bond--that of the phosphoryl oxygen--is shorter than the others.

Based on parachor determinations it was concluded that in the tertiary esters of the type $(RO)_3PO$ the three chains OR are in parallel positions.[75]

In barium diethyl phosphate, $[(C_2H_5O)_2PO]_2Ba$, the two C-O bonds were found to be in gauche position in respect to the P-O bonds.[837]

As examples of structure determination of nucleic derivatives, one can cite those of adenosine 5'-phosphate,[825b] adenosine 3'-phosphate dihydrate (adenylic acid b),[1282] β-adenosine-2-β-uridine phosphate.[1246a]

The conformation of nucleoside mono and diphosphates has been established by PMR (100 MHz).[1371]

Data on structure, conformation, and other physical properties of aryl phosphates have been collected by Gamrath et al.[590]

Cyclic esters, especially pentagonal (phospholanes) and hexagonal (phosphorinanes), present interesting problems.

In methyl ethylene phosphate the O-P-O angle in the ring is 99° (against about 106° for the (R)-O-P-O-(R) angle in non-cyclic tertiary esters). Again the P=O bond is shorter than the P-O-(R) bonds. The ring is puckered, due probably to hydrogen bonding.[1318]

Phenyl trimethylene phosphate (2-oxo-2-phenyl-1,3,2-dioxaphosphorinane) presents its ring in chair conformation with the P=O group in equatorial and the phenyl group in axial position.[597]

Examining respectively 1,2-diphenylethylene phosphate (sodium salt) and 1-phenyltrimethylene phosphate (ammonium salt), Tsuboi et al. reached similar conclusions. The conformation of the trimethylene ester (phosphorinane derivative) was found to be fixed at room temperature; the C-R bonds are considered to be rather trans with respect to the O-P bond around than gauche. The coupling constant H-C-O-P was found (in D_2O) to be J_{HP} = 6.9 Hz for the phospholane derivative, and 1.5 Hz for the phosphorinane derivative.[1372]

For 5,5-disubstituted 1,3,2-dioxaphosphorinane derivatives (free acid, bromidate, and phenyl ester) the following conformations were found in the crystal and in solution, respectively, as follows:

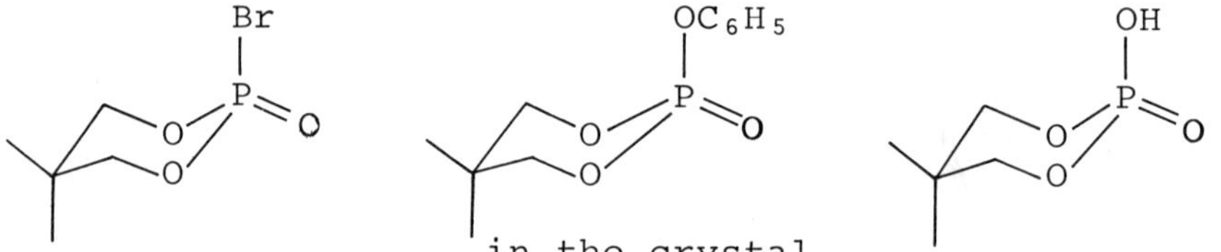

in the crystal

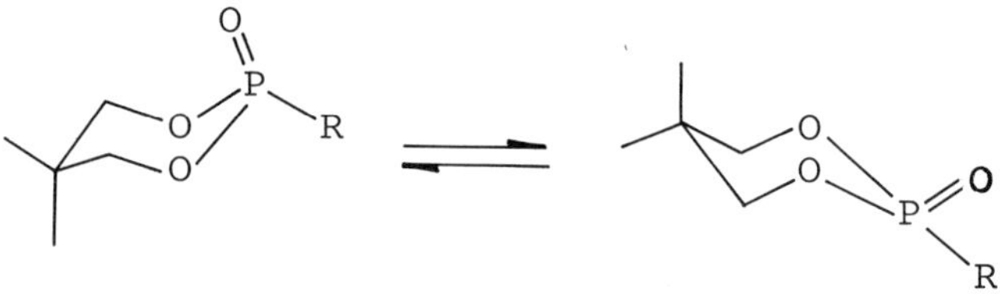

R = $O-CH_2CH_2CH_3$, $O-CH_2CH_2CH_2CH_3$, $O-C_6H_5$

in solution

The equatorial orientation of the P=O group is favored in solution as well as in the crystal.[758]

Hexacyclic phosphorohalidates have further been examined with similar results by Beinecke: 2-bromo-2-oxo-(5-methyl-5-bromomethyltrimethylen)-1,3,2-dioxaphosphorinane with the p-Br and the 5-methyl groups on the same side of the ring,[219b] and 2-bromo-2-oxo-(4-bromo-5-methyl-5-bromomethyltrimethylene) phosphate with the P-Br, the 4-bromo, and the 5-methyl groups on the same side of the ring,[219a] and by Thierry et al.: the chloridate and the bromidate of the cyclic phosphate ester of a complicated 1,3-glycol resulting from hydration-rearrangement of cyperene epoxide (cf. Section A.V.1.a.α where the developed formula is given[885b]).[1349b]

IV. COMPLEXES--USE OF ESTERS FOR SOLVENT EXTRACTION

The number of complexes derived from phosphoric esters is very large and only examples of some types will be mentioned.

With very strong alkylating agents such as stable trialkyl oxonium salts with anions of very low nucleophilicity complexes are formed as follows (in general the reverse reaction takes place):

$$R_3P{=}O + R'X \longrightarrow R_3P(OR')X$$

The reaction has been examined with the systems $(C_2H_5)_n$-$(C_2H_5O)_{3-n}PO$ + $(C_2H_5)_3OBF_4$ with n = 0, 1, 2, and 3. The complexes are prepared by boiling the components in methylene chloride or ethylene chloride during two to three hours. Thus the compound $(C_2H_5O)_4PBF_4$ has been obtained.[1034]

The complexation of tributyl phosphate with the tetrachlorides of Ti, Sn, Zr, Hf, and Th, followed by IR spectroscopy, gave the following results. Tin and titanium tetrachloride give rise to complexes in the ratios 1:2 and 1:1. In the two tin complexes and the titanium complex 1:2 the phosphate is bound by its phosphoryl oxygen. In the 1:1 complex of titanium tetrachloride the phosphate ester acts as bidentate ligand with the phosphoryl group and one P-O-C oxygen.[1076]

The adducts formed between triamyl phosphate and lanthanide salts have been investigated by Siddall et al. In the reaction with lanthanum nitrate the environment of the La ion is not affected by the solvents used (deuteriochloroform, chloroform-hexane, hexane). NMR revealed a difference between light and heavy rare earths.[1259]

In the complex between O-phosphorylserylglycine with Cu^{II} the ligand is bound mainly through the phosphoryl group.[1077]

The ease of complexation of various metal cations by phosphoric acid esters, where the phosphoryl group seems to be particularly active, has lead to the use of phosphoric esters (partial or neutral) for solvent extraction. Hereunto the publications--scientific papers or patents--are too numerous for an exhaustive enumeration. Examples of the most frequent uses shall be given.

For the extraction of hexavalent or tetravalent uranium, dibutyl phosphoric acid was found to be the best agent compared with monobutyl phosphoric acid, tributyl phosphate, and dibutyl pyrophosphate.[1389] Dibutyl phosphate in benzene solution has been successful for the extraction of cerium.[984]

Tributyl phosphate has been proposed for the extraction of uranium(VI) (solution in carbon tetrachloride used).[89] The distribution of uranyl nitrate and nitric acid between water and the ester has been investigated.[1278] As this example shows, inorganic acids may be extracted with phosphoric esters. Thus hydrofluoric acid and sulfuric acid, both present together in aqueous solution, are extractable with tributyl phosphate.[817] Boric acid has been extracted with the addition compound $H_3PO_4 \cdot 2(C_4H_9O)_3PO$.[818]

The formation of adducts between triamyl phosphate and lanthanide cations[1259] just mentioned above may be used for solvent extraction.

With monoisoamyl phosphoric acid indium sulfate has been extracted.[618]

(2-Ethylhexyl) phosphoric acid has found numerous applications: extraction of indium sulfate,[618] antimony(V) from 5 - 7N hydrochloric acid or 4 - 6N sulfuric acid (organic solvent: heptane),[873] gallium.[872]

Bis(2-ethylhexyl) phosphoric acid: for neodymium and samarium chloride, perchlorate or nitrate,[860] gallium,[872] tetravalent metal cations, particularly Ce^{IV} from dilute sulfuric acid into kerosene (no sulfate anion in the organic phase),[1343] uranium(VI) (organic solvent: carbon tetrachloride),[89a] antimony(V) from 3N hydrochloric acid or 5 - 9N sulfuric acid.[873]

(2,6,8-Trimethylnonyl) phosphoric acid for indium sulfate.[618]

(2-Ethylhexyl) pyrophosphoric acid for gallium.[872]

Dibenzyl phosphoric acid for zinc and hafnium.[439]

V. SPECTRAL PROPERTIES

1. UV Absorption Spectrum

The simple phosphoric esters have not very significant UV absorption spectra. Triethyl phosphate is almost transparent between 1260 and 1100 cm^{-1}. Molar extinctions for this compound in methanol, ethanol, hexane, and heptane from 2250 to 2700 Å have been tabulated.[650]

The UV absorption spectra of D-arabinose and D-ribose containing dinucleotide phosphates are discussed by Maurizot et al.[924]

Wada et al. have investigated the UV spectra of four crystalline forms of S-benzoylthiamine O-phosphate at various pH and found isobestic points at 239 and 277 nm.[1427]

2. IR Absorption Spectrum, Raman Spectrum

A recent review (Corbridge, 1969[432a]) presents a comprehensive survey of the known facts which may be resumed as follows.

P-F stretching frequencies in $(RO)_2POF$ 905-875 cm^{-1}; bending 500-300 cm^{-1} (approximatively)

P-Cl in $(RO)_2POCl$ 590-480 cm^{-1}, in $(RO)POCl_2$ 587-488 cm^{-1} (strong) and 544-420 cm^{-1} (less strong)

P-Br 495-320 cm^{-1}

P=O 1450-1980 cm^{-1} (overall range, very strong), 1320-1200 cm^{-1} (most compounds); hydrogen bonding resulting from interaction with P-OH produces shifts of 50-80 cm^{-1}

P-O-H in organophosphorus acids: shallow absorption at 2700-2560 cm^{-1} (disappearing on salt formation) indicative of hydrogen bonding greater than that of carboxylic acids, -C(O)OH. Six regions are connected empirically with the P-O-H group

I	3000-2525 cm^{-1}	(P)-O-H stretching
II	2400-2000	" "
III	1900-1600	P-O-H combination bands
IV	1400-1200	(P)-O-H deformation
V	1030- 820	P-O-(H) stretching
VI	540- 450	" deformation

(P)-O-C region of 1030 cm^{-1}

P-O-(C) 1242-1087 cm^{-1} (stretching)

P-O-C aliphatic: $P-O-CH_3$ 1200-1168 cm^{-1}; $P-O-CH_2$ 1170-1105 cm^{-1}; P-O-CH 1190-1087 cm^{-1}

P-O-C aromatic 1242-1110 cm^{-1} (O-C stretching); 996-905 cm^{-1} (P-O-(Aryl))

Special papers: P=O: general,[12b] 900 compounds,[1350a] frequencies between 1400-1160 cm^{-1} depending on electronegativity of the substituents (rising frequency with rising electronegativity).[377a] Esters, haloesters.[651,923a]

P-O-C stretching: combination bands at 2000-1700 cm^{-1} ∿sum of frequencies of C-O and P-O bonds.[1068] P-Cl 12 compounds.[1060] PO_2.[889,889a] PCl_2.[395b] P-O-C at ordinary temperature,[651,673b-c,1257a] at low temperature (77K).[673a] Si-O-P in (R R'SiO) POF 1015 cm^{-1} and in $(CH_3)_3SiOPOF$ 950 cm^{-1}.[487a] In cyclic esters (5,5-dimethyl-2-oxo-2-(chloro, alkoxy, aryloxy)-1,3,2-oxaphosphorinanes) P=O 1300-1290 cm^{-1}, P-O-C three bands in the region 1077-1003 cm^{-1}.[501]

Raman spectra: Diisopropyl phosphate derivatives present at least two frequencies for P=O stretching: symmetric at 720-750 cm^{-1} (strong) and asymmetric at 973-1030 cm^{-1} (weak or missing).[889a] PCl_2 compounds.[395b]

3. NMR spectra

The solvent effect on ^{1}H-NMR in aqueous solutions of nucleotides has been investigated by Prestegard et al.[1140]

The properties of the CH_3OP group are examined in Ref. 924a, those of phosphorochloridates, $(RO)_2POCl$, and of tertiary esters, $(RO)_3PO$, in Ref. 916a (Martin et al.).

Sidell et al. conclude that the phosphate esters spend an important fraction of time in a single preferred general conformation. Rotation around P-O-C (and P-C) is considered to be rapid.[1308b]

Penta and hexacoordinated phosphorus derivatives have been examined by Latscha.[853]

^{1}H-NMR data are given for numerous individual compounds.

^{13}C-NMR has been examined in CH_3-O-P derivatives by Martin et al.[916b]

In the field of ^{19}F-NMR Bystruv et al. have studied the additive relations for the chemical shifts, and the F-P spin interaction.[317a]

The P•••H spin-spin coupling in P-O-CH in various dimethyl phosphate and cyclic (phosphorinane type) derivatives has been examined by Kainosko, who found that the ring of the phosphorinanes was virtually fixed in the chair form.[757a] This conforms with the conclusions mentioned in Subsection III, above.

For the ^{31}P-NMR spectra (compilation of data, interpretation and quantum-mechanic theory of chemical shifts, coupling constants, high resolution spectra) "Topics in Phosphorus Chemistry" offer in Vol. 5 an excellent review published in 1968.[1358a]

^{31}P-NMR collected data are published in Ref. 978a, P•••H coupling constants are given by Latscha,[853] chemical shifts for halophosphate esters by Nielsen et al.[1047a]

Steric effects: in tertiary esters, the replacement of a bulky t-alkyl residue by ethyl or methyl gives a nearly linear downfield shift (in contrast to the behavior of trialkyl phosphites where the first replacement of this type results in an upfield shift, and only the second one in a downfield effect) explained by fractional bond-angle changes.[914]

The theory of interpretation of ^{31}P shifts in symmetric and unsymmetric molecules is discussed by Letcher et al.[861a]

In pentagonal cyclic esters the double bond character of the P→O group was found to be diminished.[245]

4. Mass Spectrum

This method of investigation has not given rise to generalized research work. The base peaks were found to result from rearrangement or cleavage. The rearranged ions were formed by hydrogen migration from an alkyl or aryl residue to the P-O skeleton.[460]

VI. PROPERTIES RELATED TO BIOLOGICAL ACTIVITY

1. Free Energy of Hydrolysis

The driving force for a large number of endergonic biochemical reactions is furnished by high-energy compounds, a notion introduced in biochemistry by Lipmann.[877a] Biochemists consider compounds to be "energy-rich" when their reaction with a substance commonly present in the environment is accompanied by a large negative free energy change at a physiological pH.

As far as phosphoric acid esters or anhydrides are considered, we are interested in the energy of hydrolysis of substances compatible with the living cell, i.e., phosphoric esters, compounds with P-O-P bonds, and acyl phosphates.

The limit between energy-rich and energy-poor substances is drawn arbitrarily at a free energy of hydrolysis at pH 7 of about 7000 cal/mole.

In this respect it is interesting to note the position of some of the phosphoric acid compounds of the mentioned types. The standard free energy of hydrolysis at pH 7 amounts to 3000 cal/mole for ordinary phosphoric monoesters, to 6000 for ordinary diesters; these compounds are therefore energy-poor, just as for instance ordinary acetic acid esters (5100 cal/mole). The corresponding energies (always expressed in cal/mole) are 10,500 for acetyl phosphate, 7400 for magnesium adenosine triphosphate (ATP) when split into adenosine diphosphate and inorganic phosphate, and 7600 when split into adenosine monophosphate and inorganic pyrophosphate, 13,000 for phosphoenolpyruvate (as example of vinyl phosphates). Therefore the compounds of these last types are all "energy-rich."

2. Factors Important for Insecticidal Properties

Three types of the phosphorus derivatives considered in this chapter have found application in the field of insecticides: fluorophosphoric esters, vinyl (enol) phosphates, and pyrophosphoric esters. All these compounds are characterized by their anticholinesterase activity which is very powerful with certain individual compounds.

The mode of action is considered to be based on the electrophilic phosphorylation of a nucleophilic site of the enzyme (see, e.g., Ref. 13a). Once phosphorylated, the enzyme has lost its activity, and the inhibition is of course the longer the more stable the phosphorylated derivative is.

The relation between insecticidal activity (contact or systemic action), general toxicity, and the remainder of the inhibiting molecule is still obscure. Data for insecticidal properties of numerous organophosphorus compounds have been collected, e.g., Refs. 1082a, 1418.

The relation between resistance against hydrolysis and the insecticidal power has been investigated in particular for vinyl (enol) phosphates by Schuler.[1306] A relation exists: compounds with good insecticidal activity are found mostly in the class of medium hydrolyzability in acidic medium, but no simple conclusions can be drawn.

Steric factors may be of influence. In the case of phosdrin for instance, methyl (3-dimethoxyphosphoryloxy)-but-2-enoate, the rate of hydrolysis as well as its course are different for the cis (more rapid hydrolysis) and the trans isomer, as follows:[1276]

$$
\begin{array}{c}
H_3C \quad\quad\quad H \\
\diagdown C{=}C \diagup \\
(CH_3O)_2P(O)O \diagup\quad\diagdown CO_2CH_3
\end{array}
\qquad\qquad
\begin{array}{c}
H_3C \quad\quad\quad CO_2CH_3 \\
\diagdown C{=}C \diagup \\
(CH_3O)_2P(O)O \diagup\quad\diagdown H
\end{array}
$$

hydrolysis, pH 11, 30°, 24 hr

$$
\begin{array}{c}
H_3C \quad\quad\quad H \\
\diagdown C{=}C \diagup \\
(CH_3O)_2P(O)O \diagup\quad\diagdown COOH
\end{array}
$$

and

$$
\begin{array}{c}
H_3C \quad\quad\quad H \\
\diagdown C{=}C \diagup \\
{CH_3O \atop HO}{>}P(O)O \diagup\quad\diagdown CO_2CH_3
\end{array}
$$

$(CH_3O)_2PO_2H$

(and trace of)

$$
\begin{array}{c}
H_3C \quad\quad\quad CO_2CH_3 \\
\diagdown C{=}C \diagup \\
{CH_3O \atop HO}{>}P(O)O \diagup\quad\diagdown H
\end{array}
$$

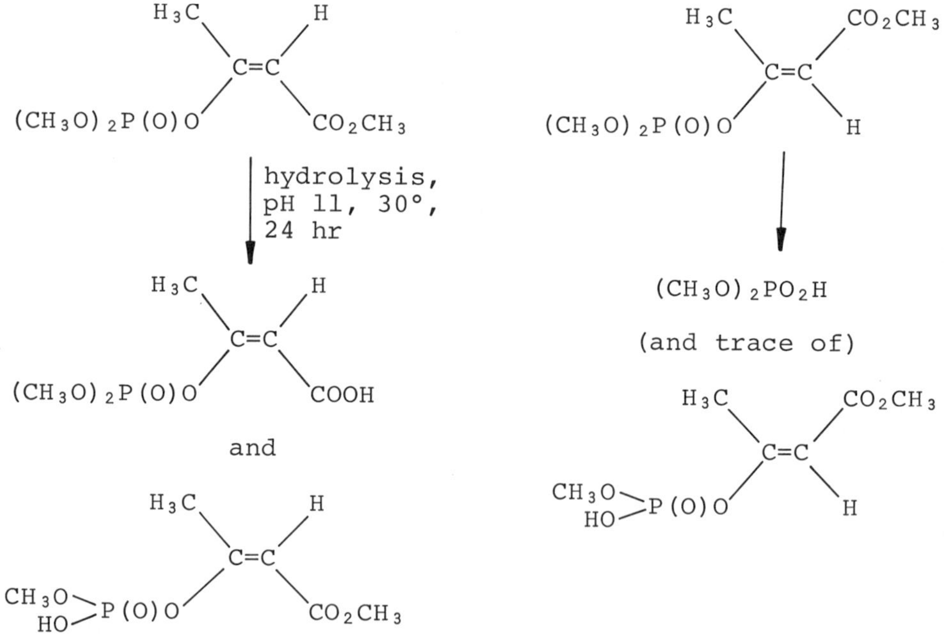

But on the other hand, in the case of the following vinyl phosphate, dimethyl 1,3-di(methoxycarbonyl)-1-propene-2-yl phosphate, $(CH_3O)_2P(O)O{-}C{=}CH(CO_2CH_3)$
$\qquad\qquad\qquad\qquad\qquad\qquad\quad | $
$\qquad\qquad\qquad\qquad\qquad\quad CH_2(CO_2CH_3)$

the two isomers cis and trans, respectively, exhibited the same toxicity for insects.[1047]

There is still a large field to explore in this problem.

VII. MISCELLANEOUS

The investigation of the polarographic half-wave potential of tertiary phosphoric esters, $(RO)_3PO$, with identical or different R radicals has shown that a 2d event takes place yielding either RH and $(RO)_2PO_2$ or ROH and $(RO)_2P(OH)$.[904]

The molecular ionization potentials of trimethyl phosphate and triethyl phosphate were found to be 10.7_7 and 10.0_6 eV, respectively.[157a]

Tolkmith has compiled data of electron group polarizability and molecular properties of organophosphorus compounds.[1356b]

The vapor pressure of tertiary phosphoric esters has been determined by Dobry et al.[483]

Physico-chemical data for fluorophosphoric derivatives have been compiled by Gutowsky et al.[644a]

D. LIST OF COMPOUNDS

This list of compounds registered until the beginning of 1970 does not pretend to be exhaustive--more than 2000 compounds are listed--but the author hopes to have omitted no important and specific item. The compounds are classed as indicated below (the finer subdivisions are not repeated here). In each final subdivision the compounds are listed in order of increasing number of carbon atoms of the skeleton or of increasing complexity; the titles of the subdivisions are self-explanatory.

Non-Cyclic Orthophosphoric Esters

I. Mono Primary Esters, $ROPO_3H_2$

A. R bound to O by an Aliphatic Carbon Atom

1. R = Hydrocarbon or Halogenated Hydrocarbon Radical

a. R Aliphatic

$MeOPO_3H_2$. I.1.a.α,[348,655,884] I.a.β,[338] I.1.b.γ-1,[1296] I.1.d.ε,[165,166] I.1.c.α,[448a] I.1.e.α-2 (from hydrobenzoin ester by hydrogenolysis),[138b] I.1.f.ζ-12,[504] I.2.b.α-6 (from H_3PO_3 with MeOH and Hg^{2+}),[1065] I.2.b.β-4 (from phosphoromorpholidite by $NCCHBrCONH_2$ and hydrolysis);[972] from $(MePO_3)_3$ by methanolysis (besides sym. $Me_2H_2P_2O_7$).[448a] Na_2 salt;[165] Ba salts;[165,166,348,448a,655] Ca salts;[165] Pb salt;[655] monoanilinium salt, m. 167-8°.[1065] Conformation of monosodium salt, J_{H-P} 10.3 Hz (in D_2O).[1372]

$EtOPO_3H_2$. I.1.a.α,[348,884,1122,1413] I.1.a.β,[388,400,842,875,1091] I.1.b.γ - I.1.e.α-2,[96] I.1.b.η (from R_4NPO_4 in EtOH in presence of $C{=}N^+$ containing compounds),[408] I.1.c.α[182,842,1014] I.1.c.β-2,[1205] I.1.d.ε,[165,166,474] I.1.e.α-1,[975] I.1.e.α-2 (hydrogenolysis of hydrobenzoin ester,[1386] or diphenyl ester[1335]), I.1.e.α-3 (from dibenzyl ester with NaBr or NaLi),[974] I.1.f.ζ-5 (with dimorpholino-POCl(Br), etc.),[988,989] I.1.f.ζ-6 (from $PhCONHPO_3H_2$),[1509] I.1.f.ι (from 2-Cl-decyl phosphonate in EtOH),[925,926] I.2.b.α-6 (from H_3PO_3 with ROH by Hg^{2+}),[1065] I.2.b.β-2 - I.1.e.α-2,[95,97] I.2.b.β-4.[971,972] From $(EtPO_3)_3$ by H_2O-pyridine (besides sym. $Et_2H_2P_2O_7$).[448a] Sirup;[988,989] Ca salts;[165,1120] Ba salts;[182,342,1091] di c-$C_6H_{11}NH_3$ salt, m. 205-6°;[926] di anilinium salt, m. 200-5° (dec.),[1109] 164-5°;[971] mono anilinium salt, m. 104-6°.[1065] Conformation of monosodium salt, J_{H-P} 6.3 Hz (in D_2O).[1372]

$ClCH_2CH_2OPO_3H_2$. I.1.a.β,[383x,547,1122] I.1.b.γ-1,[397,633] I.1.c.α,[842] I.1.e.α-1.[667] Ba salt;[383x,547,633,842] Na_2 salt.[667]

$BrCH_2CH_2OPO_3H_2$. I.1.d.ε. Ba salt.[1501]

$PrOPO_3H_2$. I.1.c.β-1,[1022] I.1.d.δ-ε,[165] I.2.α-6 (from H_3PO_3 with ROH by Hg^{2+}),[1065] I.2.β-4.[971,972] Ca salt;[165] Ba salt;[349,1122] $C_6H_5NH_3$ salt, m. 136-9°,[971] 137-8°.[1065] Conformation of $Ba_{1/2}$ salt, J_{H-P} 6.5 Hz (in D_2O).[1372]

$CH_2{=}CHCH_2OPO_3H_2$. I.1.a.α,[343,344,348] I.1.d.δ-ε,[165,344,1501] I.1.f.ζ-5,[988,989] I.1.f.ι.[925] Sirup;[989] Ca salt;[165] Ba salt;[160,343,344] di-c-$C_6H_{11}NH_3$ salt, m. 175°,[989] 200°.[925]

$CHF_2CF_2CH_2OPO_3H_2$. I.1.a.α-β. Needles, m. 65.0-68.0.[226]

$ClCH_2CH_2CH_2OPO_3H_2$. I.1.a.β. Ba salt.[383x]

$BrCH_2CH_2CH_2OPO_3OH_2$. I.1.d.δ-ε. Barium salt.[1501]

i-$PrOPO_3H_2$. I.1.d.δ-ε,[165] I.1.e.α-2 (hydrogenolysis of hydrobenzoin ester),[1386] I.2.b.α-6 (from H_3PO_3 with ROH by Hg^{2+}),[971] I.2.b.β-4.[971,972] Ca salt;[165] Ba salt;[349] mono $C_6H_5NH_3$ salt, m. 159-160°,[1065] 160-162°.[971]

$BuOPO_3H_2$. I.1.b.γ-2 - I.1.e.α-2 (use of $PhCH_2$ or Ph ester,[95] of hydrobenzoin ester[1386,1387]), I.2.b.α-6 (from H_3PO_3 with ROH by Hg^{2+}),[1065] I.2.b.β-4 (from $PhCH_2OPO_2H_2$ with ROH and bromo-amide, then hydrogenation),[971] Ba salt;[95] mono $C_6H_5NH_3$ salt, m. 138-140°,[1065] 139-141°.[971]

$ClCH_2CH_2CH_2CH_2OPO_3H_2$. I.1.a.β. Ba salt.[383a]

i-$BuOPO_3H_2$. I.1.d.δ,[165,191] I.1.e.α-2 (using phenyl ester),[1335] I.2.b.α-6 (from H_3PO_3 with ROH by Hg^{2+}).[1065] Ca salt,[165] Ba salt;[349] anilinium salt, m. 155-156°.[1065]

$Me_2C{=}CHOPO_3H_2$. I.1.f.ζ-2 (ROH with $(Et_3NH)_2HPO_4$ in MeCN). di-c-$C_6H_{11}NH_3$ salt, m. 189-190°.[1394]

t-$BuOPO_3H$. I.1.f.ζ-2,[449a] I.1.f.ι (rearrangement of 2-Cl-decyl phosphonate with t-BuOH). Di-c-$C_6H_{11}NH_3$ salt, m. 205-206° (dec.),[925,926] 191-193°.[449a]

$AmOPO_3H_2$. I.1.e.α-3 (using benzyl ester and NaI),[480] I.2.b.α-6 (from H_3PO_3 with ROH by Hg^{2+}).[1065] Cryst., m. 62-64°. Ir.[480] Anilinium salt, m. 135-137°.[1065]

i-$AmOPO_3H_2$. I.1.a.β,[644] I.1.e.α-2 (using phenyl ester),[1335] I.2.b.α-6 (from H_3PO_3 with ROH by Hg^{2+}).[1065] Cu^{2+} salt;[644] anilinium salt, m. 149-151°.[1065]

n-$C_7H_{15}OPO_3H_2$. I.1.a.β.[1033]

$H(CF_2)_6CH_2OPO_3H_2$. I.1.a.α and β. Cryst., m. 111-112°.[226]

n-$C_8H_{17}OPO_3H_2$. I.1.f.ζ-5 (using o-phenylene phosphorochloridate),[319] I.2.b.α-6 (from H_3PO_3 with ROH by Hg^{2+}).[1065] Na salt; $Ba{\cdot}2H_2O$ salt; mono c-$C_6H_{11}NH_3$ salt, m. 153-155°;[319] mono $C_6H_5NH_3$ salt, m. 129-130°.[1065]

2-$EtC_6H_{12}OPO_3H_2$. I.1.a.β.[1032,1033]

$H(CF_2)_8CH_2OPO_3H_2$. I.1.a.α and β. Cryst., m. 139-140°.[226]

n-$C_{10}H_{21}OPO_3H_2$. I.1.a.β. Cryst., m. 45°.[1033]

Geranyl phosphate. I.1.f.ζ-2 (from H_3PO_4 with ROH in MeCN + CCl_3CN).[449a] Di c-$C_6H_{11}NH_3$ salt, m. 190-192° (dec.).[449a]

Neryl phosphate. I.1.f.ζ-2 (from H_3PO_4 with ROH in MeCN + CCl_3CN). Di c-$C_6H_{11}NH_3$ salt, m. 177° (dec.).[449a]

Linalyl phosphate. I.1.f.ζ-2 (from H_3PO_4 with ROH in MeCN + CCl_3CN). $(NH_4)_2$ salt, m. 148-150° (dec.).[449a]

"Citronellal phosphate" (probably $C_8H_{15}CH{=}CHOPO_3H_2$). I.1.a. Plates, m. 203°.[484]

$H(CF_2)_{10}CH_2OPO_3H_2$. I.1.a.α and β. Cryst., m. 158°.[226]

n-$C_{12}H_{25}OPO_3H_2$. I.1.a.β,[1032,1033] I.1.f.ζ-5 (using o-phenylene phosphorochloridate).[319]

n-$C_{14}H_{29}OPO_3H_2$. I.1.a.β. Cryst., m. 68°.[1032,1033]

Farnesyl phosphate. I.1.b.f.ζ-2 (from H_3PO_4 with ROH in MeCN + CCl_3CN). Mono c-$C_6H_{11}NH_3$ salt, m. 164-165° (dec.).[449a]
Nerolidyl phosphate. I.1.f.ζ-2 (from H_3PO_4 with ROH in MeCN + CCl_3CN). $(NH_4)_2$ salt, m. 133-136°.[449a]
n-$C_{16}H_{33}OPO_3H_2$. I.1.a.α,[239] I.1.a.β,[338] I.1.b.δ-1,[395,633,1433] I.1.c.α,[1122] I.1.f.ζ-5 (using o-phenylene phosphorochloridate, followed by hydrolysis).[319] Cryst., m. 73-76°,[1433] 74-75°,[319] 72°,[239,1122] 71°;[395] Na salt,[319] NaH salt, m. 178-179°;[1433] Ba salt, not water soluble[395,633] (the indicated water-solubility[1122] is in error); piperazinium salt ($4H_2O$), m. 229-230°.[1433] Mol. weight in camphor: tetrameric.[1433]
Oleyl phosphate. I.1.a.α. Waxy solid; no definite melting point.[316]
n-$C_{20}H_{41}OPO_3H_2$. I.1.a.β. Liquid, n_D^{25} 1.4514.[1033]

b. R Containing a Cycle

Hydnocarpyl phosphate and dihydrohydnocarpyl phosphate. I.1.a.α. Waxy solids; no definite melting points.[316]
Chaulmoogryl phosphate and dihydrochaulmoogryl phosphate. I.1.a.α. Waxy solids; no definite melting points.[316]
$PhCH_2OPO_3H_2$. I.1.d.δ,[1501] I.1.c.α,[842] I.1.e.α-2 (using hydrobenzoin ester).[1386] Ba salt.[842,1501]
$PhCH_2CH_2OPO_3H_2$. I.1.f.ζ-3,[214] I.1.f.ζ-5 (using dimorpholino-POCl(Br)).[989] Sirup;[989] Ba salt;[214] mono c-$C_6H_{11}NH_3$ salt, m. 177°.[989]
$MePh_2COPO_3H_2$. I.1.b.α. $(NH_4)_2$ salt ($1H_2O$), m. 108°.[1482]

2. R Bearing Hydroxyl Groups (Free, Etherified) and Halogenated Derivatives

$HOCH_2CH_2OPO_3H_2$. I.1.a.α,[386] I.1.a.β,[329,336] I.1.d.γ,[1122] I.1.d.δ,[1122] I.1.e.α-2 (from $AcOCH_2CH_2OH$ with $(PhO)_2$-POCl)[186] I.2.b.α-2 ($P_{white\ or\ yellow}$ with $(HO)_2R$ and O-containing gas).[195] By boiling the 2-chloro analog with PbO in H_2O.[1122] Na_2 salt ($6H_2O$); m. 61°;[1122] Ba salt ($1H_2O$);[96,329,336,386] Ca salt;[336] Ag_2 salt;[175] di-c-$C_6H_{11}NH_3$ salt, m. 165-167°.[186]
MeO·$CH_2CH_2OPO_3H_2$. I.1.a.α. Ca salt; Ba salt.[175]
Myristoylglycol phosphate. I.1.e.α-2. Cryst., m. 79.5-80.5°.[1046]
Palmitoylglycol phosphate. I.1.e.α-2. Cryst., m. 85-86°.[1046]
Stearoylglycol phosphate. I.1.d.δ (from RI with AgO_2-$P(OCH_2Ph)_2$ and hydrogenolysis),[608] I.1.e.α-2 (using di Ph ester).[1046,1407] Cryst., m. 89.5-90.5°.[1046,1407]
Behenoylglycol phosphate. I.1.e.α-2 (using di Ph ester[1046] or mono Ph ester[1004]). Cryst., m. 96.5-97.5°.[1004,1046]

$H_2NCH_2C(O)OCH_2CH_2OPO_3H_2$. From $ClCH_2CH_2OPO_3H_2$ with Na glycinate. Cryst., m. 162-170° (dec.).[397]

$HOCH_2CH_2CH_2OPO_3H_2$. I.l.e.α-2 (using di Ph ester);[790] from $Cl(CH_2)_3OPO_3H_2$ by alkaline hydrolysis.[383] Ba salt.[383x]

Palmitoyl-$OCH_2CH_2CH_2OPO_3H_2$. I.l.b.γ-2 (ROH with $(PhO)_2$-POCl, then hydrogenolysis). Cryst., m. 121-122°.[654]

Me•CHOH•$CH_2OPO_3H_2$. I.l.d.γ. Silver salt.[96]

$ClCH_2CH(OH)CH_2OPO_3H_2$. I.l.d.γ.[161,167,1500] Ba salt.[167]

$ICH_2CH(OH)CH_2OPO_3H_2$. I.l.d.γ,[503] I.l.d.δ.[1500,1501] Ca salt;[503] Ba salt.[1501]

Mixed 1- and 2-glycerophosphates. I.l.a.β,[8,335,1136,1477] usual technical procedure: I.l.a.α and β (separate or combined).[8,334,388,735,1093,1127,1137] Free acid from Pb salt with H_2S.[335] Many salts.

DL-1-Glycerophosphate. I.l.a.β (impure)[158,159,630] I.l.b.β,[547] I.l.b.γ-2 then I.l.e.α-1[640] or α-2,[262,277] I.l.d.γ[164,547,1500] then I.l.e.α-1,[161] I.l.d.δ.[170] Na salt;[161] Ba salt;[547,640,1500] strychnine and quinoline salts (for resolution of racemate).[767]

L-1-Glycerophosphate. I.l.b.β (from 1,2-isopropylidene derivative). Ba salt;[135,143] Ag salt.[143]

D-1-Glycerophosphate. I.l.b.β (from 1,2.isopropylidene derivative). Ba salt; Ag salt.[144]

$MeOCH_2CH(OMe)CH_2OPO_3H_2$. By hydrolysis of the di Me ester. Ba salt.[770]

2,3-Isopropylidene glycerophosphate. From the dibenzyl-ester by hydrogenolysis. $[\alpha]_D^{17}$ -1.29° (H_2O).[127]

2,3-Isopropylidene-1-glycerophosphate. I.l.c.β-2 (with $(PhCH_2)_4P_2O_7$ followed by hydrogenolysis). Ag salt, m. 194-198°; $[\alpha]_D^{20}$ 1.40°.[262]

Batyl α-phosphoric acid. (Saponification of palmitic or stearic ester). Cryst., m. 78.5-79.5°.[1314]

2,3-Palmital-1-glycerophosphate. I.l.b.β. (Migration not excluded.)[237]

2,3-Octadecanal-1-glycerophosphate. I.l.b.β. Needles; Ag salt.[237]

(β-Palmitoylbatyl) α-phosphate. I.l.b.γ-2 (ROH with $(PhO)_2POCl$ then hydrogenolysis).[1314,1402] Cryst., m. 65-66°,[1314] 65.2-65.9°.[1402]

(β-Stearoylbatyl) α-phosphate. I.l.b.γ-2 (ROH with $(PhO)_2POCl$ then hydrogenolysis).[1314,1402] Cryst., m. 70-70.5°,[1314] 69.8-70°.[1402]

2,3-Dimyristoyl-1-glycerophosphate. I.l.e.α-2 (using dibenzyl ester). Cryst., m. 54-55°.[674]

2,3-Dimyristoyl-L-1-glycerophosphate. From ox brain phosphatide. m. 61.5-62.5°; $[\alpha]_D$ 4.3° ($CHCl_3$).[152]

2,3-Dipalmitoyl-1-glycerophosphate. I.l.b.β,[231] I.l.e.α-2 (using dibenzyl ester). Cryst., m. 62.5-63.5°.[674]

2,3-Dipalmitoyl-L-1-glycerophosphate. From ox brain phosphatides. Cryst., m. 70-71°; $[\alpha]_D$ 4° ($CHCl_3$).[152]

3-Palmitoyl-2-stearoyl-1-glycerophosphate. I.1.e.α-2 (using di Ph ester). Cryst., m. 73.5-74.5°.[1407]
2-Palmitoyl-3-stearoyl-1-glycerophosphate. I.1.e.α-2 (using di Ph ester). Cryst., m. 62.5-63.5°.[1407]
2,3-Distearoyl-1-glycerophosphate. I.1.a.α,[634] I.1.d.δ (from RI with $AgO_2P(OPh)(OCH_2Ph)$ and H_2/Pt,[1405] I.1.e.α-2 (using di Ph ester[1407] or di $PhCH_2$ ester[674]). Cryst., m. 71°,[634] 70-71°[674,1407] (not stable on standing(?)[634]).
(+)-2,3-Distearoyl-1-glycerophosphate. From ox brain phosphatide. Cryst., m. 75.5-76.8°; $[\alpha]_D$ 3.7° ($CHCl_3$).[152]
2,3-Dibehenoyl-1-glycerophosphate. I.1.d.δ (from RI with $AgO_2P(OPh)(OCH_2Ph)$, and hydrogenolysis).[1405]
2-Glycerophosphate. I.1.a.β,[168] I.1.a.β (from glycerol dichlorohydrin[1377] or diacetine - 80% transformation into 1-derivative,[1411] followed by alkaline hydrolysis), I.1.b.γ-2 - I.1.e.α-1,[277] I.1.e.α-1.[1377] By hydrolysis of $[(CH_2Cl)_2CHO]_2PO_2H$.[798] Easy separation from 1-derivative by formation of poorly soluble Ba double salt with $Ba(NO_3)_2$.[768,770] Sirup;[560,1084] Ca salt;[1377] Ba salt.[277,767,768,770]
Batyl β-phosphoric acid. From stearic or palmitic ester with CH_3ONa. Cryst., m. 59-60°.[1314]
$HOCH_2CH(OPO_3H_2)CH_2O{\cdot}C(O)CH_2CH_2CH(NH_2)CO_2H$. I.1.a.β (with H_3PO_4 at 120° in vacuo). Isolated as Fe salt.[839]
$(PhOCH_2)_2CHOPO_3H_2$. I.1.a.α,[593] I.1.b.β.[268,593] Cryst., m. 137-137.5°; Na salt (10 H_2O), m. 54°.[268]
$(2\text{-}MeC_6H_4OCH_2)_2CHOPO_3H_2$. I.1.b.β. Cryst. Na salt.[268]
$(4\text{-}MeC_6H_4OCH_2)_2CHOPO_3H_2$. I.1.b.β. Cryst.[268]
$(n\text{-}C_9H_{19}CO{\cdot}OCH_2)(HOCH_2)CHOPO_3H_2$. From 2-glycerophosphate with acyl chloride. Na salt; acid Ba salt, m. 261-263°.[85]
(Lauroyl-OCH_2)($HOCH_2$)$CHOPO_3H_2$. From 2-glycerophosphate with acyl chloride. Na salt; acid Ba salt, m. 245-255°.[85]
$[4\text{-}HO\text{-}2,5\text{-}Me_2C_6H_2C(Ph)_2CO{\cdot}OCH_2](HOCH_2)CHOPO_3H_2$. I.1.a.α. Oil.[1294]
$[MeCH(C_6H_4\text{-}4\text{-}Me)CO{\cdot}OCH_2](HOCH_2)CHOPO_3H_2$. I.1.a.α. Oil.[1294]
$[CH_2{=}C(C_6H_3\text{-}2,5\text{-}Me_2)CO{\cdot}OCH_2](HOCH_2)CHOPO_3H_2$. I.1.a.α. Oil.[1294]
$(PhCO{\cdot}OCH_2)(HOCH_2)CHOPO_3H_2$. I.1.b.β (from 1-benzoyl-3-trityl-glycerol, followed by I.1.e.α-2). K salt.[141]
(α-Palmitoylbatyl) β-phosphate. I.1.b.γ-2 (ROH with $(PhO)_2POCl$ then hydrogenolysis).[1314,1402] Cryst., m. 62-62.5°,[1314] 57.7-58.1°.[1402]
(α-Stearoylbatyl) β-phosphate. I.1.b.γ-2 (ROH with $(PhO)_2POCl$ then hydrogenolysis).[1314,1402] Cryst., m. 62-63°,[1314] 62-62.5°.[1402]

$(H_2NCH_2CO{\cdot}OCH_2)_2CHOPO_3H_2$. From $(CH_2Cl)_2CHOPO_3H_2$ with Na glycinate. Cryst., m. 169-180° (dec.).[396]

1,3-Dimyristoyl-2-glycerophosphate. I.1.b.β. Acid quinoline salt, m. 96.6-97.5°.[714]

1,3-Dipalmitoyl-2-glycerophosphate. From 2-glycerophosphate with acyl chloride. Ag salt.[767]

1-Palmitoyl-3-stearoyl-2-glycerophosphate. I.1.b.γ-2 (using di Ph ester, followed by hydrogenolysis). Cryst., m. 63.4-64.5°.[1407]

1,3-Distearoyl-2-glycerophosphate. I.1.b.γ-2 (using di Ph ester). Cryst., m. 68.5-69.5°.[1407]

1,3-Dichaulmoogryl-2-glycerophosphate. I.1.b.β. Pb salt; m. 175°; Na salt, m. 149-150°; choline salt, m. 160-165°.[1429]

$HO(CH_2)_4OPO_3H_2$. I.1.a.β,[383a] I.1.b.γ-2 (using $(PhO)_2$-POCl followed by hydrogenolysis);[790] from $Cl(CH_2)_4$-OPO_3H_2 by alkaline hydrolysis.[383a] Ba salt.[383a,790]

$(HOCH_2)_2C(NO_2)CH_2OPO_3H_2$. I.1.b.β. Ba salt.[1503]

DL-Erythritol 1-phosphate. I.1.f.ζ-10 (using o-phenylene phosphate, followed by hydrolysis).[1326a]

D-Erythritol 4-phosphate. By degradation of D-glucose phosphate. $[\alpha]_D^{25}$ 2.6°; di $c\text{-}C_6H_{11}NH_3$ salt, m. 183-186°, $[\alpha]_D^{25}$ -2.30.[897]

L-Erythritol 4-phosphate. Similarly. Di $c\text{-}C_6H_{11}NH_3$ salt, m. 186-190°; $[\alpha]_D^{23}$ 2.3°.[897]

L-Threitol 1-phosphate. By enzymic reduction of L-glycerotetrulose 1-phosphate.[1341]

$HOCH_2C(Me)_2CH_2OPO_3H_2$. From the diol with P + oxygen in presence of catalysts (Al, Sc, V, Ti, Cr, and their oxides).[195]

2-Deoxy-D-ribitol 5-phosphate. From deoxy-D-ribonic acid phosphate by KBH_4. Ba salt, $[\alpha]_D^{20}$ -16.8°.[1286]

2-Deoxy-D-threo-pentitol 5-phosphate. From deoxy-D-xylose 5-phosphate by KBH_4.[58]

L-Ribitol 1-phosphate. From D-ribose 5-phosphate by reduction with BH_4Na. Chromatographic and electrophoretic data.[124]

$(2,2\text{-}(HOCH_2)_2\text{-}BuO)PO_3H_2$. From the diol with P + oxygen in presence of catalysts (Al, Sc, V, Ti, Cr, and their oxides).[195]

Mannitol phosphate (position unknown). I.1.a.α[1267] and β.[329]

D-Mannitol 1-phosphate. I.1.f.ζ-10 (from polyol with o-phenylene phosphate, followed by hydrolysis).[1326a]

Dulcitol phosphate (position unknown). I.1.a.β. Barium salt.[337]

3. R Bearing Carbonyl Groups (Actual or Potential) without or with Hydroxyl Groups (Sugars and Aminosugars), and Halogenated Derivatives

$MeCO{\cdot}OCH_2OPO_3H_2$. I.1.d.δ. Calcium salt.[1139]

$OHCCH_2OPO_3H_2$. By oxidation of 1-glycerophosphate with HIO_4,[555,556,558a] of ribose-5-phosphate or glucose-6-phosphate with $Pb(O_2CMe)_2$,[807] of dihydrosphingosine-1-phosphate with HIO_4.[1446] Ba salt;[555,556,558a] 2,4-$(O_2N)_2C_6H_3$-hydrazone.[1446]

Glyceraldehyde 2- or 3-phosphate. I.1.b.β (from $(EtO)_2$-$CH_2CH(OH)$-CH_2OH). Ca salt.[203,548]

DL-$OHCCH(OH)CH_2OPO_3H_2$. I.1.b.β - I.1.e.α-2 (from glycerol dibenzyl ether),[549] I.1.b.γ-2 - I.1.e.α-1 and α-2 (from glyceraldehyde dimer; intermediate octaphenyl ester: m. 110-111°,[146] 108-109°[145]),[147,150] I.d.γ-2 (from epoxypropanol).[617] Na salt;[617] Ca salt.[147]

D-$OHCCH(OH)CH_2OPO_3H_2$. From Me_2 acetal on Dowex 50.[188]

$(MeO)_2CHCH(OH)CH_2OPO_3H_2$. I.1.b.γ-2 (using $(PhO)_2POCl$, followed by hydrogenolysis). c-$C_6H_{11}NH_3$ salt, m. 155-156°, $[\alpha]_D^{25}$ 8.4°.[188]

$HOCH_2C(O)CH_2OPO_3H_2$. I.1.c.α;[844] by oxidation of 1-glycerophosphate with Br_2;[630] from 2,5-$(EtO)_2$-2,5-$(H_2O_3$-$POCH_2)_2$-dioxane by acid hydrolysis;[422] from MeCO·OCH_2-$C(OMe)_2CH_2OPO_3H_2$ in situ.[189] Ba salt.[630]

$HOCH{=}C(OPO_3H_2)CHO$. From Ribitol 3-phosphate by HIO_4. Di brucine salt. Very easily hydrolyzed.[1435]

L-glycero-Tetrulose 1-phosphate ("L-erythrulose" 1-phosphate). From the Me_2 acetal.[1341]

L-glycero-Tetrulose-Me_2-acetal 1-phosphate. I.1.b.γ-2 (from 3,4-di-O-benzoyltetrulose-Me_2-acetal, using $(PhO)_2POCl$). c-$C_6H_{11}NH_3$ salt, m. 165-167°; $[\alpha]_D$ -13.0° (phosphate buffer p_H 8).[1341]

3-Deoxy-D-erythro-pentofuranose 5-phosphate. From 1,2-isopropylidene derivative. Ba salt, $[\alpha]_D^{25}$ -10.65°.[1287]

1,2-Isopropylidene-3-deoxy-D-erythro-pentofuranose 5-phosphate. I.1.e.α-2 (from Ph_2 ester). $[\alpha]_D^{25}$ -24.8°.[1287]

Arabinose 1-phosphates

D and L-arabinofuranose 1-phosphates. I.1.d.δ (from acetobromosugar and $Et_3NHOPO(OCH_2Ph)_2$, etc.). Ba salts, L-derivative $[\alpha]_D^{20}$ 6.4°, D-derivative $[\alpha]_D^{20}$ -5.7°.[1476]

D and L-arabinopyranose 1-phosphate. I.1.d.δ (from acetobromosugar and $Et_3NHOPO(OCH_2Ph)_2$, etc.). Ba salts, L-derivative $[\alpha]_D^{24}$ 45°, D-derivative $[\alpha]_D^{23}$ -44.3°.[1476]

α-D-arabinose 1-phosphate. I.1.d.δ (from acetobromosugar and AgH_2PO_4, etc.). Di c-$C_6H_{11}NH_3$ salt, m. 144-150°, $[\alpha]_D^{26}$ 30.8°.[1150]

β-L-arabinose 1-phosphate. I.1.d.δ (from acetobromosugar and Ag_3PO_4, etc.). Di c-$C_6H_{11}NH_3$ salt, m. 155-161°, $[\alpha]_D^{26}$ 91.0°.[1150]

Arabinose-5-phosphate. I.1.b.β. Ba salt; brucine salt.[864]

α-D-Ribofuranose 1-phosphate. I.1.d.δ (from 5-O-acetyl-2,3-cycl. carbonate -1-Br-sugar with $Et_3NHOPO(OCH_2Ph)_2$, etc.). $c\text{-}C_6H_{11}NH_3$ salt.[1347]

β-D-Ribofuranose 1-phosphate. I.1.d.δ (from acetobromosugar with $Et_3NHOPO(OCH_2Ph)_2$, etc.). Ba salt.[1474, 1475] β derivative assumed for enzymatic resistance reasons.[1475]

D-Ribofuranose 5-phosphate. I.1.b.γ-2 and e.α-2 (from 2,3-isopropylidene deriv.). Isolated as Ba salt. Free acid has $[\alpha]_D^{14}$ 16.5° (in dil. HCl).[959a]

1-Deoxy-1-amino-2,3-isopropylidene-ribose 5-phosphate (isopropylidene ribosylamine phosphate). By reduction of the corresponding azide.[339]

2,3-Isopropylidene-ribosylazide 5-phosphate. I.1.f.ζ-1, I.1.f.ζ-3. UV.[339]

(Formylglycinamido)-ribotide. From isopropylideneribosylamine 5-phosphate with $ClCO{\cdot}CH_2NHCHO$. Identical with natural product.[339]

Me D-ribofuranoside 2(3)-phosphate. I.1.f.ζ-10 (from ROH with hydrobenzoin cyclic phosphate activated by $(PhO)_2POCl$).[1387]

Me D-ribofuranoside 5-phosphate. I.1.f.ζ-10 (from ROH with hydrobenzoin cyclic phosphate activated by $(PhO)_2POCl$).[1387]

<u>Xylose phosphates</u>

D-Xylopyranose 1-phosphate. I.1.d.δ (from acetobromosugar). K salt; Ba salt.[927]

α Derivative. I.1.d.δ (from acetobromosugar with Ag_3PO_4 or AgH_2PO_4 (besides β derivative)[1150] or $AgO_2P(OCH_2Ph)_2$ (besides β derivative)[57] or $AgO_2P(OPh)_2$ (only α)[57]). Di $c\text{-}C_6H_{11}NH_3$ salt, m. 152-158°, $[\alpha]_D^{26}$ 58.0°.[1150]

β Derivative. I.1.d.δ (from acetobromosugar with AgH_2PO_4 (besides α derivative)). Di $c\text{-}C_6H_{11}NH_3$ salt, m. 144-150°, $[\alpha]_D^{26}$ 21.0°.[1150]

D-Xylose 3-phosphate. From 1,2-isopropylidenexylose 3,5-cyclic phosphate by alkaline hydrolysis (besides 5-phosphate). Ba salt, $[\alpha]_D^{21}$ 0°.[981]

D-Xylose 5-phosphate. I.1.b.β;[866] from 1,2-isopropylidene xylose 3,5-cyclic phosphate by alkaline hydrolysis (besides 3-phosphate).[981] Na salt, $[\alpha]_D^{21}$ 8.3° $\xrightarrow{\text{7 days}}$ -1°;[981] Ba salt.[866]

α-L-Fucopyranosyl 1-phosphate. I.1.d.δ (from acetochloro derivative with Ag_2HPO_4).[853a]

2-Deoxy-α-D-glucose 6-phosphate. From the Me pyranoside phosphate by 0.5N HBr. Ba salt.[1188]

3-Deoxy-D-glucofuranose 6-phosphate. From the isopropylidene derivative. Ba salt, $[\alpha]_D^{20}$ 6.6°.[1288]

3-Deoxy-1,2-isopropylidene-D-glucofuranose 6-phosphate. I.1.d.γ-2 (from 5,6-anhydroderivative with Na_2HPO_4). Ba salt, $[\alpha]_D^{30}$ -6.8°.[1286]

Me 2-deoxy-α-D-glucopyranoside 6-phosphate. I.1.b.γ-2 (using $(PhO)_2POCl$ - di Ph ester, sirup - followed by hydrogenation). Di $c\text{-}C_6H_{11}NH_3$ salt, m. 216-218° (dec.), $[\alpha]_D^{23}$ 42.8°.[1188]

3-Deoxy-1,2-isopropylidene-D-xylo-hexofuranose 6-phosphate. I.1.d.γ-2 (from 5,6-anhydro derivative with KH_2PO_4). Ca salt, $[\alpha]_D^{25}$ -11°.[58]

α-L-Rhamnopyranose phosphate. I.1.f.γ (from tetra-O-actylrhamnose with H_3PO_4). Di $c\text{-}C_6H_{11}NH_3$ salt (1/2 H_2O), m. 195°, $[\alpha]_D^{21}$ -21.5°; NMR.[373]

<u>Glucose phosphates</u>

Without specification: I.1.b.β, I.1.b.γ-1.[22,1040] Pb salt, m. 187°.[22] I.1.f.γ (from penta-acetyl derivative). $[\alpha]_D^{23}$ 78°.[978]

α-D-Glucopyranose 1-phosphate (Cori ester). I.1.d.δ (from acetobromosugar with AgH_2PO_4,[1150] Ag_3PO_4,[434,1150,1469] or $AgO_2P(OPh)_2$[1129]). K salt,[1469] $[\alpha]_D^{26}$ 78°;[1129] $c\text{-}C_6H_{11}NH_3$ salt, m. 163-169°, $[\alpha]_D^{26}$ 64.0° (pH 7.8);[1150] brucine salt, m. 173-178°.[1469]

β-D-Glucopyranose 1-phosphate. I.1.d.δ (from acetobromosugar with AgH_2PO_4[1150] or $AgO_2P(OCH_2Ph)_2$.[1469] $c\text{-}C_6H_{11}NH_3$ salt, m. 137-143°, $[\alpha]_D^{26}$ 0.8°; brucine salt decahydrate, m. 160-165°, anhydrous, m. 162-166°.[1469]

α-L-Glucopyranose 1-phosphate. I.1.d.δ (using Ag_3PO_4). K salt; Ba salt.[1135]

Glucofuranose 3-phosphate. I.1.b.β - I.1.e.α-1. Ba salt. The 1,2-isopropylidene and the di-isopropylidene derivatives isolated as Ba salts.[1095]

Me α-D-glucoside 3-phosphate. I.1.d.γ-2 (from 4,6-benzylidene-2,3-anhydro derivative with $(PhCH_2O)_2$-PO_2H).[659]

Glucose 4-phosphate. I.1.b.β (from tetra-O-acetyl sugar),[1179] I.1.b.γ-2 - I.1.e.α-2 (from tetra-o-acetyl sugar).[1186] Na salt, dec. 155°, $[\alpha]_D^{20}$ 51.5°; Ba salt;[1179] brucine salt,[1179] m. 173-174°.[1186]

Glucose 6-phosphate. I.1.b.γ-2 - e.α-2 (from tetra-O-acetyl sugar with $(PhO)_2POCl$,[551] from 1,2-isopropylidene derivative with $(PhCH_2O)_2POCl$[93b,94,659,865]), I.1.b.γ-2 - e.α-1 (from tetra-O-acetyl derivative with $(MeO)_2POCl$, and acid hydrolysis),[1105] I.1.d.ζ (by mouse pancreas islet cells[90]). From 6-p-toluenesulfonyl-tetra-O-acetylglucose with $(PhO)_2PO_2Li$, etc. (yield low).[374] From glucose: I.1.a.β (with polyphosphoric acid, Ba salt isolated,[1241] with HPO_3, Ba salt, $[\alpha]_D^{21}$ 32.3°, isolated[1410]), I.1.f.ζ-10 (using o-phenylene phosphate).[1326a] From starch by phosphorolysis.[1270] $[\alpha]_D^{21}$ 32.3.[1410] Ba salt,[94,1095,1270,1410] $[\alpha]_D^{15}$ 17.5°,[1105] $[\alpha]_D$ 12.2°;[93b] K salt.[551]

3-O-Me-D-glucose 6-phosphate. I.1.b.γ-2 (from 1,2-isopropylidene derivative with $(PhO)_2POCl$).[874]

1,2-Isopropylidene-glucose 6-phosphate. I.1.b.γ-2 (using $(PhCH_2O)_2POCl$). Ba salt, [α]D -5.35°.[93b]
1,2-Isopropylidene-D-glucofuranose 6-phosphate. I.1.b.γ-2 (using $PhCH_2O)_2POCl$,[659] $(PhO)_2POCl$ with 3,5-benzylidene derivative[1285]). Cryst., m. 135-140° then 195-197° (dec.), $[\alpha]_D^{25}$ -6.9°;[1285] $[\alpha]_D^{25}$ 3.8°.[659] Ba salt, $[\alpha]_D$ -5.35°.[93b]
1,2,3,4-Tetra-O-acetyl-β-D-glucopyranose 6-phosphate. I.1.b.γ-2 (using $(PhCH_2O)_2POCl$). Cryst., m. 128°,[551] 126-127°.[659]
Me α-D-glucopyranoside 6-phosphate. I.1.b.β.[546,1095] From 6-p-toluenesulfonyl-tetra-O-acetyl glucose with $(PhO)_2PO_2Li$.[374] Ba salt.[374,546,1095]
Me β-D-glucopyranoside 6-phosphate. With $(PhO)_2PO_2Li$ similarly to α derivative (above). Ba salt.[374]

Glucosamine phosphates
α-D-Glucosamine 1-phosphate. I.1.d.δ - e.α-2 (from acetobromo derivative with $(Et_3NH)O_2P(OPh)_2$). K salt, $[\alpha]_D^{23}$ 100°.[908]
D-Glucosamine 3-phosphate. From 2-N-Cbo-4,6-benzylidene-D-glucose Ph_2-phosphate by hydrogenolysis. Cryst., m. 180°; $[\alpha]_D^{20}$ ±5° (no mutarotation).[1450]
D-Glucosamine 6-phosphate. By deacetylation and dephenylation of N-anisylidene-tri-O-acetyl Ph_2 ester,[907] I.1.a.β (with metaphosphoric acid).[24] Ba salt.[24]
N-Acetyl-D-glucosamine 6-phosphate. By reaction with $(MeCO)_2O$. Cryst., m. 157-158°; $[\alpha]_D^{20}$ 24° (MeOH).[907]
1,2,3-Tri-O-acetyl-glucosamine 6-phosphate. I.1.e. -2 (from Ph_2 ester hydrochloride). Cryst., m. 106-107°; $[\alpha]_D^{20}$ 48.9° (1N HCl).[907]

α-D-Mannose 1-phosphate. I.1.d. (from acetochlorosugar with $AgO_2P(OPh)_2$ or $AgO_2P(OCH_2Ph)_2$). Ba salt, $[\alpha]_D^{23}$ 33.7°, $[\alpha]_D^{20}$ 32.7°; brucine salt, m. 179-182° (dec. 190°).[1132]
D-Mannose 6-phosphate. I.1.b.γ-2 (from tetraacetate using $(PhO)_2POCl$). Ba salt, $[\alpha]_D^{22}$ 13.3 (0.1N HCl); phenylhydrazone phenylhydrazine salt, m. 144-145° (yielding the acid with benzaldehyde).[1132]
1-Deoxy-1-Cl-2,3,4-tri-O-acetyl-D-mannose 6-phosphate. From tetra-O-acetyl diphenylphosphoryl mannose with $TiCl_4$. Cryst., m. 84-86°.[1132]
Me α-D-altroside 2-phosphate. I.1.d.γ-2 (from Me 4,6-benzylidene-2,3-anhydro-α-D-altropyranoside with $(PhCH_2O)_2POCl$). Ba salt.[659]

Galactose phosphates
Without specification. I.1.b.γ-1. Ca salt.[1039]
α-D-Galactopyranose 1-phosphate. I.1.f.γ (from penta-O-acetyl sugar),[898] I.1.d.δ (from acetobromosugar with Ag_3PO_4 or AgH_2PO_4,[1150] with $AgO_2P(OPh)_2$[1129]). K salt,

$[\alpha]_D^{22}$ 100°;[1129] di c-$C_6H_{11}NH_3$ salt, $[\alpha]_D^{26}$ 78.5°.[1150]
β-D-Galactose 1-phosphate. I.1.d. (from acetobromosugar).[1150,1185] Ba salt;[823,1185] di c-$C_6H_{11}NH_3$ salt, m. 145-151°, $[\alpha]_D^{26}$ 21°.[1150]
D-Galactose 6-phosphate. I.1.c. -2 (using $(PhCH_2)_4P_2O_7$). Ba salt.[1205]

Fructose phosphates

D-Fructose 1(?)-phosphate. I.1.b.γ-2 - e.α-2 (from 1,2;2,4-di-isopropylidene-fructose with $(PhO)_2POCl$. Ba salt.[277]
β(?)-D-fructopyranose 1-phosphate. I.1.f.ζ-10 (from 1,2-cyclic phosphate prepared in situ (DCC) with 1N NaOH). Ba salt.[1126]
Fructose 3(?)-phosphate. I.1.b.γ-2 - e.α-2 (from 1,2;3,4-di-isopropylidene-fructose with $(PhO)_2POCl$). Strongly reducing.[277]
Fructose 6-phosphate. From 1,6-di-phosphate. Ba salt.[1036]

Uncertain orientation. I.1.c.α. Ba salt. Phenylosazone, m. 158°. Reducing properties.[844]

L-Sorbose 1-phosphate. I.1.b.γ-2 - e.α-2 (from 2,3;4,6-di-isopropylidene-sorbose with $(PhO)_2POCl$). KH salt, $[\alpha]_D^{30}$ -16.5°; Ba salt, $[\alpha]_D^{20}$ -7.2° (0.1N HCl).[912]
L-Sorbose 6-phosphate. I.1.b.γ-2 - e.α-2 (via 2,3-isopropylidene 1,6-di Ph_2phosphate).[912]
Sedoheptulose 7-phosphate. By condensation of D-ribose 5-phosphate with $O_2NCH_2CH_2OH$. Ba salt.[899]
Maltose-1-phosphate. I.1.d.δ - e.α-1. Ba salt. The periodate oxidation indicates the structure of glucopyranosido-4-glucopyranose-1-phosphate. Does not decompose Fehling solution.[927]

4. R Containing Carboxyl Groups (Actual or Potential)

a. One Carboxyl Without Other Groups

$HO_2C{\cdot}CH_2OPO_3H_2$. By oxidation of the aldehyde derivative by hypoiodite.[555] Isolated either as a poorly soluble barium salt ($4H_2O$) or quinine salt, m. 148-149°. The compound is more stable to hydrolysis than the aldehyde analog.[557] By hydrolysis of carboxylic Et ester, amide, or nitrile.[384]
$EtO_2CCH_2OPO_3H_2$. I.1.a.β. Ba salt ($1H_2O$).[384]
$H_2NC(O)CH_2OPO_3H_2$. I.1.a.β (from amide with polyphosphoric acid). Ba salt ($1H_2O$).[384]
$NCCH_2OPO_3H_2$. I.1.a.β. Ba salt ($3H_2O$).[384]
$HO_2CCH_2CH_2OPO_3H_2$. By hydrolysis of the nitrile. $Ba_{1.5}$ salt.[384]
$NCCH_2CH_2OPO_3H_2$. I.1.a.β,[384] I.1.b.β (with $POCl_3$), I.1.f.ζ 2 (from $(Et_3NH)_3PO_4$ and $NCCH_2CH_2OH$ by Cl_3CCN),[1101a] I.2.b.α-6 (from H_3PO_3 and ROH with Hg^{2+}).[1065] Cryst.,

m. 165°, IR;[1344] Ba salt,[384,1101a] Ba salt ($3H_2O$);[384] mono anilinium salt, m. 154-155°.[1065]

$CH_2ClCH(CO_2H)OPO_3H_2$. I.1.b.β ($POCl_3$).[148]

$CNCH=CHOPO_3H_2$. I.1.f.β.[537]

$MeCH(CO_2Et)OPO_3H_2$. I.1.a.β. Ba salt ($2H_2O$).[384]

$CH_2=C(CO_2H)OPO_3H_2$. I.1.a.β,[388] I.1.b.β (from pyruvic acid,[795] from $ClCH_2CH(OH)CO_2H$[148]), I.1.d.ξ (from phosphoglycerate by yeast);[220] by hydrolysis of carboxylic Et ester,[384] by dehydrohalogenation of CH_2-$ClCH(CO_2H)OPO_3H_2$.[148] AgBa salt;[148] Ba salt,[148,795] with $3H_2O$.[384]

$CH_2=C(CO_2Me)OPO_3H_2$. I.1.a.β. Ba salt ($1H_2O$).[384]

b. <u>With Other Groups</u>

<u>Phosphoglyceric acids</u>

α- or/and β-phosphoglyceric acid. I.1.c.α - e.α-1. Racemic. Ba salt.[1044] Partially resolved by means of brucine salt; (+) form, $[\alpha]_D$ 2.4°.[1417] The natural product is a mixture of α- and β-phosphates.[794]

$HO_2CCH(OH)CH_2OPO_3H_2$. I.1.d.γ-2 (from epoxyacrylic acid with $Na_2(K_2)HPO_4$);[1310] by oxidation of glycerophosphate with Br_2,[794] of $OHCCH(OH)CH_2OPO_3H_2$ with I_2.[188] Ba salt;[794,1310] Ag salt.[794]

D-(-)-$HO_2CH(OH)CH_2PO_3H_2$. I.1.c.α (besides some (+) form). $[\alpha]_D^{18}$ -13.27°.[1038]

$HO_2CCH(CH_2OH)OPO_3H_2$. By oxidation of glycerophosphate with Br_2. Ba salt.[794]

L-(+)-$HO_2CH(CH_2OH)OPO_3H_2$. I.1.b.γ-2 - e.α-2 (from Me glycerate using (PhH O) POCl and final saponification). $[\alpha]^{22}$ 12.5°; Na_3 salt ($5H_2O$), $[\alpha]^{22}$ 3.6°.[187]

$HO_2CC(O)CH_2OPO_3H_2$. By oxidation of $OHCC(OMe)_2CH_2OPO_3H_2$ with $KMnO_4$ and acidic hydrolysis. Only Ba salt (easy acidic hydrolysis).[190]

$MeC(OPO_3H_2)=CHCO_2H$. By alkaline hydrolysis of carboxylic Et ester. Na_3 salt.[766]

$MeC(OPO_3H_2)=CHCO_2Et$. I.1.b.β (from $MeC(ONa)=CHCO_2Et$ with $POCl_3$). Na salt; Ba salt.[766]

D-Erythronolactone 2-phosphate. I.1.b.γ-2 - e.α-2 (from 3-O-benzoyl erythronic acid with $(PhO)_2POCl$). Di c-$C_6H_{11}NH_3$ salt, $[\alpha]_D^{22}$ -55° (1N HCl).[20x]

D-Erythronolactone 4-phosphate. I.1.b.γ-2 - e.α-2 (from 2,3-di-O-benzoyl-erythronic acid with $(PhO)_2POCl$). Cryst. ($2H_2O$) $[\alpha]_D^{22}$ -20°.[20x]

$HO_2CCH_2CH(OPO_3H_2)CO_2H$. I.1.b.γ-2 - e.α-2 (from active Et_2 malates). (-) Isomer: Ba salt, $[\alpha]_D^{22}$ -4.03° (2N HCl). (+) Isomer: Ba salt, $[\alpha]_D^{23}$ 4.08° (2N HCl).[576]

$HO_2CCH(OH)CH(OPO_3H_2)CO_2H$. (+) Isomer: I.1.b.β. Ba salt; benzidine salt.[1041]

D-(+)-$HO_2CCH(OH)CH(OPO_3H_2)CO_2H$. I.1.b.β (from Me_2 ester with $POCl_3$ followed by hydrolysis). $[\alpha]_D^{20}$ 12° (H_2O),

75.3° (8.3% $(NH_4)_2MO_4$); dibbrucine salt, with $9H_2O$ shiny prisms, m. 175-177° (dec.), anhydrous m. 183-186° (dec.); Ca, Sr, Ba, Pb salts.[887]

2-Deoxy-D-ribonic acid 5-phosphate. From deoxyribose 5-phosphate by Br_2. Ba salt, $[\alpha]_D^{25}$ 1.95°.[1286]

3-Deoxy-5-phospho-D-pentonic acid (mixture of threo and erythro forms). From D-xylose 3,5-cyclic phosphate or D-arabinose 3,5-cyclic phosphate by alkaline degradation.[741a]

D-Arabonic acid 5-phosphate. From D-fructose 6-phosphate by alkaline oxidation.[911]

2,3,4-Tri-O-acetyl-D-galacturonic acid 1-phosphate. I.1.b.γ-2 - e.α-2 (using $(PhO)_2POCl$ or $(PhCH_2O)_2$-POCl).[1108]

Me 1-phospho-2,3,4-tri-O-acetyl galacturonate. I.1.d.δ (from Me acetobromogalacturonate with $AgO_2P(OPh)_2$, I.1.b.γ-2 - I.1.e.α-2 (using $(PhCH_2O)_2POCl$). Cryst., m. 115-116°; $[\alpha]_D$ 24.2°.[1108]

6-Phosphogluconic acid. I.1.a.β. Ba salt.[1241]

α,β-Gluco-metasaccharinic acid 6-phosphate. From 3-O-Me-glucose 6-phosphate with Br_2.[874]

DL-Mevalic acid phosphate. From 3-hydroxy-3-Me-hex-5-en-1-yl phosphate by oxidation with $NaIO_4$.[902]

2-$HOCH_2$-3-deoxy-5-phospho-D-pentonic acid (mixture of threo and erythro forms). From D-glucose 4,6-cyclic phosphate by alkaline treatment.[741a]

$H_2O_3POCH_2C(Me)_2CH(OH)CO{\cdot}NH_2CH_2CH_2CN$. I.1.e.α-2 (from $(PhCH_2)_2$ ester). $[\alpha]_D^{25}$ 11.40°.[131,1027]

$H_2O_3POCH_2C(Me)_2CH(OH)CO{\cdot}NH_2CH_2CH_2CO_2H$. From nitrile (C) with $Ba(OH)_2$.[94,1027] Ba salt,[94] $[\alpha]_D^{23}$ 13.3°.[1027]

5. R (Aliphatic C Skeleton) Containing Other Groups

a. Derivatives of N Bases (Basic Function Free or Substituted) and Aminoalcohols

$H_2NCH_2CH_2OPO_3H_2$. I.1.a.β,[384,388,1123] I.1.b.γ-1,[1123] I.1.b.γ-2 (from colamine with $FPO(OR)_2$),[627] I.1.d.γ (using ethyleneimine),[395] I.1.e.α-2 (from Ph_2 ester);[1370] by rearrangement of $(i\text{-}PrO)_2P(O)NHCH_2CH_2OH$, followed by acidic hydrolysis).[1116] Cryst., m. 242°,[1116] 240° (dec.).[395]

$AcylNH{\cdot}CH_2CH_2OPO_3H_2$.

Acetyl derivative. By action of $(MeCO)_2O$. Ba salt $(2H_2O)$.[384]

Benzoyl derivative. By action of PhCOCl. Ba salt $(3H_2O)$.[384]

Tosyl derivative. By action of RSO_2Cl. Ba salt.[378]

$PhCH_2NHC(S)NHCH_2CH_2OPO_3H_2$. By action of RNCS. Na salt.[378]

4-$BrC_6H_4CH_2NHC(S)NHCH_2CH_2OPO_3H_2$. Similarly. Na salt.[378]

$3\text{-}F_3CC_6H_4CH_2NHC(S)NHCH_2CH_2OPO_3H_2$. Similarly. Na salt.[378]
$4\text{-}FC_6H_4CH_2NHC(S)NHCH_2CH_2OPO_3H_2$. Similarly. Na salt.[378]
$PhNHC(S)NHCH_2CH_2OPO_3H_2$. Similarly. Na salt.[378]
$4\text{-}FC_6H_4NHC(S)NHCH_2CH_2OPO_3H_2$. Similarly. Na salt.[378]
$4\text{-}BrC_6H_4NHC(S)NHCH_2CH_2OPO_3H_2$. Similarly. Na salt.[378]
$4\text{-}O_2NC_6H_4NHC(S)NHCH_2CH_2OPO_3H_2$. Similarly. Na salt.[378]
$3\text{-}F_3CC_6H_4NHC(S)NHCH_2CH_2OPO_3H_2$. Similarly. Na salt.[378]

Oleoyl-$NHCH_2CH_2OPO_3H_2$. From the hydroxyethylamide with H_3PO_4 in C_6H_6 at 80°.[1209a]

Stearoyl-$NHCH_2CH_2OPO_3H_2$. From the hydroxyethylamide with H_3PO_4 in C_6H_6 at 80°.[1209a]

Lauroyl-$NHCH_2CH_2OPO_3H_2$. From the hydroxyethylamide with H_3PO_4 in C_6H_6 at 80°.[1209a]

$MeHNCH_2CH_2OPO_3H_2$. I.1.a.β.[863]

$Me_3CNHCH_2CH_2OPO_3H_2$. I.1.a.β. Cryst., m. 240-241°.[378]

$Me_2NCH_2CH_2OPO_3H_2$. I.1.a.β. Ba salt.[863]

$Et_2NCH_2CH_2OPO_3H_2$. I.1.a.β.[86,356] I.1.d.δ (from Et_2NCH_2-CH_2Cl with Na_3PO_4).[86] Sirup. Ca, $HCa_{1/2}$, Ba salts.[356]

$Me_3\overset{+}{N}CH_2CH_2OPO_3^-H$. I.1.a.β,[384x,388,743,1123] I.1.b.γ-1,[1123] I.1.b.γ-2 (from choline chloride using $(PhO)_2$-$POCl$).[150] Ca salt;[743,1123,1196] Ba salt;[133] Ca-Cl salt;[384] chloroplatinate, dec. 207-208°;[743] reineckate; adduct with $HgCl_2$, m. 180-184°.[1196]

$PhCH_2NHCH_2CH_2OPO_3H_2$. I.1.a.β. Cryst., m. 250-252°.[378]

$H_2NCH_2CH_2CH_2OPO_3H_2$. I.1.a.β.[379,384] Cryst., m. 176-180°,[384] with $1H_2O$, m. 176°;[379] Ba salt ($3H_2O$).[384]

$AcylNHCH_2CH_2CH_2OPO_3H_2$
Acetyl derivative. By action of $(MeCO)_2O$. Ba salt ($3H_2O$).[384]
Benzoyl derivative. By action of PhCOCl. Ba salt ($2H_2O$).[384]
Tosyl derivative. By action of RSO_2Cl. Ba salt.[384]

$Me_3\overset{+}{N}CH_2CH_2CH_2OPO_3^-H$. I.1.b.γ-2 - I.1.e.α-2 (using $(PhO)_2$-$POCl$). Cryst., m. 235-238° (dec.).[140]

$Me_3\overset{+}{N}CH(Me)CH_2OPO_3^-H$. I.1.b.γ-2 - e.α-2 (using $(PhO)_2POCl$). Cryst., m. 200-203°.[140]

$Me_3\overset{+}{N}CH_2CH(Me)OPO_3^-H_2$. I.1.b.γ-2 - e.α-2 (using $(PhO)_2$-$POCl$). Cryst., m. 271-273° (dec.).[140]

$CH_3CH_2CH(NH_2)CH_2OPO_3H_2$. I.1.a.β Cryst., m. 235-237°.[384]

$CH_3CH_2CH(NHAcyl)CH_2OPO_3H_2$
Acetyl derivative. By action of $(MeCO)_2O$. Ba salt ($3H_2O$).[384]
Benzoyl derivative. By action of PhCOCl. Ba salt ($2H_2O$).[384]
Tosyl derivative. By action of RSO_2Cl. Ba salt.[384]

$Et_2NCH_2CH(OH)CH_2OPO_3H_2$. I.1.a.β, I.1.d.δ (from RCl with Na_3PO_4).[86]
$H_2NCH_2CH_2CH_2CH_2OPO_3H_2$. I.1.a.β. Cryst., m. 223-225°.[384]
$AcylNH(CH_2)_4OPO_3H_2$.
Acetyl derivative. By action of $(MeCO)_2O$. Ba salt $(3H_2O)$.[384]
Benzoyl derivative. By action of PhCOCl. Ba salt $(2H_2O)$.[384]
Tosyl derivative. By action of RSO_2Cl. Ba salt.[384]
$H_2N(CH_2)_5OPO_3H_2$. I.1.a.β. Cryst., m. 238-240°.[384]
$PhCONH(CH_2)_5OPO_3H_2$. By action of PhCOCl. Cryst. $(1H_2O)$, m. 147°; Ba salt.[384]
$H_2N(CH_2)_6OPO_3H_2$. I.1.a.β. Cryst., m. 245-247°.[384]
Sphingin 1-phosphate. From 3-O-Me-dihydrosphingosine 1-phosphate by energic hydrogenation.[1446]
Dihydrosphingosine 1-phosphate. I.1.b.γ-2 - e.α-2 (from N-Cbo Ph_2 ester); by acidic hydrolysis of 1,3-di-phosphate. Solid.[1446]
3-O-Me-dihydrosphigosine-II 1-phosphate. I.1.b.γ-2 - e.α-2 (from N-Cbo derivative with $(PhCH_2O)_2POCl$).[1446]
$H_2NC(NH)N(Me)CH_2CH_2OPO_3H_2$. I.1.a.β (from creatinol HBr). Cryst., m. 243-244°.[536]

b. Aliphatic Amino Acid Derivatives

Serine O-phosphate. I.1.a.β,[322,868,869,1121] I.1.b.β (from benzylidene-serine, yield poor),[868] I.1.b.γ-2 e.α-2 (from N-Cbo-serine benzyl ester with $(PhO)_2POCl$ using DL, D, and L isomers);[563] by rearrangement of $MeO_2CCH[NHPO(O\text{-}i\text{-}Pr)_2]CH_2OH$. DL isomer: cryst., m. 166-167°,[1116] 153-156° (dec.);[563] brucine salt, m. 130° (dec.), used for resolution of DL compound. L isomer: m. 168-172°; $[\alpha]_D^{19}$ 7.4° (H_2O), $[\alpha]_D^{21}$ 16.2° (2N HCl); Ba salt (identical with that of natural product), $[\alpha]_D^{25}$ 9.4°. D isomer: m. 170-173°; $[\alpha]_D^{21}$ -7.0° (H_2O), $[\alpha]_D^{21}$ -15.6° (2N HCl).[563]
(O)-Phospho-isoserine. I.1.a.β. Ba salt.[1121]
$MeO_2CCH(NH_2)CH_2OPO_3H_2$. I.1.a.β. $Ba_{1/2}$ salt.[384]
N-Benzoyl-O-phosphoserine methylamide. I.1.a.α-1. Cryst., m. 145-157° (dec.).[114]
$t\text{-}BuO_2CCH(NHCO\cdot O\text{-}t\text{-}Bu)CH_2OP(OCH_2Ph)O_2H$. I.1.d.α-1 (from ROH with $PhCH_2OPO_3H_2$ by DCC). Cryst., m. 173-175° (racemic).[646]
Glycyl-serine O-phosphate. I.1.e.α-3 - e.α-2 (from N-Cbo (C) benzyl (P) Ph_2 ester by NaI and hydrogenolysis).[113]
(O-Phospho-L-seryl)-L-glutamic acid. I.1.b.γ-2 - e.α-2 (from N-Cbo-dipeptide using $(PhCH_2O)_2POCl$).[563a]
(O-Phospho-L-seryl)-L-lysine. I.1.b.γ-2 - e.α-2 (from N-Cbo-dipeptide benzyl ester using $(PhCH_2O)_2POCl$). Cryst., $[\alpha]_D^{19}$ -4.7° (4N HCl).[563a]

Threonine O-phosphate. I.1.a.β. Plates, dec. 169°. Pb salt.[1121]

c. Sulfonic Acid Derivative

$HO_3SCH_2CH_2OPO_3H_2$. I.1.a.β. $Ba_{1.5}$ salt.[384]

6. R Containing a Cyclic Group (Unsubstituted or Substituted

a. Carbocyclic Groups

$PhCH_2OPO_3H_2$. I.1.c.α,[847] I.1.d.δ,[1501] I.2.b.α-6 (from H_3PO_3 with ROH by Hg^{2+});[1065] by oxidation of $(PhCH_2O)_2$-(4-HO-2,3-Me_2-1-$C_{10}H_4O$)PO by Br_2.[406a] Ba salt;[847,1501] di anilinium salt, m. 150-153°.[1065]

$PhCOCH_2OPO_3H_2$. I.1.e.α-2 (from Ph_2 ester). Cryst., m. 142°.[1335]

2-$C_{10}H_7$-$COCH_2OPO_3H_2$. I.1.e.α-2 (from Ph_2 ester). Cryst., m. 180° (dec.).[1335]

2-Me-4-Cl-$C_6H_3OCH_2CH_2OPO_3H_2$. I.1.b.β (using $POCl_3$). Na, K, Ag, and Ca salts.[217]

2-$HOC_6H_4C(O)OCH_2CH_2OPO_3H_2$. I.1.b.γ-2 - e.α-2 (using $(PhO)_2POCl$). Oil.[355]

Steroids

11-Deoxycorticosterone 21-phosphate. I.1.d.α-1 (from ROH with $EtSPO_3H_2$ and DCC, followed by oxidation by I_2).[1059]

Cortisone 21-phosphate. I.1.d.δ (from 3,3-ethylenedioxy-21-I-5-pregnene-17α-ol-11,20-dione with $AgO_2P(OCH_2Ph)_2$, followed by hydrogenolysis and action of N-Me-morpholine). Cryst., m. 190-193°; Na salt, $[\alpha]_D^{24}$ 117° (MeOH); Me-morpholine salt, m. 203-205°.[456]

9α-F-4-pregnene-11β,17,21-triol-3,20-dione 21-phosphate. I.1.d.δ (from 3,3-ethylenedioxy-9α-F-21-I-5-pregnene-11β,17α-diol-20-one with $AgO_2(OCH_2Ph)_2$, etc.).[1230]

Prednisolone 21(?)-phosphate. From the 17,21-cyclic phosphate by alkaline hydrolysis.[623a]

b. Heterocyclic Groups

3-Pyridylmethyl phosphate. I.1.a.β. NH_4 salt, m. 195-196°.[881]

2-Me-3,4-$(HO)_2$-5-pyridylmethyl phosphate. By oxidation of codecarboxylase with H_2O_2. Cryst., m. 229-230° (dec.).[680]

Pyridoxol 5'-phosphate. I.1.b.β (from pyridoxol·HCl with $POCl_3$). Ca salt.[677]

Pyridoxal 3-phosphate. I.1.b.β (from pyridoxal 4',5'-Et-acetal with $POCl_3$). Ba salt; UV; oxime, m. 210-211°.[678]

Pyridoxal 5'-phosphate (Codecarboxylase). I.1.a.β (from pyridoxal with polyphosphoric acid[1460] or (using the hydrazone) HPO_3;[1201] from condensation product of pyridoxal with ephedrine and polyphosphate[1300]), I.1.b.β,[677] I.2.α-9 (from bis(pyridoxal)polymethyleneimine Cu^{II}-chelate with H_3PO_3);[1028] from pyridoxol phosphate by $KMnO_4$,[677] from pyridoxamine phosphate by MnO_2 on Celite.[1098] NH_4 salt, UV;[1460] Mg, La salts;[1201] $Ca_{1/2}$ salt;[677] oxime (equally from pyridoxaloxime by $POCl_3$), m. 229-230°.[677]

1,2-Me_2-3-HO-4-OHC-5-pyridylmethyl phosphate betaine. I.1.b.β (from pyridoxalmethochloride with $POCl_3$). $Ca_{1/2}$ salt; oxime, m. 224.5°.[677]

Pyridoxamine 5'-phosphate. I.1.a.β (using polyphosphoric acid[1460] or HPO_3[1206]), I.1.b.β (from hydrochloride with $POCl_3$).[679] Hydrochloride, m. 224°;[1206] Ca salt,[679,1206] UV;[1206] NH_4 salt;[1460] 4'-N tosylate (from phosphate with 4-toluenesulfochloride), m. 190° (dec.).[679]

2-Piperidinoethyl phosphate. I.1.a.β,[86,356] I.1.d.δ (from RCl with Na_3PO_4).[86] Sirup.[356]

2-Morpholinoethyl phosphate. I.1.a.β,[86] I.1.d.δ (from RCl with Na_3PO_4).[86]

2-(β-Indolyl)-ethyl phosphate. I.1.e.α-2 (from Ph_2 or $(PhCH_2)_2$ ester). Ba salt.[670]

β-Indolyl-$CH_2CH(NH_2)CH_2OPO_3H_2$. I.1.e.α-2 (from $(PhCH_2)_2$ ester). Ba salt.[670]

2-(4-Me-5-thiazolyl)-ethyl phosphate. I.1.a.β,[1210] I.1.b.β - e.α-4 (using $(PhO)_2POCl$, then N-Me-morpholine),[857] I.1.f.ζ-5 (using dimorpholino-POBr(Cl)).[989] Cryst., m. 197-198°,[857] 162°; Ag salt;[1210] c-$C_6H_{11}NH_3$ salt, m. 190°.[989]

Thiamine phosphate. I.1.a.β,[232,1083,1210] (using H_3PO_4 at 140°).[232] Cryst., m. 200.2°,[1210] with $1H_2O$, dec. 180-182°;[232] Ag salt.[1210]

S-Benzoylthiamine O-phosphate. 4 cryst. forms; UV at various pH.[1427]

Riboflavine 5'-phosphate. I.1.a.β,[321] I.1.b.β (using $POCl_3$,[566,831] or a mixture of 1 POX_3 (X = Cl, Br, F) with 1 or $2H_2O$;[562] with O-protected riboflavine), I.1.b.γ-2 - e.α-1 (using $Me(Et)OPOCl_2$ and 1N HCl)[1231,1232,1233] I.1.b.γ-2 - e.α-2 (using $PhCH_2OPOCl_2$),[1231] I.1.f.ζ-1 (from O-protected riboflavine),[1332] I.1.f.ζ-10 (using o-phenylene phosphate),[1326a] I.2.b.α-9 (from phosphite by $KMnO_4$).[1337] Cryst., m. 195°; $[\alpha]_D^{25}$ 44.5°;[562] Na salt.[831]

Dihydroriboflavine 5'-phosphate. I.1.b.β (using mixture of 1 $POCl_3$ with 1 or $2H_2O$). Oxidizable to riboflavine 5-phosphate.[553]

4-H_2NCO-5-H_2N-1-ribofuranosylimidazole 5'-phosphate. From isopropylideneribosylamine 5-phosphate with $EtOCH{=}NCH(CN)CONH_2$.[339]

N-(5-H_2N-1-β-D-ribofuranosylimidazole-4-carbonyl)-L-aspartic acid 5'-phosphate. From (4-carbamoyl-5-H_2N-1-β-ribofuranosyl)-imidazole 5'-phosphate by enzymic conversion;[339] I.1.f.ζ-3 (from the nonphosphorylated aspartic acid Me_2 ester with $CNCH_2CH_2OPO_3H_2$ and DCC).[1308a] Spectroscopic evidence.[1308a]

1-(β-D-ribofuranosyl)-2-pyridone 2'(3')-phosphate. I.1.f.ζ-3 (from 1-(5-O-trityl-β-D-ribofuranosyl)-2-pyridone).[1113]

1-[(3(2)-O-cyanoethylphosphoryl)-5-O-trityl-D-ribofuranosyl]-2-pyridone. I.1.f.ζ-3. Characterized by Rf and enzymic reactions.[1380]

Nicotinamide N-D-ribonucleoside 5'-phosphate. From the non phosphorylated derivative with Et phosphorothioate and DCC. Ba salt.[579]

Nicotinamide N-(β-D-glucopyranosyl-6-phosphate). I.1.b.β (from 1-(tri-O-acetylglucopyranosyl)-3-carbamoylpyridinium chloride with $POCl_3$ followed by deacetylation).[1466]

Orotidine 5'-phosphate. I.1.f.ζ-3 (from orotidine Me ester, followed by saponification). Identical with natural product.[980]

Deoxyuridine 5'-phosphate. From 3'-O-acetyl derivative with $(Cl_3CCH_2O)_2POCl$, followed by treatment with Zn + $MeCO_2H$. Physicochemical evidence.[499]

Uridine 2'(3',5')-phosphate. From uridine heated with inorganic phosphate 9 months at 65°;[219] NaH_2PO_4 at 160°.[1125]

Uridine 2'(3')-phosphate. I.1.f.ζ-10 (using hydrobenzoin cyclic phosphate activated by $(PhO)_2POCl$).[1387]

Uridine 2'-phosphate. The compound reported[641] is considered to be in error, the benzylidene derivative used being 2',3' (and not 3',5' !) and the resulting product being therefore the 5' compound.[295]

Uridine 3'-phosphate. I.1.b.γ-2 - e.α-1 (from trityluridine),[273,959a] (from 2'-O-acetyl-5'-O-trityluridine),[959a] I.1.a.β.[9] Cryst., m. 192°;[959a] brucine salt, m. 195°;[273] di brucine salt, m. 182-187°.[959a]

5'-O-CH(OR)OR' uridine 3'-phosphates. By condensation of the 3'-phosphate with ROH + R'CHO. Thus: 5'-OCH-(OMe)Me derivative with MeOH + MeCHO; 5'-O-CH(OBu)Me with BuOH + MeCHO, and its 2'-O-acetyl derivative by action of $(MeCO)_2O$; 5'-O-CH(O-sec-Bu)Me with sec-BuOH + MeCHO; 5'-O-CH(O-t-Bu)Me with t-BuOH + MeCHO; 5'-O-CH(O-i-Pr)CH_2CH_2Me with i-PrOH + $MeCH_2CH_2CHO$; 5'-O-CH(OBu)$CHMe_2$ with BuOH + Me_2CHCHO; 5'-O-CH(O-i-Bu)-$CHMe_2$ with i-BuOH + Me_2CHCHO.[1240]

Uridine 5'-phosphate. I.1.b.β (from uridine;[640] 2',3'-isopropylideneuridine,[640,870,1254] using $POCl_3$ in $(MeO)_3PO$[1480,1487]), I.1.b.γ-2 - e.α-2 (from benzylidene-uridine which is the 2',3'-derivative,[295] and not the 3',5' as thought,[641] using $(PhCH_2O)_2POCl$,[954] (using $(4\text{-}O_2NC_6H_4O)_2POCl$),[982] (from 2',3'-isopropylidine derivative),[959a] I.1.f.ζ-1,[733,816,1331] I.1.f.ζ-8,[537] I.1.f.ζ-10 (using hydrobenzoin cyclic phosphate activated with $(PhO)_2POCl$),[1387] I.2.a.β (from 2',3'-isopropylidene derivative with PCl_3 heated in air stream);[698] from (2 → 5')-cyclouridine with acyl- or benzylphosphate;[977a] from 2',3'-isopropylidene derivative with $(Cl_3CCH_2O)_2POCl$, followed by treatment with Zn + $MeCO_2H$;[499] from uridine with Et phosphorothioate (DCC);[579] from 3'-phosphate by heating with uridine (besides cyclophosphate and dinucleotides).[992] Ba salt;[579,640] (described as 2'-phosphate)[641,733,959a] brucine salt,[870] m. 185-190°,[959a] with $4H_2O$ (described as 2'-phosphate derivative).[641] Protolysis const. in $HCONMe_2$: $K \times 10^4 = 3.2$.[1133]

5-Br-uridine 5'-phosphate. From uridine phosphate with N-Br-succinimide.[954]

5-O_2N-uridine 5'-phosphate. I.1.f.ζ-1 (from 2',3'-isopropylidene derivative).[1254]

6-Azauridine 5'-phosphate. I.1.f.ζ-1 (from 2',3'-isopropylidene-4-thioxo-6-azauridine, followed by treatment with NH_3),[230] I.1.e.α-2 (from mono or di Ph ester).[1142]

6-Azauridine 2',3'-di-O-benzoyl-5'-phosphate. I.1.e.α-2 (from mono Ph ester by energic hydrogenation).[1142]

3-Me-6-azauridine 5'-phosphate. I.1.a.β, I.1.f.ζ-8 (from the 2',3'-isopropylidene or benzylidene derivative).[1495a]

5-Me-6-azauridine 5'-phosphate. I.1.e.α-2 (from Ph_2 ester). $c\text{-}C_6H_{11}NH_3$ salt.[1142]

4-Thioxo-6-azauridine 5'-phosphate. I.1.f.ζ-1. Ba salt.[230]

Lyxouridine 2'(3') phosphate. I.1.f.ζ-3 (from 1-(5-O-benzoyl-β-D-lyxofuranosyl)-uracil). Ba salt.[1381]

1-(3-Deoxy-3-O_2N-β-D-glucopyranosyl)-uracil 6'-phosphate. From deoxyglucosyluracil phosphate by $NaIO_4$ - MeONO.[1333]

1-(3-Deoxy-3-H_2N-β-D-glucopyranosyl)-uracil 6'-phosphate. From O_2N derivative by Raney Ni. Cryst., m. 215°; $[\alpha]_D^{20}$ 0°; UV.[1333]

Spongouridine 5'-phosphate (β-D-arabinofuranosyluracil phosphate). I.1.f.ζ-1 (from 2',3'-isopropylidene derivative).[1254]

Cytidine 2'-phosphate. I.1.b.γ-2 - e.α-2 (from 3',5'-benzylidene derivative. The derivative used by Ref. 642 is shown by Ref. 295 to be the 2',3'-derivative yielding therefore the 5'-phosphate),[959a] I.1.f.ζ-3 (from N^6,$O^{3'}$,$O^{5'}$-tribenzoylcytidine).[1173a] Cryst., dec. 235°.[959a] Rate of hydrolysis with dil. H_2SO_4 intermediate between that of the 3' and 5' isomers,

as for other 2'-nucleotides.[959a]
Cytidine 3'-phosphate. I.1.f.ζ-3 (from N^6, $O^{2'}$, $O^{3'}$-tribenzoylcytidine).[1173a]
2',5'-Bis-O-(1-ethoxyethyl)-N^4-dimethylaminomethylenecytidine 3'-phosphate. From cytidine 3'-phosphate by reaction with $Me_2NCH(OMe)_2$, $EtOCH{=}CH_2$, and ammonia.[129,692]
Cytidine 5'-phosphate. I.1.a.β (from 2',3'-benzylidenecytidine),[954] I.1.b.β (from 2',3'-isopropylidenecytidine with $POCl_3$ in $(MeO)_3PO$),[1480,1487] I.1.b.γ-2 - e.α-2 (from 2',3'-isopropylidenecytidine).[959a] The phosphate described by Ref. 642 as 2'-derivative seems to be the 5'-phosphate as the benzylidene derivative used was shown[295] to be in reality the 2',3'-derivative yielding therefore the 5'-phosphate by I.1.b.β ($POCl_3$). I.1.f.ζ-1 ($P_2O_3Cl_4$ in 4-cresol with metaboric acid),[734,1330,1331] (from 2',3'-isopropylidenecytidine),[816] I.2.a.β (from 2',3'-isopropylidenecytidine with PCl_3 in the presence of air).[698] Plates, dec. 233°; brucine salt, dec. 215°.[959a] (The product described by Ref. 642, see above: prismes, dec. 240-242°; Pb and Ba salts).
N^4-Me-cytidine 5'-phosphate. I.1.a.β (from 2',3'-isopropylidene derivative). UV.[729]
N^4-Me_2-cytidine 5'-phosphate. I.1.a.β (from 2',3'-isopropylidene derivative). UV.[729]
D-Arabinofuranosylcytosine 5'-phosphate. I.1.f.ζ-1.[697,732]
1-(3-Deoxy-3-O_2N-β-D-glucopyranosyl)-cytosine 6'-phosphate. From glucosylcytosine 6-phosphate by $NaIO_4$ - MeONO. Cryst., m. 199-200°; $[\alpha]_D^{20}$ 49.3°; UV.[1333]
1-(3-Deoxy-3-H_2N-β-D-glucopyranosyl)-cytosine 6'-phosphate. From the 3-O_2N derivative by Raney Ni). Solid, $[\alpha]_D^{21}$ 11.2°; UV.[1333]
6-Azacytidine 5-phosphate. From 4-thioxo-6-azauridine phosphate by NH_3.[230]
Isocytidine 5'-phosphate. I.1.f.ζ-1 (from 2',3'-isopropylidene derivative).[1254]
Thymidine 3'-phosphate. I.1.b.γ-2 - e.α-2 (from 5'-O-tritylthymidine with $(PhCH_2O)_2POCl$);[960] from 5'-O-tritylthymidine with $(Cl_3CCH_2O)_2POCl$, followed by treatment with Zn (Cu) + $MeCO_2H$.[499] Ba salt, $[\alpha]_D^{20}$ 7.3°; brucine salt, m. 178°.[960]
Thymidine 5'-phosphate. I.1.b.γ-2 - e.α-2 (from 2',3'-O-benzylidenethymidine with $(PhCH_2O)_2POCl$),[960] I.1.f.ζ-1;[697,732] from 3'-O-acetylthymidine 5'-phosphate by saponification; from 3'-O-acetylthymidine with $(Cl_3CCH_2O)_2POCl$, and Zn + $MeCO_2H$.[499]
3'-O-Acetylthymidine 5'-phosphate. I.1.b.γ-2 - e.α-2 (from acetylthymidine with $(PhCH_2O)_2POCl$). Ba salt, $[\alpha]_D^{17}$ -3.0°; brucine salt, m. 175°.[960]

1-(β-D-ribofuranosyl)-thymine 5'-phosphate. I.1.b.γ-2 - e.α-2;[1254] by condensation of diacetyl-5-diphenylphosphorylribofuranosyl bromide with dithymyl mercury.[1382] Solid, $[\alpha]_D$ -15.6° (1N HCl).[1382]

1-(β-D-glucopyranosyl)-thymine 6'-phosphate. By condensation of 2,3,4-tri-O-acetyl 6-Ph_2-phosphoryl-α-D-glucopyranosyl bromide with dithymyl mercury, and hydrogenolysis. Ba salt, $[\alpha]_D^{32}$ -5.28°.[1382]

9-(β-D-ribofuranosyl)-purine 5'-phosphate. I.1.f.ζ-3 (from the non phosphorylated isopropylidene derivative with $CNCH_2CH_2OPO_3H_2$ and DCC). Spectroscopic data.[1102]

6-Mercapto-9-(β-D-ribofuranosyl)-purine 5'-phosphate. I.1.f.ζ-3 (from the non phosphorylated isopropylidene derivative with $CNCH_2CH_2OPO_3H_2$ and DCC). Yellow, amorphous.[1102]

6-Methylthio-9-(β-D-ribofuranosyl)-purine 5'-phosphate. I.1.f.ζ-3 (from the non phosphorylated isopropylidene derivative with $CNCH_2CH_2OPO_3H_2$ and DCC). Unstable.[1102]

6-Me_2N-9-β-D-Ribofuranosylpurine 5'-phosphate. I.1.f.ζ-5 (from the 2',3'-isopropylidene derivative with morpholino-$POCl_2$).[722]

Inosine 5'-phosphate. I.1.b.β (from inosine with $POCl_3$ and MeCN, yield low;[581] from 2',3'-isopropylidene derivative only with $POCl_3$,[1485] with $POCl_3$ in $(MeO)_3PO$ [1480,1487]), I.1.c.β-2 - e.α-1 (from 2',3'-di-O-acetyl inosine with $Ph_4P_2O_7$ and alkaline hydrolysis),[1117] I.1.f.ζ-1,[697,732] (in 4-cresol and metaboric acid), [734,1330] I.1.f.ζ-2 (from 2',3'-isopropylideneinosine and H_3PO_4 with malonitril or acrylonitril,[1228] or with $(BuO)_3PO$ and malonitril[1229]), I.2.a.β (from 2',3'-isopropylideneinosine with PCl_3 in the presence of air);[698] from inosine with Et phosphorothioate and DCC.[579] Na salt;[697,732] Ba salt.[1117]

6-Thioinosine 5'-phosphate. I.1.f.ζ-1. Pale yellow solid, $[\alpha]_D^{22}$ -58.5°.[732]

2-Me_2N-inosine 5'-phosphate. I.1.b.β (from 2',3'-isopropylidene derivative with $POCl_3$).[1485]

2-MeO-inosine 5'-phosphate. I.1.b.β (from 2',3'-isopropylidene derivative with $POCl_3$).[1485]

2-MeS-inosine 5'-phosphate. I.1.b.β (from 2',3'-isopropylidene derivative with $POCl_3$).[1485]

9-β-D-Xylofuranosylhypoxanthine 5'-phosphate. I.1.f.ζ-1, from the corresponding adenosine phosphate derivative by $NaNO_2$. Ba salt, $[\alpha]_D^{25}$ -21.0°.[733]

Deoxyadenosine 5'-phosphate. I.1.f.ζ-1. Yield low.[733]

Adenosine 2'(3')(5')-phosphate (mixture). I.1.f.ζ-5 (from adenosine with 2,6-lupetidino-$POCl_2$ or morpholino-$POCl_2$,[725] with morpholino-(PhO)(O)POP(O)(OPh)Cl[726]).

Adenosine 2'-phosphate (adenylic acid a). I.1.b.γ-2 - e.α-2 (from 3',5'-benzylidene derivative)[959a] (from 3',5'-di-O-acetyl derivative with $(MeO)_2POCl$ and acidic hydrolysis).[1105] Cryst., dec. 205-215°;[959a,1105] brucine salt, m. 165-175°; acridine salt, m. 215° (dec.).[959a]

Adenosine 3'-phosphate (adenylic acid b). I.l.a.β,[9] I.l.b.γ-1 - e.α-2 (from 5'-O-trityl-2'-O-acetyl derivative).[959a] Cryst., m. 194°; brucine salt, m. 177°; acridine salt, dec. 175°.[959a] Molecular and crystal structure of the free acid dihydrate.[1282]

2',5'-Bis-O-(1-ethoxyethyl)-N^6-(dimethylaminomethylene)-adenosine 3'-phosphate. From adenosine 3'-phosphate with $Me_2NCH(OMe)_2$, $EtOCH{=}CH_2$, ammonolyse).[1209]

Adenosine 5'-phosphate. I.l.b.β (from adenosine),[640,741] (from 2',3'-isopropylideneadenosine with $POCl_3$ without solvent,[1485] with tertiary base,[871] in $(MeO)_3PO$[1480,1487]), I.l.b.γ-2 - e.α-2 (from 2',3'-di-O-acetyladenosine with $(PhCH_2O)_2POCl$),[274] (similarly from 2',3'-isopropylidene derivative),[132] I.l.f.ζ-1 (from adenosine,[697,732,733,1331] in 4-cresol and metaboric acid;[734,1330] from 2',3'-isopropylideneadenosine[816]), I.l.f.ζ-5 (from 2',3'-isopropylideneadenosine with morpholino-$POCl_2$),[722] I.2.a.β (from 2',3'-isopropylideneadenosine with PCl_3 in the presence of air),[663,698] I.2.b.β-4 (from isopropylideneadenosine with $(PhCH_2O)_3P$[698] or dimorpholino-POEt[972] and NCCHBr-$CONH_2$); from adenosine with Et phosphorothioate and DCC;[579] from 2',3'-isopropylideneadenosine with $(Cl_3C\text{-}CH_2O)_2POCl$ and treatment with Zn + $MeCO_2H$.[499] Cryst., m. 198°,[733] with $2H_2O$, m. 190-192°,[722] 190°;[132] $[\alpha]_D^{20}$ -45.5°;[132] Ba salt;[722,1480] acridine salt, m. 208°.[132] UV, IR.[1480]

Adenosine N^1-oxide 5'-phosphate. From adenosine phosphate with H_2O_2. UV.[449]

9-β-D-glucopyranosyladenine 6'-phosphate. I.l.b.γ-2 - e.α-2 (from triacetylglucosyl adenosine with $(PhCH_2O)_2$-POCl).[197]

Xanthosine 5'-phosphate. I.l.b.β (from 2',3'-isopropylidene derivative with $POCl_3$ without solvent (very slow),[1485] in $(MeO)_3PO$[1480,1487]).

Guanosine 2'-phosphate. I.l.b.γ-2 and e.α-2 (from 3',5'-benzylidene deriv.). Cryst., dec. 192°.[959a]

Guanosine 3'-phosphate. I.l.b.γ-2 (from guanosine with $(PhO)_2POCl$ and $Ba(OH)_2$).[640]

2',5'-Bis-O-(1-ethoxyethyl)-N^4-(dimethylaminomethylene)-guanosine 3'-phosphate. From guanosine 3'-phosphate with $Me_2NCH(OMe)_2$, $EtOCH{=}CH_2$, ammonolyse.[1209]

Guanosine 5'-phosphate. I.l.b.β, I.l.b.γ-2 (with $(PhO)_2$-POCl and pyridine, yield is low),[640] I.l.b.β (from guanosine with $POCl_3$ in MeCN, yield low;[581] from 2',-3'-isopropylidene derivative with $POCl_3$ without solvent,[1485] in $(MeO)_3PO$[1480,1487]), I.l.b.γ-2 - e.α-2 (from isopropylidene derivative),[959a] I.l.e.α-1 (from mono or di (4-$O_2NC_6H_4$) ester with LiOH),[363] I.1.f.ζ-1 (from guanosine (yield low),[732,1331] in 4-cresol with metaboric acid;[734,1330] from isopropylidene guanosine [816]); from guanosine with NaH_2PO_4 at 160° (chromatographic evidence).[1125] Cryst., dec. 190-200°; brucine salt, dec. 210°. (Results show difference with natural guanylic acid, which therefore is the 3'-phosphate.)

2',3'-Isopropylideneguanosine 5'-phosphate. I.1.f.ζ-3. UV.[1321]

N^2-Benzoyl-2',3'-isopropylideneguanosine 5'-phosphate. From preceding compound with PhCOCl. UV.[1321]

$N^2,O^{2'},O^{3'}$-Tribenzoylguanosine 5'-phosphate. From guanosine phosphate with PhCOCl. UV.[1321]

(7-Theophyllinyl-CH_2CH_2O)PO_3H_2. I.1.b.γ-2 - e.α-1 (using $(PhO)_2POCl$ and NaOH).[355]

7-(β-D-Glucopyranosyl)-theophylline 3'-phosphate. I.1.d.γ-2 - e.α-2 (from 7-(4,6-benzylidene-2,3-anhydro-β-D-allopyranosyl)-theophylline with $(PhCH_2O)_2PO_2H$). Cryst., m. 218° (dec.).[659]

Aristeromycine 6'-phosphate. I.1.f.ζ-1. Solid, $[\alpha]_D^{24}$ -34.0°; UV.[733]

B. R = Cyclic Nucleus Bound Directly to O-P

1. <u>Carbocyclic Non-Aromatic Nuclei</u>

c-$C_6H_{11}OPO_3H_2$. I.1.f.ζ-5 (from ROH with dimorpholino-POCl(Br))[988,989] I.1.f.ι (rearrangement of 2-Cl-decyl phosphonate + c-$C_6H_{11}NH_2$ in ROH),[925] I.2.b.α-6 (from ROH with H_3PO_3 and Hg^{2+}).[1065] Cryst., m. 86°,[989] 85°; c-$C_6H_{11}NH_3$ salt, m. 208-210°;[925] mono anilinium salt, m. 168-169°.[1065]

trans-2-HO-c-$C_6H_{10}OPO_3H_2$. I.1.e.α-2 (from 2-menthyloxy-acetoxy-c-C_6H_{10}-diphenylphosphate). c-$C_6H_{11}NH_3$ salt, m. 200-205°, $[\alpha]_D^{20}$ 10.3°.[286]

(-)-Inositol 3-phosphate. I.1.b.γ-2 - e.α-2 (from 1,2;5,6-diisopropylideneinositol with $(PhO)_2POCl$). $[\alpha]_D^{25}$ -25.6°; c-$C_6H_{11}NH_3$ salt, m. > 250°.[796]

1,2;5,6-Diisopropylidene(-)inositol 3-phosphate. I.1.b.γ-2 - e.α-2 (from diisopropylideneinositol with $(PhO)_2$-POCl). $[\alpha]_D^{25}$ -2.2°; c-$C_6H_{11}NH_3$ salt, m. > 300°.[796]

Inositol 5-phosphate. I.1.b.γ-2 - e.α-2 (from penta-acetyl derivative). Na salt, m. 233-234°; free acid, cryst., m. 198-200°.[737b]

Scyllitol monophosphate. I.1.b.γ-2 and e.α-2 (from penta O-acetyl derivative). Cryst., m. 212-214°. Na_2 salt, solid.[737b]

Pinitol 5-phosphate. I.1.b.γ-2 - e.α-2 (from diisopropylidenepinitol with $(PhO)_2POCl$). c-$C_6H_{11}NH_3$ salt, m. > 250°; $[\alpha]_D^{25}$ 20.5°.[796]

Diisopropylidenepinitol 5-phosphate. I.1.b.γ-2 - e.α-2 (from diisopropylidenepinitol with $(PhO)_2POCl$). c-$C_6H_{11}NH_3$ salt, m. 177-177.5°; $[\alpha]_D^{25}$ 2.0°.[796]
Myoinositol 1-phosphate. I.1.b.γ-2 - e.α-2 (from tetraacetyl derivative with $(PhO)_2POCl$);[1114] by interconversion of 2-phosphate; by hydrolysis of 1,2-cyclic phosphate.[1114] Absolute configuration of the product resulting from monophosphoinositides;[286] $[\alpha]_D^{25}$ 3.4° (pH 9 by c-$C_6H_{11}NH_2$).[1114]
Bornyl phosphate. I.1.b.β (racemic:[747,1043]; (+) and (-) forms[1043]), I.2.b.β-4 - I.1.e.α-2 (from ROH with $(PhCH_2O)_3P$ and $NCCHBrCONH_2$).[663] Racemic form: needles, m. 155-156°.[1043] (+) Form: cryst., m. 154-156°; $[\alpha]_D$ 23.3° (EtOH).[1043] (-) Form: cryst., m. 154-156°; $[\alpha]_D$ -22.6° (EtOH).[1043]
Estradiol 17-phosphate. I.1.b.β (from estradiol;[1269] from 3-O-benzoylestradiol, followed by acidic hydrolysis[406]). Cryst., m. 216-217°.[406,1269]
19-nor-Testosterone 17-phosphate. I.1.b.β. Cryst., m. 132-134°; UV; IR.[1193,1194]
19-nor-$\Delta^{1,4}$-Androstadien-3-one-17-yl phosphate. I.1.b.β. Cryst., m. 132-134°; UV; IR.[1194]
Testosterone 17-phosphate. I.1.b.β (from ROH;[559,1193,1194] from the enol Me ether, followed by acidic hydrolysis[1016]). Cryst., m. 155-156°,[1016] 157-159° (dec.; shrinking at 135-138° then opalescent, demisolid at 138-143°),[599] with $1H_2O$ m. 140-141°;[1193] $[\alpha]_D$ 73°,[1193] $[\alpha]_D^{27.6}$ 72.6°,[1193] $[\alpha]_D^{20}$ 71.9°[1016] (all in MeOH); UV; IR.[1193,1194]
Dehydroepiandrosterone phosphate. I.1.b.β ($ROPO_3H_2$ and $(RO)_2PO_2H$ separated as Me esters (CH_2N_2)),[1193] I.1.f.ζ-5 (with dimorpholino-POCl).[1194] Cryst.; $[\alpha]_D$ 13° (MeOH),[1193,1194] $[\alpha]_D$ 5.5° ($CHCl_3$); IR;[1193,1194] UV; circular dichroism, NMR.[1194]
Androsterone-3,17-diol-17-phosphate. I.1.b.β (from the 3-acetate). Cryst., m. 218.5°; $[\alpha]_D^{20}$ -42.9° (EtOH); Na salt, cryst.[1076]
$\Delta^{1,4}$-Androstadiene-3-one-17-yl phosphate. I.1.b.β. Cryst., m. 140-141°; $[\alpha]_D$ 97°.[1194]
Cholestanyl phosphate. I.1.f.ζ-5 (using o-phenylene cyclic phosphorochloridate, followed by hydrolysis). c-$C_6H_{11}NH_3$ salt, m. 223-224°.[319]
Cholesteryl phosphate. I.1.b.β,[522,523,1122,1432,1432a] I.1.b.γ-1,[1122] I.1.b.γ-2 - e.α-1 (with $(PhO)_2POCl$, then alcoholic KOH),[1376] I.1.f.ζ-5 (with o-phenylene-POCl, followed by hydrolysis).[319] Cryst., m. 195-196°,[522] 193°,[523,1432,1432a] 175°,[1432,1432a] 162-163°,[1376] 161.5°;[319] Na salt, m. 265-270° (forms on heating with EtONa);[1432,1432a] Ba salt ($4H_2O$), plates.[1122] (The formulation as hydrated pyrophosphate, $(RO)(HO)_3$-$POP(OR)O_2H$,[1432b] is considered as erroneous.) $[\alpha]_D$

-36°; UV.[1376]

Vitamine D_3 phosphate. I.1.b.β (from ROLi with $POCl_3$). Na salt, m. 215-216° (dec.); Ca salt, m. 210-211° (dec.); Ba salt, m. 193-194°; IR.[962]

2. Aromatic Nucleus

a. Hydrocarbon Derivatives and Halogenated Compounds

$PhOPO_3H_2$. I.1.a.α,[224,1187] I.1.b.γ-1,[689,739,746,1122] I.1.e.α-3 (from (RO)$(PhCH_2O)_2PO$ by LiCl or NaBr),[974] I.1.f.ζ-5 (using dimorpholino-POCl),[988] dimorpholino-POBr(Cl)[989]), I.1.f.ι (by rearrangement of 2-Cl-decyl phosphonate + c-$C_6H_{11}NH_2$ in ROH),[925] I.2.b.β-4 (from ROH with dimorpholino-POEt and an α-Br-amide),[972] I.2.b.β-4 - I.1.e.α-2 (from ROH with $(PhCH_2O)_2PHO$ and $NCCHBrCONH_2$).[971] Scales, m. 99.5° ($CHCl_3$),[689] needles, m. 97-98° (H_2O),[1175] 94°;[989] NH_4 salt, dec. 140-150°; K salt, plates;[1122] Na salt;[575] Ca, Ba, Cu^{II} salts poorly soluble; c-$C_6H_{11}NH_3$ salt, m. 212-215°,[129,925] 211°;[988] Ag salt; complex with $(Et)_2Au$, m. 130°,[569a] anilinium salt, m. 171-175°.[971]

4-$ClC_6H_4OPO_3H_2$. I.1.b.γ-1.[776,1215,1502] Cryst., m. 80-81°,[776] 93°;[1215] Ba salt.[776]

Pentachlorophenyl phosphate. I.1.b.α (from PCl_5 by warming with Cl_4-p-benzoquinone,[202,1506] Cl_6-1,4-cyclohexadien-3-one,[202] Cl_4-o-benzoquinone,[1508] or Cl_6-1-cyclohexen-3,5-dione[1506,1508] followed by treatment with water). Plates, anhydrous, m. 203°, monohydrate, m. 224°.[1506,1508]

4-$BrC_6H_4OPO_3H_2$. By bromination of $PhOPO_3H_2$ in $CHCl_3$. Plates, m. 161°.[1502]

4-$O_2NC_6H_4OPO_3H_2$. I.1.f.ι (by rearrangement of $(EtO)_2$-P(O)CH(NHR)Ph + $SOCl_2$, with $O_2NC_6H_4OH$ + Et_3N, via chloridate),[15] I.2.b.β-4 (from ROH with dimorpholino-POEt and an α-Br-amide);[972] by nitration of $PhOPO_3H_2$ with HNO_3 (d 1.5),[1175] best procedure.[689] Cryst., m. 165-167°,[972] 153°;[689] Et_3NH salt.[15]

2,4-$(O_2N)_2C_6H_4OPO_3H_2$. I.1.b.β. Pyridinium salt, m. 156-157°.[123]

4-$MeC_6H_4OPO(OH)_2$. I.1.b.γ-1. Plates, m. 116°.[1175]

3-Me-4-$ClC_6H_3OPO_3H_2$. I.1.b.γ-1. Cryst., m. 131°.[1215]

2-Me-5-i-$PrC_6H_3OPO_3H_2$. I.1.b.γ-1. K salt, plates.[681]

2-Me-3-Cl-5-i-$PrC_6H_2OPO_3H_2$. I.1.b.γ-1. Cryst., m. 148°.[1215]

2-i-Pr-5-$MeC_6H_3OPO_3H_2$. I.1.b.γ-1. Liquid. Ba salt, plates.[481]

3-Me-6-i-Pr-2-$ClC_6H_2OPO_3H_2$. I.1.b.γ-1. Cryst., m. 142°.[1215]

2-Me-4-$BuC_6H_3OPO_3H_2$. I.1.b.γ-1. Cryst., m. 83°.[1215]

4-$(PhCMe_2)C_6H_4OPO_3H_2$. I.1.b.β. Cryst., m. 136°.[1369]

1-$C_{10}H_7OPO_3H_2$. I.1.b.γ-1.[577,835] Cryst., m. 155-157°,[577] m. 142°; diphenylhydrazine salt, m. 147-148°; monophenylhydrazine salt, m. 188°.[835]
2-$C_{10}H_7OPO_3H_2$. I.1.b.β,[577,835] I.1.b.γ-2 - e.α-2 (using $(PhO)_2POCl$).[96] Cryst., m. 167°,[835] 176-177°,[577] 172-173°;[96] $ROPO_3HNa \cdot ROPO_3H_2$, m. 202-204° (partial resolidif.), 244°;[578] diphenylhydrazine salt, m. 168°; monophenylhydrazine salt, m. 180°.[835]
6-Br-2-$C_{10}H_6OPO_3H_2$. I.1.b.β - γ-1. Cryst., m. 207-209°; $ROPO_3NaH \cdot ROPO_3H_2$.[578]

b. Aromatic Nucleus with Other Substituents

2-$HOC_6H_4OPO_3H_2$. I.1.a.α (from 1,2-$(HO)_2C_6H_4$);[598] from o-phenylene cyclophosphate by H_2O.[385] Needles, m. 139°.[598]
4-HO-2-$ClC_6H_3OPO_3H_2$. I.1.b.γ-1. Cryst., m. 98-100°.[690]
2-$MeOC_6H_4OPO_3H_2$. I.1.b.γ-1, I.1.e.α-1. Needles, m. 94°; Na salt; acid Ca salt, needles; Ca salt insoluble.[101]
4-i-$AmOC_6H_4OPO_3H_2$. I.1.b.γ-1. Cryst., m. 55-58°.[690]
2-$PhOC_6H_4OPO_3H_2$. I.1.b.γ-1. Cryst., m. 121-123°.[690]
4-$PhOC_6H_4OPO_3H_2$. I.1.b.γ-1. Cryst., m. 127-129°.[690]
4-$PhCH_2OC_6H_4OPO_3H_2$. I.1.b.γ-1. Cryst., m. 122°.[690]
2-MeO-4-CH_2=$CHCH_2 \cdot C_6H_3OPO_3H_2$. I.1.b.γ-1. Prisms, anhydrous, m. 105°, with 1H_2O 46-50°.[256]
2-MeO-4-MeCH=$CHOPO_3H_2$. From preceding compound by heating with dil. KOH. Needles (1.5 H_2O), m. 105-106°.[256]
2-Me-4-$PrCO \cdot C_6H_3OPO_3H_2$. I.1.b.γ-1. Cryst., m. 139°.[1215]
4-HO-2,3-Me_2-1-$C_{10}H_4OPO_3H_2$. I.1.e.α-2 (from Ph-$PhCH_2$ ester,[406a] $(PhCH_2)_2$ ester[123,406a]). Needles, m. 140°,[406a] m. 129°.[123]
Dihydrovitamine K_1 1-phosphate. I.2.b.γ-1 - I.1.e.α-2 (from vitamine K_1 with $(PhCH_2O)_2PHO$).[1202]
Estradiol 3-phosphate. I.1.b.β (using $POCl_3$). Cryst., m. 245-246°. UV, IR.[1193,1194]
17-O-propionylestradiol 3-phosphate. I.1.b.β. Cryst., m. 185-187°.[1269]
Estrone 3-phosphate. I.1.b.β (using $POCl_3$),[1193,1194] I.1.c.β-2 - e.α-1 (using $(PhCH_2)_4P_2O_7$).[1205] Cryst., m. 202-204°,[1205] with 1H_2O m. 214-215°;[1193,1194] $[\alpha]_D$ 129° (MeOH), UV, IR.[1193,1194]
(±)-Tocopheryl phosphate. I.1.b.β,[769,1273] I.1.c.β-2 - e.α-1 (using $(PhCH_2)_4P_2O_7$).[1205] $(NH_4)_2$ salt, m. 164-166°;[1205,1273] Na_2 salt;[769,1273] NaH, Ca, $(Et_2NH_2)_2$, $[(HOCH_2CH_2)_2NH_2]_2$ salts.[1273]
2,2,5',4"-Me_4-3',4',5',6'-tetrahydrodibenzopyran-6"-yl phosphate. I.1.b.γ-1. Viscous. Na salt.[233]
Tetrahydrocannabinyl phosphate. I.1.b.β, oil.[233]

Carboxylic Acid Derivatives

$2\text{-}HO_2CC_6H_4OPO_3H_2$. I.1.b.γ-1 (from $Cl_3CC_6H_4OH$),[40,371,438] I.1.c.β-1 (from the chloroformylderivative);[438] from $H_2NC(O)C_6H_4OPO_3H_2$ by HNO_2.[384] Cryst., m. 140-142°;[40,438] Ba salt ($Ba_{1.5}\cdot 2H_2O$).[384]

$2\text{-}MeO_2CC_6H_4OPO_3H_2$. I.1.a.β. Ba salt ($1H_2O$).[384]

$2\text{-}H_2NC(O)C_6H_4OPO_3H_2$. I.1.e.α-2 (using dibenzyl ester),[91] I.1.a.β.[384] Cryst., m. 155-156°;[91] Ba salt ($1H_2O$).[384]

$2\text{-}NCC_6H_4OPO_3H_2$. I.1.a.β. Ba salt ($4H_2O$).[384]

$2\text{-}HO_2C\text{-}4\text{-}ClC_6H_3OPO_3H_2$. I.1.b.γ-1 (from the 2-chloroformyl derivative). Powder, m. 161-162°.[33]

$3\text{-}HO_2CC_6H_4OPO_3H_2$. I.1.b.γ-1 (from the 3-chloroformyl derivative). Cryst., m. 200-201°.[46]

$4\text{-}HO_2CC_6H_4OPO_3H_2$. I.1.b.γ-1 (from the 4-chloroformyl derivative). Plates, m. 200°.[46]

$4\text{-}Et_2NC\text{-}2\text{-}MeOC_6H_3OPO_3H_2$. I.1.b.β. $c\text{-}C_6H_{11}NH_3$ salt, m. 160°; Ca salt.[1242]

$2\text{-}HO_2C\text{-}4\text{-}MeC_6H_3OPO_3H_2$. I.1.b.γ-1 (from the 2-chloroformyl derivative).[51] Cryst., m. 139.5-140.4°.[51]

2-(PhNHCO)-4-$MeC_6H_3OPO_3H_2$. I.1.b.α (from 6-hydroxy-3-methylbenzophenone oxime). Cryst., m. 187-189° (from dil. EtOH).[111]

$2\text{-}HO_2C\text{-}5\text{-}MeC_6H_3OPO_3H_2$. I.1.b.γ-1 (from the 2-chloroformyl derivative). Cryst., m. 150°.[51]

$2\text{-}HO_2C\text{-}6\text{-}Me\text{-}C_6H_3OPO_3H_2$. I.1.b.γ-1 (from the 2-chloroformyl derivative). Cryst., m. 148-149°.[52]

2,6-$(HO_2C)_2C_6H_3OPO_3H_2$. Kinetic study of hydrolysis.[1021]

$2\text{-}HO_2C\text{-}6\text{-}EtO_2CC_6H_3OPO_3H_2$. Kinetic study of hydrolysis.[1021]

O-Phospho-L-tyrosine. I.1.a.β,[1121] I.1.b.β (from tyrosine Et ester or its N-Cbo derivative),[1131] I.1.b.γ-1 (from formyl tyrosine).[640] Cryst., m. 253°,[867] 227°,[1131] 225°;[1121] $[\alpha]_D^{20}$ -8.8° (2N HCl), $[\alpha]_D$ -9.19° (2N HCl); Pb salt.[1121]

O-Phospho-L-tyrosyl-glycine. I.1.b.β (from Me ester of N-Cbo-O-acetyl-tyrosyl-glycine). Cryst., m. 178°; $[\alpha]_D^{23}$ 20.0° (1N H_2SO_4); Ba and Pb salts.[1131]

O-Phosphotyrosyl-glycyl-glycine. I.1.b.β (from N-Cbo-O-acetyl-tyrosyl-glycyl-glycine Me ester). Cryst., m. 182°, $[\alpha]_D^{23}$ 5.7° (1N H_2SO_4).[1131]

Glycyl-O-phosphotyrosine. I.1.b.β (from the N-Cbo-dipeptide). Cryst., m. 224-225°; $[\alpha]_D^{20}$ 27.9° (1N H_2SO_2); Na salt.[1131]

Glycyl-O-phosphotyrosyl-glycine. I.1.b.β (from N-Cbo-tripeptide Me ester). Cryst., m. 198°, $[\alpha]_D^{23}$ 8.0° (1N H_2SO_4); Na salt.[1131]

$2\text{-}HO_2CC_{10}H_6\text{-}1\text{-}OPO_3H_2$. I.1.b.γ-1 (from the 2-chloroformyl derivative). Needles, unstable in aqueous solution.[1468]

$1\text{-}HO_2CC_{10}H_6\text{-}2\text{-}OPO_3H_2$. I.1.b.γ-1 (from the 1-chloroformyl derivative). Needles, m. 156° (from benzene-Me_2CO).[1151]

Phenolphthalein phosphate. I.l.b.β,[797] I.l.f.ζ-3 (from Me ester, followed by double saponification).[936] Na_3 salt, m. > 400° (reddens at 90°, then went black);[936] Ca, Ba, Pb salts.[797]
N',N'-Me_2-tryptamine-6-yl phosphate. I.l.b.γ-2 (ROH with $(PhCH_2O)_2POCl$ and hydrogenolysis). Cryst., m. 180-200°.[1226]
N',N'-Me_2-tryptamine-7-yl phosphate. I.l.b.γ-2 (ROH with $(PhCH_2O)_2POCl$ and hydrogenolysis). Cryst., m. 229-231°.[1226]

Sulfonic Acid Derivative

6-$HO_3SC_{10}H_6$-2-OPO_3H_2. I.l.b.β. Ba salt, powder.[414]

3. R = Heterocyclic Nucleus, n = 2

4-Hydroxy-tetrahydrofuryl-3-phosphate. I.l.a.β. Ba salt.[329]
L-Ascorbic acid 2-phosphate. I.l.b.β (from 5,6-isopropylidene derivative). NMR.[1052]
L-Ascorbic acid 3-phosphate. I.l.b.β (from 5,6-isopropylidene derivative,,besides 2-phosphate). NMR.[1052]
L-Hydroxyproline phosphate. I.l.a.β.[869,1121] Needles, m. 130-131° (anhydrous), 115° (hydrate);[1121] Ba salt ($1H_2O$),[869,1121] $[\alpha]_D^{25}$ -13.3° (10% HCl);[869] Pb salt, insoluble powder;[1121] brucine salt, m. 180-183°.[869]
2-Pyridyl phosphate. I.l.b.β (from pyridone with $POCl_3$; no phosphorylation occurring at N). c-$C_6H_{11}NH_3$ salt; UV, NMR.[638]

II. Bis Primary Esters; Tris Primary Ester, etc. $R(OPO_3H_2)_n$, n = 2 or 3 (or more)

1. R Bound to O by an Aliphatic Carbon, n = 2

a. R Purely Aliphatic

$CH_2(OPO_3H_2)_2$. The original preparation has been disclaimed.[428]
$HOCH_2CH(OPO_3H_2)CH_2OPO_3H_2$ (?). I.l.a.β.[158]
$HOCH(CH_2OPO_3H_2)_2$ (?). I.l.d.γ-2 - I.l.a.β. Na salt.[161]
Dihydrosphingosine 1,3-di-phosphate. I.l.b.γ-2 - e.α-2 (from N-Cbo derivative using $(PhO)_2POCl$). Mono-Ba salt.[1446]
$OC\cdot(CH_2OPO_3H_2)_2$ (?). I.l.c.α. Ba salt.[844]
2,3-Diphospho-D-glyceric acid. I.l.e.α-2 (from Ph_2 ester). Sirup, $[\alpha]_D$ -1.7° → 2.3° (1N HNO_3); $Ba_{2.5}$ salt.[136]
D-glycero-Tetrulose 1,4-di-phosphate. From the Me_2-acetal. $[\alpha]_D^{25}$ -1.4°, $[\alpha]_{400}^{25}$ -11.7°.[1341,1342]
D-glycero-Tetrulose Me_2-acetal 1,4-di-phosphate. I.l.b.γ-2 - e.α-2 (using $(PhO)_2POCl$). $(c-C_6H_{11}NH_3)_3$ salt, m. 205° (dec.); $[\alpha]_D^{27}$ 13.5°.[1341,1342]

2-Deoxy-L-ribose 3,4-di-phosphate. I.1.b.γ-2 - e.α-1 (from β-Me deoxyriboside with $(PhO)_2POCl$). Sirup.[1079]

α-D-Ribofuranose 1,5-di-phosphate. I.1.d.δ - e.α-2 (from 1-Br-2,3-cyclocarbonato-D-ribofuranose using $(PhCH_2O)_2$-POCl). Ba salt.[1345]

α-D-Glucopyranose 1,6-di-phosphate. I.1.d.δ - e.α-2 (from 1-Br-2,3,4-tri-O-acetyl-β-D-glucose 6-Ph_2-phosphate with Ag_3PO_4, and deacetylation),[857] I.1.f.γ (from 1,2,3,4-tetra-O-acetyl-α-D-glucopyranose 6-phosphate with H_3PO_4, and deacetylation).[297] Ba salt, $[\alpha]_D$ 72°;[857] $(c-C_6H_{11}NH_3)$ salt, $[M]_D$ 19,560°.[297]

α-D-Mannose 1,6-di-phosphate. I.1.d.δ - e.α-2 (from 1-Cl-2,3,4-tri-O-acetyl-6-Ph_2-phosphoryl sugar with AgO_2-$P(OPh)_2$). Brucine$_4$ salt, m. 144-150°; Ba_2 salt. $[\alpha]_D^{21}$ 29.9° (solution of K_4 salt, calculated for acid).[1132]

Sedoheptulose 1,7-di-phosphate. From $OHCCH_2CH_2OPO_3H_2$ by condensation with erythrose 4-phosphate by G3P isomerase and aldolase.[807] In fresh human erythrocytes 27×10^{-9} moles/g hemoglobine.[1398]

$HO_2CCH(OPO_3H_2)CH_2OPO_3H_2$. I.1.d.ζ (from the components with chicken breast muscle extract).[631]

Pantothenic acid di-phosphate. I.1.b.γ-2 (Me pantothenate with $(PhO)_2POCl$, then alkaline hydrolysis).[799]

b. R Containing a Heterocyclic Nucleus

Uridine 2'(3'),5'-di-phosphate. I.1.a.β (from 2',3'-benzylidene uridine).[954]

Deoxycytidine 3',5'-di-phosphate. I.1.b.γ-2 - e.α-2 (using $(PhCH_2O)_2POCl$). Isolated as Ba_2 salt; brucine$_4$ salt, m. 185°. Found in acidic hydrolysate of herring sperm.[465]

Thymidine 3',5'-di-phosphate. I.1.f.ζ-3 (from thymidylic acid with $CNCH_2CH_2OPO_3H_2$ and DCC),[1293] I.1.b.γ-2 - e.α-2 (using $(PhCH_2O)_2POCl$).[465] Ba_2 salt; brucine$_4$ salt, m. 182-184°. Found in acidic hydrolysate of herring sperm.[465]

Adenosine 2'(3'),5'-di-phosphate. I.1.f.ζ-10 (from 2',3'-cyclic phosphate-5'-benzylphosphate).[956]

Thiamine-disulfide di-phosphate. By oxidation of thiamine phosphate in alkaline solution with H_2O_2, halogen, ferricyanide. UV.[582]

2. R = Aromatic Nucleus, n = 2

$1,4-(H_2O_3PO)_2C_6H_4$. I.1.a.α. Hygroscopic solid, m. 168-169°.[599]

$3,4-(H_2O_3PO)_2C_6H_3CH_2CH(NH_2)CO_2H$. I.1.a.β (from Dopa). Cryst., m. 147°; Ba salt.[1116a]

$1,4-(H_2O_3PO)_2C_6H_3$-2-$CHMe_2$. I.1.b.γ-1. Crystalline powder.[341]

4-$H_2O_3POC_6H_4CH(Et)CH(Et)C_6H_4$-4-$OPO_3H_2$. I.1.b.γ. Colorless solid. Na salt.[961]
1,4-$(H_2O_3PO)_2$-2-Me-$C_{10}H_5$. I.1.b.β,[543] I.1.b.γ-1,[544] I.1.c.β-2 - e.α-2 (using $(PhCH_2)_4P_2O_7$).[1205] Na salt ($2H_2O$);[543,544,571] Ba salt.[1205]
1,4-$(H_2O_3PO)_2$-2-phytyl-$C_{10}H_5$. I.1.b.γ-1. Waxy solid.[544]
1,4-$(H_2O_3PO)_2$-2-Me-3-phytyl-$C_{10}H_4$. I.1.b.β. Brown solid.[543]
Diphenylenedioxide 2,6-di-phosphate. I.1.b.β. Cryst., m. 236°.[1357]
3,7-Me_2-diphenylenedioxide 1,5-di-phosphate. I.1.b.β. Cryst., m. 198°.[1357]
2,5-$(EtO)_2$-2,5-$(H_2O_3POCH_2)_2$-dioxane. I.1.e.α-2 (from Ph ester). $(c\text{-}C_6H_{11}NH_3)_2$ salt, m. 201°.[422]

3. R = Alicyclic Nucleus, n = 2, 3 (or more)

(±)-Myoinositol 1,4-di-phosphate. I.1.b.γ-2 - e.α-2 (from 1,2;4,5-di-O-cyclohexylidene derivative using $(PhO)_2POCl$). $(c\text{-}C_6H_{11}NH_3)_3$ salt, m. 186-197°.[24x]
(±)-Myoinositol 1,6-di-phosphate. I.1.b.γ-2 - e.α-2 (from 1,2;5,6-di-O-cyclohexylidene derivative using $(PhO)_2POCl$). $(c\text{-}C_6H_{11}NH_3)_3$ salt, m. 161-163°.[24x]
(±)-Myoinositol 4,5-di-phosphate. I.1.b.γ-2 - e.α-2 (from 1,2;3,4-di-O-cyclohexylidene derivative using $(PhO)_2POCl$). $(c\text{-}C_6H_{11}NH_3)_3$ salt, m. 182-183°.[24x]
Myoinositol hexa-phosphate (Phytic acid). I.1.a.β (with H_3PO_4 + P_2O_5). Na_{12} salt ($38H_2O$); Ca_2Na_8 salt ($3H_2O$).[1131a] X-ray analysis of Na salt.[538a]
1,3,5-$(H_2O_3POCH_2)_3$-1,3,5-$(HO)_3$-cyclohexane. I.1.b.β (from acetol with $POCl_3$ + quinoline). UV, IR.[1420]

III. Mono Secondary Esters, $(RO)(R'O)PO_2H$

A. R and R' Bound by an Aliphatic Carbon Atom

1. R and R' with Aliphatic Carbon Skeleton

a. Hydrocarbon Derivatives and Their Halo Derivatives

$(MeO)_2PO_2H$. I.1.a.α,[346,348,655,884] I.1.b.β (using $POCl_3$ adduct with tertiary bases),[190a] I.1.b.γ-1,[1296] I.1.d.δ,[165,166] I.1.e.α-1,[489,1152] I.1.e.α-2 (from $(MeO)_2(PhO)PO$),[357] I.1.e.α-4 (using NMe_3),[357] I.2.b.α-5 (using N_2O_4).[1100b] Sirup,[348] $b_{0.0001}$ 78-80°; d_{20} 1.3451; n_D^{20} 1.4080.[1100b] Me_4N salt, m. 215°, Ag salt,[884] Ca salt,[1296] Ba salt (soluble in H_2O),[348,655] Pb salt, needles, m. 155°.[348] Conformation of $Ba_{0.5}$ salt by NMR, J_{H-P} 10.5 (in D_2O).[1372]
(MeO)(EtO)PO_2H. I.1.c.α. Ba salt.[182]

$(MeO)(ClCH_2CH_2O)PO_2H$. I.1.b.γ-2 then γ-1 (using $ClCH_2CH_2$-$OPOCl_2$).[1412]

$(EtO)_2PO_2H$. I.1.a.α,[346,348,706,1414] I.1.a.β,[842] I.1.γ-1,[248] I.1.c.α,[1086,1122] I.1.c.β-1,[1360] I.1.d.δ,[165,166] I.1.e.α-1,[489,884] I.1.e.α-2 (using Ph ester),[1335] I.1.e.α-4 (from $(MeO)(EtO)_2PO_2H$ with Me_3N;[357] from t-Bu-peroxyester with c-$C_6H_{11}NH_2$),[1274] I.2.b.α-5 (from $(EtO)_2(MeCO{\cdot}CH_2CH_2O)P$ by N oxides),[1199] I.2.b.α-7 (from $(EtO)_2PONa$),[1313] I.2.b.β-4,[931] I.2.a.α-5 (using N_2O_4),[1100b] I.1.b.β-6 (from $(EtO)_3P$ by $EtO_2CN{=}NCO_2Et$, then conc. HCl),[614a] (from $(EtO)_2PCl$ by NaO(O)N=CH-alkyl).[1011a] From $(EtO)_2PONHR$ by treatment with R'_2-CO + NaH),[1439a] revealed by chromatography in "Langheld" ester.[1453] Sirup,[1360] $b_{0.08}$ 113-116°,[971] $b_{0.01}$ 116-118°,[1360] $b_{0.0001}$ 93-95°; d_{20} 1.1041; n_D^{20} 1.4252;[1100b] n_D^{25} 1.4146-1.4152.[1360] Na salt ($3H_2O$);[165,166,1086] Me_4N salt, m. 170°;[357] c-$C_6H_{11}NH_3$ salt, m. 78-80°;[1274] Ca salt ($2H_2O$);[165] Ba salt ($6H_2O$);[165,348] Pb salt, needles, m. 180°, soluble in H_2O.[348] Conformation of $Ba_{0.5}$ salt by NMR, J_{H-P} 7.0 Hz (in H_2O).[1372]

$(EtO)(ClCH_2CH_2O)PO_2H$. I.1.b.γ-2, then γ-1 (using $ClCH_2$-CH_2OPOCl_2),[1412] I.1.c.α.[842] Ba salt.[842]

$(EtO)(Cl_2C{=}CHO)PO_2H$. I.2.b.γ-2 (from $(EtO)_2(MeCO{\cdot}CH_2CH_2$-$O)P$ with chloral, besides $(EtO)_2(Cl_2C{=}CHO)PO$).[1199]

$(F_2CH{\cdot}CH_2O)_2PO_2H$. I.1.e.α-1. NH_4 salt; Ba salt.[1284]

$(EtO)(PrO)PO_2H$. I.2.b.α-8 (from $EtOPO_2H_2$ + PrOH or from $PrOPO_2H_2$ + EtOH, by Br_2), I.2.b.β-2 (from the same components by CCl_4 + Et_3N). NH_4 salt.[246]

$(ClCH_2CH_2O)(PrO)PO_2H$. I.1.γ-2, then γ-1 (from $ClCH_2CH_2$-$OPOCl_2$ with EtOH, then H_2O).[1412]

$(ClCH_2CH_2O)(CH_2{=}CHCHO)PO_2H$. I.1.c.α, I.1.d.δ. Ba salt.[842]

$(PrO)_2PO_2H$. I.1.b.β (using $POCl_3$ with a Lewis acid, via $R_4P_2O_7$),[1408] I.1.c.α,[1122] I.1.d.δ,[165] I.1.e.α-1,[489] I.1.b.α-4 (from $(MeO)(PrO)_2PO$ with Me_3N),[357] I.2.b.α-5 (from $(PrO)_2(MeCO{\cdot}CH_2CH_2O)P$ and N oxides),[1199] I.2.b.α-8 (from $(RO)_2PHO$ and Cl_2, then H_2O).[632] Oil, n_D^{20} 1.4198;[632] Me_4N salt, m. 132°;[357] Ba salt, needles;[1122] Pb salt, needles, m. 145-147°.[349] Conformation of $Ba_{0.5}$ salt by NMR, J_{H-P} 6.4 Hz (in H_2O).[1372]

$(F_2CHCF_2CH_2O)_2PO_2H$. I.1.a.α, I.1.a.β. Liquid, $b_{0.2}$ up to 210°.[226]

$(ClCH_2CHClCH_2O)_2PO_2H$. I.1.b.γ-1. Ca salt (dihydrate), needles, m. 273°.[798]

$(BrCH_2CHBrCH_2O)_2PO_2H$. I.1.b.γ-1. Needles. Ca salt ($4H_2O$), needles, dec. 250°.[798]

$(CH_2{=}CHCH_2O)_2PO_2H$. I.1.a.α,[161,348] I.1.e.α-1,[345] I.2.b.β-4.[971] Sirup,[348] $b_{0.015}$ 122-125°;[971] K salt;[161] Ba salt, fairly soluble in H_2O; Pb salt, m. 151°.[345]

$(PrO)(i\text{-}PrO)PO_2H$. I.2.b.α-8 (from i-$PrOPO_2H_2$ with PrOH and Br_2), I.2.b.β-2 (from the same components and

CCl_4 + Et_3N). c-$C_6H_{11}NH_3$ salt, m. 141°.[246]

(i-PrO)$_2PO_2H$. I.1.e.α-4 (from (MeO)(RO)$_2$PO by Me_3N;[357] from (RO)$_2$(t-BuOO)PO and c-$C_6H_{11}NH_3$[1274]), I.1.f.α (from P_2S_5 with ROH, then hydrolysis),[774] I.2.b.α-8 (from $ROPO_2H_2$ + ROH and Br_2), I.2.b.β-2 (from the same components and CCl_4 + Et_3N).[246] NH_4 salt, m. 133°;[246] Me_4N salt, m. 128°;[357] c-$C_6H_{11}NH_3$ salt, m. 191-193°.[1274]

[($ClCH_2$)$_2$CHO]$_2PO_2H$. I.1.b.γ-1. Ca salt (4H_2O), needles, dec. 249°.[798]

(Cl_3CCHMeO)$_2PO_2H$. I.1.b.γ-1. Viscous mass.[673]

(CH_2=CHMeO)$_2PO_2H$. I.1.e.α-1 (from (RO)$_3$PO by NaOH).[778]

(EtO)(t-BuO)PO_2H. I.2.b.α-8 (from $EtOPO_2H_2$ + ROH and Br_2),[246] I.2.b.β-2 (from the same components and CCl_4 + Et_3N).[246] NH_4 salt, m. 115-116°.[246]

(PrO)(BuO)PO_2H. I.2.b.α-8 (from $PrOPO_2H_2$ + BuOH and Br_2), I.b.β-2 (from the same components and CCl_4 + Et_3N). NH_4 salt.[246]

($ClCH_2CH_2O$)(AmO)PO_2H. I.1.b.γ-2, then γ-1 (from AmOH with $ClCH_2CH_2OPOCl_2$, then H_2O).[1412]

(BuO)$_2PO_2H$. I.1.b.γ-1,[248] I.2.b.α-8 (from (RO)$_2$PHO and Cl_2, then H_2O),[632] I.2.b.β-4 (from (RO)$_2$PHO with NCCHBrCONH$_2$ + PhCH_2OH),[971] I.2.b.β-6 (from (BuO)$_3$P and EtO_2CN=NCO_2Et, then conc. HCl; yield low).[614a] By hydrolysis of (RO)$_2$PONHMe.[1308a] Oil, n_D^{20} 1.4288;[632,1308a] $b_{0.05-0.1}$ 135-138°;[971] anilinium salt, m. 68°.[1308a]

(i-BuO)$_2PO_2H$. I.1.b.β (using $POCl_3$ and a Lewis acid, via $R_4P_2O_7$ then hydrolysis),[1408] I.2.b.α-8 (from (RO)$_2$PHO with Cl_2, then H_2O).[1408] Oil, n_D^{20} 1.4248.[632]

[($ClCH_2$)$_2$C(NO_2)CH_2O]$_2PO_2H$. I.1.b.α (using PCl_5). Amorphous, thermally unstable solid.[803]

[($BrCH_2$)$_2$C(NO_2)CH_2O]$_2PO_2H$. I.1.b.α (using PCl_5). Amorphous solid.[803]

($ClCH_2CH_2O$)(n-C_7H_{15}O)PO_2H. I.1.b.γ-2, then γ-1 (from ROH with $ClCH_2CH_2OPOCl_2$, then H_2O).[1412]

(AmO)$_2PO_2H$. I.1.b.β (using $POCl_3$ and a Lewis acid, via $R_4P_2O_7$),[1408] I.1.e.α-4 (from (MeO)(RO)$_2$PO and Me_3N),[357] I.2.b.α-8 (from (RO)$_2$PHO with Cl_2, then H_2O).[632] Oil, n_D^{20} 1.4370;[632] Me_4N salt, m. 212°.[357]

[H(CF_2)$_4CH_2$O]$_2PO_2H$. I.1.a.α, I.1.b.β (using $POCl_3$). Liquid, b_5 200°.[226]

(i-AmO)$_2PO_2H$. I.1.b.γ-1,[826] I.2.b.α-8 (from (RO)$_2$PHO and Cl_2, then H_2O).[632] Oil, n_D^{20} 1.4327;[632] Ag salt; Ca salt.[826]

(n-C_6H_{13}O)$_2PO_2H$. I.1.b.γ-1,[248] I.2.b.α-5 (from (RO)$_2$PHO and N_2O_4),[1100b] I.2.b.α-8 (from (RO)$_2$PHO and Cl_2, then H_2O).[632] Oil,[632] $b_{0.0001}$ 133-138°; d_{20} 1.0180; n_D^{20} 1.4350,[1100b] 1.4421.[632]

(n-C_7H_{15}O)$_2PO_2H$. I.2.b.α-8 (from (RO)$_2$PHO and Cl_2, then H_2O),[632] I.2.b.α-5 (from (RO)$_2$PHO and N_2O_4).[1100b]

Oil,[632,1100b] d_{20} 1.0021; n_D^{20} 1.4383,[1100b] 1.4421.[632]

$(n-C_8H_{17}O)_2PO_2H$. I.l.e.α-2 (using the Ph ester),[1335] I.l.e.α-4 (from (MeO)(RO)$_2$PO and Me_3N),[357] I.2.b.α-5 (using N_2O_4),[1100b] I.2.b.α-8 (using Cl_2 then H_2O).[632] Oil;[632] cryst., m. 29-30°,[1100b] n_D^{20} 1.4464.[632] Me_4N salt, m. 173°.[357]

$(i-C_8H_{17}O)_2PO_2H$. I.l.b.γ-1,[248] I.l.f.α (from ROH with P_2S_5, then H_2O).[774]

(2-Et-hexylo)$_2PO_2H$. I.l.b.β (using $POCl_3$ with a Lewis acid, via $R_4P_2O_7$),[1408] I.2.b.α-8 (from (RO)$_2$PHO with Cl_2, then H_2O).[632] Oil, n_D^{20} 1.4448.[632] pK_a 2.85±0.05; dimerization constant in C_6H_6 and CCl_4; interaction with (BuO)$_3$PO in the same solvents.[1288]

$(n-C_9H_{19}O)_2PO_2H$. I.l.b.γ-1,[248] I.2.b.α-8 (from (RO)$_2$PHO with Cl_2, then H_2O). Oil, n_D^{20} 1.4609.[632]

$(ClCH_2CH_2O)(n-C_{16}H_{33}O)PO_2H$. I.l.b.γ-1.[395,633] Colorless needles, m. 54.5°; Ba salt.[395]

$(n-C_{10}H_{21}O)_2PO_2H$. I.l.b.β (using $POCl_3$ with a Lewis acid, via $R_4P_2O_7$),[1408] I.l.b.γ-1,[248] I.l.e.α-4 (from (MeO)-(RO)$_2$PO with Me_3N),[357] I.2.b.α-5 (using N_2O_4),[1100b] I.2.b.α-8 (from (RO)$_2$PHO with Cl_2, then H_2O).[632] Cryst., m. 50-51°,[248] 46-47°,[1100b] 37-39°;[632] Me_4N salt, m. 212°.[357]

$(n-C_{11}H_{23}O)_2PO_2H$. I.l.e.α-2 (using the Ph ester). Cryst., m. 53.5-54°.[923]

$[H(CF_2)_{10}CH_2O]_2PO_2H$. I.l.a.α, I.l.b.β. NH_4 salt, m. 182.5-190°.[226]

$(n-C_{16}H_{33}O)_2PO_2H$. I.l.b.γ-1, I.l.c.α, I.l.e.α-1,[1122] I.2.b.α-5 (using N_2O_4).[1100b] Cryst., m. 68-70°;[1100b] Na salt; Ba salt.[1122]

b. R and/or R' Bearing Hydroxy Groups (Free, Etherified, or Esterified), and Their Halo Derivatives

(MeO)$(HOCH_2CH_2O)PO_2H$. I.l.d.β (from $HOCH_2CH_2OPO_3H_2$ with CH_2N_2),[1385] I.l.d.ε (from $HOCH_2CH_2PO_3Na_2$ with Me_2SO_4).[175] Ca salt.[175]

$(HOCH_2CH_2O)_2PO_2H$. I.l.a.β,[383b] I.l.d.γ-2 (from $ROPO_3K_2$ with ethylene oxide),[1385] I.l.e.α-1 (from the Cl_3CCH_2 ester with Pb_2O in H_2O), I.l.b.γ-2, then γ-1 (from ethylene chlorohydrin with $POCl_3$, then Pb_2O in H_2O),[1122] I.2.b.α-1 (from white or yellow P with glycol, in an O_2 containing gas).[195] By alkaline hydrolysis of $(ClCH_2CH_2O)_2PO_2H$.[383x] Ba salt ($2H_2O$).[383b,1122]

(Myristoyl-$OCH_2CH_2O)_2PO_2H$. I.l.e.α-2 (using the Phe ester). Cryst., m. 86-87°.[610]

(Palmitoyl-$OCH_2CH_2O)_2PO_2H$. I.l.e.α-2 (using the Ph ester).[134,610] Cryst., m. 91.5-92.5°,[610] 92.5-93.5°; IR.[134]

(Stearoyl-$OCH_2CH_2O)_2PO_2H$. I.l.e.α-2 (using the Ph ester).[134,610] Cryst., m. 95-96°,[610] 92.5-93.5°; IR.[134]

(MeO)(MeCH(OH)CH_2O)PO_2H. I.1.d.γ-2 (from $MeOPO_3H_2$ with propylene oxide).[1385]
(MeO)[$(HOCH_2)_2CHO$]PO_2H. I.1.b.γ-1. Ca salt, m. 255°.[173]
(MeO)[MeCH(OH)CH(Me)O]PO H. I.1.d.β (from the HO-butyl phosphate with CH_2N_2).[1385]
(MeO)[MeCH(OMe)CH(OMe)O]PO_2H. From the preceding by energic treatment with CH_2N_2. Cryst., m. 134-135.5°.[1385]
($ClCH_2CH_2O$) (2,3-distearoyl-1-glycero-O)PO_2H. From ($HOCH_2CH_2O$)(RO)PO_2H and SO_2Cl_2. Cryst., m. 65-68°.[634]
(1,3-Di-O-palmitoyl-2-glycero-O)($HOCH_2CH_2O$)PO_2H. From the corr. cephalin and HNO_2. Cryst., m. 49-51°.[1210]
($HOCH_2CH_2O$)(2,3-distearoyl-1-glycero-O)PO_2H. I.1.a.α (from distearin + glycol with P_2O_5).[634]
(MeCH(OH)CH_2O)$_2PO_2H$. I.1.d.γ-2 (from $ROPO_3H_2$ with propylene oxide). Chromatographic characteristics.[1385]
(ICH_2•CHOH•CH_2O)$_2PO_2H$. I.1.d.γ-1 (from epi-iodohydrin). Ca salt.[503]
(Palmitoyl-$OCH_2CH_2CH_2O$)$_2PO_2H$. I.1.b.γ-2 (ROH with $PhOPOCl_2$, and hydrogenolysis). Cryst., m. 70.5-71.5°.[654]
($HOCH_2$CH(OH)CH_2O)$_2PO_2H$.
Rac. compound. I.1.b.β (from isopropylidene glycerol with $POCl_3$),[547 1319] I.1.d.δ (from epichlorohydrin, followed by hydrolysis of the epoxide group).[161] By permanganate oxidation of diallyl phosphate.[161] Ba salt.[547,1319]
(L-α-glyceryl-O)$_2PO_2H$. I.1.b.γ-2 - e.α-2 (from the corr. isopropylidene glycerol using $PhOPOCl_2$).[189]
($HOCH_2$•CHOH•CH_2O)(($HOCH_2$)$_2$CHO)PO_2H. I.1.d.δ (from epichlorohydrin and disodium 2-glycerophosphate in hot water). Na salt, cryst.[171,172]
[($HOCH_2$)$_2$CHO]$_2PO_2H$. I.1.a.β (crude),[1459] I.1.b.γ-1 (from ($ClCH_2$)$_2$CH(OH), followed by hydrolysis with $Ca(OH)_2$ in hot H_2O). Ca salt ($13H_2O$), needles, m. 249-250°.[1377]
(2,3-Distearoyl-1-glycero)$_2$ phosphate. I.1.a.α. Cryst., m. ~67°.[634]
(1,3-Distearoyl-2-glycero)$_2$ phosphate. I.1.b.γ-2 - e.α-2 (from ($ROCH_2$)$_2$CHOH with $PhOPOCl_2$). Cryst., m. 74-75°.[609]
1-(3-Oleoyl-2-palmitoyl-glycero-1-phosphoryl)-glycerol. From the isopropylidene derivative and $B(OMe)_3$ + $B(OH)_3$. Cryst., m. 176-179°, $[\alpha]_D^{20}$ 1.02° ($CHCl_3$).[262]
1-(3-Oleoyl-2-palmitoyl-glycero-1-phosphoryl)-2,3-isopropylidene glycerol. I.1.d.δ (from oleoyl-palmitoyl-glycerol-iodohydrin with Ag isopropylideneglycero-1-phosphate). $[\alpha]_D^{20}$ 3.44 ($CHCl_3$).[262]
1-L-(2,3-Distearoyl-L-glycero-1-phosphoryl)-glycerol. From the dioleoyl derivative and H_2/PtO_2. Cryst., m. 66.5-67.0°, $[\alpha]_D^{22}$ 2.0° ($CHCl_3$).[137]

1-L-(2,3-Dioleoyl-L-glycero-1-phosphoryl)-glycerol. I.1.b.β (from dioleine with $POCl_3$). Liquid, n_D^{23} 1.407; $[\alpha]_D^{21}$ 2.0° ($CHCl_3$), $[\alpha]_D^{22}$ 2.35° (EtOH).[137]

R-1-(2,3-Distearoyl-glycero-1-phosphoryl)-3-(L-alanyl)-glycerol. From 1-(...benzylphosphoryl)-2-benzyl-3-alanyl-glycerol and H_2/Pd. Cryst., m. partially 95°, tot. 175°, $[\alpha]_D^{20}$ 1.6°.[985]

S-1-(2,3-Distearoyl-glycero-1-phosphoryl)-3-(L-alanyl)-glycerol. From 1-(...benzylphosphoryl)-2-benzyl-3-alanyl-glycerol and H_2/Pd. Cryst., m. 70-75°, $[\alpha]_D^{10}$ 7.8°.[985]

1-O-(2-O-palmitoyl-3-O-oleoyl-1-glycerophosphoryl)-3-O-L-lysyl-glycerol. I.1.d.δ (from N^2,N^6-bis-Cbo-lysyl-2-O-t-Bu-glycerol iodohydrin with $AgO_2P(OCH_2Ph)$(O-diacylglyceryl), then elimination of the protecting groups). Identical with the product from Staphylococcus aureus.[261]

$[(PhOCH_2)_2CHO]_2PO_2H$. I.1.b.β - γ-1. Needles, m. 105°.[268]

$[(4\text{-}MeC_6H_4OCH_2)_2CHO]_2PO_2H$. I.1.b.β - γ-1. Needles, m. 160°.[268]

c. R and/or R' Bearing Carbonyl Groups (without or with Hydroxyl or Carbonyl Groups)

(EtO)(MeCO·CH(Me)O)PO_2H. I.1.f.ζ-10 (from enolacetoin cyclic phosphate by controlled hydrolysis).[624]

$(OHCCH_2O)_2PO_2H$. By $NaIO_4$ oxidation of di-1-glycerophosphate.[1319]

$(2\text{-}MeOC_6H_4OCH_2CO\cdot CH_2O)_2PO_2H$. By CrO_3 oxidation of $(2\text{-}MeOC_6H_4OCH_2CH(OH)CH_2O)_2PO_2H$. Viscous liquid.[599]

(1,2,3,4-Tetra-O-acetyl-β-D-glucopyranose-6)(2,3,4,6-tetra-O-acetyl-β-D-glucopyranose-1) phosphate. I.1.d.δ (from acetobromoglucose with Ag tetraacetylglucose 6-phosphate). Cryst., m. 204-206°.[659]

d. R and/or R' Bearing Basic Nitrogen Groups (without or with the Preceding Functions)

(MeO)$(Me_3\overset{+}{N}CH_2CH_2O)PO_2^-$. From $(MeO)_2(BrCH_2CH_2O)PO$ with Me_3N. Cryst., m. 110°.[858]

(EtO)$(HMeNCH_2CH_2O)PO_2H$. From 2-EtO-2-oxo-3-Me-1,3,2-oxazaphospholidine by controlled scission (4-toluenesulfonic acid). 4-Toluenesulfonate.[812]

(EtO)$(Me_2NCH_2CH_2O)PO_2H$. By thermic decomposition of (EtO)$(HMeNCH_2CH_2O)(Me_2NCH_2CH_2O)P$. N,N,N',N'-$Me_4$-piperazinium salt, m. 192-193°.[812]

(EtO)$(Me_3\overset{+}{N}CH_2CH_2O)PO^-$. From Me_2N derivative with MeI.[812] From (EtO)$(BrCH_2CH_2O)PO_2H$ with Me_3N. Cryst., m. 160°;[858] Reineckate, m. 175-178°.[812]

(Alkyl-O)$(Me_3\overset{+}{N}CH_2CH_2O)PO_2^-$. From (Alkyl-O)$(BrCH_2CH_2O)PO_2H$ with Me_3N. Thus:

(i-PrO)($Me_3\overset{+}{N}CH_2CH_2O$)$PO_2^-$. Cryst., m. 288°.[858]

(BuO)($Me_3\overset{+}{N}CH_2CH_2O$)$PO_2^-$. Cryst., m. 270°.[858]

(i-BuO)($Me_3\overset{+}{N}CH_2CH_2O$)$PO_2^-$. Cryst., m. 332°.[858]

(sec-BuO)($Me_3\overset{+}{N}CH_2CH_2O$)$PO_2^-$. Cryst., m. 330°.[858]

($H_2NCH_2CH_2O$)$_2PO_2H$. I.1.b.γ-2 - e.α-2 (from NCbo-ethanolamine using $PhOPOCl_2$). Perchlorate, m. 233-235°.[752]

($Me_3\overset{+}{N}CH_2CH_2O$)$_2PO_2^-$. Isolated as the dibromide, by I.1.c.α. Plates, m. 125°, dec. 166°.[1]

(Alkyl-O)($Me_3\overset{+}{N}CH_2CH_2CH_2O$)$PO_2^-$. From Alkyl-O)($Br(CH_2)_3O$)-$PO_2H$ with Me_3N. Thus:

(MeO)($Me_3\overset{+}{N}CH_2CH_2CH_2O$)$PO_2^-$. Cryst., m. 185°.[858]

(EtO)($Me_3\overset{+}{N}CH_2CH_2CH_2O$)$PO_2^-$. Cryst., m. 190°.[858]

(PrO)($Me_3\overset{+}{N}CH_2CH_2CH_2O$)$PO_2^-$. Cryst., m. 275°.[858]

(i-PrO)($Me_3\overset{+}{N}CH_2CH_2CH_2O$)$PO_2^-$. Cryst., m. 253°.[858]

(BuO)($Me_3\overset{+}{N}CH_2CH_2CH_2O$)$PO_2^-$. Cryst., m. 273°.[858]

(sec-BuO)($Me_3\overset{+}{N}CH_2CH_2CH_2O$)$PO_2^-$. Cryst., m. 314°.[858]

(n-$C_{16}H_{33}O$)($H_2NCH_2CH_2O$)PO_2H. By heating the $ClCH_2CH_2$ compound with alcoholic NH_3.[395] From the phthalimido derivative with $N_2H_4 \cdot 2H_2O$.[686] Cryst., m. 230°,[686] 226° (dec.).[395]

(n-$C_{16}H_{33}O$)(2-phthalimidoethyl-O)PO_2H. I.1.b.γ-2 (from $C_6H_4(CO)_2NCH_2CH_2OPOCl_2$). Cryst., m. 64-65°.[686]

(n-$C_{18}H_{37}O$)($H_2NCH_2CH_2O$)PO_2H. I.1.d.δ - e.α-2 (from AgO_2-$P(OPh)(OCH_2CH_2NHCbo)$. Cryst., m. 239°.[909]

Glycol Cephalins and Lecithins

(Palmitoyl-OCH_2CH_2O)($Me_3\overset{+}{N}CH_2CH_2O$)$PO_2^-$. I.1.e.α-2 (using Ph ester). Solid, sint. 68°, droplets 80°, coalescent 242-243°; IR.[134]

(Stearoyl-OCH_2CH_2O)($H_2NCH_2CH_2O$)PO_2H. I.1.d.δ - e.α-2 (using stearoylglycoliodohydrin and $AgO_2P(OCH_2CH_2$-$NHCbo)(OPh)$)). Cryst., m. 35-36°.[909]

(Stearoyl-OCH_2CH_2O)($Me_3\overset{+}{N}CH_2CH_2O$)$PO_2^-$. I.1.e.α-2 (using Ph ester). Solid, sint. 70°, droplets 80°, coalescent 239-240°; IR.[134]

($HOCH_2 \cdot CHOH \cdot CH_2O$)($Me_3\overset{+}{N}CH_2CH_2O$)$PO_2^-$. I.1.b.γ-2 - e.α-2 (using $PhOPOCl_2$ and isopropylidene glycerol). Liquid. L-Isomer: $[\alpha]_D^{23}$ -2.85°. Racemate, similarly from rac-glycerophosphate.[149]

[$(HOCH_2)_2CHO$]($Me_3\overset{+}{N}CH_2CH_2O$)$PO_2^-$. I.1.d.δ.[84,1178] Needles, m. 104-105°.[84]

Glycerol Ether Phosphatides

(2-O-Lauroyl-chimyl-O)($H_2NCH_2CH_2O$)PO_3H. I.1.d.δ - e.α-2 (from $ROCH_2CH(OR')CH_2I$ and $AgO_2P(OCH_2CH_2NHCbo)(OPh)$). Cryst., m. 207°.[216]

(2-O-Myristoyl-chimyl-O)($H_2NCH_2CH_2O$)PO_2H. Similarly. Cryst., m. 205°.[216]

(2-O-Palmitoyl-chimyl-O)($H_2NCH_2CH_2O$)PO_2H. Similarly. Cryst., m. 201°.[216]

(2-O-Lauroyl-batyl-O)($H_2NCH_2CH_2O$)PO_2H. Similarly. Cryst., m. 203°.[216]

(2-O-Myristoyl-batyl-O)($H_2NCH_2CH_2O$)PO_2H. Similarly. Cryst., m. 200°.[216]

(2-O-Palmitoyl-batyl-O)($H_2NCH_2CH_2O$)PO_2H. Similarly. Cryst., m. 197°.[216]

(1,2-Di-O-cis-9'-octadecenyl-3-glycero)($H_2NCH_2CH_2O$)PO_2H. Similarly. Chromatographic characteristics, IR.[359]

(1-O-cis-9'-octadecenyl-2-O-stearoyl-3-glycero)(H_2NCH_2-CH_2O)PO_2H. Similarly. Cryst., m. 192-193°, $[\alpha]_D^{22}$ 3.1° ($CHCl_3$-MeOH).[359]

(1-O-cis-9'-octadecenyl-2-O-stearoyl-3-glycero)($Me_3\overset{+}{N}CH_2$-CH_2O)PO_2^-. Similarly. Cryst., m. 232°, $[\alpha]_D^{22}$ 2.6 ($CHCl_3$-MeOH).[359]

(1-O-cis-9'-octadecenyl-3-O-stearoyl-2-glycero)(H_2NCH_2-CH_2O)PO_2H. Similarly. Cryst., m. 192°, $[\alpha]_D^{22}$ -3.5° ($CHCl_3$-MeOH); IR. Identical with monoether from bovin erythrocytes.[359]

(1-O-cis-9'-octadecenyl-3-O-stearoyl-2-glycero)($Me_3\overset{+}{N}CH_2$-CH_2O)PO_2^-. Similarly. Cryst., $[\alpha]_D^{22}$ -3.8° ($CHCl_3$-MeOH); IR.[359]

(2,3-Di-O-cis-9'-octadecenyl-1-glycero)($Me_3\overset{+}{N}CH_2CH_2O$)$PO_2^-$. Similarly. Cryst., m. 214-216°, $[\alpha]_D^{22}$ 1.4° ($CHCl_3$-MeOH).[359]

(2,3-O,O-Palmital-1-glycero)($H_2NCH_2CH_2O$)PO_2H. From palmital-glycero phosphate with "colamine chloride."[237]

(2,3-O,O-Stearal-1-glycero)($H_2NCH_2CH_2O$)PO_2H. Similarly from stearal-glycerophosphate.[237]

Cephalins and Related Compounds

(2-O-Stearoyl-1-glycero)($H_2NCH_2CH_2O$)PO_2H. From 3-O-benzyl N-phthaloyl derivative. From the 3-stearoylglyceryl derivative by transposition on SiO_2 with $CHCl_3$-MeOH. Cryst., m. 215-216°.[1263]

(3-O-Benzyl-2-stearoyl-1-glycero)(1,2-$C_6H_4(CO)_2NCH_2CH_2O$)-PO_2H. I.1.d.δ (using benzyl stearoyl Ag phosphate and $ICH_2CH_2N(OC)_2C_6H_4$). Cryst., m. 128-130°.[1263]

(3-O-Stearoyl-1-glycero)($H_2NCH_2CH_2O$)PO_2H. Similarly to the 2-stearoyl isomer (above). Cryst., m. 218-221°.[1263]

(2,3-Di-O-lauroyl-1-glycero)($H_2NCH_2CH_2O$)PO_2H. I.1.d.δ - e.α-2 (from $(CboNHCH_2CH_2O)(PhO)PO_2Ag$ and $ROCH_2CH(OR)$-

CH_2I). Cryst., m. 216°.[216]

(2,3-Di-O-myristoyl-1-glycero)($H_2NCH_2CH_2O$)PO_2H. Similarly. Cryst., m. 207°.[216]

(2,3-Di-O-palmitoyl-1-glycero)($H_2NCH_2CH_2O$)PO_2H. Similarly. From the N phthaloyl derivative by $N_2H_4 \cdot 2H_2O$.[686] Cryst., m. 207°,[216] 202° (sint. 130°).[686]

(2-O-Linoleoyl-3-pamlitoyl-L-1-glycero)($H_2NCH_2CH_2O$)PO_2H. I.1.d.δ - e.α-2 (from (CboNHCH$_2$CH$_2$O)(PhCH$_2$O)PO_2Ag and $ICH_2CH(OR)CH_2(OR')$. Cryst., m. 155°, $[\alpha]_D^{20}$ 6° ($CHCl_3$).[457]

(2,3-Di-O-stearoyl-1-glycero)($H_2NCH_2CH_2O$)PO_2H. I.1.d.δ - e.α-2 (from stearin-iodohydrin and (CboNHCH$_2$CH$_2$O)-(PhO)PO_2Ag).[216,909] From the dilinolenoyl derivative by H_2/Pt.[485,1258] Cryst., m. 224°,[909] 198°;[216] L-derivative, m. 180-181°.[485]

(2-O-Linolenoyl-3-O-stearoyl-1-glycero)($H_2NCH_2CH_2O$)PO_2H. I.1.b.γ-2 - e.α-1 (from $ROCH_2CH(OR')CH_2OH$ and C_6H_4-$(CO)_2NCH_2CH_2O)POCl_2$). Waxy solid.[1258]

(2-O-Linolenoyl-3-O-oleoyl-1-glycero)($H_2NCH_2CH_2O$)PO_2H. I.1.b.γ-2 - e.α-1 (similarly). Oil.[1258]

(+)-(L-2,3-Di-O-linoleoyl-1-glycero)($H_2NCH_2CH_2O$)PO_2H. I.1.b.γ-2 (using $C_6H_4(CO)_2NCH_2CH_2OPOCl_2$, then hydrazine). Oil, $[\alpha]_D^{20}$ 6°.[485]

(2,3-Di-O-linolenoyl-1-glycero)($H_2NCH_2CH_2O$)PO_2H. Similarly. Oil.[1258]

(1,3-Di-O-myristoyl-2-glycero)($H_2NCH_2CH_2O$)PO_2H. I.1.b.β - γ-2 (from myristin, $HOCH_2CH_2N(CO)_2C_6H_4$ and hydrazine). Cryst., m. 173-174°.[714]

(1,3-Di-O-palmitoyl-2-glycero)($H_2NCH_2CH_2O$)PO_2H. From the following by N_2H_4.[686] I.1.d.δ (structure ?),[757] I.1.b.β - γ-2 (from palmitin with $POCl_3$, then $HOCH_2$-CH_2N-phthaloyl or $HOCH_2CH_2NHCbo$.[1210] Cryst., m. 202° (sint. 130-139°),[686] 193-194°;[1210] N-Cbo derivative, cryst., m. 39.5-40°;[1210] N-phthaloyl derivative, cryst., m. 71-72°,[1210] wax, m. 45°; 2,4-$(O_2N)_2C_6H_4$CO-derivative, m. 62°.[686]

(1,3-Di-O-stearoyl-2-glycero)($H_2NCH_2CH_2O$)PO_2H. I.1.d.δ (from acylated glycerophosphate). Cryst., m. 175°.[757]

(1-O-Stearoyl-3-O-erucoyl-2-glycero)($H_2NCH_2CH_2O$)PO_2H. I.1.b.β - γ-2 (from erucostearin using $POCl_3$, then $HOCH_2CH_2$N-phthaloyl). Cryst., m. 163.5-164°.[714]

Lecithins and Related Compounds

(3-O-Palmitoyl-1-glycero)($Me_3\overset{+}{N}CH_2CH_2O$)$PO_2^-$. I.1.d.δ. Cryst., dec. 262°.[757]

(2,3-Di-O-palmitoyl-L-1-glycero)($Me_3\overset{+}{N}CH_3CH_3O$)$PO_3^-$. From 1-glycerophosphorylcholine with palmitoyl chloride.[1340]

(2,3-Di-O-stearoyl-1-glycero)($Me_3\overset{+}{N}CH_2CH_2O$)$PO_2^-$. I.1.a.α (stepwise; group migration is not precluded).[757] From the corr. 2-chloroethyl ester and Me_3N.[634]

Plates or needles, m. 187°, dec. 190°;[635] wax, fluid at 64-65°, opaque at 74°.[634]

(1,3-Di-O-palmitoyl-2-glycero)$(Me_3\overset{+}{N}CH_2CH_2O)PO_2^-$. From choline glycerophosphate with palmitoyl chloride. Cryst., m. 160°, dec. 185°.[1064]

(1,3-Di-O-stearoyl-2-glycero)$(Me_3\overset{+}{N}CH_2CH_2O)PO_2^-$. I.1.a.α (stepwise; group migration is not precluded). Needles or plates, m. 195°, dec. 198°.[635]

O-(2-O-Oleoyl-3-O-stearoyl-1-glycerophosphoryl)-N-glycyl-ethanolamine. I.1.d.δ (from N-Cbo and O-benzyl protected components). Cryst., m. 185-187°.[931]

O-(2,3,-Di-O-palmitoyl-1-glycerophosphoryl)-N-(DL-alanyl)-ethanolamine. I.1.d.δ - e.α-3 (from (phthaloylalanyl)-(2-I-ethyl)amine with dipalmitinyl Ag-phosphate; debenzylation by LiBr). Cryst., m. 180-182°.[929]

$[HOCH_2CH(OH)CH_2O][H_2NC(Me)_2CH_2O]PO_2H$. I.1.b.γ-2 (from isopropylidene glycerophosphoryl dichloridate and $HOCH_2C(Me)_2NHCbo$). Amorphous solid, $[\alpha]_D^{28}$ -2.0°.[154]

e. R and/or R' Bearing Carboxyl (Actual or Potential) Groups (without or with Preceding Groups)

$(2\text{-}MeC_6H_4NHCO\cdot CH_2O)_2PO_2H$. I.1.b.α. Amorphous, m. 168-170°.[244]

$(4\text{-}MeC_6H_4NHCO\cdot CH_2O)_2PO_2H$. I.1.b.α. Solid, m. 255-257°.[244]

$(MeO)(MeO_2CCH{=}C(Me)O)PO_2H$. By partial hydrolysis of phosdrin; the trans isomer from trans-phosdrin, and the cis isomer from cis-phosdrin.[1276] By Perkow reaction. Na salt, m. 190°.[257]

$(H_2NCH_2CH_2O)(DL\text{-}HO_2CCH(NH_2)CH_2O)PO_2H$. I.1.d.γ-2 (from ethylene imine and phosphoserine). Cryst., m. 179-180° (dec.).[1370]

$(H_2NCH_2CH_2O)(DL\text{-}H_2NC(CO_2Et)CH_2O)PO_2H$. I.1.b.γ-2 (from N-Cbo-serine Et ester and N-Cbo-$HOCH_2CH_2NH_2$ with $PhOPOCl_2$). Cryst., m. 180-181°.[752]

$(H_2NCH_2CH_2O)(L\text{-}H_2NC(CO_2Et)CH_2O)PO_2H$. Similarly, but using L-serine derivative. Yellow oil, $[\alpha]_D^{32}$ -4.0° (EtOH); IR; 1:1 adduct with $HCONH_2$, cryst., m. 139-141°, $[\alpha]_D^{23.5}$ 15.0.[752]

O,O-(DL-serine Et ester)$_2$ phosphate. I.1.b.γ-2 - e.α-2 (using $PhOPOCl_2$). Cryst., m. 115° (dec.).[752]

O-(2-O-Palmitoyl-3-O-oleoyl-L-1-glycerophosphoryl)-N-Cbo-DL-serine t-Bu ester. By debenzylation of the Ph ester. Cryst., m. 126-128°, $[\alpha]_D^{20}$ 5.8° ($CHCl_3$).[646]

O-(L-2,3-Di-O-stearoyl-1-glycerophosphoryl)-L-serine. I.1.e.α-2 (from the Ph-phosphate N-Cbo-serine benzyl ester derivative). Cryst., m. 159-161° (dec.; sint. ∿120°), $[\alpha]_D^{40}$ -5.2° ($CHCl_3$), $[\alpha]_D^{20}$ -23.2° → 33.0° (12.5 hr) (C_6H_6).[151]

O-(Distearoyl-L-1-glycerophosphoryl)-L-threonine. I.1.e.α-2 (from the P-Ph and C-benzyl ester). Cryst.,

m. 140-141°, insoluble in most solvents up to 60°.[142]
(+)-O-(Distearoyl-L-1-glycerophosphoryl)-L-threonine benzyl ester. I.1.b.γ-2 - b.γ-1 (from diacylglycerophosphorodichloridate and N-Cbo-threonine ester). Cryst., m. 68-69°, $[\alpha]_D$ 15° ($CHCl_3$).[142]
O-(2,3-Di-O-stearoyl-L-1-glycerophosphoryl)-L-seryl-glycylglycine. I.1.b.γ-2 - e.α-2 (using distearin successively with $PhOPOCl_2$ and the N-Cbo tripeptide). Cryst., m. 181-182°, $[\alpha]_D$ 7.6° ($CHCl_3$).[153]

2. R, or R and R' Bearing on a Carbon Atom of the Aliphatic Skeleton a Cyclic Group

a. Carbocyclic Groups

α. Only R Bearing a Cyclic Group

$(MeO)(PhCH_2O)PO_2H$. I.1.e.α-3 (from $PhCH_2$ ester by CaI_2).[454]
$(EtO)(PhCH_2O)PO_2H$. I.1.c.α. Sirup; Ba salt (sol. H_2O, insol. EtOH).[842]
$(PrO)(PhCH_2O)PO_2H$. I.1.e.α-3 (using $PhCH_2$ ester and CaI_2).[454]
$(i\text{-}PrO)(PhCH_2O)PO_2H$. Similarly.[454]
$(BuO)(PhCH_2O)PO_2H$. I.1.e.α-3 (using $PhCH_2$ ester and CaI_2),[454] I.2.b.α-8 (from $BuOPO_2H_2$ and $PhCH_2OH$ by Br_2), I.2.b.β-2 (from the same components and CCl_4 + Et_3N). c-$C_6H_{11}NH_3$ salt, m. 115-116°.[246]
$(i\text{-}AmO)(PhCH_2O)PO_2H$. I.1.b.γ-2 (using $(PhCH_2O)_2POCl$) then I.1.e.α-2[96] or I.1.e.α-4 (using N-Me-morpholine).[129] Ag salt;[96,129] c-$C_6H_{11}NH_3$ salt, m. 131-132°.[129]
((Myristoyl-OCH_2CH_2O)$(PhCH_2O)PO_2H$. I.1.e.α-3 (using $PhCH_2$ ester and NaI).[1403]
(Palmitoyl-OCH_2CH_2O)$(PhCH_2O)PO_2H$. Similarly. Na salt.[453]
$[CH_2{=}C(CO_2H)O](PhCH_2O)PO_2H$. Similarly. Na salt.[453]
2,3,4,6-Tetra-O-acetyl-β-D-1-glucosyl $PhCH_2$-phosphate. I.1.e.α-4 (using $PhCH_2$ ester and c-$C_6H_{11}NH_2$). c-$C_6H_{11}NH_3$ salt.[1204]
1,2,3,4-Tetra-O-acetyl-β-D-glucopyranose 6-$PhCH_2$-phosphate. I.1.e.α-3 (using $PhCH_2$ ester and LiCl). Cryst., m. 132-134°.[659]
(1,2-Isopropylidene-α-D-glucofuranose 6-$PhCH_2$-phosphate. I.1.e.α-4 (using $PhCH_2$ ester and N-Me-morpholine).[659]
$HO_2P(OCH_2Ph)OCH_2CH(NH{\cdot}CO_2CH_2Ph)CO_2CH_2Ph$. I.1.e.α-3 (from its $PhCH_2$ ester by NaI). Cryst., m. 108°; Na salt.[112,113]
$HO_2P(OCH_2Ph)OCH_2CH(NH{\cdot}CO_2CH_2Ph)CO{\cdot}NHMe$. Similarly. Cryst., m. 108-110°; Na salt.[112]
$(4\text{-}BrC_6H_4CH_2O)(EtO_2CCH_2O)PO_2H$. I.1.e.α-3 (using $BrC_6H_4CH_2$ ester and NaI). Na salt, m. 128-130°; Ag salt.[480]
$(4\text{-}O_2NC_6H_4CH_2O)$(2,3-di-O-palmitoyl-1-glycero)PO_2H. I.1.e.α-1 (using its $O_2NC_6H_4CH_2$ ester and Dowex 50 H^+ form). Cryst., m. 64-65°.[806]

$(4\text{-}O_2NC_6H_4CH_2O)$(Phthaloyl-$NHCH_2CH_2O)PO_2H$. I.1.e.α-3 (using its $O_2NC_6H_4CH_2$ ester and NaI). Ag salt, m. 192-194°.[929]

$(4\text{-}O_2NC_6H_4CH_2O)$(t-BuOCO·$NHCH_2CH_2O)PO_2H$. I.1.b.γ-2 - e.α-3 (using $CboNHCH_2CH_2OH$ and $(O_2NC_6H_4CH_2O)_2POCl$, then BaI_2). Ag salt, m. 300°.[457]

(MeO)(PhCH(OH)CH(Ph)O)PO_2H. I.1.e.α-1 (from hydrobenzoin cyclic H-phosphate by MeOH).[1386] Na salt; NH_4 salt.[1386]

(EtO)(PhCH(OH)CH(Ph)O)PO_2H. Similarly, using EtOH. Na and NH_4 salts.[1386]

(i-PrO)(PhCH(OH)CH(Ph)O)PO_2H. Similarly, using i-PrOH. Na and NH_4 salts.[1386]

(BuO)(PhCH(OH)CH(Ph)O)PO_2H. Similarly, using BuOH. Na and NH_4 salts.[1386]

(t-BuO)(PhCH(OH)CH(Ph)O)PO_2H. Similarly, using t-BuOH. Na and NH_4 salts.[1386]

β. R and R' Bearing Cyclic Groups

$(PhCH_2O)_2PO_2H$. I.1.a.α,[886] I.1.b.γ-1,[95,96] I.1.e.α-1 (using $(RO)_3PO$),[228,884,1355] I.1.e.α-2 (using $(RO)_3PO$ and catalytic hydrogenolysis,[1355] electrolysis[930]), I.1.e.α-3 (using $(RO)_3PO$ and LiCl,[411] NaI[1011,1013]), I.1.e.α-4 (using $(RO)_3PO$ and tertiary bases,[129,1355] N-Me-N-benzylmorpholinium-Cl[411]), (using $(MeO)(RO)_2PO$ and Me_3N),[357] I.2.b.α-8 (from $ROPO_2H_2$ and ROH + Br_2),[246] I.2.β-2 (using the same components, and CCl_4 + Et_3N),[246] I.2.b.β-4 (using $(RO)_2PHO$).[971] By heating $(RO)_3PO$ with phenol.[779] Cryst., m. 79.5°,[228] 79°,[246,1246] 79-80°,[95] 78-79°,[97,884] 78°,[971] 76-78°;[930] Me_4N salt, m. 188°;[357] $c\text{-}C_6H_{11}NH_3$ salt; m. 137°;[129] Li salt;[411] Na salt;[1355] Ag salt, dec. 216°;[96,886] Ba salt, dec. 255-261°.[886] Protolysis constant in $HCONMe_2$: K x 10^4 0.94.[1133]

$(PhCH_2O)$[PhCH(OH)CH(Ph)O]PO_2H. I.1.e.α-1 (from hydrobenzoin cyclic H-phosphate with $PhCH_2OH$).[1386]

[(6-Cl-1,3-benzodioxane-8-yl)-methyl-O]$_2PO_2H$. I.1.e.α-4 (from tris ester and N-Me-morpholine), cryst., m. 183-184°.[129]

$(4\text{-}BrC_6H_4CH_2O)_2PO_2H$. I.1.e.α-4 (from $(RO)_3PO$ and N-Me-morpholine). Needles, m. 155-156°; Ag salt.[129]

$(4\text{-}O_4NC_6H_4CH_2O)_2PO_2H$. I.1.e.α-4 (from triester and N-Me-morpholine). Cryst., m. 179-180°.[129]

b. Heterocyclic Groups

1-(3-O-Cyanoethylphosphoryl-5-O-trityl-D-ribofuranosyl)-2(1H)-pyridone. I.1.d.α-1. Characterized by chromatography and enzymic degradation.[1380]

1-(3-O-cyanoethylphosphoryl-5-O-trityl-D-ribofuranosyl)-2(1H)-pyrimidinone. I.1.d.α-1. Characterized by chromatography and enzymic degradation.[1380]

1-(3-O-cyanoethylphosphoryl-5-O-trityl-D-ribofuranosyl)-2,6(1H)-pyrimidinedione. I.l.d.α-1. Characterized by chromatography and enzymic degradation.[1380]
Uridine 2'(3')-Me-phosphate. I.l.d.ζ (from 2',3'-cyclic phosphate with MeOH and ribonuclease).[1173a]
Uridine 2'(3')-Et-phosphate. I.l.d.ζ (similarly, using EtOH).[197]
Uridine 2'(3')-Pr-phosphate. I.l.d.ζ (similarly, using PrOH).[197]
Uridine 2'(3')-Bu-phosphate. I.l.d.ζ (similarly, using BuOH).[197]
Uridine 2'(3')-i-Bu-phosphate. I.l.d.ζ (similarly, using i-BuOH).[197]
Uridine 3'-Me-phosphate. I.l.d.α-1 (from 2',5'-di-O-tetrahydro-pyranyl 3'-phosphate and MeOH + DCC).[1173a]
Uridine 5'-($HOCH_2CH_2$)-phosphate. I.l.d.α-1 (using 5'-phosphate and glycol), I.l.d.γ-2 (using 5'-phosphate and ethylene oxide).[1387a]
(Uridine 2')$_2$ phosphate. I.l.b.γ-2 - e.α-1 (from 3,5'-benzylidene-uridine using $PhOPOCl_2$). Ba salt, dec. 249-252°.[643]
Uridine-(3' → 5')uridine. I.l.d.α-1 (from 2',5'-ditrityl uridine 3'-phosphate and 2',3'-isopropylidene[649] or 2',3'-diacetyl uridine,[1388] similarly[1173a]), I.l.d.δ (from 5'-Cl-5'-deoxyuridine and Bu_3NH 3'-uridylate).[1279]
Uridine-(3' → 5')[and 2' → 5']-uridine. By heating uridine 3'(2')-phosphate with uridine at 160°.[992,993,1252]
Uridine-(3' → 5')-2'-3'-isopropylidene-uridine. I.l.f.ζ-10 (from 2',5'-diacetyl-uridine 3'-phosphate and 2',-3'-isopropylidene-O^2,5'-cyclouridine).[1497]
5'-O-(α-Butoxyethyl)-uridine-(3'(2') → x')-uridine. I.l.d.ζ (2',3'-cyclic uridine phosphate and uridine by pancreas ribonuclease). Chromatographic identification.[1239]
Uridine-(3' → 5')-6-azauridine. I.l.d.α-1 (from 2',5'-ditrityluridine phosphate and 2',3'-isopropylidene-6-azauridine).[649]
Uridine-(3' → 5')-cytidine. I.l.d.γ-1 (from 2',5'-protected uridine 3'-phosphate and 2',3'-protected cytidine),[1173a] I.l.f.ζ-10 (from 2',3'-protected uridine phosphate and 2',3'-isopropylidene O^2,5'-cyclocytidine).[1497]
5'-O-(α-Butoxyethyl)-uridine 3'(2')-phosphate-x'-cytidine. I.l.d.ζ (from 2',3'-cyclic phosphate uridine derivative and cytidine by pancreatoribonuclease). Chromatographic identification.[1239]
Uridine-(3' → 5')-2',3'-isopropylidene-adenosine, and Uridine-(2' → 5')-2',3'-isopropylidene-adenosine. I.l.f.ζ-10 (from N^6-acetyl-anhydroadenosine and uri-

dine 2',3'-di-phosphate). UV.[977]

Uridine-(3' → 5')adenosine. I.l.d.α-l (from 2',5'-protected uridine 3'-phosphate and protected nucleoside).[394,1173a,1174]

Uridine-(3' → 5')-guanosine. I.l.d.α-l (from protected uridine 3'-phosphate and protected nucleoside).[394]

5'-O-(α-Butoxyethyl)-uridine 3'(2')-phosphate-N-(Me_2N-methylene)-guanosine. I.l.d.ζ (from uridine 2',3'-cyclic phosphate derivative and guanosine derivative by pancreatoribonuclease). Chromatographic identification.[1239]

6-Azauridine 5'-Me-phosphate. I.l.d.α-l (from 5'-phosphate and MeOH), I.l.e.α-l (from Me ester by alkaline hydrolysis). UV.[1141]

6-Azauridine 5'-Et-phosphate. I.l.d.α-l (phosphate and EtOH). Cryst., m. 114-116°. UV.[1141]

6-Azauridine 5'-Am-phosphate. I.l.e.α-l (from Am ester by alkaline hydrolysis). Cryst., m. 98-100°. UV.[1141]

6-Azauridine-(5' → 5')-6-azauridine. I.l.d.α-l (from 6-azauridine with 6-azauridine 5'-phosphate and DCC). Isolated as NEt_3H salt.[695]

Deoxycytidine 5'-Et-phosphate. I.l.f.ζ-9 (from 5'-phosphate and EtOH + $NCCHBrCONH_2 \cdot Ph_3P$).[448]

Deoxycytidine-(3' → 5')-deoxycytidine. I.l.d.α-l (from N^6-anisoyl-3'-acetyl-deoxycytidine 3'-phosphate and N^6-anisoyl-5'-$(4\text{-MeO})_2$-trityl-deoxycytidine). Rf; UV.[1292b]

Deoxycytidine-(3' → 5')deoxycytidine. I.l.d.α-l (from 5'-$(4\text{-MeO})_2$-trityl-deoxycytidine and 3'-acetylcytidine 5'-phosphate).[1292a] From deoxycytidine 5'-bound to a polymer support and 3'-acetyldeoxycytidine 5'-phosphate, using tri-i-Pr-benzenesulfonyl-Cl or picryl-Cl).[928] Chromatographic and UV characteristics.[1292a]

Deoxycytidine-(3' → 5')-thymidine. I.l.d.α-l (from N^6-anisoyl-deoxycytidine and 3'-acetyl-thymidine 5'-phosphate).[1292b] From deoxycytidine 5'-bound to a polymer support and 3'-O-acetyl-thymidine 5'-phosphate, using $(i\text{-Pr})_3$-benzenesulfonyl-Cl or picryl-Cl.[928] Chromatographic and UV characteristics.[1292b]

Deoxycytidine-(3' → 5')-deoxyadenosine. From deoxycytidine 5'-bound to a polymer support and 3'-acetyl-deoxyadenosine 5'-phosphate, using $(i\text{-Pr})_3$-benzenesulfonyl-Cl or picryl-Cl).[928]

Deoxycytidine-(3' → 5')-deoxyguanosine. I.l.d.α-l (from 5'-trityl(or $(4\text{-MeO})_2$-trityl)-deoxycytidine and 3'-acetyl-deoxyguanosine 5'-phosphate).[1292b] From deoxycytidine 5'-bound to a polymer support and 3'-acetyl-deoxyguanosine 5'-phosphate, using $(i\text{-Pr})_3$-benzenesulfonyl-Cl or picryl-Cl.[928] Chromatographic and UV characteristics.[1292b]

Cytidine-(3' → 5')-uridine. I.1.d.α-1 (from N,O$^{2'}$,O$^{5'}$-triacetylcytidine 3'-phosphate and N,O$^{2'}$,O$^{3'}$-tribenzoyluridine).[879a]

Cytidine-(3' → 5')-cytidine. I.1.d.α-1 (N,O$^{2'}$,O$^{5'}$-triacetylcytidine 3'-phosphate and N,O$^{2'}$,O$^{3'}$-tribenzoylcytidine). Chromatographic characteristics.[394,879a]

Cytidine-(3' → 5')-adenosine. I.1.d.α-1 (from N^6,O$^{2'}$,O$^{5'}$-triacetylcytidine 3'-phosphate and benzhydryl-adenosine[452] or N,N',O$^{2'}$,O$^{3'}$-tetraacetyladenosine[879a]; similarly[394]), I.1.f.ζ-10 (using adenosine 5'-phosphate and O^2,3'-anhydrocytidine).[976]

Cytidine-(3' → 5')-guanosine. I.1.d.α-1 (from 2',3'-di-O(or N,O$^{2'}$,O$^{3'}$-tri)-acetylguanosine and N,O$^{2'}$,O$^{5'}$-triacetylcytidine 3'-phosphate),[879a] (similarly).[394]

Thymidine 5'-cyanoethyl-phosphate. I.1.f.ζ-9 (from 5'-phosphate and $HOCH_2CH_2CN$ + $NCCHBrCONH_2 \cdot (Ph)_3P$).[448]

Thymidine-(3' → 5')-uridine. I.1.d.α-1. Protolysis constant in $HCONMe_2$: K x 10^4 3.76.[1133]

Thymidine-(3' → 5')-deoxycytidine. I.1.d.α-1 (from N^6-anisoyl-3-O-acetyldeoxycytidine 5'-phosphate and 5'-O-tritylthymidine).[1057] From thymidine 5'-bound to a polymer support and 3'-O-acetyldeoxycytidine 5'-phosphate, using (i-Pr)$_3$-benzenesulfonyl-Cl or picryl-Cl).[928]

Thymidine-(3' → 5')-thymidine. I.1.d.α-1 (from 3'-O-acetylthymidine 5'-phosphate and 5'-O-(4-MeO)$_2$-tritylthymidine),[1292b] (from 5'-O-acetylthymidine 3'-phosphate and 3'-O-acetylthymidine, or 3'-O-acetylthymidine 5'-phosphate and 5'-O-tritylthymidine),[744] I.1.d.α-2 (from the same components, using 4-toluenesulfochloride).[744] I.1.f.ζ-9 (from 3'-phosphate and 5'-protected thymidine + $NCCHBrCONH_2 \cdot (Ph)_3P$).[446,448] Chromatographic and UV characteristics.[1292b]

(3'-[5-O-Tritylthymidine])(5'-[3'-O-acetylthymidine]) phosphate. I.1.b.γ-2 (?) (from 5'-O-tritylthymidine and 3'-O-acetylthymidine 5'-monophosphate with $SOCl_2$ in Me_2NCHO, probably •••POCl group formed as intermediary).[730]

Thymidine-(3' → 5')-deoxyadenosine. I.1.d.α-1 (from N,O$^{3'}$-diacetyl-deoxyadenosine 5'-phosphate and 5'-O-(MeO)$_2$-trityladenosine).[1292b] From thymidine 5'-bound to a polymer support and 3'-O-acetyl deoxyadenosine 5'-phosphate, using (i-Pr)$_3$-benzenesulfonyl-Cl or picryl-Cl.[928] Chromatographic and UV characteristics.[1292b]

Thymidine-(3' → 5')-deoxyguanosine. I.1.d.α-1 (from 3'-protected deoxyguanosine 5'-phosphate and 5'-protected thymidine).[1292b] From thymidine 5'-bound to a polymer support and 3'-O-acetylguanosine 5'-phosphate, using (i-Pr)$_3$-benzenesulfonyl-Cl or picryl-Cl.[928] Chromatographic and UV characteristics.[1292b]

Deoxyadenosine 5'-cyanoethyl-phosphate. I.1.f.ζ-9 (from 5'-phosphate and $HOCH_2CH_2NH_2$, using $NCCNBrCONH_2 \cdot (Ph)_3$-P).[448]

Deoxyadenosine-(3' → 5')-deoxycytidine. From deoxyadenosine 5'-bound to a polymer support and 3'-O-acetyl-deoxycytidine 5'-phosphate, using $(i\text{-}Pr)_3$-benzenesulfonyl-Cl or picryl-Cl.[928]

Deoxyadenosine-(3' → 5')-thymidine. Similarly to the preceding, using 3'-O-acetylthymidine 5'-phosphate.[928]

Deoxyadenosine-(3' → 5')-deoxyadenosine. Similarly to the preceding, using 3'-O-deoxyadenosine 5'-phosphate.[928]

Deoxyadenosine-(3' → 5')-deoxyguanosine. Similarly to the preceding, using 3'-O-acetyl-deoxyguanosine 5'-phosphate.[928]

Adenosine 3'-Me-phosphate. I.1.f.ζ-9 (from 3'-phosphate with MeOH, using picryl-Cl).[1250]

Adenosine 5'-Me-phosphate. Similarly from 5'-phosphate;[1250] using $NCCHBrCONH_2 \cdot (Ph)_3P$.[447,448]

Adenosine 2'-benzyl phosphate. I.1.d.β (from a-adenylic acid with $PhCH_2N_2$). Prisms.[395a]

Adenosine 3'-benzyl phosphate. I.1.d.β (from b-adenylic acid with $PhCH_2N_2$). Needles.[395a]

Adenosine 5'-($PhCH_2$)-phosphate. I.1.b.γ-2 - e.α-2 (from adenosine using $(PhCH_2O)POCl_2$; besides monophosphate),[132] I.1.f.ζ-6 (from adenosine phosphoroimidazolidate prepared in situ, with $PhCH_2OH$),[448a] I.1.f.ζ-9 (from adenosine phosphate and $PhCH_2OH$ using N-Me_2-formamidiniumchloride).[446] Cryst., m. 234°.[132]

<u>Adenosine 5'-(α-aminoalkyl)-phosphates</u>

Adenosine 5'-glycinyl-phosphate. I.1.d.α-1 (from N^6,$O^{2'}$,-$O^{3'}$-triacetyladenosine 5'-phosphate and N-Cbo-glycinol by DCC). Cryst., m. 185°; UV.[226a,259]

Adenosine 5'-DL-alanylylphosphate. Similarly, using N-Cbo-alaninol. Cryst., m. 193°; UV.[226a,259]

Adenosine 5'-DL-methioninylphosphate. Similarly, using N-Cbo-methioninol. Cryst., m. 180°; UV.[226a,259]

Adenosine 5'-L-valinylphosphate. Similarly, using N-Cbo-valinol. Cryst., m. 150°; UV.[226a,259]

Adenosine 5'-L-leucinylphosphate. Similarly, using N-Cbo-leucinol. Cryst., m. 116° (dec.); UV.[226a,259]

Adenosine 5'-L-isoleucinylphosphate. Similarly, using N-Cbo-isoleucinol. Cryst., m. 175°; UV.[226a,259]

Adenosine 5'-L-phenylalaninyl phosphate. Similarly, using N-Cbo-phenylalaniol. Cryst., m. 150° (dec.); UV.[226a,259]

Adenosine 5'-L-tyrosinylphosphate. Similarly, using N-Cbo-tyrosinol. Cryst., m. 190°; UV.[226a,259]

Adenosine 5'-L-prolinylphosphate. Similarly, using N-Cbo-prolinol. Cryst., m. 165°; UV.[226a,259]

(Adenosine 2')(uridine 5') phosphate. I.1.b.γ-2 (from

3',5'-di-O-acetyl-adenosine 2'-benzyl-phosphorochloridate with 2',3'-di-O-acetyl uridine, followed by deacetylation).[959]

(Adenosine 5')(uridine 5') phosphate. I.1.d.δ - e.α-2 (from 2',3'-isopropylideneadenosine 5'-benzylphosphate Ag salt and 2',3'-isopropylidene-5'-deoxy-5'-I-uridine). Amorphous solid, hygroscopic, degraded by Russel's viper venom.[507]

Adenosine-(3' → 5')-adenosine. I.1.d.α-1 (from 5'-protected adenosine 3'-phosphate and 3'-protected adenosine).[1173a]

Deoxyguanosine 5'-cyanoethyl-phosphate. I.1.f.ζ-9 (from 5'-phosphate and $HOCH_2CH_2CN$, using $NCCHBrCONH_2 \cdot (Ph)_3$-P).[448]

Deoxyguanosine-(3' → 5')-deoxycytidine. I.1.d.α-1 (from N^6-anisoyl-3'-O-acetyl-deoxycytidine 5'-phosphate and N,$O^{5'}$-bis-$(4\text{-MeO})_2$-trityl-deoxyguanosine).[1292b] From deoxyguanosine 5'-bound to a polymer support and 3'-O-acetyl-deoxycytidine 5'-phosphate, using $(i\text{-Pr})_3$-benzenesulfonyl-Cl or picryl-Cl.[928] Chromatographic and UV characteristics.[1292 b]

Deoxyguanosine-(3' → 5')-thymidine. I.1.d.α-1 (from 3'-O-acetylthymidine 5'-phosphate and N,$O^{5'}$-bis $(4\text{-MeO})_2$-trityl-deoxyguanosine).[1292b] From deoxyguanosine 5-bound to a polymer support and 3-O-acetylthymidine 5'-phosphate, using $(i\text{-Pr})_3$-benzenesulfonyl-Cl or picryl-Cl.[928] Chromatographic and UV characteristics.[1292b]

Deoxyguanosine-(3' → 5')-deoxyadenosine. From deoxyadenosine 5'-bound to a polymer support and 3'-O-acetyl-deoxyadenosine 5'-phosphate).[928]

Deoxyguanosine-(3' → 5')-deoxyguanosine. I.1.d.α-1 (from N,$O^{3'}$-diacetyl-deoxyguanosine 5'-phosphate and N,$O^{5'}$-bis-$(4\text{-MeO})_2$-trityl-deoxyguanosine).[1292b] From deoxyguanosine 5'-bound to a polymer support and 3'-O-acetyl-deoxyguanosine 5'-phosphate, using $(i\text{-Pr})_3$-benzenesulfonyl-Cl or picryl-Cl.[928] Chromatographic and UV characteristics.[1292b]

N-Benzoyl-2'-tetrahydropyranoyl-guanosine 5'-cyanethylphosphate. I.1.d.α-1 (from the N,2'-protected guanosine phosphate derivative and $HOCH_2CH_2CN$). UV characteristics. A nucleotide derivative with 3'-OH as unique free grouping.[1321]

Guanosine-(3' → 5')-uridine. I.1.d.α-1 (from N,$O^{2'}$,$O^{5'}$-triacetylguanosine 3'-phosphate and 2',3'-di-O-benzoyluridine).[879a]

Guanosine-(3' → 5')-cytidine. I.1.d.α-1 (from N,$O^{2'}$,$O^{5'}$-triacetylguanosine 3'-phosphate and N,$O^{2'}$,$O^{3'}$-tribenzoyladenosine).[879a]

Guanosine-(3' → 5')-adenosine. Similarly, but using as second component N,N',$O^{2'}$,$O^{3'}$-tetrahydrobenzoyladenosine.[879a]

Guanosine-(3' → 5')-guanosine. Similarly, but using as second component N,O$^{2'}$,O$^{3'}$-triacetylguanosine.[879a]
($PhCH_2O$)[2-(4-Me-thiazol-5-yl)ethyl-O]PO_2H. I.l.e.α-4 (from the corr. dibenzyl ester with N-Me-morpholine). Cryst., m. 82-83°.[857]
[(6-Cl-1,3-benzodioxane-8-yl)-methyl]$_2$ phosphate. I.l.e.α-4 (from $(RO)_3PO$ using N-Me-morpholine). Needles, m. 183-184°.[129]

B. R, or R and R' Bound by a Carbon Atom of a Cycle

1. Only One R Alicyclic

a. R' Aliphatic

Dehydroepiandrosterone Me-phosphate. I.l.b.β - γ-l. Cryst., m. 165-166°; IR.[1193]
Cholesteryl cyanoethyl-phosphate. From the following by $N_2H_4 \cdot 2H_2O$. Cryst., m. 270.2° (sint. 250°).[686]
Cholesteryl phthalimidoethyl-phosphate. I.l.b.γ-2 - γ-1 (using cholesterol and $ROPOCl_2$). Glass, m. 160-162°.[686]
Ergosteryl 2-chloroethyl-phosphate. I.l.b.γ-2 - γ-l. Cryst., m. 165-167°.[523]

b. R' Bearing a Cyclic Group

(c-$C_6H_{11}O$)($PhCH_2O$)PO_2H. I.l.e.α-4 (from $PhCH_2$ ester using N-Me-morpholine). Ag salt.[129]
(2-HO-c-$C_6H_{10}O$)($PhCH_2O$)PO_2H. I.l.e.α-3 (from benzyl ester using LiCl,[411,411a] using c-$C_6H_{11}NH_3SCN$ or $PhCH_2NH_3SCN$[1204]), I.l.e.α-4 (from benzyl ester using N-Me-morpholine).[129] c-$C_6H_{11}NH_3$ salt, m. 179°,[411,411a] 175-177°;[1204] $PhCH_2NH_3$ salt, m. 163-164°.[1204]
1,2,3,4-Tetra-O-acetyl-β-D-glucopyranose 6-(2-HO-c-C_6H_{10})-phosphate. I.l.e.α-2 (using the benzyl ester). c-$C_6H_{11}NH_3$ salt, m. 204-205° (dec.).[659]
3,11,20-Trioxo-4-pregnene-17α,21-diol 21-benzyl-phosphate. I.l.e.α-3 (using the benzyl ester and NaI). Cryst., m. 184-186°.[456]

2. R and R' Alicyclic

(c-$C_6H_{11}O$)$_2PO_2H$. I.l.b.β - c.β-l (from ROH using $POCl_3$ and a Lewis acid, via $R_4P_2O_7$).[1408]
Dimenthyl phosphate. I.l.b.β (using (-)-ROH and $POCl_3$). Cryst., 105°; Na salt; Ag salt; Pb salt, also obtained by heating an alleged pyrophosphate, $R_2H_2P_2O_2$, with $Pb(NO_3)_2$.[964]
Bis-(L-ascorbic acid 3) phosphate. I.l.b.γ-2 (from 5,6-isopropylidene derivative using $POCl_3$). NMR.[1052]
Bis-(dehydroepiandrosterone) phosphate. I.l.b.γ-2. $[\alpha]_D$ 2.0° ($CHCl_3$); circular dichroisme; NMR.[1194]
Dicholesteryl phosphate. I.l.b.β - γ-l,[522,523,1122] I.l.b.γ-2 (using N,N-(Ph)$_2$-amidophosphorodichloridate).[1494] Cryst., m. 208°, 204°,[1494] m. 186°;[522] Ba salt.[1122] The anhydrous ester is highly associated

and has been considered as a pyrophosphate because of its insolubility in dilute alkali and absence of reaction with Na in warm MeC_6H_5,[1432,1432a] and the existence of half salts;[1432b] this claim seems to be erroneous.

Diergosteryl phosphate. I.1.b.γ-2 - γ-1. Cryst.[523]

3. Only R Aromatic

a. R' Aliphatic

(PhO)(MeO)PO_2H. I.1.b.γ-1,[923] I.1.b.γ-2 (using adducts of $PhOPOCl_2$ with Me_3N, Et_3N, or pyridine),[190a] I.1.e.α-3 (using $(MeO)_2(PhO)PO$ and CaI_2),[454] I.1.f.ζ-6 (using imidazolium phenyl-phosphoroimidazolidate),[617a] I.1.f.ζ-6 (using $PhOPO_3H_2$ and MeOH with picryl-Cl).[1250] $c\text{-}C_6H_{11}NH_3$ salt.[1250]

(PhO)(EtO)PO_2H. I.1.b.γ-1. Sirup. Ba salt, cryst. powder; Na salt; Pb salt, cryst. powder.[994]

(PhO)(i-PrO)PO_2H. I.1.f.ζ-9 (from $PhOPO_3H_2$ and i-PrOH, using N-Me_2-formamidinium chloride). Cryst., m. 157-160°; $c\text{-}C_6H_{11}NH_3$ salt, m. 177-180°.[446]

(PhO)(BuO)PO_2H. I.1.b.γ-1. Oil.[923]

(PhO)(t-BuO)PO_2H. I.1.c.β-2 (from sym. $Ph_2H_2P_2O_7$ with t-BuOH), I.1.f.ζ-9 (from $PhOPO_3H_2$ and t-BuOH using N-Me_2-formamidinium chloride).[446]

(PhO)(n-$C_{11}H_{23}$)PO_3H_2. I.1.b.γ-1. Oil.[923]

Phenyl acylglycol phosphates

(PhO)(palmitoyl-OCH_2CH_2O)PO_2H. I.1.d.δ - e.α-2 (using $AgO_2P(OPh)(OCH_2Ph)$ (and H_2/Pd!) and acylglycol iodohydrin). Unstable, Ag salt stable.[1404]

(PhO)(stearoyl-OCH_2CH_2O)PO_2H. Similarly. Cryst., m. 70.5-71°.[1004,1404]

(PhO)(behenoyl-OCH_2CH_2O)PO_2H. Similarly.[1004,1404] Cryst., m. 78-79°; Ag salt.[1004]

(PhO)($H_2NCH_2CH_2O$)PO_2H. I.1.d.γ-2 (using ethylene imine). Cryst., m. 246.5-248.0° (dec.).[1370] N-Cbo derivative, Ag salt.[909]

(PhO)($Me_3\overset{+}{N}CH_2CH_2O$)$PO_2^-$. From (PhO)($BrCH_2CH_2O$)$PO_2H$ with Me_3N. Cryst., m. 208°.[858]

Phenyl diacylglyceryl phosphates. I.1.d.δ - e.α-2 (using $AgO_2P(OPh)(OCH_2Ph)$ (and H_2/Pd!) and diacylglycerol iodohydrine. Cryst., unstable, Ag salts stable.

(PhO)(2,3-di-O-myristoyl-glycero-1)PO_2H. Cryst, m. 48.5-49.5° (unstable modification: 39.5-40.5°).[1004]

(PhO)(2,3-di-O-palmitoyl-glycero-1)PO_2H. Cryst., m. 54.5-55.5°.[1004,1404]

(PhO)(2,3-di-O-stearoyl-glycero-1)PO_2H. Cryst., m. 51.5-52.5°.[1004,1404]

(PhO)(2,3-di-O-behenoyl-glycero-1)PO_2H.[1404]

$(PhO)((Me_3\overset{+}{N}CH_2CH_2CH_2O)PO_2^-$. From $(PhO)(BrCH_2CH_2CH_2O)PO_3H_2$ with Me_3N. Cryst., m. 260°.[858]

$(PhO)(MeO_2CCH(NH_2)CH_2O)PO_2H$. I.1.b.γ-2 - γ-1 (using N-Cbo-serine Me ester and $PhOPOCl_2$ → $HO(PhO)(O)P(OCH_2$-$CH(NHCbo)CO_2Me$ (oil), then H_2/Pd-C).[115]

<u>$(4\text{-}O_2NC_6H_4O)(alkyl\text{-}O)PO_2H$</u>

$(4\text{-}O_2NC_6H_4O)(MeO)PO_2H$. I.1.e.α-1 (from $(MeO)(ArO)_2PO$ by 1N LiOH),[982] I.1.f.ζ-9 (from $O_2NC_6H_4OPO_3H_2$ and MeOH, using $NCCHBrCONH_2 \cdot (Ph)_3P$).[447,448] Cryst., m. 123.5-124.5°,[982] 122°;[447,448] $c\text{-}C_6H_{11}NH_3$ salt, m. 150.5-151.5°.[982]

$(4\text{-}O_2NC_6H_4O)(EtO)PO_2H$. I.1.f.ζ-9 (as above but using EtOH).[447,448]

Et, Pr, i-Pr, Bu ester similarly.[447,448]

$(4\text{-}O_2NC_6H_4O)(n\text{-}C_6H_{13}O)PO_2H$. I.1.e.α-1 (from $(ArO)_2$-(Aliph.-O)PO by 1N LiOH). Cryst., m. 109-110°; $c\text{-}C_6H_{11}NH_3$ salt.[982]

$(4\text{-}O_2NC_6H_4O)(n\text{-}C_8H_{17}O)PO_2H$. I.1.f.ζ-9 (as above for Et ester).[447,448]

$(2\text{-}HO_2CC_6H_4O)(EtO)PO_2H$. I.1.b.γ-2 - f.ζ-10 (from 2-Cl-4-oxo-5,6-benzo-1,3-dioxa-phosphorinane-2-oxide and EtOH → 2-Et-O-4-oxo-5,6-1,3-dioxa-benzophosphorinane-2-oxide). Cryst., m. 120-121°.[93]

$(4\text{-}O_2NC_6H_4O)(MeO_2CCH_2CH_2O)PO_2H$. I.1.f.ζ-9 (from $O_2NC_6H_4$-OPO_3H_2 and ROH using $NCCHBrCONH_2 \cdot (Ph)_3P$).[447,448]

(2-Naphthyl-O)(MeO)PO_2H. I.1.d.β. Na salt, m. 222-223°.[578]

<u>b. R' Aliphatic but with a Cyclic C-Substituent</u>

$(PhO)(PhCH_2O)PO_2H$. I.1.e.α-3 (from $PhCH_2$ ester using LiCl,[411,411a] NaI[1004]), I.1.e.α-4 (from $PhCH_2$ ester using N-Me-morpholine),[129] I.1.f.ζ-9 (from $PhOPO_3H_2$ and $PhCH_2OH$ using N-Me_2-formamidinium chloride).[446] From $PhCH_2$ ester by heating with PhOH.[779] Oil; $c\text{-}C_6H_{11}NH_3$ salt, m. 147°,[411a] 145-147°,[446] 145°;[129] Ag salt.[1004]

Uridine 2'(3')-Ph-phosphate. I.1.d.ζ (from 2',3-cyclic phosphate and PhOH by ribonuclease).[197]

6-Azauridine 5'-Ph-phosphate. I.1.e.α-1 (using Ph ester and alkaline hydrolysis),[1141] I.1.e.α-2 (using Ph ester and moderate hydrogenolysis).[1141] Cryst., m. 142-144°; UV.[1141]

6-Azauridine 2',3'-di-O-benzoyl-5'-Ph-phosphate. I.1.e.α-2 (using its Ph ester and controlled hydrogenolysis).[1142]

$(4\text{-}O_2NC_6H_4O)(PhCH_2O)PO_2H$. I.1.f.ζ-9 (from $O_2NC_6H_4OPO_3H_2$ and $PhCH_2OH$ using $NCCHBrCONH_2 \cdot (Ph)_3P$).[447,448] I.1.e.α-1 (from the $O_2NC_6H_4$ ester and 1N LiOH). Cryst., m. 168-169°.[982]

Uridine 5'-(4-$O_2NC_6H_4$)-phosphate. I.1.e.α-1 (from the nitrophenyl ester and 1N LiOH). Ba salt ($2H_2O$).[982]
Thymidine 5'-(4-$O_2NC_6H_4$)-phosphate. I.1.e.α-1. Protolysis constant in $HCONMe_2$: K x 10^4 2.5.[1133]
(4-HO-2,3-Me_2-1-naphthyl-O)($PhCH_2O$)PO_2H. I.1.e.α-3 (from its $PhCH_2$ ester using LiCl). Needles, m. 126-127°.[406a]

c. R' Alicyclic

(4-BrC_6H_4O)(2-HO-c-$C_6H_{10}O$)PO_2H. I.1.e.α-3 (from its BrC_6H_4 ester using c-$C_6H_{11}NH_3SCN$). Cryst., m. 194-196°.[1204]
(4-$O_2NC_6H_2O$)(c-$C_6H_{11}O$)PO_2H. I.1.e.α-1 (from its nitrophenyl ester and 1N LiOH). Cryst., m. 202-203°; c-$C_6H_{11}NH_3$ salt.[982]
(4-$O_2NC_6H_4O$)(2-HO-c-$C_6H_{10}O$)PO_2H. I.1.e.α-3 (from its nitrophenyl ester using c-$C_6H_{11}NH_3SCN$). c-$C_6H_{11}NH_3$ salt, m. 187-189°.[1204]
(4-$O_2NC_6H_4O$)(bornyl-O)PO_2H. I.1.f.α-9 (from $ArOPO_3H_2$ and borneol using $NCCHBrCONH_2$·$(Ph)_3P$).[447,448]

4. R and R' Aromatic

a. Hydrocarbon Derivatives

$(PhO)_2PO_2H$. I.1.a.α,[1187] I.1.b.β,[102] I.1.b.γ-1,[236,689,746,1175] I.1.c.α,[1122] I.1.e.α-1,[616] I.1.e.α-3 (using LiCl or $CaCl_2$ and Pr, i-Pr, or c-C_6H_{11} esters;[854] CaI_2 and Me, Pr, i-Pr, or Bu esters;[454] NaI, c-$C_6H_{11}NH_3I$, or $PhNH_3SCN$ and $PhCH_2$ ester;[454] $PhNH_2$ at 180° and c-C_6H_{11} ester[1054]). I.1.e.α-4 (using its Me ester and Me_3N).[357] I.2.b.β-2 (from $(RO)_2PHO$ using CCl_4 + NaOH).[957] From amidates: hydrolysis of Me-amidate,[1181,1308] Et-amidate;[1181] from Me- and Et-amidate by action of N_2O_3 or $NOHSO_4$;[1181] from Ph-amidate and HNO_2.[307a] By pyrolysis of its c-C_6H_{11} ester.[1054] By heating of its $PhCH_2$ ester with PhOH.[779] Needles, m. 34-39° (?),[307a] plates (partial hydrates !) m. 61-62°,[102] 56° (impure),[1175] dihydrate, plates, m. 51°,[689] anhydrous, needles, m. 70°,[689] 68-70°,[854] 66-68°;[1308a] NH_4 salt, m. 130°;[689] Me_4N salt, m. 210°;[357] $PhNH_3$ salt, m. 165-166°;[689] c-$C_6H_{11}NH_3$ salt;[1054] Na salt,[957] plates ($5H_2O$), m. 70°;[689] Ag salt,[569a] needles, m. 213°;[689] Ba salt ($4H_2O$), needles;[689] complex with Et_2Au, m. 70-71°.[569a] Protolysis constant in $HCONMe_2$: K x 10^4 5.2.[1133]
(PhO)(4-MeC_6H_4O)PO_2H. I.1.b.γ-1. Needles, m. 54°.[885]
(2-MeC_6H_4O)$_2PO_2H$. I.1.b.β ($POCl_3$). c-$C_6H_{11}NH_3$ salt, m. 161-163°.[433]
(4-MeC_6H_4O)$_2PO_2H$. I.2.b.β-2 (from $(RO)_2PHO$ using CCl_4 + NaOH). Na salt.[975]

$(2\text{-i-Pr-5-MeC}_6H_3O)_2PO_2H$. I.l.b.γ-l, I.l.e.α-l. Solid; Ba salt, needles;[481] Na salt, needles.[827]

$(2\text{-Me-4-BuC}_6H_3O)_2PO_2H$. I.l.b.γ-l. K salt.[1215]

$(2\text{-Me-4-i-AmC}_6H_3O)_2PO_2H$. I.l.b.γ-l. K.salt.[1215]

$(2\text{-Me-4-}C_6H_{13}\cdot C_6H_3O)_2PO_2H$. I.l.b.γ-l. K salt.[1215]

$(2\text{-i-Pr-5-Me-6-Bu}\cdot C_6H_2O)_2PO_2H$. I.l.b.γ-l. K salt.[1215]

$(PhO)(2\text{-}C_{10}H_7O)PO_2H$. I.l.b.β. Needles, m. 92-93°.[800]

$(1\text{-}C_{10}H_7O)_2PO_2H$. I.l.b.γ-l. Fine needles, m. 137-139°.[577]

$(2\text{-}C_{10}H_7O)_2PO_2H$. I.l.b.γ-l,[835] I.l.b.β - γ-l,[102] I.l.e.α-l.[102] Prisms, m. 147-148°,[102] 142°;[835] Na salt, plates; $PhNHNH_2$ salt, m. 183°.[102]

$1\text{-}(2,4\text{-}Ph_2\text{-}C_{10}H_5O)_2PO_2H$. I.l.e.α-l. Cryst., m. 220-221°.[574]

b. Substituted Derivatives

$(PhO)(2,4\text{-}(O_2N)_2C_6H_3O)PO_2H$. I.l.d.α-l (from $PhOPO_3H_2$ and ROH). Unstable oil.[1465]

$(2\text{-ClC}_6H_4O)_2PO_2H$. I.l.b.γ-l.[276,1502] Needles, m. 121.5°,[276] m. 105-106°.[1502]

$(4\text{-ClC}_6H_4O)_2PO_2H$. I.l.b.β,[102] I.l.b.γ-l,[1215,1502] I.l.e.α-l,[102] I.2.b.β-2 (from $(RO)_2PHO$ using CCl_4 + NaOH).[975] Needles or plates, m. 133-135°,[1502] 126-127°;[102] Na salt, poorly soluble in cold water;[102,975] Fe^{III} salt, very insoluble in H_2O.[1502]

$(2,4,6\text{-}Cl_3C_6H_2O)_2PO_2H$. I.l.b.γ-l. Cryst., m. 238°.[276] [Less pure by modified I.l.b.α, using $Cl_3C_6H_2OH$ heated with PCl_3, then with PCl_5 at 200-300°, followed by H_2O; cryst., m. 230°[1490]]. NH_4 salt, poorly soluble in H_2O; Ba salt, needles.[1490]

$(4\text{-BrC}_6H_4O)_2PO_2H$. I.2.b.β-2 (from $(RO)_2PHO$ using CCl_4 + NaOH).[975] By bromination of $(PhO)_2PO_2H$. Needles, m. 199-201°.[1502] Na salt.[975]

$(4\text{-}O_2NC_6H_4O)_2PO_2H$. I.l.b.β - γ-l (using $POCl_3$),[433] I.l.e.α-l (using $(RO)_3PO$),[982,1311] I.l.e.α-3 (using $(RO)_3PO$ and KSCN),[1204] I.l.e.α-3 (using $(RO)_3PO$ and piperidine),[1204] I.l.f.ζ-5 (from ROH using 2,6-lupetidinophosphorodichloridate, then LiOH).[725] By nitration of $(PhO)_2PO_2H$ with HNO_3 (d 1.5).[689] By nitration at 0° of $(PhO)_2PONH_2$ (nitration in presence of EtOH or treatment of the reaction mixture with EtOH yields not $(RO)_2PO_2H$ as claimed[1175], the compound, m. 133.5°, being its Et ester[689]). Cryst., m. 176-177°,[1204] 175-175.5°,[982] 175°;[689] $c\text{-}C_6H_{11}NH_3$ salt, yellow, m. 174-176°;[433] piperidinium salt, m. 163-164°;[1204] Na salt, poorly soluble plates; Ag salt, needles.[689] Protolysis constant in $HCONMe_2$: K x 10^4 92.[1133]

$(2,4\text{-}(O_2N)_2C_6H_3O)_2PO_2H$. I.l.b.β. Pyridinium salt, m. 156-157°.[123] Mechanism and kinetics of hydrolysis.[303]

$(2\text{-MeOC}_6H_4O)_2PO_2H$. I.l.a.α,[495] I.l.b.γ-l,[101] I.l.b.β - γ-l,[101] I.l.e.α-l.[101] Prisms, m. 97°;[101,495] Na

salt;[101] K salt;[495] Ca salt, water soluble needles; Cu^{II} salt, water soluble needles.[101]
$(4\text{-MeOC}_6H_4O)_2PO_2H$. I.l.b.β. Cryst., m. 92-94°.[443]
$(3\text{-Me-4-ClC}_6H_3O)_2PO_2H$. I.l.b.γ-l. Cryst., m. 116°.[1215]
$(2,4,6\text{-Cl}_3\text{-3-MeC}_6H{\cdot}O)_2PO_2H$. I.l.b.α. Needles, m. 94.5°.[309]
$(2\text{-Me-5-i-Pr-6-ClC}_6H_2O)_2PO_2H$. I.l.b.γ-l. Solid.[1215]
$(2\text{-i-Pr-5-Me-6-ClC}_6H_2O)_2PO_2H$. I.l.b.γ-l. Cryst., m. 134°.[1215]
$(2\text{-Me-4-PrCO}{\cdot}C_6H_3O)_2PO_2H$. I.l.b.γ-l. K salt.[1215]
$(2\text{-i-Pr-5-Me-6-PrCO}{\cdot}C_6H_2O)_2PO_2H$. I.l.b.γ-l. K salt.[1215]
$(\text{Vanillyl})_2$ phosphate. I.l.b.β - γ-l. Ca salt.[1242]
$(1\text{-EtO}_2C\text{-3-MeO-}C_6H_3\text{-4-O})_2PO_2H$. I.l.b.β - γ-l. Cryst., m. 238°.[1242]
$(1\text{-Et}_2NCO\text{-3-MeO-}C_6H_3\text{-4-O})_2PO_2H$. I.l.b.β - γ-l. Na salt, m. 153-154°.[1242]
$(4\text{-HO}_3S\text{-2-HO}_2CC_6H_3O)_2PO_2H$. From 5-sulfosalicylic acid and Na_3PO_4 in hot H_2O. Prisms $(2H_2O)$; Na_3 salt.[204]
$2\text{-}(1\text{-ClC}_{10}H_6O)_2PO_2H$. I.l.b.β. Needles, m. 251°.[102]
Diestrone phosphate. I.l.e.α-l (from Me ester). Cryst., m. 163-165°, UV, IR.[1194]

5. R and R' Heterocyclic

Di-isopyromucyl phosphate. I.l.e.α-l (from $(RO)_3PO$). Cryst., m. 154°.[375]

IV. Bis and Poly Phosphoric Esters (at Least One Secondary Ester Group)

1,3-Bis-(2-oleoyl-3-stearoyl-L-1-glycerophosphoryl) glycerol. From the t-Bu ester of the benzyl-phosphoryl derivative by acid hydrolysis. Ba salt, m. 192-194°, $[\alpha]_D^{20}$ -6.25° $(CHCl_3)$.[645]
Hexestrol bis-(benzylphosphate). I.l.e.α-3 (from bis-(dibenzylphosphate using KSCN). K salt.[1204]
Stilbestrol bis-(benylphosphate). I.l.e.α-3 (from bis-(dibenzylphosphate using KSCN). Cryst., m. 124-125°.[1204]

Nucleotides

Uridine-(3' → 5')-uridine 3'-phosphate. I.l.d.α-l (from 2'-O-acetyluridine 3'-phosphate and DCC;[1174] from 5'-O-acetyl-2'-O-tetrahydrofuryl 3'-phosphate and benzhydryl 3'-uridylic acid with DCC).[451]
5'-Phospho-deoxycytidine-(3' → 5')thymidine. I.l.d.α-l (from 3'-O-acetylthymidine 5'-phosphate and N'-anisoyl-deoxycytidine 5'-phosphate). UV.[1057]
Deoxycytidine-(3' → 5')-deoxyadenosine 3'-phosphate.

I.l.d.α-l (from protected deoxycytidine 3'-phosphate and protected deoxyadenosine 3'-phosphate). Chromatographic and UV characteristics.[791]

Cytidine-(3' → 5')-uridine 3'-phosphate. I.l.d.α-l (from N^6,$O^{2'}$,$O^{5'}$-triacetylcytidine 3'-phosphate and benzhydryluridine 3'-phosphate),[452] I.l.d.α-l - e.α-l (from N',$O^{2'}$,$O^{5'}$-triacetylcytidine 3'-phosphate and uridine 3'(2')phosphate with DCC → 2',3'-cyclic phosphate dinucleotide derivative, followed by enzymatic scission of the cyclic ester group and alkaline treatment).[1271]

Cytidine-(3' → 5')-adenosine 3'-phosphate. I.l.d.α-l (from N^6,$O^{2'}$,$O^{5'}$-triacetylcytidine 3'-phosphate and benzhydryladenosine 3'-phosphate).[452]

Thymidine-(3' → 5')-deoxycytidine 3'-phosphate. I.l.d.α-l (from protected corr. mononucleotides). Chromatographic and UV characteristics.[452]

Thymidine-(3' → 5')-thymidine 3'-phosphate. I.l.d.α-l (from 5'-tritylthymidine 3'-phosphate and thymidine 3'-cyanoethyl-phosphate).[442b]

5'-Phospho-thymidine-(3' → 5')-deoxyadenosine. I.l.d.α-l (from protected components). Chromatographic and UV characteristics.[1293]

5'-Phospho-thymidine-(3' → 5')-N-benzoyl-deoxyadenosine. I.l.d.α-l (from protected components). Chromatographic and UV characteristics.[1293]

Thymidine-(3' → 5')-deoxyguanosine 3'-phosphate. I.l.d.α-l (from protected components). Chromatographic and UV characteristics.[791]

Inosinyl-(3' → 5')-uridine 3'-phosphate. I.l.d.α-l - e.α-l (from inosine 3'-phosphate and uridine 3'(2')-phosphate by DCC → 2',3'-cyclic phosphate dinucleotide derivative, then enzymatic scission of the cyclic ester group).[1271]

Deoxyadenosine-(3' → 5')-thymidine 3'-phosphate. I.l.f.ζ-3 (from the protected dinucleotide using $CNCH_2CH_2OPO_3H_2$). Chromatographic and UV characteristics.[1293]

Deoxyadenosine-(3' → 5')-deoxyguanosine 3'-phosphate. I.l.d.α-l (from protected mononucleotides). Chromatographic and UV characteristics.[791]

Adenosine-(3' → 5')-uridine 3'-phosphate. I.l.d.α-l - e.α-l (from protected adenosine 3'-phosphate and uridine 3'(2')-phosphate with DCC → 2',3'-cyclic phosphate dinucleotide derivative, then enzymatic scission of the cyclic ester group).[1271]

Uridine-(3' → 5')-uridine'(3' → 5')-uridine. I.l.d.α-l (from 2',5'-protected uridine 3'-phosphate and 2',3'-protected uridine-(5' → 3')-2'-protected-uridine). Identification only by paper chromatography.[450]

Uridine-(3' → 5')-uridine-(3' → 5')-uridine 3' phosphate. I.l.d.α-l (repeated condensation of 2'-O-acetyluridine 3'-phosphate).[1174]

Uridine-(3',5')-uridine-(3',5')-cytidine. III.1.b.α-1 (from N-acetyl-2',3'-protected cytidine with 2',5'-protected uridylic acid, deacetylation by NH_3 and condensation with the same uridylic derivative, and elimination of protecting groups).[987]

Cytidine-(3' → 5')-uridine-(3' → 5')-uridine. I.1.d.α-1 (from N-acetyl-2',5'-protected cytidine 3'-phosphate and 2',3'-protected-uridine-(5' → 3')-2-protected-uridine). Identification only by paper chromatography.[450]

Cytidine-(3' → 5')-uridine-(3' → 5')-cytidine. III.1.b.α-1 (from N-acetyl-2',3'-protected cytidine with 2',5'-protected uridylic acid, deacetylation with NH_3 and condensation with 2',5'-protected cytidylic acid, and elimination of protecting groups).[987]

Cytidine-(3' → 5')-uridine-(3' → 5')-adenosine. I.1.α-1 (by progressive condensation of the adequately protected components).[394]

Cytidine-(3' → 5')-cytidine-(3' → 5')-cytidine. III.1.b.α-1 (from N-acetyl-2',3'-protected cytidine with 2',5'-protected cytidylic acid, deacetylation by NH_3 and condensation with the same cytidylic derivative, and elimination of protecting groups).[987]

Cytidine-(3' → 5')-cytidine-(3' → 5')-adenosine. I.1.d.α-1 (from N^6,2',3'-protected adenosine by successive condensation with N,2',5'-protected cytidine 3'-phosphate, liberation of the 5' position of the cytidine group and with N,2',3'-protected cytidine 3'-phosphate). Chromatographic evidence; 3' → 5' linkages proved by degradability by pancreatic ribonuclease.[693]

Thymidine-(3' → 5')-deoxycytidine-(3' → 5')deoxycytidine. I.1.d.α-1 (from protected thymidine-deoxycytidine dinucleotide and N^6,3'-protected cytidine 5'-phosphate). Chromatographic and UV characteristics.[1057]

5'-Phospho-thymidine-(3' → 5')-deoxycytidine-(3' → 5')-deoxycytidine. I.1.f.ζ-3 (by phosphorylation of protected trinucleotide). Chromatographic and UV characteristics.[1057]

Thymidine-(3' → 5')-deoxycytidine-(3' → 5')-deoxycytidine. I.1.d.α-1 (from protected thymidine-deoxycytidine dinucleotide and N^6,3'-protected adenylic acid). Chromatographic and UV characteristics.[1057]

5'-Phospho-thymidine-(3' → 5')-deoxycytidine-(3' → 5')-deoxycytidine. I.1.f.ζ-3 (from the adequately protected preceding trinucleotide). Chromatographic and UV characteristics.[1057]

5'-Phospho-thymidine-(3' → 5')-deoxycytidine-(3' → 5')-deoxyadenosine. I.1.d.α-1 - f.ζ-3 (by condensation of protected deoxyadenosine 5'-phosphate and adequately protected thymidine-deoxycytidine dinucleotide, and final phosphorylation). Chromatographic and UV characteristics.[1057]

Deoxyguanosin-(3' → 5')-thymidine-(3' → 5')-thymidine. I.1.f.ξ-9 (from thymidine 5'-bound to a polymer support by repeated condensation with 3'-O-acetylthymidine 5'-phosphate (and intermediary deacetylation) by picryl-Cl).[928]

5'-Phospho-thymidine-(3' → 5')-deoxyadenosine-(3' → 5')-thymidine-(3' → 5')-deoxyadenosine. I.1.d.α-1 (by successive condensation of adequately protected mononucleotides).[1293]

Thymidine-(3' → 5')-(thymidine-(3' → 5')-)$_2$-thymidine. From 5'-bound thymidine on a polymer support by repeated condensation with 3'-O-acetylthymidine 5'-phosphate activated by picryl-Cl).[928]

Deoxycytidine-(3' → 5')-deoxyguanosine-(deoxycytidine-(3' → 5')-)$_2$-deoxycytidine. I.1.d.α-1 (by stepwise condensation of adequately protected components).[1292]

Pentathymidine 3' → 5'-tetraphosphate. By repeated condensation of thymidine 5'-bound to a polymer support with 3'-O-acetylthymidine 5'-phosphate (and intermediary deacetylation) by (i-Pr)$_3$-benzenesulfonyl-Cl or picryl-Cl. Chromatographic and UV characteristics.[928]

Thymidine-(3' → 5')-deoxyadenosine-(thymidine-(3' → 5')-)$_2$ thymidine. I.1.d.α-1 (by stepwise condensation of adequately protected components).[1292]

Uridine-(3' → 5')-(uridine-(3' → 5')-)$_2$-uridine-3'-phosphate. I.1.d.α-1 (by repeated condensation of 2'-O-acetyluridine 3'-phosphate, up to hexanucleotide: n=4).[442]

Hexathymidine 3' → 5' pentaphosphate (and higher, until hexadecathymidine pentadecaphosphate). I.1.d.α-1 (by repeated condensation of 5'-protected thymidine with 3'-acetylthymidine 5'-phosphate and intermediary liberation of the 5' position).[745]

Thymidine-(3' → 5')-deoxyadenosine-(thymidine-(3' → 5')-)$_3$-thymidine. I.1.f.ζ-3 - d.α-1 (by phosphorylation of 5'-protected thymidine-(3' → 5')-N-protected deoxyadenosine, followed by repeated condensation with 5'-protected thymidine-(3' → 5')-3'-protected thymidine, with intermediary phosphorylation of a liberated 3' position).[1293]

Thymidine-(3' → 5')-(deoxyadenosine-(3' → 5')-thymidine-(3' → 5')-)$_2$-deoxyadenosine. I.1.d.α-1 (stepwise condensation or adequately protected dinucleotides with intermediary liberation of the necessary positions).[1293]

[Deoxycytidine-(3' → 5')-deoxyadenosine-(3' → 5')-deoxyadenosine]$_2$ (no terminal phosphate !). I.1.d.α-1 (condensation of protected nucleoside with successive protected nucleotides).[791]

[Deoxyguanosine-(3' → 5')-deoxyadenosine-(3' → 5')-deoxyadenosine]$_2$ (no terminal phosphate !). I.1.d.α-1 (similarly to the preceding compound).[791]

[Thymidine-(3' → 5')-cytidine-(3'.]$_4$ 5'-phosphate. By polymerization of thymidine-(3' → 5')-cytidine 5'-phosphate.[791]

[Thymidine-(3' → 5')-deoxyadenosine-(3' → 5')-]$_3$ (no terminal phosphate !). I.1.d.α-1 (by repeated condensation of appropriated dinucleotides).[1293]

[Thymidine-(3' → 5')deoxyadenosine-(3' → 5')-]$_n$ (no terminal phosphate group !); n = 4, 5, and 6. I.1.d.α-1 (by repeated condensation of appropriated dinucleotides).[1293]

[Thymidine-(3' → 5')-thymidine-(3' → 5')-deoxycytidine-(3' → 5')-]$_4$ (no terminal phosphate group !). I.1.d.-α-1 (by repeated condensation of appropriated dinucleotides).[791]

[Thymidine-(3' → 5')-thymidine-(3' → 5')-deoxyinosine-(3' → 5')-]$_4$ (no terminal phosphate group !). I.1.d.-α-1 (by repeated condensation of appropriated dinucleotides).[791]

V. Mono Tertiary Esters, (RO)(R'O)(R"O)PO

A. R, R', and R" Aliphatic C Skeleton

1. Hydrocarbon Derivatives

$(MeO)_3PO$. I.1.b.β,[488,524,750,1415] I.1.d.δ,[348,884,1152,1440,1484] I.1.e.γ,[348] I.2.b.α-1 (from $(MeO)_3P$ using O_2 and UV irradiation,[1264] SO_3[527]), I.2.b.α-8 (from $(MeO)_3P$ in MeOH by Cl_2),[154,572] I.2.b.β-4 (from $(MeO)_3P$ using CCl_3CONRR').[1276a] From $(MeO)_3P$ by $SOMe_2$;[22a] by $(MeO)_2$ (formation of $(MeO)_3PO$ revealed by chromatography and NMR).[469] As a by-product, from MeONa and PCl_3.[66,67] Liquid, b. 197.2°,[1440] 192-193°,[66,67] b_{762} 192°, b_{60} 110°, b_{36} 97°, b_{24} 85°,[348] b_{12} 79°,[1037] b_{10} 73°,[524] b_{8-10} 72-73°,[66] b_{6-7} 67-68°,[750] b_5 62°;[1415] d_0^0 1.2365,[348] 1.2156,[66,67] 1.2148,[66] 1.218,[488] d_0^{15} 1.2195,[348] $d_0^{19.5}$ 1.1971,[66,67] d_0^{20} 1.2144,[1415] d_0^{22} 1.200,[1152] d_4^{25} 1.2052,[524] $d_4^{60.9}$ 1.1722, $d_4^{119.9}$ 1.1062;[1415] n_D^{25} 1.4708,[1276a] n_C^{20} 1.39452, n_D^{20} 1.39630, n_F^{20} 1.40049,[1415] n_C^{25} 1.3934, n_D^{25} 1.3950, n_F^{25} 1.3990.[524] Solubility in H_2O: 1:1 at 25°.[488] ^{31}P + 1 ppm.[913c]

$(MeO)_2(EtO)PO$. I.1.b.γ-2 (using $(MeO)_2POCl$),[651] I.1.d.δ.[448,884,1440] Liquid, b. 203.3°,[884,1440] 203°,[488] b_{41} 112-113°;[651] d_0^0 1.176,[488] 1.1752,[884,1440] d_0^{22} 1.161;[488] n_D^{22} 1.3984.[651] IR.[651] Solubility in H_2O: 1:1 at 25°.[488]

$(MeO)(EtO)_2PO$. I.1.d.δ,[884] I.2.b.β-4 (from $(EtO)_3P$ in MeOH using $CBr_2(CONH_2)_2$).[1013] From a salt of (MeO)-

(EtO)PO_2H with $ClCO_2$Et.[255] Liquid, b. 208.2°,[884] b_{20} 100.2°,[1013] $b_{0.01}$ 75-77°;[255] d_0^0 1.1228.[884]

$(EtO)_3PO$. I.1.b.γ-2,[330,344,488,498,524,876,1415,1452] I.1.c.α,[845] I.1.d.δ,[348,884] I.1.e.β (from $(EtO)_2(CH_2=C(Me)O)PO$ with EtOH),[662,1011,1013] I.1.e.γ (using $[(EtO)_2PO_2]_2Pb$),[1414] I.1.f.α (using HNO_3),[1111] I.2.-b.α-1 (from $(EtO)_3P$ using O_2 and UV irradiation),[13] I.2.b.α-5 (using NOCl,[72] N_2O_4[442a,b]), I.2.b.α-8 (from yellow or finely divided red P with EtOH by Cl_2),[573] I.2.b.β-1,[312] I.2.b.β-2 (from $(EtO)_3P$ using CBr_4 or CCl_4),[455] I.2.b.β-3 (from $(EtO)_3P$ using ethylene epoxide or propylene epoxide),[1237] I.2.b.β-4 (from $(EtO)_3P$ using Cl_3CONRR';[1276a] from $(EtO)_2(PhCH_2O)P$ and EtOH using $NCHBrCONH_2$[1012]), (from $(EtO)_3P$ using CBr_2-$(CONH_2)_2$),[1013] (from $(EtO)_2$PHO and EtOH using NCCHBr-$CONH_2$- besides $(EtO)_2PO_2H$),[971] I.2.b.β-5 (from $(EtO)_2$-$(PhCH_2O)P$ and EtOH using N-Br-succinimide),[1237] I.2.-b.β-6 (using $EtO_2CN=NCO_2Et$).[973] Revealed in "Lang-held"-ester by chromatography.[1453] From P_2O_5 or $EtPO_3$ and $(EtO)_2SO$.[271] From $(EtO)_3P$ by $MeCCl_2NO_2$, Cl_3CNO_2, or Me_2CBrNO_2 (besides oxime ester);[19] by $SOMe_2$.[22a] From $(EtO)_3P$ by $(EtO)_2$ (via $(EtO)_5P$), formation revealed by NMR and chromatography.[469] Liquid, b_{770} 216°,[68] b_{775} 211.5° (?),[348] b_{760} 215-216°,[66] 215°,[488] b_{745} 214°,[68] b. 203° (in H_2 atm.?),[1452] b_{50} 123°,[348] b_{30} 116°,[344] b_{25} 108°,[68] 103°,[348] b_{24} ∿110°,[1237] b_{13} 99.2°,[68] b_{12} 92-94°,[971] b_{10} 93-95°,[662,1011,1013] 90°,[524] b_{8-10} 98-98.5°,[348] b_5 75.5°,[1414] 65-67°,[1276a] $b_{3.5}$ 69-71°,[442a,b] b_2 77-79°;[61] d_0^0 1.0929,[348] 1.0897,[61] 1.0897,[66,67] $d_0^{12.5}$ 1.0785,[348] d_0^{19} 1.0725,[66] d_4^{20} 1.06817,[68] 1.0695,[1415] 1.056 (?)[488] d_4^{25} 1.0637,[524] d_0^{55} 1.0214,[348] $d_4^{60.9}$ 1.0301, $d_4^{120.1}$ 0.9708;[1415] n_D^{20} 1.4063,[61] 1.40533,[488] 1.40616,[68] n_D^{17} 1.40674,[1493] n_D^{25} 1.4948,[1276a] 1.4039,[524] n_C^{20} 1.40343, n_F^{20} 1.40983,[1415] n_C^{25} 1.4021, n_F^{25} 1.4082;[524] IR.[1237] Solubility in H_2O: 1:1 at 25°.[192] IR.[651] Molecular extinction.[650] ^{31}P + 0.9 ppm.[913c]

$(MeO)_2(CH_2=CHO)PO$. From $(MeO)_3P$ with chloroethylene cyclocarbonate. Liquid. MNR.[1244] Hydrolysis; insecticidal properties.[1306]

$(EtO)_2(CH_2=CHO)PO$. I.1.b.γ-2 (from $(EtO)_3P$ using $ClCH_2$-CHO).[20,564a] From $(EtO)_2(ClCH_2CH_2O)PO$ by dehydrochlorination.[1390] From $(EtO)_3P$ using $ClHgCH_2CHO$,[902] chloroethylene cyclocarbonate.[1244] Liquid, b_{20-30} 99-105°,[903] b_6 79°,[20] b_{5-7} 79°,[564a] b_5 79°, $b_{2.5}$ 67°;[1390] d_{35} 1.0724; n_D^{35} 1.4100,[20] n_D^{25} 1.4040;[1390] NMR.[1244]

$(MeO)_2(PrO)PO$. I.1.d.δ. Liquid, b_{15} 116°, d_0^0 1.195, d_0^{22} 1.180.[488]

(MeO)(EtO)(PrO)PO. From a salt of (MeO)(PrO)PO_2H with $ClCO_2$Et. Liquid, $b_{0.001}$ 96°.[255]

(MeO)(PrO)$_2$PO. I.1.d.δ. Liquid, b_{20} 129°, d_0^0 1.077, d_0^{22} 1.059.[488]

(EtO)$_2$(PrO)PO. I.1.d.δ,[488] I.2.b.β-4 (from (EtO)$_3$P and PrOH using CBr$_2$(CONH$_2$)$_2$,[1013] (from (EtO)$_2$(PhCH$_2$O)P and PrOH using NCCHBrCONH$_2$),[1012] I.2.b.β-5 (from (EtO)$_2$(PhCH$_2$O)P and PrOH using N-Br-succinimide). From a salt of (EtO)(PrO)PO$_2$H with ClCO$_2$Et.[255] Liquid, b_{20} 130°,[488] $b_{10.5}$ 109-114°,[1013] $b_{0.1}$ 100-101°;[255] d_0^0 1.098, d_0^{22} 1.077.[488]

(EtO)(PrO)$_2$PO. I.1.b.γ-2 (from (PrO)$_2$POCl),[651] I.1.d.δ.[488] Liquid, b_{23} 110-112°,[651] b_{20} 145°;[488] d_0^0 1.046, d_0^{22} 1.025,[488] d_{25}^{25} 1.0066; n_D^{25} 1.4044; IR.[651]

(PrO)$_3$PO. I.1.b.α,[1461] I.1.b.γ-2,[488,498,524,1395,1415] I.1.d.δ,[349,1284] I.2.b.α-1 (from (RO)$_3$P using O$_2$ and UV irradiation),[1264] I.2.b.α-5 (using N$_2$O$_4$).[442a,b] As a by-product from RONa and PCl$_3$ reaction.[66,67] Liquid, b_{47} 138°, b_{22} 133°,[349] b_{15} 131°,[488] 128-134°,[498] b_{10} 121°,[524] b_9 119°,[1395] b_{8-10} 120.5-121.5°,[66] b_8 120-121°,[67] b_5 107.5°,[1415] $b_{0.5}$ 49-51°;[442a,b] d_0^0 1.025,[488] 1.0282,[66] d_4^{20} 1.0121,[1415] d_0^{22} 1.007,[488] d_4^{25} 1.0023,[524] $d_4^{120.7}$ 0.9209;[1415] n_D^{20} 1.41646,[1415] n_d^{25} 1.4136,[524] n_C^{20} 1.4148,[1415] n_C^{25} 1.4118,[524] n_F^{20} 1.42120,[1415] n_F^{25} 1.4182.[524] ^{31}P + 0.5;+ 0.8 ppm.[913c]

(EtO)(CH$_2$=CHCH$_2$O)$_2$PO. I.1.b.γ-2 (from diallyl phosphorochloridate). Liquid, b_{1-2} 72°; n_D^{20} 1.14350.[1314a,1315]

(CH$_2$=CHCH$_2$O)$_3$PO. I.1.b.β,[1364,1451] I.1.d.δ,[348,1484] I.2.b.α-1 (from (RO)$_3$P using O$_2$ with UV irradiation[13] or a metal or metal oxide catalyst[1249]), I.2.b.α-2 (using H$_2$O$_2$).[668] Liquid, b_1 85-90°,[688] $b_{0.5}$ 80°,[1451] b_{44} 157°, b_{20} 142°.[348] ^{31}P + 1.2 ppm.[913c]

(EtO)$_2$(i-PrO)PO. I.2.b.β-4 (from (EtO)$_3$P and i-PrOH using NCCHBrCONH$_2$,[662,1011,1013] CBr$_2$(CONH$_2$)$_2$.[1013] From a salt of (EtO)(i-PrO)PO$_2$H with ClCO$_2$Et.[255] Liquid, b_{11} 91°,[1013] $b_{0.1}$ 100°.[255]

(i-PrO)$_3$PO. I.1.b.γ-2,[1415] I.1.d.δ,[349] I.2.b.α-5 (from (RO)$_3$P using N$_2$O$_4$).[442a,b] From (RO)$_3$P using Cl$_3$-CCONRR'.[1276a] As by-product in the reaction of RONa with PCl$_3$.[66,67] Liquid, b. 218-220°,[67] b_{68} 136°,[349] b_{8-10} 95-96°,[66,67] b_5 83.5°,[1415] $b_{0.45}$ 46.5°;[442a,b] d_0^0 1.0054,[66,67] d_4^{20} 0.9867, $d_4^{62.5}$ 0.9472, $d_4^{119.9}$ 0.8931; n_C^{20} 1.40376, n_D^{20} 1.40573, n_F^{20} 1.41034.[1415] ^{31}P + 3.3 ppm.[913c]

(MeO)$_2$(CH$_2$=C(Me)O)PO. I.2.b.γ-3 (from (EtO)$_3$P with MeCO•CH$_2$O-Ts).[468] Hydrolysis; insecticidal properties.[1306]

(MeO)(CH$_2$=C(Me)O)$_2$PO. I.1.b.γ-2. Liquid, $b_{0.5}$ 60°.[778]

(EtO)$_2$(CH$_2$=C(Me)O)PO. I.2.b.γ-2 (from (EtO)$_3$P with ClCH$_2$-CO•Me[20] or BrCH$_2$CO•Me[1149]). From (EtO)$_3$P with BrHgCH$_2$CO•Me.[903] Liquid, b_{12} 96°,[1149] b_1 72-73°,[20] $b_{0.1}$ 50.5-51.5°;[903] d_{20} 1.0708; n_D^{20} 1.4190.[1149] ^{31}P + 7.2 ppm.[913c]

(EtO)(CH$_2$=C(Me)O)$_2$PO. I.1.b.γ-2. Liquid, $b_{0.5}$ 66°.[778]

$(CH_2=C(Me)O)_3PO$. I.1.b.γ-2. Liquid, $b_{0.8}$ 78°; n_D^{20} 1.449.[778]

$(EtO)_2(BuO)PO$. I.1.b.γ-2,[602] I.1.e.β (from $(EtO)_3PO$ with BuOH (BuONa), besides $(EtO)(BuO)_2PO$;[1219] with BuOSOCl, besides $(EtO)(BuO)_2PO$;[270] with $BuCl \cdot Ph_3P$[672]), I.2.b.α-8 (from $(EtO)_2PHO$ and BuOH using CCl_4),[1315] I.2.b.β-4 (from $(EtO)_3P$[1013] or $(EtO)_2(PhCH_2O)P$[1012] and BuOH using $CBr_2(CONH_2)_2$[1013] or $NCCHBrCONH_2$[1012]), I.2.b.β-5 (from $(EtO)_2(PhCH_2O)P$ and BuOH using N-Br-succinimide).[1012] Liquid, b_{15} 123°,[602] b_{3-5} 82-87°,[1013] b_{3-4} 82-87°,[1219] b_{2-3} 100-101°;[1315] d_4^{10} 1.0380,[1219] d_4^{13} 1.0340,[602] d_4^{25} 1.0243; n_D^{10} 1.4170, n_D^{20} 1.4131.[1219]

$(EtO)(BuO)_2PO$. I.1.e.β (from $(EtO)_3PO$ using BuOH (BuONa)[1219] or BuOSOCl,[270] besides $(EtO)_2BuO)PO$). Liquid, b_{3-4} 95-96°; d_4^{10} 1.0112, d_4^{25} 0.9984; n_D^{10} 1.4215, n_D^{20} 1.4182.[1219]

$(i\text{-}PrO)_2(BuO)PO$. I.1.b.β-2 (from $(i\text{-}PrO)_2PHO$ and BuOH using CCl_4). Liquid, b_{2-3} 100-101°; n_D^{25} 1.4085.[1314a]

$(BuO)_3PO$. I.1.b.α (using PCl_5),[602] I.1.b.β,[74,498,524,602,717,1395,1415] (with $TiCl_3$ as catalyst),[969] I.1.b.-γ-2,[602] I.1.d.δ (using BuBr),[1484] I.2.b.α-1 (using O_2 and UV irradiation;[1264] air/SO_3 mixture[298]), I.2.-b.α-5 (using N_2O_4,[442a,b] or NOCl[72]), I.2.b.α-8 (from yellow or finely divided red P and BuOH by Cl_2;[573] from PCl_3 and BuOH by Cl_2[249]), I.2.b.β-1 (using peracetic acid),[354] I.2.b.β-6 (using $EtO_2CN{=}NCO_2Et$, then $PhCH_2OH$ or $CH_2{=}CHCH_2OH$).[973] Liquid, b_{20} 180°, b_{15} 160-162°, b_{10} 154°,[602] b_8 143-145°,[498] 148.5°,[74] b_6 143-146°,[249] 138.5°,[1415] b_5 135°,[717] $b_{0.45}$ 47°; when crystallized, m. 71-75°;[442a,b] d_4^{13} 0.9824,[602] d_0^{20} 0.9731,[74] d_4^{20} 0.9766,[1415] d_4^{25} 0.9727,[524] $d^{120.5}$ 0.8941;[1415] n_C^{20} 1.42295, n_D^{20} 1.42496,[1415] 1.4247,[74] n_F^{20} 1.42988,[602] n_C^{25} 1.4203, n_D^{25} 1.4224, n_F^{25} 1.4274.[524] IR.[442a,b] Interaction with $(2\text{-}EtC_6H_{12}O)_2PO_2H$ in C_6H_6 and in $CHCl_3$.[1288] ^{31}P + 1.0; + 0.6 ppm.[913c]

$(i\text{-}BuO)_3PO$. I.1.b.β,[488,524,717,1415,1426] I.1.d.δ,[349] I.2.b.α-1 (using O_2 and UV irradiation).[1264] By-product of the reaction of RONa and PCl_3.[66,67] Liquid b_{15} 152°,[488] b_{12} 180-200° (?),[1426] b_{10} 135-136°,[66] 138°,[524] $b_{5.5}$ 117°,[1415] b_4 112°;[717] d_0^0 0.9698,[66,67] d_4^{20} 0.9681,[1415] 0.9465 (?),[1426] d_4^{25} 0.9617,[524] $d_4^{120.3}$ 0.8818;[1415] n_C^{20} 1.41729, n_D^{20} 1.41931, n_F^{20} 1.42416,[1415] n_C^{25} 1.4152, n_D^{25} 1.4173.[524]

$(MeEtCHO)_3PO$. I.1.b.β. Liquid, b_{8-12} 119-129°.[498]

$(MeCH{=}CMeO)(EtCH{=}CHO)(CH_2{=}CHO)PO$. I.1.b.γ-2. Liquid, $b_{0.001}$ 96-100°; d_{20} 1.1096; n_D^{20} 1.4569.[619]

$(CH_2{:}C(Me)CH_2O)_2(EtO)PO$. Crude. I.1.b.β. Liquid, b_6 120-135°, n_D^{25} 1.4390.[1006]

$(CH_2{=}C(Me)CH_2O)_3PO$. I.1.b.β.[778,1006] Liquid, b_5 134.5-140°,[1006] $b_{0.1}$ 90-92°;[778] d_4^{26} 0.988;[1006] n_D^{19} 1.452,[778] n_D^{25} 1.4454.[1006]

$(MeO)_2(AmO)PO$. I.1.b.γ-2 (using $(MeO)_2POCl$). Liquid, $b_{0.4}$ 72-74°; n_D^{20} 1.4127. Pyrolysis.[215]
$(AmO)_3PO$. I.1.b.β,[498,524,717,1415] I.1.d.δ,[1484] I.2.b.β-8 (from PCl_3 and ROH by Cl_2).[249] Liquid, b_{50} 225°,[524] b_6 158-165°,[498] b_5 167°,[1415] 155-156°,[249] $b_{2.5}$ 143-144°;[717] d_4^{20} 0.9608,[1415] d_4^{25} 0.9497,[524] $d_4^{120.6}$ 0.8816;[1415] n_C^{20} 1.42975, n_D^{20} 1.43188, n_F^{20} 1.43701,[1415] n_C^{25} 1.4262, n_D^{25} 1.4283, n_F^{25} 1.4332.[524]
$(BuO)(i\text{-}AmO)_2PO$. I.1.b.γ-2. Liquid, $b_{4.5}$ 145°.[718]
$(i\text{-}AmO)_3PO$. I.1.b.β,[1395] I.1.d.δ.[664] Liquid, b_3 143°.[1395]
$(MeO)_2(n\text{-}C_6H_{13}O)PO$. I.1.b.β - γ-2 (from $C_6H_{13}OH$ using $POCl_3$, then MeONa). Liquid, $b_{1.7}$ 100°; n_D^{20} 1.4180. Pyrolysis.[215]
$(n\text{-}C_6H_{13}O)_3PO$. I.1.b.β,[75] I.1.d.δ.[1484] Liquid, b_2 187-188°; d_0^{20} 0.9396; n_D^{20} 1.4340. Viscosity indicates extended structure, in contrast to parachor.[75,76]
$(i\text{-}C_6H_{13}O)_3PO$. I.1.b.β (using $POCl_3$ with $TiCl_3$ as catalyst.[969]
$(2\text{-}Et\text{-}n\text{-}C_4H_8O)_3PO$. I.1.b.β. Liquid, $b_{2.5}$ 177-180°.[1426]
$(MeO)_2[CH_2{=}C(CMe_3)O]PO$. I.2.b.γ-3 (from $(MeO)_3P$ using $Me_3CCO{\cdot}CH_2OTs$).[468]
$(EtO)_2(n\text{-}C_7H_{15}O)PO$. I.1.e.β (from $(EtO)_3PO$ with $C_7H_{15}Br$[671] or $C_7H_{15}Cl{\cdot}Ph_3P$[672]). Liquid, $b_{0.08}$ 85°.[671,672]
$(EtO)(n\text{-}C_7H_{15}O)_2PO$. I.1.e.β (from $(EtO)_2(C_7H_{15}O)PO$ with $C_7H_{15}Br$[671] or from $(EtO)_3PO$ with $C_7H_{15}Cl{\cdot}Ph_3P$[672]). Liquid, $b_{0.003}$ 133°.[671,672]
$(BuO)_2(n\text{-}C_7H_{15}O)PO$. I.1.b.β (using $(BuO)_2POCl$). Liquid, b_2 148-150°.[248]
$(BuO)(i\text{-}AmO)(n\text{-}C_7H_{15}O)PO$. I.1.b.γ-2. Liquid, $b_{0.2}$ 129-131°; d_{20} 0.9551; n_D^{20} 1.4291.[904a]
$(n\text{-}C_7H_{15}O)_3PO$. I.1.e.β (from $(EtO)(C_7H_{15}O)_2PO$ with C_7H_{15}-Br,[671] or from $(EtO)_3PO$ with $C_7H_{15}Cl{\cdot}Ph_3P$[672]). Liquid, $b_{0.0005}$ 172°.[671,672]
$(MeO)_2(n\text{-}C_8H_{17}O)PO$. I.1.b.β (from $POCl_3$ with $C_8H_{17}OH$, then MeONa). Liquid, $b_{1.7}$ 130°; n_D^{20} 1.4236.[215]
$(MeO)(n\text{-}C_8H_{17}O)_2PO$. I.1.b.β (stepwise, similarly to the preceding compound). Liquid, b_1 162-164°; n_D^{25} 1.4362. Pyrolysis.[215]
$(EtO)_2(n\text{-}C_8H_{17}O)PO$. I.1.b.γ-2,[1335] I.1.e.β (from $(EtO)_3PO$ using $C_8H_{17}Br$).[918] Liquid, b_5 228-232°;[1335] n_D^{20} 1.4235.[918]
$(BuO)(i\text{-}AmO)(n\text{-}C_8H_{17}O)PO$. I.1.b.γ-2. Liquid, $b_{0.1}$ 131-132°; d_{20} 0.9521; n_D^{20} 1.4310.[904a]
$(n\text{-}C_8H_{17}O)_3PO$. I.1.b.β,[75,76] (using $TiCl_3$ as catalyst),[969] I.2.b.α-1 (from $(RO)_3P$ using O_2 and UV irradiation),[1264] I.2.b.α-8 (from PCl_3 and ROH by Cl_2).[249] Liquid, b_2 228-235°,[249] 225-227°;[75,76] d_0^{20} 0.9200; n_D^{20} 1.4410. Viscosity indicates extended structure, contrary to parachor.[75,76]
$(i\text{-}C_8H_{17}O)_3PO$. I.1.b.β (using $TiCl_3$ as catalyst).[969]

$(i\text{-}C_8H_{17}O)(2\text{-Et-n-}C_6H_{12}O)_2PO$. I.1.b.β (stepwise, using $TiCl_3$ as catalyst).[969]
$(2\text{-Et-n-}C_6H_{12}O)_3PO$. I.1.b.β,[969,1426] (using NH_4VO_3 or VCl_4 as catalysts),[688] I.2.b.α-1 (from $(RO)_3P$ with O_2 and UV irradiation),[1264] I.2.α-8 (from $(RO)_3P$ in presence of ROH by Cl_2).[194,572] Liquid, b_4 215°; d_{20} 0.9225.[1426]
$(BuCEtH\cdot CH_2O)_3PO$. I.1.b.β. Liquid, $b_{3.5}$ 203°, d_{20}^{20} 0.924.[340]
$(2,2,4\text{-}Me_3\text{-n-}C_5H_8O)_3PO$. I.1.b.β (using $TiCl_3$ as catalyst).[969]
$(EtO)(BuO)(t\text{-}C_8H_{17}O)PO$. I.1.b.γ-2 (ROH with (EtO)(BuO)-POCl). Liquid, $b_{0.2}$ 160°; $n_D^{20.5}$ 1.4802.[1498]
$(BuO)_2(t\text{-}C_8H_{17}O)PO$. I.1.b.γ-2 (ROH with $(BuO)_2POCl$). Liquid, $b_{0.15}$ 172-174°; $n_D^{20.5}$ 1.4790.[1498]
$(BuO)_2(n\text{-}C_9H_{19}O)PO$. I.1.b.β (from $(BuO)POCl_2$ in presence of vinylcyanide). Liquid, b_1 147-148°.[248]
$(2\text{-Bu-n-}C_8H_{16}O)_3PO$. I.1.b.β (using $TiCl_3$ as catalyst).[969]
$(n\text{-}C_{13}H_{27}O)_3PO$. I.1.b.β (using $TiCl_3$ as catalyst).[969]
$(EtO)_2(n\text{-}C_{14}H_{29}O)PO$. I.1.e.β (from $(EtO)_3PO$ using RBr).[918]
$(n\text{-}C_{16}H_{33}O)_3PO$. I.2.b.β-8 (from $(RO)_3P$ by Cl_2 in the presence of ROH),[194,572] I.1.a.α. Cryst., m. 61°.[1122]
$(EtO)_2(n\text{-}C_{18}H_{37}O)PO$. I.1.e.β (from $(EtO)_3PO$ using RBr). Liquid, n_D^{20} 1.4475.[918]
$(n\text{-}C_{18}H_{37}O)PO\cdot n\text{-}C_{18}H_{37}OH$. I.2.b.β-8 (from yellow P, ROH, and Cl_2, or from PCl_3, ROH, and Cl_2). Cryst., m. 55.5-57.5°.[573]

2. Halogen Derivatives

$(MeO)_2(ClCH_2CH_2O)PO$. I.1.b.γ-2.[1146,1189] Liquid, b_4 95-96°,[1189] b_1 93-94°; d_{20} 1.3372; n_D^{20} 1.4382.[1146]
$(MeO)_2(Cl_3CCHClO)PO$. From the dichlorovinyl derivative by Cl_2. Cryst., m. 42-43°; $b_{0.2}$ 99-102°; n_D^{25} 1.4701.[14]
$(MeO)_2(BrCH_2CH_2O)PO$. I.1.e.α-4 - I.1.d.δ (from $(MeO)_3PO$ by successive reaction with Me_3N and $BrCH_2CH_2Br$). Liquid, b_1 94°; n_D^{20} 1.4517.[858]
$(MeO)_2(Br_3CCHBrO)PO$. From the dibromovinyl derivative with Br_2. Undistillable yellowish oil, d_{21} 2.251.[1096]
$(MeO)_2(Cl_2C{=}CHO)PO$. I.2.b.γ-2 (from $(MeO)_3P$ and chloral).[14,20,205,1096,1097] By rearrangement of Cl_3-$CCH(OH)PO(OMe)_2$ in alkali.[14,205] Liquid, b_{14} 120°,[14,205] b_3 86-87°, $b_{2.5}$ 100-104°,[1096,1097] b_2 74-87°;[20] d_{20} 1.423, d_{21} 1.4243,[1096,1097] d_{25} 1.413;[14,205] n_D^{20} 1.4541, n_D^{25} 1.4524.[14,205] Hydrolysis; insecticidal properties.[1306]
$(MeO)(EtO)(BrCH_2CH_2O)PO$. I.1.e.α-4 - I.1.d.δ (from $(MeO)_2(EtO)PO$ by successive reaction with Me_3N and $BrCH_2CH_2Br$).[858] From a salt of $(MeO)(BrCH_2CH_2O)PO_2H$ with $ClCO_2Et$. Liquid, $b_{0.05}$ 81.5°,[858] $b_{0.01}$ 110-

115°;[255] n_D^{20} 1.448.[858]

(MeO)(EtO)(Cl_2C=CHO)PO. I.2.b.γ-2 (from $(MeO)_2$(EtO)P and chloral).[20] From (EtO)(Cl_2C=CHO)PO_2Na with $ClCO_2Me$.[257] Liquid, b_1 80-96°,[20] $b_{0.01}$ 75-77°.[257]

(MeO)($ClCH_2CH_2O$)(Cl_2C=CHO)PO. I.2.b.γ-2 (from Me ethylene cyclic phosphite with chloral).[1301,1304] From (MeO)-(ClC=CHO)PO_2Na with $ClCO_2CH_2CH_2Cl$.[257] Liquid, $b_{0.01}$ 97-99°,[257] 65-66°.[1301]

(MeO)($BrCH_2CH_2O$)(Cl_2C=CHO)PO. From (MeO)(Cl_2C=CHO)PO_2Na with $ClCO_2CH_2CH_2Br$. Liquid, $b_{0.1}$ 110-115°.[257]

$(EtO)_2$(FCH_2CH_2O)PO. I.1.b.γ-2. Liquid, b_{13} 123-124°.[1236]

$(EtO)_2$($ClCH_2CH_2O$)PO. I.1.b.γ-2,[462,1146,1236] I.1.d.γ (from $(EtO)_2POCl$ and ethylene oxide),[1146,1390] I.2.b.-β-2 (from $(EtO)_2PHO$ and ethylene oxide with CCl_4 and $TiCl_4$ as catalyst).[517] Liquid, b_{18} 144-145°,[1236] b_5 115-117°,[1390] $b_{4.5}$ 118-119°,[462] b_2 100°, b_1 96-97°,[1146] $b_{0.3}$ 103-106°;[517] d_{20} 1.2063, n_D^{20} 1.4300,[1146] n_D^{25} 1.4281.[1390]

$(EtO)_2$($ClCH_2CHClO$)PO. From $(EtO)_2$(CH_2=CHO)PO by Cl_2. Liquid b_1 75-92°.[564a]

$(EtO)_2$($Cl_3CCHClO$)PO. From $(EtO)_2$(CCl_2=CHO)PO with Cl_2.[14,1096] Liquid, $b_{1.2}$ 123-125°,[1096] $b_{0.4}$ 116-119°;[14] d_{21} 1.392;[1096] n_D^{25} 1.4620.[14]

$(EtO)_2$[CH_2Cl(resp. Br)CHBr(resp. Cl)O]PO. From $(EtO)_2$-(CH_2=CHO)PO with Cl_2 + Br_2. Liquid, b_1 106-112°.[564a]

$(EtO)_2$($BrCH_2CH_2O$)PO. I.2.b.γ-2 (from $(EtO)_2PHO$ and ethylene oxide with CBr_4). Liquid, b_1 124-125°.[517]

$(EtO)_2$($Br_3CCHBrO$)PO. From $(EtO)_2$(Br_2C=CHO)PO with Br_2. Undistillable yellowish oil, d_{21} 2.105.[1096]

$(EtO)_2$(ClCH=CHO)PO. I.2.b.γ-2 (from $(EtO)_3P$ with Cl_2-CCHO).[20,225,1097] By dehydrochlorination and rearrangement of $Cl_3CCH(OH)PO(OEt)_2$.[225] Liquid, b_8 112°,[225] b_{1-2} 92-105°,[20] $b_{0.5}$ 61°,[225] $b_{0.2}$ 80°;[1097] d_{20} 1.159;[1097] n_D^{25} 1.4342,[225] n_D^{35} 1.4276.[20]

$(EtO)_2$(Cl_2C=CHO)PO. I.2.b.γ-2 (from $(EtO)_3P$ with chloral),[14,20,205,564,787,1055,1096,1097a] (from $(EtO)_2$(CH≡CCHO)P with chloral),[1144] (from $(EtO)_2$(MeCO·CH_2CH_2-O)P with chloral).[205] From $Cl_3CCH(OH)PO(OEt)_2$ by dehydrochlorination-rearrangement (NaOH).[14,205] Liquid, b_{20} 146°,[787] b_{14} 132-135°,[205] 131-133°,[14] b_5 114.5-115°,[20,564] b_3 130°,[1055] 99-100°,[1144] b_2 113-115°,[1096,1097a] $b_{0.5}$ 69-71°,[787] $b_{0.04}$ 73-75°;[1199] d_{21} 1.299,[1096,1097a] d_{25} 1.295,[14,205] d_{35} 1.2821;[20] n_D^{20} 1.4498,[787] 1.4492,[1144] n_D^{25} 1.4479,[14] 1.4475,[205] n_D^{35} 1.4428.[20] Hydrolysis; insecticidal properties.[1306]

$(EtO)_2$(Cl_2C=CClO)PO. I.2.b.γ-2 (from $(EtO)_3P$ with Cl_3-CCOCl).[564]

$(EtO)_2$(Br_2C=CHO)PO. I.2.b.γ-2 (from $(EtO)_3P$ with bromal).[20,564,1096] Liquid, $b_{1.5}$ 139-140°,[1096] $b_{0.15}$ ∿100°;[20] d_{21} 1.661.[1096]

(EtO)($ClCH_2CH_2O$)(Cl_2C=CHO)PO. I.2.b.γ-2 (from ethyl

ethylene cyclic phosphite with chloral). Liquid, b_2 136°.[1304]

(EtO)(BrCH$_2$CH$_2$O)(Cl$_2$C=CHO)PO. From (EtO)(Cl$_2$C=CHO)PO$_2$Na with ClCO$_2$CH$_2$CH$_2$Br. Liquid, $b_{0.001}$ 130-139°.[257]

(FCH$_2$CH$_2$O)$_3$PO. I.1.b.β. Liquid, b_{11} 169°. d_{20}^{20} 1.365. n_D^{20} 1.4043.[809b]

(F$_2$CHCH$_2$O)$_3$PO. From ROH, bromine, and red phosphorus. Oil, b. 253-255°.[1284]

(F$_3$CCH$_2$O)$_3$PO. I.1.b.α (using PCl$_5$, POCl$_3$ is intermediate product). Liquid, b. 186-189°; m. -22°; d_{20} 1.5865; n_D^{20} 1.3198.[398] ^{31}P + 4.0 ppm.[913c]

(ClCH$_2$CH$_2$O)$_3$PO. I.1.b.β,[743,754,1122] I.1.d.γ-1 (using ethylene oxide),[721,938,1146,1487] I.2.b.α-1 (from (RO)$_3$P using O$_2$ and UV irradiation),[13] I.2.b.α-8 (from (RO)$_3$P in presence of ROH using Cl$_2$),[194,572] I.2.b.β-1 (using MeCO$_3$H).[354] From (RO)$_3$P by SOMe$_2$.[22a] Liquid, b_{40} 190° (?),[1122] b_{25} 214°, b_{15} 202°, b_{10} 194°, b_5 180°,[754] b_{2-3} 180-182°,[743] $b_{2.5}$ 173-174°,[1146] b_1 146°;[194,572] d_4^{20} 1.4256,[754] d_{20}^{20} 1.428,[743] d_{20} 1.4236;[1146] n_D^{20} 1.4725,[1146] n_C^{20} 1.4708, n_D^{20} 1.4731, n_F^{20} 1.4786.[754]

(ClCH$_2$CH$_2$O)$_2$(Cl$_3$CCHClO)PO. From (ClCH$_2$CH$_2$O)$_2$(Cl$_2$C=CHO)PO with Cl$_2$.[564a]

(Cl$_3$CCHO)$_3$PO. I.1.b.α,[466] I.1.b.β, I.2.b.α-1 (using Cl$_2$).[600] Cryst., m. 73-74°,[466] 71-72°;[600] $b_{0.2}$ 160°.[600] ^{31}P + 1.3; + 8.9 ppm.[913c]

(ClCH$_2$CH$_2$O)$_2$(CH$_2$=CHO)PO. I.2.b.γ-2 (from (ClCH$_2$CH$_2$O)$_3$P with ClCH$_2$CHO). Liquid, b_1 133-148°.[564a]

(ClCH$_2$CH$_2$O)$_2$(CHCl=CHO)PO. I.2.b.γ-2 (from (ClCH$_2$CH$_2$O)$_3$P with Cl$_2$CHCHO). Liquid, b_1 152-157°.[20]

(ClCH$_2$CH$_2$O)$_2$(Cl$_2$C=CHO)PO. I.2.b.γ-2 (from (ClCH$_2$CH$_2$O)$_3$P with chloral),[20,564] (from chloroethyl ethylene cyclic phosphite with chloral).[1304] Liquid, b_2 169°,[1304] $b_{0.1}$ 93-112°; n_D^{35} 1.4820.[20]

(ClCH$_2$CH$_2$O)(Cl$_3$CCH$_2$O)(Cl$_2$C=CHO)PO. I.2.b.γ-2 (from trichloroethyl ethylene cyclic phosphite with chloral).[1301,1304] Oil, b_2 75°.[1301]

(MeO)$_2$(ClCH$_2$CHClCH$_2$O)PO. I.1.b.γ-2 (from (MeO)$_2$POCl with ClCH$_2$CHClCH$_2$OH), I.1.d.γ (from (MeO)$_2$POCl and epichlorohydrin). Liquid, b_2 123.5-125°.[892]

(MeO)$_2$(BrCH$_2$CH$_2$CH$_2$C)PO. I.1.e.α-4 - I.1.d.δ (from (MeO)$_3$PO reacted successively with Me$_3$N and Br(CH$_2$)$_3$-Br). Liquid, $b_{0.05}$ 96°; n_D^{20} 1.4562.[858]

(MeO)$_2$[(F$_3$C)$_2$CHO]PO. I.2.b.γ-2 (from (MeO)$_2$PHO and (F$_3$C)$_2$CO). Liquid, b. 168°; n_D^{25} 1.3279.[272]

(MeO)$_2$[MeCH(CH$_2$Cl)O]PO. I.1.b.γ-2. Liquid, $b_{2.5}$ 83°; d_{20} 1.2542; n_D^{20} 1.4328.[1146]

(MeO)$_2$[(ClCH$_2$)$_2$CHO]PO. I.1.b.γ-2. Liquid, b_1 126-127°; d_{20} 1.3778; n_D^{20} 1.4560.[1146]

(MeO)$_2$[Cl$_2$C=C(Me)O]PO. I.2.b.γ-2. Hydrolysis and insecticidal properties.[1306]

$(MeO)_2[F_2C{=}C(CFCl_2)O]PO$. I.2.b.γ-2 (from $(MeO)_3P$ and $F_2ClCCOCFCl_2$. Liquid, $b_{1.5}$ 72-73°; n_D^{25} 1.4120.[1289]
$(MeO)(EtO)(BrCH_2CH_2CH_2O)PO$. I.1.e.α-4 - I.1.d.δ (from $(MeO)_2(EtO)PO$ reacted successively with Me_3N and $Br(CH_2)_3Br$). Liquid, $b_{0.05}$ 112°; n_D^{20} 1.4510.[858]
$(MeO)(BrCH_2CH_2O)(PrO)PO$. I.1.e.α-4 - I.1.d.δ (from $(MeO)_2(PrO)PO$ reacted successively with Me_3N and $BrCH_2CH_2Br$). Liquid, $b_{0.05}$ 98°; n_D^{20} 1.4485.[858]
$(MeO)(BrCH_2CH_2O)(i\text{-}PrO)PO$. I.1.e.α-4 - I.1.d.δ (from $(MeO)_2(i\text{-}PrO)PO$ reacted successively with Me_3N and $BrCH_2CH_2Br$). Liquid, $b_{0.05}$ 101°; n_D^{20} 1.4448.[858]
$(MeO)(Cl_2C{=}CHO)(PrO)PO$. From $(MeO)(Cl_2C{=}CHO)PO_2Na$ with $ClCO_2Pr$. Liquid, $b_{0.001}$ 90°.[257]
$(MeO)(Cl_2C{=}CHO)(i\text{-}PrO)PO$. From $(MeO)(Cl_2C{=}CHO)PO_2Na$ with $ClCO_2\text{-}i\text{-}Pr$). Liquid, $b_{0.05}$ 87-92°.[257]
$(EtO)_2(MeCHClCH_2O)PO$. I.2.b.β-2 (from $(EtO)_2PHO$ and propylene epoxide with CCl_4). Liquid, $b_{0.4}$ 92-100°.[517]
$(EtO)_2(ClCH_2CH_2CH_2O)PO$. I.1.d.γ (from $(EtO)_2POCl$ with epichlorohydrin. Liquid, b_2 100²; d_{20} 1.1863; n_D^{20} 1.4360.[1146]
$(EtO)_2(ClCH_2CHClCHClO)PO$. From $(EtO)_2(ClCH_2CH{=}CHO)PO$ with Cl_2. Liquid, $b_{0.6}$ 142-144°.[1097]
$(EtO)_2[MeCH(CH_2Cl)O]PO$. I.1.b.γ-2. Liquid, b_2 100°; d_{20} 1.1863; n_D^{20} 1.4360.[1146]
$(EtO)_2(ClCH_2CCl(Me)O)PO$. From $(EtO)_2(CH_2{=}C(Me)O)PO$ with Cl_2. Liquid, b_{11} 147-148°; d_{20} 1.3414; n_D^{20} 1.4580.[1085]
$(EtO)_2[(ClCH_2)_2CHO]PO$. I.1.b.γ-2 (from EtOH with $ROPOCl_2$),[1146] I.2.b.β-2 (from $(EtO)_2PHO$ and epichlorohydrin with CCl_4).[517] Liquid, $b_{1.3}$ 118-120°,[517] $b_{0.3}$ 103-104°; d_{20} 1.2729; n_D^{20} 1.4500.[1146]
$(EtO)_2[(ClCH_2)_2CClO]PO$. From $(EtO)_2[CH_2{=}C(CH_2Cl)O]PO$ with Cl_2. Liquid, b_9 150-152°; d_{20} 1.3593; n_D^{20} 1.4631.[1085]
$(EtO)_2[ClCH_2CCl(CH_2Br)O]PO$. From $(EtO)_2[CH_2{=}C(CH_2Br)O]PO$ with Cl_2. Liquid, b_9 159°; d_{20} 1.5163; n_D^{20} 1.4733.[1085]
$(EtO)_2[ClBrCHCCl(Me)O]PO$. From $(EtO)_2[CHBr{=}C(Me)O]PO$ with Cl_2. Liquid, b_{10} 154-155°; d_{20} 1.4997; n_D^{20} 1.4735.[1085]
$(EtO)_2(MeCCl{=}CHO)PO$. I.2.b.γ-2 (from $(EtO)_3P$ with $MeCCl_2$-CHO). Liquid, $b_{0.001}$ ∿80°.[20]
$(EtO)_2(ClCH_2CH{=}CHO)PO$. I.2.b.γ-2 (from $(EtO)_3P$ and $ClCH_2$-CHClCHO).[20,1097] Liquid, $b_{0.8}$ 115-118°.[1097]
$(EtO)_2(ClCH_2CCl{=}CHO)PO$. I.2.b.γ-2 (from $(EtO)_3P$ and $ClCH_2CCl_2CHO$). Liquid, b_2 140-142°.[1097]
$(EtO)_2[F_2C{=}C(CF_2Cl)O]PO$. I.2.b.γ-2 (from $(EtO)_3P$ and $(F_2ClC)_2CO$). Liquid, b_4 69-71°; n_D^{25} 1.3880; IR.[490,1456]
$(EtO)_2[F_2C{=}C(CFCl_2)O]PO$. I.2.b.γ-2 (from $(EtO)_3P$ and

$F_2ClCCO \cdot CFCl_2$). Liquid, $b_{2.5}$ 96-98°; n_D^{25} 1.4153.[1289]
$(EtO)_2[ClFC{=}C(CFCl_2)O]PO$. I.2.b.γ-2 (from $(EtO)_3P$ and $Cl_2FCCO \cdot CFCl_2$). Liquid, $b_{0.25}$ 81-82°; n_D^{25} 1.4437; IR.[1456]
$(EtO)_2[ClCH{=}C(Me)O]PO$. I.2.b.γ-2 (from $(EtO)_3P$ and $MeCO \cdot CHCl_2$). Liquid, b_{10} 116.5-117°; d_{20} 1.1833; n_D^{20} 1.4370.[1085]
$(EtO)_2[CH_2{=}C(CH_2Cl)O]PO$. I.2.b.γ-2 (from $(EtO)_3P$ and $(ClCH_2)_2CO$).[20,1085] Liquid, b_{11} 133.5-134.5°,[1085] b_1 76-89°;[20] d_{20} 1.1934; n_D^{20} 1.4435.[1085]
$(EtO)_2[Cl_2C{=}C(Me)O]PO$. I.2.b.γ-2. Hydrolysis, insecticidal properties.[1306]
$(EtO)_2[Cl_2C{=}C(CCl_3)O]PO$. I.2.b.γ-2 (from $(EtO)_3P$ and $(Cl_3C)_2CO$). Liquid, $b_{0.8}$ 131-132°; n_D^{25} 1.4954; IR.[1456]
$(EtO)_2[BrCH{=}C(Me)O]PO$. I.2.b.γ-2 (from $(EtO)_3P$ and $MeCO \cdot CHBr_2$). Liquid, b_{10} 126-127°; d_{20} 1.3643; n_D^{20} 1.4548.[1085]
$(EtO)_2[CH_2{=}C(CH_2Br)O]PO$. I.2.b.γ-2 (from $(EtO)_3P$ and $(BrCH_2)_2CO$). Liquid, b_{11} 142.5-143°; d_{20} 1.3928; n_D^{20} 1.4622.[1085]
$(EtO)(Cl_2C{=}CHO)(PrO)PO$. From $(Cl_2C{=}CHO)(EtO)PO_2Na$ and $ClCO_2Pr$. Liquid, $b_{0.1}$ 100-101°.[257]
$(EtO)(Cl_2C{=}CHO)(i\text{-}PrO)PO$. Similarly to the preceding, using $ClCO_2$-i-Pr. Liquid, b_1 100°.[257]
$(EtOCH_2CH_2O)(Cl_2C{=}CHO)(i\text{-}PrO)PO$. I.1.f.ε (from (vinyl-O)-(i-PrO)PO_2Na and $ClCO_2CH_2CH_2OEt$).[257]
$(ClCH_2CH_2O)_2(ClCH_2CHClCH_2O)PO$. I.1.d.γ-2 (using epichlorohydrin. Liquid, b_2 174-184°.[1487]
$(MeO)(PrO)(BrCH_2CH_2CH_2O)PO$. I.1.e.α-4 - I.1.d.δ (from $(MeO)_2(PrO)PO$ reacted successively with Me_3N and $Br(CH_2)_3Br$). Liquid, $b_{0.05}$ 121°; n_D^{20} 1.425.[858]
$(MeO)(i\text{-}PrO)(BrCH_2CH_2CH_2O)PO$. Similarly, but using $(MeO)_2(i\text{-}PrO)PO$. Liquid, $b_{0.05}$ 107°; n_D^{20} 1.454.[858]
$(ClCH_2CH_2O)(PrO)_2PO$. I.1.b.γ-2. Liquid, b_1 104-105°; d_{20} 1.1423; n_D^{20} 1.4345.[1146]
$(ClCH_2CH_2O)(ClCH_2CHClCH_2O)_2PO$. I.1.d.γ-2 (using ethylene oxide). Liquid, b_2 182-200°.[1487]
$(Cl_3CCHClO)(i\text{-}PrO)_2PO$. From $(Cl_2C{=}CHO)(i\text{-}PrO)_2PO$ with Cl_2. Liquid, $b_{0.6}$ 126-126°; d_{21} 1.323.[1096]
$(BrCCHBrO)(i\text{-}PrO)_2PO$. From $(Br_2C{=}CHO)(i\text{-}PrO)_2PO$ with Br_2. Yellowish undistillable oil; d_{21} 1.794.[1096]
$(ClCH{=}CHO)(PrO)_2PO$. I.2.b.γ-2 (from $(PrO)_3P$ and Cl_2-$CHCHO$). Liquid, $b_{0.1}$ 94-100°; n_D^{25} 1.4364.[20]
$(Cl_2C{=}CHO)(PrO)_2PO$. I.2.b.γ-2 (from $(PrO)_3P$ and chloral).[14,20,205,1096] From $Cl_3CH(OH)PO(OR)_2$ by dehydrochlorination and rearrangement (NaOH).[14,205] Liquid, b_3 109-123°,[20] $b_{1.1}$ 114°,[14,205] $b_{0.7}$ 118-120°, $b_{0.3}$ 107°;[14] d_{21} 1.216,[1096] d_{25} 1.215;[14,205] n_D^{25} 1.4450,[14] 1.4440.[205]
$(ClCH{=}CHO)(i\text{-}PrO)_2PO$. I.2.b.γ-2 (from $(i\text{-}PrO)_3P$ and

$Cl_2CCHO)$.[20,1097] Liquid, $b_{0.3}$ 89-91°,[1097] $b_{0.1}$ 74-87°;[20] d_{20} 1.111,[1097] n_D^{25} 1.4307.[20]

$(Cl_2C{=}CHO)(i\text{-}PrO)_2PO$. I.2.b.γ-2 (from $(i\text{-}PrO)_3P$ and chloral).[14,20,1096] From $Cl_3CCH(OH)PO(OR)_2$ by dehydrochlorination and rearrangement (NaOH).[14,20] Liquid, b_2 96.5°,[20] $b_{0.9}$ 108-111°,[14,205] $b_{0.8}$ 106-108°,[1096] $b_{0.7}$ 99-105°;[14] d_{21} 1.210,[1096] d_{25} 1.201,[14,205] d_4^{35} 1.1924;[20] n_D^{25} 1.4423(2),[14,205] n_D^{35} 1.4372.[20]

$(Br_2C{=}CHO)(i\text{-}PrO)_2PO$. I.2.b.γ-2 (from $(i\text{-}PrO)_3P$ and bromal). Liquid, $b_{2.0}$ 139-142°; d_{21} 1.484.[1096]

$(F_2HCF_2CH_2O)_3PO$. I.1.a.α,[226] I.1.b.α (using PCl_5 yielding intermediate $POCl_3$),[398] I.1.b.β.[226,615] Liquid, b. 250-255°,[398] b_2 121°,[615] $b_{0.04}$ 63-64°;[226] m. -50°; d_{20} 1.6380,[615] 1.6487;[398] n_D^{20} 1.3433,[615] 1.3450.[398]

$(MeCHClCH_2O)_3PO$. I.2.b.α-1 (from $(RO)_3P$ using O_2 and UV irradiation).[73]

$(MeCHClCH_2O)_2(ClCH_2CHClCH_2O)PO$. I.1.d.γ-2 (using epichlorohydrin.) Liquid, $b_{0.004}$ 85-90°.[1487]

$(ClCH_2CHClCH_2O)_3PO$. I.1.d.γ-1 (using epichlorohydrin),[1487] I.2.b.α-1 (using O_2 and UV irradiation).[13] Liquid, $b_{0.003}$ 81-87°.[1487]

$(PrO)_2[MeCH(CH_2Cl)O]PO$. I.1.b.γ-2. Liquid, b_2 110°; d_{20} 1.1125; n_D^{20} 1.4320.[1146]

$(F_2ClCCF_2CCl_2O)_3PO$. From $(F_2HCF_2CH_2O)_3PO$ with Cl_2 and UV irradiation. Cryst., m. 75-76°.[615]

$(MeCHClCH_2O)_3PO$. I.1.d.γ-1 (using epichlorohydrin). Liquid, $b_{0.004}$ 55-60°.[1497]

$(CH_2{=}CClCH_2O)_3PO$. I.1.b.β. Liquid, b_1 131-133°; n_D^{20} 1.4866.[1184]

$(i\text{-}PrO)_2(ClCH_2CH{=}CHO)PO$. I.2.b.γ-2 (from $(i\text{-}PrO)_3P$ and $ClCH_2CHClCHO$). Liquid, $b_{0.4}$ 124°.[1097]

$[MeCH(CH_2Cl)O]_3PO$. I.1.d.γ (from epoxide). Liquid, $b_{2.5}$ 153-154°; d_{20} 1.2885; n_D^{20} 1.4615.[1146]

$[(ClCH_2)_2CHO]_3PO$. I.1.b.β,[754] I.1.d.γ-1 (using epichlorohydrin.)[814,1146] Liquid, b_{10} 246°, b_5 236-237°; d_4^{20} 1.5182;[754] n_D^{20} 1.5029,[814] 1.5022, n_C^{20} 1.4997, n_F^{20} 1.5083.[754]

$[BrCH_2CH(CH_2Cl)O]_3PO$. I.1.d.γ-1 (from $POBr_3$ and epichlorohydrin.[814]

$[(BrCH_2)_2CHO]_3PO$. Similarly, but using epibromohydrin.[814]

$(MeO)_2[Cl_3CC(Me)_2O]PO$. From the corresponding phosphite with $COCl_2$. Oil.[529a]

$(MeO)(BrCH_2CH_2O)(i\text{-}BuO)PO$. I.1.e.α-4 - I.1.d.δ (from $(MeO)_2(i\text{-}BuO)PO$ reacted successively with Me_3N and $Br(CH_2)_2Br$). Liquid, $b_{0.05}$ 107°; n_D^{20} 1.449.[858]

$(MeO)(BrCH_2CH_2O)(sec\text{-}BuO)PO$. I.1.e.α-4 - I.1.d.δ (from $(MeO)_2(sec\text{-}BuO)PO$ reacted successively with Me_3N and $BrCH_2CH_2Br$). Liquid, $b_{0.05}$ 101°; n_D^{20} 1.4472.[858]

$(MeO)(Cl_2C{=}CHO)[Cl_3C(Me)_2O]PO$. I.2.b.γ-2 (from $(MeO)_2$-

(RO)P and chloral). Oil.[529a]

(MeO)($BrCH_2CH_2CH_2O$)(BuO)PO. I.1.e.α-4 - I.1.d.δ (from $(MeO)_2$(BuO)PO by successive reaction with Me_3N and $BrCH_2CH_2CH_2Br$). Liquid, $b_{0.04}$ 126°; n_D^{20} 1.4537.[858]

(MeO)($BrCH_2CH_2CH_2O$)(i-BuO)PO. Similarly to the preceding. Liquid, $b_{0.05}$ 122°; n_D^{20} 1.4507.[858]

(MeO)($BrCH_2CH_2CH_2O$)(sec-BuO)PO. Similarly to the preceding. Liquid, $b_{0.05}$ 123°; n_D^{20} 1.4518.[858]

$(EtO)_2$($BrCMe_2CHBrO$)PO. From $(EtO)_2$(CMe_2=CHO)PO with Br_2.[564a]

$(EtO)_2$(MeCHClCCl=CHO)PO. I.2.b.γ-2 (from $(EtO)_3P$ and $MeCHClCCl_2CHO$). Liquid, $b_{0.01}$ 80°.[20]

($ClCH_2CH_2O$)$(BuO)_2$PO. I.1.b.γ-2. Liquid, $b_{2.5}$ 132-133°; d_{20} 1.0953; n_D^{20} 1.4361.[1146]

(ClCH=CHO)$(BuO)_2$PO. I.2.b.γ-2 (from $(BuO)_3P$ and Cl_2-CHCHO). Liquid, $b_{0.5}$ 96°; n_D^{25} 1.4392.[225]

(Cl_2C=CHO)$(BuO)_2$PO. I.2.b.γ-2 (from $(BuO)_3P$ and chloral).[14,20,205] From $Cl_3CH(OH)PO(OBu)_2$ by dehydrochlorination and rearrangement.[14,205] Liquid, $b_{0.5}$ 107-121°,[20] $b_{0.35}$ 125-128°,[14] $b_{0.2}$ 124-131°,[205] 124-130°;[14] d_{25} 1.166; n_D^{25} 1.4487.[14,205]

(Br_2C=CHO)$(i\text{-}BuO)_2$PO. I.2.b.γ-2 (from $(i\text{-}BuO)_3P$ with chloral). Liquid, $b_{1.5}$ 159-160°; d_{21} 1.465.[1096]

[MeCH(CH_2Cl)O]$(BuO)_2$PO. I.1.b.γ-2. Liquid, b_1 122°; d_{20} 1.0767; n_D^{20} 1.4360.[1146]

(ClCH=C(Me)O)$(BuO)_2$PO. I.2.b.γ-2 (from $(BuO)_3P$ with MeCO•$CHCl_2$). Liquid, b_{11} 154-155°; d_{20} 1.0892; n_D^{20} 1.4400.[1085]

($ClCH_2$CHClCHClO)$(i\text{-}BuO)_2$PO. From $(i\text{-}BuO)_2$($ClCH_2$CH=CHO)PO with Cl_2. Liquid, $b_{0.3}$ 151-152°.[1097]

($ClCH_2$CH=CHO)$(i\text{-}BuO)_2$PO. I.2.b.γ-2 (from $(i\text{-}BuO)_3P$ with $ClCH_2$CHClCHO). Liquid, $b_{0.4}$ 136-137°.[1097]

$(F_3CCF_2CF_2CH_2O)_2$(BuO)PO. I.1.b.β. Liquid, b_1 80-85°.[427]

($F_3CCF_2CF_2CH_2O$)$(BuO)_2$PO. I.1.b.β. Liquid, b_1 92-100°.[427]

$(F_3CCF_2CF_2CH_2O)_3$PO. I.1.b.β,[427] I.1.a.α, I.1.b.α (using PCl_5 yielding intermediary $POCl_3$).[398] Liquid, b. 220-223°,[398] $b_{0.5}$ 65-70°;[427] m. -4°; d_{20} 1.7163; n^{20} 1.3110.[398]

$(ClCMeHCCl_2CH_2O)_3$PO. I.1.b.α. Needles, m. 85.3-85.4° (from EtOH).[1056]

$[F_2CH(CF_2)_3CH_2O]_3$PO. I.1.b.α (using PCl_5 with intermediary formation of $POCl_3$),[398] I.1.a.α,[226] I.1.b.β (using $POCl_3$).[226,615] Odorless[226] liquid, b. 295-302°,[398] b_7 161°,[226] b_5 175°;[615] m. -45°;[398] d_{20} 1.7684,[615] 1.7641,[398] d 1.6 (?);[226] n_D^{20} 1.3368,[615] 1.3340.[398]

$[Cl(CF_2)_4CCl_2O]_3$PO. From the preceding with Cl_2 and UV irradiation. Cryst., m. 94°.[615]

($ClCH_2CH_2O$)(Cl_2C=CHO)($n\text{-}C_6H_{13}$)PO. I.2.b.γ-2 (from hexyl ethylene cyclic phosphite with chloral).[1301,1303,1304] Liquid, b_2 80°.[1301]

$[F_3C(CF_2)_2CH_2O]_2$($n\text{-}C_6H_{13}O$)PO. I.1.b.β (using $POCl_3$, step-

wise). Liquid, $b_{0.5}$ 100°.[427]
$[F_2CH(CF_2)_5CH_2O]_3PO$. I.1.b.α (using PCl_5 with intermediary formation of $POCl_3$),[398] I.1.b.β.[615] Liquid, b_{100} 200-201°,[398] b_1 157-159°; m. -37°;[615] d_{20} 1.8090,[615] d 1.8246;[398] n^{20} 1.3291,[398] 1.3280.[615]
$[F_2ClC(CF_2)_5CCl_2O]_3PO$. From the preceding with Cl_2 and UV irradiation. Cryst., m. 129-130°.[615]
$(ClCH{=}CHO)(2\text{-Et-n-}C_6H_{12}O)_2PO$. I.2.b.γ-2 (from $C_8H_{17}O)_3P$ with Cl_2CHCHO). Liquid, $b_{0.1}$ 156-173°; n_D^{25} 1.4492.[20]
$(Cl_2C{=}CHO)(2\text{-Et-n-}C_6H_{12}O)_2PO$. I.2.b.γ-2 (from $C_8H_{17}O)_3P$ with chloral).[20]
$[(F_3C)_2CHO](2\text{-Et-n-}C_6H_{12}O)_2PO$. I.2.b.γ-2 (from $(C_8H_{17}O)_2$-PHO with $(F_3C)_2CO$). Liquid, n_D^{25} 1.4065.[272]
$[F_2CH(CF_2)_7CH_2O]_3PO$. I.1.b.α (using PCl_5 with intermediary formation of $POCl_3$). Cryst., m. 65-67°; b. 360-366°.[398]
$[F_2CH(CF_2)_9CH_2O]_3PO$. I.1.b.α (using PCl_5 with intermediary formation of $POCl_3$). Cryst., m. 115°.[226]

3. Derivatives with HO Groups (Free or Substituted, OH or OR)

a. With No Other Substituents

$(MeO)_2(HOCH_2CH_2O)PO$. I.1.d.β (using CH_2N_2).[383a,1385] Liquid, b_1 93-95°.[383a]
$(MeO)_2(MeOCH_2CH_2O)PO$. I.1.d.δ. Liquid, b_9 112-113°, d_4^{20} 1.1820, n_D^{20} 1.4140.[175]
$(EtO)_2(CH_2{=}C(Me)C(O)OCH_2CH_2O)PO$. I.1.b.γ-2 (ROH with $(EtO)_2POCl$). Liquid, $b_{0.5}$ 104-105°; n_D^{20} 1.4380.[1147a]
$(PrO)_2(CH_2{=}CHC(O)OCH_2CH_2O)PO$. I.1.b.γ-2 (ROH with $(PrO)_2$-POCl). Liquid, b_1 129-130°; n_D^{20} 1.4410.[1147a]
$(PrO)_2(CH_2{=}C(Me)C(O)OCH_2CH_2O)PO$. I.1.b.γ-2 (ROH with $(PrO)_2POCl$). Liquid, b_3 153-154°; n_D^{20} 1.4407.[1147a]
$(BuO)_2(EtOCH_2CH_2O)PO$. I.1.b.β (in two steps). Liquid, b_{20} 200-250°.[720]
$(BuO)_2(CH_2{=}CHC(O)OCH_2CH_2O)PO$. I.1.b.γ-2 (from ROH with $(BuO)_2POCl$). Liquid, $b_{0.5}$ 136-137°; n_D^{20} 1.4450.[1147a]
$(BuO)_2(CH_2{=}C(Me)C(O)OCH_2CH_2O)PO$. I.1.b.γ-2 (from ROH with $(BuO)_2POCl$). Liquid, b_3 159-160°; n_D^{20} 1.4460.[1147a]
$(BuO)_2(2,4\text{-}Cl_2C_6H_3OCH_2CH_2O)PO$. I.1.b.γ-2 (using $(BuO)_2$-POCl). Undistillable oil.[248]
$(i\text{-}BuO)_2(CH_2{=}C(Me)C(O)OCH_2CH_2O)PO$. I.1.b.γ-2 (from ROH with $(i\text{-}BuO)_2POCl$). Liquid, $b_{3.5}$ 159-160°; n_D^{20} 1.4415.[1147a]
$(HOCH_2CH_2O)_3PO$. I.1.d.γ-2.[376]
$(MeOCH_2CH_2O)_2(BuOCH_2CH_2O)PO$. I.1.b.β (in two steps). Liquid, b_{20} 215-220°.[720]
$(MeOCH_2CH_2O)_2(PhOCH_2CH_2O)PO$. I.1.b.β (in two steps). Liquid, b_{20} 225-235°.[720]
$(EtOCH_2CH_2O)_3PO$. I.1.b.β. Liquid, b_{20} 225°.[720]
$(BuOCH_2CH_2O)_3PO$. I.1.b.β.[248,720] Liquid, b_{10} 255°,[720] b_1

145-150°.[248]

$(n-C_{17}H_{35}CO_2\cdot CH_2CH_2O)_3PO$. I.1.b.γ-2. Structure proved by hydrolysis.[609]

$(PhOCH_2CH_2O)_3PO$. I.1.b.β. Liquid, which crystallizes and m. 142°.[720]

$(MeCH(OH)CH_2O)_3PO$. I.1.d.γ-2.[376]

$(BuOCH_2CH_2CH_2O)_3PO$. I.1.b.β. Liquid, b_{10} 248°.[720]

$(MeO)_2[MeOCH_2CH(OMe)CH_2O]PO$. From Ag glycerophosphate, MeI and Ag_2O. (+)-Form: liquid, $b_{0.7}$ 122°; $[\alpha]_D$ 2.38° (max., EtOH).[767,1153] (-)-Form: liquid, $b_{0.8}$ 125-126°, 126-128°; $[\alpha]_D$ -3.28° (max., EtOH);[767,770] $b_{0.13}$ 87°; $[\alpha]_D$ -4.78° (EtOH).[143]

$(MeO)_2$(2,3-di-O-myristoyl-L-1-glycero)PO. From the corresponding cephalin by CH_2N_2. Cryst., m. 32.3°, $[\alpha]_D$ 2.3° ($CHCl_3$).[152]

$(MeO)_2$(2,3-di-O-palmitoyl-L-1-glycero)PO. As above. Cryst., m. 42.5-43.5°; $[\alpha]_D$ 2.0° ($CHCl_3$).[152]

$(MeO)_2$(2,3-di-O-stearoyl-L-1-glycero)PO. As above.[152] I.1.f.ζ-11 (from distearoylglycerophosphoryl-L-seryl-glycylglycine).[153] Cryst., m. 52-53°; $[\alpha]_D$ 1.9° ($CHCl_3$).[152,153]

$(MeO)_2[(MeOCH_2)_2CHO]PO$. From Ag_2 2-glycerophosphate with MeI and Ag_2O. Liquid, $b_{0.8}$ 126-128°.[770]

$(EtO)_2[EtOCH_2CH(OEt)CH_2O]PO$. D-Form: from Ag_2-glycerophosphate with EtI and Ag_2O. Liquid, $b_{0.22}$ 104-105°; n_D^{19} 1.4252.[144] L-(-)-Form: from the corresponding glycerophosphate, liquid, $b_{0.13}$ 100-100.5°, $b_{0.03}$ 92-93°; $[\alpha]_D^{19}$ -5.76° (EtOH); n_D^{19} 1.4260.[143]

(2,3-Di-O-stearoyl-1-glycero)$_3$PO. I.1.a.α.[634]

$(EtO)_2[MeC(OEt)=C(Me)O]PO$. I.2.b.γ-1 (from $(EtO)_3P$ with diacetyl). Liquid, b_{10} 127-130°; n_D^{20} 1.4250.[832]

$(PrO)_2[MeC(OPr)=C(Me)O]PO$. Similarly from $(PrO)_3P$. Liquid, b_{10} 140-142°; n_D^{20} 1.4230.[832]

$(BuO)_2[MeC(OBu)=C(Me)O]PO$. Similarly from $(BuO)_3P$. Liquid, b_{10} 148-150°; n_D^{20} 1.4240.[832]

b. Halogen or Nitro Derivatives

$(MeO)_2[Cl_2C=C(OMe)O]PO$. I.2.b.γ-2 (from $(MeO)_3P$ with Cl_3CCO_2Me. Liquid, $b_{0.6}$ 115-120°.[596]

$(MeO)_2[Cl_2C=C(OEt)O]PO$. Similarly using Cl_3CCO_2Et. Liquid, b_1 116-117°.[596]

$(MeO)_2[Cl_2C=C(OBu)O]PO$. Similarly using Cl_3CCO_2Bu. Liquid, $b_{0.07}$ 140-143°.[596]

$(MeO)_2[Cl_2C=C(O\text{-}i\text{-}Bu)O]PO$. Similarly using Cl_3CCO_2-i-Bu. Liquid, $b_{0.4-0.5}$ 120-123°.[596]

$(MeO)_2[Cl_2C=C(O\text{-}i\text{-}Am)O]PO$. Similarly using Cl_3CCO_2-i-Am. Liquid, $b_{0.2}$ 116-117°.[596]

$(MeO)_2[Cl_2C=C(O\text{-}n\text{-}C_6H_{13})O]PO$. Similarly using Cl_3CCO_2-n-C_6H_{13}. Liquid, $b_{0.2}$ 140-141°.[596]

$(MeO)_2[Cl_2C=C(OCH_2CH_2OMe)O]PO$. Similarly using Cl_3CCO_2-CH_2CH_2OMe). Liquid, $b_{0.3}$ 136-138°.[596]

$(MeO)_2[Cl_2C{=}C(OCH_2CH_2OEt)O]PO$. Similarly using Cl_3CCO_2-CH_2CH_2OEt). Liquid, $b_{0.15}$ 131-133°.[596]
$(MeO)_2[Cl_2C{=}C(OCH_2CH_2OCH_2CH_2OEt)O]PO$. Similarly using $Cl_3CCO_2CH_2CH_2OCH_2CH_2OEt$). Liquid, $b_{0.2}$ 159-160°.[596]
$(MeO)(EtOCH_2CH_2O)(Cl_2C{=}CHO)PO$. I.1.f. (from vinyl-O)-$(MeO)PO_2Na$ with $ClCO_2CH_2CH_2OEt$). Liquid, $b_{0.005}$ 115-120°.[257]
$(ClCH_2CH_2O)_2(EtOCH_2CH_2O)PO$. I.1.b.β (in two steps). Liquid, n_D^{20} 1.453.[720]
$(EtO_2)[CCl_3CH(OMe)O]PO$. I.1.b.γ-2 (from $CCl_3CH(OMe)OH$ with $(RO)_2POCl$). $b_{0.03}$ 79.5-81°.[1048]
$(EtO)_2[CCl_3CH(OEt)O]PO$. I.1.b.γ-2 (from $CCl_3CH(OEt)OH$ with $(RO)_2POCl$). $b_{0.03}$ 81-83°.[1048]
$(EtO)_2[CCl_3CH(OPr)O]PO$. I.1.b.γ-2 (from $CCl_3CH(OPr)OH$ with $(RO)_2POCl$). $b_{0.03}$ 89-91°.[1048]
$(EtO)_2[CCl_3CH(O\text{-}i\text{-}Pr)O]PO$. I.1.b.γ-2 (from $CCl_3CH(O\text{-}i\text{-}Pr)OH$ with $(RO)_2POCl$). $b_{0.03}$ 88-89°.[1048]
$(BrCH_2CH_2O)(MeOCH_2CH_2O)(BuOCH_2CH_2O)PO$. I.1.b.γ-2. Liquid, b_7 205-210°.[720]
$(EtO)(EtOCH_2CH_2O)(Cl_2C{=}CHO)PO$. I.1.f.ε (from (MeO)-(vinyl-O)PO_2Na with $ClCO_2CH_2CH_2OEt$). Liquid, $b_{0.001}$ 120°.[257]
$(EtO)_2[Cl_2C{=}C(OEt)O]PO$. I.2.b.γ-2 (from $(EtO)_3P$ and Cl_3CCO_2Et).[20,564,787] Liquid, b_1 120-127°,[20,564] $b_{0.5}$ 103°;[787] n_D^{20} 1.4284,[787] n_D^{25} 1.4173.[20]
$(MeOCH_2CH_2OCH_2CH_2O)_2(Cl_2C{=}CHO)PO$. I.2.b.γ-2 (from $(RO)_3P$ and chloral).[20]
$(EtO)_2(ClCH_2CH(OH)CH_2O)PO$. From $(EtO)_2$(glycidyl-O)PO and HCl. Liquid, $b_{0.003}$ 114-116°; n_D^{20} 1.4478.[1200]
$[Cl_3CCH(OMe)O](PrO)_2PO$. I.1.b.γ-2 (from $(PrO)_2POCl$ with $Cl_3CCH(OMe)OH$). Liquid, $b_{0.025}$ 86-88°.[1048]
$[Cl_3CCH(OEt)O](PrO)_2PO$. Similarly using $Cl_3CCH(OEt)OH$. Liquid, $b_{0.03}$ 96-98°.[1048]
$[Cl_3CCH(OPr)O](PrO)_2PO$. Similarly using $Cl_3CCH(OPr)OH$. Liquid, $b_{0.025}$ 95-97°.[1048]
$[Cl_3CCH(OBu)O](PrO)_2PO$. Similarly using $Cl_3CCH(OBu)OH$. Liquid, $b_{0.025}$ 106-108°.[1048]
$(EtO)_2[(ClCH_2)_2CHO]PO$. I.1.b.γ-2. Liquid, $b_{0.3}$ 103-104°; d_{20} 1.2759; n_D^{20} 1.4527.[1146]
$[Cl_3CCH(OMe)O](BuO)_2PO$. I.1.b.γ-2 (from $(BuO)_2POCl$ with $Cl_3CCH(OMe)OH$). Liquid, $b_{0.03}$ 108-109°.[1048]
$[Cl_3CCH(OEt)O](BuO)_2PO$. Similarly using $Cl_3CCH(OEt)OH$. Liquid, $b_{0.03}$ 113-115°.[1048]
$[Cl_3CCH(OPr)O](BuO)_2PO$. Similarly using $Cl_3CCH(OPr)OH$. Liquid, $b_{0.03}$ 118-120°.[1048]
$[Cl_3CCH(OBu)O](BuO)_2PO$. Similarly using $Cl_3CCH(OBu)OH$. Liquid, $b_{0.03}$ 119-121°.[1048]
$(ClCH_2CH_2O)(MeOCH_2CH_2O)(BuO)PO$. I.1.b.γ. Liquid, b_{13} 195-205°.[720]
$[Me_2C(NO_2)CH_2O]_2PO$. I.1.b.α - b.β,[1397] I.1.b.β.[1251] Cryst., m. 155°,[1397] 151.5-152.5°.[1251]

$[MeEtC(NO_2)CH_2O]_3PO$. I.1.b.α - b.β. Undistillable solid, m. 23-25°.[1397]

c. With X-Thio Groups

$(MeO)_2(MeSCH_2CH_2O)PO$. I.1.b.γ-2 (using $(EtO)_2POCl$). Liquid, $b_{0.02}$ 73-74°.[528]

$(MeO)_2[CH_2=C(SEt)O]PO$. I.2.b.γ-2 (from $(MeO)_3P$ using $ClCH_2CO\cdot SEt$). Liquid, b_3 100°; n_D^{20} 1.4620.[621]

$(MeO)_2[ClCH=C(SEt)O]PO$. Similarly using $Cl_2CHCO\cdot SEt$. Liquid, b_2 103°; n_D^{20} 1.4885.[621]

$(MeO)_2[Cl_2C=C(SEt)O]PO$. Similarly using $Cl_3CCO\cdot SEt$. Liquid, b_1 118°; n_D^{20} 1.5015.[621]

$(MeO)_2[CH_2=C(SPh)O]PO$. I.2.b.γ-2 (from $(MeO)_3P$ using $ClCH_2CO\cdot SPh$). Oil, $b_{0.001}$ 105-110°; n_D^{25} 1.5348.[1437]

$(MeO)_2[CH_2=C(SC_6H_2Cl_3\text{-}2,4,6)O]PO$. Similarly using $ClCH_2$-$CO\cdot SC_6H_2Cl_3$. Oil; n_D^{25} 1.5610.[1437]

$(MeO)_2[CH_2=C(SC_6H_4NO_2\text{-}4)O]PO$. Similarly using $ClCH_2CO\cdot SC_6H_4NO_2$. Cryst., m. 70-71°.[1437]

$(MeO)_2[CH_2=C(S\text{-}2\text{-}C_{10}H_7)O]PO$. Similarly using $ClCH_2CO\cdot S$-2-$C_{10}H_7$. Oil; n_D^{25} 1.600.[1437]

$(MeO)_2[PhSCH=C(Me)O]PO$. I.2.b.γ-2 (from $MeO)_3P$ using PhSCHClCOMe). Liquid, b_{10} 81-86°.[83]

(MeO)(EtO)[EtSCH=C(Me)O]PO. I.1.f.ε (from (MeO)(vinyl-O)PO_2Na and $ClCO_2Et$). Liquid, $b_{0.03}$ 100-110°.[257]

$(EtO)_2(MeSCH_2CH_2O)PO$. I.1.b.γ-2 (from $MeSCH_2CH_2OH$ using $(EtO)_2POCl$). Liquid, $b_{0.03}$ 74-76°.[528]

$(EtO)_2(EtSCH_2CH_2O)PO$. Similarly from $EtSCH_2CH_2OH$. Liquid, $b_{0.02}$ 74-76°.[528]

$(EtO)_2(4\text{-}MeC_6H_4SCH_2CH_2O)PO$. Similarly from $MeC_6H_4OCH_2CH_2$-OH. Liquid, $b_{0.04}$ 116-118°.[528]

$(EtO)_2[CH_2=C(SEt)O]PO$. I.2.b.γ-2 (from $(EtO)_3P$ using $ClCH_2CO\cdot SEt$). Liquid, $b_{0.05}$ 80°; n_D^{20} 1.4600.[620]

$(EtO)_2[ClCH=C(SEt)O]PO$. Similarly using $Cl_2CHCO\cdot SEt$. Liquid, b_2 107°; n_D^{25} 1.4790.[621]

$(EtO)_2[Cl_2C=C(SEt)O]PO$. Similarly using $Cl_3CCO\cdot SEt$. Liquid, b_3 128°; n_D^{20} 1.4910.[621]

$(EtO)_2[CH_2=C(SC_6H_4NO_2\text{-}4)O]PO$. Similarly using $ClCH_2CO$. $SC_6H_4NO_2$. Oil; n_D^{25} 1.544(?).[1437]

$(EtO)_2[PhSCH=C(Me)O]PO$. I.2.b.γ-2 (from $(EtO)_3P$ using PhSCHClCOMe). Liquid, $b_{1.5}$ 65-66°.[83]

$(PrO)_2[CH_2=C(SEt)O]PO$. I.2.b.γ-2 (from $(PrO)_3P$ using $ClCH_2CO\cdot SEt$). Liquid, $b_{0.05}$ 83°; n_D^{20} 1.4652.[621]

$(PrO)_2[ClCH=C(SEt)O]PO$. Similarly using $Cl_2CHCO\cdot SEt$. Liquid, b_2 128°; n_D^{20} 1.4745.[621]

$(PrO)_2[Cl_2C=CH(SEt)O]PO$. Similarly using $Cl_3CCO\cdot SEt$. Liquid, $b_{1.5}$ 141°; n_D^{20} 1.4860.[621]

$(i\text{-}PrO)_2[EtSCH=C(Et)O]PO$. I.2.b.γ-2 (from $(i\text{-}PrO)_3P$ using $EtSCHClCO\cdot Et$).[254]

$(BuO)_2[CH_2=C(SEt)O]PO$. I.2.b.γ-2 (from $(BuO)_3P$ using $ClCH_2CO\cdot SEt$). Liquid, $b_{1.5}$ 143°; n_D^{20} 1.4560.[621]

$(BuO)_2[ClCH=C(SEt)O]PO$. Similarly using $Cl_2CHCO\cdot SEt$.

Liquid, $b_{1.5}$ 140°; n_D^{20} 1.4720.[621]
$(BuO)_2[Cl_2C=C(SEt)O]PO$. Similarly using $Cl_3CCO{\cdot}SEt$. Liquid, b_1 150°; n_D^{20} 1.4840.[621]

4. Derivatives with Carbonyl and/or Carboxyl (Actual or Potential) Groups (without or with the Preceding Functions)

$(MeO)_2(OHCCH_2O)PO$. I.2.b.γ-1 (from $(MeO)_3P$ and glyoxal via pentaoxyphosphorane). Liquid, $b_{0.05}$ 77-78°; spectral characteristics; ^{31}P -1.2 ppm.[1167]
$(MeO)_2[OHCH(COCl)O]PO$. From Me_3-ethylene pentaoxyphosphorane by $COCl_2$. Liquid, $b_{0.2}$ 82-83°; ^{31}P + 3.2 ppm.[1167]
$(MeO)_2[(EtO_2C)_2CHO]PO$. I.2.b.γ-1 (from $(MeO)_2PHO$ and $CO(CO_2Et)_2$). Liquid, $b_{1.5}$ 137-138°; n_D^{20} 1.4280.[1145a]
$(MeO)_2[OHCCH(COMe)O]PO$. From Me_3-ethylene pentaoxyphosphorane by MeCOCl. Liquid, $b_{0.2}$ 120-122°; ^{31}P + 2.6 ppm.[1167]
$(MeO)_2[OHCCH(CO{\cdot}CF_3)O]PO$. From Me_3-ethylene pentaoxyphosphorane by F_3CCOCl. Liquid, $b_{0.2}$ 83-84°; spectral characteristics; ^{31}P + 3.2 ppm.[1167]
$(MeO)_2[MeO_2CC(Me)_2O]PO$. I.2.b.γ-1 (from $(MeO)_3P$ with $MeCOCO_2Me$). Liquid, b_8 127.5-129°; n_D^{20} 1.4232.[1148]
$(MeO)_2[MeC(CO_2Et)_2O]PO$. I.2.b.γ-1 (from $(MeO)_3P$ with $CO(CO_2Et)_2$). Liquid, b_3 142-143°; d_{20} 1.2112; n_D^{20} 1.4350.[1147b]
$(MeO)_2[HO_2CCH=C(Me)O]PO$ cis. By partial hydrolysis of its Me ester (phosdrin).[1276]
$(MeO)_2[MeO_2CCH=C(Me)O]PO$. I.2.b.γ-2. cis-Isomer: liquid, b_2 114°; n_D^{20} 1.4460; IR. trans-Isomer: liquid, b_1 112°; n_D^{20} 1.4504; IR.[1276] Hydrolysis; insecticidal properties.[1306]
$(MeO)_2[EtO_2CCH=C(Me)O]PO$. I.2.b.γ-2. Hydrolysis, insecticidal properties.[1306]
$(MeO)_2[i\text{-}PrO_2CCH=C(Me)O]PO$. I.2.b.γ-2 (from $(MeO)_3P$ using $MeCO{\cdot}Cl(Br)CHCO_2$-i-Pr). Liquid, $b_{0.2}$ 145-152°.[258]
$(MeO)_2[i\text{-}BuO_2CCH=C(Me)O]PO$. Similarly using $MeCO{\cdot}Cl(Br)$-$CHCO_2$-i-Bu. Liquid, b_1 160-165°.[258]
$(MeO)_2[(Me(Et)CHO_2CH=C(Me)O]PO$. Similarly using $MeCO{\cdot}Cl(Br)CHCO_2CH(Me)(Et)$. Liquid, $b_{0.005}$ 138-140°.[258]
$(MeO)_2[AmO_2CCH=C(Me)O]PO$. Similarly using $MeCO{\cdot}Cl(Br)CH$-CO_2Am. Liquid, $b_{0.01}$ 160-165°.[258]
$(MeO)_2[MeOCH_2CH_2O_2CCH=C(Me)O]PO$. Similarly using $MeCO{\cdot}ClCHCO_2R$. Liquid, $b_{0.001}$ 100°; d 1.22; n_D^{25} 1.4537.[213]
$(MeO)_2[Me(OCH_2CH_2O)_9O_2CCH=C(Me)O]PO$. Similarly using $MeCO{\cdot}ClCHCO_2R$.[234]
$(MeO)_2[i\text{-}PrOCH_2CH_2O_2CH=C(Me)O]PO$. Similarly using $MeCO{\cdot}ClCHCO_2R$. Liquid, $b_{0.2}$ 131°.[234]
$(MeO)_2[i\text{-}BuOCH_2CH_2O_2CCH=C(Me)O]PO$. Similarly using the corresponding α-Cl-acetoacetate. Liquid, $b_{0.02}$ 127-

131°.[234]

(MeO)$_2$[PhOCH$_2$CH$_2$O$_2$CCH=C(Me)O]PO. Similarly using MeCO·ClCHCO$_2$CH$_2$CH$_2$OPh.[234]

(MeO)$_2$[MeCO$_2$·CH$_2$CH$_2$O$_2$CCH=C(Me)O]PO. Similarly using the corresponding α-Cl-acetoacetate. Liquid, $b_{0.08}$ 155-157°; d 1.23; n_D^{25} 1.4580.[213]

(MeO)$_2$[PhCO$_2$·CH$_2$CH$_2$O$_2$CCH=C(Me)O]PO. Similarly using the corresponding α-Cl-acetoacetate. Liquid, $b_{0.001}$ 160-165°; n_D^{25} 1.5079.[213]

(MeO)$_2$[EtSCH$_2$CH$_2$O$_2$CCH=C(Me)O]PO. Similarly using the corresponding α-Cl-acetoacetate. Liquid, $b_{0.1}$ 110°.[234]

(MeO)$_2$[(OCH$_2$O)CH CH$_2$CH$_2$O$_2$CCH=C(Me)O]PO. Similarly using the corresponding α-Cl-acetoacetate. Liquid, $b_{0.05}$ 152-154°.[404]

(MeO)$_2$[(2,3-isopropylidene-1-glycero)-O$_2$CCH=C(Me)O]PO. Similarly using the corresponding α-Cl-acetoacetate. Liquid, $b_{0.3}$ 150-157°.[404]

(MeO)$_2$[(MeOCH$_2$)$_2$CHO$_2$CCH=C(Me)O]PO. Similarly using the corresponding α-Cl-acetoacetate. Liquid, $b_{0.01}$ 127-134°.[234]

(MeO)$_2$[2-tetrahydrofurfuryl-O$_2$CCH=C(Me)O]PO. Similarly using the corresponding α-Cl-acetoacetate. Liquid, $b_{0.04}$ 156°.[404]

(MeO)$_2$[H$_2$NCO·CH=C(Me)O]PO. From (MeO)$_2$[ClOCCH=C(Me)O]PO with NH$_3$. Cryst., m. 102-103°.[1436]

(MeO)$_2$[MeHNCO·CH=C(Me)O]PO. I.2.b.γ-2 (from (MeO)$_3$P using MeHNCO·ClCHCO·Me). Liquid, $b_{0.003}$ 120-125°.[1247]

(MeO)$_2$[MeOCH$_2$CH$_2$NHCO·CH=C(Me)O]PO. Similarly using the corresponding α-Cl-acetoacetamide.[403]

(MeO)$_2$[MeOCH$_2$CH$_2$N(Me)CO·CH=C(Me)O]PO. Similarly using the corresponding α-Cl-acetoacetamide.[403]

(MeO)$_2$[(MeOCH$_2$CH$_2$)$_2$NCO·CH=C(Me)O]PO. Similarly using the corresponding α-Cl-acetoacetamide.[403]

(MeO)$_2$[PhOCH$_2$CH$_2$NHCO·CH=C(Me)O]PO. Similarly using the corresponding α-Cl-acetoacetamide.[403]

(MeO)$_2$[MeO(CH$_2$)$_3$NHCO·CH=C(Me)O]PO. Similarly using the corresponding α-Cl-acetoacetamide.[403]

(MeO)$_2$[MeON(Me)CO·CH=C(Me)O]PO. Similarly using the corresponding α-Cl-acetoacet-hydroxamide. Liquid, $b_{0.02}$ 115-120°.[234a]

(MeO)$_2$[EtON(Et)CO·CH=C(Me)O]PO. Similarly using the corresponding α-Cl-acetoacet-hydroxamide. Liquid, $b_{0.01}$ 115°.[234a]

(MeO)$_2$[MeON(i-Pr)CO·CH=C(Me)O]PO. Similarly using the corresponding α-Cl-acetoacet-hydroxamide.[234a]

(MeO)$_2$[NCCH(Me)O$_2$CCH=C(Me)O]PO. Similarly using the corresponding α-Cl-acetoacetate. Liquid, $b_{0.04}$ 138-140°.[258]

(MeO)$_2$[NCC(Me)(Et)O$_2$CCH=C(Me)O]PO. Similarly using the

corresponding α-Cl-acetoacetate. Liquid, $b_{0.06}$ 149-152°.[258]
$(MeO)_2[PrO_2CCCl{=}C(Me)O]PO$. Similarly using the corresponding α,α-Cl_2-acetoacetate. Liquid, $b_{0.01}$ 160-162°.[258]
$(MeO)_2[i\text{-}PrO_2CCCl{=}C(Me)O]PO$. Similarly using the corresponding α,α-Cl_2-acetoacetate. Liquid, $b_{0.03}$ 158-160°.[258]
$(MeO)_2[AmO_2CCCl{=}C(Me)O]PO$. Similarly using the corresponding α,α-Cl_2-acetoacetate. Liquid, $b_{0.15}$ 155-158°.[258]
$(MeO)_2[EtOCH_2CH_2O_2CCCl{=}C(Me)O]PO$. Similarly using the corresponding α,α-Cl_2-acetoacetate. Liquid, $b_{0.2}$ 130-140°.[234]
$(MeO)_2[i\text{-}PrOCH_2CH_2O_2CCCl{=}C(Me)O]PO$. Similarly using the corresponding α,α-Cl_2-acetoacetate. Liquid, $b_{0.07}$ 137°.[234]
$(MeO)_2[PhOCH_2CH_2O_2CCCl{=}C(Me)O]PO$. Similarly using the corresponding α,α-Cl_2-acetoacetate.[234]
$(MeO)_2[Me(OCH_2CH_2)_9O_2CCCl{=}C(Me)O]PO$. Similarly using the corresponding α,α-Cl_2-acetoacetate.[234]
$(MeO)_2[EtSCH_2CH_2O_2CCCl{=}C(Me)O]PO$. Similarly using the corresponding α,α-Cl_2-acetoacetate.[234]
$(MeO)_2$[(2,3-di-O-methylene-1-glycero)$O_2CCCl{=}C(Me)O$]PO. Similarly using the corresponding α,α-Cl_2-acetoacetate. Liquid, $b_{0.1-0.25}$ 161-171°.[404]
$(MeO)_2$[(2,3-isopropylidene-1-glycero)-$O_2CCCl{=}C(Me)O$]PO. Similarly using the corresponding α,α-Cl_2-acetoacetate. Liquid, $b_{0.25}$ 164-170°.[234]
$(MeO)_2[(MeOCH_2)_2CHO_2CCCl{=}C(Me)O]PO$. Similarly using the corresponding α,α-Cl_2-acetoacetate. Liquid, $b_{0.15}$ 158-160°.[234]
$(MeO)_2[Et_2NCO\cdot{}^{14}CCl{=}{}^{14}C(Me)O]PO$. Similarly using $Me^{14}CO\cdot{}^{14}CCl_2CONEt_2$. Liquid, $b_{0.0001}$ 120-122°; n_D^{25} 1.4718.[25a]
$(MeO)_2[MeOCH_2CH_2NHCO\cdot CCl{=}C(Me)O]PO$. Similarly using the corresponding α,α-Cl_2-acetoacetamide.[403]
$(MeO)_2[MeOCH_2CH_2N(Me)CO\cdot CCl{=}C(Me)O]PO$. Similarly using the corresponding α,α-Cl_2-acetoacetamide.[403]
$(MeO)_2[(MeOCH_2CH_2)_2NCO\cdot CCl{=}C(Me)O]PO$. Similarly using the corresponding α,α-Cl -acetacetamide.[403]
$(MeO)_2[PhOCH_2CH_2NHCO\cdot CCl{=}C(Me)O]PO$. Similarly using the corresponding α,α-Cl_2-acetoacetamide.[403]
$(MeO)_2[MeO(CH_2)_2O(CH_2)_2N(Me)CO\cdot CCl{=}C(Me)O]PO$. Similarly using the corresponding α,α-Cl_2-acetoacetamide.[403]
$(MeO)_2[MeO(CH_2)_3NHCO\cdot CCl{=}C(Me)O]PO$. Similarly using the corresponding α,α-Cl_2-acetoacetamide.[403]
$(MeO)_2[MeON(Me)CO\cdot CCl{=}C(Me)O]PO$. Similarly using the corresponding α,α-Cl_2-acetoacet-hydroxamide. Liquid, $b_{0.02}$ 130-136°.[234a]
$(MeO)_2[MeON(i\text{-}Pr)CO\cdot CCl{=}C(Me)O]PO$. Similarly using the

corresponding α,α-Cl_2-acetoacet-hydroxamide.[234a]

$(MeO)_2$[EtON(Et)CO·CCl=C(Me)O]PO. Similarly using the corresponding α,α-Cl_2-acetoacet-hydroxamide.[234a]

$(MeO)_2$[(Me)(Et)CHO_2CCBr=C(Me)O]PO. Similarly using the corresponding α,α-Br_2-acetoacetate. Liquid, $b_{0.05}$ 151-164° (dec.).[258]

$(MeO)_2$[MeCO·C(Me)(COMe)O]PO. I.2.b.γ-1 (from $(MeO)_3P$ by successive reaction with diacetyl, ketene, and HCl).[1161]

$(MeO)_2$[MeCO·C(Me)($COCH_2Br$)O]PO. Similarly but using Br_2 instead of HCl in the last phase.[1161]

(MeO)(EtO)[EtO_2CCH=C(Me)O]PO. I.1.f.ε (from (MeO)(RO)PO_2H as a salt and $ClCO_2Et$). Liquid, $b_{0.02}$ 115-117°.[255]

(MeO)(Cl_2C=CHO)(EtO_2CCH_2O)PO. I.1.f.ε (from (MeO)(vinyl-O)PO_2Na and $ClCO_2CH_2CO_2Et$. Liquid, $b_{0.001}$ 115-117°.[257]

$(EtO)_2$(CH_2=CHC(O)OCH_2CH_2O)PO. I.1.b.γ-2 (from ROH with $(EtO)_2POCl$). Liquid, $b_{0.5}$ 108-110°; n_D^{20} 1.4380.[1147a]

(PhNHCO·$CH_2O)_3$PO. I.1.b.α, I.1.b.β. Needles, m. 196° (from EtOH).[244]

(2-MeC_6H_4NHCO·$CH_2O)_3$PO. I.1.b.α. Prisms, m. 143° (from EtOH).[244]

(4-MeC_6H_4NHCO·$CH_2O)_3$PO. I.1.b.α. Needles, m. 188° (from EtOH).[244]

(2-$C_{10}H_7$NHCO·$CH_2O)_3$PO. I.1.b.α. Needles, m. 192-196° (from EtOH).[244]

$(EtO)_2$[$(EtO_2C)_2$CHO]PO. I.2.b.γ-1 (from $(EtO)_2$PHO and $CO(CO_2Et)_2$). Liquid, b_9 181-182°, n_D^{20} 1.4275.[1148a]

$(EtO)_2$[MeCO·CH(Me)O]PO. I.1.f.ζ-10 (from ethyl 1,2-Me_2-ethylene cyclic phosphate with EtOH). Liquid, b_{10} 131-133°.[624]

$(EtO)_2$($NCCMe_2O$)PO. I.1.b.γ-2 (from ROH with $(EtO)_2POCl$). Liquid, $b_{0.015}$ 67°; n_D^{20} 1.4182.[81]

$(EtO)_2$[MeO_2CCH=C(Me)O]PO. I.2.b.γ-2. Hydrolysis; insecticidal properties.[1306]

$(EtO)_2$[EtO_2CCH=C(Me)O]PO. I.1.b.β,[241] I.2.b.γ-2 (from $(EtO)_3P$ and MeCO·CClHCO_2Et).[213] Liquid, b_2 138°,[241] $b_{0.001}$ 110°; d 1.15; n_D^{25} 1.4481.[213] Hydrolysis; insecticidal properties.[1306]

$(EtO)_2$[i-PrO_2CCH=C(Me)O]PO. I.2.b.γ-2 (from $(EtO)_3P$ using MeCO·CCl(Br)HCO_2-i-Pr). Liquid, $b_{0.01}$ 140°.[258]

$(EtO)_2$[(Me)(Et)CHO_2CCH=C(Me)O]PO. Similarly using the corresponding α-Cl(Br)-acetoacetate. Liquid, $b_{0.01}$ 142-145°.[258]

$(EtO)_2$[AmO_2CCH=C(Me)O]PO. Similarly using the corresponding α-Cl(Br)-acetoacetate. Liquid, $b_{0.005}$ 136-138°.[258]

$(EtO)_2$[(Me)(i-Pr)CHO_2CCH=C(Me)O]PO. Similarly using the corresponding α-Cl(Br)-acetoacetate. Liquid, $b_{0.0-0.7}$ 158-161°.[258]

$(EtO)_2$[NCC(Me)HO_2CCH=C(Me)O]PO. Similarly using the corresponding α-Cl-acetoacetate. Liquid, $b_{0.01}$ 134-136°.[258]

$(EtO)_2[NCC(Et)HO_2CCH=C(Me)O]PO$. Similarly using the corresponding α-Cl-acetoacetate. Liquid, $b_{0.02}$ 155°.[258]

$(EtO)_2[MeOCH_2CH_2O_2CCH=C(Me)O]PO$. Similarly using the corresponding α-Cl-acetoacetate. Liquid, $b_{0.001}$ 105°; d 1.14; n_D^{25} 1.4497.[213]

$(EtO)_2[i\text{-}PrOCH_2CH_2O_2CCH=C(Me)O]PO$. Similarly using the corresponding α-Cl-acetoacetate. Liquid, $b_{0.1}$ 130°.[234]

$(EtO)_2[Me(OCH_2CH_2)_9O_2CCH=C(Me)O]PO$. Similarly using the corresponding α-Cl-acetoacetate.[234]

$(EtO)_2[PhOCH_2CH_2O_2CCH=C(Me)O]PO$. Similarly using the corresponding α-Cl-acetoacetate.[234]

$(EtO)_2[MeCO_2 \cdot CH_2CH_2O_2CCH=C(Me)O]PO$. Similarly using the corresponding α-Cl-acetoacetate. Liquid, $b_{0.001}$ 150°; d 1.2; n_D^{25} 1.4533.[213]

$(EtO)_2[EtSCH_2CH_2O_2CCH=C(Me)O]PO$. Similarly using the corresponding α-Cl-acetoacetate.[234]

$(EtO)_2$[(2,3-di-O-methylene-1-glycero)-$O_2CCH=C(Me)O$]PO. Similarly using the corresponding α-Cl-acetoacetate. Liquid, $b_{0.03}$ 161-164°.[404]

$(EtO)_2$[(2,3-isopropylidene-1-glycero)-$O_2 \cdot CCH=C(Me)O$]PO. Similarly using the corresponding α-Cl-acetoacetate. Liquid, $b_{0.03}$ 151-157°.[404]

$(EtO)_2$[(2,3-cyclocarbonato-1-glycero)-$O_2CCH=C(Me)O$]PO. Similarly using the corresponding α-Cl-acetoacetate.[404]

$(EtO)_2[(MeOCH_2)_2CHO_2CCH=C(Me)O]PO$. Similarly using the corresponding α-Cl-acetoacetate. Liquid, $b_{0.01}$ 134-138°.[234]

$(EtO)_2$[2-tetrahydrofurfuryl-$O_2CCH=C(Me)O$]PO. Similarly using the corresponding α-Cl-acetoacetate. Liquid, $b_{0.06}$ 134-140°.[404]

$(EtO)_2[MeOCH_2CH_2NHCO \cdot CH=C(Me)O]PO$. Similarly using the corresponding α-Cl-acetoacetamide.[403]

$(EtO)_2[MeOCH_2CH_2N(Me)CO \cdot CH=C(Me)O]PO$. Similarly using the corresponding α-Cl-acetoacetamide.[403]

$(EtO)_2[MeSCH_2CH_2NHCO \cdot CH=C(Me)O]PO$. Similarly using the corresponding α-Cl-acetoacetamide.[403]

$(EtO)_2[MeO(CH_2)_3NHCO \cdot CH=C(Me)O]PO$. Similarly using the corresponding α-Cl-acetoacetamide.[403]

$(EtO)_2[MeON(Me)CO \cdot CH=C(Me)O]PO$. Similarly using the corresponding α-Cl-acetoacet-hydroxamide. Liquid, $b_{0.05}$ 120-126°.[234a]

$(EtO)_2[EtON(Et)CO \cdot CH=C(Me)O]PO$. Similarly using the corresponding α-Cl-acetoacet-hydroxamide. Liquid, $b_{0.01}$ 140°.[234a]

$(EtO)_2[MeON(i\text{-}Pr)CO \cdot CH=C(Me)O]PO$. Similarly using the corresponding α-Cl-acetoacet-hydroxamide.[234a]

$(EtO)_2[EtO_2CCCl=C(Me)O]PO$. Similarly using the corres-

ponding α,α-Cl_2-acetoacetate. Liquid, b_1 120-128°.[20]
$(EtO)_2[i-PrO_2CCCl=C(Me)O]PO$. Similarly using the corresponding α,α-Cl_2-acetoacetate. Liquid, $b_{0.02}$ 155-160°.[258]
$(EtO)_2[i-PrOCH_2CH_2O_2CCCl=C(Me)O]PO$. Similarly using the corresponding α,α-Cl_2-acetoacetate. Liquid, $b_{0.04}$ 142°.[234]
$(EtO)_2[i-BuOCH_2CH_2O_2CCCl=C(Me)O]PO$. Similarly using the corresponding α,α-Cl_2-acetoacetate. Liquid, $b_{0.03}$ 138-145°.[234]
$(EtO)_2[Me(OCH_2CH_2)_9O_2CCCl=C(Me)O]PO$. Similarly using the corresponding α,α-Cl_2-acetoacetate.[234]
$(EtO)_2[EtSCH_2CH_2O_2CCCl=C(Me)O]PO$. Similarly using the corresponding α,α-Cl_2-acetoacetate.[234]
$(EtO)_2$[(2,3-isopropylidene-1-glycero)-$O_2CCCl=C(Me)O]PO$. Similarly using the corresponding α,α-Cl_2-acetoacetate. Liquid, $b_{0.1}$ 163-170°.[404]
$(EtO)_2[(MeOCH_2)_2CHO_2CCCl=C(Me)O]PO$. Similarly using the corresponding α,α-Cl_2-acetoacetate. Liquid, $b_{0.01}$ 142-150°.[234]
$(EtO)_2[MeOCH_2CH_2NHCO{\cdot}CCl=C(Me)O]PO$. Similarly using the corresponding α,α-Cl_2-acetoacetamide.[403]
$(EtO)_2[MeOCH_2CH_2N(Me)CO{\cdot}CCl=C(Me)O]PO$. Similarly using the corresponding α,α-Cl_2-acetoacetamide.[403]
$(EtO)_2[MeSCH_2CH_2NHCO{\cdot}CCl=C(Me)O]PO$. Similarly using the corresponding α,α-Cl_2-acetoacetamide.[403]
$(EtO)_2[MeO(CH_2)_3NHCO{\cdot}CCl=C(Me)O]PO$. Similarly using the corresponding α,α-Cl_2-acetoacetamide.[403]
$(EtO)_2[(MeOCH_2CH_2)_2NCO{\cdot}CCl=C(Me)O]PO$. Similarly using the corresponding α,α-Cl_2-acetoacetamide.[403]
$(EtO)_2[PhOCH_2CH_2NHCO{\cdot}CCl=C(Me)O]PO$. Similarly using the corresponding α,α-Cl_2-acetoacetamide.[403]
$(EtO)_2[MeON(Me)CO{\cdot}CCl=C(Me)O]PO$. Similarly using the corresponding α,α-Cl_2-acetoacet-hydroxamide. Liquid, $b_{0.02}$ 131-135°.[234a]
$(EtO)_2[MeON(i-Pr)CO{\cdot}CCl=C(Me)O]PO$. Similarly using the corresponding α,α-Cl_2-acetoacet-hydroxamide.[234a]
$(EtO)_2[(Me)(Et)CHO_2CCBr=C(Me)O]PO$. Similarly using the corresponding α,α-Br_2-acetoacetate.[258]
$(EtO)_2[F_2C=C(CF_2CN)O]PO$. From $(EtO)_2[F_2C=C(CF_2Cl)O]PO$ with KCN. Liquid, b_9 54-55°; n_D^{24} 1.3698.[490]
$(EtO)(MeOCH_2CH_2O)[EtO_2CCH=C(Me)O]PO$. I.2.b.γ-2 (from $(EtO)_2(MeOC_2H_4O)P$ and $MeCO{\cdot}CClHCO_2Et$). Liquid, $b_{0.001}$ 100°; d 1.15; n_D^{25} 1.4481.[213]
$(MeOCH_2CH_2O)_2[MeOCH_2CH_2O_2CCH=C(Me)O]PO$. I.2.b.γ-2 (from $(MeOCH_2CH_2O)_3P$ and methoxyethyl α-Cl-acetoacetate). Liquid, $b_{0.001}$ 150°; d 1.20; n_D^{25} 1.4535.[213]
$(MeO)(i-PrO)[MeO_2CCH=C(Me)O]PO$. I.1.f.ε (from (MeO)-(vinyl-O)PO_2Na and $ClCO_2$-i-Pr). Liquid, $b_{0.05}$ 120°.[257]
$(EtO)_2[MeCH(CO_2Me)CH_2O]PO$. I.2.b.α-5 (from the corres-

ponding phosphite by N oxides). Liquid, $b_{0.04}$ 90-92°; d 1.1070; n_D^{20} 1.4291.[1199]
$(EtO)_2[Me(Et)C(CO_2Me)O]PO$. I.2.b.γ-1 (from $(EtO)_3P$ and $MeCO \cdot CO_2Et$). Liquid, b_{13} 153-155°; n_D^{20} 1.4260.[1148]
$(EtO)_2[EtC(CO_2Et)_2O]PO$. I.2.b.γ-1 (from $(EtO)_3P$ and $CO(CO_2Et)_2$). Liquid, b_1 142-143°; d_{20} 1.1522; n_D^{20} 1.4340.[1147b]
$(PrO)_2[(EtO_2C)_2CHO]PO$. I.2.b.γ-1 (from $(PrO)_2PHO$ and $CO(CO_2Et)_2$). Liquid, $b_{1.5}$ 146-148°; n_D^{25} 1.4319.[1148a]
$(EtO_2C \cdot CHMeO)_3PO$. I.1.b.β. Undistillable oil, n_D^{18} 1.4350, d_4^{20} 1.200.[606]
$(PhNHCO \cdot CHMeO)_3PO$. I.1.b.α - β. Plates, m. 205° (from EtOH).[244]
$(2\text{-}MeC_6H_4NHCO \cdot CHMeO)_3PO$. I.1.b.α - β. Cryst., m. 177° (from EtOH).[244]
$(4\text{-}MeC_6H_4NHCO \cdot CHMeO)_3PO$. I.1.b.α - β. Cryst., m. 156° (from EtOH).[244]
$(PrO)_2[MeCO \cdot CH(Me)O]PO$. I.1.f.ζ-10 (from dipropyl 1,2-Me_2-vinylene tetraoxyphosphorane Et_2-amidate by controlled hydrolysis). Liquid, b_1 96-97°.[1223]
$(PrO)_2(NCCMe_2O)PO$. I.1.b.γ-2 (from ROH with $(PrO)_2POCl$). Liquid, $b_{0.015}$ 90-93°; n_D^{20} 1.4228.[81]
$(CH_2{=}CHCH_2O)_2(EtOC(O)CH{=}C(OEt)O)PO$. I.2.b.γ-2 (from triallyl phosphite with $BrHC(COOEt)_2$). Liquid, n_D^{23} 1.4663.[919]
$(i\text{-}PrO)_2(CNCMe_2O)PO$. I.1.b.γ-2 (from ROH with $(i\text{-}PrO)_2$-POCl). Liquid, $b_{0.015}$ 70-86°; n_D^{20} 1.4172.[81]
$(i\text{-}PrO)_2[i\text{-}PrO_2CCH{=}C(Me)O]PO$. I.2.b.γ-2 (from $(i\text{-}PrO)_3P$ using $MeCO \cdot CHCl(Br)CO_2\text{-}i\text{-}Pr$). Liquid, b_1 137-152°.[258]
$(i\text{-}PrO)_2[MeO(CH_2)_3NHCO \cdot CH{=}C(Me)O]PO$. Similarly using the corresponding α-Cl-acetoacetamide.[403]
$(i\text{-}PrO)_2[MeON(Me)CO \cdot CH{=}C(Me)O]PO$. Similarly using the corresponding α-Cl-acetoacet-hydroxamide.[234a]
$(i\text{-}PrO)_2[MeO(CH_2)_3NHCO \cdot CCl{=}C(Me)O]PO$. Similarly using the corresponding α,α-Cl_2-acetoacetamide.[403]
$(i\text{-}PrO)_2[MeON(Me)CO \cdot CCl{=}C(Me)O]PO$. Similarly using the corresponding α,α-Cl_2-acetoacet-hydroxamide.[234a]
$[(EtO_2C)_2CHO](BuO)_2PO$. I.2.b.γ-1 (from $(BuO)_2PHO$ using $CO(CO_2Et)_2$). Liquid, b_2 152-154°; n_D^{25} 1.4346.[1148a]
$[(EtO_2C)_2CHO](i\text{-}BuO)_2PO$. Similarly starting from $(i\text{-}BuO)_2PHO$. Liquid, b_2 135-136°; n_D^{25} 1.4276.[1148a]
$(BuO)_2(NCCMe_2O)PO$. I.1.b.γ-2 (using $(BuO)_2POCl$),[81] I.2.-b.α-1 (from the corresponding phosphite by O_2 at 70°). Liquid, b_{11} 162-164°,[836] $b_{0.017}$ 98-100°; n_D^{20} 1.4264.[81]
$(i\text{-}BuO)_2(NCCMe_2O)PO$. I.1.b.γ-2 (from ROH with $(i\text{-}BuO)_2$-POCl). Liquid, $b_{0.015}$ 91.5-92.5°; n_D^{20} 1.4258.[81]
$(BuO)_2[i\text{-}PrO_2CCH{=}C(Me)O]PO$. I.2.b.γ-2 (from $(i\text{-}PrO)_3P$ and $MeCO \cdot CHCl(Br)CO_2\text{-}i\text{-}Pr$). Liquid, $b_{0.01}$ 148-150°.[258]
$(PhNHCO \cdot CMe_2O)_3PO$. I.1.b.α - β. Plates, m. 158-159°.[244]
$(2\text{-}MeC_6H_4NHCO \cdot CMe_2O)_3PO$. I.1.b.α. Needles, m. 194-196°.[244]

$(4\text{-}MeC_6H_4NHCO{\cdot}CMe_2O)_3PO$. I.1.b.α. Needles, m. 160-162°.[244]

Methyl di-1-(1-chloro-D-galactosepenta-acetate) phosphate. I.1.b.γ-2. Cryst., m. 187-188° (dec.).[1470]

Ethyl di-1-(1-chloro-D-galactosepenta-acetate) phosphate. I.1.b.γ-2. Cryst., m. 156-158°.[1470]

$(EtO)_2[MeO_2CCH(C_8H_{17})O]PO$. I.1.e.β (from $(EtO)_3PO$ with RBr). n_D^{20} 1.4297.[918]

$(EtO)_2[MeO_2CH(C_{10}H_{21})O]PO$. Similarly with the corresponding RBr. n_D^{20} 1.4323.[918]

$(EtO)_2[MeO_2CCH(C_{16}H_{33})O]PO$. Similarly with the corresponding RBr. n_D^{20} 1.4337.[918]

Tri-(diacetoneglucose) phosphate. I.1.b.β. Solid, m. 55°.[1025]

Tris-(Me-tri-O-acetyl-β-D-glacturonate-1) phosphate. I.1.d.δ (from the 1-Br derivative with Ag_3PO_4). Solid; $[\alpha]_D$ 134°.[1108]

Tri-(hepta-acetyl-maltose) phosphate. I.1.d.δ. Crude solid.[927]

4. Derivatives of Nitrogen Bases (without or with Preceding Groups)

$(MeO)_2(ClMe_3NCH_2CH_2O)PO$. From the 2-chloroethyl analog and trimethylamine in toluene. Hygroscopic needles, m. 136.5-137°.[1189]

$(EtO)_2(H_2NCH_2CH_2O)PO$. From the N-Cbo derivative by H_2/Pt.[292]

$(EtO)_2(Me_2NCH_2CH_2O)PO$. I.1.b.γ-2 (using $(EtO)_2POCl$). Liquid, d_{25} 1.0486; n_D^{25} 1.4220.[1336a]

$(EtO)_2(IMe_3NCH_2CH_2O)PO$. From the Me_2N derivative with MeI. Cryst., m. 100°.[1336a]

$(EtO)_2(PhCH_2OCONHCH_2CH_2O)PO$. I.1.b.γ-2 (using $(EtO)_2$-POCl). Oil.[292]

$(EtO)(Me_2NCH_2CH_2O)_2PO$. I.1.b.γ-2. Liquid, $b_{0.06}$ 103-106°.[812]

$(ClMe_3NCH_2CH_2O)_3PO$. From the 2-Cl-ethyl analog and Me_3N. Solid, dec. 245°. Chloroaurate, needles, m. 216°.[743]

$(MeO)(BrMe_3NCH_2CH_2CH_2O)(i\text{-}BuO)PO$. From $(MeO)(BrCH_2CH_2$-$CH_2O)(i\text{-}BuO)PO$ with Me_3N. Cryst., m. 320°.[858]

$(Me_2NCH_2CH_2O)(PrO)(i\text{-}PrO)PO$. I.1.b.γ-2 (using (PrO)(i-PrO)POCl). Oil, d_{25} 1.0042; n_D^{25} 1.4215.[1336a]

$(IMe_3NCH_2CH_2O)(PrO)(i\text{-}PrO)PO$. From the preceding with MeI. Cryst., m. 172°.[1336a]

$(H_2NCH_2CH_2O)(i\text{-}PrO)_2PO$. I.1.b.γ-2 (using $(i\text{-}PrO)_2POF$). IR; NMR.[627]

$(MeO)_2[Cl_2C{=}C(N(Me)CHO)O]PO$. I.2.b.γ-2 (from $(MeO)_3P$ using $Cl_3CCON(Me)CHO$).[1226a,1306] Oil, $b_{0.1}$ 103-106°,[1226a] 104-106°.[1306]

$(MeO)_2[Cl_2C{=}C(N(Me)CO_2Me)O]PO$. Similarly using the corresponding substituted Cl_3-acetamide. Undistillable oil, n_D^{20} 1.4787,[1306] 1.4712.[1226a]

$(MeO)_2[Cl_2=C(N(Me)CO_2Et)O]PO$. Similarly using the corresponding substituted Cl_3-acetamide. Oil, $b_{0.04}$ 124-126°,[1306] n_D^{20} 1.4625.[1226a]

$(MeO)_2[Cl_2C=C(N(Me)CONHMe)O]PO$. Similarly using the corresponding substituted Cl_3-acetamide.[1226a]

$(MeO)_2[Cl_2C=C(N(Me)CONMe_2)O]PO$. Similarly using the corresponding substituted Cl_3-acetamide. Undistillable oil, n_D^{20} 1.4925.[1306]

$(MeO)_2[Cl_2C=C(N(Me)CONEt_2)O]PO$. Similarly using the corresponding substituted Cl_3-acetamide.[1226a,1306] Undistillable oil, n_D^{20} 1.4886.[1306]

$(MeO)_2[Cl_2C=C(N(Me)CO{\cdot}Me)O]PO$. Similarly using the corresponding substituted Cl_3-acetamide.[1226a,1306] Undistillable oil, n_D^{20} 1.4773.[1306]

$(MeO)_2[Cl_2C=C(N(Me)CO{\cdot}Et)O]PO$. Similarly using the corresponding substituted Cl_3-acetamide. Undistillable oil, n_D^{20} 1.4757.[1226a,1306]

$(MeO)_2[Cl_2C=C(N(Me)SO_2NMe_2)O]PO$. Similarly using the corresponding substituted Cl_3-acetamide.[1226a,1306] Undistillable oil, n_D^{20} 1.4843.[1306]

$(MeO)_2[Cl_2C=C(N(Me)SO_2NEt_2)O]PO$. Similarly using the corresponding substituted Cl_3-acetamide.[1226a,1306] Undistillable oil, n_D^{20} 1.4828.[1306]

$(MeO)_2[Cl_2C=C(N(Me)SO_2Me)O]PO$.[1306]

$(MeO)_2[Cl_2C=C(N(Et)CHO)O]PO$. I.2.b.γ-2 (from $(MeO)_3P$ using $Cl_3CCONEt(CHO)$).[1226a,1306] Oil, $b_{0.08}$ 112-115°,[1226a] 113-115°.[1306]

$(MeO)_2[Cl_2C=C(N(Et)CO_2Me)O]PO$. Similarly using the corresponding substituted Cl_3-acetamide. Oil, n_D^{20} 1.4740.[1226a]

$(MeO)_2[Cl_2C=C(N(Et)CO_2Et)O]PO$. Similarly using the corresponding substituted Cl_3-acetamide. Oil, $b_{0.02}$ 123-125°;[1226a,1306] n_D^{20} 1.4698.[1226a]

$(MeO)_2[Cl_2C=C(N(Et)CO{\cdot}Me)O]PO$. Similarly using the corresponding substituted Cl_3-acetamide.[1226a,1306] Undistillable oil, n_D^{20} 1.4732.[1306]

$(MeO)_2[Cl_2C=C(N(Et)SO_2NMe_2)O]PO$. Similarly using the corresponding substituted Cl_3-acetamide.[1226a]

$(MeO)_2[Cl_2C=C(N(Et)SO_2NEt_2)O]PO$. Similarly using the corresponding substituted Cl_3-acetamide.[1226a,1306] Undistillable oil, n_D^{20} 1.4807.[1306]

$(EtO)_2[Cl_2C=C(N(Me)CHO)O]PO$. I.2.b.γ-2 (from $(EtO)_3P$ using $Cl_3CCON(Me)CHO$). Oil, $b_{0.08}$ 105-108°,[1226a] 106-108°.[1306]

$(EtO)_2[Cl_2C=C(N(Me)CO_2Me)O]PO$. Similarly using the corresponding substituted Cl_3-acetamide. Undistillable oil, n_D^{20} 1.4712.[1306]

$(EtO)_2[Cl_2C=C(N(Me)CO_2Et)O]PO$. Similarly using the corresponding substituted Cl_3-acetamide. Oil, $b_{0.02}$ 120-122°.[1306]

$(EtO)_2[Cl_2C=C(N(Me)CONHMe)O]PO$. Similarly using the cor-

responding substituted Cl_3-acetamide.[1226a]
$(EtO)_2[Cl_2C{=}C(N(Me)CONMe_2)O]PO$. Similarly using the corresponding substituted Cl_3-acetamide.[1226a,1306] Undistillable oil, n_D^{20} 1.4831.[1306]
$(EtO)_2[Cl_2C{=}C(N(Me)CONEt_2)O]PO$. Similarly using the corresponding substituted Cl_3-acetamide.[1226a,1306] Undistillable oil, n_D^{20} 1.4810.[1306]
$(EtO)_2[Cl_2C{=}C(N(Me)CO{\cdot}Me)O]PO$. Similarly using the corresponding substituted Cl_3-acetamide.[1226a,1306] Undistillable oil, n_D^{20} 1.4705.[1306]
$(EtO)_2[Cl_2C{=}C(N(Me)CO{\cdot}Et)O]PO$. Similarly using the corresponding substituted Cl_3-acetamide. Oil, n_D^{20} 1.4686.[1226a]
$(EtO)_2[Cl_2C{=}C(N(Me)SO_2NMe_2)O]PO$. Similarly using the corresponding substituted Cl_3-acetamide.[1226a,1306] Undistillable oil, n_D^{20} 1.4746.[1306]
$(EtO)_2[Cl_2C{=}C(N(Me)SO_2NEt_2)O]PO$. Similarly using the corresponding substituted Cl_3-acetamide.[1226a,1306] Undistillable oil, n_D^{20} 1.4735.[1306]
$(EtO)_2[Cl_2C{=}C(N(Et)CHO)O]PO$. I.2.b.γ-2 (from $(EtO)_3P$ using $Cl_3CCON(Et)CHO$). Undistillable oil, n_D^{20} 1.4696.[1226a,1306]
$(EtO)_2[Cl_2C{=}C(N(Et)CO_2Me)O]PO$. Similarly using the corresponding substituted Cl_3-acetamide. Oil, n_D^{20} 1.4650.[1226a]
$(EtO)_2[Cl_2C{=}C(N(Et)CO_2Et)O]PO$. Similarly using the corresponding substituted Cl_3-acetamide. Oil, $b_{0.06}$ 132-133°;[1226a,1306] n_D^{20} 1.4648.[1226a]
$(EtO)_2[Cl_2C{=}C(N(Et)CO{\cdot}Me)O]PO$. Similarly using the corresponding substituted Cl_3-acetamide.[1226a]
$(EtO)_2[Cl_2C{=}C(N(Et)SO_2NMe_2)O]PO$. Similarly using the corresponding substituted Cl_3-acetamide.[1226a]
$(EtO)_2[Cl_2C{=}C(N(Et)SO_2NEt_2)O]PO$. Similarly using the corresponding substituted Cl_3-acetamide.[1226a,1306] Undistillable oil, n_D^{20} 1.4718.[1306]
$(EtO)_2[Cl_2C{=}C(NEt_2)O]PO$. Similarly using $Cl_3CCONEt_2$. Liquid, b_2 76-79°.[564]

B. At Least One R Bearing on Its Aliphatic Carbon Skeleton a Cyclic Structure

1. Alicyclic or/and Aromatic Cycles

a. Only on One R

$(EtO)_2[(1\text{-}EtO_2C\text{-cycloprop-2-yl})\text{-methyl-}O]PO$. From $(EtO)_2(CH_2{=}CHCH_2O)PO$ with N_2CHCO_2Et). Liquid, $b_{1.5}$ 142-143°; n_D^{20} 1.4412.[1147]
$(MeO)_2(PhCH_2O)PO$. I.1.e.β (from $(MeO)_3PO$ with $RCl{\cdot}Ph_3P$).[672]
$(MeO)_2(PhCOCH_2O)PO$. I.1.d.δ (from $(MeO)_2PO_2Na$ and α-Br-ketone). Liquid, $b_{0.02}$ 139-140°.[324]

$(MeO)_2[(4\text{-}O_2NC_6H_4)CH(OH)CH(C_6H_4NO_2\text{-}4)O]PO$. From $(MeO)_3P$ and $O_2NC_6H_4CHO$ via a pentoxyphosphorane cyclic phosphate and controlled hydrolysis.[1162]

$(MeO)_2[CH_2{=}C(Ph)O]PO$. I.2.b.γ-3 (from $(MeO)_3P$ using PhCO-CH_2O-ts).[468]

$(MeO)_2[CH_2{=}C(C_6H_3\text{-}2\text{-}OMe\text{-}5\text{-}Cl)O]PO$. I.2.b.γ-2 (from MeO)$_3$P using $ArCOCH_2Cl$).[235]

$(MeO)_2[CH_2{=}C(C_6H_2\text{-}2\text{-}OMe\text{-}4,5\text{-}Cl_2)O]PO$. Similarly using the corresponding substituted α-Cl-acetophenone. Liquid, n_D^{21} 1.5270.[235]

$(MeO)_2[CH_2{=}C(C_6H_2\text{-}4\text{-}OMe\text{-}2,5\text{-}Cl_2)O]PO$. Similarly using the corresponding substituted α-Cl-acetophenone. Liquid, n_D^{24} 1.5310.[235]

$(MeO)_2[PhSCH{=}C(Ph)O]PO$. Similarly using PhSCHClCO·Ph. Liquid, $b_{0.07}$ 103°.[83]

$(MeO)_2[ClCH{=}C(C_6H_3\text{-}2,4\text{-}Br_2)O]PO$. Similarly using 2,4-$Br_2C_6H_3CO\cdot CHCl_2$. Cryst., m. 91.5-92.5°.[1103]

<$(MeO)_2[ClCH{=}C(C_6H_2\text{-}4\text{-}OMe\text{-}2,5\text{-}Cl_2)O]PO$. Similarly using> the corresponding α,α-Cl_2-acetophenone derivative. Liquid, n_D^{21} 1.5320.[235]

$(MeO)_2[OHCCH(COPh)O]PO$. From Me_3-ethylene pentaoxyphosphorane with PhCOCl. Liquid, $b_{0.2}$ 148-150°; ^{31}P 2.6 ppm.[1167]

$(MeO)(EtO)(PhCOCH_2O)PO$. I.1.d.δ (from $(MeO)(EtO)PO_2H$ and $PhCO\cdot CH_2Br$). Liquid, $b_{0.02}$ 135°.[324]

$(MeO)(PhCH_2O)(Cl_2C{=}CHO)PO$. I.1.f.ε (from (MeO)(vinyl-O)PO_2Na and $ClCO_2CH_2Ph$). Liquid, $b_{0.01}$ 75-78°.[257]

$(EtO)_2(PhCH_2O)PO$. I.1.e.β (from $(EtO)_3PO$ and $RHal\cdot Ph_3P$). Liquid, b_{18} 183-185°,[672] b_{12} 150°.[845]

$(EtO)_2(3\text{-}Me_3Si\text{-}C_6H_4CH_2O)PO$. I.1.b.γ-2. Liquid, b_1 148-150°; n_D^{25} 1.4900.[266]

$(EtO)_2(4\text{-}Me_3Si\text{-}C_6H_4CH_2O)PO$. I.1.b.γ-2. Liquid.[266]

$(EtO)_2(Ph_2CHO)PO$. I.1.e.β (from $(EtO)_3PO$ and $RCl\cdot Ph_3P$).[672]

$(EtO)_2(Ph_3CO)PO$. Similarly using $Ph_3CCl\cdot Ph_3P$.[672]

$(EtO)_2[4\text{-}O_2NC_6H_4CH(OH)CH(C_6H_4NO_2\text{-}4)O]PO$. From $(EtO)_3P$ and $O_2NC_6H_4CHO$ via a pentaoxyphosphorane cyclic phosphate and controlled hydrolysis.[1162]

$(EtO)_2[CH_2{=}C(C_6H_2\text{-}2\text{-}OMe\text{-}4,5\text{-}Cl_2)O]PO$. I.2.b.γ-2 (from $(EtO)_3P$ using $ArCO\cdot CH_2Cl$). Liquid, n_D^{21} 1.5128.[235]

$(EtO)_2[CH_2{=}C(C_6H_2\text{-}4\text{-}OMe\text{-}2,5\text{-}Cl_2)O]PO$. Similarly using the corresponding substituted α-Cl-acetophenone. Liquid, n_D^{21} 1.5118.[235]

$(EtO)_2[CH_2{=}C(C_6H_3\text{-}2\text{-}OMe\text{-}5\text{-}Br)O]PO$. Similarly using the corresponding substituted α-Cl-acetophenone. Liquid, n_D^{21} 1.5165.[235]

$(EtO)_2[ClCH{=}C(C_6H_3\text{-}2\text{-}OMe\text{-}5\text{-}Cl)O]PO$. Similarly using the corresponding substituted α,α-Cl_2-acetophenone. Liquid, $b_{0.07}$ 155-169°.[235]

$(EtO)_2[ClCH{=}C(C_6H_2\text{-}4\text{-}OMe\text{-}2,5\text{-}Cl_2)O]PO$. Similarly using the corresponding substituted α,α-Cl_2-acetophenone.

Liquid, n_D^{21} 1.5181.[235]
$(EtO)_2[Cl_2C{=}C(Ph)O]PO$. Similarly using $PhCO{\cdot}CCl_3$. Liquid, $b_{0.001}$ 105-110°; n_D^{20} 1.5195.[787]
$(EtO)_2[PhSCH{=}C(Ph)O]PO$. Similarly using PhSCHClCO·Ph. Liquid, b_{12} 172°.[83]
$(EtO)_2[PhCH{=}C(Ph)O]PO$. Similarly using PhCHClCO·Ph. Liquid, $b_{0.25}$ 179-180°.[1336a]
$(EtO)_2(PhCHClCH{=}CHO)PO$. Similarly using PhCHClCHClCHO.[20]
$(EtO)_2[Cl_2C{=}CHCH{=}C(Ph)O]PO$. Similarly using $Cl_3CCH{=}CHC(O)Ph$.[20]
$(EtO)(PhCH_2O)[MeO(CH_2)_3NHCO{\cdot}CH{=}C(Me)O]PO$. I.2.b.γ-2 (from $(EtO)_2(PhCH_2O)P$ and the substituted α-Cl-acetoacetamide).[403]
1-(D-2,3-di-O-Stearoyl-1-glycero-benzylphosphoryl)-2-benzoyl-3-(N-Cbo-L-alanyl)-glycerol. I.1.d.δ (from Ag diacylglycero-benzylphosphate and diacylglycerol iodohydrine; yield 8%!). Cryst., m. 38-39°; $[\alpha]_D^{20}$ -3.5°.[985]
1-(L-2,3-di-O-Stearoyl-1-glycero-benzylphosphoryl)-2-benzoyl-3-N-Cbo-L-alanyl)-glycerol. I.1.d.δ (similarly to the preceding compound). Cryst., m. ~40°; $[\alpha]_D^{20}$ -2.9°.[985]
(1,2,3,4-tetra-O-acetyl-β-D-glucopyranose-6)(1,2-O-isopropylidene-α-D-glucofuranose-6) benzyl phosphate. I.1.d.γ-2 (from acetylglucose benzyl phosphate and isopropylidene-5,6-anhydroglucose). Cryst., m. 169-170°.[659]
O-[$(PhCH_2)$(2-O-Palmitoyl-3-O-oleoyl-L-1-glycero)-phosphoryl]=N-Ctbo-DL-serine t-Bu ester. I.1.d.δ (from Ag salt of benzylphosphoryl N-Ctbo-serine ester and diacylglycerol iodohydrine). $[\alpha]_D^{20}$ 1.5° ($CHCl_3$).[646]

b. On Two R

$(MeO)(PhCH_2O)_2PO$. I.1.d.δ. Liquid, d_0^0 1.2089.[884]
$(EtO)(4\text{-}O_2NC_6H_4CH_2O)_2PO$. I.1.d.δ,ε (in (MeCN). Needles, m. 65-66°.[129]
$(AmO)(4\text{-}O_2NC_6H_4CH_2O)_2PO$. I.1.b.γ-2.[480]
$(i\text{-}AmO)(PhCH_2O)_2PO$. I.1.b.γ-2.[480]
$(PhCH_2O)_2(Myristoyl\text{-}OCH_2CH_2O)PO$. I.1.d.δ (using $AgO_2P(OCH_2Ph)_2$).[1403]
$(PhCH_2O)_2(Palmitoyl\text{-}OCH_2CH_2O)PO$. Similarly to the preceding.[1403]
$(4\text{-}BrC_6H_4CH_2O)_2(EtO_2CCH_2O)PO$. I.1.d.δ (using $(ArCH_2O)_2PO_2Ag$ and ICH_2CO_2Et).[480]
$(4\text{-}BrC_6H_4CH_2O)(4\text{-}O_2NC_6H_4CH_2O)(HO_2CCH_2O)PO$. By partial saponification of the following compound. Cryst., m. 127-128°; Ag salt.[480]
$(4\text{-}BrC_6H_4CH_2O)(4\text{-}O_2NC_6H_4CH_2O)(EtO_2CCH_2O)PO$. I.1.d.δ (using $(BrC_6H_4CH_2O)(EtO_2CH_2O)PO_2Ag$ and $ICH_2C_6H_4NO_2$). Sirup.[480]
$(PhCH_2O)_2$(2,3-di-O-acyl-1-glycero-O)PO. I.1.d.δ (from

diacyl glycerol iodohydrine using $(PhCH_2O)_2PO_2Ag$).
Dimyristoyl derivative. Cryst., m. 40-41°.[674]
Dipalmitoyl derivative. Cryst., m. 49-50°.[674]
Distearoyl derivative. Cryst., m. 56-57°.[674]
$(4-O_2NC_6H_4CH_2O)_2$(2,3-di-O-palmitoyl-1-glycero-O)PO. Similarly using $(O_2NC_6H_4CH_2O)_2PO_2Ag$. Cryst., m. 70°.[806]
(-)$(PhCH_2O)_2$(2,3-O-Isopropylidene-1-glycero-O)PO. I.1.b.-γ-2 (from $(PhCH_2O)_2POCl$ and active ROH). $[\alpha]_D^{20}$ -3.65° ($CHCl_3$).[262]
$(PhCH_2O)_2[MeCO_2{\cdot}CH_2C(OMe)_2CH_2O]PO$. I.1.b.γ-2 (using $(PhCH_2O)_2POCl$). Cryst., m. 183-184°.[189]
$(PhCH_2O)_2[MeCO_2{\cdot}CH_2C(OEt)_2CH_2O]PO$. Similarly. Cryst., m. 180° (dec.).[189]
$(PhCH_2O)_2[CH_2{=}C(CO_2H)O]PO$. I.2.b.γ-2 (from $(PhCH_2O)_3P$ and $BrCH_2CO_2H$). Cryst., m. 75-78°.[453]
O-(Dibenzylphosphoryl)-N-Cbo-serine benzyl ester. I.1.d.-α-1. Cryst., m. 74°.[112]
O-(Dibenzylphosphoryl)-N-Cbo-serine methyl amide. I.1.d.-α-1. Cryst., m. 90°.[112]
$(4-O_2NC_6H_4CH_2O)_2$-2-[N-(N,N-Phthaloyl-DL-alanyl)-amino-ethyl-O]PO. I.1.b.γ-2 (using $(ArCH_2O)_2POCl$). Cryst., m. 119-121°.[929]
D-Pantothenonitrile 4-O-dibenzylphosphate. I.1.b.γ-2 (using $(PhCH_2O)_2POCl$). $[\alpha]_D^{23}$ 19.1° (EtOH).[1027]
Me 2,3-Isopropylidene-D-ribofuranoside 5-dibenzylphosphate. I.1.c.β-2.[1205]
1,2,3,4-Tetra-O-acetyl-β-D-glucose 6-dibenzylphosphate. I.1.b.γ-2. Cryst., with 1 H_2O, m. 89-90°, anhydrous m. 105-107°; $[\alpha]_D^{26}$ 24.4°.[659]
Dibenzyl 2,3,4,6-tetra-acetyl-D-glucose phosphate. I.1.-d.δ. Crystals, m. 79°.[1499]
Dibenzyl 1,2-di-O-Isopropylidene-α-D-glucofuranose 6-phosphate. I.1.d.γ-2 (from the 5,6-anhydroglucose derivative with $(PhCH_2O)_2PO_2H$). Glass.[659]
Me 1-O-Dibenzylphosphoryl-2,3,4-tri-O-acetyl-β-D-galacturonate. I.1.d.δ (using $(PhCH_2O)_2PO_2Ag$). Cryst., m. 84.5-86°; $[\alpha]_D$ 12.2°.[1108]

c. On the Three R

$(c-C_6H_{11}CH_2O)_3PO$. I.2.b.α-1 (from the phosphite by O_2 and UV irradiation).[1269]
$(PhCH_2O)_3PO$. I.1.b.β (using $POBr_3$),[539] I.1.d.δ,[276,884,1484] I.2.b.α-1 (from $(RO)_3P$ using O_2 and UV irradiation).[1264] From R_2O with P_2O_5.[713] Prisms, m. 65°,[276] 64°,[539,884] 61-63°;[713] $b_{0.005}$ 110-120°.[1246]
$(4-BrC_6H_4CH_2O)_3PO$. I.1.d.δ,ε (in MeCN),[129] I.1.d.δ and e.β (from $(AmO)(ArCH_2O)PO_2Ag$ and $BrC_6H_4CH_2I$, yield small!).[480] Cryst., m. 132-133°,[129] 130-132°.[480]
$(4-O_2NC_6H_4CH_2O)_3PO$. I.1.d.δ (from RBr and Ag_3PO_4). Pale yellow cryst., m. 127-128°.[129] ^{31}P + 0.7 ppm.[913C]
$(Ph_2CHO)(PhCH_2O)_2PO$. I.1.d.β. Cryst., m. 72-72.5°.[95]

(PhCO·CPhHO)$(PhCH_2O)_2$PO. I.1.d.β. Cryst., m. 53-54°.[95]
(+)-[MeCH(Ph)O]$_3$PO. I.1.b.β (from active ROH with $POBr_3$). $[\alpha]_D^{23}$ 24.44°.[539]
$(PhCH_2CH_2O)_3$PO. I.2.b.α-8 (from $(RO)_3P$ with Cl_2 in the presence of ROH).[194,572]
3,11,20-Trioxo-4-pregnene-17α,21-diol 21-dibenzylphosphate. I.1.d.δ (21-I derivative and $AgO_2P(OCH_2Ph)_2$). Cryst., m. 161-162°.[456]
Tris-[(6-Cl-1,3-benzodioxane-8-yl)-methyl] phosphate. I.1.d.δ. Cryst., dimorphous: m. 142-143° or 157-158°.[129]

2. Heterocycles (besides, or not, Carbocyclic Structures)

(CH$_2$CHCH$_2$O)$_2$(CH$_2$=CHCH$_2$O)PO (CH$_2$–CH epoxide, O-bridged). I.1.b.γ-2 (from glycidol using allyl-$OPOCl_2$).[1305]
(CH$_2$CHCH$_2$O)$_2$(n-$C_{18}H_{37}$O)PO (CH$_2$–CH epoxide, O-bridged). I.1.b.γ-2 (from glycidol using $ROPOCl_2$).[1305]
(CH$_2$CHCH$_2$O)$_3$PO (CH$_2$–CH epoxide, O-bridged). I.1.2.γ-2 (from glycidol with $POCl_3$).[1305]
(Epoxyoleyl-O)$_3$PO. I.2.b.β-1 (from trioleyl phosphite by $MeCO_3H$.[354]
(Furfuryl-O)$_3$PO. I.2.b.β-1 (from $(RO)_3P$ by $MeCO_3H$).[354]
$(PrO)_2$(2-Pyridyl-CH_2O)PO. I.1.b.γ-2 (using $(PrO)_2POCl$). Liquid, b_{17} 155-160°; d_{20} 1.0197; n_D^{20} 1.4388.[487]
$(i\text{-}PrO)_2$(2-Pyridyl-CH_2O)PO. Similarly. Liquid, b_4 163-164°, d_{20} 1.0568; n_D^{20} 1.4349.[487]
$(BuO)_2$(2-Pyridyl-CH_2O)PO. Similarly. Liquid, b_{10} 157-159°; d_{20} 0.9780; n_D^{20} 1.4201.[487]
$(i\text{-}BuO)_2$(2-Pyridyl-CH_2O)PO. Similarly. Liquid, b_9 160°; d_{20} 1.0065; n_D^{20} 1.4395.[487]
$(AmO)_2$(2-Pyridyl-CH_2O)PO. Similarly. Liquid, b_2 178-180°; d_{20} 0.9581; n_D^{20} 1.4330.[487]
$(n\text{-}C_6H_{13}O)_2$(2-Pyridyl-CH_2O)PO. Similarly. Liquid, b_2 189-192°; d_{20} 0.9407; n_D^{20} 1.4381.[487]
$(EtO)_2$(3-Pyridyl-CH_2O)PO. Similarly. Liquid, b_3 115-119°; d_{20} 1.0288; n_D^{20} 1.4268.[487]
$(PrO)_2$(3-Pyridyl-CH_2O)PO. Similarly. Liquid, b_3 130-132°; d_{20} 1.0064; n_D^{20} 1.4316.[487]
$(i\text{-}PrO)_2$(3-Pyridyl-CH_2O)PO. Similarly. Liquid, b_1 147-150°; d_{20} 1.0455; n_D^{20} 1.4427.[487]
$(BuO)_2$(3-Pyridyl-CH_2O)PO. Similarly. Liquid, b_2 156-158°; d_{20} 0.9992; n_D^{20} 1.4387.[487]
$(i\text{-}BuO)_2$(3-Pyridyl-CH_2O)PO. Similarly. Liquid, b_3 154-155°; d_{20} 1.0118; n_D^{20} 1.4424.[487]
$(MeO)_2$[Cl_2C=C(2-Oxopiperidino)O]PO. I.2.b.γ-2 (from $(MeO)_3P$ using Cl_3CCONX).[1226a,1306] Undistillable oil, n_D^{20} 1.4930.[1306]
$(EtO)_2$[Cl_2C=C(2-Oxopiperidino)O]PO. Similarly starting

from $(EtO)_3P$.[1226a,1306] Undistillable oil, n_D^{20} 1.4823.[1306]

(2,3-di-O-Ethylidene-1-glycero-O)$(EtO)_2PO$. I.1.b.γ-2 (using $(EtO)_2POCl$). Liquid, b_4 119°; n_D^{20} 1.4318.[77]

$(MeO)_2[Cl_2C=C$(2-Oxopyrrolidino)O]PO. I.2.b.γ-2 (from $(MeO)_3P$ using Cl_3CCONX). Undistillable oil, n_D^{20} 1.4968.[1306]

$(MeO)_2[Cl_2C=C$(2-Oxo-5-Me-pyrrolidino)O]PO. Similarly using the corresponding Cl_3-acetamide derivative.[1226a]

$(EtO)_2[Cl_2C=C$(2-Oxopyrrolidino)O]PO. I.2.b.γ-2 (from $(EtO)_3P$ using the corresponding Cl_3-acetamide derivative.[1226a,1306] Undistillable oil, n_D^{20} 1.4868.[1306]

$(EtO)_2[Cl_2C=C$(2-Oxo-5-Me-pyrrolidino)O]PO. Similarly using the corresponding Cl_3-acetamide derivative.[1226a]

$(MeO)_2[Cl_2C=C$(2-Oxo-imidazolidin-1-yl)O]PO. I.2.b.γ-2 (from $(MeO)_3P$ using the corresponding Cl_3-acetamide derivative). Undistillable oil, n_D^{20} 1.4897.[1306]

$(EtO)_2[Cl_2C=C$(2-Oxo-imidazolidin-1-yl)O]PO. Similarly starting from $(EtO)_3P$. Undistillable oil, n_D^{20} 1.4790.[1306]

$(MeO)_2[Cl_2C=C$(2-Oxazolidinone-3-yl)O]PO. I.2.b.γ-2 (from $(MeO)_3P$ using the corresponding Cl_3-acetamide derivative.[1226a]

$(EtO)_2[Cl_2C=C$(2-Oxazolidinone-3-yl)O]PO. Similarly starting from $(EtO)_3P$. Undistillable oil, n_D^{20} 1.4805.[1306]

$(MeO)_2[Cl_2C=C$(2-Oxoimidazolidinone-1-yl)O]PO. I.2.b.γ-2 (from $(MeO)_3P$ using the corresponding Cl_3-acetamide derivative.[1226a]

$(EtO)_2[Cl_2C=C$(2-Oxoimidazolidinone-1-yl)O]PO. Similarly starting from (EtO_3P).[1226a]

6-Azauridine 5'-dialkylphosphates. I.1.b.γ-2 (from 2',3'-isopropylidene derivative using $(RO)_2POCl$): Me_2, Am_2 derivatives; UV characteristics.[1141]

Adenosine 5'-dibenzylphosphate. I.1.b.γ-2. Cryst., m. 97-98°.[132]

2',3'-Isopropylidene-adenosine 5'-dibenzylphosphate. I.1.f.ζ-10 (from N^6-acetyl-isopropylidene 3-$O^{5'}$-anhydro derivative with $(PhCH_2O)_2PO_2H$). Cryst., m. 97-98°.[977]

Triquinine phosphate. I.1.b.β. Cryst., dec. 260°.[1504]

C. At Least One Ester Linkage to a Carbon Atom of a Cyclic Structure

1. Alicyclic Structures

$(MeO)_2$(2-MeO-3,3,4,4-F_4-cyclobutene-1-yl-O)PO. I.2.b.γ-2 (from 1,2-dioxo-3,3,4,4-F_4-cyclobutane). Liquid, b_1 90°; n_D^{25} 1.3950.[1457]

$(MeO)_2$(2-Me-3,3,4,4-F_4-cyclobutene-1-yl-O)PO. I.2.b.γ-2 (from $(MeO)_3P$ with 2-Me-2,3,3,4,4-F_5-cyclobutanone).[1457]

$(EtO)_2(2\text{-}EtO_2C\text{-cyclopentene-1-yl-O})PO$. I.1.b.γ-2. Distillable oil.[636]

$(MeO)_2(Cl_2C{=}CHO)(2\text{-}Cl_3C\text{-}c\text{-}C_5H_8O)PO$. I.2.b.γ-2 (from $(MeO)_2(RO)P$ and chloral). Yellow oil.[529a]

$(MeO)_2(c\text{-}C_6H_{11}O)PO$. I.1.b.β - γ-2 (from $POCl_3$ stepwise). Liquid, $b_{1.5}$ 106°; n_D^{20} 1.4417. Pyrolysis.[215]

$(EtO)_2(c\text{-}C_6H_{11}O)PO$. I.2.b.β-4 (from the phosphite by $CBr_2(CONH_2)_2$). Liquid, b_4 59-60°.[1013]

$(BuO)_2(c\text{-}C_6H_{11}O)PO$. I.2.b.β-2 (from $(BuO)_2PHO$ and ROH by CCl_4). Liquid, $b_{0.0001}$ 90°; n_D^{20} 1.4462.[1101]

$(i\text{-}AmO)_2(c\text{-}C_6H_{11}O)PO$. I.1.b.γ-2,[718] I.2.b.β-2 (from $(RO)_2PHO$ and R'OH with CCl_4).[1101] Liquid, $b_{0.5}$ 110°,[718] $b_{0.0001}$ 110°; n_D^{20} 1.4461.[1101]

$(n\text{-}C_6H_{13}O)_2(c\text{-}C_6H_{11}O)PO$. I.2.b.β-2 (similarly to the preceding compound). Liquid, $b_{0.0001}$ 120°; n_D^{20} 1.4472.[1101]

$(i\text{-}C_8H_{17}O)_2(c\text{-}C_6H_{11}O)PO$. Similarly to the preceding. Liquid, $b_{0.0001}$ 140°; n_D^{20} 1.4534.[1101]

$(2\text{-}Et\text{-}n\text{-}C_6H_{12}O)_2(c\text{-}C_6H_{11}O)PO$. Similarly to the preceding. Liquid, $b_{0.0001}$ 120°; n_D^{20} 1.4472.[1101]

$(EtO)_2(2\text{-Oxo-cyclohexyl-O})PO$. I.2.b.γ-1 (from $(EtO)_3P$ and 1,2,dioxocyclohexane, followed by controlled hydrolysis). Liquid, b_8 165.6°; n_D^{20} 1.4510.[833]

$(PrO)_2(2\text{-Oxo-cyclohexyl-O})PO$. By controlled hydrolysis of Pr c-tetramethylene-ethylene cyclic phosphoro-diethylamidate. Liquid, $b_{0.018}$ 78°.[1223]

$(EtO)_2(2\text{-}EtO_2C\text{-}c\text{-hexen-1-yl-O})PO$. I.1.d.γ-2 (from c-RONa and $(EtO)_2POCl$). Oil.[636]

$(MeO)(Cl_2C{=}CHO)(2\text{-}Cl_3C\text{-}c\text{-}C_6H_{12}O)PO$. I.2.b.γ-2 (from $(MeO)_2(c\text{-}RO)P$ with chloral). Cryst., m. 48°.[529a]

$(PhCH_2O)_2(2\text{-}HO\text{-}c\text{-}C_6H_{10}O)PO$. I.1.d.δ,ε. I.1.d.γ-2. Needles, m. 78-80°.[129]

$(4\text{-}BrC_6H_4CH_2O)_2(2\text{-}HO\text{-}c\text{-}C_6H_{10}O)PO$. I.1.d.δ,ε, I.d.γ-2. Cryst., m. 104-105° or 112-113°.[129]

$(4\text{-}O_2NC_6H_4CH_2)_2(2\text{-}HO\text{-}c\text{-}C_6H_{10}O)PO$. I.1.d.δ (from $IC_6H_{10}OH$ and $AgO_2P(OAr)_2$). Cryst., m. 112-113°.[129]

O-(2,3-Di-O-stearoyl-glyceryl 1-myoinositylphosphoryl)-ethanolamine alanyl-glycine. From the following by H_2/Pd. Cryst., m. 162-162.5°.[888]

O-(2,3-Di-O-stearoyl-glyceryl 1-myoinositylphosphoryl)-ethanolamine alanyl-N-Cbo-glycine. From the diacylglyceryl myoinositylphosphoryl ethanolamine and CbO-glycylalanylazide. Cryst., m. 136.5-137°.[888]

1,2;5,6-Diisopropylidene-(-)-inositol 3-diethylphosphate. I.1.b.γ-2. Needles, m. 135°; $[\alpha]_D^{25}$ -49.7° (EtOH).[796]

1,2,3,4-Tetra-O-acetyl-β-D-glucopyranose 6-(benzyl-2-HO-$c\text{-}C_6H_{10}O$-phosphate). I.1.d.γ-2 (from the benzyl H phosphate and cyclohexene-epoxide). Cryst., m. 76-77°.[659]

$(EtO)_2(3\text{-Br-Born-2-ene-2-yl-O})PO$. I.2.b.γ-2 (from $(EtO)_3P$ and Br_2-camphor).[20]

Dimethyl 17-[(5,6)androsterone-3,17-diol] phosphate. I.l.d.β. Cryst., m. 192-193°.[1016]
Diethyl 17-[(5,6)androsterone-3,17-diol] phosphate. I.l.d.β. Cryst., m. 165° (from benzene).[1016]
Dipropyl 17-[(5,6)androsterone-3,17-diol] phosphate. I.l.d.β. Cryst., m. 101° (from benzene-ligroin).[1016]
Dibutyl 17-[(5,6)androsterone-3,17-diol] phosphate. I.l.d.β. Very unstable substance.[1016]
Dimethyl 17-testosterone phosphate. I.l.b.β (from $POCl_3$, stepwise),[1193] I.l.d.β,[1016] I.2.b.β-4 (from $(MeO)_3P$ and testosterone by $NCCHBrCONH_2$). By oxidation of the corresponding androsterone-3,17-diol phosphate with CrO_3.[1016] Cryst., m. 152-153°,[1016] 151-152°; IR.[1193] Oxime, dec. 185°.[1016]
Diethyl 17-testosterone phosphate. I.l.d.β. Oil, b. 160° (in high vacuum).[1016]
Dimethyl dehydroepiandrosterone phosphate. I.l.b.β (using $POCl_3$ followed by MeOH[1194] or by CH_2N_2[1193]). Cryst. m. 152-154°; IR;[1193] $[\alpha]_D$ 1.4°; circular dichroism; NMR.[1194]
Methyl 2-cyanoethyl dehydroepiandrosterone phosphate. I.l.b.β (stepwise with $POCl_3$). IR.[1193]
$(1\text{-}Cl_3C\text{-}c\text{-}C_5H_8O)(1\text{-}Cl_3C\text{-}c\text{-}C_6H_{10}O)(Cl_2C{=}CHO)PO$. I.2.b.γ-2 (from (MeO)(c-R)(c-R'O)P and chloral). Yellow oil.[529a]
Di-(dehydroepiandrosterone) methyl phosphate. I.l.b.β - d.γ (stepwise using $POCl_3$ and CH_2N_2).[1193,1194] Cryst., m. 210-212°; IR;[1193] $[\alpha]_D$ 1.3° ($CHCl_3$); circular dichroism; NMR.[1194]
Dicholesteryl methyl phosphate. I.l.2.β (stepwise using $POCl_3$). Cryst., m. 159-161°; IR.[1193]
Dicholesteryl 2-chloroethyl phosphate. I.l.b.γ-2. Cryst., m. 158°.[523]
Diergosteryl 2-chloroethyl phosphate. I.l.b.γ-2. Cryst., m. 165-167°.[523]
$(2\text{-}Cl\text{-}3\text{-}(CH_2{=}CH)\text{-}c\text{-}C_6H_9O)_3PO$ (?). From PCl_3 and 3-vinyl-1,2-epoxy-cyclohexene.[545]
Trifenchyl phosphate. Levo derivative. I.l.b.β. Cryst., m. 160°.[707]
Trimenthyl phosphate. I.l.b.α,[965] I.l.b.β (best $POCl_3$ and Na c-alkoxide). Plates, m. 86°.[964,965] (-)-Derivative: $[\alpha]_D$ -100° (MePh).[965]
Tri-trans-2-decahydronaphthyl phosphate. I.l.b.β. Cryst., m. 159°.[707]

2. Aromatic Structures

a. One Aromatic Nucleus

α. Hydrocarbon Derivatives

$(MeO)_2(PhO)PO$. I.1.b.γ-2 (using $PhOPOCl_2$).[923] From $(MeO)_2PHO$ by successive reaction with Cl_2 and MeOH.[1072] Liquid, $b_{2.0}$ 114°,[1072] $b_{0.1}$ 92.5°.[923]

$(MeO)(EtO)(PhO)PO$. I.1.f.ε (from a salt of $(MeO)(PhO)PO_2$-H with $ClCO_2Et$). Liquid, $b_{0.02}$ 110-125°.[255]

$(EtO)_2(PhO)PO$. I.1.b.β,[994,995] I.1.b.γ-2,[699,1335] I.1.e.β[996] (from $(EtO)_2(CH_2{=}CMeO)PO$ with PhOH),[1011,1013] I.1.b.α-8 (from $(EtO)_2PHO$ with CCl_4 then with PhOH),[421] I.1.b.β-4 (from $(EtO)_3P$ and PhOH with CBr_2-$(CONH_2)_2$).[1013] Liquid, b_{70} 200-230°, b_{18} 146-162° (apparently impure),[994,995,996] b_5 130-135°,[1335] $b_{1.6}$ 90-108°,[1013] b_1 90-103°,[1011,1013] $b_{0.1}$ 118°,[421] $b_{0.01}$ 90°.[699]

$(EtO)(PrO)(PhO)PO$. I.1.e.β. Liquid.[996]

$(BuO)_2(PhO)PO$. I.1.b.γ-2,[718,923] I.2.b.α-8 (from $(BuO)_2$-PHO with Cl_2 then with PhOH).[1072] Liquid, b_{15} 183-185°,[718] $b_{0.3}$ 136-140°,[1072] $b_{0.1}$ 125-125.5°.[923]

$(MeO)(n\text{-}C_{11}H_{23}O)(PhO)PO$. I.1.b.γ-2 (from $(C_{11}H_{23}O)(PhO)$-POCl using MeOH).[923]

$(BuO)(n\text{-}C_{11}H_{23}O)(PhO)PO$. Similarly using BuOH.[923]

$(2\text{-Et-}n\text{-}C_6H_{12}O)_2(PhO)PO$. I.2.b.α-1 (from the phosphite by O_2 and UV irradiation).[1264]

$(n\text{-}C_{11}H_{23}O)_2(PhO)PO$. I.1.b.γ-2 (from $PhOPOCl_2$).[923]

$(PhCH_2O)_2(PhO)PO$. I.1.b.γ-2 (using $(PhCH_2O)_2POCl$).[129,1004] Cryst., m. 42°.[129]

$(EtO)_2(2\text{-}MeC_6H_4O)PO$. I.1.b.γ-2. Liquid, b_{15} 119°; d_{20} 1.1310; n_D^{20} 1.4812.[770a]

$(MeO)_2(3\text{-}MeC_6H_4O)PO$. I.1.b.γ-2 (using $ArOPOCl_2$),[1072] I.2.b.α-8 (from $MeO)_2PHO$ with Cl_2 then PhOH).[518,1072] Liquid, $b_{1.0}$ 113-115°,[1072] $b_{0.7}$ 107-110°.[518]

$(EtO)_2(3\text{-}MeC_6H_4O)PO$. I.1.b.γ-2 (using $(EtO)_2POCl$). Liquid, $b_{0.002}$ 85°.[699]

$(PrO)_2(3\text{-}MeC_6H_4O)PO$. I.2.b.α-8 (from $PrO)_2PHO$ with Cl_2 then ArOH). Liquid, $b_{1.0}$ 146-148°.[1072]

$(MeO)_2(4\text{-}MeC_6H_4O)PO$. I.2.b.α-8 (from $(MeO)_2PHO$ with Cl_2 then ArOH). Liquid, $b_{0.5}$ 114°.[518,1072]

$(EtO)_2(4\text{-}MeC_6H_4O)PO$. I.2.b.β-2 (from PCl_3 with CCl_4 stepwise). Liquid, $b_{0.01}$ 131-133°.[1050]

$(MeO)_2(3,5\text{-}Me_2C_6H_3O)PO$. I.2.b.α-8 (from $(MeO)_2PHO$ with Cl_2 then ArOH).[518,1072] Liquid, b_{20} 177°,[518] $b_{1.2}$ 125°.[1072]

$(MeO)_2(3\text{-t-Bu-4-}MeC_6H_3O)PO$. I.2.b.γ-2 (from $(MeO)_3P$ with 3-t-Bu-4-Br-2,5-cyclohexadienone). Cryst., m. 55-57°.[1017]

$(EtO)_2(2\text{-}EtC_6H_4O)PO$. I.1.b.γ-2 (using $(EtO)_2POCl$). Liquid, $b_{0.015}$ 96°.[699]

$(EtO)_2(4\text{-}EtC_6H_4O)PO$. I.1.b.γ-2 (using $(EtO)_2POCl$). Liquid, $b_{0.02}$ 118°.[699]
$(EtO)_2(2\text{-}i\text{-}PrC_6H_4O)PO$. I.1.b.γ-2 (using $(EtO)_2POCl$). Liquid, $b_{0.01}$ 106°.[699]
$(EtO)_2(4\text{-}i\text{-}PrC_6H_4O)PO$. I.1.b.γ-2.[699,770a] Liquid, b_{10} 174.5-175.5°,[770a] $b_{0.3}$ 121°; d_{20} 1.0852; n_D^{20} 1.4770.[770a]
$(EtO)_2(4\text{-}(Me_2Et)C_6H_4O)PO$. I.1.b.γ-2. Liquid, $b_{0.5}$ 144-146°; d_{20} 1.0640; n_D^{20} 1.4838.[770a]
$(EtO)_2(4\text{-}Me_2C_4H_7\cdot C_6H_4O)PO$. I.1.b.γ-2. Liquid, b_1 164-167°; d_{20} 1.1380; n_D^{20} 1.4812.[770a]
$(EtO)_2(4\text{-}n\text{-}C_7H_{15}C_6H_4O)PO$. I.1.b.γ-2. Liquid, b_2 178-179°; d_{20} 1.0328; n_D^{20} 1.4761.[770a]
$(MeO)_2[4\text{-}Me_3CCH_2C(Me)_2\text{-}C_6H_4O)PO$. I.2.b.α-8 (from $(MeO)_2$-PHO with Cl_2 then ArOH). Liquid, $b_{0.7}$ 160-165°.[1072]
$(EtO)_2(4\text{-}n\text{-}C_9H_{19}C_6H_4O)PO$. I.1.b.γ-2. Liquid, b_1 181.5-184.0°; d_{20} 1.0125; n_D^{20} 1.4765.[770a]
$(EtO)_2(4\text{-}n\text{-}C_{12}H_{25}C_6H_4O)PO$. I.1.b.γ-2. Liquid, $b_{0.5}$ 204-207°; d_{20} 0.9673; n_D^{20} 1.4750.[770a]
$(EtO)_2(1\text{-}C_{10}H_7O)PO$. I.1.b.γ-2.[835] Liquid, $b_{0.5}$ 151-154°;[770a] d_0^{18} 1.0441;[835] n_D^{20} 1.5245.[770a]
$(MeO)_2(2\text{-}C_{10}H_7O)PO$. I.1.d.β (from $ArOPO_3H_2$). Pale yellow oil, $b_{0.5}$ 150-160°; n_D^{25} 1.5610.[578]
$(EtO)_2(2\text{-}C_{10}H_7O)PO$. I.1.b.γ-2.[770a,835] Liquid, b_1 170°; d_{20} 1.1792;[770a] d_0^{18} 1.0439;[835] n_D^{20} 1.5250.[770a]
$(PhCH_2O)_2(4\text{-}HO\text{-}2,3\text{-}Me_2\text{-}1\text{-}C_{10}H_4O)PO$. I.2.b.γ-1 (from $(PhCH_2O)_2PHO$ and the naphthoquinone). Needles, m. 94-95°.[406a]
$(MeO)_2(10\text{-anthryl-O})PO$. I.2.b.γ-2 (from $(MeO)_3P$ using 10-Br-anthrone). Cryst., m. 111.5-112.5°; IR.[82]
$(EtO)_2(10\text{-anthryl})PO$. Similarly starting from $(EtO)_3P$. Cryst., m. 84.5-85.5°; IR.[82]
$(i\text{-}PrO)_2(10\text{-anthryl})PO$. Similarly starting from $(i\text{-}PrO)_3$-P. Cryst., m. 78-79°; IR.[82]
$(i\text{-}BuO)_2(10\text{-anthryl})PO$. Similarly starting from $(i\text{-}BuO)_3$-P. Cryst., m. 66-70°; IR.[82]

β. Halogen Derivatives

$(MeO)(ClCH_2CH_2O)(PhO)PO$. I.1.b.γ-2 (from $(MeO)(PhO)POCl$[1031] or $(ClCH_2CH_2O)(PhO)POCl$[1]). Liquid, b_4 163-165°,[1] $b_{0.01}$ 130-131°.[1031]
$(MeO)(BrCH_2CH_2O)(PhO)PO$. I.1.e.α-4 - d.δ (from $(MeO)_2$-$(PhO)PO$ reacted with Me_3N then $BrCH_2CH_2Br$). Liquid, $b_{0.05}$ 124°; n_D^{20} 1.514.[858]
$(MeO)(Cl_2C{=}CHO)(PhO)PO$. I.1.f.ε (from $(MeO)(vinyl\text{-}O)PO_2Na$ with $ClCO_2Ph$). Liquid, $b_{0.01}$ 110-125°.[257]
$(MeO)(BrCH_2CH_2CH_2O)(PhO)PO$. I.1.e.α-4 - d.δ (from $(MeO)_2$ $(PhO)PO$ reacted with Me_3N then $Br(CH_2)_3Br$).[858]
$(EtO)(ClCH_2CH_2O)(PhO)PO$. I.1.b.γ-2 (using $(EtO)(PhO)$-$POCl$). Liquid, $b_{0.01}$ 123-124°.[1031]
$(EtO)(Cl_2C{=}CHO)(PhO)PO$. I.1.f.ε (from $(EtO)(vinyl\text{-}O)PO_2Na$

and $ClCO_2Ph$),[257] I.2.b.γ-2 (from $(EtO)_2(PhO)P$ and chloral).[20] Liquid, $b_{0.5}$ 116-130°,[20] $b_{0.04}$ 100-110°.[257]
$(ClCH_2CH_2O)_2(PhO)PO$. I.1.b.γ-2,[355] I.1.d.γ (from $PhOPOCl_2$ and ethylene oxide).[1220] Liquid, $b_{0.5}$ 161-163°,[1220] $b_{0.2-0.4}$ 160-163°;[355] d_{20} 1.399; n_D^{20} 1.5108.[1220]
$(BrCH_2CH_2O)_2(PhO)PO$. I.1.b.γ-2. Liquid, $b_{0.3-0.5}$ 182-185° (dec.).[355]
$(ClCH_2H_2O)(PrO)(PhO)PO$. I.1.b.γ-2 (using $(PrO)(PhO)POCl$). Liquid, $b_{0.01}$ 128-130°.[1031]
$(ClCH_2CH_2O)(BuO(PhO)PO$. I.1.b.γ-2 (using $(BuO)(PhO)POCl$). Liquid, $b_{0.01}$ 133-135°.[1031]
$(ClCH_2CHClCH_2O)_2(PhO)PO$. I.1.2.γ-2 (using $PhOPOCl_2$ and epichlorohydrine). Liquid, $b_{0.5}$ 199-203°; d_{20} 1.4211; n_D^{20} 1.5218.[1220]
$(F_3CCF_2CF_2CH_2O)_2(PhO)PO$. I.1.b.β (from $POCl_3$ stepwise). Liquid, b_1 93-96°.[427]
$(MeO)_2(2\text{-}ClC_6H_4O)PO$. I.2.b.α-8 (from $(MeO)_2PHO$ with Cl_2 then ArOH). Liquid, $b_{2.5}$ 134°.[518]
$(EtO)_2(2\text{-}ClC_6H_4O)PO$. I.1.b.γ-2 (using $(EtO)_2POCl$),[699] I.2.b.α-8 (from $(EtO)_2PHO$ with Cl_2 then ArOH).[1072] Liquid, $b_{2.5}$ 134°,[1072] b_2 140°,[241] $b_{0.005}$ 110°.[699]
$(ClCH_2CH_2O)_2(2\text{-}ClC_6H_4O)PO$. I.1.d.γ (from ethylene oxide using $ArOPOCl_2$). Liquid, $b_{1.5}$ 177-179°; d_{20} 1.4221; n_D^{20} 1.5198.[1220]
$(ClCH_2CHClCH_2O)_2(2\text{-}ClC_6H_4O)PO$. Similarly starting from epichlorohydrine. Liquid, $b_{0.5}$ 205-207°; d_{20} 1.4691; n_D^{20} 1.5280.[1220]
$(MeO)_2(3\text{-}ClC_6H_4O)PO$. I.2.b.α-8 (from $(MeO)_2PHO$ with Cl_2 then ArOH). Liquid, $b_{2.5}$ 139°.[1072]
$(EtO)_2(3\text{-}ClC_6H_4O)PO$. I.1.b.γ-2 (using $(EtO)_2POCl$),[699] I.2.b.α-8 (from $(EtO)_2PHO$ with Cl_2 then ArOH).[1072] Liquid, $b_{1.0}$ 124-126°,[1072] $b_{0.002}$ 95°.[699]
$(MeO)_2(4\text{-}ClC_6H_4O)PO$. I.2.b.α-8 (from $(MeO)_2PHO$ with Cl_2 then ArOH). Liquid, $b_{0.5}$ 108-110°.[518,1072]
$(MeO)(i\text{-}PrO)(4\text{-}ClC_6H_4O)PO$. I.2.b.α-3 (from $(MeO)(i\text{-}PrO)$-PHO with SO_2Cl_2 then ArOH). Liquid, $b_{0.01}$ 94°.[530]
$(EtO)_2(4\text{-}ClC_6H_4O)PO$. I.1.b.γ-2 (using $(EtO)_2POCl$).[699,1225] Liquid, b_2 142°,[241] $b_{0.6}$ 146-150°,[1225] $b_{0.01}$ 114°.[699]
$(ClCH_2CH_2O)(Cl_2C{=}CHO)(4\text{-}ClC_6H_4O)PO$. I.2.b.γ-2 (from aryl ethylene cyclic phosphite and chloral).[1301,1303,1304] Liquid, b_2 100°.[1301]
$(EtO)_2(2\text{-}BrC_6H_4O)PO$. I.1.b.γ-2 (from (alkyl-O)$POCl_2$). Liquid, $b_{0.04}$ 112°.[699]
$(EtO)_2(3\text{-}BrC_6H_4O)PO$. Similarly. Liquid, $b_{0.001}$ 104°.[699]
$(EtO)_2(4\text{-}BrC_6H_4O)PO$. Similarly. Liquid, $b_{0.005}$ 119°.[699]
$(EtO)_2(2\text{-}IC_6H_4O)PO$. Similarly, Liquid, $b_{0.02}$ 131°.[699]
$(EtO)_2(4\text{-}IC_6H_4O)PO$. Similarly. Liquid, $b_{0.005}$ 144°.[699]
$(MeO)(i\text{-}PrO)(2,4\text{-}Cl_2C_6H_3O)PO$. I.2.b.α-3 (from $(MeO)(i\text{-}PrO)PHO$ with SO_2Cl_2 then ArOH). Liquid, $b_{0.01}$ 99°.[530]

$(EtO)_2(2,4\text{-}Cl_2C_6H_3O)PO$. I.1.b.γ-2 (using $(EtO)_2POCl$). Oil, $b_{0.2}$ 142-146°; d_{15} 1.320.[1224]
$(ClCH_2CH_2O)(Cl_2CH{=}CHO)(2,4\text{-}Cl_2C_6H_3O)PO$. I.2.b.γ-2 (from aryl ethylene cyclic phosphite and chloral).[1301,1303,1304] Liquid, b_2 135°.[1301]
$(MeO)_2(2,4,5\text{-}Cl_3C_6H_2O)PO$. I.1.b.γ-2 (using $ArOPOCl_2$). Liquid, $b_{0.4}$ 151-154°; d_{32} 1.44; n_D^{35} 1.5335.[1005]
$(MeO)(i\text{-}PrO)(2,4,5\text{-}Cl_3C_6H_2O)PO$. I.2.b.α-3 (from (MeO)(i-PrO)PHO using SO_2Cl_2 then ArOH),[530] I.2.b.α-8 (similarly using Cl_2).[526] Liquid, $b_{0.01}$ 118°.[526,530]
$(EtO)_2(2,4,5\text{-}Cl_3C_6H_2O)PO$. I.1.b.γ-2 (using $ArOPOCl_2$). Liquid, $b_{0.7}$ 157-158°; d_{32} 1.38; n_D^{35} 1.5129.[1005]
$(MeO)_2(2,4,6\text{-}Cl_3C_6H_2O)PO$. I.1.b.γ-2 (using $ArOPOCl_2$). Cryst., m. 69-71°.[486]
$(EtO)_2(2,4,6\text{-}Cl_3C_6H_2O)PO$. Similarly. Cryst., m. 40-41°.[486]
$(BuO)_2(C_6F_5O)PO$. I.1.b.γ-2 (using $ArOPOCl_2$).[265a,1243] Oil, $b_{0.12(0.15)-0.16}$ 102-106°;[265a,1243] $d_{77°F}$ 1.24; kinetic viscosity at -40°F 263 cStokes.[265a]
$(MeO)_2(C_6Cl_5O)PO$. I.1.b.γ-2 (using $(MeO)_2POCl$),[1481] I.2.-b.α-8 (from $(MeO)_2PHO$ with Cl_2 then ArOH).[1312] Cryst., m. 137-139°,[1481] 124-127°.[1312]
$(EtO)_2(C_6Cl_5O)PO$. I.1.b.γ-2 (using $(EtO)_2POCl$),[1481] I.2-b.α-8 (from $(EtO)_2PHO$ with Cl_2 then ArOH).[1312] Cryst., m. 115-116°,[1481] 110-112°.[1312] ^{31}P + 7.6 ppm.[913c]
$(i\text{-}PrO)_2(C_6Cl_5O)PO$. I.1.b.γ-2 (using $(alkyl\text{-}O)_2POCl$). Cryst., m. 81°.[1481]
$(BuO)_2(C_6Cl_5O)PO$. Similarly. Cryst., m. 50-52°[1481]
$(MeO)_2(3\text{-}Me\text{-}4\text{-}Cl\text{-}C_6H_3O)PO$. I.1.b.γ-2 (using $ArOPOCl_2$),[1072] I.2.b.α-8 (from $(MeO)_2PHO$ with Cl_2 then ArOH).[518] Cryst., m. 131°.[518,1072]
$(ClCH_2CHClCH_2O)_2(4\text{-}cumyl\text{-}C_6H_4O)PO$. I.1.d.γ-1 (from $ArOPOCl_2$ and epichlorohydrine). Liquid, $b_{0.5}$ 230° (dec.); d_{10} 1.2452; n_D^{20} 1.5403.[1220]
$(EtO)_2(2\text{-}Cl_3C\text{-}C_{10}H_6O)PO$. I.1.b.γ-2. Cryst., m. 63°.[1468]
$(ClCH_2CH_2O)_2(3\text{-}Cl\text{-}4\text{-}Me\text{-}7\text{-}cumarinyl\text{-}O)PO$. I.2.b.β-2 (from $(alkyl\text{-}O)_2PHO$ with CCl_4 and the substituted hydroxy-cumarin). Cryst., m. 91°.[431]

γ. Derivatives with Nitro, Alkylthio, or Silyl Groups

$(EtO)_2(2\text{-}O_2NC_6H_4O)PO$. Liquid, b_2 176°.[241]
$(EtO)_2(3\text{-}O_2NC_6H_4O)PO$. Liquid, b_1 172°.[241]
$(MeO)_2(4\text{-}O_2NC_6H_4O)PO$. I.1.b.γ-2 (using $ArOPOCl_2$[352] or $(MeO)_2POCl$[471a,1209]), I.2.b.α-8 (from $(MeO)_2PHO$ with CCl_4 then ArOH).[421] Liquid, b_{10} 164°,[352] $b_{0.52}$ 151°; n_D^{20} 1.5203;[471a,1209] IR.[1209]
$(MeO)(i\text{-}PrO)(4\text{-}O_2NC_6H_4O)PO$. I.2.b.α-3 (from (MeO)(i-PrO)PHO with SO_2Cl_2 then ArOH).[530]
$(EtO)_2(4\text{-}O_2NC_6H_4O)PO$. I.1.b.β,[241,630] I.1.b.γ-2,[241,352,471a,630,699,770a,1209,1224] I.2.b.β-2 (stepwise, from PCl_3 and EtOH with CCl_4 then ArOH[1050] or from $(EtO)_2$-

PHO with CCl_4 then ArOH[421]), I.2.b.β-4 (from $(EtO)_3P$ and ArOH by $CBr_2(CONH_2)_2$).[1013] By nitration (fuming HNO_3, 0°) of $(EtO)_2(PhO)PO$.[241,630,1479] Liquid, b_8 186°,[352] b_1 178-9°,[770a] 175°,[1479] 173°,[241] $b_{0.86-0.90}$ 163.5-5.5°,[471a,1209] $b_{0.6}$ 174-7°,[1224] $b_{0.45}$ 160-5°,[1013] $b_{0.4}$ 153°,[1050] $b_{0.05-0.2}$ 145-60°, $b_{0.2-0.3}$ 150-4°,[421] $b_{0.01}$ 150°;[699] d_{20} 1.2782,[770a] 1.2736,[1050] d_{15} 1.288;[1224] n_D^{20} 1.5105,[1050] 1.5086,[471a] 1.5080;[770a,1209] IR.[1209]

(MeO)(sec-BuO)$(4-O_2NC_6H_4O)PO$. I.2.b.α-3 (from (MeO)(BuO)-PHO with SO_2Cl_2 then ArOH). Oil.[530]

(EtO)(i-PrO)$(4-O_2NC_6H_4O)PO$. Similarly, starting from (EtO)(i-PrO)PHO. Oil.[530]

$(PrO)_2(4-O_2NC_6H_2O)PO$. I.1.b.γ-2,[471a,1209] I.2.b.β-2 (from PCl_3 with CCl_4 then ArOH). Liquid, $b_{0.7}$ 176-7°,[1050] $b_{0.35-0.40}$ 156.5-9°; n_D^{20} 1.5013;[471a,1209] IR.[1209]

$(i-PrO)_2(4-O_2NC_6H_4O)PO$. I.1.b.γ-2,[352,471a,1209] I.2.β-2 (from PCl_3 and i-PrOH with CCl_4 then ArOH;[1050] from $(i-PrO)_2PHO$ with CCl_4 then ArOH[421]). Liquid, b_8 190° (dec.),[352] $b_{0.6}$ 165-6°,[1050] $b_{0.25-0.50}$ 149-50°; n_D^{20} 1.4938;[471a,1209] IR.[1209]

$(BuO)_2(4-O_2NC_6H_4O)PO$. I.1.b.γ-2.[352,471a,1209] Liquid, $b_{0.02}$ 155-6°; n_D^{20} 1.4977.[471a,1209]

$(i-BuO)_2(4-O_2NC_6H_4O)PO$. I.1.b.γ-2. Liquid, $b_{0.05}$ 139-9.5°; n_D^{20} 1.4954.[471a,1209]

$(sec-Bu)_2(4-O_2NC_6H_4O)PO$. I.1.b.γ-2. Liquid, n_D^{20} 1.4959.[471a,1209]

$(AmO)_2(4-O_2NC_6H_4O)PO$. I.1.b.γ-2. Liquid, $b_{0.05-0.09}$ 165-76°; n_D^{20} 1.4944.[471a,1209]

$(i-AmO)_2(4-O_2NC_6H_4O)PO$. I.1.b.β-2 (using $ArOPOCl_2$). Undistillable oil.[352]

(MeO)$(c-C_6H_{11}O)(4-O_2NC_6H_4O)PO$. I.2.b.α-3 (from (MeO)(c-alkyl-O)PHO with SO_2Cl_2 then ArOH). Oil.[530]

$(EtO)_2(3-Me-4-O_2N-C_6H_3O)PO$. Liquid, b_2 149°.[241]

(MeO)(i-PrO)$(4-MeS-C_6H_4O)PO$. I.2.b.α-3 (from (MeO)(i-PrO)PHO with SO_2Cl_2 then ArOH). Liquid, $b_{0.01}$ 109°.[530]

$(EtO)_2(4-MeS-C_6H_4O)PO$. I.1.b.γ-2 (using $(EtO)_2POCl$). Liquid, $b_{0.01}$ 156°.[699]

(MeO)(i-PrO)$(3-Me-4-MeS-C_6H_3O)PO$. I.2.b.α-3 (from (MeO)-(i-PrO)PHO with SO_2Cl_2 then ArOH). Liquid, $b_{0.01}$ 116°.[530]

$(EtO)_2(4-Me_3Si-C_6H_4O)PO$. I.1.b.γ-2. Liquid, $b_{2-2.5}$ 145-50°; n_D^{25} 1.4840.[266]

δ. Derivatives with Other Substituents (without or with Preceding Functions)

$(MeO)(MeCOCH_2O)(PhO)PO$. I.1.d.δ (from $AgO_2P(OMe)(OPh)$ and $BrCH_2CO{\cdot}Me$). Liquid, $b_{0.02}$ 133°.[324]

$(Myristoyl-OCH_2CH_2O)_2(PhO)PO$. I.1.b.γ-2 (from $PhOPOCl_2$ using the corr. acyl glycol). Cryst., m. 42.5-3.5°.[610]

$(Palmitoyl-OCH_2CH_2O)_2(PhO)PO$. Similarly.[134,610] Cryst.,

m. 52-3°,[610] 51.5-2°; IR.[134]
(Stearoyl-$OCH_2CH_2O)_2$(PhO)PO. Similarly.[134,610] Cryst., m. 60.5-1.5°,[610] 59-60°; IR.[134]
(2-$HOC_6H_4CO_2$·$CH_2CH_2O)_2$(PhO)PO. Similarly. Oil.[355]
(Palmitoyl-OCH_2CH_2O)($HOMe_3NCH_2CH_2O$)(PhO)PO. I.1.b.γ-2 (from monoacyl glycol using $PhOPOCl_2$ and choline chloride). Isolated as reineckate, m. 141-2°; sulfate; IR.[134]
(Stearoyl-OCH_2CH_2O)($HOMe_3NCH_2CH_2O$)(PhO)PO. Similarly. Reineckate, m. 144.5-5°; sulfate; IR.[134]
(2,3-Isopropylidene-1-glycero-O)($HOMe_3NCH_2CH_2O$)(PhO)PO. I.1.b.γ-2 (from isopropylidene glycero-phenyl-phosphorochloridate (not isolated) with choline chloride. Reineckate, m. 137-7.5°.[149]
O-(2,3-Di-O-stearoyl-L-1-glycero-P(O)(OPh)-)-N-Cbo-L-serine benzyl ester. I.1.b.γ-2 (from diacylglycero-OP(OPh)OCl and N-Cbo-serine ester). Cryst., m. 58.5-9.5°; $[\alpha]_D^{20}$ 8.45 ($CHCl_3$).[151]
(1,3-Di-O-stearoyl-2-glycero-O$)_2$(PhO)PO. I.1.b.γ-2 (using $PhOCOCl_2$). Cryst., m. 72.2-3.5°.[609]
(Epoxyoleyl-O$)_2$(PhO)PO. From the dioleyl derivative by $MeCO_3H$.[354]
(MeO)($PhCH_2CO$·CH_2O)(PhO)PO. I.1.d.δ (from AgO_2P(OMe)(OPh) and the α-Br-ketone). Liquid, $b_{0.05}$ 150°.[324]
(Stearoyl-OCH_2CH_2O)($PhCH_2O$)(PhO)PO. I.1.d.δ (from AgO_2P-(OPh)(OCH_2Ph) and acyl-OCH_2CH_2I). Cryst., m. 29-30° (unstable modification: m. 22.5-23.5°).[1004]
(Behenoyl-OCH_2CH_2O)($PhCH_2O$)(PhO)PO. Similarly. Cryst., m. 44-5°.[1004]
(2,3-Di-O-myristoyl-1-glycero-O)($PhCH_2O$)(PhO)PO. I.1.d.δ (from diacylglycerol iodohydrine and $AgO_2P(OCH_2Ph)$-(OPh)). Cryst., m. 34-5°.[1004]
(2,3-Di-O-palmitoyl-1-glycero-O)($PhCH_2O$)(PhO)PO. Similarly. Cryst., m. 41.5-2.5°,[1004] 52-3°.[1404]
(2,3-Di-O-stearoyl-1-glycero-O)($PhCH_2O$)(PhO)PO. Similarly. Cryst., m. 48-9°.[1004]
Methyl phenyl dehydroepiandrosterone phosphate. I.1.b.β (from $POCl_3$ stepwise). IR.[1193]
(Dipropyleneglycol$)_2$(2,4,6-$Br_3C_6H_2O$)PO. I.2.b.α-2 (from the phosphite by H_2O_2).[851]
($EtOCH_2CH_2O)_2$(4-$O_2NC_6H_2O$)PO. I.1.b.γ-2 (using $(RO)_2POCl$). Liquid; d_{20} 1.2720; n_D^{20} 1.5050.[585]
(MeO)(pinacolyl-O)(4-$O_2NC_6H_4O$)PO. I.2.b.α-3 (from (MeO)-(alkyl-O)PHO with SO_2Cl_2 then ArOH). Oil.[530]
$(MeO)_2$(4-HOC_6H_4O)PO. I.2.b.α-1 (from $(MeO)_3P$ and p-benzoquinone).[1163] ^{31}P + 4.1 ppm.[913c]
$(EtO)_2$(2-$MeOC_6H_4O$)PO. I.1.b.γ-2.[699,770a] Liquid, $b_{1.5}$ 144°,[770a] $b_{0.001}$ 108°;[699] d_{20} 1.1872; n_D^{20} 1.4977.[770a]
$(EtO)_2$(3-$MeOC_6H_4O$)PO. I.1.b.γ-2.[699,770a] Liquid, $b_{1.5}$ 154-6°,[770a] $b_{0.005}$ 104°;[699] d_{20} 1.1750; n_D^{20} 1.4984.[770a]
$(EtO)_2$(4-$MeOC_6H_4O$)PO. I.1.b.γ-2.[699,770a] Liquid, b_1

148°,[770a] $b_{0.25}$ 119°;[699] d_{20} 1.1757; n_D^{20} 1.4865.[770a]
(2,3-Epoxypropyl-O)$_2$(4-MeOC_6H_4O)PO. I.2.b.γ-1 (from Me diglycidyl phosphite and p-benzoquinone). Liquid, $b_{0.006}$ 167-70°.[1200]
(MeO)$_2$(2-HO-3,4,5,6-Cl_4-phenyl-O)PO. From Me_3 Cl_4-1,2-phenylene cyclic pentaoxyphosphorane by controlled hydrolysis. NMR; ^{31}P + 2.7 ppm.[1159]
(MeO)$_2$(4-HO-2,3,5,6-Cl_4-phenyl-O)PO. I.2.b.γ-1 (from (MeO)$_2$PHO using chloranil). Cryst., m. 236-8°; IR.[1164]
(EtO)$_2$(4-HO-2,3,5,6-Cl_4-phenyl-O)PO. Similarly starting from (EtO)$_2$PHO. Cryst., m. 180-1°; IR.[1164]
(EtO)$_2$(4-EtO-2,3,5,6-Cl_4-phenyl-O)PO. I.2.b.α-1 (from (EtO)$_3$P and chloranil). Cryst., m. 38-40°; IR.[1165]
($PhCH_2O$)$_2$(4-HOC_6H_4O)PO. I.2.b.α-1 (from ($PhCH_2O$)$_2$PHO using p-benzoquinone). Needles, m. 110°.[406a]
(EtO)(glycidyl-O)(4-EtOC_6H_4O)PO. I.2.b.α-1 (from (EtO)$_2$-(glycidyl-O)P using p-benzoquinone). Liquid, $b_{0.004}$ 142-3°; n_D^{20} 1.4979.[1200]
(EtO)(glycidyl-O)(4-glycidyl-C_6H_4O)PO. As above, secondary product. Liquid, $b_{0.003}$ 149-50°.[1200]
(PrO)(glycidyl-O)(4-glycidyl-C_6H_4O)PO. Similarly starting from (PrO)$_2$(glycidyl-O)P. Liquid, $b_{0.004}$ 155-7°; n_D^{20} 1.4410.[1200]
(i-PrO)(glycidyl-O)(4-i-PrC_6H_4O)PO. Similarly starting from (i-PrO)$_2$(glycidyl-O)P. Liquid, $b_{0.005}$ 154-6°; n_D^{20} 1.4909.[1200]
(Glycidyl-O)$_2$(4-PrC_6H_4O)PO. Similarly starting from (PrO)(glycidyl-O)$_2$P. Liquid, $b_{0.004}$ 188-90°; n_D^{20} 1.5030.[1200]
(Glycidyl-O)$_2$(4-i-PrC_6H_4O)PO. As above, with i-Pr derivative. Liquid, $b_{0.004}$ 180-2°.[1200]
(EtO)$_2$(2-$Me_2NC_6H_4$O)PO. I.1.b.γ-2 (using (EtO)$_2$POCl). Liquid, $b_{0.02}$ 129°.[699]
(MeO)$_2$(3-$Me_2NC_6H_4$O)PO. I.1.b.γ-2 (using (MeO)$_2$POCl). Liquid, $b_{0.05}$ 120-3°.[1203]
(EtO)$_2$(3-$Me_2NC_6H_4$O)PO. Similarly using (EtO)$_2$POCl. Liquid $b_{0.0003}$ 116-20°; n_D^{22} 1.5112.[1203]
(i-PrO)$_2$(3-$Me_2NC_6H_4$O)PO. Similarly using (i-PrO)$_2$POCl. Liquid, $b_{0.0001}$ 110-4°; n_D^{22} 1.5102.[1203]
(EtMeCHO)$_2$(3-$Me_2NC_6H_4$O)PO. Similarly using (EtMeCHO)$_2$POCl. Liquid, $b_{0.00002-3}$ 119-24°.[1203]
(EtO)$_2$(4-$H_2NC_6H_4$O)PO. From the nitro derivative by Raney Ni and H_2. Cryst., m. 57°.[699]
(EtO)$_2$(4-$Me_2NC_6H_4$O)PO. I.1.b.γ-2. Liquid, $b_{0.02}$ 132°.[699]
(EtO)$_2$(OHC(2-?)-C_6H_4O)PO. Liquid, b_1 146°.[241]
(EtO)$_2$(OHC(4-?)-C_6H_4O)PO. Liquid, $b_{0.8}$ 160°.[241]
(EtO)$_2$(2-MeO-4-OHC-C_6H_3O)PO. I.2.b.β-2 (from PCl_3 and EtOH with CCl_4 then ArOH). Liquid, $b_{1.8}$ 190°.[1050]
(EtO)$_2$(2-MeO-5-OHC-C_6H_3O)PO. I.2.b.β-2 (from PCl_3 and EtOH with CCl_4 then ArOH).[1050]
(EtO)$_2$(4-MeCO·C_6H_4O)PO. I.1.b.γ-2 (using (EtO)$_2$POCl).

Liquid, $b_{0.001}$ 125°.[699]
$(EtO)_2(2\text{-}EtO_2C{\cdot}C_6H_4O)PO$. Liquid, b_1 156°.[241]
$(EtO)_2(2\text{-}EtO_2C\text{-}4\text{-}O_2N{\cdot}C_6H_3O)PO$. Liquid, b_4 190°.[241]
$(PhCH_2O)_2(2\text{-}H_2NCO{\cdot}C_6H_4O)PO$. I.1.c.β (from salicylamide and $R_4P_2O_7$). Cryst., m. 102-4°.[91]
$(EtO)_2(4\text{-}HO_2C{\cdot}C_6H_4O)$ PO. I.1.b.γ-2 (using $(RO)_2POCl$).[699]
$(EtO)_2(4\text{-}EtO_2C{\cdot}C_6H_4O)$. Liquid, b_2 175°.[241]
(MeO)(i-PrO)$(4\text{-}CN{\cdot}C_6H_4O)PO$. I.2.b.α-3 (from (MeO)(i-PrO)-PHO with SO_2Cl_2 then ArOH). Oil.[530]
$(EtO)_2(4\text{-}CN{\cdot}C_6H_4O)PO$. I.1.b.γ-2 (using $(EtO)_2POCl$),[699] I.2.b.β-2 (from PCl_3 and EtOH with CCl_4 then ArOH).[1050] Liquid, $b_{0.001}$ 105°.[699]
(MeO)(i-PrO)$(3\text{-}Me\text{-}4\text{-}MeO_2S{\cdot}C_6H_4O)PO$. I.2.b.α-3 (from (MeO)(i-PrO)PHO with SO_2Cl_2 then ArOH). Oil.[530]
$(MeO)_2(4\text{-}MeO\text{-}1\text{-}C_{10}H_6O)PO$. I.2.b.γ-1 (from $(MeO)_3P$ and α-naphthoquinone. NMR; ^{31}P + 3.3 ppm.[1159]
$(EtO)_2(1\text{-}HO_2C\text{-}2\text{-}C_{10}H_6O)PO$. I.1.b.γ-2. Cryst., m. 113°.[451]
$(PhCH_2O)_2(4\text{-}HO\text{-}2,3\text{-}Me_2\text{-}1\text{-}C_{10}H_4O)PO$. I.2.b.γ-1 (from $(RO)_2PHO$ and the naphthoquinone derivative). Needles, m. 110°.[406a]
$(MeO)_2$(10-MeO-9-anthryl-O)PO. I.2.b.γ-1 (from $(MeO)_3P$ and anthraquinone). NMR; ^{31}P + 2.8 ppm.[1159]
Dimethyl estrone phosphate. I.1.b.β (stepwise from $POCl_3$). Cryst., m. 151-2°; IR.[1193]
$(MeO)_2$(4-HO-5-pyrenyl-O)PO. From Me pyrenylidene cyclic pentaoxyphosphorane by controlled hydrolysis. IR.[1159]
$(EtO)_2$(xanthone-1-yl) phosphate. I.1.b.γ-2 (from ROH with $(EtO)_2POCl$). Cryst., m. 76-7°.[1026]
$(i\text{-}PrO)_2$(xanthone-1-yl) phosphate. I.1.b.γ-2 (from ROH with $(i\text{-}PrO)_2POCl$). Cryst., m. 93-4°.[1026]
$(EtO)_2$(xanthone-2-yl) phosphate. I.1.b.γ-2 (from ROH with $(EtO)_2POCl$). Cryst., m. 72-4°.[1026]
$(EtO)_2$(xanthone-4-yl) phosphate. I.1.b.γ-2 (from ROH with $(EtO)_2POCl$). Cryst., m. 92-3°.[1026]
$(i\text{-}PrO)_2$(xanthone-4-yl) phosphate. I.1.b.γ-2 (from ROH with $(i\text{-}PrO)_2POCl$). Cryst., m. 89-90°.[1026]

b. Two R Aromatic

α. Hydrocarbon Derivatives

(EtO)$(PhO)_2PO$. I.1.b.β,[994,995] I.1.e.β[996] (both methods are unsatisfactory: impure products); I.1.b.γ-2 (using $(PhO)_2POCl$).[651,1335] Liquid, b_{70} 250-63°, b_{18} 211-21°,[994-6] b_5 187-90°,[1335] b_2 181-5°;[651] d_0^0 1.2113,[994-6] d_{25}^{25} 1.185; n_D^{30} 1.5250.[651]
(PrO)$(PhO)_2PO$. I.1.b.γ-2 (using $(PhO)_2POCl$). Liquid, $b_{2.1}$ 148-50°; $n_D^{15.5}$ 1.5942.[854]
(i-PrO)$(PhO)_2PO$. I.1.b.γ-2 (using $(PhO)_2POCl$).[854,1335] Liquid, $b_{10\text{-}14}$ 98-100°,[854] b_5 192-7°;[1335] $n_D^{15.5}$ 1.5232.[854]
(i-BuO)$(PhO)_2PO$. I.1.b.γ-2 (from $(PhO)_2POCl$). Liquid,

b_5 194-200°.[1335]

$(CH_2{:}CMe{\cdot}CH_2O)(PhO)_2PO$. I.1.b.β. Undistillable liquid, n_D^{25} 1.5242.[1006]

(i-AmO)$(PhO)_2PO$. I.1.b.γ-2 (from $(PhO)_2POCl$). Liquid, b_5 205-10°.[1335]

$(n\text{-}C_8H_{17}O)(PhO)_2PO$. I.1.b.γ-2 (from $(PhO)_2POCl$). Liquid, b_5 233-7°.[1335]

$(2\text{-}Et\text{-}n\text{-}C_6H_{12}O)(PhO)_2PO$. I.1.b.β (stepwise from $POCl_3$ with alkyl-OH then ArOH). Liquid, b_5 ∿232°; d_{25}^{25} 1.090; n_D^{25} 1.510.[587,588]

$(n\text{-}C_{12}H_{25}O)(PhO)_2PO$. I.1.b.β (stepwise from $POCl_3$ with alkyl-OH and ArOH). Oil; $d_{15.5}^{26.5}$ 1.038; n_D^{25} 1.4979.[587]

$(PhCH_2O)(PhO)_2PO$. I.1.b.γ-2 (from $(PhO)_2POCl$). Liquid, b_5 250-60°.[1335]

$(c\text{-}C_6H_{11}O)(PhO)_2PO$. I.1.b.γ-2 (from $(PhO)_2POCl$). Cryst., m. 34-5°.[854]

$(2\text{-}MeC_5H_{10}O)(3\text{-}MeC_6H_4O)_2PO$. I.1.b. (stepwise from $POCl_3$ with alkyl-OH then ArOH). Oil; d_{25}^{25} 1.081; n_D^{25} 1.511.[588]

$(Me_2CHCH_2CH_2CH_2O)(3\text{-}MeC_6H_4O)_2PO$. Similarly. Oil; d_{21}^{21} 1.093; n_D^{25} 1.508.[588]

$(2\text{-}Et\text{-}n\text{-}C_6H_{12}O)(3\text{-}MeC_6H_4O)PO$. Similarly (the m-cresol used contained some p-cresol). Oil, b_5 ∿243°; d_{25}^{25} 1.064; n_D^{25} 1.5072.[587,588]

$(EtO)(1\text{-}C_{10}H_7O)_2PO$. I.1.b.γ-2. Plates, m. 31-2°.[835]

$(6\text{-}Me\text{-}n\text{-}C_7H_{14}O)(1\text{-}C_{10}H_7O)_2PO$. I.1.b.γ-2 (from alkyl-$OPOCl_2$ and ArOH). Oil; d_{25}^{25} 1.1554; viscosity.[986]

$(2\text{-}Me\text{-}n\text{-}C_5H_{10}O)(2\text{-}C_{10}H_7O)_2PO$. Similarly.[986]

$(6\text{-}Me\text{-}n\text{-}C_7H_{14}O)(2\text{-}C_{10}H_7O)_2PO$. Similarly. Oil, d_{25}^{25} 1.1407; n_D^{25} 1.5820.[986]

$(2\text{-}Et\text{-}n\text{-}C_6H_{12}O)(2\text{-}C_{10}H_7O)_2PO$. Similarly. Oil, d_{25}^{25} 1.1415; n_D^{25} 1.5821; viscosity.[986]

β. Halogen and Nitro Derivatives

$(ClCH_2CH_2O)(PhO)_2PO$. I.1.b.γ-2 (from $(PhO)_2POCl$),[323,355,567] I.1.d.γ-1 (from $(PhO)_2POCl$ and ethylene oxide),[1220] I.2.b.β-2 (from $(PhO)_2PHO$ with CCl_4 then ethylene oxide).[517] Sirup, $b_{1.5}$ 201-4°,[1220] $b_{0.5-1}$ 175-8°,[517] $b_{0.5-0.1}$ 165-73°,[667] $b_{0.2-0.5}$ 160-3°,[323] $b_{0.1-0.5}$ 168-73°;[355] d_{20} 1.2911; n_D^{20} 1.5380.[1220]

$(BrCH_2CH_2O)(PhO)_2PO$. I.1.b.γ-2 (from $(PhO)_2POCl$).[323,355] Oil, $b_{0.3-0.5}$ 182-5°.[323]

$(ClCH_2CHClCH_2O)(PhO)_2PO$. I.1.d.γ-1 (from $(PhO)_2POCl$ and epichlorohydrine). Liquid, $b_{0.5}$ 191-7°; d_{20} 1.3263; n_D^{20} 1.5438.[1220]

$[(F_3C)_2CHO](PhO)_2PO$. I.1.b.γ-2 (from $(PhO)_2PHO$ and $(F_3C)_2CO$). Cryst., m. 35-6°; $b_{1.75}$ 140°.[272]

$(F_3CCF_2CF_2CH_2O)(PhO)_2PO$. I.1.b.β (stepwise from $POCl_3$ beginning with PhOH). Liquid, b_1 127-30°.[427]

$(4\text{-}O_2NC_6H_4CH_2O)(PhO)_2PO$. I.1.c.β-2 (from $Ph_4P_2O_7$ with p-nitrobenzyl alcohol). Solid, m. 60-2°.[921]

(BuO)(C_6F_5O)$_2$PO. I.1.b.γ-2 (from (ArO)$_2$POCl).[1243]
(MeO)(2,4,6-$Cl_3C_6H_2O$)$_2$PO. I.1.d.δ. Cryst., m. 132-3°.[1490,1491]
(EtO)(C_6Cl_5O)$_2$PO. I.1.b.γ-2 (from ArOH with EtOPOCl_2). Cryst., m. 158-9°.[1481]
(BuO)(C_6Cl_5O)$_2$PO. I.1.b.γ-2 (from ArOH with BuOPOCl_2). Cryst., m. 143°.[1481]
(EtO)(4-BrC_6H_4O)$_2$PO. I.1.b.γ-2 (from (ArO)$_2$POCl).[975]
(ClCH_2CH_2O)(4-$O_2NC_6H_4O$)$_2$PO. I.1.b.γ-2 (from alkyl-OPOCl_2). Cryst., m. 86-7°.[585]
(MeO)(4-$O_2NC_6H_4O$)$_2$PO. I.1.b.β (stepwise from POCl_3 beginning with MeOH),[786] I.1.d.α-1 (from (ArO)$_2PO_2$H and MeOH by di-p-tolylcarbodiimide).[982] Cryst., m. 142-3°,[982] 141-2.5°.[786]
(EtO)(4-$O_2NC_6H_4O$)$_2$PO. I.1.b.β (stepwise from POCl_3 beginning with EtOH).[786] I.1.e.β (by prolonged boiling of (Ar)$_3$PO in EtOH- result questioned![689]).[1175] Cryst., m. 132-6°,[786] 135°.[1175]
(PhCH_2O)(4-$O_2NC_6H_4O$)$_2$PO. I.1.d.γ-1 (from (ArO)$_2PO_2$H and ROH by di-p-tolylcarbodiimide). Cryst., m. 101-2°.[982]
($F_3CCF_2CF_2CH_2O$)(MeC_6H_4O)$_2$PO. I.1.b.β (stepwise from POCl_3; type of cresol not given). Liquid, b_1 132-6°.[427]
(Cl_2C=CHO)(4-t-Bu·C_6H_4O)$_2$PO. I.2.b.γ-2 (from EtO)(ArO)$_2$P with chloral).[20]
(ClCH_2CHClCH_2O)(4-cumyl-C_6H_4O)$_2$PO. I.1.d.γ-1 (from (ArO)$_2$POCl and epichlorohydrine). Undistillable oil; d_{20} 1.3218; n_D^{20} 1.5431.[1220]

γ. Derivatives with Other Substituents (without or with Preceding Functions)

(Myristoyl-OCH_2CH_2O)(PhO)$_2$PO. I.1.b.γ-2 (from (PhO)$_2$POCl using the monoacyl glycol). Cryst., m. 23.5-4.5°.[1046]
(Palmitoyl-OCH_2CH_2O)(PhO)$_2$PO. Similarly. Cryst., m. 31.5-2.5.[1046]
(Stearoyl-OCH_2CH_2O)(PhO)$_2$PO. Similarly. Cryst., m. 38.5-9.5°,[1046] 37.5-8.5°.[1407]
(Behenoyl-OCH_2CH_2O)(PhO)$_2$PO. Similarly. Cryst., m. 49-50°.[1046]
2-O-Palmitoylbatyl Ph_2 phosphate. I.1.b.γ-2 (from (PhO)$_2$-POCl and monoacyl batyl alcohol). Cryst., m. 46-6.5°.[1314,1402]
2-O-Stearoylbatyl Ph_2 phosphate. Similarly. Cryst., m. 48-9°.[1314,1402]
1-O-Palmitoylbatyl Ph_2 phosphate. Similarly. Cryst., m. 42.5-3°.[1314,1402]
1-O-Stearoylbatyl Ph_2 phosphate. Similarly. Cryst., m. 45.5-6.0°.[1314,1402]
(HO$CH_2CH_2CH_2O$)(PhO)$_2$PO. I.1.b.γ-2 (from (PhO)$_2$POCl).[790]
3-O-Palmitoyl-2-O-stearoyl-1-glycero Ph_2 phosphate. I.1.b.γ-2 (using (PhO)$_2$POCl and diacyl glycerol).

Cryst., m. 50.5-1.5°.[1407]

3-O-Stearoyl-2-O-palmitoyl-1-glycero Ph_2 phosphate. Similarly. Cryst., m. 53-4°.[1407]

2,3-Di-O-stearoyl-1-glycero Ph_2 phosphate. Similarly. Cryst., m. 58-8.5°.[1407]

1-O-Pamitoyl-3-O-stearoyl-2-glycero Ph_2 phosphate. Similarly. Cryst., m. 35.5-40°.[1407]

1,3-Di-O-stearoyl-2-glycero Ph_2 phosphate. Similarly. Cryst., 50.5-1.5°.[1407]

1,2,3-Tri-O-benzoyl-D-erythritol 4-Ph_2-phosphate. I.1.b.γ-2. Cryst., m. 128-32°; $[\alpha]_D^{25}$ -1.1° ($CHCl_3$).[201]

1,2-Isopropylidene-xylofuranose 5-Ph_2-phosphate. I.1.b.γ-2. Cryst., m. 102.2-2.4°; $[\alpha]_D^{20}$ 10.8° ($CHCl_3$).[201]

1,2-Isopropylidene-D-erythro-pentofuranose 5-Ph_2-phosphate. I.1.b.γ-2. Cryst., m. 74°; $[\alpha]_D^{25}$ -13.6°.[1287]

Benzyl β-D-ribofuranoside 2,3-cyclic carbonate 5-Ph_2-phosphate. From benzyl furanoside with $(PhO)_2POCl$ then $COCl_2$. Cryst., m. 86-6.5°.[1345]

3,4,5-Tri-O-acetyl-2-deoxyglucose diethylmercaptal 6-Ph_2-phosphate. I.1.b.γ-2. Sirup, $[\alpha]_D^{17}$ 32° ($CHCl_3$).[196]

3,4,5-Tri-O-acetyl-2-deoxy-D-galactose diethylmercaptal 6-Ph_2-phosphate. I.1.b.γ-2. Sirup; $[\alpha]_D^{15}$ 12° ($CHCl_3$).[196]

1,2,3,4-Tetra-O-acetyl-D-glucose 6-Ph_2-phosphate. I.1.b.γ-2. Cryst., m. 68°.[551]

3,4,6-Tri-O-acetyl-α-D-glucosamine•HCl 1-Ph_2-phosphate. I.1.d.δ (from 1-Br derivative and $Et_3HNO_2P(OPh)_2$). Cryst., m. 137-8°; $[\alpha]_D^{23}$ 110° (MeOH).[908]

4,6-Benzylidene-2-Cboamino-2-deoxy-D-glucose 3-Ph_2-phosphate. I.1.b.γ-2. α-D-Anomer: cryst., m. 98°; $[\alpha]_D^{20}$ 37.5°. β-D-Anomer: cryst., m. 125°; $[\alpha]_D^{20}$ -39.0°.[1450]

2,3,4,5-Tetra-O-benzoyl-aldehydo-D-galactose 6-Ph_2-phosphate. From diethylmercaptal by $CdCO_3$ or $HgCl_2$. Cryst., m. 60-2°; $[\alpha]_D^{17}$ -4° ($CHCl_3$).[196]

2,3,4,5-Tetra-O-acetyl-aldehydo-D-galactose 6-Ph_2-phosphate. From the following by $CdCO_3$ or $HgCl_2$. Cryst., m. 97°; $[\alpha]_D^{18}$ 24° ($CHCl_3$).[196]

2,3,4,5-Tetra-O-acetyl-aldehydo-D-galactose diethylmercaptal 6-Ph_2-phosphate. I.1.b.γ-2. Cryst., m. 83°; $[\alpha]_D^{18}$ -6° ($CHCl_3$).[196]

Methyl 2,3,4-Tri-O-acetyl-β-D-galacturonate 1-Ph_2-phosphate I.1.b.γ-2. Cryst., m. 115-6°; $[\alpha]_D$ 24.2° ($CHCl_3$).[1108]

1,2,3,4-Tetra-O-acetyl-β-D-mannose 6-Ph_2-phosphate. I.1.b.γ-2. Cryst., m. 113-5°; $[\alpha]_D^{20}$ -9.0° ($CHCl_3$).[1132]

1,2-Isopropylidene-fructose 3(?)-Ph_2-phosphate. From the following by 70% acetic acid. Cryst., m. 136°.[277]

1,2;4,5-Di-isopropylidene-fructose 3(?)-Ph_2-phosphate. I.1.b.γ-2. Cryst., m. 71-2°.[277]

L-Sorbose 1-Ph_2-phosphate. I.1.b.γ-2 (starting from the

2,3;4,6-diisopropylidene sugar). $[\alpha]_D^{22}$ -11.7° ($CHCl_3$).[912]
($MeCO{\cdot}CH_2O$)(PhO)$_2$PO. I.1.d.δ (from $(PhO)_2PO_2Na$ and α-Br-ketone). Liquid, $b_{0.02}$ 153°.[324]
($PhCO{\cdot}CH_2O$)(PhO)$_2$PO. I.1.b.γ-2,[1335] I.1.d.δ (from $(PhO)_2$-PO_2Na and α-Br-ketone). Cryst., m. 43°.[324]
($2\text{-}C_{10}H_7CO{\cdot}CH_2O$)($PhO$)$_2$PO. I.1.b.γ-2.[1335]
($HOMe_3NCH_2CH_2O$)(PhO)$_2$PO. I.1.b.γ-2.[150] Reineckate, m. 162-4°,[133] chloroaurate, m. 122°.[150]
N-Cbo-Dihydrosphingosine 1-Ph_2-phosphate. I.1.b.γ-2. Cryst., m. 55°.[1446]
Sterol Derivatives. I.1.b.γ-2 (using $(PhO)_2POCl$).
Dehydroepiandrosterone Ph_2-phosphate. Cryst., m. 94-6°.[1376]
Calciferyl Ph_2-phosphate. Gum; UV.[1376]
Cholesteryl Ph_2-phosphate. Cryst., m. 114-6°; $[\alpha]_D$ -9.2°, UV.[1376]
7-Dehydrocholesteryl Ph_2-phosphate. Cryst., m. 75-7°; $[\alpha]_D$ -31°.[1376]
5-Dehydroergosteryl Ph_2-phosphate. Cryst., m. 74-6°; $[\alpha]_D$ -12°.[1376]
7-Dehydroergosteryl Ph_2-phosphate. Cryst., m. 82-4°; $[\alpha]_D$ -31°, UV.[1376]
Stigmasteryl Ph_2-phosphate. Cryst., m. 99-101°; $[\alpha]_D$ -22°; UV.[1376]
Fucosteryl Ph_2-phosphate. Cryst., m. 80-1°; $[\alpha]_D$ -24°; UV.[1376]
6-Azauridine 5'-Ph_2-phosphate. I.1.b.γ-2 (from 2',3'-isopropylidene uridine with $(PhO)_2POCl$). UV.[1140,1141]
5-Me-6-Azauridine Ph_2-phosphate. I.1.b.γ-2. UV.[1142]
2',3'-di-O-Benzoyl-5-Me-6-azauridine 5'-Ph_2-phosphate. I.1.b.γ-2.[1142]
6-Azauracil 1-[2,3-di-O-benzoyl-5-O-(Ph_2-phosphoryl)-β-D-ribofuranose]. I.1.d.δ (from Br-ribose derivative and Ag salt of 6-aza-3-Ph_3C-uracil). Cryst., m. 192-3°; $[\alpha]_D$ -50° ($CHCl_3$).[1141]
[2-(7-Theophyllinyl)-ethyl] Ph_2-phosphate. I.1.d.δ (from the theophylline salt and $BrCH_2CH_2OP(O)(OPh)_2$). Cryst., m. 165°.[355]
$(PhO)_2$(4-Me-5-$HOCH_2CH_2$-thiazolyl-O)PO. I.2.b.β-2. Liquid, n_D^{20} 1.5620.[857]
(trans-2-Menthyloxyacetoxy-c-C_6H_{10}-O)$(PhO)_2$PO. I.1.b.γ-2. Cryst., m. 76.5°; $[\alpha]_{589}^{20}$ -14.4° (EtOH).[286]
1,2;5,6-Diisopropylidene-(-)-inositol 3-Ph_2-phosphate. I.1.b.γ-2. Solid, m. 150-200°.[796]
Diisopropylidene-pinitol 3-Ph_2-phosphate. I.1.b.γ-2. Cryst., m. 69°; $[\alpha]_D^{25}$ 45.3° (EtOH).[796]
($Et\text{-}OCH_2CH_2O$)($4\text{-}O_2NC_6H_4O$)$_2$PO. I.1.b.γ-2 (from alkyl-$OPOCl_2$). Cryst., m. 72-3°.[585]
1,2,3,4-Tetra-O-acetyl-β-D-glucopyranose 6-($4\text{-}O_2NC_6H_4$)$_2$-phosphate. I.1.b.γ-2. Cryst., m. 146-7°; $[\alpha]_D^{23}$ 28.7°

($CHCl_3$).[1382]

2,3,4-Tri-O-acetyl-[6-O-(4-$O_2NC_6H_2O$)$_2$PO]-α-D-glucopyranosyl bromide. From the sugar derivative with HBr. Cryst., m. 110-1°; $[\alpha]_D^{33}$ 167° ($CHCl_3$).[1382]

Uridine 5'-(4-$O_2NC_6H_4$)$_2$-phosphate. I.1.c.β-2 (from uridine 5'-phosphate with $Ar_4P_2O_7$). Cryst., m. 118.5-20°.[982]

2',3'-Isopropylidene-uridine 5'-(4-$O_2NC_6H_4$)$_2$-phosphate. I.1.d.α-1 (using di-p-tolylcarbodiimide). Needles, m. 118.5-20.5°.[982]

Guanosine 5'-(4-$O_2NC_6H_4$)$_2$-phosphate. I.1.c.β-2 (using $Ar_4P_2O_7$). Amorphous. Isopropylidene derivative: cryst., m. 161-3°, then dec. at 263-4°.[363]

($PhCH_2O$)(4-$H_2NC_6H_4O$)$_2$PO. I.1.e.β (from ($PhCH_2O$)$_3$PO and $H_2NC_6H_4OH$ with metal hydrides as catalyst). Cryst., m. 137-43°.[405]

[($Me_3\overset{+}{N}CH_2CH_2O$)(PhO)(4-$MeC_6H_4$)PO]$_2SO_4^{2-}$. I.1.b.γ-2 (from (ArO)(Ar'O)POCl and choline chloride). Reineckate, m. 138-40°.[480]

c. The Three R Aromatic

α. Hydrocarbon Derivatives

$(PhO)_3PO$. I.1.b.β,[11,102,105,276,669,746] (using $ZnCl_2$ as catalyst),[1295] ($AlCl_3$ as catalyst);[377] from $(RO)_3P$: I.2.b.α-1 (by O_2 and UV irradiation - yield low),[318] I.2.b.α-3 (by SO_3),[298] I.2.b.α-5 (by N_2O_4),[442a,b] I.2.b.β-1 (by $MeCO_3H$),[354] I.2.b.γ-1 (by chloranil then H_2O - no intermediate isolated),[1200] I.2.b.γ-2 (by $BrCH_2CO{\cdot}Ph$, intermediate in C_6H_6: $(RO)_3\overset{+}{P}OC(Ph)=CH_2Br^-$).[1149] From R_2O heated with P_2O_5 (I_2 as catalyst) or $(RO)_2PO_2H$;[712,713] from $(RO)_2PO_2H$ and ROH at 300°;[1138] by distillation of $(RO)_3PCl_2$;[1099] from $(RO)_3P$ and SO_2 or $SOCl_2$ (very complex reactions)[1128] or with $SOMe_2$;[22] from PCl_5 and $(RO)_2SO$;[338] from red P and ROH at 200-300°.[1421] Cryst., m. 49-51°,[442a,b] needles (from Et_2O-ligroin), prisms (from EtOH), m. 50°,[1280] 49°;[276,689] b_{11} 245°.[41] ^{31}P + 18 ppm.[913c]

$(PhO)_2$(2-CH_2=$CHCH_2{\cdot}C_6H_4O$)PO. I.1.b.γ-2. Liquid, $b_{6.5}$ 250-60°, n_D^{25} 1.5640.[282]

$(PhO)_2$(4-t-Bu-C_6H_4O)PO. I.1.b.γ-2. Liquid, b_6 261°.[210]

$(PhO)_2$[2,6-(CH_2=$CHCH_2$)$_2C_6H_3O$]PO. I.1.b.γ-2. Liquid, b_5 254-8°, n_D^{25} 1.5637.[282]

$(PhO)_2$(2,6-t-Bu_2-4-MeC_6H_2O)PO. From $(PhO)_3P$ and 2,6-t-Bu-4-Me-2,5-cyclohexadienone. Cryst., m. 138-9°.[1017]

$(PhO)_2$(2-$Ph{\cdot}C_6H_4O$)PO. I.1.b.γ. Oil, b_{11} 289-90°.[208]

$(PhO)_2$(3-$Ph{\cdot}C_6H_4O$)PO. I.1.b.γ. Liquid, b_8 293°.[278]

$(PhO)_2$(4-$Ph{\cdot}C_6H_4O$)PO. I.1.b.γ. Liquid, b_{10} 302-309°.[279]

(PhO)(4-MeC_6H_4O)$_2$PO. I.1.b.β. Plates, m. 54.[885]

(PhO)(2-CH_2=$CHCH_2{\cdot}C_6H_4O$)$_2$PO. I.1.b.γ. Liquid, b_6 254-62°

n_D^{25} 1.5669.[282]

(PhO)(4-t-BuC$_6$H$_4$O)$_2$PO. I.1.b.γ. Liquid, $b_{5.5}$ 281°.[210]

(PhO)(2-CH$_2$=CMeCH$_2$·C$_6$H$_4$O)$_2$PO. I.1.b.γ. Liquid, $b_{7.5}$ 267-9°, n_D^{25} 1.5647.[282]

(PhO)(2-Me-4-t-BuC$_6$H$_3$O)$_2$PO. I.1.b.γ. Liquid, b_8 280-5°, n_D^{60} 1.5272.[1007]

(PhO)(4-t-BuC$_6$H$_4$O)(2-C$_6$H$_{11}$·C$_6$H$_4$O)PO. I.1.b.γ. Liquid, b_6 293-300°.[210]

(PhO)(2-Ph·C$_6$H$_4$O)$_2$PO. I.1.b.γ. Liquid, $b_{0.5}$ 273-5°.[208]

(PhO)(2-MeC$_6$H$_4$O)(2-Ph·C$_6$H$_4$O)PO. I.1.b.γ. Oil, b_{11} 286-8°.[208]

(PhO)(3-Ph·C$_6$H$_4$O)$_2$PO. I.1.b.γ. Liquid, b_8 345°.[278]

(PhO)(4-Ph·C$_6$H$_4$O)$_2$PO. I.1.b.γ. Liquid, b_{6-7} 360-1°; freezing on cooling; m. 88-90°.[279]

(PhO)(2-C$_6$H$_{11}$-4-Ph-C$_6$H$_3$O)$_2$PO. I.1.b.γ. Liquid, $b_{9.5}$ 371-95°.[1009]

(PhO)(2-Ph-4-C$_6$H$_{11}$·C$_6$H$_3$O)$_2$PO. I.1.b.γ. Liquid, $b_{7.5}$ 340-4°.[1009]

(PhO)(2-C$_{10}$H$_7$O)$_2$PO. I.1.b.β (with AlCl$_3$ catalyst). Oil, b_9 300° (dec.).[277]

(MeC$_6$H$_4$O)$_3$PO (position not given). I.1.a.α (from P$_2$O$_5$ with Al(OR)$_3$ at 320°),[1197] I.1.b.α (with TiCl$_3$ as catalyst).[969]

(2-MeC$_6$H$_4$O)$_3$PO. I.1.b.β.[11,276,669,828a,1175] Cryst., m. 90-1°;[828a] b_{760} 410°, b_3 275-80°;[828a] d_4^{25} 1.183.[276] ^{31}P + 17 ppm.[913c]

(3-MeC$_6$H$_4$O)$_3$PO. I.1.b.β. Liquid, b_{41} 258-63°; m. 25-6°.[276] ^{31}P + 17 ppm.[913c]

(4-MeC$_6$H$_4$O)$_3$PO. I.1.b.α,[104,1471] I.1.b.β.[276,669,1175] Needles, m. 77°,[276] 77.5-78°,[669] 76°.[104,1175] ^{31}P + 17 ± 1 ppm.[913c]

(2-MeC$_6$H$_4$O)$_2$(2-Ph·C$_6$H$_4$O)PO. I.1.b.γ. Oil, b_{23} 315°.[208]

(2-MeC$_6$H$_4$O)$_2$(4-t-BuC$_6$H$_4$O)$_2$PO. I.1.b.γ. Liquid, b_{20} 284°.[210]

(2-MeC$_6$H$_4$O)$_2$(3-Ph·C$_6$H$_4$O)PO. I.1.b.γ. Liquid, b_5 284-98°.[278]

(2-MeC$_6$H$_4$O)$_2$(4-Ph·C$_6$H$_4$O)PO. I.1.b.γ. Liquid, b_{6-7} 303-5°.[279]

(2-MeC$_6$H$_4$O)(4-Ph·C$_6$H$_4$O)$_2$PO. I.1.b.γ. Liquid, b_6 353°.[279]

(2,4-Me$_2$C$_6$H$_3$O)$_3$PO. I.1.b.β. Viscous oil.[827]

(2,5-Me$_2$C$_6$H$_3$O)$_3$PO. I.1.b.β. Cryst., m. 77°.[276]

(3,4-Me$_2$C$_6$H$_3$O)$_3$PO. I.1.b.β. Viscous oil.[827]

(2-CH$_2$=CHCH$_2$·C$_6$H$_4$O)(4-t-BuC$_6$H$_4$O)$_2$PO. I.1.b.γ-2. Liquid, b_8 291-7°.[282]

(2-CH$_2$=CHCH$_2$·C$_6$H$_4$O)$_2$(2-Ph·C$_6$H$_4$O)PO. I.1.b.γ-2. Oil, b_8 293-6°; n^{25} 1.5872.[282]

(2-Me-5-i-PrC$_6$H$_3$O)$_3$PO. I.1.b.α. I.1.b.β. Prisms or plates, m. 75°,[827] 71.5-2°.[749]

(2-i-Pr-5-MeC$_6$H$_3$O)$_3$PO. I.1.b.β. Prisms, m. 59°.[509,827]

(4-t-BuC$_6$H$_4$O)$_3$PO. I.1.b.β. Crystals, m. 101°, b_4 315-7°.[210]

$(4\text{-t-BuC}_6H_4O)_2(2\text{-Me-4-t-BuC}_6H_3O)PO$. I.1.b.γ-2. Liquid, b_8 314-8°.[1007]
$(4\text{-t-BuC}_6H_4O)_2(2\text{-Ph-4-t-BuC}_6H_3O)PO$. I.1.b.γ-2. Liquid, b_5 300-25°.[1009]
$(4\text{-t-BuC}_6H_4O)(2\text{-Me-4-t-BuC}_6H_3O)$ PO. I.1.b.γ. Liquid, b_8 310-4°.[1007]
$(4\text{-t-BuC}_6H_4O)(4\text{-C}_6H_{11}\cdot C_6H_4O)_2PO$. I.1.b.γ. Cryst., m. 81.5°.[210]
$(4\text{-t-BuC}_6H_4O)(2\text{-Ph}\cdot C_6H_4O)(2\text{-Ph-4-t-Bu}\cdot C_6H_3O)PO$. I.1.b.γ-2. Oil, b_6 323-45°.[1009]
$(2\text{-Me-4-t-BuC}_6H_3O)_3PO$. I.1.b.β. Liquid, b_8 304-8°; n_D^{60} 1.5182.[1007]
$(2\text{-Me-4-t-BuC}_6H_4)(2\text{-Ph}\cdot C_6H_4O)_2PO$. I.1.b.γ-2. Liquid, b_8 340-5°.[1007]
$(2\text{-Me-4-t-BuC}_6H_4O)(4\text{-Ph}\cdot C_6H_4O)_2PO$. I.1.b.γ-2. Liquid, b_8 378-85°.[1007]
$(4\text{-t-AmC}_6H_4O)_3PO$. I.1.b.β. Viscous oil.[827]
$(2\text{-i-Pr-4-t-Am-5-MeC}_6H_2O)_3PO$. I.1.e.β. (from triphenyl phosphate). Oil.[809]
$(Me_3CCH_2CMe_2\cdot C_6H_4O)_3PO$ (Para isomer, presumably). I.1.b.β. Oil, b_{12} 361-5°.[212]
$(2\text{-C}_6H_{11}\text{-4-t-BuC}_6H_3O)_2(4\text{-PhCMe}_2\cdot C_6H_4O)PO$. I.1.b.γ. Liquid, b_{10} 378-90°.[1009]
$(2\text{-Ph}\cdot C_6H_4O)_3PO$. I.1.b.β. (using $MgCl_2$ catalyst). Cryst., m. 114°.[206]
$(3\text{-Ph}\cdot C_6H_4O)_3PO$. I.1.b.β. Cryst., m. 84-6°, b_{10} 384°.[278]
$(4\text{-Ph}\cdot C_6H_4O)_3PO$. I.1.b.β. Crystals, m. 137.5°.[279]
$(4\text{-PhCH}_2\cdot C_6H_4O)_3PO$. I.1.b.α. Needles, m. 93-4°.[1088]
$(2\text{-Ph}\cdot C_6H_4O)(2\text{-Ph-4-t-BuC}_6H_3O)_2PO$. I.1.b.γ. Liquid, $b_{7.6}$ 356-60°.[1009]
$(2\text{-Ph}\cdot C_6H_4O)(2\text{-C}_{10}H_7O)_2PO$. I.1.b.γ. Undistillable liquid.[208]
$(2\text{-Ph-4-t-BuC}_6H_3O)_3PO$. I.1.b.β. Liquid, $b_{7.5}$ 344-6°.[1009]
$(4\text{-PhCMe}_2\cdot C_6H_4O)_3PO$. I.1.b.β. Cryst., m. 144°.[1368]
$(1\text{-C}_{10}H_7O)_3PO$. I.1.b.α,[42,413,1290] I.1.b.β.[11,102,276,669,835] Cryst., m. 149-50°,[1191] 148-9°,[11] 145°,[276,835,1290] 144.5-5°.[669]
$(2\text{-C}_{10}H_7O)_3PO$. I.1.b.α.[42,102,1290] I.1.b.β.[102,276,377,669] Needles, m. 111°,[102,276] 110-1°,[42,377,669] 108°.[1290]
$(2,4\text{-Ph}_2\cdot C_{10}H_5O)_3PO$. I.1.b.α. Cryst., m. 130° (from benzene), m. 198-199° (from CCl_4-EtOH).[574]
Tri-3-phenanthryl phosphate. I.1.b.β. Plates, m. 180-2°.[1447]

β. Halogen or Nitro Derivatives

$(PhO)_2(2\text{-ClC}_6H_4O)PO$. I.1.b.γ-2. Liquid, b_4 236°.[207]
$(PhO)_2(2,4\text{-Br}_2C_6H_3O)PO$. I.1.b.γ-2. Liquid, b_8 273-83°, n_D^{25} 1.5992.[281] ^{31}P + 18.6 ppm.[913c]
$(PhO)(2\text{-ClC}_6H_4O)_2PO$. I.1.b.γ-2. Liquid, b_4 254°.[211]
$(C_6F_5O)_3PO$. I.1.b.β (from $POCl_3$). Viscous oil, b_1 164-8°.[1243]

$(2\text{-}ClC_6H_4O)_3PO$. I.1.b.β,[207] I.2.b.α-8 (from $(PhO)_3P$ by Cl_2, besides the 4-Cl derivative).[762] Cryst., m. 37°; $b_{17.5}$ 309°,[207] b_3 228-9°.[762]
$(2\text{-}ClC_6H_4O)(4\text{-}ClC_6H_4O)_2PO$. I.1.b.γ-2. Liquid, $b_{11.5}$ 290°.[207]
$(4\text{-}ClC_6H_4O)_3PO$. I.1.b.β,[11,102,110,276,1015,1502] I.2.b.α-8 (from $(PhO)_3P$ and Cl_2, besides the 2-Cl derivative).[762] By heating $(PhO)_3PO$ with SO_2Cl_2 and Fe as catalyst.[496] Needles, m. 117°,[276] 113°,[496] 112-3°,[1015] 112°;[11] b_{18} 292-5°,[1015] b_3 235-6°.[762]
$(2,4\text{-}Cl_2C_6H_3O)_3PO$. I.1.b.β, I.2.b.α-1 (from $(RO)_3P$ by O_2). Cryst., m. 87°.[760]
$(4\text{-}BrC_6H_4O)_3PO$. Modified I.1.b.α (from PBr_5 with PhO_2CMe; poor yield).[616] By bromination of $(PhO)_3PO$ at 180°.[616] Scales, m. 109°,[276] 101°.[109,616]
$(PhO)(2\text{-}MeC_6H_4O)(2,4\text{-}Br_2C_6H_3O)PO$. I.1.b.γ-2. Liquid, b_8 270-85°; n_D^{25} 1.5934.[281]
$(2\text{-}MeC_6H_4O)_2(2\text{-}ClC_6H_4O)PO$. I.1.b.γ-2. Liquid, b_7 272-3°.[211]
$(2\text{-}MeC_6H_4O)_2(2,4\text{-}Br_2C_6H_3O)PO$. I.1.b.γ-2. Liquid, b_6 290-300°; n_D^{25} 1.5939.[281]
$(3\text{-}MeC_6H_4O)_2(2\text{-}ClC_6H_4O)PO$. I.1.b.γ-2. Liquid, b_{11-2} 279-80°.[211]
$(4\text{-}MeC_6H_4O)_2(2\text{-}ClC_6H_4O)PO$. I.1.b.γ-2. Cryst., m. 52-3°, b_{9-10} 282-4°.[211]
$(2\text{-}MeC_6H_4O)(2\text{-}ClC_6H_4O)_2PO$. I.1.b.γ-2. Liquid, b_{15} 268-70°,[211] b_{15} 312-15° (?).[207]
$(3\text{-}F_3C\cdot C_6H_4O)_3PO$. I.1.b.γ-2. Liquid, dec. on distillation at 14 Torr.[997]
$(2\text{-}Cl_2CH\cdot C_6H_4O)_3PO$. I.1.b.α (from salicylaldehyde).[1325] By chlorination of $(2\text{-}MeC_6H_4O)_3PO$ at 160-80°.[1176,1177] Needles, m. 78°.
$(2\text{-}Me\text{-}4,6\text{-}Cl_2C_6H_2O)_3PO$. I.1.b.α. Needles, dec. 248°.[308]
$(2,4,6\text{-}Cl_3\text{-}3\text{-}MeC_6HO)_3PO$. I.1.b.α. Needles, m. 230°.[309]
$(3\text{-}Me\text{-}4(?)BrC_6H_3O)_3PO$. By bromination of $(3\text{-}MeC_6H_4O)_3PO$. Cryst., m. 90°.[276]
$(4\text{-}Me\text{-}2,6(?)\text{-}Br_2C_6H_2O)_3PO$. By bromination of $(4\text{-}MeC_6H_4)_3PO$. Cryst., m. 90°.[276]
$(2\text{-}MeC_6H_4O)_2(2\text{-}CH_2{=}CClCH_2\cdot C_6H_4O)PO$. I.1.b.γ-2. Liquid, b_7 258-67°.[282]
$(2,4\text{-}Br_2C_6H_3O)(2\text{-}Me\text{-}5\text{-}i\text{-}PrC_6H_3O)_2PO$. I.1.b.γ-2. Liquid, b_8 295-305°.[281]
$(2\text{-}i\text{-}Pr\text{-}4\text{-}Br\text{-}5\text{-}MeC_6H_2O)_3PO$. I.1.b.α. Needles, m. 94-5°.[1115]
$(4\text{-}ClC_6H_4O)_2(4\text{-}t\text{-}BuC_6H_4O)PO$. I.1.b.γ-2. Liquid, b_4 287°.[210]
$(2\text{-}ClC_6H_4O)_2(4\text{-}C_6H_{11}\cdot C_6H_4O)PO$. I.1.b.γ-2. Liquid, b_5 315-25°.[207]
$(2,6\text{-}Cl_2\text{-}4\text{-}PhCMe_2\cdot C_6H_4O)_3PO$. I.1.b.β. Cryst., m. 187°.[1369]
$(2\text{-}Br\text{-}4\text{-}PhCMe_2\cdot C_6H_4O)_3PO$. I.1.b.β. Cryst., m. 87°.[1369]

$(2\text{-ClC}_6H_4O)_2(2\text{-C}_{10}H_7O)PO$. I.1.b.γ-2. Liquid, b_{7-10} 315-30°.[207]

$(1\text{-Cl-2-naphthyl-O})_3PO$. I.1.b.α,[418] I.1.b.β.[102] Needles, m. 152°.[102,418]

$(1,6\text{-Br}_2\text{-2-naphthyl-O})_3PO$. I.1.b.α. Cryst., dec. 200-1°.[42]

$(2\text{-O}_2N\cdot C_6H_4O)_3PO$. I.1.b.α. Needles, m. 126°.[510]

$(4\text{-O}_2NC_6H_4O)_3PO$. I.1.b.α,[510] I.1.b.β.[786,982,1175] By nitration of $(PhO)_3PO$ with fuming HNO_3 below 0° (best).[689,1175] Some 2-O_2N isomer is formed (not isolated).[689] Cryst., m. 156°,[786] 155-6°,[982] 155°,[689,1175] 148°.[510]

$(2\text{-O}_2N\text{-4-PhCMe}_2\cdot C_6H_3O)_3PO$. I.1.b.β. Cryst., m. 96°.[1369]

γ. Derivatives with Hydroxy, Carbonyl and/or Carboxyl Groups (Actual or Potential)

$(2\text{-MeOC}_6H_4O)_3PO$. I.1.b.α,[478] I.1.b.β.[276,828a] Cryst., m. 91°,[276,478] 90-1°; b_3 275-80°.[828a]

$(2\text{-EtOC}_6H_4O)_3PO$. I.1.b.β. Cryst., m. 131-2°.[935]

$(3\text{-HOC}_6H_4O)_3PO$. I.1.b.α. (from resorcinol). Monohydrate, cryst., m. 75°.[1238]

$(4\text{-HOC}_6H_4O)_3PO$. I.1.b.α. (from hydroquinone). Needles, m. 149°.[1238]

$(2\text{-MeO-4-CH}_2\text{=CHCH}_2C_6H_3O)_3PO$. I.1.b.β. Yellow undistillable oil.[505]

$(3\text{-t-Bu-4-HOC}_6H_3)_3PO$. I.1.b.β (from $POCl_3$ with t-Bu-hydroquinone). Cryst., m. 232-4°.[596]

$[3,5\text{-(t-Bu)}_2\text{-4-HOC}_6H_2O]_3PO$. Similarly with $2,4\text{-(t-Bu)}_2$-hydroquinone. Cryst., m. 204.5-6°.[596]

$(PhO)_2(2\text{-MeO}_2C\cdot C_6H_4O)PO$. I.1.b.γ-2. Liquid, $b_{7.6}$ 254-65°.[1008]

$(PhO)(2\text{-MeO}_2C\cdot C_6H_4O)_2PO$. I.1.b.γ-2. Liquid, b_5 274-8°.[1008]

$(2\text{-ClC}_6H_4O)(2\text{-MeO}_2C\cdot C_6H_4O)_2PO$. I.1.b.γ-2. Liquid, $b_{7.6}$ 304-15°.[1008]

$(2\text{-MeO}_2C\cdot C_6H_4O)_3PO$. I.1.b.β. Liquid, $b_{7.6}$ 294-309°.[1008]

$(4\text{-EtO}_2C\text{-2-MeOC}_6H_3O)_3PO$. I.1.b.β. Cryst., m. 99°.[1242]

$(4\text{-t-BuC}_6H_4O)(2\text{-MeO}_2C\cdot C_6H_4O)_2PO$. I.1.b.γ-2. Liquid, b_{7-8} 304-315°.[1008]

$(2\text{-MeO}_2C\cdot C_6H_4O)(2\text{-PhC}_6H_4O)_2PO$. I.1.b.γ-2. Liquid, b_8 340-51°.[1008]

$(2\text{-MeOC}_6H_4O)_2(4\text{-PhC}_6H_4O)PO$. I.1.b.γ-2. Liquid, b_{12} 327°.[279]

$(2\text{-i-BuO}_2C\cdot C_6H_4O)(2\text{-PhC}_6H_4O)_2PO$. I.1.b.γ-2. Cryst., m. 99-102°, b_7 333-6°.[1008]

$(4\text{-ClSO}_2\cdot C_6H_4O)_3PO$. From $(PhO)_3PO$ and $ClSO_3H$ (substitution chiefly in 4 position). Cryst., m. 119-21°; IR; NMR.[521]

$(PhO)_2(4\text{-HO-2,3-Me}_2\text{-1-naphthyl-O})PO$. I.2.b.γ-1 (from $(PhO)_2PHO$ and Me_2-α-naphthoquinone). Cryst., m. 49-5°.[406a]

δ. Derivatives with Nitrogen Groups other than Nitro

(PhO)(4-$H_2N{\cdot}C_6H_4O)_2PO$. I.1.e.β (from $(PhO)_3PO$ and $H_2NC_6H_4OH$). Cryst., m. 118-23°.[405]

(2-$Me_2N{\cdot}C_6H_4O)_3PO$. I.1.b.β. Liquid; n_D^{25} 1.5792.[552]

(3-$H_2N{\cdot}C_6H_4O)_2$(3-$Me_2N{\cdot}C_6H_4O)PO$. I.1.e.β (from $(PhO)_3PO$ reacted successively with $H_2NC_6H_4OH$ and $Me_2NC_6H_4OH$). Cryst., m. 146-51°.[405]

(3-$Me_2N{\cdot}C_6H_4O)_3PO$. I.1.b.β,[502] I.1.e.β (from $(PO)_3PO$ and $Me_2N{\cdot}C_6H_4OH$ in presence of metal hydrides). Cryst., m. 48-9°; $b_{0.001}$ 200°.[552]

(3-$IMe_3N{\cdot}C_6H_4O)_3PO$. From the preceding with MeI. Cryst., m. 207-12° (dec.).[552]

(3-$Et_2N{\cdot}C_6H_4O)_3PO$. I.1.b.β (from $POCl_3$ and $Et_2N{\cdot}C_6H_4OH$). Liquid; n_D^{25} 1.5796.[552]

(3-$IEt_3N{\cdot}C_6H_4O)_3PO$. From the preceding with MeI. Cryst. (5 H_2O), m. 90-103.[552]

(4-$H_2N{\cdot}C_6H_4O)_3PO$. I.1.e.β (from $(PhO)_3PO$ and $Me_2N{\cdot}C_6H_4OH$). Cryst., m. 152-3°.[405]

(4-$Me_2N{\cdot}C_6H_4O)_3PO$. I.1.b.β (from $POCl_3$ and $Me_2N{\cdot}C_6H_4OH$). Cryst., m. 92.5-3°.[552]

(4-$IMe_3N{\cdot}C_6H_4O)_3PO$. From the preceding with MeI. Cryst. (2 H_2O), m. 160-1°.[552]

(4-$H_2N{\cdot}C_6H_4O)_2$(4-t-$BuC_6H_4O)PO$. I.1.e.β (from $(PhO)_2$(4-t-$BuC_6H_4O)PO$ and $H_2N{\cdot}C_6H_4OH$). Oil.[405]

(4-$H_2N{\cdot}C_6H_4O)_2(Ph{\cdot}C_6H_4O)PO$. I.1.e.β (from $(PhO)_2(PhC_6H_4$-O)PO and $H_2N{\cdot}C_6H_4OH$). Cryst., m. 136°.[405]

(4-$PhN{=}NC_6H_4O)_3PO$. I.1.b.α. Yellow needles, m. 148° (from Me_2CO).[676]

[4-(2-$MeC_6H_4N{=}N)C_6H_4O]_3PO$. I.1.b.α. Orange-red needles, m. 116°.[1082]

[4-(4-$MeC_6H_4N{=}N)C_6H_4O]_3PO$. I.1.b.α. Needles, m. 140° (from Me_2CO).[1082]

(2-$Me_2NCH_2{\cdot}C_6H_4O)_3PO$. I.1.b.β (from $POCl_3$ and $Me_2NCH_2{\cdot}C_6$-H_4OH). Oil; n_D^{25} 1.5443.[552]

(2-$IMe_3NCH_2{\cdot}C_6H_4O)_3PO$. From the preceding with MeI. Cryst. (2 H_2O), m. 232-2.5° (dec.).[552]

(4-$Me_2NCH_2{\cdot}C_6H_4O)_3PO$. I.1.b.β (from $POCl_3$ and $Me_2NCH_2{\cdot}C_6$-H_4OH). Oil; n_D^{25} 1.5419; trioxalate, m. 114-5°.[552]

(4-$IMe_3NCH_2{\cdot}C_6H_4O)_3PO$. From the preceding with MeI. Cryst. (3 H_2O), m. 164-6° (dec.).[552]

(1-Me_2N-2-naphthyl-O$)_3PO$. I.1.b.β (from $POCl_3$ and the aminonaphthol). Cryst., m. 106-7°.[552]

(2-Piperidino-1-naphthyl-O$)_3PO$. I.1.b.β (from $POCl_3$ and the naphthylamine). Cryst., m. 169-71°; salt: 3 HCl, 2 H_2O, m. 235-8°.[552]

[2-(I·N-Me·Piperidino)-1-naphthyl-O$]_3PO$. From the preceding with MeI. Cryst., m. 235-8° (dec.).[552]

(1-Piperidinomethyl-2-naphthyl-O$)_3PO$. I.1.b.β (from $POCl_3$ and the naphthylamine). Cryst., m. 206-7°.[552]

$(PhO)_2$(8-Quinolyl-O)PO. I.1.b.γ-2 (from $(PhO)_2POCl$). Cryst., m. 58-9°; $b_{0.12}$ 211-2°; $n_D^{35.5}$ 1.6040.[1155]

c. Heterocyclic Structures

(EtO)(Tetrahydrofuryl)[$MeOCH_2CH_2CH_2NHCO{\cdot}CCl{=}C(MeO)O$]PO. I.2.b.γ-2 (from $(EtO)_2$(c-RO)P and the corr. α,α-Cl_2-acetoacetamide).[404]

Tri-i-pyromucyl phosphate. I.1.b.α, I.1.b.β. Prisms, m. 138°.[375]

$(EtO)_2$(4-Me-1,3-Dioxacyclohexyl-O)PO. I.1.b.γ-2 (from $(EtO)_2POCl$ and 1,3-ethylidene glycerol). Liquid, b_1 113°; n_D^{20} 1.4386.[77]

$(EtO)_2$(3-pyridyl) phosphate. I.1.b.γ-2 (from ROH with $(EtO)_2POCl$). $b_{0.0025}$ 95-8°; n_D^{22} 1.4752.[999]

$(i\text{-}PrO)_2$(3-pyridyl) phosphate. I.1.b.γ-2 (from ROH with $(i\text{-}PrO)_2POCl$). $b_{0.0015}$ 83-6°; n_D^{22} 1.4687.[999]

Diethyl (2,6-Me_2-4-pyridyl) phosphate. I.1.b.γ-2 (from ROK with $(EtO)_2POCl$). Liquid, $b_{0.2}$ 118-20°.[592]

Dibutyl (2,6-Me_2-4-pyridyl) phosphate. I.1.b.γ-2 (from ROK with $(BuO)_2POCl$). Liquid, $b_{0.1}$ 150°.[592]

$(PhO)_2$(3-Pyridyl-O)PO. I.1.b.γ-2. Cryst., m. 33°; $b_{0.1}$ 183-5°; n_D^{22} 1.5672.[1155]

$(1\text{-}C_{10}H_7\text{-}O)_2$(3-Pyridyl-O)PO. I.1.b.γ-2. Oil, $b_{0.11}$ 243-8°; n_D^{27} 1.6372.[1155]

$(EtO)_2$(2-EtO-4-Me-6-Pyrimidyl-O)PO. I.1.b.γ-2 (from ROK with $(EtO)_2POCl$). Liquid, $b_{0.3}$ 145-7°.[592]

$(MeO)_2$[2,4,6-$(MeO)_2$-5-Pyrimidyl-O]PO. From $(MeO)_2$(6-MeO-5-uracilyl-O)PO with CH_2N_2. Cryst., m. 69-71°; ^{31}P + 2.2 ppm.[1160]

$(EtO)_2$(2-MeS-5-Me-6-Pyrimidyl-O)PO. I.1.b.γ-2 (from ROK with $(EtO)_2POCl$). Liquid, $b_{0.05}$ 152-4°.[592]

$(EtO)_2$(2-EtS-4-Me-6-Pyrimidyl-O)PO. I.1.b.γ-2 (from ROK with $(EtO)_2POCl$). Liquid, $b_{0.15}$ 150-3°.[592]

$(EtO)_2$(2,4-Me_2-6-Pyrimidyl-O)PO. I.1.b.γ-2 (from $(EtO)_2$-POCl and RONa,[80] ROAg,[79,80] ROK[592]). Liquid, $b_{5.5}$ 153-7° (some decomposition),[79,80] $b_{0.1}$ 117-9°;[592] d_0^{20} 1.1581; n_D^{20} 1.4720.[79,80]

$(i\text{-}PrO)_2$(2-Et-4-Me-6-Pyrimidyl-O)PO. I.1.b.γ-2 (from ROK with $(i\text{-}PrO)_2POCl$). Liquid, $b_{0.08}$ 120-2°.[592]

$(2\text{-}Et\text{-}n\text{-}C_6H_{12}O)_2$(2-Et-4-Me-6-Pyrimidyl-O)PO. I.1.b.γ-2 (from ROK and $(Et{\cdot}C_6H_{12}O)_2POCl$). Liquid, $b_{0.05}$ 180-3°.[592]

$(EtO)_2$(2-Pr-4-Me-6-Pyrimidyl-O)PO. I.1.b.γ-2 (from ROK and $(EtO)_2POCl$). Liquid, $b_{0.06}$ 128-30°.[592]

$(i\text{-}PrO)_2$(2-Pr-4-Me-6-Pyrimidyl-O)PO. I.1.b.γ-2 (from ROK and $(i\text{-}PrO)_2POCl$). Liquid, $b_{0.1}$ 122-4°.[592]

$(EtO)_2$(2-i-Pr-4-Me-6-Pyrimidyl-O)PO. I.1.b.γ-2 (from ROK and $(EtO)_2POCl$). Liquid, $b_{0.3}$ 123-5°.[592]

$(EtO)_2$(2-Bu-4-Me-6-Pyrimidyl-O)PO. I.1.b.γ-2 (from ROK and $(EtO)_2POCl$). Liquid, $b_{0.2}$ 125-7°.[592]

$(EtO)_2$(2-Am-4-Me-6-Pyrimidyl-O)PO. I.1.b.γ-2 (from ROK and $(EtO)_2POCl$). Liquid, $b_{0.2}$ 140-2°.[592]

$(EtO)_2$(2-Ph-4-Me-6-Pyrimidyl-O)PO. I.1.b.γ-2 (from $(EtO)_2$-POCl and RONa,[80] or ROAg[79,80]). Cryst., m. 69-70°.[79,80]

(i-BuO)$_2$(2-Ph-4-Me-6-Pyrimidyl-O)PO. I.1.b.γ-2 (from (i-BuO)$_2$POCl and RONa,[80] or ROAg[79,80]). Cryst., m. 64-5°.[79,80]
(MeO)$_2$(6-HO-5-Uracilyl-O)PO. I.2.b.γ-4 (from (MeO)$_3$P or (MeO)$_2$PHO and alloxane hydrate, besides the 6-MeO derivative). Cryst., m. 180-3°; IR; ^{31}P -1.3 ppm.[1160]
(MeO)$_2$(6-MeO-5-Uracilyl-O)PO. Compound equally produced in the reaction yielding the preceding compound. Cryst., m. 195-7°; IR; ^{31}P + 4.1 ppm.[1160]
(EtO)$_2$(2-Imido-4-Me-6-uracilyl-O)PO. I.1.d.δ (from the corr. uracil Na salt[80] or Ag salt[79] using (EtO)$_2$POCl). Cryst., m. 110-1°.[79,80]
(BuO)$_2$(2-Imido-4-Me-6-uracilyl-O)PO. Similarly using (BuO)$_2$POCl. Cryst., m. 106-7°.[79,80]
(i-BuO)$_2$(2-Imido-4-Me-6-uracilyl-O)PO. Similarly using (i-BuO)$_2$POCl. Cryst., m. 114-6°.[79,80]
(MeO)$_2$(3-Quinolyl-O)PO. I.1.b.γ-2 (from (MeO)$_2$POCl and RONa). Oil.[1000]
(EtO)$_2$(3-Quinolyl-O)PO. Similarly from (EtO)$_2$POCl. Oil, $b_{0.000011}$ 120-4°; n_D^{21} 1.5360. Methosulfate, cryst., m. 106-7°.[1000]
(i-PrO)$_2$(3-Quinolyl-O)PO. Similarly from (i-PrO)$_2$POCl. Oil, $b_{0.000023}$ 120-4°. Methosulfate, oil.[1000]
Diethyl (5,7-Me$_2$-pyrazolo[1.5-a]pyrimidine-2-yl) phosphate. I.1.b.γ-2 (hydroxy derivative with (EtO)$_2$POCl). Cryst., m. 79-81°.[531]
Diethyl (5-Me-6-EtOC(O)-pyrazolo[1.5-a]pyrimidine-2-yl) phosphate. I.1.b.γ-2 (hydroxy derivative with (EtO)$_2$-POCl). Red-brown oil.[531]

VI. Bis and Tris Tertiary Esters

A. Bis Esters

[-CH$_2$OPO(OBu)$_2$]$_2$. I.1.b.γ-2 (from (BuO)$_2$POCl and glycol). Liquid, b_{10} 164°; d_{20} 0.9910; n^{20} 1.4264.[1003]
O[CH$_2$CH$_2$OPO(OCH$_2$CH$_2$Cl)$_2$]$_2$. I.1.b.β (from POCl$_3$ stepwise), I.1.b.γ-2 (from (alkyl-O)$_2$POCl). Liquid, d_{20} 1.408; n_D^{20} 1.4820.[1003]
O[CH$_2$CH$_2$OPO(OBu)$_2$]$_2$. Similarly. Liquid, d_{20} 1.001; n_D^{20} 1.4418.[1003]
O[CH$_2$CH$_2$OPO(O-n-C$_6$H$_{13}$)$_2$]$_2$. Similarly. Liquid, d_{20} 1.037; n_D^{20} 1.4500.[1003]
O[CH$_2$CH$_2$OPO(O-n-C$_8$H$_{17}$)$_2$]$_2$. Similarly. Liquid, n_D^{20} 1.4530.[1003]
O[CH$_2$CH$_2$OPO(O-c-C$_6$H$_{11}$)$_2$]$_2$. Similarly. Liquid, d_{20} 1.148; n_D^{20} 1.4894.[1003]
O[CH$_2$CH$_2$OPO(OPh)$_2$]$_2$. Similarly. Liquid, d_{20} 1.262; n_D^{20} 1.5470.[1003]
O[CH$_2$CH$_2$OCH$_2$CH$_2$OPO(OBu)$_2$]$_2$. I.1.b.γ-2 (from (BuO)$_2$POCl and HOROH). Oil, n_D^{20} 1.4535.[1003]
MeN[CH$_2$CH$_2$OPO(OBu)$_2$]$_2$. Similarly. Liquid, d_{20} 1.058; n_D^{20}

1.4434.[1003]

$S[CH_2CH_2OPO(OBu)_2]_2$. Similarly. Liquid, d_{20} 1.060; n_D^{20} 1.4410.[1003]

$O_2NCH[CH_2OPO(OBu)_2]_2$. Similarly.[1003]

$ClCH_2CH[OPO(OEt)_2]CH_2OPO(OEt)_2$. I.1.b.γ-2 (from $(EtO)_2$-POCl and $ClCH_2CH(OH)CH_2OPO(OEt)_2$). Liquid, $b_{0.002}$ 148-50°; n_D^{20} 1.4396.[1200]

1,3-Bis-(2-O-oleoyl-3-O-stearoyl-L-1-glycero-benzylphosphoryl)glycerol t-butyl ether. I.1.d.δ (from diacylglycero-benzylphosphate Ag salt and $(ICH_2)_2CHO$-t-Bu). $[\alpha]_D^{20}$ 2.15°.[645]

(D)-$(PhO)_2P(O)OCH_2CH[OPO(OPh)_2]CO_2Me$. I.1.b.γ-2 (from $(PhO)_2POCl$ and D-$HOCH_2CH(OH)CO_2Me$). Liquid; n_D^{26} 1.5540; $[\alpha]_D$ -0.18 ± 0.03° (in substance).[136]

$[-CH_2CH_2OPO(OBu)_2]_2$. I.1.b.γ-2 (from $(PhO)_2POCl$ and HOROH). Liquid, b_8 190-2° (dec.); d_{20} 1.064; n_D^{20} 1.4410.[1003]

$CH_2[CH_2CH_2OPO(O\text{-glycidyl})_2]_2$. I.1.b.γ-2 (from $(CH_2)_5$-$(OPOCl_2)_2$) and glycidol).[1305]

1,3-Bis-(diphenylphosphoryl)-N-Cbo-dihydrosphingosine. I.1.b.γ-2 (from $(PhO)_2POCl$). Cryst., m. 55°.[1446]

1,3-Bis-(diphenylphosphoryl)-N-Cbo-sphingosine. I.1.b.γ-2 (from $(PhO)_2POCl$). Sirup.[1446]

$Me_2C[C_6H_4\text{-}4\text{-}OPO(OPh)_2]_2$. I.1.b.γ-2.[405]

$Me_2C[C_6H_4\text{-}4\text{-}OPO(OC_6H_4\text{-}4\text{-}NH_2)_2]_2$. I.1.e.β (from the preceding and $H_2NC_6H_4OH$).[405]

Hexestrol bis-(dibenzylphosphate). I.1.b.γ-2.[1205]

Stilbestrol bis-(dibenzylphosphate). I.1.b.γ-2. Cryst., m. 109-10°.[1205]

1,3-Bis-(dibenzylphosphoryloxy)-chalcone. I.1.b.γ-2. Cryst., m. 89.5-90.5°.[1205]

2,5-$(EtO)_2$-2,5-$[(PhO)_2P(O)OCH_2]_2$-Dioxane. I.1.b.γ-2 (from $(PhO)_2POCl$ and 2,5-$(HOCH_2)_2$-2,5-$(EtO)_2$-dioxane). Cryst., m. 78°.[422]

Thiamine disulfide bis-(diethylphosphate). I.1.b.γ-2 (from $(RO)_2POCl$ and thiamine disulfide). By oxidation of thiamine dialkylphosphate. Cryst., m. 113-4°.[1024]

Thiamine disulfide bis-(diisopropylphosphate). Similarly. Cryst., m. 131-2°.[1024]

Thiamine disulfide bis-(dibutylphosphate). Similarly. Cryst., m. 92-4°.[1024]

Thiamine disulfide bis-(diphenylphosphate). Similarly. Cryst., m. 153-4°.[1024]

1,3-$[(MeO)_2P(O)O]_2$-C_6H_4. I.1.b.γ-2 (from resorcinol using $(MeO)_2POCl$). Liquid, $b_{0.025}$ 110-5°; d_{20} 1.2720; n_D^{20} 1.4997.[910]

1,3-$[(EtO)_2P(O)O]_2$-C_6H_4. Similarly using $(EtO)_2POCl$.[808, 910] Liquid, $b_{0.13}$ 155-6°; d_{20} 1.2048; n^{20} 1.4856.[910]

1,3-$[(i\text{-}PrO)_2P(O)O]_2$-$C_6H_4$. Similarly using $(i\text{-}PrO)_2POCl$. Liquid, $b_{0.035}$ 110-2 ; d_{20} 1.1069; n_D^{20} 1.4500.[910]

1,4-$[(MeO)P(O)O]_2$-C_6H_4. I.1.b.γ-2 (from hydroquinone using

$(MeO)_2POCl$). Liquid, $b_{0.02}$ 95-8°; d_{20} 1.3883; n_D^{20} 1.4738.[910]

1,4-$[(EtO)_2P(O)O]_2$-C_6H_4. Similarly using $(EtO)_2POCl$.[770a, 808,910] Liquid, b_1 184°,[770a] $b_{0.15}$ 118-21°;[910] d_{20} 1.2789 (?),[770a] 1.1913;[910] n_D^{20} 1.4725,[770a] 1.4575.[910]

1,4-$[(i\text{-}PrO)_2P(O)O]_2$-$C_6H_4$. Similarly using $(i\text{-}PrO)_2POCl$. Liquid, $b_{0.024}$ 102-10°; d_{20} 1.2250; n_D^{20} 1.440.[910]

1,4-$[(EtO)_2P(O)O]_2$-2,3,5,6-Cl_4-Benzene. I.2.b.γ-1 (from $(EtO)_3P$ and chloranil). Cryst., m. 170-1°; IR.[1165]

4,10-$(HO)_2$-5,9-$[(MeO)_2P(O)O]_2$-Pyrene (or 4,9-$(HO)_2$-5,10-$(Ph_2phosphoryl)_2$-pyrene). From the Me_3 4,5;9,10-dicyclo pentaoxyphosphorane derivative by controlled hydrolysis. NMR; ^{31}P + 1.0 ppm.[1159]

B. Tris Ester

1,3,5-Cl_3-2,4,6-$[(MeO)_2P(O)O]_3$-Benzene. I.2.b.γ-2 (from $(MeO)_3P$ and 1,1,3,3,5,5-Cl_6-2,4,6-trioxocyclohexane). Cryst., m. 155°; IR.[1456]

Cyclic Esters

A. Esters with Pentagonal Cycle (Phospholanes)

1. Mono Cyclophosphoric Esters

$(C_2H_4O_2)PO_2H$: 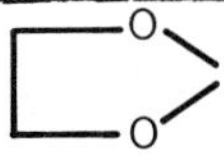 PO_2H. II.1.c (from diol with polyphosphoric acid),[383b] II.2.a (from $BrCH_2CH_2OPO_3H_2$ at pH 7),[833a] II.1.b (from $HOCH_2CH_2PO_3H_2$ at 80°,[383a,383b] or by ClCOOEt[958]); by controlled hydrolysis of the corresponding pyrophosphate.[1456] Sirup; Ba salt;[383a,383b] c-$C_6H_{11}NH_3$ salt.[833a] NMR.[1456]

$(C_2H_4O_2)P(O)OMe$. From the acid with CH_2N_2. Liquid, b_1 85-6°.[383a]

$(C_2H_4O_2)P(O)OCH{=}CH_2$. II.3.a (from Et ethylene phosphite with $ClCH_2CHO$).[998]

$(C_2H_4O_2)P(O)OCH{=}CCl_2$. II.3.a (from Me ethylene phosphite with chloral,[564] similarly from Et ethylene phosphite[20]). Liquid, b_1 110-7°,[564] $b_{0.5}$ 110-7°.[20]

$(C_2H_4O_2)P(O)OPr$. II.3.a (oxidation with N_2O_4).[1100a]

$(C_2H_4O_2)P(O)OCH{=}CHCH_2Cl$. II.1.a (glycol with $POCl_3$ then H_2O and $HC{\equiv}CCH_2Cl$ and $HgSO_4$ as catalyst).[998]

$(C_2H_4O_2)P(O)OBu$. II.3.a (oxidation with N_2O_4).[1100a]

$(C_2H_4O_2)P(O)O$-i-Bu. II.3.a (oxidation with N_2O_4).[1100a]

$(C_2H_4O_2)PO(O)OAm$. II.3.a (oxidation with N_2O_4).[1100a]

$(C_2H_4O_2)P(O)O$-n-C_6H_{13}. II.3.a (oxidation with N_2O_4).[1100a]

$(C_2H_4O_2)P(O)OPh$. II.3.a (oxidation with N_2O_4),[442a,b,1100a] II.1.b (glycol and $PhOPOCl_2$). Cryst., m. 28-9°, $b_{0.3}$ 151-3°[502a], $b_{1.5}$ 106-8°.[442a,b]

$[(CH_2{=}CHCH_2O)C_2H_3O_2]P(O)OCH{=}CH_2$. II.3.a (from the corr. phosphite and H_2O_2).[998]

$[(CH_2{=}CHCH_2O)C_2H_3O_2]P(O)OCH{=}CCl_2$. II.3.a (from the corr. Et ethylene phosphite with chloral).[998]
$(MeC_2H_3O_2)P(O)OMe$. From the corr. cyclophosphorochloridate. Liquid, b_5 114-8°; d_{20} 1.273; n_D^{20} 1.4252.[70]
$(MeC_2H_3O_2)P(O)OEt$. From the corr. cyclophosphorochloridate. Liquid, b_3 105-8°; d_{20} 1.208; n_D^{20} 1.4265.[70]
$(MeC_2H_3O_2)P(O)OCH{=}CH_2$. II.3.a (from Et (Me-ethylene) phosphite with $ClCH_2CHO$).[998]
$(MeC_2H_3O_2)P(O)OCH{=}CCl_2$. II.3.a (from Me propylene phosphite with chloral,[20] similarly from Et propylene phosphite[998]). Liquid, b_2 118-20°,[20] $b_{1.7}$ 121°.[998]
$(MeC_2H_3O_2)P(O)OPr$. From the corr. cyclophosphorochloridate. Liquid, b_3 116-8°; d_{20} 1.1623; n_D^{20} 1.4290.[70]
$(MeC_2H_3O_2)P(O)OBu$. From the corr. cyclophosphorochloridate. Liquid, b_3 127-30°; d_{20} 1.121; n_D^{20} 1.4312.[70]
$(MeC_2H_3O_2)P(O)O$-i-Bu. From the corr. phosphorochloridate. Liquid, b_3 122-4°; d_{20} 1.131; n_D^{20} 1.4310.[70]
$(MeC_2H_3O_2)P(O)OCH_2CH(Et)Bu$. From $(RO_2)POCl(Br)$ with R'OH.[846]
$(MeC_2H_3O_2)P(O)OPh$. From $(PhO)_3PO$ with propylene glycol (Na-alkoxide as catalyst). Liquid, $b_{0.007}$ 123°; n_D^{20} 1.5068.[122a]
$(MeC_2H_3O_2)P(O)OC_6H_4Me$-4. From $(RO_2)POCl(Br)$ with R'OH.[846]
$[(HOCH_2)C_2H_3O_2]PO_2H$. II.1.e (from glycerol),[1328] II.2.b (from β-glycerol phosphate and $(PhO)_2POCl$,[955] by $ClCOOEt$[958]). Liquid, polymerized by $(PhO)_2POCl$ or $Ph_4P_2O_7$.[955]
$[(EtOCH_2)C_2H_3O_2]PO_2H$. II.3.a (from the corr. phosphorochloridite with H_2O_2 and controlled hydrolysis). Liquid, b_{25} 125-30°.[998]
$[(EtOCH_2)C_2H_3O_2]P(O)OCH{=}CCl_2$. II.3.a (from Me 3-ethoxypropylene phosphite with chloral).[20]
$[(CH_2{=}CHCH_2OCH_2)C_2H_3O_2]P(O)OCH{=}CH_2$. II.3.a (from the phosphite and H_2O_2).[998]
Pinacol cyclic phosphate, $(Me_4C_2O_2)PO_2H$. II.1.b (with $POCl_3$).[978] Cryst., m. 186-7°; anilinium salt, m. 138-9°; c-$C_6H_{11}NH_3$ salt, m. 234-6°; IR.[978]
$(Me_4C_2O_2)P(O)OPh$. II.1.b (from pinacol with $PhOPOCl_2$). Cryst., m. 188-8.5°; IR.[978]

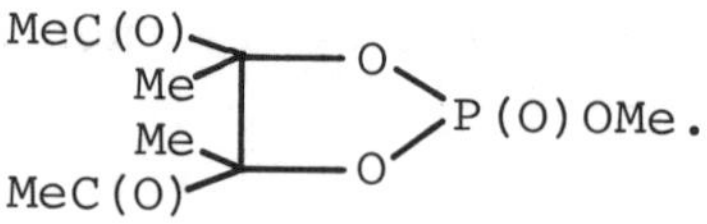

II.3.b.α (from $(MeO)_3P$ with 2 biacetyl then 1 H_2O). Cryst., m. 106-12°. NMR data,[1166] the 2 Me cis.
3,4-Tetrahydrofurylene phosphate. II.1.c (from erythritol heated in vacuo with H_3PO_4), II.2.b (from 4-HO-tetrahydrofuryl 3-phosphate by heating). Prisms, dec. 205°.[329] Identity of this compound (publication of 1862!) uncertain.

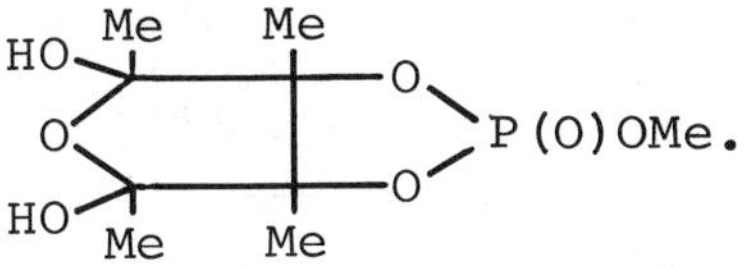
From the preceding with MeOH. Spectroscopic evidence.[1166]

Myoinositol 1,2-cyclic phosphate. II.2.b (from 1- or 2-phosphate by DCC). Solid, m. 130-40° (foaming). Hydrolysis yields mostly 1-phosphate.[1114]

Inositol 1,2-cyclic phosphate. From cycitidine diphosphate inositol by action of 1N NH_3.[863]

1,2:5,6-Diisopropylidene inositol 3,4-cyclic phosphate. II.2.b (from 3-Et_2-phosphate by NH_3). c-$C_6H_{11}NH_3$ salt, $[\alpha]_D^{25}$ 16.7°.[796]

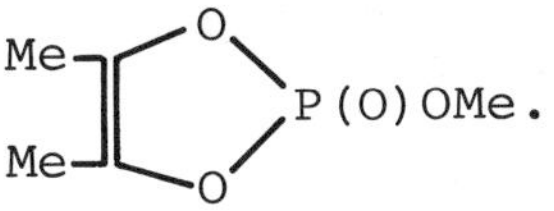
II.3.b.α (from the $(MeO)_3$-phosphorane derivative by controlled hydrolysis). Cryst. m. 42-3°, $b_{0.06}$ 48-9°; n_D^{25} 1.4416; IR.[1170]

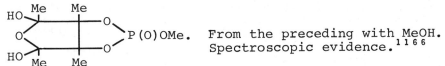

L-Ribitol 1,2-cyclic phosphate. II.b. (from L-ribitol 1-phosphate by action of $(CF_3CO)_2O$. Chromatographic and electrophoretic data.[124]

DL-erythro-(3-Me-hexane-3,4-diol-2-one) 3,4-cyclic Me-phosphate. II.3.b.α (from the corr. $(MeO)_3$-phosphorane by controlled hydrolysis). IR.[1158]

DL-erythro-(3-Me-hexane-3,4-diol-2-one) 3,4-cyclic ethylphosphate. II.3.b.α (from the corr. $(EtO)_3$-phosphorane by controlled hydrolysis). Liquid, $b_{0.2}$ 78-80°; n_D^{25} 1.4381; IR.[1158]

DL-erythro-(3-Me-hexane-3,4-diol-2-one) 3,4-cyclic propylphosphate. II.3.b.α (from the corr. $(PrO)_3$-phosphorane by controlled hydrolysis). Liquid, $b_{0.2-0.1}$ 84-6°; n_D^{25} 1.4402; IR.[1158]

DL-erythro-(3-Me-hexane-3,4-diol-2-one) 3,4-cyclic hexylphosphate. II.3.b.α (from the corr. $(C_6H_{13})_3$-phosphorane by controlled hydrolysis). Liquid, $b_{0.1}$ 109-12°; n_D^{25} 1.4441; IR.[1158]

Glucose 1,2-cyclic phosphate. From adenosine diphosphate glucose by alkali.[1180a]

1,2,3,4-Tetra-O-acetyl-β-D-glucose 6-hydrobenzoin-cyclic-phosphate. From the corresponding phosphorochloridate by controlled hydrolysis, Cryst., m. 168-9°; $[\alpha]_D^{23}$ 18.8° ($CHCl_3$).[1380a]

Galactose 1,2-cyclic phosphate. From adenosine diphosphate galactose by alkali.[1180a]

1-(β-D-ribofuranosyl)-2-pyridone 2',3'-cyclic phosphate. II.2.b (from the 2'(3') phosphate and DCC).[1113]

Uridine 2',3'-cyclic phosphate. II.2.b (from yeast uridylic acid and $(F_3CCO)_2O$),[296] (from 2'(3')-phosphate by heating in Me_2NCHO at 180°,[991] or with c-hexyl-β-(N-Me-morpholino)ethyl carbodiimide 4-toluenesulfonate),[991] (from 5'-O protected 2'(3') phosphate and

$(PhO)_2P(O)OP(O)$(morpholino)Cl, followed by alkaline treatment).[727] Ba salt, very hygroscopic; UV.[296] Protolysis constant in $HCONMe_2$: $K \times 10^4 = 10$.[1133]

5'-O-(1-BuO-ethyl)-uridine 2',3'-cyclic phosphate. From uridine cyclic phosphate with $BuOCH{=}CH_2$.[1240]

5-Br-uridine 2',3'-cyclic phosphate. II.3.a (from the cyclic phosphite and $CO(CCl_3)_2$.[694]

Lyxouridine 2',3'-cyclic phosphate. II.2.b (from 2'(3')-phosphate and DCC).[1381]

Cytidine 2',3'-cyclic phosphate. II.2.b (from cytidine-2'- or 3'-phosphate and $(F_3CCO)_3O$,[296] or ClCOOEt,[958] or DCC[1352]). Ba salt;[296] $(c\text{-}C_6H_{11}NH_3)_2$ salt;[1352] UV.[296]

N^4-(Dimethylaminomethylene)-cytidine 2',3'-cyclic phosphate. From the cyclic cytidine phosphate with Me_2-$NCH(OMe)_2$.[129]

Adenosine 2',3'-cyclic phosphate. II.2.b (from adenylic acid and $2,4\text{-}(O_2N)_2C_6H_3F$,[1465] or $(F_3CCO)_2O$[296]), (from 5'-O protected 3'-phosphate and $(PhO)_2P(O)OP(O)$(morpholino)Cl, followed by alkaline treatment).[727] Ba salt;[296] UV.[296,727]

N^6-(Dimethylaminomethylene)-adenosine 2',3'-cyclic phosphate. From adenosine cyclic phosphate with Me_2-$NCH(OMe)_2$.[129]

2',3'-O-Isopropylidene-adenosine 5'-hydrobenzoin phosphate. II.1.b (using hydrobenzoin phosphorochloridate). $c\text{-}C_6H_{11}NH_3$ salt, m. 208-9°.[1380a]

Guanosine 2',3'-cyclic phosphate. II.2.b (from 2'(3')-phosphate by c-hexyl-β-(N-Me-morpholino)ethyl carbodiimide 4-toluenesulfonate). Isolated by chromatography.[1354]

N^2-(Dimethylaminomethylene)-guanosine 2',3'-cyclic phosphate. From guanosine cyclic phosphate with Me_2-$NCH(OCH_2CMe_3)$.[129]

Adenyl-(3'→5')-cytidine 2',3'-cyclic phosphate. II.2.b (from 2'(3') phosphate and c-hexyl-β-(N-Me-morpholine)-ethyl carbodiimide 4-toluenesulfonate).[1354] Isolated by chromatography.[1354]

Adenyl-(3'→5')-uridine 2',3'-cyclic phosphate. II.2.b (from 2'(3')-phosphate and c-hexyl-β-(N-Me-morpholino)-ethyl carbodiimide 4-toluenesulfonate);[1354] from N,3',5'-triacetyladenosine 3'-phosphate with uridine 2',3'-cyclic phosphate and DCC.[773] Chromatographic data.[1354]

Guanyl-(3'→5')-uridine 2',3'-cyclic phosphate. II.2.b (from 2'(3')-phosphate and c-hexyl-β-(N-Me-morpholino)-ethyl carbodiimide 4-toluenesulfonate). Isolated by chromatography.[1354]

Riboflavine 4',5'-cyclic phosphate. II.2.b (from 5'-phosphate with $POCl_3$[566] or with $Ph_4P_2O_7$ or $(PhCH_2)_4$-P_2O_7[565]). From Ag 2',3'-isopropylidene-adenosine 5'-benzylphosphate and riboflavine 5'-phosphate with

$(PhO)_2POCl$ (breakdown of intermediate flavin-adenosine dinucleotide) or from natural flavin-adenosine dinucleotide by NH_3.[566] Microcrist. solid.[566]

Hydrobenzoin cyclic phosphate. II.1.a (from meso-hydrobenzoin with $POCl_3$).[1327] Cryst. with 1 H_2O, m. 190-2° (dec.); Na salt, m. 262°, $c\text{-}C_6H_{11}NH_3$ salt, m. 190-2°.[1327,1386] Conformation (fixed at room temp.); NMR. ($J_{H\text{-}P}$ 6.9 Hz).[1372]

1,2-Phenylene phosphate. II.1.c (from pyrocatechol with P_2O_5). Glassy solid, b_{15} 244-70°, b_3 225°.[385]

Me 1,2-phenylene phosphate. II.1.d. Liquid, b_{11} 148°.[31]

Et 1,2-phenylene phosphate. II.1.d. Liquid, b_{12} 157°.[31]

3,4,5,6-Br_4-1,2-$C_6O_2P(O)OPh$. II.1.b (using $PhOPOCl_2$). Resinous solid, m. ∿110.[1500]

2. Bis Phosphoric Esters (One or Both Ester Groups Cyclic)

D-Ribofuranose 1,2-cyclic phosphate 5-phosphate. II.2.b (from ribose 1,5-di-phosphate and DCC).[1345]

Guanosine 2',3'-cyclic phosphate 5'-phosphoromorpholidate. II.2.b (from 2'(3'),5'-di-phosphate and DCC in morpholine).[1252]

$(C_2H_4O_2)P(O)OCH_2CH_2OP(O)(O_2H_4C_2)$. From ethylene oxide and P_2O_5 (secondary product besides ethylene pyrophosphate.[1456]

1,2-$C_6H_4[OP(O)O(1,2\text{-}O_2\text{-}C_6H_4)]_2$. From pyrocatechol using o-phenylene cyclic phosphorochloridate. Plates, m. 230°; b_{12} 300°.[37,808]

1,2-[4-t-Bu-1,2-$C_6H_3O_2P(O)O]_2$-4-t-BuC_6H_3. II.1.a. Solid; b_{10} 335-7°,[474] b_{10} 333-7°.[280]

B. Esters with Hexagonal Cycle (Phosphorinanes)

1. Mono Cyclophosphoric Esters

$[(CH_2)_3O_2]PO_2H$: 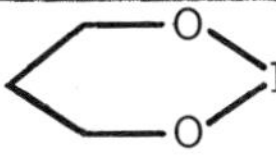 PO_2H. II.2.a (from $ClCH_2CH_2CH_2OH$ with $H_4P_2O_7$ via $ClCH_2CH_2CH_2$-OPO_3H_2),[383a] II.2.b (from $HOCH_2CH_2CH_2OPO_3H_2$ and DCC).[790] Cryst., m. 102-5°,[790] 91-2°;[383a] Ba salt;[383a] $c\text{-}C_6H_{11}NH_3$ salt, m. 177.5-8°.[790]

$[(CH_2)_3O_2]P(O)OMe$. From the acid with N_2CH_2. Liquid, b_1 120°.[383a]

$[(CH_2)_3O_2]P(O)OPh$. II.2.b (from $HOCH_2CH_2CH_2OPO(OPh)_2$ on standing in pyridine). Cryst., m. 76-6.5°.[790] Conformation in the crystal.[758]

$[(CH_2)_3O_2]P(O)OPh$. From 1,3-propanediol with $(PhO)_3PO$ (Na alkoxide as catalyst). Cryst., m. 76-7°.[122a]

2-HO-1,3-propylidene cyclic phosphate. II.2.a (from epichlorohydrine with Na_2HPO_4 via $ClCH_2CH(OH)CH_2PO_3H_2$ production proven but substance not isolated),[159a] (from $ClCH_2CH(OH)CH_2OPO_3Na_2$ by heating in H_2O).[161,167] From CH_2=$CHCH_2OPO_3Na_2$ with Br_2 in H_2O, then evaporation.[160] Na salt.[160,161,167]

1,3-Butanediol cyclic $ClCH_2CH_2$-phosphate. II.1.a (from the diol with $POCl_3$, then ethylene oxide).[847]

1,3-Butanediol cyclic CCl_2=CH-phosphate. II.3.a (from 1,3-butanediol cyclic alkyl phosphite with chloral). Liquid, b_1 132-40°.[20]

1,3-Butanediol cyclic phenyl-phosphate. II.1.a (from the diol with $POCl_3$ and then NaOR). Oil; d_{25}^{25} 1.2507; n_D^{25} 1.5163.[589]

1,3-Butanediol cyclic $4\text{-}ClC_6H_4$-phosphate. II.1.a (from the diol with $POCl_3$ and then NaOR). Oil; d_{25}^{25} 1.3495; n_D^{25} 1.5238.[589]

$2\text{-}O_2N\text{-}2\text{-}HOCH_2$-1,3-propylene phenylphosphate. II.1.b (using $PhOPOCl_2$). Yellow oil, not distillable under 11 Torr.[1503]

2,4-Pentanediol cyclic $(c\text{-}C_6H_{11})$-phosphate. II.1.a (from the diol with $POCl_3$ then ROH).[846]

2,2-$(Me)_2$-1,3-propanediol cyclic phosphate: 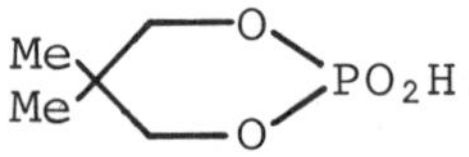

From the corr. phosphorochloridate by controlled hydrolysis. Cryst., m. 174-6°; c-hexylamine salt, m. 240-3°.[500]

2,2-$(Me)_2$-1,3-propanediol cyclic methyl-phosphate. II.1.a (from the diol with $POCl_3$ then NaOMe), II.1.b (from the diol with $MeOPOCl_2$). Cryst., m. 94°, $b_{0.5}$ 118-21°; IR.[501]

2,2-$(Me)_2$-1,3-propanediol cyclic ethyl-phosphate. II.1.b (from the diol with $EtOPOCl_2$). Liquid, $b_{0.3}$ 94-5°; IR.[501]

2,2-$(Me)_2$-1,3-propanediol cyclic Pr-phosphate. II.1.a. Conformation: P=O equatorial both in solution and in solid state.[758]

2,2-$(Me)_2$-1,3-propanediol cyclic i-Pr-phosphate. II.1.b (from the diol with $i\text{-}PrOPOCl_2$). Cryst., m. 57-8°, $b_{3.5}$ 136-9°; IR.[501]

2,2-$(Me)_2$-1,3-propanediol cyclic Bu-phosphate. II.1.a. Conformation: P=O equatorial both in solution and in solid state.[758]

2,2-$(Me)_2$-1,3-propanediol cyclic Ph-phosphate. II.1.a. Conformation: P=O equatorial both in solution and in solid state.[758]

2,2-$(Me)_2$-1,3-propanediol cyclic $4\text{-}O_2NC_6H_4$-phosphate. II.1.b (from the diol with $4\text{-}O_2NC_6H_4OPOCl_2$). Cryst., m. 124-5.5; IR.[501]

2,2-$(ICH_2)_2$-1,3-propylidene phosphate. II.1.a. Cryst., m. 102-3°; Ca salt.[503]

2-Me-1,3-pentanediol cyclic $(CCl_2$=CH)-phosphate. II.3.a (from 2-Me-1,3-pentanediol cyclic alkyl phosphate with chloral).[20]

2-Me-2,4-pentanediol cyclic $(MeCHClCH_2)$-phosphate. II.1.a (from the diol with $POCl_3$ then propylene oxide).[847]

P(O)O-i-Pr. II.1.a (from the diol with $POCl_3$ then ROH).[846]

P(O)OEt. II.1.a (from the diol with $POCl_3$ then EtOH).[846]

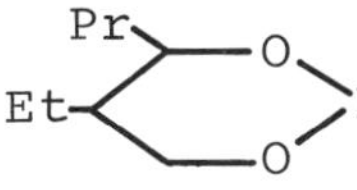P(O)OCH_2CH_2Cl. II.1.a (from the diol with $POCl_3$ then ethylene oxide).[847]

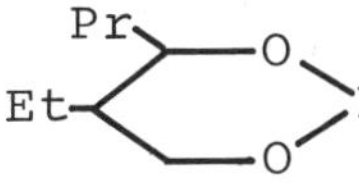P(O)$OCH_2CH_2OCH_2CH(Et)C_4H_9$. II.1.a (from the diol with $POCl_3$ then ROH).[846]

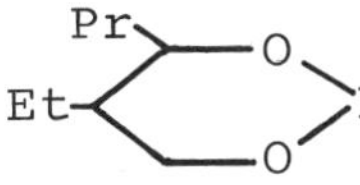P(O)$OCH_2CH_2OC_6H_3Cl_2$-2,4. II.1.a (from the diol with $POCl_3$ then ROH).[846]

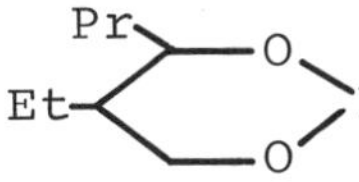P(O)$OCH_2CHClMe$. II.1.a (from the diol with $POCl_3$ then propylene oxide.[847]

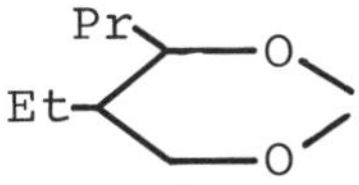P(O)$OCH_2CHClCH_2Cl$. II.1.a (from the diol with $POCl_3$ then epichlorohydrine.[847]

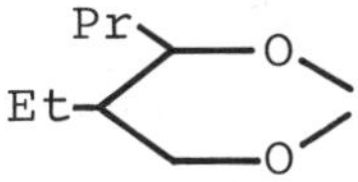P(O)OPh. II.1.a (from the diol with $POCl_3$ and HOAr[846] or NaOAr[589]). Oil; d_{25}^{25} 1.1539; n_D^{25} 1.5017.[589]

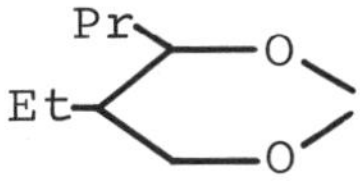P(O)OC_6H_4Me-4. II.1.a (from the diol with $POCl_3$ then ROH).[846]

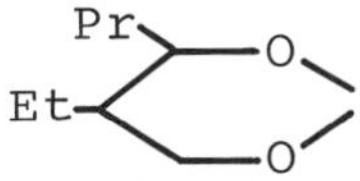P(O)O-x-cresyl. II.1.a (from the diol with $POCl_3$ and NaOR). Oil; d_{25}^{25} 1.1298; n_D^{25} 1.5019.[589]

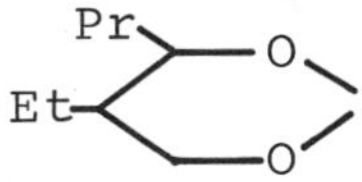P(O)$OC_6H_4CMe_3$-4. II.1.a (from the diol with $POCl_3$ then ROH).[846]

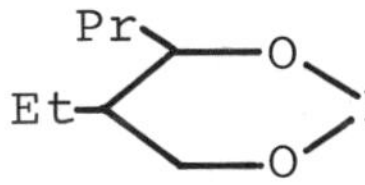P(O)O-2-naphthyl. II.1.a (from the diol with $POCl_3$ and NaOAr). Oil; d_{25}^{25} 1.1814; n_D^{25} 1.5543.[589]

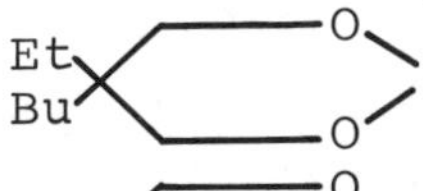P(O)OCH_2CCl_3. II.1.a (from the diol with $POCl_3$ then ROH).[846]

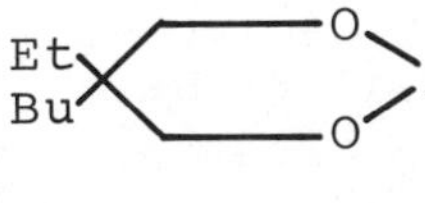P(O)$OCH_2CHBrMe$. II.1.a (from the diol with $POBr_3$ then propylene oxide).[847]

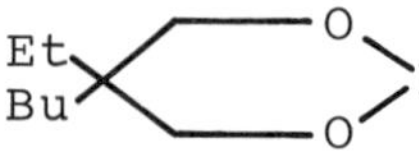

P(O)OPh. II.1.a (from the diol with $POCl_3$ then ROH).[846]

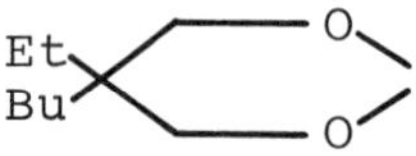

$P(O)OC_6H_3Cl_2$-2,4. II.1.a (from the diol with $POCl_3$ then ROH).[846]

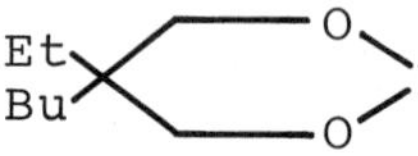

$P(O)OC_6H_4NO_2$-4. II.1.a (from the diol with $POCl_3$ then ROH).[846]

1-Ph-1,3-propanediol cyclic phosphate. II.1.a (via the chloridate). NH_4 salt; conformation (fixed at room temp.); NMR, J_{H-P} 1.5 Hz.[1372]

Chloromycetine 1,3-cyclic phosphate. II.1.a (with $POCl_3$ and controlled hydrolysis of the phosphorochloridate), II.1.c (with tetraphosphoric acid). Needles, m. 134-7°; $[\alpha]_D^{27}$ -16.2°; IR.[1002] No more antibiotic.[1002]

D-Arabinose 3,5-cyclic phosphate. II.1.b (with $PhOPOCl_2$ and dephenylation).[741a]

D-Xylose 3,5-cyclic phosphate. II.1.b.[741a,981] Ba salt.[981]

1,2-O-Isopropylidene-D-xylose 3,5-cyclic phosphate. From the phenyl ester by hydrogenolysis, or 4-nitrophenyl ester by controlled hydrolysis. c-$C_6H_{11}NH_3$ salt, dec. 217-9°, $[\alpha]_D^{21}$ 5.5°.[981]

1,2-O-Isopropylidene-D-xylose 3,5-cyclic Ph-phosphate. II.1.b. Sirup, $b_{0.01}$ (short path) 170°.[981]

1,2-O-Isopropylidene-D-xylose 3,5-cyclic 4-$O_2NC_6H_4$-phosphate. II.1.b. Cryst., m. 148°.[981]

1-Deoxy-1-cyclohexylamino-keto-fructose 4,6-cyclic phosphate. From phenyl β-D-glucoside 4,6-cyclic Ph-phosphate by hydrogenation and heating with c-$C_6H_{11}NH_2$.[128]

D-Glucose 4,6-cyclic phosphate. II.1.b (with $PhOPOCl_2$ and dephenylation).[741a] From phenyl ester of Ph or benzyl glucoside cyclic phosphate.[128,1221] Ba salt, $[\alpha]_D^{20}$ 16.1°.[128,1221]

Me α-D-glucoside 4,6-cyclic phosphate. From the phenyl ester by hydrogenolysis. Cryst., dec. 228-30°, $[\alpha]_D^{20}$ 83.1°.[128]

Me α-D-glucoside 4,6-cyclic phenyl-phosphate. II.1.b (with $PhOPOCl_2$). Cryst., m. 196-7°; $[\alpha]_D^{20}$ 100.8° (EtOH).[128]

Ph β-D-glucoside 4,6-cyclic phenyl-phosphate. II.1.b (with $PhOPOCl_2$). Cryst., m. 193-4°; $[\alpha]_D^{20}$ -86.3°.[128]

Benzyl β-D-glucoside 4,6-cyclic phsophate. From the Ph ester. Cryst., m. 230-4° (dec.); $[\alpha]_D^{24.5}$ -53.0°.[1221]

Benzyl β-D-glucoside 4,6-cyclic Ph-phosphate. II.1.b (with $PhOPOCl_2$). Cryst., m. 202-3°; $[\alpha]_D^{27.5}$ -7.9° (EtOH).[1221]

D-Gluconic acid 4,6-cyclic phosphate. From β-D-glucose 4,6-cyclic phosphate by Br_2[1221] or $NaIO_4$.[741a] Ba salt.[1221] No [α] indicated.[741a,1221]

Deoxyuridine 3',5'-cyclic phosphate. II.2.b (from 5'-phosphate by N,N'-di-c-hexyl-4-morpholinocarboxamidine).[580]

Uridine 3',5'-cyclic phosphate. II.2.b (from 5'-phosphate by N,N'-di-c-hexyl-4-morpholinocarboxamidine).[580]

9-β-D-Xylofuranosyl-hypoxanthine 3',5'-cyclic phosphate. II.2.b (from xylosyladenine 5'-phosphate with $P_2O_3Cl_4$, then with $NaNO_2$). Ba salt, $[\alpha]_D^{25}$ -44.3°; UV.[733]

Deoxyadenosine 3',5'-cyclic phosphate. II.2.b (from 5'-phosphate and N,N'-di-c-hexyl-4-morpholinocarboxamidine).[580]

Adenosine 3',5'-cyclic phosphate. II.2.a (from 5'-fluophosphate and $Ba(OH)_2$),[1465] II.2.b (from 5'-phosphate and N,N'-di-c-hexyl-4-morpholinocarboxamidine),[580] (from N-benzoyl 5'-phosphate and c-hexyl isocyanate).[458]   Isolated from Escherichia coli.[905]

9-β-D-Glucopyranosyl-adenine 4',6'-cyclic phosphate.   From Ph ester and 80% acetic acid. Ba salt.[197]

9-β-D-Glucopyranosyl-adenine 4',6'-cyclic Ph-phosphate. II.1.b (from glucosyladenine using $PhOPOCl_2$).   Needles, m. 272-5° (dec.).[197]

Guanosine 3',5'-cyclic phosphate. II.2.b (from 5'-phosphate by DCC).[1320]

N-Benzoyl guanosine 3',5'-cyclic phosphate. II.2.b (from the 5'-phosphate and DCC or N,N-di-c-C_6H_{11}-4-morpholinocarboxamidine).   UV.[1320]

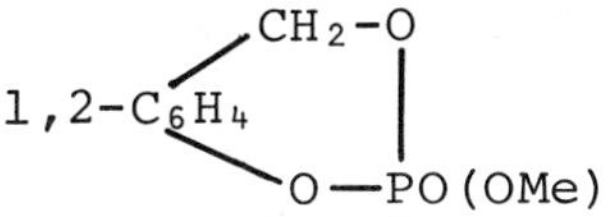

1,2-C_6H_4 $\begin{array}{c} CH_2-O \\ | \\ O-PO(OMe) \end{array}$. II.1.b. Liquid, $b_{0.01}$ 110-2°, n_D^{25} 1.5155.[1281]

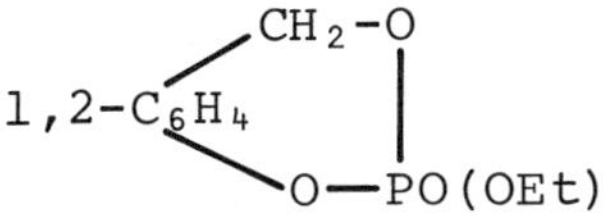

1,2-C_6H_4 $\begin{array}{c} CH_2-O \\ | \\ O-PO(OEt) \end{array}$. II.1.b. Liquid, $b_{0.1}$ 137-9°; n_D^{25} 1.5032.[1075,1281]

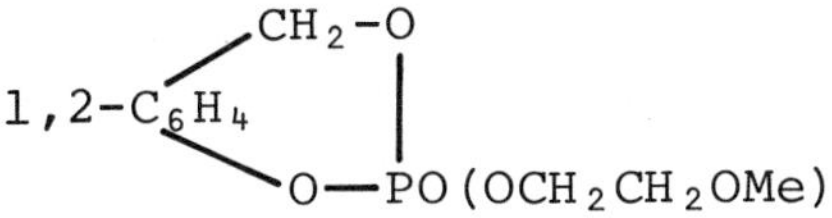

1,2-C_6H_4 $\begin{array}{c} CH_2-O \\ | \\ O-PO(OCH_2CH_2OMe) \end{array}$. II.1.b. Liquid, $b_{0.09}$ 158-60°.[809a]

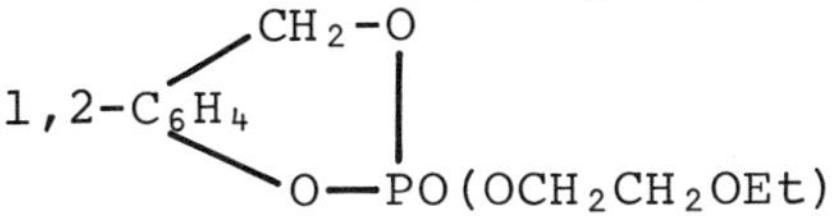

1,2-C_6H_4 $\begin{array}{c} CH_2-O \\ | \\ O-PO(OCH_2CH_2OEt) \end{array}$. II.1.b. Liquid, $b_{0.04}$ 155-8°.[809a]

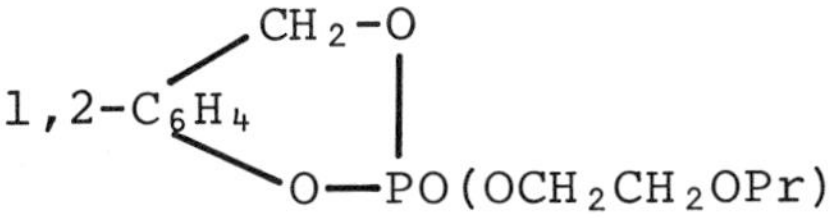

1,2-C_6H_4 $\begin{array}{c} CH_2-O \\ | \\ O-PO(OCH_2CH_2OPr) \end{array}$. II.1.b. Liquid, $b_{0.06}$ 160-1°.[809a]

1,2-C_6H_4 $\begin{array}{c} CH_2-O \\ | \\ O-PO(OCH_2CH_2O-i-Bu) \end{array}$. II.1.b. Liquid, $b_{0.04}$ 169-70°.[809a]

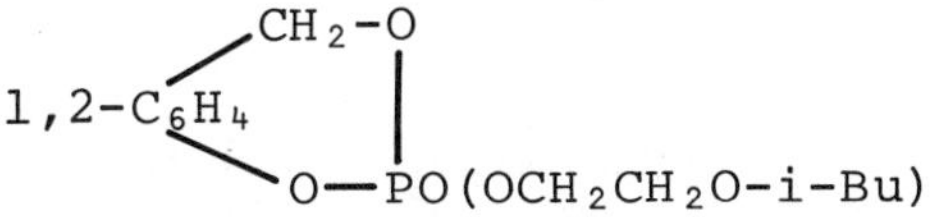

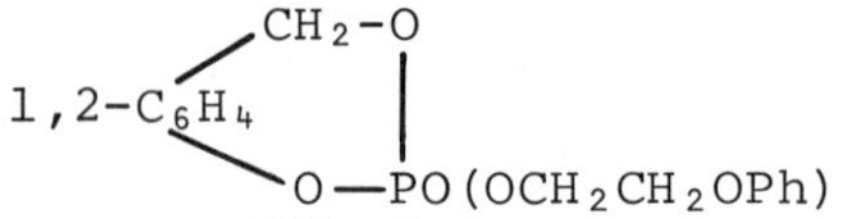

$1,2\text{-}C_6H_4$ $\begin{array}{c} CH_2-O \\ | \\ O-PO(OCH_2CH_2OPh) \end{array}$ • II.1.b. Liquid, $b_{0.05}$ 180-5°.[809a]

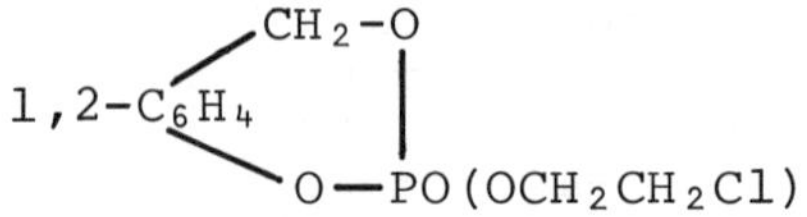

$1,2\text{-}C_6H_4$ $\begin{array}{c} CH_2-O \\ | \\ O-PO(OCH_2CH_2Cl) \end{array}$ • II.1.b. Liquid, $b_{0.06}$ 160-7°.[809a]

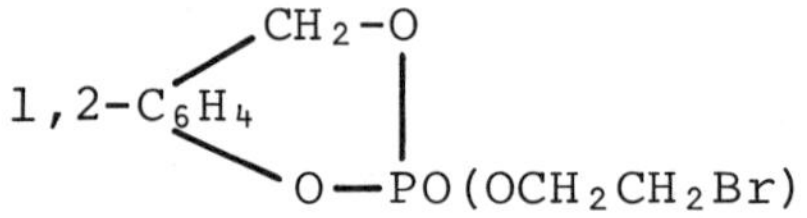

$1,2\text{-}C_6H_4$ $\begin{array}{c} CH_2-O \\ | \\ O-PO(OCH_2CH_2Br) \end{array}$ • II.1.b. Liquid, $b_{0.05}$ 162-4°.[809a]

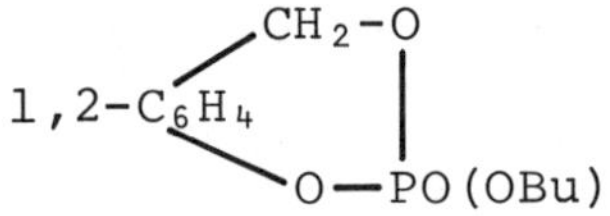

$1,2\text{-}C_6H_4$ $\begin{array}{c} CH_2-O \\ | \\ O-PO(OBu) \end{array}$ • II.1.b. Liquid, $b_{0.05}$ 150-4°; n_D^{25} 1.5020.[1075]

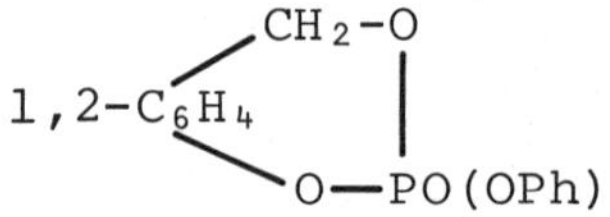

$1,2\text{-}C_6H_4$ $\begin{array}{c} CH_2-O \\ | \\ O-PO(OPh) \end{array}$ • II.1.b (from the diol with PhOPO-Cl_2),[519,1075] II.2.b (from $(2\text{-}HOCH_2\text{-}C_6H_4)(PhO)_2PO$ enzymatically by various plasmata (albumine fraction)--liver homogenate is inactive).[520] Cryst., m. 77-9°.[519,1075] Nucleophilic agents split P-O-Ar bond.[519]

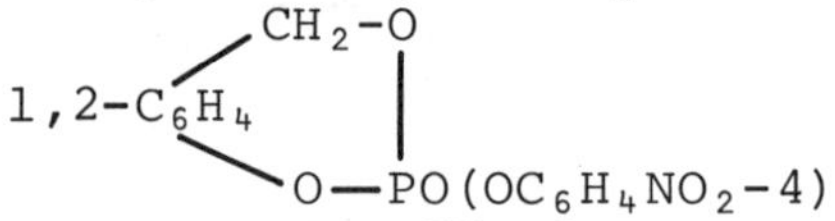

$1,2\text{-}C_6H_4$ $\begin{array}{c} CH_2-O \\ | \\ O-PO(OC_6H_4NO_2\text{-}4) \end{array}$ • II.1.b. Liquid, $b_{0.1}$ 137-9° (?).[1075]

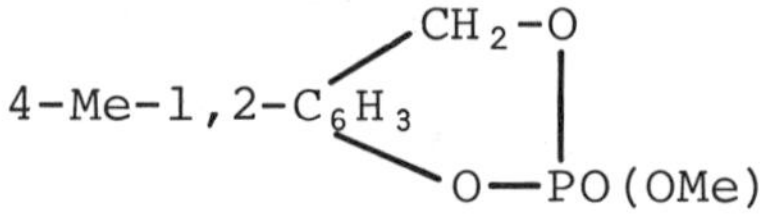

$4\text{-}Me\text{-}1,2\text{-}C_6H_3$ $\begin{array}{c} CH_2-O \\ | \\ O-PO(OMe) \end{array}$ • II.1.b. Liquid, $b_{0.05}$ 109°; n_D^{25} 1.5164.[1075,1281]

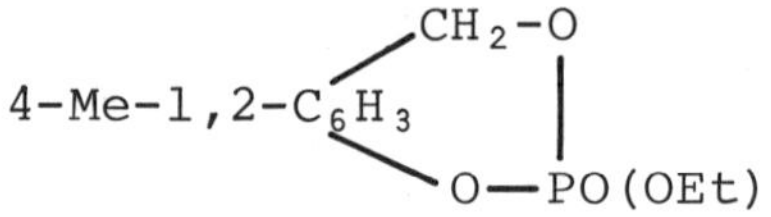

$4\text{-}Me\text{-}1,2\text{-}C_6H_3$ $\begin{array}{c} CH_2-O \\ | \\ O-PO(OEt) \end{array}$ • II.1.b. Liquid, $b_{0.05}$ 112-8°;[1075,1281] n_D^{25} 1.5080.[1281]

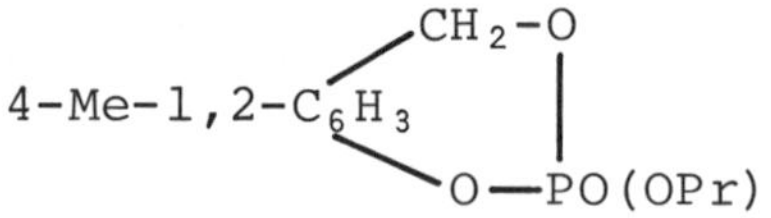

$4\text{-}Me\text{-}1,2\text{-}C_6H_3$ $\begin{array}{c} CH_2-O \\ | \\ O-PO(OPr) \end{array}$ • II.1.b. Liquid, $b_{0.05}$ 140-7°; n_D^{25} 1.5027.[1075]

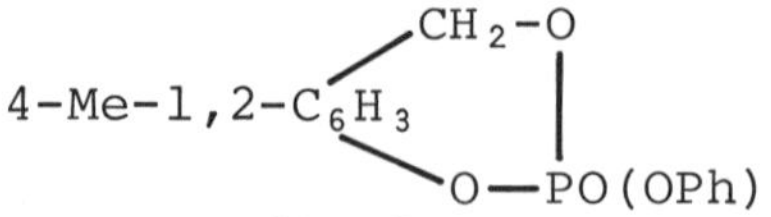

$4\text{-}Me\text{-}1,2\text{-}C_6H_3$ $\begin{array}{c} CH_2-O \\ | \\ O-PO(OPh) \end{array}$ • II.1.b.[1075]

$1,2\text{-}C_6H_4$ $\begin{array}{c} CO-O \\ | \\ O-PO(OEt) \end{array}$ • From the corr. phosphorochloridate with EtOH. Cryst., m. 126.6-7.5°.[93]

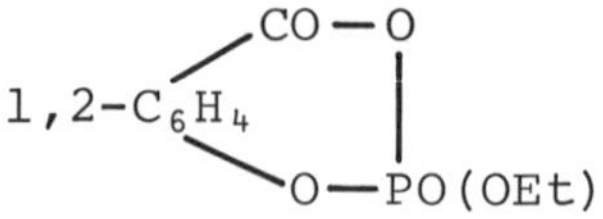

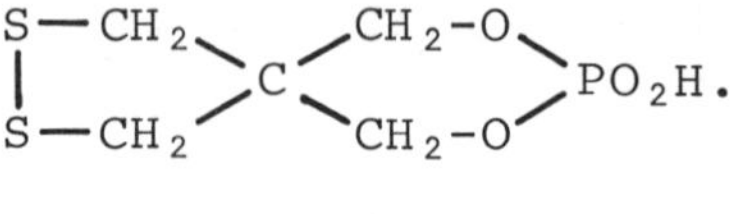

$\begin{array}{c} S-CH_2 \\ | \qquad\quad C \\ S-CH_2 \end{array} \begin{array}{c} CH_2-O \\ \diagdown \\ CH_2-O \end{array} PO_2H.$ From the corr. phosphorochloridate and aqueous NaOH. Solid, dec. ∿200°. Morpholidate (from the chloridate with morpholine), cryst., m. 231-33.5°.[915a]

Prednisolone 17,21-cyclic phosphate. From the (2-HOC_6H_4)-phosphate cyclic phosphate by controlled hydrolysis.[623a]

Prednisolone (2-HOC_6H_4)-phosphate 17,21-cyclic phosphate. II.1.e. Cryst., m. 177-80° (dec.).[623a]

CCl_2=CHOPO  OPOCH=CCl_2. II.3.a (from pentaerythritol dialkylphosphate with chloral). Cryst., m. 130-2°.[20]

2. <u>Bis Phosphoric Ester (One Ester Group Non Cyclic)</u>

Hydrocortisone (2-HOC_6H_4)-phosphate 17,21-cyclic phosphate. II.1.e.[623a]

C. Esters with Heptagonal Cycle (or Larger)

[$(CH_2)_4O_2$]PO_2H. II.1.c (from diol with polyphosphoric acid),[383a] II.2.b (from HO$(CH_2)_4OPO_2H$ and DCC).[790] Cryst., m. 152-3°;[790] Ba salt.[383a]

D-Glucose 3,6-cyclic phosphate. From the 1,2-O-isopropylidene derivative. $[\alpha]_D^{21}$ 20.7°.[1285]

1,2-O-Isopropylidene-α-D-glucose 3,6-cyclic phosphate. II.2.b (from the 3-phosphate derivative with DCC). c-$C_6H_{11}NH_3$ salt, m. 195-200°; $[\alpha]_D^{25}$ 13.4°.[1285]

2,2'-(1,1'-dinaphthylene) phosphate. From the chloridate by controlled hydrolysis. Na salt.[915]

5-O_2N-Uridine 5',4-cyclic phosphate. From the 2',3'-isopropylidene derivative with $P_2O_3Cl_2$, etc.[1254]

Cyclo-[→5'-uridylyl-(3'→5')-uridylyl-(3'→]. II.2.b (from the protected dinucleotide phosphate and DCC). Identification by enzymatic degradation.[1174]

Cyclo-[→5')-thymidylyl-(3'→5')-adenylyl-(3'→]. II.2.b (from the protected mononucleotides and DCC).[1293]

Cyclo-[→5')-uridylyl-(3'→5')-uridylyl-(3'→5')-uridylyl-(3'→]. II.2.b (from the protected trinucleotide phosphate and DCC). Identification by enzymic degradation.[1174]

D. Esters with Condensed Cycles on the P Atom

1,2,3-Glyceryl phosphate. II.1.b (from glycerol heated with polyphosphate). Amorphous mass, generally insoluble.[335] The tricyclic monomolecular structure proposed appears highly improbable.

$MeC(CH_2O)_3PO$: 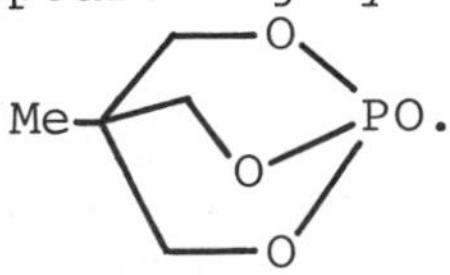 II.3.a (from the corr. phosphite by O_2 or H_2O_2).[700] Cryst., m. 245°.[700]

$HOCH_2C(CH_2O)_3PO$. II.3.a (from the corr. phosphite with

O_2 or H_2O_2).[700]
$O_2NC(CH_2O)_3PO$. II.1.a (from $(HOCH_2)_3CNO_2$ with $POCl_3$). Prisms, m. 243°.[1503]
$EtC(CH_2O)_3PO$. II.3.a (from the corr. phosphite and O_2 or H_2O_2,[700,1045a,1477a] by O_2 with UV irradiation[1264]).
n-$C_6H_{13}C(CH_2O)_3PO$. II.3.a (from the corr. phosphite and O_2 or H_2O_2).[700]
n-$C_{16}H_{33}C(CH_2O)_3PO$. II.3.a (from the corr. phosphite and O_2 or H_2O_2).[700]
1-Oxo-1-phospha-2,8,9-trioxa-adamantane. II.3.a (from the corr. phosphite and H_2O_2).[1318a]

Esters of Anhydro Acids with Inorganic Oxy Acids

A. Esters of Polyphosphoric Acids

1. Pyrophosphoric Esters

a. Monoester

$EtH_3P_2O_7$. III.1.a.β-2. Presence revealed by chromatography in "Langheld" ester.[1453]
t-$BuH_3P_2O_7$. III.1.b.α-1 (from ROH with H_3PO_4 by DCC + CCl_3CN, besides orthophosphate). c-$C_6H_{11}NH_3$ salt, m. 184-5° (dec.).[449a]
3-Me-3-buten-1-yl pyrophosphate (isopentenyl pyrophosphate). III.1.b.δ (from isopentenyl phosphoroamidate). Isolated as tri-(c-$C_6H_{11}NH_3$) salt.[1488]
Geranyl pyrophosphate. III.1.b.α-1 (using H_3PO_4 with DCC + Cl_3CCN, besides orthophosphate). Bis S-benzylthiuronium salt, m. 165-7° (dec.).[449a]
Neryl pyrophosphate. III.1.b.α-1 (using H_3PO_4 with DCC + Cl_3CCN, besides orthophosphate). Bis S-benzylthiuronium salt, m. 150.1° (dec.).[449a]
Linalyl pyrophosphate. III.1.b.α-1 (using H_3PO_4 with DCC + Cl_3CCN, besides orthophosphate). Bis S-benzylthiuronium salt, m. 140-1° (dec.).[449a]
Farnesyl pyrophosphate. III.1.b.α-1 (using H_3PO_4 with DCC + Cl_3CCN, besides orthophosphate). Bis S-benzylthiuronium salt, m. 142-3° (dec.).[449a]
Farnesyl pyrophosphate (tritium and ^{14}C labelled). Enzymatic condensation of geranyl pyrophosphate with isopentenyl pyrophosphate.[1196]
Nerolidyl pyrophosphate. III.1.b.α-1 (using H_3PO_4 with DCC + Cl_3CCN, besides orthophosphate). Bis S-benzylthiuronium salt, m. 134° (dec.).[449a]
Presqualene(tritium labelled), $C_{30}H_{49}OP_2O_6H_2$. From farnesyl pyrophosphate by enzymic action. NMR data.[1196] The non-labelled compound intermediate of biological squalene synthesis.[1196]
(2,3-Distearoyl-1-glyceryl) pyrophosphoric acid.

III.1.a.β-2 (from ROH with P_2O_5). Cryst., m. 63°.[634]

DL-Mevalic acid pyrophosphate. III.1.b (from 3-hydroxy-3-Me-hex-5-en-1-yl phosphoromorpholidate or amidate, and $NaIO_4$ oxidation). Dibrucine salt with 6 H_2O, cryst., m. 173-5°.[902]

L-Ascorbic acid 3-pyrophosphate. III.1.b.γ-1 (from 5,6-di-O-isopropylidene ascorbic acid). NMR.[1052]

L-Ascorbic acid 5-pyrophosphate. (Similarly to 3-pyrophosphate) NMR.[1052]

L-Leucyl-(O-pyrophosphoryl)-L-serine. III.1.2.γ-2 (from the Cbo-dipeptide benzyl ester with $(PhCH_2O)_2POCl$, and hydrogenolysis).[116]

O-Pyrophosphoryl-L-seryl-L-leucine. III.1.b.γ-2 (from the Cbo-dipeptide benzyl ester with $(PhCH_2O)_2POCl$, and hydrogenolysis).[116]

(α-L-Aspartyl-(O-pyrophosphoryl)-L-serine. III.1.b.γ-2 (from the Cbo-dipeptide benzyl ester with $(PhCH_2O)_2$-POCl, and hydrogenolysis).[116]

γ-L-Glutamyl-(O-pyrophosphoryl)-L-serine. III.1.b.γ-2 (from the Cbo-dipeptide benzyl ester with $(PhCH_2O)_2$-POCl, and hydrogenolysis).[116]

2-(4-Me-5-thiazolyl)-ethyl pyrophosphate. III.1.a.β-1 (from dehydrated phosphoric acid at 150°). Ag_3 salt, needles.[1442]

(3-Creatinyl-4-Me-5-($HOCH_2CH_2$)thiazolyl) pyrophosphate. III.1.c (from methyl-hydroxethyl-thiazole with the reaction product of creatine ethyl ester + P_{red} + Br_2 and saponification of the ethyl ester). Cryst., m. 236-40°.[815]

Thiamine chloride pyrophosphate (cocarboxylase). III.1.a.α (from bromothiamine and Ag pyrophosphate),[1443] III.1.a.β-1 (from thiamine chloride and HPO_3[1442] or polyphosphoric acid--besides mono and triphosphate[1083]). Cryst., m. 238-40°.[1441,1442]

Cytidine 5'-diphosphate. III.1.b.α-1 (from monophosphate with H_3PO_4 by DCC),[1265] (from 2',3'-benzylidene derivative of mono phosphate with $(PhCH_2O)_2POCl$, and hydrogenolysis).[954]

Deoxycytidine diphosphate. III.1.b.α-1 (from monophosphate with H_3PO_4 by DCC).[1265]

6-Azacytidine 5'-diphosphate. III.1.b. (from azacytidine 5'-phosphoromorpholidate).[230] Spectroscopic data.[230]

Uridine 5'-diphosphate. III.1.a. (from 2',3'-isopropylidene-5'-deoxy=5'-I-uridine with $Ag(PhCH_2)_3P_2O_7$, and hydrogenolysis),[23] III.1.b.α-1 (from monophosphate with H_3PO_4),[648] III.1.b.γ-2A (from 2',3'-isopropylidene monophosphate with $(PhO)_2POCl$, followed by hydrogenolysis,[648,953] similarly with $(PhCH_2O)_2POCl$,[648] also from 2',3'- benzylidene uridine monophosphate[954]), III.1.b.ε (from 2',3'-isopropylidene monophosphate with dibenzyl N-Phe-benzimidazolyl phosphate, and debenzylation),[95] (from monophosphate with cyanovinyl phosphate),[537,538] III.1.c. (2',3'-isopropylidene uridine 5'-benzyl phosphite with N-Cl-succinimide and $(PhCH_2O)_2POCl$, and hydrogenolysis).[782] Ba salt.[648]

6-Azauridine 5'-pyrophosphate. III.1.b.δ (from 5' phosphoromorpholidate with phosphate). Ba salt; UV.[230]

3-Me-6-azauridine 5'-pyrophosphate. III.1.b.γ-2A (from the monophosphate with $(PhCH_2O)_2POCl$), III.1.b.δ (from the uridine phosphoromorpholidate with H_3PO_4). Constants.[1495a]

Thymidine 5'-diphosphate. III.1.b.γ-2A (from 2',3'-isopropylidene monophosphate with $(PhO)_2POCl$, and hydrogenolysis,[953] 2',3'-benzylidene derivative with $(PhCH_2O)_2POCl$ etc.,[954] 3'-O-benzylthymidine 5'-benzylphosphorochloridate with $PhCH_2OPO_3H_2$, and debenzylation[628]).

Adenosine 5'-diphosphate. III.1.b.α-1 (from 2',3'-isopropylidene adenosine with morpholinophosphorodichloridate, hydrolysis into the monophosphate and reaction with H_3PO_4 and DCC),[721a] (from the monophosphate with H_3PO_4 and DCC),[1265] III.1.b.γ-2A (from adenosine 5'-benzylphosphate Ag salt with $(PhCH_2O)_2POCl$, and hydrogenolysis),[132] (from 2',3'-isopropylidene adenosine phosphate with $(PhO)_2POCl$, and hydrogenolysis),[953] (from 2',3'-benzylidene adenosine phosphate with $(PhCH_2O)_2POCl$ etc.),[954] (from adenosine phosphate Ag salt with $(PhCH_2O)_2POCl$, and hydrogenolysis),[420,954] III.1.b.δ (from monophosphate with $NH_2PO_3H_2$),[360,362,409] (from adenosine phosphoroimidazolidate prepared _in situ_, with H_3PO_4),[448a] III.1.b.η (from monophosphate by oxidation in the presence of 2,3-diMe naphthohydroquinone phosphate),[406b] III.1.b.ζ (from monophosphate + triphosphate by catalytic exchange),[702] III.1.c (from monophosphate with Na_2HPO_3 and Br_2, besides the triphosphate; chromatographic evidence).[1464] From triphosphate by action of $(Bu_3NH)_3PO_4$.[722] Ba salt; acridine salt, m. 215.[132]

Adenosine 5'-diphosphate N^1-oxide. From diphosphate with H_2O_2. UV.[449]

8-Acetoxyadenosine 5'-diphosphate. III.1.b.δ (from the phosphoroamidate).[983]

Guanosine diphosphate. III.1.b.α-1 (from guanosine monophosphate with H_3PO_4 by DCC).[361,1265] Ba salt.[361]

Deoxyguanosine diphosphate. III.1.b.α-1 (from monophosphate with H_3PO_4 by DCC).[1265]

Thiamine diphosphate. III.1.a.β-1 (besides mono- and triphosphate). Chromatographic evidence.[232,1083]

$PhH_3P_2O_7$. III.1.b.η-3 (Oxidation of $PhOPO_3H_2$ in pres. of phosphate and $C{=}N^+$ containing compounds).[408]

b. Diester

[(HO(MeO)P(O)]$_2$O. III.l.a.β-l (from $(MeOPO_2)_3$ by methanolysis, besides $MeOPO_3H_2$).[448a]

$(EtO)_2P(O)OPO_3H_2$. From di$(3\text{-}O_2NC_6H_4)$ ester by alkaline hydrolysis. Na salt (hygroscopic).[963]

[(HO(EtO)P(O)]$_2$O. III.l.a.β-l (from $(EtOPO_2)_3$ by controlled hydrolysis, besides $EtOPO_3H_2$),[448a] III.l.a.β-2 (revealed by chromatography in "Langheld" ester),[1453] III.l.b.α-2 (from $EtOPO_3H_2$ by 2,4-$(O_2N)_2C_6H_3F$),[1464a] III.l.b.γ-2A (from $EtOPOCl_2$ with 1 H_2O),[1020] III.l.c (from $EtOPO_2H_2$ with Br_2 in the absence of EtOH).[246] Revealed or isolated by chromatography.

[(HO)(PrO)P(O)]$_2$O. III.l.c (from $ROPO_2H_2$ with Br_2 in pyridine).[246]

$(PrO)_2P(O)OP(O)(OH)_2$(?). Claimed to be formed from ethyl metaphosphate and propanol. Isolated as barium salt.[1122]

[(CH_2=$CHCH_2O$)(HO)PO]$_2$O. III.l.b.ζ (from $(RO)PO_3H_2$ with $(EtO)_2POP(O)(OR)_2$). Isolated as di-(c-hexylammonium) salt, m. 193-6°.[919]

[(HO)(BuO)P(O)]$_2$O. III.l.c (from $ROPO_2H_2$ with Br in pyridine). $(NH_4)_2$ salt.[246]

[(HO)(CH-CMe=CHO)P(O)]$_2$O. III.l.a.α (from ROH and $(Et_3NH)_2HPO_4$ with MeCN, besides $(RO)_3PO$). Di-c-$C_6H_{11}NH_3$ salt, m. 172-3°.[1394]

Dimenthyl pyrophosphate. $(RO)_2P(O)OP(O)(OH)_2$(?). III.l.b.γ-2. Cryst., m. 198°; Pb, Ag, Ca salts.[964]

$(CH_2)_3$<O,O>P(O)(OH)OP(O)(OH)<O,O>$(CH_2)_3$. III.l.b.α-2 (by $CNCH_3$). Cryst., m. 137°.[410a]

[(HO)($PhCH_2O$)P(O)]$_2$O. III.l.b.δ (from $PhCH_2OP(O)(OH)NHNH_2$ with I_2 or N-Br-succinimide).[287] By oxidation of $(PhCH_2O)_2$(4-HO-2,3-Me_2-1-$C_{10}H_4$)PO with Br_2, besides other products;[406a] from $(PhCH_2O)_2PO_2H_2$ with Br_2 in pyridine.[246]

(HO)($PhCH_2O$)P(O)OP(O)(OPh)(OH). III.l.b.δ (from benzyl phosphoroguanidate with $(PhO)PO_3H_2$). Diammonium salt, m. 260° (dec.).[412]

$(PhO)_2P(O)OPO_3H_2$. III.l.b.γ-2 (from phenol and 1.4 moles $POCl_3$ in pyridine (?)),[1042] III.l.b.ε (from $(RO)_2(CH_2$=CMeO)PO with H_3PO_4).[1011,1013] K_2 salt, needles;[1042] Ba salt.[1011,1013]

[(HO)(PhO)P(O)]$_2$O. III.l.α-2 (from $PhOPO_3H_2$ and 2,4-$(O_2N)_2C_6H_3F$),[1465] (from $PhOPO_3H_2$ and ketene dimer),[88,1011] (from $PhOPO_3H_2$ and MeCN, $MeOCH_2Cl$, or (best) CCl_3CN),[407] (from $PhOPO_3H_2$ and PhOCN),[916] (from $PhOPO_3H_2$ and Cl_2CPh_2, Ph_3PHal_2, Ph_3P + polyhalogenomethanes, or Ph_2SHal_2 + Ph_3P),[775] (from PhOP(O)(OMe)SMe with PhOP(O)(OMe)OK),[685] III.l.b.δ (from PhOP(O)(OH)-$NHNH_2$ and I_2 or N-Br-succinimide),[267] III.l.b.η (from $Ph(Bu_4N)_2PO_4$ and $C{=}N^+$ containing compounds + Br_2 or $KMnO_4$),[408] III.l.b.ζ (from $PhOPO_3H_2$ with $(EtO)_2POP(O)$-

$(OCH_2CH{=}CH_2)_2$ or $MeC(O)OP(O)(OCH_2CH{=}CH_2)_2$).[919] Ba salt,[88] $(c\text{-}C_6H_{11}NH_3)_2$ salt, m. 254-6°.[916,919,1465]

$(HO)(PhO)P(O)OP(O)(O\text{-}4\text{-}ClC_6H_4)(OH)$. III.1.b.δ (from imidazolium phenyl phosphoroimidazolidate with $(4\text{-}ClC_6H_4O)PO_3H_2$). Oil.[617a]

$(4\text{-}ClC_6H_4O)_2P(O)OPO_3H_2$. III.1.b.ε (from $(RO)_2(CH_2{=}CMeO)PO$ with H_3PO_4). Ba salt.[1011,1013]

$[(HO)(4\text{-}ClC_6H_4O)P(O)]_2O$. III.1.b.α-2 (from $ROPO_3H_2$ and ketene dimer),[88] III.1.b.γ-2A (from $ROPOCl_2$ with $ROPO_3H_2$).[1248] Liquid; Ba salt;[88] $(c\text{-}C_6H_{11}NH_3)_2$ salt, m. 253°.[1248]

$[(HO)(4\text{-}O_2NC_6H_4O)P(O)]_2O$. III.1.b.α-2 (from $ROPO_3H_2$ and ketene dimer). Ba salt.[88]

$[(HO)(2\text{-}MeC_6H_4O)P(O)]_2O$. III.1.b.γ-1 (from ROH with $POCl_3$). Sirup. K salt, needles.[1038]

P^1,P^2-bis 2,3-bis(palmitoyloxy)-1-propyl pyrophosphate. III.1.b.γ-1 (from dipalmitin with $POCl_3$ then H_2O + pyridine).[231]

P^1,P^2-Bis[2,3-bis(palmitoyl)-1-glyceryl] pyrophosphate. III.b.γ-1 (using $POCl_3$ + 2 H_2O + 4 pyridine).[231]

Cytidine 5'-pyrophosphate glycerol. III.1.b.α-1 (from cytidine 5'-phosphate with isopropylidene glycerol 1-phosphate by DCC, followed by elimination of the isopropylidene group).[127] III.1.b.δ (from morpholidate with $CH_2OHCHOHCH_2OPO_3H_2$).[1211] Found in Lactobacillus arabinosus, split by dil. NH_3 into cytidine 5'-phosphate and 1,2-glycerol cyclic phosphate.[126]

Cytidine 5'-diphosphate 2-inositol. III.1.b.ζ (from cytidine 5'-phosphate with $(PhO)_2POCl$, then inositol phosphate and hydrogenolysis).[863]

Cytidine 5'-diphosphate 1-ribitol. Found in Lactobacillus arabinosus, split by dil. NH_3 into cytidine 5'-phosphate and ribitol 1,2-cyclic phosphate.[124]

Cytidine 5'-diphosphate α-D-glucopyranose. III.1.b.δ (from cytidine 5'-phosphoromorpholidate with α-D-glucopyranosyl phosphate). Characterized by electrophoresis and R_f (paper).[300]

Cytidine 5'-diphosphate choline. From cytidine 5'-P^2-Ph_2-pyrophosphate with phosphocholine).[863]

Cytidine 5'-diphosphate-serine. From cytidine 5'-P^2-Ph_2-pyrophosphate with phosphoserine).[863]

Isocytidine 5'-diphosphate glucose. III.1.b.δ (from isocytidine 5'-phosphoromorpholidate with α-D-glucose phosphate).[811]

Nicotinamide 5'-diphosphate 5'-(9-β-D-ribofuranosyl)-purine (NAD). III.1.b.α-1 (from the 5'-monophosphorylated components with DCC). Spectroscopic data.[1102]

Nicotinamide 5'-diphosphate 5'-[6-mercapto-9-(β-D-ribofuranosyl)purine]. III.1.b.α-1 (from the 5'-monophosphorylated components with DCC). Spectroscopic data.[1102]

Nicotinamide 5'-diphosphate 5'-[6-methylthio-9-(β-D-ribofuranosyl)purine]. III.1.b.α-1 (from the 5'-monophosphorylated components with DCC). Spectroscopic data.[1102]

Uridine 5'-phenylpyrolphosphate. III.1.b. (From uridine phosphoroamidate with $PhOPO_3H_2$).[983]

Uridine 5'-(3-deoxy-α-D-ribohexopyranosyl) pyrophosphate. III.1.b.δ (from uridine 5'-phosphoromorpholidate with the glycosyl phosphate).[810]

Uridine 5'-(4-deoxy-α-D-ribohexopyranosyl) pyrophosphate. III.1.b.δ (from uridine 5'-phosphoromorpholidate with the glycosyl phosphate).[810]

Uridine 5'-diphosphate 1-glucose. III.1.b.γ-2A (from 2',3'dibenzyluridine 5'-benzylphosphorochloridate with α-D-glucose 1-phosphate, and hydrogenolysis),[960] III.1.b.δ (from uridine phosphoromorpholidate with glucose phosphate),[1211] (from uridine phosphoroamidate with glucose phosphate).[983] Identical with a natural coenzyme.[960] Found in fresh human erythrocytes, 29×10^{-9} moles/g homoglobine.[1398]

Uridine 5'-diphosphate N-acetylglucosamine. III.1.b.δ (from the uridine phosphoromorpholidate with the sugar phosphate).[1211] Isolated from mung bean (Phaseolus aureus) seedlings.[1272]

Uridine 5'-diphosphate galactose. III.1.b.δ (from the uridine morpholidate with the sugar phosphate).[1211]

Uridine 5'-diphosphate glucuronic acid. III.1.b. (from the uridine phosphoromorpholidate with the sugar phosphate).[1211] Isolated from mung bean (Phaseolus aureus) seedlings.[1272]

2-Thiouridine 5'-diphosphate glucose. III.1.b.δ (from the uridine phosphoromorpholidate with α-D-glucose phosphate).[811]

4-Thiouridine 5'-diphosphate glucose. III.1.b.δ (from the uridine phosphoromorpholidate with α-D-glucose phosphate).[811]

3-Me-uridine 5'-diphosphate glucose. III.1.b.δ (from the uridine 5'-phosphoromorpholidate with α-D-glucose phosphate).[811]

P^1,P^2-Bis-(6-azauridylyl) 5'-pyrophosphate. III.1.b.α-1 (from 6-azauridine 5'-phosphate by DCC). Isolated as Na salt.[695]

6-Azauridine 5'-(α-D-glucopyranoysyl) pyrophosphate. III.1.b.δ (from the uridine 5'-phosphoromorpholidate with the glucosyl phosphate).[301,811]

5'-(3'-O-Acetylthymidine)$_2$ P^1,P^2-pyrophosphate. III.1.b.α-3 (from the 5'-monophosphate and $SOCl_2$ in Me_2NCHO).[730]

Deoxyadenosine 5'-diphosphate β-nicotinamide nucleotide. III.1.b.α-1 (from deoxyadenosine 5'-phosphate and nicotinamide mononucleotide with DCC). $[\alpha]_D$ -20.5°; spectroscopic data.[696]

P^1-(Adenosine 5')-P^2-glucose. III.1.b.α-1 (from adenylyl piperidate and sugar phosphate with DCC).[1180a]
P^1-(Adenosine 5')-P^2-glucosamine. III.1.b.α-1 (from adenylyl piperidate and sugar phosphate with DCC).[1180a]
P^1-(Adenosine 5')-P^2-galactose. III.1.b.α-1 (from adenylyl piperidate and sugar phosphate with DCC).[1180a]
Adenylyl 5'-(HO)(O)POP(O)(OH)OCH_2CH(OH)COOH. III.1.b.δ (from adenosine 5'-phosphoromorpholidate with 3-phosphoglyceric acid).[646a]
P^1-Adenosine-5'-P^2-riboflavine-5' pyrophosphate (= natural coenzyme ADF) from 2',3'-isopropylideneadenosine 5'-benzylphosphate + 2',3'-adenosine 5'-benzylphosphorochloridate. Spectrum 240-470nm; Ag_2 salt.[395a]
Nicotinamide-D-glucose adenosine dinucleotide. III.1.b.α-1 (from nicotinamide-D-glucose 6'-phosphate and adenosine monophosphate by DCC).[1466]
Guanosine 5'-diphosphate mannose. III.1.b.δ (from guanosine phosphoromorpholidate with the sugar phosphate).[1211]
Sym. Diriboflavin pyrophosphate. III.1.a.β-1. Di-Na salt.[275]
P^1,P^2-Di(N-benzoylserinemethylamide) O-pyrophosphate. III.1.b.α-1 (from N-benzoyl-O-phosphoserinemethylamide with DCC).[114]
P^1,P^2-Diseryl O-pyrophosphate. III.1.b.α-1 (from N-benzyloxycarbonyl-O-monobenzylphosphorylserine benzyl ester with DCC and simultaneous hydrogenolysis).[114]
P^1,P^2-Di(glycylseryl) O-pyrophosphate. III.1.b.α-1 (from N-benzyloxycarbonyl-Glycyl-(O-monobenzylphosphorylserine) benzyl ester with DCC and simultaneous hydrogenation).[114]
P^1,P^2-Cholesteryl pyrophosphate. Double formula proposed for "Cholesteryl phosphate."[1432b]
Di(estrone) pyrophosphate (sym.). III.1.b.γ-2A (from the steroid phosphate, $ROPO_3H_2$, with $(PhO)_2POCl$ via $(PhO)_2P$(O)OP(O)(OR)OH)). Cryst., m. 163-5°. Spectroscopic data.[1194]
Di(dehydroepiandrosterone) pyrophosphate (sym.). III.1.b.γ-2A (from the steroid phosphate, $ROPO_3H_2$, with $(PhO)_2POCl$ via $(PhO)_2P$(O)OP(O)(OR)(OH)). Cryst., m. 135-6°; spectroscopic data; $[\alpha]_D$ 4.3°($CHCl_3$).[1194]
P^1-(Prednisolone-21-yl),P^2-(adenosine 5'-yl) pyrophosphate. III.1.b.δ (from prednisolone phosphoromorpholidate with adenosine 5'-monophosphate).[1058]
[MeC(O)OCH_2CH_2HgOP(O)(OH)$]_2O$. From MeC(O)CH_2CH_2OHgOC(O)Me with H_3PO_4.[429]

c. Triester

$Et_3HP_2O_7$. III.1.a.β-2. Revealed by chromatography in "Langheld" ester.[1453]
$(PhCH_2O)_2P$(O)OP(O)(OCH_2Ph)(OH). III.1.b.γ-2 (tetrabenzyl

ester debenzylated with N-methylmorpholine).[129 1355] Sirup; monohydrate, m. 65°. Silver salt soluble in $CHCl_3$.[129]

(HO)[2-(4-Me-thiazol-5-yl)ethyl-O]P(O)OP(O)($OCH_2Ph)_2$. III.1.b.γ-2A (from the thiazolylethyl phosphoric acid with $(PhCH_2O)_2POBr$ prepared in situ). c-$C_6H_{11}NH_4$ salt.[857]

d. Tetraester

α. Non-Cyclic, Non-Aromatic Ester Links

$Me_4P_2O_7$. III.1.α,[1213] III.1.b.α-3 (from R_3PO_4 with $SOCl_2$),[221] III.1.b.β-2,[6] III.1.b.γ-2A,[1090,1360,1360a] III.1.c (from $(MeO)_2PHO$ and N_2O_4, contrary to Petrov et al., C.A. 55, 23312h (1961)).[637] Liquid, $b_{0.5}$ 114-6°,[1360] $b_{0.3}$ 106-8°,[1360] $b_{0.05}$ 105°;[1090] d_4^{25} 1.3609; n_D^{25} 1.4121.[1360]

$(MeO)_2P(O)OP(O)(OEt)_2$. III.1.b.γ-2A (from $(MeO)_2POCl$ with $(EtO)_2PO_2H$ or a salt),[1090,1361] III.1.ε (from $(EtO)_2$-$(CH_2$=CMeO)PO with $(MeO)_2PO_2H$),[1011,1013] III.1.c (from $(EtO)_3P$ + $(MeO)_2PO_2H$ and $CBr_2(CONH_2)_2$).[1013] Liquid, $b_{0.1}$ 111°,[1090] $b_{0.06}$ 125-6°,[1011,1013] $b_{0.003}$ 100-5°; d_4^{25} 1.259; n_D^{25} 1.4156.[1361]

$Et_4P_2O_7$. III.1.a.α,[350,417,436,647,1067,1213] III.1.a.β-1,[647,1360] III.1.a.β-2 (revealed by chromatography in "Langheld" ester),[1483] (from HC$(OEt)_3$ with P_2O_5),[269] III.1.b.α-2 (from $(EtO)_2PO_2H$ with EtOCH=CH_2, $(EtO)_2$-POCl, $Et_4P_2O_7$),[1011] (similarly with $(MeCO)_2O$);[479] with EtOCH=CH_2 and HgO·BF_3 catalyst[661]), III.1.b.α-3 (from $(EtO)_2PO_2H$ with $SOCl_2$),[221] III.1.b.β-1 (?),[179] III.1.-b.β-2,[6,657] III.1.b.γ-1 (from $(EtO)_3PO$),[647,656,658,1334] (from EtOH),[1124] III.1.b.γ-2,[1360] III.1.b.γ-2A (from $(EtO)_2POCl$ with 1/2 H_2O and CH_2=CHCN),[248,1362] III.1.b.ε (from $(EtO)_2(CH_2$=CMeO)PO with $HO_2P(OEt)_2$),[444,662,1011,1013] III.1.b.ζ (from $(EtO)_2POCl$ with $(EtO)_2$-POP(O)$(OEt)_2$,[952a] III.1.c (from $(EtO)_2PONa$ with $(EtO)_2POCl$ and oxidation,[1061] with Cl_2,[60,63,65] with SO_2[1313]), (from $(EtO)_3P$ with $(EtO)_2PO_2H$ and CBr_2-$(CONH_2)_2$[1013]), (from $Et_4P_2O_5$ or $Et_4P_2O_6$ and O_2),[820] (from $(EtO)_2PHO$ + H_2O or + $(EtO)_2POCl$, with CCl_4).[1314a] From P_2O_5 or $EtPO_3$ with $(EtO)_2SO$ (yield low, besides more $(EtO)_3PO$);[271] from $(EtO)_3PO$ with PCl_5.[980] Liquid, b_8 166-70°,[1061] b_5 155-5.5°,[63] b_3 144-5°,[63] $b_{2.5}$ 133-4°,[820] b_{1-2} 140°,[1061] b_1 135-8°,[1360] 134-8°,[1362] 128-30°,[248] $b_{0.5}$ 125-30°,[1360] 124-6°,[444] $b_{0.4-0.5}$ 125-30° (impure),[269] $b_{0.08}$ 104-10°,[647] $b_{0.01}$ 95-7°;[952a] d_0^0 1.2040[60,63] (?), d (?) 1.190,[1124] d_4^{24} 1.1901,[1360,1362] d_4^{20} 1.172,[1061] d_4^{17} 1.1978,[1360,1362] d_0^{20} 1.1847,[60] d_{25}^{25} 1.1845;[647] n_D^{25} 1.4513 (?),[444] 1.4185,[1362] 1.4182,[952a,1360] 1.4170,[647] n_D^{20} 1.4222 (high ?),[60,63] 1.417,[6] n (?) 1.418.[1124] Evolution of ethylene begins at 208°.[647]

$(EtO)_2P(O)OP(O)(OEt)(OCH_2CH_2Cl)$. III.1.b.γ-2A (from equimol. mixture of the corresponding chloridates + 1 H_2O + base). Liquid, $b_{0.02}$ 136-7°.[1049]

$[(EtO)(ClCH_2CH_2O)P(O)]_2O$. III.1.b.γ-2A (from (RO)(R'O)POCl + 1/2 H_2O with base). Liquid, $b_{0.02}$ 145-5.5°.[1049]

$(EtO)_2P(O)OP(O)(OCH_2CH_2Cl)_2$. III.1.b.γ-2A (from equimol. mixture of the corr. chloridates + 1 H_2O + base). Liquid, $b_{0.02}$ 164-5°.[1049]

$(EtO)(ClCH_2CH_2O)P(O)OP(O)(OCH_2CH_2Cl)_2$. III.1.b.γ-2A (from equimol. mixture of the corr. chloridates + 1 H_2O + base). Liquid, no constants.[1049]

$(ClCH_2CH_2O)_4P_2O_7$. III.1.b.γ-2A (from $(RO)_2POCl$ + 1/2 H_2O + base). Liquid, non-distillable.[1049]

$(MeO)_2P(O)OP(O)(OPr)_2$. III.1.b.γ-2A (from $(MeO)_2POCl$ with $(PrO)_2PO_2H$). Liquid, $b_{0.002}$ 105-10°; d_4^{25} 1.192; n_D^{25} 1.4210.[1361]

$(MeO)_2P(O)OP(O)(O\text{-}i\text{-}Pr)_2$. III.1.b.γ-2A (from $(MeO)_2POCl$ and $(i\text{-}Pr\text{-}O)_2PO_2H$). Liquid, $b_{0.005-0.008}$ 105° (bath); d_4^{25} 1.178; n_D^{25} 1.4165.[1361]

$(EtO)_2P(OPOP(O)(OPr)_2$. III.1.b.γ-2 (from $(RO)_2POCl$ and $(PrO)_2PO_2H$). Liquid, $b_{0.005-0.007}$ 118-20° (bath); d_4^{25} 1.141; n_D^{25} 1.4210.[1361]

$(EtO)_2P(O)OP(O)(O\text{-}i\text{-}Pr)_2$. III.1.b.γ-2 (from $(EtO)_2POCl$ and $(i\text{-}Pr\text{-}O)_2PO_2H$). Liquid, $b_{0.01-0.006}$ 105° (bath); d_4^{25} 1.132; n_D^{25} 1.4175.[1361]

$(MeO)_2P(O)OP(O)(OBu)_2$. III.1.b.γ-2A (from $(MeO)_2PO_2NMe_4$ with $(BuO)_2POCl$). Liquid, $b_{0.05}$ 115°.[1090]

$(PrO)_2P(O)OP(O)(OPr)(OCH_2CH_2Cl)$. III.1.b.γ-2A (from equimol. mixture of the corr. chloridates + 1/2 H_2O + base). Liquid, $b_{0.03}$ 162.5-3.0°.[1049]

$Pr_4P_2O_7$. III.1.a.α,[350] III.1.b.γ-2A (from $(RO)_2POCl$, controlled hydrolysis in pres. of pyridine),[1360] III.1.b.ζ (from $(RO)_2PO_2H$ with $R_4P_2O_6$).[952a] Liquid, b_4 178-9.5,[71] $b_{0.01}$ 112-6°,[1360] 115°;[952a] d_0^0 1.1211,[71] d_4^{25} 1.1037;[1360] n_D^{25} 1.4248,[1360] 1.4238,[952a] $n_D^{17.5}$ 1.4300.[71]

$(PrO)_2P(O)POP(O)(O\text{-}i\text{-}Pr)_2$. III.1.b.γ-2A (from $(PrO)_2POCl$ and $(i\text{-}PrO)_2PO_2H$). Liquid, $b_{0.003-0.009}$ 110-4°; d_4^{25} 1.095; n_D^{25} 1.4210.[1361]

$(i\text{-}PrO)_4P_2O_7$. III.1.γ-2 (using pyridine). Liquid, $b_{0.01-0.02}$ 92-5°; d_4^{25} 1.0854; n_D^{25} 1.4170.[1360]

$[(CH_2{=}CHCH_2O)_2PO]_2O$. III.1.b.γ-2 (from $(RO)_2POCl$ and $(RO)_2PO_2H$ or $(RO)_3PO)$, III.1.b.ζ (from $(RO)_2PO_2H$ and $(EtO)_2POP(O)(OR)_2$). Liquid, n_D^{23} 1.4555.[919]

$(EtO)_2P(O)OP(O)(OBu)_2$. III.1.b.γ-2A (from $(EtO)_2POCl$ with $(BuO)_2PO_2H$),[1361] III.1.b.ε (from $(EtO)_2(CH_2{=}CMeO)PO$ with $HOPO(OBu)_2$),[1011,1013] III.1.c (from $(BuO)_3P$ + $(EtO)_2PO_2H$ with $CBr_2(CONH_2)_2$).[1013] Liquid, $b_{0.01}$ 108-14°,[1011,1013] $b_{0.007}$ 115-8°; d_4^{25} 1.107; n_D^{25} 1.4245.[1361]

$(i\text{-}PrO)_2P(O)OP(O)(OBu)_2$. III.1.b.γ-2 (from $(i\text{-}PrO)_2POCl$ and $(BuO)_2PO_2H$). Liquid, $b_{0.002}$ 118-20°; d_4^{25} 1.068; n_D^{25} 1.4235.[1361]

$Bu_4P_2O_7$. III.1.a.α,[350] III.1.b.β-2,[6] III.1.b.γ-2A (from R_2PO_3Cl with $R_2PO_4NMe_4$[1090] or using pyridine[1360]), (from R_2POCl + 1/2 H_2O with CH_2=CHCN).[248] Liquid, b_1 150-2°,[248] $b_{0.01}$ 115°,[1090] 143-6°; d_4^{25} 1.0533; n_D^{25} 1.4296.[1360]

$(i\text{-}Am)_4P_2O_7$. III.1.a. Liquid.[350]

$(EtO)_2P(O)OP(O)(O\text{-}2\text{-}Et\text{-}C_6H_{12})_2$. III.1.b.γ-2B (from $(EtO)_3PO$ with $(2\text{-}Et\text{-}C_6H_{12}O)_2POCl$). Liquid, n_D^{25} 1.4390.[821]

$(PhCH_2)_4P_2O_7$. III.1.b.α-2 (from R_2PO_4H with $4\text{-}MeC_6H_4SO_2$-Cl,[433] with ketene dimer,[88] with 2,6-$(O_2N)_2C_6H_3OH$,[1465] Ag salt with CS_2[98]), III.1.b.α-3 (R_2PO_4H with $SOCl_2$),[921] III.1.b.γ-2A (from $(RO)_2POCl$ with dil. KOH),[97] III.1.b.ε (from $(RO)_2PO_2H$ with dibenzyl imidoyl phosphate prepared in situ),[95] III.1.b.ζ (from R_2PO_4H with $(PhO)_2P(O)OP(O)(OCH_2Ph)_2$ + base),[433] III.1.c (from $(RO)_2PHO$ with CCl_4 + KOH).[97] From $(PhO)_2P(O)OP(O)(OCH_2Ph)_2$ in presence of NH_3;[433] from bis(dibenzylphosphoryl)oxalate by thermic decomposition.[921] Cryst., m. 60-1,[97] 59-60°,[476] 62°.[921]

$[(PhCH_2OC(O)CH(NHCOCH_2Ph)CH_2O)(PhCH_2O)(O)P]_2O$. III.1.b.α-1 (from N-benzyloxycarbonyl-O-(monobenzylphosphoryl) serine benzyl ester by DCC). Cryst., m. 95° (on standing rising to 107°).[112]

$[(PhCH_2O_2CCH(NHCOCH_2NHCbo)CH_2O)(PhCH_2O)P(O)]_2O$. III.1.b.α-1 (from (RO)(R'O)PO_2H and DCC).[113]

β. Ester Links Partially or Totally Aromatic

$(EtO)_2P(O)OP(O)(OPh)_2$. III.1.b.α-2 (from $(EtO)_2PO_2H$ and $(PhO)_2PO_2H$ with $EtOCH=CH_2$, poor),[661] III.1.b.γ-2B (from $(EtO)_3PO$ with $(PhO)_2POCl$).[821] Liquid, $b_{0.01}$ 125-8°.[821]

$(EtO)_2P(O)OP(O)(OC_6H_4NO_2\text{-}3)_2$. III.1.b.γ-2 (from $(EtO)_2POCl$ with $(RO)_2PO_2H$).[963]

$(PhO)_2P(O)OP(O)(OCH_2Ph)_2$. III.1.b.γ-2A (from $(PhO)_2POCl$ with $(PhCH_2O)_2PO_2H$),[433] III.1.b.δ (from $(PhO)_2PO_2Ag$ with dibenzyl N-Ph-imidazolyl phosphate),[95] III.1.b.ε (from $(PhO)_2PO_2H$ with dibenzyl imidoyl phosphate prepared in situ).[781] Viscous resin, not distillable.[433]

$Ph_4P_2O_7$. III.1.b.α-2 (from Ph_2PO_4H with ketene dimer),[88] (from Ph_2PO_4Ag with CS_2),[98] III.1.b.α-3 (from $(RO)_3PO$ with $SOCl_2$),[221] III.1.b.γ-2A.[921] Liquid, $b_{0.1}$ 165° (bath temp.).[921]

$(PhO)_2P(O)OP(O)(O\text{-}4\text{-}MeOC_6H_4)_2$. III.1.b.γ-2 (from $(PhO)_2$-POCl with $(MeOC_6H_4O)_2PO_2H$). Viscous oil.[433]

$(4\text{-}ClC_6H_4)_4P_2O_7$. III.1.b.γ-2A (from $(RO)_2POCl$ with $(RO)_2PO_2H$). Cryst., m. 271°.[1248]

$(4\text{-}O_2NC_6H_4)_4P_2O_7$. III.1.b.α-1,[363] III.1.b.α-2 (by ketene dimer). Cryst., m. 126-7° then 146-8°.[88]

γ. Cyclic Esters

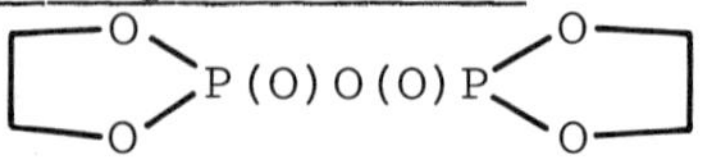

. From $(CH_2)_2O$ with P_2O_5. Very easily hydrolyzed.[1456]

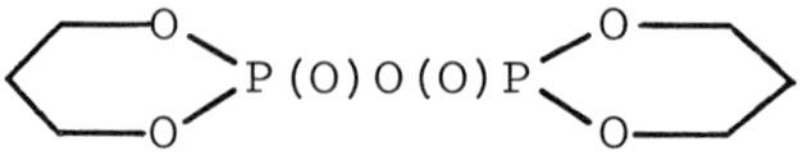

. III.1.α-1,[790] III.b.α-2 (by $CNCH_3$).[410a] Cryst., m. 137-7.5°,[790] 137°.[410a]

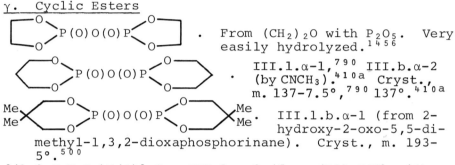

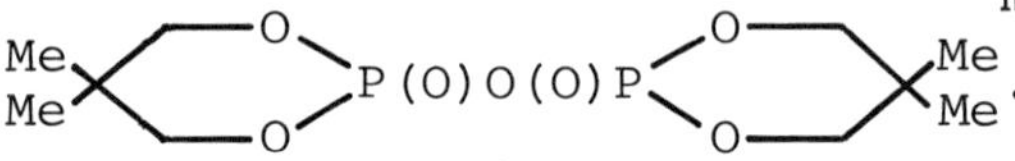

. III.1.b.α-1 (from 2-hydroxy-2-oxo-5,5-dimethyl-1,3,2-dioxaphosphorinane). Cryst., m. 193-5°.[500]

$[(1,2\text{-}C_6H_4O_2)P(O)]_2O$. III.1.γ-2 (from $(RO)_2PCl_3$ with $(COOH)_2$). Cryst., m. 136-8°; b_{12} 222°.[32]

2. Pyrophosphoric Esters with a Supplementary Phosphoric Ester Group

α-D-Ribofuranose 1-pyrophosphate 5-phosphate. III.1.a.α (from benzyl 5-dibenzyl phosphoryl-β-D-ribofuranoside 2,3-cyclic carbonate with $(PhCH_2)_3P_2O_7NHEt_3$, followed by alkaline treatment. Spectral evidence.[1345,1346]

Cytidine 2'-phosphate 5'-dipnosphate. III.1.b.δ (from cytidine 2'(3'), 5'-di-phosphate via the 2',3'-cyclic phosphate 5'-phosphoromorpholidate(anisidate) and H_3PO_4).[1401]

Adenylyl 5'-O-(HO)(O)POP(O)(OH)$OCH_2CH(OPO_3H_2)CO_2H$. III.1.b.δ (from adenosine 5'-phosphoromorpholidate with 2,3-diphosphoglyceric acid Bu_3NH salt). Pyrophosphate link location in 3 position of diphosphoglycerate probable.[646a]

α-D-Ribofuranose 5-phosphate 1-pyrophosphate. III.1.b.ζ (from 1-Br-2,3-cyclocarbonato-ribose 5-Ph_2-phosphate with $(PhCH_2)_3Et_3NHP_2O_7$, followed by hydrogenolysis).[1345]

Uridine 2'-3'-cyclophosphate 5'-pyrophosphate. III.1.b.γ-2A (from 2',3'-benzylidene uridine 2'(3'),5'-di-phosphate).[954]

3. Triphosphoric Esters (Non-Cyclic)

(HO)(EtO)P(O)OP(O)(OEt)OP(O)(OEt)(OH). III.1.a.β-2. Revealed by chromatography in "Langheld" ester.[1453]

$Et_5P_3O_{10}$. III.1.c (from 2 $(EtO)_2PONa$ + $EtOPOCl_2$ heated in O_2).[820]

Ribonucleosido and deoxyribonuclosido 5'-triphosphates have been synthesized following III.1.b.δ by reacting the nucleosido phosphoromorpholidates with pyrophosphate[979] or the phosphoroamidates with pyrophosphate.[983a]

Cytidine triphosphate. III.1.b.α-1 (from diphosphate with H_3PO_4 and DCC),[1265] (from monophosphate with di-c-hexylguanidine).[1329]

Deoxycytidine triphosphate. III.l.b.α-1 (from diphosphate with H_3PO_4 and DCC).[1265]
Uridine 5'-triphosphate. III.l.b.α-1 (from uridine 5'-phosphate with pyrophosphate),[1329] (from monophosphate by repeated condensation with H_3PO_4).[1265]
6-Azauridine 5'-triphosphate. III.l.b.δ (from the 5'-phosphoromorpholidate with pyrophosphate).[1496]
Ca salt. Spectroscopic data.
Thymidine 5'-triphosphate. III.l.γ-2A (from 3'-O-benzylthymidine 5'-benzylphosphorochloridate with $(PhCH_2O)_3$-P_2O_3(OH) followed by debenzylation).[628]
Inosine triphosphate. Found in fresh human erythrocytes, 19×10^{-9} moles/g hemoglobine.[1398]
Adenosine 5'-triphosphate. III.l.b.α-1 (by phosphorylation of 2',3'-isopropylideneadenosine with morpholinophosphorodichloridate followed by hydrolysis with dilute acid and condensation with H_3PO_4 and DCC),[721a] (from adenosine diphosphate with H_3PO_4 and DCC),[1265] III.b.γ-2A (from Ag dibenzyl adenosine pyrophosphate or--yield better--Ag_2 adenosine phosphate),[959b] (from Ag adenosine benzylphosphate with repeated use of $(PhCH_2O)_2POCl$),[130] (from 2',3'-isopropylidene diphosphate with morpholino-$POCl_2$),[722] (from Ag adenosine benzyl pyrophosphate with $(PhCH_2O)_2POCl$ and hydrogenolysis),[420] (from 2',3'-benzylidene adenosine diphosphate with $(PhCH_2O)_2POCl$, etc.),[954] III.l.b.δ (from adenosine diphosphate with $H_2NPO_3H_2$),[362,409] III.l.b.η (oxidation of adenosine mono or diphosphate in presence of H_3PO_4 with thiohydroquinone derivatives and other compounds),[155,1454,1455b] III.l.c (from adenosine monophosphate with Na_2HPO_4 + Br_2, besides adenosine diphosphate; chromatographic evidence).[1464] Ba salt;[130] triacridine salt, m. 209°,[130] with dec.,[129a] acridine (2:5) salt, m. 218° (dec.).[129a]
Adenosine triphosphate N^1-oxide. From triphosphate by H_2O_2.[449]
9-D-Erythrityladenine 4'-triphosphate. III.l.b.γ-2A (from 9-(2,3-di-O-acetyl-D-erythrityl)-adenylic acid with $(PhO)_2P(O)OP(O)$(morpholino)Cl followed by dephenylation and deacetylation).[723]
Guanosine triphosphate. III.l.b.α-1 (from guanosine monophosphate with H_3PO_4 and DCC).[361,1265] Ba salt.[361]
Deoxyguanosine triphosphate. III.l.b.α-1 (from diphosphate with H_3PO_4 and DCC).[1265]
Thiamine triphosphate. III.l.a.β-1 (besides mono- and diphosphate). Chromatographic evidence.[232,1083]
Oxythiamine triphosphate. From monophosphate with P_2O_5. Light yellow microcryst. powder, m. 215°.[1393]
Prednisolone 21-triphosphate. III.l.b.δ (from the 21-phosphoromorpholidate with pyrophosphate).[1058]

4. Esters of Higher Polyphosphoric Acids and of Metaphosphoric Acid

$(MeOPO_2)_3$. III.1.a.α (from $(AgPO_3)_3$ and MeI). Oil.[448a] Yields Me pyrophosphate and orthophosphate by methanolysis.[448a]

$(EtOPO_2)_3$. III.1.a (from $(AgPO_3)_3$ with EtI),[448a] III.1.b. α-1 (from $EtOPO_3H_2$ with DCC),[1444] III.1.α-2 (revealed by chromatography in "Langheld" ester).[310] Hexagonal cyclic structure.[310]

$Et_6P_4O_{13}$. III.1.a.β-2 (revealed by chromatography in "Langheld" ester and in "ethyl metaphosphate").[310] III.1.b.β-1 (from $EtPO_3$ by thermal decomposition). Liquid (impure).[7]

$Et_4P_4O_{12}$. III.1.a.β-2 (from P_2O_5 with Et_2O). Octagonal cycle. Chromatographic evidence.[310]

$Et_2P_4O_{11}$, 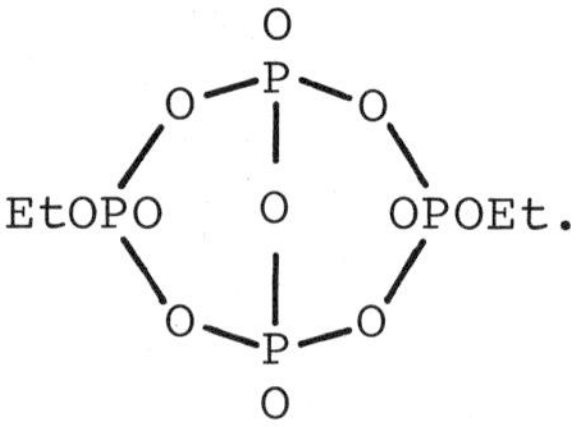 III.1.a.β-2 (from P_2O_5 with Et_2O). Chromatographic evidence.[310]

$Et_8P_6O_{19}$. III.1.b.γ-1. Liquid, d 1.272; n_D 1.424.[1124]

$Bu_8P_6O_{19}$. III.1.b.γ-1. Liquid, d 1.22.[1124]

Metaphosphoric esters. "Metaphosphoric acid" is a generic name for compounds corresponding to, or approaching the general formula HPO_3. Monomolecular HPO_3 is unknown, its anion, PO_3^-, is a very reactive, not isolable intermediate, e.g. during hydrolysis of phosphoric esters. "Metaphosphoric acid" may be cyclic, e.g. $(HPO_3)_3$, or higher condensed linear, branched or unbranched polyphosphoric acids. Among their esters pure compounds have been, if ever, rarely isolated.

$(MeOPO_2)_x$. III.1.a.α (from $(AgPO_3)_3$ with CH_3I),[448a,1439] III.1.b.β-1,[822] III.1.b.β-2,[6] III.1.b.γ-1.[822] The product prepared from $(AgPO_3)_3$ is considered to be a mixture of linear polyphosphate esters (?).[1439] Oil.

$(EtOPO_2)_x$. III.1.a.α (from $(AgOPO_2)_3$ with EtI),[448,843,1439] III.1.a.β-1 (from $EtOPO_3H_2$ and DCC),[1444] III.1.a. β-2 ("Langheld" ester),[310,842,843,1081,1439,1453] best procedure,[1317] III.1.b.β-1 (by heating $(EtO)_2PO_2H$ (?),[843] III.1.b.γ-1.[179] Indistillable sirup (the old indication (1861!) that the ester prepared by III.1.a.α from $PbPO_3$[334] is distillable (b. ∿100°) seems to be erroneous). Reprecipitation from solution in halogenated solvents by ether (old purification method) yielded a product of d_4^{25} 1.42 and n_D^{25} 1.438[6] but this product is certainly not homogeneous. By chromatographic analysis, the "Langheld" ester gave contradictory results; mixture of linear polyphosphate esters[1439,1453] or mixture of ethyl tetraphosphate, cyclic esters (see the three above mentioned P_4 derivatives and $(EtOPO_2)_3$) and simpler esters.[310] Molecular weight: hexameric by freezing point method in naphthalene, dimeric by boiling point method in $CHCl_3$.[1081,1122]

$(BuOPO_2)_x$. III.1.b.β-2. Unstable liquid, d_4^{25} 1.227, n_D^{25} 1.445.[6]

$(C_8H_{17}OPO_2)_x$. III.1.b.β-2. Unstable at room temperature; d_4^{25} 1.151, n_D^{25} 1.45.[6]

2-ClCO-$C_6H_4OPO_2$. III.1.b.γ-2A (from $ROPOCl_2$ with $(COOH)_2$), III.1.b.γ-1. Cryst., m. 95°; b_{11} 170-1°.[28,437] (Structure doubtful.)

2-ClCO-4-Me$C_6H_3OPO_2$. III.1.b.γ-2A (from $ROPOCl_2$ with $(COOH)_2$). Plates, m. 88; b_{14} 185-6.[53] (Structure doubtful.)

2-ClCO-5-Me$C_6H_3OPO_2$. III.1.b.γ-2A (from $ROPOCl_2$ with $(COOH)_2$). Plates, m. 77°; b_{14} 195.4-6.2°.[53] (Structure doubtful.)

4-$ClSO_2$-$C_6H_3OPO_2$. III.1.b.γ-2A (from $ROPOCl_2$ with $(COOH)_2$). Cryst., m. 150.1.[45] (Structure doubtful.)

B. Esters of Anhydro Acids with Other Inorganic Oxy Acids

1. With Sulfurous, Sulfuric and Ethansulfonic Acid

$[(4\text{-}O_2NC_6H_4O)_2P(O)O]_2SO$. From $(RO)_2PO_2H$ with $SOCl_2$. Identified only by its reaction products with c-$C_6H_{11}NH_2$.[433]

Adenosine 5'-phosphosulfate. From adenosine 5'-phosphate with Et-O-CH=$CHOSO_3H$ or $NCCH_2CH_2OSO_3H$, followed by catalytic dealkylation;[193] or condensation with H_2SO_4 (DCC).[125,1183] No pure substance; spectroscopic (IR) evidence,[193,1183] Li Salt, IR abs.[125]

$EtS(O_2)OP(O)(OEt)_2$. From $(RO)_2POP(O)(OR)_2$ with $EtSO_3H$. Liquid, $b_{0.005}$ 77°; n_D^{25} 1.4296.[952a]

2. Silyl and Oxysilyl Phosphate Esters

$Me_3SiOP(O)(OMe)_2$. From Me_3SiCl with $NaOP(O)(OMe)_2$. Oil, disproportionated when heated.[1211a]

$(Me_3SiO)_3PO$. From Me_3SiCl and H_3PO_4;[1224] Me_3SiOEt with Na_2HPO_4 (poor yield) or H_3PO_4;[1211a] Me_3SiOMe with H_3PO_3 in an air stream;[1423] $(Me_3Si)_2O$ with P_2O_5.[1234] Liquid, b_{750} 231.8°, b_6 97°,[1422] 98-100°,[1423] b_5 91-3°,[1424] b_4 85-7°;[1234] d_{20} 0.959;[1422] n_D^{20} 1.4090,[1234] 1.4089;[1422] $n_D^{20} - n_C^{20}$ 0.0076.[1422]

$(MeEt_2SiO)_3PO$. From $MeEt_3SiOMe$ and H_3PO_4. Liquid, b_1 145-7°.[1424]

$(Et_3SiO)_3PO$. From Et_3SiCl (with trace of $FeCl_3$) or Et_3SiBr and $(EtO)_3PO$;[1424] Et_3SiOMe with H_3PO_4[1422] or with H_3PO_3 in an air stream;[1423] $(Et_3Si)_2O$ with P_2O_5.[1422] Liquid, b_{11} 200.5°,[1422] b_3 183-6°,[1424] b_1 166.5°;[1422] d_{20} 0.9670; n_D^{20} 1.4457.[1422]

$(MePr_2SiO)_3PO$. From $MePr_2SiOMe$ and H_3PO_4. Liquid, b_{10} 221-4°.[1424]
$(Pr_3SiO)_3PO$. From $(Pr_3Si)_2O$ and P_2O_5. Liquid, b_5 215-25° (not pure).[1422]
$(Bu_3SiO)_3PO$. From $(Bu_3Si)_2O$ and P_2O_5. Liquid, b_5 260-70° (not pure).[1422]
$Me_2Si(OPO_3H_2)_2$. From Me_2SiCl_2 and H_3PO_4 in equimol. proportions. Cryst., m. 60°.[785]
$Me_2Si[OP(O)(OMe)_2]_2$. From Me_2SiCl_2 and $NaOPO(OMe)_2$. Liquid, not quite pure.[1211a]
$Me_2Si[OPO(OEt)_2]_2$. From Me_2SiCl_2 and $(EtO)_3PO$. Liquid, b_1 140-2° (not quite pure).[1211a]
$Ph_2Si(OPO_3H_2)_2$. From Ph_2SiCl_2 and H_3PO_4 in equimol. proportions. Solid.[785]
$(Me_2Si)_3(PO_4)_2$. From 3 Me_2SiCl_2 and 2 H_3PO_4. Viscous polymer, structure probably cyclic.[1424]
$(EtO)_3SiOPO(OEt)_2$. From $Si(OEt)_4$ and $(EtO)_2POCl$. Liquid, $b_{0.001}$ 102-3° (dec.).[1211a]

3. Stannyl and Germanyl Phosphate Esters

$Bu_3SnOPO(OEt)_2$. From Bu_3SnCl and $NaOPO(OEt)_2$. Liquid.[828]
$Ph_3SnOPO(OPh)_2$. From Ph_3SnCl and $(PhO)_2PO_2NH_4$. Cryst., m. 170°.[828]
$Ph_3SnOPO(OEt)_2$. From Ph_3SnCl and $NaOPO(OEt)_2$. Cryst., m. 196°.[828]
$(Me_3GeO)_3PO$. From $Me_3GeOGeMe_3$ and P_2O_5. Liquid, b_1 50°.[1438]

4. Imidoyl Phosphate Esters (Nitrosyl Phosphates)

$(MeO)_2(O_2NCF_2CF{=}NO)PO$. From $(MeO)_3P$ and $O_2NCF_2CFClNO$. Liquid, b_3 84°; d_{20} 1.5160; n_D^{20} 1.3963.[917a]
$(EtO)_2(CFCl{=}NO)PO$. From $(EtO)_3P$ or $EtOPCl_2$ and FCl_2CNO. Liquid, b_3 89°; d_{20} 1.3096; n_D^{20} 1.4246.[917a]
$(EtO)_2(CCl_2{=}NO)PO$. From $(EtO)_3P$ and Cl_3CNO_2.[18,19] Liquid b_1 90-103°,[19] $b_{0.2}$ 80-85°.[18]
$(EtO)_2(CF_3CF{=}NO)PO$. From $(EtO)_3P$ or $(EtO)_2PHO$ and CF_3-$CFClNO$. Liquid, b_3 57.5°; d_{20} 1.3239; n_D^{20} 1.3700.[607]
$(EtO)_2(O_2NCF_2CF{=}NO)PO$. From $(EtO)_3P$ and $O_2NCF_2CFClNO$. Liquid, b_2 83-4°, d_{20} 1.3781; n_D^{20} 1.4002.[607]
$(EtO)_2(CF_3CCl{=}NO)PO$. From $(EtO)_3P$ and CF_3CCl_2NO. Liquid, b_4 79-80°; d_{20} 1.3988; n_D^{20} 1.3600.[607]
$(EtO)_2(CH_2ClCH{=}NO)PO$. From $(EtO)_3P$ and $CH_2ClCHClNO_2$. Liquid, $b_{0.3}$ 103°; IR.[18]
$(EtO)_2(MeCCl{=}NO)PO$. From $(EtO)_3P$ and $MeCCl_2NO_2$ (besides $(EtO)_3PO$). Liquid, b_1 105-15°.[19]
$(EtO)_3(Me_2C{=}NO)PO$. From $(EtO)_3P$ and Me_2CBrNO_2[18,19] or Me_2CClNO_2[18] (besides $(EtO)_3PO$); from $(EtO)_2PCl$ with $Me_2C{=}NO_2Na$;[1011a] from $(EtO)_2POCl$ with $Me_2C{=}NOH$.[18,19] Liquid, $b_{0.2}$ 95-8°,[18] 95-8.5°,[19] 95-7°,[1011a] $b_{0.18}$ 93-6°,[1011a] $b_{0.1}$ 88-90;[18] d_{35} 1.0990; n_D^{25} 1.4358. IR.[1011a]

$(EtO)_2[(F_3C)_2C=NO]PO$. From $(EtO)_3P$ and $(F_3C)_2CClNO$, or from $(EtO)_2POCl$ and $(F_3C)_2C=NOH$. Liquid, b_7 73.5-4.5°.[607]

$(EtO)_2(PhCH=NO)PO$. From $(EtO)_2POCl$ and $PhCH=NOH$. Undistillable oil.[18,19]

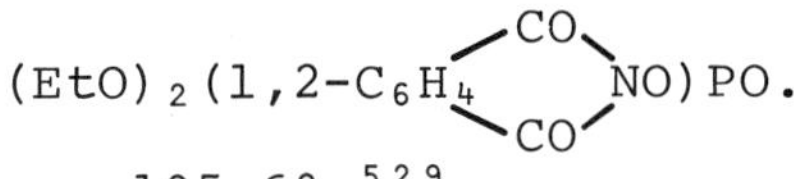

$(EtO)_2(1,2-C_6H_4\overset{CO}{\underset{CO}{\diagdown}}NO)PO$. From $(EtO)_2POCl$ and Naphthalyl hydroxylamine. Cryst., m. 105-6°.[529]

Anhydro Acids with Carboxylic Acids (Acyl Phosphates) and Their Esters

A. Derivatives of Orthophosphoric Acid

1. Derivatives of Carbonic Acid Monoesters

$MeOC(O)OP(O)(OMe)_2$. IV.2.a (phosphoryl chloride and K salt). Liquid, b_2 99°.[1245] Unstable at room temp., storable at 0°.[1245]

$MeOC(O)OP(O)(OEt)_2$. IV.2.a (phosphoryl chloride and $MeKCO_3$). Liquid, b_1 108°; unstable at room temp., storable at 0°.[1245]

$MeOC(O)OP(O)(OPr)_2$. IV.2.a (from $(RO)_2POCl$ and $PrKCO_3$). Liquid, b_1 114-6°; unstable at room temp., storable at 0°.[1245]

$MeOC(O)OP(O)(OBu)_2$. IV.2.a (from $(RO)_2POCl$ and $MeKCO_3$). Liquid, b_1 132-4°; d_{20} 1.1142; n_D^{20} 1.4247; unstable at room temperature.[1245]

$MeOC(O)OP(O)(O-i-Bu)_2$. IV.2.a (from $(RO)_2POCl$ and $MeKCO_3$). Liquid, b_2 132°; d_{20} 1.1021; n_D^{20} 1.4238; unstable at room temperature.[1245]

$EtOC(O)OP(O)(OMe)_2$. IV.2.a (phosphoryl chloride and $EtKCO_3$). Liquid, b_1 100°; unstable at room temp., storable at 0°.[1245]

$EtOC(O)OP(O)(OEt)_2$. IV.1.a (from $(EtO)_2PO_2Na$ and acyl chloride),[840] IV.2.a (from $(RO)_2POCl$ and $EtKCO_3$).[1245] Liquid, b_1 85-6°,[1245] $b_{0.02}$ 89°; n_D^{25} 1.4106;[840] unstable at room temperature, storable at 0°.[1245]

$EtOC(O)OP(O)(OPr)_2$. IV.2.a (from $(RO)_2POCl$ and $EtKCO_3$). Liquid, b_5 148°; unstable at room temperature; d_{20} 1.1320; n_D^{20} 1.4191.[1245]

$EtOC(O)OP(O)(OBu)_2$. IV.2.a (from $(RO)_2POCl$ and $EtKCO_3$). Liquid, b_1 140-1°; d_{20} 1.0830; n_D^{20} 1.4233; unstable at room temperature.[1245]

$EtOC(O)OP(O)(O-i-Bu)_2$. IV.2.a (from $(RO)_2POCl$ and $EtKCO_3$). Liquid, b_1 124°; d_{20} 1.0724; n_D^{20} 1.4243; unstable at room temperature.[1245]

$EtOC(O)OP(O)(OCH_2C_6H_5)_2$. IV.1.a (from $EtOCOCl$ and Ag salt). Liquid; IR bands.[849]

$PrOC(O)OP(O)(OMe)_2$. IV.2.a (from $(RO)_2POCl$ and $PrKCO_3$). Liquid, b_1 111°; unstable at room temperature.[1245]

$PrOC(O)OP(O)(OEt)_2$. IV.2.a (from $(RO)_2POCl$ and $PrKCO_3$). Liquid, b_2 138°; unstable at room temperature.[1245]
$PrOC(O)OP(O)(OPr)_2$. IV.2.a (from $(RO)_2POCl$ and $PrKCO_3$). Liquid, b_{5-6} 144°; unstable at room temperature.[1245]
$PrOC(O)OP(O)(OBu)_2$. IV.2.a (from $(RO)_2POCl$ and $PrKCO_3$). Liquid, b_1 142-3°; d_{20} 1.0673; n_D^{20} 1.4265; unstable at room temperature.[1245]
$PrOC(O)OP(O)(O\text{-}i\text{-}Bu)_2$. IV.2.a (from $(RO)_2POCl$ and $PrKCO_3$). Liquid, b_1 137-8°; d_{20} 1.0641; n_D^{20} 1.4256; unstable at room temperature.[1245]

2. Derivatives of Aliphatic or Araliphatic, Non-nitrogenous, Monocarboxylic Acids

$HC(O)OPO_3H_2$. IV.1.a (from formyl fluoride and K_2HPO_4). Isolated as Li salt (93% pure); Ag salt; free acid oil, unstable at 0°.[748]
$MeC(O)OPO_3H_2$. IV.1.a (from "mono" silver phosphate),[879] (from Ag dibenzyl phosphate),[886] IV.1.b,[288] IV.1.c (from vinyl acetate and H_3PO_4 with Hg^{2+} as catalyst, isolated as Li salt),[139] (from isopropenyl acetate and H_3PO_4, isolated as Li salt).[1309] Ag salt, needles,[228,886] lead salt, poorly soluble needles.[886] Free acid (from Pb salt) sirup;[228] hydrolysis (traced by ^{18}O) by P-O cleavage in acid and by C-O cleavage in alkaline medium;[229] half life at 38° at pH 7.4 is 3 hr.[886]
$MeC(O)OP(O)(OEt)_2$. IV.1.c (from RCOOH and $Cl_2C{=}C(OEt)$-$OP(O)(OEt)_2$[443] or $EtOC(O)CH{=}C(OEt)OP(O)(OEt)_2$[444]), IV.2.b (from RCOOH and $(RO)_2POP(O)(OEt)_2$[952a]). Liquid, $b_{0.2}$ 70°,[444] $b_{0.01}$ 60°;[952a] n_D^{25} 1.4110,[952a] 1.4115.[444]
$MeC(O)OP(O)(OCH_2CH{=}CH_2)_2$. IV.2.b (from $(EtO)_2POP(O)(OCH_2$-$CH{=}CH_2)_2$ and MeCOOH). Liquid, $b_{0.05}$ 75°; n_D^{25} 1.4382.[919]
$MeC(O)OP(O)(OBu)_2$. IV.1.a (from acyl chloride and t-amine salt).[432]
$MeC(O)OP(O)(OCH_2Ph)_2$. IV.1.a (from acyl chloride and Ag salt[629] or t-amine salt[432] of dibenzylphosphate).
$(MeCO)_2P(OH)_3$(?). This derivative of ortho structure is claimed to be produced by IV.1.a (from silver phosphate).[332] It is probably crude triacetyl phosphate.
$(MeCO{\cdot}O)_3PO$. IV.1.a (from silver phosphate).[886] Plates, m. 59-61°. Very hygroscopic and readily hydrolyzed, probably to the diacetyl derivative, which readily passes on to phosphoric acid, as the hydrolysis rates of mono- and diacetyl derivatives appear to be nearly identical.[886]
$CF_3C(O)OP(O)(OEt)_2$. IV.2.a (phosphoryl chloride + Na salt). Liquid, $b_{0.01}$ 30°, m. -10°.[840]
$EtC(O)OPO_3H_2$. IV.1.a (from "mono" silver phosphate). Silver salt.[879]

$EtC(O)OP(O)(OEt)_2$. IV.2.c (from RCOOH and $EtOC(O)CH{=}C(OEt)OP(O)(OEt)_2$),[444] IV.2.b (from RCOOH and $(EtO)_2POP(O)(OEt)_2$.[952a] Liquid, $b_{0.05}$ 68°,[444,952a] $b_{0.001}$ 65.7°;[952a] n_D^{25} 1.4140.[444,952a]

$EtC(O)OP(O)(OBu)_2$. IV.3.c (from phosphoro-diethylamidate and acid anhydride). Liquid, b_1 76-8°; n_D^{20} 1.4329.[525]

$EtC(O)OP(O)(OC_6H_{13})_2$. IV.3.c (from phosphoro-diethylamidate and acid anhydride). Liquid, b_1 106-7°; n_D^{20} 1.4398.[525]

$PrC(O)OPO_3H_2$. IV.1.a (from "mono" silver phosphate). Silver salt.[879]

$PrC(O)OP(O)(OEt)_2$. IV.2.c (from $EtOC(O)CH{=}C(OEt)OP(O)(OEt)_2$ and RCOOH). Liquid, $b_{0.03}$ 76-77°; n_D^{25} 1.4172.[444]

$PrC(O)OP(O)(OBu)_2$. IV.3.c (from phosphoro-diethylamidate and acid anhydride). Liquid, b_1 100-3°; n_D^{20} 1.4343.[525]

$PrC(O)OP(O)(OC_6H_{13})_2$. IV.3.c (from phosphoro-diethylamidate and acid anhydride). Liquid, b_1 120-1°; n_D^{20} 1.4411.[525]

$BuC(O)OP(O)(OEt)_2$. IV.2.c (from $EtOC(O)CH{=}C(OEt)OP(O)(OEt)_2$ and RCOOH). Liquid, $b_{0.05}$ 79°; n^{25} 1.4186.[444]

$Me_3CC(O)OP(O)(OEt)_2$. IV.1.a (from $NaO(EtO)_2PO$ and acyl chloride). Liquid, $b_{0.05}$ 64°; n_D^{25} 1.4144.[840]

$AmC(O)OP(O)(OEt)_2$. IV.2.c (from $EtOC(O)CH{=}C(OEt)OP(O)(OEt)_2$ and RCOOH). Liquid, $b_{0.01}$ 82-4°; n_D^{25} 1.4202.[444]

$AmC(O)OP(O)(OBu)_2$. IV.1.a (from acyl chloride and t-amine salt).[432]

$C_7H_{15}C(O)OPO_3H_2$. IV.1.a (from "mono" silver phosphate). Silver salt, solid, by precipitation with silver nitrate from slightly acid solution. Sodium, barium, calcium, and strychnine salts prepared by metathesis. Half life at 37° and pH 7.4 is 12 hr.[856]

$C_{15}H_{31}C(O)OPO_3H_2$. IV.1.a (from "mono" silver phosphate).[856] Hygroscopic plates, dec. 61-3° (from benzene-ligroin). Metathesis was used to isolate the following poorly soluble salts: silver, calcium, barium, and strychnine (useful for the separation from inorganic matter). Half life is comparable to the C_8 derivative.

$C_{15}H_{31}C(O)OP(O)(OCH_2C_6H_4NO_2\text{-}p)_2$. IV.1.a (from palmitoyl chloride and $AgOP(O)(OR)_2$. Cryst., m. 70-1°.[1089]

$Ph_3CC(O)OP(O)(OEt)_2$. IV.1.a ($NaO(EtO)_2PO$ and acyl chloride). Cryst., m. 71-2°.[840]

3. Derivatives of α-Amino Acids

β-Aspartyl phosphate, γ-glutamyl phosphate, leucine phosphate. These derivatives have been prepared as extremely reactive and hygroscopic oils--but not obtained pure--by treating the Ag salts of the benzoxycarbonyl derivatives with $(PhCH_2O)_2POCl$ [IV.2.a] and debenzylating the dibenzyl esters.[771]

Glycyl phosphate. By hydrogenolysis (Pd, H) of the dibenzylester. Ag_2 salt, m. 254-8°; Ba salt.[227]

Glycyl dibenzyl phosphate. By controlled hydrogenation (Ni, H) of the azidoacetyl derivative. Prisms, m. 62°.[227]

$N_3CH_2C(O)OP(O)(OCH_2C_6H_5)_2$. IV.1.a (acyl chloride and Ag salt). Needles, m. 152° (dec.); dec. at room temp. in several hr.[227]

$MeCH(NH_2)C(O)OPO_3H_2$. By hydrogenolysis (Pd, H) of the dibenzyl ester. Pale oil, undistillable; Ag_2 salt, m. 295-300°, stable in the dark.[227]

$MeCH(NH_2)C(O)OP(O)(OCH_2C_6H_5)_2O$. By selective hydrogenation (Ni, H) of the α-azidopropionyl derivative. Oil.[227]

$MeCH(N_3)C(O)OP(O)(OCH_2C_6H_5)_2$. IV.1.a (acyl chloride and Ag salt). Oil, unstable.[227]

$C_6H_5CH_2CH(NH_2)C(O)OPO_3H_2$. By hydrogenolysis (Pd, H) of the dibenzyl ester. Solid; Ag_2 salt, m. >320°, stable in the dark; Ba salt, needles.[227]

$C_6H_5CH_2CH(NH_2)C(O)OP(O)(OCH_2C_6H_5)_2$. By selective hydrogenation of the α-azido-phenylalanyl derivative). Cryst., m. 104-106° (dec.).[227]

$C_6H_5CH_2CH(N_3)C(O)OP(O)(OCH_2C_6H_5)_2$. IV.1.a (acyl chloride and Ag salt). Cryst., m. 98°, unstable.[227]

4. Derivatives of Aromatic Monoacids

$PhC(O)OPO_3H_2$. IV.1.a (from "mono" silver phosphate, Ag salt isolated).[368] Ag salt.[368a]

$(PhC(O)O)(PhO)PO_2H$. IV.1.a. Silver salt.[368b]

$(PhC(O)O)(PhCH_2O)PO_2H$. From its benzyl ester by controlled hydrogenolysis. Bu_3NH salt, IR.[977a]

$PhC(O)OP(O)(OEt)_2$. IV.2.c (from $EtOC(O)CH{=}C(OEt)OP(O)(OEt)_2$ and RCOOH). Liquid, $b_{0.01}$ 110-1°; n_D^{25} 1.4922.[444]

$(PhC(O)O)(PhCH_2O)_2PO$. IV.1.a (from $(RO)_2PO_2Ag$ and ClCOPh). Prisms, m. 62-5°; IR.[977a]

$C_6H_2Cl_3C(O)OP(O)(OEt)_2$. IV.1.a ($NaO(EtO)_2PO$ and acyl chloride). Liquid, $b_{0.004}$ 81°; n_D^{25} 1.5139.[840]

$C_6HCl_4C(O)OP(O)(OEt)_2$. IV.1.a ($NaO(EtO)_2PO$ and acyl chloride). Cryst., m. 76-8°.[840]

$C_6Cl_5C(O)OP(O)(OEt)_2$. IV.1.a ($NaO(EtO)_2PO$ and acyl chloride). Cryst., m. 72-4°.[840]

$4\text{-}O_2NC_6H_4C(O)OP(O)(OEt)_2$. IV.2.c (from $EtOC(O)CH{=}C(OEt)\text{-}OP(O)(OEt)_2$ and RCOOH). Liquid, $b_{0.05}$ 139-40°; n_D^{25} 1.5088.[444]

$4\text{-}O_2NC_6H_4C(O)OP(O)(OCH_2C_6H_5)_2$. IV.1.a. Cryst., m. 67-9°; IR.[629]

$(PhC(O)O)_2PO_2H$. IV.1.a (from silver phosphate). Sodium salt isolated. Half life at 37° and pH 7.4 is 45 hr. One benzoyl group is readily available for N-benzoylations in water solution.[368] Silver salt.[368a]

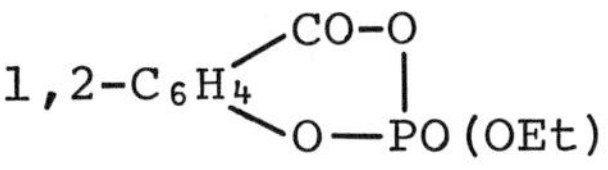

$1,2\text{-}C_6H_4\diagdown\begin{smallmatrix}CO-O\\|\\O-PO(OEt)\end{smallmatrix}$. From the corr. phosphorochloridate and ethanol. Cryst.[93a], m. 126.7-7.5°[93]

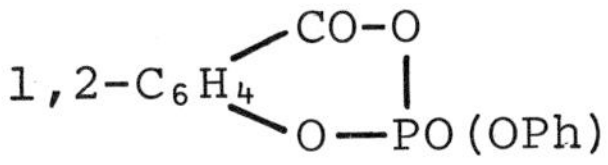

$1,2\text{-}C_6H_4\diagdown\begin{smallmatrix}CO-O\\|\\O-PO(OPh)\end{smallmatrix}$. From the corr. phosphorochloridate with phenol. Cryst., m. 85-6°.[93a]

5. Derivatives of Dicarboxylic Acids

$[-C(O)OP(O)(O\text{-benzyl})_2]_2$. IV.1.a (from acyl chloride and free acid). Cryst., m. 107-8° (dec.).[921]

$O[CH_2C(O)OP(O)(OBu)_2]_2$. IV.1.a (from acyl chloride and t-amine salt).[432]

$(HO_2C \cdot CH_2CH_2CO)OPO_3H_2$. IV.1.a (from "mono" silver phosphate) results in formation of both mono- (illustrated) and the diphosphate derivatives. Isolated as silver salts.[879]

$[-CH_2CH_2C(O)OP(O)(OBu)_2]_2$. IV.1.a (from acyl chloride and t-amine salt). n_D^{25} 1.4438.[432]

$[-CH_2CH_2C(O)OP(O)(O\text{-benzyl})_2]_2$. IV.1.a (from acyl chloride and t-amine salt). Solid.[432]

$1,4\text{-}C_6H_4[C(O)OP(O)(OBu)_2]_2$. IV.1.a (from acyl chloride and t-amine salt).[432]

B. Derivatives of Pyrophosphoric Acid

$MeCO \cdot OP(O)(OH)OP(O)(OH)_2$ (?). IV.3.b. Barium salt isolation is claimed: acetyl pyrophosphite (Chapter 13) is oxidized by BaO_2.[933] The slowness of reported hydrolysis by hot acids or bases makes this structure dubious.

Esters of Halophosphoric Acids

I. Phosphoric Acid Derivatives

A. Monoesters

1. $ROP(F)O_2H$

$MeOP(F)O_2H$. V.3.a.α-4 (from $MeOPO_3H_2$ + HF (DCC or CCl_3-CN))[933a] and α-5 (from $MeOPO_3H_2$ by 2,4-$(O_2N)_2C_6H_3$-F).[1465] Na salt, m. 88°;[933a] K salt, m. 166°.[1465]

$EtOP(F)O_2H$. V.3.a.α-4 (from F_2PO_2H). Liquid, by short path vacuum dist., not pure, Na and K salt.[698a] V.3.a.β-3 ($EtPO_3$ + HF),[698b] (from $[HO_2P(OEt)]_2O$ with HF). $b_{0.001}$ 55-8°.[840a] V.3.a.β-5 (from $EtOPO_3H_2$ with 2,4-$(O_2N)_2C_6H_4F$).[1464a] By controlled hydrolysis of $EtOPOFCl$. $b_{0.0005}$ 50-5°; d_2^{20} 1.3129; n_D^{20} 1.3680.[1319a]

$i\text{-}PrOP(F)O_2H$. V.3.a.α-4 (from F_2PO_2H). Liquid, by short path vacuum dist., not pure, Na and K salt.[698a]

$PhOP(F)O_2H$. V.3.β-4. Na salt from $PhOPO_3H_2$ + HF (DCC or CCl_3CN).[933a] V.3.a.β-5 (from $PhOPO_3H_2$ by fluoranil). NH_4 salt, m. 208-10°.[410a] V.3.a.β-5 (from

$PhOPO_3H_2$ by 2,4-$(O_2N)_2C_6H_3F$). K salt, m. 214-6°; c-$C_6H_{11}NH_3$ salt, m. 148-9°.[1465]

(4-$MeC_6H_4O)P(F)O_2H$. V.3.a.β-5 (from $ROPO_3H_2$ by 2,4-$(O_2N)_2C_6H_3F$). K salt, m. 235-6°; c-$C_6H_{11}NH_3$ salt, m. 109-11°.[1465]

(4-$ClC_6H_4O)P(F)O_2H$. V.3.a.β-5 (from $ROPO_3H_2$ by 2,4-$(O_2N)_2C_6H_3F$). K salt, m. 238-9°; c-$C_6H_{11}NH_3$ salt, m. 138°.[1465]

Adenosyl 5'-monofluorophosphate, Na salt. V.3.a.β-4. From adenosine phosphate + HF (DCC or CCl_3CN);[933a] with 2,4-$(O_2N)_2C_6H_3F$ and strong base; NH_4 salt, m. 178°.[1465]

Thymidine 5'-fluophosphate. V.3.a.β-5 (with 2,4-$(O_2N)_2$-C_6H_3F). Chromatographic and UV evidence.[1465]

Uridine 5'-fluophosphate. V.3.a.β-5 (with 2,4-$(O_2N)_2$-C_6H_3F). Chromatographic and UV evidence.[1465]

2. $ROPOF_2$

$MeOPOF_2$. V.3.b.α (from $(MeO)_3P$ and PF_5 besides other products).[285a]

$EtOPOF_2$. V.3.a.β-1.[616a,1236] V.3.a.α-5.[1206a] V.3.a.β-3 (neutral Et polyphosphates + HF).[840a] Liquid, b. 82-5°,[1206a] b. 84-5°,[616a] b. 85-6°;[1236] NMR,[1206a] ^{31}P 20.9 ppm.[913c]

$Me_3SiOPOF_2$. V.3.a.α-4 (from Me_3SiCl).[239a,826a] Liquid, b_{36} 36°; d_{20} 1.1139; n_D^{20} 1.3428;[826a] IR.[487a,826a] [1206a] V.3.a.α-5 (from Me_3SiCl). b. 118°; NMR.[1206a]

$Me_2EtSiOPOF_2$. V.3.a.α-4 (from $Me_2EtSiCl$). Liquid, b_{30} 52°; d_{20} 1.1356; n_D^{20} 1.3581; IR.[826a]

$Et_3SiOPOF_2$. V.3.a.α-4 (from Et_3SiCl). Liquid, b_{20} 72°, b.176°; d_{20} 1.0635; n_D^{20} 1.3890; IR.[826a]

$Et_2PrSiOPOF_2$. V.3.a.α-4 (from $Et_2PrSiCl$). Liquid, b_{14} 75°; d_{20} 1.0569; n_D^{20} 1.3901; IR.[826a]

$Bu_3SiOPOF_2$. V.3.a.α-5 (from Bu_3SiCl). Yellow brownish oil, not distillable; IR; NMR.[1206a]

$Me_2PhSiOPOF_2$. V.3.a.α-4 (from $Me_2PhSiCl$). Liquid, b_3 80°; d_{20} 1.1833; n_D^{20} 1.4541; IR.[826a]

$MePh_2SiOPOF_2$. V.3.a.α-4 (from $MePh_2SiCl$). Liquid, b_1 123°; d_{20} 1.1776; n_D^{20} 1.5102; IR.[826a]

$Me_2FSiOPOF_2$. V.3.a.α-4 (from Me_2SiCl_2). Liquid, b_{55} 105°; d_{20} 1.3692; n_D^{20} 1.3270.[826b]

$Et_2FSiOPOF_2$. V.3.a.α-4 (from Et_2SiCl_2). Liquid, b_{52} 53°; d_{20} 1.2592; n_D^{20} 1.3510.[826b]

$(CH_2Cl)MeSiF(OPOF_2)$. V.3.a.α-4 (from $CH_2ClMeSiCl_2$). Liquid, b_{92} 38°; d_{20} 1.3910; n_D^{20} 1.3610.[826b]

$PhMeSiF(OPOF_2)$. V.3.a.α-4 (from $PhMeSiCl_2$). Liquid, b_7 57°; d_{20} 1.3042; n_D^{20} 1.4392.[826b]

(p-$FC_6H_4SiMe_2O)POF_2$. V.3.a.α-4 (from p-$FC_6H_4Me_2SiCl$). Liquid, b_4 93°; d_{20} 1.2682; n_D^{20} 1.4467; IR.[826a]

3. ROPOFCl and ROPOFBr

MeOPOFCl. V.3.a.α-2 (from $POFCl_2$). Liquid, b_{75} 55°; d_{20}^{20} 1.4365; n_D^{20} 1.3671.[1319a]
EtOPOFCl. V.3.a.α-2 (from $POFCl_2$).[893a,1319a] Liquid, b. 171°,[893a] b_{50} 50°; d_{20}^{20} 1.3184; n_D^{20} 1.3750.[1319a]
BuOPOFCl. V.3.a.α-2 (from $POFCl_2$). Liquid, b_{15} 54°; d_{20}^{20} 1.2054; n_D^{20} 1.3843.[1319a]
PhOPOFCl. V.3.a.α-2 (from $POFCl_2$).[264a,1069a] Liquid, $b_{0.10}$ 81-2°,[264a] b_5 66-70°,[1069a] b. 205°; d_{20}^{20} 1.3756; n_D^{20} 1.4793.[264a]
EtOPOFBr. V.3.a.α-2 (from $POFBr_2$). Liquid, b_{15} 40-2°; d_{20}^{20} 1.699; n_D^{20} 1.413.[1319a]

4. $ROPOCl_2$

a. Ester Bond to Non-Aromatic Carbon

$MeOPOCl_2$. V.1.a.β;[351] (from diazomethane, yield small);[478a] V.1.c.[636b] Liquid, b_{15} 49°,[636a,b] b_{15} 62-4°;[351] d_{25} 1.4878;[636a] ^{31}P - 5.6 ppm.[913c]
$EtOPOCl_2$. V.1.a.β (from diazoethane, yield small);[478a] V.1.a.α;[809b,946,1236,1434,1452] V.1.a.β;[364] V.2.a (from Et_2O)[1270b] (from ethane);[625] V.2.b.α.[1452] Liquid, b. 167°,[946,1452] b_{19} 63°,[1236] b_{13} 58°,[636a] b_{10} 64-5°;[1434] d_4^{19} 1.353,[1434] d_{25} 1.3819;[636a] ^{31}P - 6.4; - 3.4 ppm.[913c]
$FCH_2CH_2OPOCl_2$. V.1.a.α. Liquid, b_{30} 106-7°, d_{20}^{20} 1.5367, n_D^{20} 1.4400.[809b]
$ClCH_2CH_2OPOCl_2$. V.1.a (from epoxide),[1146] V.1.a.α,[397,462,633,1123,1146,1189] V.2.a,[903a] V.2.b.α-1 (from ethylene phosphorochloridite or Me ethylene phosphite by Cl_2).[1217] Liquid, b_{15} 108-10°,[1189] b_{20} 101°,[1146] b_{15} 109-11°,[397,1146] b_{12} 103°,[462,1123] b_5 81.5°,[1217] $b_{2.2}$ 71.5°,[1217] $b_{0.8}$ 96°;[633] d_{20}^{20} 1.5430,[1146] 1.5560;[1217] n_D^{20} 1.4960,[1217] 1.4694;[1146] n_D^{22} 1.4689.[397] ^1H, ^{31}P - 5.9 ppm.[903a]
$BrCH_2CH_2OPOCl_2$. V.1.a.α. Liquid, b_2 70-1°.[749a]
Phthalimido-$CH_2CH_2OPOCl_2$. V.1.a.α. Cryst., m. 72-3°.[686]
(2-Me-4-Cl-$C_6H_3OCH_2CH_2O$)$POCl_2$. V.1.a.α. Cryst., m. 90-2°.[217]
CH_2=$CHOPOCl_2$. V.1.a.β. Liquid, b_{30} 36-40°; d_{20} 1.4273; n_D^{20} 1.4429, IR.[619]
$PrOPOCl_2$. V.1.c. Liquid, b_{13} 72°; d_{25} 1.3075.[636a,b]
$(CF_2HCF_2CH_2O)POCl_2$. V.2.b.α-1 (from $(RO)_2PCl$ with Cl_2). Liquid, b_1 65°.[859a]
i-$PrOPOCl_2$. V.1.c. Liquid, b_{13} 60°; d_{25} 1.2814.[636a,b]
$MeCH(CH_2Cl)OPOCl_2$. V.1.a.β (from epoxide). Liquid, $b_{1.5}$ 67-8°; d_{20} 1.4582; n_D^{20} 1.4640.[1146]
$(CH_2Cl)_2CHOPOCl_2$. V.1.a.α;[836a] V.1.a.β (from epoxide).[1146] Liquid, b_{12} 122-4°,[836a] b_3 103-4°;[1146] d_{20} 1.5800,[836a] 1.5844;[1146] n_D^{20} 1.4885,[836a] 1.4900.[1146]
$Cl_3CCH(COCl)OPOCl_2$. V.1.b.α (from $Cl_3CH(OH)COOH$). Liquid,

b_2 101-2°; d_{20} 1.7682.[857a]
CH_2=CMeOPOCl_2. V.1.a.β. Liquid, b_{30} 69-70°; d_{20} 1.3398; n_D^{20} 1.4492; IR.[619]
BuOPOCl_2. V.1.a.α, V.1.a.β (poor),[601] V.1.a.β (from Bu acetate),[1153a] V.1.c,[636a,b] V.2.a (from Bu_2O).[1270b] Liquid, b_{17} 90°, b_{13} 85°,[601] b_{13} 84°,[636a,b] b_2 60.5-1.5°;[1153a] d_4^{11} 1.2711, d_4^{25} 1.2560,[601] d_{25} 1.2562,[1153a] 1.2554;[636a,b] n_D^{11} 1.4453,[601] n_D^{20} 1.4420.[1153a]
n-$C_{18}H_{37}$OPOCl_2. V.1.a.α. Oil.[1305]
Cl_2OPO$(CH_2)_5$OPOCl_2. V.1.a.α.[1305]

b. Ester Bond to Aromatic Carbon

α. Aryl Phosphorodihalidates

PhOPOCl_2. V.1.a.α,[259a,689,746,770a,1072,1175] V.1.f.δ (dist. in vacuo of PhOC(O)N=PCl_3),[803] (from PhKSO_4 heated with PCl_5),[1316] V.2.b.α-2 (O_2 with UV).[1264] Liquid, b. 241-3°,[746,1175] 240°,[575,1222] b_{21} 130-4°,[575] b_{14} 138-40°,[103] b_{12} 119-21°,[355] b_{11} 121-2°,[746,1175] b_7 106-7.5,[770a] b_5 99.5-101.5°,[1072] $b_{0.1}$ 73°;[259a] d_4^{20} 1.41214;[747,1175] n_D^{25} 1.5215;[259a] IR;[259a] ^{31}P -1.5; -1.8 ppm.[913a]
C_6F_5OPOCl_2. V.1.a.α. Liquid, b_{20} 112-25°.[265a]
(2-ClC_6H_4O)POCl_2. V.1.a.α,[259a] V.1.b.α.[30] Liquid, b_{12} 135-7°,[30] b_8 205;[259a] IR.[259a]
(4-ClC_6H_4O)POCl_2. V.1.a.α,[1072,1215,1502] V.1.b.β,[776] V.1.b.α.[45,776] Liquid, b. 265°, b_{12} 141°,[45] b_{11} 142°,[1502] $b_{1-1.5}$ 96-104°,[1072] $b_{0.1}$ 95-115°.[1215]
(2,4,6-$Cl_3C_6H_2$O)POCl_2. V.1.a.α. Cryst., m. 126-7°; IR.[259a]
(2-$O_2NC_6H_4$O)POCl_2. V.1.a.α. Liquid, $b_{2.8}$ 150-2°.[1072]
(4-$O_2NC_6H_4$O)POCl_2. V.1.a.α.[12a,302,770a] Liquid, b_2 154-5°;[770a] cryst., m. 39-41°, b_2 145-8°.[12a]
2,4-$(O_2N)_2C_6H_3$OPOCl_2. V.1.a.α. Oil, $b_{0.05}$ 170-80° (bath temperature; short path distillation).[123]
2-MeOC_6H_4OPOCl_2. V.1.a.α.[770a,828a] Liquid, b_3 126-9°,[828a] b_2 120°.[770a]
(3-MeOC_6H_4O)POCl_2. V.1.a.α.[770a,835a] Liquid, b_5 132 3°,[835a] $b_{1.5}$ 118-20°;[770a] d_{19} 1.4378; n_D^{19} 1.5330.[835a]
(4-MeOC_6H_4O)POCl_2. V.1.a.α.[770a,835a] Liquid, b_6 140°,[835a] $b_{1.5}$ 118-20°;[770a] d_{23} 1.4338; n_D^{23} 1.5305.[835a]
(2-MeC_6H_4O)POCl_2. V.1.a.α.[208,770a] Liquid, b_{19} 135-6°,[211] b_{15} 127°,[208] b_8 118°.[770a]
(3-MeC_6H_4O)POCl_2. V.1.a.α. Liquid, b_{10} 124.5-5.5°; $b_{1.0}$ 90-7°.[1072]
(4-MeC_6H_4O)POCl_2. V.1.a.α.[259a,1072,1175] Liquid, b. 255°,[1175] b_{12} 145-50°,[103] b_1 90-1°,[1072] $b_{0.5}$ 87-8°; n^{25} 1.5296; IR.[259a]
(4-Cl-3-MeC_6H_3O)POCl_2. V.1.a.α. Liquid, $b_{0.1}$ 95°,[1215] 79-82°.[1072]
(3,5-$Me_2C_6H_3$O)POCl_2. V.1.a.α. Liquid, $b_{0.5}$ 82-3°.[1072]

$(4\text{-}i\text{-}PrC_6H_4O)POCl_2$. V.l.a.α. Liquid, b_{11} 145-6°.[770a]

$(4\text{-}(Me_2Et)C_6H_4O)POCl_2$. V.l.a.α. Liquid, b_{11} 155-9°.[770a]

$(4\text{-}t\text{-}BuC_6H_4O)POCl_2$. V.l.a.α. Liquid, b_{10} 176°, b_6 150-3°; n_D^{20} 1.244.[209]

$(2\text{-}i\text{-}Pr\text{-}5\text{-}MeC_6H_3O)POCl_2$. V.l.a.α. Liquid, b_{300} 246-9°.[481]

$(2\text{-}Me\text{-}6\text{-}PrCO\text{-}C_6H_3O)POCl_2$. V.l.a.α. $b_{0.2}$ 167°.[1215]

5-Chlorocarvacrylphosphorodichloridate. V.l.a.α. Liquid, $b_{0.6}$ 123-5°.[1215]

$(2\text{-}i\text{-}Pr\text{-}6\text{-}Cl\text{-}5\text{-}MeC_6H_2O)POCl_2$. V.l.a.α. Liquid, b_{12} 168°.[1215]

$(4\text{-}t\text{-}AmC_6H_4O)POCl_2$. V.l.a.α. Liquid, b_{10} 174°; d_4^{25} 1.159.[209]

$(4\text{-}Bu\text{-}2\text{-}MeC_6H_3O)POCl_2$. V.l.a.α. Liquid, $b_{0.4}$ 128-33°.[1215]

$(4\text{-}n\text{-}C_6H_{13}\text{-}2\text{-}MeC_6H_4O)POCl_2$. V.l.a.α. Liquid, $b_{0.15}$ 140-5°.[1215]

$(4\text{-}Me_2BuC_6H_4O)POCl_2$. V.l.a.α. Liquid, $b_{0.5}$ 123-5°.[770a]

$(2,6\text{-}(i\text{-}Pr)_2C_6H_3O)POCl_2$. V.l.a.α. Liquid, $b_{0.7}$ 118°; d_{25} 1.183; IR.[259a]

$(4\text{-}i\text{-}Am\text{-}2\text{-}MeC_6H_3O)POCl_2$. V.l.a.α. Liquid, $b_{0.3}$ 125-33°.[1215]

$(4\text{-}n\text{-}C_7H_{15}C_6H_4O)POCl_2$. V.l.a.α. Liquid, $b_{1.5}$ 167-9°.[770a]

$(6\text{-}Bu\text{-}2\text{-}i\text{-}Pr\text{-}5\text{-}MeC_6H_2O)POCl_2$. V.l.a.α. Liquid, $b_{0.2}$ 138-41°.[1215]

$(6\text{-}Bu\text{-}2\text{-}i\text{-}Pr\text{-}5\text{-}MeC_6H_2O)POCl_2$. V.l.a.α. Liquid, $b_{0.2}$ 138-41°.[1215]

$(4\text{-}t\text{-}C_8H_{17}\text{-}C_6H_4O)POCl_2$. V.l.a.α. Liquid, b_{13} 197-203°.[209]

$(4\text{-}n\text{-}C_9H_{19}\text{-}C_6H_4O)POCl_2$. V.l.a.α. Liquid, $b_{0.5}$ 167°.[770a]

$(4\text{-}Me\text{-}2,6\text{-}(t\text{-}Bu)_2C_6H_2O)POCl_2$. V.l.a.α. Liquid, b_{1-5} 120-35°; IR.[259a]

$(4\text{-}n\text{-}C_{12}H_{25}\text{-}C_6H_4O)POCl_2$. V.l.a.α. Liquid, b_1 204-8°.[770a]

$(2\text{-}PhOC_6H_4O)POCl_2$. V.l.a.α. Liquid, b_{11} 195-8°.[690]

$(2\text{-}Cl\text{-}4\text{-}PhOC_6H_3O)POCl_2$. V.l.a.α. Oil, b_{11} 216-9°.[690]

$(1\text{-}C_{10}H_7O)POCl_2$. V.l.a.α.[577,770a,835] Liquid, b. 325-7°,[835] b_{20} 199-201°,[577] 198-200°,[835] $b_{0.5}$ 138-40°;[770a] n_D^{27} 1.596.[577]

$(2\text{-}C_{10}H_7O)POCl_2$. V.l.a.α.[577,770a,835] Cryst., m. 34-5°,[577] 39°,[835] b_{20} 204-5°,[835] b_1 150-5°,[577] 155-6°.[770a]

$(2\text{-}PhC_6H_4O)POCl_2$. V.l.a.α. Liquid, b_{47} 228°.[208]

$(3\text{-}PhC_6H_4O)POCl_2$. V.l.a.α. Liquid, b_9 218-21°.[278]

$(4\text{-}PhC_6H_4O)POCl_2$. V.l.a.α. Liquid, b_{12-3} 211-23°.[279]

6"-(2,2,5',4"-Tetramethyl-3',4',5',6'-tetrahydro-dibenzopyran)-phosphorodichloridate. V.l.a.α. Viscous mass, $b_{0.15}$ 170°.[233]

Tetrahydrocannabinyl phosphorodichloridate. V.l.a.α. Liquid, $b_{0.1}$ 185°.[233]

Cholesteryl phosphorodichloridate. V.l.a.α. Cryst., m. 122°.[523]

Tocopheryl phosphorodichloridate. V.l.a.α. Liquid, $b_{0.06}$

204°.[1028a]

β. Derivatives Prepared from Hydroxysulfonic and Hydroxybenzoic Acids

$(2\text{-}ClSO_2C_6H_4O)POCl_2$. V.1.b.α. Liquid.[30]
$(4\text{-}ClSO_2C_6H_4O)POCl_2$. V.1.b.α. Needles, m. 87-8°, $b_{13.5}$ 203°.[29]
$(4\text{-}ClSO_2\text{-}2,6\text{-}Br_2C_6H_2O)POCl_2$. V.1.b.α. Cryst., m. 76-8°,[29] m. 69-70°.[45]
$(4\text{-}ClSO_2\text{-}6\text{-}Br\text{-}4\text{-}MeC_6H_2O)POCl_2$. V.1.b.α. Prisms, m. 147°.[1507]
$(2\text{-}ClCOC_6H_4O)POCl_2$. V.1.b.α,[27,41,46,438] V.2.b.α.[40] Liquid, b_{11} 168°; d_4^{20} 1.55873;[46] ^{31}P -2.3 ppm.[913c]
$(3\text{-}ClCOC_6H_4O)POCl_2$. V.1.b.α. Liquid, b. 315-22°, b_{11} 168-70°; d_4^{20} 1.54844.[46]
$(4\text{-}ClCOC_6H_4O)POCl_2$. V.1.b.α. Liquid, b_{13} 176°; d_4^{20} 1.54219.[46]
$(2\text{-}ClCO\text{-}4\text{-}ClC_6H_3O)POCl_2$. V.1.b.α.[26,33] V.2.b.α.[33] Liquid, b_{13} 183-4°.[33]
$(2\text{-}ClCO\text{-}6\text{-}ClC_6H_3O)POCl_2$. V.2.b.γ. Liquid, b_{13} 195-6°.[33]
$(2\text{-}ClCO\text{-}4\text{-}MeC_6H_3O)POCl_2$. V.1.b.α, V.2.b.γ. Liquid, b_{12} 185°.[51]
$(2\text{-}ClCO\text{-}5\text{-}MeC_6H_3O)POCl_2$. V.1.b.α, V.2.b.γ. Liquid, b_{12} 184.6-5.4°.[51]
$(2\text{-}ClCO\text{-}6\text{-}MeC_6H_3O)POCl_2$. V.2.b.γ. Liquid, b_{12} 185.6-6.2°.[35,52]
$1\text{-}ClCO\text{-}2\text{-}Cl_2P(O)O\text{-}C_{10}H_6$. V.1.b.α. Needles, m. 38°.[1151]
$3\text{-}ClCO\text{-}2\text{-}Cl_2P(O)O\text{-}C_{10}H_6$. V.1.b.α. Needles, m. 63°.[704]
Derivatives $Cl_3CArOPOCl_2$ which may be $ClCOArOPCl_4$:
$(2\text{-}Cl_3CC_6H_4O)POCl_2$. V.1.b.α. Liquid, b_{11} 178-9°, d_4^{20} 1.62019.[46] ^{31}P -8.7 ppm.[913c]
$(3\text{-}Cl_3CC_6H_4O)POCl_2$. V.1.b.α. Liquid, b_{11} 178°.[46]
$(2\text{-}Cl_3C\text{-}4\text{-}ClC_6H_3O)POCl_2$. V.1.b.α. Cryst., m. 59-60°, b_{15} 197°.[33]
$(2\text{-}Cl_3C\text{-}4,6\text{-}Cl_2C_6H_2O)POCl_2$. V.1.b.α. Cryst., m. 102-4°.[43]
$(2\text{-}Cl_3C\text{-}4,6\text{-}Br_2C_6H_2O)POCl_2$. V.1.b.α. Plates, m. 129-30°.[48]
$(2\text{-}Cl_3C\text{-}4,6\text{-}I_2C_6H_2O)POCl_2$. V.1.b.α. Cryst., m. 126°.[50]
$(2\text{-}Cl_3C\text{-}6\text{-}MeC_6H_3O)POCl_2$. V.1.b.α. Plates, m. 80°, b_{13} 199.4-9.8°.[51]
$2\text{-}Cl_3C\text{-}1\text{-}Cl_2P(O)O\text{-}C_{10}H_6$. V.1.b.α. Prisms, m. 115°.[1467,1468]

γ. Bis-Phosphorodihalidates

$1,3\text{-}[Cl_2P(O)O]_2C_6H_4$. V.1.a.α.[770a,808] Liquid, b_{115} 263°, b_{75} 216°,[808] b_1 157°;[770a] d_{15} 1.643.[808]
$2\text{-}ClC_6H_3\text{-}1,3\text{-}(OPOCl_2)_2$. Liquid, b_{10} 196-8°.[1140a]
$1,4\text{-}[Cl_2P(O)O]_2C_6H_4$. V.1.a.α.[770a,808] Cryst., m. 123°, b_{70} 270°,[808] liquid, b_{20} 212°.[770a]
$1,4\text{-}(Cl_2OPO)_2\text{-}2,3,5,6\text{-}(PhCCl_2)_4$-benzene. V.1.b.α (from

tetra-C-benzoylhydroquinone). Cryst., m. 196-7°.[1149b]

c. Ester Bond to Other Atoms

CF_2=NOPOCl_2. V.2.b.β (EtOPCl_2 with F_2ClCNO). Liquid, b_{20} 65°; d_{20} 1.6360; n_D^{20} 1.4231.[917a]

CFCl=NOPOCl_2. V.2.b. (EtOPCl_2 with FCl_2CNO). Liquid, b_6 61-2°; d_{20} 1.7080; n_D^{20} 1.4565.[917a]

HClC=N(Me)$_2$OPOCl_2. From HCO·NH_2 with POCl_3.[85a]

Me_3SiOPOCl_2. V.1.a.β. Liquid, $b_{0.1}$ 40°,[539a] b_{11} 62-7°,[1297b] b_{10} 63-4°, m. -22°; d_{25} 1.210; n_D^{20} 1.4288.[1297a]

B. Diesters

1. (RO)(R'O)POF

a. Aliphatic and Alicyclic

(MeO)$_2$POF. V.3.a.α-1,[1069b] V.3.a.β-1,[617c,654a,819,1070] V.3.a.β-5 (from (MeO)$_3$P with $CF_3CF_2CCl_2CF_3$);[194a] from (RO)$_2$POSH by picryl fluoride[265b,c] or 2,4-$(O_2N)_2C_6H_3$-F;[265b] from (MeO)$_3$P by $(F_5C_2)_2$CO.[1456] Liquid, b. 149-50°,[617c] b_{20} 58-9°,[1069b] 60-65°.[1070]

(MeO)(EtO)POF. From EtOP(F)O_2H by CH_2N_2. Liquid, b_{15} 53-4°; n_D^{20} 1.3650.[1319b]

(MeO)($Me_2NCH_2CH_2O$)POF. From (MeO)POFCl with ROH. Liquid, $b_{0.2}$ 40°; unstable, but yielding with MeI the following stable product:[1336]

(MeO)($IMe_3NCH_2CH_2O$)POF. Cryst., m. 84°.[1336]

(EtO)$_2$POF. V.1.a.α,[369,517a,1070,1236] V.1.3.a.α-1,[1069b] V.3.a.α-2 (from POFCl_2),[369] V.3.a.β-1[241,517a,617b,1070,1235] (by K fluosulfinate),[1240a] (from (EtO)$_2$-morpholinoPOCl or (EtO)$_2$POCl with EtOH and NaF),[906b] V.3.a.α-4,[841,1235] V.3.a.β-3,[840a] V.3.a.β-5 (from picryl fluoride and (EtO)$_2$POSH),[265b,c] (from 2,4-$(O_2N)_2C_6H_3$F and (EtO)POSH),[265b] V.3.a.γ (from (EtO)$_2$-P(O)CCl_3 with KF),[236a] V.3.b.β (from (EtO)$_3$P with $(CF_3)_2$CFNO).[498a] Liquid, b. 171°,[369] 170-1°,[840a] 168-71°,[1236] b_{45} 88-90°, b_{25} 76-7°, b_{23} 74-6°,[1235] b_{20} 74.5-5.5°,[617b] 71-3°,[1070] b_{18} 70-2°,[369,1235] b_{16} 66-7°,[241] b_{12} 63°,[1235] b_{11} 61-2°,[241] b_{10} 48-9° (?),[1069b] b_9 54.5°.[517a]

(EtO)(ICH_2CH_2O)POF. V.3.a.α-3 (from EtOPOF_2 and $HOCH_2CH_2I$). Liquid, $b_{0.10}$ 62-3°.[616a]

(EtO)(CCl_3CCl_2O)POF. From (EtO)(CCl_2=CClO)POF and Cl_2. Liquid, $b_{0.001}$ 83-4°; d_{20} 1.6320; n_D^{20} 1.4650; IR.[1270a]

(FCH_2CH_2O)$_2$POF. V.3.a.α,[369] V.3.a.β-1 (from the chloridate).[617c] Liquid, b_{13} 125-7°,[369] $b_{0.8}$ 111-2°.[617c]

($ClCH_2CH_2O$)$_2$POF. V.3.a.α,[369] V.3.a.α-4,[430a] V.3.a.β-1.[429a,617c] Liquid, b_{23} 159-60°,[617c] b_{15} 142-4°,[369] 142°,[430a] $b_{0.8}$ 107°.[429a]

(CH_2=$CHCH_2O$)$_2$POF. V.3.a.β-1 (from (RO)$_2$POCl). Liquid,

b_{23} 99-100°.[617c]
(EtO)(CCl_2=CClO)POF. V.3.b.γ (from $(EtO)_2PF$ with CCl_3-COCl). Liquid, $b_{0.05}$ 40°; d_{20} 1.5120; n_D^{20} 1.4447; IR.[1270a]
(MeO)[$Me_2NCH(Me)CH_2O$]POF. From MeOPOFCl with ROH. Liquid, $b_{0.1}$ 40°.[1336]
(MeO)[$IMe_3NCH(Me)CH_2O$]POF. From the preceding with MeI. Cryst., m. 84°.[1336]
(EtO)(PrO)POF. Liquid, b_{11} 72-5°; n_D^{20} 1.3814.[1319b]
$(PrO)_2POF$. V.3.a.α-2 (from $POFCl_2$). Liquid, b_{20} 98-100°.[369]
$(i\text{-}PrO)_2POF$. V.1.a.α,[517a,1070] V.1.a.α-1,[1069b] V.3.a.α-2 (from $POFCl_2$),[369,893a] V.3.α-4 (from Ag_2PO_3F),[1235] V.3.a.β-1,[517a,564b,617c,654a,819,1069c,1070,1235] (from (EtO)morpholinoPOCl or (EtO)(Et_2N)POCl with i-PrOH and NaF),[906b] V.3.a.β-5 (from $(RO)_2POSH$ with picryl fluoride or 2,4-$(O_2N)_2C_6H_3F$).[265b,c] Liquid, b. 183° (calc.),[1235] b_{25} 84-5°,[369,893a] b_{22} 83°,[617c] b_{20} 84-6°,[1070] b_{17} 80-2°,[1069c] 74-5°,[819] b_{16} 73°,[1235] b_{10} 62-3°,[517a] 60-1°,[1069b] b_9 62°, b_5 46°,[819] b_2 42-3°;[564b] m. -82°;[1235] n_D^{20} 1.3814; vapor pressure; viscosity.[564b] ^{31}P + 11.2 ppm.[913c]
[$(ClCH_2)_2CHO$]$_2$POF. V.3.a.α-4, V.3.a.β-1. Liquid, $b_{0.7}$ 163-5°.[430a]
$(EtOC(O)CHMeO)_2POF$. V.3.a.α-4, V.3.a.β-1 (with AgF),[430a] V.3.a.β-1 (with NH_4F).[617c] Liquid, b_1 128-30°,[617c] $b_{0.6}$ 126-8°.[430a]
(EtO)(BuO)POF. V.3.a.α. Liquid, b_{18} 83°.[241]
$(BuO)_2POF$. V.3.a.α-4 (from Ag_2PO_3F),[430a] V.3.b.β (from $(RO)_3P$ with $(CF_3)_2CFNO$),[498a] (from $(RO)_3P$ with $F(CF_2CFCl)_nF$, $CF_3CF_2CCl_2CF_3$, or $CF_3CFClCCl_2CF_3$).[571a,194a] Liquid, b_{30} 128°,[430a] b_{10} 85-7°.[194a]
$(i\text{-}BuO)_2POF$. V.3.a.α-4 (from Ag_2PO_3F). Liquid, b_{30} 135-8°.[430a]
$(MeEtCHO)_2POF$. V.3.a.β-1. Liquid, $b_{0.8}$ 62-4°, $b_{0.15}$ 64.5°.[430a]
$(AmO)_2POF$. V.3.a.α-4 (from Ag_2PO_3F). Liquid, b_{30} 143-4°.[430a]
$(i\text{-}AmO)_2POF$. V.3.a.α-4 and β-1. Liquid, b_{28} 142°, b_{23} 135-8°.[430a]
$(Et_2CHO)_2POF$. V.3.a.α-4, V.3.a.β-1. Liquid, b_2 97-8°.[430a]
$(Me_2CHCH_2CHMeO)_2POF$. V.3.a.α-4, V.3.a.β-1,[430a] V.3.β-1 (from $(RO)_2POCl$--from $(RO)_2PHO$ by N-Cl-succinimide (V.2.b.β-3), not isolated--by NH_4F).[617c] Liquid, $b_{2.7}$ 102-3°,[430a] b_1 105-6°.[617c]
(i-PrO)(c-C_6H_{11}-O)POF. V.3.a.β-1 (from the corr. chloridate by KF). Liquid, $b_{0.2}$ 93-4°; d_{20} 1.113; n_D^{20} 1.4349.[904b]
(i-BuO)(c-C_6H_{11}O)POF. V.3.a.β-1 (from the corr. chloridate by KF). Liquid, $b_{0.3}$ 98-100°; d_{20} 1.033; n_D^{20}

1.4352.[904b]

$(c\text{-}C_6H_{11}O)_2POF$. V.3.a.α-2 (from $POFCl_2$),[369,893a] V.3.a.β-1 (from $(RO)_2POCl$ by NH_4F).[1353a] Liquid, $b_{0.3}$ 116°,[369] $b_{0.02}$ 90-6°,[369] 96°;[893a] n_D^{20} 1.4785.[1353a]

$(2\text{-Me-}c\text{-}C_6H_{10}O)_2POF$. V.3.a.α-2. Liquid, $b_{0.15}$ 137°, $b_{0.1}$ 120°.[369]

b. Aromatic-Aliphatic and Purely Aromatic

(BuO)(PhO)POF. V.3.a.β-1 (from the corr. chloridate by KF). Liquid, $b_{0.1}$ 78-80°; d_{20} 1.062; n_D^{20} 1.4300.[904b]

$(PhO)_2POF$. V.3.a.α-2 (from $POFCl_2$)[369] (an earlier preparation, V.3.a.β-1,[623] is in error), V.3.β-1 (from $(RO)_2POCl$ with NaF in polar medium),[1368] V.3.a.β-4 (from $(RO)_3PO$ with HF).[698b] Liquid, $b_{0.4}$ 115-8°,[369] $b_{0.3}$ 118-20°,[264a] $b_{0.07}$ 106-8°;[369] d^{20} 1.2722; n_D^{20} 1.5203.[264a]

$(PhO)(4\text{-}ClC_6H_4O)POF$. V.3.a.α-2 (from (PhO)POFCl). Liquid, $b_{1.5}$ 133-4°.[1069a]

$(PhO)(4\text{-}MeC_6H_4O)POF$. V.3.a.α-2 (from (PhO)POFCl). Liquid, $b_{1.5}$ 129-30°.[1069a]

Di-tetrahydrofurfuryl phosphorofluoridate. V.3.a.β-1 (from the corr. chloridate with NH_4F). Liquid, $b_{0.005}$ 112-6°; d_{20} 1.2496; n_D^{20} 1.4495.[1353a]

c. Ester Bonds to Other Atoms than C

$(MeO)(CF_2{=}NO)POF$. From (MeO) PF with F_2ClCNO. Liquid, b_{20} 58°; d_{20} 1.4240; n_D^{20} 1.3578.[917a]

$(MeO)(CFCl{=}NO)POF$. From $(MeO)_2PF$ with FCl_2CNO. Liquid, b_2 70°; d_{20} 1.5070; n_D^{20} 1.3910.[917a]

$(EtO)(CFCl{=}NO)POF$. From $(EtO)_2PF$ with FCl_2CNO. Liquid, b_4 75°; d_{20} 1.4450; n_D^{20} 1.3940.[917a]

$(MeO)(O_2NCF_2CF{=}NO)POF$. From $(MeO)_2PF$ with $O_2NCFCFClNO$. Liquid, b_3 76°; d_{20} 1.6390; n_D^{20} 1.3940.[917a]

$(Et_3PbO)_2POF$. V.3.α-4. Colorless solid, m. ∿260°.[1236]

2. $(RO)_2POCl$, $(RO)(R'O)POCl$, $R{<}^{O}_{O}{>}POCl$

a. R and R' Bound with Aliphatic C

$(MeO)(FCH_2CH_2O)POCl$. V.1.e (from $FCH_2CH_2OPOCl_2$). Liquid, b_1 61.5°.[756a]

$(MeO)_2POCl$. V.2.b.α,[819] V.2.b.α-1 (from PCl_3 + MeOH then, in one operation, Cl_2[654a] or SO_2Cl_2[551a]), (from $(RO)_3P$ by Cl_2).[471a,765,1209] V.2.b.α-2 (from $(RO)_3P$ by $COCl_2$,[756,1149a] or $Me_4P_2O_6$ by SO_2Cl_2[1099b]), (from $(RO)_2PCl$ by N_2O_4),[775a] V.2.b.β-1 (from $(RO)_2PHO$ by CCl_4),[1314a] V.2.b.β-3 (from $(RO)_3P$ by $CHCl_2CCl_2CN$,[1083a] from $(RO)_2PHO$ by S_2Cl_2[520a] or by $ClSCCl_3$[520b]). Liquid, $b_{20\text{-}25}$ 75-80°,[819] b_{17} 70-1°,[1209] b_{15} 70-1°,[471a] b_{11} 65-7°,[756] b_{10} 70°,[1314a] b_4 60°;[819] n_D^{25} 1.4107,[551a] n_D^{20} 1.4119;[471a,1209] ^{31}P -6.4 ppm.[913c]

(MeO)($ClCH_2CH_2O$)POCl. V.1.e (from $ROPOCl_2$ with MeOH),[1412] V.2.b.α-1.[1217] Liquid, b_3 88.5°,[1217] $b_{0.1}$ 83°;[1412] d_4^{20} 1.4135; n_D^{20} 1.4468.[1217]

$(EtO)_2POCl$. V.1.a,[922,1061,1434,1452] V.1.f.β,[1262a] V.2.a (with MeOH then SO_2Cl_2 in one operation),[551a] V.2.b.α-1,[471a,894,1452] V.b.α-2 (from $(RO)_2PHO$ by Cl_2,[1209] by SO_2Cl_2,[94] by $CuCl_2$;[1266,1308b] from $(RO)_3P$ by SCl_2),[1128] V.2.b.β-1 (from $(RO)_2PHO$ by CCl_4,[1314a] from $(RO)_3P$ by CCl_3NO_2,[760a] from $Et_4P_2O_5$ by CCl_4[760,761]), V.2.b.β-2 (from $(RO)_3P$ by hexachlorocyclopentadiene,[913a] by perchlorofulvene[913b]), V.2.b.β-3 (from $(RO)_2PHO$ by $ClSCCl_3$,[520b] by $PhSO_2NCl_2$;[402] from $(RO)_3P$ by CH_2ClCCl_2CN[1083b]). Liquid, b_{18} 93-5°,[402,894] b_{12} 86°,[551a] b_{10} 93-4°,[1434] b_8 76-8°,[471a] 70-1°,[1209] 80-2°,[1061] $b_{2.5}$ 61-3°,[922] $b_{0.4}$ 39°;[913a] n_D^{25} 1.4253,[913a] 1.4162;[551a] n_D^{20} 1.463,[471a] 1.4163;[1209] ^{31}P -2.8 ppm.[913c]

(EtO)($ClCH_2CH_2O$)POCl. V.1.e (from (RO)POCl and EtOH),[1412] V.2.b.α-1 (from ethyl ethylene phosphite). Liquid, $b_{4.5}$ 105.7°, $b_{0.1}$ 87°;[1412] d_4^{20} 1.3184; n_D^{20} 1.4426.[121]

$(ClCH_2CH_2O)_2POCl$. V.1.a. (from epoxide),[938,1146] V.2.a (with ROH then SO_2Cl_2 in one operation),[551a] V.2.b.α-1,[430a,755] (from $HOP(OCH_2CH_2O)_2POH$ [?, cf.[333a]] by Cl_2).[429a] Liquid, b_5 137-9°,[755] b_2 139°,[430a] 116-7°,[1146] $b_{0.6}$ 122-4°,[429a] $b_{0.45}$ 132-4°;[551a] d_4^{20} 1.4623,[755] d_{20} 1.4588;[1146] n_D^{25} 1.4711,[551a] n_D^{20} 1.4742,[755] 1.4710.[1146]

$(Cl_3CCH_2O)_2POCl$. V.1.a.α. Cryst., m. 37-42°.[499]

$(CH_2{=}CHO)_2POCl$. V.1.a.β',[619] V.2.b.α-2 (from $(RO)_2(BuO)P$ by PCl_5).[885a] Liquid, b_{12} 60-1°,[885a] b_{11} 58-9°,[619] b_4 58-9°;[619] d_{20} 1.2408,[619] 1.2442;[885a] n_D^{20} 1.4319,[619] 1.4350.[885a]

(EtO)$CH_2{=}CMeO$)POCl. V.2.b.α-1 (from $(EtO)_2(CH_2{=}CHMeO)P$ by PCl_5). Liquid, b_4 68-70°; d_{20} 1.1811; n_D^{20} 1.4332.[885a]

(EtO)[MeCH(CH_2Cl)O]POCl. V.1.a.α. Liquid, b_1 79-80°; d_{20} 1.3164; n_D^{20} 1.4500.[1146]

($ClCH_2CH_2O$)(PrO)POCl. V.1.e (PrOH with $(RO)POCl_2$). Liquid, $b_{0.1}$ 94°.[1412]

$(PrO)_2POCl$. V.2.b.α-1 (from $(RO)_2PHO$ by Cl_2),[632] (from $(RO)_3P$ by Cl_2),[471a] V.2.b.β-1 (from $(RO)_2PHO$ by CCl_4,[1314a] from $(RO)_3P$ by C_2Cl_6,[760b] from $R_4P_2O_5$ by CCl_4[760,761]), V.2.b.β-3 (from $(RO)_2PHO$ by $ClSCCl_3$).[520] Liquid, b_{13} 107-8°,[760b] b_{12} 106-7°,[760,761] b_{3-4} 65° (?),[1314a] b_2 78° (?),[471a,1209] b_{1-2} 93-5°,[632] $b_{1.5}$ 76°;[551a] n_D^{25} 1.4334,[632] 1.4236,[551a] n_D^{20} 1.4256,[471a] 1.426,[1314a] 1.4258.[1209]

$(CH_2{=}CHCH_2O)_2POCl$. V.2.b.β-1 (from $(RO)_2PHO$ by CCl_4),[1314] V.2.b.β-2 (from $(RO)_3P$ by perchlorofulvalene),[913b] V.2.b.β-3 (from $(RO)_2PHO$ by N-Cl-succinimide).[617c] Liquid, $b_{0.9}$ 89-90°,[617c] $b_{0.27}$ 45°,[913b] $b_{0.001}$

35°;[1314a] n_D^{20} 1.4504,[1314a] n_D^{21} 1.4410.[913b]
$(i\text{-}PrO)_2POCl$. V.2.a (with ROH and SO_2Cl_2 in one operation),[551a] V.2.b.α-1,[471a,654a,819,894,1209] V.2.b.α-2,[94,564b] V.2.b.β-3;[520b] from $(RO)_2PHO$ by $PhSO_2NCl_2$.[402] Liquid, b_{14} 95-6°,[94] 92°,[402] b_{12} 92-4°,[551a] $b_{7-7.5}$ 78-80°,[471a] b_2 78°,[1209] $b_{0.08}$ 41°;[894] n_D^{25} 1.4159,[551a] n_D^{20} 1.4178.[471a,1209]
$(MeCH(CH_2Cl)O)_2POCl$. V.1.a.β (from epoxide). Liquid, $b_{1.5}$ 131-2°; d_{20} 1.3422; n_D^{20} 1.4625.[1146]
$[(ClCH_2)_2CHO]_2POCl$. V.1.a.α,[836a] V.1.a. (from epoxide),[1146] V.2.b.α-1.[430] Liquid, b_2 182-6°,[430] 180-2° (dec.),[836a] $b_{0.03}$ 125-6°;[1146] d_{20} 1.5301; n_D^{20} 1.4955.[1146]
$(Cl_3CCHMeO)_2POCl$. V.1.b.α. Undistillable oil.[673]
$(CH_2{=}CMeO)_2POCl$. V.2.b.α-1 (from $(CH_2{=}CHMeO)_2(BuO)P$ by PCl_5). Liquid, b_1 51-2°; d_{20} 1.1746; n_D^{20} 1.4425.[885a]
$(EtOC(O)CHMeO)_2POCl$. V.2.b.α-1,[430,436] V.2.b.β-3 (from (RO)PHO by N-Cl-succinimide).[617c] Liquid, b_1 158-60°,[617c] $b_{0.02}$ 128°,[430] 120°.[436]
(EtO)(BuO)POCl. V.2.b.α-2 (from (EtO)(BuO)PHO by SO_2Cl_2). Liquid, b_1 86-7°; d_{20} 1.137; n_D^{20} 1.4206.[904a]
$(CH_2{=}CHO)(BuO)POCl$. V.2.b.α-1 (from $(CH_2{=}CHO)(BuO)_2P$ by PCl_5). Liquid, $b_{3.5}$ 69°; d_{20} 1.1529; n_D^{20} 1.4360.[885a]
(EtO)(AmO)POCl. V.2.b.α-2 (from (RO)(R'O)PHO by SO_2Cl_2). Liquid, $b_{0.8}$ 108°; d_{20} 1.112; n_D^{20} 1.4267.[904b]
(EtO)(i-AmO)POCl. V.2.b.α-2 (from (RO)(R'O)PHO by SO_2Cl_2). Liquid, $b_{0.2}$ 95°; d_{20} 1.109; n_D^{20} 1.4269.[904b]
$(ClCH_2CH_2O)(AmO)POCl$. V.1.e (AmOH with $ROPOCl_2$). Liquid, $b_{0.1}$ 104°.[1412]
$(BuO)_2POCl$. V.1.a.β,[601] V.2.b.α-1,[471a,601,632,1209] V.2.b.β-1 (from $(RO)_3P$ by CCl_3NO_2,[760a] by C_2Cl_6[760b]), V.2.b.β-3 (from $(RO)_2PHO$ by $ClSCCl_3$,[520b] by CCl_4[1314a]). Liquid, b_{15} 132-5°,[601] b_{12} 132-4°,[551a] b_6 110-3°,[1314a] b_1 112-4°,[632] $b_{0.5}$ 80.5-1.0°;[471a,1209] d_4^{14} 1.0822;[601] n_D^{25} 1.4378,[632] 1.4289,[551a] n_D^{20} 1.4280,[471a] 1.4288,[1209] 1.4308,[1314a] n_D^{15} 1.4335;[601] ^{31}P -3.4 ppm.[913a]
$(i\text{-}BuO)_2POCl$. V.2.a (with ROH and SO_2Cl_2 in one operation),[551a] V.2.b.α-1,[430,436,471a,632,1209] V.2.b.β-3 (from $(RO)_2PHO$ by $ClSCCl_3$).[520b] Liquid, b_{12} 122-4°,[551a] b_1 100-5°,[632] $b_{0.18}$ 62°,[471a,1209] $b_{0.1}$ 57°;[430] n_D^{25} 1.4249,[551a] 1.4358,[632] n_D^{20} 1.4215,[471a] 1.4265.[1209]
$(sec\text{-}BuO)_2POCl$. V.2.b.α-1.[430a,471a,1209] Liquid, $b_{0.8}$ 84°,[471a,1209] 92-4°;[430a] n_D^{20} 1.4280,[471a] 1.4286.[1209]
$(c\text{-}Propyl\text{-}CH_2O)_2POCl$. V.2.b.β-3 (from $(c\text{-}propyl\text{-}CH_2O)_2PHO$ with N-Cl-succinimide). Undistillable oil; IR; NMR.[1299]
(BuO)(i-AmO)POCl. V.2.b.α-2 (from (RO)(R'O)PHO by SO_2Cl_2). Liquid, $b_{0.05}$ 73-4°; d_{20} 1.150; n_D^{20} 1.4400.[904b]
$(EtO)(n\text{-}C_7H_{15}O)POCl$. V.2.b.α-2 (from (RO)R'O)PHO by SO_2Cl_2). Liquid, $b_{0.4}$ 125-6°; d_{20} 1.069; n_D^{20}

1.4336.[904b]

$(ClCH_2CH_2O)(n\text{-}C_7H_{15}O)POCl$. V.1.e (ROH with $ClCH_2CH_2$-$OPOCl_2$). $b_{0.1}$ 116°.[1412]

$(AmO)_2POCl$. V.2.b.α-1,[471a,632] V.2.b.β-3 (N-Cl-succinimide).[1209] Liquid, b_1 131-3°,[632] $b_{0.04}$ 71-2°;[471a,1209] n_D^{20} 1.4356,[471a,1209] n_D^{25} 1.4375.[632]

$(i\text{-}AmO)_2POCl$. V.2.b.α-1.[430,632] Liquid, b_1 122-4°,[632] $b_{0.02}$ 74°;[430] n_D^{25} 1.4341.[632]

$(Et_2CHO)_2P(O)Cl$. V.2.b.α-1. Liquid, $b_{0.1}$ 73.5°.[430]

Di-tetrahydrofurfuryl phosphorochloridate. V.2.b.β-3 (from $(RO)_2PHO$ by N-Cl-succinimide). Liquid, d_{20} 1.1672; n_D^{20} 1.4750.[1353a]

$(i\text{-}AmO)(n\text{-}C_6H_{13}O)POCl$. V.2.b.α-1 (from (RO)(R'O)PHO by Cl_2). Liquid, $b_{0.2}$ 122-5°; d_{20} 1.0452; n_D^{20} 1.4397.[904a]

$(n\text{-}C_6H_{13}O)_2POCl$. V.2.b.α-1 (from $(RO)_2PHO$ by Cl_2). Liquid, n_D^{25} 1.4392.[632]

$(1,3\text{-}(Me)_2BuO)_2POCl$. V.2.b.α-1 (from the corr. secondary phosphite). Liquid, $b_{0.01}$ 72.5-3.5°.[430]

$(i\text{-}AmO)(n\text{-}C_7H_{15}O)POCl$. V.2.b.α-2 (from (RO)(R'O)PHO by SO_2Cl_2 or Cl_2). Liquid, $b_{0.3}$ 129-32°; d_{20} 1.056; n_D^{20} 1.4500.[904a]

$(i\text{-}AmO)(n\text{-}C_8H_{17}O)POCl$. V.2.b.α-2 (from (RO)(R'O)PHO by Cl_2). Liquid, $b_{0.3}$ 132-4°; d_{20} 1.035; n_D^{20} 1.452.[904a]

$(n\text{-}C_7H_{15}O)_2POCl$. V.2.b.α-1 (from $(RO)_2PHO$ by Cl_2). Liquid, n_D^{25} 1.4410.[632]

$(n\text{-}C_{16}H_{33}O)_2POCl$. V.1.a.α. Crude solid.[1222]

$(PhCH_2O)_2POCl$. V.1.b.β,[476] V.2.b.α-1,[94,96] (from $(RO)_2PHO$ with Cl_2),[93b] V.2.b.β-3 (from $(RO)_2PHO$ with N-Cl-succinimide).[1027] Undistillable liquid; ^{31}P -4.7 ppm.[913c]

Di-(1-Cl-D-galactose-penta-O-acetate)-phosphorochloridate. V.1.b.α. Solid.[1470]

2',3'-O-isopropylidene adenosine 5'-(benzyl phosphorochloridate). V.2.b.β-3 (from the corr. 5'-benzyl phosphite by N-Cl-succinimide).[395a]

b. R Bound with Aliphatic and R' with Alicyclic C, or Both Bound with Alicyclic C

$(EtO)(c\text{-}C_6H_{11}O)POCl$. V.2.b.α-2 (from (RO)(R'O)PHO by SO_2Cl_2). Liquid, $b_{0.7}$ 115-7°; d_{20} 1.080; n_D^{20} 1.4305.[904b]

$(i\text{-}PrO)(c\text{-}C_6H_{11}O)POCl$. V.2.b.α-2 (from (RO)(R'O)PHO by SO_2Cl_2). Liquid, b_1 114-5°; d_{20} 1.076; n_D^{20} 1.4521.[904b]

$(c\text{-}C_6H_{11}O)_2POCl$. V.2.b.β-3 (from $(RO)_2PHO$ by N-Cl-succinimide). Liquid, d_{20} 1.1691; n_D^{20} 1.4772.[1353a]

c. R' Bound with Aromatic C; R Bound with Aliphatic or with Aromatic C

(MeO)(PhO)POCl. V.1.e. Liquid, $b_{0.1}$ 90-1°.[1031]

(EtO)(PhO)POCl. V.1.a.α,[994] V.1.e,[1031] V.2.b.α-2 (from (RO)(R'O)PHO by SO_2Cl_2). Liquid, b_1 96-7°,[904b] $b_{0.01}$

94-5°;[1031] d_{20} 1.231; n_D^{20} 1.4950.[904b]
($ClCH_2CH_2O$)(PhO)POCl. V.1.e (ROH with $PhOPOCl_2$). Liquid, b_3 139-41°.[1,2]
(CH_2=CMeO)(PhO)POCl. V.2.b.α-1 (from (CH_2=CMeO)$_2$(PhO)P by PCl_5). Liquid, b_1 114-4.5°; d_{20} 1.2510; n_D^{20} 1.5040.[885a]
(PrO)(PhO)POCl. V.1.e. Liquid, $b_{0.01}$ 105-8°.[1031]
(i-PrO)(PhO)POCl. V.2.b.α-2 (from (RO)(R'O)PHO by SO_2Cl_2). Liquid, $b_{0.1}$ 80-2°; d_{20} 1.188; n_D^{20} 1.4685.[904b]
(EtO)(4-MeC_6H_4O)POCl. V.2.b.α-2 (from (RO)(R'O)PHO by SO_2Cl_2). Liquid, b_1 136-8°; d_{20} 1.229; n_D^{20} 1.5060.[904b]
(BuO)(PhO)POCl. V.1.e, V.2.b.α-2 (from (RO)(R'O)PHO by SO_2Cl_2). Liquid, $b_{0.8}$ 102-5°,[904b] $b_{0.01}$ 120-2°;[1031] d_{20} 1.118; n_D^{20} 1.4615.[904b]
(n-$C_{11}H_{23}O$)(PhO)POCl. V.1.e (from ROH with (PhO)$POCl_2$).[923]
$(PhO)_2POCl$. V.1.a.α,[259a,1175] V.2.b.α-1 and α-2 (from $(RO)_2PHO$ by $SOCl_2$ or Cl_2;[975,1434a] from $(RO)_2PCl$ by O_2 and UV[1264]). Liquid, b_{272} 314-6°, b_{216} 275°,[1175] b_{21} 212-5°,[514,575,689] b_{13-4} 195°,[41] b_1 141°,[1434a] $b_{0.1}$ 134°;[259a] d_4^{20} 1.29604,[689] n_D^{20} 1.5490,[1434a] n_D^{25} 1.5475; IR;[259a] ^{31}P + 6.2 ppm.[913c]
$(C_6F_5O)_2POCl$. V.1.a.α. Liquid, b_{20} 181-3°.[265a]
(4-$ClC_6H_4O)_2POCl$. V.1.a.α,[1015,1215,1502] V.2.b.α-2 (from $(RO)_2PHO$ with $SOCl_2$).[975] Needles, m. 53-4°[1015] b_{15} 225-6°,[1015] $b_{0.1}$ 164-76°.[1215]
(2,4,6-$Cl_3C_6H_2O)_2POCl$. V.1.b.α. Cryst., m. 126-9°.[31]
(4-$BrC_6H_4O)_2POCl$. V.2.b.α-2 (from $(RO)_2PHO$ with $SOCl_2$).[975]
(PhO)(m-$MeOC_6H_4O$)POCl. V.1.a.α. Liquid, b_{3-4} 193-4°; d_{18} 1.3231; n_D^{18} 1.5545.[835a]
(PhO)(p-$MeOC_6H_4OP$)POCl. V.1.a.α,[835a] V.1.e.[480] Liquid, b_{5-6} 209-10°;[835a] b_2 186°;[480] d_{23} 1.3163; n_D^{23} 1.5515.[835a]
(2-$MeOC_6H_4O)_2POCl$. V.1.a.α,[101] V.1.d.[494] Liquid, b_{15} 258°;[101] cryst., m. 65-7°, b_3 213-5°.[828a]
(4-$O_2NC_6H_4O)_2POCl$. V.1.b.α,[1382] V.2.b.α-2 (from $(RO)_2PHO$ by S_2Cl_2).[562a] Cryst., m. 97-7.5°,[1382] 107-8°.[562]
(PhO)(2-MeC_6H_4O)POCl. V.1.a.α.[208,835a] Liquid, b_{35} 214-8°,[835a] b_{11} 200-12°;[208] d_{22} 1.2629; n_D^{22} 1.5490.[835a]
(PhO)(4-MeC_6H_4O)POCl. V.1.a.α.[835a,885] Liquid, b_{35} 224-8°;[885] d_{22} 1.2612; n_D^{22} 1.5471.[835a]
(PhO)(2,4-(i-PrO)$_2C_6H_3O$)POCl. V.1.a.α. Liquid, b_5 186-90°; d_{21} 1.1836; n_D^{21} 1.5385.[835a]
(2-$MeC_6H_4O)_2POCl$. V.1.a.α. Liquid, b_{13} 239-41°; d_{22} 1.3297; n_D^{22} 1.5592.[835a]
(3-$MeC_6H_4O)_2POCl$. V.1.a.α. Liquid, $b_{1.7}$ 173°, $b_{1.0}$ 160-4°.[1072]
(4-$MeC_6H_4O)_2POCl$. V.1.a.α.[259a,835a,1072] Liquid, b_5 232-4°,[835a] $b_{1.25}$ 166-70°,[1072] $b_{0.15}$ 142°;[259a] d_{23} 1.3265; n_D^{23} 1.5540,[835a] n_D^{25} 1.5420;[259a] IR.[259a]
(4-Cl-3-$MeC_6H_3O)_2POCl$. V.1.a.α.[1072,1215] Liquid, $b_{0.1}$ 170°,[1215] 155-7°.[1072]

$(3,5\text{-}Me_2C_6H_3O)_2POCl$. V.1.a.α. Liquid, $b_{0.6}$ 162-3°.[1072]
$(4\text{-}t\text{-}BuC_6H_4O)_2POCl$. V.1.a.α. Cryst., m. 100-1.5°, b_{10} 190-310° (crude).[210]
$(2\text{-}Me\text{-}4\text{-}i\text{-}PrC_6H_3O)_2POCl$. V.1.a.α. Liquid, b_{3-4} 184-6°; d_{22} 1.1300; n_D^{22} 1.5293.[835a]
$(2\text{-}i\text{-}Pr\text{-}5\text{-}MeC_6H_3O)_2POCl$. V.1.a.α. Liquid, b_{320} 330-40°.[481]
$(2\text{-}i\text{-}Pr\text{-}5\text{-}Me\text{-}6\text{-}ClC_6H_2O)_2POCl$. V.1.a.α. Liquid, $b_{0.2}$ 185-95°.[1215]
Di-(5-chlorocarvacryl) phosphorochloridate. V.1.a.α. Liquid, $b_{0.6}$ 190-2°.[1215]
Di-(6-Bu-thymyl) phosphorochloridate. V.1.a.α. Liquid, $b_{0.2}$ 218-30°.[1215]
$(4\text{-}Bu\text{-}2\text{-}MeC_6H_3O)_2POCl$. V.1.a.α. Liquid, $b_{0.4}$ 219-23°.[1215]
$(4\text{-}i\text{-}Am\text{-}2\text{-}MeC_6H_3O)_2POCl$. V.1.a.α. Liquid, $b_{0.2}$ 215-22°.[1215]
$(4\text{-}t\text{-}C_8H_{17}C_6H_4O)_2POCl$. V.1.a.α. Viscous mass, b_{10-3} 203-30°.[209]
$[4\text{-}Me\text{-}(2,6\text{-}(t\text{-}Bu)_2C_6H_2O]_2POCl$. V.1.a.α. Liquid, b_8 ∿350°; IR.[259a]
$(PhO)(2\text{-}C_{10}H_7O)POCl$. V.1.a.α. Liquid, b_{29} 286°.[800]
2,2'-(1,1'-Dinaphthylene) phosphorochloridate. V.1.a.α. Cryst.[915]

d. One R Bound with an Other Atom than C to O

$(MeO)(CFCl{=}NO)POCl$. From $(MeO)_2PCl$ with FCl_2CNO. Liquid, b_6 95°; d_{20} 1.5790; n_D^{20} 1.4390.[917a]
$(i\text{-}PrO)(CF_2{=}NO)POCl$. From $(i\text{-}PrO)_2PCl$ with F_2ClCNO. Liquid, b_2 61°; d_{20} 1.3720; n_D^{20} 1.4079.[917a]

e. Cyclic Esters

α. Five-Membered Rings

2-Cl-2-oxo-4-Me-1,3,2-dioxaphosphorochloridate (propylene phosphorochloridate). V.1.a.α. Liquid, b_4 102.5°; d_{20} 1.4260; n_D^{20} 1.4520.[70]
2-Cl-2-oxo-4,5-Me_2-1,3,2-dioxaphospholane. V.1.b.β. Liquid, $b_{8.5}$ 90-1.5°; d_{20} 1.3429; n_D^{20} 1.4444.[623a]
$1,2\text{-}C_6H_4O_2POCl$. V.1.a.β,[37,808] V.1.d,[31,37,631b,670a] V.2.b.α-1.[631b] Needles, m. 59-60°,[37] 58-9°;[31] liquid;[631b] b_{55} 162°,[808] b_{12} 122°,[31] 119-20°,[631b] b_9 150° (?);[37] ^{31}P -18 ppm.[913c]

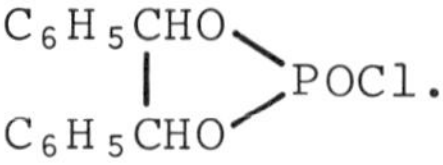

V.1.a.α. Cryst., m. 160-2°.[1380a]

2 OH
1 CH_3

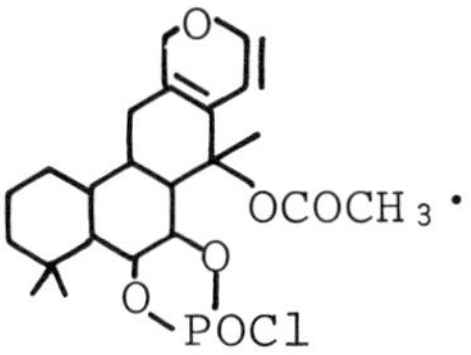

V.1.a (caesalpine derivative). Cryst., m. 130° (dec.); IR, UV; NMR; $[\alpha]_D^{25}$ 16° (c = 1, dioxane).[322a]

β. Six-Membered Rings

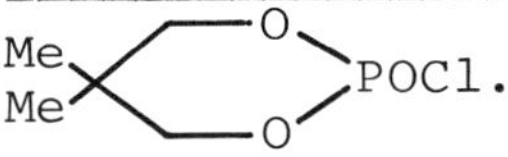

II.1.a. Cryst., m. 104.5-6°.[500]

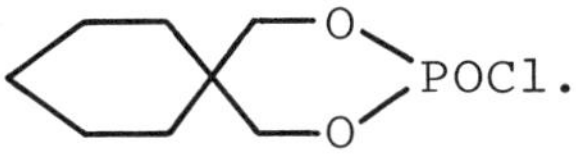

V.1.a.α. Cryst., m. 121-25°.[848a]

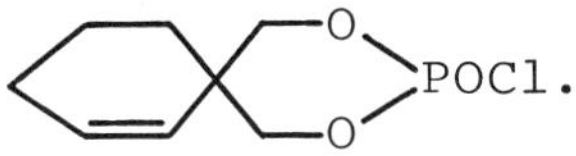

V.1.a.α. Cryst., m. 94-7°.[848a]

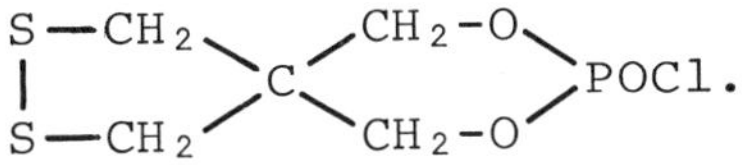

V.1.a.α. Cryst., m. 146-8°.[915a]

V.1.a.α. (from pentaerythritol).[9a,848a,1071a] Cryst., m. 233-5°,[1071a] 243-5°.[9a]

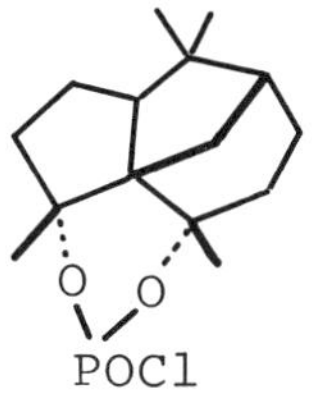

($C_{15}H_{24}O_2$)POCl. V.1.a.α. Cryst., m. 134°; $[\alpha]_D$ -41°.[885b]

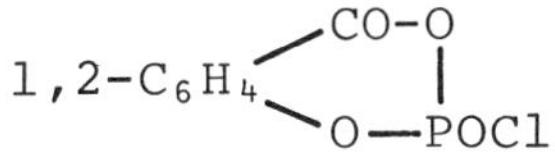

V.1.a.α (from salicyclic acid),[93,93a] V.2.b.α-2 (from the phosphorochloridite by O_2). Cryst. m. 95.5-6.0°; $b_{0.5}$ 145-8°; n_D^{20} 1.5544;[1107b] ^{31}P + 9.5 ppm.[913c]

3-Cl-3-oxo-5-Me-2,4,3-dioxaphospha-bicyclo[4.4.0]decane. V.1.a. (from 2-(1-HO-ethyl)-cyclohexanol). Liquid, n_D^{30} 1.4883.[848]

γ. Seven-Membered Rings

2,2'-(1,1'-dinaphthylene) phosphorochloridate. V.1.a.α. Undistillable oil (yields perylene on distillation).[915]

3. $(RO)_2POBr$, $(RO)(R'O)POBr$, $R{<}^{O}_{O}{>}POBr$

a. No Aromatic Ester Bond

$(MeO)_2POBr$. V.2.b.β-3 (from $(RO)_2PHO$ by N-Br-succinimide),[617b] (from $(RO)_3P$ by $CH_2BrCBrCBr_2CN$);[1083a] Liquid, $b_{0.8}$ 45-7°.[617b]

$(EtO)_2POBr$. V.1.f.β,[1262a] V.2.b.α-2 (from $(RO)_2PHO$ by $CuBr_2$),[1266] V.2.b.β-1 (from $(RO)_3P$ by $CHBr_3$),[760a] V.2.b.β-3 (from $(RO)_2PHO$ by N-Br-succinimide),[617b] (from $(RO)_3P$ by CH_2BrCBr_2CN).[760a] Liquid, $b_{1.5}$ 75°,[617b] b_1 138-9°; n_D^{20} 1.4319.[760a]

$(CH_2{=}CHO)_2POBr$. V.2.b.α-1 (from $(CH_2{=}CHO)_2(BuO)P$ by Br_2). Liquid, b_{12} 75.6°; d_{20} 1.5203; n_D^{20} 1.4650.[885a]

$(PrO)_2POBr$. V.2.b.β-3 (from $(RO)_2PHO$ by N-Br-succinimide) Liquid, $b_{0.4}$ 88-90°.[617b]

$(i\text{-}PrO)_2POBr$. V.2.b.β-3 (from $(RO)_2PHO$ by N-Br-succinimide). Liquid, $b_{0.4}$ 77-8°.[617b]

$[(CCl_3)_2CHO]_2POBr$. V.1.a.α. Cryst., m. 120°.[605]

$(CH_2{=}CMeO)_2POBr$. V.2.b.α-1 (from $(CH{=}CMeO)_2(BuO)P$ by Br_2). Liquid, $b_{0.3}$ 71.5-2°; d_{20} 1.3993; n_D^{20} 1.4702.[885a]

$(t\text{-}BuO)_2POBr$. V.2.b.α-1 (from $(t\text{-}BuO)_3P$). Liquid, b_1 90°.[606]

$(C_6H_5CH_2O)_2POBr$. V.2.b.α-1 (from $(RO)_2PHO$ with Br_2).[93b]

b. One or Two Aromatic Ester Bonds

$(MeO)(PhO)POBr$. V.2.b.β-1 (from $(MeO)(PhO)_2P$ by CBr_4). Liquid, b_2 115-23°.[761a]

$(BrCH_2CH_2O)(PhO)POBr$. V.2.b.α-1 (from ethylene phenyl phosphite and Br_2).[122a]

$(CH_2{=}CMeO)(PhO)POBr$. V.1.d (from $CH_2{=}CMeO)_2(PhO)PBr_2$ and moist air). Liquid, $b_{0.5}$ 120°; d_{20} 1.4748; n_D^{20} 1.5261.[885a]

$(BrCH_2CH_2CH_2O)(PhO)POBr$. V.2.b.α-1 (from trimethylene phenyl phosphite and Br_2).[122a]

$(PhO)_2POBr$. V.1.f.β,[1262a] V.2.b.β-1 (from $(RO)_3P$ by CBr_4). Cryst., m. 46-7°, b_3 198-200°.[761a]

c. Cyclic Esters

2-Br-2-oxo-4-Me-4-$BrCH_2$-1,3,2-dioxaphosphorinane. V.1.a. Configuration.[219a,758]

1,2-$C_6H_4{<}^{O}_{O}{>}POBr$. V.1.d (from O-phenylenedioxy-Br_3-phosphorane and $(COOH)_2$),[631b] V.2.b.α-1 (from O-phenylene methyl phosphite)[631b] V.2.b.β-3 (from bromidite and RO_2CH).[631a] Cryst., m. 35-9°, b_{15} 132-5°;[631a] liquid, b_{11} 134-8°;[631b] ^{31}P -3.6 ppm.[913c]

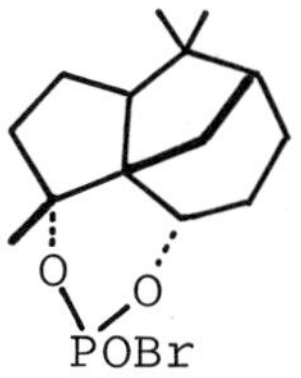

$(C_{15}H_{24}O_2)POBr$. V.1.a. Two epimers at the P atom. Cryst.; m. 126° and $[\alpha]_D$ -34°; m. 128° and $[\alpha]_D$ -40°;[885b] structure.[1349b]

4. $(RO)_2POI_n$

$(PhO)_2POI_4$. V.2.b.α-1 (from $(PhO)_3P$ by I_2, besides $(PhO)_2POI_9$). Golden brown plates, unstable, → brown oil.[567]

$(PhO)_2POI_9$. V.2.b.α-1 (from $(PhO)_2$(c-hexyl-O)P by I_2); perhaps: $(PhO)_2PO_7$ + $(PhO)_2POI_{10}$? Emerald prisms, m. 89-90°.[567]

B. Pyrophosphorofluoridates

$(EtO)_2P(O)OP(O)(OEt)F$. V.3.a.α-6 (from CH_2=CHO)$(EtO)_2PO$ + EtOP(F)O_2H, or $(EtO)_2PO_2H$ + EtOPOFCl, or EtOP(F)O_2H + ClPO$(OEt)_2$). Oil (by short path dist.; is diproportionated by vacuum dist.); d_{20}^{20} 1.248; n_D^{20} 1.4017.[1319b]

$[(MeO)PF(O)]_2O$. V.3.a.α-6 (from MeOPOFCl and MeOP(F)O_2H). Liquid, $b_{0.3}$ 44-6°; d_{20}^{20} 1.5181; n_D^{20} 1.3626.[1319b]

$[(EtO)PF(O)]_2O$. V.3.a.α-6 (from EtOPOFCl and EtOP(F)O_2H). Liquid, $b_{0.002}$ 61-4°; d_{20}^{20} 1.3512; n_D^{20} 1.3725.[1319b]

$[(BuO)PF(O)]_2O$. V.3.a.α-6 (from BuOPOFCl and BuOP(F)O_2H). Liquid, $b_{0.01}$ 70-80° (dec.); d_{20}^{20} 1.2016; n_D^{20} 1.3946.[1319b]

Esters of Peroxyphosphoric Acids

$(MeO)_2PO(OOCMe_3)$. VI.1. Cryst., m. 24-5°,[1192a] 26°;[1275] $dist._{0.001}$ (bath, 70°);[1192a] n_D^{25} 1.4160; NMR.[1275]

$(EtO)_2PO(OOCMe_3)$. VI.1. Liquid, $b_{0.03}$ 53-4°,[1192,1192a] $b_{0.15}$ 66-7°;[1275] n_D^{20} 1.4208,[1192] n_D^{25} 1.4169.[1275]

$(PrO)_2PO(OOCMe_3)$. VI.1. Liquid, $b_{0.3}$ 85-7°; n_D^{25} 1.4219; NMR.[1275]

$(i\text{-}PrO)_2PO(OOCMe_3)$. VI.1. Liquid, $b_{0.1}$ 64-7°; n_D^{25} 1.4148.[1275]

$(BuO)_2PO(OOCMe_3)$. VI.1. Liquid, non-distillable; n_D^{25} 1.4247; NMR.[1275]

$(i\text{-}BuO)_2PO(OOCMe_3)$. VI.1. Liquid, $b_{0.4}$ 99.5-100°; n_D^{25} 1.4200; NMR.[1275]

$(n\text{-}C_8H_{17}O)_2PO(OOCMe_3)$. VI.1. Oil; n_D^{25} 1.4389; NMR.[1275]

$(C_6H_5CH_2O)_2PO(OOCMe_3)$. VI.1. Unstable oil; NMR.[1275]

$(PhO)_2PO(OOCMe_3)$. VI.1. Oil, n_D^{25} 1.5133.[1275]

$(4\text{-}MeC_6H_4O)_2PO(OOCMe_3)$. VI.1. Unstable oil.[1192a]

^{32}P Labelled Esters

$BuO^{32}PO_3H_2$. VI.2.a (from $(BuO)_3PO$ with $H_3{}^{32}PO_4$ above 100°; besides the two following).[683]
$(BuO)_2{}^{32}PO_2H$. See above.[683]
$(BuO)_2{}^{32}PO$. See above.[683]
^{32}P-Phosphoryl choline. VI.2.a - (indirectly) I.1.a.β (from choline chloride and $H_3{}^{32}PO_4$ at 100°/1-2 Torr).[1260]
$(MeO)_2{}^{32}POF$. VI.2- V.3.a.β-1. Liquid, b_{3-4} 57°. n_D 1.3748.[253a]
$(PrO)_2{}^{32}POF$. VI.2-V.3.a.β-1. Liquid, b_{3-4} 90°. n_D 1.3875.[253a]
$(i\text{-}PrO)_2{}^{32}POF$. VI.2- V.3.a.β-1. Liquid, b_{3-4} 65°. n_D 1.3850.[253a]
$(MeO)_2{}^{32}POCl$. VI.2-V.2.b.α-1. Liquid, b_{3-4} 64°. n_D 1.4165.[253a]
$(PrO)_2{}^{32}POCl$. VI.2-V.2.b.α-1. Liquid, b_{3-4} 97°. n_D 1.4290.[253a]
$(i\text{-}PrO)_2{}^{32}POCl$. VI.2-V.2.b.α-1. Liquid, b_{3-4} 94°. n_D 1.4200.[253a]

REFERENCES

(received August 31, 1972)

1. Abderhalden, E., H. Paffrath, H. Sickel, Arch. Physiol., 207, 249 (1925).
2. Acker, L., Chem. Ber., 88, 376 (1955).
3. Adams, C. E., and B. H. Shoemaker, U. S. Patent 2,372,244 (1945; to Standard Oil Co.).
4. Adler, H., and H. B. Gottlieb, U. S. Patent 1,983,588 (1934; to Victor Chemical Works).
5. Adler, H., and W. H. Woodstock, Chem. Industries, 51, 516 (1942).
6. Adler, H., and W. H. Woodstock, Chem. Industries, 51, 516 (1942); Woodstock, W. H., U. S. Patent 2,402,703 (1946; to Victor Chemical Works).
7. Adler, H. (to Victor Chem. Works), U. S. 2,462,057 (Feb. 22, 1949) [C.A. 43, 3836g (1949)].
8. Adrian, Trillat, Bull. Soc. Chim. (3), 19, 266 (1898)
9. Agarwal, K. L., and M. M. Dhar, Ind. J. Chem., 1, 451 (1963) [C.A. 60, 5615 (1964)].

9a. Agfa A.-G., Belg. Patent 635,167 (1964) [C.A. 62, 1567a (1965)].

10. Agfa A.-G., Fr. Patent 1,394,564 (1965) [C.A. 63, 4163d (1965)].
11. Agfa A.-G., German Patent 246,871 (1912).
12. Ainley, A. D., L. A. Elson, and W. A. Sexton, J. Chem. Soc., 1946, 776.

12a. Ajinmoto Co. Inc., Japan. 108('65) (1965) [C.A. 62, 11737g (1965)].

12b. Aksnes, D., and G. Aksnes, Acta Chem. Scand., 17 1262 (1962).
12c. Aksnes, G., and S. I. Snaprud, Acta Chem. Scand., 15, 457 (1961).
13. Albright & Wilson (Manufg.) Ltd., Br. Patent 937,609 (1963) [C.A. 60, 1592e (1964)].
13a. Aldridge, W. N., Ann. Rep. Progr. Chem., 53, 294 (1956).
14. Alexander, B. H., P. A. Giang, and S. A. Hall, J. Am. Chem. Soc., 77, 2424 (1955).
15. Alexander, B. H., L. S. Hafner, M. V. Garrison, and J. E. Brown, J. Org. Chem., 28, 3499 (1963).
16. Alimov, P. I., M. A. Zvereva, and O. N. Fedorova, Khim. i Primenenie Fosfororgan. Soedinenii, Akad. Nauk SSSR, le Konf., 1955, 164 (1957) [C.A. 52, 244a (1958)].
17. Nil.
18. Allen, J. F., J. Am. Chem. Soc., 79, 3071 (1957).
19. Allen, J. F., U. S. Patent 2,816,128 (1957) [C.A. 52, 5449e (1958)].
20. Allen, J. F., and O. H. Johnson, J. Am. Chem. Soc., 77, 2871 (1955).
21. Allen, C. M., Jr., E. Richelson, and M. E. Jones, Curr. Aspects Biochem. Energ., 1966, 401 [C.A. 68, 104286 (1968)].
22. Amato, D., Gazz. Chim. Ital., 1, 56 (1871).
22a. Amonoo-Neizer, E. H., S. K. Ray, R. A. Shaw, and B. C. Smith, J. Chem. Soc., 1962, 4296.
23. Anand, N., V. M. Clark, R. A. Hall, and A. R. Todd, J. Chem. Soc., 1952, 3665.
24. Anderson J. M., and E. Percival, Chem. and Ind., 1954, 1018.
25. Anglo-Iranian Oil Co. Ltd., McLean, Oldham & Wirth, Br. Patent 696,219 (1953) [C.A. 48, 12790c (1954)].
25a. Anliker, R., E. Beriger, and K. Schmid, Experientia, 17, 492 (1961).
26. Anilinfarben A. G., German Patent 89,556 (1897).
27. Anschütz, R., Ann., 228, 308 (1885).
28. Anschütz, R., Ann., 228, 308 (1885); 346, 286 (1906).
29. Anschütz, R., Ann., 358, 92 (1908).
30. Anschütz, R., Ann., 415, 64 (1918).
31. Anschütz, L., Ann., 454, 106 (1927).
32. Anschütz, L., H. Boedeker, W. Broeker, and F. Wenger, Ann., 454, 71 (1927).
33. Anschütz, R., and R. Anspach, Ann., 346, 312 (1906).
34. Anschütz, R., Ber., 30, 221 (1897).
35. Anschütz, R., Ber., 30, 223 (1897).
35a. Anschütz, L., H. Boodeker, W. Broeker, and F. Wenyer, Ann. Chem., 454, 71 (1927).
36. Anschütz, L., and W. Broeker, Ber., 61, 1265 (1928).
37. Anschütz, L., and W. Broeker, J. Prakt. Chem., (2),

115, 379 (1927).
37a. Anschütz, L., and W. Broeker, Ber., 59B, 2848 (1926).
38. Anschütz, L., and W. Broeker, Ber., 77, 439 (1944).
39. Anschütz, R., and W. O. Emery, Ann., 239, 301 (1887).
40. Anschütz, R., and W. O. Emery, Ann., 239, 304 (1887).
41. Anschütz, R., and W. O. Emery, Ann., 253, 105 (1889).
42. Anschütz, L., H. Kraft, and K. Schmidt, Ann., 542 14 (1939).
43. Anschütz, R., and H. Mehring, Ann., 346, 300 (1906).
44. Anschütz, R., and H. Mehring, Ann., 346, 311 (1906).
45. Anschütz, R., and E. Molineus, Ann., 415, 51 (1918).
46. Anschütz, R., and G. D. Moore, Ann., 239, 314 (1887).
47. Anschütz, R., and A. Robitsek, Ann., 346, 311 (1906).
48. Anschütz, R., and A. Robitsek, Ann., 346, 323 (1906).
49. Anschütz, R., A. Robitsek, and F. Schmitz, Ann., 346, 335 (1906).
50. Anschütz, R., A. Robitsek, and F. Schmitz, Ann., 346, 330 (1906).
51. Anschütz, R., and E. Schroeder, Ann., 346, 341 (1906)
52. Anschütz, R., and E. Schroeder, Ann., 346, 349 (1906)
53. Anschütz, R., and E. Schroeder, Ann., 346, 300 (1906)
54. Anschütz, R., E. Schroeder, E. Weber, and R. Annspach, Ann., 346, 300 (1906).
55. Anschütz, L., and H. Walbrecht, J. Prakt. Chem., 133, 65 (1932).
56. Anschütz, L., and F. Wenger, Ann., 482, 25 (1930).
57. Antia, N. J., and R. W. Watson, J. Am. Chem. Soc., 80, 6134 (1958).
58. Antonakis, K., A. Dowgiallo, and L. Szabo, Bull. Soc. Chim. France, 1962, 1355.
59. Arbuzov, A. E., and V. S. Abramov, Trudy Kazan. Khim. Tekhnol. Inst., 1, 28 (1935).
60. Arbuzov, A. E., and B. A. Arbuzov, J. Prakt. Chem., 130, 103 (1931); Ber., 63, 195 (1932).
61. Arbuzov, A. E., and B. A. Arbuzov, J. Prakt. Chem., 130, 110 (1931).
62. Arbuzov, A. E., and B. A. Arbuzov, J. Prakt. Chem., 131, 337 (1931).
62a. Arbuzov, A. E., and B. A. Arbuzov, J. Obshch. Khim., 2, 345 (1932); 7, 1762 (1937).
63. Arbuzov, A. E., and B. A. Arbuzov, Zhur. Obshch. Khim., 2, 348 (1932).
64. Arbuzov, A. E., and B. A. Arbuzov, Zhur. Obshch. Khim., 2, 368 (1932).
65. Arbuzov, A. E., and B. A. Arbuzov, Zhur. Obshch. Khim., 2, 371 (1932).
66. Arbuzov, A. E., Ber., 38, 1172 (1903).
67. Arbuzov, A. E., Dissertation, Kazan (1905); J. Russ. Phys. Chem. Soc., 38, 161 (1906).
68. Arbuzov, A. E., and E. Ivanov, J. Russ. Phys. Chem. Soc., 47, 2015 (1915).

69. Arbuzov, A. E., and B. P. Lugovkin, Zhur. Obshch. Khim., 6, 394 (1936).
70. Arbuzov, A. E., K. V. Nikonorov, and Z. G. Shishova, Izv. Akad. Nauk SSSR, Otdel. Khim. Nauk, 1954, 823 [C.A. 49, 13891b (1955)].
71. Arbuzov, A. E., and A. I. Razumov, Zhur. Obshch. Khim., 7, 1762 (1937).
72. Arbuzov, B. A., and E. N. Ukhvatura, Izv. Akad. Nauk SSSR, Otdel Khim. Nauk, 1955, 1395 [C.A. 53, 6988f (1959)].
73. Arbuzov, B. A., and V. S. Vinogradova, Dokl. Akad. Nauk SSSR, 54, 787 (1946).
74. Arbuzov, B. A., and V. S. Vinogradova, Izv. Akad. Nauk SSSR, Otdel. Khim. Nauk, 1947, 459.
75. Arbuzov, B. A., and V. S. Vinogradova, Izv. Akad. Nauk SSSR, Otdel. Khim. Nauk, 1951, 733 [C.A. 46, 7515e (1952)].
76. Arbuzov, B. A., and V. S. Vinogradova, Izv. Akad. Nauk SSSR, Otdel. Khim. Nauk, 1952, 865 (C.A. 47, 10485e (1953)].
77. Arbuzov, B. A., and D. Kh. Yarmukkametova, Izv. Akad. Nauk SSSR, Otdel. Khim. Nauk, 1957, 292 [C.A. 51, 14542b (1957)].
78. Arbuzov, A. E., V. M. Zoroastrova, and N. I. Rizpolozhenskii, Izv. Akad. Nauk SSSR, Otdel. Khim. Nauk, 1948, 208.
79. Arbuzov, B. A., and V. M. Zoroastrova, Izv. Akad. Nauk SSSR, Otdel. Khim. Nauk, 1958, 1331 [C.A. 53, 7182h (1959)].
80. Arbuzov, B. A., and V. M. Zoroastrova, Izv. Akad. Nauk SSSR, Otdel. Khim. Nauk, 1958, 1331 [C.A. 53, 7182i (1959)].
81. Arbuzov, B. A., V. M. Zoroastrova, and N. D. Ibragimova, Izv. Akad. Nauk SSSR, Ser. Khim., 1964, 656 [C.A. 61, 2959c (1964)].
82. Arbuzov, B. A., V. M. Zoroastrova, and N. D. Ibragimova, Izv. Akad. Nauk SSSR, Ser. Khim., 1967, 704 [C.A. 67, 108477 (1967)].
83. Arcoria, A., and S. Fisichella, Ann. Chim. (Roma), 56, 1504 (1966).
84. Arnold, H., Ber., 73, 87 (1940).
85. Arnold, H., Ber., 74, 1736 (1941).
85a. Arnold, Z., and A. Holy, Coll. Czech. Chem. Commun., 27, 2886 (1962).
86. Aron-Samuel, J. M. D., P. Chabrier, and A. E. G. Desjobert, Fr. Patent 992,802 (1951).
87. Arvin, J. A., U. S. Patent 2,058,394 (1936; to E. I. du Pont Co.).
88. Asahi Chemical Industry Co., Ltd., Fr. Patent 1,361,963 (1964); [C.A. 61, 14586b (1964)].
89. Asano, M., Tech. Rep. Eng. Res. Inst. Kyoto Univ.,

1968, No. 138 [C.A. 69, 100071 (1968)].
89a. Asano, M., and N. Kohsaka, Tech. Rep. Eng. Res. Inst. Kyoto Univ., 1968, nr. 140 [C.A. 69, 100073 (1968)].
90. Askroft, S. J. H., and P. J. Randle, Biochem. J., 107, 599 (1968).
91. Atherton, F. R., Br. Patent 789,563 (1958) [C.A. 52, 12913h (1958)].
92. Atherton, F. R. (to Roche Products Ltd.), Br. Patent 793,722 (1958) [C.A. 52, 20063f (1958)].
93. Atherton, F. R. (to Roche Products Ltd.), Br. Patent 793,017 (1958) [C.A. 53, 2109b (1959)].
93a. Atherton, F. R. (to Roche Products Ltd.), Br. Patent 806,879 [C.A. 53, 14125i (1959)].
93b. Atherton, F. R., F. Bergel, A. Cohen, J. W. Haworth, H. T. Openshaw, and A. R. Todd, U. S. Patent 2,490,-573 (Dec. 1949) [C.A. 44, 3525a (1950)].
94. Atherton, F. R., H. T. Howard, and A. R. Todd, J. Chem. Soc., 1948, 1106.
95. Atherton, F. R., A. L. Morrison, J. W. Cremlyn, G. W. Kenner, A. R. Todd, and R. F. Webb, J. Chem. Soc., 1955, 1183.
96. Atherton, F. R., Openshaw, and A. R. Todd, J. Chem. Soc., 1945, 382.
97. Atherton F. R., and A. R. Todd, J. Chem. Soc., 1947, 674.
98. Atkinson, R. E., and J. I. G. Cadogan, J. Chem. Soc., C, 1967, 1356.
99. Audrieth, L. F., and R. Steinman, J. Am. Chem. Soc., 63, 2115 (1941).
100. Audrieth, L. F., and A. D. E. Toy, J. Am. Chem. Soc., 63, 2117 (1941).
101. Auger, V., and P. Dupuis, Compt. Rend., 146, 1152 (1908).
102. Autenrieth, W., Ber., 30, 2369 (1897).
103. Autenrieth, W., and E. Bölli, Ber., 58, 2144 (1925).
104. Autenrieth, W., and A. Geyer, Ber., 41, 146, 4256 (1908).
105. Autenrieth, W., and H. Hefner, Ber., 58, 2156 (1925).
106. Autenrieth, W., and O. Hildebrand, Ber., 31, 1094 (1898).
107. Autenrieth, W., and O. Hildebrand, Ber., 33, 2111 (1900).
108. Autenrieth, W., and W. Meyer, Ber., 58, 840 (1925).
109. Autenrieth, W., and P. Mühlinghaus, Ber., 40, 747 (1907).
110. Autenrieth, W., and Z. v. Vamossy, Zeitschr. Phys. Chem., 25, 440 (1898).
111. Auwers, K., and H. Czerny, Ber., 31, 2697 (1898).
112. Avaeva, S. M., M. M. Botvinik, and I. F. Syromyatnikova, Zh. Obshch. Khim., 34, 1749 (1964) [C.A.

61, 8393d (1964)].
113. Avaeva, S. M., M. M. Botvinik, I. F. Syromyatnikova, and V. I. Grigorovich, Vestn. Mosk. Univ., Ser. II, Khim., 20, 78 (1965) [C.A. 63, 8476g (1965)].
114. Avaeva, S. M., M. M. Botvinik, and I. F. Syromyatnikova, Sintez Privodn. Soedin. Ikh Analagov; Fragmentov, Akad. Nauk SSSR, Otd. Obshch. i Tekhn. Khim., 1965, 154 [C.A. 65, 9008f (1966)].
115. Avaeva, S. M., M. M. Botvinik, M. G. Vafina, and L. F. Matyazh, Zh. Obshch. Khim., 34, 1754 (1964) [C.A. 61, 8393g (1964)].
116. Avaeva, S. M., G. Folsch, L. Strid, and O. Mellander, Acta Chem. Scand., 17, 2718 (1963).
117. Avaeva, S. M., N. V. Ras'kova, O. V. Koval'chuk, and M. M. Botvinik, Khim. Org. Soedin. Fosfora, Akad. Nauk SSSR, Otd. Obshch. Tekh. Chim., 1967, 230 [C.A. 69, 36429 (1968)].
118. Avaeva, S. M., and Y. A. Sklyankina, Vestn. Mosk. Univ., Khim., 23, 119 (1968) [C.A. 70, 11964 (1969)].
119. Avaeva, S. M., V. A. Sklyankina, and M. M. Botvinik, Zh. Obshch. Khim., 39, 591 (1969) [C.A. 71, 39362 (1969)].
120. Avaeva, S. M., V. A. Sklyankina, L. V. Ermolenko, and M. M. Botvinik, Zh. Obshch. Khim., 38, 2363 (1968) [C.A. 70, 47811 (1969)].
121. Avaeva, S. M., I. F. Syromyatnikova, and M. M. Botvinik, Zh. Obshch. Khim., 38, 2231 (1968) [C.A. 70, 38081 (1969)].
122. Avaeva, S. M., I. F. Syromyatnikova, V. M. Grishchenko, and M. M. Botvinik, Khim. Prir. Soedin., 3, 126 (1967) [C.A. 67, 64671 (1967)].
122a. Ayres, D. C., and H. N. Rydon, J. Chem. Soc., 1957, 1109.
123. Azerad, R., D. Gautheron, and M. Vilkas, Bull. Soc. Chim. France, 1963, 2078.
124. Baddiley, J., J. G. Buchanan, B. Carss, and A. P. Mathias, J. Chem. Soc., 1956, 4583.
125. Baddiley, J., J. G. Buchanan, and R. Letters, J. Chem. Soc., 1957, 1067.
126. Baddiley, J., J. G. Buchanan, A. P. Mathias, and A. R. Sanderson, J. Chem. Soc., 1956, 4186.
127. Baddiley, J., J. G. Buchanan, and A. R. Sanderson, J. Chem. Soc., 1958, 3107.
128. Baddiley, J., J. G. Buchanan, and L. Szabó, J. Chem. Soc., 1954, 3826.
129. Baddiley, J., V. M. Clark, J. Michalski, and A. R. Todd, J. Chem. Soc., 1949, 815.
129a. Baddiley, J., A. M. Michelson, and A. R. Todd, J. Chem. Soc., 1949, 582.
130. Baddiley, J., A. M. Michelson, and A. R. Todd, Nature, 161, 761 (1948).

130a. Baddiley, J., and M. Thain, J. Chem. Soc., 1953, 903.
131. Baddiley, J., and E. M. Thain (to Nat. Research Devt. Corp.), Br. Patent 749,512 (1958) [C.A. 51, 1252 (1957)].
132. Baddiley, J., and A. R. Todd, J. Chem. Soc., 1947, 648.
133. Baer, E., J. Am. Chem. Soc., 69, 1253 (1947).
134. Baer, E., J. Am. Chem. Soc., 75, 5533 (1953).
135. Baer, E., Biochem. Preparations, 2, 31 (1952).
136. Baer, E., J. Biol. Chem., 185, 763 (1950).
137. Baer, E., and D. Buchner, J. Biol. Chem., 232, 895 (1958).
138. Baer, E., and D. Buchner, Can. J. Biochem. Physiol., 36, 243 (1958).
139. Baer, E., L. T. Cipliijanskas, and T. Visser, J. Biol. Chem., 234, 1 (1959).
140. Baer, E., and P. Cooper, Can. J. Biochem., 43, 1343 (1963).
141. Baer, E., I. B. Cushing, and H. O. L. Fischer, Can. J. Research, 21, 119 (1943).
142. Baer, E., and F. Eckstein, J. Biol. Chem., 237, 1449 (1962).
143. Baer, E., and H. O. L. Fischer, J. Biol. Chem., 128, 491 (1939).
144. Baer, E., and H. O. L. Fischer, J. Biol. Chem., 135, 321 (1940).
145. Baer, E., and H. O. L. Fischer, J. Biol. Chem., 143, 563 (1942).
146. Baer, E., and H. O. L. Fischer, J. Biol. Chem., 150, 214 (1943).
147. Baer, E., and H. O. L. Fischer, J. Biol. Chem., 150, 223 (1943).
148. Baer, E., and H. O. L. Fischer, J. Biol. Chem., 180, 145 (1949).
149. Baer, E., and M. Kates, J. Am. Chem. Soc., 70, 1394 (1948).
150. Baer, E., and C. S. McArthur, J. Biol. Chem., 154, 451 (1944).
151. Baer, E., and T. Maurukas, J. Biol. Chem., 212, 25 (1955).
152. Baer, E., and T. Maurukas, J. Biol. Chem., 212, 39 (1955).
153. Baer, E., J. Maurukas, and D. D. Clarke, J. Biol. Chem., 228, 181 (1957).
154. Baer, E., and G. Venkat, Can. J. Biochem., 42, 1547 (1964).
155. Baeuerlein, E., and Th. Wieland, Chem. Ber., 102, 1299 (1969).
156. Baeyer & Co., German Patent 280,000 (1914).
157. Baeyer, H. v., and K. A. Hofmann, Ber., 30, 1973

(1897).
157a. Bafus, D. A., E. J. Gallegos, and R. W. Kiser, J. Phys. Chem., 70, 2614 (1966).
158. Bailly, O., Ann. Chim., (9), 6, 96 (1916).
159. Bailly, O., Ann. Chim., (9), 6, 259 (1916).
159a. Bailly, O., Bull. Soc. Chim. Fr., 29, 274 (1921).
160. Bailly, O., Bull. Soc. Chim. (4), 29, 280 (1921).
161. Bailly, O., Bull. Soc. Chim. (4), 31, 848 (1922).
162. Bailly, O., Bull. Soc. Chim., 9, 314, 340, 405, 421 (1942).
163. Bailly, O., Compt. Rend., 160, 663 (1915).
164. Bailly, O., Compt. Rend., 161, 677 (1915); Ann. Chim. (9), 6, 133 (1916).
165. Bailly, O., Compt. Rend., 168, 561 (1919); Bull. Soc. Chim., 25, 244 (1919).
166. Bailly, O., Compt. Rend., 170, 1062 (1920).
167. Bailly, O., Compt. Rend., 172, 689 (1921).
168. Bailly, O., Compt. Rend., 206, 1902 (1938).
169. Bailly, O., Compt. Rend., 208, 443 (1939).
170. Bailly, O., and J. Gaumé, Compt. Rend., 178, 1192 (1924); Bull. Soc. Chim., 35, 590 (1924).
171. Bailly, O., and J. Gaumé, Bull. Soc. Chim. (4), 39, 1420 (1926).
172. Bailly, O., and J. Gaumé, Compt. Rend., 183, 67 (1926).
173. Bailly, O., and J. Gaumé, Compt. Rend., 199, 793 (1934); 198, 2258 (1934).
174. Bailly, O., and J. Gaumé, Bull. Soc. Chim. (5), 2, 354 (1935).
175. Bailly, O., and J. Gaumé, Bull. Soc. Chim. (5), 3, 1396 (1936).
176. Baina, N. F., R. V. Kokunova, and L. Z. Soborovskii, USSR Patent 207,900 (Dec. 29, 1967) [C.A. 69, 52296 (1968)].
177. Nil.
178. Balarew, D., Z. Anorg. Chem., 88, 133 (1914).
179. Balarew, D., Z. Anorg. Chem., 99, 187, 190 (1917); 101, 225 (1917).
180. Balarew, D., Z. Anorg. Chem., 99, 101, 191 (1917).
181. Balarew, D., Z. Anorg. Chem., 99, 187 (1917).
182. Balarew, D., Z. Anorg. Chem., 100, 355 (1917).
183. Balarew, D., Z. Anorg. Chem., 101, 225, 227 (1917).
184. Balarew, D., J. Prakt. Chem., 104, 368 (1922).
185. Balarew, D., Z. Anorg. Allg. Chem., 158, 105 (1926).
186. Ballou, C. E., Arch. Biochem. Biophys., 78, 328 (1958).
187. Ballou, C. E., and H. O. L. Fischer, J. Am. Chem. Soc., 76, 3188 (1954).
188. Ballou, C. E., and H. O. L. Fischer, J. Am. Chem. Soc., 77, 3329 (1955).
189. Ballou, C. E., and H. O. L. Fischer, J. Am. Chem.

Soc., 78, 1659 (1956).
190. Ballou, C. E., and R. Hesse, J. Am. Chem. Soc., 78, 3718 (1956).
191. Bamann, E., W. Stadelmann, and J. Riehl, Arch. Pharm., 292, 36 (1959).
192. Bamann, E., H. Trapmann, and A. Schnegraf, Chem. Ber., 88, 1726 (1955).
193. Banks, G. R., and D. Cohen, J. Chem. Soc., 1965, 6209.
193a. Bany, T., Bull. Intern. Acad. Polon. Sci. Classe Sci. math. et nat., Ser. A, 1951, 1 [C.A. 47, 3791c (1953)].
194. Baranauckas, C. F., and A. W. Frank, U. S. Patent 3,340,333 (Sept. 5, 1967) [C.A. 67, 116557 (1967)].
194a. Baranauckas, C. F., and A. W. Frank, U. S. Patent 3,349,151 (Oct. 1967) [C.A. 68, 39098 (1968)].
195. Baranauckas, C. E., and J. J. Hodan, U. S. Patent 3,445,547 (May 20, 1969) [C.A. 71, 22587 (1969)].
196. Barclay, J. L., A. B. Foster, and W. G. Overend, J. Chem. Soc., 1955, 2505.
197. Barker, G. R., M. D. Montague, R. J. Moss, and M. A. Parsons, J. Chem. Soc., 1957, 3756.
198. Barker, G. T., and G. E. Toll, J. Chem. Soc., 1957, 3794, 3798.
199. Barnard, P. W. C., C. A. Bunton, D. Kellerman, M. M. Mhala, B. Silver, C. A. Vernon, and V. A. Welch, J. Chem. Soc., B, 1966, 227.
200. Barnard, P. W. C., C. A. Bunton, D. R. Llewellyn, V. G. Oldham, B. L. Silver, and C. A. Vernon, Chem. and Ind., 1955, 760.
201. Barnwell, J. L., W. A. Saunders, and R. W. Watson, Chem. and Ind., 1953, 173.
202. Barral, E., Bull. Soc. Chim. (3), 13, 420 (1895).
203. Barrenscheen, H. K., and L. Klebermass-Messiner, Biochem. Z., 265, 157 (1933).
204. Barthe, L., Compt. Rend., 150, 401 (1910).
205. Barthel, W. F., B. H. Alexander, P. A. Giang, and S. A. Hall, J. Am. Chem. Soc., 77, 2224 (1955).
206. Bass, S. L., U. S. Patent 1,858,659 (1932; to Dow Chemical Co.).
207. Bass, S. L., U. S. Patent 2,033,916 (1936; to Dow Chemical Co.).
208. Bass, S. L., U. S. Patent 2,033,918 (1936; to Dow Chemical Co.).
209. Bass, S. L., U. S. Patent 2,071,017 (1937; to Dow Chemical Co.).
210. Bass, S. L., U. S. Patent 2,071,323 (1937; to Dow Chemical Co.).
211. Bass, S. L., U. S. Patent 2,117,283 (1938; to Dow Chemical Co.).
212. Bass, S. L., U. S. Patent 2,228,222 (1941; to Dow

Chemical Co.).
213. N. V. de Bataafsche Petroleum Maatschappij, Br. Patent 807,119 (1959) [C.A. 53, 13498c (1959)].
214. Bauer, S., and G. Avigad, Israel J. Chem., 5, 171 (1967).
215. Baumgarten, H. E., and R. A. Setterquist, J. Am. Chem. Soc., 79, 2605 (1958).
216. Bayliss, R. L., H. Bevan, and T. Malkin, J. Chem. Soc., 1958, 2962.
217. Becchey, R. B., A. M. Roberton, and C. T. Holloway, Biochemistry, 6, 3867 (1961).
218. Beck, A., R. Lohrmann, and L. E. Orgel, Science, 157, 952 (1967).
219. Beineke, T. A., Acta Crystallogr., B, 25, 413 (1969) [C.A. 70, 119167 (1969)].
219a. Beineke, T. A., Chem. Commun., 1966, 860.
220. Belitser, V. A., and T. V. Saenko, Ukrain. Biokhim. Zhur., 18, 211 (1946) [C.A. 49, 866b (1955)].
221. Bell, A. (to Eastman Kodak Co.), Br. Patent 652,632 (1951) [C.A. 46, 1025a (1952)].
222. Bel'skii, V. E., N. N. Bezzubova, and I. P. Gozman, Zh. Obshch. Khim, 38, 1330 (1968) [C.A. 69, 85866 (1968)].
223. Bel'skii, V. E., and I. P. Gozman, Zh. Obshch. Khim., 37, 2730 (1967) [C.A. 69, 58646 (1968)].
224. Belugou, G., Compt. Rend., 126, 1575 (1898).
225. Bengelsdorf, J. Org. Chem., 21, 475 (1956).
226. Benning, A. F. (to E. I. Dupont de Nemours & Co.), U. S. Patent 2,597,702 (1952) [C.A. 47, 2196b (1953)].
227. Bentler, M., and H. Netter, Z. Physiol. Chem., 295, 362 (1953).
228. Bentley, R., J. Am. Chem. Soc., 70, 2183 (1948).
229. Bentley, R., J. Am. Chem. Soc., 71, 2765 (1949).
230. Beranek, J., and F. Sorm, Collection Czech. Chem. Commun., 28, 469 (1963).
231. Berecoechea, J., M. Faure and J. Anatol, Bull. Soc. Chim. Biol., 50, 1561 (1968).
232. Berezovskii, V. M., V. I. Koltunova, E. A. Shlimovich, and V. A. Devyatnin, Zh. Obshch. Khim., 32, 3890 (1962) [C.A. 58, 13953b (1963)].
233. Bergel, F., A. L. Morrison, H. Rinderknecht, A. R. Todd, A. D. MacDonald, and G. Woolfe, J. Chem. Soc., 1943, 286.
234. Beriger, E. (to CIBA), U. S. Patent 3,014,955 (1961) [C.A. 57, 11028d (1962)].
234a. Beriger, E., U. S. Patent 3,016,326 (1962) [C.A. 57, 12328h (1962)].
235. Beriger, E., and L. Pinter (CIBA Ltd.), S. African Patent 68 00,124 (1968) [C.A. 70, 67884 (1969)].
236. Bernton, A., Ber., 55, 3365 (1922).

236a. Berry, J. P., J. R. Arnold, and A. F. Isbell, J. Org. Chem., 33, 1644 (1968).
237. Bersin, T., Moldtmann, et al., Z. Physiol. Chem., 269, 241 (1941).
238. Bevan, T. H., and T. Malkin, Biochem. Problems Lipids, Proc. 2nd. Int. Conf. Ghent, 1955 [C.A. 52, 12759i (1958)].
239. Biehringer, J., Ber., 38, 3974 (1905).
239a. Biermann, U., and O. Glemser, Chem. Ber., 102, 3342 (1969).
240. Billman, J. H., A. Radike, and B. W. Mundy, J. Am. Chem. Soc., 64, 2977 (1942).
241. B.I.O.S. Final Report 714.
242. B.I.O.S. Final Report 1095.
243. B.I.O.S. Final Report 1480.
244. Bischoff, C. A., and P. Walden, Ann., 279, 45 (1894)
245. Blackburn, G. M., J. S. Cohen, and A. R. Todd, Tetrahedron Lett., 1964, 2873.
246. Blackburn, G. M., J. S. Cohen, and Lord Todd, J. Chem. Soc., C, 1966, 239.
247. Blackburn, G. M., et al., J. Chem. Soc., C, 1969, 676.
248. Bliznyuk, N. K., P. S. Khokhlov, R. V. Strel'tsov, Z. N. Kvasha, and A. F. Kolomiets, Zh. Obshch. Khim., 37, 1119 (1967) [C.A. 68, 105296 (1968)].
249. Bliznyuk, N. K., and A. F. Kolomiets, Zh. Obshch. Khim., 34, 1169 (1964) [C.A. 61, 1747g (1964)].
250. Bliznyuk, N. K., A. F. Kolomiets, L. P. Ivershina, and S. G. Zhemchuzhin, USSR Patent 178,819 (1966) [C.A. 65, 2126c (1966)].
251. Bliznyuk, N. K., and 5 others, USSR Patent 184,863 (1966) [C.A. 66, 65083 (1967)].
252. Bliznyuk, N. K., A. F. Kolomiets, and Z. N. Kvasha, USSR Patent 196,817 (May 31, 1967) [C.A. 68, 104717 (1968)].
253. Blumenthal, E., and J. B. M. Herbert, Trans. Faraday Soc., 41, 611 (1945).
253a. Bocquel, J. R., Ann. Soc. Roy. Sc. Med. et Nat. Bruxelles, 9, 161 (1956) [C.A. 51, 7296 (1957)].
254. Boehringer Ingelheim GmbH., Br. Patent 1,020,809 (1966) [C.A. 64, 17425b (1966)].
255. C. H. Boehringer Sohn, Belg. Patent 660,365 (1965) [C.A. 64, 1960d (1966)].
256. Boehringer & Söhne, German Patent 98,522 (1897).
257. C. H. Boehringer Sohn, Ger. Patent 1,232,961 (1967) [C.A. 66, 94698 (1967)].
258. C. H. Boehringer, Sohn, Fr. Patent 1,518,255 (1963) [C.A. 71, 2991 (1969)].
259. Boissonnas, R., and E. Sandrin (to SANDOZ), Swiss Patent 447,185 (1968) [C.A. 70, 4548 (1969)].
259a. Bojars, N., Tech. Apskats. nr. 52, 7 (1965) [C.A.

68, 50951 (1968)].
260. Bond, A., and H. S. Mason, Biochem. Biophys. Res. Commun., 9, 574 (1962).
261. Bonsen, P. P. M., G. H. de Haas, L. L. M. van Deenen, Biochemistry, 6, 1114 (1961).
262. Bonsen, P. P. M., G. H. de Haas, L. L. M. von Deener, Chem. Phys. Lipids, 1, 33 (1966) [C.A. 66, 65027 (1967)].
263. Booth, H. S., D. R. Martin, and F. E. Kendall, J. Am. Chem. Soc., 70, 2523 (1948).
264. Boozalez, T. S., and R. B. McKecver, U. S. Patent 3,146,254 (1964) [C.A. 61, 13241g (1964)].
264a. Börner, K. B., and C. Stölzer, Chem. Ber., 96, 1328 (1963).
265. Borowitz, I. J., M. Anschel, and J. Firstenberg, J. Org. Chem., 32, 1723 (1965).
265a. Boschan, R. H., and J. P. Holder, U. S. Patent 3,341,630 (1967) [C.A. 68, 77961 (1968)].
265b. Boter, H. L., NASA Accession no N66-31913, Rept. No TDCK-45558 [C.A. 67, 21404 (1967)].
265c. Boter, H. L., and G. R. vanden Berg, Rec. Trav. Chim., 85, 1099 (1966).
266. Bott, R. W., B. F. Dowden, and C. Eaborn, Intern. Symp. Organosilicon Chem., Sci. Commun., Prague, 1965, 290 [C.A. 65, 10606b (1966)].
267. Boyd, D. B., J. Am. Chem. Soc., 91, 1200 (1969).
268. Boyd, D. R., and D. E. Ladhams, J. Chem. Soc., 1928, 215.
269. Brannock, K. C., J. Am. Chem. Soc., 73, 4953 (1951).
270. Brannock, K. C. (to Eastman Kodak Co.), U. S. Patent 2,639,292 (1953) [C.A. 48, 3393a (1954)].
271. Brannock, K. C., and R. L. McConnell (to Eastman Kodak Co.), U. S. Patent 2,865,947 (1958) [C.A. 53, 9056i (1959)].
272. Braun, R. A. (to du Pont de Nemours, E. I., & Co.), U. S. Patent 3,412,181 (1968) [C.A. 70, 57131 (1969)].
273. Bredereck, H., and E. Berger, Ber., 73, 1124 (1940).
274. Bredereck, H., E. Berger, and J. Ehrenberg, Ber., 73, 269 (1940).
275. Breivogel, P. J. (to White Laboratories, Inc.), U. S. Patent 2,535,385 (1950) [C.A. 45, 3428d (1950)].
276. Breusch, F. L., and H. Keskin, Rev. Facult. Sci. Univ. Istanbul, 7A, 182 (1942) [C.A. 38, 1483_2 (1944)].
277. Brigl, P., and H. Müller, Ber., 72, 2121 (1939).
278. Britton, E. C., and S. L. Bass, U. S. Patent 2,117,290 (1938; to Dow Chemical Co.).
279. Britton, E. C., and S. L. Bass, U. S. Patent 2,117,291 (1938; to Dow Chemical Co.).
280. Britton, E. C., and C. L. Moyle, U. S. Patent

2,151,680 (1939; to Dow Chemical Co.).
281. Britton, E. C., and C. L. Moyle, U. S. Patent 2,182,309 (1939; to Dow Chemical Co.).
282. Britton, E. C., and C. L. Moyle, U. S. Patent 2,275,041 (1942; to Dow Chemical Co.).
282a. Britton, E. C., and H. R. Slagh (to Dow Chemical Co.), U. S. Patent 2,829,131 (1958) [C.A. 52, 13776c (1958)].
283. Brock, F. H., J. Org. Chem., 22, 1114 (1957).
284. Broeker, W., J. Prakt. Chem. (2), 118, 287 (1928).
285. Brooks, B. T., J. Am. Chem. Soc., 34, 492 (1912).
285a. Brown, D., K. D. Crosbie, G. W. Fraser, and D. W. Sharp, J. Chem. Soc., 1969, 872.
286. Brown, D. M., and B. F. C. Clark, J. Chem. Soc., 1963, 1475.
287. Brown, D. M., J. A. Flint, and N. K. Hamer, J. Chem. Soc., 1964, 326.
288. Brown, D. M., and M. J. Frearson, Chem. Commun., 1968, 1342.
289. Brown, D. M., G. E. Hall, and H. M. Higson, J. Chem. Soc., 1958, 1360.
290. Brown, D. M., F. Hayes, and A. R. Todd, Chem. Ber., 90, 936 (1957).
291. Brown, D. M., and H. M. Higson, J. Chem. Soc., 1957, 2034.
292. Brown, D. M., and G. O. Osborne, J. Chem. Soc., 1957, 2590.
293. Brown, D. M., A. R. Todd, and S. Varadarajan, J. Chem. Soc., 1956, 2388.
294. Brown, D. M., and D. A. Usher, Proc. Chem. Soc., 1963, 309.
295. Brown, D. M., L. J. Haynn, and A. R. Todd, J. Chem. Soc., 1950, 3299.
296. Brown, D. M., D. I. Magrath, and A. R. Todd, J. Chem. Soc., 1952, 2708.
297. Buch, K. W., Carbohydr. Res., 6, 247 (1968).
298. Buchheim, K., U. S. Patent 2,059,084 (1936; to Chemische Fabrik von Heyden); British Patent 398,659 (1933).
299. Budoviskii, E. I., et al., Izv. Akad. Nauk SSSR, Ser. Khim., 1964, 1236 [C.A. 61, 13398d (1964)].
300. Budovskii, E. I., V. N. Shibaev, G. I. Eliseeva, and N. K. Kochetkov, Izv. Akad. Nauk SSSR, Otd. Khim. Nauk, 1962, 1491 [C.A. 58, 3500a (1963)].
301. Budovskii, E. I., and V. N. Shibaev, Khim. Prir. Soedin., 4, 293 (1968) [C.A. 70, 78303 (1969)].
302. Bunter, C. A., D. R. Llewellyn, K. G. Oldham, and C. A. Vernon, J. Chem. Soc., 1958, 3574.
303. Bunton, C. A., and S. J. Farber, J. Org. Chem., 34, 767 (1969).
304. Bunton, C. A., S. J. Farber, and E. J. Peneller, J.

Org. Chem., 33, 29 (1968).
305. Bunton, C. A., E. J. Fendler, E. Humeres, and K.-U. Yang, J. Org. Chem., 32, 2806 (1967).
306. Bunton, C. F., E. J. Fendler, G. L. Sepulveda, and K.-U. Yang, J. Am. Chem. Soc., 90, 5512 (1968).
307. Bunton, C. A., D. Kellerman, K. G. Oldham, and C. A. Vernon, J. Chem. Soc., B, 1966, 292.
307a. Bunyan, P. J., and J. I. Cadogan, J. Chem. Soc., 1962, 1304.
308. Bures, E., Chem. Listy, 21, 148 (1927).
309. Bures, E., Chem. Listy, 21, 221 (1927).
310. Burkhardt, G., M. P. Klein, and M. Calvin, J. Am. Chem. Soc., 87, 591 (1965).
311. Burn, A. J., and I. G. Cadogan, Chem. and Ind., 1963, 736.
312. Burn, A. J., J. I. G. Cadogan, and P. J. Bunyan, J. Chem. Soc., 1963, 1527.
313. Butcher, W. W., and F. H. Westheimer, J. Am. Chem. Soc., 77, 2420 (1955).
314. Butz, K., U. S. Patent 2,053,653 (1936; to American Hyalsol Corp.).
315. Butz, K., U. S. Patent 2,119,523 (1938; to H. Th. Böhme A.G.).
316. Buu-Hoi, and P. Cagniant, Bull. Soc. Chim., (5) 10, 135 (1943).
317. Bykova, L. V., Z. V. Mikkailova, P. Z. Li, L. K. Rubtsova, and O. S. Osadchuk, Plast. Massy, 1967, 31 [C.A. 68, 69371 (1968)].
317a. Bystruv, V. F., A. A. Neimyskeva, A. U. Stepanyants, and I. L. Knunyants, Dokl. Akad, Nauk SSSR, 156, 637 (1964) [C.A. 61, 6548c (1964)].
317b. Cade, J. A., and W. Gerrard, Chem. and Ind., 1954, 402; J. Chem. Soc., 1960, 1249.
318. Cadogan, J. I., M. Cameron-Wood, and W. R. Foster, J. Chem. Soc., 1963, 2549.
319. Calderón, J. Anales Real. Soc. Españ. Fis. y Quim., 53B, 69 (1957) [C.A. 51, 11273f (1957)].
320. Cambi, L., Chimica e Industr., 26, 97 (1944).
321. Canali, V., and G. Casciotti, Boll. Soc. Ital. Biol. Sper., 27, 1478 (1951) [C.A. 47, 11208d (1953)].
322. Canali, V., G. Casciotti, R. De Luca, and A. Baldantoni, Boll. Soc. Ital. Biol. Sper., 27, 1479 (1951) [C.A. 47, 11130i (1953)].
322a. Canonica, L., G. Jommi, P. Manitto, U. M. Pagnoni, and F. Pelizzoni, Gazz. Chim. Ital., 96, 662 (1966).
323. Carayón-Gentil, A., Compt. Rend., 241, 1785 (1955).
324. Carayon-Gentil, A., N. T. Thuong, G. Gonzy, and P. Chabrier, Bull. Soc. Chim. France, 1967, 1616.
325. Carayon-Gentil, A., and Van-Thoai, Compt. Rend., 239, 1051 (1954).
326. Carius, L., Ann., 112, 190 (1851).
327. Carius, L., Ann., 119, 289 (1861); Jahresber., 1861,

583.
328. Carius, L., Ann., 119, 298 (1861).
329. Carius, L., Ann., 124, 57 (1862).
330. Carius, L., Ann., 137, 121 (1866).
331. Carius, L., Jahresber., 1861, 585
332. Carius, L., and H. Kämmerer, Ann., 131, 170 (1864).
333. Carré, P., Compt. Rend., 136, 456 (1903); Ann. Chim. Phys., (8) 5, 345 (1905).
333a. Carré, P., Compt. Rend., 136, 756 (1903).
334. Carré, P., Compt. Rend., 137, 1070 (1903).
335. Carré, P., Compt. Rend., 138, 47 (1904).
336. Carré, P., Compt. Rend., 138, 374 (1904).
337. Carré, P., Compt. Rend., 139, 638 (1904).
338. Carré, P., and D. Liberman, Compt. Rend., 196, 864 (1933).
339. Carrington, R., G. Shaw, and D. V. Wilson, Tetrahedron Lett., 1964, 2861.
340. Carruthers, T. F., U. S. Patent 2,406,802 (1946; to Carbide & Carbon Chemicals Corp.).
341. Causse, H., Bull. Soc. Chim. (3), 7, 565 (1892).
342. Cavalier, J., Compt. Rend., 118, 1275 (1894).
343. Cavalier, J., Compt. Rend., 122, 69 (1895).
344. Cavalier, J., Bull. Soc. Chim., 13, 885 (1895); 15, 932 (1896).
345. Cavalier, J., Compt. Rend., 124, 91 (1897).
346. Cavalier, J., Compt. Rend., 126, 1214 (1898).
347. Cavalier, J., Compt. Rend., 127, 114 (1898).
348. Cavalier, J., Bull. Soc. Chim. (3), 19, 883 (1898); Ann. Chim. Phys. (7), 18, 449 (1899).
349. Cavalier, J., and E. Proust, Bull. Soc. Chim. (3), 23, 678 (1900).
350. Cavalier, J., Compt. Rend., 142, 885 (1906).
351. Caven, R. M., J. Chem. Soc., 81, 1368 (1902).
352. Cebrián, G. R., Anales Real Soc. Españ. Fis. y Quim., 47B, 841 (1951) [C.A. 46, 11140c (1952)].
353. Celanese Corp. of America, Br. Patent 985,752 (1965) [C.A. 62, 14571d (1965)].
354. Celanese Corp. of America, Br. Patent 999,793 (1965) [C.A. 64, 1961c (1966)].
355. Chabrier, P., and A. Carayon-Gentil, Bull. Soc. Chim. France, 1957, 639.
356. Chabrier, P., and A. Desjobert, Br. Patent 684,977 (1952) [C.A. 49, 1773f (1955)].
357. Chabrier, P., and M. Selim, Compt. Rend., 244, 2730 (1957).
358. Chabrier, P., N. T. Thuong, and J. Favre-Bonvin, Compt. Rend., 258, 5004 (1964).
359. Chacko, G. K., and D. J. Hanakan, Biochim. Biophys. Acta, 164, 252 (1968).
360. Chambers, R. W., and H. G. Khorana, Chem. and Ind., 1956, 1022.

361. Chambers, R. W., and H. G. Khorana, J. Am. Chem. Soc., 79, 3752 (1957).
362. Chambers, R. W., and H. G. Khorana, J. Am. Chem. Soc., 80, 3749 (1958).
363. Chambers, R. W., J. G. Moffatt, and H. G. Khorana, J. Am. Chem. Soc., 79, 3747 (1957).
364. Chambon, E., Jahresber., 1876, 205.
365. Chanley, J. D., and E. Feageson, J. Am. Chem. Soc., 85, 1181 (1963).
366. Chanley, J. D., and E. Feageson, J. Am. Chem. Soc., 77, 4002 (1955).
367. Chanley, J. D., E. M. Gindler, and H. Sobotka, J. Am. Chem. Soc., 74, 4347 (1952).
368. Chantrenne, H., Nature, 158, 603 (1947).
368a. Chantrenne, H., Compt. Rend. Trav. Lab. Carlsberg, Ser. Chim., 26, 297 (1948).
368b. Chantrenne, H., Biochim. Biophys. Acta, 2, 286 (1948).
369. Chapman, N. B., and B. C. Saunders, J. Chem. Soc., 1948, 1010.
369a. Chargaff, E., J. Biol. Chem., 144, 455 (1942).
370. Chargaff, E., J. Am. Chem. Soc., 60, 1700 (1938).
370a. Chargaff, E., J. Biol. Chem., 144, 455 (1942).
371. Chasanowitsch, J., Ber., 20, 1165 (1887).
372. Chase, B. H., G. W. Kenner, A. R. Todd, and R. E. Webb, J. Chem. Soc., 1956, 1371.
373. Chatterjee, A. K., and D. L. MacDonald, Carbohydr. Res., 6, 253 (1968).
374. Chatterjee, A. K., and D. L. MacDonald, J. Org. Chem., 33, 1584 (1968).
375. Chavanne, G., Compt. Rend., 134, 1439 (1902); Bull. Soc. Chim. (3), 29, 398 (1903); Ann. Chim. Phys. (8), 3, 529 (1904).
376. Chemicals and Phosphates Ltd., Israeli Patent 15,861 (1963) [C.A. 59, 8595e (1963)].
377. Chemische Fabrik Griesheim, German Patent 367,954 (1923).
377a. Chen, W.-C., Hua Hsueh Pao, 31, 29 (1965) [C.A. 63, 4128e (1965)].
378. Cherbuliez, E., Br. Baehler, J. Jindra, and J. Rabinowitz, Helv. Chim. Acta, 48, 1069 (1965).
378a. Cherbuliez, E., Br. Baehler, J. Marszalek, G. Weber, and J. Rabinowitz, Helv. Chim. Acta, 52, 2676 (1969).
379. Cherbuliez, E., and M. Bouvier, Helv. Chim. Acta, 36, 1200 (1953).
380. Cherbuliez, E., J. P. Leber, and M. Bouvier, Helv. Chim. Acta, 36, 1200 (1953).
380a. Cherbuliez, E., S. Čolak-Antić, R. Prince, and J. Rabinowitz, Helv. Chim. Acta, 47, 1659 (1964).
381. Cherbuliez, E., C. Gandillon, Arch. Sci., 9, 103 (1956) [C.A. 51, 2366i (1957)].

382. Cherbuliez, E., and J. P. Leber, Helv. Chim. Acta, 35, 644 (1952).
383. Cherbuliez, E., and J. P. Leber, Helv. Chim. Acta, 35, 2589 (1952).
383a. Cherbuliez, E., H. Probst, and J. Rabinowitz, Helv. Chim. Acta, 42, 1377 (1959); 43, 464 (1960).
383x. Iidem, Helv. Chim. Acta 41, 1693 (1958).
384. Cherbuliez, E., and J. Rabinowitz, Helv. Chim. Acta, 39, 1455 (1956).
384a. Cherbuliez, E., and J. Rabinowitz, Helv. Chim. Acta, 39, 1844 (1956); 40, 526 (1957).
385. Cherbuliez, E., M. Schwarz, and J. P. Leber, Helv. Chim. Acta, 34, 841 (1951).
386. Cherbuliez, E., and M. Schwarz, Helv. Chim. Acta, 36, 1189 (1953).
386a. Cherbuliez, E., G. Weber, and J. Rabinowitz, Helv. Chim. Acta, 46, 2464 (1963).
387. Cherbuliez, E., and H. Weniger, Helv. Chim. Acta, 28, 1584 (1945).
388. Cherbuliez, E., and H. Weniger, Helv. Chim. Acta, 29, 2006 (1946); Bull. Soc. Chim. Biol., 29, 256 (1947).
389. Chevrier, Jahresber., 1869, 344
390. Chevrier, Z. Chem., 1869, 413; Compt. Rend., 68, 924 (1869).
391. Cheymol, J., P. Chabries, M. Selim, and P. Leduc, Fr. Patent 1,356,967 (1964) [C.A. 61 9402h (1964)].
392. Chigirev, V. S., V. I. Shvets, and E. N. Bezinger, Dokl. Akad. Nauk SSSR, 181, 747 (1968) [C.A. 69, 107046 (1968)].
393. Chin, T. H., and D. S. Feingold, Methods Enzymol., 9, 464 (1966).
394. Chladek, S., and J. Smrt, Collection Czech. Chem. Commun., 29, 214 (1964).
395. Christensen, H. N., J. Biol. Chem., 135, 399 (1940).
395a. Christie, S. M. H., G. W. Kenner, and A. R. Todd, J. Chem. Soc., 1954, 46.
395b. Christol, C., and W. Christol, J. Chim. Phys., 62, 246 (1965).
396. Chrzaszczewska, I. A., and J. Kaczan, Lodz. Towarz. Nauk., Wydzial III, Acta Chim., 9, 213 (1964) [C.A. 63, 1862e (1965)].
397. Chrzaszczewsky, A., and J. Kaczan, Lodz. Towarz. Nauk., Wydzial III, Acta Chim., 9, 227 (1964) [C.A. 62, 7625f (1965)].
398. Chugunov, V. S., and E. V. Sapgir, Zh. Obshch. Khim. 38, 416 (1968) [C.A. 69, 86258 (1968)].
399. Church, J. M. (to Metal & Thermit Corp.), U. S. Patent 2,630,436 (1953) [C.A. 48, 1420e (1954)].
400. Church, J. M., Proc. Roy. Soc., 13, 520 (1864).
401. Church, J. M., H. E. Ramsden, H. Hirschland, and

H. W. Buchanan (to Metal & Thermit Corp.), U. S. Patent 2,630,442 (1953) [C.A. 48, 1420c (1954)].
402. Chwalinski, S., and W. Rypinska, Roczniki Chem., 31, 539 (1957) [C.A. 52, 5284a (1958)].
403. CIBA Ltd., Ger. Patent 1,140,776 (1962) [C.A. 58, 11225e (1963)].
404. CIBA Ltd., Br. Patent 940,033 (1962) [C.A. 60, 5340d (1964)].
405. CIBA Ltd., Br. Patent 1,027,059 (1966) [C.A. 65, 2172e (1966)].
406. CIBA, Swiss Patent 236,233 (1948) [C.A. 43, 7056f (1949)].
406a. Clark, V. M., D. W. Hutchinson, G. W. Kirby, and A. Todd, J. Chem. Soc., 1961, 715.
406b. Clark, V. M., D. W. Hutchinson, and A. Todd, J. Chem. Soc., 1961, 722.
407. Clark, V. M., D. W. Hutchinson, and P. E. Varey, J. Chem. Soc., C, 1969, 74.
408. Clark, V. M., D. W. Hutchinson, and D. E. Wilson, Angew. Chem., 77, 259 (1965).
409. Clark, V. M., G. N. Kirby, and A. Todd, J. Chem. Soc., 1957, 1497.
410. Clark, V. M., G. W. Kirby, and A. Todd, J. Chem. Soc., 1958, 3039.
410a. Clark, V. W., D. W. Hutchinson, A. R. Lyons, and R. K. Roschnik, J. Chem. Soc., C, 1969, 233.
411. Clark, V. M., and A. R. Todd, J. Chem. Soc., 1950, 2023.
411a. Clark, V. M., and A. R. Todd, J. Chem. Soc., 1950, 2030.
412. Clark, V. M., Lord Todd, and S. G. Warren, Biochem. Z., 338, 591 (1963).
413. Claus, A., and H. Oehler, Ber., 15, 312 (1882).
414. Claus, A., and O. Zimmermann, Ber., 14, 1482 (1881).
415. Clemmensen, E., U. S. Patent 1,931,056-9 (1933; to Monsanto Chemical Co.).
416. Clemmensen, E., U. S. Patent 1,945,183 (1934; to Monsanto Chemical Co.).
417. Clermont, P. de, Ann., 91, 375 (1854); Compt. Rend., 39, 338 (1854); Ann. Chim. Phys., 44, 330 (1855).
418. Cleve, P. T., Ber., 21, 896 (1888).
419. Cloez, Jahresber., 1847-8, 695.
420. Closse, A., Chemiker-Ztg., 81, 103 (1957).
421. Coates, H. (to Albright & Wilson Ltd.), Br. Patent 687,596 (1953) [C.A. 48, 3392g (1954)].
422. Colbran, R. L., J. K. N. Jones, N. K. Matheson, and I. Rozema, Carbohydr. Res., 4, 355 (1967).
423. Collin, R. L., J. Am. Chem. Soc., 88, 3281 (1966).
424. Combustion Utilities Corp., U. S. Patent 1,869,312 (1932).
425. Commercial Solvents Corp., U. S. Patent 1,589,608

(1926).
426. Commercial Solvents Corp., U. S. Patent 1,799,349 (1931).
427. Conly, J. C. (to Douglas Aircraft Co.), U. S. Patent 2,754,316 (1956) [C.A. 51, 1244a (1957)].
428. Contardi, A., Gazz. Chim. Ital., 51, I, 109 (1921).
429. Convevy, R. J., U. S. Patent 3,193,660 (1964) [C.A. 62, 1688h (1965)].
429a. Cook, H. G., H. McCombie, and B. C. Saunders, J. Chem. Soc., 1949, 2921.
430. Cook, H. G., H. McCombie, and B. C. Saunders, J. Chem. Soc., 1945, 873.
430a. Cook, H. G., B. C. Saunders, and F. E. Smith, J. Chem. Soc., 1949, 635.
431. Cooper, McDougall & Robinston, Ltd., Br. Patent 1,007,332 (1965) [C.A. 64, 2061d (1966)].
432. Cope, A. C., and H. L. Jackson (E. I. du Pont de Nemours & Co.), U. S. Patent 2,748,153 (1956) [C.A. 51, 498f (1957)].
432a. Corbridge, D. E. C., in "Topics in Phosphorus Chemistry," Vol. 6, p. 235 (Interscience Publishers, 1969).
433. Corby, N. S., G. W. Kenner, and A. Todd, J. Chem. Soc., 1952, 1234.
433a. Corbridge, D. E. C., in "Topics in Phosphorus Chemistry," Vol. 3, p. 57 (Interscience Publishers, 1966).
434. Cori, C. F., S. P. Colowick, and G. T. Cori, J. Biol. Chem., 121, 470 (1937).
435. Coulon-Morelec, M. J., and D. Giraud, Bull. Soc. Chim. Biol., 41, 47 (1965).
436. Council of Scient. & Industr. Research, Indian Patent 46,405 (1953) [C.A. 48, 10050b (1954)].
437. Couper, A. S., Compt. Rend., 46, 1108 (1858).
438. Couper, A. S., Compt. Rend., 46, 1107 (1859); Ann., 109, 369 (1859).
439. Courtemanche, P., and J. C. Merlin, Bull. Soc. Chim. France, 1968, 836.
440. Courtois, J., and R. Barré, Bull. Soc. Chim. Biol., 28, 854 (1946).
441. Courtois, J., and P. Biget, Compt. Rend., 213, 192 (1941).
442. Coutsogeorgopoulos, C., and H. G. Khorana, J. Am. Chem. Soc., 86, 2926 (1964).
442a. Cox, J. R., Jr., R. E. Wall, and F. H. Westheimer, Chem. and Ind., 1959, 929.
442b. Cox, J. R., Jr., and F. H. Westheimer, J. Am. Chem. Soc., 80, 5441 (1958).
442c. Cox, J. R., Jr., and F. H. Westheimer, J. Am. Chem. Soc., 80, 5441 (1958).
442d. Cox, J. R., Jr., and O. B. Ramsay, Chem. Rev., 64, 324 (1964).

442e. Cramer, F., Ann. Chem., 683, 199 (1965).
442f. Cramer, F., H. J. Baldauf, and H. Kuentzel, Angew. Chem., 74, 77 (1962).
442g. Cramer, F., and W. Böhm, Angew. Chem., 71, 775 (1959).
443. Cramer, F., and K. G. Gärtner, Angew. Chem., 68, 649 (1956).
444. Cramer, F., and K.-G. Gärtner, Chem. Ber., 91, 704 (1958).
445. Cramer, F., and K.-G. Gärtner, Chem. Ber., 91, 1562 (1958).
446. Cramer, F., Ger. Patent 1,179,551 (1964) [C.A. 62, 626c (1965)].
447. Cramer, F., Ger. Patent 1,251,322 (1965) [C.A. 68, 104729 (1968)].
448. Cramer, F., and T. Hata, Ann. Chem., 692, 22 (1966).
448a. Cramer, F., and H. Hettler, Chem. Ber., 91, 1181 (1958).
448b. Cramer, F., and H. Neunhoeffer, Chem. Ber., 95, 1664 (1962).
449. Cramer, F., and K. Randerath, Angew. Chem., 70, 571 (1958).
449a. Cramer, F., W. Rittersdorf, and W. Böhm, Ann. Chem., 654, 180 (1962).
450. Cramer, F., and S. Rittner, Tetrahedron Lett., 1964, 107.
450a. Cramer, F., and H. Schaller, Chem. Ber., 94, 1634 (1961).
451. Cramer, F., and K. H. Scheit, Angew. Chem., 74, 717 (1962).
452. Cramer, F., and G. Schneider, Ann. Chem., 717, 193 (1968).
453. Cramer, F., and D. Voges, Chem. Ber., 92, 952 (1959).
454. Cremlyn, R. J. W., G. W. Kenner, J. Mather, and A. Todd, J. Chem. Soc., 1958, 258.
454a. Cremlyn, R. J. W., G. W. Kenner, and A. Todd, J. Chem. Soc., 1960, 4511.
455. Crofts, P. C., and I. M. Downie, J. Chem. Soc., 1963, 2559.
456. Cutler, F. A., Jr., J. P. Conbere, et al., J. Am. Chem. Soc., 80, 6300 (1958).
457. Daemen, F. J., G. H. de Haas, and L. L. M. van Doenen, Rec. Trav. Chim., 82, 487 (1963).
458. Daiichi Seiyoku Co., Ltd., Japan. Patent 9036 (1965) (May 11, 1962) [C.A. 63, 7095a (1965)].
459. Daly, A. J., and W. G. Lowe, U. S. Patent 2,157,164 (1939; to Celanese Corp.).
460. Damico, J. N., J. Assoc. Offic. Anal. Chemists, 49, 1027 (1966) [C.A. 66, 1834 (1967)].
461. Darlix, J. L., H. P. M. Fromageot, and P. Fromageot,

Biochem. Biophys. Acta, 145, 517 (1967).
462. Darmon, M., Bull. Soc. Chim. France, 1947, 262.
463. Danison, R. M. C., M. Ford, and J. Eichberg, Biochem. J., 95, 104.
463a. Datskevich, L. A., V. D. Maiboroda, and I. P. Losev, Vysokmolekul. Soedin. Geterotsepnye Vysokomolekul. Soedin., 1964, 243 [C.A. 61, 5779f (1964)].
464. Daul, G. C., and J. C. Reid (to the United States of America, Secy. of Agr.), U. S. Patent 2,583,549 (1952) [C.A. 46, 7584i (1952)].
464a. David, W. A. L., and B. A. Kilby, Nature, 164, 522 (1949).
465. Dekker, C. A., A. M. Michelson, and A. R. Todd, J. Chem. Soc., 1953, 947.
466. Delacre, M. Bull. Soc. Chim. (2), 48, 787 (1887).
467. Delepine, M., Bull. Soc. Chim. (4), 11, 577 (1912).
468. Denney, D. B., N. Gershman, and J. Giacin, J. Org. Chem., 31, 2833 (1966).
469. Denney, D. B., and H. M. Relles, J. Am. Chem. Soc., 86, 3897 (1964).
470. Denney, D. B., and J. W. Hanifin Jr., Tetrahedron Lett., 1963, 2177.
471. Dennis, E. A., and F. H. Westheimer, J. Am. Chem. Soc., 88, 3432 (1966).
471a. deRoos, A. M., and H. J. Toet, Rec. Trav. Chim., 77, 946 (1958).
471b. De Selms, R. C., and T.-W. Lin, J. Org. Chem., 32, 2023 (1961).
472. Desjobert, A., Bull. Soc. Chim. France, 1963, 683.
473. Desjobert, A., Colloqu. Nat. CNRS, 1965, 311 (publ. 1966).
474. Desjobert, A., Compt. Rend., 224, 575 (1947); Bull. Soc. Chim. 1947, 809.
475. Deutsch. Akad. Wiss. zu Berlin, Ger. Patent 1,119,278 (1958) [C.A. 57, 2316f (1962)].
476. Deutsch, A., and O. Fernö, Nature, 156, 604 (1945).
477. Diamond, M. J., T. H. Applewhite, R. E. Knowles, and C. A. Goldblatt, J. Am. Oil Chemists Soc., 41, 9 (1964) [C.A. 60, 7908e (1964)].
478. DiBoscogrande, S., Atti Real Accad. Lincei (5), 6, II, 35 (1897).
479. Dickey, J. B., and H. W. Coover, Jr. (to Eastman Kodak Co.), U. S. Patent 2,600,378 (1952) [C.A. 46, 8322c (1952)].
480. Dilaris, I., and G. Eliopoulos, J. Org. Chem., 30, 686 (1969).
480a. Dimroth, K., and B. Lerch, Angew. Chem., 72, 751 (1960).
480b. DiSabato, G., and W. P. Jenks, J. Am. Chem. Soc., 83, 4400 (1961).
481. Discalzo, G., Gazz. Chim. Ital., 15, 278 (1885).

482. Divinskii, A. F., M. I. Kabachnik, and V. V. Sidorenko, Dokl. Akad. Nauk SSSR, 60, 999 (1948).
483. Dobry, A., and R. Keller, J. Phys. Chem., 61, 1448 (1958).
484. Dodge, F. D., Am. Chem. J., 12, 555 (1890).
485. Dorofeeva, L. T., O. N. Tolkachev, N. A. Preobrazhenskii, Zh. Obshch. Khim., 33, 2880 (1963) [C.A. 60, 1579e (1964)].
485a. Dostrovsky, I., and M. Halmann, Bull. Research Council Israel, 1, 142 (1951) [C.A. 46, 3376a (1952)]; J. Chem. Soc., 1953, 502; 1956, 1004.
485b. Drago, R. S., V. A. Mode, J. G. Kay, and D. L. Lydy, J. Am. Chem. Soc., 87, 5210 (1965).
486. Drake, L. R., and A. J. Erbel, U. S. Patent 2,599,375 (1959) [C.A. 46, 8321h (1952)].
487. Dregval, G. F., and N. A. Rybak, Khim. Geterotsikl. Soedin., Sb. 1: Azotsoderzhashch. Geterotsikly, 1967, 227 [C.A. 70, 87501 (1969)].
487a. Drozdo, V. A., G. S. Karetnikov, and I. Y. Orlova, Zh. Fiz. Khim., 40, 695 (1966) [C.A. 65, 4851g (1966)].
488. Drushel, W. A., Am. J. Sci. (4), 40, 643 (1915).
489. Drushel, W. A., and A. R. Felty, Am. J. Sci. (4), 43, 57 (1918).
490. Drysdale, J. J., and H. S. Simmons, U. S. Patent 3,149,142 (1964) [C.A. 61, 14530f (1964)].
491. Dudek, G. O., and F. H. Westheimer, J. Am. Chem. Soc., 81, 2641 (1959).
492. Dulova, V. I., and N. R. Molchanova, Izv. Vysst. Ucheb. Zaved., Khim. Khim. Tekhnol., 11, 980 (1968) [C.A. 70, 32091 (1968)].
493. Dünhaupt, F., J. Prakt. Chem., 61, 399 (1854).
494. Dupuis, P., Compt. Rend., 150, 622 (1910).
495. Dupuis, P., Bull. Soc. Chim. (4), 7, 846 (1910).
496. Durrans, G. J., J. Chem. Soc., 121, 48 (1922).
497. Dursch, H. R., and F. J. Reithel, J. Am. Chem. Soc., 74, 850 (1952).
498. Dutton, G. R., and C. R. Noller, J. Am. Chem. Soc., 55, 424 (1933).
498a. Dyatkin, B. L., E. P. Mochalina, Y. S. Konstantinov, S. R. Sterlin, and I. L. Knunyants, Izv. Akad. Nauk SSSR, Ser. Khim., 1967, 2297 [C.A. 68, 77632 (1968)].
499. Eckstein, F., and K. H. Scheit, Angew. Chem. Internat. Ed., 6, 362 (1967).
500. Edmundson, R. S., and A. J. Lambic, J. Chem. Soc., B, 1967, 577.
500a. Edmundson, R. S., Chem. and Ind., 1962, 1770.
501. Edmundson, R. S., Tetrahedron, 20, 2781 (1964).
502. Edmundson, R. S., Tetrahedron, 21, 2379 (1965).
502a. Edmundson, R. S., and A. J. Lambie, J. Chem. Soc., B, 1967, 577.

503. Eidebenz, E., and M. Depner, Arch. Pharm., 280, 227 (1942).
504. Eideberg, J., and R. M. C. Dawson, Biochem. Biophys. Acta, 93, 425 (1964).
505. Einhorn, A., and C. Frey, Ber., 27, 2456 (1894); German Patent 74,748 (1895).
506. Elektrochemische Industrie A. G., U. S. Patent 2,046,031 (1936).
507. Elmore, D. T., and A. R. Todd, J. Chem. Soc., 1952, 3681.
508. Emmett, W. G., and H. O. Jones, J. Chem. Soc., 99, 715 (1911).
509. Engel'hardt, A., and P. Lachinov, Z. Chem., 1869, 44.
510. Engel'hardt, A., and P. Lachinov, J. Russ. Phys. Chem. Soc., 2, 116 (1870).
511. Engelke, E. F., U. S. Patent 2,260,303 (1941; to Cities Service Corp.).
512. Engelke, E. F., U. S. Patent 2,260,304-5 (1941; to Cities Service Corp.).
513. Engelke, E. F., U. S. Patent 2,373,670 (1945; to Cities Service Corp.).
514. Ephraim, F., Ber., 44, 633 (1911).
515. Ephraim, F., and M. Sackheim, Ber., 44, 3422 (1911).
516. Ephraim, F., and R. Stein, Ber., 44, 3405 (1911).
517. Etablissements Kuhlmann, Fr. Patent 1,413,874 (1965) [C.A. 64, 4941h (1966)].
517a. Etat Francais, Fr. Patent, 1,282,501 (1962) [C.A. 57, 16402h (1962)].
518. Ethyl Corp., Br. Patent 791,526 (1958) [C.A. 53, 1253e (1959)].
519. Eto, M., and Y. Oshima, Agr. Biol. Chem., 26, 452 (1962) [C.A. 59, 3269f (1963)].
520. Eto, M., Y. Oshima, and J. E. Casider, Biochem. Pharmacol., 16, 295 (1967) [C.A. 66, 75231 (1967)].
520a. Ettel, V., and M. Zbirovsky, Chem. Listy, 50, 1261 (1956) [C.A. 50, 16025f (1956)].
520b. Ettel, V., and M. Zbirovsky, Chem. Listy, 50, 1265 (1956) [C.A. 50, 16025h (1956)].
521. Etterweh, J., J. Org. Chem., 31, 2422 (1966).
522. Euler (von), H., and A. Bernton, Ber., 60, 1720 (1927).
523. Euler (von), H., A. Wolf, and H. Hellström, Ber., 62, 2451 (1929).
524. Evans, D. P., W. C. Davies, and W. J. Jones, J. Chem. Soc., 1930, 1310.
525. Evdakov, V. P., and E. I. Alipova, Zh. Obshch. Khim., 35, 1587 (1965) [C.A. 63, 18142e (1965)].
526. Farbenfabriken Bayer A. G., Br. Patent 1,044,000 (1966) [C.A. 66, 28522 (1967)].
527. Farbenfabriken Bayer, A. G., Ger. Patent 872,040 (1953) [C.A. 52, 16197g (1958).

528. Farbenfabriken Bayer, A. G., Ger. Patent 936,037 (1955) [C.A. 50, 4448d (1956)].
529. Farbenfabriken Bayer A. G., Ger. Patent 962,608 (1957) [C.A. 51, 15588a (1957)].
529a. Farbenfabriken Bayer A. G., Ger. Patent 1,178,420 (1964) [C.A. 62, 445e (1965)].
530. Farbenfabriken Bayer, A. G., Neth. Appl. 6,514,070 (1966) [C.A. 66, 75811 (1967)].
531. Farbwerke Hoechst A. G., Neth. Appl. 6,602,131 (1966) [C.A. 66, 37946 (1967)].
532. Faure, M., and J. LeCocq, Compt. Rend., 248, 2252 (1959).
533. Faure, M., J. Marechal, and J. Troester, Compt. Rend., 259, 941 (1969).
534. Fawaz, G., and U. Zeile, Z. Physiol. Chem., 263, 175 (1940).
535. Fernández, O., and E. Martínez, Farm. Nueva (Madrid), 10, 597 (1945) [C.A. 43, 1526 (1949)].
536. Ferrari, G., and C. Casagrande, Farmaco (Pavia), Ed. Sci., 20, 879 (1965) [C.A. 64, 19404c (1966)].
537. Ferris, J. P., Science, 161, 53 (1968).
538. Ferris, J. P., Science, 161, 3836 (1968).
539. Fertig, J., and W. Gerrard, Chem. and Ind., 1958, 1957.
539a. Fertig, J., W. Gerrard, and H. Herbst, J. Chem. Soc., 1957, 1488.
540. Feulgen, A., and T. Bersin, Z. Physiol. Chem., 260, 217 (1939).
541. F.I.A.T. Final Report 949.
542. Fiera, W. L., and W. J. Sandner, U. S. Patent 3,228,998 (1966) [C.A. 64, 12549h (1966)].
543. Fieser, L. F., U. S. Patent 2,407,823 (1946; to Research Corp.).
544. Fieser, L. F., and E. M. Fry, J. Am. Chem. Soc., 62, 228 (1940).
545. Filippenko, D. M., V. N. Gorbunov, and E. L. Gefter, USSR Patent 192,807 (1967) [C.A. 68, 114163 (1968)].
546. Fischer, E., Ber., 47, 3193 (1914).
547. Fischer, E., and E. Pfähler, Ber., 53, 1606 (1920).
548. Fischer, H. O. L., Ber., 65, 1040 (1932).
549. Fischer, H. O. L., and E. Baer, Ber., 65, 337 (1932).
550. Fischer, H. O. L., and E. Baer, Ber., 65, 337, 1040 (1932).
551. Fischer, H. O. L., and H. A. Lardy, J. Biol. Chem., 164, 513 (1946).
551a. Fiszer, B., and J. Michalski, Roczniki Chem., 26, 688 (1952) [C.A. 49, 2306c (1955)].
552. Fitch, H. M., U. S. Patent 2,759,961 (1956) [C.A. 51, 482b (1957)].
553. Flexser, L. A., and W. G. Farkas (to Hoffmann-La

Roche Inc.), U. S. Patent 2,604,470 (1953) [C.A. 47, 3354c (1953)].
554. Fletcher, J. H., J. C. Hamilton, I. Hechenbleikner, E. I. Hoegberg, B. J. Sertl, and J. T. Cassaday, J. Am. Chem. Soc., 70, 3943 (1948).
555. Fleury, P., and J. Courtois, Bull. Soc. Chim. France, 8, 69 (1941).
556. Fleury, P., and J. Courtois, Bull. Soc. Chim. France, 8, 397 (1941).
557. Fleury, P., and J. Courtois, Bull. Soc. Chim. France, 9, 570 (1942).
558. Fleury, P., J. Courtois, and A. Desjobert, Compt. Rend., 226, 801 (1948).
558a. Fleury, P., J. Courtois, and A. Desjobert, Bull. Soc. Chim. France, 1948, 694.
559. Fleury, P., J. Courtois, and A. Desjobert, Bull. Soc. Chim. France, 1952, 458.
560. Fleury, P., and J. Marque, Compt. Rend., 188, 1688 (1929).
561. Fleury, P., and R. Paris, J. Pharm. Chim. (8), 18, 470 (1933).
562. Flexser, L. A., and W. G. Farkas (to Hoffmann-La Roche Inc.), U. S. Patent 2,610,178 (1952) [C.A. 47, 8781d (1953)].
562a. Fölsch, G., Acta Chem. Scand., 10, 686 (1956).
563. Fölsch, G., and O. Mellander, Acta Chem. Scand., 11, 1232 (1957).
563a. Foelsch, G., and R. Oesterberg, Acta Chem. Scand., 15, 1963 (1961).
564. Food Machinery and Chemical Corp., Br. Patent 783,69 (1957) [C.A. 52, 8194h (1958)].
564a. Food Machinery and Chemical Corp., Br. Patent 784,98 (1957) [C.A. 52, 8195e (1958)].
564b. Ford-Moore, A. H., L. J. Lermit, and C. Stratford, J. Chem. Soc., 1953, 1776.
565. Forrest, H. S., H. S. Mason, and A. R. Todd, J. Chem. Soc., 1952, 2530.
566. Forrest, H. S., and A. R. Todd, J. Chem. Soc., 1950, 3295.
567. Forsman, J. P., and D. Lipkin, J. Am. Chem. Soc., 75, 3145 (1953).
568. Foss, O., Acta Chem. Scand., 1, 8 (1947); Kgl. Norske Videnskab Selskabs. Forh., 15, 119 (1942).
569. Foss, O., Acta Chem. Scand., 1, 307 (1947).
569a. Foss, M. E., and C. S. Gibson, J. Chem. Soc., 1949, 3075.
570. Fosse, R., Compt. Rend., 136, 1007 (1903).
571. Foster, R. H. K., J. Lee, and U. V. Solmssen, J. Am. Chem. Soc., 62, 453 (1940).
571a. Frank, A. W., and C. F. Baranauckas, J. Org. Chem., 30, 3970 (1965).

572. Frank, A. W., and C. F. Baranauckas, J. Org. Chem., 31, 872 (1966).
573. Frank, A. W., and C. F. Baranauckas, J. Org. Chem., 31, 1644 (1966).
574. Franssen, A., Bull. Soc. Chim. (4), 49, 550 (1931).
575. Freeman, H. F., J. Am. Chem. Soc., 60, 750 (1938).
576. Friedkin, M., and A. L. Lehninger, J. Biol. Chem., 169, 183 (1947).
577. Friedman, O. M., and A. M. Seligman, J. Am. Chem. Soc., 72, 624 (1950).
578. Friedman, O. M., and A. M. Seligman, J. Am. Chem. Soc., 73, 5292 (1951).
579. Fujimoto, S., Japan. Patent 68 25,495 (1968) [C.A. 70, 68706 (1969)].
580. Fujimoto, Y., and M. Naruse, Japan. Patent 68 16,988 (Jul. 17, 1968) [C.A. 70, 68705 (1969)].
581. Fujimoto, Y., and M. Naruse, Yakugaku Zasshi, 87, 270 (1967) [C.A. 67, 11691k (1967)].
582. Fujita, T., and Y. Mushika, Bitamin, 32, 51 (1965) [C.A. 63, 7221a (1965)].
583. Fukuto, T. R., and R. L. Metcalf, J. Med. Chem., 8, 759 (1965).
584. Gabov, N. I., and I. A. Korshunov, Radiokhimiya, 10, 64 (1968) [C.A. 68, 110566 (1968)].
584a. Gabov, N. I., and A. I. Shapirov, Metody Analiza Radioaktivn. Preparatov, 1965, 115 [C.A. 64, 6029g (1966)].
585. Galashina, M. L., I. L. Vladimirova, Ya. A. Mandel'-baum, and N. N. Mel'nikov, Zh. Obshch. Khim., 23, 433 (1953) [C.A. 48, 3887g (1954)].
586. Gamrath, H. R. (to Monsanto Chemical Co.), U. S. Patent 2,504,121 (1950) [C.A. 44, 6435h (1950)].
587. Gamrath, H. R., and J. K. Craver (to Monsanto Chemical Co.) Br. Patent 656,471 (1951) [C.A. 46, 7589h (1952)].
588. Gamrath, H. R., and J. K. Craver (to Monsanto Chemical Co.), U. S. Patent 2,596,140 (1952) [C.A. 47, 4367f (1953)].
589. Gamrath, H. R., and R. E. Hatton (to Monsanto Chemical Co.), U. S. Patent 2,661,366 (1953) [C.A. 49, 1098h (1955)].
590. Gamrath, H. R., R. E. Hatton, and W. E. Weesner, Ind. Eng. Chem., 46, 208 (1954).
591. Gefter, E. L., and I. A. Rogacheva, Zh. Obshch. Khim., 36, 1712 (1966) [C.A. 66, 54759 (1967)].
592. J. R. Geigy A. G., Br. Patent 713,278 (1954) [C.A. 50, 1092i (1956)].
593. J. R. Geigy A. G., Br. Patent 1,140,811 (Jan. 22, 1969) [C.A. 70, 69054 (1969)].
594. J. R. Geigy A. G., Swiss Patent 232,459 (Sept. 1, 1944) [C.A. 43, 6232i (1949)].

595. J. R. Geigy A. G., Swiss Patent 245,537 (July 16, 1947) [C.A. 43, 6231g (1949)].
596. J. R. Geigy A. G., Swiss Patent 326,948 (1958) [C.A. 52, 16197f (1958)].
597. Geise, H. J., Rec. Trav. Chim. 86, 362 (1967).
598. Gensler, W. J., and A. P. Mahadevan, J. Am. Chem. Soc., 76, 6192 (1954).
599. Genvresse, P., Compt. Rend., 127, 523 (1898).
600. Gerrard, W., J. Chem. Soc., 1940, 218.
601. Gerrard, W., J. Chem. Soc., 1940, 1464.
602. Gerrard, W., J. Chem. Soc., 1945, 106.
603. Gerrard, W., J. Chem. Soc., 1946, 741.
604. Gerrard, W., and M. J. Richmond, J. Chem. Soc., 1945, 853.
605. Gerrard, W., and B. K. Howe, J. Chem. Soc., 1955, 505.
606. Gerrard, W., A. Nechvatal, and B. Wilson, J. Chem. Soc., 1950, 2088.
606a. Gerrard, W., and R. J. Phillips, Chem. and Ind., 1952, 540.
607. Gevorkyan, A. A., B. L. Dyatkin, and I. L. Knunyants, Izv. Akad. Nauk SSSR, Ser. Khim., 1965, 1599 [C.A. 64, 1944g (1966)].
608. Gielkens, J. W., M. A. Hoefnagel, L. J. Stegerhoek, and P. E. Verkade, Rec. Trav. Chim., 77, 656 (1958).
609. van Gijzen, J., M. A. Hoefnagel, and P. E. Verkade, Rec. Trav. Chim., 83, 267 (1964).
610. van Gijzen, J., and P. E. Verkade, Rec. Trav. Chim., 73, 496 (1954).
611. Gilham, P. T., and G. M. Tener, Chem. and Ind., 1959, 542.
612. Gillis, R. G., J. F. Horwood, and G. L. White, J. Am. Chem. Soc., 80, 2969 (1958).
613. Gilman, H., and J. D. Robinson, Rec. Trav. Chim., 48, 328 (1929).
614. Gilman, H., and C. C. Vernon, J. Am. Chem. Soc., 48, 1063 (1926).
614a. Ginsburg, V. A., M. N. Vasil'eva, S. S. Dubov, and A. Y. Yakubovich, Zh. Obshch. Khim., 30, 2854 (1960) [C.A. 55, 17477b (1961)].
614b. Ginsburg, V. A., and A. Y. Yakubovich, Zh. Obshch. Khim., 30, 3979 (1960) [C.A. 55, 22099c (1961)].
615. Gitel, P. O., L. F. Osipova, O. P. Solovova, and A. Y. Yakubovich, Zh. Obshch. Khim., 39, 301 (1969) [C.A. 71, 2907 (1969)].
616. Glutz, C., Ann., 143, 181 (1867).
616a. Gold, A. M., J. Org. Chem., 26, 3991 (1961).
617. Gold, J., U. S. Patent 3,070,620 (1962) [C.A. 58, 8905h (1963)].
617a. Goldman, L., J. W. Marsico, and G. W. Anderson, J. Am. Chem. Soc., 82, 2969 (1960).

617b. Goldwhite, H., and B. C. Saunders, J. Chem. Soc., 1955, 2040.
617c. Goldwhite, H., and B. C. Saunders, J. Chem. Soc., 1955, 3564.
618. Golinskii, M., Przem. Chem., 46, 730 (1967) [C.A. 68, 117546 (1968)].
619. Golobolov, Y. G., T. F. Dimitreva, Y. M. Zinov'ev, and L. Z. Soborovskii, Zh. Obshch. Khim., 35, 1460 (1965) [C.A. 63, 14689c (1965)].
620. Golobolov, Y. G., and L. Z. Soborovskii, Zh. Obshch. Khim., 33, 2955 (1963) [C.A. 60, 1572e (1964)].
621. Golobolov, Y. G., I. A. Zaishlova, and A. S. Buntyakov, Zh. Obshch. Khim., 35, 1240 (1965) [C.A. 63, 13061g (1965)].
622. Gottlieb, H. B., J. Am. Chem. Soc., 54, 748 (1932).
623. Gottlieb, H. B., J. Am. Chem. Soc., 58, 532 (1936).
623a. Gould, D. H., and E. C. Shapiro, U. S. Patent 3,045,033 (1962) [C.A. 57, 16718g (1962)].
623b. Gozman, I. P., Izv. Akad. Nauk SSSR, Ser. Khim., 1968, 2362 [C.A. 70, 37090 (1969)].
624. Gozman, I. P., and V. A. Kukhtin, Zh. Obshch. Khim., 37, 1644 (1967) [C.A. 68, 59036 (1968)].
625. Graf, R., Chem. Ber., 85, 9 (1952).
625a. Green, M., and R. F. Hudson, J. Chem. Soc., 1963, 540, 3883.
626. Green, M., and D. M. Thorp, J. Chem. Soc., 1965, 466.
627. Greenhalgh, R., and M. A. Weinberger, Can. J. Chem., 45, 495 (1967).
628. Griffin, B. E., and A. Todd, J. Chem. Soc., 1958, 1389.
629. Griffith, D. L., and M. Stiles, J. Am. Chem. Soc., 87, 3710 (1965).
630. Grimbert, L., and O. Bailly, Compt. Rend., 160, 207 (1915).
631. Grisolia, S., and B. K. Joyce, J. Biol. Chem., 233, 18 (1958).
631a. Gross, H., and U. Karsch, J. Prakt. Chem., 29, 315 (1965).
631b. Gross, H., S. Katzwinkel, and J. Gloede, Chem. Ber., 99, 2631 (1966).
632. Grosse-Ruyken, H., and K. Uhlig, J. Prakt. Chem., 18, 287 (1962).
633. Grün, A., Stiasny Festschr., 1937, 88.
634. Grün, A., and F. Kade, Ber., 45, 3362 (1912).
635. Grün, A., and R. Limpächer, Ber., 59, 1355 (1926); 60, 151 (1927).
636. Grundmann, C. J., and J. A. Nakagiri, Am. J. Enol., 8, 1 (1957) [C.A. 51, 10831g (1957)].
636a. Grunze, H., Chem. Ber., 92, 850 (1959).
636b. Grunze, H., Angew. Chem., 71, 70 (1959).

637. Gryskiewicz-Trochimowski, E., Mém. Poudres, 44, 133 (1962) [C.A. 62, 11676g (1965)].
638. Guibe-Sampel, E., and M. Wakselman, J. Chem. Soc., D, 1969, 720.
639. Guichard, F., Ber., 32, 1572 (1899).
640. Gulland, J. M., and G. I. Hobday, J. Chem. Soc., 1940, 746.
641. Gulland, J. M., and H. Smith, J. Chem. Soc., 1947, 338.
642. Gulland, J. M., and H. Smith, J. Chem. Soc., 1948, 1527.
643. Gulland, J. M., and H. Smith, J. Chem. Soc., 1948, 1532.
644. Guthrie, F., Ann., 99, 57 (1856); J. Chem. Soc., 9, 131 (1857).
644a. Gutowsky, H. S., D. W. McCall, and C. P. Slichter, J. Chem. Phys., 21, 279 (1953).
645. de Haas, G. H., and L. L. M. van Deenen, Rec. Trav. Chim., 84, 436 (1965).
646. de Haas, G. H., H. v. Zutphen, P. P. M. Bonsen, and L. L. van Deenen, Rec. Trav. Chim., 83, 99 (1964).
647. Hall, S. A., and M. Jacobson, Ind. Eng. Chem., 40, 694 (1948); Agric. Chem., 1948, 30.
648. Hall, R. H., and H. G. Khorana, J. Am. Chem. Soc., 76, 5056 (1954).
649. Hall, R. H., and R. Thedford, J. Org. Chem., 28, 1506 (1963).
650. Halmann, M., J. Chem. Soc., 1954, 2158.
650a. Halmann, M., J. Chem. Soc., 1959, 305.
651. Halmann, M., and S. Pinchas, J. Chem. Soc., 1953, 626.
652. Halmann, M., and I. Platzner, J. Chem. Soc., 1965, 5380.
653. Hanson, M. W., and J. B. Bouck, J. Am. Chem. Soc., 79, 5631 (1957).
654. Hara, I., and H. Haneko, Bull. Soc. Chim. Biol., 46, 339 (1964).
654a. Hardy, E. E., and G. M. Kosolapoff, U. S. Patent 2,409,039 (1946) [C.A. 41, 1233 (1947)].
655. Harlay, V., J. Pharm. Chim., 20, 160 (1934).
656. Harris, J. S. (to Monsanto Chemical Corp.), U. S. Patent 2,572,806 (1951) [C.A. 46, 2234d (1952)].
657. Harris, J. S. (to Monsanto Chemical Corp.), U. S. 2,702,299 (1955) [C.A. 50, 1074h (1956)].
657a. Harris, S. A., and S. P. Hammond, J. Am. Chem. Soc., 78, 2568 (1956).
658. Harris, L. W., G. R. Sanders, and C. C. Cassil (to California Spray. Chemical Corp.), U. S. Patent 2,641,606 (1953) [C.A. 48, 4582f (1954)].
659. Harvey, W. E., J. Michalski, and A. R. Todd, J. Chem. Soc., 1951, 2271.

659a. Hashimoto, T., M. Tachibara, Y. Ishii, and H. Yoshikawa, J. Biochem. (Tokyo), 50, 548 (1961) [C.A. 57, 16615f (1962)].
660. Hashizume, T., and H. Iwamura, Tetrahedron Lett., 1965, 3095.
661. Hata, T., and K. Ashida, Japan. Patent 19,436 (1964) [C.A. 62, 10339c (1965)].
662. Hata, T., and T. Mukaiyama, Bull. Chem. Soc. Japan, 35, 1106 (1962).
663. Hata, T., and T. Mukaiyama, Bull. Chem. Soc. Japan, 37, 103 (1964).
664. Hatt, H. H., J. Chem. Soc., 1933, 776.
665. Havinga, E., R. O. de Jongh, and W. Dorst, Rec. Trav. Chim., 75, 378 (1956).
666. Hawthorne, J. N., and J. Hawthorne, Biochem. Problems Lipids, Proc. 2nd Intern. Conf. Ghent, 1955, 104 (1956) [C.A. 52, 20317d (1958)].
667. Hazard, R., P. Chabrier, A. Carayon-Gentil, and Y. Fjévet, Compt. Rend., 240, 986 (1955).
668. Hechenbleikner, I. (to Shea Chemical Corp.), U. S. Patent 2,851,476 (1958) [C.A. 53, 1146i (1959)].
669. Heim, R., Ber., 16, 1764 (1883).
670. Hellmann, H., F. Lingens, and H. J. Burkhardt, Chem. Ber., 91, 2290 (1958).
670a. Henning, H. G., Z. Chem., 5, 103 (1965).
671. Henning, H. C., and G. Busse, Angew. Chem., 77, 963 (1965).
672. Henning, H. G., G. Hilgetag, and G. Busse, J. Prakt. Chem., 33, 188 (1966).
673. Henry, L., Rec. Trav. Chim., 24, 343 (1905).
673a. Herail, F., Compt. Rend., 261, 3375 (1965).
673b. Herail, F., Compt. Rend., 262 C, 1624 (1966).
673c. Herail, F., and V. Viosset, Compt. Rend., 259, 4629 (1964).
674. Hessel, L. W., I. D. Morton, A. R. Todd, and P. E. Verkade, Rec. Trav. Chim., 73, 150 (1954).
675. Hetzer, C. D., and L. Szabo, Colloq. Nat. CNRS., 1965, 327 (publ. 1966).
676. Heumann, K., and R. Paganini, Ber., 23, 3552 (1890).
677. Heyl, D., E. Luz, S. A. Harris, and K. Folkers, J. Am. Chem. Soc., 73, 3430 (1951).
678. Heyl, D., and S. A. Harris, J. Am. Chem. Soc., 73, 3434 (1951).
679. Heyl, D., E. Luz, S. A. Harris, and K. Folkers, J. Am. Chem. Soc., 73, 3436 (1951).
680. Heyl, D., E. Luz, and S. A. Harris, J. Am. Chem. Soc., 73, 3437 (1951).
681. Heymann, B., and W. Königs, Ber., 19, 3310 (1886).
682. Heymann, H. E., and T. Rosenberg, Nature, 165, 317 (1950).
683. Higgins, C. E., and W. H. Baldwin, J. Org. Chem., 21, 1156 (1956).

684. Higgins, C. E., and W. H. Baldwin, J. Org. Chem., 30, 3173 (1965).
685. Hilgetag, G., M. Krenger, and H. Teichmann, Z. Chem., 5, 180 (1965).
686. Hirt, R., and R. Berchtold, Helv. Chim. Acta, 40, 1928 (1957).
687. Hochwalt, C. A., J. H. Lum, J. E. Malowan, and C. P. Dyer, Ind. Eng. Chem., 34, 20 (1942).
688. Hochwalt, C. A., U. S. Patent 3,155,710 (Nov. 3, 1964) [C.A. 62, 445g (1965)].
689. Hoeflake, J. M. A., Rec. Trav. Chim., 36, 24 (1916).
690. Hoffman La Roche & Co., U. S. Patent 1,960,184 (1934).
691. Holbrook, K. A., and L. Ouellet, Can. J. Chem., 35, 1496 (1957).
692. Holy, A., S. Chladek, and J. Zemlicka, Collect. Czech. Chem. Commun., 34, 253 (1969).
693. Holy, A., and J. Smrt, Coll. Czech. Chem. Commun., 31, 1528 (1966).
694. Holy, A., J. Smrt, and F. Sorm, Collection Czech. Chem. Commun., 33, 3809 (1968).
695. Holy, A., N. C. Spasovska, and J. Smrt, Collection Czech. Chem. Commun., 29, 2567 (1964).
696. Honjo, M., Y. Furukawa, H. Moriyama, and K. Tanaka, Chem. Pharm. Bull., 16, 73 (1962).
697. Honjo, M., K. Imar, and S. Yoshikawa, Japan. Patent 68 25,500 (1968) [C.A. 70, 68696 (1969)].
698. Honjo, M., R. Marumoto, K. Kobayashi, and Y. Yoskioka, Tetrahedron Lett., 1966, 3851.
698a. Hood, A., U. S. Patent 2,712,548 (1955) [C.A. 51, 459a (1957)].
698b. Hood, A., and W. Lange, J. Am. Chem. Soc., 72, 4956 (1950).
699. van Hooidonk, C., and L. Ginjaar, Rec. Trav. Chim., 86, 449 (1961).
700. Hooker Chemical Corp., Br. Patent 889,338 (1962) [C.A. 58, 8906e (1963)].
701. Hooker Chemical Corp., Br. Patent 937,560 (1963) [C.A. 60, 15733g (1964)].
702. Hopkins, E. A. H., and J. H. Wang, J. Am. Chem. Soc. 87, 4391 (1965).
703. Hori, T., O. Itasaka, H. Inone, and M. Akai, Japan. J. Exptl. Med., 35, 81 (1965) [C.A. 63, 10225c (1965)].
704. Hosaeus, H., Ber., 26, 667 (1893).
705. Houssa, A. H. J., and H. Phillips, J. Chem. Soc., 1931, 108.
706. Hove (van), T., Bull. Acad. Roy. Belg., 1909, 282.
707. Hückel, W., O. Neunhoeffer, A. Gercke, and E. Frank, Ann., 477, 99 (1929).
708. Hückel, B. W., and H. Pietrzok, Ann., 540, 250 (1939

709. Hudson, R. F., Ann. Chim. (Roma), 53, 47 (1953) [C.A. 58, 13742d (1963)].
709a. Hudson, R. F., "Structure and Mechanism in Organo-Phosphorus Chemistry," Academic Press, 1965.
709b. Hudson, R. F., and R. Greenhalgh, J. Chem. Soc., B, 1969, 325.
709c. Hudson, R. F., and D. C. Harper, J. Chem. Soc., 1958, 1356.
709d. Hudson, R. F., and L. Keay, J. Chem. Soc., 1956, 2463.
710. Hull, D. E., and A. H. Agett, U. S. Patent 2,430,569 (1947; to Eastman Kodak Co.).
711. Hull, D. E., and J. R. Snodgrass, U. S. Patent 2,407,279 (1946; to Eastman Kodak Co.).
712. Hull, D. E., and J. R. Snodgrass (to Eastman Kodak Co.), Br. Patent 649,584 (1951) [C.A. 45, 9074c (1951)].
713. Hull, D. C., and J. R. Snodgrass (to Eastman Kodak), U. S. Patent 2,518,692 (Aug. 15, 1950) [C.A. 45, 692g (1951)].
714. Hunter, J. R., R. L. Roberts, and E. B. Kester, J. Am. Chem. Soc., 70, 3244 (1948).
715. Hurd, C. D., and M. F. Dull, J. Am. Chem. Soc., 54, 3427 (1932).
716. I. G. Farbenindustrie A. G., German Patent 517,538 (1931).
717. I. G. Farbenindustrie A. G., U. S. Patent 1,766,720-1 (1930).
718. I. G. Farbenindustrie A. G., U. S. Patent 1,869,768 (1932).
719. I. G. Farbenindustrie A. G., U. S. Patent 1,936,985 (1933).
720. I. G. Farbenindustrie A. G., U. S. Patent 1,944,530 (1934).
721. I. G. Farbenindustrie A. G., Swiss Patent 152,767 (1932).
721a. Ikehara, M., and E. Ohtsuka, Chem. Pharm. Bull., 10, 536 (1962).
722. Ikehara, M., and E. Ohtsuka, Chem. Pharm. Bull., 11, 435 (1963).
723. Ikehara, M., and E. Ohtsuka, Chem. Pharm. Bull., 11, 1095 (1963).
724. Ikehara, M., and E. Ohtsuka, Chem. Pharm. Bull., 11, 1353 (1963).
725. Ikehara, M., E. Ohtsuka, and Y. Kodama, Chem. Pharm. Bull., 11, 1456 (1963).
726. Ikehara, M., E. Ohtsuka, and Y. Kodama, Chem. Pharm. Bull., 12, 145 (1964).
727. Ikehara, M., and I. Tazawa, J. Org. Chem., 31, 819 (1966).
728. Ikehara, M., I. Tazawa, and T. Fukui, Chem. Pharm.

Bull., 17, 1019 (1969).
729. Ikehara, M., T. Ueda, and K. Ikeda, Chem. Pharm. Bull., 10, 767 (1962).
730. Ikehara, M., and H. Uno, Chem. Pharm. Bull., 12, 742 (1964).
731. Imaev, M. G., and A. M. Shakirova, Dokl. Akad. Nauk SSSR, 163, 656 (1965) [C.A. 63, 11338b (1965)].
732. Imai, K., S. Fujii, K. Takanohashi, Y. Furukawa, T. Masuda, and M. Honjo, J. Org. Chem., 34, 1547 (1969).
733. Imai, K., S. Fujii, K. Takanohashi, Y. Furakawa, T. Masuda, and M. Honjo, J. Org. Chem., 34, 1947 (1969).
734. Imai, K., T. Hiraba, and M. Honjo, Takeda Kenkyusho Nempo, 23, 1 (1964) [C.A. 63, 4553g (1965)].
735. Imbert, H., and G. Belugou, Bull. Soc. Chim. (3), 21, 935 (1899); Compt. Rend., 125, 1041 (1897).
736. Imhausen & Co., German Patent 302,501 (1919).
737. Ipat'ev, V., U. S. Patent 2,062,312 (1936; to Universal Oil Products Corp.).
737a. Ishikazu, O., J. Pharm. Soc. Japan, 73, 568 (1953).
737b. Iselin, B. M., J. Am. Chem. Soc., 71, 3825 (1949).
738. Ito, A., O. Amaksu, K. Okmoto, W. Yamanaka, Sankyo Kenkyusho Nempo, 15, 65 (1963) [C.A. 60, 12006f (1964)].
739. Iwatsuru, R., Biochem. Z., 173, 349 (1926).
740. Iwei, H., A. Kiyomoto, and M. Takeshita, Biochim. Biophys. Acta, 162, 302 (1968).
741. Jachimowitz, T., Biochem. Z., 292, 356 (1937).
741a. Jachymczyk, W., L. Ménager, and L. Szabó, Tetrahedron, 21, 2049 (1965).
742. Jacini, G., and G. de Zotti, Ind. Parfums et Cosmét. 12, 389 (1951) [C.A. 52, 6505c (1958)].
743. Jackson, E. L., J. Am. Chem. Soc., 57, 1903 (1935).
744. Jacob, T. M., and H. G. Khorana, J. Am. Chem. Soc., 86, 1630 (1964).
745. Jacob, T. M., and H. G. Khorana, J. Am. Chem. Soc., 87, 368, 2971 (1965).
746. Jacobsen, G., Ber., 8, 1521 (1875).
747. Jacobsohn, H. J., and M. Jacobsohn, Compt. Rend. Soc. Biol., 105, 154 (1930).
748. Jaenicke, L., and J. Koch, Ann. Chem., 663, 50 (1963).
749. Jahns, E., Arch. Pharm., 215, 11 (1879).
749a. Jean, H., Bull. Soc. Chim. France, 1957, 783.
750. Jenkins, R. L., U. S. Patent 2,426,691 (1947; to Monsanto Chemical Co.).
751. Johnson, D. I. O., and G. L. Milward, Br. Patent 778,081 (1957) [C.A. 51, 17980h (1957)].
752. Jones, E. E., and D. Lipkin, J. Am. Chem. Soc., 78, 2408 (1956).
753. Jones, W., G. O. Osborne, G. J. Sutherland, R. D. Topson, and J. Vaughan, Chem. Commun., 1966, 18.

754. Jones, W. J., L. H. Thomas, E. H. Pritchard, and S. T. Bowden, J. Chem. Soc., 1946, 824.
755. Kabachnik, M. I., and M. P. A. Rossiiskaya, Izv. Akad. Nauk SSSR, Otdel Khim. Nauk, 1946, 403.
756. Kabachnik, M. I., and M. P. A. Rossiiskaya, Izv. Akad. Nauk SSSR, Otdel Khim. Nauk, 1958, 1398 [C.A. 53, 6988e (1959)].
756a. Kabachnik, M. I., et al., Zh. Obshch. Khim., 29, 1680 (1959) [C.A. 54, 8594i (1960)].
757. Kabashima, I., Ber., 71, 76, 1071 (1938).
757a. Kainosko, M., et al., Bull. Chem. Soc. Japan, 42, 1713 (1962).
758. Kainosko, M., and T. Shimozeva, Tetrahedron Lett., 1965, 865.
759. Kaiser, E. T., and K. Kudo, J. Am. Chem. Soc., 89, 6725 (1967).
760. Kamai, G., Dokl. Akad. Nauk SSSR, 70, 233 (1950) [C.A. 45, 5611f (1951)].
760a. Kamai, G., Dokl. Akad. Nauk SSSR, 79, 795 (1951) [C.A. 46, 6081f (1952)].
760b. Kamai, G., Izv. Akad. Nauk SSSR, Otdel Khim. Nauk, 1952, 923 [C.A. 47, 10461a (1953)].
761. Kamai, G., and A. P. Bogdanov, Trudy Kazan. Khim. Tekhnol. Inst. im. S. M. Kirova, 1913, nr. 18, 22 [C.A. 51, 5721b (1957)].
761a. Kamai, G., and F. M. Kharrasova, Trudy Kazan. Khim. Tekhnol. Inst., 23, 127 (1957) [C.A. 52, 9946a (1958)].
761b. Kamai, G., and F. M. Kharrasova, Vysshikh Ucheb. Zavedenii, Khim. i Khim. Tekhnol., 4, 229 (1961) [C.A. 55, 21762e (1961)].
762. Kamai, G., and E. S. Koshkina, C.A. 50, 6346i (1956).
763. Kanaoka, Y., O. Yonemitsu, K. Tanizawa, and Y. Ban, Chem. Pharm. Bull., 12, 773 (1964).
764. Kanaoka, Y., O. Yonemitsu, K. Tanizawa, K. Matsukaki, and Y. Ban, Chem. and Ind., 1964, 2102.
765. Kanaoka, Y., et al., Tetrahedron Lett., 1964, 2419.
766. Karrer, P., and H. Bendas, Helv. Chim. Acta, 19, 98 (1936).
767. Karrer, P., and P. Benz. Helv. Chim. Acta, 9, 24, 599 (1926).
768. Karrer, P., and P. Benz, Helv. Chim. Acta, 10, 89 (1927).
769. Karrer, P., and G. Bussmann, Helv. Chim. Acta, 23, 1137 (1940).
770. Karrer, P., and H. Salomon, Helv. Chim. Acta, 9, 13 (1926).
770a. Katyshkina, V. V., and M. Y. Kraft, Zh. Obshch. Khim., 26, 3060 (1956) [C.A. 51, 8028a (1957)].
771. Katchalsky, A., and M. Paecht, J. Am. Chem. Soc., 76, 6042 (1954).

772. Katzman, M. B., U. S. Patent 2,128,946 (1938; to Emulsol Corp.).
773. Kavunenko, A. P., and N. S. Tikhomirova-Sidorova, Zh. Obshch. Khim., 38, 2368 (1968) [C.A. 70, 68678 (1969)].
774. Kawahara, F. K., U. S. Patent 3,228,925 (Jan. 11, 1966) [C.A. 64, 15744a (1966)].
775. Kaye, H., and Lord Todd, J. Chem. Soc., C, 1967, 1420.
775a. Keay, L., and E. M. Crook, J. Chem. Soc., 1961, 710.
776. Kekule, A., and G. A. Barbaglia, Ber., 5, 876 (1872); Kekule, A., Ber., 6, 944 (1873).
777. Kelso et al., U. S. Patent 2,315,529; 2,316,078-91 (1943; to Standard Oil Co.); Mixon, U. S. Patent 2,375,315 (1945; to Standard Oil Co.); Noland, U. S. Patent 2,379,453 (1945; to Standard Oil Co.).
778. Kennedy, J., E. S. Lane, and B. K. Robinson, J. Appl. Chem., 8, 459 (1958).
779. Kenner, G. W., and J. Mather, J. Chem. Soc., 1956, 3524.
779a. Kenner, G. W., C. B. Rees, and A. R. Todd, J. Chem. Soc., 1958, 546.
780. Kenner, G. W., and A. R. Todd, J. Chem. Soc., 1956, 1231.
781. Kenner, G. W., A. R. Todd, and R. F. Webb, J. Chem. Soc., 1956, 1231.
782. Kenner, G. W., A. R. Todd, and F. J. Weymouth, J. Chem. Soc., 1952, 3675.
783. Kenyon, J., A. G. Lipscomb, and H. Phillips, J. Chem. Soc., 1931, 2275.
784. Kenyon, J., H. Phillips, and F. M. H. Taylor, J. Chem. Soc., 1931, 382.
785. Kerger, K., and R. Kohlhaas, Z. Anorg. Allg. Chem., 354, 44 (1965).
786. Kevelaar, J. A. A., and H. R. Gersmann, J. Am. Chem. Soc., 72, 5777 (1950).
787. Kharasch, M. S., and I. S. Bengelsdorf, J. Org. Chem. 20, 1356 (1955).
788. Kharlampovich, G. D., and T. F. Aksenova, USSR Patent 176,895 (Dec. 1, 1965) [C.A. 64, 12600d (1966)].
789. Kharlampovic, G. D., and T. F. Aksenova, ref. in Zh. Khim., 1966, II, Abstr. 23P93 [C.A. 66, 104766 (1967)].
790. Khorana, H. G., G. M. Tener, R. S. Wright, and J. D. Mofatt, J. Am. Chem. Soc., 79, 430 (1957).
791. Khorana, H. G., T. M. Jacob, M. W. Moon, S. A. Narang, and E. Ohtsuka, J. Am. Chem. Soc., 87, 2954, 2956, 2971, 2981, 2988 (1965).
792. Kiely, D. E., and H. G. Fletcher, Jr., J. Org. Chem. 33, 3723 (1968).

793. Kiessling, W., Biochem. Z., 273, 103 (1934).
794. Kiessling, W., Ber., 68, 243 (1935).
795. Kiessling, W., Ber., 69, 2331 (1936); 68, 597 (1935).
796. Kilgour, G. L., and C. E. Ballou, J. Am. Chem. Soc., 80, 3956 (1958).
797. King, E. J., J. Pathol. Bacteriol., 55, 311 (1943).
798. King, H., and F. L. Pyman, J. Chem. Soc., 105, 1247 (1914).
799. King, T. E., and F. H. Strong, Science, 112, 562 (1950).
800. Kipping, F. S., and F. Challenger, J. Chem. Soc., 99, 626 (1911).
801. Kirby, A. J., and A. G. Varvoglis, J. Am. Chem. Soc., 89, 415 (1967).
802. Kirillova, K. M., and V. A. Khutin, Zh. Obshch. Khim., 35, 544 (1965) [C.A. 63, 523d (1965)].
803. Kirsanov, A. V., and M. S. Marenets, Zh. Obshch. Khim., 31, 1605 (1961) [C.A. 55, 23339f (1961)].
804. Kleinfeller, H., Ber., 62, 1585 (1929).
805. Klenk, E., and H. Debuch, Z. Physiol. Chem., 296, 179 (1954).
806. Klepp, M., and L. Schmid, Nahrung, 1968, 12, 41 [C.A. 69, 10072 (1968)].
807. Klybas, V., M. Schramm, and E. Racker, Arch. Biochem. Biophys., 80, 229 (1959).
808. Knauer, W., Ber., 27, 2565 (1894).
809. Knoll & Co., German Patent 201,369 (1908).
809a. Kobayashi, K., T. Hirano, S. Wakamori, M. Eto, and Y. Oshima, Bochu-Kagu, 34, 66 (1969) [C.A. 71, 112896 (1969)].
809b. Knunyants, I. L., O. V. Kil'disheva, and E. Bykhovskaya, Zh. Obshch. Khim., 19, 101 (1949).
810. Kochetkov, N. K., and 6 others, Carbohydr. Res., 10, 152 (1969).
811. Kochetkov, N. K., E. I. Budovskii, V. N. Shibaev, G. I. Yeliseeva, M. A. Grachev, and B. P. Demushkin, Tetrahedron, 19, 1207 (1963).
812. Kodaira, Y., and T. Mukaiyama, J. Org. Chem., 31, 2903 (1966).
813. Koefoed, J., and A. H. Jensen, Acta Chem. Scand., 5, 23 (1951).
814. Kojima, T., and R. Ito, C.A. 69, 43693 (1968).
814a. Kolka, A. J., U. S. Patent 2,866,809 (1958) [C.A. 53, 19879f (1959)].
815. Kominato, K., Japan. Patent 9076 (1957); 9077 (1957) [C.A. 52, 14699d (1958)].
816. Koransky, W., H. Grunze, and G. Muench, Z. Naturforsch., 17 B, 291 (1962).
817. Korovin, S. S., Y. I. Kol'tsov, I. A. Apraksin, and A. M. Reznik, Zh. Neorgan. Khim., 11, 948 (1966) [C.A. 65, 1472f (1966)].
818. Korovin, S. S., V. G. Yurkin, and A. P. Mironenko,

Zh. Neorgan. Khim., 11, 1910 (1966) [C.A. 65, 12913h (1966)].
818a. Koshland, D. E., Jr., J. Am. Chem. Soc., 73, 4103 (1951).
818b. Koshland, D. E., Jr., J. Am. Chem. Soc., 74, 2286 (1952).
819. Kosolapoff, G. M., and E. E. Hardy, U. S. Patent 2,409,039 (to Monsanto Chemical Co.).
820. Kosolapoff, G. M. (to Monsanto Chemical Co.), U. S. Patent 2,503,204 (1950) [C.A. 44, 5897i (1950)].
821. Kosolapoff, G. M. (to Monsanto Chemical Co.), U. S. Patents 2,634,226 and 2,634,227 (1953) [C.A. 47, 7154a,c (1953)].
822. Kosolapoff, G. M., Science, 108, 485 (1948).
822a. Kosolapoff, G. M., C. K. Arpke, R. W. Lamb, and H. Reich, J. Am. Chem. Soc., 90, 815 (1968).
823. Kosterlitz, H. W., Biochem. J., 37, 321 (1943).
824. Kovalevskii, A., Ann., 119, 303 (1861).
825. Kozlova, N. Y., I. V. Mel'nichenko, and A. A. Yasnikov, Ukr. Khim. Zh., 39, 1041 (1968) [C.A. 70, 86743 (1969)].
825a. Kraft, M. Y., and V. V. Katyshkina, Dokl. Akad. Nauk SSSR, 86, 725 (1952) [C.A. 47, 8032b (1953)].
825b. Kraut, J., and L. H. Jensen, Acta Cryst., 16, 79 (1963).
826. Kraut, K., Ann., 118, 102 (1861).
826a. Kreshkov, A. P., V. A. Drozdov, and I. Y. Orlova, Zh. Obshch. Khim., 36, 525 (1966) [C.A. 65, 735h (1966)].
826b. Kreshkov, A. P., V. A. Drozdov, and I. Y. Orlova, Zh. Obshch. Khim., 36, 2014 (1966) [C.A. 66, 76068 (1967)].
826c. Kreutzkamp, N., and H. Kayser, Ann. Chem., 609, 39 (1957).
827. Kreysler, E., Ber., 18, 1701 (1885).
828. Kubo, H., Agr. Biol. Chem. (Tokyo), 29, 43 (1965) [C.A. 63, 7032c (1965)].
828a. Kucherov, V. F., Zh. Obshch. Khim., 19, 126 (1949).
829. Kugel, L., and M. Halmann, J. Am. Chem. Soc., 88, 3566 (1966).
830. Kugel, L., and M. Halmann, J. Org. Chem., 32, 642 (1967).
831. Kuhn, R., H. Rudy, and F. Weygand, Ber., 69, 1543 (1936).
832. Kukhtin, V. A., Dokl. Akad. Nauk SSSR, 121, 466 (1958) [C.A. 53, 1105a (1959)].
833. Kukhtin, V. A., T. N. Voskoboeva, and K. M. Kyrillova, Zh. Obshch. Khim., 32, 2333 (1962) [C.A. 58, 9127g (1963)].
833a. Kumamoto, J., J. Cox, and F. H. Westheimer, J. Am. Chem. Soc., 78, 4858 (1956).

833b. Kumamoto, J., and F. H. Westheimer, J. Am. Chem. Soc., 77, 2515 (1955).
834. Kumler, W. D., and J. J. Eiler, J. Am. Chem. Soc., 65, 2355 (1943).
835. Kunz, P., Ber., 27, 2559 (1894).
835a. Kuz'menko, I. I., and L. B. Rapp, Zh. Obshch. Khim., 38, 158 (1968) [C.A. 69, 95863 (1968)].
836. Kuznetsov, E. V., and R. K. Valendinov, Trudy Kazan. Khim. Tekhnol. Inst., 23, 161 (1957) [C.A. 52, 8938i (1958)].
836a. Kuznetsov, E. V., and R. K. Valendinov, Zh. Obshch. Khim., 29, 235 (1959) [C.A. 53, 21648d (1959)].
837. Kyogoku, Y., and Y. Iitaka, Acta Cryst., 21, 49 (1966).
838. Kyrides, L. P. (to Monsanto Chemical Co.), U. S. Patent 2,510,033 (1950) [C.A. 44, 8361b (1950)].
839. Laboratoires Le Brun, Fr. Patent 1,072,327 (1954) [C.A. 53, 4134a (1959)].
840. Lambic, A. J., Tetrahedron Lett., 1966, 3709.
840a. Lange, W., and A. Hood, U. S. Patent 2,614,116 (1952) [C.A. 47, 8771g (1953)].
841. Lange, W., and G. V. Krüger, Ber., 65, 1598 (1932).
842. Langheld, K., Ber., 43, 1857 (1910); 44, 2077 (1911).
843. Langheld, K., Ber., 44, 2076 (1911).
844. Langheld, K., Ber., 45, 1126 (1912).
845. Langheld, K., F. Oppmann, and E. Meyer, Ber., 45, 3753 (1912); German Patent 248,956 (1912).
846. Lanham, W. M. (to Union Carbide & Carbon Corp.), Br. Patent 762,125 (1956) [C.A. 51, 11399d (1957)].
847. Lanham, W. M. (to Union Carbide & Carbon Corp.), Br. Patent 762,147 (1956) [C.A. 51, 11399i (1957)].
848. Lanham, W. M. (to Union Carbide & Carbon Corp.), Br. Patent 791,739 (1958) [C.A. 53, 2263i (1959)].
848a. Lanham, W. M., Br. Patent 807,896 (1959) [C.A. 53, 14027a (1959)].
849. Lapidot, A., and M. Halmann, J. Org. Chem., 28, 1394 (1963).
850. Lapidot, A., and D. Samuel, J. Chem. Soc., 1964, 1931.
851. Larrison, M. S., U. S. Patent 3,333,027 (July 1967) [C.A. 67, 81935 (1967)].
851a. Larsson, L., Acta Chem. Scand., 12, 723 (1958).
852. Lassaigne, J. L., Ann. Chim., 3, 294 (1820).
853. Latscha, H. P., Z. Naturforsch., B, 23, 137 (1968).
853a. Leaback, D. L., E. C. Heath, and S. Roseman, Biochemistry, 8, 1351 (1969).
854. Lecocq, J., and A. Todd, J. Chem. Soc., 1954, 2381.
855. Lee, Y. C., and C. E. Ballou, Biochemistry, 4, 1395 (1965).
856. Lehninger, A. L., J. Biol. Chem., 162, 333 (1946).

857. Leichssenring, G., and J. Schmidt, Chem. Ber., 95, 767 (1962).
857a. Leinen, R., and P. Ronkainen, Acta Chem. Scand., 8, 591 (1954).
858. Leloir, L. F., O. M. Repetto, C. E. Cardini, A. C. Paladini, and R. Caputto, Anales Asoc. Quím. Argentina, 37, 187 (1949) [C.A. 44, 4425g (1950)].
859. Le Maire, D., and M. Perat, Compt. Rend., C, 267, 732 (1968).
859a. Lenton, M. V., and B. Lewis, Chem. and Ind., 1965, 946.
860. Lenz, T. G., and M. Smutz, J. Inorg. Nucl. Chem., 30, 621 (1968).
861. Leplawy, M., J. Michalski, and J. Zabrocki, Chem. and Ind., 1964, 835.
861a. Letcher, J. H., and J. R. Van Wazer, J. Chem. Phys., 45, 2916 (1966).
862. Letters, R., and A. M. Michelson, J. Chem. Soc., 1962, 71.
863. Letters, R., and A. M. Michelson, Bull. Soc. Chim. Biol., 45, 1353 (1963).
864. Levene, P. A., and C. C. Christman, J. Biol. Chem., 123, 607 (1938).
865. Levene, P. A., and A. L. Raymond, J. Biol. Chem., 92, 757 (1931).
866. Levene, P. A., and A. L. Raymond, J. Biol. Chem., 107, 75 (1934).
867. Levene, P. A., and A. Schormüller, J. Biol. Chem., 100, 583 (1933).
868. Levene, P. A., and A. Schormüller, J. Biol. Chem., 105, 547 (1934).
869. Levene, P. A., and A. Schormüller, J. Biol. Chem., 106, 595 (1934).
870. Levene, P. A., and R. S. Tipson, J. Biol. Chem., 106, 113 (1934).
871. Levene, P. A., and R. S. Tipson, J. Biol. Chem., 121, 131 (1937).
872. Levin, I. S., N. A. Bustovaya-Balakireva, and I. A. Vorsina, Zh. Neorg. Khim., 13, 857 (1968) [C.A. 68, 117559 (1968)].
873. Levin, I. S., A. A. Shatalova, and I. A. Vorsina, Izv. Sib. Otd. Akad. Nauk SSSR, Ser. Khim. Nauk, 1967, 89 [C.A. 69, 90363 (1968)].
874. Lewak, S., R. Derache, and L. Szabó, Compt. Rend., 248, 1837 (1959).
874a. Lichtenthaler, F. W., and F. Cramer, Chem. Ber., 95, 1971 (1962).
875. Liebig, J. v., Ann. Pharm., 6, 149 (1833).
876. Limpricht, H., Ann., 134, 347 (1865).
877. Lindner, K. (to Hall Laboratories), U. S. Patent 2,218,582 (1940).

877a. Lipmann, F., Adv. Enzymol., 1, 99 (1941).
878. Lippert, A., and E. E. Reid, J. Am. Chem. Soc., 60, 2370 (1938).
879. Lippman, F., and L. C. Tuttle, J. Biol. Chem., 153, 571 (1944).
879a. Lohrmann, R., and H. G. Khorana, J. Am. Chem. Soc., 86, 4188 (1964).
880. Lomonosov, M. V., USSR Patent 186,469 (1966) [C.A. 66, 76150 (1967)].
881. Long, R. F., and A. L. Morrisson (Hoffmann-La Roche Inc.), U. S. Patent 2,777,851 (1957) [C.A. 51, 12984c (1957)].
882. Longinotti, L., and F. Pocchiari, Gazz. Chim. Ital., 84, 1171 (1954).
882a. Lorenz, W., A. Henglein, and G. Schrader, J. Am. Chem. Soc., 77, 2554 (1955).
883. Lorentz, K., Mikrochim. Acta, 1969, 358.
883a. Loshadkin, N. A., et al., Zh. Obshch. Khim., 36, 1105 (1966) [C.A. 65, 13467f (1966)].
884. Lossen, W., and A. Köhler, Ann., 262, 209 (1891).
885. Luff, B. D. W., and F. S. Kipping, J. Chem. Soc., 95, 1993 (1909).
885a. Lutsenko, I. F., Z. S. Kraits, and M. V. Proskurina, Dokl. Akad. Nauk SSSR, 148, 846 (1963) [C.A. 59, 3759d (1963)].
885b. Luu, B., M. A. Diaz-Parra, and G. Ourisson, Tetrahedron Lett., 1969, 227.
886. Lynen, F., Ber., 73, 367 (1940).
887. Lynen, F., and H. Bayer, Chem. Ber., 85, 905 (1952).
888. Lyutik, A. I., A. V. Luk'yanov, E. S. Zhdanovich, and N. A. Preobrazhenskii, Zh. Obshch. Khim., 38, 2251 (1968) [C.A. 70, 58186 (1969)].
889. Maarsen, J. W., and M. C. Smit, Rec. Trav. Chim., 76, 724 (1957).
889a. Maarsen, J. W., M. C. Smit, and J. Matze, Rec. Trav. Chim., 76, 713 (1957).
890. MacAfee, U. S. Patent 2,060,815 (1936).
891. MacAfee, U. S. Patent 2,198,915 (1940; to Dow Chemical Co.).
892. McBee, E. T., and J. D. Damrath, U. S. Patent, 3,206,495 (Sept. 14, 1965) [C.A. 63, 14705a (1965)].
893. McCombie, H., and B. E. Saunders, Nature, 157, 287, 776 (1946).
893a. McCombie, H., B. C. Saunders, N. B. Chapman, and R. Heap, U. S. Patent 2,489,917 (1949) [C.A. 44, 3005g (1950)].
894. McCombie, H., B. C. Saunders, and G. J. Stacey, J. Chem. Soc., 1945, 380.
895. McCombie, H., B. C. Saunders, and G. J. Stacey, J. Chem. Soc., 1945, 921.
896. McCready, R. M., and W. Z. Hassid, J. Am. Chem. Soc.,

66, 560 (1944).
897. Mac Donald, D. L., H. O. L. Fischer, and C. E. Ballou, J. Am. Chem. Soc., 78, 3720 (1956).
898. Mac Donald, D. L., Methods Enzymol., 8, 121 (1966).
899. McFadden, B. A., and L. J. Sick, J. Am. Chem. Soc., 87, 5505 (1965).
900. McGilvery, J. Biol. Chem., 200, 835 (1953).
901. McKenzie, B. A., and G. W. Clough, J. Chem. Soc., 103, 687 (1913).
902. Machleidt, H., E. Cohnen, and R. Tschesche, Ann., Chem., 672, 215 (1964).
903. Magee, P. S., Tetrahedron Lett., 1965, 3995.
903a. Maier, L., Helv. Chim. Acta, 52, 1337 (1969).
904. Mairanovskii, V. G., N. Fokina, L. A. Vakulova, and G. F. Samokhvalov, Zh. Obshch. Khim., 36, 1345 (1968) [C.A. 65, 16502b (1966)].
904a. Maklyaev, F. L., M. I. Druzin, and I. V. Palagina, Zh. Obshch. Khim., 31, 1312 (1961) [C.A. 57, 27012f (1961)].
904b. Maklyaev, F. L., et al., Zh. Obshch. Khim., 32, 3421 (1962) [C.A. 58, 8887g (1963)].
905. Makman, R. S., and E. W. Sutherland, J. Biol. Chem., 240, 1309 (1965).
906. Malatesta, P., and R. Pizzotti, Chimica e Industr., 27, 6 (1945).
906a. Malatesta, P., and B. D'Atri, Farmaco, Ed. Sci., 8, 398 (1953) [C.A. 48, 9312f (1954)].
906b. Malatesta, P., and B. D'Atri, Farmaco, Ed. Sci., 8, 470 (1953) [C.A. 48, 9901i (1954)].
907. Maley, Fr., and H. A. Lardy, J. Am. Chem. Soc., 78, 1393 (1956).
908. Maley, F., G. F. Maley, and H. A. Lardy, J. Am. Chem. Soc., 78, 5303 (1956).
909. Malkin, T., R. L. Baylis, T. H. Bevan, and D. J. Wetley, C. A. 51, 1835i (1957).
910. Mandel'baum, Ya. A., I. L. Vladimirova, and N. N. Mel'nikov, Zh. Obshch. Khim., 23, 429 (1953) [C.A. 48, 3887d (1954)].
911. Mandl, I., and C. Neuberg, Arch. Biochem. Biophys., 37, 83 (1952).
912. Mann, J. J., and H. A. Lardy, J. Biol. Chem., 187, 379 (1950).
913. Marcus, H., and W. B. Eliott, J. Am. Chem. Soc., 80, 4287 (1958).
913a. Mark, V., Tetrahedron Lett., 1961, 295.
913b. Mark, V., U. S. Patent 3,328,472 (1967) [C.A. 67, 118098 (1967)]; Org. Synth. 46, 93 (1966).
913c. Mark, V., C. H. Dungan, M. M. Crutchfield, and J. R. Van Wazer, Topics in Phosphorus Chemistry, 5, 227 (1967).
914. Mark, V., and J. R. van Wazer, J. Org. Chem., 32,

1187 (1967).
914a. Markov, S. M., et al., Zh. Obshch. Khim., 36, 1098 (1966) [C.A. 65, 13467c (1966)].
915. Marschalk, C., Bull. Soc. Chim. France (4), 43, 1397 (1928).
915a. Martensson, O., and E. Nilsson, Acta Chem. Scand., 19, 1256 (1965).
916. Martin, D., Chem. Ber., 98, 3286 (1965).
916a. Martin, G., and A. Besnard, Compt. Rend., 257, 2463 (1963).
916b. Martin, G., and G. Marvel, Proc. Colloque Ampere, 11, 577 (1962) [C.A. 61, 138d (1964)]; Compt. Rend., 257, 1703 (1964).
917. Martynov, I. V., Yu. L. Kruglyak, Z. I. Khromova, and S. I. Malekin, USSR Patent 280,813 (1968) [C.A. 70, 67601k (1969)].
917a. Martynov, I. V., Y. L. Kruglyak, and N. F. Privezentseva, Zh. Obshch. Khim., 37, 1125 (1967) [C.A. 67, 116497 (1967)].
917b. Martynov, I. V., N. F. Privezentseva, and Y. L. Kruglyak, USSR Patent 184,847 (Jul., 1966) [C.A. 66, 65081 (1967)].
918. Maruishi, I., and S. Tsuda, Osaka Furitsu Kogyo-Shoreikan Hokoku, 1962, nr. 28, 86 [C.A. 61, 15968c (1964)].
919. Maruszewska-Wieczorkowska, E., and J. Michalski, Rozniki Chem., 37, 1579 (1963) [C.A. 60, 9134f (1964)].
920. Marvel, C. S., M. J. Copley, and E. Ginsberg, J. Am. Chem. Soc., 62, 3109 (1940).
921. Mason, H. S., and A. R. Todd, J. Chem. Soc., 1951, 2267.
922. Mastin, T. W., G. R. Norman, and E. A. Weilmuenster, J. Am. Chem. Soc., 67, 1662 (1945).
923. Mathieson, D. W., and D. W. Russell, J. Pharm. Pharmacol., 9, 612 (1917) [C.A. 52, 3716c (1958)].
923a. Mathis, R., M. T. Boisdon, J. P. Vives, and F. Mathis, Compt. Rend., 257, 402 (1963).
924. Maurizot, J. C., W. J. Wechter, J. Brahms, and C. Sadron, Nature, 219, 377 (1968).
924a. Mavel, G., and G. Martin, J. Chem. Phys., 59, 762 (1962).
925. Maynard, S. A., and J. M. Swan, Australian J. Chem., 16, 596 (1963).
926. Maynard, J. A., and J. M. Swan, Proc. Chem. Soc., 1963, 61.
926a. Mazur, A., J. Biol. Chem., 164, 271 (1946).
927. Meagher, W. R., and W. Z. Hassid, J. Am. Chem. Soc., 68, 2135 (1946).
928. Melby, L. R., and D. R. Strobach, J. Org. Chem., 34, 421 (1969).

929. Mel'nik, S. Y., M. A. Miropol'skaya, and G. I. Samokhvalov, Zh. Obshch. Khim., 36, 1905 (1966) [C.A. 67, 44067 (1967)].
930. Mel'nik, S. Ya., V. G. Mairanovskii, M. A. Miropol'skaya, and G. I. Samokhvalov, Zh. Obshch. Khim., 38, 1495 (1968) [C.A. 69, 95890 (1968)].
931. Mel'nik, S. Ya., et al., Zh. Obshch. Khim., 38, 1495 (1968) [C.A. 69, 95894 (1968)].
932. Menschutkin, N., Compt. Rend., 59, 295 (1864).
933. Menschutkin, N., Bull. Soc. Chim. (2), 3, 269 (1865).
933a. E. Merck A.-G., Neth. Appl. 6,516,242 (1966) [C.A. 66, 11162 (1967)].
934. E. Merck A. G., Ger. Patent 1,134,075 (1962) [C.A. 58, 8009h (1963)].
935. Merck, E., Jahresber., 1898, 25.
936. Merrill, E. J., and A. L. Babson, U. S. Patent 3,331,862 (July 18, 1967) [C.A. 68, 21704 (1968)].
937. Messens, E., and M. Van Montagu, FEBS Lett., 1, 326 (1968).
938. Metallgesellschaft A. G., Br. Patent 692,633 (1953) [C.A. 48, 10031f (1954)].
939. Metallgesellschaft A. G., Br. Patent 706,410 (1954) [C.A. 49, 6988g (1955)].
940. Meyerhof, O., and K. Lohmann, Biochem. Z., 185, 113 (1927).
941. Mhala, M. M., and M. D. Patwardhan, Indian J. Chem., 6, 704 (1968).
942. Mhala, M. M., M. D. Patwardhan, and T. R. Kastari, Indian J. Chem., 7, 145 (1969).
943. Michaelis, A., Ann., 164, 30 (1872).
944. Michaelis, A., Ann., 293, 193 (1896); 294, 1 (1896).
945. Michaelis, A., Ann., 315, 43 (1901).
946. Michaelis, A., Ann., 326, 129 (1903).
947. Michaelis, A., Ber., 5, 4 (1872).
948. Michaelis, A., and A. Flemming, Ber., 34, 1291 (1901).
949. Michaelis, A., and R. Kaehne, Ber., 31, 1048 (1898).
950. Michaelis, A., and G. L. Linke, Ber., 40, 3419 (1907).
951. Michaelis, A., and F. Rothe, Ber., 25, 1747 (1892).
952. Michaelis, A., and F. Wegner, Ber., 48, 316 (1915).
952a. Michalski, J., and T. Modro, Chem. Ber., 95, 1629 (1962).
952b. Michalski, J., T. Modro, and A. Zwierzak, J. Chem. Soc., 1961, 4904.
952c. Michalski, J., and A. Zwierzak, Rocznicki Chem., 38, 97 (1962) [C.A. 57, 12298h (1962)].
953. Michelson, A. M., Chem. and Ind., 1957, 1669.
954. Michelson, A. M., J. Chem. Soc., 1958, 1957.
955. Michelson, A. M., Chem. and Ind., 1958, 1147.
956. Michelson, A. M., J. Chem. Soc., 1958, 2055.

957. Michelson, A. M., J. Chem. Soc., 1959, 1371.
958. Michelson, A. M., J. Chem. Soc., 1959, 3655.
959. Michelson, A. M., L. Szabo, and A. R. Todd, J. Chem. Soc., 1956, 1546.
959a. Michelson, A. M., and A. R. Todd, J. Chem. Soc., 1949, 2476.
959b. Michelson, A. M., and A. R. Todd, J. Chem. Soc., 1949, 2487.
960. Michelson, A. M., and F. R. Todd, J. Chem. Soc., 1953, 951.
960a. Michelson, A. M., and A. Todd, J. Chem. Soc., 1956, 3459.
961. Miescher, K., and J. Heer, U. S. Patent 2,395,934 (1946; to Ciba Pharmaceutical Products, Inc.).
962. Milas, N. A., Davis, P., and Chiang Li-Chin, J. Am. Chem. Soc., 77, 1640 (1955).
963. Miller, D. L., and T. Ukeno, J. Am. Chem. Soc., 91, 3050 (1969).
964. Milobendzki, T., and W. Janezak, Roczniki Chem., 11, 840 (1931).
965. Milobendzki, T., and J. H. Kolitowska, Roczniki Chem., 6, 70, (1926).
966. Milobendzki, T., and J. Walczynska, Roczniki Chem., 8, 486 (1928).
967. Mineree Corp., U. S. Patent 2,434,357 (1948).
968. Miranda, M., An. Acad. Brasil. Cienc., 1967, 39 [C.A. 70, 3208 (1969)].
969. Mitchell, R. S., U. S. Patent 3,205,251 (1965) [C.A. 64, 595f (1966)].
970. Mitsui and Yamamoto (to Nippon Kagaku K.K.), Japan. Patent 2723 (1953) (June 16) [C.A. 49, 2482e (1955)].
971. Mitsunobu, O., T. Obata, and T. Mukaiyama, J. Org. Chem., 30, 1071 (1965).
972. Mitsunobu, O., T. Obata, Y. Sagaya, and T. Mukaiyama, Bull. Chem. Soc. Japan, 38, 2100 (1965).
973. Mitsunobu, O., M. Yamada, and T. Mukaiyama, Bull. Chem. Soc. Japan, 40, 935 (1961).
974. Miyano, M., J. Am. Chem. Soc., 77, 3524 (1955).
975. Miyano, M., and S. Funahashi, J. Am. Chem. Soc., 77, 3522 (1955).
976. Mizuno, Y., and T. Sasacki, Tetrahedron Lett., 1965, 4579.
977. Mizuno, Y., and T. Sasaki, J. Am. Chem. Soc., 88, 863 (1966).
977a. Mizuno, Y., T. Sasaki, T. Kanai, and H. Igarashi, J. Org. Chem., 30, 1533 (1965).
978. Modro, T., and J. Sokolowski, Zeszyty Nauk. Mat. Fiz. Chem. Wyzsza Szkola Pedagog. Gdansk, 5, 77 (1965) [C.A. 66, 2153 (1967)].
978a. Moedrizer, W., L. Maier, and L. C. D. Groenweghe, J. Chem. Eng. Data, 7, 307 (1962) [C.A. 57, 308f (1962)].
979. Moffatt, J. G., Can. J. Chem., 42, 599 (1964).

980. Moffatt, J. G., J. Am. Chem. Soc., 85, 1118 (1963).
981. Moffatt, J. G., and H. G. Khorana, J. Am. Chem. Soc., 79, 1194 (1957).
982. Moffatt, J. G., and H. G. Khorana, J. Am. Chem. Soc., 79, 3741 (1957).
983. Moffatt, J. G., and H. G. Khorana, J. Am. Chem. Soc., 80, 3756 (1958).
983a. Moffatt, J. G., and H. G. Khorana, J. Am. Chem. Soc., 83, 649 (1961).
984. Mohai, M., and E. Upor, Magy. Kem. Foly., 73, 488 (1967) [C.A. 68, 35562 (1968)].
984a. Moll, H., P. W. Schneider, and H. Brintzinger, Helv. Chim. Acta, 47, 1937 (1964).
985. Molotkovskii, Y. G., and L. D. Bergel'son, Izv. Akad. Nauk SSSR, Ser. Khim., 1966, 1098 [C.A. 65, 10656 (1966)].
986. Monsanto Chemical Co., Br. Patent 677,171 (Aug. 13, 1952) [C.A. 47, 10006h (1953)].
987. van Montagu, M., J. Smrt, F. Sorm, Arch. Intern. Physiol. Biochem., 72, 705 (1964).
988. Montgomery, H. A. C., and J. H. Turnbull, Proc. Chem. Soc., 1957, 178.
989. Montgomery, H. A. C., and J. H. Turnbull, J. Chem. Soc., 1958, 1963.
990. Montignie, E., Bull. Soc. Chim. (4), 49, 73 (1931).
991. Moravek, J., J. Kopecky, and J. Skoda, Collect. Czech. Chem. Commun., 33, 4120 (1968).
992. Moravek, J., J. Kopecky, and J. Skoda, Collect. Czech. Chem. Commun., 33, 960 (1968).
993. Moravek, J., Tetrahedron Lett., 1967, 1707.
994. Morel, A., Bull. Soc. Chim. (3), 21, 402 (1899).
995. Morel, A., Compt. Rend., 127, 1024 (1898).
996. Morel, A., Compt. Rend., 128, 508 (1899).
997. Moreton, D. H., and B. Keilin (to Douglas Aircraft Co., Inc.), U. S. Patent 2,694,083 (1954) [C.A. 50, 2667i (1956)].
998. Morris, R. C., and J. L. Van Winkle, U. S. Patent 2,744,128 (1956) [C.A. 52, 1208g (1958)].
999. Morrison, A. L., and F. R. Atherton (to Roche Products Ltd.), Br. Patent 666,596 (1952) [C.A. 46, 11249a (1952)].
1000. Morrison, A. L., F. R. Atherton, and K. J. M. Andress (to Roche Products Ltd.), Br. Patent 684,045 (1952) [C.A. 48, 742b (1954)].
1001. Morrison, A. L., and R. F. Long, J. Chem. Soc., 1958, 211.
1002. Mosher, H. S., J. Reinhart, and H. C. Prosser, J. Am. Chem. Soc., 75, 4899 (1953).
1003. Moshkina, I. M., and A. N. Pudovik, Zh. Obshch. Khim., 32, 1671 (1962) [C.A. 58, 5496b (1963)].
1004. Mostert, S., L. J. Stegerhoek, and P. E. Verkade,

Rec. Trav. Chim., 77, 133 (1958).
1005. Moyle, C. L. (to Dow Chemical Co.), U. S. Patent 2,599,515 (June 3, 1952) [C.A. 46, 8322a (1952)].
1006. Moyle, C. L. (to Dow Chemical Co.), U. S. Patent 2,176,416 (1939).
1007. Moyle, C. L. (To Dow Chemical Co.), U. S. Patent 2,182,817 (1939).
1008. Moyle, C. L. (to Dow Chemical Co.), U. S. Patent 2,223,329 (1940).
1009. Moyle, C. L. (to Dow Chemical Co.), U. S. Patent 2,225,285 (1940).
1010. Moyle, C. L. (to Dow Chemical Co.), U. S. Patent 2,250,049 (1941).
1011. Mukaiyama, T., T. Hata, and O. Mitsunobu, J. Org. Chem., 27, 1815 (1962).
1011a. Mukaiyama, T., T. Hata, and O. Mitsunobe, J. Org. Chem., 28, 481 (1963).
1011b. Mukaiyama, T., and H. Nambu, J. Org. Chem., 27, 2201 (1962).
1012. Mukaiyama, T., T. Obata, and O. Mitsunobo, Bull. Chem. Soc. Japan, 38, 1088 (1965).
1013. Mukiyama, T., T. Hata, and K. Tahasaka, J. Org. Chem., 28, 481 (1963).
1014. Müller, O., Ber., 58, 2106 (1944).
1015. Müller, O., Dissertation, Stuttgart, 1944.
1016. Müller, E., A. Langerbeck, and W. Riedel, Z. Physiol. Chem., 281, 29 (1944).
1017. Müller, B., J. Org. Chem., 30, 1964 (1965).
1018. Müller, D. L., and F. H. Westheimer, J. Am. Chem. Soc., 88, 1507 (1966).
1019. Munoz, A., Double Liaison, nr. 126, 185 (1966) [C.A. 66, 55547 (1967)].
1020. Munoz, A., and J. P. Vives, Compt. Rend., 256, 4017 (1963).
1021. Murakami, Y., and A. E. Martell, J. Am. Chem. Soc., 86, 2119 (1964).
1022. Murakami, M., and T. Nashima, Nippon Kagaku Zasshi, 75, 443 (1954) [C.A. 51, 11242d (1957)].
1023. Murakami, Y., and M. Tagaki, Bull. Chem. Soc. Japan, 40, 2724 (1967).
1024. Mushika, Y., and T. Fujita, Yakagaku Zasshi, 87, 33 (1967) [C.A. 67, 3057 (1967)].
1025. Muskat, I. E., J. Am. Chem. Soc., 56, 2449 (1934).
1026. Mustafa, A., M. M. Sidky, and F. M. Soliman, Ann. Chem., 698, 109 (1966).
1027. Nagase, O., Chem. Pharm. Bull., 15, 648 (1967).
1028. Nakagawa, K., and M. Matsui, Nippon Nogei Kagaku Kaishi, 40, 300 (1968) [C.A. 70, 57582 (1969)].
1028a. Nakagawa, T., and Y. Mori, J. Pharm. Soc. Japan, 75, 1322 (1955).
1029. Naranez, S. A., M. T. Jacob, and H. G. Khorana, J.

Am. Chem. Soc., 87, 2988 (1965).
1030. Naranez, S. A., and H. G. Khorana, J. Am. Chem. Soc., 87, 2981 (1965).
1030a. Navech, J., Ann. Fac. Sci. Univ. Toulouse, Sc. Math. Sc. Phys., 20, 9 (1962) [C.A. 65, 5387g (1966)].
1031. Navech, J., and F. Matnis, Compt. Rend., 246, 2001 (1958).
1032. Nelson, A. K., and A. D. F. Toy, Inorg. Chem., 2, 775 (1963).
1033. Nelson, A. K., and A. D. F. Toy, U. S. Patent 3,146,255 (1964) [C.A. 61, 14530e (1964)].
1034. Nesterov, L. V., and R. I. Mutalapova, Tetrahedron Lett., 1968, 51.
1035. Neuberg, C., Arch. Biochem., 3, 105 (1943).
1036. Neuberg, C., Biochem. Z., 88, 432 (1917).
1037. Neuberg, C., and K. P. Jacobsohn, Biochem. Z., 199, 503 (1928).
1038. Neuberg, C., and K. P. Jacobsohn, Biochem. Z., 199, 510 (1928).
1039. Neuberg, C., and E. Kretschmer, Biochem. Z., 36, 10 (1911).
1040. Neuberg, C., and H. Pollack, Biochem. Z., 23, 515 (1910); 26, 523 (1910); Ber., 43, 2060 (1910).
1041. Neuberg, C., and T. Schuchardt, Enzymolog., 1, 39 (1936).
1042. Neuberg, C., and J. Wagner, Biochem. Z., 171, 489 (1927).
1043. Neuberg, C., and J. Wagner, Jacobsohn, Biochem. Z., 188, 230 (1927).
1044. Neuberg, C., F. Weinman, and M. Vogt, Biochem. Z., 199, 250 (1928).
1045. Neumann, H., Rec. Trav. Chim., 67, 101 (1948).
1045a. Neunhoeffer, O., and W. Maiwald, Chem. Ber., 95, 108 (1962).
1046. van der Neut, J. H., J. H. Uhlenbrock, and P. E. Verkade, Rec. Trav. Chim., 72, 365 (1953).
1047. Newallis, P. E., P. Lombardo, E. E. Gilbert, E. Y. Spencer, and A. Morello, J. Agr. Food Chem., 15, 940 (1967).
1047a. Nielsen, M. L., J. V. Pustinger, Jr., and J. Strobel, J. Chem. Eng. Data, 9, 167 (1967) [C.A. 61, 2618a (1964)].
1048. Nikonorov, K. V., and E. A. Gurylev, Izv. Akad. Nauk SSSR, Ser. Khim., 1967, 1872 [C.A. 68, 29191 (1968)].
1049. Nikonorov, K. V., G. M. Vinokurova, and Z. G. Speranskaya, Khim. i Primenenie Fosfororgan. Soedinii, Akad. Nauk SSSR, Trudy 1[e] Konf., 1955, 223 (publ. 1957) [C.A. 52, 240e (1958)].
1050. Nissan Chemical Ind., Ltd., Japan. Patent 22,066

(1965) (Sept. 30) [C.A. 64, 4996d (1966)].
1051. Nomura, J. Japan. Chem., 3, 145 (1949) [C.A. 46, 3962c (1952)].
1052. Nomura, H., et al., Chem. Pharm. Bull., 17, 387 (1969).
1053. Nomura, H., and Y. Minami, Japan. Patent 68 09,219 (1968) [C.A. 69, 107017 (1968)].
1054. Noone, T. M., Chem. and Ind., 1958, 1512.
1055. Norddeutsche Affinerie und C. F. Spiess & Sohn, Ger. Patent 944,430 (1956) [C.A. 52, 16199h (1958)].
1056. Norton, L. M., and A. A. Noves, Am. Chem. J., 10, 430 (1888).
1057. Nussbaum, A. L., G. Scheuerbrand, and A. M. Duffield, J. Am. Chem. Soc., 86, 102 (1964).
1058. Nussbaum, A. L., and R. Tiberi, Tetrahedron, 20, 2467 (1964).
1059. Nussbaum, A. L., and R. Tiberi, J. Am. Chem. Soc., 87, 2513 (1965).
1060. Nyquist, P. A., Appl. Spectroscopy, 11, 161 (1957).
1061. Nylen, P., Dissertation, Uppsala, 1930.
1062. Nylen, P., and O. Stelling, Z. Anorg. Allgem. Chem., 212, 169 (1933); Nylen, P., ibid., 212, 182 (1933).
1063. Nylen, P., Tids. Kjemi Bergvesen, 18, 59 (1938).
1064. Obata, Y., Bull. Inst. Phys. Chem. Research, Tokyo, 22, 115 (1943).
1065. Obata, T., and T. Mukaiyama, J. Org. Chem., 32, 1063 (1967).
1066. O'Connor, M. J., and W. K. Detweiler (to Union Carbide Corp.), U. S. 2,868,827 (1959) [C.A. 53, 11224e (1959)].
1067. Oestreich, C. H., and M. M. Jones, Biochemistry, 5, 2926 (1966).
1068. Ohwada, K., Appl. Spectrosc., 22, 209 (1968).
1069. Okada, S., O. Nagase, and M. Shimizu, Chem. Pharm. Bull., 15, 713 (1961).
1069a. Oláh, G. A., and A. A. Oswald, Can. J. Chem., 40, 1917 (1962).
1069b. Oláh, G., and A. Oswald, J. Org. Chem., 24, 1568 (1959).
1069c. Oláh, G., and A. Pavláth, Acta Chim. Acad. Sci. Hung., 3, 191 (1953) [C.A. 48, 7533b (1954)].
1070. Oláh, G., A. Pávlath, and G. Hosszáng, Acta Chim. Acad. Sci. Hung., 8, 41 (1955) [C.A. 52, 8139e (1958)].
1071. Olezzo, S., R. Moratti, and M. L. Speranzo, Farmaco (Pavia), Ed. Prat., 21, 319 (1966) [C.A. 65, 7525h (1966)].
1071a. Olin Mathieson Chemical Corp., Belg. Patent 628,795 (1963) [C.A. 60, 9297d (1964)].
1072. Orloff, H. D., C. J. Worrel, and F. X. Markley, J. Am. Chem. Soc., 80, 727 (1958).

1073. Osborne, D. W., J. Org. Chem., 39, 3570 (1964).
1074. Oshima, T., Arch. Biochem. Biophys., 126, 478 (1968).
1075. Oshima, Y., and M. Eto, Japan. Patent 17,395 (1965) [C.A. 63, 17971d (1965)].
1076. Osipov, O. A., V. I. Gaivoronskii, and A. A. Shvets, Primen. Mol. Spektrosk. Khim., Sb. Dokl. Sob. Soveshch., 3rd Krasnoyarsk, USSR, 1964 (publ. 1966) [C.A. 69, 6592 (1968)].
1077. Osterberg, R., Arkiv Kemi, 25, 177 (1965).
1078. Outhouse, E. L., Biochem. J., 30, 197 (1936); 31, 1459 (1937).
1079. Overend, W. G., and M. Stacey, Chem. and Ind., 1952, 952.
1080. Pacák, J., and M. Cerny, Chem. Listy, 51, 1165 (1957) [C.A. 51, 13762g (1957)].
1081. Paecht, M., and A. Katchalsky, Israel. J. Chem., 1, 483 (1963) [C.A. 60, 11872b (1964)].
1082. Paganini, R., Ber., 24, 365 (1891).
1082a. Paikin, D. M., M. P. Shabanova, N. M. Gamper, and L. F. Efimova, Soedinnenii Akad. Nauk SSSR, Kazana Filial, Trudy ler Konf., 1955, 408 [C.A. 52, 4096c (1958)].
1083. Pakula, R., and B. Szczycinski, Acta. Polon. Pharm., 22, 49 (1965) [C.A. 63, 4558e (1965)].
1083a. Pande, K. C., and G. Trampe, J. Org. Chem., 35, 1169 (1970).
1084. Paolini, V., Gazz. Chim Ital., 42, I, 63 (1912).
1085. Nil.
1086. Pascal, P., Bull. Soc. Chim. (4), 33, 1617 (1923).
1087. Pascal, P., Compt. Rend., 176, 1398 (1923).
1088. Paterno, E., and M. Fileti, Gazz. Chim. Ital., 3, 125 (1873).
1089. Parfenov, E. A., G. A. Serebrennikova, and N. A. Preobrazhenskii, Zh. Obshch. Khim., 37, 2363 (1967) [C.A. 68, 86934 (1968)].
1090. Péchiney-Progil, Fr. Patent 1,381,206 (1964) [C.A. 62, 16056c (1965)].
1091. Pelouze, J., Ann. Pharm., 6, 129 (1833).
1092. Pelouze, J., Ann. Chim., 52, 37 (1833).
1093. Pelouze, J., J. Prakt. Chem., 36, 257 (1845).
1094. Peppard, D. F., J. R. Ferraro, and G. W. Mason, J. Inorg. Nucl. Chem., 7, 231 (1959).
1095. Percival, E. E., and E. G. V. Percival, J. Chem. Soc., 1945, 874.
1096. Perkow, W., Chem. Ber., 87, 755 (1954).
1097. Perkow, W., E. W. Krockow, and K. Knoevenagel, Chem. Ber., 88, 662 (1955).
1097a. Perkow, W., K. Ullerich, and F. Meyer, Naturw., 39, 353 (1952).
1098. Peterson, E. A., A. Sober, and A. Meister, Biochem.

Preparations, 3, 34 (1955) [C.A. 48, 11410i (1954)].
1099. Petkovic, D. M., J. Inorg. Nucl. Chem., 30, 603 (1968).
1099a. Petrov, G., A. Sokol'skii, and B. M. Polees, Zh. Obshch. Khim., 26, 3381 (1956) [C.A. 51, 9473i (1956)].
1099b. Petrov, K. A., N. K. Bliznyuk, and V. E. Burygin, Zh. Obshch. Khim., 29, 1486 (1959) [C.A. 54, 8806a (1960)].
1099c. Petrov, K. A., A. I. Gavrilova, and V. P. Korotkova, Zh. Obshch. Khim., 36, 853 (1966) [C.A. 65, 10638g (1966)].
1100. Petrov, K. A., L. V. Khorkhoyanu, V. V. Pazdnev, et al., USSR Patent 210,861 (Feb. 8, 1968) [C.A. 69, 35415 (1968)].
1100a. Petrov, K. A., E. E. Nifant'ev, and L. A. Fedorchuck, Vysokomolekulyarnoye Soedineniya, 2, 417 (1960) [C.A. 54, 24372i (1960)].
1100b. Petrov, K. A., E. E. Nifant'ev, and T. N. Lysenko, Zh. Obshch. Khim., 31, 1709 (1961) [C.A. 55, 23312h (1961)].
1100c. Petrov, K. A., E. E. Nifant'ev, and R. F. Nikitina, Zh. Obshch. Khim., 31, 1705 (1961) [C.A. 55, 22199f (1961)].
1101. Petrov, K. A., E. E. Nifant'ev, I. I. Sopikova, and L. P. Levitan, Zh. Prikl. Khim., 37, 1132 (1964) [C.A. 61, 4237b (1964)].
1101a. Pfitzner, K. E., and J. G. Moffatt, Biochem. Biophys. Res. Commun., 17, 146 (1964).
1102. Pfleiderer, G., C. Woenckhaus, K. Scholz, and H. Feller, Ann. Chem., 675, 205 (1964).
1103. Philipps, D. D., U. S. Patent 3,297,798 (1967) [C.A. 66, 55212 (1967)].
1104. Phillips, D. R., and T. H. Five, J. Am. Chem. Soc., 90, 6803 (1968).
1105. Phytochemical Research Assoc., Japan. Patent 10,911 (1964) [C.A. 61, 14770c (1964)].
1106. Pierre, J., and E. Puchot, Ann., 163, 253 (1872).
1107. Pilgram, K., and H. Ohse, Angew. Chem., 78, 820 (1966).
1107a. Pinkus, A. G., and P. G. Waldrep, Chem. and Ind., 1962, 302.
1107b. Pinkus, A. G., P. G. Waldrep, and W. J. Collie, J. Org. Chem., 26, 682 (1961).
1108. Pippen, E. L., and R. M. McCready, J. Org. Chem., 16, 262 (1951).
1109. Pishchimuka, P., Ber., 41, 3854 (1908).
1110. Pishchimuka, P., J. Prakt. Chem. (2), 84, 746 (1911).
1111. Pishchimuka, P., J. Russ. Phys. Chem. Soc., 44, 1406 (1912).

1112. Pishchimuka, P., J. Russ. Phys. Chem. Soc., 56, 11 (1925).
1113. Pischel, H., and A. Holy, Collect. Czech. Chem. Commun., 34, 89 (1969).
1114. Pizer, F. L., and C. E. Ballou, J. Am. Chem. Soc., 81, 915 (1959).
1115. Plancher, G., Gazz. Chim. Ital., 23, II, 70 (1893).
1116. Plapinger, R. E., and Th. Wagner-Jauregg, J. Am. Chem. Soc., 75, 5757 (1953).
1116a. Plant, H., U. S. Patent 3,132,171 (1964) [C.A. 61, 4482c (1964)].
1117. Plant Science Research Assoc., Japan. Patent 15,999 (1964) [C.A. 62, 7856c (1965)].
1118. Plets, V. M., Zh. Obshch. Khim., 6, 1198 (1936).
1119. Plets, V. M., Zh. Obshch. Khim., 8, 1296 (1938).
1120. Plimmer, R. H. A., Biochem. J., 7, 75 (1913).
1121. Plimmer, R. H. A., Biochem. J., 35, 461 (1941).
1122. Plimmer, R. H. A., and W. J. N. Burch, J. Chem. Soc., 1929, 279, 292.
1123. Plimmer, R. H. A., and W. J. N. Burch, Biochem. J., 31, 398 (1937).
1124. Plueddemann, E. P. (to Food Machinery & Chemical Corp.), U. S. Patent 2,558,380 [C.A. 46, 1024i (1952)].
1125. Ponnamperuma, C., and R. Mack, Science, 148, 124 (1965).
1126. Pontis, H. G., and C. L. Fischer, Methods Enzymol., 8, 125 (1966).
1127. Portes, L., and G. Prunier, Bull. Soc. Chim. France (3), 13, 96 (1895).
1127a. Poshkus, A. C., and J. E. Herweh, J. Am. Chem. Soc., 79, 6127 (1957).
1128. Poshkus, A. C., and J. E. Herweh, J. Am. Chem. Soc., 84, 555 (1962).
1129. Posternak, Th., J. Am. Chem. Soc., 72, 4824 (1950).
1130. Posternak, Th. J. Biol. Chem., 188, 317 (1951).
1131. Posternak, Th., and S. Grafl, Helv. Chim. Acta, 28, 1258 (1945).
1131a. Posternak, Th., and H. Pollaczek, Helv. Chim. Acta, 24, 1190 (1941).
1132. Posternak, Th., and T. P. Rosselet, Helv. Chim. Acta, 36, 1614 (1953).
1133. Potajov, V. K., A. P. Ryskov, and Z. A. Shabarova, Vestn. Mosk. Univ., Ser. II, 21, 115 (1966) [C.A. 66, 38196 (1967)].
1134. Potter, C., Nature, 163, 379 (1949).
1135. Potter, A. L., J. C. Sowden, W. Z. Hassid, and D. Doudoroff, J. Am. Chem. Soc., 70, 1751 (1948).
1136. Poulenc Frères, German Patent 208,700 (1907).
1137. Power, F. B., and F. Tutin, J. Chem. Soc., 87, 251 (1905).

1138. Prahl, W. H., U. S. Patent 2,805,240 (1957) [C.A. 52, 2066d (1958)].
1139. Pratesi, P., Ber., 72, 1459 (1939).
1140. Prestegard, J. H., and S. I. Chan, J. Am. Chem. Soc., 91, 2843 (1969).
1140a. Protsenka, L. D., and K. A. Kornev, Ukr. Khim. Zh., 28, 719 (1962) [C.A. 59, 489a (1963)].
1141. Prystas, M., and F. Sorm, Collection Czech. Chem. Commun., 28, 3113 (1963).
1142. Prystas, M., and F. Sorm, Collection Czech. Chem. Commun., 30, 537 (1965).
1143. Puchir, M., and S. Tryon (to Allied Chemical & Dye Corp.), U. S. Patent 2,562,244 (1951) [C.A. 46, 2563e (1952).
1144. Pudovik, A. N., I. M. Aladzheva, and L. N. Yakovenko, Zh. Obshch. Khim., 35, 1210 (1965) [C.A. 63, 11689f (1965)].
1145. Pudovik, A. N., and L. G. Biktimirova, Zh. Obshch. Khim., 27, 2104 (1957) [C.A. 52, 6156i (1958)].
1146. Pudovik, A. N., E. M. Faizullin, and I. Zhuravlev, Zh. Obshch. Khim., 36, 1454 (1966) [C.A. 66, 10516 (1967)].
1147. Pudovik, A. N., and R. D. Gareev, Zh. Obshch. Khim., 34, 3942 (1964) [C.A. 62, 7191g (1965)].
1147a. Pudovik, A. N., E. I. Kashevarova, and G. L. Goloven'kin, Zh. Obshch. Khim., 34, 3240 (1964) [C.A. 62, 3922b (1965)].
1147b. Pudovik, A. N., and I. U. Konovalova, Zh. Obshch. Khim., 34, 3848 (1964) [C.A. 62, 6390d (1965)].
1148. Pudovik, A. N., and I. V. Konovalova, Zh. Obshch. Khim., 35, 1591 (1965) [C.A. 63, 17887b (1965)].
1148a. Pudovik, A. N., I. V. Konovalova, and L. V. Banderova, Zh. Obshch. Khim., 35, 1206 (1965) [C.A. 63, 13064h (1965)].
1149. Pudovik, A. N., and N. M. Lebedeva, Doklady Akad. Nauk SSSR, 101, 889 (1955) [C.A. 50, 3219e (1956)].
1149a. Pudovik, A. N., and R. N. Platonova, Zh. Obshch. Khim., 29, 505 (1959) [C.A. 54, 254f (1960)].
1149b. Pudovik, A. N., and L. G. Salekhova, Zh. Obshch. Khim., 26, 1431 (1956) [C.A. 50, 14513e (1956)].
1149c. Pummerer, R., E. Buchta, and E. Deimler, Chem. Ber., 84, 583 (1951).
1150. Putman, E. W., and W. Z. Hassid, J. Am. Chem. Soc., 79, 5057 (1957).
1151. Rabe, H., Ber., 22, 392 (1889).
1152. Rabzevich-Subkovskii, I., J. Russ. Phys. Chem. Soc., 44, 153 (1912).
1153. Rae, J. J., H. D. Kay, and E. J. King, Biochem. J., 28, 148 (1934).
1153a. Rafikov, S. R., G. N. Chelnokova, and Y. V. Artemova, Zh. Obshch. Khim., 35, 591 (1965) [C.A. 63,

477h (1965)].
1154. Rakuzin, Arsen'ev, J. Russ. Phys. Chem. Soc., 53, 376 (1921); Chem.-Ztg., 47, 178 (1923).
1155. Raley, C. F., Jr. (to U.S.A.), U. S. Patent 2,844,582 (1958) [C.A. 53, 3250i (1959)].
1156. Ralph, R. K., W. J. Connor, H. Schaller, and H. G. Khorana, J. Am. Chem. Soc., 85, 1983 (1963).
1156a. Ramirez, F., Pure Appl. Chem., 9, 337 (1964); Composés organiques du phosphore (Colloques nationaux du CNRS), Ed. du CNRS, Paris 1966, p. 185.
1157. Ramirez, F., A. V. Patwardhan, N. Ramanathan, N. B. Desai, C. V. Greco, and S. R. Heller, J. Am. Chem. Soc., 87, 543 (1965).
1158. Ramirez, F., A. V. Patwardhan, N. B. Desai, and S. R. Heller, J. Am. Chem. Soc., 87, 549 (1965).
1159. Ramirez, F., S. B. Bhatia, A. V. Patwardhan, E. H. Chen, and C. P. Smith, J. Org. Chem., 33, 20 (1968).
1160. Ramirez, F., S. B. Bhatia, and C. P. Smith, J. Org. Chem., 31, 4105 (1966).
1161. Ramirez, F., S. B. Bhatiz, and C. P. Smith, J. Am. Chem. Soc., 89, 3026 (1967).
1162. Ramirez, F., S. B. Bhatia, and C. P. Smith, Tetrahedron, 23, 2067 (1967).
1163. Ramirez, F., and S. Dershowitz, J. Org. Chem., 22, 856 (1951).
1164. Ramirez, F., and S. Dershowitz, J. Org. Chem., 22, 1282 (1957).
1165. Ramirez, F., and S. Dershowitz, J. Am. Chem. Soc., 81, 587 (1959).
1165a. Ramirez, F., and N. B. Desai, J. Am. Chem. Soc., 82, 2652 (1960); 85, 3252 (1963).
1166. Ramirez, F., N. D. Desai, and N. Ramanathan, J. Am. Chem. Soc., 85, 1874 (1963).
1167. Ramirez, F., S. L. Glaser, A. J. Bigler, and J. P. Pilet, J. Am. Chem. Soc., 91, 496 (1969).
1168. Ramirez, F., B. Hansen, and N. B. Desai, J. Am. Chem. Soc., 84, 4588 (1962).
1169. Ramirez, F., H. J. Kugler, A. V. Patwardhan, and C. P. Smith, J. Org. Chem., 33, 1185 (1968).
1170. Ramirez, F., O. P. Madan, and C. P. Smith, J. Am. Chem. Soc., 87, 670 (1965).
1171. Ramirez, F., A. V. Patwardhan, and C. P. Smith, J. Org. Chem., 30, 2575 (1965).
1172. Ramirez, F., N. Ramanathan, and N. B. Desai, J. Am. Chem. Soc., 84, 1317 (1962).
1173. Ramirez, F., N. Ramanathan, and N. B. Desai, J. Am. Chem. Soc., 85, 3465 (1963).
1173a. Rammler, D. H., and H. G. Khorana, J. Am. Chem. Soc., 84, 3112 (1962).

1174. Rammler, D. H., Y. Lapidot, and H. G. Khorana, J. Am. Chem. Soc., 85, 1989 (1963).
1175. Rapp, M., Ann., 224, 156 (1884).
1176. Raschig, F., German Patent 223,684 (1910).
1177. Raschig, F., German Patent 233,631 (1911).
1178. Ravazzoni, C., and A. Fenaroli, Ann. Chimi. Applicata, 30, 318 (1940).
1179. Raymond, A. L., J. Biol. Chem., 113, 375 (1936).
1180. Razumov, A. I., and Sim-Do-Khen, Zh. Obshch. Khim., 26, 2233 (1956) [C.A. 51, 1818g (1956)].
1180a. Recondo, E., M. Dankert, and S. Passeron, Biochem. Biophys. Acta, 107, 129 (1965).
1181. Regan, P. D., J. A. Stock, and W. J. Hopwood, J. Chem. Soc., C, 1966, 637, 6407.
1182. Reich, W. S., Nature, 157, 133 (1946).
1183. Reichard, P., and N. R. Ringertz, J. Am. Chem. Soc., 79, 2025 (1957).
1184. Reinhardt, J. Am. Chem. Soc., 74, 1093 (1952).
1185. Reithel, F. J., J. Am. Chem. Soc., 67, 1056 (1945).
1186. Reithel, F. J., and C. K. Claycomb, J. Am. Chem. Soc., 71, 3669 (1949).
1187. Rembold, O., Z. Chem., 1866, 651.
1188. Remizov, A. L., Zh. Obshch. Khim., 31, 3769 (1961) [C.A. 57, 9932e (1962)].
1189. Renshaw, R. R., and C. Y. Hopkins, J. Am. Chem. Soc., 51, 953 (1929).
1190. Revel, M., A. Munoz, and J. Navech, Compt. Rend., C, 265, 1053 (1967).
1190a. Revel, M., J. Navech, and J. P. Vives, Bull. Soc. Chim. France, 1963, 2327.
1191. Reverdin, F., and A. Kaufmann, Ber., 28, 3054 (1895).
1192. Rieche, H., G. Hilgetag, and G. Schramm, Angew. Chem., 71, 285 (1959).
1192a. Rieche, H., G. Hilgetag, and G. Schramm, Ger. (East) Patent 21489 (1959) [C.A. 56, 9967a (1962)]; Ger. Patent 1,082,895 (1960) [C.A. 55, 16422c (1961)]; Chem. Ber., 95, 381 (1962).
1193. Riess, J., Bull. Soc. Chim. France, 1965, 18.
1194. Riess, J., Bull. Soc. Chim. France, 1965, 18.
1195. Riley, R. F., J. Am. Chem. Soc., 66, 512 (1944).
1196. Rilling, H. C., and W. W. Epstein, J. Am. Chem. Soc., 91, 1041 (1969).
1197. Rinse, J. (to J. W. Ayers & Co.), U. S. Patent 2,886,586 (1959) [C.A. 53, 17903f (1959)].
1197a. Ripper, W. E., R. M. Greenslade, and L. A. Lickerish, Nature 163, 787 (1949).
1198. Rivaille, P., and L. Szabo, Bull. Soc. Chim. France, 1963, 712.
1199. Rizpolozhenskii, N. I., and F. S. Mukhametov, Izv. Akad. Nauk SSSR, Ser. Khim., 1968, 2755 [C.A. 70,

77229 (1969)].
1200. Rizpoloshenskii, N. I., L. V. Stepashkina, and R. R. Shagidullin, Izv. Nauk SSSR, Ser. Khim., 1967, 2006 [C.A. 68, 29789 (1968)]; ibid., 2013 [C.A. 68, 29790 (1968)].
1201. Roche Products Ltd., Br. Patent 711,442 (1954) [C.A. 49, 11024f (1955)].
1202. Roche Products Ltd., Br. Patent 893,172 (1962) [C.A. 57, 11131e (1962)].
1203. Roche Products Ltd., Morrison and Atherton, Br. Patent 665,315 (1952) [C.A. 46, 11239d (1952)].
1204. Roche Products Ltd., A. L. Morrison, and F. R. Atherton, Br. Patent 675,779 (1952) [C.A. 47, 4911d (1953)].
1205. Roche Products Ltd., A. R. Todd, and F. R. Atherton, Br. Patent 674,087 (Sept. 9, 1952) [C.A. 47, 6436d (1953)].
1206. Roche Products Ltd., Br. Patent 715,183 (1954) [C.A. 49, 13299i (1955)].
1206a. Roesky, H. W., Chem. Ber., 100, 2147 (1967).
1207. Rokhlin, E. M., Y. V. Zeifman, Y. A. Cheburkov, N. P. Gambaryan, and I. L. Knunyants, Dokl. Akad. Nauk SSSR, 161, 1356 (1965) [C.A. 63, 4153g (1965)].
1208. Romieux, C. J., and K. D. Ashley, U. S. Patent 2,266,514 (1941; to American Cyanamid Co.).
1209. de Roos, A. M., and H. J. Toet, Rec. Trav. Chim., 77, 946 (1958).
1209a. Ropuszynski, S., Przem. Chem., 1969, 48 [C.A. 71, 2934 (1969)].
1210. Rose, W. G., J. Am. Chem. Soc., 69, 1384 (1947).
1211. Roseman, S., J. J. Distler, J. G. Moffatt, and H. G. Khorana, J. Am. Chem. Soc., 83, 659 (1961).
1211a. Rosenberg, E. S., Dissertation, Geneva (Switzerland), 1956.
1212. Rosenheim, A., and J. Pinsker, Ber., 43, 2007 (1910).
1213. Rosenheim, A., and M. Pritze, Ber., 41, 2708 (1908).
1214. Rosenheim, A., W. Stadler, and F. Jacobsohn, Ber., 39, 2837 (1906).
1215. Rosenmund, K. W., and H. Vogt, Arch. Pharm., 281, 317 (1943).
1216. Rosnati, L., Gazz. Chim. Ital., 76, 272 (1946).
1217. Rossiiskaya, P. A., and M. I. Kabachnik, Izvestiya Akad. Nauk SSSR, Otdel. Khim. Nauk, 1947, 509.
1218. Rubber Service Laboratories, U. S. Patent 1,840,335 (1932).
1219. Rueggeberg, W. H. C., and J. Chernack, J. Am. Chem. Soc., 70, 1802 (1948).
1220. Rubtsova, I. K., and S. M. Shner, Plasticheskie Massy, 1962, 23 [C.A. 59, 1512e (1963)].

1221. Sachymczyk, W., L. Ménager, and L. Szabó, Tetrahedron, 21, 2049 (1965).
1222. Salzberg, P. L., and J. H. Werntz, U. S. Patent 2,063,629 (1936; to E. I. du Pont Co.).
1223. Sandalova, L. Y., L. I. Mizrakh, and V. P. Evdakov, Zh. Obshch. Khim., 36, 1451 (1966) [C.A. 66, 10627 (1966)].
1224. Sandoz Ltd., Swiss Patent 257,649 (1949) [C.A. 44, 3020g (1950)].
1225. Sandoz Ltd., Swiss Patent 261,806 (1949) [C.A. 44, 3527 (1950)].
1226. Sandoz Ltd., Swiss Patent 375,010 (1964) [C.A. 61, 3075b (1964)].
1226a. Sandoz Ltd., Swiss Patent 422,743; Fr. Patent 1,339,156 (1963) [C.A. 60, 4154h (1964)].
1226b. Sandrin, E., and R. A. Boissonnas, Helv. Chim. Acta, 49, 76 (1966).
1227. Sänger, A., Ann. Chem., 232, 1 (1886).
1228. Sanno, Y., and A. Nohara, Chem. Pharm. Bull., 16, 2056 (1968).
1229. Sanno, Y., A. Nohara, and Y. Kanai, Japan. Patent 67 21,832 (1967) [C.A. 69, 27743 (1968)].
1230. Sarett, L. H. (to Merck & Co., Inc.), U. S. Patent 2,779,775 (1957) [C.A. 51, 8817f (1957)].
1231. Sato, T., and H. Yoshimura, Japan. Patent 3675 (1956) [C.A. 51, 13943h (1957)].
1232. Sato, T., and T. Yoshimura, Japan. Patent 8745 (1955) [C.A. 52, 3875e (1958)].
1233. Sato, T., J. Yoshimura, and T. Takaoka, Proc. Japan Acad., 29, 260 (1953) [C.A. 49, 1063i (1955)].
1234. Sauer, R. O., J. Am. Chem. Soc., 66, 1707 (1944).
1235. Saunders, B. C., and G. J. Stacey, J. Chem. Soc., 1948, 695.
1236. Saunders, B. C., G. J. Stacey, F. Wild, and I. G. E. Wilding, J. Chem. Soc., 1948, 699.
1237. Scott, C. B., J. Org. Chem., 22, 1118 (1957).
1238. Secretant, Bull. Soc. Chim. (3), 15, 361 (1896).
1239. Sedel'nikova, E. A., M. Z. Khabarova, and S. M. Zhenodarova, Izv. Akad. Nauk SSSR, Ser. Khim., 1969, 196 [C.A. 70, 115463 (1969)].
1240. Sedel'nikova, E. A., and S. M. Zhenodarova, Zh. Obshch. Khim., 38, 2254 (1968) [C.A. 70, 68682 (1969)].
1240a. Seel, F., Ger. Patent 1,000,355 (1957) [C.A. 54, 4501d (1960)].
1241. Seegmiller, J. E., and B. L. Horecker, J. Biol. Chem., 192, 175 (1951).
1242. Segre, H., and A. Larizza, Gazz. Chim. Ital., 87, 519 (1957).
1243. Seil, C. A., R. H. Boschan, and J. R. Holder, U. S. Patent 3,341,631 (Sept. 12, 1967) [C.A. 68, 95520

(1968)].
1245. Shamshurin, A. A., O. E. Krivoshchekova, and M. Z. Krimer, Zh. Obshch. Khim., 35, 1877 (1965) [C.A. 64, 1945a (1966)].
1245a. Shaw, G., and D. V. Wilson, J. Chem. Soc., 1963, 1077.
1246. Sheehan, J. C., and V. S. Frank, J. Am. Chem. Soc., 72, 1312 (1950).
1246a. Shefter, E., M. Barlow, R. Sparks, and K. Trueblood, J. Am. Chem. Soc., 86, 1872 (1964).
1247. Shell Internationale Maatschappij N. V., Neth. Appl. 6,403,122 (1964) [C.A. 62, 7638g (1965)].
1247a. Shell Internationale Research Maatschappij, N. V., Ger. Patent 1,111,147 (1959) [C.A. 56, 8637a (1962)].
1248. Shell Internationale Research Maatschapij N. V., Ger. Patent 1,153,373 (1963) [C.A. 60, 2830h (1964)].
1249. Shell Internationale Research Maatschappij, N. V., Neth. Patent 110,895 (1965) [C.A. 63, 1701c (1965)].
1250. Shell Internationale Research Maatschappij, N. V., Ger. Patent 1,212,100 (1966) [C.A. 64, 15799c (1966)].
1251. Shelver, W., M. I. Blake, and C. E. Miller, J. Am. Pharm. Assoc., 47, 72 (1958).
1252. Shimizu, M., O. Nagase, and S. Okada, Japan. Patent 68 24,190 (1968) [C.A. 70, 68701 (1969)].
1253. Shinike, T., J. Shiokawa, and T. Ishino, Nippon Kagaku Zasshi, 88, 944 (1967) [C.A. 68, 83891 (1968)].
1254. Shinskii, N. G., N. N. Preobrazhenskaya, M. G. Ivanovskaya, Z. A. Shabarova, and M. A. Prokof'er, Dokl. Akad. Nauk SSSR, 184, 622 (1969) [C.A. 70, 115478 (1969)].
1255. Shionogi & Co., Ltd., Japan. Patent 4148 (1962) (June 11) [C.A. 61, 8392d (1964)].
1256. Shlyakhtenko, Lebedev, and R. Mandel, Novosti Tekh., 1938, No. 20, 42 [C.A. 32, 8383 (1938)].
1257. Shuman, R. L., U. S. Patent 2,133,310 (1938; to Celluloid Corp.).
1257a. Shvedov, V. P., Y. F. Orlov, and A. M. Shevyakov, Zh. Prikl. Spektroskopii, Akad. Nauk Belorussk. SSR, 2, 59 (1965) [C.A. 63, 5991a (1965)].
1258. Shvets, V. I., L. V. Volkova, S. F. Ryzhenkova, E. E. Lukashenko, and N. A. Preobrazhenskii, Zh. Obshch. Khim., 33, 2876 (1963) [C.A. 61, 1579c (1964)].
1258a. Siddall, T. H., and C. A. Prohaska, J. Am. Chem. Soc., 84, 3467 (1962).
1259. Siddall, T. H., W. E. Stewart, and D. G. Karraker, Inorg. Nucl. Chem., 3, 479 (1967).

1260. Silaev, A. B., A. N. Nesmeyanov, and V. M. Kutuyurin, Zh. Obshch. Khim., 27, 1283 (1957) [C.A. 52, 2749e (1958)].
1261. Simon, A., and C. Stölzer, Chem. Ber., 89, 2253 (1956).
1262. Skoda, J., and J. Moravek, Tetrahedron Lett., 1966, 4167.
1262a. Skrowaczewska, Z., and P. Mastalerz, Rozniki Chem., 29, 415 (1955) [C.A. 50, 7075a (1956)].
1263. Slotboom, A. J., G. H. de Haas, and L. L. M. van Deenen, Rec. Trav. Chim., 82, 469 (1963).
1264. Smeykal, K., H. Baltz, and H. Fischer, J. Prakt. Chem., 22, 186 (1963).
1265. Smith, M., and H. G. Khorana, J. Am. Chem. Soc., 80, 1141 (1958).
1266. Smith, T. D., J. Chem. Soc., 1961, 1122.
1267. Société des Usines chimiques Rhône-Poulenc, German Patent 484,356 (1925).
1268. Société des Usines chimiques Rhône-Poulenc, U. S. Patent 1,598,370 (1926).
1269. Société pour l'industrie chimique à Bâle, Br. Patent 578,551 (1946).
1270. Soc. franc. des Lab. Labaz, Fr. M. 5,782 (1968) [C.A. 71, 3626 (1969)].
1270a. Soborovskii, L. Z., and Y. G. Golobolov, Zh. Obshch. Khim., 34, 1142 (1964) [C.A. 61, 1747b (1964)].
1270b. Soborovskii, L. Z., Y. M. Zinov'ev, and M. A. Englin, Dokl. Akad. Nauk SSSR, 73, 333 (1950) [C.A. 45, 2854c (1951)].
1271. Soell, D., and H. G. Khorana, Angew. Chem., 70, 435 (1964).
1272. Solms, J., and W. Z. Hassid, J. Biol. Chem., 228, 357.
1273. Solmssen, U. V., and J. Lee (to Hoffmann-LaRoche, Inc.), U. S. Patent 2,457,932 (1949) [C.A. 43, 4702d (1949)].
1274. Sosnovsky, G., and E. H. Zaret, Chem. and Ind., 1967, 1297.
1275. Sosnovsky, G., and E. H. Zaret, J. Org. Chem., 34, 968 (1969).
1276. Spener, E. Y., A. Todd, and R. F. Webb, J. Chem. Soc., 1958, 2968.
1276a. Speziale, A. J., and R. C. Freeman, J. Am. Chem. Soc., 82, 903 (1960).
1277. Spiegel, A., Ann. Chem., 219, 1 (1883); Ber., 13, 2219 (1880).
1278. Sraier, V., Jaderna Energie, 11, 373 (1965) [C.A. 64, 1572c (1966)].
1279. Srivastava, P. C., K. L. Nagpal, and M. M. Dhar, Experientia, 25, 356 (1969).
1280. Sugden, S., J. B. Reed, and H. Wilkins, J. Chem.

Soc., 127, 1539 (1925).
1281. Sumitomo Chemical Co., Ltd., Belg. Patent 633,481 (Nov. 4, 1963) [C.A. 61, 9513h (1964)].
1282. Sundaralingam, M., Acta Cryst., 21, 495 (1966).
1283. Nil.
1284. Swarts, F., Rec. Trav. Chim., 28, 166 (1909).
1285. Szabó, P., and L. Szabó, J. Chem. Soc., 1961, 448.
1286. Szabó, P., and L. Szabó, J. Chem. Soc., 1964, 5139.
1287. Szabó, P., and L. Szabó, J. Chem. Soc., 1965, 2944.
1288. Szabo, E., and J. Szabo, Acta Chim. Acad. Sci. Hung., 48, 299 (1966) [C.A. 65, 11411h (1966)].
1289. Szabo, K., U. S. Patent 3,288,889 (1966) [C.A. 66, 37624 (1967)].
1290. Schaeffer, L., Ann., 152, 279 (1869).
1291. Schaller, H., and H. G. Khorana, J. Am. Chem. Soc., 85, 3828 (1963).
1292. Schaller, H., G. Weimann, and H. G. Khorana, J. Am. Chem. Soc., 85, 355 (1963).
1292a. Schaller, H., G. Weimann, B. Lerch, and H. G. Khorana, J. Am. Chem. Soc., 85, 3828 (1963).
1292b. Schaller, H., and H. G. Khorana, J. Am. Chem. Soc., 85, 3828 (1963).
1293. Schaller, H., G. Weimann, B. Lerch, and H. G. Khorana, J. Am. Chem. Soc., 85, 3821 (1963).
1294. Schering Akt. Ges., Ger. Patent 1,007,310 (1957) [C.A. 53, 18917b (1959)].
1295. Schiaparelli, C., Gazz. Chim. Ital., 11, 69 (1881).
1296. Schiff, H., Ann., 102, 334 (1857).
1297. Schmid, H. H. O., and T. Takahaski, Z. Physiol. Chem., 349, 1673 (1968).
1297a. Schmidt, M., and H. Schmidbaur, Angew. Chem., 71, 553 (1959).
1297b. Schmidt, M., H. Schmidbaur, and H. Binger, Chem. Ber., 93, 872 (1960).
1297c. Schmidt, M., H. Schmidbaur, and I. Ruidisch, Angew. Chem., 73, 408 (1961).
1298. Schneider, P. W., and H. Brintzinger, Helv. Chim. Acta, 47, 1717 (1964).
1299. Schoffstall, A. M., and H. Tieckelmann, Tetrahedron, 22, 399 (1966).
1300. Schorre, G., U. S. Patent 3,124,587 (1964) [C.A. 61, 3078g (1964)].
1301. Schrader, G. (to Farbenfabriken Bayer Akt.-Ges.), Br. Patent 806,638 (1958) [C.A. 53, 13497i (1959)].
1302. Schrader, G., Ger. Patent 720,577 (1942); U. S. Patent 2,336,302 (1943); (to I. G. Farbenindustrie A. G.).
1303. Schramm, G., Origin. Prebiol. Systems, Their Mol. Matrices, Proc. Conf. Wakulla Springs, Fla., 1963, 299 [C.A. 63, 3026f (1965)].
1304. Schroder, G., Israeli Patent 8839 (1956) [C.A. 52,

11890i (1958)].
1305. Schroeder, E. W., and E. L. Shokal, U. S. Patent 2,826,592 (1958) [C.A. 52, 12895g (1958)].
1306. Schuler, M., Chimia, 21, 342 (1967).
1307. Schwarze, J. Prakt. Chem. (2), 10, 227 (1874).
1308. Schwartz, A., and C. Ponnamperuma, Nature, 218, 443 (1968).
1308a. Staab, H. A., Ann. Chem., 609, 75 (1957); Staab, H. A., H. Schaller, and F. Cramer, Angew. Chem., 71, 736 (1959).
1309. Stadtmann, E. R., and F. Lipmann, J. Biol. Chem., 185, 549 (1950).
1310. Stamicarbon N. V., Neth. Appl. 67,04,361 (1968) [C.A. 70, 46839 (1969).
1311. Stansly, P. G., J. Org. Chem., 23, 148 (1958).
1312. Starkov, A. V., I. A. Shenkman, M. P. Bogomolov, and Y. P. Volkov, Zh. Obshch. Khim., 35, 352 (1965) [C.A. 62, 13070f (1965)].
1313. Stee, W., A. Zwierzak, and J. Michalski, Tetrahedron Lett., 1968, 5873.
1314. Stegerhoek, L. J., and P. E. Verkade, Rec. Trav. Chim., 75, 467 (1956).
1314a. Stein, S. S., and D. E. Koshland, Arch. Biochem. Biophys., 39, 229 (1952).
1314b. Steinberg, G. M., J. Org. Chem., 15, 637 (1950).
1315. Steinberg, G. M. (to the U.S.A.), U. S. Patent 2,666,778 (1954) [C.A. 49, 6300e (1955)].
1316. Steinkopf, W., W. Mieg, and J. Herold, Ber., 53, 1145 (1920).
1317. Steinkopf, W., and I. Schubart, Ann., 424, 1 (1921).
1318. Steitz, T. A., and W. N. Lipscomb, J. Am. Chem. Soc., 87, 2488 (1965).
1318a. Stetter, H., and K. H. Steinacker, Chem. Ber., 85, 451 (1952).
1318b. Stock, J. A., W. J. Hepwood, and P. D. Regan, J. Chem. Soc., 1966, 637.
1319. Stockx, J., and L. Vandendriessche, Bull. Soc. Chim. Belges, 65, 919 (1956).
1319a. Stölzer, C., and A. Simon, Chem. Ber., 93, 1323 (1960).
1319b. Stölzer, C., and A. Simon, Chem. Ber., 93, 2578 (1960).
1319c. Stölzer, C., and A. Simon, Chem. Ber., 96, 288 (1963).
1320. Strauss, D. B., and J. R. Fresco, J. Am. Chem. Soc., 87, 1364 (1965).
1321. Strauss, D. B., J. Am. Chem. Soc., 87, 1375 (1965).
1322. Strecker, W., and C. Grossmann, Ber., 49, 63 (1916).
1323. Strecker, W., and H. Heuser, Ber., 57, 1364 (1924).
1324. Strecker, W., and R. Spitaler, Ber., 59, 1772 (1926).

1325. Stuart, C. M., J. Chem. Soc., 53, 403 (1888).
1326. Taban, R. L., and J. Fondarai, Bull. Soc. Chim. France, 1956, 459.
1326a. Takeda, Chemical Industries, Ltd., Japan. Patent 23,014 (1961) [C.A. 57, 16729d (1962)].
1327. Takeda Chemical Industries, Ltd., Japan. Patent 3868 (1961) [C.A. 58, 10123c (1963)].
1328. Takeda Chemical Industries Ltd., Japan. Patent 5916 (1962) [C.A. 59, 1489b (1963)].
1329. Takeda Chemical Industries, Ltd., Japan. Patent 4636 (1964) [C.A. 61, 7092d (1964)].
1330. Takeda Chemical Ind. Ltd., Fr. Patent 1,428,186 (1966) [C.A. 64, 10654b (1966)].
1331. Takeda Chemical Industries, Neth. Appl. 6,610,578 (1967) [C.A. 67, 117222 (1967)].
1332. Takeda Chemical Industries Ltd., Fr. Patent 1,504,374 (1967) [C.A. 69, 109820 (1968)].
1333. Takei, S., and Y. Kuwada, Chem. Pharm. Bull., 16, 944 (1968).
1334. Takizaki, K., et al. (to Nitto Chemical Ind. Co.), Japan. Patent 2273 (1952) [C.A. 48, 2084b (1954)].
1335. Tamayo, M. L., and G. Ramon, Anales Real Soc. Españ. Fis. y Quim., 45B, 1123 (1949) [C.A. 44, 8882i (1950)].
1336. Tammelin, L. E., Acta Chem. Scand., 11, 859 (1957).
1336a. Tammelin, L. E., Acta Chem. Scand., 11, 1340 (1957)
1337. Tanaka, T., Yakugaku Zasshi, 79, 437 (1959) [C.A. 53, 16932c (1959)].
1338. Tarasov, V. V., Y. S. Arbisman, N. S. Rylyakova, and Y. A. Kondrat'ev, Zh. Fiz. Khim., 42, 2720 (1968) [C.A. 70, 77083 (1969)].
1339. Tasker, H. S., and H. O. Jones, J. Chem. Soc., 95, 1910 (1909).
1340. Tattrie, and C. S. McArthur, Can. J. Biochem. Biophysiol., 35, 1165 (1957).
1341. Taylor, G. A., and C. E. Ballou, Biochemistry, 2, 47 (1963).
1342. Taylor, G. A., and C. L. Ballou, Biochemistry, 2, 553 (1963).
1343. Tedesso, P. H., V. B. de Rumi, J. A. Gonzalez Quintana, J. Inorg. Nucl. Chem., 29, 2813 (1967).
1344. Tener, G. M., J. Am. Chem. Soc., 83, 159 (1961).
1345. Tener, G. M., and H. G. Khorana, Chem. and Ind., 1957, 562.
1346. Tener, G. M., and H. G. Khorana, J. Am. Chem. Soc., 80, 1999 (1958).
1347. Tener, G. M., R. S. Wright, and H. G. Khorana, J. Am. Chem. Soc., 79, 441 (1957).
1348. Termens, L., AEC Accession No 15413, Rept. no KFK-246 [C.A. 63, 11321c (1965).
1349. Termens, L., and K. H. Schweer, Radiochim. Acta,

9, 125 (1968).
1349a. Thadford, R., et al., J. Med. Chem., 8, 486 (1966).
1349b. Thierry, J. C., R. Weiss, et al., Tetrahedron Lett., 1969, 3757.
1350. Thomas, D. G., J. H. Billman, and C. E. Davis, J. Am. Chem. Soc., 68, 895 (1946).
1350a. Thomas, L. C., and R. A. Chittenden, Spectrochim. Acta, 20, 467 (1964).
1351. Dr. Karl Thomas G.m.b.H. Chemisch-pharmazeutische Fabrik, Ger. Patent 945,236 (1956) [C.A. 52, 16202e (1958)].
1352. Thomas, H. J., K. Hewson, and J. A. Montgomery, J. Org. Chem., 27, 192 (1962).
1353. Tian, A., Bull. Soc. Chim. France, 1946, 407.
1353a. Tichý, V., and S. Truchlík, Chem. Zvesti, 12, 345 (1958) [C.A. 52, 18258a (1958)].
1354. Tikhomirova-Sidorova, N. S., and A. P. Kavunenko, Zh. Obshch. Khim., 38, 1403 (1968) [C.A. 69, 106996 (1968)].
1355. Todd, A. R., Angew. Chem., 60A, 69 (1948).
1356. Todd, A. R., J. Chem. Soc., 1946, 647.
1356a. Todd, A. R., Chem. and Ind., 1956, 802; 1958, 170; Angew. Chem., 70, 527 (1958).
1356b. Tolkmith, H., Ann. N. Y. Acad. Sci., 79, Art. 7, 189 (1959) [C.A. 54, 11607h (1960)].
1357. Tomita, M., J. Pharm. Soc. Japan, 54, 490 (1936).
1358. Tomlinson, R. V., and G. M. Tener, J. Org. Chem., 29, 493 (1964).
1358a. "Topics in Phosphorus Chemistry" (Van Wazer, J. R.; Letcher, J. N.; Mark, V.; Cruchfield, M. M.; Dungan, C. H.), Interscience Publishers, 1968.
1359. Toy, A. D. F., J. Am. Chem. Soc., 66, 499 (1944).
1360. Toy, A. D. F., J. Am. Chem. Soc., 70, 3882 (1948).
1360a. Toy, A. D. F., J. Am. Chem. Soc., 71, 3561 (1949).
1361. Toy, A. D. F., J. Am. Chem. Soc., 72, 2065 (1950).
1362. Toy, A. D. F. (to Victor Chemical Works), U. S. Patent 2,504,165 (1950) [C.A. 44, 5898c (1950)].
1363. Toy, A. D. F. (to Victor Chemical Works), U. S. Patent 2,643,265 (1953) [C.A. 47, 9672g (1953)].
1364. Toy, A. D. F., and Costello, J. R. (Victor Chemical Works), U. S. Patent 2,754,315 (1956) [C.A. 51, 1244e (1957)].
1365. Toy, A. D. F. (to Victor Chemical Works), U. S. Patent 2,683,733 (1954) [C.A. 49, 6989b (1955)].
1366. Trapmann, H., and M. Devani, Naturwissenschaften, 52, 208 (1965).
1367. Trapmann, H., and M. Devani, Arch. Pharm., 298, 253 (1965).
1368. Truchlík, S., Chem. Zvesti, 12, 256 (1958) [C.A. 52, 15458a (1958)].
1369. Tsivanin, V. S., R. G. Ivanova, and G. Kh. Kamai,

Zh. Obshch. Khim., 38, 1062 (1968) [C.A. 69, 106076 (1968)].
1370. Tsizin, Y. S., and N. A. Preobrazhenskii, Zh. Obshch. Khim., 33, 2873 (1963) [C.A. 60, 1579a (1964)].
1371. Ts'O, P. O. P., N. S. Kondo, and M. P. Schweizer, Biochemistry, 8, 997 (1968).
1372. Tsuboi, M., F. Kuriyagawa, K. Matsuo, and K. Koygoku, Bull. Chem. Soc. Japan, 40, 1813 (1967).
1373. Tung, T., Hua Hsueh Hsueh Pao, 27, 160 (1961) [C.A. 59, 12628c (1963)].
1374. Turbak, A. F., and J. R. Livingston, Jr., Ind. Eng. Chem. Prod. Res. Develop. 2, (1963) [C.A. 59, 8581h (1963)].
1375. Turnbull, J. H., Chem. and Ind., 1956, 350.
1376. Turnbull, J. H., and W. Wilson, J. Chem. Soc., 1954, 2301.
1377. Tutin, F., and A. C. O. Hann, J. Chem. Soc., 89, 1749 (1906).
1378. Ueda, T., et al., Japanese Patent 99,857 (1933).
1379. Uhlenbroek, J. H., and P. E. Verkade, Rec. Trav. Chim., 72, 558 (1953).
1379a. Ukita, T., N. A. Bates, and H. E. Carter, J. Biol. Chem., 216, 867 (1955).
1380. Ukita, T., R. Funakoshi, and Y. Hirose, Chem. Pharm Bull., 12, 828 (1964).
1380a. Ukita, C., A. Hamada, and A. Kobata, Chem. Pharm. Bull., 9, 363 (1961).
1381. Ukita, C., and H. Hayatsu, Chem. Pharm. Bull., 9, 1000 (1961).
1382. Ukita, C., and H. Hayatsu, J. Am. Chem. Soc., 84, 1879 (1962).
1383. Ukita, C., K. Nagasawa, and M. Iric, Pharm. Bull., 5, 121, 127 (1957).
1384. Ukita, C., K. Nagasawa, and M. Iric, Pharm. Bull., 5, 208 (1957).
1385. Ukita, C., K. Nagasawa, and M. Iric, Pharm. Bull., 5, 215 (1957).
1386. Ukita, C., K. Nagasawa, and M. Iric, J. Am. Chem. Soc., 80, 1373 (1958).
1387. Ukita, C., N. Imura, K. Nagasawa, and N. Aimi, Chem Pharm. Bull., 10, 1113 (1963).
1387a. Ukita, C., H. Okuyama, and H. Hayatsu, Chem. Pharm. Bull., 11, 1399 (1963).
1388. Ukita, C., Y. Takede, and H. Hayatsu, Chem. Pharm. Bull., 12, 1503 (1964).
1389. Upor, E., and L. Gorbicz, Magy. Kem. Foly., 73, 484 (1967) [C.A. 68, 35535 (1968)].
1390. Upson, R. W. (to E. I. du Pont de Nemours & Co.), U. S. Patent 2,557,805 [C.A. 45, 8298a (1951)].
1391. U. S. Department of Commerce, OTS, P. B. 60,890;

F.I.A.T. Final Report 949.
1392. U. S. Department of Commerce, OTS, P. B. 73,754.
1393. Usines chimiques des Lab. francais, Br. Patent, 697,107 (1952) [C.A. 49, 4824i (1955)].
1394. Vakulova, A., L. N. Fokina, T. S. Fradkina, L. V. Luk'yanova, and G. I. Samokhvalov, Dokl. Akad. Nauk SSSR, 147, 103 (1962) [C.A. 58, 8919b (1963)].
1395. Vanderbilt, B. M., and H. B. Gottlieb, U. S. Patent 2,008,478 (1935; to Victor Chemical Works).
1396. Vanderbilt, B. M., U. S. Patent 2,084,361 (1937; to Victor Chemical Works).
1397. Vanderbilt, B. M., U. S. Patent 2,177,757 (1939; to Commercial Solvents Corp.).
1398. Vanderheiden, B. S., Biochem. Biophys. Res. Commun., 21, 265 (1965).
1398a. Vanderheiden, B. S., Analyt. Biochem., 8, 1 (1964).
1399. Vanderheiden, B. S., and I. Boszormenyi-Nagy, Analyt. Biochem., 13, 496 (1965).
1400. VEB Leuna Werke "Walter Ulbricht", Ger. Patent 1,135,459 (1962) [C.A. 58, 1346h (1963)].
1401. Verheyden, J. P. H., and J. G. Moffatt, J. Am. Chem. Soc., 86, 1236 (1964).
1402. Verkade, P. E., and L. J. Stegerhoek, Ned. Akad. Wetenschap., Proc., 57B, 494 (1954) [C.A. 50, 176g (1956)].
1403. Verkade, P. E., L. J. Stegerhoek, M. A. Hoefnagel, and J. W. Gielkens, Koninkl. Ned. Akad. Wetenschap., Ser. B. 60, 308 (1957) [C.A. 52, 6156a (1958)].
1404. Verkade, P. E., L. J. Stegerhoek, and S. Mostert, Koninkl. Ned. Akad. Wetenschap, Proc., Ser. B, 60, 147 (1957) [C.A. 52, 297e (1958)].
1405. Verkade, P. E., L. J. Stegerhoek, and S. Mostert, Croat. Chem. Acta, 29, 413 (1951) [C.A. 53, 16037i (1959)].
1406. Verkade, P. E., J. C. Stoppelenburg, and W. Cohen, Rec. Trav. Chim., 59, 886 (1940).
1407. Verkade, P. E., and J. H. Uhlenbrock, Proc. Kon. Nederland. Akad. Wetenschap. Ser. B, 55, no 110 (1950) [C.A. 47, 3234f (1953).
1408. Vilsmeier, W., Ger. Patent 1,272,283 (1968) [C.A. 69, 76632 (1968)].
1409. Viscontini, M., M. Bonetti, C. Ebnöther, and P. Karrer, Helv. Chim. Acta, 34, 1384, 1388 (1951).
1410. Viscontini, M., and C. Olivier, Helv. Chim. Acta, 36, 466 (1953).
1411. Viscontini, M., and J. Pudles, Helv. Chim. Acta, 33, 594 (1950).
1412. Vives, J. P., and F. Mathis, Compt. Rend., 246, 1879 (1958).
1413. Vögeli, F., Jahresber., 1847-8, 694.
1414. Vögeli, F., Ann., 69, 180 (1849).

1415. Vogel, A. I., and D. M. Cowan, J. Chem. Soc., 1943, 16.
1416. Vogel, A. I., J. Chem. Soc., 1948, 1833.
1417. Vogt, M., Biochem. Z., 211, 8 (1929).
1418. Volkova, M. I., Khim. i Primenenie Fosfororg. Soedinn. Akad. Nauk SSSR, Kazana Filial, Trudy le Konf., 1955, 420 (publ. 1957) [C.A. 52, 4095g (1958)].
1419. Volkova, N. V., I. I. Semenyuk, and A. A. Yasnikov, Ukr. Khim. Zh., 33, 712 (1967) [C.A. 68, 49908 (1968)].
1420. Volkova, N. V., and A. Yasnikov, Ukr. Khim. Zh., 31, 119 (1965) [C.A. 62, 14516d (1965)].
1421. Vol'kovich, Kuskov, & Koroteeva, Akad. Nauk. SSSR, Otdel Khim. Nauk, 1954, 5 [C.A. 49, 6839c (1955)].
1422. Voronkov, M. G., Zh. Obshch. Khim., 25, 469 (1955) [C.A. 50, 2418d (1956)].
1423. Voronkov, M. G., V. A. Kolesova, and V. N. Zgonnik, Izv. Akad. Nauk SSSR, Otdel Khim. Nauk, 1957, 1363 [C.A. 52, 7128e (1958)].
1424. Voronkov, M. G., and V. N. Zgonnik, Zh. Obshch. Khim., 27, 1483 (1957) [C.A. 52, 3569g (1958)].
1425. Vvedenskii, W., J. Russ. Phys. Chem. Soc., 20, 29 (1888).
1426. Wacker-Chemie G.m.b.H., Ger. Patent 900,091 (1955) [C.A. 52, 16198a (1958)].
1427. Wada, T., H. Takagi, S. Miyazawa, Y. Suzuki, and H. Minakami, Bitamin, 22, 342 (1961) [C.A. 62, 2956b (1965)].
1427a. Wadsworth, W. S., and W. D. Emmons, J. Am. Chem. Soc., 84, 610 (1962).
1427b. Wadsworth, W. S., and W. D. Emmons, J. Org. Chem., 29, 2816 (1964).
1428. Wagner-Jauregg, T., Ber., 68, 670 (1935).
1429. Wagner-Jauregg, T., and H. Arnold, Ber., 70, 1459 (1937).
1430. Wagner-Jauregg, T., and H. Griesshaber, Ber., 70, 1 (1937).
1431. Wagner-Jauregg, T., and H. Griesshaber, Ber., 70, 8 (1937).
1432. Wagner-Jauregg, T., and T. Lennartz, Kothny, Ber., 74, 1513 (1941).
1432a. Wagner-Jauregg, T., and T. Lennartz, Ber., 75, 178 (1942).
1433. Wagner-Jauregg, T., and A. Wildermuth, Ber., 77, 481 (1944).
1434. Walczynska, J., Roczniki Chem., 6, 110 (1926).
1434a. Walsh, E. N., J. Am. Chem. Soc., 81, 3023 (1959).
1435. Walter-Levy, L., Bol. Inf. cient. nucl., 9, no 77, 54 (1958) [C.A. 51, 219g (1957)].
1436. Ward, L. F., Jr., U. S. Patent 3,369,062 (1968)

[C.A. 68, 95336 (1968)].
1437. Ward, L. F., Jr., and D. D. Phillips, U. S. Patent 3,069,313 (1962) [C.A. 59, 7432b (1963)].
1438. Wasag-Chemie A. G., Ger. Patent 1,194,858 (1965) [C.A. 63, 13315d (1965)].
1439. Van Wazer, J. R., and S. Nerval, J. Am. Chem. Soc., 88, 4415 (1966).
1440. Weger, F., Ann., 221, 61 (1883).
1441. Weijlard, J., J. Am. Chem. Soc., 63, 1160 (1941); Weijlard, J., and H. Tauber, J. Am. Chem. Soc., 60, 2263 (1938).
1442. Weijlard, J., J. Am. Chem. Soc., 64, 2279 (1942).
1443. Weil-Malherbe, H., Biochem. J., 34, 980 (1940).
1444. Weimann, G., and H. G. Khorana, J. Am. Chem. Soc., 84, 4329 (1962).
1444a. Weimann, G., and H. G. Khorana, Chem. and Ind., 1962, 271.
1445. Weimann, G., H. Schaller, and H. G. Khorana, J. Am. Chem. Soc., 85, 3828 (1963).
1446. Weiss, B., J. Am. Chem. Soc., 79, 5553 (1957).
1447. Werner, A., Ann., 321, 248 (1902).
1448. Werntz, J. H. (to E. I. du Pont de Nemours & Co.), U. S. Patent 2,563,506 (1951) [C.A. 46, 1579i (1951)].
1449. Wertyporoch, E., and H. Kiekenberg, Biochem. Z., 268, 8 (1934).
1449a. Westheimer, F. H., Accounts Chem. Res., 1, 70 (1968).
1450. Westphal, O., and R. Stadler, Angew. Chem., 75, 452 (1963).
1451. Whitehill, L. N., and R. S. Barker, U. S. Patent 2,394,829 (1946; to Shell Oil Co.).
1452. Wichelhaus, H., Ann. Suppl., 6, 257 (1868).
1453. Wicker, W., H. P. Walter, and E. Thilo, Chem. Ber., 97, 2385 (1964).
1454. Wieland, Th., and H. Aquila, Chem. Ber., 101, 3031 (1968).
1455. Wieland, Th., and H. Aquila, Chem. Ber., 102, 2285 (1969); Angew. Chem., Int. Ed., 7, 213 (1968).
1455a. Wieland, Th., and E. Bäuerlein, Mh. Chem., 98, 1381 (1967); Naturwiss., 54, 80 (1967).
1455b. Wieland, Th., and E. Bäuerlein, Chem. Ber., 100, 3869 (1967).
1456. Wilcox, R. D., G. H. Harris, and R. S. Olson, Tetrahedron Lett., 1968, 6001.
1457. Wiley, D. W., and H. E. Simmons, J. Org. Chem., 29, 1876 (1964).
1458. Williamson, A. G., Proc. Roy. Soc., 7, 18 (1854).
1459. Willstätter, R., and K. Lüdecke, Ber., 37, 3753 (1904).
1460. Wilson, A. N., and S. A. Harris, J. Am. Chem. Soc.,

73, 4693 (1951).
1461. Winssinger, C., Bull. Soc. Chim., 48, 111 (1887).
1462. Witt, E. R., U. S. Patent 3,363,033 (Jan. 9, 1968) [C.A. 69, 10232 (1968)].
1463. Witten, B., and J. I. Miller, J. Am. Chem. Soc., 70, 3886 (1948).
1464. Wittersheim, H., G. Dirheimer, and J. P. Ebel, Bull. Soc. Pharm. Strasbourg, 7, 23 (1964) [C.A. 64, 17908d (1966)].
1464a. Wittmann, R., Angew. Chem., 74, 214 (1962).
1465. Wittmann, R., Chem. Ber., 96, 771 (1963).
1466. Woenckhaus, C., M. Volz, and G. Pfleiderer, Z. Naturforsch., 19b, 467 (1964).
1467. Wolffenstein, R., Ber., 20, 1966 (1887).
1468. Wolffenstein, R., Ber., 21, 1186 (1888).
1469. Wolfrom, M. L., and D. E. Pletcher, J. Am. Chem. Soc., 63, 1050 (1941); Wolfrom, M. L., C. S. Smith, D. E. Pletcher, and A. E. Brown, J. Am. Chem. Soc., 64, 23 (1942).
1470. Wolfrom, M. L., and D. I. Weisblat, J. Am. Chem. Soc., 62, 1149 (1940).
1471. Wolkow, A., Z. Chem., 1870, 322.
1472. Woodstock, W. H., and P. E. Pelletier, U. S. Patent 2,410,118 (1946; to Victor Chemical Co.).
1473. Working, E. B., and A. C. Andrews, Chem. Rev., 29, 245 (1941).
1474. Wright, R. S., and H. G. Khorana, J. Am. Chem. Soc., 77, 3423 (1955).
1475. Wright, R. S., and H. G. Khorana, J. Am. Chem. Soc., 78, 811 (1956).
1476. Wright, R. S., and H. G. Khorana, J. Am. Chem. Soc., 80, 1994 (1958).
1477. Wülfing, J. A., Ger. Patent 205,579 (1908).
1478. Yagi, T., S. A. El-Kinaroy, and A. A. Benson, J. Am. Chem. Soc., 85, 3462 (1963).
1478a. Yakubovich, A. Y., and V. A. Ginsburg, Zh. Obshch. Khim., 22, 1534 (1952) [C.A. 47, 9254 (1953)].
1479. Yamamoto, T., Japan. Patent 3269 (1951) [C.A. 47, 4898b (1953)].
1480. Yamazaki, A., J. Kumashiro, and T. Takenishi, J. Org. Chem., 33, 2585 (1968).
1481. Yarmukhametova, D. K., and I. V. Cheplanova, Izv. Akad. Nauk SSSR, Ser. Khim., 1964, 1998 [C.A. 62, 7667g (1965)].
1482. Yasnopol'skii, V. D., Zh. Obshch. Khim., 39, 582 (1969) [C.A. 71, 38251 (1969)].
1483. Yonemitsu, O., K. Miyashita, Y. Ban, and Y. Kanaoka Tetrahedron, 25, 95 (1968).
1484. Yoshida, Z., S. Yoneda, A. Nakamura, and K. Fukui, Kogyo Kagaku Zasshi, 69, 644 (1966) [C.A. 66, 10538 (1967)].

1485. Yoshikawa, M., and T. Kato, Bull. Chem. Soc. Japan, 40, 2849 (1967).

1486. Yoshikawa, M., T. Kato, and T. Takenishi, Tetrahedron Lett., 1967, 5065.

1487. Yoshino, M., F. Monden, T. Ikeda, and S. Mukai, Kogyo Kagaku Zasshi, 68, 1689 (1963) [C.A. 64, 4925a (1966)].

1488. Yuan, C., and K. Blcoh, J. Biol. Chem., 234, 2605 (1959).

1489. Yvernault, Th., and Ph. Couillaut, Compt. Rend., C, 265, 652 (1967).

1490. Zaharia, A. J., Bulet, Soc. Sci. Bucuresci, 4, 131 (1895).

1491. Zaharia, A. J., Jahresber., 1896, 1176.

1492. Zal'kind, Yu. S., and E. G. Dmitrieva, Plastich. Massy, 1934, No. 4, 28.

1493. Zecchini, F., Gazz. Chim. Ital., 24, I, 34 (1894).

1494. Zeile, K., and W. Kruckenberg, Ber., 75, 1127 (1942).

1495. Zemlicka, J., and S. Chadek, Tetrahedron Lett., 1969, 715.

1495a. Zemlicka, J., J. Smrt, and F. Sorm, Tetrahedron Lett., 1962, 392.

1496. Zemlicka, J., J. Smrt, and F. Sorm, Collection Czech. Chem. Commun., 28, 241 (1963).

1497. Zemlicka, J., and J. Smrt, Tetrahedron Lett., 1964, 2081.

1497a. Zenftman, H., Am. Chem. Soc. Div. Paints etc., Preprints, 18, No 2, 361 (1958) [C.A. 57, 6110g (1962)].

1498. Zenftman, H., E. Whitworth, and H. R. Wright (to ICI Ltd.), Br. Patent 713,757 (1954) [C.A. 49, 15961g (1955)].

1499. Zervas, L., Naturwiss., 27, 317 (1939).

1500. Zetsche, F., and F. Aeschlimann, Helv. Chim. Acta, 9, 708 (1926).

1501. Zetsche, F., and M. Nachmann, Helv. Chim. Acta, 8, 943 (1925).

1502. Zetsche, F., and M. Nachmann, Helv. Chim. Acta, 9, 420 (1926).

1503. Zetsche, F., and E. Zurbrugg, Helv. Chim. Acta, 9, 297 (1926).

1504. Zimmer & Co., Ger. Patent 115,920 (1900).

1505. Zimmermann, C., Ann., 175, 1 (1875).

1506. Zincke, T., and O. Fuchs, Ann., 267, 1 (1892).

1507. Zincke, T., and J. Kempf, Ber., 44, 418 (1911).

1508. Zincke, T., and F. Küster, Ber., 24, 927 (1891).

1509. Zioudrou, C., Tetrahedron, 18, 197 (1962).

1510. Zwierzak, A., Bull. Acad. Polon. Sci., Ser. Sci. Chim., 11, 333 (1963) [C.A. 61, 1571h (1964)].

Chapter 16. Phosphorus(V)-Nitrogen Compounds with Phosphorus in Coordination Number 4

E. FLUCK AND W. HAUBOLD

Institut für Anorganische Chemie der Universität Stuttgart, Germany

A. SYNTHETIC ROUTES*

Reactions which lead merely to a change in the nitrogen containing ligand on the phosphorus atom are, in general, not included in this chapter. Reactions of this type are described, for example, in Refs. 267, 893. However, the list of compounds does include compounds prepared by such reactions.

I. $PX_5 + N\lessgtr$ (X = hal, OAr)

Reactions of compounds containing phosphorus in coordination number 5 with amines, amides, imides and nitriles:

$$PCl_5 + H_2NAr \xrightarrow{-HCl} ArHN\text{-}PCl_4 \xrightarrow{-HCl} ArN{=}PCl_3$$

$$(ArO)_5P + H_2NAr \longrightarrow (ArO)_3P{=}NAr + 2\ HOAr$$

$$(ArO)_3PCl_2 + 3\ H_2NAr \longrightarrow (ArO)_3P{=}NAr + 2\ [H_3NAr]Cl$$

The reaction of equi-molar quantities of an aromatic amine or its hydrochloride with PCl_5 in carbon tetrachloride results in the formation of N-aryl-trichlorophosphinimides. At lower temperatures the reaction can be accomplished with a two-molar amount of the amine. With weakly basic amines the products are monomeric N-aryl-trichlorophosphinimides, whereas with strongly basic amines dimeric compounds are formed.[935,936] N-alkyl-trichlorophosphinimides can also be prepared by heating phosphorus pentachloride and the alkylamine hydrochloride in tetrachloroethane to reflux temperature.[154a] On heating pentaaryloxyphosphorane with an equivalent amount of a primary aromatic amine to 140-150 °C, N-aryl-triaryloxyphosphinimide is formed.[933] N-Phenyl-triphenoxyphosphinimide is prepared by heating triphenoxyphosphorus dichloride with three moles of aniline to 120-130 °C.[933]

*For nomenclature used, see J. Chem. Soc. 1952, 5122.

$$PCl_5 + RCO\text{-}NH_2 \longrightarrow RCO\text{-}N{=}PCl_3 + 2\ HCl$$

$$PCl_5 + RC(OR){=}NH \cdot HCl \longrightarrow RCO\text{-}N{=}PCl_3 + 2\ HCl + RCl$$

$$PCl_5 + RO\text{-}CO\text{-}NH_2 \longrightarrow RO\text{-}CO\text{-}N{=}PCl_3 + 2\ HCl$$

$$PCl_5 + R\text{-}SO_2\text{-}NH_2 \longrightarrow R\text{-}SO_2\text{-}N{=}PCl_3 + 2\ HCl$$

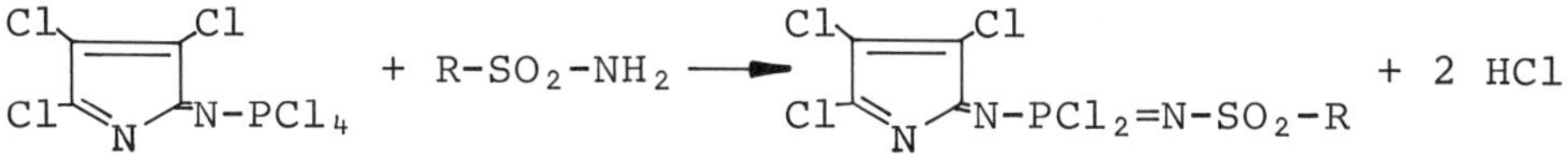

$$P(OAr)_5 + R\text{-}SO_2\text{-}NH_2 \longrightarrow R\text{-}SO_2\text{-}N{=}P(OAr)_3 + 2\ ArOH$$

$$Cl_2P(OAr)_3 + R\text{-}SO_2\text{-}NH_2 \longrightarrow R\text{-}SO_2\text{-}N{=}P(OAr)_3 + 2\ HCl$$

$$PCl_5 + R_2N\text{-}CO\text{-}NH_2 \longrightarrow Cl_3P{=}N\text{-}CO\text{-}NR_2 + 2\ HCl$$

$$2\ PCl_5 + H_2N\text{-}CO\text{-}NH_2 \longrightarrow Cl_3P{=}N\text{-}CO\text{-}N{=}PCl_3 + 4\ HCl$$

$$2\ Cl_2P(OAr)_3 + H_2N\text{-}SO_2\text{-}NH_2 \longrightarrow (ArO)_3P{=}N\text{-}SO_2\text{-}N{=}P(OAr)_3 + 4\ HCl$$

Carboxylic acid amides react with phosphorus pentachloride to form N-acyl-trichlorophosphinimides.[120,435,452] While aromatic carboxylic acid amides always react according to the above equation, aliphatic carboxylic acid amides only react in this way when an electronegative group, e.g., Cl^-, NO_2^-, or an aryl-group is bonded to the carbon atom and to the carbonyl group.[226] The same compounds are also obtained when carboxylic acid imide alkylesters of the respective hydrochlorides are allowed to react with PCl_5 under reduced pressure.[202a] Equimolar amounts of carbamic acid alkyl ester and PCl_5 react nearly quantitatively within 24 hr at lower temperatures to give the thermally unstable N-carbalkoxy-trichlorophosphinimides[455a] (see Section XV). In a similar manner to that used for carboxylic acid amides, practically all compounds which contain an unsubstituted sulfamide group can be obtained. With the cleavage of hydrogen chloride, N-sulfonyl-trichlorophosphinimides[442,518] or N-sulfonyl-amidodichlorophosphinimides are formed with equivalent amounts of phosphorus pentachloride or amido-phosphorus tetrachloride.[797] The reactions can be carried out, mostly in very good yield, without a solvent or in chlorinated hydrocarbons. If pentaphenoxyphosphorane is used instead of phosphorus pentachloride, the triphenoxy derivative is formed with the splitting off of phenol. The same products are obtained when triphenylphosphorus dichloride is allowed to react with sulfonic acid amides at 160 °C.[256,933] N,N-diphenylurea and PCl_5 react nearly quantitatively at 70-80 °C to give N,N-diphenyl-carbamyl-trichloro-phosphinimide.[203] In urea[461] and in sulfamide[605a] both amino groups are available for analogous reactions.

$$PCl_5 + (ArO)_2P(O,S)NH_2 \longrightarrow (ArO)_2P(O,S)\text{-}N{=}PCl_3 + 2\ HCl$$

$$P(OAr)_5 + (RO)_2P(O,S)NH_2 \longrightarrow (RO)_2P(O,S)\text{-}N{=}P(OAr)_3 + 2\ ArOH$$

$$PCl_5 + R\text{-}SO_2\text{-}NH\text{-}P(O)(OAr)_2 \longrightarrow R\text{-}SO_2\text{-}N{=}P(OAr)_2Cl + OPCl_3 + HCl$$

$$PCl_5 + R\text{-}SO_2\text{-}NH\text{-}P(O)Cl_2 \longrightarrow R\text{-}SO_2\text{-}N{=}PCl_3 + HCl + OPCl_3$$

$$PCl_5 + R\text{-}SO_2\text{-}NH\text{-}P(O)Cl_2 \longrightarrow R\text{-}S(O)(Cl){=}N\text{-}P(O)Cl_2 + HCl + OPCl_3$$

N-unsubstituted diaryl phosphoramidates and phosphoramidothionates react with phosphorus pentachloride to give N-diaryloxyphosphoryl (or N-diaryloxythiophosphoryl)-trichloro-phosphinimides.[469] Pentaphenoxyphosphorane reacts in an identical way at 140-150 °C under reduced pressure.[933] Equi-molar quantities of diaryl N-sulfonyl-phosphoramidate and phosphorus pentachloride react smoothly at higher temperatures to produce N-sulfonyl-diaryloxychloro-phosphinimide.[525] N-aryl-sulfonyl-phosphoramidic dichloride forms, on heating with an equi-molar amount of phosphorus pentachloride, N-aryl-sulfonyl-trichlorophosphinimide when the aryl group contains a second-order substituent. N-phenyl-, N-methyl-phenyl-, or N-chlorophenyl-sulfonylamide-dichloride, however, gives compounds of the type $RS(O)Cl{=}N\text{-}P(O)Cl_2$.[520,523]

$$ArCH_2CN + 3\ PCl_5 \longrightarrow Ar\text{-}CCl_2\text{-}CCl_2\text{-}N{=}PCl_3 + 2\ HCl + 2\ PCl_3$$

Nitriles react with excess phosphorus pentachloride to give trichlorophosphinimide compounds in which all aliphatic hydrogen atoms are replaced by chlorine atoms. The chlorination can be partially or completely repressed by using smaller amounts of PCl_5.[312a,787,799]

IIa. $OPX_3 + N\leqslant$ (X = hal)

Reactions of phosphoryl trihalides with ammonia, amines, aldimines and hydrazine:

$$OPCl_3 + 2\ R_2NH \longrightarrow R_2N\text{-}P(O)Cl_2 + [R_2NH_2]Cl$$

$$OPCl_3 + 4\ R_2NH \longrightarrow (R_2N)_2P(O)Cl + 2\ [R_2NH_2]Cl$$

$$OPCl_3 + 6\ R_2NH \longrightarrow OP(NR_2)_3 + 3\ [R_2NH_2]Cl$$

Phosphorus oxychloride reacts with two moles of a primary or secondary amine to give phosphoramidic dichlorides. The reaction is generally vigorous and so is carried out in an inert solvent such as benzene, ether, or chloroform.[573,821] The reactions with four moles of a primary or secondary amine to give phosphorodiamidic halides and with six moles of amine to give phosphoric triamides are similar. Only aromatic amines can be used without a solvent.

For the preparation of phosphoric triamides it is necessary to heat the reaction mixture under increased pressure for a long time in order to complete the reaction. In all of the forementioned reactions it is possible to use a tertiary base, e.g., pyridine or triethylamine, in place of half of the amine. For the reactions with lower aliphatic amines, ammonia, which does not react as rapidly with phosphorus oxychloride as the amine, can be used as acid acceptor. Finally, in the preparation of aromatic phosphoric triamides, NaOH or Na_2CO_3 can be used as acid acceptor. The solutions are used in excess and the NaOH solution must be at least 25%, the sodium carbonate at least 30-40%.

Phosphoramidic dibromides can be synthesized in analogous ways to those used for phosphoramidic dichlorides. In contrast to the chlorides, the bromides are, in general, not distillable without decomposition.

$$OPCl_2F + 2\ R_2NH \longrightarrow R_2N\text{-}P(O)ClF + [R_2NH_2]Cl$$

$$OPF_3 + 2\ R_2NH \longrightarrow R_2N\text{-}P(O)F_2 + [R_2NH_2]F$$

$$OPF_3 + 4\ R_2NH \longrightarrow (R_2N)_2P(O)F + 2\ [R_2NH_2]F$$

$$OPClF_2 + 2\ RNH_2 \longrightarrow RNH\text{-}P(O)F_2 + [RNH_3]Cl$$

$$OPBrF_2 + 2\ RNH_2 \longrightarrow RNH\text{-}P(O)F_2 + [RNH_3]Br$$

Phosphorus oxyfluoride reacts, even at room temperature with 2 or 4 moles of a primary or secondary amine to give phosphoramidic difluoride and phosphorodiamidic fluoride, respectively.[656,658] Also, mixed phosphoryl chloride fluorides react rapidly at 0° to 20 °C with primary or secondary amines. For the preparation of phosphoramidic chloride fluorides and also bromide fluorides by the reactions of phosphoryl-dichloride-fluoride or phosphoryl-dibromide-fluoride it is necessary to use very dilute solutions.[656,821] In more concentrated solutions, phosphorodiamidic fluoride is formed simultaneously. N-aryl phosphoramidic difluorides can be prepared in good yields by the reaction of two moles of an aromatic amine with phosphoryl-chloride-difluoride or bromide-difluoride.[656]

$$OPCl_3 + 6\ H_2N-NHAr \longrightarrow OP(NH-NHAr)_3 + 3\ H_2N-NHAr\cdot HCl$$

Phosphorus oxychloride forms with 6 moles of an aryl-hydrazine, in ether or chloroform, phosphoric-tri-[N'-aryl-hydrazide]; pyridine can also be used as hydrogen chloride acceptor.[69]

$$OPCl_3 + R-N=CH-CH_2-R' \longrightarrow \underset{\underset{O=PCl_2}{|}}{R-N-\overset{\overset{Cl}{|}}{C}H-CH_2-R'} \overset{-HCl}{\longrightarrow} \underset{\underset{O=PCl_2}{|}}{R-N-CH=CH-R'}$$

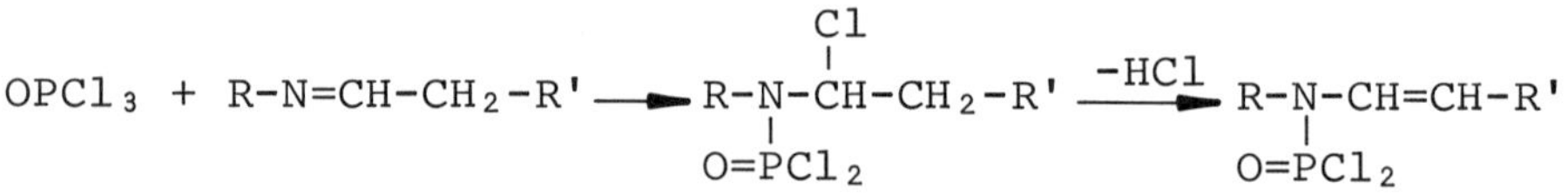

In the presence of a tertiary amine, aldimines react with phosphoryl chloride to form N,N-dialkyl phosphoramidic dichlorides, in which a vinyl group is bonded to nitrogen.[121,122]

IIb.   SPX$_3$ + N$\leqslant$ (X = hal)

Reactions of thiophosphoryltrihalides with ammonia, amines and hydrazines:

$$SPF_3 + 2\ R_2NH \longrightarrow (R_2N)P(S)F_2 + [R_2NH_2]F$$

$$SPCl_3 + 2\ R_2NH \longrightarrow (R_2N)P(S)Cl_2 + [R_2NH_2]Cl$$

$$SPBr_3 + 2\ R_2NH \longrightarrow (R_2N)P(S)Br_2 + [R_2NH_2]Br$$

$$SPF_2Cl + 2\ R_2NH \longrightarrow (R_2N)P(S)F_2 + [R_2NH_2]Cl$$

$$SPCl_2F + 2\ R_2NH \longrightarrow (R_2N)P(S)ClF + [R_2NH_2]Cl$$

$$SPBr_2F + 2\ R_2NH \longrightarrow (R_2N)P(S)BrF + [R_2NH_2]Br$$

$$SPCl_3 + 2\ H_2NR + 4\ NaOH \longrightarrow (RNH)_2P(O)SNa + 3\ NaCl + 3\ H_2O$$

Thiophosphoryl halides react with primary and secondary aliphatic amines primarily to give N-alkyl or N,N-dialkyl phosphoramidic dihalides.[658,821] With secondary amines the reactions can be performed generally without a solvent. With primary aliphatic amines it is preferable to use very dilute solutions otherwise the formation of triamides is preferred. It is possible to use a tertiary base such as pyridine or triethylamine in place of half of the amine. Primary aromatic amines also react in ether solution when the reaction mixture is allowed to stand at room temperature for a long time.[658] In mixed thiophosphoryl chloride-fluorides or bromide-fluorides a chlorine or bromine atom reacts first when the reaction with 2 moles of a primary or secondary amine is carried out in benzene solution below 0 °C.[657,658,821] In this way phosphoramidothioic chloride fluorides or bromide fluorides can be obtained. Also aromatic amines react with thiophosphoryl dichloride fluoride, when left at room temperature for several days. In an analogous manner phosphoramidothioic bromide fluorides can be prepared from thiophosphoryl dibromide fluoride. In excess dilute sodium hydroxide solution, thiophosphoryl chloride reacts with primary aromatic amines to give the sodium salts of the N,N'-diaryl phosphorodiamidothionic acid, from which the free compounds can be obtained upon acidification. N,N',N"-triaryl phosphorothioic triamides are formed as by-products.

$$SPCl_3 + 4\ HNR_2 \longrightarrow (R_2N)_2P(S)Cl + 2\ [R_2NH_2]Cl$$

$$SPCl_2F + 4\ HNR_2 \longrightarrow (R_2N)_2P(S)F + 2\ [R_2NH_2]Cl$$

As previously stated, thiophosphoryl chloride reacts smoothly with 2 moles of a primary or secondary amine to give, in the first stage, phosphoramidothioic dichlorides. By further treatment with amines the remaining two chlorine atoms react but less selectively, so that with four moles of amine, as well as the expected phosphorodiamidothioic chloride, phosphoramidothioic dichloride and thiophosphoric triamide are obtained.[513] With primary aromatic amines especially at higher temperatures, diaminodithiocyclodiphosphazanes are easily formed. In contrast thiophosphoryl-dichloride fluoride reacts with both aliphatic and aromatic amines to give phosphorodiamidothioic fluorides in very good yields.[657]

$$SPCl_3 + 6\ HNR_2 \longrightarrow SP(NR_2)_3 + 3[R_2NH_2]Cl$$

Symmetrical thiophosphoric triamides are formed by the reaction of at least 6 moles of a primary or secondary amine with thiophosphoryl chloride.[69,513] The substitution of the first halogen atom is rapid but in order to achieve full substitution the mixture must be heated for a long time.

Primary aliphatic amines generally react in ether solution. Secondary aliphatic and primary aromatic amines are mostly made to react by heating to 100 °C without a solvent.[513] In the last case diaminodithiocyclodiphosphazanes are easily formed by prolonged heating. Their formation can be achieved when the reaction of thiophosphoryl chloride with 3 moles of amine is carried out in excess 10% sodium hydroxide solution. N,N'-diaryl-phosphorodiamidothioic acid in the form of its water soluble salt is, of course, present as a by-product.[77]

$$SPCl_3 + 6\ H_2N\text{-}NHR \longrightarrow SP(NH\text{-}NHR)_3 + 3\ RNH\text{-}NH_2\cdot HCl$$

Thiophosphoryl chloride reacts with 6 moles of a monosubstituted hydrazine to give thiophosphoric tris(-N'-substituted hydrazide).[580,858]

$$\text{III.}\quad OPX_3 + [-\overset{|}{\underset{|}{N}}-]^+ \quad (X = \text{hal})$$

Reactions of phosphoryl trihalides with salts of ammonia and amines:

$$OPCl_3 + [R_2NH_2]Cl \longrightarrow (R_2N)P(O)Cl_2 + 2\ HCl$$

$$OPCl_3 + 2\ [R_2NH_2]Cl \longrightarrow (R_2N)_2P(O)Cl + 4\ HCl$$

$$OPCl_3 + 3\ [R_2NH_2]Cl \longrightarrow (R_2N)_3P(O) + 6\ HCl$$

In particular, lower monoalkylammonium chlorides and lower dialkylammonium chlorides react with excess phosphoryl chloride, on long boiling, to give N-alkyl phosphoramidic dichlorides and N,N-dialkyl phosphoramidic dichlorides, respectively, in very good yields.[573,601] Also monoarylammonium chlorides react readily in a similar manner.[529,573] Excess phosphoryl chloride or a solvent such as benzene or xylene is used as reaction medium. The most suitable method for the preparation of phosphorodiamidic chlorides is as follows: Phosphoryl chloride is first allowed to react with 2 moles of amine, to give the phosphoramidic dichloride and then after removal of the solvent the reaction mixture is allowed to react with more amine at a higher temperature of about 200 °C.

Like phosphoryl chloride, phosphoryl bromide reacts with aniline; the product, phosphorodiamidic bromide, is not distillable and is very acid sensitive.[611]

Finally, phosphoric triamides can also be synthesized by the reaction of alkyl- or arylammonium chlorides with phosphorus pentachloride when the reaction temperature is high enough and the reaction time long enough. The reaction is accelerated by the addition of concentrated phosphoric acid.[127]

$$2 \ OPCl_3 \ + \ 4 \ ArNH_2 \cdot HCl \longrightarrow \ \begin{array}{c} O \diagdown \quad \diagup NHAr \\ \quad P \\ Ar-N \diagup \quad \diagdown N-Ar \\ \quad P \\ ArHN \diagup \quad \diagdown O \end{array} \ + \ 10 \ HCl$$

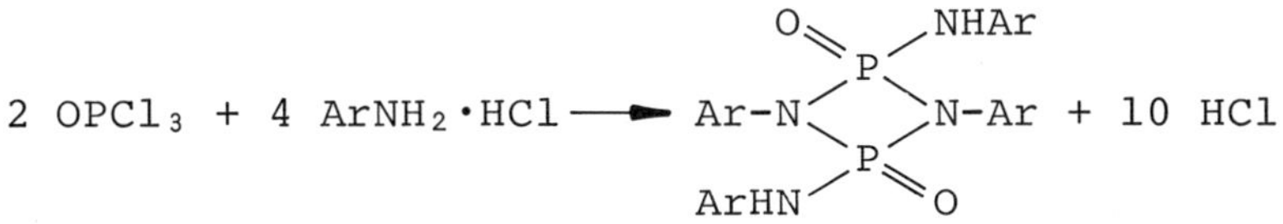

By the prolonged heating of equi-molar amounts of phosphoryl chloride and a primary aromatic amine hydrochloride, a dichlorodioxocyclodiphosphazane is formed which eventually, by higher temperatures and a further quantity of a primary or secondary amine hydrochloride, is converted to a diaminodioxocyclodiphosphazane.[660]

$$SPCl_3 \ + \ [R_2NH_2]Cl \longrightarrow R_2N-P(S)Cl_2 \ + \ 2 \ HCl$$

$$2 \ SPCl_3 \ + \ 2 \ [ArNH_3]Cl \longrightarrow \ \begin{array}{c} S \diagdown \quad \diagup Cl \\ \quad P \\ Ar-N \diagup \quad \diagdown N-Ar \\ \quad P \\ Cl \diagup \quad \diagdown S \end{array} \ + \ 6 \ HCl$$

$$2 \ SPCl_3 \ + \ 4 \ [ArNH_3]Cl \longrightarrow \ \begin{array}{c} S \diagdown \quad \diagup NHAr \\ \quad P \\ Ar-N \diagup \quad \diagdown N-Ar \\ \quad P \\ ArHN \diagup \quad \diagdown S \end{array} \ + \ 10 \ HCl$$

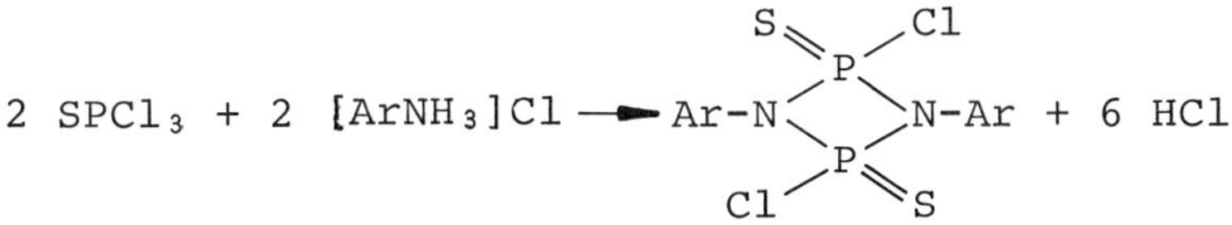

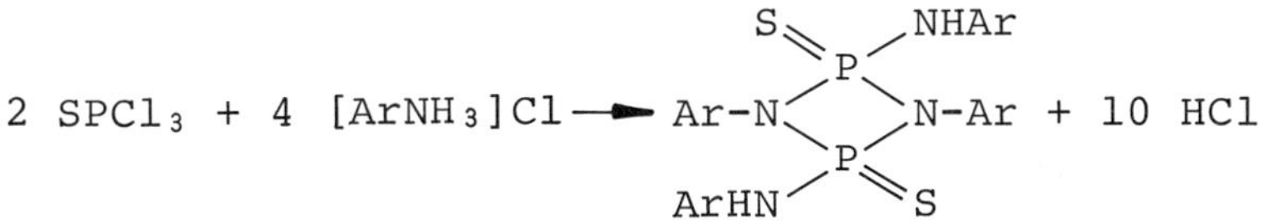

Thiophosphoryl chloride reacts with the hydrochlorides of primary or secondary aliphatic amines, dissolved in excess thiophosphoryl chloride, in the warm to give N-alkyl or N,N-dialkyl phosphoramidothioic dichlorides in good yields.[658] Also, N-aryl phosphoramidothioic dichlorides can be synthesized in this manner. With the hydrochlorides of primary aromatic amines on several days heating to reflux temperature thiophosphoryl chloride forms cyclic 1,3-diaryl-2,4-dichloro-2,4-dithio-2,4,1,3-cyclodiphosphazane. 1,3-diaryl-2,4-diarylamino-2,4-dithio-2,4,1,3-cyclodiphosphazane is present as a by-product even when thiophosphoryl chloride is used in excess.[576]

IVa. $OP(X_2)- \ + \ N\leqslant$ $(X = hal)$

Reactions of phosphoryl dihalo compounds with ammonia, amines, hydrazine and substituted hydrazines:

$$(RO)P(O)Cl_2 \ + \ 2 \ HNR_2' \longrightarrow (RO)P(O)Cl(NR_2') \ + \ [R_2'NH_2]Cl$$

$$(RO)P(O)ClF \ + \ 2 \ HNR_2' \longrightarrow (RO)P(O)F(NR_2') \ + \ [R_2'NH_2]Cl$$

Esters of phosphorodichloridic acid are very reactive and, with two moles of a primary or secondary amine in the presence of ether or benzene as reaction medium, give esters of phosphoramidochloridic acid.[370]   In esters of phosphorochloridofluoridic acid, the chlorine atom is much more reactive than the fluorine atom so that exclusively phosphoramidofluoridates are formed.[821]

$$(RO)P(O)Cl_2 + 4 R_2'NH \longrightarrow (RO)P(O)(NR_2')_2 + 2 [R_2'NH_2]Cl$$

Phosphorodichloridates react with ammonia and primary or secondary amines to give phosphorodiamidates.[51] For the preparation of unsubstituted diamides it is preferable to conduct the reaction in liquid ammonia.[333]   Aryl phosphorodichloridates can also be used in concentrated aqueous ammonia, but the yields are frequently lower.[826] In the reactions with primary or secondary amines a tertiary base such as triethylamine or pyridine can be used instead of one half of the amount of amine.

Unsymmetrical phosphorodiamidates are prepared from amines or an amine and ammonia and the dichloride in two stages (see Section Va).

$$2 \ (ArO)P(O)Cl_2 + 6 \ H_2NR' \longrightarrow ArO-P\underset{\underset{R'}{\overset{|}{N}}}{\overset{\overset{R'}{\overset{|}{N}}}{\diagdown}}P-OAr + 4 \ [R'NH_3]Cl$$

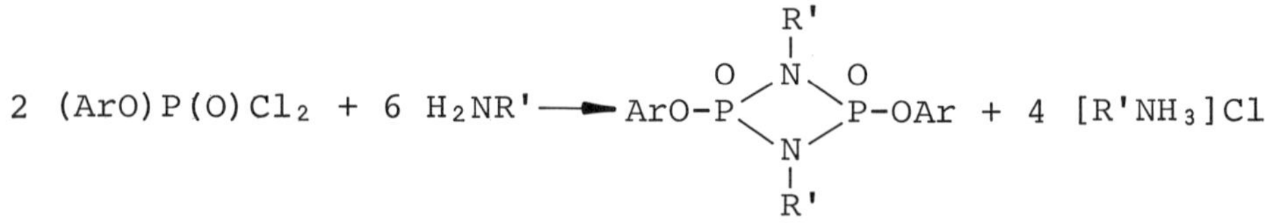

Aryl phosphorodichloridates react, at higher temperatures, with 3 moles of a primary aliphatic or aromatic amine to yield 2,4-aryloxy-2,4-dioxo-2,4,1,3-cyclodiphosphazanes. Half of the requisite amine can be replaced by a tertiary base. In some cases the application of pressure is necessary to force the reaction to go to completion.[100a]

$$(R_2N)P(O)ClF + 2 R_2'NH \longrightarrow (R_2N)(R_2'N)P(O)F + [R_2'NH_2]Cl$$

Unsymmetrical phosphorodiamidofluoridates can be obtained by treatment of phosphoramidic chloride fluorides with amines.[821] Also, in phosphoramidic dicyanides a cyanide group can be replaced by an amino group by reaction with a primary or secondary amine in an inert solvent at about 40 °C.[767]

$$R-CO-NH-P(O)Cl_2 + 4 H_2NR' \longrightarrow$$
$$R-CO-NH-P(O)(NHR')_2 + 2 [R'NH_3]Cl$$

$$RO\text{-}CO\text{-}NH\text{-}P(O)Cl_2 + 2\ HNR'_2 \longrightarrow RO\text{-}CO\text{-}NH\text{-}P(O)(NR'_2)_2 + 2\ HCl$$

$$R\text{-}NH\text{-}CO\text{-}NH\text{-}P(O)Cl_2 + 2\ HNR'_2 \longrightarrow R\text{-}NH\text{-}CO\text{-}NH\text{-}P(O)(NR'_2)_2 + 2\ HCl$$

$$Ar\text{-}C(Cl)\text{=}N\text{-}P(O)Cl_2 + 6\ H_2NAr' \longrightarrow Ar\text{-}C(NHAr')\text{=}N\text{-}P(O)(NHAr')_2 + 3\ [Ar'NH_3]Cl$$

N-acyl-phosphoramidic dichlorides form, on treatment with 4 moles of a primary or secondary amine, N-acyl-phosphoric triamides.[222] N-(dichlorophosphoryl)-carbamic acid esters react with ammonia and primary or secondary amines to give N-(diaminophosphoryl)-carbamic acid esters.[83a,661] N-(dichlorophosphoryl)-ureas react in an analogous way with amines to give N-(diaminophosphoryl)-ureas.[294,485,661]

Finally N-(dichlorophosphoryl)-benzimidochlorides react with 6 moles of an aromatic amine to yield N-(bis-arylaminophosphoryl)-N'-aryl-benzamidines.[211]

$$R\text{-}SO_2\text{-}NH\text{-}P(O)Cl_2 + 5\ NH_3 \longrightarrow [R\text{-}SO_2\text{-}N\text{-}P(O)(NH_2)_2]NH_4 + 2\ NH_4Cl$$

$$R\text{-}SO_2\text{-}NH\text{-}P(O)Cl_2 + 4\ HNR'_2 \longrightarrow R\text{-}SO_2\text{-}NH\text{-}P(O)(NR'_2)_2 + 2\ [R'_2NH_2]Cl$$

N-arylsulfonyl-phosphoramidic dichlorides are converted by gaseous ammonia in an inert solvent to the ammonium salts of N-arylsulfonyl phosphoric triamides, which, on acidification, give the free N-arylsulfonyl phosphoric triamides (see Section XVII).[429a] Secondary aliphatic amines react to give N,N,N',N'-tetraalkyl-N"-arylsulfonyl phosphoric triamides.[447]

$$Cl_2P(O)\text{-}O\text{-}P(O)Cl_2 + 8\ HNR_2 \longrightarrow (R_2N)_2P(O)\text{-}O\text{-}P(O)(NR_2)_2 + 4\ [R_2NH_2]Cl$$

$$ClP(O)F\text{-}O\text{-}P(O)FCl + 4\ R_2NH \longrightarrow R_2N\text{-}P(O)F\text{-}O\text{-}P(O)F(NR_2) + 2\ [R_2NH_2]Cl$$

$$(ArO)_2P(O)\text{-}NH\text{-}P(O)Cl_2 + 4\ H_2NAr' \longrightarrow (ArO)_2P(O)\text{-}NH\text{-}P(O)(NHAr')_2 + 2\ [Ar'NH_3]Cl$$

Diphosphoryl tetrachloride reacts almost quantitatively with water-free secondary aliphatic amines, under cooling, to give tetraamides of diphosphoric acid.[332] The chlorine atoms in diphosphoryl dichloride-difluoride are much more reactive than the fluorine atoms so that, in the first step, diphosphoric acid-P,P'-diamide-P,P'-

difluoride is formed exclusively.[822] Imido-diphosphoric acid-P,P-diarylester-P',P'-dichlorides react with amines to give imido-diphosphoric acid-P,P-diarylester-P',P'-diamides.[469]

$$(ArO)P(O)Cl_2 + 4 H_2N-NH_2 \longrightarrow$$
$$(ArO)P(O)(NH-NH_2)_2 + 2 H_2N-NH_2 \cdot HCl$$

$$2 (ArO)P(O)Cl_2 + 6 H_2N-NH_2 \longrightarrow$$

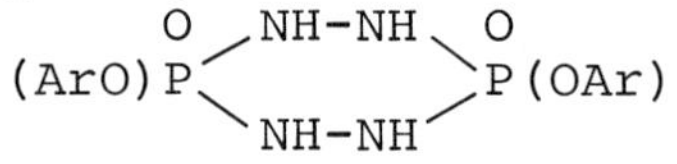

$$(ArO)P \underset{NH-NH}{\overset{O \quad NH-NH \quad O}{<\qquad>}} P(OAr) + 4 H_2N-NH_2 \cdot HCl$$

$$(R_2N)P(O)Cl_2 + 4 H_2N-NHAr \longrightarrow$$
$$(R_2N)P(O)(NH-NHAr)_2 + 2 ArNH-NH_2 \cdot HCl$$

Aryl phosphorodichloridates react only with water-free hydrazine in ether or with 80% hydrazine hydrate in alcohol to give aryl phosphorodihydrazates, when the hydrazine is used in a sufficient excess.[475] With 2 moles of hydrazine in an aqueous medium, however, 3,6-diaryloxy 3,6-dioxo-3,6,1,2,4,5-diphospha-tetrazane is obtained. Phosphoramidic dichlorides also react with aqueous hydrazine to give cyclic products.[123] In contrast, phenylhydrazine produces in good yields phosphoramidic di (N'-phenyl) hydrazides.

IVb. $SP(X_2)^- + N\lesseqgtr$ (X = hal)

Reactions of thiophosphoryl dihalo compounds with ammonia, amines, hydrazine and substituted hydrazines:

$$(ArO)P(S)Cl_2 + 3 NH_3 + H_2O \longrightarrow$$
$$(ArO)P(O)(SH)(NH_2) + 2 NH_4Cl$$

$$(R_2N)P(S)Cl_2 + 4 NH_3 \longrightarrow (R_2N)P(S)(NH_2)_2 + 2 NH_4Cl$$

With excess aqueous ammonia, O-aryl phosphorodichloridothioates give N-unsubstituted O-aryl phosphoramidothiolates when the reaction mixture is heated.[75] In the cold the products are, in contrast, O-aryl phosphordiamidothionates. By addition of phosphoramidothioic dichlorides to liquid ammonia the products are phosphorothioic triamides with two unsubstituted amido groups.[333] The separation of the reaction mixture is achieved by extraction of the phosphorus compound with chloroform or extraction of the ammonium chloride with liquid ammonia.

$$(RO)P(S)Cl_2 + 2 HNR_2' \longrightarrow (RO)P(S)(NR_2')Cl + [R_2NH_2]Cl$$

$$(RO)P(S)FCl + 2\ HNR'_2 \longrightarrow (RO)P(S)F(NR'_2) + [R'_2NH_2]Cl$$

O-alkyl and O-aryl esters of phosphorodichloridothionic acid produce O-alkyl or O-aryl esters of phosphoramidochloridothionic acid on treatment with a primary or secondary amine in ether or benzene solution at about 0° C.[105] When O-alkyl or O-aryl phosphorochloridofluoridates are used only the more reactive chlorine atoms are substituted.[655] With dimethylamine the reaction can even be carried out in aqueous acetone solution.[657]

$$(RO)P(S)Cl_2 + 4\ HNR'_2 \longrightarrow (RO)P(S)(NR'_2)_2 + 2[R'_2NH_2]Cl$$

$$(R_2N)P(S)Cl_2 + 4\ HNR'_2 \longrightarrow (R_2N)P(S)(NR'_2)_2 + 2[R'_2NH_2]Cl$$

$$(R_2N)_2P(O)\text{-}O\text{-}P(S)Cl_2 + 4\ HNR'_2 \longrightarrow (R_2N)_2P(O)\text{-}O\text{-}P(S)(NR'_2)_2 + 2[R'_2NH_2]Cl$$

Phosphorodichloridothionates react more slowly and less vigorously with ammonia and primary or secondary amines than the respective sulfur-free phosphorodichloridates. However, the reactions do proceed smoothly to give phosphorodiamidothionates. Half of the amount of amine shown in the above equations can also be replaced by a tertiary amine or ammonia. For the preparation of N-unsubstituted amides, concentrated aqueous ammonia at lower temperatures,[75] gaseous ammonia in water-free solvents or liquid ammonia is allowed to interact with the dichloride.[333] Primary or secondary aliphatic amines are usually allowed to react in an inert solvent. By stepwise reaction with different amines, unsymmetrical O-aryl phosphorodiamidothionates can also be synthesized. With primary aromatic amines in the absence of a further acid acceptor only one chlorine atom can be substituted by an amino group, while, in the presence of aqueous alkali, disubstitution occurs. With 1,2-diamines such as ethylenediamine cyclic products are formed.[76] Phosphoramidothioic dichlorides react with amines to give phosphorothioic triamides. This reaction type is especially useful for the synthesis of unsymmetrical phosphorothioic triamides. Diphosphorothioic acid bis-(dialkylamide)-dichlorides react readily with primary or secondary aliphatic amines to produce diphosphorothioic acid tetraamides with, if necessary, different types of amide groups.[863]

$$(ArO)P(S)Cl_2 + 2\ H_2N\text{-}NR_2 \longrightarrow (ArO)P(S)Cl(NH\text{-}NR_2) + H_2N\text{-}NR_2\cdot HCl$$

$$(ArO)P(S)Cl_2 + 4\ H_2N\text{-}NH_2 \longrightarrow (ArO)P(S)(NH\text{-}NH_2)_2 + 2\ H_2N\text{-}NH_2 \cdot HCl$$

$$(RNH)P(S)Cl_2 + 2\ H_2N\text{-}NHR' \longrightarrow (RNH)P(S)(NH\text{-}NHR')_2 + 2\ HCl$$

O-aryl phosphorodichloridothionates react at or below 0 °C with 2 moles of an N-mono-substituted or an N,N-disubstituted hydrazine to give solely the O-aryl phosphorohydrazidothionic chloride.[105] Only by the reaction of at least 4 moles of hydrazine or an N-substituted hydrazine the respective phosphorodihydrazidothionates can be synthesized.[105,475a] Both chlorine atoms in phosphoramidothioic dichlorides react clearly with hydrazine compounds.

Va. OP(X)= + N$\lessgtr$ (X = hal)

Reactions of phosphoryl monohalo compounds with ammonia, amines, hydrazine, N-substituted hydrazines, amides and imides:

$$(RO)_2P(O)X + 2\ HNR^1_2 \longrightarrow (RO)_2P(O)(NR^1_2) + [R^1_2NH_2]X \quad X = hal$$

Diesters of phosphorochloridic, -bromidic and -iodidic acid react readily with ammonia, primary or secondary amines to give diesters of phosphoramidic acid.[746] Diesters of phosphorofluoridic acid are less reactive but can, in some cases, be converted with strongly basic amines. The reaction with ammonia is most conveniently performed by passing gaseous ammonia through a solution of the phosphorochloridate in an inert solvent. Similarly, the reactions with primary or secondary amines are carried out in an inert solvent as reaction medium. Instead of one mole of the amine a tertiary base such as pyridine or, also, ammonia can be used.

$$(RO)P(O)(SR')Cl + R''_2NH \longrightarrow (RO)P(O)(SR')(NR''_2) + HCl$$

O,S-diesters of phosphorochloridothiolic acid react with amines to give O,S-diesters of phosphoramidothiolic acid.[137]

$$2\ (R'O)_2P(O)Cl + H_2N\text{-}P(O)(OR)_2 \longrightarrow (R'O)_2P(O)\text{-}N[P(O)(OR')_2]\text{-}P(O)(OR)_2 + 2\ HCl$$

$$2\ (R'O)_2P(O)Cl + H_2N\text{-}P(O)(NR_2)_2 \longrightarrow (R'O)_2P(O)\text{-}N[P(O)(OR')_2]\text{-}P(O)(NR_2)_2 + 2\ HCl$$

$$2\ (R'_2N)_2P(O)Cl + H_2N\text{-}P(O)(NR_2)_2 \longrightarrow (R'_2N)_2P(O)\text{-}\underset{(R'_2N)_2\text{-}P(O)}{\underset{|}{N}}\text{-}P(O)(NR_2)_2 + 2\ HCl$$

$$(R''_2N)_2P(O)Cl + HNR'\text{-}P(O)(OR)_2 \longrightarrow (R''_2N)_2P(O)\text{-}NR'\text{-}P(O)(OR)_2 + HCl$$

$$(RO)(R''_2N)P(O)Cl + HNR'\text{-}P(O)(OR)_2 \longrightarrow (RO)(R''_2N)P(O)\text{-}NR'\text{-}P(O)(OR)_2 + HCl$$

$$(RO)_2P(O)Cl + HNR'\text{-}P(O)(NR_2)_2 \longrightarrow (RO)_2P(O)\text{-}NR'\text{-}P(O)(NR_2)_2 + HCl$$

$$(RO)(R'_2N)P(O)Cl + HNR''\text{-}P(O)(OR)(NR'_2) \longrightarrow (RO)(R'_2N)P(O)\text{-}NR''\text{-}P(O)(OR)(NR'_2) + HCl$$

N-unsubstituted phosphoramidic acid derivatives give, on treatment with a two molar quantity of a phosphorochloridic derivative, the nitrilo-triphosphoric acid. The reaction proceeds only slowly in the presence of pyridine as acid acceptor even at higher temperatures, while in the presence of sodium or sodium hydride the reaction proceeds more rapidly at lower temperatures.[737a] N-monosubstituted phosphoramidic acid derivatives react with a molar quantity of a phosphorochloridate to give derivatives of imidodiphosphoric acid. The yield from these reactions is not good and seldom exceeds 50%.[26]

$$(RO)_2P(S)Cl + NaNR'\text{-}P(O)(OR)_2 \longrightarrow (RO)_2P(S)\text{-}NR'\text{-}P(O)(OR)_2 + NaCl$$

$$(RO)_2P(O)Cl + NaNR'\text{-}P(S)(OR)_2 \longrightarrow (RO)_2P(O)\text{-}NR'\text{-}P(S)(OR)_2 + NaCl$$

$$(R''O)_2P(O)Cl + NaNR'\text{-}P(O)(OR)_2 \longrightarrow (R''O)_2P(O)\text{-}NR'\text{-}P(O)(OR)_2 + NaCl$$

$$(R_2N)_2P(O)Cl + NaNR'\text{-}P(O)(NR_2)_2 \longrightarrow (R_2N)_2P(O)\text{-}NR'\text{-}P(O)(NR_2)_2 + NaCl$$

N-monosubstituted diesters of phosphoramidic acid react, in inert solvents such as petroleum, ether, or benzene in the presence of a tertiary amine, pyridine, sodium, or sodium hydride to form N-substituted imido diphosphoric acid tetraesters. In an analogous manner imido diphosphoric acid-tetraesters, in which one half of the two phosphoric acid groups has a sulfur atom bonded to it, or imido diphosphoric acid tetraamides can be obtained.[46]

$$(ArO)_2P(O)Cl + H_2N\text{-}CS\text{-}NH_2 \longrightarrow (ArO)_2P(O)NH\text{-}CS\text{-}NH_2 + HCl$$

$$(ArO)_2P(O)Cl + NaNH\text{-}SO_2\text{-}R \longrightarrow (ArO)_2P(O)\text{-}NH\text{-}SO_2\text{-}R + NaCl$$

Thiourea can be phosphorylated by diarylphosphorochloridates in aqueous sodium hydroxide solution. The sodium salts of sulfonic acid amides react to give diaryl N-sulfonyl phosphoramidates.[710]

$$(RO)_2P(O)Cl + 2\ H_2N\text{-}NHR \longrightarrow (RO)_2P(O)\text{-}NH\text{-}NHR + RNH\text{-}NH_2\cdot HCl$$

$$(RO)_2P(O)Cl + H_2N\text{-}NH\text{-}P(O)(OR)_2 \longrightarrow (RO)_2P(O)\text{-}NH\text{-}NH\text{-}P(O)(OR)_2 + HCl$$

$$(RO)(R_2N)P(O)Cl + 2\ H_2N\text{-}NR'_2 \longrightarrow (RO)(R_2N)P(O)\text{-}NH\text{-}NR'_2 + H_2N\text{-}NR'_2\cdot HCl$$

$$(R_2N)_2P(O)Cl + H_2N\text{-}NHR \longrightarrow (R_2N)_2P(O)\text{-}NH\text{-}NHR + HCl$$

Phosphorochloridates react with hydrazine or substituted hydrazines to form phosphorohydrazid. The reaction can be carried out in water[274a,475,858] and sodium carbonate can be used in place of half of the necessary of hydrazine. Further treatment of the diaryl phosphorohydrazates with diaryl phosphorochloridates in the presence of pyridine leads to the formation of N,N'-bis(diaryloxyphosphoryl)-hydrazine.[858] Also the chlorine atoms in alkyl phosphoramidochloridates are easily replaced by hydrazine groups.[646] Phosphorodiamidochloridates give phosphoroamidohydrazid with substituted hydrazines.

$$(RO)_2P(O)Cl + HN{=}CR(NH_2) \longrightarrow (RO)_2P(O)\text{-}N{=}C(NH_2)\text{-}R + HCl$$

$$(RO)_2P(O)Cl + HN{=}C(OR')(NH_2) \longrightarrow (RO)_2P(O)\text{-}N{=}C(OR')(NH_2) + HCl$$

$$(RO)_2P(O)Cl + HN{=}C(SR')(NH_2) \longrightarrow (RO)_2P(O)\text{-}N{=}C(SR')(NH_2) + HCl$$

$$(RO)_2P(O)Cl + 2\ HN{=}C(NR'_2)(NH_2) \longrightarrow (RO)_2P(O)\text{-}N{=}C(NR'_2)(NH_2) + (H_2N)(R'_2N)C{=}NH\cdot HCl$$

When an aqueous solution of an amidine salt is treated simultaneously with sodium hydroxide solution and a solution of a dialkyl or diaryl phosphorochloridate in benzene N-dialkyloxy phosphoryl- or N-diaryloxy phosphorylamidine is formed in very good yield.[174] O-alkylisoureas, in contrast to N-substituted ureas, react directly with halogen compounds in the presence of an acid acceptor.[174] S-alkylisothioureas react similarly. Free guanidines, or free guanidines obtained from guanidinium salts and sodium hydroxide solution, react with dialkyl or diaryl phosphorochloridates to form dialkoxy- (or diaryloxy)-phosphoryl guanidines which are generally formed in this way. The free guanidine is used mostly as an alcohol solution.[174]

Vb. SP(X)= + N$\lessgtr$ (X = hal)

Reactions of thiophosphoryl monohalo compounds with ammonia, amines, hydrazine, and N-substituted hydrazines:

$$(RO)_2P(S)Cl + 2\ HNR'_2 \longrightarrow (RO)_2P(S)NR'_2 + [R'_2NH_2]Cl$$

$$(RO)P(S)Cl(NH\text{-}NHR') + 2\ HNR'_2 \longrightarrow (RO)P(S)(NR'_2)(NH\text{-}NHR') + [R'_2NH_2]Cl$$

O,O-diesters of phosphorochloridothionic acids are converted smoothly to O,O'-diesters of phosphoramidothionic acids by treatment with 2 moles of ammonia, primary or secondary amines.[75] N-unsubstituted derivatives are best prepared by passing gaseous ammonia through the dialkylester-chloride,[568] diarylester compounds can also be synthesized by using aqueous,[75] or liquid ammonia. Lower aliphatic amines can be employed as concentrated aqueous solutions. Phosphorochloridohydrazidothionic acid O-esters react with 2 moles of ammonia, primary amine or secondary amine to give the respective phosphoramidothionic acid O-esters.[105]

$$(RO)_2P(S)Cl + 2\ H_2N\text{-}NHR' \longrightarrow (RO)_2P(S)\text{-}NH\text{-}NHR' + R'NH\text{-}NH_2\cdot HCl$$

$$(R'S)_2P(O)Cl + 2\ H_2N\text{-}NHR \longrightarrow (R'S)_2P(O)\text{-}NH\text{-}NHR + RNH\text{-}NH_2\cdot HCl$$

$$(RO)P(S)Cl(NHR') + 2\ H_2N\text{-}NHR \longrightarrow (RO)P(S)(NHR')(NH\text{-}NHR) + RNH\text{-}NH_2\cdot HCl$$

$$(R_2N)_2P(S)Cl + 2\ H_2N\text{-}NH_2 \longrightarrow (R_2N)_2P(S)(NH\text{-}NH_2) + H_2N\text{-}NH_2\cdot HCl$$

O,O-diesters of phosphorochloridothionic acid are converted to O,O-diesters of phosphorohydrazidothionic acid by treatment with a two molar amount of hydrazines or a substituted hydrazine.[858,924] S,S-diesters of phosphorochloridodithiolic acid react similarly.

O-esters of phosphoramidochloridothionic acid react with 2 moles of a hydrazine to give O-esters of phosphoramidohydrazidothionic acid.[858] In contrast, N,N,N',N'-tetraalkyl phosphorodiamidothioic chlorides react only with excess hydrazine to form N,N,N',N'-tetraalkyl phosphorodiamidothioic hydrazides.[858] This last reaction is carried out in benzene solution.

$$ClP(S)(OR')_2 + (RO)_2P(S)\text{-}NH\text{-}NH_2 \longrightarrow (RO)_2P(S)\text{-}NH\text{-}NH\text{-}P(S)(OR')_2 + HCl$$

$$ClP(S)(OR')_2 + (R_2N)_2P(S)\text{-}NH\text{-}NH_2 \longrightarrow (R_2N)_2P(S)\text{-}NH\text{-}NH\text{-}P(S)(OR')_2 + HCl$$

N,N'-bis-(thiophosphoryl)-hydrazines are formed when O,O-diesters of phosphorochloridothionic acid are allowed to react with O,O-diesters of phosphorohydrazidothionic acid or phosphorodiamidothioic hydrazides in the presence of a tertiary amine:

VI. $Cl_3P{=}N\text{-} + N{\lessgtr}$

Reactions of trichlorophosphinimide derivatives with ammonia and amines:

$$R\text{-}CO\text{-}N{=}PCl_3 + H_2NAr + 2\ R'_3N \longrightarrow [R\text{-}CO\text{-}N{=}PCl_2NAr][R'_3NH] + [R'_3NH]Cl$$

$$R\text{-}CO\text{-}N{=}PCl_3 + 6\ H_2NAr \longrightarrow R\text{-}CO\text{-}N{=}P(NHAr)_3 + 3[ArNH_3]Cl$$

When one mole of aniline and two moles of triethylamine are allowed to react with N-acyl trichlorophosphinimide the triethylammonium salt of N-aryl-N'-acyl-phosphoramido dichloridimidate is formed;[222] with 6 moles of aniline N-acyl trianilido-phosphinimide is obtained.[211] With ethyleneimine in the presence of triethylamine the product is N-acyl triethylenimido-phosphinimide.[695]

$$R\text{-}SO_2\text{-}N{=}PCl_3 + 6\ NH_3 \longrightarrow R\text{-}SO_2\text{-}N{=}P(NH_2)_3 + 3\ NH_4Cl$$

$$R\text{-}SO_2\text{-}N{=}PCl_3 + 4\ H_2NAr \longrightarrow R\text{-}SO_2\text{-}N{=}PCl(NHAr)_2 + 2[ArNH_3]Cl$$

N-arylsulfonyltrichlorophosphinimide reacts with gaseous ammonia in an inert solvent in the absence of water to give N-arylsulfonyl triamido-phosphinimide. The N-unsubstituted triamides are not stable and decompose within hours. Triamides are only formed from aromatic amines when excess amine is used and the reaction mixture is heated for a long time. In the cold two chlorine atoms are substituted.[429a,432]

VIIa. $\geqslant$P-O-P$\leqslant$ + N$\leqslant$
+ HX, X^- (X = hal, pseudohal)

Cleavage of P-O-P-bridges:

$$(RO)_2P(O)\text{-}O\text{-}P(O)(OR)_2 + 2\ R'NH_2 \longrightarrow (RO)_2P(O)(NHR') + (RO)_2P(O)(O^-)[R'NH_3^+]$$

$$(RO)_2P(O)\text{-}O\text{-}P(O)(OH)(OR') + H_2NR'' \longrightarrow (RO)_2P(O)(OH) + R''NH\text{-}P(O)(OH)(OR')$$

Diphosphoric acid tetraalkylesters are cleaved by ammonia or amines to dialkylphosphoramidate and the respective ammonium salt of the O,O-dialkylhydrogenphosphate.[65] Unsymmetrical diphosphoric acid tetraalkylesters are so cleaved that the amide of the weaker acid is formed.[166] These reactions are best performed in an inert solvent. Diphosphoric acid triesters are also cleaved by amines to give esters of phosphoramidic acid and salts of phosphate.

$$[\text{-}O\text{-}P(O)(OR)\text{-}]_3 + 3\ H_2NAr \longrightarrow 3\ ArNH\text{-}P(O)(OH)(OR)$$

Trimetaphosphoric acid ethyl ester reacts with aniline to produce O-ethyl N-phenyl phosphoramidate.[768]

$$[\text{-}O\text{-}P(O)(OR)\text{-}]_3 + 3\ H_2N\text{-}NHAr \longrightarrow 3\ RO\text{-}P(O)(OH)(NH\text{-}NHAr)$$

Arylhydrazines split trimetaphosphoric acid triethylester to give O-ethyl-phosphoro-(N'-aryl)hydrazates.

$$(RO)_2P(O)\text{-}O\text{-}P(O)(OR)_2 + HN{=}C(NH_2)(NHR') \longrightarrow (RO)_2P(O)\text{-}N{=}C(NH_2)(NHR') + P(O)(OH)(OR)_2$$

Diphosphoric acid esters are cleaved by guanidine to form N-(dialkyloxy-phosphoryl)-N'-arylguanidines.

$$(ArO)_2P(O)\text{-}O\text{-}P(O)(OAr)_2 + KSCN \longrightarrow (ArO)_2P(O)(OK) + SCN\text{-}P(O)(OAr)_2$$

Potassium isothiocyanate splits diphosphoric acid-tetraarylester to form O,O-diaryl phosphorate and diarylphosphorisothiocyanatidate.[425]

VIIb. $\geqslant$P-S-P$\leqslant$ + N$\leqslant$

Cleavage of P-S-P-bridges:

$$1/2\ P_4S_{10} + 2\ HNR_2 \longrightarrow R_2N\text{-}P(S)(SH)_2 + R_2N\text{-}PS_2$$

$$1/2 \ P_4S_{10} + 4 \ H_2NR \longrightarrow 2 \ (RNH)_2P(S)SH + H_2S$$

$$1/2 \ P_4S_{10} + 6 \ RNH_2 \longrightarrow 2 \ SP(HNR)_3 + 3 \ H_2S$$

Phosphorus V sulfide reacts with two moles of a primary aromatic amine to give undefinable products. Secondary aliphatic amines react to produce trithiophosphoric acid alkyl amides in low yield. By the reaction with primary aromatic amines in molar ratio 4:1, phosphorus V sulfide is converted to dithiophosphoric acid-bis-arylamide, at low temperature. With 6 moles of a primary aromatic amine and a temperature of approximately 150 °C phosphorothioic triamide is formed.[131,913] Low-boiling primary aliphatic amines can also be forced to react under pressure, but good yields cannot be obtained.

VIIc. $\overset{|}{\geq}P-\overset{|}{N}-P\leq$ + $N\leq$

Cleavage of P-N-P-bridges:

$$+ \ 2 \ RNH_2 \longrightarrow 2 \ (RNH)_3P(S)$$

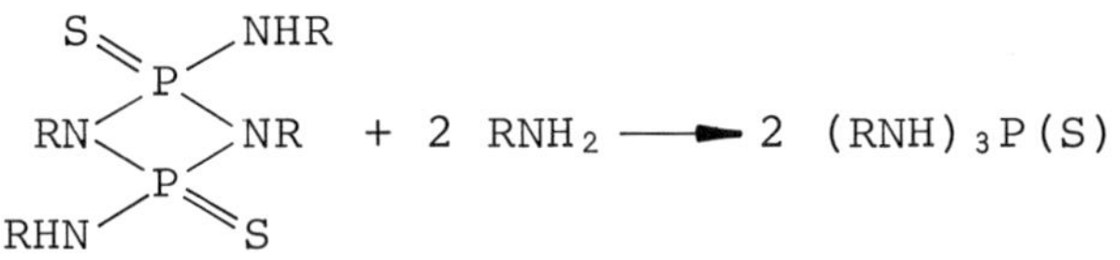

Diamidodithioic-cyclodiphosphazanes react with amines to give phosphorothioic triamides. Only symmetrical phosphorothioic triamides can be prepared by this route.[131]

VIIIa. $OP(X_2)-$ + H_2O
          $OP(X)=$ + H_2O

Hydrolysis of phosphoryl dihalo- and phosphoryl monohalo- compounds:

$$OPCl_2(NHAr) + 2 \ H_2O \longrightarrow OP(OH)_2(NHAr) + 2 \ HCl$$

N-aryl phosphoramidic dichlorides, especially those which contain a halogen atom in the aromatic nucleus, can be, at least partially, converted to N-aryl phosphoramidic acids by aqueous sodium hydroxide solution followed by acidification. The hydrolysis proceeds more easily on refluxing for three hours with the stoichiometrical amount of formic acid in benzene. As well as N-aryl phosphoramidic acids, N-alkyl phosphoramidic acids can also be obtained from the dichlorides.[529]

$$OPCl_3 \xrightarrow{RNH_2/H_2O} R\text{-}NH\text{-}P(O)(OH)_2$$

$$R\text{-}NH\text{-}P(O)Cl_2 \xrightarrow{NH_3/H_2O} R\text{-}NH\text{-}P(O)(NH_2)(OH)$$

N-phospho-amino acids are obtained directly by the reaction of an amino acid such as glycine, alanine, or glutamic acid with phosphorus oxytrichloride in aqueous solution; phosphoramidic dichlorides are intermediates. Magnesium oxide is used as HCl acceptor so that the N-phospho-amino acids are formed and isolated as magnesium salts. For the synthesis of the hydrolytically stable phosphoric acid monoguanidine, concentrated hydroxide solution is used as HCl acceptor instead of MgO.[629] From the hydrolysis of N-aryl-phosphoramidic dichlorides with aqueous ammonia N-aryl phosphorodiamidic acids are obtained.[734]

$$Ar\text{-}CO\text{-}NH\text{-}P(O)Cl_2 + 2\ H_2O \longrightarrow Ar\text{-}CO\text{-}NH\text{-}P(O)(OH)_2 + 2\ HCl$$

$$RO\text{-}CO\text{-}NH\text{-}P(O)Cl_2 + 2\ H_2O \longrightarrow RO\text{-}CO\text{-}NH\text{-}P(O)(OH)_2 + 2\ HCl$$

$$(C_5H_4N)NH\text{-}CO\text{-}NH\text{-}P(O)Cl_2 + 2\ H_2O \longrightarrow (C_5H_4\overset{\oplus}{N}H)NH\text{-}CO\text{-}NH\text{-}P(O)(O^{\ominus})OH + 2\ HCl$$

N-acyl phosphoramidic dichlorides with aromatic carboxylic acid rests can, by careful hydrolysis, be converted to the N-acyl phosphoramidic acids.[856] The corresponding aliphatic compounds cannot be prepared in the same way as the dichloride is immediately hydrolyzed to phosphoric acid and carboxylic acid amide. N-phosphocarbamic acid esters are obtained from the hydrolysis of N-dichlorophosphorylcarbamic acid ethyl ester. The reaction is slow with water but more rapid with sodium hydroxide solution; aqueous dioxan in the presence of silver carbonate is especially suitable.[509] N-dichlorophosphoryl ureas are, in general, not hydrolyzed to the free N-phospho-urea with water but are split to give urea and phosphoric acid. Only when intramolecular stabilization through betain formation with a nitrogen-containing heterocyclic substituent is possible, can N-phosphoureas be isolated.[450] N-phospho-N'-aryl-ureas are also not hydrolytically stable; they can, however be isolated by treatment of the dichloride with a stoichiometric amount of formic acid.[461]

$$R\text{-}SO_2\text{-}NH\text{-}P(O)Cl_2 + HCOOH \longrightarrow R\text{-}SO_2\text{-}NH\text{-}P(O)Cl(OH) + CO + HCl$$

$$R\text{-}SO_2\text{-}NH\text{-}P(O)Cl_2 + 2\ HCOOH \longrightarrow R\text{-}SO_2\text{-}NH\text{-}P(O)(OH)_2 + 2CO + 2HCl$$

N-sulfonyl phosphoramidic dihalides can be stepwise hydrolyzed with stoichiometric amounts of formic acid. Benzene is generally used as reaction medium and the reaction is usually complete after approximately 4 hr at reflux temperature.[429a,446] With excess formic acid sulfonic acid amide and phosphoric acid are formed. Also N-dialkylaminosulfonylphosphoramidic acids ($R\text{-}NR'_2$) can be obtained by this route.[459]

$$2\ (R_2N)P(O)FCl + H_2O \longrightarrow (R_2N)P(O)F\text{-}O\text{-}P(O)F(NR_2) + 2\ HCl$$

$$2\ (R_2N)_2P(O)Cl + H_2O \longrightarrow (R_2N)_2P(O)\text{-}O\text{-}P(O)(NR_2)_2 + 2\ HCl$$

Diphosphoric acid-P-P'-diamide-P,P'-difluoride is obtained by the partial hydrolysis of phosphoramidic chloride fluoride in the presence of pyridine.[824] Similarly, phosphoramidic chlorides react with a limited amount of water in the presence of pyridine or another tertiary base to give diphosphoric acid tetraamides. Instead of the tertiary base, concentrated aqueous solutions of NaOH, $Ba(OH)_2$, or Na_2CO_3, or finely divided solid alkali hydroxides can be used.[683a,876]

$$(R_2N)P(O)Cl(OR') + H_2O \longrightarrow R_2N\text{-}P(O)(OR')(OH) + HCl$$

$$(R_2N)_2P(O)Cl + 2\ NaOH \longrightarrow (R_2N)_2P(O)(ONa) + NaCl + H_2O$$

In some cases phosphoramido halidates can be converted to phosphoramidic acids by hydrolysis.[542] In most cases the hydrolysis must be carried out in barium hydroxide or potassium carbonate solutions, and the corresponding salt is isolated.[149] The hydrolytically stable aryl N-aryl phosphoramidates can be obtained from their alkali salts by treatment with acids. Phosphorodiamidates are rapidly decomposed in acidic solutions. They can, however, still be obtained from the solution of the alkali salt by neutralization under cooling. The solution is prepared by hydrolysis of phosphoric acid-diamide-halide in aqueous alkalihydroxide or carbonate solution.[582]

$$R\text{-}CO\text{-}NH\text{-}P(O)(Cl)(OR') + H_2O \longrightarrow R\text{-}CO\text{-}NH\text{-}P(O)(OR')(OH) + HCl$$

N-acyl phosphoramidochloridates can be converted in good yield to N-acyl phosphoramidates.[943]

VIIIb. $SP(X_2)- + H_2O$
$SP(X)= + H_2O$

Hydrolysis of thiophosphoryl dihalo- and thiophosphoryl monohalo-compounds:

$$(R_2N)P(S)ClF + 2\ KOH \longrightarrow (R_2N)P(O)(SK)F + KCl + H_2O$$

N,N-dialkyl phosphoramidothioic chloride fluorides and bromide fluorides can be selectively hydrolyzed to the salts of N,N-dialkyl phosphoramidothioic fluoride by treatment with an equivalent amount of potassium hydroxide in aqueous dioxan.[821]

$$2\ (R_2N)P(S)FCl + H_2O \longrightarrow (R_2N)P(S)F\text{-}O\text{-}P(S)F(NR_2) + 2\ HCl$$

On partial hydrolysis of N,N-dialkyl phosphoramidothioic chloride fluorides in the presence of pyridine dithiodiphosphoric acid-P,P'-diamide-P,P'-difluoride is obtained.[824]

$$(R_2N)P(S)(OR')Cl + 2\ NaOH \longrightarrow (R_2N)P(S)(OR')(ONa) + NaCl + H_2O$$

By treatment with a two molar quantity of alkali salts of O-esters of hydrogen, phosphoramidothionic acid can be obtained from O-esters of phosphoramidochloridothionic acids.[764]

IXa. $OP(X_2)- + ROH, RSH$
$+ ROM, RSM$ (M = alkalimetal)
$OP(X)= + ROH, RSH$
$+ ROM, RSM$

Reactions of phosphoryl dihalo- and phosphoryl monohalo- compounds with alcohols, phenols and thiols or their alkali metal derivatives:

$$(R_2N)P(O)Cl_2 + 2\ HOR' \longrightarrow (R_2N)P(O)(OR')_2 + 2\ HCl$$

$$(R_2N)P(O)Cl_2 + NaOR' \longrightarrow (R_2N)P(O)Cl(OR') + NaCl$$

$$(RHN)P(O)ClF + NaOR' \longrightarrow (RHN)P(O)(OR')F + NaCl$$

The not very reactive phosphoramidic dichlorides can be converted to diesters of phosphoramidic acid by the reaction of alcohols in the presence of a hydrogen chloride acceptor.[927] With polyfunctional alcohols cyclic diester-amides are formed. The reaction with phenols and cresols in the absence of an acid acceptor occurs only at higher temperatures, approximately 150-200 °C.[582] Phosphoramidic dichlorides are converted in high yield to alkyl phosphoramidochloridates by treatment with alcoholates.[573] Phosphoramidic dichloride reacts with phenols in the presence of a tertiary amine to form aryl phosphoramidochloridates;[574] alkali phenolates react similarly.[573] Because of the differing reactivities of the halogen atoms in phosphoramidic chloride fluorides treatment with an equimolar amount of alcoholate yields predominantly alkyl phosphoramidofluoridates.[821]

$$R\text{-}CO\text{-}NH\text{-}P(O)Cl_2 + HOR' \longrightarrow R\text{-}CO\text{-}NH\text{-}P(O)(Cl)(OR') + HCl$$

$$R\text{-}CO\text{-}NH\text{-}P(O)Cl_2 + 3\ NaOCH_3 \longrightarrow R\text{-}CO\text{-}N(Na)\text{-}P(O)(OCH_3)_2 + 2\ NaCl + CH_3OH$$

N-acyl phosphoramidic dichlorides can be converted to the phenyl N-acyl-phosphoramidochloridates by the reaction of phenol in the presence of pyridine. With methanol or ethanol the dialkyl N-acyl phosphoramidates are formed even on standing. The reaction proceeds more rapidly when sodium alcoholate in excess alcohol, followed by acidification of the resultant mixture, is employed.[816] Similarly with sodium phenolate diaryl N-acyl phosphoramidates are obtained.[943]

$$RO\text{-}CO\text{-}NH\text{-}P(O)Cl_2 + 2\ NaOR' \longrightarrow RO\text{-}CO\text{-}NH\text{-}P(O)(OR')_2 + 2\ NaCl$$

N-dichlorophosphoryl-carbamic esters react with sodium alcoholates or phenolates to form N-dialkoxy-phosphoryl and N-diaryloxy-phosphoryl carbamic esters respectively.[456,467]

$$R\text{-}NH\text{-}CO\text{-}NH\text{-}P(O)Cl_2 + 3\ NaOR' \longrightarrow R\text{-}NH\text{-}CO\text{-}NNa\text{-}P(O)(OR')_2 + 2\ NaCl + R'OH$$

N'-substituted-N-dichlorophosphoryl ureas and sodium alcoholates interact to produce the respective N-dialkyl-oxyphosphoryl ureas. The yields are only satisfactory with methoxy compounds; with longer chain alcohols trialkyl phosphates are preferentially formed.[450]

$$(ArO)_2P(O)\text{-}NH\text{-}P(O)Cl_2 + 2\ NaOR' \longrightarrow (ArO)_2P(O)\text{-}NH\text{-}P(O)(OR')_2 + 2\ NaCl$$

$$(ArO)_2P(S)\text{-}NH\text{-}P(O)Cl_2 + 2\ NaOR' \longrightarrow (ArO)_2P(S)\text{-}NH\text{-}P(O)(OR')_2 + 2\ NaCl$$

$$R\text{-}SO_2\text{-}NH\text{-}P(O)Cl_2 + 3\ NaOR' \longrightarrow R\text{-}SO_2\text{-}NNa\text{-}P(O)(OR')_2 + 2\ NaCl + R'OH$$

Imido-diphosphoric acid P,P-diphenylester-P',P'-dichlorides react with alcoholates or phenolates to give symmetrical or unsymmetrical imido-diphosphoric acid tetraesters.[469] Unsymmetrical imido-thiodiphosphoric-acid-O,O-diarylester-dichlorides react analogously.[404] With sodium alcoholates N-sulfonyl phosphoramidic dichlorides form dialkyl N-sulfonyl-phosphoramidates, the respective alcohol, or, e.g., benzene are used as solvents. At least three moles of alcoholates must be employed as one mole is used in the formation of the sodium salt. The free compounds are liberated from the sodium salts by subsequent acidification.[429a,462]

$$(R_2N)P(O)Cl(OR') + HOR'' \longrightarrow (R_2N)P(O)(OR')(OR'') + HCl$$

$$(R_2N)_2P(O)Cl + NaOR' \longrightarrow (R_2N)_2P(O)(OR') + NaCl$$

$$(R_2N)_2P(O)Cl + NaS\ R' \longrightarrow (R_2N)_2P(O)(SR') + NaCl$$

$$(R_2N)P(O)Cl(OR') + NaS\ R'' \longrightarrow (R_2N)P(O)(SR'')(OR') + NaCl$$

$$(R_2N)P(O)(SR')Cl + NaOR'' \longrightarrow (R_2N)P(O)(SR')(OR'') + NaCl$$

Esters of phosphoramidochloridic acid also react with alcohols. In many cases the acid acceptor such as triethylamine, pyridine, or potassium tert. butylate can be omitted; in other cases they are necessary.[316] Phosphordiamidic halides react with alcohols only in the presence of a tertiary amine. In contrast esters of phosphoro-diamidic acid are easily formed by the reaction with sodium alcoholates in benzene, xylene, or alcohol.[574] Phenols react at about 200 °C to give aryl phosphorodiamidates with evolution of hydrogen chloride.[573] This reaction also proceeds more readily when a phenolate in an inert solvent is used instead of phenol.[582] S-esters of phosphoro-diamidothiolates are formed in moderate yields from reactions with alkali salts of thiols.[680] O,S-dialkyl N,N-dialkyl-phosphoramidothiolates are formed clearly by the reaction of alkyl N,N-dialkyl phosphoramidochlorides with alkali mercaptides.[763] The same products are also obtained from the reaction of S-esters of phosphoramidochloridothiolic acid with alcoholates.[776]

$$R\text{-}C(Cl)\text{=}N\text{-}P(O)Cl_2 + 3\ NaOR' \longrightarrow R\text{-}C(OR')\text{=}N\text{-}P(O)(OR')_2 + 3\ NaCl$$

N-dialkoxyphosphoryl- or N-diaryloxyphosphoryl-carboxylic acid-imide-esters with similar alkoxy or aryloxy groups on phosphorus ana carbon are preparea from 3 moles of an alcoholate or phenolate and N-dichlorophosphoryl-carboxylic-acid-imide-chloride.[436]

$$\text{Ar-N} \underset{\underset{Cl}{P}}{\overset{\overset{O}{P}\diagdown^{Cl}}{\diagup}} \text{N-Ar} + 4\ \text{NaOR} \longrightarrow 2\ \text{Ar-NNa-P(O)(OR)}_2 + 2\ \text{NaCl}$$

$$\text{Ar-N} \underset{\underset{Cl}{P}}{\overset{\overset{S}{P}\diagdown^{Cl}}{\diagup}} \text{N-Ar} + 2\ \text{NaOR} \longrightarrow \text{Ar-N} \underset{\underset{RO}{P}}{\overset{\overset{S}{P}\diagdown^{OR}}{\diagup}} \text{N-Ar} + 2\ \text{NaCl}$$

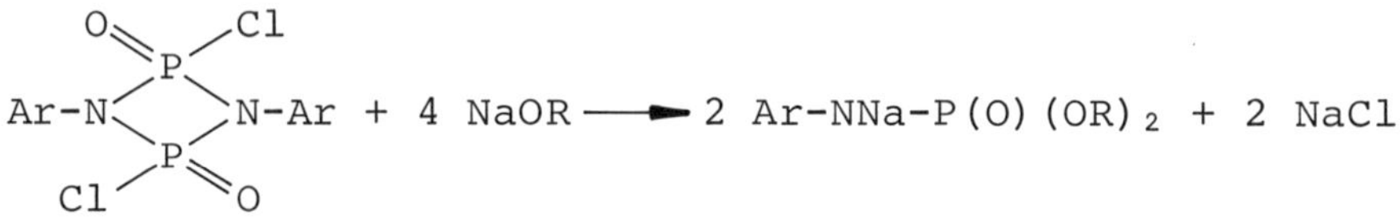

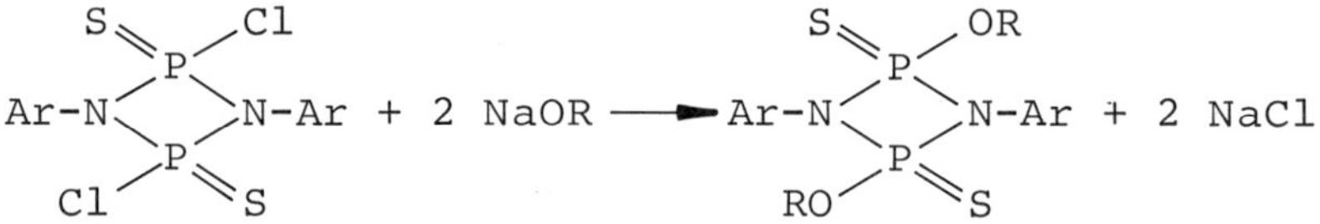

Depending upon conditions, one or two phosphorus-nitrogen bonds in 1,3-diphenyl-2,4-dichloro-2,4-dioxo-2,4,1,3-cyclodiphosphazane are cleaved by water-free sodium phenolate in benzene; in the latter case diphenyl phosphoramidate is formed.[574] Using alcoholates or phenolates the corresponding dithio-cyclophosphazanes are converted to 1,3-diphenyl-2,4-dialkoxy (or diaryloxy-)-2,4-dithio-2,4,1,3-cyclodiphosphazanes.[576]

IXb. SP(X$_2$)- + ROH, RSH
           + ROM, RSM (M = alkalimetal)
      SP(X)= + ROH, RSH
           ROM, RSM

Reactions of thiophosphoryl dihalo- and thiophosphoryl monohalo- compounds with alcohols and phenols or their alkali metal derivatives:

$$(R_2N)P(S)Cl_2 + NaOR' \longrightarrow (R_2N)P(S)(OR')Cl + NaCl$$

$$(R_2N)P(S)Cl_2 + 2\ NaOR' \longrightarrow (R_2N)P(S)(OR')_2 + 2\ NaCl$$

$$(R_2N)P(S)(OR'')Cl + NaOR' \longrightarrow (R_2N)P(S)(OR'')(OR') + NaCl$$

$$(R_2N)P(S)FCl + NaOR' \longrightarrow (R_2N)P(S)F(OR') + NaCl$$

Phosphoramidothioic dichlorides react with equi-molar amounts of alcoholates or phenolates, or with the corresponding amount of alcohols or phenols in the presence of a tertiary amine, to form the O-esters of phosphoramidochloridothionic acid.[255] With 2 moles of a sodium alcoholate O,O'-dialkylesters of phosphoramidothionic acid are formed.[573] If O-alkyl phosphoramidochloridothionates are allowed to react with alcoholates or alcohols in the presence of dry alkali hydroxide, O,O'-dialkyl phosphoramidothionates are obtained, in which the two ester groups are different. With 1,2-, 1,3- or 1,4-glycols used in the presence of a tertiary amine, 5- to 7-membered ring compounds are obtained.[573] Sodium phenolate reacts smoothly at 50 °C to yield N-alkyl (or N,N-dialkyl)O,O'-diaryl phosphoramidothionates,[619] whereas phenol itself only reacts on heating under pressure.[573] Phosphoramidothioic chloride fluorides react with equivalent amounts of sodium alcoholate in carbon tetrachloride at 0 °C to give O-alkyl-phosphoramidofluoridothionates in very high yields.[821]

$$(R_2N)_2P(S)Cl + NaOR' \longrightarrow (R_2N)_2P(S)(OR') + NaCl$$

$$(R_2N)P(S)Cl(OR') + NaSR'' \longrightarrow (R_2N)P(S)(SR'')(OR') + NaCl$$

The difficultly accessible phosphorodiamidothioic chlorides react with alcoholates and phenolates to form alkyl or aryl phosphorodiamidothionates.[395] Also O-alkyl phosphoramidochloridothionates react with alkali mercaptides to give O,S-dialkyl phosphoramidothiolothionates.[395]

X. $Cl_3P{=}N- + H_2O$

Hydrolysis of trichlorophosphinimide derivatives:

$$R\text{-}CO\text{-}N{=}PCl_3 + H_2O \longrightarrow R\text{-}CO\text{-}NH\text{-}P(O)Cl_2 + HCl$$

$$R\text{-}CO\text{-}N{=}PCl_3 + HCOOH \longrightarrow R\text{-}CO\text{-}NH\text{-}P(O)Cl_2 + HCl + CO$$

$$RO\text{-}CO\text{-}N{=}PCl_3 + H_2O \longrightarrow RO\text{-}CO\text{-}NH\text{-}P(O)Cl_2 + HCl$$

$$Ar_2N\text{-}CO\text{-}N{=}PCl_3 + HCOOH \longrightarrow Ar_2N\text{-}CO\text{-}NH\text{-}P(O)Cl_2 + CO + HCl$$

N-acyl-trichloro-phosphinimides are hydrolyzed by water, or preferably, to avoid further hydrolysis, by an equivalent amount of formic acid in benzene or carbon tetrachloride at room temperature, to N-acyl-phosphoramidic dichlorides.[934] N-dichlorophosphoryl-carbamic acid alkyl (or aryl) esters are obtained analogously from the respective carbalkoxy and carbaryloxy derivatives. With formic acid N-diarylcarbamyl-trichloro-phosphinimides give N-dichlorophosphoryl-N',N'-diarylurea.[461]

$$R\text{-}SO_2\text{-}N{=}PCl_3 + H_2O \longrightarrow R\text{-}SO_2\text{-}NH\text{-}P(O)Cl_2 + HCl$$

$$R\text{-}SO_2\text{-}N{=}PCl_3 + 5\ KF + 2\ H_2O \longrightarrow R\text{-}SO_2\text{-}NK\text{-}P(O)F(OK) + 3\ KCl + 4\ HF$$

N-sulfonyl-trichlorophosphinimides are very readily hydrolyzed. Even in moist air they form N-sulfonyl phosphoramidic dichlorides. The hydrolysis is preferably carried out with formic acid, acetic acid, or also benzoic acid.[429a,443] From the reaction with a saturated aqueous solution of potassium fluoride the potassium salt of N-arylsulfonyl phosphoramidofluoridic acid is obtained. The reaction proceeds via the also isolated N-potassium N-arylsulfonyl-phosphoramide difluoridate, $R\text{-}SO_2N(K)P(O)F_2$.[451]

$$(ArO)_2P(O)\text{-}N{=}PCl_3 + HCOOH \longrightarrow (ArO)_2P(O)\text{-}NH\text{-}P(O)Cl_2 + HCl + CO$$

$$(ArO)_2P(O)\text{-}N{=}PCl_3 + 2\ HCOOH \longrightarrow (ArO)_2P(O)\text{-}NH\text{-}P(O)(OH)Cl + 2\ HCl + 2\ CO$$

$$(ArO)_2P(O)\text{-}N{=}PCl_3 + 3\ H_2O \longrightarrow (ArO)_2P(O)\text{-}NH\text{-}P(O)(OH)_2 + 3\ HCl$$

$$(ArO)_2P(S)\text{-}N{=}PCl_3 + HCOOH \longrightarrow (ArO)_2P(S)\text{-}NH\text{-}P(O)Cl_2 + HCl + CO$$

N-phosphoryl-trichlorophosphinimides are also readily hydrolyzed with an equimolar quantity of formic acid, imido diphosphoric acid P,P-diarylester-P',P'-dichloride can be obtained from N-diaryloxyphosphoryl-trichlorophosphinimide.[469] The respective monochloride can also be prepared by using 2 moles of formic acid. On careful hydrolysis with water imido-diphosphoric acid-P,P-diarylesters can be prepared.[469] The respective thio-compounds are also partially saponified by formic acid.[469]

XI. $Cl_3P{=}N\text{-} + ROH$

Reactions of trichlorophosphinimide derivatives with alcohols:

$$Ar\text{-}N{=}PCl_3 + 3\ NaOAr' \longrightarrow Ar\text{-}N{=}P(OAr')_3 + 3\ NaCl$$

Monomeric and dimeric N-aryl-trichlorophosphinimides react with sodium phenolate to give exclusively monomeric N-aryl-triphenoxyphosphinimides.[939]

$$R\text{-}CO\text{-}N{=}PCl_3 + 3\ NaOR' \longrightarrow R\text{-}CO\text{-}N{=}P(OR')_3 + 3\ NaCl$$

$$R\text{-}CO\text{-}N{=}PCl_3 + 3\ ROH \longrightarrow R\text{-}CO\text{-}NH\text{-}P(O)(OR)_2 + 2\ HCl + RCl$$

$$RO\text{-}CO\text{-}N{=}PCl_3 + 3\ R'OH \longrightarrow RO\text{-}CO\text{-}NH\text{-}P(O)(OR')_2 + R'Cl + 2\ HCl$$

The reactions of 3 moles of an alcoholate or phenolate with N-acyl-trichlorophosphinimides result in the formation of N-acyl-trialkoxy-(or triaryloxy)-phosphinimides. The yields are mostly small as the products, especially the trialkyloxy compounds, are easily cleaved by alkali to give N-acyl-O,O'-dialkyl-phosphoramidates. The N-acyl-triaryloxy-phosphinimides are obtainable in better yields. When 4 moles of alcoholate, followed by acidification, are used, N-acyl-O,O'-dialkyl phosphoramidates are obtained. The same compounds can also be prepared by using 3 moles of the alcohol. N-carbalkoxy-trichloro-phosphinimides react with excess alcohol even below 10° C to form the N-dialkoxy-phosphoryl-carbamic acid esters in very good yields.[430]

$$R\text{-}SO_2\text{-}N{=}PCl_3 + R'OH \longrightarrow R\text{-}SO_2\text{-}N{=}P(OR')Cl_2 + HCl$$

$$R\text{-}SO_2\text{-}N{=}PCl_3 + R'OH \longrightarrow R\text{-}SO_2\text{-}NH\text{-}P(O)Cl_2 + R'Cl$$

$$R\text{-}SO_2\text{-}N{=}PCl_3 + 3\ R'OH \longrightarrow R\text{-}SO_2\text{-}N{=}P(OR')_3 + 3\ HCl$$

$$R\text{-}SO_2\text{-}N{=}P(OR')_3 + HCl \longrightarrow R\text{-}SO_2\text{-}NH\text{-}P(O)(OR')_2 + R'Cl$$

Some N-arylsulfonyl-trichloro-phosphinimides, such as the biphenyl- and naphthyl-sulfonyl compounds, react with equimolar amounts of methanol to produce methyl N-arylsulfonyl-phosphorodichloridimidate, when the evolved hydrogen chloride is removed by passing CO_2 through the reaction mixture.[464] If the hydrogen chloride is not removed or more hydrogen chloride is added during the reaction, N-sulfonyl phosphoramidic dichloride is obtained.[464] When excess alcohol is used a mixture of N-sulfonyl-trialkoxy-phosphinimides and dialkyl N-sulfonyl phosphoramidates results, due to the evolved hydrogen chloride causing partial ester cleavage. On allowing the reaction mixture to stand for a long time only the latter products remain. The formation of these products is suppressed when 3 moles of alcoholate are used instead of the alcohol.[462] Triaryloxy-phosphinimides are best synthesized by using phenolates.[444] The reactions with alcoholates and phenolates are carried out using, e.g., benzene as solvent.[429a,462,464]

$$(ArO)_2P(O,S)\text{-}N{=}PCl_3 + 3\ NaOR \longrightarrow (ArO)_2P(O,S)\text{-}N{=}P(OR)_3 + 3\ NaCl$$

N-diphenoxyphosphoryl-trichloro-phosphinimides and the corresponding thiophosphoryl compounds react with 3 moles of sodium phenolate to give N-diphenoxy-phosphoryl and N-diphenoxythiophosphoryl-triphenoxy-phosphinimides, respectively.[469]

XII. P-NCO + H_2O, ROH, RSH, NH_3, NH_2R, NHR_2, H_2NNH_2, H_2NNHR
P-NCS + H_2O, ROH, RSH, NH_3, NH_2R, NHR_2, H_2NNH_2, H_2NNHR, Cl_2

Reactions of phosphorus isocyanates and isothiocyanates with water, alcohols, thiols, ammonia, amines, hydrazines, and chlorine:

$$OP(NCS)_3 + RNH_2 \longrightarrow RNH\text{-}CS\text{-}NH\text{-}P(O)(NCS)_2$$

$$OP(NCS)_3 + 2\ RNH_2 \longrightarrow (RNH\text{-}CS\text{-}NH)_2P(O)(NCS)$$

Phosphoryl-triisothiocyanate reacts with amines in stages; with ammonia phosphoryl-trithiourea is formed.[250]

$$Cl_2P(O)(NCO) + H_2O \longrightarrow Cl_2P(O)(NH\text{-}COOH)$$

$$Cl_2P(O)(NCO) + ROH \longrightarrow Cl_2P(O)(NH\text{-}COOR)$$

$$Cl_2P(O)(NCO) + RSH \longrightarrow Cl_2P(O)\text{-}NH\text{-}CO\text{-}SR$$

$$Cl_2P(O)(NCO) + R_2NH \longrightarrow Cl_2P(O)\text{-}NH\text{-}CO\text{-}NR_2$$

$$F_2P(O)(NCO) + R_2NH \longrightarrow F_2P(O)\text{-}NH\text{-}CO\text{-}NR_2$$

The hydrolysis of phosphorisocyanatic dichloride with concentrated hydrochloric acid at -20° C gives N-dichlorophosphoryl-carbamic-acid.[471] The alkyl and aryl esters are obtained by the reaction with an equimolar amount of an alcohol or phenol, respectively. In the latter case it is preferable to work in a solvent such as benzene or petroleum ether.[456,467] The analogous thiocarbamic acid derivatives can be prepared by using mercaptans. Phosphorisocyanatidic dichlorides and also difluorides react, in ethereal solution in the cold, with equimolar amounts of a primary or secondary aromatic amine to produce N-dihalophosphorylureas.[450]

$$P(O)(OR)_2(NCO) + H_2O \longrightarrow P(O)(OR)_2(NH_2) + CO_2$$

$$P(O)(OR)_2(NCO) + R'OH \longrightarrow P(O)(OR)_2(NH\text{-}CO\text{-}OR')$$

$$P(O)(OR)_2(NCO) + NH_3 \longrightarrow P(O)(OR)_2(NH\text{-}CO\text{-}NH_2)$$

Dialkyl (and diaryl) phosphorisocyanatidates give on hydrolysis dialkyl (or diaryl) phosphoramidates, on alcoholysis, N-dialkoxy (or diaryl)-phosphorylcarbamic acid esters,[430,468] and on ammonolysis or aminolysis N-dialkoxy-(or diaryloxy)-phosphorylureas.[468]

$$OCN\text{-}P(O)(NR_2)_2 + R'SH \longrightarrow R'S\text{-}CO\text{-}NH\text{-}P(O)(NR_2)_2$$

$$OCN\text{-}P(O)(NR_2)_2 + R'NH_2 \longrightarrow R'\text{-}NH\text{-}CO\text{-}NH\text{-}P(O)(NR_2)_2$$

With mercaptans, phosphorodiamidic isocyanates give N-diamino-phosphoryl-thiocarbamic acid S-esters, while with amines the respective ureas are produced.

$$OCN\text{-}P(O)Cl_2 + RNH\text{-}NH_2 \longrightarrow R\text{-}NH\text{-}NH\text{-}CO\text{-}NH\text{-}P(O)Cl_2$$

$$OCN\text{-}P(O)(OR)_2 + RNH\text{-}NH_2 \longrightarrow R\text{-}NH\text{-}NH\text{-}CO\text{-}NH\text{-}P(O)(OR)_2$$

Substituted hydrazines react analogously as amines to form semicarbazides.[470]

$$SCN\text{-}P(O)(OR)_2 + R'OH \longrightarrow R'O\text{-}CS\text{-}NH\text{-}P(O)(OR)_2$$

$$SCN\text{-}P(O)(NR_2)_2 + R'OH \longrightarrow R'O\text{-}CS\text{-}NH\text{-}P(O)(NR_2)_2$$

$$SCN\text{-}P(O)(OR)_2 + R'SH \longrightarrow R'S\text{-}CS\text{-}NH\text{-}P(O)(OR)_2$$

$$SCN\text{-}P(S)(OR)_2 + R'OH \longrightarrow R'O\text{-}CS\text{-}NH\text{-}P(S)(OR)_2$$

$$SCN\text{-}P(S)(NR_2)_2 + R'OH \longrightarrow R'O\text{-}CS\text{-}NH\text{-}P(S)(NR_2)_2$$

Isothiocyanates also react with alcohols or mercaptans in the warm to produce thiocarbamic acid esters or dithiocarbamic acid esters, respectively.[268] The reactions with alcoholates are especially mild, the sodium salt of the ester being formed; the free compound is readily obtained by treatment of the salt with an equimolar amount of an acid.[524]

$$SCN\text{-}P(O)(OR')_2 + RNH_2 \longrightarrow R\text{-}NH\text{-}CS\text{-}NH\text{-}P(O)(OR')_2$$

$$SCN\text{-}P(O)(OR')(NR''_2) + RNH_2 \longrightarrow R\text{-}NH\text{-}CS\text{-}NH\text{-}P(O)(OR')(NR''_2)$$

$$SCN\text{-}P(O)(NR'_2)_2 + RNH_2 \longrightarrow R\text{-}NH\text{-}CS\text{-}NH\text{-}P(O)(NR'_2)_2$$

$$SCN\text{-}P(S)(OR')_2 + RNH_2 \longrightarrow R\text{-}NH\text{-}CS\text{-}NH\text{-}P(S)(OR')_2$$

$$SCN\text{-}P(S)(NR'_2)_2 + RNH_2 \longrightarrow R\text{-}NH\text{-}CS\text{-}NH\text{-}P(S)(NR'_2)_2$$

Ammonia, primary aliphatic, and primary aromatic amines are cleanly added to phosphoric isothiocyanate derivatives to produce thiourea derivatives.[268,587,589,760]

$$(RO)_2P(O)NCS + 2\ Cl_2 \longrightarrow (RO)_2P(O)(N{=}CCl_2) + SCl_2$$

$$(RO)P(O)(NCS)_2 + 4\ Cl_2 \longrightarrow (RO)P(O)(N{=}CCl_2)_2 + 2\ SCl_2$$

$$(RO)P(S)F(NCS) + 4\ Cl_2 \longrightarrow ClP(O)F(N{=}CCl_2) + 2\ SCl_2 + RCl$$

The isothiocyanate groups of dialkyl phosphorisothiocyanatidates and alkyl phosphorodiisothiocyanatidates are converted by elemental chlorine to isocyanatodichloride groups. In other cases the chlorination can go further.[192,495]

XIII. $\geqslant$P + oxidans

Oxidation of phosphorus(III) compounds:

$$P(NCO)_3 + O_3 \longrightarrow OP(NCO)_3 + O_2$$

$$P(NR_2)_3 + H_2O_2 \longrightarrow OP(NR_2)_3 + H_2O$$

$$P(NR_2)_3 + Cl_2 \longrightarrow Cl_2P(NR_2)_3 \xrightarrow[-HCl]{+\ H_2O} OP(NR_2)_3$$

$$(R'O)P(NR_2)_2 + H_2O_2 \longrightarrow (R'O)P(O)(NR_2)_2 + H_2O$$

Some examples of the oxidation of phosphorus(III) nitrogen compounds are shown in the above equations. Ozonated oxygen[378] or hydrogen peroxide[829] are used as oxidizing agents; also oxidation with chlorine and subsequent hydrolysis is successful in many cases.[737b]

$$(R_2N)_3P + S \longrightarrow (R_2N)_3PS$$

$$(R_2N)_2P\text{-}OR' + S \longrightarrow (R_2N)_2P(S)(OR')$$

$$R_2N\text{-}P(OR')_2 + S \longrightarrow R_2N\text{-}P(S)(OR')_2$$

$$R_2N\text{-}PCl_2 + S \longrightarrow R_2N\text{-}P(S)Cl_2$$

$$(R_2N)_2P\text{-}SR' + S \longrightarrow (R_2N)_2P(S)(SR')$$

$$(RS)_2P\text{-}NR'_2 + S \longrightarrow (RS)_2P(S)(NR'_2)$$

Derivatives of phosphorothioic acid are obtained from phosphorus(III) nitrogen compounds by the addition of sulfur. Some of the reactions take place at room temperature, especially with diesters of phosphoramidous acids, in other cases temperatures as high as 200 °C are necessary.[47,573,673,829]

$$(RO)_2P(O)H + (SCN)_2 \longrightarrow (RO)_2P(O)\text{-}NCS + HSCN$$

$$(RO)_2P(S)H + (SCN)_2 \longrightarrow (RO)_2P(S)\text{-}NCS + HSCN$$

Dialkylesters of phosphorous and phosphorothious acids react with dithiocyanogen in a similar manner to the reactions with halogens, to give dialkyl phosphorisothiocyanatidates or phosphor-isothiocyanatido-thionates. The reactions are performed in benzene or cyclohexane; they are relatively slow but give good yields.[268,587]

$$P(NCO)_3 + SO_3 \longrightarrow OP(NCO)_3 + SO_2$$

$$P(NCO)_3 + NO_2 \longrightarrow OP(NCO)_3 + NO$$

$$(PhO)P(NCO)_2 + NO_2 \longrightarrow (PhO)P(O)(NCO)_2 + NO$$

Other oxidizing agents used with phosphorus(III)isocyanates are sulfur trioxide,[471] nitrogen dioxide, or a nitrogen dioxide/oxygen mixture.[304a]

$$P(OR)_3 + HN_3 \longrightarrow (RO)_2P(O)(NHR) + N_2$$

$$P(OR)_3 + ArN_3 \longrightarrow (RO)_3P{=}N\text{-}Ar + N_2$$

$$(RO)_2P\text{-}NH\text{-}COR + ArN_3 \longrightarrow (RO)_2P(NHAr)({=}N\text{-}COR) + N_2$$

$$HP(O)(OR)_2 + ArN_3 \longrightarrow (RO)_2P(O)\text{-}N{=}N\text{-}NH\text{-}Ar \quad \text{or} \quad (RO)_2P(O)(NHAr) + N_2$$

$$P(NR_2)_3 + ArN_3 \longrightarrow (R_2N)_3P{=}N\text{-}N{=}N\text{-}Ar \quad \text{or} \quad (R_2N)_3P{=}N\text{-}Ar + N_2$$

Trialkylphosphites react with hydrazoic acid to produce, essentially, dialkyl N-alkyl-phosphoramidates, while phenylazide in ether solution at 0 °C is added with nitrogen evolution to give N-phenyl-trialkoxy-phosphinimides.[403] This latter reaction sometimes gives very good yields. Dialkyl N-acyl-phosphoramidites are equally easily converted to N-acyl-dialkylanilido-phosphinimides. Dialkyl phosphites in the presence of triethylamine, or the sodium or ammonium salts of the dialkyl phosphites react with phenylazide to form dialkylesters of phenylamidazophosphoric acid.[324] In many cases diesters of

N-phenyl-phosphoramidic acid are produced directly by the reaction between diesters of phosphorous acid and phenylazide with the splitting off of nitrogen. Phosphorous triamides form triaminophosphazides with aryl and also acylazides; these products are stable at normal temperatures,[290] but at higher temperatures evolve nitrogen.

$$P(OR')_3 + R\text{-}CO\text{-}N_3 \longrightarrow (R'O)_3P{=}N\text{-}CO\text{-}R + N_2$$

$$(RO)_2P\text{-}NHR' + R\text{-}CO\text{-}N_3 \longrightarrow (RO)_2P(NHR')({=}N\text{-}CO\text{-}R) + N_2$$

Triesters of phosphorous acid are oxidized on boiling with carbonic acid azides to form trialkyl or triaryl N-acyl-phosphorimidates.[440] Dialkylesters of phosphoramidous acid react accordingly.

$$P(OR')_3 + R\text{-}SO_2\text{-}N_3 \longrightarrow (R'O)_3P{=}N\text{-}SO_2\text{-}R + N_2$$

$$P(SR')_3 + R\text{-}SO_2\text{-}N_3 \longrightarrow (R'S)_3P{=}N\text{-}SO_2\text{-}R + N_2$$

When trialkyl or triaryl esters of phosphorous acid are allowed to react with sulfonylazides triesters of N-sulfonyl-phosphorimidic acid are formed. Triesters of phosphorotrithious acids behave similarly to give S,S,S-triesters of N-sulfonyl-phosphorimido-trithioic acid.[334]

$$PHal_3 + R\text{-}SO_2\text{-}N(Cl)Na \longrightarrow Hal_3P{=}N\text{-}SO_2\text{-}R + NaCl$$

$$P(OR')_3 + R\text{-}SO_2\text{-}N(Cl)Na \longrightarrow (R'O)_3P{=}N\text{-}SO_2\text{-}R + NaCl$$

$$(ArO)PCl_2 + R\text{-}SO_2\text{-}N(Cl)Na \longrightarrow (ArO)Cl_2P{=}N\text{-}SO_2\text{-}R + NaCl$$

$$(ArO)_2PCl + R\text{-}SO_2\text{-}N(Cl)Na \longrightarrow (ArO)_2ClP{=}N\text{-}SO_2\text{-}R + NaCl$$

The sodium salts of N-chloroarylsulfonylamides are strong oxidizing agents for phosphorus(III) compounds. With phosphorus III halides the reactions are strongly exothermic, in the absence of a solvent even explosive; in carbon tetrachloride they are slower and give N-sulfonyl-trihalophosphinimides. With triesters of phosphorous acid in the absence of water, e.g., in boiling CCl_4, triesters of N-sulfonyl-phosphorimidic acid are formed.[140] Diarylesters of phosphorochloridous acid and arylesters of phosphorodichloridious acid react similarly.[464] Alkylesters of phosphorochloridous and phosphorodichloridous acid do not react in this way but give unknown products.

$$(RO)_2P(O)\text{-}N_3 + P(OR)_3 \longrightarrow (RO)_2P(O)\text{-}N{=}P(OR)_3 + N_2$$

Dialkylesters of phosphorazidic acid react with trialkyl phosphites to give trialkylesters of dialkoxyphosphoryl phosphorimidic acid.[404]

$$P(OR)_3 + N_2CH_2 \longrightarrow (RO)_3P{=}N{-}N{=}CH_2$$

$$(RO)_2P(O)H + N_2CH_2 \longrightarrow (RO)_2P(O){-}N{=}N{-}CH_3$$

$$P(NR_2)_3 + N_2CR_2 \longrightarrow (R_2N)_3P{=}N{-}N{=}CR_2$$

$$(RO)_2P(O)H + [R'N{\equiv}N]X \longrightarrow (RO)_2P(O){-}N{=}N{-}R' + HX$$

Diazomethane and its derivatives can also be used to oxidize compounds of phosphorous acid. Diazomethane is added to trialkylphosphites to form trialkoxyphosphaformaldazines. Dialkylphosphites are not alkylated at the oxygen atom by diazomethane in the presence of triethylamine, but, through direct addition at phosphorus, are converted to dialkylesters of methylazophosphoric acid.[324] From hexaalkyl phosphorous triamides and aliphatic diazocompounds are formed triaminophosphaketazines.[283a] Finally the reactions of dialkyl phosphites with diazonium salts of aromatic and heterocyclic compounds should be mentioned; these give, in weakly acidic to neutral solutions at room temperature, N-aryl,N'-dialkoxyphosphorylazo-compounds.[838]

$$P(OR)_3 + R'_2NCl \longrightarrow (RO)_2P(O)NR'_2 + RCl$$

$$HP(O)(OR)_2 + R'_2NCl \longrightarrow (RO)_2P(O)NR'_2 + HCl$$

$$P(OAr)_3 + R'_2NCl \longrightarrow (ArO)_3P{=}NR' + R'Cl$$

Trialkylphosphites react with N-chloride compounds of secondary aliphatic amines to produce dialkyl phosphoramidates.[678] The reaction can be carried out in carbon tetrachloride solution. Also dialkyl and diaryl phosphites react readily and in good yields to form dialkyl or diaryl phosphoramidates. N-unsubstituted compounds are prepared by using aqueous chloramine, NH_2Cl, solutions.[679] By heating the adduct of triphenyl phosphite and N-chlorodiethylamine to 100° C in vacuum triphenylphosphorimidate is obtained by cleavage of ethyl chloride.

$$P(OR)_3 + (R'O)_2C{=}N{-}Cl \longrightarrow (RO)_2P(O){-}N{=}C(OR')_2 + RCl$$

$$P(OR)F_2 + (R'O)_2C{=}N{-}Cl \longrightarrow F_2P(O){-}N{=}C(OR')_2 + RCl$$

$$Cl_2P(OAr) + (RO)_2C{=}N{-}Cl \longrightarrow (ArO)Cl_2P{=}N{-}CO{-}OR + RCl$$

$$ClP(OAr)_2 + (RO)_2C{=}N{-}Cl \longrightarrow (ArO)_2ClP{=}N{-}CO{-}OR + RCl$$

Trialkyl phosphites, alkyl phosphorodichloridites, and dialkyl phosphorochloridites react with dialkylesters of N-chloroimino-carbonic acid to form N-phosphoryl-iminocarbamic acid dialkyl esters and alkyl chlorides.[672] Esters of phosphorofluoridous and phosphorodifluoridous acids react in a similar way.[672] Upon reaction of dialkylesters of N-chloroimino-carbonic acid with phenyl-phosphorodichloridite or diphenyl-phosphorochloridite phenyl-N-acyl-phosphorodichloridimidate or diphenyl-N-acyl-phosphorochloridimidate and alkyl chlorides are formed. On heating to about 100 °C these compounds lose a further molecule of alkyl chloride to form phenyl-phosphorisocyanatidic chloride or diphenyl-phosphoriso-cyanatidate, respectively.

$$P(OR)_3 + R''\text{-}C(OR')\text{=}NCl \longrightarrow (RO)_2P(O)\text{-}N\text{=}C(OR')R'' + RCl$$

$$P(OAr)_3 + R\text{-}C(OR')\text{=}NCl \longrightarrow (ArO)_3P\text{=}N\text{-}CO\text{-}R + R'Cl$$

$$P(OR)_3 + R'\text{-}C(NH_2)\text{=}N\text{-}Cl \longrightarrow (RO)_2P(O)\text{-}N\text{=}C(NH_2)R' + RCl$$

Trialkyl phosphites and triaryl phosphites react differently with esters of N-chlorocarboxylic acid imides. Trialkyl phosphites form N-(dialkoxy-phosphoryl)carboxylic acid imide esters,[672] while triaryl phosphites give triaryl N-acyl-phosphorimidates.[193] N-chlorbenzamidine reacts with phosphites, which contain at least one alkyl group, to form N-dialkoxy-(or N-diaryloxy-)-phosphoryl benzamidines with elimination of alkyl chloride.

$$(RO)_2P(O)H + CCl_4 + 2\ HNR_2^1 \longrightarrow (RO)_2P(O)(NR_2^1) + HCCl_3 + [R_2^1NH_2]Cl$$

$$(RO)_2P(O)H + CCl_4 + 2\ H_2N\text{-}NH_2 \longrightarrow (RO)_2P(O)\text{-}NH\text{-}NH_2 + HCCl_3 + H_2N\text{-}NH_2\cdot HCl$$

$$(PhCH_2O)_2P(O)H + CCl_4 + HN\text{=}C(NH_2)\text{-}R \longrightarrow (PhCH_2O)_2P(O)\text{-}N\text{=}C(NH_2)R + CHCl_3 + HCl$$

$$(RO)_2P(O)H + CCl_4 + 2\ HN\text{=}C(NH_2)(NHR') \longrightarrow (RO)_2P(O)\text{-}N\text{=}C(NH_2)(NHR') + CHCl_3 + (H_2N)(R'NH)C\text{=}NH\cdot HCl$$

Esters of phosphorous acid react with primary and secondary amines, and also with ammonia in halogenated hydrocarbon solvents, to give diesters of phosphoramidic acid. Carbon tetrachloride is used most frequently; $CHBr_3$, CBr_4, and others are sometimes used. With the reactions of aromatic amines a strongly basic tertiary amine, such as triethylamine or also solid alkali hydroxide, must be added. Dialkylphosphites give, on treatment with 2 moles of hydrazine in carbon tetrachloride,

dialkyl phosphorohydrazates.[475] Under similar conditions
amidines give N-phosphorylamidines. Similarly diesters
of phosphorous acid react with guanidine in the presence
of carbon tetrachloride or another perhalo-hydrocarbon
to afford dialkyloxy-phosphoryl-guanidines.[174]

$$HP(O)(OR)_2 + RO-CO-N=N-CO-OR \longrightarrow RO-CO-N-P(O)(OR)_2$$
$$| \\ RO-CO-NH$$

$$P(OR)_3 + RO-CO-N=N-CO-OR \longrightarrow RO-CO-N-P(O)(OR)_2$$
$$| \\ RO-CO-N-R$$

Esters of phosphorous acid combine with azo-bis(ethylfor-
mate) to give derivatives of phosphoric acid hydrazides.[614]

$$PCl(OR)_2 + R_2'N-S-R'' \longrightarrow (RO)P(O)(NR_2')(SR'') + RCl$$

$$PCl_2(OR) + R_2'N-S-R'' \longrightarrow (R''S)-P(O)(NR_2')Cl + RCl$$

$$P(NR_2)_2(OR') + Cl-SCCl_3 \longrightarrow (R_2N)_2P(O)-SCCl_3 + R'Cl$$

Dialkyl phosphorochloridites react with dialkylamino
alkyl sulfides on heating to give an alkyl chloride and
O,S-diesters of phosphoramidothiolic acid. Analogously
alkyl phosphorodichloridites form S-esters of phosphoramidochloridothiolic acid. Alkyl phosphorodiamidites,
finally, undergo a type of Michaelis-reaction with sulfur
chlorides. Alkyl chlorides are split off and S-esters of
phosphorodiamido-thiolic acids are formed.[561]

$$P(NR_2)(OR)_2 + Cl_3C-CHO \longrightarrow (RO)P(O)(NR_2)(O-CH=CCl_2) + RCl$$

$$\begin{matrix} H_2C-O \\ | \\ H_2C-O \end{matrix} > P-NR_2 + CH_3-CO-CH_2Cl \longrightarrow$$

$$ClCH_2-CH_2-O-P(O)(NR_2)[OC(CH_3)=CH_2]$$

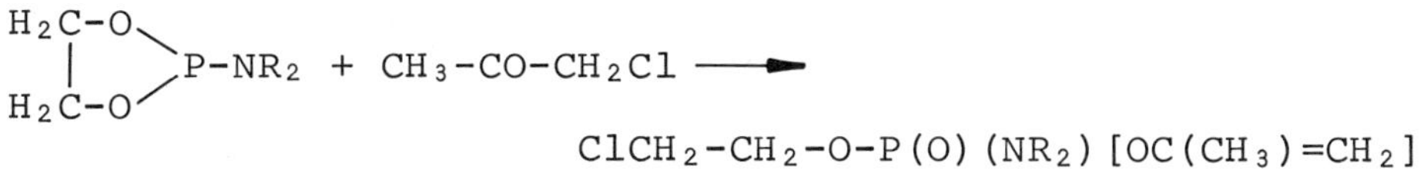

Dialkyl-N,N-dialkyl-phosphoramidites react in the
warm with α-halocarbonyl compounds with the expulsion of
alkyl halide to form O-alkyl-O-vinyl N,N-dialkyl-phos-
phoramidates, while with cyclic dialkylaminodioxa-phos-
pholanes, ring cleavage occurs to give analogous products.

$$2\ P(OR)_3 + \begin{matrix} F \\ \diagdown \\ F \end{matrix} C \diagup \begin{matrix} N \\ \| \\ N \end{matrix} \longrightarrow (RO)_3PF_2 + (RO)_3P=N-C\equiv N$$

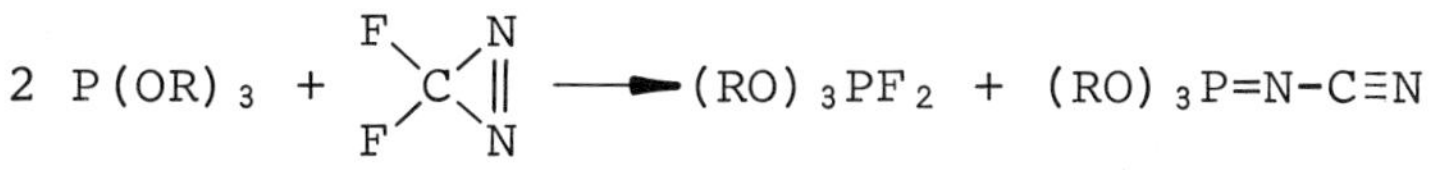

$$2 \ P(NR_2)_3 + \underset{F}{\overset{F}{>}}C\overset{N}{\underset{N}{<}}\| \longrightarrow (R_2N)_3PF_2 + (R_2N)_3P=N-C\equiv N$$

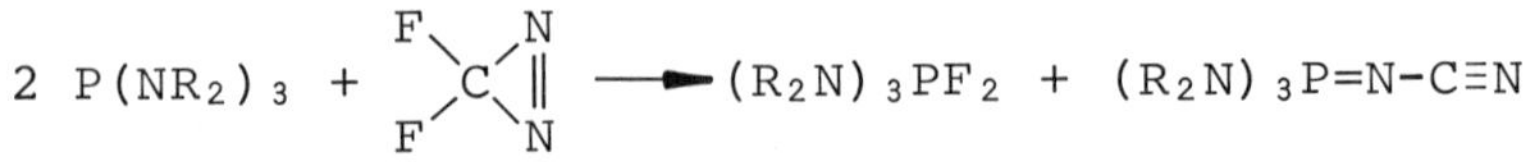

Difluorodiazirine oxidizes trialkyl phosphites, triaryl phosphites, phosphorous triamides, or trialkyl phosphoro-trithioites to give triesters or triamides of N-cyano-phosphorimidic acid and fluorophosphoranes.[599]

$$3 \ P(OR)_3 + o\text{-}MeC_6H_4NO_2 \longrightarrow 2 \ (RO)_3PO + o\text{-}MeC_6H_4N=P(OR)_3$$

$$3 \ P(OR)_3 + Ar\text{-}NSO \longrightarrow (RO)_3PS + (RO)_3PO + (RO)_3P=NAr$$

$$3 \ P(OR)_3 + (R'O)_2P(O)\text{-}NSO \longrightarrow$$
$$(RO)_3PS + (RO)_3PO + (RO)_3P=N-P(O)(OR')_2$$

Some further reactions in which phosphorus III compounds are oxidized to phosphorus(V)-nitrogen compounds are shown in the above equations.[847,874,906]

XIVa. OP-X + Y (X, Y = hal, pseudohal)

Exchange of halogen and pseudohalogen ligands in phosphoryl-nitrogen compounds:

$$R_2N\text{-}P(O)Cl_2 + 2 \ NaF \longrightarrow R_2N\text{-}P(O)F_2 + 2 \ NaCl$$

$$R_2N\text{-}P(O)Cl(OR') + KF \longrightarrow R_2N\text{-}P(O)F(OR') + KCl$$

Phosphoramidic dichlorides react exothermically with alkali fluoride, if necessary mixed with SbF_3, producing chlorine/fluorine exchange.[374] Also potassium fluoride, potassium hydrogen fluoride, $[R_3NH][HF_2]$, or zinc fluoride can be used in inert solvents such as halogenated hydrocarbons, benzene, or toluene to excecute chlorine/fluorine exchange. Esters of phosphoroamidochloridic acid or phosphorodiamidic chlorides are also converted to the respective fluorine compounds by these routes. Esters of phosphoramidocyanidic acid on similar treatment undergo cyanide/fluorine exchange.[753]

$$3 \ OPCl_2(NCO) + 2 \ SbF_3 \longrightarrow 3 \ OPF_2(NCO) + 2 \ SbCl_3$$

In many cases the use of antimony trifluoride allows the readily replacement of the chlorine atoms in chlorides of phosphoric acid by fluorine atoms. For example, phos-phorisocyanatidic dichloride gives phosphorisocyanatidic difluoride.[497]

$$(RO)_2P(O)Cl + NaN_3 \longrightarrow (RO)_2P(O)\text{-}N_3 + NaCl$$

$$(R_2N)_2P(O)Cl + NaN_3 \longrightarrow (R_2N)_2P(O)\text{-}N_3 + NaCl$$

Phosphoric acid chlorides are converted to the respective azides on treatment with sodium azide in acetone; the yields are usually very good.[771]

$$R_2N\text{-}P(O)Cl(OR') + NaCN \longrightarrow R_2N\text{-}P(O)(CN)(OR') + NaCl$$

$$R_2N\text{-}P(O)Cl_2 + 2\ NaCN + HOR' \longrightarrow R_2N\text{-}P(O)(CN)(OR') + 2\ NaCl + HCN$$

Esters of phosphoroamidochloridic acid exchange a chlorine atom for a cyanide group on treatment with an alkali cyanide. The same products are obtained from phosphoramidic dichloride when the reaction with alkali cyanide is carried out either in an inert solvent with 1 mole of alcohol or in excess alcohol at higher temperatures.[374]

$$OPCl_3 + 3\ AgOCN \longrightarrow OP(NCO)_3 + 3\ AgCl$$

Phosphoric triisocyanate is obtained from the reaction of phosphoryl chloride with silver cyanate; the yield, however, is only 11%.[37]

$$OPCl_3 + 3\ NH_4SCN \longrightarrow OP(NCS)_3 + 3\ NH_4Cl$$

$$R_2N\text{-}P(O)Cl_2 + 2\ AgSCN \longrightarrow R_2N\text{-}P(O)(NCS)_2 + 2\ AgCl$$

$$(RO)_2P(O)Cl + KSCN \longrightarrow (RO)_2P(O)(NCS) + KCl$$

$$(RO)P(O)Cl(NR_2') + KSCN \longrightarrow (RO)P(O)(NR_2')(NCS) + KCl$$

$$(R_2N)_2P(O)Cl + KSCN \longrightarrow (R_2N)_2P(O)(NCS) + KCl$$

The chlorine atoms in compounds of phosphoric acid are especially easily replaced by isothiocyanate groups.[589] The reaction can be carried out with potassium or ammonium isothiocyanate without a solvent or in acetone, benzene or acetonitrile. Liquid sulfur dioxide can also be used as reaction medium, e.g., in the reaction of phosphoryl chloride with ammonium thiocyanate;[837] silver thiocyanate can also be employed.[482C]

XIVb. SP-X + Y (X, Y = hal, pseudohal)

Exchange of halogen and pseudohalogen ligands in thiophosphoryl-nitrogen compounds:

$$3\ R_2N\text{-}P(S)Cl_2 + 2\ SbF_3 \longrightarrow 3\ R_2N\text{-}P(S)F_2 + 2\ SbCl_3$$

Chlorides of phosphorothioic acid are converted to the corresponding fluorides by alkali fluorides in dry solvents or with SbF_3 without a solvent.[754a,756]

$$(RO)_2P(S)Cl + NaN_3 \longrightarrow (RO)_2P(S)\text{-}N_3 + NaCl$$

In order to replace the chlorine atoms in phosphorothioic acid chlorides by azide groups, alkali azides in solvents such as butanone are used. From O,O'-dialkyl phosphorochloridothionate is thus obtained O,O'-dialkyl phosphorazidothionate, which is distillable under vacuum.[756]

$$R_2N\text{-}P(S)Cl_2 + 2\ NaCN + HOR' \longrightarrow R_2N\text{-}P(S)(CN)(OR') + HCN + 2\ NaCl$$

The chlorine atoms in O-esters of phosphoroamidochloridothionic acid are readily exchanged by cyanide groups by treatment with alkali cyanides at slightly elevated temperatures. The same products are also obtained from the reaction of phosphoramidothioic dichlorides with alkali cyanides in excess alcohol or in an inert solvent, such as acetonitrile, and the equivalent amount of alcohol.[755]

$$(R_2N)_2P(S)Cl + KOCN \longrightarrow (R_2N)_2P(S)NCO + KCl$$

Phosphorodiamidothioic chlorides are converted to phosphorodiamidothioic isocyanates by reaction with potassium cyanate.[661]

$$SPCl_3 + 3\ NH_4SCN \longrightarrow SP(NCS)_3 + 3\ NH_4Cl$$

$$(RO)_2P(S)Cl + KSCN \longrightarrow (RO)_2P(S)NCS + KCl$$

$$(R_2N)(R'O)P(S)Cl + KSCN \longrightarrow (R_2N)(R'O)P(S)NCS + KCl$$

$$(R_2N)_2P(S)Cl + KSCN \longrightarrow (R_2N)_2P(S)NCS + KCl$$

Thiophosphoryl chloride reacts with ammonium thiocyanate in acetonitrile or liquid SO_2 in very good yield. Similarly, the respective isothiocyanate derivatives can be obtained from the corresponding thiophosphoryl chloride by treatment with potassium thiocyanate in acetone or acetonitrile.[425,524,587,837]

$$SPCl_3 + SP(NCS)_3 \rightleftharpoons SP(NCS)_2Cl + SP(NCS)Cl_2$$

Thiophosphoryl halides and phosphorothioic triisothiocyanates undergo ligand exchange at higher temperatures.[309]

XV. Pyrolysis Reactions:

$$RO\text{-}CO\text{-}N{=}PCl_3 \longrightarrow OCN\text{-}P(O)Cl_2 + RCl$$

$$RO\text{-}CO\text{-}N{=}PCl_2(OPh) \longrightarrow OCN\text{-}P(O)Cl(OPh) + RCl$$

$$RO\text{-}CO\text{-}NH\text{-}P(O)(OR')_2 \longrightarrow OCN\text{-}P(O)(OR')_2 + ROH$$

$$RO\text{-}CO\text{-}NH\text{-}P(O)(NR_2)_2 \longrightarrow OCN\text{-}P(O)(NR_2)_2 + ROH$$

N-carbalkoxy-trichlorophosphinimide decomposes slowly at lower temperatures; at higher temperatures the decomposition, with evolution of alkyl chloride, to give phosphorisocyanatidic dichloride is explosive.[430] Similarly, phenyl-N-carbalkoxy-phosphorodichloridimidates yield phenyl-phosphorochloridisocyanatidate.[195,597] Carbamic acid alkylesters decompose to alcohols and isocyanates on heating.[468] The pyrolysis can, for example, be carried out in boiling toluene.[661]

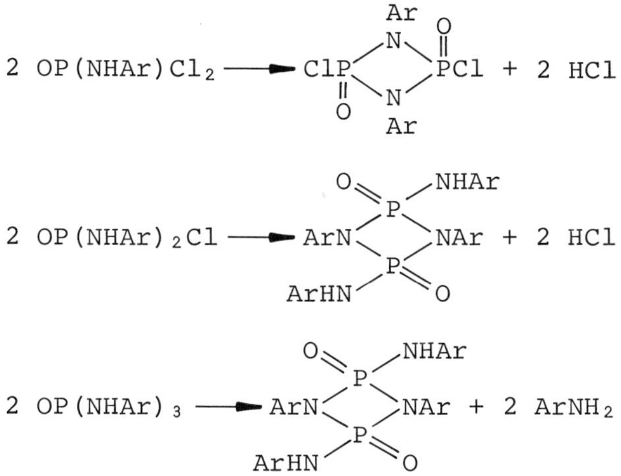

N-aryl-phosphoramidic dichlorides undergo self-condensation upon heating to form dichloro-dioxo-cyclodiphosphazanes with evolution of hydrogen chloride,[574] N, N'-diaryl phosphorodiamidic chlorides and phosphoric triamides of primary aliphatic and aromatic amines split off hydrogen chloride or amine, respectively, to produce diamino-dioxo-cyclodiphosphazanes.[583]

$$3\ (R_2N)P(O)Cl(OR) \longrightarrow [(R_2N)PO_2]_3 + 3\ RCl$$

Alkyl N,N-dialkyl-phosphoramidochloridates decompose on heating into alkyl chlorides and trimeric metaphosphoric acid dialkylamides.[573]

$$2\ (ArO)_2P(O)\text{-}NH\text{-}NH_2 \longrightarrow (ArO)_2P(O)\text{-}NH\text{-}NH\text{-}P(O)(OAr)_2 + H_2N\text{-}NH_2$$

Hydrazine is split off from diphenyl phosphorohydrazidate on heating above its melting point to form N,N'-bis(diphenoxyphosphoryl)-hydrazine.[475]

$$(EtO)_2P(O)\text{-}NHAr \longrightarrow (HO)_2P(O)\text{-}NHAr + 2\ C_2H_4$$

$$(t\text{-}BuO)_2P(O)NHAr \longrightarrow (HO)_2P(O)NHAr + 2\ C_4H_8$$

Dialkyl N-aryl-phosphoramidates, on heating with aniline in a high boiling solution solvent such as toluene or xylene, are converted to N-aryl-phosphoramidic acid,[771] on heating the corresponding di-t-butylesters N-aryl-phosphoramidic acid and iso-buten are formed.[336,337]

$$R_2N\text{-}P(S)(OR')_2 \longrightarrow R_2N\text{-}P(O)(OR')(SR')$$

$$(R_2N)_2P(S)(OR') \longrightarrow (R_2N)_2P(O)(SR')$$

A series of O,O'-diesters of phosphoramidothioic acid and O-esters of phosphorodiamidothioic acid rearrange on heating above 150° C to the corresponding thiolates.[395]

$$OPCl_3 + 2\ OP(NR_2)_3 \longrightarrow 3\ OPCl(NR_2)_2$$

Phosphoryl chloride and hexaalkyl phosphoric triamides, in molar ratio 1:2 react together at about 150° C to form N,N,N',N'-tetraalkyl-phosphorodiamidic chlorides in good yields.

$$R'\text{-}SO_2\text{-}N{=}P(OR)_3 \longrightarrow R'\text{-}SO_2\text{-}NR\text{-}P(O)(OR)_2$$

Trialkyl N-sulfonyl-phosphorimidates rearrange, on heating to about 200° C, to dialkyl N-alkyl-N-sulfonyl-phosphoramidates.[802]

XVI. $\geqslant P(OH) + ClP\leqslant$
$\geqslant P(OR) + ClP\leqslant$

Reactions of phosphoric acid and alkyl phosphates with phosphoric chlorides:

$$(RO)_2P(O)(OH) + Cl_2P(O)(NR_2) \longrightarrow (RO)_2P(O)\text{-}O\text{-}P(O)Cl(NR_2) + HCl$$

Dialkyl phosphates react with phosphoramidic dichlorides in the presence of a tertiary amine to form P,P-diesters of diphosphoric P'-amide-P'-chloride.[376a]

$$(RO)_2P(S)(OH) + ClP(S)(OR)(NR_2) \longrightarrow (RO)_2P(S)\text{-}O\text{-}P(S)(OR)(NR_2) + HCl$$

$$(RO)_2P(S)(OH) + ClP(S)(NR_2)_2 \longrightarrow (RO)_2P(S)\text{-}O\text{-}P(S)(NR_2)_2 + HCl$$

$$(RO)_2P(S)(OH) + ClP(O)(NR_2)_2 \longrightarrow (RO)_2P(S)\text{-}O\text{-}P(O)(NR_2)_2 + HCl$$

$$(R_2N)(RO)P(S)(OH) + ClP(S)(NR_2)_2 \longrightarrow (R_2N)(RO)P(S)\text{-}O\text{-}P(S)(NR_2)_2 + HCl$$

Dialkyl phosphorothionates and monoalkyl phosphoramidothionates react with esters of phosphoramidochloridic or phosphoramidochloridothioic acids or phosphorodiamidic or phosphorodiamidicthioic chlorides to the respective derivatives of diphosphoric acid.[824] Triethylamine is usually used as HCl-acceptor.[637] Very many amino-derivatives of diphosphoric and higher polyphosphoric acids, and also of diphosphorothioic and diphosphorodithioic acids can be synthesized using the above shown condensation principle. The P-O-P bridging bonds are also formed by elimination of alkyl chloride from an alkyl ester and a chloride of phosphoric acid. Diphosphoric acid monoamide-triesters, symmetrical and unsymmetrical diamide-diesters, triamide-esters, tetramides, ester-amide-halides and the corresponding derivatives of diphosphor monothioic and diphosphor dithioic acids can be prepared by this route. Also the amino derivatives of triphosphoric and higher polyphosphoric acids are obtainable in this way.

$$(R_2N)P(O)(OR)Cl + OP(OR)_3 \longrightarrow (R_2N)P(O)(OR)\text{-}O\text{-}P(O)(OR)_2 + RCl$$

$$(R_2N)P(O)(OR)_2 + ClP(O)(OR)_2 \longrightarrow (R_2N)P(O)(OR)\text{-}O\text{-}P(O)(OR)_2 + RCl$$

Alkyl phosphoramidochloridates react with trialkyl phosphates on heating to form trialkyl diphosphoramidates.[754] The same products are also obtained from the reaction of dialkylphosphoramidates with dialkyl phosphorchloridates.[754]

$$(R_2N)_2P(O)(OR) + OPCl_3 \longrightarrow (R_2N)_2P(O)\text{-}O\text{-}P(O)Cl_2 + RCl$$

$$(R_2N)_2P(O)Cl + OP(OR)_3 \longrightarrow (R_2N)_2P(O)\text{-}O\text{-}P(O)(OR)_2 + RCl$$

$$(R_2N)_2P(O)(OR) + ClP(O)(OR)_2 \longrightarrow (R_2N)_2P(O)\text{-}O\text{-}P(O)(OR)_2 + RCl$$

Methyl phosphorodiamidate reacts with excess phosphorylchloride on heating to form methyl chloride and diphosphoric acid-P,P-diamide-P',P'-dichloride.[863] Diphosphoric acid-P,P-diester -P',P'-diamides are prepared by the reaction of phosphorodiamidic chloride with trialkyl phosphates at 130-140 °C in the presence of a copper powder catalyst.[754] This last product is also obtained from alkyl phosphorodiamidates and dialkyl phosphorochloridates.[743a]

$$(R_2N)P(O)(OR)Cl + (RO)P(O)(OR)(NR_2) \longrightarrow (R_2N)P(O)(OR)\text{-}O\text{-}P(O)(OR)(NR_2) + RCl$$

$$(R_2N)P(O)(OR)_2 + ClP(O)F(NR_2) \longrightarrow (R_2N)P(O)(OR)\text{-}O\text{-}P(O)F(NR_2) + RCl$$

$$(R_2N)P(O)FCl + (RO)P(O)F(NR_2) \longrightarrow (R_2N)P(O)F\text{-}O\text{-}P(O)F(NR_2) + RCl$$

Dialkyl phosphoramidates and alkyl phosphoramidochloridates together form symmetrical diesters of diphosphorodiamidic acid and alkyl chloride.[754] Analogously derivatives of diphosphoric acid are obtained from dialkyl N,N-dialkyl-phosphoramidates and N,N-dialkylphosphoramidic chloride fluorides or from phosphoramidic chloride fluorides and alkyl phosphoramidofluoridates.[824]

$$(R_2N)_2P(O)(OR) + ClP(O)(NR_2)(OR) \longrightarrow (R_2N)_2P(O)\text{-}O\text{-}P(O)(NR_2)(OR) + RCl$$

$$(R_2N)_2P(O)Cl + (RO)_2P(O)(NR_2) \longrightarrow (R_2N)_2P(O)\text{-}O\text{-}P(O)(NR_2)(OR) + RCl$$

$$(R_2N)_2P(O)(OR) + Cl_2P(O)(NR_2) \longrightarrow (R_2N)_2P(O)\text{-}O\text{-}P(O)(NR_2)Cl + RCl$$

$$(R_2N)_2P(O)(OR) + ClP(O)F(NR_2) \longrightarrow (R_2N)_2P(O)\text{-}O\text{-}P(O)F(NR_2) + RCl$$

$$(R_2N)_2P(O)Cl + (RO)P(O)F(NR_2) \longrightarrow (R_2N)_2P(O)\text{-}O\text{-}P(O)F(NR_2) + RCl$$

The above equations show other combinations of alkyl esters of phosphorodiamidic acids and phosphoric chlorides which give, on heating, derivatives of diphosphoric acid and alkyl halides.[754,767,824]

$$(R_2N)_2P(O)(OR) + ClP(O)(NR_2)_2 \longrightarrow (R_2N)_2P(O)\text{-}O\text{-}P(O)(NR_2)_2 + RCl$$

Diphosphoric tetramides are prepared in good yields by the reaction between phosphorodiamidic chlorides and alkyl phosphorodiamidates.[754]

$$(R_2N)_2P(O)(OR) + SPCl_3 \longrightarrow (R_2N)_2P(O)\text{-}O\text{-}P(S)Cl_2 + RCl$$

$$(R_2N)P(S)FCl + (RO)P(O)(NR_2)_2 \longrightarrow (R_2N)P(S)F\text{-}O\text{-}P(O)(NR_2)_2 + RCl$$

$$(R_2N)P(S)FCl + (RO)P(O)(NR_2)F \longrightarrow (R_2N)P(S)F\text{-}O\text{-}P(O)(NR_2)F + RCl$$

$$(R_2N)_2P(O)Cl + (RO)P(S)(NR_2)_2 \longrightarrow (R_2N)_2P(O)\text{-}O\text{-}P(S)(NR_2)_2 + RCl$$

$$(R_2N)_2P(S)Cl + (RO)P(O)(NR_2)_2 \longrightarrow (R_2N)_2P(S)\text{-}O\text{-}P(O)(NR_2)_2 + RCl$$

Methyl phosphorodiamidates react on warming with thiophosphoryl chloride to form diphosphoromonothioic acid P',P'-diamide-P,P-dichlorides by splitting off methyl chloride.[863] If the ester and/or the chloride part contain fluorine atoms bonded to phosphorus, these are not affected by the reaction and diphosphoromonothioic acid amide fluorides are formed.[824] Diphosphoromonothioic tetraamides result from the treatment of phosphorodiamidic chlorides with alkyl phosphorodiamidothionates or of alkyl phosphorodiamidates with phosphorodiamidothioic chlorides. Because of the low reactivity of alkyl phosphorodiamidates, the latter reaction proceeds slowly at higher temperatures.[639]

XVII. Miscellaneous

a. Formation of Phosphoric Acid Amides, Ester-Amides and Their Salts

$$(Ar_2N)_3P(O) + KOH \longrightarrow [(Ar_2N)_2P(O)(O^-)]K^+ + Ar_2NH$$

The amine groups of hexaaryl-phosphoric triamide are successively split off on boiling with potassium hydroxide solution. First potassium phosphorodiamidate is formed.[172]

$$(R_2N)_2P(O)(OMe) + NMe_3 \longrightarrow [(R_2N)_2P(O)(O^-)][NMe_4{}^+]$$

$$(R_2N)P(O)(OR')(OMe) + NMe_3 \longrightarrow [(R_2N)P(O)(OR')(O^-)][NMe_4{}^+]$$

Esters of phosphoramidic and phosphorodiamidic acid which contain at least one O-methyl group are converted by trimethylamine in acetone solution to tetramethylammonium salts of the respective acid.[155]

$$(R_2N)P(O)(OR')_2 \xrightarrow{Ba(OH)_2} (R_2N)P(O)(OR')(O^-)[Ba^{2+}/2] + R'OH$$

$$(RHN)P(O)(OR')_2 + LiCl \longrightarrow (RHN)P(O)(OR')(OLi) + R'Cl$$

$$(H_2N)P(S)(OR)_2 + NaOH \longrightarrow (H_2N)P(O)(OR)(SNa) + ROH$$

Diesters of phosphoramidic acid are saponified by strong alkalis such as barium hydroxide to the salts of the monoesters of phosphoramidic acid.[712,825] Ammonia can also be used to cleave diaryl phosphoramidates.[825] Dibenzyl phosphoramidates are partially debenzylated by warming with sodium iodide, lithium chloride, or barium iodide to form O-salts of benzyl phosphoramidate.[161] The equilibrium reaction can be performed in ketones, e.g., acetone. The free acid can be easily isolated by acidification of the salt solution. O,O'-diesters of phosphoramidothioic acid are partially saponified by treatment with a one or two molar amount of alcoholic alkali hydroxide solution to N-unsubstituted salts of O-esters of phosphoramidothiolic acid.[272,417]

$$(ArO)_2\overset{\overset{\displaystyle O}{\|}}{P}\text{-NH-CH}_2\text{-COOEt} + 3/2\ Ba(OH)_2 \longrightarrow [^-O\text{-}\underset{\underset{\displaystyle O^-}{|}}{\overset{\overset{\displaystyle O}{\|}}{P}}\text{-NH-CH}_2\text{-COO}^-][3/2\ Ba^{++}] + EtOH + 2\ ArOH$$

$$(RNH)\overset{\overset{\displaystyle O}{\|}}{P}(OCH_2\text{-}Ph)_2 + 2\ H_2(Pd) \longrightarrow (RNH)\overset{\overset{\displaystyle O}{\|}}{P}(OH)_2 + 2\ MePh$$

$$(R_2N)P(O)(OR')(OCH_2Ph) \xrightarrow[-\ MePh]{+\ H_2(Pd)} (R_2N)P(O)(OR')(OH)$$

Diaryl phosphoramidates of amino-acid esters can, in some cases, be converted to N-phospho-amino-acid salts with strong alkali; but this method does not have any general significance, as the saponification of esters of phosphoramidic acids often leads to cleavage of the phosphorus-nitrogen bonds.[926] A very general method for the synthesis of phosphoramidic acids, however, utilizes the hydrogenolytic cleavage of dibenzyl phosphoramidates.[63] Generally, palladium in the form of PdO is used as catalyst. In unsymmetrical diesters of phosphoramidic acid which contain only one O-benzyl group, only the O-benzyl group is split off. Analogously, with platinum as catalyst, monoesters of phosphoramidic acid can be obtained by hydrogenolysis of diesters of phosphoramidic acid which contain one O-benzyl group.

$$(RO)_2P(O)\text{-}N_3 + H_2NPh \longrightarrow (RO)_2P(O)\text{-}NH\text{-}Ph + N_2$$

Dialkyl phosphorazidates, on boiling with aniline, are converted to N-phenyl-phosphoramidic acid, i.e., as well as aminolysis cleavage of the ester bonds also occurs.[771]

$$(H_2N\text{-}NH)P(O)(OR)_2 + 2\ NaOH \longrightarrow (H_2N\text{-}NH)P(O)(OR)(ONa) + RONa + H_2O$$

Diaryl phosphorohydrazidates are hydrolyzed to salts of aryl phosphorohydrazidates by warm concentrated solutions of sodium hydroxide, ammonia, or barium hydroxide.[274a,475,838]

b. Formation of Esters of Phosphoramidic Acids from the Acids of Their Salts

$$[(R_2N)P(O)(OMe)(O^-)][N(Me)_4^+] + R'X \longrightarrow (R_2N)P(O)(OMe)(OR') + [NMe_4]X$$

$$(R_2N)P(O)(OR')(SNa) + R''X \longrightarrow (R_2N)P(O)(OR')(SR'') + NaX \quad (X = hal)$$

Unsymmetrical O-methyl-O'-alkyl N,N-dialkyl-phosphoramidates result from the reaction between tetramethylammonium salts of methyl phosphoramidates and alkyl halides.[155] S-alkali salts of O-esters of phosphoramidothiolic acids can be alkylated by alkyl halides to form O,S-diesters of phosphoramidothiolic acid.[419]

c. Transamination

$$(RNH)_3PS + 3\ R'NH_2 \rightleftharpoons (R'NH)_3P(S) + 3\ R\text{-}NH_2$$

$$\overline{NC_3H_3N}\text{-}P(O)(OR)_2 + R'NH_2 \longrightarrow (R'NH)P(O)(OR)_2 + \overline{NC_3H_3NH}$$

Phosphorthioic triamides of primary amines are transaminated by different primary amines to give different phosphorothioic triamides. Alkylamino groups are more strongly bonded to phosphorus atoms than arylamino groups. The reaction equilibrium can be forced to the right by distilling off the lower boiling amine. Diesters of N-imidazole-phosphoramidic, acid and, less easily, diesters of N-benzimidazole phosphoramidic acid are transaminated on treatment with primary or secondary amines. The reactions must be done under anhydrous conditions.[735,748,814]

d. Transesterification

$$(R_2N)_2P(S)(OR') + R''I \longrightarrow (R_2N)_2P(O)(SR'') + R'I$$

O-esters of phosphorodiamidothionic acid are easily converted to the corresponding S-esters of phosphoroamidothiolic acid on warming with alkyl halides. Solvents with high dielectric constants are used as reaction medium.[137]

e. Reaction of Aminotetrachlorophosphoranes with SO_2

$$(R_2N)PCl_4 + SO_2 \longrightarrow (R_2N)P(O)Cl_2 + SOCl_2$$

Dialkylaminotetrachlorophosphoranes are converted to N,N-dialkyl phosphoramidic dichlorides on treatment with sulfur dioxide.[573]

f. Sulfurization

$$(R_2N)P(O)Cl_2 \xrightarrow{P_2S_5} (R_2N)P(S)Cl_2$$

$$(R_2N)_2P(O)Cl \xrightarrow{P_2S_5} (R_2N)_2P(S)Cl$$

On heating with phosphorus pentasulfide at about 150 °C phosphoramidic dichlorides are converted into the corresponding phosphoramidothioic dichlorides in reasonable yields.[331,395]

g. Introduction of Amino Groups

$$(RO)P(O)(OH)_2 + NH_3 + C_6H_{11}N{=}C{=}NC_6H_{11} \longrightarrow (RO)P(O)(OH)(NH_2) + C_6H_{11}\text{-NH-CO-NH-}C_6H_{11}$$

$$\overline{NC_3H_3N}\text{-CO-}\overline{NC_3H_3N} + (RO)P(O)(OH)_2 \longrightarrow [\overline{NC_3H_3N}\text{-}P(O)(O^-)(OR)][H\overline{NC_3H_3N}H^+] + CO_2$$

$$(RO)_2P(S)(OR) + 2\ HNR'_2 \longrightarrow (RO)_2P(S)(NR'_2) + [RO][R'_2NH_2]$$

Monoesters of dihydrogen phosphoric acid are converted to monoesters of hydrogen N-unsubstituted phosphoramidic acid on treatment with ammonia in the presence of excess dicyclohexylcarbodiimide. Methanol can be used as the reaction medium but solvent mixtures, e.g., formamide, water, and t-butanol, are better.[337] The reaction of carbonyl diimidazole with monoesters of dihydrogen phosphoric acid gives rise to the formation of imidazolium salts of monoesters of N-imidazolephosphoramidic acid.[172,814] O,O',O"-triaryl phosphorothionates lose one molecule of phenol on reaction with 2 moles of a primary or secondary amine at room temperature and form O,O'-diaryl phosphoramidothionates.[567]

$$(RO)_3P{=}NAr + H_2O \longrightarrow (RO)_2P(O){-}NHAr + ROH$$

$$(RO)_3P{=}N{-}SO_2Ar + H_2O \longrightarrow (RO)_2P(O){-}NH{-}SO_2Ar + ROH$$

$$(R_2N)_3P{=}N{-}SO_2Ar + H_2O \longrightarrow (R_2N)_2P(O){-}NH{-}SO_2Ar + R_2NH$$

With water or also acids such as HCl or benzoic acid trialkyl N-aryl phosphorimidates are converted to dialkyl N-aryl phosphoramidates. With alkyl halides the reaction proceeds similarly, i.e. there is additionally alkylation at nitrogen, and the corresponding N-alkyl-N-aryl derivative is formed.[403] Triesters of N-sulfonyl-phosphorimidic acid react with about 1 N sodium hydroxide solution to form diesters of N-sulfonyl-phosphoramidic acid.[429a,444] Similarly N,N,N',N'-tetraalkyl-N"-sulfonyl phosphoric triamides are formed from hexaalkyl N-sulfonyl phosphorimidic triamides on treatment with aqueous alcoholic sodium hydroxide solution.[447]

$$R{-}N{=}N{-}P(O)(OR')_2 + H_2 \longrightarrow R{-}NH{-}NH{-}P(O)(OR')_2$$

N-alkyl-N'-dialkoxyphosphoryl-azo-compounds are catalytically hydrogenated or are reduced with, e.g., Cu(I) compounds, tin(II)chloride, sulfurous acid, or sulfites to dialkyl phosphorohydrazidates. With bisulfites, salts of the type $RN(SO_3Na){-}NH{-}P(O)(OR')_2$ are mostly formed.[838] N-dialkoxyphosphoryl,N'-aryl-azo-compounds are formed by oxidation of the dialkyl N'-arylphosphorohydrazidates with mercury(II) oxide.[838]

$$(R_2N)P(OR')_2 + BrCN \longrightarrow (R_2N)P(O)(OR')(CN) + R'Br$$

Dialkyl,N,N-dialkyl phosphoramidites react with cyanogen bromide or iodide to give alkyl halides and alkyl N,N-dialkyl phosphoramidocyanidates.[753]

$$R{-}\underset{}{\overset{Cl}{\overset{|}{C}}}{=}N{-}P(O)(OR')_2 + H_2S \longrightarrow R{-}\overset{S}{\overset{\|}{C}}{-}NH{-}P(O)(OR')_2 + HCl$$

$$R{-}\overset{Cl}{\overset{|}{C}}{=}N{-}P(O)(OR')_2 + HSAr \longrightarrow R{-}\overset{SAr}{\overset{|}{C}}{=}N{-}P(O)(OR')_2 + HCl$$

$$R\text{-}C(=O)\text{-}NH\text{-}P(O)Cl_2 + PCl_5 \longrightarrow R\text{-}C(Cl){=}N\text{-}P(O)Cl_2 + OPCl_3 + HCl$$

$$R\text{-}C(=O)\text{-}NH\text{-}P(O)(OR')_2 + PCl_5 \longrightarrow R\text{-}C(Cl){=}N\text{-}P(O)(OR')_2 + OPCl_3 + HCl$$

N- Dialkoxy-(or N-diaryloxy-) phosphoryl-carbonic acid imide chlorides react with H_2S in the presence of a tertiary amine such as triethylamine to form diesters of N-thioacyl-phosphoramidic acid[466] and with thiophenol to form phosphoryl thiocarboxylic acid imide aryl esters.[465] N-acyl phosphoramidic dichlorides form, on treatment with equimolar amounts of PCl_5 at 120-140 °C, N-dichlorophosphoryl-carboxylic acid imide chlorides.[430] On similar treatment dialkyl or diaryl N-acyl phosphoramidates give N-dialkoxy-(or N-diaryloxy)-phosphoryl-carboxylic acid imide chlorides.[190]

$$(RO)_2P(O)NH_2 + COCl_2 \longrightarrow (RO)_2P(O)NCO + 2\ HCl$$

$$(RO)_{3-n}P(O)(NH_2)_n + n\ COCl_2 \longrightarrow (RO)_{3-n}P(O)(NCO)_n + 2n\ HCl;\ n = 1,2$$

$$(RO)_{3-n}P(S)(NH_2)_n + n\ COCl_2 \longrightarrow (RO)_{3-n}P(S)(NCO)_n + 2n\ HCl;\ n = 1,2$$

$$(RO)_2P(O)NH_2 + SOCl_2 \longrightarrow (RO)_2P(O)NSO + 2\ HCl$$

Diesters of phosphoramidic acid, which contain an N-unsubstituted NH_2 group, react in dry solvents at low temperature with phosgene in the presence of an organic base, such as pyridine, triethylamine or trimethylamine, as HCl acceptor to form diesters of phosphorisocyanatidic acid. Similarly isocyanate derivatives are obtained from phosphonamidates and phosphinamidates.[740,741] In an analogous reaction, diesters of phosphoramidic acid and thionyl chloride give rise to diesters of N-thionyl-phosphoramidic acid.[905]

$$(R_2N)P(S)(OR)_2 + Cl\text{-}CO\text{-}R' \longrightarrow (R_2N)P(S)(OR)(OCOR') + RCl$$

$$(R_2N)_2P(S)(OR) + Cl\text{-}CO\text{-}R' \longrightarrow (R_2N)_2P(S)(OCOR') + RCl$$

O-alkyl esters of phosphoramidothionic and phosphorodiamidothionic acids form, on heating with acyl chlorides, alkyl chlorides and O-acyl derivatives of the respective acid.[357]

$$(R_2N)P(O)Cl_2 + 2\ RCOOK \longrightarrow (R_2N)P(O)(OCOR)_2 + 2\ KCl$$

$$(R_2N)_2P(O)Cl + RCOOAg \longrightarrow (R_2N)_2P(O)(OCOR) + AgCl$$

$$(R_2N)P(O)Cl_2 + 2\ RCOONa + R'OH \longrightarrow (R_2N)P(O)(OR')(OCOR) + 2NaCl + RCOOH$$

$$(R_2N)P(S)Cl_2 + 2\ RCOOK \longrightarrow (R_2N)P(S)(OCOR)_2 + 2\ KCl$$

N,N-dialkylphosphoramidic dichlorides and alkali salts of fatty acids react together on heating to produce O,O'-diacyl N,N-dialkylphosphoramidates.[908a] Similarly, phosphorodiamidic chlorides react with silver salts of the acids to give O-acyl phosphorodiamidates.[677a] If the alkali salt of a carboxylic acid, e.g., sodium acetate, in absolute alcohol is allowed to react with phosphoramidic dichloride, O-acyl-O'-alkyl phosphoramidate is formed.[170a] The respective sulfur containing phosphoric acid derivatives react with alkali salts of fatty acids analogously. N,N-dimethylphosphoramidothioic dichloride, for example, forms O,O'-diacetyl N,N-dimethylphosphoramidothionate with potassium acetate.[357]

$$(R_2N)P(O)Cl_2 + R'O\text{-}CS\text{-}SK \longrightarrow (R_2N)P(O)Cl(SCSOR') + KCl$$

N,N-dialkylphosphoramidic dichloride reacts with potassium xanthate to produce S-thiocarboxy N,N-dialkylphosphoramidochloridothiolate.[861]

$$[(R_2N)P(O)(OR)(O^-)][NMe_4^+] + (RO)_2P(O)Cl \longrightarrow (R_2N)P(O)(OR)\text{-}O\text{-}P(O)(OR)_2 + [NMe_4]Cl$$

$$[(RO)_2P(O)(O^-)][NMe_4^+] + ClP(O)(NR_2)_2 \longrightarrow (RO)_2P(O)\text{-}O\text{-}P(O)(NR_2)_2 + [NMe_4]Cl$$

By condensation of salts of alkyl phosphoramidates with dialkyl phosphorochloridates diphosphoric acid triester amides are synthesized.[155a] Similarly salts of diesters of phosphoric acid combine with N,N,N',N'-tetraalkylphosphorodiamidic chlorides to produce diphosphoric acid-P,P-diester-P',P'-diamides.[155a]

$$2\ (R_2N)_2P(O)(OR) + R'CO\text{-}O\text{-}COR' \longrightarrow (R_2N)_2P(O)\text{-}O\text{-}P(O)(NR_2)_2 + 2\ R'COOR$$

Alkyl phosphorodiamidates undergo self-condensation at temperatures between 170 and 180 °C under the influence of acetic anhydride or another acid anhydride to give diphosphoric acid tetrakis-(dialkyl)amides. $BF_3 \cdot O(Et)_2$ or sulfuric acid, for example, serve as catalysts.[164]

$$2 \ (ArNH)P(O)(OR)(OH) \longrightarrow$$
$$(ArNH)P(O)(OR)-O-P(O)(NHAr)(OR) \ + \ H_2O$$

Monoalkyl or monoaryl hydrogen phosphoramidates split off water in the presence of a water-acceptor such as dicyclohexylcarbodiimide and form diphosphoric acid-P,P'-diester-P,P'-diamides.[115]

$$(RO)_2P(O)-N\underset{CH_2}{\overset{CH_2}{\diagup\diagdown}} \ + \ R_2NH \longrightarrow (RO)_2P(O)-NH-CH_2-CH_2-NR_2$$

$$(RO)P(O)\left[-N\underset{CH_2}{\overset{CH_2}{\diagup\diagdown}}\right]_2 \ + \ 2 \ R_2NH \longrightarrow (RO)P(O)(NH-CH_2-CH_2-NR_2)_2$$

$$(RO)_2P(O)-N\underset{CH_2}{\overset{CH_2}{\diagup\diagdown}} \ + \ RCOOH \longrightarrow (RO)_2P(O)-NH-CH_2-CH_2-O-CO-R$$

$$(RO)_2P(O)-N\underset{CH_2}{\overset{CH_2}{\diagup\diagdown}} \ + \ ClR \longrightarrow (RO)_2P(O)-N\underset{R}{\overset{CH_2-CH_2Cl}{\diagup\diagdown}}$$

$$(RO)_2P(O)-N\underset{CH_2}{\overset{CH_2}{\diagup\diagdown}} \ + \ Cl_2 \longrightarrow (RO)_2P(O)-\overset{\overset{Cl}{|}}{N}-CH_2-CH_2Cl$$

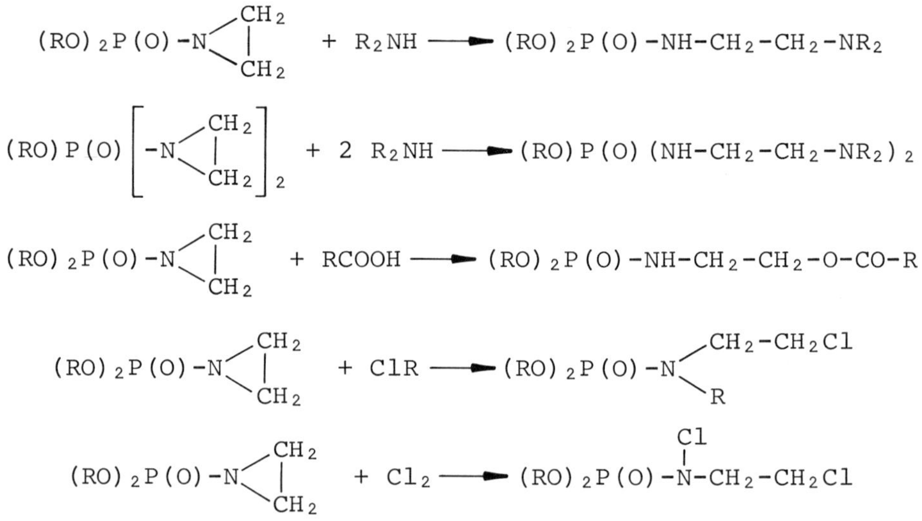

The ring of the ethyleneimine group in diesters of N-ethyleneimine phosphoramidic acids is cleaved by reagents such as amines, carboxylic acids, lower alkyl halides at high temperatures and, if necessary under pressure, or by chlorine even at 0 °C, to give diesters of N-ethyl phosphoramidates, in which the N-ethyl group is β-substituted.[836]

h. Alkylations

$$(RO)_2P(S)(NHR') \ + \ MeI \longrightarrow (RO)_2P(S)(NMeR') \ + \ HI$$

O,O'-diesters of N-monosubstituted phosphoramidothionates can be alkylated on nitrogen in the presence of potassium t-butoxide.[593]

$$(R_2N)P(O)(OR')(SNa) \ + \ R''X \longrightarrow (R_2N)P(O)(OR')(SR'') \ + \ NaX$$

$$(RO)_2P(S)(NR_2') \ + \ R''I \longrightarrow (RO)P(O)(NR_2')(SR'') \ + \ RI$$

$$(RO)P(S)(NR_2')_2 \ + \ R''I \longrightarrow (R''S)P(O)(NR_2')_2 \ + \ RI$$

O-esters of the S-alkali salts of phosphoramidothiolates are easily alkylated by treatment with alkyl halides to form O,S-diesters of phosphoramidothiolic acids.[417] The same products are obtained by isomerization of O,O'-diesters of phosphoramidothionic acid on treatment with alkyl halides in nitromethane or methyl acetamide.[137] O-esters of phosphorodiamidothionic acid react analogously.[137]

B. BASIC CHEMISTRY

$RNHP(O)X_2$

Phosphoramidic dihalides are easily cleaved to the amine and phosphoric acid in acidic and alkaline media. In contrast, the phosphorus-nitrogen bonds in N-alkyl and N-aryl phosphoramidic dihalides are largely unaffected, when the hydrolysis is carried out with water-free formic acid; more hydrolytically stable are the phosphorus-nitrogen bonds in monoarylamide derivatives of phosphoric acid, in which the aromatic group contains a halogen ligand.[527] Partial hydrolysis of phosphoramidic dihalides is only possible for chlorides, fluorides and bromide fluorides.[821] Monoaryl-phosphoramidic dihalides are converted by aqueous ammonia to N-aryl-phosphordiamidic acids.[852b,874a] With alcohols and phenols, or better alcoholates and phenolates, both monoesters of phosphoramidochloridic acid and diesters of phosphoramidic acid are obtained. On heating N-aryl-phosphoramidic dichloride, dichloro-dioxo-cyclodiphosphazanes are formed.[574] With Grignard or aryl-lithium compounds, phosphinic amides $R_2P(O)(NR_2^1)$ are produced.[482a,613a] Also, partial reactions to form phosphonamidic chlorides can be observed.[482b]

$(R_2N)P(O)(OR)_2$

Although the phosphorus-nitrogen bonds in diesters of phosphoramidic acids are rather unstable, they are so stable in alkaline media that, with strong alkalis, the ester groups can be saponified.[712] The alcoholysis is, as yet, only observed in diesters of N-imidazole-phosphoramidic acid.[80] Diphenyl phosphoramidates react with carboxylic acids at temperatures of about 200 °C to form diphenyl hydrogen phosphates and amides.[771a] Diesters of N-monoalkyl-(or N-monoaryl)-phosphoramidic acid react with elemental sodium or sodium hydride to form N-sodium salts.

$OP(OR)_2(NH-NH_2)$

Diesters of phosphorohydrazidic acid which are not substituted on nitrogen add hydrogen halides to form hydrohalides.[475] In strongly acidic media the phosphorus-nitrogen bond is cleaved, while in strongly alkaline media they are stable.[475] N-unsubstituted diesters of phosphorohydrazidic acid are oxidized by iodine in aqueous solutions to diesters of phosphoric acid and nitrogen.[124a]

$(R_2N)P(O)(OR')Cl$

Esters of phosphoramidochloridic acid are relatively easily cleaved in the warm by water to give monoesters of phosphoric acid and the respective ammonium chloride. With aqueous alkali, the salts of monoesters of phosphoramidic acid are formed; with alcohols or phenols in the presence of an acid acceptor the products are unsymmetrical diesters of phosphoramidic acid and with mercaptides, O,S-diesters of phosphoramidothiolic acid.[763] Alkylesters of N,N-dialkyl-phosphoramidochloridates split off alkyl chloride on heating to about 150 °C and form dialkylamides of metaphosphoric acid.[573]

$R_2NP(O)(OR')(OH)$

Monoesters of phosphoramidic acid can undergo self-condensation in the presence of a water acceptor such as dicyclohexylcarbodiimide to form diphosphoric acid-P,P'-diester-P,P'-diamides.[115] With phosphoric acid the condensation products are diphosphoric acid monoesters and with monoesters of phosphoric acid, diphosphoric acid-P,P'-diesters are formed.[160]

$OP(OR)_2(N{=}N-R')$

Dialkylesters of alkylazophosphoric acid are cleaved in strongly acidic media to form diazonium compounds and esters of phosphorous On reduction dialkyl phosphorohydrazidates are formed,[838] in strongly alkaline media are formed among others, diesters of phosphoric acid-(N,N- and N,N'-diarylhydrazide). Thermally N-dialkoxyphosphoryl-N'-alkyl-azo-compounds are stable.

$(R_2N)P(S)X_2$
$(R_2N)P(S)(OR)(OH)$ or
$(R_2N)P(O)(OR)(SH)$

Phosphoramidothioic dichlorides are sometimes so hydrolytically stable that they can be steam-distilled. They are also indifferent to alcohols and phenols. The derivatives of primary aliphatic and aromatic amines are cleaved on heating to give hydrogen chloride and dichlorodithio-cyclodiphosphazanes.[576] O-monoesters of phosphoramidothionic acid in the form of their salts are converted to O,S-diesters of phosphoramidothiolic acid by alkyl halides.[419] With iodine the products are bis(alkoxy-amino-phosphoryl)-disulfides or bis(aryloxy-amino phosphoryl)-disulfides, respectively.[76]

$(R_2N)_2P(O)X$

Phosphordiamidic chlorides are completely hydrolyzed by water or more rapidly by acids. In contrast, the phosphorus-nitrogen bonds are much more stable in alkaline media.[582,584] Treatment with water[359a,875] or hydrogen sulfide[586] in the presence of pyridine or another tertiary amine leads to the formation of the tetraamides of diphosphoric or thiodiphosphoric acid. With alcohols or thiols, O-esters of phosphorodiamidic acid and S-esters of phosphorodiamidothiolic acid, respectively, are formed. Grignard or aryl-lithium compounds give rise to phosphonic diamides[613a]; with silver salts of carboxylic acids, aryl phosphorodiamidates are formed.[677a] On heating phosphorodiamidic chlorides of primary aliphatic and aromatic amines split off hydrogen chloride and form diamino-dioxo-cyclodiphosphazanes.[573,583]

$(R_2N)_2P(O)(OR)$

The phosphorus-nitrogen bonds in esters of phosphorodiamidic acids are easily cleaved in acidic media; a morpholine rest is especially easily split off.[611] While the expulsion of an amine group is, in most cases, connected with the saponification of an ester group, this reaction can be carried out in such a way that monoesters of phosphoric acid are formed. Phenyl phosphorodiamidate can be catalytically hydrogenated by platinum to give the free diamide.[93a,316]

$(R_2N)_2P(O)(OH)$
$(R_2N)_2P(S)(OH)$

Alkali salts of phosphorodiamidic acids are converted into the respective acid chlorides on treatment with phosphorus pentachloride.[77] The tetramethylammonium salts are alkylated to the esters of phosphorodiamidic acid with alkyl halides. Phosphorodiamidothionic acid decomposes on heating with water.[77]

$(R_2N)_2P(O){-}O{-}P(O)(NR_2)_2$

The P-O-P bonds in diphosphoric acid tetraamides are easily cleaved on hydrolysis. At the same time at least some of the amino groups are removed. On cleavage of the P-O-P-bridge with potassium bifluoride phosphorodiamidic fluoridates are formed as well as phosphorodiamidic acid.

$OP(NR_2)_3$
$SP(NR_2)_3$

Phosphoric triamides are completely hydrolized in warm acidic media. They are, however, soluble in cold concentrated hydrochloric acid and can be precipitated out undecomposed with alkali. On heating derivatives of primary amines, the amine is split off and diamino-dioxo-cyclodiphosphazanes are formed.[77] Phosphorothioic triamides are exceptionally stable to hydrolysis. Even on heating with dilute acid or alkali solutions they are not cleaved. With alkyl halides they react to form triamido-alkylmercapto-phosphonium halides.[137] The derivatives, on heating, lose amine and give diamino-dithio-cyclodiphosphazanes.[573]

$Cl_3P{=}NR$ or 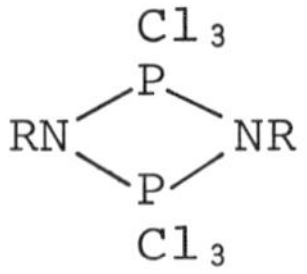

Trichlorophosphinimides are monomeric when they are derivatives of weakly basic amines. The monomers are well soluble in non-polar solvents. The derivatives of strongly basic amines, e.g., the N-methyl derivative, are dimeric and poorly soluble. The dimeric N-aryl-trichlorophosphinimides dissociate on heating in suitable solvents to give the monomers.[933,935,936]

$(RO)_3P{=}NR'$

Concentrated hydrochloric acid cleaves the phosphorus-nitrogen bonds in triaryl N-alkyl-phosphorimidates to form triaryl phosphates. The corresponding bonds in triesters of N-aryl-phosphorimidic acids are considerably more stable to hydrolysis. The trialkylesters are cleaved by water to form alcohol and dialkyl N-aryl phosphoramidates. The same products are obtained on treatment with dry hydrogen chloride or benzoic acid.[403] Triaryl N-aryl phosphorimidates react especially easily in the same way.[939] Alkyl halides and trialkyl N-aryl phosphorimidates react together to produce dialkyl N-alkyl-N-aryl phosphoramidates.[403] Worthy of note is also the cleavage of phosphorus-nitrogen bonds with CS_2 to give trialkyl phosphorothionates and isothiocyanates.

$Cl_3P=N-CO-R$

N-acyl-trichlorophosphinimides are very sensitive to hydrolysis. On careful hydrolysis with water or formic acid N-acyl-phosphoramidic dichlorides are formed[454] and, on alcoholysis, dialkyl N-acyl phosphoramidates. On heating the compounds decompose to phosphoryl chloride and nitriles.

$(RO)_3P=N-CO-R$

The phosphorus-nitrogen bonds in triesters of N-acyl phosphorimidic acid are stable to hydrolysis in alkaline or neutral solutions. In alkaline media one ester group is saponified to form diesters of N-acyl-phosphoramidic acid.[234,436] Dry hydrogen chloride cleaves the compounds to produce triesters of phosphoric acid and nitriles. The same products are obtained on pyrolysis of triesters of N-acyl phosphorimidic acid.[430,440]

$Cl_3P=N-SO_2-R$
$(RO)_3P=N-SO_2-R$
$(R_2N)_3P=N-SO_2-R$

The chlorine atoms in N-sulfonyl-trichlorophosphinimide can be hydrolyzed stepwise.[429a] With excess water the phosphorus-nitrogen bond is finally cleaved. The chlorine atoms can be substituted by alkyl or aryl groups on treatment with Grignard reagents. The compounds are, in general, thermally very stable. The phosphorus-nitrogen bonds in triesters of N-arylsulfonyl phosphorimidic acid are very stable against hydrolysis. On treatment with hot water or, preferably, alcoholic sodium hydroxide solution, one ester group is saponified to form diesters of N-sulfonyl-phosphoramidic acid. Arylsulfonyl phosphorimidic triamides react slowly with cold water to form ammonium salts of N-arylsulfonyl phosphoric triamides.[429a]

$OPX_2(NH-SO_2-R)$
$OP(OH)_2(NH-SO_2-R)$
$OP(OR)_2(NH-SO_2-R)$
$SP(OR)_2(NH-SO_2-R)$

N-sulfonyl phosphoramidic dihalides are partially hydrolyzed by formic acid in benzene to the monohalide and the free acid.[446] N-sulfonyl phosphoramidic acids are soluble in water. On heating in neutral or acidic solutions they are, however, cleaved to form phosphoric acid and sulfonamide. Diesters of N-sulfonyl phosphoramidic acid need a long time boiling in alcoholic hydrogen chloride before being split to the diesters of phosphoric acid and sulfonamide. They are, like diesters of N-sulfonyl phosphoramidothionic acid, moderately strong acids which form salts with alkalis. The pK_a values of $p\text{-}X\text{-}C_6H_4\text{-}SO_2\text{-}NH\text{-}P(S)(OR)_2$ were well correlated with the Hammett substituent constants.[28,944] The sodium and silver salts of dialkyl N-arylsulfonyl phosphoramidates are alkylated on nitrogen by treatment with alkyl halides.[791a]

$OPZ_2(NCO)$ (Z = hal, OR)
$OPZ_2(NCS)$

Phosphorisocyanatidic dichloride is cleaved by water/CO_2, HCl, and ammonium phosphate (see also Ref. 471). Dialkyl and diaryl phosphorisocyanatidates react with water to split off CO_2 and form N-unsubstituted diesters of phosphoramidic acid. Dialkyl phosphorisothiocyanatidates are cleaved by warm water to give HSCN and dialkyl hydrogen phosphates.[496] On pyrolysis they give alkyl isothiocyanates and esters of metaphosphoric acid.

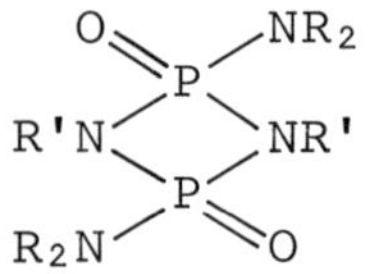

Diamino-dioxo-cyclodiphosphazanes, especially the aromatic derivatives are relatively stable to hydrolysis. Alcohols or phenols cleave the four-membered ring to form esters of phosphorodiamidic acid. In the case of aromatic derivatives it is possible to isolate imidodiphosphoric acid ester triamides as the first reaction products.

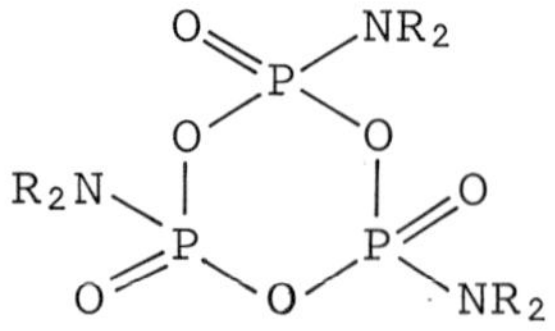

Metaphosphoric acid amides are stable for a short time in neutral solutions. In acidic solutions they are very rapidly cleaved to phosphoric acid and ammonium salts.

They are inert to alcohols. With phosphorus pentachloride they react to form phosphoramidic dichlorides.

C. GENERAL DISCUSSION OF PHYSICAL PROPERTIES

C.1. NMR Spectroscopy

The ^{31}P NMR chemical shifts of compounds which contain four-coordinate phosphorus(V) span a relatively small region. When ligands bonded through oxygen to phosphorus are replaced stepwise by other elements or element groups, with the exception of fluorine, a shift of the resonance signal to lower field is almost always observed. It follows, therefore, that the shielding of the phosphorus nucleus by other ligands is weaker than that by oxygen-bonded ligands. The shielding of the phosphorus increases in sequence with the increasing electronegativity of the ligands, when a series of exceptions are omitted. The substitution of an oxygen-bonded ligand by a nitrogen-bonded ligand, in general, causes a change in the chemical shift of -11 ± 3 ppm, i.e., a movement within relatively small limits to lower field. This regularity is connected with the fact that the bonding systems in phosphorus-nitrogen and phosphorus-oxygen compounds are very similar. Considerable variations from regularity, however, occur when an OH group is replaced by an NH_2 group.

The substitution of an oxygen atom by a sulfur atom in O-esters of phosphoric acid and also in amine derivatives of phosphoric acid or esters leads to a considerable shift to lower field strengths. The replacement of an oxygen bridging atom by sulfur is characterized by a fairly constant shift of -25 to -30 ppm, whereas the conversion of a phosphoryl- to a thiophosphoryl-compound can lead to strongly differing changes of chemical shift. For literature on the ^{31}P NMR spectra of phosphorus nitrogen compounds see Refs. 309a,311,977.

Proton magnetic resonance studies of dialkylamido derivatives of phosphoric acid and of phosphoric chlorides as well as of the corresponding phosphorthioic derivatives have shown that the covalency of the phosphorus atom, which is increased by replacement of R_2N groups with halogens, increases the magnitude of $^3J_{HCNP}$ and simultaneously decreases the τ value of the N-methyl or N-methylene protons.[169] Examples of long-range proton-phosphorus coupling were observed in a number of phosphorus-nitrogen compounds containing the system P-N-C-C-H but not, however, in any containing the P-N-N-C-H system. The major contribution to the coupling is supposed to be through bonding electrons, although "through-space" coupling cannot be excluded.[410] Tabulated ^{1}H-chemical shifts see Refs. 79,84,136,423,811,836.

C.2. IR and Raman Spectroscopy

Infrared and Raman spectroscopy help greatly in the clarification of the structures of the here-mentioned compound types, they are also used in the qualitative or half-quantitative analysis of mixtures of compounds. For compounds containing P-O-, P-S-, and P-N- bonds, the valence vibrations of these groups, on the grounds of their characteristic and, by manifold comparisons, determined frequency ranges, are especially meaningful. In addition to the types of ligand on phosphorus, their hybridizations are of decisive influence on the positions of the valence vibrations. For the compounds described in this chapter, the phosphorus-nitrogen bond absorptions occuring between 600 and 1450 cm^{-1} are of interest. The higher frequency absorptions in this region, i.e., 1200-1350 cm^{-1} are due to P=N double bonds and those in the lower range, i.e., about 600-700 cm^{-1} are due to P-N single bonds. The symmetrical and unsymmetrical valence movements of P-N ring systems, P=N-P and P-N-P groups are observed between these two systems. The unsymmetrical vibrations are expected to be more intense in the infrared spectra and the symmetrical vibrations more intensive in the Raman spectra.

For literature on the infrared and Raman spectroscopy of phosphorus-nitrogen compounds see Refs. 158,339,811a, 854.

C.3. P-N Bond Lengths

Using Pauling covalent radii (P 1.10; N 0.74 Å) and the Schomaker-Stevenson correction[752a] (-0,09(3.0 [E.neg. N] - 2.1 [E.neg. P])) the value of 1.77 Å was calculated for the length of a pure P-N single bond. The value was observed, for example, for the P-N bond in the phosphoramidate anion.[264a] π-bonding, depending on its strength, can shorten this bond length by as much as 0.3 Å. Thus, for the exocyclic P-N bond in $[NP(NMe_2)_2]_4$ a value of 1.68 Å was found.[133a] In $(H_2N)_3PBH_3$ the P-N bond length is 1.65 Å,[646a] in the 4-membered ring $[MeNP(S)Cl]_2$ it is 1.67 Å,[903a] while in cyclic phosphornitrilic compounds the distance between the phosphorus and nitrogen atoms in the ring is observed to be between 1.51 and 1.60 Å (average 1.56 Å)[251a] (see also Chapter 19). Furthermore bond distances were measured in $[MeO(O)PNMe]_3$.[38a]

C.4. Miscellaneous

For general information on thermochemical properties of phosphorus compounds see Ref. 360a, on paper chromatography see Refs. 258a,258b,258c,349a,411a,852a; on analytical chemistry of P-N compounds see Ref. 309b. For biological aspects and uses see Refs. 754,902.

D. LIST OF COMPOUNDS

Comments on List of Compounds

Phosphorus-nitrogen compounds with phosphorus in oxidation number 5 and coordination number 4 are presented in the list of compounds in this chapter. Compounds containing P-C bonds are to be found in the chapters concerning derivatives of phosphonic (Chapter 12) or phosphinic acids (Chapter 13). Likewise compounds with phosphonium structures (Chapter 4) and cyclotriphosphazatrienes (Chapter 19) are not included in this chapter.

Although $>\overset{\|}{P}-C\equiv N$ compounds are not strictly within the sphere of this chapter, these compounds are included as the $-C\equiv N$ group is considered to be a pseudo-halide. -N=C=O, -N=C=S, and $-N_3$, in contrast, are classified as nitrogen ligands bonded to phosphorus.

All $>\overset{\|}{P}-\overset{|}{N}-\overset{|}{N}-$, $>\overset{\|}{P}-N=N-$ and $>\overset{\|}{P}=N-N<$ groups, including ring compounds, are classified under hydrazine derivatives.

Compounds which contain two or more different phosphorus atoms, all of which belong to this chapter, are listed under the first heading in the list of compounds; e.g., (i) for $(RO)_2P(O)NHP(O)Cl_2$ see section D.1.1.4. [$AcNHP(O)X_2$] not D.1.5.7. (ii) for $(RO)_2P(O)NHP(O)(NHR)_2$ see section D.1.5.7. not section D.3.3. However, there is a special section for di-(and poly-)phosphoric acid amides.

References in the table to Section XVII (Miscellaneous) indicate that the synthetic routes are to be found in this section or also in the original literature.

The comment "for other compounds see Ref. XXX means that further analogous compounds can be found in this reference, but for these, especially in patent literature, often no definite characterization is given.

Within one section the compounds are arranged in order of increasing numbers of carbon atoms. Where general structural formulas are given, the order is only valid for the first of the various compounds.

In contrast to other chapters, for n_D^{20} and d^{20} only n and d, respectively, are written; but, e.g., n_D^{15} is written in the accepted manner.

A general formula in the index of the list of compounds does not mean that an example for it is in the list.

The symbols to the left of the list of compounds C_1, C_2... give the number of carbon atoms in the formula registered and all sucessive formulae up to the next symbol. C_x indicates that the number of carbon atoms is larger than in the immediately preceeding symbol. In short sections this indexing system is omitted. Heterocyclic compounds such as thiophene, pyrrole, pyrimidine or $HN-N=N-N=C-NH_2$ as ligands directly bonded to phosphorus are listed as Ar, e.g., for $HN-N=N-N=C-NH-P(O)(N\leq)_2$, $ArNHPO(NR_2)_2$ see D.3.3.6., however, for $R'-N=C(R)-NH-P(O)-(N\leq)_2$, $AcNHPO(NR_2)_2$ see D.3.3.5.

Contents

D.1. Phosphoric and Thiophosphoric Acid Monoamides and Hydrazides (Phosphoramidic Acid Derivatives, Phosphoramido[Mono,Di,Tri]Thioic Acid Derivatives, etc.

D.1.1. $>$N-P(O)X_2, X = Halogen

D.1.1.1. RNH-P(O)X_2, X = Halogen, R = Alky

C_1 $MeNHPOCl_2$. III. b_{27} 132°,[573] b_{12} 120°,[753] ^{31}P -17.7.[873]
$MeNHPOF_2$. XIV. b_{14} 70-2°.[753]

C_2 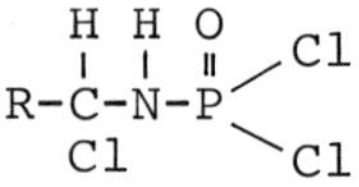 X.[937] R=CCl_3, m. 136-8°; R=$MeCCl_2$, m. 112-5°; R=$EtCCl_2$, m. 64-6°; R= Me_2CCl, m. 115-8° dec.
$EtNHPOCl_2$. III. b_{22} 140°.[573,601]
$EtNHPOF_2$. II. b_8 35-6°.[656,658]
$ClCOCHClNHPOCl_2$. X. m. 84-5°.[932]

C_3 CCl_2=C($CHCl_2$)$NHPOCl_2$. X. m. 141-2°.[932]
CCl_2=C(CCl_3)$NHPOCl_2$. X. m. 158-61°.[932]
$CHCl_2CCl$(COCl)$NHPOCl_2$. X. m. 90-1° dec.[932]
$(CH_2Cl)_2CHNHPOCl_2$. III. $b_{0.6}$ 121°, n_D^{22} 1.5163.[910]
$(CF_3)_2CHNHPOCl_2$. X. m. 106-8°.[484]
CH_2=$CHCH_2NHPOCl_2$. III. $b_{0.6}$ 104-6°,[295] $b_{0.001}$ 50°,[677] n 1.4840,[677] d 1.354.[677]
$PrNHPOCl_2$. III. b_{16} 146°.[573,601]
i-$PrNHPOF_2$. II, XIV. b_{10} 62-3°,[656] b_9 85°.[753]

C_4 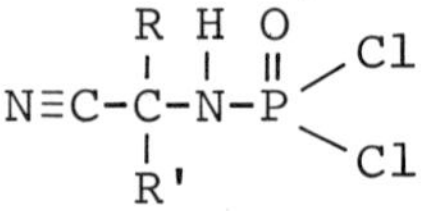 X.[687] R=R'=Me, m. 92-3°; R=R'=Et, $b_{0.05}$ 127-8°, n 1.4837, d 1.2950; R=Me, R'=Et, m. 42°, b_1 141-2°; R=R' = $(CH_2CH_2)_2$, m. 82-3°; R=R' = $CH_2(CH_2CH_2)_2$, m. 108-10°.
MeCH=$CHCH_2NHPOCl_2$. III. b_2 124°.[295]
$BuNHPOCl_2$. III. b_2 115-7°.[601]
i-$BuNHPOCl_2$. III. b_{14} 141°.[573]
t-$BuNHPOCl_2$. X. m. 110-1°.[932]
$Me_2CClCH_2NHPOCl_2$. X. m. 58-60°.[932]
$Me_2CClCHClNHPOCl_2$. X. m. 115-8° dec.[932]
$BuNHPOF_2$. II. b_8 64°.[656,658]
t-$BuNHPOF_2$. II. b_8 50-1°.[656,658]

C_5 $Me_3CCH_2NHPOCl_2$. X. m. 67-9°.[932]
$Et_2CHNHPOCl_2$. X. m. 30°, b_8 134-5°.[932]
$AmNHPOCl_2$. III. b_3 127-8°,[601] b_{17} 159°.[573]

C_6 $Me_3CCCl_2CHClNHPOCl_2$. X. m. 88-91°.[932]
$CH_3(CH_2)_5NHPOCl_2$. III. b_{3-4} 135-9°.[601]
i-$C_6H_{13}NHPOF_2$. XIV. b_{15} 115°.[753]

C_7 $PhCH_2NHPOCl_2$. II. Oil.[573,529]
$C_7H_{15}NHPOCl_2$. III. b_4 154-5°.[601]

C_8 $BuEtCHCH_2NHPOCl_2$. II. Oil.[507]
$C_8H_{17}NHPOCl_2$. III. b_3 150-4°.[601]

C9 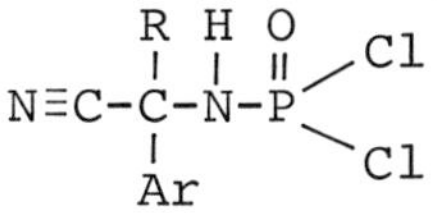 X.[686]

Ar	R	m.
Ph	Me	76-7°
Ph	Et	106-8°
Ph	Pr	107.5-9°
Ph	i-Pr	121-3°
Ph	Bu	101-2°
p-MeC_6H_4	Me	96-8°
p-ClC_6H_4	Me	126-7°
2,4-$Me_2C_6H_3$	Me	131-2.5°
2,5-$Me_2C_6H_3$	Me	128-30°

$C_9H_{19}NHPOCl_2$. III. $b_{1.5}$ 133-7°.[601]

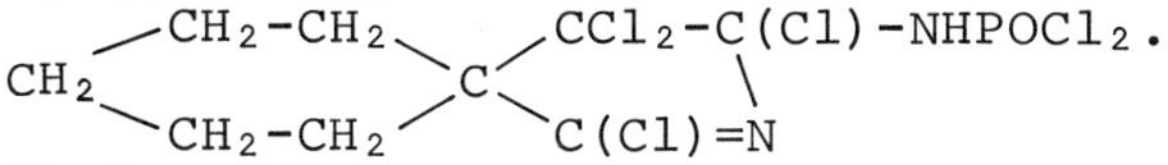 X. m. 139-40°.[787]

C10 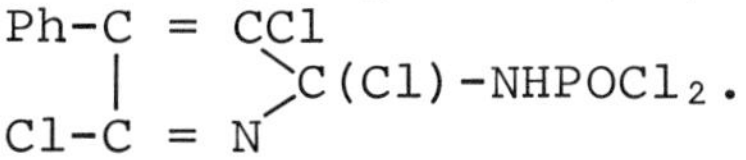 X. m. 220-1°.[787]

$C_{10}H_{21}NHPOCl_2$. III. $b_{1.5}$ 166-9°.[601]

Cx $C_{11}H_{23}NHPOCl_2$. III. b_1 175-7°.[601]

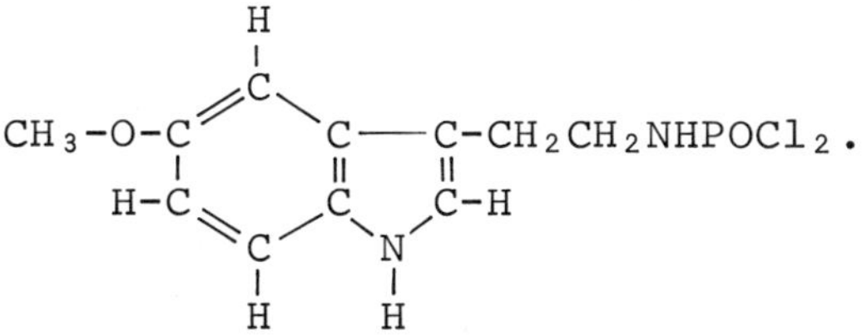 II. m. 100-2°.[613]

$C_{12}H_{25}NHPOCl_2$. III. b_2 180-6°.[601]

$Ph_2CHNHPOCl_2$. X. m. 169-70°.[938]

D.1.1.2. ArNH-P(O)X_2, X = Halogen, Ar = Aryl

C3 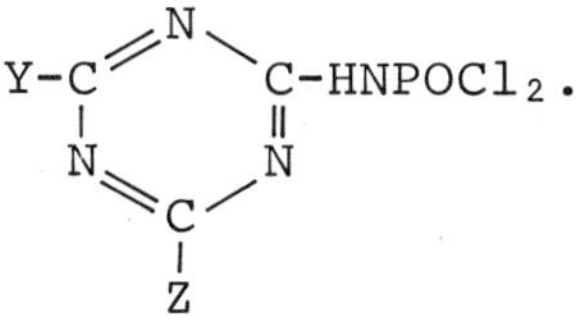 X. Y=Z=Cl, m. 168-9°.[202]
Y=Cl, Z=$NHPOCl_2$, m. 174-6°.[202]
Y=Z=$NHPOCl_2$, m. 250-5°.[202]

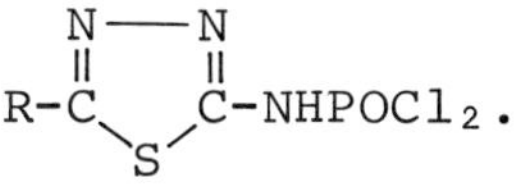 II. R=Me, m. 170-2°; R=Et, m. 122-4°; R=Pr, m. 117-9°; R=Ph, m. 199-202°.[841]

C4 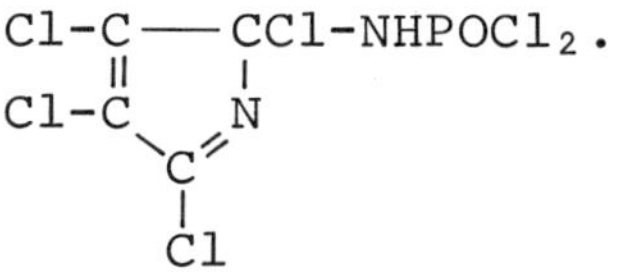 X. m. 194-5°.[787]

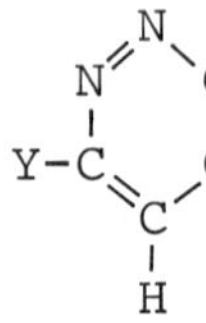

$C-N(R)POCl_2$. II.[840] R=H, Y=Cl, m. 156-9°; R=H, Y=H, m. 122-4°; R=Me, Y=Cl, m. 71-3°; R=Et, Y=Cl, m. 66-9°.[840]

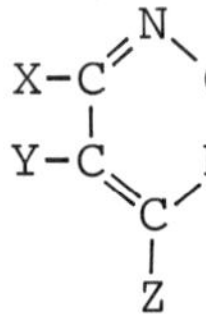

$C-NHPOCl_2$. III.[489,491] X.[490] X=Y=Z=H, m. 171-2°,[489] m. 182-4°;[49 Y=X=H, Z=Cl, m. 167 9°,[490] m. 163-4°;[48 X=Z=H, Y=Cl, m. 163 5°,[490] m. 163-3.5°;[489] X=Z=H, Y=Br, m. 180-1°;[490] X=Z=Cl, Y=H, m. 184 6°;[490] X=Cl, Y=H, Z=MeO, m. 150-2°,[49 Y=H, X=Z=MeO, m. 133.5-5.5°;[490] X=Cl Y=H, Z=Me, m. 166-8°;[490] X=Me, Y=Br, Z=H, m. 173-4°;[490] X=Me, Y=Z=H, m. 16 7°.[489,490] X=MeO, Y=Z=H, m. 190°,[489] m. 199-200°;[490] X=M Y=H, Z=MeO, m. 168-71°,[490] X=Z=Me, Y=H, m. 161-3°,[490] $X=Et_2N$, Y=Z=H, m. 212-5°;[490] $X=C_5H_{10}N$ Y=Z=H, m. 226-8°;[49 $X=O(CH_2CH_2)_2N$, Y=Z=H, m. 228-30°.[490]

C_6 $2,4,6-Br_3C_6H_2NHPOCl_2$. II. m. 148°,[573] m. 186-8°.[934]

$2,6,4-Cl_2(O_2N)C_6H_2NHPOCl_2$. X. m. 171-3°.[934]

$2,4,6-Cl_3C_6H_2NHPOCl_2$. II. m. 128°,[573] m. 156-8°.[934]

$2,4-Br_2C_6H_3NHPOCl_2$. III.[573] X.[936] m. 134°,[573] m. 96-8°.[936]

$2,4-(O_2N)_2C_6H_3NHPOCl_2$. X. m. 121-2°.[934]

$2,4-Cl_2C_6H_3NHPOCl_2$. III. m. 126°,[573] m. 98-100°.[934]

$m-BrC_6H_4NHPOCl_2$. III. m. 87°.[573]

$p-BrC_6H_4NHPOCl_2$. III. m. 98°.[573]

$p-IC_6H_4NHPOCl_2$. II. m. 105-7°.[487]

$o-O_2NC_6H_4NHPOCl_2$. X. m. 40-2°.[934]

$m-O_2NC_6H_4NHPOCl_2$. II.[487] III.[573] m. 85-6°,[487] m. 148°.[573]

$p-O_2NC_6H_4NHPOCl_2$. III, X. m. 155-6°,[529,934] m. 156°.[57

$o-ClC_6H_4NHPOCl_2$. X. m. 77-9°.[934]

m-$ClC_6H_4NHPOCl_2$. X. m. 60-2°.[428]
p-$ClC_6H_4NHPOCl_2$. III,[529] X.[936] m. 104-6°,[936] m. 105-7°,[734] m. 107°,[573,660] m. 107-8°.[529]
$PhNHPOCl_2$. II,III,X. m. 84°,[73,573,581] m. 87°,[927] m. 89-91°,[936] m. 93-4°,[529,149] ^{31}P -7.6.[633,634]
o-$(Cl_2PONH)C_6H_4SO_2NHPOCl_2$. X. m. 127-8° dec.[941]
$PhNHPOF_2$. II. m. 48-9°, b_5 103-4°.[656]
o-$(H_2NSO_2)C_6H_4NHPOCl_2$. II. m. 175-6°.[941]
p-$(H_2NSO_2)C_6H_4NHPOCl_2$. II. m. 230-1° dec.[529]

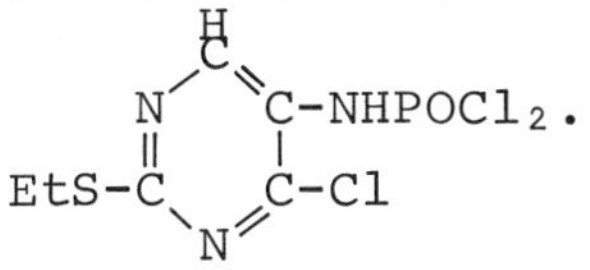
II. m. 247-50° dec.[393]

C_7 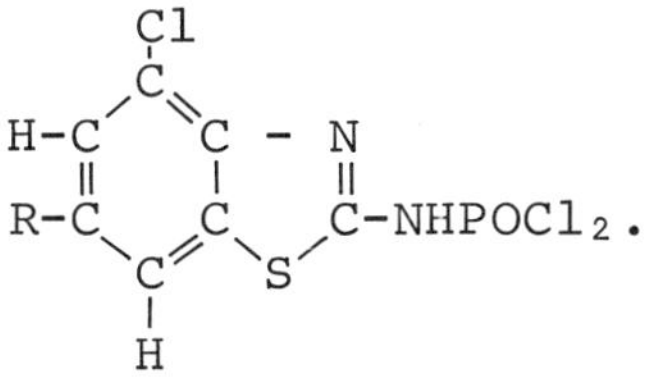
II. R=H, m. 241-3°; R=Me, m. 240-3°; R=EtO, m. 232-3°.[843]

o-$ClCOC_6H_4NHPOCl_2$. I. m. 62°.[882]
m-$ClCOC_6H_4NHPOCl_2$. I. m. 109-10°.[487,573]
p-$ClCOC_6H_4NHPOCl_2$. I. m. 165-8°,[529] m. 168°.[573]
p-$HOCOC_6H_4NHPOCl_2$. II. m. 240-1°.[529]
2,4-Br,$MeC_6H_3NHPOCl_2$. III.[573]
o-$MeC_6H_4NHPOCl_2$. III,[582] X.[936] m. 91°,[582] m. 68-70°.[936]
p-$MeC_6H_4NHPOCl_2$. II,[149] III,[573,581] X.[936] m. 110-1°,[149] m. 104°,[573,581] m. 108-10°,[936] m. 108-11°.[529]
o-$MeOC_6H_4NHPOCl_2$. III. Oil,[529] m. 83°.[374]
p-$MeOC_6H_4NHPOCl_2$. II. m. 71-2°.[487]
$MeC_6H_3(NHPOCl_2)_2$-2,4. II.[487]
o-$MeC_6H_4NHPOF_2$. II. m. 40°, b_6 98-100°.[656]
m-$MeC_6H_4NHPOF_2$. II. b_6 124-5°.[656]
p-$MeC_6H_4NHPOF_2$. II. m. 79-80°, b_6 118°.[656]
o-$MeOC_6H_4NHPOF_2$. II. m. 61°, b_6 105-6°.[656]
C_8 2,4-$Me_2C_6H_3NHPOCl_2$. II,III. m. 79°.[573]
2,5-$Me_2C_6H_3NHPOCl_2$. II,III. m. 119°.[573]
3,4-$Me_2C_6H_3NHPOCl_2$. II,III. m. 76°.[573]
p-$EtOC_6H_4NHPOCl_2$. II. m. 100-2°.[487]
o-$Me_2NSO_2C_6H_4NHPOCl_2$. X. m. 85-8° dec.[941]
C_9 $Me_3C_6H_2NHPOCl_2$. II,III. m. 122°.[573]
1-$C_{10}H_7NHPOCl_2$. III. m. 183° dec.[529]
2-$C_{10}H_7NHPOCl_2$. II. m. 115-7°.[487]
$[(Cl_2PONH)C_6H_4]_2$-p. X. ^{31}P -5.8.[512a]
[p-$(Cl_2PONH)C_6H_4]_2SO_2$. II.[389]
p-$(PhCH_2OCOCH_2)C_6H_4NHPOCl_2$. II. m. 99.5-100.5°.[947]

D.1.1.3. R_2N-P(O)X_2, RR'N-P(O)X_2, R or R' = Alkyl or Aryl, X = Halogen

C2 Me_2NPOCl_2. II,[744] XVII.[1] b. 194-5°,[573] b_{22} 90-1°,[573,74] b_{10} 72°,[370] b_{10} 76°,[1] b_1 45°,[898] n_D^{25} 1.4610,[898] n 1.4637,[370] n 1.4650,[1] $d_{15.5}^{15.5}$ 1.369,[898] d 1.3630,[370] ^1H,[169,644] ^{31}P -16,[633] -18.1,[873] elect.diff.[889]

Me_2NPOF_2. II,[101] XIV.[753] b_{10} 28°,[101] b_{12} 47-9°,[656,658] b_{20} 40-2°,[753] ^1H.[644]

$(CF_3)_2NPOF_2$. XVII. b. 31°, ^{19}F, mass spectrum.[269]

C3 $MeEtNPOCl_2$. II. b_{10} 84°.[753]

$MeEtNPOF_2$. XIV. b_{10} 35°.[753]

C4 $Me_2NPO(CN)_2$. XIV.[767]

$(FCH_2CH_2)_2NPOCl_2$. III. $b_{0.25}$ 75-6°.[662]

$(ClCH_2CH_2)(FCH_2CH_2)NPOCl_2$. III. $b_{0.4-0.5}$ 97-9°, n_D^{25} 1.4870.[662]

$O(CH_2CH_2)_2NPOCl_2$. II. b_5 110°,[60] $b_{0.1}$ 65-8°,[563] d 1.430,[563] n 1.4975,[563] m. 6-7.5°.[563]

$(ClCH_2CH_2)_2NPOCl_2$. II. m. 56°,[123] $b_{0.2}$ 127°,[123] m. 54-6°,[316,421] $b_{0.6}$ 123-5°.[316]

$Cl_2PON(CH_2CH_2)_2NPOCl_2$. III. m. 170-2°,[773] m. 175°.[488]

Et_2NPOBr_2. II. Oil.[573]

Et_2NPOCl_2. II. b. 220°, b_{15} 100°,[573] b_{15} 99°,[376] ^1H.[169]

Et_2NPOF_2. II,[658] VII,[820] XIV.[753] b_9 46°,[753] b_{11} 42-3°,[820] b_{13} 45-6°,[376] b_8 30-1°.[656] n 1.3730.[820] ^{31}P +3.6.[644]

C5 $(ClCH_2CH_2)[(ClCH_2)_2CH]NPOCl_2$. III. $b_{0.25}$ 126°.[910]

$C_5H_{10}NPOCl_2$. III. b. 257°, b_{11} 124°, d_0^{18} 1.323.[573]

CH_2=CH-CH_2(R)N-P(O)Cl_2. III.[295] R=Et, b_{10} 101°; R=(CH_2=$CHCH_2$-), b_9 102-3°; R=$PhCH_2$, b_5 148-9°; R=PhCH=CH, b_1 164-5°; R=Ph, b_2 140-2°; R= p-ClC_6H_4, $b_{1.5}$ 152-4°; R= p-MeC_6H_4, b_1 141-3°; R= p-$MeOC_6H_4$, $b_{2.5}$ 169-70°

C6 $MeOCON(CH_2CH_2)_2NPOCl_2$. II. m. 74-6°.[488]

$C_6H_{11}NHPOCl_2$. II. m. 81-2°.[60]

Pr(CH_3CH=CH)$NPOCl_2$. II. $b_{0.6}$ 78-80°,[121] n 1.4861.[121] $b_{0.5}$ 71-6.[122]

$(CH_3CHClCH_2)_2NPOCl_2$. III. m. 90-2°.[57]

Pr_2NPOBr_2. II. Oil.[573]

Pr_2NPOCl_2. II, III. b. 243-4°, b_{20} 123°.[573,579]

i-Pr_2NPOCl_2. XVII. ^{31}P -15,3.[873]

Pr_2NPOF_2. XIV. b_{21} 84°.[753]

C7 $PhMeNPOCl_2$. II, III. b. 282, b_{10} 150-1°.[573] ^{31}P -13,6.[634]

p-$H_2NSO_2C_6H_4NMePOCl_2$. II. Powder.[366]

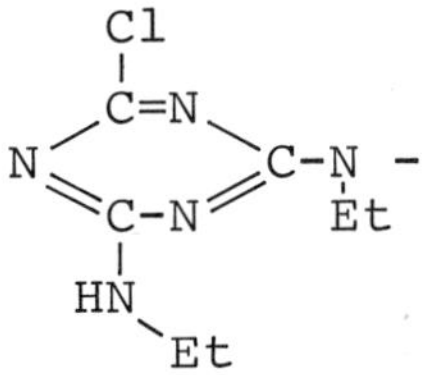 $- POCl_2$. II. m. 114-5°.[242]

$(ClCH_2CH_2)[ClCH_2(CH_2)_2CHClCH_2]NPOCl_2$. II. $b_{0.1}$ 170-5°.[316]

C_8 $PhEtNPOCl_2$. II, III. b_{16} 159°.[573]

$i\text{-}Bu_2NPOBr_2$. II. m. 68°.[573]

$i\text{-}Bu_2NPOCl_2$. II, XIII. m. 54°.[573,579]

$i\text{-}Bu_2NPOF_2$. XIV. b_{43} 114-6°,[384] b_{17} 98°,[753] n 1.4020,[384] d 1.0689.[384]

C_9 $Et_2NCON(CH_2CH_2)_2NPOCl_2$. II. b_2 185°.[488]

$C_9H_{10}NPOCl_2$. II. m. 79°, tetrahydroquinoline derivative.[573]

C_x $i\text{-}Am_2NPOCl_2$. II. b_{12} 150°. d_0^{13} 1.0804.[573,579]

Ph_2NPOCl_2. II. m. 57°.[573,660] ^{31}P -8,2.[634]

$(i\text{-}Hexyl)_2NPOF_2$. XIV. b_4 103°.[753]

D.1.1.4. $AcNH\text{-}P(O)X_2$, Ac = R-CO-, $ArSO_2$-, etc. Any Acid Group, X = Halogen

$HOCONHPOCl_2$. XII. Dec. 73-4°.[471]

C_7 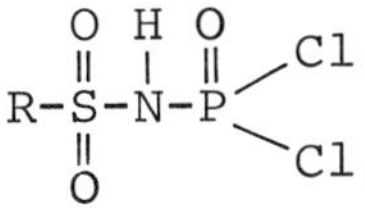 X.

R(alkyl)	m.
$ClCH_2$	101-3°[523]
Me	78-81°[443]
Et	65-8°[443]
Pr	89-91°[523]
i-Pr	114-6°[443]
Bu	70-2°[443]
i-Bu	86-8°[523]
Am	95-6°[523]
i-Am	76-8°[523]
C_6H_{11}	90-2°[443]
C_6H_{13}	79-81°[523]
$PhCh_2$	123-5°[443]
$PhCh_2CH_2$	111-2°[523]

C_2 $Cl_2BrCCONHPOCl_2$. X. m. 147°.[816,817]

$Br_3CCONHPOCl_2$. X. m. 105-6°.[816,817]

$CF_3CONHPOCl_2$. X. m. 83-4.5°.[551,399] IR.[551]

$Cl_2(O_2N)CCONHPOCl_2$. X. Dec. 165°.[816,817]

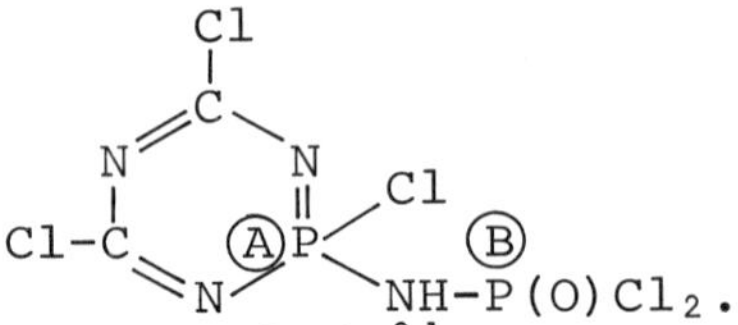
X. m. 164-6°.[91] ^{31}P A: -51.5; B: -3.8.[91]

$CCl_3CONHPOCl_2$. X. m. 146-8°,[63,816,817] m. 148-50°.[435]
$ClBrCHCONHPOCl_2$. X. Crystals.[816]
$Cl_2CHCONHPOCl_2$. X. m. 113°,[120] m. 112-3°.[816,817]
$ClCH_2CONHPOCl_2$. X. Crystals.[816]

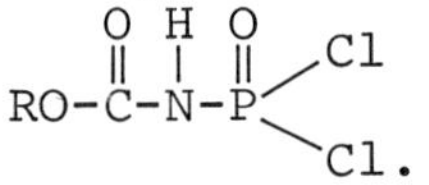
X, XII.

R	m.
Me	47-59°[430]
Et	23-5°,[430] 20-5°[467,509]
i-Pr	75-7°,[430] 74-8°[467]
Pr	49-50°[430]
C_6H_{11}	110-1°[430]
Ph	dec. 102-4°[467]
o-ClC_6H_4	100-2°[467]
p-ClC_6H_4	115-7°[467]
p-BrC_6H_4	124-6°[467]
p-$MeOC_6H_4$	90-6°[467]
$PhCH_2$	85-6°[456]
5,2-i-Pr,MeC_6H_3	72-4°[456]
m-HOC_6H_4	132-3°[456]
p-HOC_6H_4	126-8°[456]
o-$O_2NC_6H_4$	40-2°[456]
p-$O_2NC_6H_4$	88-90°[456]

$MeOCONHPOF_2$. XII. b_5 95°, m. 51-2°.[497]
$Me_2NSO_2NHPOCl_2$. X. m. 110-2°.[459]
MeO-$NHCONHPOCl_2$. XII. m. 91-2°.[144]
$CCl_3C(=NPOCl_2)NHPOCl_2$. X. m. 139-41°.[232]
$Cl_2PONHCONHNHCONHPOCl_2$. XII. $[N_2H_5]^+$-salt: m. 213-5° dec.[470]

C_3 $EtOCONHPOF_2$. XII. b_3 91-2°, m. 35-6°.[497]
$HCF_2CF_2CONHPOCl_2$. XVII. m. 65-9°.[270]
$HCClFCF_2CONHPOCl_2$. XVII. m. 82-4°.[270]
$HCCl_2CF_2CONHPOCl_2$. XVII. m. 104-6°.[270]
$CH_2(CN)CONHPOCl_2$. X. m. 140-2°.[800]
$ClCH_2CHClCONHPOCl_2$. X. m. 74-6°.[198]
$ClCH_2CCl_2CONHPOCl_2$. X. m. 121-3°.[198]
$MeCCl_2CONHPOCl_2$. X. m. 119-21°,[231] m. 127-8°.[816]
$(C_2H_4N)CONHPOCl_2$. XII. m. 70°.[144]
$EtNHCONHPOF_2$. XII. m. 66-7°.[497]
$Me_3SiNHPOCl_2$. II. m. 94-5°.[94]
$Me_2NNHCONHPOCl_2$. XII. $[Me_2NNH_3]^+$-salt: m. 110-3° dec.[47]

C_4 i-PrOCONHPOF$_2$.   XII.   b$_6$ 101-2°. m. 54-5°.[497]

$Et_2NSO_2NHPOCl_2$.   X.   m. 70-2°.[459]

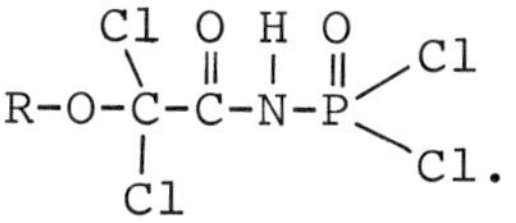

$$\underset{\underset{Cl}{|}}{\overset{\overset{Cl}{\diagup}}{R-O-C}}-\overset{\overset{O}{\|}}{C}-\overset{\overset{H}{|}}{N}-\overset{\overset{O}{\|}}{P}\diagdown_{Cl}^{Cl}$$ X.

R	m.
Et	124-6°,[458] m. 128-30°[895] (gives wrong structure)
Bu	114-6°[458]
Ph	155-6°[458]
o-MeC$_6$H$_4$	128-30°[458]
m-MeC$_6$H$_4$	161-3°[458]
p-MeC$_6$H$_4$	144-6°[458]
1-C$_{10}$H$_7$	140-2°[458]
2-C$_{10}$H$_7$	163-5°[458]

$C_3F_7CONHPOCl_2$.   X.   m. 89-90°.[551]   IR.[551]

$CCl_3C(=NCOCCl_3)NHPOCl_2$.   m. 166-8°.[232]

C_5 $Cl_2PONHCO(CF_2)_3CONHPOCl_2$.   X.   m. 145-8°.[201]   For similar compounds see Ref. 201.

$Me_2C(CN)CONHPOCl_2$.   X.   m. 149-50°.[788]   For similar compounds see Ref. 788.

C_6 $\left[Ar-\overset{\overset{O}{\|}}{\underset{\underset{O}{\|}}{S}}-N-\overset{\overset{O}{\|}}{P}\diagdown_{F}^{F} \right] K.$ X.[451]

Ar	m.
Ph	230-2°
p-MeC$_6$H$_4$	232-5°
p-ClC$_6$H$_4$	265-8°
p-FC$_6$H$_4$	225-8°
o-O$_2$NC$_6$H$_4$	181-2°
m-O$_2$NC$_6$H$_4$	217-20°
p-O$_2$NC$_6$H$_4$	225-8°
1-C$_{10}$H$_7$	264-5°
p-BrC$_6$H$_4$	269-70°

$PhSO_2NHPOBr_2$.   X.   m. 122-4°.[473]

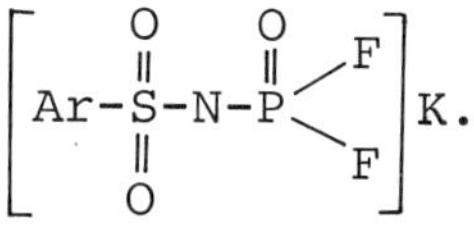

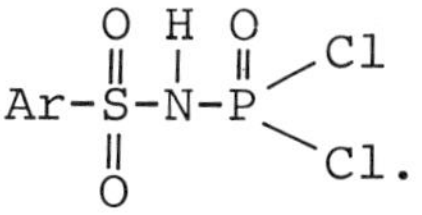

$$Ar-\overset{\overset{O}{\|}}{\underset{\underset{O}{\|}}{S}}-\overset{\overset{H}{|}}{N}-\overset{\overset{O}{\|}}{P}\diagdown_{Cl}^{Cl}$$ X.

Ar	m.
Ph	130-1°,[573,904] dec. 110°[431] K-salt: 195-6°,[385] 197-8° dec.[451,518] [PhNH$_3$]$^+$-salt: 107-10°[451,518]
o-MeC$_6$H$_4$	115° dec.[431]
p-MeC$_6$H$_4$	110° dec.,[431] 115-8°[919] K-salt: 210-3°[451,518]
o-O$_2$NC$_6$H$_4$	149-51°[446] K-salt: 199-200°[451,518]
m-O$_2$NC$_6$H$_4$	145-7°[446] K-salt: 205-6°[451,518]

Ar	m.
$p\text{-}O_2NC_6H_4$	143-5°[446]
	K-salt: 190-2°[451] [518]
$1\text{-}C_{10}H_7$	120° dec.[421]
$p\text{-}FC_6H_4$	135-6°[391]
$m\text{-}CF_3C_6H_4$	82-3°[391]
$p\text{-}ClC_6H_4$	K-salt: 207-8° dec.[451] [518]
$m\text{-}C_6H_4(SO_2NHPOCl_2)$	dec. 152-4°[448]
$p\text{-}C_6H_4(SO_2NHPOCl_2)$	dec. 220°[448]

$CCl_3C(=NBu)NHPOCl_2$.   X.   m. 49-51°.[232]
$PhNHSO_2NHPOCl_2$. ^{31}P +2.5.[512a]
C_7 $PhOCONHPOF_2$.   XII.   m. 80-3°.[497]
$PhNHCONHPOF_2$.   XII.   m. 111-2°.[497]
$o\text{-}MeC_6H_4SO_2NHPOBr_2$.   X.   m. 123-6°.[473]
$p\text{-}MeC_6H_4SO_2NHPOBr_2$.   X.   m. 123-6°.[473]

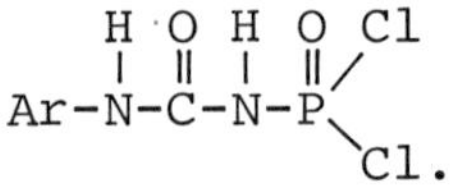

$$\text{Ar-N-C-N-P} \begin{array}{c} \text{H O H O Cl} \\ | \; \| \; | \; \| \diagup \\ \phantom{\text{Ar-N-C-N-P}} \diagdown \text{Cl} \end{array} \quad \text{XII.}$$

Ar	dec.
Ph	124-5°.[450] m. 121-2°[71]
$p\text{-}MeC_6H_4$	120°[450]
$o\text{-}MeC_6H_4$	117-8°[450]
$p\text{-}EtC_6H_4$	127-8°[450]
$p\text{-}(i\text{-}Pr)C_6H_4$	128-30°[450]
$p\text{-}MeOC_6H_4$	118-20°[450]
$p\text{-}EtOC_6H_4$	122-4°[450]
$o\text{-}O_2NC_6H_4$	146-7°[450]
$m\text{-}O_2NC_6H_4$	161-2°[450]
$p\text{-}O_2NC_6H_4$	160°[450]
benzthiazolyl-2	152-4°[450]
thiazolyl-2	solid[450]
pyridyl-2	dec. in air[450]
$m\text{-}ClC_6H_4$	130-2°[240]
$p\text{-}IC_6H_4$	140-1°[240]
$p\text{-}CF_3C_6H_4$	159-61°[240]

$2,4\text{-}Cl(O_2N)C_6H_3CONHPOCl_2$.   X.   m. 121-2°.[452]
$O\text{-}ClC_6H_4CONHPOCl_2$.   X.   m. 92-3°.[452]
$m\text{-}ClC_6H_4CONHPOCl_2$.   X.   m. 109-9.5°.[699]
$p\text{-}ClC_6H_4CONHPOCl_2$.   X.   m. 106-7°.[452]
$2,4\text{-}Cl_2C_6H_3CONHPOCl_2$.   X.   m. 106-7°.[452]
$3,5\text{-}Cl_2C_6H_3CONHPOCl_2$.   X.   m. 129.5-30°.[699]
$2,3,5\text{-}Cl_3C_6H_2CONHPOCl_2$.   X.   m. 129.5°.[699]
$3,4,5\text{-}Cl_3C_6H_2CONHPOCl_2$.   X.   m. 151.5-2°.[699]
$2,3,6\text{-}Cl_3C_6H_2CONHPOCl_2$.   X.   m. 128-32°,[832] with impurities.
$PhCONHPOCl_2$.   X.   m. 92.5-5°, rapid heating 96-114°,[452] m. (rapid heating) 115°, (slow heating) dec. 110°.[856,852]

o-$IC_6H_4CONHPOCl_2$. X. m. 91-2°.[700]
m-$IC_6H_4CONHPOCl_2$. X. m. 111-3°.[700]
2,5-$I_2C_6H_3CONHPOCl_2$. X. m. 141-2°.[700]
3,5-$I_2C_6H_3CONHPOCl_2$. X. m. 99-9.5°.[700]
2,4,5-$I_3C_6H_2CONHPOCl_2$. X. m. 148-50°.[700]
2,4-$I_2C_6H_3CONHPOCl_2$. X. m. 89-91°.[700]
3,4-$I_2C_6H_3CONHPOCl_2$. X. m. 155-6°.[700]
2,3,5-$I_3C_6H_2CONHPOCl_2$. X. m. 148-50°.[700]
3,4,5-$I_3C_6H_2CONHPOCl_2$. X. m. 173-75°.[700]

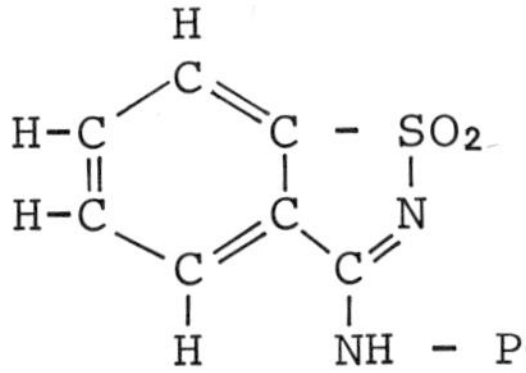

NH - $POCl_2$. X. m. 213-4°.[207]
o-$O_2NC_6H_4CONHPOCl_2$. X. m. 117-8°.[452]
m-$O_2NC_6H_4CONHPOCl_2$. X. m. 121-2°.[452]
p-$O_2NC_6H_4CONHPOCl_2$. X. m. 130-1°.[452]
2,4-$(O_2N)_2C_6H_3CONHPOCl_2$. X. m. 102-4°.[452]
3,5-$(O_2N)_2C_6H_3CONHPOCl_2$. X. m. 167-8°.[452]
o-$BrC_6H_4CONHPOCl_2$. X. m. 83-4°.[693]
m-$BrC_6H_4CONHPOCl_2$. X. m. 108-10°.[693]
2,5-$Br_2C_6H_3CONHPOCl_2$. X. m. 138-40°.[693]
3,5-$Br_2C_6H_3CONHPOCl_2$. X. m. 184-5°.[693]
2,3,5-$Br_3C_6H_2CONHPOCl_2$. X. m. 177-9°.[693]
3,4,5-$Br_3C_6H_2CONHPOCl_2$. X. m. 146-8°.[693]
p-$FC_6H_4CONHPOCl_2$. X. m. 103-4°,[698] m. 111-2°.[235]
m-$FC_6H_4CONHPOCl_2$. X. m. 104-5°,[698] m. 110-1°.[235]
o-$FC_6H_4CONHPOCl_2$. X. m. 82.5-4.5°,[698] m. 100-1°.[235]
$PhNHNHCONHPOCl_2$. XII. dec. 160-3°.[470]
p-$O_2NC_6H_4NHNHCONHPOCl_2$. XII. m. 188-90° dec.[470]
2,4-$(O_2N)_2C_6H_3NHNHCONHPOCl_2$. XII. m. 175-7° dec.[470]
C_8 p-$MeC_6H_4NHCONHPOF_2$. XII. m. 97°.[497]
$PhCF_2CONHPOCl_2$. X. m. 89-91°.[698]
$CF_3C(=NPh)NHPOCl_2$. X. m. 85-7°.[232]
p-$CF_3C_6H_4CONHPOCl_2$. X.[698,235] m. 126-8°.[698]
m-$CF_3C_6H_4CONHPOCl_2$. X. m. 102-3°.[698]
p-$MeC_6H_4NHNHCONHPOCl_2$. XII. m. 114-7° dec.[470]
$PhCONHNHCONHPOCl_2$. XII. m. 126-8° dec.[470]
m-$C_6H_4(CONHPOCl_2)_2$. X. m. 130-1°.[438]
$CCl_3C(=NPh)NHPOCl_2$. X. m. 121-3°.[197] For other $CCl_3C(=NAr)NHPOCl_2$ compounds see Ref. 197.
o-$MeC_6H_4CONHPOCl_2$. X. m. 76-7°.[692]
m-$MeC_6H_4CONHPOCl_2$. X. m. 70-1°.[692]
p-$MeC_6H_4CONHPOCl_2$. X. m. 99-101°.[438]
m-$MeOC_6H_4CONHPOCl_2$. X. m. 83-5°.[692]
p-$MeOC_6H_4CONHPOCl_2$. X. m. 182-3°.[692]
C_9 m-$EtOC_6H_4CONHPOCl_2$. X. m. 108-10°.[692]
p-$EtOC_6H_4CONHPOCl_2$. X. m. 126-8°.[692]
$PhEtNCONHPOCl_2$. XII. m. 113-4°.[450]

C_{10} 1-$C_{10}H_7SO_2NHPOBr_2$. X. m. 131-4°.[473]
2-$C_{10}H_7SO_2NHPOBr_2$. X. m. 161-3°.[473]
m-CH_2=$CHCH_2$-$C_6H_4CONHPOCl_2$. X. m. 86-7°.[692]
p-CH_2=$CHCH_2$-$C_6H_4CONHPOCl_2$. X. m. 77-8°.[692]
3,4,5-$(MeO)_3C_6H_2CONHPOCl_2$. X. m. 87-9°.[692]
C_{11} 1-$C_{10}H_7CONHPOCl_2$. X. m. 99-101°.[438]
2-$C_{10}H_7CONHPOCl_2$. X. m. 111-3°.[438]
C_{12} $(PhO)_2PONHPOCl_2$. X. Oil.[469]
$(PhO)_2PSNHPOCl_2$. X. m. 102-4°.[469]
C_{13} p-$O_2NC_6H_4C$(=NSO_2Ph)$NHPOCl_2$. X. m. 161-3°.[207]
PhC(=NSO_2Ph)$NHPOCl_2$. X. m. 123-5°.[203]
PhC(=$NSO_2C_6H_4$Cl-p)$NHPOCl_2$. X. m. 166-7°.[203]
PhC(=$NSO_2C_6H_4NO_2$-p)$NHPOCl_2$. X. m. 183-4°.[203]
$Ph_2NCONHPOCl_2$. X, XII. m. 136-7°.[450,461]
C_{14} m-$PhCH_2OC_6H_4CONHPOCl_2$. X. m. 95-6°.[692]
p-$PhCH_2OC_6H_4CONHPOCl_2$. X. m. 110-2°.[692]
PhC(=$NSO_2C_6H_4$Me-p)$NHPOCl_2$. X. m. 174°.[203]
$Ph_2CClCONHPOCl_2$. X. m. 122-3°.[433,816]
C_x PhC(=$NSO_2C_{10}H_7$-1)$NHPOCl_2$. X. m. 177-8°.[203]
PhC(=$NSO_2C_{10}H_7$-2)$NHPOCl_2$. X. m. 162-3°.[203]
PhC(=NPO$(OPh)_2$)$NHPOCl_2$. X. m. 159-61°.[203]
$Ph_3CCONHPOCl_2$. X. m. 128-30°.[433]

D.1.1.5. =N-P(O)X_2, N Double Bond to Any Other Atom or Substituent, X = Halogen

C_1 $SCNPOCl_2$. VII. b_1 40°,[752] IR,[752] Raman,[752] ^{31}P +21.[309,311]
$OCNPOCl_2$. I. b_5 19.7-20°,[430] d^{15} 1.649,[430] n_D^{15} 1.470,[430] n_D^{21} 1.4682,[661] ^{31}P +9.2,[309] b_8 36-7°.[661]
$Cl_2CNPOCl_2$. XVII. b_8 75-6°,[254] b_{15} 82-4°,[483] b_{25} 83-6°,[223] d 1.7359,[223] d^{20} 1.7520,[254] n 1.5252,[254,483] n 1.5248,[223] ^{31}P +1.3.[312]
$Cl_3P[\alpha]$=NCCl=$NP[\beta]OCl_2$. ^{31}P [α] -23.8, [β] -0.1.[91]
$(Cl_3P[\alpha]=N)_2C$=$NP[\beta]OCl_2$. XVII. m. 93-4°, ^{31}P [α] -14.7, [β] -0.6.[91]
$SCNPOF_2$. VII. b_{87} 35-6°,[729] ^{19}F,[729] ^{31}P +36.4.[729]
$OCNPOF_2$. VII,[728] XIV.[497] b. 69°,[728] b. 68-68.5°,[497] n 1.3381,[497] ^{19}F,[728] ^{31}P +29.5.[728]
MeSO(=$NPOCl_2$)Cl. XVII. $b_{0.001}$ 91-2°, b_8 140-2°, m. 42-3°.[523]
C_2 EtSO(=$NPOCl_2$)Cl. XVII. $b_{0.002}$ 104-5°, b_3 148-9°.[523] For other alkyl-SO(=$NPOCl_2$)Cl (oils, n_D^{20}, d_{20}) see Ref. 523.
$Cl_2BrCCCl$=$NPOCl_2$. I. m. 68°.[816]
CBr_3CCl=$NPOCl_2$. I. Liquid.[816]
CF_3CCl=$NPOCl_2$. XVII. b_6 37-9°, n 1.4360.[200]
$Cl_2(O_2N)CCCl$=$NPOCl_2$. I. m. 55-60°.[816]
CCl_3CCl=$NPOCl_2$. I, XVII. b. 255-9°?,[895] b_{11} 140°,[120] m. 80°?,[120] m. 40-1°.[435] b_2 95-6°,[799] m. 45-6°,[799] for further RCCl=$NPOCl_2$ compounds see Ref. 799.
Cl_3P=NC(CCl_3)=$NPOCl_2$. XVII. $b_{0.03}$ 76-8°.[232]
ClHBrCCCl=$NPOCl_2$. I. Oil.[816]
Cl_2CHCCl=$NPOCl_2$. I. m. 39°,[120] $b_{0.3}$ 92-4°.[895]
$BrCH_2CCl$=$NPOCl_2$. I. Oil.[816]

ClCH$_2$CCl=NPOCl$_2$.   I.   b$_{0.2}$ 100°.[120]

MeNHCCl=NPOCl$_2$.   I.   m. 103-5°.[226]

C$_3$ EtNHCCl=NPOCl$_2$.   I.   m. 98-100°.[226]

MeCCl$_2$CCl=NPOCl$_2$.   I.   m. 80°.[816]

MeCCl$_2$CH=NPOCl$_2$.   XVII.   b$_3$ 82-5°, d 1.5502, n 1.5150.[937]

MeNHC(CCl$_3$)=NPOCl$_2$.   XVII.   m. 141-4°.[230]   For 8 similar compounds see Ref. 230.

ClCOCCl$_2$CCl=NPOCl$_2$.   X.   b$_{0.02}$ 78-80°, d 1.7555, n 1.5339.[793]

C$_4$ EtCCl$_2$CH=NPOCl$_2$.   XVII.   b$_3$ 78-80°, d 1.5255, n 1.5150.[973]

MeCCl(CN)CCl=NPOCl$_2$.   X. b$_{0.035}$ 84-6°, n 1.5119, d 1.5366.[792]

Me(EtO)C=NPOCl$_2$.   XIII.   b$_{2.5}$ 78°, n$_D^{22}$ 1.4765, d$_{22}$ 1.3281.[672]

MeCCl(COCl)CCl=NPOCl$_2$.   X.   b$_{0.2}$ 92-4°, n$_D^{25}$ 1.5238, d$_{25}$ 1.6260.[794]

EtOC(CN)=NPOCl$_2$.   XII.   b$_{20}$ 36-8°, n 1.4482, d 1.3365.[209]

EtCCl$_2$CCl=NPOCl$_2$.   X.   b$_{0.08}$ 76-7°, n 1.5200.[796]

Me(EtO)C=NPOF$_2$.   XIII.   b$_7$ 51°, n 1.3895, d$_{22}$ 1.2501.[672]

Me$_3$SiN=C=NPOF$_2$.   II.   m. -70°, b$_{16}$ 83.5°.[630]

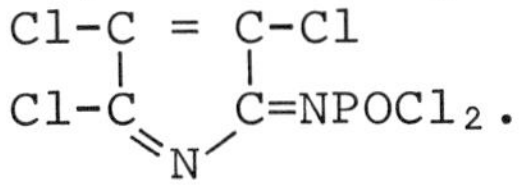

Cl-C = C-Cl
Cl-C$\diagdown$_N$\diagup$C=NPOCl$_2$.   XVII.   m. 193-4°.[797]

C$_5$ (MeO)$_2$C=NPO(CN)$_2$.   XIII.   b$_3$ 118-20°, n 1.4435, d 1.2260.[672]

EtCCl(CN)CCl=NPOCl$_2$.   X.   b$_{0.02}$ 83-5°, n 1.5095, d 1.4859.[792]

EtCCl(COCl)CCl=NPOCl$_2$.   X.   b$_{0.05}$ 85-8°, n 1.5213, d$_{25}$ 1.5977.[794]

PrCCl$_2$CCl=NPOCl$_2$.   X.   b$_3$ 123-5°, n 1.5144.[796]

(EtO)$_2$C=NPOCl$_2$.   XIII.   b$_{0.4}$ 86-90° dec. n 1.4715, d 1.2800.[672]

(EtO)$_2$C=NPOF$_2$.   XIII.   b$_{0.6}$ 92-4°.[672]

C$_6$ PrCCl(COCl)CCl=NPOCl$_2$.   X.   b$_{0.1}$ 93-6°, n 1.5208, d 1.5508.[794]

i-PrCCl(COCl)CCl=NPOCl$_2$.   X.   b$_{0.2}$ 95-8°, n 1.5219, d 1.5262.[794]

PrCCl(CN)CCl=NPOCl$_2$.   X.   b$_{0.015}$ 83-6°, n 1.5081, d 1.4441.[792]

i-PrCCl(CN)CCl=NPOCl$_2$.   X.   b$_{0.02}$ 87-90°, n 1.5073, d 1.4439.[792]

Cl$_2$PON=CClCCl$_2$CH$_2$CH$_2$CCl$_2$CCl=NPOCl$_2$.   X.   m. 144-6°.[798]

PhSO(=NPOCl$_2$)Cl.   XVII.   b$_{0.001}$ 141-2°, n$_D^{17}$ 1.5769, d$_{17}$ 1.5704.[523]

p-FC$_6$H$_4$SO(=NPOCl$_2$)Cl.   XVII.   b$_{0.001}$ 134-5°, n$_D^{17}$ 1.5669, d$_{17}$ 1.6459.[523]

p-ClC$_6$H$_4$SO(=NPOCl$_2$)Cl.   XVII.   m. 71-2°.[520]   For other aryl derivatives see Ref. 520.

C$_7$ o-MeC$_6$H$_4$SO(=NPOCl$_2$)Cl.   XVII.   m. 70-1°.[520]

BuCCl(CN)CCl=NPOCl$_2$.   X.   b$_{0.018}$ 96-9°, n 1.5046, d 1.3896.[792]

(EtO)$_2$C=NPO(CN)$_2$.   XIII.   b$_2$ 133-4°, n 1.4321, d 1.1806.[672]

p-O$_2$NC$_6$H$_4$CCl=NPOCl$_2$.   XVII.   m. 121-4°.[438]

3,5-(O$_2$N)$_2$C$_6$H$_3$CCl=NPOCl$_2$.   XVII.   m. 82-3°,[438] for other ArCCl=NPOCl$_2$ compounds (oils) see Ref. 438.

PhCCl=NPOCl$_2$.   XV,[203] XVII.[438]   Oil,[438] b$_{0.1}$ 146-7°, n 1.6035, d 1.468.[203]

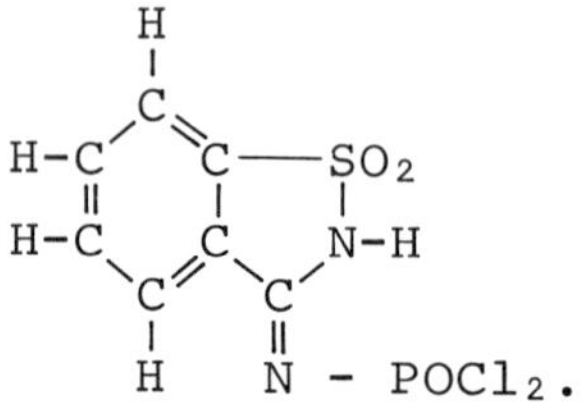

X. m. 213-4°.[207]

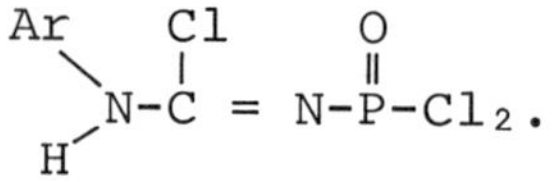

I,[228] XVII.[240]

Ar	m.
Ph	138-41°[228,240]
o-MeC_6H_4	105-7°[240]
p-MeC_6H_4	152-4°[228.240]
p-$CF_3C_6H_4$	138-40°[240]
m-ClC_6H_4	135-6°[240]
p-ClC_6H_4	153-4°[240]
p-BrC_6H_4	147-9°[228,240]
p-IC_6H_4	139-41°[240]
o-$O_2NC_6H_4$	98-100°[240]
m-$O_2NC_6H_4$	147-9°[228,240]
p-$O_2NC_6H_4$	164-5°[240]
4,3-Me(O_2N)C_6H_3	136-7°[240]
p-$MeOC_6H_4$	128-30°[228]

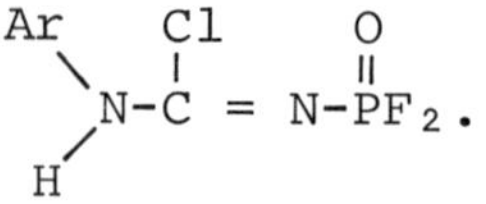

XVII.[227]

Ar	m.
Ph	88-9°
p-ClC_6H_4	130-2°
p-BrC_6H_4	127-9°
p-$O_2NC_6H_4$	158-60°
m-$O_2NC_6H_4$	129-31°
p-MeC_6H_4	120-2°
p-$MeOC_6H_4$	108-10°
2-$C_{10}H_7$	127-9°

C_8 $Cl_2PON=CClCCl_2(CH_2)_4CCl_2CCl=NPOCl_2$. X. m. 122-3.[798]
p-$C_6H_4[CCl(=NPOCl_2)]_2$. XVII. m. 118-20°, after storage m. 82-5°.[695]
$PhBrCHCCl=NPOCl_2$. I. Liquid.[816]

C_x $(Cl_2PON=CClCCl_2(CH_2)_6CCl_2CCl=NPOCl_2$. X. m. 84-6°.[798]

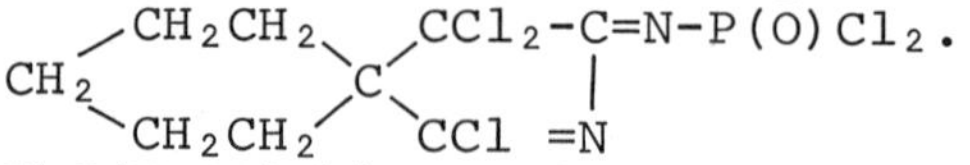

XVII. m. 96-8°.[787]

$PhC(N=CClPh)=NPOCl_2$. I. $b_{0.1}$ 202-4°.[203]
p-$BrC_6H_4C(N=CClPh)=NPOCl_2$. I. m. 96-9°.[203]
$Ph_2CClCCl=NPOCl_2$. I, XVII. m. 81-3°,[433] liquid.[816]

D.1.2. >N-P(S)X_2, X = Halogen

D.1.2.1. $RNH-P(S)X_2$, $AcNH-P(S)X_2$, $=N-P(S)X_2$, R = Alkyl or Aryl, X = Halogen

C_1 $SCNPSBr_2$. XIV. ^{31}P +62.8.[311]
$OCNPSCl_2$. XIV.[661]
$SCNPSCl_2$. XIV. ^{31}P -16.[309]
$SCNPSF_2$. XIV.[728,729] b. 89.5-90°,[728] ^{31}P +28.6,[728] ^{19}F.[728]
$MeNHPSCl_2$. II, III. b_{33} 115°.[573]
$MeNHPSF_2$. XIV. b_{12} 40°.[753]
C_2 $EtNHPSCl_2$. II, III. b. 216°, b_9 105°, b_{20} 115°.[573]
$EtNHPSF_2$. II. b_8 28-9°.[658]
C_3 $CH_2=CHCH_2NHPSCl_2$. III. $b_{1.5}$ 86°,[677] b_3 83-5°.[295] n 1.5535, d 1.369.[677]
$PrNHPSCl_2$. III,[573] II.[113] b_{17} 121°.[573,113]
C_4 $BuNHPSCl_2$. II. b_{17} 115°.[113]
i-$BuNHPSCl_2$. II, III. b_9 116°, b_{15} 123°, b. 251°.[573]
$BuNHPSF_2$. II. b_8 46-7°.[658]
C_5 $AmNHPSCl_2$. III. b_{16} 140°.[573]
C_6 $PhNHPSCl_2$. II. b_8 133-4°.[658]
$PhNHPSF_2$. II. b_1 86°. n 1.5414.[656]
$C_6H_{11}NHPSCl_2$. II. m. 70°. b_4 135-43°.[618]
$C_6H_{11}NHPSF_2$. II. b_8 73-4°.[658]
C_7 $PhCH_2NHPSCl_2$. II.[573]
m-$MeC_6H_4NHPSCl_2$. II. b_8 148-9°.[658]
p-$MeC_6H_4NHPSCl_2$. II. b_{11} 165-6°.[658]
o-$MeOC_6H_4NHPSCl_2$. II. b_{10} 161-2°.[658]
o-$MeC_6H_4NHPSF_2$. II. $b_{0.9}$ 76°, n 1.5223.[656]
m-$MeC_6H_4NHPSF_2$. II. $b_{0.9}$ 85.5, n 1.5361.[656]
p-$MeC_6H_4NHPSF_2$. II. b_1 98°, 1.5358.[656]
o-$MeOC_6H_4NHPSF_2$. II. b_1 93°, m. 46-7°.[656]

C_x 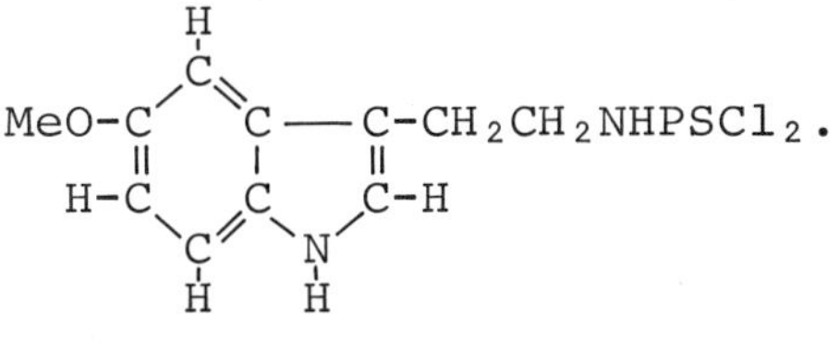 II. m. 95-6°.[613]

D.1.2.2. $RR'N-P(S)X_2$, R or R' = Alkyl or Aryl, X = Halogen

C_2 Me_2NPSCl_2. II,[744,573] XIII,[47] XVII.[331] b_{10} 77.5°,[47] b_{10} 75-6°,[331] b_{16} 89-90°,[744] b_{16} 85-90°.[573] n 1.5404,[331] n 1.5390.[47] d 1.3761,[331] d 1.3731.[47]
Me_2NPSF_2. II,[658] XIV.[753] b_{11} 22-5°,[753] b_{30} 33-4°.[658] ^{31}P -75.7.[371] ^1H, ^{19}F.[371]
C_4 $(ClCH_2CH_2)_2NPSCl_2$. III. $b_{0.007}$ 92-4°,[251] b_2 117°.[421] m. 32-3°,[251] m. 30-2°.[421]
$Cl_2PSN(CH_2CH_2)_2NPSCl_2$. II. m. 160°,[488] m. 158.5-160.5°.[773]
$(C_4H_8N)PSF_2$. II. ^{31}P -6.5.[371] ^1H, ^{19}F.[371]
Et_2NPSBr_2. III. Oil.[573]
Et_2NPSCl_2. II, XIII, XVII. b_1 63°,[331] b_{13} 103°, b_{14} 107°.[376,573] n 1.5262.[331] d_0^{15} 1.105,[376,573] d 1.2673.[331] ^{31}P -59.8,[873] ^1H.[169]

Et_2NPSF_2. II,[658] XIV.[376,753] b_{12} 49°,[658] b_{12} 50°,[753] b_{12} 50-1°,[376] ^{31}P -71.3,[371] ^1H, ^{19}F.[371]

C_5 $(C_5H_{10}N)PSCl_2$. II, III. b_{21} 146-9°, d_0^{15} 1.3092.[573]

C_6 $(CH_2{=}CHCH_2)_2NPSCl_2$. III. b_8 111-2°.[295]

$MeOCON(CH_2CH_2)_2NPSCl_2$. II. m. 53°.[488]

$(MeCHClCH_2)_2NPSCl_2$. II. m. 72°.[57]

Pr_2NPSBr_2. III. Oil.[573]

Pr_2NPSCl_2. II, III. b_{15} 132-4°, b. 240-5°,[573] b_{17} 123°.[113] d_0^{15} 1.077.[573]

Pr_2NPSF_2. II. ^{31}P -70.0.[371] ^1H, ^{19}F.[371]

C_7 $MePhNPSCl_2$. II, XIII. d_0^{22} 1.357.[573]

C_8 $EtPhNPSCl_2$. II, XIII. Liquid.[573]

$(i\text{-}Bu)_2NPSBr_2$. II. m. 66°.[573]

Bu_2NPSCl_2. II. b_{17} 143°.[113]

$(i\text{-}Bu)_2NPSCl_2$. II, III, XIII. m. 36°. b_{10} 150°.[573]

C_x $Et_2NCON(CH_2CH_2)_2NPSCl_2$. II. m. 72°.[488]

$(i\text{-}Am)_2NPSBr_2$. II. Oil.[573]

$(i\text{-}Am)_2NPSCl_2$. II, III. b_{10} 160-3°. d_0^{15} 1.0288.[573,579]

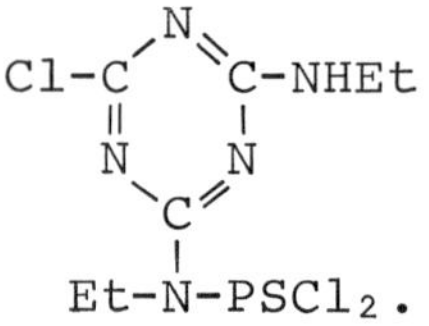

II.[242]

D.1.3. >N-P(O)(OH)$_2$

C_1 $HC(NH_2){=}NPO(OH)_2$. XVII. m. 137°.[174]

$(NH_2)_2C{=}NPO(OH)_2$. XVII. m. 132°,[174] Ca-salt.[922]

C_2 $MeC(NH_2){=}NPO(OH)_2$. XVII. m. 161°.[174]

$MeNHC(NH_2){=}NPO(OH)_2$. XVII. m. 86°.[174]

$MeOC(NH_2)NHPO(OH)_2$. XVII. m. 155-6°.[174]

$HOCH_2CH_2NHPO(OH)_2$. XVII. m. 242°.[689]

$Me_2NSO_2NHPO(OH)_2$. X. dec. 187-9°.[459]

C_3 $EtOCONHPO(OH)_2$. XVII. m. 125°.[509]

C_4 $Et_2NSO_2NHPO(OH)_2$. X. m. 80-95°.[459]

C_5 $(HO)_2PONHCO(CF_2)_3CONHPO(OH)_2$. VIII. m. 149-51°.[201]

C_6 $2,4\text{-}Cl_2C_6H_3NHPO(OH)_2$. VIII. m. 167°.[573]

$p\text{-}BrC_6H_4NHPO(OH)_2$. VIII. m. 158°.[573]

$p\text{-}ClC_6H_4NHPO(OH)_2$. VIII. m. 292°,[527,529] [m. 155°,[573] not $-PO(OH)_2$ but $-PO(OH)(NH_2)$], XV.[336,337]

$p\text{-}FC_6H_4SO_2NHPO(OH)_2$. VIII. m. 148-9°.[391]

$PhNHPO(OH)_2$. XVII,[163] VIII,[529] XV.[336,337] m. 276-7°,[163] 276-7.[529] ^{31}P +12.[634]

$PhSO_2NHPO(OH)_2$. X. m. 148-9°,[431] 147-8°.[473]

$o\text{-}O_2NC_6H_4SO_2NHPO(OH)_2$. X. m. 188-90°.[446]

$m\text{-}O_2NC_6H_4SO_2NHPO(OH)_2$. X. m. 159-61°.[446]

$p\text{-}O_2NC_6H_4SO_2NHPO(OH)_2$. X. m. 173-5°.[446]

$o\text{-}H_2NSO_2C_6H_4NHPO(OH)_2$. VIII. m. 207-8°.[941]

C_7 $o\text{-}MeC_6H_4SO_2NHPO(OH)_2$. X. m. 139-141°.[431,473]

$p\text{-}MeC_6H_4SO_2NHPO(OH)_2$. X. m. 150-3°.[431,473]

$p\text{-}H_2NSO_2C_6H_4NHPO(OH)_2$. VIII. m. 233-34° dec.[529]
$p\text{-}FC_6H_4CONHPO(OH)_2$. X. m. 83-5°.[235]
$PhCONHPO(OH)_2$. X. m. 157-8°,[856] m. 136-8° dec.,[454] m. 136-57°.[943]
$m\text{-}CF_3C_6H_4SO_2NHPO(OH)_2$. VIII. m. 135-6°.[391]
$o\text{-}O_2NC_6H_4CONHPO(OH)_2$. VIII. m. 148-9° dec.[454]
$m\text{-}O_2NC_6H_4CONHPO(OH)_2$. VIII. m. 135-7° dec.[454]
$2,4\text{-}Cl,O_2NC_6H_3CONHPO(OH)_2$. VIII. m. 135-7°.[454]
$3,5\text{-}(O_2N)_2C_6H_3CONHPO(OH)_2$. VIII. m. 149-51° dec.[454]
$2,4\text{-}(O_2N)_2C_6H_3CONHPO(OH)_2$. VIII. m. 153-5° dec.[454]
$o\text{-}ClC_6H_4CONHPO(OH)_2$. VIII. m. 134-6°.[454]
$p\text{-}ClC_6H_4CONHPO(OH)_2$. VIII. m. 128-9°, resolidifies m. 188-9°.[454]
$2,4\text{-}Cl_2C_6H_3CONHPO(OH)_2$. VIII. m. 140-2° dec.[454]
$p\text{-}O_2NC_6H_4CONHPO(OH)_2$. VIII. $[C_6H_{11}NH_3]^+$ salt m. 194-6°.[943]
$p\text{-}HOCOC_6H_4NHPO(OH)_2$. VIII. Darkens at 239°.[529]
$2,4\text{-}Br,MeC_6H_3NHPO(OH)_2$. VIII. m. 142°.[573]
$PhC(NH_2)\text{=}NPO(OH)_2$. XVII. m. 180-5°.[174]
$PhCH_2NHPO(OH)_2$. VIII. m. 165-70°.[529]
$p\text{-}MeC_6H_4NHPO(OH)_2$. VIII,[529] XV,[336,337] XVII.[925] m. 250-2°.[529] Ba-salt.[925]
$o\text{-}MeOC_6H_4NHPO(OH)_2$. VIII. m. 130° dec.[529]
$PhNHC(NH_2)\text{=}NPO(OH)_2$. XVII. m. 165°.[173,174]
$C_6H_{11}NHC(\text{=}NH)NHPO(OH)_2$. XVII. m. 164°.[173]
C_8 $p\text{-}MeC_6H_4NHC(NH_2)\text{=}NPO(OH)_2$. XVII. m. 158°.[174]
$CCl_3C(\text{=}NPh)NHPO(OH)_2$. VIII. m. 155-6° dec.[197] For other $CCl_3C(\text{=}NAr)NHPO(OH)_2$ compounds see Ref. 197.
$p\text{-}MeC_6H_4CONHPO(OH)_2$. VIII. m. 158-9°.[438]
$PhCH_2NHC(\text{=}NH)NHPO(OH)_2$. XVII. m. 158°.[173]
C_{10} $1\text{-}C_{10}H_7NHPO(OH)_2$. VIII. m. 230-5° dec.[529]
$1\text{-}C_{10}H_7SO_2NHPO(OH)_2$. X. m. 159-62°.[431,473]
$2\text{-}C_{10}H_7SO_2NHPO(OH)_2$. X. m. 174-5°.[431,473]
C_{11} $1\text{-}C_{10}H_7CONHPO(OH)_2$. VIII. m. 139-41°.[438]
$2\text{-}C_{10}H_7CONHPO(OH)_2$. VIII. m. 155-7°.[438]
$Ph_2NPO(OH)_2$. VIII.[660] Crystals.
$Ph_2NCONHPO(OH)_2$. X. m. 110° dec.[461]
$PhC[\text{=}NPO(OPh)_2]NHPO(OH)_2$. VIII. m. 175-8°.[203]

C_X 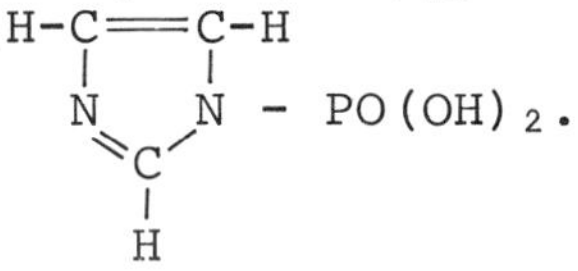 XVII. Na-, Na_2-salt.[172] Imidazolium-salt m. 164°,[85] m. 135-60°.[172]

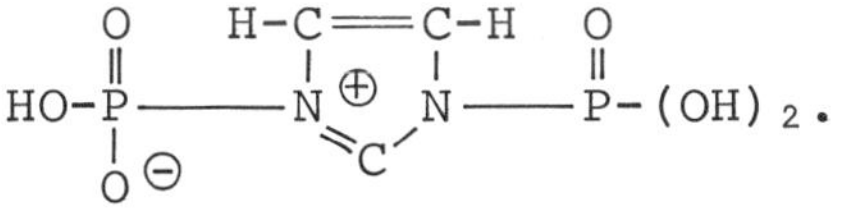 XVII. Ca-salt.[735]

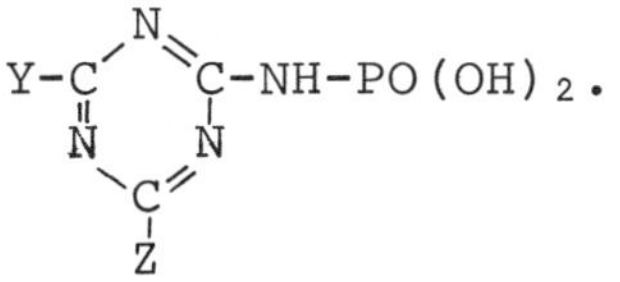 VIII. Y=Z=Cl, m. 155-61°.[203]
Y=Cl, Z=$NHPO(OH)_2$, m. 236°.[202]
Y=Z=$NHPO(OH)_2$, m. 160-5°.[202]

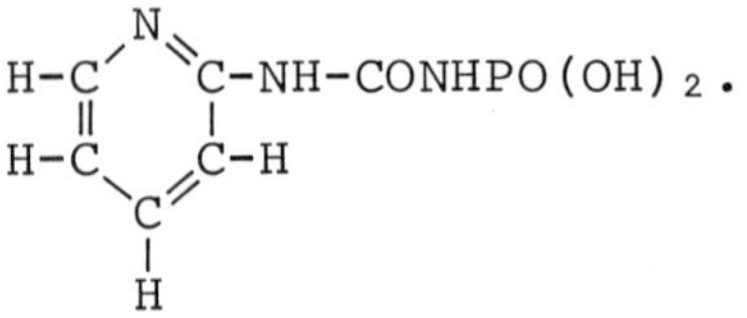

VIII. m. 160-2° dec.[450]

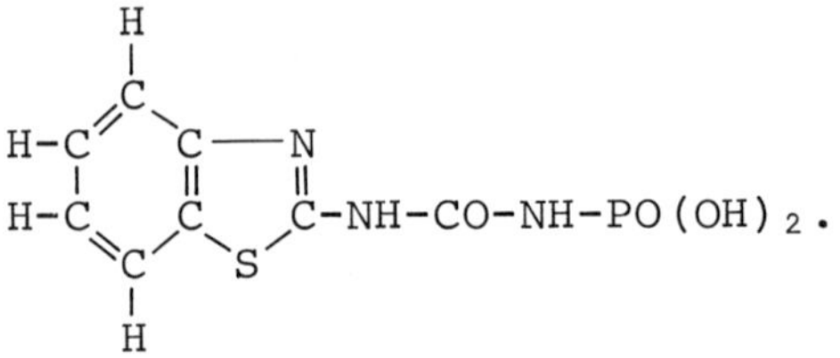

VIII. m. 305-10° dec.[450]

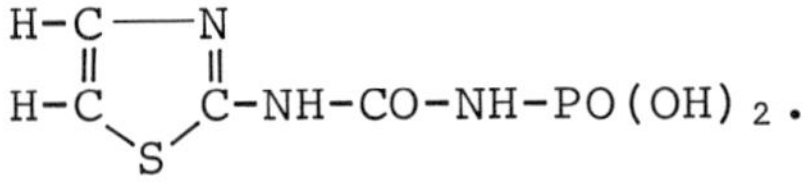

VIII. m. 156-8° dec.[450]

N-Phospho-Glycine. $HOCOCH_2NHPO(OH)_2$. II,[912] XVII.[926] Mg-salt,[912] Ba-salt.[926]

N-Phospho-Alanine. $HOCOCHMeNHPO(OH)_2$. II. Mg-salt.[66,912]

N-Phospho-Serine. $HOCOCH(CH_2OH)NHPO(OH)_2$. XVII. m. 166-7°.[689]

N-Phospho-Threonine. $MeCH(OH)CH(COOH)NHPO(OH)_2$. XVII. m. 184°.[689]

N-Phospho-Glycocyamine. $HOCOCH_2CH(NH_2)NHPO(OH)_2$. XVII.[175] Ca-salt.[297]

N-Phospho-Glycyl-glycine. $HOCOCH_2NHCOCH_2NHPO(OH)_2$. XVII Ba-salt.[926]

N-Phospho-Creatine. $HOCOCH_2NMeC(=NH)NHPO(OH)_2$. XVII.[36] Ca-salt,[175,921] Na-salt.[175,271]

N-Phospho-Proline. $HOCOCH(CH_2)_3NPO(OH)_2$ (ring CH–N). XVII.[735]

N-Phospho-L-Arginine. $HOCOCH(NH_2)(CH_2)_3NHC(=NH)NHPO(OH)_2$. II.[151,851] m. 175-80°.[173]

N-Phospho-Glutaminicacid. $HOCOCH_2CH_2CH(COOH)NHPO(OH)_2$. II. Mg-salt.[66,912]

N-Phospho-L-Tyrosine. $p\text{-}HOC_6H_4CH_2CH(COOH)NHPO(OH)_2$. XVII. Ba-salt.[735,926]

N,O-Diphospho-Tyrosine. $(HO)_2P(O)NHCH(COOH)CH_2\text{-}C_6H_4\text{-}[OP(O)(OH)_2]\text{-}p$. II. Mg-salt.[629]

Triphospho-Tyrosine. II. Ba-salt.[705]

N-Phospho-Phenylalaninemethylester. $PhCH_2CH(COOMe)NHPO(OH)_2$. XVII. m. 143-5°.[528]

N-Phospho-Tryptophane. XVII.[528]

N-Phospho-Tryptophane-methylester. XVII. m. 130-2°.[528]

N-Phospho-Glycyl-L-tyrosine. $HOCOCH_2NHCOCH(CH_2C_6H_4OH\text{-}p)NHPO(OH)_2$. XVII. Ba-salt.[926]

Phospho-Lombricine. XVII. D and L form. (2-amino-2-carboxyethyl 2-N'-phosphono-guanidinoethyl-hydrogen phosphate).[89]

D.1.4. $>N\text{-}P(S)(SH)_2$

$Pr_2NPS(SH)_2$. V. m. 167-70°, moisture sensitive.[913]
$Bu_2NPS(SH)_2$. V. m. 191-3°, moisture sensitive.[913]
$(C_5H_{11})_2NPS(SH)_2$. V. m. 151-4°, moisture sensitive.[913]

D.1.5. $(RO)_2P(O)-N<$, R = Alkyl or Aryl or Acyl

D.1.5.1. $(RO)_2P(O)-NH_2$

C_2 $(MeO)_2PONH_2$. XII, XVII. m. 40-2°, b_1 136-8°.[940] ^{31}P -15.2.[634]
C_4 $(ClCH_2CH_2O)_2PONH_2$. II,[834] XIII.[897] m. 75-6°,[834] m. 77°.[897]
$(EtO)_2PONH_2$. V,[333,746] XIII.[64,404] m. 46-7°,[545] m. 50-1°,[64] m. 52°,[333] m. 53.5-4.5°,[404] m. 54.5°,[712,746] $b_{0.2}$ 131-8°,[746] very hygroscopic,[545] ^{31}P -12.8,[873] -11.1.[633]
C_6 $(CH_3CHClCH_2O)_2PONH_2$. XIII. n_D^{25} 1.4725, d_{25} 1.3177.[897]
$(PrO)_2PONH_2$. XVII. $b_{0.6}$ 115-6°, m. 39-42°.[139]
$(i\text{-}PrO)_2PONH_2$. XIII. m. 56-7°,[64] m. 57°.[404]
C_8 $(BuO)_2PONH_2$. XIII. b_8 172-4°,[679] $b_{0.1}$ 117°,[878] n 1.4353,[679] n_D^{30} 1.4345,[878] d 1.0363.[679]
$(t\text{-}BuO)_2PONH_2$. V. m. 122-5° dec.[337]
C_{10} $(i\text{-}AmO)_2PONH_2$. XIII. $b_{0.2}$ 142-3°, n 1.4392, d 1.0061.[679]
C_{12} $(p\text{-}BrC_6H_4O)_2PONH_2$. XII. m. 186-8°.[468]
$(p\text{-}ClC_6H_4O)_2PONH_2$. V. m. 152°.[72]
$(PhO)_2PONH_2$. V,[67,272,333,897,825] XII,[468] XIII,[65] XV.[940] m. 145-6°,[67] m. 148°,[272,333,825] m. 148.7-9.3°,[154] m. 148-50°,[468] m. 151-2°,[70] dec. 180°,[825] ^{31}P -2.2,[634] -2.8.[633]
$(C_6H_{13}O)_2PONH_2$. XIII. $b_{0.2}$ 168-9°, n 1.4452, d 0.9955.[679]
C_{14} $(p\text{-}BrC_6H_4CH_2O)_2PONH_2$. XIII. m. 130-1°.[600]
$(p\text{-}ClC_6H_4CH_2O)_2PONH_2$. XIII. m. 136-7°.[600]
$(PhCH_2O)_2PONH_2$. V,[63] XIII.[64] m. 103.5-4.5°,[63,64,61] ^{31}P -12.0.[634]
$(p\text{-}MeC_6H_4O)_2PONH_2$. V. m. 146°.[72]
C_x $(2\text{-}C_{10}H_7O)_2PONH_2$. V. m. 215°.[72]
$(Ph_2CHO)_2PONH_2$. XIII. m. 145°.[62]

D.1.5.2. $(RO)_2P(O)-NHR'$, R' = Alkyl

C_3 $(MeO)_2PONHMe$. XIII. b_1 81°, n 1.4260, d 1.2002.[404]
C_4 $(MeO)_2PONHCH(OH)CCl_3$. XVII. m. 103-4°.[10]
$(MeO)_2PONHCH_2CH_2Cl$. XVII. $b_{1.5}$ 114-5°, n 1.4537, d 1.2960.[340]
$(MeO)_2PONHCH_2CONH_2$. V. m. 112-3°.[19]
$(MeO)_2PONHEt$. V. b_{10} 122-4°, n 1.4275, d 1.1438.[638]
C_5 $(MeO)_2PONHCHMeCONH_2$. V. m. 111-2°,[530] D,L alanine amide derivative.
$(MeO)_2PONHPr$. V. b_{11} 132-4°, n 1.4290, d 1.1101.[638]
$(EtO)_2PONHMe$. V. b_{17} 135°,[559] b_{15} 130°.[746]
C_6 $(EtO)_2PONHCH_2CONH_2$. V. m. 73-6° dec.[19]
$(ClCH_2CH_2O)_2PONHCH(OH)CCl_3$. XVII. m. 91-2°.[539]
$(EtO)_2PONHCH(OH)CCl_3$. XVII. m. 96-7°.[10]
$(MeO)_2PONHCH_2COOEt$. V. m. 55-6°.[10]
$(EtO)_2PONHCH_2CONHOH$. V. Oil. n 1.4455, d 1.1840.[20]

$(EtO)_2PONHCH_2CO_2H$.   V.   Oil.   n 1.4510, d 1.2579.[20]

$(EtO)_2PONHCH_2CH_2Cl$.   XVII. b_5 148°, n 1.4473, d 1.1865.[340]

$(MeO)_2PONHBu$.   V. b_9 143-54°, n 1.4326, d 1.0782.[638]

$(EtO)_2PONHEt$.   V,[638] IX,[574] XIII.[404] $b_{0.03}$ 79-81°,
    n 1.4254, d 1.0583,[638] b_1 92-4°, n 1.4250,
    d 1.0590,[404] $b_{1.5}$ 91°, b_{25} 135°,[574] ^{31}P -10.7.[873]

$(EtO)_2PONHCH_2CH_2OH$.   XVII.   Yellow sirup, dec. 140°.[182]

C_7 $(EtO)_2PONHCH_2CH_2CN$.   XVII. b_{10} 176-7°, n 1.4400,
    d 1.1260.[709]

$(EtO)_2PONHCH_2CONHMe$.   V. b_1 180°, n 1.4695, d 1.2126.[20]

$(MeO)_2PONHCH(CONH_2)CH_2CH_2CONH_2$.   V.   m. 117-20°.[530]

$CH_2(CH_2O)_2PONHBu-i$.   IX.   m. 82.5-3°.[261]

$CH_2(CH_2O)_2PONHBu-t$.   IX.   m. 131-3.5°.[261]

$(EtO)_2PONHPr$.   V,[139,638] IX.[574] $b_{0.5}$ 94-5°,[139]
    n_D^{21} 1.4266,[139] b_{10} 140°,[638] n 1.4278,[638] d 1.0338,[638]
    b_8 112°.[574]

C_8 $(EtO)_2PONHC(CF_3)_3$.   XVII. b_8 105-6°, m. 31.5-2°.[603]

$(ClCH_2CH_2O)_2PONHBu$.   IX. n_D^{30} 1.4699.[507]

$Me_2C(CH_2O)_2PONHPr$.   V.   m. 90.5-1.5°.[261]

$Me_2C(CH_2O)_2PONHPr-i$.   V.   m. 141-2°.[261]

$(MeO)_2PONHC_6H_{11}$.   XIII.   m. 65.5-6.2°.[88]

$(CH_2O)_2PONHC_6H_{11}$.   V.[605]

$(EtO)_2PONHCH_2COOEt$.   XVII. b_1 135.5°, n 1.4390,
    d 1.1495.[26]

$(PrO)_2PONHCH(OH)CCl_3$.   XVII.   m. 81-2°.[10]

$(i-PrO)_2PONHCH(OH)CCl_3$.   XVII.   m. 138-9°.[10]

$(EtO)_2PONHCH_2CONHCH(OH)Me$.   V.   m. 111-2°.[20]

$(PrO)_2PONHCH_2CONH_2$.   V.   m. 68-73° dec.[19]

$(i-PrO)_2PONHCH_2CONH_2$.   V.   m. 91-4°,[19] m. 81-4°.[894]

$(PrO)_2PONHEt$.   V. $b_{0.03}$ 85-9°, n 1.4295, d 1.0160.[638]

$(EtO)_2PONHCH_2CH_2SCOMe$.   XVII. $b_{0.09}$ 132-3°, n 1.4757,
    d 1.1716.[650]

$(EtO)_2PONHBu$.   V. $b_{0.035}$ 101-3°, n 1.4300, d 1.0194,[638]
    $b_{0.35}$ 107°, n_D^{25} 1.4260.[893]

$(EtO)_2PONHBu-i$.   IX. b_{14} 146°.[574]

$(EtO)_2PONHBu-t$.   V. $b_{1.0}$ 95°, n_D^{25} 1.4290.[893]

$(i-PrO)_2PONHCH_2CH_2OH$.   V. $b_{0.8}$ 151°, n_D^{23} 1.4400.[689]

$(MeO)_2PONHC_6H_{13}$.   XIII. $b_{0.5}$ 119°, n_D^{25} 1.4337.[88]

$$\underset{EtO}{\overset{EtO}{\diagdown}}\underset{}{\overset{O\;\;H\quad\quad O}{\underset{\|\;\;\;|\quad\quad\;\|}{P-N-CH_2CH_2COR}}}$$

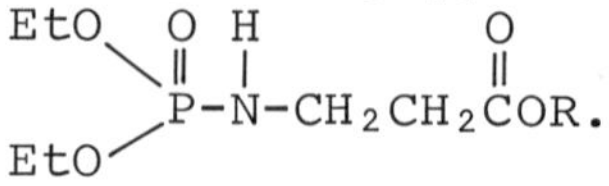

XVII.
R=Me: b_{45} 152-3°, n 1.4392,
d 1.1509.[342]
R=Et: b_2 147-50°, n 1.4998,
d 1.1245.[342]
R=Pr: b_1 145-6°, n 1.4401,
d 1.1009.[342]
R=i-Pr: b_1 134-6°, n 1.4384,
d 1.0935.[342]
R=i-Bu: b_1 140-2°, n 1.4398,
d 1.0727.[342]

R=Am: b_1 153-4°, n 1.4418, d 1.0660.[342]
R=Ph: m. 44-7°.[342]

$(EtO)_2PONHCH_2COOEt$. XVII. $b_{0.3}$ 123-8°, n_D^{30} 1.4338.[531]

$(EtO)_2PONHCH_2COOH$. XVII. Guanidiniumsalt: m. 159-60°,[531] Ba-salt.[531]

$(PrO)_2PONHCH_2CH_2Br$. IX. $b_{0.05}$ 101-3°.[143]

$(PrO)_2PONHCH_2CH_2Cl$. XVII. $b_{0.7}$ 123-4°, n 1.4478, d 1.1297.[340]

$(i\text{-}PrO)_2PONHCH_2CH_2Cl$. XVII. b_1 104-6°, n 1.4404, d 1.1090.[340]

$(EtO)_2PONHCH_2SCHMe_2$. XVII. b_1 130-1°, n 1.4620, d 1.1220.[21]

C9 $CH_2(CH_2O)_2PONHC_6H_{11}$. IX. m. 152-3°.[261]

$(EtO)_2PONHAm\text{-}i$. IX. b_{25} 185°.[574]

$(BuO)_2PONHMe$. V. $b_{0.25}$ 126-8°, n_D^{27} 1.4315.[819]

$(PrO)_2PONHPr$. V. $b_{0.03}$ 96-8°, n 1.4310, d 1.0035,[638] $b_{0.1}$ 92°.[139]

$(EtO)_2PONHCH_2CH_2OCOCH{=}CH_2$. XVII. $b_{0.05}$ 138-9°, n 1.4526, d 1.1502.[342]

$(EtO)_2PONHCH_2CH_2NHCH_2CH{=}CH_2$. XVII. $b_{0.02}$ 125-7°, n 1.4546, d 1.0588.[342]

$Me_2C(CH_2O)_2PONHBu$. V. m. 81-2°.[261]

$Me_2C(CH_2O)_2PONHBu\text{-}i$. V. m. 116-7°.[261]

$Me_2C(CH_2O)_2PONHBu\text{-}t$. V. m. 166°.[261]

$(i\text{-}PrO)_2PONHPr\text{-}i$. XIII. b_1 86-9°, n 1.4220, d 0.9902.[404]

$(MeO)_2PONHCH_2C_6H_4NO_2\text{-}o$. V. m. 90-2°.[355]

$(MeO)_2PONHCH_2C_6H_4NO_2\text{-}p$. V. m. 95-6°.[355]

C10 $(EtO)_2PONHC_6H_{11}$. V, XIII. m. 71-2°,[894] m. 101-2°.[893]

$(i\text{-}PrO)_2PONHCH_2COOEt$. V. $b_{0.5}$ 115-28°, m. 28-9°.[894]

$(i\text{-}PrO)_2PONHCH(COOMe)CH_2OH$. V. m. 48-50°, waxy,[689] D,L-serine deriv.

$(i\text{-}PrO)_2PONHCH(COOMe)CH_2SH$. V.[689] L-cysteine derivative.

$(i\text{-}PrO)_2PONHCH(COOMe)CH(OH)Me$. V.[689] m. 54-6°,[894] D,L-threonine deriv.

$(BuO)_2PONHCH_2CH_2Cl$. XVII. $b_{1.7}$ 148-50°, n 1.4499, d 1.0847.[340]

$(i\text{-}BuO)_2NHCH_2CH_2Cl$. XVII. $b_{1.7}$ 136.5-8.0°, n 1.4458, d 1.0761.[340]

$(PrO)_2PONHBu$. V. $b_{0.03}$ 105-7°, n 1.4330, d 0.9910.[638]

$(MeO)_2PONHCH_2CHEtBu$. XIII. $b_{1.5}$ 142.5, n_D^{25} 1.4403.[88]

$(BuO)_2PONHEt$. V. $b_{0.03}$ 98-100°, n 1.4335, d 0.9911.[638]

$(i\text{-}BuO)_2PONHEt$. V. $b_{0.025}$ 96-8°, n 1.4295, d 0.9796.[638]

$(MeO)_2PONH(CH_2)_7CH_3$. XIII. b_1 127°, n_D^{25} 1.4283.[88]

$(EtO)_2PONHCH_2CH_2NEt_2$. XVII. b_1 104-5°, n 1.4439, d 1.0517.[340]

$(BuO)_2PONHCH_2CONH_2$. V. m. 77-9°.[19]

$(i\text{-}BuO)_2PONHCH_2CONH_2$. V. m. 93-5°.[19]

$(PrO)_2PONHCH_2CO_2Et$. V. b_1 141-2°, n 1.4375, d 1.0926.[10]

$(i\text{-}PrO)_2PONHCH_2CO_2Et$. V. b_1 129-30°, n 1.4332, d 1.0856.[10]

$(BuO)_2PONHCH(OH)CCl_3$. XVII. m. 79-80°.[10]
$Me_2C(CH_2O)_2PONHAm$. V. m. 71-2°.[261]
$(EtO)_2PONHCH_2CH_2OCOCMe{=}CH_2$. XVII. $b_{0.015}$ 138-9°, n 1.4542, d 1.1315.[342]
$(PrO)_2PONHCH_2CH_2SC(O)Me$. XVII. $b_{0.1}$ 143-4°, n 1.4706, d 1.1179.[650]

C_{11} $(MeO)_2PONHCH(CONH_2)CH_2Ph$. V. m. 148-9°,[530] phenylalanin amide deriv.
$Me_2C(CH_2O)_2PONHC_6H_{13}$. V. m. 60-1°.[261]
$Me_2C(CH_2O)_2PONHC_6H_{11}$. V. m. 187-8°.[262]
$(PrO)_2PONH(CH_2)_3NMe_2$. V. b_2 136°, n 1.442.[182]
$(PrO)_2PONHCH_2CH_2OCOCH{=}CH_2$. XVII. $b_{0.4}$ 147-8°, n 1.4518, d 1.0997.[342]
$(BuO)_2PONHPr$. V. $b_{0.03}$ 112-3°, n 1.4360, d 0.9838.[638]

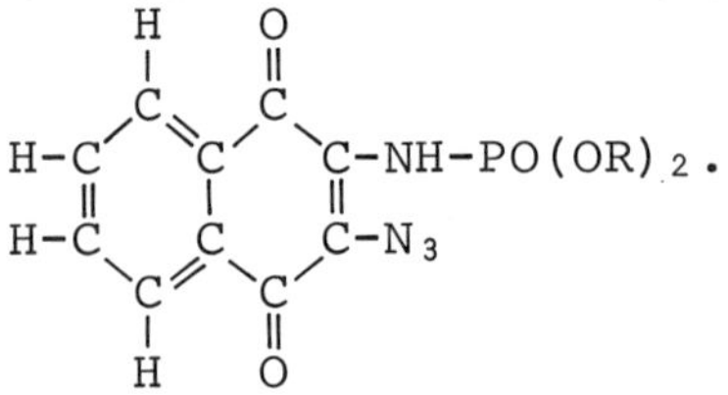
XVII. R=Me: m. 129-31°,[615] m. 126-7°.[616] R=Et: m. 130-1°.[616]

C_{12} $(EtO)_2PONHCH_2CH_2SPh$. XVII. b_2 195-8°, n 1.5288, d 1.1655.[340]
$(CH_2{=}CHCH_2O)_2PONHC_6H_{11}$. V. $b_{0.05}$ 102-3°, m. 37-8°.[554]
$(EtO)_2PONHCH_2CH_2N(CH_2CH{=}CH_2)_2$. XVII. $b_{0.2}$ 130-1°, n 1.4617, d 1.0202.[347] For other $(RO)_2PONHCH_2CH_2N(CH_2CH{=}CH_2)_2$ see Ref. 347.
$(PrO)_2PONHCH_2CH_2OCOCMe{=}CH_2$. XVII. $b_{0.05}$ 148-9°, n 1.4557, d 1.0882.[342]
$(BuO)_2PONHCH_2CH_2SC(O)Me$. XVII. $b_{0.05}$ 146-7°, n 1.4700, d 1.0861.[650]
$(i\text{-}BuO)_2PONHCH_2CH_2SC(O)Me$. XVII. $b_{0.03}$ 137-8°, n 1.466, d 1.0818.[650]
$(i\text{-}AmO)_2PONHCH_2CH_2Cl$. XVII. $b_{1.4}$ 152-3.5°, n 1.4495, d 1.0552.[340]
$(BuO)_2PONHBu$. V. $b_{0.03}$ 128-30°,[638] $b_{0.3}$ 145-6°,[819] b_1 145°,[182] n 1.4381,[182,638] n^{25} 1.4351,[819] d 0.9740.[638]
$(BuO)_2PONHBu\text{-}i$. V. $b_{0.7}$ 139-40°, n_D^{25} 1.4351.[819]
$(BuO)_2PONHBu\text{-}sec$. V. $b_{4.5}$ 160°, n_D^{25} 1.4354.[819]
$(BuO)_2PONHBu\text{-}t$. V. $b_{7.5}$ 152-4°, n_D^{25} 1.4356.[819]
$(ClCH_2CH_2O)_2PONHCH_2CHEtBu$. IX. n_D^{30} 1.4649.[507]
$(PrO)_2PONHC_6H_{11}$. V. m. 53°.[182]
$(i\text{-}PrO)_2PONHC_6H_{11}$. V. m. 53.5-5.0°.[894]
$(BuO)_2PONHCH_2CO_2Et$. V. b_1 160-0.5°, n 1.4408, d 1.0660.[10]
$(i\text{-}BuO)_2PONHCH_2CO_2Et$. V. b_1 145-6°, n 1.4375, d 1.0578.[10]

C_{13} $(PhO)_2PONHMe$. V. m. 95°,[68] m. 95°,[94] m. 95-6°,[819] ^{31}P 1.2.[634]

$(EtO)_2PONHCH_2COOCH_2Ph$. IX. b_1 135.5°, n 1.4390, d 1.1495.[46]

C_{14} $(i\text{-}PrO)_2PONHCH_2Ph$. V. b. 238°, m. 48.5-50.0°.[894]

$(p\text{-}ClC_6H_4O)_2PONHCH(OH)CCl_3$. XVII. m. 142-3°.[539]

$(PhO)_2PONHEt$. IX,[574] V.[819] b_{11} 205°,[574] $b_{0.9}$ 186-8°, m. 49-50°,[819] ^{31}P -1.[901]

$[(ClCH_2CH_2O)_2PONHCH_2CH_2CH_2\text{-}]_2$ V. m. 50°.[737]

$[(EtO)_2PONHCH_2CH_2CH_2\text{-}]_2$. ^{31}P -10.3.[873]

$[(PrO)_2PONHCH_2\text{-}]_2$. V. m. 81°.[182]

C_{15} $(PhO)_2PONHCH(CF_3)_2$. IX. m. 173-4°.[484]

$(PhO)_2PONHCH_2CH{=}CH_2$. V. m. 54-4.5°.[53]

$(PhO)_2PONHCH_2COOMe$. V. m. 93°.[926]

$(PhO)_2PONHPr$. IX,[574] V.[819] b_8 208°,[574] m. 52-4°,[139] m. 56-7°.[819]

$(PhO)_2PONHPr\text{-}i$. V. m. 75-6°.[819]

$(t\text{-}BuO)_2PONHCH_2Ph$. V. m. 97-8°.[336,337]

$[(PrO)_2PONHCH_2\text{-}]_2CH_2$. V. Oil, n 1.4405.[182]

C_{16} $(p\text{-}IC_6H_4CH_2O)_2PONHCH_2COOH$. V. m. 115°. dec. 175-8°.[926]

$(p\text{-}O_2NC_6H_4CH_2O)_2PONHCH_2COOH$. V. m. 149°.[926]

$(PhO)_2PONHCH_2COOEt$. V. m. 77-8°,[770] m. 76°.[750]

$(PhCH_2O)_2PONHCH_2CONH_2$. V. m. 103-4.5°.[530]

$(PhO)_2PONHBu\text{-}sec$. V. m. 62-3°.[819]

$(PhO)_2PONHBu\text{-}i$. IX,[574] V.[819] m. 69-70,[819] m. 58°, b_{11} 218°.[574]

$(PhO)_2PONHBu$. V. m. 59-60,[819] m. 58°.[70]

$(PhO)_2PONHBu\text{-}t$. V,[70] IX,[932] V.[819] m. 114-5°,[819] m. 111-2,[932] m. 116°.[70]

$(EtO)_2PONHC_{12}H_{25}$. IX. b_5 180-90°.[308]

$[(PrO)_2PONHCH_2CH_2\text{-}]_2NH$. V. n_D^{18} 1.4470.[183]

C_{17} $(p\text{-}IC_6H_4CH_2O)_2PONHCH_2COOMe$. V. m. 124-5°.[926]

$(PhCH_2O)_2PONHCHMeCONH_2$. V. m. 97-9°.[530] Alanine amide derivative.

$(p\text{-}O_2NC_6H_4CH_2O)_2PONHCH_2COOMe$. V. m. 89°.[926]

$(p\text{-}O_2NC_6H_4CH_2O)_2PONHCHMeCOOH$. XVII. m. 60-2°.[167] Alanine derivative.

C_{18} $(p\text{-}O_2NC_6H_4CH_2O)_2PONHCHMeCOOMe$. XVII. m. 88-9°.[167] Alanine derivative. For the analogous L-leucine derivative and other aminoacid derivatives see Ref. 167.

$(PhO)_2PONHC_6H_{11}$. V. m. 104-5°,[68] m. 102.5°,[87] m. 103-4.[819] ^{31}P +0.4.[634]

$(PhCH_2O)_2PONHCH_2COOEt$. XVII. m. 43-5°.[528]

$(PhCH_2O)_2PONHCHMeCOOMe$. XVII. m. 40-1°,[528] alanine derivative.

$(PhCH_2O)_2PONHCH(COOMe)CH_2OH$. V.[530] Serine derivative.

$(PhO)_2PONHC_6H_{13}$. V. m. 48°.[87]

C_{19} $(PhO)_2PONHCH_2Ph$. V, XVII. m. 104-5°,[573] m. 101-2.5°,[493] m. 107-8°.[70]

C_{20} $(p\text{-}O_2NC_6H_4CH_2O)_2PONHCH_2CONHCH_2COOEt$. V. m. 112-3°.[926] Glycyl-glycin deriv.

$(p\text{-}BrC_6H_4CH_2O)_2PONHC_6H_{11}$. XIII. m. 114-5°.[600]

$(p\text{-}ClC_6H_4CH_2O)_2PONHC_6H_{11}$. XIII. m. 97.5-8.5°.[600]
$(PhCH_2O)_2PONHC_6H_{11}$. V,[63] XIII.[65] m. 79-80°.[63,65]
$(PhO)_2PONHCH_2CHEtBu$. V. m. 68.5°.[87]
$(PhO)_2PONHC_8H_{17}\text{-}t$. V. m. 86°.[70]
$[(BuO)_2PONHCH_2CH_2\text{-}]_2NH$. V. Oil, n 1.4540.[183]

C_{27} $(PhCH_2O)_2PONHCH_2Ph$. V. m. 84-5°,[63] m. 96-8°,[337] m. 81-3°.[782] ^{31}P -10.0.[394]
$(PhO)_2PONHCH(COOEt)CH_2CH_2COOEt$. V. m. 73.5-4°.[770]
$(m\text{-}Me_2NC_6H_4O)_2PONHCH_2CH_2CH_2NMe_2$. V. $b_{0.001}$ 175°. n_D^{25} 1.5678.[306]

C_{22} $(PhCH_2O)_2PONHCHMePh$. V. m. 81-2°.[63]
$(p\text{-}MeC_6H_4CH_2O)_2PONHC_6H_{11}$. XIII. m. 98.5-9.5°.[600]
$(PhO)_2PONHC_{10}H_{21}$. V. m. 52°.[87]
$[(BuO)_2PONHCH_2CH_2CH_2\text{-}]_2$. V. n 1.4490.[182]

C_{23} $(p\text{-}IC_6H_4CH_2O)_2PONHCH_2COOCH_2Ph$. V. m. 89°.[926]
$(p\text{-}IC_6H_4CH_2O)_2PONHCH(COOH)CH_2C_6H_4OH\text{-}p$. V. m. 80-5°.[926] L-Tyrosine-deriv.
$(PhCH_2O)_2PONHCH_2COOCH_2Ph$. V.[530]
$(PhO)_2PONHCH(CH_2Ph)COOEt$. V. m. 78-9°,[770] phenylalanine deriv.
$(PhO)_2PONHCH(COOEt)CH_2C_6H_4OH\text{-}p$. V. m. 93-4°,[926] L-Tyrosine-derivative.

C_{24} $(p\text{-}O_2NC_6H_4CH_2O)_2PONHCH(CH_2Ph)COOMe$. XVII. m. 111°,[167] L-phenylalanine deriv.
$(PhCH_2O)_2PONHCH(CH_2Ph)COOMe$. XVII. m. 82-3,[528] phenylalanine deriv.
$(PhCH_2O)_2PONHCHMeCOOCH_2Ph$. V.[530] Alanine derivative.
$(PhCH_2O)_2PONHCH(COOMe)CH(OH)Ph$. V. m. 115-6° dec.,[530] phenylserine deriv.
$(PhO)_2PONHC_{12}H_{25}$. ^{31}P in benzene -1.[901]

C_{25} $(PhO)_2PONHCHPh_2$. IX. m. 132-3°.[938]
$(p\text{-}IC_6H_4CH_2O)_2PONHCH(COOEt)CH_2C_6H_4OH\text{-}p$. V. m. 143°.[926]
$(PhO)_2PONHCH(CH_2C_6H_4OH\text{-}p)CONHCH_2COOEt$. V. m. 123-4°.[92] Glycyl-L-Tyrosine-derivative.

C_x 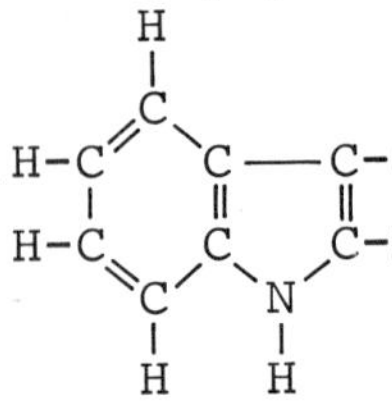 $C\text{-}CH_2CH(COOMe)NHP(O)(OCH_2Ph)_2$. XVII. m. 104.5°,[528] Tryptophane derivative.

$[(PhO)_2PONHCH_2\text{-}]_3$. V. m. 130°.[182]
$(p\text{-}IC_6H_4CH_2O)_2PONHCH(CH_2C_6H_4OH\text{-}p)CONHCH_2COOEt$. V. m. 127-8°.[926] Glycyl-L-Tyrosine-derivative.
$(PhCH_2O)_2PONHCH(COOCH_2Ph)CH_2Ph$. V. m. 67-9°,[530] phenyl alanine derivative.
$(PhCH_2O)_2PONHCH(COOCH_2Ph)CH_2C_6H_4OH\text{-}p$. V. m. 54-5°,[530] L-tyrosine derivative.
$[(PhO)_2PONHCH_2CH_2CH_2\text{-}]_2$. V. m. 114°.[182]

$(p\text{-}O_2NC_6H_4CH_2O)_2PONHCH(CH_2Ph)CONHCH_2COOCH_2Ph$. XVII. m. 135°.[167] $[\alpha]_D^{20}$ -8°. c=1.5 in $CHCl_3$.
$(Ph_2CHO)_2PONHC_6H_{11}$. XIII. m. 101-2°.[62]
$(Ph_2CHO)_2PONHCH_2Ph$. XIII. m. 104-5°.[62]
$(PhCH_2O)_2PONHCH(COOCH_2Ph)CH_2CH_2COOCH_2Ph$. V. m. 45-7°.[530]

D.1.5.3. $(RO)_2P(O)$-NHAr, Ar = Aryl

C_7 $(MeO)_2PONH(C_5H_4N)$-2. XIII. m. 108-9°, monopicrate m. 149-50°.[50]
C_8 $(MeO)_2PONHC_6H_4Cl$-p. IX. m. 97.5°.[279]
$(MeO)_2PONHPh$. V,[559] XVII,[403] XIII.[880] m. 88-8.5°,[559] m. 88°,[403] m. 84.5-5.0°.[880]
$(CH_2O)_2PONHPh$. V.[605]
C_9 $(CH_2O)_2PONHC_6H_4Me$-p. V.[605]
$(EtO)_2PONH(C_5H_4N)$-2. V. m. 87°,[146] m. 86-8°,[50] for substituted pyridine derivatives see Ref. 146.
C_{10} $(ClCH_2CH_2O)_2PONHC_6H_4Br$-p. ^{31}P -1.8.[873]
$(EtO)_2PONHC_6H_3Br_2$-2,4. IX. m. 114°.[573]
$(EtO)_2PONHC_6H_3Cl_2$-2,4. IX. m. 106°.[573]
$(FCH_2CH_2O)_2PONHPh$. V. m. 68-70°.[163]
$(MeO)_2PONHC_6H_4COOMe$-m. IX. b. 184-6°(?).[573]
$(MeO)_2PONHC_6H_4COOMe$-p. IX. b. 166-7°(?).[573]
$(EtO)_2PONHC_6H_4Cl$-p. IX. m. 76°.[279,660]
$(EtO)_2PONHC_6H_4NO_2$-m. IX. m. 120°.[573]
$(EtO)_2PONHPh$. V,[62,582] IX,[559] XII,[880] XVII.[403] m. 93°,[582] m. 93-4.5°,[880] m. 94-5°,[403] m. 95-6°,[62,893] m. 96.5°,[559,560] b_{15} 142-4°.[745]
$(MeO)_2PONHC_6H_4SO_2NMe_2$-o. V. m. 70-2°.[941] Meta isomer, m. 128-30°;[941] para isomer, m. 192-4°.[941]
C_{11} $(MeO)_2PONH(C_9H_6N)$-2. V. m. 94-6°.[42] Quinoline derivative.
$(EtO)_2PONHC_6H_3Cl,(SCN)$-3,4. XIII. m. 86-8°.[643]
$(EtO)_2PONHC_6H_3Me,Br$-2,4. IX. m. 102°.[573]
$(EtO)_2PONHC_6H_3Me,Cl$-3,4. V. m. 85-6°.[145]
$(EtO)_2PONHC_6H_3Me,F$-3,4. V. m. 91-2°.[145]
$(EtO)_2PONHC_6H_4Me$-o. IX. m. 95°,[582] m. 94.5°.[374]
$(EtO)_2PONHC_6H_4Me$-p. IX. m. 98°,[582] m. 96°.[374]
$(PrO)_2PONH(C_5H_4N)$-2. V. m. 59-62°.[42]
$(i\text{-}PrO)_2PONH(C_5H_4N)$-2. V. m. 135-6°,[50] m. 109-9.5°.[42]

C_{12} 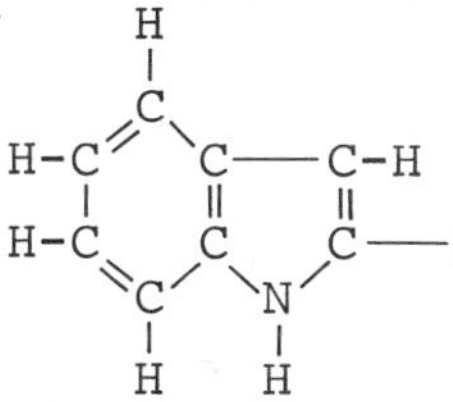 —NH-PO-$(OEt)_2$. V. m. 228-9°.[570]

$[(ClCH_2)_2CHO]_2PONHPh$. V. m. 81°.[162]
$(PrO)_2PONHC_6H_4Cl$-p. IX. m. 70°.[279]
$(i\text{-}PrO)_2PONHC_6H_4Cl$-p. IX. m. 137-8°.[279]
$(PrO)_2PONHPh$. V,[182] XIII.[880] m. 55°,[182] m. 52-3.5°.[880]

$(i\text{-}PrO)_2PONHPh$. V,[62,559,560] XIII.[880] m. 121-1.5°,[559,56 m. 120-1°,[62] m. 119-20°.[880]
$(EtO)_2PONHC_6H_3Me_2$-2,4. IX. m. 96°.[573]
$(EtO)_2PONHC_6H_4NMe_2$-p. V. m. 94°.[163]
C_{13} $(i\text{-}BuO)_2PONH(C_5H_4N)$-2. V. m. 50-3°,[42,146] m. 75-6°.[50]
$(EtO)_2PONH(C_9H_6N)$-2. V. m. 87-7.5°,[42] quinoline derivative.
$(PrO)_2PONHC_6H_3Cl,(SCN)$-3,4. XIII. m. 32-3°.[643]
$(PrO)_2PONHC_6H_4(SCN)$-p. XIII. m. 65-6°.[643]
$(EtO)_2PONHC_6H_4CO_2Et$-m. IX. b. 232-4°, b_{35} 130-5°.[573]
$(EtO)_2PONHC_6H_4CO_2Et$-p. IX. b. 206-7°, b_{45-50} 113-8°.[573]
$(i\text{-}PrO)_2PONHC_6H_4Me$-o. V. m. 84-5°.[182]
C_{14} $(EtO)_2PONHC_{10}H_7$-2. V. $b_{0.3}$ 126-9°.[745]
$[Me_2C(CN)O]_2PONHPh$. XIII. m. 87°.[409]
p-$[(ClCH_2CH_2O)_2PONH\text{-}]_2C_6H_4$. V. m. 181°.[737]
$(i\text{-}BuO)_2PONHC_6H_4Cl$-p. IX. m. 70-1°.[279]
$(sec\text{-}BuO)_2PONHPh$. XIII. b_{14} 201°, m. 42-4°, n 1.4950.[88
$(BuO)_2PONHPh$. XIII. b_{14} 222°. n 1.4940.[880]
$(i\text{-}BuO)_2PONHPh$. V,[162] XIII.[880] m. 43.5-5.0°,[162] m. 43-4°.[880]

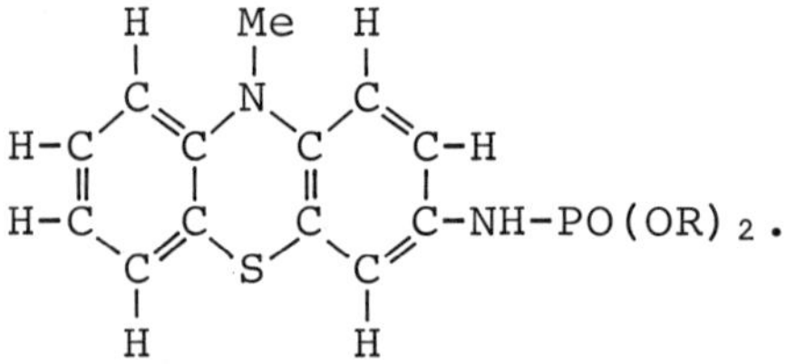

V. R=Me: m. 176-7°, and other R.[494]

C_{15} $(PrO)_2PONH(C_9H_6N)$-2. V. m. 114-5°,[42] quinoline derivative.
C_{16} $(EtOCOCHMeO)_2PONHPh$. V. m. 91°.[162]
$(AmO)_2PONHPh$. XIII. b_{10} 218°, n 1.4821.[880]
C_{17} $(PhO)_2PONH(C_5H_4N)$-2. V.[42,146] m. 151-2°.[42]
$(BuO)_2PONH(C_9H_6N)$-2. V. m. 102-4°,[42] quinoline derivative.
C_{18} $(PhO)_2PONHC_6H_2Br_3$-2,4,6. XVII. m. 165-6°.[939]
$(PhO)_2PONHC_6H_2Cl_2(NO_2)$-2,6,4. XVII. m. 159-61°.[939]
$(PhO)_2PONHC_6H_2Cl_3$-2,4,6. XVII. m. 142-3°.[939]
$(PhO)_2PONHC_6H_3Br_2$-2,4. IX,[573] XVII.[939] m. 141°,[573] m. 86-8°.[939]
$(PhO)_2PONHC_6H_3Cl_2$-2,4. IX,[573] XVII.[939] m. 132°,[573] m. 71-3°.[939]
$(PhO)_2PONHC_6H_3Cl_2$-3,5. XVII. m. 115-6°.[939]
$(PhO)_2PONHC_6H_3(NO_2)_2$-2,4. XVII. m. 144-6°.[939]
$(PhO)_2PONHC_6H_4Br$-o. XVII. m. 122-3°.[939]
$(PhO)_2PONHC_6H_4Br$-m. XVII. m. 117-9°.[939]
$(PhO)_2PONHC_6H_4Br$-p. IX,[573] XVII.[939] m. 112°,[573] m. 110-2°.[939]
$(PhO)_2PONHC_6H_4Cl$-o. XVII. m. 120-2°.[939]
$(PhO)_2PONHC_6H_4Cl$-m. XVII. m. 100-2°.[939]

$(PhO)_2PONHC_6H_4Cl$-p. IX,[660] XVII.[939] m. 117°,[660] m. 117-8°.[939]
$(PhO)_2PONHC_6H_4NO_2$-o. V,[493] XVII.[939] m. 108-9°.[939]
$(PhO)_2PONHC_6H_4NO_2$-m. XVII. m. 127-8°.[939]
$(PhO)_2PONHC_6H_4NO_2$-p. V,[493] XVII.[939] m. 146-8°,[939] m. 146.5-7.5°.[493]
$(PhO)_2PONHPh$. V, IX, XVII. ^{31}P +7.1,[634] ^{31}P +6.3.[633,631] m. 129-30°,[68] m. 129°,[582] m. 127-9°,[896] m. 129-30°,[939] m. 131-1°.[631]
$(PhO)_2PONHC_6H_4SO_2NH_2$-p. V. m. 201°.[772]
p-$[(PrO)_2PONH-]_2C_6H_4$. V. m. 174°.[183]
19 $(PhO)_2PONHC_6H_3$Me,Br-2,4. IX. m. 126°.[573]
$(PhO)_2PONHC_6H_3$Me,Cl-3,4. V. m. 139-40°.[145]
$(PhO)_2PONHC_6H_3$Me,F-3,4. V. m. 115-6°.[145]
$(PhO)_2PONHC_6H_4$Me-o. IX,[582] XVII.[939,134] m. 176°,[582] m. 121-3°,[939] m. 119-20°.[134]
$(PhO)_2PONHC_6H_4$Me-m. XVII. m. 122-3°.[939]
$(PhO)_2PONHC_6H_4$Me-p. IX,[582] XVII.[939,134] m. 134°,[582] m. 138-40°,[939] m. 135°.[134]
$(PhO)_2PONHC_6H_4$OMe -p. XVII. m. 139-41°.[939]
$(PhO)_2PONHC_6H_4$COOH-p. V. m. 194-5°.[148]
20 $(p\text{-}BrC_6H_4CH_2O)_2PONHPh$. V. m. 119-20°.[600]
$(p\text{-}MeC_6H_4O)_2PONHC_6H_3Br_2$-2,4. IX. m. 158°.[573]
$(p\text{-}ClC_6H_4CH_2O)_2PONHPh$. V. m. 103-4°.[600]
$(p\text{-}MeC_6H_4O)_2PONHC_6H_3Cl_2$-2,4. IX. m. 162°.[573]
$(p\text{-}IC_6H_4CH_2O)_2PONHPh$. V. m. 160-2°.[249]
$(PhO)_2PONHC_6H_4CH{=}CH_2$-m. V. m. 96°.[53]
$(p\text{-}NO_2C_6H_4CH_2O)_2PONHPh$. V. m. 151-2°.[249]
$(p\text{-}MeC_6H_4O)_2PONHC_6H_4$Br-p. IX. m. 138°.[573]
$(PhCH_2O)_2PONHC_6H_4$Cl-p. XIII. m. 93-111°.[527]
$(o\text{-}MeC_6H_4O)_2PONHPh$. V,[493] I.[74] m. 129-30°,[493] m. 126-7°.[74]
$(m\text{-}MeC_6H_4O)_2PONHPh$. I. m. 82°.[74]
$(p\text{-}MeC_6H_4O)_2PONHPh$. I,[74] IX.[582] m. 125°,[74] m. 133°.[582]
$(PhCH_2O)_2PONHPh$. V,[63] XIII.[61,62,64] m. 91-2.5°,[63,64] m. 91-2°.[62] ^{31}P -3.8.[394]
$(PhO)_2PONHC_6H_3Me_2$-2,4. IX. m. 115°.[573]
$(PhO)_2PONHC_6H_3Me_2$-3,5. XVII. m. 132-3°.[939]
$(PhO)_2PONHC_6H_4$Et-o. V. m. 105-6°.[134]
$(PhO)_2PONHC_6H_4$OEt-p. XVII. m. 109-10°.[939]
$(o\text{-}MeOC_6H_4O)_2PONHPh$. V. m. 129-30°.[493]
$(PhO)_2PONHC_6H_4SO_2NMe_2$-o. V. m. 174-77°.[941]
$(PhCH_2O)_2PONHC_6H_4SO_2NH_2$-p. V. 174°.[61]
1 $(p\text{-}IC_6H_4CH_2O)_2PONHC_6H_4$Me -p. V.[925]
$(PhO)_2PONHC_6H_4$COOEt-o. V. m. 148.5-9.0°.[493]
$(PhO)_2PONHC_6H_4$COOEt-p. V. m. 151-2°.[148]
$(p\text{-}O_2NC_6H_4CH_2O)_2PONHC_6H_4$Me-p. V. m. 155°.[925]
$(p\text{-}MeC_6H_4O)_2PONHC_6H_3$Br,Me-2,4. IX. m. 154°.[573]
$(o\text{-}MeC_6H_4O)_2PONHC_6H_4$Me-o. IX. m. 161°.[582]
$(p\text{-}MeC_6H_4O)_2PONHC_6H_4$Me-p. IX. m. 161°.[582]
$(PhCH_2O)_2PONHC_6H_4$Me-p. XIII. m. 89.5-90.5°.[65]

C_{22} $(PhCH_2O)_2PONHC_6H_4NMe_2$-p. V. m. 123-4°.[163]
p-$[(BuO)_2PONH-]_2C_6H_4$. V. m. 144°.[183]
p-$[(i\text{-}BuO)_2PONH-]_2C_6H_4$. V. m. 195°.[183]

C_x (o-$MeOC_6H_4O)_2PONHC_6H_4COOEt$-p. V. m. 108-9°.[493]
$(PhO)_2PONHC_6H_4[SO_2NH(C_5H_4N)-2]$-p. V. m. 215-6°.[493]
$(PhCH_2O)_2PONHC_{10}H_7$-2. XIII. m. 75.5-6.5°.[65]
$[(PrO)_2PONHC_6H_4-]_2$-p. V. m. 211°,[183] benzidine derivative.
$(PhO)_2PONHC_6H_4COOPh$-o. IX. m. 94°.[882]
$(2\text{-}C_{10}H_7O)_2PONHC_6H_4Br$-m. IX. m. 166.5°.[573]
$(2\text{-}C_{10}H_7O)_2PONHPh$. I. m. 193-5°.[74]

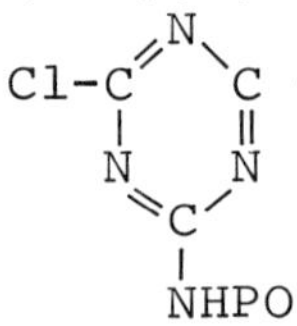
- $NHPO(OPh)_2$. IX. m. 214-5°.[202]

$[(PhO)_2PONH]_2C_{10}H_6$-1,8. V. m. 208°, pink crystals; 1,8-Diamino-naphthaline derivative.[183]
[p-$(PhO)_2PONHC_6H_4-]_2SO_2$. V. m. 232-4°.[493]
[p-$(PhO)_2PONHC_6H_4-]_2CH_2$. V. m. 185°.[183]
$(C_{27}H_{45}O)_2PONHPh$. IX. m. 196-7°.[927]

D.1.5.4. $(RO)_2P(O)$-NR'R", R' or R" = Alkyl or Aryl, R' = R" or R' ≠ R"

C_4 $(CH_2O)_2PONMe_2$. XIII. b_1 113-4°, m. 47.5-8.5°.[2]
$(MeO)_2PO(NC_2H_4)$. V. b_1 50°, n 1.4963, d 1.213,[130] b_{10} 99.5-100°, n 1.4375, d 1.2212.[340]
$(MeO)_2PONMe_2$. V, IX. b_1 78-9°, n_D^{25} 1.4175.[156] b_{11} 72-2.5°, n 1.4175, d 1.1317.[408] b_{15} 78-9°.[138] b_{15} 80°.[883]

C_5 $CH_2(CH_2O)_2PO(NC_2H_4)$. XIII. m. 82-3°.[345]
$(MeO)_2PO(NC_3H_6)$. V. b_{11} 108-9°, n 1.4447, d 1.1801.[346]

C_6 $(CH_2O)_2PON(CH_2CH_2Cl)_2$. IX. m. 57-8°.[59]
$(ClCH_2CH_2O)_2PO(NC_2H_4)$. V. $b_{0.01}$ 130°, n 1.4837.[737]
$(CHMeO)_2PO(NC_2H_4)$. XIII. $b_{0.09}$ 132°, n 1.4624, d 1.223 Similar cyclic esters.[345]
$(CH_2O)_2PON(CH_2CH_2)_2O$. XIII. m. 117-9°.[2]
$(MeO)_2PON(Et)CH(OH)CCl_3$. XVII. m. 128-9°. Similar N-substituted compounds.[638]
$(EtO)_2PO(NC_2H_4)$. V. $b_{0.03}$ 60-1°.[572] b_1 55°, n 1.4841, d 1.119.[130] $b_{9.5}$ 108.5°, n 1.4362, d 1.1148.[340]
$(CH_2O)_2PONEt_2$. XIII. b_2 117-7.5°, n 1.4565, d 1.1666.[2] b_6 132°, n 1.4522, d 1.1735.[49]
$(MeO)_2PON(CH_2CH_2)_2O$. V. b_1 96°, n_D^{25} 1.4530.[155,156]
$(MeO)_2PONEt_2$. V,[156] IX,[883] XIII.[408,678,879] b_8 82.5-3.0°, n 1.4265, d 1.0664.[408] b_8 85-6°, n 1.4284, d 1.0740.[678] b_{15} 92°, n_D^{25} 1.4240.[156] b_{16-17} 98°, n 1.4225.[879] b_{22} 100°.[883]

$(EtO)_2PONMe_2$. V, IX. b_5 85-90°.[573] b_{10} 93°, n 1.4231,
    d 1.0473.[857] b_{12} 84°, n 1.4180, d 1.0469.[408]
    b_{13} 92°.[883] ^{31}P -11.1.[873]

C_7 $CH_2(CH_2O)_2PON(CH_2CH_2Cl)_2$. IX. m. 49-50°.[59]
    $(CH_2O)_2PO(NC_5H_{10})$. V.[605]
    $Me_2C(CH_2O)_2PONMe_2$. V. b_6 64-6°.[84] ^1H.[84]
    $(MeO)_2PO(NC_5H_{10})$. V, IX. b_1 88°, n_D^{25} 1.4517.[88]
        b_{13} 125°.[883] b_{15} 119°, n_D^{25} 1.4528.[156,155]
    $(EtO)_2PO(NC_3H_6)$. V. b_8 109°, n 1.4410, d 1.0933.[346]
    $(EtO)_2PONMeEt$. XVII. b_1 56-8°, n 1.4210, d 1.0239.[26]
    $(EtO)_2PON(Et)CH_2OH$. XVII. b_1 147-50°, n 1.4442,
        d 1.1163.[21]

C_8 $(o-C_6H_4O_2)PO(NC_2H_4)$. V. m. 60° dec.[846]
    $(CH_2CH_2O)_2PON(CH_2CH_2Cl)_2$. IX. m. 72-3°.[59]
    $(ClCH_2CH_2O)_2PON(CH_2CH_2)_2O$. V. $b_{0.01}$ 143°, n 1.4851.[737]
    $(ClCH_2CH_2O)_2PONEt_2$. IX. $b_{0.2}$ 78°, n_D^{30} 1.4592.[507]
    $(EtO)_2PON(CH_2CH_2)_2$. V. $b_{0.6}$ 82°, n 1.4451.[180]
    $(CH_2O)_2PONPr_2$. XIII. b_3 127-9°, n 1.4576, d 1.1096.[2]
    $(PrO)_2PO(NC_2H_4)$. V. b_{10} 127-8°, n 1.4382, d 1.0584.[340]
    $(i-PrO)_2PO(NC_2H_4)$. V. b_{12} 112-3°,[340] b_{1-2} 76-8°.[894]
        n 1.4310, d 1.0439.[340]
    $(EtO)_2PON(CH_2CH_2)_2O$. V. b_{11} 137°.[746]

$(EtO)_2PON(Me)CH_2-\overset{O}{\overset{\diagup\diagdown}{CH}}-CH_2$. XVII. b_3 118-9°, n_D^{25} 1.4365.[51]
    $(MeO)_2PONPr_2$. V. b_{15} 112°.[155,156] n_D^{25} 1.4300.[156]
    $(EtO)_2PONEt_2$. IX,[573,883] XIII.[879,408,678] b_6 91°,[330]
        b_{13} 101°,[408] b_{18} 113-4°,[532] b_{20} 106-8°,[138] b_{25} 116-
        9°.[678] n 1.4242,[408] n 1.4318.[678] d 1.0113,[408]
        d 1.0401.[678] ^{31}P -9.0.[873]
    $(PrO)_2PONMe_2$. XIII. b_{11} 125-6°, n 1.4280, d 0.9826.[408]
    $(i-PrO)_2PONMe_2$. XIII. b_{12} 84-84.5°, 1.4160, d 0.9892.[408]
    $(EtO)_2PONEtCH_2OMe$. XVII. $b_{0.5}$ 65-6°,[298] b_1 80-0.5°,[26]
        n 1.4330,[298] n 1.4260.[26] d 1.0580,[298] d 1.0599.[26]
    $(EtO)_2PON(Pr)CH_2OH$. XVII. b_1 158-60°, n 1.4510,
        d 1.0898.[21,298]
    $(EtO)_2PON(CH_2CH_2OH)_2$. ^{31}P -13.7.[873]
    $(HOCH_2CH_2O)_2PONEt_2$. IX.[883]
    $(MeOCH_2CH_2O)_2PONMe_2$. IX. b_{13} 155-6°.[883]

C_9 $(MeO)_2PON(CF_3)C_6H_4Cl-p$. V, XIII. $b_{0.1}$ 95-105°.[285]

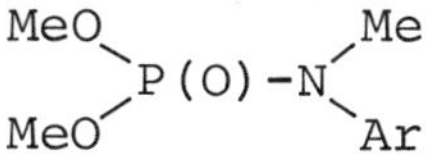

$\overset{\displaystyle MeO\diagdown}{\underset{\displaystyle MeO\diagup}{P(O)-N}}\overset{\diagup Me}{\diagdown Ar}$ XV, XVII.[326]

Ar	b.	n	d
$p-BrC_6H_4$	$b_{0.04}$ 118-20°	1.5461	1.4880
$p-ClC_6H_4$	$b_{0.25}$ 101-2°	1.5275	1.2856
$m-ClC_6H_4$	$b_{0.04}$ 90-1°	1.5289	1.2869
$m-FC_6H_4$	b_3 135-6°	1.4991	1.2600
$p-MeC_6H_4$	$b_{0.1}$ 100-1°	1.5124	1.1611
$m-MeC_6H_4$	$b_{0.04}$ 93.5-4°	1.5163	1.1627
$p-MeOC_6H_4$	$b_{0.04}$ 110-2°	1.5150	1.2148

$(EtO)_2PON(Bu)CH_2OH$. XVII. $b_{0.5}$ 161-2°, n 1.4500, d 1.0599.[21,298]
$(EtO)_2PON(Pr)CH_2OMe$. XVII. $b_{0.5}$ 78-9°, n 1.4370, d 1.0618.[21,298]
$(EtO)_2PON(Et)CH_2CONHMe$. V. $b_{0.5}$ 128-30°, n 1.4550, d 1.1234.[20]
$(EtO)_2PON(Et)CH_2OC(O)Me$. $b_{0.5}$ 88-90°, n 1.4310, d 1.1051.[21,298]
$(SCH_2)_2C(CH_2O)_2PON(CH_2CH_2Cl)_2$. IX. m. 113-5°.[553]
$(SCH_2)_2C(CH_2O)_2PON(CH_2CH_2)_2O$. IX. m. 221-9°.[553]
$CH_2(CH_2O)_2PON(CH_2CH=CH_2)_2$. XIII. $b_{0.8}$ 132-3°, n 1.4804 d 1.1363.[345]
$(HOCH_2)_2C(CH_2O)_2PON(CH_2CH_2Cl)_2$. IX. Penta-erythrite derivative.[59]
$(EtO)_2PONEt(CH_2CH=CH_2)$. XVII. b_1 78°, n 1.4349, d 1.0136.[26]
$(EtO)_2PO(NC_5H_{10})$. V.[573,180] $b_{0.3}$ 80°, n 1.4480.[180]

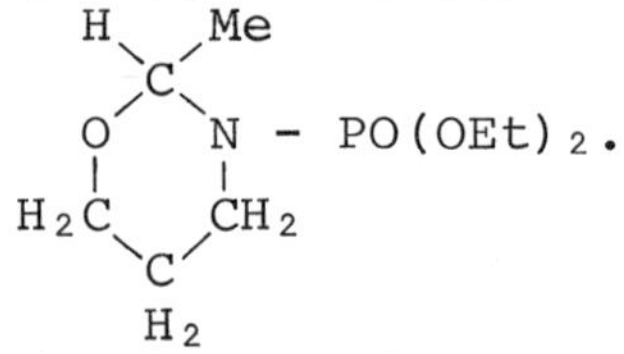
V. Eight similar five- and six-membered ring-compounds.[564]

$(EtO)_2PONMe(CH_2COOEt)$. XVII. b_2 112-4°, n 1.4295, d 1.1048.[26,46]
$(PrO)_2PO(NC_3H_6)$. V. b_{10} 133-4°, n 1.4423, d 1.0469.[346]
$(i\text{-}PrO)_2PO(NC_3H_6)$. V. b_{15} 123°, n 1.4353, d 1.0310.[346]
$(EtO)_2PONEtPr$. XVII. b_5 97-8°, n 1.4260, d 0.9963.[26]
C_{10} $(EtO)_2PON(Et)CH_2SPr$. XVII. $b_{0.5}$ 99-101°, n 1.4680, d 1.0539.[21,298]
$(EtO)_2PON(Bu)CH_2OMe$. XVII. $b_{0.5}$ 83-4°, n 1.4370, d 1.0250.[21,298]
$(EtO)_2PON(Et)CH_2COOEt$. XVII. b_1 106-7°, n 1.4340, d 1.0903.[26]
$(EtO)_2PON(CH_2CH_2CN)_2$. XVII. b_{12} 197-9°, n 1.4430, d 1.0932.[709]
$(i\text{-}BuO)_2PO(NC_2H_4)$. V. $b_{0.5}$ 135.5-6°, n 1.4367, d 1.0174.[340]
$(ClCH_2CH_2O)_2PON(CH_2CH=CH_2)_2$. V. $b_{0.4}$ 145°, n 1.4780.[73]

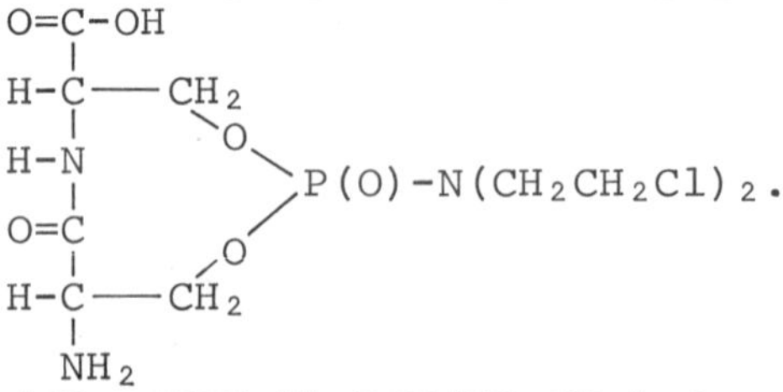
IX. L,L form, m. 96°, $[\alpha]_D^{20}$ 4.16°; D,D form, m. 96°, $[\alpha]_D^{20}$ -4.22°; D,L form, m. 88°.[850]

$(CH_2=CHCH_2O)_2PON(CH_2CH_2)_2O$. IX. $b_{0.1}$ 84-7°, n_D^{19} 1.4716.[563]
$(EtO)_2PON(CH_2CH=CH_2)_2$. V. b_9 118°.[330]

$(CH_2O)_2PONBu_2$. XIII. b_5 168-9°, n 1.4564, d 1.0607.[2]
$(EtO)_2PO[N(CH_2)_6]$. V. $b_{0.3}$ 87°, n 1.4550.[180]
$(BuO)_2PO(NC_2H_4)$. V. $b_{0.03}$ 88°.[572] b_{11} 146-7°, n 1.4407, d 1.0260.[340]
$(MeO)_2PONBu_2$. V. b_{15} 96°, n_D^{25} 1.4355.[155,156]
$(EtO)_2PONPr_2$. IX. $b_{0.5}$ 73-5°,[139] b_{12} 105-10°.[573] n_D^{22} 1.4280.[139] d_0^{15} 0.975.[573]
$(EtO)_2PONEtBu$. XVII. b_2 93-4°, n 1.4286, d 0.9891.[26]
$(i\text{-}PrO)_2PONEt_2$. XIII. b_{20} 115°,[408] b_1 78-9°.[543] n 1.4205, d 0.9660.[408]
$(MeOCH_2CH_2O)_2PONEt_2$. IX. b_{13} 165°.[883]
11 $(EtO)_2PONPhMe$. IX, XVII. b_1 91-2°,[324,403] b_1 109°.[26] n 1.5030,[324,403] n 1.5020.[26] d 1.1233,[324,403] d 1.1216.[26]
$(MeCH_2)(BrCH_2)C(CH_2O)_2PO(NC_5H_{10})$. V. m. 157-8°.[891]
$(EtO)_2PONMe(C_6H_{11})$. XIII. b_1 110°, n_D^{25} 1.4511.[88]
$(BuO)_2PO(NC_3H_6)$. V. b_{10} 158-9°, n 1.4446, d 1.0134.[346]
$(i\text{-}BuO)_2PO(NC_3H_6)$. V. b_{10} 143°, n 1.4404, d 1.0090.[346]
$(i\text{-}PrO)_2PONMe(CH_2COOEt)$. XVII. b_1 107-8°, n 1.4266, d 1.0591.[26]
$[(EtO)_2PONMe\text{-}]_2CH_2$. XVII. b_1 149-50°, n 1.4394, d 1.1309.[46]
12 $(C_3F_7CH_2O)_2PON(Me)CH_2CH_2CN$. V. $b_{0.1-0.3}$ 93-5°, n_D^{25} 1.3520.[608]

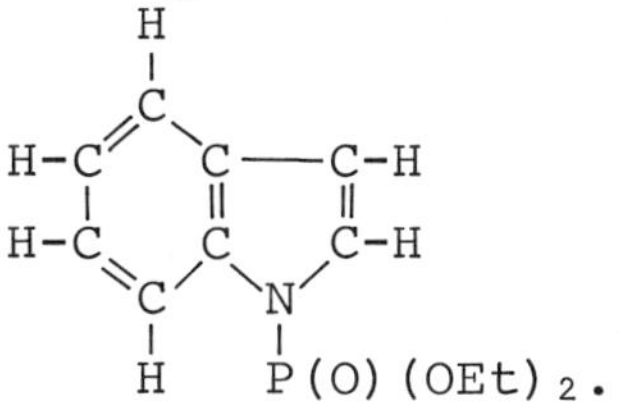
V. $b_{0.01}$ 110-2°.[570]

$(EtO)_2PON(Et)CH_2CONEt_2$. V. b_1 131-2°, n 1.4569, d 1.0668.[20]
$(EtO)_2PONEtPh$. XVII. $b_{0.5}$ 91°, n 1.4972, d 1.0917.[324,403]
$(PrO)_2PON(CH_2COOMe)_2$. V. b. 162°, n 1.4419.[182]
$(BuO)_2PO(NC_4H_8)$. V. $b_{0.3}$ 106°, n 1.4471.[180]
$(i\text{-}C_5H_{11}O)_2PO(NC_2H_4)$. V. $b_{10.5}$ 155.5-6.5°, n 1.4415, d 1.0004.[340]
$(BuO)_2PON(CH_2CH_2)_2O$. V,[182] IX.[563] b_2 143°.[182] $b_{0.03}$ 94°.[563] n 1.4522.[182]
$(EtO)_2PONBu_2$. V. b_3 115°.[182]
$(EtO)_2PON(Bu\text{-}i)_2$. IX. d_0^{14} 0.9663.[573]
$(PrO)_2PONPr_2$. XVII. $b_{0.07}$ 78-9°, n 1.4314.[139]
$(i\text{-}PrO)_2PONPr_2$. ^{31}P -8.4.[873]
$(BuO)_2PONEt_2$. XIII. b_9 133.5°,[408] b_8 136-7°,[679] n 1.4320,[408] n 1.4326.[679] d 0.9641,[408] d 0.9659.[679]
$(i\text{-}BuO)_2PONEt_2$. V. b_1 104°, n 1.4243.[182]
$(PrO)_2PONBu(CH_2CH_2OH)$. V. b_2 150-60°, n 1.4353.[182]
$(EtOCH_2CH_2O)_2PONEt_2$. IX. b_{10} 167-70°.[246]

C_{13} $(EtO)_2PON(C_9H_{10})$. V. b_8 155°.[573] Tetrahydroquinoline-derivative.
$(EtO)_2PON(CH_2Ph)(CH_2CH_2Cl)$. XVII. $b_{1.7}$ 151-1.5°, n 1.5013, d 1.1621.[341]
$(EtO)_2PONEt(CH_2Ph)$. XVII. b_1 109°, n 1.4871, d 1.0745.[2]
(i-AmO)$_2$PO(NC_3H_6). V. b_{10} 164-5°, n 1.4453, d 0.9956.[34]
$(EtOCH_2CH_2O)_2PO(NC_5H_{10})$. IX. b_{22} 210-3°.[246]
C_{14} $(C_6Cl_5O)_2PONMe_2$. IX. m. 180-3°.[916]
(p-$H_2NC_6H_4O)_2PONMe_2$. XVII. m. 101-3°.[291]
$[CHF_2(CF_2)_3CH_2O]_2PON(Me)CH_2CH_2CN$. V. $b_{0.5}$ 161°, n_D^{25} 1.3670.[608]
$(CH_2{=}CHCH_2O)_2PONBu_2$. V. $b_{0.05}$ 91-2°.[554]
(i-BuO)$_2$PONEt(CH_2COOEt). XVII. $b_{1.5}$ 134-5°, n 1.4320, d 1.0208.[26]
(i-BuO)$_2PONPr_2$. V. $b_{0.5}$ 112°, n 1.4342.[182]
(i-AmO)$_2PONEt_2$. XIII. b_8 144-6°, n 1.4320, 0.9472.[679]
$(BuO)_2PON(Bu)CH_2CH_2OH$. V. $b_{2.5}$ 143°, n 1.4370.[182]
$(MeOCH_2CH_2O)_2PONBu_2$. IX. b_5 172-8°.[834]
C_{15} 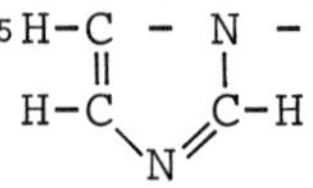 - PO(OPh)$_2$. V. m. 73-80°.[172] ^{31}P +15.7.[634]
$(PhO)_2PO(NC_3H_6)$. V. b_6 203-4°, n 1.5548, d 1.2186.[346]
C_{16} (m-$CF_3C_6H_4O)_2PONMe_2$. V, IX. $b_{2.9}$ 151-2°, n_D^{25} 1.4651.[60]
$(PhO)_2PON(CH_2CH_2Cl)_2$. V. m. 104-5°,[716] and cresyl-esters.[716]
$(PhO)_2PON(CH_2CH_2)_2O$. V. m. 71-2,[819] m. 72.5-3.5.[68] b_5 240-50°.[68]
$(PhO)_2PONEt_2$. V,[819] IX.[573] m. 61-2°.[819]
$(EtO)_2PONPh_2$. IX. m. 175°.[660]
$[(EtO)_2PONMe]_2C_6H_4$-p. V. m. 94°.[745]
$(EtO)_2PON(C_6H_{11})_2$. V. m. 140°.[182]
$(BuO)_2PONBu_2$. V. b_6 158-61°,[834] b_1 122°.[182] n 1.4406.[182]
(BuO)$_2$PON(Bu-i)$_2$. V. b_1 111°, n 1.4384.[182]
(i-BuO)$_2PONBu_2$. V. b_2 139-40°, n 1.4382.[182]
(i-BuO)$_2$PON(Bu-i)$_2$. V. b_3 127°, n 1.4381.[182]
C_{17} $(PhO)_2PO(NC_5H_{10})$. V. m. 70°,[573] m. 75-6°,[819,493] m. 78-9°.[70]
$(EtO)_2PONPh(CH_2Ph)$. XVII. $b_{0.1}$ 130°, n 1.5390, d 1.1397.[326]
$(MeO)_2PON(CH_2Ph)(CH_2CH_2NHPh)$. V. m. 70-1°.[355]
C_{18} (p-$CF_3C_6H_4O)_2PONMe(CH_2CH_2CN)$. V. $b_{0.07}$ 173°, n_D^{25} 1.4758.[608]
(i-PrO)$_2PON(Et)CH_2CONBu_2$. V. b_1 146-7°, n 1.4525, d 0.9923.[20]
$(PhCHO)_2PON(CH_2CH_2Cl)_2$. IX. Meso-hydrobenzoine, two isomers; m. 149.5-150° and m. 116.5-7.0°.[300]
$(PhO)_2PON(CH_2CH{=}CH_2)_2$. V. m. 37°.[53]
$(PhCH_2O)_2PON(CH_2CH_2)_2O$. ^{31}P -8.4.[394]
$(PhCH_2O)_2PON(CH_2CH_2Cl)_2$. XIII. n_D^{28} 1.5452.[315]

$(o\text{-}MeOC_6H_4O)_2PON(CH_2CH_2Cl)_2$. V. m. 85-6°.[502]
$(p\text{-}MeOC_6H_4O)_2PON(CH_2CH_2Cl)_2$. V. m. 101-2°.[502]
$(PhCH_2O)_2PON(CH_2CH_2)_2O$. XIII. m. 71-2°.[64]
$(PhO)_2PONPr_2$. V. m. 82-4°.[70]
$(PhCH_2O)_2PON(CH_2CH_2OH)_2$. XIII. Pale yellow oil; n_D^{25} 1.5465.[315]
$(m\text{-}Me_2NC_6H_4O)_2PONMe_2$. V. $b_{0.001}$ 200-25°, n_D^{25} 1.5760.[306]
$(MeO)_2PON(C_8H_{17})_2$. XVII. b_1 156-7°, n_D^{25} 1.4420.[156]
$(m\text{-}CF_3C_6H_4O)_2PONMe(CH_2CF_2CF_2CF_3)$. V. $b_{0.025}$ 123°, n_D^{25} 1.4240, NMR.[607]
$(m\text{-}ClC_6H_4O)(m\text{-}CF_3C_6H_4O)PONMe(CH_2CF_2CF_2CF_3)$. V. $b_{0.05}$ 133°, n_D^{25} 1.4500.[607] Similar compounds.[607]
19 $(PhO)_2PO(NPhMe)$. IX. m. 50°.[573]
$(p\text{-}H_2NC_6H_4O)_2PONMePh$. XVII. m. 110-1°.[291]
$(o\text{-}MeOC_6H_4O)_2PO(NC_5H_{10})$. V. m. 55-6°.[493]
20 $[3,5(CF_3)_2C_6H_3O]_2PONEt_2$. V. IX. $b_{0.020-0.030}$ 108-10°, n_D^{25} 1.4286.[609] Similar compounds.[609]
$(C_6Cl_5O)_2PONBu_2$. V. m. 144-5°.[532]
$(PhO)_2PON(Bu\text{-}i)_2$. V,[70] IX.[573] m. 139-41°,[70] m. 56°.[573]
$(PhO)_2PON(Bu\text{-}sec)_2$. V. m. 118-9°.[70]
$(BuOCH_2CH_2O)_2PONBu_2$. IX. b_4 198-205°.[834]
21 $(PhO)_2PO(NC_9H_{10})$. V. Tetrahydroquinoline-derivative.[573]
$(PhCH_2O)_2PONMePh$. XIII. m. 86-7°.[65] ^{31}P -6.2.[394]
x $(m\text{-}Et_2NC_6H_4O)_2PONMe_2$. V. $b_{0.001}$ 225°, n_D^{25} 1.5595.[306]
$(PhO)_2PONPh_2$. IX. m. 180°.[660]
$(o\text{-}MeC_6H_4O)_2PONPh_2$. IX. m. 178°.[660]
$(PhO)_2PON(CH_2Ph)(CH_2CH_2NHPh)$. V. m. 91-2°.[355]

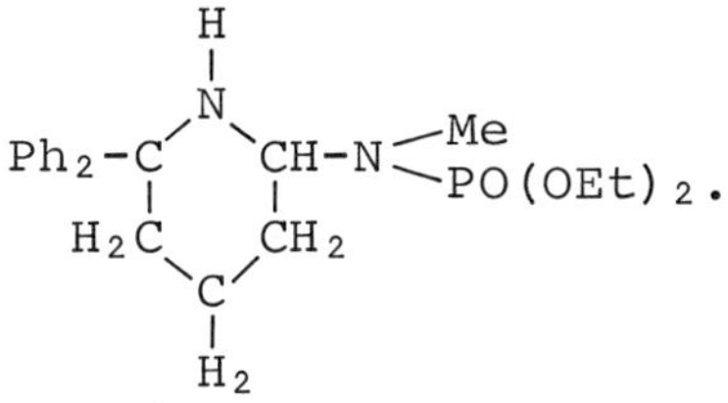
V. m. 162-5°.[855]

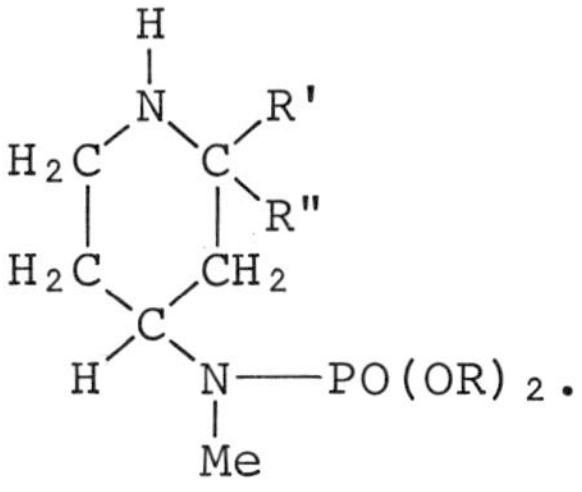
V.[855] R=Et, R'=Ph, R"=C_6H_4Me-o: m. 229-31°.
R=Et, R'=Ph, R"=C_6H_4Cl-p: m. 202-4°.
R=Et, R'=Ph, R"=C_6H_4OEt-p: m. 200-4°.
R=Et, R'=R"=C_6H_4Me-o: m. 185-6°.
R=Bu, R'=R"=Ph: m. 162-3°.

$R=C_8H_{17}$, R'=R"=Ph: m. 89-91°.
R=Et, R'=$PhCH_2$, R"=Ph: m. 188-9°.

D.1.5.5. $(RO)_2P(O)$-NH-C(=Z)-, Z=O,S,=NR'
Note: >P(O)NHC(=NH)- compounds may be listed as >P(O)N=C(NH_2)- in the next section.

C_3 $(MeO)_2PONHCONH_2$. XIII. m. 186-7°.[909]
C_4 $(MeO)_2PONHCOCBrCl_2$. IX. m. 107°.[816,817]
$(MeO)_2PONHCOCHCl_2$. IX. m. 81-3°.[805] For other $(RO)_2PONHCOR'$ see Ref. 805.
$(MeO)_2PONHCOCCl_3$. IX, XI. m. 105-7°.[436,816,817]
$(MeO)_2PONHC(=NH)CCl_3$. V. m. 112-3°.[232]
$(MeO)_2PONHCOOMe$. IX, XII. m. 63-5°,[467] m. 63-4°.[430]
C_5 $(MeO)_2PONHCOCBr_2OMe$. IX. m. 92-3°.[816,817]
$(MeO)_2PONHCOCCl_2CH_2Cl$. IX. m. 141-2°.[198]
$(MeO)_2PONHCOCHClCH_2Cl$. IX. m. 96-9°.[198]
$(MeO)_2PONHCO(NC_2H_4)$. XII. m. 70-3°.[233]
$(EtO)_2PONHCHO$. XVII. $b_{0.25}$ 134.5-5.0°, n_D^{24} 1.4368.[95]
$(MeO)_2PON(Me)COMe$. XVII. Oil.[669]
$(EtO)_2PONHC(S)NH_2$. XII. m. 146-7°.[524]
$(EtO)_2PONHCONH_2$. XIII. m. 208-9°.[909]
C_6 $(EtO)_2PONHCOCBrCl_2$. IX. m. 76-7°.[816,817]
$(EtO)_2PONHCOCCl_2NO_2$. IX. m. 56°.[816,817]
$(EtO)_2PONHCOCCl_3$. IX, XI. m. 47-8°,[816,817] m. 46-7°,[43] m. 55-6°.[399] $[K^+][CCl_3CONPO(OEt)_2^-]$: m. 177° dec.[55]
$(EtO)_2PONHCOCF_3$. IX, XI. $b_{0.4}$ 88.5-9.5°, n 1.3958, d 1.3215, IR.[399] Na-salt: m. 157.5-9.5°, IR, UV.[558] Ag-salt: dec. 120-1.5°.[400]
$(EtO)_2PONHCOCHBrCl$. IX. m. 67-8°.[816]
$(EtO)_2PONHCOCHCl_2$. XI. m. 73-4°,[805] m. 72-3°.[399]
$(EtO)_2PONHC(=NH)CCl_3$. V. m. 51-3°.[232]
$(EtO)_2PONHCOCHCl_2$. IX, XI. m. 72-3°,[816,817] m. 65-6.5°.[399]

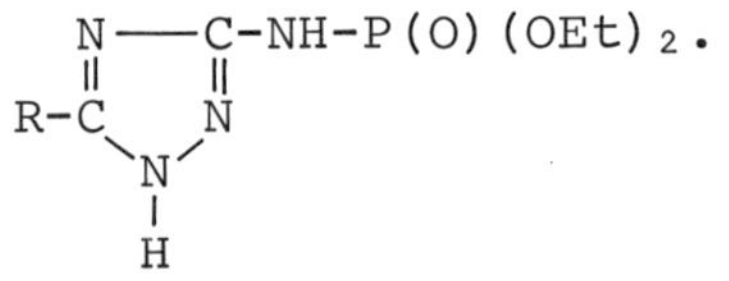
V.[570] R=H: m. 82-5°. R=Pr-i: m. 53-4°.

$(EtO)_2PONHCOMe$. IX. m. 49-9.5°.[404] Na-salt: m. 106.5-8.0°.[558] Ag-salt: m. 81.5-3.0°.[558]
$(EtO)_2PONHCSOMe$. XII. Na-salt: dec. 144-5°.[524]
$(MeO)_2PONHCONHNHCONHPO(OMe)_2$. XII. m. 183-4°.[470]
C_7 $(EtO)_2PONHCOCCl_2CH_2Cl$. XVII. m. 64-6°.[198]
$(MeO)_2PONHCOOC(Me_2)CN$. IX. m. 75-6°.[456]

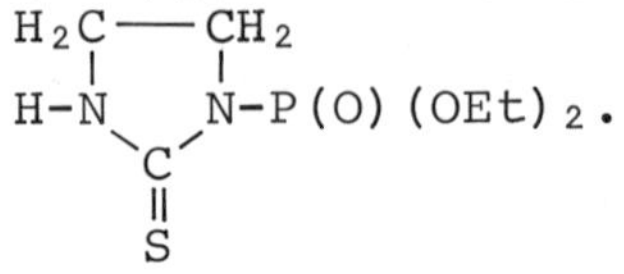
XVII. m. 77°.[513]

$(MeO)_2PONHCON(CH_2CH_2)_2O$.   XII.   m. 118-9°.[233]
$(EtO)_2PONHCSOEt$.   XII.   Na-salt: dec. 177-8°.[524]
$(EtO)_2PONHCSNHEt$.   XII.   m. 61-2°.[268]
$(i-PrO)_2PONHCSNH_2$.   XII.   m. 148-9°.[524]
$(PrO)_2PONHCONH_2$.   XIII.   m. 196-7°.[909]
$(MeO)_2PONHCONEt_2$.   XII.   m. 128°.[627]

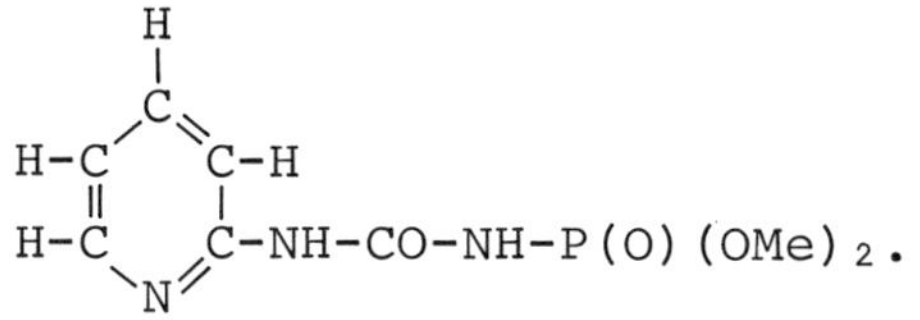

$$\underset{N}{\overset{H}{\underset{\|}{C}}}$$

H-C   C-H
H-C   C-NH-CO-NH-P(O)(OMe)$_2$.   IX.   m. 154-5°.[450]
         N   3-pyridyl-isomer: m. 160-1°.

$(EtO)_2PONHCOCBr_2OEt$.   IX.   m. 91°.[816]
$(i-PrO)_2PONHCOCHCl_2$.   IX.   m. 94-6°,[805] m. 86-8°.[231]
$(i-PrO)_2PONHC(=NH)CCl_3$.   V.   m. 70-2°.[232]
$(EtO)_2PON[C(O)Me]_2$.   XVII.   Oil.[182]
$(PrO)_2PONHC(O)Me$.   XVII.   b$_2$ 106°, n 1.4199.[182]
$(EtO)_2PONEtC(O)Me$.   XVII.   b$_{0.4}$ 76-8°, n 1.4330,
     d 1.0910.[558]
$(i-PrO)_2PONHCSOMe$.   XII.   Na-salt: dec. 183-4°.[524]
$(EtO)_2PONMeCOOEt$.   XVII.   b$_{1.5}$ 95-6°, n 1.4299, 1.1311.[46]
$(i-PrO)_2PONHCONHMe$.   IX.   m. 144-6°.[226]
$(MeO)_2PONHCSC_6H_4NO_2$-p.   XVII.   m. 147-8°.[465]

$$\underset{MeO}{\overset{MeO}{>}}P(O)-\underset{H}{N}-\underset{\|}{\overset{O}{C}}-Ar.$$ IX.

Ar	m. (Lit)
$2,3,6-Cl_3C_6H_2$	149-52° (impure)[832]
$p-FC_6H_4$	122-3°[235]
$m-FC_6H_4$	97-8°[235]
$o-FC_6H_4$	98-9°[235]
Ph	116-8°[208,455]
$p-ClC_6H_4$	125-6°[455]
$p-O_2NC_6H_4$	153-4°[455]
$m-O_2NC_6H_4$	124-6°[234,208]
$3,5-(O_2N)_2C_6H_3$	189-91°[439]

$(MeO)_2PONHCOOPh$.   IX.   m. 78-80°.[456]
$(MeO)_2PONHCO(OC_6H_4OH-p)$.   IX.   m. 121-4°.[456]
$(MeO)_2PONHCO(OC_6H_4NO_2-o)$.   IX.   m. 54-6°.[456]
$(MeO)_2PONHCO(OC_6H_4OH-m)$.   XII.   m. 115-8°.[237]
$(MeO)_2PONHCO(OC_6H_4OH-p)$.   XII.   m. 118-9°.[237] For other
    $(RO)_2PONHCOOAr(OH)$ compounds see Ref. 237.

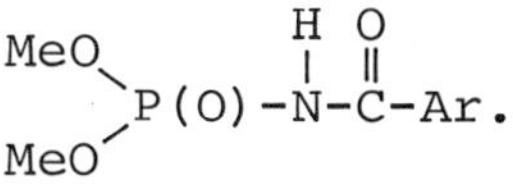

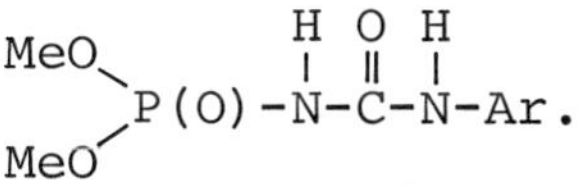

$$\underset{MeO}{\overset{MeO}{>}}P(O)-\underset{H}{N}-\underset{\|}{\overset{O}{C}}-\underset{H}{N}-Ar.$$ IX, XII.[450]

Ar	m.[450]
Ph	135-6°, 136-7°[468]
$p-MeC_6H_4$	135-8°
$o-MeC_6H_4$	116-7°
$p-EtC_6H_4$	126-7°
$p-i-PrC_6H_4$	145-8°

Ar	m.[450]
p-MeOC$_6$H$_4$	148-9°
p-EtOC$_6$H$_4$	147-9°
p-O$_2$NC$_6$H$_4$	176-7°
m-O$_2$NC$_6$H$_4$	165-7°
o-O$_2$NC$_6$H$_4$	165-6°
2,4-(O$_2$N)$_2$C$_6$H$_3$	180-1°
p-BrC$_6$H$_4$	144-6°
p-ClC$_6$H$_4$	139-40°
2,4,6-Cl$_3$C$_6$H$_2$	159-61°
1-C$_{10}$H$_7$	155-6°
2-C$_{10}$H$_7$	164-5°

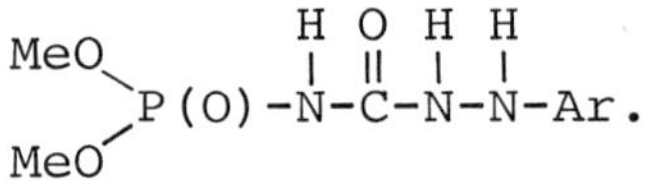

$$\text{MeO} \diagdown \atop \text{MeO} \diagup P(O)-\overset{H}{\underset{|}{N}}-\overset{O}{\underset{||}{C}}-\overset{H}{\underset{|}{N}}-\overset{H}{\underset{|}{N}}-\text{Ar. \quad IX, XII.}^{470}$$

Ar	m.
Ph	166-7°
p-MeC$_6$H$_4$	147-9°
PhCO	149-50°
p-O$_2$NC$_6$H$_4$CO	dec. 164-6°
2,4-(O$_2$N)$_2$C$_6$H$_3$	dec. 198-200
Me$_2$NNH-derivative	109-10°

[(MeO)$_2$PONHCO-]$_2$(CF$_2$)$_3$. IX. m. 170-2°.[201]

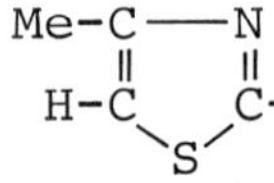

$$\begin{matrix} \text{Me-C} & \text{---} & \text{N} \\ \| & & \| \\ \text{H-C} & & \text{C-NHCSNHPO(OEt)}_2. \\ & \diagdown \text{S} \diagup & \end{matrix} \quad \text{XII. \quad m. 136-7°.}^{22}$$

(i-PrO)$_2$PONHCOCCl$_2$CH$_2$Cl. IX. m. 116-8°.[198]
(i-PrO)$_2$PONHCOOC(Me$_2$)CN. IX. m. 168-70°.[456]
(i-PrO)$_2$PONHCOCHClCH$_2$Cl. XVII. m. 150-2°.[198]
(i-PrO)$_2$PONHCOCCl$_2$Me. IX. m. 102-4°.[231]
(EtO)$_2$PONHCSNHCH$_2$COOEt. XII. m. 72-3°.[22]
(BuO)$_2$PONHCHO. XVII. b$_2$ 168.5-70°, n$_D^{24}$ 1.4430.[95]
(BuO)$_2$PONHCSNH$_2$. XII. m. 79-80°,[524] m. 78-9°.[709]
(BuO)$_2$PONHCONH$_2$. XIII. m. 175-6°.[909]
(i-BuO)$_2$PONHCSNH$_2$. XII. m. 116-7°.[524]
(i-PrO)$_2$PONHCSOEt. XII. Na-salt: dec. 182-3°.[524]
(EtO)$_2$PONHCSNHBu-t. XII. m. 108-9°.[268]
(EtO)$_2$PONHCSNEt$_2$. XII. m. 70-2°.[496]

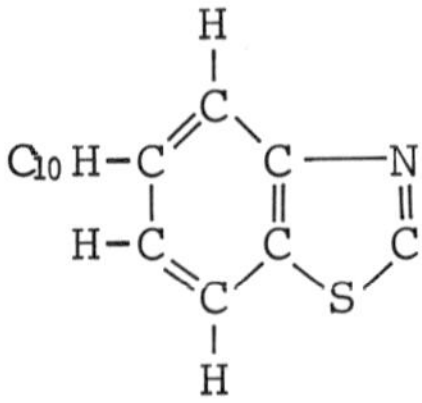

$$\begin{matrix} & & \overset{H}{\underset{|}{C}} & & \\ & \diagup & \| & \diagdown & \\ C_{10}\text{H-C} & & \text{C} & \text{---N} & \\ \| & & | & & \diagdown \\ \text{H-C} & & \text{C} & & \text{C-NH-CO-NH-P(O)(OR)}_2. \\ & \diagdown & | & \diagup \text{S} & \\ & & \underset{H}{\underset{|}{C}} & & \end{matrix} \quad \text{IX.}^{450}$$

R=Me: m. 157-8°
R=Et: m. 153-4°
R=Ph: m. 165-6°

(MeO)$_2$PONHCONHC$_6$H$_4$COOH-p. XII. m. 179-80°.[237]
(i-BuO)$_2$PONHCOCCl$_3$. IX. m. 181.5-3.0°.[399]
(i-BuO)$_2$PONHCOCF$_3$. IX. m. 87.5-8.5°.[399]

(i-PrO)$_2$PONHCSOPr-i. XII. Na-salt: dec. 173-4°.[524]
(i-BuO)$_2$PONHCSOMe. XII. Na-salt: dec. 197°.[524]
(i-PrO)$_2$PONHCOOPr-i. IX. m. 128-9°.[456]
(EtO)$_2$PONMeCONEt$_2$. XVII. b$_1$ 112°, n 1.4478, d 1.0849.[46]
C$_{11}$ (EtO)$_2$PONHCOC$_6$H$_2$Cl$_3$-2,3,6. IX. m. 180-3°.[832]
(EtO)$_2$PONHCOOPh. IX. m. 124-6°.[456]
(EtO)$_2$PONHCONHPh. IX, XII. m. 125-7°.[450]

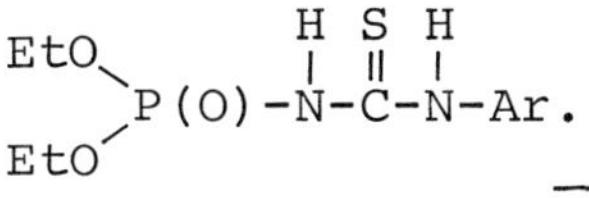

$$\begin{array}{c} \text{EtO} \\ \diagdown \\ \text{EtO} \diagup \end{array} \overset{\substack{H \quad S \quad H \\ | \quad \| \quad |}}{P(O)-N-C-N-Ar.} \quad XII.$$

Ar	m. (Lit)
Ph	107-8°,[524,496] 108-9°[709]
p-ClC$_6$H$_4$	129-30°[496]
3,4-(MeO)$_2$C$_6$H$_3$	124-5°[496]
2,4-Cl$_2$C$_6$H$_3$CH$_2$	133-4°[496]
2-C$_{10}$H$_7$	117-8°[496]
C$_6$H$_{11}$	84-5°,[496] 91-2°[268]
PhCH$_2$CH$_2$	80-1°[496]
C$_6$Cl$_5$NHCH$_2$CH$_2$	134-5°[496]

(i-BuO)$_2$PONEtCHO. XIII. b$_{2.5}$ 122-4°, n 1.4365,
    d 1.0386.[24]
(i-BuO`$_2$PONHCSOEt. XII. dec. 192-3°.[524]
[(EtO)$_2$PONMe-]$_2$CO. XVII. b$_{0.5}$ 145-6.5°, n 1.4505,
    d 1.2019.[46]
C$_{12}$ (EtO)$_2$PONPh(CO-Me). XVII.[324,403] b$_{0.01}$ 84.5°, n 1.5038,
    d 1.1670.[324]
(BuO)$_2$PON(COCH$_2$)$_2$. XIII. m. 45-7°.[878]
(BuO)$_2$PONHCSOPr-i. XII. Na-salt: dec. 190-1°.[524]
(i-BuO)$_2$PONHCSOPr-i. XII. Na-salt: dec. 202-3°.[524]
C$_{13}$ (PhO)$_2$PONHCN. XIII.[175,920] m. 136°.[175]
(PhO)$_2$PONHCONH$_2$. XII. m. 199-200°.[468]
(i-PrO)$_2$PONHCOC$_6$H$_2$Cl$_3$-2,3,6. IX. m. 187-9°.[832]
(EtO)$_2$PONHCSN(Ph)CH$_2$COOH. XII. m. 108-9°.[22]
(i-PrO)$_2$PONHCOPh. XIII. m. 107-9°.[208]
(MeO)$_2$PONHCO(OC$_6$H$_3$Me,Pr-i-2,5). IX. m. 110-2°.[456]
(i-PrO)$_2$PONHCOOPh. IX. m. 122-3°.[456]
(PrO)$_2$PONHCSNHPh. XVII. m. 101-2°.[709]
(i-PrO)$_2$PONHCSNHPh. XII. m. 124-5°.[524]
(i-PrO)$_2$PONHCONHPh. IX, XII. m. 138-40°.[450]
(BuO)$_2$PONHCSOBu. XII. Na-salt: dec. 169-70°.[524]
(i-BuO)$_2$PONHCSOBu-i. XII. Na-salt: dec. 181-90°.[524]
(o-ClC$_6$H$_4$O)$_2$PONHCOCCl$_3$. XI. m. 132-3°.[436]
C$_{14}$ (o-O$_2$NC$_6$H$_4$O)$_2$PONHCOCCl$_3$. XI. m. 175-7°.[436]
(p-O$_2$NC$_6$H$_4$O)$_2$PONHCOCCl$_3$. XI. m. 164-6°.[436]
(PhO)$_2$PONHCOCOCl. XVII. dec. 67°.[739]
(PhO)$_2$PONHCOCONH$_2$. XVII. m. 192-3°.[739]
(PhO)$_2$PONHCOCCl$_3$. XI. m. 101-2°.[436]
(PhO)$_2$PONHCOCF$_3$. IX. m. 87-9°.[200]
(p-BrC$_6$H$_4$O)$_2$PONHCOCF$_3$. IX. m. 183-5°.[200]
(PhO)$_2$PONHCOCN. XI. m. 114-5°.[209]

$(PhO)_2PONHCOCHCl_2$. IX. m. 114-6°.[231]
$(2,4\text{-}Cl_2C_6H_3O)_2PONHCOCHCl_2$. IX. m. 126-8°.[231]
$(PhO)_2PONHC(=NH)NHCN$. V. m. 177°.[100]
$(p\text{-}ClC_6H_4O)_2PONHC(=NH)NHCN$. V. m. 177°.[100]
$(PhO)_2PONHCSSMe$. XII. m. 133-4°.[268]
$(PhO)_2PONHCOOMe$. IX, XII. m. 109-11°.[467]
$(PhO)_2PONHCSOMe$. XII. m. 132-4°,[524] m. 129-30°.[268]
$(p\text{-}BrC_6H_4O)_2PONHCOOMe$. IX, XII. m. 152-4°.[467]
$(o\text{-}ClC_6H_4O)_2PONHCOOMe$. IX, XII. m. 116-8°.[467]
$(EtO)_2PONHCONHC_6H_4COOEt\text{-}p$. XII. m. 100-2°.[237]
$(t\text{-}BuCH_2O)_2PON(COCH_2)_2$. XIII. m. 167-8°.[878]
C_{15} $(p\text{-}ClC_6H_4O)_2PONHCOCF_2CF_2H$. IX. m. 108°.[270]
$(2,4\text{-}Cl_2C_6H_3O)_2PONHCOCCl_2Me$. IX. m. 105-7°.[231]
$(p\text{-}ClC_6H_4O)_2PONHCOCCl_2Me$. IX. m. 141-3°.[231]
$(p\text{-}ClC_6H_4O)_2PONHCOCHClCH_2Cl$. XI. m. 162-4°.[198]
$(PhO)_2PONHCOCHClCH_2Cl$. XI. m. 190-2°.[198]
$(PhO)_2PONHCOCCl_2CH_2Cl$. XI. m. 124-6°.[198]
$(p\text{-}ClC_6H_4O)_2PONHCOCCl_2CH_2Cl$. XI. m. 134-6°.[198]
$(PhO)_2PONHCO(NC_2H_4)$. XII. m. 97-9°.[233]
$(PhO)_2PONHCOCOOMe$. XVII. m. 111-3°.[739]
$(PhO)_2PONHCOOEt$. IX, XII. m. 94-6° dec.[467]
$(PhO)_2PONHCSOEt$. XII. m. 69-71°.[524]
$(p\text{-}BrC_6H_4O)_2PONHCOOEt$. IX, XII. m. 98-100°.[467]
$(o\text{-}ClC_6H_4O)_2PONHCOOEt$. IX, XII. 78-80°.[467]
$(EtO)_2PONHCONHC_{10}H_7$(isomer unstated). XII. m. 88-9°.[70]
$[Me_2C(CN)O]_2PONHCOOPh$. IX. m. 178-80°.[456]
$(PhO)_2PONHCONHNMe_2$. XII. m. 129-30°.[470]
$(BuO)_2PONHCSNHPh$. XII. m. 72-4°.[524]
$(i\text{-}BuO)_2PONHCSNHPh$. XII. m. 127-8°.[524]
$(BuO)_2PONHCSNHC_6H_4Cl\text{-}p$. XII. m. 65-6°.[496]
C_{16} $(MeO)_2PONHCOCClPh_2$. IX. m. 104-6°.[816,817]
$(p\text{-}MeC_6H_4O)_2PONHC(=NH)NHCN$. V. m. 199°.[100]
$(PhO)_2PONHCOOPr\text{-}i$. IX, XII. m. 79-81° dec.[467]
$(PhO)_2PONHCSOPr\text{-}i$. XII. m. 82-4°.[524]
C_{17} $(p\text{-}BrC_6H_4O)_2PONHCOOPr\text{-}i$. IX, XII. m. 114-6°.[467]
$(p\text{-}MeC_6H_4O)_2PONHCOOMe$. IX, XII. m. 149-51°.[467]
$(o\text{-}MeC_6H_4O)_2PONHCOOMe$. IX, XII. m. 135-6°.[467]
$(p\text{-}MeOC_6H_4O)_2PONHCOOMe$. IX, XII. m. 103-5°.[467]
C $(p\text{-}BrC_6H_4O)_2PONHCOOBu$. XII. m. 61-3°.[468]
$(PhO)_2PONHCON(CH_2CH_2)_2O$. XII. m. 182-3°.[233]
$(p\text{-}MeC_6H_4O)_2PONHCOOEt$. IX, XII. m. 100-1°.[467]
$(PhO)_2PONHCOOBu$. XII. m. 161-4°.[468]
$(i\text{-}PrO)_2PONHCO(OC_6H_3Me,Pr\text{-}i\text{-}2,5)$. IX. m. 131-2°.[456]
C_{18} $(PhO)_2PONHCSN(Me)CH_2COOEt$. XII. m. 124.5-5.5°.[268]
C_{19} $(p\text{-}ClC_6H_4O)_2PONHCSPh$. XVII. m. 92-3°.[465]
$(p\text{-}O_2NC_6H_4O)_2PONHCSPh$. XVII. m. 97-8°.[465]

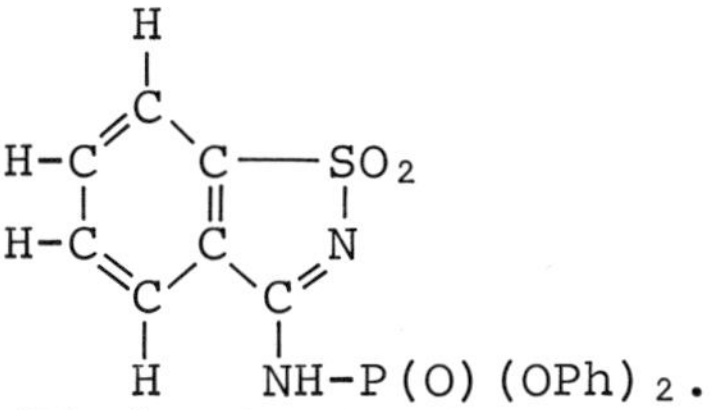
IX. m. 257-9°.[207]

$(PhO)_2PONHCONHC_6H_4Br$-p. XII. m. 182-3°.[233]
$(PhO)_2PONHCONHC_6H_4Cl$-p. XII. m. 177-8°.[233]
$(PhO)_2PONHCSNHC_6H_4NO_2$-p. XII. m. 139-40°.[268]

$(PhO)_2P(O)$-N(H)-C(=O)-Ar. IX.[832,235]

Ar	m. (Lit)
Ph	147-9°,[193,455,440,439] 148°[943]
2,3,6-$Cl_3C_6H_2$	236-9°[832]
o-BrC_6H_4	111-2°[234]
p-BrC_6H_4	132-3°,[234] 137-9°[193]
o-$O_2NC_6H_4$	121-2°[234]
p-MeC_6H_4	131-3°[234]
p-FC_6H_4	116-7°[235]
m-FC_6H_4	142-3°[235]
2,4-$Cl_2C_6H_3$	136-8°[439]
m-$O_2NC_6H_4$	166-7°[439,193]
3,5-$(O_2N)_2C_6H_3$	197-9°[439,440]
2,4-Cl,$O_2NC_6H_3$	107-9°[439,440]
p-ClC_6H_4	142-3°[455]
p-$O_2NC_6H_4$	151-2°[455,439]
1-$C_{10}H_7$	127-9°[439,440]
o-ClC_6H_4	102-4°[439]

$(ArO)_2P(O)$-N(H)-C(=O)-Ar'. Ar≠Ph

Ar	Ar'	m. (Lit)
p-ClC_6H_4	p-BrC_6H_4	152-3°[193]
p-ClC_6H_4	o-$O_2NC_6H_4$	142-4°[193]
p-ClC_6H_4	3,5-$(O_2N)_2C_6H_3$	216-7°[439,440]
p-ClC_6H_4	o-ClC_6H_4	179-80°[439]
p-ClC_6H_4	2,4-$Cl_2C_6H_3$	152-4°[439]
p-ClC_6H_4	m-$O_2NC_6H_4$	171-3°[439]
p-ClC_6H_4	Ph	131-2°[455,439,193]
p-$O_2NC_6H_4$	p-$O_2NC_6H_4$	175-6°[193,455]
p-ClC_6H_4	p-ClC_6H_4	155-6°[455]
p-$O_2NC_6H_4$	Ph	151-2°[455]
p-$O_2NC_6H_4$	p-ClC_6H_4	167-8°[455]
p-$O_2NC_6H_4$	p-$O_2NC_6H_4$	179-80°[455,439]
p-$O_2NC_6H_4$	o-$O_2NC_6H_4$	168-9°[234]
p-$O_2NC_6H_4$	m-$O_2NC_6H_4$	176-8°[439]

Ar	Ar'	m. (Lit)
p-$O_2NC_6H_4$	2,4-$Cl_2C_6H_3$	160-2°[439]
o-$O_2NC_6H_4$	p-ClC_6H_4	179-80°[455]
o-$O_2NC_6H_4$	Ph	154-5°[455],
o-$O_2NC_6H_4$	p-$O_2NC_6H_4$	178-9°[440];[439]
o-$O_2NC_6H_4$	2,4-$Cl_2C_6H_3$	170-1°[439]
p-MeC_6H_4	Ph	146-7°[455],[193]
p-MeC_6H_4	p-ClC_6H_4	139-40°[455]
p-MeC_6H_4	p-$O_2NC_6H_4$	165-7°[193],[455]
p-MeC_6H_4	p-BrC_6H_4	133-5°[193]
m-MeC_6H_4	p-$O_2NC_6H_4$	131-3°[439]
1-$C_{10}H_7$	Ph	195-6°[439],[455]
1-$C_{10}H_7$	p-ClC_6H_4	165-7°[440],[439]

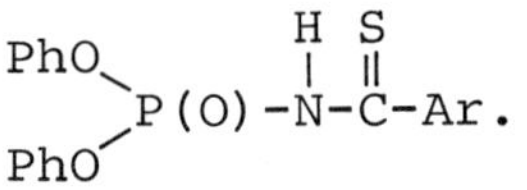

PhO
  $\backslash$ H S
   P(O)-N-C-Ar.   XVII.[456]
PhO$^{/}$

Ar	m.
Ph	88-9°
p-ClC_6H_4	106-7°
o-$O_2NC_6H_4$	127-8°
p-$O_2NC_6H_4$	127-8°
1-$C_{10}H_7$	111-2°

$(PhO)_2PONHCONHPh$.   IX, XII.   m. 155-6°.[468]
$(p-BrC_6H_4O)_2PONHCONHPh$.   IX, XII.   m. 170-1°.[468]
$(PhO)_2PONHCOOPh$.   IX.   m. 121-3°.[456]
$(C_6H_{11}O)_2PONHCOOPh$.   IX.   m. 111-3°.[456]
$(p-O_2NC_6H_4O)_2PONHCOOPh$.   IX.   m. 149-50°.[456]
$(p-O_2NC_6H_4O)_2PONHCO(OC_6H_4NO_2-p)$.   IX.   m. 114-5°.[456]

PhO
  $\backslash$ H O H H
   P(O)-N-C-N-N-Ar.   IX, XII.[470]
PhO$^{/}$

Ar	m.
Ph	167-8°
p-MeC_6H_4	176-7°
PhCO	180-1°
p-$O_2NC_6H_4$	211-3° dec.
2,4-$(O_2N)_2C_6H_3$	202-3°

C_{20} $(PhO)_2PONHC(=NPh)CCl_3$.   XI.   m. 105-6°.[197]   For other
      $(RO)_2PONHC(=NR')CCl_3$ see Ref. 197.
   $(PhO)_2PONHCOCONHPh$.   XVII.   m. 140-1°.[739]
   $(PhO)_2PONHCSNHC_6H_4COOH-p$.   XII.   m. 128-30°.[268]
   $(PhO)_2PONHCSNH(C_6H_3Cl,Me-2,4)$.   XII.   m. 114-5°.[146]   For
      11 other $(PhO)_2PONHCSNHAr$ see Ref. 146.
   $(PhO)_2PONHCONHC_6H_4Me-p$.   XII.   m. 155-7°.[233]
   $(PhO)_2PONHCOOCH_2Ph$.   XII.   m. 79-81°.[468]
   $(p-BrC_6H_4O)_2PONHCOOCH_2Ph$.   XII.   m. 113-5°.[468]
C_{21} $(PhCH_2O)_2PONHCO(OC_6H_4NO_2-p)$.   IX.   m. 124-6°.[456]
   $(PhCH_2O)_2PONHCSNHPh$.   XII.   m. 107-8°.[268]

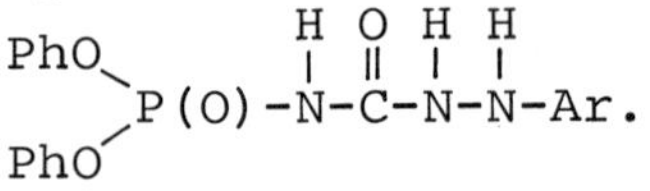

$(PhCH_2O)_2PONHCOPh$. XI. m. 123-4°.[943]
$(PhCH_2O)_2PONHCOOPh$. IX. m. 124-6°.[456]
$[1\text{-}(4\text{-}ClC_{10}H_6)O]_2PONHCOCCl_3$. XVII. m. 180-2°.[437]
C22 $(1\text{-}C_{10}H_7O)_2PONHCOCCl_3$. XVII. m. 139-41°.[437]
$(PhO)_2PONHCSNHC_6H_4COOEt\text{-}p$. XII. m. 126-7°.[268]
Cx $(PhO)_2PONHCSN(Me)CH_2CONHC_6H_4Me\text{-}p$. XII. m. 141.5-2.0°.[268]
$p\text{-}C_6H_4[NHCSNHPO(OBu)_2]_2$. XII. m. 149-50°.[496]
$(PhO)_2PONHCONPh_2$. XI. m. 147-9°.[461]
$(p\text{-}ClC_6H_4O)_2PONHCONPh_2$. XI. m. 151-2°.[461]
$(o\text{-}ClC_6H_4O)_2PONHCONPh_2$. XI. m. 63-5°.[461]
$(p\text{-}O_2NC_6H_4O)_2PONHCONPh_2$. XI. m. 148-50°.[461]
$(o\text{-}O_2NC_6H_4O)_2PONHCONPh_2$. XI. m. 182-3°.[461]
$(PhO)_2PONHCOCONHPO(OPh)_2$. XVII. dec. 175-81°.[739,740]
$(PhO)_2PONHCONHNHCONHPO(OPh)_2$. XII. m. 188-90°.[470]
$(PhO)_2PONHC(=NSO_2Ph)C_6H_4NO_2\text{-}p$. XVII. m. 154-6°.[207]

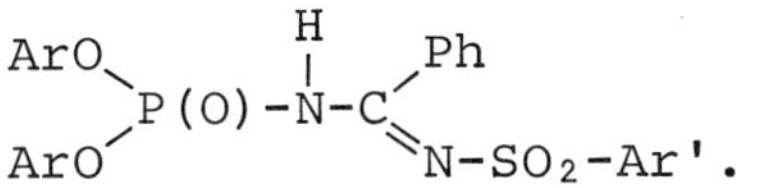
. XVII.[203]

Ar	Ar'	m.
Ph	Ph	153-4°
Ph	$p\text{-}MeC_6H_4$	143-4°
Ph	$p\text{-}ClC_6H_4$	165-6°
Ph	$p\text{-}O_2NC_6H_4$	181-2°
Ph	$1\text{-}C_{10}H_7$	154-6°
Ph	$2\text{-}C_{10}H_7$	177-8°
$p\text{-}ClC_6H_4$	Ph	143-4°
$p\text{-}ClC_6H_4$	$1\text{-}C_{10}H_7$	189-90°
$p\text{-}ClC_6H_4$	$2\text{-}C_{10}H_7$	173-4°
$p\text{-}ClC_6H_4$	$p\text{-}O_2NC_6H_4$	185-6°
$p\text{-}ClC_6H_4$	$p\text{-}ClC_6H_4$	165-6°
$p\text{-}ClC_6H_4$	$p\text{-}MeC_6H_4$	155-6°

$(1\text{-}C_{10}H_7O)_2PONHCOC_6H_4NO_2\text{-}p$. XVII. m. 188-90°.[433]
$(1\text{-}C_{10}H_7O)_2PONHCOCClPh_2$. XVII. m. 168-70°.[433]

D.1.5.6. $(RO)_2P(O)\text{-}N=C<$

C3 $(MeO)_2PO(NCO)$. XV. b_4 75-6°, n 1.4280, d 1.3136.[468]
C5 $(MeO)_2PON=C(OMe)CF_3$. IX. $b_{0.03}$ 69-71°, n 1.4161.[200]
$(EtO)_2PO(N=CCl_2)$. XVII. $b_{0.08}$ 75-8°,[284] b_1 81.5-2.0°,[22,23] $b_{0.1}$ 81-2°.[441] n 1.4540,[441] n 1.4646.[23] d 1.2182,[441] d 1.3181.[23] ^{31}P +8.5.[361a]
$(MeO)_2PON=C(SCH_2)_2$. V. $b_{0.001}$ 100-5°, n_D^{25} 1.5683;[8] for further compounds see Ref. 8.
$(EtO)_2PO(NCS)$. XVI. b_{13} 40°,[746] b_1 80-2°,[163] b_7 94°,[524] b_{12} 112-3°,[496] $b_{0.6}$ 70-2°,[268] $b_{0.07}$ 80-2°,[287] n 1.4749,[524] n_D^{23} 1.4772,[496] n_D^{19} 1.4751,[268] d 1.1694,[524] ^{31}P +18.9.[873]
$(EtO)_2PO(NCO)$. XVII. b_{14} 91-3°, n 1.4175, d 1.1820.[741]
$(MeO)_2PON=C(NH_2)NHCH_2COOH$. V. m. 125-6°.[175]

$(MeO)_2PON{=}C(OMe)NHMe$. IX. $b_{0.6}$ 100-3°.[226]

C_6 $(EtO)_2PO(N{=}CClCF_3)$. XVII. b_6 82-5°, n 1.4020, d 1.3434.[558]

$(MeO)_2PON{=}C(CN)OEt$. IX, XIII. b_{10} 148-50°, n 1.4442, d 1.2234.[209]

$(MeO)_2PON{=}C(OMe)CCl_2Me$. XI. $b_{0.06}$ 97-8°, n 1.4768, d 1.3636;[796] for other $(RO)_2PON{=}C(OR')R''$ see Ref. 79

$(EtO)_2PON{=}C(Me)NH_2$. XIII. $b_{0.08}$ 49-50°, n 1.4110, d 1.0738.[217]

$(MeO)_2PON{=}C(OMe)NHEt$. IX. $b_{0.5}$ 110-2°.[226]

C_7 $(SCN)P(O)(OCH_2)_2C(CH_2O)_2PO(NCS)$. XIV. m. 197°.[711]

$(EtO)_2PON{=}C(SCH_2)_2$. V. $b_{0.001}$ 100-5°, n_D^{25} 1.5463.[9]

$(i\text{-}PrO)_2PO(NCO)$. XIV,[524] XVII.[740] b_5 105-7°,[524] b_{10} 97-9°.[740] n 1.4635,[524] n 1.4294.[740] d 1.1039.[52]

$(i\text{-}PrO)_2PO(N{=}CCl_2)$. XVII. $b_{0.05}$ 62-3°,[441] $b_{1.5}$ 83-4°,[23] n 1.4556,[23] d 1.2182.[23]

$(PrO)_2PO(N{=}CCl_2)$. XVII. b_1 94-6°, n 1.4630, d 1.2293.[23]

$(EtO)_2PON{=}C(OCH_2)_2$. XVII. b_1 134-5°, n 1.4520, d 1.2629.[11]

$(EtO)_2PON{=}C(Me)OMe$. XVII. $b_{0.6}$ 75-7°, n 1.4397, d 1.1140.[400]

$(EtO)_2PON{=}C(NH_2)NHCH_2COOH$. V. m. 122-3°.[175]

$(EtO)_2PON{=}C(OMe)_2$. XVII. $b_{0.1}$ 90-5°,[223] b_1 88-90°.[25] n_D^{25} 1.4309,[223] n 1.4353.[25] d_{25} 1.1682,[223] d 1.1724.[25] For other $(RO)_2PON{=}C(OR')_2$ compounds see Ref. 223.

$(EtO)_2PON{=}C(SEt)NH_2$. XVII. m. 85-6°.[526]

$(EtO)_2PON{=}CHNMe_2$. XVII. $b_{0.15}$ 125-7°, n_D^{25} 1.4691.[911]

C_8 $(EtO)_2PON{=}C(CN)OEt$. IX, XIII. b_{10} 148°, n 1.4418, d 1.1366.[209]

$(i\text{-}PrO)_2PON{=}C(CCl_3)NH_2$. XIII. m. 71-3°.[217]

$(EtO)_2PON{=}CHOCHMe_2$. XVII. b_2 109-11°, n_D^{31} 1.4290.[95] For other $(EtO)_2PON{=}CH\text{-}OR$ compounds see Ref. 95.

$(EtO)_2PON{=}C(Me)OEt$. XIII, XVII. b_1 71°,[400] b_5 111-2°.[67] n 1.4425,[400] n 1.4405.[672] d 1.0862,[400] d 1.0895.[672]

$(EtO)_2PON{=}C(SPr\text{-}i)NH_2$. XVII. m. 73-4°.[526]

C_9 $(MeO)_2PON{=}C(Cl)C_6H_4NO_2\text{-}p$. XVII. m. 107-10°.[190]

$(MeO)_2PON{=}C(Cl)C_6H_4NO_2\text{-}m$. XVII. m. 20-5°.[190]

$(MeO)_2PON{=}C(Cl)C_6H_3(NO_2)_2\text{-}3,5$. XVII. m. 125-7°.[190]

$(MeO)_2PON{=}C(C_6H_4NO_2\text{-}p)NH_2$. XVII. m. 134-6°.[221]

$(MeO)_2PON{=}C(C_6H_4Br\text{-}p)NH_2$. XIII. m. 120-2°.[215]

$(MeO)_2PON{=}C(Ph)NH_2$. XIII. m. 26-8°.[215]

$(BuO)_2PO(NCS)$. XIV. b_2 135-7°,[524] b_{12} 153-4°.[496] n 1.4712,[524] n 1.4728.[496] d 1.0792[524]

$(BuO)_2PO(N{=}CCl_2)$. XVII. b_1 111-3°,[23] $b_{0.08}$ 104-8°.[284] n 1.4622,[23] d 1.1737.[23]

$(i\text{-}BuO)_2PO(NCS)$. XIV. b_{18} 145-6°, d 1.0771, n 1.4680.[5]

$(BuO)_2PO(NCO)$. XIV. b_1 116°.[379]

$(EtO)_2PON{=}C(SEt)_2$. XVII. $b_{1.5}$ 147-8°,[11] $b_{0.05}$ 147-9°.[2] n 1.5260, d 1.1700,[11] and others.

$(EtO)_2PON{=}C(OEt)_2$. XIII. b_1 133-4°, n_D^{18} 1.4430, d 1.1400.[672]
$(i\text{-}PrO)_2PON{=}C(OMe)_2$. XVII. $b_{0.04}$ 40-5°, n 1.4210.[223]
$(EtO)_2PON{=}C(OEt)NMe_2$. XVII. b_1 116-8°, n 1.4628, d 1.1050.[25]
$(EtO)_2PON{=}C(OEt)NHEt$. IX. $b_{0.6}$ 120-2°.[226]
$(EtO)_2PON{=}C(SPr)NHMe$. XVII. b_1 128°, n 1.4954, d 1.1329.[526]
$(i\text{-}BuO)_2PON{=}C(NH_2)_2$. V. m. 171°.[174]
$(EtO)_2PON{=}C(NMe_2)_2$. XVII. b_1 122-4°, n 1.4823, d 1.1046.[25]
$(EtO)_2PON{=}C(NHEt)_2$. XVII. b_2 160-2°, n 1.4698, d 1.0866.[25]
$(EtO)_2PON{=}C(OEt)_2$. XVII. b_2 123-4°, n 1.4432, d 1.1097.[25]
C_{10} $(MeO)_2PON{=}C(OMe)C_6H_4NO_2\text{-}p$. IX. m. 94-5°.[221]
$(i\text{-}PrO)_2PON{=}C(CN)OEt$. IX, XIII. b_1 154-6°, n 1.4372, d 1.0829.[209]
$(EtO)_2PON{=}C(SPr)N(C_2H_4)$. XVII. b_1 129-31°, n 1.5005, d 1.1445.[11]
$(EtO)_2PON{=}C(SPr)OEt$. XVII. $b_{0.1}$ 112-4°, n 1.4880, d 1.1300.[526]
$(MeO)_2PON{=}C(OEt)C(OEt){=}NPO(OMe)_2$. XIII. b_{10} 205°, n 1.4596, d 1.2657.[209]
$(EtO)_2PON{=}C(Pr\text{-}i)NMe_2$. XVII. b_1 132-4°, ^1H.[97]
$(EtO)_2PON{=}C(SBu)NHMe$. XVII. b_1 116-8°, n 1.4938, d 1.1135.[526]
$(EtO)_2PON{=}C(SPr)NMe_2$. XVII. b_1 132-3°, n 1.4979, d 1.1230.[25]
C_{11} $(EtO)_2PON{=}C(C_6H_4Br\text{-}p)NH_2$. XVII. m. 102-4°.[215]

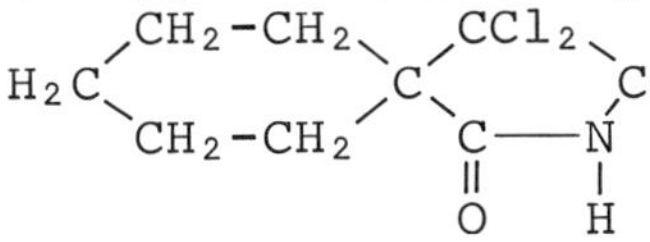 $C{=}N{-}P(O)(OMe)_2$. IX. m. 135-6°.[787]

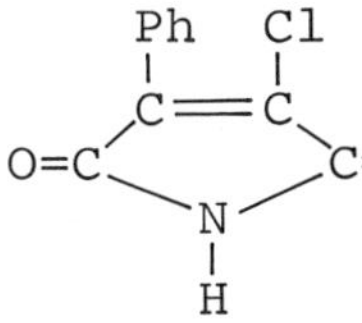 $C{=}N{-}P(O)(OMe)_2$. IX. m. 122-4°.[787]

$(EtO)_2PON{=}C(Ph)NH_2$. XVII. m. 24-6°.[215]
$(i\text{-}AmO)_2PO(N{=}CCl_2)$. XIV, XVII. b_1 112-3°, n 1.4595, d 1.1402.[12]
$(EtO)_2PON{=}C(SPr)Pr$. XVII. b_1 140-1°, n 1.5028, d 1.1235.[11]
$(EtO)_2PON{=}C(SPr)_2$. XVII. $b_{0.1}$ 134°,[526] b_1 142-4°.[11] n 1.5162,[526] n 1.5168.[11] d 1.1257.[11]
$(EtO)_2PON{=}C(SPr\text{-}i)_2$. XVII. $b_{1.5}$ 141-3°, n 1.5127, d 1.1201.[11]
$(EtO)_2PON{=}C(SBu)NHEt$. XVII. b_1 120-1.5°, n 1.4882, d 1.0929.[526]

$(EtO)_2PON{=}C(SBu)NMe_2$. XVII. b_1 140-2°, n 1.4978, d 1.1091.[25]

C_{12} $(EtO)_2PON{=}C(SPr)NEt_2$. XVII. b_1 130-3°, n 1.5002, d 1.0939.[526]

C_{13} $(p\text{-}ClC_6H_4O)_2PO(NCO)$. XIII. b_3 186-7°, m. 41-3°.[233]

$(p\text{-}BrC_6H_4O)_2PO(NCO)$. XV. b_{10} 250°, m. 72-4°.[468]

$(PhO)_2PO(N{=}CCl_2)$. XVII. $b_{0.05}$ 157°,[441] n 1.5603.[671]

$(PhO)_2PO(NCS)$. XIV. $b_{0.1}$ 105°, $b_{0.05}$ 90°,[425] $b_{0.001}$ 130-40°.[268] n 1.5851,[425] n 1.5840.[268] ^{31}P +29.3°.[634,873]

$(PhO)_2PO(NCO)$. XV, XVII. $b_{0.1}$ 140°,[741] $b_{0.5}$ 142-4°,[818] b_5 184-6°.[468] n 1.5473,[818] n 1.5470.[468] d 1.2828.[468]

$(p\text{-}ClC_6H_4O)_2PON{=}CHNH_2$. V. m. 124-7°.[288]

$(PhO)_2PON{=}C(NH_2)_2$. V. m. 118°.[174,750]

$(EtO)_2PON{=}C(OEt)C_6H_3(NO_2)_2$-3,5. IX. m. 89-91°.[221]

$(i\text{-}PrO)_2PON{=}C{=}N\text{-}Ph$. XVII. $b_{0.1}$ 115°.[223]

$(C_6H_{13}O)_2PO(N{=}CCl_2)$. XVII. b_1 143-4°, n 1.4602, d 1.04.[12]

$(EtO)_2PON{=}C(SBu)_2$. XVII. b_1 168-70°, n 1.5117, d 1.0958.[11]

C_{14} $(PhO)_2PON{=}C(Me)NH_2$. V. m. 86-7°.[174]

$(PhO)_2PON{=}C(CCl_3)NH_2$. XVII. m. 79-82°.[230]

$(PhO)_2PON{=}C(SMe)NH_2$. V. m. 78°.[174]

$(PhO)_2PON{=}C(OMe)NH_2$. V. m. 80°.[174]

$(PhO)_2PON{=}C(NH_2)NHMe$. V. m. 90°.[174]

$(EtO)_2PON{=}C(OEt)\text{-}C(OEt){=}NPO(OEt)_2$. XIII. $b_{0.1}$ 141-3°, n 1.4512, d 1.1636.[209]

$(BuO)_2PON{=}C(Pr\text{-}i)NMe_2$. XVII. b_1 176°, ^1H.[97]

C_{15} $(p\text{-}O_2NC_6H_4CH_2O)_2PON{=}C{=}NH$. XVII. Na-salt: m. 218° dec.[175]

$(PhCH_2O)_2PO(NCS)$. XIII. Not isolated.[268]

$(MeO)_2PON{=}C(Ph)OC_6H_4Cl\text{-}p$. XVII. m. 85-7°.[434]

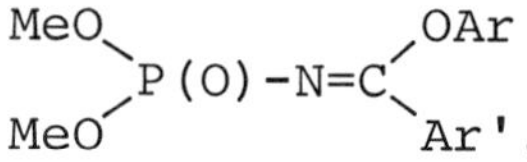

XVII.[434]

Ar	Ar'	m.
$p\text{-}O_2NC_6H_4$	Ph	131-3°
$p\text{-}ClC_6H_4$	$p\text{-}ClC_6H_4$	112-4°
$p\text{-}O_2NC_6H_4$	$p\text{-}ClC_6H_4$	131-3°
$p\text{-}ClC_6H_4$	$p\text{-}O_2NC_6H_4$	140-2°
$p\text{-}O_2NC_6H_4$	$p\text{-}O_2NC_6H_4$	128-30°
$p\text{-}ClC_6H_4$	$m\text{-}O_2NC_6H_4$	145-7°
$p\text{-}O_2NC_6H_4$	$m\text{-}O_2NC_6H_4$	126-8°
$p\text{-}ClC_6H_4$	3,5-$(O_2N)_2C_6H_3$	100-3°

$(PhO)_2PON{=}C(OMe)_2$. XVII. $b_{0.05}$ 78-80°, n 1.4940.[223]

$(PhO)_2PON{=}C(NH_2)NHCH_2COOH$. V. m. 144°.[175]

$(p\text{-}O_2NC_6H_4CH_2O)_2PON{=}C(NH_2)_2$. XIII. m. 177-8°.[174]

$(PhO)_2PON{=}CHNMe_2$. XVII. m. 92-3.5°.[911]

$(PhO)_2PON{=}C(SEt)NH_2$. V. m. 46°.[174]

$(PhCH_2O)_2PON{=}C(NH_2)H$. XIII. m. 116°.[174]
$(PhO)_2PON{=}C(NH_2)NMe_2$. V. m. 76-7°.[174]
$(PhO)_2PON{=}C(NH_2)NHEt$. V. m. 74°.[174]
$(PhCH_2O)_2PON{=}C(NH_2)_2$. XIII. m. 123°,[174] m. 166.5-7.5°.[241]
$(i\text{-}BuO)_2PON{=}C(NH_2)NHPh$. XIII. m. 108-9°.[174]
$(EtO)_2PON{=}C(NHPh)NEt_2$. XVII. $b_{0.2}$ 140°.[223]
$(C_7H_{15}O)_2PO(N{=}CCl_2)$. XVII. b_1 160-2°, n 1.4620, d 1.0788.[12]
C_{16} $(p\text{-}O_2NC_6H_4CH_2O)_2PON{=}C(SMe)NH_2$. V. m. 128-9°.[174]
$(PhO)_2PON{=}C(NH_2)N(Me)CH_2COOH$. XVII. m. 140-1°.[175]
$(PhCH_2O)_2PON{=}C(NH_2)Me$. XIII. m. 59°.[174]
$(PhCH_2O)_2PON{=}C(NH_2)NHMe$. XIII. m. 63°.[174]
$(PhCH_2O)_2PON{=}C(SMe)NH_2$. XIII. m. 89°.[174]
$(PhCH_2O)_2PON{=}C(OMe)NH_2$. XIII. m. 57-8°.[174]
$(i\text{-}PrO)_2PON{=}C(OPr\text{-}i)C_6H_3(NO_2)_2\text{-}3,5$. IX. m. 166-8°.[221]
C_{17} $(PhCH_2O)_2PON{=}C(NH_2)NHCH_2COOH$. V. m. 103-4°.[175]
$(EtO)_2PON{=}C(Ph)NHPh$. XIII. m. 119-20°.[217]
$(PhCH_2O)_2PON{=}C(SEt)NH_2$. XVII. m. 46°.[174]
$(PhCH_2O)_2PON{=}C(NH_2)NMe_2$. XIII. m. 65°.[174]
$(EtO)_2PON{=}C(NHPh)_2$. XVII. m. 64-6°.[223]
$(C_8H_{17}O)_2PO(N{=}CCl_2)$. XVII. b_1 182-3°, n 1.4607, d 1.0542.[12]
C_{18} $(p\text{-}O_2NC_6H_4CH_2O)_2PON{=}C(NH_2)NHCH_2COOMe$. V. m. 120-1°.[175]
C_{19} $(PhO)_2PON{=}C{=}NC_6H_4Br\text{-}p$. XVII. m. 83-5°.[223]
$(PhO)_2PON{=}C(Ph)Cl$. XVII. m. 74-6°.[190]

$(PhO)_2P(O){-}N{=}C(Ar)Cl$. XVII.[190]

Ar	m.
$p\text{-}ClC_6H_4$	55-7°
$p\text{-}O_2NC_6H_4$	87-9°
$m\text{-}O_2NC_6H_4$	127-9°
$3,5\text{-}(O_2N)_2C_6H_3$	115-8°

$(ArO)_2P(O){-}N{=}C(Ar')Cl$. XVII.[190,434] Ar ≠ Ph

Ar	Ar'	m.
$p\text{-}ClC_6H_4$	Ph	69-71°
$p\text{-}ClC_6H_4$	$p\text{-}O_2NC_6H_4$	150-2°
$p\text{-}ClC_6H_4$	$p\text{-}ClC_6H_4$	109-11°
$p\text{-}O_2NC_6H_4$	Ph	157-60°
$p\text{-}O_2NC_6H_4$	$p\text{-}O_2NC_6H_4$	161-3°
$p\text{-}O_2NC_6H_4$	$p\text{-}ClC_6H_4$	137-9°
$p\text{-}O_2NC_6H_4$	$m\text{-}O_2NC_6H_4$	130-1°
$p\text{-}ClC_6H_4$	$3,5\text{-}(O_2N)_2C_6H_3$	128-30°
$p\text{-}O_2NC_6H_4$	$3,5\text{-}(O_2N)_2C_6H_3$	55-7°

$(PhO)_2PON{=}C(NHPh)Cl$. XVII. m. 96-8°.[198]
$(PhO)_2PON{=}C(NHC_6H_4Cl)Cl$. XVII. m. 126-7°.[198]

$(PhO)_2PON{=}C(NHC_6H_4Br)Cl$. XVII. m. 129-32°.[198]
$(PhO)_2PON{=}C(NHC_6H_3Cl_2\text{-}2,4)Cl$. XVII. m. 95-8°.[198]
$(PhO)_2PON{=}C(Ph)NH_2$. III,[174] XIII,[215] XVII.[214] m. 103°,[174] m. 99-101°.[214]
$(PhO)_2PON{=}C(C_6H_4Cl\text{-}p)NH_2$. XVII. m. 147-9°,[214] m. 132-4°.[221]
$(PhO)_2PON{=}C(C_6H_4NO_2\text{-}p)NH_2$. XVII. m. 168-70°.[214]
$(PhO)_2PON{=}C(C_6H_4NO_2\text{-}m)NH_2$. XVII. m. 128-30°.[214]
$(PhO)_2PON{=}C[C_6H_3(NO_2)_2\text{-}3,5]NH_2$. XVII. m. 182-3°.[214]

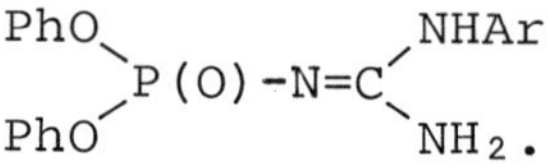
V.[174]

Ar	m.
Ph	139°
$p\text{-}MeC_6H_4$	154°
$o\text{-}ClC_6H_4$	108°
$p\text{-}ClC_6H_4$	164°
$o\text{-}MeOC_6H_4$	131°
$p\text{-}MeOC_6H_4$	127°

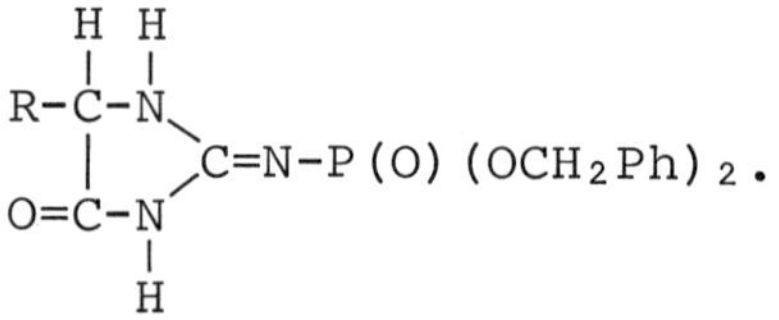
XVII.[276] R=CH_2COOH: m. 161-3°. R=CH_2CH_2COOH: m. 153°

$(PhO)_2PON{=}C(NH_2)NHC_6H_{11}$. V. m. 88°.[174]
$(p\text{-}O_2NC_6H_4CH_2O)_2PON{=}C(NH_2)NHBu$. V. m. 110-1°.[174]
$(EtO)_2PON{=}C(SCH_2Ph)_2$. XVII. $b_{1.5}$ 235-7°, n 1.5912, d 1.2109.[11]
$(i\text{-}PrO)_2PON{=}C(NHPh)_2$. XVII. m. 108°.[223]
$(i\text{-}PrO)_2PON{=}C(SCH_2CH_2NEt_2)_2$. XVII. $b_{0.04}$ 91°.[223]

C_{20} $(PhO)_2PON{=}C(OPh)CCl_3$. IX. n 1.561, d 1.34.[436]
$(PhO)_2PON{=}C(OPh)CF_3$. IX. $b_{0.04}$ 105-7°, n 1.5130.[200]
$(PhO)_2PON{=}C(OMe)Ph$. XVII. m. 144-6°.[221]
$(PhO)_2PON{=}C(OMe)C_6H_4Cl\text{-}p$. XVII. m. 140-2°.[221]
$(PhO)_2PON{=}C(OMe)C_6H_4NO_2\text{-}p$. XVII. m. 80-2°.[221]
$(p\text{-}ClC_6H_4O)_2PON{=}C(OMe)C_6H_4Cl\text{-}p$. XVII. m. 105-6°.[221]
$(p\text{-}O_2NC_6H_4O)_2PON{=}C(OMe)C_6H_3(NO_2)_2\text{-}3,5$. XVII. m. 199-201°.[221]
$(PhO)_2PON{=}C(OPh)NHMe$. IX. m. 86-7°.[226]
$(p\text{-}ClC_6H_4O)_2PON{=}C(OC_6H_4Cl\text{-}p)NHMe$. IX. m. 98-100°.[226]
$(p\text{-}O_2NC_6H_4CH_2O)_2PON{=}C(NH_2)N(Me)CH_2COOEt$. XVII. m. 72-3°.[175]

C_{21} $(PhO)_2PON{=}C(CCl_3)NHCH_2Ph$. XVII. m. 80-2°.[230] Other $(RO)_2PON{=}C(CCl_3)NHR'$ compounds see Ref. 230.
$(p\text{-}ClC_6H_4O)_2PON{=}C(OC_6H_4Cl\text{-}p)NHEt$. IX. m. 88-90°.[226]
$(PhO)_2PON{=}C(OEt)Ph$. XVII. m. 138-40°.[221]
$(PhO)_2PON{=}C(OEt)C_6H_3(NO_2)_2\text{-}3,5$. XVII. m. 111-3°.[221]
$(p\text{-}O_2NC_6H_4CH_2O)_2PON{=}C(NH_2)NHPh$. V. m. 135-6°.[174]
$(PhCH_2O)_2PON{=}C(NH_2)Ph$. XIII. m. 120°.[174]
$(PhO)_2PON{=}C(OPh)NHEt$. IX. m. 98-100°.[226]

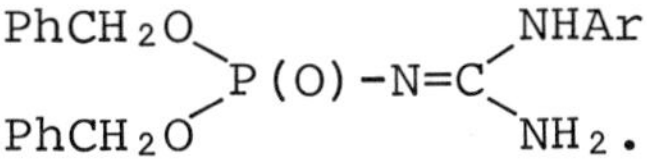

XIII.[174]

Ar	m.
Ph	82°
p-ClC_6H_4	95°
p-MeC_6H_4	90°
p-$EtOC_6H_4$	127°

$(PhCH_2O)_2PON{=}C(NH_2)NHC_6H_{11}$. XIII. m. 72°.[174]

$(EtO)_2PON{=}C(NPh_2)NEt_2$. XVII. $b_{0.1}$ 100-10°.[223]

C23 $(PhO)_2PON{=}C(C_{10}H_7\text{-}2)Cl$. XVII. m. 107-9°.[198]

$(PhO)_2PON{=}C(C_{10}H_7\text{-}1)NH_2$. XVII. m. 119-21°.[214]

$(PhO)_2PON{=}C(NH_2)N(Me)CH_2COOCH_2Ph$. XVII. m. 103°.[175]

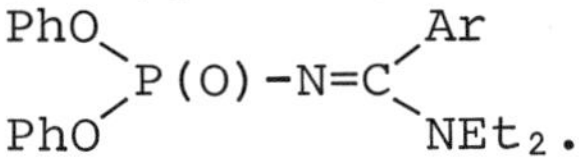

XVII.[214]

Ar	m.
Ph	77-9°
p-ClC_6H_4	69-71°
p-$O_2NC_6H_4$	102-4°
m-$O_2NC_6H_4$	100-2°
3,5-$(O_2N)_2C_6H_3$	99-100°

C24 $(p\text{-}O_2NC_6H_4CH_2O)_2PON{=}C(NH_2)NHCH_2COOCH_2Ph$. V. m. 140°.[175]

C25 $(p\text{-}O_2NC_6H_4O)_2PON{=}CPh(OC_6H_4Cl\text{-}p)$. XVII. m. 180-2°.[434]

$(p\text{-}O_2NC_6H_4O)_2PON{=}C(C_6H_4Cl\text{-}p)OC_6H_4Cl\text{-}p$. XVII. m. 139-41°.[434]

$(p\text{-}O_2NC_6H_4O)_2PON{=}C(C_6H_4NO_2\text{-}p)OC_6H_4Cl\text{-}p$. XVII. m. 217-9°.[434]

$(p\text{-}O_2NC_6H_4O)_2PON{=}C(C_6H_4NO_2\text{-}m)OC_6H_4Cl\text{-}p$. XVII. m. 149-51°.[434]

$(p\text{-}O_2NC_6H_4O)_2PON{=}C[C_6H_3(NO_2)_2\text{-}3,5]OC_6H_4Cl\text{-}p$. XVII. m. 200-2°.[434]

$(p\text{-}ClC_6H_4O)_2PON{=}CPh(OC_6H_4NO_2\text{-}p)$. XVII. m. 142-3°.[434]

$(p\text{-}ClC_6H_4O)_2PON{=}C(C_6H_4Cl\text{-}p)OC_6H_4NO_2\text{-}p$. XVII. m. 151-3°.[434]

$(p\text{-}ClC_6H_4O)_2PON{=}C(C_6H_4NO_2\text{-}p)OC_6H_4NO_2\text{-}p$. XVII. m. 162-4°.[434]

$(p\text{-}ClC_6H_4O)_2PON{=}C(C_6H_4NO_2\text{-}m)OC_6H_4NO_2\text{-}p$. XVII. m. 133-5°.[434]

$(p\text{-}ClC_6H_4O)_2PON{=}C[C_6H_3(NO_2)_2\text{-}3,5]OC_6H_4NO_2\text{-}p$. XVII. m. 209-10°.[434]

$(PhO)_2PON{=}C(Ph)OC_6H_4Cl\text{-}p$. XVII. m. 124-6°.[434]

$(PhO)_2PON{=}C(Ph)OC_6H_4NO_2\text{-}p$. XVII. m. 157-9°.[434]

$(PhO)_2PON{=}C(C_6H_4Cl\text{-}p)OC_6H_4Cl\text{-}p$. XVII. m. 143-5°.[434]

$(PhO)_2PON{=}C(C_6H_4Cl\text{-}p)OC_6H_4NO_2\text{-}p$. XVII. m. 155-7°.[434]

$(PhO)_2PON{=}C(C_6H_4NO_2\text{-}p)OC_6H_4Cl\text{-}p$. XVII. m. 118-20°.[434]

$(PhO)_2PON{=}CPh_2$. XVII. m. 184-5°.[210]

```
ArO         SAr'
   P(O)-N=C
ArO         AR".
```
XVII.[466]

Ar	Ar'	Ar"	m.
Ph	Ph	Ph	104-5°
Ph	p-ClC_6H_4	Ph	122-3°
Ph	p-$O_2NC_6H_4$	Ph	133-4°
p-ClC_6H_4	Ph	Ph	142-3°
p-ClC_6H_4	p-ClC_6H_4	Ph	169-70°
p-ClC_6H_4	p-$O_2NC_6H_4$	Ph	177-8°
p-$O_2NC_6H_4$	Ph	Ph	179-80°
p-$O_2NC_6H_4$	p-ClC_6H_4	Ph	192-3°
Ph	p-ClC_6H_4	p-$O_2NC_6H_4$	166-7°
p-$O_2NC_6H_4$	p-$O_2NC_6H_4$	p-$O_2NC_6H_4$	208-9°

and 42 more compounds with different Ar, Ar', Ar" combinations.

```
ArO         OAr
   P(O)-N=C
ArO         Ar'.
```
IX.[439]

Ar	Ar'	m.
Ph	Ph	100-2°
Ph	1-$C_{10}H_7$	97-8°
Ph	o-ClC_6H_4	119-20°
Ph	2,4-$Cl_2C_6H_3$	99-101°
Ph	m-$O_2NC_6H_4$	122-3°
Ph	p-$O_2NC_6H_4$	139-41°
Ph	3,5-$(O_2N)_2C_6H_3$	105-8°
Ph	2,4-Cl,$(O_2N)C_6H_3$	112-4°
m-MeC_6H_4	p-$O_2NC_6H_4$	113-5°
1-$C_{10}H_7$	Ph	146-8°
2-$C_{10}H_7$	p-ClC_6H_4	115-7°
p-ClC_6H_4	Ph	107-9°
p-ClC_6H_4	o-ClC_6H_4	161-3°
p-ClC_6H_4	2,4-$Cl_2C_6H_3$	107-9°
p-ClC_6H_4	m-$O_2NC_6H_4$	119-21°
p-ClC_6H_4	p-$O_2NC_6H_4$	152-3°
p-ClC_6H_4	3,5-$(O_2N)_2C_6H_3$	151-3°
o-$O_2NC_6H_4$	2,4-$Cl_2C_6H_3$	107-9°
o-$O_2NC_6H_4$	p-$O_2NC_6H_4$	162-3°
p-$O_2NC_6H_4$	p-$O_2NC_6H_4$	220-2°
p-$O_2NC_6H_4$	2,4-$Cl_2C_6H_3$	147-9°
p-$O_2NC_6H_4$	m-$O_2NC_6H_4$	167-9°

```
ArO         NHPh
   P(O)-N=C
ArO         Ar'.
```
IX, XIII.[214]

Ar	Ar'	m.
Ph	Ph	153-5°, 187-8°[218]
Ph	p-ClC$_6$H$_4$	192-3°
Ph	p-O$_2$NC$_6$H$_4$	195-6°
Ph	m-O$_2$NC$_6$H$_4$	204-6°
Ph	3,5-(O$_2$N)$_2$C$_6$H$_3$	164-5°
p-ClC$_6$H$_4$	Ph	181-2°[218]
p-O$_2$NC$_6$H$_4$	Ph	153-5°
p-O$_2$NC$_6$H$_4$	p-ClC$_6$H$_4$	225-8°
p-O$_2$NC$_6$H$_4$	p-O$_2$NC$_6$H$_4$	232-3°
p-O$_2$NC$_6$H$_4$	m-O$_2$NC$_6$H$_4$	172-4°
p-O$_2$NC$_6$H$_4$	3,5-(O$_2$N)$_2$C$_6$H$_3$	210-2°
p-MeC$_6$H$_4$	Ph	167-8°[218]

$$\begin{array}{c} ArO \\ \diagdown \\ P(O)-N=C \\ \diagup \\ ArO \end{array} \begin{array}{c} OAr' \\ \diagup \\ \diagdown \\ NHAr''. \end{array} \quad IX.$$

Ar	Ar'	Ar"	m. (Lit)
Ph	Ph	Ph	97-8°[229]
Ph	2-C$_{10}$H$_7$	Ph	88-90°[198]
Ph	2,4-Cl$_2$C$_6$H$_3$	Ph	89-90°[198]
p-ClC$_6$H$_4$	p-ClC$_6$H$_4$	Ph	119-20°[229]
p-ClC$_6$H$_4$	p-ClC$_6$H$_4$	p-MeC$_6$H$_4$	118-9°[229]
Ph	Ph	p-MeOC$_6$H$_4$	81-3°[229]

For other aryl-combinations see Refs. 198, 229.

(PhO)$_2$PON=C(NHPh)NHC$_6$H$_4$Br-p. XVII. m. 106-8°.[198]

C$_{26}$ (PhO)$_2$PON=C(Ph)NHCSNHPh. XVII. m. 145-6°.[203] For further (PhO)$_2$PON=C(Ar)NHCSNHAr' compounds see Ref. 203.

(PhO)$_2$PON=C(NHC$_6$H$_4$Me-p)Ph. XVII. m. 184-5°.[466]

(PhO)$_2$PON=C(Ph)NHCH$_2$Ph. XVII. m. 137-8°.[466] For other (ArO)$_2$PON=C(Ar')NHR compounds see Ref. 466.

C$_X$ (1-C$_{10}$H$_7$O)$_2$PON=C(OC$_{10}$H$_7$-1)CCl$_3$. XVII. m. 60-4°.[437]

(PhO)$_2$PON=C(Ph)N=C(Ph)OPh. IX. Oil.[203] For similar compounds see Ref. 203.

(1-C$_{10}$H$_7$O)$_2$PON=C(C$_6$H$_4$NO$_2$-p)OC$_{10}$H$_7$-1. IX. m. 147-8°.[433]

(1-C$_{10}$H$_7$O)$_2$PON=C(OC$_{10}$H$_7$-1)CClPh$_2$. XVII. m. 42-7°.[433]

D.1.5.7.   (RO)$_2$P(O)-NAB, A = Non-Carboxylic Acid Rests
(ArSO$_2$-, etc.), B Any Atom or Ligand; Also
(RO)$_2$P(O)N$_3$

C$_2$ (MeO)$_2$PON$_3$. XIV. b$_{10}$ 79.5-81.0°, n 1.4276, d 1.3013.[404]

C$_4$ (MeO)$_2$PON(Me)SCCl$_3$. XVII. Oil, n 1.495, d 1.356.[668]
For other (RO)$_2$PON(R')SR" compounds see Ref. 668.

(EtO)$_2$PO(NSO). XVII. b$_1$ 77°, n 1.4576.[905]

(EtO)$_2$PON$_3$. XIV.[404,771] b$_{4.5}$ 76-7°, n 1.4260, d 1.1672.[404]

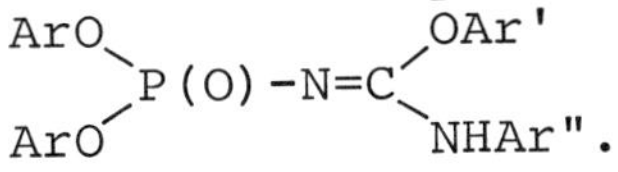

C_5 $(EtO)_2PONHSO_2Me$. m. 96-6.5°.[404] Deuterium substituted N-atom: m. 95-5.5°.[404]

$(MeO)_2PON(Me)SO_2Et$. XVII. Oil.[669] For other $(MeO)_2PON(R)SO_2R'$ compounds see Ref. 669.

$(MeO)_2PON{=}P(OMe)_3$. XVII. $b_{0.2}$ 104-6°, n 1.4430, d 1.3118.[404] The compound may be the N-alkylated isomer.[807]

$(MeO)_2PONMePO(OMe)_2$. XIII, XV. $b_{0.002}$ 78-9°, $b_{0.9}$ 122-5°, n 1.4426, d 1.3153.[807]

$(EtO)_2PON(OH)CF_3$. XIII. b_5 121-2°, n_D^{18} 1.3790, d_{18} 1.2882.[477]

$(EtO)_2PONH(OMe)$. XVII. $b_{0.5}$ 107°, n_D^{25} 1.4275.[893]

C_6 $(MeO)_2PON(OH)C(CF_3)_3$. XIII. m. 68-9°, b_2 94-6°.[477]

$(EtO)_2PON(Cl)CH_2CH_2Cl$. XVII. b_1 99-100°, n 1.4574, d 1.2710.[341]

$(i\text{-}PrO)_2PON_3$. XIV. b_2 62-4°, n 1.4265, d 1.0824.[328]

$(PrO)_2PON_3$. XIV. $b_{0.25}$ 64°, n 1.4367, d 1.1029.[328]

$(PrO)_2PO(NSO)$. XVII. $b_{0.05}$ 71°, n 1.4562.[905]

$(EtO)_2PON(SCl)Et$. XVII. b_2 105-6°, n 1.4718, d 1.2107.[13]

$(EtO)_2PONHSO_2Et$. IR, acidity constant for NH-proton.[31]

C_7 $(EtO)_2PON(Cl)CH_2CH_2CH_2Cl$. XVII. b_2 124-5°, n 1.4565, d 1.2174.[346]

$(MeO)_2PONMePO(OEt)_2$. XIII, XV. m. 34-6.5°.[305] $b_{0.002}$ 85-7°, n 1.4376, d 1.2328.[807]

$(EtO)_2PON{=}P(OMe)_3$. XIII. $b_{0.04}$ 85-8°, n 1.4528, d 1.1711.[305]

C_8 $(MeO)_2P(O)NH\text{-}SO_2\text{-}Ar$. IX.

Ar	m. (Lit)
Ph	106-8°,[451,518] 108-9°[462]
p-ClC_6H_4	128-9°[451,518]
p-$O_2NC_6H_4$	177-9°,[451,518] 175-7°[447]
m-$O_2NC_6H_4$	149-50°,[451,518] 145-8°[447]
o-$O_2NC_6H_4$	134-6°[447]
o-MeC_6H_4	145°[462]
p-MeC_6H_4	110-1°,[462] ^{31}P +3.5[873]
1-$C_{10}H_7$	164-5°[462,464]
2-$C_{10}H_7$	144-5°[462]
o-$O_2NC_6H_4$	Na-salt: m. 156-8°[447]
m-$O_2NC_6H_4$	Na-salt·$2H_2O$: m. 195-8°[447]
p-$O_2NC_6H_4$	Na-salt·$2H_2O$: m. 188-90°[447]
o-MeC_6H_4	Ag-salt: m. 155-7°[791a]
1-$C_{10}H_7$	Ag-salt: m. 155-7°[791a]
2-$C_{10}H_7$	Ag-salt: m. 159-60°[791a]

$(MeO)_2PON{=}N\text{-}NHPh$. XIII. m. 94.5°.[324]

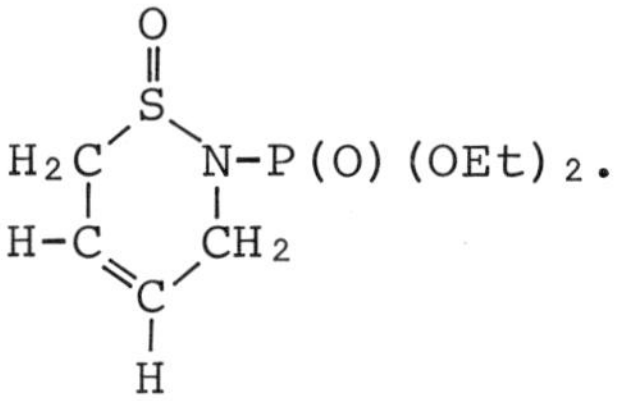
XVII. $b_{0.05}$ 122°, n 1.4945.[905] ^1H.[412]

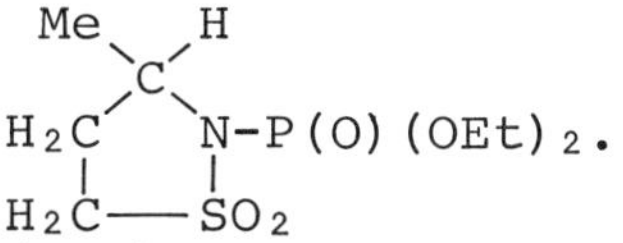
n 1.4665, d_4^{27} 1.2689.[299]

$(BuO)_2PON_3$. XIV. $b_{0.5}$ 86-7°, n 1.4372, d 1.0713.[328]
$(BuO)_2PO(NSO)$. XVII. $b_{0.1}$ 84°, n 1.4552.[905]
$(i\text{-}PrO)_2PON(Cl)CH_2CH_2Cl$. XVII. b_2 101-2°, n 1.4513, d 1.1867.[341]
$(EtO)_2PON(Et)PS(OCH_2)_2$. XIII. m. 52-3°, $b_{1.5}$ 144-5°.[11]
$(EtO)_2PON(Et)PO(OCH_2)_2$. XIII. b_1 150-2°, m. 64°.[11]

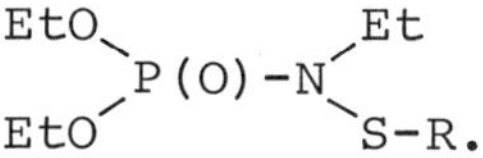
XVII.[13,15]

R	b.	n	d
$-(NC_2H_4)$	b_1 100-2°	1.4736	1.1426
$-NMe_2$	$b_{0.5}$ 96.5-8.0°	1.455	1.0894
$-NEt_2$	$b_{0.5}$ 103-4°	1.458	1.0618
-NHPh	[m. 62-3°]		
$-N(CH_2CH=CH_2)_2$	b_2 108-9°	1.4776	1.0646
-OEt	b_1 85-6°	1.4494	1.1072
-OPr-i	b_1 90-1°	1.4469	1.079
-OBu-i	b_1 98-9°	1.4490	1.0701
$-OCH_2CH_2Cl$	b_1 105°	1.4670	1.1914
$-OCH_2CH_2Br$	b_2 126-8°	1.4789	1.3413
$-OCH_2CH=CH_2$	b_2 116-7°	1.7640	1.1524
$-OCH_2CH_2NEt_2$	$b_{1.5}$ 115-6°	1.4600	1.0678

$(EtO)_2PONHPS(OEt)_2$. XVII. $b_{0.0001}$ 78-80°, n 1.4728, d 1.2055.[328]
$(MeO)_2PON=P(OEt)_3$. XIII. $b_{0.002}$ 86-7°, n 1.4368, d 1.1867.[807]
$(EtO)_2PONHPO(OEt)_2$. XVII. $b_{0.001}$ 79-80°, n 1.4415, d 1.2029.[404] See also Ref. 328. Deuterium substituted N-atom: $b_{0.001}$ 77-8°, n 1.4412.[404]
$(MeO)_2PON(Me)SO_2Ph$. XIII,[808] XV.[802] m. 37-8°, b_3 174-5°.
$(MeO)_2PON(Me)SO_2C_6H_4Cl\text{-}p$. XIII,[808] XV.[802] m. 61-2°, b_1 179-81°.
$(MeO)_2PON(Me)SO_2C_6H_4NO_2\text{-}p$. XIII,[808] XV.[802] m. 133-5°.
$(EtO)_2PON(Me)PO(OEt)(OCH=CCl_2)$. XIII. b_1 152-3°, n 1.4606, d 1.3077.[45]
$(EtO)_2PON(Me)PS(OEt)_2$. V. b_1 120°, n 1.4656, d 1.0420.[46]

$(EtO)_2PON(Me)PO(OEt)_2$. V. $b_{0.3}$ 105-6°,[404] b_3 138-9°,[46] $b_{1.3}$ 147-53°.[165] n 1.4372,[404] n 1.4353,[46] d 1.1665.[404,46]

$(MeO)_2PON(Me)SnEt_3$. XIII. $b_{0.1}$ 72-4°.[533] ^1H.[533]

$(EtO)_2PON(Me)PO(OEt)NMe_2$. V. $b_{1.5}$ 139-41°, n 1.4453, d 1.3831.[46]

$(EtO)_2PON(Me)PO(NMe_2)_2$. V. b_3 147-9°, n 1.4621, d 1.0452.[46]

C10 $(EtO)_2P(O)NH-SO_2-Ar$. V,[710] IX.[462]

Ar	m. (Lit)
Ph	111-2°[462,710]
$p-ClC_6H_4$	141-2°[710]
$p-O_2NC_6H_4$	139°[710]
$p-H_2NC_6H_4$	157.5-8.5°[710]
$o-MeC_6H_4$	97-8°[462]
$p-MeC_6H_4$	105-6°,[462] 104°[710]
$1-C_{10}H_7$	154-5°[462]
$2-C_{10}H_7$	161-2°[462]

$(EtO)_2PON=N-NHPh$. XIII. m. 85°.[324]

$(MeO)_2PON(Me)SO_2C_6H_4Me$-p. XIII. m. 57-8°.[808]

$(EtO)_2PON(Et)PO(OCH_2CH_2Cl)(OCH=CCl_2)$. XIII. b_2 170-2°.[1

$[(EtO)_2PO]_2NCH_2CH_2Cl$. XIII. b_1 133-5°, n 1.4467, d 1.2114.[341]

$(i-BuO)_2PON(Cl)CH_2CH_2Cl$. XVII. b_3 132-3°, n 1.4539, d 1.1416.[341]

$(EtO)_2PON=P(OEt)_2(NHCOMe)$. XIII. $b_{0.00005}$ 93-5°, n 1.4602, d 1.1862.[398]

$(EtO)_2PON(Et)SPO(OEt)_2$. XIII. b_2 134-6°, n 1.4582, d 1.1801.[15]

$(EtO)_2PON=P(OEt)_3$. XIII. $b_{0.03}$ 79-83°,[305] $b_{0.5}$ 121°,[404 $b_{0.1}$ 98°,[404] $b_{0.003}$ 87-8°,[807] $b_{0.05}$ 95°.[906] n 1.4459,[305] n 1.4346,[807] n 1.4372,[404] n 1.4369.[906] d 1.0744,[305] d 1.1309,[807] d 1.129.[404]

$(EtO)_2PON(Et)PO(OEt)_2$. V. b_1 123-4°,[46] b_2 125-50°,[184] $b_{0.001}$ 100-2°.[11] n 1.4338,[184] n 1.4342,[11] n 1.4359.[4 d 1.1466.[46,11]

$(EtO)_2PON(Me)SSN(Me)PO(OEt)_2$. XVII. b_2 161-3°, n 1.460(d 1.1908.[46]

$(EtO)_2PON(Et)PO(NMe_2)_2$. V. b_2 138-9°, n 1.4609, d 1.1345.[46]

C11 $(EtO)_2PON(Me)SO_2C_6H_4I$-p. V. m. 77°.[32]

$(EtO)_2PON(Me)SO_2Ph$. XIII. $b_{0.8}$ 158-60°.[808]

C12 $(PhO)_2PON=PCl_3$. I.[469] Oil. ^{31}P α,β= 11.2.[634]

$(PhO)_2P[α]ONHP[β]O(OH)_2$. VIII. m. 170-2°.[469]
^{31}P [α] +8.1, [β] +2.5.[633]
^{31}P K-salt: [α] +5.6, [β] +4.0.[633]
^{31}P K_2-salt: [α] +3.3, [β] +2.1.[634]
^{31}P K_3-salt [α] 5.7, [β] -2.2.[633]

^{31}P see also 632.
$(CH_2{=}CBrCH_2O)_2PONHSO_2Ph$. XVII. m. 42-4°.[801]
$(CH_2{=}CBrCH_2O)_2PONHSO_2C_6H_4Cl$-p. XVII. m. 80-3°.[801]
$(CH_2{=}CBrCH_2O)_2PONHSO_2C_6H_4NO_2$-p. XVII. m. 78-80°.[801]
$(CH_2{=}CHCH_2O)_2PONHSO_2Ph$. XI. m. 48-50°.[803]
$(CH_2{=}CHCH_2O)_2PONHSO_2C_6H_4Cl$-p. XI. m. 71-2°.[803]
$(CH_2{=}CHCH_2O)_2PONHSO_2C_6H_4NO_2$-o. XI. m. 72-3°.[803]
m-isomer: m. 89-90°.
p-isomer: m. 97-8°.
$(BrCH_2CHBrCH_2O)_2PONHSO_2Ph$. XVII. m. 69-72°.[801]
$(BrCH_2CHBrCH_2O)_2PONHSO_2C_6H_4Cl$-p. XVII. m. 107-10°.[801]
$(BrCH_2CHBrCH_2O)_2PONHSO_2C_6H_4NO_2$-p. XVII. m. 123-8°.[801]
$(EtO)_2PON(Et)SO_2Ph$. XV. $b_{0.05}$ 150-3°.[802]
$(EtO)_2PON(Et)SO_2C_6H_4NO_2$-p. XV. m. 75-6°.[802]
$(i\text{-}PrO)_2PON{=}N{-}NHPh$. XIII. m. 70°.[324]
$(EtO)_2PON(Me)SO_2C_6H_4Me$-p. XIII. $b_{0.6}$ 168-70°.[808]
p-$C_6H_4[SO_2NHPO(OMe)_2]_2$. XI. m. 206-7°.[448]
$(AmO)_2PON(Cl)CH_2CH_2Cl$. XVII. b_2 149-50°, n 1.4586, d 1.1144.[341]
$(EtO)_2PONHPO(OBu)_2$. XI. $b_{0.0001}$ 118-20°, n 1.4445, d 1.1171.[328] For other $(RO)_2PONHPO(OR')_2$ compounds see Ref. 328.
$[(EtO)_2PO]_3N$. V.[188,165] n 1.4368.[188]

C_{13} $(PhO)_2PONMeCl$. XVII. n 1.5557, d 1.2894.[671] Other $(RO)_2PONR'Cl$ compounds (n, d) see Ref. 671.

$(PhO)_2P(O)NH{-}SO_2{-}R$. XI.[444]

R	m.
Me	96-8°
Et	117-9°
i-Pr	140-2°
Bu	109-10°
$PhCH_2$	147-8°

$(MeO)_2PON(Me)SO_2C_{10}H_7$-2. XIII, XV. m. 82-3°.[802,808]
(p-$ClC_6H_4O)_2PONHSO_2Me$. XI. m. 58-60°.[444]
$(CH_2{=}CHCH_2O)_2PONHSO_2C_6H_4Me$-p. XI. m. 63-5°.[803]
$(BrCH_2CHBrCH_2O)_2PONHSO_2C_6H_4Me$-p. XVII. m. 77-9°.[801]
$(EtO)_2PON(Et)SO_2C_6H_4Me$-p. XV. b_1 172-5°.[802]
$(i\text{-}PrO)_2PON(Me)PS(OPr\text{-}i)_2$. XIII. b_1 126-7.5°,[45] n 1.4535, d 1.078.[26]

$(RO)_2P(O){-}N(R'){-}P(O)(OR)_2$. V.[26]

R	R'	b.	n	d
Pr	Me	b_2 159-60°	1.4365	1.0867
i-Pr	Me	b_2 128-9°	1.4292	1.0686
Pr	Et	b_1 160-1°	1.4385	1.0746
i-Pr	Et	b_1 129-30°	1.4288	1.0612

R	R'	b.	n	d
i-Bu	Me	b_2 162-3°	1.4375	1.0335
Bu	Et	b_1 173-4°	1.4411	1.0398
i-Bu	Et	b_1 157-8°	1.4370	1.0264

C_{14}(p-$O_2NC_6H_4O$)$_2$PONHSO$_2$Et. XI. m. 166-7°.[444]

(MeO)$_2$PONHPS(OPh)$_2$. V. m. 115-7°.[469]

(PhO)$_2$P[α]ONHP[β]O(OEt)OH. ^{31}P [α] +8.6, [β] +1.1.[634]

(PhO)$_2$PONHSO$_2$NMe$_2$. IX. m. 143-4°.[460]

(o-$O_2NC_6H_4O$)$_2$PONHSO$_2$NMe$_2$. IX. m. 163-4°.[460]

(p-ClC$_6$H$_4$O)$_2$PONHSO$_2$NMe$_2$. IX. m. 167-8°.[460]

(o-ClC$_6$H$_4$O)$_2$PONHSO$_2$NMe$_2$. IX. m. 137-8°.[460]

(p-ClC$_6$H$_4$O)$_2$PONHSO$_2$Et. XI. m. 133-5°.[444]

(BuO)$_2$PONHSO$_2$Ph. XI. m. 145-7°.[462] Na-salt: m. 146-7°.[804]

(EtO)$_2$PON(Ph)PS(OEt)$_2$. V. Oil. n 1.5025.[189]

(EtO)$_2$PON=P(OPh)(OEt)$_2$. XIII. $b_{0.015}$ 134-9°, n 1.4804, d 1.1745.[810]

[(EtO)$_2$PO]$_2$NSO$_2$Ph. V.[710]

p-C$_6$H$_4$[SO$_2$NHPO(OEt)$_2$]$_2$. XI. m. 198-200°.[448]

(C$_6$H$_{13}$O)$_2$PON(Cl)CH$_2$CH$_2$Cl. XVII. b_2 160-2°, n 1.4590, d 1.0827.[341]

(EtO)$_2$PON(CH$_2$COOEt)PS(OPr-i)$_2$. XIII. b_1 152-4°, n 1.4570, d 1.1434.[18]

(PrO)$_2$PON(Et)PS(OPr)$_2$. XIII. b_1 142-4°, n 1.4570, d 1.7044.[26,45]

(BuO)$_2$PON=P(OEt)$_3$. XIII. $b_{0.005}$ 104°,[906] $b_{0.0001}$ 80-1°.[328] n 1.4386,[906] n 1.4426.[328] d 1.0738.[328] For 14 other (RO)$_2$PON=P(OR')$_3$ compounds see Ref. 328.

C_{15} (BuO)$_2$PONHSO$_2$C$_6$H$_4$Me-o. XI. m. 153-7°.[462]

C_{16} (CH$_2$=CBrCH$_2$O)$_2$PONHSO$_2$C$_{10}$H$_7$-1. XVII. m. 105-7°.[801] The 2-C$_{10}$H$_7$ isomer: m. 78-81°.[801]

(BrCH$_2$CHBrCH$_2$O)$_2$PONHSO$_2$C$_{10}$H$_7$-1. XVII. m. 90-2°.[801] The 2-C$_{10}$H$_7$ isomer: m. 124-5°.[801]

(CH$_2$=CHCH$_2$O)$_2$PONHSO$_2$C$_{10}$H$_7$-1. XI. m. 138-9°.[803] The 2-C$_{10}$H$_7$ isomer: m. 89-91°.[803]

(EtO)$_2$PONHPS(OPh)$_2$. V. m. 67-9°.[469]

(PhO)$_2$P[α]ONHP[β]O(OEt)$_2$. ^{31}P [α] +9.2, [β] +1.7.[873]

(PhO)$_2$PONHSO$_2$NEt$_2$. IX. m. 122-3°.[460]

(o-$O_2NC_6H_4O$)$_2$PONHSO$_2$NEt$_2$. IX. m. 142-3°.[460]

(p-ClC$_6$H$_4$O)$_2$PONHSO$_2$NEt$_2$. IX. m. 110-2°.[460]

(o-ClC$_6$H$_4$O)$_2$PONHSO$_2$NEt$_2$. IX. m. 148-50°.[460]

(EtO)$_2$PON(Et)SO$_2$C$_{10}$H$_7$-1. XV. m. 77-9°.[802]

(EtO)$_2$PON=P(OBu)$_3$. XIII. $b_{0.01}$ 113°, n 1.4398.[906]

C_{18} (o-ClC$_6$H$_4$O)$_2$PONHSO$_2$Ph. XI. m. 149-51°.[453]

p-ClC$_6$H$_4$O\
        P(O)NH-SO$_2$-Ar. IX, XI.\
p-ClC$_6$H$_4$O/

	Ar	m. (Lit)
	Ph	155-7°[453]
	p-$O_2NC_6H_4$	163-5°[447]
	m-$O_2NC_6H_4$	153-5°[447]

Ar	m. (Lit)
$o\text{-}O_2NC_6H_4$	159-61°[447]
$p\text{-}MeC_6H_4$	165-7°[453]
$o\text{-}MeC_6H_4$	139-40°[453]

$(o\text{-}O_2NC_6H_4O)_2P(O)NH\text{-}SO_2\text{-}Ar$. IX, XI.

Ar	m. (Lit)
Ph	150-1°[462]
$o\text{-}O_2NC_6H_4$	171-4°[447]
$m\text{-}O_2NC_6H_4$	175-8°[447]
$p\text{-}O_2NC_6H_4$	167-70°[447]
$o\text{-}MeC_6H_4$	161-2°[462,453]
$1\text{-}C_{10}H_7$	172-3°[462]
$p\text{-}MeC_6H_4$	163-4°[453]

$(p\text{-}O_2NC_6H_4O)_2P(O)NH\text{-}SO_2\text{-}Ar$. IX, XI.

Ar	m. (Lit)
Ph	199-200°[453]
$p\text{-}O_2NC_6H_4$	165-8°[447]
$o\text{-}O_2NC_6H_4$	188-91°[447]
$m\text{-}O_2NC_6H_4$	191-4°[447]
$o\text{-}MeC_6H_4$	189-92°[453]
$p\text{-}MeC_6H_4$	173-4°[453]

$(PhO)_2P(O)NH\text{-}SO_2\text{-}Ar$. IX, XI.

Ar	m. (Lit)
Ph	145°[462,256]
$p\text{-}BrC_6H_4$	170°[256]
$p\text{-}ClC_6H_4$	168°[256]
$p\text{-}O_2NC_6H_4$	192-3°[447]
$m\text{-}O_2NC_6H_4$	146-7°[447]
$o\text{-}O_2NC_6H_4$	156-8°[447]
$p\text{-}MeC_6H_4$	188-9°,[256] 186-7°[140]
$o\text{-}MeC_6H_4$	160-1°[462]
$2,4\text{-}Me_2C_6H_3$	155-6°[256]
$2,5\text{-}Cl_2C_6H_3$	126-7°[256]
$1\text{-}C_{10}H_7$	137-8°[462]
$2\text{-}C_{10}H_7$	178-9°[462]
$p\text{-}H_2NC_6H_4$	221°[772]

$(PhO)_2PON(NO)Ph$. XVII.[134]

$(PhO)_2PONHPO(OPh)OH$. ^{31}P (both P-atoms) +9.7.[634]

$(PhO)_2P[\alpha]ONHP[\beta]O(OPh)NH_2$. XVII. m. 119-20°.[631] ^{31}P [α] +8.4, [β] −3.8.[633]

$(EtO)_2PON{=}P(OPh)_2(OEt)$. XIII. $b_{0.01}$ 155°, n 1.5170, d 1.2142.[810]

$(BuO)_2PONHSO_2C_{10}H_7$-1. XI. m. 187-9°, m. 75-7°.[462] May be the Na-salt. 2-$C_{10}H_7$-isomer: m. 175-80°.[462]

$(C_6H_{13}O)_2PONHSO_2Ph$. XI. Na-salt: m. 112-3°.[804] For Na-salts of the analogous C_7H_{15}, C_8H_{17}, C_9H_{19}, $C_{10}H_{21}$ compounds see Ref. 804.

$(PrO)_2PON(Ph)PO(OPr)_2$. V. n 1.4770.[184]

$(p\text{-}O_2NC_6H_4O)_2P(O){-}N{=}S(O)(R){-}OC_6H_4NO_2\text{-}p$. IX.[523]

R	m.
Me	121-2°
Et	97-9°
Pr	114-5°
Bu	86-7°
i-Bu	115-6°
Am	97-9°
i-Am	90-1°
C_6H_{13}	88-90°
$PhCH_2CH_2$	97-8°

C_{19} $(o\text{-}ClC_6H_4O)_2PONHSO_2C_6H_4Me$-o. XI. m. 130-2°.[453]

$(o\text{-}ClC_6H_4O)_2PONHSO_2C_6H_4Me$-p. XI. m. 172-4°.[453]

$(p\text{-}ClC_6H_4O)_2PONHSO_2CH_2Ph$. XI. m. 149-51°.[444]

C_{20} $(MeC_6H_4O)_2P(O)NH{-}SO_2{-}Ar$. XI.[462]

MeC_6H_4	Ar	m.
ortho	Ph	136-7°
ortho	1-$C_{10}H_7$	134-6°
meta	Ph	107-9°
para	Ph	172-3°
para	o-MeC_6H_4	116-7°
para	1-$C_{10}H_7$	168-9°
para	2-$C_{10}H_7$	161-2°

$[(BuO)_2PO]_2N[PO(OEt)_2]$. V. n 1.4402.[188]

C_{22} $(1\text{-}C_{10}H_7O)_2PONHSO_2NMe_2$. IX. m. 190-1°.[460]

C_{24} $(p\text{-}O_2NC_6H_4O)_2PON{=}S(OC_6H_4NO_2\text{-}p)_2$. IX. m. 166-7°.[521]

$(PhO)_2PON{=}SO(OPh)Ph$. IX. m. 101-2°, for similar compounds see Ref. 520.

$(PhO)_2PONHPS(OPh)_2$. IX. m. 100-2°.[469]

$(p\text{-}O_2NC_6H_4O)_2PONHPS(OPh)_2$. IX. m. 174-6°.[469]

$(1\text{-}C_{10}H_7O)_2PONHSO_2NEt_2$. IX. m. 171-2°.[460]

$(PhO)_2PONHPO(OPh)_2$. XVII. m. 110-2°,[469] m. 111-2°.[631] ^{31}P +10.7,[631,633] ^{31}P +11.3.[634]

$(PhO)_2PONHSO_2NHPO(OPh)_2$. XVII. Dec. 188-9°, Na-, K-, NH_4^+-salt.[457]

$(p\text{-}ClC_6H_4O)_2PONHSO_2NHPO(OC_6H_4Cl\text{-}p)_2$. XVII. Dec. 183-5°.[457]

$(p\text{-}O_2NC_6H_4O)_2PONHSO_2NHPO(OC_6H_4NO_2\text{-}p)_2$. XVII. Dec. 218-21°.[457]

$(PhO)_2PONHPO(NHPh)_2$. IV. m. 197-8°.[469]

$(PhO)_2P[\alpha]ONHP[\beta]O(OPh)NHP[\gamma]O(OPh)NH_2$. XV. m 158-61°.[631] ^{31}P [α] +8.4, [β] +6,3, [γ] -0.8.[633]

25 $(PhO)_2PON(Me)PO(OPh)_2$. V,[184,165] XVII.[631] m. 92-2.5,[631] m. 68°[184](?), ^{1}H, ^{31}P +7.3.[631]

x $[(i\text{-}PrO)_2PO]_2NCH_2CH_2N[PO(OPr\text{-}i)_2]_2$. V. n 1.4232.[186]

$(PhO)_2PON(Bu)PS(OPh)_2$. V. n 1.5662.[189]

$(o\text{-}MeC_6H_4O)_2PONHPO(OC_6H_4\text{-}o)_2$. ^{31}P +11.0.[634] p-isomer: ^{31}P +11.0.[634]

$(o\text{-}MeC_6H_4O)_2PONHSO_2NHPO(OC_6H_4Me\text{-}o)_2$. XVII. dec. 173-6°.[457] meta-isomer: dec. 170-2°.[457] para-isomer: dec. 184-5°.[457]

$(p\text{-}O_2NC_6H_4O)_2PON{=}P(OC_6H_4NO_2\text{-}p)_3$. XI. m. 205-6°.[521]

$(PhO)_2PON(Ph)PO(OPh)_2$. ^{31}P +11.6.[634]

$(PhO)_2PON{=}P(OPh)_3$. XVII. m. 74-4.5°.[667]

$(PhO)_2PON(Ph)PO(OPh)(NHPh)$. VII. m. 185°.[574]

$p\text{-}C_6H_4[SO_2NHPO(OPh)_2]_2$. XI. m. 125-8°.[448]

$(PhO)_2P[\alpha]ONHP[\beta]O(OPh)NHP[\alpha]O(OPh)_2$. XV. m. 126-7°.[631] ^{1}H.[631] ^{31}P [α] +12.5, [β] +5.9.[633]

$(1\text{-}C_{10}H_7O)_2PONHSO_2NHPO(OC_{10}H_7\text{-}1)_2$. XVII. dec. 213-5°.[457] $2\text{-}C_{10}H_7$-isomer: dec. 195-8°.[457]

$[(PhO)_2PO]_2N(CH_2)_6N[PO(OPh)_2]_2$. V. m. 79-82°.[186]

D.1.6. $(RO)_2P(S)\text{-}N\lessdot$, $(RS)_2P(O)\text{-}N\lessdot$, $(RS)_2P(S)\text{-}N\lessdot$, R = Alkyl or Aryl or Acyl

D.1.6.1. $(RO)_2P(S)\text{-}NH_2$, $(RO)_2P(S)\text{-}NHAc$ Carboxylic Acid Rest, $(RO)_2P(S)\text{-}N{=}C\lessdot$; also $(RO)_2P(S)N_3$

2 $(MeO)_2PSN_3$. XIV. $b_{0.5}$ 38-9°.[756]

$(MeO)_2PSNH_2$. V. b_{10} 105-8°, n 1.4982, d 1.2649.[568]

3 $(MeO)_2PSN{=}CCl_2$. XVII.[284]?

$(MeO)_2PS(NCS)$. XVII. b_2 65°.[759]

$(MeO)_2PS(NCO)$. XVII. b_8 58-9°, n 1.4718, d 1.2706.[741]

$(MeO)_2PSNHCSNH_2$. XII. m. 82°.[760]

4 $(EtO)_2PSN_3$. XIV,[756,328] XIII.[753] b_7 80°,[328] $b_{1.5}$ 60°,[756] $b_{0.5}$ 55°,[753] d 1.1647,[328] n 1.4758.[328]

$(EtO)_2PSNH_2$. V. b_9 112-5°,[568] b_{20} 132°,[137] n 1.4828,[568] n_D^{26} 1.4842,[137] d 1.1456.[568]

5 $N_3PS(OCH_2)_2C(CH_2O)_2PSN_3$. XIV. m. 177-8°.[109]

$(EtO)_2PS(NCS)$. XIII,[587] XIV,[524,587] XVII.[759] $b_{0.7}$ 75-6°,[587] b_2 82-3°,[759] b_{10} 108-10°,[524] n_D^{25} 1.5231,[587] ^{31}P -45.5.[873,361a]

$(MeO)_2PS[N{=}C(SCH_2)_2]$. V. $b_{0.001}$ 100-5°, n_D^{25} 1.5683.[535] 27 similar compounds.

$(EtO)_2PS(NCO)$. XVII. b_{20} 78-9°, n 1.4598, d 1.1606.[740]

$(EtO)_2PSNHCSNH_2$. XII. m. 89-90°,[524] m. 91°.[760]

6 $(EtO)_2PSNHCOCCl_3$. IX. n_D^{25} 1.500.[870]

$(EtO)_2PSNHCOOMe$. XII. n 1.4790, d 1.2032.[740]

$$\begin{array}{c} \text{EtO} \\ \hphantom{x} \\ \text{EtO} \end{array} \!\!\!\! \overset{\displaystyle \overset{\text{S H S H}}{\underset{||\ \ |\ \ ||\ \ |}{}}}{\text{P-N-C-N-R.}}$$

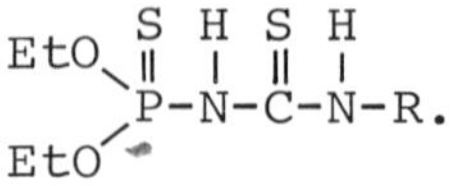

XII.   R=Me: m. $94°$.[760]
R=Et: m. $73°$.[760]
R=Ph: m. $98°$,[760] m. $95°$.[524]
R=C_6H_4OMe-p: m. $109°$.[760]
R=2-pyridyl: m. $108°$.[760]
R=NH_2: m. $109°$,[760] semicarbazide derivative.

$(PrO)_2PSN_3$.   XIV. b_1 77-8°.[756]
$(PrO)_2PSNH_2$.   V. $b_{0.3}$ 82°, n 1.4798, d 1.0883.[568]
$(i-PrO)_2PSNH_2$.   V. $b_{0.025}$ 82°, n 1.4670, d 1.0563.[568]
C_7 $(PrO)_2PS(NCS)$.   XIII, XIV,[587] XVII.[759] $b_{1.8}$ 99-102°,[587] b_2 96°.[759]
$(i-PrO)_2PS(NCS)$.   XVII. b_2 82°.[759]
$(PrO)_2PSNHCSNH_2$.   XII.   m. 77°.[760]
$(i-PrO)_2PSNHCSNH_2$.   XII.   m. 112°.[760]
C_8 $(EtO)_2PS[N=C(SCH_2)_2CH_2]$.   V. $b_{0.003}$ 100°, n_D^{25} 1.600.[9]
$(EtO)_2PSN=\overline{C-SCHMe}CH_2S$.   V. $b_{0.001}$ 114-6°, n_D^{25} 1.5814.[9]
$(BuO)_2PSNH_2$.   V. $b_{0.1}$ 93-6°, n 1.4715, d 1.0407.[568]
C_9 $(BuO)_2PS(NCS)$.   XIII, XIV. $b_{0.5}$ 109°.[587]
$(EtO)_2PSNHCSNEtCH_2CH_2OH$.   XII.   m. 59°, other thiourea compounds, n_D^{20} given.[691]
$(MeO)_2PSNHCSNHPh$.   XII.   m. 93-4°,[524] m. 96°.[760]
C_{10} $(MeO)_2PSNHCSNHC_6H_4OMe$-p.   XII.   m. 104°.[760]
$(MeO)_2PS[N(CO)_2C_6H_4-o]$.   V, XVII.   m. 126.5-8.0°, other alkylesters.[659]
$(MeO)_2PSN=C(CH=CH)_2C=NPS(OMe)_2$.   XVII.   m. 127-8°.[257]
C_{11} $(EtO)_2PSNHCOPh$.   V, XVII.   m. 113°.[33]
$(EtO)_2PSNHCONHPh$.   XII.   m. 121-2°.[740]
$(EtO)_2PSNHCOC_6H_4Cl$-p.   V.   m. 104°.[34]
C_{12} $(EtO)_2PS[N(CO)_2C_6Cl_4-o]$.   IX.   m. 162-5°.[870]
$(p-ClC_6H_4O)_2PSNH_2$.   V.   m. 96°,[75] m. 92-4°.[830]
$(o-BrC_6H_4O)_2PSNH_2$.   V.   m. 92-3°.[830]
$(p-IC_6H_4O)_2PSNH_2$.   V.   m. 156-7°.[830]
$(PhO)_2PSNH_2$.   V.   m. 115°,[75,272] m. 112°.[272]
$(EtO)_2PSNHCOC_6H_4Me$-p.   V.   m. 90°.[34]
C_{13} $(PhO)_2PS(NCS)$.   XIV.   m. 30-2°.[524]
$(PhO)_2PS(NCO)$.   XVII. $b_{0.08}$ 128°,[740,741] m. 50-2°.[740,7]
$(p-ClC_6H_4O)_2PSN=CHNH_2$.   V.   m. 86-90°.[288]
$(p-BrC_6H_4O)_2PSN=CHNH_2$.   V.   m. 89-91°.[288]
$(PhO)_2PSNHCSNH_2$.   XII.   m. 136-7°.[524]
$(PrO)_2PSNHCOC_6H_4Cl$-p.   V.   m. 77°.[34]
$(i-PrO)_2PSNHCOC_6H_4Cl$-p.   V.   m. 118°.[34]
$(PrO)_2PSNHCOPh$.   V, XVII.   m. 90.5°.[33]
$(i-PrO)_2PSNHCOPh$.   V, XVII.   m. 99°.[33]
C_{14} $(PrO)_2PSNHCOC_6H_4Me$-p.   V.   m. 66°.[34]
$(i-PrO)_2PSNHCOC_6H_4Me$-p.   V.   m. 84°.[34]
$(p-ClC_6H_4S)_2PONHCOCHCl_2$.   IX.   m. 160-2°.[231]
$(PhO)_2PSNHC(=NH)NHCN$.   V.   m. 168°.[100]
$(PhO)_2PSNHC(=NH)NHCONH_2$.   V.   m. 218°.[100]

(PhO)$_2$PSNHCOOMe. XII. m. 87-8°.[740]
(p-ClC$_6$H$_4$O)$_2$PSNHC(=NH)NHCN. V. m. 172°.[100]
(PhO)$_2$PSNHC(=NH)NHCSNH$_2$. V.[100]
(PhO)$_2$PSNHCS(OMe). XII. m. 75-7°.[524]
(p-MeC$_6$H$_4$O)$_2$PSNH$_2$. V. m. 131°,[75] m. 134-7°.[830]
(p-MeOC$_6$H$_4$O)$_2$PSNH$_2$. II, IX. m. 136-8°.[255]
(o-MeOC$_6$H$_4$O)$_2$PSNH$_2$. II, IX. m. 135-8°.[255]
(PhO)$_2$PSNHCOCCl$_3$. IX. d_4^{25} 1.4801.[870]
(p-ClC$_6$H$_4$S)$_2$PONHCOCF$_3$. IX. m. 134-6°.[200]
C$_{15}$ (EtO)$_2$PS[N(SPh)CSNEt$_2$]. XII. n 1.5631.[654]
(PhO)$_2$PSNHCS(OEt). XII. m. 47-9°.[524]
(p-ClC$_6$H$_4$S)$_2$PONHCOCCl$_2$Me. IX. m. 169-71°.[231]
(p-MeC$_6$H$_4$O)$_2$PS(NCO). XVII. $b_{0.04}$ 137-8°,[740] $b_{0.07}$ 137-8°,[741] n 1.5640,[740] d 1.2220.[740] By-product: [(p-MeC$_6$H$_4$O)$_2$PSNHCO-]$_2$ m. 143-4°.[740]
(BuO)$_2$PSNHCOPh. V, XVII. m. 53°.[33]
(i-BuO)$_2$PSNHCOPh. V, XVII. m. 76°.[33]
(PrO)$_2$PSNHCSNHC$_6$H$_4$OEt-p. XII. m. 91°.[760]
Cx (p-Me$_2$NC$_6$H$_4$O)$_2$PSNH$_2$. V. m. 162-3°.[306]
(p-MeC$_6$H$_4$O)$_2$PSNHCOOMe. XII. m. 61-3°.[740]
(PhO)$_2$PSNHCSNHPh. XII. m. 84-5°.[524]
(PhO)$_2$PSNHCOPh. V, XVII. m. 124°.[33]
(PhO)$_2$PSNHCONHPh. XII. m. 166-8°.[740]
(2-C$_{10}$H$_7$O)$_2$PSNH$_2$. V. m. 215°.[75]
(p-MeC$_6$H$_4$O)$_2$PSNHCONHPh. XII. m. 159-60°.[740]
O,O-diethyl O-7-oxabicyclo[2,2,1] hept-5-ene-2,3 dicarboximido phosphorothioate. XVII. m. 80.5-2°.[725]

D.1.6.2. (RO)$_2$P(S)-NHR', R' = Alkyl or Aryl

C$_6$ (EtO)$_2$PSNHEt. IX. b_{12} 94°,[573] $b_{(?)}$ 67-70°,[474] n 1.4716.[474]
(EtO)$_2$PSNHCH$_2$CH$_2$SH. V. n 1.5166, d 1.159.[128]
(EtO)$_2$PSNHCH$_2$CH$_2$OH. V. b_3 155°, n 1.4935.[181]
C$_7$ (EtO)$_2$PSNHCH$_2$CHMeSH. V. n 1.5092, d 1.128.[128]
(EtO)$_2$PSNHPr. IX. b_{11} 98°, d_0^{15} 1.005.[573]
C$_8$ (MeO)$_2$PSNH(C$_5$H$_3$NMe-5)-2. V. m. 95°.[146] For other (RO)$_2$PSHN- pyridines see Ref. 146.
(MeO)$_2$PSNHC$_6$H$_4$Cl-p. V. m. 78-80°.[390]
(MeO)$_2$PSNHPh. XIII. b_3 129.5-30.5°, d 1.2333, n 1.5730.[401]
(MeO)$_2$PSNHC$_6$H$_{11}$. V. m. 91°.[354]
Me$_2$C(CH$_2$O)$_2$PSNHPr. V. m. 70-70.5°, n_D^{18} 1.4463.[261]
Me$_2$C(CH$_2$O)$_2$PSNHPr-i. V. m. 97.5-8.5°.[261]
(EtO)$_2$PSNHCH$_2$CH$_2$-S-COMe. XVII. $b_{0.08}$ 125°, n 1.5102, d 1.1742.[650]
(i-PrO)$_2$PSNHCH$_2$CH$_2$-S-COMe. XVII. $b_{0.04}$ 114-5°, n 1.4952, d 1.1167.[650]
(EtO)$_2$PSNHCH$_2$COOEt. XVII. b_1 116°, n 1.4720, d 1.1451.[26]
(EtO)$_2$PSNHBu. V. b_7 128°, n 1.4738.[181]
(PrO)$_2$PSNHEt. IX. $b_{(?)}$ 84-7°, n 1.4695.[474]
(i-PrO)$_2$PSNHEt. IX. $b_{(?)}$ 72-5°, n 1.4617.[474]

C_9 $Me_2C(CH_2O)_2PSNHBu$. V. b_2 106-8°, n_D^{21} 1.4995.[261]
$Me_2C(CH_2O)_2PSNHBu$-i. V. $b_{0.05}$ 90°, n_D^{23} 1.500.[261]
C_{10} $(EtO)_2PSNHC_6H_4Cl$-p. V. $b_{0.3}$ 153-56°.[390]
$(EtO)_2PSNHC_6H_4Cl$-m. V. $b_{0.3}$ 157-60°.[390]
$(EtO)_2PSNHPh$. XVII,[405] V,[181] XIII.[401] $b_{0.2}$ 110-1°,[405] b_2 140-50°,[181] b_2 134°.[187] n 1.5482,[187,181] n 1.5502.[405] d 1.1526,[405] d 1.1537.[401]
$(EtO)_2PSNHC_6H_4SO_2NH_2$-p. V. m. 159.5°.[29]
$[(MeO)_2PSNH]_2C_6H_4$-p. V. m. 156°.[257]
$(EtO)_2PSNHC_6H_{11}$. V. $b_{0.75}$ 121°,[137] b_2 146,[181] n_D^{25} 1.4920,[137] n 1.4962.[181]
$(BuO)_2PSNHEt$. IX. $b_{(?)}$ 112-6°, n 1.4690.[474]
i-Bu: $b_{(?)}$ 90-2°, n 1.4640.[474]
sec-Bu: $b_{(?)}$ 87-90°, n 1.4658.[474]
t-Bu: $b_{(?)}$ 78-81°, n 1.4662.[474]
C_{11} $(EtO)_2PSNHC_6H_3Cl,Me$-3,4. V. m. 52-4°.[145]
$(EtO)_2PSNHC_6H_4Me$-p. V. m. 38°.[137]
$Me_2C(CH_2O)_2PSNHC_6H_{11}$. m. 98-9°.[262]

C_{12} 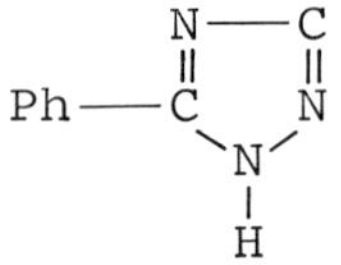 -NH-PS-$(OEt)_2$. V. m. 63-5°.[119,570]

$(PrO)_2PSNHPh$. XIII. b_2 148-9°, n 1.5388, d 1.1146.[401]
$(i\text{-}PrO)_2PSNHPh$. XIII. m. 89-90°.[401]
$(PrO)_2PSNHC_6H_4SO_2NH_2$-p. V. m. 168°.[29]
C_{13} $(p\text{-}ClC_6H_4O)_2PSNHMe$. IX. d^{26} 1.26, n_D^{35} 1.5356.[619]
$(EtO)_2PSNHCH_2CH_2SCH_2Ph$. V. n 1.5550, d 1.151.[128]
$(EtO)_2PSNH(C_9H_6N$-2). V. Picrate m. 174-5° (quinoline derivative).[42]
C_{14} $(C_6Cl_5O)_2PSNHEt$. IX. n_D^{35} 1.5536, d^{30} 1.36.[619]
$(2,3,4,6\text{-}Cl_4C_6HO)_2PSNHEt$. IX. n_D^{35} 1.5608, d^{30} 1.22.[619]
$(p\text{-}ClC_6H_4O)_2PSNHEt$. IX. n_D^{35} 1.5290, d^{28} 1.15.[619]
$(p\text{-}O_2NC_6H_4O)_2PSNHEt$. XVII. m. 152.8-3.8°.[567]
$(PhO)_2PSNHEt$. IX. Liquid.[573]
$(BuO)_2PSNHPh$. XIII. $b_{2.5}$ 160-3°, n 1.5315, d 1.0801.[40]
$(BuO)_2PSNHC_6H_4SO_2NH_2$-p. V. m. 118°.[29]
$[(PrO)_2PSNHCH_2\text{-}]_2$. V. n 1.4821.[181]
$[(EtO)_2PSNH\text{-}]_2C_6H_4$-p. V. m. 177°.[181]
C_{15} $(4,2,6\text{-}BrCl_2C_6H_2O)_2PSNHPr$-i. IX. n_D^{35} 1.5103, d^{30} 1.23.[619]
$(p\text{-}ClC_6H_4O)_2PSNHPr$-i. IX. n_D^{35} 1.5250, d^{19} 1.13.[619]
$(p\text{-}O_2NC_6H_4O)_2PSNHPr$-i. XVII. m. 154.5-6.0°.[567]
C_{16} $(AmO)_2PSNHC_6H_4SO_2NH_2$-p. V. m. 83-4°.[29]
$(2,4,6\text{-}Cl_3C_6H_2O)_2PSNHAm$. IX. n_D^{35} 1.4956, d^{31} 1.07.[619]
$(p\text{-}Me_2NC_6H_4O)_2PSNHMe$. V. m. 112.5-4.0°.[306]
$[(m\text{-}Me_2NC_6H_4O)_2PSNH]_2CH_2$. V. m. 156-64°, other diamine derivatives, oils.[307]
C_{18} $(m\text{-}Me_2NC_6H_4O)_2PSNHCH_2CH_2NH_2$. V. m. 121-2.5°.[307]
$[(o\text{-}C_6H_4O_2)PSNH]_2C_6H_4$-p. XIII. m. 187° dec.[420]
$(2,4,5\text{-}Cl_3C_6H_2O)_2PSNHC_6H_{11}$. IX. m. 66-72°.[619]

(PhO)$_2$PSNHPh.   V.[75,117]   m. 92°.[75]
(PhO)$_2$PSNHC$_6$H$_4$SO$_2$NH$_2$-p.   V.   m. 191°.[29]
C$_{19}$ (PhO)$_2$PSNHC$_6$H$_4$OMe-p.   V.   m. 83-5°.[117]
(PhO)$_2$PSNHC$_6$H$_4$Me-o.   V.   m. 68-70°.[117]
(PhO)$_2$PSNHC$_6$H$_4$Me-p.   V.   m. 94-6°.[117]
(o-CH$_2$=CHCH$_2$C$_6$H$_4$O)$_2$PSNHMe.   IX.   n$_D^{35}$ 1.5302, d^{26} 1.09.
    Also: NHPr-i, NMe$_2$, and NEt$_2$ derivatives.[618]
C$_{20}$ (p-MeC$_6$H$_4$O)$_2$PSNHPh.   V.[75,117]   m. 106°.[75]
C$_x$ (p-Me$_2$NC$_6$H$_4$O)$_2$PSNHPh.   V.   m. 177-80°.[306]

m-Me$_2$NC$_6$H$_4$O
$\quad\quad\quad\overset{\displaystyle S}{\underset{\displaystyle \Vert}{}}\quad$ H
$\quad\quad\quad\quad$P−N
m-Me$_2$NC$_6$H$_4$O$\quad\quad\quad$R

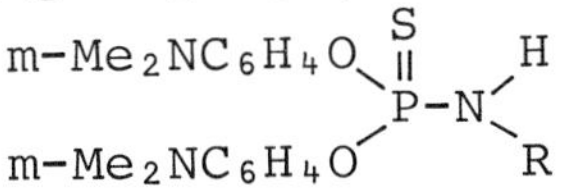

V.[306]   R=Bu: n$_D^{25}$ 1.591.
R=CH$_2$CH$_2$CH$_2$NMe$_2$: n$_D^{25}$ 1.5873,
    oxalate m. 122-3°.
R=(C$_5$H$_4$N-2): m. 130-1°.
R=C$_6$H$_{11}$: m. 59.5-61.0.
R=Ph: m. 97-8°.
R=C$_6$H$_4$NMe -p: m. 124-6°.
R=CH$_2$Ph: n$_D^{25}$ 1.6212.
R=CHMePh: n$_D^{25}$ 1.6165.

(EtO)$_2$PS- D,L-methionine-Et-ester.   b$_{0.25}$ 145-7°,
    n 1.4950.[474]
(EtO)$_2$PS- L-proline-Et-ester.[474]
6-[(MeO)$_2$PSNH-]penicillanic acid.[781]   Alkyl- and Aryl-
    analogs.[781]

D.1.6.3.   (RO)$_2$P(S)-NR'R", R' or R" = Alkyl or Aryl,
           R' = R" or R' ≠ R"

C$_4$ (CH$_2$O)$_2$PS(NC$_2$H$_4$).   XIII.   m. 61-2°.[344]
(CH$_2$O)$_2$PSNMe$_2$.   XIII.   m. 42.5-4°,[264] m. 46-6.5°,[2]
    b$_2$ 102-3°.[2]
(MeO)$_2$PSNMe$_2$.   XIII,[47,883] V.[568]   b$_9$ 77-8°,[568] b$_{11}$ 80°,[47]
    b$_{15}$ 90°,[883] d 1.1712,[568] d 1.1303,[47] n 1.4766,[568]
    n 1.4728.[47]
C$_6$ (CH$_2$O)$_2$PSN(CH$_2$CH$_2$)$_2$O.   XIII.   m. 72-2.5°.[2]
(MeCOO)$_2$PSNMe$_2$.   IX.[101,883]
(MeCHO)$_2$PSNMe$_2$.   XIII.   b$_{0.2}$ 103-4°, n$_D^{21}$ 1.5018.[264]
(CH$_2$O)$_2$PSNEt$_2$.   XIII.   b$_3$ 133-4.5°,[49] b$_2$ 129-30°,[2]
    m. 18.5-9.0°,[2] n 1.5050,[49] n 1.5075,[2] d 1.1825,[49]
    d 1.1781.[2]
(EtO)$_2$PS(NC$_2$H$_4$).   V.   b$_{0.01}$ 55°.[572]
(MeO)$_2$PSNEt$_2$.   V.   b$_9$ 88-91°, n 1.4780, d 1.0905.[568]
(EtO)$_2$PSNMe$_2$.   V,[568] IX,[573] XIII.[47]   b$_3$ 97-100°,[568]
    b$_{13}$ 93°,[47] b$_{45}$ 107°.[573] n 1.4634.[47,568]   d 1.0622,[568]
    d 1.0576,[47] ^{31}P -76.6.[873]
(MeO)$_2$PSN(CH$_2$CH$_2$OH)$_2$.   V.   b$_{0.15}$ 136°, n 1.5090,
    d 1.2470.[568]
C$_7$ CH$_2$(CH$_2$S)$_2$PON(CH$_2$CH$_2$Cl)$_2$.   IX.   m. 112-4°.[314]
Me$_2$C(CH$_2$O)$_2$PSNMe$_2$.   V.   m. 73-3.5°, ^{1}H.[84]
C$_8$ (Me$_2$CO)$_2$PSNMe$_2$.   XIII.   m. 32-3°.[264]
(MeOCOCH$_2$S)$_2$PONMe$_2$.   IX.   b$_{0.1}$ 140°,[380] similar com-
    pounds.[380]

$(CH_2O)_2PSNPr_2$. XIII. b_2 135.5-6.5°, n 1.5020, d 1.1254.
$(EtO)_2PSN(CH_2CH_2)_2O$. V. b_2 105°, n 1.4917.[181]
$(EtO)_2PSNEt_2$. IX. b_{20} 110°, d_0^{15} 1.0056.[573] ^{31}P -75.4.[873]
$(PrO)_2PSNMe_2$. V,[568] XIII.[47] $b_{0.36}$ 79-82°,[568] b_{10} 93-4°.[47] n 1.4640,[568] n 1.4500.[47] d 1.0222,[568] d 1.0013.[47]
$(i\text{-}PrO)_2PSNMe_2$. V,[568] XIII.[47] $b_{0.6}$ 71°, n 1.4585, d 1.0070,[568] ^{31}P -74,6.[873]
$(EtO)_2PSN(CH_2CH_2OH)_2$. V. $b_{0.035}$ 143-5°, n 1.4945,[568] n 1.4940,[181] d 1.1742.[568]

C_9 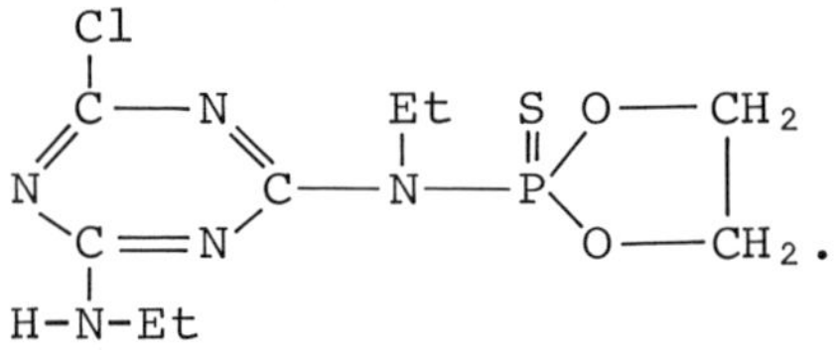 IX. m. 160-3°.[243]

$(EtO)_2PSNMeCH{=}CHCOMe$. XVII. b. 112-3°.[592]
$(EtO)_2PS(NC_5H_{10})$. IX. b_{10} 138°, d_0^{16} 1.0433.[573]

C_{10} $(MeO)_2PSNPhEt$. V. $b_{0.05}$ 87-93°, n 1.5565, d 1.1904.[568]
$(CH_2O)_2PSNBu_2$. XIII. b_2 160-1°, n 1.4960, d 1.0802.[2]
$(EtNHCOCH_2S)_2PONMe_2$. IX. n_D^{24} 1.4792.[380] Similar compounds.[380]
$(PrO)_2PSNEt_2$. V. $b_{0.45}$ 81-2°, n 1.4620, d 1.0029.[568]
$(BuO)_2PSNMe_2$. XIII,[47] V.[568] b_9 132°,[568] b_{10} 128-30°,[47] n 1.4629,[47] n 1.4622.[568] d 0.9975,[47] d 0.9958.[568]
$(i\text{-}BuO)_2PSNMe_2$. XIII. b_{10} 121.5-2.0°, n 1.4550, d 0.9842.[47]
$(t\text{-}BuO)_2PSNMe_2$. XIII. $b_{0.5}$ 76°, n_D^{19} 1.463, d_{18}^{18} 1.0075 ^1H.[136]
$(PrO)_2PSN(CH_2CH_2OH)_2$. V. $b_{0.03}$ 149-51°, n 1.4900, d 1.1313.[568]
$(i\text{-}PrO)_2PSN(CH_2CH_2OH)_2$. V. $b_{0.05}$ 103-110°, n 1.4820, d 1.1292.[568]

C_{12} 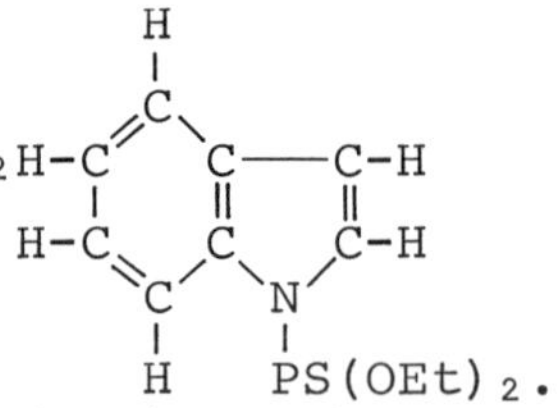 V. $b_{0.005}$ 112-4°.[570]

$(EtO)_2PSNPhEt$. IX,[573] V.[568] $b_{0.7}$ 99-107°, n 1.5499, d 1.1378.[568]

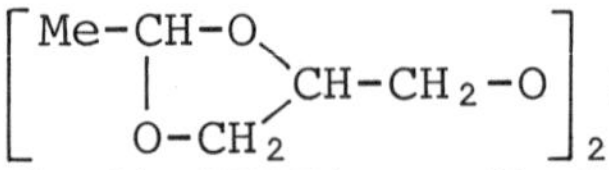 PS-NMe2. XIII. $b_{0.5}$ 153°, n 1.4848, d 1.1997.[48]

$(BuO)_2PSNEt_2$. V, XIII. $b_{0.3}$ 91-3°,[568] $b_{0.3}$ 108-10°,[673] n 1.4610.[568,673]
$(MeO)_2PSN(Am\text{-}i)_2$. IX. b_{13} 118-21°, d_0^{15} 1.0024.[573]

(t-AmO)$_2$PSNMe$_2$.   XIII.   b$_{0.05}$87-9°, n 1.466, d 0.9946.[135]
(BuO)$_2$PSN(CH$_2$CH$_2$OH)$_2$.   V.   b$_{0.05}$ 156-63°, n 1.4850,
    d 1.0916.[568]

C$_{14}$ (C$_6$Cl$_5$O)$_2$PSNMe$_2$.   IX.   m. 216-7°.[916]
(2,4,5-Cl$_3$C$_6$H$_2$O)$_2$PSNMe$_2$.   IX.   n$_D^{35}$ 1.5737, d$_{31}$ 1.35.[619]
(p-ClC$_6$H$_4$O)$_2$PSNMe$_2$.   IX.   n$_D^{35}$ 1.5533, d$_{24}$ 1.16.[619]
(PhO)$_2$PS(NC$_2$H$_4$).   V.[248]
(PhS)$_2$PSNMe$_2$.   IX.   b$_{0.05}$ 110°.[380]
(p-ClC$_6$H$_4$S)$_2$PONMe$_2$.   m. 75-6°.[380]
(PrO)$_2$PSNPhEt.   V.   b$_{0.01}$ 104-8°, n 1.5272, d 1.0864.[568]
(i-PrO)$_2$PSNPhEt.   V.   m. 81°.[568]
(EtNHCOCH$_2$S)$_2$PSNPr$_2$.   XVII.   m. 91-3°.[549]   Similar com-
    pounds.[549]

C$_{16}$ (p-ClC$_6$H$_4$O)$_2$PSNEt$_2$.   IX,[619] XVII.[567]   m. 151-2°.[567]
    n$_D^{35}$ 1.5458.   d$_{28}$ 1.16.[619]
(p-MeC$_6$H$_4$S)$_2$PSNMe$_2$.   b$_{0.01}$ 130-40°.[380]
(p-O$_2$NC$_6$H$_4$O)$_2$PSNEt$_2$.   XVII.   m. 173-3.5°.[567]

C$_x$ (p-O$_2$NC$_6$H$_4$O)$_2$PSNC$_5$H$_{10}$.   XVII.   m. 158-9°.[567]
(PhO)$_2$PSNEt$_2$.   V,[75] IX,[573] XVII.[567]   m. 58°,[75] m. 70°,[573]
    m. 114-4.5°.[567]
(PhS)$_2$PSNEt$_2$.   IX,[380] XIII.[673]   b$_{0.005}$ 190-205°,[673]
    b$_{0.01}$ 125-35°.[380]   n 1.6405, d 1.1886.[673]
(PhS)$_2$PONEt$_2$.   IX.   b$_{0.025}$ 125-40°.[380]   For -(NC$_5$H$_{10}$),
    -N(CH$_2$CH$_2$)$_2$O and imidazole analog see Ref. 380.
(BuO)$_2$PSNPhEt.   V.   b$_{0.025}$ 115°, n 1.5245, d 1.0679.[568]
(PhO)$_2$PS(NC$_5$H$_{10}$).   V.   Liquid.[573]   m. 49-51°.[117]
(p-ClC$_6$H$_4$S)$_2$PSNPr$_2$.   IX.   b$_{0.5}$ 95-100°.[380]
(p-Me$_2$NC$_6$H$_4$O)$_2$PSNMe$_2$.   V.   m. 103-5.5°.[306]
(PhO)$_2$PSN(Am-i)$_2$.   IX.   m. 64°.[573]
(p-Me$_2$NC$_6$H$_4$O)$_2$PSNBu$_2$.   V.   b$_{0.001}$ 225°, n$_D^{25}$ 1.5716.[306]

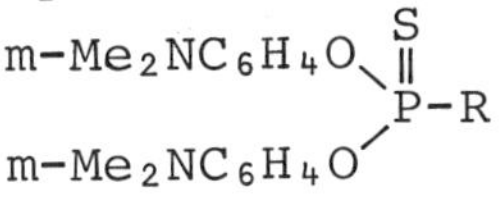

m-Me$_2$NC$_6$H$_4$O$\diagdown$
    $\overset{\text{S}}{\underset{}{\|}}$P-R   V.[306]
m-Me$_2$NC$_6$H$_4$O$\diagup$

R=NMe$_2$: n$_D^{25}$ 1.599.
R=NPr$_2$: m. 108-9°.
R=NBu$_2$: n$_D^{25}$ 1.5689.
R=N(Bu-i)$_2$: n$_D^{25}$ 1.5696.
R=N(C$_7$H$_{15}$)$_2$: n$_D^{25}$ 1.5527.
R=NPhMe: n$_D^{25}$ 1.623.
R=NBuPh: n$_D^{25}$ 1.6048.
R=NMe(CH$_2$Ph): n$_D^{25}$ 1.614
R=[N(C$_2$H$_4$)$_2$O]: m. 78.5-79.5°.
R=NBu(CH$_2$Ph): b$_{0.001}$ 175°.
n$_D^{25}$ 1.5986.

R=N$\overset{\text{(CH$_2$)$_3NMe_2$}}{\underset{\text{CH$_2C_6H_4NMe_2$-p}}{\diagdown}}$: n$_D^{25}$ 1.603.

R=N(CH$_2$Ph)$_2$: n$_D^{25}$ 1.623.

D.1.6.4.   (RO)$_2$P(S)-NAB, A,B Any Ligands Except
Those Listed Before

C$_7$
$$\begin{array}{c} H_2C\text{----}CH_2 \\ | \quad\quad | \\ H\text{-}N \quad N - P(S)(OEt)_2. \quad XII, XVII. \quad m. \ 83\text{-}4°.^{513} \\ \diagdown C \diagup \\ \| \\ S \end{array}$$

C$_8$ (EtO)$_2$PSN(Et)SO$_2$CH$_2$CH$_2$Cl.   XVII.   n 1.475.[669]
C$_9$ (i-PrO)$_2$PSNEtCHO.   XIII.   b$_3$ 103.5-5.0°, n 1.4620,
    d 1.0755.[24]
  (PrO)$_2$PSN(Me)SEt.   XVII.   n 1.51, d 1.175.[668]
  [(EtO)$_2$PS]$_2$NMe.   V.   b$_{1.5}$ 134-5°, n 1.4940, d 1.1662.[26,4]
C$_{10}$ [(EtO)$_2$PS]$_2$NEt.   V.   b$_1$ 133-4°, n 1.4913, d 1.1476.[26,45]
  (EtO)$_2$PSNHSO$_2$C$_6$H$_4$Br-p.   V.   IR.[31]   Ag-salt.[946]
  (EtO)$_2$PSNHSO$_2$C$_6$H$_3$Cl$_2$-2,5.   V.   m. 128°.[30]
  (EtO)$_2$PSNHSO$_2$C$_6$H$_3$Cl$_2$-3,4.   V.   m. 66-7°.[30]
  (EtO)$_2$PSNHSO$_2$C$_6$H$_4$I-p.   Diss. constant.[28]
C$_{11}$ (EtO)$_2$P(S)N=P(OEt)$_3$.   XIII, XVII.   b$_{0.1}$ 112-4°, n 1.4703,
    d 1.1445.[328]
  (EtO)$_2$PSN(Me)SO$_2$C$_6$H$_4$I-p.   XVII.   m. 70°.[30]
  (i-BuO)$_2$PSNEtCHO.   XIII.   b$_1$ 107-8°, n 1.466, d 1.0428.[2]
C$_{12}$ (PhO)$_2$P(S)N=PCl$_3$.   I.[469]
  (PrO)$_2$PSNHSO$_2$C$_6$H$_4$I-p.   V.   m. 48°.[30]
  [(EtO)$_2$PS]$_2$NCH$_2$COOEt.   XIII.   b$_1$ 156-7°, n 1.4835,
    d 1.1807.[18]
C$_x$ (EtO)$_2$PSN[SN(CH$_2$CH$_2$)$_2$O][C(S)N(CH$_2$CH$_2$)$_2$O].   XII.
    m. 88°.[280]
  (EtO)$_2$PSN(Bu)SO$_2$C$_6$H$_4$Cl-p.   XVII.[669]
  (PhO)$_2$P(S)N=P(NC$_2$H$_4$)$_3$.   VI.   m. 60-2°.[830]
  (o-BrC$_6$H$_4$O)$_2$P(S)N=P(NC$_2$H$_4$)$_3$.   VI.   m. 80-2°,[830] other
    aryl derivatives.[830]
  (PhO)$_2$PSNHPO(NHPh)$_2$.   V.   m. 198-200°.[469]
  (PhO)$_2$P(S)N=P(OPh)$_3$.   IX, XI.   m. 96-8°.[92]

D.1.7. $\geq$N-P(O)AB, A $\neq$ B, A,B = Halogen, -OH, -OR,
    -OAc

D.1.7.1. $\geq$N-P(O)XY, X,Y = Halogen
OCN-POClF.   XIV.   b. 101-3°, n$_D^{25}$ 1.4024.[497]
Cl$_2$C=N-POClF.   XVII.   b$_{15}$ 52-4°, n 1.4717, d 1.6999.[386]
EtNHPOClF.   II.   b$_{0.3}$ 56-7°, n 1.4098, d 1.3162.[821]
Et$_2$NPOBrF.   II.   b$_{0.5}$ 46°, n 1.4468, d 1.4671.[821]
Et$_2$NPOClF.   II.   b$_{12}$ 67°, n 1.4168, d 1.1962.[821]
C$_6$H$_{11}$NHPOClF.   II.   b$_{0.3}$ 98-100°, m. 52°.[820]
PhNHPOClF.   II.   b$_{0.5}$ 116-9°, m. 65°.[821] ^{31}P -2.4.[361a]
PhNHC(=NPOClF)Cl.   XVII.   m. 95-8°.[227]
p-O$_2$NC$_6$H$_4$NHC(=NPOClF)Cl.   XVII.   m. 160-2°.[227]
Ph$_2$NC(=NPOClF)Cl.   XVII.   m. 118-20°.[227]

D.1.7.2.   N-P(O)X(OH), X = Halogen
C$_2$ EtNHPO(F)OK.   XVII.   Hygroscopic solid.[821]

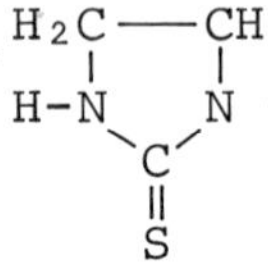

C_3 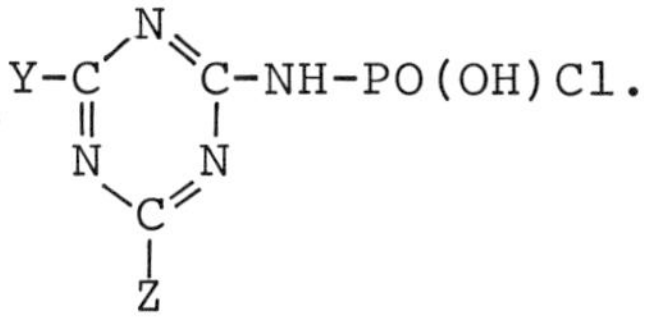 X.[202] Y=Z=Cl: m. 140-2°. Y=Cl, Z=NH-PO(OH)Cl: m. 260-5°. Y=Z=NH-PO(OH)Cl: m. 310-1°.

C_4 $Et_2NPOF(OK)$. XVII. K-salt.[821] Na-salt.[820]

C_6 $[p\text{-}ClC_6H_4SO_2NPO_2F]K_2$. X. dec. 312-6°.[451]

$[o\text{-}O_2NC_6H_4SO_2NPO_2F]K_2$. X. dec. 235-7°.[451]

$[m\text{-}O_2NC_6H_4SO_2NPO_2F]K_2$. X. dec. 302-6°.[451]

$[p\text{-}O_2NC_6H_4SO_2NPO_2F]K_2$. X. dec. 299-302°.[451]

$p\text{-}FC_6H_4SO_2NHPOCl(OH)$. VIII. m. 123-4°.[391]

PhNHPOF(OK). XVII. Solid.[821]

$PhSO_2NHPOBr(OH)$. X. m. 147-50°.[473]

$PhSO_2NHPOCl(OH)$. X. dec. 120°.[431]

$o\text{-}O_2NC_6H_4SO_2NHPOCl(OH)$. X. dec. 123-5°.[446]

$m\text{-}O_2NC_6H_4SO_2NHPOCl(OH)$. X. dec. 71-3°.[446]

$p\text{-}O_2NC_6H_4SO_2NHPOCl(OH)$. X. dec. 45-6°.[446]

C_7 $o\text{-}MeC_6H_4SO_2NHPOBr(OH)$. X. m. 126-9°.[473]

$p\text{-}MeC_6H_4SO_2NHPOBr(OH)$. X. m. 137-40°.[473]

$o\text{-}MeC_6H_4SO_2NHPOCl(OH)$. X. dec. 120°.[431]

$p\text{-}MeC_6H_4SO_2NHPOCl(OH)$. X. dec. 115°.[431]

$3,4\text{-}Cl,MeC_6H_3NHPOCl(OH)$. II, VIII. m. 218-20°.[145]

$2,4\text{-}Cl,MeC_6H_3NHPOCl(OH)$. II, VIII. m. 88-90°.[145]

$PhC(NH_2)=NPO(OH)Cl$. XIII. dec. 108-10°.[218]

C_8 $2,4\text{-}Me_2C_6H_3NHPOCl(OH)$. II, VIII. m. 178-9°.[145]

C_x $1\text{-}C_{10}H_7SO_2NHPOBr(OH)$. X. m. 127-30°.[473]

$2\text{-}C_{10}H_7SO_2NHPOBr(OH)$. X. m. 163-5°.[473]

$1\text{-}C_{10}H_7SO_2NHPOCl(OH)$. X. dec. 120°.[431]

$Ph_2NCONHPOCl(OH)$. X. m. 171-4°.[461]

D.1.7.3. >N-P(O)X(OR), X = Halogen, R = Alkyl or Aryl or Acyl

All ROPO(N<)CN compounds are poisons!

C_2 MeOPOCl(NCO). XVII. b_{30} 78-80°, n 1.4365, d 1.4033.[627]

$\overline{OCH_2CH_2NHP}OCl$. II.[98]

C_3 EtOPOCl(NCO). XVII. b_{55} 100-2°, n 1.4320, d 1.3478.[627]

$EtOPOF(N=CCl_2)$. XVII. $b_{1.5}$ 40-2°, n 1.4490, d 1.4459.[387]

$EtOPOCl(N=CCl_2)$. XVII. b_1 73-4°, n 1.4845, n 1.4842.[12] ^{31}P +3.7.[361a]

EtOPOF(NCO). XIV. b_8 39°, n 1.3900, d 1.3213.[387]

EtOPOF(NCS). XIV. b_2 34-6°, n 1.4710, d 1.2963.[387]

$\overline{OCH_2CH_2CH_2NHP}OCl$. II. m. 80-3°.[388]

C_4 PrOPOF(NCO). XIV. b_{13} 59°, n 1.3980, d 1.2405.[387]

PrOPOF(NCS). XIV. b_2 45-7°, n 1.4650, d 1.2455.[387]

$PrOPOCl(NCCl_2)$. XVII. b_1 77.5-8°, n 1.4810, d 1.4088.[12]

$CH_2=CHOPO(NMe_2)Cl$. IV. b_1 60-6°, n 1.4502, d 1.1487.[338]

$MeOPO(NMe_2)CN$. XIV. b_1 65°,[369] b_{13} 89-91°.[755]

$ClCH_2CH_2OPO(NMe_2)F$. XVII. b_3 69°.[101]

EtOPO(NMe$_2$)Cl. IV.[163,370] b$_{18}$ 98-100°.[163]
EtOPO(NMe$_2$)F. IV,[163] XIV.[753] b$_{18}$ 76-8°,[163] b$_{14}$ 72°.[753]
C$_5$ CCl$_3$CH$_2$OPO(NMe$_2$)CN. XIV. m. 57°.[753]
CF$_3$CH$_2$OPO(NMe$_2$)CN. XIV. b$_2$ 67°.[753]
BrCH$_2$CH$_2$OPO(NMe$_2$)CN. XIV.[758]
ClCH$_2$CH$_2$OPO(NMe$_2$)CN. XIV. b$_{14}$ 159°.[753]
FCH$_2$CH$_2$OPO(NMe$_2$)CN. XIV. b$_2$ 110°.[753]
EtOPO(NMe$_2$)CN. XIII, XIV. b$_2$ 95-6°,[766] b$_6$ 92°,[753] b$_9$ 100-8°.[755] n 1.4250, d_4^{24} 1.077.[755] See also Refs. 71, 370. ^{31}P +10.7.[394] Poison, nerve gas, "tabun."
ClCH$_2$CH$_2$OPO(NMe$_2$)CN. XIV. b$_{12}$ 160°.[755]
EtOPO[N=C(OMe)$_2$]F. XIII. b$_4$ 129-30°, n 1.4265, d$_{22}$ 1.2917.[672]
EtOPO(NHCSNHEt)F. XII. m. 104-5°.[387]
i-PrOPO(NMe$_2$)Cl. IV.[119]
C$_6$ i-AmOPOCl(N=CCl$_2$). XVII. b$_2$ 89-90°, n 1.4782, d 1.3121.[12]
EtOPO(CN)N=CMe(OMe). XIII, XIV. b$_3$ 108-10°, n_D^{21} 1.4389 d$_{21}$ 1.1716.[672]
ClCH$_2$CH$_2$OPO[N(CH$_2$CH$_2$)$_2$O]F. XIV. b$_1$ 101°, n_D^{10} 1.4379.[54
ClCH$_2$CH$_2$OPO[N(CH$_2$CH$_2$)$_2$O]Cl. XVII. b$_1$ 123°.[542]
ClCH$_2$CH$_2$OPO[N(CH$_2$CH$_2$Cl)$_2$]Cl. XVII. Oil, ClO_4^--salt: m. 167°.[542]
EtOPO[N(CH$_2$CH$_2$)$_2$O]Cl. XVII. b. 95°.[542]
EtOPO[N(CH$_2$CH$_2$Cl)$_2$]Cl. XVII. Oil, ClO_4^--salt: m. 163-4°.[542]
EtOPO[N(CH$_2$CH$_2$)$_2$O]F. XIV. b$_7$ 105°, n 1.4384.[542]
PrOPO(NMe$_2$)CN. XIV. b$_{15}$ 123°.[755]
MeOCH$_2$CH$_2$OPO(NMe$_2$)CN. XIV. b$_{11}$ 155°.[755]
i-PrOPO(NMe$_2$)CN. XIV. b$_2$ 95°,[766] b$_4$ 93-4°.[369]
EtOPO(NMeEt)CN. XIV. b$_{10}$ 117°.[753]
EtOPO(N=CMeOEt)F. XIII. b$_{25}$ 86°, n 1.4015, d 1.1518.[67
i-PrOPO[N=C(OMe)$_2$]F. XIII. b$_4$ 134-5°, n_D^{21} 1.4242, d$_{21}$ 1.2179.[672]
ClCH$_2$CH$_2$OPO(NEt$_2$)F. XIV. b$_1$ 88-9°.[541]
ClCH$_2$CH$_2$OPO(NEt$_2$)Cl. XVII. b$_1$ 114°.[541]
EtOPO(NEt$_2$)Cl. IV.[573,370] b$_{18}$ 113°, n 1.4437, d 1.1200.[573] ^{31}P -14.7.[873]
EtOPO(NEt$_2$)F. IV. b$_{11}$ 85°, n 1.4050, d 1.0708.[821]
C$_7$ PhOPO(NCO)Cl. XIII. b$_7$ 100-2°.[233] n 1.5167, d 1.401.[1
p-ClC$_6$H$_4$OPO(NCO)Cl. XIII. b$_3$ 125-9°, n 1.5335, d 1.508.[233]
i-PrOPO(CN)N=CMe(OMe). XIII, XIV. b$_{0.4}$ 94-5°, n 1.4353 d 1.1625.[672]
MeOPO(NHPh)Cl. IV. m. 82-3°.[149]
MeOPO(NHSO$_2$Ph)F. X, XIII. K-salt: m. 202-5°. [$PhNH_3^+$]-salt: m. 109-11°.[385] Similar compounds see Ref. 385
EtOCOCH$_2$OPO(NHCH$_2$CH=CH$_2$)Cl. IV. b$_{0.0001}$ 100-5°, n 1.4528, d 1.222.[676]
EtOPO[N(CH$_2$CH$_2$)$_2$O]CN. XIV. b$_7$ 138°.[753]

i-PrOPO(N=CMeOMe)CN. XIII. $b_{0.4}$ 94-5°, n 1.4353, d 1.1625.[672]
$C_6H_{13}OPOCl(N{=}CCl_2)$. XVII. b_1 99-101°, n 1.4781, d 1.2864.[12]
EtOPO(CN)N=CMe(OEt). XIII, XIV. b_3 105-7°, n_D^{19} 1.4520, d_{19} 1.1867.[672]
$BuOPO(NMe_2)CN$. XIV. b_{14} 130°.[755]
$i\text{-}PrOPO[N(CH_2CH_2)_2O]Cl$. XVII. b_1 108°.[542]
$EtOPO[N{=}C(OEt)_2]Cl$. XIII. $b_{0.5}$ 96-8°, n 1.4520, d 1.2381.[672]
$EtOPO[N{=}C(OEt)_2]F$. XIII. b_3 130-3°, n_D^{18} 1.4250, d_{18} 1.2050.[672]
i-PrOPO[N=CMe(OEt)]Cl. XIII. $b_{0.2}$ 80-2°, n 1.4520, d 1.1600.[672]
$EtOPO[N(CH_2CH_2)_2O]CN$. XIV. b_9 140°.[755]
i-PrOPO[N=CMe(OEt)]F. $b_{1.5}$ 74°, n_D^{21} 1.4200, d_{21} 1.1157.[672]
$i\text{-}PrOPO[N(CH_2CH_2)_2O]F$. XIV. b_1 95°, n 1.4392.[542]
$i\text{-}BuOPO(NMe_2)CN$. XIV. b_2 105°.[766]
$EtOPO(NEt_2)CN$. XIV. b_2 95°,[369] b_{17} 124°.[755,883]
$i\text{-}PrOPO(NEt_2)Cl$. IV. b_1 100°, n 1.4276, d 1.0866.[543]
$i\text{-}PrOPO(NEt_2)F$. XIV. b_1 82-3°, d 1.0802, n 1.4366.[543]
$MeOPO(NHC_6H_4Cl\text{-}p)Cl$. IV. m. 115-6°.[149]
C_8 EtOPO(NHPh)Cl. IV. m. 61-2°.[149]
EtOPO(NHPh)F. XIV. $b_{0.6}$ 137°,[821] $b_{0.2}$ 100-50°.[163] m. 50°,[163] m. 49-50°.[821]
$EtOPO(NC_5H_{10})CN$. XIV. b_{17} 129°.[755]
$C_7H_{15}OPOCl(N{=}CCl_2)$. XVII. b_1 112-4°, n 1.4768, d 1.2358.[12]
$i\text{-}PrOPO[N{=}C(OEt)_2]F$. XIII. b_2 129-31°, n_D^{21} 1.4242, d_{21} 1.2179.[672]
i-PrOPO[N=CMe(OEt)]CN. XIII. $b_{1.2}$ 98-100°, n 1.4435, d 1.1357.[672]
$EtOPO(NPr_2)Cl$. IV. b_{10} 240°.[573]
C_9 EtOPO(NHCONHPh)F. XII. m. 94-5°.[387] For similar compounds see Ref. 387.
$EtOPO(NHC_6H_4Me\text{-}p)Cl$. IV. m. 74-5°.[149]
EtOPO(NHCONHNHPh)Cl. XII. m. 135-7°.[627]
$C_6H_{11}OPO(NMe_2)CN$. XIV. b_2 140°.[766]

C_{10} 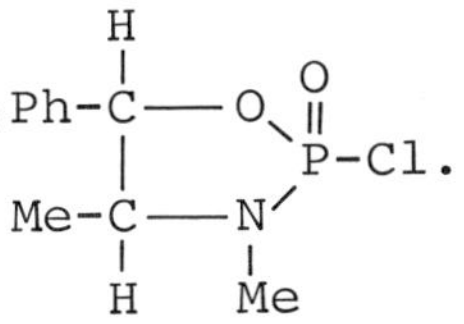 II. Pseudo-ephedrine-derivative.[300]

$PhOPO[N(CH_2CH_2Cl)_2]Cl$. IX. $b_{0.2}$ 167-9°.[316] m. 52-4°.[503]
$o\text{-}IC_6H_4OPO[N(CH_2CH_2Cl)_2]Cl$. IX. m. 78.5-9°.[501,503] For further I or Cl substituted aryl derivatives (n_D^{20} given) see Refs. 501,503.
$PhOPO(NEt_2)Cl$. IX. $b_{0.4}$ 118°, n_D^{25} 1.507.[914]
$EtOPO[N(Bu\text{-}i)_2]Cl$. IV. Liquid.[573]

C_{11} o-$MeC_6H_4OPO[N(CH_2CH_2Cl)_2]Cl$. IX. m. 42-3°.[501,503] For the meta and para isomers (n_D^{20} given) see Refs. 501, 503.
C_{12} $PhOPO(NHC_6H_4Br$-p$)Cl$. IX.[573]
PhOPO(NHPh)Cl. IV, IX. m. 131.5-2.0°,[716] m. 137°.[573]
$PhOPO(NHSO_2Ph)Cl$. X, XIII. K-salt: m. 174-6°.[385]
C_{13} PhOPO(N=CClPh)Cl. XVII. m. 61-2°.[220]
$PhOPO[N{=}CClC_6H_3(NO_2)_2$-3,5]Cl. XVII. m. 134-6°.[220]
PhOPO(NHCOPh)Cl. X. m. 123-4°.[220]
$PhOPO(NHCOC_6H_4NO_2$-p$)Cl$. m. 142-3°.[220]
$PhOPO[NHCOC_6H_3(NO_2)_2$-3,5]Cl. X. 162-3°.[220]
PhOPO(NHCONHPh)Cl. XII. m. 140-1°, similar compounds with substituted aryls.[233]
$PhOPO(NHC_6H_4Me$-p$)Cl$. IV. m. 77°.[573]
o-$MeC_6H_4OPO(NHPh)Cl$. IX. m. 97-8°.[716]
m-$MeC_6H_4OPO(NHPh)Cl$. IX. m. 91-1.5°.[716]
p-$MeC_6H_4OPO(NHPh)Cl$. IX. m. 108-10°.[716]
C_{14} $MeOPO(NHCONPh_2)Cl$. XII. m. 133-5°.[627] Other aryls.[627]
$ClCH_2CH_2OPO[N{=}C(NHC_6H_4Me$-p$)_2]F$. XVII. m. 122-4°.[391]
For ROPO(N<)NCS, ROPO(N<)NCO see $ROPO(N<)_2$ compounds.

D.1.7.4. >N-P(O)(OH)(OR), R = Alkyl or Aryl or Acyl

C_1 $MeOPO(N_3)OH$. XVII. Na-salt: m. 146-8° dec.[356]
C_2 $EtOPO(NH_2)OH$. XVII. K-salt, $[NH_4^+]$-salt.[712]
C_3 $MeOPO(NMe_2)OH$. XVII. (Me_4N^+)-salt: m. 221°.[156]
C_4 $[EtOPO(OH)]_2NH$. XVII. $[NH_4^+]_2$-salt.[505]
C_5 $Me_2C{=}CHCH_2OPO(NH_2)OH$. XVII. $[(C_6H_{11}NH)_2CNH_2^+]$-salt.[538]
$MeOPO[N(CH_2CH_2)_2O]OH$. XVII. $[Me_4N^+]$-salt: m. 177°.[156]
$MeOPO(NEt_2)OH$. XVII. $[Me_4N^+]$-salt: m. 231-2°.[156]
C_6 $PhOPO(NH_2)OH$. XVII. $[C_6H_{11}NH_3^+]$-salt: m. 186-8°,[748] m. 220-7°.[176] sint. 179-84°. recryst. m. 237-240°.[176] ^{31}P $[NH_4^+$-salt$]$ -0.5.[634] Cinchonine-salt: m. 194°.[272,825] Ag-salt.[825] $[NH_4^+]$-salt.[826]
$PhOPO(OH)NHPO(OH)_2$. XVII. Na-salt. ^{31}P (single peak) ±0.[632]
$MeOCH_2CH_2OPO(NC_3H_3N)OH$. XVII. imidazole derivative. Na-salt: m. > 250°.[172]
$EtOPO[N(CH_2CH_2Cl)_2]OH$. XVII. $(C_6H_{11}NH_3^+]$-salt: m. 146-7°.[316]
$MeOPO(NC_5H_{10})OH$. XVII. $[Me_4N^+]$-salt: m. 212°.[156]
$EtOPO[N(CH_2CH_2)_2O]OH$. XVII. $[Me_4N^+]$-salt.[155]
$EtOPO[N(CH_2CH_2Cl)(CH_2CH_2OH)]OH$. VIII. $HClO_4$-salt: m. 147-8°.[542]
$ClCH_2CH_2OPO[N(CH_2CH_2Cl)(CH_2CH_2OH)]OH$. VIII. $HClO_4$-salt m. 147-8°.[542]
$EtOPO(NEt_2)OH$. XVII. $[Me_4N^+]$-salt.[155]
C_7 MeOPO(NHPh)OH. VIII. Ba-salt.[149]
$PhCH_2OPO(NH_2)OH$. XVII. Li-salt.[161]
$MeOPO(NHC_6H_{11})OH$. XVII. m. 91-2°.[353]
$EtOPO(NC_5H_{10})OH$. XVII. $[Me_4N^+]$-salt.[155]

$H_2N(CH_2)_3OPO[N(CH_2CH_2Cl)_2]OH$. XVII. m. 154-5°, betaine-structure.[56]
$MeOPO(OH)OPO(NHC_6H_{11})OH$. XVII. Na_2-salt.[353]
$MeOPO(NPr_2)OH$. XVII. $[Me_4N^+]$-salt: m. 245°.[156]
C_8 $PhCH_2OPO(NHCN)OH$. XVII. Na_2-salt.[173]
$PhOPO(NHCH_2COOH)OH$. IV, VIII. Ba-salt.[926,424]
$MeOPO(NHC_6H_4Me\text{-}p)OH$. VIII. K-salt, Ba-salt.[149]
$EtOPO(NHPh)OH$. VIII. Ba-salt.[149]
$MeOPO(NHCH_2C_6H_4NH_2\text{-}o)OH$. XVII. Li-salt, Pb-salt.[355]
$PhOPO(NHCH_2CH_2NH_2)OH$. IV, VIII. m. 195-6°. monohydrate: m. 219-21°. Other aryls.[263]
t-$BuOPO[N(CH_2CH_2Cl)_2]OH$. XVII. $[C_6H_{11}NH_3^+]$-salt: m. 131-3°.[316]
$BuOPO(NEt_2)OH$. XVII. $[Me_4N^+]$-salt.[155]
C_9 p-$ClC_6H_4OPO(NC_3H_3N)OH$. XVII. Imidazolium-(NC_3H_3NH)-salt: m. 87°.[85] Na-salt: m. > 250°.[172]
$PhOPO(NC_3H_3N)OH$. XVII. Imidazole derivative. Na-salt: m. > 250°, $[C_6H_{11}NH_3^+]$-salt: m. 112°,[172] m. 115°.[170]
$MeOPO[N{=}C(OMe)C_6H_3(NO_2)_2\text{-}3,5]OH$. XVII. m. 230-1°.[221]
$PhOPO(NHCHMeCOOH)OH$. IV, VIII. Ba-salt.[424]
$EtOPO(NHC_6H_4Me\text{-}p)OH$. VIII. Ba-salt.[149]
$C_5H_{11}OPO[N(CH_2CH_2Cl)_2]OH$. XVII. $[C_6H_{11}NH_3^+]$-salt: m. 160-1°.[316]
$MeOPO(NBu_2)OH$. XVII. $[Me_4N^+]$-salt: m. 243-4°.[156]
p-$IC_6H_4OPO(NHCH_2COOMe)OH$. XVII. Ba-salt.[249]
p-$O_2NC_6H_4OPO(NHCH_2COOMe)OH$. XVII. Ba-salt.[249]
C_{10} $PhCH_2OPO(NC_3H_3N)OH$. XVII. imidazole derivative, Na-salt.[80,172]
$PhOPO[N(CH_2CH_2Cl)_2]OH$. XVII. $[C_6H_{11}NH_3^+]$-salt: m. 184-5°.[316]
t-$BuOPO(NHPh)OH$. V. m. 253-6° dec. aniline-salt: m. 140-1°. Resolidifies, melts again 264-70°.[337]
$H(CH_2CMe{=}CHCH_2)_2OPO(NH_2)OH$. XVII. Na-salt,[266] geranyl ester.
C_{11} $MeOPO(NHSO_2C_{10}H_7\text{-}1)OH$. X. m. 95-9°.[464]
$PhOPO(NHCHPrCOOH)OH$. XVII. Ba-salt.[424]
C_{12} $PhOPO(NHC_6H_4Br\text{-}p)OH$. VIII. m. 164°.[573]
$PhOPO(NHPh)OH$. VIII, XVII. m. 134°.[573] ^{31}P +4.1.[634] $[C_6H_{11}NH_3^+]$-salt: m. 200-3°,[783] m. 190-5°.[748] $[Et_2NH_2^+]$-salt: ^{31}P +6.0.[873]
$PhOPO(NHSO_2C_6H_4NH_2\text{-}p)OH$. V, VIII. m. 198°.[772]
p-$ClC_6H_4OPO(NHC_6H_{11})OH$. XVII. $[C_6H_{11}NH_3^+]$-salt: m. 195-8°.[748]
$PhOPO(NHC_6H_{11})OH$. XVII.[138,814] $[C_6H_{11}NH_3^+]$-salt: m. 192-3°.[176]
C_{13} 2-$C_{10}H_7OPO(NC_3H_3N)OH$. XVII. Imidazole derivative. Na-salt: m. > 250°, $[C_6H_{11}NH_3^+]$-salt: m. 110°.[172]
$PhOPO(NHCOPh)OH$. VIII. m. 141-2°,[220] m. 170°.[943]
$PhOPO[NHCOC_6H_3(NO_2)_2\text{-}3,5]OH$. VIII. m. 195-6°.[220]
p-$MeC_6H_4OPO(NHC_6H_4Br\text{-}p)OH$. VIII. m. 230°.[573]
p-$IC_6H_4CH_2OPO(NHPh)OH$. XVII. Ba-salt.[249]

$PhOPO[N(NO)CH_2Ph]OH$. XVII. Na-salt,[352] also $PhOPO[N(NO)Et]OH$ and $PhOPO[N(NO)C_6H_{11}]OH$.
$p\text{-}O_2NC_6H_4CH_2OPO(NHPh)OH$. XVII. Ba-salt.[249]
$PhOPO(NHCH_2Ph)OH$. VIII. Na-salt.[352]
$PhCH_2OPO(NHPh)OH$. XVII. Li-salt,[161] Ba-salt,[249] Na-salt.[160]
$PhOPO(NHSO_2C_6H_4Me\text{-}p)OH$. IV, VIII. m. 185-6°.[263]
$PhCH_2OPO(NHC_6H_{11})OH$. XVII. m. 101-4°.[161] m. 99°.[170] m. 98-100°.[176]
C_{14} $p\text{-}IC_6H_4CH_2OPO(NHC_6H_4Me\text{-}p)OH$. XVII. Ba-salt.[249]
$p\text{-}O_2NC_6H_4CH_2OPO(NHC_6H_4Me\text{-}p)OH$. XVII. Ba-salt.[249]
$PhCH_2OPO(NHCH_2Ph)OH$. XVII. m. 98-100°.[161,176]
$PhCH_2OPO(NHC_6H_4Me\text{-}p)OH$. XVII.[160]
$PhOPO(NHCH_2CH_2NHPh)OH$. XVII. m. 148°.[355]
C_{15} $PhOPO[NHCH(CH_2C_6H_4OH\text{-}p)COOH]OH$. IV, VIII. L-tyrosine derivative, Ba-salt.[926]
$PhOPO[NHCH(CH_2Ph)COOH]OH$. IV, VIII. Ba-salt, phenylalanine derivative.[424]
$PhCH_2OPO(NHCH_2CH_2Ph)OH$. XVII.[160]
$PhCH_2OPO(NHC_6H_4OEt\text{-}p)OH$. XVII. m. 101-3°.[176]
C_{17} $PhOPO[NHCH(CH_2C_6H_4OH\text{-}p)(CONHCH_2COOH)]OH$. IV, VIII. Ba-salt, glycyl-tyrosine derivative.[926]
$MeOPO[N(C_8H_{17})_2]OH$. XVII. $[Me_4N^+]$-salt: m. 186°.[156]
C_x $PhCH_2OPO(NHCSNHPh)OH$. XII. $[C_6H_{11}NH_3^+]$-salt: m. 151-2°.[268] aniline-salt.[268]
$C_{27}H_{45}OPO(NPh_2)OH$. IX. m. 173°.[923] Cholesteryl derivative.
$[C_{15}H_{31}COOCH_2CH(OCOC_{15}H_{31})CH_2O]PO[NEt(CH_2CH_2Cl)]OH$. XVII m. 65-6°.[86]

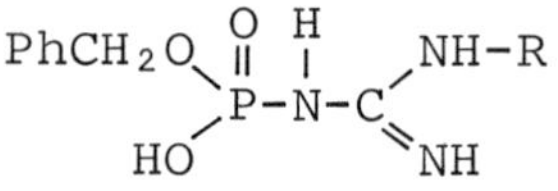

XVII.[173] Betaine structure.
R=$PhCH_2$: m. 161°.
R=$HOOCCH_2$: m. 198°.
R=C_6H_{11}: m. 245°.
R=Ph.

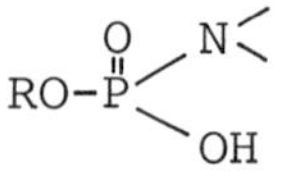

R=Adenosine-5', $-NH_2$,[590] Li-salt,[154] $[(C_6H_{11}NH)_2CNH_2^+]$-salt: m. 240-1°.[152,153]
R=Adenosine-5', $-NC_5H_{10}$.[606]
R=Adenosine-5', $-N(CH_2CH_2)_2O$.[606]
R=Adenosine-5', $-NHC_6H_4OMe\text{-}p$.[606]
R=Adenosine-5', $-[NC_3H_3N]$, imidazole derivative.[172]
R=Adenosine-5', NHC_6H_{11}.[171]
R= 2',3'-O-isopropylidene adenosine-5', $-NHCH(COOMe)CH_2Ph$. $[Et_3NH^+]$-salt.[38]
R=Uridine-5', $-NH_2$.[152,153]
R=Uridine-5', $-N(CH_2CH_2)_2O$.[606]
R= 3-Me-uridine-5', $-N(CH_2CH_2)_2O$.[478]
R=Cytidine-5', $-N(CH_2CH_2)_2O$.[606]
R=Guanosine-5', $-N(CH_2CH_2)_2O$.[606]

R=Thymidine-5', $-N(CH_2CH_2)_2O$.[606]
R=Desoxycytidine-5', $-N(CH_2CH_2)_2O$.[606]
R= 2,3,4,6-tetra-O-acetyl-β-D-glucose-1, $-N(CH_2CH_2Cl)_2$.[885]

D.1.7.5. >N-P(O)(OR)(OR'), R,R' = Alkyl or Aryl or Acyl

C_4 (MeO)(EtO)PO(NCO). XV. b_7 81°.[236]

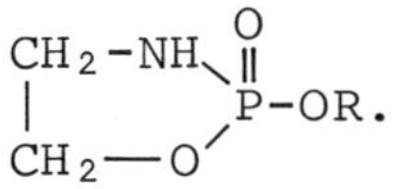
IV.[55] R= Et, $ClCH_2CH_2$, δ-ClC_4H_8, C_8H_{17}, Ph. Oils.

C_5 (MeO)(CCl_2=CHO)$PONMe_2$. XIII. b_2 98-8.5°, n 1.4650, d 1.3352.[17]

$\overline{OCHMeCH_2OP}O(NC_2H_4)$. XIII. $b_{0.06}$ 95-6°, n 1.4644, d 1.2828.[345]

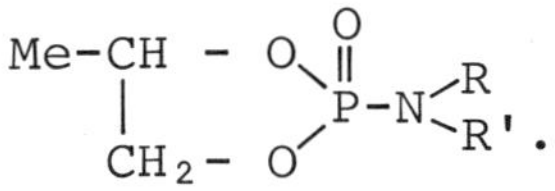
XIII.[2] R=R'=Me: b_3 110-1°, n 1.4512, d 1.1926.
R=R'=Et: b_3 126.5-7.5°, n 1.4558, d 1.1162.
R=R'=Pr: b_4 134-4.5°, n 1.4570, d 1.0785.
R=R'=Bu: b_5 156-7°, n 1.4547, d 1.0444.
XVII.[59] R=R'= $-CH_2CH_2Cl$
V.[604] R=H, R'=Ph: m. 152-4°.
R=H, R'=p-MeC_6H_4: m. 155-7°.
R=H, R'=$PhCH_2$: m. 125-6°.
R=H, R'=C_6H_{11}: m. 129-31°.

(MeO)(FCH_2CH_2O)$PONMe_2$. V. b_1 65-6°, n 1.4170, d 1.1954.[396]

(EtO)(PrO)$PONH_2$. V. b_1 119°, m. 30-2°.[139]

C_6 ($ClCH_2CH_2O$)(CCl_2=CHO)$PONMe_2$. XIII. b_2 137°.[761]

(EtO)(CCl_2=CHO)$PONMe_2$. XIII. b_3 90-1°, n 1.4580, d 1.2801.[17]

(EtO)(FCH_2CH_2O)$PONMe_2$. V. $b_{0.5}$ 62°, n 1.4178, d 1.1449.[396]

C_7 (EtO)(CCl_2=CHO)PONEt(HCO). XIII. b_1 107-9°, n 1.4706, d 1.3099.[24]

(EtO)(CCl_2=CHO)PONHCOOEt. XIII. b_2 65-115°.[39]

(EtO)(CCl_2=CHO)PONEtCOOEt. XIII. b_1 130-2°, n 1.4700, d 1.2791.[39]

(MeO)(CCl_2=CHO)$PONEt_2$. XIII. $b_{0.5}$ 88-9°,[17] b_3 116-7°.[4] n 1.4655,[17] n 1.4646.[4] d 1.2518.[17]

(i-PrO)(CCl_2=CHO)$PONMe_2$. XIII. $b_{0.5}$ 77°, n 1.4545, d 1.2237.[17]

(MeO)(EtO)$PON(CH_2CH_2)_2O$. XVII.[155]

(i-PrO)(FCH_2CH_2O)$PONMe_2$. V. b_1 68°, n 1.4169, d 1.1059.[396]

(MeO)(EtO)$PONEt_2$. XVII. b_1 59-60°.[156]

C_8 (MeO)(2,4-$Cl_2C_6H_3O$)PONHMe. V. m. 54-5°,[253] similar compounds.[253]

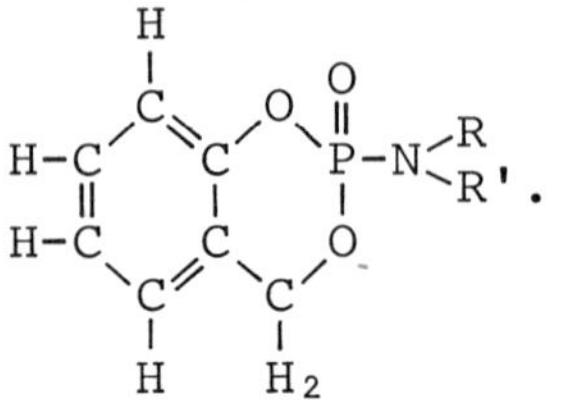

IX.[844] R=H, R'=Me: m. 87°.[277,845,842]
R=H, R'=Et: m. 68°.[842,845]
R=H, R'=Pr: $b_{0.5}$ 135-40°.[277,842,845]
R=H, R'=Pr-i: m. 84°.[842,845]
R=H, R'=Bu: m. 46-7°.[842,845]
R=H, R'=Ph: m. 131-2.5°.[842,84
R=R'=Me: m. 121°.[842,845]
R=R'=Et: $b_{0.5}$ 133-6°.[845] m. 44°.[842,845]

(EtO)(PhO)$PONH_2$. V. m. 133°.[612]
(MeO)(p-$MeSC_6H_4O$)$PONH_2$. V. m. 91°.[418]
(EtO)(CCl_2=CHO)$PONEt_2$. XIII. b_1 85-7°,[17] b_3 112-4°.[4] n 1.4620,[17] n 1.4600.[4] d 1.2132.[17]
(EtO)(CCl_2=CHO)PON(CH_2CH_2Cl)$_2$. XIII. b_3 143-6°, n 1.4920, d 1.3862.[39]
(EtO)(CCl_2=CHO)PONMe(COOEt). XIII. b_1 121-2°, n 1.4750 d 1.3150.[39]
($ClCH_2CH_2O$)(CCl_2=CHO)$PONEt_2$. XIII. b_1 147-8°,[6] $b_{0.2}$ 113-23°,[27] b_2 147-9°.[377] n 1.4839,[6] n 1.4771.[37 d 1.3087,[6] d 1.2852.[377]
(i-PrO)(CCl_2=CHO)PONEt(HCO). XIII. b_1 114-6°, n 1.4662, d 1.2667.[24]
(BuO)(CCl_2=CHO)$PONMe_2$. XIII. b_2 99°, n 1.4610, d 1.2044.[17]
(EtO)($EtOCOCH_2O$)PO(NC_2H_4). V. $b_{0.0001}$ 95-105°, n 1.441[illegible] d 1.170.[676]
(MeO)(EtO)PO(NC_5H_{10}). XVII. b_1 79°.[156]
(FCH_2CH_2O)(i-BuO)$PONMe_2$. V. b_1 75°, n 1.4225, d 1.0882.[396]
(EtO)(PrO)PONHPr. V. $b_{0.2}$ 85-6°, n_D^{21} 1.4270.[139]

C_9 (MeO)[4,2-(NC),ClC_6H_3O]PONHMe. IX. m. 115-6°.[107]
(MeO)(PhO)PONHCH(OH)CCl_3. XVII. m. 122-3°.[14]

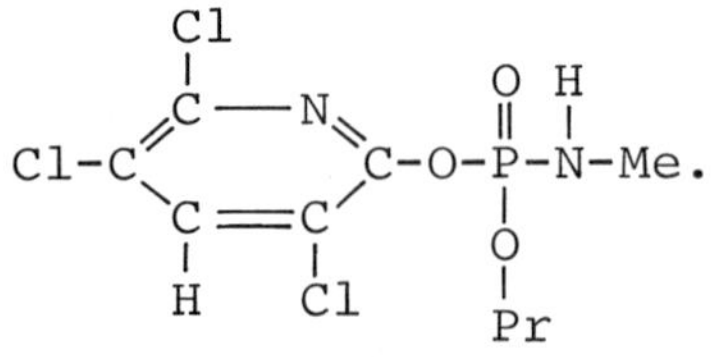

IX. m. 68-71°.[252] For other pyridine ester substituent see Ref. 252.

(EtO)(PhO)PONHMe. XVII. $b_{0.015}$ 120°, n 1.5416, m. 76-8°.[674]
(EtO)(p-$MeSC_6H_4O$)$PONH_2$. V. m. 78-9°.[418]
(MeO)(4,2-MeS,MeC_6H_3O)$PONH_2$. V. m. 80-1°.[418]

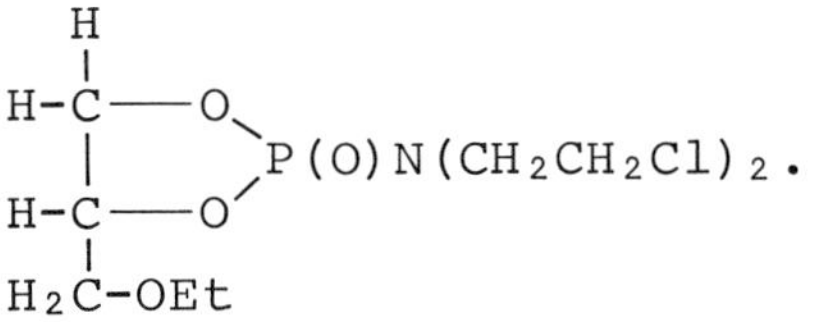

$$\begin{array}{c} H \\ | \\ H-C-O \\ | \quad\quad\quad >P(O)N(CH_2CH_2Cl)_2. \quad IX.[59] \\ H-C-O \\ | \\ H_2C-OEt \end{array}$$

(EtO)[CH$_2$=C(CH$_2$Cl)O]PONEt$_2$.   XVII.   b$_3$ 120-2°, n 1.4545.[4]
(MeO)(BuO)PONEt$_2$.   XVII.[155]
(EtO)(CCl$_2$=CHO)PONHPh.   XIII.   m. 80-1°.[39]
(BuO)(CCl$_2$=CHO)PONEt$_2$.   XIII.'   b$_1$ 109-10°,[17] b$_3$ 131-3°.[4]
    n 1.4605,[17] n 1.4590.[4]   d 1.1495.[17]
(BuO)(CCl$_2$=CHO)PONHBu.   XIII.   b$_1$ 134-5°, n 1.4570,
    d 1.1528.[6]

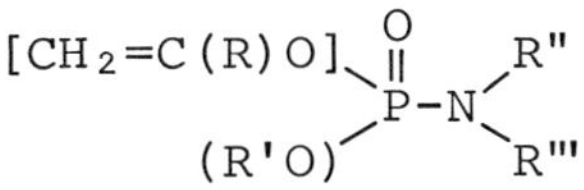

$$\begin{array}{c} \quad\quad\quad\quad O \\ \quad\quad\quad\quad || \quad R'' \\ [CH_2=C(R)O] \quad P-N \\ \quad\quad\quad\quad \quad R''' \\ (R'O) \end{array}$$ XIII. For compounds with R = Me,
OMe, CH$_2$Cl, R' = alkyl, R" = alkyl
or H, R''' = alkyl or H see Refs. 3,4,
6,7,27,345,377.

(EtO)(PhO)PONHCH(OH)CCl$_3$.   XVII.   m. 89-90°.[14]
(EtO)(p-ClC$_6$H$_4$O)PONHCH(OH)CCl$_3$.   XVII.   m. 104-5°.[14]
(MeO)(EtO)PONHCONHPh.   XII.   90-2°.[236]
(EtO)(o-O$_2$NC$_6$H$_4$O)PONMe$_2$.   b$_1$ 156°.[101]
(EtO)(p-O$_2$NC$_6$H$_4$O)PONMe$_2$.   b$_2$ 185°.[101]
(EtO)(PhO)PONMe$_2$.   b$_1$ 156°.[101]
(EtO)(4,3-MeS,MeC$_6$H$_3$O)PONH$_2$.   V.   m. 73-5°.[418]
(i-PrO)(p-MeSC$_6$H$_4$O)PONH$_2$.   V.   m. 75°.[418]

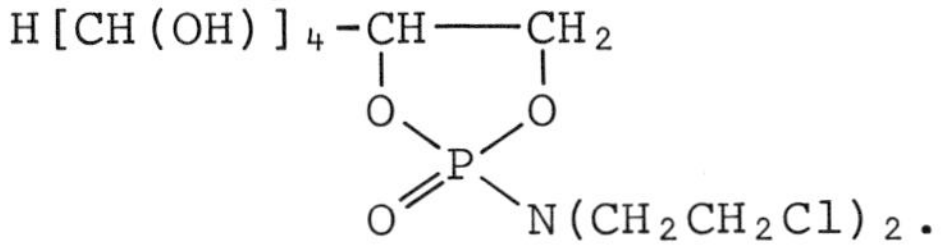

$$\begin{array}{c} H[CH(OH)]_4-CH-CH_2 \\ \quad\quad\quad | \quad\quad | \\ \quad\quad\quad O \quad\quad O \\ \quad\quad\quad\quad >P< \\ \quad\quad\quad O \quad\quad N(CH_2CH_2Cl)_2. \end{array}$$ IX.   D-sorbit derivative.[59]

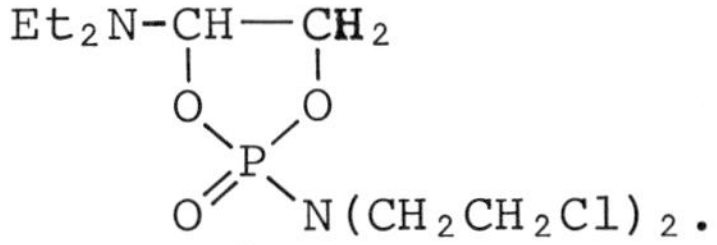

$$\begin{array}{c} Et_2N-CH-CH_2 \\ \quad\quad\quad | \quad\quad | \\ \quad\quad\quad O \quad\quad O \\ \quad\quad\quad\quad >P< \\ \quad\quad\quad O \quad\quad N(CH_2CH_2Cl)_2. \end{array}$$ V.[59]

$$\begin{array}{c} H \quad Ph \\ | \quad\quad | \quad\quad\quad\quad\quad\quad Cl \\ H-C-N \quad\quad\quad\quad\quad\quad | \\ | \quad\quad\quad >P(O)-O-CH_2-C-CH_2-Cl. \\ H-C-O \quad\quad\quad\quad\quad\quad | \\ | \quad\quad\quad\quad\quad\quad\quad\quad NO_2 \\ H \end{array}$$ IV.   m. 98-9°.[318]   p-ClC$_6$H$_4$
    instead of Ph: m. 135-6°.[318]

(i-PrO)(PhO)PONHCH(OH)CCl$_3$.   XVII.   m. 107-8°.[14]
(i-PrO)(o-ClC$_6$H$_4$O)PONHCH(OH)CCl$_3$.   XVII.   m. 118-9°.[14]
(MeO)[o-(NC)C$_6$H$_4$O]PONHPr.   IX.   m. 170-9°.[107]
(MeO)(i-PrO)PONHCONHPh.   XII.   m. 134-5°.[236]
(BuO)(HC≡CCH$_2$O)PONEt$_2$.   IX.   Oil.[110]
(EtO)(PrO)PONPr$_2$.   V.   b$_{0.5}$ 76-8°, n 1.4290.[139]
(EtO)(PhO)PON(CH$_2$CH$_2$Cl)$_2$.[316]   For other (RO)(PhO)PON
    (CH$_2$CH$_2$Cl)$_2$ compounds, no data listed, see Ref. 316.
(EtO)(i-PrO)PONHCONHPh.   XII.   m. 120-2°.[236]
(i-PrO)[4,2-(NC)MeC$_6$H$_3$O]PONHMe.   IX.   m. 74-80°.[107]

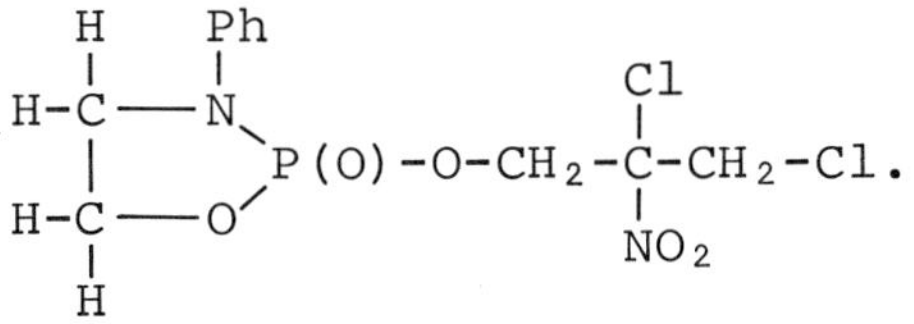

(EtO)(PhO)PONEt$_2$. XVII. $b_{0.03}$ 170°, m. 49-50°.[674]
(MeO)(PhCH$_2$O)PONEt$_2$. XVII. b_1 114°.[156]

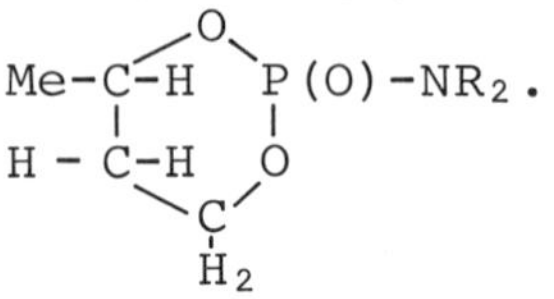

XIII.[345] -NBu$_2$ $b_{0.5}$ 119-20°, n 1.4566 d 1.0346. -N(CH$_2$CH=CH$_2$)$_2$ b_1 133-4°, n 1.4770, d 1.0972. -(NC$_2$H$_4$) b_1 126-7°, n 1.4739, d 1.2504.

C_{13} 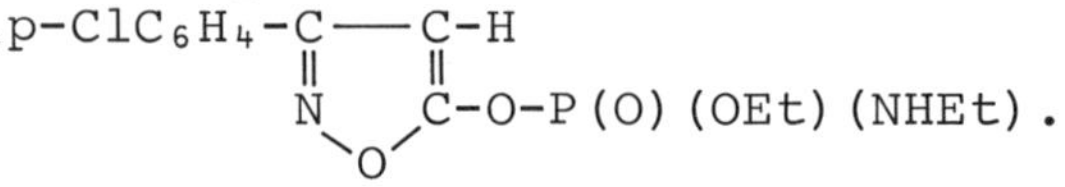

V.[724] Other heterocyclic substituents.[72]
(MeO)(p-t-BuC$_6$H$_4$O)PONHCH(OH)CCl$_3$. XVII. m. 140-1°.[14]
(EtO)(o-EtOCOC$_6$H$_4$O)PONMe$_2$. b_2 160°.[101]
C_{14} (MeO)(PhO)PONHCOC$_6$H$_4$NO$_2$-p. IX. m. 114-6°.[220]
(MeO)(PhO)PONHCOC$_6$H$_3$(NO$_2$)$_2$-3,5. IX. m. 192-3°.[220]
(BrCH$_2$CH$_2$O)(PhO)PONHPh. V. m. 128°.[78]
(MeO)(PhCH$_2$O)PONHC$_6$H$_4$Cl-p. XVII. m. 111-3°.[529]
(EtO)(PhO)PONHPh. V, IX. m. 143°,[612] m. 120°.[573]

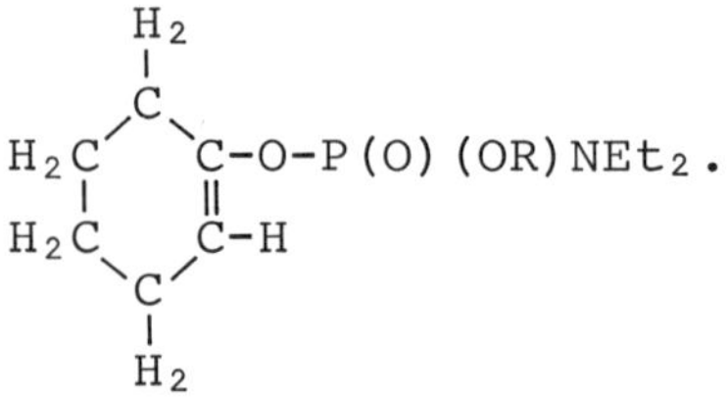

XIII. R=Bu: b_6 160-1°, n 1.4644.[4] R=CH$_2$CH$_2$Cl: b_3 149-50°,[3] n 1.4797.[3] d 1.1477.[3]

C_{15} 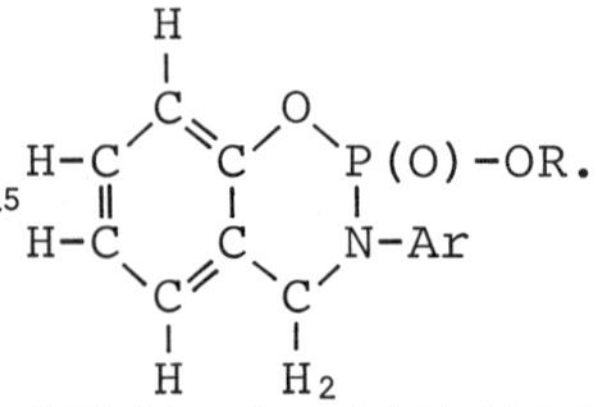

R=Et, Ar=m-O$_2$NC$_6$H$_4$;[278] and R=Ph, Ar=Ph.[278]

[Cl(CH$_2$)$_3$O](PhO)PONHPh. V. m. 65°.[78]
(PrO)(PhO)PONHPh. V. m. 122°.[628]
C_{16} (PhO)(2-C$_{10}$H$_7$O)PONH$_2$. V. m. 152-3°.[428]
C_{17} [F$_2$CH(CF$_2$)$_5$CH$_2$O](PhO)PONMe(CH$_2$CH$_2$CN). V. $b_{0.05}$ 163°, n_D^{25} 1.4179.[608] Other fluorinated ester substituents.[608]
(PhO)(o-MeC$_6$H$_4$O)PON(CH$_2$CH$_2$Cl)$_2$. III. m. 98-9°. Other arylester substituents.[502]
C_{19} (PhO)(p-MeC$_6$H$_4$O)PONHPh. IX. m. 106°.[573]
(PrO)(PhO)PONHC$_{10}$H$_7$-2. V. m. 101°.[628]
C_{22} (BuO)[HC≡CC(MeEt)O]PONPh$_2$. V. Oil.[110]
C_x (EtO)PO(NHPh)-O-tetra-acetyl-glucose. IX. m. 116-7°.[92]

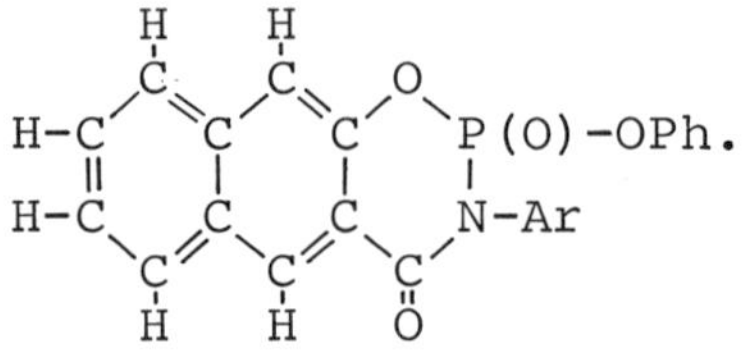

II. Ar=Ph: m. 173°.[670] Ar=C$_{10}$H$_7$-2: m. 185°.[670]

(PhO)(2-$C_{10}H_7O$)PONHC_6H_4Me-p. V. m. 126-7°.[428]
(BuO)(HC≡CCH_2O)PONH(CH_2)$_8$CH=CH(CH_2)$_7CH_3$. V. Oil.[110]
($C_8H_{17}O$)(HC≡CCHMeO)PONH$C_{18}H_{35}$. V. Oil.[110]
(PhO)[$C_{15}H_{31}$COOCH_2CH(OCO$C_{15}H_{31}$)CH_2O]PON(CH_2CH_2Cl)$_2$. XVII. m. 35.5-37°.[86]

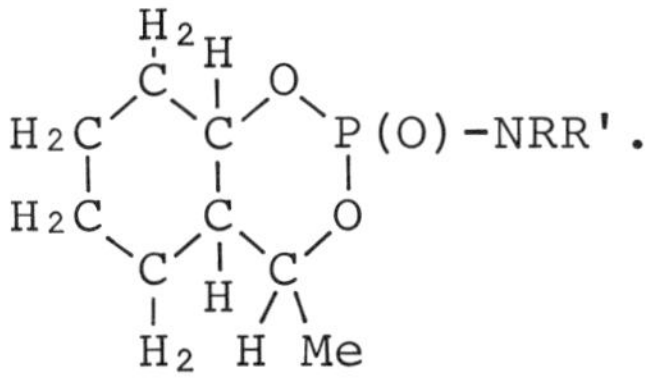

V.[506] R=R'=H: m. 153-5°.
R=H, R'=Ph: m. 89°.
R=H, R'=Bu: n_D^{30} 1.4810.
R=R'=Me: n_D^{30} 1.4808.

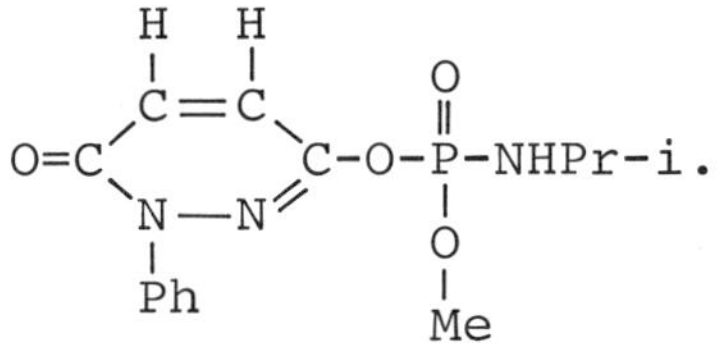

n_D^{25} 1.5247;[722] for similar compounds see Ref. 722.

3,6-[Ph_2N-P(O)O,O]-α-methylglucoside. IX. m. 251°,[923] the 2,4-diacetyl-methylglucosid m. 138°.[923]
N-menthyl-phosphoramidicacid phenyl, 2-naphtyl ester. V. (From levo form of the amine), two forms: m. 135-6° and m. 94-6°.[428]
N-menthyl-phosphoramidicacid phenyl, p-tolyl ester. V. (From levo form of the amine), two forms: m. 109-10° and m. 85-6°.[536]
N-hydrindyl-phosphoramidicacid phenyl, p-tolyl ester. V. Inactive form: m. 98-100°; active form (from d-amine) two isomers: m. 127°, m. 82-6°.[536]
Adenosine-(2',3'-isopropylidene)-5'-O-P(O)(OR)(N<).

R	(N<)
$PhCH_2$-	-NH_2 dec. 65-7°[812]
$PhCH_2$-	-NHEt dec. 57-60°[812]
$PhCH_2$-	-NEt_2,[812] picrate: m. 141-3°[914]
Et-	-NEt_2 picrate: m. 169-70°[914]
$PhCH_2$-	phenylalanine-Me-ester dec. 79-81°[38]
$PhCH_2$-	glycine-Me-ester dec. 60-3°[38]
$PhCH_2$-	phenylalanine-Et-ester dec. 88-90°[38]
$PhCH_2$-	leucine-Me-ester dec. 68-72°[38]
$PhCH_2$-	N-carbobenzoxy-lysine-Me-ester dec. 50-5°[38]
$PhCH_2$-	glutamicacid-di-Et-ester dec. 70-6°[38]

(RO)(EtO)PONEt_2. R = adenosine-5', picrate: m. 138-40° dec., sint. 125°.[914]
(RO)(PhO)PONH_2. R = 2',3'-O-diacetyl-adenosine-5'.[154]
(RO)(PhCH_2O)PONHC$H_2C_6H_4$Br-p. R = 2',3'-O-isopropylidene-uridine-5'.[166]
(RO)(PhCH_2O)PON(CH_2CH_2Cl)$_2$. R = 2,3,4,6-tetra-O-acetyl-β-D glucose-1.[885]

D.1.8. >N-P(S)AB, A ≠ B, A,B = Halogen, -OH,-OR,-SR -OAc, Also >N-P(O)(SR)B

D.1.8.1. >N-P(S)XY, X,Y = Halogen

$(ClCH_2CH_2)_2$NPSClF. II. b_{4-5} 110-5°, n 1.5219.[831]
Et_2NPSBrF. II. $b_{0.5}$ 47°, n 1.5088, d 1.4576.[821]
Et_2NPSClF. II. b_9 67°,[821] b_{12} 60-6_°.[657]
n 1.4769.[657,821] d 1.2082.[821]
BuNHPSClF. II. b_{10} 81-3°.[657] n 1.4737.[657]
PhNHPSClF. II. b_{3-5} 100-3°,[831] b_9 111-2°.[658]
n 1.5474,[831] n 1.5846.[658]
p-ClC_6H_4NHPSClF. II. b_{3-5} 95-100°, n 1.5639.[831]
p-FC_6H_4NHPSClF. II. b_{3-5} 74-5°, n 1.5645.[831]
C_6H_{11}NHPSClF. II. b_8 95-7°, n 1.5148.[657]
o-MeC_6H_4NHPSClF. II. b_9 128-9°, n 1.5722.[658]
m-MeC_6H_4NHPSClF. II. b_9 126-7°, n 1.5758.[658]
p-MeC_6H_4NHPSClF. II. b_9 123-4°,[658] b_{4-5} 85-7°.[831]
n 1.5758,[658] n 1.5523.[831]
o-$MeOC_6H_4$NHPSClF. II. b_9 123-4°, n 1.5782.[658]

D.1.8.2. >N-P[(O,S)H]X, >N-P(S)X(OR), >N-P(O)X(SR), >N-P(S)X(SR), X = Halogen, R = Alkyl or Aryl or Acyl

C_2 MeOPS(NCS)F. XIV. b_{15} 53-4°, n 1.5320, d 1.3573.[386]
MeOPS(NHMe)Cl. IV. $b_{1.5}$ 68.5°.[381]
C_3 EtOPS(NCS)F. XIV. b_9 59-60°.[386] ^{31}P -40, ^{19}F.[361a]
EtOPS(NHMe)Cl. IV. b_2 75°,[381] b_2 74-7°.[641]
MeOPS(NHEt)Cl. IV. $b_{0.5}$ 65°,[381] b_1 70-3°.[641]
MeOPS(NMe_2)Cl. IV. $b_{1.5}$ 50-2°.[641]
MeSPS(NMe_2)F. IV. $b_{0.01}$ 38-40°, IR, ^1H, ^{19}F, ^{31}P -106,
mass spectrum.[731]
C_4 PrOPS(NCS)F. XIV. b_{10} 73-5°.[386]
i-PrOPS(NCS)F. XIV. b_8 64-6°.[386]
MeOPS(NHPr)Cl. IV. $b_{1.5}$ 62-5°,[381] $b_{1.5}$ 87°.[641]
EtOPS(NHEt)Cl. IV. $b_{1.5}$ 85°.[641]
EtOPS(NMe_2)Cl. IV. b_3 72-3°,[395] b_2 58-60°,[641]
n 1.4972,[395] d 1.2008.[395]
HOPS(NEt_2)F. XVII. K-salt.[821]
EtOPS(NMe_2)F. IV. b_6 41-2°, n 1.4478.[657]
EtSPS(NMe_2)F. IV. $b_{0.02}$ 33.5-5.5°, IR, ^1H, ^{31}P -103.8,
mass spectrum.[731]
C_5 EtOPS(NMe_2)CN. XIV. b_2 94°,[766] b_{12} 107-9°.[755]
MeOPS(NHBu-i)Cl. IV. $b_{0.5}$ 80-2°.[381]
EtOPS(NHPr-i)Cl. IV. $b_{1.5}$ 80°.[641]
MeOPS(NHBu)F. IV. b_{10} 91-3°, n 1.4660.[657]
MeSPS(NEt_2)F. IV. $b_{0.02}$ 45-7°, IR, ^1H, ^{19}F, ^{31}P -102.6,
mass spectrum.[731]
EtOPS(NHPr)SH. m. 91°.[283]

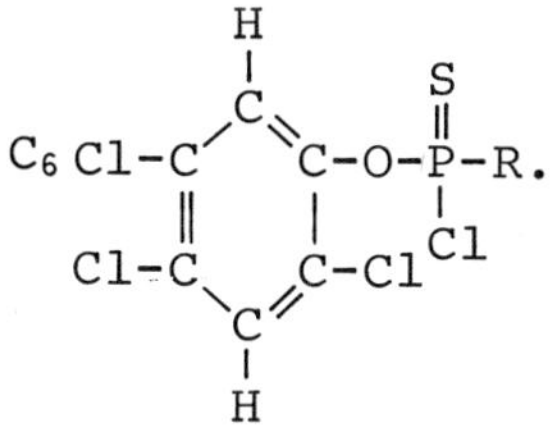

IV.[105]

R	m
NH_2	55-6°
NHMe	61-2°
NHPr-i	51-3°
NHPh	n_D^{25} 1.6386, d_{25} 1.5140
NHC_6H_{11}	88-9°
$NHCH_2Ph$	73-3.5°
NMe_2	49-50°
(NC_5H_{10})	40-3°
$N(CH_2CH_2)_2O$	93-5°
NHC_6H_4Cl-m	n_D^{25} 1.6318, d_4^{25} 1.5470
(NC_4H_8)	66-8°

$ClCH_2CH_2OPS(NEt_2)F$. XIV. b_1 77-8°.[541]
$ClCH_2CH_2OPS(NEt_2)Cl$. XIII. b_{10} 104-5°.[541]
$EtOPS(NEt_2)Cl$. IV. b_{10} 105-6°, n 1.4931, d 1.1352.[395] ^{31}P -61.1.[873]
i-$BuSPO(NMe_2)Cl$. XIII. b_6 123-4°, n 1.4970, d 1.1536.[675]
EtOPS(NHBu)Cl. IV. $b_{1.5}$ 102-4°,[641] $b_{0.08}$ 73-7°, n 1.4940, d 1.1433.[566] Similar alkyl substituted compounds.[566]
$EtSPS(NEt_2)F$. IV. $b_{0.03}$ 53-6°, IR, ^1H, ^{31}P -101.3, mass spectrum.[731]
$EtOPS(NEt_2)F$. IV. b_{10} 82°,[821] b_{10} 70-1°.[657] n 1.4551,[821] n 1.4548.[657] d 1.1020.[821]
C_7 2,4,6-$Cl_3C_6H_2OPS(NHMe)Cl$. IV. m. 88-91°.[105]
$C_6Cl_5OPS(NHMe)Cl$. IV. m. 114.5-6.5°.[918]
PhOPS(NHMe)Cl. IV. m. 22-2.5°.[105]
$EtOPS[N(CH_2CH_2)_2O]CN$. XIV. b_{12} 152°.[755]
$EtOPS(NEt_2)CN$. XIV. b_{12} 125°.[755]
$[EtOC(S)S]PO(NEt_2)Cl$. IX. d_{18} 1.195.[861]
i-$PrOPS(NEt_2)F$. IV. b_8 60-1°, n 1.4482.[657]
C_8 $C_6Cl_5OPS(NHEt)Cl$. IV. m. 120.5-2.0°.[918]
$C_6Cl_5OPS(NMe_2)Cl$. IV. m. 94-5°.[917]
EtOPS(NHPh)F. IV. $b_{0.01}$ 90-5°, ^{31}P -64, ^{19}F.[361a]

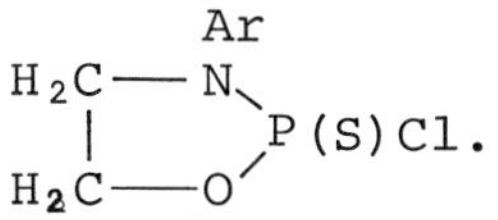

XIII.[318] Ar = Ph, b_1 165-80°.
Ar = p-ClC_6H_4, b_{23} 218°.

$PhOPS(NMe_2)F$. IV. b_3 93-3.5°.[655]
o-$MeC_6H_4OPS(NHMe)F$. IV. b_3 110-0.5°.[655]
$PhSPS(NMe_2)F$. IV. $b_{0.3}$ 103-4°, ^1H, ^{19}F, ^{31}P -94.[730]
$EtOC(S)SPO(NEt_2)CN$. XIV. Oil. d_{18} 1.187.[816]
$EtOPS(NC_5H_{10})CN$. XIV. b_{13} 152°.[755]
$EtOPS(NHC_6H_{11})F$. IV. b_{12} 125-7°, n 1.4878.[657]
$BuSPO(NEt_2)Cl$. XIII. b_4 128-9°, n 1.4962, d 1.1059.[675]
C_9 $C_6Cl_5OPS(NHCH_2CH{=}CH_2)Cl$. IV. m. 104-7°.[918]

$C_6Cl_5OPS(NHPr)Cl$. IV. m. 109-13°.[917]
2,4-$Cl_2C_6H_3OPS(NHPr$-i)Cl. IV. m. 50.5-1.0°.[105]
EtOPS(NHCSNHPh)F. XII. m. 78-9°.[386] For other ROPS(NHR')F compounds see Ref. 386.
EtOPS(NPr_2)CN. XIV. b_{13} 137°.[755]
C_{10} $C_6Cl_5OPS(NHBu)Cl$. IV. m. 93-6°.[917]
$C_6Cl_5OPS(NEt_2)Cl$. IV. m. 74-6°.[917]
PhSPS(NEt_2)F. IV. $b_{0.3}$ 127-8°, ^1H, ^{19}F, ^{31}P -94.[730]
C_{11} EtOPS(NBu_2)CN. XIV. b_{13} 146°.[755]
C_{12} PhOPS(NHPh)Cl. IV. m. 153°,[76] n_D^{25} 1.6242, d_4^{25} 1.3091.[10
2-$C_{10}H_7OPS(NMe_2)Cl$. IV. m. 49-51°, $b_{0.03-0.05}$ 178-80°.[12
C_{13} 2-$C_{10}H_7OPS(NHPr)Cl$. IV. $b_{0.03-0.1}$ 230-50°, n 1.6178.[129
2-$C_{10}H_7OPS(NHPr$-i)Cl. IV. m. 73-5°.[129]
C_{14} 2-$C_{10}H_7OPS(NEt_2)Cl$. IV. $b_{0.5}$ 163-6°, n 1.6002.[129]
C_x 4,2,6-Me$(C_6H_{11})_2C_6H_2OPS(NHMe)Cl$. IV.[106]
RR'N-PS(OR")Cl. Not isolated intermediates, no data given.[618]

D.1.8.3. >N-P[(O,S)H](OR), R = Alkyl or Aryl or Acy

MeOPS($NHCH_2CF_3$)OH. XVII. [$C_6H_{11}NH_3^+$]-salt: m. 83-4°.[354]
MeOPS(NMe_2)OH. XVII. [Et_3MeN^+]-salt: m. 85-90°.[738] For similar compounds see Ref. 738.
PhOP(OSH)NH_2. m. 127-8°.[75] Na-salt.[272]
MeOPS(NHPh)OH. XVII. $[C_6H_{11}NH_3]^+$-salt: m. 144-5°. (+)form: $[\alpha]_D$ -27° in ethanol.[354]
MeOPS(NHC_6H_{11})OH. XVII. m. 115-6°. quinine-salt: (+)form: m. 176°; (-)form: m. 158°.[354]
i-PrOPS(NHBu)SH. m. 88°.[283]

D.1.8.4. >N-P(O)(OR)(SR'), >N-P(S)(OR)(OR'), >N-P(S)(OR)(SR'), R,R' = Alkyl or Aryl or Acyl

C_2 (MeO)(MeS)$PONH_2$. XVII. m. 54°.[292]
C_3 (MeS)(EtO)$PONH_2$. XVII. m. 70°.[292]
C_4 (MeO)(EtO)PS(NCO). XVII. b_{19} 68-70°, n 1.4660, d 1.2233.[740]
(MeO)(Cl_3CSSS)PONHEt. XVII. m. 84-9°.[142] Similar S-compounds.[142]
(MeO)(CH_2=$CHCH_2S$)$PONH_2$. XVII. m. 36-7°.[293]
(MeO)($ClCH_2CH_2O$)PSNHMe. IX. n_D^{25} 1.5012, many similar compounds as oils.[126]
(MeO)(EtS)PSNHMe. IX. $b_{0.08}$ 65.5-71.5°.[777]
(MeS)(i-PrO)$PONH_2$. XVII. m. 68-70°.[292]
(EtO)(EtS)$PONH_2$. XVII. m. 51°.[137]

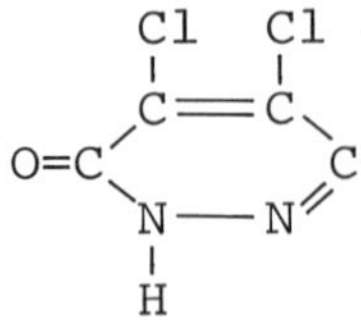

C-O-P(S)(OMe)(NHMe). IX. m. 120-2°, similar pyridazinylesters.[722]

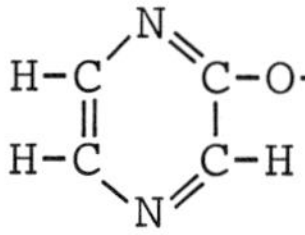
-O-P(S)(OMe)(NHMe). V. m. 57-8°.[35]

C_5 (MeO)(CH_2=CClCH$_2$S)PONHMe. IX. n_D^{25} 1.5410.[776] Other (H_2C=CClCH$_2$S)(RO)PONHR', R = alkyl, n_D^{25} given.[776]

(EtO)(CH_2=CHCH$_2$S)PONH$_2$. XVII. $b_{0.2}$ 128-30°, m. 45-6°.[293]

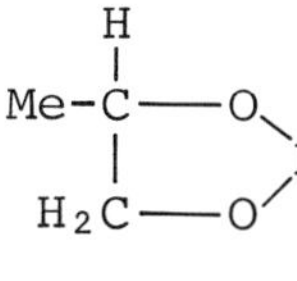
P(S)(N<). XIII.

-NMe$_2$: b_2 107-8°,[2] $b_{0.2}$ 98-102°.[264] m. 43-4°.[2] n_D^{21} 1.5066.[264]

-NEt$_2$: b_1 118-9°, n 1.5000, d 1.1388.[2]

-NPr$_2$: b_2 125-6°, n 1.4921, d 1.0906.[2]

-NBu$_2$: b_3 158-9°, n 1.4884, d 1.0485.[2]

-N(CH_2CH_2)$_2$O: b_7 174-5°, n 1.5212, d 1.2747.[2]

(MeO)(EtO)PSNHCOOMe. XII. n 1.4868, d 1.2614.[740]

(MeO)(PrS)PSNHMe. IX. $b_{0.09}$77-84°.[777]

(EtO)(PrS)PONH$_2$. XVII. $b_{0.05}$ 122-8°.[137]

(EtO)(EtSCH$_2$S)PONH$_2$. XVII. $b_{0.01}$ 60°.[417]

C_6 (MeO)(CCl_3C≡CCH$_2$O)PSNHMe. IX. n_D^{25} 1.543, similar compounds, n_D^{25} given.[778]

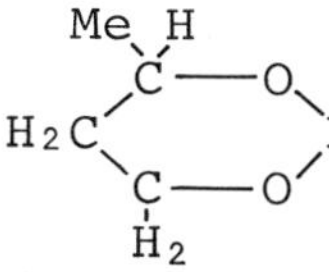
P(S)(NC_2H_4). XIII. m. 76-7°.[344]

(EtO)(CH_2=CHCH$_2$S)PONHMe. XVII. b_1 60°.[293]

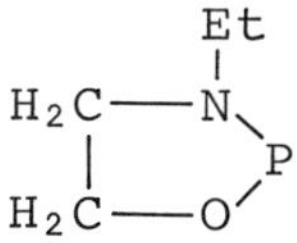
P(S)OEt. IV. b_2 108-10°, $b_{0.5-1}$ 80°, n 1.498.[318]

(MeO)(HOCH$_2$CH=CHCH$_2$O)PSNHMe. IX. b_{30} 45°, n_D^{25} 1.521, d_{24} 1.270.[779] Similar compounds.[779]

(EtO)(FCH$_2$CH$_2$O)PSNMe$_2$. V. b_1 68°, n 1.4665, d 1.1482.[396]

(EtO)(ClS)PONEt$_2$. XVII. $b_{0.3}$ 64°, n_D^{22} 1.4770.[588]

(MeO)(BuS)PSNHMe. IX. $b_{0.07}$ 81-7°.[777]

(MeO)(sec-BuS)PSNHMe. IX. $b_{0.15-0.45}$ 78-99°.[777]

(MeO)(t-BuS)PSNHMe. IX. $b_{0.15-0.45}$ 78-99°.[777]

(EtO)(PrS)PSNHMe. IX. $b_{0.06}$ 96-103°.[777]

(MeO)(EtSCH$_2$CH$_2$O)PSNHMe. IX. n_D^{25} 1.5250, similar compounds, n_D^{25} given.[102]

(EtO)(EtSCH$_2$S)PONHMe. XVII. $b_{0.05}$ 145°.[419]

C_7 (MeO)(2,4,5-$Cl_3C_6H_2$O)PSNH$_2$. IX. m. 57-9°.[103]

(MeO)(2,4-Cl(O_2N)C_6H_3O)PSNH$_2$. V. m. 41-3°.[868]

(MeO)(3,4-$Cl_2C_6H_3$O)PSNH$_2$. V. n_D^{25} 1.5769, d_{25} 1.439.[414]

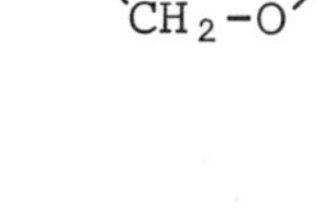
P(O)N(CH_2CH_2Cl)$_2$. IX. m. 102-4°.[314]

(EtO)(N≡CCH$_2$CH$_2$S)PONMe$_2$. XVII. $b_{0.01}$ 85°.[764]
(MeO)(EtO)PSNHCOOPr-i. XII. n 1.4730, d 1.1773.[740]
(EtO)(BuS)PSNHMe. IX. $b_{0.08}$ 92-7°.[777]
(EtO)(EtSCH$_2$CH$_2$S)PONHMe. XVII. $b_{0.05}$ 95-100°.[419]
(MeO)(Et$_2$N-S)PONMe$_2$. XVII. b_1 94-5°, n 1.4730, d 1.0785.[16]

C$_8$ (MeO)(p-ClC$_6$H$_4$O)PS(NCS). IX. n_D^{25} 1.6032.[124]
(MeO)(C$_6$Cl$_5$O)PSNHMe. IX. m. 132-34°.[918]
(EtO)(2,4,5-Cl$_3$C$_6$H$_2$O)PSNH$_2$. IX. m. 59-63°.[255] For other (RO)(ArO)PSNH$_2$ compounds (as oils) see Ref. 255
(MeO)(3,4-Cl(O$_2$N)C$_6$H$_3$O)PSNHMe. IX. n_D^{20} 1.5709.[868]
(MeO)(3,4-Cl$_2$C$_6$H$_3$O)PSNHMe. V. n_D^{25} 1.5662,[414] n_D^{25} 1.5562.[103] d_{25} 1.3914,[414] d_{25} 1.3914.[103]
(MeO)(2,5-Cl$_2$C$_6$H$_3$S)PSNHMe. IX. m. 67-9°.[383]

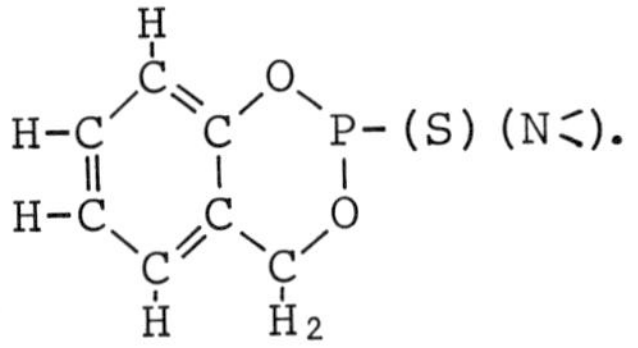

IX.
-NHMe: $b_{0.2}$ 120-3°.[277,842,844,845]
-NHEt: $b_{0.2}$ 110-5°.[842,844,845]
-NMe$_2$: $b_{0.2}$ 118-22°.[842,844,845]
-NEt$_2$: $b_{0.2}$ 110°.[842,844,845]

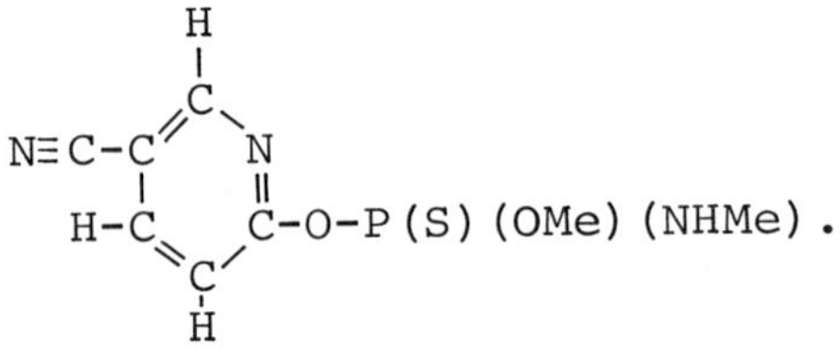

IX. m. 69-71°.[726]

(MeO)(p-O$_2$NC$_6$H$_4$CH$_2$S)PONH$_2$. XVII. m. 129-30°.[417]
(MeO)(PhS)PSNHMe. IV, V, IX. $b_{0.25}$ 127-8°, n 1.6110, d 1.2431.[547] For 60 other (RO)(ArS)PSNHR' compounds see Ref. 547.
(MeO)(o-MeOC$_6$H$_4$O)PSNH$_2$. V. n_D^{35} 1.5636, d_{23} 1.49.[255]

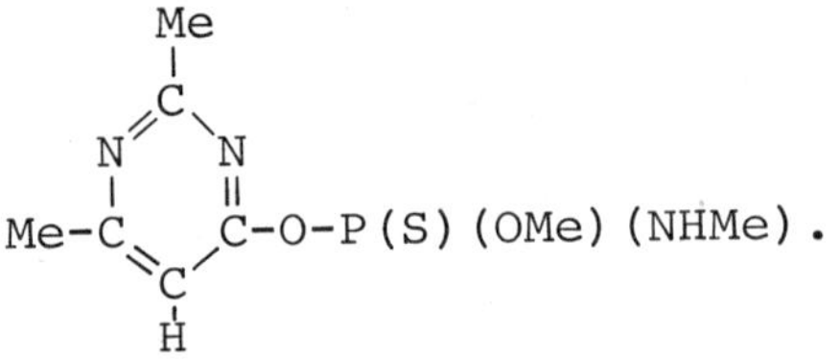

IX. n 1.5433.[382] For other pyrimidine-ol-esters, n given, see Ref. 382.

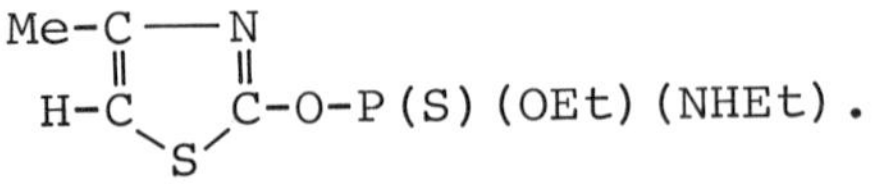

IX. n_D^{25} 1.5192.[723] For other thiazolyl esters see Ref. 723.

(EtO)(ClCH$_2$CH$_2$S)PONEt$_2$. XVII. $b_{0.05}$ 76°, n_D^{26} 1.4850.[58]
(MeO)(MeNHCOCH$_2$S)PSNEt$_2$. IX. n 1.5458, d 1.1855.[835] For other (RO)(R'NHCOCH$_2$S)PSNR$_2$" compounds see Ref. 835.
(EtO)(EtSCH$_2$CH$_2$S)PONMe$_2$. IX, XVII. $b_{0.01}$ 88°,[763] $b_{2.5}$ 122-3°, n 1.5100, d 1.1272.[395]

$(EtO)(EtSCH_2CH_2O)PSNMe_2$. IX. $b_{2.5}$ 112-4°, n 1.5015, d 1.1091.[395]

$(EtO)(EtSCH_2CH_2S)PSNMe_2$. IX. $b_{2.5}$ 136-7°, n 1.5501, d 1.1409.[395]

C_9 $(2,4\text{-}Cl_2C_6H_3O)(CF_3CH_2O)PSNHMe$. IX. n_D^{25} 1.5260, for other $(RO)(2,4\text{-}Cl_2C_6H_3O)PSNHR'$ compounds (R=fluoro-alkyls) see Ref. 717.

$(MeO)(2,4,5\text{-}Cl_3C_6H_2O)PS(NC_2H_4)$. V. m. 50-2°.[864]

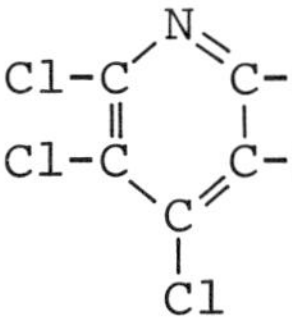
-O-P(S)(OMe)(NHPr-i). IX. m. 77-81°. For other pyridine-ol-esters see Ref. 252.

$(MeO)(3,4\text{-}Cl_2C_6H_3O)PSNHEt$. V. n_D^{25} 1.5599, d_{25} 1.3315.[414]

$(MeO)(2,4,5\text{-}Cl_3C_6H_2O)PSNHEt$. IX. m. 37-9°.[103]

$(EtO)(2,5\text{-}Cl_2C_6H_3S)PSNHMe$. IX. m. 65-8°. For other $(RO)(2,5\text{-}Cl_2C_6H_3S)PSNHR'$ compounds, n given, see Ref. 383.

$(MeO)(p\text{-}ClC_6H_4O)PSNHEt$. IX. n_D^{25} 1.5478, d_{25} 1.2489.[103]

$(EtO)(p\text{-}ClC_6H_4O)PSNHMe$. IX. d_{26} 1.13, n_D^{35} 1.5147.[618] For other (RO)(ArO)PSNR'R" compounds see Ref. 618.

$(EtO)(p\text{-}ClC_6H_4S)PSNHMe$. IX. b_1 85-100°, n 1.5933.[383]

$(EtO)(p\text{-}O_2NC_6H_4S)PSNHMe$. IX. m. 55°. Other (RO)(ArS) PSNHR' compounds see Ref. 383.

(EtO)(PhS)PSNHMe. II, V, IX. $b_{0.2}$ 130-2°.[548] For other (RO)(ArS)PSNR'R" compounds see Ref. 548.

$p\text{-}MeSC_6H_4OPS(MeO)NHMe$. IX. b_{15} 100-10°, n 1.5909.[381] For other (RO)(ArO)PSNHR', n given, see Ref. 381.

$(EtO)(PhSCH_2S)PONH_2$. XVII. m. 52-4°.[417]

$(EtO)(p\text{-}ClC_6H_4CH_2S)PONH_2$. XVII. m. 85°.[417] Similar compounds.[417]

$(EtO)(p\text{-}MeOC_6H_4O)PSNH_2$. V. m. 113-5°.[255]

$(EtO)(MeOCOCH{=}CMeO)PSNMe_2$. XIII. $b_{0.5}$ 104-6°, n 1.4970, d 1.1456.[397]

$(EtO)[EtOC(S)S]PONEt_2$. IX. d 1.054.[861]

$(MeO)(Et_2NS)PONEt_2$. XVII. $b_{1.5}$ 102-3°, n 1.4880, d 1.0553.[16]

C_{10} $(EtO)(C_6Cl_5O)PSNHEt$. IX. m. 115.5-7.0°; for other $(RO)(C_6Cl_5O)PSNHR'$ compounds see Ref. 918.

$(MeS)(EtO)PONHC_6H_4Me\text{-}p$. XVII. m. 88.5-91.5°.[137]

$(MeO)(2,4,5\text{-}Cl_3C_6H_2CH_2CH_2O)PSNHMe$. IX. m. 75-6°.[125]

$(MeS)(EtO)PONHCOC_6H_4Cl\text{-}p$. XVII. m. 128°.[34]

$(MeO)(3,4\text{-}Cl_2C_6H_3O)PSNHPr$. V. n_D^{25} 1.5508, d_{25} 1.3096.[414]

(MeS)(EtO)PONHCOPh. XVII. m. 99°.[34]

(MeO)(EtO)PSNHCONHPh. XII. m. 122-3°.[740]

$(EtO)(p\text{-}ClC_6H_4O)PSNMe_2$. IX. n_D^{35} 1.5375, d_{23} 1.17.[618]

$(EtO)(p\text{-}O_2NC_6H_4O)PSNMe_2$. IX. $b_{0.2}$ 153-5°, n 1.5460, d 1.2468.[565]

$(EtO)(p\text{-}O_2NC_6H_4CH_2S)PONHMe$. XVII. m. 80-2°.[419]

(EtO)(PhS)$PSNMe_2$. IX. $b_{0.01}$ 95°; for other (EtO)(RS) $PSNMe_2$ compounds see Ref. 281.

(MeO)(p-$Me_2NSO_2C_6H_4O$)PSNHMe. IX. m. 106-8°. For other (RO)(ArO)PSNHR' see Ref. 640.

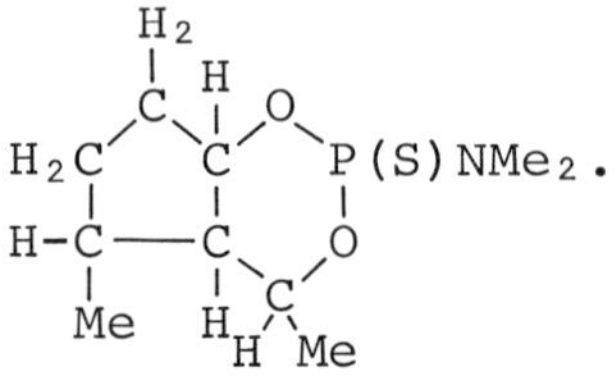

V. n_D^{30} 1.5128.[506]

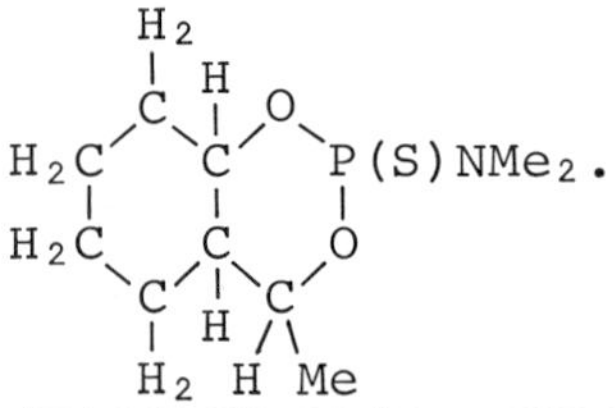

V. n_D^{30} 1.5230.[506]

($EtOCOCH_2O$)(Me_2CHCH_2S)PON(C_2H_4). V. $b_{0.01}$ 106-8°, n 1.5070, d 1.1250.[676]

(EtO)(EtS)$PONHC_6H_{11}$. V, XVII. $b_{0.25}$ 140°,[137] $b_{0.25}$ 130°.[137] m. 53-4°,[137] m. 61°.[137]

(EtO)(PrS)PON(C_5H_{10}). XVII. $b_{0.05}$ 106-10°.[137]

(EtO)(BuS)$PONEt_2$. XIII. b_4 125-6°, n 1.4698, d 1.0210.[67]

(EtO)($EtSCH_2CH_2S$)$PONEt_2$. XVII. b_3 142-3°, n 1.5046, d 1.0884.[395]

(EtO)($EtSCH_2CH_2O$)$PSNEt_2$. IX. b_2 118-20°, n 1.4972, d 1.0779.[395]

(EtO)($EtSCH_2CH_2S$)$PSNEt_2$. IX. b_2 133-4°, n 1.5430, d 1.1113.[395]

(EtO)($C_6H_{13}S$)$PONMe_2$. XVII. $b_{0.01}$ 68°.[764]

(EtO)(MeNH)P(S)OCH_2CH=$CHCH_2OP$(S)(OEt)(NHMe). IX. n_D^{25} 1.5136°.[775] For other [(RO)(R'NH)$PSOCH_2CH$=]$_2$ (no data) see Ref. 775.

C_{11} (MeO)(2,4,5-$Cl_3C_6H_2O$)PS(NC_4H_8). V. d_{25} 1.3447.[864]

(MeO)(2,4,5-$Cl_3C_6H_2O$)PSN(CH_2CH_2)$_2$O. V. m. 66-8°.[864]

(EtO)[o-C_6H_4(CO)$_2NCH_2S$]$PONH_2$. XVII. m. 151-2°.[417]

(MeO)[(C_9H_6N-8)O]PSNHMe. IX. m. 74-5.5°.[727] For other (RO)(8-quinolyl-O)PSNHR' compounds see Ref. 727.

(MeO)(p-ClC_6H_4O)PS(NC_4H_8). V. n_D^{25} 1.5592, d_{25} 1.277.[864]

(MeS)(PrO)PONHCOPh. XVII. m. 75°.[34]

(MeS)(i-PrO)PONHCOPh. XVII. m. 137°.[34]

(MeS)(EtO)$PONHCOC_6H_4$Me-p. XVII. m. 105°.[34]

(MeS)(PrO)$PONHCOC_6H_4$Cl-p. XVII. m. 53°.[34]

(MeS)(i-PrO)$PONHCOC_6H_4$Cl-p. XVII. m. 155°.[34]

(MeO)(2,4-Cl,t-BuC_6H_3O)$PSNH_2$. V. n_D^{25} 1.5562, d_{25} 1.241.[413]

(MeO)(2,4-Br,t-BuC_6H_3O)$PSNH_2$. V. n 1.5708, d 1.3976.[413]

(MeO)(p-$O_2NC_6H_4O$)$PSNEt_2$. IX. n 1.5700, d 1.3086.[565]

(EtO)($PhCH_2S$)$PONMe_2$. XVII. $b_{0.01}$ 89°.[764]

(EtO)(3,4-Me,MeSC$_6$H$_3$O)PSNHMe. V. b_{10} 123-5°.[418]
(EtO)(PrS)PONHC$_6$H$_{11}$. XVII. m. 30-2°.[137]
(EtO)(EtS)PONHC$_6$H$_4$Me-p. XVII. m. 59-61°, $b_{0.025}$ 137-40°.[137]
(EtO)(Pr$_2$NS)PSNHPr. XVII. m. 74-5°.[283]
C$_{12}$ (p-ClC$_6$H$_4$O)(2,4,5-Cl$_3$C$_6$H$_2$O)PSNH$_2$. V. n_D^{35} 1.5985, d_{22} 1.685.[619]
(p-BrC$_6$H$_4$O)(2,4,5-Cl$_3$C$_6$H$_2$O)PSNH$_2$. V. n_D^{35} 1.6062, d_{22} 1.715.[619]
(EtO)(C$_6$Cl$_5$O)PSNHBu. IX. m. 67-9°.[917]

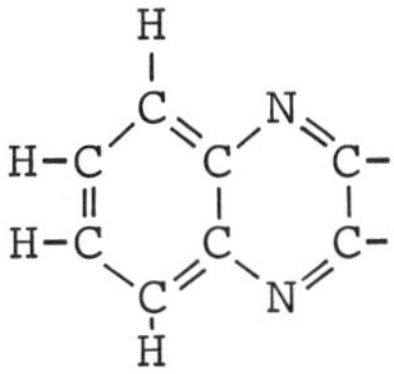
-O-P(S)(OEt)(NMe$_2$). IX. $b_{0.0003}$ 106°.[743]

(MeS)(PrO)PONHCOC$_6$H$_4$Me-p. XVII. m. 74°.[34]
(MeS)(i-PrO)PONHCOC$_6$H$_4$Me-p. XVII. m. 135°.[34]
(EtO)(p-O$_2$NC$_6$H$_4$O)PSNEt$_2$. IX. $b_{0.1}$ 158-64°, m. 34°.[565]
(EtO)(p-MeC$_6$H$_4$SCH$_2$S)PONMe$_2$. IX. $b_{0.01}$ 120-4°.[763]
(MeO)(2,4-Cl,t-BuC$_6$H$_3$O)PSNHMe. V. n 1.5468, d 1.1963.[413]
(EtO)(PhO)PSNEt$_2$. IX. $b_{0.4}$ 120-2°, n 1.5238, d 1.1475.[565]
(EtO)(3,4-Me,MeSC$_6$H$_3$O)PSNMe$_2$. V. $b_{0.01}$ 113°.[418]
C$_{13}$ (MeO)(2,4,5-Cl$_3$C$_6$H$_2$O)PSNHC$_6$H$_4$Cl-m. V. n 1.6172.[867]
(MeO)(3,4-Cl,PhC$_6$H$_3$O)PSNH$_2$. V. n_D^{25} 1.6356.[103] For other bi-phenylylesters see Ref. 103.
(MeO)(2-C$_{10}$H$_7$O)PSNMe$_2$. IX. m. 71-2°.[129]
(MeO)(o-C$_6$H$_{11}$C$_6$H$_4$O)PSNH$_2$. III. n_D^{25} 1.5582, d_{25} 1.179.[103,106]
(MeO)(2,4-Cl,C$_6$H$_{11}$C$_6$H$_3$O)PSNH$_2$. III. m. 74-5°.[106]
(MeO)(2,4-Cl,t-BuC$_6$H$_3$O)PSNHEt. V. n 1.5371, d 1.1602.[413]
(MeO)(2,4-Br,t-BuC$_6$H$_3$O)PSNHEt. V. n 1.5479, d 1.272.[103,413]
(MeO)[2,4-Cl(EtOCH$_2$CH$_2$S)C$_6$H$_3$O]PSNHEt. IX. m. 27.5°, n 1.5400.[642] For other (RO)(ArO)PSNHR' see Ref. 642.
(EtO)(p-MeOC$_6$H$_4$O)PSNEt$_2$. V. n_D^{35} 1.5422, d_{23} 1.385.[255] For other (RO)(ArO)PSNR'R" compounds see Ref. 255.
(PrO)(Et$_2$NCH$_2$CH$_2$O)PSNEt$_2$. IX. b_1 147-50°, n 1.475, d 1.008.[544]
(i-PrO)(Et$_2$NCH$_2$CH$_2$O)PSNEt$_2$. IX. b_4 136-7°, n 1.473, d 0.999.[544]
C$_{14}$ (PhO)(p-MeC$_6$H$_4$O)PS(NCO). XVII. $b_{0.07}$ 127-9°, n 1.5688, d 1.2388.[741]
(MeO)(3,4-Cl$_2$C$_6$H$_3$O)PSNHCH$_2$Ph. V. n_D^{25} 1.5867, d_{25} 1.3268.[867]
(MeO)(2,4,5-Cl$_3$C$_6$H$_2$O)PSNHCH$_2$Ph. V. m. 48-50°.[867]
(EtO)(2,4,5-Cl$_3$C$_6$H$_2$O)PSNHC$_6$H$_4$Cl-m. V. n 1.600.[867]
(EtO)(p-O$_2$NC$_6$H$_4$O)PSNHPh. IX. $b_{0.2}$ 165-70°, n 1.5720, d 1.2663.[565]
(MeO)(2-C$_{10}$H$_7$O)PSNHPr. IX. m. 88-9°.[129]

(EtO)(2-$C_{10}H_7O$)PSNMe$_2$. IX. m. 53-5°.[129]

(sec-BuO)(2,4,5-$Cl_3C_6H_2O$)PSN(CH_2CH_2)$_2$O. V. n_D^{25} 1.5623,
   d_{25} 1.3829.[864]

$$\begin{array}{c} Ph\text{-}C\ \text{-}\ C\text{-}H \\ \parallel\qquad\parallel \\ N\qquad C\text{-}O\text{-}P(S)(OMe)(NHBu). \\ \diagdown\ \diagup\ O \end{array}$$

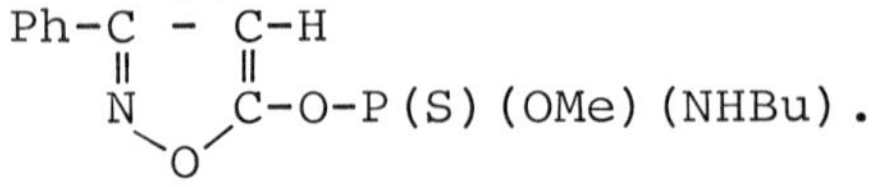

IX. n_D^{25} 1.5366.[724] For
similar heterocyclic esters
see Ref. 724.

(EtO)(Bu_2NS)PSNHBu. XVII. m. 85-6°.[283]

C_{15} (MeO)(2-$C_{10}H_7O$)PSNEt$_2$. IX. $b_{0.04-0.7}$ 230-50°, n 1.5937.[12]

$$\begin{array}{c} H \\ O \qquad C \\ O\text{=}C\diagdown\diagup\ \diagup C\diagdown \\ \ C\qquad C\text{-}O\text{-}P(S)(OMe)(NEt_2). \\ H\text{-}C\ \parallel\qquad\ C\text{-}H \\ C\diagup\ \diagdown C \\ \ \ \mid\qquad\mid \\ \ \ Me\quad H \end{array}$$

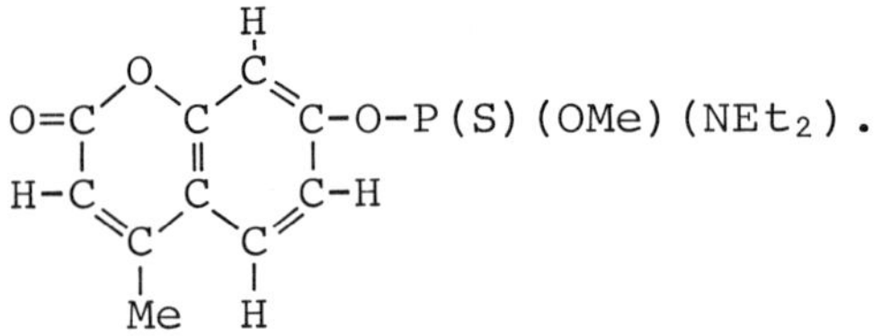

V. m. 106-8°. For simi-
lar heterocyclic compound
see Ref. 721.

(i-PrO)(Bu_2NS)PSNHBu. XVII. m. 61°.[283]

C_{16} [(MeO)(2,4,5-$Cl_3C_6H_2O$)PSNHCH$_2$-]$_2$. V. m. 118-23°.[104]

[(MeO)(p-ClC_6H_4O)PSNHCH$_2$-]$_2$. V. n_D^{25} 1.5844.[104]

[(MeO)(2,4-$Cl_2C_6H_3O$)PSNHCH$_2$-]$_2$. V. n_D^{25} 1.5918.[104]

(EtO)(2-$C_{10}H_7O$)PSNEt$_2$. IX. $b_{0.1-0.2}$ 230-40°, n 1.5806.[12]

(i-PrO)(p-t-BuC_6H_4S)PSNHPr-i. V, IX. n_D^{25} 1.5020.[102]
   For other (RO)(ArS)PSNHR' see Ref. 102.

C_{17} (2,4,5-$Cl_3C_6H_2O$)(2,4,5-$Cl_3C_6H_2OCH_2CH_2O$)PSNHPr. V.
   m. 78°.[866]

C_{18} (PhS)(PhSCH$_2$CH$_2$O)PONEt$_2$. XIII. dec.$_{0.05}$ 170°.[784] For
similar compounds see Ref. 784.

(PhO)(Bu_2NS)PSNHBu. XVII. m. 99°.[283]

C_{19} [(EtO)(2,4,5-$Cl_3C_6H_2O$)PSNHCH$_2$-]$_2$CH$_2$. V. n_D^{25} 1.4200.[104]

(p-ClC_6H_4O)(p-t-BuC_6H_4S)PSNHPr-i. IX. n_D^{25} 1.5737,
   d_{25} 1.1491.[416] For other (ArO)(Ar'S)PSNHR see
   Ref. 416.

C_X (PhS)(PhSCH$_2$CHMeO)PON(C_5H_{10}). XIII. $b_{0.1}$ 205°.[685]

(2,3-Me,$O_2NC_6H_3S$)(2,4-Me,$O_2NC_6H_3SCH_2CH_2O$)PONEt$_2$. XIII.
   m. 82-83°. For other (ArS)(ArSCH$_2$CH$_2$O)PONR$_2$ compound
   see Ref. 685.

(MeO)[4,2,6-Me(C_6H_{11})$_2C_6H_2O$]PSNHMe. IX. m. 154-6°.[106]

(MeO)[4,2,6-Me(C_6H_{11})$_2C_6H_2O$]PSNHEt. IX. m. 148-9°.[106]

(2,4-Me,$ClC_6H_3OCH_2CH_2O$)(2,4,5-Cl_2,MeC_6H_2O)PSNHCH$_2$Ph. IX.
   n 1.5869.[866]

PhOP(O)(NHPh)S-SP(O)(NHPh)OPh. XVII. m. 165°.[76]

D.1.9. $>$N-$\overset{\mid}{N}$-P(O)AB, -N=N-P(O)AB, A = B, A $\neq$ B,
   A,B = Halogen,-OH,-OR, R = Alkyl or Aryl or
   Acyl

C_2 (MeO)$_2$PONHNH$_2$.[622]

C_3 Me$_2$NN(Me)POF$_2$. II. b_{12} 54-8°, m. 28.5°, ^{1}H, ^{19}F,
   ^{31}P +9.2, IR, mass spectrum.[731a]

C_4 (MeO)$_2$PONHN=CHCH$_2$Cl. XVII. $b_{0.03}$ 176°.[622]

$(EtO)_2PONHNH_2$.[622]

C_5 $(EtO)_2PON{=}NMe$. XIII. $b_{0.5}$ 107.5°.[324]

C_6 $(EtO)_2PONHNMe_2$. V. $b_{0.55}$ 102°, n_D^{25} 1.4300.[893]

$PhOPO(OH)NHNH_2$.[274,475,742]

$(MeO)_2PONHN{=}CHCOOEt$. XIII. $b_{0.5}$ 128-31°, n 1.4470, d 1.2369.[708]

$(MeO)_2PONHN(CH_2CH_2)_2O$. V.[624]

C_7 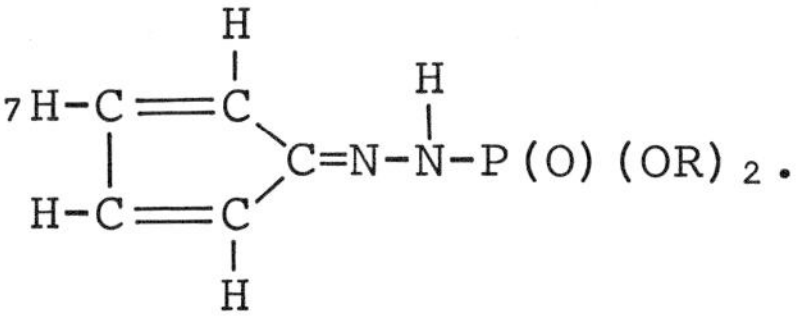 XVII. R=Me: m. 174-5°.[903] R=Et: m. 177-8°.[903]

$(PhCH_2O)PO(OH)NHNH_2$. XVII. Na^+-, K^+-salt.[475]

$(PrO)_2PONHNHMe$. V. b_3 114°, n 1.4430.[185]

C_8 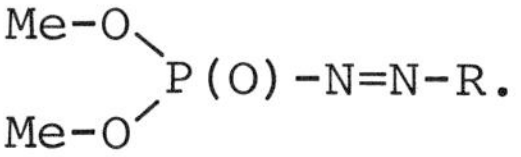 XIII.[839,838]
R=p-$O_2NC_6H_4$: m. 119° red.
R=p-ClC_6H_4: m. 0°.
R=3,5-$Cl_2C_6H_3$: m. 67° orange.
R=p-$MeCONHC_6H_4$: m. 134°.
R=p-$EtOCOC_6H_4$: m. 34-5°.
R=p-$PhN{=}NC_6H_4$: m. 91°.

For 12 further compounds see Refs. 839, 838.

$(MeO)_2PON{=}NPh$. XIII. $b_{0.0001}$ 122°.[111] Oil.[838,839]

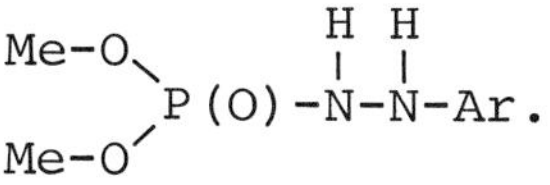 XIII, XVII.
Ar=Ph: m. 132-4°,[880] m. 142°.[838]
Ar=p-$O_2NC_6H_4$: m. 206°.[838]
Ar=p-$PhSO_2OC_6H_4$: m. 125°.[838]
Ar=5,4,2-Cl,(O_2N),Me,C_6H_2: m. 191°.[838]
Ar=4,2-Cl,(O_2N)C_6H_3: m. 187°.[838]
Ar=2,3-$Cl_2C_6H_3$: m. 106°.[838]
Ar=m-MeC_6H_4: m. 109-10°.[838]

$(EtO)_2PONHN{=}CHCOOEt$. XIII. $b_{0.5}$ 136-8°, n 1.4440, d 1.1115.[708]

$(MeO)_2PONHN(CH_2CH_2)_2NNHPO(OMe)_2$. V. 107°dark, 210°black.[245]

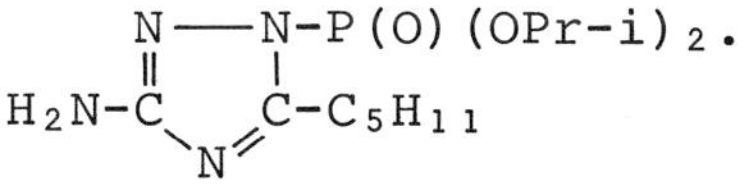 V. m. 53-4°.[119] Diethylester, oil.[119]

C_{10} $(EtO)_2PON{=}NC_6H_4NO_2$-p. XIII. m. 48-9°.[838,839]

$(ClCH_2CH_2O)_2PONHNHPh$. V. m. 93°.[737]

$(EtO)_2PONHNHC_6H_4NO_2$-p. XIII, XVII. m. 186°.[838]

$(EtO)(ClCH_2CH_2O)PONHNHPh$. V. m. 76°.[628,890]

$(PrO)_2PONHN{=}CHCOOEt$. XIII. $b_{0.2}$ 132-5°, n 1.4471, d 1.1370.[708]
$(EtO)_2PONHNHPh$. V, XIII. m. 113-4°,[555] m. 113°,[185] m. 124°,[838] m. 113.5-5.0.[880]
$(EtO)_2PON(COOEt)NH(COOEt)$. XVII. b_3 178-82°.[329,614]
$(EtO)_2PONHN{=}CHPh$. XVII. m. 77°.[622] For other $(RO)_2PONHN{=}CR_2^{\prime}$ compounds see Ref. 623.
C12 $PhOPO(OH)N{=}NPO(OH)OPh$. XVII. di-K-salt dec. 245°.[112]
$(PhO)_2PONHNH_2$. V. m. 112°,[274] m. 116°,[475] m. 113°.[185]
$PhOPO(OH)NHNHPO(OH)OPh$. K-salt.[112]
$(CH_2{=}CHCH_2O)_2PONHNHPh$. XIII. m. 85-7°.[880]
$(EtO)(BuO)PONHNHPh$. XIII. m. 77.5-80.5°.[880]
$(i\text{-}PrO)_2PONHNHPh$. XIII. m. 154.5-6°.[880]
$(PrO)_2PONHNHPh$. V. m. 112°.[185]
$(EtO)_2PON(CH_2CH_2Cl)N(CH_2CH_2Cl)PO(OEt)_2$. XVII. $b_{1.3}$ 109° n 1.4484, d 1.1912.[341]

C13 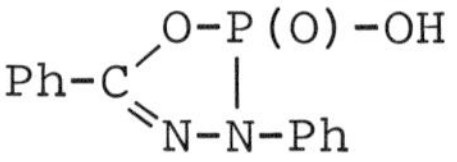 XVII. m. 161°.[666]

$(EtO)(AmO)PONHNHPh$. XIII. m. 76-9.5°.[880]
$(PhO)_2PONHNHMe$. V. m. 51°.[185]
C14 $(MeO)_2PONPhNHPh$. XVII. m. 140-1°.[838]
$(PhCH_2O)_2PONHNH_2$. XIII. m. 71° (from water), 73° (from ethanol).[475]

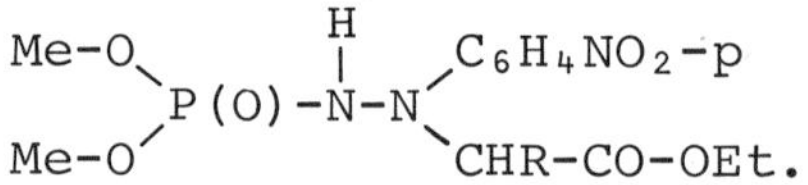 XVII.[617] R=MeCO: m. 155°
R=EtCOO: m. 135-6°.
R=CN: m. 133-4°.

For similar compounds see Ref. 617.
$(i\text{-}BuO)_2PONHNHPh$. XIII, XVII. m. 114.5-6.0°.[880]
C15 $(PhO)_2PONHNHSiMe_3$. XVII. m. 53-4°.[94]
C16 $(EtO)_2PONHNPh_2$. V. n 1.5642, d 1.169.[41]
$(AmO)_2PONHNHPh$. XIII. m. 108.5-10°.[880]
$(i\text{-}AmO)_2PONHNHPh$. XIII. m. 113-4°.[880]
C18 $(PhO)_2PON{=}NPh$. XVII. m. 10-2° (red).[111]
$(PhO)_2PONHNHPh$. V. m. 155.5°,[185] m. 148-9°.[111]
$(i\text{-}PrO)_2PONHNPh_2$. XVII.[704]

C19 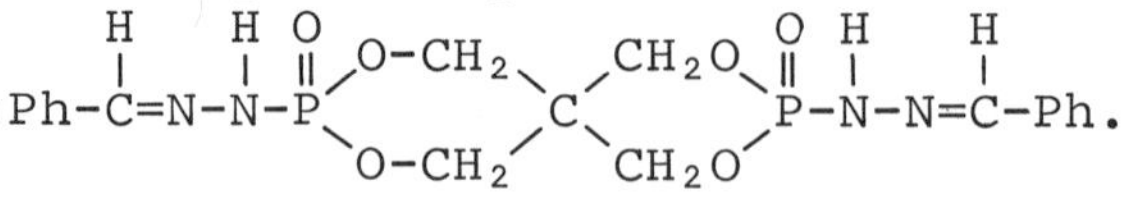 V. m. 241-5° dec.[10] For similar compounds see Ref. 108.

Cx $(MeO)_2PON{=}NC_6H_2(Bu\text{-}t)_3\text{-}2,4,6$. XIII. m. 93-4°.[720]
$(MeO)_2PONHNHC_6H_2(Bu\text{-}t)_3\text{-}2,4,6$. XVII. m. 147°.[720]
$(PhO)_2PON{=}NPO(OPh)_2$. XVII. m. 97° dec. (violet).[112]
$(PhO)_2PONHNPh_2$. V. m. 133-4°.[41]
$(PhO)_2PONHNHPO(OPh)_2$. V. m. 146-8°,[112] m. 148°.[475]
See also Ref. 274.

D.1.10. $>$N-$\overset{|}{N}$-P(S)AB, -N=N-P(S)AB, A = B, A $\neq$ B,
A,B = Halogen,-OH,-OR,-SR, R = Alkyl or Aryl
or Acyl, Also $>$N-$\overset{|}{N}$-P(O)(SR)B

C_2 (MeO)$_2$PSNHNH$_2$.   V.   m. 103-9°.[568] n_D^{25} 1.5428.[858]
    d_{24} 1.368.[858]

C_4 (EtO)$_2$PSNHNH$_2$.   V.   m. 83-6°.[568] b_2 110°,[753] n_D^{25} 1.5010,
    d_{25} 1.185.[858]

(MeO)$_2$PSNHNHPS(OMe)$_2$.   V.   m. 68°.[858]

C_5 (EtO)$_2$PSNHNHMe.   V. b_2 78°, n 1.4857.[187]

C_6 PhNHNHPSF$_2$.   II.   m. 44-6°.[372] ^{19}F, ^{31}P -65.3.[371,372]

(EtO)$_2$PSNHN=CHCCl$_3$.   XVII. $b_{0.08}$ 200° dec.[622]   For simi-
    lar compounds see Refs. 622, 623.

(MeO)$_2$PSNHN(CH$_2$CH$_2$)$_2$O.   V.[624]   Similar compounds see
    Ref. 624.

(PrO)$_2$PSNHNH$_2$.   V. $b_{0.04}$ 76-80°, n 1.4896, d 1.1031.[568]

(i-PrO)$_2$PSNHNH$_2$.   V.   m. 123-9°.[568]

(MeO)$_2$PSNHNHPS(OEt)$_2$.   V.   m. 52-3°.[858]

C_7 Ar-O$\diagdown$
    P(S)-$\overset{H}{N}$-NH$_2$.   V.   Ar=Ph: n 1.5875, d 1.3114.[924]
Me-O$\diagup$ Ar=p-O$_2NC_6H_4$: m. 100°,[924] m. 103°.[106]
    Ar=2,4,5-Cl$_3C_6H_2$: m. 88-8.5°.[106]
    Ar=2,4-Br,t-BuC$_6H_3$: m. 92.5-3.5°.[106]
    Ar=o-PhC$_6H_4$: m. 76-7°.[106]
    Ar=p-MeOC$_6H_4$: n_D^{25} 1.5761,
    d_{25} 1.302.[106]
    Ar=4,2-Cl(C$_6H_{11}$)C$_6H_3$: m. 85-7°.[106]
    Ar=2,5-Me$_2C_6H_3$: n_D^{25} 1.5698,
    d_{25} 1.2447.[106]

C_8 (2,4,5-Cl$_3$C$_6$H$_2$O)PS(NHNMe$_2$)Cl.   V.   m. 76-8°.[105]

(EtO)(2,4,5-Cl$_3$C$_6$H$_2$O)PSNHNH$_2$.   V.   m. 56-8°.[106]

(EtO)(2,4,-Cl$_2$C$_6$H$_3$O)PSNHNH$_2$.   V.   n 1.5852, d 1.4174.[924]

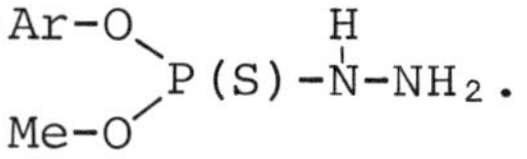

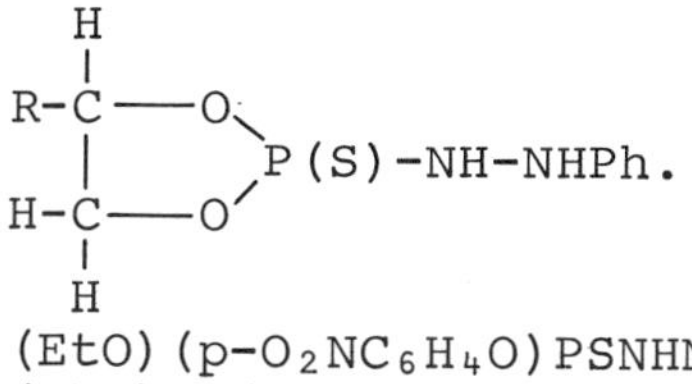

R-$\overset{H}{\underset{|}{C}}$——O$\diagdown$
    P(S)-NH-NHPh.   V.[915]   R=H: m. 187°.
H-$\overset{|}{\underset{|}{C}}$——O$\diagup$ R=Me: m. 139°.
    $\overset{|}{H}$ R=MeOCH$_2$: m. 104°.

(EtO)(p-O$_2$NC$_6$H$_4$O)PSNHNH$_2$.   V.   m. 80°.[924]

(PhO)PS(NHNMe$_2$)Cl.   V. n_D^{25} 1.5663, d$_{25}$ 1.2464.[105]   See
    also Refs. 374, 475.

(o-ClC$_6$H$_4$O)PS(NHNMe$_2$)Cl.   V. n_D^{25} 1.5701, d$_{25}$ 1.3237.[105]

(2,4-Cl$_2$C$_6$H$_3$O)PS(NHNMe$_2$)Cl.   V.   m. 74-5°.[105]

(3,4-Cl$_2$C$_6$H$_3$O)PS(NHNMe$_2$)Cl.   V.   m. 76-8°.[105]

(EtO)(p-ClC$_6$H$_4$S)PSNHNH$_2$.   V.   m. 53°.[547]

(MeO)$_2$PSNHNHPh.   V.   m. 83-5°.[568]

(EtO)(PhO)PSNHNH$_2$.   V. $b_{0.3}$ 147-50°, n 1.5600, d 1.2242.[924]

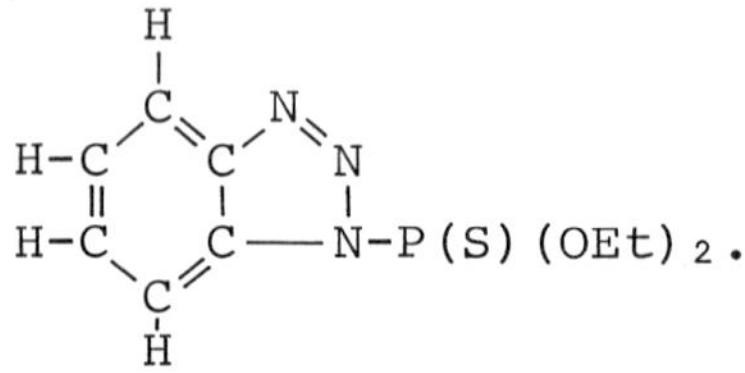

XIII. Oil, substituted benztri-azoles.[259,762]

$(MeO)_2PSN(COMe)N(COMe)_2$. XVII. m. 82-4°.[858,869]

$(BuO)_2PSNHNH_2$. V. $b_{0.04}$ 110-5°, n 1.5410, d 1.0598.[56

$(EtO)_2PSNHNHPS(OEt)_2$. V. m. 83-4°.[858]

$(EtO)_2PSNHNHPS(NMe_2)_2$. V. m. 100-2°.[858]

C_9 $(MeO)_2PSNHN{=}CHC_6H_4Cl$-o. XVII. m. 51-2°.[869]

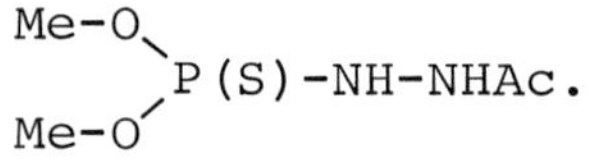

XVII.[858,869] Ac:PhNHCS: m. 104-5°. Ac:PhNHCO: m. 101-2°,[86 m. 101.5-2.0°.[858] Ac:$CH_2{=}CHCH_2NHCS$: m. 110-1° Ac:MeCO: m. 96-7°. Ac:PhCO: m. 84-5°.

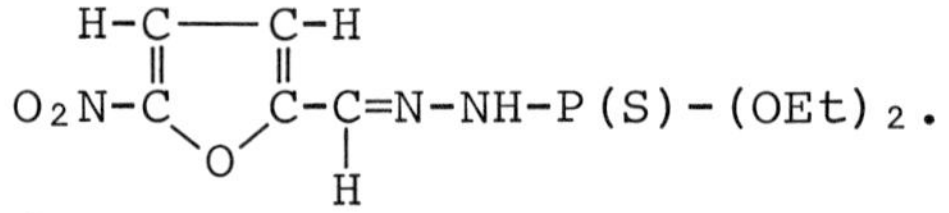

XVII. m. 101.5-2.5°.[833]

$(PrO)_2PSNHN{=}CHCH_2CN$. XVII. $b_{0.03}$ 160°.[622]

C_{10} $(EtS)_2PONHNHPh$. V. m. 106.5-7.0°.[555]

$(EtO)_2PSNHNHPh$. V. m. 72°,[187] m. 68-9°,[555] m. 62-5°.[56

$(EtO)_2PSN(COMe)N(COMe)_2$. XVII. m. 78-80°.[858,869]

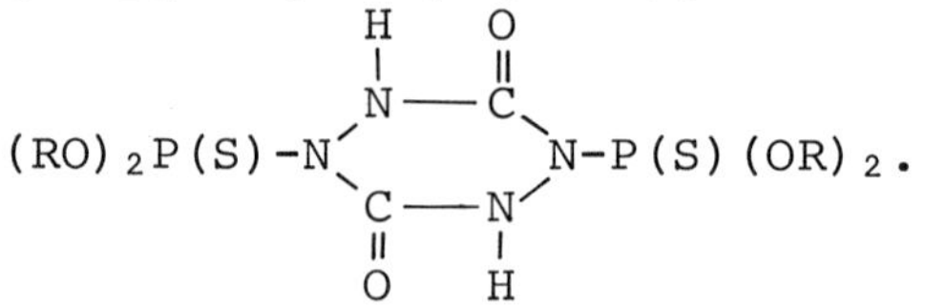

XVII.[357] R = Et: m. 154-5°. R = Me: m. 201°.

C_{11} $(EtO)_2PSNHN{=}CHC_6H_4Cl$-o. XVII. m. 52-3°.[869] For 10 further compounds see Ref. 869.

$(MeO)(2,4\text{-}Cl,t\text{-}BuC_6H_3O)PSNHNH_2$. V. m. 60-70°.[858]

$(MeO)(p\text{-}t\text{-}BuC_6H_4O)PSNHNH_2$. V. m. 51-2°.[858]

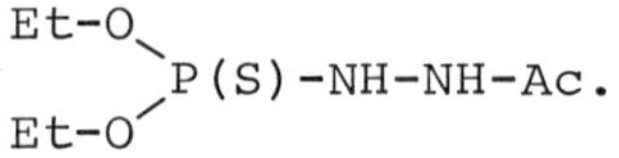

XVII.[858,869] Ac:PhNHCO: m. 145°. Ac:PhNHCS: m. 121-2°. Ac:$C_6H_{11}NHCO$: m. 83-4°. Ac:$CH_2{=}CHCH_2NHCS$: m. 114-5°. Ac:PhCO: m. 102-3°. Ac:MeO: m. 75-6°.

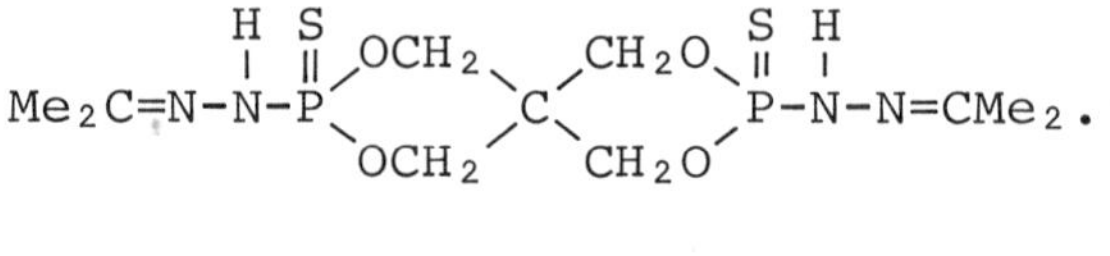

V. m. 221-2°.[108] For similar compounds see Ref. 108.

C_{12} 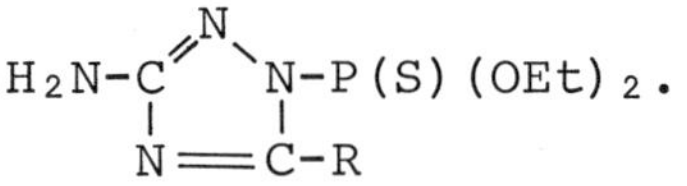 V.[119] R = Ph: m. 63-5°, R = H: m. 82-5°. R = Me,Et,Am: oils. Uncertain which N atom is connected with the phosphoryl group.

$(2,4,5\text{-}Cl_3C_6H_2O)PS(NHNHPh)Cl$. V. m. 120-2°.[105] See also Refs. 374, 475.

$(PhO)_2PSNHNH_2$. V. m. 63°,[76,858] m. 62-3°,[827] m. 64°.[475]

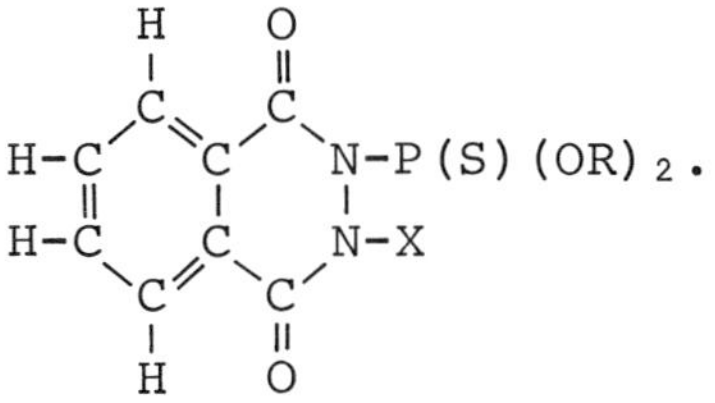 V.[858] R=Me, X=P(S)(OMe)$_2$: m. 113-4°. R=Et, X=P(S)(OEt)$_2$: m. 106-7°. R=Me, X=COMe: m. 112-3°. R=Et, X=COMe: m. 97-8°.

$(PrO)_2PSNHNHPh$. V. m. 57-62°.[568]

$(i\text{-}PrO)_2PSNHNHPh$. V. m. 42-5°.[568]

C_{13} $(MeO)(2,4,5\text{-}Cl_3C_6H_2O)PSNHNHPh$. V. m. 134-5°.[106]

$(MeO)(PhO)PSNHNHPh$. V. m. 80°.[924]

$(MeO)(p\text{-}O_2NC_6H_4O)PSNHNHPh$. V. m. 125-7°.[924]

C_{14} $(p\text{-}MeC_6H_4O)_2PSNHNH_2$. V. m. 141°.[827]

$(EtO)(p\text{-}ClC_6H_4S)PSNHNHPh$. V. m. 110°.[547]

$(EtO)(PhO)PSNHNHPh$. V. m. 57°.[924]

$(EtO)(p\text{-}O_2NC_6H_4O)PSNHNHPh$. V. m. 68°.[924]

$(EtO)(2,4\text{-}Cl_2C_6H_3O)PSNHNHPh$. V. m. 62°.[924]

$(EtO)(2,4,5\text{-}Cl_3C_6H_2O)PSNHNHPh$. V. m. 108-10°.[924]

$(BuO)_2PSNHNHPh$. V. m. 19-21°.[568]

C_{15} $(PhO)_2PSNHN{=}CMe_2$. XVII. m. 79°.[475]

C_x $(EtO)_2PSNHNPh_2$. V. m. 69-70°.[43]

$(m\text{-}Me_2NC_6H_4O)_2PSNHNH_2$. V. m. 117-8°.[306,307]

$(m\text{-}Me_2NC_6H_4O)_2PSNHNHPh$. V. m. 146-7.5°.[306]

D.2. Phosphoric and Thiophosphoric Acid Diamides, Amide-Hydrazides and Trihydrazides (Phosphorodiamidic Acid, Phosphorodiamido[Mono,Di]Thioic Acid Derivatives)

D.2.1. $(>N)_2P(O)X$, X = Halogen (Including Ring Compounds

D.2.1.1. (RR'N)(R"NH)P(O)X, All R's Equal or Different, R,R',R" = H, Alkyl, Aryl, Acyl, Also (>C=N-) Ligands

C_2 $(OCN)_2POBr$. XIII. b_{20} 83-4°, n 1.5060, d 1.9306.[238]

$(SCN)_2POCl$. XV. ^{31}P +41.5.[309,311]

$(OCN)_2POCl$. XIII. b_{12} 66-8°, n 1.4720, d 1.5983.[238] ^{31}P +26.6.[309]

$(SCN)_2POF$. XIV. $b_{3.5}$ 64-5°, ^{31}P +47.7, ^{19}F.[728,729]

$(OCN)_2POF$. XIV. b. 144-6°, n 1.4160, d 1.6021.[238]

C_3 $(Me_2N)(MeNH)POF$. b_3 110°.[101]

C_4 $(MeOCONH)_2POBr$. XII. m. 112-4°.[238]

$(MeOCONH)_2POCl$. XII. m. 117-9°.[238]

$(MeOCONH)_2POF$. XII. m. 107-9°.[238]

$(Me_2N)(MeNH)PO(CN)$. XIV.[767]

$(EtNH)_2POCl$. IV. m. 74°.[574] Hydrolysis.[177]

C_5 $(i-PrNH)(Me_2N)POCl$. IV. $b_{1.5}$ 118°. m. 30°.[360]

C_6 $(EtOCONH)_2POBr$. XII. m. 110-2°.[238]

$(CH_2=CHCH_2NH)_2POCl$. II. m. 51-3°.[677]

$(EtOCONH)_2POCl$. XII. m. 126-8°.[238]

$(EtOCONH)_2POF$. XII. m. 120-1°.[238]

$(PrNH)_2POCl$. II. m. 104.2-4.8°,[877] m. 88°,[573] ^1H.[877]

$(i-PrNH)_2POF$. II. m. 59°.[706]

C_7 $(Et_2N)(i-PrNH)POCl$. IV. $b_{0.01}$ 142°.[113]

C_8 $(i-PrOCONH)_2POBr$. XII. m. 107-8°.[238]

$(i-PrOCONH)_2POCl$. XII. m. 125-6°.[238]

$(i-PrOCONH)_2POF$. XII. m. 117-8°.[238]

$(i-BuNH)_2POCl$. II. m. 86°.[573]

$(BuNH)_2POF$. II. m. 59.5°, $b_{2.5}$ 177°.[362]

C_{10} $(Et_2N)(C_6H_{11}NH)POF$. IV. m. 52.5°.[821]

C_{12} $(p-ClC_6H_4NH)_2POCl$. II.[660]

$(PhNH)(p-ClC_6H_4NH)POCl$. IV. m. 133-4°.[149]

$(PhNH)_2POCl$. II, III. m. 176°,[927] m. 174°,[573,583] m. 171-2°,[652] m. 159°,[77] 167°.[163] ^{31}P -4.1.[633,634]

$(PhNH)_2POF$. II. m. 145°,[362] m. 144°.[163] Poison.

$(C_6H_{11}NH)_2POF$. II. m. 127°.[362] Poison.

C_{13} $(Ph_2N)(OCN)POCl$. XIII. b_6 83-4°, n 1.4592, d 1.248.[233]

$(PhNH)(PhCCl=N)POCl$. XVII. m. 84-5°.[220]

$(PhNH)(p-O_2NC_6H_4CCl=N)POCl$. XVII. m. 72-3°.[220]

$(PhNH)[3,5-(O_2N)_2C_6H_3CCl=N]POCl$. XVII. m. 51-2°.[220]

$$
\begin{array}{c}
\text{H} \\
| \\
\text{Ph-N} \\
\diagdown \\
\quad\quad \text{P(O)-Cl.} \\
\diagup \\
\text{Ar-C-N} \\
\| \quad | \\
\text{O} \quad \text{H}
\end{array}
$$

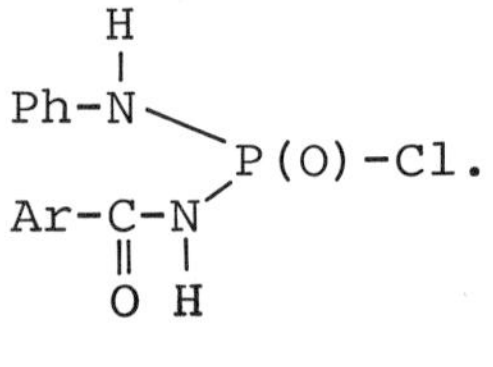

X. Ar=Ph: m. 155-7°,[222] m. 176°.[856]

Ar=$p-BrC_6H_4$: m. 148-50°.[222]

Ar=$m-O_2NC_6H_4$: m. 130-2°.[222]

Ar=$p-O_2NC_6H_4$: m. 140-2°.[222]

Ar=$3,5-(O_2N)_2C_6H_3$: m. 153-5°.[222]

$(PhNH)(PhMeN)POCl$. ^{31}P -10.[634]

$(PhNH)(p-MeC_6H_4NH)POCl$. IV. m. 133-4°.[149]

C_{14} $(PhNHCONH)_2POBr$. XII. m. 155-7°.[238]

$(PhNHCONH)_2POCl$. XII. m. 142-4°.[238]

$[3,4-Cl,MeC_6H_3NH]_2POCl$. II. m. 154-6°.[145]

$(PhNHCONH)_2POF$. XII. m. 163-5°.[238]

$(o-MeC_6H_4NH)_2POCl$. II, III. m. 190°.[573,582] See also Ref. 736.

$(p-MeC_6H_4NH)_2POCl$. II. m. 210°.[573,582]

$(PhCH_2NH)_2POF$. II. m. 96°.[362]

$(PhEtN)(PhNH)POCl$. IV. m. 113°.[573]

C_{15} $(C_9H_{10}N)(PhNH)POCl$. IV. m. 174-5°.[573] Tetrahydroquinoline derivative.

Cx $(C_9H_{10}N)(o\text{-}MeC_6H_4NH)POCl$. IV. m. 122°,[573] tetrahydroquinoline derivative.

$(p\text{-}EtOOCC_6H_4NH)_2POCl$. II. m. 172-3°.[148]

$[(PhNH)_2C{=}N](PhNH)POF$. XVII. m. 148-51°.[386]

$[(p\text{-}ClC_6H_4NH)_2C{=}N](p\text{-}ClC_6H_4NH)POF$. XVII. m. 206-7°.[386]

$[(p\text{-}MeC_6H_4NH)_2C{=}N](p\text{-}MeC_6H_4NH)POF$. XVII. m. 149-51°.[386]

$[(p\text{-}MeOC_6H_4NH)_2C{=}N](p\text{-}MeOC_6H_4NH)POF$. XVII. m. 171-3°.[386]

D.2.1.2. (RR'N)(R"R'''N)P(O)X, All R's Equal or Different, R,R',R",R''' = Alkyl, Aryl, Acyl

C_4 $(OCNMe)_2POCl$. ^{31}P -8.0.[512a]

$[CH_2N(Me)]_2POCl$. II. m. 213-5°.[884]

$(Me_2N)_2POBr$. II. b_{17} 123-6°,[684] b_{12} 122°.[887]

$(Me_2N)_2POCl$. II. b_{10} 110°,[857] b_6 97°,[320] b_6 102°,[179] b_6 98-105°,[330] $b_{0.6}$ 79-82°,[163] b_{30} 133-4°,[163] b_{6-7} 96-102°,[369] b_{16} 118°,[515] b_6 93-9°,[744] b_{11} 109°,[887] b_{10} 108°.[1] n 1.4670,[857] n 1.4695,[369] n 1.4677,[1] n_D^{25} 1.4661.[887] d 1.1823,[857] d 1.1980.[369] ^1H.[169,644] ^{31}P -30.3,[394] -29.6.[633] Hydrolysis.[177,877]

$(Me_2N)_2POF$. II, IV, XIV. b_{4-6} 70-6°,[370] b_{17} 96°,[101] b_{15} 86°,[362] b_{10} 86°,[101] b_5 69-70°,[101] b_4 67°,[101] b_{15} 92-3°.[163] n 1.4267.[370] d 1.1151,[370] d_{25} 1.1.[362] ^1H.[644] ^{31}P -40,[715] mass spectrum.[715] Poison.

$(Me_2N)_2POI$. XIV. dec. 100°.[887]

C_5 $MeN[CON(Me)]_2POCl$. XVII. m. 77-9°.[510] ^{31}P -3.6.[512]

$(Me_2N)_2PO(CN)$. XIV. b_2 106°.[767]

$(Me_2N)(MeEtN)POF$. b_{11} 90°.[101]

C_6 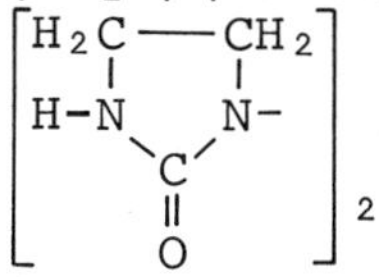 P(O)Cl. I.[625]

$(Me_2N)(Et_2N)POCl$. II, IV. b_5 105-8°.[757]

$(Me_2N)(Et_2N)POF$. b_{10} 102°.[101]

$(MeEtN)_2POCl$. II. b_{15} 133°.[757]

$(MeEtN)_2POF$. b_9 105°.[101]

C_8 $[O(CH_2CH_2)_2N]_2POF$. II. m. 40°.[362]

$[O(CH_2CH_2)_2N]_2POBr$. II.[611]

$[O(CH_2CH_2)_2N]_2POCl$. II. m. 80°,[757] m. 81°,[611] m. 81°.[60] $b_{0.02}$ 137-40°.[611]

$(Et_2N)_2POCl$. II, XIII. $b_{0.02}$ 69-70°,[948] b_{10} 136°,[757] $b_{0.03}$ 79°.[757] n 1.4662,[948] n 1.4655.[757] ^1H.[169] ^{31}P -25.7.[873]

$(Et_2N)_2POF$. II. b_8 113°,[101] b_{20} 124.5-5.5°, b_{22} 127-8°.[362]

C_9 $(Me_2N)(MePhN)POF$. b_6 77°.[101]

$[(C_5H_{10})N][O(CH_2CH_2)_2N]POCl$. IV. b_5 170°.[757]

C_{10} 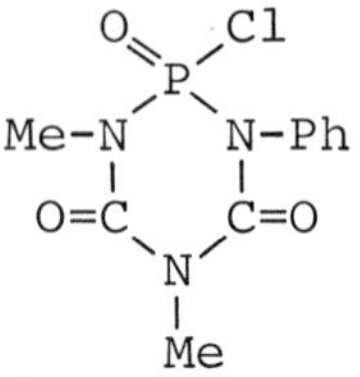 . XVII. ^{31}P -4.5.[512a]

$(Me_2N)(EtPhN)POF$. b_4 110-5°.[101]
$(C_5H_{10}N)_2POCl$. IV. b_{12} 184°.[573]
$[(C_5H_{10})N]_2POF$. II. $b_{0.3}$ 145°.[362]
$(Me_2N)[Et(C_6H_{11})N]POF$. XIV. b_2 100°.[706]
$(Me_2N)(Bu_2N)POCl$. IV. n 1.4658, d 1.0420.[863]
C_{14} $(PhMeN)_2POCl$. II.[573,611] $b_{0.03}$ 149-51°.[611]
$(MePhN)_2POF$. IV. $b_{0.08}$ 163-5°.[362]
C_x $[CH_2N(C_6H_4COOMe\text{-}p)]_2POCl$. II. m. 240-2°.[499]
$(Bu_2N)_2POCl$. II. n 1.4661, d 1.0257.[863]
$(Ph_2N)_2POCl$. ^{31}P -8.1.[633,873]
$[MeNPOCl]n$. II. m > 314°.[1]

D.2.1.3. $[RNP(O)X]_n$, R = Alkyl, Aryl, or Acyl, Ring Compounds

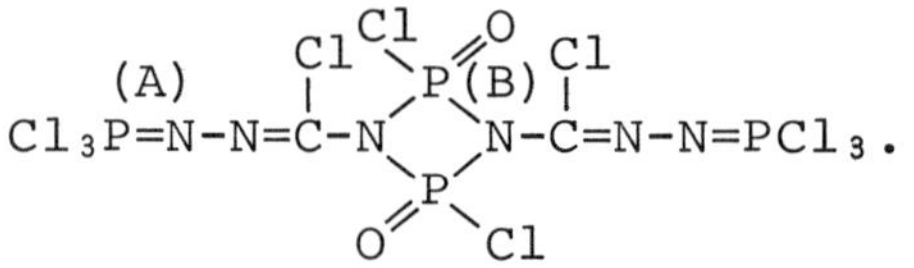 I. Oil. ^{31}P A -14.2, B -5.1.[512a]

$[MeNPOCl]_2$. XVII. m. 101-3°,[93] m. 100-2°.[348a] ^1H,[348a] IR.[348a] ^{31}P +5.3,[93] ^{31}P(cis/trans) +3.0 and 0.0.[348a]
$[EtNPOCl]_2$. XVII. m. 65-8°,[348a] IR, ^1H, ^{31}P(cis/trans) +3.1 and +5.7.[348a]
$[PrNPOCl]_2$. XVII. $b_{0.1}$ 110°, IR, ^1H, ^{31}P(cis/trans) +1.1 and +2.4.[348a]
$[PhN{=}POCl]_2$. II,[574] XVII.[348a] m. 228°,[574] m. 144.5-5.0.[348a]
$[p\text{-}MeC_6H_4NPOCl]_2$. II. m. 336°.[574]

D.2.2. $(>N)_2P(S)X$, X = Halogen (Including Ring Compounds)

$(SCN)_2PSBr$. ^{31}P +28.4.[311]
$(SCN)_2PSCl$. ^{31}P -3.4.[309]
$(SCN)_2PSF$. XIV. $b_{1.5}$ 66-7°,[728,729] b_1 46-7°.[386] n 1.6390, d 1.4827.[386] ^{19}F.[729] ^{31}P -12.2.[309,728]
$(MeNH)_2PSF$. b_2 105°.[101]
$(Me_2N)(MeNH)PSF$. b_6 93°.[101]
$(Me_2N)_2PSBr$. II, XIII. $b_{0.1}$ 68°, m. 32°.[887]
$(Me_2N)_2PSCl$. II,[744] XVII.[331,395] $b_{0.8}$ 52-3°, b_1 57-8°,[3] b_2 80-2°,[744] b_4 87°,[887] b_{10} 108-10°.[119] m. 22°,[331,395] m. 21.5°.[395,744] n 1.5217,[887] n_D^{23} 1.5229,[395] n_D^{25} 1.5209.[744] ^1H.[169] ^{31}P -91.4.[873]

$(Me_2N)_2PSF$. $b_{1.5}$ 58°.[101]
$MeN(CONMe)_2PSCl$. ^{31}P -53.0.[512a]
$(CH_2{=}CHCH_2NH)_2PSCl$. II. $b_{0.001}$ 70-80°, n 1.5528, d 1.208.[677]
$(PhNH)(C_2H_4N)PSF$. IV. n 1.5813, d 1.2358.[831] For other arylamine compounds, n, d listed, see Ref. 831.
$(Et_2N)_2PSCl$. XIII. b_1 117-9°, n 1.5130, d 1.0873.[649] ^1H.[169] ^{31}P -61.4.[873]
$(C_5H_{10}N)_2PSCl$. II. m. 98°.[573]
$(PhNH)_2PSF$. II. m. 123°.[657]
$(C_6H_{11}NH)_2PSF$. II. m. 97°.[657]

Ring compounds:
$[MeNPSCl]_2$. XVII. m. 120-2°.[93] X-ray.[903a] ^{31}P -51.5.[93]
$[PhNPSCl]_2$. XV. m. 149°, b_{80} 280-90°.[576] ^{31}P -39.1.[512]
$[p\text{-}ClC_6H_4NPSCl]_2$. XV. m. 188°, b_{16} 230°.[576]
$[o\text{-}MeC_6H_4NPSCl]_2$. XV. m. 260°, b_{28} 290°.[576]
$[p\text{-}MeC_6H_4NPSCl]_2$. XV. m. 170°.[576]
$[2,4,5\text{-}Me_3C_6H_2NPSCl]_2$. XV. m. 257°.[576]

D.2.3. $(\gt N)_2P(O)OH$

2 $(MeNH)_2POOH$. XVII.[742]
4 $(ClCH_2CH_2)_2NPO(NH_2)OH$. XVII. $[C_6H_{11}NH_3^+]$-salt: m. 125-6°.[316]
$[Me_2N]_2POOH$. XVII. $[Me_4N^+]$-salt: m. 180°.[156]
$(Et_2N)PO(NH_2)OH$. VII. m. 144°.[573]
$(MeNHSO_2NMe)_2POOH$. ^{31}P +3.7.[512a]
6 $[EtNP(O)OH]_3$. ^{31}P -5.3.[634]

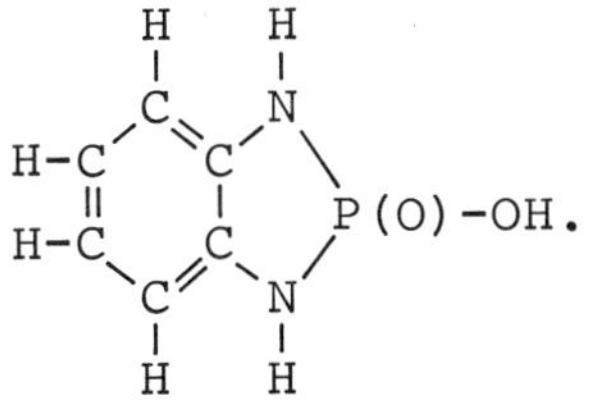

Na-salt: ^{31}P -21.2.[634]

$(C_3H_3N_2)_2POOH$. XVII. Na-salt: m. > 250°.[172] Imidazole derivative.
$p\text{-}ClC_6H_4NHPO(NH_2)OH$. IV, VIII. m. 156-7°.[734]
$p\text{-}ClC_6H_4SO_2NHPO(NH_2)OH$. VI, VIII. m. 161-2°.[517]
$PhNHPO(NH_2)OH$. IV, VIII. m. 157-8°.[149]
$PhSO_2NHPO(NH_2)OH$. VI, VIII. m. 155-6°.[432]
$p\text{-}H_2NSO_2C_6H_4NHPO(NH_2)OH$. IV, VIII. m. 167-9°.[367,899]
$(PrNH)_2POOH$. VIII. m. 135-5.5°.[877]
7 $p\text{-}MeC_6H_4NHPO(NH_2)OH$. IV, VIII. m. 159°.[149]
$MePhNPO(NH_2)OH$. IV, VIII. m. 125°,[573] m. 137°.[499]
$o\text{-}MeC_6H_4SO_2NHPO(NH_2)OH$. VI, VIII. m. 168-9°.[432]
$p\text{-}MeC_6H_4SO_2NHPO(NH_2)OH$. VI, VIII. m. 148-9°.[432]
$Me(p\text{-}H_2NSO_2C_6H_4)NPO(NH_2)OH$. IV, VIII.[367]
8 $[O(CH_2CH_2)_2N]_2POOH$. XVII. $[Me_4N^+]$-salt: m. 257-9°.[156]
$(Et_2N)_2POOH$. XVII. $[Me_4N^+]$-salt: m. 222°.[156]

C_{10} 1-$C_{10}H_7SO_2NHPO(NH_2)OH$. VI, VIII. m. 167-9°.[432]
$[C_5H_{10}N]_2POOH$. XVII. $[Me_4N^+]$-salt: m. 300-3°.[156]

C_{12} $(p\text{-}ClC_6H_4NH)_2POOH$. II, VIII. dec. 218°.[77] m. 126°.[660]

$(PhNH)_2POOH$. II, VII, VIII. m. 214-6°,[77] m. 123°,[573, 582, 584] m. 199-200°.[132]
$(p\text{-}H_2NSO_2C_6H_4NH)_2POOH$.[900]
$(Pr_2N)_2POOH$. XVII. $[Me_4N^+]$-salt: m. 308-9°.[156]
C_{13} $PhCONHPO(NHPh)OH$. X. dec. 136-8°.[222]
$p\text{-}BrC_6H_4CONHPO(NHPh)OH$. X. dec. 171-3°.[222]
$m\text{-}O_2NC_6H_4CONHPO(NHPh)OH$. X. dec. 147-9°.[222]
$p\text{-}O_2NC_6H_4CONHPO(NHPh)OH$. X. dec. 176-8°.[222]
$3,5\text{-}(O_2N)_2C_6H_3CONHPO(NHPh)OH$. X. dec. 134-6°.[222]
$p\text{-}MeC_6H_4NHPO(NHPh)OH$. VIII. shrink at 134°, solidify and m. 195-6°.[149]
C_{14} $(p\text{-}HOOCC_6H_4NH)_2POOH$. II, VIII. m. 200° dec.[148]
$(o\text{-}MeC_6H_4NH)_2POOH$. VIII. m. 120°,[582] m. 95°.[736]
$(p\text{-}MeC_6H_4NH)_2POOH$. VIII. soften at 148°, m. 193-4°,[149] m. 195°,[77] m. 124°.[736]
C_{15} $(C_9H_{10}N)PO(NHPh)OH$. VIII.[573] Tetrahydroquinoline derivative.
$[CH_2\text{-}N(C_6H_4COOH\text{-}p)]_2POOH$. VIII. m. 350°.[499]
$(p\text{-}EtOOCC_6H_4NH)PO(NHC_6H_4COOH\text{-}p)OH$. II, VIII. m. 154-7°.[148]
$(p\text{-}EtOC_6H_4NH)_2POOH$. II, VIII. m. 202°.[77]
$(Bu_2N)_2POOH$. XVII. $[Me_4N^+]$-salt: m. 257-8°.[156]
C_x $(p\text{-}EtOOCC_6H_4NH)_2POOH$. VIII. m. 200-2° dec.[148]
$(1\text{-}C_{10}H_7NH)_2POOH$. VIII. m. 197°.[736]
$(2\text{-}C_{10}H_7NH)_2POOH$. VIII. m. 150°.[736]
$(Ph_2N)_2POOH$. Na-salt: ^{31}P +3.8.[634]
$[(C_8H_{17})_2N]_2POOH$. XVII. $[Me_4N^+]$-salt: m. 320°.[156]
$p\text{-}[H_2NC_6H_4SO_2C_6H_4NHPO(OH)NHC_6H_4SO_2C_6H_4NH]_2POOH$. VIII.[38]

D.2.4. $(\gt N)_2P[(O,S)H]$, $(\gt N)_2P(S)SH$

$OC(NH)_2PS(SH)$. VII. Dihydrate dec. 78°.[479] $[NH_4^+]$-salt dec. 252°.[479] Ba-, Ag-salt.[363] For urea or biuret derivative see Ref. 374.
$(MeNH)_2PS(SH)$. VII. $[MeNH_3^+]$-salt: m. 175°.[310] ^{31}P -92.[310]
$(Me_2N)_2P(OS)H$. XVII. b_6 98-100°.[330]
$(BuNH)_2PS(SH)$. VII. $[BuNH_3^+]$-salt: m. 140-1°.[90] ^{31}P -82.5°.[512a]
$(Pr_2N)(PrNH)PS(SH)$. VII. m. 194-6°.[913]
$(C_5H_{10}N)_2PS(SH)$. VII. $[C_5H_{10}NH_2^+]$-salt: m. 145°, m. 138°.[90] ^{31}P -100.2.[512a]
$(o\text{-}ClC_6H_4NH)_2PS(SH)$. VII. m. 183-90°.[131]
$(PhNH)_2PS(SH)$. VII. m. 161-3°.[131]
$(C_6H_{11}NH)_2PS(SH)$. VII. $[C_6H_{11}NH_3^+]$-salt: m. 197-200°.[90]
$(Bu_2N)(BuNH)PS(SH)$. VII. m. 215-7°.[596,913]
$(Am_2N)(AmNH)PS(SH)$. VII. m. 212-4°.[913]

$(p\text{-}EtOC_6H_4NH)_2P(OS)H$. II. Unstable.[77]
$(Bu_2N)_2PS(SH)$. VII. $[Bu_2NH_2^+]$-salt.[90] ^{31}P -101.5.[512a]
$(Ph_2N)_2PS(SH)$. ^{31}P -54.8.[873]

D.2.5. $(\gt N)_2P(O)OR$, R = Alkyl or Aryl or Acyl (Including Ring Compounds)

D.2.5.1. $(H_2N)_2P(O)OR$, $(AcNH)_2P(O)OR$, $(\gt C{=}N)_2P(O)OR$, Ac = Acyl

C_2 $EtOPO(NH_2)_2$. IV. m. 110°.[333]
C_4 $EtOPO(NCS)_2$. XIV. $b_{1.5}$ 96-8°, n 1.5697, d 1.330,[22] ^{31}P +38.6.[361a]
C_5 $PrOPO(NCS)_2$. XIV. b_1 111-2°, n 1.5680, d 1.2985.[22]
$MeOPO(NHCOOMe)_2$. XII. m. 115-6°.[238]
C_6 $PhOPO(NH_2)_2$. IV. m. 183-5°.[67] dec. 185-90°,[826] m. 185°,[333] m. 183°.[552] Polycondensation[552]! ^{31}P -15.2.[633]
$BuOPO(NCS)_2$. XIV. b_1 101-2°.[22] n 1.5532, d 1.2523.[287]
$(i\text{-}BuO)PO(NCS)_2$. XIV. $b_{0.5}$ 97-9°, n 1.5523, d 1.2450.[22]
$BuOPO(NCO)_2$. XIV. b_3 80°.[379]
$MeN[CON(Me)]_2POOMe$. IX. m. 70-2°.[510]
C_7 $p\text{-}MeSC_6H_4OPO(NH_2)_2$. IV. m. 184-5°.[418]
C_8 $PhOPO(NCS)_2$. ^{31}P +43.[361a]
$PhOPO(NCO)_2$. XIII.[303,818] b_2 104°, n 1.5201.[303]
$PhOPO(NMeCl)_2$. XVII. n 1.5380, d 1.3688.[671]
$4,3\text{-}MeS,MeC_6H_3OPO(NH_2)_2$. IV. m. 137-8°.[418]
$p\text{-}EtSC_6H_4OPO(NH_2)_2$. IV. m. 184-5°.[418]
$EtOPO(NHCOOEt)_2$. XII. m. 133-5°.[238]
C_{10} $PhOPO(NEtCl)_2$. XVII. n 1.5330, d 1.2908.[671]
C_{11} $p\text{-}t\text{-}AmC_6H_4OPO(NH_2)_2$. IV. m. 160°,[610] m. 160°.[482]
$(i\text{-}PrO)PO(NHCOOPr\text{-}i)_2$. XII. m. 128-30°.[238]
C_{12} $o\text{-}PhC_6H_4OPO(NH_2)_2$. IV. m. 151°.[610]
$EtOPO(NHCSNHCH_2CO_2Et)_2$. XII. m. 139-40°.[22]
$PhOPO(NHSiMe_3)_2$. XVII. m. 119-20°.[94]

C_x 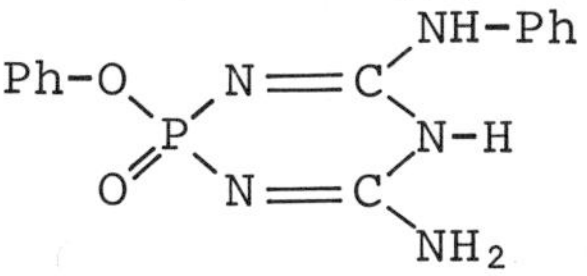 . Solid.[100]

$EtOPO(NHCSNHC_6H_4Cl)_2$. XII. m. 127-8°.[22]
$EtOPO(NHSO_2C_6H_4Me\text{-}p)_2$. m. 188°.[263]
$BuOPO(NHCSNHC_6H_4Cl)_2$. XII. m. 115-6°.[22]
$EtOPO(N{=}CPh_2)_2$. IV. m. 130.5-1.0°.[375]

D.2.5.2. $(R'NH)_2P(O)OR$, R' = Alkyl or Aryl

C_4 $(MeNH)_2POOEt$. IV. b_3 132-6°, n_D^{25} 1.4489.[51]
C_8 $(CH_2NH)_2POOPh$. IV. m. 196°.[73]
$(MeNH)_2PO(OC_6H_4NO_2\text{-}p)$. IV. m. 46-8°.[168] H_2O adduct.[168]
$(MeNH)_2POOPh$. IV. m. 103°,[552] m. 103-5°.[68] Polycondensation.[552] ^{31}P -16.0.[634]
$(PrNH)_2POOEt$. IX. m. 108°.[574]

C_9 $CH_2(CH_2NH)_2POOPh$. XV. m. 144.5-6.5°.[263]
$(CH_2NH)_2PO(OC_6H_4Me$-p). IV. m. 204°.[73]
$(MeNH)_2PO(OC_6H_4SMe$-p). IV. m. 96-7°.[418]
C_{10} $(EtNH)_2POOPh$. IV. $b_{0.7}$ 179°,[671] b_1 150°,[552] n 1.5152, d 1.1104.[671] Polycondensation.[552]
$(MeNH)_2PO(OC_6H_3Me,SMe$-3,4). IV. m. 92-4°.[418]
$(i\text{-}BuNH)_2POOEt$. IX. m. 123°.[574]
C_{11} $(MeNH)_2PO(OC_6H_2SMe,Me_2$-4,3,5). IV. m. 90-2°.[550]
C_{12} o-$C_6H_4(NH)_2POOPh$. IV. m. 185°.[73] ^{31}P -18.1.[633]
$(CH_2{=}CHCH_2NH)_2POOPh$. IV. m. 63°.[53]
$(i\text{-}PrNH)_2POOPh$. IV. m. 61-2°.[361]
C_{13} o-$C_6H_4(NH)_2PO(OC_6H_4Me$-p). IV. m. 158°.[73]
$(PhNH)_2POOMe$. IX. m. 108-9°,[147] m. 123°.[279]
C_{14} $(PhNH)_2POOEt$. IV, IX. m. 114°,[573] m. 114.5°.[134] m. 118°.[279]
$(PhNH)_2PO(OCH_2CH_2Cl)$. IV. m. 102-4°.[159]
$(BuNH)_2POOPh$. IV. m. 54-5°.[361]
$(Et_2NCH_2CH_2NH)_2POOEt$. XVII. $b_{1.3}$ 168°, n 1.4655, d 0.9895.[343]
C_{15} $(PhNH)_2POOPr$. IX. m. 101°,[279] m. 101-3°.[147]
$(PhNH)_2POOPr$-i. IX. m. 149-50°,[147] m. 154°.[279]
$(PhNH)_2PO(OCH_2CH_2OCONH_2)$. IV. m. 170-80°.[159]
C_{16} $(PhNH)_2POOBu$. IX. m. 103°.[279]
$(PhNH)_2POOBu$-i. IX. m. 155°.[279]
$(PhNH)_2POOBu$-sec. IX. m. 124.5°.[279]
(p-$MeC_6H_4NH)_2POOEt$. IV. m. 108°.[573]
(o-$MeC_6H_4NH)_2POOEt$. IV, IX. m. 115°.[573]
$(Et_2NCH_2CH_2NH)_2POOBu$. XVII. b_1 176°, n 1.4695, d 0.9761.[343]
C_{17} (o-$MeOCOC_6H_4NH)_2POOMe$. IX. m. 174°.[882]
$(PhNH)_2POOAm$. IX. m. 101.5°.[279]
C_{18} (2,4-$Cl_2C_6H_3NH)_2POOPh$. VII. m. 227°.[583]
$(PhNH)_2PO(OC_6H_4Cl$-p). IV, IX. m. 167-8°.[573]
$(PhNH)_2PO(OC_6H_4NO_2$-p). IV. m. 159°.[168]
$(PhNH)_2POOPh$. IV, VII, IX. m. 179-80°,[68] m. 179.5°,[573] m. 169°,[573] m. 165°,[583] m. 126°,[76] m. 125°,[582] m. 169°.[279] ^{31}P +2.2.[633]
(p-$H_2NSO_2C_6H_4NH)_2POOPh$. IV. m. 243-5°.[772]
$(PhNH)_2PO(OCH_2CH_2NEt_2)$. IX. m. 115°.[147]
$(C_6H_{11}NH)_2POOPh$. IV. m. 124-5°.[68] ^{31}P -10.7.[634]
C_{20} (2,4-Br,$MeC_6H_3NH)_2POOPh$. VII. m. 221°.[583]
(3,4-Cl,$MeC_6H_3NH)_2POOPh$. m. 149°.[145]
(o-$MeC_6H_4NH)_2POOPh$. IV, IX. m. 157.5°.[573]
(p-$MeC_6H_4NH)_2POOPh$. IV. m. 147-8°.[428]
$(PhCH_2NH)_2POOPh$. IV. m. 114°.[573]
C_{22} $(PhNH)_2PO[OCH_2\text{-}CH(OH)\text{-}CH_2OOCC_6H_4NO_2$-p] (?). IX. m. 220°.[927]
$(EtNH)_2PO(OSnPh_3)$. IX. m. 218-21°.[492]
C_x $(PrNH)_2PO(OSnPh_3)$. IX. m. 153°.[492]
$(PhNH)_2PO(OC_6H_4CO_2Ph$-o). IV. m. 174-5°.[577]
$(PhNH)_2POOCH_2CH_2OPO(NHPh)_2$. IX. m. 180°.[927]

$(p\text{-}MeC_6H_4NH)_2PO(OC_6H_4CO_2Ph\text{-}o)$. IV. m. 146°.[577]
$(PhNH)_2PO(OSnPh_3)$. IX. m. 187-9°.[492]
$o\text{-}C_6H_4O_2[PO(NHPh)_2]_2$. IX. m. 192°.[927]
1,2,3-Tri(N,N'-diphenyl-diamidophosphoryl)glycerol. IX. m. 206°.[927]
Cholesteryl N,N'-diphenyl-diamidophosphate. IX. m. 182°.[927]
Octa-(N,N'-diphenyl-diamidophosphoryl)sucrose. IX. m. 219-20°.[927]
$[RNP(O)(OR')]_n$ polymers.[361]

D.2.5.3. $(R'R''N)_2P(O)OR$, R' = R" or R' ≠ R", R',R" = Alkyl or Aryl

C_5 $(C_2H_4N)_2POOMe$. IV. b_3 98-100°.[602]
$(Me_2N)_2POOMe$. IV. b_1 49-50°,[156] b_{10-15} 76-82°.[532] n_D^{25} 1.4368,[156] n_D^{27} 1.4359,[532] n 1.4385.[863] d 1.0647.[863] Ph_3SnCl adduct m. 97-102°.[114]
$(Me_2N)_2PO(OSO_2Me)$. IX. b_2 125-30°.[767]
C_6 $(Me_2N)_2PO(OCH{=}CCl_2)$. XIII. b_1 90-5°, n_D^{35} 1.4708, d_{35} 1.2432.[27]
$(C_2H_4N)_2POOEt$. IV. $b_{1.3}$ 95-7°,[99] b_2 91-2.5°,[340] b_3 105°.[602] n 1.4712.[340] d 1.1688.[340]
$(Me_2N)_2PO(OCH_2CH_2Cl)$. IV. b_9 126-30°.[330]
$(Me_2N)_2POOEt$. IV, IX. b_6 82°,[753] b_8 93.5°,[179] b_{18} 103-4°.[330] ^{31}P -18,[901] -16.3.[394]
$(MeNHSO_2NMe)_2POOEt$. ^{31}P +3.3.[512a]
C_7 $(C_2H_4N)_2POOPr$. IV. b_1 103.5-4°,[343] b_3 108-9°.[602] n 1.4681, d 1.1263.[343]
$(Me_2N)_2PO(OCH_2CH_2CN)$. IX. Red oil.[81] Pr_2N- and $(CH_2{=}CHCH_2)_2N$- derivatives: brown oils.
$(Me_2N)_2PO(OCH_2CH_2SMe)$. IX. $b_{1.5}$ 120-5°.[571]
C_8 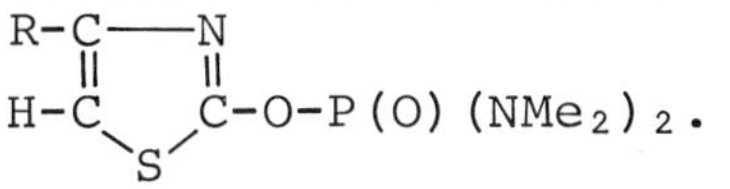 IX.[682] R=Me: $b_{0.2}$ 119-9.5°. R=Ph: m. 56-9°. R=p-BrC_6H_4: m. 61-2°.
$(C_2H_4N)_2POOBu$. IV. b_3 119-20°.[602]
$(C_3H_6N)_2POOEt$. IV. $b_{1.2}$ 82-3°, n 1.4746, d 1.1209.[346]
$(Me_2N)_2POOBu$. IV. b_{15} 123-5°,[532] n_D^{26} 1.4360.[330]
$(Me_2N)_2PO(OCH_2CH_2NMe_2)$. IX. b_1 114-6°.[571]
$(C_2H_4N)_2PO(OCH_2COOEt)$. V. $b_{0.0001}$ 100-10°, n 1.4745, d 1.232.[676]
C_9 $(C_2H_4N)_2POOAm$. IV. b_3 124-8°.[602]
$(C_3H_6N)_2POOPr$. IV. $b_{1.3}$ 102-2.5°, n 1.4735, d 1.0936.[346]
$[O(CH_2CH_2)_2N]_2POOMe$. IV. m. 45-7°.[156]
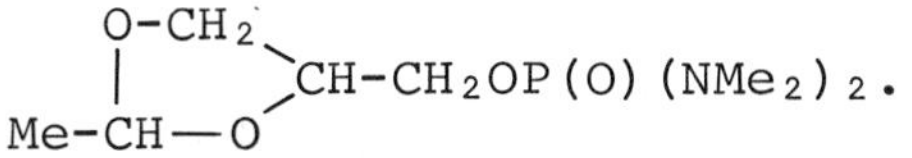 IX. b_1 113°, n 1.4550, d 1.1241.[48]
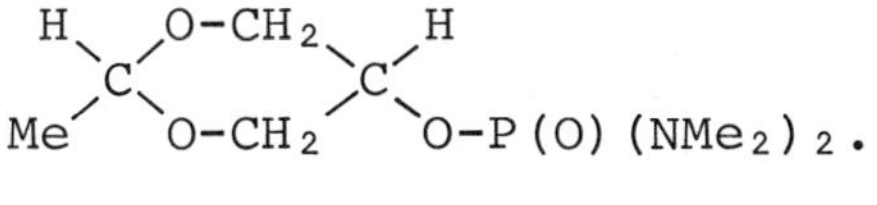 IX. b_2 104-5°, n 1.4600, d 1.1366.[48]

$(Et_2N)_2POOMe$. IV. b_1 96°, n_D^{25} 1.4410.[156] n_D^{27} 1.4519.[532]

C_{10} $(Me_2N)_2PO(OC_6Cl_5)$. IX. m. 144°,[680,681] m. 143-4°.[916]

$(C_2H_4N)_2PO(OC_6H_4NO_2\text{-}p)$. IV. m. 73.5-4.0°.[487]

$(Me_2N)_2PO(OC_6HCl_4\text{-}2,3,4,6)$. IX. m. 122°.[680,681]

$(Me_2N)_2PO(OC_6H_2Cl_3\text{-}2,4,6)$. IX. $b_{0.4}$ 146-7°.[680,681]

$(C_2H_4N)_2POOPh$. IV. Oil.[846] m. 58-9°.[487]

$(Me_2N)_2PO(OC_6H_4NO_2\text{-}o)$. $b_{0.5}$ 166°.[101]

$(Me_2N)_2PO(OC_6H_4NO_2\text{-}m)$. b_1 175°.[101]

$(Me_2N)_2PO(OC_6H_4NO_2\text{-}p)$. IV, IX.[82] b_1 192°.[101]

$(Me_2N)_2POOPh$. $b_{0.5}$ 158-60°.[101]

$(Me_2N)_2PO(OC_6H_4NH_2\text{-}p)$. b. 136°.[101]

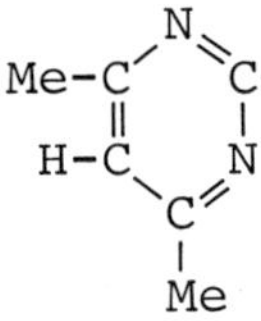 -O-P(O)$(NMe_2)_2$. IX. m. 60-75°.[683]

$[O(CH_2CH_2)_2N]_2PO(OCH_2CH_2Cl)$. IV. m. 60-2°.[159]

$(Et_2N)_2PO(OCH{=}CCl_2)$. XIII. b_1 115-25°.[27] b_2 131-2°, n 1.4700, d 1.1493.[7] For similar esters see Ref. 7.

$(C_2H_4N)_2PO(OC_6H_{13})$. IV. b_3 138-40°.[602]

$(C_3H_6N)_2POOBu$. IV. b_1 103°, n 1.4723, d 1.0720.[346]

$[O(CH_2CH_2)_2N]_2POOEt$. IV, XVII. b_1 155°.[542] m. 48°.[611] See also Ref. 155.

$(Et_2N)_2POOEt$. IV, IX. $b_{0.8}$ 79-80°,[532] b_{15} 140°,[573] b_{17-23} 125-34°.[330] n_D^{27} 1.4380,[532] ^{31}P -16.7.[873]

C_{11} 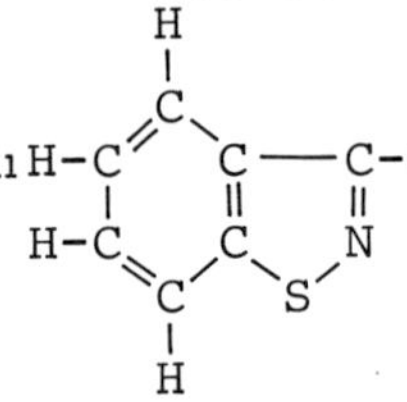 -O-P(O)$(NMe_2)_2$. IX. m. 66-7°.[534]

$[O(CH_2CH_2)_2N]_2PO(OCH_2CH{=}CH_2)$. IV.[611]

$(Me_2N)_2PO(OCOPh)$. IX. b_2 110°.[767]

$(Me_2N)_2PO(OC_6H_4SMe\text{-}p)$. IV. Oil.[418]

$(C_2H_4N)_2PO(OC_7H_{15})$. IV. b_3 153-5°.[602]

$(C_5H_{10}N)_2POOMe$. IV. b_1 125°, n_D^{25} 1.4872.[156]

$(Et_2N)_2PO[OC(OMe){=}CH_2]$. XIII. b_{10} 176-7°, n 1.4675, d 1.0522.[5,6]

C_{12} $(NC_3H_3N)_2POOPh$. IV. m. 90-2°.[172] Imidazole derivative

$(Me_2N)_2PO(OC_6H_3Me,SMe\text{-}3,4)$. IV. Oil.[418]

$(C_2H_4N)_2PO(OC_8H_{17})$. IV. $b_{2.5}$ 157-60°.[602]

$(Et_2N)_2PO[OC(OEt){=}CH_2]$. XIII. b_7 176-7°, n 1.4652, d 1.0312.[5,6]

$(Et_2N)_2POOBu$. IX. $b_{0.6}$ 111-3°, n_D^{26} 1.4413.[532]

$(Me_2N)_2PO(OC_8H_{17})$. IX. $b_{3.5}$ 178-80°, n_D^{27} 1.4414.[532]

$(C_5H_{10}N)_2POOEt$. IV. b_{20} 176-80°.[573]

$(Et_2N)_2PO(OCH_2CH_2SEt)$. IX. n_D^{27} 1.4720.[532]

$(Et_2N)_2PO(OCH_2CH_2OEt)$. IV. b_{15} 170-80°.[246]

C13 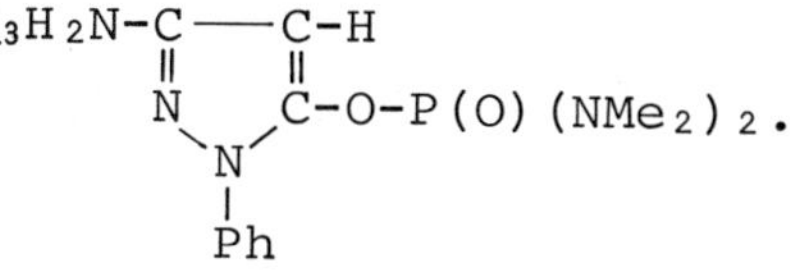
IX. m. 142-3°.[682] 11 other substituted pyrazole derivatives.[682]

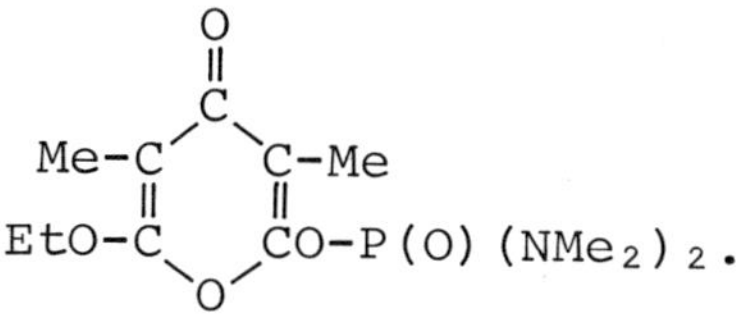
IX. m. 66-8°.[881] For similar pyrone, furane, and cumarin derivatives see Ref. 881.

$(C_2H_4N)_2PO(OC_9H_{19})$. IV. $b_{2.5}$ 168-72°.[602]
$(Pr_2N)_2NPOOMe$. IV. b_1 107°.[156]
C14 $(C_2H_4N)_2PO(OC_{10}H_7\text{-}1)$. IV. Oil.[846]
$(C_2H_4N)_2PO(OC_{10}H_7\text{-}2)$. IV. m. 66-7°.[846]
$2,1,4\text{-}C_6H_3Cl[PO(NC_2H_4)_2]_2$. IV. m. 62-4°.[697]

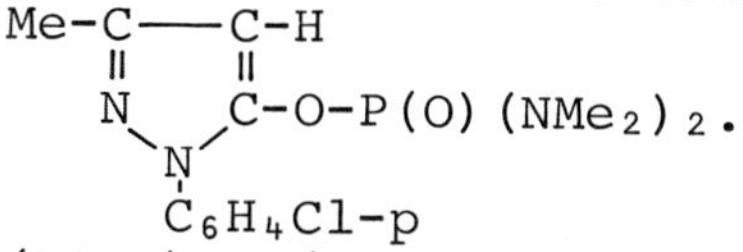
IX. m. 84-5°.[683] For benzoxazole and thiazole derivatives (without data) see Ref. 683.

$(Et_2N)_2PO(OC_6Cl_5)$. IX. m. 152°.[680,681]
$1,3\text{-}C_6H_4[PO(NC_2H_4)_2]_2$. IV.[697,846] m. 190°.[846]
$1,4\text{-}C_6H_4[PO(NC_2H_4)_2]_2$. IV. m. 106-8°,[846] m. 104-6°.[697]
$[O(CH_2CH_2)_2N]_2POOPh$. IV. m. 85-6°, b_2 220°.[68]
$[(CH_2{=}CHCH_2)_2N]_2POOEt$. IV. $b_{1.5}$ 121-5°.[82] See also Ref. 374.
$(Et_2N)_2POOPh$. IV. $b_{0.015}$ 84-5°.[108] n 1.5100,[108] n_D^{24} 1.5215.[532] d 0.9886.[108]
$[O(CH_2CH_2)_2N]_2PO(OC_6H_{11})$. IX. m. 57°.[611]
$(C_2H_4N)_2PO(OC_{10}H_{21})$. IV. b_2 163°.[602]
$(Pr_2N)_2POOEt$. XIII. b_{20} 164-6°.[573]
15 $(Et_2N)_2PO(OC_6H_4Me\text{-}o)$. IV. $b_{0.03}$ 94-5°,[108] b_2 153-5°.[532] n_D^{26} 1.4990.[532]
$(Et_2N)_2PO(OC_6H_4Me\text{-}m)$. IV. $b_{0.027}$ 89-90°.[108]
$(Et_2N)_2PO(OC_6H_4Me\text{-}p)$. IV. $b_{0.02}$ 89-90°.[108]
$(C_2H_4N)_2PO(OC_{11}H_{23})$. IV. $b_{2.5}$ 178-81°.[602]
16 $(Me_2N)_2PO(OC_6Cl_5)\cdot PhOH$. XVII. m. 101-2°.[569] 1:1 adduct. Similar complexes.[569]
$(PhMeN)_2POOEt$. IV,[532] XVII.[405] $b_{0.001}$ 101-3°.[405] n 1.5605,[405] n_D^{27} 1.5545.[532] d 1.1518.[405]
$(C_5H_{10}N)_2POOPh$. IV, IX. b_{10} 215-6°.[573]
$[O(CH_2CH_2)_2N]_2PO(OCH_2CH_2Ph)$. IV.[611]
$(Et_2N)_2PO(OCH_2CH_2OC_6H_3Cl_2\text{-}2,4)$. IV. n_D^{27} 1.5240.[532]
$(C_2H_4N)_2PO(OC_{12}H_{25})$. IV. b_2 184-6°.[602]
$(Et_2N)_2PO(OC_8H_{17})$. IX. $b_{0.5}$ 168-75°, n_D^{26} 1.4450.[532]
$(Pr_2N)_2POOBu$. IV, IX.[82]
$(Me_2N)_2PO(OC_{12}H_{25})$. IV.[330]
$(Me_2N)_2PO(OSnBu_3)$. IX. m. 158°.[492]
$(Bu_2N)_2POOMe$. IV. dec.[156]

C_{18} $[(CH_2{=}CHCH_2)_2N]_2POOPh$. IV. $b_{0.25}$ 164-8°.[53]
$(Bu_2N)_2POOEt$. IV, IX.[82,532] $b_{0.5}$ 150-5°, n_D^{27} 1.4455.[532]
C_{20} $(Bu_2N)_2POOBu$. IV. n_D^{21} 1.4500.[532]
$(Me_2N)_2PO(OC_{16}H_{35})$. ^{31}P -18.9.[873]
$(Et_2N)_2PO(OSnBu_3)$. IX. m. 151-3°.[492]
C_{22} $(Me_2N)_2PO(OSnPh_3)$. IX. m. > 250°.[492]
$(Me_2N)_2PO(OC_{18}H_{37})$. ^{31}P -18.2.[873]
C_x $(C_2H_4N)_2PO(OCH_2CH_2COOC_{17}H_{35})$. IV. m. 56-8°.[86]
$(Bu_2N)_2PO(OC_8H_{17})$. IX. n_D^{27} 1.4524.[532]
$(Et_2N)_2PO(OSnPh_3)$. IX. m. 250°.[492]
$[(C_8H_{17})_2N]_2POOMe$. IV. dec.[156]
$[O(CH_2CH_2)_2N]_2PO(OC_{27}H_{45})$. IV. m. 159°.[611] Cholesteryl ester.[611]
$(C_2H_4N)_2PO[OCH_2CH(OCOC_{17}H_{35})CH_2OCOC_{17}H_{35}]$. IV. m. 59-60°.[86]
$[(ClCH_2CH_2)_2N]_2PO[OCH_2CH(OCOC_{17}H_{35})CH_2OCOC_{17}H_{35}]$. IV. m. 53-4°.[86]

D.2.5.4. (>N)(>N)'P(O)OR

C_5 $(Me_2N)(SCN)POOEt$. XIV. $b_{0.12}$ 59-61°, n 1.4952, d 1.1601.[587,589]

C_6 $[CH_2]_n \begin{smallmatrix} CH_2-O \\ \\ CH_2-NH \end{smallmatrix} P(O)(N<)$. 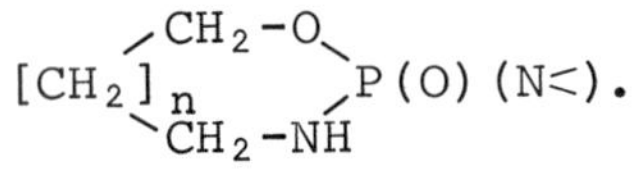 IV.
n=1, $-NH-CH(CH_2Cl)_2$ m. 133-4°.[910]
n=0, $-N(CH_2CH_2Cl)_2$ m. 99.4°.[55,5
n=1, $-N(CH_2CH_2Cl)_2$ m. 48-9°.[305a] monohydrate m. 41-5°,[55,59] "endoxan" reg. trade mark.
n=2, $-N(CH_2CH_2Cl)_2$ m. 76-7°.[55,59]
n=3,4, $-N(CH_2CH_2Cl)_2$ oils.[55,5
n=1, $-N(CH_2CH_2F)_2$ n_D^{25} 1.4760.[66
n=1, $-N(CH_2CH_2Cl)(CH_2CH_2F)$ n_D^{25} 1.4925.[662]

$(Me_2N)(SCN)POOPr$. XIV. $b_{0.18}$ 69-72°, n 1.4962.[589]

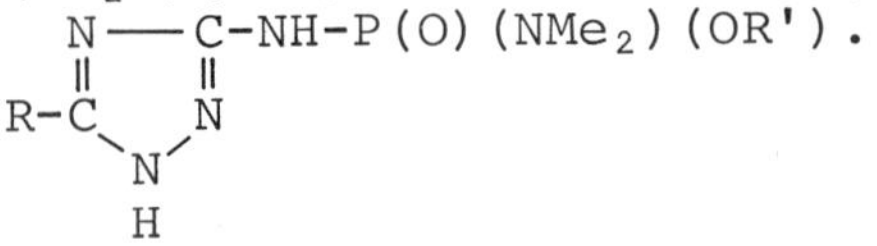 V.[119,570]
R=H, R'=Et, m. 57-64°.
R=H, R'=Pr-i, m. 108-10°.
R=Ph, R'=Pr-i, m. 104-5°.
R=Ph, R'=Pr-i, 1/2 H_2O m. 93-4°.

$[O(CH_2CH_2)_2N](H_2N)POOEt$. V. m. 195°.[542]
$(Me_2N)(H_2NCOCH_2NH)POOEt$. V. m. 62-5°.[19]

C_7 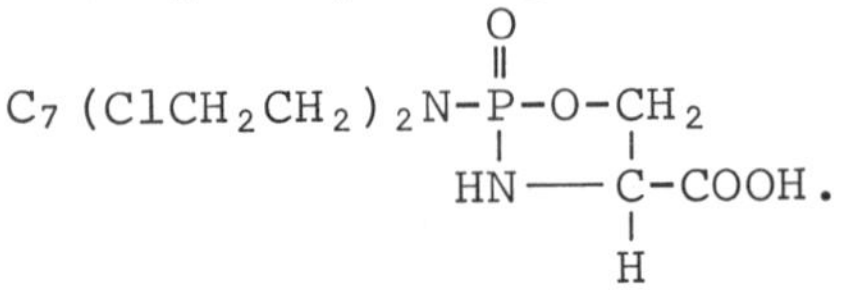 IV, XVII. m. 161°.
L form m. 168°, $[\alpha]_D^{20}$ 5.5°.[849,85
D form m. 168°, $[\alpha]_D^{20}$ -5.8°.[849,8

For threonine derivative see Refs. 849, 850.

$[O(CH_2CH_2)_2N](H_2N)POOPr\text{-}i.$ V. m. 215°.[542]

C_8 $(PhNH)(H_2N)POOEt.$ IV, V. m. 127°.[149]

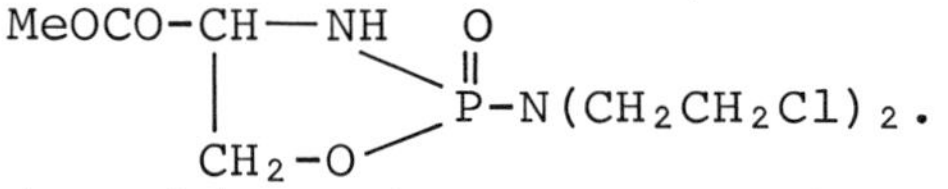

MeOCO-CH—NH O
$\quad\quad|\quad\quad\backslash\quad\|$
$\quad\quad CH_2\text{-}O\quad P\text{-}N(CH_2CH_2Cl)_2.$ IV. m. 96°.[421] For other ring compounds see Ref. 421.

$(Et_2N)(C_2H_4N)POOEt.$ V. $b_{0.01}$ 62°.[572]

$(Me_2N)(EtOCOCH_2NH)POOEt.$ V. b_2 141-2°, n 1.4492, d 1.1293.[19]

C_9 $(p\text{-}MeC_6H_4NH)(H_2N)POOEt.$ IV, V. m. 125°.[149]

$(CH_2=CHCH_2NH)(C_2H_4N)PO(OCH_2COOEt).$ V. $b_{0.0001}$ 100-10°, n 1.4700, d 1.1504.[676]

C_{10}
Ph-C N O
$\quad\quad\quad\|$
$\quad\quad\quad P\text{-}OEt.$ XII, XIII. m. 236-8°.[199]
H-N N-H
$\quad\quad C$
$\quad\quad\|$
$\quad\quad O$

$[(ClCH_2CH_2)_2N](H_2N)POOPh.$ V. m. 57-9°.[316]

$[(EtO)(Me_2N)PO]_2NEt.$ V. b_2 140-2°, n 1.4560, d 1.1214.[26,45] See also Ref. 165.

C_{11} $(Et_2N)(C_5H_{10}N)POOEt.$ IV, V. b_{15} 150°.[573]

$(Me_2N)(Bu_2N)POOMe.$ IV, V. n 1.4465.[863]

C_{12} $(ClCH_2CH_2)_2N\quad O$
$\quad\quad\quad\quad\quad P$
$\quad\quad (C_2H_4N)\quad OAr$. V.[503]

Ar	n	d
Ph	1.5353	1.2834
$o\text{-}MeC_6H_4$	1.5331	1.2625
$m\text{-}MeC_6H_4$	1.5329	1.2470
$p\text{-}MeC_6H_4$	1.5319	1.2580
$o\text{-}IC_6H_4$	1.5772	1.6420
$m\text{-}IC_6H_4$	1.5788	--
$p\text{-}IC_6H_4$	1.5759	1.6215

C_{13} $(Et_2N)(PhNH)POOPr\text{-}i.$ IV. m. 170°.[543]

C_{14} $(PhNH)(p\text{-}O_2NC_6H_4CCl=N)POOMe.$ XVII. m. 103-4°.[220]

$(PhNH)(PhCONH)POOMe.$ XI. m. 169-70°.[222,449]

$(PhNH)(p\text{-}BrC_6H_4CONH)POOMe.$ XI. m. 182-3°.[222,449]

$(PhNH)(m\text{-}O_2NC_6H_4CONH)POOMe.$ XI. m. 181-2°.[222,449]

$(PhNH)(p\text{-}O_2NC_6H_4CONH)POOMe.$ XI. m. 148-50°.[222]

$(PhNH)[3,5\text{-}(O_2N)_2C_6H_3CONH]POOMe.$ XI. m. 186-7°.[222]

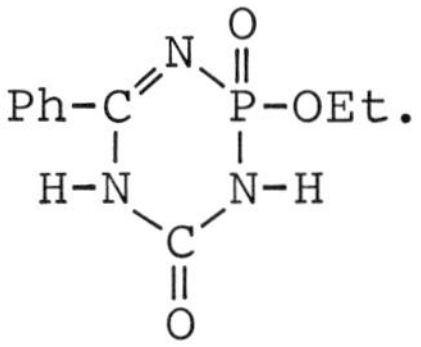

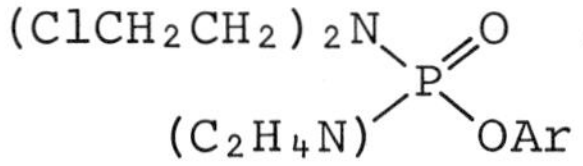

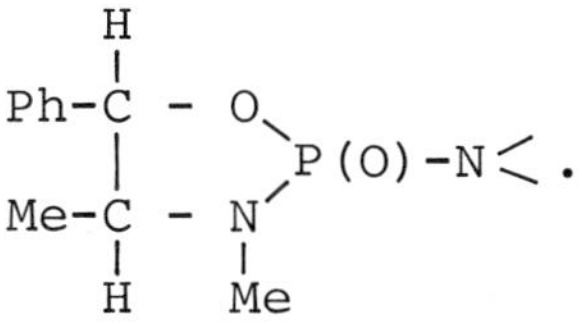

H
$|$
Ph-C - O
$|\quad\quad\quad\backslash$
Me-C - N $P(O)\text{-}N{<}$. II, V.[300] d-pseudo ephedrine.
$|\quad\quad|\quad\quad$ $-N(CH_2CH_2Cl)_2$: m. 106.5-7°. $[\alpha]_D^{26}$ 34°.
H   Me$\quad\quad$ -NHPh: m. 191.5-3.0°. $[\alpha]_D^{21}$ 43,8°.
$\quad\quad\quad\quad\quad$ For other isomers see Ref. 301.

C_{15} $(PhNH)(p\text{-}O_2NC_6H_4CCl=N)POOEt.$ XVII. m. 103-4°.[220]

$(PhNH)(PhCONH)POOEt.$ XI. m. 182-3°.[222,449]

$(PhNH)(p\text{-}BrC_6H_4CONH)POOEt.$ XI. m. 189-90°.[222,449]

$(PhNH)(m\text{-}O_2NC_6H_4CONH)POOEt.$ XI. m. 166-7°.[222,449]

$(PhNH)(p\text{-}O_2NC_6H_4CONH)POOEt.$ XI. m. 176-7°.[222]

(PhNH)[3,5-$(O_2N)_2C_6H_3$CONH]POOEt. XI. m. 182-3°.[222]
(PhNH)(p-MeC_6H_4NH)POOEt. IV, V. m. 116-7°.[149]

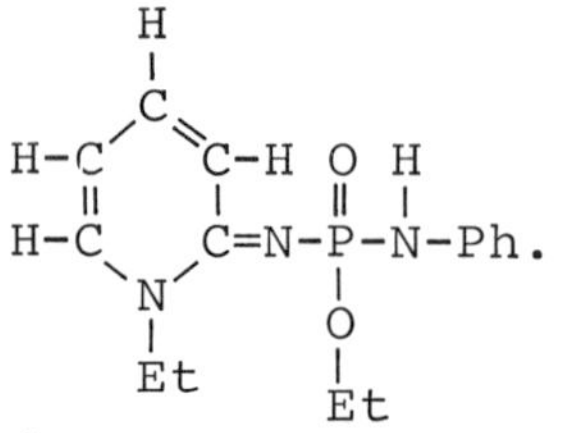
XVII. m. 170-1°.[406]

$(C_5H_{10}N)[(ClCH_2CH_2)_2N]$POOPh. V. m. 51-2°.[502]
(Et_2N)(PhNHCSNH)POOBu. XII.[587]

C_{16} (PhNHCONH)(m-$O_2NC_6H_4$CONH)POOEt. XII, XIII. m. 183-4°.[19]

PhNH(O)P-OAr with $(ClCH_2CH_2)_2N$. V.[501]
Ar=p-IC_6H_4: m. 114-5°.
Ar=m-IC_6H_4: m. 126-7°.
Ar=o-IC_6H_4: m. 169-70°.
Ar=p-ClC_6H_4: m. 107-8°.
Ar=m-ClC_6H_4: m. 133-4°.
Ar=o-ClC_6H_4: m. 102.5-3.5°.

(PhNH)$[(ClCH_2CH_2)_2N]$POOPh. V. m. 87-8.5°.[716] For cresyl esters see Ref. 716.

C_{19} (PhNH)(PhCONH)POOPh. XI. m. 199-201°.[222,449]
(PhNH)[3,5-$(O_2N)_2C_6H_3$CONH)POOPh. XI. m. 193-4°.[222,449]
(PhNH)(p-MeC_6H_4NH)POOPh. IV, V. m. 136-7°.[573]
(PhNH)(EtO)PONPhPO$(NHPh)_2$. VII. m. 220°.[583]
(PhNH)(PhO)PONPhPO$(NHPh)_2$. VII. m. 240°.[583]
$[(ClCH_2CH_2)_2N](C_6H_{11}NH)$PO-O-$C_6H_4$-CEt=CEt-$C_6H_4$-O-PO $(NHC_6H_{11})[N(CH_2CH_2Cl)_2]$. XVII. m. 206-8°.[645] Para substitution.

D.2.5.5. $[R'NP(O)OR]_n$, R' = Alkyl, Aryl, or Acyl, Ring Compounds

$[MeOPONMe]_3$. XV. m. 127-7.5°.[307a] X-ray.[38a]
$[MeOPONMe]_4$. XV. m. 235°.[307a]
$[EtOPONEt]_3$. XV.[307a,714a] m. 74-5°.[307a] ^{31}P -5.3.[633]
$[MeNPOOPh]_2$. XV.[100a]
$[EtOPONPr\text{-}i]_3$. XVII. m. 150-2°.[307b]
$[EtOPONEt]_4$. XV. Oil.[714a] m. 208-10°.[307a]
$[PrOPONPr]_3$. XV. m. 21-3°.[307a]
$[i\text{-}PrOPONPr\text{-}i]_3$. XV. m. 179-80°.[307a]
$[HOPONCH_2Ph]_3$. XVII. m. 185° dec.[307c] Na-salt: m. 230°.[307c]
$[PhOPONPh]_2$. XV. m. 185°.[574]
$[EtOPO(NC_6H_4NO_2\text{-}p)]_3$. XVII. m. 168-9°.[307b]

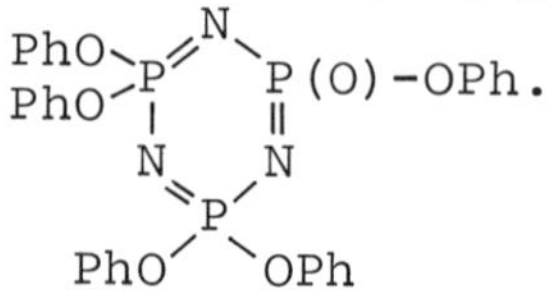
XVII. m. 163-5°, IR, ^1H.[307c]

$[o\text{-}PhOCOC_6H_4OPONPh]_2$.   XV.   m. 152°.[577]
$[PhCH_2OPONCH_2Ph]_3$.   XV.   m. 88-9°.[307a]

D.2.6. $(\geq N)_2P(S)OR$, $(\geq N)_2P(O)SR$, $(\geq N)_2P(S)SR$,
R = Alkyl or Aryl or Acyl, (Including Ring
Compounds)

C_4 $(SCN)_2PS(OEt)$.   XIV. $b_{0.2}$ 62-5°.[361a] ^{31}P -18.4.[361a]
C_5 $(Me_2N)(SCN)PS(OEt)$.   XIV. $b_{0.5}$ 52-4°, n 1.5481,
    d 1.1771.[587,589]
  $(Me_2N)_2PO(SCCl_3)$.   XIII.[561]   For other $(R_2N)_2PO(SCCl_3)$
    compounds (no data listed) see Ref. 561.
C_6 $(H_2N)_2PS(OC_6Cl_5)$.   IV.   m. 172-4°.[255]
  $(H_2N)_2PS(OC_6H_2Cl_3\text{-}2,4,5)$.   IV.   m. 143.5-4.5°.[105]
    n_D^{35} 1.5982.[255]
  $(H_2N)_2PS(OC_6H_2Br_3\text{-}2,4,6)$.   IV.   m. 145-60°.[255]
  $(H_2N)_2PS(OC_6H_2Cl_3\text{-}2,4,6)$.   IV. n_D^{35} 1.6080.[255]
  $(H_2N)_2PS(OC_6H_3Cl_2\text{-}3,4)$.   IV.   m. 129-30°.[103]
  $(H_2N)_2PS(OC_6H_3Cl_2\text{-}2,4)$.   IV.   m. 102-4°.[103] n_D^{35} 1.5961.[255]
  $(H_2N)_2PS(OPh)$.   IV.   m. 119°,[75]   m. 118°,[273]   m. 119°.[333]
    ^{31}P -68.6.[634]
  $(C_2H_4N)_2PS(OCH_2CH_2Cl)$.   IV. $b_{0.03}$ 110°, n 1.5407,
    d 1.292.[130]

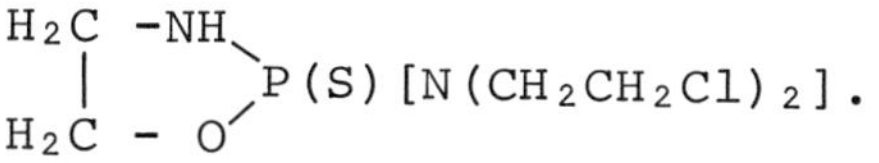

H_2C -NH
  | $P(S)[N(CH_2CH_2Cl)_2]$.   IV, IX.   m. 66-7°.[57]   For
H_2C - O   other amino-alcohol deriva-
                                             tives see Ref. 57.
  $(C_2H_4N)_2PS(OEt)$.   IV. $b_{0.4}$ 96-8°.[248] $b_{0.2}$ 90-1°.[343]
    n 1.5203,[130] 1.5198.[343]   d 1.166,[130] d 1.1663.[343]
  $(Me_2N)_2PS(OCOCH_2Cl)$.   XVII.   Red oil.[357]   For other
    $(R_2N)_2PS(OCOCH_2Cl)$ compounds (oils) see Ref. 357.

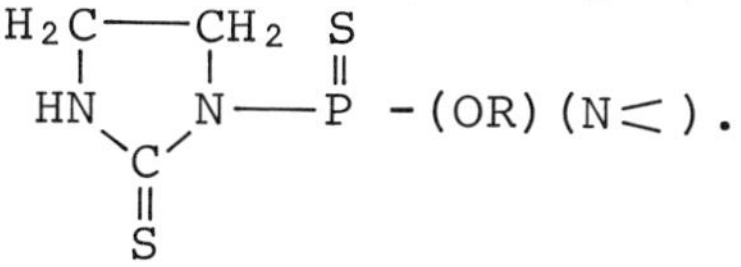

H_2C——CH_2 S
  | | ||
HN   N——P -(OR)(N$\leq$).   XVII.[513]   R=Me, -NMe$_2$: m. 160.5°.
    \C/   R=Et, -NEt$_2$: m. 99°.
     || R=Me, -NEt$_2$: m. 129°.
     S   R=Et, -NMe$_2$: m. 140.5°.
                                            R=Et, -NHMe: m. 70°.
  $(Me_2N)_2PS(OCH_2CH_2Cl)$.   XIII. b_2 89-90°, n 1.5040,
    d 1.1730.[407]
  $(HSCH_2CH_2NH)_2PS(OEt)$.   IV.   n 1.5730, d 1.227.[178]
$_7$ $(MeNH)(H_2N)PS(OC_6H_2Cl_3\text{-}2,4,5)$.   V.   m. 113-4°.[105]
  $(H_2N)_2PS(OC_6H_4Me\text{-}p)$.   IV.   m. 84°.[828]
  $(H_2N)_2PS(OC_6H_4OMe\text{-}o)$.   IV.   m. 149-50°.[255]
  $(H_2N)_2PS(OC_6H_4OMe\text{-}m)$.   IV.   m. 87-8°.[255]
  $(H_2N)_2PS(OC_6H_4OMe\text{-}p)$.   IV.   m. 120-3°.[255]
  $(H_2N)_2PS(OC_6H_4SMe\text{-}p)$.   IV.   m. 140-1°.[418]
  $(C_2H_4N)_2PS(OPr)$.   IV. $b_{0.4}$ 103.5-4.5°,[343] $b_{0.05}$ 65°.[130]
    n 1.5129,[343] n 1.5139.[130]   d 1.1300,[343] d 1.135.[130]
  $(C_2H_4N)_2PS(OPr\text{-}i)$.   IV. $b_{0.1}$ 79-80°, n 1.5098,
    d 1.1237.[343]

$(Me_2N)_2PS(OCH_2CH_2CN)$. IX.[81]
$(Me_2N)_2PS(OCONMe_2)$. XIII. $b_{0.15}$ 108-10°,[654] $b_{0.15}$ 109°.[282] n 1.5035.[654]
$(Me_2N)_2PO(SCH_2SEt)$. IX. $b_{0.01}$ 90°.[763]
C_8 $(OCN)_2PS(OPh)$. XVII. $b_{0.8}$ 85°, n 1.5640, d 1.3562.[741]
$(CH_2NH)_2PS(OC_6H_2Br_3\text{-}2,4,6)$. IV. m. 221-3°.[865,871]
$(CH_2NH)_2PS(OC_6H_2Cl_3\text{-}2,4,6)$. IV. m. 206-8°.[865,871]
$(CH_2NH)_2PS(OC_6H_2Cl_3\text{-}2,4,5)$. IV. m. 156-8°.[865,871]
$(CH_2NH)_2PS(OC_6H_3NO_2,Cl\text{-}4,3)$. IV. m. 108-9°.[871]
$(MeNH)_2PS(OC_6Cl_5)$. IV. m. 216-8°.[917]
$(Me_2N)(H_2N)PS(OC_6H_2Cl_3\text{-}2,4,5)$. V. m. 119-20°.[105]
$(MeNH)_2PS(OC_6H_2Cl_3\text{-}2,4,5)$. IV. m. 116-7.5°.[105]
$(CH_2NH)_2PS(OC_6H_4F\text{-}p)$. IV. m. 63-4°.[871]
$(MeNH)_2PS(OC_6H_3Cl_2\text{-}3,4)$. IV. m. 60-1°.[103]
$(C_2H_4N)_2PS(OBu)$. IV. $b_{0.5}$ 106-8°,[248] $b_{0.2}$ 113.5-4.5°,[34] $b_{0.018}$ 83°.[130] n 1.5101.[130] n 1.5087.[343] d 1.111,[130] d 1.1066.[343]
$(CH_2NH)_2PS(OPh)$. IV. m. 90°,[263] m. 189°,[76] m. 185.5-6.5°.[871]

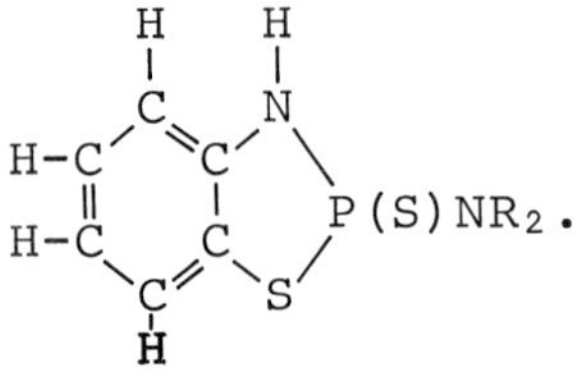

IV.[318] R=Me: m. 113.5-4.5°.
R=Et: m. 146-8°.

$(EtNH)(H_2N)PS(OC_6H_4NO_2\text{-}p)$. V, IX. n_D^{35} 1.5039, d_{24} 1.15.[618] For other $(R_2N)_2PS(OC_6H_4NO_2\text{-}p)$ compounds see Ref. 618. n_D^{35}, d_{24} listed.
$(H_2N)_2PS(OC_6H_3SMe,Me\text{-}4,3)$. IV. m. 112-3°.[418]

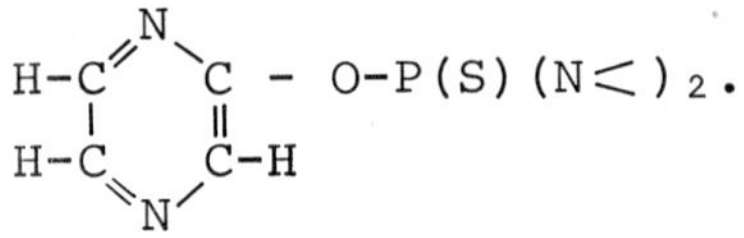

IV. $-NMe_2$: m. 54-5°.[35,591]
$-NEt_2$: m. 43°.[35]

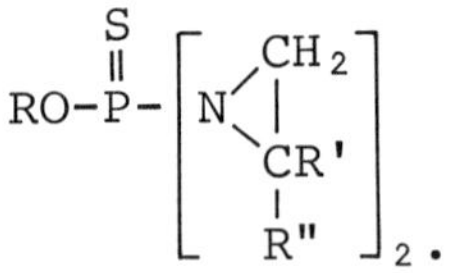

IV.[130] R=Et, R'=Me, R"=H: $b_{0.07}$ 40°, n 1.5000, d 1.087.
R=Pr, R'=Me, R"=H: $b_{0.035}$ 70°, n 1.4981, d 1.076.
R=Ph, R'=Me, R"=H: $b_{0.013}$ 63°, n 1.5560, d 1.162.
R=Et, R'=R"=Me: $b_{0.175}$ 63°, n 1.4983, d 1.064.
R=Pr, R'=R"=Me: $b_{0.05}$ 78°, n 1.4940, d 1.042.
R=Ph, R'=R"=Me: m. 64°.

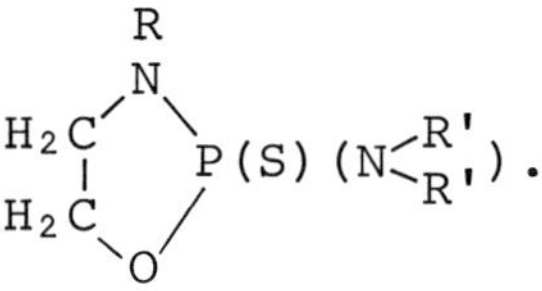

IV.[318] R=R'=Et: b_1 90-2°. R=C_6H_{11}, R'=Et: b_1 135-8°. n 1.521. R=Ph, R'=Et: m. 105-6°. R=Ph, R'=Me: m. 124-6°. R=p-ClC_6H_4, R'=Me: m. 130.5-1.0°.

$(i\text{-}PrNH)_2PS(OEt)$. IV. m. 25-8°.[780] For 46 other (RNH)(R'NH)PS(OR") compounds see Ref. 780.

$(Me_2N)_2PS(OBu\text{-}t)$. XIII. $b_{0.5}$ 73-4°, n 1.480, d 1.081, ^1H.[136]

$(Me_2N)_2PS(OCH_2CH_2SEt)$. IX. $b_{1.5}$ 105-6°, n 1.5151, d 1.1008.[395]

$(Me_2N)_2POSCH_2CH_2SEt$. IX,[763] XV.[395] $b_{0.01}$ 88°,[763] b_1 103-4°.[395] n 1.5208, d 1.1165.[395]

$(Me_2N)(MeNH)PS(OC_6H_2Cl_3\text{-}2,4,5)$. V. m. 106-7.5°.[105]

$(MeNH)(EtNH)PS(OC_6H_2Cl_3\text{-}2,4,5)$. V. m. 91-2.5°.[105]

$(MeNH)(EtNH)PS(OC_6H_2Cl_3\text{-}2,4,6)$. V. m. 106-7°.[105]

$(MeNH)(EtNH)PS(OC_6H_3NO_2,Cl\text{-}4,2)$. V. m. 75-7°.[105]

$CH_2(CH_2NH)_2PS(OPh)$. IV. m. 51-2°.[263]

(MeNH)(PhNH)PS(OEt). V. m. 38-40°.[566] For other (RNH)(ArNH)PS(OR') compounds see Ref. 566.

$(MeNH)_2PS(OC_6H_4SMe\text{-}p)$. IV. m. 63-4°.[418]

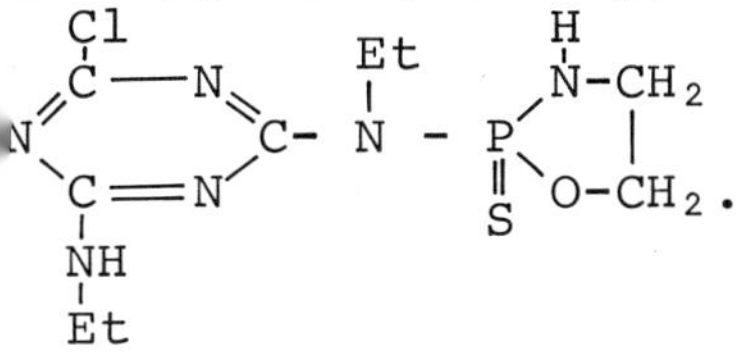

IV. m. 125-7°.[243]

$(C_2H_4N)_2PS(OCH_2CH_2CHMe_2)$. IV. $b_{0.045}$ 73°, n 1.5055, d 1.080.[130]

$(C_2H_4N)_2PS(OAm)$. IV. $b_{0.9}$ 120-3°.[248]

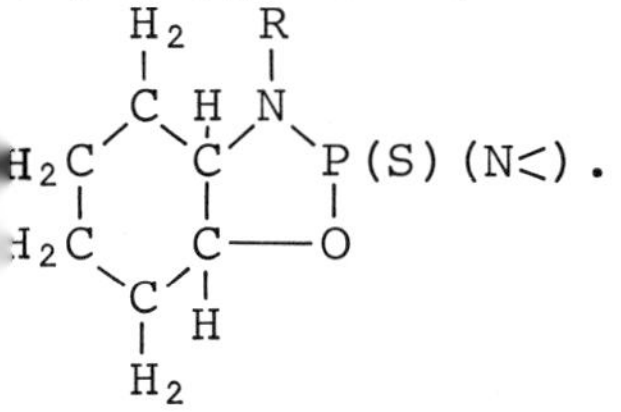

IV.[318] R=Me, $-NMe_2$: b_2 135-7°. n_D^{29} 1.523. R=i-Pr, $-N(CH_2CH_2)_2O$: b_2 150-5°. n_D^{29} 1.536. 2-Dimethylamino-3-methyl-4, 5-benzo-2-thiono-2-phospha-1,3-oxazolidine: m. 110-0.5°.

$(Me_2N)_2PS(OCMe{=}CHCOOMe)$. XIII. $b_{0.5}$ 112-4°, n 1.5075, d 1.1381.[397]

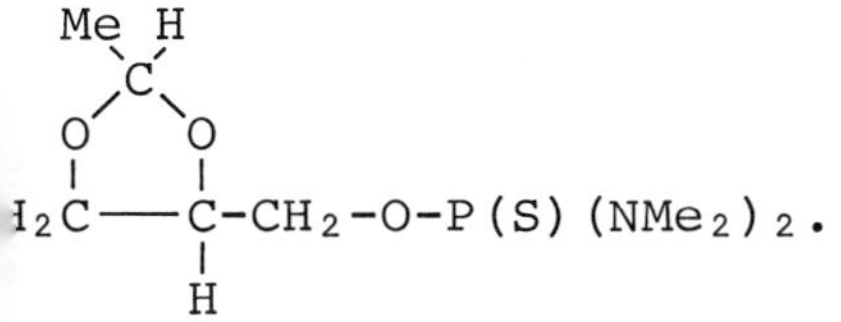

XIII. $b_{0.2}$ 118-8.5°, n 1.4907, d 1.1317.[48]

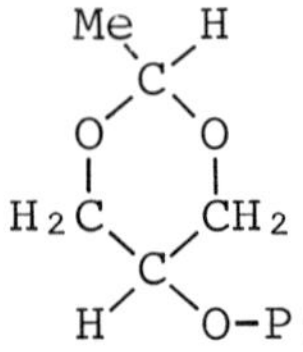
$O-P(S)(NMe_2)_2$. XIII. b_1 120°, n 1.4980, d 1.1446.[48]

C_{10} $(H_2N)_2PS(OC_{10}H_7-2)$. IV. m. 184-5°,[129] m. 176°.[75]

$(EtNH)_2PS(OC_6Cl_5)$. IV. m. 138-40°.[917]

$(Me_2N)_2PO(SC_6Cl_5)$. IX. m. 133°.[680,681]

$(Me_2N)_2PO(SC_6H_2Cl_3-2,4,5)$. IX. $b_{0.4}$ 150-60°.[680,681]

$(Me_2N)_2PO(SC_6H_2Cl_3-2,4,6)$. IX. m. 63°.[680,681]

$(Me_2N)_2PO(SC_6H_3Cl_2-2,5)$. IX. $b_{1.2}$ 170°, m. 40°.[680,681]

$(Me_2N)_2PO(SC_6H_4Cl-p)$. IX. b_1 173-5°.[680,681]

$(C_2H_4N)_2PS(OPh)$. IV.[130,248] $b_{0.008}$ 85°, n 1.5801, d 1.235.[130]

$(MeNH)(i-PrNH)PS(OC_6H_2Cl_3-2,4,5)$. V. m. 88-9°.[105]

$(EtNH)_2PS(OC_6H_2Cl_3-2,4,5)$. IV. m. 79-80°.[105]

$(ClCH_2CH_2NH)_2PS(OC_6H_4Br-p)$. XVII. m. 63-5°.[702]

$(ClCH_2CH_2NH)_2PS(OC_6H_4Cl-p)$. XVII. m. 83-5°.[702]

$(ClCH_2CH_2NH)_2PS(OC_6H_4I-p)$. XVII. m. 67-99°.[702]

$(ClCH_2CH_2NH)_2PS(OC_6H_4NO_2-p)$. XVII. m. 58-60°.[702]

$(Me_2N)_2PS(OC_6H_2Cl_3-2,4,5)$. IV. m. 62.5-3.0°.[105] n_D^{35} 1.5823.[255] d_{27} 1.36.[255] For other $(R_2N)_2PS(OC_6H_2Cl_3-2,4,5)$ compounds (n,d listed) see Ref. 255.

$(EtNH)_2PS(OC_6H_4Cl-p)$. IV. m. 99-100°.[103]

$(C_2H_4N)_2PS(OC_6H_4Cl-p)$. IV. $b_{0.02}$ 145°, n 1.5857, d 1.312.[130]

$(C_2H_4N)_2PS(OC_6H_{11})$. IV. $b_{0.7}$ 128°.[248]

$(MeNH)_2PS(OC_6H_3SMe,Me-4,3)$. IV. m. 51-2°.[418]

$(Me_2N)_2PS(SPh)$. XIII.[765]

$(Me_2N)_2PS(SC_6H_4Cl-p)$. XIII.[765]

$(Et_2N)_2PS(OCH_2CH_2Cl)$. XIII. b_3 128-9°, n 1.4972, d 1.1019.[407]

$(Et_2N)_2PS(OEt)$. XIII. b_{20} 149-51°.[573] ^{31}P -61.4.[873]

$(Me_2N)_2PS[SC(S)NHNHC(S)S]PS(NMe_2)_2$. XVII.[621]

C_{11} $[O(CH_2CH_2)_2N](MeNH)PS(OC_6H_2Cl_3-2,4,5)$. V. m. 149-50°.[10]

$(C_2H_4N)_2PS(OC_6H_4Me-p)$. IV. $b_{0.02}$ 100°, n 1.5769, d 1.213.[130]

$(C_2H_4N)_2PS(OC_6H_3Me,Cl-3,4)$. IV. $b_{0.02}$ 130°, n 1.5819, d 1.286.[130]

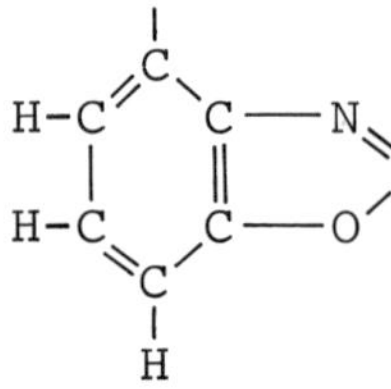
$-S-P(O)(NMe_2)_2$. IX. m. 109-11°.[682,683] For similar triazole and thiadiazole derivatives see Refs. 683, 682.

$(ClCH_2CH_2NH)_2PS(OC_6H_4Me\text{-}p)$. XVII. m. 72-3°.[702]
$(ClCH_2CH_2NH)_2PS(OC_6H_4Me\text{-}o)$. XVII. n 1.5769.[702]
$(Me_2N)_2PS(OC_6H_4SMe\text{-}p)$. IV. m. 66-7°.[418]
$(Et_2N)(H_2N)PS(OC_6H_4OMe\text{-}p)$. IV, V. m. 122-4°.[255] For other $(R_2N)_2PS(OC_6H_4OMe$-o, m, p) compounds see Ref. 255.
$(Me_2N)(C_6H_{11}NHCSNH)PS(OEt)$. XII. m. 125-6°.[589] See also Ref. 374.
$(Et_2N)_2PS[SC(S)OEt]$. IV. d 1.1625.[861]

C12 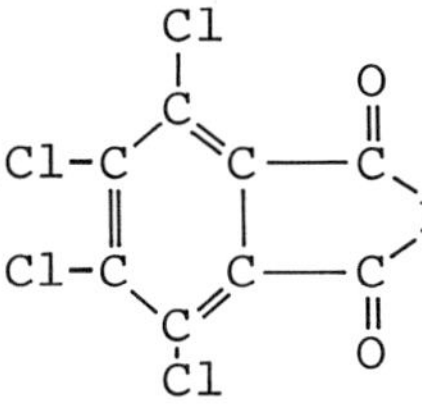 $N\text{-}P(S)(OEt)(NMe_2)$. V. m. 170°.[870] The same compound with all Cl substituted by H: m. 83-4°.[870]

$o\text{-}C_6H_4(NH)_2PS(OPh)$. IV. m. 185°,[75] m. 176°.[263]
$(H_2N)_2PS(OC_6H_3Br,Ph\text{-}3,4)$. IV. m. 120-2°.[103]
$(CH_2{=}CHCH_2NH)_2PS(OC_6Cl_5)$. IV. m. 144-5°.[917]
$(PrNH)_2PS(OC_6Cl_5)$. IV. m. 148-50°.[917]
$(i\text{-}PrNH)_2PS(OC_6H_2Cl_3\text{-}2,4,5)$. IV. m. 90-1°.[105]

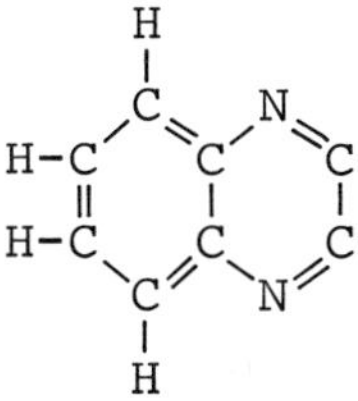 $\text{-}O\text{-}P(S)(NMe_2)_2$. IX. m. 77-8°.[743]

$(H_2N)_2PS(OC_6H_4C_6H_{11}\text{-}p)$. IV. m. 158-9°.[103,106]
$(EtNH)_2PS(OC_6H_4OEt\text{-}p)$. IV. m. 74-5°.[103]
$(Me_2N)_2PS(OC_6H_3SMe,Me\text{-}4,3)$. IV. m. 60-1°.[418]
$(C_5H_{10}N)_2PS(OEt)$. IX. b_{22} 198-210°, b_{10} 191°, d^{15} 1.0633.[573]
$(Et_2N)_2PS(OCH_2CH_2SEt)$. IX,[532] XIII.[317] $b_{0.005}$ 77-80°, n 1.5076,[317] n_D^{21} 1.5130.[532] d 1.0475.[317]
$(Et_2N)_2PO(SNEt_2)$. XVII. b_1 120-2°, n 1.5000, d 1.0145.[16]
C13 $(MeNH)(p\text{-}ClC_6H_4NH)PS(OC_6H_2Cl_3\text{-}2,4,5)$. V. m. 132-3°.[105,867]
$o\text{-}C_6H_4(NH)_2PS(OC_6H_4Me\text{-}p)$. IV. m. 147°.[76]
$(C_6H_{11}NH)_2PO(SMe)$. XVII. m. 175°.[137]
$(C_4H_8N)_2PS(OC_6H_2Cl_3\text{-}2,4,5)$. IV. n_D^{25} 1.5936.[864]
C14 $(Me_2N)_2PS(OC_{10}H_7\text{-}2)$. IV. m. 93-4°.[129]
$p[(C_2H_4N)_2P(S)O]_2C_6H_4$. IV. m. 108-10°.[846]
$(BuNH)_2PS(OC_6Cl_5)$. IV. m. 131-3°.[917]
$(C_4H_8N)_2PS(OC_6H_2Cl_3\text{-}2,4,5)$. IV. m. 99-101°.[105]
$[O(CH_2CH_2)_2N]_2PS(OC_6H_2Cl_3\text{-}2,4,5)$. IV. m. 156-7°.[105]
$(Et_2N)_2PS(OPh)$. XVII. $b_{0.0001}$ 95°, n 1.5330, d 1.0717.[133]
$(C_6H_{11}NH)_2PS(OEt)$. IV. m. 76°.[137]

$(C_2H_4N)_2PS(OC_{10}H_{21})$. IV. $b_{0.05}$ 130°, n 1.4940, d 1.012.[130]

$(Pr_2N)_2PS(OEt)$. XIII. b_{22} 178-80°.[573]

C_{15} $(H_2N)_2PS(OC_6H_4-C_6H_4Pr$-i-p). IV. m. 110-3°.[103]

(i-PrNH)(PhNH)PS(OPh). V. m. 96-7°, n_D^{25} 1.6242, d_4^{25} 1.3091.[105]

$(Et_2N)_2PS(OC_6H_4Me$-o). XVII. $b_{0.0001}$ 109°, n 1.5340, d 1.0694.[133]

$(Et_2N)_2PS(OC_6H_4Me$-m). XVII. $b_{0.0001}$ 111°, n 1.5305, d 1.0598.[133]

$(Et_2N)_2PS(OC_6H_4Me$-p). XVII. $b_{0.0001}$ 120°.[133]

$(C_6H_{11}NH)_2PO(SPr)$. XVII. m. 158°.[137]

C_{16} $(p-MeC_6H_4NH)_2PS(OEt)$. IV. m. 107°.[137]

$(C_5H_{10}N)_2PS(OPh)$. IX. m. 108°.[573]

$(EtNH)_2PS(OC_6H_4C_6H_{11}$-p). IV. m. 81-2°.[106]

C_{17} $(p-MeC_6H_4NH)_2PO(SPr)$. XVII. m. 161°.[137]

C_{18} $(PhNH)_2PS(OC_6Cl_5)$. IV. m. 120-1°.[917]

$(C_6H_{11}NH)_2PS(OC_6H_2Cl_3$-2,4,5). IV. m. 124-4.5°.[105]

$(PhNH)_2PS(OPh)$. IV. m. 126°,[75,76] m. 122-3°.[718]

C_{20} $(PhCH_2NH)_2PS(OPh)$. IV. m. 73°.[573]

$(PhCH_2SCH_2CH_2NH)_2PSOEt$. IV. n 1.6029, d 1.179.[178]

$(EtNH)[(C_6H_{11})_2N]PS(OC_6H_2Cl_3$-2,4,5). V. m. 144-5°.[105]

$(EtNH)(C_{12}H_{25}NH)PS(OC_6H_4NO_2$-o). V, IX. n_D^{35} 1.4862, d_{30} 1.03.[618]

C_{22} $(p-EtOC_6H_4NH)_2PS(OPh)$. IV. monohydrate: m. 145°.[76]

$(Et_2N)[(C_6H_{11})_2N]PS(OC_6H_4NO_2$-o). V, IX. n_D^{36} 1.4926, d_{31} 1.05.[618]

C_x $(Bu_2N)_2PS(SCH_2Ph)$. XIII. n 1.5931.[829]

$(BuNH)(Bu_2N)PS(SCH_2CH_2COOCH_2CHEtBu)$. XVII.[596] For similar compounds see Ref. 596.

$(Ph_2C=N)_2PS(OMe)$. IV, XIII. m. 79-81°.[375]

$(Ph_2C=N)_2PS(OEt)$. IV, XIII. m. 149°.[375]

$(Ph_2C=N)_2PS(OBu)$. IV, XIII. m. 126°.[375]

$(Ph_2C=N)_2PS(OBu$-i). IV, XIII. m. 150°.[375]

$(Ph_2C=N)_2PS(OPh)$. IV, XIII. m. 137°.[375]

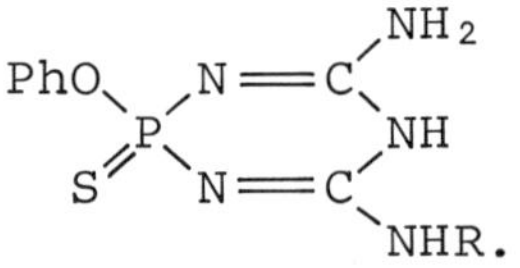

XV.[100]

Ring compounds

$[MeNPO(SMe)]_2$. IX. m. 176-8°. IR. ^1H.[348a]

$[PhNPS(OEt)]_2$. IX. m. 206°.[576]

$[p-ClC_6H_4NPS(OEt)]_2$. IX. m. 91°.[576]

$[o-MeC_6H_4NPS(OEt)]_2$. IX. m. 176°.[576]

$[p-MeC_6H_4NPS(OEt)]_2$. IX. m. 176°.[576]

$[2,4,5-Me_3C_6H_2NPS(OEt)]_2$. IX. m. 201°.[576]

$[o-MeC_6H_4NPS(OPh)]_2$. IX. m. 236°.[576]

$[o-MeC_6H_4NPS(OC_6H_4Me-o)]_2$. IX. m. 247°.[576]

D.2.7. $(\geq N)(\geq N\text{-}\overset{|}{N})P(O)A$, A = Halogen,-OH,-OR,
R = Alkyl or Aryl or Acyl

$H_2C \begin{smallmatrix} CH_2 \text{——} NH \\ \\ CH_2 \text{——} O \end{smallmatrix} P(O)NHN(CH_2CH_2Cl)_2$.   IV.   m. 149-50°.[690]

[(ClCH$_2$CH$_2$)$_2$N](Me$_2$NNH)POOPh.   XVII.[646]   Possibly cyclic
structure.
(Et$_2$N)(PhNHNH)PO(OPr-i).   V.   m. 194-5°.[543]

D.2.8. $(\geq N)(\geq N\text{-}\overset{|}{N})P(S)A$, A = Halogen,-OH,-OR,-SR,
R = Alkyl or Aryl or Acyl

MeOPS(NHMe)(NHNH$_2$).   V. n_D^{25} 1.5604, d_6^{25} 1.308.[858]
(MeNH)(MeO)PS(NHNH)PS(OMe)(NHMe).   V.   m. 99-100°.[858]
(2,4,5-Cl$_3$C$_6$H$_2$O)PS(NH$_2$)NHNH$_2$.   V.   m. 86-7°.[105]
(2,4,5-Cl$_3$C$_6$H$_2$O)PS(NHNH$_2$)(NHMe).   V.   m. 104-5°.[105]
(2,4,5-Cl$_3$C$_6$H$_2$O)PS(NHNMe$_2$)(NHMe).   V.   m. 130-1°.[105]
(2,4,5-Cl$_3$C$_6$H$_2$O)PS(NHNHPh)(NH$_2$).   V.   m. 137-9°.[105]
(2,4,5-Cl$_3$C$_6$H$_2$O)PS(NHNHPh)(NHMe).   V.   m. 112-3°.[105]

D.2.9. $(\geq N\text{-}\overset{|}{N})_2P(O)A$, A = Halogen,-OH,-OR, R =
Alkyl or Aryl or Acyl

PhOPO(NHNH$_2$)$_2$.   IV.   m. 100°.[828]   m. 103°.[475]
PhOPO(NHNH)$_2$PO(OPh).   IV.   m. 132°,[73]   m. 210-1°.[871]
p-MeC$_6$H$_4$OPO(NHNH)$_2$PO(OC$_6$H$_4$Me-p).   IV.   m. 168°.[73]
1-C$_{10}$H$_7$OPO(NHNHPh)$_2$.   IV.   m. 168-9°.[498]
2-C$_{10}$H$_7$OPO(NHNHPh)$_2$.   IV.   m. 198°.[498]
o-PhOC(O)C$_6$H$_4$OPO(NHNHPh)$_2$.   IV.   m. 170°.[577]

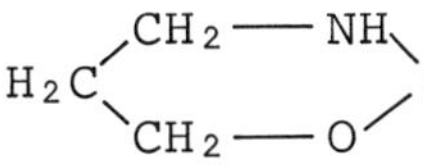

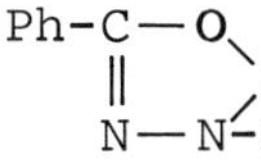

$\begin{smallmatrix} Ph\text{-}C \text{——} O \\ \| \\ N \text{——} N\text{-}Ph \end{smallmatrix} P(O)\text{-}NPh\text{-}NH\text{-}CO\text{-}Ph$.   I, VIII.   m. 164.5°.[666]

(PhCONHNPh)$_2$PO(OH).   VIII.   m. 131-2°.[666]

D.2.10. $(\geq N\text{-}\overset{|}{N})_2P(S)A$, A = Halogen,-OH,-OR,-SR,
R = Alkyl or Aryl or Acyl

(2,4,5-Cl$_3$C$_6$H$_2$O)PS(NHNH$_2$)$_2$.   IV.   m. 152-3°.[105,106]
(2,5-Cl,BrC$_6$H$_3$O)PS(NHNH$_2$)$_2$.   IV.   m. 190-1°.[105,106]
(o-ClC$_6$H$_4$O)PS(NHNH$_2$)$_2$.   IV.   m. 145-7°.[105,106]
(p-O$_2$NC$_6$H$_4$O)PS(NHNH$_2$)$_2$.   IV.   m. 142°.[924]
PhOPS(NHNH$_2$)$_2$.   IV.   m. 95°,[76,828]   m. 96°.[475]
(p-MeC$_6$H$_4$O)PS(NHNH$_2$)$_2$.   IV.   m. 106°.[828]
(2,4,5-Cl$_3$C$_6$H$_2$O)PS(NHNMe$_2$)$_2$.   IV.   m. 123-5°.[105,106]
PhOPS(NHNH)$_2$PS(OPh).   IV.   m. 183°,[76]   m. 184-5°.[871]
PhOPS(NHN=CMe$_2$)$_2$.   XVII.   m. 121°.[475]
(2,4,5-Cl$_3$C$_6$H$_2$O)PS(NHNHPh)$_2$.   IV.   m. 158-9°.[105]
(2,4-Cl$_2$C$_6$H$_3$O)PS(NHNHPh)$_2$.   IV.   m. 156-7°.[105,106]
(p-O$_2$NC$_6$H$_4$O)PS(NHNHPh)$_2$.   IV.   m. 176°.[924]
PhOPS(NHNHPh)$_2$.   IV.   m. 136°.[75]

PhOPS(NHN=CHPh)$_2$. XVII. m. 115°.[76]
(2,4-Cl,t-BuC$_6$H$_3$O)PS(NHNHPh)$_2$. IV. m. 151-3°.[105]

D.3. Phosphoric and Thiophosphoric Acid Triamides, Amide-Hydrazides, and Trihydrazides. (Phosphoric Triamides, Phosphorothioic Triamides, etc.)

D.3.1. (>N)$_3$PO

D.3.1.1. (AcNH)$_3$PO, (>C=N)$_3$PO, Ac = Acyl

(Cl$_2$C=N)$_3$PO. XVII. b$_{0.08}$ 83-5°.[495]?
(SCN)$_3$PO. XIV. b$_{0.07}$ 117-9°,[287] b$_{0.5}$ 106-7°,[837] b$_2$ 120-2°,[311] b$_{5-6}$ 145°.[427] b. 300°.[37] m. 13.8°.[37] IR.[833] Vapor pressure.[37] d 1.484.[37] ^{31}P +61.[309,31
(OCN)$_3$PO. XIII. b$_{10}$ 88°.[303,378] IR.[594] ^{31}P +40.9.[309]
[H$_2$NC(S)NH]$_3$PO. XII.[250]
[(C$_2$H$_4$N)CONH]$_3$PO. XII. m. 285-7° dec.[144]

D.3.1.2. (RNH)$_3$PO, R = Alkyl or Aryl

C$_3$ (MeNH)$_3$PO. II. m. 102-3°.[51] IR, ^{1}H.[368] No pure product could be isolated.[368]
C$_6$ (HSCH$_2$CH$_2$NH)$_3$PO. II. n 1.5877.[178]
(H$_2$NCH$_2$CH$_2$NH)$_3$PO. XVII.[815]
C$_9$ (CH$_2$=CHCH$_2$NH)$_3$PO. II. b$_{0.0001}$ 110-5°.[677] n 1.5028, d 1.071.[677]
(PrNH)$_3$PO. II. Liquid.[573]
C$_{12}$ (t-BuNH)$_3$PO. II. m. 246-7°, ^{1}H.[368]
(i-BuNH)$_3$PO. II. m. 46-7°,[573] m. 57°.[113] Thermolysis.[11
C$_{15}$ (i-AmNH)$_3$PO. II. Oil.[573,574]
C$_{18}$ [2,4(?)-Br$_2$C$_6$H$_3$NH]$_3$PO. XVII. m. 252-3°.[584]
[4-Cl,2(or 3)-O$_2$NC$_6$H$_3$NH]$_3$PO. XVII. m. 249°.[660]
[p-ClC$_6$H$_4$NH]$_3$PO. II,[77] III.[660] m. 248-50°,[77] m. 230°.[660]
[PhNH]$_3$PO. II, III, VII. m. 213-5°,[132] m. 212-5°,[77] m. 211-4°,[69] m. 212°,[652] m. 208-10°,[323] m. 208°.[574] ^{31}P +4.8.[633]
(C$_6$H$_{11}$NH)$_3$PO. IV. dec. 245-6°.[69]
C$_{21}$ [2-Me,(?)BrC$_6$H$_3$NH]$_3$PO. XVII. m. 253°.[736]
(2,4-Br,MeC$_6$H$_3$NH)$_3$PO. VII. m. 268°.[583]
(2-Me,(?)-BrC$_6$H$_3$NH)$_3$PO. XVII. m. 253°.[736]
[4-Me,(?)-BrC$_6$H$_3$NH]$_3$PO. XVII. m. 221°,[736] m. 222°.[77]
[4-Me,(?)-(O$_2$N)C$_6$H$_3$NH]$_3$PO. XVII. m. 247°.[736]
(PhCH$_2$NH)$_3$PO. II. m. 98-9°,[69] m. 98°.[573]
(o-MeC$_6$H$_4$NH)$_3$PO. II. m. 236°,[573] m. 229-30° dec.[69] m 225°.[736]
(p-MeC$_6$H$_4$NH)$_3$PO. II, VII. m. 198-9°,[69] m. 192-4°,[77] m. 192°.[736]
C$_x$ (2,4-O$_2$N,EtOC$_6$H$_3$NH)$_3$PO. XVII. m. 126°.[77]
(2,4-Me$_2$C$_6$H$_3$NH)$_3$PO. II. m. 225°,[514] m. 198°.[573]
(2,5-Me$_2$C$_6$H$_3$NH)$_3$PO. II. m. 247°.[573]
(3,4-Me$_2$C$_6$H$_3$NH)$_3$PO. II. m. 183°.[573]
(p-EtOC$_6$H$_4$NH)$_3$PO. II. m. 172-3°,[69] m. 168°.[77]

$(2,4,5\text{-}Me_3C_6H_2NH)_3PO$. II. m. 217°.[573]
$(PhCH_2SCH_2CH_2NH)_3PO$. II. m. 76-7°.[178]
$(1\text{-}C_{10}H_7NH)_3PO$. II. m. 216°.[736]
$(2\text{-}C_{10}H_7NH)_3PO$. VII. m. 170°,[736] m. 168-70°.[131]
$(C_{12}H_{25}NH)_3PO$. II. m. 75°.[308]

D.3.1.3. $(RR'N)_3PO$, R = R' or R ≠ R', R,R' = Alkyl or Aryl

C_6 $(C_2H_4N)_3PO$. II. $b_{0.3}$ 90-1°, m. 41°.[99]
$(Me_2N)_3PO$. II. b_1 76°,[481] b_2 80°,[757] b_6 94-6°,[744] b_6 97-9°.[330] b_1 68-70°, b_{15} 115°, b. 235°.[648] m. 7.2°.[648] Vapor pressure,[648] n_D^{25} 1.4570,[481,648] n 1.4582.[648] d 1.0253.[648] Dielectric const. ε= 30(20°),[648] dipolmoment.[648] ^1H,[169] ^{31}P -27,[901] -23.4,[620] -22.[715] As solvent in organometall. chem. see Ref. 647, chemical reactivity see summary article.[648] Ph_3SnCl adduct m. 158-60°.[114]
C_9 $(C_3H_3N_2)_3PO$. II. m. 135-7.[172] ^{31}P +16.3.[633] Imidazole derivative.

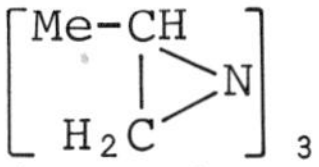 PO. ^{31}P -38.8.[873]

$(C_3H_6N)_3PO$. II. $b_{2.5}$ 124°.[346]
C_{12} $[O(CH_2CH_2)_2N]_3PO$. II. m. 191-2°,[69] m. 192°.[127]
$(Et_2N)_3PO$. II,[127,573] XIII.[829] b_{12} 135°,[127] $b_{0.3}$ 105-10°.[829] ^{31}P -23.5,[873] ^1H.[169]
C_{15} $(C_5H_{10}N)_3PO$. II. m. 75-6°,[573,578,585] m. 76°.[127] $HgCl_2$ double salt m. 105°. Ph_3SnCl adduct m. 161-3°.[114]

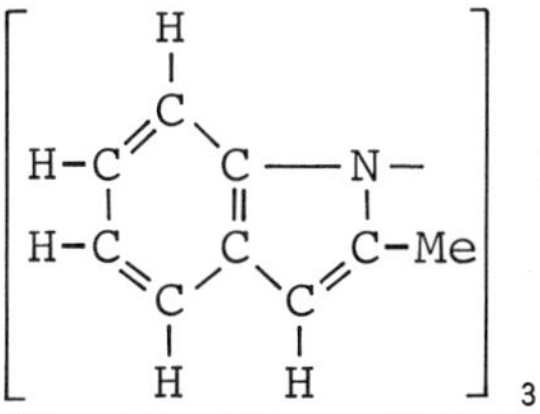 PO. II. m. 140-2°.[598]

C_{18} $(Pr_2N)_3PO$. XIII. $b_{0.05}$ 131-5°.[829] Liquid.[573]
C_x $(MePhN)_3PO$. II. m. 162°,[573] m. 162°.[127] ^{31}P -13.0.[633]
$(EtPhN)_3PO$. II. m. 182°.[573]
$(i\text{-}Bu_2N)_3PO$. II. Liquid.[573]
$(Bu_2N)_3PO$. XIII. $b_{0.06}$ 153-7°.[829] ^{31}P -23.[394,901]
$(C_9H_{10}N)_3PO$. II. m. 90-1°.[575] Tetrahydroquinoline derivative.
$[Ph_2N]_3PO$. ^{31}P -1.7.[633]

D.3.2. $(\geq N)_3PS$

D.3.2.1. $(RNH)_3PS$, $(\geq C{=}N)_3PS$, R = Alkyl or Aryl or Acyl

C_3 $(SCN)_3PS$. XIV. $b_{0.1}$ 124°,[311] $b_{0.3}$ 121-3°.[837] IR.[837] ^{31}P +9.2,[309,728] +10.0.[311]

$(OCN)_3PS$. XIII. m. 8.8°, [b. 215°], n_D^{25} 1.5116, d_{25} 1.538.[313]
$(MeNH)_3PS$. II,[51,368] VII.[310] m. 108°,[113] m. 105-6°,[310] m. 105-7°,[51] m. 65°.[368] pyrolysis,[368] ^{1}H.[859] ^{31}P -70.3,[310] -68.8.[859]
C6 $(HSCH_2CH_2NH)_3PS$. II. n_5 1.6280, d 1.307.[178]
$(EtNH)_3PS$. II. m. 68°.[573]
C9 $(CH_2{=}CHCH_2NH)_3PS$. II. $b_{0.0001}$ 80-90°.[677] n 1.5461, d 1.091.[677]
$(PrNH)_3PS$. II,[573] VII.[131] m. 73°,[131,573] m. 75°.[113]
C12 $(BuNH)_3PS$. II, VII. m. 54°.[913]
$(i\text{-}BuNH)_3PS$. II. m. 78-8.5°,[573] m. 78°.[113]
C18 $(p\text{-}ClC_6H_4NH)_3PS$. VII. m. 225-6°.[131]
$(PhNH)_3PS$. II,[77] VII,[131,476] XVII.[132,736] m. 153-4°,[77,131] m. 153.5°,[476] m. 150-1°.[718]
$(C_6H_{11}NH)_3PS$. II. m. 143-4.5°.[69] metal ion extraction.[359] ^{31}P -58.4.[859]
Cx $(PhCH_2NH)_3PS$. II,[573] VII,[131,913] XVII.[132] m. 127°,[573] m. 125-6°,[131] m. 122-3°.[913]
$(o\text{-}MeC_6H_4NH)_3PS$. II. m. 134.5°.[736]
$(p\text{-}MeC_6H_4NH)_3PS$. II. m. 185°,[736] m. 186°.[77]
$(p\text{-}EtOC_6H_4NH)_3PS$. II. m. 152°.[77]
$(PhCH_2SCH_2CH_2NH)_3PS$. II. m. 47-8.5°.[178]

D.3.2.2. $(RR'N)_3PS$, R = R' or R ≠ R', R,R' = Alkyl or Aryl or Acyl

$(C_2H_4N)_3PS$. II. m. 53-4°.[302]
$(Me_2N)_3PS$. XIII. $b_{1.2}$ 63°, m. 29°.[888] ^{1}H.[169,644] ^{31}P -81.6.[859,873]
$[O(CH_2CH_2)_2N]_3PS$. II. m. 145.5-6.0°.[69]
$(Et_2N)_3PS$. II,[573] XIII.[673,829] $b_{0.025}$ 104°,[137] $b_{0.12}$ 121-2°.[829] $b_{0.3}$ 129-31°.[673] n_D^{23} 1.5040,[137] n 1.5051.[673,829] ^{1}H.[169] ^{31}P -77.8.[394,620]
$(C_5H_{10}N)_3PS$. II, XIII. m. 121-2°,[69] m. 120°.[573]
$(Pr_2N)_3PS$. XIII. $b_{0.01}$ 108-12°, n 1.4783.[829]
$(i\text{-}Bu_2N)_3PS$. II,[573] XIII.[829] m. 19-20°,[829] n 1.4869,[829] d_0^{15} 0.9965,[573] metal ion extraction.[359]
$(C_9H_{10}N)_3PS$. II,[575] VII.[131] m. 192°,[575] m. 190-2°.[131] Tetrahydroquinoline derivative.

D.3.3. $(\geq N)_2P(O)(N\leq)'$ (Including Ring Compounds)

D.3.3.1. $(H_2N)_2P(O)(N\leq)$, $(\geq C{=}N)_2P(O)(N\leq)$, $(AcNH)_2P(O)(N\leq)$, Ac = Acyl

C2 $Me_2NPO(NH_2)_2$. IV. m. 119°.[333]
C4 $(ClCH_2CH_2)_2NPO(NH_2)_2$. m. 126-7°.[316]
$O(CH_2CH_2)_2NPO(NH_2)_2$. IV. m. 158-61°.[563]
$BuNHPO(NH_2)_2$. IV. b_2 120°.[333]
$Et_2NPO(NH_2)_2$. IV. m. 81°.[333]
C6 $p\text{-}ClC_6H_4SO_2NHPO(NH_2)_2$. IV. m. 195-200°.[919]
$Et_2NPO(NCS)_2$. XIV.[573,883] m. 151°.[883]

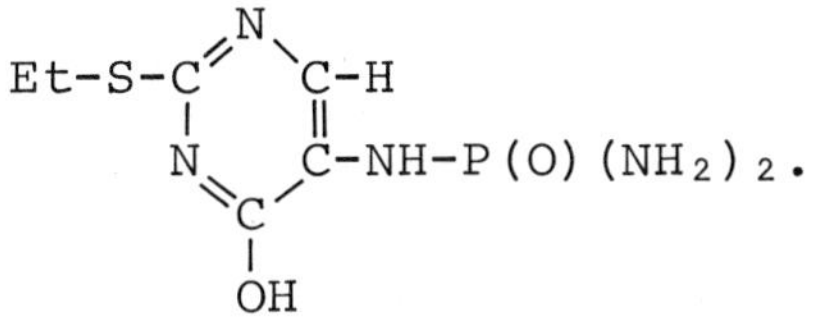

dec. 290-300°.[393]

$Et_2CBrCONHPO(NH_2)_2$. m. 133-5°.[319]

C_7 $p-MeC_6H_4SO_2NHPO(NH_2)_2$. IV. m. 178-9°.[919]

C_9 $PhNHCSNHPO(NCS)_2$. XII.[250]

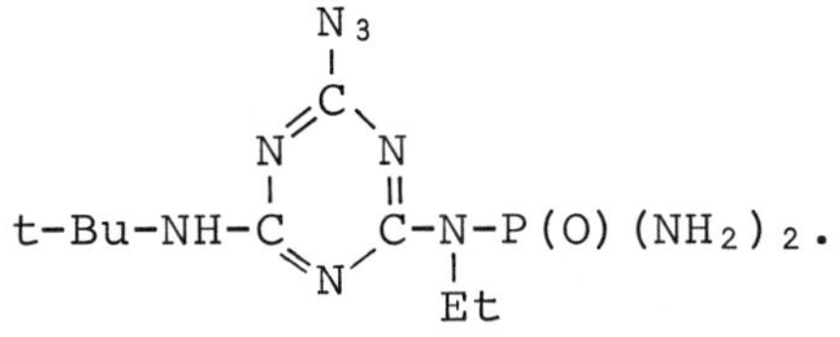

IV. m. 150-1°.[244] For other substituted triazine derivatives see Ref. 244.

C_{10} $2,4,6-Cl_3C_6H_2NHPO(NClCH_2CH_2Cl)_2$. XVII. m. 102-3°.[702]

$2,4-Cl_2C_6H_3NHPO(NClCH_2CH_2Cl)_2$. XVII. m. 87-9°.[702]

$p-BrC_6H_4NHPO(NClCH_2CH_2Cl)_2$. XVII. m. 133-4°.[702]

$p-ClC_6H_4NHPO(NClCH_2CH_2Cl)_2$. XVII. m. 98-100°.[702]

$PhNHPO(NClCH_2CH_2Cl)_2$. XVII. m. 93-5°.[702]

C_{11} $p-MeC_6H_4NHPO(NClCH_2CH_2Cl)_2$. XVII. m. 89-91°.[702]

$C_{18}H_{37}NHPO(NH_2)_2$. IV. m. 95-6°.[296]

$(C_{18}H_{37})(Me)NPO(NH_2)_2$. IV. m. 103-4°.[296]

$PhNHPO(NHCONHPh)_2$. XII. m. 179-81°.[238]

D.3.3.2. $(RNH)_2P(O)(NH_2)$, $(RNH)_2P(O)(NHAc)$, $(RNH)_2P(O)(N{=}C\lessdot)$, R = Alkyl or Aryl, Ac = Acyl

C_8 $Et_2CBrCONHPO(NHMe)_2$. IV. m. 146-8°.[319]

C_{11} $MeCCl_2CONHPO(NHBu)_2$. IV. m. 72-3°.[231]

$ClCH_2CHClCONHPO(NHBu)_2$. IV. m. 133-6°.[198]

C_{12} $H_2NPO(NHPh)_2$. V. m. 210-2°,[471] m. 212-3°.[521]

C_{13} $HOCONHPO(NHPh)_2$. XII. aniline salt: m. 160-90°.[471]

$p-MeC_6H_4SO_2NHPO(NHPr-i)_2$. IV. m. 136-8°.[919]

C_{14} $CF_3CONHPO(NHC_6H_4Br-p)_2$. IV. m. 245-6°.[200]

$CF_3CONHPO(NHC_6H_4Cl-p)_2$. IV. m. 253-5°.[200]

$CF_3CONHPO(NHC_6H_4I-p)_2$. IV. m. 274-6°.[200]

$CF_3CONHPO(NHPh)_2$. IV. m. 224-5°.[200]

$CCl_3CONHPO(NHPh)_2$. IV. m. 194-5°.[816,817]

$Cl_2CHCONHPO(NHPh)_2$. IV. m. 219-20°,[816,817] m. 214-5°.[231] For derivatives with substituted arylamines see Ref. 231.

$EtSO_2NHPO(NHPh)_2$. X. m. 127-30°.[445]

$p-ClC_6H_4SO_2NHPO(NHBu)_2$. IV. m. 96-8°.[919]

C_{15} $CCl_3C(NHMe){=}NPO(NHC_6H_4Br-p)_2$. IV. m. 189-90°.[230]

$ClCH_2CCl_2CONHPO(NHPh)_2$. IV. m. 191-3°.[198] For other arylamine compounds see Ref. 198.

$ClCH_2CHClCONHPO(NHPh)_2$. IV. m. 214-6°.[198] For other arylamine compounds see Ref. 198.

$$Me-CCl_2-CO-NH-P \begin{subarray}{l} O\ NHAr. \\ \| \\ NHAr \end{subarray}$$ IV.[231]

Ar	m.
Ph	170-2°
PhCH$_2$	110-2°
o-MeC$_6$H$_4$	140-1°
m-MeC$_6$H$_4$	177-9°
p-MeC$_6$H$_4$	204-6°
o-MeOC$_6$H$_4$	197-9°
m-MeOC$_6$H$_4$	178-81°
p-MeOC$_6$H$_4$	186-9°
o-ClC$_6$H$_4$	119-21°
m-ClC$_6$H$_4$	147-9°
p-ClC$_6$H$_4$	216-9°
p-IC$_6$H$_4$	188-9°
p-EtOC$_6$H$_4$	187-9°
2-C$_{10}$H$_7$	214-6°

(C$_2$H$_4$N)CONHPO(NHPh)$_2$. XII. m. 150-2°.[144] For other (ArNH)$_2$PONHCOR compounds see Ref. 144.

p-MeC$_6$H$_4$SO$_2$NHPO(NHBu)$_2$. IV. m. 95-7°.[919]

C$_{16}$ CF$_3$CONHPO(NHCH$_2$Ph)$_2$. IV. m. 182-4°.[200]

CF$_3$CONHPO(NHC$_6$H$_4$Me-o)$_2$. IV. m. 189-91°.[200]

CF$_3$CONHPO(NHC$_6$H$_4$Me-p)$_2$. IV. m. 243-4°.[200]

CF$_3$CONHPO(NHC$_6$H$_4$OMe-m)$_2$. IV. m. 169-72°.[200]

CF$_3$CONHPO(NHC$_6$H$_4$OMe-p)$_2$. IV. m. 229-30°.[200]

CHCl$_2$CONHPO(NHCH$_2$Ph)$_2$. IV. m. 175-6°.[231]

BuSO$_2$NHPO(NHPh)$_2$. X. m. 188-9°.[445]

C$_{17}$ Me$_2$C(CN)CONHPO(NHPh)$_2$. IV. m. 220-1°.[788] For other arylamine compounds see Ref. 788.

C$_{18}$ PhN=S=NPO(NHPh)$_2$. IV. m. 193-4°.[521]

PhSO$_2$NHPO(NHPh)$_2$. X. m. 205-7°.[432]

Et$_2$CBrCONHPO(NHPh)$_2$. IV. m. 185-6°.[319]

p-ClC$_6$H$_4$SO$_2$NHPO(NHC$_6$H$_{11}$)$_2$. IV. m. 170°.[919]

C$_{19}$ $$\begin{subarray}{l} ArNH\ \ \ O \\ \ \ \ \ \|\ \ \ \ \\ P-NH-CO-C_6H_2Cl_3-2,3,6. \\ ArNH \end{subarray}$$ IV.[832]

Ar	m.
Ph	204-6°
p-BrC$_6$H$_4$	159-62°
m-ClC$_6$H$_4$	156-9°
o-MeC$_6$H$_4$	171-3°
m-MeC$_6$H$_4$	214-7°
p-MeC$_6$H$_4$	214-6°
m-MeOC$_6$H$_4$	184-5°
p-MeOC$_6$H$_4$	224-6°
2-C$_{10}$H$_7$	208-10°

$$\begin{subarray}{l} PhNH\ \ O \\ \ \ \ \ \|\ \ \\ P-NH-CO-Ar. \\ PhNH \end{subarray}$$ IV.[219]

Ar	m.
Ph	215-6°
p-ClC$_6$H$_4$	215-6°
o-O$_2$NC$_6$H$_4$	237-8°

Ar	m.
$m-O_2NC_6H_4$	213-4°
$p-O_2NC_6H_4$	225-6°
$3,5-(O_2N)_2C_6H_3$	226-8°
$p-MeOC_6H_4$	208-10°

$o-MeC_6H_4SO_2NHPO(NHPh)_2$. X. m. 198-200°.[432]
$p-MeC_6H_4SO_2NHPO(NHPh)_2$. X. m. 201-2.5°.[432]
$MeSO[=NPO(NHPh)_2]NHPh$. IV. m. 81-2°.[523]
$p-MeC_6H_4SO_2NHPO(NHC_6H_{11})_2$. IV. m. 171°.[919]

C_{20} $CF_3C(NHC_6H_4Cl-p)=NPO(NHC_6H_4Cl-p)_2$. IV. m. 214-6°.[200]
$CF_3C(NHPh)=NPO(NHPh)_2$. IV. m. 115-7°,[200] m. 116-7°.[232]
$PhNHCH_2CONHPO(NHPh)_2$. IV. m. 156°.[120]
$PhSO_2NHPO(NHC_6H_4Me-p)_2$. X. m. 203-4°.[432]
$MeNHC(NHPh)=NPO(NHPh)_2$. IV. m. 149-50°.[226]
$MeNHC(NHC_6H_4Br-p)=NPO(NHC_6H_4Br-p)_2$. X. m. 145-8°.[226]
$EtSO[=NPO(NHPh)_2]NHPh$. IV. m. 78-80°.[523]
$Et_2CBrCONHPO(NHCH_2Ph)_2$. IV. m. 122-3°.[319]

C_{21}
$p-MeC_6H_4NH$, $\overset{O}{\overset{\|}{P}}$–NH–CO–Ar. IV.[219]
$p-MeC_6H_4NH$

Ar	m.
Ph	218-20°
$o-O_2NC_6H_4$	252-3°
$m-O_2NC_6H_4$	229-30°
$p-O_2NC_6H_4$	216-9°
$3,5-(O_2N)_2C_6H_3$	238-9°
$p-MeOC_6H_4$	202-4°

$o-MeC_6H_4SO_2NHPO(NHC_6H_4Me-p)_2$. X. m. 195-7°.[432]
$p-MeC_6H_4SO_2NHPO(NHC_6H_4Me-p)_2$. X. m. 207-8°.[432]
$EtNHC(NHPh)=NPO(NHPh)_2$. X. m. 113-5°.[226]
$EtNHC(NHC_6H_4Br-p)=NPO(NHC_6H_4Br-p)_2$. X. m. 111-4°.[226]

C_{22} $CF_3CONHPO(NHC_{10}H_7-1)_2$. IV. m. 257-9°.[200]
$1-C_{10}H_7SO_2NHPO(NHPh)_2$. X. m. 180-2°.[432]
$2-C_{10}H_7SO_2NHPO(NHPh)_2$. X. 210-1°.[789]

C_{23} $CCl_3C(=NCH_2Ph)NHPO(NHC_6H_4OMe-p)_2$. IV. m. 86-8°.[232]
$Cl_2CHC(=NC_6H_4Me-p)NHPO(NHC_6H_4Me-p)_2$. IV. m. 227-9°.[230]
For other arylamine substituted compounds see Ref. 230.

C_{24} $1-C_{10}H_7SO_2NHPO(NHC_6H_4Me-p)_2$. X. m. 184-5°.[432]
C_{25} $p-O_2NC_6H_4C(=NSO_2Ph)NHPO(NHPh)_2$. IV. m. 221-2°.[207]

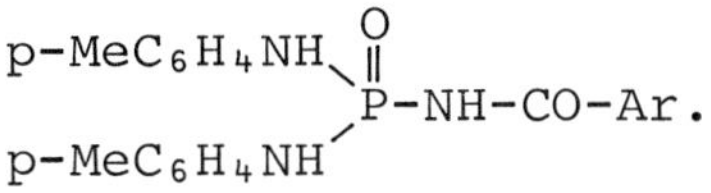

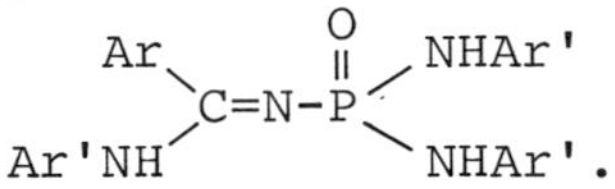

$\begin{matrix} Ar \\ & \diagdown \\ & C=N-\overset{O}{\overset{\|}{P}} \\ & \diagup \\ Ar'NH \end{matrix} \begin{matrix} NHAr' \\ \\ NHAr'. \end{matrix}$ IV.[211]

Ar	Ar'	m.
Ph	Ph	93-5°
Ph	$p-MeC_6H_4$	94-6°
$p-BrC_6H_4$	Ph	190-1°
$p-BrC_6H_4$	$p-MeC_6H_4$	145-6°
$m-O_2NC_6H_4$	Ph	102-3°
$m-O_2NC_6H_4$	$p-MeC_6H_4$	116-7°

$$\underset{Ph}{\overset{Ar'SO_2N}{>}}C\text{-}NH\text{-}\underset{\underset{NHAr}{|}}{\overset{\overset{O}{\|}}{P}}\text{-}NHAr. \quad IV.^{203}$$

Ar	Ar'	m.
p-O$_2$NC$_6$H$_4$	Ph	198-9°
p-O$_2$NC$_6$H$_4$	p-MeC$_6$H$_4$	220-1°

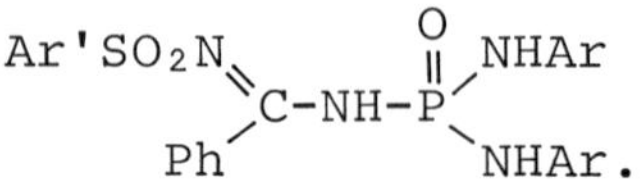

Ar'	Ar	m.
Ph	Ph	210-1°
Ph	p-MeC$_6$H$_4$	197-8°
Ph	p-ClC$_6$H$_4$	197-8°
Ph	p-BrC$_6$H$_4$	200-1°
p-MeC$_6$H$_4$	Ph	204-5°
p-MeC$_6$H$_4$	p-MeC$_6$H$_4$	206-8°
p-ClC$_6$H$_4$	Ph	205-6°
p-O$_2$NC$_6$H$_4$	Ph	217-8°
1-C$_{10}$H$_7$	Ph	244-5°
2-C$_{10}$H$_7$	Ph	202-3°

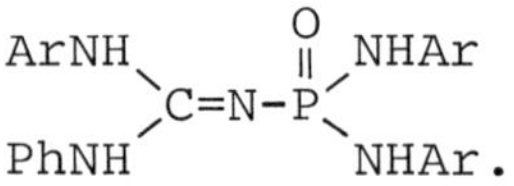

$$\underset{PhNH}{\overset{ArNH}{>}}C=N\text{-}\underset{\underset{NHAr}{|}}{\overset{\overset{O}{\|}}{P}}\text{-}NHAr. \quad IV.^{228}$$

Ar	m.
Ph	187-8°
p-MeC$_6$H$_4$	181-2°
p-MeOC$_6$H$_4$	166-8°
p-ClC$_6$H$_4$	205-7°
p-BrC$_6$H$_4$	225-6°
m-O$_2$NC$_6$H$_4$	280-1°

For other Ar-groups see Ref. 22&

C$_x$ p-BrC$_6$H$_4$C[=NPO(NHPh)$_2$]N=C(NHPh)Ph. IV. m. 220-2°.[203]

PhC[=NPO(NHPh)$_2$]N=C(NHPh)C$_6$H$_4$Br-p. IV. 270-4°.[203]

PhC[=NPO(NHPh)$_2$]N=C(NHPh)Ph. IV. m. 228-30°.[203]

p-MeC$_6$H$_4$C[=NPO(NHPh)$_2$]N=C(NHPh)Ph. IV. m. 248-9°.[203]

o-ClCOC$_6$H$_4$N[PO(NHC$_6$H$_4$COCl-o)$_2$]$_2$. I. m. 148-53°.[882]

D.3.3.3.   (RNH)$_2$P(O)(NHR'), R $\neq$ R', R,R' = Alkyl or
            Aryl

C$_8$ PhNHPO(NHCH$_2$)$_2$. IV. m. 232°.[73]

C$_{10}$ (2,4,6-Cl$_3$C$_6$H$_2$NH)PO(NHCH$_2$CH$_2$Cl)$_2$. XVII. m. 140-2°.[702]

(2,4-Cl$_2$C$_6$H$_3$NH)PO(NHCH$_2$CH$_2$Cl)$_2$. XVII. m. 75-7°,[702]
   m. 77-8°.[703]

(p-BrC$_6$H$_4$NH)PO(NHCH$_2$CH$_2$Cl)$_2$. XVII. m. 62-3°,[702]
   m. 63-5°.[703]

(p-ClC$_6$H$_4$NH)PO(NHCH$_2$CH$_2$Cl)$_2$. XVII. m. 84.5-5.0°,[703]
   m. 85-7°.[702]

PhNHPO(NHCH$_2$CH$_2$Cl)$_2$. XVII. m. 45-6°,[702] m. 61-3°.[703]

C11 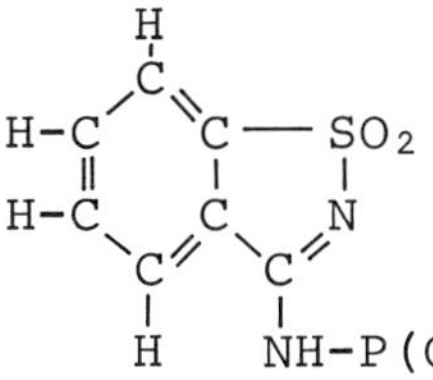
NH-P(O)(NHR)$_2$. IV.[207] R=Et: m. 219-20°.
R=Ph: m. 260-1°.
R=p-ClC$_6$H$_4$: m. 249-53°.

p-MeC$_6$H$_4$NHPO(NHCH$_2$CH$_2$Cl)$_2$. XVII. m. 60-2°.[702]

C12 PhNHPO(NH)$_2$C$_6$H$_4$-o. IV. m. 214°.[73]

C14 2,3or4-H$_2$N,MeC$_6$H$_3$NHPO(NH)$_2$C$_6$H$_3$-1,2-Me-3. I. m. 200°.[364]

EtNHPO(NHPh)$_2$. IV, V. m. 147°.[573]

PhNHPO(NHCH$_2$CH$_2$OCOCH$_2$Cl)$_2$. XVII. n 1.3786.[703]

p-BrC$_6$H$_4$NHPO(NHCH$_2$CH$_2$OCOCH$_2$Cl)$_2$. XVII. n 1.3810.[703]

p-ClC$_6$H$_4$NHPO(NHCH$_2$CH$_2$OCOCH$_2$Cl)$_2$. XVII. n 1.3777.[703]

p-FC$_6$H$_4$NHPO(NHCH$_2$CH$_2$OCOCH$_2$Cl)$_2$. XVII. n 1.3802.[703]

C15 p-MeC$_6$H$_4$NHPO(NHCH$_2$CH$_2$OCOCH$_2$Cl)$_2$. XVII. n 1.3821.[703]

PrNHPO(NHPh)$_2$. IV, V. m. 146°.[573]

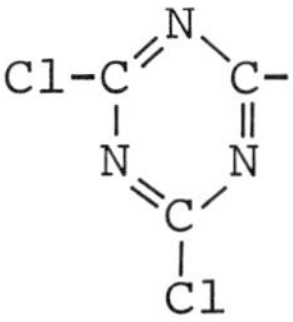
-NH-P(O)(NHPh)$_2$. IV. m. 223-5°.[202] For other triazine derivatives see Ref. 202.

C16 (i-BuNH)PO(NHPh)$_2$. IV, V. m. 207°.[573]

C17 2-(C$_5$H$_4$N)NHPO(NHPh)$_2$. V. m. 240-1°.[42] m. 217-8°.[50] Pyridine derivative.

i-AmNHPO(NHPh)$_2$. IV, V. m. 117°.[573]

C18 PhNHPO(NHC$_6$H$_4$Br-m)$_2$. VII. m. 165°.[583]

2,4-Br$_2$C$_6$H$_3$NHPO(NHPh)$_2$. IV, V. m. 228°.[573]

p-O$_2$NC$_6$H$_4$NHPO(NHPh)$_2$. IV, V. m. 242°.[573]

m-O$_2$NC$_6$H$_4$NHPO(NHPh)$_2$. IV, V. m. 177°.[573]

PhNHPO(NHC$_6$H$_{11}$)$_2$. ^{31}P -8.0.[634]

C19 o-MeC$_6$H$_4$NHPO(NHPh)$_2$. IV, V. m. 175°.[582]

p-MeC$_6$H$_4$NHPO(NHPh)$_2$. IV, V. m. 168°.[582]

C20 (2,4-Br$_2$C$_6$H$_3$NH)PO(NHC$_6$H$_4$Me-p)$_2$. IV, V. m. 214°.[573]

PhNHPO(NHC$_6$H$_4$Me-o)$_2$. IV, V. m. 201°.[582]

PhNHPO(NHC$_6$H$_4$Me-p)$_2$. IV, V. m. 168°.[582]

(x-O$_2$NC$_6$H$_4$NH)PO[NHC$_6$H$_3$(NO$_2$-x)Me-P]$_2$. XVII. m. 220°.[582]

Cx PhNHPO(NHCH$_2$CH$_2$OSO$_2$C$_6$H$_4$Me-p)$_2$. XVII. m. 89-91°.[703]

p-MeC$_6$H$_4$NHPO(NHCH$_2$CH$_2$OSO$_2$C$_6$H$_4$Me-p)$_2$. XVII. m. 61-3°.[703]

p-ClC$_6$H$_4$NHPO(NHCH$_2$CH$_2$OSO$_2$C$_6$H$_4$Me-p)$_2$. XVII. m. 88-90°.[703]

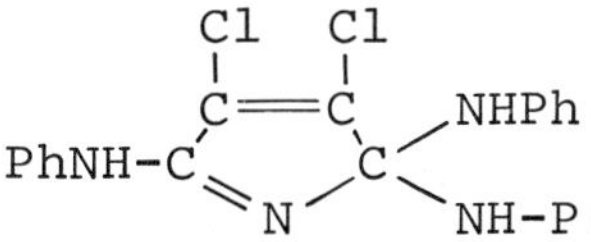
NH-P(O)(NHPh)$_2$. XVII. m. 181-2°.[787]

D.3.3.4. $(RNH)_2P(O)(NR'R'')$, R' = R" or R' ≠ R", R,R',R" = Alkyl or Aryl

C_6 $(ClCH_2CH_2)_2NPO(NHMe)_2$. IV. m. 86-7°.[54]

$Me_2NPO(NHCH_2CH_2SH)_2$. IV. n 1.5548, d 1.233.[178]

C_8 $(C_2H_4N)PO(NHCH_2CH{=}CH_2)_2$. V. $b_{0.001}$ 120-5°, n 1.5017, d 1.10.[677]

C_9 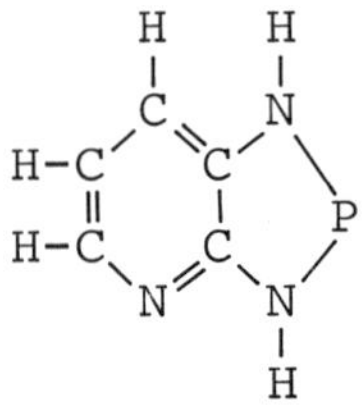 $P(O)-N(CH_2CH_2Cl)_2$. IV. m. 210°. Softening 188-91°.[942]

C_{10} 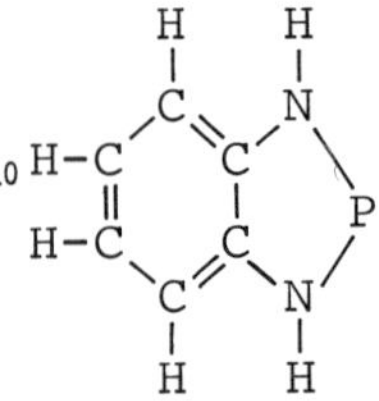 $P(O)-N(CH_2CH_2Cl)_2$. IV. m. 168-9°.[942]

$Et_2NPO(NHPr\text{-}i)_2$. IV. $b_{0.01}$ 137°.[113]

C_{12} $(ClCH_2CH_2)_2NPO(NHCH_2COOEt)_2$. IV. m. 88-90°,[848] m. 87°.[500] For $(ClCH_2CH_2)_2NPO(NHCHRCOOEt)_2$ compounds (n_D listed) see Ref. 848.

$Et_2NPO(NHBu)_2$. IV.[113]

C_{14} $Me_2NPO(NHPh)_2$. IV. m. 196°,[573] m. 198-200°.[879] ^{31}P -6.4.[634]

$Me_2NPO(NHC_6H_4SO_2NH_2\text{-}p)_2$. XVII. m. 235°.[772]

$Me_2NPO(NHC_6H_{11})_2$. ^{31}P -15.7.[634]

C_{16} $(ClCH_2CH_2)_2NPO(NHPh)_2$. IV. m. 185-6°.[716]

$Et_2NPO(NHPh)_2$. IV, V. m. 150°.[573]

C_{18} $(ClCH_2CH_2)_2NPO(NHC_6H_4Me\text{-}p)_2$. IV. m. 154-5°.[716]

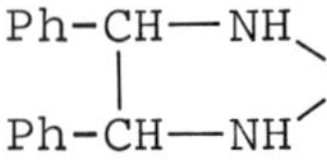 $P(O)NR_2$. IV.[300] R=Et: two isomers (cis/trans) m. 201.5-3.0°, m. 162.4° R=CH_2CH_2Cl: two isomers (cis/trans): m. 180-1° m. 150-2°.

$Pr_2NPO(NHPh)_2$. IV, V. m. 220°.[573]

C_{19} $MePhNPO(NHPh)_2$. IV, V. m. 192°.[573]

$MePhNPO(NHC_6H_{11})_2$. ^{31}P -11.0.[634]

C_{20} $Me_2NPO(NHCH_2CH_2SCH_2Ph)_2$. IV. m. 46-7.5°.[178]

$(i\text{-}Bu)_2NPO(NHPh)_2$. IV, V. m. 202°.[573]

$Pr_2NPO(NHC_6H_4Me\text{-}p)_2$. IV, V. m. 168°.[573]

C_{21} $(C_9H_{10}N)PO(NHPh)_2$. IV. m. 176°.[573] Tetrahydroquinoline derivative.

$MePhNPO(NHC_6H_4Me\text{-}p)_2$. IV, V. m. 232°.[573]

C_{22} $Et_2NPO(NHC_6H_4COOEt\text{-}p)_2$. IV. m. 233-5°.[141]

$(i\text{-}Bu)_2NPO(NHC_6H_4Me\text{-}p)_2$. IV, V. m. 180°.[573]

$(ClCH_2CH_2)_2NPO[NHCH(CH_2CH_2COOEt)COOEt]_2$. IV. m. 85-6.5°.[848]

C_x $Ph_2NPO(NHPh)_2$. IV, V. m. 232°.[660]
$Ph_2NPO(NHC_6H_{11})_2$. ^{31}P -6.8.[634]
$Ph_2NPO(NHC_6H_4Me\text{-}o)_2$. IV, V. m. 219°.[660]
$(ClCH_2CH_2)_2NPO[NHCH(CH_2Ph)COOEt]_2$. IV. m. 113-5°.[848]

D.3.3.5. $(RR'N)_2P(O)(NHAc)$, $(RR'N)_2P(O)(N{=}C\lt)$, $(RR'N)_2P(O)(N_3)$, R = R' or R ≠ R', R,R' = Alkyl or Aryl, Ac = Acyl

C_4 $N_3PO(NMe_2)_2$. XIV.[360,887] b_2 93-4°, n_D^{21} 1.4673.[887]
C_5 $(SCN)PO(NMe_2)_2$. XIV. $b_{0.3}$ 83-5°, n 1.5318, d 1.1575.[587,589] ^{31}P -5.5.[361a]
C_6 $MeOCONHPO(NC_2H_4)_2$. IV, XII. m. 119-21°.[52,661,663]
$MeONHCONHPO(NC_2H_4)_2$. XVII. m. 99-101°.[144]
C_7 $EtOCONHPO(NC_2H_4)_2$. IV, XII. m. 89-90°.[52,661]
C_9 $(C_2H_4N)_2P(O)\text{-}NH\text{-}CO\text{-}NH\text{-}R$. IV, XII. See also Ref. 714.

R	m. (Lit)
Bu	148-9°[294]
i-Bu	157-8°[294]
$CH_2{=}CHCH_2$	94-6°[294]
C_6H_{11}	191°,[294] 250-2° dec.[661,663]
$PhCH_2$	182°[294]
$PhCH_2CH_2$	150°[294]
Ph	184°,[713] 209°,[714] 250°[663]
$4\text{-}Me\text{-}C_6H_{10}$	196°[294]
-CH(Me)COOEt	126-8°[485]
$4\text{-}MeOC_6H_{10}$	192-4°[294]
$p\text{-}BrC_6H_4$	188-9°[485]
$p\text{-}ClC_6H_4$	195-7°,[485] dec. 220°,[294] 211°[713]
$o\text{-}ClC_6H_4$	159-60°,[485] 177°[713]
$p\text{-}MeC_6H_4$	174-5°,[485] 170°[294]
$o\text{-}MeC_6H_4$	167-9°[485]
$p\text{-}MeOC_6H_4$	149-50°,[485] 166°[294]
$o\text{-}MeOC_6H_4$	153-4°[485]
$2\text{-}C_{10}H_7$	147-9°[485]
$2,4\text{-}Cl,MeC_6H_3$	180°[294]
$2,5\text{-}Cl_2C_6H_3$	171-2°[713]
$3,4\text{-}Cl_2C_6H_3$	190-2°[713]
$2,4,6\text{-}Cl_3C_6H_2$	202°[713]
$2,4\text{-}O_2N,ClC_6H_3$	187-8°[713]
$p\text{-}O_2NC_6H_4$	199-201°[713]
$o\text{-}O_2NC_6H_4$	156.5°[713]
$p\text{-}NCC_6H_4$	197-8°[713]

$(OCN)PO(NEt_2)_2$. XV. b_1 112-4°, n 1.4629, d 1.0790.[809]
$(SCN)PO(NEt_2)_2$. XIV. $b_{0.1}$ 82-4°, n 1.5030, d 1.0727.[587,589]

$(Me_2N)_2PON(Me)PO(NMe_2)_2$. V. b_1 145-9°,[165] b_2 160-70),[184] m. 58°.[184] For metal complexes see Ref. 508.

C_{10} 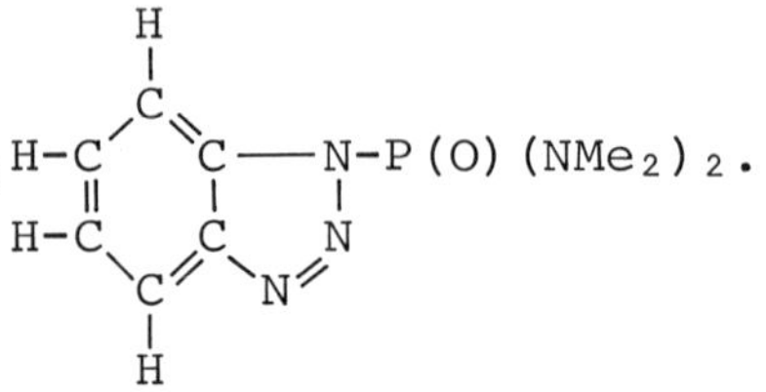 V. m. 80-1°.[118]

$o-O_2NC_6H_4SO_2NHPO(NMe_2)_2$. IV. m. 161-4°.[447]
$m-O_2NC_6H_4SO_2NHPO(NMe_2)_2$. IV. m. 145-7°.[447]
$p-O_2NC_6H_4SO_2NHPO(NMe_2)_2$. IV. m. 169-70°.[447]
$CHCl_2CONHPO(NEt_2)_2$. IV. m. 126-8°.[231]
$(Me_2N)_2PON(Et)PO(NMe_2)_2$. V. b_2 155-63°, m. 48°.[184]
$(Me_2N)_3P{=}NPO(NMe_2)_2$. XIII. b_1 135, n 1.4847.[887]

C_{11} 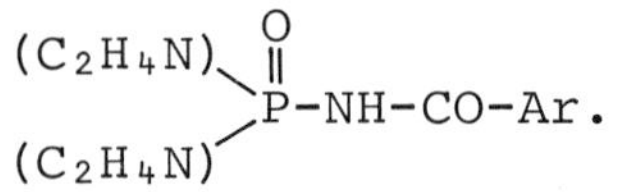 IV.

Ar	m. (Lit)
Ph	93-4°[696]
$p-BrC_6H_4$	109-11°[696]
$m-BrC_6H_4$	148°[693,694]
$o-BrC_6H_4$	111-2°[693,694]
$2,5-Br_2C_6H_3$	166°[693]
$3,5-Br_2C_6H_3$	175°[693]
$2,3,5-Br_3C_6H_2$	185°[693]
$3,4,5-Br_3C_6H_2$	122°[693]
$p-ClC_6H_4$	106-7°[696]
$m-ClC_6H_4$	139-40°[699]
$o-ClC_6H_4$	124-5°[699]
$2,4-Cl_2C_6H_3$	154-5°[699]
$3,5-Cl_2C_6H_3$	137-8°[699]
$2,3,5-Cl_3C_6H_2$	142-3°[699]
$3,4,5-Cl_3C_6H_2$	232-3°[699]
$p-FC_6H_4$	103-4°,[696] 141-3°[698]
$m-FC_6H_4$	142-4°[698]
$o-FC_6H_4$	135-6.5°[698]
$p-CF_3C_6H_4$	143-4°[698]
$m-CF_3C_6H_4$	158-9°[698]
$p-IC_6H_4$	112-4°[696]
$m-IC_6H_4$	111-2°[700]
$o-IC_6H_4$	112-4°[700]
$2,4-I_2C_6H_3$	211-3°[700]
$2,5-I_2C_6H_3$	dec. 192-3°[700]
$3,4-I_2C_6H_3$	228-30°[700]
$3,5-I_2C_6H_3$	212-3°[700]
$2,3,4-I_3C_6H_2$	208-10°[700]
$2,4,5-I_3C_6H_2$	218-9°[700]
$3,4,5-I_3C_6H_2$	218-9°[700]

Ar	m. (Lit)
p-$O_2NC_6H_4$	129-30°[696]
p-MeC_6H_4	126-8°[696]
m-MeC_6H_4	145-7°[692]
o-MeC_6H_4	132-4°[692]
p-$MeOC_6H_4$	215-7°[692]
m-$MeOC_6H_4$	122-4°[692]
p-$EtOC_6H_4$	171-3°[692]
m-$EtOC_6H_4$	106-7°[692]
$3,4,5$-$(MeO)_3C_6H_2$	172-4°[692]
p-$CH_2=CHCH_2OC_6H_4$	140-2°[692]
m-$CH_2=CHCH_2OC_6H_4$	108-10°[692]
p-$PhCH_2OC_6H_4$	155-7°[692]
m-$PhCH_2OC_6H_4$	138-40°[692]

$ClCH_2CCl_2CONHPO(NEt_2)_2$. IV. m. 106-8°.[198]

$C_6H_{11}NHCSNHPO(NMe_2)_2$. XII.[587]

$_{12}$ $PhCF_2CONHPO(NC_2H_4)_2$. IV. dec. 180-1°.[698]

$PhCH_2OCONHPO(NC_2H_4)_2$. IV. m. 134-5°.[52,661,663]

p-$O_2NC_6H_4CH_2OCONHPO(NC_2H_4)_2$. IV. m. 106-8°.[52]   Meta-
    analog m. 106-8°.[52]

$[(Me_2N)_2PO]_3N$. IV.[165]

$_{13}$ $PhC(NC_2H_4)=NPO(NC_2H_4)_2$. IV. m. 78-80°.[486]

p-$BrC_6H_4C(NC_2H_4)=NPO(NC_2H_4)_2$. IV. m. 92-3°.[486]

p-$ClC_6H_4C(NC_2H_4)=NPO(NC_2H_4)_2$. IV. m. 78-80°.[486]

$$\left[\begin{array}{c} R \diagdown C \\ R' \diagup \\ | \\ H_2C \end{array} N \right]_2 \overset{O}{\overset{\|}{P}}\text{-NH-CO-NHR''}.$$

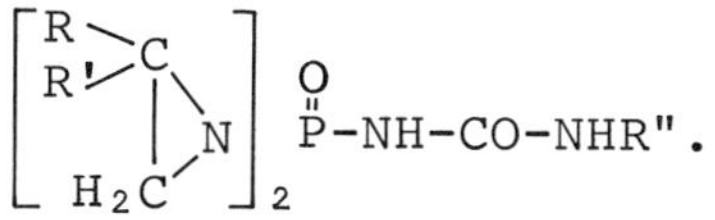

IV.   R=H, R'=Me, R''=p-ClC_6H_4:
    m. 177°.[294]
    R=H, R'=Me, R''=o-$O_2NC_6H_4$:
    m. 138-40°.[713]
    R=R'=Me, R''=i-Bu: m. 128°[294]

$(C_2H_4N)_2PONHCO(CF_2)_3CONHPO(NC_2H_4)_2$. IV. m. 216-9°.[201]
    For similar compounds see Ref. 201.

$PhNHC(NC_2H_4)=NPO(NC_2H_4)_2$. IV. m. 115-6°.[194]

p-$BrC_6H_4NHC(NC_2H_4)=NPO(NC_2H_4)_2$. IV. m. 59-60°.[194]

p-$ClC_6H_4NHC(NC_2H_4)=NPO(NC_2H_4)_2$. IV. m. 51-3°.[194]

p-$IC_6H_4NHC(NC_2H_4)=NPO(NC_2H_4)_2$. IV. m. 116-7°.[194]

$Et_2NCONHPO(NEt_2)_2$. IV, XII. m. 50-2°.[627]

$_{14}$ p-$O_2NC_6H_4SO_2NHPO(NEt_2)_2$. IV. m. 109-11°.[447]

m-$O_2NC_6H_4SO_2NHPO(NEt_2)_2$. IV. m. 125-7°.[447]

$Et_2CBrCONHPO[N(CH_2CH_2)_2O]_2$. IV. m. 119-20°.[319]

$_{15}$ $\begin{array}{c} Et_2N \diagdown \\ \\ Et_2N \diagup \end{array} \overset{O}{\overset{\|}{P}}\text{-NH-CO-Ar}.$ IV.[219]

Ar	m.
Ph	99-101°
p-ClC_6H_4	142-4°
o-$O_2NC_6H_4$	151-2°
m-$O_2NC_6H_4$	119-21°
p-$O_2NC_6H_4$	187-9°

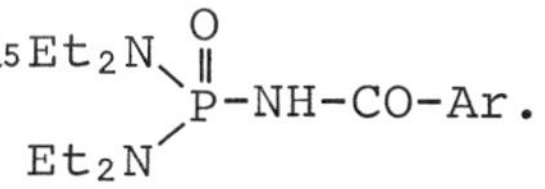

Ar	m.
3,5-$(O_2N)_2C_6H_3$	159-60°
p-$MeOC_6H_4$	128-30°
2,3,6-$Cl_3C_6H_2$	208-10°[832]

C_x p-$MeC_6H_4SO_2NHPO(NC_5H_{10})_2$. IV. m. 171°.[919]
$(Et_2N)_2PON(Et)PO(NEt_2)_2$. V. b_2 185-200°.[184]
p-$C_6H_4[C(=NPO(NC_2H_4)_2)(NC_2H_4)]_2$. IV. m. 108-9°.[695]

D.3.3.6. $(RR'N)_2P(O)(NHR'')$, R = R' or R ≠ R', R,R',R" = Alkyl or Aryl

C_5 $MeNHPO(NC_2H_4)_2$. IV. m. 105.5-6.5°.[601]

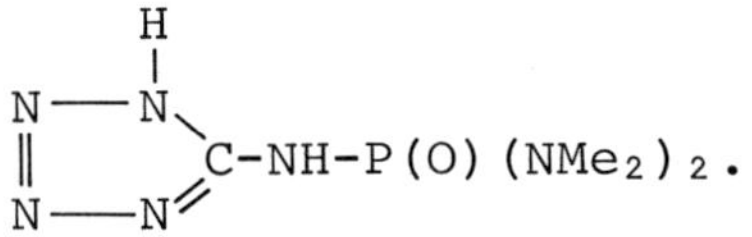
V. dec. 185°.[682]

$MeNHPO(NMe_2)_2$. V. b_3 137-43°, n_D^{25} 1.4551.[51]
C_6 $EtNHPO(NC_2H_4)_2$. IV. m. 57-61°, b_5 144°.[601]

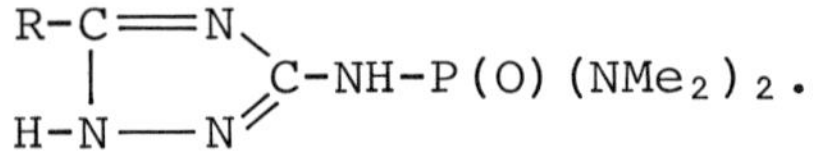
V.[119,480] The phosphoryl group may also be attached to a ring atom; see Refs. 119,570

R	m.
H	136-8°
Me	91-2°
Et	92-5°
Pr	62-3°
i-Pr	105-6°
EtOCO	151-4°
i-PrOCO	181-3°
Bu	55-6°
i-Bu	91-3°
Am	44-7°, 52-3°
$C_6H_{11}OCO$	105-8°
$C_{11}H_{23}$	47-8°
$PhCH_2$	130-1°, 107-8°
Ph	167-8°
p-ClC_6H_4	170-1°
p-$MeOC_6H_4$	173-4°
PhCH=CH	151-2°

The corresponding $(C_2H_2N_3)NHPO(NEt_2)_2$, m. 105-6°.

$H_2NCOCH_2NHPO(NMe_2)_2$. V. m. 128-9°.[19]

C_7 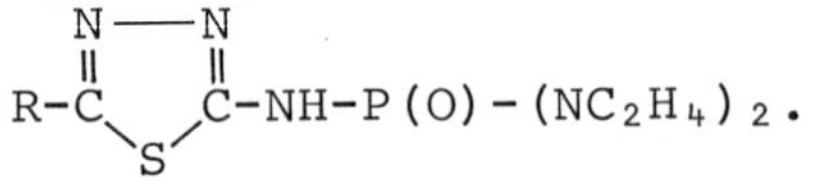
IV.[841] R=Me: m. 143-5°; R=Et: m. 124-6°; R=Pr: m. 104-5°.

$CH_2=CHCH_2NHPO(NC_2H_4)_2$. IV. m. 42-3°,[247,295] m. 40°.[677]

$PrNHPO(NC_2H_4)_2$. IV. b_2 125-6°.[601]

$MeOCH_2CH_2NHPO(NMe_2)_2$. XVII. $b_{0.25}$ 107-17°.[836]

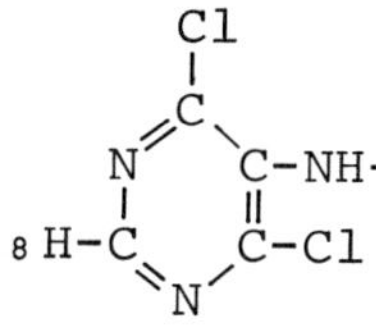

$-NH-P(O)(NC_2H_4)_2$. IV. dec. 161.5°.[546]

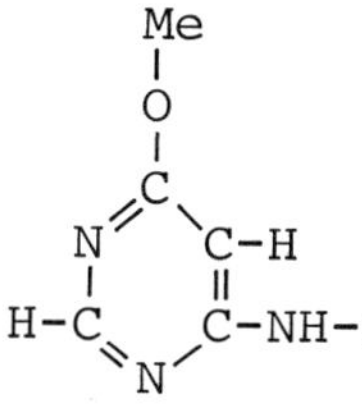

$-NH-P(O)(NC_2H_4)_2$. IV. m. 158-8.5°.[747]

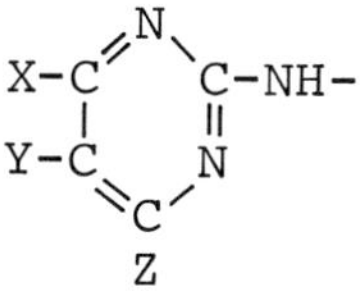

$-NH-P(O)(NC_2H_4)_2$. IV.

X	Y	Z	m. (Lit)
H	H	H	128-9°[489]
Cl	H	Cl	136-7°[491]
H	Br	H	153-4°[491]
Cl	H	H	121-2°[489]
H	Cl	H	157-8°[489]
Me	H	H	123-4°[489]
MeO	H	H	128-9°[489]
Cl	H	Me	123-4°[491]
Me	Br	H	130-1°[491]
Cl	H	MeO	121-2°[491]
Me	H	Me	128-9°[489]
Et_2N	H	H	150-0.5°[489]
MeO	H	Me	116.5-7.5°[491]
MeO	H	MeO	108-9°[491]
Me_2N	H	Cl	136°[491]
$(C_5H_{10}N)$	H	H	172-2.5°[491]

$MeCH=CHCH_2NHPO(NC_2H_4)_2$. IV. $b_{0.03}$ 131°.[247,295]

$BuNHPO(NC_2H_4)_2$. IV. m. 67-70°.[601]

$EtOCOCH_2NHPO(NMe_2)_2$. V. m. 41-4°.[19]

$Me_2CClCH_2NHPO(NMe_2)_2$. XVII. m. 117-9°.[836] For $RCMe_2CH_2NHPO(NMe_2)_2$ compounds see Ref. 836.

$AmNHPO(NC_2H_4)_2$. IV. b_2 146-8°.[601]

$2,5-Cl_2C_6H_3NHPO(NMe_2)_2$. V. m. 91-5°.[275] For other chloroarylamine derivatives see Ref. 275.

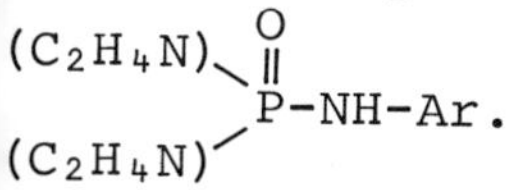

IV.

Ar	m. (Lit)
Ph	143-4°[487]
$p-ClC_6H_4$	170-1.5°[487]
$p-FC_6H_4$	150-2°[698]

Ar	m. (Lit)
p-IC_6H_4	176-7°[487]
m-$O_2NC_6H_4$	166-7°[487]
p-$MeOC_6H_4$	102-3°[487]
p-$CF_3C_6H_4$	172.5-4.5°[698]
m-$CF_3C_6H_4$	172.5-4°[698]
p-$EtOOCC_6H_4$	151-2°[487]
m-[(C_2H_4N)CO]C_6H_4	131-2.5°[487]
2-$C_{10}H_7$	148°[487]
p-$PhCH_2COOCH_2C_6H_4$	104-5°[947]

For further arylamine derivatives see Ref. 947.

p-$H_2NSO_2C_6H_4NHPO(NMe_2)_2$. IV. m. 243°.[772]

$C_6H_{11}NHPO(NC_2H_4)_2$. IV. m. 104-4.7°.[947]

C_{13} $O_2NC_6H_4COOCH_2CH_2NHPO(NMe_2)_2$. XVII. o, m, p-aryl compounds.[836]

C_{14} 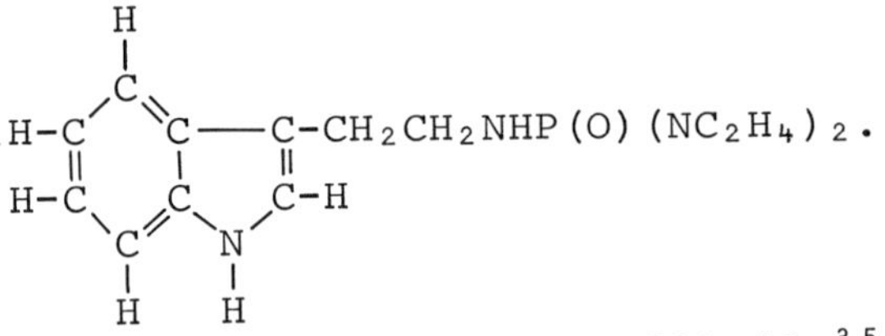 IV. m. 139-41°.[613] For substituted indole derivatives see Ref. 613.

$PhNHPO(NEt_2)_2$. IV. m. 113-5°.[350]

C_{16} 2,4-$Br_2C_6H_3NHPO(NC_5H_{10})_2$. IV, V. m. 186°.[573]

p-$BrC_6H_4NHPO(NC_5H_{10})_2$. IV, V. m. 169°,[573] m. 169°.[51]

p-$ClC_6H_4NHPO(NC_5H_{10})_2$. IV, V. m. 175°.[660]

$PhNHPO(NC_5H_{10})_2$. IV, V. m. 159°.[573]

C_x o-$MeC_6H_4NHPO(NC_5H_{10})_2$. IV, V. m. 146°.[573]

p-[$(C_5H_{10}N)_2PO(NHC_6H_4-)]_2SO_2$. IV. m. 221-2°, m. 190-1°.[389]

D.3.3.7. $(RR'N)_2P(O)(NR''R''')$, R,R',R'',R''' = Alkyl or Aryl, R,R' or R'',R''' Equal or Different

C_6 $Me_2NPO(NC_2H_4)_2$. IV.[757]

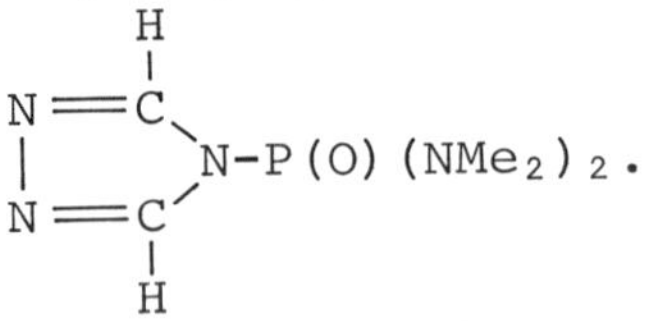 V. $b_{0.45}$ 109-12°.[682] For substituted triazole derivatives see Ref. 682.

$(C_2H_4N)PO(NMe_2)_2$. V. $b_{0.15}$ 42-6°, ^1H.[836]

C_7 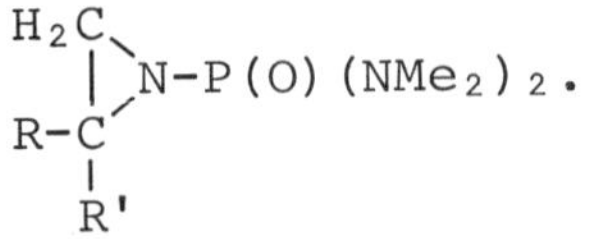 V.[836] ^1H. R=H, R'=Me: $b_{0.05}$ 54-9°. R=R'=Me: $b_{0.04}$ 51-4°.

C_8 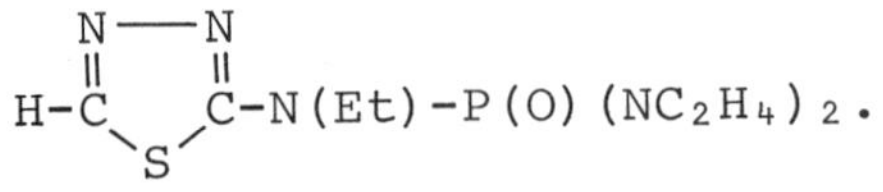 V. m. 95-6.5°.[774]

$(ClCH_2CH_2)_2NPO(NC_2H_4)_2$. IV. m. 31-6°.[58]
$Et_2NPO(NC_2H_4)_2$. IV. b_1 98-100°.[99]

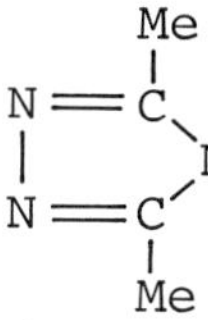
-P(O)$(NMe_2)_2$. V. m. 44.5-6.0°.[118]

C9 $(C_5H_{10}N)PO(NC_2H_4)_2$. IV. $b_{0.4-0.5}$ 103-4°.[664]
$[(CH_2=CHCH_2)EtN]PO(NC_2H_4)_2$. V. b_4 114°.[295]
C10 $(CH_2=CHCH_2)_2NPO(NC_2H_4)_2$. V. $b_{0.5}$ 132-3°.[247,295] $b_{0.0001}$ 81-2.5°, n 1.4954, d 1.0752.[347]
$MeOOCN(CH_2CH_2)_2NPO(NC_2H_4)_2$. IV. m. 106-7°.[488]
$(C_2H_4N)PO[N(CH_2CH_2)_2O]_2$. V. m. 40-3°.[572]
$(C_2H_4N)PO(NEt_2)_2$. V. $b_{0.03}$ 83°.[572]

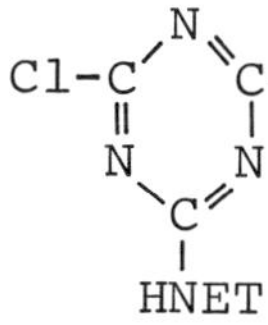
-N(Et)-P(O)$(NMe_2)_2$. IV. m. 155-6°.[242] For other triazine derivatives see Ref. 242.

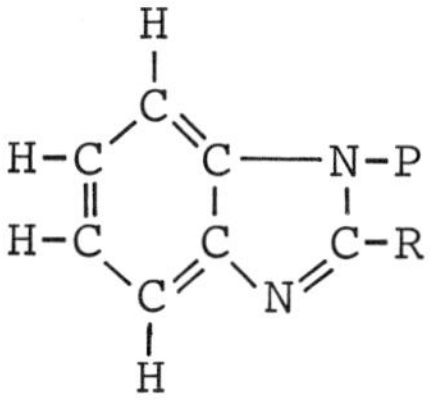
-P(O)$(NMe_2)_2$. V. R=H: b_1 178-9°.[118,682] R=Me: b_1 175-6°.[118] For other heterocyclic compounds see Ref. 682.

C12 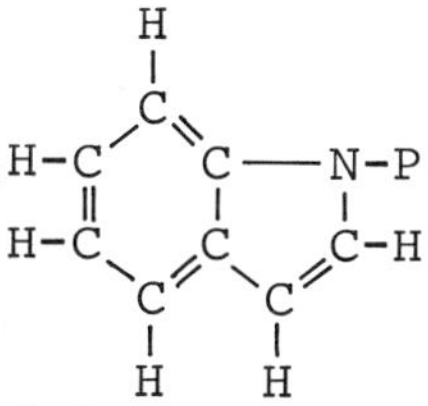
-P(O)$(NMe_2)_2$. V. m. 121-3°.[118]

$(ClCH_2CH_2)_2NPO[N(CH_2CH_2)_2O]_2$. IV. m. 134.5-5.5°.[251]
$(C_2H_4N)PO(NC_5H_{10})_2$. V. $b_{0.01}$ 110°.[118]
$(C_2H_4N)_2PON(CH_2CH_2)_2NPO(NC_2H_4)_2$. IV. m. 185.5-7.5°.[488]
C13 $MePhNPO(NC_3H_3N)_2$. ^{31}P +7.6.[633] Diimidazole derivative.

$[(CH_2{=}CHCH_2)PhN]PO(NC_2H_4)_2$. V. m. 39°.[295]
$[(CH_2{=}CHCH_2)(p\text{-}ClC_6H_4)N]PO(NC_2H_4)_2$. V. m. 73°.[295]
$[(CH_2{=}CHCH_2)(C_6H_{11})N]PO(NC_2H_4)_2$. V. n_D^{27} 1.5042.[295]
$Et_2NCON(CH_2CH_2)_2NPO(NC_2H_4)_2$. IV. m. 59°.[488]
C14 $[(CH_2{=}CHCH_2)(PhCH_2)N]PO(NC_2H_4)_2$. V. $b_{0.8}$ 156-8°.[295]
$[(CH_2{=}CHCH_2)(p\text{-}MeC_6H_4)N]PO(NC_2H_4)_2$. V. m. 45°.[295]
$Et_2NPO(NC_5H_{10})_2$. IV, V. Liquid.[573]

C16 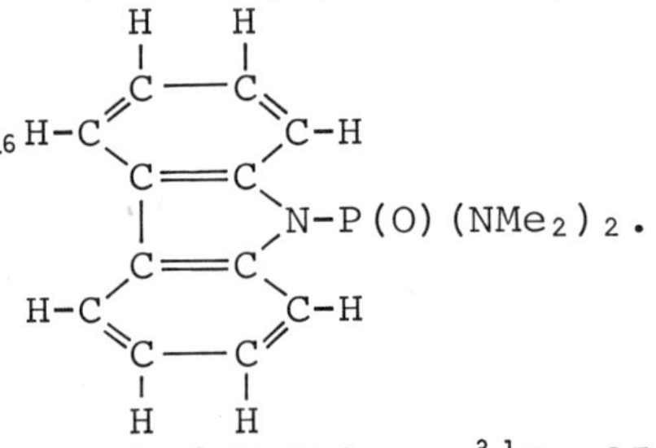
V. m. 103-4.[118] Carbazole derivative.

$Me_2NPO(NMePh)_2$. ^{31}P -25.5.[634]
Cx $MePhNPO(NC_5H_{10})_2$. IV, V. m. 86°.[51,573]
$Ph_2NPO(NC_5H_{10})_2$. IV, V. m. 200°.[660]
$Me_2NPO(NPh_2)_2$. ^{31}P -8.9.[634]

D.3.3.8. $[RNP(O)(NR'R'')]_n$, R,R',R" = Alkyl or Aryl or Acyl, R's Equal or Different, Ring Compounds

$[MeNPONMe_2]_2$. V,[348a] XV.[51] m. 168-70°.[51] m. 172-3°.[348a] IR, ^1H.[348a] ^{31}P -7.0.[348a]
$[EtNPONMe_2]_2$. V. m. 73-5°, IR.[348a]
$[PrNPONMe_2]_2$. V. m. 54-5°, IR.[348a]
$[PrNPONHPr]_2$. XV. m. 213°.[574]
$[i\text{-}PrNPONHPr\text{-}i]_2$. XV. m. 270°.[113]
$[i\text{-}PrNPONEt_2]_2$. XV. m. 92°.[113]
$[i\text{-}BuNPONHBu\text{-}i]_2$. XV. m. 271°,[574] m. 279°.[113]
$[PhNPO(NC_5H_{10})]_2$. XV. m. 233°.[574]
$[4\text{-}Cl,2(\text{or } 3)\text{-}O_2NC_6H_3NPONHC_6H_3Cl,NO_2\text{-}4,2(\text{or } 3)]_2$. XVII. m. > 300°.[660]
$[m\text{-}BrC_6H_4NPONHC_6H_4Br\text{-}m]_2$. XV. m. 329°.[583]
$[p\text{-}ClC_6H_4NPONHC_6H_4Cl\text{-}p]_2$. XVII. m. > 300°.[660]
$[PhNPONHPh]_2$. XV. m. 357-9°,[132,574] m. 225-6°(?).[150] ^{31}P (dimer) +12.0.[634]
$[o\text{-}MeC_6H_4NPO(NC_5H_{10})]_2$. XV. m. 195°.[574]
$[p\text{-}MeC_6H_4NPO(NC_5H_{10})]_2$. XV. m. 294°.[574]
$[PhNPONHCOPh]_2$. II. m. 226°.[856]
$[PhNPONMePh]_2$. XV. m. 234°.[574]
$[p\text{-}MeC_6H_4NPONHPh]_2$?. II. m. 188°.[150]
$[p\text{-}MeC_6H_4NPONHC_6H_4Me\text{-}o]_2$. XV. m. 309°.[574]
$[p\text{-}MeC_6H_4NPONHC_6H_4Me\text{-}p]_2$. XV. m. 226-8°,[150] m. 328°.[574,583]
$[o\text{-}MeC_6H_4NPONMePh]_2$. XV. m. 191°.[574]
$[o\text{-}MeC_6H_4NPONHC_6H_4Me\text{-}o]_2$. XV. m. 309°.[583]
$[p\text{-}MeC_6H_4NPONMePh]_2$. XV. m. 251°.[574]

$[2,4,5\text{-}Me_3C_6H_2NPONHC_6H_2Me_3\text{-}2,4,5]_2$. XV. m. 217°.[583]
$[2,4,6\text{-}Me_3C_6H_2NPONHC_6H_2Me_3\text{-}2,4,6]_2$. XV. m. 240°.[583]
[PhNPO(N=CPhNHPh)]$_2$. I, II. m. 227-8°.[856]

D.3.4. $(\geq N)_2P(S)(N\leq)'$, Including Ring Compounds

C_2 $Me_2NPS(NH_2)_2$. IV. m. 107°.[333]
C_4 $N_3PS(NMe_2)_2$. XIV. $b_{1.5}$ 80°, n_D^{21} 1.5177.[887]
$Et_2NPS(NH_2)_2$. IV. m. 64°.[333]
$BuNHPS(NH_2)_2$. IV. m. 54°.[333]
C_5 $(SCN)PS(NMe_2)_2$. XIV. $b_{0.3}$ 84°,[887] $b_{0.2}$ 66-7°.[587,589] n_D^{25} 1.5704,[887] n 1.5752.[587,589] d 1.1739.[587,589]
$H_2NCSNHPS(NMe_2)_2$. XII. m. 118°.[887]
$(MeNH)_2PSN(Me)PS(NHMe)_2$. VII. m. 92°, ^{31}P -73.[93]
C_6 $MeOCONHPS(NC_2H_4)_2$. XII, XIV. m. 85-7°.[83]

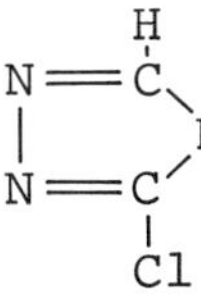
N-P(S)(NMe$_2$)$_2$. V. m. 40-3°, $b_{0.005}$ 103-6°.[682]

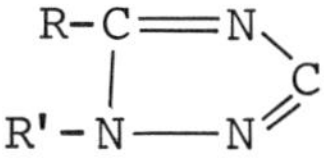
C-NH-P(S)(NMe$_2$)$_2$. V.[119,480] The P(S)(NMe$_2$)$_2$-group may be attached to a ring N atom.[119,570]
R=R'=H: m. 121-2°.[119,480]
R=C_5H_{11}, R'=H: m. 139-40°.[480]
R=Me, R'=H: two isomers.[119]
R=H, R'=P(S)(NMe$_2$)$_2$: m. 203-5°.[119]

$MeOCSNHPS(NMe_2)_2$. XII. m. 98-9°.[887]
$(C_2H_4N)PS(NMe_2)_2$. V. $b_{0.005}$ 51°.[572]
C_7 $Cl_2C(CO)_2NPS(NMe_2)_2$. V. m. 110-2°.[659]
$EtOCONHPS(NC_2H_4)_2$. IV, XII.[661] For other $RCONHPS(NC_2H_4)_2$ compounds see Ref. 661
$CH_2{=}CHCH_2NHPS(NC_2H_4)_2$. IV. $b_{0.001}$ 100-10°,[677] b_1 119-20°.[247] n 1.5492,[677] d 1.1426.[677]

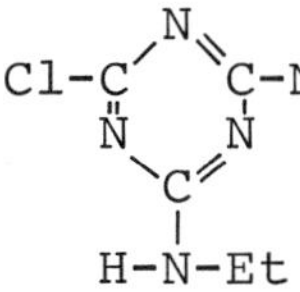
C-N(Et)-P(S)(NH$_2$)$_2$. IV. m. 115°.[242]

C_8 $(C_2H_4N)PS(NH\text{-}CH_2CH{=}CH_2)_2$. XIII. $b_{0.001}$ 100-10°, n 1.5474, d 1.11.[677]
$MeC(O)SCH_2CH_2NHPS(NMe_2)_2$. XVII. $b_{0.5}$ 149-50°, n 1.5453, d 1.1587.[650]
$Me_3SiNMePS[N(Me)CH_2]_2$. XIII. $b_{0.1}$ 98°, m. 37-8°, ^1H.[749]

C_9 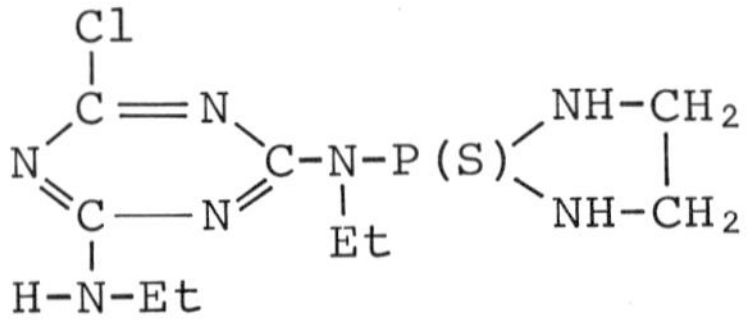 IV. m. 185-8°.[243]

$MeN(CH_2CH_2)_2NPS(NC_2H_4)_2$. IV. m. 65.5-6.5°.[322]

$[CH_2N(Me)]_2PSN(Me)PS[N(Me)CH_2]_2$. XIII. m. 115-7°, ^{1}H.[74]

C_{10} $PhNHPS(NC_2H_4)_2$. IV. m. 96.5-8.0°.[701]

$[C(O)N(Et)]_2PS[N(Et)COMe]$. V. m. 111-2°.[870]

$p\text{-}ClC_6H_4NHPS(NC_2H_4)_2$. IV. m. 153-4°.[701]

$p\text{-}FC_6H_4NHPS(NC_2H_4)_2$. IV. m. 128-30°.[701]

$p\text{-}IC_6H_4NHPS(NC_2H_4)_2$. IV. m. 118-20°.[701]

$PhNHPS(NHCH_2CH_2Cl)_2$. XVII. m. 165-6°.[701]

$p\text{-}ClC_6H_4NHPS(NHCH_2CH_2Cl)_2$. XVII. m. 127-9°.[701]

$p\text{-}FC_6H_4NHPS(NHCH_2CH_2Cl)_2$. XVII. m. 155-7°.[701]

$p\text{-}IC_6H_4NHPS(NHCH_2CH_2Cl)_2$. XVII. m. 195-7° dec.[701]

$MeOCON(CH_2CH_2)_2NPS(NC_2H_4)_2$. IV. m. 96-7°.[488]

$C_6H_{11}NHPS(NC_2H_4)_2$. IV. m. 96.5-7.0.[947]

$(CH_2{=}CHCH_2)_2NPS(NC_2H_4)_2$. IV. b_7 138°.[247]

$EtNHPS(NHBu\text{-}i)_2$. IV. m. 48.5°.[573]

$(Me_2N)_3P{=}N\text{-}PS(NMe_2)_2$. XIII. $b_{0.2}$ 132°, m. 58-9°.[887]

C_{11} $p\text{-}MeC_6H_4NHPS(NC_2H_4)_2$. IV. m. 91.5-2.5°.[701]

$p\text{-}MeC_6H_4NHPS(NHCH_2CH_2Cl)_2$. XVII. m. 155-6°.[701]

$p\text{-}MeOC_6H_4NHPS(NHCH_2CH_2Cl)_2$. XVII. m. 130-2°.[701]

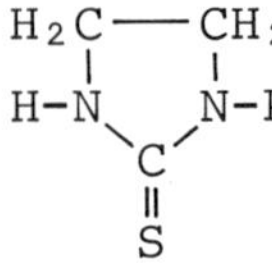 $\text{-}P(S)(NEt_2)_2$. XVII. m. 96°.[513]

$C_6H_{11}NHCSNHPS(NMe_2)_2$. XII. m. 130°.[589] See also Ref. 374.

C_{12} $o\text{-}C_6H_4(CO)_2NPS(NMe_2)_2$. V. m. 168.5-70°.[659]

$p\text{-}EtOC_6H_4NHPS(NC_2H_4)_2$. IV. m. 111-2°.[701]

$p\text{-}EtOC_6H_4NHPS(NHCH_2CH_2Cl)_2$. XVII. m. 110-2°.[701]

$(ClCH_2CH_2)_2NPS(NHCH_2CO_2Et)_2$. IV. m. 146-7°.[421]

$(C_2H_4N)_2PSN(CH_2CH_2)_2NPS(NC_2H_4)_2$. IV. m. > 300°,[773] m. 203°.[488]

$EtNHPS(NC_5H_{10})_2$. IV, V. m. 95°.[573]

$Pr_2NPS(NHPr)_2$. IV, V. oil.[113] For similar Bu and Pr derivatives (oils) see Ref. 113.

C_{13} $Et_2NCON(CH_2CH_2)_2NPS(NC_2H_4)_2$. IV. m. 78-80°.[488]

C_{14} $Me_2NPS(NHPh)_2$. IV. m. 209-10°.[573]

$EtNHPS(NHPh)_2$. IV, V. m. 106°.[573]

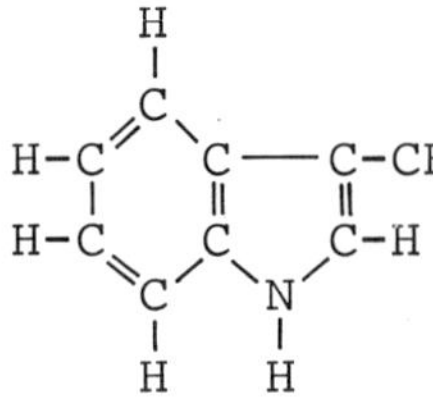 $\text{-}CH_2CH_2NHP(S)(NC_2H_4)_2$. IV. m. 145-7°.[613] For other indole derivatives see Ref. 613.

$Et_2NPS(NC_5H_{10})_2$. IV. m. 126°.[573]
i-$BuNHPS(NC_5H_{10})_2$. IV, V. m. 106°.[573]
C_{15} $PrNHPS(NHPh)_2$. IV, V. m. 116°.[573]
C_{16} i-$BuNHPS(NHPh)_2$. IV, V. m. 118°.[573]
$EtNHPS(NHC_6H_4Me\text{-}p)_2$. IV, V. m. 140°.[573]
$Et_2NPS(NHPh)_2$. IV, V. m. 192°.[573]
$PhNHPS(NC_5H_{10})_2$. IV, V. m. 112°.[573]
$(ClCH_2CH_2)_2NPS(NHC_6H_{11})_2$. IV. m. 184-5°.[251]
$HN(CH_2CH_2)_2NPS(NC_2Me_4)_2$. IV. m. 68-9°.[322]
C_{17} $(C_5H_{10}N)PS(NHPh)_2$. IV, V. m. 199°.[573]
p-$MeC_6H_4NHPS(NC_5H_{10})_2$. IV, V. m. 157°.[573]
C_{18} $Et_2NPS(NHC_6H_4Me\text{-}p)_2$. IV. m. 166-7°.[573]
$Pr_2NPS(NHPh)_2$. IV. m. 145°.[573]
i-$BuNHPS(NHC_6H_4Me\text{-}p)_2$. IV, V. m. 152°.[573]
C_{19} $(C_5H_{10}N)PS(NHC_6H_4Me\text{-}p)_2$. IV. m. 190°.[573]
i-$AmNHPS(NHC_6H_4Me\text{-}p)_2$. IV, V. m. 129°.[573]
C_{20} $EtPhNPS(NHPh)_2$. IV, V. m. 140°.[573]
$(ClCH_2CH_2)_2NPS(NHCH_2CH_2Ph)_2$. IV. m. 141-2°.[251]
$Et_2NPS[N(Bu\text{-}i)_2]_2$. IV, XIII. d_0^{15} 1.0023.[573]
C_x $EtPhNPS(NHC_6H_4Me\text{-}p)_2$. IV. m. 158°.[573]
$(i\text{-}Am)_2NPS(NHPh)_2$. IV. m. 141°.[573]
$O(CH_2CH_2)_2NPS(N{=}CPh_2)_2$. IV. m. 86-8°.[375]
$Et_2NPS(N{=}CPh_2)_2$. IV. m. 138°.[375]
$H_2N[P(S)(NRR')NH\text{-}]_n\text{-}H$.[289] $RNHPS(NR_2')_2$. R,R' = C_6, C_8, C_{16}, C_{18}, C_{31}. Carbon-hydride ligands.[116]

<u>Ring compounds</u>

$[MeNPSNHMe]_2$. V. m. 220-5°,[51] m. 224°.[93] ^{31}P -60.[93]
$[EtNPSNHEt]_2$. XV. m. 169°.[574]
$[MeNPSNEt_2]_2$. V. m. 169°, ^{31}P -62.[93]
$[PrNPSNHPr]_2$. XV. m. 152°,[574] m. 157°.[113]
$[MeNPSNHPh]_2$. V. m. 146°, ^{31}P -54.[93]
$[i\text{-}BuNPSNHBu\text{-}i]_2$. XV. m. 150°,[574] m. 151°.[113]
$[i\text{-}AmNPSNHAm\text{-}i]_2$. XV. m. 90°.[574]
$[PhNPSNHPh]_2$. XV. m. 233-5°,[131] m. 226-7°.[576]
$[o\text{-}MeC_6H_4NPSNHPh]_2$. V. m. 162°.[576]
$[o\text{-}MeC_6H_4NPS(NC_5H_{10})]_2$. XV. m. 236°.[576]
$[p\text{-}MeC_6H_4NPS(NC_5H_{10})]_2$. V. m. 275°.[576]
$[o\text{-}MeC_6H_4NPSNHC_6H_4Me\text{-}o]_2$. V. m. 258°.[576]
$[p\text{-}MeC_6H_4NPSNHC_6H_4Me\text{-}p]_2$. XV. m. 182°.[576]
$[PhCH_2NPSNHCH_2Ph]_2$. XV. m. 197-9°.[131]

D.3.5. (>N)(>N)'(>N)"PO

$(C_5H_{10}N)[O(CH_2CH_2)_2N]PO(NC_2H_4)$. V. $b_{0.01}$ 118-20°.[572]
$(PhNH)PO(NEtPh)(NC_2H_4)$. V. m. 137°.[572]

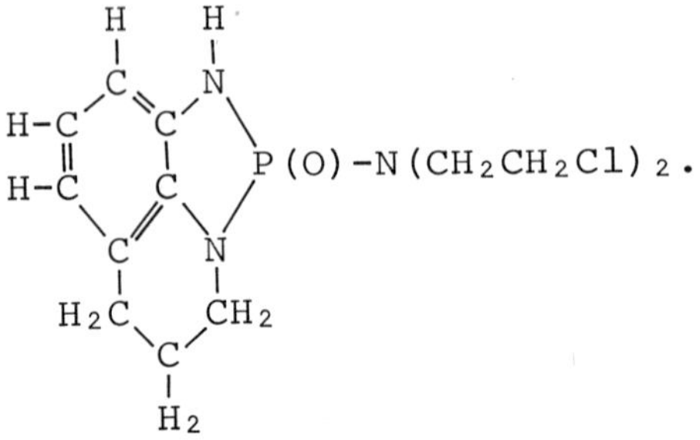

IV. m. 189-90°.[942]

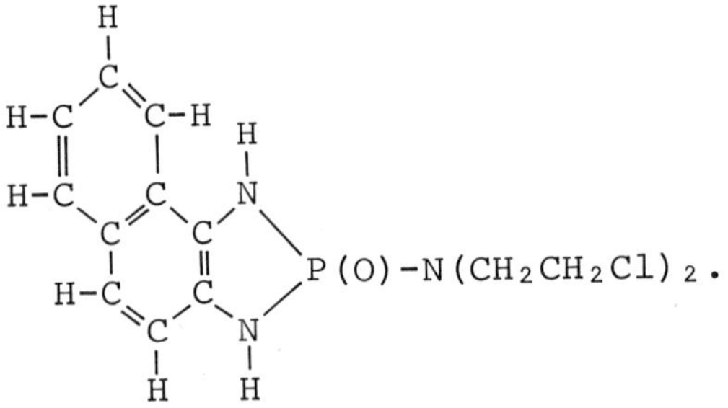

IV. m. 189-90°.[942]
1,8-naphtalene derivative: m. 183°.[942]
2,3-fluorene derivative: m. 238-9°.[942]

D.3.6. $(\geqslant N)_2P(O)(\overset{|}{N}-N\leqslant)$, $(\geqslant N)_2P(S)(\overset{|}{N}-N\leqslant)$, Also (N=N-) Groups

$H_2NNHPS(NMe_2)_2$. V. m. 41°.[858]
$(Me_2N)_2PSNHNHPS(NMe_2)_2$. V. m. 90-1°.[858]

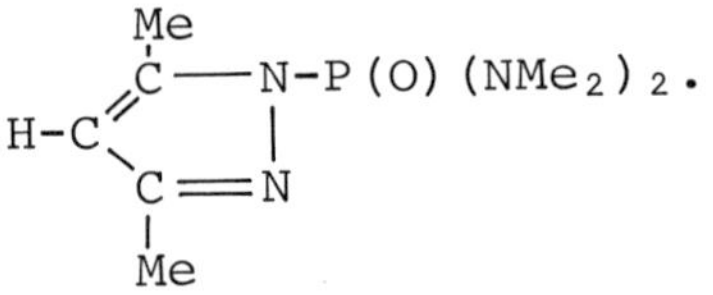

V. $b_{0.2}$ 84-6°.[118]

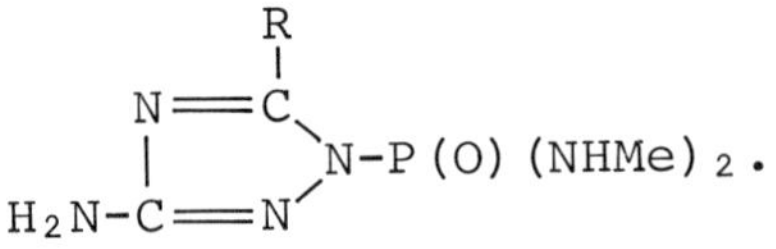

IV.[119] R=Ph: m. 182-3°.
R=$PhCH_2$: m. 177-80°.
The $-P(O)(NHMe)_2$-group may be attached to another ring N-atom,[119] but also to the $-NH_2$-group.[570]

$PhN{=}NPO(NMe_2)_2$. XVII. $b_{0.0001}$ 100°.[111]
$PhNHNHPO(NMe_2)_2$. V. m. 187-9°.[111]

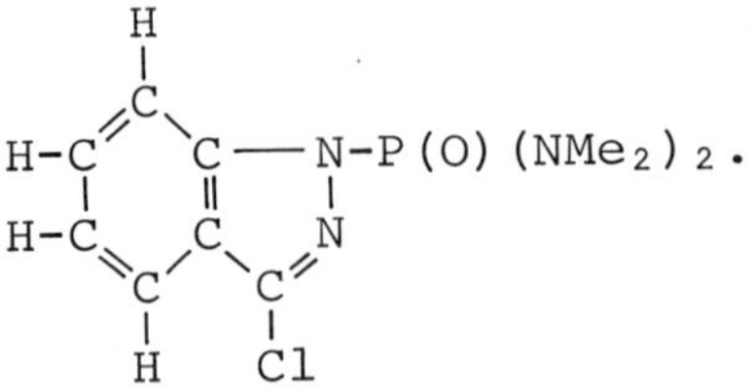

V. m. 56-64.[118]

$PhNHCSNHNHPS(NMe_2)_2$. XVII. m. 134-4.5°.[858]
$PhNHNHPO(NEt_2)_2$. V. m. 152°.[185]
$PhNHNHPO(NC_5H_{10})_2$. V. m. 155°.[573]

D.3.7. $(\geqslant N)P(O)(\overset{|}{N}-N\leqslant)_2$, $(\geqslant N)P(S)(\overset{|}{N}-N\leqslant)_2$, Also (N=N-) Groups

$(ClCH_2CH_2)_2NPO(NHNMe_2)_2$. IV.[646]
$PhNHPO(NHNH)_2PO(NHPh)$. IV. m. 208-10°.[73]
$CCl_3CONHPO(NHNHPh)_2$. IV. m. 237-8°.[816,817]
$Cl_2CHCONHPO(NHNHPh)_2$. IV. m. 190°.[816,817]
$Me_2NPO(NHNHPh)_2$. IV. m. 194-5°.[573]
$EtNHPO(NHNHPh)_2$. IV. m. 153°.[573]
$PrNHPO(NHNHPh)_2$. IV. m. 151°.[573]
$(ClCH_2CH_2)_2NPO(NHNHPh)_2$. IV, V. m. 158°.[123]
$(i\text{-}BuNH)PO(NHNHPh)_2$. IV. m. 141°.[573]
$Et_2NPO(NHNHPh)_2$. IV. m. 184-5°.[573]
$Et_2NPS(NHNHPh)_2$. IV.[573]
$(i\text{-}BuNH)PS(NHNHPh)_2$. IV. m. 129°.[573]
$(C_5H_{10}N)PS(NHNHPh)_2$. IV. m. 158°.[573]
$(i\text{-}AmNH)PO(NHNHPh)_2$. IV, V. m. 122°.[573]
$Pr_2NPS(NHNHPh)_2$. IV. m. 196°.[573]
$Pr_2NPO(NHNHPh)_2$. IV. m. 164°.[573]
$MePhNPO(NHNHPh)_2$. IV. m. 148°.[573]
$(i\text{-}Bu)_2NPO(NHNHPh)_2$. IV. m. 168°.[573]

D.3.8. $(\geqslant N-\overset{|}{N})_3PO$, $(\geqslant N-\overset{|}{N})_3PS$, Also (N=N-) Groups

$OP(NMeNMe)_3PO$. XIII. m. 320-5°.[665]
$SP(NMeNMe)_3PS$. XIII. m. > 360°.[665]
$(Me_2NNH)_3PO$. II. m. 193.5-5.0°, ^{31}P -12.5, ^{1}H.[635]
$(Me_2NNH)_3PS$. II. m. 79-80°,[858] m. 75.5-8.0°.[635] ^{31}P -12.5.[635]
$(PhNHNH)_3PO$. II. dec. 204°,[852] dec. 185-7°,[69] m. 196°,[580] m. 188°.[858]
$(PhNHNH)_3PS$. II. m. 154°,[580] m. 156°.[858]
$(p\text{-}MeC_6H_4NHNH)_3PO$. II. m. 189°.[580]

D.4. Compounds with P-O-P and P-S-P Bridging Bond

D.4.1. ABP(O)-O-P(O)CD, A,B,C,D = Halogen,-OH,-OR, -NRR', at Least One of the Ligands Must Be an Amino Group

$(Me_2N)_2PO\text{-}O\text{-}POCl_2$. XVI. n 1.4669.[863]
$(SCN)_2PO\text{-}O\text{-}PO(NCS)_2$. XIV. red oil.[426] polymer? see Ref. 752.
$(Me_2N)_2PO\text{-}O\text{-}PO(NMe_2)Cl$. XVI. b_2 122-4°.[767]
$(Me_2N)_2PO\text{-}O\text{-}PO(NMe_2)F$. XVI. b_2 118°.[767]
$(Me_2N)_2PO\text{-}O\text{-}PO(NHMe)_2$. IV. n 1.4711.[863]

C_8 [O(CH$_2$CH$_2$)$_2$N]PO(OH)-O-PO(OH)[N(CH$_2$CH$_2$)$_2$O]. VIII. [NH$_4^+$]-salt: m. 138-41°.[563]
(Et$_2$N)(F)PO-O-PO(F)(NEt$_2$). V, VIII, XVI. $b_{0.001}$ 93-7° and 95-100°. n 1.4190, 1.4194. d 1.1944.[822,824]
(Et$_2$N)(F)PO-O-PO(NMe$_2$)$_2$. VIII,[820] XVI.[824] $b_{0.001}$ 108-13°,[820] $b_{0.001}$ 100-12°.[824] n 1.4415, d 1.1687.[820]
Me$_2$N(EtO)PO-O-PO(OEt)NMe$_2$. VIII.[370]
(Me$_2$N)$_2$PO-O-PO(OEt)(NMe$_2$). XVI. b_3 145°.[767]
(Me$_2$N)$_2$PO-O-PO(NMe$_2$)$_2$. IV,[332,863] VIII,[684,875] XVI,[320,753] XVII.[164,863] $b_{0.003}$ 110-2°,[320] $b_{0.35}$ 130-2°,[684] $b_{0.5}$ 120-2°,[876] $b_{0.65}$ 119-20°,[857] b_1 134-5°,[44] b_1 126-8°,[753] $b_{1.5}$ 139-40°,[863] b_2 142°.[767] n_D^{25} 1.4620,[857,863] 1.4612.[753] d_4^{25} 1.1360,[857] 1.1343.[753] ^{31}P -12.0.[901] Poison. "OMPA" registered name. Metal complexes.[392]
C_9 MeO(EtO)PO-O-PO(NMe$_2$)NEt$_2$. VIII. $b_{1.5-2}$ 138-40°, n 1.4448, d 1.1395.[636] For other >PO-O-PO< compounds see Ref. 636.
(Me$_2$N)$_2$PO-O-PO(NMe$_2$)NMeEt. XVI. b_3 151°.[767]
C_{10} (Me$_2$N)(Bu$_2$N)PO-O-POCl$_2$. XVI. n 1.4728.[863]
Et$_2$N(EtO)PO-O-POF(NEt$_2$). XVI. $b_{0.001}$ 115-8°, n 1.4323.[820,824]
(Me$_2$N)$_2$PO-O-PO(NMe$_2$)(NEt$_2$). XVI, VIII. $b_{0.04}$ 132-3°, n 1.4625, d 1.1067.[639]
C_{12} PhNH(HO)PO-O-PO(OH)NHPh. XVII. [C$_6$H$_{11}$NH$_3^+$]-salt: m. 223-4°, K-salt.[115]
C$_6$H$_{11}$NH(F)PO-O-PO(F)NHC$_6$H$_{11}$. VIII. m. 83-4°.[820]
Et$_2$N(ClCH$_2$CH$_2$O)PO-O-PO(OCH$_2$CH$_2$Cl)NEt$_2$. VIII. $b_{0.03}$ 170-2°, n 1.4640, d 1.2270.[637]
(Et$_2$N)(F)PO-O-PO(NEt$_2$)$_2$. VIII, XVI. $b_{0.001}$ 133-8°, n 1.4479, d 1.1053.[822,824]
Et$_2$N(EtO)PO-O-PO(OEt)NEt$_2$. XVI. b_2 153-3.5°, n_D^{18} 1.4402, d_0^{18} 1.2102.[677a]
(Me$_2$N)$_2$PO-O-PO(NEt$_2$)$_2$. VIII, XVI. b_2 156-7°, n 1.4640, d 1.0851.[639]
Me$_2$N(Et$_2$N)PO-O-PO(NMe$_2$)NEt$_2$. VIII, XVI. b_2 153-5°, n 1.4630, d 1.0831.[639]
C_{13} (Me$_2$N)(Bu$_2$N)PO-O-PO(NMe$_2$)(NHMe). IV. n 1.4750.[863]
C_{14} [O(CH$_2$CH$_2$)$_2$N](CH$_2$=CHCH$_2$O)PO-O-PO(OCH$_2$CH=CH$_2$)[N(CH$_2$CH$_2$)$_2$O]. VIII. $b_{0.01}$ 130-40°, n 1.4387.[563]
Me$_2$N(Et$_2$N)PO-O-PO(NEt$_2$)$_2$. VIII, XVI. $b_{1.5}$ 159-60°, n 1.4645, d 1.0627.[639]
C_x (Et$_2$N)$_2$PO-O-PO(NEt$_2$)$_2$. VIII,[639,875] XVII.[164] b_1 95-110°,[875] b_3 150°,[767] b_1 174-6°,[639] n 1.4650,[639] n_D^{25} 1.4668,[875] d 1.0444.[639]
(PhNH)$_2$PO-O-PO(NHPh)$_2$. VIII. m. 222°.[923]
[(CH$_2$=CHCH$_2$)$_2$N]$_2$PO-O-PO[N(CH$_2$CH=CH$_2$)$_2$]$_2$. II. n 1.4985.[5
(i-Pr$_2$N)$_2$PO-O-PO[N(Pr-i)$_2$]$_2$. VIII.[707]
[(PhCH$_2$O)P(O)(NHPh)]$_2$O. XVII. m. 142-3°.[115]

D.4.2. ABP(Z)-Z-P(Z)CD, Z = S or O, at Least One Z = S, A,B,C,D = Halogen,-OH,-OR,-SR,-NRR', at Least One of the Ligands Must Be an Amino Group

C_4 $(Me_2N)_2PO$-O-$PSCl_2$. XVI. n 1.5132.[863]

C_6 (MeNH)(EtO)PS-O-PS(OEt)NHMe. VIII. n_D^{25} 1.5232, d_{25} 1.2499.[415]

C_8 (Et_2N)(F)PO-O-PS(F)NEt_2. XVII. $b_{0.001}$ 95-100°, n 1.450, d 1.200.[820]

Et_2N(F)PS-O-PS(F)NEt_2. VIII, XVI. $b_{0.001}$ 95-100°.[820,824] n 1.4815, d 1.2041.[824]

(Me_2N)(EtO)PO-O-PS$(OEt)_2$. XVII. b_1 120°, n 1.4593, d 1.1784.[40]

(Me_2N)(EtO)PS-O-PS$(OEt)_2$. XVII. b_2 135°, n 1.4921, d 1.1829.[40]

(i-PrNH)(MeO)PS-O-PS(OMe)NHPr-i. VIII. m. 91-3°.[415]

$(Me_2N)_2PO$-O-PS$(OEt)_2$. XVII. b_1 127-8°, n 1.4719, d 1.1667.[40]

$(Me_2N)_2PO$-O-PS(NHMe)(NHPr-i). IV. n 1.5139.[863]

$(Me_2N)_2PO$-S-PO$(NMe_2)_2$. XVII. $b_{0.03}$ 94°, n_D^{25} 1.4675, d_{25} 1.1443.[586]

$(Me_2N)_2PO$-O-PS$(NMe_2)_2$. XVI. b_3 125-30°.[767]

C_{10} (Et_2N)(EtO)PO-O-PS$(OEt)_2$. XVII. b_1 143-5°, n 1.4595, d 1.1402.[40]

$(Me_2N)_2PO$-O-PS$(OPr)_2$. XVII. $b_{3.5}$ 159-60°, n 1.4700, d 1.1186.[637]

$(Me_2N)_2PO$-O-PS$(OPr\text{-}i)_2$. XVII. b_3 150-2°, n 1.4665, d 1.1237.[637]

C_{12} $(Et_2N)_2PO$-O-PS(F)NEt_2. VIII, XVI. $b_{0.001}$ 132-6°, n 1.4736, d 1.1142.[824]

(Et_2N)(F)PO-O-PS$(NEt_2)_2$. VIII, XVI. $b_{0.001}$ 95-100°, n 1.450, d 1.200.[824]

$(Et_2N)_2PS$-O-PS(OEt_2). XVII. b_3 169-70°,[40] b_{1-2} 153-6°,[637] n 1.500,[637] d 1.1097.[637]

$(Me_2N)(Et_2N)$PO-O-PS$(OPr)_2$. XVII. $b_{1.5}$ 149-52°, n 1.4710, d 1.0950.[637]

$(Et_2N)_2PO$-O-PS$(OEt)_2$. XVII. b_2 163-6°, n 1.4669, d 1.0871.[40]

$(Me_2N)_2PO$-O-PS$(OBu)_2$. XVII. b_1 154-7°, n 1.4695, d 1.0883.[637]

$(Me_2N)_2PO$-O-PS$(OBu\text{-}i)_2$. XVII. $b_{2.5}$ 159-60°, n 1.4652, d 1.082.[637]

$(Me_2N)_2PO$-O-PS$(NEt_2)_2$. XVI. $b_{0.05}$ 128-31°, n 1.4928, d 1.0951.[639]

C_{14} $(Me_2N)(Et_2N)$PS-O-PO$(NEt_2)_2$. XVI. $b_{1.5}$171-2°, n 1.4910, d 1.0754.[639]

$(Et_2N)_2PS$-O-PS$(OPr)_2$. XVII. b_{2-3} 167-70°, n 1.4970, d 1.0890.[637]

$(Et_2N)_2PS$-O-PS$(OPr\text{-}i)_2$. XVII. b_{2-3} 165-8°, n 1.4912, d 1.0746.[637]

C_x $(Et_2N)_2PO\text{-}O\text{-}PS(NEt_2)_2$. XVI. $b_{0.02}$ 159-61°, n 1.4900, d 1.0611.[639]

$(Et_2N)_2PS\text{-}O\text{-}PS(OBu)_2$. XVII. b_2 172-5°, n 1.4960, d 1.0670.[637]

$(Et_2N)_2PS\text{-}O\text{-}PS(OBu\text{-}i)_2$. XVII. b_{2-3} 173-7°, n 1.4880, d 1.0567.[637]

D.4.3. Molecules with More Than Two P Atoms, (However see D.2.2.5. and D.3.3.5.)

$Me_2NPO[OPO(NMe_2)(OMe)]_2$. XVII.[164]

$Me_2NPO[OPO(OEt)_2]_2$. XVI. n 1.4345, d 1.239.[862]

$Me_2NPO[OPO(NMe_2)(OEt)]_2$. XVI. n 1.4471, d 1.2141.[862]

$Me_2NPS[OPO(NMe_2)_2]_2$. XVI. $b_{1.3}$ 175-6°, n 1.4880, d 1.2072.[44]

$Me_2NPO[OPO(NMe_2)_2]_2$. XVI. $b_{0.5}$ 190-200°,[258] b_1 170-1°,[4] n_D^{25} 1.4660,[258] n 1.4644,[857] n 1.4685,[44] d 1.1931,[857] n 1.2353.[44]

$[Et_2NP(O)O]_3$. XVI. m. 103°.[573]

$[(EtO)(Me_2N)P(O)O]_3PO$. XVI. n 1.4470, d 1.2671.[862]

$[(Me_2N)_2P(O)O]_3PO$. XVI. n 1.4685,[857,860] d 1.2474,[860] d 1.2478.[857]

$(Me_2N)_2P(O)\text{-}[OP(O)(NMe_2)]_3\text{-}NMe_2$. XVI. n 1.4719, d 1.2561.[857]

$[Pr_2NP(O)O]_3$. XVI. b_{10} 240°.[573]

$[(i\text{-}Bu)_2NP(O)O]_3$. XVI. m. 79°, b_{15} 255°.[573]

$[(Me_2N)(Bu_2N)P(O)O]_3PO$. XVI. n 1.4622.[860]

$[(Bu_2N)_2P(O)O]_3PO$. XVI. n 1.4600.[860]

$[-OP(O)(NMe_2)-]$ unit. ^{31}P +12.[769]

$[-OP(O)(NMe_2)_2]$ end group. ^{31}P -11.5.[769]

D.5. Derivatives of Phosphorimidic Acid, [-N=P⩽], Also [>N-N=P⩽] Compounds

D.5.1. $[-N=PX_3]$, X = Halogen

C_1 $ClCH_2N=PCl_3$. I. m. 133-5°.[937]

```
  O   Cl
  ‖   |
R-S-N=P-Cl.
  ‖   |
  O   Cl
```

I. R = Alkyl.

R	m. (Lit)
$ClCH_2$	50°, 54°[523]
Me	47-50°[442]
Et	18-21°[442]
Pr	60°[523]
i-Pr	19-22°[442]
Bu	48-51°[442]
i-Bu	6.5°[523]
Am	35°[523]
i-Am	2°[523]
C_6H_{11}	39-41°[442]

R	m. (Lit)
C_6H_{13}	10°[523]
$PhCH_2$	77-82°[442]
$PhCH_2CH_2$	57°[523]

$MeN(SO_2N{=}PCl_3)_2$. I. m. 228-9°.[626]

$CCl_3N{=}PCl_3$. XVII. Monomeric. $b_{0.02}$ 56°.[312] b_{12} 102-3°, n 1.5502, d 1.7877.[483] ^{31}P +16.[312]

$C_2ClCOCHClN{=}PCl_3$. I. b_2 84-6°, n 1.5408, d 1.7167.[930,931]

$CCl_3CHClN{=}PCl_3$. I. b_3 106-7°, n 1.5500, d 1.7754.[937]

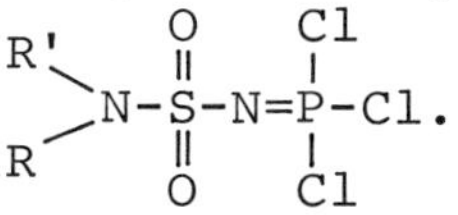 I.

R	R'	
Me	Me	m. 73-5°,[459] m. 74°[886]
Et	Et	$b_{0.2}$ 113°, n 1.5072[886]
Pr	Pr	$b_{0.005}$ 113°, n 1.5010[886]
Bu	Bu	$b_{0.05}$ 140°, n 1.4952[886]
[RR'N]		
[$O(CH_2CH_2)_2N$-]		m. 94°[886]
[PhNH-]		^{31}P -3.8[512a]
[p-ClC_6H_4NH-]		^{31}P -4.2[512a]

$BrCl_2CCON{=}PCl_3$. I. m. 68°.[816]

$O_2NCCl_2CON{=}PCl_3$. I. m. 55-60°.[816]

$O_2NCBr_2CON{=}PCl_3$. I. m. 65°.[816]

$CBr_3CON{=}PCl_3$. I. m. 115-7°.[484]

$ClCOCCl_2N{=}PCl_3$. I. b_2 85-7°, n 1.5445, d 1.7610.[931]

$CCl_2{=}CClN{=}PCl_3$. I. b_2 92-4°,[511,790] b_3 101-8°.[791] n 1.5725, d 1.757.[790] ^{31}P +23.8.[511]

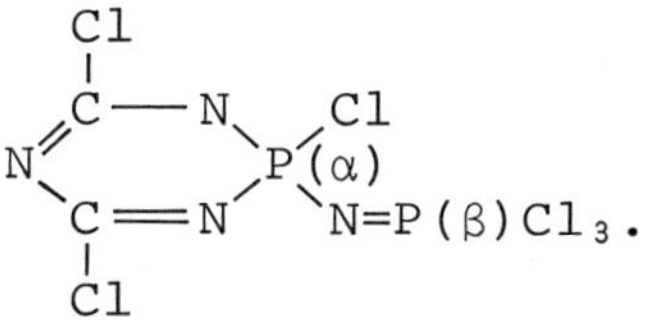 I. m. 108-111°. ^{31}P [α]: -57.0. [β]: -22.7.[91]

$CCl_3CCl_2N{=}PCl_3$. I. b_2 100-5°,[931] b_3 103-4°.[790,791] m. 17.5-8.0°,[931] m. 20-3°.[790,791] n_D^{24} 1.5608, d_{24} 1.823.[790,791] ^{31}P +11.1.[511]

$C_3(CF_3)_2CHN{=}PCl_3$. I. b. 136-8°, n 1.3921, d 1.6760.[484]

$(CF_3)_2CClN{=}PCl_3$. I. b_{10} 47-8°, n 1.4125, d 1.7486.[484]

$CCl_2{=}C(CHCl_2)N{=}PCl_3$. I. b_3 107-10°.[938]

$CCl_3CCl(CHCl_2)N{=}PCl_3$. I. m. 45-6°.[938]

$Cl_2CHCCl(COCl)N{=}PCl_3$. I. $b_{1.5}$ 100-2°, n 1.5492, d 1.7598.[930,931]

$ClCH{=}C(COCl)N{=}PCl_3$. I. b_1 81-2°, n 1.5759, d 1.6885.[930]

NC-$CH_2CON{=}PCl_3$. I. $b_{0.18}$ 96°, n 1.5260, d 1.700.[800]

$ClCH_2CCl_2CCl_2N{=}PCl_3$. I. $b_{0.1}$ 104-6°, n 1.5650, d 1.7750.[785]

$MeCCl_2CON{=}PCl_3$. I. m. 82-5°.[231]

$ClCH_2CHClCON{=}PCl_3$. I. m. 36-9°.[198]
$ClCH_2CCl_2CON{=}PCl_3$. I. m. 64-7°.[198]
$MeOCCl_2CON{=}PCl_3$. I. m. 55-7°.[458]
$Cl_3CC({=}NMe)N{=}PCl_3$. I. m. 128-31°.[230]
$MeCCl_2CCl_2N{=}PCl_3$. I. b_2 110-1°, n 1.5509, d 1.7071.[799]
$(ClCH_2)_2CHN{=}PCl_3$. I. b_3 93-5°, n 1.5335, d 1.5616.[940]
$NC\text{-}CCl{=}CClN{=}PCl_3$. I. $b_{0.02}$ 84-6°, n 1.5919, d 1.6880.[78]
$NC\text{-}CCl_2CCl_2N{=}PCl_3$. I. $b_{0.03}$ 76-8°, m. 39-40°.[786]
$NC\text{-}CCl_2CON{=}PCl_3$. I. m. 50-1°.[800]

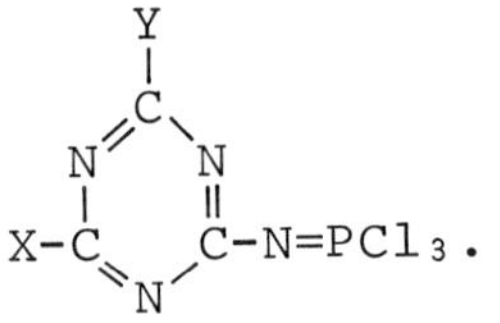

I.[202] X=Y=Cl: m. 108-9°;
X=Cl, Y= $N{=}PCl_3$: m. 110-4°;
X=Y= $N{=}PCl_3$: m. 113-4°;

$ClCOCCl{=}CClN{=}PCl_3$. I. $b_{0.015}$ 92-3°, n 1.5711, d 1.7553.[793]
$ClCOCCl_2CCl_2N{=}PCl_3$. I. $b_{0.015}$ 88-91°, n 1.5611, d 1.8186.[793]
$CCl_2{=}C(CCl_3)N{=}PCl_3$. I. b_{18} 164-5°, n 1.5850, d 1.8084.[930]
$Cl_2C(CON{=}PCl_3)_2$. I. m. 165-6°.[800]
$Br_2C(CON{=}PCl_3)_2$. I. m. 165°.[800]
$ClCH(CON{=}PCl_3)_2$. I. m. 96-7°.[800]
$O_2NCBr(CON{=}PCl_3)_2$. I. m. 121°.[800]

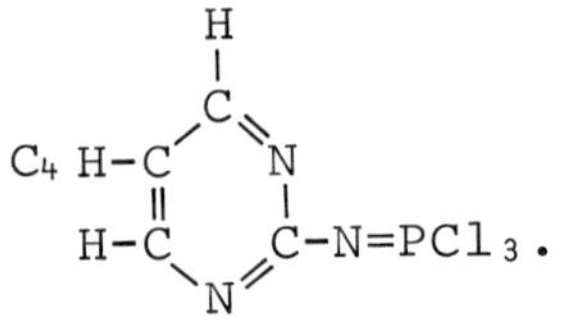

I.[490] For substituted 2-pyrimidinyl compounds see Ref. 490.

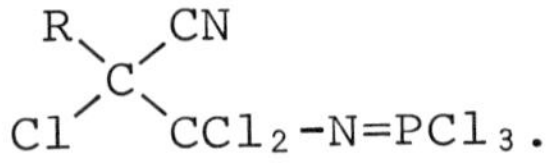

I.[792]

R	b.	n	d
Me	$b_{0.03}$ 80-2°	1.5472	1.649
Et	$b_{0.02}$ 87-9°	1.5364	1.555
Pr	$b_{0.025}$ 89-100°	1.5330	1.526
i-Pr	$b_{0.015}$ 109-10°	1.5352	1.533
Bu	$b_{0.015}$ 106-9°	1.5272	1.473

$MeCCl(COCl)CCl_2N{=}PCl_3$. I. $b_{0.1}$ 94-7°, n_D^{25} 1.5461, d_{25} 1.6684.[794]
$MeCHClCCl(COCl)N{=}PCl_3$. I. b_1 96-8°, n_D^{25} 1.5352, d_{25} 1.6202.[930]
$Me_2C(CN)N{=}PCl_3$. I. Monomer: b_2 62-3°, n 1.4880, d 1.3300.[687] Dimer: m. 211-2°.[687]
$Me_2C{=}CCl\text{-}N{=}PCl_3$. I. $b_{0.3}$ 122-4°, n 1.5309, d 1.578.[795]

$Me_2CClCCl_2N=PCl_3$. I. $b_{0.1}$ 75-6°, m. 30-1°.[795]
$MeCCl(CH_2Cl)CCl_2N=PCl_3$. I. $b_{0.07}$ 93-4°.[795]
$MeCCl(CH_2Br)CCl_2N=PCl_3$. I. $b_{0.05}$ 107-8°.[795]
$Me_2CClCHClN=PCl_3$. I. b_2 82-4°.[928]
$Me_2CClCH_2N=PCl_3$. I. b_2 63-4°, n 1.5042, d 1.3864.[928]
$Me_3CN=PCl_3$. I. b. 153-4°, d 1.2190.[928]
$(CF_3)_3CN=PCl_3$. I. b. 134-5°, n 1.3723, d 1.7600.[484]
$CCl_3C(=NCOCCl_3)N=PCl_3$. I. m. 94-6°.[232]
C_5 $Cl_3P=NCCl_2CCl_2CH_2CCl_2CCl_2N=PCl_3$. I. m. 62-5°.[798] For
    other $[Cl_3P=N-CCl_2CCl_2]_2(CH_2)_n$ compounds (m. listed)
    see Ref. 798.
$EtCCl(COCl)CCl_2N=PCl_3$. I. $b_{0.05}$ 100-2°, n_D^{25} 1.5371,
    d_{25} 1.6161.[794]
$Me_2C(CN)CON=PCl_3$. I. $b_{0.06}$ 77-9°, m. 48°.[788] For simi-
    lar compounds see Ref. 788.
$MeEtCClCCl_2N=PCl_3$. I. $b_{0.1}$ 82-6°.[795]
$Me_3CCH_2N=PCl_3$. I. Monomer: b_{56} 99-100°, n 1.4766,
    d 1.1980.[940]
$Et_2CHN=PCl_3$. I. b_{30} 94-5°, n 1.4855, d 1.2280.[940]
C_6 $Cl_3P=NC_6H_4SO_2N=PCl_3$. I.[941] o-: m. 90-3°. m-: m. 149-52°.
    p-: m. 162-5°.
$Cl_3P=NSO_2C_6H_4SO_2N=PCl_3$. I.[448] p-: m. 153-7°. m-:m. 101-4°.
$PhSO_2N=PBr_3$. XIII. m. 94-7°.[472]
$PrCCl(COCl)CCl_2N=PCl_3$. I. $b_{0.04}$ 90-3°, n_D^{25} 1.5246,
    d_{25} 1.5635.[794]
$i-PrCCl(COCl)CCl_2N=PCl_3$. I. $b_{0.04}$ 90-2°, n_D^{25} 1.5271,
    d_{25} 1.5417.[794]

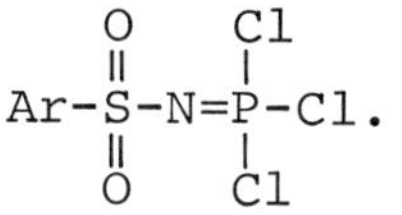

$$Ar-\overset{\overset{O}{\|}}{\underset{\underset{O}{\|}}{S}}-N=\overset{\overset{Cl}{|}}{\underset{\underset{Cl}{|}}{P}}-Cl.$$ I, XIII.

Ar	m. (Lit)
Ph	50-3°,[519] 51-3°,[431] 54.5°[429]
$p-ClC_6H_4$	69-71°,[519] 71-3°[518]
$p-FC_6H_4$	72-3°[391]
$o-O_2NC_6H_4$	73-5°,[446] 69-71°[525]
$m-O_2NC_6H_4$	82-4°,[446] 70-80°,[519] 72-3°[525]
$p-O_2NC_6H_4$	115-7°,[525] 118-9°[446]
$m-CF_3C_6H_4$	52-4°[391]
$p-ClC(O)C_6H_4$	82°[719,733]
$o-MeC_6H_4$	53.5°,[429] 51-3°[431]
$p-MeC_6H_4$	106°,[429] 104-6°,[431] 106-8°[519] 103-4°[907]
$1-C_{10}H_7$	110-2°,[462] 117-9°[431]
$2-C_{10}H_7$	130-2°,[462] 131-3°[431]

$Me_3CCCl_2CHClN=PCl_3$. I. b_1 112-4°, n 1.5367, d 1.4901.[940]
$Et_2C(CN)N=PCl_3$. I. Monomer: b_1 84-5°, n 1.4909,
    d 1.2680.[687] Dimer: m. 225-6°.[687]
$(CH_2CH_2)_2C(CN)N=PCl_3$. II. Monomer: b_3 96-7°, n 1.5173,
    d 1.3700.[687] Dimer: m. 204-5°.[687] For similar com-
    pounds see Ref. 687.

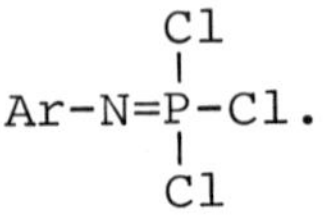

$$\text{Ar-N=P-Cl} \begin{array}{c} \text{Cl} \\ | \\ | \\ \text{Cl} \end{array}$$ I.[935,936]

Ar	m.
Ph	180-2°
o-ClC$_6$H$_4$	127-8°
p-ClC$_6$H$_4$	181-3°, ^{31}P +11.0[512a]
2,4-Cl$_2$C$_6$H$_3$	116-8°
o-O$_2$NC$_6$H$_4$	109-11°
p-O$_2$NC$_6$H$_4$	140-1°
o-MeC$_6$H$_4$	124-6°
p-MeC$_6$H$_4$	198-200°
o-MeOC$_6$H$_4$	124-6°
p-MeOC$_6$H$_4$	196-8°

$$C_7 \text{ Ar-C-N=P-Cl} \begin{array}{c} \text{O} \quad \text{Cl} \\ \| \quad | \\ | \\ \text{Cl} \end{array}$$ I.

Ar	m. (Lit)
2,3,6-Cl$_3$C$_6$H$_2$	52-6°[832]
p-ClC$_6$H$_4$	62-3°[452]
p-BrC$_6$H$_4$	64-6°[452]
p-FC$_6$H$_4$	50-1°[235]
m-FC$_6$H$_4$	11-2°[235]
o-FC$_6$H$_4$	49-50°[235]
o-O$_2$NC$_6$H$_4$	115-20° dec.[452]
m-O$_2$NC$_6$H$_4$	103-5°[452]
p-O$_2$NC$_6$H$_4$	121-3°[452]
3,5-(O$_2$N)$_2$C$_6$H$_3$	125-6°[452]
2,4-Cl,O$_2$NC$_6$H$_3$	62-4°[438]
p-[Cl$_3$P=NSO$_2$C$_6$H$_4$]	72-5°[695]
1-C$_{10}$H$_7$	66-8°[438]
Ph	60-1°[430,452]

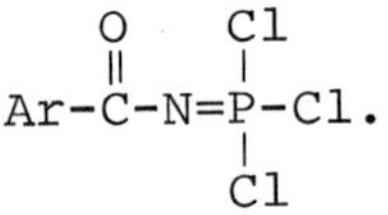

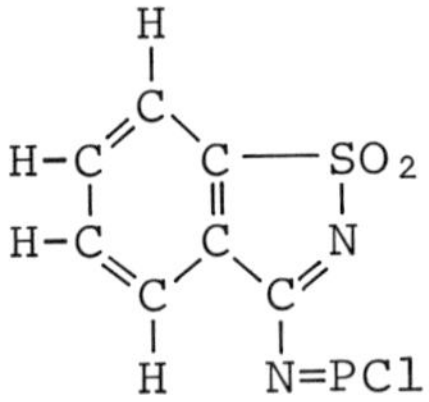

N=PCl$_3$. I. m. 165-9°.[207]

p-MeC$_6$H$_4$SO$_2$N=PBr$_3$. XIII. m. 138-41°.[472]
o-MeC$_6$H$_4$SO$_2$N=PBr$_3$. XIII. m. 148-51°.[472]
p-ClC$_6$H$_4$OCCl$_2$N=PCl$_3$. I. m. 52-5°.[458]
C$_8$ PhCCl(COCl)N=PCl$_3$. I. b$_{0.085}$ 117-20°, n 1.5813,
        d 1.5592.[938]
PhOCCl$_2$CON=PCl$_3$. I. m. 49-50°.[458]
CCl$_3$C(=NPh)N=PCl$_3$. I. m. 56-8°.[197]
CF$_3$C(=NPh)N=PCl$_3$. I. m. 49-51°.[232]
Me$_2$NSO$_2$C$_6$H$_4$N=PCl$_3$. I.[941] o-: m. 88-93°; m-: m. 158-62°
        p-: m. 138-41°.
C$_9$ o-MeC$_6$H$_4$OCCl$_2$CON=PCl$_3$. I. m. 49-51°.[458]
CCl$_3$C(=NCH$_2$Ph)N=PCl$_3$. I. m. 72-6°.[230]

$p\text{-}MeC_6H_4OCCl_2CON{=}PCl_3$. I. m. 74-6°.[458]
$MePhC(CN)N{=}PCl_3$. I. $b_{0.045}$ 92-4°, n 1.5517, d 1.3436.[686]
For similar compounds see Ref. 686.

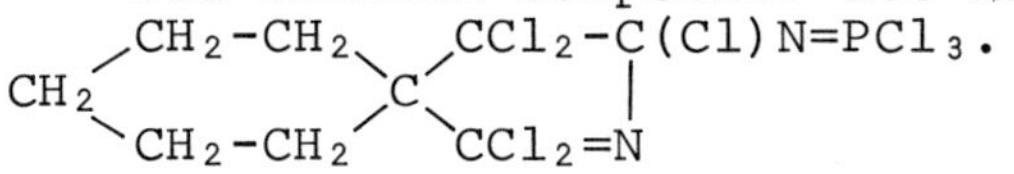

I. m. 87-9°, $b_{0.04}$ 163-6°.[787]

C_{10} 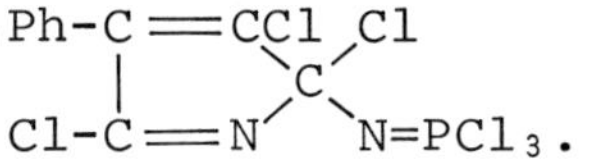
I. m. 95-70°, $b_{0.03}$ 145-7°.[787]
$1\text{-}C_{10}H_7SO_2N{=}PBr_3$. XIII. 157-9°.[472]
$2\text{-}C_{10}H_7SO_2N{=}PBr_3$. XIII. m. 150-3°.[472]

C_x $Ph_2NCON{=}PCl_3$. I. m. 112-4°.[461]
$PhSO_2N{=}C(Ph)N{=}PCl_3$. I. m. 59-61°.[204] For similar compounds see Ref. 204.
$Ph_2CHN{=}PCl_3$. I. $b_{0.05}$ 110-5°, n 1.6000, d 1.3240.[938]
$Ph_2CClCON{=}PCl_3$. I. m. 60-2°.[433,816]
$(CF_2)[C_6H_4(CON{=}PCl_3)\text{-}p]_2$. I. m. 82-5°, m. 127-9°(?).[201]
$Ph_3CCON{=}PCl_3$. I. m. 123-5°.[433]

D.5.2. [$-N{=}P(OR)_3$], [$-N{=}P(SR)_3$], R = Alkyl or Aryl or Acyl

C_4 $(MeO)_3P{=}NCN$. XIII. $b_{0.16}$ 69°. IR.[599]

C_7 $(EtO)_3P{=}NN{=}CH_2$. XIII. b_1 55-7°.[402]
$(EtS)_3P{=}NSO_2Me$. XIII. m. 34-5°.[334]
$(EtO)_3P{=}NSO_2Me$. XIII. $b_{0.05}$ 102-4°, n_D^{25} 1.4446, d 1.2194.[334]

C_8 $(EtO)_3P{=}NCOCCl_3$. XI. n 1.598, d 1.24.[436]
$(EtO)_3P{=}NCOCH_2Cl$. XIII. $b_{0.45}$ 90°, n 1.4559, d 1.1820.[399]
$(EtO)_3P{=}NCOCHCl_2$. XIII. $b_{0.0006}$ 53-3.6°, n 1.4646, d 1.2485.[399]
$(EtO)_3P{=}NCOCCl_3$. XIII. $b_{0.0006}$ 66-6.5°, n 1.4718, d 1.3139.[399]
$(EtO)_3P{=}NCOCF_3$. XIII. $b_{0.08}$ 50-3°, n 1.4010, d 1.2178.[399]
$(EtO)_3P{=}NSO_2Et$. XIII. $b_{0.1}$ 115-6°, n_D^{25} 1.4447, d 1.2194.[334]

C_9 $(MeO)_3P{=}NPh$. XIII. b. 150-4°,[907] $b_{0.0002}$ 44-5°,[403] $b_{0.0001}$ 46-7°.[326] n 1.5231.[326,403] d 1.1604,[326] d 1.1713.[403] For substituted aryl derivatives see Ref. 326.

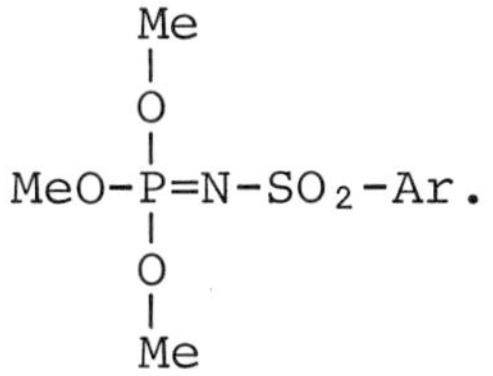

$MeO\text{-}P(OMe)(OMe){=}N\text{-}SO_2\text{-}Ar$. XI.

Ar	m. (Lit)
Ph	38-40°[462]
$p\text{-}ClC_6H_4$	51-2°[518]
$p\text{-}O_2NC_6H_4$	94-6°[447]

Ar	m. (Lit)
m-$O_2NC_6H_4$	57-9°[447]
p-MeC_6H_4	44-5°,[462] 40-1°[907]
o-MeC_6H_4	74-5°[462]
p-FC_6H_4	45-6°[391]
1-$C_{10}H_7$	84-5°[462]
2-$C_{10}H_7$	93-4°[462]
p-[$(MeO)_3P$=NSO_2]C_6H_4	117-9°[448]
m-[$(MeO)_3P$=NSO_2]C_6H_4	96-8°[448]

$(EtO)_3P$=NCOEt. XIII. b_1 103-4°, n 1.4360, d 1.0557.[398]

C10 $(MeO)_3P$=NCOPh. XIII. $b_{0.2}$ 135-7°, n 1.5102, d 1.2165.[208]

$(MeO)_3P$=$NCOC_6H_4NO_2$-m. XIII. m. 34-6°.[208]

$(MeO)_3P$=$NCOC_6H_4NO_2$-p. XIII. m. 56-8°.[208]

$(MeO)_3P$=NCONHPh. XIII. m. 42-4°.[225]

$(MeO)_3P$=$NCONHC_6H_4Br$-p. XIII. m. 72-4°.[225]

$(EtO)_3P$=$NC(CF_3)_3$. XIII. $b_{6.5}$ 70-1°, n 1.3540, d 1.3582.[603]

$(PrO)_3P$=NN=CH_2. XIII. b_1 74.5-6.0°.[402]

(i-PrO)$_3P$=NN=CH_2. XIII. b_1 55°.[402]

$(EtO)_3P$=NSO_2Bu. XIII. $b_{0.15}$ 120-1°.[334]

C11 $(PrO)_3P$=NCOMe. XIII. $b_{0.25}$ 69°, n 1.4419, d 1.0213.[398]

C12 $(EtO)_3P$=NPh. XIII. b_1 107°,[597] b_2 119-21°,[874] $b_{3.5}$135°.[403]

$(EtO)_3P$=NC_6H_4Cl-o. XIII. $b_{0.5}$ 117°.[597]

$(EtO)_3P$=NSO_2Ph. XIII. $b_{0.3}$ 162-4°.[334]

$(EtO)_3P$=$NSO_2C_6H_4F$-p. XI. b_{10} 202-4°.[391]

$(EtO)_3P$=NNHPh. XIII. b_2 113°.[874]

C13 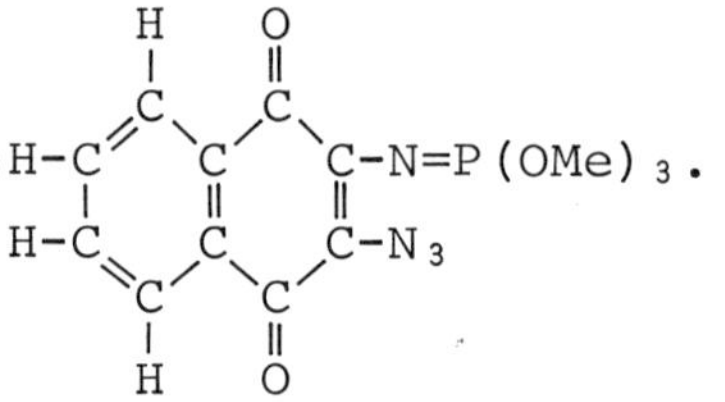
XIII. m. 99.5-100.5° dec.[615,616] For similar compounds see Refs. 615, 616.

$(EtO)_3P$=$NSO_2C_6H_4Me$-p. XIII. $b_{0.01}$ 159-70°, n_D^{25} 1.5041.[140]

$(EtO)_3P$=NCOPh. XIII. $b_{0.5}$ 146-7°, n 1.5019, d 1.1289.[208]

$(EtO)_3P$=$NCOC_6H_4NO_2$-m. XIII. m. 19-21°, n 1.5132, d 1.2377.[208]

$(EtO)_3P$=$NCOC_6H_4NO_2$-p. XIII. m. 21-3°, n 1.5170, d 1.2341.[208]

$(BuS)_3P$=NCN. XIII.[599]

$(BuO)_3P$=NN=CH_2. XIII. b_1 101-2°.[402]

$(EtO)_3P$=NC_6H_4Me-o. XIII. $b_{1.2}$ 127-33°.[874]

$(EtO)_3P$=NC_6H_4Me-m. XIII. $b_{0.5}$ 115-7°.[597]

$(EtO)_3P$=NC_6H_4Me-p. XIII. $b_{1.5}$ 132.2°.[874]

C15 $(PrO)_3P$=NPh. XIII. b_1 122°.[403]

$(CH_2$=$CHCH_2O)_3P$=NSO_2Ph. XI. n 1.5170, d 1.2069.[803] For substituted aryl derivatives (n, d given) see Ref. 803.

(i-PrO)$_3$P=NPh.   XIII.   b$_{0.0002}$ 49-50°.[403]
(EtO)$_3$P=NSO$_2$(CH$_2$)$_3$SO$_2$N=P(OEt)$_3$.   XIII.   m. 47°.[334]
(EtS)$_3$P=NSO$_2$(CH$_2$)$_3$SO$_2$N=P(SEt)$_3$.   XIII.   m. 40-1°.[334]

C$_{16}$ (EtO)$_3$P=NSO$_2$C$_{10}$H$_7$-1.   XI.   m. 94-5°.[462]
(EtO)$_3$P=NSO$_2$C$_{10}$H$_7$-2.   XI.   m. 51-2°.[462]

O=C——C-N=P(OMe)$_3$
| ‖
PhN   C-N=P(OMe)$_3$.   XIII.   m. 75-85°.[615]
  \ C /
    ‖
    O

(i-PrO)$_3$P=NCOPh.   XIII.   b$_{0.3}$ 143-4°, n 1.4863, d 1.0605.[208]
(i-PrO)$_3$P=NCOC$_6$H$_4$NO$_2$-p.   XIII.   n 1.5005, d 1.1597.[208]
    Meta-isomer: n 1.5320, d 1.2213.[208]

C$_{18}$ (EtO)$_3$P=NC$_6$H$_4$-Ph-o.   XIII.   b$_{0.4}$ 160°.[597]
(BuO)$_3$P=NPh.   XIII.   b$_1$ 144-5°.[403]
(EtO)$_3$P=NSO$_2$C$_6$H$_4$[SO$_2$N=P(OEt)$_3$]-p.   XIII.   m. 104.5°.[334]
(EtS)$_3$P=NSO$_2$C$_6$H$_4$[SO$_2$N=P(SEt)$_3$]-p.   XIII.   m. 79.5°.[334]

C$_{19}$
p-O$_2$NC$_6$H$_4$O
p-O$_2$NC$_6$H$_4$O ——P=N-SO$_2$-R.   XI.

R	m. (Lit)
Me	163-4°[523]
Pr	120-1°[523]
Bu	109-10°[523]
i-Bu	129-30°[523]
Am	134-5°[523]
i-Am	138-40°[523]
C$_6$H$_{13}$	93-6°[523]
PhCH$_2$CH$_2$	166-7°[523]
ClCH$_2$	131-2°[523]
Ph	123-4°[453]
NMe$_2$	144-6°[460]

(PhO)$_3$P=N-CN.   XIII.   m. 92.5-3.5°.[599]
(PhO)$_3$P=NSO$_2$Me.   XI, XIII.   m. 90.2°,[444] m. 89.5°.[334]

C$_{20}$
PhO
PhO-P=N-CO-R.   XI, XIII.
PhO

R	
CCl$_3$	n 1.638, d 1.31[436]
CF$_3$	b$_{0.035}$ 158-60°, n 1.5278[200]
CN	m. 58-60°[209]
Me	oil[193]
Ph	m. 74-6°[193,220,440]
m-O$_2$NC$_6$H$_4$	m. 96-7°[193]
3,5-(O$_2$N)$_2$C$_6$H$_3$	m. 87-8°[220,440]
1-C$_{10}$H$_7$	m. 60-83°[440]
Ph$_2$N	m. 83-4°[461]

(p-ClC$_6$H$_4$O)$_3$P=NCO-CN.   XIII.   n 1.4065, d 1.5780.[209]
(PhO)$_3$P=NEt.   XIII.   b$_2$ 171-3°, m. 33-4°.[678]
(PhO)$_3$P=NSO$_2$Et.   XI.   m. 100-1°.[444]

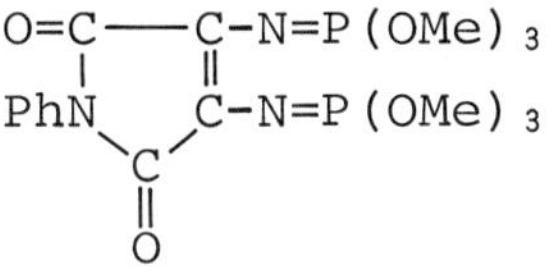

$(PhO)_3P{=}NSO_2NMe_2$. XI. m. 88-90°.[460]
$(o\text{-}ClC_6H_4O)_3P{=}NSO_2NMe_2$. XI. m. 69-70°.[460]
$(p\text{-}ClC_6H_4O)_3P{=}NSO_2NMe_2$. XI. m. 119-20°.[460]
$(o\text{-}O_2NC_6H_4O)_3P{=}NSO_2NMe_2$. XI. m. 141-2°.[460]
$(EtO)_3P{=}N\text{-}P({=}NPh)(NEt_2)_2$. XIII. $b_{0.0001}$ 116-7°, n 1.5120.[327] For similar P=N-P(=N-)N compounds see Ref. 327.

C_{21} 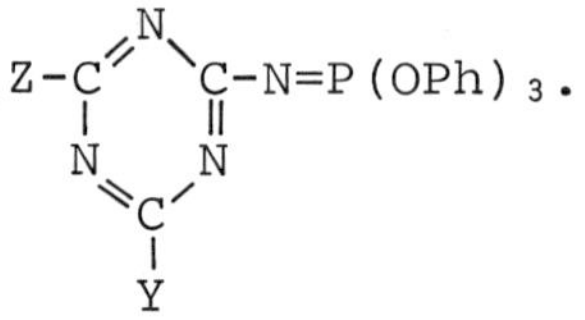 XI.[202] Y=Z=Cl: m. 49-50°. Y=Cl, Z= $N{=}P(OPh)_3$: m. 230°. Y=Z= $N{=}P(OPh)_3$: m. 300°.

$(PhO)_3P{=}NSO_2Pr\text{-}i$. XI. m. 66-8°.[444]
$(i\text{-}PrO)_3P{=}N\text{-}N{=}CPh_2$. XIII. m. 66-7°.[704]
C_{23} $(p\text{-}MeC_6H_4O)_3P{=}NCO\text{-}CN$. XIII. n 1.5459, d 1.1874.[209]
C_{24} $(PhS)_3P{=}NSO_2Ph$. XIII. m. 102.5°.[334]

PhO
|
PhO-P=N-Ar. I, XI,[939] XIII.[933]
|
PhO

Ar	
Ph	$b_{0.05}$ 222-4°, n 1.6005, d 1.2085
$o\text{-}O_2NC_6H_4$	oil
$m\text{-}O_2NC_6H_4$	n 1.6032, d 1.2665
$p\text{-}O_2NC_6H_4$	m. 76-8°
$o\text{-}ClC_6H_4$	n 1.6048, d 1.2605
$m\text{-}ClC_6H_4$	n 1.6035, d 1.2606
$p\text{-}ClC_6H_4$	n 1.5980, d 1.2599
$o\text{-}BrC_6H_4$	n 1.6069, d 1.3591
$m\text{-}BrC_6H_4$	n 1.6060, d 1.3342
$p\text{-}BrC_6H_4$	n 1.6099, d 1.3714
$2,4\text{-}(O_2N)_2C_6H_3$	m. 78-80°
$2,4\text{-}Cl_2C_6H_3$	n 1.6082, d 1.3113
$2,5\text{-}Cl_2C_6H_3$	n 1.6030, d 1.3155
$2,4\text{-}Br_2C_6H_3$	n 1.6255, d 1.5222
$2,6,4\text{-}Cl_2(O_2N)C_6H_2$	m. 90-2°
$2,4,6\text{-}Cl_3C_6H_2$	m. 68-9°
$2,4,6\text{-}Br_3C_6H_2$	m. 84-6°
$o\text{-}MeC_6H_4$	n 1.6006, d 1.2080
$m\text{-}MeC_6H_4$	n 1.5942, d 1.2023
$p\text{-}MeC_6H_4$	n 1.5969, d 1.2032
$p\text{-}MeOC_6H_4$	n 1.5891, d 1.2093
$3,5\text{-}Me_2C_6H_3$	n 1.5935, d 1.1809
$p\text{-}EtOC_6H_4$	n 1.5862, d 1.2069

PhO
|
PhO-P=N-SO_2-Ar. XI, XIII.
|
PhO

Ar	m. (Lit)
Ph	85-6°,[463,462,933] m. 89.5°[334]
$p\text{-}FC_6H_4$	69-70°[391]
$p\text{-}O_2NC_6H_4$	101-4°[447]
$m\text{-}O_2NC_6H_4$	56-8°[447]
$o\text{-}O_2NC_6H_4$	78-9°[447]
$p\text{-}MeC_6H_4$	72-5°[140]
$o\text{-}MeC_6H_4$	79-80°[462]
$1\text{-}C_{10}H_7$	94-5°[462]
$2\text{-}C_{10}H_7$	90-1°[463]

$(p\text{-}ClC_6H_4O)_3P{=}NSO_2Ph$. XI. m. 112-4°.[453]

$(o\text{-}ClC_6H_4O)_3P{=}NSO_2Ph$. XI. m. 72-3°.[453]

$(o\text{-}O_2NC_6H_4O)_3P{=}NSO_2Ph$. XI. m. 94-5°.[453]

C25 $(PhO)_3P{=}NSO_2CH_2Ph$. XI. m. 89-90°.[444]

$(p\text{-}ClC_6H_4O)_3P{=}NCOC_6H_4Br\text{-}p$. XIII. m. 74-6°.[193]

$(o\text{-}O_2NC_6H_4O)_3P{=}NCOC_6H_4NO_2\text{-}p$. XI. m. 130-3°.[440]

$(PhO)_3P{=}NC({=}NPh)CCl_3$. XI. m. 83-4°.[197] For similar compounds with different N-Ar groups see Ref. 197.

Cx $(p\text{-}MeC_6H_4O)_3P{=}NCOPh$. XIII. m. 84-6°.[193]

$(o\text{-}MeC_6H_4O)_3P{=}NSO_2Ph$. XI. m. 90-1°.[462]

$(p\text{-}ClC_6H_4O)_3P{=}NCONPh_2$. XI. m. 128-30°.[461] Ortho-isomer: m. 40-3°. For other o-aryl compounds see Ref. 461.

$(1\text{-}C_{10}H_7O)_3P{=}NSO_2NMe_2$. XI. m. 123-4°.[460]

$(1\text{-}C_{10}H_7O)_3P{=}NCOCCl_3$. XI. m. 114-6°.[437]

$(PhO)_3P{=}NC({=}NSO_2C_6H_4Me\text{-}p)Ph$. XI. m. 127-9°.[205] For analogous compounds with different aryls see Ref. 205.

$(PhO)_3P{=}NSO_2N{=}P(OPh)_3$. I, XI. m. 132-3°.[457,933]

$(p\text{-}ClC_6H_4O)_3P{=}NSO_2N{=}P(OC_6H_4Cl\text{-}p)_3$. XI. m. 111-3°.[457]

$(p\text{-}O_2NC_6H_4O)_3P{=}NSO_2N{=}P(OC_6H_4NO_2\text{-}p)_3$. XI. m. 198-200°.[457]

$[(PhO)_3P{=}NSO_2\text{-}]_2NMe$. XI. m. 91-2°.[626]

$(1\text{-}C_{10}H_7O)_3P{=}NCOC_6H_4Cl\text{-}p$. XI. m. 156-8°.[440]

$(1\text{-}C_{10}H_7O)_3P{=}NCOC_6H_4NO_2\text{-}p$. XI. m. 83-6°.[433]

$[(PhO)_3P{=}NSO_2N(Me)\text{-}]_2SO_2$. XI. m. 90°.[626]

$(PhO)_3P{=}NSO_2(CH_2)_3SO_2N{=}P(OPh)_3$. XIII. m. 70.5°.[334]

$(PhO)_3P{=}NSO_2C_6H_4[SO_2N{=}P(OPh)_3]\text{-}p$. XI. m. 132-3°.[448] Meta-isomer: m. 100-2°.[448]

$(o\text{-}MeC_6H_4O)_3P{=}NSO_2N{=}P(OC_6H_4Me\text{-}o)_3$. XI. m. 127-9°.[457] Para-analogue: m. 80-1°.[457]

$(1\text{-}C_{10}H_7O)_3P{=}NCOCClPh_2$. XI. m. 121-2°.[433]

D.5.3. $[\text{-}N{=}P(NRR')_3]$, R = R' or R ≠ R', R,R' = H, Alkyl or Aryl

C12 $(C_2H_4N)_3P{=}NPh$. XIII. m. 65-6°.[224] For substituted aryl derivatives see Ref. 224.

$(Me_2N)_3P{=}NSO_2C_6H_4NO_2\text{-}p$. VI. m. 109-11°.[447] Ortho isomer: m. 160-3°.[447] Meta isomer: m. 66-8°.[447]

$(Me_2N)_3P=NPh$. XIII. $b_{0.4}$ 127°, m. 11.5°, n 1.5537, d 1.054.[888]
$(C_2H_4N)_3P=NSO_2N=P(NC_2H_4)_3$. IV. m. 94-5°.[695]
$(Me_2N)_3P=N-SnEt_3$. XIII. $b_{0.1}$ 81-4°, ^1H.[533]
C13 $(C_2H_4N)_3P=NCOPh$. VI. m. 57-8°.[486]
$(C_2H_4N)_3P=NCOC_6H_4Cl$-p. VI. m. 72-4°.[486]
$(C_2H_4N)_3P=NCOC_6H_4Br$-p. VI. m. 80-2°.[486]
$(C_2H_4N)_3P=NCOC_6H_4NO_2$-p. VI. m. 80-2°.[486]

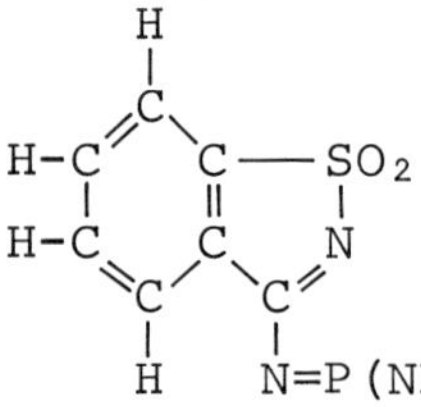

N=P(NHR)$_3$. VI.[207] R=Et: m. 171-2°.
R=Ph: m. 214-5°.

C15 $[O(CH_2CH_2)_2N]_3P=NCOOEt$. XIII. m. 127-8°.[196]
C17 $(Et_2N)_3P=N-N=C(COOMe)COMe$. XIII. m. 41-2°.[374]
C18 $(C_2H_4N)_3P=NSO_2C_6H_4[SO_2N=P(NC_2H_4)_3]$-p. VI. dec. 175°.[695]
Meta isomer: dec. 130-2°.
$[O(CH_2CH_2)_2N]_3P=N-N=NPh$. XIII. m. 129°.[290]
$[O(CH_2CH_2)_2N]_3P=N-N=NC_6H_4NO_2$-p. XIII. m. 145-7°.[290]
$(Et_2N)_3P=NPh$. VI. m. 28-30°, ^1H, ^{31}P -14.2.[350]
$(Me_2N)_3P=N-SnBu_3$. XIII. $b_{0.1}$ 138-40°.[533]

C19 $(PhNH)_3P=N-SO_2R$. VI.

R	m. (Lit)
Me	265-6°[523]
Et	251-2°[445]
Pr	241-2°[523]
Bu	232-3°[445]
i-Bu	253-4°[523]
Am	228-9°[523]
i-Am	254-5°[523]
C_6H_{13}	221-2°[523]
$PhCH_2CH_2$	252-3°[523]
$ClCH_2$	246-7°[523]
Ph	256-7°[432]
p-ClC_6H_4	255°[522]
p-BrC_6H_4	258-9°[522]
p-FC_6H_4	244-5°[522]
o-ClC_6H_4	232-3°[522]
m-$CF_3C_6H_4$	250-1°[522]
o-MeC_6H_4	191-1.5°[432]
p-MeC_6H_4	253-4°[432]
p-$MeOC_6H_4$	265-6°[522]

$[O(CH_2CH_2)_2N]_3P=N-N=NCOPh$. XIII. m. 86-7°.[290]
Cx $(C_2H_4N)_3P=NCOC_6H_4]CON=P(NC_2H_4)_3]$-p. VI. m. 136-8°.[695]
Meta isomer: m. 130-2°.[695]
$[O(CH_2CH_2)_2N]_3P=N-N=NC_{10}H_7$-2. XIII. m. 136-8°.[290]

$(Pr_2N)_3P{=}N{-}N{=}NPh$. XIII. m. 55°.[290]
$(PhNH)_3P{=}NCOPh$. VI. m. 112-5°.[211]
$(PhNH)_3P{=}NCOC_6H_4NO_2$-p. VI. m. 207-9°.[211] Meta isomer: m. 81-3°.[211]
$[O(CH_2CH_2)_2N]_3P{=}N{-}N{=}CPh_2$. XIII. m. 104-5°.[374]
$(p\text{-}MeC_6H_4NH)_3P{=}NCOPh$. VI. m. 101-3°.[211]
$(PhNH)_3P{=}NC({=}NSO_2Ph)Ph$. VI. m. 225-6°.[206] For analogous compounds with different aryls see Ref. 206.

D.5.4. [-N=PABC], A,B,C = Halogen,-OR,-SR,-NRR'

C_3 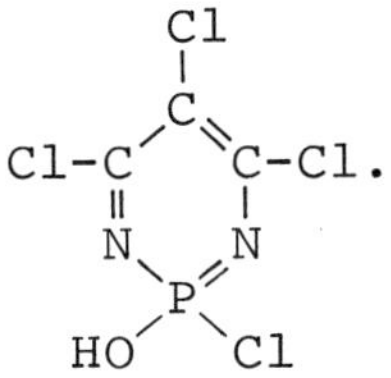 X. m. 182-3°.[786]

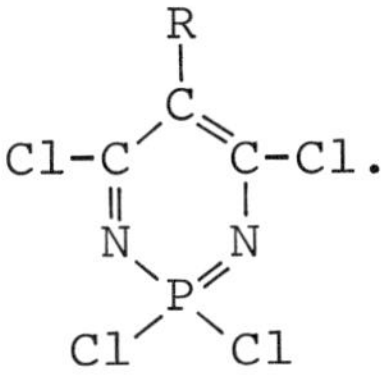 XVII.

R	m. (Lit)
Cl	150-2°[786]
H	40-5°. b_{13} 108-9°[786]
Me	116-7°[792]
Et	59-65°[792]
Pr	52-5°[792]
i-Pr	62-4°[792]
Bu	48-50°[792]

C_8 $(EtO)_2(MeCONH)P{=}NCOMe$. XIII. $b_{0.0004}$ 62-4°, n 1.4680, d 1.1860.[398]
C_{10} 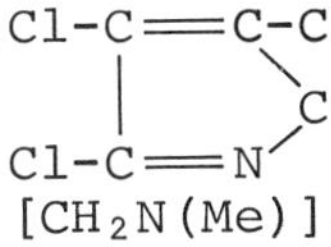 C=N-P(Cl_2)=NSO_2Ph. I. m. 145-7°.[797]
$[CH_2N(Me)]_2ClP{=}NPh$. XIII. m. 34-6°.[884]
$(Et_2N)Cl_2P{=}NPh$. VI. $b_{0.1}$ 96-8°, ^{31}P +21.6.[350]
$(Me_2N)_2ClP{-}NPh$. XIII. $b_{0.05}$ 92-3°, n 1.5650, d 1.1544.[327]
$(Me_2N)_2(N_3)P{=}NPh$. XIII. $b_{0.04}$ 88-9°, n 1.5580, d 1.1270.[327]
C_{11} $(MeO)Cl_2P{=}NSO_2C_{10}H_7$-1. XI. m. 82-3°.[464]
$(EtO)_2(MeS)P{=}NSO_2C_6H_3Br,Cl$-2,5. XVII. m. 54-5°.[30]
$(EtO)_2(MeS)P{=}NSO_2C_6H_4I$-p. XVII. m. 75-6°.[30]
C_{12} $(MeO)_2ClP{=}NSO_2C_{10}H_7$-1. XI. m. 138-40°.[464]
$[CH_2N(Me)]_2(Me_2N)P{=}NPh$. XIII. m. 30-1°, ^{31}P -7.6.[884]
$(EtO)_2(MeS)P{=}NCOPh$. XVII. $b_{0.15}$ 136°, n 1.5425, d 1.1763.[33,34] For other $(RO)_2(MeS)P{=}NCOAr$ compounds (b, n, d) see Refs. 33,34.
$(EtO)_2(PhNH)P{=}NCOMe$. XIII. $b_{0.007}$ 85-6°, n 1.5200, d 1.1356.[398]
C_{13} $(PhO)Cl_2P{=}NCOC_6H_4NO_2$-p. XI. m. 72-5°.[220]

$(PhO)Cl_2P{=}NCOC_6H_3(NO_2)_2$-3,5. XI. m. 81-3°.[220]
$(PhO)F_2P{=}NC({=}NH)Ph$. XIII. $b_{0.03}$ 180-2°, n 1.6102.[218]
$(EtO)_2(PhMeN)P{=}NCOMe$. XIII. $b_{0.0001}$ 52-4°, n 1.5135, d 1.1295.[398]
C_{14} $(PhO)_2ClP{=}NCOOMe$. XIII. m. 36-7°.[195,196]
$(Et_2N)_2ClP{=}NPh$. VI. $b_{0.1}$ 110-5°, ^{31}P -9.8.[350]
C_{16} $(EtO)_2(p\text{-}ClC_6H_4NH)P{=}NPh$. XIII. m. 107.5-8.5°.[405]
$(EtO)_2(PhNH)P{=}NC_6H_4NO_2$-p. XIII. m. 103-4.5°.[398]
$(Et_2N)_2(EtO)P{=}NPh$. XIII. $b_{0.09}$ 110-1°, n 1.5150, d 1.0059.[327]
C_{17} $(EtO)_2(p\text{-}MeC_6H_4NH)P{=}NC_6H_4Cl$-p. XIII. m. 124-5°.[325] For $(EtO)_2(ArNH)P{=}NAr'$ compounds see Ref. 325.
$(EtO)_2(MePhN)P{=}NPh$. XIII, XVII. $b_{0.003}$ 150-70°, n 1.5632, d 1.1126.[405]
$(EtO)_2(PhNH)P{=}NC_6H_4Me$-p. XIII. m. 112°.[405] Meta isomer m. 88-9°.[405] It may be the tautomerie form $(EtO)_2(p\text{-}MeC_6H_4NH)P{=}NPh$.

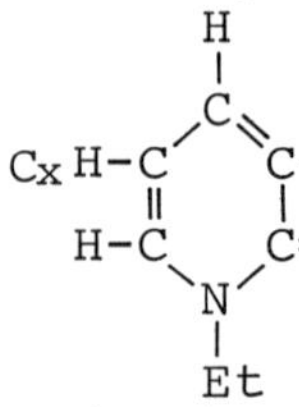
$=N-P(OEt)_2(=NPh)$. XIII. m. 68.5-70°.[406]

$(PhO)_2ClP{=}NSO_2Ph$. XIII. m. 66-9°.[464]
$[O(CH_2CH_2)_2N]_2(C_5H_{10}N)P{=}N{-}N{=}NPh$. XIII. m. 129°.[290]
$(PhO)_2(PhNH)P{=}NCOPh$. IX. m. 125-6°.[222]
$(PhO)_2(PhNH)P{=}NCOC_6H_3(NO_2)_2$-3,5. IX. m. 128-30°.[222]
$(o\text{-}C_6H_4O_2)(PhN{=})P{-}NHC_6H_4[NHP({=}NPh)(O_2C_6H_4\text{-}o)]$-p. XVII. dec. 132-3°.[420]

(received June 14, 1971)

REFERENCES

1. Abel, E. W., D. A. Armitage, and G. R. Willey, J. Chem. Soc., 1965, 57.
2. Abramov, V. S., and Z. S. Druzhina, Zh. Obshch. Khim., 36, 923 (1966); C. A. 65, 10581 a (1966).
3. Abramov, V. S., and Z. S. Druzhina, Zh. Obshch. Khim., 37, 718 (1967); C. A. 67, 64490 a (1967).
4. Abramov, V. S., and N. A. Il'ina, Dokl. Akad. Nauk SSSR, 132, 823 (1960); C. A. 54, 22329 g (1960).
5. Abramov, V. S., and N. A. Il'ina, Dokl. Akad. Nauk SSSR, 155, 112 (1964); C. A. 60, 13131 g (1964).
6. Abramov, V. S., and N. A. Il'ina, Nekotorye Vopr. Organ. Khim., 1964, 256; C. A. 65, 3902 h (1966).
7. Abramov, V. S., and N. A. Il'ina, Khim. Org. Soedin. Fosfora Akad. Nauk SSSR, Otd. Obshch. Tekh. Khim. 1967, 119; C. A. 69, 43339 j (1968).
8. Addor, R. W., U. S. 3,197,481 (1965); C. A. 64, 2089 a (1966).

9. Addor, R. W., and J. B. Lovell, Belg. 618,155 (1962); C. A. 59, 10066 f (1963).
10. Alimov, P. I., Izv. Akad. Nauk SSSR, 1961, 61; C. A. 55, 18577 e (1961).
11. Alimov, P. I., O. N. Fedorova, and L. N. Levkova, Izv. Akad. Nauk SSSR, Ser. Khim., 1965, 1208 and 1298; C. A. 63, 13059 f and 13060 h (1965).
12. Alimov, P. I., and M. P. Alimov, Izv. Akad. Nauk SSSR, Ser. Khim., 1967, 1344; C. A. 68, 48958 z (1968).
13. Alimov, P. I., and L. A. Antokhina, Izv. Akad. Nauk SSSR, 1963, 1132; C. A. 59, 8576 h (1963).
14. Alimov, P. I., and L. A. Antokhina, Izv. Akad. Nauk SSSR, Ser. Khim., 1963, 2204; C. A. 60, 9180 g (1964).
15. Alimov, P. I., and L. A. Antokhina, Izv. Akad Nauk SSSR, Ser. Khim., 1964, 1316; C. A. 61, 11882 h (1964).
16. Alimov, P. I., and L. A. Antokhina, Izv. Akad Nauk SSSR, Ser. Khim., 1966, 1486; C. A. 66, 54956 x (1967).
17. Alimov, P. I., and I. V. Cheplanova, Izv. Kazansk. Akad. Nauk SSSR, 1961, 61; C. A. 59, 9775 e (1963).
18. Alimov, P. I., and O. N. Fedorova, Izv. Akad. Nauk SSSR, 1960, 1985; C. A. 55, 13297 g (1961).
19. Alimov, P. I., and O. N. Fedorova, Izv. Kazansk. Filiala Akad. Nauk SSSR, Ser. Khim. Nauk, 1961, 48; C. A. 59, 9782 (1963).
20. Alimov, P. I., and O. N. Fedorova, Izv. Akad. Nauk SSSR, Ser. Khim., 1966, 1370; C. A. 66, 65818 e (1967).
21. Alimov, P. I., and O. N. Fedorova, Izv. Akad. Nauk SSSR, Ser. Khim., 1966, 1461; C. A. 67, 63664 y (1967).
22. Alimov, P. I., and L. N. Levkova, Izv. Akad. Nauk SSSR, Ser. Khim., 1964, 187; C. A. 60, 9144 h (1964).
23. Alimov, P. I., and L. N. Levkova, Izv. Akad. Nauk SSSR, Ser. Khim., 1964, 932; C. A. 61, 5502 (1964).
24. Alimov, P. I., and N. L. Levkova, Izv. Akad. Nauk SSSR, Ser. Khim., 1964, 1801; C. A. 62, 2702 (1965).
25. Alimov, P. I., and L. N. Levkova, Izv. Akad. Nauk SSSR, Ser. Khim., 1964, 1889; C. A. 62, 3921 f (1965).
26. Alimov, P. I., M. A. Zvereva, and O. N. Fedorova, Khim. i. Prim., 1955, 164; C. A. 52, 244 a (1958).
27. Allen, J. F., and O. H. Johnson, J. Am. Chem. Soc., 77, 2871 (1955).
28. Almasi, L., M. Giurgiu, and A. Hantz, Akad. Rep. Pop. Romine, Fil. Cluj. Studii Cercetari Chim., 14, 271 (1963); C. A. 62, 15490 e (1965).

29. Almasi, L., and A. Hantz, Acad. Rep. Pop. Romine, Studii Cercetari Chim., 11, 297 (1960); C. A. 58, 5556 (1963).
30. Almasi, L., and A. Hantz, Rev. Roemaine Chim., 9, 433 (1964); C. A. 62, 6417 d (1965).
31. Almasi, L., A. Hantz, and E. Hamburg, Chem. Ber., 96, 3148 (1963), and Rev. Roumaine Chim., 11, 273 (1966); C. A. 65, 5334 b (1966).
32. Almasi, L., and A. Hantz, Chem. Ber., 98, 617 (1965).
33. Almasi, L., and L. Paskucz, Chem. Ber., 99, 3293 (1966).
34. Almasi, P. I., and L. Paskucz, Chem. Ber., 100, 2625 (1967).
35. American Cyanamid, Brit. 948,522 (1964); C. A. 60, 12028 a (1964).
36. Anatol, J., Fr. 1,211,099, 1,210,435 (1960); C. A. 55, 14323 i, 23563 c (1961).
37. Anderson, H. H., J. Am. Chem. Soc., 64, 1757 (1942).
38. Andronova, L. G., Z. A. Shabarova, T. S. Ryabova, and M. A. Prokof'ev, Zh. Obshch. Khim., 31, 3243 (1961); C. A. 57, 7365 a (1962).
38a. Ansell, G. B., and G. J. Bullen, Chem. Comm., 1965, 493.
39. Antokhina, L. A., and P. I. Alimov, Izv. Akad. Nauk SSSR, Ser. Khim., 1966, 2135; C. A. 66, 95132 u (1967).
40. Arbuzov, B. A., P. I. Alimov, M. A. Zvereva, I. D. Neklesova, and M. A. Kudrina, Izv. Akad. Nauk SSSR, Otdel. Khim. Nauk, 1954, 1038; C. A. 50, 161 b (1956).
41. Arbuzov, A. E., F. G. Valitova, A. V. Il'yasov, B. M. Kozyrev, and Yu. V. Yablokov, Dokl. Akad. Nauk SSSR, 147, 839 (1962); C. A. 58, 8535 d (1963).
42. Arbuzov, B. A., V. M. Zoroastrova, and S. P. Shcherbak, Zh. Org. Khim., 1, 2190 (1966); C. A. 66, 94896 j (1967).
43. Arbuzov, A. E., S. Yu. Baigil'dina, F. G. Valitova, and R. R. Shagidullin, Izv. Akad. Nauk SSSR, Ser. Khim., 1967, 1966; C. A. 68, 39722 r (1968).
44. Arbuzov, B. A., P. I. Alimov, and O. N. Fedorova, Izv. Kazan, Filiala, Akad. Nauk SSSR, Ser. Khim. Nauk, 1955, Nr. 2,25; C. A. 52, 243 f (1958).
45. Arbuzov, B. A., P. I. Alimov, and O. N. Fedorova, Izv. Akad. Nauk SSSR, 1956, 932; C. A. 51, 4932 d (1957).
46. Arbuzov, B. A., P. I. Alimov, and M. A. Zvereva, Izv. Akad. Nauk SSSR, 1954, 1042, 1047; C. A. 50, 215 i and 216 e (1956).
47. Arbuzov, B. A., and D. C. Jarmuchametova, Dokl. Akad. Nauk SSSR, 101, 675 (1955); C. A. 50, 3214 g (1956).

48. Arbuzov, B. A., and D. C. Jaramuchametova, Izv. Akad. Nauk SSSR, Ser. Khim., 1957, 292; C. A. 51, 14542 c (1957).
49. Arbuzov, A. E., and V. M. Soroatrova, Izv. Akad. Nauk SSSR, 1952, 789; C. A. 47, 10461 c (1953).
50. Arbuzov, B. A., V. M. Soroatrova, and M. P. Osipova, Izv. Akad. Nauk SSSR, 1961, 2163; C. A. 57, 8536 f (1962).
51. Arceneaux, R. L., et. al., J. Org. Chem., 24, 1419 (1959).
52. Armour Pharmaceutical Co., Fr. 1,397,493 (1965); C. A. 63, 5605 f (1965).
53. Arni, P. C., and E. J. Jones, J. Appl. Chem. (London), 14, 221 (1964).
54. Arnold, H., F. Bourseaux, and N. Brock, Arzneimittelforschung, 11, 143 (1961).
55. Arnold, H. and F. Bourseaux, Angew. Chem., 70, 539 (1958).
56. Arnold, H., and F. Bourseaux, Arzneimittelforsch., 13, 927 (1963); C. A. 60, 3993 b (1964).
57. Arnold, H., F. Bourseaux, and N. Brock, Ger. 1,054,997 (1956); Brit. 853,044; C. A. 55, 19793 d (1961); C. 1960, 228.
58. Arnold, H., and N. Brock, Ger. 1,046,621 (1956); C. 1959, 10001.
59. Arnold, H., N. Brock, and F. Bourseaux, Ger. 1,057,119 (1956); C. 1960, 3312.
60. Artemkina, R. V., and V. M. Berezovski, Zh. Obshch. Khim., 36, 823 (1966); C. A. 65, 12276 e (1966).
61. Atherton, F. R., U. S. 2,490,573 (1947); C. A. 44, 3525 a (1950).
62. Atherton, F. R., H. T. Howard, and A. R. Todd, J. Chem. Soc., 1948, 1106.
63. Atherton, F. R., H. T. Openshaw, and A. R. Todd, J. Chem. Soc., 1945, 382.
64. Atherton, F. R., H. T. Openshaw, and A. R. Todd, J. Chem. Soc., 1945, 660.
65. Atherton, F. R., and A. R. Todd, J. Chem. Soc., 1947, 674.
66. Aubel and Reich, Compt. Rend., 195, 183 (1932).
67. Audrieth, L. F., and A. D. F. Toy, J. Am. Chem. Soc., 63, 2117 (1941).
68. Audrieth, L. F., and A. D. F. Toy, J. Am. Chem. Soc., 64, 1337 (1942).
69. Audrieth, L. F., and A. D. F. Toy, J. Am. Chem. Soc., 64, 1553 (1942).
70. Audrieth, L. F., H. Zimmer, and F. Zimmer, J. Pr. [4], 8, 117 (1959).
71. Augustinsson, K. B., Svensk. Kem. Tidskr., 1952, 87.
72. Authenrieth, W., Ber. Dtsch. Chem. Ges., 30, 2369 (1897).

73. Authenrieth, W., and E. Bölli, Ber. Dtsch. Chem. Ges., 58, 2144 (1925).
74. Authenrieth, W., and A. Geyer, Ber. Dtsch. Chem. Ges., 41, 146 (1908).
75. Authenrieth, W., and O. Hildebrand, Ber. Dtsch. Chem. Ges., 31, 1094, 1111 (1898).
76. Authenrieth, W., and W. Meyer, Ber. Dtsch. Chem. Ges., 58, 840, 848 (1925).
77. Authenrieth, W., and P. Rudolph, Ber. Dtsch. Chem. Ges., 33, 2099, 2112 (1900).
78. Ayres, D. C., and H. N. Rydon, J. Chem. Soc., 1957, 1109.
79. Babad, H., and W. Herbert, Anal. Chim. Acta, 41, 259 (1968).
80. Baddiley, J., J. G. Buchanan, and R. Letters, J. Chem. Soc., 1956, 2812.
81. Baker, J. W., U. S. 2,957,018 (1957); C. A. 55, 9280 b (1961).
82. Baker, J. W., U. S. 2,993,775 (1956); C. A. 55, 25146 h (1961).
83. Bardos, T. J., et al., Chem. Ind., 35, 1464 (1963).
83a. Bardos, T. J., et al., Nature, 183, 399 (1959).
84. Bartle, K. D., R. S. Edmundson, and D. W. Jones, Tetrahedron, 23, 1701 (1967).
85. Bataafsche (Shell), Ger. 1,113,937 (1959); C. 1962, 10278.
86. Batrakov, S. G., et al., Zh. Obshch. Khim., 37, 426 (1967); C. A. 67, 99595 r (1967).
87. Baumgarten, H. E., and R. E. Allen, J. Org. Chem., 26, 1533 (1961).
88. Baumgarten, H. E., and R. A. Setterquist, J. Am. Chem. Soc., 81, 2132 (1959).
89. Beatty, I. M., D. I. Magrath, and A. H. Ennor, J. Chem. Soc., 1965, 12.
90. Becke-Goehring, M., and H. Hofmann, Z. Anorg. Allg. Chem., 369, 73 (1969).
91. Becke-Goehring, M., and D. Jung, Z. Anorg. Allg. Chem., 372, 233 (1970).
92. Becke-Goehring, M., and W. Lehr, Z. Anorg. Allg. Chem., 325, 287 (1963).
93. Becke-Goehring, M., L. Leichner, and B. Scharf, Z. Anorg. Allg. Chem., 343, 154 (1966).
94. Becke-Goehring, M., and G. Wunsch, Chem. Ber., 93, 326 (1960).
95. Berlin, K. D., and M. A. R. Khayat, Tetrahedron, 22, 975, 987 (1966).
96. Berlin, K. D., and L. A. Wilson, Chem. Commun., 1965, 280.
97. Berlin, K. D., and L. A. Wilson, Chem. Ind., 1965, 1522.

98. Bersin, T., et al., Z. Physiol. Chem. Hoppe Seylers, 269, 241 (1941).
99. Bestian, H., Ann., 566, 210 (1950).
100. Beyer, H., T. Pyl, and H. Lemke, J. Pr. [4], 16, 132, 137 (1962).
100a. Binder, H., and R. Heinle, Ger. 1,075,611 (1957); C. A. 55, 14383 (1961); Ger. 139,497 (1957); C. A. 60, 426 (1964).
101. B. I. O.S. Final Report, 714.
102. Blair, E. H., U. S. 3,165,545 (1965); C. A. 62, 10371 g (1965); U. S. 3,159,665 (1964); C. A. 62, 3939 b (1965).
103. Blair, E. H., K. C. Kauer, and E. C. Britton, U. S. 2,875,233/234 (1959); C. A. 53, 16076 (1959).
104. Blair, E. H., H. R. Slagh, and E. C. Britton, U. S. 2,870,189 (1956); C. A. 53, 11305 g (1959).
105. Blair, E. H., and H. Tolkmith, J. Org. Chem., 25, 1620 (1960).
106. Blair, E. H., H. Tolkmith, E. C. Britton, and K. C. Kauer, U. S. 2,855,423-426 (1958); C. A. 53, 4210, 4211 (1959).
107. Blair, E. H., and J. L. Wasco, U. S. 3,281,504 (1966); C. A. 66, 10777 r (1967).
108. Bliss, A. D., M. I. Gruber, and R. Raetz, U. S. 3,355,521 (1967); C. A. 68, 39657 y (1968).
109. Bliss, A. D., and R. Raetz, U. S. 3,358,004 (1967); C. A. 68, 39661 v (1968).
110. Blum, D., G. Notles, and H. Pasedach, Ger. 1,085,524 (1959); C. 1961, 4142.
111. Bock, H., and E. Baltin, Chem. Ber., 98, 2844 (1965).
112. Bock, H., and G. Rudolph, Chem. Ber., 94, 1457 (1961).
113. Bock, H., and W. Wiegräbe, Chem. Ber., 99, 377 (1966).
114. Boehringer, C. H. and Sohn, Neth. 6,612,312 (1967); C. A. 67, 73687 x (1967).
115. Bogner, E., and O. M. Friedman, J. Am. Chem. Soc., 80, 2583 (1958).
116. Boisselet, L., and G. Avico, Fr. 1,140,878 (1955); C. 1959, 3972.
117. Borke, M. L., and E. R. Kirch, J. Am. Pharm. Assoz., 47, 461 (1958); C. A. 51, 2910 (1957).
118. Bos, B. G. van den, C. J. Schoot, M. J. Koopmans, and J. Meltzer, Rec. Trav. Chim., 80, 1040 (1961); 79, 836,1129 (1960).
119. Bos, B. G. van den, M. J. Koopsmanns, and H. O. Huisman, Rec. Trav. Chim., 79, 807 (1960); C. A. 55, 16522 i (1961).
120. Braun, J., and W. Rudolph, Ber. Dtsch. Chem. Ges., 67, 1762 (1934).

121. Breederveld, H., U. S. 3,102,910 (1963); C. A. 60, 2756 a (1964).
122. Breederveld, H., Ger. 1,129,946 (1960); C. A. 57, 12328 e (1962).
123. Brintzinger, H., K. Pfannenstiel, and H. Koddebusch, Chem. Ber., 82, 389 (1949).
124. Britton, E. C., and E. H. Blair, U. S. 2,874,178 (1956); C. A. 54, 2258 a (1960).
124a. Brown, D. M., and N. K. Hamer, Proc. Chem. Soc., 1960, 212.
125. Brust, H. F., U. S. 2,971,976 (1959); C. A. 55, 12358 e (1961).
126. Brust, H. F., U. S. 3,087,955 (1963); C. A. 59, 12644 g (1963).
127. Bub, L., W. Franke, and A. Holz, Ger. 1,005,963 (1955); C. 1957, 11716.
128. Buchner, B., and G. G. Curtius, U. S. 3,286,001 (1966); C. A. 66, 18508 u (1967).
129. Buchner, B., and E. Jacoves, U. S. 3,328,494 (1967); C. A. 67, 73440 m (1967).
130. Buchner, B., G. G. Kertesz, and A. F. Jackson, J. Org. Chem., 27, 1051 (1962).
131. Buck, A. C., J. D. Bartleson, and H. P. Lankelma, J. Am. Chem. Soc., 70, 744 (1948).
132. Buck, A. C., and H. P. Lankelma, J. Am. Chem. Soc., 70, 2396, 2398 (1948).
133. Buina, N. A., I. A. Nuretdinov, and N. P. Grechkin, Izv. Akad. Nauk SSSR, Ser. Khim., 1967, 1606; C. A. 68, 12594 p (1968).
133a. Bullen, A. C., J. Chem. Soc., 1962, 3201.
134. Bunyan, P. J., and J. I. G. Cadogan, J. Chem. Soc., 1962, 1304.
135. Burgada, R., Bull. Soc. Chim. France, 1964, 1735.
136. Burgada, R., G. Martin, and G. Mavel, Bull. Soc. Chim. France, 1963, 2154.
137. Burn, A. J., and J. I. G. Cadogan, J. Chem. Soc., 1961, 5532.
138. Cadogen, J. I. G., J. Chem. Soc., 1957, 1079.
139. Cadogan, J. I. G., R. K. Mackie, and J. A. Maynard, J. Chem. Soc. C., 1967, 1356.
140. Cadogan, J. I. G., and H. N. Moulden, J. Chem. Soc., 1961, 3079.
141. Caldwell, J. R., and J. C. Martin, U. S. 2,882,294 (1956), 2,976,267 (1958); C. A. 54, 1443 (1960).
142. California Research Corp., Neth. 6,501,006; C. A. 64, 6560 h (1966).
143. Carayon-Gentil, A., et al., Bull. Soc. Chim. Biol., 49, 873 (1967); C. A. 68, 22010 j (1968).
144. Cates, L. A., J. Med. Chem., 10, 924 (1967).
145. Cates, L. A., and N. M. Ferguson, J. Pharm. Sci., 53, 973 (1964).

146. Cates, L. A., and N. M. Ferguson, J. Pharm. Sci., 54, 331, 465 (1965).
147. Cates, L. A., and T. E. Jones, J. Am. Pharm. Assoc., 48, 547 (1959).
148. Cates, L. A., and T. E. Jones, J. Pharm. Sci., 53, 691 (1964).
149. Caven, R. M., J. Chem. Soc., 81, 1362 (1902).
150. Caven, R. M., J. Chem. Soc., 83, 1045 (1903).
151. Chabrier, P., and A. Carayon-Gentil, Produit. Pharmac., 17, 397 (1962).
152. Chambers, R. W., J. G. Moffatt, and H. G. Khorana, J. Am. Chem. Soc., 79, 4240 (1957).
153. Chambers, R. W., and J. G. Moffatt, J. Am. Chem. Soc., 80, 3752 (1958).
154. Chambers, R. W., and H. G. Khorana, J. Am. Chem. Soc., 80, 3749 (1958).
154a. Chapman, A. C., W. S. Holmes, N. L. Paddock, and H. T. Searle, J. Chem. Soc., 1961, 1825.
155. Cheymol, J., P. Chabrier, M. Selim, and T. N. Thanh, Compt. Rend. Hebdomadaires de l'Acad. des Sci., 249, 1240 (1959), and 251, 550 (1960).
155a. Cheymol, J., P. Chabrier, and M. Selim, Compt. Rend., 251, 1171 (1960).
156. Cheymol, J., Fr. 1,239,989 (1959); C. 1964, 19-2274; C. A. 58, 1345 h (1963).
157. Chittenden, R. A., and L. C. Thomas, Spectrochim. Acta, 20, 1679 (1964).
158. Chittenden, R. A., and L. C. Thomas, Spectrochim. Acta, 22, 1449 (1966).
159. Chrzaszczewska, A., and J. Kaczan, Lodz. Towarz. Nauk Wydzial III, Acta Chim., 9, 227 (1964); C. A. 62, 7625 f (1965).
160. Clark, V. M., G. W. Kirby, and A. R. Todd, J. Chem. Soc., 1957, 1497; 1958, 3039.
161. Clark, V. M., and A. R. Todd, J. Chem. Soc., 1950, 2030.
162. Cook, H. G., H. McCombie, and B. C. Saunders, J. Chem. Soc., 1945, 873.
163. Cook, H. G., J. D. Ilett, B. C. Saunders, G. J. Stacey, J. G. Watson, I. G. E. Wilding, and S. J. Woodcock, J. Chem. Soc., 1949, 2921.
164. Coover, H. W., Jr., U. S. 2,756,250 (1953); C. A. 51, 1275 e (1957).
165. Coover, H. W., Jr., and R. L. McConnell, U. S. 2,798,086 (1956); C. A. 51, 16535 a (1957).
166. Corby, N. S., G. W. Kenner, and A. R. Todd, J. Chem. Soc., 1952, 1234, 3669.
167. Cosmatos, A., I. Photaki, and L. Zervas, Chem. Ber., 94, 2644 (1961).
168. Coult, D. B., and M. Green, J. Chem. Soc., 1964, 5478.

169. Cowley, A. H., and R. P. Pinnell, J. Am. Chem. Soc., 87, 4454 (1965).
170. Cramer, F., Angew. Chem., 72, 236 (1960).
170a. Cramer, F. D., and K. G. Gärtner, Chem. Ber., 91, 1562 (1958).
171. Cramer, F., and H. Neunhoeffer, Chem. Ber., 95, 1664 (1962).
172. Cramer, F., H. Schaller, and H. A. Staab, Chem. Ber., 94, 1612 (1961).
173. Cramer, F., E. Scheiffele, and A. Vollmer, Chem. Ber., 95, 1670 (1962).
174. Cramer, F., and A. Vollmer, Chem. Ber., 91, 911, 919 (1958).
175. Cramer, F., and A. Vollmer, Chem. Ber., 92, 392 (1959).
176. Cramer, F., and R. Wittmann, Chem. Ber., 94, 322, 328, 1634 (1961).
177. Crunden, E. W., and R. F. Hudson, J. Chem. Soc., 1962, 3591.
178. Curtis, G. G., and B. Buchner, U. S. 3,286,002 (1966); C. A. 66, 28390 p (1967).
179. David, W. A., and B. A. Kilby, Nature, 164, 522 (1949).
180. Debo, A., Brit. 858,453 (1950); C. 1962, 11736.
181. Debo, A., Ger. 1,033,201/202 (1957); C. 1959, 1324 and 2967.
182. Debo, A., Ger. 1,033,200 and 1,034,173 (1956); C. 1959, 301, 3641.
183. Debo, A., Ger. 1,040,027 and 1,041,956 (1957); C. 1959, 6636, 6958.
184. Debo, A., Ger. 1,041,044/1,050,766 (1957); C. 1959, 12387 and 1960, 994; Ger. 1,042,582 (1956); C. 1959, 10064.
185. Debo, A., Ger. 1,042,581 (1956); C. 1959, 4307.
186. Debo, A., Ger. 1,047,781 (1957); C. 1959, 13361.
187. Debo, A., Ger. 1,048,917 (1956); C. 1959, 11069.
188. Debo, A., Ger. 1,048,918 (1957) and 1,048,919 (1958); C. 1959, 15506 and 11732.
189. Debo, A., Ger. 1,067,432 (1957); C. 1960, 7984.
190. Derkach, G. I., Zh. Obshch. Khim., 29, 241 (1959); C. A. 53, 21750 d (1959).
191. Derkach, G. I., Angew. Chem., 81, 407 (1969).
192. Derkach, G. I., Z. Chem., 9, 369 (1969).
193. Derkach, G. I., E. S. Gubnitskaya, V. A. Shokol, and A. V. Kirsanov, Zh. Obshch. Khim., 32, 1201 (1962); C. A. 58, 5721 a (1963).
194. Derkach, G. I., L. D. Protsenko, L. P. Zhuravleva, and A. V. Kirsanov, Zh. Obshch. Khim., 32, 2992 (1962); C. A. 58, 9003 e (1963).

195. Derkach, G. I., L. I. Samarai, A. S. Shtepanek, and A. V. Kirsanov, Zh. Obshch. Khim., 32, 3759 (1962); C. A. 58, 13985 e (1963).
196. Derkach, G. I., E. S. Gubnitskaya, L. I. Samarai, and V. A. Shokol, Zh. Obshch. Khim., 33, 557 (1963); C. A. 59, 2684 (1963).
197. Derkach, G. I., V. P. Rudavskii, and G. F. Dregval, Zh. Obshch. Khim., 34, 3959 (1964); C. A. 62, 9046 (1965).
198. Derkach, G. I., A. V. Narbut, and A. V. Kirsanov, Probl. Organ. Sinteza Akad. Nauk SSSR, Otd. Obshch. i. Tekhn. Khim., 1965, 278 and 268; C. A. 64, 12580 b, 12577 c (1966).
199. Derkach, G. I., E. S. Gubnitskaya, M. V. Kolotilo, and A. G. Matyusha, Zh. Obshch. Khim., 36, 2215 (1966); C. A. 66, 75980 r (1967).
200. Derkach, G. I., V. P. Rudavskii, A. A. Koval, and V. I. Shevchenko, Zh. Obshch. Khim., 37, 445 (1967); C. A. 67, 32742 t (1967).
201. Derkach, G. I., V. P. Rudavskii, and B. F. Malichenko, Khim. Org. Soedin, Fosfora, Akad. Nauk SSSR, Otd. Obshch. Tekh. Khim., 1967, 75; C. A. 69, 10169 a (1968).
202. Derkach, G. I., M. I. Bukovskii, S. N. Solodushenkov, and A. I. Mosiichuk, Khim. Org. Soedin. Fosfora, Akad. Nauk SSSR, Otd. Obshch. Tekh. Khim., 1967, 89, 93; C. A. 69, 36082 s, 36083 t (1968).
202a. Derkach, G. I., V. A. Shokol, L. I. Samarai, and A. V. Kirsanov, Zh. Obshch. Khim., 32, 159 (1962); C. A. 57, 12379 (1962).
203. Derkach, G. I., G. F. Dregval, and A. V. Kirsanov, Zh. Obshch. Khim., 30, 3402 (1960); 31, 2385 (1961); 32, 154,3002 (1962); C. A. 55, 21029 b (1961); 56, 3405 d (1962), 57, 13668 f (1962); 58, 8964 f (1963). Derkach, G. I., V. A. Shokal, and A. V. Kirsanov, Zh. Obshch. Khim., 31, 2275 (1961); C. A. 56, 4667 i (1962).
204. Derkach, G. I., G. F. Dregval, and A. V. Kirsanov, Zh. Obshch. Khim., 30, 3402 (1960); C. A. 55, 21029 b (1961).
205. Derkach, G. I., G. F. Dregval, and A. V. Kirsanov, Zh. Obshch. Khim., 31, 2385 (1961); C. A. 56, 3405 d (1962).
206. Derkach, G. I., G. F. Dregval, and A. V. Kirsanov, Zh. Obshch. Khim., 32, 154 (1962); C. A. 57, 13668 f (1962).
207. Derkach, G. I., G. F. Dregval, and A. V. Kirsanov, Zh. Obshch. Khim., 32, 1878 (1962); C. A. 58, 4535 a (1963).
208. Derkach, G. I., and E. S. Gubnitskaya, Zh. Obshch. Khim., 34, 604 (1964); C. A. 60, 13268 h (1964).

209. Derkach, G. I., and E. S. Gubnitskaya, Zh. Obshch. Khim., 35, 1009 (1965); C. A. 63, 9801 b (1965).
210. Derkach, G. I., E. S. Gubnitzkaya, and A. V. Kirsanov, Zh. Obshch. Khim., 31, 3679 (1961); C. A. 57, 9876 g (1962).
211. Derkach, G. I., E. S. Gutnitzkaya, and A. V. Kirsanov, Zh. Obshch. Khim., 31, 3746 (1961); C. A. 57, 9735 e (1962).
212. Derkach, G. I., E. S. Gutnitzkaya, V. A. Shokal, and A. A. Kisilenko, Zh. Obshch. Khim., 34, 82 (1964); C. A. 60, 11502 d (1964).
213. Derkach, G. I., Zh. M. Ivanova, and N. I. Liptuga, Khim. Org. Soedin. Fosfora, Akad. Nauk SSSR, Otd. Obshch. Tekh. Khim., 1967, 72; C. A. 69, 35212 x (1968).
214. Derkach, G. I., A. V. Kirsanov, Zh. Obshch. Khim., 29, 3424 (1959); C. A. 54, 15277 b (1960).
215. Derkach, G. I., and A. V. Kirsanov, Zh. Obshch. Khim., 32, 2254 (1962); C. A. 58, 9126 a (1963).
216. Derkach, G. I., and A. A. Kisilenko, Zh. Obshch. Khim., 34, 3060 (1964); C. A. 62, 2692 (1965).
217. Derkach, G. I., and M. V. Kolotilo, Zh. Obshch. Khim., 36, 82 (1966); C. A. 64, 14127 h (1966).
218. Derkach, G. I., and M. V. Kolotilo, Zh. Obshch. Khim., 36, 1437 (1966); C. A. 66, 2292 a (1967).
219. Derkach, G. I., V. V. Kruzement, and A. V. Kirsanov, Zh. Obshch. Khim., 31, 2391 (1961); C. A. 56, 2374 h (1962).
220. Derkach, G. I., and A. M. Lepesa, Zh. Obshch. Khim., 34, 525 (1964); C. A. 60, 13179 d (1964); C. A. 61, 1784 g (1964).
221. Derkach, G. I., A. M. Lepesa, and A. V. Kirsanov, Zh. Obshch. Khim., 31, 3424 (1961); C. A. 57, 3353 d (1962).
222. Derkach, G. I., A. M. Lepesa, and A. V. Kirsanov, Zh. Obshch. Khim., 32, 2600 (1962); C. A. 58, 7864 e (1963).
223. Derkach, G. I., and N. I. Liptuga, Zh. Obshch. Khim., 36, 461 (1966); C. A. 65, 634 h (1966).
224. Derkach, G. I., and S. K. Mikhailik, Khim. Org. Soedin. Fosfora, Akad. Nauk SSSR, Otd. Obshch. Tekh. Khim., 1967, 59; C. A. 69, 2757 a (1968).
225. Derkach, G. I., and A. V. Narbut, Zh. Obshch. Khim., 35, 1006 (1965); C. A. 63, 9857 g (1965).
226. Derkach, G. I., and A. V. Narbut, Zh. Obshch. Khim., 36, 322 (1966); C. A. 64, 15738 c (1966).
227. Derkach, G. I., and A. V. Narbut, Zh. Obshch. Khim., 37, 1364 (1967); C. A. 67, 108401 d (1967).
228. Derkach, G. I., A. V. Narbut, and A. V. Kirsanov, Zh. Obshch. Khim., 33, 1584 (1963); C. A. 59, 12669 g (1963).

229. Derkach, G. I., A. V. Narbut, and A. V. Kirsanov, Zh. Obshch. Khim., 33, 3062 (1963); C. A. 60, 1622 g (1964).
230. Derkach, G. I., and V. P. Rudavskii, Zh. Obshch. Khim., 35, 1202 (1965); C. A. 63, 11398 a (1965).
231. Derkach, G. I., and V. P. Rudavskii, Zh. Obshch. Khim., 35, 2200 (1965); C. A. 64, 12580 g (1966).
232. Derkach, G. I., and V. P. Rudavskii, Zh. Obshch. Khim., 37, 1893 (1967); C. A. 68, 21489 k (1968).
233. Derkach, G. I., and L. I. Samarai, Zh. Obshch. Khim., 33, 1587 (1963); C. A. 59, 12669 d (1963).
234. Derkach, G. I., V. A. Shokol, and A. V. Kirsanov, Zh. Obshch. Khim., 30, 3393 (1960); C. A. 55, 22220 i (1961).
235. Derkach, G. I., and E. I. Slyusarenko, Zh. Obshch. Khim., 35, 532 (1965); C. A. 63, 531 f (1965).
236. Derkach, G. I., and E. I. Slyusarenko, Zh. Obshch. Khim., 35, 2220 (1965); C. A. 64, 11082 b (1966).
237. Derkach, G. I., and E. I. Slyusarenko, Zh. Obshch. Khim., 36, 1639 (1966); C. A. 66, 55135 r (1967).
238. Derkach, G. I., and E. I. Slyusarenko, Zh. Obshch. Khim., 37, 2069 (1967); C. A. 68, 59664 b (1968).
239. Derkach, G. I., I. N. Zhmurova, A. V. Kirsanov, V. I. Shevchenko, and A. S. Shtepanek, Phosphazo-Compounds, Izdatelstro "Naukova Dumka", Kiev, 1965.
240. Derkach, G. I., L. P. Zhuraleva, and A. V. Kirsanov, Zh. Obshch. Khim., 32, 879 (1962); C. A. 58, 2388 b (1963).
241. Deutsch, A., and O. Fernö, Nature, 156, 604 (1945).
242. Deutsche Gold-und Silberscheideanstalt, Belg. 635,690 (1963); C. A. 61, 10693 e (1964).
243. Deutsche Gold-und Silberscheideanstalt, Belg. 665,127 (1965); C. A. 64, 19681 f (1966).
244. Deutsche Gold-und Silberscheideanstalt, Neth. 6,402,886 (1964); C. A. 62, 14704 h (1965); Neth. 6,403,732 (1965); C. A. 62, 16275 d (1965); Ger. 1,182,667 (1965); C. A. 62, 7781 g (1965).
245. Dick, J., M. Neascu, and A. Lupea, Rev. Roumaine Chim., 11, 517 (1966); C. A. 65, 5460 d (1966).
246. Dickey, J. B., and J. B. Normington, U.S. 2,299,535 (1942).
247. Diery, H., Ger. 1,083,267 (1958); C. 1960, 17540.
248. Diery, H., and J. Heyna, Ger. 1,048,583 (1959); C. 1960, 17207.
249. Dilaris, I., Chem. Ber., 91, 833 (1958).
250. Dixon, A. E., J. Chem. Soc., 79, 541 (1901); 93, 2148 (1908).
251. Dorn, H., Monatsber. Deutsch. Akad. Wiss. Berlin, 3, 683 (1961); C. A. 58, 5666 f (1963).

251a. Dougill, M. W., J. Chem. Soc., 1961, 5471; 1963, 3211; Wilson, A., and D. F. Carroll, J. Chem. Soc., 1960, 2548; Hazekamp, R., T. Migchelsen, and A. Vos, Acta Crystallogr. [Copenhagen], 15, 539 (1962).
252. Dow Chemical Co., Fr. 1,360,901 (1964); C. A. 61, 16052 (1964).
253. Dow Chemical Co., Ger. 1,224,312 (1966); C. A. 65, 10057 (1966).
254. Drach, B. S., and A. D. Sinitsa, Zh. Obshch. Khim., 38, 1325 (1968); C. A. 69, 66827 f (1968).
255. Drake, L. R., et al., U. S. 2,552,536/537/538/539/540/541 (1951); C. A. 46, 136 b (1952).
256. Dubina, V. L., and S. I. Burmistrov, Ukr. Khim. Zh., 33, 719 (1967); C. A. 67, 108366 w (1967).
257. Dunbar, J. E., U. S. 2,907,775 and 2,950,308 (1958/59); C. A. 54, 2260 i (1960); 55, 2590 i (1961).
258. Dye, W. T. Jr., U. S. 2,610,139 (1950); C. A. 47, 5424 h (1953).
258a. Ebel, J. P., Compt. Rend., 234, 621 (1952).
258b. Ebel, J. P., Soc. Chim. France, 10, 991 (1953).
258c. Ebel, J. P., and Y. Volmar, Compt. Rend., 233, 415 (1951).
259. Eckelmann, A., H. J. Renner, J. Weissflog, and H. Damm, Ger. East 15,457 (1956); C. 1960, 2327.
260. Edmundson, R. S., Chem. Ind., 1963, 784.
261. Edmundson, R. S., Tetrahedron, 20, 2781 (1964).
262. Edmundson, R. S., Tetrahedron, 21, 2379 (1965).
263. Edmundson, R. S., J. Chem. Soc., C. 1969, 2730.
264. Edmundson, R. S., and A. J. Lambie, J. Chem. Soc., C 1966, 1997.
265. Edmundson, R. S., and A. J. Lambie, J. Chem. Soc., B 1967, 577.
266. Eggerer, H., Chem. Ber., 94, 174 (1961).
267. Eiden, F., and E. Schoenduve, Tetrahedron Letters, 1967, 2119.
268. Elmore, D. T., J. R. Ogle, J. Chem. Soc., 1959, 2286.
269. Emeleus, H. J., and T. Onak, J. Chem. Soc., 1966, 1291.
270. England, D. C., R. V. Lindsey Jr., and L. R. Melby, J. Am. Chem. Soc., 80, 6442 (1958).
271. Ennor, A. H., and L. A. Stocken, Biochem. Prepar., 5, 9 (1957).
272. Ephraim, F., Ber. Dtsch. Chem. Ges., 44, 631 (1911).
273. Ephraim, F., Ber. Dtsch. Chem. Ges., 44, 3414 (1911).
274. Ephraim, F., and M. Sackheim, Ber. Dtsch. Chem. Ges., 44, 3416 (1911).
274a. Ephraim, F., and M. Sackheim, Ber. Dtsch. Chem. Ges., 57, 1364 (1924).

275. Erbel, A. I., and E. E. Kenaga, U. S. 2,776,311 (1957); C. A. 51, 6690 g (1957).
276. Etablissements Kuhlmann, Fr. M4258 (1966); C. A. 68, 68984 h (1968).
277. Eto, M., K. Kobayashi, T. Kato, K. Kojima, and Y. Oshima, Agr. Biol. Chem., 29, 243 (1965); C. A. 63, 4298 h (1965).
278. Eto, M., T. Eto, and Y. Oshima, Agr. Biol. Chem., 26, 630 (1962); C. A. 59, 3269 f (1963).
279. Ettel, V., J. Myska, and J. Bestova-Zavorkova, Sb. Vysoke Skoly. Chem. Technol. Praze Org. Technol., 5, 211 (1961); C. A. 61, 14561 h (1964).
280. Farbenfabriken Bayer A. G., Belg. 628,173 (1963); C. A. 61, 1874 e (1964).
281. Farbenfabriken Bayer A. G., Brit. 913,996 (1962); C. A. 59, 514 b (1963).
282. Farbenfabriken Bayer A. G., Brit. 991,979 (1965); C. A. 64, 757 c (1966).
283. Farbenfabriken Bayer A. G., Fr. 1,377,118 (1964); C. A. 62, 9011 f (1965).
283a. Farbenfabriken Bayer A. G., Ger. 1,158,971 (1961).
284. Farbenfabriken Bayer A. G., Ger. 1,173,469 (1964); C. A. 61, 11892 h (1964).
285. Farbenfabriken Bayer A. G., Ger. 1,187,612 (1965); C. A. 62, 16119 c (1965), or C. 1965, 43-2387.
286. Farbenfabriken Bayer A. G., Ger. 1,195,750 (1965); C. A. 63, 11371 e (1965).
287. Farbenfabriken Bayer A. G., Ger. 1,215,144 (1966); C. A. 65, 3908 c (1966).
288. Farbenfabriken Bayer A. G., Ger. 1,220,411 (1966); C. A. 65, 8820 b (1966).
289. Farbenfabriken Bayer A. G., Ger. 1,021,526 (1956); and 1,025,086 (1956); C. 1958, 12287, and 1959, 4350.
290. Farbenfabriken Bayer A. G., Ger. 1,104,958 (1960); C. A. 56, 10041 e (1962).
291. Farbenfabriken Bayer A. G., Ger. 1,122,075 (1958); C. 1962, 11733.
292. Farbenfabriken Bayer A. G., Neth. 6,508,556 (1966); C. A. 65, 7058 b (1966); and Belg. 666,143 (1965); C. A. 65, 16864 e (1966).
293. Farbenfabriken Bayer A. G., Neth. 6,600,465 (1966); C. A. 66, 10578 v (1967).
294. Farbwerke Hoechst, Belg. 610,695 (1961); C. A. 58, 505 b (1963).
295. Farbwerke Hoechst, Brit. 906,428 (1962); C. A. 58, 11329 c (1963).
296. Farbwerke Hoechst, Fr. 1,331,110 (1963); C. A. 60, 407 d (1964).
297. Fazaw and Seraidarian, J. Biol. Chem., 165, 97 (1946).

298. Federova, O. N., and P. I. Alimov, Izv. Akad. Nauk SSSR, Ser. Khim., 1967, 1348; C. A. 67, 108125 s (1967).
299. Feichtinger, H., and H. Tummes, Ger. 930,210 (1953); C. 1956, 5948.
300. Feldman, I. K., and A. I. Berlin, Zh. Obshch. Khim., 32, 1604, 3379 (1962); C. A. 58, 12563 e (1963).
301. Feldman, I. K., and A. I. Berlin, Zh. Obshch. Khim., 32, 575 (1962); C. A. 58, 6820 e (1963).
302. Ferrari, G., Brit. 894,820 (1959); C. A. 57, 15074 f (1962).
303. Fielding, H. C., Brit. 891,861 (1959); C. A. 57, 7170 e (1962).
304. Fielding, H. C., Brit. 916,279 (1963); C. A. 59, 2714 e (1963).
304a. Fielding, H. C., Ger. 1,135,456 (1960); C. A. 57, 7170 (1962).
305. Filatova, I. M., et al., Zh. Obshch. Khim., 38, 1304 (1968); C. A. 69, 87091 b (1968).
305a. Fischer, F., and W. Kapitza, Ger. East 36,523 (1965); C. A. 63, 11362 e (1965).
306. Fitch, H. M., U. S. 2,895,981 (1952); C. A. 54, 416 f (1960).
307. Fitch, H. M., U. S. 2,901,503 (1956); C. A. 54, 1431 b (1960).
307a. Fitzsimmons, B. W., C. Hewlett, and R. A. Shaw, J. Chem. Soc., 1964, 4459.
307b. Fitzsimmons, B. W., C. Hewlett, and R. A. Shaw, J. Chem. Soc., 1965, 7432.
307c. Fitzsimmons, B. W., C. Hewlett, K. Hills, and R. A. Shaw, J. Chem. Soc., (A) 1967, 679.
308. Flint and Salzberg, U. S. 2,151,380 (1939).
309. Fluck, E., Z. Naturforsch., 19b, 869 (1964).
309a. Fluck, E., "Die kernmagnetische Resonanz und ihre Anwendung in der anorganischen Chemie," Springer-Verlag, Heidelberg, 1963.
309b. Fluck, E., and H. Binder, in "Analytical Chemistry of Phosphorus and its Compounds," ed. M. Halman, Interscience Publ., New York, 1971.
310. Fluck, E., and H. Binder, Z. Anorg. Allg. Chem., 359, 102 (1968).
311. Fluck, E., H. Binder, and F. L. Goldmann, Z. Anorg. Allg. Chem., 338, 58 (1965).
312. Fluck, E., and F. L. Goldmann, Z. Anorg. Allg. Chem., 356, 307 (1968).
312a. Fluck, E., and W. Steck, unpublished.
313. Forbes, G. S., and H. H. Anderson, J. Am. Chem. Soc., 65, 2271 (1943).
314. Friedman, O. M., F. B. Papanastassiou, R. S. Levi, H. R. Till, Jr., and W. M. Whaley, J. Med. Chem., 6, 82 (1963); C. A. 58, 13958 a (1963).

315. Friedman, O. M., D. L. Klass, and A. M. Seligman, J. Am. Chem. Soc., 76, 916 (1954).
316. Friedman, O. M., and A. M. Seligman, J. Am. Chem. Soc., 76, 655, 658 (1954).
317. Furdi, K. M., and J. Masek, Acta Fac. Rerum Nat. Univ. Comenianae. Chimia, 6, 611 (1961); C. A. 59, 8578 b (1963).
318. Fusco, R., Fr. 1,098,798 (1954); C. 1958, 1126; Chimica e Ind., 37, 849 (1955).
319. Gadekar, S. M., and E. Ross, J. Org. Chem., 26, 606 (1961).
320. Gardiner, J. E., and B. A. Kilby, J. Chem. Soc., 1950, 1769.
321. Garrison, A. W., and Ch. E. Boozer, J. Am. Chem. Soc., 90, 3486 (1968).
322. Gauss, W., S. Petersen, and C. Hackmann, Ger. 1,020,631 (1954); C. 1958, 13305.
323. Gilpin, Am. Chem. J., 27, 444 (1902).
324. Gilyarov, V. A., Akad. Nauk SSSR, Trudy 1.oi Konferent, 1955, 275; C. A. 52, 243 c (1958).
325. Gilyarov, V. A., and M. I. Kabachnik, Zh. Obshch. Khim., 36, 282 (1966); C. A. 64, 15790 b (1966).
326. Gilyarov, V. A., R. V. Kudryavtsev, and M. I. Kabachnik, Zh. Obshch. Khim., 36, 708 (1966); C. A. 65, 8798 e (1966).
327. Gilyarov, V. A., R. V. Kudryavtsev, and M. I. Kabachnik, Zh. Obshch. Khim., 38, 352 (1968); C. A. 69, 96097 k (1968).
328. Gilyarov, V. A., E. N. Tsvetkov, and M. I. Kabachnik, Zh. Obshch. Khim., 36, 274 (1966); C. A. 64, 17408 e (1966).
329. Ginsburg, V. A., Zh. Obshch. Khim., 30, 2854 (1960); C. A. 55, 17477 b (1961).
330. Godfrey, K. L., U. S. 2,852,550 (1958); C. A. 53, 4131 i (1959).
331. Godovikov, N. N., and M. I. Kabachnik, Zh. Obshch. Khim., 31, 1628 (1961); C. A. 55, 22200 f (1961).
332. Goehring, M., and K. Niedenzu, Angew. Chem., 68, 704 (1956).
333. Goehring, M., and K. Niedenzu, Chem. Ber. 89, 1768 (1956).
334. Goerdeler, J., and H. Ullmann, Chem. Ber., 94, 1067 (1961).
335. Goetz, T., Ger. 855,248 (1952); C. A. 48, 11841 b (1954).
336. Goldwhite, H., and B. C. Saunders, Chem. and. Ind., 1956, 663.
337. Goldwhite, H., and B. C. Saunders, J. Chem. Soc., 1957, 2409.

338. Gololobov, Y. G., T. F. Dmitrieva, Yu. M. Zinov'ev, and L. Z. Soborovskii, Zh. Obshch. Khim., 35, 1460 (1965); C. A. 63, 14689 c (1965).
339. Goubeau, J., Angew. Chem., 78, 565 (1966) and 81, 343 (1969).
340. Grechkin, N. P., Izv. Akad. Nauk SSSR, 1956, 538; C. A. 51, 1933 a (1957); C. A. 52, 241 f (1958).
341. Grechkin, N. P., Izv. Akad. Nauk SSSR, 1957, 1053; C. A. 52, 6157 d (1958).
342. Grechkin, N. P., and R. R. Shagidullin, Izv. Akad. Nauk SSSR, 1960, 2135, and 1962, 295; C. A. 55, 16409 (1961), and C. A. 57, 11135 d (1962).
343. Grechkin, N. P., and G. S. Bobchenko, Dokl. Akad. Nauk SSSR, 129, 569 (1959); C. A. 54, 7671 (1960).
344. Grechkin, N. P., and L. N. Grishina, Dokl. Akad. Nauk SSSR, 146, 1333 (1962); C. A. 58, 90004 f (1955).
345. Grechkin, N. P., and L. N. Grishina, Izv. Akad. Nauk SSSR, Ser. Khim., 1967, 2120; C. A. 68, 39381 d (1968).
346. Grechkin, N. P., and R. N. Khamitov, Dokl. Akad. Nauk SSSR, 162, 1063 (1965); C. A. 63, 6939 e (1965).
347. Grechkin, N. P., and I. A. Nuretdinov, Izv. Akad. Nauk SSSR, 1963, 302; C. A. 59, 5194 (1963).
348. Green, B. S., D. B. Sowerby, and K. J. Wihksne, Chem. and Ind. 1960, 1306; Sowerby D. B., J. Inorg. Nucl. Chem., 22, 205 (1961).
348a. Green, M., R. N. Haszeldine, and G. S. A. Hopkins, J. Chem. Soc. A, 1966, 1766.
349. Greenhalgh, R., and J. R. Blanchfield, Can. J. Chem., 44, 501 (1966).
349a. Grunze, H., and E. Thilo, "Die Papierchromatographie der kondensierten Phosphate," Akademie-Verlag, Berlin, 1954.
350. Gutmann, V., Ch. Kemenater, and K. Utvary, Mh. Chem., 96, 836 (1965).
351. Halman, M., A. Lapidot, and D. Samuel, J. Chem. Soc., 1963, 1299.
352. Hamer, N. K., J. Chem. Soc., 1964, 1961.
353. Hamer, N. K., J. Chem. Soc., 1965, 46.
354. Hamer, N. K., J. Chem. Soc., 1965, 2731.
355. Hamer, N. K., J. Chem. Soc., C 1966, 404.
356. Hamer, N. K., Chem. Comm., 1967, 758.
357. Hamm, P. C., and G. A. Saul, U. S. 3,020,141 (1958); C. A. 57, 2075 e (1962).
358. Handley, Th. H., Talanta, 12, 983 (1965).
359. Handley, Th. H., Anal. Chem., 36, 2467 (1964).
359a. Hartley, G. S., and D. W. Pound, U. S. 652,981 (1948); C. A. 46, 1025 (1952).

360. Hartley, G. S., and D. W. Pound, U. S. 2,649,464 (1950); C. A. 48, 10050 d (1954).
360a. Hartley, S. B., W. S. Holmes, J. K. Jacques, M. F. Mole, and J. C. McCoubrey, Quart. Rev., 17, 204 (1963).
361. Hashimoto, Sh., I. Furukawa, and Sh. Tanibuchi, Doshisha Daigaku Rikogaku Kenkyu Kokoku, 8, 68-77 (1967); C. A. 68, 40154 p (1968).
361a. Haubold, W., and E. Fluck, unpublished.
362. Heap, R., and B. C. Saunders, J. Chem. Soc., 1948, 1313.
363. Hemmelmayr, F. von., Monatsh., 26, 772 (1905).
364. Hinsberg, O., Ber. Dtsch. Chem. Ges., 27, 2178 (1894).
364a. Hobbs, E., D. E. C. Corbridge, and B. Raistrick, Acta Crystallogr. [Copenhagen], 6, 621 (1953); Cruickshank, D. W. J., J. Chem. Soc., 1961, 5486.
365. Ho Chien and R. J. Kurland, J. Biol. Chem., 241, 3002 (1966).
366. Hoffmann La Roche Co., Fr. 853,666 (1938); Swiss. 211,297, 213,196, 214,335 (1938); U. S. 2,287, 154/59 (1939); C. A. 36, 26886, 36329, 46733, 49749 (1942); C. A. 37, 2312 (1943).
367. Hoffmann La Roche Co., Swiss. 206,549 (1931) and 214,335 (1941).
368. Holmes, R. R., and J. A. Forster, Inorg. Chem., 2, 380 (1963), and 1, 89 (1962).
369. Holmstedt, B., Acta Physiol. Scand., 25, Suppl. 90, 26 (1951).
370. Holmstedt, B., Acta Physiol. Scand., 25, Suppl. 90, 120 (1951).
371. Horn, H. G., Z. Naturforsch., 21 b, 617 (1966).
372. Horn, H. G., and O. Glemser, Chem. Ber., 100, 2258 (1967).
373. Horn, H. G., and A. Mueller, Z. Naturforsch., 21 b, 729 (1966).
374. Houben-Weyl, Methoden der Organischen Chemie, XII/2, Georg-Thieme Verlag, Stuttgart 1963.
375. Hunger, K., Brennst.-Chem., 49, 201 (1968); C. A. 69, 106813 g (1968).
376. I. G. Farbenindustrie A. G., Fr. 807,769 (1937), Ger. 664,438 (1938); Winthrop Chemical Co., U. S. 2,146,356 (1939).
376a. Ikehara, M., and E. Ohtsuka, Chem. Pharm. Bull. Tokyo, 10, 997 (1962); C. A. 59, 4021 (1963).
377. Il'ina, N. A., A. P. Lukovnikowa, and V. S. Abramov, Tr. Kazansk. Khim. Tekhnol Inst., 30, 40 (1962); C. A. 60, 7908 g (1964).
378. Imperial Chemical Ind. Ltd., Belg. 591,707 (1960), and 612,359 (1962); C. A. 57, 13411 g (1962).

379. Imperial Chemical Ind. Ltd., Brit. 968,886; C. A. 61, 13194 e (1964).
380. Imperial Chemical Ind. Ltd., Fr. 1,378,035 (1964); C. A. 62, 10371 b (1965).
381. Ishida, H., S. Asaka, and Y. Kawamura, Japan. 16,465 (1963) and 16,469 (1963); C. A. 60, 2828 e (1964).
382. Ishida, H., S. Asaka, and Y. Kawamura, Japan. 23,181 (1963); C. A. 60, 5520 e (1964).
383. Ishida, H., S. Asaka, and Y. Kawamura, Japan. 2476 (1964); C. A. 60, 15781 c (1964).
384. Ivanova, Zh. M., Zh. Obshch. Khim., 35, 164 (1965); C. A. 62, 13026 h (1965).
385. Ivanovan, Zh. M., E. S. Levchenko, and A. V. Kirsanov, Zh. Obshch. Khim., 35, 1607 (1965); C. A. 63, 17949 h (1965).
386. Ivanova, Zh. M., E. A. Stukalo, and G. I. Derkach, Zh. Obshch. Khim., 37, 1144 (1967); C. A. 68, 2664 x (1968).
387. Ivanova, Zh. M., E. A. Stukalo, and G. I. Derkach, Zh. Obshch. Khim., 38, 551 (1968); C. A. 69, 67029 j (1968).
388. Iwamoto, R. H., E. M. Acton, L. Goodman, and B. R. Baker, J. Org. Chem., 26, 4743 (1961).
389. Jackson, E. L., J. Org. Chem., 9, 457 (1944).
390. Jaeger, A., and H. Werres, Ger. 1,154,669 (1960); C. 1964, 242167.
391. Jagupolskij, L. M., and V. I. Troitskaja, Zh. Obshch. Khim., 29, 552 (1959); C. A. 54, 356 (1960).
392. Joesten, M. D., and K. M. Nykerk, Inorg. Chem., 3, 548 (1964).
393. Johnson, T. B., Am. Chem. J., 34, 200 (1905).
394. Jones, R. A. Y., and A. R. Katritsky, Angew. Chem., 74, 60 (1962).
395. Kabachnik, M. I., Zh. Obshch. Khim., 29, 2182 (1959); C. A. 54, 10830 c (1960).
396. Kabachnik, M. I., E. I. Golubeva, D. M. Paikin, M. P. Shabanova, N. M. Gamper, and L. F. Efimova, Zh. Obshch. Khim., 29, 1680 (1959), C. A. 54, 8594 i (1960).
397. Kabachnik, M. I., P. A. Rossiiskaya, M. P. Shabanova, D. M. Paikin, L. F. Efimova, and N. M. Gamper, Zh. Obshch. Khim., 30, 2218 (1960); C. A. 57, 4531 f (1962).
298. Kabachnik, M. I., V. A. Gilyarov, and E.M. Popov, Zh. Obshch. Khim., 32, 1598 (1962); C. A. 58, 4401 b (1963).
399. Kabachnik, M. I., V. A. Gilyarov, C. T. Chang, and E. I. Matrosov, Izv. Akad. Nauk SSSR, 1962, 1589; C. A. 58, 6673 g (1963).

400. Kabachnik, M. I., V. A. Gilyarov, and C. T. Chang, Izv. Akad. Nauk SSSR, Ser. Khim., 1965, 665; C. A. 63, 2888 f (1965).
401. Kabachnik, M. I., and V. A. Gilyarov, Dokl. Akad. Nauk SSSR, 96, 991 (1954); C. A. 49, 8842 (1955).
402. Kabachnik, M. I., and V. A. Gilyarov, Dokl. Akad. Nauk SSSR, 106, 473 (1956); C. A. 50, 13723 b (1956).
403. Kabachnik, M. I., and V. A. Gilyarov, Izv. Akad. Nauk SSSR, 1956, 790; C. A. 51, 1823 b (1957).
404. Kabachnik, M. I., and V. A. Gilyarov, Izv. Akad. Nauk SSSR, 1961, 816, 819, 1022; C. A. 55, 27014 d (1961).
405. Kabachnik, M. I., V. A. Gilyarov, and R. V. Kudryavtsev, Tetrahedron Letters, 1965, 2691; Zh. Obshch. Khim., 35, 1476 (1965); C. A. 63, 14733 b (1965); and Zh. Obshch. Khim., 36, 57 (1966); C. A. 64, 14118 (1966).
406. Kabachnik, M. I., V. A. Gilyarov, and M. M. Yusupov, Dokl. Akad. Nauk SSSR, 160, 1079 (1965); C. A. 62, 14650 h (1965).
407. Kabachnik, M. I., and T. Ya. Medved, Izv. Akad. Nauk SSSR, Ser. Khim., 1966, 1365; C. A. 66, 54802 u (1967).
408. Kamai, G., and F. M. Kharrasova, Zh. Obshch. Khim., 27, 3064 (1957); C. A. 52, 8039 i (1958).
409. Kamai, G., E. V. Kuznetsov, and R. K. Valetdinov, Trudy Kazan. Khim. Tekhnol., 23, 190 (1957); C. A. 52, 8940 f (1958).
410. Kaplan, F., G. Singh, and H. Zimmer, J. Phys. Chem., 67, 2509 (1963).
411. Kaplan, F., and C. O. Schulz, Chem. Comm., 1967, 376.
411a. Karl-Kroupa, E. F., Anal. Chem., 28, 1091 (1956).
412. Kataev, E. G., V. V. Plemenkov, and V. V. Markin, Dokl. Akad. Nauk SSSR, 165, 1313 (1965); C. A. 64, 9727 c (1966).
413. Kauer, K. C., and E. C. Britton, U. S. 2,836,612 (1958); C. A. 52, 16291 f (1958).
414. Kauer, K. C., and E. C. Britton, U. S. 2,855,422 (1958); C. A. 53, 8077 e (1959).
415. Kauer, K. C., U. S. 3,095,438 (1963); C. A. 59, 12644 f (1963).
416. Kauer, K. C., U. S. 3,162,666 (1964); C. A. 62, 10372 b (1965).
417. Kayser, H., W. Lorenz, and G. Schrader, Ger. 1,077,215 (1958); C. 1960, 16911.
418. Kayser, H., and G. Schrader, Ger. 1,121,882 (1958); C. A. 56, 8635 b (1962).
419. Kayser, H., and G. Schrader, Ger. 1,135,905 (1962); C. A. 59, 3836 h (1963).

420. Kazimirchik, I. V., G. F. Bebikh, F. S. Denisov, and M. I. Kabachnik, Zh. Obshch. Khim., 36, 1226 (1966); C. A. 65, 16889 h (1966).
421. Kaz'mina, N. B., O. V. Kil'disheva, and I. L. Knunyants, Izv. Akad. Nauk SSSR, Ser. Khim., 1964, 117; C. A. 60, 9352 c (1964).
422. Keat, R., and R. A. Shaw, J. Chem. Soc., A 1968, 703.
423. Keith, L. H., A. W. Garrison, and A. L. Alford, J. Ass. Offic. Anal. Chem., 51, 1063 (1968).
424. Keller, H., H. Netter, and B. Niemann, Hoppe Seyler, 313, 244 (1958).
425. Kenner, G. W., H. G. Khorona, and R. J. Stedman, J. Chem. Soc., 1953, 673.
426. Kenney, C. N., Brit. 872,832 (1959); C. A. 56, 2153 i (1962).
427. Kenney, C. N., Brit. 883,488 (1959); C. A. 56, 15156 d (1962).
428. Kipping, F. S. and F. Challenger, J. Chem. Soc., 99, 626 (1911).
429. Kirsanov, A. V., Zh. Obshch. Khim., 22, 269 (1952); C. A. 46, 11135 f (1952).
429a. Kirsanov, A. V., Izv. Akad. Nauk SSSR, 1952, 710; C. A. 47, 9904 f (1953).
430. Kirsanov, A. V., Zh. Obshch. Khim., 24, 1033 (1954), 29, 2256 (1959); C. A. 49, 8787 b (1955); C. A. 54, 10855 e (1960); Izv. Akad. Nauk SSSR, 1954, 646; C. A. 49, 13161 h (1955).
431. Kirsanov, A. V., and E. A. Abrazhnova, Sbornik Statei Obshch. Khim., 2, 1048 (1953); C. A. 49, 3052 a (1955).
432. Kirsanov, A. V., and E. A. Abrazhanova, Sbornik Statei Obshch. Khim., 2, 1059, 1370 (1953); C. A. 49, 4600, 5405 i (1955).
433. Kirsanov, A. V., and G. I. Derkach, Zh. Obshch. Khim., 27, 3248 (1957); C. A. 52, 8997 d (1958).
434. Kirsanov, A. V., and G. I. Derkach, Zh. Obshch. Khim., 29, 600 (1959); C. A. 54, 382 d (1960).
435. Kirsanov, A. V., and G. I. Derkach, Zh. Obshch. Khim., 26, 2009 (1956); C. A. 51, 1821 i (1957).
436. Kirsanov, A. V., and G. I. Derkach, Zh. Obshch. Khim., 26, 2631 (1956); C. A. 51, 1821 b (1957).
437. Kirsanov, A. V., and G. I. Derkach, Zh. Obshch. Khim., 27, 1080 (1957); C. A. 52, 3715 e (1958).
438. Kirsanov, A. V., and G. I. Derkach, Zh. Obshch. Khim., 28, 1887 (1958); C. A. 53, 1208 d (1959).
439. Kirsanov, A. V., and G. I. Derkach, Zh. Obshch. Khim., 28, 2247 (1958); C. A. 53, 2134 g (1959).
440. Kirsanov, A. V., G. I. Derkach, and R. G. Makitra, Zh. Obshch. Khim., 28, 1227 (1958); C. A. 52, 20002 g (1958).

441. Kirsanov, A. V., G. I. Derkach, and N. I. Liptuga, Zh. Obshch. Khim., 34, 2812 (1964); C. A. 61, 14514 b (1964).
442. Kirsanov, A. V., and N. L. Egorova, Zh. Obshch. Khim., 25, 187 (1955); C. A. 50, 1647 i (1956).
443. Kirsanov, A. V., and N. L. Egorova, Zh. Obshch. Khim., 25, 1140 (1955); C. A. 50, 3215 b (1956).
444. Kirsanov, A. V., and N. L. Egorova, Zh. Obshch. Khim., 28, 1052 (1958); C. A. 52, 18290 d (1958).
445. Kirsanov, A. V., and N. L. Egorova, Zh. Obshch. Khim., 28, 1587 (1958); C. A. 53, 1203 a (1959).
446. Kirsanov, A. V., and N. G. Feshchenko, Zh. Obshch. Khim., 27, 2817 (1957); C. A. 52, 8069 (1958).
447. Kirsanov, A. V., and N. G. Feshchenko, Zh. Obshch. Khim., 28, 339 (1958), 29, 4085 (1959); C. A. 52, 13663 d (1958); C. A. 54, 20948 f (1960).
448. Kirsanov, A. V., and N. A. Kirsanova, Zh. Obshch. Khim., 29, 1802 (1959); C. A. 54, 8693 e (1960).
449. Kirsanov, A. V., A. M. Lepesa, and G. I. Derkach, Dopovidi Akad. Nauk Ukr. RSR, 1962, 384; C. A. 58, 2402 (1963).
450. Kirsanov, A. V., and E. S. Levchenko, Zh. Obshch. Khim., 26, 2285 (1956); 27, 2585 (1957); C. A. 51, 1875 f (1957), 52, 7190 i (1958).
451. Kirsanov, A. V., and E. S. Levchenko, Zh. Obshch. Khim., 28, 1589 (1958); C. A. 53, 1219 f (1959).
452. Kirsanov, A. V., and R. G. Makitra, Zh. Obshch. Khim., 26, 905,907 (1956); C. A. 50, 14600 i and 14633 d (1956).
453. Kirsanov, A. V., and R. T. Makitra, Zh. Obshch. Khim., 27, 245 (1957); C. A. 51, 12844 h (1957).
454. Kirsanov, A. V., and R. G. Makitra, Zh. Obshch. Khim., 27, 450 (1957); C. A. 51, 15443 d (1957).
455. Kirsanov, A. V., and R. G. Makitra, Zh. Obshch. Khim., 28, 35 (1958); C. A. 52, 12787 b (1958).
455a. Kirsanov, A. V., and M. S. Marenets, Z. Obshch. Khim., 29, 2256 (1959); C. A. 54, 10855 (1960).
456. Kirsanov, A. V., and M. S. Marenets, Zh. Obshch. Khim., 31, 1605, 1607, (1961); C. A. 55, 23339 f and 25752 a (1961).
457. Kirsanov, A. V., and L. L. Matveenko, Zh. Obshch. Khim., 28, 1892 (1958); C. A. 53, 1265 b (1959).
458. Kirsanov, A. V., and V. P. Molosnova, Zh. Obshch. Khim., 25, 772 (1955), 28, 347 (1958); C. A. 50, 2416 g (1956); C. A. 52, 13663 h (1958).
459. Kirsanov, A. V., and Z. D. Nekrasova, Zh. Obshch. Khim., 27, 1253 (1957); C. A. 52, 3666 g (1958).
460. Kirsanov, A. V., and Z. D. Nekrasova, Zh. Obshch. Khim., 27, 3241 (1957); C. A. 52, 8996 i (1958).
461. Kirsanov, A. V., and Z. D. Nekrasova, Zh. Obshch. Khim., 28, 1595 (1958); C. A. 53, 1206 f (1959).

462. Kirsanov, A. V., and V. I. Shevchenko, Zh. Obshch. Khim., 24, 474, 882, 1980 (1954); C. A. 49, 61640, 8168, 13927 c (1955).
463. Kirsanov, A. V., and V. I. Shevchenko, Zh. Obshch. Khim., 26, 74 (1956); C. A. 50, 13786 h (1956).
464. Kirsanov, A. V., and V. I. Shevchenko, Zh. Obshch. Khim., 26, 250, 504 (1956); C. A. 50, 13785 g and 13783 e (1956).
465. Kirsanov, A. V., and V. A. Shokol, Zh. Obshch. Khim., 30, 3031 (1960); C. A. 55, 18636 i (1961).
466. Kirsanov, A. V., and V. A. Shokol, Zh. Obshch. Khim., 31, 582 (1961); C. A. 55, 23407 c (1961).
467. Kirsanov, A. V., and I. N. Zhmurova, Zh. Obshch. Khim., 26, 2642 (1956); C. A. 51, 1820 g (1957).
468. Kirsanov, A. V., and I. N. Zhmurova, Zh. Obshch. Khim., 27, 1002 (1957); C. A. 52, 3715 h (1958).
469. Kirsanov, A. V., and I. N. Zhmurova, Zh. Obshch. Khim., 28, 2478 (1958), C. A. 53, 3118 i (1959).
470. Kirsanov, A. V., and L. P. Zhuravleva, Zh. Obshch. Khim., 31, 210 (1961); C. A. 55, 22129 f (1961).
471. Kirsanov, A. V., and L. P. Zhuravleva, Zh. Obshch. Khim., 31, 598 (1961); C. A. 55, 25751 f (1961).
472. Kirsanov, A. V., and Y. M. Zolotov, Zh. Obshch. Khim., 24, 122 (1954); C. A. 49, 3052 h (1955).
473. Kirsanov, A. V., and Y. M. Zolotov, Zh. Obshch. Khim., 25, 571 (1955); C. A. 50, 2521 d (1956).
474. Klee, F. C., and E. R. Kirch, J. Pharm. Sci., 51, 423 (1962); C. A. 58, 3503 h (1963).
475. Klement, R., and K. O. Knollmüller, Chem. Ber., 93, 834,840 (1960).
475a. Klement, R., and K. O. Knollmüller, Chem. Ber., 93, 1088 (1960).
476. Knop, A. Ber. Dtsch. Chem. Ges., 20, 3352 (1887).
477. Knunyants, I. L., E. G. Bykhovskaya, B. L. Dyatkin, V. N. Frosin, and A. A. Gevorkyan, Zh. Vses. Khim. Obshch. im D. I. Mendeleeva, 10, 472 (1965); C. A. 63, 14687 a (1965).
478. Kochetkov, N. K., E. I. Budovskii, and V. N. Shibaer, Izv. Akad. Nauk SSSR, 1962, 1035; C. A. 58, 571 g (1963).
479. Kohn, Monatsh., 9, 407 (1888).
480. Koopmans, M. J., et al., Belg. 567,166 (1958) = Fr. 1,205,582 = Austr. 230,960; C. 1963, 5187.
481. Kosolapoff, G. M., "Organo-phosphorus Compounds," Wiley, New York, 1950.
482. Kosolapoff, G. M., U. S. 2,385,713 (1944); C. A. 40, 608 g (1946).
482a. Kosolapoff, G. M., J. Am. Chem. Soc., 71, 369 (1949).
482b. Kosolapoff, G. M., J. Am. Chem. Soc., 72, 5508 (1950).

482c. Kovache, A., H. Jean, and G. Garnier, Chim. et Ind., 64, 287 (1950).
483. Kozlov, E. S., and B. S. Drach, Zh. Obshch. Khim., 36, 760 (1966); C. A. 65, 8742 g (1966).
484. Kozlov, E. S., A. A. Kisilenko, A. I. Sedlov, and A. V. Kirsanov, Zh. Obshch. Khim., 37, 1611 (1967).
485. Kropacheva, A. A., G. I. Derkach, L. P. Zhuravleva, N. V. Sazonov, and A. V. Kirsanov, Zh. Obshch. Khim., 32, 1540 (1962); C. A. 58, 3375 e (1963).
486. Kropacheva, A. A., G. I. Derkach, and A. V. Kirsanov, Zh. Obshch. Khim., 31, 1601 (1961); C. A. 55, 22274 g (1961).
487. Kropacheva, A. A., and V. A. Parshina, Zh. Obshch. Khim., 29, 556 (1959); C. A. 54, 472 e (1960).
488. Kropacheva, A. A., V. A. Parshina, and S. I. Sergievskaja, Zh. Obshch. Khim., 30, 3584 (1960); C. A. 55, 18695 a (1961).
489. Kropacheva, A. A., and N. V. Sasonov, Zh. Obshch. Khim., 31, 3601, (1961); C. A. 57, 8573 h (1962).
490. Kropacheva, A. A., and N. V. Sasonov, Khim. Geterotsikl. Soedin., Akad. Nauk Latv. SSSR, 1965, 433; C. A. 63, 14859 a (1965).
491. Kropacheva, A. A., N. V. Sasonov, and S. Sergievskaja, Zh. Obshch. Khim., 32, 3796 (1962); C. A. 58, 12552 h (1963).
492. Kubo, H., Agr. Biol. Chem. (Tokyo), 29, 43 (1965); C. A. 63, 7032 c (1965).
493. Kucherov, V. F., Zh. Obshch. Khim., 19, 126 (1949); C. A. 43, 6178 d (1949).
494. Kudryavtsev, B. V., and D. Kh. Yarmukhametova, Izv. Akad. Nauk SSSR, Ser. Khim., 1968, 1412 (1968); C. A. 69, 96620 a (1968).
495. Kühle, E., B. Anders, and G. Zumach, Angew. Chem., 76, 663 (1967).
496. Kulka, M., Can. J. Chem., 37, 525 (1959).
497. Kuhn, S. I., and G. A. Olah, Can. J. Chem., 40, 1951 (1962).
498. Kunz, Ph., Ber. Dtsch. Chem. Ges., 27, 2559 (1894).
499. Kutzback, C., and L. Jaenicke, Ann., 692, 26 (1966).
500. Kuz'menko, I. I., Fiziol. Aktiv. Veshchestva Akad. Nauk Ukr. SSR, 1966, 78; C. A. 67, 117242 e (1967).
501. Kuz'menko, I. I., and L. B. Rapp, Zh. Obshch. Khim., 35, 1221 (1965); C. A. 63, 11406 f (1965).
502. Kuz'menko, I. I., and L. B. Rapp, Zh. Obshch. Khim., 38, 158 (1968); C. A. 69, 95863 v (1968).
503. Kuz'menko, I. I., and L. B. Rapp, Probl. Organ. Sinteza, Akad. Nauk SSSR, Otd. Obshch. i. Tekhn. Khim., 1965, 316; C. A. 64, 6597 b (1966).
504. Laar, C., J. Pr. [2], 20, 250 (1879).
505. Langheld, K., Ber. Dtsch. Chem. Ges., 44, 2076 (1911).

506. Lanham, W. M., U. S. 2,875,235 (1959); C. A. 53, 15110 g (1959).
507. Lanham, W. M., U. S. 2,610,978 (1952) = Ger. 848,946 (1950); C. A. 47, 6975 b (1953).
508. Lannert, K. P., and M. D. Joesten, Inorg. Chem., 7, 2048 (1968).
509. Lapidot, A., and M. Halman, J. Chem. Soc., 1958, 1713.
510. Latscha, H. P., Z. Anorg. Allg. Chem., 346, 166 (1966), and 359, 78,81 (1968).
511. Latscha, H. P., Z. Anorg. Allg. Chem., 367, 40 (1969).
512. Latscha, H. P., Z. Naturforsch., 23 b, 139 (1968).
512a. Latscha, H. P., P. B. Hormuth, and H. Vollmer, Z. Naturforsch., 24 b, 1237 (1969).
513. Leber, J. P., Swiss. 439,306 (1967); C. A. 69, 43913 s (1968).
514. Lemoult, P., Compt. Rend., 139, 206 (1904).
515. Lester, P., U. S. 2,678,335 (1953); C. A. 49, 6300 g (1958).
516. Letcher, H., and J. van Wazer, J. Chem. Phys., 45, 2916 (1966).
517. Levchenko, E. S., N. Y. Derkach, and A. V. Kirsanov, Zh. Obshch. Khim., 32, 1212 (1962); C. A. 58, 4456 h (1963).
518. Levchenko, E. S. and A. V. Kirsanov, Zh. Obshch. Khim., 27, 3078 (1957); C. A. 52, 8070 f (1958).
519. Levchenko, E. S., and A. V. Kirsanov, Zh. Obshch. Khim., 29, 1813 (1959); C. A. 54, 8694 i (1960).
520. Levchenko, E. S., and A. V. Kirsanov, Zh. Obshch. Khim., 30, 1553 (1960); C. A. 55, 3484 g (1961).
521. Levchenko, E. S., and I. E. Sheinkman, Zh. Obshch. Khim., 34, 1145 (1964); C. A. 61, 1787 c (1964).
522. Levchenko, E. S., I. E. Sheinkman, and A. V. Kirsanov, Zh. Obshch. Khim., 30, 1941 (1960); C. A. 55, 6426 b (1961).
523. Levchenko, E. S., I. E. Sheinkman, and A. V. Kirsanov, Zh. Obshch. Khim., 33, 3315 (1963); C. A. 60, 4043 (1964).
524. Levchenko, E. S., and I. N. Zhmurova, Ukrain. Khim. Zhur., 22, 623 (1956); C. A. 51, 5719 e (1957).
525. Levchenko, E. S., I. N. Zhmurova, and A. V. Kirsanov, Zh. Obshch. Khim., 29, 2262 (1959); C. A. 54, 10926 a (1960).
526. Levkova, L. N., and P. I. Alimov, Izv. Akad. Nauk SSSR, Ser. Khim., 1966, 2218; C. A. 66, 75644 j (1967).
527. Li, S. O., Acta Chem. Scand., 4, 610 (1950).
528. Li, S. O., J. Am. Chem. Soc., 74, 5959 (1952).
529. Li, S. O., and C. P. Chang, Acta Chim. Sinica, 23, 99 (1957); C. A. 52, 14552 c (1958).

530. Li, S. O., and R. E. Eakin, J. Am. Chem. Soc., 77, 1866 (1955).
531. Lies, T., R. E. Plapinger, and T. Wagner-Jauregg, J. Am. Chem. Soc., 75, 5755 (1953).
532. Loev, B., and J. T. Massengale, J. Org. Chem., 22, 1186 (1957).
533. Lorberth, J., H. Krapf, and H. Nöth, Chem. Ber., 100, 3511 (1968).
534. Lorenz, W., Ger. 1,160,440 (1964); C. A. 60, 10718 e (1964).
535. Lovell, J. B., U. S. 3,197,365 (1965); C. A. 66, 37931 t (1967).
536. Luff, B. D. W., and F. S. Kipping, J. Chem. Soc., 895, 1993 (1909).
537. Lutskii, A. E., Z. A. Shevchenko, L. I. Samarai, and A. M. Pinchuk, Zh. Obshch. Khim., 37, 2034 (1967); C. A. 68, 34386 z (1968).
538. Lynen, F., B. W. Agranoff, H. Eggerer, U. Henning, and E. M. Möslein, Angew. Chem., 71, 657 (1959).
539. Maeder, A., Ger. 1,052,984 (1959); C. 1959, 15191.
540. Major, R. T., and R. J. Hedrick, J. Org. Chem., 30, 1268 (1965).
541. Malatesta, P., and B. D'Atri, Farmaco (Pavia), Ed. Sci., 8, 398 (1953); C. A. 48, 9312 f (1954).
542. Malatesta, P., and B. D'Atri, Farmaco (Pavia), Ed. Sci., 8, 470 (1953); C. A. 48, 9901 (1954).
543. Malatesta, P., Farmaco (Pavia), Ed. Sci., 8, 193 (1953); C. A. 47, 12077 f (1953).
544. Malinovskij, M. S., Z. F. Solomko, and L. M. Jurilina, Zh. Obshch. Khim., 30, 3454 (1960); C. A. 55, 19775 e (1961).
545. Malowan, J. E., Inorg. Synth., 4, 77 (1953).
546. Mamaeva, I. E., et al., Khim. Farm. Zh., 2, 31 (1968); C. A. 69, 77204 b (1968).
547. Mandel'baum, Ya. A., G. L. Abramova, N. N. Mel'nikov, and L. M. Golovleva, Zh. Obshch. Khim., 37, 2540 (1967); C. A. 68, 104511 n (1968).
548. Mandel'baum, Ya. A., G. L. Abramova, T. M. Pivovarova, and Z. M. Bakanova, Khim. v. Sel'sk Khot., 3, 42 (1965); C. A. 63, 16240 h, 16261 a (1965).
549. Mandel'baum, Ya. A., R. S. Soifer, N. N. Mel'nikov, and L. A. Belova, Zh. Obshch. Khim., 37, 2287 (1967); C. A. 68, 86790 w (1968).
550. Mannes, K., K. Wedemeyer, and G. Unterstenhofer, Ger. 1,160,436 (1964); C. A. 60, 10601 a (1964).
551. Mao, T. J., R. D. Dresdner, and J. A. Young, J. Inorg. Nucl. Chem., 24, 53 (1962).
552. Marcus, P., Compt. Rend., Ser. C., 262, 1048 (1966).
553. Martensson, O., and E. Nilsson, Acta Chem. Scand., 19, 1256 (1965).

554. Maruszewska, L., E. Wieczerkowska, J. Michalski, and A. Zwierzak, Chem. Ind., 1961, 1668.
555. Mastin, T. W., G. R. Norman, and E. A. Weilmuenster, J. Am. Chem. Soc., 67, 1662 (1945).
556. Matrosov, E. I., Zh. Srukt. Khim., 7, 708 (1966); C. A. 66, 23974 x (1967).
557. Matrosov, E. I., A. Gilyarov, and M. I. Kabachnik, Izv. Akad. Nauk SSSR, Ser. Khim., 1965, 16262 e (1966).
558. Matrosov, E. I., V. A. Gilyarov, and M. I. Kabachnik, Izv. Akad. Nauk SSSR, Ser. Khim., 1967, 1465; C. A. 68, 38831 p (1968).
559. McCombie, H., W. C. Saunders, and G. J. Stacey, J. Chem. Soc., 1945, 921.
560. McCombie, H., W. C. Saunders, and G. J. Stacey, J. Chem. Soc., 1945, 380.
561. McConnell, R. L., and H. W. Coover, Jr., U. S. 2,861,092 (1956); C. A. 54, 1297 d (1960).
562. McKinley, W. P., Proc. Can. Soc. Forensic Sci., 2, 433 (1964); C. A. 61, 7638 c (1964).
563. Meise, W., and H. Machleidt, Ann., 693, 76 (1966).
564. Melamed, S., U. S. 2,843,586 (1954); C. A. 53, 12311 d (1959).
565. Mel'nikov, N. N., Zh. Obshch. Khim., 29, 3286 (1959); C. A. 54, 14215 g (1960).
566. Mel'nikov, N. N., A. F. Prokof'eva, T. P. Krylova, and I. L. Vladimirova, Khim. Org. Soedin. Fosfora, Akad. Nauk SSSR, Otd. Obshch. Tekh. Khim., 1967, 267; C. A. 69, 2629 k (1968).
567. Mel'nikov, N. N., B. A. Khaskin, and K. D. Shvetsova-Shilovskaja, Zh. Obshch. Khim., 31, 3605 (1961); C. A. 57, 11071 f (1962); Zh. Obshch. Khim., 32, 1836 (1962); C. A. 58, 4456 a (1963).
568. Mel'nikov, N. N., and A. G. Zenkevich, Zh. Obshch. Khim., 25, 828 (1955); C. A. 50, 2415 d (1956).
569. Meltzer, J., B. G. van den Bos, and C. H. Schoot, U. S. 3,089,808 (1963); C. A. 59, 11341 c (1963).
570. Meltzer, J., K. Weelinga, and B. G. van den Bos, U. S. 3,111,525 (1963); C. A. 60, 9285 e (1964).
571. Metivier, J., Brit. 761,104 (1954); C. A. 51, 11371 g (1957).
572. Meyer, F., and C. Hackmann, Ger. 963,976 (1953); C. 1958, 2815.
573. Michaelis, A., Ann. 326, 129, 172, 201, 223 (1903).
574. Michaelis, A., Ann. 407, 290, 294 (1915).
575. Michaelis, A., Ber. Dtsch. Chem. Ges., 31, 1037 (1898).
576. Michaelis, A., and W. Kaersten, Ber. Dtsch. Chem. Ges., 28, 1237 (1895).
577. Michaelis, A., and W. Kerkhof, Ber. Dtsch. Chem. Ges., 31, 2172 (1898).

578. Michaelis, A., and K. Luxembourg, Ber. Dtsch. Chem. Ges., 28, 2205 (1895).
579. Michaelis, A., and K. Luxembourg, Ber. Dtsch. Chem. Ges., 29, 710 (1896).
580. Michaelis, A., and F. Oster, Ann. 270, 126, 136 (1892).
581. Michaelis, A., and G. Schulze, Ber. Dtsch. Chem. Ges., 26, 2937 (1893).
582. Michaelis, A., and G. Schulze, Ber. Dtsch. Chem. Ges, 27, 2572 (1894).
583. Michaelis, A., and E. Silberstein, Ber. Dtsch. Chem. Ges., 29, 716 (1896).
584. Michaelis, A., and H. v.Soden, Ann. 229, 295, 334 (1885).
585. Michaelis, A., Ber. Dtsch. Chem. Ges., 28, 1017 (1895).
586. Michalski, J., Roczniki Chem., 29, 960 (1955); C. A. 50, 10614 h (1956).
587. Michalski, J., and J. Wieczorkowski, Roczniki Chem., 31, 585, 879 (1957); C. A. 52, 5283 i, 8037 i (1958).
588. Michalski, J., B. P. Krawiecka, and A. Skowronska, Roczniki Chem., 37, 1479 (1963); C. A. 60, 9135 f (1964).
589. Michalski, J., H. Strzelecka, and J. Wieczorkowski, Roczniki Chem., 31, 879 (1957); C. A. 52, 8037 h (1958).
590. Michelson, A. M., Biochim. Biophys. Acta, 91, 1 (1964).
591. Miller, B., and D. W. Long, Fr. 1,282,807 (1961).
592. Miller, B., and T. P. O'Leary Jr., U. S. 3,113,958 (1963); C. A. 60, 5337 a (1964).
593. Miller, B., and T. P. O'Leary Jr., J. Org. Chem., 27, 3382 (1962).
594. Miller, F. A., Pure Appl. Chem., 7, 125 (1963), and Spectrochim. Acta 18, 1311 (1963). See also C. A. 58, 7515 f (1963).
595. Miller, J. K., and J. N. Lomonte, U. S. 3,084,190 (1963); C. A. 59, 11250 d (1963).
596. Millikan, A. F., and G. W. Crosby, U. S. 2,996,532 (1957); C. A. 56, 319 i (1962).
597. Minami, T., H. Miki, H. Matsumoto, Y. Ohshiro, and T. Agawa, Tetrahedron Letters, 1968, 3049.
598. Mingoia, Gazz. Chim. Ital., 62, 333 (1932).
599. Mitsch, R. A., J. Am. Chem. Soc., 89, 6297 (1967).
600. Miyano, M., and S. Funahashi, J. Am. Chem. Soc., 77, 3522 (1955).
601. Mizuma, T., Y. Minaki, and S. Toyoshima, J. Pharm. Soc. Japan, 81, 48 (1961); C. A. 55, 13403 f (1961).
602. Mizuma, T., Y. Minaki, and S. Toyoshima, J. Pharm. Soc. Japan, 81, 51 (1961); C. A. 55, 12379 g (1961).

603. Mochalina, E. P., B. L. Dyatkin, and I. L. Knunyants, Izv. Akad. Nauk SSSR, Ser. Khim., 1966, 2247; C. A. 66, 76089 u (1967).
604. Modro, T., and J. Sokolowski, Roczniki Chem., 40, 1203 (1966); C. A. 66, 28713 w (1967).
605. Modro, T., and J. Sokolowski, Bull. Acad. Polon. Sci. Ser. Sci. Chim., 13, 687 (1965); C. A. 64, 15705 a (1966).
605a. Moeller, T., and A. Vandi, J. Org. Chem., 27, 3511 (1962).
606. Moffatt, J. G., and H. G. Khorana, J. Am. Chem. Soc., 83, 649 (1961).
607. Monsanto Co., Belg. 672,659 (1966); C. A. 67, 32458 e (1967).
608. Monsanto Co., Belg. 670,693 (1966); C. A. 65, 8819 g (1966).
609. Monsanto Co., Fr. 1,456,182 (1966); Belg. 672,658 (1966); C. A. 67, 21648 r, 21649 s (1967).
610. Monsanto Co., U. S. 2,385,713 (1945).
611. Montgomery, H. A. C., and J. H. Turnbull, J. Chem. Soc., 1958, 1963; Proc. Chem. Soc., 1957, 178.
612. Morel, A., Bull. Soc. Chim., (3), 21, 494 (1889).
613. Morozovskaja, L. M., et al., Khim. Org. Soedin, Fosfora, Akad. Nauk SSSR, Otd. Obshch. Tekh. Khim., 1967, 223; C. A. 69, 77047 c (1968).
613a. Morrison, D. C., J. Am. Chem. Soc., 73, 5896 (1951).
614. Morrison, D. C., J. Org. Chem., 23, 1072 (1958).
615. Mosby, W. L., and M. L. Silva, J. Chem. Soc., 1965, 2727.
616. Mosby, W. L., and M. L. Silva, U. S. 3,387,004 (1968); C. A. 69, 77481 g (1968).
617. Moshkina, T. M., and A. N. Pudovik, Zh. Obshch. Khim., 35, 2042 (1965); C. A. 64, 6680 g (1966).
618. Moyle, C. L., and E. E. Kenaga, U. S. 2,552,574/577 (1951); C. A. 45, 9079 (1951).
619. Moyle, C. L., and L. R. Drake, U. S. 2,615,037/38/39 (1952); C. A. 47, 9343 (1953).
620. Muller, N., P. C. Lauterbur, and J. Goldenson, J. Am. Chem. Soc., 78, 3557 (1956).
621. Nagasawa, M., Y. Imamiya, and H. Sugiyama, Japan, 17211 (1962); C. A. 59, 11254 b (1963).
622. Nagasawa, M., and Y. Jmamiya, Japan, 11,013 (1962); C. A. 59, 3833 (1963).
623. Nagasawa, M., and T. Tsuboi, Japan, 1,398 (1963); C. A. 59, 11337 d (1963).
624. Nagasawa, M., K. Takida, and S. Okada, Japan, 6,198 (1963); C. A. 60, 4158 (1964).
625. Najer, H., R. Giudicelli, and J. Sette, Bull. Soc. Chim. France, 1961, 2114.
626. Nannelli, P., A. Failli, and T. Moeller, Inorg. Chem., 4, 560 (1965).

627. Narbut, A. V., and G. I. Derkach, Zh. Obshch. Khim., 38, 1321 (1968); C. A. 69, 86260 u (1968).
628. Navech, J., Ann. Fac. Sci. Univ. Toulouse Sci. Phys., 20, 9 (1962).
629. Neuberg, C., and W. Oertel, Bio. Z., 60, 491 (1914).
630. Niecke, E., and J. Stenzel, Z. Naturforsch., 22 b, 785 (1967).
631. Nielsen, M. L., Inorg. Chem., 3, 1760 (1964).
632. Nielsen, M. L., R. R. Ferguson, and W. S. Coakley, J. Am. Chem. Soc., 83, 99 (1961).
633. Nielsen, M. L., and J. V. Pustinger, Jr., J. Phys. Chem., 68, 152 (1964).
634. Nielsen, M. L., J. V. Pustinger, Jr., and J. Strobel, J. Chem. Eng. Data, 9, 167 (1964).
635. Nielsen, R. P., and H. H. Sisler, Inorg. Chem., 2, 753 (1963).
636. Nikonorov, K. V., and Z. G. Speranskaya, Izv. Akad. Nauk SSSR, 1963, 1697; C. A. 59, 15166 d (1963).
637. Nikonorov, K. V., G. M. Vinokurova, and Z. G. Speranskaya, Khim. i. Primenenie, 1955, 223; C. A. 52, 240 e (1958); Izv. Akad. Nauk SSSR, 1957, 1059; C. A. 52, 6163 a (1958).
638. Nikoronov, K. V., E. A. Guryler, and A. V. Chernova, Izv. Akad. Nauk SSSR, Ser. Khim., 1968, 587; C. A. 69, 95864 w (1968).
639. Nikoronov, K. V., and Z. G. Speranskaya, Izv. Akad. Nauk SSSR, 1958, 964; C. A. 53, 1118 b (1959).
640. Nippon, Kayaku Co., Japan., 18,131 (1964); C. A. 62, 13087 h (1965).
641. Nippon Kayaku Co., Japan., 19, 151 (1964); C. A. 62, 9011 d (1965).
642. Nippon Kayaku Co., Japan., 6,737 (1967); C. A. 67, 73355 n (1967).
643. Nitantera, L. V., N. N. Shvetsova, and O. N. Vlasow, Zh. Obshch. Khim., 38, 1450 (1968); C. A. 69, 86495 z (1968).
644. Nixon, J. F., and R. Schmutzler, Spectrochim. Acta, 20, 1835 (1964), and 22, 565 (1966).
645. Nogrady, T. F., K. M. Vagi, and V. W. Adamkiewicz, Can. J. Chem., 40, 2126 (1962).
646. Nogrady, T., and K. M. Vagi, J. Org. Chem., 27, 2270 (1962).
646a. Nordman, C. E., Acta Crystallogr. [Copenhagen], 13, 535 (1960).
647. Normant, H., J. Bull. Soc. Chim. France, 1963, 1868; C. A. 59, 15154 g (1963).
648. Normant, H., Angew. Chem., 79, 1029 (1967).
649. Nuretdinova, O. N., Zh. Obshch. Khim., 35, 1880 (1965); C. A. 64, 2122 a (1966).

650. Nuretdinov, I. A., and N. P. Grechkin, Izv. Akad. Nauk SSSR, Ser. Khim., 1966, 1466; C. A. 66, 75641 f (1967).
651. Nyquist, R. A., E. H. Blair, and D. W. Osborne, Spectrochim. Acta, 23A, 2505 (1967).
652. Oddo, G., Gazz. Chim. Ital., 29, 330 (1899).
653. Öney, I., and M. Caplow, J. Am. Chem. Soc., 89, 6972 (1967).
654. Oertel, G., H. Malz, and H. Holtschmidt, Chem. Ber., 97, 891 (1964).
655. Olah, G. A., and A. A. Oswald, Can. J. Chem., 38, 2053 (1960).
656. Olah, G. A., and A. A. Oswald, J. Org. Chem., 24, 1443 (1959).
657. Olah, G. A., A. A. Oswald, and A. Mlinko, Ann. 602, 123 (1957).
658. Olah, G. A., A. A. Oswald, and S. Kuhn, Ann. 625, 88, 92 (1959).
659. Osborne, D. W., H. O. Senkbeil, and J. L. Wasco, J. Org. Chem., 31, 192, 197 (1966).
660. Otto, P., Ber. Dtsch. Chem. Ges., 28, 613 (1895).
661. Papanastassiou, Z. B., and Th. J. Bardos, Med. Pharm. Chem., 5, 1000 (1962); C. A. 58, 5599 f (1963); Armour Co., Brit. 877,671 (1959); C. A. 56, 11572 a (1962).
662. Papanastassiou, Z. B., R. J. Bruni, F. P. Fernandes, and P. L. Levins, J. Med. Chem., 9, 357 (1966); C. A. 64, 19395 g (1966).
663. Papanastassiou, Z. B., and Th. J. Bardos, Fr. 1,220,583, and 1,220,655 (1959); C. 1962, 9569.
664. Parker, R. P., D. R. Seeger, and E. Kuk, U. S. 2,663,705 (1951); C. A. 48, 13709 e (1954).
665. Payne, D. S., H. Noeth, and G. Henniger, Chem. Comm., 1965, 327.
666. Pechmann, H. von, and L. Seeberger, Ber. Dtsch. Chem. Ges. 27, 2121 (1894).
667. Penert, J. C., and R. B. Tideswell, U. S. 3,331,893 (1967); C. A. 67, 64040 d (1967).
668. Perkow, W., Ger. 1,022,587 (1957); C. 1958, 12516.
669. Perkow, W., Ger. 1,067,433; C. 1960, 16232.
670. Petitcolas, P., A. P. Richard, and R. Goupil, Fr. 1,009,369 (1948); C. A. 51, 18632 h (1957).
671. Petrov, K. A., F. L. Maklyaev, A. A. Meimyaheva, and N. K. Bliznyuk, Zh. Obshch. Khim., 30, 4060 (1960); C. A. 55, 21012 e (1961).
672. Petrov, K. A., A. A. Neimysheva, M. G. Fomenko, L. M. Chernushevich, and A. D. Kuntsevich, Zh. Obshch. Khim., 31, 516 (1961); C. A. 55, 22111 a (1961).

673. Petrov, K. A., V. P. Evdakov, G. I. Abramtsev, and A. K. Strautman, Zh. Obshch. Khim., 32, 3070 (1962); C. A. 58, 11395 (1963).
674. Petrov, K. A., V. A. Kravchenko, V. P. Evdakov, and L. I. Mizrakh, Zh. Obshch. Khim., 34, 2586 (1964); C. A. 61, 14706 f (1964).
675. Petrov, K. A., N. K. Bliznjuk, and V. A. Savostenok, Zh. Obshch. Khim., 31, 1361 (1961); C. A. 55, 23317 e (1961).
676. Petrov, K. A., A. I. Gavrilova, and M. M. Butilov, Zh. Obshch. Khim., 35, 1856 (1965); C. A. 64, 2120 b (1966).
677. Petrov, K. A., A. I. Gavrilova, and V. P. Korotkova, Zh. Obshch. Khim., 32, 915 (1962); C. A. 58, 1485 g (1963).
677a. Petrov, K. A., and A. A. Nejmysheva, Z. Obshch. Khim., 29, 1822 (1959); C. A. 54, 8600 c (1960).
678. Petrov, K. A., and G. A. Sokolskij, Zh. Obshch. Khim., 26, 3378 (1956); C. A. 51, 8029 b (1957).
679. Petrov, K. A., and O. S. Urbanskaja, Zh. Obshch. Khim., 30, 1233 (1960); C. A. 55, 352 g (1961).
680. Philips' Gloeilampenfabrieken, Austr. 214,941 (1959); C. 1964, 18-2163.
681. Philips' Gloeilampenfabrieken, Brit. 920,117 (1963); C. A. 59, 5077 d (1963).
682. Philips' Gloeilampenfabrieken, Brit. 940,921 (1963); C. A. 60, 2951 e (1964); Brit. 947,485; C. A. 60, 14513 h (1964).
683. Philips' Gloeilampenfabrieken, Belg. 588,331 (1960).
683a. Pianka, M., J. Appl. Chem., 5, 109 (1955).
684. Pianka, M., and V. H. Chambers, Ger. 899,433 (1951) = Brit. 703,160; C. A. 49, 4011 a (1955).
685. Pilgram, K., D. D. Philips, and F. Korte, J. Org. Chem., 29, 1844 (1964).
686. Pinchuk, A. M., I. M. Kosinskaya, and V. I. Shevchenko, Zh. Obshch. Khim., 37, 2693 (1967); C. A. 69, 67069 x (1968).
687. Pinchuk, A. M., I. M. Kosinskaya, and V. I. Shevchenko, Zh. Obshch. Khim., 37, 852, 856 (1967); C. A. 67, 108169 j, 108407 k (1967).
688. Pishchimuka, P., J. Russ. Phys. Chem. Soc., 44, 1406 (1912).
689. Plapinger, R. E., and T. Wagner-Jauregg, J. Am. Chem. Soc., 75, 5757 (1953).
690. Preussmann, R., Arzneimittelforschung, 12, 260 (1962); C. A. 58, 3389 h (1963).
691. Price, G. R., E. N. Walsh, and J. T. Hallett, U. S. 3,083,135 (1963); C. A. 59, 8602 a (1963).
692. Protsenko, L. D., Fiziol. Aktiv. Veshchestva, Akad. Nauk Ukr., SSSR, 1966, 74; C. A. 67, 53935 y (1967).

693. Protsenko, L. D., Zh. Obshch. Khim., 35, 368 (1965); C. A. 62, 13107 (1965).
694. Protsenko, L. D., P. V. Rodionov, N. Yq. Skul'skaya, Yu. I. Bogodist, P. Ya. Sologub, M. I. Tarnavskaya, and S. V. Nikolaeva, Puti Sin. Izyskaniya Protivoopukholevykh Prep., Tr. Somp. 2nd, Moskow, 1965, 100; C. A. 69, 96630 d (1968).
695. Protsenko, L. D., G. I. Derkach, and A. V. Kirsanov, Zh. Obshch. Khim., 31, 3433 (1961); C. A. 57, 3357 g (1962).
696. Protsenko, L. D., and K. A. Kornev, Ukr. Chem. Z., 24, 636 (1958); C. A. 53, 12266 b (1959).
697. Protsenko, L. D., and K. A. Kornev, Ukr. Khim. Zh., 28, 719 (1962); C. A. 59, 489 a (1963).
698. Protsenko, L. D., K. A. Kornev, and Y. I. Bogodist, Tr. Simpoziuma po Khim., Moscow, 1960, 182; C. A. 58, 4491 g (1963).
699. Protsenko, L. D., and L. A. Negievich, Ukr. Khim. Zh., 30, 1328 (1964); C. A. 62, 9085 e (1965).
700. Protsenko, L. D., and L. A. Negievich, Zh. Obshch. Khim., 35, 1564 (1965); C. A. 63, 18003 c (1965).
701. Protsenko, L. D., and N. Y. Skul'skaya, Zh. Obshch. Khim., 33, 2284 (1963); C. A. 59, 12666 c (1963).
702. Protsenko, L. D., and N. Y. Skul'skaya, Zh. Obshch. Khim., 35, 715 (1965); C. A. 63, 4195 b (1965).
703. Protsenko, L. D., and E. A. Svyatnenko, Zh. Obshch. Khim., 37, 2517 (1967); C. A. 69, 26945 m (1968).
704. Poshkus, A. C., and J. E. Herweh, J. Org. Chem., 27, 2700 (1962).
705. Posternak, T., and S. Grafl, Helv. Chim. Acta, 28, 1258 (1945).
706. Pound, D. W., Brit. 688,760, and 688,787 (1949); C. A. 47, 7717 and 7728 (1953).
707. Pound, D. W., Brit. 690,386 (1949); C. A. 48, 4582 d (1954).
708. Pudovik, A. N., and R. D. Gareev, Zh. Obshch. Khim., 34, 3942 (1964); C. A. 62, 7791 h (1965).
709. Pudovik, A. N., and G. P. Krupnov, Zh. Obshch. Khim., 33, 1654 (1963); C. A. 59, 12639 e (1963).
710. Raetz, R., J. Org. Chem., 22, 372 (1957).
711. Raetz, R. F. W., U. S. 3,090,800 (1963); C. A. 59, 11294 e (1963).
712. Raetz, R., and L. Engelbrecht, Z. Anorg. Allg. Chem., 272, 326 (1953).
713. Raetz, R. F. W., and M. J. Gruber, U. S. 3,314,848 (1967); C. A. 67, 54030 m (1967).
714. Raetz, R., and M. Gruber, J. Med. Chem., 9, 144 (1966).
714a. Raetz, R., and M. Hess, Chem. Ber., 84, 889 (1951).
715. Ramirez, F., and L. P. Smith, Tetrahedron, 30, 3651 (1966).

716. Rapp, L. B., and I. I. Kuz'menko, Zh. Obshch. Khim., 33, 2277 (1963); C. A. 59, 12676 g (1963).
717. Raths, F. W., U. S. 2,988,566 (1961); C. A. 55, 25862 (1961).
718. Reist, E. J., I. G. Junga, and B. R. Baker, J. Org. Chem., 25, 666 (1960).
719. Remsen, Hartman, and Mückenfürs, Am. Chem. J., 18, 150 (1896).
720. Rigandy, J., and J. C. Vernieres, C. R. Akad. Sci. Paris, Ser. C., 266, 828 (1968); C. A. 69, 2617 e (1968).
721. Rigterink, R. H., U. S. 3,064,009 (1962); C. A. 58, 7912 b (1963).
722. Rigterink, R. H., U. S. 3,100,206 (1963); C. A. 60, 1775 d (1964).
723. Rigterink, R. H., U. S. 3,159,645 (1964); C. A. 62, 4032 c (1965).
724. Rigterink, R. H., U. S. 3,166,565 (1965); C. A. 62, 10439 h (1965).
725. Rigterink, R. H., U. S. 3,228,960 (1966); C. A. 64, 17542 g (1966).
726. Rigterink, R. H., U. S. 3,326,752 (1967); C. A. 67, 72806 y (1967).
727. Rigterink, R. H., U. S. 3,268,536 (1966); C. A. 65, 18566 g (1966).
728. Roesky, H. W., Angew. Chem., Int. Ed., 6, 90 (1967).
729. Roesky, H. W., Chem. Ber., 100, 2142 (1967).
730. Roesky, H. W., Chem. Ber., 101, 636 (1968).
731. Roesky, H. W., Chem. Ber., 101, 2977 (1968).
731a. Roesky, H. W., Z. Naturforsch., 24 b, 818 (1969).
732. Roesky, H. W., and A. Müller, Z. Anorg. Allg. Chem., 353, 265 (1967).
733. Rodionov, V. M., and E. V. Yavorskaya, Zh. Obshch. Khim., 18, 110 (1948); C. A. 42, 4976 h (1948).
734. Rorig, K., J. Am. Chem. Soc., 71, 3561 (1949).
735. Rosenberg, E. T., Ger. 1,139,839 (1957) = U. S. 2,917,517; C. A. 54, 8853 d (1960).
736. Rudert, P., Ber. Dtsch. Chem. Ges., 26, 565 (1893).
737. Rudy, H., and A. Debo, Brit. 865,848 (1959); C. A. 55, 22136 a (1961); C. 1962, 16797.
737a. Rudy, H., and A. Debo, Ger. 1,048,919 (1958); C. 1959, 11732.
737b. Rydon, H. N., and B. L. Tonge, J. Chem. Soc., 1957, 4682.
738. Rymareva, T. G., K. D. Shvetsova-Shilovskaya, and N. N. Mel'nikov, Zh. Obshch. Khim., 37, 2074 (1967); C. A. 68, 77881 z (1968).
739. Samarai, L. I., and G. I. Derkach, Zh. Obshch. Khim., 35, 755 (1965); C. A. 63, 5548 c (1965).
740. Samarai, L. I., and G. I. Derkach, Zh. Obshch. Khim., __, 1433 (1966); C. A. 66, 2297 p (1967).

741. Samarai, L. I., O. I. Kolodyazhnii, and G. I. Derkach, Angew. Chem. Int. Ed., 7, 618 (1968).
742. Sambeth, J., and M. Becke-Goehring, Angew. Chem., 70, 594 (1958).
743. Sandoz Ldt., Neth. 6,611,511 (1967); C. A. 67, 73620 y (1967).
743a. Saul, G. A., Brit. 732,384 (1952); C. A. 50, 7122 (1956).
744. Saul, G. A., and K. L. Godfrey, Brit. 744,484 (1953); C. A. 50, 16825 c (1956).
745. Saunders, B. C., et al., J. Chem. Soc., 1949, 2921.
746. Saunders, B. C., G. J. Stacey, F. Wild, and I. G. E. Wilding, J. Chem. Soc., 1948, 699.
747. Sazonov, N. Y., and H. A. Kropacheva, Khim.-Farm. Zh., 1, 32 (1967); C. A. 68, 11541 m (1968).
748. Schaller, H., H. A. Staab, and F. Cramer, Chem. Ber., 94, 1621 (1961).
749. Scherer, O. J., and J. Wokulat, Z. Anorg. Allg. Chem., 361, 296 (1968).
750. Schering-Kahlbaum, A. G., Ger. 556,145 (1930).
751. Schiff, H., Ann. 101, 302 (1857).
752. Schmitt, R., and K. Dehnicke, Z. Anorg. Allg. Chem., 358, 38 (1968).
752a. Schomaker, V., and D. P. Stevenson, J. Am. Chem. Soc., 63, 37 (1941).
753. Schrader, G., results published in Houben-Weyl, see Ref. 374.
754. Schrader, G., Die Entwicklung neuer insektizider Phosphorsäure-Ester, Verlag Chemie, Weinheim/Bergstr., 1963.
754a. Schrader, G., and O. Bayer, Ger. 664,438 (1935); C. 1937, I, 4550.
755. Schrader, G., Ger. 767,511 (1937); C. A. 49, 14795 g (1955); C. 1953, 2020.
756. Schrader, G., Ger. 880,443 (1951); C. 1954, 10570.
757. Schrader, G., Ger. 900,814 (1942); C. 1954, 5852.
758. Schrader, G., Ger. 949,650 (1952); C. 1957, 6878.
759. Schrader, G., Ger. 952,085 (1955); C. 1957, 6265.
760. Schrader, G., Ger. 952,712 (1955); C. 1957, 3631.
761. Schrader, G., Ger. 968,486 (1952); C. 1959, 9369.
762. Schrader, G., Ger. 1,025,882 (1956); C. 1959, 272.
763. Schrader, G., Ger. 1,032,247 (1954); C. 1959, 17391.
764. Schrader, G., Ger. 1,080,109 (1960); C. 1961, 2421.
765. Schrader, G., Ger. 1,142,866 (1963); C. A. 59, 1533 h (1963).
766. Schrader, G., and H. Gebhardt, Ger. 767,830 (1939); C. 1954, 4039.
767. Schrader, G., and H. Kükenthal, Ger. 918,603 (1941); C. 1955, 4197.
768. Schramm, G., and H. Wissmann, Chem. Ber., 91, 1073 (1958).

769. Schwarzmann, E., and J. R. van Wazer, J. Am. Chem. Soc., 82, 6009 (1960).
770. Sciarini, L. J., and J. S. Fruton, J. Am. Chem. Soc., 71, 2940 (1949).
771. Scott, F. L., R. Riordan, and P. D. Morton, J. Org. Chem., 27, 4255 (1962).
771a. Scrowaczewska, Z., and P. Mastallerz, Roczniki. Chem., 31, 531 (1957); C. A. 52, 2798 (1958).
772. Scrowaczewska, Z., and R. Tyka, Roczniki Chem., 33, 51 (1959); C. A. 53, 16045 f (1959).
773. Seeger, D. R., Ger. 1,014,543 (1955); C. A. 54, 18564 b (1960); C. 1958, 14176.
774. Seeger, D. R., and A. S. Tomcufcik, Brit. 885,370 (1960); C. A. 58, 7949 e (1963).
775. Senkbeil, H. O., U. S. 3,005,001 (1960); C. A. 56, 7136 c (1962).
776. Senkbeil, H. O., U. S. 3,038,925 (1960); C. A. 57, 13614 c (1962).
777. Senkbeil, H. O., U. S. 3,072,702 (1963).
778. Senkbeil, H. O., U. S. 3,087,957 (1963); C. A. 59, 12644 d (1963).
779. Senkbeil, H. O., U. S. 3,088,966 (1963); C. A. 59, 9794 h (1963).
780. Senkbeil, H. O., U. S. 3,251,675 (1966); C. A. 65, 3749 h (1966).
780a. Serfontein, W. J., and J. H. Jordaan, J. Org. Chem., 27, 3332 (1962).
781. Seto, T. A., and B. K. Koe, Belg. 608,777 (1961); C. A. 57, 7276 d (1962).
782. Sheehan, J. C., and V. S. Frank, J. Am. Chem. Soc., 72, 1312 (1950).
783. Shell. Dev., Ger. 1,111,187 (1959); C. A. 56, 8637 a (1962).
784. Shell Internat. Research Maatschappi j., Ger. 1,194,854 (1965); C. A. 63, 11515 e (1965).
785. Shevchenko, V. I., V. P. Kukhar, and A. V. Kirsanov, Zh. Obshch. Khim., 36, 467 (1966); C. A. 65, 615 h (1966).
786. Shevchenko, V. I., P. P. Kornuta, N. D. Bodnarchuk, and A. V. Kirsanov, Zh. Obshch. Khim., 36, 730 (1966); C. A. 65, 8912 (1966).
787. Shevchenko, V. I., V. P. Kukhar, A. A. Koval, and A. V. Kirsanov, Zh. Obshch. Khim., 38, 1270 (1968); C. A. 69, 106391 t (1968).
788. Shevchenko, V. I., M. El Dik, and A. M. Pinchuk, Zh. Obshch. Khim., 38, 1527 (1968); C. A. 69, 95898 k (1968).
789. Shevchenko, V. I., and V. V. Bondarev, Zh. Obshch. Khim., 26, 270 (1956); C. A. 50, 13787 b (1956).

790. Shevchenko, V. I., N. D. Bondarchuk, and A. V. Kirsanov, Zh. Obshch. Khim., 33, 1342 and 1591 (1963); C. A. 59, 11240 a and 12839 d (1963).
791. Shevchenko, V. I., and N. D. Bodnarchuk, Zh. Obshch. Khim., 36, 1645 (1966); C. A. 66, 54962 w (1967).
791a. Shevchenko, V. I., and G. I. Derkach, Zh. Obshch. Khim., 28, 1085 (1958); C. A. 52, 16264 b (1958).
792. Shevchenko, V. I., and P. P. Kornuta, Zh. Obshch. Khim., 36, 1254 (1966); C. A. 65, 15381 b (1966).
793. Shevchenko, V. I., and P. P. Kornuta, Zh. Obshch. Khim., 36, 1642 (1966); C. A. 66, 65032 u (1967).
794. Shevchenko, V. I., P. P. Kornuta, and A. V. Kirsanov, Zh. Obshch. Khim., 35, 1598 (1965); C. A. 63, 17891 h (1965).
795. Shevchenko, V. I., P. P. Kornuta, and A. V. Kirsanov, Zh. Obshch. Khim., 35, 1970 (1965); C. A. 64, 8027 d (1966).
796. Shevchenko, V. I., A. A. Koval, and A. V. Kirsanov, Zh. Obshch. Khim., 38, 555 (1968); C. A. 69, 26701 d (1968).
797. Shevchenko, V. I., and V. P. Kukhar, Zh. Obshch. Khim., 36, 735 (1966); C. A. 65, 8858 e (1966).
798. Shevchenko, V. I., and V. P. Kukhar, Zh. Obshch. Khim., 36, 1260 (1966); C. A. 65, 16845 g (1966).
799. Shevchenko, V. I., E. E. Nizhnikova, N. D. Bodnarchuk, and P. P. Kornuta, Zh. Obshch. Khim., 37, 1358 (1967); C. A. 68, 39055 a (1968).
800. Shevchenko, V. I., and V. I. Stadnik, Probl. Org. Sinteza, Akad. Nauk SSSR, Otd. Obshch. i. Tekhn. Khim., 1965, 281; C. A. 64, 4933 g (1966).
801. Shevchenko, V. I., V. P. Tkach, and A. V. Kirsanov, Zh. Obshch. Khim., 34, 624 (1964); C. A. 60, 13171 a (1964).
802. Shevchenko, V. I., V. P. Tkach, and A. V. Kirsanov, Zh. Obshch. Khim., 35, 1224 (1965); C. A. 63, 11411 c (1965).
803. Shevchenko, V. I., V. P. Tkach, and A. V. Kirsanov, Zh. Obshch. Khim., 32, 4047 (1962) and 33, 562 (1963); C. A. 58, 13832 c (1963) and C. A. 59, 2688 (1963).
804. Shokol, A. V., L. I. Fedotova, A. N. Frolova, and A. V. Kirsanov, Zh. Obshch. Khim., 35, 534 (1965); C. A. 63, 524 f (1965).
805. Shokol, A. V., and G. I. Derkach, Zh. Obshch. Khim., 35, 1468 (1965); C. A. 63, 14750 a (1965).
806. Shokol, V. A., G. I. Derkach, and A. A. Kisilenko, Zh. Obshch. Khim., 33, 2660 (1963); C. A. 60, 2455 g (1964).

807. Shokol, A. V., N. K. Mikhailyuchenko, and G. I. Derkach, Khim. Org. Soedin, Fosfora, Akad. Nauk SSSR, Otd. Obshch. Tekh. Khim., 1967, 78; C. A. 69, 10065 p (1968).
808. Shokol, V. A., L. I. Molyarko, and G. I. Derkach, Zh. Obshch. Khim., 36, 930 (1966); C. A. 65, 12229 g (1966).
809. Shtepanek, A. S., Yu. V. Piven, and G. I. Derkach, Khim. Org. Soedin. Fosfora, Akad. Nauk SSSR, Otd. Obshch. Tekh. Khim., 1967, 61; C. A. 68, 114695 q (1968).
810. Shvetsov, N. I., I. V. Lebedeva, and I. M. Filatova, Zh. Neorgan. Khim., 10, 993 (1965); C. A. 63, 1812 g (1965).
811. Siddall, T. H. III., and W. E. Stewart, Spectrochim. Acta, 24 A, 81 (1968).
811a. Siebert, H., Anwendung der Schwingungsspektroskopie in der anorganischen Chemie, Springer-Verlag, Heidelberg, 1966.
812. Silaeva, S. A., L. A. Kazitsyna, and M. A. Prokof'ev, Zh. Obshch. Khim., 35, 80 (1965); C. A. 62, 14799 h (1965).
813. Simon, W., and G. Gassl., Ger. 935,206 (1954); C. 1956, 11881.
814. Staab, H. A., H. Schaller, and F. Cramer, Angew. Chem., 71, 736 (1959).
815. Steinhauer, R. C., Ger. 1,047,228 (1959); C. A. 57, 2464 e (1962).
816. Steinkopf, W., Ber. Dtsch. Chem. Ges., 41, 3571 (1908).
817. Steinkopf, W., J. Prakt. Chem., 81, 97, 193 (1910).
818. Steyermark, P. R., U. S. 3,177,239 (1965); C. A. 62, 16129 b (1965).
819. Stock, J. A., W. J. Hopwood, and P. D. Regan, J. Chem. Soc. (C), 1966, 637.
820. Stölzer, C., Chem. Ber., 96, 881, 902 (1963); C. A. 58, 13777 e (1963).
821. Stölzer, C., and A. Simon, Chem. Ber., 93, 1323 (1960).
822. Stölzer, C., and A. Simon, Chem. Ber., 94, 1976 (1961).
823. Stölzer, C., and A. Simon, Z. Anorg. Allg. Chem., 339, 30, 38 (1965).
824. Stölzer, C., and A. Simon, Naturwiss., 46, 377 (1959); see also Stölzer, C., A. Simon, and H. Frömmel, Fr. 1,266,490 (1960); C. A. 57, 2075 g (1962).
825. Stokes, H. N., Amer. Chem. J., 15, 198 (1893).
826. Stokes, H. N., Amer. Chem. J., 16, 123,154 (1894).
827. Strecker, W., and Ch. Grossmann, Ber. Dtsch. Chem. Ges., 49, 63 (1916).

828. Strecker, W., and H. Heuser, Ber. Dtsch. Chem. Ges., 57, 1368 (1924).
829. Stuebe, C., and H. P. Lankelma, J. Am. Chem. Soc., 78, 976 (1956).
830. Skul'skaya, N. Ya., Fiziol. Aktiv. Veshchestva. Akad. Nauk Ukr. SSR, 1966, 83; C. A. 67, 53938 b (1967).
831. Skul'skaya, N. Ya., and L. D. Protsenko, Zh. Obshch. Khim., 37, 2724 (1967); C. A. 69, 2754 x (1968).
832. Smolina, A. I., G. N. Tsybul'skaya, V. P. Rudavskii, and G. I. Derkach, Fiziol. Aktiv. Veshchestva. Akad. Nauk Ukr. SSR, 1966, 96; C. A. 67, 53861 w (1967).
833. Snyder, H. R., Jr., Belg. 630,616 (1963); C. A. 60, 14474 f (1964).
834. Société Usines Rhône-Poulenc, Fr. 729,966 (1932).
835. Soifer, R. S., Ya. A. Mandel'baum, N. N. Mel'nikow, and L. A. Belova, Zh. Obshch. Khim., 37, 2291 (1967); C. A. 68, 86793 z (1968).
836. Sonnet, P. E., and A. B. Borkovec, J. Org. Chem., 31, 2962 (1966).
837. Sowerby, D. B., J. Inorg. Nucl., Chem., 22, 205 (1961).
838. Suckfüll, F., and H. Haubrich, Angew. Chem., 70, 238 (1958); Ger. 1,011,432 (1955); C. 1958, 3169.
839. Suckfüll, F., and H. Haubrich, Ger. 1,008,313 (1955); C. 1958, 1990.
840. Sumitomo Chemical Co., Japan. 25,052 (1964); C. A. 62, 13158 h (1965).
841. Sumitomo Chemical Co., Belg. 640,798 (1964); C. A. 63, 1767 h (1965).
842. Sumitomo Chemical Co., Belg. 650,729 (1964); C. A. 63, 13152 b (1965).
843. Sumitomo Chemical Co., Japan. 10,574 (1965); C. A. 63, 11569 c (1965).
844. Sumitomo Chemical Co., Fr. 1,408,408 (1965); C. A. 64, 4995 h (1966); U. S. 3,235,448 (1966); C. A. 64, 17485 h (1966).
845. Sumitomo Chemical Co., Brit. 1,040,734 (1966); C. A. 65, 15272 d (1966).
846. Sunagawa, G., Y. Sato, and H. J. Nakano, J. Pharm. Soc. Japan, 77, 1176 (1957); C. A. 52, 6304 c (1958).
847. Sundberg, R. J., Tetrahedron Letters, 1966, 477.
848. Sung, Wei-Liang, S. C. Hou, H. F. Chao, C. H. Yang, and H. H. Ku, Yao Hsueh Pao, 13, 126 (1966); C. A. 65, 7261 h (1966).
849. Szekerke, M., J. Csaszar, and V. Brückner, Chem. Ind., 1964, 1385.
850. Szekerke, M., J. Csaszar, and V. Brückner, Acta Chim. Acad. Sci. Hung., 46, 379 (1965); C. A. 64, 12787 b (1966).

851. Thiem, N. van., N. van Thoai, and J. Roche, Bl. Soc. Chim. Biol., 44, 285 (1962).
852. Thieme, B., Ann., 272, 212 (1893).
852a. Thilo, E., and H. Grunze, Z. Anorg. Allg. Chem., 281, 262 (1955).
852b. Thomas, D. G., J. H. Billmann, and C. E. Davis, J. Am. Chem. Soc., 68, 895 (1946).
853. Thomas, L. C., and R. A. Chittenden, Spectrochim. Acta, 20, 467 (1964).
854. Thomas, L. C., and R. A. Chittenden, Spectrochim. Acta, 26A, 781 (1970).
855. Tilford, C. H., F. J. McCarty, and M. G. van Campen Jr., U. S. 2,832,786 (1956); C. A. 52, 14705 g (1958).
856. Titherly, A. W., and E. Worrall, J. Chem. Soc., 95, 1143 (1909); Proc. Chem. Soc., 25, 150.
857. Tolkmith, H., J. Am. Chem. Soc., 75, 5270, 5276 (1953).
858. Tolkmith, H., J. Am. Chem. Soc., 84, 2097 (1962).
859. Tolkmith, H., J. Am. Chem. Soc., 85, 3248 (1963).
860. Tolkmith, H., U. S. 2,620,355 (1952); C. A. 47, 9342 (1953).
861. Tolkmith, H., U. S. 2,668,825/831/832/833 (1954); C. A. 49, 5517 ff (1955).
862. Tolkmith, H., U. S. 2,668,835/836/837 (1954); C. A. 49, 5518 f (1955).
863. Tolkmith, H., U. S. 2,654,780/781/782/783/784 (1953); C. A. 48, 10050 f (1954).
864. Tolkmith, H., E. C. Britton, and C. F. Holoway, U. S. 2,802,823/824 (1957); C. A. 52, 2065, 2083 (1958).
865. Tolkmith, H., and E. C. Britton, U. S. 2,805,256 (1957); C. A. 52, 2086 b (1958).
866. Tolkmith, H., U. S. 2,831,015 (1956); C. A. 54, 3315 f (1960).
867. Tolkmith, H., and K. C. Kauer, U. S. 2,879,286 (1956); C. A. 53, 15009 d (1959).
868. Tolkmith, H., and E. C. Britton, U. S. 2,908,706 (1959); C. A. 54, 5573 d (1960).
869. Tolkmith, H., U. S. 2,965,667/668 (1960); C. A. 55, 14381 (1961).
870. Tolkmith, H., U. S. 3,336,419 (1967); C. A. 69, 77479 v (1968).
871. Tolkmith, H., and E. C. Britton, J. Org. Chem., 24, 705 (1959).
872. "Topics in Phosphorus Chemistry," Vol. 4; Interscience Publishers, J. Wiley and Sons, New York, 1967.
873. "Topics in Phosphorus Chemistry," Vol. 5; Interscience Publishers, J. Wiley and Sons, New York, 1967.

874. Toru Minami, Hisaya Miki, and Toshio Agawa, Kogyo Kagarku Zasshi, 70, 1829 (1967); C. A. 68, 38988 (1968).
874a. Toy, A. D. F., J. Am. Chem. Soc., 66, 499 (1944).
875. Toy, A. D. F., and J. R. Costello Jr., U. S. 2,717,249 (1950); C. A. 50, 8747 c (1956).
876. Toy, A. D. F., and E. N. Walsh, Inorg. Synth., 7, 73 (1963).
877. Traylor, P. S., and F. H. Westheimer, J. Am. Chem. Soc., 87, 553 (1965).
878. Tsolis, A. K., W. E. McEwen, and C. A. Vander, Tetrahedron Letters, 1964, 3217.
879. Tzeng-Shou Tung, Acta Chim. Sinica, 23, 307 (1958); C. A. 52, 19903 g (1958).
880. Tzeng-Shou Tung and Shyh-Tsang Chern, Acta Chim. Sinica, 24, 30 (1959); C. A. 53, 3113 g (1959).
881. Uhlenbrock, J. H., and B. G. van Bos, U. S. 3,051,615 (1960); C. A. 58, 1436 c (1963).
882. Uhlfelder, E., Ber. Dtsch. Chem. Ges., 36, 1824 (1903).
883. U. S. Department of Commerce, P. B. Report, 73, 754.
884. Utvary, K., V. Gutmann, and Ch. Kemenater, Inorg. Nucl. Chem. Letters, 1, 75 (1965).
885. Vagi, K. M., V. M. Adamkiewicz, and T. Nogrady, Can. J. Chem., 40, 1049 (1962).
886. Vandi, A., and Th. Moeller, Inorg. Synth., 8, 116 (1966); and Moeller, T., and A. Vandi, J. Org. Chem., 27, 3511 (1962).
887. Vetter, H. J., Z. Naturforsch., 19 b, 72, 167 (1964).
888. Vetter, H. J., and H. Noeth, Chem. Ber., 96, 1308 (1963).
889. Vilkov, L. V., and L. S. Khaikin, Dokl. Akad. Nauk SSSR, 168, 810 (1966); C. A. 65, 8724 a (1966).
890. Vives, J. P., and F. Mathis, Compt. Rend., 246, 1879 (1958).
891. Wadsworth, W. S., and W. D. Emmons, J. Am. Chem. Soc., 84, 610 (1962).
892. Wadsworth, W. S., and W. D. Emmons, J. Am. Chem. Soc., 84, 1316 (1962).
893. Wadsworth, W. S., and W. D. Emmons, J. Org. Chem., 29, 2816 (1964).
894. Wagner-Jauregg, T., J. J. O'Neill, and W. H. Summerson, J. Am. Chem. Soc., 73, 5202 (1951).
895. Wallach, O., Ann., 184, 1 (1877).
896. Wallach, O., and T. Heymer, Ber. Dtsch. Chem. Ges., 8, 1235 (1875).
897. Walsh, E. N., J. Am. Chem. Soc., 81, 3023 (1959).
898. Walsh, E. N., and A. D. F. Toy, Inorg. Synth., 7, 69 (1960).
899. Warnat, K., U. S. 2,305,751 (1942).

900. Warnat, K., U. S. 2,316,908 (1939); C. A. 37, 5832 (1943); cf. C. A. 36, 4974 (1942).
901. Wazer, J. R. van, C. F. Callis, J. N. Shoolery, and R. C. Jones, J. Am. Chem. Soc., 78, 5715 (1956).
902. Wegler, R., Chemie der Pflanzenschutz- und Schädlingsbekämpfungsmittel, Band I und II, Springer-Verlag Heidelberg 1970/71.
903. Weil, Th., J. Org. Chem., 28, 2472 (1963).
903a. Weiss, J., and G. Hartmann, Z. Naturforsch. 21b, 891 (1966).
904. Wichelhaus, H. E., Ber. Dtsch. Chem. Ges., 2, 502 (1869).
905. Wieczorkowski, J., J. Chem. Ind., 1963, 825.
906. Wieczorkowski, J., Bull. Acad. Polon. Sci. Ser. Sci. Chim., 13, 155 (1965); C. A. 63, 6837 h (1965).
907. Wiegraebe, W., and H. Bock, Chem. Ber., 101, 1414 (1968).
908. Wiegraebe, W., H. Bock, and W. Luettke, Chem. Ber., 99, 3737 (1966).
908a. Wieland, T., and H. Bernhard, Ann. 572, 190 (1951).
909. Wiesboeck, R. A., J. Org. Chem., 30, 3161 (1965).
910. Williamson, C. E., T. J. Sayers, A. M. Seligman, and B. Witten, J. Med. Chem., 10, 511 (1967).
911. Winberg, H. E., U. S. 3,121,084 (1964); C. A. 60, 13197 f (1964).
912. Winnick, T., and E. M. Scott, Arch. Biochem., 12, 201 (1947).
913. Wise, G., and H. P. Lankelma, J. Am. Chem. Soc., 74, 529 (1952).
914. Wolff, M. E., and A. Burger, J. Am. Chem. Soc., 79, 1970 (1957).
915. Yamasaki, T., and T. Sato, Sci. Rep. Tokoku Univ., 6 A, 384 (1954); C. A. 50, 314 d (1956).
916. Yarmukhametova, D. K., and I. V. Cheplanova, Izv. Akad. Nauk SSSR, Ser. Khim., 1964, 1998; C. A. 62, 7667 h (1965).
917. Yarmukhametova, D. K., and I. V. Cheplanova, Izv. Akad. Nauk SSSR, Ser. Khim., 1967, 602; C. A. 67, 99773 x (1967).
918. Yarmukhametova, D. K., and I. V. Cheplanova, Izv. Akad. Nauk SSSR, Ser. Khim., 1967, 1511; C. A. 68, 12603 r (1968).
919. Yoshitomi, Pharmaceutical Ind. Ldt., Japan. 7379 (1967); C. A. 67, 81947 x (1967).
920. Zeile, K., Hoppe Seylers, Z. f. Physiol. Chem., 236, 263 (1935).
921. Zeile, K., and G. Fawaz, Z. Physiol. Chem., 256, 193 (1938).
922. Zeile, K., and G. Fawaz, Z. Physiol. Chem., 263, 175 (1940).
923. Zeile, K., and W. Kruckenberg, Ber. Dtsch. Chem. Ges., 75, 1127 (1942).

924. Zenkevich, A. G., et al., Zh. Obshch. Khim., 30, 2317 (1960); C. A. 55, 9320 g (1961).
925. Zervas, L., and I. Dilaris, J. Am. Chem. Soc., 77, 5354 (1955).
926. Zervas, L., and P. G. Katsoyannis, J. Am. Chem. Soc., 77, 5351 (1955).
927. Zetsche, F., and W. Büttiker, Ber. Dtsch. Chem. Ges., 73, 47 (1940).
928. Zhmurova, I. N., and B. S. Drach, Zh. Obshch. Khim., 34, 1441 (1964); C. A. 61, 5499 f (1964).
929. Zhmurova, I. N., and B. S. Drach, Zh. Obshch. Khim., 34, 3055 (1964); C. A. 62, 14480 e (1965).
930. Zhmurova, I. N., B. S. Drach, and A. V. Kirsanov, Ukr. Khim. Zh., 31, 223 (1965); C. A. 63, 1689 d (1965); Zh. Obshch. Khim., 35, 718 (1965); C. A. 63, 4149 a (1965).
931. Zhmurova, I. N., B. S. Drach, and A. V. Kirsanov, Zh. Obshch. Khim., 35, 344 (1965); C. A. 62, 13030 b (1965).
932. Zhmurova, I. N., B. S. Drach, and A. V. Kirsanov, Zh. Obshch. Khim., 35, 1018 (1965); C. A. 63, 9796 h (1965).
933. Zhmurova, I. N., and A. V. Kirsanov, Zh. Obshch. Khim., 29, 1687 (1959); C. A. 54, 8688 i (1960).
934. Zhmurova, I. N., and A. V. Kirsanov, Zh. Obshch. Khim., 30, 4048 (1960); C. A. 55, 22197 a (1961).
935. Zhmurova, I. N., and A. V. Kirsanov, Zh. Obshch. Khim., 30, 3044 (1960); C. A. 55, 17551 c (1961).
936. Zhmurova, I. N., and A. V. Kirsanov, Zh. Obshch. Khim., 32, 2576 (1962); C. A. 58, 7848 f (1963).
937. Zhmurova, I. N., and A. P. Martynyuk, Khim. Org. Soedin, Fosfora, Akad. Nauk SSSR, Otd. Obshch. Tekh. Khim., 1967, 191; C. A. 69, 2442 n (1968).
938. Zhmurova, I. N., and A. P. Martynyuk, Zh. Obshch. Khim., 37, 896 (1967); C. A. 68, 29350 d (1968).
939. Zhmurova, I. N., Y. Voitsekhovskaya, and A. V. Kirsanov, Zh. Obshch. Khim., 31, 3741 (1961); C. A. 57, 9702 g (1962).
940. Zhuravleva, L. P., M. A. Grinyuk, and A. V. Kirsanov, Zh. Obshch. Khim., 35, 998 (1965); C. A. 63, 9799 f (1965).
941. Zhuravleva, L. P., and A. V. Kirsanov, Zh. Obshch. Khim., 32, 3752 (1962); C. A. 58, 11253 h (1963).
942. Zimmer, H., and A. D. Sill, Naturwiss., 49, 256 (1962).
943. Zioudrou, C., Tetrahedron, 18, 197 (1962).
944. Zsako, I., L. Almasi, M. Giurgiu, and A. Hantz, Zh. Obshch. Khim., 35, 1866 (1965); C. A. 64, 3331 d (1966).

945. Zsako, I., M. Giurgiu, L. Almasi, and A. Hantz, Rev. Roumaine Chim., 11, 1019 (1966); C. A. 66, 41123 y (1967).
946. Zsako, J., M. Giurgin, L. Almasi, and A. Hantz, Z. Anorg. Allg. Chem., 359, 220 (1968).
947. Zubrabyan, S. E., S. S. Keblas, and I. L. Knunyants, Izv. Akad. Nauk SSSR, Ser. Khim., 1964, 2036; C. A. 62, 9230 c (1965).
948. Zwiezak, A., Bull. Akad. Polon. Sci. Chim., 13, 609 (1965); C. A. 64, 9576 e (1966).

Notes added in proof concerning the problem:

a) Phosphorothiocyanatidate/Phosphoroisothiocynatidate. The first synthesis of a phosphorothiocyanatidate $(Me_3CCH_2O)_2P(O)\text{-}S\text{-}Cl + AgCN \longrightarrow AgCl + (Me_3CCH_2O)_2P(O)\text{-}SCN$ was achieved by Lopusinski A., and J. Michalski, Angew. Chem., 84, 896 (1972).

b) Influence of Y in $Y\text{-}\overline{N}{=}PX_2(OR)$ upon eventual rearrangement to N-alkyl compounds.

$$Y\text{-}N{=}PX_2(OR) \longrightarrow Y\text{-}NR\text{-}P(O)X_2$$

see Roesky, H. W., and W. Grosse Bowing, Z. Anorg. Allg. Chem., 386, 191 (1971).

c) Chain lengthening.

$$R\text{-}N\text{-}PCl_3 + (Me_3Si)_2NH \longrightarrow R\text{-}N{=}PCl_2\text{-}NH\text{-}SiMe_3 + Me_3SiCl$$

$$R\text{-}N{=}PCl_2\text{-}NH\text{-}SiMe_3 + PX_5 \longrightarrow R\text{-}N{=}PCl_2\text{-}N{=}PX_3 + HX + Me_3SiX$$

see Roesky, H. W., Chem. Ber., 105, 1439 (1972).

d) On structure and purity of reaction products of nitriles with PCl_5.
see Fluck, E., and W. Steck, Z. Anorg. Allg. Chem., 387, 349 (1972).

Chapter 17. Cyclophosphazenes and Related Ring Compounds

R. KEAT

Chemistry Department, University of
Glasgow, W.2., Scotland

R. A. SHAW

Department of Chemistry, (Birkbeck College)
University of London, Malet Street, WC1E 7HX
England

Contents

The products of the reaction of ammonium chloride with phosphorus pentachloride, the phosphazenes, a group of compounds containing the $-X_2P{=}N-$ unit, have proved a source of fascination and inspiration for chemists for well over one hundred years. Even those pioneers of chemical thought in the nineteenth century, such as Davy,[129] Liebig,[298] and Stokes,[477-480] who were originally associated with this reaction, would surely have found it difficult to envisage the formation of well over one thousand derivatives of $N_3P_3Cl_6$ (I) and its homologues, which are described in this chapter. Various aspects of cyclophosphazene chemistry have been reviewed in recent years,[35,117,119,209,217,219a,258,365,366,367,415,419,432-435,437,439,515] but the chemistry of the linear polymeric phosphazenes, $-(N{=}PX_2)_n-$, is not as well documented.

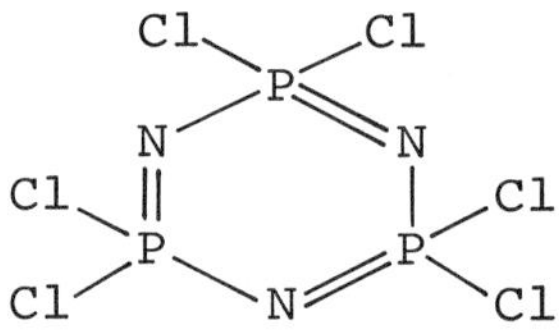

(I)

The properties of these polymers vary widely according to the method of preparation, and for this reason, are not included here, but the reader is referred to a recent authoritative review[263] of this topic. Phosphazene chemistry is now reviewed annually.[258]

The nomenclature used in this chapter is that described previously,[437] where the names are based on the presence of the recurring phosphazene linkage, $-N{=}P\lessgtr$.

A. SYNTHETIC ROUTES

A.1. Cyclophosphazenes, $(NPX_2)_n$, and Their Derivatives by Ring-Forming Reactions

I. X = HALOGEN

The halogenocyclotriphosphazatrienes are the precursors of most of the cyclophosphazene derivatives to be described and accordingly, considerable effort has been expended in the perfection of synthetic routes, especially those to the chloro-derivatives. Hexachlorocyclotriphosphazatriene, $N_3P_3Cl_6$, was first identified by Liebig,[298] and obtained by the reaction of phosphorus pentachloride with ammonia or ammonium chloride. Later, Stokes[477-478] found that a homologous series of cyclophosphazenes may be obtained, as expressed by the equation:

$$PCl_5 + NH_4Cl \longrightarrow (1/n)(NPCl_2)_n + 4HCl$$

The reaction was carried out in sealed tubes, or in sym-tetrachloroethane, and the homologs n = 3-6 were isolated. It is essentially the latter method[300,350] that is now generally used for the preparation of the chlorides. Improved fractional distillation techniques have enabled the homologs n = 7 and 8 to be separated also.[300]

In the preparation of the chlorides, phosphorus pentachloride and ammonium chloride are boiled under reflux in sym-tetrachloroethane for periods up to 20 hr. Unreacted ammonium chloride is filtered off and the solvent removed. The cyclic oligomers are generally obtained with a mixture of linear salt-like phosphazenes (the proportions of which vary with the reaction conditions). The linear products are generally insoluble in petrol, so that the cyclic products may be extracted with this solvent. It has recently been shown that nearly 100% of cyclic products may be obtained from the reaction of phosphorus pentachloride with ammonia. The initial product isolated from the reaction in nitrobenzene is $[Cl_3P{-}N{=}PCl_3]^+[PCl_6]^-$,[42] and it has been suggested[42] that this salt is formed via PCl_4NH_2 and $PCl_3{=}NH$. More credence for the formation of the first intermediate, PCl_4NH_2, was recently supplied by a study of the PCl_5/NH_3 reaction at low temperatures, where $(H_2N)_4P^+Cl^-$ was isolated.[415] Growth of the linear phosphazene chains may then occur by steps such as

$$[Cl_3P{-}N{=}PCl_3]^+\ X^- + NH_4Cl \longrightarrow$$

(a)

$$[Cl_3P{-}N{=}PCl_2{-}N{=}PCl_3]^+\ X^- + 4HCl$$

$$\Big\downarrow PCl_3NH \quad \text{(b)}$$

$$[Cl_3P(N{=}PCl_2)_2N{=}PCl_3]^+\ X^- + HCl$$

(c)

$$(X^- = Cl^- \text{ or } PCl_6^-)$$

Further chain lengthening may then occur by condensation of these cations;[153] alternatively, intramolecular cyclization of (c) could give $N_3P_3Cl_6$.[153] The formation of higher homologs may, in general, proceed by intramolecular condensation of linear phosphazenes with one more phosphorus atom in the linear chain than in the cyclic products. Salts derived from the cations (a),[42] (b),[42] and (c)[179a] have been isolated and their formations during the reaction have been carefully followed by ^{31}P NMR.[153] Yields of cyclic chlorophosphazenes are improved when finely ground ammonium chloride is used,[153] and also when certain metallic salts are present.[153]

Other preparative routes to chlorophosphazenes include the reaction of ammono-basic mercuric chloride,[425a] or silicon-nitrogen compounds[517a] with phosphorus pentachloride, the chlorination of phosphorus thionitride,[342] and the reaction between tetrasulphur tetranitride, thionyl chloride, and phosphorus trichloride.[143]

Only two direct syntheses of fluorocyclophosphazenes have been reported,[310] which involve the reaction of NF_3 or of CF_3SF_5 with N_5P_3 at 700° --routes unsuitable for routine synthesis.

Cyclic bromophosphazenes, $(NPBr_2)_n$, are generally obtained by routes related to those for the chlorophosphazenes. Thus phosphorus pentabromide and ammonium bromide in refluxing sym-tetrachloroethane[242-244] give the cyclic homologues where n = 3,4. Separation of these homologs has been achieved by fractional crystallization from light petroleum. The use of sym-tetrabromoethane as a solvent for the reaction appears to be useful when preferential formation of the tetramer, n = 4, is required. Higher homologs (n = 5,6) have also been isolated from reactions in this solvent.[115,116] The halides PCl_5 or PCl_3Br_2 and ammonium bromide have provided a convenient route to mixed chloro-bromocyclotriphosphazatrienes, $N_3P_3Cl_{6-n}Br_n$ (n = 1-5)[111,391,472] which were separated by GLC. A related series of nongeminal chloride-bromides were obtained by chlorination of $N_3P_3Br_6$ by mercuric chloride.[112]

Attempts to prepare iodo-derivatives of cyclophosphazenes by direct iodination have been unsuccessful.[367-371]

II. X = ORGANO-GROUP OR HALOGEN

The reaction of ammonium chloride with organo-substituted derivatives of phosphorus pentachloride or pentabromide (phosphoranes) provides a versatile route to nongeminal homologues, $(NPXR)_{3,4}$ (X = Cl, R = alkyl,[62] aryl,[199,200,201,215,441,442] NMe_2,[416] X = Br, R = aryl,[332,344] and, to a lesser extent, derivatives of the type, $(NPR_2)_{3,4}$ (R = alkyl, aryl).[218,272] The phosphoranes are hydrolytically unstable and are best prepared _in situ_, e.g., $PhPCl_2 + Cl_2 \longrightarrow PhPCl_4$. With the aryl phosphoranes, tetrameric, rather than trimeric ring systems are generally favored:[215,235,442]

e.g., $4PhPCl_4 + 4NH_4Cl \longrightarrow N_4P_4Cl_4Ph_4 + 16HCl$
(3 geometrical isomers obtained)

A comprehensive study of the reaction of phenyltetrachlorophosphorane with ammonium chloride[215,216] shows that the proportion of trimer to tetramer in the $(NPClPh)_n$ series can be raised by the use of chlorobenzene as a solvent for the reaction. The analogous bromides, $PhPBr_4$ and NH_4Br, generally provide larger yields of the trimer $N_3P_3Br_3Ph_3$ (two isomers identified)[332,344] than the tetramer $N_4P_4Br_4Ph_4$ (one isomer identified).[344]

Yields of the fully phenylated trimer, $N_3P_3Ph_6$, and tetramer, $N_4P_4Ph_8$, from the reaction of ammonium chloride with diphenyltrichlorophosphorane in sym-tetrachloroethane are generally poor (< 10%),[218,272] but can be improved by the use[272] of ortho-dichlorobenzene as a solvent. The intermediate, $[Ph_2PCl\text{-}N\text{=}PClPh_2]^+Cl^-$ (cf. $[Cl_3P\text{=}N\text{-}PCl_3]^+[PCl_6]^-$ from PCl_5 and NH_4Cl) has been isolated[181] in good yield from Ph_2PCl_3 and NH_4Cl, which suggests that the initial stages of the reaction may proceed in a manner closely related to that of PCl_5 and NH_4Cl. Treatment of the phenylated intermediate with ammonia gives $[Ph_2P(NH_2)\text{=}N\text{-}PPh_2NH_2]^+Cl^-$, an air-stable solid, which has proven[450] to be a useful intermediate in cyclisation reactions. Pyrolysis of this salt, and its reaction with Ph_2PCl_3 both provide good yields of $N_4P_4Ph_8$, or $N_3P_3Ph_6$ with $N_4P_4Ph_8$, respectively.[450] With PCl_5, the salt gives the geminal phenylchloro-derivatives, $N_3P_3Cl_2Ph_4$[421] and $N_4P_4Cl_4Ph_4$.[227] In the same way, ring closure by $PhPCl_4$ provides pentaphenyl and hexaphenyl-derivatives, $N_3P_3ClPh_5$[227,423] and $N_4P_4Cl_2Ph_6$,[227] the latter containing the -PClPh=N-PPhCl- grouping. The product of ring closure by dimethyltrichlorophosphorane appears to be more basic, for only the hydrochloride shown has been isolated.[51a]

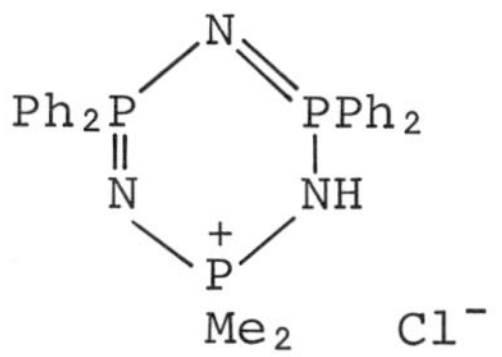

The synthesis of alkyl-derivatives of cyclophosphazenes has attracted much less attention than that of the aryl derivatives. Nevertheless, it is known that dialkyltrichlorophosphoranes and ammonium chloride give the fully alkylated derivatives in refluxing sym-tetrachloroethane:

$$R_2PCl_3 + NH_4Cl \longrightarrow (1/n)(NPR_2)_n + 4HCl$$
(R = Me,[364,426,427] Et,[62] n = 3,4)

The addition of triethylamine increases the rate of hydrogen chloride abstraction and consequently, an improved yield of cyclophosphazenes is obtained (when R = Me).[426] Diethyltrichlorophosphorane gave an adduct with liquid ammonia, which gave $(NPEt_2)_{3,4}$ on pyrolysis.[62] A similar reaction with $(CF_3)_2PCl_3$ gives $(CF_3)_2P(NH_2)Cl_2$, which may be dehydrochlorinated by trimethylamine to give $[N_3P_3(CF_3)_2]_{3,4}$, and a polymer, $\{NP(CF_3)_2\}_n$.[482]

These methods have been extended to prepare nongeminal dimethylamino-derivatives of $N_3P_3Cl_6$:

$$3Me_2NPCl_4 + 3NH_4Cl \longrightarrow N_3P_3Cl_3(NMe_2)_3 + 12HCl.$$

No tetrameric derivative was isolated and the trimers were obtained as a mixture of cis- and trans-isomers.[415] Attempts to cyclise bis(amino)trichlorophosphoranes, e.g., $(Me_2N)_2PCl_3$, have been unsuccessful.[415]

III. OTHER RING-FORMING REACTIONS

A number of other routes to $N_3P_3Ph_6$ and $N_4P_4Ph_8$, have been established, all of which involve the pyrolysis of phenylated mononuclear phosphorus compounds:

$Ph_2PN_3 \longrightarrow N_3P_3Ph_6$ (Refs. 273,362)
$Ph_2PCl + NH_2Cl \longrightarrow N_3P_3Ph_6 + N_4P_4Ph_8$ (Refs. 191,198)
$Ph_2PNHNMe_2 \longrightarrow N_3P_3Ph_6$ (Ref. 511)
$Ph_2P(S)NHR \longrightarrow N_3P_3Ph_6$ (R = H, Me) (Ref. 438)

In all cases, the phenyl-phosphorus bond remained intact.

Linear diphosphazadienes have been used as the precursors of an extensive series of cyclodiphosphazatrienes, compounds with P, N, and C atoms in a formally unsaturated ring system, e.g.,

$$4H_2NC(R){:}NH + [ClP(X_2){=}N{-}PX_2Cl]^+Cl^-$$
(or hydrochloride)

$\xrightarrow[20^\circ]{CH_2Cl_2}$ 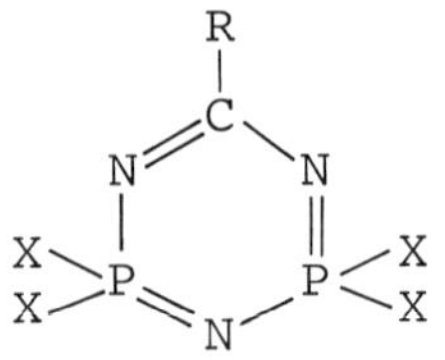 + $3RC(NH_2)_2Cl$

(R and X are alkyl or aryl substituents)

$$\text{PhC}\begin{array}{c}\diagup\text{N}=\text{PCl}_3\\ +\\ \diagdown\text{N}-\text{PCl}_3 \ \ \text{SbCl}_6^-\end{array} + 2\text{NH}_4\text{Cl} \longrightarrow$$

$$\begin{array}{c}\text{Ph}\\ |\\ \text{C}\\ \diagup\diagdown\\ \text{N} \qquad \text{N}\\ | \qquad\quad ||\\ \text{Cl}_2\text{P} \qquad \text{PCl}_2\\ \diagdown\quad\diagup\\ \text{N}\end{array} \qquad + \ 4\text{HCl} + \text{NH}_4^+\text{SbCl}_6^-$$

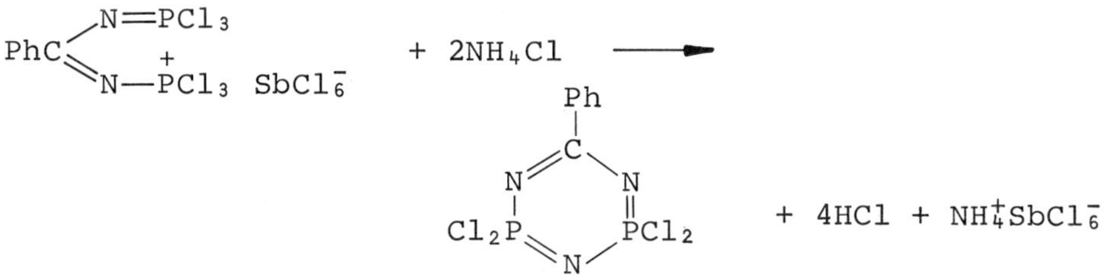

Several cyclophosphazenes in which one or more atoms of the ring have been replaced by another group have been described in the literature. In particular, there are several systems known in which one or two phosphorus or nitrogen atoms have been replaced by the following groups: $S=O$, $C-Cl$, $C=O$, $B-R$, $S-R$, SR_2, SO_2, CH_2, O, S, $SbCl_3$, and metals. Details of the synthesis of these rings will be found in Section D.3.

A.2.   Reactions Involving Chlorine-Atom Displacement from Chlorocyclophosphazenes

IV.   FLUORINATION

Fluorocyclophosphazenes, $(NPF_2)_n$ ($n = 3-9$), are generally obtained by reaction of a metal fluoride, such as NaF[425] suspended in acetonitrile or KSO_2F in nitrobenzene,[95,428,429] with the corresponding chlorocyclophosphazene. Heavy metal fluorides may also be used to effect fluorination, but they have the disadvantage that fluorination is not always complete and that changes in ring size may occur.[417,418] Cyclic homologs up to $n = 17$ have been identified[95] as products of the reaction of KSO_2F with a mixture of chlorides $(NPCl_2)_n$ The fluoride-chlorides, $N_3P_3Cl_{6-n}F_n$ ($n = 1-5$) are best made[151] from reactions of $N_3P_3Cl_6$ with NaF in refluxing nitromethane ($n = 1-3$), or with NaF in nitrobenzene ($n = 4-5$).[151] Tetrameric fluoride-chlorides, $N_4P_4Cl_{8-n}F_n$ have been obtained by reactions with KSO_2F in the absence of solvent.[23,151] Fluorination of $N_4P_4Cl_8$ proceeds very rapidly, so that the mixed halides are best obtained by removal from the reaction mixture as soon as they are formed. Fluoride-chlorides (bromides) have also been obtained from the reaction of dimethylamino-derivatives, e.g., $N_3P_3F_{6-n}(NMe_2)_n$ with HCl or HBr[100,193] (see Section B).

V. ALKYLATION AND ARYLATION

The products of alkylation of halogenocyclophosphazenes by organometallic reagents are generally difficult to characterise because the reaction is often accompanied by phosphazene ring cleavage and polymerization. The reactions are often exceedingly complex (cf. the more thoroughly investigated phenylation reactions below). This ring cleavage appears to be minimized[328,336] in the case of the fluoride $N_3P_3F_6$, where alkylation by n-butyl lithium under mild conditions gave $N_3P_3Cl_{6-n}(n\text{-}Bu)_n$ (n = 1,2). It has been found that the presence of an amino-group at the phosphorus atom undergoing alkylation ensures a smoother reaction:[485]

$$N_3P_3Cl_3(NMe_2)_3 + 3MeMgI \longrightarrow N_3P_3Me_3(NMe_2)_3 + 3MgICl$$

The amino groups can then be removed again by reaction with hydrogen chloride to give $N_3P_3Cl_3Me_3$.

Arylation by organometallic reagents has been more thoroughly studied. $N_3P_3Cl_6$ and phenyl magnesium bromide generally gives small quantities of $N_3P_3Ph_6$, and a complex mixture of linear phenylated phosphazenes which may include $Ph_3P{=}N{-}PPh_2{=}N{-}PPh_2{=}NMgBr$.[59] With $N_4P_4Cl_8$ the reaction is slower and two major products, both formulated $N_4P_4Cl_4Ph_4$, which have the structures (2) and (3) are obtained,[60] together with small quantities of $N_4P_4Ph_8$.

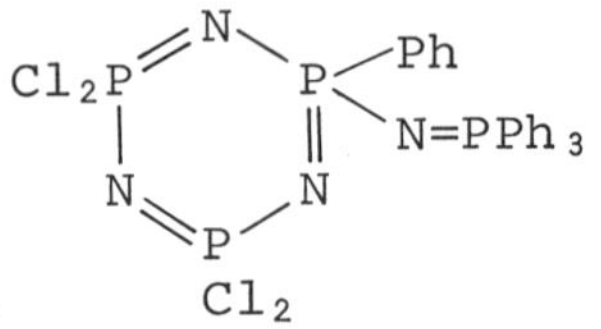

(2)

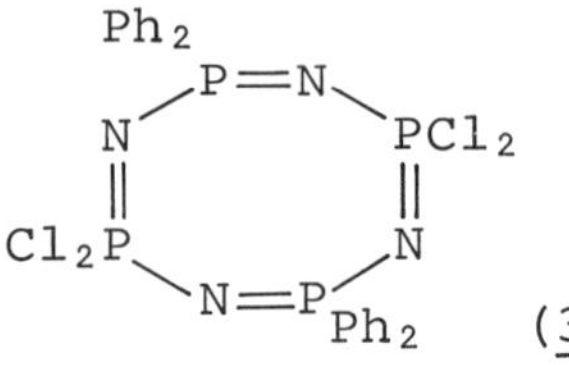

(3)

The formation of linear structures is less important in this case, possibly because recyclization is better accomplished with a longer chain. Diphenyl magnesium and $N_3P_3Cl_6$ in dioxan give the unusual products (4)[61] and (5)[61] in addition to $N_3P_3Ph_6$.

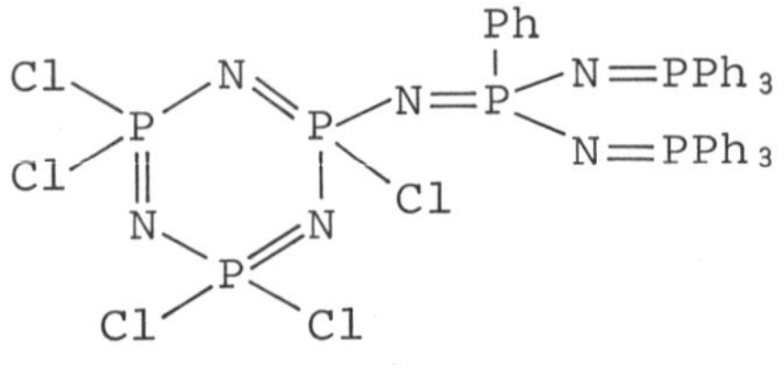

(4)

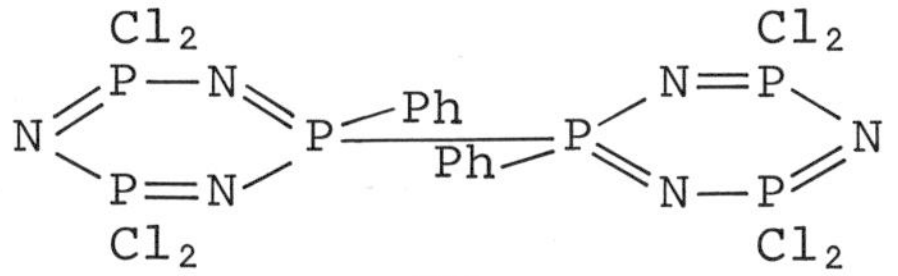

(5)

Phenyl lithium and n-butyl lithium, like phenyl magnesium bromide cause extensive ring cleavage in $N_3P_3Cl_6$.[439] However, analogous reactions of $N_3P_3F_6$ with phenyl lithium and phenyl magnesium bromide give good yields of arylfluorocyclotriphosphazatrienes, $N_3P_3F_{6-n}Ph_n$ (n = 1-4). When n = 2, phenylation by phenyl lithium[20] gives a non-geminal product, but with phenyl magnesium bromide the product obtained is a geminal isomer.[22]

VI. FRIEDEL-CRAFTS REACTIONS

Friedel-Crafts reactions of chloro- and fluorocyclophosphazenes provide a route to geminally substituted derivatives, e.g., $N_3P_3Cl_4Ph_2$ and $N_3P_3Cl_2Ph_4$.[1,45,65,320a,345,444] Yields of these derivatives, for example,

$$N_3P_3Cl_6 + 2PhH \xrightarrow{AlCl_3} N_3P_3Cl_4Ph_2 + 2HCl$$

may be improved by the use of a tertiary aliphatic amine as hydrogen halide acceptor.[320a] The hexaphenyl-derivative, $N_3P_3Ph_6$, has been obtained in low yield by this route.[1,320a,345] Similar reactions with $N_4P_4Cl_8$ are complex[137] and only the ring contracted derivative, $N_3P_3Cl_4Ph(N{=}PPh_3)$, and pentaphenyl-derivative, $N_4P_4Cl_3Ph_5$, have been identified. Arylation may proceed by initial formation of a complex:

$$N_3P_3Cl_6 + AlCl_3 \longrightarrow N_3P_3Cl_5^+AlCl_4^-$$

This could then be followed by nucleophilic attack of benzene at the $\equiv\overset{+}{P}Cl$ centre. Alternatively, the $AlCl_3$ may complex a ring nitrogen atom, cf. ref. 113, thus making the phosphorus atoms more electrophilic. Both routes would be consistent with the observed geminal replacement pattern, where heterolysis of a P-Cl bond is more likely to occur at the $\equiv$PClPh group than at the $\equiv PCl_2$ group. Also in agreement with this suggestion is the observation that phenylation of nongeminal amino-derivatives, such as $N_3P_3Cl_4(NMe_2)_2$ or $N_3P_3Cl_5 \cdot N{=}PPh_3$ proceeds preferentially at the $\equiv PCl \cdot NMe_2$ or $\equiv PCl \cdot N{=}PPh_3$ groups, rather than at the $\equiv PCl_2$ groups.[55,182]

VII. AMMONOLYSIS AND AMINOLYSIS

Partial ammonolysis of cyclophosphazenes can readily be accomplished by passage of ammonia through a solution of the cyclophosphazene: $N_3P_3Cl_5NH_2$,[166,169] $N_3P_3F_5NH_2$,[393] and $N_4P_4Cl_4(NH_2)_4$[44,46,166,169,255,294,338,478] are, however, the only aminohalogenocyclophosphazenes that have so far been isolated, the former by reaction of $N_3P_3Cl_4(NH_2)_2$ with HCl.[166] Complete ammonolysis is readily effected by reaction of chlorocyclophosphazenes with liquid ammonia.[34,53,325]

The aminolysis of halogenocyclophosphazenes has probably been studied more than any other aspect of phosphazene chemistry. Reactions with ca. 70 different types of primary and secondary amines have been carried out, generally by one of the following methods:

The addition of the amine to a solution of the halogenocyclophosphazene in an inert solvent such as ether is the commonest route employed. Reaction with the lower aliphatic primary and secondary amines, RNH_2 (R = Me, Et, n-Pr) R_2NH (R = Me, Et), is generally strongly exothermic and cooling to ca. -78° is necessary not only to control the reaction, but also to ensure that the product does not contain compounds with more than one degree of chlorine replacement. Complete replacement of the chlorine atoms in $N_3P_3Cl_6$, for example, with all amines requires refluxing chloroform solutions of the amine, or even the use of sealed tubes if the amine is weakly nucleophilic or has a large steric requirement. The reaction with less nucleophilic amines is assisted by the presence of triethylamine to abstract the elements of hydrogen chloride.

The addition of an aqueous solution of the amine to an ether solution of the chlorophosphazene sometimes is employed. The amine hydrochloride formed dissolves in the aqueous layer. In some cases, the amine may be generated in situ by addition of sodium hydroxide solution to an aqueous amine hydrochloride solution.

VIII. REACTION WITH SILYLAMINES

Only the fluorocyclophosphazenes undergo ready reaction with silylamines under normal conditions and, even then, complete aminolysis has not been achieved:[100,102]

$$\text{e.g., } N_3P_3F_6 + nMe_3SiNMe_2 \longrightarrow N_3P_3F_{6-n}(NMe_2)_n + nMe_3SiF$$

In a few cases, reactions of functional groups with silylamines have been achieved, e.g., $N_3P_3Cl_5(NHR) + (Me_3Si)_2NH \longrightarrow N_3P_3Cl_5NRSiMe_3 + Me_3SiNH_2$

(R = Me, Et)[315]

IX. PHOSPHAZENYL DERIVATIVES BY THE KIRSANOV REACTION

An extension of the ammonolysis reactions is the preparation of phosphazenyl-derivatives by the Kirsanov reaction.[166,167,249,250,294,393]

$$RNH_2 + X'_3PX_2 \longrightarrow RN{=}PX'_3 + 2HX$$

$$\left[\begin{array}{l} R = \text{phosphazenyl} \\ X' = Cl, Br, Ph \\ X = Cl, Br \end{array}\right]$$

These reactions are discussed in detail in Section B.

X. ALCOHOLYSIS OF CHLOROPHOSPHAZENES

Two general methods are in common use,[179] the reaction of a chlorophosphazene with an alcohol in the presence of a hydrogen chloride acceptor (B) such as pyridine or triethylamine:

e.g., $N_3P_3Cl_6 + nROH \xrightarrow{nB} N_3P_3Cl_{6-n}(OR)_n + 6nBHCl$

or the reaction with a metal alkoxide:

e.g., $N_3P_3Cl_6 + nNaOR \longrightarrow N_3P_3Cl_{6-n}(OR)_n + nNaCl$

In most cases, the product of complete alcoholysis has been isolated and relatively little is known of the products of the mixed chloroalkoxy(aryloxy)-derivatives. There appears to be little to choose between the two methods, except that it is easier to exclude traces of water (which may result in decomposition reactions) in the case of the metal alkoxides. Reactions are generally carried out in ether cooled to 0° where R = Me, Et, n-Pr. More forcing conditions for the reactions have been provided by carrying out reactions in tetrahydrofuran, dioxan, or high boiling hydrocarbons, at their reflux temperatures. Sodium chloride or amine hydrochloride is filtered off at the completion of the reaction. The solution is generally then washed with water, dried, (Na_2SO_4), the solvent removed and the product purified by crystallization or vacuum distillation.

XI. THIOALCOHOLYSIS OF CHLOROPHOSPHAZENES

Few investigations have been carried out on this topic and where well authenticated products have been obtained, sodium thioalkoxides(aryloxides) have been employed,[88,89,445] e.g.,

$$N_3P_3Cl_6 + nNaSR \longrightarrow N_3P_3Cl_{6-n}(SR)_n + nNaCl$$

(R = various alkyl and aryl substituents)

The sodium thiolates are almost insoluble in suitable reaction media (ether, tetrahydrofuran, benzene, etc.), resulting in heterogeneous reactions. Because of this, an excess of the reagent is usually employed. Given degrees of chlorine replacement are usually achieved by suitable variation of reaction conditions such as reaction solvent, temperature and time. A preliminary report of the thioethanolysis of $N_3P_3F_6$ in ether has appeared,[348] and in this case, the reaction appears to proceed more smoothly and derivatives with all degrees of fluorine atom replacement, $N_3P_3F_{6-n}(SEt)_n$ (n = 1-6) were reported.

B. BASIC CHEMISTRY

Studies of the chemistry of the cyclophosphazenes have largely been confined to reactions which involve halogen-atom replacement by other atoms or groups, and where the P-N ring system is usually, but not always, retained. Most of these studies have been confined to ammonolysis and aminolysis with the result that most structural and mechanistic information pertains to this area. Correlations may be drawn between the reactivity to halogen-atom displacement in phosphoryl- and thiophosphoryl-compounds and to that in the cyclophosphazenes, but the latter, with more than one available reaction site, can provide more information about the ways in which electronic effects can be transmitted over P-N bonds. These differences are often manifested in the occurrence of two halogen-atom displacement schemes, which are illustrated diagrammatically for the trimeric ring system:

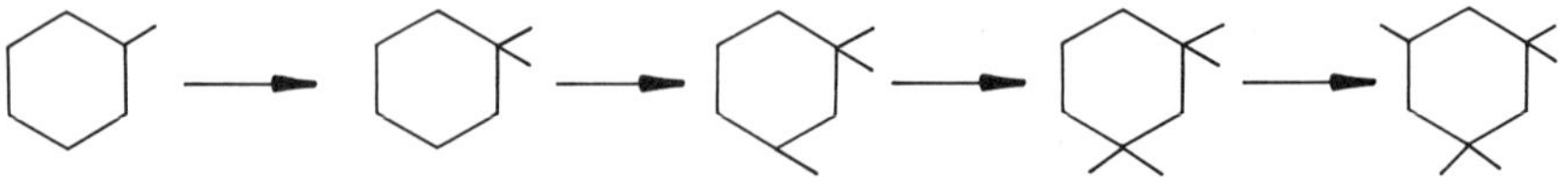

Geminal substitution*

*P- and N- atoms are omitted; the positions marked on the rings represent the incoming substituents only.

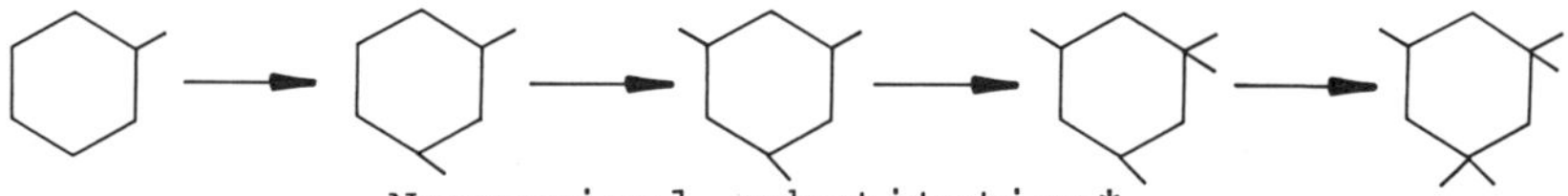

Nongeminal substitution*

More often than not, it is a mixture of the two reaction patterns that is observed rather than exclusive geminal or nongeminal replacement of chlorine atoms.

The fluorination of chlorophosphazenes by metal fluorides invariably proceeds[96,151] by a geminal route as might be expected on simple electrostatic grounds, in which F^- ion, or a closely related nucleophile, readily approaches the most electrophilic phosphorus atom, i.e., that in the ≡PFCl group rather than that in the ≡PCl_2 group. The fact that the fluorination of $N_4P_4Cl_8$ proceeds via (6) rather than (7) is also consistent with these ideas.

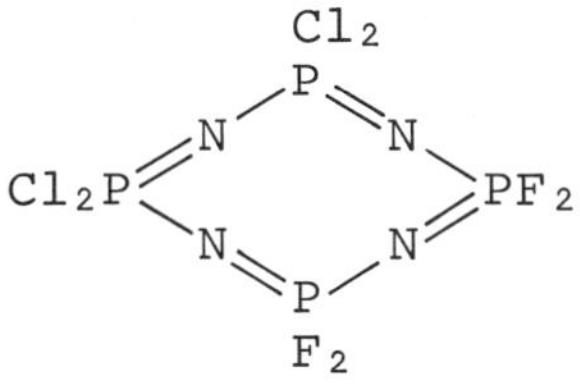

(6)

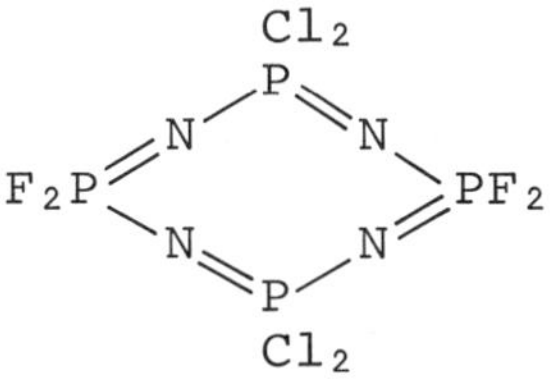

(7)

However, preliminary kinetic studies[153] with KSO_2F have shown that the ratio of the rate of formation of a monofluoride (K_1) to the rate of formation of a difluoride (K_2) is ca. 8 for $N_3P_3Cl_6$ and ca. 100 for $N_4P_4Cl_8$; this rather unexpected result may show that the π-donor effects of fluorine at phosphorus are important in determining the replacement pattern.[151] There have also been a few studies of the fluorination of aminochlorocyclophosphazenes,[206,207,208] and these showed[206,208] that the pattern of chlorine atom replacement is dependent on the nature of the fluorinating agent.

*P- and N- atoms are omitted; the positions marked on the rings represent the incoming substituents only. The possibilities for geometrical isomerism have been omitted.

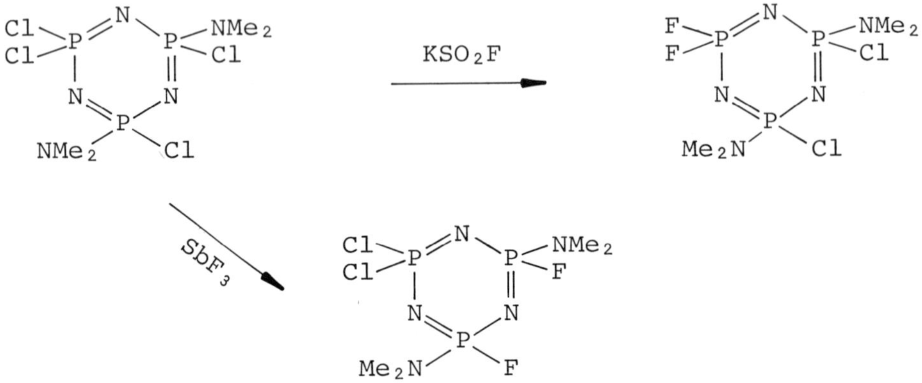

It was suggested that the unexpected route observed for SbF_3 may be due to initial complex formation with the ring N-atom between the two $\equiv PCl \cdot NMe_2$ groups as the donor site. This would be expected to make the phosphorus atoms of the $\equiv PClNMe_2$ groups more electrophilic than in the uncomplexed species. Chloride ion exchange in chlorocyclophosphazenes $(NPCl_2)_n$, n = 3,4,5,6, has been followed by tracer techniques and shown to obey a second-order rate law.[457] The order of reactivity is

$$n = 4 > n = 5 > n = 6 > n = 3$$

In contrast to the fluorination of chlorophosphazenes, the chlorination of bromocyclophosphazenes by mercuric chloride proceeds by a predominantly nongeminal route.[112]

The formation and reactions of pseudohalogeno compounds of cyclophosphazenes are typical of those of mononuclear phosphorus compounds. Substituents have included N_3, NCO, NCS, NSO, and are generally introduced via sodium salts[92,102,212,361,484] in solvents such as acetonitrile or by suitable reaction with an amino-derivative; for example,

$$N_3P_3F_5NH_2 + SOCl_2 \longrightarrow N_3P_3F_5NSO + 2HCl \quad \text{(Ref. 349)}$$

or

$$N_3P_3Cl_4(NH_2)_2 + COCl_2 \longrightarrow N_3P_3Cl_4(NH_2)NCO + 2HCl \quad \text{(Ref. 486)}$$

The thiocyanate groups readily react with amines and alcohols as expected. For example,

$$N_3P_3(NCS)_6 + 6RNH_2 \longrightarrow N_3P_3(NHCSNHR)_6$$

(R = various alkyl and aryl groups)

Few pseudohalogeno derivatives with halogen atoms remaining on the phosphorus atoms have been detected,[466] probably because the polarisable pseudohalide ions undergo ready reaction at phosphorus, possibly because of increased soft-soft interaction.

The reactions of alkylchloro- and arylchloro-cyclophosphazenes are closely related to those of the fully chlorinated analogues, e.g., see aminolysis below. The nongeminal cis- and trans-derivatives, $N_3P_3Cl_3Ph_3$, may be interconverted by aluminium chloride or ferric chloride [214] in a manner analogous to that observed for nongeminal $N_3P_3Cl_3(NMe_2)_3$, with aluminium chloride.[260] Fully alkylated derivatives form complexes with metal carbonyls;[149] for example,

$$N_4P_4Me_8 + Mo(CO)_6 \longrightarrow N_4P_4Me_8Mo(CO)_4 + 2CO$$

In this case, infrared spectroscopy suggested that molybdenum was coordinated by two nitrogen atoms at cis-positions at the molybdenum atom. Titanium and tin tetrachlorides also form adducts with $N_3P_3Me_6$.[291] Numerous weak adducts are formed by $N_3P_3Ph_6$ and acceptor species such as $HICl_2$, ICl, etc.

The established replacement patterns of ammonolysis and aminolysis of halogenocyclophosphazenes have proved so varied that it is very difficult to make generalizations on their possible mechanisms. Table 1 below summarizes the predominant chlorine atom replacement routes for a number of amines. It may be emphasized that in very few cases have derivatives corresponding to all degrees of chlorine atom replacement been obtained. Detailed kinetic studies in this area have been confined to measurements[84,85] on the rate of formation of $N_3P_3Cl_5{\cdot}NC_5H_{10}$, $N_3P_3Cl_4(NC_5H_{10})_2$ (nongeminal) and $N_3P_3Cl_5{\cdot}NEt_2$, and these exhibit in toluene solution a mixture of second- and third-order kinetics. If it is assumed that amino-groups are capable of supplying (conjugatively) electron-density to the adjacent phosphorus atom, then a nongeminal chlorine atom replacement pattern might be expected when an $S_N(2)(P)$-type mechanism is operative. The true situation is never as simple as this, but, at the same time, no evidence for an ionisation mechanism has yet been reported (see below). Geminal replacement patterns are followed by ammonia[166,294] and some primary amines, e.g., NH_2Ph, NH_2-t-Bu. The fact that t-butylamine also follows a geminal pattern suggests that the size of the amine is not the most important factor.

To account for this mode of replacement in the case of ammonia and primary amines, a proton abstraction mechanism has been proposed[123] but it may be that the

Table 1. Predominant Chlorine Atom Replacement Patterns in $N_3P_3Cl_6$ by Ammonia and Other Amines

Geminal		Nongeminal	
NH_3	(2,6)[a]	NH_2Me	(1,2,6)[a]
NHC_2H_4	(1,2,3,4,5,6)	$NHMe_2$	(1,2,3,4,6)
NH_2-i-Pr	(1,2,4,6)	NH_2Et	(1,2,3,4,6)
NH_2-t-Bu	(1,2,4,6)	$NHEt_2$	(1,2,3,6)
NH_2CH_2COOEt	(1,2,4,6)	NHC_4H_8	(1,2,3,4,5,6)
NH_2Ph	(1,2,3,4,6)	NHC_4H_8O	(1,2,3,4,5,6)
		NHC_5H_{10}	(1,2,3,4,6)
		NH_2NHPh	(2,6)

[a]Numbers in parenthese indicate the degrees of ammonolysis (aminolysis) observed; in several cases geometric and positional isomers have been identified.

nucleophile is more important than the substituent in determining the course of the reaction. This suggestion comes from the following observations.[257]

$$N_3P_3Cl_5NH\text{-t-Bu} \xrightarrow{NH_2Et} N_3P_3Cl_4(NH\text{-t-Bu})(NHEt) \text{ (nongeminal)}$$

$$N_3P_3Cl_5NHEt \xrightarrow{NH_2\text{-t-Bu}} N_3P_3Cl_4(NHEt)(NH\text{-t-Bu}) \text{ (geminal)}$$

If the substituent present on the phosphazene ring is important in determining the course of the aminolysis (as in the proton abstraction mechanism), the first reaction would be expected to give a geminal product since $N_3P_3Cl_6$ with t-butylamine gives geminal products. This is not the case. $N_3P_3Cl_6$ and ethylamine give nongeminal bis-derivatives, and consistent with the nucleophile being important in determining the course of the reaction, the second reaction gives a geminal product. Although the secondary amines, dimethylamine,[253,271] diethylamine,[121] pyrrolidine,[274-276] morpholine,[283,340] and piperidine[255] all follow predominantly nongeminal patterns, detailed studies[253,255] with dimethylamine and piperidine show that geminal bis- and tris- derivatives are favored as reaction temperatures are increased. A predominantly nongeminal route is favored by these amines:

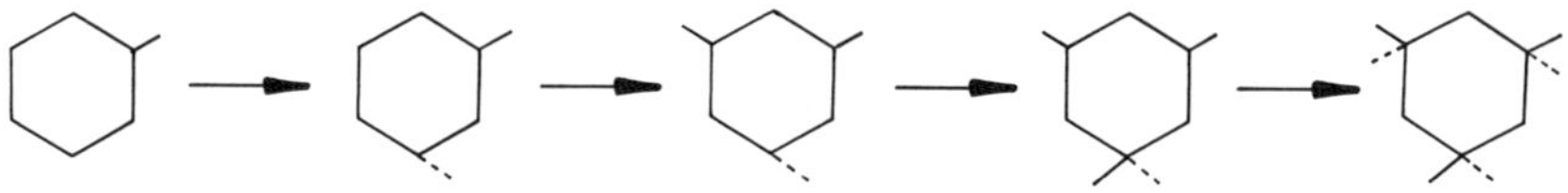

(⁄ represents an NMe_2 groups above the plane of the ring and ⋰ represents the same group below the plane of the ring. Other nongeminal bis- and tris- derivatives are isolated in small quantities).

This route, which also appears to be predominant with pyrrolidine and morpholine, led to the suggestion[255] that a "cis-effect" is important in determining the reaction path in these cases. The idea underlying the "cis-effect" is that an electron-supplying substituent, such as an $-NMe_2$ group, may transfer electron-density preferentially to a chlorine atom which is cis- to the dimethylamino-group (on an adjacent P-atom). This would have the effect of making the chlorine atom(s) cis- to the dimethylamino-group more labile so that intermediates such as those in an idealized S_N2 (P) mechanism would be favored when the incoming dimethylamino-group approaches trans- to the first. Such a suggestion would also account for the formation of a cis-tetrakisamino-derivative. There appears to be little doubt that the latter structure is thermocynamically favored.

A few pentakisamino-derivatives have been fully authenticated. A pentakispiperidino-derivative has been reported,[285,286] but the preparations could not be repeated elsewhere. It is possible that there may be a change to a relatively facile ionization mechanism of aminolysis when a hexakisamino-derivative is formed from a pentakisamino-derivative, thus accounting for the absence of the latter. It is worth noting that enhanced reactivity of cyclophosphazenes to aminolysis (or hydrolysis) may be encountered when the phosphorus atom is part of a strained five-membered ring (exocyclic to the phosphazene ring). This follows from the unusually rapid degradation of halogenocyclophosphazenes by o-aminophenol,[10] although it is not clear whether aminolysis or phenolysis is the most important reaction.

Large numbers of mixed amino-derivatives of cyclophosphazenes as well as amino (alkyl, aryl, alkoxy, aryloxy- or thioalkoxy)-derivatives have been prepared. Very often, these derivatives can give an indication of the structure of the precursors of unknown structure. Thus, structural assignments can be made[255] to the three isomeric trispiperidino-derivatives m. 114.5, 190, and -17, from a knowledge of the structures of the complimentary trisdimethylamino-derivatives.

$N_3P_3Cl_3(NC_5H_{10})_3$, m. 114.5 $\xrightarrow{NHMe_2}$ $N_3P_3(NC_5H_{10})_3(NMe_2)_3$, m. 113 $\xleftarrow{NHC_5H_{10}}$ $N_3P_3Cl_3(NMe_2)_3$, m. 105

$N_3P_3Cl_3(NC_5H_{10})_3$, m. 190 $\longrightarrow$ $N_3P_3(NC_5H_{10})_3(NMe_2)_3$, m. 156 $\longleftarrow$ $N_3P_3Cl_3(NMe_2)_3$, m. 152

$N_3P_3Cl_3(NC_5H_{10})_3$, m. -17 $\longrightarrow$ $N_3P_3(NC_5H_{10})_3(NMe_2)_3$, m. 87-88 $\longleftarrow$ $N_3P_3Cl_3(NMe_2)_3$, m. 71

The formation of all three possible geometrical and positional isomers leads to conclusive structural assignments for the trispiperidino-derivatives. However, it is not often that all isomers corresponding to a given degree of chlorine replacement are obtained so that structural assignments under such conditions are not as conclusive, unless taken in conjunction with supplementary data, e.g., ^1H NMR. Consideration of solvent effects is also important.[124] The fact that amine hydrochlorides can effect a cis-trans-isomerization[254] of some nongeminal bis- and tris-amino-derivatives also needs to be borne in mind when such structural assignments are made. Aluminium chloride can also isomerize[260] some nongeminal dimethylamino- and piperidino-cyclotriphosphazatrienes.

Fully aminolyzed cyclophosphazenes have base strengths comparable with that of the amines from which they are derived, so that it is not surprising that numerous adducts with hydrogen halides have been isolated, particularly in the case of primary amine derivatives. These adducts are formed in the presence of excess of amine and are generally believed to contain a protonated ring nitrogen atom.[160,162,330] In the case of $N_3P_3Cl_2(NH\text{-}i\text{-}Pr)_4 \cdot HCl$, unambiguous x-ray crystallographic evidence[305] has been obtained for this. Hydrogen halides are also effective in the removal of amino-groups from cyclophosphazenes and the hexakisdimethylamino-derivative, $N_3P_3(NMe_2)_6$, with hydrogen chloride gives[343] chlorodimethylamino derivatives, $N_3P_3(NMe_2)_nCl_{6-n}$ ($n = 3,4$), with structures identical to those obtained by dimethylaminolysis of $N_3P_3Cl_6$. Analogous results have been obtained with hydrogen bromide.[343] These reactions may also be used[100,193,194] to synthesize mixed halogeno-derivatives; for example,

$$N_3P_3F_5NMe_2 + 2HCl \longrightarrow N_3P_3F_5Cl + H_2\overset{+}{N}Me_2Cl^-$$

Hexakisaziridinocyclotriphosphazatriene undergoes ready reactions[384] with protic species as might be expected:

$$N_3P_3(NC_2H_4)_6 + 6HX \longrightarrow N_3P_3(NHC_2H_4X)_6$$
$$(X = Cl,\ SO_3H,\ SH,\ SR)$$

Aminocyclotriphosphazatrienes form adducts with a variety of metal halides.[330,384] Quaternization by $Me_3O^+BF_4^-$ generally, but not invariably, occurs[383] at the exocyclic nitrogen-atoms:

$$N_3P_3Cl_{6-n}(NMe_2)_n + Me_3O^+BF_4^- \longrightarrow N_3P_3Cl_{6-n}(NMe_2)_{n-1}(\overset{+}{N}Me_3)\ BF_4^-$$
$$(n = 2,3,4,6)$$

In contrast,[383] $N_3P_3Cl_2(NH\text{-}i\text{-}Pr)_4$ alkylates at a ring N-atom to give a structure analogous to the adduct $N_3P_3Cl_2(NH\text{-}i\text{-}Pr)_4,HCl$, mentioned above.

The range of reactions at the amino-groups, and not involving the cyclophosphazene rings is so far rather limited. Ammonolysis products undergo the Kirsanov reaction with phosphorus pentachloride in which exocyclic trichlorophosphazenyl-groups are introduced:

$$\text{e.g., } N_3P_3Cl_4(NH_2)_2 + 2PCl_5 \longrightarrow N_3P_3Cl_4(N{=}PCl_3)_2 + 4HCl$$ [166,294]

Dichlorotriphenylphosphorane also undergoes this reaction,[167,249,250] and triphenylphosphazenyl-groups are introduced, but the presence of a tertiary base (usually triethylamine) is necessary to remove the second molecule of hydrogen chloride. A novel ring closure reaction[297] occurs with the bis(trichlorophosphazenyl)-derivative and heptamethyldisilazane:

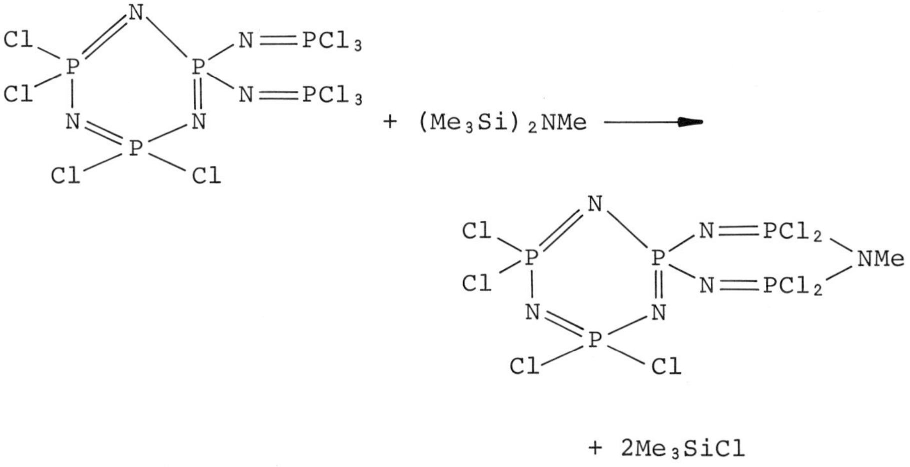

Silicon-nitrogen compounds do not generally react with the $\equiv PCl_2$ groups of the phosphazene ring, but reactions of amine substituents are possible,[316] for example,

$$N_3P_3Cl_4(NH_2)_2 + (Me_3Si)_2NH \longrightarrow N_3P_3Cl_4(NHSiMe_3)_2 + NH_3$$

In a few cases it has been shown that aminocyclophosphazenes add to isocyanates and isothiocyanates, for example,

$$N_3P_3(NMe_2)_4(NH_2)_2 + RNCX \longrightarrow N_3P_3(NMe_2)_4(NH_2)NHCXNR$$

(X = O, R = n-Bu, Ph[447,483]
X = S, R = n-Bu)

The course of ammonolysis and aminolysis of arylchlorocyclophosphazatrienes has been studied in some detail.[138,230] The course of aminolysis of the geminal diphenyl derivative, $N_3P_3Cl_4Ph_2$, is closely related to that observed for the parent hexachloro compound, $N_3P_3Cl_6$. Thus, ammonia and aniline effect geminal replacement of chlorine, but with dimethylamine and piperidine mainly nongeminal products are obtained on replacement of two chlorine atoms. Complementary results[409,410] are also obtained with the diphosphazadiene (8) and dimethylamine, piperidine, and t-butylamine. The first two amines replace chlorine by a nongeminal route and the latter by a geminal route.

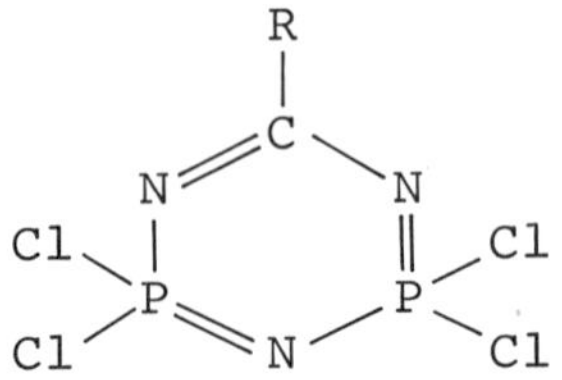

(R = Me,Ph)

(8)

The aminolysis of higher homologues of chlorocyclophosphazenes has not been studied in detail with the exception[139,389] of reaction of $N_4P_4Cl_8$ and dimethylamine. A complex mixture of the geminal and nongeminal products was obtained, the latter predominating.

By contrast with structures and reactions of the aminocyclophosphazenes, relatively little is known of the corresponding alkoxy- and aryloxy-derivatives. In the alkoxy-series, the chlorine atom displacement path appears to be mainly nongeminal, as might be expected on the basis of an S_N2 (P)-type mechanism and the well-established ability of an alkoxy group to increase the electron-density at phosphorus by a conjugative interaction. The only alkoxychloro-derivatives, $N_3P_3Cl_{6-n}(OR)_n$ (n = 2,3,), that have been well characterized are

$N_3P_3Cl_3(OMe)_3$,[70,171] $N_3P_3Cl_4(O\text{-}n\text{-}Bu)_2$,[455] and $N_3P_3Cl_3$-$(O\text{-}n\text{-}Bu)_3$,[455] all of which have nongeminal structures.

A feature of the chemistry of some of the aryloxy-derivatives is their ability to form molecular inclusion compounds.[12,13,75,454] For example,

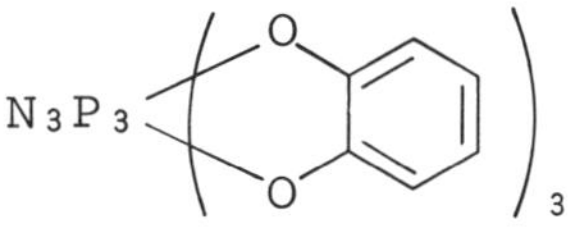

occludes benzene[448,454] and bromobenzene[448] and the crystal structures of these adducts have been established.

More detailed information is available on the structure of the aryloxy-derivatives. It has been shown[131,132] that the replacement of chlorine atoms by phenoxy-groups follows an exclusively nongeminal pattern and the same is probably true with p-bromophenoxy-groups.[131,132] These structures were established by the preparation of a series of mixed dimethylamino (phenoxy)-derivatives, $N_3P_3(NMe_2)_n$-$(OPh)_{6-n}$ ($n = 1$-5) starting from both the chloro(phenoxy)- and chlorodimethylamino-derivatives. Several amino (phenoxy)-derivatives, $N_3P_3(NH_2)_{6-n}(OPh)_n$ ($n = 3,4,5$) have been obtained.[320]

An interesting feature of the chemistry of the alkoxy-cyclophosphazenes is their ability to rearrange under the influence of heat[174] or of an alkyl halide. The thermal rearrangement may be represented as follows:

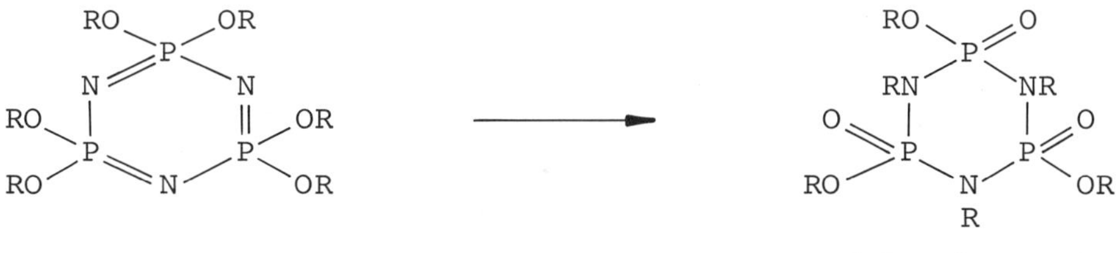

(phosphazane)

(R = Et, CH_2Ph)

It has been suggested that rearrangement proceeds by inter- or intra-molecular attack of a ring nitrogen atom on the α-carbon atom, e.g. (for the sake of simplicity intramolecular mechanisms are illustrated),

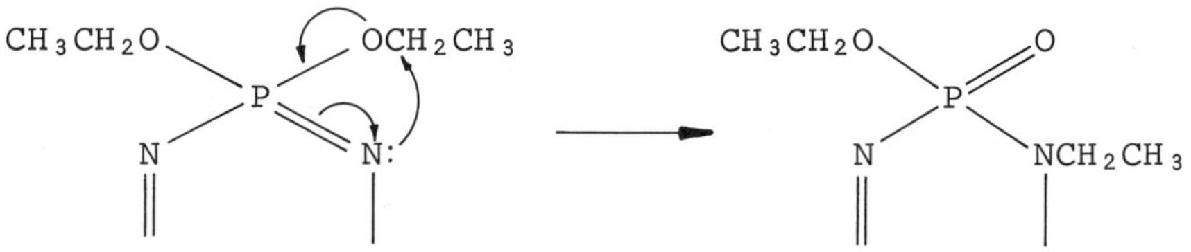

Such a mechanism would be closely analogous to that occurring in the second (fast) stage of the Arbuzov reaction.

The preparative scope of the reaction may be extended[176] by the use of an alkyl halide in the reaction

$N_3P_3(OR)_6 \xrightarrow{3R'I}$ 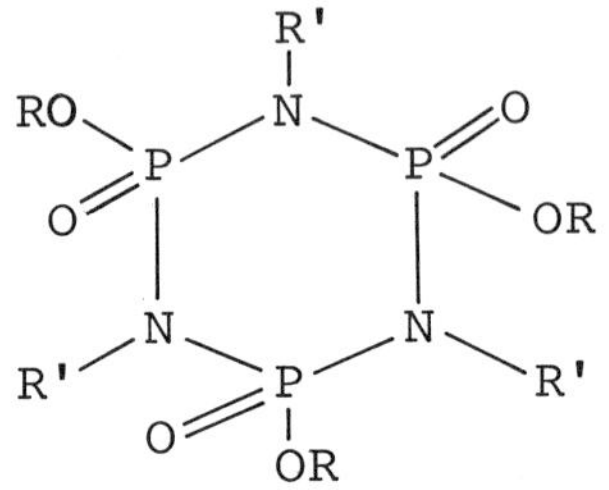 $+ 3RI$

$$\left[\begin{array}{l} R = R' = Me \\ R = Et,\ R' = i\text{-}Pr \\ R = Et,\ R' = O_2N\text{-}p\text{-}C_6H_4{\cdot}CH_2Br \end{array}\right]$$

In some cases, products in which rearrangement has occured at only one of the phosphorus atoms may be isolated.[172] With benzoyl bromide, the alkoxyphosphazane intermediate (R = Et, R' = COPh) is unstable and eliminates benzonitrile[173,175] which in turn trimerises to triphenyl-s-triazine, $N_3C_3Ph_3$.

Other reactions of the alkoxy-derivatives that have been studied include those with chlorosilanes,[49,50] which provides a convenient method of introducing $-OSiR_3$ groups:

$$\text{e.g.,}\ N_3P_3Cl_{6-n}(O\text{-}n\text{-}Bu)_n + PhMe_2SiCl \longrightarrow$$

$$N_3P_3Cl_{6-n}(O\text{-}n\text{-}Bu)_{n-1}(OSiPhMe_2) + n\text{-}BuCl$$

$$(n = 1,6)$$

The chlorination of alkoxy-side chains has also been accomplished:[385]

$$[NP(OCH_2R_F)_2]_n + 2nCl_2 \longrightarrow [NP(OCCl_2R_F)_2]_n + 2nHCl$$

$$\left[\begin{array}{ll} R_F = (CF_2)_mX, & m = 1,3;\ n = 3,4;\ X = F \\ & m = 2,4,6,8,10;\ n = 3,4;\ X = F,Cl \end{array}\right]$$

The base catalyzed hydrolysis of $N_3P_3(OCH_2CF_3)_6$ to $N_3P_3(ONa)(OCH_2CF_3)_5$ and $N_3P_3(ONa)_2(OCH_2CF_3)_4$, has been followed[16] and it has been shown that in the latter compound OCH_2CF_3-groups are displaced from different phosphorus atoms.

Most of the cyclophosphazenes are hydrolytically stable except the bromide, but the chlorides and fluorides may be hydrolyzed under acidic or basic conditions.

The hydrolysis of chlorophosphazenes has been studied in detail and the results for $N_3P_3Cl_6$ may be summarized:

$$N_3P_3Cl_6 \longrightarrow N_3P_3Cl_4O_2H_2{}^{475} \longrightarrow N_3P_3Cl_2O_4H_4{}^{377} \longrightarrow N_3P_3O_6H_6{}^{351,352}$$

Infrared and other measurements suggest that the hydrolysis products exist in the "phosphazane," $-NH-\overset{\overset{O}{\|}}{\underset{\underset{OH}{|}}{P}}-$, form, rather than the "phosphazene" form, $-N{=}P(OH)_2-$ (assuming geminal replacement of chlorine). Several metal salts of the "hexahydroxy" product have been isolated.[351,352] Further hydrolysis proceeds through a complex series of condensed phosphates with -NH- and -O- bridges, to simple phosphate salts. The hydrolysis of the higher homologs $(NPCl_2)_n$,[351,376] $(NPF_2)_{3,4}$,[428] and $N_3P_3Cl_4(NH_2)_2$[143] has also been studied.

Two monohydroxy-derivatives have been isolated: $N_3P_3Cl_5OH$, from the hydrolysis of $N_3P_3Cl_5OSiMe_2Ph$,[49] and $N_3P_3Ph_5OH$, by hydrolysis[424] of the analogous chloride.

C. GENERAL DISCUSSION OF PHYSICAL PROPERTIES

C.1. Molecular Structures

Many structures of cyclophosphazenes have now been determined accurately by diffraction methods (see Table 2). The equality of P-N bond lengths in homogeneously substituted derivatives and the shortness of these bonds compared with the sum of the single bond covalent radii for phosphorus and nitrogen (1.76 Å) has provided very good evidence for a conjugative interaction involving nitrogen and phosphorus. However, the controversy over the best description of the bonding in the π-electron systems continues.[119,140]

Table 2. Compounds Whose Detailed Molecular Structures Have Been Established by Diffraction Methods

Compound	Ref.
$N_3P_3F_6$	145
$N_3P_3Cl_5F$	356
$N_3P_3Cl_6$	76,128

Table 2 (Continued)

Compound	Ref.
$N_3P_3Br_6$	190,519
$N_3P_3(NCS)_6$	158
$N_3P_3F_4Ph_2$ (2,2,4,4:6,6)	21
$N_3P_3Cl_4Ph_2$ (2,2,4,4:6,6)	303
$N_3P_3Cl_2Ph_4$ (2,2:4,4,6,6)	304
$N_3P_3Cl_2(NH\text{-}i\text{-}Pr)_4 \cdot HCl$ (2,2:4,4,6,6)	305
N_3P_3 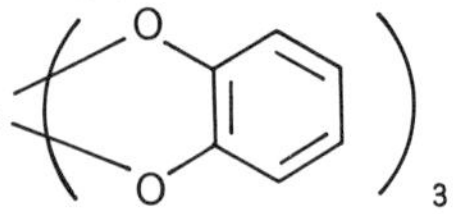 $\cdot C_6H_6$	448,454
N_3P_3 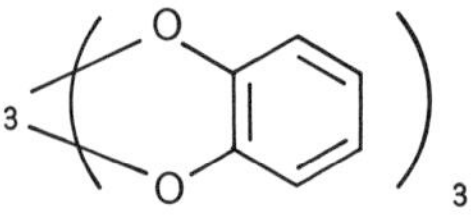 $\cdot C_6H_5Br$	448
N_3P_3 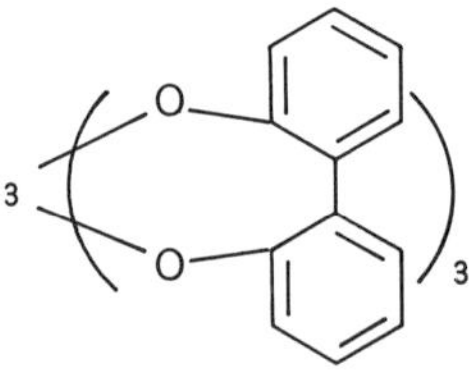	14
$N_4P_4F_8$	322
$N_4P_4Cl_8$ (K form)	222,262
$N_4P_4Cl_8$ (T form)	502
$N_4P_4F_6Me_2$ (2,2,4,4,6,6:8,8)	313
$N_4P_4F_4Me_4$ (2,2,6,6:4,4,8,8)	313
$[(NPMe_2)_4 \cdot H]CuCl_3$	492
$[(NPMe_2)_4H]_2CoCl_4$	493
$N_4P_4Cl_4Ph_4$ (2-cis-4-cis-6-cis-8:2,4,6,8)	80
$N_4P_4Cl_4Ph_4$ (2-cis-4-trans-6-cis-8:2,4,6,8)	79
$N_4P_4(NMe_2)_8$	78
$N_4P_4Ph_4(NHMe)_4$ (2-cis-4-trans-6-cis-8:2,4,6,8)	79
$N_4P_4(OMe)_8$	28
$N_5P_5Cl_{10}$	400
$N_6P_6(NMe_2)_{12}$	503
$[N_6P_6(NMe_2)_{12}CuCl]^+CuCl_2^-$	312
$N_8P_8(OMe)_{16}$	368
Miscellaneous cyclophosphazenes	103,104, 264,499

Much recent interest has centered on the variation in P-N bond lengths encountered on changing the substituents on the cyclophosphazene rings, particularly where they are heterogeneously substituted. It has been pointed out[2] that the P-N bond lengths in a derivative $N_3P_3X_6$ decrease with increasing electron-negativity of X. This is also paralleled by a decrease in the angle subtended by the exo-substituents at phosphorus, an increase in the ring NPN angle, and a decrease in the ring PNP angle. In the series $N_3P_3Cl_{6-n}Ph_n$ (n = 0,2,4,6[2,303,304,570]) the P-Cl and P-C bond lengths also follow regular trends. The structure of the geminal derivative, $N_4P_4F_6Me_2$, (9), is particularly interesting.[313]

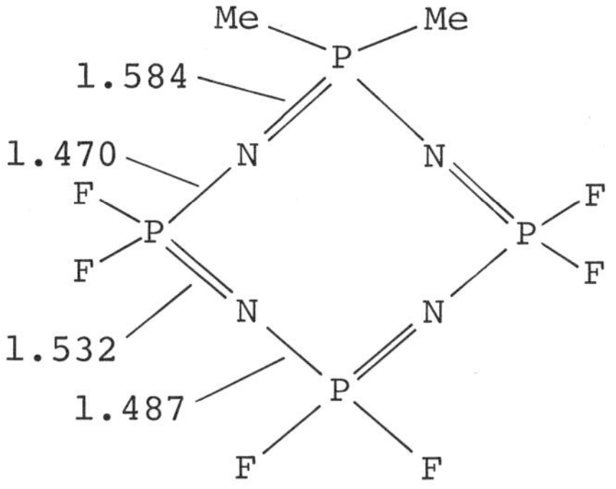

(9)

There is an alternation in P-N bond lengths on moving away from the ≡PMe_2 group and the P-N bond of 1.470 Å (σ = 0.006) is the shortest known. This result has been interpreted in terms of a delocalized π-system, within the limits of simple Hückel theory.[313] A similar effect is significant in other structures such as geminal $N_3P_3Cl_4Ph_2$[303] and $N_4P_4F_4Me_4$.[313]

C.2. Nuclear Magnetic Resonance Spectroscopy

There is little doubt that NMR spectroscopy provides the most powerful method of identification, estimation of purity, and establishing molecular structures of cyclophosphazenes in solution. The greater availability of ^1H NMR facilities has meant that more information is available in this area than from ^{31}P NMR spectroscopy.

The ^1H NMR spectra of simple mononuclear phosphorus compounds with an organo-group such as Me, OMe, NMe_2, or SMe bonded to phosphorus generally, comprises a simple doublet of separation $J_{P....H}$. The ^1H spectrum of $(Me_2N)_3PO$ consists of a doublet of separation ~10 Hz.,[256] which corresponds to $J_{P-N-C-H}$. Similar spectra might initially be expected from the ^1H spectra of organo-substituted cyclophosphazenes and, in general, a doublet is observed but, in addition, extra lines of an unresolved "hump" between the two components of the doublet appear. These are the result of the fact that many of these compounds provide a second-order spin system involving more than one ^{31}P nucleus in the phosphazene ring.[171,220] Thus, the cis-nongeminal isomer, $N_3P_3Cl_3(OMe)_3$ contains nuclei which may be represented as a $AA'A''X_3X'_3X_3''$ spin system.[171] These spin systems can give rise to a very large number of signals and, in the case of the ^1H spectra, satisfactory agreement between observed and computed spectra has yet to be achieved.[171] One difficulty lies in understanding why the signals comprising the "hump" are not resolved. In spite of these complications, valuable information can be obtained on a first-order basis. Thus, the appearance of two chemically shifted doublets in a 1 : 2 ratio (Figure 1) establishes the structure of the compound, $N_3P_3Cl_3(NMe_2)_3$, m. 105°, as nongeminal trans-.[155,182,251,253,256,271] The use of benzene as a solvent is often useful[256] in distinguishing chemically shifted signals from those which are spin coupled and in many cases increases chemical shift differences relative to those in CCl_4 or $CDCl_3$ solution.

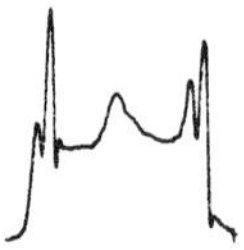

(a)

(b)

Figure 1. (a) ^1H spectrum of nongeminal trans-$N_3P_3Cl_3$-$(NMe_2)_3$.
(b) Same spectrum with ^{31}P decoupled.

The cis- and trans-bisdimethylamino-derivatives, $N_3P_3Cl_4(NMe_2)_2$, each give a single doublet with a "hump," but their structures can be assigned on the basis of empirical correlations.[251] The separation of the doublet structure varies considerably with the type of grouping, e.g., $\equiv PClNMe_2$, ca. 16 Hz., $\equiv P(NMe_2)_2$, ca. 12 Hz., so providing a means of distinguishing positional isomers.

The appearance of the "hump" between the doublets in the ^1H spectra is dependent on the relative magnitudes of $J_{P....H}$ and J_{P-N-P} and on there being a small or zero chemical shift between the ^{31}P nuclei.[171,172] Thus, when a considerable chemical shift between the ^{31}P nuclei occurs, e.g., in (10), the $-NMe_2$ signal is a clear doublet, but in (11) where there is no chemical shift between the $^{31}P(NH_2)(NMe_2)$ nuclei, a doublet enclosing a hump is observed.[138] This effect has provided a useful method of distinguishing between positional isomers but it may be noted that this is not the case in certain halogeno-derivatives. Thus,

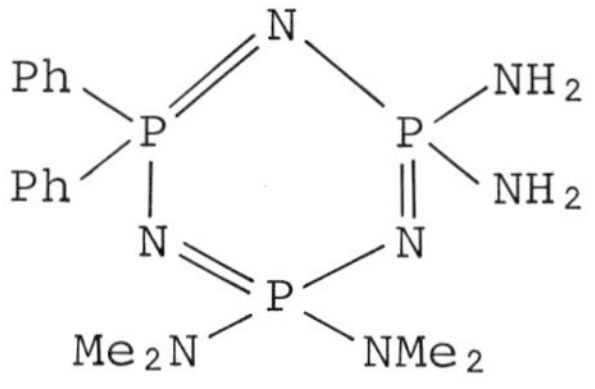

(10)

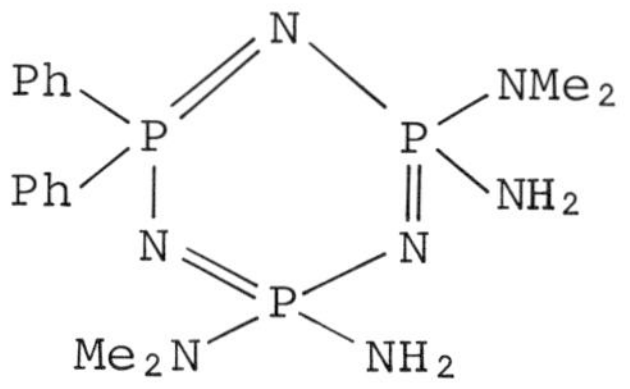

(11)

e.g., $N_3P_3X_5NMe_2$ (X = F,[193] Cl,[223] Br[155]) have a ^1H spectra consisting of a doublet enclosing a "hump;" possibly this is related to the relatively large J_{P-N-P} values operative in the halogeno-derivatives. A related effect is observed[229] in the ^1H spectra of the series $N_3P_3(OEt)_{6-n}Ph_n$ (n = 0, 2, 4). The CH_3 spin decoupled CH_2 spectrum is simplest when n = 4 and most complex when n = 0. It has also been shown that in homologous series, such as $\{NP(NMe_2)_2\}_n$, the intensity of the central "hump" increases relative to that of the doublet with increase in n.[220] The ^{19}F spectra of fluorocyclophosphazenes have received less attention but provide examples of second-order spin systems also. Thus, a doublet of separation not very different in magnitude from J_{P-F} appears[151,224] in the ^{19}F spectrum of $N_3P_3Cl_4F_2$ (geminal) but, because of the relative magnitudes and signs of the coupling constants, some of the signals appear outside the main doublet. A full analysis of the ^{19}F spectrum of $N_3P_3F_6$ has not yet appeared but this, in principle, should give J_{P-N-P}.

The ^{31}P chemical shifts of the cyclophosphazenes are closely related to those of related mononuclear P(V) compounds and the effect of substituents is similar. Thus, the chemical shifts of $N_3P_3X_6$ follow the order:

$N_3P_3Cl_6$	<	$N_3P_3F_6$	<	$N_3P_3Br_6$	
-20[300]		+13.9[22]		+49.5[243]	ppm

The phosphoryl halides follow the same order with change of halogen atom. Unfortunately, very little is yet known of the factors that determine the ^{31}P chemical shifts in the cyclophosphazenes.

An attempt[170] has been made to correlate the magnitudes of P-N-P coupling constants in the cyclophosphazenes with the properties of the substituents. Wide variations in J_{P-N-P} are apparent (known values for cyclotriphosphazatrienes are given in Table 3), but they only bear an approximate relationship to the electronegativities of the substituents.

Table 3. Across-Ring Phosphorus-Phosphorus Coupling Constants in Cyclophosphazatrienes

Compound	J_{P-N-P} (Hz)	Ref.
$N_3P_3Cl_5F$	78	223
$N_3P_3Cl_4F_2$ (2,2,4,4:6,6)	100	96,224
$N_3P_3Cl_2F_4$ (2,2:4,4,6,6)	100	96
$N_3P_3F_4Ph_2$ (2,2,4,4:6,6)	86	22
$N_3P_3F_4Ph_2$ (2,2,4-cis-6:4,6)	58	20
$N_3P_3F_4Ph_2$ (2,2,4-trans-6:4,6)	65	20
$N_3P_3F_3Ph_3$ (2,2,4:4,6,6)	44,25,57	22
$N_3P_3F_2Ph_4$ (2,2:4,4,6,6)	32.9	22
$N_3P_3Cl_5NH_2$	46.52	166
$N_3P_3Cl_4(NH_2)_2$ (2,2,4,4:6,6)	48.45	166,294
$N_3P_3Cl_5NMe_2$	49.7	223
$N_3P_3Cl_4(NMe_2)_2$ (2,2,4-cis-6:4,6)	~48	251
$N_3P_3Cl_4(NMe_2)_2$ (2,2,4-trans-6:4,6)	~44	251
$N_3P_3Cl_2$(NH-i-Pr)$_4$ (2,2:4,4,6,6)	48	148
$N_3P_3Cl_4$(NHPh)$_2$ (2,2,4,4:6,6)	48	293
N_3P_3(NHPh)$_2$(OCH_2CF_3)$_4$ (2,2:4,4,6,6)	72	293
$N_3P_3Cl_5$OMe	66.3	223
$N_3P_3Cl_5OCH_2CF_3$	66.2	223
$N_3P_3Cl_5$O-i-Pr	62.7	223
$N_3P_3Cl_5(N{=}PCl_3)$	58.36	166
$N_3P_3Cl_5(N{=}PPh_3)$	47	148
$N_3P_3Cl_4Ph(N{=}PPh_3)$ (2,2,4,4:6:6)	21	148
$N_3P_3Cl_4(N{=}PCl_3)_2$ (2,2,4,4:6,6)	60.99	166,294
$N_3P_3Cl_4(NHPOCl_2)_2$ (2,2,4,4:6,6)	46	294
$N_3P_3Cl_4(NH_2)(N{=}PPh_3)$ (2,2,4,4:6:6)	46	294
$N_3P_3Cl_4(N{=}PCl_3)(N{=}PPh_3)$ (2,2,4,4:6:6)	~27	294
$N_3P_3Cl_4$(SEt)$_2$ (2,2,4,4:6,6)	4.8	68
$N_3P_3Br_4(NMe_2)_2$ (2,2,4-cis-6:4,6)	~18	155
$N_3P_3Br_4(NMe_2)_2$ (2,2,4-trans-6:4,6)	~18	155
$N_3P_3Br_5(NMe_2)$	~18	155

A recent development in this area is the measurement of the dynamic nuclear enhancement of ^{31}P signals.[147,148] This effect is generally observed in the presence of a free radical, which is irradiated at the resonance frequency of the unpaired electron with sufficient power to cause saturation. Enhancement of ^{31}P signal intensities varies with the nature of the substituents[147,148] at phosphorus and with the nature of the free radical.[148]

C.3. Vibrational Spectra

Infrared spectroscopy (IR) is largely used as a fingerprint for a given cyclophosphazene. The dominant absorption band in the IR spectra in this class of compounds is associated with the presence of the $\geqslant$P=N- linkage and generally occurs in the range 1150-1450 cm^{-1}. This band has been shown to arise from a degenerate ring stretching vibration,[94] and as expected, the band generally moves to higher wavenumbers with an increase in the electronegativity of the substituents. Typical values for cyclotriphosphazatrienes, $N_3P_3X_6$, are

X =	F	OMe	Cl	NMe_2	Me	Br	SEt
ν(cm.$^{-1}$)	1297	1240	1221	1195	1180	1171	1170
Ref.	e.g. 95	70	e.g. 95	432	426	472	445

It is also observed that ν(P-N) increases with increase in the size of the cyclophosphazene ring[95] up to $N_5P_5X_{10}$ (X = F, Cl), after which ν(P=N) decreases again.

IR spectroscopy was first suggested as a means of distinguishing positional isomers in the case of the mixed fluoride chlorides, $N_3P_3Cl_{6-n}F_n$ (n = 2,3,4).[96] The degeneracy of the P=N stretching mode is lower in the case of the geminal than in the nongeminal isomers, with the result that ν(P=N) appears as two bands in each case with separation of 24 and 7 cm^{-1}, respectively.[96] (See also Ref. 152.) The same effects are observed in the case of positional isomers of dimethylamino-[259,461] and piperidino-derivatives,[259] but there is little difference between geometrical isomers. Relatively large splittings in ν(P=N) were apparent in the geminal isomers $N_3P_3Cl_4Ph_2$, (36 cm^{-1}), $N_3P_3Cl_2Ph_4$ (34 cm^{-1}),[259] but no splitting appeared in the spectra of the alkanethiol derivatives, $N_3P_3Cl_4(SEt)_2$ and $N_3P_3Cl_2(SEt)_4$.[445] The geminal structures of the latter two compounds are well established, so that the reasons for absence of any splitting in ν(P=N) remain obscure.

It has also been shown that positional isomers in the series $N_3P_3Cl_{6-n}(NMe_2)_n$ (n = 2,3) may be distinguished by examination of the P-N vibrations of the exocyclic (dimethylamino) groups.[461]

Considerable interest has been shown in the symmetry of homologous halogenocyclophosphazenes in the solid, liquid, and solution states. Several detailed studies have been made generally using IR and Raman spectroscopy. In the solid state, $N_3P_3Cl_6$[231] and $N_3P_3Br_6$[306] have structures with C_{3v} symmetry in agreement with the results of x-ray analyses. In solution, $N_3P_3X_6$ (X = F, Cl, Br[94,231,306]), all have higher (D_{3h}) symmetry. The symmetry types become less distinct as the homologous series for each halide is ascended,[94,95,114,115,116,300,468,470,471] but in the solid state, there is now reasonable agreement between the structures $N_4P_4X_8$ predicted by x-ray methods and those predicted by vibrational spectroscopy.

C.4. Other Spectroscopic Methods

The ultraviolet spectra of many cyclophosphazene-derivatives have been studied,[140,187,247,289,290,300,444] but no absorptions related to those in an "aromatic" ring system have been detected. In the chlorides, the lowest energy absorption (near 250 mμ) is probably related to an electron excitation on the chlorine atoms. Bromophosphazenes absorb at slightly lower energies and fluorophosphazenes at higher energies than the chlorides. These compounds do not absorb in the visible region. More information is available from photoelectron spectroscopy,[71] and vertical ionization energies of the series $(NPF_2)_n$ (n = 3-8), have been measured by this means. Ionization energies have also been obtained by mass spectrometry for the compounds $(NPX_2)_n$ (X = Cl, n = 3-7; X = OCH_2CF_3, OMe, OPh, NMe_2, or Me, n = 3,4).[71] An alternation of ionization potentials for the fluorides with increase in ring size is detected[71] and, in the other cases, the derivatives where n = 3 have ionization energies higher than when n = 4. The results have been discussed[71] in terms of possible bonding schemes involving the phosphazene rings. Several other studies of the fragmentation patterns of halogenocyclophosphazenes have been made,[73,94,110,111,115,420] but the most useful aspect of the technique is the identification of the masses of parent ions giving accurate molecular weights. ^{35}Cl nuclear quadrupole resonance studies have been made on some of the chlorophosphazenes and their derivatives.[142,248,299,346,347,489,508]

Bond energy values for cyclophosphazenes are sparse. Evaluation of the standard heats of formation led to the following P-N bond energies, E:

Compound	$N_3P_3Cl_6$	$N_4P_4Cl_8$	$N_3P_3Me_6$	$N_4P_4Ph_8$	$N_3P_3(OC_6H_{11})_6$
E (Kcal/mole)	72.3	72.5	68	85.6	83.0
Ref.	48a,221	221	48a	48a	48a

These values do not suggest that the P-N bonds in cyclophosphazenes are significantly stronger than those in formally saturated P-N bonds.

C.5. Basicity Measurements

Basicity measurements have been made on a large number of cyclophosphazene derivatives in nitrobenzene solution and to a lesser extent, in aqueous solution.[159-165] These give a relative measure of base strength in $C_6H_5NO_2$ termed a pK'_a value (for definition see Ref. 159). It has been suggested that protonation of the cyclophosphazenes occurs on a ring nitrogen atom and this is substantiated by the crystal structure of $N_3P_3Cl_2(NH\text{-}i\text{-}Pr)_4 \cdot HCl$.[305] These measurements revealed that different substituents have a very marked effect on the base strength of the ring nitrogen atoms. Selected values are shown in Table 4.

Table 4. Basicities of Cyclotriphosphazatrienes, $N_3P_3R_6$[161]

R	NMe_2	Et	Ph	OEt	SEt	OPh	Cl
pK'_a	7.6 (Ref. 160)	6.4	1.5	-0.2	-2.8	-5.8	<-6.0

A series of derivatives of $N_4P_4Cl_8$ follow a closely related trend.[160,161] Even more interesting results were obtained from pK'_a measurements on compounds with related structures, particularly positional isomers, which lead to structural assignments based solely on pK'_a data.[164,165] These assignments are in good agreement with the results obtained from other physical techniques and, in some cases, have lead to the correction of erroneous assignments (mostly based on the results of chemical reactions). In general, it is found that isomers with geminal structures are more basic than those with nongeminal structures:[162] The basicity of geometrical isomers is not significantly different. This is exemplified by the results[162] obtained for the following triamino-derivatives:

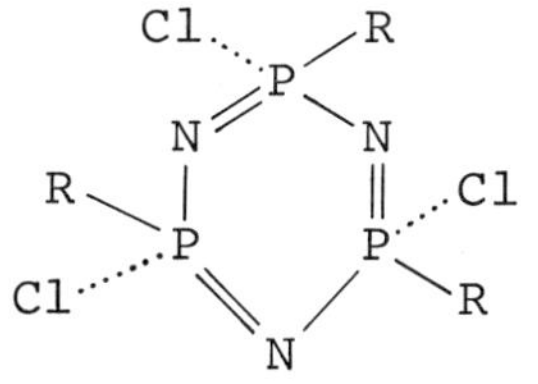

cis-

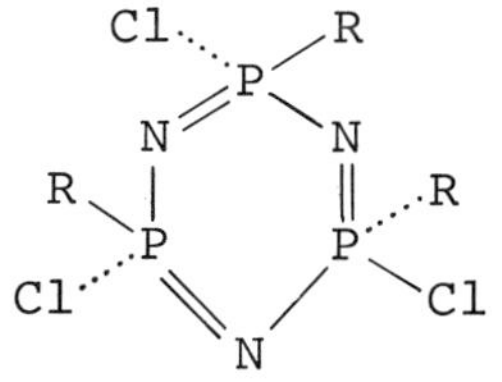

trans-

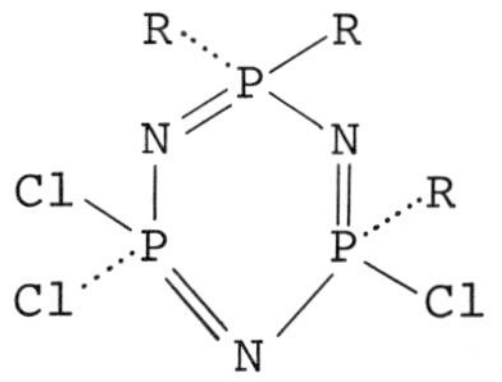

gem

pK$'_a$ values:

R = NMe_2	-5.5	-5.4	-4.4
R = NC_5H_{10}	-5.1	-5.3	-3.9

Even when different amino-substituents are present, pK$'_a$ values indicate differences in structure. For example, the pK$'_a$ values of $N_3P_3Cl_2(NMe_2)_4$, $N_3P_3Cl_2(NC_5H_{10})_4$, and $N_3P_3Cl_2(NH\text{-}t\text{-}Bu)_4$ are -1.4, -0.9, and +3.5, respectively. Accordingly, the first two derivatives have nongeminal structures and the third has a geminal structure. From measurements carried out on a wide range of derivatives, a series of substituent constants have been evaluated for given groups.[163,164,165] These constants are, naturally, different for ring nitrogen atoms α- and γ- to the group being studied. These substituent constants provide a reliable method of calculating the pK$'_a$ value of a cyclophosphazene with one or more substituents, provided the relevant substituent constants are known. The application of these values to the identification of cyclophosphazenes of given structure may be widely employed.

The basicities of the chlorocyclophosphazenes, $(NPCl_2)_n$, have been compared[72] by measurement of the quantity of HCl that they absorb. The basicities are greater when n = 4, 6, and 8 than when n = 3, 5, and 7, and these values bear a close relation to the first ionization potentials of the chlorides as measured by mass spectrometric methods.

The literature contains a vast amount of other physical data relating to the cyclophosphazenes. Since comparisons based on this data are not always very instructive, the reader is referred to Section D of this chapter.

D. LIST OF COMPOUNDS

Notes to Accompany List of Compounds

1. Where the possibility of isomers arises and where the structures of these have been established, the nomenclature is that described previously, e.g.,

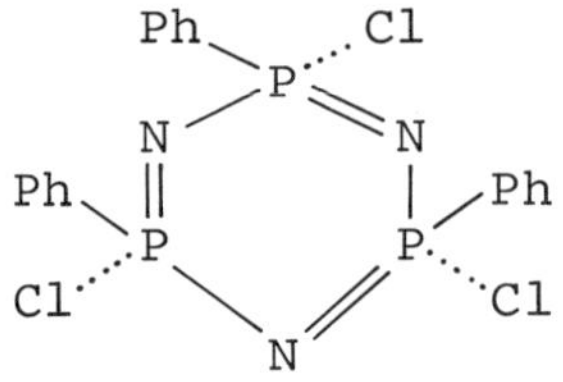

is $N_3P_3Cl_3Ph_3$ (2-cis-4-cis-6:2,4,6).

The 2-phosphorus atom is taken as the reference point for distinguishing cis- and trans-groupings so that in a tetramer derivative the naming is, e.g.,

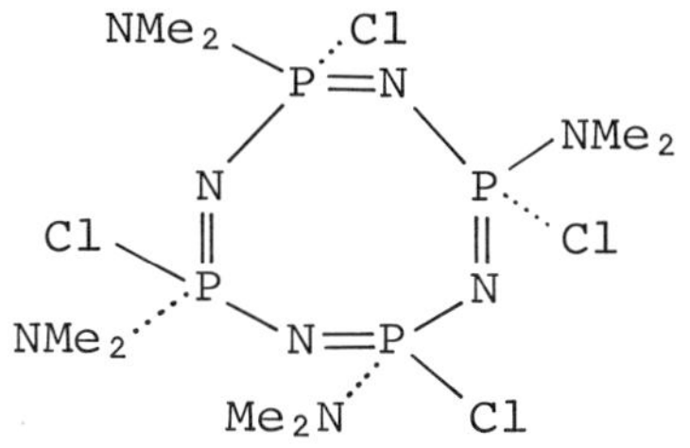

$N_4P_4Cl_4(NMe_2)_4$ (2-cis-4-trans-6-trans-8).

2. ^1H chemical shifts are measured relative to TMS in $CDCl_3$ or CCl_4 solution. Where other solvents have been employed, their identity is noted.

3. ^{19}F chemical shifts are all relative to CCl_3F (generally internal).

4. ^{31}P chemical shifts are all relative to H_3PO_4 (external).

5. Negative shifts in the ^1H, ^{19}F, and ^{31}P spectra are downfield from the respective reference compounds.

D.1. Cyclotriphosphazatrienes

D.1.1. Halogeno-derivatives

$N_3P_3F_6$. IV.[95,219,329,425,428,429] m. 27.8,[95] b_{760} 50.9,[95] crit. temp. 187.7,[98] d^{20} 2.237,[429] log p = -1670/T + 8.04 (m. -50°)[429] $\Delta H_{vap.}$ 7.60 Kcal./mole.[429] $\Delta S_{vap.}$ 23.5 cal./mole/deg.,[429] $\Delta H_{fus.}$ 5.3 Kcal./mole,[429] thermodynamic functions (50-2000°F).[307] X-ray structure[145] established planar ring with three crystallographically non-equivalent P-N bonds, 1.546, 1.563, and 1.572, Av. P-F 1.521, $P\hat{N}P$ 119.6 and 121.1°, $N\hat{P}N$ 119.5 and 119.3°, isomorphous with $N_3P_3Cl_6$. Bonding in,[326] IR,[94,152,468,471] bond stretching and interaction constants,[93] Raman,[94,152,470,471] ^{31}P + 13.9,[22] ^{19}F + 71.9,[102,151] mass spect.,[73] photoelectron spectr.,[71] first ionization potl. 11.4 eV,[71] UV[187,290] hydrolysis.[428] Forms adducts: $N_3P_3F_6 \cdot CsF$,[146] IR[146] conductivity,[146] $N_3P_3F_6 \cdot 2SbF_5$,[101] m. 143-145,[101] IR,[101] has F-bridged structure,[101] and $N_3P_3F_6 \cdot 2HF \cdot 2H_2O$ with HF/H_2O.[418]

$N_3P_3ClF_5$. IV.[151] [$N_3P_3F_5 \cdot NMe_2$ + HCl],[100,193] m. -32,[151] b_{760} 82,[96,100,151,193] d^{20} 1.7594,[151] $\Delta H_{vap.}$ 8.7 Kcal./mol.,[151] $\Delta S_{vap.}$ 24.5 cal./mole/deg.,[151] IR,[152] Raman,[152] ^{19}F + 70.2 (PF_2), + 30.7 (PFCl).[102,151]

$N_3P_3Cl_2F_4$ (2,2:4,4,6,6). IV.[96,151] m. -46,[151] b_{760} 115,[96,151] d^{20} 1.7650,[151] $\Delta H_{vap.}$ 9.6 Kcal./mole,[151] $\Delta S_{vap.}$ 24.7 cal./mole/deg.,[151] IR,[152] Raman,[152] ^{19}F + 71.9 (PF_2).[151]

$N_3P_3Cl_2F_4$ (2,4:2,4,6,6). [$N_3P_3F_4(NMe_2)_2$ + HCl],[194] b_{760} 112,[194] ^{19}F + 31.5 (PFCl).[194]

$N_3P_3Cl_3F_3$ (2,2,4:4,6,6). IV.[96,151] m. -35,[96,151] b_{760} 150,[151] d^{20} 1.7588,[151] $\Delta H_{vap.}$ 10.2 Kcal./mole,[151] $\Delta S_{vap.}$ 24.1 cal./mole/deg.[151] IR,[152] Raman,[152] ^{19}F + 71.7 (PF_2), + 31.9 (PFCl).[151]

$N_3P_3Cl_3F_3$ (2,4,6:2,4,6). [$N_3P_3F_3(NMe_2)_3$ + HCl],[206] b_{760} 140.[206]

$N_3P_3Cl_4F_2$ (2,2,4,4:6,6). IV.[96,151] m. 18,[151] b_{760} 182,[96,151] d^{20} 1.7851,[151] $\Delta H_{vap.}$ 11.0 Kcal./mole,[151] $\Delta S_{vap.}$ 24.4 cal./mole/deg.,[151] IR,[152] Raman,[152] ^{31}P, ^{19}F + 72.3 (PF_2).[151,224]

$N_3P_3Cl_4F_2$ (2,2,4,6:4,6). [$N_3P_3Cl_2F_2(NMe_2)_2$ + HCl],[206] b_{760} 170.[206]

$N_3P_3Cl_5F$. IV.[151] m. 50,[151] b_{760} 215,[151] d^{20} 1.981,[151] $\Delta H_{vap.}$ 12.5 Kcal./mole,[151] $\Delta S_{vap.}$ 25.5 cal./mole/deg.[151] X-ray structure[356] established nearly planar ring with all P-N bond lengths equal within experimental error, Av. P-N 1.563 (0.007), Av. $N\hat{P}N$ 118.5 (0.5), disordered structure prevented location of F. IR,[152] Raman,[152] ^{31}P -14.4 (PFCl), -23.0 (PCl_2),[223] ^{19}F + 31.9.[151]

$N_3P_3BrF_5$. [$N_3P_3F_5{\cdot}NMe_2$ + HBr],[100,193] b_{760} 86,[100,193,482] IR,[467] ^{19}F +19.6 (PFBr), +70.0 (PF_2).[100,102,193]

$N_3P_3BrClF_4$ (2:4:2,4,6,6). [$N_3P_3F_4Cl{\cdot}NMe_2$ + HBr],[194] $b_{0.4}$ 54,[194] ^{19}F + 30.5 (PFCl), + 20 (PFBr).[194]

$N_3P_3Br_2F_4$ (2,4:2,4,6,6). [$N_3P_3F_4(NMe_2)_2$ + HBr][194] b_{760} 145,[194] ^{19}F + 20 (PFBr).[194]

$N_3P_3Br_nF_{6-n}$ (n = 2-5). IR.[467]

$N_3P_3Cl_6$. I.[153,300,350,477,479] m. 112.8,[153,300,350,477,479] b_{760} 256,[479] d^{20} 1.99,[300] log p = 3978/T + 11.187 (75° - m),[473] log p = 2880/T + 8.357 (m - 190°),[473] log p = 2830/T + 8.215 (m - b_{760}),[339] $\Delta H_{sub.}$ 18.2[473] 19.1[106] Kcal./mole, $\Delta H_{vap.}$ 13.2[236,300] 12.9[300] Kcal./mole, $\Delta H_{fus.}$ 5.0 Kcal./mole,[300] $\Delta S_{vap.}$ 24.9 (m - 190) 23.9 (m - b_{760}) cal./mole/deg.[300] $\Delta H_f°$ (c) -194.1 Kcal./mole,[221] E (N-P) 72.3 Kcal./mole[221] thermodynamic functions (50-2000°F).[307] Heat of polymerization.[236]

Electron diffraction data in agreement with planar ring[76] in vapor phase and, more recently,[128] slightly puckered ring (chair conformation) P-N 1.585 (0.01), P-Cl 2.006 (0.007), av. ring angle 119.7 (0.3). Two x-ray structure determinations[189,510] reveal almost planar ring, but in one case[189] P-N bonds of unequal length reported. Av. values[510] P-N 1.59 (0.017), P-Cl 1.98 (0.013), $P\hat{N}P$ 120.38 (0.75), 118.48 (1.25) $N\hat{P}N$ 120.93 (1.10) 118.33 (1.30) $Cl\hat{P}Cl$ 101.77 (0.52) 102.05 (0.32), isomorphous with $N_3P_3F_6$,[145] bonding in.[118,119,140,326,369]

IR,[82,83,94,120,210,231,247,306,460,468,471] bond stretching and interaction constants,[93] Raman,[82,94,210,231,471] ^{31}P -20,[300,500] -19.0.[46] Dynamic nuclear polarization (^{31}P),[147,148] NQR.[248,299,346,347,489,508] Mass spectr.,[71,74,110,402] first ionization potl. 10.26 eV,[71,72,110] 2p binding energy,[369] UV,[187,289,290,300,490] dipole moment <0.14,[105] conductivity,[6,150,390] magnetic anisotropy,[118] molecular polarizability,[32] chromatographic methods of separation,[192,376,495] rel. basicity (to HCl),[72] Cl^- exchange,[457] detn. of Cl in,[476] hydrolysis,[351,376,377,514] with Cl_2O gives (PO_2Cl),[130] activator for carboxylic acids.[81] Forms adducts:

$N_3P_3Cl_6 \cdot 2AlCl_3$.[65]

$N_3P_3Cl_6 \cdot AlBr_3$. m. 174,[113] IR.[113]

$N_3P_3Cl_6 \cdot NO_2$.[52]

$N_3P_3Cl_6 \cdot 3SO_3$.[43,196]

$N_3P_3Cl_6 \cdot HClO_4$.[66] NQR.[508]

$N_3P_3Cl_6 \cdot 2C_6Me_6$. m. 149.[127]

$N_3P_3Cl_6 \cdot$naphthalene. Formation constant.[302]

$N_3P_3Cl_6 \cdot 6\ C_5H_5N$. m. ~90.[323,474,475]

$N_3P_3Cl_5Br$. I.[111,112,391,469] [$N_3P_3Br_6$ +$HgCl_2$][112], m. 120,[112] IR,[112,462,469,472] Raman,[462] ^{31}P -17.7 (PCl_2), + 7.8 (PBrCl).[156] Mass spectr.,[110,111] ΔH°_f -168.6 Kcal./mole,[110] first ionization potl. 9.83 eV.[110]

$N_3P_3Cl_4Br_2$ (2,2,4,6:4,6). I.[111,112,391] [$N_3P_3Br_6$ +$HgCl_2$][112], m 131,[112] IR,[112,462,469,472] Raman,[462] ^{31}P -16.1 (PCl_2), +8.7 (PBrCl),[156] mass spectr.,[110,111] first ionization potl. 9.80 eV.[110]

$N_3P_3Cl_3Br_3$ (2,4,6:2,4,6). I.[111,112,391,469] [$N_3P_3Br_6$ +$HgCl_2$][112], m 143,[112] IR,[112,462,469,472] Raman,[462] ^{31}P +9.8,[112] mass spectr.[110,111] First ionization potl. 9.72 eV.[110]

$N_3P_3Cl_3Br_3$ (2,2,4:4,6,6). I.[156] ^{31}P -14.0 (PCl_2), +10.0 (PBrCl), +39.8 (PBr_2).[156]

$N_3P_3Cl_2Br_4$ (2,4:2,4,6,6). I.[111,112,391] [$N_3P_3Br_6$ +$HgCl_2$],[112] m. 143,[112] IR,[112,462,472] Raman,[462] ^{31}P +12.1 (PBrCl), +41.3 (PBr_2),[156] mass spectr.,[110,111] first ionization potl. 9.60 eV.[110]

$N_3P_3ClBr_5$. I.[111,112,391] [$N_3P_3Br_6$ +$HgCl_2$][112], m. 175,[112] IR,[112,462,472] Raman,[462] ^{31}P +14.0 (PClBr), +42.5 (PBr_2),[156] mass spectr.[110,111] First ionization potl. 9.47.[110]

$N_3P_3Br_6$. I.[39,116,237,242,243,244] m. 192,[39,116,242,243,244] log p = 11.15 - 5066/T (54-99°),[106] $\Delta H_{sub.}$ 23.2 Kcal./mole,[106] solubility in various solvents at 25°.[243] Four x-ray studies,[64,190,397,519] two using 3-dimensional data,[190,519] show that[519] the ring has a slight chair conformation, (av.) P-N 1.576 (0.008), P-Br 2.162 (0.004), $P\hat{N}P$ 122.4 (0.5), 119.3 (0.6), $N\hat{P}N$ 118.5 (0.5). X-ray powder data,[116,243]

IR,[116,306,460,471] thermodynamic functions,[307] Raman,[307,470,471] ^{31}P +49.5,[243] +45.4,[155] mass spectr.,[109] UV,[187,290] hydrolyses slowly in moist air. Forms adducts:

$N_3P_3Br_6 \cdot AlBr_3$. m. 180-2,[113] IR.[113]

$N_3P_3Br_6 \cdot 2AlBr_3$. m. 180.[113]

D.1.2. Pseudohalogeno-derivatives

$N_3P_3(N_3)_6$. ($N_3P_3Cl_6$ + NaN_3 in acetone),[212] m. < -20,[212] readily detonated by friction.[212]

$N_3P_3(NH_2)_2(N_3)_4$. (2,2:4,4,6,6). (Structure reassigned by authors of chapter on basis of later evidence): [$N_3P_3Cl_4(NH_2)_2$ + NaN_3 in H_2O],[92] m. 81-82.[92]

$N_3P_3Ph_5 \cdot N_3$. ($N_3P_3ClPh_5$ + NaN_3),[430] m. 153.5-155.5.[430]

$N_3P_3Br_n(NCO)_{6-n}$. ($N_3P_3Br_6$ + AgOCN in $MeNO_2$),[466] IR.[466]

$N_3P_3Cl_4(NH_2)(NCO)$. (2,2,4,4:6:6):($N_3P_3Cl_4(NH_2)_2$ + $COCl_2$),[486] m. 105,[486] IR,[486] with EtOH forms $N_3P_3Cl_4(NH_2)$-(NHCOOEt) m. 132.[486]

$N_3P_3F_5 \cdot NCS$. ($N_3P_3F_5Cl$ + KSCN in MeCN),[102] IR,[102] ^{19}F +49.6 (PF·NCS), +67.9 (PF_2).[102]

$N_3P_3(NCS)_6$. ($N_3P_3Cl_6$ + KSCN),[33,361,484] m. 42,[33,361,484] IR,[463] Raman,[463] ^{31}P +26.8,[180,353] solubility.[484] X-ray crystal structure[158] shows that five-ring atoms lie in a plane, but one N atom is 0.15 Å out of the plane. Av. P-N 1.58, av. P-N (exo) 1.63, av. $N\hat{P}N$ 119°, av. $P\hat{N}P$ 121°, av. $P\hat{N}C$ 152°. Forms the following derivatives:

$N_3P_3(NHCSNH_2)_6$. (With NH_3), m. 190(d),[33,353] ^{31}P +10.8.[33,353]

$N_3P_3[NHCSO(CH_2)_2OH]_6$. [With $HO(CH_2)_2OH$], m. 110.[33]

$N_3P_3(NHCSNH\text{-}n\text{-}Bu)_6$. (With $n\text{-}BuNH_2$), m. 155.[33,484]

$N_3P_3[NHCSN\text{-}(n\text{-}Bu)_2]_6$. (With $n\text{-}Bu_2NH$), $b_{0.005}$ 120-130.[33]

$N_3P_3(NHCSNHPh)_6$. (With $PhNH_2$), m. 151,[33,484] ^{31}P +2.2.[353]

$N_3P_3[NHCSNH(CH_2)_6NH_2]_6$. [With $NH_2(CH_2)_6NH_2$][33], m. 200-260 (d).[33]

$N_3P_3(NHCSNHNHPh)_6$. (With $PhNHNH_2$), m. 165 (d).[484]

$N_3P_3(NHCSOEt)_6$. (With EtOH), m. 174.[484]

$N_3P_3(NMe_2)_4(NCS)_2$ (2,2,4,6:4,6):($N_3P_3Cl_2(NMe_2)_4$ + KSCN,[447] $b_{0.1}$ 175.[447]

$N_3P_3F_5 \cdot NSO$. ($N_3P_3F_5 \cdot NH_2$ + $SOCl_2$),[349] b_{760} 143,[349] IR,[349] ^{19}F +70 (PF_2), +57 (PF).[349] With pyridine gives ($N_3P_3F_5$)-NSN-$N_3P_3F_5$, m. 52,[349] IR,[349] ^{19}F +70 (PF_2), +55.5 (PF).[349] With PCl_5 gives $N_3P_3F_5 \cdot NSCl_2$, b_{760} 188,[349] IR,[349] ^{19}F +69 (PF_2), +51.5 (PF).[349]

D.1.3. Alkyl-derivatives

$N_3P_3F_5\text{-}n\text{-}Bu$. V.[328,336] b_{8-10} 73-76.[328,336]

$N_3P_3F_4(n\text{-}Bu)_2$. V.[328] b_1 87-89.[328]

$N_3P_3Cl_4Me_2$ (2,2,4,4:6,6). [$Me_2\overset{+}{P}(NH_2)_2Cl^-$ + PCl_5],[188] m. 174-176,[188] IR.[188]
$N_3P_3Cl_3Me_3$ (2,4,6:2,4,6). [$N_3P_3Cl_3(NMe_2)_3$ + MeMgI, then HCl],[485] m. 156-157,[485] IR.[485]
$N_3P_3Me_6$. II.[364,426,427] m. 195-196,[364,426,427] IR,[426] ^{1}H -1.28,[27] pK$_a'$ 5.17.[341] First ionization potl. 8.35 eV,[71] forms adducts:
$[N_3P_3Me_7]^+I^-$. ^{1}H -2.91 (NMe), -1.92 (PMe), -0.70 (PMe),[24] ring N quaternized,[24] $[N_3P_3Me_6Et]^+I^-$,[24] $[N_3P_3Me_7]^+$-HgI_3^-,[24] $2[N_3P_3Me_7]^+[Zn(NCS)_4]^{2-}$.[24]
$N_3P_3Me_6{\cdot}TiCl_4$. m. 315 (d),[291] IR,[291] and $N_3P_3Me_6{\cdot}SnCl_4$, m. 320,[291] IR.[291]
$N_3P_3(CF_3)_6$. [Heat $(CF_3)_2PN_3$],[482a] [heat $(CF_3)_2PCl_2NH_2$][482] m. 64,[482] IR,[482] pK$_a'$ < -6.0.[161]
$N_3P_3Et_6$. (II), m. 117.5-119,[62] pK$_a'$ 6.4.[161]
$N_3P_3(C_3F_7)_6$. [Heat $(C_3F_7)_2PN_3$],[482] m. 85,[482] IR.[482]
$N_3P_3(n\text{-}Bu)_6$. [Pyrolysis of $(n\text{-}Bu)_2\overset{+}{P}(NH_2)_2Cl^-$],[451] m. 45-50,[451] IR.[451]

D.1.4. Aryl-derivatives

$N_3P_3F_5Ph$. V,[18,20,336] b_{760} 183-184,[18,20,336] IR.[20]
$N_3P_3F_5{\cdot}C_6F_5$. V,[99] ^{19}F.[99]
$N_3P_3F_5{\cdot}C_6H_5$-m-Me. V,[336] $b_{0.12}$ 79.[336]
$N_3P_3F_5{\cdot}C_6H_4$-p-Me. V,[336] $b_{0.12}$ 83.[336]
$N_3P_3F_4Ph_2$ (2,2,4,4:6,6). VI,[22,336] ($N_3P_3F_6$ + PhMgBr),[18] m. 68.5-69.5,[18,22,336] x-ray structure[19,21] shows that five of the phosphazene ring atoms lie in a plane and that the phenylated phosphorus atom is 0.2 Å out of this plane. Three sets of P-N bond lengths 1.618 (0.005), 1.558 (0.004), and 1.539 (0.005), $N\hat{P}Ph_2N$ 115.5 (0.3), $N\hat{P}F_2N$ 120.6 (0.3), $Ph_2P\hat{N}PF_2$, 120.5 (0.2), $F_2P\hat{N}PF_2$ 120.1 (0.2), IR,[22] ^{31}P -30.4 (PPh_2), -12.1 (PF_2),[22] ^{19}F +69.7.[22]
$N_3P_3F_4Ph_2$ (2,2,4,6:4-cis-6). ($N_3P_3F_6$ + PhLi),[20] m. 74-75,[20] IR,[20] ^{31}P -38.4 (PFPh), -12.4 (PF_2),[20] ^{19}F +49.3 (PFPh), +69.3, +65.2 (PF_2),[20] dipole moment 4.3D.[20]
$N_3P_3F_4Ph_2$ (2,2,4,6:4-trans-6). ($N_3P_3F_6$ + PhLi),[20] m. 78-79,[20] IR,[20] ^{31}P -38.3 (PFPh), -10.8 (PF_2),[20] ^{19}F +51.5 (PFPh), +67.7 (PF_2),[20] dipole moment 2.5D.[20]
$N_3P_3F_4(C_6H_4\text{-o-Me})_2$ (2,2,4,6:2-cis-6). ($N_3P_3F_6$ + Me-o-C_6H_4Li),[20] m. 64-65,[20] IR,[20] dipole moment 4.4D.[20]
$N_3P_3F_3Ph_3$ (2,2,4:4,6,6). [$N_3P_3F_4Ph_2$ (2,2,4,4:6,6) + PhLi],[22] m. 79.5-80.5,[22] IR,[22] ^{31}P -31.2 (PFPh), -27.3 (PPh_2), -8.6 (PF_2),[22] ^{19}F +49.1 (PFPh), +66.0, +68.6 (PF_2).[22]

$N_3P_3F_2Ph_4$ (2,2:4,4,6,6). [$N_3P_3F_4Ph_2$ (2,2,4,6:4,6) + $AlCl_3$ + PhH],[22] m. 137-138,[22] IR,[22] ^{31}P -27.3 (PPh_2), -6.2 (PF_2).[22]

$N_3P_3Cl_5(C_6H_4$-p-Cl). VI.[1] m. 88-90.[1]

$N_3P_3Cl_4Ph_2$ (2,2,4,4:6,6). VI.[1,45,65,320a,345,444] m. 95,[1,45,65,320a,345,444] ^{31}P -20.6 (PPh_2), -18.3 (PCl_2),[148,214] NQR (^{35}Cl),[508] pK$_a'$ < 6.0,[164] x-ray structure[303] shows that P-N ring has slight chair conformation P-N 1.555 ($Ph_2PN-PCl_2$), 1.578 ($Cl_2P-N-PCl_2$), 1.615 ($Ph_2P-NPCl_2$) (all 0.005), $Ph_2P\hat{N}PCl_2$ 121.7, 122.4 (0.3), $Cl_2P\hat{N}PCl_2$ 119.2 (0.3), $N\hat{P}Ph_2N$ 115.2 (0.2), $N\hat{P}Cl_2N$ 119.8 (0.3).

$(N_3P_3Cl_4Ph)_2$. (Linked by PhP PPh unit), ($N_3P_3Cl_6$ + Ph_2Mg),[58] m. 250,[58] P-P bond of unusual stability.

$N_3P_3Cl_4(C_6H_4$-p-Me$)_2$ (2,2,4,4:6,6). VI.[1] m. 124-126.[1]

$N_3P_3Cl_4(C_6H_4$-p-Cl$)_2$ (2,2,4,4:6,6). VI.[1] m. 138-139.[1]

$N_3P_3Cl_4(C_6H_4$-p-Br$)_2$. VI.[1] m. 133-135.[1]

$N_3P_3Cl_3Ph_3$ (2-trans-4,6:2,4,6). II.[199,200,201,215] m. 158-159,[199,200,201,215] 161-163,[235] IR,[215] ^{31}P -30.3,(2), -32.7 (1)[215] pK$_a'$ < -6.0,[164] isomerized by $FeCl_3$ or $AlCl_3$.[214]

$N_3P_3Cl_3Ph_3$ (2-cis-4-cis-6:2,4,6). II.[199,200,201,215] m. 191-192,[199,200,201,215] 188,[441,442] IR,[215] ^{31}P -29.4,[215] isomerized by $AlCl_3$.[214]

$N_3P_3Cl_2Ph_4$ (2,2:4,4,6,6). VI.[1,227,320a,345] [$(Ph_2PNH_2NPPh_2NH_2)^+Cl^-$ + PCl_5][421] m. 142-143,[1,227,320a,345,421] IR,[421] ^{31}P -19.1 (PPh_2), -16.6 (PCl_2),[148,214] pK$_a'$ -3.9,[164] x-ray structure[304] shows that P-N ring has a slight boat conformation with (av.) P-N 1.554 (0.010) ($Cl_2P-NPPh_2$), 1.578 (0.010) ($Ph_2P-N-PPh_2$), 1.609 (0.010) ($Ph_2P-NPCl_2$), $Ph_2P\hat{N}PCl_2$ 120.3 (0.5), and 121.6 (0.6), $N\hat{P}Ph_2N$, 116.1 (0.4), 115.0 (0.4), $NPCl_2N$, 121.1 (0.5), 120.4 (0.5), two molecules in asymmetric unit. Forms adduct, $N_3P_3Cl_2Ph_4$·MeCN.[421]

$N_3P_3Cl_2Ph_4$ (2-trans-4:2,4,6,6). [$N_3P_3Cl_3Ph_3$ (2,4,6:2,4,6) + $AlCl_3$ + PhH],[214] m. 169,[214] ^{31}P -19.0 (1), -28.5 (2),[214] x-ray powder diffraction data.[214]

$N_3P_3Cl_2Ph_4$ (2-cis-4:2,4,6,6). [$N_3P_3Cl_3Ph_3$ (2,4,6:2,4,6) + $AlCl_3$ + PhH],[214] m. 195-196,[214] ^{31}P -19.0 (1), -30.2 (2),[214] x-ray powder diffraction data.[214]

$N_3P_3Cl_2(C_6H_4$-p-Cl$)_4$ (2,2:4,4,6,6). VI.[1] m. 189-191.[1]

$N_3P_3ClPh_5$. [$N_3P_3Cl_3Ph_3$ (2,4,6:2,4,6) + $AlCl_3$ + PhH],[201,214] {$[Ph_2P(NH_2)NPPh_2(NH_2)]^+Cl^-$ + $PhPCl_4$},[227,423] m. 151-152,[201,214,227,423] ^{31}P -17.6 (PPh_2), -28.7 (PPhCl),[214] x-ray powder diffraction data,[214] kinetics of hydrolysis to $N_3P_3Ph_5OH$,[424] m. 275-276.[424] Forms $(N_3P_3Ph_5)_2O$, m. 238-239 with pyridine/water mixture.[423]

$N_3P_3Ph_6$. II,[218,272] V,[59,65,67,381,395,396] VI.[1,320a,345] {$[Ph_2P(NH_2)NPPh_2(NH_2)]^+Cl^-$ + Ph_2PCl_3}[202,227] (pyrolysis of product from Ph_2PCl + NH_2Cl[191,198]) (pyrolysis of Ph_2PN_3[273,362]), (pyrolysis of $[Ph_2P(NH_2)NPPh_2$-$(NH_2)]^+Cl^-$[54,56,450]), (pyrolysis of product from Ph_2PCl + $H_2NNH_2 \cdot HCl$[337]), (pyrolysis of $Ph_2P(S)(NHMe)$,[438] (pyrolysis of $Ph_2PNHNMe_2$[511]), m. 232-233,[1,54,56,59,65,67,191,198,199,200,202,218,227,272,273,320a,337,345,362,381,395,396,438,450,511] ^{31}P -15.2,[214] ESR,[7,97] pK_a' 1.5,[161] x-ray structure[2] shows that P-N ring has slight chair form, av. P-N 1.597 (0.006), av. $P\hat{N}P$ 122.1 (0.4), av. $N\hat{P}N$ 117.8 (0.3). Forms adducts: $N_3P_3Ph_6 \cdot 3C_2H_2Cl_4$ (Clathrate),[506] thermodynamics of decomp.,[506] $N_3P_3Ph_6 \cdot HICl_2$: m. 197-199,[507] $N_3P_3Ph_6 \cdot 2ICl$: m. 181-183 (d).[507]

$N_3P_3(C_6F_5)_6$: [heat $(C_6F_5)_2PCl_2 \cdot NH_2$].[301]

$N_3P_3Br_3Ph_3$ (2-trans-4,6:2,4,6). II.[332,344] m. 152-153,[332,344] 162-163,[332,344] IR,[344] ^{31}P -20.1, -18.0,[333] solubility,[344] x-ray powder data,[344] dipole moment 2.36 D.[344]

$N_3P_3Br_3Ph_3$ (2-cis-4-cis-6:2,4,6). II.[332,344] m. 194-195,[332,344] ^{31}P -16.4,[333] solubility,[344] x-ray powder data,[344] dipole moment 5.27 D.[344]

$N_3P_3Ph_4(H)Me$ (2,2,4,4:6:6). [$H_2NP(Ph_2)$=N-$P(Ph_2)$=NH + $MeP(OPh)_2$],[405] m. 152,[405] IR,[405] ^{31}P -12.9 (PPh_2), -6.9 (PHMe),[405] ^1H -7.36 (PH), -1.57 (PMe).[405]

$N_3P_3Ph_4(H)Et$ (2,2,4,4:6:6). [$H_2NP(Ph_2)$=N-$P(Ph_2)$=NH + $EtP(OPh)_2$],[405] m. 152,[405] IR,[405] ^{31}P -15.5 (PPh_2), -12.5 (PHEt),[405] ^1H -7.58 (PH), -1.72 (CH_2).[405]

$N_3P_3Ph_4Me_2 \cdot HCl$ (2,2,4,4:6,6). {$[Ph_2P(NH_2)NPPh_2NH_2]^+Cl^-$ + Me_2PCl_3},[51a] m. 180-183,[51a] IR,[51a] ^{31}P -39.6 (PMe_2), -18.8 (PPh_2),[51a] ^1H -1.98 (PMe_2),[51a] ring N atom adjacent to PMe_2 group protonated.[51a]

D.1.5. Borohydride Derivatives

$N_3P_3Br_4(B_{10}H_{13})_2$. ($N_3P_3Br_6$ + $NaB_{10}H_{13}$),[288] IR.[288]

$N_3P_3Br_3(B_{10}H_{13})_3$. ($N_3P_3Br_6$ + $NaB_{10}H_{13}$).[287]

$N_3P_3(B_{10}H_{13})_6$ + ($N_3P_3Br_6$ + $NaB_{10}H_{13}$).[288] IR.[288]

D.1.6. Amino-Derivatives

$N_3P_3F_5 \cdot NH_2$. VII.[393] m. 59.5,[393] IR,[393] mass spectr.,[393] with SF_4 gives

$N_3P_3F_5 \cdot NSF_2$. b_{102} 75-76,[349] IR,[349] ^{19}F +70 (PF_2), +51.5 (PF), -59.5 (NSF_2).[349]

$N_3P_3F_5 \cdot NHMe$. VII.[193] b_{760} 151,[193] IR,[193] ^{19}F +62.5 (PF), +70.7 (PF_2),[193] ^1H -2.75 (NMe),[193] -3.70 (NH).[193]

$N_3P_3F_4(NHMe)_2$ (2,2,4,6:4,6). VII.[194] $b_{0.05}$ 78,[194] ^{19}F +59 (PF·NHMe).[194]

$N_3P_3F_5 \cdot NMe_2$. VII,[100,102,193] VIII.[100,102] b_{760} 140,[100,193] b_{200} 100,[102] IR,[102,193] ^{31}P -24.4 (PF), -10.6 (PF_2),[102] ^{19}F + 63.0 ($PFNMe_2$), +70.9 (PF_2),[102,193] ^1H -2.78.[193]

$N_3P_3F_4Cl \cdot NMe_2$ (2,2,4,6:4:6). [$N_3P_3F_4(NMe_2)_2$ + HCl],[194] b_{50} 92,[194] ^{19}F +63 ($PF \cdot NMe_2$), +29 (PFCl).[194]

$N_3P_3F_4Br \cdot NMe_2$ (2,2,4,6:4:6). [$N_3P_3F_4(NMe_2)_2$ + HBr],[194] b_{10} 70,[194] ^{19}F +62 ($PFNMe_2$), +18.5 (PFBr).[194]

$N_3P_3F_4(NMe_2)_2$ (2,2,4,4:6,6). [$N_3P_3Cl_2F_4$(2,2:4,4,6,6) + $NHMe_2$],[100,102] b_{18} 96-110,[102] IR,[102] ^{31}P -29.4 [$P(NMe_2)_2$], -14.2 (PF_2),[102] ^{19}F +70.1,[102] ^1H -2.59.[102]

$N_3P_3F_4(NMe_2)_2$ (2,2,4-trans-6:4,6). VII.[194] [$N_3P_3Cl_4(NMe_2)_2$ + NaF or KSO_2F],[206,208] $b_{0.5}$ 64 (cis + trans isomers),[194] IR,[208] ^{19}F +70.1[208] (PF_2), +63.5,[208] +63,[194] ($PF \cdot NMe_2$), H -2.70.[208]

$N_3P_3F_4(NMe_2)_2$ (2,2,4-cis-6:4,6). [$N_3P_3Cl_4(NMe_2)_2$ + NaF or KSO_2F],[206,208] $b_{0.5}$ (see above), IR,[208] ^{19}F +61.7 ($PFNMe_2$), ca. + 69.9, + 68.5 (PF_2),[208] ^1H -2.69.[208]

$N_3P_3F_2Cl_2(NMe_2)_2$ (2,2:4-trans-6:4,6). [$N_3P_3Cl_4(NMe_2)_2$ + KSO_2F],[206,208] m. 84-85,[208] IR,[208] ^{19}F +69.0,[208] ^1H -2.71.[208]

$N_3P_3F_2Cl_2(NMe_2)_2$ (2,2:4-cis-6:4,6). [$N_3P_3Cl_4(NMe_2)_2$ + KSO_2F],[206,208] m. 60-61,[208] IR,[208] ^{19}F +71, +66.9,[208] ^1H -2.68.[208]

$N_3P_3F_2Cl_2(NMe_2)_2$ (2-trans-4:6,6:2,4). [$N_3P_3Cl_4(NMe_2)_2$ + SbF_3],[206,208] liquid sepd. by GLC,[208] IR,[208] ^{19}F +63.0,[208] ^1H -2.71.[208]

$N_3P_3F_2Cl_2(NMe_2)_2$ (2-cis-4:6,6:2,4). [$N_3P_3Cl_4(NMe_2)_2$ + SbF_3],[206,208] m. 54-55,[208] IR,[208] ^{19}F +61.6,[208] ^1H -2.70.[208]

$N_3P_3F_3(NMe_2)_3$ (2,4,6:2,4,6). VII.[194] $b_{0.05}$ 68,[194] ^{19}F +63.[194]

$N_3P_3Cl_5 \cdot NH_2$. [Pyrolysis of $N_3P_3Cl_4(NH_2)_2$],[169] [$N_3P_3Cl_4(NH_2)_2$ + HCl],[166] [$N_3P_3Cl_4(NH_2)_2$ + $COCl_2$],[486] m. 138-139,[166,169,486] IR,[166,486] ^{31}P -20.4 (PCl_2), -19.0 ($PCl \cdot NH_2$).[166]

$N_3P_3Cl_4(NH_2)_2$ (2,2,4,4:6,6). VII.[44,46,166,169,255,294,338,478] m. ca. 164,[44,46,166,169,255,294,338,478] IR,[166] ^{31}P -18.3 (PCl_2), -9.03 [$P(NH_2)_2$],[166,294] hydrolysis,[143] formation of fluoro-alkoxy-derivatives.[297a]

$N_3P_3(NH_2)_6$. ($N_3P_3Cl_6$ + liq. NH_3),[34,53,325] m. 220 (d),[34,53,325] ^{31}P -15.3,[327] pK_a 7.70 (in water),[160] d^{20} 1.68,[325] pyrolysis of,[324,325,440] electrolysis of,[154] hydrolysis of,[143] forms adduct $N_3P_3(NH_2)_6 \cdot H_2O$: heat cap. 11.768 Kcal./mol.,[503a] entropy 70.98 cal./deg./mol. at 298 K.[503a]

$N_3P_3(NHNH_2)_6$. VI.[360] m. > 360,[360] forms $N_3P_3(NHN{=}CHPh)_6$: m. 250 (d)[360] with benzaldehyde, and $N_3P_3(NHN{=}CHC_6H_4OH)_6$: m. 225 (d)[360] with salicaldehyde.

$N_3P_3Cl_5 \cdot NHMe$. VII.[295,334,335] m. 67,[295,334,335] $b_{0.1}$ 92-94,[295] IR,[295,335] ^{31}P -22.8 (one signal),[295] ^1H -2.89 (NMe),[295] dipole moment 3.41 D.[295]

$N_3P_3Cl_4(NHMe)_2$ (2,2,4,4:6,6). VII.[184,295] m. 128,[184,295] IR,[295] ^{31}P -20.1, -12.3,[295] ^1H -2.62 (NMe),[184,295] dipole moment 4.33 D,[295] only obtained in small quantities compared with nongeminal isomers below.[295]

$N_3P_3Cl_4(NHMe)_2$ (2,2,4-trans-6:4,6). VII.[44,46,66,184,295,387] m. 103,[44,46,66,184,295,387] IR,[295] ^{31}P -22.2 (one signal),[295] ^1H -2.76 (NMe),[295] dipole moment 2.88 D.[295]

$N_3P_3Cl_4(NHMe)_2$ (2,2,4-cis-6:4,6). VII.[295] m. 149,[295] IR,[295] ^{31}P -21.6 (one signal),[295] ^1H -2.73 (NMe),[295] dipole moment 4.12 D.[295]

$N_3P_3(NHMe)_6$. VII.[46,184,386,387] m. 259,[46,184,386,387] IR,[184] ^{31}P -24,[46] ^1H -2.57,[184] pK_a' 8.8,[160] forms adduct:

$N_3P_3(NHMe)_6 \cdot 2C_6H_2(NO_2)_3OH$. m. 190-192 and 210-212.[126]

$N_3P_3Cl_5 \cdot NC_2H_4$. VII.[265,282,357] m. 69-70,[265,282,357] ^{31}P,[357] ^1H -2.37,[357] dipole moment 3.07 D.[269]

$N_3P_3Cl_4(NC_2H_4)_2$ (2,2,4,4:6,6). VII.[265,282,357,358] m. 105,[265,282,357,358] ^{31}P -22 (PCl_2), -34 [$P(NC_2H_4)_2$],[265] ^1H -2.22.[357]

$N_3P_3Cl_3(NC_2H_4)_3$ (2,2,4:4,6,6). VII.[282,357] m. 68-70,[282,357] ^1H -2.27, -2.20, -2.15.[357]

$N_3P_3Cl_2(NC_2H_4)_4$ (2,2:4,4,6,6). VII.[265,282,357,358] m. 131,[265,282,357,358] ^{31}P -24 (PCl_2), -35 [$P(NC_2H_4)_2$],[265] ^1H -2.14.[357]

$N_3P_3Cl_2(NC_2H_4)_4$. VII.[282] m. 99-101; 114-115.[282]

$N_3P_3Cl(NC_2H_4)_5$. VII.[265,282,357] m. 122,[265] 108-109,[282] ^1H -2.22 (1), -2.14 (4).[357]

$N_3P_3(NC_2H_4)_6$. VII.[265,282,384] m. 149-150,[265,282,384] IR,[458] ^1H -2.08,[357] forms complexes $N_3P_3(NC_2H_4)_6$, $3AgNO_3$,[384] $N_3P_3(NC_2H_4)_6 \cdot 2CuSO_4$,[384] $N_3P_3(NC_2H_4)_6 \cdot 3ZnCl_2$,[384] and picrate: m. 234-236.[384] With HCl gives $N_3P_3(NHCH_2CH_2Cl)_6$: m. 82-82.5.[384]

$N_3P_3Cl_5 \cdot NHEt$. VII.[121,257,335,372] m. 33-34,[372] 34-35,[121] $b_{0.05}$ 105,[335,372] IR,[335] n_D^{25} 1.5414,[335] d. 1.065.[335]

$N_3P_3Cl_4(NHEt)_2$ (2,2,4-trans-6:4,6). VII.[121] m. 85.[121]

$N_3P_3Cl_4(NHEt)_2$ (2,2,4-cis-6:4,6). VII.[121] m. 98.[121]

$N_3P_3Cl_3(NHEt)_3$ (2,4,6:2,4,6). VII.[121] m. 131,[121] pK_a' -4.9.[162]

$N_3P_3Cl_2(NHEt)_4$ (2,2:4,4,6,6). VII.[121] m. 126,[121] pK_a' 3.2.[162]

$N_3P_3(NHEt)_6$. VII.[121,268,386,387] m. 118-119,[121,268,386,387] IR,[268] pK_a' 8.2,[160] forms adduct, $N_3P_3(NHEt)_6 \cdot HCl$: m. 202.[121]

$N_3P_3Cl_5 \cdot NMe_2$. VII.[55,182,197,253,269,271,335,389,481] m. 12-14,[55,182,197,253,269,271,387] $b_{0.01}$ 81-83,[55,182,197,253,269,271,335,387,481] IR,[271,459,461] Raman,[461] ^{31}P -22.7 ($PCl \cdot NMe_2$), -20.6 (PCl_2),[182,223]

^{1}H -2.75,[155,182,223,251,256,271] dipole moment 3.12 D,[271] n_D^{25} 1.5377,[335] Sp. gr. 1.63.[335]

$N_3P_3Cl_4(NMe_2)_2$ (2,2,4,4:6,6). VII.[253,271] m. 62,[253,271] IR,[271] ^{1}H -2.71.[251,253,256,271]

$N_3P_3Cl_4(NMe_2)_2$ (2,2,4-trans-6:4,6). VII.[44,46,55,182,186,253,271,386,387] m. 103,[44,46,55,182,186,253,271,386,387] IR,[271,459,461] Raman,[461] ^{31}P -21.6 (PCl_2), -24.9 ($PCl{\cdot}NMe_2$),[251] ^{1}H -2.73,[155,182,251,256,271] dipole moment. 2.61. D.[271]

$N_3P_3Cl_4(NMe_2)(\overset{+}{N}Me_3)BF_4^-$ (2,2,4,6:4:6). (Above deriv. + $Me_3O^+BF_4^-$),[383] m. 120 (d),[383] ^{1}H -2.80 (NMe_2), -3.45 (NMe_3),[383] hydrolysis,[383] conductivity.[383]

$N_3P_3Cl_4(NMe_2)_2$ (2,2,4-cis-6:4,6). VII.[253,271] m. 86,[253,271] IR,[271,459,461] Raman,[461] ^{31}P -20.5 (PCl_2), -24.5 ($PCl{\cdot}NMe_2$),[251] ^{1}H -2.71,[251,256,271] dipole moment 4.3 D.[271]

$N_3P_3Cl_3(NMe_2)_3$ (2,2,4:4,6,6). VII.[253,271] m. 71,[253,271] IR,[271,459,461] Raman,[461] ^{31}P ca. -22.0, -27.5,[148,259] ^{1}H -2.69 [$P(NMe_2)_2$] (two signals), -2.72 ($PCl{\cdot}NMe_2$),[251,256,271] pK_a' -4.4.[162]

$N_3P_3Cl_3(NMe_2)_3$ (2-trans-4,6:2,4,6). VII.[44,46,182,253,273,387,416,459] [$N_3P_3(NMe_2)_6$ + HCl],[343] m. 105,[44,46,182,253,273,387,416,459] IR,[271,459,461] Raman,[461] ^{31}P -29.0,[259] ^{1}H -2.68(2), -2.72(1),[155,182,251,253,256,271] dipole moment 2.01 D,[271] pK_a' -5.4.[162]

$N_3P_3Cl_3(NMe_2)_2(\overset{+}{N}Me_3)BF_4^-$ (2,4,6:2,4:6). (Above deriv. + $Me_3O^+BF_4^-$),[383] m. 114-139(d),[383] ^{1}H -2.76 (NMe_2), -3.42 (NMe_3),[383] hydrolysis,[383] conductivity.[383]

$N_3P_3Cl_3(NMe_3)_3$ (2-cis-4-cis-6:2,4,6). VII.[182,253,416] [$N_3P_3(NMe_2)_6$ + HCl][343] (isomerization of trans-isomer above)[254,260] m. 152,[182,253,343,416] ^{31}P -27.6,[251] ^{1}H -2.68,[171,182,251,256] pK_a' -5.5.[162]

$N_3P_3Cl_2(NMe_2)_4$ (2-cis-4:2,4,6,6). VII.[44,46,182,253,386,387] [$N_3P_3(NMe_2)_6$ + HCl],[343] m. 104,[44,46,182,253,386,387] IR,[459,461] Raman,[461] ^{31}P -23.9 ($PCl{\cdot}NMe_2$), -30.5 [$P(NMe_2)_2$],[251] ^{1}H -2.52 (2), -2.64 (1), -2.65 (1),[179,182,183] pK_a' -1.4.[162]

$N_3P_3Cl_2(NMe_2)_3(\overset{+}{N}Me_3)BF_4^-$ (2,4:2,4,6:6). (Above deriv. + $Me_3O^+BF_4^-$),[383] m. 167-168 (d),[383] ^{1}H -2.77, -3.02 (NMe_2), -3.12 (NMe_3),[383] hydrolysis,[383] conductivity.[383]

$N_3P_3(NMe_2)_6$. VII.[46,182,253,386,387] m. 104,[46,182,253,386,387] IR,[432] UV,[447] ^{31}P -25,[46] ^{1}H -2.55,[27,182,183,220,251] pK_a' 7.6,[160] first ionization potl. 7.85 eV.,[71] forms adducts,

$N_3P_3(NMe_2)_6{\cdot}HCl$. m. 198-200,[343]

$N_3P_3(NMe_2)_6{\cdot}HBr$. m. 216-217,[343]

$N_3P_3(NMe_2)_6{\cdot}HI$. m. 220-222,[343]

$N_3P_3(NMe_2)_6{\cdot}HI_3$. m. 144-145,[343]

$N_3P_3(NMe_2)_5(\overset{+}{N}Me_3)BF_4^-$. (With $Me_3O^+BF_4^-$),[383] m. 181-200 (d),[383] hydrolysis,[383] conductivity,[383]

$[N_3P_3(NMe_2)_4(\overset{+}{N}Me_3)_2]\ 2BF_4^-$. (With $Me_3O^+BF_4^-$),[382,383] m. 207-209,[382,383] ^1H,[382] conductivity,[382,383] exocyclic N-quaternized,

$N_3P_3(NMe_2)_6 \cdot C_6H_2(NO_3)_3OH$. m. 195-197,[126]

$N_3P_3(NMe_2)_6 \cdot C_6H_3(NO_2)_2OH$. m. 44.5-46, 112,[126]

$N_3P_3(NMe_2)_6 \cdot C_2(CN)_4$. m. 113-114.[126]

$N_3P_3Cl_4(NHCH_2)_2$ (2,2,4,4:6,6). VII.[41] m. 199,[41] IR.[41]

$N_3P_3Cl_4(NHCH_2)_2$. VII.[87] m. 180.5-181.5.[87]

$N_3P_3Cl_4(NHCH_2CH_2NH_2)_2$. VII.[66] m. 188.[66]

$N_3P_3Cl_4(NHCH_2CH_2O)$. VII.[87,505] m. 147-147.5,[87] 150.5.[505]

$N_3P_3Cl_5(NH\text{-}n\text{-}Pr)$. VII.[279,335] m. 54.5-55.5,[279] $b_{1.75}$ 137,[335] n_D^{25} 1.5362,[335] Sp. gr. 1.58,[335] IR,[335] rate of formation.[331]

$N_3P_3Cl_2(NH\text{-}n\text{-}Pr)_4$ (2,2:4,4,6,6). VII.[387] m. 93,[387] pK_a' 3.1.[162]

$N_3P_3(NH\text{-}n\text{-}Pr)_6$. VII.[268,387] m. 59,[268,387] IR,[268] ^1H,[330] pK_a' 7.9,[160] forms adducts:

$N_3P_3(NH\text{-}n\text{-}Pr)_6 \cdot HCl$. m. 184,[330,388] conductivity,[330] IR,[330] ^1H,[330]

$N_3P_3(NH\text{-}n\text{-}Pr)_6 \cdot 2HCl$. m. 179,[133]

$N_3P_3(NH\text{-}n\text{-}Pr)_6 \cdot HBr$. m. 154,[330] conductivity,[330] IR,[330] ^1H.[330] UV evidence[330] for formation of $\{M^{II}[N_3P_3(NH\text{-}n\text{-}Pr)_6]_x\}^{2+}$ (x = 1,2,6; M^{II} = Co^{2+}, Ni^{2+}, Cu^{2+}).

$N_3P_3Cl_5 \cdot NH\text{-}i\text{-}Pr$. VII.[124,279] m. 54-55.[124,279]

$N_3P_3Cl_4(NH\text{-}i\text{-}Pr)_2$ (2,2,4,6:4,6). VII.[124] m. 114-115,[124] forms adduct, $N_3P_3Cl_4(NH\text{-}i\text{-}Pr)_2 \cdot N_3P_3Cl_2(NH\text{-}i\text{-}Pr)_4$ m. 80-81,[122] which with HCl gives $N_3P_3Cl_2(NH\text{-}i\text{-}Pr)_4 \cdot HCl$: m. 220-225,[122]

$N_3P_3Cl_2(NH\text{-}i\text{-}Pr)_4$ (2,2:4,4,6,6). VII.[124,387] m. 126,[124,387] ^{31}P -9.9 [$P(NH\text{-}i\text{-}Pr)_2$], -23.4 ($PCl_2$),[148] pK_a' 3.4,[162] forms hydrochloride $N_3P_3Cl_2(NH\text{-}i\text{-}Pr)_4 \cdot HCl$. m. 195,[124] x-ray crystal structure[305] shows that ring N protonated between aminolyzed P-atoms, where P-N-P 132 (0.4) and P-N 1.67 (0.005) Cl_2PN-P 1.58 (0.005), Cl_2P-NP 1.56 (0.005). Ring has slight "boat" conformation. Exocyclic P-N 1.60-1.62 (0.005).

$[N_3(Me)P_3Cl_2(NH\text{-}i\text{-}Pr)_4]^+BF_4^-$ (5:2,2:4,4,6,6). (Above deriv. + $Me_3O^+BF_4^-$),[383] m. 205-208,[383] ^1H -2.91 (NMe),[383] conductivity,[383] hydrolysis.[383]

$N_3P_3(NH\text{-}i\text{-}Pr)_6$. VII.[124,387] m. 81,[124,387] pK' 8.4,[160] forms adduct $N_3P_3(NH\text{-}i\text{-}Pr)_6 \cdot HCl$. m. 203.[124,388]

$N_3P_3Cl_5(NH\text{-}cyclo\text{-}Pr)$. VII.[319] m. 58,[319] $b_{0.1}$ 130-160.[319]

$N_3P_3Cl_4(NHCH_2)_2CH_2$ (2,2,4,6:4,6). VII.[41] m. 166,[41] IR.[41]

$N_3P_3Cl_4(NHCH_2)_2CH_2$ (2,2,4,4:6,6). VII.[86,87] m. 157.5-158.5,[86,87] ^{31}P,[86] ^1H.[86]

$N_3P_3Cl_{6-n}(NHCH_2CH{=}CH_2)_n$. n = 1,3 [(2,2,4:4,6,6) and (2,4,6:2,4,6)] and 5 reported,[15] VII.[15]

$N_3P_3(NHCH_2CH{=}CH_2)_6$. VII.[15] m. 90-91.[15]

$N_3P_3Cl_4(NHCH_2CH_2CH_2O)$. VII.[505] m. 135.[505]
$N_3P_3Cl_5(NH\text{-}n\text{-}Bu)$. VII.[335] $b_{0.05}$ 110,[335] b_2 148-149,[279] IR,[335] n_D^{25} 1.5312,[335] sp. gr. 1.54,[335] forms complexes with $CuCl_2$ and $CoCl_2$.[241]
$N_3P_3(NH\text{-}n\text{-}Bu)_6$. VII.[268,387] m. 48,[268,387] IR,[268] pK$_a'$ 7.9,[160] forms adducts $N_3P_3(NH\text{-}n\text{-}Bu)_6 \cdot HCl$: m. 132,[330,388] conductivity,[330] IR,[330] ^1H;[330] $N_3P_3(NH\text{-}n\text{-}Bu)_6 \cdot 2HCl$: m. 164.[133] Forms complexes with metal ions as for $N_3P_3(NH\text{-}n\text{-}Pr)_6$.[241]
$N_3P_3Cl_2(NH\text{-}sec\text{-}Bu)_4$. VII.[387] m. 71.[387]
$N_3P_3Cl_5(NH\text{-}t\text{-}Bu)$. VII.[123,257] m. -10 to -11.[123,257]
$N_3P_3Cl_4(NH\text{-}t\text{-}Bu)_2$ (2,2,4,4:6,6). VII.[123] m. 120-122.[123]
$N_3P_3Cl_2(NH\text{-}t\text{-}Bu)_4$ (2,2:4,4,6,6). VII.[123,387] m. 156,[123,387] pK$_a'$ + 3.5,[162] forms adduct $N_3P_3Cl_2(NH\text{-}t\text{-}Bu)_4 \cdot HCl$: m. 295 (d).[123]
$N_3P_3(NH\text{-}t\text{-}Bu)_6$. VII.[123] m. 280-282,[123] pK$_a'$ 8.0.[160]
$N_3P_3Cl_5 \cdot NEt_2$. VII.[121a,335] m. 15,[121a] $b_{0.05}$ 97-100,[335] IR,[335] n_D^{25} 1.5287,[335] sp. gr. 1.54.[335]
$N_3P_3Cl_4(NEt_2)_2$ (2,2,4,6:4,6). VII.[121a] Differentiated by GLC and ^1H n.m.r. from isomer below.
$N_3P_3Cl_4(NEt_2)_2$ (2,2,4,6:4,6). VII.[121a] See compound above.
$N_3P_3Cl_3(NEt_2)_3$ (2-trans-4,6:2,4,6). VII.[121a] m. 43,[121a] pK$_a'$ -5.2.[162]
$N_3P_3Cl_3(NEt_2)_3$ (2-cis-4-cis-6:2,4,6). VII.[121a] m. 115,[121] pK$_a'$ -5.2.[162]
$N_3P_3(NEt_2)_6$. VII.[121a,387] m. 205,[121a,387] pK$_a'$ 8.5.[160]
$N_3P_3Cl_4(NHCH_2CH_2)_2$ (2,2,4,6:4,6). VII.[41] m. 187,[41] IR.[41]
$N_3P_3[NH(CH_2)_2NH(CH_2)_2NH_2]_6$. VII.[354] n_D^{20} 1.5565.[354]
$N_3P_3Cl_5(NC_4H_8)$. (Pyrrolidino). VII.[274,276] m. 65.5-66.5,[274,276] dipole moment 3.74 D.[269]
$N_3P_3Cl_4(NC_4H_8)_2$ (2,2,4-trans-6:4,6). VII.[274,276] m. 56-56.5,[274,276] dipole moment 3.28 D.[269]
$N_3P_3Cl_4(NC_4H_8)_2$ (2,2,4-cis-6:4,6). VII.[274,276] m. 94.5-95.5,[274-276] dipole moment 5.02 D.[269]
$N_3P_3Cl_3(NC_4H_8)_3$ (2-trans-4,6:2,4,6). VII.[274-276] m. 82.5-83.5,[274-276] dipole moment 2.44 D.[269]
$N_3P_3Cl_3(NC_4H_8)_3$ (2-cis-4-cis-6:2,4,6). VII.[274-276] m. 174-175,[274-276] dipole moment 5.64 D.[269]
$N_3P_3Cl_3(NC_4H_8)_3$ (2,2,4:4,6,6). VII.[274-276] m. 74-75,[274-276] dipole moment 4.34 D.[269]
$N_3P_3Cl_2(NC_4H_8)_4$ (2-trans-4:2,4,6,6). VII.[274-276] m. 137.5-138.[274-276]
$N_3P_3Cl_2(NC_4H_8)_4$ (2-cis-4:2,4,6,6). VII. m. 121.5-122,[274-276] dipole moment 4.19 D.[269]
$N_3P_3Cl(NC_4H_8)_5$. VII.[274-276] m. 183-185,[274-276] generally obtained in very low yield.
$N_3P_3(NC_4H_8)_6$. VII.[274-276] m. 211-213,[274-276] dipole moment 1.75 D.[269]

$N_3P_3Cl_5(NC_4H_8O)$ (morpholino). VII.[283] m. 92-93.5,[283] dipole moment 1.91 D.[269]
$N_3P_3Cl_4(NC_4H_8O)_2$ (2,2,4-trans-6:4,6). VII.[283,340] m. 106-108,[283,340] dipole moment 1.85 D.[269]
$N_3P_3Cl_4(NC_4H_8O)_2$ (2,2,4-cis-6:4,6). VII.[284,340] m. 202-203,[284,340] dipole moment 2.7 D.[269]
$N_3P_3Cl_3(NC_4H_8O)_3$. VII.[283,340] m. 101.5-102.5.[283,340]
$N_3P_3Cl_3(NC_4H_8O)_3$. VII.[283,340] m. 117-118,[283,340] the two tris derivatives are positional isomers.[340]
$N_3P_3Cl_2(NC_4H_8O)_4$ (2,4:2,4,6,6). VII.[284,340] m. 145-147.[284,340]
$N_3P_3Cl_2(NC_4H_8O)_4$ (2,4:2,4,6,6). VII.[283,340] m. 158-159.[283,340]
$N_3P_3Cl(NC_4H_8O)_5$. VII.[283] m. 171-172.[283]
$N_3P_3(NC_4H_8O)_6$. VII.[283] m. 293.[283]
$N_3P_3Cl_5(NHCH_2COOEt)$. VII.[281,285,286] m. 74-75.[281,285,286]
$N_3P_3Cl_4(NHCH_2COOEt)_2$ (2,2,4,4:6.6). VII.[277,281,285,286] m. 83-84.[277,281,285,286]
$N_3P_3Cl_2(NHCH_2COOEt)_4$ (2,2:4,4,6,6). VII.[277,281,285,286] m. 73-74.[277,281,285,286]
$N_3P_3(NHCH_2COOEt)_6$. VII.[277,281,285,286] m. 72-72.5.[277,281,285,286]
$N_3P_3Cl_5[NH(CH_2)_2COOMe]$. VII.[279,285] m. 59-60.[279,285]
$N_3P_3Cl_4[NH(CH_2)_2COOMe]_2$. VII.[285] m. 59-60.[285]
$N_3P_3Cl_2[NH(CH_2)_2COOMe]_4$. VII.[285] m. 42.[285]
$N_3P_3[NH(CH_2)_2COOMe]_6$. VII.[285] Oil.[285]
$N_3P_3Cl_5(NMe{\cdot}SiMe_3)$. $[N_3P_3Cl_5{\cdot}NHMe + (Me_3Si)_2NH]$,[315] $[N_3P_3Cl_6 + (Me_3Si)_2NMe]$,[259] m. 90.5-91,[315] ^1H -2.68 (NMe), -0.25 ($SiMe_3$),[259] ^{31}P -20.5.[259]
$N_3P_3Cl_5{\cdot}NH(CH_2)_4Me$. VII.[335] $b_{0.05}$ 115,[335] IR,[335] n_D^{25} 1.5265,[335] sp. gr. 1.63.[335]
$N_3P_3[NH(CH_2)_4Me]_6$. VII.[268] m. 38,[268] IR,[268] forms adduct $N_3P_3[NH(CH_2)_4Me]_6{\cdot}3HCl$: m. 120.[133]
$N_3P_3Cl_5{\cdot}$NH-cyclo-pentyl. VII.[319] b_2 195-9.[319]
$N_3P_3Cl_5{\cdot}NC_5H_{10}$. VII.[255,285,286] m. 68,[255,285,286] ^1H -3.19 (α-H), -1.63 (β- and γ-H),[255] dipole moment 3.67 D,[269] kinetics of formation.[84,85]
$N_3P_3Cl_4(NC_5H_{10})_2$ (2,2,4-trans-6:4,6). VII.[255,285,286] m. 104-105,[255,285,286] ^1H -3.19 (α-H), -1.62 (β- and γ-H),[255] dipole moment 3.02 D,[269] kinetics of formation.[84,85]
$N_3P_3Cl_4(NC_5H_{10})_2$ (2,2,4-cis-6:4,6). VII.[255] m. 129,[255] ^1H -3.18 (α-H), -1.61 (β- and γ-H),[255] dipole moment 4.61 D.[269]
$N_3P_3Cl_3(NC_5H_{10})_3$ (2-trans-4,6:2,4,6). VII.[255,285,286] m. 114.5,[255,285,286] ^1H -3.15 (α-H), -1.59 (β- and γ-H),[255] pK_a' -5.3.[162]
$N_3P_3Cl_3(NC_5H_{10})_3$ (2-cis-4-cis-6:2,4,6). VII.[255] m. 190,[255] ^1H -3.11 (α-H), -1.63 (β- and γ-H),[255] pK_a' -5.1.[162]

$N_3P_3Cl_3(NC_5H_{10})_3$ (2,2,4:4,6,6). VII.[255] m. -17,[255] ^{1}H -3.12 (α-H), -1.58 (β- and γ-H),[255] pK_a' -3.9.[162]

$N_3P_3Cl_2(NC_5H_{10})_4$ (2-cis-4:2,4,6,6). VII.[255,285,286] m. 111-112,[255,285,286] ^{1}H -3.10 (α-H), -1.56 (β- and γ-H),[255] pK_a' -0.9,[162] dipole moment 3.99 D.[269]

$N_3P_3Cl(NC_5H_{10})_5$. VII.[285,286] m. 121-123.5.[285,286]

$N_3P_3(NC_5H_{10})_6$. VII.[108,233,255,285,286] m. 266,[108,233,255,285,286] ^{1}H -3.05 (α-H), -1.49 (β- and γ-H),[255] pK_a' +8.4,[160] dipole moment 1.16 D.[269]

$N_3P_3Cl_5 \cdot NEt \cdot SiMe_3$. [$N_3P_3Cl_5 \cdot NHEt$ + $(Me_3Si)_2NH$],[315] $b_{1.5}$ 135,[315] n_D^{18} 1.5281.[315]

$N_3P_3Cl_5$(NH-n-Hex). VII.[69] IR,[69] ^{1}H.[69]

N_3P_3(NH-n-Hex)$_6$. VII.[268] m. 49,[268] IR.[268]

$N_3P_3Cl_5$(NH-cyclo-Hex). VII.[319] m. 63,[319] $b_{0.02}$ 190-200.[319]

N_3P_3(NH-cyclo-Hex)$_6$. VII.[282,386] m. 166,[282,386] pK_a' 7.9.[160]

$N_3P_3Cl_5 \cdot N\text{-}(n\text{-}Pr)_2$. VII.[335] $b_{0.01}$ 122,[335] IR,[335] n_D^{20} 1.5218,[335] sp. gr. 1.47.[335]

$N_3P_3Cl_5(NC_4H_6\text{-}\alpha,\alpha\text{-}Me_2)$ (α,α-dimethylpyrrolidino). VII.[278] m. 67-68.5.[278]

$N_3P_3Cl_4(NC_4H_6\text{-}\alpha,\alpha\text{-}Me_2)_2$. VII.[278] m. 101-103.[278]

$N_3P_3Cl_4(NH \cdot SiMe_3)_2$ (2,2,4,4:6,6). (Structure assigned by authors of chapter in the light of subsequent findings),[166,253,294] [$N_3P_3Cl_4(NH_2)_2$ + $(Me_3Si)_2NH$],[316] m. 173.5-174.2.[316]

$N_3P_3Cl_5 \cdot NHPh$. VII.[136,319] m. 85,[319] 98,[136] $b_{0.06}$ 140-180,[319] rate of formation.[38]

$N_3P_3Cl_4(NHPh)_2$ (2,2,4,4:6,6). VII.[136,293] m. 207.5-208.5,[293] 212,[136] ^{31}P -20.4 (PCl_2), -2.3 [$P(NHPh)_2$].[293]

$N_3P_3Cl_4(NHPh)_2$ (2,2,4,6:4,6). VII.[136] m. 165.[136]

$N_3P_3Cl_3(NHPh)_3$ (2,2,4:4,6,6). VII.[136] m. 172.[136]

$N_3P_3Cl_2(NHPh)_4$ (2,2:4,4,6,6). VII.[66,136] m. 191,[66] 198,[136] pK_a' -2.3,[162] forms adduct $2N_3P_3Cl_2(NHPh)_4 \cdot HCl$: m. 217.[66]

$N_3P_3(NHPh)_6$. VII.[10,136,233] m. 270-272,[10,136,233] pK_a' 1.2,[162] forms adduct $N_3P_3(NHPh)_6 \cdot HCl$: m. 221.[66]

$N_3P_3Cl_5(NHC_6H_4\text{-}o\text{-}Cl)$. VII.[319] m. 75,[319] $b_{0.09}$ 175.[319]

$N_3P_3Cl_5(NHC_6H_4\text{-}m\text{-}Cl)$. VII.[319] m. 103,[319] $b_{0.06}$ 145.[319]

$N_3P_3Cl_4(NHC_6H_4\text{-}m\text{-}Cl)_2$. VII.[293] (Not isolated, but converted to derivative.)

$N_3P_3Cl_5(NHC_6H_4\text{-}p\text{-}Cl)$. VII.[319] m. 77,[319] $b_{0.06}$ 155.[319]

$N_3P_3Cl_4(NHNHPh)_2$ (2,2,4-trans-6:4,6). VII.[374] m. 183-184.[374]

$N_3P_3(NHNHPh)_6$. VII.[108] m. 200.[108]

$N_3P_3Cl_4(NHC_6H_4\text{-}o\text{-}NH)$ (2,2,4,4:6,6). VII.[9,66] m. 325,[9,66] IR,[9] UV,[9] ^{31}P -20 (one signal only),[9] reaction with o-aminophenol.[10]

$N_3P_3Cl_2(NHC_6H_4\text{-}o\text{-}NH)_2$ (2,2:4,4,6,6). VII.[9] m. > 300,[9] IR,[9] UV.[9]

$N_3P_3(NHC_6H_4\text{-o-}NH)_3$. VII.[9] m. > 300,[9] IR,[9] UV,[9] ^{31}P -23,[9] forms inclusion compounds,[75] reaction with o-aminophenol.[10]
$N_3P_3Cl_5 \cdot NH$-cyclo-heptyl. VII.[319] $b_{0.1}$ 170-190.[319]
$N_3P_3Cl_5 \cdot NH$-cyclo-hexyl-2-Me. VII.[319] m. 44,[319] $b_{0.1}$ 130-140.[319]
$N_3P_3Cl_5 \cdot NH$-cyclo-hexyl-4-Me. VII.[314] m. 50,[319] $b_{0.1}$ 172.[319]
$N_3P_3Cl_5 \cdot NHC_6H_4$-p-Me. VII.[319] $b_{0.1}$ 150-160.[319]
$N_3P_3Cl_4(NHC_6H_4\text{-p-Me})_2$. VII.[265] m. 138.[265]
$N_3P_3Cl_2(NHC_6H_4\text{-p-Me})_4$ (2,2:4,4,6,6). (Structure assigned by authors of chapter). VII.[66] m. 174,[66] pK_a' -1.8.[162]
$N_3P_3(NHC_6H_4\text{-p-Me})_6$. VII.[233,387] m. 249,[233,387] pK_a' 3.0.[162]
$N_3P_3(NHC_6H_4\text{-o-Me})_6$. VII.[108] m. 241-242.[108]
$N_3P_3Cl_5 \cdot NHCH_2Ph$. VII.[319] m. 58-59,[319] $b_{0.01}$ 200-210.[319]
$N_3P_3Cl_5 \cdot NHC_6H_4$-p-OMe. VII.[319] m. 71,[319] $b_{0.09}$ 180.[319]
$N_3P_3Cl_5 \cdot NHCH_2C_6H_4$-p-Cl. VII.[319] m. 63-65,[319] $b_{0.06}$ 195-200.[319]
$N_3P_3(NMePh)_6$. VII.[436] m. 141-142.[436]
$N_3P_3Cl_4(NHC_6H_3Me\text{-o-}NH)_2$ (2,2,4,4:6,6). VII.[66] m. 211.[66]
$N_3P_3Cl_5$(NH-n-octyl). VII.[318] $b_{0.01}$ 130-135.[318]
N_3P_3(NH-n-octyl)$_6$. VII.[64] m. 58-59,[69] IR,[69] ^1H.[69]
$N_3P_3Cl_5 \cdot N\text{-}(n\text{-}Bu)_2$. VII.[334] $b_{0.01}$ 130-131,[334] IR,[334] n_D^{25} 1.5148,[334] sp. gr. 1.36.[334]
$N_3P_3Cl_5$-NH-1-Me-heptyl. VII.[318] b_1 180.[318]
$N_3P_3Cl_5 \cdot NHCHMePh$. VII.[279] b_3 150.[279]
$N_3P_3Cl_5 \cdot NHC_6H_4\text{-p-}NMe_2$. VII.[69] m. 128-129,[69] IR,[69] ^1H.[69]
$N_3P_3(NHC_6H_4\text{-p-}NMe_2)_6$. VII.[69] m. 140 (d),[69] IR,[69] ^1H.[69]
$N_3P_3Cl_5(NHCH_2C_6H_4\text{-o-Me})$. VII.[319] $b_{0.09}$ 186.[319]
$N_3P_3Cl_5$(NH-n-nonyl). VII.[318] $b_{0.35}$ 205.[318]
$N_3P_3Cl_5(NH\text{-6-heptyl-2,6-}Me_2)$. VII.[318] m. 25,[318] $b_{0.08}$ 140-150.[318]
$N_3P_3Cl_5(NH\text{-3,5,5-}Me_3\text{-hexyl})$. VII.[318] b_1 189-194.[318]
$N_3P_3Cl_5(NH\text{-}CH_2C_6H_3\text{-2,6-}Me_2)$. VII.[319] m. 82,[319] b_1 220.[319]
$N_3P_3Cl_5(NHCH_2C_6H_3\text{-3,4-}Me_2)$. VII.[319] $b_{0.8}$ 210-220.[319]
$N_3P_3Cl_5(NHCHMeCH_2Ph)$. VII.[319] $b_{0.1}$ 150-160.[319]
$N_3P_3Cl_5$(NHCHPhCOOMe). VII.[279] m. 100.5-102.[279]
$N_3P_3Cl_5 \cdot NH$-n-decyl. VII.[318] $b_{0.15}$ 210.[318]
$N_3P_3Cl_5 \cdot NH\text{-3,7-}Me_2$-octyl. VII.[318] $b_{0.05}$ 170-180.[318]
$N_3P_3Cl_5(NC_4H_7\text{-}\alpha\text{-Ph})$ (α-phenylpyrrolidino). VII.[280] m. 82.5-83.5.[280]
$N_3P_3Cl_4(NC_4H_7\text{-}\alpha\text{-Ph})_2$. VII.[280] m. 120-121.[280]
$N_3P_3Cl_3(NC_4H_7\text{-}\alpha\text{-Ph})_3$. VII.[280] m. 135-137.[280]
$N_3P_3Cl_2(NC_4H_7\text{-}\alpha\text{-Ph})_4$. VII.[280] m. 171-173.[280]
$N_3P_3Cl_5$(NH-n-undecyl). VII.[318] $b_{0.3}$ 210.[318]
$N_3P_3Cl_5$(NH-n-tridecyl). VII.[318] $b_{0.05}$ 170-190.[318]
$N_3P_3Cl_5$(NH-n-tetradecyl). VII.[318] m. 25,[318] $b_{0.08}$ 180-190.[318]
$N_3P_3Br_4(NH_2)_2$. VII.[211]

$N_3P_3Br_5 \cdot NMe_2$. VII.[459,464] m. 112.5-113.5,[459,464] IR,[459,461] Raman,[461] ^{31}P +39.3 (PBr_2), -4.5 ($PBr \cdot NMe_2$),[155] ^1H -2.68.[155,256]

$N_3P_3Br_4(NMe_2)_2$ (2,2,4,4:6,6). VII.[461] IR.[461]

$N_3P_3Br_4(NMe_2)_2$ (2,2,4-trans-6:4,6). VII.[459,464] m. 135-136.[459,464] IR,[459,461] Raman,[461] ^{31}P +36.6 (PBr_2), -10.0 ($PBr \cdot NMe_2$),[155] ^1H -2.67.[155,256]

$N_3P_3Br_4(NMe_2)_2$ (2,2,4-cis-6:4,6). VII.[459,464] m. 138-139.5,[459,464] Raman,[461] ^{31}P +36.8 (PBr_2), -9.2 ($PBr \cdot NMe_2$),[155] ^1H -2.63.[155,256]

$N_3P_3Br_3(NMe_2)_3$ (2,2,4:4,6,6). VII.[464] m. 87-88,[464] IR,[461,464] Raman,[461] ^1H -2.32, -2.48, -2.53 (in C_6H_6).[155]

$N_3P_3Br_3(NMe_2)_3$ (2-trans-4,6:2,4,6). VII.[459,464] [$N_3P_3(NMe_2)_6$ + HBr],[343] m. 120-121[343,459,464] (two crystal modifications[464]), IR,[459,461] Raman,[461] ^{31}P -15.4,[155] ^1H -2.62 (2), -2.65 (1).[155,256]

$N_3P_3Br_3(NMe_2)_3$ (2-cis-4-cis-6:2,4,6). [$N_3P_3(NMe_2)_6$ + HBr][343]. m. 170-172,[343] ^1H -2.57.[256]

D.1.7. Amino-Derivatives with Mixed Substituents

$N_3P_3Cl_2(NH_2)_2(NHMe)_2$ (2,2:4,6:4,6).* VII.[46] m. 140,[46] ^{31}P.[46]

$N_3P_3(NH_2)_2(NHMe)_4$ (2,2:4,4,6,6).* VII.[44,46] m. 161.5,[44,46] IR.[46]

$N_3P_3(NH_2)_2(NHMe)_4$ (2,4:2,4,6,6).* VII.[44,46] m. 204,[44,46] IR.[46]

$N_3P_3(NH_2)_2(NMe_2)_4$ (2,2:4,4,6,6).* VII.[46,182,255,447] m. 155,[182,255,447] 144,[46] ^{31}P,[46] ^1H -2.61 (NMe_2)[182,255] IR,[447] UV,[447] gives the following products with the reactants (stated).

$N_3P_3(NMe_2)_4(NHCOMe)_2$ (2,2,4,4:6,6). [$(MeCO)_2O$],[447] m. 191-192.[447]

$N_3P_3(NMe_2)_4(NHCOPh)_2 \cdot PhCOOH$ (2,2,4,4:6,6). [$(PhCO)_2O$],[447] m. 161.5 (d).[447]

$N_3P_3(NMe_2)_4(NH_2)(NHCONHPh)$ (2,2,4,4:6:6). (PhNCO),[447,483] m. 185,[483] 180 (d).[447]

$N_3P_3(NMe_2)_4(NH_2)(NHCONH$-n-Bu) (2,2,4,4:6:6). (n-BuNCO),[483] m. 132.[483]

$N_3P_3(NMe_2)_4(NH_2)(NHCSNHPh)$ (2,2,4,4:6:6). (PhNCS),[483] m. 145-146,[483] 155-156 (isomer?),[483] 137.5,[447] IR.[447]

$N_3P_3(NMe_2)_4(NH_2)$NHCONH 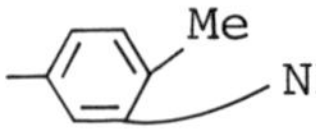

NHCONH $N_3P_3(NMe_2)_4NH_2$:

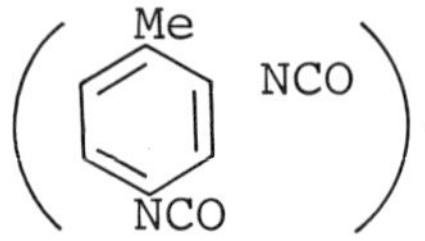

,[483] m. 216-217.[483]

$N_3P_3(NH_2)_2(NMe_2)_4$ (2-cis-4:2,4,6,6). VII.[255] m. 117,[255] ^1H -2.60, -2.63, -2.70 (all NMe_2).[255]

*Structures reassigned by the authors of this chapter on the basis of later results.[166,294,295]

$N_3P_3(NH_2)_2(NHMe)_2(NMe_2)_2$ (2,4:2,4:6,6).* VII.[46] m. 76.[46]
$N_3P_3(NH_2)_2(NHPh)_4$ (2,2:4,4,6,6). VII.[66] m. 256-257.[66]
$N_3P_3(NHMe)_2(NMe_2)_4$ (2,4:2,4,6,6). VII.[46] m. 125.[46]
$N_3P_3(NHMe)_4(NMe_2)_2$ (2,2,4,6:4,6). VII.[46] m. 124.[46]
$N_3P_3(NHMe)_3(NMe_2)_3$ (2,4,6:2,4,6). VII.[46] m. 119.[46]
$N_3P_3(NHMe)_2(NH\text{-}t\text{-}Bu)_4$ (2,2:4,4,6,6). VII.[46] m. 199.[46]
$N_3P_3Cl_2(NC_2H_4)(NMe_2)_3$ (2,4:6:2,4,6). VII.[265,358] m. 90-91,[265,358] ^1H -2.65 (NMe_2), -2.13 (NC_2H_4).[357]
$N_3P_3Cl_2(NC_2H_4)_2(NMe_2)_2$ (2,4:6,6:2,4). VII.[265,358] m. 121,[265] 128,[463] ^1H -2.63 (NMe_2), -2.13 (NC_2H_4).[357]
$N_3P_3(NC_2H_4)(NMe_2)_5$. VII.[357] m. 80-80.5,[357] ^1H -2.60, -2.62, -2.70 (NMe_2), -1.90 (NC_2H_4).[357]
$N_3P_3(NC_2H_4)_2(NMe_2)_4$ (2,2:4,4,6,6). VII.[357] m. 93,[357] ^1H -2.60 (NMe_2), -2.03 (NC_2H_4).[357]
$N_3P_3(NC_2H_4)_3(NMe_2)_3$ (2,4,6:2,4,6). VII.[357] m. 83-84,[357] ^1H -2.68 (NMe_2), -1.90 (NC_2H_4).[357]
$N_3P_3(NC_2H_4)_4(NMe_2)_2$ (2,2,4,4:6,6). VII.[265] m. 129-130,[265] ^1H -2.75 (NMe_2), -2.03 (NC_2H_4).[357]
$N_3P_3(NC_2H_4)_4(NMe_2)_2$ (2,2,4,6:4,6). VII.[265] m. 99-101,[265] ^1H -2.85 (NMe_2), -2.03, -1.90 (NC_2H_4).[357]
$N_3P_3(NC_2H_4)_5(NH_2)$. VII.[384] m. 132-133,[384] IR.[384]
$N_3P_3(NC_2H_4)_5(NHNH_2)$. VII.[357] m. 157-158,[357] ^1H -2.03, -2.07 (NC_2H_4),[357] ^{31}P -37.6 (2) [$P(NC_2H_4)_2$], -30.6 (1) [$P(NC_2H_4)(NHNH_2)$].[357]
$N_3P_3(NC_2H_4)_5(NHMe)$. VII.[357] m. 128-129.5,[257] ^1H -2.53 (NMe), -2.07, -2.03, -1.97 (NC_2H_4),[357] ^{31}P -37.3 (2) [$P(NC_2H_4)_2$], -29.7 (1) [$P(NC_2H_4)(NHMe)$].[357]
$N_3P_3(NC_2H_4)_5(NMe_2)$. VII.[357] m. 116.5-117.5,[357] ^1H -2.85 (NMe_2), -1.90, -2.03 (NC_2H_4).[357]
$N_3P_3Cl_2(NC_2H_4)(NC_4H_8)_3$. VII.[278] m. 65-65.5.[278]
$N_3P_3Cl_2(NC_2H_4)_2(NC_4H_8)_2$ (2,4:6,6:2,4). VII.[278] m. 117-119.[278]
$N_3P_3Cl_2(NC_2H_4)_2(NC_4H_8)_2$ (2,4:6,6:2,4). VII.[278] m. 133-135.[278]
$N_3P_3(NC_2H_4)_2(NC_4H_8)_4$ (2,2:4,4,6,6). VII.[278] m. 97-98.[278]
$N_3P_3(NC_2H_4)_2(NC_4H_8)_4$ (2-trans-4:2,4,6,6). VII.[278] m. 118-119.[278]
$N_3P_3(NC_2H_4)_3(NC_4H_8)_3$. VII.[278] m. 85-86.[278]
$N_3P_3(NC_2H_4)_3(NC_4H_8)_3$ (2-trans-4,6:2,4,6). VII.[278] m. 103-105.[278]
$N_3P_3(NC_2H_4)_4(NC_4H_8)_2$ (2,2,4,4:6,6). VII.[278] m. 89-90.[278]
$N_3P_3(NC_2H_4)_4(NC_4H_8)_2$ (2-trans-4,6,6:2,4). VII.[278] m. 69-70.[278]
$N_3P_3(NC_2H_4)_4(NC_4H_8)_2$ (2-cis-4,6,6:2,4). VII.[278] m. 96-98.[278]
$N_3P_3(NC_2H_4)_5(NC_4H_8)$. VII.[278] m. 104.5-105.5.[278]
$N_3P_3Cl_2(NC_2H_4)_2(NC_4H_6\text{-}\alpha,\alpha\text{-}Me_2)_2$. VII.[278] m. 120.5-122.[278]
$N_3P_3(NC_2H_4)_5(NC_4H_6\text{-}\alpha,\alpha\text{-}Me_2)$. VII.[278] m. 53-54.[278]

*Structures reassigned by the authors of this chapter on the basis of later results.[166,294,295]

$N_3P_3(NC_2H_4)_5(NC_4H_7$-α-Ph). VII.[278] m. 88-89.[278]
$N_3P_3Cl(NC_2H_4)_2(NC_4H_8O)_3$ (2:4,4:2,6,6). VII.[340] m. 144-146.[340]
$N_3P_3(NC_2H_4)_3(NC_4H_8O)_3$ (2,2,4:4,6,6). VII.[340] m. 161-163.5.[340]
$N_3P_3(NC_2H_4)_2(NC_4H_8O)_4$ (2,4:2,4,6,6). VII.[340] m. 163-165.[340]
$N_3P_3(NC_2H_4)_2(NC_4H_8O)_4$ (2,4:2,4,6,6). VII.[340] m. 161.5-163.5.[340]
$N_3P_3(NC_2H_4)_2(NHCH_2COOEt)_4$ (2,2:4,4,6,6). VII.[277,281] m. 77-78.[277,281]
$N_3P_3(NC_2H_4)_3(NHCH_2COOEt)_3$ (2,2,4:4,6,6). VII.[281] m. 93-94.5.[281]
$N_3P_3(NC_2H_4)_4(NHCH_2COOEt)_2$ (2,2,4,4:6,6). VII.[277,281] m. 99-100.[277,281]
$N_3P_3(NC_2H_4)_5(NHCH_2COOEt)$. VII.[281] m. 93.5-95.[281]
$N_3P_3Cl_2(NC_2H_4)_2(NHNHPh)_2$. VII.[374] m. 173-174.[374]
$N_3P_3Cl_4$(NHEt)(NH-t-Bu) (2,2,4,4:6:6). VII.[257] m. 104-105.[257]
$N_3P_3Cl_4$(NHEt)(NH-t-Bu) (2,2,4,6:4:6). VII. m. 68-70.[257]
$N_3P_3Cl_2(NMe_2)_2(NH\text{-i-Pr})_2$ (2,2:4,6:4,6). VII.[124] m. 120,[124] ^1H -2.62 (NMe_2),[124] pK$'_a$ 2.5.[165]
$N_3P_3(NMe_2)_2(NH\text{-i-Pr})_4$ (2,4:2,4,6,6). VII.[124] m. 55,[124] ^1H -2.66 (NMe_2),[124] forms hydrochloride, $N_3P_3(NMe_2)_2(NH\text{-i-Pr})_4\cdot HCl$: m. 112.[124]
$N_3P_3(NMe_2)_2(NH\text{-i-Pr})_4$ (2,4:2,4,6,6). VII.[124] m. 64,[124] ^1H -2.65 (NMe_2),[124] forms hydrochloride, $N_3P_3(NMe_2)_2(NH\text{-i-Pr})_4\cdot HCl$: m. 123-124,[124] ^1H -2.74 (NMe_2).[124]
$N_3P_3(NMe_2)_2(NH\text{-i-Pr})_4$ (2,2:4,4,6,6). VII.[124] m. 42,[124] ^1H -2.59 (NMe_2).[124] forms hydrochloride $N_3P_3(NMe_2)_2(NH\text{-i-Pr})_4\cdot HCl$: m. 168-169,[124] ^1H -2.67 (NMe_2).[124]
$[N_3(Me)P_3(NMe_2)_2(NH\text{-i-Pr})_4]^+BF_4^-$ (5:2,2:4,4,6,6). (Above deriv. + $Me_3O^+BF_4^-$),[383] m. 178-180.5,[383] ^1H -2.85 (NMe),[383] conductivity.[383]
$[N_3(Me)P_3(NMe_2)(NMe_3)(NH\text{-i-Pr})_4]^{2+}2BF_4^-$ (5:2:2:4,4,6,6). (As above),[383] m. 211-213,[383] ^1H -2.83 (NMe), -2.98 (NMe_2), -3.03 (NMe_3),[383] conductivity.[383]
$N_3P_3(NMe_2)_4(NH\text{-i-Pr})_2$ (2-trans-4,6,6:2,4). VII.[124] m. 79-80,[124] ^1H -2.52, -2.55 (NMe_2),[124] forms hydrochloride, $N_3P_3(NMe_2)_4(NH\text{-i-Pr})_2\cdot HCl$: m. 162,[124] ^1H -2.70, -2.75 (NMe_2).[124]
$N_3P_3(NMe_2)_4(NH\text{-i-Pr})_2$ (2-cis-4,6,6:2,4). VII.[124] m. 45,[124] ^1H -2.53, -2.55, -2.57 (all NMe_2),[124] forms hydrochloride $N_3P_3(NMe_2)_4(NH\text{-i-Pr})_2\cdot HCl$: m. 182-183,[124] ^1H -2.71, -2.71, -2.76, -2.80 (all NMe_2).[124]
$N_3P_3(NMe_2)_5$(NH-i-Pr). VII.[124] m. 74-75,[124] ^1H -2.66, -2.66, -2.67 (NMe_2).[124]
$N_3P_3(NMe_2)_2(NH\text{-t-Bu})_4$ (2,2:4,4,6,6). VII.[211] m. 67,[211] forms hydrochloride $N_3P_3(NMe_2)_2(NH\text{-t-Bu})_4\cdot HCl$: m. 259.[211]

$N_3P_3(NMe_2)_2(NH\text{-}t\text{-}Bu)_4$ (2,4:2,4,6,6). VII.[211] m. 106,[211] forms hydrochloride $N_3P_3(NMe_2)_2(NH\text{-}t\text{-}Bu)_4 \cdot HCl$: m. 243.[211] A second isomer of this hydrochloride m. 240[211] has also been obtained.

$N_3P_3(NMe_2)_4(NH\text{-}t\text{-}Bu)_2$ (2,2,4,4:6,6). VII.[211] m. 72.[211]

$N_3P_3(NMe_2)_4(NH\text{-}t\text{-}Bu)_2$ (2,2,4,6:4,6). VII.[211] $b_{0.001}$ 110.[211]

$N_3P_3(NMe_2)_2(NC_5H_{10})_4$ (2-trans-4:2,4,6,6). VII.[255] m. 141,[255] ^1H -2.57 (NMe_2).[255]

$N_3P_3(NMe_2)_2(NC_5H_{10})_4$ (2-cis-4:2,4,6,6). VII.[255] m. 136,[255] ^1H -2.56 (NMe_2).[255]

$N_3P_3(NMe_2)_3(NC_5H_{10})_3$ (2-trans-4,6:2,4,6). VII.[255] m. 113,[255] ^1H -2.56 (NMe_2).[255]

$N_3P_3(NMe_2)_3(NC_5H_{10})_3$ (2-cis-4-cis-6:2,4,6). VII.[255] m. 156,[255] ^1H -2.55 (NMe_2).[255]

$N_3P_3(NMe_2)_3(NC_5H_{10})_3$ (2,2,4:4,6,6). VII.[255] m. 87-88,[255] ^1H -2.54 (NMe_2).[255]

$N_3P_3(NMe_2)_4(NC_5H_{10})_2$ (2-trans-4,6,6:2,4). VII.[255] m. 107,[255] ^1H -2.55 (NMe_2).[255]

$N_3P_3(NMe_2)_4(NC_5H_{10})_2$ (2-cis-4,6,6:2,4). VII.[255] m. 96,[255] ^1H -2.55 (NMe_2).[255]

$N_3P_3(NMe_2)_5(NC_5H_{10})$. VII.[255] m. 56-57,[255] ^1H -2.55 (NMe_2).[255]

$N_3P_3(NHPh)(NMe_2)_5$. VII.[136] m. 98,[136] ^1H -2.48, -2.56, -2.60, (NMe_2).[136]

$N_3P_3Cl_2(NHPh)(NMe_2)_3$ (2,2:4:4,6,6). VII.[136] m. 149,[136] ^1H -2.53, -2.67, -2.70 (NMe_2).[136]

$N_3P_3(NHPh)_2(NMe_2)_4$ (2,2:4,4,6,6). VII.[136] m. 165,[136] ^1H -2.61 (NMe_2).[136]

$N_3P_3(NHPh)_2(NMe_2)_4$ (2,4:2,4,6,6). VII.[136] m. 86-88,[136] ^1H -2.58, -2.59,[136] forms hydrochloride $N_3P_3(NHPh)_2(NMe_2)_4 \cdot HCl$: m. 110, 210,[136] ^1H -2.67, -2.72 (NMe_2).[136]

$N_3P_3(NHPh)_3(NMe_2)_3$ (2,2,4:4,6,6). VII.[136] m. 142,[136] ^1H -2.50, -2.62, -2.65 (NMe_2).[136]

$N_3P_3(NHPh)_3(NMe_2)_3$ (2,4,6:2,4,6). VII.[136] m. 63,[136] ^1H -2.51 (1), -2.66 (2) (NMe_2),[136] forms hydrochloride $N_3P_3(NHPh)_3(NMe_2)_3 \cdot HCl$: m. 195,[136] ^1H -2.35, -2.82 (NMe_2).[136]

$N_3P_3(NHPh)_4(NMe_2)_2$ (2,2,4,4:6,6). VII.[136] m. 222,[136] ^1H -2.54 (NMe_2).[136]

$N_3P_3(NHPh)_5(NMe_2)$. VII.[136] m. 195,[136] ^1H -2.55 (NMe_2).[136]

D.1.8. Phosphazenyl-Derivatives

$N_3P_3F_5(N{=}PCl_3)$. IX.[393] b_1 63,[393] IR,[393] mass spect.[393]

$N_3P_3Cl_5(N{=}PCl_3)$. IX.[166] m. 39.5-41,[166] IR,[166] ^{31}P -20.5 (PCl_2), +2.2 $(PCl \cdot N{=}PCl_3)$, +3.3 $(N{=}PCl_3)$.[166]

$N_3P_3Cl_4(N{=}PCl_3)_2$ (2,2,4,4:6,6). IX.[166,294] m. 57.5-59,[166,994] IR,[166] ^{31}P -17.5 (PCl_2), +20.4. [$P(N{=}PCl_3)_2$], +13.5, ($N{=}PCl_3$).[166,294] With formic acid gives $N_3P_3Cl_4(NHPOCl_2)_2$ (2,2,4,4:6,6): m. > 360,[294] ^{31}P -15.6 (PCl_2), +1.2 [$P(NHPOCl_2)_2$], -10.6 ($NHPOCl_2$).[294] With $(Me_3Si)_2NMe$ gives

$N_3P_3Cl_4$() (2,2,4,4:6,6): m. 228 (d),[297] ^{31}P -20.1 (ring PCl_2), -6.1 ($=PCl_2{\cdot}NMe$), +15.8 [$P(N{=})_2$],[297] ^{1}H -3.42.[297]

$N_3P_3Cl_5(N{=}PPh_3)$. ($N_3P_3Cl_5{\cdot}N{=}PCl_3$ + PhMgBr),[167] [$N_3P_3Cl_4(NH_2)_2$ + Ph_3PCl_2],[250] m. 214-125,[167,250] IR.[167] ^{31}P -20.1 (PCl_2), -15.3 ($N{=}PPh_3$), -10.5 ($PCl{\cdot}N{=}PPh_3$).[167]

$N_3P_3Cl_4(NH_2)(N{=}PPh_3)$ (2,2,4,4:6:6). IX.[167,249,250] m. 173-174,[167,249,250] IR,[167] ^{31}P -17.0 (PCl_2), +1.8 to +4.1, [$P(NH_2)(N{=})$], -12.1 ($N{=}PPh_3$).[148,167]

$N_3P_3Cl_4(N{=}PCl_3)(N{=}PPh_3)$ (2,2,4,4:6:6). (Above deriv. + PCl_5),[294] ^{31}P -18.5 (PCl_2), +20.3 [$P(N{=})_2$], +11.8 ($N{=}PCl_3$), ~ -19 ($N{=}PPh_3$).[294]

$N_3P_3Cl_4Ph(N{=}PPh_3)$ (2,2,4,4:6:6). [$N_3P_3Cl_5(N{=}PPh_3)$ + $AlCl_3$ + PhH],[250] [$N_3P_3Cl_5(N{=}PCl_3)$ + PhMgBr],[167] ($N_4P_4Cl_8$ + PhMgBr),[57,60] ($N_4P_4Cl_8$ + PhH + $AlCl_3$),[137] m. 181,[57,60,137,167,250] IR,[167] ^{31}P -15.8 (PCl_2), ~ -14.6 [$PPh(N{=})$], -2.4 ($N{=}PPh_3$),[148,167] mass spectr.[60]

$N_3P_3Cl_4(N{=}PPh_2NH_2)_2$ (2,2,4,4:6,6). [$N_3P_3Cl_4(NH_2)_2$ + Ph_2PCl_3, followed by NH_3],[249,250] m. 208-209,[249,250] with Me_2NH gives $N_3P_3(NMe_2)_4(N{=}PPh_2NH_2)_2{\cdot}HCl$: m. 223-224.[250]

$N_3P_3Cl_4(N{=}PPh_3)_2$ (2,2,4,4:6,6). IX.[250] [$N_3P_3Cl_4(N{=}PCl_3)_2$ + PhMgBr],[167] m. 199-200,[167,250] IR,[167] ^{31}P -13.4 (PCl_2), +11.6 [$P(N{=})_2$], -5.5 ($N{=}PPh_3$),[167] mass spectr.[167]

$N_3P_3Cl_4[N{=}P(C_6H_4\text{-p-}Me)_3]_2$ (2,2,4,4:6,6). [$N_3P_3Cl_4(N{=}PCl_3)_2$ + Me-p-C_6H_4MgBr],[167] m. 214-214.5,[167] IR.[167]

$N_3P_3Cl_5{\cdot}N{=}PPh(N{=}PPh_3)_2$. ($N_3P_3Cl_6$ + Ph_2Mg),[61] m. 152,[61] ^{31}P -19.8 (PCl_2), -7.6 [$PCl(N{=})$], +5.1 [$=PPh(N{=})_2$],[61] +9.6 ($N{=}PPh_3$),[61] nuclear-electron double resonance,[61] forms $N_3P_3(OCH_2CF_3)_5{\cdot}N{=}PPh(N{=}PPh_3)_2$ with $NaOCH_2CF_3$,[61] ^{19}F.[61]

$N_3P_3(NMe_2)_4(NH_2)(N{=}PPh_3)$ (2,2,4,4:6:6). [$N_3P_3Cl_4(NH_2)(N{=}PPh_3)$ + Me_2NH],[249,250] m. 180-181.[249,250]

$N_3P_3(NMe_2)_4Ph(N{=}PPh_3)$ (2,2,4,4:6:6). [$N_3P_3Cl_4Ph(N{=}PPh_3)$ + Me_2NH],[60] m. 133,[60] ^{1}H -2.49 (NMe_2),[60] forms hydrochloride $N_3P_3(NMe_2)_4Ph(N{=}Ph_3){\cdot}HCl$. m. 204.[60]

$N_3P_3(NMe_2)_5(N{=}PPh_3)$. [$N_3P_3Cl_5(N{=}PPh_3)$ + Me_2NH],[250] m. 149-150.[249,250]

$N_3P_3(NMe_2)_4(N{=}PPh_3)_2{\cdot}HBr$ (2,2,4,4:6,6). [$N_3P_3(NH_2)_2(NMe_2)_4$ + Ph_3PBr_2],[250] m. 219-220.[250]

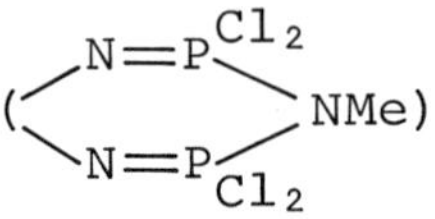

D.1.9. Alkyl and Aryl(Amino)-Derivatives

$N_3P_3PhCl_3(NHMe)_2$ (2:4,6,6:2,4). [$N_3P_3Cl_4(NHMe)_2$ + $AlCl_3$ + PhH],[184] m. 138,[184] ^1H -2.70, -2.63 (both NMe).[184]

$N_3P_3Ph_2Cl_2(NH_2)_2$ (2,2:4,4:6,6). VII.[41,44,45,66,138] m. 165,[41,44,45,66,138] pK'a 0.0.[163]

$N_3P_3Ph_2(NH_2)_4$ (2,2:4,4,6,6). VII.[45] m. 106, 275 (two crystalline forms),[45] IR,[45] forms adduct $N_3P_3Ph_2(NH_2)_4 \cdot HCl$: m. 120 (d).[138]

$N_3P_3Ph_2(NH_2)_2(NHMe)_2$ (2,2:4,4:6,6). VII.[45] m. 140.[45]

$N_3P_3Ph_2(NH_2)_2(NMe_2)_2$ (2,2:4,4:6,6). VII.[44,45,138] m. 137,[44,45,138] ^1H -2.58 (NMe_2).[138]

$N_3P_3Ph_2(NH_2)_2(NMe_2)_2$ (2,2:4,6:4,6). VII.[138] m. 171,[138] ^1H -2.59 (NMe_2).[138]

$N_3P_3Ph_2(NH_2)_2(HNCH_2CH_2CH_2NH)$. VII.[41] m. 185,[41] IR.[41]

$N_3P_3Ph_2Cl_3 \cdot NHMe$ (2,2:4,4,6:6). VII.[230] m. 151.[230]

$N_3P_3Ph_2Cl_2(NHMe)_2$ (2,4:6,6:2,4). VII.[184] m. 145,[184] ^1H -2.67 (NMe).[184]

$N_3P_3Ph_2Cl_2(NHMe)_2$ (2,2:4,6:4,6). VII.[230] m. 171,[230] pK'a -2.8.[163]

$N_3P_3Ph_2Cl_2(NHMe)_2$ (2,2:4,4:6,6). VII.[230] m. 103,[230] pK'a -0.4.[163]

$N_3P_3Ph_2(NHMe)_4$ (2,2:4,4,6,6). VII.[45,184,230] m. 174,[45,184,230] ^1H -2.52 (NMe),[184] pK'a 7.5,[163] forms adduct $N_3P_3Ph_2(NHMe)_4 \cdot HCl$: m. 195.[230]

$N_3P_3Ph_2(NHMe)_4$ (2-cis-4:2,4,6,6). [$N_3P_3Ph_2Cl_2(NHMe)_2$ m. 145 + NH_2Me],[184] m. 125,[184] ^1H -2.62 (NMe).[184]

$N_3P_3Ph_2Cl_3(NHEt)$ (2,2:4,4,6:6). VII.[230] m. 96-97.[230]

$N_3P_3Ph_2Cl_2(NHEt)_2$ (2,2:4,4:6,6). VII.[230] m. 93,[230] pK' -0.5.[163]

$N_3P_3Ph_2(NHEt)_4$ (2,2:4,4,6,6). VII.[230] m. 121-122,[230] pK' 7.4,[163] forms adduct $N_3P_3Ph_2(NHEt)_4 \cdot HCl$: m. 217.[230]

$N_3P_3Ph_2Cl_3(NMe_2)$ (2,2:4,4,6:6). VII.[138,230] m. 110,[138] ^1H -2.59 (NMe_2),[138] pK'a -5.8.[165]

$N_3P_3Ph_2(NMe_2)(NHPh)_3$ (2,2:4:4,6,6). VII.[138] ^1H -2.61 (NMe_2).[138]

$N_3P_3Ph_2Cl_2(NMe_2)(NHPh)$ (2,2:4,4:6:6). VII.[138] m. 143,[138] pK'a -2.0,[165] ^1H -2.63 (NMe_2).[138]

$N_3P_3Ph_2Cl_2(NMe_2)_2$ (2,2:4-trans-6:4,6). VII.[55,138,182,230] m. 144,[55,138,182,230] pK'a -4.0,[163] ^1H -2.70 (NMe_2).[55,138,182]

$N_3P_3Ph_2Cl_2(NMe_2)_2$ (2-cis-4:6,6:2,4). [$N_3P_3Cl_4(NMe_2)_2$ + $AlCl_3$ + PhH],[55,182] m. 99,[55,182] ^1H -2.63 (NMe_2).[55,182]

$N_3P_3Ph_2(NMe_2)_2(NHPh)_2$ (2,2:4-trans-6:4,6). VII.[138] ^1H -2.53 $(NMe)_2$.[138]

$N_3P_3Ph_2(NMe_2)_2(NHPh)_2$ (2,2:4-cis-6:4,6). VII.[138] ^1H -2.63 (NMe_2).[138]

$N_3P_3Ph_2(NMe_2)_2(NHPh)_2$ (2,2:4,4:6,6). VII.[138] m. 153,[138] ^1H -2.48 (NMe_2).[138]

$N_3P_3Ph_2(NMe_2)_3(NHPh)$ (2,2:4,4,6:6). VII.[138] m. 122-123,[138] ^1H -2.44, -2.57, -2.64 (all NMe_2).[138]

$N_3P_3Ph_2(NMe_2)_4$ (2,2:4,4,6,6). VII.[45,55,138,182,230] m. 122,[45,55,138,182,230] pK_a' + 6.2,[163] ^{1}H -2.52 (NMe_2).[55,138,171,182]

$N_3P_3Ph_2(NMe_2)_4$ (2-cis-4:2,4,6,6). [$N_3P_3Ph_2Cl_2(NMe_2)_2$, m. 99 + Me_2NH],[187] m. 83-86,[55,182] av. ^{1}H -2.46 (NMe_2).[55,182]

$N_3P_3Ph_2Cl_2(NHCH_2CH_2NH)$ (2,2:4,6:4,6). VII.[41] m. 142-5,[41] IR.[41]

$N_3P_3Ph_2Cl_2(NHCH_2CH_2CH_2NH)$ (2,2:4,6:4,6). VII.[41] m. 152,[41] IR.[41]

$N_3P_3Ph_2Cl_2(NH\text{-}t\text{-}Bu)_2$ (2,2:4,4:6,6). VII.[230] m. 108,[230] pK_a' 0.2.[163]

$N_3P_3Ph_2(NH\text{-}t\text{-}Bu)_4$ (2,2:4,4,6,6). VII.[230] m. 132-133,[230] pK_a' 7.2,[163] forms adduct $N_3P_3Ph_2(NH\text{-}t\text{-}Bu)_4 \cdot HCl$: m. 290.[230]

$N_3P_3Ph_2Cl_2(NEt_2)_2$ (2,2:4,6:4,6). VII.[230] m. 66,[230] pK_a' -3.2.[163]

$N_3P_3Ph_2(NEt_2)_4$ (2,2:4,4,6,6). VII.[230] m. 76-77,[230] pK_a' 6.8.[163]

$N_3P_3Ph_2Cl_3(NC_5H_{10})$ (2,2:4,4,6:6). VII.[230] m. 126.[230]

$N_3P_3Ph_2Cl_2(NC_5H_{10})_2$ (2,2:4,6:4,6). VII.[230] m. 146,[230] pK_a' -3.6.[163]

$N_3P_3Ph_2Cl_2(NC_5H_{10})_2$ (2,2:4,4:6,6). VII.[230] m. 131,[230] pK_a' -1.0.[163]

$N_3P_3Ph_2(NC_5H_{10})_4$ (2,2:4,4,6,6). VII.[230] m. 195-197,[230] pK_a' 6.7.[163]

$N_3P_3Ph_2Cl_3(N_2C_5H_{11})$ (4-methylpiperazin-1-yl) (2,2:4,4,6:6). VII.[230] m. 102.5-103.5,[230] forms HCl adduct, m. 211-213,[230] and MeI adduct, m. 235 (d).[230]

$N_3P_3Ph_2Cl_3(NHC_6H_{11})$ (2,2:4,4,6:6). VII.[230] m. 111-112,[230] pK_a' -5.5.[163]

$N_3P_3Ph_2Cl_2(NHC_6H_{11})_2$ (2,2:4,4:6,6). VII.[230] m. 144-145,[230] pK_a' 0.4.[163]

$N_3P_3Ph_2(NHC_6H_{11})_4$ (2,2:4,4,6,6). VII.[230] m. 138,[230] pK_a' 7.6,[163] forms adduct $N_3P_3Ph_2(NHC_6H_{11})_4 \cdot HCl$: m. 243.[230]

$N_3P_3Ph_2Cl_3 \cdot NHPh$ (2,2:4,4,6:6). VII.[138] m. 118.[138]

$N_3P_3Ph_2Cl_2(NHPh)_2$ (2,2:4,4:6,6). VII.[66,138] m. 198,[66,138] pK_a' -2.7.[165]

$N_3P_3Ph_2(NHPh)_4$ (2,2:4,4,6,6). VII.[66,138] m. 204-205,[66,138] pK_a' 1.7.[165]

$N_3P_3Ph_2(NHCH_2Ph)_4$ (2,2:4,4,6,6). VII.[230] m. 92.5,[230] pK_a' 6.6.[165]

$N_3P_3(C_6H_3Me_2)_2Cl_2(NMe_2)_2$ (2-cis-4:6,6:2,4). [$N_3P_3Cl_4(NMe_2)_2$ + $AlCl_3$ + $C_6H_4Me_2$],[182] m. 122,[182] ^{1}H -2.52 (NMe_2).[182]

$N_3P_3Me_3(NMe_2)_3$ (2,4,6:2,4,6). [$N_3P_3Cl_3(NMe_2)_3$ + MeMgBr],[485] b_3 146-147.[485]

$N_3P_3Ph_3(NH_2)_3$ (2-trans-4,6:2,4,6). VII.[344] m. 202-203,[344] IR.[344]

$N_3P_3Ph_3(NH_2)_3$ (2-cis-4-cis-6:2,4,6). VII.[344] m. 274,[344] IR,[344] pyrolysis.[205]

$N_3P_3Ph_3(NHMe)_3$ (2-trans-4,6:2,4,6). VII.[344] m. 112-113,[344] IR.[344]

$N_3P_3Ph_3(NHMe)_3$ (2-cis-4-cis-6:2,4,6). VII.[344,442] m. 167,[344,442] IR.[344]

$N_3P_3Ph_3(NMe_2)_3$ (2-trans-4,6:2,4,6). VII.[214] m. 93.5-94.5,[214] ^1H -2.26(1), -2.47(2) (both NMe_2).[214]

$N_3P_3Ph_3(NMe_2)_3$ (2-cis-4-cis-6:2,4,6). VII.[214] m. 113.5-115,[214] ^1H -2.55 (NMe_2).[214]

$N_3P_3Ph_3(NHEt)_3$ (2-trans-4,6:2,4,6). VII.[344] m. 91-92,[344] IR.[344]

$N_3P_3Ph_3(NHEt)_3$ (2-cis-4-cis-6:2,4,6). VII.[344] m. 126-127,[344] IR.[344]

$N_3P_3Ph_3(NH\text{-}n\text{-}Pr)_3$ (2-trans-4,6:2,4,6). VII.[332,344] m. 92-93,[332,344] IR.[344]

$N_3P_3Ph_3(NH\text{-}n\text{-}Pr)_3$ (2-cis-4-cis-6:2,4,6). VII.[332,344] m. 128-129,[332,344] IR.[344]

$N_3P_3Ph_3(NH\text{-}n\text{-}Bu)_3$ (2-trans-4,6:2,4,6). VII.[332,344] m. 78-79,[332,344] IR.[344]

$N_3P_3Ph_3(NH\text{-}n\text{-}Bu)_3$ (2-cis-4-cis-6:2,4,6). VII.[332,344] m. 97-98,[332,344] IR.[344]

$N_3P_3Ph_3(NC_5H_{10})_3$ (2-cis-4-cis-6:2,4,6). VII.[442] m. 187.[442]

$N_3P_3Ph_4(NH_2)_2$ (2,2,4,4:6,6). VII.[320] ^1H.[320]

$N_3P_3Ph_4(NH_2)_2$ (2,2,4-trans-6:4,6). VII.[201] m. 149-150.[201]

$N_3P_3Ph_4(NH_2)_2$ (2,2,4-cis-6:4,6). VII.[201] m. 191-192.[201]

$N_3P_3Ph_4Cl{\cdot}NHMe$ (2,2,4,4:6:6). VII.[230] m. 165-166,[230] pK_a' -0.75.[163]

$N_3P_3Ph_4(NHMe)_2$ (2,2,4,4:6,6). VII.[230] m. 190-192,[230] pK_a' 4.65.[163]

$N_3P_3Ph_4(NHEt)_2$ (2,2,4,4:6,6). VII.[230] m. 165.[230]

$N_3P_3Ph_4Cl{\cdot}NMe_2$ (2,2,4,4:6:6). VII.[230] m. 164-165.[230]

$N_3P_3Ph_4(NMe_2)_2$ (2,2,4,4:6,6). VII.[230] m. 145,[230] pK_a' 3.8.[165]

$N_3P_3Ph_4(NMe_2)_2$ (2,2,4-trans-6:4,6). VII.[214] m. 123-124,[214] ^1H -2.59 (NMe_2).[214]

$N_3P_3Ph_4(NMe_2)_2$ (2,2,4-cis-6:4,6). VII.[214] m. 145-145.5,[214] ^1H -2.38 (NMe_2).[214]

$N_3P_3Ph_4Cl{\cdot}NEt_2$ (2,2,4,4:6:6). VII.[230] m. 147.5,[230] pK_a' -1.2.[163]

$N_3P_3Ph_4(NEt_2)_2$ (2,2,4,4:6,6). VII.[230] m. 122-123,[230] pK_a' 4.0.[163]

$N_3P_3Ph_4(NH\text{-}t\text{-}Bu)_2$ (2,2,4,4:6,6). VII.[230] m. 162-163,[230] pK_a' 4.7.[163]

$N_3P_3Ph_4Cl(NC_5H_{10})$ (2,2,4,4:6:6). VII.[230] m. 175.[230]

$N_3P_3Ph_4(NC_5H_{10})_2$ (2,2,4,4:6,6). VII.[230] m. 180-181,[230] pK_a' 4.3.[163]

$N_3P_3Ph_4(NHC_6H_{11})_2$ (2,2,4,4:6,6). VII.[230] m. 122-123,[230] pK_a' 5.0.[163]

$N_3P_3Ph_4(NHCH_2Ph)_2$ (2,2,4,4:6,6). VII.[230] m. 133,[230] pK_a' 3.8.[165]

$N_3P_3Ph_5{\cdot}NH_2$. VII.[201] m. 170-175.[201]

$N_3P_3Ph_5 \cdot NMe_2$. VII.[442] m. 142-144.[442]

D.1.10. Alkoxy- and Aryloxy-Derivatives

$N_3P_3Cl_5OH$. (Heat $N_3P_3Cl_5OSiMe_2Ph$ and hydrolyze product),[49] m. 133-134,[49] IR.[49]
$N_3P_3Cl_5 \cdot OMe$. X.[70] m. 30-32,[70] ^1H -3.84,[223] ^{31}P -16.7.[223] (PCl·OMe), -22.5 (PCl_2),[223] kinetics of formation.[456,518]
$N_3P_3Cl_3(OMe)_3$ (2-cis-4-cis-6:2,4,6). X.[70] ^1H -3.88.[171]
$N_3P_3Cl(OMe)_5$. X.[70] m. 17.5,[70] $b_{0.1}$ 117,[70] pK$_a'$ -4.2.[164]
$N_3P_3(OMe)_6$. X.[144,177,178] m. 48,[144,177,178] ^1H -3.94,[27] pK$_a'$ -1.9,[161] first ionization potl. 9.29 eV,[71] with MeI rearranges to $N_3Me_3P_3O_3(OMe)_3$: m. 127-127.5,[174] with HCl gives $N_3H_2P_3O_2(OMe)_4$: m. 182;[172,176] with $PhCH_2Br$ gives $N_3Me(CH_2Ph)_2P_3O_3(OMe)_3$: amorphous solid.[176]
$N_3P_3Cl_5 \cdot OEt$. ^{31}P -13.6 (PCl·OEt), -21.3 (PCl_2),[223] ^1H -1.46 (CH_3), -4.22 (CH_2).[223]
$N_3P_3(OEt)_6$. X.[144,177,178,179,512] $b_{0.1}$ 115-116,[144,177,178,179,512] ^{31}P -17.9,[353] ^1H -1.3 (CH_3), -3.92 (CH_2),[299] pK$_a'$ -0.2,[161] rearranges thermally to $N_3Et_3P_3O_3(OEt)_3$: m. 74-75,[174] ^{31}P -5.3,[229] reacts with benzoyl chloride to give $N_3C_3Ph_3$,[173,175] with alkyl halides to give N_3-i-$Pr_3P_3O_3(OEt)_3$ (i-PrI): m. 150-152,[176] $N_3(CH_2 \cdot C_6H_4$-p-$NO_2)_3P_3O_3(OEt)_3$ (O_2N-p-$C_6H_4 \cdot CH_2Br$): m. 168-169,[176] $N_3EtP_3O(OEt)_5$(EtI): $b_{0.01}$ 100,[176] which gives N_3P_3Et(i-Pr)$_2P_3O_3(OEt)_3$ (i-PrI): $b_{0.01}$ 130.[176]
$N_3P_3Cl_5 \cdot OCH_2CF_3$. ^{31}P -16.5 ($PCl \cdot OCH_2CF_3$), -22.7 (PCl_2),[223] ^1H -4.58.[223]
$N_3P_3(OCH_2CF_3)_6$. X.[27,385] m. 48,[385] b_3 115-166,[385] IR,[385] ^1H -4.25,[27] pK$_a'$ < -6.0,[161] first ionization potl. 10.43 eV,[71] base catalyzed hydrolysis to $N_3P_3(ONa)_2(OCH_2CF_3)_4$ and $N_3P_3(ONa)_2(OCH_2CF_3)_4$ (2,4:2,4,6,6).[16]
$N_3P_3Cl_5(OCH_2CH_2O)N_3P_3Cl_5$. X.[516] m. 102-103.[516]
$N_3P_3Cl_4(OCH_2CH_2O)$ (2,2,4,4:6,6). X.[505] m. 163.[505]
$N_3P_3(OCH_2CH_2O)_3$. X.[317,380] (d) < 200.[317,380]
$N_3P_3(OCH_2CH_2OH)_6$. X.[355] m. 112 (d),[355] d^{20} 1.111,[355] n_D^{20} 1.4821.[355]
N_3P_3(O-n-Pr)$_6$. X.[178] $b_{0.1}$ 146-148,[178] n_D^{25} 1.4494.[178]
$N_3P_3Cl_5$(O-i-Pr). X.[505] $b_{0.01}$ 95,[505] ^{31}P -12.6 (PCl·O-i-Pr), -21.7 (PCl_2),[223] ^1H -1.34 (CH_3), -4.66 (CH),[223] n_D^{20} 1.5165,[505] d_{20} 1.5523.[505]
N_3P_3(O-i-Pr)$_6$. X.[178] m. ~28,[178] $b_{0.005}$ 104-105,[178] pK$_a'$ 1.4.[161]
$N_3P_3Cl_5(OCH_2CH{=}CH_2)$. X.[505] $b_{0.01}$ 93,[505] n_D^{20} 1.5352,[505] d_{20} 1.6359.[505]
$N_3P_3(OCH_2CH_2CH_2O)_3$. X.[380] m. 280.[380]
$N_3P_3[OCH_2CH(CH_3)O]_3$. X.[380] 200 (d).[380]

$N_3P_3(OCH_2CF_2CF_3)_6$. X.[385] m. 16-18,[385] b_6 136.5,[385] IR.[385]

$N_3P_3(OCCl_2CF_2CF_3)_6$. X.[385] m. 147-148,[385] b_5 176.[385]

$N_3P_3Cl_5(ON{=}CMe_2)$. X.[373] m. 91.[373]

$N_3P_3Cl_5(O\text{-}n\text{-}Bu)$. X.[455] $b_{0.0002}$ 58-59,[455] IR,[455] n_D^{20} 1.5130,[455] d_{20} 1.5212,[455] kinetics of formation,[456,518] with $PhMe_2SiCl$ gives $N_3P_3Cl_5(OSiMe_2Ph)$: n_D^{20} 1.5149,[49] reactions with chlorosilanes.[49,50]

$N_3P_3Cl_4(O\text{-}n\text{-}Bu)_2$ (2,2,4,6:4,6). X.[455] $b_{0.0004}$ 109-110,[455] n_D^{20} 1.4936,[455] d_{20} 1.3936,[455] reactions with chlorosilanes.[50]

$N_3P_3Cl_3(O\text{-}n\text{-}Bu)_3$ (2,4,6:2,4,6). X.[455] $b_{0.0003}$ 143-145,[455] n_D^{20} 1.4806,[455] d_{20} 1.2427,[455] reactions with chlorosilanes.[50]

$N_3P_3(O\text{-}n\text{-}Bu)_6$. X.[49,144,177,178] $b_{0.01}$ 162-164,[49,144,177,178] n_D^{25} 1.4330,[178] pK_a' 0.1,[161] with $PhMe_2SiCl$ gives $N_3P_3(O\text{-}n\text{-}Bu)_5(OSiPhMe_2)$: n_D^{20} 1.4719[49] with $Ph_2MeSiCl$ gives $N_3P_3(O\text{-}n\text{-}Bu)_5(OSiPh_2Me)$: n_D^{20} 1.4959,[49] with Ph_3SiCl gives $N_3P_3(O\text{-}n\text{-}Bu)_5(OSiPh_3)$: n_D^{20} 1.5151.[49]

$N_3P_3(OCH_2CF_2CF_2CF_3)_6$. X.[385] b_3 154.[385]

$N_3P_3(OCF_2CF_2CF_2CF_2H)_6$. X.[50]

$N_3P_3(OCCl_2CF_2CF_2CF_3)_6$. X.[385] m. 94,[385] b_4 194-197.[385]

$N_3P_3[O(CH_2)_4OH]_6$. X.[355] (d) 122,[355] n_D^{20} 1.4960,[355] d_{20} 1.218.[355]

$N_3P_3[O(CH_2)_2O(CH_2)_2OH]_6$. X.[355] (d) 119,[355] n_D^{20} 1.4921,[355] d_{20} 1.224.[355]

$N_3P_3[O(CH_2)_nO]_3$ (n = 4, 5, 6). X.[317]

$N_3P_3[OCH_2(CF_2)_2CH_2O]_3$. X.[385] m. 300.[385]

$N_3P_3[OCH(CH_3)CH(CH_3)O]_3$. X.[380] m. 201.[380]

$N_3P_3[OCH_2(CF_2)_3CH_2O]_3$. X.[385] m. 242-244.[385]

$N_3P_3(OC_6H_{11})_6$. E(P-N) 83.0 Kcal/mole.[48a]

$N_3P_3[OCH_2(CF_2)_5CF_2H]_6$. X.[311,385] m. 33-34,[385] b_2 258-260.[385]

$N_3P_3[OCH_2(CF_2)_9CF_2H]_6$. X.[385] m. 103-105,[385] b_2 320-324.[385]

$N_3P_3Cl_5(OPh)$. X.[132,487] m. 48,[132] $b_{0.05}$ 60,[132] $b_{0.00005}$ 83-83.5.[487]

$N_3P_3Cl_4(OPh)_2$ (2,2,4,6:4,6). X.[132,455] $b_{0.02}$ 130 (mixture of isomers),[132] $b_{0.00008}$ 129.5-130.5,[455,487] n_D 1.568,[132] d^{20} 1.4973.[455]

$N_3P_3Cl_3(OPh)_3$ (2,4,6:2,4,6). X.[132,487] $b_{0.05}$ 160 (mixture of isomers),[132] $b_{0.00005}$ 148-149.5,[487] oil.[183]

$N_3P_3Cl_3(OPh)_3$ (2-cis-4-cis-6:2,4,6). X.[132,497] m. 68,[132] 93-95.[497]

$N_3P_3Cl_2(OPh)_4$ (2,4:2,4,6,6). X.[132,487] Oil (mixture of isomers),[132] $b_{0.000085}$ 165-166.[487]

$N_3P_3Cl_2(OPh)_4$ (2-cis-4:2,4,6,6). X.[132,320] m. 75.[132,320]

$N_3P_3Cl(OPh)_5$. X.[132,178,320,497] m. 67-68,[132,178,320,497] with KOH hydrolyzes to $N_3HP_3O(OPh)_5$: m. 163-165.[172]

$N_3P_3(OPh)_6$. X.[178,320] m. 110-111,[178] $b_{0.1}$ 280,[320] ^{31}P -10.0,[9] dipole moment 3.0 D,[4] first ionization potl. 8.83,[71] rate of hydrolysis to $N_3P_3(ONa)(OPh)_5$.[17]

$N_3P_3Cl_5 \cdot OC_6H_4$-p-Br. X.[132] m. 60,[132] $b_{0.01}$ 100.[132]

$N_3P_3Cl_4(OC_6H_4$-p-Br$)_2$ (2,2,4,6:4,6). X.[132] $b_{0.01}$ 210.[132]

$N_3P_3Cl_3(OC_6H_4$-p-Br$)_3$ (2,4,6:2,4,6). X.[132] m. 110-130 (mixture of isomers).[132]

$N_3P_3Cl_2(OC_6H_4$-p-Br$)_4$ (2,4:2,4,6,6). X.[132] m. 87.[132]

$N_3P_3Cl(OC_6H_4$-p-Br$)_5$. X.[132] m. 106.5-108.[132]

$N_3P_3(OC_6H_4$-p-Br$)_6$. X.[132] m. 175.[132]

$N_3P_3Cl_5(OC_6H_4$-p-$NO_2)$. X.[266] m. 95-97.[266]

$N_3P_3Cl_3(OC_6H_4$-p-$NO_2)_3$ (2,4,6:2,4,6). X.[266,267] m. 225-226,[266,267] ^{31}P -17.4.[266,267]

$N_3P_3Cl(OC_6H_4$-p-$NO_2)_5$. X.[266,267] m. 183-184.[266,267]

$N_3P_3(OC_6H_4$-p-$NO_2)_6$. X.[266,267] m. 263-264,[266,267] 212-214,[513] ESR,[7] rate of hydrolysis to $N_3P_3(ONa)(OC_6H_4$-p-$NO_2)_5$.[17]

$N_3P_3(OC_6H_4$-o-$NO_2)_6$. X.[17] rate of hydrolysis to $N_3P_3(ONa)(OC_6H_4$-o-$NO_2)_5$.[17]

$N_3P_3(OC_6H_4$-p-$NH_2)_6$. (Reduction of analogous p-NO_2 deriv. with H_2),[267,359] m. 189-190.[267,359]

$N_3P_3Cl_5(OC_6H_4$-m-O$)N_3P_3Cl_5$. X.[516,517] m. 71-72,[516,517] IR.[516]

$N_3P_3Cl_5(OC_6H_4$-p-O$)N_3P_3Cl_5$. X.[516] 114-116,[516] IR.[516]

$N_3P_3(OC_6H_4$-o-O$)_3$. X.[3,4,379] m. 244-245,[3,4,379] IR,[4] UV,[4] ^{31}P -11.8,[9] ^1H -7.32 (singlet),[4] dipole moment 1.9 D,[4] ESR,[7] forms molecular inclusion compounds with organic molecules,[12,13,75,454] crystal structures of inclusion compounds with benzene,[448,454] and bromobenzene[448] show that the P-N ring is planar, with, in the latter case, P-N 1.575, av. $O\hat{P}O$ 97°. Reactions with o-aminophenol.[10]

$N_3P_3(OCH_2Ph)_6$. X.[178] m. 51.5,[178] same reaction also gives $N_3(CH_2Ph)_3P_3O_3(OH)_3$: m. 185 (d),[172] which with NaOH, gives trisodium salt, m. 230 (d).[172]

$N_3P_3(OC_6H_4$-p-Me$)_6$. Rate of hydrolysis to $N_3P_3(ONa)(OC_6H_4$-p-Me$)_5$.[17]

$N_3P_3(OC_6H_4$-m-$CF_3)_6$. X.[378] $b_{0.2}$ 250,[378] n_D^{25} 1.490,[378] d^{25} 1.50.[378]

$N_3P_3(OC_6H_4$-p-OMe$)_6$. X.[178] m. 103-104.[178]

$N_3P_3(OC_6H_4$-p-NCO$)_6$. X.[359] m. 149-150.[359]

$N_3P_3(OC_6H_4$-p-NH$\cdot$COOMe$)_6$. (Above deriv. + MeOH),[359] m. 148.[359]

$N_3P_3Cl_5 \cdot OSiPhMe_2$. ($N_3P_3Cl_5 \cdot$O-n-Bu + $PhMe_2SiCl$),[49] oil,[49] n_D^{20} 1.5149.[49]

$N_3P_3(OSiPhMe_2)_6$. [N_3P_3(O-n-Bu$)_6$ + $PhMe_2SiCl$],[49] oil,[49] n_D^{20} 1.5421.[49]

N_3P_3 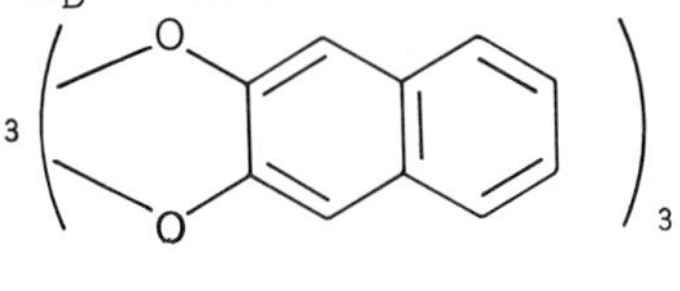 X.[9] m. 335-336,[9] IR,[9] UV,[9] ESR,[7] reaction with o-aminophenol.[10]

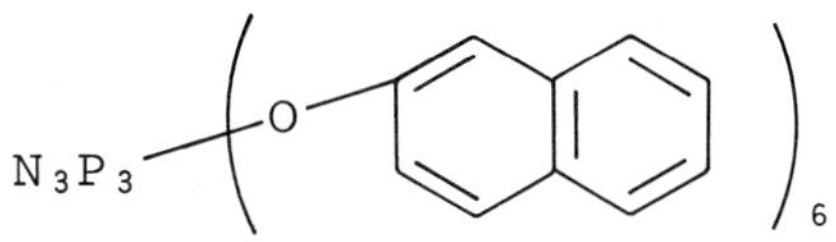

X.[5,278] m. 168-169.[5,278]

$N_3P_3(OC_6H_4$-p-NH·COO-n-Bu$)_6$. [$N_3P_3(OC_6H_4$-p-NCO$)_6$ + n-BuOH],[359] m. 163.5-164.[359]

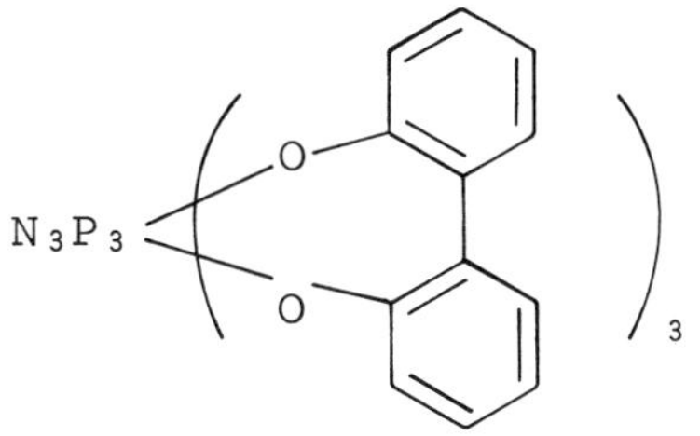

X.[4,5] m. > 300,[4,5] reaction with o-aminophenol,[10] ESR,[7] x-ray crystal structure[14] shows that P-N ring has slightly distorted boat conformation av. P-N 1.572, av. P-O 1.584 av. P-$\hat{N}$-P 121.0, av. N$\hat{P}$N 118.5.

D.1.11. Aryl(Alkoxy or Aryloxy)-Derivatives

$N_3P_3Ph_2(OMe)_4$ (2,2:4,4,6,6). X.[172] m. 99-99.5.[172]

$N_3P_3Ph_2(OEt)_4$ (2,2:4,4,6,6). X.[172] m. 63-64,[172] ^1H -1.2 (CH_3), -3.88 (CH_2),[229] formation accompanied by $N_3HP_3(O)Ph_2(OEt)_3$: m. 116-117,[172] ^1H,[172] IR.[172]

$N_3P_3Ph_2Cl_2(OCH_2CH_2NH)$ (2,2:?). X.[505] m. 135.[505]

$N_3P_3Ph_2(O$-n-Pr$)_4$ (2,2:4,4,6,6). X.[504] $b_{0.1}$ 220,[504] d^{20} 1.1443,[504] $N_3P_3Ph_2Cl_4$ + NaO-n-Pr gives $N_3HP_3(O)Ph_2(O$-n-Pr$)_3$: m. 139-140,[172] ^1H,[172] IR.[172]

$N_3P_3Ph_2(O$-i-Pr$)_4$ (2,2:4,4,6,6). X.[172] m. 75.[172]

$N_3P_3Ph_3(OCH_2CF_3)_3$ (2,4,6:2,4,6). X.[7] ESR.[7]

$N_3P_3Ph_4(OMe)_2$ (2,2,4,4:6,6). X.[172] m. 109-110,[172] pK'_a 0.4,[164] accompanied by $N_3HP_3(O)Ph_4(OMe)$: m. 253,[172] ^1H,[172] IR.[172]

$N_3P_3Ph_4(OEt)_2$ (2,2,4,4:6,6). X.[172] m. 131-133,[172] ^1H -1.1 (CH_3), -3.79 (CH_2),[172] formation accompanied by $N_3HP_3(O)Ph_4(OEt)$: m. 244,[172] IR [also formed from $N_3P_3Ph_4(OEt)_2$ + HCl].[172]

$N_3P_3Ph_5OH$. ($N_3P_3Ph_5Cl$ + pyridine/H_2O),[423] m. 275-276,[423] pK_a 9.92 (in 77% EtOH),[423] forms $(N_3P_3Ph_5)_2O$ in refluxing CH_3CN,[423] IR.[423]

D.1.12. Alkoxy(Amino)- and Aryloxy(Amino)-Derivatives

$N_3P_3(OMe)(NC_2H_4)_5$. X.[357] m. 125-126,[357] ^{31}P -38.2 [$P(NC_2H_4)_2$], -29.6[$P(NC_2H_4)(OMe)$],[357] ^1H -3.63 (OMe), -2.08, -2.03 (NC_2H_4).[357]

$N_3P_3(OMe)_2(NC_2H_4)_4$ (2,2:4,4,6,6). X.[357] m. 119-120,[357] ^{31}P -38.0 [$P(NC_2H_4)_2$], -20.1 [$P(OMe)_2$],[357] ^1H -3.73 (OMe), -2.07 (NC_2H_4).[357]

$N_3P_3(OMe)_2(NMe_2)_4$ (2,4:2,4,6,6). VII.[375] $b_{0.15}$ 106,[375] n_D^{20} 1.4888,[375] d^{20} 1.1620.[375]

$N_3P_3(OMe)_3(NMe_2)_3$ (2,4,6:2,4,6). VII.[504] $b_{0.6}$ 135,[504] n_D^{20} 1.4888.[504]

$N_3P_3(OEt)_2(NMe_2)_4$ (2,4:2,4,6,6). VII.[375] $b_{0.15}$ 110-111,[375] n_D^{20} 1.4811,[375] d_4^{20} 1.1176.[375]

$N_3P_3(OEt)_3(NMe_2)_3$ (2,4,6:2,4,6). VII.[504] b_1 145,[504] n_D^{20} 1.4780,[504] d^{20} 1.1246.[504]

$N_3P_3(OCH_2CF_3)_4(NHPh)_2$ (2,2,4,4:6,6). X.[293] m. 98-100,[293] $b_{0.2-0.7}$ 209-214,[293] ^{31}P -16.38 [$P(OCH_2CF_3)_2$], -6.37 [$P(NHPh)_2$],[293] ^1H -4.14 (CH_2).[293]

$N_3P_3(OCH_2CF_3)_4$ $(NHC_6H_4$-m-Cl$)_2$ (2,2,4,4:6,6). X.[293] m. 102-104,[293] ^{31}P -16.2 [$P(OCH_2CF_3)_2$], -6.94 [$P(NHC_6H_4$-m-Cl$)_2$],[293] ^1H -4.24 (CH_2).[293]

$N_3P_3(OCH_2CF_3)_4$ $(NHC_6H_4$-m-Me$)_2$ (2,2,4,4:6,6). X.[293] m. 102.5-104,[293] $b_{0.1}$ 186-191,[293] ^1H -4.08 (CH_2), -2.19 (CH_3).[293]

N_3P_3(O-n-Pr$)_2(NMe_2)_4$ (2,4:2,4,6,6). VII.[375] $b_{0.05}$ 112,[375] n_D^{20} 1.4781,[375] d^{20} 1.0957.[375]

N_3P_3(O-i-Pr$)_2(NMe_2)_4$ (2,4:2,4,6,6). VII.[375] m. 29-30,[375] $b_{0.05}$ 104.[375]

$N_3P_3(OCH_2CH{=}CH_2)_2(NMe_2)_4$ (2,4:2,4,6,6). VII.[504] $b_{0.08}$ 112,[504] n_D^{20} 1.4931,[504] d^{20} 1.1231.[504]

N_3P_3(O-n-Pr$)_3(NMe_2)_3$ (2,4,6:2,4,6). VII.[504] $b_{0.22}$ 149,[504] n_D^{20} 1.4730,[504] d^{20} 1.0845.[504]

$N_3P_3(OCH_2CH{=}CH_2)_4(NMe_2)_2$ (2,2,4,6:4,6). X.[504] $b_{0.01}$ 133,[504] n_D^{20} 1.4929,[504] d^{30} 1.1505.[504]

$N_3P_3(OCH_2CF_2CF_2H)_4(NHPh)_2$ (2,2,4,4:6,6). X.[293] m. 98-100.[293]

$N_3P_3Cl(ON{=}CMe_2)(NMe_2)_4$ (2:4:2,4,6,6). X.[373] $b_{0.1}$ 135-137.[373]

$N_3P_3(NCS)(ON{=}CMe_2)(NMe_2)_4$ (2:4:2,4,6,6). (Above deriv. + KSCN),[373] m. 127-128.[373]

$N_3P_3(OR)(ON{=}CMe_2)(NMe_2)_4$ (2:4:2,4,6,6). X.[373] R = Me, $b_{0.08}$ 122-123,[373] n_D^{20} 1.4970,[373] d^{20} 1.4118.[373]

R = Et. $b_{0.15}$ 126-128, n_D^{20} 1.4953, d^{20} 1.1307.[373]

R = n-Pr. $b_{0.1}$ 128-130, n_D^{20} 1.4929, d^{20} 1.1056.[373]

R = i-Pr. $b_{0.08}$ 118-119, n_D^{20} 1.4918, d^{20} 1.1138.[373]

R = $-CH_2CH{=}CH_2$. $b_{0.1}$ 123-124, n_D^{20} 1.4979, d^{20} 1.1267.[373]

R = n-Bu. $b_{0.08}$ 124-125, n_D^{20} 1.4908, d^{20} 1.1085.[373]

R = i-Bu. $b_{0.1}$ 132-133, n_D^{20} 1.4911, d^{20} 1.1030.[373]

R = sec-Bu. $b_{0.08}$ 124-125, n_D^{20} 1.4929, d^{20} 1.1131.[373]

R = n-Pent. $b_{0.1}$ 133-136, n_D^{20} 1.4895, d^{20} 1.0988.[373]

R = n-Hex. $b_{0.05}$ 133-135, n_D^{20} 1.4843, d^{20} 1.0756.[373]

R = i-Hex. $b_{0.15}$ 130-133, n_D^{20} 1.4873, d^{20} 1.0869.[373]

R = n-Hept. $b_{0.1}$ 140-143, n_D^{20} 1.4889, d^{20} 1.0829.[373]

N_3P_3(O-n-Bu$)_2(NMe_2)_4$ (2,4:2,4,6,6). VII.[375] $b_{0.05}$ 121,[375] n_D^{20} 1.4778,[375] d^{20} 1.0729.[375]

N_3P_3(O-i-Bu$)_2(NMe_2)_4$ (2,4:2,4,6,6). VII.[375] $b_{0.05}$ 117-119,[375] n_D^{20} 1.4741,[375] d^{20} 1.0685.[375]

$N_3P_3(O\text{-}sec\text{-}Bu)_2(NMe_2)_4$ (2,4:2,4,6,6). VII.[375] $b_{0.05}$ 113-115,[375] n_D^{20} 1.4793,[375] d^{20} 1.0718.[375]

$N_3P_3(O\text{-}n\text{-}Bu)_3(NMe_2)_3$ (2,4,6:2,4,6). VII.[504] $b_{0.05}$ 165,[504] n_D^{20} 1.4729,[504] d^{20} 1.0519.[504]

$N_3P_3(O\text{-}n\text{-}Bu)_4(NMe_2)_2$ (2,2,4,6:4,6). VII.[504] $b_{0.01}$ 164-165,[504] n_D^{20} 1.4700,[504] d^{20} 1.0701.[504]

$N_3P_3(O\text{-}n\text{-}Pent)_2(NMe_2)_4$ (2,4:2,4,6,6). VII.[375] $b_{0.03}$ 133-135,[375] n_D^{20} 1.4769,[375] d^{20} 1.0513.[375]

$N_3P_3(O\text{-}iso\text{-}Pent)_2(NMe_2)_4$ (2,4:2,4,6,6). VII.[375] $b_{0.05}$ 136-137,[375] n_D^{20} 1.4742,[375] d^{20} 1.0481.[375]

$N_3P_3(O\text{-}n\text{-}Hex)_2(NMe_2)_4$ (2,4:2,4,6,6). VII.[375] $b_{0.05}$ 144-145,[375] n_D^{20} 1.4751,[375] d^{20} 1.0354.[375]

$N_3P_3(O\text{-}n\text{-}Hept)_2(NMe_2)_4$ (2,4:2,4,6,6). VII.[375] $b_{0.08}$ 172-176,[375] n_D^{20} 1.4750,[375] d^{20} 1.0282.[375]

$N_3P_3Cl_3(OPh)(NMe_2)_2$ (2-trans-4,6:2:4,4,6). X.[186] m. 55,[186] ^1H.[186]

$N_3P_3(OPh)(NMe_2)_5$. VII.[131] m. 52,[131] ^1H -2.39(2), -2.54(2), -2.73(1) (all NMe_2),[131] pK'_a 5.6.[164]

$N_3P_3Cl(OPh)_2(NMe_2)_3$. VII.[131] m. 94.[131]

$N_3P_3(OPh)_2(NMe_2)_4$. VII,[131] X.[131] m. 69,[131] ^1H -2.53(1), -2.60(1), (all NMe_2),[131] pK'_a 3.3.[164]

$N_3P_3(OPh)_2(NMe_2)_4$ (2-cis-4:2,4,6,6). VII,[131] X.[131] $b_{0.01}$ 200,[131] ^1H -2.11(1), -2.53(1), -2.73(2) (all NMe_2).[131]

$N_3P_3(OPh)_3(NH_2)_3$ (2,2,4:4,6,6). X.[320] m. 135.5-137,[320] ^1H.[320]

$N_3P_3(OPh)_3(NH_2)_3$ (2-trans-4,6:2,4,6). X.[320] m. 210-212,[320] ^1H.[320]

$N_3P_3(OPh)_3(NH_2)_3$ (2-cis-4-cis-6:2,4,6). X.[320] m. 189-190,[320] ^1H.[320]

$N_3P_3Cl(OPh)_3(NMe_2)_2$. X.[131] $b_{0.01}$ 200.[131]

$N_3P_3(OPh)_3(NMe_2)_3$ (2,2,4:4,6,6). X.[131] ^1H -2.24(1), -2.56(1), -2.78(1) (all NMe_2).[131]

$N_3P_3(OPh)_3(NMe_2)_3$ (2-trans-4,6:2,4,6). VII,[131,183] X.[131,183,504] m. 90,[131,183,504] ^1H -2.32(1), -2.56(2) (all NMe_2),[131,183] pK'_a 1.05.[164]

$N_3P_3(OPh)_3(NMe_2)_3$ (2-cis-4-cis-6:2,4,6). VII,[131] X.[131] m. 82,[131] ^1H -2.70 (NMe_2),[131] pK'_a 1.05.[164]

$N_3P_3(OPh)_4(NH_2)_2$ (2,2,4,4:6,6). X.[320] m. 104-106,[320] ^1H.[320]

$N_3P_3(OPh)_4(NH_2)_2$ (2,2,4-trans-6:4,6). VII.[320] m. 110-111.5,[320] ^1H.[320]

$N_3P_3(OPh)_4(NH_2)_2$ (2,2,4-cis-6:4,6). VII.[320] m. 118-120,[320] ^1H.[320]

$N_3P_3(OPh)_4(NMe_2)_2$ (2,2,4-trans-6:4,6). VII,[131] X.[131] m. 65-66,[131] ^1H -2.36 (NMe_2).[131]

$N_3P_3(OPh)_4(NMe_2)_2$ (2,2,4-cis-6:4,6). VII,[131] X.[131] m. -30,[131] ^1H -2.59 (NMe_2).[131]

$N_3P_3(OPh)_5(NH_2)$. VII.[320] m. 66-67.[320]

$N_3P_3(OPh)_5(NMe_2)$. VII.[131] m. -11,[131] ^1H -2.42[131] (NMe_2).

$N_3P_3(OC_6H_4\text{-p-Br})_2(NMe_2)_4$ (2,4:2,4,6,6). VII, X. m. 100-101.[131]

$N_3P_3(OC_6H_4\text{-p-Br})_2(NMe_2)_4$ (2,4:2,4,6,6). VII, X. m. 88-89.[131]

$N_3P_3(OC_6H_4\text{-p-Br})_3(NMe_2)_3$ (2,4,6:2,4,6). VII, X. m. 174-175.[131]

$N_3P_3(OC_6H_4\text{-p-Br})_3(NMe_2)_3$ (2,4,6:2,4,6). VII, X. m. 180-181.[131]

$N_3P_3(OC_6H_4\text{-p-Br})_4(NMe_2)_2$ (2,2,4,6:4,6). VII, X. m. 148-149.[131]

$N_3P_3(OC_6H_4\text{-p-Br})_4(NMe_2)_2$ (2,2,4,6:4,6). VII, X. m. 101-102.[131]

$N_3P_3(OCH_2Ph)_2(NMe_2)_4$ (2,4:2,4,6,6). X.[375] m. 67-68,[375] $b_{0.15}$ 201.[375]

D.1.13. Phenyl(Phenoxy)- and Phenyl(Phenoxy)(Amino)-Derivatives

$N_3P_3Cl_2Ph_2(OPh)_2$ (2-cis-4:6,6:2,4). X.[185,320] m. 140.5-143,[185,320] ^{31}P -16.6 (PCl·OPh), -21.9 (PPh_2),[185] ^1H.[320]

$N_3P_3Cl_2Ph_2(OPh)_2$ (2-trans-4:6,6:2,4). X.[185] m. 104,[185] ^{31}P -16.6 (PCl·OPh), -21.9 (PPh_2).[185]

$N_3P_3Ph_2(OPh)_2(NH_2)_2$ (2,2:4-trans-6:4,6). VII.[320] m. 161-163,[320] ^1H.[320]

$N_3P_3Ph_2(OPh)_2(NH_2)_2$ (2,2:4-cis-6:4,6). VII.[320] m. 156-157.5,[320] ^1H.[320]

$N_3P_3Ph_2(OPh)_2(NH_2)_2$ (2,2:4,4:6,6). X.[320] m. 150-152,[320] ^1H.[320]

$N_3P_3ClPh_2(OPh)_2(NHMe)$ (2:4,4:2,6:6). VII.[185] m. 110.[185]

$N_3P_3Ph_2(OPh)_2(NHMe)_2$ (2,2:4,6:4,6). VII.[185] m. 140,[185] ^1H -2.47 (NMe).[185]

$N_3P_3Ph_2(OPh)_2(NHMe)_2$ (2,2:4,6:4,6). VII.[185] m. 162,[185] ^1H -2.67 (NMe).[185]

$N_3P_3ClPh_2(OPh)_2(NMe_2)$ (2:4,4:2,6:6). VII.[185] m. 76.[185]

$N_3P_3Ph_2(OPh)_2(NMe_2)_2$ (2,2:4,6:4,6). VII.[185] m. 96,[185] ^1H -2.74 (NMe_2).[185]

$N_3P_3Ph_2(OPh)_2(NMe_2)_2$ (2,2:4,6:4,6). VII.[185] m. 108,[185] ^1H -2.52 (NMe_2).[185]

$N_3P_3Ph_4H(OPh)$ (2,2,4,4:6:6). [H_2NPPh_2=N-P(Ph_2)=NH + $(PhO)_3P$], m. 153,[405] IR,[405] ^{31}P -14.5 (PPh_2), -5.8 [P(H)OPh],[405] ^1H -7.36 (PH).[405]

D.1.14. Alkylthio- and Arylthio-Derivatives

$N_3P_3F_5$·SEt. XI. b_{760} 153, IR, ^{31}P -6 (PF_2), -50 (PF·SEt), ^{19}F + 70.5 (PF_2), + 35.5 (PF·SEt), ^1H -1.51 (CH_3), -3.10 (CH_2), n_D^{20} 1.413, d^{20} 1.560.[348]

$N_3P_3F_4(SEt)_2$ (2,2,4,4:6,6). XI. $b_{3.3}$ 89, IR, ^{31}P -5 (PF_2), -59 [$P(SEt)_2$], ^{19}F + 69.3, ^1H -1.56 (CH_3), -3.08 (CH_2), n_D^{20} 1.478, d^{20} 1.449.[348]

$N_3P_3F_3(SEt)_3$ (2,2,4:4,6,6). XI. $b_{1.5}$ 139, IR, ^{31}P -1 (PF_2), -40 (PF·SEt), -55 [$P(SEt)_2$], ^{19}F + 68.2 (PF_2), + 33.5 (PF·SEt), ^1H -1.61 (CH_3), -3.20 (1), -3.14 (2) (CH_2), n_D^{20} 1.522, d^{20} 1.372.[348]

$N_3P_3F_2(SEt)_4$ (2,2:4,4,6,6). XI. b_1 175, IR, ^{31}P +1 (PF_2), -52 [$P(SEt)_2$], ^{19}F + 66.9, ^1H -1.67 (CH_3), -3.17 (CH_2), n_D^{20} 1.555, d^{20} 1.314.[348]

$N_3P_3F(SEt)_5$. XI. b_1 215, IR, ^{31}P -33 (PF·SEt), -48 [$P(SEt)_2$], ^{19}F + 27.5, ^1H -1.71 (CH_3), -3.25(1), -3.20(2) (CH_2), n_D^{20} 1.595, d^{20} 1.283.[348]

$N_3P_3[SC(=NH)NH_2]_6$. [$N_3P_3Cl_6$ + $NH_2C(NH)SH$],[239] m. 156.[239]

$N_3P_3Cl_4(SEt)_2$ (2,2,4,4:6,6). XI.[88-90] $b_{0.2}$ 128,[88-90] ^{31}P -51.7 [$P(SEt)_2$], -17.7 (PCl_2).[68]

$N_3P_3Cl_3(SEt)_3$ (2,2,4:4,6,6). XI.[88,89] $b_{0.1}$ 148-150.[88,89]

$N_3P_3Cl_2(SEt)_4$ (2,2:4,4,6,6). XI.[88,89] $b_{0.07}$ 174.[88,89]

$N_3P_3(SEt)_6$. XI.[88-90] m. 35.5,[88-90] $b_{0.06}$ 196,[88,89] pK_a' -2.8,[161] ^{31}P -45.7.[68]

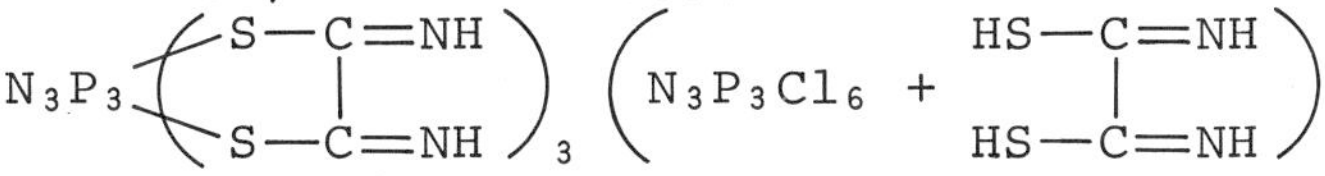
,[239] m. > 300.[239]

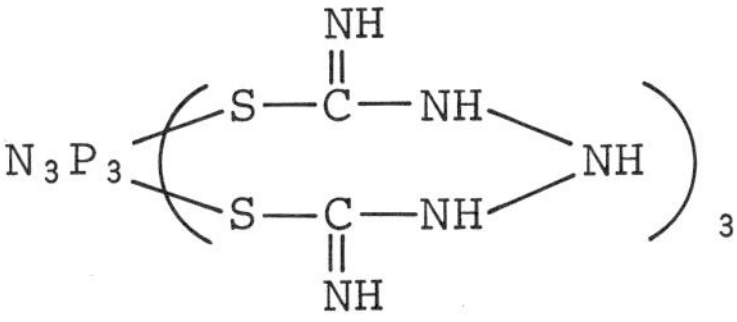
($N_3P_3Cl_6$ + HSC(=NH)NHNH$_2$),[239] m. 275 (d).[239]

$N_3P_3Cl_4(S\text{-}n\text{-}Pr)_2$ (2,2,4,4:6,6). XI.[88,89] $b_{0.1}$ 124.[88,89]

$N_3P_3(S\text{-}n\text{-}Pr)_6$. XI.[88,89] $b_{0.02}$ 209,[88,89] pK_a' -2.6,[161] n_D^{20} 1.5813.[89]

$N_3P_3Cl_4(S\text{-}i\text{-}Pr)_2$ (2,2,4,4:6,6). XI.[88,89] m. 53-54,[88,89] $b_{0.03}$ 116-118.[88,89]

$N_3P_3Cl_2(S\text{-}i\text{-}Pr)_4$ (2,2:4,4,6,6). XI.[88,89] m. 47-49,[88,89] $b_{0.09}$ 160.[88,89]

$N_3P_3(S\text{-}i\text{-}Pr)_6$. XI.[88,89] $b_{0.03}$ 175.[88,89]

$N_3P_3Cl_4(S\text{-}n\text{-}Bu)_2$ (2,2,4,4:6,6). XI.[88,89] $b_{0.08}$ 150.[88,89]

$N_3P_3(S\text{-}n\text{-}Bu)_6$. XI.[88,89] $b_{0.04}$ 238,[88,89] n_D^{20} 1.5645.[89]

$N_3P_3Cl_4(S\text{-}i\text{-}Bu)_2$ (2,2,4,4:6,6). XI.[88,89] $b_{0.07}$ 138.[88,89]

$N_3P_3Cl_4(SPh)_2$ (2,2,4,4:6,6). XI.[88-90] m. 107,[88-90] $b_{0.01}$ 180,[88,89] ^{31}P -46.8 [$P(SPh)_2$], -18.6 (PCl_2).[68]

$N_3P_3(SPh)_6$. XI.[88-90] m. 153-5,[88-90] pK_a' -4.8.[161]

$N_3P_3Cl_2(S\text{-}cyclo\text{-}Hex)_4$ (2,2:4,4,6,6). XI.[88,89] m. 80-81.[88,89]

$N_3P_3(S\text{-}cyclo\text{-}Hex)_6$. XI.[88,89] m. 99-100,[88,89] pK_a' -2.2.[161]

$N_3P_3(SC_6H_4\text{-}p\text{-}Cl)_6$. XI.[445] m. 164.[445]

$N_3P_3(SCH_2Ph)_6$. XI.[88,89] m. 77.5-78.5,[88,89] pK_a' -4.2.[161]

$N_3P_3Cl_4(SC_6H_4\text{-}p\text{-}Me)_2$ (2,2,4,4:6,6). XI.[445] m. 129,[445] ^{31}P -18.9 (PCl_2), -47.4 [$P(SC_6H_4\text{-}p\text{-}Me)_2$].[148]

$N_3P_3(SC_6H_4\text{-}p\text{-}Me)_6$ XI.[445] m. 147.[445]

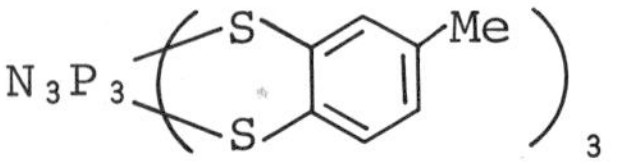

XI.[10] m. 256-263,[10] reaction with o-aminophenol.[10]

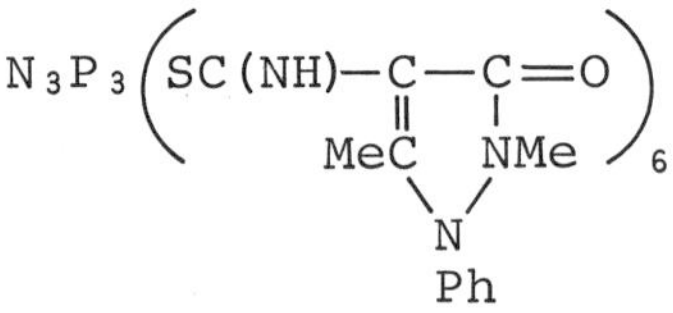

XI.[239] m. 175,[239] use in quant. determination of Ag and Cu.[240]

D.1.15. Thio(Aryl)- and Thio(Amino)-Derivatives

$N_3P_3Ph_4(SMe)(SPh)$ (2,2,4,4:6:6). [$H_2NP(Ph_2)$=N-PPh_2=NH + 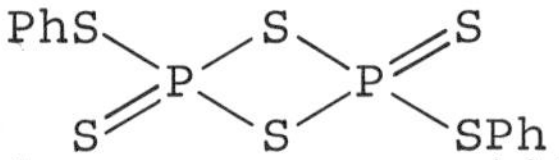 followed by MeI],[414] m. 100-103,[414] ^{1}H -2.37 (SMe).[414]

$N_3P_3Ph_4(SMe)(C_6H_4$-p-OMe) (2,2,4,4:6:6). [$H_2NP(Ph_2)$=N-PPh_2=NH + 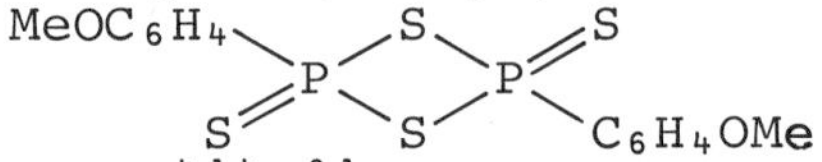 followed by MeI],[414] m. 129-130,[414] ^{31}P -27.0 [P(SMe)(C_6H_4-p-OMe)], -14.1 (PPh_2),[414] ^{1}H -1.86 (SMe).[414]

$N_3P_3(SMe)(NC_2H_4)_5$. XI.[357] m. 106.5-107.5.[357]

$N_3P_3(SEt)_2(NMe_2)_4$ (2,4:2,4,6,6). VII.[91] $b_{0.015}$ 133,[91] pK_a' 3.8.[164]

$N_3P_3(SEt)_2(NMe_2)_4$ (2,2:4,4,6,6). XI.[91] m. 62,[91] pK_a' 4.7.[164]

$N_3P_3(SEt)_3(NMe_2)_3$. VII.[91] $b_{0.02}$ 152.[91]

$N_3P_3(SPh)_2(NMe_2)_4$ (2,2:4,4,6,6). VII.[445] m. 87,[445] pK_a' 4.0.[164]

$N_3P_3(SPh)_2(NMe_2)_4$ (2,4:2,4,6,6). XI.[91] m. 84,[91] pK_a' 2.8.[164]

$N_3P_3(SC_6H_4$-p-Cl$)_4(NMe_2)_2$ (2,2,4,4:6,6). VII.[445] m. 114.[445]

$N_3P_3Cl_3(SC_6H_4$-p-Me$)_2(NMe_2)$ (2,2,4:6,6:4). IX.[445] m. 118.[445]

$N_3P_3(SC_6H_4$-p-Me$)_2(NMe_2)_4$ (2,2:4,4,6,6). VII.[445] m. 99.[445]

$N_3P_3(SC_6H_4$-p-Me$)_2(NMe_2)_4$ (2-cis-4:2,4,6,6). XI.[445] m. 120.[445]

$N_3P_3(SC_6H_4$-p-Me$)_3(NMe_2)_3$ (2-trans-4,6:2,4,6). XI.[445] m. 98.[445]

D.2. Cyclotetraphosphazatetraenes

D.2.1. Halogeno-Derivatives

$N_4P_4F_8$. IV.[95,219,329,425,428,429,494] (P_3N_5+ CF_3SF_5 or NF_3),[310] m. 30.4,[95,219,329,425,428,429,494] b_{760} 89.7,[429] crit. temp. 223.2,[98] n_D^{20} 2.239,[429] d^{20} ~1.7,[310] molar vol. 149.6 ml.[95] log p = -1952/T + 8.26 (m - 90°),[219,429] log p = -3013/T + 11.76 (0° - m.)[219,429] ΔHvap 8.91 Kcal./mole,[219,429] ΔS vap. 24.6

cal./mole/deg.,[219,429] $\Delta H_{sub.}$ 13.8 Kcal./mole.[219,429] X-ray structure[322] shows that ring is planar with P-N 1.51 (0.02), P-F 1.51 (0.02), $P\hat{N}P$ 147.2(1.4), $N\hat{P}N$ 122.7(1.0), $F\hat{P}F$ 99.9(1.0), prelim. crystal data,[237,238,321] bonding in,[326] IR,[94,95,471] Raman,[94,95,470,471] ^{31}P + 17.0,[327] ^{19}F + 72.3,[23] + 71.9,[102] mass spectr.,[73] photoelectron spectr.,[71] first ionization potl. 10.7 eV,[71] UV,[187,290] dipole moment < 0.1 D,[105] hydrolysis,[428] forms adduct $N_4P_4F_8 \cdot 2SbF_5$: m. 143-145,[101] IR.[101]

$N_4P_4ClF_7$. IV.[23,151] [$N_4P_4F_7(NMe_2)$ + HCl],[100,102] m. -6,[23,151] b_{760} 117,[23,151] ΔH vap. 9.5 Kcal./mole,[151] ΔS vap. 24.4 cal./mole/deg.,[151] d^{20} 1.8020,[151] IR,[152] ^{19}F + 70.3 (PF_2), + 38.2 (PFCl).[102,151]

$N_4P_4Cl_2F_6$ (2,2:4,4,6,6,8,8). IV.[23,151] m. -22,[23,151] -12,[418] b_{760} 147,[23,151] log p = 1911/T + 7.923 (60-77°),[418] ΔH vap. 10.0,[151] 8.74,[418] Kcal./mole, ΔS vap. = 23.9,[151] 23.07,[418] cal./mole/deg., d^{20} 1.8061,[151] IR,[23,151] ^{19}F + 72.4(1), + 70.5(2).[23,151]

$N_4P_4Cl_3F_5$ (2,2,4:4,6,6,8,8). IV.[23,151] m. -29,[23,151] b_{760} 178,[23,151] ΔH vap. 11.0 Kcal./mole,[151] ΔS vap. 24.5 cal./mole/deg.,[151] d^{20} 1.8197,[151] IR,[152] ^{19}F + 71.9 (PF_2), + 39.4 (PFCl).[23,151]

$N_4P_4Cl_3F_5$ (nongeminal). [$N_4P_4F_5(NMe_2)_3$ + HCl].[207]

$N_4P_4Cl_4F_4$ (2,2,4,4:6,6,8,8). IV.[23,151,417] m. -23,[23,151] b_{760} 205,[23,151] ΔH vap. 12.2,[151] 8.74 Kcal./mole,[417] ΔS vap. 25.6,[151] 21.66[417] cal./mole/deg., d^{20} 1.8343,[151] IR,[152] ^{19}F + 72.2.[23,151]

$N_4P_4Cl_5F_3$ (2,2,4,4,6:6,8,8). IV.[23,151] m. -11,[23,151] b_{760} 232,[23,151] ΔH vap. 13.0 Kcal./mole,[151] ΔS vap. 25.8 cal./mole/deg.,[150] d^{20} 1.8420,[151] IR,[152] ^{19}F + 72.2 (PF_2), + 39.4 (PFCl).[23,151]

$N_4P_4Cl_6F_2$ (2,2,4,4,6,6:8,8). IV.[23,151] m. 23,[23,151] b_{760} 267,[23,151] ΔH vap. 13.6 Kcal./mole,[151] ΔS vap. 25.1 cal./mole/deg.,[151] d^{20} 2.060,[151] IR,[152] ^{19}F + 72.8.[23,151]

$N_4P_4Cl_7F$. IV.[23,151] m. 63,[23,151] b_{760} 301,[23,151] ΔH vap. 14.0 Kcal./mole,[151] ΔS vap. 24.5 cal./mole/deg.,[151] d^{20} 2.049,[151] IR,[152] ^{19}F + 41.9.[151]

$N_4P_4BrF_7$. ($N_4P_4F_7 \cdot NMe_2$ + HBr),[100,102] b_{760} 120,[100,102] ^{19}F + 30.6 (PFBr), + 70.6 (PF_2).[100,102]

$N_4P_4Cl_8$. I.[262,300,350,398,477,480] m. 122.8,[262,300,350,398,477,480] b_{10} 185,[262] b_{760} 382.5,[477,480] d^{20} 2.18, log p mm. = 13.79-5593/T (28.2-81.4°),[106] $\Delta H_{sub.}$ 25.6 Kcal./mole.[106] log p mm. = 3400/T + 8.560 (m - 275°),[339] ΔH vap. 15.6 Kcal./mole,[339] log p mm. = 8.498 - 3358/T (m - 306°),[236] ΔH vap. = 15.35 Kcal./mole,[236] $\Delta H_f°(c)$ = -259.2 Kcal./mole,[221] $\bar{E}$(N-P) 72.5 Kcal./mole,[221] exists in two crystalline forms, x-ray study of low temp. (K) form[222,262] shows that ring has "tub" conformation with all P-N bonds equal in length

1.570 (0.01), P-Cl 1.989 (0.004), $N\hat{P}N$ 121.2 (0.5), $P\hat{N}P$ 131.3 (0.6), the T form[502] has rings in a "chair" conformation with P-N 1.559 (0.012), P-Cl 1.992 (0.004), $N\hat{P}N$ 120.5 (0.7), $P\hat{N}P$ 133.6 (0.8), and 137.6 (0.8), IR,[94,120,210,247,300,306,308,460,468,471] Raman,[94,210,232,470,471] ^{31}P + 7,[300] NQR,[142,248,346,347,489,508] dynamic nuclear polarisation,[147,148] dipole moment < 0.2 D,[105] mass spectr.,[71,74,399] first ionization potl. 9.80 eV,[71,72] magnetic anisotropy,[118] UV,[247,289,290,300,490] relative basicity (to HCl),[72] electrical conductivity,[150] Cl^- exchange,[457] chromatographic separation,[376,495] hydrolysis,[351] forms adducts $N_4P_4Cl_8{\cdot}C_6Me_6$: m. 136,[127] $N_4P_4Cl_8{\cdot}2C_6Me_6$: m. 152,[127] $N_4P_4Cl_8$·naphthalene: formation constant.[302]

$N_4P_4Br_8$. I.[39,64,116,242,243,244] m. 202,[39,64,116,242,243,244] log p mm. = 12.78 - 6478/T (87.6 - 99.6°),[107] ΔH sub. 29.4 Kcal./mole,[107] crystal data,[64] x-ray powder data,[116] IR,[116,306,460,471] Raman,[470,471] ^{31}P + 71.8,[243,327] mass spectr.,[109] UV,[290] hydrolyzes slowly in moist air.

D.2.2. Pseudohalogeno-Derivatives

$N_4P_4F_7{\cdot}NCS$. ($N_4P_4F_8$ + KSCN).[102] IR,[102] ^{19}F 53.9 (PF·NCS), 69.9 (PF_2).[102]

$N_4P_4(NCS)_8$. ($N_4P_4Cl_8$ + KSCN),[484] m. 90,[484] ^{31}P +49.1,[353] solubility,[484] forms derivatives with the following protic species (stated).

$N_4P_4(NHCSNH_2)_8$. (NH_3), m. 120 (d),[33,484] ^{31}P + 21.2.[353]

$N_4P_4(NHCSNH\text{-}n\text{-}Bu)_8$. ($NH_2$-n-Bu), m. 105.[33,484]

$N_4P_4(NHCSNHPh)_8$. (NH_2Ph), m. 139,[484] ^{31}P + 23.2.[353]

$N_4P_4(NHCSNHNHPh)_8$. (NH_2NHPh), m. 153 (d).[33,484]

$N_4P_4(NHCSOEt)_8$. (EtOH), m. 189.[33,484]

$N_4P_4[NHCSNH(CH_2)_2NH(CH_2)_2NH_2]_8$. [$H_2N(CH_2)_2NH(CH_2)_2NH_2$], m. 140-149 (d).[33]

$N_4P_4Ph_6(N_3)_2$ (2,2,4,6,6,8:4-trans-8). [$N_4P_4Cl_2Ph_6$ + LiN_3],[431] m. 206-208,[431] IR,[431] dipole moment 0.2D,[431] kinetics of decomp.,[431] N_3 derivative partially converts to cis-isomer,[431] with Ph_3P gives $N_4P_4Ph_6(N{=}PPh_3)_2$: (2,2,4,6,6,8:4-trans-8), m. 240-241,[431] with $Ph_2PCF_2CF_2CF_3$ gives $N_4P_4Ph_6(N{=}PPh_2CF_2CF_2CF_3)_2$: (2,2,4,6,6,8:4-trans-8), m. 104-106.[431]

$N_4P_4Ph_6(N_3)_2$ (2,2,4,6,6,8:4-cis-8). ($N_3P_3Cl_2Ph_6$ + LiN_3),[431] m. 147.5-148,[431] IR,[431] dipole moment 3.7 D,[431] kinetics of decomp.,[431] with $Ph_2P(CH_2)_4PPh_2$ gives $N_4P_4Ph_6[N{=}PPh_2(CH_2)_4PPh_2{=}N]$: (2,2,4,6,6,8:4-cis-8), m. 314-317.[431]

D.2.3. Alkyl-Derivatives

$N_4P_4F_6Me_2$ (2,2,4,4,6,6:8,8). X-ray study[313] shows that P-N ring has slight "saddle" conformation with P-N bonds of alternating lengths P_8-N_1 1.586, N_1-P_2 1.470, P_2-N_3 1.532, N_3-P_4 1.487 (all 0.006), $N\hat{P}_8N$ 116.9, $N\hat{P}_2N$ 126.1, $P\hat{N}_1P$ 146.7, $P\hat{N}_3P$ 143.3°.

$N_4P_4F_4Me_4$ (2,2,6,6:4,4,8,8). X-ray study[313] shows that ring again has slight "saddle" conformation with P_2-N_3 1.54, N_3-P_4 1.59 (both 0.01) $N\hat{P}_2N$ 125.9, $N\hat{P}_4N$ 117.5 $P\hat{N}P$ 134.6°.

$N_4P_4Me_8$. I.[427] m. 163-164,[427] IR,[427] ^1H -1.31,[27] pK_a 5.75,[149,341] first ionization potl. 7.99 eV.,[71] x-ray study[144] shows that molecule has a "saddle" shaped ring with P-N 1.591 and 1.601 (0.003), P-C 1.805 (0.004), av. $N\hat{P}N$ 119°48', av. $P\hat{N}P$ 131°57', with $Mo(CO)_6$ forms $N_4P_4Me_8Mo(CO)_4$ in which Mo is most likely cis-coordinated by ring N-atoms,[149] with MeI gives $[N_4P_4Me_9]^+I^-$: ^1H -3.02 (NMe), -1.94 (PMe),[24] which, in turn, forms $[N_4P_4Me_9]^+[M(CO)_5I]^-$ (M = Cr,Mo) [149] with $M(CO)_6$. Salt-like adducts $N_4P_4Me_8\cdot 2HCl$: m. 282,[341] $[N_4P_4Me_9]^+[HgI_3]^-$,[24] and $[N_4P_4Me_9H]^{2+}$ $[Zn(NCS)_4]^{2-}$ [24] also known. X-ray study[491,492] of $[(NPMe_2)_4H]\ CuCl_3$ shows that proton and $CuCl_3$ group are covalently bonded to N-atoms on opposite sides of ring where P-N is 1.63, 1.60, 1.56, and 1.67. Structure of $[(NPMe_2)_4H]_2CoCl_4$ shows[491,493] two types of protonated (at N) rings and $CoCl_4^-$ species with P-N 1.70, 1.54, 1.61, and 1.58.

$N_4P_4(CF_3)_8$. $[(CF_3)_2P(NH_2)Cl_2 + NMe_3]$,[482] m. 109,[482] IR.[482]

$N_4P_4Et_8$. II.[62] Oil,[62] pK_a 6.47.[341]

$N_4P_4Cl_7$-n-Bu. ($N_4P_4Cl_8$ + n-BuLi),[328] $b_{0.6}$ 43-46.[328]

$N_4P_4Cl_6(n\text{-}Bu)_2$. ($N_4P_4Cl_8$ + n-BuLi),[328] $b_{2.5}$ 113.[328]

D.2.4. Aryl-Derivatives

$N_4P_4F_7\cdot C_6F_5$. ^{19}F.[99]

$N_4P_4F_2Ph_6$ (2,6:2,4,4,6,8,8). IV.[63] m. 176-177.[63]

$N_4P_4Cl_4Ph_4$ (2-cis-4-cis-6-trans-8:2,4,6,8). II.[203,204,216,441,442,453] m. 148-150.5.[203,204,216,441,442,453]

$N_4P_4Cl_4Ph_4$ (2-cis-4-cis-6-cis-8:2,4,6,8). II.[203,204,213,216,441,442,453] (Pyrolysis of $PhPClN_3$),[225] m. ca. 202,[225,441,442,453] 225-226[203,204,213,216] (probably two crystalline modifications),[80] dielectric const.,[213] x-ray study[80] shows that ring has flattened "crown" conformation with av. P-N 1.570 (0.008) av. P-C 1.783 (0.010), av. P-Cl 2.042 (0.004), $P\hat{N}P$ 133.2-141.8 (0.5), $N\hat{P}N$ 119.6-122,4° (0.5).

$N_4P_4Cl_4Ph_4$ (2-cis-4-trans-6-trans-8:2,4,6,8). II.[203,204,213,441,442,453] m. 260-263,[203,204,213] 248,[441,442,453] dielectric constant,[213] x-ray study[79] shows that ring has "chair" conformation and that the molecule is centrosymmetric with av. P-N 1.57 (≤ 0.1), av. P-Cl 2.03 (≤ 0.1), av. P-Ph 1.79 (≤ 0.1), av. $N\hat{P}N$ 120, $P\hat{N}P$ 132, 139° (0.6).

$N_4P_4Cl_4Ph_4$ (2,2,4,4:6,6,8,8). {$[Ph_2P(NH_2)\cdots N\cdots PPh_2NH_2]^+Cl^-$ + PCl_5},[227] m. 135.[227]

$N_4P_4Cl_4Ph_4$ (2,2,6,6:4,4,8,8). ($N_4P_4Cl_8$ + PhMgBr),[57,60,67] m. 212.5,[57,60] mass spectr.,[60] ^{31}P -4.3 (PPh_2), + 2.0 (PCl_2).[148]

$N_4P_4Cl_2Ph_6$ (2,6:2,4,4,6,8,8). (NaN_3 + $PhPCl_2$ + Ph_2PCl),[228] [$N_4P_4(OH)_2Ph_6$ + PCl_5],[63] m. 294,[228] 303-304,[63] reactions with diols.[228]

$N_4P_4Cl_2Ph_6$ (2,4:2,4,6,6,8,8). [$Ph_2(NH_2)\cdots N\cdots PPh_2NH_2]^+Cl^-$ + $PhPCl_4$],[227] m. 192-193.[227]

$N_4P_4Cl_2Ph_6$ (2,4:2,4,6,6,8,8). (As isomer above),[227] m. 154-155.[227]

$N_4P_4Ph_8$. II.[218,272,496] V.[57,60,67] {$[Ph_2P(NH_2)\cdots N\cdots PPh_2NH_2]^+Cl^-$ + Ph_2PCl_3}, (pyrolysis of pdt. from Ph_2PCl + NH_2Cl),[191,198] (pyrolysis of Ph_2PN_3),[226] (pyrolysis of pdt. from Ph_2PCl + $H_2NNH_2\cdot HCl$),[450] [$N_4P_4Cl_4Ph_4$ (2,4,6,8:2,4,6,8) + PhMgBr],[442] m. 318-319,[57,60,67,191,198,218,226,272,442,450,496] pK_a' 2.2,[161] ESR,[7] E(P-N) 85.6 Kcal/mole.[48a]

$N_4P_4Br_4Ph_4$ (2,4,6,8:2,4,6,8). II.[344] m. 227-229.[344]

$N_4P_4Br_2Ph_6$ (2,6:2,4,4,6,8,8). [$N_4P_4Ph_6(OH)_2$ + PBr_5],[63] [$N_4P_4Ph_6(NMe_2)_2$ + HBr],[63] m. 254.[63]

D.2.5. Amino-Derivatives

$N_4P_4F_7\cdot NH_2$. VII.[392] sublimes 30° in vacuo,[392] IR,[392] ^1H -5.0 (NH_2),[392] with $SOCl_2$ gives $N_4P_4F_7\cdot NSO$: b_{13} 51,[392] IR.[392]

$N_4P_4F_7\cdot NMe_2$. VII,[100,102] VIII.[100,102] b_{20} 72-73,[102] IR,[102] ^{31}P -3.5 ($PF\cdot NMe_2$), +15.4 (PF_2),[102] ^{19}F +59.2 ($PF\cdot NMe_2$), +69.9 (PF_2),[102] ^1H -2.78 (NMe_2).[102]

$N_4P_4F_6(NMe_2)_2$ (2,2,4,6,6,8:4,8). VII,[100,102] VIII.[100,102] b_{20} 125,[102] IR,[102] ^{31}P -4.5 ($PF\cdot NMe_2$), +15.3 (PF_2),[102] ^{19}F +60.0 ($PF\cdot NMe_2$), +69.4 (PF_2),[102] ^1H -2.79 (NMe_2).[102]

$N_4P_4F_5(NMe_2)_3$. IV.[207] Liquid (isomeric mixture).[207]

$N_4P_4F_4(NMe_2)_4$. IV.[2] m. ~138 (isomeric mixture).[207]

$N_4P_4Cl_4(NH_2)_4$ (2,2,6,6:4,4,8,8). VII.[452]

$N_4P_4(NH_2)_8$. VII.[34,325] m. ca. 220 (d),[34,325] pK_a' 7.50,[160] pyrolysis.[325]

$N_4P_4Cl_7\cdot NHMe$. VII.[314] m. 67-67.5.[314]

$N_4P_4Cl_6(NHMe)_2$. VII.[245,314] m. 135-136.5,[314] 143,[245] (av.) ^{31}P +3.8.[327]

$N_4P_4(NHMe)_8$. VII.[246,389] m. 206,[246,389] pK'a 8.2,[160] pKa 7.95.[160]
$N_4P_4(NC_2H_4)_8$. VII.[384,422] m. 262.[384,422]
$N_4P_4Cl_7(NHEt)$. VII.[314] $b_{0.3}$ 133-134.[314]
$N_4P_4Cl_6(NHEt)_2$. VII.[245,314] 114-114.5,[245,314] (av.) ^{31}P +6.0.[327]
$N_4P_4(NHEt)_8$. VII.[246,389] m. 116,[389] 122,[246] pK'a 8.1,[160] pKa 8.70.[160]
$N_4P_4Cl_7(NMe_2)$. VII.[296,389,465] m. 52,[296,389,465] (av.) ^{31}P -0.7.[148]
$N_4P_4Cl_6(NMe_2)_2$ (2,4,4,6,8,8:2-trans-6). VII.[245,296,389,465] m. 170.[245,296,389,465]
$N_4P_4Cl_6(NMe_2)_2$ (2,2,4,4,6,6:8,8). VII.[296] m. 67.[296]
$N_4P_4Cl_6(NMe_2)_2$ (2,2,4,4,6,8:6-trans-8). VII.[296] m. 92.[296]
$N_4P_4Cl_6(NMe_2)_2$ (2,2,4,4,6,8:6-cis-8). VII.[296] m. 87.[296]
$N_4P_4Cl_6(NMe_2)_2$ (2,4,4,6,8,8:2-cis-6). VII.[296] m. 85.[296]
$N_4P_4Cl_5(NMe_2)_3$ (2,2,4,6,6:4,8,8). VII.[139,296] m. 103.[139,296]
$N_4P_4Cl_5(NMe_2)_3$ (2,2,4,6,8:4-trans-6-trans-8). VII.[139,296] m. 106-107.[139,296]
$N_4P_4Cl_4(NMe_2)_4$ (2,2,6,6:4,4,8,8). VII.[139] m. 110.[139]
$N_4P_4Cl_4(NMe_2)_4$. VII.[139] m. 148.[139]
$N_4P_4Cl_4(NMe_2)_4$ (2,4,6,8:2-cis-4-trans-6-trans-8). VII.[139,389] m. 200.[139,389]
$N_4P_4Cl_3(NMe_2)_5$ (2,2,6:4,4,6,8,8). VII.[139,389] m. 146.[139,389]
$N_4P_4Cl_3(NMe_2)_5$ (2-cis-4-trans-6:2,4,6,8,8). VII.[139] m. 99.5.[139]
$N_4P_4Cl_2(NMe_2)_6$ (2-trans-6:2,4,4,6,8,8). VII.[139,389] m. 168,[139,389] ^{31}P -9.2 [$P(NMe_2)_2$], -5.6 ($PCl\cdot NMe_2$).[148]
$N_4P_4Cl_2(NMe_2)_6$. VII.[139] m. 83.[139]
$N_4P_4(NMe_2)_8$. VII.[139,386,389] m. 220-238(d),[78,139,386,389] ^1H -2.86,[27] pK'a 8.3,[160] x-ray study[77,78] shows that ring has "saddle" shape with P-N 1.60 (0.01), exo P-N 1.67, 1.69 (0.01), PN̂P 133(0.6), av. NP̂N 120°(0.5), first ionization potl. 7.45 eV.[71]
$N_4P_4Cl_7\cdot$NH-n-Pr. VII.[314] b_1 157.[314]
$N_4P_4Cl_6$(NH-n-Pr)$_2$. VII.[314] m. 100-100.5.[314]
N_4P_4(NH-n-Pr)$_8$. VII.[246,389] m. 98,[246,389] pK'a 8.3.[160]
$N_4P_4Cl_7$(NH-i-Pr). VII.[314] b_1 128-129.[374]
$N_4P_4Cl_6$(NH-i-Pr)$_2$. VII.[314] m. 120.5-121.5.[314]
$N_4P_4Cl_4$(NH-i-Pr)$_4$. VII.[389] m. 187.[389]
N_4P_4(NH-i-Pr)$_8$. VII.[389] m. 170,[389] pK'a 8.1.[160]
$N_4P_4Cl_6$(NH-t-Bu)$_2$. VII.[245,389] m. 168,[389] 124,[245] (av.) ^{31}P +10.0.[327]
N_4P_4(NH-n-Bu)$_8$. VII.[246,389] m. 81,[389] 86,[246] pK'a 7.6.[160]
N_4P_4(NH-sec.-Bu)$_8$. VII.[389] m. 111.[389]
N_4P_4(NH-i-Bu)$_8$. VII.[389] m. 94,[389] pK'a 8.0.[160]
N_4P_4(NH-t-Bu)$_8$. VII.[389] m. > 300 (d),[389] pK'a 8.8.[160]

$N_4P_4Cl_4(NEt_2)_4$. VII.[389] m. 172.[389]
$N_4P_4(NEt_2)_8$. VII.[389] m. 200,[389] pK$_a'$ 8.3.[160]
$N_4P_4Cl_6(NC_4H_8O)_2$. VII.[245] m. 205,[245] (av.) ^{31}P +3.5.[327]
$N_4P_4(NC_4H_8O)_8$. VII.[246] m. 230(d).[246]
$N_4P_4Cl_6(NC_5H_{10})_2$. VII.[245] m. 194,[245] (av.) ^{31}P +4.1.[327]
$N_4P_4Cl_4(NC_5H_{10})_4$ (2,4,6,8:2,4,6,8). VII.[246] m. 204,[246] ^{31}P -1.7.[246]
$N_4P_4(NC_5H_{10})_8$. VII.[246] m. > 230 (d),[246] pK$_a'$ 8.4.[160]
N_4P_4(NH-n-Pent)$_8$. VII.[246] m. 79.[246]
N_4P_4(NH-n-Hex)$_8$. VII.[246] m. 70.[247]
$N_4P_4Cl_6(NHPh)_2$ (2,2,4,6,6,8:4,8). VII.[245] m. 166,[245] ^{31}P +4.6, +13.5.[245]
$N_4P_4Cl_6(NMePh)_2$. VII.[245] m. 146,[245] (av.) ^{31}P +5.3.[327]
$N_4P_4Cl_4(NMePh)_4$ (2,4,6,8:2,4,6,8). VII.[246] m. 146,[246] (av.) ^{31}P +1.1.[246]
$N_4P_4(NHCH_2Ph)_8$. VII.[246] m. 121.5.[246]
$N_4P_4Cl_6(NHC_6H_4$-o-Me$)_2$ (2,2,4,6,6,8:4,8). VII.[245] m. 172,[245] ^{31}P +5.9, +11.6.[245]
$N_4P_4(NHC_6H_4$-o-Me$)_8$. VII.[246] m. 229(d).[246]
$N_4P_4Cl_6(NHC_6H_4$-m-Me$)_2$ (2,2,4,6,6,8:4,8). VII.[245] m. 204,[245] ^{31}P +5.6, +13.5.[245]
$N_4P_4(NHC_6H_4$-m-Me$)_8$. VII.[246] m. 199.[246]
$N_4P_4Cl_6(NHC_6H_4$-p-Me$)_2$. VII.[245] m. 196,[245] (av.) ^{31}P +9.1.[327]
$N_4P_4(NHC_6H_4$-p-Me$)_8$. VII.[246] m. 207.[246]
$N_4P_4Cl_4(NMe{\cdot}C_6H_{11})_4$ (2-trans-4-trans-6-cis-8:2,4,6,8). VII.[51] m. 126.5-128,[51] ^1H.[51]

D.2.6. Amino-Derivatives with Mixed Substituents

$N_4P_4(NH_2)_6(NHNH_2)_2$ (2,2,4,4,6,6:8,8). ^{31}P -10.0.[327]
$N_4P_4(NMe{\cdot}C_6H_{11})_4(NHMe)_4$ (2-trans-4-trans-6-cis-8:2,4,6,8). VII.[51] m. 176-177,[51] ^1H.[51]
$N_4P_4(NMe{\cdot}C_6H_{11})_4(NHMe)_3$(NMeCONHPh) (2-trans-4-trans-6-cis-8:?). (Above deriv. + PhNCO),[51] m. 134-135,[51] ^1H.[51]
$N_4P_4(NMe{\cdot}C_6H_{11})_4(NHMe)_2(NMeCONHPh)_2$ (2,4,6,8:?). (As above),[51] m. 216-217.[51]

D.2.7. Phosphazenyl-Derivatives

$N_4P_4F_7(N{=}PCl_3)$. IX.[392] $b_{0.01}$ 42,[392] IR.[392]
$N_4P_4Cl_4(N{=}PCl_3)_4$ (2,2,6,6:4,4,8,8). IX.[452] IR,[452] ^{31}P +11.4 (PCl_2), +18.0 (PCl_3), +34.1 [$P(N{=}PCl_3)_2$].[452]
$N_4P_4Cl_6(N{=}VCl_3)_2$ (2,2,4,4,6,6:8,8). (Reaction of above deriv. with $VOCl_3$),[452] m. 60(d),[452] IR,[452] ^{31}P +7.3 (PCl_2), +3.2 [$P(N{=}VCl_3)_2$].[452]
$N_4P_4Cl_7(N{=}PPh_3)$. ^{31}P +18.6 [$PCl(N{=}PPh_3)$], +10.5 (PCl_2) (2), +6.7 (PCl_2) (1), -14.9 (PPh_3).[148]

D.2.8. Aryl(Amino)-Derivatives

$N_4P_4Ph_4(NH_2)_4$ (2-cis-4-trans-6-trans-8:2,4,6,8). VII. (From $N_4P_4Cl_4Ph_4$ m. 248),[442] m. 229.[442]

$N_4P_4Ph_4(NHMe)_4$ (2-cis-4-trans-6-trans-8:2,4,6,8). VII. (As above),[213,442] m. 130-154,[442] 151-152,[213] IR,[442] conductivity,[213] x-ray study[79] shows that P-N ring has "chair" conformation with endo P-N 1.59 ($\leqslant$ 0.1), exo P-N 1.68 ($\leqslant$ 0.1), $N\hat{P}N$ 120 ($\leqslant$ 0.6). With Ph_2PCl and Et_3N forms $N_4P_4Ph_4(NHMe)_3(NMePPh_2)$: m. 220-222,[213] ^1H,[213] and with PhNCO gives $N_4P_4Ph_4(NHMe)_3$ (NMe·CONHPh): m. 165-176,[213] ^1H.[213]

$N_4P_4Ph_4(NHEt)_4$. m. 101.[442] Structure and prepn. as for NH_2 deriv. above.

$N_4P_4Ph_4(NMe_2)_4$. m. 176-178,[213,442,453] ^1H -2.38 (NMe_2),[313,453] conductivity.[213] Structure and prepn. as for NH_2 deriv. above.

$N_4P_4Ph_4$(NH-n-Pr)$_4$. m. 98.[442] Structure and prepn. as for NH_2 deriv. above.

$N_4P_4Ph_4$(NH-i-Pr)$_4$. m. 140.[442] Structure and prepn. as for NH_2 deriv. above.

$N_4P_4Ph_4$(NH-t-Bu)$_4$. m. 252.[442] Structure and prepn. as for NH_2 deriv. above.

$N_4P_4Ph_4(NEt_2)_4$. m. > 320.[442] Structure and prepn. as for NH_2 deriv. above.

$N_4P_4Ph_4(NC_5H_{10})_4$. m. 255.[442] Structure and prepn. as for NH_2 deriv. above.

$N_4P_4Ph_4(NHC_6H_{11})_4$. m. 132.[442] Structure and prepn. as for NH_2 deriv. above.

$N_4P_4Ph_4(NHPh)_4$. m. 241.[442] Structure and prepn. as for NH_2 deriv. above.

$N_4P_4Ph_4(NHNHPh)_4$. m. 176.[442] Structure and prepn. as for NH_2 deriv. above.

$N_4P_4Ph_4(NHMe)_4$ (2,4,6,8:2-cis-4-cis-6-cis-8). VII.[213,442] (From $N_4P_4Cl_4Ph_4$, m. 202[442] or 225.5-226,[213]) m. 131,[213,442] IR,[442] conductivity,[213] with PhNCO gives $N_4P_4Ph_4(NHMe)_3$(NMe·CONHPh). m. 78-80.5.[213]

$N_4P_4Ph_4(NHEt)_4$. m. 122.[442] Structure and prepn. as NHMe deriv. above.

$N_4P_4Ph_4(NMe_2)_4$. m. 154,[213,216,442,453] ^1H -2.65 (NMe_2).[213,453] Structure and prepn. as NHMe deriv. above.

$N_4P_4Ph_4$(NH-i-Pr)$_4$. m. 133.[442] Structure and prepn. as NHMe deriv. above.

$N_4P_4Ph_4(NC_5H_{10})_4$. m. 224.[442] Structure and prepn. as NHMe deriv. above.

$N_4P_4Ph_4(NHMe)_4$ (2,4,6,8:2-trans-4-trans-6-trans-8). VII. (From $N_4P_4Cl_4Ph_4$, m. 148), m. 113,[442] 119-120,[213] conductivity,[213] IR.[442]

$N_4P_4Ph_4(NMe_2)_4$. m. 139-141,[453] 136-138,[213,216] ^1H -2.17 (1), -2.50(2), -2.58(1) (all NMe_2).[213,216,453] Structure and prepn. as for NHMe deriv. above.

$N_4P_4Ph_4(NC_5H_{10})_4$. m. 240.[442] Structure and prepn. as for NHMe deriv. above.
$N_4P_4Ph_4(NMe_2)_4$ (2,2,6,6:4,4,8,8). VII.[60] m. 178,[60] ^{1}H -2.30 (NMe_2),[60] pK$_a'$ +5.6.[60]
$N_4P_4Ph_5Cl_2(NMe_2)$. (VI, followed by reaction with $NHMe_2$),[137] m. 171,[137] ^{1}H,[137] IR,[137] mass spectr.[137]
$N_4P_4Ph_6(NH_2)_2$ (2,2,4,6,6,8:4,8). VII.[63] m. 205-207.[63]
$N_4P_4Ph_6(NHMe)_2$ (2,2,4,6,6,8:4,8). VII.[63] m. 192-194.[63]
$N_4P_4Ph_6(NMe_2)_2$ (2,2,4,6,6,8:4,8). VII.[63] m. 189-190.[63]
$N_4P_4Ph_6(NEt_2)_2$ (2,2,4,6,6,8:4,8). VII.[63] m. 260-265 (sublimes).[63]
$N_4P_4Ph_6(NC_5H_{10})_2$ (2,2,4,6,6,8:4,8). VII.[63] m. 290-292.[63]
$N_4P_4Ph_6(NMePh)_2$ (2,2,4,6,6,8:4,8). VII.[63] m. 183-184.[63]

D.2.9. Alkoxy- and Aryloxy-Derivatives

$N_4P_4(OMe)_8$. X.[177,178] m. 41,[177,178] b_1 125,[178] IR,[432] ^{1}H -3.52,[27] pK' -1.0.[161] First ionization potl. 8.83 eV,[71] x-ray study[28] shows that ring has "saddle" shape with P-N 1.57, P-O 1.60, P$\hat{N}$P 132, N$\hat{P}$N 122°, rearranges to $N_4Me_4P_4O_4(OMe)_4$: m. 235[174] with MeI.
$N_4P_4(OEt)_8$. X.[177-179] m. 45-47,[177-179] $b_{0.001}$ 128,[178,179] IR,[179] ^{31}P +0.6,[229] pK$_a'$ 0.6,[161] rearranges thermally to $N_4Et_4P_4O_4(OEt)_4$: m. 208-210, catalyzed with EtI,[174] and reacts with benzoyl chloride[175] to give $N_3C_3Ph_3$.
$N_4P_4(OCH_2CF_3)_8$. X.[385] m. 65,[385] b_3 139-140,[385] ^{1}H -4.26,[27] pK$_a'$ < -6.0,[161] first ionization potl. 10.01 eV.[71]
$N_4P_4(O\text{-}n\text{-}Pr)_8$. X.[177,178] m. 38-38.5,[177,178] $b_{0.01}$ 176-178.[178]
$N_4P_4(OCH_2CH_2CF_3)_8$. X.[385] b_3 142-144,[385] n_D^{27} 1.3530.[385]
$N_4P_4(O\text{-}i\text{-}Pr)_8$. X.[177,178] m. 74-75(d),[177,178] $b_{0.01}$ ~130,[178] pK$_a'$ 2.1,[161] rearranges to $N_4Me_4P_4O_4(O\text{-}i\text{-}Pr)_4$ with MeI.[174]
$N_4P_4(O\text{-}n\text{-}Bu)_8$. X.[177,178] $b_{0.005}$ 196-198,[177,178] n_D^{25} 1.4565,[178] pK$_a'$ 0.7.[161]
$N_4P_4[OCH_2(CF_2)_2CF_3]_8$. X.[385] m. 107,[385] b_3 169.5-171.5.[385]
$N_4P_4(OPh)_8$. X.[177,178,179] m. 85-86,[177,178,179] IR,[179] pK$_a'$ -6.0,[161] first ionization potl. 8.70 eV.[71]
$N_4P_4[OCH_2(CF_2)_5CF_2H]_8$. X.[385] b_1 265,[385] n_D^{27} 1.3441.[385]
$N_4P_4(OCH_2Ph)_8$. X.[177,178] m. 38-39,[177,178] pK$_a'$ -1.6.[161]
$N_4P_4[OCH_2(CF_2)_7CF_2H]_8$. Mass spectr.[157]
$N_4P_4(OC_6H_4\text{-}p\text{-}OMe)_8$. X.[177,178] m. 73-74,[177,178] pK$_a'$ -5.2.[161]
$N_4P_4[OCH_2(CF_2)_9CF_2H]_8$. X.[385] m. 102-105,[385] $b_{0.1}$ 310-312.[385]

D.2.10. Aryl(Alkoxy- or Aryloxy)-Derivatives

$N_4P_4Ph_4(OMe)_4$ (2,4,6,8:2-cis-4-trans-6-trans-8). X. (From $N_4P_4Cl_4Ph_4$, m. 248),[442] m. 142.[442]

$N_4P_4Ph_4(OEt)_4$ (2,4,6,8:2-cis-4-trans-6-trans-8). X. (From $N_4P_4Cl_4Ph_4$, m. 248),[442] m. 104.[442]
$N_4P_4Ph_4(OEt)_4$ (2,4,6,8:2-cis-4-cis-6-cis-8). X. (From $N_4P_4Cl_4Ph_4$, m. 202),[442] m. 122.[442]
$N_4P_4Ph_4(OPh)_4$ (2,4,6,8:2-cis-4-cis-6-cis-8). X. (From $N_4P_4Cl_4Ph_4$, m. 202),[442] m. 133.5.[442]
$N_4P_4Ph_6(OH)_2$ (2,2,4,6,6,8:4,8). ($N_4P_4Cl_2Ph_6$, m. 294 + aq. pyridine),[228] m. 272.[228]
$N_4P_4Ph_6(OEt)_2$ (2,2,4,6,6,8:4,8). X. (From $N_4P_4Cl_2Ph_6$, m. 294),[228] m. 183-185.[228]

D.2.11. Alkylthio- and Arylthio-Derivatives

$N_4P_4Cl_4(SEt)_4$ (2,2,6,6:4,4,8,8). XI.[88,90] m. 105-106.5,[88,90] $b_{0.2}$ 148-150,[88,90] ^{31}P -28.6 [$P(SEt)_2$], +9.7 (PCl_2).[68]
$N_4P_4Cl_4(S\text{-}n\text{-}Pr)_4$. XI.[88] m. 74.5-75,[88] $b_{0.1}$ 144.[88]
$N_4P_4Cl_4(S\text{-}n\text{-}Bu)_4$. XI.[88] $b_{0.2}$ 148-150.[88]
$N_4P_4Cl_5(S\text{-}i\text{-}Bu)_3$. XI.[88] $b_{0.04}$ 175-180.[88]
$N_4P_4Cl_4(SPh)_4$ (2,2,6,6:4,4,8,8). XI.[88,90] m. 156.[88,90]
$N_4P_4Cl_4(SC_6H_4\text{-}p\text{-}Cl)_4$ (2,2,6,6:4,4,8,8). XI.[445] m. 180,[445] ^{31}P +10.3 (PCl_2), -19.9 [$P(SC_6H_4\text{-}p\text{-}Cl)_2$].[443]
$N_4P_4Cl_4(SC_6H_4\text{-}p\text{-}Me)_4$ (2,2,6,6:4,4,8,8). XI.[445] m. 192.[445]

D.2.12. Thio(Amino)-Derivatives

$N_4P_4(SEt)_4(NMe_2)_4$ (2,2,6,6:4,4,8,8). VII.[445] m. 39.[445]
$N_4P_4(SPh)_4(NMe_2)_4$ (2,2,6,6:4,4,8,8). VII.[445] m. 192.[445]
$N_4P_4(SC_6H_4\text{-}p\text{-}Cl)_4(NMe_2)_4$ (2,2,6,6:4,4,8,8). VII.[445] m. 154.[445]
$N_4P_4(SC_6H_4\text{-}p\text{-}Me)_4(NMe_2)_4$ (2,2,6,6:4,4,8,8). VII.[445] m. 145.[445]

D.2.13. Cyclopentaphosphazapentaenes

$N_5P_5F_{10}$. IV.[95] m. -50,[95] b_{760} 120.1,[95] $\log_{10} p(mm.) = -2141.7/T + 8.3373$ (25-120°), ΔH vap 9.8 Kcal./mole,[95] ΔS vap 24.9 cal./mole/deg.,[95] crit. temp. 250.8,[98] d^{20} 1.8259,[95] n_D^{20} 1.3482,[95] viscosity,[95] IR,[94,95,468] Raman,[94] ^{19}F +69.0,[102] mass spectr.,[73] photoelectron spectr.,[71] first ionization potl. 11.4 eV.,[71] forms complex $N_5P_5F_{10}\cdot 2SbF_5$: m. 80-81,[101] IR.[101]
$N_5P_5F_9Cl$. ($N_5P_5F_9\cdot NMe_2$ + HCl), b_{760} 144-145,[100,102] ^{19}F +33.9 (PFCl), +67.4 (PF_2).[100,102]
$N_5P_5F_9Br$. ($N_5P_5F_9\cdot NMe_2$ + HBr), b_{146} 107-109,[100,102] ^{19}F +25.0 (PFBr), +67.5 (PF_2).[100,102]
$N_5P_5Cl_{10}$. I.[300,479] m. 41.3,[300,479] b_{13} 223-224.3,[479] b_{760} 371.5,[236] $\log p\ mm = 9.10 - 3973/T$ (170-371°),[236] ΔH vap 18.15 Kcal./mole,[236] heat of polymerization,[236] d^{20} 2.02,[300] dynamic nuclear polarization (^{31}P),[147,148] NQR,[508] UV,[300] IR,[94,300,468] Raman,[94,468]

^{31}P +17,[300] mass spectr.,[74] first ionization potl. 9.83 eV,[71,72] magnetic anisotropy,[118] Cl^- exchange,[457] rel. basicity to HCl,[72] x-ray study[399,400] shows nearly planar ring with (av.) P-N 1.521 (0.014), P-Cl 1.961 (0.009), P$\hat{N}$P 148.6, N$\hat{P}$N 118.4°, forms adduct $N_5P_5Cl_{10}\cdot 2C_6Me_6$: m. 146.[127]

$N_5P_5Br_{10}$. I.[116] m. 104.5,[116] IR,[114,116] Raman,[114] x-ray powder data,[116] mass spectr.[109]

$N_5P_5F_9\cdot NCS$. ($N_5P_5F_{10}$ + KSCN),[102] IR,[102] ^{19}F +52.3 (PF·NCS), +67.1 (PF_2).[102]

$N_5P_5F_9\cdot C_6F_5$. ^{19}F.[99]

$N_5P_5F_9\cdot NMe_2$. VII,[100,102] VIII.[100,102] b_{77} 116-117,[102] IR,[102] ^{31}P -4.0 (PF·NMe_2), +16.1 (PF_2),[102] ^{19}F +57.3 (PF·NMe_2), +67.2 (PF_2),[102] ^1H -2.78.[102]

$N_5P_5F_8(NMe_2)_2$ (nongeminal). VII,[100,102] VIII.[100,102] b_{90} 175-176,[102] IR,[102] ^{31}P -4.2 (PF·NMe_2), +16.1 (PF_2),[102] ^{19}F +58.9 (PF·NMe_2), +66.9 (PF_2),[102] ^1H -2.77.[102]

$N_5P_5(NMe_2)_{10}$. VII.[27] m. 230-236,[27] ^1H -3.62.[27]

$N_5P_5(OMe)_{10}$. X.[27] b_1 > 200,[27] ^1H -3.62.[27]

$N_5P_5(OCH_2CF_3)_{10}$. X.[27] b_{30} 206,[27] ^1H -4.27.[27]

$N_5P_5(OPh)_{10}$. X.[27] m. 129-130.[27]

D.2.14. Cyclohexaphosphazahexaenes

$N_6P_6F_{12}$. IV.[95] m. -45.5,[95] b_{760} 147.2,[95] d^{20} 1.8410,[95] n_D^{20} 1.3533,[95] IR,[95,468] ^{19}F +68.6,[102] photoelectron spectr.,[71] mass spectr.,[73] first ionization potl. 10.9 eV,[71] forms adduct $N_6P_6F_{12}\cdot 2SbF_5$: m. 92-94,[101] IR.[101]

$N_6P_6F_{11}Cl$. ($N_6P_6F_{11}\cdot NMe_2$ + HCl),[100,102] b_{25} 78,[102] ^{19}F +32.8 (PFCl), + 67.1 (PF_2).[100,102]

$N_6P_6F_{11}Br$. ($N_6P_6F_{11}\cdot NMe_2$ + HBr),[100,102] b_{25} 84-86,[102] ^{19}F +23.4 (PFCl), +67.0 (PF_2).[100,102]

$N_6P_6Cl_{12}$. I.[300,479] m. 92.3,[300] b_{26} 281-282,[479] log p mm = 9.576 - 4533/T (205-349°),[236] ΔH vap 20.7 Kcal./mole,[236] log p mm. = 11.66 - 5533/T (59.1-91.3°),[107] ΔH sub. 25.3 Kcal./mole,[107] heat of polymerization,[236] d^{20} 1.96,[300] IR,[300,468] Raman,[468] ^{31}P +16,[300] dynamic nuclear polarization (^{31}P),[147,148] NQR,[508] UV,[300] mass spectr.,[74] first ionization potl. 9.81 eV.,[71,72] magnetic anisotropy,[118] Cl^- exchange,[457] relative basicity to HCl,[72] forms adduct $N_6P_6Cl_{12}\cdot C_6Me_6$: m. 110.[127]

$N_6P_6Br_{12}$. I.[115] m. 153-154,[115] IR,[115] x-ray powder data,[115] mass spectr.[109]

$N_6P_6F_{11}\cdot NCS$. ($N_6P_6F_{12}$ + KSCN),[102] b_{20} 90-92,[102] IR,[102] ^{19}F +51.7 (PF·NCS), +67.1 (PF_2).[102]

$N_6P_6F_{11}\cdot C_6F_5$. ^{19}F.[99]

$N_6P_6F_{11}{\cdot}NMe_2$. VII,[100,102] VIII.[100,102] b_{14} 90-91,[102] IR,[102] ^{31}P -3.9 ($PF{\cdot}NMe_2$), +16.8 (PF_2),[102] ^{19}F +58.3 ($PF{\cdot}NMe_2$), +66.9 (PF_2),[102] ^{1}H -2.80.[102]

$N_6P_6F_{10}(NMe_2)_2$ (nongeminal). VII,[100,102] VIII.[100,102] b_{17} 151,[102] IR,[102] ^{31}P -0.9 ($PF{\cdot}NMe_2$), +20.5 (PF_2),[102] ^{19}F +59.0 ($PF{\cdot}NMe_2$), +66.4 (PF_2),[102] ^{1}H -2.78.[102]

$N_6P_6F_9(NMe_2)_3$ (nongeminal). VII,[100,102] VIII.[100,102] $b_{0.001}$ 112-115,[102] IR,[102] ^{31}P -4.2 ($PF{\cdot}NMe_2$), +16.5 (PF_2),[102] ^{19}F +59.5 ($PF{\cdot}NMe_2$), +66.3 (PF_2),[102] ^{1}H -2.77.[102]

$N_6P_6(NMe_2)_{12}$. VII.[27] m. 255-260,[27] ^{1}H -2.90,[27] x-ray study[501,503] at -170° shows puckered ring (analogous to boat form) with P-N (endo) 1.563 (0.010), P-N (exo) 1.669 (0.010), $P\hat{N}P$ 148, $N\hat{P}N$ 120°, forms adducts $[N_6P_6(NMe_2)_{12}CuCl]^+CuCl_2^-$, in which an x-ray study[312] shows that four endocyclic N-atoms are coordinated to Cu.

$N_6P_6(OMe)_{12}$. X.[27] m. 48,[27] ^{1}H -3.67.[27]

$N_6P_6(OCH_2CF_3)_{12}$. X.[27] m. 189,[27] ^{1}H -4.26.[27]

$N_6P_6(OPh)_{12}$. X.[27] m. 320-324.[27]

D.2.15. Cycloheptaphosphazaheptaenes

$N_7P_7F_{14}$. IV.[95] m. -61,[95] b_{760} 170.7,[95] ΔH vap 11.6 Kcal./mole,[95] ΔS_{vap} 26.2 cal./mole/deg.,[95] d^{20} 1.8496,[95] $n_D{}^{20}$ 1.3644,[95] IR,[95] mass spectr.,[73] photoelectron spectr.,[71] first ionization potl. 11.3 eV.[71]

$N_7P_7F_{13}{\cdot}C_6F_5$. ^{19}F.[99]

$N_7P_7Cl_{14}$. I.[300,479] m. 8-12,[300,479] b_{13} 289-294,[479] log p mm. = 10 - 5060/T (226-305°),[236] ΔH_{vap} 23.1 Kcal./mole,[236] heat of polymerization,[236] d^{20} 1.890,[300] IR,[300] ^{31}P +18,[300] dynamic nuclear polarization (^{31}P),[147,148] mass spectr.,[74] first ionization potl. 9.80 eV.,[71] magnetic anisotropy,[118] relative basicity to HCl.[72]

$N_7P_7(NMe_2)_{14}$. VII.[27] m. 230-233,[27] ^{1}H -2.92.[27]

$N_7P_7(OMe)_{14}$. X.[27] b_1 > 200,[27] ^{1}H -3.72.[27]

$N_7P_7(OCH_2CF_3)_{14}$. X.[27] Wax,[27] ^{1}H -4.26.[27]

$N_7P_7(OPh)_{14}$. X.[27] m. 340-344.[27]

D.2.16. Cyclooctaphosphazaoctaenes

$N_8P_8F_{16}$. IV.[95] m. -16.9,[95] b_{760} 192.8,[95] ΔH_{vap} 12.0 Kcal./mole,[95] ΔS_{vap} 25.8 cal./mole/deg.,[95] d^{20} 1.8567,[95] $n_D{}^{20}$ 1.3570,[95] mass spectr.,[73] photoelectron spectr.,[71] first ionization potl. 10.9 eV.[71]

$N_8P_8F_{15}{\cdot}C_6F_5$. ^{19}F.[99]

$N_8P_8Cl_{16}$. I.[300] m. 57-58,[300] d^{20} 1.99,[300] IR,[300] ^{31}P +18,[300] mass spectr.[74]

$N_8P_8(NMe_2)_{16}$. VII.[27] Gum,[27] ^{1}H -2.90.[27]

$N_8P_8(OMe)_{16}$. X.[27] m. 81,[27] ^1H -3.70,[27] x-ray study[368] shows that P-N ring consists of two six-atom segments joined by a "step"; all P-N bonds equal 1.561 (0.005), P-O 1.576 (0.005), PN̂P 136.7 (1.0), NP̂N 116.7° (0.4).
$N_8P_8(OCH_2CF_3)_{16}$. X.[27] b_{30} 254,[27] ^1H -4.21.[27]
$N_8P_8(OPh)_{16}$. X.[27] b_{42} 338.[27]

D.2.17. Higher Fluorocyclophosphazenes

$N_9P_9F_{18}$. IV. m. < -78, b_{760} 214.4, ΔHvap 12.7 Kcal/mole, ΔS vap. 26.1 cal./mole/deg., d^{20} 1.8589, n_D^{20} 1.3622, viscosity, IR.[95]
$N_{10}P_{10}F_{20}$. IV. m. -51, b_{760} 230.8, ΔHvap 13.5 Kcal./mole, ΔSvap. 26.9 cal./mole/deg., d^{20} 1.8638, n_D^{20} 1.3633, IR.[95]
$N_{11}P_{11}F_{22}$. IV. m. < -78, b_{760} 246.7, ΔHvap. 14.6 Kcal./mole, ΔSvap. 28.0 cal./mole/deg., d^{20} 1.8644, n_D^{20} 1.3645, IR.[95]
$N_{12}P_{12}F_{24}$. IV. $b_{12.65}$ 143-148,* IR.[95]
$N_{13}P_{13}F_{26}$. IV. $b_{12.65}$ 154-159,* IR.[95]
$N_{14}P_{14}F_{28}$. IV. $b_{12.65}$ 166-170,* IR.[95]
$N_{15}P_{15}F_{30}$. IV. $b_{1.7}$ 136-142,* IR.[95]
$N_{16}P_{16}F_{32}$. IV. $b_{1.85}$ 143-148,* IR.[95]
$N_{17}P_{17}F_{34}$. IV. $b_{1.85}$ 152-157,* IR.[95]

D.3. Miscellaneous Cyclophosphazenes

D.3.1. Cyclic Monophosphazenes

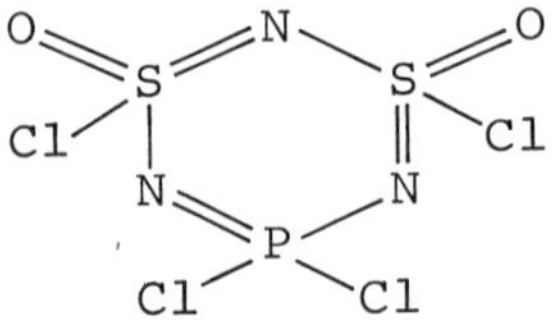

(Heat $Cl_3P{=}NSO_2Cl$ and irradiate with UV light),[498] m. 94.9-95.9,[498] ^{31}P -28.6,[292] x-ray study[499] shows that ring has a distorted chair conformation with P-N 1.585 (0.013), P-Cl 1.957 (0.006), NP̂N 115.3 (0.7).

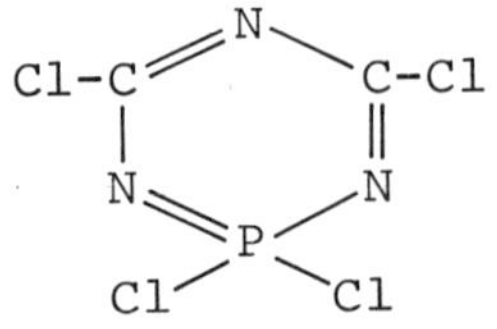

$[Na^+N(CN)_2^- + PCl_5]$[47] m. 141-143,[47] ^{31}P -55.6.[47]

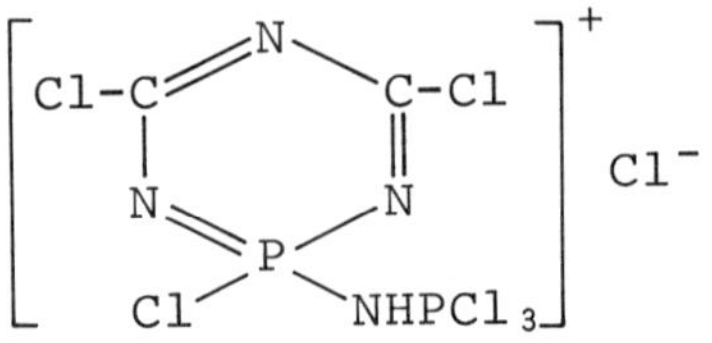

$[(H_2N)_2C{=}NCN + PCl_5$ at 65°],[47] m. 123-125,[47] ^{31}P -34.1 (PCl_3), -54.4 [$PCl(NHPCl_3)$],[47] with BCl_3 forms BCl_4^- salt,[47] m. 168-170.[47]

*Not pure samples.

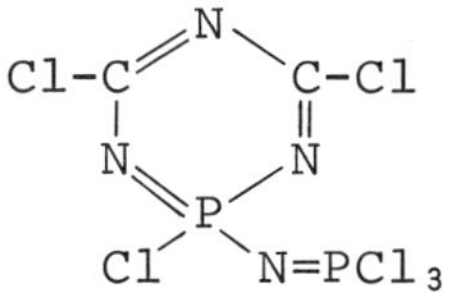

(As above at 90°),[47] m. 108-111,[47] ^{31}P -5.70 ($PCl{\cdot}N{=}PCl_3$), -22.7 (PCl_3).[47]

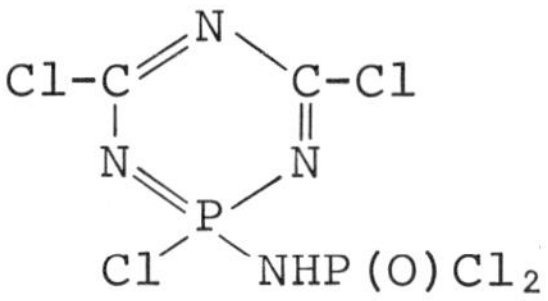

(Above deriv. + HCOOH),[47] m. 164-166,[47] ^{31}P -51.5 [$PCl(NHPOCl_2)$], -3.8 ($P(O)Cl_2$).[47]

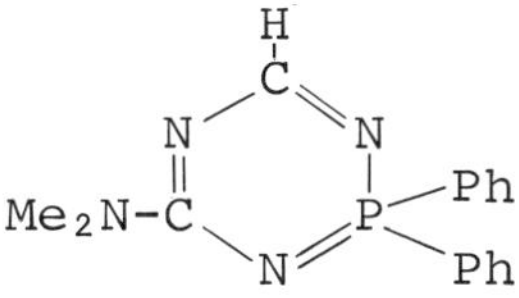

[$Me_2N(H_2N)C{=}N{-}PPh_2{=}NH$ + PhNHCH=NPh],[404] m. 146.5-147,[404] ^{31}P -26.4.[404] ^1H -3.07 (CH_3), -8.04 (CH).[404]

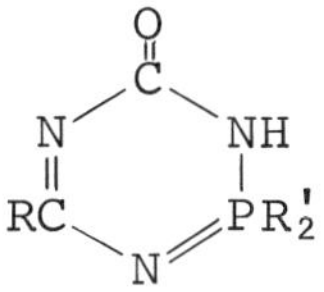

[$RC(NH_2){=}NCl$ + R'_2PNCO, and subsequent reaction with Et_3N].[270]
R = Ph, R' = OPh, m. 139-195.[270]
R = Ph, R' = Ph, m. 218-220.[270]
R = C_6H_4-p-Me, R' = OPh, m. 215-216.[270]
R = C_6H_4-p-Me, R' = Ph, m. 241-243.[270]

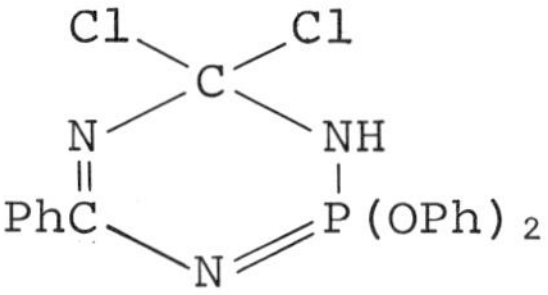

[Above deriv. (R = Ph, R' = OPh) with $POCl_3$],[270] viscous liquid, n_D^{20} 1.5963.[270]

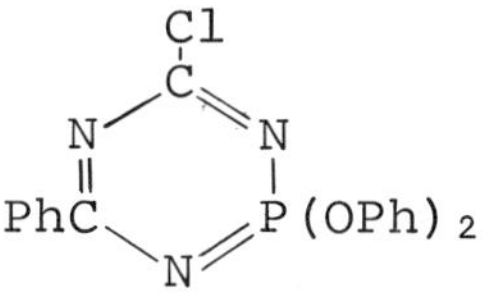

[Above deriv. + Et_3N],[270] m. 147-149.[270]

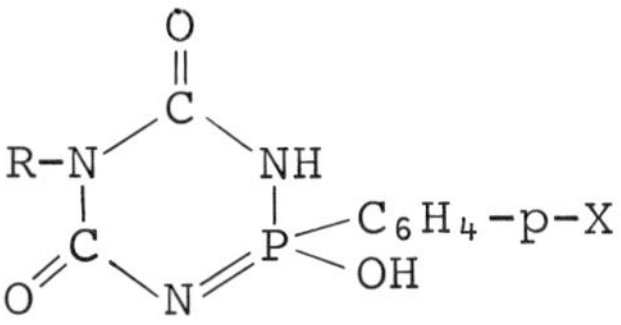

[Heat X-p-$C_6H_4P(O)(NCO)_2$ + RNH_2],[488]
X = H, R = H; m. 250, pK_a 4.89, IR.[488]
X = H, R = Me; m. 236.5, pK_a 5.45, IR, UV.[488]
X = H, R = Et; m. 225-226, IR.[488]
X = H, R = i-Pr; m. 202-203.[488]
X = H, R = n-Bu; m. 224-225.[488]
X = H, R = Ph; m. 202-203.[488]
X = Me, R = H; m. 244-245, pK_a 5.35, UV.[488]

X = Me, R = Me; m. 226-227, pK_a 5.50.[488]
X = NO_2, R = H; m. 194-196.[488]
X = NO_2, R = Me; m. 220-221, pK_a 4.75.[488]
X = OMe, R = Me; m. 238-239, pK_a 5.70.[488]
X = Cl, R = Me; m. 223-224.[488]

D.3.2. Cyclic Diphosphazadienes

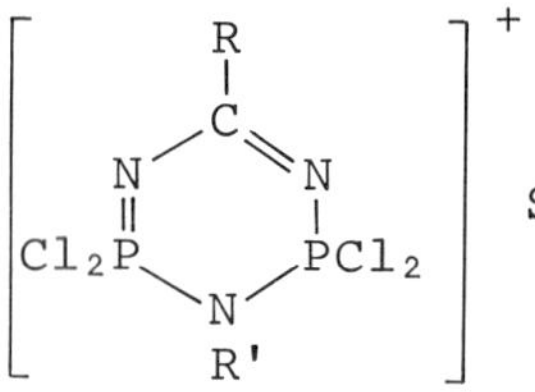

$SbCl_6^-$ {[RC(N=PCl_3)$_2$] $SbCl_6^-$ + R'NH_3Cl^-}. R = Me, R' = Me; m. 183-191 (d), ^1H (in PhCN) -3.72 (NMe), -2.45 (CMe).[401]

R = Me, R' = Ph; not purified, ^1H -2.52 (CMe).[401]
R = Ph, R' = Me; m. 188-192 (d), ^1H (in MeCN) -3.67 (NMe), forms adduct with CH_2Cl_2.[401]
R = Ph, R' = Ph; m. 201-215.[401]

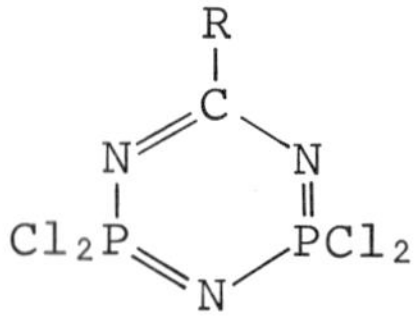

{[RC(N=PCl_3)$_2$]$^+$$SbCl_6^-$ + NH_4Cl} (R = Me, Ph).[401] [(Cl_3P=N-PCl_3)$^+$$SbCl_6^-$ + $(NH_2)_2Cl^-$] (R = Me, Ph, NMe_2).[403] R = Me, m. 39-40,[403] ^{31}P -39.3,[403] ^1H -2.26 (CMe).[403] R = Ph, m. 92-93,[401,403] ^{31}P -41.6,[403] hydrolyzes to $PhCN_3P_2Cl_3OH$: m. 240-280 (d),[403] pyrolysis gives $(NPCl_2)_n$ + PhCN.[403] R = NMe_2, m. 54-55,[403] ^{31}P -3.60,[403] ^1H -3.08 (NMe_2).[403]

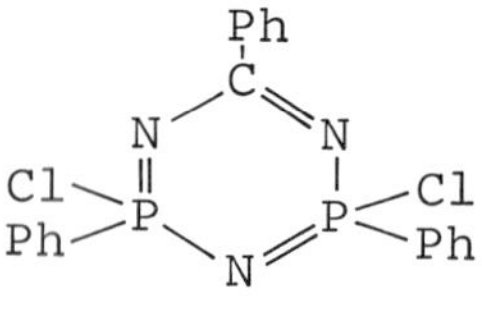

{[$PhPCl_2N$=$PPhCl_2$]$^+$$Cl^-$ + Ph$\overset{+}{C}$$(NH_2)_2Cl^-$}.[408] m. 127-132 (mixture of geometrical isomers),[408] ^{31}P -43.0, -41.5.[408]

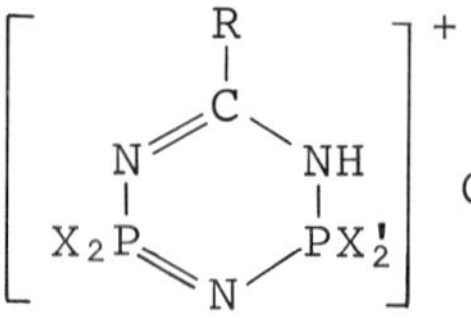

Cl^- [R$\overset{+}{C}$$(NH_2)_2Cl^-$ + (X_2PClN=$PClX_2'$)$^+$$Cl^-$].[403,407]

R = Me; X = X' = Me. m. 241-244,[407] ^{31}P -36.4,[407] ^{1}H -2.30 (CMe), -1.88 (PMe), -12.2 (NH).[407]
R = Me; X = Me, X' = Ph. m. 208-211,[407] ^{31}P -16.7 (PPh_2), -40.3 (PMe_2),[407] ^{1}H -2.43 (CMe), -2.00 (PMe).[407]
R = Me; X = X' = Ph. m. 192-195,[403,407] ^{31}P -18.9,[403,407] ^{1}H -2.63 (CMe), -15.2 (NH),[403,407] forms BF_4^- salt, m. 114-116 (d),[403,407] ^{31}P -20.4,[403,407] ^{1}H -1.80.[403]
R = Ph; X = X' = Ph. m. 220-227 (d),[403,407] ^{31}P -21.9,[403,407] ^{1}H -13.3 (NH).[403,407]
R = CH_2Ph, X = X' = Ph. m. 182-184,[403,407] ^{31}P -19.9,[403,407] ^{1}H -4.22 (CH_2Ph), -14.2 (NH).[403,407]

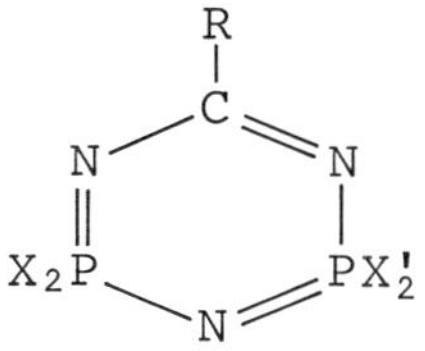

Synthesis from (a) [hydrochlorides (above) + NH_3],[407] (b) {$H_2NC(R)=NH$ + [$ClP(X_2)=N-P(X_2)Cl]^+Cl^-$}.[407]

R = Me; X = X' = Me. (a),[407] m. 120-122,[407] ^{31}P -29.9,[407] ^{1}H -1.93 (CMe), -1.43 (PMe).[407]
R = Me; X = Me, X' = Ph. (a),[407] m. 148-150,[407] ^{31}P -14.7 (PPh_2), -31.3 (PMe_2),[407] ^{1}H -2.08 (CMe), -1.48 (PMe).[407]
R = Me; X = X' = Ph.[404,407] (a), or with NaOMe,[404,407] m. 175-177,[404,407] ^{31}P -15.6, ^{1}H -2.17 (CMe).[404,407]
R = Ph; X = X' = Ph. (a), (b),[404,407] m. 239-241,[404,407] ^{31}P -17.3.[404,407]
R = CH_2Ph; X = X' = Me. (b),[407] m. 119-120 (sublimes),[407] ^{31}P -31.1,[407] ^{1}H -3.45 (CH_2), -1.42 (PMe).[407]
R = CH_2Ph; X = X' = Ph. (a), (b),[404,407] m. 113-114,[404,407] ^{31}P -15.9,[404,407] ^{1}H -3.67 (CH_2).[404,407]
R = NH_2; X = X' = Ph. (b),[406,407] m. 214-216,[406,407] ^{31}P -21.7,[406,407] ^{1}H -1.85 (NH_2).[406,407]
R = NMe_2; X = X' = Me. (b),[407] m. 132 (sublimes), ^{31}P -31.3,[407] ^{1}H -2.86 (NMe_2),[407] -1.31 (PMe).[407] X-ray study shows ring containing P is planar, with C-atoms in NMe_2 group in same plane.[264]
R = NMe_2; X = X' = Ph. (b),[406,407] m. 190,[406,407] ^{31}P -19.3,[406,407] ^{1}H -3.17 (NMe_2).[406,407]
R = H, X = X' = Me. [$H_2NP(Me_2)=N-PMe_2=NH$ + PhNH-CH=NPh],[404,407] m. 121-122,[404,407] ^{31}P -27.1,[404,407] ^{1}H -1.40 (PMe), -7.57 (CH).[404,407]
R = H, X = X'= Ph. [$H_2NP(Ph_2)=NH$ + PhNH-CH=NPh],[404.407] m. 242-243,[404,407] ^{31}P -14.4,[404,407] ^{1}H -8.43 (CH).[404,407]

D.3.3. Amino Derivatives of

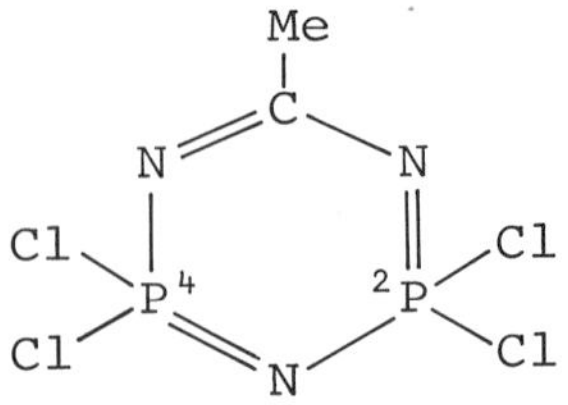

$N_3CMeP_2(NC_2H_4)_4$. VII.[409] m. 106-108,[409] ^{31}P -41.1,[409] ^1H -2.06 (CH_2), -1.98 (CMe).[409]
$N_3CMeP_2(NMe_2)_4$. VII.[409] m. 54-55,[409] ^{31}P -24.3,[409] ^1H -1.95 (NMe_2), -1.72 (CMe).[409]
$N_3CMeP_2Cl_2(NH\text{-}t\text{-}Bu)_2$ (2,2:4,4). VII.[410] m. 140-145 (d),[410] ^{31}P -34.9 (PCl_2), -13.6 [$P(NH\text{-}t\text{-}Bu)_2$],[410] ^1H -2.24 (CMe), -1.32 (t-Bu), -3.52 (NH).[410]
$N_3CMeP_2Cl(NH\text{-}t\text{-}Bu)_3$. VII.[410] m. 130-135 (mixed with deriv. below),[410] ^{31}P -12.4,[410] ^1H -2.02 (CMe), -1.26 (t-Bu), -2.84 (NH).[410]
$N_3CMeP_2(NH\text{-}t\text{-}Bu)_4$. VII.[410] m. 130-135[410] (mixed with deriv. above), ^{31}P -13.6,[410] ^1H -1.90 (CMe), -1.26 (t-Bu), -2.30 (NH).[410]
$N_3CMeP_2(NH\text{-}t\text{-}Bu)_2(OMe)_2$ (2,2:4,4). X.[410] m. 102-104,[410] ^{31}P -19.6 [$P(NH\text{-}t\text{-}Bu)_2$], -26.6 [$P(OMe)_2$],[410] ^1H -2.00 (CMe), -1.29 (t-Bu), -2.54 (NH), -3.62 (OMe).[410]

D.3.4. Amino-Derivatives of

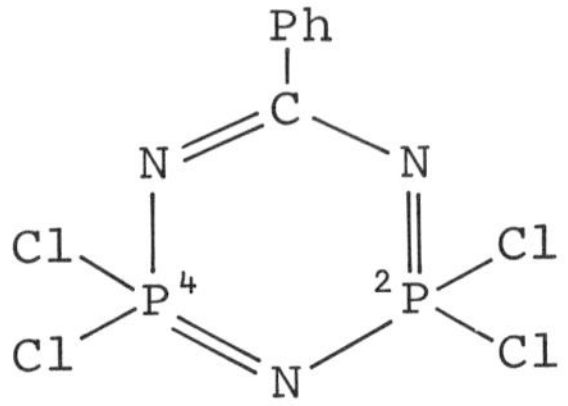

$N_3CPhP_2(NH_2)_4$. VII. m. 70 (d), ^{31}P -18.7, ^1H (in Me_2SO) -4.10 (NH_2).[409]
$N_3CPhP_2(NC_2H_4)_4$. VII. m. 128-130, ^{31}P -43.5, ^1H -2.10 (NC_2H_4).[409]
$N_3CPhP_2(NMe_2)_4$. VII. m. 102-104, ^{31}P -32.3, ^1H -2.45 (NMe_2).[409]
$N_3CPhP_2(NC_4H_8)_4$. VII. m. 168-171, ^1H -1.72 (β-CH_2), -2.10 (α-CH_2).[409]
$N_3CPhP_2(NC_4H_8O)_4$. VII. m. 130, ^{31}P -25.8, ^1H -3.12, -3.60 (CH_2).[409]
$N_3CPhP_2Cl_2(NH\text{-}t\text{-}Bu)_2$ (2,2:4,4). VII. m. 140-142, ^{31}P -40.6 (PCl_2), -13.2 [$P(NH\text{-}t\text{-}Bu)_2$], ^1H -1.32 (t-Bu), -2.62 (NH).[410]
$N_3CPhP_2Cl_2(NC_5H_{10})_2$ (2,4:2,4). VII. m. 155-157. ^{31}P -38.3, ^1H -3.29 (α-CH_2), -1.62 (β,γ-CH_2).[410]

$N_3CPhP_2(NH\text{-}t\text{-}Bu)_2(NMe_2)_2$ (2,2:4,4). VII. m. 125-127, ^{31}P -18.8 [$P(NH\text{-}t\text{-}Bu)_2$], -32.3 [$P(NMe_2)_2$], ^1H -1.32 (t-Bu), -2.47 (NH), -2.63 (NMe_2).[410]
$N_3CPhP_2(NC_5H_{10})_2(NMe_2)_2$ (2,4:2,4). VII. m. 150-151, ^{31}P -30.1, ^1H -3.07 (α-CH_2), -1.55 (β,γ-CH_2), -2.62 (NMe_2).[410]

D.3.5. Amino-Derivatives of

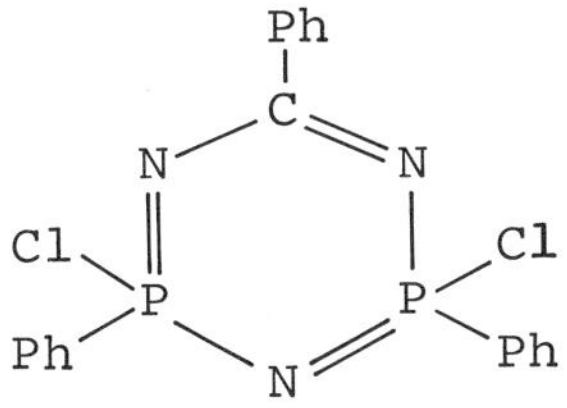

(cis- and trans-Isomer Mixture)

$N_3CPhP_2Ph_2(NH_2)_2$ (2,4:2,4). VII. m. 110-115,* ^{31}P -23.3, -21.3, ^1H -3.45 (NH_2), polymerizes on heating, hydrolyzes to $H_2NC(Ph)=N\text{-}P(NH_2)(Ph)=N\text{-}P(O)(Ph)(NH_2)$ m. 152-153.[409]
$N_3CPhP_2Ph_2(NMe_2)_2$. VII. m. 111-116,* ^{31}P -28.3 (broad), ^1H -2.63, -2.43 (NMe_2), gave one pure isomer m. 127, ^1H -2.63 (NMe_2).[409]
$N_3CPhP_2Ph_2(NEt_2)_2$. VII. m. 113-121,* ^{31}P -26.0, -24.8, ^1H -3.00 (NCH_2).[409]

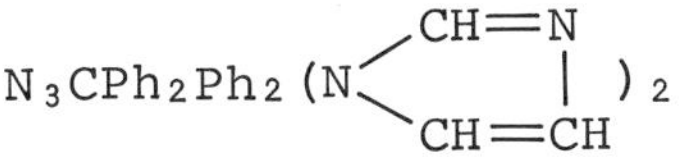

VII. m. 172-177, ^{31}P -18.6.[409]

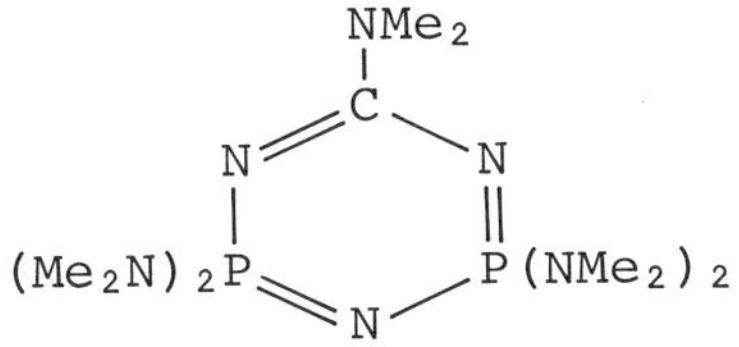

VII. m. 79-81, ^{31}P -32.9, ^1H -2.98 ($CNMe_2$), -2.55 ($PNMe_2$).[409]

D.3.6. Alkoxy- and Aryloxy-Derivatives of

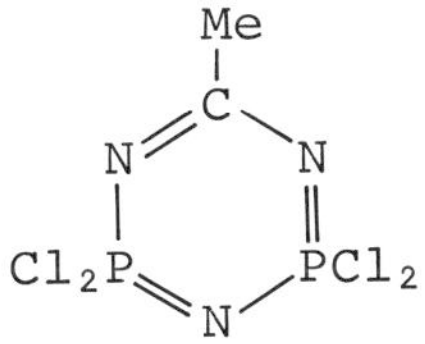

$N_3CMeP_2Cl(OMe)_3$. X. $b_{0.01-0.001}$ 85-105,† ^{31}P -39.4, -33.3.[411]

*Mixture of isomers.
†Not obtained pure.

$N_3CMeP_2(OMe)_4$. X. $b_{0.001}$ 90-100,† ^{31}P -26.6, ^1H -3.62 (OMe), -2.10 (CMe).[411]
$N_3CMeP_2Cl(O\text{-}n\text{-}Pr)_3$ and $N_3CMeP_2(O\text{-}n\text{-}Pr)_4$. X. Mixture $b_{0.001}$ 130-135, ^{31}P -38.4, -22.9 (both from tri-n-Pr-deriv.), -24.2 (tetra-n-Pr-deriv.).[411]
$N_3CMeP_2Cl(O\text{-}n\text{-}Bu)_3$. X. $b_{0.01-0.001}$ 170-190, ^{31}P -39.4, -21.0, contains deriv. below.[411]
$N_3CMeP_2(O\text{-}n\text{-}Bu)_4$. X. $b_{0.001}$ 155-160, ^{31}P -24.4, ^1H -1.98 (CMe).[411]
$N_3CMeP_2(OC_6H_4\text{-}p\text{-}Me)_4$. X. m. 90-91, ^{31}P -17.3, ^1H 2.26 (tolyl-Me), -2.12 (CMe).[411]

D.3.7. Alkoxy- and Aryloxy-Derivatives of

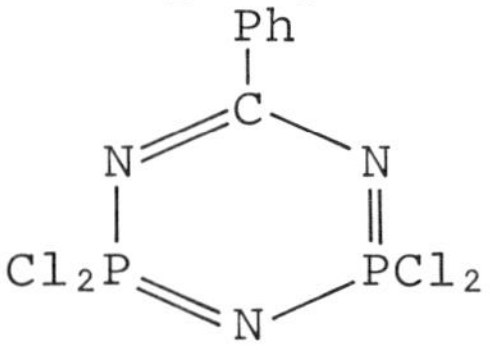

$N_3CPhP_2(OMe)_4$. X. m. 61-62, $b_{0.001}$ 170-175, ^{31}P -28.9, ^1H -3.67 (OMe).[411]
$N_3CPhP_2Cl(OEt)_3$. X. $b_{0.01-0.001}$ 150-175, ^{31}P -39.5, -22.6, contains deriv. below.[411]
$N_3CPhP_2(OEt)_4$. X. $b_{0.001}$ 160-170, ^{31}P -25.8, ^1H -4.04 (CH_2).[411]
$N_3CPhP_2Cl(O\text{-}n\text{-}Pr)_3$. X. ^{31}P -38.4, -22.8, contains deriv. below.[411]
$N_3CPhP_2(O\text{-}n\text{-}Pr)_4$. X. $b_{0.01}$ 175-190, ^{31}P -26.4, ^1H.[411]
$N_3CPhP_2(O\text{-}n\text{-}Bu)_4$. X. $b_{0.01}$ 195-205, ^{31}P -24.4, ^1H.[411]
$N_3CPhP_2(OC_6H_4\text{-}p\text{-}Me)_4$. X. m. 121-122, ^{31}P -22.2, ^1H -2.25 (tolyl-Me).[411]

D.3.8. Alkoxy-Derivatives of

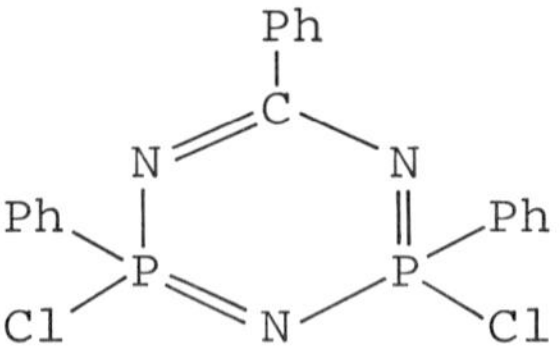

$N_3CPhP_2Ph_2(OMe)_2$. X. m. 55-70 (mixture of isomers), ^{31}P -33.4, -30.8, ^1H -3.72, -3.45.[411]
$N_3CPhP_2Ph_2(O\text{-}n\text{-}Bu)_2$. X. $b_{0.001}$ 210-215 (mixture of isomers), ^{31}P -31.2, -28.8, ^1H.[411]

†Not obtained pure.

D.3.9. Alkoxy-Derivatives of

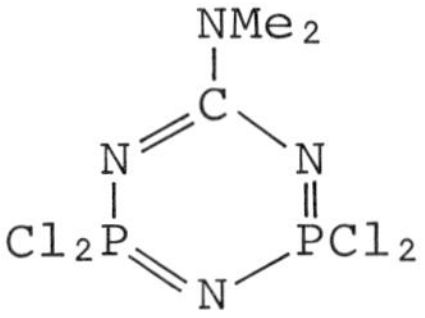

$N_3C(NMe_2)P_2Cl(OMe)_3$. X. m. 48-57, ^{31}P -41.4, -26.3, ^{1}H -3.86, -3.62 (OMe), -3.05 (NMe_2),[411] contains derivative below.
$N_3C(NMe_2)P_2(OMe)_4$. X. m. 68-69, ^{31}P -30.6, ^{1}H -3.58 (OMe), -3.02 (NMe_2).[411]
$N_3C(NMe_2)P_2Cl(OEt)_3$. X. $b_{0.01}$ 100-130, ^{31}P -39.2, -23.3,[411] contains derivative below.
$N_3C(NMe_2)P_2(OEt)_4$. X. $b_{0.01}$ 120-125, ^{31}P -26.2, ^{1}H -3.88 (CH_2), -3.00 (NMe_2).[411]

D.3.10. Cyclophosphazenes Containing Boron or Sulfur Ring Atoms.

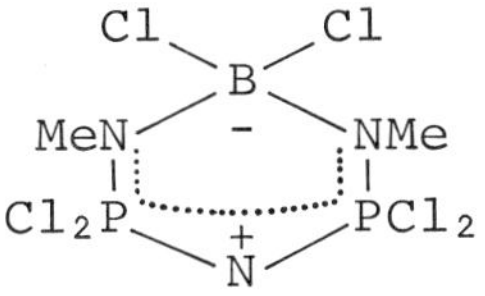

$[(Cl_3P{=}N{-}PCl_3)^+Cl^- + Me\overset{+}{N}H_3Cl^- + BCl_3]$,[48] $(PCl_5 + Me\overset{+}{N}H_3Cl^- + BCl_3)$,[48] m. 140-145,[48] ^{31}P -28,[48] conductivity.[48]

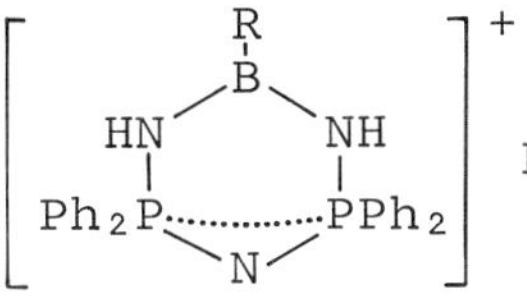 RBX_3^-

$[2RBX_2 + (H_2NPPh_2{\cdots}N{\cdots}PPh_2NH_2)^+ Cl^-]$.[446]
R = Cl, X = Cl. m. 204,[446] IR.[446]
R = Ph, X = Cl. m. > 340,[446] IR.[446]
R = Br, RBX_3^- = $BClBr_3^-$. m. 204,[446] IR.[446]

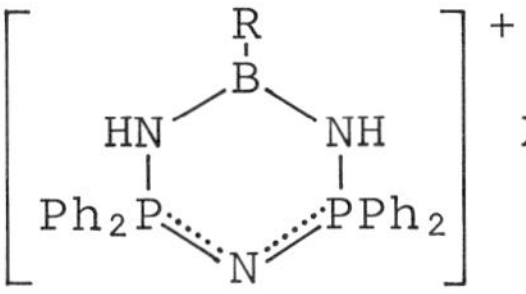 X^-

R = Cl, X = Cl. (pyrolysis of BCl_4^- salt above)[446] m. 302-304,[446] IR.[446]

R = Ph, X = Cl. $[PhBCl_2 + (H_2NPPh_2{\cdots}N{\cdots}PPh_2NH_2)^+Cl^-]$,[446] m. 323-325,[446] IR.[446]
R = Br, X = Cl. $[BBr_3 + (H_2NPPh_2{\cdots}N{\cdots}PPh_2NH_2)^+Cl^-]$,[446] m. 201 (d),[446] IR.[446]

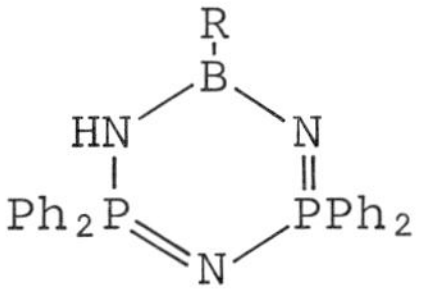

$[RBCl_2 + (H_2NPPh_2{\cdots}N{\cdots}PPh_2NH_2)^+Cl^-]$.[446]
R = Cl. m. 304-306,[446] IR.[446]
R = Ph. m. 210,[446] IR.[446]
R = Br. $[BBr_3 + (H_2NPPh_2{\cdots}N{\cdots}PPh_2NH_2)^+ Cl^-]$,[446] m. 328-330,[446] IR.[446]

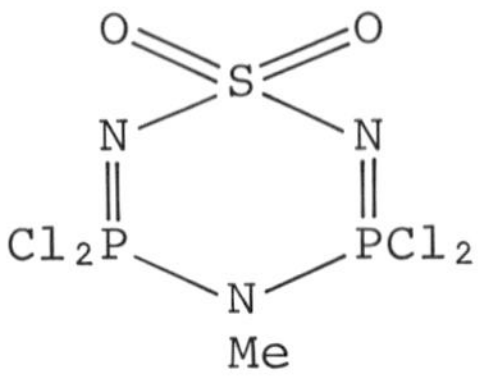

$[SO_2(N{=}PCl_3)_2 + (Me_3Si)_2NMe]$,[40] m. 177,[40] ^{31}P -15.1.[40]

NH_4^+ 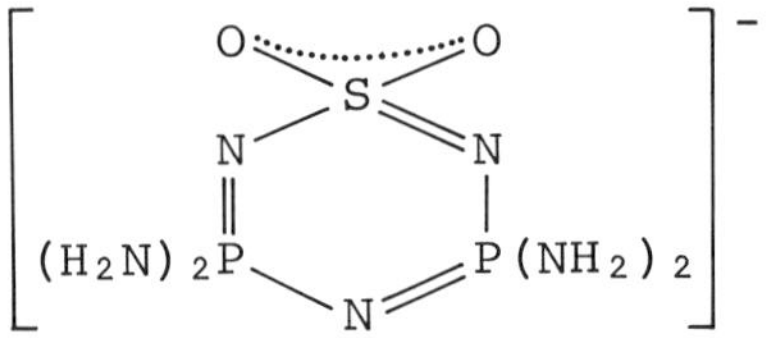

$[SO_2(N{=}PCl_3)_2 + NH_3]$.[40]

Ag^+ and K^+ salts also isolated.[40]

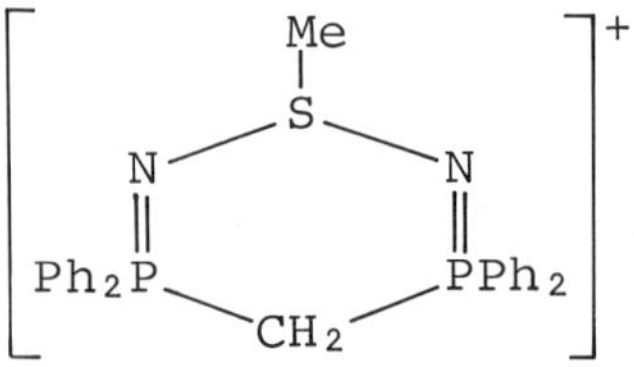 Br^-

{Pyrolysis of 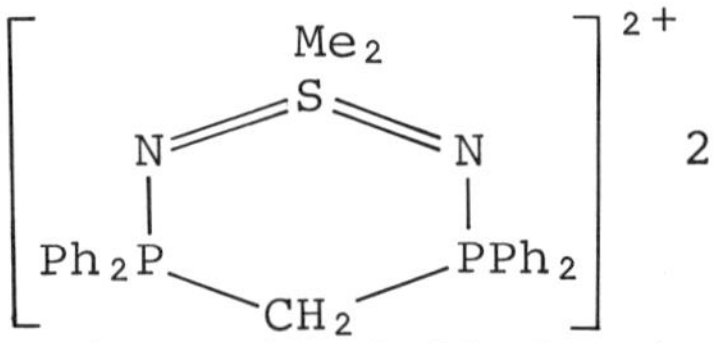 $2Br^-$}.[30,31]

m. 255-257(d),[30,31] ^1H (Me_2SO) -3.20 (SMe), -4.85 (CH_2).[449] with NH_3 gives

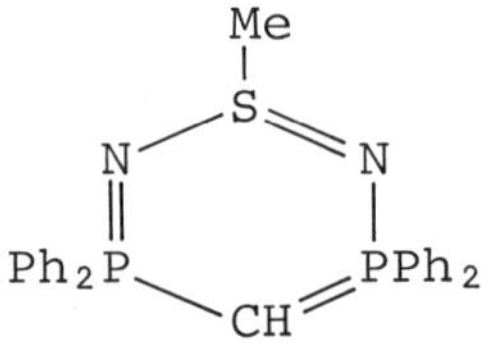

m. 140-142.[30,31] ^1H -2.80 (SMe), -1.71 (CH).[449]

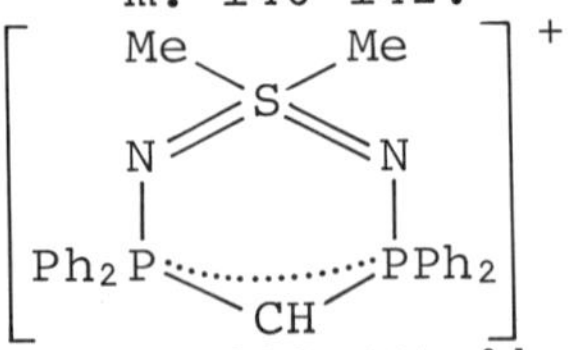 Br^- {NH_3 + 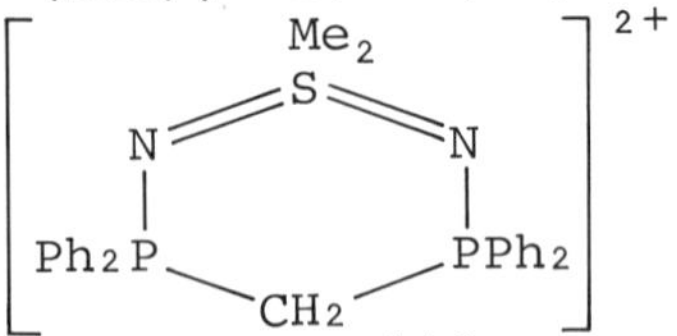 $2Br^-$},[31]

m. 200 (d),[31] ^1H (Me_2SO) -3.76 (SMe).[449]

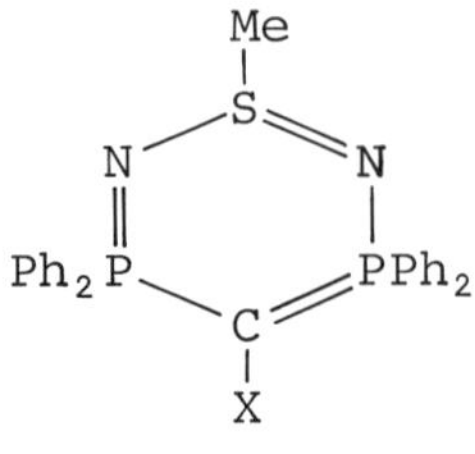

X = Br. ^1H -2.81 (SMe).[449]
X = Me. ^1H -2.57 (SMe), -1.69 (CMe).[449]

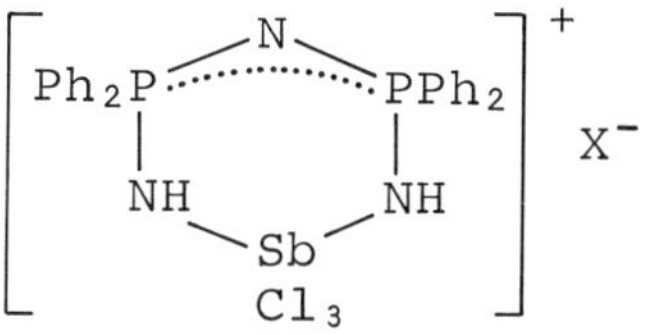

{[$Ph_2P(NH_2)\cdots N\cdots P(Ph_2)NH_2$] Cl + $SbCl_5$}.[422]

X = Cl. m. 252-254,[422] IR,[422] ^{31}P -2.81,[422] conductivity.[422]
X = $SbCl_6$. m. 143-144,[422] IR,[422] conductivity.[422]

D.3.11. Cyclophosphazenes Which Form Part of a Chelate Structure

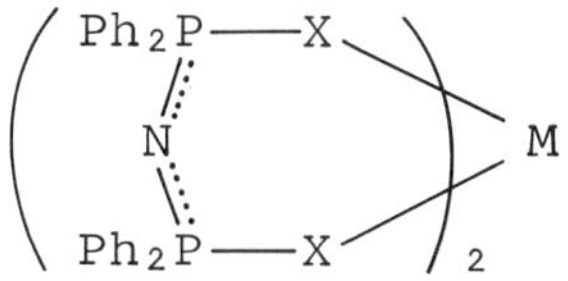

X = NH, M = Cu. [$(Ph_2PNH_2)_2\overset{+}{N}Cl^-$ + $CuCl_2$ + NaOMe, then reaction with H_2O],[261 402] m. 178-180 (d),[402] ESR.[261]

X = O, M = Cu. [$(Ph_2PO)_2NH$ + $Cu(OOCCH_3)_2$],[261] m. 271-272,[261] ESR,[261] forms adduct with two moles of pyridine.[261]
X = O, M = Sn. [$Sn(OOCCH_3)_2$ + $(Ph_2PO)_2NH$],[412] m. 215-217,[412] ^{31}P -17.9.[412]
X = O, M = Zn. [$ZnCl_2$ + $(Ph_2PO)_2NH$],[402] m. 228-229.5.[402]
X = S, M = Ni. [$(Ph_2PS)_2\bar{N}K^+$ + $NiCl_2$],[402] m. 265-266 (d).[402]
X = S, M = Cd. [$(Ph_2PS)_2\bar{N}K^+$ + $CdCl_2$],[402] m. 309-311 (d).[402]

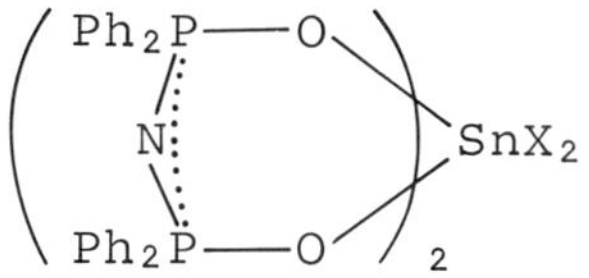

X = Cl. [boil $(Ph_2PO)_2NH\cdot SnCl_4$ in water], [melt $(Ph_2PO)_2NH\cdot SnCl_4$ + $(PhPO)_2NH$], [$(Ph_2PO)_2NNa$ + $SnCl_4$], {Cl_2 + [$(Ph_2PO)_2N]_2Sn$}, m. 325-327, ^{31}P -28.8.[412]

X = Br. (By methods analogous to X = Cl), m. 283-285, ^{31}P -28.4.[412]
X = I. [$(Ph_2PO)_2NH$ + SnI_4], m. 258-260 (d), ^{31}P -28.0, -26.3.[412]

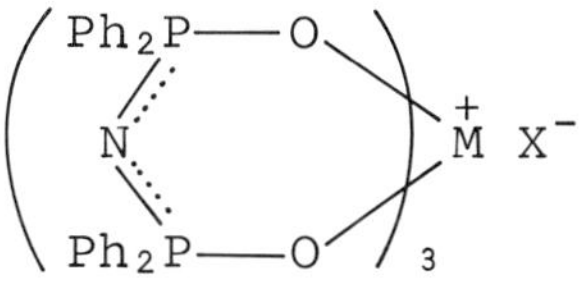

M = Si, X = Cl. [$SiCl_4$ + $(Ph_2PO)_2NH$], m. 300, ^{31}P -29.4, Cl^- may be replaced to give the following derivatives: X = Br_3, m. 360 (d); X = $SbCl_6$, m. 355 (d); X = $CuCl_3$, m. 305 (d); X = NO_3, m. 250.[413]
M = Ge, X = Cl. [$GeCl_4$ + $(Ph_2PO)_2NH$], m. 264, ^{31}P -31.3, Cl^- replacement gives X = $SbCl_6$, m. 315 (d), X = $CuCl_3$, m. 320 (d).[413]

$$M = Sn, X = Cl. \quad \left[\left(\begin{array}{c} Ph_2P\text{---}O \\ N \\ Ph_2P\text{---}O \end{array} \right)_2 SnCl_2 + NaN(Ph_2PO)_2 \right],$$

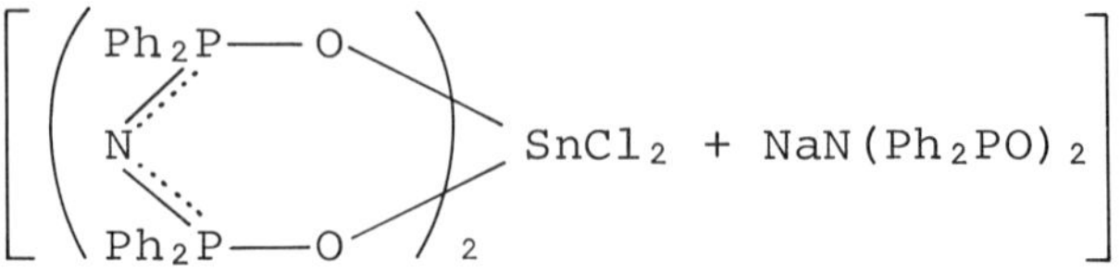

m. 178, ^{31}P -35.7,[413] readily disproportionates to

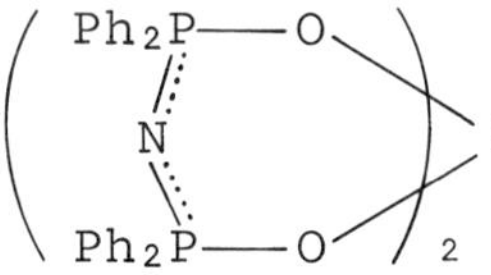

$$\left(\begin{array}{c} Ph_2P\text{---}O \\ N \\ Ph_2P\text{---}O \end{array} \right)_2 SnCl_2, \quad Cl^- \text{ replacement gives } X = I,$$

m. 380, X = ICl_2, m. 279, X = I_3, m. 365 (d), X = $I_8/2$, m. 370, X = Br_3, m. 284, X = $SbCl_6$, m. 270, X = $CuCl_3$, m. 322 (d), X = HgI_3, X = NO_3, m. 332, X = ClO_4, m. 400 (d), X = MnO_4, X = OH, m. 188.[413]

$$\begin{array}{c} Ph_2P\text{---}O \quad\quad O\text{---}C\text{---}Me \\ N \quad Be \quad\quad CH \\ Ph_2P\text{---}O \quad\quad O\text{---}C\text{---}Me \end{array}$$

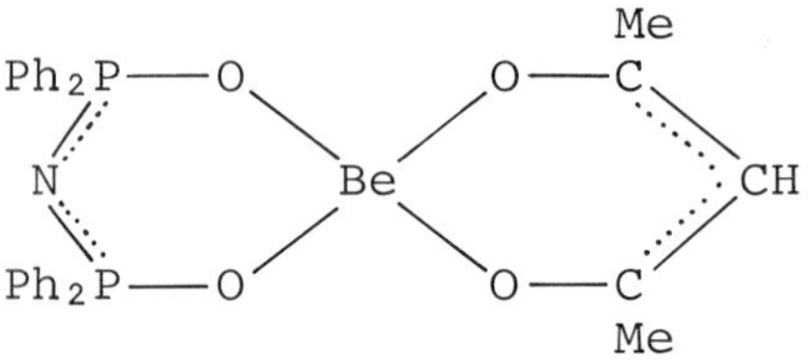

[Be(acac)$_2$ + (Ph$_2$PO)$_2$NH], m. 193-193.5, IR, ^{31}P -28.8.[363]

$$\left(\begin{array}{c} Ph_2P\text{---}O \\ N \\ Ph_2P\text{---}O \end{array} \right)_2 Be$$

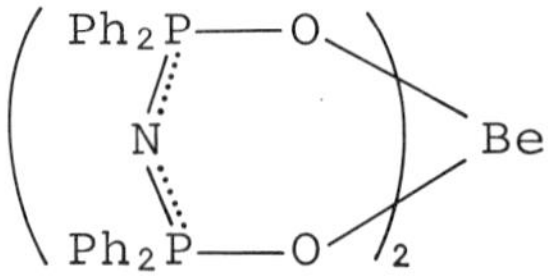

[Above deriv. + (Ph$_2$PO)$_2$NH], m. 237-238, IR.[363]

$$\begin{array}{c} Ph_2PO \\ | \\ Ph_2P\text{---}N \quad\quad O\text{---}C\text{---}Me \\ N \quad Be \quad\quad CH \\ Ph_2P\text{---}N \quad\quad O\text{---}C\text{---}Me \\ | \\ Ph_2PO \end{array}$$

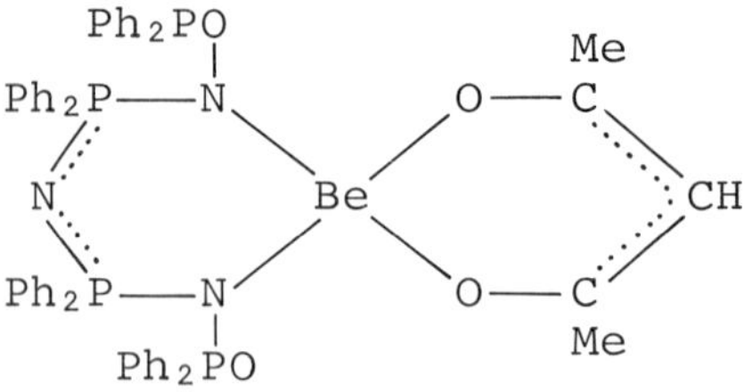

[Be(acac)$_2$ + Ph$_2$PO(NPPh$_2$)$_3$OH], m. 188-189, IR, ^{31}P -10.6, O.[363]

$$\left(\begin{array}{c} Ph_2PO \\ | \\ Ph_2P\text{---}N \\ N \\ Ph_2P\text{---}N \\ | \\ Ph_2PO \end{array} \right)_2 Be$$

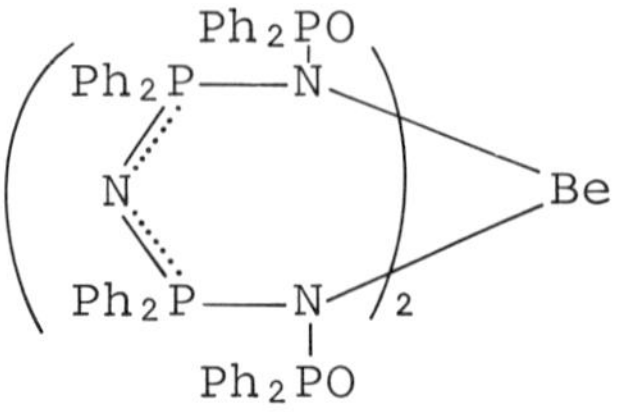

[Above deriv. + PhPO(NPPh$_2$)$_3$OH], m. 359-361, IR, ^{31}P -6.3, +2.6.[363]

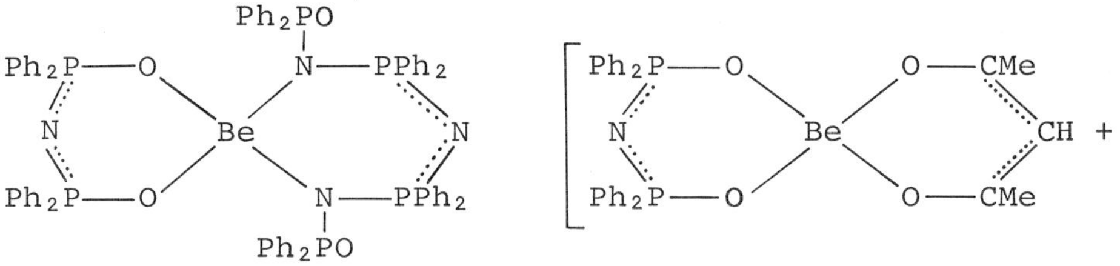

$Ph_2PO(NPPh_2)_3OH]$, m. 357-359, IR.[363]

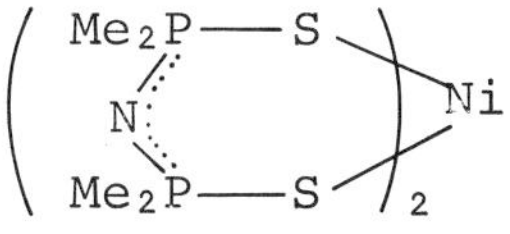

$[Et_4N^+NiCl_4^- + NaN(Me_2PS)_2]$,[103] m. 194-195,[103] paramagnetic (3.40 BM),[103] x-ray study[103] shows that Ni is tetrahedrally coordinated, and that P-N 1.606 (0.035), P$\hat{N}$P 127.3 (1.5).

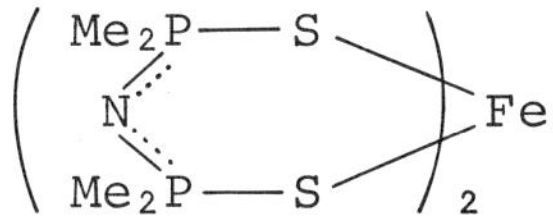

X-ray study[104] shows that Fe is tetrahedrally coordinated, and that P-N 1.591 (0.010), P$\hat{N}$P 132.3 (3.8)°.

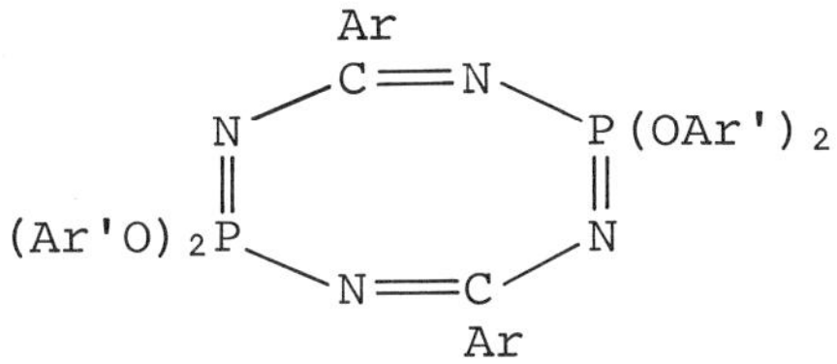

$[ArC(:NCl)NH_2 + P(OAr')_3$, then $Et_3N]$,[135]

Ar	Ar'	m.	
Ph	Ph	139-140	IR, all ref. 135.
Ph	C_6H_4-p-Me	121-122	
Ph	C_6H_4-p-Cl	124-125	
C_6H_4-p-Cl	Ph	156-157	
C_6H_4-p-Cl	C_6H_4-p-Me	154-155	
C_6H_4-p-Br	Ph	160-161	
C_6H_4-p-Me	Ph	156-158	

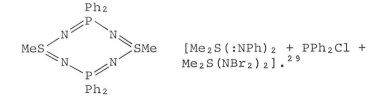

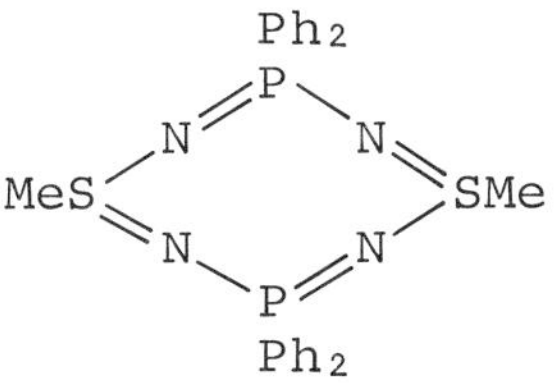

$[Me_2S(:NPh)_2 + PPh_2Cl + Me_2S(NBr_2)_2]$.[29]

ACKNOWLEDGEMENTS

We are grateful to Dr. M. Woods for his help in checking the content of the chapter, and to Mrs. B. M. Salmon for typing the manuscript.

(received April 5, 1971)

REFERENCES

1. Acock, K.G., R.A. Shaw, and F. B. G. Wells, J. Chem. Soc., 1964, 121.
2. Ahmed, F.R., P. Singh, and W. H. Barnes, Acta Cryst., 25B, 316 (1969).
3. Allcock, H. R., J. Am. Chem. Soc., 85, 4050 (1963).
4. Allcock, H. R., J. Am. Chem. Soc., 86, 2591 (1964).
5. Allcock, H. R., U. S. Pat. 3,356,769; C. A. 69, 192,324, (1968).
6. Allcock, H. R., and R. J. Best, Can. J. Chem., 42, 447 (1964).
7. Allcock, H. R., and W. J. Birdsall, J. Am. Chem. Soc., 91, 7541 (1969).
8. Allcock, H. R., G. F. Konopski, R. L. Kugel, and E. G. Stroh, Chem. Comm., 1970, 985.
9. Allcock, H. R., and R. L. Kugel, Inorg. Chem., 5, 1016 (1966).
10. Allcock, H. R., and R. L. Kugel, J. Am. Chem. Soc., 91, 5452 (1969).
11. Allcock, H. R., and D. P. Mack, Chem. Comm., 1970, 685.
12. Allcock, H. R., and L. A. Siegel, J. Am. Chem. Soc., 86, 5140 (1964).
13. Allcock, H. R., and L. A. Siegel, U. S. Pat. 3,356,768; C. A. 69, 86,996 (1968).
14. Allcock, H. R., M. T. Stein, and J. A. Stanko, Chem. Comm., 1970, 944.
15. Allcock, H. R., and W. M. Thomas, U. S. Pat. 3,329,663; C. A. 67, 64,883 (1967).
16. Allcock, H. R., and E. J. Walsh, J. Am. Chem. Soc., 91, 3102 (1969).
17. Allcock, H. R., and E. J. Walsh, Chem. Comm., 1970, 580.
18. Allen, C. W., Chem. Comm., 1970, 152.
19. Allen, C. W., J. B. Faught, T. Moeller, and I. C. Paul, Inorg. Chem., 8, 1719 (1969).
20. Allen, C. W., and T. Moeller, Inorg. Chem., 7, 2177 (1968).
21. Allen, C. W., I. C. Paul, and T. Moeller, J. Am. Chem. Soc., 89, 6361 (1967).
22. Allen, C. W., F. Y. Tsang, and T. Moeller, Inorg. Chem., 7, 2183 (1968).

23. Allen, G., M. Barnard, J. Emsley, N. L. Paddock, and R. F. M. White, Chem. and Ind. (London), 1963, 952.
24. Allen, G., J. Dyson, and N. L. Paddock, Chem. and Ind. (London), 1964, 1832.
25. Allen, G., C. J. Lewis, and S. M. Todd, Polymer, 11, 31 (1970).
26. Allen, G., C. J. Lewis, and S. M. Todd, Polymer, 11, 44 (1970).
27. Allen, G., D. J. Oldfield, N. L. Paddock, F. Rallo, J. Serregi, and S. M. Todd, Chem. and Ind. (London), 1965, 1032.
28. Ansell, G. B., and G. J. Bullen, Chem. Comm., 1966, 430.
29. Appel, R., and W. Eichenhofer, Z. Naturforsch., 24B, 1659 (1969).
30. Appel, R., and D. Häussgen, Angew. Chem., 6, 560 (1967).
31. Appel, R., D. Häussgen, and B. Ross, Z. Naturforsch., 22B, 1354 (1967).
32. Aroney, M. J., R. J. W. LeFevre, D. S. N. Murphy, G. J. Peacock, and J. D. Saxby, J. Chem. Soc. (B), 1966, 657.
33. Audrieth, L. F., and R. J. A. Otto, U. S. Pat. 3,048,631; C. A. 59, 8654 (1963).
34. Audrieth, L. F., and D. B. Sowerby, Chem. and Ind. (London), 1959, 749.
35. Audrieth, L. F., R. S. Steinmann, and A. D. F. Toy, Chem. Rev., 32, 109 (1943).
36. Azhikina, Yu. V., M. Ya Koroleva, B. M. Maslenikov, and L. Ya Kulikova, Izv. Akad. Nauk SSSR Neorg. Mater., 4, 1711 (1968); C. A. 70, 53,572 (1969).
37. Babaeva, A. V., and G. V. Derbisher, Zh. Neorgan. Khim., 10, 296 (1965); Russ. J. Inorg. Chem., 10, 156 (1965).
38. Bailey, J. V., and R. E. Parker, Chem. and Ind. (London), 1962, 1823.
39. Bean, N. E., and R. A. Shaw, Chem. and Ind. (London), 1960, 1189.
40. Becke-Goehring, M., K. Bayer, and T. Mann, Z. Anorg. Allgem. Chem., 346, 143 (1966).
41. Becke-Goehring, M., and B. Boppel, Z. Anorg. Allgem. Chem., 322, 239 (1963).
42. Becke-Goehring, M., and E. Fluck, Angew. Chem. Int. Edn. Egl., 1, 281 (1962).
43. Becke-Goehring, M., Chem. Soc. Spec. Publ., 12, 297 (1958).
44. Becke-Goehring, M., and K. John, Angew. Chem., 70, 657 (1958).
45. Becke-Goehring, M., and K. John, Z. Anorg. Allgem. Chem., 304, 126 (1960).
46. Becke-Goehring, M., K. John, and E. Fluck, Z. Anorg. Allgem. Chem., 302, 103 (1959).

47. Becke-Goehring, M., and D. Jung, Z. Anorg. Allgem. Chem., 372, 233 (1970).
48. Becke-Goehring, M., and H. J. Muller, Z. Anorg. Allgem. Chem., 362, 51 (1968).
48a. Bedford, A. F., and C. T. Mortimer, J. Chem. Soc., 1960 4649.
49. Belykh, S. I., S. M. Zhivukhin, V. V. Kireev, and G. S. Kolesnikov, Zh. Neorgan. Khim., 14, 1278 (1969); Russ. J. Inorg. Chem., 14, 668 (1969).
50. Belykh, S. I., S. M. Zhivukhin, V. V. Kireev, and G. S. Kolesnikov, Zh. Obshch. Khim., 39, 799 (1969); J. Gen. Chem. USSR, 39, 761 (1969).
51. Berlin, A. J., B. Grushkin, and L. R. Moffet, Inorg. Chem., 7, 589 (1968).
51a. Berman, M., and K. Utvary, J. Inorg. Nucl. Chem., 31, 271 (1969).
52. Besson, A., and H. Rosset, Compt. Rend., 143, 37 (1906).
53. Besson, A., and H. Rosset, Compt. Rend., 146, 1149 (1908).
54. Bezman, I. I., U. S. Pat. 3,098,871; C. A. 59, 14,024 (1963).
55. Bezman, I. I., and C. T. Ford, Chem. and Ind. (London), 1963, 163.
56. Bezman, I. I., and J. H. Smalley, Chem. and Ind. (London), 1960, 839.
57. Biddlestone, M., and R. A. Shaw, Chem. Comm., 1965, 205.
58. Biddlestone, M., and R. A. Shaw, Chem. Comm., 1968, 407.
59. Biddlestone, M., and R. A. Shaw, J. Chem. Soc. (A), 1969, 178.
60. Biddlestone, M., and R. A. Shaw, J. Chem. Soc. (A), 1970, 1750.
61. Biddlestone, M., R. A. Shaw, and D. Taylor, Chem. Comm., 1969, 320.
62. Bilbo, A. J., Z. Naturforsch., 15B, 330 (1960).
63. Bilbo, A. J., C. M. Grieve, D. L. Herring, and D. E. Saltzbrun, Inorg. Chem., 7, 2671 (1968).
64. Bode, H., Angew Chem., 61, 438 (1949).
65. Bode, H., and H. Bach, Ber., 75B, 215 (1942).
66. Bode, H., K. Bütow, and G. Lienau, Chem. Ber., 81, 547 (1948).
67. Bode, H., and R. Thamer, Ber., 76B, 121 (1943).
68. Boden, N., J. W. Emsley, J. Feeney, and L. H. Sutcliffe, Chem. and Ind. (London), 1962, 1909.
69. Bogeat, G., and G. Cauquis, Bull. Soc. Chim. France, 1966, 2735.
70. Bond, M. R., C. Hewlett, K. Hills, and R. A. Shaw, unpublished results.
71. Branton, G. R., C. E. Brion, D. C. Frost, K. A. R. Mitchell, and N. L. Paddock, J. Chem. Soc. (A), 1970, 151.

72. Brion, C. E., D. J. Oldfield, and N. L. Paddock, Chem. Comm., 1966, 226.
73. Brion, C. E., and N. L. Paddock, J. Chem. Soc. (A), 1968, 392.
74. Brion, C. E., and N. L. Paddock, J. Chem. Soc. (A), 1968, 338.
75. British Petroleum Co., Fr. Pat., 1,525,419, C. A. 71, 38,561 (1969).
76. Brockway, L. O., and W. M. Bright, J. Am. Chem. Soc., 65, 1551 (1943).
77. Bullen, G. J., Proc. Chem. Soc., 1960, 425.
78. Bullen, G. J., J. Chem. Soc., 1962, 3193.
79. Bullen, G. J., P. R. Mallinson, and A. H. Burr, Chem. Comm., 1969, 691.
80. Bullen, G. J., and P. A. Tucker, Chem. Comm., 1970, 1185.
81. Caglioti, L., M. Poloni, and G. Rosini, J. Org. Chem., 33, 2979 (1968).
82. Califano, S., J. Inorg. Nucl. Chem., 24, 483 (1962).
83. Califano, S., and A. Ripamonti, J. Inorg. Nucl. Chem., 24, 491 (1962).
84. Capon, B., K. Hills, and R. A. Shaw, Proc. Chem. Soc., 1962, 390.
85. Capon, B., K. Hills, and R. A. Shaw, J. Chem. Soc., 1965, 4059.
86. Cardillo, B., G. Mattogno, A. Melera, and F. Tarli, Atti Accad. Nazl. Lincei. Rend., Classe Sci., Fis. Mat. Nat., 37, 194 (1964); C. A. 62, 13,030 (1965).
87. Cardillo, B., G. Mattogno, and F. Tarli, Atti. Accad. Nazl. Lincei. Rend., Classe Sci., Fis. Mat. Nat., 35, 328 (1963); C. A. 61, 4357 (1964).
88. Carroll, A. P., and R. A. Shaw, Chem. and Ind. (London), 1962, 1908.
89. Carroll, A. P., and R. A. Shaw, J. Chem. Soc. (A), 1966, 914.
90. Carroll, A. P., and R. A. Shaw, Inorg. Syn., 8, 84 (1966).
91. Carroll, A. P., and R. A. Shaw, unpublished results.
92. Chang, M. S., and A. J. Matuszko, J. Am. Chem. Soc., 82, 5796 (1960).
93. Chapman, A. C., and D. F. Carroll, J. Chem. Soc., 1963, 5005.
94. Chapman, A. C., and N. L. Paddock, J. Chem. Soc., 1962, 635.
95. Chapman, A. C., N. L. Paddock, D. H. Paine, H. T. Searle, and D. R. Smith, J. Chem. Soc., 1960, 3608.
96. Chapman, A. C., D. H. Paine, H. T. Searle, D. R. Smith, and R. F. M. White, J. Chem. Soc., 1961, 1768.
97. Chapman, D., S. H. Glarum, and A. G. Massey, J. Chem. Soc., 1963, 3140.

98. Cheng, D. C-H., and J. C. McCoubrey, J. Chem. Soc., 1963, 4993.
99. Chivers, T., and N. L. Paddlock, Chem. Comm., 1968, 704.
100. Chivers, T., and N. L. Paddlock, Chem. Comm., 1969, 337.
101. Chivers, T., and N. L. Paddock, J. Chem. Soc. (A), 1969, 1687.
102. Chivers, T., R. T. Oakley, and N. L. Paddock, J. Chem. Soc. (A), 1970, 2324.
103. Churchill, M. R., J. Cooke, J. Wormald, A. Davison, and E. S. Switkes, J. Am. Chem. Soc., 91, 6518 (1969).
104. Churchill, M. R., and J. Wormald, Chem. Comm., 1970, 703.
105. Corfield, G., J. Chem. Soc., 1962, 4258.
106. Cotson, S., and K. A. Hodd, J. Inorg. Nucl. Chem., 27, 335 (1965).
107. Cotson, S., and K. A. Hodd, J. Inorg. Nucl. Chem., 31, 245 (1969).
108. Couldridge, W., J. Chem. Soc., 53, 398 (1888); Bull. Soc. Chim. France, 50, 535 (1888).
109. Coxon, G. E., T. F. Palmer, and D. B. Sowerby, J. Chem. Soc. (A), 1967, 1568.
110. Coxon, G. E., T. F. Palmer, and D. B. Sowerby, J. Chem. Soc. (A), 1969, 358.
111. Coxon, G. E., T. F. Palmer, and D. B. Sowerby, Inorg. Nucl. Chem. Letters, 2, 215 (1966).
112. Coxon, G. E., and D. B. Sowerby, Inorg. Chim. Acta, 1, 381 (1967).
113. Coxon, G. E., and D. B. Sowerby, J. Chem. Soc. (A), 1969, 3012.
114. Coxon, G. E., and D. B. Sowerby, Spectrochim. Acta, 24A, 2145 (1968).
115. Coxon, G. E., and D. B. Sowerby, J. Chem. Soc. (A), 1967, 1566.
116. Coxon, G. E., D. B. Sowerby, and G. C. Tranter, J. Chem. Soc. (A), 1965, 5697.
117. Cragg, R. H., Essays in Chemistry, 1, 77 (1970).
118. Craig, D. P., M. L. Heffernan, R. Mason, and N. L. Paddock, J. Chem. Soc., 1961, 1376.
119. Craig, D. P., and K. A. R. Mitchell, J. Chem. Soc., 1965, 4682.
120. Daasch, L. W., J. Am. Chem. Soc., 76, 3403 (1954).
121. Das, R. N., R. A. Shaw, B. C. Smith, and M. Woods, J. Chem. Soc. (Dalton), 1973, 709.
121a. Das, R. N., D. J. Lingley, R. A. Shaw, B. C. Smith, H. S. Yu, and M. Woods, unpublished results.
122. Das, S. K., D. Feakins, W. A. Last, S. N. Nabi, R. A. Shaw, and B. C. Smith, J. Chem. Soc. (A), 1970, 616.
123. Das, S. K., R. Keat, R. A. Shaw, and B. C. Smith, J. Chem. Soc., 1965, 5032.

124. Das, S. K., R. Keat, R. A. Shaw, and B. C. Smith, J. Chem. Soc. (A), 1966, 1677.
125. Das, S. K., R. A. Shaw, and B. C. Smith, Chem. Comm., 1965, 176.
126. Das, S. K., R. A. Shaw, B. C. Smith, W. A. Last, and F. B. G. Wells, Chem. and Ind. (London), 1963, 866.
127. Das, S. K., R. A. Shaw, B. C. Smith, and C. P. Thakur, Chem. Comm., 1966, 33.
128. Davis, M. I., and J. W. Paul, Acta Cryst. Supp., 25A, S116 (1969).
129. Davy, H., Ann. Physik, 39, 3 (1811); Schweiggers J. Chem. Phys. 3, 79 (1811).
130. Dehnicke, K., Chem. Ber., 97, 3358 (1964).
131. Dell, D., B. W. Fitzsimmons, R. Keat, and R. A. Shaw, J. Chem. Soc. (A), 1966, 1680.
132. Dell, D., B. W. Fitzsimmons and R. A. Shaw, J. Chem. Soc., 1965, 4070.
133. Denny, K., and S. Lanoux, J. Inorg. Nucl. Chem., 31, 1531 (1969).
134. Derbisher, G. V., and A. V. Babaeva, Zh. Neorgan. Khim., 10, 2194 (1965); Russ. J. Inorg. Chem., 10, 1194 (1965).
135. Derkach, G. I., and M. V. Kolotilo, Zh. Obshch. Khim., 35, 1001 (1965); J. Gen. Chem. U.S.S.R., 35, 1007 (1965).
136. Desai, V. B., B. C. Smith, and R. A. Shaw, J. Chem. Soc. (A), 1970, 2023.
137. Desai, V. B., R. A. Shaw, and B. C. Smith, Angew. Chem., Int. Edn. Egl., 7, 887 (1968).
138. Desai, V. B., R. A. Shaw, and B. C. Smith, J. Chem. Soc. (A), 1969, 1977.
139. Desai, V. B., R. A. Shaw, B. C. Smith, and D. Taylor, Chem. and Ind. (London), 1969, 1177.
140. Dewar, M. J. S., E. A. C. Lucken, and M. A. Whitehead, J. Chem. Soc., 1960, 2423.
141. Dishon, B. R., J. Am. Chem. Soc., 71, 2251 (1949).
142. Dixon, M., H. D. B. Jenkins, J. A. S. Smith, and D. A. Tong, Trans. Farad. Soc., 63, 2852 (1967).
143. Dostal, K., M. Kouril, and J. Novak, Z. Chem., 4, 353 (1964).
144. Dougill, M. W., J. Chem. Soc., 1961, 5471.
145. Dougill, M. W., J. Chem. Soc., 1963, 3211.
146. Douglas, W. M., M. Cooke, M. Lustig, and J. K. Ruff, Inorg. Nucl. Chem. Letters, 6, 409 (1970).
147. Dwek, R. A., N. L. Paddock, J. A. Potenza, and E. H. Poindexter, J. Am. Chem. Soc., 91, 5436 (1969).
148. Dwek, R. A., R. E. Richards, D. Taylor, and R. A. Shaw, J. Chem. Soc., 1970, 1173.
149. Dyson, J., and N. L. Paddock, Chem. Comm., 1966, 191.

150. Eley, D. D., and M. R. Willis, J. Chem. Soc., 1963, 1534.
151. Emsley, J., and N. L. Paddock, J. Chem. Soc. (A), 1968, 2590.
152. Emsley, J., and N. L. Paddock, J. Chem. Soc. (A), 1970, 109.
153. Emsley, J., and P. B. Udy, J. Chem. Soc. (A), 1970, 3025.
154. Engelbrecht, A., E. Meyer, and Chr. Pupp, Monatsh., 95, 633 (1964).
155. Engelhardt, G., E. Steger, and R. Stahlberg, Z. Naturforsch., 21B, 586 (1966).
156. Engelhardt, G., E. Steger, and R. Stahlberg, Z. Naturforsch., 21B, 1231 (1966).
157. Fales, H. M., Analyt. Chem., 38, 1058 (1966).
158. Faught, J. B., T. Moeller, and I. C. Paul, Inorg. Chem., 9, 1656 (1970).
159. Feakins, D., W. A. Last, and R. A. Shaw, J. Chem. Soc., 1964, 2387.
160. Feakins, D., W. A. Last, and R. A. Shaw, J. Chem. Soc., 1964, 4464.
161. Feakins, D., W. A. Last, N Neemuchwala, and R. A. Shaw, J. Chem. Soc., 1965, 2804.
162. Feakins, D., W. A. Last, S. N. Nabi, and R. A. Shaw, J. Chem. Soc. (A), 1966, 1831.
163. Feakins, D., S. N. Nabi, R. A. Shaw, and P. Watson, J. Chem. Soc. (A), 1968, 10.
164. Feakins, D., W. A. Last, S. N. Nabi, R. A. Shaw, and P. Watson, J. Chem. Soc. (A), 1969, 196.
165. Feakins, D., R. A. Shaw, P. Watson, and S. N. Nabi, J. Chem. Soc. (A), 1969, 2468.
166. Feistel, G. R., and T. Moeller, J. Inorg. Nucl. Chem., 29, 2731 (1967).
167. Feldt, M. K., and T. Moeller, J. Inorg. Nucl. Chem., 30, 2351 (1968).
168. De Ficquelmont, A. M., Compt. Rend., 202, 423 (1936).
169. De Ficquelmont, A. M., Ann. Chim., 12, 169 (1939).
170. Finer, E. G., J. Molecular Spectroscopy, 23, 104 (1967).
171. Finer, E. G., R. K. Harris, M. R. Bond, R. Keat, and R. A. Shaw, J. Molecular Spectroscopy, 33, 72 (1970).
172. Fitzsimmons, B. W., C. Hewlett, K. Hills, and R. A. Shaw, J. Chem. Soc. (A), 1967, 679.
173. Fitzsimmons, B. W., C. Hewlett, and R. A. Shaw, Proc. Chem. Soc., 1962, 340.
174. Fitzsimmons, B. W., C. Hewlett, and R. A. Shaw, J. Chem. Soc., 1964, 4459.
175. Fitzsimmons, B. W., C. Hewlett, and R. A. Shaw, J. Chem. Soc., 1965, 4799.
176. Fitzsimmons, B. W., C. Hewlett, and R. A. Shaw, J. Chem. Soc., 1965, 7432.

177. Fitzsimmons, B. W., and R. A. Shaw, Chem. and Ind. (London), 1961, 109.
178. Fitzsimmons, B. W., and R. A. Shaw, J. Chem. Soc., 1964, 1735.
179. Fitzsimmons, B. W., and R. A. Shaw, Inorg. Synth., 8, 77 (1966).
179a. Fluck, E., Z. Anorg. Allgem. Chem., 320, 64 (1963).
180. Fluck, E., Z. Naturforsch., 19B, 869 (1964).
181. Fluck, E., and F. L. Goldmann, Chem. Ber., 96, 3091 (1963).
182. Ford, C. T., F. E. Dickson and I. I. Bezman, Inorg. Chem., 3, 177 (1964).
183. Ford, C. T., F. E. Dickson, and I. I. Bezman, Inorg. Chem., 4, 419 (1965).
184. Ford, C. T., F. E. Dickson, and I. I. Bezman, Inorg. Chem., 4, 890 (1965).
185. Ford, C. T., J. M. Barr, F. E. Dickson, and I. I. Bezman, Inorg. Chem., 5, 351 (1966).
186. Ford, C. T., F. E. Dickson, and I. I. Bezman, Inorg. Chem., 6, 1594 (1967).
187. Foster, R., L. Mayor, P. Warsop, and A. D. Walsh, Chem. and Ind. (London), 1960, 1445.
188. Frazier, S. E., and H. H. Sisler, Inorg. Chem., 5, 925 (1966).
189. Giglio, E., Ricerca Sci. Rend., 30, 721 (1960).
190. Giglio, E., and R. Puliti, Acta Cryst., 22, 304 (1967).
191. Gilson, I. T., and H. H. Sisler, Inorg. Chem., 4, 273 (1965).
192. Gimblett, F. G. R., Chem. and Ind. (London), 1958, 365.
193. Glemser, O., E. Niecke, and H. W. Roesky, Chem. Comm., 1969, 282.
194. Glemser, O., E. Niecke, and H. Thamm, Z. Naturforsch., 25B, 754 (1970).
195. Goehring, M., and J. Heinke, Z. Anorg. Allgem. Chem., 278, 53 (1955).
196. Goehring, M., H. Hohenschutz, and R. Appel, Z. Naturforsch., 9B, 678 (1954).
197. Goldschmidt, J. M. E., and J. Weiss, J. Inorg. Nucl. Chem., 26, 2023 (1964).
198. Grace, W. R., and Co., Fr. Pat., 1,339,384; C. A. 60, 3012 (1964).
199. Grace, W. R., and Co., Brit. Pat., 992,377; C. A. 63, 14,908 (1965).
200. Grace, W. R., and Co., Brit. Pat., 1,023,415; C. A. 64, 17,639 (1966).
201. Grace, W. R., and Co., Fr. Pat., 1,393,692; C. A. 64, 8238 (1966).
202. Grace, W. R., and Co., Brit. Pat., 1,016,467; C. A. 64, 17,639 (1966).

203. Grace, W. R., and Co., Brit. Pat., 1,031,170; C. A. 65, 5488 (1966).
204. Grace, W. R., and Co., U.S. Pat., 3,394,177 (see Ref. 203).
205. Grace, W. R., and Co., Fr. Pat., 1,461,442; C. A. 67, 12,243 (1967).
206. Green, B., and D. B. Sowerby, Chem. Comm., 1969, 628.
207. Green, B., and D. B. Sowerby, Inorg. Nucl. Chem. Letters, 5, 989 (1969).
208. Green, B., and D. B. Sowerby, J. Chem. Soc. (A), 1970, 987.
209. Gribova, I. A., and U. Ban-Yuan, Russ. Chem. Rev. (English Transl.) 30, 1 (1961).
210. Griffith, W. P., and K. J. Rutt, J. Chem. Soc. (A), 1968, 2331.
211. Grimme, W., Dissertation, Münster, 1926.
212. Grundmann, C., and R. Rätz, Z. Naturforsch., 10B, 116 (1955).
213. Grushkin, B., A. J. Berlin, J. L. McClanahan, and R. G. Rice, Inorg. Chem., 5, 172 (1966).
214. Grushkin, B., M. Gali-Sanchez, M. V. Ernest, J. L. McClanahan, G. E. Ashby and R. G. Rice, Inorg. Chem., 4, 1538 (1965).
215. Grushkin, B., M. Gali-Sanchez, and R. G. Rice, Inorg. Chem., 3, 623 (1964).
216. Grushkin, B., J. L. McClanahan, and R. G. Rice, J. Am. Chem. Soc., 86, 4204 (1964).
217. Haber, C. P., Inorganic Polymers, Chemical Society Special Publication, No. 15, p. 115, 1961.
218. Haber, C. P., D. L. Herring, and E. A. Lawton, J. Am. Chem. Soc., 80, 2116 (1958).
219. Haber, C. P., and R. K. Uenishi, Chem. Eng. Data Ser., 3, 232 (1958).
219a. Haiduc, I., "The Chemistry of Inorganic Ring Systems," Part 2, Wiley Interscience, New York, 1970, p. 624.
220. Harris, R. K., Inorg. Chem., 5, 701 (1966).
221. Hartley, S. B., N. L. Paddock, and H. T. Searle, J. Chem. Soc., 1961, 430.
222. Hazekamp, R., T. Migchelson, and A. Vos, Acta Cryst., 15, 539 (1962).
223. Heatley, F., and S. M. Todd, J. Chem. Soc. (A), 1966, 1152.
224. Heffernan, M. L., and R. F. M. White, J. Chem. Soc., 1961, 1382.
225. Herring, D. L., unpublished results, quoted in Ref. 228.
226. Herring, D. L., Chem. and Ind. (London), 1960, 717.
227. Herring, D. L., and C. M. Douglas, Inorg. Chem., 3, 428 (1964).

228. Herring, D. L., and C. M. Douglas, Inorg. Chem., 4, 1012 (1965).
229. Hewlett, C., and R. A. Shaw, J. Chem. Soc., 1966, 56.
230. Hills, K., and R. A. Shaw, J. Chem. Soc., 1964, 130.
231. Hisatsune, I. C., Spectrochim. Acta, 21, 1899 (1965).
232. Hisatsune, I. C., Spectrochim. Acta, 25A, 309 (1969).
233. Hofmann, A. W., Ber., 17, 1909 (1884).
234. Horn, H.-G., Chem.-Zeitung, 93, 241 (1969).
235. Humiec, F. S., and I. I. Bezman, J. Am. Chem. Soc., 83, 2210 (1961).
236. Jacques, J. K., M. F. Mole, and N. L. Paddock, J. Chem. Soc., 1965, 2112.
237. Jagodzinski, H., J. Langer, I. Oppermann, and F. Seel, Z. Anorg. Allgem. Chem., 302, 81 (1959).
238. Jagodzinski, H., and I. Oppermann, Z. Krist., 113, 241 (1960).
239. Janik, B., V. Zeshutko, and T. Pelćzar, Zh. Obshch. Khim., 36, 1444 (1966); J. Gen. Chem. USSR, 36, 1451 (1966).
240. Janik, B., V. Zeshutko, and T. Pelćzar, Zh. Anal. Khim., 22, 1103 (1967); C. A. 67, 104,884 (1967).
241. Jenkins, R. W., and S. Lanoux, J. Inorg. Nucl. Chem., 32, 2429 (1970).
242. John, K., and T. Moeller, J. Am. Chem. Soc., 82, 2647 (1960).
243. John, K., and T. Moeller, J. Inorg. Nucl. Chem., 22, 199 (1961).
244. John, K., and T. Moeller, Inorg. Synth., 7, 76 (1963).
245. John, K., T. Moeller, and L. F. Audrieth, J. Am. Chem. Soc., 82, 5616 (1960).
246. John, K., T. Moeller, and L. F. Audrieth, J. Am. Chem. Soc., 83, 2608 (1961).
247. Jurinski, N. B., and P. A. D. DeMaine, J. Inorg. Nucl. Chem., 27, 1571 (1965).
248. Kaplansky, M., and M. A. Whitehead, Can. J. Chem., 45, 1669 (1967).
249. Keat, R., M. C. Miller, and R. A. SHaw, Proc. Chem. Soc., 1964, 137.
250. Keat, R., M. C. Miller, and R. A. Shaw, J. Chem. Soc. (A), 1967, 1404.
251. Keat, R., S. K. Ray, and R. A. Shaw, J. Chem. Soc., 1965, 7193.
252. Keat, R., and R. A. Shaw, Chem. and Ind. (London), 1964, 1232.
253. Keat, R., and R. A. Shaw, J. Chem. Soc., 1965, 2215.
254. Keat, R., and R. A. Shaw, J. Chem. Soc., 1965, 4067.

255. Keat, R., and R. A. Shaw, J. Chem. Soc., 1966, 908.
256. Keat, R., and R. A. Shaw, J. Chem. Soc. (A), 1968, 703.
257. Keat, R., and R. A. Shaw, Angew. Chem. Int. Edn. Egl., 7, 212 (1968).
258. Keat, R., and R. A. Shaw, Chemical Society Specialist Reports on Organophosphorus Chemistry, 1, 214 (1969).
259. Keat, R., and R. A. Shaw, unpublished results.
260. Keat, R., R. A. Shaw, and C. Stratton, J. Chem. Soc., 1965, 2223.
261. Keller, H. J., and A. Schmidpeter, Z. Naturforsch., 22B, 231 (1967).
262. Ketelaar, J. A. A., and J. A. de Vries, Rec. Trav. Chim., 58, 1081 (1939).
263. Kireev, V. V., G. S. Kolesnikov, and I. M. Raigorodskii, Russ. Chem. Rev. (Egl. Transl.), 38, 667 (1969).
264. Klement, U., and A. Schmidpeter, Z. Naturforsch., 23B, 1610 (1968).
265. Kobayashi, Y., L. A. Chasin, and L. B. Clapp, Inorg. Chem., 2, 212 (1963).
266. Kober, E., H. F. Lederle, and G. Ottmann, Inorg. Chem., 5, 2239 (1966).
267. Kober, E., H. F. Lederle, and G. Ottmann, U.S. Pat. 3,280,222; C. A. 66, 38,047 (1967).
268. Kokalis, S. G., K. John, T. Moeller, and L. F. Audrieth, J. Inorg. Nucl. Chem., 19, 191 (1961).
269. Yu-Kokoreva, I., Ya K. Syrkin, A. A. Kropacheva, and N. M. Kashnikova, Dokl. Akad. Nauk SSSR, 166, 155 (1966); C. A. 64, 12,522 (1966).
270. Kolotilo, M. V., A. G. Matyusha, and G. I. Derkach, Zh. Obshch. Khim., 39, 188 (1969); J. Gen. Chem. USSR, 39, 173 (1969).
271. Koopman, H., F. J. Spruit, F. van Deursen, and J. Bakker, Rec. Trav. Chim., 84, 341 (1965).
272. Korshak, V. V., I. A. Gribova, T. V. Artamonova, and A. N. Bushmarina, Vysokmol. Soedineniya, 2, 377 (1960); C. A. 54, 24,479 (1960).
273. Kratzer, R. H., and K. L. Paciorek, Inorg. Chem., 4, 1767 (1965).
274. Kropacheva, A. A., and N. M. Kashnikova, Zh. Obshch. Khim., 32, 652 (1962); J. Gen. Chem. USSR, 32, 645 (1962).
275. Kropacheva, A. A., and N. M. Kashnikova, Zh. Obshch. Khim., 33, 1046 (1963); J. Gen. Chem. USSR, 33, 1036 (1963).
276. Kropacheva, A. A., and N. M. Kashnikova, Zh. Obshch. Khim., 35, 1988 (1965); J. Gen. Chem. USSR, 35, 1978 (1965).

277. Kropacheva, A. A., and N. M. Kashnikova, Zh. Obshch. Khim., 35, 2229 (1965); J. Gen. Chem. USSR, 35, 2219 (1965).
278. Kropacheva, A. A., and N. M. Kashnikova, Zh. Obshch. Khim., 38, 136 (1968); J. Gen. Chem. USSR, 38, 135 (1968).
279. Kropacheva, A. A., and N. M. Kashnikova, Khim. Org. Soedin Fosfora Akad. Nauk USSR, Otd. Obshchei Tekh. Khim., 1967, 186; C. A. 69, 10,066 (1968).
280. Kropacheva, A. A., and N. M. Kashnikova, Khim. Org. Soedin Fosfora Akad. Nauk SSSR,Otd. Obshchei Tekh. Khim., 1967, 188; C. A. 69, 10,331 (1968).
281. Kropacheva, A. A., N. M. Kashnikova, and V. A. Parshina, Zh. Obshch. Khim., 34, 530 (1964); J. Gen. Chem. USSR, 34, 532 (1964).
282. Kropacheva, A. A., and L. E. Mukhina, Zh. Obshch. Khim., 31, 2437 (1961); J. Gen. Chem. USSR, 31, 2274 (1961).
283. Kropacheva, A. A., and L. E. Mukhina, Zh. Obshch. Khim., 32, 521 (1962); J. Gen. Chem. USSR, 32, 512 (1962).
284. Kropacheva, A. A., and L. E. Mukhina, Zh. Obshch. Khim., 33, 706 (1963); J. Gen. Chem. USSR, 33, 699 (1963).
285. Kropacheva, A. A., L. E. Mukhina, and N. M. Kashnikova, Put. Sin. i Izy. Prot. Prep. Tr. Simp. po Khim., Prot. Vesh., Moscow, 1960, 174; C. A. 58, 5663 (1963).
286. Kropacheva, A. A., L. E. Mukhina, N. M. Kashnikova, and V. A. Parshina, Zh. Obshch. Khim., 31, 1036 (1961); J. Gen. Chem. USSR, 31, 957 (1961).
287. Kuznetsov, N. T., Izv. Akad. Nauk SSSR Neorg. Mater, 2, 2258 (1966); C. A. 66, 61,396 (1967).
288. Kutznetsov, N. T., and G. S. Klimchuk, Izv. Akad. Nauk SSSR, Neorg. Mater, 3, 587 (1967); C. A. 67, 73,635 (1967).
289. Lakatos, B., A. Hesz, S. Holly, and G. Horvath, Naturwissenschaften, 49, 493 (1962).
290. Lakatos, B., A. Hesz, Z. Vetessy, and G. Horvath, Acta Chim. (Budapest), 66, 309 (1969); C. A. 71, 55,090 (1969).
291. Lappert, M. F., and G. Srivasteva, J. Chem. Soc., 1966, 210.
292. Latscha, H. P., Z. Naturforsch., 236, 139 (1968).
293. Lederle, H. F., G. F. Ottmann, and E. H. Kober, Inorg. Chem., 5, 1818 (1966).
294. Lehr, W., Z. Anorg. Allgem. Chem., 350, 18 (1967).
295. Lehr, W., Z. Anorg. Allgem. Chem., 352, 27 (1967).
296. Lehr, W., Naturwissenschaften, 56, 214 (1969).
297. Lehr, W., Z. Anorg. Allgem. Chem., 371, 225 (1969).

297a. Lenton, M. V., and B. Lewis, J. Chem. Soc. (A), 1966, 665.
298. Liebig, J., and F. Wöhler, Ann., 11, 139 (1834).
299. Lucken, E. A. C., Proc. Colloq. Ampere, 1962, 678; C. A. 60, 142 (1964).
300. Lund, L. G., N. L. Paddock, J. E. Proctor, and H. T. Searle, J. Chem. Soc., 1960, 2542.
301. Magnelli, D. D., G. Tesi, J. U. Lowe, and W. E. McQuiston, Inorg. Chem., 5, 457 (1966).
302. de Maine, P. A. D., and R. D. Srivastava, J. Miss. Acad. Sci., 10, 67 (1964); C. A. 62, 216 (1965).
303. Mani, N. V., F. R. Ahmed, and W. H. Barnes, Acta Cryst., 19, 693 (1965).
304. Mani, N. V., F. R. Ahmed, and W. H. Barnes, Acta Cryst., 21, 375 (1966).
305. Mani, N. V., and A. J. Wagner, Chem. Comm., 1968, 658.
306. Manley, T. R., and D. A. Williams, Spectrochim. Acta, 23A, 149 (1967).
307. Manley, T. R., and D. A. Williams, Spectrochim. Acta, 23A, 1221 (1967).
308. Manley, T. R., and D. A. Williams, Spectrochim. Acta, 24A, 166 (1968).
309. Manley, T. R., and D. A. Williams, Polymer, 10, 307 (1969).
310. Mao, T. J., R. D. Dresdner, and J. A. Young, J. Am. Chem. Soc., 81, 1020 (1959).
311. Mao, T. J., R. D. Dresdner, and J. A. Young, J. Inorg. Nucl. Chem., 24, 53 (1962).
312. Marsh, W. C., N. L. Paddock, C. J. Stewart, and J. Trotter, Chem. Comm., 1970, 1190.
313. Marsh, W. C., T. N. Ranganthan, J. Trotter, and N. L. Paddock, Chem. Comm., 1970, 815.
314. Mattogno, G., and A. Monaci, Ric. Sci. Rend. Sez. A., 8, 1139 (1965); C. A. 64, 14,078 (1966).
315. Mattogno, G., A. Monaci, and F. Tarli, Ann. Chim. (Rome), 55, 599 (1965); C. A. 63, 13,307 (1965).
316. Mattogno, G., and F. Tarli, Ric. Sci. Rend. Sez. A, 4, 487 (1964); C. A. 61, 14,702 (1964).
317. Matuszko, J., and M. S. Chang, J. Org. Chem., 31, 2004 (1966).
318. May and Baker Ltd., Belg. Pat., 660,454; C. A. 64, 591 (1966).
319. May and Baker Ltd., Belg. Pat., 660,562; C. A. 64, 591 (1966).
320. McBee, E. T., K. Okuhara, and C. J. Morton, Inorg. Chem., 5, 450 (1966).
320a. McBee, E. T., K. Okuhara, and C. J. Morton, Inorg. Chem., 4, 1672 (1965).
321. McGeachin, H. M., and F. R. Tromans, Chem. and Ind. (London), 1960, 1131.

322. McGeachin, H. M., and F. R. Tromans, J. Chem. Soc., 1961, 4777.

323. Migachev, G. I., and B. I. Stepanov, Z. Neorgan. Khim., 11, 1739 (1966); C. A. 65, 18,132 (1966).

324. Miller, M. C., D. W. Rhys, and R. A. Shaw, Ind. Chemist, 40, 183 (1964).

325. Miller, M. C., and R. A. Shaw, J. Chem. Soc., 1963, 3233.

326. Mitchell, K. A. R., J. Chem. Soc. (A), 1968, 2683.

327. Moeller, T., unpublished results referred to in Topics in Phosphorus Chemistry, 5, 406 (1967).

328. Moeller, T., A. Failli, and F. Y. Tsang, Inorg. Nucl. Chem. Letters, 1, 49 (1965).

329. Moeller, T., K. John, and F. Y. Tsang, Chem. and Ind. (London), 1961, 347.

330. Moeller, T., and S. G. Kokalis, J. Inorg. Nucl. Chem., 25, 875 (1963).

331. Moeller, T., and S. G. Kokalis, J. Inorg. Nucl. Chem., 25, 1397, (1963).

332. Moeller, T., and P. Nannelli, Inorg. Chem., 1, 721 (1962).

333. Moeller, T., and P. Nannelli, Inorg. Chem., 2, 659 (1963).

334. Moeller, T., and S. Lanoux, J. Inorg. Nucl. Chem., 25, 229 (1963).

335. Moeller, T., and S. Lanoux, Inorg. Chem., 2, 1061 (1963).

336. Moeller, T., and F. Y. Tsang, Chem. and Ind. (London), 1962, 361.

337. Moran, E. F., and D. P. Reider, Inorg. Chem., 8, 1551 (1969).

338. Moureu, H., and A. M. de Ficquelmont, Compt. Rend., 198, 1417 (1934).

339. Moureu, H., and A. M. de Ficquelmont, Compt. Rend., 213, 306 (1941).

340. Mukhina, L. E., and A. A. Kropacheva, Zh. Obshch. Khim., 38, 313 (1968); J. Gen. Chem. USSR, 38, 314 (1968).

341. Munch, R. M. and S. V. Dighe, 155th A.C.S. meeting, 1968, Abstract M97.

342. Nabi, S. N., S. N. Nabi, and N. K. Das, J. Chem. Soc., 1965, 3857.

343. Nabi, S. N., R. A. Shaw, and C. Stratton, Chem. and Ind. (London), 1969, 166.

344. Nannelli, P., and T. Moeller, Inorg. Chem., 2, 896 (1963).

345. National Research and Development Corporation (London), Brit. Pat. 965,424; C. A. 61, 9528 (1964).

346. Negita, H., and S. Satou, Bull. Chem. Soc. Japan, 29, 426 (1956).

347. Negita, H., and S. Satou, J. Chem. Phys., 24, 621 (1956).
348. Niecke, E., O. Glemser, and H. W. Roesky, Z. Naturforsch., 24B, 1187 (1969).
349. Niecke, E., O. Glemser, and H. Thamm, Chem. Ber., 103, 2864 (1970).
350. Nielson, M. L., and G. Cranford, Inorg. Synth., 6, 94 (1960).
351. Nielson, M. L., and T. J. Morrow, Inorg. Synth., 6, 97 (1960).
352. Nielson, M. L., and T. J. Morrow, Inorg. Synth., 6, 99 (1960).
353. Nielson, M. L., J. V. Pustinger, and J. Strobel, J. Chem. Eng. Data, 9, 167 (1964).
354. Nikolaev, A. F., and Eŕ-Ten Wan, Zh. Obshch. Khim., 34, 1831 (1964); J. Gen. Chem. USSR, 34, 1843 (1964).
355. Nikolaev, A. F., and Eŕ-Ten Wan, Zh. Obshch. Khim., 34, 1833 (1964); J. Gen. Chem. USSR, 34, 1846 (1964).
356. Olthoff, R., Acta Cryst., 25B, 2040 (1969).
357. Ottmann, G. F., H. Agahigian, H. Hooks, G. D. Vickers, E. H. Kober, and R. F. W. Raetz, Inorg. Chem., 3, 753 (1964).
358. Ottmann, G. F., H. Hooks, E. H. Kober, R. F. W. Raetz, and S. S. Ristich, U.S. Pat., 3,136,754; C. A. 61, 4312 (1964).
359. Ottmann, G. F., H. Lederle, H. Hooks, and E. H. Kober, Inorg. Chem., 6, 394 (1967).
360. Otto, R. J. A., and L. F. Audrieth, J. Am. Chem. Soc., 80, 3575 (1958).
361. Otto, R. J. A., and L. F. Audrieth, J. Am. Chem. Soc., 80, 5894 (1958).
362. Paciorek, K. L., and R. H. Kratzer, Inorg. Chem., 3, 594 (1964).
363. Paciorek, K. L., and R. H. Kratzer, Inorg. Chem., 5, 538 (1966).
364. Paciorek, K. L., and R. H. Kratzer, U.S. Pat. 3,297,751; C. A. 66, 76,152 (1967).
365. Paddock, N. L., Endeavour, 19, 134 (1960).
366. Paddock, N. L., Quart. Rev., 18, 168 (1964).
367. Paddock, N. L., and H. T. Searle, in "Advances in Inorganic Chemistry and Radiochemistry," Vol. 1, Emeléus, H. J., and A. G. Sharpe, eds., Academic Press, New York, 1959, p. 347.
368. Paddock, N. L., J. Trotter, and S. H. Whitlow, J. Chem. Soc. (A), 1968, 2227.
369. Palavin, M., D. N. Henrickson, J. M. Hollander, and W. L. Jolly, J. Phys. Chem., 74, 1116 (1970).
370. Parrod, J., and R. Pornin, Compt. Rend., 258, 3022 (1964).

371. Parrod, J., and R. Pornin, Compt. Rend., 260, 1438 (1965).
372. Philips Gloeilampenfabrieken, N. V., Fr. Pat. 1,372,097; C. A. 62, 444 (1965).
373. Pitina, M. R., T. M. Ivanova, and N. I. Shvetsov-Shilovskii, Zh. Obshch. Khim., 37, 2076 (1967); J. Gen. Chem. USSR, 37, 1968 (1967).
374. Pitina, M. R., V. V. Nebrebskii, and N. I. Shvetsov-Shilovskii, Zh. Obshch. Khim., 39, 1216 (1969); J. Gen. Chem. USSR, 39, 1186 (1969).
375. Pitina, M. R., and N. I. Shvetsov-Shilovskii, Zh. Obshch. Khim., 36, 498 (1966); J. Gen. Chem. USSR, 36, 517 (1966).
376. Pollard, F. H., G. Nickless, K. Burton, and J. Hubbard, Microchem. J., 10, 131 (1966); C. A. 64, 13,366 (1966).
377. Pollard, F. H., G. Nickless, and R. W. Warrender, J. Chromatog., 9, 485 (1962).
378. du Pont de Nemours, E. I., and Co., Fr. Pat. 1,356,145; C. A. 61, 1796 (1964).
379. Pornin, R., Bull. Soc. Chim. France, 1966, 2861.
380. Pornin, R., and J. Parrod, Compt. Rend., 260, 1198 (1965).
381. Ramain, R., Y. Runavot, and P. Schneebeli, J. Chim. Phys., 56, 659 (1959).
382. Rapko, J. N., and G. R. Feistel, Chem. Comm., 1968, 474.
383. Rapko, J. N., and G. Feistel, Inorg. Chem., 9, 1401 (1970).
384. Raetz, R., E. Kober, C. Grundmann, and G. Ottmann, Inorg. Chem., 3, 757 (1964).
385. Raetz, R., H. Schroeder, H. Ulrich, E. Kober, and C. Grundmann, J. Am. Chem. Soc., 84, 551 (1962).
386. Ray, S. K., and R. A. Shaw, Chem. and Ind. (London), 1959, 53.
387. Ray, S. K., and R. A. Shaw, J. Chem. Soc., 1961, 872.
388. Ray, S. K., and R. A. Shaw, Chem. and Ind. (London), 1961, 1173.
389. Ray, S. K., R. A. Shaw, and B. C. Smith, J. Chem. Soc., 1963, 3236.
390. Rencroft, P. J., P. L. Kronick, H. Scott, and M. M. Labes, Nature, 201, 609 (1964).
391. Ritzsche, H., R. Stahlberg, and E. Steger, J. Inorg. Nucl. Chem., 28, 687 (1966).
392. Roesky, H. W., and W. Grosse-Bowing, Inorg. Nucl. Chem. Letters, 6, 781 (1970).
393. Roesky, H. W., and E. Niecke, Inorg. Nucl. Chem. Letters, 4, 463 (1968).
394. Rose, S. H., J. Poly Sci., Poly. Letters, 6b, 837 (1968).

395. Rosset, H., Compt. Rend., 180, 750 (1925).
396. Rosset, H., Bull. Soc. Chim. France, 37, 518 (1925).
397. de Santis, P., E. Giglio, and A. Ripamonti, J. Inorg. Nucl. Chem., 24, 469 (1962).
398. Schenck, R., and G. Römer, Ber., 57B, 1343 (1924).
399. Schlueter, A. W., and R. A. Jacobson, J. Am. Chem. Soc., 88, 2051 (1966).
400. Schlueter, A. W., and R. A. Jacobson, J. Chem. Soc. (A), 1968, 2317.
401. Schmidpeter, A., and R. Böhm, Z. Anorg. Allgem. Chem., 362, 65 (1968).
402. Schmidpeter, A., R. Böhm, and H. Groeger, Angew. Chem. Intl. Edn. Egl., 3, 704 (1964).
403. Schmidpeter, A., and J. Ebeling, Angew. Chem. Intl. Edn. Egl., 6, 87 (1967).
404. Schmidpeter, A., and J. Ebeling, Angew. Chem. Intl. Edn. Egl., 6, 565 (1967).
405. Schmidpeter, A., and J. Ebeling, Angew. Chem. Intl. Edn. Egl., 7, 209 (1968).
406. Schmidpeter, A., and J. Ebeling, Chem. Ber., 101, 2602 (1968).
407. Schmidpeter, A., and J. Ebeling, Chem. Ber., 101, 3883 (1968).
408. Schmidpeter, A., and N. Schindler, Z. Anorg. Allgem. Chem., 362, 281 (1968).
409. Schmidpeter, A., and N. Schindler, Z. Anorg. Allgem. Chem., 367, 130 (1969).
410. Schmidpeter, A., and N. Schindler, Chem. Ber., 102, 856 (1969).
411. Schmidpeter, A., and N. Schindler, Z. Anorg. Allgem. Chem., 372, 214 (1970).
412. Schmidpeter, A., and K. Stoll, Angew. Chem. Int. Edn. Egl., 6, 252 (1967).
413. Schmidpeter, A., and K. Stoll, Angew. Chem. Int. Edn. Egl., 7, 549 (1968).
414. Schmidpeter, A., and C. Weingand, Z. Naturforsch., 24B, 177 (1969).
415. Schmidpeter, A., and C. Weingand, Angew. Chem. Int. Edn. Egl., 8, 615 (1969).
416. Schmidpeter, A., C. Weingand, and E. Hafner-Roll, Z. Naturforsch., 24B, 799 (1969).
417. Schmitz-Dumont, O., and A. Braschos, Z. Anorg. Allgem. Chem., 243, 113 (1939).
418. Schmitz-Dumont, O., and H. Külkens, Z. Anorg. Allgem. Chem., 238, 189 (1938).
419. Schmulbach, C. D., "Progress in Inorganic Chemistry," Vol. 4, F. A. Cotton, ed., Interscience, New York, 1962, 275.
420. Schmulbach, C. D., A. G. Cook, and V. R. Miller, Inorg. Chem., 7, 2463 (1968).

421. Schmulbach, C. D., and C. Derderian, J. Inorg. Nucl. Chem., 25, 1395 (1963).
422. Schmulbach, C. D., and C. Derderian, J. Inorg. Nucl. Chem., 32, 3397 (1970).
423. Schmulbach, C. D., and V. R. Miller, Inorg. Chem., 5, 1621 (1966).
424. Schmulbach, C. D., and V. R. Miller, Inorg, Chem., 7, 2189 (1968).
425. Schmutzler, R., T. Moeller, and F. Tsang, Inorg. Synth., 9, 75 (1967).
426. Searle, H. T., Proc. Chem. Soc. (London), 1959, 7.
427. Searle, H. T., Brit. Pat. 924,620; C. A. 59, 10,124 (1963).
428. Seel, F., and J. Langer, Angew. Chem., 68, 461 (1956).
429. Seel, F., and J. Langer, Z. Anorg. Allgem. Chem., 295, 316 (1958).
430. Sharts, C. M., U.S. Pat., 3,347,876; C. A. 67, 116,942 (1967).
431. Sharts, C. M., A. J. Bilbo, and D. R. Gentry, Inorg. Chem., 5, 2140 (1966).
432. Shaw, R. A., Chem. and Ind. (London), 1959, 54.
433. Shaw, R. A., New Scientist, 8, 1603 (1960).
434. Shaw, R. A., Endeavour, 27, 74 (1968).
435. Shaw, R. A., Rec. Chem. Prog., 28, 243 (1967).
436. Shaw, R. A., unpublished results.
437. Shaw, R. A., B. W. Fitzsimmons, and B. C. Smith, Chem. Rev., 62, 247 (1962).
438. Shaw, R. A., and E. H. M. Ibrahim, Angew. Chem. Int. Edn. Egl., 6, 556 (1967).
439. Shaw, R. A., R. Keat, and C. Hewlett, "Preparative Inorganic Reactions," Vol. 2, W. L. Jolly, ed., Interscience, New York, 1965, p. 1.
440. Shaw, R. A., and T. Ogawa, J. Poly. Sci., 3A, 3343 (1965).
441. Shaw, R. A., and C. Stratton, Chem. and Ind. (London), 1959, 52.
442. Shaw, R. A., and C. Stratton, J. Chem. Soc., 1962, 5004.
443. Shaw, R. A., and D. Taylor, unpublished results.
444. Shaw, R. A., and F. B. G. Wells, Chem. and Ind. (London), 1960, 1189.
445. Shaw, R. A., and M. Woods, unpublished results.
446. Sherif, F. G., and C. D. Schmulbach, Inorg. Chem., 5, 322 (1966).
447. Shvetsov, N. I., K. A. Nuriozhanyan, A. Ya Yakubovich, and F. F. Sukhov, Zh. Obshch. Khim., 33, 3936 (1963).
448. Siegel, L. A., and J. H. van den Hende, J. Chem. Soc. (A), 1967, 817.

449. Siekmann, L., H. O. Hoppen, and R. Appel, Z. Naturforsch., 23B, 1156 (1968).
450. Sisler, H. H., H. S. Ahuja, and N. L. Smith, Inorg. Chem., 1, 84 (1962).
451. Sisler, H. H., and S. E. Frazier, Inorg. Chem., 4, 1204 (1965).
452. Slawisch, A., and J. Pietschmann, Z. Naturforsch., 25B, 321 (1970).
453. Smalley, J. H., F. E. Dickson, and I. I. Bezman, Inorg. Chem., 3, 1780 (1964).
454. Smith, G. W., and D. Wood, Nature, 210, 520 (1966).
455. Sorokin, M. F., and V. K. Latov, Zh. Obshch. Khim., 35, 1471, (1965); J. Gen. Chem. USSR, 35, 1472 (1965).
456. Sorokin, M. F., and V. K. Latov, Kinetika i Kataliz., 7, 42 (1966); C. A. 64, 19,348 (1966).
457. Sowerby, D. B., J. Chem. Soc., 1965, 1396.
458. Spell, H. L., Anal. Chem., 39, 185 (1967).
459. Stahlberg, R., and E. Steger, J. Inorg. Nucl. Chem., 28, 684 (1966).
460. Stahlberg, R., and E. Steger, Spectrochim. Acta, 23A, 627 (1967).
461. Stahlberg, R., and E. Steger, Spectrochim. Acta, 23A, 2005, (1967).
462. Stahlberg, R., and E. Steger, Spectrochim. Acta, 23A, 2057, (1967).
463. Stahlberg, R., and E. Steger, Spectrochim. Acta, 23A, 2185, (1967).
464. Stahlberg, R., and E. Steger, J. Inorg. Nucl. Chem., 29, 961 (1967).
465. Stahlberg, R., and E. Steger, J. Inorg. Nucl. Chem., 30, 737 (1968).
466. Steger, E., and G. Bachmann, Z. Chem., 10, 306 (1970).
467. Steger, E., and D. Clemm, J. Inorg. Nucl. Chem., 29, 1812 (1967).
468. Steger, E., and G. Mildner, Z. Naturforsch., 16B, 836 (1961).
469. Steger, E., and J. Rost, J. Inorg. Nucl. Chem., 25, 732 (1963).
470. Steger, E., and R. Stahlberg, Z. Naturforsch., 17B, 780 (1962).
471. Steger, E., and R. Stahlberg, Z. Anorg. Allgem. Chem., 326, 243 (1964).
472. Steger, E., and R. Stahlberg, J. Inorg. Nucl. Chem., 28, 688 (1966).
473. Steinman, R., F. B. Schirmer, and L. F. Audrieth, J. Am. Chem. Soc., 64, 2377 (1942).
474. Stepanov, B. I., and G. I. Migachev, Zh. Obshch. Khim., 35, 2254 (1965); J. Gen. Chem. USSR, 35, 2245 (1965).

475. Stepanov, B. I., and G. I. Migachev, Zh. Obshch. Khim., 36, 1447 (1966); J. Gen. Chem. USSR, 36, 1454 (1966).
476. Stepanov, B. I., and G. I. Migachev, Zavodsk. Lab., 32, 416 (1966); C. A. 66, 25,854 (1967).
477. Stokes, H. N., Am. Chem. J., 17, 275 (1895).
478. Stokes, H. N., Ber., 28, 437 (1895).
479. Stokes, H. N., Am. Chem. J., 18, 780 (1896).
480. Stokes, H. N., Am. Chem. J., 19, 782 (1897).
481. Tarli, F., Ric. Sci. Rend. Sez. A., 3, 761 (1963); C. A. 62, 7624 (1965).
482. Tesi, G., and C. M. Douglas, J. Am. Chem. Soc., 84, 549 (1962).
482a. Tesi, G., C. P. Haber, and C. M. Douglas, Proc. Chem. Soc., 1960, 219.
483. Tesi, G., A. J. Matuszko, R. Zimmer-Galler, and M. S. Chang, Chem. and Ind. (London), **1964**, 623.
484. Tesi, G., R. J. A. Otto, F. G. Sherif, and L. F. Audrieth, J. Am. Chem. Soc., 82, 528 (1960).
485. Tesi, G., and P. J. Slota, Proc. Chem. Soc. (London), 1960, 404.
486. Tesi, G., and R. Zimmer-Galler, Chem. and Ind. (London), 1964, 1916.
487. Tolstoguzov, V. B., V. V. Pisarenko, and V. V. Kireev, Z. Neorgan. Khim., 10, 712 (1965); Russ. J. Inorg. Chem., 10, 382 (1965).
488. Tomaschewskii, G., C. Berseck, and G. Hilgetag, Chem. Ber., 101, 2037 (1968).
489. Torizuka, K., J. Phys. Soc. Japan, 11, 84 (1956); C. A. 50, 9149 (1956).
490. Trieber, E., W. Berndt, and H. Taplak, Angew. Chem., 67, 69 (1955).
491. Trotter, J., N. L. Paddock, and S. H. Whitlow, Chem. Comm., 1969, 695.
492 Trotter, J., and S. H. Whitlow, J. Chem. Soc. (A), 1970, 455.
493. Trotter, J., and S. H. Whitlow, J. Chem. Soc. (A), 1970, 460.
494. Tullock, C. W., and D. D. Coffmann, J. Org. Chem., 25, 2016 (1960).
495. Uhlig, E., and H. Eckert, Z. Anal. Chem., 204, 332 (1964).
496. Ul Haque, R., and B. ud Din, Pakistan J. Sci. Ind. Res., 9, 121 (1960); C. A. 66, 28,850 (1967).
497. United States Rubber Co., Neth. Appl., 6,608,987; C. A. 67, 21,612 (1967).
498. Van de Grampel, J. C., and A. Vos, Rec. Trav. Chim., 82, 246 (1963).
499. Van de Grampel, J. C., and A. Vos, Acta Cryst., B25, 651 (1969).

500. Van Wazer, J. R., C. F. Callis, J. N. Shoolery, and R. C. Jones, J. Am. Chem. Soc., 78, 5715 (1956).
501. Wagner, A. J., and A. Vos, Rec. Trav. Chim., 84, 603 (1965).
502. Wagner, A. J., and A. Vos, Acta Cryst., B24, 707 (1968).
503. Wagner, A. J., and A. Vos, Acta Cryst., B24, 1423 (1968).
503a. Wakefield, Z. T., B. B. Luff, and J. J. Kohler, J. Chem. Eng. Data, 15, 241 (1970).
504. Wende, A., and D. Joel, Z. Chem., 3, 466 (1963).
505. Wende, A., and D. Joel, Z. Chem., 3, 467 (1963).
506. Whitaker, R. D., A. J. Barreiro, P. A. Furman, W. C. Guida, and E. S. Stallings, J. Inorg. Nucl. Chem., 30, 2921 (1968).
507. Whitaker, R. D., J. C. Carlton, and H. H. Sisler, Inorg. Chem., 2, 420 (1963).
508. Whitehead, M. A., Can. J. Chem., 42, 1212 (1964).
509. Wilson, A., and D. F. Carroll, Chem. and Ind. (London), 1958, 1558.
510. Wilson, A., and D. F. Carroll, J. Chem. Soc., 1960, 2548.
511. Winyall, M., and H. H. Sisler, Inorg. Chem., 4, 655 (1965).
511a. Wunsch, G., R. Scheidermier, V. Kiener, E. Fluck, and G. Heckmann, Chem. Zeitung, 94, 832 (1970).
512. Yokoyama, M., J. Chem. Soc. Japan, 80, 1192 (1959).
513. Yokoyama, M., J. Chem. Soc. Japan, 81, 481 (1960).
514. Yokoyama, M., H. Cho, and M. Sakuma, Kogyo Kagaku Zasshi, 66, 422 (1963); C. A. 60, 1322 (1964).
515. Yvernault, T., and G. Casteignau, Bull. Soc. Chim. France, 1966, 1464.
516. Zhivukhin, S. M., and V. V. Kireev, Zh. Obshch. Khim., 34, 3126 (1964); J. Gen. Chem. USSR, 34, 3169 (1964).
517. Zhivukhin, S. M., V. V. Kireev, N. V. Aulova, and L. T. Gerasimenko, Dokl. Akad. Nauk USSR, 158, 896 (1964); C. A. 62, 2685 (1965).
517a. Zhivukhin, S. M., V. V. Kireev, V. G. Kolesnikov, V. P. Popolin, and G. S. Kolesnikov, Otkrytiya. Izobret. Rom. Obraztsy Tovarnye Znaki, 46, 24 (1969); C. A. 72, 123,535 (1970).
518. Zhivukhin, S. M., V. B. Tolstoguzov, and Z. Lukuszewski, Zh. Neorgan. Khim., 10, 1653 (1965); Russ. J. Inorg. Chem., 10, 901 (1965).
519. Zoer, H., and A. J. Wagner, Acta Cryst., B26, 1812 (1970).